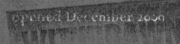

Guide to Standard Floras of the World

Guide to Standard Floras of the World is a selective annotated bibliography of the principal floras and related works of inventory for vascular plants. This new edition has been completely revised, updated and expanded to take into account the substantial literature of the late twentieth century, and features a more fully developed review of the history of floristic documentation. The works covered are principally specialist publications, encompassing descriptive floras and checklists, distribution atlases, systematic iconographies, and enumerations or catalogues. A relatively few more popularly oriented books are also included. The *Guide* is organized into 10 geographical divisions, with these successively divided into regions and units. Each geographical unit or larger region is prefaced with a historical review of floristic studies, including references to key literature as well as to more specialized area bibliographies. The bibliography itself is preceded by three general chapters on botanical bibliography, the history of floras, and general principles and current trends, and the book concludes with an appendix on bibliographic searching, a lexicon of serial abbreviations, and author and geographical indices.

DAVID FRODIN is a researcher in the Herbarium at the Royal Botanic Gardens, Kew, having previously held positions in the Department of Biology at the University of Papua New Guinea and the Department of Botany at the Academy of Natural Sciences of Philadelphia. His botanical interests focus on the systematics of the family Araliaceae, particularly the genus *Schefflera*, while his more general interests include tropical biology and the history of biology. He is also an acknowledged authority on botanical bibliography, documentation and informatics, having compiled the first edition of *Guide to Standard Floras of the World* (1984) and co-authored several volumes in the Kew series *World Checklists and Bibliographies of Seed Plants*, including *Magnoliaceae* (1996), *Fagales* (1998), *Euphorbiaceae (with Pandaceae)* (2000), *Sapotaceae* (2000) and *Araceae and Acoraceae* (2000).

Guide to
Standard Floras
of the World

...

An annotated, geographically arranged systematic
bibliography of the principal floras, enumerations,
checklists and chorological atlases of different areas

SECOND EDITION

David G. Frodin

Royal Botanic Gardens, Kew

PUBLISHED BY THE PRESS SYNDICATE OF THE UNIVERSITY OF CAMBRIDGE
The Pitt Building, Trumpington Street, Cambridge, United Kingdom

CAMBRIDGE UNIVERSITY PRESS
The Edinburgh Building, Cambridge CB2 2RU, UK
40 West 20th Street, New York, NY 10011-4211, USA
10 Stamford Road, Oakleigh, VIC 3166, Australia
Ruiz de Alarcón 13, 28014 Madrid, Spain
Dock House, The Waterfront, Cape Town 8001, South Africa

http://www.cambridge.org

First published 1984
Second edition 2001

Printed in the United Kingdom at the University Press, Cambridge

Typeface Monotype Ehrhardt 10/12pt   *System* QuarkXpress®   [SE]

A catalogue record for this book is available from the British Library

ISBN 0 521 79077 8 hardback

To Sidney Fay Blake and Alice Cary Atwood
authors of *Geographical guide to floras of the world*

Phyllis I. Edwards and Rudolf Schmid
advocates for biological documentation

and E. J. H. Corner
who eloquently reminded us of what
Floras are *for*

Contents

Contents

No branch of botanical literature is more useful, and at the same time more neglected than [floras] . . . For a beginner [the Flora] is the first, and one of the most important aids for obtaining botanical knowledge.

> de Candolle and Sprengel, *Elements of the philosophy of plants* (Edinburgh, 1821).

Quatenus bibliotheca in omni scientia primum à studioso evolvi debeat, ita etiam est Botanico *maxime necessaria*, quum multiplex usus inde deducitur ...

> Linnaeus, *Bibliotheca botanica* (Amsterdam, 1736; reprinted Halle 1747).

Now, there are two different attitudes towards learning from others. One is the dogmatic attitude of transplanting everything, whether or not it is suited to our conditions. This is no good. The other attitude is to use our heads and learn those things which suit our conditions, that is, to absorb whatever experience is useful to us. That is the attitude we should adopt.

> Mao Tse-tung, 27 February 1957, in *Quotations from Chairman Mao Tse-tung* (New York, 1967).

Of all forms of human activity related to plants, that of knowing the kinds, properties and uses of such plants as grow in one's *Landschaft*, or 'parish', is perhaps the longest-established. Most, if not all, 'traditional' cultures centered on the land possess, or once possessed, a comparatively detailed knowledge of the local flora, in many cases recognizing the same species (and sometimes genera) as would a modern professional botanist; and, in like manner to many 'advanced' societies, this knowledge is best developed amongst a comparatively small circle of *savants*.

It is thus not surprising that, in those civilizations which achieved literacy in the pre-Columbian era, this kind of botanical knowledge should have come to be recorded at an early date. However, such works as are now known were largely compilations of what was common knowledge, considerable though this might have been, and for long were conceptually pragmatic. This obtains, for instance, in the oldest known extant geographically oriented botanical works, the treatises of Theophrastus reporting discoveries on Alexander's campaigns in the fourth century B.C.E. and *Nan-fang ts'ao mu chuang* of the fourth century C.E. on south China and Indochina, whose post-Columbian semantic descendants include the Swedish surveys of Linnaeus, the many geographical accounts of distant lands of the nineteenth century containing substantial botanical information, and, with greater impersonality, the Australian and Pacific-area terrain studies by various military and civilian agencies in the mid-twentieth century. Not, however, until the rise of the Western European tradition of independent scientific enquiry and the consequent development of a systematics based on the *nature* of plants *in themselves* rather than on traditionally pragmatic values did the compilation of organized area floristic accounts and plant lists become a distinct activity, which after the Linnean revolution and during the century of European colonial expansion came to constitute a significant proportion of what in some lands was to be called 'special botany' (*spezieller Pfanzenkunde, bijzondre plantkunde*) or, in more modern parlance, systematic botany in a broad sense. Floristic studies and flora- and checklist-writing have ever since constituted an important part of the work of this subdiscipline and the published (and, increasingly, semi-published) results cumulatively contain an immense amount of botanical information. To the non-specialist, these works, along with the provision 'on demand' of identification and information services, perhaps represent the most easily comprehensible aspect of the systematist's work.

The relative importance of area floras and check-lists in the world of systematic botany has varied over nearly four centuries, but since the 1930s, and especially since 1945 and the advent of liberal but often short-term state support, floras, checklists, and related area works and contributions thereto have come to pre-dominate. Stafleu[1] has termed the present cycle as 'an age of floras and floristic work', but at the same time notes that this has partly come about at the expense of serious monographic and revisionary work, a trend strongly aggravated by the virtual destruction of the Berlin Herbarium and the German systematics profession by the end of World War II. Current indications are, as Jäger[2] has noted, that this pattern will continue, causing the collective mass of floristic works, especially the more significant ones, to become the single most important source of modern taxonomic knowledge, and thus by default supplanting the great synthetic works of the mid-nineteenth to early-twentieth centuries such as the *Prodromus systematis naturalis regni vegetabilis*, *Monographiae Phanerogamarum*, *Die natür-lichen Pflanzenfamilien*, and *Das Pflanzenreich*. Of a verity have floras expanded in scope far beyond what was originally conceived: from simple inventory (and diagnosis) they have successively assumed the roles of identification manual and taxonomic encyclopedia, in the latter case now often also accounting for current notions on the classification of a given group above, as well as at, species level. In too many instances, however, their *effective* role has been lost sight of.

That floras and checklists had a distinctive place in botanical literature was already recognized from late in the eighteenth century and by 1820 had become canonical. Classified lists of those which were indepen-dently published appear in general bibliographies from Linnaeus onwards, but by 1879 their importance had become so recognized that a separate list was deemed

necessary. This first list, a slender but closely printed pamphlet of twelve pages, was *The floras of different countries* by G. L. Goodale of Harvard University. Two others have followed since: the Lloyd Library's *Bibliography of the floras* by W. Holden and E. Wycoff (1911–14), rather more comprehensive than Goodale's list but like it compiled 'in-house' and (following tradi-tional practice) limited to independently published works, and the original and critical *Geographical guide to floras of the world* by S. F. Blake and A. C. Atwood (1942–61, not completed). Several regional and national lists have also been produced.

These earlier general guides to floras, however, were produced when the totality of botanical literature, even accounting for that published in periodicals and serials, was far less and overall bibliographical control more satisfactory (particularly before World War I). These conditions no longer existed by the 1960s. Twenty years alone were required by Blake for distilla-tion of volume 2 of the *Geographical Guide* from the vast mass of Western European floristic literature, and by the end of that period, volume 1 was already in need of revision, although the flow of new literature had not yet taken on the proportions of the 1970's flood. Developments since 1960 have been such that, although Blake was said to be well aware of the magni-tude of his task,[3] it is likely that in the years after his death, even had the will and the means existed, comple-tion of the work on the original plan would, for a variety of reasons, have been very difficult if not impossible.

At the present time the climate for a revised and completed version of the *Geographical Guide* along its original lines is even less favorable, however much it may be desired in some quarters. The exponential growth of biological literature in the 1980s, and its control, is but one factor: others include the effects of the disruption and fragmentation of the world botani-cal information system due to two World Wars, trans-Atlantic isolationism in the inter-war period, additional centers of botanical activity and publication, changes in scientific fashions, and political and social develop-ments of recent decades including the currently chang-ing relationship between science and society in a more austere economic climate.[4]

Moreover, much current retrospective biblio-graphical work has been directed elsewhere: *Index nominum genericorum*, *Taxonomic literature* (and its 2nd edition referred to as *TL-2*), *Bibliographia Huntiana*, etc. With respect to floristic bibliography, the fragmentation and partial disintegration of the botanical *referat* system alone has posed significant obstacles which only sophisticated organization and large financial expenditure can overcome. A number of the principal sources utilized by Blake and Atwood no longer exist; these include their primary source, the USDA Botany Subject Catalogue, terminated in mid 1952 (fortunately for others, it appeared in book form in 1958) and current literature coverage, especially of independently published items which comprise the majority of significant contributions, is more diffuse and uneven and less complete than in the past, although since 1950 two indexing journals specifically dealing with systematic botany have come into being: *Excerpta Botanica*, A (from 1959) and *Kew Record of Taxonomic Literature* (from 1971). Lately, 'semi-publication' has presented an increasing problem to bibliographers as inexpensive, comparatively permanent modes of offset printing have become widely diffused. Another approach was needed if the heterogeneous flood of floristic literature, which had increased greatly during the 1950s and later came to be considered a key contributor to what Heywood[5] has termed the contemporary 'crisis' in taxonomy, was ever to be mastered and meaningful world-wide coverage once more provided.

The actual stimulus for the present book, from which grew its basic idea, arose from a conversation in the summer of 1962 with a fellow student at the University of Michigan Biological Station in the northernmost part of the state's Lower Peninsula. As an invertebrate zoologist planning to participate in the 1963 International Indian Ocean Expedition, he desired to obtain some basic references on the vascular plants of the islands in the region. A search through the first volume of the *Geographical Guide* revealed a goodly number of titles, but upon reflection it became apparent that many were too specialized or restricted in scope for the kinds of information sought. Ultimately it was found that a comparatively limited selection of floras and enumerations would provide, within a reasonable compass, a proportionately high degree of useful information about the region's vascular plants; in other words, these works could be viewed as 'standard' floras.

From this beginning there developed the idea that such a selective process could, with variations, be applicable world-wide, and that this would in due time enable the preparation of a one-volume annotated general bibliography of 'standard' floristic works on vascular plants which would cover the entire world, region by region. I also came to believe that such a work would be of particular interest to non-botanists as well as to botanists without a detailed knowledge of regional floristic literature outside their own sphere. Other factors contributing to a decision to prepare such a bibliography were the limited nature of lists of 'useful' floras provided in systematics textbooks as well as the unlikely prospect, noted above, that the *Geographical Guide* would ever be completed, especially considering the death of its senior author in 1959 (as it stands, it does not cover central or eastern Europe or the continent of Asia). Furthermore, in addition to Part I becoming increasingly out of date, the size of Part II appeared likely to daunt all those not having some familiarity with the vast corpus of western European floristic literature.

During 1962–3, various experiments in relation to depth of coverage were attempted, but the main catalyst proved to be in the pair of 'Green Books'[6] published by the Flora Europaea Organization which came to my attention in March 1963 on a visit to the University of Michigan Herbarium. Therein was given a list, with supplement, of 'standard' floras of Europe deemed most significant for the preparation of *Flora Europaea*. (The 'standard' flora concept had itself evidently been formulated by the Organization in the mid-1950s.)[7]

The final result, for which work originally began in a substantial way during the summer of 1963 at the Field Museum of Natural History, Chicago, is represented by the present book. However, lack of experience as well as time suggested that the *Guide* be first written up and distributed in short-title form without annotations or commentary. That effort materialized as the mimeographed booklet written largely at the University of Tennessee, Knoxville, and issued from its Department of Botany in 1964.

The consequent strong and continuing demand for that booklet, even to the time of final revision of this preface, ultimately led me to consider an expanded, more definitive edition. For various reasons, however, no serious research was begun until the end of 1967 when, encouraged by representations from colleagues all over the world as well as a publication proposal from the University of Tennessee Press, I felt compelled to undertake the task, one which would be greatly facilitated by being at the time at the University of Cambridge. Primary compilation of the necessary material was undertaken largely in Cambridge and London, with additions from Australian libraries in 1971 following my move to an academic position at Port Moresby (Papua New Guinea), but short visits were made to libraries in several other centers.

It was a basic tenet of both the preliminary and the present versions of this book that as far as possible all titles selected for inclusion should be examined and annotated at first hand. To a very large degree, this has been achieved, in a few cases with the aid of photo-copied extracts. Where an entry has had to be based upon a secondary source, that source has been indicated.

The original selection of titles was made by sys-tematic browsing along the shelves of the Botany Library in the Field Museum. Additions were made through work in the University of Tennessee (Knoxville) Libraries and short visits to some special botanical libraries in the central and eastern United States. Guidelines for the selection process were also provided by a number of secondary sources as well as advice from colleagues. For the present version, the botanical libraries at Cambridge (England), the Royal Botanical Gardens, Kew, and the British Museum (Natural History) were extensively utilized, along with the working library of the Flora Europaea Secretariat (at Liverpool, later at Reading), the library of the Komarov Botanical Institute, Leningrad, and the libraries of the New York Botanical Garden and the Arnold Arboretum/Gray Herbarium at Harvard University. Small amounts of work were done at addi-tional special botanical libraries as opportunities arose. Advice was also sought from a great number of other botanists, both in person and in writing. It may here be noted that the number of botanical libraries in which a substantial primary search for floras and related works may be carried out efficiently is comparatively small: five in the United States (in four centers) and three in Europe (in two centers). It is in London that the most substantial collections of these works exist, and it has been my good fortune to have been able to make exten-sive use of them over the years.

As might be expected, the coverage of material in the periodical literature has presented the greatest problems, both in ferreting out references and in seeing the articles concerned. No good cumulative classified index is currently available and extensive searches of the various abstracting and indexing journals would have been tedious and very time-consuming. Furthermore, floristic material is found in a wide and scattered range of biological, zoological, general scientific, and other periodicals as well as in those more specifically concerned with botany. In more recent years, material published or 'semi-published' in various kinds of technical series or runs of 'occasional papers' emanating from a plethora of university departments, institutes, and other organizations has proliferated to an inordinate degree. A great misfor-tune has been the above-mentioned discontinuance of the botany subject catalogue in what is now the United States National Agricultural Library; this provides the best classified source for the first half of the twentieth century. Its suspension without an adequate replace-ment can only be deplored. Fortunately, in more recent decades there has been a marked rise in the number of regional compilations of botanical literature (both bib-liographies and indices), and much use was made of them; they are now available over many parts of the world though variable in scope and quality. Some of these provided their own selections of key floristic works. Lists of references in major floras themselves were searched for periodical material. It must be con-fessed, however, that a goodly number of items were yet found 'by chance'. In all respects, having made a sys-tematic study of a world-wide tropical and subtropical genus, *Schefflera* (Araliaceae), which followed earlier work on *Cytisus* and allied genera (Leguminosae–Genisteae), proved a considerable asset.

Principal secondary sources utilized included, above all, the two volumes of Blake and Atwood's *Geographical Guide*. Another useful but older general source was *Bibliography relating to the Floras* (1911–14) in the *Bibliographical Contributions* series of the Lloyd Library. Other key works were, in the main, regional: among them were the bibliographies in Hultén's *The amphi-Atlantic plants* (1958) and *The circumpolar plants*, II (1971); *Bibliography of eastern Asiatic botany* by Merrill and Walker (1938) and its *Supplement* by Walker (1960); the two volumes of *Island bibliographies* by Sachet and Fosberg (1955, 1971); *Botanical bibliog-raphy of the islands of the Pacific* by Merrill, with subject index by Walker (1947); *Vvedenie v botaničeskuju literaturu* SSSR by Lebedev (1956) and *Literaturnye istočniki po flore SSSR* by Lipschitz (1975); the *Guide for contributors to 'Flora Europaea'* and its *Supplement*, both by Heywood (1958, 1960), otherwise known as the 'Green Books'; *History of botanical researches in India, Burma, and Ceylon*, II:

Systematic botany of angiosperms by Santapau (1958), and *A guide to selected current literature on vascular plant floristics for the contiguous United States, Alaska, Canada, Greenland, and the US Caribbean and Pacific Islands* by Lawyer, Miller, Morse, and Kartesz (in press). Some individual library or union catalogues were useful, particularly the *Botany Subject Index* (1958), which constitutes the above-mentioned former USDA botany subject catalogue of 1906–52 in book form, the *Catalogue of the Library, Royal Botanic Gardens, Kew* (1974) with both author and classified divisions, the *Catalogue of the Library of the Arnold Arboretum* (1914–33), and, for bibliographic control, the *National Union Catalog* [USA]: *Pre-1956 imprints* and its retrospective and post 1956 supplements together with the *Botany Subject Index* and *Biological Abstracts*. Major indices used from time to time included *Excerpta Botanica* and *Kew Record of Taxonomic Literature* and, at regional level, *Index to American Botanical Literature* (and the former *Taxonomic Index*), *AETFAT Index*, *Flora Malesiana Bulletin*, and the European and Australasian indices published through the International Association for Plant Taxonomy in the 1960s. For search purposes, however, only occasional use was made of *Biological Abstracts*, however, and with the advent of the many regional botanical bibliographies now in existence there proved relatively little need to consult the older general indices, even had they been readily available. Of general current awareness lists, those extensively utilized included the *referat* sections in *Taxon* and *Progress in Botany* [Fortschritte der Botanik] as well as the 'semi-published' accession lists from the New York Botanical Garden (now defunct) and Kew libraries (the latter classified); these were supplemented by a range of dealers' catalogues (mainly Antiquariat Junk, Koeltz, Krypto, Scientia, Stechert-Hafner, and Wheldon and Wesley) and trade announcements (the latter sometimes providing descriptions). None of these, however, acted as substitutes for examination of the originals save when no other opportunity was available, but nevertheless they prove especially valuable whilst working in a relatively remote country such as Papua New Guinea.

The actual preparation of the *Guide*, although undertaken in 1970, was unfortunately considerably prolonged on account of my many university responsibilities as well as the attractions of a tropical flora, and only in late 1975 could it be terminated. The remote-

ness of Port Moresby was also a handicap, but on account of circumstances perhaps less so than might be imagined. More importantly, it enabled the work to be written from the point of view of a botanist attempting to cope with an imperfectly known tropical flora and actively involved in teaching. Much of the writing was accomplished during spells in remote outstations and camps while 'on patrol', often when waiting for airplanes or sitting out the rain. Following submission of the manuscript, a variety of technical difficulties led to a long delay in publication and in January 1979 it was formally transferred to Cambridge University Press. Accumulating additions and other changes as well as ideological refinements necessitated complete revision of the manuscript and this was largely carried out in Papua New Guinea and Australia during study leave from July 1979 to February 1980. Overseas visits in 1973, 1975–6, 1976–7, and 1978–9 enabled coverage of new or overlooked works. As far as possible 1980 is taken as the 'cut-off' year, with some indication of likely future developments and publications given in the various regional commentaries.

It is hoped that the *Guide* as now presented will meet the needs of a wide range of users, both botanical and non-botanical. It has been written in the belief that, since a thorough revision and completion of Blake and Atwood's *Geographical Guide* is not likely in the foreseeable future (and in any case would have to be an institutional project), a simpler one-volume analytical work would serve as a practical and more easily realizable alternative which would yet suffice for a majority of interested persons.

The work as it stands, though, is also intended to draw attention to the need for developments in floristic and other botanical bibliography comparable with what Heywood[8] and some other authors have called for with regard to floras generally. Although the necessity for various kinds of functional articulation and resource redeployment was long ago recognized in bibliographic science through sheer force of circumstances, it has been slow to come to systematic botany: the dream of the definitive, hard-cover, omnibus work has been long-persistent. Yet two (or more) functions are served in both floras and botanical bibliography – chiefly the archival and the practical – which in most cases can no longer usefully be combined within a single work and now require separation in publication. It is here suggested, for instance, that comparable selectivity with articulation is as necessary for flora-bibliography as for floras themselves, and that this is but part of a continuing process in information handling with implications for all fields of knowledge.[9] A work such as the *Geographical Guide* – considered in its day as 'selective' in relation to the general corpus of systematic botanical literature, and representative of the 'new trend' of scholarly bibliography which arose out of World War I[10] – is seen here as to a marked extent now archival, whereas the present *Guide*, though less extensive in its coverage, should prove useful at a more practical level whilst still remaining a meaningful indicator. It thus continues to stand for the 'state of the art' in Malclès' sense. Put in another way, it represents a level of selection twice removed from the coverage spread represented by the last of the great retrospective subject bibliographies (Pritzel, 1871–7; Rehder, 1911–18). 'Standard' floras may be viewed as having a place in floristic literature comparable to the head of a comet; the rest forms the gradually thinning tail.[11] As with floras themselves, the overall view is that there is room for both kinds of bibliographical works.

Any deliberate abridgement of the kind represented by the present work, though, always involves subjective decisions over inclusion or exclusion of particular titles, even though they be based upon heuristic criteria. Many items inevitably will have a 'borderline' status, even given the intuitively recognized 'point of balance' which limits this work. Such items may show a 'shift' in that they possess rather more importance in a local as opposed to a global context. All that can be said is that all care has been taken in such decisions, using the only computer available. Nonetheless, I shall always welcome any reasonable suggestions for addition (or deletion) of titles (within limits) with appropriate arguments. It should also be noted that the actual preparation of this work has to a considerable extent been carried out at locations remote from large botanical libraries, making quick rechecking or reinterpretation of sources difficult or impossible; unintentional errors may, therefore, have crept in. Any technical omissions or errors or misleading statements should, if possible, be brought to my attention. All changes accepted would be incorporated in a supplement contemplated for publication in the late 1980s.

Finally, it should be noted that whereas earlier bibliographies of floras have been largely empirical or descriptive, the present work attempts as well to be analytical and interpretative, essaying also some integration on historical principles. The belief has, latterly, grown in my mind that a classified subject bibliography should not only present and describe titles but also reach outwards: to act as a *Spiegelbild der Forschungsergebnisse*, a mirror on the progress of the subject,[12] as well as to guide – in the words of an earlier promotor of bibliographic science – 'a young man [who], instead of wasting months getting lost in unimportant reading . . . would be [thus] directed toward the best works and more easily and quickly attain a better education'.[13] It is hoped that this *Guide*, at least to some extent, fulfills these ideals, which with variations are of long standing in bibliography. Modern methods of bibliographical analysis moreover, indicate that a literature cross-section of the kind presented here can be about as meaningful as a comprehensive bibliography in revealing patterns of development in the subject, in this case floristic botany. Further research on the themes embodied here might (1) utilize citation analysis of a wide range of floristic articles as a means of quantifying the selection criteria and the 'point of balance', and (2) estimate patterns of usage through time by analysis along similar lines of a series of historical cross-sections of the literature. Both could serve as contributions to the history of systematic botany; and other insights might also be obtained in ways not yet suspected.

D. G. Frodin
Port Moresby, Papua New Guinea
August 1980/January 1982

Notes

1 Stafleu, F. A., 1959. The present status of plant taxonomy. *Syst. Zool.* 8: 59–68.
2 Jäger, E. J., 1978. Areal- und Florenkunde (Floristische Geobotanik). *Prog. Bot.* 40: 413–28.
3 Schubert, B. G., 1960. Sidney Fay Blake. *Rhodora*, 62: 325–38.
4 Drucker, P. F., 1979. Science and industry, challenges of antagonistic interdependence. *Science*, 204: 806–10.
5 Heywood, V. H., 1973. Taxonomy in crisis? or taxonomy is the digestive way of biology. *Acta Bot. Acad. Sci. Hung.* 19: 139–46.
6 Heywood, V. H., 1958. *The presentation of taxonomic information: a short guide for contributors to Flora Europaea.* 24 pp. Leicester: Leicester University Press; *idem*, 1960. *Supplement*. 20 pp. Coimbra, Portugal.
7 Heywood, V. H., 1957. A proposed flora of Europe. *Taxon*, 6: 33–42.

8 Heywood, V. H., 1973. Ecological data in practical taxonomy. In *Taxonomy and ecology* (ed. V. H. Heywood), pp. 329–47. London: Academic Press.

9 Garfield, E., 1979. *Citation indexing*. New York: Wiley.

10 Malclès, L. N., 1961. *Bibliography* (Trans. T. C. Hines). New York: Scarecrow (reprinted 1973). [Originally publ. 1956, Paris, as *La bibliographie*.]

11 Garfield, E., 1980. Bradford's Law and related statistical patterns. *Current Contents/Life Sciences* 23 (19): 5–12.

12 Simon, H.-R., 1977. *Die Bibliographie der Biologie*, p. 75. Stuttgart: Hiersemann.

13 Napoléon I to Finkestein, 19 April 1807; quoted in Maclès, 1961, p. 75 (see n. 10).

Prologue to the second edition

The reception of this book since its original publication some 15 years ago, and the frequent questions put to the author over the past decade about a revised edition, suggest that it has found a place amongst the tools of working botanists as well as of reference librarians. I hope this revision will find a similar reception, in spite of – inevitably – an increase in bulk.

In the nearly 20 years since coverage was closed for the original edition, floras and related works have continued by and large to gush forth. The need for them remains, although it may be driven more by practical than by academic considerations. The renewal and increasing prominence of the environmental and conservation movements, the associated promulgation of international treaties such as the Convention on International Trade in Endangered Species (CITES) and the Convention on Biological Diversity (CBD), and the consequent requirement to have a better understanding of national biotas have moreover created new 'markets' for floristic information. This is all in addition to natural cycles of renewal as scientific knowledge expands and deepens, best expressed in more developed countries. Altogether, many more new floras and enumerations have been published than superseded, improving coverage for many parts of the world – sometimes well beyond what was the case in 1979 (Map I). They retain an important place within the botanical literary warrant, and continue to be one of the most important points of contact between user and producer.

At the same time, however, the nature of floras may undergo change, driven in particular by the increasing power and flexibility of the Internet as an information source over the past five years or so as well as changes in their organization and the way users interact with them. Five points seem apparent: (1) floras should be backed by an information system; (2) manual-floras for identification should become less 'academic'; (3) enumerations and checklists are valuable 'interim' tools but should be backed as far as possible by specialist advice; (4) the elements of floras should be analyzed and, where possible, the works built up using widely available routines; and (5) the financial and human needs of flora projects should be worked out in such a way that their real costs become more transparent.

That said, what are some of the advances of the past two decades? Among large-scale floras, one may count the launch of *Flora of North America* (1993), *Flora fanerogámica argentina* (1994), *Flora iberica* (1989) and *Flora hellenica* (1996) as well as the progress of *Flora of Australia*, *Flora reipublicae popularis sinicae*, *Flora of Thailand*, *Flora of tropical East Africa* and *Flora Zambesiaca* and the near-completion of *Flora iranica*. Successful large-scale enumerations – all accounting for 15 000 species or more – include, in the Americas, *Catalogue of the flowering plants and gymnosperms of Peru* (1993) and *Catalogue of the vascular plants of Ecuador* (1999) and, in Africa, *Énumération des plantes à fleurs d'Afrique tropicale* (1991–97) and the various editions of what is now *Plants of southern Africa: names and distribution* (most recently in 1993). Many floras, manuals and enumerations of lesser extent have been completed or are in progress, some of them – such as *The Jepson Manual: higher plants of California* (1993), *Michigan flora* (1972–86), *Flora of Egypt* (1999–), *Standardliste der Farn- und Blütenpflanzen der Bundesrepublik Deutschland* (1998), *Manual of the flowering plants of Hawai'i* (1990; revised 1999), and *Flora vitiensis nova* (1979–91) – 'successors' to earlier works, and others – such as *Flora of the Lesser Antilles* (1974–89), *Flora of Bhutan* (1983–), *Flora of Orissa* (1994–96), *Flore analytique du Togo* (1984), and *Flora of central Australia* (1981) – 'breaking new ground', i.e., accounting for areas never previously covered or only by rather older, larger-scale works. There has also been – as suggested in the previous edition of this book – a further growth in floras and enumerations of relatively small 'target' areas. Although by their nature 'local', they may serve clearly defined areas or needs and, significantly, have come to be seen as realistically feasible in the three- to six-year terms of many project grants (which have largely succeeded the relatively open-ended financial commitments more common in the years after World War II). Four recent examples include *Flora of Pico das Almas* (1995) and *Flora da Reserva Ducke* (1999) in Brazil, *Flórula de las Reservas Biológicas de Iquitos, Perú* (1997) in Peru, and *The plants of Mount Cameroon: a conservation checklist* (1998) in Cameroon; several more could be mentioned. In the *Guide*, I have accounted for such 'local' works in areas for which there is little or no larger-scale coverage, or such is significantly out of date.

It is clear from the above that publication of floras and related works has continued apace. As a literary warrant they have continued to be prominent in the literature of plant biodiversity, if perhaps not quite as pervasive as 30–40 years ago when Frans Stafleu spoke of an 'age of floras'. It is likely that they will survive the advance of the Internet: in time a balance may be reached between traditional and virtual media in a likely larger market; moreover, paper remains a primary symbol of professional achievement. Nevertheless, as in the print world of the past, divergence in the kinds of data stored and presented will occur. The dictionary, manual-key, concise descriptive manual, and enumeration seem most likely to continue in their present forms; larger, more scholarly works will metamorphose into monograph-series or virtual publications (or information systems) on the Web or CD-ROMs or will be presented as differentiated print and virtual products (as have already some works of more limited scope). Large tropical floras can be, and are being, broken down into more manageable units for presentation. Whatever their form of presentation in future, however, floras and related works remain one of the most important forms of interaction between specialist and user; indeed, with respect to botany and other plant sciences in general they would figure prominently in any renewal of the question of the accountability of 'normal' science.[1]

The considerable number of floras and related works published since 1980 – along with a felt need to develop a fuller historical perspective – has meant that the *Guide* as presented here is some one-half again as large as its predecessor. I have, however, attempted to maintain its presentation as a practical as well as analytical introduction to the literature of identification and floristic documentation, occupying a level below, though hopefully more comprehensible than, the often title-rich overall, subdisciplinary, or regional bibliographies. It must be said, though, that the *Guide* has about reached the limit of what is feasible in a one-volume work. In future, a re-analysis of its principles may be required, with the possibility of new directions including (1) rendition of historical analyses and detailed descriptions in electronic form, with the printed text limited to titles and brief annotations, or (2) a bringing forward of the 'base line' from 1840 to 1940, with the possible creation of two temporally limited volumes (though with some overlap where deemed essential). In the latter instance, a differentiation would be made between works still 'standard' for a practising systematist or regional specialist and those of more immediate value for identification, fact-checking or basic documentation. Whatever path is followed, however, it would seem imperative that the present text – now in electronic form as its predecessor was not – is in the first instance converted into a structured database or marked up in XML; this would aid the development of differentiated products encompassing print, the Web and other media.

For it is in a variety of formats that the way forward in botanical information lies. In spite of the staggering growth of the World Wide Web as a source, its content is inevitably very uneven. Moreover, products with serious input and editorial control – whatever the medium – cost money which will have to be identified, allocated, and often recouped. Though the importance of data and information management in the progress of botanical research and dissemination may be undervalued, I remain convinced that there will always be a place for analytical reference works in print of the kind presented here. Indeed, within five years of publication a new edition of the original work was called for. Availability of text in electronic form, with the possibility of new kinds of products, will not only facilitate the process of future revision and dissemination but also enable new kinds of links to current data – including those available only in virtual form – not readily possible in the past. The *Guide* has in its field become, and will hopefully continue to be, a key tool not only for botanists in general but also for what is an increasingly important profession, that of reference librarians.[2]

A few final words should be said about preparation of the present edition. I have by and large followed the scope and methods of its predecessor, though important advantages in recent years have included regular access to the Library of the Royal Botanic Gardens, Kew as well as the advent of searchable remote-access library catalogues – the latter helpful for the checking of holdings as well as bibliographic details. Account was taken of the many additional references given in *Plants in danger: what do we know?* by S. B. Davis *et al.* (1986, Gland/Cambridge). A number of visits were made to other libraries as opportunities permitted, and extracts from some additional items obtained by post. A slightly more liberal view was taken of partial and local floras, especially where more general works were not available or significantly out of date. The original text – not available in electronic form – was optically scanned in 1990 but then was entirely rewritten as well as checked for errors. Unit areas and vascular or seed plant flora sizes have been incorporated as far as possible, and opening commentaries have been expanded to account, in running form, for references to floras and related works of historical interest. Where possible, works published in 1999 have been included but beyond that a line has been drawn. No attempt was made to create a database at the risk of further delay to what was in the end becoming – in the face of other commitments – a long drawn-out effort. Portions of the text, including the general chapters and Appendix A, were read by others before being worked up in final form.

David G. Frodin
Kew
4 July 2000

Notes

1 Cf. R. Schmid, 2000. An excellent flora of New York City and its easterly and northerly environs *sensu latissimo* [review]. *Taxon* **49**: 353–355. On the general question of science and society, see for example J. R. Ravetz, 1996. *Scientific knowledge and its social problems*. 2nd edn. New Brunswick, N.J.; and Z. Sardar, 2000. *Thomas Kuhn and the science wars*. Duxford, England: Icon Books; New York: Totem Books.

2 The growth in electronic sources has not so far displaced books or reference librarians, nor has this been seen by professionals as likely in spite of words to the contrary. See, for example, M. Runkle (then-director of the University of Chicago Libraries) in *University of Chicago Magazine* **77**(2): 19 (1985); and S. C. Sutter (acting assistant director for humanities and social sciences, Joseph Regenstein Library, University of Chicago Libraries) in *ibid.*, **92**(4): 3–5 (2000). Indeed, Sutter notes that the rise in the variety and complexity of on-line (and other) resources has *increased* the need for reference librarians, with four new posts being created in the library system.

Acknowledgments for the first edition

The preparation of both the preliminary version and this present edition of the *Guide to standard floras of the world*, especially the latter, has necessitated the consultation over several years of a great many sources, as noted in the Preface, and furthermore has involved the assistance in various ways of numerous individuals and institutions. These latter must now be acknowledged formally, for without their aid this book could not have appeared in its present form, or indeed at all. The author wishes here to express his deep appreciation to all the support and assistance given him over the nearly two decades required for gestation of the work in its present form.

For the original version (1963–4), the author wishes to record his sincere gratitude to all those in charge of library collections for granting him access to them, particularly the late John Millar, then Chief Curator of Botany at the Field Museum of Natural History, Chicago, and those in charge of the University of Tennessee (Knoxville) libraries. Thanks are also due to the authorities of the Biology Library of The University of Chicago; the Lloyd Library, Cincinnati, Ohio; the Missouri Botanical Garden Library, St Louis; the library of the Department of Botany, Smithsonian Institution; the New York Botanical Garden Library; and the libraries of the Arnold Arboretum and Gray Herbarium of Harvard University, Cambridge, Massachusetts.

Advice and assistance was also given by many individuals, but the author is particularly indebted to the following: E. G. Voss, University Herbarium, University of Michigan, for an introduction to the *Flora Europaea* 'Green Books', L. B. Smith, Washington, for assistance with South American references; and above all to A. J. Sharp and other staff and students at the Department of Botany at the University of Tennessee, Knoxville, for their continuing interest in and support of the project. It was Prof. Sharp who made it possible for the preliminary edition to be reproduced and circulated around the world.

The preparation for and writing of the present version has unfortunately extended over a much longer period (late 1967 to mid-1980), owing to the considerably expanded format, changes in the philosophy of the work, publication difficulties, and the author's many other responsibilities while at Cambridge and in Port Moresby. A major contributing factor to the time span was naturally the decision to annotate, as far as possible, all floristic works included in the *Guide*. This made it necessary to examine personally, or obtain full notes upon, the contents, style, and philosophy of each title, and the author considers himself fortunate to have been able to carry out much of the work of compilation in Europe and especially in London. For completeness of world-wide coverage and for convenience of access and usage, the libraries of the Royal Botanic Gardens, Kew, and the Department of Botany, British Museum (Natural History), are perhaps without peer for research on a work of this kind; and it was, as noted in the Prologue, at these two libraries that the greater part of the materials for the present edition was compiled during 1968–70 and in short intervals in the succeeding ten years. Special thanks are therefore due to R. G. C. Desmond and V. T. H. Parry, successively Librarians at the Royal Botanic Gardens, Kew, and their assistants, and to Miss P. I. Edwards, formerly Botany Librarian, Department of Botany, British Museum (Natural History), and her successors, for their help (and patience!) during my extended visits to their libraries.

A significant amount of compilation was also carried out in 1971 and again in 1979–80 at the library of the Royal Botanic Gardens and National Herbarium of Victoria in Melbourne. This resource is perhaps the most extensive of its kind in Australasia, despite past neglect, and proved of great value at a time when substantial work on the general chapters and area commentaries was necessary but owing to circumstances beyond my control could not be done in Europe or the United States. My thanks are due to the director and staff of that institution, but especially to J. H. Ross, Senior Botanist, and Miss Olwyn Evans, assistant in the library. The help of J. Ashworth, Assistant Secretary, Department of Lands and Environment of Victoria, in resolving an unforeseen crisis over access to the facilities is also hereby acknowledged. During the second period in Melbourne, much use was also made of the Baillieu Library and of the branch library in the Department of Botany in the University of Melbourne,

and the opportunity to make use of these well-endowed resources is much appreciated.

Extensive use was naturally made of the University Library, the Scientific Periodicals Library, and the Libraries of the Department of Botany and of the Botanic Garden in the University of Cambridge whilst the author was in residence as a Research Student from 1967 to 1970. As one of the centers for preparation of *Flora Europaea*, the Department of Botany housed a fine collection of major European floras, ably cared for in the Herbarium by P. D. Sell and (at the time) S. M. Walters. It was under the guidance of Prof. E. J. H. Corner, however, that the varying worth of tropical floras came to be appreciated through research into the large genus *Schefflera* (Araliaceae), a stimulus enriched by subsequent personal experience. This augmented earlier experience at Liverpool in 1964–5 when a study was made of *Cytisus* and its allies (including preparation of an account for *Flora Europaea*) under the direction of Prof. V. H. Heywood.

Other resources substantively utilized include the libraries of the Royal Botanic Gardens, Sydney, and the Commonwealth Scientific and Industrial Research Organization, Black Mountain, Canberra; the library of the Komarov Botanical Institute, Academy of Sciences of the USSR, Leningrad; the library of the Conservatoire et Jardin Botaniques, City of Geneva; the libraries of the Institut für systematische Botanik, Universität Zürich, the Botanische Staatssammlung, München, the Rijksherbarium, Leiden University, and the Botaniska Avdeling, Naturhistoriska Riksmuseet, Stockholm; the library of the New York Botanical Garden; the libraries of the Arnold Arboretum and Gray Herbarium of Harvard University, Cambridge, Massachusetts; and the libraries of the Linnean Society of London and the Commonwealth Forestry Institute, Oxford. Use was also made of the library of the Flora Europaea Secretariat, both in Liverpool and in Reading, and of a number of private collections. The author is much indebted to all those persons in charge of institutional libraries as well as private owners for permission to consult the collections in their care and for their assistance in locating needed references.

As with the earlier version of this work, the author is indebted to all those who freely gave assistance during the various stages of preparation and writing of the present edition. The difficult task of searching out, selecting, and locating the various

European floras and manuals occupied a goodly amount of attention in the early stages; in this connection, particular thanks are due to the Flora Europaea Organization (and especially to S. M. Walters) for arranging to have a draft of the bibliographic text of Division 6 (Europe) typed, mimeographed, and sent from Reading to all regional advisers for comment. To all those who replied, many thanks. Thanks are also due to Prof. V. H. Heywood, now at Reading, for advice on European Floras generally, and especially A. O. Chater, London (formerly Leicester), for assistance over several years (mainly before 1977) in locating and annotating obscure works and for arranging contacts with Soviet botanists.

The very exacting and time-consuming task of selecting titles and preparing text for those sections of the book covering the Soviet Union was considerably eased through the generous assistance of M. E. Kirpicznikov of the Komarov Botanical Institute, Leningrad. Not only did he prepare extracts and sample pages from a goodly number of works scarcely available outside the Soviet Union but he also sent a copy of S. J. Lipschitz' *Literaturnye istočniki po flore SSSR*, mentioned above, by air post to New Guinea immediately upon its publication. Moreover, during my visit to Leningrad in the summer of 1975, he graciously read through the completed manuscript for those portions covering the USSR and made many valuable suggestions. In addition, V. I. Grubov of the same Institute gave advice on his special region, central Asia (i.e., from Tibet to Mongolia). The author is also indebted to Prof. Al. A. Fëdorov, Director of the Institute, for permission to make use of the Institute library as well as the collections for a period of several days following the International Botanical Congress, as well as during the Congress itself.

Other botanists in Europe who gave assistance in various forms and to whom the author is likewise indebted include K. Browicz, Zakład Dendrologii, PAN, Kórnik, Poland (Poland and adjacent countries); H. M. Burdet, Geneva (Corsica and other parts of the Mediterranean, as well as general advice); Prof. E. Hultén, Stockholm (Eurasia in general); L. A. Lauener, Edinburgh (China); Prof. C. G. G. J. van Steenis and other staff members of the Rijksherbarium, Leiden (Malesia and adjacent regions); F. White, Oxford, and Prof. J. Léonard, Brussels (Africa); and especially Prof. F. A. Stafleu, Utrecht, for his general advice, criticism, and support.

Botanists in Asia who rendered assistance include R. I. Patel, Baroda, Gujarat State, India (India); Profs. H. Hara and H. Inoue, Tokyo (works published in Japan; Korea); K.-C. Oh (Korea); Prof. Pham Hoang Hô, Ho Chi Minh City (Indo-China); Stella Thrower and Y. S. Lau, Hong Kong (China, Hong Kong); H. Keng, Singapore (China and other areas); and B. C. Stone, Kuala Lumpur (miscellaneous).

In Australia, advice was received from J. H. Ross, Melbourne (Africa and Australia) and Hj. Eichler and A. Kanis as well as the late Nancy Burbidge, Canberra (Australia, Europe, and in general). Casual comments were advanced by many other colleagues on that continent in the course of a number of visits over the past decade while pursuing this and other, perhaps too ambitious, projects.

Assistance from Africa was received from E. A. C. L. E. Schelpe, Cape Town, and Prof. H. P. van der Schijff, Pretoria (southern Africa). Some pointers regarding the Tethyan side of that continent were suggested by Marie-Thèrese Misset, Oran, Algeria.

As with Europe and the Soviet Union, sifting through the mass of more recent floristic literature on North America was a not inconsiderable task. Special thanks accrue to E. L. Little, Jr., Washington, DC, for his general advice and for information on woody floras and on the Americas in general, and to L. E. Morse, New York, for general advice and for sending a copy, in advance of publication, of the typewritten manuscript of *A guide to selected current literature* by Lawyer and others, mentioned in the Prologue. Other advice was received from A. Cronquist and N. Holmgren, New York; P. F. Stevens, Cambridge, Mass.; and S. G. Shetler and C. R. Gunn, Washington.

Botanists in the United States who gave advice for other parts of the world included L. B. Smith, Washington, DC (Brazil); In-cho Chung, Mansfield, Pa. (Korea); Prof. A. Löve, formerly of Boulder, Colo. (miscellaneous, but especially the Arctic regions); E.H. Walker, formerly of Washington (China); and especially F. R. Fosberg, Washington (general advice and support, and for steering me through the scattered shoals of the insular literature).

During the years the author was resident in England and on subsequent visits, many staff members of both the Royal Botanic Gardens, Kew, and the Department of Botany, British Museum (Natural History), gave freely of their time and knowledge to answer my questions regarding the floristic literature of many different parts of the world, thus enabling a kind of collective picture to be formed. My 1970 sojourn at Kew furthermore coincided with a series of staff briefings related to an internal reorganization of responsibilities in the Herbarium; I am indebted to Prof. J. P. M. Brenan, then Keeper, for copies of the area circulars produced for these briefings.

Special assistance on China and Korea was received from J. Needham, Cambridge, and E. Wu, Harvard-Yenching Institute, Cambridge, Mass. Advice on bibliographic matters was given by W. T. Stearn, London, and by the staffs of the University of Tennessee Press and Cambridge University Press.

Members of the botanical sodality in New Guinea, past and present, as well as professional and personal visitors, assisted in various ways and gave moral support. Among them were J. Croft, J. Dodd, V. Demoulin, Elizabeth Gagné, Lord Alistair Hay, M. Heads, Camilla Huxley, R. J. Johns, I. M. Johnstone, P. Kores, G. Leach, P. van Royen, B. Verdcourt, Estelle van der Watt, and A. Wheeler. Margaret O'Grady, formerly on the staff of the University of Papua New Guinea Library, assisted in tracking down some obscure items, one of which proved not to exist as cited.

The author also owes much to the late J. L. Gressitt, and to his associates H. Sakulas and A. Allison for enabling accommodation to be made available at Wau Ecology Institute, in the mountains south of Lae in Papua New Guinea, for an extended period in 1979. This meant that much of the final revision of this book could be accomplished in comparative comfort. Some chapters were rewritten at their nearby Mt Kaindi branch house, where near-total isolation at 2362 m in a diurnal temperature range of 9 to 23 degrees C acted as strong incentives! All associated with the Institute gave support and encouragement to the work. In Melbourne over a two-month period from December 1979 to February 1980, residence at Graduate House, University of Melbourne, was kindly granted by the Warden, W. E. F. Berry. Several residents, botanists and otherwise, provided conversation and moral support over this period; particular thanks are due to P. Bernhardt and D. Fleming. A further short visit to Melbourne and Canberra was made just before completion of the manuscript, which had been unavoidably delayed; at this time A. Kanis, Hj. Eichler and P. Bernhardt kindly undertook to go through the general chapters. A debt of gratitude is also owed to the University of Papua New Guinea and to its Department of Biology for the grant of six months' study leave to carry out the task of thorough revision of the manuscript which, as noted in the Preface, had originally been completed in 1975 but owing to technical and other difficulties was not published as intended.

Finally, I wish to thank my father, Reuben Frodin, for advice and encouragement at all times and for assistance in locating some obscure references and arranging for notes and copies of sample pages to be sent; to the late L. T. Iglehart, of The University of Tennessee Press, for early financial assistance, much advice and encouragement, and above all patience with the long drawn-out initial period of preparation of this book and sympathy when publication arrangements had to be terminated; to A. Winter and M. Walters at Cambridge University Press for advice, encouragement, and gentle nagging; to D. J. Mabberley, Oxford, for assistance at a critical stage in 1978, at a time when the author also suffered severe losses in an office and herbarium fire; to the Society of the Sigma Xi, USA, for a grant-in-aid in 1970 to enable visits to botanical libraries in Australia and elsewhere; to the Research Committee, the University of Papua New Guinea, for a grant-in-aid towards expenses associated with replication of the manuscript and carriage of two copies by air to the United Kingdom; to A. Butler, Librarian, and his staff in the University of Papua New Guinea Library, for the opportunity to utilize their extensive general bibliographical resources during final corrections to the manuscript in November and December 1981; to Prof. E. J. H. Corner, Cambridge and Great Shelford, for general encouragement over many years; to C. J. Humphries, British Museum (Natural History), London, and to R. Wetherbee in Melbourne, G. J. Leach in Port Moresby, and M. Heads, formerly in Bulolo, for real support during the final stages of the project; and lastly (but no less importantly) to the staff of CUP for a thorough editing of a manuscript written under unconventional circumstances to say the least. Full responsibility for the text, including the onerous task of typing and retyping some 1800 pages of manuscript is, nevertheless, mine and mine alone.

Chapter 3 of the general introduction to this book is based upon the author's essay of the same title which appeared in *Gardens' Bulletin, Singapore* **29**: 239–50 (1976 (1977)). Acknowledgment is hereby made to the Government of Singapore for permission to reuse this material.

The map on p. 20 in Chapter 2 of the General introduction, depicting the relative state of present floristic knowledge for different parts of the world, was kindly supplied by E. J. Jäger, Halle/Saale, German Democratic Republic. It is a revised version of that which appeared in *Progress in Botany*, 38: 317 (1976).

The not inconsiderable task of proofreading, carried out at Port Moresby, was assisted by Nancy Birge, G. J. and Amanda Leach, P. Osborne, and N. V. C. Polunin. Preparation of the indices was much facilitated by the use of 'Profile II', a Radio Shack (Tandy Corporation) proprietary file management package run on a Tandy TRS-80 Model II microcomputer; access to this machine was kindly granted by E. D'Sa and J. C. Renaud of the Mathematics Department, University of Papua New Guinea.

Acknowledgments for the second edition

As with the first edition, I am indebted to many bodies and individuals for their interest and assistance during the preparation of this book.

Firstly, I would like to thank the Royal Botanic Gardens, Kew, its successive Directors, Sir Ghillean Prance and Peter Crane, successive Keepers of the Herbarium, Grenville Lucas and Simon Owens, as well as David Hunt, Keith Ferguson, R. K. Brummitt and Alan Paton for leave to include this project amongst my official duties while on the Herbarium staff from 1993 onwards. I am very grateful for their interest and support. I would also like to thank my many other colleagues at that institution for questions answered, assistance rendered and general discussions over the years, and particularly Sylvia Fitzgerald, John Flanagan and other staff of the Library for their unstinting help in the face of many other demands on their time.

Elsewhere, I would like particularly to thank Malcolm Beasley and his staff (Natural History Museum), Gina Douglas (Linnean Society of London), John Reed (New York Botanical Garden), Connie Wolf (Missouri Botanical Garden), Ruth Schallert (Botany Library, Smithsonian Institution), Bernadette Callery (now Carnegie Museum, Pittsburgh), Judy Warnement and other staff of the botanical libraries at Harvard University, Kees Lut (National Herbarium of the Netherlands, Leiden Branch), Raymond Clarysse (National Botanic Garden, Belgium), Hervé Burdet (Conservatoire et Jardin Botaniques, Geneva, Switzerland), Walter Lack and Norbert Kilian (Botanischer Gärten und Botanisches Museum, Berlin, Germany), Helen Cohn (Royal Botanic Gardens, Melbourne, Australia), and – last but not least – Carol Spawn and her staff at the library of the Academy of Natural Sciences, Philadelphia, as well as staff at the Philadelphia Free Library, where work was carried out on this revision prior to 1993.

Many thanks are also due to Maria Murphy, Cambridge University Press, for her interest and patience during what became a protracted undertaking due to other commitments, and to Erica Schwarz, acting on behalf of the Press, for critically working over the text and putting up with late changes in an inevitably futile attempt to 'fix' what is, after all, a moving frontier.

Finally, I would like to thank my former supervisor, the late E. J. H. Corner, my sister, Joanna Frodin and my father, Reuben Frodin, as well as Bill Baker, Martin Cheek, Sarah Darwin, Aljos Farjon, Mark Griffiths, Chris Humphries, Frances Livingstone-Ra, David Mabberley, Bob Makinson, Rudi Schmid, Sy Sohmer, Camilla Speight, William T. Stearn, and the late B. C. Stone as well as others not specifically named for their moral and other support as well as in some cases for reading and advising on parts of the text. I am most grateful for their consistent support over the years, seeing me through medical and other problems.

Part I

• • •

General introduction

1
• • •

An analytical–synthetic systematic bibliography of 'standard' floras: scope, sources and structure

Primarius noster scopus hic est ad redigendos auctores in ordinem, seu libros botanicos in methodum naturalem, ut tyrones sciant quos libros eligere debeant, auctoresque noscant, qui in hac vel illa scientiae nostrae partae scripserint.

> Linnaeus, *Bibliotheca botanica* (1736).

Die Bibliographie ist in ihrem weiteren Umfange der Codex diplomaticus der Literar-Geschichte, der sicherste Grad- und Höhenmesser der literarischen Kultur und Tätigkeit.

> Ebert, *Allgemeines bibliographisches Lexikon* (1821); quoted from Simon, *Die Bibliographie der Biologie* (1977).

The difficulty in publishing an extended list of floras is to know where to stop.

> Turrill, 'Floras'; in *Vistas in Botany* (ed. Turrill), vol. 4 (1964).

Definition and scope of the work

The aim of the present work, a revised and expanded version of that first published in 1984, is to furnish in bibliographic form a geographically arranged one-volume guide to the most useful nominally complete floras, checklists and related works dealing with the vascular plants of the world.[1] Also included are concise historically oriented reviews of the state of floristic knowledge in different parts of the world, geographical conspectuses, and references to local and general bibliographies and indices. The work attempts as far as possible to account for titles up through 1999 that fall within its scope. The sequence of geographical units is, with slight modifications, that devised for the first edition.

In contrast to *Geographical guide to floras of the world* by Sidney F. Blake and Alice C. Atwood (vol. 1, 1942; vol. 2 by Blake alone, 1961) only one to a few 'standard' works are listed for each recognized geographical unit. With some exceptions, no detailed coverage of florulas and lists of comparatively local scope has been attempted, and only limited attention has been given to works on weeds and poisonous or useful plants. Such limitations have made it possible to cover, in an approximately uniform fashion and within a single volume, a well-tempered selection of floristic works for the student and general reader as well as the specialist. For those interested in more information on any given unit, region or ecological synusia, the work provides references to local, regionally or topically specialized bibliographies, guides and indices. As with Linnaeus's *Bibliotheca botanica* (1736; 2nd edn., 1751), our aim is to furnish not only a bibliography but also an introductory digest.

Sources and the historical background

General

Since the seventeenth century, various world-wide botanical bibliographies and indices have been produced; with the passage of time these have become increasingly specialized, more or less automated, or absorbed into biological information systems. More recently they have been supplemented by numerous local, regional and supraregional bibliographies. The following paragraphs review the most significant of these works, starting with general botanical bibliographies and followed by those specifically relating to floras.[2]

Botanical bibliography effectively began, as did bibliography in general, with the work of the sixteenth-century Swiss natural historian and polymath Conrad Gesner (1516–65). His *Bibliotheca universalis*, a general compendium of some 12000 items in Latin, Greek or Hebrew arranged by authors' forenames, appeared in 1545 as an attempt to bring some order into the rapidly increasing range of literature consequent to the Renaissance and the introduction of printing. A classified index, the *Pandectarum*, followed in 1548 and a supplement, *Appendix bibliothecae C. Gesneri*, with 2000 additional works, in 1555. Further editions of the *Bibliotheca* appeared from time to time after the author's death, the last in the 1720s. In Italy, the Bologna professor of medicine and natural history Ulisses Aldrovandi (1522–1605) essayed a similar work in 12 volumes; unfortunately, this remained unpublished. Gesner himself contributed bibliographical chapters to the Kyber edition of Hieronymus Bock's *De stirpium* (1552) as well as his own edition of Valerius Cordus's *Historia stirpium* et *Sylva* (1561). Caspar Bauhin – whose elder brother Johannes had been a student of Gesner's – continued this tradition of a special bibliographical supplement with the *Recensio* in his *Pinax theatri botanici* (1623).[3] Such supplements (or sections) have ever since remained a feature of serious textbooks; recent examples include Woodland's *Contemporary plant systematics* (1997) and *Plant systematics: a phylogenetic approach* (1999) by Walter Judd *et al*.[4]

With the gradual differentiation of botany as a distinct scientific discipline in the seventeenth century, it is not surprising that at some time there would appear a botanical bibliography. This was first achieved by Ovidio Montalbani (1601–72), like Aldrovandi at Bologna University. His *Biblioteca botanica* (1657, published under the pseudonym of J. A. Bumaldi), a chronologically arranged duodecimo work, covered literature through 1652. With its reissue in 1740 (and again in 1762) as an appendix to Séguier's *Bibliotheca botanica*, it became more widely disseminated.[5] In Switzerland, the Gesnerian tradition was for natural history maintained through the work of his fellow-*Zürcher* Johann Jakob Scheuchzer (1672–1733). Scheuchzer's key published contribution was *Bibliotheca scriptorum historiae naturalis* (1716; reissued 1751), written preliminary to a fuller study of Swiss natural history. Its primary arrangement was therefore geographical; titles were arranged chronologically under authors in each section. As such, it was the first worldwide geographical guide to natural history works – including floras.[6]

It is to Carl Linnaeus that credit must go for the first botanical bibliography arranged by subject: his didactic, somewhat baroque *Bibliotheca botanica* (1736; 2nd edn., 1751). This was first written during his sojourn in Holland and put forward as part of his comprehensive botanical reform campaign.[7] Here, titles were arranged hierarchically into 16 *classes* or chapters – each with one or more *ordines* or sections – based on the author's perception of their contents, as outlined in the brief introduction, and often furnished with sometimes pointed commentary. Principal sources (*historici litterarii*), including the already-mentioned works of Gesner, Montalbani and Scheuchzer, are listed on pp. 2–3. His class VIII, 'Floristae', is in the present context significant: it is in effect a geographically arranged world guide to regional and local floristic literature. Here, country subdivisions became in effect 'genera' and countries 'orders' (with all extra-European works being grouped together in a single 'order', *Extranei*).[8]

That Linnaeus could thus apply his so-called *methodus naturalis* to books – and people – in the same way as fauna and flora was a mark of his 'scholastic' view of the world. As Cain and Stearn have pointed out, Linnaeus's approach, while containing some elements of empiricism, was primarily based upon Aristotelian logic.[9] Later 'universal' systems of knowledge, such as the Dewey Decimal System (DDC) with its common geographical denominators, were, however, seldom adopted in botanical bibliography. Most subsequent classifications of botanical literature, including geographical entities, would be more or less *empirically* based. Such differences in approach not unnaturally reflect the divergent outlooks of specialists and generalists. They also highlight a recurrent conflict among essentialism, empiricism, nominalism and other doctrines in the theory and practice of any kind of classification.[10]

With empirical or more strictly historical principles being considered more desirable, Linnaeus's *methodus naturalis* was accordingly rejected as impractical by other compilers. Among them were the authors of the two other major botanical bibliographies of the mid-eighteenth century: the homonymic *Bibliotheca botanicae* of Jean François Séguier (1740; supplement, 1745; 2nd edn., 1760) and Albrecht von Haller (1771–72; revised index by J. C. Bay, 1908). Linnaeus drew upon the former for the 1751 edition of his own *Bibliotheca*, while in the latter the last of its 10 'books' or primary divisions was named after him. Both were very critical as well as more complete than that of Linnaeus. Séguier adopted but three main subject divisions (botany proper, *materia medica* and agriculture and horticulture), while within his historically based classes from 'Book 1' (the Greeks and Romans) through 'Book 10' von Haller arranged authors chronologically from the date of their first publication.[11] Neither author recognized floras and related works as a separate class.

In the wider world of the natural sciences – corresponding to the three kingdoms of Linnaeus – there appeared two other key works before the final years of the century. These comprised a suite prepared by L. T. Gronovius including the second edition of Séguier's *Bibliotheca botanica* (1760) as well as his own *Bibliotheca regni animalis atque lapidei* (1760) and, a quarter-century later, *Bibliotheca scriptorum historiae naturalis* (1785–89) by G. R. Boehmer. The latter, a relatively massive work of some 65 000 partly annotated titles in five nominal 'volumes' or *Bände*, physically running to eight volumes, was arranged in the first instance by discipline; Bd. 3 (in 2 vols.) covered botany. Bd. 5 includes an expanded table of contents and author indices. As in von Haller's work, the internal arrangement of titles under subheadings was chronological, and – likewise – the lack of a subject index rendered the work difficult to use.[12]

The concept of a didactic subject classification comparable to that adopted in Linnaeus's *Bibliotheca botanica*, but in a more empirical and rational form, nevertheless gained more general currency by the end of the eighteenth century. This is an important feature of Jonas Dryander's *Catalogus bibliothecae historico-*

naturalis Josephi Banks (1796–1800), which accounts for some 25 000 items.[13] The third volume (1798), on botany, includes the first significant listing of floras and related works through and after Linnaeus's time. Although based upon a single book collection, this dry but very scholarly catalogue, though limited to independently published books and papers, was of such a quality and completeness as to be called at the time an *opus aureum*, or 'golden standard'.[14] Though in general lacking deep structure, the approach of the *Catalogus* gives the user a quick impression of the kinds of botanical studies then being undertaken. Floras, arranged geographically but without a hierarchy of areas, encompass classes 126 through 163 over 63 pages.[15]

The Banksian catalogue as a whole marks the beginning of the tradition of monographic subject bibliographies in the natural sciences which, although inevitably becoming more specialized, reached its fullest development in the century after 1815.[16] In spite of its limitation to independent works, it remained a standard reference for the first half of the nineteenth century.[17] It was afterwards for systematic biology largely superseded by *Bibliotheca historico-naturalis* (1846) by Wilhelm Engelmann, *Thesaurus literaturae botanicae* (1847–52; 2nd edn., 1871–77) by George A. Pritzel, and *Bibliographia zoologiae et geologiae* (1848–54) by Louis Agassiz. Of these, only the *Thesaurus* will be further considered here.[18]

The two editions of Pritzel's *Thesaurus*, both highly critical and based as far as possible on personal observations, are with respect to systematic botany the apogee of the broadly based nineteenth-century bibliographic tradition. Both were much praised in their time as well as afterwards.[19] They respectively encompass 11 906 and 10 871 entries, with some classes of works being eliminated for the second edition. While the primary arrangement of titles in the *Thesaurus* is by author, it shows historical sensibility in its chronological arrangement of multiple works by a given writer along with, in many cases, concise biographical notes. As in Dryander's work, each entry is bibliographically fully described. In the classified index, all entries appear in short-title form. In both editions several of the index classes deal with regional and local floristic literature. These, along with the work's quarto format, provide a good visual overview of the state of progress in description and analysis of the world's flora.

The second edition of the *Thesaurus* was soon followed by Benjamin Daydon Jackson's *Guide to the literature of botany* (1881).[20] Although offered as a companion to the *Thesaurus*, it is effectively an independent work. With some 10000 entries organized by empirically derived subject classes, it may be directly compared to the index of the *Thesaurus*; entries are in short-title format and there is no alphabetical author section. A substantial portion (over 180 pages) in Jackson's *Guide* is devoted to geographically arranged classes of regional and local floras, enumerations and lists. The level of geographical subdivision therein, especially for regions outside Europe, is more precise than in Pritzel's work. This arguably acknowledges the rapid development of 'overseas' literature (notably in North America and South Asia).

In neither of these works is there extensive commentary. Annotations are few and for the most part strictly bibliographic, although in the *Thesaurus* brief critical notes do appear here and there. As in the Banksian *Catalogue*, only independently published works are covered. The already significant periodical literature was for the most part bypassed; this was done not only for reasons of economy but also in recognition of the advent (in 1867) of the Royal Society of London's *Catalogue of Scientific Papers*. Pritzel himself acknowledged the latter with volume and page cross-references from each author entry in the *Thesaurus*.[21] To these criteria might be added a not-uncommon contemporary scholarly view that periodical papers were 'ephemeral' or at least precursory compared with monographic works.[22]

The final major monographic botanical bibliography largely to appear before World War I, and – save for the late twentieth-century *Taxonomic Literature-2* – the only real successor to the tradition set by Pritzel and Jackson, is the *Bradley Bibliography* (1911–18) by Alfred Rehder. This is a five-volume guide to literature on woody plants published through 1900 and encompassing 145000 entries. A total of 75000 (more than half) are concerned with dendrology, with a large proportion of them taxonomic. An innovation in the 'Bradley' is the inclusion of papers in serials. In the first volume (Dendrology, I) is a classified list of woody floras and 'tree books'.

All these nineteenth and early twentieth century works combine various traditions of earlier bibliographers but they are also the final more or less general botanical bibliographies.[23] World War I with its attendant disruption and loss of resources as well as changes in fashion and technology led to what has become a permanent fragmentation in the coverage of systematic and related botanical literature. The manyfold expansion in the number of titles alone (let alone potential technical problems) would now render all but impossible the compilation of a full retrospective botanical bibliography. To cope with the increasing volume as well as specialization of the literature – clearly evident by the mid-nineteenth century – three main directions have been pursued: (1) monographic subject or thematic bibliographies, including world guides to floras; (2) national and regional bibliographies, beginning as early as 1831 but most notably after World War II; and (3) periodical surveys of new literature, initially in more general journals but by the mid-nineteenth century in specialized bibliographic journals and, from the 1960s, computerized information retrieval services. To these may be added the catalogues of major libraries, especially those specialized in botany or natural history, as well as alternative professional or commercial outlets. All these are in turn considered in the sections that follow.

World guides to floras

The publication of Pritzel's *Thesaurus* led directly to the first known separate guide to floras of the world, namely George L. Goodale's *The floras of different countries* (1879), originally published by the Harvard University Library in its *Bulletin* and then separately as one of its 'Bibliographical Contributions'. This selective compilation of 12 pages, with about 400 entries, is comparable to the present work in scope although by and large it was limited to independently published works available within Harvard University. The primary arrangement of titles is as in the *Pars systematica* of the *Thesaurus*: geographical and then chronological. The brief annotations are mainly bibliographical. Noteworthy is the omission of the great majority of the smaller local floras, already very numerous in Europe and elsewhere increasing in number, both inside and outside North America. At the end of the list is an appendix entitled 'Botanical Handbooks for Tourists'. In his brief foreword, Goodale indicated that his list was 'simply an attempt to answer questions frequently asked respecting the systematic treatises upon the vegetation of different countries'.[24]

Goodale's list was followed in 1911–14 by a rather more substantial compilation, a mostly unannotated series of contributions by William Holden and Edith Wycoff entitled 'Bibliography relating to the

Floras'. With some 7750 entries, it comprised most of volume 1 of *Bibliographical Contributions from the Lloyd Library*.[25] More than a mere library catalogue, however, the series was an attempt to list all known independently published floras; those actually present in the Library were especially indicated. The work is divided into major geographical units comparable to those in the *Thesaurus* or Jackson's *Guide*; however, within each the arrangement of titles is alphabetical by author. As with Goodale's list, the series was produced in the interest of service to the public. Though seemingly not well known, it remained for long the only substantial guide to floras completely covering the earth, and is still useful for some parts.[26]

As the twentieth century progressed, critical bibliographic scholarship filtered through to more specialized biological fields including vascular plant floristics. In both Europe and North America several key monographic bibliographies were produced.[27] Among these was the next bibliography of floras: *Geographical guide to floras of the world* by Sidney F. Blake and Alice C. Atwood (vol. 1, 1942; vol. 2 by Blake alone, 1961). The first volume, completed by 1940, covers Africa, the Americas, Australasia, and the islands of the Atlantic, Indian and Pacific Oceans; the second volume provides detailed coverage for most of western Europe (save the German states). Based upon a wide range of primary and secondary sources and many years of critical research and experience on the part of its authors, it was in its time the most comprehensive and original contribution of its kind to be published.[28] Unfortunately, the work, left incomplete upon the death of Blake in 1959, does not cover the rest of Europe and the continent of Asia. No official plans were ever made to complete it,[29] although in a posthumous contribution a leading Kew botanist, William B. Turrill, considered this to be a task of high priority.[30]

The arrangement of the *Geographical guide* is fairly simple, with continents and their subdivisions arranged alphabetically in volume 1 and the countries and their administrative subdivisions similarly arranged in volume 2. Coverage extends to local florulas and checklists as well as encompassing the more important larger works and – appropriately to an agricultural research branch – works on applied botany (medicinal and poisonous plants, useful plants, and weeds) are also included. Each primary citation contains extensive bibliographic details and is briefly annotated; associated with these are many secondary

citations (supplements, reviews, related or superseded works, etc.). Like the *Bradley Bibliography* but in contrast to the works of Goodale and of Holden and Wycoff, it features detailed coverage of floristic contributions in periodical and serial literature. Geographical and author indices are also provided. The *Geographical guide*, an *opus aureum* like those of Dryander and Pritzel, was a primary source for the original edition of the present work.

Following publication of the first edition of the present *Guide*, there appeared *Plants in danger: what do we know?* (1986) by S. D. Davis *et al.*, published by the International Union for the Conservation of Nature and Natural Resources (IUCN) with support from the World Wide Fund for Nature (WWF) and its Plant Conservation Programme. Exemplifying the collective approach feasible within an established organization, this work was a response to the needs of the rapidly growing environment and conservation movements and the requirements imposed by the Convention on International Trade in Endangered Species (CITES), promulgated in 1973. Organized by countries, it lists in addition to 'standard floras' other useful works as well as references on threatened plants.[31] *Plants in danger* has been of great value for the revision of this *Guide*.

Other, more or less abridged, lists of floras have appeared in a wide variety of references. Among these are textbooks of systematic botany, notably *Taxonomy of vascular plants* by G. H. M. Lawrence (1951), *Taxonomy of flowering plants* by C. L. Porter (1959; 2nd edn., 1967), *Vascular plant systematics* by A. E. Radford *et al.* (1974), and *Contemporary plant systematics* by D. W. Woodland (1997) (see also Appendix A). There is also a compact list in *Biodiversity assessment: field manual 1* (1996), published by HMSO in the United Kingdom.

Regional and national floristic bibliographies

In addition to the world guides just described, there have been since the mid-nineteenth century many lists of floristic publications with a regional or local scope. These have been published either independently or as parts of more general national and regional botanical (or biological) bibliographies. Only the more salient aspects of this now rather extensive literature will be dealt with here.

The earliest regional bibliography in North America devoted exclusively to floras appears to be *A list of state and local floras of the United States and*

British America by N. L. Britton (1890; in *Annals of the New York Academy of Sciences* 5: 237–300). Its main feature was a geographically arranged listing of 791 works.[32] Partial successors included *State and local floras* (1930; in *Bull. Wild Flower Preserv. Soc.* 1: 1–16) by A. C. Atwood and S. F. Blake and, more fully, the North American section of Blake and Atwood's *Geographical guide*, with coverage through 1939. Canada (along with Alaska, Greenland and Newfoundland) was through 1945 very thoroughly documented in the nine installments of *Bibliography of Canadian plant geography* (1928–51) by J. Adams, M. H. Norwell and H. A. Senn.

Since about 1950, however, continent-wide lists of floras in North America have been limited to the most significant works. Short lists were published by Charles Gunn in 1956 for the United States and by Stanwyn Shetler in 1966 for North America north of Mexico. More substantial was a list by Lawyer *et al.*, announced for *Torreya* in the late 1970s but never published. Popular floras of the United States, including 'wild-flower books', were covered in some detail by Blake in 1954 and later, but less thoroughly, by Elaine Shetler in 1967. United States tree books have similarly been rather fully covered, firstly by Dayton in 1952 and subsequently by Little and Honkala in 1976.

Of more import, particularly in the twentieth century, have been bibliographies for states, provinces, or other more or less limited areas in the continent. A notable pre-1950 contribution was *Bibliography of botany of New York State, 1751–1940* (1942) by then-state botanist Homer D. House. Others were incorporated into floras and enumerations. There have since been numerous – some of them quite substantial – additions to this range; as far as possible they have been accounted for in the present book.

In Europe, national or regional bibliographies or indices have been produced more or less in tandem with the growth of interest in local floristics, beginning as early as 1831 with *Conspectus litteraturae botanicae in Suecicae* by Stockholm professor Johann Wikström but becoming more numerous only after 1860.[33] Now available in one or another form in most countries, they have become a significant source for literature on floristics. There have also been some more general botanical bibliographies, sometimes the work of specialist librarians. Literature has also been cumulated, at least partly, within national floras or enumerations; an example is Erwin Janchen's treatment of seed plants in *Catalogus*

florae austriae (1956–60). Perhaps not surprisingly, the only comprehensive work for nearly a century following Pritzel and Jackson was the second volume of Blake and Atwood's *Geographical guide* (1961). Even then, it does not cover Germany or its predecessors, the rest of Central Europe, the Balkans, or the European part of the former Soviet Union.

The first modern European lists of floras dealing with the whole of that continent did not make their appearance until after the initiation of the *Flora Europaea* project in the 1950s.[34] As with the lists of Gunn and Shetler in North America, these latter were limited to what their authors considered to be the most significant and/or generally useful works, thus obtaining a depth of coverage comparable to that in the present *Guide*. Heywood's list appeared, with successive revisions, in every volume of *Flora Europaea* (1964–80) and in the first volume of its second edition (1993). With respect to individual countries, two sets of listings were published under the aegis of the Flora Europaea Organisation, firstly in 1963 following their second international symposium and again in 1974–75 following the seventh; these were important sources for the present *Guide* (see **Division 6**). Significant floras in Europe – and, less thoroughly, other parts of the Holarctic zone – were listed in a botanical bibliography for Central Europe published (initially in 1970, with a second edition in 1977 but not since revised) to accompany *Illustrierte Flora von Mitteleuropa*.[35] Literature for countries surrounding the Mediterranean was listed in 1975 in *La flore du bassin méditerranéen*.[36]

Biological literature in the former Soviet Union has been the subject of surveys since 1847 but only in 1968–69 were floras, at least in part, separately reviewed. This critical study by M. E. Kirpicznikov, however, never covered more than Russia-in-Europe, Belarus, Moldova and Ukraine as well as the Baltic States. Good coverage can also be had in Lebedev's historico-didactic but selective *Vvedenie v botaničeskuju literaturu SSSR* (1956) as well as in Lipschitz's empirical but more complete *Literaturnye istočniki po flore SSSR* (1975). There are also many national, republican and regional bibliographies. With economic, social, political and technological changes since 1991, new works in that genre have, however, become scarce.

For other parts of the world, there are now a considerable number of botanical bibliographies, many published since 1981. Important supranational works include those by Merrill and Walker for eastern Asia

(1938; supplement by Walker, 1960) and van Steenis for Malesia and adjacent areas (1955), the Field Research Projects' bibliography for southwestern Asia (1953–72), Hultén's excellent source bibliographies (1958, 1971) covering the whole of the north temperate and polar zones, that by Yudkiss and Heller for the *Flora orientalis* arca (1987), and three bibliographies for southern Africa (1988, 1990, 1997). Many national bibliographies have also appeared; some, like those of Langman for Mexico (1964), Kanai for Japan (1994) and Strid for Greece (1996), are extremely detailed. That by Nayar and Giri (1988–) for India is geographically arranged. There are also some brief continental or subcontinental literature surveys; among them are those by Léonard for Africa and the islands of the southwestern Indian Ocean (1965; in *Webbia* 16: 869–876) and Zohary for southwestern Asia and adjacent areas (1966, in the first volume of *Flora palaestina*). With respect to floras, these latter cover 'standard' works and thus, like Heywood's lists for Europe or those in North America, provide a level of coverage comparable to this *Guide*.

The majority of printed bibliographies discussed here are arranged in the first instance by author, the entries sometimes being numbered. Any classification is limited to the indices, which generally are confined to a numerical or author cross-reference. In some cases there may be a limited regional or subject breakdown within the primary listing. Rarely are the indices themselves in short-title form – a recent example being D. M. C. Fourie's *Guide to publications on the southern African flora* (1990) – or even inclusive of keywords (used by Egbert H. Walker among others) which might offer clues. Where cross-referencing is skeletal, subject-related searches may potentially be time-consuming, requiring much copying and page-turning. Far less common are classified bibliographies, which for well-established topics (including taxa and regions) have been much easier to use.

Until relatively recently, all bibliographies and catalogues perforce were published in print (after World War II sometimes also, or only, in microform). Electronic dissemination became possible from the 1960s but, though gradually increasing its penetration, remained relatively limited until the 1980s. With the advent of less costly and more convenient storage media such as the CD-ROM, as well as the introduction of the World Wide Web, such material has begun also – or even exclusively – to appear in electronic form,

with increasingly enhanced searchability.[37] These developments and their consequences will be more fully discussed in Chapters 2 and 3.

Periodical indices and other current awareness services

From the seventeenth century, timely coverage of new literature had been a regular feature of many scientific journals.[38] The first botanical periodical began publication in 1787, and in 1840 a weekly newsletter, *Botanische Zeitung*, was established. Specialized bibliographic journals made their appearance mainly after 1860, although the Swedish Academy published an annual *Öfversigt af botaniska arbeten* from 1825 to 1843/44 (again the work of Wikström) and, in Berlin, the *Archiv für Naturgeschichte* from its foundation in 1837 had included a second, purely bibliographic section.[39] From 1864 through 1871 the well-known German journal *Flora* carried in its *Beiblättern* listings of new literature. In the decade of the 1870s there were founded four serials – all German – which would find wide use in general as well as systematic botany: *Repertorium annuum literature botanicae periodicae* (1873–86), covering literature for 1873 through 1879, *Just's Botanischer Jahresbericht* (established in 1874), *Naturae Novitates* (from 1879), and the relatively timely *Botanisches Centralblatt* (from 1880). From 1902 they were joined by the *International Catalogue for Scientific Literature*, section M: *Botany* (established as one of the coordinated successors to the *Catalogue of Scientific Papers*).[40] In the Americas, the Torrey Botanical Club in 1886 initiated the *Index to American Botanical Literature* as part of their *Bulletin* and, in 1918, a group of interested botanists led by the physiological ecologist B. E. Livingston of Johns Hopkins University founded *Botanical Abstracts* (in 1926 expanded into *Biological Abstracts*).[41] *Biological Abstracts*, and its sister journal *Biological Abstracts/RRM* (as well as, since 1968, the on-line *BIOSIS Previews*), are now (along with *Bibliography of Agriculture* and *CAB Abstracts* and their electronic counterparts) among the leading information sources for new biological literature. These and others are further described and evaluated in Appendix A. However, no botanical counterpart to *Zoological Record* (begun in 1864) was established until the advent of *Kew Record for Taxonomic Literature* in 1971.

As time progressed, however, the continuing and indeed exponential growth of biological literature along with the increasingly lesser percentage accounted for by systematics, floristics and related subjects have resulted in changes which have not necessarily been favorable either to effective coverage in these fields or to easy retrieval. Until the advent of on-line electronic dissemination and indexing in the late 1960s an inevitable failing of abstracting and indexing services was, over time, their relative inflexibility in relation to the kinds of deeply retrospective searches required in systematics or, indeed, any history-dependent or encyclopedic area. Already in the latter part of the nineteenth century, therefore, classified taxonomic-bibliographic card catalogues were established in some botanical institutions.[42] The catastrophes of the two world wars of the twentieth century would also leave their mark. The *International Catalogue of Scientific Literature* network of bureaux was disrupted by World War I and its aftermath and, in spite of efforts at revival, ceased operations in the 1920s – the United States in particular having chosen not to assume a greater share of support.[43] *Botanisches Centralblatt* also became less truly international, its coverage being reduced from 1922 – concomitantly with the rise of *Botanical Abstracts* in the United States. More serious were the effects of World War II, especially the physical destruction and subsequent division of Germany (including in particular the loss of the library of the Berlin Botanical Museum) which put an end to *Botanisches Centralblatt* (renamed *Botanisches Zentralblatt* in the 1930s), *Just's Botanischer Jahresbericht*, and *Naturae Novitates*. Nothing would succeed them until the late 1950s and indeed by then in some respects their time had passed. The institutional card catalogues would also, one by one, cease to grow as costs rose and scientific fashions as well as technologies changed; that in Washington, for example – a major source for Blake's *Geographical guide* – was closed in 1952.[44]

The place of the former journals would eventually be taken by two new works: *Excerpta Botanica*, sectio A, begun in 1959 by Gustav Fischer Verlag (the publishers of the defunct *Zentralblatt*) under an agreement with the International Association for Plant Taxonomy, and *Kew Record of Taxonomic Literature*, which initially absorbed certain regional indices including the *Index to European Taxonomic Literature* (begun in 1965) and *Index to Australasian Taxonomic Literature* (begun in 1968).[45] The former, edited at first from Berlin but later from Kassel and finally Cologne before its termination in 1998, included short summaries for each title, prepared by a network of collaborators. In this fashion it continued the tradition of its Central European predecessors but inevitably there developed a time lag ultimately reaching some 2–3 years. It also to the end remained purely a paper product. The initially annual *Kew Record* became a quarterly in the mid-1980s – at the same time going 'on-line' – and remains timely. It is now the only worldwide indexing serial of its kind in the field.[46]

Apart from these sources, reliance – especially for more up-to-date coverage – has customarily had to be placed upon more general botanical and biological abstracting and indexing journals (and their electronic counterparts), worldwide and regional newsletters with literature lists, booksellers' catalogues, advertising leaflets, and announcements and reviews in professional journals. Summary lists of new floras and related works have appeared from time to time in the annual *Progress in Botany* (formerly *Fortschritte der Botanik*), begun in 1932.[47] Rudolf Schmid as book review editor of *Taxon* since the mid-1980s has created a detailed and well-indexed section for new literature in that journal which carries some of the flavor of the old *Botanisches Zentralblatt*. *Biological Abstracts* along with *Referativnyj Žurnal* (established in 1954) and *Bulletin Signalétique* comprise the main group of more general abstracting and indexing journals useful for systematics and floristics; they focus, however, on journal articles and are not as broad in their coverage as *Excerpta Botanica* (through 1998) or *Kew Record*. By contrast, *Current Contents (Agriculture, Biology, and Environmental Sciences)*, a widely consulted commercial publication begun in 1970, is with respect to systematic botany more useful for developing areas such as molecular systematics, phylogenetic reconstruction and biodiversity analyses rather than floristics.[48] Its emphasis has not unnaturally been on more widely used journals (as measured through citation analysis)[49] as well as more prominent symposium reports. The relative strengths and weaknesses of the various periodical indices are considered along with other general sources in Appendix A.

Various indices have also functioned at national or regional level. In North America, the *Taxonomic Index*, based on the *Index to American Botanical Literature*, was conducted (partly in *Brittonia*) by the American Society of Plant Taxonomists from 1939 through 1967. From 1996, however, it was in effect revived – again in *Brittonia* – with the restriction of the larger *Index* to systematics and related fields. With other changes, it has now become a continent-wide index to floristic literature, and moreover is also (and, from 1999, exclusively) available on-line.[50] Apart from the *Index*, recourse must be had to *Biological Abstracts* (and *BIOSIS Previews*) or *Kew Record for Taxonomic Literature*. In Europe, the country reports prepared for the second *Flora Europaea* symposium gave rise to an interest in ongoing documentation of new literature. Initially this was realized in *Index to European Taxonomic Literature* (1966–71, 1977), covering the years 1965 through 1970; afterwards, coverage was absorbed into *Kew Record*. At a later date came the 'European Floristic, Taxonomic and Biosystematic Documentation System' (more commonly known as the 'European Science Foundation/European Documentation System' or, for short, ESFEDS). This was first proposed in 1977 as a means of continuing the integrative processes in European taxonomic botany set in motion by *Flora Europaea*.[51] Due to technical and conceptual difficulties, however, an initially projected bibliographic module had not been developed by the close of the project in 1987.[52] Current documentation of European botanical literature, where undertaken, is – apart from *Kew Record* (and, through 1998, *Excerpta Botanica*) – presently at national or regional level. In the Russian Federation, indexing of new literature on any scale has since the 1950s been concentrated in *Referativnyj Žurnal*, although *Botaničeskij Žurnal* remains useful for reviews and notices. Elsewhere, recent outlets for continuing documentation have included *Flora Malesiana Bulletin* (1947–), *AETFAT Index* (1952–86, afterwards absorbed into *Kew Record*), and *Bibliografia Brasileira de Botânica* (1957–75).

Progress reports and reviews

In recent decades, the publication of review articles and reports in plant systematics and geography has extended to include reports on the state of floristic knowledge for different parts of the world. This is, in part, related to the growth of the conservation movement as well as to increased general awareness of the tropical biota. Such reports vary considerably in scope and quality, and range from isolated articles to sometimes elaborate surveys covering large areas; more or less extensive bibliographies may be included.

Examples of these reports include the previously mentioned surveys of European and Mediterranean floristics; the reviews of the state of tropical floristic inventory firstly by Prance and later by Prance and Campbell and Campbell and Hammond,[53] the many articles in Verdoorn's *Plants and plant science in Latin America*,[54] and reviews presented at the congresses of AETFAT (Association pour l'Étude Taxonomique de la Flore d'Afrique Tropicale), Flora Malesiana, the Pacific Science Association, the Inter-American Botanical Association, and elsewhere.[55] In recent years, there has also been floristic reporting at International Botanical Congresses.

All these sources collectively constitute a valuable source of information on the progress of floristic research and (where applicable) the institutional background. They are, however, scattered far and wide through the literature and could potentially be overlooked.[56] They have sometimes been intertwined with historical surveys of botanical exploration or biographical sketches.[57] Valuable also are the introductory portions or volumes of many floras and checklists.[58] On the other hand, as Jonsell has warned, the user should take note of the standard of these reviews and surveys; many are not well documented and in addition may be unreliable.[59] It is also important to distinguish levels of floristic documentation from mere botanical inventory, as E. J. Jäger (see below) has done.

The best periodical worldwide surveys of progress in floristics were those produced from 1976 through 1993 by Jäger in the already-mentioned *Fortschritte der Botanik/Progress in Botany*.[60] The initial survey included a world map depicting floristic progress based upon four criteria.[61] A revised version of this map was presented as Map II in the original edition of this book and, in the absence of a successor, is reproduced here (as Map I). Much progress has since been made in hitherto imperfectly known parts of the Americas, Asia, Malesia and Australia, but in others advance has been slower and in some polities civil disturbances and other factors have all but prevented field and other studies. Prolonged economic recession, slow development, and a relative reduction generally in public funds have also limited progress. Nevertheless, the many additional floras and related works published since 1980 have certainly, if nothing else, helped towards the construction of improved world species richness maps.[62]

Major library catalogues

A final – and by no means inconsequential – major source of floristic references are printed library

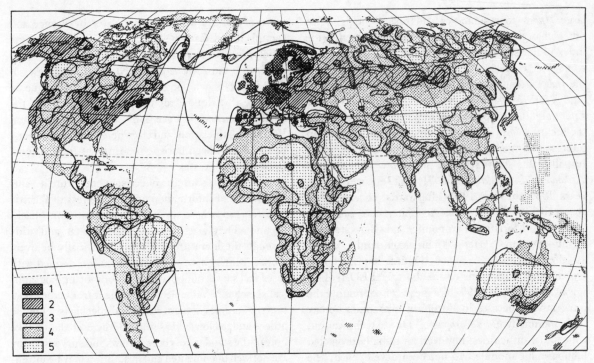

Map I. Five-grade map of the approximate state of world floristic knowledge as of 1979. Based upon (1) quantity, quality, age and completeness of floras, (2) collecting density, (3) an estimation of the percentage of undescribed and/or unreported species, and (4) status of distribution mapping. [From E. J. Jäger in *Progress in Botany* **38**: 317 (1976); revised by him for the first edition of this *Guide*. No subsequent version has been published.]

catalogues (and their on-line successors). That of floras issued by the Lloyd Library in 1911–14 has already been discussed. Other principal printed catalogues from before 1950 include those of the Royal Botanic Gardens, Kew (1899; supplement, 1919), the present Natural History Museum, London (1903–15; supplement, 1922–40), and the Arnold Arboretum of Harvard University (1914–17; supplement, 1933). In the third quarter of the twentieth century the Boston (Mass.) firm G. K. Hall produced numerous catalogues in book form reproduced from library cards; among those covered was the Kew Library (1974). A catalogue, with supplement, of the U.S. National Agricultural Library through 1970 was published in 1967–73 by a New York (later Totowa, N.J.) firm, Rowman and Littlefield. In that decade and the next, however, the application of computer-based information technology in libraries – already initiated for production purposes in the 1950s – began to spread widely. Since then, major developments have included the rise of network services such as OCLC and RLN and on-line access to individual catalogues – including most of those referred to above – via Telnet or the World Wide Web. Further details appear in Appendix A.

Plan and philosophy of the present work
Definition of a 'standard' flora

For the purposes of this *Guide*, a 'standard' flora (or corresponding manual, manual-key, enumeration, or list) is considered to be a current scientific work which yields the maximum information about the vascular plants of a given geographical unit within parameters set by the nature and style of the work and available resources. It thus saves the enquirer an extensive (and often time-consuming) search in the more detailed (and usually very scattered) taxonomic and floristic literature. Put in another way, standard floras are generally those which one turns to first for information about the plants of a given region, state or country; in many instances they may be the only ones consulted, as they are likely to suffice for the query in hand. They represent among floristic literature an optimum ratio of information to effort. Ideally a 'standard' flora should contain descriptions, keys for identification, and supporting documentation, but often only an enumeration or checklist is available for a given area. Further evaluations of the different kinds of floristic writing appear in Chapters 2 and 3.

The concept of a standard flora as expressed herein is by no means original. Its initial formulation appears to have been by Vernon Heywood in his 1957 report on the organization of the *Flora Europaea* project.[63] His suggestion therein was that a list of about 100 titles had to be considered in obtaining a general overview on any given European taxonomic or floristic problem. The value of the concept was shortly afterwards reiterated by Thomas G. Tutin in his foreword to the *Flora Europaea* 'Green Book' of 1958: 'It is our belief that the list of Standard Floras . . . will be generally welcome. These floras, as far as we can ascertain, are the ones most generally acknowledged by botanists in the countries concerned'.[64] Although originally developed in a European context, the author believes the standard flora concept to be, with variations, applicable worldwide.[65] Indeed, a satisfactory paraphrase of Tutin's words might read as follows: 'Standard floras, as far as can be ascertained, are the ones most generally acknowledged by botanists in, or working on, the countries or other regions concerned'. As indicated in the previous section, the concept was reflected directly or indirectly in various continental and subcontinental lists of floras published in the 1950s and 1960s and moreover has passed into other languages.[66]

'Standard' floras contrast with, but should relate to, works which are less geographically comprehensive, such as county or provincial floras or checklists. These latter normally deal only with areas of relatively limited extent and are, comparatively speaking, of more interest to specialists on local floristics, local amateurs, and persons engaged on detailed monographic, revisionary or chorological work. They should also as far as possible include references to taxonomic monographs and revisions and other key contributions. For some parts of the world – above all Europe – the regional and systematic literature is very large indeed; as already related, there is room for improvement in the ease of extraction of desired information.

Selection and coverage of standard floras
The preparation of a comprehensive list of standard floras, no matter what definitions or guidelines are

available or may be evolved, necessarily entails a difficult process of evaluation and selection. It is also essential that a reasonably uniform standard of coverage be adhered to throughout the bibliography. The nature, quantity and quality of the corpus of regional literature, however, varies greatly from one part of the world to another. Many tropical areas, such as the island of New Guinea, have no general floras or enumerations of relatively recent date and the student or non-specialist is faced with an ill-digested mass of florulas, expedition reports, and scattered 'contributions', revisions, notes, and the occasional monograph of varying scope. By contrast, the bulk of Europe is covered for the most part by a plethora of local, national and regional floras and lists of varying dates from which it was necessary to make a careful and limited choice. These areas and others have also become blanketed with more or less widely used 'popular' works.

Fortunately, the exacting tasks of selection and establishment of an approximately uniform standard of coverage were for the 1984 edition greatly facilitated by the existence of some useful guidelines. These were (1) the regional lists of floras already referred to (including the 'Green Books' and the lists of Shetler, Lawyer, Léonard, van Steenis, and Zohary); (2) the selected lists in the standard textbooks referred to on p. 7; and (3) two lists of works considered to be of 'greatest general utility' in Blake and Atwood's *Geographical guide*.[67] Other reference points have included a series of unpublished memoranda on various regions prepared in 1970 for internal use in the Kew Herbarium as part of a major reorganization;[68] a 1979 list prepared at Geneva for the projected 'Med-Checklist'; published 'state of knowledge' reports for a wide variety of countries and geographical areas; and verbal and written advice from a number of specialists and others with local knowledge. Similar surveys and sources have been consulted for the present edition.

The *Guide* is modeled on Blake and Atwood's *Geographical guide* but features historically oriented unit prologues along with more detailed commentary. As far as possible, every primary entry in this book has been provided with an annotation describing its style and contents. These have been as far as possible based upon personal examination of the works concerned. For those not seen, my annotations have been based on notes and/or extracts supplied by correspondents, who have been acknowledged in the text, or published or circulated secondary sources. Any material not seen at first hand has been so indicated. Subsidiary and historical titles – i.e., those not given separate entries – appear in the unit prologues unless they are direct extensions of or closely related to a primary work.

Some works covering only parts of basic geographical units as delineated in this work have been included. Such works are seen as bridging gaps left by the absence, relative antiquity, or inadequacy of a general work or works. They may also be of an exceptionally high standard or of acknowledged value well beyond their nominal circumscription.[69] Amelioration of the limitations on coverage has also been applied with respect to sets of 'contributions' and/or expedition reports covering imperfectly known areas where these appear to be of exceptional importance or are otherwise often routinely consulted.

Provision has also been made for certain kinds of ancillary works. Atlases of illustrations, if of major importance, have usually been accorded the status of primary entries, unless they are clearly companions to descriptive works. Separate subheadings have been set aside under a given unit heading if there are separate keys to families (and genera) and/or dictionaries, but in practice this has been done only at regional level and above. The same has been done with atlases of distribution maps and like chorological works, save for a few such as *Pacific plant areas* (given under **001** as they are not readily referable elsewhere).

Under unit headings, any 'local' or 'partial' work deemed important enough for inclusion has been treated as a 'secondary' work and its citation and commentary appear in smaller type, usually following a subheading. The same procedure has been adopted with respect to works on the woody flora (including 'tree books'), the ferns and fern-allies, and (in a very few cases) the grasses, groups also accounted for in the *Guide* due to general interest or where these groups are not well accounted for in available floras.

Schedule of geographical entities

The arrangement of titles is, as already noted, geographically systematic in accordance with a three-tier hierarchical decimal scheme devised especially for the original edition of this book. Development of this scheme was begun in the belief that existing special schedules in standard library classification schemes or other, more specialized works – though sometimes with a wealth of detail – were obsolete or not particularly suited to the material in hand.[70] Moreover, many existing schedules were largely rooted in nineteenth-century 'Eurocentric' notions of history and geography, past and present. A new scheme was also seen as useful not only for floras but, by extension, for any geographically oriented systematic biological (and earth sciences) literature.

The possibility that universal geographical schemes as used in major library classifications were unworkable appears first to have been raised by de Grolier in 1953.[71] With respect to history and geography, de Grolier argued that a schedule suitable for physical geography would not suit economic geography, and even less would it suit history (upon which most general schemes had been based). Likewise, following de Grolier, it is argued here that the regional literature of botany (and zoology) is more closely related to that of physical and 'political' geography (and geology) than to history or economic geography. However, apart from two recent proposals discussed below, no geopolitical scheme rooted in the biological or earth sciences regional literature and at the same time potentially compatible with one or more of the existing widely used classifications (particularly the Universal Decimal Classification or UDC, which formally allows for specialized schedules) has been seen.[72]

The first of these proposals, published some time prior to the 1984 edition of this book, was – as will be further noted below – S. W. Gould's *Geo-code*.[73] Purely geographical, it was based on latitudinally and longitudinally founded sectors similar to those used for the 1:1 000 000 *Map of the World* and related products. Such a rigid structuring, however, negated any sense of geographical continuity as well as any relationship to existing (and likely) publication patterns; its adoption for the present book was impossible. The second scheme is that of the Taxonomic Databases Working Group, first published in 1992 under the authorship of S. Hollis and R. K. Brummitt as *World geographical scheme for recording plant distributions*.[74] Its basic hierarchy is similar to that in the UDC and the present book but lacks a first-level 'zero' element (corresponding to our 'World floras, isolated oceanic islands, and polar regions'). In addition, for its third level it uses more or less mnemonic triplets of letters in place of a single digit.[75] Its geographical progression at the first and second levels is 'Eurocentric'; such a methodology requires major sequential 'retracings' and moreover fragments the temperate parts of the Southern Hemisphere. It is also wholly politically based, being, as its title suggests, primarily intended for precision in recording the sovereign geographical distribution of biota.

In summary, what best suited this work was a representative and uniform geographical schedule suitable in the first instance for floristic (and, by extension, faunistic) literature. It was early evident that the structural pattern – or what is known in librarianship as the 'literary warrant' – of existing (and expected) floristic literature was such that it could be grouped into successive hierarchical arrays, thus enabling construction of a 'decimal' system in form resembling the UDC.[76] In comparison with those systems, however, our actual geographical arrangement of divisions, regions and polities is quite different. In constructing a necessarily linear schedule of geographical units, primary concerns have been logic, practicality, mnemonic value, and physical and biogeographical relationships.[77]

Common auxiliaries

A purely geographic schedule is, however, not enough for current floristic literature. It is also necessary to formulate an adequate classification of physiographic and synusial isolates such as alpine zones and wetlands. Many key floras meeting our criteria as 'standard' already existed for these isolates by 1981; more have appeared since. At the time of writing of the 1984 edition, no logical schedules or sets of common auxiliaries suited to floristics and faunistics appeared to exist.[78] Following a first empirical attempt at listing works not conveniently included in a geopolitical unit, a system of nine common auxiliaries based upon those used for the UDC was developed.[79] As revised for the present edition, it features the following structure:

- –01 Vague areas (e.g., Patagonia, tropical Africa)
- –02 Major uplands or highlands (e.g., the Guayana Highland, the Ural)
- –03 Alpine and upper montane areas (e.g., the Andes, the Alps, the Pamir)
- –04 Ectopotrophic areas (e.g., serpentine and limestone formations)
- –05 Steppes and deserts (e.g., the Sahara, the Gobi, the North American Great Plains)
- –06 Rivers and riverbanks
- –07 Great lakes and their littoral (e.g., Lake Baikal, Victoria Nyanza, the Great Lakes of North America)
- –08 Wetlands
- –09 Oceans and the oceanic littoral; islands

The nine auxiliaries are in theory definable in all 10 divisions of the *Guide*'s geographical system; in practice they do not appear unless there are appropriate works to be covered.

Usage of these auxiliaries has been comparatively sparing, save for –03 and –08. For these two the opportunity has been taken to refer to them all (or most) such works covered in the *Guide*, even where their geographical compass fell wholly within one third-level polity (as in *Rocky Mountain flora* (**103**) and *Alpine flora of New Guinea* (**903**)). Wetland floras of subregional level or below have, however, largely been omitted. Auxiliary –09 in particular has the potential for coverage of marine and littoral non-vascular as well as vascular taxa.

The system hierarchy

The highest category in the system adopted here is the *division*. These are numbered from **0** through **9**; general floristic works with a division-wide coverage are designated by the numbers **100, 200**, etc., up to **900**. The category below is the *region*. These are numbered from **01** through **99**, according to the division into which they fall (**00** being used notionally for worldwide floras, world synusial works (such as *Rheophytes of the world* by C. G. G. J. van Steenis, here under **006**), and (under **001**) certain major chorological works such as *The amphi-Atlantic plants* by E. Hultén). Some regions are grouped together into *superregions*, with separate principal headings; these are designated by hyphenated figures, such as **14–19, 42–45**, or **91–93**, indicative of the regions they encompass. Very large single regions comprising more than nine units (among them the northeastern U.S.A., Brazil, and eastern Europe) are designated by a stroke between two figures, such as **14/15, 35/36**, or **68/69**. Individual regional floras, enumerations, etc., are always given a three-digit number ending in a single zero, viz. **160, 220, 560, 830**, or **990**, except that floras of superregions, such as *Flora orientalis* or *Index florae sinensis*, are designated by 'inclusive' unit numbers such as **770–90, 910–30**, etc.

The lowest category – the 'species' of the system – is the *unit*. These are designated by figures running from **001** through **999** (excluding those ending in a zero). Units as recognized here generally correspond to geographical areas such as states, countries of small or medium size, large provinces, or significant islands or island groups. It is for these that the bulk of 'standard' floras have been written. By contrast, *regions* comprise large countries (or natural groups of smaller countries or states) or comparable areas of large size; while *divisions* consist of continents, parts of continents, giant aggregates of islands, or combinations of these. No category has been devised for the relatively small number of local or partial floras included in the *Guide*; they are set off from principal works by subheadings.

Examples of divisions are North America, Europe, or Greater Malesia and Oceania. The polar zones beyond the 'tree-lines' of north and south, together with some isolated oceanic islands, have been allocated to Division 0. Representative superregions include the West Indies, South Asia, Greater Malesia, and Australia (with Tasmania). Areas such as the southeastern United States, Argentina, South Central Africa, Madagascar, Western Australia, Central Europe, the British Isles, the Russian Far East, Southeast Asia, Papuasia, and the Hawaiian Islands constitute regions. At the unit level are areas such as Macquarie Island, St. Helena, Alberta (Canada), New York State (U.S.A.), Puerto Rico, Mato Grosso (Brazil), Buenos Aires Province (Argentina), South Australia, Mauritius, KwaZulu/Natal (South Africa), Nigeria, France, Finland, Ukraine, Sakha, Iraq, Uttar Pradesh (India), Nepal, Korea, Sichuan Province (China), Java, the Solomon Islands, and the Marquesas.

Physiographically, ecologically or synusially defined standard floras, or those covering broad but vague geographical areas, are classified according to the 'common auxiliaries' introduced under the previous subheading. The resulting three-digit numbers feature a *middle* zero, e.g., 201, 703. Examples of the areas covered are the Sonoran Desert, the Andes, the Afroalpine zone, and the Altai and Sayan Mountains. In general, this class comprises areas which are too awkward to fit into geopolitical regions, or which otherwise deserve special emphasis. As already noted, under these auxiliaries are included *all* appropriate works for a given division; thus, *Alpenfloren* should not be sought for under a country or region, but under x03 where x is any number from 0 through 9.

The 10 primary divisions are all listed in the table of contents, but for ready reference are repeated below:

Division 0: World floras, isolated oceanic islands and polar regions
Division 1: North America (north of Mexico)
Division 2: Middle America
Division 3: South America
Division 4: Australasia and islands of the southwest Indian Ocean (Malagassia)[80]

Division 5: Africa
Division 6: Europe
Division 7: Northern, central and southwestern (extra-monsoonal) Asia
Division 8: Southern, eastern and southeastern (monsoonal) Asia
Division 9: Greater Malesia and Oceania

The full classification scheme for each division appears as a conspectus under the respective main heading. The spread and limits of the primary divisions are depicted in Map II.

Bibliographies and indices

A special feature of this *Guide* is the systematic inclusion of references to more detailed local, regional, and general botanical and floristic bibliographies. Anyone seeking more detailed information on any given area will thus learn where to turn. These references are included under their appropriate headings. For general bibliographies (such as those of Blake and Atwood, Hultén, or Jackson) and indices (such as *Excerpta Botanica* or *Kew Record*), abbreviated references or mnemonic devices appear throughout the text at divisional and regional levels; full citations of these works are given in the **General bibliographies** and **General indices** lists located under **Conventions and abbreviations** at the beginning of Part II, the *Guide* proper.

Under the appropriate headings are also included references to reviews of the state of floristic knowledge for given major geographical entities; no attempt is made, however, at exhaustive coverage of such literature.

Limitations

In order to make this *Guide* as compact and practical as possible, various limitations have been imposed. These are:

1. The *Guide* is limited to works covering **vascular plants**, either exclusively or as part of their total scope. Extension of coverage to non-vascular plants and fungi would have unduly increased the size of the work. There is, however, certainly scope for similar guides to these groups.

2. **Superseded** floras or enumerations are covered only in regional or unit introductions or, in some cases, as subsidiary titles. This is part of an attempt to place current listings in a historical perspective. Such works generally appear only in

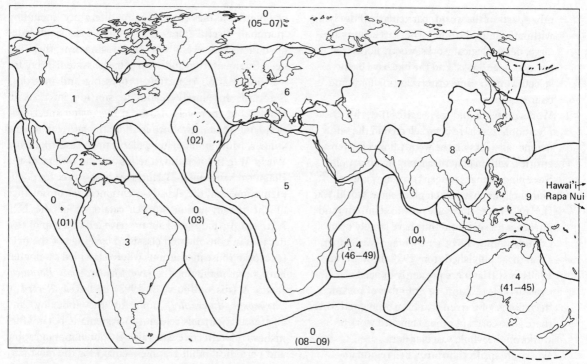

Map II. The spread of Divisions 0–9 as used in this book. For explanation see text. (Antarctica has for technical reasons not been depicted.)

short-title form; fuller details may be had elsewhere, including the sources listed in Appendix A.

3. With but few exceptions, no literature dating from **before 1840** appears as primary entries. As discussed in Chapter 2, only from about this time did the format of descriptive floras begin consistently to be recognizably 'modern' (as exemplified by W. J. Hooker's *Flora boreali-americana* (1829–40), *Flora brasiliensis* (begun in 1840), Torrey's *Flora of the state of New-York* (1843), J. D. Hooker's *Flora antarctica* (1843–47), and Grenier and Godron's *Flore de France* (1848–56)). The year 1840 moreover marks, with rare exceptions, the demise in floras of the Linnaean system of classification.[81]

4. No purely **popular** works are included, nor does coverage extend to lexica and other works on vernacular names. To do so again would greatly increase the bulk of the *Guide*. In recent decades, however, the distinction between 'scientific' and 'popular' floras has become less clear. Exceptions have consequently had to be made, especially for

areas for which no good recent standard floras exist. The European Alps furnish a good example of a compromise.[82] In addition, many more or less popular works on trees (and woody plants in general) have also been included as explained under §5 below.

5. With regard to works dealing only with **trees** (or **woody plants**), coverage varies according to the importance of these life-forms in the total vascular flora. Speaking generally with respect to trees alone, within the largely Holarctic divisions 1, 6 and 7 only works which cover areas the size of regions or larger have been fully listed. Wherever the whole woody flora is accounted for, however, works covering smaller units are included. In addition, where the shrub flora is substantial (as, for example, in California), separate works on this synusia are also listed and described. Many works dealing with the woody flora (or the trees) in Europe and northern Asia also include a substantial number of introduced park and garden trees, reflecting a long interest in dendrology and landscape improvement. For

[17]

other parts of the world, particularly those lying within the humid tropics where tree floras are large, dendrological works, woody floras, semi-popular 'tree books' and the like have been selected on the same criteria as full floras and enumerations.

6. Works on **ferns** and **fern-allies** (i.e., the pteridophytes) have been selected in the same manner as works on the woody flora of given entities, with in general a later 'starting-point'. Sweeping changes to fern taxonomy and nomenclature have taken place since World War II.[83] Older fern floras are now to all intents and purposes obsolete and thus have largely been excluded unless no other coverage is available. Even those published from 1939 through the 1960s or still later are presently in need of considerable revision. In a number of instances, 'fern floras' of a given area have been cited where there is no corresponding standard work or works on the whole vascular flora.

7. Works on **applied botany**, i.e., regional treatises on economic, medicinal or poisonous plants and on weeds have generally been omitted. It is the author's belief that, important though many of these works are, they should not come within the scope of a basic guide to floras. Moreover, as with other classes of regional works referred to above, their inclusion would greatly increase the size of this work. There is, however, scope for a separate topical guide along similar lines to the present work.

8. With few exceptions, no works covering **single families** of seed plants are included. It should be noted, though, that for the Poaceae, Fabaceae and Orchidaceae (and for some other groups such as Cactaceae in the New World and Dipterocarpaceae in Malesia) a more or less extensive canon of regional works exists, which might merit the preparation of separate bibliographies.[84]

Summary remarks

During the preparation of the original edition of this work, the author sometimes was asked to defend the preparation of a selective rather than a comprehensive treatment. In response to this question, two major points should be considered.

Firstly, it seems evident that as in all other fields

of botany the mass of taxonomic literature, including 'nominally useful' floras, has within the last six decades or so increased severalfold. At the same time, there has been fragmentation and change in the system of botanical information reporting, processing and indexing. Some of this surely relates to shifting interests in biology but there has also been increasing specialization and regionalization in floristic and taxonomic studies. More immediately, disruptions resulting from World War II (including the loss of the library of the Botanical Museum in Berlin, a leading source for documentation) and the already-mentioned discontinuance of the botany union subject catalogue of the U.S. National Agricultural Library have led to a gap of two decades in consolidated classified coverage of the field (except in those institutions where classified catalogues have been maintained). Save for *Excerpta Botanica*, sectio A (discontinued in 1998) and *Kew Record of Taxonomic Literature*, none of the indexing and abstracting journals relevant to systematic botany furnishes truly effective coverage. Regional monographic and periodical bibliographies remain for the most part only in print mode. Retrospective coverage on the scale necessary for a renewal of comprehensive coverage of floristic or revisionary literature would require substantial institutional support, financing and personnel, and could sensibly be realized at but few locations.[85] It was thus unfortunate – but perhaps understandable – that the appropriate authorities made no provision for completion of the *Geographical guide* after Blake's death.[86]

The second point, less obvious but perhaps more important, revolves around the *need* or *desire* for such a work, especially when measured against the mechanics involved. With increasing specialization and changing interests and methodologies, there is a necessity from time to time to review the *scope* and *style* of publications, including reference works, with regard to function and efficiency. This has been done for floras: in the 1960s and 1970s by Aymonin, Heywood, and the author,[87] in the 1980s by Heywood and by Morin *et al.*,[88] and in the 1990s by Jarvie and van Welzen, Palmer *et al.*, and Schmid.[89] Some of these writers, in particular Jarvie and van Welzen, believe that floristic works serve two or more functions; what is required are publications of differing scope rather than all-purpose works. Similarly, in bibliographical compilation and writing in the field of botany it has long been evident that functional differentiation is necessary.[90] Apart from the sheer volume of literature to be assessed, much of the material that would perforce be included through simple extensions of the older general works is likely to be of relatively local or specialized interest. Thus, a single, comprehensive work covering floras of the world as conceived by Blake and Atwood – while perhaps still conceptually valid as a statement of knowledge – is very likely not now satisfactory or even desirable as a methodological, let alone practical, solution.[91]

Given these limiting factors, in the late 1960s there still seemed clearly to be a need for a convenient general-interest guide to floras in a single volume. Application of the 'standard flora' concept first suggested by Heywood and the development of relatively objective criteria for inclusion, along with the provision of pointers to more extensive source bibliographies and indices, allowed for the separation of the functions of comprehensiveness and general utility. This is not dissimilar to post-World War I directions in information handling as described by Malclès[92] and, in taxonomy, to the distinction between 'general-purpose' and 'special-purpose' classifications strongly advocated by Gilmour in the mid-twentieth century.[93] Such a distinction is also *a posteriori* a measure of the principle of parsimony[94] and moreover is broadly congruent with the bibliometric Bradford 'law' (actually an axiom) of 'scatter' and its inverse, Garfield's 'law' of 'concentration'.[95] Quantitative testing of patterns of usage in floristic literature by recognized procedures is a task which remains;[96] it is, however, likely that these will merely confirm the perceived pattern of usage and its broad conformity with the above-mentioned bibliometric 'laws', already demonstrated in many different contexts.[97]

The final result as originally presented had a number of advantages. With a more limited scope than the comprehensive treatment customarily considered as ideal in systematic botany, the use of 'pointers' to detailed sources, and with the formal listings supplemented by historical and other commentary related to the genesis of the standard works selected, it has been possible to fashion this *Guide* as a kind of analytico-synthetic systematic bibliography. It thus recalls the bibliographic styles of Linnaeus and von Haller in being more communicative than a purely 'empirical' work and thus more 'open' to the student and non-specialist – the 'tyrones' of Linnaeus's *Bibliotheca botanica*. Rather than a mere list of books, perhaps the

Guide could serve a *codex diplomaticus* as advocated by F. A. Ebert in the first volume of his *Allgemeines bibliographisches Lexikon* (1821).[98] The value of critical selectivity has been well demonstrated in other fields, as, for example, in the studies of Leonard Webb and others on rain forest vegetation.[99] Where the means exist, quantitative procedures, including the use of information technology, can (and should) be used in support of the overall study, but never so mindlessly that they dominate the final form and thrust of the work.[100]

A similar philosophy has guided preparation of the present edition with additional features being increased attention to the historical setting of current literature as well as a somewhat deeper coverage of national and regional bibliographies and dendrological manuals. Many items accorded full entries in the 1984 edition have been superseded and are therefore treated as historical. The sheer increase over the last two decades in the number of current works meeting the original criteria has, however, brought about a considerable expansion of the work. In addition, the author has thought it necessary to expand somewhat on the history of floras in general; this now forms the subject of the next chapter.

Notes

1 Both works are successors to a preliminary version (Frodin, 1964).
2 For a chronological sequence of major biological bibliographies, see table 17 in Simon, 1977, pp. 185–187. They are also listed alphabetically therein (pp. 12–23).
3 For Gesner, see Wellisch, 1984, and Heller, 1983 [originally publ. 1970], p. 171; for Aldrovandi, see Simon, 1977, pp. 28–30; for Bauhin, see Heller, 1983 [originally publ. 1970], p. 171. Simon makes reference to Aldrovandi's contributions to bibliographical scholarship in general, and notes that his *Bibliothecarum thesaurus* of 1583 remains extant in the Bologna University Library.
4 Woodland, 1997; Judd *et al.*, 1999.
5 For Montalbani, see Heller, 1983 [originally publ. 1970], pp. 171–172, and Simon, 1977, p. 30. Simon suggests that Montalbani may well have drawn upon Aldrovandi's work. Although the *Bibliotheca* was said by Linnaeus in his own *Bibliotheca botanica* to be very rare – he himself had not seen it – Ewan (1970) has recorded that the late seventeenth-century English priest, explorer and natural historian John Banister possessed a copy. Ewan

further notes that it was subsequently acquired by the Virginian planter William Byrd II, whose library was before 1750 one of the two or three most important collections in North America.

6 For Scheuchzer, see Simon, 1977, pp. 30–35. Scheuchzer also prepared more detailed bio-bibliographies in botany and zoology; these were never published but remain in the Zürich *Zentralbibliothek*.

7 Stearn, 1957.

8 For commentaries, see Heller, 1983 [originally publ. 1970], pp. 146–204, and Simon, 1977, pp. 36–39.

9 Cain, 1958; Stearn, 1959.

10 Davis and Heywood, 1963, p. 18; Ghiselin, 1997.

11 As already noted, Séguier included Montalbani's catalogue as an appendix to his main work.

12 Simon, 1977, pp. 43–44.

13 Besterman, 1965–66. The Banks Library was willed to the British Museum; it is now part of the British Library, London. The *Catalogus* was reissued in 1966 by Johnson (as *Sources of science* 22).

14 Heller, 1983 [originally publ. 1970], p. 202; from F. J. Cole, *A history of comparative anatomy* (1944, London). The historian of botany E. Meyer would in 1849 call the whole work 'ein Muster bibliographischer Genauigkeit' (*Bot. Zeit.* (Berlin) 7: 290–292); Heller himself regarded it as stylistically and intellectually a great advance on Linnaeus's *Bibliotheca botanica*.

15 The *Catalogus* as a whole is more fully described by Heller, 1983 [originally publ. 1970], pp. 201–202.

16 Simon, 1977, pp. 44–45, 184, 186–187.

17 Periodical literature to 1800 was covered in *Repertorium commentationum a societatibus litterariis editarum* (1801–02, in 2 vols.) by J. D. Reuss, with botany in vol. 2. Its successor was the Royal Society *Catalogue of Scientific Papers* (1867–1925).

18 Engelmann's *Bibliotheca historico-naturalis* was originally intended to comprise three volumes, with the second and third devoted respectively to botany and geology; these latter, however, were never published. Its two successors were exclusively zoological.

19 For a modern commentary, see Stafleu, 1973. Pritzel, trained as a botanist, was librarian of the Prussian State Library, Berlin. Completion of the second edition had to be supervised by his associate K. F. W. Jessen (author of *Botanik der Gegenwart und Vorzeit* (1864), an important and culturally oriented history of botany) on account of Pritzel's debilitating illness and (in 1874) death.

20 Jackson was for many years librarian of the Linnean Society of London. He was also managing editor of the original *Index Kewensis* (1893–95) and of its first supplement (1901–06).

21 The *Catalogue* is more fully discussed under **General indices** in Appendix A. It was fortunate for Pritzel that, with the substantial growth of serial literature, this critical reference had come into being.

22 Cf. Malclès, 1961. In today's scholarly world, monographs, especially by a single author, are comparatively rare.

23 A successor to the *Thesaurus*, to cover the period from 1870 through 1899, was planned by J. Christiaan Bay, in the early twentieth century librarian of the John Crerar Library, Chicago, Ill., U.S.A. (now part of the University of Chicago Libraries). However, all that he published was a list of bibliographies (1909; see **General bibliographies** in Appendix A).

24 Although largely derivative, Goodale's little bibliography was an early example of the life-long interest in public relations and popular education on the part of the creator of the Harvard Botanical Museum and its famous 'glass flowers' (Sutton, 1970, pp. 171–172; see also B. L. Robinson, 1926. *Biographical memoir: George Lincoln Goodale, 1839–1923*. Washington, D.C.: U.S. Government Printing Office. (Mem. Natl. Acad. Sci. 21(6).))

25 The Lloyd Library was established in the late nineteenth century as a private foundation by the Lloyd family (including the mycologist C. G. Lloyd) in Cincinnati, Ohio, U.S.A. Its specialities have been in systematic botany, mycology and pharmacognosy. The authors of the bibliography were at the time respectively chief librarian and assistant (later chief) librarian.

26 Some omissions were, however, unavoidable; as acknowledged by the compilers, its external sources were largely secondary. No special trips outside Cincinnati were essayed and much use had thus to be made of such works as the *Thesaurus* and Jackson's *Guide* as well as the available volumes of the catalogue of the library of the British Museum (Natural History), *Botanisches Centralblatt*, and the *Index to American Botanical Literature*.

27 Simon, 1977, pp. 68ff.

28 Blake was a botanist with the Crops Division of the Agricultural Research Service of the United States Department of Agriculture; Atwood, a librarian and bibliographer with the departmental library (now the National Agricultural Library). The latter had also been responsible for the library's botany subject union card catalogue, a prime source for the *Guide* until its discontinuance in 1952 (for description, see Atwood, 1911).

29 Elbert L. Little, Jr., personal communication.

30 Turrill, 1964.

31 The archives for this work are presently housed in the World Conservation Monitoring Centre near Cambridge, England.

32 Britton's list was arguably comparable to Linnaeus's *Bibliotheca botanica* in being part of an overall research

programme. For the author this was the reform of North American taxonomy and floristics including the development of a nominalistic (but for a time influential) 'American' school of taxonomy at once more 'scientific' and less reliant on 'tradition'.

33 Wikström's work is the first purely bibliographic national literature survey. Other contemporary works were primarily historical or bio-bibliographic, including those of Sternberg for Bohemia (1817–18), Adamski for the Polish lands (1825), Haberle for the Hungarian lands (1830), and Trautvetter for the Russian Empire (1837).

34 Heywood, 1958, 1960; Lawalrée, 1960.

35 Hamann and Wagenitz, 1977.

36 Heywood (coord.), 1975.

37 Indeed, it is arguably one of the most important uses for the Web and its search engines.

38 Simon, 1977, pp. 82ff.

39 The Swedish Academy also published a zoological review (1826–42). The Berliner *Archiv* accounted for new botanical literature only through 1855, with geographical botany contributed by the noted plant geographer August Grisebach. In later years it became all but a 'house organ' of the Berlin Zoological Museum.

40 The *International Catalogue* is described more fully in Appendix A.

41 *Botanical Abstracts* was established as a direct response to the entry of the U.S.A. into World War I and the consequent disruption to *Botanisches Centralblatt*.

42 Such subject catalogues existed in St. Petersburg, Brussels, Geneva, Washington, and perhaps elsewhere. In zoology, however, some institutionalization of information handling took place with the formation in 1895 of the *Concilium bibliographicum* in Zürich (Simon, 1977, pp. 145–152). This body published author and classified cards as well as annual indices (*Bibliographia zoologica*) until the mid-1930s. There was, however, no comparable contemporary movement in botany. Classified bibliography – though not limited to the sciences – was also an objective of the *Institut International de Bibliographie* in Brussels. Organized in the same year as the *Concilium* and a participant in the ICSL, it moreover effectively introduced the Dewey Decimal Classification (DDC) to Europe and other parts of the world through its sponsorship of a derivative, the Universal Decimal Classification (UDC), first published in full in 1904–07.

43 The efforts of the U.S. bureau are recorded in contemporary annual reports of the Smithsonian Institution. Also effectively interrupted or altered were the activities of both the *Concilium bibliographicum* and the *Institut International de Bibliographie*. The latter was in 1924 reorganized as an international federation of documentation organizations (now known as *Fédération Internationale d'Information et de Documentation*) while the former, after a partial revival in the 1920s and 1930s, was liquidated in 1941. By this time, of course, *Biological Abstracts* was well established.

44 It was partly succeeded by *Bibliography of Agriculture* (Blake, 1961).

45 In the 1980s *Kew Record* would also absorb the *AETFAT Index*.

46 The *Kew Record* database may be consulted within the Royal Botanic Gardens, Kew, and on-demand lists of titles generated. It has also been available through bibliographic search services. All queries, however, have hitherto been command-line based. In 1999–2000, though, a World Wide Web 'client' interface was developed and, after internal release, was made generally available to the public in September 2000 (at http://www.rbgkew.org.uk/kr/KRHomeExt.html).

47 Cf. Jäger, 1976 *et seq.*

48 For *Biological Abstracts* and ornithology, see R. Mengel in Buckman, 1966, pp. 121–130; for *Biological Abstracts*, *Current Contents* and systematic botany, see Delendick, 1990.

49 Garfield, 1979.

50 Available through the New York Botanical Garden website (http://www.nybg.org/bsci/iabl.html).

51 The project was mounted under the aegis of the Committee of the European Science Research Councils and financially supported by the European Science Foundation (European Science Foundation, 1978–81). The ESFEDS itself was described in some detail in Heywood and Derrick, 1984; a further summary appears in Heywood, 1989. The project itself ran for five years from November 1981. A successor initiative (currently known as 'Euro+Med PlantBase') received substantial support from the European Union in 1999 after a decade of discussion, meetings, and proposals to funding agencies beginning in 1988. A succinct summary appears in *Linnean Society Annual Report 1998*, pp. 17–18 (1999).

52 The capability of the computer hardware in use at the time was by current standards quite limited. A basic taxonomic database was, however, realized; it is maintained at Edinburgh and may be accessed through the World Wide Web (http://www.rbge.org.uk/forms/fe/).

53 Prance, 1977 (publ. 1978); Prance and Campbell, 1988; Campbell and Hammond, 1989.

54 Verdoorn, 1945.

55 Also of value is the already-mentioned *Plants in danger* (Davis *et al.*, 1986).

56 This category of botanical literature is difficult to survey and may be seen as one example of the inadequacy of parts of the present biological information system (cf. Wyatt, 1997). Fortunately, the area has to a considerable extent been covered by the periodic studies of plant

geographical literature by Jäger, 1976 *et seq.*, in *Progress in Botany* – a review annual not, however, mentioned in Wyatt's book. The surveys of Davis *et al.*, 1986, and Campbell and Hammond, 1989, are also valuable.

57 For a good survey of this material, see Bridson and Forman, 1998.

58 Examples include *Flora Malesiana* and *Flora of the Venezuelan Guayana*.

59 Jonsell, 1979. Chapter 4 (pp. 91–111) in the UNESCO synthesis report *Tropical Forest Ecosystems* (1978, Paris) can serve as an example.

60 Jäger, 1976 *et seq.*

61 Jäger, 1976, p. 317.

62 Barthlott, Lauer and Placke, 1996; this follows on from a first attempt by Malyschev, 1975.

63 Heywood, 1957.

64 Heywood, 1958, 1960.

65 The high level of congruence between the selections in the original edition of the *Guide* and in *Plants in danger* (Davis *et al.*, 1986) seems to support this view.

66 The French and German equivalents are, for example, respectively 'flore de base' and 'Standardflora'.

67 Blake and Atwood, 1942, pp. 15–16; Blake, 1961, pp. 27–28.

68 Sections in the Kew Herbarium responsible for collections were at that time reorganized on a systematic rather than a geographical basis as had been in place since the nineteenth century.

69 An exception has been made for the district floras of India; these have consistently been included as until recent years there have been few state floras.

70 Among those then (and still) in wide use were the purely enumerative geographical units within the QK (Botany) section of the Library of Congress (U.S.A.) Classification (1901 onwards) and the common or universal geographical auxiliaries in the Dewey Decimal Classification (DDC; 1876 onwards) and its derivative, the Universal Decimal Classification (UDC; 1895 onwards). Within natural history several schemes were available; those seen included, for floras, the broadly geographically arranged Lloyd Library scheme (Holden and Wycoff, 1911–14) and the alphabetical schemes of Blake and Atwood (1942) and Blake (1961) – also for floras – and Travis *et al.* (1962) for entomological literature.

71 Cf. Vickery, 1975, pp. 46–47.

72 I have not here attempted a fuller examination of the development of this aspect of bibliographic classification. A potential source is E. L. Schamurin, 1967. *Geschichte der bibliothekarisch-bibliographischen Klassification*, 1. Munich: Dokumentation.

73 Gould, 1968–72.

74 Hollis and Brummitt, 1992. This scheme evolved partly from work done by the International Legume Database and Information Service (ILDIS); see S. Hollis, 1990. *ILDIS type one data: geography*. Version 4. 35 pp. Southampton. A revision of the 1992 scheme is in preparation.

75 The first two of the letters in each triplet embody the ISO-3166 country code.

76 For a discussion of the concept of the 'literary warrant', see Kumar, 1979, pp. 266–267, 283.

77 Relatively few changes have been made for the present edition. Among them are subdivision of the Arabian Peninsula (Region 78), the shift of the Baltic republics to Region 67 and of Slovenia to Region 64, and renumbering of some other units in Regions 63, 64 and 68/69.

78 Among possible alternatives was a 'symmetrical' scheme proposed by Ranganathan (1957). Its basic principle became a partial basis for the common auxiliaries adopted here.

79 The UDC standard consulted was British Standard (B.S.) 1000, 5th edn. (1961).

80 'Malagassia' is here introduced as a portmanteau word for the islands and reefs of the southwest Indian Ocean. It is based on 'Malagasy', after the inhabitants of Madagascar, by far the largest island, and 'Thalassia', referring to their oceanic location.

81 It was also expected that *Bibliographia Huntiana* would provide a detailed review of all pre-1840 botanical literature, inclusive of floristic works; however, as of writing the project has effectively been abandoned (Sylvia FitzGerald, personal communication, 1999). Photocopies of the master list of this project exist in some botanical libraries.

82 For this important and well-studied physiographic unit no separate complete modern flora is available – a lacuna already noted in the 1950s by the Innsbruck botanist Helmut Gams (Gams, 1954). Various more or less popular works, notably *Unsere Alpenflora* by Elias Landolt (available in four languages), have perforce been included. A new general flora for the European Alps has, however, been projected.

83 Pichi-Sermolli, 1973; Wagner, 1974.

84 This is in fact being addressed in two ways: through independent family bibliographies or, since 1996, within the *World Checklists and Bibliographies* series of the Royal Botanic Gardens, Kew. Useful selections also appear in the Springer series *Families and genera of vascular plants*, edited by K. Kubitzki (1990–).

85 Consideration was, however, being given by the Library of the Royal Botanic Gardens, Kew, to extension of coverage by *Kew Record* to pre-1971 literature (Sylvia FitzGerald, personal communication, 1998).

86 Apart from the sheer length of time required – 20 years were required by Blake for vol. 2 – technological advances have been such that the need for such a work

may be largely satisfied in other ways including simultaneous Web searches.

87 Aymonin, 1962; Heywood, 1973*a,b*; Frodin, 1976 (publ. 1977).

88 Heywood, 1984; Morin *et al.*, 1989.

89 Jarvie and van Welzen, 1994; Palmer, Wade and Neal, 1995; Schmid, 1997.

90 Cf. Malclès, 1961.

91 The standards of coverage adopted for the *Geographical guide*, while perhaps relatively satisfactory as an index of the status of knowledge in entities such as Europe, North America, and a scattering of others elsewhere where good floras are more or less numerous, may also on the other hand fail to reflect accurately actual standards of floristic knowledge over a great part of the earth's surface. In such areas, there may exist a considerable 'literature' but comparatively few substantial floras or checklists (as in much of Latin America, where until recently at least publication of floristic and taxonomic records has been very much more in journals than in books). Enumerative bibliography is now only a part of the wider field of information science, and the whole approach towards fields of knowledge – and the questions asked – have become more systemic. The already-mentioned proposal for retrospective extension of coverage by *Kew Record* represents, however, an important first step.

92 Malclès, 1961, pp. 109–110.

93 Gilmour, 1952.

94 Ziman, 1968, p. 125.

95 Bradford, 1953, pp. 144–159; Garfield, 1979, pp. 21–23. Garfield (1980) later likened it to a comet.

96 cf. Leimkuhler, 1967; Bulick, 1978.

97 Garfield, 1980. This law of 'scatter' is actually a manifestation of the Zipf distribution, of which another is J. C. Willis's 'law' of distribution of subordinate ranks. See Nalimov, 1985, pp. 13–14.

98 Simon, 1977, p. 1.

99 Webb *et al.*, 1970, 1976.

100 An analytical bibliography may also be looked upon as a kind of scientific monograph or treatise, a vehicle for communication eloquently defended by Ziman (1968). Paradoxically, however, such works often are seen as not 'orthodox'. As a result, scholarly bibliographies, even of comparatively restricted scope (when compared with the major artisan-bibliographies of the eighteenth, nineteenth and early twentieth centuries), are now – as with major monographic studies in general – less often attempted (a notable recent exception being *Taxonomic Literature-2* and its supplements). This reflects present patterns of funding and management as well as widespread short-term thinking; but on a deeper plane may be related to a lessened interest in intellectual values. This work is nevertheless offered in the hope that some scope remains in the canons of science for serious monographs, bibliographies and similar treatises.

2

• • •

The evolution of floras

As concerns the flowering plants we may say that we live again in an age of floras and floristic work. [It is part of] a cyclic development [with several phases].

> Stafleu, *Syst. Zool.* **8**: 66 (1959).

Seit der Mitte des Jahrhunderts hält eine Epoche des Florenschreibens an.

> Jäger, *Prog. Bot.* [*Fortschr. Bot.*] **40**: 413 (1978).

Floras should always be regarded only as a stage, although an important one, in understanding plants and contributing to botanical knowledge. . . . They can never be definitive: new facts and information are always coming to light.

> Hedge, in *Contributions selectae ad floram et vegetationem Orientis* (eds. Engel *et al.*), p. 312 (1991).

Introduction and general considerations

The preparation and publication of floras and related works has been a constant feature of systematic botany since late in the sixteenth century. In that time, this activity in a formal sense has spread from central and western Europe to other parts of the world. In some cases, however – and particularly in eastern Asia – it absorbed, and was to an extent influenced by, autochthonous floristic traditions. At different times – and particularly in the twentieth century – floristic work has prevailed over other approaches to plant diversity. This trend, first noted in 1959 by Stafleu[1] and later by Thorne[2] and Gómez-Pompa,[3] was affirmed in an extensive review by Jäger in the late 1970s.[4] From the 1980s, floristics (and its products) have received renewed emphasis from conservation and biodiversity interests, and the scope, strengths and weaknesses of floras have been analyzed in symposia[5] and individual articles.[6]

The rapid development of computer technology and the recent spread of the Internet with its tools for information access and exchange are bringing, and will continue to bring, further change in the ways floristic information is presented; this will be further examined in Chapter 3. While critical remarks have already been heard,[7] it is nevertheless likely that geographically based taxonomy related to state, national and regional flora and checklist production will continue to predominate in systematics. Such floristic studies, along with research on medicinal plants, figure among the oldest classes of botanical literature; they will remain essential if not representative of more truly original study and thought.

This chapter is concerned with the general development of floras and floristic documentation over some four centuries or so, with some attention given to particular regions. It is an area of enquiry which until recent years attracted relatively little attention from historians of botany. A. G. Morton in his *History of botanical science* (1981), writing primarily from the general-botanical point of view notably championed in the late nineteenth century by Julius von Sachs (depicted in the frontispiece of Morton's book), does not single it out;[8] neither did Sachs himself in his *Geschichte der Botanik von 16. Jahrhundert bis 1860* (1875) beyond ascribing (wrongly) the introduction of the term 'flora' to Linnaeus.[9] The diversity of aims, styles and content in floras – already recognized in the 1730s by Linnaeus[10] and, almost two and a half centuries later, by Brenan[11] – will be examined with an attempt to relate these to philosophical, methodological and historical movements in botany. Orthodox and alternative styles will alike be examined, with reference to both professional respectability and user experience.

Floras in Europe and by Europeans before 1805

The floras of the sixteenth, seventeenth and eighteenth centuries – up to the appearance of the Candollean edition of *Flore française* in 1805 – fall into two relatively discernible periods: those published before the suite of methodological reforms introduced by Linnaeus in the 1730s and those of the mid- and later eighteenth century. These are considered in succession.

Early writers

The effective renewal in the Renaissance of botany as a field of enquiry began firstly with commentaries on Theophrastus, Dioscorides, Pliny and other classical writers; it then gradually extended its reach to encompass native plants, medicinal and otherwise.[12] To these were added the findings of voyages of exploration from the second half of the fifteenth century onwards and of survey commissions, notably in the Americas but also in the Far East. By 1536 Antonio Musa Brasavola (of Ferrara) noted in his *Examen omnium simplicium medicamentorum* that 'not a hundredth part of the herbs existing in the whole world was described by Dioscorides, not a hundredth part by Theophrastus or Pliny, but we add more every day and [thus] the art of medicine advances'. By the middle of the century there arose a serious appreciation of the diversity of plant life worldwide, a phenomenon reduced only by drought or by cold.

A first chair of *materia medica* including botany was established in Padua in the then-Venetian Republic in 1533. In 1534, Luca Ghini became reader in *materia medica* at Bologna University and in 1538, professor; later he moved to Pisa in Tuscany. In this period and particularly in the 1540s Ghini revolutionized the study of plant diversity with his innovations of the herbarium for the preservation of plants and plant parts and, at both Pisa and Florence, the botanical garden for the cultivation and observation of living plants. It was in Central Europe, however, that a lasting tradition of encyclopedic plant documentation first became established, aided by the spread of plant portraiture based on the wood-cut.[13] The earliest such works are those of Otto Brunfels (*Herbarium vivae eicones*, 1530), Leonhart Fuchs (*De historia stirpium*, 1542) and Hieronymus Bock (Tragus) (*Historia stirpium*, 1552, with later editions to 1630). The works of these so-called 'German fathers of botany'[14] were followed by those of Rembertus Dodoens (*Historia stirpium*, 1554; 2nd edn., 1563, with reissues to 1644; French version by Carolus Clusius, 1557; English version (translated by H. Lyte), 1578, with reissues to 1619), Valerius Cordus (*Historia stirpium*, 1561), Petrus Matthioli (*Commentaria in VI. libros Dioscoridis*, 1554 and later editions, and especially *Compendium de plantis omnibus*, 1571), Lobelius (*Plantarum seu stirpium historia*, 1576, with revisions under other titles to 1655), and Jacob Tabernaemontanus (*Historia plantarum tomi tres*, 1588).[15]

These works were all arranged according to classifications based on gross form or simply by plant names; the main developments were in descriptions and iconography. A distinct philosophical movement arose in 1583 – two years before the Council of Trent – with publication of *De plantis libri XVI* by Andrea Caesalpino, a student of Ghini. The 1500 plants accounted for therein were arranged according to a system based directly on plant observation, particularly of fruits, and perceived relationships. With this work systematic botany could be said to have become separate from the study of *materia medica*. The question of plant relationships as an area of enquiry could be pursued for its own sake; indeed, already evident in Caesalpino's classification were groups corresponding to several now generally accepted families. By its nature, however, Caesalpino's work would have relatively little direct influence on the developing literature of regional floristics.

By the mid-sixteenth century, as already noted, it was accepted that there was a considerable diversity of plants around the world. It was also well understood that plants were not uniformly distributed. Bock in particular in his *Historia stirpium* had already made note of this here and there – and likewise with respect to habitats and altitude. As Möbius has indicated, this marks a transition in emphasis from herbal to local flora.[16] At the same time, exploration of Europe and the rest of the world was continuing, and travel and survey reports including many references to animal and plant life were being published.

This rapid increase in information about the natural world and its 'wonders' not unnaturally stimulated interest in a total documentary approach. Notable exponents of this method were two polymaths, the Swiss Conrad Gesner and the Fleming Carolus Clusius (Charles de l'Écluse). Gesner focused on Central Europe, especially Switzerland; after his death his wider botanical interests were diffused mainly through his student Johannes Bauhin and, in turn, Bauhin's younger brother Caspar.[17] By contrast, Clusius, forced out of the greater Netherlands by the struggle for independence from Spain, traveled extensively, firstly in that country and then in Pannonia (the Hungarian Basin). His reports of these travels, respectively published as *Rariorum aliquot stirpium per Hispanias observatarum historia* (1576) and *Rariorum aliquot stirpium per Pannoniam, Austriam et vicinas quasdam provincias observatarum historia* (1583), along with editions of the *Coloquios dos simples* by the Portuguese Garcia da Orta (a work on Indian and other medicinal plants), engendered an interest in 'exotick' botany. This would be translated into his subsequent career in the newly independent United Provinces, where he became professor of botany and head of the botanical garden at Leiden University (1593–1609). Clusius also contributed a digest of his work on Pannonia, *Nomenclator pannonicus* (1584) – the first known 'checklist' of a distant land.

It was the Bauhins who, along with Jacques Daléchamps at Lyon, were effectively the founders of world floristic documentation. Johannes Bauhin firstly contributed to Daléchamps's *Historia generalis plantarum* (1586–87; reissues 1615, 1653) and then, with Johann Cherler, produced *Historia plantarum universalis* (1650–51, posthumously), covering 4000 plants. Not to be outdone and aware of the many errors in the *Historia generalis*, Caspar Bauhin firstly produced a preliminary checklist, ΦΥΤΟΠΙΝΑΞ (*Phytopinax*) *seu Enumeratio Plantarum* (1596) and then, late in life, published his most famous work, ΠΙΝΑΞ (*Pinax*) *theatri botanici* (1623; reissue 1671). Subdivided into 12

'books' corresponding to his ideas on primary classification, it covered some 6000 species and was based on collections, literature records and personal information. Its universality also set a new contemporary standard for botanical nomenclature. Truly the first 'world list', the *Pinax* was not unnaturally much used by Linnaeus, as his own annotated copy testifies.[18]

Conrad Gesner himself also had had a considerable influence on floristic botany through his work on the Alps; various observations on individual mountains were made and lists of plants compiled by him along with others.[19] He may thus well be considered the founder of 'topographical botany', a principal basis for the local flora tradition. One of Gesner's students was Anton Schneeberger, later of the Jagiellonian University in Krakow; while there he published *Catalogus stirpium quarundum* (1557), an alphabetically arranged, annotated enumeration of 432 native and exotic plants intended for student use. This work may be considered a precursor to the earliest true local flora, which appeared in Germany in 1588. This was *Sylva harcynia* by Johannes Thal; covering the Harz Mountains, it featured localities, vernacular names, and scientific and economic commentary for some 600 plants. As would remain customary for another century or more, the plants were arranged alphabetically by their botanical names.[20]

With Thal's work the basic conventions of local floras were established; from then on such works began to proliferate, particularly after 1600. The earliest are those of Johannes Wigand on Prussia, *De herbis in Borussia nascentibus* (1590), and Caspar Schwenckfelt on Silesia, *Stirpium et fossilium Silesiae enumeratio* (1601), while the best is *Catalogus plantarum circa Cantabrigiam nascentium* (1660) by John Ray. Others particularly recommended by Linnaeus are those of Rupp on Jena (1718), Dillenius on Giessen (1719), and Vaillant on Paris (1723, 1727). Improvements to the arrangement of the plants, nomenclature, content and organization along with the provision (sometimes) of descriptions and keys gradually took place. Of particular note was the adoption of the Caesalpinio system by Christoph Knauth for his *Enumeratio plantarum circa Halem Saxonium* (1687); however, alphabetical, phenological or gross-morphological arrangements remained usual until the spread of the Linnaean system in the mid-eighteenth century, beginning with that author's own *Flora lapponica* (1737).

During the seventeenth century, the geographical horizons of the local flora began to widen, particularly in those countries with something of a national consciousness. Not unnaturally, the idea of a national flora took root most quickly in the smaller polities. The first of these was *Flora danica* (1648) by Simon Pauli; this was followed by *Plantae in Borussia sponte nascentium* (1655) by Johannes Loeselius. In 1670 Ray extended his Cambridge flora to cover all England, as *Catalogus plantarum Angliae*; later, this was revised and expanded as *Synopsis methodica stirpium britannicarum* (1690). For some time this remained standard, undergoing revisions in 1696 and 1724 (the latter by Dillenius who had moved to Oxford). Dillenius's edition was regarded by Linnaeus as the best among a small number of good floras. In the Low Countries, the first were *Herbarius belgicus* (1670) by Petrus Nyland and *Catalogus plantarum indigenarum Hollandiae* (1683) by Jan Commelijn (a revision, also well thought of by Linnaeus, appeared in 1709). In France, *Prodromus botanici parisiensis* (1723) and its revision, *Botanicum parisiense* (1727), both by S. Vaillant, were influential, with the latter containing many observations. Few, if any, others of real scope would appear until the middle of the eighteenth century. The last major 'pre-Linnaean', though transitional, work was *Flora helvetica* by Albrecht von Haller (1742). While not following the Linnaean precepts of classification, it nevertheless called itself a 'flora' in the manner suggested by the rising Swedish master.[21]

The relative lack of good floristic works would leave some scope for Linnaeus and his students to fill gaps, as will be noted in the next section. Already in 1736 – but with more lasting effect in 1751 – he had commented unfavorably on Spain: 'dolendum est quod in locis Europae cultioribus, tanta existat nostro tempore barbaries botanices!' Yet already some catalogue-floras had been published for extra-European lands; among them were *Flora malabarica* (1696) by Jan Commelijn (based on *Hortus malabaricus* by van Rheede), *Catalogus plantarum insulae Jamaicae* (1696) by Hans Sloane, *Catalogus plantarum americanarum* (1703) by Charles Plumier (1703), and *Museum zeylanicum* (1717) by Paul Hermann. Many florulas were also published by James Petiver; these are collected in *Musei petiveriani cent. I–X* [1692–1703] and *Gazophylacii . . . decades X* (1702–09).

General compendia continued to appear subsequent to Bauhin's *Pinax*. Among these were 'Le petit Bauhin', *Histoire des plantes de l'Europe* (1670, with many reissues and a revision by Gilibert in 1798 and again in 1806), *Historia plantarum universalis* by Robert Morison (1680–99, not completed), and Ray's magisterial *Historia plantarum* (1686–1704). The last-named was the ultimate solo pre-Linnaean descriptive world flora.[22]

The Linnaean and post-Linnaean eras

It is generally appreciated that by the 1730s the world of botany was in some disorder, ripe for new proposals in management of its information. In particular, the number of known plants was increasing to the point where the old Latin polynomial diagnostic nomenclature had become a hindrance; moreover, there was an increasing interest in new approaches to classification as the potential of additional plant parts became recognized. Into this milieu came the confident young Linnaeus, who within the space of a few years would furnish what we might today call a 'total package'. Set out in a variety of works including the original edition of his *Systema naturae* (1735), his strongly pragmatic scheme was for botany first expounded in *Fundamenta botanica* and *Bibliotheca botanica* (1736) and in *Critica botanica* (1737). These include clear expositions of the concepts of a flora and of 'florists', with an indication of possible aliases and synonyms as well as ideas about the form and contents of floras.[23]

Linnaeus's ideas were first realized in a practical fashion in his *Flora lapponica* (1737), the first proto-modern flora. Covering 534 species and arranged according to its author's system of classification, it included geographical distribution along with taxonomic notes.[24] Augustin-Pyramus de Candolle in 1813 credited Linnaeus with our present concept of a flora and *Flora lapponica* as being an excellent model,[25] while in 1905 E. L. Greene wrote of this work that it was 'the most classic and delightful' of Linnaeus's writings.[26] *Flora lapponica* was soon followed by the first (and so far only!) flora of Virginia in North America, *Flora virginica* (1739; 2nd edn., 1762) by Jacob Gronovius (with whom Linnaeus had been well acquainted in Leiden) and then by Haller's already-mentioned *Flora helvetica* (1742) and Gmelin's *Flora sibirica* (1747–69). Linnaeus's next flora was *Flora suecica* (1745; 2nd edn., 1755, with binomial nomenclature). Its two editions respectively covered 1140 and 1297 species inclusive of cryptogams, and incorporated diagnostic features, synonymy, vernacular names, and notes on habitat, distribution, uses and properties. A version in checklist form appeared as *Chloris suecica*.[27] From these, Linnaeus – by 1753 having added species epithets to his other methodological reforms – turned his attention to the flora of other parts of Europe and the world in a series of theses, all except the first featuring binomials (their respondents in parentheses):

Plantae rariores camschatcenses (Jonas P. Halenius, 1750)

Flora anglica (Isaac O. Grufberg, 1754)

Herbarium amboinense (Olof Stickman, 1754)

Flora palaestina (Bengt J. Strand, 1756)

Flora monspeliensis (Theophil E. Nathhorst, 1756)

Flora alpina (Nicol N. Åmann, 1756)

Prodromus florae danicae (Georg T. Holm, 1757)

Flora jamaicensis (Carl G. Sandmark, 1759)

Flora capensis (Carl H. Wännmann, 1759)

Flora belgica (Christian F. Rosenthal, 1760)

Plantae rariores africanae (Jacob Printz, 1760)

Necessitatum promovendae historiae naturalis in Rossia (Flora sibirica) (A. Karamyschew, 1766)

Flora Akeröensis (Carl J. Luut, 1769)

Pandora et Flora Rybyensis (Daniel H. Söderberg, 1771)

Plantae surinamenses (Jacob Alm, 1774)

On his own account, Linnaeus had already ventured outside Europe with *Flora zeylanica* (1747),[28] and in 1758 he edited and published *Iter hispanicum* by his former student Pehr Löfling. The various editions of his *Genera plantarum* (beginning in 1737) set precedents for generic floras (a genre with yet few examples in spite of some strong advocacy in the past). Finally, his and later editions of *Species plantarum* are in effect world floras, while the later editions of *Systema naturae* and *Systema vegetabilium* are world enumerations.

The work of Linnaeus brought about nomenclatural and methodological changes which for the most part were relatively soon adopted, if in some quarters his classification system was not.[29] Most floras after the 1750s would adopt Linnaean binomial nomenclature along with his preferred format. Among the first by an author other than Linnaeus was *Flora carniolica* by J. A. Scopoli (1760), covering the major part of modern Slovenia; others quickly followed, among them those of Crantz in Austria (1762–67), Gunner in Norway (1766–72), and Gorter in the Netherlands (1767–77). Works adhering to earlier forms of nomenclature, such as *Flora gallo-provencialis* by Gérard (1761), soon were seen as old-fashioned, with any merits overshadowed.[30]

The spread of Linnaean methodology has been well documented by Stafleu;[31] it will suffice here to record that by the end of the eighteenth century much of Europe was covered by 'modern' floras. For other parts of the world, several 'Linnaean' floras and check-lists of varying quality had also been produced; leading authors included Nicolaas Burman, Johann Reinhold and Georg Forster, Pehr Forskål, Olof Swartz, and Carl Thunberg. A preliminary account of the flora of the Spanish viceroyalty of Peru, *Flora peruviana, et chilensis prodromus* by Hipolito Ruiz and José Pavón, appeared in 1794 as the first of the few contemporary fruits of the series of expeditions in the Americas initiated in the 1770s under the Spanish king Carlos III.

A need for economy as well as convenience had forced many floras to be relatively compact. Linnaeus himself from time to time faced paper shortages; moreover, he was concerned with making information affordable as well as widely available.[32] Elsewhere, however, liberal patronage – as well as prestige – led to the presentation of geographical and natural history reports of new or little-known lands in a generous, often lavishly illustrated fashion, beginning already in the Renaissance. Examples of such works prior to Linnaeus's time – some of them already mentioned – include *Primera parte de la historia natural y general de las Indias* (1535) by Gonzalo Fernández de Oviedo, the two *Historiae* of Clusius (1576 and 1583) and his *Exoticorum libri decem* (1605),[33] *De plantis Aegypti liber* (1592) by the Venetian Prosper Alpinus,[34] *Historiae rerum naturalium Brasiliae libri* (1648) by Georg Marggraf, *Historiae naturalis et medicae Indiae orientalis libri sex* (1658) by Jacob Bondt, *A voyage to the islands Madera, Barbados, Nieves, S. Christophers and Jamaica* (1707–25) by Hans Sloane, *Amoenitatium exoticarum* (1712) by Engelbert Kaempfer, and *The natural history of Carolina, Florida and the Bahama Islands* (1730–47) by Mark Catesby. By the latter part of the seventeenth century, however, more methodical approaches to description and documentation of exotic natural history had emerged. The best-known works in this genre include *Hortus indicus malabaricus* (1678–93) by Hendrik van Rheede, *Herbarium amboinense* (1741–50; supplement, 1755) by Georg Rumpf, *Flora peruviana, et chilensis* by Hipolito Ruiz and José Pavón (1798–1802, not completed), and – methodologically closer to our own time – *Flora brasiliensis* (1840–1906), begun by Carl Philip von Martius and further considered below. Expedition reports would also continue to appear until well into the twentieth century, some with significant florulas and, from around 1900, vegetatiological accounts.[35]

The European fashion for large-scale descriptive works of natural history was not limited to exotica. Increasing wealth, a growth in national consciousness, improved technology and more widespread connoisseurship, along with the Linnaean reforms and the growth of the encyclopedic tradition, led to the appearance after 1750 of similar large-scale illustrated floras in the old subcontinent. The examples set by Denis Diderot's *Encyclopédie* and its contemporaries also led to a development of an interest in phytography as a literary rather than purely documentary form (while still retaining the rational organization promoted by Linnaeus), while public sensibility was increased with the introduction of fine illustrations.

The first of the 'large' works was *Flora danica*, begun by Johann C. Oeder and published from 1762 through 1883; ultimately it featured 3240 plates of phanerogamic and non-phanerogamic plants and fungi from Denmark and other territories then under Danish rule.[36] *Flora danica* was followed by the similar *Flora austriaca* (1773–78) by Nicolas Jacquin (limited to plants not figured in *Flora danica*), *Flora rossica* by Peter S. Pallas (1784–88, not completed), *English botany* (1790–1814; supplements, 1829–66) by James Sowerby and James Edward Smith, *Flora batava* (1800–1940), begun by Jan Kops and running to 2240 plates, *Svensk Botanik* (1801–43) begun by Johan W. Palmstruch and C. W. Venus, *Plantes de la France* (1805–22) by Jean Henri Jaume de Saint-Hilaire, *Flora graeca* (1806–40) begun by John Sibthorp and J. E. Smith, and finally – late in the day, and fulfilling Oeder's wish for a major all-German flora – *Icones florae germanicae et helveticae* (1834–1914), begun by Ludwig Reichenbach.[37]

By the end of the eighteenth century, floristic documentation had settled into three forms: the large-scale descriptive flora, the smaller national or regional flora with synonymy, diagnoses and notes, and the enumeration or checklist. The smaller works, the vast majority essentially Linnaean in method and system, were in a later generation termed *diagnostic documentary enumerations*.[38] To these had recently been added a new form: the analytical, strictly dichotomous key, introduced by J. B. de Lamarck in his *Flore françoise* (1778; 2nd edn., 1783 or 1788). Lamarck, strongly critical of parts of the Linnaean methodology, saw floras as

tools for identification and not as documentary works in their own right, aimed largely towards the scholarly world. To him, the latter function was best addressed through such works as *Encyclopédie méthodique: Botanique*, begun by himself in 1783 and, after his move to zoology in 1793, continued until 1817 by Jean Louis Marie Poiret. This featured an alphabetical arrangement of entries including individual genera and species. This last work was equivalent to a 'world flora', and may be compared with Caspar Bauhin's unfinished *Theatri botanici*. It had the added merit of extensively documented sources, whether literature or specimens.

Lamarck's views on floras, as well as his analytical method, were expressed in the 'Discours préliminaire' of *Flore françoise*. The work was designed almost entirely as a handy means of plant identification and, although species were included under their respective genera (as in the works of Linnaeus), no families or other suprageneric categories were introduced.[39] While there were antecedents to analytical keys in, for example, Morison's *Plantarum umbelliferarum distributio nova* (1672) and Ray's *Historia piscium* (1686), Lamarck has to be credited with their application in a consistent fashion. At the same time he should be credited for his projection of floras into one of their lasting and still-relevant roles.[40]

The work of Lamarck, however, did not effectively question the Linnaean order, which continued to prevail – even in France[41] – until well into the nineteenth century and changes in sensibility within as well as without.[42] Antoine-Laurent de Jussieu made no attempt to popularize his revolutionary *Genera plantarum secundum ordines naturales disposita* (1789); indeed, no version in French appeared until 1824. Similarly, the development and exposition of Linnaeus's own ideas on 'natural' suprageneric groups by Paul Giseke in his *Praelectiones in ordines naturales plantarum* (1792) only slowly made themselves felt. With respect to world compendia, successive editions of Linnaeus's *Systema vegetabilium* retained their vogue, while the Berlin botanist Carl Willdenow went so far as to prepare a wholly revised edition of *Species plantarum* (1797–1805, not completed). The latter moreover improved on Linnaean practice through the provision of more complete descriptive phrases. Only Christiaan Persoon (better known to posterity as a mycologist) endeavored to bring a more cursive, though still concise, uniform and truly comparative, style to bear on plant description with his *Synopsis plantarum* (1805–06;

revised as *Species plantarum*, 1817–21). Though Linnaean in arrangement, in phytography it clearly reflects the influence of Lamarck and the *Encyclopédie*. A transition to a 'natural' system and a more definitive phytography were to be the next steps, part of the effective entry of empiricism into systematics.

The emergence of 'conventional' floristic styles, 1805–70

The 'Lamarckian' phytographic style

The first half of the nineteenth century saw the evolution of the descriptive flora into more or less its present format. This development began in France after 1800 with the third edition of *Flore françoise* (1805–15, as *Flore française*) and would be expressed as principles in *Théoire élémentaire de la botanique* (1813; 2nd edn., 1819). Both works, as we shall see, were by the Genevan botanist Augustin-Pyramus de Candolle, then resident in France. Like the first, this third edition introduced radical changes which would make largely obsolete the Linnaean method in flora-writing and set in train the development of new 'schools'. Both were part of the broader range of developments in French botany in the latter part of the eighteenth century during which much of Linnaean philosophy was rejected and modern systematics founded.[43]

Chief among the new precepts and methods as applied to floras was the concise paragraph-length descriptive 'plant portrait' with necessary supporting information including in particular distribution, properties and uses. Such a style of phytography had first been developed around 1782 by Lamarck for the botanical volumes of Panckoucke's epochal *Encyclopédie méthodique* (1783–1808; supplement, 1810–17), the large-scale revision of Diderot's *Encyclopédie* and a definitive work of the Revolutionary and Napoleonic eras.[44] Lamarck evidently devised it as an alternative not only to the diagnostic but telegraphic Linnaean style but also to the rambling 'herbalist' mode of documentation and writing of many pre-Linnaean authors which here and there was undergoing a revival.[45] The 'Lamarckian' style, maintained by Lamarck's successor Poiret in the project after 1791, in time became very influential. Apart from *Flore française*, an important early application was by Aimé Bonpland and Carl S. Kunth in the botanical volumes of Humboldt and Bonpland's Spanish-American expedition of 1799–1804, including the latter's *Nova genera et species plantarum* (1815–25).

The excellence of this last work, along with de Candolle's own writings, effectively further diffused the new style. Time, resources, and democratization here and there forced condensation of this 'French style' both in Europe and elsewhere. In other cases, however, generous financial and other support, project institutionalization, prestige, and a continuing sense of 'elitism' led to a sometimes substantial degree of stylistic expansion (or 'inflation'), as will be seen in the next section. Nevertheless, even with the addition of many new classes of information and the appearance of all kinds of variations, the Lamarckian phytographic formula, a product of reasoned essentialism framed in a nominalist superstructure, has continued to the present time as the accepted standard for descriptive floras of all kinds from the concise to the elaborate.[46] Over nearly two centuries, few significant modifications have been incorporated. Among these have been increasingly extensive taxonomic and biological commentary, greater ecological and geographical detail, and – significantly – another of Lamarck's innovations, the now-commonplace analytical keys for identification.

Flore française *and the* Prodromus

Flore française may be considered the first 'modern' descriptive flora. It was published under the nominal authorship of Lamarck and the Genevan botanist Augustin-Pyramus de Candolle. Lamarck, however, played no direct part in this work, having ceded botany to proponents of natural classification: de Jussieu, Poiret, Jaume de Saint-Hilaire (whose *Exposition des familles naturelles* also appeared in 1805) and René Desfontaines. Instead, the whole was, as already noted, written by de Candolle, then 32 years old. Added to Lamarck's keys, retained from the earlier editions, were cursive plant descriptions with notes on distribution, habitat and special features similar in conception to those of Persoon's *Synopsis*, if less concise. The work was so arranged that all the keys were in one volume, the descriptions in the others. Moreover, the work was also the first following a 'natural' arrangement – that of de Jussieu.[47]

The *Flore* was a ground-breaking work. It represented a melding of the flora as a medium for regional plant documentation, the plant name and a diagnostic phrase, the descriptive, methodological phytographic style, indication of distribution and habitat along with pertinent notes, keys for identification, application of a natural system of classification, and the use of a vernacular language with vernacular names. It was at once scholarly and popularly oriented. By contrast, its synonymy was relatively limited and citation of specimens was omitted. These lacunae would draw some criticism from the academically inclined; however, they were (and are) the province of the more elaborate 'research' flora or of monographs and revisions, the latter a genre which would rapidly develop during the first half of the nineteenth century. A current work such as *Jepson's Flora of California* (1993) – covering a flora comparable in size to that of France – remains true to the principles of *Flore française*. What has increased is mainly the range and sophistication of data now considered apropos to a manual-flora; there have also been stylistic changes including the interleaving of keys and descriptions under generic and family headings, a mid-nineteenth century development indicative of a renewal of 'academic' influence.

With the success of *Flore française*, de Candolle began to look for new challenges. In the decade or so following, he conceived a wide-ranging reform of general and systematic botany, much as had Linnaeus 80 years before. This would be firstly expressed in the form of a textbook and then crowned by a new world flora written according to his concept of the natural system. The latter was undertaken in the seemingly firm belief that such a task could still be achieved largely single-handed (the world flora being then thought to encompass some 25000–30000 species). Seven years after *Flore française* he published the first edition of his *Théoire élémentaire de la botanique*; in 1818 came the first volume of *Regni vegetabilis systema naturae*, with a second in 1821.

Floras form the topic of sections 217–219 (pp. 269–274) of *Théoire élémentaire*. Section 217 comprises a general critique of contemporary works as well as the still-prevalent Linnaean canons influencing them. Section 218 contains guidelines for a 'good' flora, a summary of which follows:

1) An introductory account, inclusive of such topics as physical features and vegetation
2) A systematic arrangement, in the first instance by families according to a 'natural' system
3) Name and diagnosis [or 'specific phrase']
4) Essential synonymy
5) Vernacular names
6) A sufficient description

7) Local variability
8) Distribution with detail in inverse proportion to a plant's ubiquity, localities to be given only for the uncommon or rare; in mountainous areas the altitudinal range to be included. For small areas documentation to be relatively strict; larger areas to make use of 'authorities'
9) Notes on uses, medicinal values, etc.

Selected works are cited under each rubric. The final section cautions against omission of commonly cultivated plants. Disagreeing with purist opinion, de Candolle in particular indicated that not to account for the olive in Provence, or the 'trèfle' in the Palatinate, would be quite wrong. All that was needed was clearly to indicate their status. A flora so enhanced would be of far more value to land economy and other fields.[48]

These nine principles reflect the author's experience with *Flore française*, and not unnaturally they would also guide *Regni vegetabilis systema naturae*. The definitive approach taken in the latter, however, called for additional synonymy and documentation. It was this greater detail which, after two volumes, caused de Candolle to realize that to continue in that vein would require 80–100 years more. He then initiated a more concise work in the manner of Persoon's *Synopsis*: *Prodromus systematis naturalis regni vegetabilis*. Despite inevitable limitations, the *Prodromus* (1824–73) became *the* definitive work of the first two-thirds of the nineteenth century: 'le régulateur de la botanique descriptive'.[49] As such, it will always remain a standard research reference. For many plant groups, ground was often seriously broken for the first time – and too often never since in such a fashion.[50] By the time of its termination following completion of the dicotyledons it encompassed 5134 genera and 58975 species. The latter figure amounted to *twice* what had been estimated for the *whole plant kingdom* in the early 1820s and more than *ten times* what Linnaeus had accounted for in the second edition of his *Species plantarum* (1762–63). From a solo undertaking when begun it gradually developed – notably under de Candolle's son Alphonse, editor from 1841 – into a collaborative enterprise, with ultimately 33 contributors. It also – perhaps inevitably – became more like the original *Regni vegetabilis systema naturae*, with longer descriptions and more details.

As the nineteenth century progressed, stylistic refinements to the elder de Candolle's principles of flora-writing were put into place by both de Candolles as well as by their contemporaries William J. Hooker, George Bentham, Joseph D. Hooker, and others. Rooted in empiricism as well as faith, these principles were succinctly summarized by Alphonse de Candolle[51] and, for the English-speaking world, by George Bentham.[52] They have remained more or less current to the present. Until relatively recently they were rarely seriously questioned, not only in absolute terms but possibly also because of increasing conservatism in the profession.[53]

Arguments have since been made from time to time for other forms of presentation, but apart from the manual-key (itself derived from Lamarck's *Flore françoise*) few found wide acceptance. The professional 'standard of excellence' has thus by general consent been the critical descriptive or 'phytographic' flora with the species as the main working unit: essentially the form so eloquently re-advocated in the 1950s by van Steenis.[54]

'Encyclopedic' floras and Central Europe

The above-mentioned ideas of Lamarck and de Candolle on the writing of floras would in time find wide acceptance, displacing most of the Linnaean methodology and style. However, there remained during much of the nineteenth century a belief that a descriptive flora, particularly of a new 'exotic' area, should continue to act as a detailed compendium and repository of information about its plants – in short, a specialized encyclopedia. Like the growing genre of critical monographs and revisions of 'natural' families and lesser taxa, the sometimes submonographic accounts presented in such works had to contain detailed descriptions, synonymy, specimen citations, extensive notes, and (often) illustrations in large plates. Tradition called for publication in a sumptuous format in the manner of the 'scientific results' of most contemporary voyages and expeditions.[55]

This concept of a flora seems to have taken hold most strongly in the Central European intellectual sphere, and cannot fail to have been influenced by the Germanic predilection for detail rather than conciseness. This trait could be viewed as having descended directly from the herbals and other botanical compilations of preceding centuries, the very works against whose frequent verbosity and mindlessness Lamarck so strongly reacted. For many writers, the Linnaean sexual system had simply furnished a new and improved framework for the preparation of general

compendia of the plant kingdom which seemingly enjoyed a continuing popularity.[56]

It thus comes as little surprise that, with exceptions, the Linnaean system of classification, along with much of its methodology, persisted as dogma longer in the German Confederation than elsewhere. This reflected the strengths of scholastic traditions and, in universities, the authoritarian professorial system. The marked differences which existed between Central European and French botany in the early nineteenth century are exemplified in the philosophy and styles of the respective treatments of Humboldt and Bonpland plants by Kunth for *Nova genera et species plantarum* – already mentioned – and by Roemer and Schultes for their version of Linnaeus's *Systema vegetabilium*.[57] The first major systematic work in Central Europe professing a 'natural' system was *Ordines naturales plantarum* (1830) by Friedrich G. Bartling in Göttingen – more than a generation after Giseke's proposals – yet use of the Linnaean system persisted here and there for another decade or more.[58] With the rise of the 'new botany' over the next generation or so taxonomy would come under attack for its seeming mindlessness – particularly from the influential physiologist Julius von Sachs.[59]

Following *Nova genera et species plantarum*, a number of large-scale semi-monographic descriptive floras written on the French model and following 'natural' systems were commenced. In format they approximated to their sumptuous Linnaean-era predecessors including *Flora peruviana, et chilensis*. Among them were Blume and Fischer's *Flora Javae* (1828–51), Webb and Berthelot's *Phytographia canariensis* (1836–50), Moris's *Flora sardoa* (1837–59, not completed), Torrey's *Flora of the state of New-York* (1843), and, notably, the Reichenbachs' *Icones florae germanicae* (1834–1914). These were, however, outdone by the regally sponsored and, symbolically, king-sized *Flora brasiliensis* (1840–1906) begun by the Bavarian botanist and Amazonian explorer Carl Philip von Martius. At the same time it was the first comprehensive, wide-area 'tropical' flora given the sheer size of the country covered and its geographical position. Contemporary reviews greeted the first fascicles of this work as a major step forward in floristic phytography, and it soon became widely influential as the best Central European systematic work of the period.[60]

This greatest of comprehensive nineteenth-century floras was to drag its detailed pages slowly on for what would ultimately be 66 years. Together with the de Candolles' *Prodromus* and (after 1870) *Monographiae phanerogamarum* it was long a prominent influence in contemporary European phytography. Like those undertakings it was a collaborative work – an approach now standard for most flora projects. Martius's project, however, featured – as would *Flora Europaea* a century later – an organization comprising general and managing editors, professional co-workers or flora-writers (*Privatassistenten*), and ultimately a total of 65 specialist contributors. Nearly all the leading botanists of the day were in one way or another involved, so fostering development of the spirit of international collaboration ever since characteristic of most of systematic botany. Through it – as well as other works – the Central European predilection for large compendia was shifted into new and fruitful channels, with in later decades lasting results – notably under Adolf Engler, a *Privatassistent* in Munich to the second general editor, August Eichler, and later ordinary professor successively in Kiel and Breslau (Wrocław) and (from 1889) Eichler's successor in Berlin (in the last-named with the notable support of Ignatz Urban, the *Flora*'s third and last general editor).

Flora brasiliensis thus established the tradition – still with us – of large-scale, multi-volume, descriptive regional floras, although these are now, like the British colonial floras, usually published in octavo format. They came to be seen as suitable vehicles for submonographic studies (often by students in their sponsoring institute or institutes). Most remain more or less encyclopedic, and as well retain an aura of prestige: a form of institutional 'cachet'. For many botanists, they represent the ideal in floristic phytography, particularly for areas not intensively well known. Their merits and inadequacies are further discussed in Chapter 3.

The concise, critical descriptive flora: British and other examples

While large-scale works such as *Flora brasiliensis*, *Phytographia canariensis* and *Flora sardoa* were (and are) representative of one phytographic ideal, they were – as Linnaeus had already complained in 1753 – necessarily costly and relatively limited in their distribution.[61] The powers of the conservative 'Holy Alliance' may have been victorious in 1815, but the next decades would see the effective rise of a middle class in much of Europe, the propagation of more liberal social and political ideas, and the progress of real or imagined nationalism.[62]

The impact of the 'Scottish Enlightenment' and a fashion for utilitarianism meant that the ideas on conciseness in floras and cursive descriptions with supporting data espoused by Lamarck, Persoon and Augustin-Pyramus de Candolle relatively early found favor among some British botanists, in spite of the continued strength of Linnaean traditions. Much of the credit for propagating these ideas should go to William J. Hooker, author of *Flora scotica* (1821) and until 1841 professor of botany at Glasgow University, George A. Walker-Arnott, a later incumbent of that chair and author (with Robert Wight) of *Prodromus florae peninsulae Indiae orientalis* (1834), and Robert Brown, a Scot who, after his Australian explorations, was resident in London as Joseph Banks's last librarian/assistant, botanist to the British Museum, discoverer of the biological nucleus, and author of the first Australian flora, *Prodromus florae Novae Hollandiae* (1810, not completed).[63] Further refinements and a more formal expression of the 'utilitarian' philosophy were the work of George Bentham and Hooker's son, Joseph D. Hooker.[64]

Following *Flora scotica*, the elder Hooker first employed 'concise' stylistic principles in an overseas flora, *Flora boreali-americana* (1829–40). Yet, while contemporaneous with the early editions of his *British Flora* (an octavo work first published in 1830), in its quarto format the North American work remained faithful to the tradition of 'prestige' overseas floras like Kunth's already-mentioned *Nova genera et species plantarum*. Similarly in quarto were three major expedition reports, two of them collections of floras: the *Botany of the Antarctic Voyage of H.M. Discovery Ships* Erebus *and* Terror (1843–59) by the younger Hooker (with colored illustrations by that great mid-century botanical artist Walter H. Fitch), *Botany of the voyage of H.M.S. Sulphur* (1844) by Bentham, and *Botany of the voyage of H.M.S. Herald* (1852–57) by Berthold Seemann.[65] By the late 1850s, however, the senior Hooker was to argue that the proposed colonial floras – discussed below – should be in octavo, 'botany [not being] what it once was, a science confined to the learned, and of little or no benefit to the people at large'.[66] Precedents had been set by the already-mentioned *Prodromus florae peninsulae Indiae orientalis* and by the first volume of J. D. Hooker and Thomas Thomson's *Flora indica* (1855), both of them in an octavo format. The latter was, however, aborted due to other commitments of the authors and finally by the dissolution of the East India Company. Its style also proved too detailed for expeditious completion; its successor, *Flora of British India*, would be more concise.[67]

As the mid-nineteenth century progressed, some differences of opinion were expressed concerning the use of analytical keys in floras. With experience of those in *Flore française* while living in France as a young man (1817–26), Bentham, who termed them 'indexes', naturally was an advocate.[68] He introduced them into his *Handbook of the British Flora* (1858 and subsequent editions), and with this and his later colonial floras may be credited with their effective integration into floras, now a standard practice and used in place of the contrasting synoptic statements characteristic of the *Prodromus* and many other contemporary works. By contrast, the Hookers continued to use such synoptic statements, both in the *British Flora* of the elder (through the 8th and last edition in 1860) and by the younger in his *Student's Flora of the British Islands* (1870; 3rd edn., 1884). The latter evidently believed that such keys made things too easy: students might well pay insufficient attention to diagnoses and descriptions.[69] Instead, in his *Student's Flora* he gave particular emphasis to geographical distribution and habitat.[70]

The relative popular success of the above-mentioned British floras and an unquestioned, characteristically Victorian confidence in their governing precepts caused them to be adopted as standards for the growing British Empire by the elder Hooker – by now director of Kew – upon the commencement of his colonial floras scheme.[71] Conceived in 1857 and launched three years later, it has continued – with modifications – to the present as part of the work of the Kew Herbarium.[72] Indeed, J. D. Hooker was to write in the preface to his *Flora of British India* in 1872 that in style and phraseology he was specifically following 'my *Flora of the British Islands*'. Like his father's *British Flora*, the latter work was originally written with a view to the requirements of the Scottish universities; with characteristic singlemindedness, however, the younger Hooker believed it a suitable model for a much longer work for a very different part of the world. Bentham's experience with his own *Handbook* doubtless similarly influenced his two major contributions to the imperial botanical survey: *Flora hongkongensis* (1861) and *Flora australiensis* (1863–78) as well as his *Outlines of Botany*, further considered below.

So influential were these colonial floras and such

was the spirit of the era within which most of them were published that not until the first Imperial Botanical Conference of 1924 did the approach and style represented by these works begin to be questioned.[73] Stronger criticisms would appear in the 1930s;[74] these will be further discussed in Chapter 3.

Outside Britain, the concise styles of the *Prodromus* and the British floras were adopted by Karl F. Ledebour for his *Flora altaica* (1821–34) and *Flora rossica* (1842–53), Charles Grenier and D. A. Godron in their *Flore de France* (1848–56), F. A. W. Miquel in his *Flora indiae batavae* (1855–59) covering the then-Dutch East Indies and neighboring lands, Moritz Wilkomm and Johan Lange in their *Prodromus florae hispanicae* (1861–80; supplement, 1893), and Edmond Boissier in his *Flora orientalis* (1867–88).[75] Analytical keys are, however, absent or were but partially employed in these works; as in the *Prodromus* and the Hookerian manuals, separation in larger groups was achieved through synoptic devices, necessitating close reading of descriptions to achieve identification. Even then, without authentically named specimens one could not always be certain, especially if, as was very often the case, the flora was imperfectly known.[76]

In North America, floras and identification manuals first appeared in any numbers only after 1810. A democratic tradition and relatively limited means in the United States surely contributed towards the relative utilitarianism and conciseness of most early work, including Amos Eaton's *Manual of botany for the northern states* (1817; 8th edn., 1840), Thomas Nuttall's *Genera of North American plants* (1818), and John Torrey's *Compendium of the flora of the northern and middle states* (1826, not completed) as well as floras of lesser areas such as Jacob Bigelow's *Florula bostoniensis* (1814; 2nd edn., 1824; 3rd edn., 1840), Stephen Elliott's *A sketch of the botany of South Carolina and Georgia* (1816–24), Constantin Rafinesque's *Flora ludoviciana* (1817), and William Darlington's *Flora cestrica* (1826; 2nd edn., 1837). The publication of *Flora boreali-americana* as well as the advent of state surveys in the 1830s furnished new opportunities and challenges. Torrey with Asa Gray commenced a continental flora on the natural system, *Flora of North America* (1838–43, not completed), while the well-financed New York survey provided Torrey with the opportunity to write his already-mentioned large-scale *Flora of the state of New-York* (1843). The needs of teaching at Harvard, as well as competition from an entrepreneur,

Alphonso Wood (author of *Class-book of botany*, first published in 1845), then turned Gray towards a concise regional work, his *Manual of botany of the northern United States* (1848). The works of Torrey and Gray by and large established the manual-flora format characteristic of state and regional works in North America; with variations, this has remained standard. Large-scale semi-monographic works remained few until the twentieth century.

The principles of the concise flora have scarcely been put better than in Bentham's *Outlines of Botany*. They first appeared in 1861 in that author's *Flora hongkongensis* (as pp. i–xxxvi) and would do so again in nearly all of the other 'Kew floras' as well as some other contemporary anglophone works.[77] They are embodied in the first five of the 247 aphorisms of the *Outlines*. The first three of these are particularly apropos and I repeat them here:

1) The principal object of a Flora of a country, is to afford the means of determining (i.e. ascertaining the name of) any plant growing in it, whether for the purpose of ulterior study or of intellectual exercise.

2) With this view, a Flora consists of descriptions of all the wild or native plants contained in the country in question, so drawn up and arranged that the student may identify with the corresponding description any individual specimen which he may so gather.

3) These descriptions should be clear, concise, accurate, and characteristic, so as that each one should be readily adapted to the plant it relates to, and to no other one; they should be as nearly as possible arranged under natural divisions, so as to facilitate the comparison of each plant with those nearest allied to it; and they should be accompanied by an artifical key or index, by means of which the student may be guided step by step in the observation of such peculiarities or characters in his plant, as may lead him, with the least delay, to the individual description belonging to it.

The second part of the fifth aphorism is also of some interest and is likewise quoted:

The botanist's endeavours should always be, on the one hand, to make as near an approach to precision as circumstances will allow, and, on the other hand, to avoid that prolixity of detail and overloading with technical terms which tends rather to confusion than clearness. In this he will be more or less successful. The aptness of a botanical description, like the beauty of a work of imagination, will always vary with the style and genius of the author.

Analytical keys and the 'manual-key' flora

The cleavage in the function of systematic works – especially floras – effectively espoused by Lamarck before 1793[78] gradually found practical expression outside France, though rarely in anglophone circles – very likely on account of, as we have seen, Bentham's effective integration in his manuals of analytical keys with concise floristic text.[79] As the century progressed, the pure analytical key was, however, modified to provide some of the elements of descriptive floras. Ultimate leads (those ending in a given taxon) were furnished with diagnoses, essential synonymy, and very concise (often coded) indication of distribution, habitat and other classes of information along with vernacular and accepted botanical names. The resulting form, which I am here calling a 'manual-key', thus – as Lamarck would have wished – came largely to supplement or complement larger descriptive works. These latter continued to be viewed as belonging to the herbarium, library or salon and, as we have seen, were ideally of a more or less encyclopedic (and sometimes 'prestigious') character. In this manner, for example, in Denmark the manuals and field-guides firstly of Johan Lange (*Haandbog i den Danske flora*, 1851; 4th edn., 1886–88) and later of Rostrup and Raunkiær came to complement *Flora danica*,[80] while in Central Europe Wilhelm Koch's *Synopsis florae germanicae et helveticae* (1837, with further editions to 1903) similarly complemented the Reichenbachs' *Icones*.[81] The real spread of the 'manual-key' in Europe came, however, only after 1870 as travel greatly increased, secondary school education became much more widespread, and recreational field-botany gained popularity. In 1878 the method made its first substantial appearance in Australia in Spicer's *Handbook of the plants of Tasmania*; von Mueller's larger but still compact *Key to the system of Victorian plants* followed a decade later.[82]

Enumerations and checklists

In spite of the new approaches to flora-writing espoused from early in the nineteenth century, economic forms of presenting systematic information, dating back to Bauhin's *Pinax* and exemplified by the many editions of Linnaeus's *Systema naturae* and its successor, *Systema vegetabilium*, yet continued to find favor. These included Carl Kunth himself who wrote two major lists, *Synopsis plantarum quas in itinere ad plagam aequinoctialem orbis novi Alexander de Humboldt et Amatus Bonpland* (1822–23) and the ambitious *Enumeratio plantarum omnium hucus cognitarum* (1833–50, not completed). His contemporary Carl Blume began a similar work for Java, also not completed: *Enumeratio plantarum Javae* (1827–28). These had followed such eighteenth-century works as the already-mentioned *Flora malabarica* by Caspar Commelijn, *Enumeratio systematica plantarum* (1760) by Nicholas Jacquin (a preliminary list of the West Indian flora), and J. R. Forster's *Florae americae septentrionalis* (1771). In such works, only essential information of greater or lesser brevity is provided, and descriptions and (for the most part) keys are lacking. They were prepared with an eye towards rapid and convenient publication of results, but at first they were regarded as summaries of, or precursors to, larger descriptive undertakings. Gradually, however, the synopsis or enumeration (and its even more telegraphic relative, the checklist – a genre in existence since the sixteenth century) developed into an independent area of floristic writing.

Although the greater part of the self-contained floristic enumerations – which from the mid-nineteenth century onwards appeared in considerably increased numbers – were for local or insular areas of relatively limited extent, a number were written for whole countries, groups of countries, or even subcontinents. Many earlier enumerations and checklists were more or less uncritical compilations and contained numerous errors, a problem of which even before 1600 Caspar Bauhin was well aware. Those based wholly or largely upon personal research by the author were not unnaturally more reliable. An important step forward towards respectability for the floristic enumeration in the mid-nineteenth century was Thwaites's *Enumeratio plantarum Zeylaniae* (1858–64). Its major development as a form, however, took place after 1870 as the great floristic richness of, for example, China as well as many tropical areas came to be realized.

While enumerations (and checklists) have been sharply criticized by some writers, such works, particularly if an attempt has been made at critical evaluation of taxa, should be regarded as better than no consolidated work at all. In many instances they represent the only serious work for botanically poorly known areas, especially in the tropics, and more than once have fared, or may well fare, better than semi-monographic floras. Even in extra-tropical areas use has sometimes been made of checklists where floras are large, as in Natal and Western Australia. Their authors and/or editors have often lacked the means and/or the time to prepare full descriptive works but believed some kind of consolidated publication, even if imperfect, to be necessary.[83]

The 'imperial' era and its aftermath, 1870–1930

By the time of the Franco-Prussian War in 1870–71 and the coeval foundation of the Second German Empire, the main genres of floristic writing had by and large assumed the forms which would remain standard for the next 100 or more years. While floristic works became more sophisticated through critical research and inclusion of new classes of information, they also grew intrinsically more 'remote' from wider audiences – in spite of the general substitution of modern languages for Latin in the majority of works. This probably reflects the relative stasis which had come over the practice of taxonomy from the latter part of the nineteenth century and the concomitant growth of interest in 'general' or 'causal' botany.[84] More particularly, the period after 1880 – and even more so after 1900, with the rise of genetics and ecology – was characterized by an evident lessening of interest in the production of major descriptive floras (save in certain circles in North America, western Europe, and the Russian Empire) as many key projects of the mid-nineteenth century were completed or became far advanced.[85] This trend was, however, offset by a rise in the publication of state and regional floras in increasingly better-known areas, among them parts of North America,[86] European Russia and adjacent areas, South Asia,[87] the Japanese Empire, southern Africa, and Australia where strong demand of one or another kind prevailed. The era also saw the production of the first tropical forest floras, beginning in the British Indian domains but by the first half of the twentieth century also in Malesia and Africa.[88]

Centers of influence: Kew and Berlin

Fortunately for posterity, the period 1870–1930 may be viewed as one of outstanding progress in synthetic systematics. In no small measure credit for this should go to contemporary patterns of geographical, economic and political development, particularly before World War I. This in turn gave rise in some countries to the emergence of large, mainly state-supported taxonomic centers with half a dozen or more specialist staff. The most important of these before 1900 were the Royal Botanic Gardens, Kew, and the Botanischer Garten und Botanisches Museum, Berlin. Several more such centers attained variously comparable positions of strength in the following two to three decades, mainly in Europe and North America; for most, either Berlin or Kew or both were models. Each of these institutions was active in both flora preparation and monographic studies; many were also concerned with 'lower plants' as well as 'economic botany'. A major impetus for the development of these centers was the need for effective knowledge of the flora of recently acquired or opened territories.[89] Strong personalities or social and cultural factors also played a role.

In the last third of the nineteenth century Kew was perhaps the most influential center. This was in no small measure due to the several colonial floras published or in preparation along with *Genera plantarum*, completed in 1883, and two major enumerations: one comprising the botanical part of *Biologia centrali-americana*, the other, *Index florae sinensis*. Yet Kew's taxonomic work overall was more practical than theoretical;[90] moreover, by the end of the nineteenth century decentralization and local scientific development had led to changes in priorities in flora-writing. In the author's opinion, the absence of an early successor to *Genera plantarum* on avowedly evolutionary principles – and a very marked emphasis on flora-writing as well as imperial consultancy related to what may be seen as a narrowly conceived remit – could be said to have placed Kew scientifically at a disadvantage.[91] Ultimately more influential for twentieth-century practice was the work of the integrated 'Englerian school' of taxonomy, phytogeography and comparative morphology in Berlin whose 'heyday' about spans our 60-year period.[92] As a research institute of the then-University of Berlin, it was more closely associated with the academic world than was Kew, until 1984 under the direct control of successive government departments.

This leadership from Germany was no isolated phenomenon: in the five decades before World War I the 'teuton' had come to dominate most branches of science and scholarship as well as to assume leadership in scientific bibliography.[93] This not unnaturally reflected the drive towards German unity – and later world influence – led initially by Prince Otto von Bismarck and, later, Kaiser Wilhelm II. What has come to be called the 'Englerian school' of systematic botany itself had in effect been founded in the 1870s by Eichler upon his appointment at Berlin; but because of his long tenure as Eichler's successor (1889–1921) and organizational ability this 'school' will always bear Engler's name.[94] Together with Ignatz Urban, who as already noted had been, like Eichler and Engler, seriously involved with *Flora brasiliensis* and in Berlin had connections in high places, and Ludwig Diels – Engler's later associate and successor and 'einer der letzten grossen, in der ganzen Welt geachteten deutschen Pflanzensystematiker'[95] – Engler was largely responsible for his 'school' becoming imbued with the scholarship and *Weltanschauung* which were to make it so influential.

With its leaders' intellectual interests and formal academic links, the *Berliner Kreis* was able to specialize in large-scale monographic works rather than floras. Among their lasting contributions were the monumental *Die natürlichen Pflanzenfamilien* (1887–1915; 2nd edn., 1926– , not yet completed) and *Das Pflanzenreich* (1900–), detailed series of regional revisions and other studies, notably for Africa and the western Pacific (*Beiträge zur Flora von Afrika, Monographien afrikanischer Pflanzen-Familien und -Gattungen, Beiträge zur Flora von Papuasien*, etc.), and plant geographical and vegetatiological studies (especially the monographic series *Die Vegetation der Erde*). On the other hand, apart from the continuation and completion of *Flora brasiliensis* their only major floristic work was the very detailed *Synopsis der mitteleuropäischen Flora* of Paul Ascherson and Paul Graebner. In their endeavors, the Berlin group was strongly supported by the botanical circle at Breslau, from 1884 until the end of the period led successively by Engler and (after 1889) Ferdinand Pax. Elsewhere in Germany there was rather less activity, save in Hamburg where the younger Reichenbach and his circle were completing *Icones florae germanicae et helveticae* (1837–1914), and in Munich where a successor of Eichler, Gustav Hegi, was responsible for another leading encyclopedic work, *Illustrierte Flora von Mitteleuropa* (1906–31; 2nd and 3rd edns., 1935– , not yet completed). On the whole, however, there were relatively few concise floras outside of the *Heimat*; the

only significant colonial work was Karl Schumann and
Karl Lauterbach's privately sponsored *Flora der deutschen Schutzgebiete in der Südsee* (1900; *Nachträge*, 1905)
for German New Guinea, Micronesia and Samoa.[96]
Indeed, floras as a whole perhaps were seen – particularly in Berlin – as secondary in relation to revisions,
monographs and taxonomic studies; Diels in his treatise on the methodology of systematics makes little or
no mention of them.[97]

The Russian Empire and the Soviet Union

Close links continued to exist between German
and Russian botanists into the 'imperial' era, although
under Czars Alexander III (1881–95) and Nicholas II
(1895–1917) more nationalistic policies came into
effect.[98] For many years in the latter part of the nineteenth century the head of the botanic garden in St.
Petersburg had been Eduard Regel who was also
editor of a leading German horticultural periodical,
Gartenflora. From the last decade of the century,
however, the prominent figures in floristics and plant
geography were all Russians. Expansion and development of the Russian Empire was vigorously pursued
and floristic exploration also promoted until World
War I as well as afterwards in the early Soviet period.
Significant works included, in the 1890s, *Flora srednej i
juznoj Rossii, Kryma, i Severnogo Kavkaza* (1895–97)
by I. F. Schmal'hausen and *Tentamen florae Rossiae
orientalis* (1898) by Sergei Korshinsky; after 1900 there
came *Flora caucasica critica* (1901–18) led by Nikolai I.
Kuznetsov in Jurjew, *Flora Altaja i Tomskoj gubernii*
(1901–14) by Porphyry N. Krylov in Tomsk, *Plantae
asiae mediae* (1906–16) by Olga A. and Boris A.
Fedtschenko, *Jakutskaja Flora* (1907) by B. A.
Fedtschenko, *Flora Evropejskoj Rossii* (1908–10) by
B. A. Fedtschenko and Alexander F. Flerow, the same
authors' *Flora asiatskoj Rossii* (1912–24), and, finally,
Flora Sibiri i Dal'nego Vostoka (1913–31) led by I.
Borodin and Dimitri I. Litwinow. Russian botanists
were also active in Korea, Manchuria and northern
China prior to 1910. The first Korean flora was
Conspectus florae Koreae by Ivan Palibin (1898–1901);
however, it had been able to make use of the already-
mentioned *Index florae sinensis*. Manchuria, then partly
under Russian influence, was the subject of the young
Vladimir Komarov's *Flora Man'čžurii* (1901–07); on
the other hand, his *Flora poluostrova Kamčatki*, prepared around 1910, was not published until 1927–30,
well into the Soviet era.

Komarov's Kamchatka flora was contemporane-
ous with the last of the 'primary' regional floras, *Flora
Rossiae austro-orientalis* by B. A. Fedtschenko and B. K.
Shishkin (1927–36; index, 1938), and the early volumes
of Krylov's *Flora zapadnoj Sibiri* (also begun in 1927).
Most of the rest of Siberia and the Far East was concisely covered in two works by Komarov, *Malyj opredelitel' rastenij Dal'nevostočnogo kraja* (1925, with
Evgenija Klobukova-Alisova), and *Vvedenie v izučenie
rastitel'nosti Jakutii* (1926). The former was shortly
afterwards expanded as *Opredelitel' rastenij
Dal'nevostočnogo kraja* (1931–32, with Klobukova-
Alisova), just as major organizational changes were
overtaking the botanical institutions in St. Petersburg
(by now Leningrad) and elsewhere. This very considerable effort over four decades was to be a cornerstone
for *Flora SSSR* which, however, belongs to the
'modern' era.[99]

Austria-Hungary and the Balkans

Austro-Hungarian botanists were also active in
this period, particularly in exploration and documentation of the flora of the Balkans and Greece. During the
late nineteenth century, most of their documentation
took the form of 'expedition reports', but after 1890
some notable floras were published: *Flora bulgarica* by
Josef Velenovský (1891; supplement, 1898), *Conspectus
florae graecae* by E. von Halácsy (1900–08; supplement,
1912), and *Flora Bosne, Hercegovine i Novopazarskog
sandžaka* by G. Beck von Mannagetta (begun 1903 with
three of four volumes by 1930). Serious summaries of
the Romanian flora likewise began in the 1870s, initially
with *Plantae Romaniae hucusque cognitas* by August
Kanitz (1879–81). After World War I, there were fewer
new initiatives but one major synthesis: *Prodromus
florae peninsulae balcanicae* by August von Hayek
(1924–33). Locally based activity also further developed, a notable contribution being the first edition of
Flora na Bălgarija (1924–25) by Nikolai Stojanov and
Boris Stefanov.

France

Among circles of floristic writing less directly
influenced by Central Europe, notice may be taken
firstly of those in other parts of that continent, followed by those in the United States and elsewhere.
Through the middle of the nineteenth century systematic botany declined in France along with much else in
national life.[100] Grenier and Godron's well-regarded

Flore de France – to which reference was made in the last section – lacked any significant successor until the 1890s; similarly, for some decades comparatively few overseas floras followed the example of Weddell's *Chloris andina* (1855–61). The suppression from 1853 to 1873 of any chair at the Muséum d'Histoire Naturelle in Paris specifically responsible for plant systematics, along with the not unrelated transfer by legacy of the Delessert Herbarium to Geneva in 1869, were serious setbacks.[101] Associated with this development was the continued formation and maintenance of several large private herbaria by wealthy amateurs including, in addition to the Delesserts, Ernest Cosson, Emmanuel Drake del Castillo, and Albert de Francqueville. It was upon them (and others) that French general systematic botany, with its strong tradition of monographic writing, largely rested in the mid- to late nineteenth century. They also contributed considerably to what overseas floras were produced in the last three decades of the nineteenth century.[102] At the same time, much of the abundant local and regional floristic work and publication within France was, as in other countries, the work of clerics and schoolmasters, one of whom, Abbé Hippolyte Coste, contributed a national flora of the greatest importance about which more will later be said.[103] The universities in this period became largely concerned with 'general botany', the school of comparative anatomy at the Sorbonne under Philippe Édouard van Tieghem enjoying particular prominence. No major floras were, however, associated with the Université de Paris prior to the advent of Gaston Bonnier in the 1880s. Montpellier, situated within the Mediterranean Basin and moreover with a long botanical tradition of its own, retained some distinctiveness as a place for the study of plant diversity under J.-É. and L. Planchon. In the early twentieth century under their successor, Charles Flahault, it began to emerge as an important center for the new disciplines of ecology and phytosociology.

Professional revival in systematics in Paris, led by Henri Baillon, L. Éduard Bureau and, somewhat later, Bonnier and Henri Lecomte, was gradual, Bureau having in his time at the Muséum but few resources at his disposal.[104] A strong stimulus was, not unnaturally, an awareness of and a felt need to improve knowledge of the extensive botanical resources of the developing Second French Empire. Until early in the twentieth century, however, the writing of overseas floras remained in the hands of individuals, sometimes with grants-in-aid from metropolitan or colonial authorities.[105] The concept of organization and teamwork in flora-writing already manifest elsewhere was first effectively implemented at the Muséum by Lecomte in 1906 following his succession to Bureau's chair. Influenced in particular by *Flora of British India* and perhaps also *Flora brasiliensis*, the latter completed in the same year, he initiated, under the sponsorship of the colonial administration in Saigon, *Flore générale de l'Indochine*.[106] With this work began the 'Paris school' of semi-monographic floristic writing which has continued to the present and has influenced the conduct and format of similar works in Belgium, Portugal and even Britain.

Floristic botany in France itself in the late nineteenth and early twentieth centuries, also pervaded by a renewed spirit of optimism, was marked by the production of three major national floras, two by nonprofessionals and the third by the already-mentioned Sorbonne professor Gaston Bonnier. The *Flore descriptive et illustrée de la France* by Coste was perhaps the most influential. It was in part based on the style established by Britton and Brown in their *Illustrated flora of the northern United States, Canada and the British possessions* (1896–98) and enjoyed considerable input from Flahault in Montpellier. Bonnier's manual (1909) and illustrated flora (1911–35), the latter doubtless inspired by Hegi's contemporary *Illustrierte Flora*, have also enjoyed lasting popularity.[107]

Belgium, the Netherlands, Italy and the Iberian Peninsula

Detailed national floras were also published in Belgium, Holland, Italy and Spain and Portugal at this period, likewise largely through the efforts of amateur and para-professional botanists. In Belgium the advent of the Congo Free State after 1884 gave its botanists an important new opportunity for tropical studies. Rather generous support was furnished by Léopold II for explorations, institutions and publications, but in the time of Théophile Durand and Émile De Wildeman the extent of knowledge was considered premature for a descriptive flora. Their main synthesis was *Sylloge florae congolanae* (1909), published shortly after transfer of responsibility for the Congo to the Belgian government. Italian overseas floristic contributions were mostly *ad hoc* undertakings resulting from individual needs and interests rather than as parts of any common strategy. Emilio Chiovenda's main publications on Somalia were moreover prepared in some isolation; like the French in Indochina, many of his novelties, particularly genera, have gradually fallen into synonymy. Local undertakings in Spain and Portugal were relatively limited. In the latter part of the nineteenth century the main development therein was *Prodromus florae hispanicae* (1861–80; supplement, 1893) by Moritz Willkomm and Johann Lange. This laid a foundation for various national and regional undertakings, the most notable being *Flora de Catalunya* (1913–37) by Juan Cadevall i Diars and his associates. By contrast, the remaining tropical possessions were largely neglected, save in Mozambique where T. R. Sim was contracted to write a forest flora (1909).

With respect to the Netherlands, the increasing wealth of the East Indies possessions and the strong directorships of Rudolph Scheffer and Melchior Treub enabled the *'s Lands Plantentuin* in Buitenzorg (Bogor) near Batavia (Jakarta) to flourish as a biological center, not unnaturally including considerable contributions to regional floristic botany. It was, however, becoming clear that the Indies had, like Brazil and China, an extremely rich flora and Treub accordingly took the view that a new descriptive work in succession to that of Miquel was premature. An interim solution was *Handleiding tot de kennis der flora van Nederlandsch-Indië* by J. G. Boerlage (1890–1900), a 'generic flora' with lists of species (also accounting for published records in neighboring territories). More detailed documentary efforts became focused on Java, one of which ultimately was realized as *Flora of Java* (1963–68).

The British Empire: later developments

Returning to the United Kingdom, certain changes in interests developed after 1900. As already mentioned, by this time Kew had attained a peak in its range of floristic undertakings. Even the Herbarium itself was administered on a geographical basis, its 'sections' remaining so organized until 1970. Subsequent research, however, became strongly centered on Africa, with an emphasis on two very large works: *Flora of tropical Africa* and *Flora capensis*. With these largely completed by 1920 or so, efforts were directed towards a partial revision of the former as *Flora of west tropical Africa* (1927–36). With respect to the Americas, projected floras for what are now Guyana and Belize never

materialized; and Jamaica, not surprisingly, became with the rest of the Caribbean an interest of the British Museum (Natural History). As for Asia, following the completion of *Flora of British India* in 1897 and the *Handbook to the Flora of Ceylon* in 1900 responsibility for territories east of the Suez was largely devolved to Calcutta and Singapore. The several Indian regional floras of the post-1900 period, while in part prepared at Kew, were not directly a charge on that institution.

United States of America and its 'spheres of influence'

Across the Atlantic in the United States, where until 1890 or so the botanical frontier was largely domestic, the works of Torrey and Gray (and their associates and students) had caused the manual-flora to become firmly established as a genre for most U.S. descriptive regional (and, later, state and some overseas) works. This style has for the most part remained conventional, with a gradual increase in the number of illustrations as the twentieth century has progressed. Until about 1930, however, the majority of states were covered only by checklists or documentary enumerations, sometimes with keys. Several of the accounts were part of a programme instituted by Frederick Coville, for many years chief botanist to the U.S. Department of Agriculture and head curator of the U.S. National Herbarium.[108]

The efforts of the Torrey/Gray 'school' of botanists to establish a lasting tradition of sound scholarship (as well as to complete Gray's *Synoptical flora of North America* (1878–97), itself a continuation of Torrey and Gray's original *Flora of North America* project begun 40 years before) were, however, to fail for a time. One factor – not surprisingly – was regionalism, notably in California; of greater import, however, was the rise to a dominant position of the Britton 'school' in New York.[109] From the New York Botanical Garden in Bronx Park (founded in 1891 with assistance from some of the newly wealthy 'robber barons' of the era) Britton and his group authored numerous works both for different parts of the mainland United States (and southern Canada) and for Bermuda and the Caribbean. Among them were Britton's own *Manual of the flora of the northern states and Canada* (1901; 2nd edn., 1905, 3rd edn., 1907) and *Flora of Bermuda* (1918), J. K. Small's *Flora of the southeastern United States* (1903; 2nd edn., 1913) and *Manual of the southeastern flora* (1933), and P. A.

Rydberg's *Flora of the Rocky Mountains and adjacent plains* (1917; 2nd edn., 1922) and *Flora of the prairies and plains of central North America* (1932). All were written not only with a view towards the geographically and numerically rapidly increasing commercial market but also as part of an overall strategy, including propagation of the 'American Code' of nomenclature. This code, distinct from the 'gentlemen's agreement' international code used in other parts of the world as well as by Gray and his circle, was introduced in 1892 amidst a rising tide of nationalism and widely used in the United States until 1930.[110] More ambitiously, Britton also sought to prepare a complete flora of the continent; following preliminary planning in the 1890s[111] this commenced publication in 1905 as *North American flora*. He also introduced the systematically illustrated flora, producing *An illustrated flora of the northern United States, Canada, and the British possessions* (1896–98; 2nd edn., 1913; both co-authored with Addison Brown). This inspired several comparable works, both at home (notably *Illustrated flora of the Pacific States* (1923–60) by Leroy Abrams and Roxana Ferris) and abroad (notably Coste's French flora).

In addition to preparing works on different parts of North America, botanists from the United States were also active in floristic writing in Middle America, the Philippines and, in the last decade or so of the period, in China and British India. Major contributions, all published after World War I, include *Flora of the Bahama Archipelago* (1920) by N. L. Britton and C. F. Millspaugh, the already-mentioned *Trees and shrubs of Mexico* by P. C. Standley (1920–26), *Enumeration of Philippine flowering plants* by E. D. Merrill (1923–26), and *Botany of Porto Rico and the Virgin Islands* (1923–30) by Britton and P. Wilson. In China, botanists at Lingnan University in Canton (Guangzhou) focused on little-explored Hainan, with a first contribution, *An enumeration of Hainan plants* (1927), by Merrill.

Other areas

Elsewhere in the Americas, political instability, limited resources and social factors discouraged serious flora-writing save in Uruguay, Argentina and Chile. In these countries relatively extensive exploration, interested botanists, and developing local institutions led to a number of floras, among them *Flora de Chile* by K. Reiche (1896–1911, not completed) and *Flora uruguaya* by José Arechavaleta (1898–1911). In Brazil, four parts of a *Flora paulista* by A. Löfgren and G. Edwall

appeared between 1897 and 1905; it then ceased, far from complete. Interest in floristic botany in most of Latin America generally remained relatively limited and publications few until the second half of the twentieth century.

The situation in Canada, southern Africa, Australasia and northeast Asia was broadly more favorable. In Australia, Ferdinand von Mueller published two editions of a continent-wide census in 1882 and 1889, after which emphasis switched largely to the individual colonies (or, from 1901, states) as they variously acquired the capacity to prepare basic state floras or checklists. Several such appeared between 1890 and 1931, between them covering the whole country. As the twentieth century progressed, however, interest in biology was becoming more diversified with relatively less emphasis on taxonomy and floristics; moreover, the fashion for museums had passed.[112] In spite of representations as early as 1909, the formation of the Commonwealth did not soon lead to the institution of a biological survey, a national natural history museum, or sponsorship of major floristic or faunistic works. In Canada, by contrast, a differing political history, the needs of geology, settlement and agricultural development, and strong representations on the part of the pioneer paleobotanist John W. Dawson resulted in the formation of a national museum as well as an agricultural center at Ottawa in the decades after federation in 1867. A botanical survey was also established, one major result of which was John Macoun's *Catalogue of Canadian plants* (1883–1902). Similarly, a national botanical institute was established in South Africa relatively soon after formation of the Union in 1910, and the later parts of *Flora capensis* featured several contributions from local botanists. A national herbarium was also formed in New Zealand, though only near the end of our period and following the development of substantial collections in Auckland and Wellington. In northeast Asia, the foundation of Tokyo Imperial University and its science faculty in 1877 enabled the formation of a cadre of botanists who, with the aid of Franchet and Savatier's *Enumeratio* and other 'Western' works – notably those of Karl Maximowicz as well as Siebold and Zuccarini – were able by 1910 to furnish primary floras for Japan, Taiwan and Korea along with the Ryukyu and Kurile Islands. By 1925 most of the flora of the Japanese Empire could be incorporated into a large manual, *Nihon shokubutsu sôran* (1925; 2nd edn., 1931).[113]

Overall trends

Altogether, the 'imperial' or 'post-Darwinian' phase of systematics, though long scarcely considered in most general histories of science or even biology and until recently seldom the subject of retrospective studies,[114] was one of continuing great publication efforts. This reflected a 'silver age' of exploration and natural history which accompanied imperial expansion; within the field it featured attempts at global influence through major efforts in consolidation (as exemplified by the Englerian school), international competition, chauvinism in the United States and elsewhere. In broader scientific terms, there were some attempts to merge evolutionary theory with taxonomy; systematic studies began to include speculations about putative ancestors and patterns of diversification over time. Nevertheless, floras (and other works) became increasingly stylized and academic; their role in communication gradually was lost sight of through misapplication of old formats to new different situations along with philosophical drift. At the same time, the interests of systematics and other areas of botany gradually diverged, with the latter seen as more 'progressive'. I see this here as forming a botanical parallel to some of the social, political and architectural developments of the period. There were, nevertheless, some rays of light, among them the appearance of a relatively forward-looking textbook, *Methods and principles of systematic botany* (1925) by the U.S. Department of Agriculture agrostologist and protagonist of nomenclatural reform Albert S. Hitchcock, and the emergence of 'experimental taxonomy' (or, as it would later be known, biosystematics) in Europe, North America and elsewhere.[115]

The 'modern' era

In considering developments after 1930 two sub-themes – which to an extent overlap – manifest themselves. The first centers around a marked renewal of flora-writing as well as accompanying intellectual and methodological developments, while the second concerns the rise of information technology in relation to floristic documentation.

Mid-twentieth century developments and intellectual renewal

The 'modern' era of flora-writing is for convenience here dated from 1930. Apart from it being a time of general economic recession, a notable event was the Fifth International Botanical Congress at Cambridge (England) at which a unified nomenclatural code was agreed. In that year also came the death of Engler (with that of Urban following early in 1931) and, in the Soviet Union, the organization of the *Flora SSSR* project. A year previously, Britton had retired from the directorship of the New York Botanical Garden. From this time forward, in part stimulated by the appearance (from 1934) of the successive volumes of the Soviet flora, renewed interest was shown in flora-writing. Old schemes were reviewed and new ideas floated; among the latter were proposals for modern floras of Indonesia, Europe and different parts of Africa.

This trend in floristics intensified after World War II with additional factors being proximally the loss in 1943 of most of the Berlin Herbarium, the political division of Germany, and the effective end of *Das Pflanzenreich* but also a greatly increased interaction of science with the public accompanied by changes in the nature of research funding. There was also a consummation of a shift, already apparent before 1930, of the center of gravity in the scientific world from Central Europe to the United States and its anglophone associates. Without such stimuli as *Pflanzenreich*, serious monographic work tended to lose ground to floristics except where the new cycle of floras provided suitable outlets or opportunities.[116] For some, including the author, this has been a regrettable but perhaps inevitable development.[117] By 1958 systematic botany was well and truly into a new 'Age of Floras', with activity extending over most of the earth's land surface; this 'boom' has continued to the present time.[118] From the 1970s it has also been influenced by rising public concern with the environment and the implementation of international conventions; in the same period, however, the climate for basic research has become more difficult.

The 1930s also saw the entry of major philosophical and other changes to systematic biology which have largely continued to the present. Of particular importance was the definitive formulation of the 'evolutionary synthesis' – which in effect was a manifestation in biology of the interwar Viennese 'unity of science' movement – as well as advances in the methodologies and techniques now collectively grouped under 'biosystematics'.[119] Under the impulse of what came to be known as the 'New Systematics', plant systematics – in principle at least – became more dynamic and outward-looking. Hitherto it had been for the most part a static typological exercise based largely upon comparative morphology, even retaining some elements of creationism and essentialism; indeed, for the best part of a century the 'Darwinian revolution' of 1859 had little effective influence on much of taxonomic practice.[120] It thus began to interact with ecology, karyology, biochemistry and molecular biology; much more attention was paid to variability, evolutionary dynamics, and the potential for parallelism, convergence and neoteny as expressions of change. Indeed, as a whole, considerations of time became more integral to taxonomic thinking. In an increasingly technological age, the 'biosystematics' movement improved the credibility of systematics in the eyes of other, largely laboratory-based scientists, notably in the United States or elsewhere where reductionism has been a serious force.[121]

Among notable developments of the 'modern' period may be mentioned the following:

1. The successful prosecution of *Flora SSSR* and the publication of an all-but-comprehensive range of regional floras and manuals for the then-Soviet Union (see Region 68/69 and Superregion 71–75).[122] In the People's Republic of China, a similar comprehensive programme of floristic documentation at national and provincial level has been in place since the 1950s. Heading this has been the large-scale *Flora reipublicae popularis sinicae*, publication of which commenced in 1959 (with an interruption in the late 1960s and early 1970s).[123]

2. The initiation (in the 1950s) and successful completion of *Flora Europaea*, of which the last of five volumes appeared in 1980 after 25 years of work. The history of the project, the most important of its kind in Europe for a century or more, has been well described.[124] A number of corollary projects were then or afterwards initiated, some of which are continuing.[125]

3. Commencement of new floras of Australia and North America. In Australia, the *Flora of Australia* project was formally approved in 1979 after 20 years of agitation and the first volumes were published in 1981–82. Publication has continued more or less steadily until the present, if more slowly than originally envisaged.[126] The path to a new flora of North America, a continent with no complete coverage since the early nineteenth century, has been as long if not

longer. A first modern initiative was the Flora North America Program of 1967–73; this, however, was abruptly terminated before any definitive results had appeared. The present *Flora of North America north of Mexico* is the result of a new initiative dating from 1982; formal support from the U.S. National Science Foundation was achieved in 1989 and publication commenced in 1993.[127]

4. The initiation of a goodly number of large-scale serial floras for various parts of the tropical zone as well as some extra-tropical areas (apart from Australia and North America). Among them are *Flore de Madagascar et des Comores* (1936–), *Flora of Peru* (1936–71; 2nd series, 1980–), *Flora of Panama* (1943–81), *Flora of Guatemala* (1946–77), *Flora Malesiana* (1948–), *Flora of tropical east Africa* (1952–), *Flore du Cambodge, du Laos, et du Viêt-Nam* and *Flora Zambesiaca* (1960–), *Flora Iranica* (1963– , now all but completed), *Flora of Turkey* (1964–88), *Flora Neotropica* (1966–), *Flora of Ecuador* (1966–), *Flore des Mascareignes* (1976–), *Flora de Colombia* (1983–), *Flora of India* (1993–), and *Flora Mesoamericana* (1994–). Many of these works represent a renewal and continuation of 'territorial flora' traditions which arose in many of the large taxonomic centers in Europe and North America during the latter part of the nineteenth century and beyond. Others have been entirely new initiatives. Progress, however, has often been slow, particularly where necessary materials have become widely scattered or where means or interest are limited.[128]

5. The publication of numerous modern, relatively concise, and more or less critical floras and manuals for many different parts of North America and Europe, with a more scattered output of such works in and for other parts of the world. Although viewed as 'routine' by van Steenis and others, there is a steady or even increasing market and, as van Steenis himself admitted, their quality has generally improved.[129]

6. The increasing development of inter-institutional and international links among botanists, largely replacing the old nationally oriented schools (and one-time colonial rivalries). Apart from the umbrella International

Association for Plant Taxonomy (established in 1950), these links include AETFAT (1951, for Africa south of the Tropic of Cancer), the Flora Europaea Organisation (1955–80; partially succeeded by the European Science Research Councils (ESRC) Ad Hoc Group on Biological Recording, Systematics and Taxonomy and, more recently, by the Euro–Mediterranean Initiative in Plant Systematics (now styled Euro+Med PlantBase), the Flora Malesiana Foundation (1950), the Organization for Flora Neotropica (1966), and OPTIMA (1975, for the Mediterranean Basin). To these should perhaps be added the Global Taxonomy Initiative (GTI). Forms of communication and gathering have varied, but most have conducted different kinds of serial publications along with periodic symposia; in more recent years electronic media including Web sites and newsgroups have been established. Similar groups may exist within more inclusive associations, for example the Committee on Pacific Botany of the Pacific Science Association. Official and unofficial agreements aimed at coordination of activities and reducing or eliminating duplication of floristic coverage also exist.

Offsetting these progressive developments have been a number of setbacks. Some projects have appeared just to 'run out of steam' on account of the death (or retirement) of their moving spirits, loss of institutional interest, lack of finance, or for other reasons.[130] Others have been variously affected by political or other external circumstances, either temporarily or permanently. As already indicated, the initiation of some projects has been more or less long drawn-out due to factionalism, government attitudes, and other factors.[131] A lack of firm organization and management, with consequent dissipation of available resources, is a probable major cause of many delays.[132]

Of all of the setbacks in the period after 1930 the most notable was perhaps the sudden termination of the first Flora North America Program in early 1973 shortly after it became 'operational'.[133] A first attempt at revival in the latter part of the 1970s was less ambitious; it soon, however, sank without trace at a time of political paralysis. A second attempt, initiated by the American Society of Plant Taxonomists in 1982, would in the end be more lasting. Ironically, its first task was, in the words of one contributor to a precursory 1988

workshop, seen as '*not* to design computer databases of floristic information'.[134] While a role for information technology was clearly envisaged, there was an equal concern for a more traditional presentation of results, i.e., in a series of volumes comparable to *Flora Europaea* or the *Flora of Australia*. The new plan also called for less of a superstructure and more producer and user participation (including the establishment of a newsletter). As already noted, this approach was successful and *Flora of North America north of Mexico* is now a reality.[135]

All in all, however, the mid-twentieth century – the years of World War II excepted – should be seen as one of marked progress in world floristic documentation. The somewhat bleak picture painted by Blake in 1939 (Blake and Atwood, 1942) and Lawrence (1951) – the latter making particular reference to southwestern Asia – became rather less so by the 1970s. Sufficient advances had been made for some serious estimates, in cartographic form, of world levels of plant species richness by Malyschev[136] as well as thoroughness of floristic knowledge.[137] With additions, these would over the next two decades contribute to overall knowledge of world biodiversity as well as to conservation needs assessment. Yet major gaps in published knowledge remained, notably in parts of the tropics where much alpha-taxonomic research was still required; in many other cases available coverage was aging or otherwise unsatisfactory.

The rise of information technology

Of profound importance in recent decades has been the introduction into systematics and floristic botany of information systems technology. Acceptance of the new methodology was at first slow, and from time to time criticism was voiced. Among its earliest advocates was the International Business Machines (IBM) engineer and botanical enthusiast Sydney W. Gould.[138] In the 1950s and 1960s he successively developed his 'International Plant Index' (1962) and 'Geo-code' (1968–72) as well as a standardized list (with A. C. Noyce) of authors of plant generic names (1965).[139] From 1953, automatic data processing was also introduced for the mapping scheme later published as *Atlas of the British Flora* (1962).[140] Computers moreover came into use in the then-developing field of 'numerical taxonomy', and in addition began to be explored as potential tools in museum documentation.

The development of database and other more

advanced software tools, along with introduction in the mid-1960s of the first 'third-generation' computer systems, notably the 360-series of IBM, represented, however, decisive steps forward. Projects for their application in systematics were mounted by, among others, Theodore J. Crovello at Notre Dame University, the original Flora North America team under Stanwyn Shetler, and a group in Mexico City (later Xalapa, Veracruz) under Arturo Gómez-Pompa. Integrated floristic systems were developed by both the Washington and Xalapa teams, the latter with a slight edge and in the end continuing to the present, with *Flora de Veracruz* (1979–) one of its key products.

Interest in the more advanced technology also developed rapidly outside North America wherever there was forward-looking leadership. In Pretoria, South Africa, an important regional database, PRECIS, was created; in time this developed into a full taxonomic information system.[141] From it a first complete checklist of the vascular flora of southern Africa, *List of species of Southern African plants*, was produced in 1984; two further editions have since appeared. At the end of the 1970s, support was found for development of a prototype European floristic information system.[142] A database was also created for *Med-Checklist*. Routines for key-writing and truly comparative descriptions also began their development in the 1970s.[143]

The high overheads and custom software of the 'big iron' era did, however, for several years militate against more extensive application of computer-based information technology, although in the 1970s this was offset by the spread of 'minicomputers'.[144] Effective penetration of the new technology began only in the late 1970s with the widespread introduction of visual terminals, firstly for word processing and then also for data with the rapid spread of personal computers; this will be discussed more fully in Chapter 3. The publication of *Databases in systematics* (1984), based on a 1982 symposium, establishes a useful point of reference encompassing the penetration of information technology (I.T.) from the 1960s through the advent of personal computers in the 1970s to less expensive mass storage for data, the appearance of the CD-ROM, networking, and portables in the 1980s.[145] As we come nearer the present, the use of I.T. in floristic work as well as in other areas of systematics has become all but routine, although effective standardization of packages, data libraries, etc., is far from achieved.[146]

Less obviously, but just as significantly, there have also in recent years been reassessments of several larger flora projects. Not only has progress often been slower than projected, but in some cases the estimates of the numbers of species involved have increased with continuing botanical exploration.[147] At the same time, changing circumstances as well as new technologies have had to be taken into account. This has led to decisions to accelerate production of certain works, notably *Flora of tropical East Africa* and *Flora Zambesiaca* at Kew (with completion of both envisaged by 2005) and *Flora Malesiana* (through simplification of its presentation and recruitment of a greater range of collaborators). At the same time, in at least one institution an increasing emphasis may be placed on monographic studies, including contributions towards a new *Species plantarum*. The 'Age of Floras' may for a time continue, but necessary change may eventually as such limit its run in its present form. What does seem very probable is that in the twenty-first century floras as such will be not merely also on-line but entirely absorbed into floristic information systems, with the more traditional forms comprising just some of its possible products. Already such is the case with some basic checklists as, for example, *Queensland plants* (**435**) or *Plants of southern Africa: names and distribution* (**510**). A more elaborate example is *Atlas of endemics of the Western Ghats (India)* (**802**), with a CD-ROM supplementing the printed work.

Summary

Floristics as a distinct area of botanical enquiry began with the work of Clusius and Thal in the last third of the sixteenth century. Early floras were largely enumerative or took the form of elaborate documentation of 'exotica'. A more systematic approach, already evident in some late seventeenth century works, became general as the eighteenth century progressed; many basic principles and practices were 'codified' by Linnaeus. Analytical keys were introduced beginning with Lamarck's *Flore française* in 1778; phytography itself gradually became more sophisticated and formal in a style first developed, also by Lamarck, for his *Encyclopédie* (1783 onwards) and in 1805 applied in *Flore française* by Augustin-Pyramus de Candolle. Arrangements of taxa initially followed 'folk' tradition, later progressing through the Linnaean system (which proved very effective at increasing access) to systems based on 'natural' affinities within and among major taxa.[148] Rapid progress in primary documentation of the flora of northern, central and parts of southern Europe characterized the eighteenth and early nineteenth centuries, and initial lists were compiled for various other parts of the world. In the hundred or so years from 1815 to 1930 most of the rest of the globe became 'covered' with floras, although outside Europe, eastern North America and some other largely temperate regions their basis was relatively sketchy. For the most part, however, they remained purely phytographic. A nineteenth-century innovation derived from the Lamarckian key form and applied mainly in Europe was the field manual-key, a very concise form of a flora aimed at identification and provision of basic information.

The latter half of the twentieth century has seen a great rise in the production of major floras. The bulk of the vascular floras of the north temperate and the two polar zones and, increasingly, the south temperate zone have by and large become reasonably well known with respect to inventory, the available information now variously consolidated into more or less readily accessible forms. The same applies for scattered areas in tropical and subtropical zones, often where a substantial history of local botanical endeavor has existed. On the other hand, over the bulk of the tropics and subtropics, including tropicmontane zones, all or most of the greatest importance to a proper understanding of the earth's vascular flora, floristic progress has been uneven. Efforts by individual persons, institutions, or other organizations have played an exceptional role, more often than not in the absence of general movements as well as official indifference or suspicion. Large areas still remain imperfectly studied and documented. Moreover, what literature is available is often so out of date as to be all but valueless for anything save professional revisionary work. Even so, in these zones significant progress has been made, notably in Africa, various parts of Asia and Malesia, Middle America, and the Pacific but also in Australia (although there it was gradual until the 1970s) and South America. This contrasts very positively with the opinion of Blake and Atwood that, as of 1939, outside of Europe and parts of northern Asia only Greenland, Australasia (in my view partly mistakenly), and some islands could be considered floristically relatively well known.[149] This is all the more so as standards of knowledge and documentation have increased substantially in the past six decades compared with the previous century or more.

Given this situation, as well as for economic, technological and intellectual reasons, it is perhaps not surprising that quite recently voices have been heard advocating a renewed emphasis on monography as a primary basis for floristic accounts. Certainly, in the last decade or so relatively few major new flora projects have been initiated when compared with the post-World War II era; in addition, there have also been reassessments of several of those still current. If, as was earlier suggested, floras do become parts of information systems, then – in addition to on-line access – 'traditional' printed materials can be generated on demand or formally published from time to time, complemented or supplemented by CD-ROMs with, for example, additional illustrations, maps, or interactive keys. This and other issues will be further considered in Chapter 3.

Notes

1 Stafleu, 1959.
2 Thorne, 1971.
3 Gómez-Pompa and Butanda, 1973; Gómez-Pompa and Nevling, 1973.
4 Jäger, 1978.
5 Morin *et al.*, 1989.
6 Ng, 1988; Funk, 1993; Jarvie and van Welzen, 1994; Palmer, Wade and Neal, 1995.
7 For example, Stace, 1989, pp. 224–226. There has also been a recent decision at a major institution to move away from large-scale floras once current commitments are completed.
8 Morton, 1981.
9 Sachs, 1875; Wein, 1937. Indeed, Wein argues that historically the meaning and scope of the word 'flora' have varied widely, only taking on more definitively the sense used here following its adoption by Linnaeus for his account of the plants of Sami, *Flora lapponica* (1737). A further attempt to redress the historiographical gap was made by Max Möbius in his paper 'Entstehung und Entwicklung der Floristik' (Möbius, 1938). Written as a supplement to his 1937 book *Geschichte der Botanik* (Jena: Fischer), it has been of considerable assistance in working up the present chapter.
10 Linnaeus, 1736 (see also Heller, 1983 [originally publ. 1970], pp. 146–204).
11 Brenan, 1979.
12 Morton, 1981, pp. 117ff. For early botany see also Reeds, 1991.

13 Wood blocks provided a relatively inexpensive means of illustration and became widely used in the sixteenth century (Blunt and Stearn, 1994, especially chapters 4–6).

14 Sivarajan, 1991. Of these works, the illustrations (by Johann Weidlitz) in *Herbarium vivae eicones* constitute a significant technical and artistic advance in botanical illustration (Blunt and Stearn, 1994, pp. 61–63).

15 The later works of Dodoens, along with those of Lobelius (De L'Obel) and Clusius (De L'Écluse), were all published through the famed Antwerp house of Plantin-Moretus which ensured a wide distribution. Their illustrations were drawn from a common pool – still extant – developed by the firm (Blunt and Stearn, 1994; personal observations, 1997).

16 Möbius, 1938, p. 297.

17 Gesner's published writings are largely non-botanical. In 1555, however, he produced *De raris et admirandis herbis* which included plants from Mt. Pilatus near Luzern. Over the last 10 years of his life – cut short by the plague in 1565 – he initiated a *Historia plantarum*, an illustrated work on all known plants similar to his *Historia animalium*. Never published in his lifetime, some fragments were subsequently incorporated in *Opera botanica* (1751–59, edited by C. C. Schmiedel). Additional illustrations, advanced for their time, were lost sight of after his death and only rediscovered in Erlangen in 1929. In 1973–80 a facsimile edition of some 1500 drawings in eight volumes appeared in Zürich as *Conradi Gesneri Historia Plantarum Facsimile-Ausgabe*, edited under the direction of H. Zoller, M. Steinmann and K. Schmidt. For further details on Gesner, see Wellisch, 1984.

18 For a fuller discussion of the Bauhins and their work, see Reeds, 1991. Of the *Pinax*, William Sherard in Oxford possessed an annotated copy to which Dillenius, as part of his duties, had earlier contributed. Never worked up for publication, this copy remains in Oxford University (Ewan, 1970, p. 27).

19 Möbius, 1938.

20 The work appeared posthumously as an appendix to *Hortus medicus et philosophicus* by Joachim Camerarius the Younger. For an appreciation, see Greene, 1905.

21 As a flora it was highly praised by A.-P. de Candolle (de Candolle, 1813).

22 The next attempts would be those of Lamarck in *Encyclopédie méthodique* and A.-P. de Candolle, who in 1812–13 conceived his *Regni vegetabilis systema naturae*.

23 Möbius, 1938.

24 His personal 'symbol', the twinflower *Linnaea borealis*, appears on the title-page.

25 de Candolle, 1813, p. 269.

26 Greene, 1905, p. 115.

27 First published as pp. xv–xxxii in the second edition of *Flora suecica*.

28 This was based on a herbarium left by Paul Hermann, but different from that used for Hermann's own *Museum zeylanicum* (1717).

29 A leading opponent was von Haller.

30 Williams, 1988. Meritorious features of *Flora gallo-provincialis* included the adoption of Bernard de Jussieu's proto-natural system and a fuller treatment of geographical distribution than customary in Linnaeus's works. Like *Flora helvetica*, it would also be noticed by Augustin-Pyramus de Candolle, who commented favorably on its topographical features, i.e., handling of plant distribution (de Candolle, 1813).

31 Stafleu, 1971a.

32 Linnaeus, 1753 (see also Heller, 1983 [originally publ. 1976], pp. 239–267).

33 *Exoticorum libri decem* contains one of the first known illustrations as well as a good contemporary description of the great banyan fig of South Asia, *Ficus bengalensis*.

34 *De plantis Aegypti liber* contains perhaps the earliest illustration of a banana plant in European botanical literature.

35 Examples include the island florulas by W. B. Hemsley in the British *Challenger* reports and Heinrich Schenck in the reports of the German *Valdivia* expedition.

36 *Flora danica* remains famous, not the least on account of the porcelain based on its illustrations.

37 All these were part of the great 'age of botany' which coincided with the Romantic era. Egerton has argued that a catalyst was J. J. Rousseau's letters to Mme. Delessert (posthumously published in 1781, but written from 1771 to 1773 and circulated in contemporary salons). See Williams, 1988.

38 de Candolle, 1873.

39 Voss, 1952; Stafleu, 1971b. While Lamarck had rejected the suprageneric taxa of Linnaeus, he may have judged the introduction of families, still a relatively new concept, premature or even irrelevant in the context.

40 Pankhurst, 1978, 1991; Stace, 1989. A possible motive for Lamarck was the relative size of the French flora, at least double those of, for example, Britain, Ireland and the Netherlands.

41 Duris, 1993.

42 Cf. de Candolle, 1873. Soon after the publication of *Flore françoise*, the first edition of Rousseau's *Lettres* appeared which, as already indicated, opened botany to a truly wide popular audience. This promoted a demand for popular floras as well as the already-mentioned fine illustrated works for the salon. The Linnaean system remained widely used for these; it would be another generation before 'natural' arrangements of plants made their way into floras.

43 Stafleu, 1971*a,b*.

44 Charles Panckoucke was in 1768–69 a 'rising young Paris publisher'. His first proposal was for a reprint of the Diderot edition along with a multi-volumed supplement; the latter appeared in four volumes. Though part of the plan, due to a personal offence – expressed in the end in dots – Diderot took no part in the enterprise. See A. M. Wilson, 1972. *Diderot*. New York: Oxford University Press (especially pp. 578–579). His own *Encyclopédie méthodique* followed, an entirely new work – begun just within Diderot's lifetime.

45 de Wit, 1949, p. xcvi.

46 This apparent persistence of essentialism has also been discussed in more general terms by Sivarajan (1991, pp. 17–18) and earlier by Hull (1965).

47 In the next decade de Candolle would develop his own, rather differently conceived system.

48 Curiously, however, de Candolle did not mention analytical keys, in spite of their presence in *Flore française*.

49 de Candolle, 1873.

50 Boissier's arrangement of *Euphorbia* in volume 15(2) (1862) has, for example, never been entirely supplanted (M. Gilbert, personal communication).

51 de Candolle, 1873, 1880.

52 Bentham, 1861, 1874.

53 Stevens, 1994.

54 van Steenis, 1954.

55 These as a genre enjoyed their zenith in the period 1770–1850 with a new wave in the late nineteenth and early twentieth centuries.

56 Bentham, 1874.

57 McVaugh, 1955.

58 Bartling's work was followed by *Plantarum vascularium genera* (1836–43) by Carl Meisner of Basel and particularly *Genera plantarum* (1836–40; supplements, 1842–50) by Stephan Endlicher in Vienna. The last-named, as head of the Austrian state herbarium, would become the first managing editor for *Flora brasiliensis*. Of these works, Endlicher's was surely the most influential.

59 Sachs, 1875. In many ways this critique was to have an influence which, with reiterations, has in one or another form continued to the present.

60 An extended discussion of the work's merits and defects is presented in Anonymous, 1861.

61 Linnaeus, 1753 (see also Heller, 1983 [originally publ. 1976], pp. 239–267).

62 These latter found expression in various ways, notably in 1830, 1832, 1848, 1867 and 1871. The process has also been manifest from time to time throughout the twentieth century.

63 It was also in Scotland that a professional interest in geographical botany developed (Fletcher and Brown, 1970, pp. 168–171).

64 de Wit, 1949.

65 A large format was also employed for the results of the U.S. Exploring Expedition.

66 W. J. Hooker to Colonial Office, 1857 (quoted in Thistleton-Dyer, 1906). Similar points were made by Anonymous (1863) and with respect to horticulture by Fletcher (1969). They continue to be heard in one or another form to the present day (Hickman and Duncan, 1989; Smocovitis, 1992).

67 Desmond, 1999.

68 Bentham, 1874.

69 Such a view may well have been shared by Alphonse de Candolle who as late as 1880 failed to mention them in his *Phytographie* (de Candolle, 1880). His father likewise had made no reference to keys in his *Théoire élémentiare* (de Candolle, 1813).

70 It is noteworthy that no real successor to these works appeared until after World War II. Even in recent years they have been appreciated in some quarters for their method – attributes also to be found in some measure in *New flora of the British Isles* by Clive Stace (1992; 2nd edn., 1997).

71 Anonymous, 1863; Thistleton-Dyer, 1906.

72 Commitments to *Flora of tropical East Africa* and *Flora Zambesiaca* are scheduled to continue until completion of these works about 2005 (Royal Botanic Gardens, Kew. *Corporate Strategic Plan 1996–2001* and *Kew 2020*). A recent single-volume work in the tradition is *Flora of Pico das Almas* (1995), edited by Brian Stannard.

73 Burtt-Davy *et al.*, 1925.

74 Symington, 1943; Corner, 1946.

75 The style was also followed by Walpers in his *Repertorium* and *Annales*, designed as supplements to the *Prodromus* (Stafleu, 1967, p. 75). Most of these works were in their time well received and long remained standard. Miquel's Indonesian flora was, however, 'premature' given the still-low overall level of botanical exploration therein.

76 Unavailability of authentically named material has resulted, and continues to result, in many misapplications of names.

77 Among the non-Kew works was William Hillebrand's flora of Hawaii (1888).

78 Lamarck was also among the first to challenge traditional 'linearity', with particular respect to classification and 'progression' of living things (Légée and Guédès, 1981; Stevens, 1994). This challenge has been extended in modern times to intellectual progress in general: not all of it is linearly based, nineteenth and early twentieth century precepts to the contrary (cf. Dolby, 1979).

79 In the United States, a partial cleavage was achieved, with a conventional manual-flora such as *Gray's Manual of botany* containing only relatively limited synonymy in

comparison with a contemporary work such as *Flora australiensis*. Gray, together with his associate Sereno Watson, evidently believed the full documentation of synonymy was more appropriate to a specialized work. Watson accordingly prepared a synonymized checklist, *Bibliographical index to North American botany* (1878), to accompany Gray's *Synoptical flora*. [Unfortunately only the first volume (Polypetalae) was ever published. Its main functional successor, Kartesz and Kartesz's *Synonymized checklist*, appeared only in the late twentieth century.]

80 Lange also published a synonymized 'nomenclator' for *Flora danica*.

81 Similar twentieth-century 'pairs' include *Opredel'itel' rastenij Ukrainii / Flora URSR* of the Ukrainian Institute of Botany, *Bestimmungsschlüssel zur Flora der Schweiz / Flora der Schweiz* by Hess, Landolt, and Hirzel and, from outside Europe, *Iconographia cormophytorum sinicarum / Flora reipublicae popularis sinicae* from the Academia Sinica Institute of Botany and *Flora of the Pacific Northwest / Vascular plants of the Pacific Northwest* principally by C. Leo Hitchcock and Arthur Cronquist.

82 An extensive exposition on the form and merits of analytical keys appears in von Mueller, 1888. He similarly calls attention to their relatively slow adoption.

83 A brief review of checklists and enumerations appears in Wisskirchen and Haeupler, 1998.

84 Davis and Heywood, 1963; Stevens, 1994.

85 de Wit, 1949, p. cxviii.

86 Most North American state and provincial floras in this period were enumerations or checklists. Descriptive floras or manuals were written mainly for larger regions or linguistically distinct areas such as Québec.

87 As in North America, descriptive works within British India were effectively regional in scope.

88 Forest floras in tropical American lands were, however, rare until after World War II. Standley's *Trees and shrubs of Mexico*, though completed before 1930, is more botanical than dendrological.

89 Limoges, 1980; Timler and Zepernick, 1987; Zepernick and Timler, 1990.

90 Cf. Desmond, 1995.

91 Neither the systems of Hutchinson (at Kew; 1926–34) nor Rendle (at the then-British Museum (Natural History); 1904–25) gained paramount influence, although the original version of Hutchinson's system would be adopted for a number of floras.

92 Davis and Heywood, 1963, p. 33; Stafleu, 1981.

93 Simon, 1977; Malclès, 1961, p. 84.

94 Eckardt, 1966, p. 168; Stafleu, 1981.

95 Eckardt, 1966, p. 173.

96 Lauterbach, a landowner near Breslau but trained as a botanist under Engler at Breslau University and in the 1890s a New Guinea explorer, was the sponsor; he also completed the *Nachträge* after Schumann's death in 1904.

97 Diels, 1921; see also Mildbraed, 1948.

98 Anderson, 1991, p. 87. The first 'minorities' to suffer Russification were the Baltic Germans. Russian became compulsory in schools and in 1893 the German-language University of Dorpat (at present-day Tartu, Estonia) was closed, afterwards reopening as the Russian-language University of Jurjew. Russians were also preferred in appointments to positions at the botanical institutions in St. Petersburg.

99 Much information on the history of Russian floras may be had from Shetler, 1967.

100 This was certainly apparent at the Muséum National d'Histoire Naturelle. The Muséum would enjoy a renewal in the late nineteenth and early twentieth centuries in relation to France's renewed imperial presence, but it never really regained the scientifically prestigious position it had enjoyed in the early part of that century (Limoges, 1980).

101 Leandri, 1967.

102 These included *Enumeratio plantarum in Japonia sponte crescentium* (1873–79) by A. Franchet and L. Savatier, *Compendium florae atlanticae* (1881–87) by Ernest Cosson, *Flore forestière de Cochinchine* (1881–99, not completed) by L. Pierre, *Flore de l'Algérie* (1888–95; supplement, 1910) by J. A. Battandier and L. Trabut, and *Illustrationes florae insularum maris Pacifici* (1886–92) and *Flore de la Polynésie française* (1893) of E. Drake del Castillo (who also contributed in the same period to the botanical volumes of Grandidier's encyclopedic *Histoire physique, naturelle et politique de Madagascar*). Some of these works enjoyed grants-in-aid from metropolitan or colonial authorities, but all were individually sponsored.

103 Jovet, 1954.

104 This shortage of resources hampered Adrien Franchet's research on the flora of East Asia at a time when vast collections were accruing from the work of the many French missionary priests and others active there after 1860. After his death in 1900, they were to a large extent neglected as official interest shifted elsewhere (Cox, 1945). Resources for tropical floristic inventory and research were also limited (Gagnepain, 1913), though in fact Indochina fared relatively well in the two decades or so before the economic slump of the 1920s and changing scientific interests.

105 These included most if not all of the major overseas floras to which reference has already been made.

106 Leandri, 1962; Vidal, 1984. Vidal notes that the order and division of the *Flore* into volumes were those of the *Flora of British India*, save for the inclusion of pteridophytes in the last volume.

107 The work was revised and reissued with considerable success by Belin in 1990.

108 In the late nineteenth and early twentieth centuries the U.S. National Herbarium was officially a joint undertaking of the Department of Agriculture and the Smithsonian Institution.

109 Dupree, 1959; Shinners, 1962, pp. 14, 19. The pre-eminence of Boston (and New England) as a botanical center was at this time beginning its decline relative to other parts of the country. Moreover, at Harvard itself botanical 'disunity' had become established (Morison, 1937; Hall, 1990).

110 The U.S. federal government also adopted the American Code as official for botanical names. This resulted in its use in a great many standard works, including the already-mentioned state and other floras (as well as Paul Standley's *Trees and shrubs of Mexico*) published through the National Herbarium as well as more applied botanical works issued directly by the Department of Agriculture (and the Forest Service) and in publications of the first U.S. Biological Survey.

111 Britton, 1894, 1895.

112 Sheets-Pyenson, 1988.

113 A supplement, *Nihon shokubutsu sôran-hoi* by Nemoto alone, appeared in 1936.

114 Within commonly accepted but oversimplified notions of progress in science it evidently has been thought to be of little significance. See Wheeler, 1923; Ravetz, 1975; Stafleu, 1981; Sheets-Pyenson, 1988; Gould, 1989; Stevens, 1994.

115 Other historical accounts of this period include Hagen, 1984, Briggs, 1991, and Reveal, 1991. Hagen deals in particular with the rise of 'experimental taxonomy'. With respect to more general contemporary trends in botanical research, see Smocovitis, 1992.

116 As successor to the Prussian Academy of Sciences, the rights to the work passed to the [East] German Academy of Sciences. However, its actual and likely contributors were mostly in the 'West'. One fascicle was reissued and two new ones published after 1945, wholly devoted to a revision of Campanulaceae-Lobelioideae by the Austrian cleric and botanist Franz Wimmer. Nothing since has been published. However, there do remain, or have arisen, some good outlets for major monographs or revisions; among them are *Bibliotheca botanica*, *Boissiera*, and *Systematic Botany Monographs*.

117 Cf. Corner, 1961; Jacobs, 1973; Mabberley, 1979, p. 274; Toledo and Sosa, 1993. The situation might be eased by improvements in research tools, better access to sources including digital imagery of collections, and methodological changes including a more 'industrial' philosophy.

118 Stafleu, 1959; see especially the map therein, p. 67. See also Thorne, 1971; Gómez-Pompa and Butanda, 1973; Gómez-Pompa and Nevling, 1973; Jäger, 1978.

119 Smocovitis, 1992. The 'watershed' for the evolutionary synthesis was *Genetics and the origin of species* (1937) by Theodosius Dobzhansky. At the same time, the synthesis provided an intellectual framework for Ernst Mayr's defenses firstly of the study of biological diversity (Mayr, 1982) and – paradoxically – of the autonomy of biology (Mayr, 1997). Good reviews of the penetration of population biology, experimental taxonomy, karyology and molecular biology into plant systematics include Huxley, 1940 (for plants see chapter by Turrill, pp. 47–71); Stebbins, 1950; Davis and Heywood, 1963; Raven, 1974, 1977; Hagen, 1984, 1986; Stuessy, 1990; and Soltis, Soltis and Doyle, 1992. Standard textbooks may also be consulted (e.g., Radford *et al.*, 1974; Stace, 1989; Sivarajan, 1991; Woodland, 1997; Judd *et al.*, 1999).

120 Heywood, 1974; Stevens, 1994; Ghiselin, 1997.

121 Shinners, 1962, p. 22; also Rosenberg, 1985; Briggs, 1991; Reveal, 1991; Ghiselin, 1997. In the United States, the biosystematics movement developed most significantly in the West and Midwest through the work of Frederic E. Clements and Harvey M. Hall, the Carnegie Institution group under Jens Clausen, Edgar Anderson, and G. Ledyard Stebbins (Hagen, 1984). Notable exponents elsewhere included W. B. Turrill in Britain, H. H. Allan in New Zealand, G. W. Turesson in Denmark, and B. H. Danser in Indonesia and the Netherlands.

122 The greater part of an English translation was also realized and published.

123 Ma and Liu, 1998.

124 For example, Heywood, 1964, 1989.

125 These include *Atlas florae europaeae*, *Med-Checklist*, the aborted European Floristic, Taxonomic and Biosystematic Documentation System (ESFEDS), and the Euro-Mediterranean Initiative in Plant Systematics (now Euro+Med PlantBase). A revision of the first volume of *Flora Europaea* was published in 1993.

126 Progress is documented in the ABRS newsletter *Biologue* (now bi-annual) and in the *Australian Systematic Botany Society Newsletter* (quarterly).

127 The *Flora of North America Newsletter* (nominally quarterly) furnishes reports of progress. Prior to award of its first grant, the National Science Foundation had sponsored a participants' workshop. Held in 1988, the proceedings appeared the following year as *Floristics for the 21st century* (Morin *et al.*, 1989).

128 Prosecution of these works continues to be advocated by individuals and in symposia. They are seen as contributing not only to advances in documentation and knowledge

but also to the formation of young systematists. They may also be the best we can hope for in situations when monographic studies are not possible or practicable. Some decades ago van Steenis (1954) called them 'creative' as opposed to the largely 'routine' floras of Europe (and parts of North America). Their merits or otherwise will be further examined in Chapter 3.

129 van Steenis, 1954.

130 De Wolf, 1963, 1964; Polhill, 1990.

131 Cf. Department of Science, Australia, 1979; Morin *et al.*, 1989.

132 Polhill, 1990.

133 Shetler *et al.*, 1973; Shetler and Read, 1973. A fuller description of the events is given in these references and in the original edition of this *Guide*. It needs only to be added that a proximal cause was a refusal on the part of the Department of Botany of the host institution any longer to accommodate the project. It should also be said, however, that by then six or more years had already elapsed without the appearance, in the public view, of definitive 'product': the first volume of a 'conventional' printed flora.

134 Rodman, 1989.

135 The primary emphasis has since remained on text production although over time efforts have been directed towards database development and in communication through the World Wide Web. On the other hand, organizational overheads began again to increase, a factor contributing – along with a slower than expected rate of publication – to the loss in 1999 of renewed National Science Foundation funding.

136 Malyschev, 1975; Jäger, 1976.

137 Jäger, 1976. A revision of Jäger's map appears as Map I of this *Guide*.

138 Gould, 1958.

139 Gould, 1962, 1968–72.

140 Walters, 1954. A significant part of the equipment used for this project is now in the Computer Museum, Boston, Mass., U.S.A. The records remain available through the Environmental Information Centre of the U.K. Natural Environment Research Council. For the *Atlas* see **660**.

141 PRECIS has been documented in several contributions, e.g., Gibbs Russell and Gonsalves, 1984.

142 Heywood and Derrick, 1984; Heywood, 1989. The European Floristic, Taxonomic and Biosystematic Documentation System (usually known as ESFEDS) was executed in a basic form between 1981 and 1987 but did not become permanently established.

143 Pankhurst, 1975, 1978, 1991.

144 A strong sense of caution was advanced by Shetler (1974*b*) with particular reference to the then-recently terminated Flora North America Program.

145 Allkin and Bisby, 1984. Reports from more recent symposia on biological information handling include Bisby, Russell and Pankhurst, 1993, and Hawksworth, Kirk and Dextre Clarke, 1997.

146 Among recent developments are a greater demand for supporting data as well as more color imagery and artwork. There is also an increasing interest in interactive keys. For practical and financial reasons, these demands are best accommodated by electronic media.

147 Polhill, 1990.

148 More recently there has been an increasing, somewhat fundamentalist emphasis on the monophyly of major taxa; with this have come debates on the relationship of phylogenetic approaches to the traditional, hierarchically based 'Linnaean' structure of biological classification. In floras, however, the impact of these developments has so far been relatively limited, reflected mainly in the adoption of more narrowly defined 'Jussieuan' families (a notable feature of the latest version of the Takhtajan system, as opposed to that of Cronquist). No totally worked-out cladistically based system of families of flowering plants has, however, yet been published (that in Judd *et al.*, 1999, pp. 166–167, does not cover all families). Floras, at least in the near future, have still perforce to follow one or another of the more traditional Gisekian genealogical/geographical ('Dahlgrenogrammic') or Steiner-tree (Haeckelian) schemes (cf. Judd *et al.*, 1999, pp. 37–38).

149 Blake and Atwood, 1942.

3

...

Floras at the end of the twentieth century: philosophy, progress and prospects

It happens that nearly every tropical flora is fundamentally unsuited to its subject . . . they not merely discourage the aspirant by so aggravating his difficulties but they expose their authors to unlearned ridicule.

   Corner, *New Phytol.* **45**: 187 (1946).

The Flora of the future will be a standardized data bank. It will be open-ended, dynamic and ever-growing. . . . Thus [it] will become a huge memory or series of linked memories available on-line to all users at any place and time.

   Shetler, 'Flora North America as an information system'; *BioScience* **21**: 524–532 (1971).

The whole question of the design of Floras requires considerable attention. Little advance has been made in practice during the last century.

   Heywood, 'European floristics: past, present and future'; in *Essays in plant taxonomy* (ed. Street), p. 288 (1978).

A regional flora is not the place to propose a new family classification, but the Flora writer serves botanical knowledge well if attention is drawn to the pitfalls that can result from a misuse of available characters.

   Hedge, in *Contributions selectae ad floram et vegetationem Orientis* (eds. Engel *et al.*), p. 313 (1991).

The current method of Flora writing employed . . . satisfies only specialists, who are generally not living in the geographical areas where information is quickly required for practical purposes. . . . Floras currently leave it to the end users to do as best they can in interpreting their highly technical content.

   Jarvie and van Welzen, *Taxon* **43**: 444 (1994).

In the late 1990s a new Flora is not simply a book, but a means of structuring and delivering botanical information.

   Blackmore, 'Aspects of the design for a Flora of Nepal'; in *International seminar-cum-workshop on the flora of Nepal* (coord. Jha, Adhikari and Shrestha), p. 4 (1997).

Introduction

In the preceding chapter emphasis was placed on substantive developments in flora-writing. Here an attempt is made to consider the range of aims, styles and content which are encompassed by the term 'flora' – a diversity with many stages[1] – and to relate this to methodological, philosophical and historical movements in botany and beyond. Both orthodox and alternative styles will be examined, with reference to both professional respectability and user experience, and some suggestions put forward. Some themes advanced here are: (1) that different needs call for different approaches; (2) that too many key data have been 'lost' in translation from research to publication; (3) that many so-called 'research' floras are sometimes as much political statements as they are substantive contributions to knowledge; (4) that effective but judicious use should be made of available information technology and current methods of communication; and (5) that documentary treatises apart, floristic works should be written in non-technical language and well illustrated.

As indicated in Chapter 2, a major focus in plant taxonomy in the last half-century has been the writing of floristic works. With the spread of professionalism from late in the eighteenth century, floras became one of the principal forms of contact between *savants* and users. Indeed, such has been their predominance since 1945 that several commentators have been moved to speak of the post-World War II era and beyond as an 'Age of Floras'.[2] The activities of most larger botanical research groups have featured one or more 'big', often tropical flora projects – be they for countries, lesser polities or supranational regions – along with 'domestic' national, state or local floras or manuals. Very often these projects have involved extensive collaboration among botanists, so facilitating the writing of individual family accounts by specialists. Management of flora projects has taken various forms; sometimes they have been the work of one (or more) individuals while others have been the collective work of one or more institutions, often over many years or even decades. For long, however, relatively little was said about how they should be written, the way in which their message can best be spread, and their relationship to the political and social environment as well as the needs of the different countries concerned.

Yet since publication of the original edition of this work in 1984, more attention has indeed been given to the principles and practice of flora-writing and presentation. This arose from four general developments: firstly, a greater awareness of, and intercommunication with, potential users; secondly, an increasing need effectively to justify projects in relation to changing funding and management practices; thirdly, internal reviews of existing and projected programmes, and finally – but not the least – the great spread and rapidly falling relative costs of information technology.

Floras have also acquired a more distinct warrant within systematic botany. They formed, for example, a distinct area of enquiry at International Botanical Congresses in 1987 and 1993.[3] In 1988 a workshop, 'Floristics in the 21st century', was held in the United States as part of the development process for the new *Flora of North America*.[4] In 1986 and, more fully, in 1989, floristic inventory in the tropics was extensively reviewed,[5] and in the latter year a workshop was held at Leiden as part of a re-examination of the *Flora Malesiana* programme.[6] In 1990 a conference on plant species information handling took place at Delphi, Greece.[7] Individual contributions have appeared with respect to such issues as accessibility, presentation of information, new electronically based techniques, and actual or potential applications of floras.[8] Changes in institutional planning and agency support patterns have also taken place. Nevertheless, there remains scope for a continuing examination of the content, form and purpose of floras. These topics form the main focus of the present chapter.

Philosophy and purpose of floras

The design and writing of floras and related works has long been a rather conservative sphere of activity. Often there has been an unquestioned acceptance of stereotyped formats and sets of questions, regardless of the real worth of current floristic content and styles. As discussed in Chapter 2, these formats largely became crystallized between 1805 and 1870 although the descriptive flora as such dates back to Linnaeus. The subsequent passage of time has been marked chiefly by changes in content, more illustrations, and improved keys. Some classes of works such as manual-keys, excursion guides, works on trees and shrubs, and 'applied' floras evolved to meet particular practical needs. At another level, the large-scale enumeration was introduced to meet particular needs, such as the inventorying and classification of the often large floras of humid tropical lands within limited means and times.

An examination of the relevant literature as well as personal observations suggest the existence of two views – both of long standing and to some extent at odds – concerning the central purpose of floras and related works. This dichotomy, prompting the belief by van Steenis[9] that most floras were 'dualistic' in nature, is a result of contrasting aims. One is communication (through keys, descriptions, illustrations and commentary), while the other is archival or encyclopedic.[10] Van Steenis believed the problem could be resolved in north temperate regions but, due to the sheer size of the flora in most areas, not so in the tropics.[11] The question is more general, however, as was recognized long ago in Europe and, to an increasing extent, in North America and elsewhere.[12] Indeed, it may be regarded as a natural intellectual development as a flora becomes better known and documented.

The first philosophy – one which sees floras as tools for communication (through identification and basic information) – recalls the first aphorism of Bentham quoted in Chapter 2: 'The principal object of a Flora of a country, is to afford the means of determining (i.e. ascertaining the name of) any plant growing in it, whether for the purpose of ulterior study or of intellectual exercise'. The relative value of this philosophy has several times been emphasized by Heywood as well as by other authors.[13] Heywood argued that floras were not necessarily intended to serve as sources of strictly comparative data; their main function was to address themselves to certain questions about the plants of a given area: (1) what there is, (2) how they may be recognized, and (3) where they may be found. To this end floras should include keys, descriptions, necessary auxiliary information, and essential nomenclature, synonymy, and citations. With modifications – notably the use of illustrations – several other authors have adopted this point of view, among them Shinners and (more recently) Palmer, Wade and Neal in North America, Brenan and Jacobs for, respectively, Africa and Southeast Asia, and, in Malaysia, the editors of *Tree flora of Sabah and Sarawak*.[14]

The second philosophy – in which floras are seen as essentially archival or encyclopedic – has its modern origins in such works as *Nova genera et species plantarum* (1815–25) and *Flora brasiliensis* (1840–1906). This approach rests in turn on the herbalist tradition (and predilection for detail) of Central Europe – manifest in such works as Fuchs' *De historia stirpium commentarii insignes* (1542) – wherein also lie the philosophical origins of such late seventeenth century works as Rheede's *Hortus malabaricus* and Rumphius's *Herbarium amboinense*.[15] Current manifestations include many recent and current large-scale flora projects, their justification being, according to Shetler, that floras should be 'a physical repository of descriptive data about plants which are organized and formatted, usually in book form, so as to answer a time-tested set of prescribed questions'.[16]

The differences between these two philosophies as related to developments in the nineteenth and twentieth centuries have been discussed, with many examples, in Chapter 2. It was there noted that for floristically lesser-known parts of the world the encyclopedic flora has been much favored, but that the realization of such works – commonly still by traditional means – involved a great investment of time and manpower, with gradually falling productivity (often as larger or more 'difficult' families, such as the Rubiaceae, were addressed). In the words of one commentator, such works were veritable botanical 'booby-traps', with progress measurable in decades, not years.[17] Later commentators have spoken of *centuries*.[18] By contrast, projects with more limited objectives stood a better chance of successful completion within reasonable time-spans, and more often than not were in spirit 'tighter'. As David Webb noted in his *Flora Europaea* valediction, close editorial control was at all times essential.[19]

Sometimes, though, the two philosophies have been confused. In the 'introductory notes' to one flora project underway at the time in the region where these words were originally written, it was stated that, in order to make available 'information' on the flora (then very scattered), the sponsoring institutions have 'embarked on a project to produce, in a handbook format, a *concise* Flora' [italics added]. By contrast, the first volume of the flora concerned suggests that the work, even with some information having been relegated to 'technical supporting papers', is somewhat expansive. In one small family, four pages of text are required to deal with three relatively easily recognized species. Rather than being 'concise' this flora brings to mind larger-scale works such as *Flora of Panama* or *Flore d'Afrique centrale*. Other examples demonstrating confusion of objectives or the more underlying effects of academic tradition could be given. Such a situation may have (at least partly) contributed to an impression that there is an 'uninterrupted continuum' among floras.[20]

These criticisms notwithstanding, it remains true that without a strong taxonomic and documentary foundation the achievement of conciseness in floras of tropical areas has been difficult and but rarely overcome. Works such as the second edition of *Flora of west tropical Africa, Flore des plantes ligneuses du Rwanda, Flowering plants of Jamaica, Flora of Java, Tree flora of Malaya, An excursion flora of central Tamilnadu, India, Flora of Pico das Almas* and *Flora of central French Guiana*, all published since World War II, have generally depended upon antecedents. These latter comprise major parent works or a long tradition of externally based or local botanical work or any combination thereof.[21] Yet even with an actual or potential higher impact the concise works have remained fairly traditional or feature traditional elements. Awkward compromises have not rarely been apparent, with the result that key information is lost or very scattered.[22]

Content, style and methodology of floras

In Chapter 2 the substantive evolution of floras and related works was described, with examples. It was there noted that their basic principles and practice had largely been formulated by Linnaeus[23] and afterwards Lamarck and, in the nineteenth century, by A.-P. de Candolle, the Hookers, Bentham, and A. de Candolle.[24] Further contributions were made in the mid-twentieth century by, among others, van Steenis, Brenan, Turrill, and Heywood.[25] For North American workers important early precedents were set by Torrey, Gray, Britton, Fernald, A. S. Hitchcock, S. F. Blake, Lawrence, and Radford.[26] The last few decades have seen the introduction and spread of information technology, a major phenomenon more appropriately dealt with in the next section.

General developments

The influence of the many discoveries and developments from so-called 'general botany' (*allgemeiner Botanik*) and other areas of biology from the mid-nineteenth century onwards on the actual writing of floras was only gradual. The inclusion of analytical keys (save in most enumerations and checklists) became a norm, along with more critical commentary and a greater emphasis on geographical and ecological data. This reflected a growth of interest in field botany as well as in plant geography and ecology.[27] Twentieth-century refinements have included additional categories of biological and other data (notably chromosome numbers), simplification of nomenclature, discussion of phyletic relationships, and more explicit expositions of taxonomic philosophy along with other topics such as increased consideration of habitats, variability, hybridism, introgression, clines, and conservation status.[28] Information from pollen analysis, comparative phytochemistry, anatomy, and reproductive biology have also been introduced but these are as yet mainly at generic and family level, especially in the larger 'research' floras.

A notable development of the last four decades has been the provision of 'guidelines' for contributors, such as the *Flora Europaea* 'Green Books' and *Tree flora of Sabah and Sarawak: guide to contributors.*[29] The writing of floras having now become largely a collaborative process, such handbooks are now essential.[30] Guidelines on the content and writing of floras have also appeared in textbooks of taxonomy. A noteworthy exposition of general principles appeared in Davis and Heywood's 1963 textbook, *Principles of angiosperm taxonomy.*[31] For 'traditional' manual-flora writing in North America (and other areas under North American intellectual influence) a number of the previously mentioned authorities published definitive and widely used treatises.[32]

What was prescribed in the most recent of these (Radford's *Vascular plant systematics*, 1974) would, it is fair to say, have been clearly recognizable to the leading nineteenth-century writers on floristics. A similar point was made by Meikle in a consideration of contemporary British floristic writing.[33] Indeed, until comparatively recently many floras could be seen as not so very different from those written by Torrey, Gray, Bentham, J. D. Hooker and their contemporaries.[34] The main differences are, as already indicated, the almost universal use of analytical keys and the inclusion of many more classes of data. Notable among the latter

was the indication of detailed distribution, particularly in U.S. state floras where the use of county dot maps for all species – first introduced by Deam in his *Flora of Indiana* (1940) – became widely popular.[35]

It is thus evident that, consciously or not, the 'classical' formula was one that was widely accepted, even after more than 150 years. It represented a successful combination of scientific objectivity and literary style, and its keys provided means of identification. Even by the 1950s it was passionately defended by van Steenis who argued that 'besides . . . ecology and distribution, and nomenclature, nothing needs to be added to Bentham's classic exposition of a purely descriptive Flora'.[36] Variations introduced by the Britton 'school' in North America, especially in *North American Flora* (1905–57, series II, 1954–), never gained lasting popularity and after 1930 the older precepts imparted in the name of the nineteenth-century masters enjoyed renewed support.[37]

For the larger 'research' floras more or less amplified versions of the classical formula, following the precepts of *Flora brasiliensis* but now presented in convenient octavo or quarto formats, have become standard.[38] Examples of mostly post-World War II works include the greater part of those written for the tropics, both by European authors (e.g., *Flore du Gabon, Flora of Thailand, Flora Zambesiaca, Flora de Cabo Verde*) and by Americans (e.g., *Flora of Guatemala, Flora of the Lesser Antilles, Flora of Panama, Flora de Colombia*), as well as several temperate-zone works. Among the latter are *Flora SSSR, Flora reipublicae popularis sinicae* (in style and philosophy modeled after *Flora SSSR*), former Soviet-republican and Chinese provincial floras, *Flora of Japan*, the regional floras of the Argentine, and *Flora of Southern Africa*. Among the more concise of this genre – and at the same time very true to the Benthamian tradition – are *Flora of Turkey* by Davis and collaborators (1964–88), *Flora of Australia* (1981–), *Flora of Somalia* edited by Thulin (1993–), *Flora Mesoamericana* (1994–), and *Flora of the Venezuelan Guayana* (1995–).

Enumerations, checklists and 'manual-keys'

Apart from those already discussed, other longestablished and stereotyped formulae of flora-writing have also remained in vogue despite some criticism, especially of the enumeration.[39] Practical considerations, however, dictate that for many areas enumerations will continue to be produced. These moreover lend themselves particularly well to electronic methodology. Among notable recent works in this genre are *Prodromus einer Flora von Südwestafrika* (1968–72) edited by Merxmüller, *Enumeration of the flowering plants of Nepal* (1978–82) by Hara *et al.*, and the electronically supported *Vascular plants of British Columbia: a descriptive resource inventory* (1977) by Taylor and MacBryde, *Provisional checklist of species for Flora North America (revised)* by Shetler and Skog (1978), *Med-Checklist* (1984–), *Énumération des plantes à fleurs d'Afrique tropicale* (1991–97) by Lebrun and Stork, *Catalogue of the flowering plants and gymnosperms of Peru* (1993) by Brako and Zarucchi, *Catalogue of the vascular plants of Ecuador* (1999) by Jørgensen and León-Yánez, and *Plants of southern Africa: names and distribution* (1993) by Arnold and de Wet. The first-named among the above contains keys, a feature unusual in an enumeration but much to be commended, while the last two respectively account for 16087 and 21397 species. Less critical works also continue to appear, some merely based on source extraction. A list of desirable criteria has been furnished by Wisskirchen and Haeupler in their *Standardliste der Farn- und Blütenpflanzen Deutschlands* (1998).[40]

With respect to 'manual-keys', this formula likewise remains in wide use for works of identification. Examples include the greater part of the 'routine' floras of Europe, the *opredel'itely* of Russia and neighboring states, and similar works in other continents such as *Flora of the Sydney region* by Beadle *et al.* (1963; 2nd edn., 1972; 4th edn., 1994), *Flora of the Pacific Northwest: an illustrated manual* by Hitchcock and Cronquist (1973), and the unillustrated three-volume *Flora of Java* by Backer and Bakhuizen van den Brink, Jr. (1963–68). These should not, however, be regarded as substitutes for more definitive descriptive floras or information systems.

'Biosystematic' information and 'critical' floras

Among the most important developments in biology of the first half of the twentieth century were the rise of ecology (plant geography), population biology, experimental taxonomy and 'biosystematics'.[41] From the 1920s onwards they brought in a return to broader species concepts (as well as renewed efforts towards 'objective' definitions of genera). A prevailing conservative approach to floras, however, evidently delayed or limited the consistent incorporation of 'biosystematic' information in some parts of the world.

This was particularly evident in the United States, a seemingly curious phenomenon to which Raven drew attention in the 1970s in a review of progress in these subdisciplines.[42] Among the first manual-floras demonstrating consistent use of the new classes of information associated with experimental taxonomy and biosystematics were *Flora of the British Isles* by Clapham, Tutin and Warburg (1952; 2nd edn., 1962) and the first volume of *Flora of New Zealand* by Allan (1961). With its substantial commentary covering hybridism, introgression, and other patterns of variability, the latter became almost a 'critical' flora. An early North American work of this type was *A California flora* by Munz and Keck (1959; supplement, 1968). More widespread incorporation of 'biosystematic' data in that continent came only around the end of the 1960s, with representative works of the period being *Flora of the Queen Charlotte Islands* by Calder and Taylor (1968), *Manual of the vascular flora of the Carolinas* by Radford *et al.* (1968), *Manual of the vascular plants of Texas* by Correll and Johnston (1970), *A flora of tropical Florida* by Long and Lakela (1971), and *Utah plants: Tracheophyta* by Welsh and Moore (1973). However, only the Queen Charlotte Islands flora was truly a 'critical' work along the lines of *Flora of New Zealand*.

It was in Central Europe, however, that the first real 'critical' floras appeared, some as complements to field-manuals. Although not representing a distinct stylistic class, such floras give particular emphasis to variation, infraspecific forms, and agamic 'microspecies', and usually also make reference to more detailed treatments of taxonomically 'difficult' groups.[43] Among the earliest was *Flora der Schweiz* by Schinz and Keller (1900), whose *Kritische Flora* had by 1914 appeared as a separate volume; this latter remains something of a landmark.[44] A similar, though more recent, key work is Rothmaler's *Exkursionsflora von Deutschland*. Its part IV (*Kritischer Band*), based upon extensive cooperative research from the 1930s to the 1950s (and continuing to the present), first appeared in 1963 (with several subsequent revisions). In 1996, the first of the five projected volumes of Sell and Murrell's *Flora of Great Britain and Ireland* was published, reviving an idea first expressed with Moss's *Cambridge British Flora* (1914–20). Some floras have also incorporated a significant amount of chorological and phytosociological data; notable among these is Oberdorfer's *Pflanzensoziologische Exkursionsflora* (7th edn., 1994). A complementary, but specialized, genre was the series of cytotaxonomic conspectuses prepared and published for various groups and areas by Áskell Löve and collaborators beginning in the late 1940s. On the whole, however, the influence on floristic writing of what Merritt L. Fernald once termed 'camp-ing'[45] has come through gradual diffusion into traditional flora formats. Radical departures have been uncommon.

'Generic' floras

A less common approach to the problems of time and resources in areas of high diversity has been the 'generic' flora. From a historical point of view, they had antecedents in the *genera plantarum* of the eighteenth (Linnaeus, de Jussieu, Lamarck and Poiret) and nineteenth (de Candolle, Endlicher, Bentham and Hooker, and later Engler and Prantl) centuries. Such works were developed as a form of communication about the floras of continents where a full account (to species level) was not practicable but where a desire for a descriptive account existed. Important early examples were *Genera of North American plants* (1818) by Thomas Nuttall, *Genera florae americae boreali-orientalis illustrata* (1848–49, not completed) by A. Gray (with drawings by I. Sprague), and *Genera of South African plants* by W. H. Harvey (1838; 2nd edn., 1868, by J. D. Hooker). Afterwards, in the wake of *Genera plantarum* and *Die natürlichen Pflanzenfamilien* generic floras largely fell from favor. Save for one major Asian work (the incomplete *Handleiding tot de Kennis der Flora van Nederlandsch-Indië* of 1890–1900 by J. Boerlage), the genre became largely Afro-Malagasy. The leading African exemplar of the first half of the twentieth century is a successor to Harvey's work, *The genera of South African flowering plants* by E. P. Phillips (1926; 2nd edn., 1951; 3rd edn., 1975–76, by R. A. Dyer), while in Madagascar after World War II the French forest botanist R. Capuron produced *Introduction à l'étude de la flore forestière de Madagascar* (1957). However, interest continued elsewhere, with the Austrian-American botanist Theodor Just in the 1950s making a strong plea for more 'genera' in place of (or in addition to) big semi-monographic 'research' floras.[46] This may have at least partly inspired what became the *Generic flora of the southeastern United States*, initiated at the Arnold Arboretum of Harvard University and published in installments since 1958 (currently in *Harvard Papers in Botany*).

In more recent years, a renewed role for 'generic' floras has emerged with the realization that they are suitable vehicles for illustrated keys to floras of lesser-known regions. This is manifest in *A field guide to the families and genera of woody plants of Northwest South America (Colombia, Ecuador, Peru) with supplementary notes on herbaceous taxa* by A. Gentry (1993). More recently, this approach has been extended to Madagascar where a revision and expansion of Capuron's manual to cover all vascular genera is in progress. Inclusion of selected references to monographs, revisions and other pertinent works as 'aids' would, however, make the works more useful for those interested in seeking further information.

Illustrations and maps

While among recent floras the *Flora of Java* was unusual in being without illustrations, most modern works have at least some figures; those billed as 'illustrated floras' attempt to depict every species covered. As already noted, technological changes from late in the nineteenth century made possible, at a lower unit cost, a greater dissemination of figures (although color reproduction in print remains more expensive).[47] Illustrations are in fact often more effective than descriptions in conveying information about plants and their characters.[48] For the humid tropics this is of particular import in view of the presence of so many different kinds of plants (relative to the cool-temperate zones where most botanical thought has been shaped) – and where the perception of most people is rather more visually than literarily oriented. This was early recognized by Asian writers, notably Tomitaro Makino who, after some uncompleted series of plant portraits (1888–1911), many critical taxonomic papers, and (with K. Nemoto) a conventional manual (1925), produced in 1940 a fully illustrated work, *Nippon shoku-butsu zukan (Illustrated flora of Japan)*.[49] This and similar East Asian works are atlas-floras comprising small figures with parallel text, any analytical keys playing a supporting role.[50] The more costly early floras by European authors, like the sixteenth- and seventeenth-century herbals, also were well illustrated, the figures lending an element of prestige, and following the introduction of lithography in the early nineteenth century figures gradually became more common in humbler guides and reference works, good examples being the illustrated edition of Bentham's *Handbook of the British flora* (1865) and Bonnier and Layens' *Nouvelle flore du nord de la France et de la Belgique* (about 1887) as well as the latter's slightly later

Tableaux synoptiques des plantes vasculaires de la flore de France, this last with 5291 small figures.

It thus remains surprising that Cornelius Backer, a leading early twentieth century Dutch authority on the Javanese flora, was so opposed to them in a formal flora.[51] Paradoxically, however, his manuals for weed identification are abundantly and well illustrated. Backer, however, was not alone; in Macbride's *Flora of Peru* figures similarly are wanting and in Ridley's *Flora of the Malay Peninsula* they are incidental. Gradually, however, their value became more widely recognized and, as the costs of reproduction fell in the face of further technological developments, their use became widespread. By the 1970s some tropical as well as temperate works illustrated most if not all species. Among the former are *Flora del Ávila* (1978) by Steyermark (Venezuela) and Huber and *Flora of the Rio Palenque Science Center* (1978) by Dodson and Gentry (Ecuador). This trend has by and large continued to the present; even major works such as *Flora of North America* now feature illustrations (as well as distribution maps) of all species, while in *Flora da Reserva Ducke* (1999), from Amazonian Brazil, all the illustrations are in color. On the other hand, they are absent from *Flora Mesoamericana* – potentially a serious problem in attempting to key out closely related taxa in areas of high diversity such as Mexico and Costa Rica. They were also at first omitted from *Flora of China*; however, pressure from users has since led to the issue of companion atlas-volumes. Good illustrations are in the author's view very important to field botanists and others without ready access to a large herbarium. With the spread of virtual media, their consistent use becomes even more feasible.[52]

Another feature of floras which has now become widespread is the use of distribution maps. Their application has paralleled the growth of interest in biogeography and chorology, but until the 1930s they were largely confined to monographs and revisions or specifically biogeographical papers. Moreover, they had not yet attained recognizably modern forms. For this reason, Lebrun and Stork in their index of distribution maps of African plants (see **Division 5**) took 1935 as a base line. In North America, the publication of Deam's *Flora of Indiana* (1940) marks the beginning of their continuing use there, although a first uniform application of county-based distribution maps dates from the end of the nineteenth century (in *Flora of Kansas*, 1898–99, by A. S. Hitchcock). The spectacular devel-

opment of computer-assisted mapping technology in stages since World War II as well as in printing has also manifested itself in many ways, both through the production of national and state cartographic atlases and within more conventional floristic documentation. Distribution maps have now become widely used, even in large-scale works. Indeed, in *Flora of Australia* and *Flora of North America* all species are being mapped. There have also been attempts to combine layers of information in print (with more possibilities using color), by the use of overlays, or virtually as a product of a computerized geographic information system (GIS). The latter is among the facilities offered by FloraBase in Western Australia.[53]

Floras and information technology

While there has been much progress in recent decades with respect to the style and substance of floras and related works – particularly following their re-examination as potential sources of comparative data – it is the development of information technology which can now be seen as having had the greatest impact, particularly by the 1990s with the large-scale spread of public networks with graphical interfaces. Earlier, it had come to be seen not only as offering new ways of storing and handling information but also as having the potential to provide some relief from what was coming to be seen as its version of the 'information crisis'. The introduction, growth and spread of information systems as aids to the handling of floristic information, and – more recently – as vehicles for its dissemination, furnishes the theme of this section; it will be argued that in effect traditional floras and checklists have now become manifestations in space and time of what is a continuously changing source of data – a view first espoused by S. G. Shetler in 1969 and quoted at the beginning of this chapter.

The 'information crisis'

As we have seen in the preceding section, the manifold developments in biology in the twentieth century not unnaturally led to a veritable explosion in the kinds of data appropriate to a flora. Indeed, by 1971 systematics as a whole was seen by Heywood as facing an 'information crisis'.[54] All this made the question of what to include in a given work still more acute. A 'universal flora' came to be seen as no longer practical if ever it was. The introduction of computers and advances in information technology over the previous decade or so also effectively made possible more quantitative approaches to classification which, in addition to supposedly more 'objective' philosophies, also required – as far as possible – uniform setting-out of characters in descriptions. With coverage of the plant world in revisions and monographs very spotty, attention turned to floras as supposedly being rich sources of comparative data. Only then did the question of objectives in floras again impinge seriously on systematics: were they documentary, or were they – as Lamarck, Bentham and Hooker had seen them – primarily a means for identification?

It was Leslie Watson who, also in 1971 – after exercises with the large genus *Salvia* (Lamiaceae) as well as higher taxa in Lamiaceae and other families – first seriously argued that floras – even the more elaborate ones – were, after all, objectively limited (and ideally should remain so).[55] For this reason, he called for a return to the Benthamian tradition of 'concise' works for practical use; the kind of information which went into elaborate 'archival' floras was in reality more appropriate to other kinds of taxonomic publication or for storage and retrieval through data banks or other non-print media.[56] Seriously comparative systematic studies really required more elaborate, specialized data sets. Similar ideas were mooted in the original Flora North America (FNA) Program, which planned for the establishment of an information system along with publication of a relatively concise conventional flora.[57]

Necessary choices of information on the part of individual scholars or groups for floras were for the most part pragmatic, depending on means and circumstances, although, as we have seen, tradition was often followed. Some kinds of potentially useful data, however, would inevitably be omitted and at a later date might not even be retrievable.[58] Watson believed that information selection and presentation should be more explicit and defined, in a not dissimilar fashion to Britton 75 years before. In arguing for a start that the two main philosophies of flora-writing – the 'archival' and the 'practical' – should be separated and that a given work should follow one or the other, he considered that confusion of objectives in current floras was frequent and that many represented unhappy compromises. They were neither definitive sources of comparative data nor practical tools for identification. There was little recognition of the desirability under modern conditions to separate these functions. Watson's closing challenge was: 'we have all these advantages [computerization, philosophical analysis, masses of data, etc.], yet have more difficulty in getting to grips with real problems than Bentham did'.[59] This conundrum has in various forms continued to the present notwithstanding the great advances in information technology over the intervening three decades. Paradoxically, however, with the functional separation of print and information system now seriously possible it may become less of an issue. Indeed, with floristic descriptions and other information organized in the form of defined fields, blocks or hypertext-links, automation of such features as key generation, data analysis and map generation, and the provision of (as far as possible) consistent content, the design and implementation of the information system along with 'report generation' (i.e., the formulation of virtual or printed products) are arguably now of greater importance than mere choice of styles or, as we shall see, concern for an 'ideal flora'.

Technological and methodological developments

Increased exploitation of computers and software – here referred to as 'information technology' – was not unsurprisingly seen by its advocates as the best way forward in the face of Heywood's postulated 'information crisis'. But, as will be seen, its real promise has taken more than a generation to become manifest. Not only was progress inevitably constrained by questions of cost-effectiveness – which became more of an issue as the relative worth of science budgets began to fall in the 1970s, a time generally of high inflation – as well as technical developments but, as was contemporaneously suggested by Heywood, 'until very recently any suggestion that information processing was a major role of taxonomy was vigorously repudiated as being a pursuit unworthy of scientists in this evolutionary age'.[60] The termination in 1973 of the original FNA Program shortly after commencement of its operational phase not unnaturally also caused some soul-searching.[61] Since then, advances in information technology have played an important if not always leading role in the field, given financial as well as conceptual and other constraints.[62]

The first major computer-assisted flora programmes were not mounted until after the introduction of 'third-generation' mainframe computers (with enhanced central memory and greater speed but still using punch-card and paper tape input and magnetic tape storage). As well as the already-mentioned FNA, there were *Flora de Veracruz* and *Flora of British Columbia*. The original FNA (1966–73) was the most ambitious but ultimately was premature and not cost-effective; as already mentioned, it was terminated very early in its operational phase. Its two major offshoots in Canada and Mexico continued, and additional systems began their development in other parts of the world. While the British Columbian project was ultimately reduced in scope to a checklist, *Vascular plants of British Columbia*,[63] that in Veracruz, with a vascular flora of some 8000 species, became a serious floristic information system – the first in a tropical polity as well as among the earliest worldwide.[64] Its principal goal – a descriptive *Flora de Veracruz* – began publication in 1978.[65] About the same time, the concept of an information system as a complement to *Flora Europaea* began to be explored; a central idea was the application of the then-new 'Videotex' system of passive visual display pages running on the more compact and less costly 'minicomputers' by then entering the market.[66] A generally closer relationship between floras and information technology would, however, not develop more seriously until the 1980s.

Significant advances in such areas as key generation and computer-assisted identification,[67] collections data,[68] taxonomic text compilation, storage, retrieval and generation using custom routines,[69] and geographical information systems nevertheless continued, while the formation of biological data banks has remained one of the priorities for systematic biology.[70] The advent of desktop and, somewhat later, portable microcomputers along with greatly increased storage capacity including magnetic 'hard' disks and, subsequently, CD-ROMs promised further advances which began to be exploited as the 1980s progressed.[71] With the spread of database, word-processing and text-formatting software the production in particular of documented enumerations and checklists became much easier, and by the mid-1980s several examples, some substantial, had appeared.[72] But until the spread of networking, floristic information systems perforce remained solitary entities; external communication was largely limited to paper or transferable magnetic or optical media.

Floras on-line: an ontological revolution

An entirely new dimension, however, opened after 1986 with the advent and expansion of the modern Internet and the appearance of hypertext, markup language, navigation protocols, browser software and, in 1992, the World Wide Web. The arrival shortly afterwards of browsers with graphical user interfaces and interactive capability (firstly Mosaic, then Netscape and Internet Explorer) has since enabled the 'Net' to develop into its present form and attract worldwide interest.[73] Continuing evolution, including falling relative costs and wide availability of suitable media and their operating devices, ensures that the technical possibilities for information structuring, exchange and integration have, and will become, effectively unlimited.

With respect to floristic and systematic botany, these latest developments mean in effect that computer- and network-based floristic information systems ontologically are now definitive, as Shetler had predicted. In the space of only a few years, they arguably have overtaken and enveloped traditional approaches to floristic information and documentation – including discussions of the relative advantages and disadvantages of particular formats.[74] The adoption of this or that style for a given floristic publication is now purely a matter of choice. Any printed materials – which for many purposes remain essential – can be based upon, and are produced from, an information system. Moreover, the advent of the Convention on Biological Diversity and the formation of the Biodiversity Information Network/Agenda 21 (BIN21) have also brought into being the formation or enhancement of national networks which may include floristic databases.[75]

There have not unnaturally been technical, social, legal, intellectual and, especially, financial obstacles: some new, some continuing. Technology continues to evolve rapidly, methodologies are still developing, expectations have increased, issues of authorship, copyright and intellectual property remain open, and – last but not least – the interested community still includes 'unrepentant supporters of print'.[76] These have all complicated provision of, and access to, on-line floristic information. In addition, there may be questions of how new technologies and approaches are perceived among decision-makers in institutions and agencies.[77] Not surprisingly, the creation – and maintenance – of data banks requires long-term institutional commitment as well as substantial funding.[78] Finally, there may remain – as we have seen – resistance to the idea that effective information handling in systematics and floristics is as important an intellectual goal when compared with more traditional directions such as analyses, revisions, monographs and floras. Yet taxonomic information systems *are* important, not only because potential sources for a given problem or situation are now far more fragmented than in the so-called 'golden age' of Bentham and his contemporaries, but also as under Articles 7 and 17 of the Convention on Biological Diversity contracting polities now have standing obligations respectively to identify components of their biological diversity and to facilitate exchange of information relevant to its conservation, management and sustainable use.

There are at present perhaps three possible levels of commitment to 'on-line floras'. The first, and simplest, is a Web page (or pages) advertising printed publications with only a limited amount of information conveyed on-line in basic text format. The second is represented by the provision of on-line material which attempts to be comprehensive but essentially remains processed text which has also appeared, or will appear, in print. The third, and most advanced, is the presentation of on-line information in its own right. Text, maps and graphics are carefully organized and presented and are based on links to structured data. Keys are interactive, descriptions are truly comparative, nomenclature is lexical (allowing entry to data via synonyms), and supporting information (including specimen records) is searchable. The underlying data are regularly updated and authorship is distributed. Associated print materials, while having their uses as well as status, become in effect temporal cross-sections of a dynamic system, or 'virtual flora', rather than independent entities.[79]

Paper or virtual, or paper and virtual? that is the question

The advent of the virtual flora not unnaturally raises the question: what future for printed floras and checklists? There remains considerable public demand for 'hard copy', not only for practical and symbolic value but also for citation records and personal *curricula vitarum*. Printed matter – which remains independent of extrinsic modes of delivery such as software and is classically measurable – continues to carry weight as a record of progress.[80] Attempts at moving some larger flora projects entirely to electronic delivery have so far been successfully countered and, at least with respect to *Flora of North America* and *Flora iberica*, parallel dissemination has become current practice. This pattern

could, however, change in future with new projects developed without primary reference to printed formats, such as FloraBase in Western Australia.

A significant factor favoring a shift to virtual floras, be they delivered on-line or on transferable media, is increasing public interest in illustrations and maps. In addition, a greater range of questions is now being asked of floras, which require inclusion of more classes of comparative data or *Zeigerworte*.[81] Inclusion of all these features would, however, greatly increase the bulk of printed floras (of which *Illustrierte Flora von Mitteleuropa* is perhaps an extreme example) and may in some cases, such as interactive keys or geographic information system (GIS)-based maps, simply be impossible. As already indicated, the effective limit for a traditional, single-volume, concise manual-flora is about 5000 species. Choices have thus to be made if demand continues for some form of printed publication, whether for practical or symbolic reasons.

Various approaches to this problem suggest themselves. The first is to continue primarily with print, with preparation by electronic means. This may remain the only possible option in the absence of a suitable framework for a floristic information system, or where time is limited.[82] A second is to develop mixtures of hard copy and electronic media wherein the latter would contain material and functions not available in the former.[83] Both these options would be compatible with situations where preparation of a large-scale flora forms part of bilateral or multilateral cooperation programmes, although given the requirement for a floristic information system the latter has advantages over the longer term.[84] A third is to adopt different styles and kinds of content relative to the number of taxa covered. Thus, for full coverage of a large national or regional flora one would produce a checklist or enumeration – of which revisions at relatively frequent intervals would be feasible – while smaller areas would become the focus of manuals.[85]

It is in pursuing the possibilities of simultaneous use of different media and forms of dissemination that the issues of 'common standards' become important and thus the choice of information for a given situation. This in turn relates to the recognition respectively of 'core' and 'non-core' data as proposed by Wilken *et al.*[86] The availability of an information system, whether on-line or on a transferable medium, makes economically feasible the inclusion of valuable but specialized information along with color illustrations and sophisticated maps which otherwise would not have been possible. Manuals which are derived from an information system might then be limited largely to 'core' material such as botanical and common vernacular names, key synonyms, diagnoses or descriptions, distribution and ecology, and pertinent notes as well as keys and (where necessary) critical figures. 'Non-core' data and materials – as well as the 'core' – would reside on-line or on disk along with software which would make possible searches and data combinations as well as interactive identification simply not feasible in print. The introduction of slimline, lightweight 'virtual books' with enhanced battery life may further facilitate the spread of virtual floras as a primary information source on vascular plants along with monographs, revisions and world checklists which, independent of political boundaries, represent the intellectual ideal.

I thus foresee, at least over the next decade, some coexistence of media and methodologies, both traditional and advanced, but believe that in time the 'virtual flora' or floristic information system will ultimately offer more flexibility in acquisition, processing, dissemination and maintenance of what is essentially *dynamic* data. The plant world, as is well recognized, is never static and currently is under more pressure than ever; treaty obligations also stipulate that information has as far as possible to be kept up to date. Yet a place is likely to remain for print, to furnish historical cross-sections as well as visible records of personal and collective progress. Dynamism and stasis are indeed contrasting concepts, but the former is not necessarily always 'better'.

Critiques of floras

As already noted, floristic works, large and small, have been a major feature of taxonomic botany since World War II. They remain as a whole the most recognized means for conveyance of botanical information to the public; in addition to their primary roles of inventory and identification they include data which have been applied in many different ways both within and without the biological sciences.[87] There have been, however, a number of criticisms of their content and presentation, particularly in recent years as the number and diversity of their users has increased.[88] I shall focus here on two questions: (1) the relative merits and difficulties of large- and small-scale floras, and (2) the relationship of floras with their users.

Large- vs. small-scale floras

Large-scale floras, some with more or less natural phytogeographical limits, remain important in the programmes of many botanical research groups. They are sometimes also seen in terms of the fulfillment of cultural or national goals.[89] On the other hand, at the same time – and in spite of past criticism – there remains a strong demand from professionals and the public for concise one-volume country, state or local manuals for reference or field use.[90] Each has had their advantages and disadvantages, as I shall discuss below.

It is the smaller-scale, one- (or two-) volume works with which most users of floras are by and large likely to come into contact. As recounted in Chapter 2, such inventories have been part of the botanical canon in one or another form for more than 400 years, and currently comprise a wide range of checklists, enumerations, manuals or floras. They are designed to be more or less readily portable and are oriented to students as well as professionals and other users. Not unnaturally, however, they vary considerably in quality, and several proposals with respect to content and style have been made.[91]

The **smaller-scale** flora, manual or checklist is here taken as covering up to 5000 species of vascular plants. With compact writing and tight editing, it is possible to encompass such a number within a single volume. Several good examples have appeared in recent decades, all with 3000 or more species: *Manual of the vascular plants of Texas* (1970), *Flowering plants of Jamaica* (1972), *Exkursionsflora von Deutschland: Gefässpflanzen* (15. Aufl., 1990), *The Jepson Manual: higher plants of California* (1993), and *Flore de la Suisse et des territoires limitrophes: le nouveau Binz* (2nd edn., 1994). Many more covering lesser numbers could be cited; here, compactness is easier to achieve even with the inclusion of figures for many or most species. Of these the *Jepson Manual* is perhaps the most ambitious: although in format less compact than others, it features keys, descriptions and much ancillary information including cultivation potential.

It is perhaps fortunate that in many or most temperate regions, recognized political or physical limits of nations or states generally encompass fewer than 5000 species. This has facilitated more or less regular revisions of manuals.[92] In many parts of the tropics, however, species numbers are simply too great relative to human and other resources and their taxonomy moreover often not or but poorly known. Lesser areas have therefore come to be seen as suitable foci for floras and manuals, with national or state inventories taking the form of checklists such as Brako and Zarucchi's *Catálogo de las angiospermas y gimnospermas del Perú* (1993), Arnold and de Wet's *Plants of southern Africa: names and distribution* (1993), or Turner's *A catalogue of the vascular plants of Malaya* (1996–97). Several significant 'focused' tropical florulas have in such wise appeared; among them are *Flora of Barro Colorado Island* (Croat, 1979), *Flora of Pico das Almas* (Stannard, 1995), *Flora of Central French Guiana* (Mori *et al.*, 1997), and *Flora da Reserva Ducke* (1999), with several others in preparation.[93] Smaller tropical islands or island groups provide natural foci; for these an outstanding recent work has been *Manual of the flowering plants of Hawai'i* (Wagner, Herbst and Sohmer, 1990).[94] Yet none of these covers more than 2000 species, with all having required several years of work as well as a solid organization. Even then, the time required for their preparation has sometimes been seriously underestimated; intensive research at 'focus' sites has often led to numerous range extensions as well as the discovery of new taxa.[95] Alternative objectives in temporally circumscribed projects on tropical 'focus sites' or small polities have been the tree flora, as was partially done in *Los árboles del Arborétum 'Jenaro Herrera'* (1989–90) by Spichiger *et al.* in Peru or, more recently and fully, in *Flora da Reserva Ducke* (1999) by Hopkins *et al.*, or well-documented enumerations such as *A checklist of the flowering plants and gymnosperms of Brunei Darussalam* (1996) by Coode *et al.* and *Plants of Mount Cameroon: a conservation checklist* (1998) by Cable and Cheek.

Small-scale works in themselves have certain scientific as well as practical advantages. A work covering a relatively small but key area can encompass a significant percentage of the known species in a given polity, even in Africa.[96] Moreover, geographically focused work can itself be rewarding, with the discovery of new species or significant range extensions.[97] Another 'local' flora, *Flora of the Río Palenque Science Center* by Gentry and Dodson (1978), is now one of the few records we have of the much-destroyed lowland wet forest flora of western Ecuador.[98] For Barro Colorado Island in Gatun Lake in Panama, created by the construction of the Canal early in the twentieth century, there have been two florulas (1933 and 1979), thus creating a record of actual or potential changes to the flora over time. A similar record – but with its two poles nearly

100 years apart – is now available for Singapore, clearly showing the great losses as well as survivors in the face of massive urban and other development over the last century. There is thus little doubt that for well-circumscribed areas they can be valuable management as well as informational tools.

There are, however, also disadvantages with such smaller-scale, relatively tightly circumscribed works. Most objections have been legitimately scientific. Taxonomically, not all families (and genera) may benefit from specialist or other forms of first-hand review, leading to perpetuation of long-standing misidentifications as well as taxa which should really be in synonymy (where a recent revision or monograph is not available). More generally, the ratio of species to genera is usually relatively low, making more difficult any attempt at building a picture of diversity in a larger or widely distributed genus where good separate revisions or reviews are wanting; at the same time, they may be of less value than a large-scale flora or enumeration for some kinds of biogeographic analyses. And finally, there is an element of sentiment as well as challenge: can – in spite of the seemingly ever-greater presence of micro-management in science and the inherent anonymity of the virtual world – the present generation create new, identifiable large-scale 'classics'? Or are such 'transcendental' notions of no value in a world of sometimes protracted negotiations along with decisions by committee?

It is **large-scale** floras which have long represented a *beau ideal* in plant taxonomy: at once a major contribution to knowledge, a form of social expression, and a source of recognition – sometimes permanent – for their authors or organizations. Some, such as *Hortus malabaricus* in the seventeenth, *Herbarium amboinense* and *Flora danica* in the eighteenth, *Flora australiensis*, *Flora brasiliensis* and *Flora orientalis* in the nineteenth, and *Flora SSSR*, *Flora Europaea*, *Flora of Turkey* and the almost-complete *Flora iranica* in the twentieth centuries achieved all of these aims in greater or lesser degree and have become recognized classics in the genre.[99] Others were grandly planned but in their time not or but little realized; such was the fate of the floras of the late eighteenth century Spanish royal botanical expeditions (including *Flora peruviana, et chilensis*) or, in the twentieth century, *Genera et species plantarum argentinarum* and the original FNA Program. Still others have been seemingly too elaborate, particularly in proportion to the area or size of flora covered – such as *Flora de Cabo Verde* – or too ambitious in relation to available resources or the state of existing knowledge and documentation (examples being *Flore générale de l'Indo-Chine*, *Flora of Panama*, and *Flora of Southern Africa*).

Project times and taxonomic productivity

Where the flora is large, the area great, means limited, or much detail or a luxurious format demanded, long project times have been usual.[100] Examples initiated in the first half of the twentieth century include the first series of *North American Flora* (1905–57, not completed), *Illustrierte Flora von Mitteleuropa* (1st edn., 1906–31; 2nd (and 3rd) edns., 1935–), the already-mentioned *Flore générale de l'Indochine* (1907–51) and *Flora of Panama* (1943–81), *Flora of Suriname* (1932–84, not completed), *Flora SSSR* (1934–64), and *Flora de Madagascar et des Comores* (1936–). Post-World War II examples include *Flora Malesiana* (1948– , now some 20 percent complete), what is now *Flore d'Afrique centrale* (1948– , presently about 50 percent complete), *Flora reipublicae popularis sinicae* (1959– , now more than two-thirds complete), *Flore du Cambodge, du Laos, et du Viêt-Nam* (1960– , yet still not that far advanced), *Flora Zambesiaca* (1960– , now about 65 percent complete), *Flora of Southern Africa* (1963, now 10–20 percent complete), *Flora iranica* (1963– , now all but complete), *Flora Europaea* (1964–80), and the monographic series *Flora Neotropica* (1966– , about 10 percent complete). In the 1970s and 1980s came *Flora of Ecuador* (1973–), *Flore des Mascareignes* (1976–), *Flora of Australia* (1981– , now some 45 percent complete), *Flora of the Guianas* (1985–), and the successors to *Flora SSSR* in eastern Europe, Siberia and the Far East. The 1990s have seen the commencement of *Flora of North America* (1993–), *Flora Mesoamericana* (1994–), *Tree flora of Sabah and Sarawak* (1995–) and *Flora of the Venezuelan Guayana* (1995–), although most of these projects had actually been initiated in the previous decade – and all of them before the rise of the Web.

Such prolonged times inevitably have raised questions – notably in more recent years – about financing as well as institutional and individual motivation.[101] Note was taken in Chapter 2 of the 66 years taken by *Flora brasiliensis*, but against this must be set its coverage of over 22000 species at a time when available manpower was very much less than today. Even the energetically pursued *Flora SSSR*, with something

over 17000 species, required over 30 years for completion, and *Flora reipublicae popularis sinicae*, faced additionally with a decade of political disruption, will have taken at least half a century by its conclusion.[102] The time necessary to realize *Flora of Australia* will perhaps be double the 20 years originally estimated. *Flora of Southern Africa* (1963–) was in the 1990s even suspended for a time.[103] Some projects, among them *Flora Malesiana* and *Flora reipublicae popularis sinicae*, have furthermore been faced with significant increases in estimates of the total number of species to be covered in comparison to when they began.[104] Only those works with sharply limited objectives, defined parameters, or relatively well-known floras in terms of size have been fully realized within a generation or less. Among these are, besides *Flora Europaea*, *Flora of west tropical Africa* (1927–36; 2nd edn., 1954–72), *Flora of Turkey* (1964–88), and *Prodromus einer Flora von Südwestafrika* (1966–72). It is likely that *Flora of the Venezuelan Guyana* (of which five volumes have appeared) and *Flora iberica* (now about a third complete) will also make relatively rapid progress; with rather more variables, however, such a pace may be harder for *Flora of North America* (of which just three of 30 volumes had appeared by 1999, with one more in press).

A corollary to the increasing length of time taken per species in most 'research' floras is an evident decline in average taxonomic productivity. Whereas in the nineteenth century perhaps 250 species per year could be written up for a handbook-flora by one author (Bentham achieving a still higher rate), by the 1930s the optimum for critical semi-monographic floristic work was estimated at 80 species per man-year.[105] This had by 1963 further declined to 50 species,[106] and by 1979 was down to 15–20.[107] More recently, it has been suggested that the 160 species of Malesian Sapindaceae consumed the efforts of 15–20 people over 20 years.[108] Among the few mid-twentieth century works with comparably high productivity levels covering only partly known ground was *Flora SSSR*, but its symbolic status as well as its founders' advocacy ensured a full deployment over a generation of its host institution's taxonomic resources.

It is not to be wondered, then, that De Wolf and Jacobs[109] and, more recently, a workshop at Leiden[110] have questioned the wisdom of many large-scale projects, suggesting that more attention be paid to the preparation of 'target' as well as 'concise' works. More attention is now being paid to these suggestions, partly through force of circumstances but also on account of changing needs and interests.[111] Indeed, in the world at large relatively few new 'grand' flora projects have been commenced in the last 15–20 years as compared with the mid-twentieth century, and some have been abandoned or modified (e.g., *Genera et species plantarum argentinarum*, *Flora brasilica*, *Flora Malesiana* and *Flora of Southern Africa*).[112] Moreover, as we have seen, the advent of the Web has made possible 'virtual floras' such as FloraBase in Western Australia in which varying degrees of completeness or reliability are recognized and accepted in the interests of availability of information – much more of a priority or even requirement than in the past.

Yet for many botanically lesser-known areas (below stage 3 on Jäger's map) where existing documentation is poor and/or scattered, any general flora would in Symington's sense[113] represent 'first' or 'second' coverage and should therefore possess adequate documentation and references. Some would still argue that such works should always have priority. On the other hand, the objection would also be raised that for 'first' floras, a substantial amount of basic monographic and revisionary work is required and this must be expressed in some way in the published work, because there may be no alternative. Thus, with any advocacy of conciseness in floristic publication, careful consideration must be given to the satisfactory disposal of what is not included; too often in the past this has been relegated to a plethora of scattered outlets or even lost.

With respect to 'traditional' large-scale tropical floras a good compromise has been struck by *Flora Zambesiaca* with its relatively concise descriptions, selected specimen citations, and supporting data and references. By contrast, the inadequate specimen, distributional and bibliographical data characteristic of *Flora Malesiana* – originally conceived as a 'final' product – necessitates access to, ideally, a considerable file of associated precursory papers and other records for effective use.[114] At a more local level, *Flora of Java*, a 'second' flora like *Flowering plants of Jamaica* and *Flora of west tropical Africa*, similarly is not well documented; it notably lacks any consistent citation of taxonomic sources (standard revisions, monographs and other precursory papers). By contrast – and usually with a stronger historical foundation – conciseness is more easily achieved in extra-tropical works, although a larger-scale form of presentation may be desired for

reasons of prestige or thoroughness, as in *Flora boreali-americana*, *Flora of Texas* or *Illustrierte Flora von Mitteleuropa*. Even so, however, many manuals have been in the past deficient in documentation; one of the better postwar ones was Willis's two-volume *A handbook to plants in Victoria* (1960–72).[115] More recent manuals have, however, become less 'authoritarian'; among more 'user-friendly' features now becoming more common is the inclusion of standard taxonomic references in family and generic headings. All this is part of the question of information content and handling in floras and related works: a question never seriously considered at the beginning of the 'modern' era but which, as Heywood pointed out already in the 1970s, has grown in the last 60 years to acute proportions. The advent of the Web, the realization of 'virtual floras', and changing social demands will also, as has been suggested, bring about further discontinuities in practice, with current arguments relating to content, style and presentation becoming obsolete or taking new forms.

Floras and the user

The relationship between floristic publications and their users is among the most important in plant taxonomy and thus has frequently been a topic for discussion, with a variety of viewpoints being expressed.[116] In the more distant past, however, the content and style of floras were largely the prerogative of the *savant*, with generally a one-way relationship with users. The works were moreover often viewed in terms of a pedagogical setting.[117] Indeed, innovations such as analytical keys, diagnoses or concise descriptions, critical figures, and the use of vernacular names were all devised as study aids.[118] Works for purely popular consumption were derived from the more formal floras, often by enthusiasts, and did not necessarily enjoy a comparable status.

In the twentieth century, increasing demands on curricular time as well as changes in fashion have led to a relative decline or even disappearance in classroom and excursion use of floras with the result that people may be less conversant with the language of botany. What was once a relatively common currency over several generations has, like Latin, become rarer. This has for many made the more traditional floras harder to use and thus less accessible. Yet, in the same century, the growth of plant geography (or 'ecology' in the anglophone world) and its applications (including land

and wildlife research and management, environmental impact surveys, and conservation assessments) has made new demands on floras and checklists. Field data are seen as particularly important not only as an aid to applied work but also for their intrinsic value as biology has become less exclusively museum-oriented. Nevertheless, to a considerable degree the new classes of data were simply incorporated into the formats inherited from the eighteenth and nineteenth centuries. It was mainly in dendrological works where an atlas-format was gradually adopted, with considerable use of figures and photographs.

A relatively close relationship between producer and users generally existed in temperate regions, including those colonized subsequent to the Columbian and da Gama voyages. The more enterprising writers made efforts to ensure that floristic information was effectively presented, sometimes themselves preparing works for extra-scientific consumption. In recent years, collaboration among different constituencies has – as already noted – become more common; such a course governed preparation of *The Jepson Manual: higher plants of California* (1993). Similar directions have guided the development of some contemporary floristic information systems, including FloraBase in Western Australia.

It was the spread of more intensive studies of tropical floras, including the diversity and potential of forest trees, that led to problems with the use of the run of floristic literature. Until the late nineteenth century (and the advent of relatively convenient surface travel along with an improved understanding of tropical diseases) the investigation of tropical floras was largely seen as an extension of metropolitan activities; conventional styles of organization and presentation in floristic works were deemed suitable, and indeed to a considerable extent have remained so. The first potential users of tropical floras beyond their traditional 'Banksian' botanical constituency[119] were the functional plant ecologists of the late nineteenth century.[120] Their primary interest was, however, in processes rather than floristic diversity as such. In only a few areas of enquiry was there the potential for comparative observations which could usefully be incorporated into floras.[121] With respect to vegetation, their approach was largely physiognomic; the taxonomic identity of constituent species generally was of secondary interest. It was the growth and spread of community plant ecology ('synecology') and an increasing interest in

tropical forestry in the four decades prior to World War II – along with the beginnings of the tropical conservation movement and, not least, the greater ease of travel offered by the automobile as well as improved bus and rail services – which led to a greater appreciation of tropical floristics, particularly after 1920.[122] Yet much if not most of academic botany remained temperate in its orientation, notwithstanding contributions by such tropical pioneers as Adanson, the Banksian 'circle', Griffith and other Indian botanists, Beccari, and Warming as well as the 'functional ecologists'.[123] Curricula remained conservative,[124] while fashions in research lay elsewhere – exemplified by the rise of population genetics and experimental taxonomy.[125] Indeed, it was not until after World War II and publication of Paul Richards' *The tropical rain forest* that a renewed awareness of the tropics began to manifest itself in the field at large, at least in plant ecology.[126] As for the herbaria, they appeared to some still all but Linnaean and – in spite of their undoubtedly great contributions – thus a world apart, remote from contemporary concerns.[127]

With respect to floras, there was – perhaps not surprisingly – relatively little innovation. 'Official' British thought on tropical flora handbooks, as exemplified in recommendations for African research by E. B. Worthington, remained traditional;[128] they would largely be produced by botanists at the then-Imperial Forestry Institute in Oxford, at Kew, or in Edinburgh, with local collaboration where possible.[129] In France and its territories, tropical floristic research remained almost entirely centered on the Muséum National d'Histoire Naturelle, with comparatively few initiatives elsewhere.[130] Only in temperate North Africa were significant autochthonous resources built up through the work of R. Maire and others, continuing activities begun before World War I. Few if any significantly distinctive floras or manuals arose in other cultural spheres, though Backer's well-illustrated weed floras for Java deserve mention.

It was thus not from academia or the herbarium but from the rain forests of Malaya, Africa and the Americas that the first radical critiques of traditional flora formats arose. From experience obtained during careers respectively in the Federated Malay States and the Straits Settlements – both beginning in 1929 – C. F. Symington and E. J. H. Corner[131] independently argued that 'standard' handbook-floras were in fact of comparatively little use to persons in the field.[132]

Indeed, Symington claimed that much of the admittedly extensive tropical botanical (and ecological work) of the interwar period – some of which had attracted official criticism – lacked definition, while Corner believed that the existing corpus of publications only contributed to what he called 'the enormous humbug of tropical botany'.[133] Similar problems were faced by other pioneer forest botanists such as André Aubréville in Ivory Coast and elsewhere in French Africa, Adolfo Ducke in Brazil, and W. D. Francis and (later) B. P. M. ('Bernie') Hyland in Australia.[134]

While progress in the design and production of floras has indeed been made in the past half-century or more,[135] problems remain. One is the continuing need, sometimes by default, for accounts to be written by outside specialists, at times with limited field opportunities (although this is now less the case than formerly). Another is the great length of time now required by major flora projects, which often have had to follow formats established a generation or more before. Finally, there remains the question of effective access to floristic information, still often problematic for technical, linguistic, conceptual or financial reasons.[136] Symington's remark that with respect to flora schemes 'the main defect was to imagine that the European herbarium worker could solve the field biologist's problem without the latter's full co-operation' thus to a certain extent still holds.[137]

Particular criticisms made by Corner of contemporary works included (1) ignorance of vegetative characters, (2) ambiguous descriptions, (3) faulty nomenclature, and (4) errors resulting from repeated copying and/or lack of critical investigation.[138] Similar situations have been faced by other writers of tropical field-manuals (although the growing number of smaller 'local' herbaria has improved the availability of reference material, reducing the possibility of misidentifications and furnishing a foundation for description-writing and local key construction). There is now, however, an increased acceptance of the value of 'field floras' complementary to the larger, scholarly works, and more support has been forthcoming. Yet, as with *Wayside trees of Malaya* and, more recently, *Flora da Reserva Ducke*, these are not necessarily compilations; they should (and do) incorporate substantial additional field work and research. For the team of Jarvie, Ermayanti and Mahyar in central Borneo, preciseness, comparability and transparency of information – necessary for multi-access or interactive keys –

have been paramount. Close attention was, however, at the same time paid to potential users as Symington or Corner would have wished. Four points were in particular expressed: (1) availability of the local flora in Bahasa Indonesia as well as English; (2) understandable text, with inclusion of a glossary; (3) adequate illustrations; and (4) orientation to the needs of each user group with the information required.[139] These authors also found that there were practical limits to the number of dichotomies in a single analytical key.

Beyond the tropics, there have also been significant recent contributions on the relationship of floras and users, as much on substance as on style and presentation. Wilken *et al.*, with reference to numerous examples, reiterated that floristic information was widely used beyond the world of pure biology and that this had to be taken into account in flora projects.[140] Palmer, Wade and Neal, in a study of an extensive sample of floras large and small in North America, renewed long-standing arguments for common standards and with respect to content made explicit proposals including a distinction between the 'essential' and the 'desirable'.[141] Schmid, from the viewpoint of a user and reviewer, has given a detailed list of desiderata with respect to content and style along with a plea for statistical information.[142] To this latter I would like to add some indication of the area covered with, for floras of larger entities, an internal breakdown covering individual provinces, states, counties or similar smaller polities with their numbers of species. Given modern database and spreadsheet capabilities this should be feasible. As has long been recognized, such statistical and geographical data are invaluable to ecological and biogeographical theory and, in more recent years at least, to public interest and awareness.[143] Indeed, they could be considered among the most important by-products of any flora project.

Is there an 'ideal flora'?

The preceding sections have reviewed various aspects of contemporary floras including recent methodological and other developments and have drawn attention to some continuing as well as recent problems with their writing and use. Emphasis has been given to the rapid development of information technology over the last decade which has at last – from initial attempts in the 1960s and 1970s – effectively allowed floristic information and tools to be developed and maintained as a 'virtual system' without reference to print.[144] Thus, traditional printed floras, manuals and enumerations – and their content – have ontologically become matters of essential or discretionary choice rather than absolute necessities. The parameters of what constitutes an 'ideal' flora, long a matter for debate and reviewed in the first edition of this book, have therefore changed and now may be seen as largely functional.[145] The issues relate as much to presentation as to content: is the floristic work in question useful for identification or essential information and is it comprehensible to a wide range of users? Moreover, given the existence of some form of floristic information system, what selection of tools and information is best also presented in print, relatively a more expensive medium? Finally, what scope is there for a traditional 'ideal', the 'classical' descriptive flora?

The 'classical' descriptive flora of a country or region has for some two centuries been perceived as a key intellectual or practical goal of many a systematic or floristic botanist, and – as outlined in Chapter 2 – has at least until recently largely followed parameters set by leading eighteenth- and nineteenth-century writers. However, there has been over the last decade or so an increasing recognition of the need on the part of floras and related works for more effective communication with a wider audience along with marked changes in methodologies.[146] It thus may be (and has been) argued that the 'classical' descriptive flora has no real place in the contemporary world, and that users are best served with checklists, keys and illustrated guides. The latter is often all that may be possible given current project-oriented funding practices. Moreover, the descriptions in many 'standard' floras are not truly comparative, nor were they so intended.[147]

Yet, as some authors have pointed out, there remains merit in the 'classical style' of flora-writing as an art form.[148] An elegant 'effect' may still be achieved in presentation through discerning phytography, carefully chosen synonymy, references, vernacular names, selected *exsiccatae*, concise indications of status, distribution, phenology, habitat, altitudinal range, substrates and sociology, and well-thought-out commentary. A 'classical' description should be individually prepared and be thus distinctly diagnostic while still embodying a 'feel' for a plant; this is more than can be achieved from mere application of a computer-driven descriptive language system.[149] Its making and user comprehension does, however, require a certain amount of formal education and experience. Indeed, it may now

be best suited to situations where a local, state or national flora is relatively well known and alternative 'research' sources (including critical floras or monographs) are available.[150] Other 'traditional' formats, among them the 'manual-key' with its heavy use of abbreviations and symbols, may similarly at first be rather hard to use, particularly if one has had no training. Many authors of such works have, however, recognized that illustrations are essential aids to identification. Such is not new; Bonnier and Layens did this consistently over 100 years ago, initially in their already-mentioned *Nouvelle flore du nord de la France et de la Belgique*. Among contemporary manuals, at least one is now also issued as a CD-ROM (making potentially possible links to more comprehensive floristic information therein or elsewhere).[151]

With respect to the function of floras, it has gradually become more widely recognized that for a given polity or region no single 'ideal' flora was possible.[152] Multiple products or systems were seen as necessary, with some kinds of information best represented in 'information banks'. This gave rise to discussions – now partially academic – over what was best included in printed works and what best was stored electronically.[153] In the late 1980s, 'core fields' were seen as including scientific names, authors, concise descriptions, notes on relationships, indication of phenology, habitat and distribution, literature citations, and illustrations along with general ease of use. Keys were not specifically mentioned, nor were synonymy, vernacular names or ethnobiology. More appropriate to a 'data bank' were: a GIS, supplementary morphological data, bibliographic references, specimen-based data, illustration sources and cartographic information. Yet this distinction could be seen as analogous to that long made between 'synoptic' and 'comprehensive' forms of floristic communication – a question which, as discussed in Chapter 2, has existed since the late sixteenth century.[154] Moreover, the 'core fields' are comparable to those of A.-P. de Candolle and other nineteenth-century advocates of concise floras.

It is the increasing range – and greater complexity of – 'extended' material which came to be seen by the 1980s as best handled in an information retrieval system. The advent of new technologies and their deployment have since effectively ended any *primary* need for such distinctions. Output can be designed and produced as matters of record or for particular functions. If there is any single 'ideal' national or regional flora, it has ultimately to be in the form of a dynamic, accessible and generally intelligible information system or 'metaflora'.[155] How completely this happens depends, however, on human resources and commitment, and thus decisions over objectives and priorities are still required. It should moreover not be overlooked that many large-scale scientific works – though in practice read only by a few – traditionally also have had humanistic functions; indeed, they are, or have become, expressions of social, cultural and political identity.[156]

Prospects

With the transmutation of floras into information systems, perhaps the major issues now – in contrast with 20 years ago – are less with products *per se* but more about effective access to and dissemination of particular functions centering on inventory, information and identification. The 'classical' parameters and styles, long generally accepted as definitive,[157] have become in themselves rather less adequate in the face of expanded horizons, audiences and requirements. Change has, however, sometimes come but slowly; questions raised already in the 1940s were still being asked in the mid-1990s.[158] Much more attention has to be, and is being, paid to users; floristic works – particularly the larger ones now requiring extensive collaboration – cannot now be prepared in isolation. In addition, larger projects must consider shorter temporal parameters, with one response being identification of the plant groups of greatest difficulty (or for which there was no specialist) and making some information on them available – if in a less definitive form than a 'final product'.

As for the functions of floras, several key areas remain as important as ever irrespective of the form of delivery (save for dynamic interaction, not possible in print, and family arrangement, less relevant in the virtual world). These include identification, nomenclature, descriptions, documentation, illustrations and maps, commentary, arrangement of taxa, links, and presentation. Each of these is considered below.

Identification is best addressed through artificial dichotomous, multi-access or (increasingly) interactive keys, preferably with figures or key characters and states. In large floras or in well-represented taxa with scores or even hundreds of species in a given area, illustrations are essential; a similar consideration applies where, for example, a given leaf arrangement or flower type is widespread. The more traditional synoptic key, even in analytical form, has to be seen as primarily an academic exercise although it has merits in being able to focus on principal differentiating characters.

With respect to **nomenclature**, floras show wide variation. In the more documentary undertakings accepted names, synonymy, references and applications may be rather fully treated. In more or less concise works only selected synonymy may be given, with or without references. The publication of two worldwide synonymized species checklists in addition to many at supranational and national level may bring about an increasing amount of control. This may well lessen the need for elaborate synonymies in floras, although this goes against those who see a flora as 'self-contained'.

Descriptions have their place in a concise flora but should be largely diagnostic, as aids to identification, and emphasize key features. Elaborate 'primary' descriptions, characteristic of many large documentary floras, may be seen as important components of training of young systematists but are more the province of the monograph or revision of a given taxon or part of an information system. The importance of truly *comparative* descriptions was long ago emphasized by Watson[159] but floras are not the place for them.[160]

Documentation is an extremely important part of systematics, both for the record and for analysis, but is only partially congruent with the main functions of a flora (or related work). Enumerations and checklists fundamentally are documentary, and should be encouraged.[161] By their nature, however, they can cover only a portion of the potential range of published or unpublished material; they should therefore contain 'pointers' to sources, or be part of an information system.[162] With the rise of the latter, however, the issue of choices of data for dissemination becomes relevant only for printed products; past problems such as loss of locality and specimen data as well as more extended taxonomic and biological observations due to space constraints in theory cease to exist.[163]

Illustrations and **maps** are increasingly sought after and should now be seen as essential in a flora. Not only have people become more visually oriented, but with this has come a greater appreciation of the value of practical as well as aesthetic botanical art. Maps are also extremely important visual tools. Provision of both should be systematic, and for identification illustrations have to be well thought out. There are presently few technical limitations on the storage and retrieval of images.[164] Floras (and related works) should be, and are being, accompanied by CD-ROMs with color images otherwise collectively too costly for print.[165] If maps are also available on CD-ROM, there should be software for interrelating them as part of a GIS.

It is hard to suggest definitive guidelines on **topographical**, **taxonomic** and **biological commentary** including distribution, ecology, phenology, karyotypes, chorological classes, fidelity (in a phytosociological sense), variability, and biology as well as opinions on taxonomic limits and relationships. The scope of such material varies widely. Basic distribution, habitat and phenology are generally seen as 'core data' and will always have a place in a concise flora, enumeration or checklist. Likewise, where distinct differences of opinion exist with regard to taxa, such should be noted. Where a flora is poorly known or available materials are scattered, representative specimens should be cited.[166] 'Peripheral' data are, however, better left to specialized works or to information systems.

The **arrangement** of taxa in printed floras varies widely. In part this is because of differences in opinion over higher-level relationships among flowering plants. For strictly practical purposes alphabetical or artificial arrangements may suffice; but with differing ideas about the limits of many families care has to be taken. Intellectually a systematic order may be more informative, but here again a choice among different, sometimes widely diverging, schemes has to be made.

All of the above leads to the importance of **links to and among sources**. A good flora today should as far as possible include reference to monographs, revisions and supporting papers on a consistent basis. Among other things, this acknowledges the idea of a flora as a statement of current knowledge and not an all-encompassing 'authority'. In recent years, this practice has become more common. Source references are also important in introductory general keys to state, national and supranational floras.

Last but not least is **presentation**. The appearance and ease of use of a work have become important factors in their acceptance. Though the latter problem was early addressed by writers of forest floras, the result was mainly the additional provision of keys based on vegetative features rather than a serious answer to arguments that conventional floras were to many all but impenetrable. In defense of the latter, however, it should be said that such works have always required some training; indeed, floras originated in, and were in many ways designed for, the academic world. The loss or reduction in many universities (let alone secondary schools) of course modules in identification and taxonomy, and the greater diversity of users, have now brought the 'language' of floras into greater relief.

Concluding remarks

In summarizing the above – and closing this chapter – the author believes that the best roles for a 'traditional' flora (or enumeration or checklist) in the present age are practical: inventory, identification and provision of essential related data. To Bentham's 'essential data' should be added ecological information and a good range of illustrations, as well as a clear indication of where taxonomic problems exist. Where botanically lesser-known areas are concerned there is room for some extra data and commentary.[167] Systematic provision of references to sources should be standard practice. Documentary floras – unless possessing biologically or biogeographically based parameters – are, by contrast, basically symbolic or humanistic. Much of their content – and the effort that goes into them – ideally should be directed elsewhere, for example into revisions or monographs.[168] Their data, while in scientific terms essential, are to a goodly extent of relatively specialized interest, and – as Stephen Blackmore has suggested in the last of our opening quotations – are in modern terms best presented as part of an information system. That said, however, it should be admitted that without sponsored larger-scale floras a great deal of raw data might never be synthesized. Much of the relatively limited support for plant systematics as a whole is ultimately public and often is related to specific national, developmental or environmental goals.

With respect to the place of print in a world of rapidly spreading electronic media, I believe that with respect to floras a place remains for handbooks for practical use. They should be concise and communicative, with defined parameters but associated with a floristic information system (either on-line or distributed on a CD-ROM or other electronic medium) which may carry additional, more specialized data, images or such features as interactive keys and GIS capability. All this presupposes a team effort, the necessity for which is now widely recognized. Such a multidimensional approach calls for organization, effective cooperation and use of limited human resources (particularly family specialists), adequate funding, and an enlightened use of information technology and electronic communication.[169] Similar approaches could be (and from time to time have been, with varying degrees of integration) taken with respect to advancing basic taxonomic knowledge of, for example, large families or genera.[170]

In the original edition of this book, I concluded this chapter with four points: (1) floras and manuals should be written in a way as will have a wide public appeal; (2) floras should be the products of collective effort; (3) monographs and revisions should be promoted, with larger efforts becoming similarly collegial; and (4) the language of floras should 'not be permeated with learned and prosy dullness'. Taking a lead from the historian and philosopher of science Jerry Ravetz, I wrote at the time of the need for a 'new ethic' for floras.[171] Progress on these fronts has since been varied, perhaps more in method than in language, but in the last half-decade or so all else has been overtaken by an ontological revolution: the effective transmutation of floras into information systems, with print an option rather than a necessity. Nevertheless – and whatever the medium – standards of content and presentation in any future floristic works should ultimately, as ever, be guided by 'correctness and clearness of method and language [as] the first qualities requisite',[172] with moreover an eye to 'care and coordination, and a clear sense of the priorities both from the producer's and user's point of view'.[173] Twenty years on, users are even more important; their interests should be firmly accounted for within any contemporary philosophy of floras, with a 'stress on access'.

Notes

1 For example Stafleu, 1959; Brenan, 1979.

2 Jäger, 1978, p. 413.

3 The 1999 Congress (St. Louis), however, featured a session entitled 'Species diversity information systems on the Internet'. This represents a possible future direction for large-scale floras, as will be shown later in this chapter.

4 Morin *et al.*, 1989.

5 Prance and Campbell, 1988; Campbell and Hammond, 1989.

6 Baas, Kalkman and Geesink, 1990; Polhill, 1990. Further workshops on *Flora Malesiana* have since been held, most recently in 1998.

7 Bisby, Russell and Pankhurst, 1993.

8 Gómez-Pompa and Plummer, 1990; Sivarajan, 1991, p. 5; Funk, 1993; Jarvie and van Welzen, 1994.

9 van Steenis, 1962.

10 Similar themes have recurred in subsequent discussions; see in particular Heywood, 1984; Sivarajan, 1991, p. 5; Jarvie and van Welzen, 1994. Sivarajan in particular pointed out the potential problems which might arise from the creation of 'non-uniform matrices', the result of developments oriented to given projects. With the spread of often uncoordinated, grant-based 'biodiversity surveys', this phenomenon has in the author's opinion become a real concern.

11 van Steenis, 1962.

12 Twentieth-century examples include, in Europe, the Hegi *Illustrierte Flora* vs. national manuals and, in North America, the two floras of the Pacific Northwest (**191**) coordinated by C. L. Hitchcock and A. Cronquist. In tropical countries with large floras, the focus of some recent one-volume manuals and checklists has been on relatively small areas, often reserves with research stations.

13 Heywood, 1973*a*, 1984, 1988; Watson, 1971; Palmer, Wade and Neal, 1995.

14 Shinners, 1962; Palmer, Wade and Neal, 1995; Brenan, 1963; Jacobs, 1973. The editors of *Tree flora of Sabah and Sarawak* adopted *Tree flora of Malaya* and *Flora Malesiana* as models; see Forest Research Institute of Malaysia, 1992.

15 *Herbarium amboinense* was, as indicated in Chapter 2, not actually published until the mid-eighteenth century.

16 Shetler, 1971.

17 Jacobs, 1973.

18 For example Polhill, 1990.

19 Webb, 1978.

20 Palmer, Wade and Neal, 1995; p. 339.

21 Symington, 1943. *Flora of central French Guiana* had, probably, the fewest antecedents; indeed, very few collections had been made around Saül before 1965 (Mori and Gracie, 2000). This likely contributed to the underestimate of the time needed for the project, as indicated by Mori and Gracie.

22 The *Flora of Java* is a noteworthy example. A more satisfactory solution would have been the production of a work like *Flore française*, with a volume of analytical keys (accompanied by illustrations) and two (or three) enumerative volumes with diagnoses.

23 In *Flora lapponica*, *Flora suecica* and *Philosophia botanica*.

24 de Candolle, 1813; Bentham, 1861, 1874; de Candolle, 1873, 1880.

25 van Steenis, 1954; Brenan, 1963; Turrill, 1964; Heywood, 1964.

26 Britton, 1896–98 (2nd edn., 1913); Hitchcock, 1925; Blake, in Blake and Atwood, 1942; Lawrence, 1951; Radford, in Radford *et al.*, 1974.

27 The growth of interest in field botany, perhaps in large part because of greater ease of transportation, led to an increased appreciation of plant variation and of the distribution of plants in relation to habitat factors as well as to each other. The latter area of enquiry became known in continental Europe as plant geography, with ecology – a term first introduced by Ernst Haeckel in 1866 as *Ökologie* and which came into English in 1873 – referring more narrowly to plant–habitat relations, including the study of morphological adaptations. Its extension in scope to the whole of plant geography (except floristics) initially took place in Britain and North America at the end of the nineteenth century.

28 All these reflect successive developments of new areas of botanical enquiry, their application to systematics, and evaluation of their worth in floras.

29 Heywood, 1958, 1960; Forest Research Institute of Malaysia, 1992.

30 Radford *et al.*, 1974; Yü, 1979 (publ. 1980).

31 Davis and Heywood, 1963.

32 Hitchcock, 1925; Lawrence, 1951; Radford *et al.*, 1974.

33 Meikle, 1971. Meikle's own *Flora of Cyprus* (1977–85; see 772) is a good example of a 'modern' flora still very much in the nineteenth-century tradition.

34 In comparing some mid- to late twentieth century North American floras with Torrey's *Flora of the state of New-York* (1843) one might almost think that the modern works had been written by descendants of Rip van Winkle.

35 With advancing technology, some more recent works, such as *The atlas of the vascular plants of Utah* by B. J. Albee, L. M. Schutz and S. Goodrich (1988), have been able to use localized dot maps.

36 van Steenis, 1954. For another view of the art of description, see Jacobs, 1980.

37 Cf. Britton, 1896, pp. v–xii; 1898, pp. iv–v. There was no clear philosophy in the Britton 'school' except perhaps its attempts at more or less uniform descriptions and, more particularly, its narrow generic concepts and nomenclatural ideas. Its seeming façade of 'authority' was also of a piece with the architecture of the era. With respect to nomenclature, the *North American Flora* and other works of the 'school' were not unnaturally important vehicles for propagating use of the so-called Rochester Code (also known as the 'American Code'). Formulated in 1892, it was applied in Britton's own works and those of many others but finally consolidated with the 'International Rules' in 1930, at a congress which also saw a serious consideration of plant population biology – with the noted California botanist H. M. Hall active in both movements.

38 A 'quarto' format was used for *Illustrierte Flora von Mitteleuropa* and its contemporaries on account of their many plates. It is also more amenable to two-column text, used in, for example, *Flora Europaea*, *Flora Mesoamericana*, and the English-language *Flora of China*.

39 van Steenis, 1954; Davis and Heywood, 1963, p. 299.

40 Wisskirchen and Haeupler, 1998, p. 12.

41 Hagen, 1984.

42 Raven, 1974*a*. This was in spite of efforts in the 1930s and 1940s by W. H. Camp and others to emphasize the importance of 'biosystematic' information.

43 Examples include *Alchemilla*, *Rubus*, *Sorbus*, *Taraxacum* and *Hieracium*.

44 Endress, 1977; Hj. Eichler, personal communication, 1980.

45 After the already-mentioned Wendell H. Camp.

46 Just, 1953.

47 Indeed, current technology enables incorporation of line drawings and gray-scale figures at little or no additional cost beyond an inevitable rise in the number of pages.

48 Fisher, 1968.

49 This work was fully revised in 1961 as *New Illustrated flora of Japan*.

50 For example *Dai shokubutsu zukan* (1928) by M. Murakoshi; *Cay-co Viêt-Nam* by Pham Hoàng Hô (1960; 2nd edn., 1970–72; 3rd edn., 1991–93); and *Iconographia cormophytorum sinicorum* from the Botanical Institute, Beijing (1972–76; supplements, 1982–83). While the form of these manuals may have been influenced by 'Western' models, notably (for Makino) Britton and Brown's *Illustrated flora*, the author believes them also to reflect something of the autochthonous East Asian botanical tradition (and, in Japan, its particular development in the Edo period). At the same time, their conciseness owes something to the traditions of Bentham and his contemporaries.

51 Backer, a former secondary-school teacher, saw his floras in pedagogical terms. Like J. D. Hooker and his reaction to analytical keys (see Chapter 2), he believed that figures distracted students from careful reading of descriptions.

52 With the introduction of more floras and identification works at least partly on CD-ROMs there will be additional opportunities for illustrations, particularly in color.

53 Chapman and Richardson, [1998].

54 Heywood, 1973*a*.

55 Watson, 1971.

56 Watson was by then at the Australian National University; there he had access to a mainframe computer, statistical and other software, and peripherals including tape-writers, keypunches and card- and tape-readers. With the CSIRO entomologist Mike Dallwitz he would in the later 1970s develop DELTA (DEscriptive Language for TAxonomy), a program

designed to ensure (and generate) uniformly comparable taxonomic descriptions; this led to the preparation of 'uniform' accounts of genera (and species) in selected families.

57 Shetler, 1971; Krauss, 1973. The FNA Program had a database software package as well as a mainframe computer at its disposal.

58 Such information might include, for example, detailed geographical records for a species or infraspecific taxon, useful to local workers.

59 Watson, 1971. The issue posed was subsequently examined by several writers, among them Cutbill, 1971*a*; Heywood, 1972, 1973*a*, 1974; Morse, 1975; and Raven, 1977.

60 Heywood, 1976, p. 267.

61 Shetler, 1974*b*.

62 Useful landmarks may be found in Cutbill, 1971*b*; Shetler, 1974*a*,*b*; Brenan, Ross and Williams, 1975; Allkin and Bisby, 1984; Bisby, Russell and Pankhurst, 1993; and Hawksworth, Kirk and Dextre Clarke, 1997.

63 Taylor and MacBryde, 1977 (see **124**). The projected descriptive flora never materialized.

64 Shetler, 1971; Gómez-Pompa and Butanda, 1973; Gómez-Pompa and Nevling, 1973.

65 By 1999 some 100 fascicles had appeared and a complementary 'ecological flora' was also well advanced.

66 Heywood, 1978. Major commercial or parastatal applications, still in use, were Ceefax (United Kingdom) and Minitel (France).

67 Pankhurst, 1975, 1978.

68 Brenan, Ross and Williams, 1975; Forero and Pereira, 1976; Gibbs Russell and Gonsalves, 1984; Johnson, 1991.

69 Pankhurst, 1991. Current taxonomic information packages on the market include ALICE (Kew), BRAHMS (University of Oxford), LINNAEUS-II (Free University of Amsterdam/ETI), LucID (CSIRO and University of Queensland, Australia) and SysTax (Universität Ulm).

70 Raven, 1974*b*; Systematics Agenda 2000, 1994*a*,*b*.

71 Turning points were the introduction in 1979 of the Tandy TRS-80 Model II, the first personal computer with sufficient disk capacity for a modest floristic database, the advent of dBASE II for CP-M and subsequent operating systems, and, in 1981, the arrival of the IBM PC (with the AT following in 1984, effectively ensuring the universal spread of the 'PC-standard').

72 These include *Vascular plants of British Columbia: a descriptive resource inventory* (1977) by Taylor and MacBryde (see **124**), *List of species of Southern African plants* (1st edn., 1984; see **510**), and *Med-Checklist* (1984– ; see **601**). Additional examples are reviewed in Allkin and Bisby (1984).

73 Literature on the Internet and its tools is now abundant. Worthy of note are A. J. Kennedy, 1996. *The Internet and World Wide Web: The Rough Guide.* Version 2.0. London (and its successors); and B. M. Leiner *et al.*, 1998. *A brief history of the Internet.* The Internet Society (electronic publication; at http://www.isoc.org/internet/history/brief.html). Its original basis was ARPANET, established in 1969 by the U.S. Department of Defense Advanced Research Projects Agency to connect universities and research institutes. Other major networks followed and 'Internet' accordingly came into use around 1974 as a 'collective' term. In 1986 the U.S. National Science Foundation, in funding a major 'supercomputer initiative', created the open-access NSFNet and so laid the foundation for the Internet's exponential expansion. From 100 networks in 1985 (L. R. Shannon, *New York Times*, 26 October 1993) the number of networks worldwide by 1996 was thought to have risen to at least 50 000 (Canhos, Manfio and Canhos, 1997); increase continues to be rapid.

74 A useful analogy here is the subsumption of Newtonian mechanics by relativity theory.

75 Canhos, Manfio and Canhos, 1997. Examples of such networks are ERIN (Australia), CBIN (Canada), NBII and ITIS (United States), and BIN-BR (Brazil); they may be governmental, non-governmental or collaborative.

76 All this reflects a still-imperfect grasp of the intellectual and cybernetic aspects of taxonomic information systems, with further research and modeling needed (cf. Bisby, Russell and Pankhurst, 1993).

77 Maxted (1992) has written that 'the full potential of contemporary computing technology is, however, not used in taxonomy, although its use in a revision . . . would seem to be appropriate'. Computers, he noted at the time, were 'commonly used' in fairly specific contexts.

78 Not only are there as yet relatively few full floristic databases, but the number of wholly databased herbaria remains small. Yet some have viewed specimen databases in terms not merely of a single collection but of a *network*: functional rather than institutional. These are in effect floristic information systems in the making. North American examples include SMASCH in California and SERFIS in the southeastern United States.

79 I thank Alex Chapman (Perth, Western Australia) for suggesting these 'levels of commitment'. It may be additionally noted that the third level corresponds to that aimed for by the original FNA Program, while the present *Flora of North America* has operated mainly at the second.

80 Alex Chapman, personal communication.

81 Palmer, Wade and Neal, 1995.

82 An example is *Flora of Somalia*.

83 An example is furnished by the *Flora of Ecuador* programme, where for two decades an information system has been developed as a complement to the published *Flora*. See P. Frost-Olsen and L. B. Holm-Nielsen, 1986. *A brief introduction to the AAU-Flora of Ecuador information system*. 39 pp., illus. Aarhus. (Reports Bot. Inst., Univ. Aarhus 14.)

84 It should also be noted that availability of CD-ROM players continues to spread, and newer storage technologies – such as the 'e-book' – will inevitably follow. This consideration may have guided the authors of the revision (as *Australian tropical rain forest trees and shrubs*) of *Australian tropical rainforest trees* (**430**), issued in 1999 as a wholly electronic publication (disseminated on a CD-ROM). Apart from an accompanying 60-page manual, no print version was planned.

85 This situation in effect exists in Western Australia and is developing in southern Africa, to give two examples. In Western Australia FloraBase is complemented by a state checklist and two regional manuals, *Flora of the Kimberley region* (1992; see **452**) by Wheeler *et al.*, covering 2085 species, and *Flora of the Perth region* (1987; see **455**) by Marchant *et al.*, covering 2057 species. In South Africa there are two very concise regional floras complementary to the national checklist and the PRECIS information system: *Plants of the northern provinces of South Africa: keys and diagnostic characters* by Retief and Herman (1997; see **515**), covering 5700 species, and *Plants of the Cape flora* by Bond and Goldblatt (1984; see **511**), covering 8579 species. (The latter is currently under revision.)

86 Wilken *et al.*, 1989b.

87 Palmer, Wade and Neal, 1995.

88 For example Jarvie and van Welzen, 1994; Palmer, Wade and Neal, 1995; Schmid, 1997.

89 Anderson, 1991. A corollary has been their inclusion as a part of some development assistance programmes.

90 With respect to Malesia, however, van Steenis (1949, 1954) was, at least initially, a notable critic of 'local' floras. Nevertheless many botanists therein have strongly supported their preparation (Ng, 1988; Rifai, 1997 (publ. 1998)). By the 1960s, even van Steenis lent his encouragement in a call for a new flora of Taiwan (van Steenis, 1967).

91 For example Lawrence, 1951; Radford *et al.*, 1974; Palmer, Wade and Neal, 1995.

92 Cf. Garnock-Jones and Breitwieser, 1998.

93 Preparation of such 'focused' works has also for some time enjoyed official or peri-official encouragement (cf. Committee on Research Priorities in Tropical Biology, 1980).

94 Like Meikle's *Flora of Cyprus* (1977–85), this work is in two volumes, but in addition to relatively substantial descriptions it features a considerable number of illustrations. It furthermore does not account for the large naturalized flora.

95 Mori and Gracie, 2000.

96 M. Cheek, personal communication; based on the experience of Cheek, S. Cable and others with the flora of Korup National Park, western Cameroon.

97 M. Hopkins, personal communication; based on the work of Hopkins and others on *Flora da Reserva Ducke* (near Manaus, Brazil).

98 This work, along with *Flora of Jauneche* by the same authors (with F. de M. Valverde), was part of a series of florulas projected by Gentry to illustrate his ideas on the relationship of species diversity to rainfall levels and distribution.

99 *Flora of Australia* and *Flora reipublicae popularis sinicae* are among current projects likely to become future 'classics'.

100 Polhill, 1990; Jäger, 1993.

101 De Wolf, 1963, 1964; Polhill, 1990; Jäger, 1993. For *Flora Neotropica* a period of more than 300 years was projected; however, it also encompasses non-vascular plants and fungi.

102 The *Flora* was begun in the mid-1950s and, after the interruption of the Cultural Revolution and its aftermath, was originally projected for completion in 1985 (Yü, 1979 (publ. 1980)). Progress is continuing steadily, however; when completed it will be the largest of descriptive floras. Many large as well as smaller families have now been published, and four volumes of an English edition (accompanied by a series of volumes of illustrations) have also appeared.

103 Limited manpower and means were given as the reason.

104 A notable example is the *Flora Malesiana* area, for which a total of 40 000 vascular plants is now estimated, substantially higher than the estimate of 25 000 made by van Steenis in the 1940s (van Steenis, 1949, and in *Flora Malesiana*). A similar increase is known to have occurred in China. Such estimates are, however, dependent upon whatever definition of a species is generally accepted at any given time.

105 van Steenis, 1938; recalled in van Steenis, 1979, p. 73.

106 De Wolf, 1963.

107 van Steenis, 1979, p. 73.

108 Jarvie and van Welzen, 1994.

109 De Wolf, 1963; Jacobs, 1973.

110 George, Kalkman and Geesink, 1990.

111 Mori, 1992.

112 The sumptuous *Genera et species plantarum argentinarum* (1943–56) was physically even larger than *Flora brasiliensis*. After losing its patronage in the wake of political changes, it was succeeded by several 'regional' flora pro-

jects – though such was then the state of floristic documentation that even these were often 'primary', with no antecedents. A new national flora, in octavo fascicles, commenced publication in the 1990s.

113 Symington, 1943.

114 Based on the author's own experience in Papua New Guinea.

115 This has now been complemented by *Flora of Victoria* by Foreman and Walsh (1993–), with three of the planned four volumes published by 1999.

116 It has also been a topic for textbooks of taxonomy, e.g., in chapter 9, 'Presentation of taxonomic data', of Davis and Heywood, 1963. Most floras also feature expressions of their philosophy and presentation in their general sections.

117 As students required them, they had to be more or less compact. Indeed, Linnaeus (1753; see also Heller 1983 [originally publ. 1976], pp. 239–267) strongly criticized 'sumptuous' books, published in relatively small editions and affordable only by the wealthy. On the other hand, however, such works were themselves sources for more compact floras as Linnaeus himself acknowledged.

118 Not always, however, were such developments looked upon favorably; indeed, learning the naming and characteristics of different plants as well as the associated terminology was seen as a form of mental exercise and so intellectually beneficial. It thus comes as no real surprise that many, if not most, of those responsible for advances in phytography – including the writing and presentation of floras – were teachers. Linnaeus, Lamarck, A.-P. de Candolle, W. J. Hooker, Asa Gray, Sergei Korshinsky, Hans Schinz, Nathaniel L. Britton, Alexander Tolmachev and C. G. G. J. van Steenis, to name but some, all held university lectureships or professorships.

119 A 'Banksian' botanist here refers to the kind of eighteenth- or nineteenth-century *savant* responsible for all areas, pure and applied. They were usually responsible for a botanical garden, and could refer questions to metropolitan establishments for advice. Notable representatives included Roxburgh, Griffith, Jenman, Mueller and Ridley. The advent of professional specialization reduced or changed their scope of activities.

120 Cittadino, 1990; see also McIntosh, 1985, pp. 30–38.

121 Nevertheless, Eugen Warming, one of the founders of tropical plant ecology (and, indeed, plant ecology in general) was faced while in Brazil in the 1860s with an imperfectly known flora; at the time, *Flora brasiliensis* was not yet far advanced. He thus prepared, with the aid of specialists, a long series of floristic contributions. These formed a partial basis for his classic monograph, *Lagoa Santa*, which appeared only in 1892. The functional botanists who traveled to South and Southeast Asia were somewhat more fortunate as the local network

of literature and resident botanists was somewhat better developed; their plants could be named at Calcutta, Peradeniya, Singapore, Bogor or (after 1900) Manila.

122 Natural history societies, along with handbooks, had already come into existence in the nineteenth century, particularly in India with its large potential audience, and the 'hill stations' not unnaturally became centers of local investigations by enthusiasts and others. However, *Flora simlensis* and similar works from before World War I – including the most ambitious, *Exkursionsflora von Java* – followed traditional formats.

123 For Adanson, see Stafleu, 1971*a*, pp. 310–320; for the Banksian 'circle', see McCracken, 1997, and Frodin, 1998; for Griffith, see Burkill, 1965; for Beccari, see Pichi-Sermolli and van Steenis, 1983; for Warming, see Cittadino, 1990.

124 Wheeler, 1923.

125 Hagen, 1983.

126 McIntosh, 1985, p. 38; see also Janzen, 1977 (publ. 1978). Richards' book had in turn been inspired by A. F. W. Schimper's great monograph, *Plant geography upon a physiological basis* (1898; English edn., 1903), a summation of the work of the 'functional ecology' school.

127 Such a reaction had been expressed by the pioneer Swedish experimental taxonomist Göte Turesson, who further suggested that in botany field and herbarium workers had become more widely separated than in zoology (G. Turesson, 1922. The genotypical response of the plant species to the habitat. *Hereditas* 3: 211–350; cited by E. M[ayr], 1980 [1998].).

128 Cf. Worthington, 1938, p. 197.

129 Cf. Hill *et al.*, 1925.

130 For example Aubréville in West Africa, Perrier de la Bâthie in Madagascar, Pételot in Indochina, and the Stehlés in the Antilles. The most notable work was Aubréville's *Flore forestière de la Côte d'Ivoire* (1938–39).

131 Symington, 1943; Corner, 1946.

132 From a Nigerian perspective, Symington's immediate example was *Flora of west tropical Africa* (1927–36); at the same time, it was to him the best of its kind. Among forest floras, the major anglophone example then available was *Indian trees* (1906) by Dietrich Brandis.

133 Corner's response was *Wayside trees of Malaya* (1940; 2nd edn., 1952; 3rd edn., 1988).

134 Aubréville wrote two forest floras (one in two editions) while Ducke was responsible for a lengthy stream of Amazonian novelties in *Arquivos do Jardim Botânico do Rio de Janeiro*.

135 Heywood, 1984.

136 Jarvie and van Welzen, 1994; Jarvie, Ermayanti and Mahyar, 1997 (publ. 1998); Rifai, 1997 (publ. 1998).

137 Symington, 1943, p. 16. This view in fact went back to the Hookers, particularly J. D. Hooker; see P. F. Stevens,

1997. J. D. Hooker, George Bentham, Asa Gray and Ferdinand Mueller on species limits in theory and practice: a mid-nineteenth-century debate and its repercussions. *Hist. Rec. Austral. Sci.* 11: 345–370.

138 Corner, 1946.

139 Jarvie, Ermayanti and Mahyar, 1997 (publ. 1998), p. 130.

140 Wilken *et al.*, 1989*a*.

141 Palmer, Wade and Neal, 1995.

142 R. Schmid, 'Some desiderata to make floras and other types of works user (and reviewer) friendly', in Schmid, 1997, pp. 179–194; R. Schmid and A. R. Smith, 'Some new floras that are statistically deficient: why this unrelenting annoyance?', in Schmid, 1997, pp. 168–178.

143 For example Williams, 1964, especially fig. 38; Barthlott, Lauer and Placke, 1996. The attractive colored map in the latter furnishes an example of effective translation of research into public perception.

144 Shetler, 1971; Gómez-Pompa and Butanda, 1973; Gómez-Pompa and Nevling, 1973; Keller and Crovello, 1973; Heywood, 1984; Heywood and Derrick, 1984; Bisby, Russell and Pankhurst, 1993; Hawksworth and Mibey, 1997.

145 Cf. Heywood, 1984.

146 Cf. Heywood, 1988.

147 Watson, 1971; Funk, 1993.

148 van Steenis, 1954; Jacobs, 1980.

149 The differences between descriptions written directly from observation and synthesis and those formed from a computer-driven descriptive language generation system soon become evident, at least to the author.

150 It is these, or information systems, which can (and should) contain, for example, fuller synonymy, citations of nomenclatural usage, and all specimen and other records.

151 The work in question is *Flora helvetica* by Lauber and Wagner (see **649**); this moreover has a 'pocket' version containing only the keys. From here, it becomes possible to visualize floras as 'e-books' for the new eBook readers (B. Macintyre, 'Is that a library in your pocket?', *The Times*, 27 November 1999, p. 22).

152 Davis and Heywood, 1963; Frodin, 1976 (publ. 1977); Heywood, 1984, 1988.

153 Cf. Wilken *et al.*, 1989*b*. What is 'central' and what 'peripheral' in a floristic information system continues to be much discussed, though more now in terms of priorities.

154 In 1583–84 Charles de l'Écluse had, for his pioneer account of the flora of the Hungarian basin, produced both a larger-scale illustrated work (through the great Antwerp firm of Plantin-Moretus) and a simple, handily sized enumeration of plants.

155 Cf. *Biologue* (ABRS, Environment Australia) **20**: 7 (1999).

156 Cf. Anderson, 1991. The eighteenth-century Spanish royal botanical expeditions, which included published floras in their programmes along with exploration, education and extension, contributed for example to the formation of national identity in Mexico, Colombia, Peru and other Iberoamerican countries. Other projects, such as *Flora SSSR* and *Flora iberica*, may be said to represent *expressions* of identity or renewal.

157 Cf. de Candolle, 1880; Hitchcock, 1925; Lawrence, 1951; Benson, 1962; Radford *et al.*, 1974.

158 Symington, 1943; Corner, 1946; Jarvie and van Welzen, 1994; Palmer, Wade and Neal, 1995; Jarvie, Ermayanti and Mahyar, 1997 (publ. 1998). Practically oriented works have, however, had a long tradition in forestry. Some key early examples are *Geslachtstabellen voor Nederlandsch-Indische boomsoorten naar vegetatieve kenmerken* (1928, 1953; English edn., 1956) by F. H. Endert (**910–30**) as well as Allen's *Rain forests of Golfo Dulce* (**236**). Others have since followed, including *La forêt dense d'Afrique centrale: identification pratique des principaux arbres* (1990) by Y. Talifer and, as already noted, *Australian tropical rainforest trees* (1994) by B. P. M. Hyland and T. Whiffin.

159 Watson, 1971.

160 A prime role, as Watson suggested, is in taxonomic and phylogenetic analysis. They also have a role in a floristic or taxonomic information system, as their role is therein less circumscribed.

161 As discussed in Chapter 2, their importance has received renewed recognition with the growth of concern for the environment. Information technology has greatly aided their preparation and maintenance, even where numbers of 10 000 or more taxa are involved.

162 Several countries (or lesser polities) now have official botanical information systems with greater or lesser detail. FloraBase in Western Australia (Chapman and Richardson, [1998]) is among the most advanced.

163 An information system can, for example, include not only specimen data but also 'intermediate documentation' of the kind proposed for the *Flora Malesiana* programme in the absence of more definitive family treatments.

164 Their capture to a good standard, however, takes time, manpower, and electronic capacity.

165 Some works are already appearing exclusively in CD-ROM format.

166 *Flora of the Venezuelan Guayana*, in addition to representative specimens in the main work, is also issuing a parallel series of documentation featuring all *exsiccatae* with their localities.

167 Good modern, relatively 'concise' conventional floras include Flora Zambesiaca, *Flora of Turkey*, *Flora of Australia*, *Flora of the Venezuelan Guayana*, and *Flora of Somalia*.

168 Grimes, 1998. An ideal would be the development of 'preferred outlets' for monographs and revisions with established parameters. In the past, series such as *Monographiae phanerogamarum* and *Das Pflanzenreich* served in this capacity; modern equivalents include *Flora Neotropica* and *Systematic Botany Monographs*.

169 Cf. Allkin, 1997; Chapman and Richardson, [1998].

170 Jacobs, 1969. Since then 'family conferences' or symposia have become relatively numerous, and the growth of electronic mail, specialized newsletters, and news groups has promoted communication among specialists of like interests.

171 Ravetz, 1975.

172 Bentham, 1874.

173 Brenan, 1979, p. 57.

References

ALLKIN, R., 1997. Data management within collaborative groups. In J. Dransfield, M. J. E. Coode and D. A. Simpson (eds.), *Plant diversity in Malesia*, III: 5–24. Kew: Royal Botanic Gardens.

ALLKIN, R. and F. A. BISBY, 1984. *Databases in systematics*. London: Academic Press. (Systematics Association Spec. Vol. 26.)

ANDERSON, B., 1991. *Imagined communities*. 2nd edn. London: Verso. (1st edn., 1983.)

ANONYMOUS, 1861. Flora brasiliensis. *Nat. Hist. Rev.*, N.S. **1**: 1–5.

ANONYMOUS, 1863. Colonial Floras. *Nat. Hist. Rev.*, N.S. **3**: 255–256.

ATWOOD, A. C., 1911. *Description of the comprehensive catalogue of botanical literature in the libraries of Washington* (USDA Bureau of Plant Industry Circular 87). Washington, D.C.: U.S. Government Printing Office.

AYMONIN, G., 1962. Où en sont les flores européennes? *Adansonia*, II, **2**: 159–171.

BAAS, P., C. KALKMAN and R. GEESINK (eds.), 1990. *The plant diversity of Malesia; proceedings of the Flora Malesiana Symposium commemorating Professor Dr. C. G. G. J. van Steenis*. Dordrecht: Kluwer.

BARTHLOTT, W., W. LAUER and A. PLACKE, 1996. Global distribution of species diversity in vascular plants: towards a world map of phytodiversity. *Erdkunde* **50**: 317–327, col. map.

BENSON, L., 1962. *Plant taxonomy: methods and principles*. New York: Ronald Press. (A Chronica Botanica publication.)

BENTHAM, G., 1861. Outlines of botany. In G. Bentham, *Flora hong-kongiensis*, pp. i–xxxvi. London: Reeve.

BENTHAM, G., 1874. On the recent progress and present state of knowledge of systematic botany. *Reports Brit. Assoc. Adv. Sci.* (1874): 27–54.

BESTERMAN, T., 1965–66. *A world bibliography of bibliographies*. 5 vols. Lausanne: Societas Bibliographica.

BISBY, F. A., G. F. RUSSELL and R. J. PANKHURST (eds.), 1993. *Designs for a global plant species information system*. Oxford: Clarendon/Oxford University Press. (Systematics Association Spec. Vol. 48.)

BLAKE, S. F., 1961. *Geographical guide to floras of the world*, 2 (USDA Misc. Publ. 797). Washington, D.C.: U.S. Government Printing Office. (Reprinted 1974, Königstein/Ts., Germany: Koeltz.)

BLAKE, S. F. and A. C. ATWOOD, 1942. *Geographical guide to floras of the world*, 1 (USDA Misc. Publ. 401). Washington, D.C. U.S. Government Printing Office. (Reprinted 1963, New York: Hafner; 1974, Königstein/Ts., Germany: Koeltz.)

BLUNT, W. and W. T. STEARN, 1994. *The art of botanical illustration*. 2nd edn., by W. T. Stearn. Woodbridge, Suffolk: Antique Collectors' Club (in association with the Royal Botanic Gardens, Kew). (1st edn., 1950 (publ. 1951), London: Collins.)

BRADFORD, S. C., 1953. *Documentation*. 2nd edn., with introduction by J. H. Shera and M. E. Egan. London: Crosby Lockwood. (1st edn., 1948.)

BRENAN, J. P. M., 1963. The value of floras to underdeveloped countries. *Impact* (Paris) **13**: 121–145.

BRENAN, J. P. M., 1979. The flora and vegetation of tropical Africa today and tomorrow. In K. Larsen and L. B. Holm-Nielsen (eds.), *Tropical botany*, pp. 49–58. London: Academic Press.

BRENAN, J. P. M., R. ROSS and J. T. WILLIAMS (eds.), 1975. *Computers in botanical collections*. London: Plenum.

BRIDSON, D. and L. L. FORMAN (eds.), 1998. *Herbarium handbook*. 3rd edn. Kew: Royal Botanic Gardens. (1st edn., 1989; 2nd edn., 1992.)

BRIGGS, B. G., 1991. One hundred years of plant taxonomy, 1889–1989. *Ann. Missouri Bot. Gard.* **78**: 19–32.

BRITTON, N. L., 1894. A new systematic botany of North America. *Bull. Torrey Bot. Club* **21**: 230–231.

BRITTON, N. L., 1895. The systematic botany of North America. *Bot. Gaz.* **20**: 177–179.

BRITTON, N. L., 1896–98. Prefaces to volumes I and III. In N. L. Britton and A. Brown, *An illustrated flora of the northern United States, Canada, and the British possessions*, 1: v–xii; 3: iv–v. New York. [2nd edn., 1913, in 1: v–xiii.]

BROOKS, F. T. (ed.), 1925. *Imperial Botanical Conference* (London, 1924): *report of proceedings*. Cambridge: Cambridge University Press.

BUCKMAN, T. R. (ed.), 1966. *Bibliography and natural history*. Lawrence, Kans.: University of Kansas Libraries.

BULICK, S., 1978. Book use as a Bradford-Zipf phenomenon. *College Res. Libraries* **39**: 215–219.

BURKILL, I. H., 1965. *Chapters on the history of botany in India*. Calcutta: Manager of Publications, Government of India.

BURTT-DAVY, J. *et al.*, 1925. Correlation of taxonomic work in the dominions and colonies with work at home. In F. T. Brooks, *op. cit.*, pp. 214–239.

CAIN, A. J., 1958. Logic and memory in Linnaeus's system of taxonomy. *Proc. Linn. Soc. London* **169**: 144–163.

CAMPBELL, D. G. and H. D. HAMMOND (eds.), 1989. *Floristic inventory of tropical countries: the status of plant systematics, collections and vegetation, plus recommendations for the future.* New York: The New York Botanical Garden.

CANDOLLE, A. DE, 1873. Réflexions sur les ouvrages généraux de botanique descriptive. *Arch. Sci. Phys. Nat. Genève* **48**: 185–209.

CANDOLLE, A. DE, 1880. *La phytographie.* Paris: Masson.

CANDOLLE, A.-P. DE, 1813. *Théoire élémentaire de la botanique.* Paris.

CANHOS, V. P., G. P. MANFIO and D. A. L. CANHOS, 1997. Networks for distributing information. In D. L. Hawksworth, P. M. Kirk and S. Dextre Clarke, *op. cit.*, pp. 147–156.

CHAPMAN, A. R. and B. P. RICHARDSON, [1998.] About FloraBase. Electronic publication, on-line at http://florabase.calm.wa.gov.au/about.html.

CITTADINO, E., 1990. *Nature as the laboratory.* Cambridge: Cambridge University Press.

COMMITTEE ON RESEARCH PRIORITIES IN TROPICAL BIOLOGY, NATIONAL RESEARCH COUNCIL, U.S.A., 1980. *Research priorities in tropical biology.* Washington, D.C.: National Research Council, [U.S.] National Academy of Sciences.

CORNER, E. J. H., 1946. Suggestions for botanical progress. *New Phytol.* **45**: 185–192.

CORNER, E. J. H., 1961. Evolution. In A. M. MacLeod and L. S. Cobley (eds.), *Contemporary botanical thought*, pp. 95–114. Edinburgh: Oliver and Boyd.

COX, E. H. M., 1945. *Plant hunting in China.* London: Collins. (Reissued 1986, Hong Kong: Oxford University Press.)

CUTBILL, J. L., 1971*a*. New methods for handling biological information. *Biol. J. Linn. Soc.* **3**: 253–260.

CUTBILL, J. L. (ed.), 1971*b*. *Data processing in biology and geology.* London: Academic Press. (Systematics Association Spec. Vol. 3.)

DAVIS, P. H. and V. H. HEYWOOD, 1963. *Principles of angiosperm taxonomy.* Edinburgh: Oliver and Boyd. (Reprinted with corrections, 1973, Huntington, N.Y.: Krieger.)

DAVIS, S. D. *et al.* (comp.), 1986. *Plants in danger: what do we know?* Gland, Switzerland: International Union for the Conservation of Nature and Natural Resources. (IUCN conservation library.)

DELENDICK, T. J., 1990. Citation analysis of the literature of systematic botany: a preliminary survey. *J. Amer. Soc. Inform. Sci.* **41**: 535–543.

DEPARTMENT OF SCIENCE, AUSTRALIA, 1979. *Australian Biological Resources Study 1973–78* (Commonwealth Parliamentary Paper 354/1978). Canberra: Australian Government Publishing Service.

DESMOND, R., 1995. *Kew: the history of the Royal Botanic Gardens.* London: Harvill (in association with the Royal Botanic Gardens, Kew).

DESMOND, R., 1999. *Sir Joseph Dalton Hooker: traveller and plant collector.* Woodbridge, Suffolk: Antique Collectors' Club (in association with the Royal Botanic Gardens, Kew).

DE WOLF, G. P., JR., 1963. On the *Flora Neotropica. Taxon* **12**: 251–253.

DE WOLF, G. P., JR., 1964. On the sizes of floras. *Taxon* **13**: 149–153.

DIELS, L., 1921. Die Methoden der Phytographie und der Systematik der Pflanzen. In E. Abderhalden (ed.), *Handbuch der biologischen Arbeitsmethoden.* 2nd edn., 11(1): 67–190. Berlin: Urban and Schwarzenberg.

DOLBY, R. G. A., 1979. Classification of the sciences: the nineteenth-century tradition. In R. F. Ellen and D. Reason (eds.), *Classifications in their social context*, pp. 167–193. London: Academic Press.

DUPREE, A. H., 1959. *Asa Gray.* Cambridge, Mass.: Belknap/Harvard University Press.

DURIS, P., 1993. *Linné et la France (1780–1850).* Geneva: Droz.

ECKARDT, T., 1966. 150 Jahre Botanisches Museum Berlin (1815–1965). *Willdenowia* **4**: 151–182.

ENDRESS, P. K., 1977. Das Institut für Systematische Botanik der Universität Zürich. *Vierteljahrsschr. Naturf. Ges. Zürich* **122**: 143–150. [In H. Wanner and P. K. Endress (eds.), 'Die Botanischen Institute der Universität Zürich'.]

EUROPEAN SCIENCE FOUNDATION, 1978–81. *ESF Reports 1977–80.* Strasbourg.

EWAN, J., 1970. Plant collectors in America: backgrounds for Linnaeus. In P. Smit and R. J. C. V. ter Laage (eds.), *Essays in biohistory*, pp. 19–54. Utrecht: International Bureau for Plant Taxonomy and Nomenclature. (Regnum Veg. 71.)

FISHER, F. J. F., 1968. The role of geographical and ecological studies in taxonomy. In V. H. Heywood (ed.), *Modern methods in plant taxonomy*, pp. 241–259. London: Academic Press.

FLETCHER, H. R., 1969. *The story of the Royal Horticultural Society, 1804–1968.* London: Oxford University Press (for the Royal Horticultural Society).

FLETCHER, H. R. and W. H. BROWN, 1970. *The Royal Botanic Garden Edinburgh 1670–1970.* Edinburgh: HMSO.

FORERO, E. and F. J. PEREIRA, 1976. EDP-IR in the National Herbarium of Colombia (COL). *Taxon* **25**: 85–94.

FOREST RESEARCH INSTITUTE OF MALAYSIA, 1992. *Tree flora of Sabah and Sarawak: guide to contributors.* 28 pp. Kepong.

FRODIN, D. G., 1964. *Guide to the standard floras of the world (arranged geographically).* Knoxville, Tenn.: Department of Botany, The University of Tennessee.

FRODIN, D. G., 1976 (1977). On the style of floras: some general considerations. *Gardens' Bull. Singapore* **29**: 239–250.

FRODIN, D. G., 1998. Tropical biology and research institutions in South and Southeast Asia since 1500: botanic gardens and scientific knowledge to 1870. *Pacific Sci.* **52**: 276–286.

FUNK, V., 1993. Uses and misuses of floras. *Taxon* **42**: 761–772.

GAGNEPAIN, F., 1913. Les sciences naturelles auxiliaires de la colonisation. *Bull. Soc. Hist. Nat. Autun* **26**: 109–133.

GAMS, H., 1954. Flores européennes modernes: resultats acquis, objectifs à atteindre. In *Huitième Congrès International de Botanique* (Paris, 1954): *rapports et communications parvenus avant le Congrès aux sections 2, 4, 5 et 6*, p. 101. Paris: SEDES.

GARFIELD, E., 1979. *Citation indexing – its theory and application in science, technology, and humanities*. New York: Wiley.

GARFIELD, E., 1980. Bradford's Law and related statistical patterns. *Current Contents/ Life Sci.* **23**(19): 5–12.

GARNOCK-JONES, P. J. and I. BREITWIESER, 1998. New Zealand floras and systematic botany: progress and prospects. *Austral. Syst. Bot.* **11**: 175–184.

GEORGE, A. S., C. KALKMAN and R. GEESINK (coord.), 1990. The future of *Flora Malesiana*. 60 pp. Leiden: Rijksherbarium/Hortus Botanicus. (Fl. Males. Bull., Suppl. 1.)

GHISELIN, M. T., 1997. *Metaphysics and the origin of species*. Albany, N.Y.: State University of New York Press. (SUNY series in philosophy and biology.)

GIBBS RUSSELL, G. E. and P. GONSALVES, 1984. PRECIS – a curatorial and biogeographic system. In R. Allkin and F. A. Bisby, *op. cit.*, pp. 137–153.

GILMOUR, J. S. L., 1952. The development of taxonomy since 1851. *Adv. Sci.* (UK) **9**: 70–74.

GÓMEZ-POMPA, A. and A[RMANDO] BUTANDA C., 1973. *El uso de computadoras en la 'Flora de Vera Cruz'*. Mexico City: Instituto de Biología, Universidad Nacional Autónoma de México. (Colleccíon de trabajos publicados o en proceso de publicación del Programa Flora de Veracruz 1.)

GÓMEZ-POMPA, A. and L. I. NEVLING, 1973. The use of electronic data-processing methods in the *Flora of Veracruz* program. *Contr. Gray Herb.*, N.S. **203**: 49–64.

GÓMEZ-POMPA, A. and O. E. PLUMMER, 1990. Video floras: a new tool for systematic botany. *Taxon* **39**: 576–585.

GOULD, S. J., 1989. *Wonderful life: the Burgess Shale and the nature of history*. New York: Norton.

GOULD, S. W., 1958. Punched cards, binomial names and numbers. *Amer. J. Bot.* **45**: 331–339.

GOULD, S. W., 1962. *Family names of the plant kingdom*. New Haven, Conn.: Connecticut Agricultural Experiment Station. (International Plant Index 1.)

GOULD, S. W., 1968–72. Geo-code. 2 vols. Maps. New Haven, Conn.: Gould Fund. [Vol. 1, *Western hemisphere*; vol. 2, *Eastern hemisphere*.]

GOULDEN, C. E. (ed.), 1977. *Changing scenes in the natural sciences, 1776–1976*. Philadelphia. (Academy of Natural Sciences of Philadelphia Spec. Publ. 12.)

GREENE, E. L., 1905. The earliest local flora. *Plant World* **8**(5): 115–121.

GRIMES, J. W., 1998. Before the floras – monographs. *Austral. Syst. Bot.* **11**: 243–249.

HAGEN, J. B., 1983. The development of experimental methods in plant taxonomy, 1920–1950. *Taxon* **32**: 406–416.

HAGEN, J. B., 1984. Experimentalists and naturalists in twentieth-century botany: experimental taxonomy, 1920–1950. *J. Hist. Biol.* **17**(2): 249–269.

HAGEN, J. B., 1986. Ecologists and taxonomists: divergent traditions in twentieth-century plant geography. *J. Hist. Biol.* **19**(2): 197–214.

HALL, M., 1990. The coming rejuvenation of botany. *Harvard Mag.* **93**(1): 39–49.

HAMANN, U. and G. WAGENITZ, 1977. *Bibliographie zur Flora von Mitteleuropa*. 2nd edn. 374 pp. Berlin: Parey. (1st edn., 1970, Munich.)

HAWKSWORTH, D. L. and R. K. MIBEY, 1997. Information needs of inventory programmes. In D. L. Hawksworth, P. M. Kirk and S. Dextre Clarke, *op. cit.*, pp. 55–68.

HAWKSWORTH, D. L., P. M. KIRK and S. DEXTRE CLARKE (eds.), 1997. *Biodiversity information: needs and options*. Wallingford, Oxon.: CAB International.

HEDBERG, I. (ed.), 1979. *Systematic botany, plant utilization and biosphere conservation*. Stockholm: Almqvist and Wiksell.

HELLER, J. L., 1983. *Studies in Linnaean method and nomenclature*. Frankfurt: Peter Lang. (Marburger Schriften zur Medizingeschichte 7.) [Incorporates as chapter 5 (pp. 146–204) *idem*, 1970. Linnaeus's *Bibliotheca botanica*. *Taxon* **19**: 363–411, and as chapter 7 (pp. 239–267) *idem*, 1976. Linnaeus on sumptuous books. *Taxon* **25**: 33–52.]

HEYWOOD, V. H., 1957. A proposed flora of Europe. *Taxon* **6**: 33–42.

HEYWOOD, V. H., 1958. *The presentation of taxonomic information: a short guide to contributors to 'Flora Europaea'*. Leicester: Leicester University Press.

HEYWOOD, V. H., 1960. *The presentation of taxonomic information: supplement*. Alcobaça, Portugal: Flora Europaea Organization.

HEYWOOD, V. H., 1964. *Flora Europaea* and the problems of organization of floras. *Taxon* **13**: 48–51.

HEYWOOD, V. H., 1972. The new Hegi Compositae. *Taxon* **19**: 937–938.

HEYWOOD, V. H., 1973*a*. Ecological data in practical taxonomy. In V. H. Heywood (ed.), *Taxonomy and ecology*, pp. 329–347. London: Academic Press.

HEYWOOD, V. H., 1973*b*. Taxonomy in crisis? or Taxonomy is the digestive system of biology. *Acta Bot. Acad. Sci. Hung.* **19**: 139–146.

HEYWOOD, V. H., 1974. Systematics – the stone of Sisyphus. *Biol. J. Linn. Soc.* **6**: 169–178.

HEYWOOD, V. H. (coord.), 1975. Données disponibles et lacunes de la connaissance floristique des pays méditerrannéens. In Centre National de la Recherche Scientifique, France, *La flore du bassin méditerranéen: essai de systématique synthétique* (Montpellier, 1974), pp. 15–142. Paris. (Colloques internationaux du CNRS 235.)

HEYWOOD, V. H., 1976. Contemporary objectives in systematics. In G. F. Estabrook (ed.), *Proceedings, Eighth International Conference on Numerical Taxonomy*, pp. 258–283. San Francisco: Freeman.

HEYWOOD, V. H., 1978. European floristics: past, present and future. In H. E. Street (ed.), *Essays in plant taxonomy*, pp. 275–288. London: Academic Press.

HEYWOOD, V. H., 1984. Designing floras for the future. In V. H. Heywood and D. M. Moore (eds.), *Current concepts in plant taxonomy*, pp. 397–410. London: Academic Press. (Systematics Association Spec. Vol. 25.)

HEYWOOD, V. H., 1988. The structure of systematics. In D. L. Hawksworth (ed.), *Prospects in systematics*, pp. 44–56. Oxford: Clarendon/Oxford University Press. (Systematics Association Spec. Vol. 36.)

HEYWOOD, V. H., 1989. *Flora Europaea* and the European Documentation System. In N. R. Morin *et al.*, *op. cit.*, pp. 8–10.

HEYWOOD, V. H. and L. N. DERRICK, 1984. The European Taxonomic, Floristic and Biosystematic Information System. *Norrlinia* **2**: 33–54.

HICKMAN, J. and T. DUNCAN, 1989. *Jepson's Manual* revision: a user-oriented publication. In N. R. Morin *et al.*, *op. cit.*, pp. 37–38.

HILL, A. W. *et al.*, 1925. The best means of promoting a complete botanical survey of the different parts of the Empire. In F. T. Brooks, *op. cit.*, pp. 196–213.

HITCHCOCK, A. S., 1925. *Methods of descriptive systematic botany*. New York: Wiley.

HOLDEN, W. and E. WYCOFF, 1911–14. [Bibliography relating to the floras.] *Bibliogr. Contr. Lloyd Library* **1**(2–13): 1–513.

HOLLIS, S. and R. K. BRUMMITT, 1992. *World geographical scheme for recording plant distributions*. Pittsburgh, Pa.: Hunt Institute for Botanical Documentation (for the International Working Group on Taxonomic Databases).

HULL, D. L., 1965. The effect of essentialism on taxonomy – two thousand years of stasis. *British J. Philos. Sci.* **15**: 314–326; **16**: 1–18.

HUXLEY, J. S. (ed.), 1940. *The new systematics*. Oxford: Clarendon/Oxford University Press.

JACOBS, M., 1969. Large families – not alone! *Taxon* **18**: 253–262.

JACOBS, M., 1973. Flora-projecten als intellectuele uitdaging. *Vakbl. Biol.* **53**(15): 252–255.

JACOBS, M., 1980. Revolutions in plant description. In J. C. Arends *et al.* (eds.), *Liber gratulatorius in honorem H. C. D. de Wit*, pp. 155–181. Wageningen, the Netherlands. (Misc. Papers Landbouwhogeschool Wageningen 19.)

JÄGER, E. J., 1976 *et seq.* Areal- und Florenkunde (floristische Geobotanik) (*later* Plant geography). *Prog. Bot.* [*Fortschr. Bot.*] **38**: 314–330, 2 maps. Continued as *idem*, 1978 (see below); 1979 in *ibid.*, **41**: 310–323; 1980 in *ibid.*, **42**: 331–345; 1983 in *ibid.*, **45**; 1985 in *ibid.*, **47**; 1987 in *ibid.*, **49**; 1990 (see below); 1993 (see below); 1995 in *ibid.*, **56**: 396–415.

JÄGER, E. J., 1978. Areal- und Florenkunde (floristische Geobotanik). *Ibid.*, **40**: 413–428.

JÄGER, E. J., 1990. Areal- und Florenkunde (floristische Geobotanik). *Ibid.*, **51**: 313–332.

JÄGER, E. J., 1993. Plant geography. *Ibid.*, **54**: 428–447.

JANZEN, D., 1977 (1978). Promising directions of study in tropical animal–plant interactions. *Ann. Missouri Bot. Gard.* **64**: 706–736.

JARVIE, J. K. and P. VAN WELZEN, 1994. What are tropical Floras for in SE Asia? *Taxon* **43**: 444–448.

JARVIE, J. K., ERMAYANTI and U. W. MAHYAR, 1997 (1998). Techniques used in the production of an electronic and hardcopy flora for the Bukit Baka-Bukit Raya area of Kalimantan. In J. Dransfield, M. J. E. Coode and D. A. Simpson (eds.), *Plant diversity in Malesia*, III: 129–143. Kew: Royal Botanic Gardens.

JOHNSON, R. W., 1991. HERBRECS: the Queensland Herbarium records system – its development and use. *Taxon* **40**: 285–300.

JONSELL, B., 1979. Europe. In I. Hedberg, *op. cit.*, pp. 34–40.

JOVET, P., 1954. Flore et phytogéographie de la France. In A. Davy de Virville (coord.), *Histoire de la botanique en France*, pp. 243–268. Paris: SEDES.

JUDD, W. S., C. S. CAMPBELL, E. A. KELLOGG and P. F. STEVENS, 1999. *Plant systematics: a phylogenetic approach*. Sunderland, Mass.: Sinauer Associates.

JUST, T., 1953. Generic synopses and modern taxonomy. *Chron. Bot.* **14**(3): 103–114.

KELLER, C. and T. J. CROVELLO, 1973. Procedures and problems in the incorporation of data from floras into a computerized data bank. *Proc. Indiana Acad. Sci.* **82**: 116–122.

KRAUSS, H. M., 1973. The information system design for the Flora North America program. *Brittonia* 25: 119–134.

KUMAR, K., 1979. *Theory of classification*. New Delhi: Vikas.

LAWALRÉE, A., 1960. Indication des principaux ouvrages de floristique européenne (plantes vasculaires). *Natura Mosana* 13(2–3): 29–68.

LAWRENCE, G. H. M., 1951. *Taxonomy of vascular plants*. New York: Macmillan.

LEANDRI, J., 1962. Deux grands artisans de la floristique tropicale: Henri Lecomte (1856–1934) et Achille Finet (1863–1913). *Adansonia*, II, 2: 147–158.

LEANDRI, J., 1967. La fin de la dynastie des Jussieu et l'éclipse d'une chaire au Muséum (1853 à 1873). *Adansonia*, II, 7: 443–450.

LÉGÉE, G. and M. GUÉDÈS, 1981. Lamarck botaniste et évolutioniste. *Hist. et Nat.* 17/18: 19–31.

LEIMKUHLER, F. F., 1967. The Bradford distribution. *J. Documentation* 23: 197–207.

LIMOGES, C., 1980. The development of the Muséum d'Histoire Naturelle of Paris, 1800–1914. In R. Fox and G. Weisz (eds.), *The organization of science and technology in France 1808–1914*, pp. 211–240. Cambridge: Cambridge University Press.

LINNAEUS, C., 1736. *Bibliotheca botanica*. Amsterdam: Schouten. (2nd edn., 1751, Amsterdam: Schouten.) [For commentary, see Heller, 1983.]

LINNAEUS, C., 1753. *Incrementa botanices proxime praeterlapsi semisaeculi* (resp. J. Biuur). Stockholm. [Reissued in *Amoenitates Academicae* 3: 377–393 (1756). For commentary, see Heller, 1983.]

MA JIN-SHUANG and LIU QUAN-RU, 1998. The present situation and prospects of plant taxonomy in China. *Taxon* 47: 67–74.

MABBERLEY, D. J., 1979. Pachycaul plants and islands. In D. Bramwell (ed.), *Plants and islands*, pp. 259–277. London: Academic Press.

MALCLÈS, L.-N., 1961. Bibliography (transl. T. C. Hines). New York: Scarecrow. (Reprinted 1973. Originally published 1956, Paris, as *La bibliographie* in the series 'Que sais-je'.)

MALYSCHEV, L. I., 1975. (Količestvennyj analiz flory: prostranstvennoe raznoobrazie, uroven vidovog bogatstva i reprezentativnost' učastkov obsledovanija (The quantitative analysis of flora: spatial diversity, the level of specific richness and representativity of sampling areas).) *Bot. Žurn. SSSR* 60: 1537–1550, map.

MAXTED, N., 1992. Towards defining a taxonomic revision methodology. *Taxon* 42: 653–660.

M[AYR], E., 1980 [1998]. Botany (introduction). In E. Mayr and W. Provine (eds.), *The evolutionary synthesis: perspectives on the unification of biology*, pp. 137–138. Cambridge, Mass.: Harvard University Press.

MAYR, E., 1982. *The growth of biological thought*. Cambridge, Mass.: Belknap/Harvard University Press.

MAYR, E., 1997. *This is biology*. Cambridge, Mass.: Belknap/Harvard University Press.

McCRACKEN, D. P., 1997. *Gardens of empire*. London: Leicester University Press.

McINTOSH, R. P., 1985. *The background of ecology: concept and theory*. Cambridge: Cambridge University Press.

McVAUGH, R., 1955. The American collections of Humboldt and Bonpland as described in the 'Systema Vegetabilium' of Roemer and Schultes. *Taxon* 4: 78–86. (Reprinted in W. T. Stearn (ed.), *Humboldt, Bonpland, Kunth and tropical American botany*, pp. 32–43. Lehre, Germany: Cramer.)

MEIKLE, R. D., 1971. Co-ordination of floristic work and how we might improve the situation. In P. H. Davis *et al.* (eds.), *Plant life of South-West Asia*, pp. 313–331. Edinburgh: Botanical Society of Edinburgh.

MILDBRAED, J. 1948. Ludwig Diels. *Bot. Jahrb. Syst.* 74: 173–198.

MÖBIUS, M., 1938. Entstehung und Entwicklung der Floristik. *Bot. Jahrb. Syst.* 69: 295–317.

MORI, S. A., 1992. Neotropical floristics and inventory: who will do the work? *Brittonia* 44: 372–375.

MORI, S. A. and C. A. GRACIE, 2000. The making of a flora. *Brittonia*, in press.

MORIN, N. R., R. D. WHETSTONE, D. WILKEN and K. L. TOMLINSON (eds.), 1989. *Floristics for the 21st century*. St. Louis: Missouri Botanical Garden. (Monogr. Syst. Bot. Missouri Bot. Gard. 28.)

MORISON, S. E., 1937. *Three centuries of Harvard, 1636–1936*. Cambridge, Mass.: Harvard University Press.

MORSE, L. E., 1975. Recent advances in the theory and practice of biological specimen identification. In R. J. Pankhurst (ed.), *Biological identification with computers*, pp. 11–52. London: Academic Press. (Systematics Association Spec. Vol. 7.)

MORTON, A. G., 1981. *History of botanical science*. London: Academic Press.

MUELLER, F. VON, 1888. Considerations of phytographic expressions and arrangements. *Proc. Roy. Soc. New South Wales* 22: 187–204.

NALIMOV, V. V., 1985. *Space, time and life: the probabilistic pathways of evolution*. Philadelphia: ISI Press.

NG, F. S. P., 1988. Problems in the organization of plant taxonomic work. *Fl. Males. Bull.* 10(1): 39–44.

PALMER, M. W., G. L. WADE and P. NEAL, 1995. Standards for the writing of floras. *BioScience* 45: 339–345.

PANKHURST, R. J. (ed.), 1975. *Biological identification with computers: proceedings of a meeting held at King's College, Cambridge, 27 and 28 September, 1973*. London: Academic Press. (Systematics Association Spec. Vol. 7.)

PANKHURST, R. J., 1978. *Biological identification: the principles and practice of identification methods in biology.* London: Arnold.

PANKHURST, R. J., 1991. *Practical taxonomic computing.* Cambridge: Cambridge University Press.

PICHI-SERMOLLI, R. E. G., 1973. Historical review of the higher classification of the Filicopsida. In A. C. Jermy *et al.* (eds.), *The phylogeny and classification of the ferns*, pp. 11–40. London: Academic Press.

PICHI-SERMOLLI, R. E. G. and C. G. G. J. VAN STEENIS, 1983. Dedication [to O. Beccari]. In C. G. G. J. van Steenis (ed.), *Flora Malesiana*, I, 9: (6)–(44). The Hague: Nijhoff/Junk.

POLHILL, R., 1990. Production rates of major regional floras. In A. S. George, C. Kalkman and R. Geesink, *op. cit.*, pp. 11–20.

PRANCE, G. T., 1977 (1978). Floristic inventory of the tropics: where do we stand? *Ann. Missouri Bot. Gard.* 64: 659–684.

PRANCE, G. T. and D. G. CAMPBELL, 1988. The present state of tropical floristics. *Taxon* 37: 519–548.

RADFORD, A. E. *et al.*, 1974. *Vascular plant systematics.* New York: Harper and Row.

RANGANATHAN, S. R., 1957. *Prolegomena to library classification.* 2nd edn. London: Library Association.

RAVEN, P. H., 1974*a*. Plant systematics 1947–1972. *Ann. Missouri Bot. Gard.* 61: 166–178.

RAVEN, P. H. (ed.), 1974*b*. Trends, priorities, and needs in systematic and evolutionary biology. *Syst. Zool.* 23: 416–439/*Brittonia* 26: 421–444.

RAVEN, P. H., 1977. The systematics and evolution of higher plants. In C. E. Goulden, *op. cit.*, pp. 59–83.

RAVETZ, J. R., 1975. '. . . et augebitur scientia'. In R. Harré (ed.), *Problems of scientific revolution: progress and obstacles to progress in the sciences*, pp. 42–57, 1 pl. Oxford: Oxford University Press.

REEDS, K. M., 1991. *Botany in medieval and Renaissance universities.* New York: Garland.

REVEAL, J. L., 1991. Botanical explorations in the American West, 1889–1989: an essay on the last century of a floristic frontier. *Ann. Missouri Bot. Gard.* 78: 65–80.

RIFAI, M., 1997 (1998). *Flora Malesiana*: a user's experience and view. In J. Dransfield, M. J. E. Coode and D. A. Simpson (eds.), *Plant diversity in Malesia*, III: 219–223. Kew: Royal Botanic Gardens.

RODMAN, J. E., 1989. Workshop goals. In N. R. Morin *et al.*, *op. cit.*, p. 1.

ROSENBERG, A., 1985. *The structure of biological science.* Cambridge: Cambridge University Press.

SACHS, J., 1875. *Geschichte der Botanik von 16. Jahrhundert bis 1860.* Munich. (English version, 1890, as *History of botany 1530–1860.* Oxford: Oxford University Press; reprinted 1906, 1967.)

SCHMID, R., 1997. Reviews and notices of publications. *Taxon* 46: 147–214. [Inclusive of R. Schmid and A. R. Smith, 'Some new floras that are statistically deficient: why this unrelenting annoyance?', pp. 168–174, and R. Schmid, 'Some desiderata to make floras and other types of works user (and reviewer) friendly', pp. 179–194.]

SHEETS-PYENSON, S., 1988. *Cathedrals of science: the development of colonial natural history museums during the late 19th century.* Kingston: McGill-Queens University Press.

SHETLER, S. G., 1967. *The Komarov Botanical Institute: 250 years of Russian research.* Washington, D.C.: Smithsonian Institution Press.

SHETLER, S. G., 1971. Flora North America as an information system. *BioScience* 21: 524, 529–532.

SHETLER, S. G., 1974*a*. Information systems and data banking. In A. E. Radford *et al.*, *op. cit.*, pp. 791–821.

SHETLER, S. G., 1974*b*. Demythologizing biological data banking. *Taxon* 23: 71–100.

SHETLER, S. G. and R. W. READ, 1973. *Flora North America/International index of current research projects in plant systematics*, 7. Washington, D.C. (Flora North America Report 71.)

SHETLER, S. G. *et al.*, 1973. *A guide for contributors to 'Flora North America' (FNA).* Provisional edn. Washington, D.C. (Flora North America Report 65.)

SHINNERS, L. H., 1962. Evolution of the Gray's and Small's manual ranges. *Sida* 1: 1–31.

SIMON, H. R., 1977. *Die Bibliographie der Biologie.* Stuttgart: Hiersemann.

SIVARAJAN, V. V., 1991. *Introduction to the principles of plant taxonomy.* 2nd edn. Cambridge: Cambridge University Press.

SMOCOVITIS, V. B., 1992. Disciplining botany: a taxonomic problem. *Taxon* 41: 459–470.

SOLTIS, P. S., D. E. SOLTIS and J. J. DOYLE (eds.), 1992. *Plant molecular systematics.* New York: Chapman and Hall.

STACE, C. A., 1989. *Plant taxonomy and biosystematics.* 2nd edn. London: Arnold. (1st edn., 1980.)

STAFLEU, F. A., 1959. The present status of plant taxonomy. *Syst. Zool.* 8: 59–68, map.

STAFLEU, F. A., 1967. *Adanson, Labilliardiere, de Candolle.* Lehre, Germany: Cramer.

STAFLEU, F. A., 1971*a*. *Linnaeus and the Linnaeans.* Utrecht: Oosthoek. (Regnum Veg. 79.)

STAFLEU, F. A., 1971*b*. Lamarck: the birth of biology. *Taxon* 20: 397–442.

STAFLEU, F. A., 1973. Pritzel and his *Thesaurus. Taxon* 22: 119–130.

STAFLEU, F. A., 1981. Engler und seine Zeit. *Bot. Jahrb. Syst.* 102: 21–38.

STEARN, W. T., 1957. *An introduction to the 'Species plantarum' and cognate botanical works of Carl Linnaeus.* London. (Reprinted in C. Linnaeus, 1957. *'Species Plantarum': a facsimile of the first edition 1753*, 1: v–xiv, 1–176. London: Ray Society.)

STEARN, W. T., 1959. The background of Linnaeus' contribution to the nomenclature and methods of systematic biology. *Syst. Zool.* 8: 4–22.

STEBBINS, G. L., 1950. *Variation and evolution in plants.* New York: Columbia University Press. (Columbia biological series 16.)

STEENIS, C. G. G. J. VAN, 1938. Recent progress and prospects in the study of the Malaysian flora. *Chron. Bot.* 4: 392–397.

STEENIS, C. G. G. J. VAN, 1949. De *Flora Malesiana* en haar betekenis voor de Nederlandse botanici. *Vakbl. Biol.* 29(2): 24–33.

STEENIS, C. G. G. J. VAN, 1954. General principles in the design of Floras. In *Huitième Congrès International de Botanique* (Paris, 1954): *rapports et communications parvenus avant le Congrès aux sections 2, 4, 5 et 6*, pp. 59–66. Paris: SEDES.

STEENIS, C. G. G. J. VAN, 1962. The school-flora as a medium for instruction in the tropics. In *Ninth Pacific Science Congress, Proceedings*, 4 (Botany): 139–140. Bangkok: Secretariat, Ninth Pacific Science Congress [at Chulalongkorn University].

STEENIS, C. G. G. J. VAN, 1967. The herb flora of Taiwan, or how to master a flora without types and with only a few books. *Fl. Males. Bull.* 22: 1562–1567.

STEENIS, C. G. G. J. VAN, 1979. The Rijksherbarium and its contribution to the knowledge of the tropical Asiatic flora. *Blumea* 25: 57–77.

STEVENS, P. F., 1994. *The development of biological systematics: Antoine-Laurent de Jussieu, nature and the natural system.* New York: Columbia University Press.

STUESSY, T., 1990. *Plant taxonomy: the systematic evaluation of comparative data.* New York: Columbia University Press.

SUTTON, S. B., 1970. *Charles Sprague Sargent and the Arnold Arboretum.* Cambridge, Mass.: Harvard University Press.

SYMINGTON, C. F., 1943. The future of colonial forest botany. *Empire For. J.* 22: 11–23.

SYSTEMATICS AGENDA 2000, 1994a. *Systematics Agenda 2000: charting the biosphere.* [New York.]

SYSTEMATICS AGENDA 2000, 1994b. *Systematics Agenda 2000: charting the biosphere. Technical report.* [New York.]

THISTLETON-DYER, W. T., 1906. Botanical survey of the Empire. *Bull. Misc. Inform.* (Kew) **1905**: 9–43.

THORNE, R. F., 1971. Introduction: North temperate floristics and the *Flora North America* project. *BioScience* 21: 511–512.

TIMLER, F. K. and B. ZEPERNICK, 1987. German colonial botany. *Ber. Deutsch. Bot. Ges.* **100**: 143–168. [For German version, see Zepernick and Timler, 1990.]

TOLEDO, V. M. and V. SOSA, 1993. Floristics in Latin America and the Caribbean: an evaluation of the numbers of plant collections and botanists. *Taxon* 42: 355–364.

TRAVIS, B. V. *et al.*, 1962. *Classification and coding system for compilations from the world literature on insects and other arthropods that affect the health and comfort of man* (U.S. Army Quartermaster Research and Engineering Center, Technical Report ES-4). Natick, Mass.

TURRILL, W. B., 1964. Floras. In W. B. Turrill (ed.), *Vistas in botany*, 4: 225–238. Oxford: Pergamon.

VERDOORN, F., 1945. *Plants and plant science in Latin America.* Waltham, Mass.: Chronica Botanica.

VICKERY, B. C., 1975. *Classification and indexing in science.* 3rd edn. London: Butterworths.

VIDAL, J.-E., 1984. Problèmes et perspectives des études floristiques relatives à la Peninsule Indochinoise. *Peninsule* 8/9: 11–22.

VOSS, E. G., 1952. The history of keys and phylogenetic trees in systematic biology. *J. Sci. Labs. Denison Univ.* **43**: 1–25. (Quoted from R. J. Pankhurst, 1978, *op. cit.*, pp. 82–87.)

WAGNER, W. H., 1974. Pteridology 1947–1972. *Ann. Missouri Bot. Gard.* **61**: 86–111.

WALTERS, S. M., 1954. The Distribution Maps Scheme. *Proc. Bot. Soc. British Isles* 1: 121–130.

WATSON, L., 1971. Basic taxonomic data: the need for organisation over presentation and accumulation. *Taxon* **20**: 131–136.

WEBB, D. A., 1978. Flora Europaea – a retrospect. *Taxon* 27: 3–14.

WEBB, L. J. *et al.*, 1970. Studies in the numerical analysis of complex rainforest communities, V. A comparison of the properties of floristic and physiognomic–structural data. *J. Ecol.* **58**: 203–232.

WEBB, L. J. *et al.*, 1976. The value of structural features in tropical forest typology. *Australian J. Ecol.* 1: 3–28.

WEIN, K., 1937. Die Wandlungen in Sinne des Wortes „Flora". *Repert. Spec. Nov. Regni Veg., Beih.* **66**: 74–87.

WELLISCH, H., 1984. *Conrad Gesner: a bio-bibliography.* 2nd edn. Zug, Switzerland: Inter-Documentation Company. (1st edn. in *J. Soc. Bibliogr. Nat. Hist.* 7: 151–247 (1975).)

WHEELER, W. M., 1923. The dry-rot of our academic biology. *Science*, N.S. **52**(1464): 61–71. (Reprinted 1970 in *BioScience* 20: 1008–1013.)

WILKEN, D., D. WHETSTONE, K. TOMLINSON and N. MORIN, 1989a. How floristic information is used. In N. R. Morin *et al.*, *op. cit.*, pp. 58–70.

WILKEN, D., D. WHETSTONE, K. TOMLINSON and N. MORIN, 1989*b*. Considerations for producing the ideal flora. In N. R. Morin *et al.*, *op. cit.*, pp. 71–89.

WILLIAMS, C. B., 1964. *Patterns in the balance of nature and related problems in quantitative ecology.* London: Academic Press.

WILLIAMS, R. L., 1988. Gérard and Jaume: two neglected figures in the history of Jussiaean classification, 1–3. *Taxon* **37**: 2–34, 233–271.

WISSKIRCHEN, R. and H. HAEUPLER, 1998. *Standardliste der Farn- und Blütenpflanzen Deutschlands.* Stuttgart: Ulmer.

WIT, H. C. D. DE, 1949. Short history of the phytography of Malaysian vascular plants. In C. G. G. J. van Steenis (ed.), *Flora Malesiana*, I, 4: lxxi–clxi. Batavia (Jakarta): Noordhoff-Kolff.

WOODLAND, D. W., 1997. *Contemporary plant systematics.* 2nd edn. Berrien Springs, Mich.: Andrews University Press. (With CD-ROM.)

WORTHINGTON, E. B., 1938. *Science in Africa.* London: Oxford University Press. (Revised edn., 1958, London, as *Science in the development of Africa.*)

WYATT, H. V. (ed.), 1997. *Information sources in the life sciences.* 4th edn. London: Bowker/Saur. (Guides to information sources. 1st edn., 1966, by R. T. Bottle and H. V. Wyatt as *The use of biological literature.* London: Butterworths; 2nd edn., 1971; 3rd edn., 1987, by Wyatt alone as *Information sources in the life sciences.*) [Plant sciences by H. V. Wyatt, pp. 237–245; botany, plant ecology and the environment by H. J. Peat, pp. 247–257.]

YÜ, TE-TSUN, 1979 (1980). Special report: status of the *Flora of China. Syst. Bot.* **4**: 257–260.

ZEPERNICK, B. and F. K. TIMLER, 1990. Beiträge zur botanischen Erforschung aussereuropäischer Länder. In C. Schnarrenberger and H. Scholz (eds.), *Geschichte der Botanik in Berlin*, pp. 319–355. Berlin: Colloquium. (Wissenschaft und Stadt 15.)

ZIMAN, J., 1968. *Public knowledge: the social dimension of science.* Cambridge: Cambridge University Press.

Part II
• • •
Systematic bibliography

•••

Conventions and abbreviations

Bibliographic conventions
Books
Author(s) or editor(s), date, title, edition, collation, place, and publisher; previous edition (if any);
reprint (if any); publisher's series (if any), bibliographic notes (if required). The author may be defined
as an institution, or other collective body.

Irregular serials
Author, date, title, edition, collation, serial reference (in parentheses), place, and publisher (if obscure).
Authorship convention as above; Roman numera-
tion/letters for series and Arabic numeration for
volumes. [A distinct convention has been adopted here
as many book-length floras and enumerations are published in various kinds of irregular serials, the handling
and cataloguing of which varies among libraries and
other book collections.]

Periodicals
Author, date, title, serial reference, and collation.
Authorship convention as above; numeration the same
as for irregular serials. [Used for more or less regularly
periodic serials, which are always treated as such in
libraries.]

Symposia, congresses, etc.
Author of article, date, title of article, 'in', title of
whole work, year and place of meeting (if known and
applicable), editor(s), collation of article, serial reference (if applicable), place, and publisher. Authorship
and numeration conventions as above.

Publication
As far as possible, only the primary point of publication (assumed to be the first or only city or town
given) is indicated, and superfluous portions of the
titles of printeries or publishing houses (where relevant) are omitted, including *izdatel'stvo* (Russian for

'publishing house') in Russian (and from 1924–91, Soviet) books consequent to the general adoption, from 1964, of acronyms or cognomina (or where such designations had previously been used). Where necessary for clarity, the term 'press' has been used in the sense of a publishing house, as in Cambridge University Press or Academia Sinica Press. In some cases, precisely who has been the publisher has been hard to ascertain. Where this is so, recourse has been had as far as possible to the seven-volume second edition of *Taxonomic Literature* by F. A. Stafleu and R. A. Cowan (1976–88, Utrecht) [*TL-2*] and its five supplements as so far published (1992–98).

The designation of a work as *mimeographed* indicates that, as far as can be ascertained, it has been reproduced from stencils with the use of a duplicating machine. In recent years, though, the introduction of more sophisticated means of reproduction, especially the laser printer, has made the application of this designation somewhat arbitrary. Whether or not such works are properly published has long been a matter for argument among bibliographers; but to the author effective circulation, usage, and citation irrespective of reproductive process constitutes a form of publication – although not necessarily *definitive* publication in the sense of the International Code of Botanical Nomenclature.

Transcription and transliteration

All principal titles in this work have been rendered in the original for languages in Roman, Cyrillic and modern Greek scripts (in some cases, as with Vietnamese, using simplified diacritics). Where the titles are in non-Roman alphabets or in ideographs or in both, any Roman alternatives furnished have been adopted. If no such alternatives were given, the originals have been transcribed or transliterated into their nearest equivalents in the modern Roman alphabet (with diacritical marks used as necessary). Transcription of Cyrillic characters follows ISO recommendation R9 (1954; see also *Unesco Bulletin for Libraries* **10**: 136–137 (1956)), except that X is rendered as *kh* and Ц as *ts* as being more familiar to anglophone users. In addition, personal names are as far as possible rendered according to the given author's own preference or to conventional usage. For Mandarin (Chinese), the system used in *Bibliography of Eastern Asiatic Botany* (1938) and its *Supplement* (1960) has been followed as far as possible; this applies also to Japanese and Korean. For Chinese works, no standard

set of transliterations on the *pinyin* system has been available and the author has therefore had to use his best judgment. Most works published since 1972 have, however, had an alternative title, usually in Latin, and further guidance has been available from S. C. Chen *et al.* (eds.), *Bibliography of Chinese systematic botany (1949–1990)* (1993). Comparatively few problems have been presented by other languages. Where works have been written in the Jawi script (Arabic or Farsi; none has been published in Malay), alternative titles in a European language have nearly always been provided.

Periodical and serial abbreviations

The system of periodical and serial abbreviations here followed is based upon that in G. H. M. Lawrence *et al.* (eds.), *Botanico-Periodicum Huntianum*. Pittsburgh, Pa.: Hunt Botanical Library (1968), and G. D. R. Bridson (ed.), *Botanico-Periodicum-Huntianum/Supplementum*. Pittsburgh (1991). Abbreviations of titles not included in that work have as far as possible been formulated according to its guidelines. The use of *Botanico-periodicum huntianum* (or *B-P-H*) brings the present work into line with the second edition of *Taxonomic Literature* previously referred to. Full titles are given in Appendix B.

General bibliographies and indices: explanation of abbreviations

Only short titles are given here. For fuller details see Appendix A.

General bibliographies

Bay, 1909. Bay, *Bibliographies of botany.*
Blake and Atwood, 1942. Blake and Atwood, *Geographical guide to floras of the world*, 1.
Blake, 1961. Blake, *Geographical guide to floras of the world*, 2.
Frodin, 1964. Frodin, *Guide to the standard floras of the world (arranged geographically).*
Frodin, 1984. Frodin, *Guide to standard floras of the world.*
Goodale, 1879. Goodale, *The floras of different countries.*
Holden and Wycoff, 1911–14. Holden and Wycoff, *Bibliography relating to the floras.*
Hultén, 1971. Hultén, *The circumpolar plants*, II [bibliography].
Jackson, 1881. Jackson, *Guide to the literature of botany.*
Pritzel, 1871–77. Pritzel, *Thesaurus literaturae botanicae.*
Rehder, 1911. Rehder, *The Bradley bibliography*, 1: *Dendrology*, part 1.
Sachet and Fosberg, 1955, 1971. Sachet and Fosberg, *Island bibliographies* and *Island bibliographies: supplement.*
USDA, 1958. US Department of Agriculture, *Botany subject index.*

General indices

The dates give years of coverage, not publication (in print). Dates of on-line electronic coverage (where available) are furnished in Appendix A; only their acronyms appear here (in square brackets).

BA, 1926. *Biological Abstracts* (and *Biological Abstracts/RRM*). [BIOSIS]

BotA, 1918–26. *Botanical Abstracts*.

BC, 1879–1944. *Botanisches Centralblatt* (from 1938 *Botanisches Zentralblatt*).

BS, 1940– . *Bulletin Signalétique* (through 1955 *Bulletin Analytique*; from 1984 *Bibliographie Internationale*). [PASCAL]

CSP, 1800–1900. *Catalogue of Scientific Papers*.

EB, 1959–98. *Excerpta Botanica*, sectio A: *Taxonomia et chorologia*.

FB, 1931– . *Progress in Botany/Fortschritte derBotanik*.

IBBT, 1963–69. *Index Bibliographique de Botanique Tropicale*.

ICSL, 1901–14. *International Catalogue of Scientific Literature*, section M: *Botany*.

JBJ, 1873–1939. *Just's Botanischer Jahresbericht*.

KR, 1971– . *Kew Record of Taxonomic Literature*.

NN, 1879–1943. *Naturae Novitates*.

RŽ, 1954– . Реферативный журнал (*Referativnyj Žurnal*): (04) Биология (*Biologija*).

General references

The following list comprises articles and other references relating to inventory, state-of-knowledge, etc., to which repeated reference in the main part of this book is made. A few additional general references of interest have been added. Material clearly identifiable with a geographic unit used in this book is listed in the appropriate place(s), with materials for the Neotropics and the Pacific appearing respectively at Division 3 (South America) and Superregion 94–99 (Oceania).

BARTHLOTT, W., W. LAUER and A. PLACKE, 1996. Global distribution of species diversity in vascular plants: towards a world map of phytodiversity. *Erdkunde* **50**: 317–327, col. map.

BLAKE, S. F., 1961. *Geographical guide to floras of the world*, 2 (USDA Misc. Publ. 797). Washington, D.C.: U.S. Government Printing Office.

BLAKE, S. F. and A. C. ATWOOD, 1942. *Geographical guide to floras of the world*, 1 (USDA Misc. Publ. 401). Washington, D.C.: U.S. Government Printing Office.

BRAMWELL, D., 1979. *Plants and islands*. London: Academic Press.

CAMPBELL, D. G. and H. D. HAMMOND (eds.), 1989. *Floristic inventory of tropical countries*. New York: The New York Botanical Garden.

CRABBE, J. A., A. C. JERMY and J. T. MICKEL, 1975. A new generic sequence for the pteridophyte herbarium. *Fern Gaz.* **11**: 141–162.

CRONQUIST, A., 1981. *An integrated system of classification of flowering plants*. xviii, 1262 pp., illus. New York: Columbia University Press.

DAVIS, S. D. *et al.* (comp.), 1986. *Plants in danger: what do we know?* xlv, 461 pp., 2 maps. Gland, Switzerland, and Cambridge, England: IUCN. [Abbreviated as PD.]

DAVIS, S. D. *et al.* (eds.), 1994–97. *Centres of plant diversity: a guide and strategy for their conservation*. 3 vols. Illus., maps. Cambridge, England: IUCN Publications Unit (for WWF/IUCN).

FOSBERG, F. R., 1985. Present state of knowledge of the floras and vegetation of emergent reef surfaces. In *Proceedings of the Fifth International Coral Reef Congress* (Tahiti, 1985), 5: 107–112. Moorea, French Polynesia: Antenne Muséum/École Pratique des Hautes Études (ÉPHÉ).

GENTRY, A. H., 1978. Floristic needs in Pacific tropical America. *Brittonia* **30**: 134–153.

GOOD, R. D'O., 1974. *The geography of the flowering plants*. 4th edn. xvi, 557 pp., illus., maps. London: Longman.

GROOMBRIDGE, B. (ed.), 1992. *Global biodiversity: status of the earth's living resources*. xviii, 585 pp., illus., maps. London: Chapman and Hall. [Abbreviated as GB.]

HEMSLEY, W. B., 1885. Report on the present state of knowledge of various insular floras. In W. Thomson and J. Murray (eds.), *Reports on the scientific results of the voyage of HMS Challenger during the years 1873–76, Botany*, 1 (Introd.): 1–75. London: HMSO.

HEYWOOD, V. H. (exec. ed.) and R. T. WATSON (chair), 1995. *Global biodiversity assessment*. x, 1140 pp., illus. Cambridge: Cambridge University Press.

MORIN, N. R., R. D. WHETSTONE, D. WILKEN and K. L. TOMLINSON (eds.), 1989. *Floristics for the 21st century*. xiii, 163 pp., illus. St. Louis: Missouri Botanical Garden. (Monogr. Syst. Bot. Missouri Bot. Gard. 28.)

PRANCE, G. T., 1977 (1978). Floristic inventory of the tropics: where do we stand? *Ann. Missouri Bot. Gard.* **64**: 659–684; [addendum]: 1–2.

PRANCE, G. T. and D. G. CAMPBELL, 1988. The present state of tropical floristics. *Taxon* **37**: 519–548.

STAFLEU, F. A. and R. S. COWAN, 1976–88. *Taxonomic Literature*. 2nd edn. 7 vols. Utrecht: Bohn, Scheltema and Holkema. (Regnum Veg. 94, 98, 105, 110, 112, 115, 116.)

TAKHTAJAN, A., 1986. *Floristic regions of the world*. xxii, 522 pp., 4 maps in text, 2 end-paper maps. Berkeley, Calif.: University of California Press.

THISTLETON-DYER, W. T., 1906. Botanical survey of the Empire. *Bull. Misc. Inform.* (Kew) **1905**: 9–43.

• • •

Conspectus of divisions and superregions

Division 0   World floras, isolated oceanic islands and polar regions
  Superregion 01–04 Isolated oceanic islands
  Superregion 05–07 North Polar regions
  Superregion 08–09 South Polar regions

Division 1   North America (north of Mexico)
  Superregion 11–13 Boreal North America
  Superregion 14–19 Conterminous United States

Division 2   Middle America
  Superregion 21–23 Mexico and Central America
  Superregion 24–29 The West Indies

Division 3   South America

Division 4   Australasia and islands of the southwest Indian Ocean (Malagassia)
  Region (Superregion) 41 New Zealand and
   surrounding islands
  Superregion 42–45 Australia
  Superregion 46–49 Islands of the southwest Indian
   Ocean (Malagassia)

Division 5   Africa

Division 6   Europe

Division 7   Northern, central and southwestern (extra-monsoonal) Asia
  Superregion 71–75 CIS-in-Asia
  Region (Superregion) 76 Central Asia
  Superregion 77–79 Southwestern Asia

Division 8   Southern, eastern and southeastern (monsoonal) Asia
  Superregion 81–84 South Asia (Indian subcontinent)
  Region (Superregion) 85 Japan, Korea and associated
   islands
  Superregion 86–88 China (except Chinese central
   Asia)
  Region (Superregion) 89 Southeastern Asia

Division 9   Greater Malesia and Oceania
  Superregion 91–93 Greater Malesia
  Superregion 94–99 Oceania

Division 0

• • •

World floras, isolated oceanic islands and polar regions

It may appear paradoxical, at first sight, to associate the plants of Kerguelen's Land with those of Fuegia, separated by 140 degrees of longitude, rather than with those of Lord Auckland's Group, which is nearer by about 50 degrees. But the features of the Flora of Kerguelen's Land are similar to, and many of the species identical with, those of the American continent, constraining me to follow the laws of botanical affinity in preference to that of geographical position.

> J. D. Hooker, *Flora antarctica* (1846).

[Our main objective is] to assist the tyro in the verification of genera and species . . . natural habit is often a safer guide than minute microscopic characters. [Such structural details are unimportant.]

> W. J. Hooker, *Species filicum*, vol. 3 (1859).

[There is a] widely held, but erroneous, idea that the floras and vegetations of coral atolls are very much alike everywhere and therefore not much worth studying.

> F. R. Fosberg, 'Coral island vegetation'; in O. A. Jones and R. Endean, *Biology and geology of coral reefs*, vol. 3 (Biology, 2): 260 (1976).

Apart from general world floras (**000**), major chorological atlases (**001**, including *Die Pflanzenareale* by E. Hannig and H. Winkler), and worldwide works relating to the vascular plants of particular physiographically defined habitats (**002–009**), this division comprises three geographical superregions: **01–04**, Isolated oceanic islands; **05–07**, North Polar regions; and **08–09**, South Polar regions. The geographical limits and the bio-history of each are considered beneath their respective headings.

Unit **000** contains descriptions of the two principal post-Linnaean *species plantarum*, the *Prodromus systematis naturalis regni vegetabilis* of the Candolles and *Das Pflanzenreich* of Engler and his school, as well as W. J. Hooker's *Species filicum*, Carl S. Kunth's *Enumeratio plantarum* (for monocotyledons), and the 'suites au *Prodromus*', the *Monographiae phanerogamarum* edited by A. and C. de Candolle. Also included are the principal vascular *genera plantarum*. The standard general keys, indices and dictionaries relating to vascular plants are, however, merely listed; these, along with the various *syllabi* or compendia of families, are more or less fully discussed in textbooks of taxonomy.[1]

Bibliographies. See under supraregional headings. Those relating to **000** are listed on p. 90 and described in Appendix A.

Indices. See under supraregional headings. Those relating to **000** are listed on p. 91 and described in Appendix A.

Conspectus

000 World: general works
001 World: chorological works
 Mappae mundi
 Regiones polarium
 Atlantica
 Eurasia
 Indo-Pacifica
002 World: 'old' upland and montane regions
003 World: 'alpine' regions
004 World: ectopotrophic areas
005 World: drylands (steppes and deserts)
006 World: rivers
007 World: lakes and their littoral
008 World: wetlands
009 World: oceans, islands, reefs and the oceanic littoral
 Sea-grasses
 Mangroves
 Salt marshes
 Strand plants
 Islands

Superregion 01–04 Isolated oceanic islands
Region 01 Islands of the eastern Pacific Ocean
011 Guadalupe Island
012 Rocas Alijos
013 Revillagigedo Islands
014 Clipperton Island
015 Cocos Island
016 Malpelo Island
017 Galápagos Islands
018 Desventuradas Islands
019 Robinson Crusoe (Juan Fernández) Islands
Region 02 Macaronesia
020 Region in general
021 Azores
022 Madeira Islands
023 Salvage Islands
024 Canary Islands
025 Cape Verde Islands
Region 03 Islands of the Atlantic Ocean (except
 Macaronesia)
030 Region in general
031 The Bermudas
032 St. Paul Rocks
033 Fernando de Noronha (with Rocas)
034 Ascension Island
035 St. Helena
036 Trindade (South Trinidad) and Martin Vaz Islands

037 Tristan da Cunha Islands
038 Gough Island
Region 04 Islands of the central and eastern Indian Ocean
041 Laccadive Islands (Lakshadweep)
042 Maldive Islands
043 Chagos Archipelago
044 Amsterdam and St. Paul Islands
045 Cocos (Keeling) Islands
046 Christmas Island

Superregion 05–07 North Polar regions
050–70 Superregion in general
Region 05 Islands of the Arctic Ocean
051 Jan Mayen Island
052 Bear Island (Bjørnøya)
053 Spitsbergen (Svalbard)
054 Franz Josef Land (Zemlja Frantsa Iosifa)
055 Novaja Zemlja
056 Severnaja Zemlja
057 New Siberian Islands (Novosibirskije Ostrova)
058 Wrangel Island (Ostrov Vrangelja)
Region 06 Palearctic mainland region
060 Region in general
061 Arctic Fennoscandia
062 Kola Peninsula (Arctic zone)
063 Northern and northeastern Russia (Arctic zone)
064 The Ural (Arctalpine zone)
065 Western Siberia (Arctic and arctalpine zones)
066 Central Siberia (Arctic and arctalpine zones)
067 Eastern Siberia (Arctic and arctalpine Sakha)
068 Anadyr and Chukotia
Region 07 Nearctic region
071 Bering Sea Islands
072 Alaska and Yukon (Arctic and arctalpine zones)
073 Nunavut and Northwest Territories of Canada (Arctic
 mainland zone)
074 Canadian Arctic Archipelago
075 Ungava (far northern Québec)
076 Greenland

Superregion 08–09 South Polar regions
080–90 Superregion in general
Region 08 Circum-Antarctic islands
081 Macquarie Island
082 McDonald (Heard) Islands
083 Kerguelen Archipelago
084 Crozet (Possession) Islands
085 Marion (Prince Edward) Islands
086 Bouvetøya (Bouvet Island)
087 South Georgia
088 South Sandwich Islands
Region 09 Antarctica
090 Region in general

000

World: general works

The descriptive works on species and genera deemed most appropriate to a selective guide to floras are here preceded by a choice of the most significant general dictionaries, indices, identification keys and descriptive compendia relating to vascular plants. One or another of these works has influenced the systematic style and arrangement of the majority of the floras and other floristic works included in the present work. It is hoped that this selection will be of particular value to the non-specialist. Works covering families alone are, however, not here accounted for unless they feature keys.

General keys to vascular plants

CRONQUIST, A., 1979. *How to know the seed plants.* vii, 153 pp., 7 lvs., 337 text-figs. Dubuque, Iowa: Brown. [Analytical students' key covering the majority of seed plant families, notably those wild or widely cultivated in North America. Terminal leads in the key are illustrated by small figures depicting representatives of the families concerned.]

DAVIS, P. H. and J. CULLEN, 1997. *The identification of flowering plant families: including a key to those native and culti-vated in north temperate regions.* 4th edn. by J. Cullen. xii, 215 pp., 8 figs. Cambridge: Cambridge University Press. (1st edn., 1965, Edinburgh; 2nd and 3rd edns., 1979, 1989, Cambridge.) [A pocket-sized set of keys to northern extra-tropical angiosperm families, with (pp. 4–44) an introductory section – careful reading of which is advised – on terminology and approaches to observation and use of the keys. The latter part of the work encompasses concise descriptions of the families (with geographical range), an annotated bibliogra-phy of 36 key references with guidelines (pp. 191–194), a glossary, and an index.]

HUTCHINSON, J., 1967. *Key to the families of flowering plants of the world.* viii, 117 pp., 8 text-figs. Oxford: Oxford University Press. (Reproduced with new index, 1979, Königstein/Ts., Germany: Koeltz.) [Worldwide in scope, based upon the general keys in his *Families of flowering plants.*]

THONNER, F. (transl. and revised by R. Geesink, A. J. M. Leeuwenberg, C. E. Ridsdale and J.-F. Veldkamp), 1981. *Thonner's analytical key to the families of flowering plants.* xxvi, 231 pp., portr. Wageningen: PUDOC; Leiden: Leiden University Press. (Leiden Botanical Series 5.) [Analytical key to seed plant families on a worldwide basis, with an abun-dance of alternative leads to take account of the many taxa with features aberrant for a given family. A glossary, index, and bio-bibliography of Franz Thonner are also included. – Based upon *idem*, 1917. *Anleitung zum Bestimmen der Familien der Blütenpflanzen (Phanerogamen).* 2nd edn. vi, 280 pp. Berlin: Friedländer. (1st edn., 1895.)][2]

WATSON, L. and M. J. DALLWITZ, 1994. *The families of flowering plants: interactive identification and information retrieval.* 1 CD-ROM; booklet of 40 pp. Collingwood, Vic.: CSIRO Publishing. [Includes interactive keys, accepted and synonymous names, descriptions, notes on distribution and habitat, commentary, representative illustrations, and refer-ences. Some query facilities are also available.][3]

General dictionaries and indices of vascular plants

CHRISTENSEN, C., 1905–06. *Index filicum.* ix, 744 pp. Copenhagen: Hagerup. Continued as *idem*, 1913–34. *Index filicum*, Suppls. 1–3. Copenhagen; R. E. G. PICHI-SERMOLLI et al., 1965. *Index filicum*, Suppl. 4. xiv, 370 pp. Utrecht: International Association for Plant Taxonomy (Regnum Veg. 37); F. M. JARRETT and collaborators, 1985. *Index filicum: supplementum quintum pro annis 1961–1975.* 245 pp. Oxford: Oxford University Press; R. J. JOHNS, 1996. *Index filicum: supplementum sextum pro annis 1976–1990.* vii, 414 pp. Kew: Royal Botanic Gardens; and *idem*, 1997. *Index filicum: supple-mentum septimum pro annis 1991–1995.* ix, 124 pp. [Homosporous and heterosporous Filicinae through 1995 (the fifth and subsequent supplements also including the 'fern-allies'). – From 1971 through 1980, new names were also published in *Kew Record of Taxonomic Literature* (1974–); and from 1997, it was intended that future supple-ments would be quinquennial, like *Index Kewensis*.][4]

FARR, E. R., J. A. LEUSSINK and F. A. STAFLEU (eds.), 1979. *Index nominum genericorum (plantarum).* 3 vols. Utrecht: Bohn, Scheltema & Holkema/International Association for Plant Taxonomy (Regnum veg. 100). Continued as E. R. FARR, J. A. LEUSSINK and G. ZIJLSTRA (eds.), 1986. *Index nominum genericorum (plantarum): Supplementum I.* xv, 126 pp. Utrecht (Regnum Veg. 113). [Authoritatively documented index to all generic names of plants (in traditional sense) published from 1754 onwards, with indication of place of publication and, where desig-nated, a type species. In part originally published (1955–72) as index cards (not, however, entirely superseded). – The thoroughness of coverage varies; inclusive of Supplement I vascular cryptogams are fairly complete through 1984 but phanerogams less so after 1975.]

GRAY HERBARIUM OF HARVARD UNIVERSITY, 1894– . *Gray Herbarium [card] index.* Cards, issued quarterly (from 1983 as annually reissued and updated microfiche and since 1993 purely as entries in an electronic database). Cambridge, Mass. (Contents to 1977 reprinted in book form as *idem*, 1968. *Gray Herbarium index*, with preface by R. C.

Rollins. 10 vols. Boston: Hall, and *idem*, 1981. *Supplement*. Boston.) [Index to all new plant names published in the Western Hemisphere from 1886, including those in infraspecific categories. In more recent years, retrospective coverage of infraspecific names to 1753 has been undertaken (a task not yet assumed at *Index Kewensis* for names prior to 1971). The complete work is now accessible via the World Wide Web and wholly available only in that format.][5]

HOOKER, J. D. and B. DAYDON JACKSON, 1893–95. *Index Kewensis plantarum phanerogamarum*. 2 vols. Oxford: Oxford University Press. Continued as T. DURAND and B. DAYDON JACKSON, 1902–06. *Index Kewensis*: Supplementum I. Brussels: Castaigne (for T. Durand; subsequently reissued by Oxford University Press), and W. T. THISTLETON-DYER *et al.*, 1904– . *Index Kewensis*: Supplementum II– . Oxford: Oxford University Press (from Supplementum XX (1996), Kew: Royal Botanic Gardens). (Both index and supplements reprinted from 1977, Königstein/Ts., Germany: Koeltz. Consolidated CD-ROM version first released 1993, Oxford; 2nd edn., 1997.) [The standard index to names of seed plant species and genera from 1753, with indication of known geographical range when first described (save for names treated as synonyms in the original work and in early supplements) as well as place of publication. Twenty supplements in all have been published, providing coverage through 1995; these quinquennial supplements will be continued. Annual coverage from 1986 through 1989 was also provided by *Kew Index* (Oxford: Oxford University Press) and continues in *Kew Record* (quarterly; see Appendix A). For an index to generic entries throughout the work, see E. ROULEAU, 1981. *Guide to the generic names appearing in the* Index Kewensis *and its fifteen supplements*. 512 pp. Cowansville, Quebec: Chatelain. (1st edn., 1970, Montreal.)][6]

MABBERLEY, D. J., 1997. *The plant-book*. xvi, 858 pp. Cambridge: Cambridge University Press. (1st edn., 1987; corrected reprint, 1989.) [Index to valid, currently used generic and family names of vascular plants, including *more commonly encountered* synonyms; key revisionary and monographic literature, with emphasis on recent works; philosophy on circumscription of genera. Also includes many common English vernacular or vernacularized names. Family limits are based in the first instance on the Cronquist system (A. Cronquist, 1981. *An integrated system of flowering plants*. New York: Columbia University Press (see **General references**)), as well as that of Kubitzki (see below under **Generic encyclopedias**).

ØLLGAARD, B., 1989. *Index of the Lycopodiaceae*. 135 pp. (Kongel. Danske Vidensk. Selsk., Biol. Skr. 34). Copenhagen: Munksgaard. [Alphabetical nomenclator for *Lycopodium* and allies from 1753; also includes a systematic conspectus and 17-page bibliography. Succeeds W. G. HERTER, 1949. *Index Lycopodiorum*. 120 pp. Basel: The author.]

REED, C. F., 1953. *Index Isoetales*. 72 pp. Alcobaça, Portugal/Baltimore, Md.: The author. (Also in *Bol. Soc. Brot.*, II, 27.) [Index to names of Isoetales.]

REED, C. F., 1966. *Index Psilotales*. 30 pp. Alcobaça, Portugal/Baltimore, Md.: The author. (Also in *Bol. Soc. Brot.*, II, 40.) [Index to names in *Psilotum*, *Tmesipteris*, and their allies.]

REED, C. F., 1966. *Index selaginellarum*. 287 pp. Alcobaça, Portugal/Baltimore, Md.: The author. (Also published as *Mem. Soc. Brot.* 18.) [Index to names of *Selaginella* sensu lato.]

REED, C. F., 1971. *Index equisetorum: index to Equisetophyta*. 2 vols. (Contr. Reed Herb. 19). Baltimore, Md.: The author (Reed Herbarium). [Vol. 2, Extantes, indexes names in the modern genus *Equisetum*.]

WILLIS, J. C., 1973. *A dictionary of the flowering plants and ferns*. 8th edn., revised by H. K. Airy-Shaw. xxii, 1245, lxvi pp. Cambridge: Cambridge University Press. (1st edn. in 2 vols., 1895–97; 6th edn., 1931.) [A standard guide to generic and family names and their synonyms, with accepted equivalents and/or brief notes on one or more lines about each, the latter including overall distribution and approximate number of species. The two most recent editions are less broad in scope than those of 1931 and before, but remain of value as a one-volume nomenclator. – Partly succeeded by D. J. MABBERLEY, 1997. *The plant-book* (described above).]

World floras (*Florae cosmopolitanae*)

The greater part of this *oeuvre*, written in the late eighteenth and early nineteenth centuries, is now chiefly of historical and nomenclatural interest. Many works were compiled in succession to, or under the inspiration of, the encyclopedic treatises of Linnaeus, and followed his sexual system of classification. Moreover, before 1840 such coverage of the plant kingdom (or at least the higher plants) to species level was still possible for one or two men, without many other encumbrances. The one real exception is the Candollean *Prodromus*, which covers the dicotyledons. Its distinguishing features include adoption of a natural system of classification, of which it was planned to be an exponent, and (as far as possible) freshly prepared descriptions. It is regarded here as the last 'world flora' (along with the contemporaneous works by Kunth on monocotyledons and W. J. Hooker on the vascular cryptogams). Completion, however, required some 50 years and the aid of several specialists; contemplated coverage of the monocotyledons was abandoned.

Later 'world floras' took the form of monograph-series. The two outstanding exemplars are *Monographiae phanerogamarum* and *Das Pflanzenreich*.

The latter, prominent in the early twentieth century, lost some momentum after World War I and was effectively terminated by World War II. Complete modern coverage of the plant kingdom to the species level has seemed an impossible dream.[7]

CANDOLLE, A.-P. DE and A. DE CANDOLLE (eds.), 1824–73. *Prodromus systematis naturalis regni vegetabilis*. 17 vols. Paris: Treuttel & Würtz (vols. 1–7); Fortin, Masson (vols. 8–9); Masson (vols. 10–17). (Reprint of vols. 1–7, 1966, Lehre, Germany: Cramer, in connection with an offer of the remaining stock of vols. 8–17.) Accompanied by H. W. BUEK, 1842–74. *Genera, species, et synonyma Candolleana alphahetico ordine disposita*. 4 parts. Berlin: Nauck (parts 1–2); Hamburg: Perthes-Besser & Nauck (part 3); Gräfe (part 4). (Reprinted 1967, Amsterdam: Asher.)

Concise descriptive formal systematic account of the families, genera and species of Dicotyledoneae, arranged according to the Candollean system; includes notes on distribution, synonymy, some citations of (usually representative) *exsiccatae*, and taxonomic commentary. Pp. 303–314 in vol. 17 comprise a historical analysis of the work, and pp. 323–493 in the same volume, an abridged index.[8]

CANDOLLE, A. DE and C. DE CANDOLLE (eds.), 1878–96. *Monographiae phanerogamarum*. Vols. 1–9. Paris: Masson.

'Prodromi nunc continuatio, nunc revisio.' A series of formal descriptive family monographs, issued without regard to systematic order but intended as an extension and revision of the *Prodromus*. All treatments were by specialists, including the Candolles themselves. The accounts include synonymy, summaries of distribution, citations of *exsiccatae*, and taxonomic commentary, as well as synoptic keys. [The series was terminated a few years after the death of Alphonse de Candolle in 1893.][9]

ENGLER, A. *et al.* (eds.), 1900– . *Das Pflanzenreich*. Heft 1– . Illus. Leipzig: Engelmann. (Heft 1–106 published 1900–43; Heft 106 reprinted 1956, Berlin: Akademie-Verlag; Heft 107–108 published 1953, 1968, Berlin; Heft 1–105 reprinted 1957–60, Weinheim, Germany: H. R. Engelmann/Cramer, and in part by Koeltz in more recent years.)

Comprises an illustrated series of monographs of families (or major parts thereof) of the plant kingdom, numbered according to the Engler and Prantl system (thus retaining some semblance of the 'world flora' concept) but not issued in sequence. The accounts include brief descriptions, synonymy, indication of distribution, citation of *exsiccatae*, and taxonomic commentary, as well as analytical keys. – The work may also be viewed as an extension of *Die natürlichen Pflanzenfamilien* as well as a successor to *Monographiae phanerogamarum*. The series is far from complete; progress on the work diminished after Engler's death in 1930 and since 1943 only two new parts have appeared. For a detailed bibliographic summary see M. DAVIS, 1957. A guide and analysis of Engler's *Das Pflanzenreich*. *Taxon* 6: 161–182 (reprinted 1967 in *Taxonomic Literature* by F. A. Stafleu; 1976 in vol. 1 of *Taxonomic Literature-2* by F. A. Stafleu and R. S. Cowan).[10]

FRODIN, D. G., R. GOVAERTS *et al.*, 1996– . [*World checklists and bibliographies of orders and families.*] In volumes. Illus. Kew: Royal Botanic Gardens.

Synonymized checklists of species and infraspecific taxa in selected orders and families of (so far) seed plants, with indication of distribution (including Taxonomic Databases Working Group country codes), usual habit, and (occasionally) further notes; commentaries on genera with selected, annotated standard references as well as family introductions inclusive of general, infrafamilial and regional revisionary and other literature. Some treatments are partly or wholly by specialists. [The first of its kind since that of Kunth (see below). As of writing, installments have appeared for Magnoliaceae (1996), Fagales (1998) and Coniferae (1998); installments for Euphorbiaceae, Sapotaceae and Araceae are expected to appear in 2000.]

HOOKER, W. J., 1844–64. *Species filicum*. 5 vols., 304 pls. London: Pamplin (vol. 5: Dulau). Complemented by *idem*, 1874. *Synopsis filicum* (revised by J. G. Baker). xiv, 559 pp., 9 pls. London. (1st edn., 1865–68.)

Species filicum is a copiously illustrated descriptive account of known fern species, with special reference to the contents of the 'Hookerian Herbarium' (now at Kew). It is noteworthy for its broad concept of fern families, particularly Polypodiaceae. *Synopsis filicum* is a 'condensed', enumerative version. The latter was supplemented in J. G. BAKER, 1891. A summary of the new ferns which have been discovered or described since 1874. *Ann. Bot.* 5: 181–222, 301–332, 455–500, pl. 14. (Reprinted separately, 1892, London. v, 119 pp.)[11]

KUNTH, C. S., 1833–50. *Enumeratio plantarum omnium hucusque cognitarum*. Vols. 1–5 (in 6). Stuttgart, Tübingen: Cotta.

Systematic enumeration, with diagnoses, of a large part of the known monocotyledons (the only major group covered). [Not completed owing to the author's suicide in 1850; such large orders as the Zingiberales and Orchidales are wanting. Bentham in 1874 estimated that it covered 'little more than half the class (i.e., sub-class)'. Although now largely of historical value, it is accounted for here as being in some respects complementary to the *Prodromus* (see above).]

Generic encyclopedias (*Generae plantarum*)

These works provide descriptions of individual families and genera of Tracheophyta (except those purely on pteridophytes; for these, see the next section) and are in effect specialized encyclopedias, although with their keys and indications of distribution they also act to some degree as 'world floras'. As a literary canon they have been in existence since Linnaeus's *generae plantarum* (1737; 5th edn., 1754; 6th edn., 1764). An influential work in the mid-nineteenth century, covering the whole of the plant kingdom as then understood, was *Genera plantarum* (1836–41; supplements, 1842–50) by Stephan Endlicher. Of subsequent works in this genre the most comprehensive is easily Engler and Prantl's *Die natürlichen Pflanzenfamilien* (1887–1915; 2nd edn., 1924–). The published parts of the second edition are particularly notable for their extensive organized lists of literature covering each major taxon. This tradition has to an extent been maintained in the most recent undertaking, *Families and genera of vascular plants* (1990–) by Klaus Kubitzki and collaborators, also described below.

BAILLON, H., 1866–95. *Histoire des plantes.* Vols. 1–13, illus. Paris: Morgand/Hachette. English edn.: *idem*, 1871–88. *The natural history of plants.* Transl. M. M. Hartog. Vols. 1–8, illus. London: Reeve.

A systematically arranged, partly derivative work containing descriptions of families and genera with extensive commentary and numerous fine illustrations, but without keys. Extensive references are given in the footnotes. The family arrangement follows that of A.-P. de Candolle, but with a very wide conception of genera. A fourteenth and final volume would have included Musaceae, Zingiberaceae and Orchidaceae, but the author's death in 1895 precluded its realization. The incomplete English version covers the Polypetalae and part of the Gamopetalae.[12]

BENTHAM, G. and J. D. HOOKER, 1862–83. *Genera plantarum.* 3 vols. London: distributed by Black, Pamplin, Reeve, and Williams & Norgate (vol. 1, part 1); Reeve and Williams & Norgate. (Reprinted 1965, Codicote near Hitchin, England: Wheldon & Wesley, and Weinheim, Germany: Cramer.)

This nineteenth-century botanical *tour de force* comprises an unillustrated, formal descriptive account, in Latin, of all families and genera of seed plants then known, arranged on a modified version of the Candollean system (now known as the Bentham and Hooker system) and including synonymy, notes on distribution and special features, and limited but pointed critical commentary. The generic accounts in each family are preceded by a concise synopsis, but no analytical keys are provided. [Now long out of date; a revision was partly realized by John Hutchinson as *The genera of flowering plants* (1964–67; see below).][13]

ENGLER, A. and K. PRANTL (eds.), 1887–1915. *Die natürlichen Pflanzenfamilien.* Teil I–IV; Nachträge zu Teil I/2; Nachträge I–IV zu Teil II–IV. 33 parts in 23 vols. Illus. Leipzig: Engelmann. Partly succeeded by A. ENGLER *et al.* (eds.), 1924– . *Die natürlichen Pflanzenfamilien.* 2. Aufl. Vols. 1b–28bI, *passim.* Illus. Leipzig: Engelmann (1924–43); Berlin: Duncker & Humblot (1953–). (Partly reprinted 1959–61, Berlin.)

One of the greatest of all plant taxonomic treatises, this work – in both its editions – is a copiously illustrated, scholarly, encyclopedic systematic account of the plant kingdom (as traditionally defined), covering all hierarchical levels down to infrageneric categories with mention or discussion of a great number of individual species. Analytical keys to all genera are provided, and much attention is paid to distribution, habitat, special features, properties and uses. The general family accounts include lists of references, extensive considerations of general taxonomy, morphology, anatomy, significant attributes, distribution, flower biology and dispersal, properties, uses, etc. The supplements in the first edition cover new data up through 1912. The second edition, commenced in 1924 and continuing, is even more extensive in scope than the first. Its *Band* 14a (1926) contains Engler's important general essay on flowering plant classification. [Many floras throughout the world were arranged according to the original edition of this work. An extensive review and analysis was prepared by F. A. Stafleu (*Taxon* **21**: 501–511 (1972); see also pp. 769–783 in vol. 1 of *Taxonomic Literature-2* (*TL-2*) by F. A. Stafleu and R. S. Cowan (1976)). Full details of coverage by the second edition through the 1970s also appear in *TL-2*.][14]

HUTCHINSON, J., 1964–67. *The genera of flowering plants*. Vols. 1–2. Oxford: Oxford University Press.

A relatively concise but formal descriptive systematic account of the families and genera of flowering plants, based partly on Bentham and Hooker's *Genera plantarum* but arranged according to the author's so-called phylogenetic system as enunciated in his *Families of flowering plants* (2nd edn., 1959). Generic accounts include synonymy, type species, key references, the number of species, and overall distribution (but little if any critical discussion, extensive infrageneric synopses, or mention of species of botanical interest). Family accounts include sections on phylogeny, unusual features, and uses as well as lists of references. Analytical keys are given to all genera. With the author's death in 1972, work on this series lapsed, although extensive materials are preserved at the Kew Herbarium.[15]

KUBITZKI, K. (gen. ed.), 1990– . *The families and genera of vascular plants*. Vols. 1– . Illus. Berlin: Springer.

Concise, moderately well-illustrated treatment, with descriptions of families and commentaries on features of their morphology, anatomy, embryology, pollen morphology, karyology, biology, phytochemistry, distribution and habitats, paleohistory, economic botany, and systematics, followed in each family by keys to and descriptions of genera (including, for each, place and date of publication as well as significant references, if any) and a select list of references covering all aspects; index to names of genera and higher taxa at end of each volume. Each volume opens with a general part on systematics and phylogeny, with discussion of the various patterns of thought. [As of writing in 1999, four volumes had been published. Volume 1 covers pteridophytes and gymnosperms; vol. 2 (1993), the magnoliid, hamamelid and caryophyllid families (based on the broad subclasses introduced in the systems of Takhtajan and Cronquist), and vols. 3–4, most monocotyledon families.]

LEMÉE, A., 1929–59. *Dictionnaire descriptif et synonymique des genres des plantes phanérogames*. 10 vols. (in 11). Brest: Imprimerie Commerciale et Administrative (for the author); Paris: Lechevalier (vol. 10).

Comprises an extensive but largely compiled descriptive account of all genera of seed plants, with indication of their geographical distribution. The arrangement of genera in volumes 1–7 and 8b is *alpha-betical*; volumes 9 and 10 are supplements. Volume 8a includes keys to the genera in each family.

Generic encyclopedias, pteridophytes (*Generae pteridophytarum*)

Included here is Copeland's *Genera filicum* (1947). – A successor to Copeland's now-outdated classification, at the same time offered as a scheme for a pteridophyte herbarium, is R. E. G. PICHI-SERMOLLI, 1977. Tentamen pteridophytarum genera in taxonomicum ordinem redigendi. *Webbia* 31: 313–512.

COPELAND, E. B., 1947. *Genera filicum: the genera of ferns*. xvi, 247 pp., 10 pls. (Annales cryptogamici et phytopathologici 5). Waltham, Mass.: Chronica Botanica.

A systematic descriptive treatment of the families and genera of the Filicinae, with keys, synonymy, references and citations, indication of types, geographical distribution, statistics, and taxonomic commentary; references to monographs, revisions, etc., under each entry; index. The most recent comprehensive survey of the group, but now in need of a thorough revision in the light of marked advances in fern taxonomy. [For corrections, see *idem*, 1951. Additions and corrections to the *Genera Filicum*. *Amer. Fern J.* 50: 16–21.]

001

World: chorological works

Included here are some key works dealing with transcontinental and transoceanic distributions. For convenience, Hannig and Winkler's *Pflanzenareale*, the only work devoted to world plant distribution maps, is also incorporated here under **Mappae mundi**.

General index to distribution maps

TRALAU, H. *et al.* (eds.), 1969– . *Index holmiensis: a world phytogeographic index*. Vols. 1– . Zurich: The Scientific Publishers (vols. 1–4); Stockholm: Publishing House of the Swedish Research Councils (vol. 5); Swedish Museum of Natural History, Department of Phanerogamic Botany (vols. 6–). [Index to all known published distribution maps of vascular plants, both recent and fossil, arranged alphabetically within major classes and subclasses. Twelve volumes have been projected, with references to over

250 000 maps; as of 1999 nine (the latest in 1998) had been published (covering lower vascular plants, gymnosperms, monocotyledons, and (in vols. 5–9) dicotyledons from A to P. Based in the first instance upon data accumulated at Stockholm (by Eric Hultén) and Halle/Saale (by Hermann Meusel).][16]

Mappae mundi

HANNIG, E. and H. WINKLER (eds.), 1926–40. *Die Pflanzenareale.* Vols. 1–4, 5(1–2). Maps. Jena: Fischer.

A series of world distribution maps of recent and fossil families, genera and species, with explanatory text and references giving sources. [Not continued after 1940.]

Regiones polarium

HULTÉN, E., 1964–71. *The circumpolar plants*, I–II. 228 pp., 301 maps (Kongl. Svenska Vetenskapsakad. Handl., IV, 8(5), 13(1)). Stockholm.

A two-part, large-format critical atlas of distribution maps of Arctic vascular plants, with inclusion of essential synonymy and appropriate literature references along with taxonomic commentary and notes on local ranges. The work is concluded (in part 2) with a comprehensive, geographically arranged bibliography of floristic and related works on the Arctic zone and regions to the south. [This atlas forms a valuable complement to the two principal circum-Arctic works treated in the *Guide* (see **050–70**), in which distribution is more sketchily indicated. Its coverage is markedly extended by the author's complementary works: *The amphi-Atlantic plants and their phytogeographical connection* and *Atlas of North European vascular plants north of the Tropic of Cancer* (Hultén, 1958; Hultén and Fries, 1986: for both, see below).]

Atlantica

HULTÉN, E., 1958. *The amphi-Atlantic plants and their phytogeographical connection.* 340 pp., 279 maps (Kongl. Svenska Vetenskapsakad. Handl., IV, 7(1)). Stockholm. (Reprinted 1973, Königstein/Ts., Germany: Koeltz.)

A series of annotated distribution maps (mainly of species or groups of species) distributed on both sides of the Atlantic Ocean, organized to show the ranges of presumed vicariant species and provided with critical commentary including synonymy, taxonomic notes, etc. (partly compiled from other literature). Arrangement of the species follows a phytogeographical classification. In the general part there is given an overall review of transatlantic phytogeography and its problems, in which among other matters the question of long-distance dispersal is considered. At the end (pp. 298–330) there is an extensive bibliography with a remarkable map showing the areal coverage of nearly every work cited. [Complementary to the same author's *The circumpolar plants* (see above).]

Eurasia

HULTÉN, E. and M. FRIES (eds.), 1986. *Atlas of North European vascular plants north of the Tropic of Cancer.* 3 vols. xvi, 1172 pp., 1936 distribution maps (in two colors). Königstein/Ts., Germany: Koeltz.

This primarily chorological work represents in large part a consolidation and synthesis of the senior author's three earlier atlases, beginning with *Atlas över växternas utbredning i Norden* (**670**) and including *The amphi-Atlantic plants* and *The circumpolar plants* (both described above). Here the *total* ranges of Scandinavian species are examined, along with map and text references to related taxa; altogether about 4475 taxa are covered. The maps, which fill the first two volumes, remain in the style of the earlier works (rather than adopting a grid system); however, the arrangement of taxa is systematic (after Engler) rather than phytogeographic as in the works on Atlantic and circumpolar plants. All maps are on a scale of 1:88 000 000. Volume 3 includes all commentaries, each of which includes descriptive remarks, critical taxonomic notes, and sources of information.

MEUSEL, H., E. JÄGER, S. RAUSCHERT and E. WEINERT (eds.), 1965–92. *Vergleichende Chorologie der zentral-europäischen Flora.* 3 Bde. (each in 2 vols.). Maps. Jena: Fischer.

Comprises a systematically arranged atlas of distribution maps, with explanatory text, of all vascular plants native to Central Europe but depicting in each case their worldwide range. The text entries comprise descriptive and tabular commentary covering details of distribution, ecology, and chorological classification or *Arealtypen*. The introductory section to the first volume of commentary (Bd. 1a) includes chapters on the principles and approaches to the study of plant areas and their classification, sources of data, and the floristic regions and provinces of extra-tropical Eurasia. Text and maps are in separate volumes within each *Band*, the arrangement of species following the traditional Englerian system. Bd. 3b includes a complete list of references (pp. 561–607) and a general index. [In contrast to any other distribution atlas (save for those by Hultén and, to a lesser extent, that by van Steenis and his associates and successors given below), here the habitat preferences, life-forms, floristic background, and phytogeography of each species are discussed in detail.]

Indo-Pacifica

Bibliography
The successive issues of *Pacific plant areas* include concisely annotated bibliographies, the first four (1963–84) by M. J. van Steenis-Kruseman, the fifth and last by M. M. J. van Balgooy.

STEENIS, C. G. G. J. VAN and M. M. J. VAN BALGOOY (eds.), 1963–93. *Pacific plant areas*. Vols. 1–5. Maps 1–375. Manila: National Institute of Science and Technology, Philippines (vol. 1); Leiden: Rijksherbarium (vols. 2–5). (Vol. 1 published as *Monogr. Philipp. Inst. Sci. Tech.* 8(1); vol. 2 (1966) as *Blumea, Suppl.* 5; vols. 3–5 separately published. Volumes 3–5 edited by van Balgooy alone.)

A series of annotated distribution maps of genera and some species partly or wholly occurring within the Indo-Pacific region, with special reference to the Pacific Basin. Each map (none gridded) depicts distribution by lines, dots or solid areas, or combinations of these. Associated text includes systematic position of the plant(s) concerned, synonymy, taxonomic background, notes on habit, habitat, ecology, diaspore type, and modes of dispersal, and indication of sources of information, signed by the contributor. No fixed sequence was followed.

002
World: 'old' upland and montane regions

Some significant works have been published on major individual non-alpine uplands and mountain systems, but any overall studies have been geographical and taxonomic rather than floristic. Regional works, including coverage of such systems as the Appalachians, the Guayana Highland and the Ural, appear under **n02** where **n** is any number from 1 to 9.

003
World: 'alpine' regions

No single work yet covers the floras of the world's high mountains, in particular those above the timberline – of which 'true alpines' are estimated to encompass some 10000 species (Körner, 1999, p. 13; see below). A useful introduction to *páramo* and other 'tropicalpine' plants, with much emphasis on their growth forms, is W. RAUH, 1988. *Tropische Hochgebirgspflanzen: Wuchs- und Lebensformen.* 206 pp., 39 figs., 212 photographs (part col.). Berlin: Springer. More general and functional is C. KÖRNER, 1999. *Alpine plant life: functional plant ecology of high mountain ecosystems.* ix, 338, [5] pp., illus. (part col.). Berlin: Springer; this includes, however, on pp. 13–17 a brief consideration of the origins and diversity of alpine plants. An attractive recent, rather extensive encyclopedia oriented towards gardeners and enthusiasts is K. BECKETT (ed.), 1993. *Encyclopaedia of alpines.* 2 vols. xxi, 1411, [3] pp., col. frontisp., numerous text-figs., 543 col. illus. Pershore, Worcs.: AGS [Alpine Garden Society] Publications. Regional and national literature appears under **n03** where **n** is any number from 1 to 9.

004

World: ectopotrophic areas

Increasing interest has been shown in the floras of areas characterized by 'unusual' substrates, often with nutrient limitations or potentially toxic compounds. Only for serpentines and other ultramafic formations is there, however, a good introductory survey: R. R. BROOKS, 1987. *Serpentine and its vegetation: a multidisciplinary approach.* 454 pp., illus., 8 col. pls. Portland, Oreg.: Dioscorides Press. (Ecology, phytogeography and physiology series 1.) Regional literature appears under **n04** where **n** is any number from **1** to **9**.

005

World: drylands (steppes and deserts)

Floristic documentation of drylands (steppes and deserts) has usually been national or regional; these appear under **n05** where **n** is any number from **1** to **9**. One world 'sourcebook' exists for deserts: W. G. MCGINNIES, B. J. GOLDMAN and P. PAYLORE, 1968. *Deserts of the world: an appraisal of research into their physical and biological environments.* xxviii, 788 pp., 7 maps. [Tucson, Ariz.]: University of Arizona Press. [The maps (at the beginning of the book) show the geographical extent of arid and semi-arid lands, while chapter 6 (pp. 381–566) considers vegetation and flora, with a compiled basic species list (sources on p. 473), status-of-knowledge regional reviews, and a 66-page bibliography.]

006

World: rivers

No single work covers river plants of the world; however, many tropical species are covered by van Steenis (1981; see below). *River plants* by S. M. Haslam (1978, Cambridge) is, despite its title, limited to Great Britain apart from three chapters on eastern and midwestern North America; it is also purely an ecosystem work, without a taxonomic section. A similar work is *idem*, 1987. *River plants of western Europe.* xii, 512 pp., illus. Cambridge.

STEENIS, C. G. G. J. VAN, 1981. *Rheophytes of the world.* xv, 407 pp., 47 text-figs., 23 photographs. Alphen aan den Rijn (Netherlands): Sijthoff & Noordhoff.

This work includes a systematic census, without keys, of known vascular rheophytes (riverbed and riverbank plants growing within reach of riverine flash floods); for each species are given a concise description, synonymy with references, literature citations, generalized indication of distribution and altitudinal range (including for some species, chiefly Malesian, citations of *exsiccatae*), habitat and biological notes (based on available information), and taxonomic commentary. In the general chapters preceding the census appear discussions of habitats, morphology and autecology, regional floristics and environmental factors, phytogeography, cultivation, the phenomenology of the habitat and the willow-leaf (the latter characteristic of the land-rheophytes), and their place in the author's theory of 'autonomous evolution', which latter is considered at some length in chapter 8. A systematic conspectus appears on pp. 70–72 preceding the phytogeographical treatment, while a list of references and a glossary are given at pp. 143–148. All plant names as well as concepts are indexed at the end. – For additions, see *idem*, 1987. Rheophytes of the world: supplement. *Allertonia* 4: 267–330.

007

World: lakes and their littoral

No general works are available. Some studies, however, exist for individual great lakes and lake systems; these latter are arranged under unit headings designated as **n07** (where **n** is any number from **1** to **9**).

008

World: wetlands

Our selections from the considerable regional and local floristic literature on aquatic and wetland plants are arranged under unit headings designated as **n08** (where **n** is any number from 1 to 9). Resources for the identification of wetland plants have been summarized in M. WADE, 1987. A review of the provision made for the identification of wetland macrophytes. In J. Pokorný, O. Lhotský, P. Denny and E. G. Turner (eds.), *Waterplants and wetland processes*, pp. 105–113. Stuttgart: Schweizerbart. (Ergeb. Limnol. 27.) A useful though now aging general academic introduction to aquatic plant biology remains C. D. SCULTHORPE, 1967. *The biology of aquatic vascular plants.* xviii, 610 pp., illus. London. Many more or less popular works for enthusiasts in a wide variety of languages are also available.

COOK, C. D. K., 1990. *Aquatic plant book.* [vi], 228 pp., 408 text-figs. The Hague: SPB Academic Publishing. (Reissued 1993. Originally published as C. D. K. COOK *et al.*, 1974. *Water plants of the world.* viii, 561 pp., 261 text-figs. The Hague: Junk.)

An alphabetically arranged manual for identification of the genera (and families) of herbaceous and rhizomatous fresh-water and marine tracheophytes (helophytes, hydrophytes and sea-grasses); includes analytical keys, descriptions, brief indication of distribution, approximate number of species, general ecology, representative illustrations, and references to appropriate taxonomic treatments; index to genera and families at end. The introductory section includes directions for use, explanations of terms, a 'Dahlgrenogram' indicating in phenetic terms the systematic distribution of aquatic angiosperms (with associated statistics), references (p. 7), and a general key based in the first instance upon vegetative attributes.

009

World: oceans, islands, reefs and the oceanic littoral

See also **008** (COOK). **009** here encompasses works on *marine* plants (sea-grasses, mangroves, tidal marshes and, by extension, marine algae) as well as those of atolls and cays and the oceanic strand. The two first-named have benefited from recent monographs, described below, but for tropical strand plants no general works have appeared since those of Schimper and Booberg and no similar treatise on the temperate strand has ever been published. Many works of more restricted scope, however, have appeared; selections appear under **n09** where **n** is any number from 1 to 9. For atolls and cays (as well as higher islands) their very diversity has precluded general floras, but several introductory works and 'progress papers' provide useful leads.

Sea-grasses

Active research on these plants in the last quarter-century or more necessitates a complete revision of Den Hartog's monograph.

DEN HARTOG, C., 1970. *The sea-grasses of the world.* 275 pp., 63 text-figs. (including 10 maps), 31 halftone pls. (Verh. Kon. Ned. Akad. Wetensch., II (Afd. Natuurk.) 59(1)). Amsterdam: North-Holland.

Conservatively styled monographic treatment of all marine Hydrocharitaceae and Potamogetonaceae *ss. ll.*, with keys, full synonymy including references and literature citations, generalized indication of overall range, citations of *exsiccatae* with localities, critical taxonomic commentary, and ecological notes including biology, reproduction, spread, habitat and variability. All species are illustrated. The introductory section includes maps, a key for sterile specimens, and a bibliography (pp. 36–38), and the work concludes with addenda and an index to all scientific names.

Mangroves

In addition to Tomlinson's work, described below, reference may also be made to V. J. CHAPMAN,

1976. *Mangrove vegetation*. viii, 447 pp., 298 illus. (including maps). Vaduz: Cramer/Gantner. [Pp. 380–402 of this work comprise a largely compiled 'taxonomic appendix', with species keys, synonymy, some illustrations, and literature citations but no commentary or indication of geographical distribution.]

TOMLINSON, P. B., 1986. *The botany of mangroves*. xii, 413 pp., illus. Cambridge: Cambridge University Press. (Cambridge Tropical Biology Series.) (Reissued in soft cover, 1994.)

The second (pp. 173–382) of the two parts of this book comprises an illustrated systematic treatment covering 36 families; included are family descriptions, brief characterizations of genera, keys to genera and species where appropriate, and detailed species descriptions with limited synonymy and notes (sometimes collective) on growth, vegetative morphology (and variability), reproductive biology, and taxonomy and nomenclature. References and a complete index (taxonomic entries in bold face) follow.[17]

Salt marshes

No systematic treatment on a world scale has been published. A partial introduction to the subject is, however, available in V. J. CHAPMAN, 1960. *Salt marshes and salt deserts of the world*. 392 pp., illus., col. frontisp. London: Leonard Hill. This is, however, largely oriented towards temperate marshes and especially those of northwestern Europe. Some manuals are available for individual areas, for example that of L. N. Eleuterius for the Gulf coast of North America (**109**).

Strand plants

Although strand plants have been the subject of many regional and local publications, no worldwide systematic treatment has ever been published. Those of the Indo–Pacific region (**809, 909**) are perhaps the best documented along with those from eastern North America (**109**). A successor to the classic works of Schimper and Booberg on Indomalesian strand plants – described below for convenience – is, however, very much to be desired. Some accounts are available for western Europe (**609**) and Australasia (**409**), while the *restingas* of eastern South America (**309**) have been partially documented in a series of revisions. Yet no

significant works exist for such key areas as the Mediterranean and the Caribbean among others, many currently or potentially under threat from resort development as well as other destructive changes.

SCHIMPER, A. F. W., 1891. *Die indo-malayische Strandflora*. xii, 204 pp., 7 pls., map. Jena: Fischer. (Botanische Mitteilungen aus den Tropen 3.)

Part III (pp. 100–151) of this work features a systematic checklist with distribution and, for each family, commentary with mention of additional species and local occurrences (unfortunately without citation of vouchers or references); the latter often are from outside the Indo–Pacific region. The list is followed by a general summary with emphasis on chemical properties; there is also a discussion of orders and families and their representatives in the shore flora. The remainder of the work is on physiological (functional) ecology and morphology (part I), description of formations with characteristic species (part II), dispersal (part IV) and dynamics (part V). Sources appear on pp. 1–5. For additions and corrections based on new observations and documented sources, see G. BOOBERG, 1933. Die malayische Strandflora: eine Revision der Schimperschen Artenliste. *Bot. Jahrb. Syst.* **66**: 1–38.

Islands

Most oceanic islands may be seen as being either 'low', in which sand and/or limestone are the principal components, or 'high' and then usually of continental, terrane, plate-interactive or volcanic origin. The emphasis of island botanical research, and, more recently, of conservation and 'rescue' efforts, has understandably been on the latter. Only in recent decades has it come to be realized that, collectively, the ecosystems of 'low' islands, especially atolls and emergent reef islands, are likewise distinctive and do not consist merely of 'waifs and strays'. This insight furnishes a conceptual foundation for a more thorough and integrative approach to the study of the biota of 'low' islands and island groups.

From the eighteenth and through much of the nineteenth century, islands were studied on a collective basis, using the materials and information gathered by the voyages of exploration by older and newer maritime nations and by scientific expeditions such as the *Challenger* voyage in 1873–76 and the subpolar and polar voyages of the two decades either side of 1900.

Studies of individual islands or island groups gradually succeeded this collective approach, and prevailed until World War II. The advent of an extensive air transport network, improved methodologies, a greater concern for an understanding of processes, increased support for island research, conferences, and publication outlets such as *Atoll Research Bulletin* enabled individual studies again to be placed in a broader perspective.

The physical features and biota of islands have been the subject of many general books, among them *Island life* by A. R. Wallace (1880, London), and *Island life* and *Island biology* by S. A. Carlquist (1965, New York: Natural History Press; 1974, New York: Columbia University Press). Classical botanical contributions include J. D. HOOKER, 1867. *Lecture on insular floras.* 12 pp. London (reprinted from *Gardeners' Chronicle* **1867**: 6–7, 27, 50–51, 75–76); and W. B. HEMSLEY, 1885. Report on the present state of knowledge of various insular floras. In W. Thomson and J. Murray (eds.), *Reports on the scientific results of the voyage of HMS* Challenger *during the years 1873–76, Botany,* 1 (Introd.): 1–75. London: HMSO.[18] A 'modern classic' is D. BRAMWELL (ed.), 1979. *Plants and islands.* x, 459 pp., illus. London: Academic Press. For tropical islands in general, see F. R. FOSBERG, 1979. Tropical floristic botany – concepts and status – with special attention to tropical islands. In K. Larsen and L. B. Holm-Nielsen (eds.), *Tropical botany*, pp. 89–105. London/New York: Academic Press. Useful introductions to the flora and vegetation of 'low' islands include *idem*, 1976. Phytogeography of atolls and other coral islands. In *Proceedings of the Second Coral Reef Symposium* (Brisbane, 1974), 1: 389–396; *idem*, 1976. Coral island vegetation. In O. A. Jones and R. Endean, *Biology and geology of coral reefs*, 3 (Biology, 2): 255–277. New York: Academic Press; and *idem*, 1985. Present state of knowledge of the floras and vegetation of emergent reef surfaces. In *Proceedings of the Fifth International Coral Reef Congress* (Tahiti, 1985), 5: 107–112. Moorea, French Polynesia: Antenne Muséum/École Pratique des Hautes Études (ÉPHÉ).

Superregion 01–04

Isolated oceanic islands

Within this superregion are included all those islands of the Atlantic, Indian and Pacific Oceans which can neither be grouped into a large quasi-continental series (such as the West Indies) nor be associated very readily with a nearby continent. Moreover, botanical work on these islands (as well as those of the far Southern Hemisphere) has often been carried out independently from similar work relating to continents or parts thereof – because many, if not most, of these islands are of considerable biological and biogeographic interest. In recent decades, actual and perceived threats to their ecosystems have stimulated new research, publication, scientific meetings and the establishment of conservation and management measures.

The history of botanical exploration and documentation of the islands included here is to a large extent associated with voyages of discovery, exploration, and oceanic documentation and research and/or specialized field work. Some have been the target of specific expeditions, but for the greater part an element of 'chance' has been involved. Logistics, season, and human factors have all played their part.

Nevertheless, by the late nineteenth century, most islands had become tolerably well known, especially through the efforts of the *Challenger* expedition and the use by William Botting Hemsley, an associate of J. D. Hooker at Kew, of its naturalist Henry N. Moseley's collections as a basis for a comprehensive report on oceanic island botany.[19] Of the remainder, Christmas Island was studied by a British Museum expedition at the end of the century, while several of those in the eastern Pacific were from the 1870s visited by North Americans, with the California Academy of Sciences particularly active after 1900. The Percy Sladen Trust-sponsored *Sealark* expedition in 1905 under John Stanley Gardiner pioneered extensive atoll research on the Indian Ocean islands, including those from the Laccadives to the Chagos Archipelago; and Tristan da Cunha, Gough, and Amsterdam and St. Paul were visited by some of the many Antarctic expeditions of 1890–1914 (cf. **Region 08–09**).

This phase of primary floristic exploration and writing of checklists or floras was followed by more intensive study on single islands or island groups. Examples from the first half of the nineteenth century include the work of Webb and Berthelot in the Canaries, Charles Darwin in the Galápagos and Johann A. Schmidt in the Cape Verdes and, in the second half, Edward Palmer in Guadalupe, Henry Ridley in Fernando Noronha, William Trelease in the Azores and Friedrich Johow in the Robinson Crusoe (Juan Fernández) group. In the early twentieth century, Nathaniel L. Britton was active in Bermuda (and elsewhere in the West Indies), Charles Pitard in the Canaries, and Carl Skottsberg in the Robinson Crusoes (and other islands, including the Falklands (Malvinas) and the Hawaiian archipelago).

Since World War II, detailed floristic and ecological exploration has been carried out in most individual islands and groups, and much attention has been given to evolutionary pathways and processes. The presence in several of field stations or botanical gardens has greatly facilitated such work. Revised floras and/or checklists are now available for many of the island units in the four regions accounted for here. Outstanding among them is *Flora of the Galapagos Islands* (1971) by Ira Wiggins and Duncan Porter. A *Flora of the Robinson Crusoe Islands* is also in prospect. Significant lacunae, however, remain; for several units available accounts are more or less antiquated, some more than 100 years old. Good modern treatments are, for instance, needed for Bermuda, Fernando Noronha, Trindade (with Martin Vaz), Amsterdam and St. Paul, and all of Macaronesia.

Progress

For general progress reviews, see **009**.

Bibliographies. Bay, 1910; Blake and Atwood, 1942; Frodin, 1964, 1984; Rehder, 1911; Sachet and Fosberg, 1955, 1971; USDA, 1958.

Regional bibliography

Literature to the mid-1880s is thoroughly covered by HEMSLEY (**009**).

Indices. BA, 1926– ; BotA, 1918–26; BC, 1879–1944; BS, 1940– ; CSP, 1800–1900; EB, 1959–98; FB, 1931– ; IBBT, 1963–69; ICSL, 1901–14; JBJ, 1873–1939; KR, 1971– ; NN, 1879–1943; RŽ, 1954– .

Region 01

Islands of the eastern Pacific Ocean

This region incorporates the following islands and island groups: Guadalupe, Rocas Alijos, the Revillagigedos, Clipperton, Cocos, Malpelo, the Galápagos, the Desventuradas, and the Robinson Crusoe (Juan Fernández) group.

Most eastern Pacific islands or island groups are small and scattered, with floras more or less closely related to those of the nearest continental areas. Some, such as Clipperton, are very impoverished and consist almost entirely of oceanic 'wides', while others, such as the Robinson Crusoes, have a high percentage of endemism including many 'relicts' and show more generalized biogeographical affinities. The best-known unit is perhaps now the Galápagos, serious botanical exploration of which began with the plants collected in 1835 by Charles Darwin.[20]

Post-1939 treatments are available for Guadalupe, Clipperton, Cocos (checklist only), and the Galápagos. Skottsberg's floras of the Desventuradas (1937) and the Robinson Crusoe group (1922) were supplemented in the 1950s. No treatments are available for Rocas Alijos, in any case wanting in vascular plants, and Malpelo, an island almost as barren.

Bibliographies. General bibliographies as for Superregion 01–04.

Indices. General indices as for Superregion 01–04.

011

Guadalupe Island

See also **198** (California Channel Islands); **211** (Baja California). – Area: 250 km². Vascular flora: 156 native species and additional infraspecific taxa, of which 34 endemic (at least 5 now extinct); at least 45 species introduced (Moran, 1996).

Though of volcanic origin and in part relatively

high (reaching 1295 m), this island is tectonically and biotically related to the California Channel Islands; some 65 percent of its species are also found therein (the strongest connection being with Santa Catalina). The land biota has suffered greatly from an overabundance of cats and especially goats, already long present when the first significant collections were made by Edward Palmer in 1875; it was upon these that the first florula, *On the flora of Guadalupe Island, Lower California* (1876) by Sereno Watson, was based. This was succeeded by *Studies in the flora of Lower California and adjacent islands: list of the plants originally recorded from Guadalupe Island, Mexico* by Alice Eastwood (1929; in *Proc. Calif. Acad. Sci.*, IV, **18**: 394–420, pls. 33–34). Eastwood's work remained current until the publication of Reid Moran's illustrated *Flora of Guadalupe Island*, also published by the California Academy of Sciences.

Bibliography
See **Region 21/22** (JONES; LANGMAN).

MORAN, R., 1996. *The flora of Guadalupe Island, Mexico.* viii, 190 pp., [1] p. pls. (col. frontisp.), 79 text-figs. (incl. 2 maps) (Mem. Calif. Acad. Sci. 19). San Francisco.

Photographically illustrated enumeration of vascular plants; entries include synonymy, overall distribution, descriptions, and more or less extensive taxonomic, ecological and biological commentary including reports of observations by the author (over nearly five decades) and his predecessors and contemporaries; some *exsiccatae* are cited. The general part comprises an extended account of physical features, climate, flora and vegetation, extinct and nearly extinct plants, biogeography, and (pp. 44–52) collectors; following the enumeration are a bibliography and index to all botanical names.

012

Rocas Alijos

No vascular plants, so far as known, have been recorded from this very low-lying group of barren rocks, which lie about halfway between Guadalupe and the Revillagigedo group (see below).

013

Revillagigedo Islands

Area: Socorro, 210 km^2; no information seen on San Benedicto, Clarión or Roca Partida (all smaller than Socorro, the last being, in the words of Levin and Moran (1989; see below) 'a mere pinnacle without vascular plants'. Vascular flora: 117 native (30 endemic and 9 subendemic) and 47 naturalized species on Socorro; 41 on Clarión; 6 on San Benedicto (Levin and Moran, 1989).

Prior to the work of Herbert Mason and Ivan Johnston in the 1920s, the main basis for the latter's 1931 paper, reports had been published by Vasey and Rose in 1889 and Brandegee in 1900. Additional background on Socorro may be found in J. ADEM *et al.*, 1960. *La isla Socorro, archipelago de la Revillagigedo.* 234, [1] pp., illus. (Mongr. Univ. Nac. Auton. Méx., Inst. Geofis. 2). Mexico, D.F. [Chapter 6 (pp. 129–152) of this work comprises an account (by Faustino Miranda) of the vegetation of the island, along with additions to the flora (considered relatively poor).]

Bibliography
See **Region 21/22** (JONES; LANGMAN).

JOHNSTON, I. M., 1931. The flora of the Revillagigedo Islands. *Proc. Calif. Acad. Sci.*, IV, **20**: 9–104.

Systematic enumeration of vascular plants, with descriptions of new taxa, synonymy (with references and pertinent citations), indication of localities with *exsiccatae*, general summary of extralimital range (if applicable), critical commentary, and extensive notes on life-form, habitat, occurrence, etc. The plant list is preceded by accounts of the geography and physical features of the several islands (with tabular summaries of their floras), the origin and general features of the flora as a whole, plant geographical relationships, and past botanical exploration.

LEVIN, G. A. and R. MORAN, 1989. *The vascular flora of Isla Socorro, Mexico.* 71 pp., 18 text-figs. and halftones (incl. map) (Mem. San Diego Soc. Nat. Hist. 16). San Diego.

Annotated descriptive flora, with keys, relevant synonymy, indication of types, local distribution (with *exsiccatae* where appropriate, especially for new or less common taxa), often extensive critical commentary, and brief indication of habitat, frequency, etc.; acknowledgments, excluded species and (pp. 67–69) references along with appendices on the floras of Clarión and San Benedicto Islands (based on Johnston's data, with corrections). The introductory part covers previous work, the basis for the present treatise, physical features and climate, vegetation, human and other biotic impacts, biogeography (including an analysis of the flora and its dispersal mechanisms and dynamics), and probable evolutionary history.

014

Clipperton Island

Area: 5 km². Vascular flora: 31 angiosperm species (Sachet, 1962; see below). – The island, a French possession without permanent human habitation, is an atoll: the only one in the eastern Pacific. For long the flora was not thought worth attention, and prior to the mid-twentieth century few if any studies were available. For a relatively recent general study, see M.-H. SACHET, 1962. Geography and land ecology of Clipperton Island. 115 pp. (Atoll Res. Bull. 86). Washington.

SACHET, M.-H., 1962. Flora and vegetation of Clipperton Island. *Proc. Calif. Acad. Sci.*, IV, **31**: 249–307, 12 halftones, map.

Includes a systematic enumeration of native and introduced non-vascular and vascular plants, with localities, citations of *exsiccatae* and general summary of extralimital range, along with possible methods of inward migration and notes on habitat, status, biology, etc. Other portions of the work deal with prior botanical investigation, the history and present condition of the vegetation, plant geographical relationships, dispersal mechanisms of the individual species, and drift seeds collected.

015

Cocos Island

See also **236** (STANDLEY). – Area: 24 km². Vascular flora: 155 native and introduced species, with 10 percent endemism (Fournier, 1966).

This 'high' island, of volcanic origin and presently a Costa Rican possession, is distinguished by its high rainfall (about 7000 mm/annum) and, in some parts, almost continuous cloud cover, which no doubt accounts for a reportedly high percentage of pteridophytes. These have been separately documented by Luis D. Gómez, but the rest of the vascular flora seems still to be imperfectly known and too often has not been considered in its own right. For relationships of the biota, see L. G. HERTLEIN, 1963. Contribution to the biogeography of Cocos Island, including a bibliography. *Proc. Calif. Acad. Sci.*, IV, **32**: 219–289.

FOSBERG, F. R. and W. L. KLAWE, 1966. Preliminary list of plants from Cocos Island. In R. I. Bowman (ed.), *The Galapagos: Proceedings of the Galapagos International Scientific Project*, pp. 187–189. Berkeley, Calif.: University of California Press. Complemented by L. A. FOURNIER, 1966. Botany of Cocos Island, Costa Rica. *Ibid.*, pp. 183–186.

Fosberg and Klawe's paper is a compiled checklist of all known non-vascular (except algae) and vascular plants, with essential synonymy. Fournier's paper comprises a general description of main features of the island and its vegetation, an account of the origin of the flora, and notes on visiting scientific expeditions, along with a list of references.

STEWART, A., 1912. Notes on the botany of Cocos Island. *Proc. Calif. Acad. Sci.*, IV, **1**: 375–404, pls. 31–34.

Systematic enumeration of mosses and vascular plants, based mainly on collections made by the author; includes synonymy (with references), localities and citations of *exsiccatae*, general summary of extralimital range, critical commentary, and notes on habitat, special features, etc. The plant list is preceded by accounts of the physical features, vegetation, and plant-geographical relationships of the island and its flora. [For additions, see H. K. SVENSON, 1935. Plants of the Astor expedition, 1930 (Galapagos and Cocos Islands). *Amer. J. Bot.* **22**: 208–277, 9 illus.]

Pteridophytes
These were earlier treated by Svenson (1938; see remarks under **017**).

GÓMEZ P., L. D., 1975. Contribuciones a la pteridología costarricense, VII. Pteridófitos de la Isla de Cocos. *Brenesia* **6**: 33–48.

Briefly descriptive pteridoflorula (60 species), with synonymy, references, citations of *exsiccatae* with locations, and notes on habitat, frequency and (sometimes) associates. A floristic analysis and general notes on vegetation associations conclude the work. [For English summary, see *idem*, 1975. The ferns and fern-allies of Cocos Island, Costa Rica. *Amer. Fern J.* **65**: 102–104.]

016

Malpelo Island

No vascular plants seem as yet to have been recorded for this mile-long, steep and nearly barren island, although 'scrub' has been reported. For a general description of this Colombian possession, see NAVAL INTELLIGENCE DIVISION, UNITED KINGDOM, 1943. *Pacific Islands*, 2 (Eastern Pacific): 21. [London.]

017

Galápagos Islands

Area: 7844 km². Vascular flora: 736 species and non-nominate infraspecific taxa, with 541 native (Porter, 1990; see Note 22).

The high scientific interest of the Galápagos Islands is manifest in the several floristic accounts published since the *Beagle* visit in 1835. The current definitive *Flora of the Galápagos Islands* succeeded, in turn, *Plants of the Galapagos Archipelago* (1845–46) and *An enumeration of the plants of the Galapagos Archipelago* (1851; both in *Transactions of the Linnean Society*) by J. D. Hooker, based in part on a partial set of Darwin's collections; *Enumeratio plantarum in insulis Galapagensibus hucusque observatorum* (1861) by Nils Andersson, a result of the Swedish 'Eugénie' expedition of the early 1850s; *Flora of the Galapagos Islands* (1902) by the asterologist Benjamin L. Robinson of Harvard University; and *A botanical survey of the Galapagos Islands* (1912) by Alban Stewart, embodying the results of a 1905–06 California Academy of Sciences expedition (a visitor also to Cocos).[21] Phanerogams were further revised in *Plants of the Astor expedition* by Henry K. Svenson (1935; in *Amer. J. Bot.* **22**); the pteridophytes shortly afterwards were fully revised by Svenson in his *Pteridophyta of the Galapagos*

and Cocos Islands (1938; in *Bull. Torrey Bot. Club* **65**). Both were based in the first instance on field work in 1930. All these were succeeded by the definitive *Flora* in 1971. However, continuing taxonomic studies, along with an increasing interest in the biology and ecology of the plants, the needs of visitors, continuing introductions of sometimes invasive biota, and problems of conservation and management, have made new works necessary. Towards this end, a new checklist appeared in 1987 and a field-manual in 1999.[22]

Bibliography

SCHOFIELD, E. K., 1973. Annotated bibliography of Galápagos botany, 1836–1971. *Ann. Missouri Bot. Gard.* **60**: 461–477. [286 references, arranged by author within major divisions of the plant kingdom.]

LAWESSON, J. E., J. ADSERSEN and P. BENTLEY, 1987. *An updated and annotated check list of the vascular plants of the Galápagos Islands*. 74 pp. (Rep. Bot. Inst. Aarhus Univ. 16). Risskov.

Tabular checklist, showing origin, status, occurrence by island, and references; includes an introduction (in Spanish and English), a list of nomenclatural changes since 1971 and a key to references (pp. 17–21).

McMULLEN, C. K., 1999. *Flowering plants of the Galápagos*. 384 pp., 41 text-figs., 383 col. photographs, map. Ithaca, N.Y.: Cornell University Press. (A Comstock Book.)

Illustrated semi-popular field-guide covering 436 species and infraspecific taxa; includes keys, descriptions, indication of distribution and habitat, and commentary including any special features.

WIGGINS, I. L. and D. M. PORTER, 1971. *Flora of the Galápagos Islands*. xx, 998 pp., 2768 text-figs., 16 col. pls., maps. Stanford: Stanford University Press.

Illustrated descriptive flora of native, naturalized and adventive vascular plants (702 species and non-nominate subspecies); includes keys to all taxa, full synonymy (with references), detailed indication of local range, citation of some *exsiccatae* and inclusion of many maps, general summary of extralimital distribution (if applicable), critical taxonomic commentary, and notes on variability, habitat, life-form, ecology, etc. The introductory section includes accounts of the geography, physical features, geology, climate, soils, vegetation, and animal life of the archipelago, together with historical reviews of botanical exploration and human influences. A glossary, bibliography and index to all botanical names conclude the work.[23]

018

Desventuradas Islands

See also **390** (MARTICORENA and QUESADA; MUÑOZ PIZARRO). – Area: 5 km². Vascular flora: 19 phanerogam species (Skottsberg, 1937).

These small, rather dry islands of Mediterranean climate, though credited with 12 species in 1875 by the Chilean botanist Federico Philippi (in *Anales Univ. Chile* **47**(1)) and – independently – with eight by Hemsley in the botanical reports of the *Challenger* expedition (1(3): 97–100; for full reference see **019**), remained botanically imperfectly known until the twentieth century. The only full summary remains that of Skottsberg (1937) although lately Hoffmann and Teillier (1991) have added considerably to our knowledge of San Félix. The islands have been notable for their difficulty of access as well as for a strongly seasonal climate, and while much has been accomplished it remains possible that the native flora, although doubtless limited, has yet entirely to be accounted for.[24]

Bibliography
See **Region 39** (MARTICORENA).

SKOTTSBERG, C. J. F., 1937. *Die Flora der Desventuradas-Inseln (San Felix und San Ambrosio)*. 87 pp., 46 text-figs. (incl. map) (Göteborgs Kungl. Vetensk. Vitterh. Samhälles Handl., V/B (Mat. Naturvetensk. Skr.), 5(6)). Göteborg. Spanish edn.: *idem*, 1949. *Flora de las islas San Félix y San Ambrosio* (transl. A. Horst). 64 pp., 39 text-figs. (incl. map) (Bol. Mus. Nac. Hist. Nat. (Santiago) 24). Santiago.

Critical illustrated descriptive flora, without keys; includes synonymy, references and citations, detailed indication of local occurrences with citations of *exsiccatae* or other pertinent sources, general indication of overall distribution (if applicable), extensive taxonomic commentary, and occasional notes on habitat, biology, etc.; general summary of the flora and its relationships. All species are illustrated. An introductory section gives a general description of the islands and their topography and climate together with an account of botanical exploration. [For additions, revised summaries and further references covering both islands, see *idem*, 1949. Eine kleine Pflanzen-

Sammlung von San Ambrosio (Islas Desventuradas, Chile). *Acta Horti Gothob.* **17**: 49–57; *idem*, 1950. Weitere Beiträge zur Flora der Insel San Ambrosio (Islas Desventuradas, Chile). *Ark. Bot.*, II, **1**: 453–469, illus.; *idem*, 1963. Zur Naturgeschichte der Insel San Ambrosio (Islas Desventuradas, Chile). *Ibid.*, **4**: 465–488; and A. J. HOFFANN and S. TEILLIER, 1991. La flora de la isla de San Félix (Archipiélago de la Desventuradas, Chile). *Gayana, Bot.* **48**: 89–99, 3 text-figs.][25]

019

Robinson Crusoe (Juan Fernández) Islands

See also **390** (MARTICORENA and QUESADA; MUÑOZ PIZARRO; RODRÍGUEZ, QUESADA and MATTHEI). The islands were also covered in detail by Hemsley in the *Challenger* expedition reports. – Area: 93 km². Vascular flora: 143 species including 54 pteridophytes, with 118 endemic taxa (PD; see **General references**).

The three Robinson Crusoes, home to a significant endemic plant family, the Lactoridaceae, and several endemic woody Compositae, were first collected only in 1823, with Alexander Selkirk (Masafuera) hardly visited even by 1884. The first account was *Report on the botany of Juan Fernandez* by Hemsley (1884, in *Report on the scientific results of the voyage of HMS* Challenger, *Botany*, 1(3): 1–96). This was followed by *Estudios sobre la flora de las islas de Juan Fernández* (1896) by Frederico Johow, covering non-vascular as well as vascular plants and based in the first instance on his own field work, and then by Skottsberg's contributions. Günther Kunkel revised the pteridophytes in 1965. In recent decades the evolution, conservation and management of the flora have been of particular concern. Six expeditions have been in particular made by teams from Concepción University in Chile and Ohio State University in the United States, and in 1993 their investigators announced plans for a new manual.[26]

Bibliography
See **Region 39** (MARTICORENA).

SKOTTSBERG, C. J. F., 1922 (1921). The phanerogams of the Juan Fernandez Islands. In *idem* (ed.), *Natural history of Juan Fernandez and Easter Islands*, 2 (Botany): 95–240, 39 text-figs., pls. 10–20 (1 col.). Uppsala: Almqvist & Wiksell. Complemented by C. CHRISTENSEN and C. J. F. SKOTTSBERG, 1920. The Pteridophyta of the Juan Fernandez Islands. *Ibid.*: 1–46, 7 text-figs., pls. 1–5. Uppsala; and C. J. F. SKOTTSBERG, 1951. A supplement to the pteridophytes and phanerogams of Juan Fernandez and Easter Island. *Ibid.*: 763–792, pls. 55–57. Uppsala.

The two initial contributions respectively constitute critical systematic enumerations of native and introduced phanerogams and pteridophytes, with a limited number of keys and including full synonymy, references and citations, fairly detailed indication of local distribution (with citations of *exsiccatae* and literature sources), general summary of extralimital range (where applicable), taxonomic commentary, and notes on habitat, biology, occurrence, etc., as well as discussions on the general features of the phanerogam and pteridophyte floras and their indigenous and introduced components. Bibliographies and indices to botanical names are also included. The supplement is in the same style. [The three contributions form part of a comprehensive monograph of the group, together with Easter Island (988).]

Pteridophytes

KUNKEL, G., 1965. Catalogue of the pteridophytes of the Juan Fernandez Islands (Chile). *Nova Hedwigia* 9: 245–284.

Nomenclatural checklist, with complete synonymy, references and all relevant citations; also includes geographical, habitat and frequency data and some critical but mostly compiled commentary.[27]

relatively closely associated with nearby continents or island systems. – Vascular flora: 3106 native and naturalized species (Eriksson *et al.*, 1993).

Macaronesia, initially examined by Georg Forster in the second section of his *De plantis magellanicis et atlanticis commentationes* (1787), was first so named by Philip Barker Webb in 1845; this was a consequence of its recognition in the 1820s as a distinct botanical–geographical unit by the Danish plant geographer Heinrich Schouw.[28] It has since been fully covered by floras and checklists of individual islands and groups, but only in 1974 did a modern inclusive account appear: *Flora of Macaronesia: checklist of vascular plants*, by O. Eriksson, Alfred Hansen and Per Sunding. The work, by 1993 three times revised, is a consolidation of earlier lists for the individual groups by the three authors and, along with associated bibliographies, effectively serves to bring together a great deal of scattered information. It also reflects the considerable amount of new floristic work accomplished, particularly in the Canaries (where there are now at least four herbaria) and the Cape Verdes (especially through the work of Wilhelm Lobin and collaborators). The checklist and bibliographies also represent steps towards the realization of a descriptive *Flora Macaronesica*, plans for which were announced in the United Kingdom more than two decades ago.[29]

Bibliographies. General bibliographies as for Superregion 01–04.

Regional bibliography

SCHÜTZ, J. F., 1929. *Baustein zu einer Bibliographie der Canarischen, Madeirischen un Capverdischen Inseln und der Azoren*. 144 pp. Graz.

Indices. General indices as for Superregion 01–04.

Region 02

Macaronesia

Included herein are all the islands of Macaronesia, i.e., the Azores, Madeira, the Salvages, the Canaries, and the Cape Verdes, which lie in the eastern part of the North Atlantic. Other Atlantic Ocean islands are grouped as Region 03, save for those

020

Region in general

The checklist described below, to date the only result of the 'Flora of Macaronesia' project, is the sole modern overall account for the five groups of islands.

ERIKSSON, O., A. HANSEN and P. SUNDING, 1993. *Flora of Macaronesia: checklist of vascular plants*.

4th edn., by A. Hansen and P. Sunding. 295 pp. (Sommerfeltia 17). Oslo: Botanical Garden and Museum, University of Oslo. (1st edn., 1974, Umeå (Sweden); 2nd and 3rd edns., 1979 and 1985, Oslo.)

Annotated tabular geographical checklist of vascular plants, arranged alphabetically within the major classes; against the accepted name of each species is a notation of its distribution by archipelago and island (or island group) and an indication of endemicity in the region (if applicable). The checklist proper is followed by a synonym lexicon (pp. 232–282, with 2327 names) and an index to accepted genera (pp. 284–295). [The original edition represented a consolidation of earlier lists for each of the Macaronesian archipelagoes by the various authors. The smaller format in the current version makes it more convenient for field use; unfortunately, a bibliography is wanting.]

021

Azores

See also **600** (TUTIN *et al.*); **611** (FRANCO). – Area: 2235 km². Vascular flora: about 600 native species, of which 55 are endemic (PD; see **General references**). The archipelago encompasses nine islands, two of them very small and with scanty floras.

The best previous accounts are *Flora azorica* (1844) by Moritz Seubert, based on collections by C. Hochstetter in 1838, *Botany of the Azores* by Hewett C. Watson (1870, in *Natural history of the Azores* by F. du C. Godman) and *Botanical observations on the Azores* by William Trelease (1897, in *Annual Rep. Missouri Bot. Gard.* 8). These along with Palhinha's catalogue are, however, all based on mutually more or less differing sources of data.

Bibliography
HANSEN, A., 1970. *A botanical bibliography of the Azores.* 9 pp. Copenhagen: [Botanical Museum]. Continued as *idem*, 1975. *A botanical bibliography of the Azores: additions 1975.* 6 pp. Copenhagen. [Both parts reissued as *idem*, 1976. A botanical bibliography of the Azores. *Bol. Mus. Munic. Funchal* **30**: 46–56.]

FERNANDES, A. and R. B. FERNANDES (eds.), 1980–87. *Iconographia selecta florae Azoricae.* Vols.

1(1–2), 2(1). Pls. 1–83. Coimbra, Portugal: Secretary for Culture, Azorean Autonomous Region.

A cultural/documentary work, comprising illustrations of selected Azorean species with detailed descriptive text including full synonymy, references and annotations. [Only pteridophytes, gymnosperms and dicotyledons from Salicaceae through Phytolaccaceae (on the Englerian system) have been accounted for.]

PALHINHA, R. T., 1966. *Catálogo das plantas vasculares dos Açores* (revised and edited by A. R. Pinto da Silva). xi, 186 pp., portr., maps (end-papers). Lisbon: Sociedade de Estudos Açorianos 'Alfonso Chaves'.

Systematic enumeration of native, naturalized and commonly cultivated vascular plants, with synonymy, references and citations, vernacular names, more or less copious citations of *exsiccatae* (mostly from Portuguese herbaria), general indication of local distribution (with some particular localities), summary note of extralimital range, and brief notes on habitat, occurrence, etc.; indices to botanical and to vernacular names. The introductory section includes a list of references. [For additions to the flora (many of them new introductions), see E. SJÖGREN, 1973. Vascular plants new to the Azores and to individual islands in the Archipelago. *Bol. Mus. Munic. Funchal* **27**: 94–120. Pteridophytes were earlier revised as R. T. PALHINHA, 1943. Pteridófitos do Arquipélago dos Açores. *Bol. Soc. Brot.*, II, **17**: 215–249.]

022

Madeira Islands

Area: 796 km². Vascular flora: 1163 species (Press and Short, 1994), of which 131 endemic (PD; see **General references**); introduced species (some naturalized) about 380 (PD). – The group in a strict sense comprises Madeira, Porto Santo and the Desertas.

Significant collections began to be made from the mid-eighteenth century and the first accounts, based on forays made respectively in 1823 and 1827, were furnished by Sarah Bowditch (1825) and Friedrich Holl (1830, in *Flora* 13). The first effective, if somewhat rambling, account was *Manual flora of Madeira* by the Rev. Richard T. Lowe, resident as English chaplain in Madeira from 1832 to 1854. Owing to the author's death by drowning in 1874 following a shipwreck, it was left unfinished and for the next 120 years the only complete treatments of vascular plants were the 1914 enumeration, *Flora de archipelago da Madeira (phanero-gamicas e cryptogamicas vasculares)* by Carlos Azevedo de Menezes and its supplements of 1923 and 1926, the checklist of 1969 by A. Hansen, and the provisional manual of 1970, *Oversigt over Madeiras flora* by T. B. Christensen *et al.* All these have now been succeeded by a descriptive flora compiled at the Natural History Museum in London under the direction of J. R. Press and M. J. Short (1994; see below).

Bibliography

HANSEN, A., 1975. *A botanical bibliography of the Madeira Archipelago.* 31 pp. Copenhagen: [Botanical Museum]. (Mimeographed.) Reissued as *idem*, 1976. A botanical bibliography of the archipelago of Madeira. *Bol. Mus. Munic. Funchal* 30: 26–45. [523 titles.]

HANSEN, A., 1969. Checklist of the vascular plants of the archipelago of Madeira. *Bol. Mus. Munic. Funchal* 24: 1–62, 2 maps.

Systematic list of native, naturalized and commonly cultivated vascular plants with synonymy (relating to Macaronesia); references and index to families at end. [For additions, see the author's papers of 1970–74 in *Bocagiana* 25, 27, 32 and 36.]

PRESS, J. R. and M. J. SHORT, 1994. *Flora of Madeira.* xvii, 574 pp., 9 text-figs., 57 pls. London: HMSO.

Concise manual-flora of vascular plants with keys to all taxa, accepted names, places of publication, vernacular names where known, descriptions, phenology, karyotypes, distribution, habitat and altitudinal range, biogeographical status, and distribution code; brief descriptions for families and genera with citations of key literature; illustrations and indices to scientific and vernacular (Portuguese) names at end. The general part includes maps plus sections on geography and geology, climate, vegetation and floristics as well as on the organization of the manual and (pp. 13–28) a key to families. Covers the Madeira group proper and the Salvages (**023**).

Pteridophytes

ROMARIZ, C., 1953. Flora da Ilha da Madeira: Pteridófitas. *Rev. Fac. Ci. Univ. Lisboa*, sér. 2, C (Ci. Nat.) 3(1): 53–112, 7 pls., 4 figs.

Descriptive treatment, with keys, geographical range,

indication of local occurrences, and vernacular names (Portuguese and English). The systematic part is preceded by sections on sources, background, regional and local geography, and vegetation, along with an analysis and synthesis of the pteridophyte flora.

023

Salvage Islands

Area: <3 km^2 (283.2 ha). Vascular flora: 105 species, of which 11 are endemic (Press and Short, 1994; see **022**). – The Salvages are a group of three small, relatively low islands between Madeira (of which they administratively form part) and the Canaries. In addition to the works below, the islands earlier were covered by R. T. Lowe in *Florulae salvagicae tentamen* (1869), a work commemorated by Monod.

MONOD, T., 1990. *Conspectus florae salvagicae.* 79 pp., 142 figs., col. frontisp. (Bol. Mus. Munic. Funchal, Suppl. 1). Funchal.

Alphabetically arranged enumeration of vascular plants with references, synonymy, *exsiccatae* with localities, Macaronesian distribution, and commentaries with key taxonomic literature; annotated bibliography and illustrations at end. The work is preceded by a description of the islands, a review of collectors and some notes on the flora and its biogeography.[30]

PÉREZ DE PAZ, P. L. and J. R. ACEBES GINOVÉS, 1978. Las Islas Salvajes: contribución al conocimiento de su flora y vegetación. In Museo de Ciencias Naturales del Cabildo Insular de Santa Cruz de Tenerife, *Contribución al estudio de la historia natural de las Islas Salvajes: resultados de la expedición científica "Agamenón 76"*, pp. 79–105, 3 text-figs., 1 pl. Santa Cruz de Tenerife: Aula de Cultura de Tenerife.

Annotated enumeration (pp. 93ff.) with indication of occurrences; preceded by accounts of topography, climate, general features of the flora, plant geography, and the vegetation and its zonation. References at end (pp. 104–105). [Among other chapters in this expedition report are treatments of the algal and lichen floras.]

PICKERING, C. H. C. and A. HANSEN, 1969. List of higher plants and cryptogams known from the Salvage Islands. *Bol. Mus. Munic. Funchal* 24: 63–71, map.

Systematic list of non-vascular and vascular plants, with limited synonymy, indication of earlier records, and brief statement of local range; short list of references.

024

Canary Islands

Area: 7273 km^2. Vascular flora: 2000 native and introduced species, including more than 500 endemic species and infraspecific taxa (PD; see **General references**). – The Canaries, with seven main islands, are the largest and highest of the Macaronesian groups, reaching 3718 m at Pico de Teide on Tenerife, and have the richest and most distinctive flora in all the islands.

The first separate account of the flora is *Allgemeine Uebersicht der Flora auf den Canarischen Inseln* by the geologist Leopold von Buch (1819, in *Abh. Königl. Akad. Wiss. Berlin* (1816–17)), but this, and its more complete successor in the author's *Physicalische Beschreibung der Canarischen Inseln* (1825), have been overshadowed by the *Phytographia canariensis* (1835–50) of Philip Barker Webb and Sabin Berthelot, the second part of the third 'tome' of their *Histoire naturelle des îles Canaries*. This renowned work, with no modern successors (except for the enumeration of Pitard and Proust, the accounts by Heinrich Schenck (1907), Herman Knoche (1923) and Leonhard Lindinger (1926; see below), and scattered post-World War II contributions including *Floristic botany of the Canary Islands* by Kornelius Lems), remains of great value, although their relatively narrow generic concepts were not accepted by some contemporaries. In 1993 a first field-manual, by A. Hohenester and W. Weiss (see below), was published. A database on the flora has been under development at the Jardín Botánico 'Viera y Clavijo' in Las Palmas de Gran Canaria.[31]

A well-written and well-illustrated popular introduction to the flora, with special attention to the many endemic species, is D. BRAMWELL and Z. BRAMWELL, 1974. *Wild flowers of the Canary Islands*. 261 pp., 118 text-figs., 205 col. illus., 16 pls. London: Stanley Thornes. (Spanish edn., 1990, as *Flores silvestres de la Islas Canarias*. 376 pp., illus. Madrid.)[32] Also of value is EDIRCA, *Flora y Vegetación del Archipelago Canario* (1986–92, as part of the larger *Gran Biblioteca Canarina* of islands publisher Edirca); this incorporates a well-illustrated introduction to families and genera comparable to that of Marloth for South Africa (**510**).

Bibliography

SUNDING, P., 1973. *A botanical bibliography of the Canary Islands*. 2nd edn. 46 pp. Oslo: Oslo [University] Botanic Garden.

ERIKSSON, O., 1971. *Check-list of vascular plants of the Canary Islands*. [i], 35 pp. Umeå, Sweden: Section of Ecological Botany, University of Umeå.

Systematic list, with essential synonymy and tabular summary of local range. Additions subsequent to Lems's *Floristic botany* (see below) are especially noted.

HOHENESTER, A. and W. WEISS, 1993. *Exkursionsflora für die Kanarischen Inseln*. 374 pp., 438 text-figs., 96 col. photographs. Stuttgart: Ulmer.

Concise field manual-key to vascular plants, the leads inclusive of diagnoses, limited synonymy and some vernacular names, status, overall distribution, habitat, phytosociological categories (where understood), and a row of boxes indicative of presence or absence in each of the seven main islands; indices to all names at end. The introductory part encompasses in brief geography, climate, vegetation, communities with a summary of their phytosociological classification, references (pp. 23–24), directions for use, abbreviations and conventions, and a brief glossary. [Selected plants from other Macaronesian islands are also included.]

LEMS, K., 1960. *Floristic botany of the Canary Islands*. 94 pp. (Sarracenia 5). Montreal: University of Montreal. (Mimeographed.)

Alphabetical enumeration of vascular plants (within their major classes); incorporates limited synonymy, citations of the author's *exsiccatae*, some taxonomic commentary, notes on habitat and biology, and symbolic indication of distribution, life-form, etc.; addenda and references at end. The general part includes a list of abbreviations, an analysis of the flora and its origin, a list of abbreviations, and statistical tables. [Provisional in nature and incomplete, the work is now of most use for its physiognomic, ecological and biological notes.]

PITARD, J. and L. PROUST, 1908. *Les Canaries: flore de l'archipel*. 502 pp., 19 pls. Paris: Klincksieck. (Reprinted 1972, Königstein/Ts., Germany: Koeltz.)

Systematic enumeration of native and naturalized bryophytes and vascular plants, with descriptions
of new taxa; entries include limited synonymy (with
references and citations of major source works), fairly
detailed summary of local range (but without *exsiccatae*) and, where applicable, extralimital distribution and
notes on habitat, phenology, etc. The introductory
section contains accounts of prior botanical exploration, physical geography, climate, soils, vegetation
zones and formations, characteristic features of the
flora and its probable history and geographic relationships (with statistical tables), and endemic taxa. An
index to all botanical names concludes the work. [For
substantial additions to the flora, see L. LINDINGER,
1926. Flora der Kanarischen Inseln. Berichtigungen
und Nachträge zu Pitard und Proust . . . . *In idem,
Beiträge zur Kenntnis von Vegetation und Flora der
Kanarischen Inseln*, pp. 135–350 (Abh. Auslandsk.
(Hamburg) 21 (Reihe C, Naturwiss. 8)). Hamburg.]

WEBB, P. B. and S. BERTHELOT, 1836–50.
Phytographia canariensis. In *idem, Histoire naturelle des
îles Canaries*, tome III, 2. ptie. 4 sects. in 6 vols. 288 pls.
(part col.). Paris: Béthune.

Finely executed large-scale systematically
arranged descriptive atlas-flora of non-vascular and
vascular plants, with synonymy, vernacular names, and
indication of local and extralimital range; indices. [Of
the four sections, the last, by C. Montagne, is devoted
to the non-vascular plants.]

Partial work: Gran Canaria

KUNKEL, M. A. and G. KUNKEL, 1974–79. *Flora de
Gran Canaria*. Vols. 1–4. Pls. 1–200. Las Palmas: Ediciones
del Excmo. Cabildo Insular de Gran Canaria.

Large-scale color-plate atlas of vascular plants, with
detailed descriptive text (including references, vernacular
names, distribution, ecology, phenology, nature of propa-
gules, variability, and source of the material together with
plant description) facing each plate; introductory matter and
indices in each volume. The species selected are allocated to
volumes organized by synusiae (woody plants, lianas, succulents, ferns, etc.) rather than in any systematic order, with 50
in each volume. [Valuable for its illustrations, but not a flora.
Eleven volumes were originally projected, but only four actually published.]

Woody plants (including trees): guide

An introductory work is G. KUNKEL, 1981. *Arboles y arbustos
de las Islas Gran Canarias: guia de campo*. 138 pp., illus. Las
Palmas: Edirca.

025

Cape Verde Islands

Area: 4033 km². Vascular flora: about 659 native
and introduced species, of which 92 are endemic (PD;
see **General references**). – Formerly also known as
the Gorgonas and with their highest peak reaching
2829 m, these 11 islands are now an independent state.
Being at Sahelian latitudes, they are relatively arid and
furthermore have suffered greatly from vegetation
destruction. The Macaronesian (and Mediterranean)
relationship is evident mostly in the mountain flora.
Geologically, the flora is but 'a few hundred thousand
years old' (Brochmann *et al.*, 1997); the endemics are
mostly 'neoendemics', only slightly distinct from their
sister taxa (mainly in northwest Africa).

Several collections, beginning with the Forsters
in 1773, were made by visiting naturalists – Charles
Darwin among them – in the late eighteenth and early
nineteenth centuries, but the first separate flora was
Spicilegia gorgonea by Philip B. Webb (1849, in *Niger
Flora* by W. J. Hooker), covering 293 taxa. This was
soon followed by *Beiträge zur Flora der Cap Verdischen
Inseln* (1852) by an Austrian naturalist, Johan A.
Schmidt. Covering all known native and introduced
plants and accounting for all previous publications, this
latter encompassed 435 vascular species. For more than
80 years it remained standard, being succeeded only in
the mid-twentieth century by the works of August
Chevalier and, after World War II, Per Sunding.

Since independence in 1975 much new collecting has been accomplished, with many new records
and range extensions; several botanical contributions
have been published, particularly by Werner Lobin,
and a series of symposia on all aspects of Cape Verdean
biodiversity organized through the Forschungsinstitut
und Naturmuseum Senckenberg in Frankfurt-am-
Main (Germany). In 1987, a flora project was organized under an agreement between Cape Verde and
Portugal; publication of *Flora de Cabo Verde* commenced in 1995. A detailed treatment of the 82
endemic species and subspecies has also appeared: C.
BROCHMANN, Ø. H. RUSTAN, W. LOBIN and N.
KILIAN, 1997. *The endemic vascular plants of the Cape
Verde Islands*. 356 pp., 188 figs. (incl. maps)
(Sommerfeltia 24). Oslo.[33]

Bibliographies

FERNANDES, A., 1982. See **Division 5, Divisional bibliographies**. [Cape Verde, pp. 49–81; 207 titles in pure and 133 in applied botany.]

SUNDING, P., 1977. A botanical bibliography of the Cape Verde Islands. *Bol. Mus. Munic. Funchal* **31**: 100–109. [213 titles.]

CHEVALIER, A., 1935. *Les îles du Cap Vert. Géographie, biogéographie, agriculture, flore de l'archipel.* 358 pp., 11 text-figs. (incl. map), 16 pls. Paris: Muséum National d'Histoire Naturelle, Laboratoire d'Agronomie Coloniale. (Also published as *Rev. Int. Bot. Appl. Agric. Trop.* **15**: 733–1090, figs. 22–32, pls. 1–16.)

Of the three parts of this work, the third comprises a systematic enumeration of non-vascular and vascular plants, with descriptions of new or noteworthy taxa and limited synonymy (with citations of significant literature), vernacular names, detailed indication of local range with *exsiccatae*, general summary of extra-limital distribution, taxonomic commentary, and notes on special features, habitat, occurrence, etc.; indices to vernacular and to generic and family names. Parts 1–2 are introductory, and cover geography, physical features, climate, geology, soils, other biota, effects of human disturbance, agriculture, vegetation and floristics, and botanical exploration; a bibliography is also included (pp. 138–140 (870–872 in journal version)).

LOBIN, W., 1986. Katalog der von den Kapverdischen Inseln beschriebenen Taxa höherer Pflanzen (Pteridophyta und Phanerogamae). *Cour. Forschungsinst. Senckenb.* **81**: 93–164, 33 halftones.

An index to 229 taxa first described from the archipelago (with indication by a large bullet if in current opinion thought to be endemic); entries include synonymy, typifications, isotypes, references to text and illustrations, distribution, localities with the author's *exsiccatae*, and commentary where appropriate. No keys, vernacular names or citations of other *exsiccatae* are included. References, addenda and indices (by names and by families) appear at the end. [For additions, see *idem*, 1988. Ergänzungen und Verbesserungen zu 'Katalog der von den Kapverdischen Inseln beschriebenen Taxa höherer Pflanzen'. *Ibid.*, **105**: 145–147; and *idem*, 1993. 2. Ergänzungen und Verbesserungen . . . *Ibid.*, **159**: 87–90.]

PAIVA, J. (ed.), 1995– . *Flora de Cabo Verde*. In fascicles. Illus. Lisbon/Praia: Centro de Documentação e Informação do IICT.

Descriptive flora with keys, synonymy, citations, localities with *exsiccatae*, vernacular names, distribution and habitat, and commentary; index in each fascicle. [In progress. Fascicles, one each to a family, are numbered following a predetermined taxonomic system but are appearing only as families are ready. As of 2000, 53 families had been published (in 1995–96); a further 34 (including all non-graminoid or cyperoid monocotyledons) were awaiting publication, and eight, along with the pteridophytes, were under study. The remaining eight include Asteraceae, Cyperaceae, Poaceae and Fabaceae. No general part has yet been published.][34]

SUNDING, P., 1973. *Check-list of the vascular plants of the Cape Verde Islands*. 36 pp. Oslo: Oslo [University] Botanic Garden.

Systematic list of vascular plants, with essential synonymy and tabular summary of local range; bibliography. Additions to the flora since the enumeration of Chevalier (see above) are specially indicated. [For supplement, see *idem*, 1974. Additions to the vascular flora of the Cape Verde Islands. *Garcia de Orta*, Sér. Bot. 2(1): 5–29.]

Pteridophytes

LOBIN, W., E. FISCHER and J. ORMONDE, 1998. *The ferns and fern-allies (Pteridophyta) of the Cape Verde Islands, West-Africa.* iii, 115 pp., 36 figs. (incl. maps) (Nova Hedwigia, Beih. 115). Berlin/Stuttgart: Cramer/Borntraeger.

Detailed revision of ferns and fern-allies (32 species, mostly found on islands over 1000 m) with synonymy, references, types, citations, concise descriptions, localities with *exsiccatae*, indication of local and overall distribution and habitat along with actual or potential threats (if any), and taxonomic commentary along with illustrations and (pp. 98–108) maps; 'red lists' (pp. 89–96), references (pp. 109–112), summary (in Portuguese, English and German) and complete index. The general part encompasses botanical history, habitats, phytogeography, a table of distribution by islands (p. 5), and a discussion of threats (a significant factor being the gradual drying of the islands on account of disturbance and climatic change over the past century or more; 65.6 percent of the species are considered rare to very rare); general key, pp. 15–17.

Region 03

Islands of the Atlantic Ocean (except Macaronesia)

This region is construed to encompass all isolated Atlantic islands not normally considered part of Macaronesia, i.e., the Bermudas, St. Paul Rocks, Fernando Noronha (and Rocas), Ascension, St. Helena, Trindade (South Trinidad) and Martin Vaz, the Tristan da Cunha group, and Gough.

Nearly all islands in this region were visited on the voyage of HMS *Challenger* in the 1870s. From the expedition's collections as well as other sources, Hemsley worked up critical florulas which appeared in 1884 in the expedition reports. For some islands, these represented the first consolidated account. Subsequent studies have generally centered around individual islands or groups. Bermuda was examined in more detail by Nathaniel L. Britton and various collaborators from New York and Philadelphia early in the twentieth century as part of Britton's West Indian programme, and his still-current but now-obsolete manual, *Flora of Bermuda*, appeared in 1918. Fernando Noronha was studied by Henry N. Ridley in 1887, with his still-definitive *Notes* appearing three years later. John C. Melliss's 1875 *St. Helena* (1875) covers, as part of that long-time resident's overall study of the island, the native and naturalized plants (with colored illustrations). However, as documentation was completed, attention largely lapsed for more than half a century; in the interim only the Tristan da Cunha group and Gough were to receive much attention.

From about 1960, a new cycle of interest in the more southerly islands (especially Trindade, St. Helena and the Tristan group) developed, partly in conjunction with the 'Southern Zone' initiative of the Royal Society as well as other British and Brazilian research programmes and latterly out of concern for conservation and effective management. Trindade was restudied by A. Curt Brade and Richard Eyde, the latter addressing the problem of its 'dead trees'. St. Helena has received, and continues to receive, much attention on account of its unique native (and critically endangered) flora; and, in the Tristan group and Gough, Nigel Wace and James Dickson carried out extensive

biogeographic (vegetatiological) and ecological studies (including a detailed plant list) and Eric Groves has produced a critical enumeration, consolidating collection records and published information. St. Paul Rocks was re-examined in some detail by a Cambridge University party in 1979. Among current projects are a full flora (by Quentin Cronk) and a field-guide (by Simon Goodenough), both for St. Helena.

A noteworthy feature of the native vascular floras of all the higher islands (Ascension, Trindade, St. Helena and Tristan da Cunha) is a relatively high percentage of pteridophytes, particularly ferns; as elsewhere in the world, the uplands of such oceanic islands furnish a highly favorable environment.

Bibliographies. General bibliographies as for Superregion 01–04.

Indices. General indices as for Superregion 01–04.

030

Region in general

HEMSLEY, W. B., 1885 (1884). Report on the botany of the Bermudas and various other islands of the Atlantic and Southern Oceans, part 2. In W. Thomson and J. Murray (eds.), *Reports on the scientific results of the voyage of HMS* Challenger *during the years 1873–76*, *Botany*, 1(2): 1–299, pls. 14–53. London: HMSO.

This stately work includes individual annotated enumerations of the vascular and non-vascular plants of the following islands or island groups: St. Paul Rocks (pp. 1–7); Fernando Noronha (pp. 9–30, pls. 14–15, 47); Ascension (pp. 31–48, pls. 16–17); St. Helena (pp. 49–122, pls. 18–22, 48–51); Trindade (South Trinidad) (pp. 123–132, pls. 23–24), and the Tristan da Cunha group (pp. 133–185, pls. 25–39, 46). Each area account includes synonymy, references and citations, descriptions of new taxa, generalized indication of local and extralimital range (with citations of *exsiccatae* or other sources), extensive critical commentaries, and often ample notes on habitats, special features, biology, uses, etc. (in part compiled from other sources). The amply scaled plates depict new or noteworthy plants. Each treatment is prefaced by remarks on physical features, botanical exploration, general and special features of the flora, and vegetation formations along with tabular phytogeographic analyses. [The remainder of part 2 deals with Amsterdam and St. Paul in the Indian Ocean (**Region 04**) and with sub-Antarctic islands (**Region 08**). Part 1, separately paged and encompassing plates 1–13, is devoted to the Bermudas (see **031** below).]

031

The Bermudas

Area: 54 km^2; the group comprises 100 closely associated limestone islands. Vascular flora: 165 native species, of which 8.7 percent endemic (PD; see **General references**).[35]

Earlier field work included Michaux's extended visit (1805–06) but the great bulk of collections dates from after 1830. Published flora accounts prior to that of Britton include two by the zoologist John Matthew Jones, firstly in *The naturalist in Bermuda* (1859) and then in *On the vegetation of the Bermudas* (1873); by Johannes Rein in *Über die Vegetations-Verhältnisse der Bermudas-Inseln* (1873); and finally by Hemsley in *Botany of the Bermudas* (1884, in the botanical reports of the *Challenger* expedition (1(1): 1–135; for fuller reference see **030**).

BRITTON, N. L., 1918. *Flora of Bermuda.* 585 pp., numerous text-figs., col. frontisp. New York: Scribners. (Reprinted 1965, New York: Hafner.)

Descriptive manual-flora of non-vascular and vascular plants; includes keys to all taxa, limited synonymy, generalized indication of local and extralimital range, vernacular names, occasional critical remarks, and notes on habitat, phenology, etc. A glossary, short cyclopedia of collectors, and indices to all botanical and vernacular names conclude the work. Figures for most species of vascular plants and bryophytes are provided to aid identification.

032

St. Paul Rocks

See **030** (HEMSLEY, pp. 1–7). No vascular or non-vascular land plants have so far been recorded from this group, a Brazilian possession consisting largely of barren rocks no more than 18.3 m above sea level. For a general and historical account, see A. J. EDWARDS, 1985. Saint Paul's Rocks: a bibliographical review of the natural history of a Mid-Atlantic island. *Arch. Nat. Hist.* **12**: 31–49.

033

Fernando de Noronha (with Rocas)

See also **030** (HEMSLEY, pp. 9–30, pls. 14–15, 47). – Area: *c.* 19 km^2. Vascular flora: no data, but endemism low and bryophyte and pteridophyte species scarcely represented. The Fernando de Noronha group, currently under the jurisdiction of the Brazilian state of Pernambuco, is relatively low in elevation and climatically strongly seasonal; it comprises one large and several smaller islands and extends to the atoll of Rocas, nearer the mainland. Long a penal colony – during which the woods were deliberately all but destroyed – part of the main island is now a national park.

Hemsley's account, the first for the islands, was soon succeeded by that of Ridley. This, based on a visit of five and a half weeks at the end of the 1887 rainy season, remains the only substantial treatment of the flora. Rocas is 'almost unknown' (Fosberg, 1986; see **002**).[36]

RIDLEY, H. N., 1890. Notes on the botany of Fernando Noronha. *J. Linn. Soc., Bot.* **27**: 1–95, pls. 1–4.

Systematic enumeration of non-vascular and vascular plants; includes descriptions of new or noteworthy species, synonymy (with references and citations), generalized indications of local and extralimital range (with localities for less common plants), and notes on habitat, occurrence, special features, biology, properties and uses. The introductory section includes accounts of physical features, geography, history of human contact and settlement, origin of the flora, pollination and dispersal, fresh-water life, past botanical work on the island, and the author's itinerary; an account of local geology (by T. Davies) concludes the work.

034

Ascension Island

See **030** (HEMSLEY, pp. 31–48, pls. 16–17). – Area: 94 km². Vascular flora: *c.* 25 native species (11 endemic, 6 of them pteridophytes).

Hemsley's account, summarizing nearly two centuries of scattered collections and reports (the earliest being by James Cunninghame in 1698 and published in 1699), was for long the only critical treatment. It has been succeeded after a fashion by a list of native and introduced plants in John Packer's *Ascension handbook.* Additional information may be found in E. DUFFEY, 1964. The terrestrial ecology of Ascension Island. *J. Appl. Ecol.* **1**: 219–251.

PACKER, J. E., 1974. *Ascension handbook: a concise guide to Ascension Island, South Atlantic.* 2nd edn. Unpaginated. Illus. Georgetown, Ascension I.: The author. (1st edn., 1968; 68 pp.)

This general work includes (in its third chapter) a checklist of the flora (including known non-vascular plants and lichens), with line drawings; for each species is given the date of first record, known occurrences, and other notes (unfortunately without status symbols). Vernacular and scientific names are given in a single sequence. Chapter 9 comprises an extensive bibliography.

035

St. Helena

See **030** (HEMSLEY, pp. 49–122, pls. 18–22, 48–51). – Area: 121 km². Vascular flora: *c.* 60 native species, 50 of them endemic; about 260 other plants are naturalized (PD; see **General references**).

The flora of St. Helena has long been recognized as one of the most peculiar in the world and is also among the most endangered. Indeed, the island was witness to an early manifestation of environmental consciousness. The first relatively thorough (though regrettably somewhat late) collections, inclusive of the endemic species, were made by William J. Burchell (1805–10) and William Roxburgh (1813–14). The latter's plants were first published in *Tracts relative to the island of St Helena* (1816) by Alexander Beatson, a former governor, and separately reprinted in the island's capital, Jamestown, in 1843 as *Botany of St Helena.* The best of the older works, however, remains J. C. MELLISS, 1875. *St Helena.* xiv, 426 pp., pls., maps. London: Reeve. [Botany, pp. 221–383, with illustrations of nearly all the endemic species.]

The main accounts since 1875 have been the checklist of Brown (1982) and an attempt at a revision of Melliss's botany by Williams (1989). Neither can be considered as truly definitive. With respect to the endemic plants, a useful work is M. D. HOLLAND (ed.), 1986. *The endemic flora of St Helena: a struggle for survival.* 44 pp., illus. [Jamestown]: Departments of Agriculture and Forestry and of Education, St. Helena.[37]

BROWN, L. C., 1982. *The flora and fauna of St Helena.* 88 pp. Surbiton, Surrey: Land Resources Centre, Overseas Development Administration, United Kingdom. (Project record S.HEL-01-12/REC-59-82.)

Includes (as chapter 4, pp. 19–26) a plant checklist, with vernacular names and status. Additional coverage of the endemic vascular plants is presented in chapter 3 (pp. 13–17) and background information and historical data in chapters 1 and 2.

WILLIAMS, R. O., 1989. *Plants on St Helena.* xxiii, 157 pp. Scotland, St. Helena: Department of Agriculture and Forestry. (Mimeographed.)

A revision of the botanical part of Melliss's *St Helena* (pp. 1–136) presented in dictionary format, with classified lists at the beginning and ancillary topics at end; references, pp. 156–157.[38]

036

Trindade (South Trinidad) and Martin Vaz Islands

See **030** (HEMSLEY, pp. 123–132, pls. 23–34). – Area: Trindade, *c.* 12 km². Vascular flora: no data, but native species apparently are few, with a goodly percentage of them being pteridophytes.

Prior to Hemsley's somewhat sketchy treatment, the islands had been little visited notwithstanding the saga of the 'dead trees' of Trindade (now established as *Colubrina glandulosa* var. *reitzii*, Rhamnaceae). Official hostility and bad luck prevented work by two French expeditions before 1850. In the three decades after the *Challenger* landing, visits were made by an astronomer, Ralph Copeland (1874), E. F. Knight (1889) and George Murray (1901). Serious collecting, particularly by J. Becker, began only in the second half of the twentieth century.

Recent contributions include A. C. BRADE, 1969. Filices novae brasilienses, XII. Algunas especias novae Pteridophyta de Ilha brasileira Trindade coletadas por J. Becker em 1965/1966. *Bradea* 1: 3–9, 5 pls.; R. H. EYDE and S. L. OLSON, 1983. The dead trees of Ilha da Trindade. *Bartonia* 49: 32–51; and, more thoroughly, R. J. VÁLKA ALVES, 1998. *Ilha de Trindade e arquipélago Martin Vaz: um ensaio geobotânico.* 144 pp., 40 col. pls., maps. Rio de Janeiro: Serviço de Documentação da Marina.[39]

see N. M. WACE and J. H. DICKSON, 1965. The terrestrial botany of the Tristan da Cunha Islands. *Phil. Trans.*, B, **249**: 273–360, figs. 14–16, pls. 35–42.

GROVES, E. W., 1981. Vascular plant collections from the Tristan da Cunha group of islands. *Bull. Brit. Mus. (Nat. Hist.), Bot.* 8(4): 333–420, 33 figs. (incl. maps).

Illustrated documentary systematic enumeration (covering the Tristan group and Gough and treating 212 native and introduced species), with synonymy, references and citations, indication of localities with *exsiccatae*, critical taxonomic commentary, and notes on status, frequency, and patterns of occurrence. The introductory sections deal with the plan of the work, history of the Tristan group and its geography, physical features and climate, and collectors. Concluding parts treat statistics, biogeographic elements, the history of work on the flora, and spread of alien plants, and end with acknowledgments, references (pp. 415–417) and an index to botanical names.

037

Tristan da Cunha Islands

See also **030** (HEMSLEY, pp. 133–185, pls. 25–39, 46). – Area: 102 km² (Tristan, 86 km²; Inaccessible, 12 km²; Nightingale, 4 km²). Vascular flora (native species): Tristan, 80; Inaccessible, 64; Nightingale, 37. Eighty-three additional species are naturalized (PD; see **General references**).[40] All the islands are of volcanic origin, with Tristan reaching 2060 m.

The first known collections of plants were obtained by Louis-Marie Aubert du Petit Thouars in 1793, before any settlement had been established, and published by him in 1811. By 1884 Hemsley declared the flora, then with 55 known vascular species, 'fully exhausted'. Nevertheless, additions continued to be made over the next century; of particular importance was the work done in the 1930s by Erling Christophersen and other Norwegians, written up by Christophersen in a series of four papers from 1937 to 1968, and in the 1950s and 1960s by Nigel Wace and James Dickson. All known records have now been documented by Eric Groves; but for additional information (including an alternative checklist) as well as a treatment of vegetation and historical biogeography,

038

Gough Island

See **037** (GROVES; WACE and DICKSON). – Area: 57 km². Vascular flora: 80 taxa, of which 63 native (Groves); two-fifths are pteridophytes.

This rainy (3225 mm/annum) southernmost of the Atlantic islands, with a vascular flora allied to that of the Tristan group, was not treated by Hemsley for want of any collections. The first plants were obtained by R. N. Rudmose Brown in 1904 during a brief visit by the Scottish National Antarctic Expedition in the *Scotia*; they were published in 1905 (and reprinted in the expedition reports in 1912) as *The botany of Gough Island*. Subsequent contributions include *Plants of Gough Island (Diego Alvarez)* (1934) by Erling Christophersen, based on a 1933 collection by Lars Christensen, and that by Nigel Wace, who remained on the island for more than four months in 1955–56 as a member of the Gough Island Scientific Survey. Wace's results, which include an unannotated list of vascular plants, appeared as N. M. WACE, 1961. The vegetation of Gough Island. *Ecol. Monogr.* **31**: 337–367, 22 figs. All collections have been accounted for by GROVES (1981; see **037**).

Region 04

Islands of the central and eastern Indian Ocean

Islands incorporated under this heading include the Laccadives (Lakshadweep), the Maldives, the Chagos Archipelago (including Diego Garcia), Amsterdam and St. Paul, the Cocos (Keeling) group, and Christmas Island.

Few new treatments for these scattered and mostly very small islands have appeared in the last 60 years; indeed, several standard accounts date from before 1914. One cited here was based on work by the Percy Sladen Trust Expedition led by John Stanley Gardiner in 1905 in HMS *Sealark* (see also Superregion 46–49).[41] Charles Darwin's collections in the Cocos have been a foundation for all subsequent lists of its flora. New florulas for parts of the Maldives and the Chagos group have appeared under the auspices of the Atoll Research Programme but do not provide complete coverage of the units concerned; and field knowledge also remains uneven (Fosberg, 1986; see **009**). Dennis Adams has, however, completed a revised enumeration for the Maldives; and new floras of the Cocos group and Christmas were published in 1993 in one of the two 'outer island' volumes of *Flora of Australia* (see also **420–50**). Significant additions have been made to our knowledge of Amsterdam and St. Paul with the advent there of a more continuous scientific presence, but no revised florula has yet appeared.

Bibliographies. General bibliographies as for Superregion 01–04.

Indices. General indices as for Superregion 01–04.

041

Laccadive Islands (Lakshadweep)

See also **042** (WILLIS and GARDINER). – Area: 32 km². Vascular flora: 348 species reported by Raghavan (see below); the figure of 121 native species given by Prain (1893–94) is certainly too low, a point made also by Renvoize (1979, p. 115).[42] This Union Territory of India comprises 27 islands (8 inhabited, 5 of them botanically explored) as well as reefs, cays and sunken banks; altogether about 300 km². No part is more than 9 m above sea level.[43]

The islands are for the Botanical Survey part of its Western Circle. Recent work on the flora has been reviewed by Raghavan (in BOTANICAL SURVEY OF INDIA, 1977, pp. 95–108; see **Superregion 81–84, Progress**). Sivadas, Narayana and Sivaprasad in their treatment of Kavaratti note the need for studies of each individual island, using as a basis the work of Prain, Willis and Gardiner (see below and **042**). In 1995, a new checklist by Rao and Ellis appeared (see below).

PRAIN, D., 1893–94. Botany of the Laccadives. *J. Bombay Nat. Hist. Soc.* **7**: 268–295, 460–486; **8**: 57–86. (Reprinted 1894, Calcutta, as part of the author's *Memoirs and memoranda, chiefly botanical*.)

Includes a systematic enumeration of the known native, naturalized and commonly cultivated non-vascular and vascular plants, with synonymy, references and citations, English vernacular names, localities (with collectors), overall range outside the archipelago, and sometimes copious notes on habitat, special features, variability, uses, etc.; statistics and characteristic features of the flora; remarks on phytogeography. The first installment comprises extensive remarks on local geography, physiography, vegetation, and methods of plant introduction as well as an account of botanical contributions and a list of references.

RAO, T. ANANDA and J. L. ELLIS, 1995. Flora of Lakshadweep Islands off the Malabar Coast, Peninsular India, with emphasis on phytogeographical distribution of plants. *J. Econ. Taxon. Bot.* **19**: 235–250, illus.

Name list with indication of overall distribution (pp. 241–248) but not otherwise annotated. The considerable general part is on geomorphology, soils, climate, phytogeography and vegetation.[44]

Local works

Kavaratti encompasses the territorial headquarters and Minikoi is the most southerly of the islands.

SIVADAS, P., B. NARAYANA and K. SIVAPRASAD, 1983. *An account of the vegetation of Kavaratti Island, Laccadives.* 9 pp., 4 halftones, 3 maps (Atoll Res. Bull. 266). Washington.

Includes a checklist of 117 vascular plants, with sources (collections or sight records) and zones of occurrence (cf. maps 2 and 3). The general part includes an introduction as well as sections on climate, soil and vegetation.

WILLIS, J. C. 1901. Notes on the flora of Minikoi. *Ann. Roy. Bot. Gard. Peradeniya* 1: 39–43.

Includes a list of 134 species supplementing the records in Prain's *Botany of the Laccadives.*

042

Maldive Islands

Area: 298 km^2. Vascular flora: 583 native, introduced and cultivated species, of which Adams (1983; see below) believes 260 to be 'native or naturalized'. The Maldives, comprising 1201 islands in 19 atolls stretching across the equator, are an independent state. The flora not surprisingly shows strong affinities with those of the coral islands of the Pacific, with many species in common; the West Indian connection is less immediate.

Willis and Gardiner's 1901 account (see below) has long been standard but is now well out of date. Field coverage remains for the greater part confined to Malé and Addu Atolls, the latter including Gan, the site of a former British Royal Air Force base. For a review of the flora and vegetation, see C. D. ADAMS, 1983. *Report to the Government of the Maldive Islands on flora identification.* 41 pp. (FAO Project RAS 79/123.) Rome. A more popular introduction by the same author may be found on pp. 106–120 of P. A. WEBB, 1988. *Maldives: people and environment.* 120 pp., col. illus. Bangkok: Media Transasia.[45] For another recent report (relating to work in Malé Atoll), see L.-C. STUTZ, 1982. Herborisation 1981 aux îles Maldives. *Candollea* 37: 599–631.[46]

Bibliography

ANONYMOUS, 1966. Bibliography of the Maldive Islands. In D. R. Stoddart (ed.), *Reef studies at Addu Atoll, Maldive Islands: preliminary results of an expedition to Addu*

Atoll in 1964, pp. 107–122 (Atoll Res. Bull. 116). Washington, D.C. [Unannotated list, arranged by author; not limited to botany.]

FOSBERG, F. R., 1957. *The Maldive Islands, Indian Ocean.* 37 pp. (Atoll Res. Bull. 58). Washington.

Briefly annotated systematic checklist (based in part on a two-day visit to Malé and adjacent islands in Malé Atoll in 1956 and accounting for 324 species), with vernacular names, revised nomenclature, citations of the author's *exsiccatae*, and pointers to Willis and Gardiner's work; references (p. 37). An introductory general part is also included.[47]

WILLIS, J. C. and J. S. GARDINER, 1901. The botany of the Maldive Islands. *Ann. Roy. Bot. Gard. Peradeniya* 1: 45–164, pl. 2.

Systematic enumeration of native, naturalized and commonly cultivated vascular plants (accounting for 284 species, with 160 thought to be truly wild) with vernacular names, detailed indication of local range including *exsiccatae*, generalized summary of overall distribution including the Laccadives (**041**), the Chagos Archipelago (**043**), and beyond, and notes on habitat, ecology, uses, etc. The introductory section includes a brief general description of the geography, physical features and climate of the archipelago, while a lengthy concluding section includes notes on vegetation of various Maldivian atolls, the origin of the flora, order of plant succession, economic products, and a discussion of floras of oceanic islands in general.

Local work: Addu

FOSBERG, F. R., E. W. GROVES and D. C. SIGEE, 1966. List of Addu vascular plants. In D. R. Stoddart (ed.), *Reef studies at Addu Atoll, Maldive Islands: preliminary results of an expedition to Addu Atoll in 1964*, pp. 75–105 (Atoll Res. Bull. 116). Washington, D.C.

Annotated, systematically arranged enumeration of vascular plants with citation of *exsiccatae*, status, distribution, occurrence and habitat; new records specially indicated.

043

Chagos Archipelago

See also **042** (WILLIS and GARDINER); **490** (HEMSLEY). – Area: 60 km^2 (Diego Garcia, 47 km^2). Vascular flora: about 150 species, of which some 100 native (FOSBERG and BULLOCK; see below). The group, at present forming the British Indian Ocean Territory, includes the atoll of Diego Garcia with its strategic air base.

Fosberg (1986; see **009**) has noted that of the Chagos only Diego Garcia, the largest unit, is relatively well studied. The current accounts are antedated by a list for that atoll incorporated by Hemsley in the *Challenger* report on the southeastern Moluccas (1885; see **928/III**).

WILLIS, J. C. and J. S. GARDINER, 1931. Flora of the Chagos Archipelago. *Trans. Linn. Soc. London*, II, Zool. **19**: 301–306.

Systematic list of native, naturalized, adventive and widely cultivated plants, with brief indication of local range and notes on habitat, ecology, taxonomic problems, etc. The introductory section includes notes on previous botanical work in the group and on the origin and general features of the flora.

Local work: Diego Garcia

FOSBERG, F. R. and A. A. BULLOCK, 1971. List of Diego Garcia vascular plants. In D. R. Stoddart and J. D. Taylor (eds.), *Geography and ecology of Diego Garcia Atoll, Chagos Archipelago*, pp. 143–160 (Atoll Res. Bull. 149). Washington.

Systematic checklist of 142 species, with localities and brief notes.

044

Amsterdam and St. Paul Islands

Area: Amsterdam, 55 km^2; St. Paul, 7 km^2. Vascular flora: HEMSLEY (1884; see below) listed 38 native species (9 endemic) and SCHENCK (1905) 36, but recent exploration consequent to establishment of a permanent base has raised the total to 55 species for Amsterdam alone (J. Jérémie in PD; see **General references**). These two very isolated islands, their names sometimes in the past interchanged in the literature, are part of the French Southern and Antarctic Territory (cf. Kerguelen; **083**).[48]

Collections to 1870, including those from the Austrian *Novara* circumnavigation of 1857, were treated by Heinrich W. Reichardt in his *Ueber die Flora der Insel St. Paul* (1871). Hemsley in 1884 also

accounted for, as far as possible, known collections and reports. Additional information has been furnished in the two chapters by Schenck. A more recent flora is, however, not available.[49]

HEMSLEY, W. B., 1885 (1884). Amsterdam and St Paul Islands. In his Report on the botany of the Bermudas and various other islands of the Atlantic and Southern Oceans, part 2. In W. Thomson and J. Murray (eds.), *Reports on the scientific results of the voyage of HMS* Challenger *during the years 1873–76*, *Botany*, **1**(2): 259–281, pls. 39, 41–45, 52. London: HMSO.

Systematic enumeration of vascular and non-vascular plants, with descriptions of new taxa; synonymy, with references and citations; details of local distribution for each island with indication of *exsiccatae*, etc.; general summary of extralimital range; critical taxonomic commentary and notes on habitat, biology, etc.; figures of new or noteworthy species. The introductory section includes accounts of physical features, botanical exploration, vegetation formations, and introduced plants together with a tabular analysis of the flora.

SCHENCK, H., 1905. Über Flora und Vegetation von St Paul und Neu-Amsterdam. In C. Chun (ed.), *Wissenschaftliche Ergebnisse der deutschen Tiefsee-Expedition auf dem Dampfer 'Valdivia', 1898–1899*, **2**(1): 179–218, 14 text-figs., pls. 11–15. Jena: Fischer. Supplemented by *idem*, 1906. Die Gefässpflanzen der deutschen Südpolar-Expedition 1901–1903 gesammelt auf der Possession-Insel (Crozet-Gruppe), Kerguelen, Heard-Insel, St Paul und Neu-Amsterdam. In E. von Drygalski (ed.), *Deutsche Südpolar-Expedition 1901–1903*, **8** (Botanik): 97–123, 10 text-figs. Berlin: Reimer.

Schenck's *Valdivia* work contains a comprehensive illustrated account of the islands' floristic phytogeography and vegetation, with tabular lists of vascular and non-vascular plants including essential synonymy and distributional details. Other parts of this work treat botanical exploration (with references), physiography, climate, vegetation formations, ecological/morphological adaptations (reflecting the work of A. F. W. Schimper, a member of this expedition), and comparisons with the flora and vegetation of the Tristan da Cunha group and Gough (**037, 038**). Schenck's second work incorporates additional observations and includes on pp. 120–123 an annotated list of vascular plants collected on St. Paul and Amsterdam during the visit of the expedition ship *Gauss*.

045

Cocos (Keeling) Islands

Area: 14 km². Vascular flora: 121 species of which 64 are considered native (ABRS, 1993; see below).[50] – These atolls are now part of the Northern Territory of Australia following earlier, partly nominal, administrations under the Straits Settlements, Singapore and the Commonwealth of Australia.

For a background survey of this island group, in which reference is made to groves of *Pisonia grandis* (Nyctaginaceae) originally described by Charles Darwin, see C. A. GIBSON-HILL, 1950. A note on the Cocos-Keeling Islands. *Bull. Raffles Mus. Singapore* 22: 11–28. The plants collected by Darwin formed the basis of the first floristic account, by J. S. Henslow in *Florula keelingensis* (1838). The group was later visited by the *Challenger* (cf. Hemsley, 1885, under 928/III), H. O. Forbes (who furnished a 'List of the Keeling Atoll plants' in his *A naturalist's wanderings in the Eastern Archipelago* (1885)) and H. B. Guppy (reported in his Dispersal of plants as illustrated by the flora of the Keeling or Cocos Islands. *J. Trans. Victoria Inst., London* 24: 267–301 (1890)). The last list also appeared in *Coral and atolls* (1912) by F. Wood-Jones. In the 1970s and 1980s, renewed collecting added considerably to the known flora. All earlier accounts have now been supplanted by the *Flora of Australia* volume cited below.

AUSTRALIAN BIOLOGICAL RESOURCES STUDY, 1993. *Flora of Australia* 50: *Oceanic Islands* 2. xxvi, 606 pp., 97 figs. (part col. and incl. maps). Canberra: Australian Government Publishing Service.

Pages 30–42 comprise an introduction to the Cocos (Keeling) Islands (by I. R. H. Telford), encompassing background information, vegetation formations, floristics, land use, history and collectors and key references along with a key to families and a concisely annotated list of species with indication of their status and whether or not they were recorded in the works of Forbes and Wood-Jones (see above). All species are further treated in the descriptive part – which follows the format of other *Flora* volumes – together with those from other islands or groups (Christmas, 046; Coral Sea Islands, 949; Ashmore and Cartier, 451; Macquarie, 081; and the McDonald (Heard) group, 082).

046

Christmas Island

See also 910–30 (VAN STEENIS). – Area: 137 km². Vascular flora: 411 species, of which 237 are considered native with 7 percent endemic (ABRS, 1993).[51] The island is a dependency of Australia.

Biological survey of this previously uninhabited island began in 1887 in connection with its examination for workable phosphate deposits and subsequent settlement. The most considerable efforts were those of the *Egeria* officers, reported by Hemsley in 1889; a British Museum (Natural History) party under C. W. Andrews, written up in his *A monograph of Christmas Island, Indian Ocean* (1900); and Henry N. Ridley, accounted for by him (with previous records) in *The botany of Christmas Island* (1906). Transfer of the islands to Australia in 1958 was followed some time afterwards by renewed interest in the biota; much new field work was then carried out. All known materials have now been accounted for by the *Flora of Australia* volume cited below. For a modern popular treatment of the island's natural history, see H. S. GRAY, 1995. *Christmas Island – naturally.* 2nd edn., revised by R. Clark. Christmas Island Natural History Association (1st edn., 1981); and for an introduction to the forest flora, covering formations, species composition, and the (fairly severe) impact of mining and other human activities (with references), see B. A. MITCHELL, 1974. The forest flora of Christmas Island. *Commonw. Forest. Rev.* 53: 19–29.

AUSTRALIAN BIOLOGICAL RESOURCES STUDY, 1993. *Flora of Australia* 50: *Oceanic Islands* 2. xxvi, 606 pp., 97 figs. (part col. and incl. maps). Canberra: Australian Government Publishing Service.

Pages xxvi and 1–30 comprise an introduction to Christmas Island (by David J. Du Puy), encompassing background information, vegetation formations and characteristic species (the latter grouped by synusiae), dispersal and endemic species, volant fauna, the National Park, and key references along with a key to families and a concisely annotated list of species with indication of their status and whether or not they were recorded in the works of Andrews and Ridley (see above). All species are further treated in the descriptive part – which follows the format of other *Flora* volumes

– together with those from other islands or groups (Cocos (Keeling), **045**; Coral Sea Islands, **949**; Ashmore and Cartier, **451**; Macquarie, **081**; and the McDonald (Heard) group, **082**).

Superregion 05–07

North Polar regions

The southern limits of the Arctic region as delimited for the purposes of this book may, generally speaking, be considered to correspond to the 'tree-line', except that the Aleutian and Commander Islands as well as Iceland are excluded. These limits, with the exception of Iceland, closely conform with those proposed by YOUNG (see **071**). They also are broadly congruent with those proposed by LÖVE and LÖVE (1975), but are wider than those of POLUNIN (1959; for both references see **050–70**); the latter omits most of the partly shrub-dominated 'low-Arctic' subzone. For reasons botanical (trees having once been present) as well as practical, Iceland is incorporated with the rest of Scandinavia (Region 67). Nevertheless, it is recognized that argument continues over delimitation of the Arctic zone (Takhtajan, 1986, pp. 12–15; see **General references**).

Arctic botany has a lengthy history, but until the present century exploration was fairly sporadic and superficial or localized and literature sparse. Early travelers included Linnaeus in Lapland, Bering and Steller around the Bering Strait, and Sir Joseph Banks in Iceland. During the nineteenth century, further contributions to our knowledge resulted from such activities as the search for the Northwest Passage (and the missing explorer, John Franklin), the penetration of Greenland and Alaska, exploration along the great rivers of northern Asia, and the push towards the North Pole. One of the earliest areas to become reasonably well known was Lapland, for which the first modern general flora was produced in 1812 by Georg Wahlenburg, a successor to Thunberg and Linnaeus in Uppsala.[52] By the late nineteenth century exploration in Greenland had also progressed to a degree worthy of a general flora: this appeared in 1880–94 (J. Lange et al., *Conspectus florae groenlandicae*). What was known of Arctic North America as a whole was also incorporated by W. J. Hooker into his *Flora boreali-americana* (1829–40) and, later, accounted for by John Macoun in his *Catalogue of Canadian plants* (1883–1902). But, both here and elsewhere, local or regional floras in any quantity were not available until the twentieth century.

The large collections emanating from the mid-century Arctic explorations in British North America led J. D. Hooker to compose a first synthesis: *Outlines of the distribution of Arctic plants* (1861, in *Trans. Linn. Soc. London* **23**: 251–348). Herein was a complete tabular checklist covering 805 vascular plants as well as non-vascular plants and fungi and including extensive taxonomic commentary. This was gradually augmented over the next several decades by much collecting in Arctic Eurasia as exploration proceeded in earnest. Among the most active in Russia were Alexander Middendorff in the 1840s and, later, E. R. von Trautvetter (author of some 10 papers in the 1870s and 1880s). The final quarter of the nineteenth century saw two additional Arctic voyages: the *Vega* of 1878–80 under A. E. van Nordenskiöld, which effectively pioneered the northern sea route from Europe to East Asia, and the German-led International Polar Year expedition of 1882–83.

The Danish interest in Greenland, the *Vega* voyage, and a natural Scandinavian interest in polar regions prompted, at the end of the century, Eugen Warming in Copenhagen to call for a new *Flora arctica*. Carl H. E. Ostenfeld was charged with the project and, in collaboration with Otto Gelert, a first volume appeared in 1902 covering vascular plants apart from dicotyledons. Unfortunately, its completion was frustrated by the death of Gelert in 1899 and other commitments on the part of Ostenfeld. Primary botanical exploration, however, continued, with in particular closer examination of the Arctic islands from Jan Mayen to Novaja Zemlja and the 1913–18 Canadian Arctic Expedition. The botany of the latter appeared in 1922 under the authorship of Theodor Holm, a student of Warming by then working in the United States; and a still-standard flora by Bernt Lynge was one result of a 1921 Norwegian trip to Novaja Zemlja.

The Canadian expedition was land- as well as sea-based, and set the pattern for work after 1920 in Alaska, Canada and the new Soviet Union. Prominent floristic botanists in this period were Eric Hultén, Nicholas Polunin, Alf Erling Porsild and Alexander Tolmatchev. Local accounts first began to appear in the

1930s; among the more significant may be mentioned Tolmatchev's *Flora tsentral'noj časti Vostočnogo Tajmyra* (1932–35) and Polunin's *Botany of the Canadian Eastern Arctic* (1940).

Exploration has continued since World War II, with major contributions coming from Ira Wiggins and B. A. Yurtsev among others. The work of all botanical explorers was greatly facilitated by improved transport, communications and logistical support as the region became more significant in terms of resources and strategic value. Several standard works have been published, covering Alaska (and the Alaskan Arctic), the Canadian Arctic, and the whole of the Soviet Arctic. Within the last-named, further contributing factors were the importance of the northern shipping route and the fossil mammoth excavations.

The period from 1946 to 1975 also saw the publication of three more or less comprehensive works, widely varying in scope. These include *Circumpolar arctic flora* (POLUNIN, 1959; see below), *The circumpolar plants* (HULTÉN, 1964–71; see **001**) and *Cytotaxonomical atlas of the Arctic flora* (LÖVE and LÖVE, 1975; see below). None of these represents a critical descriptive flora, still a desideratum. Polunin's work includes descriptions, keys and good illustrations but employs a narrower definition of the Arctic and lacks any real documentation. Its species concept is, furthermore, rather wider than the admittedly narrow one of the Löves. The Löves' work is a checklist with karyotypes and indication of distribution. Hultén's work is a critical distribution atlas with extensive commentary.

At a more regional and local level, the Arctic is now comparatively well documented. The largest project, Tolmatchev's *Arktičeskaja flora* (1960–87), is now complete; an English version commenced publication in 1995. A flora of the mainland Northwest Territories in Canada by Porsild and W. J. Cody appeared in 1980, and modern floras of Greenland and Alaska/Yukon are also available. However, the depth of coverage, including local floras, varies widely, being greatest by far in Greenland and followed by Alaska, Lapland, and the Kola Peninsula.

A useful introduction to Arctic vegetation is V. D. ALEKSANDROVA, 1983. Растительность польаных пустынь СССР (*Rastitel'nost' pol'anykh pustyn' SSSR*). Leningrad: 'Nauka'. (English edn.: *idem*, 1988. *Vegetation of the Soviet polar deserts*, transl. D. Löve. Cambridge: Cambridge University Press).

Progress
The only reviews dealing with large stretches of the North Polar zone, neither less than 30 years old, are A. I. TOLMATCHEV, 1956. К изучению Арктической флоры СССР (K izučeniju Arktičeskoj flory SSSR). *Bot. Žurn. SSSR* **41**: 783–796; and B. MAGUIRE, 1958. Highlights of botanical exploration in the New World. In W. C. Steere (ed.), *Fifty years of botany*, pp. 209–246. New York (particularly 'Boreal America', pp. 210–216). [In the former article the preparation of more 'regional' floras for different parts of the Arctic was advocated.] A more recent prospect is represented by the papers, read to the XII International Botanical Congress (Leningrad, 1975), in B. A. YURTSEV (ed.), 1978. Арктическая флористическая область (*The Arctic floristic region*). 166 pp., illus., maps. Leningrad: 'Nauka'.[53]

Bibliographies. Bay, 1910; Blake and Atwood, 1942; Frodin, 1964, 1984; Goodale, 1879; Holden and Wycoff, 1911–14; Hultén, 1958; Jackson, 1881; Pritzel, 1871–77; Rehder, 1911; USDA, 1958.

Regional bibliographies
Löve and Löve, *Cytotaxonomical atlas of the Arctic flora* (**050–70** below) also includes a substantial bibliography.

ARCTIC INSTITUTE OF NORTH AMERICA, 1953–75. *Arctic bibliography*. 16 vols. Washington, D.C.: United States Department of Defense (vols. 1-11); Montreal: McGill University Press (vols. 12-16). [108 723 abstracts, all arranged alphabetically and covering the years 1947–70 inclusive; indices.].

HULTÉN, E., 1971. [Bibliography.] In *idem, The circumpolar plants*, II: 405–446 (Kongl. Svenska Vetenskapsakad. Handl., IV, 13(1)). Stockholm. [Geographically arranged, unannotated list of floristic literature for the entire circum-Arctic floristic and vegetational zone, with some extension southwards (emphasizing upper-montane and alpine areas). Compiled in Stockholm from sources also used for *Index holmiensis* and to be viewed as an extension of the bibliography in the same author's *The amphi-Atlantic plants* (1958; see Appendix A under **General bibliographies**); in contrast to that work, however, there is no map showing areal coverage of the works cited.]

Indices. BA, 1926– ; BotA, 1918–26; BC, 1879–1944; BS, 1940– ; CSP, 1800–1900; EB, 1959–98; FB, 1931– ; ICSL, 1901–14; JBJ, 1874–1939; KR, 1971– ; NN, 1879–1943; RŽ, 1954– .

050–70

Superregion in general

In addition to the two major circum-Arctic works accounted for below, users should be aware of Hultén's two-volume *The circumpolar plants* (**001**). There is also a semi-popular illustrated guide of relatively restricted scope, in Italian: T. ZUCCOLI, 1973. *Flora arctica*. 209 pp., 96 text-figs., 6 col. pls. Bologna: Edagricole.

LÖVE, Á. and D. LÖVE, 1975. *Cytotaxonomical atlas of the Arctic flora*. xxiii, 598 pp., map. Vaduz, Liechtenstein: Cramer.

Detailed systematic list of vascular plants, with abbreviated indication of distribution together with chromosome numbers. Synonyms and taxa without known karyotypes are also included, so in effect this work constitutes a complete enumeration of the known Arctic flora (1629 species in 404 genera). The introductory section includes considerations of geographical limits, distribution of taxa, taxonomic concepts, botanical progress, and sources of information (including references); comprehensive bibliography (pp. 507–594) and index given at end.

OSTENFIELD, C. H. (ed.), 1902. *Flora arctica*. Part 1: *Pteridophyta, Gymnospermae and Monocotyledoneae* (by O. Gelert and C. H. Ostenfield). xi, 134, [2] pp., 95 text-figs., map. Copenhagen: Nordiske Forlag.

Illustrated descriptive flora, with synonymy, references, and extensive literature citations and generalized indication of internal and extralimital range; no biological or ecological notes or index. The introductory part gives the scope and plan of the work along with a bibliography (pp. vii–xi). [Not continued.]

POLUNIN, N., 1959. *Circumpolar arctic flora*. xxvii, 514 pp., 900 text-figs., map. Oxford: Oxford University Press.

Illustrated descriptive manual-flora of vascular plants, with keys to all families and genera (and to species in the larger genera), essential synonymy (without authorities, although the latter appear in the picture captions), generalized indication of circum-Arctic distribution, as well as regional and local occurrences, vernacular names, and some taxonomic commentary but only limited information on habitats, ecology, biology, etc.; glossary and general index to all botanical and vernacular names. The introductory section includes a definition of the limits of the Arctic zone for the purposes of this flora and a delimitation of botanical regions (with map). [Some 55 percent fewer species are covered in comparison to Löve and Löve's work due to the author's adoption of a broader species concept along with more restrictive geographical limits (the latter explained under **Superregion 05–07**).]

Region 05

Islands of the Arctic Ocean*

This region incorporates all the major islands and island groups of the Arctic Ocean from Jan Mayen eastwards to Wrangel, thus also including Bear Island, Spitsbergen, Franz Josef Land, Novaja Zemlja, Severnaja Zemlja, and the New Siberian Islands.

Floristic coverage of these scattered island groups is relatively complete for the European islands, but less so for those north of Asia. No useful accounts have been seen for Severnaja Zemlja. The remaining units each have separate floras, although that for Novaja Zemlja (Lynge, 1923) is now relatively incomplete. The level of botanical exploration parallels that of floristic coverage.

Bibliographies. General bibliographies as for Superregion 05–07.

Indices. General indices as for Superregion 05–07.

051

Jan Mayen Island

Area: 380 km². Vascular flora: 62 species (Lid, 1964; see below). This partially glaciated island, rising to a height of 2277 m, has from 1929 been Norwegian territory.

Lid's 1964 florula, incorporating results of several expeditions mounted from 1930 onwards,

---

*For a lexicon of Russian words used in titles see Table 1, p. 618.

succeeds *Flora der Insel Jan Mayen* (1886) by Heinrich
W. Reichardt and *Contribution à la flore de l'île Jan-
Mayen* (1897) by C. H. Ostenfield. The first-named
was one result of the establishment on the island of an
Austrian polar station, a part of the International Polar
Year expeditions of 1882–83; the second was based on a
French polar expedition of 1892 and a Danish expedi-
tion of 1896.[54]

LID, J., 1964. *The flora of Jan Mayen.* 108 pp., 26
text-figs., 1 col. pl., 64 maps (Skr. Norsk Polarinst.
130). Oslo.

Briefly descriptive illustrated flora of vascular
plants, with keys to all taxa, limited synonymy, detailed
indication of local range with citations and dot maps,
summary of overall range, and notes on habitat, special
features, etc., as well as some critical remarks; summary
of major features of the flora and its phytogeographic
relationships; lists of species present in selected sample
plots; references and an index to species. The introduction includes an account of botanical exploration
together with a synopsis of the flora.

052
Bear Island (Bjørnøya)

Area: 179 km². Vascular flora: 63 species and a
few additional infraspecific taxa are accounted for by
Rønning (1959; see below). Bjørnøya is administered
by Norway as part of Svalbard.

Botanical exploration began in 1827, with by
1833 some 28 species recorded. Later accounts are *Om
Beeren-Islands fanerogam-vegetation* (1870) by Th. M.
Fries and *The vascular plants of Bear Island* (1925) by O.
Hanssen and J. Holmboe. All vascular plant species are
now included in *Svalbards flora* (see **053**). More
detailed information on local range, ecology, variability,
etc., may be found in O. I. RØNNING, 1959. *Vascular
flora of Bear Island.* 62 pp., maps (Acta Borealia, A, Sci.
15). Tromsø: Tromsø Museum.

053
Spitsbergen (Svalbard)

Area: 62 000 km² (much of it glaciated). Vascular
flora: 150–175 species (Webb, 1978; see **Division 6,
Progress**). Some 169 were accounted for by Rønning
in 1964, of which 162 were considered native. The
islands officially became part of Norway in 1925.

Spitsbergen has been in recent decades the most
accessible of the Eurasian Arctic islands and so has
attracted many travelers. Rønning's 1979 revision of
the first guide, *Svalbards flora* by H. Resvoll-Holmsen
(1927, Oslo), has long been standard for the higher
plants. This latter in turn was derived from several
earlier expedition reports, including an initial enumeration, *Öfversigt af Spetsbergens fanerogamflora* (1862) by
A. I. Malmgren.

RØNNING, O. I., 1979. *Svalbards flora.* 2nd edn.
128 pp., 60 text-figs., 8 col. pls. (with 36 photographs),
maps (end-papers) (Polarhåndbok 1). Oslo: Norsk
Polarinstitutt. (1st edn., 1964.)

Briefly descriptive illustrated manual-flora of
native and introduced vascular plants of Spitsbergen
and Bear Island, with keys to all taxa, limited synonymy, vernacular names, generalized indication of local
range, and concise notes on habitat, ecology, special
features, etc.; summary of plant protection regulations;
index to vernacular and generic names. The introductory section includes an illustrated descriptive account
of the vegetation as well as a review of physical features,
climate, etc.; plant geography and floristic affinities are
also surveyed.

054
Franz Josef Land
(Zemlja Frantsa Iosifa)

See also **060** (TOLMATCHEV); **680** (KOMAROV *et
al.*; STANKOV and TALIEV); **688** (PERFIL'EV). – Area:
20 720 km² (in 167 islands, 90 percent glaciated).
Vascular flora: 36 species, all phanerogams (Hanssen
and Lid, 1932; see below); a few additions were
reported by Tolmatchev and Shukhtina in 1974. From

1926 the group became Soviet territory, passing in 1991 to the Russian Federation.

The first plant collections were made by Julius Payer on an Austro-Hungarian exploring expedition in 1873–74. The work of Hanssen and Lid is based primarily on a Norwegian survey of 1930 but accounts for all previous work.

HANSSEN, O. and J. LID, 1932. *Flowering plants of Franz Josef Land.* 42 pp., 5 halftones, map (Skr. Svalbard Ishavet 39). Oslo: Norsk Polarinstitutt.

Enumeration of known seed plants, with limited synonymy, detailed indication of local range (including citations of *exsiccatae*), some critical remarks, and notes on habitat, ecology, special features, etc.; list of references. An introductory section gives an account of botanical exploration, while an appendix provides a summary and analysis of general features of the flora. [For additions, see A. I. TOLMATCHEV and G. G. SHUKHTINA, 1974. Новые дание о флоре Земли Франца Иосифа (Novye danie o flore Zemli Frantsa Iosifa). *Bot. Žurn. SSSR* **59**: 275–279.]

055

Novaja Zemlja

See also **060** (TOLMATCHEV); **680** (KOMAROV *et al.*; STANKOV and TALIEV); **688** (PERFIL'EV). – Area: 82 600 km² (a considerable proportion glaciated). Vascular flora: 208 species (Tolmatchev, 1936; see below).

The first enumeration, covering 105 species, is *Conspectus florae insularum Nowaya-Semlja* (1871) by E. R. Trautvetter; this followed studies on climate and vegetation published in 1837–38 by the biologist and embryologist Karl von Baer, the first collector in the group. Tolmatchev (1936; see below) gives a list of known collectors to 1933.

LYNGE, B., 1923(–24). Vascular plants from Novaya Zemlya. In O. Holtedahl (ed.), *Report on the scientific results of the Norwegian expedition to Novaya Zemlya in 1921*, 13: *Botany*. 151 pp., 47 pls. (incl. maps). Kristiania [Oslo]: Videnskapsselskapet i Kristiania.

Descriptive account, based principally on the expedition's collections but also incorporating earlier records; includes localities with citations of *exsiccatae* and other records along with dot maps, taxonomic com-

mentary, and discursive notes on habitat, life-forms, ecology, etc.; photographs of vegetation and interesting plants; tabular summary of the flora, list of references and index to botanical names at end. The introduction includes an account of previous botanical exploration along with the itinerary of the 1921 expedition.

TOLMATCHEV, A. I., 1936. Обзор флоры Новой Земли (Obzor flory Novoj Zemli). *Arctica* **4**: 143–178.

Includes a systematically arranged tabular checklist of 208 vascular species with indication of presence or absence in six phytogeographical units on the two islands (three on the south, three on the north). The rest of the paper is concerned with vegetation, biogeography, paleohistory and dynamics; a list of collectors is also given.

056

Severnaja Zemlja

See **060** (TOLMATCHEV); **710–50** (KOMAROV *et al.*). – Area: 37 560 km². Vascular flora: no data. This partially glaciated group, its highest point (in October Revolution Island) rising to 675 m, was discovered only in 1913.

Although scattered studies of the individual islands (Bolshevik, October Revolution, Pioneer, Komsomolets, and Shmidt) exist, no separate general floras or lists appear to have been published. Hultén, however, records in his *Amphi-Atlantic Plants* (see **001**) the existence of an unpublished account.[55]

057

New Siberian Islands (Novosibirskie Ostrova)

See also **060** (TOLMATCHEV); **710–50** (KOMAROV *et al.*); **727** (all works). – Area: 38 400 km². Vascular flora: 123 species (Egorova, 1981). The group comprises three main islands: Bolshoj Ljakhovskij, Kotel'nyj/Faddeevskij (the largest and most northerly), and Novoe Sibir', as well as several scattered smaller ones. All are relatively low and uniform in relief. Administratively they are part of the republic of Sakha (Yakutia).

Botanical progress to 1980 was reviewed in A. A. EGOROVA, 1981. Флора Новосибирских островов (Якутия) (Flora Novosibirskikh ostrovov (Jakutija)). In *Biol. problemy Severa, 9-j Sympos.* (Syktyvkar, 1981), *Plenarn. dokl., Bot.*: 22. (Abstr., RŽ, 1981/10(2), V576.) Only Kotel'nyj has a checklist.

SAFRANOVA, I. N., 1980. К флоре острова Котельный (Новосибирские острова) (K flore ostrova Kotel'nyj (Novosibirskie ostrova)). *Bot. Žurn. SSSR* **65**: 544–551.

The included checklist (covering 89 species) features for each species notes on habitat, chorological type, and ecology (including associates); a floristic analysis, summary and four references conclude the work. The list proper is preceded by a brief introduction to the area as well as notes on previous botanical work.

058

Wrangel Island (Ostrov Vrangelja)

See also **060** (TOLMATCHEV); **710–50** (KOMAROV *et al.*). – Area: 7270 km². Vascular flora: 394 species and infraspecific taxa (Yurtsev, 1994, p. 257); Petrovsky in 1983 recorded 348 species. The height, situation and paleohistory of this island (1096 m) is reflected in the relative size of its vascular flora, greater than that of the whole Canadian Arctic Archipelago.

The elaborately documented К флоре острова Брангеля (*K flore ostrova Vrangelja*) by M. I. Nazarov (*Trudy Gosud. Okeanogr. Inst.* (Moscow) 3(4): 3–21 (1933)) is now very incomplete. For vegetation, see V. V. PETROVSKY, 1985. Очерк растительности острова Врангеля (Oček rastitel'nosti ostrova Vrangelja). *Bot. Žurn. SSSR* **70**: 742–751. Recent research, including a florula of the polar station at Somnitel'naja Bay, is documented in B. A. YURTSEV, 1994. Арктические тундры Острова Врангеля (Arktičeskie tundry Ostrova Vrangelja). 289 pp., illus., maps (Trudy Bot. Inst. RAN 6). St. Petersburg.

PETROVSKY, V. V., 1973. Список сосудистых растений острова Врангеля (Spisok sosudistykh rastenij ostrova Vrangelja). *Bot. Žurn. SSSR* **58**: 113–126.

Systematic enumeration of 312 vascular plant species confirmed as occurring on Wrangel, with brief descriptive notes on habitat, occurrence, etc., for each; followed by a supplementary list of species reported in the literature as being on the island but not supported by herbarium records. A list of references is also given. [The progress of exploration is documented: 42 species were known to Nazarov in 1933, 185 in 1964, and about 330 in the present work. For additions, see *idem*, 1983. Дополнение к списку растений флоры острова Врангеля (Dopolnenie k spisku rastenij flory ostrova Vrangelja). *Nov. Sist. Vysš. Rast.* **20**: 196–199.]

Region 06

Palearctic mainland region

Within this region are incorporated the Arctic and sub-Arctic zones of Eurasia north of the 'tree-line', comprising lands stretching from Lapland eastwards to the Chukotsk Peninsula. Almost all the area lies within the Russian Federation, and therefore the *Arktičeskaja flora SSSR*, initiated by A. I. Tolmatchev, has been listed under **060** as if it were a 'regional' work.

Until the completion in 1987 of Tolmatchev's great Arctic flora (now also appearing in an English edition), the only overall work had been *Rossiae arcticae plantas quasdam a peregrinatoribus variis in variis locis lectas enumeravit* by Trautvetter (1880). For particular sectors of the Eurasian Arctic, few separate floras are available – an exception being the recent *Nordisk fjallflora* by Ö. Nilsson (1986; see **061**) – and in both Scandinavia and Russia recourse usually has had to be made to the regional and national floras. The Anadyr-Chukotsk Region (**068**) is, in addition to local literature, also largely covered by Hultén's *Flora of Alaska and neighboring territories* (see **110**). The European sectors, notably west of the White Sea, are botanically the best explored and documented. Elsewhere, coverage is patchy, with the central Taimyr Peninsula, Tiksi Bay in Sakha (Yakutia) (a port of call on the northern shipping route near the Lena delta), and Anadyr-Chukotia having the most substantial literature. – The best introduction to the flora and vegetation, whose chapters include a number of systematic florulas (two of which are described in this section), is A. I. TOLMATCHEV (ed.), 1966. Растения Севера Сибири и Дадьнего Востока (*Rastenija Severa Sibiri i Dal'nego Vostoka*). 223 pp. Moscow/Leningrad: 'Nauka'. (Растительность Краинего Севера СССР и ее освоение (Rastitel'nost' Krainego Severa SSSR i ee osvoenie) (ed. B. A. Tikhomirov), vyp. 6.) English edn.: *idem*, 1969, as *Vascular plants of the Siberian north and the northern Far East*. Transl. L. Phillips. 340 pp., map. Jerusalem: Israel Program for Scientific Translations.

Bibliographies. General bibliographies as for Super-region 05–07. Bibliographies covering the former Soviet Union (**680**) should also be consulted.

Indices. General indices as for Superregion 05–07.

060
Region in general

See also **680** (KOMAROV et al.).
SEKRETAREVA, N. A., 1999. *Vascular plants of Russian Arctic and adjacent territories.* 170 pp. Sofia/Moscow: Pensoft. (Pensoft, Series Floristics 1.)

A checklist with distribution and indication of habitat, life-forms and karyotypes; covers 1478 species.
TOLMATCHEV, A. I. and B. A. YURTSEV, 1960–87. Арктическая флора СССР (*Arktičeskaja flora SSSR/Flora arctica URSS*). Vyp. 1–10 (vyp. 8–9 each in 2 fascicles). Illus., maps. Moscow/Leningrad: AN SSSR Press (later Leningrad: 'Nauka'). English edn.: A. I. TOLMATCHEV, 1995– . *Flora of the Russian Arctic* (transl. G. C. D. Griffiths; ed. J. G. Packer). Vols. 1– . Maps. Edmonton: University of Alberta Press.

Detailed, critical enumeration of vascular plants (1640 species and 122 additional infraspecific taxa) of the Russian Arctic, with descriptive keys to all taxa, full synonymy, references and citations (including illustrations), very full accounting of internal range and of distribution inside and outside the Arctic zone as a whole, individual dot maps for each species, critical taxonomic commentary, and extensive notes on habitat, ecology, biology, karyotypes, etc.; tables of distribution, lists of references, and indices to botanical names at end of each part (but no general index in vyp. 10). A brief introduction to the whole work is given in vyp. 1. The arrangement of families generally follows the Englerian system, save that families from Droseraceae through Leguminosae appeared in vyp. 9, out of sequence. [The English edition will comprise six volumes. As of 1999, two had been published (respectively in 1995 and 1996), corresponding to fascicles 1–4 of the Russian edition.]

061
Arctic Fennoscandia

See also **670, 674, 675, 676** (all works). The area, the heart of Sami, is here considered to encompass the treeless parts of northern Norway, Sweden and Finland, but includes for convenience the arctalpine 'enclaves' of the Scandinavian Mountains.

Notable works in addition to those listed below include the celebrated *Flora lapponica* by Carl Linnaeus (1737; revision by J. E. Smith, 1792), the first 'modern' flora (and the first to use its author's sexual system of classification), as well as *Norges Arktiske Flora/Florae arcticae norvegicae* by I. M. Norman (1894–1901; in 2 parts) and *Alpine Plants* by O. Gjærevoll (1990, published as vol. 2 of *Maps of distribution of Norwegian vascular plants*) (see **674**). The last-named includes for the 109 species covered sketch maps of Fennoscandian distribution as well as dot maps for Norway alone.

NILSSON, Ö., 1986. *Nordisk fjällflora.* 272 pp., illus. (mostly col.), map. [Stockholm]: Bonniers. (3rd edn., 1991.) (Norwegian edn.: *idem*, 1987. *Norges fjellflora*. Transl. B. Grenager. 272 pp. Oslo: Cappelen; 2nd edn., 1995.)

Field manual-flora of the mainly treeless parts of Fennoscandia; includes keys, concise descriptions, principal synonymy, Swedish vernacular names, altitudinal and geographical distribution, extralimital range, habitat occurrence, biology, special features, and some critical notes; appendix ('Arter i fjällens grannskap', or borderline species) and index to vernacular and generic names. The introductory section covers general aspects of the region and its flora, a map (p. 10), conventions, abbreviations, and an illustrated glossary as well as the general key. [The lower altitudinal limits range from (50–)200(-400) m to (280–)600–950(-1200) m, with the highest records ranging to 2370 m (on Jotunheimen); 65 species occur above 1500 m.]

WAHLENBERG, G., 1812. *Flora lapponica*. lxvi, 550 pp., 30 pls., 2 tables, map. Berlin: Reimer. Continued as S. C. SOMMERFELT, 1826. *Supplementum 'Florae lapponicae'*. xii, 331 pp., 3 col. pls. Christiania (Oslo): Borg.

Wahlenberg's work comprises an annotated descriptive account of non-vascular and vascular plants, without keys; localities and many critical remarks are included. The 1826 supplement is in essentially the same format.

Local work: Swedish Lapland

ROWECK, H., 1981. *Die Gefässpflanzen von Schwedisch-Lappland*. ix, 804 pp., map. Vaduz: Cramer/Gantner. (Flora et vegetatio mundi 8.)

An ecologically oriented vascular flora in the form of a detailed enumeration, without keys; species entries include Swedish and German vernacular names, notes on races, karyotypes, patterns of distribution, habitat requirements, and frequency, all with literature references. Hybrids are also accounted for. The introductory section accounts for sources and the plan of the work (nomenclature mainly after *Flora Europaea* (see **600**) and *Liste der Gefässpflanzen Mitteleuropas* by Ehrendorfer *et al.* (see **640/I**)) while at the end are the bibliography (pp. 738–789) and index to scientific names. Phytogeographic districts are given throughout, but any formal phytosociological units are omitted and the species are not individually mapped. [The scope of this work extends to Norway and south to 64°N but apparently does not include Finmark.]

062

Kola Peninsula (Arctic zone)

See **687** (GORODKOV and POJARKOVA). The only markedly elevated areas center around Khibiny Gora (1191 m), for which a floristic and geobotanical account has appeared: B. A. MISCHKIN, 1953. Флора Хибинских гор, ес анализ и история (*Flora Khibinskikh Gor, es analiz i istorija*). 114 pp. Moscow/Leningrad: AN SSSR Press.

063

Northern and Northeastern Russia (Arctic zone)

See also **688** (PERFILEV; TOLMATCHEV). The offshore islands of Kolgu'ev and Vajgač (Vaigach) are included within this area, all either in Murmansk Oblast' or Nenets National Territory (the latter being east of the White Sea).

Early works include *Flores Samojedorum Cisuralensium* (1845) by F. J. Ruprecht and *Reise nach dem Nordosten des europäischen Russlands durch die Tundren der Samojeden zum arktischen Uralgebirge* (1848–53) by A. G. Schrenk. No separate overall flora is available, but the following provide useful partial accounts.

Partial works

LESKOV, A. I., 1937. Флора Малоземельской тундры (*Flora Malozemel'skoj tundry*). 105 pp. (Trudy Severnoj bazy AN SSSR 2). Moscow/Leningrad.

A systematic enumeration of 392 vascular plant species in an area south of Kolgu'ev Island and west of the northern Urals, with indication of localities and *exsiccatae* (including earlier records) and notes on habitat, taxonomy, etc. An introductory section includes sources for the work, a survey of collectors, and a list of references.

SERGIENKO, V. G., 1986. Флора полуострова Канин (*Flora poluostrova Kanin*). 146, [2] pp. Leningrad: 'Nauka'.

Includes an enumeration of 474 species in Kanin Peninsula (18000 km², situated in western Nenets), the entries covering life-form, habitat, chorological class(es), phenology, local distribution, and other notes; lengthy floristic analysis and typology (following the so-called 'concrete floras' approach), followed by a list of references (pp. 137–146). Botanical history, topographical features, climate and vegetation are covered in the general part. [Some two-thirds of the area is floristically boreal (in the *taïga* or transitional); only the northern one-third is truly arctic.]

064

The Ural (Arctalpine zone)

See also **602** (GOVORUCHIN). The work by Igoshina, a concise floristic checklist of the arctalpine and transitional zones, succeeds *Flora boreali-uralensis* (1854) by F. J. Ruprecht, which recorded 301 species and was based on field work by its author in 1847–48.

IGOSHINA, K. N., 1966. Флора горных и равнинны тундр и редколесий Урала (Flora gornykh i ravninny tundr i redkolesij Urala). In A. I. Tolmachev (ed.), Растения Севера Сибири и Дадьнего Востока (*Rastenija Severa Sibiri i Dal'nego Vostoka*), pp. 135–223. Moscow/Leningrad: 'Nauka'. (Растительность Краинего Севера СССР и ее освосние (Rastitel'nost' Krainego Severa SSSR i ee osvoenie) (ed. B. A. Tikhomirov), vyp. 6.) English edn.: *idem*, 1969. Flora of the mountain and plain tundras and open forests of the Urals. In *Vascular plants of the Siberian north and the northern Far East* (ed. A. I. Tolmatchev; transl. L. Phillips), pp. 182–340. Jerusalem: Israel Program for Scientific Translations.

Analytical floristic checklist of vascular plants (769 species in 245 genera; 682 species exclusive of *Hieracium*) of the semi- to non-forested arctalpine areas of the Ural, including the tundra (*goltsy*) which to the south becomes gradually discontinuous; details in brief of ecology and occurrence are given. The main list is preceded by a historical account, and pp. 221–223 (Russian edition) comprise a bibliography.[56]

065

Western Siberia (Arctic and arctalpine zones)

See also **710** (KRYLOV). – This area, traversed by the Ob' estuary and its branches, is topographically almost featureless. No separate general works appear to exist, and relatively few special studies. Among the latter, a relatively substantial contribution is F. R. KJELLMAN, 1882. Sibirska nordkustens fanerogamflora. In A. E. Nordenskiöld (ed.), *Vega-expeditionens vetenskapliga iakttagelser*, 1: 247–296. Stockholm. (German edn., 1883, pp. 94–139.)

066

Central Siberia (Arctic and arctalpine zones)

See also **721** (both works); for Severnaja Zemlja, see also **056**. – This area is here understood to comprise the Arctic zone of Krasnojarsk Krai, including the Taimyr Peninsula (with elevations to 1146 m), as well as the arctalpine 'enclave' of the Putorany Plateau, a late geographical discovery which rises to a maximum of 1701 m from the *taiga* of northern Krasnojarsk (**723**).

Apart from general works on the Arctic or on Krasnojarsk Krai, only scattered florulas or lists are available. The two most useful, described below, are *Flora tsentral'noj časti Vostočnogo Tajmyra* (Tolmatchev, 1932–35) and *Flora Putorana* (Malyschev, 1976). Early knowledge of the area is embodied in *Flora taimyrensis phaenogama* (1847), *Plantae jenisseiensis* (1856) and *Flora riparia Kolymensis* (1877), all by Trautvetter and based on expeditionary work organized by Alexander Middendorff in the 1840s, and *Florula jenisseensis arctica* (1872) by Friedrich Schmidt.

Partial works

MALYSCHEV, L. I. (ed.), 1976. Флора Путорана. Материалы к познанию особеностей состава и генезиса горных субарктических флор Сибири (*Flora Putorana. Materialy k poznaniju osobennostej sostava i genezisa gornykh subarktičeskikh flor Sibiri*). 248 pp., 6 figs., 19 tables. Novosibirsk: 'Nauka'.

Incorporates on pp. 40–162 an annotated systematic enumeration (*Sostav flory Putorana*, by S. Ju. Andrulajtis and others) of 569 vascular plant species, with synonymy and notes on habitat and distribution; no index. Additional chapters deal with the flora in general, including its origin, character and size, and with approaches and methodologies.[57]

TOLMATCHEV, A. I., 1932–35. Флора центральной части Восточного Таймыра (*Flora tsentral'noj časti Vostočnogo Tajmyra*). 3 parts (Trudy Poljarn. Komiss. AN SSSR 8, 13, 25). Leningrad.

Sections 5 and 6 of this work include a systematic checklist (194 species) of the vascular flora of the central part of the eastern Taimyr, with synonymy, localities, critical commentary, and notes on habitat, biology, etc. Based mainly on collections made on a 1928 expedition.[58]

067

Eastern Siberia (Arctic and arctalpine Sakha)

See **728** for general works on Sakha and **057** for the New Siberian Islands.

Only scattered florulas and checklists otherwise exist covering the plants of this area, which is here taken as extending to include arctalpine 'enclaves' on the mountains of Sakha (particularly the Verkhoyansk Ranges, the Suntar-Khayat Range and the Chersk Range, respectively with elevations to 2389 m, 2959 m and 3147 m). The collection of papers on the far north of the former USSR published in 1966 under Tolmatchev's editorship (see IGOSHINA under **064**) includes two contributions from along the coast of Sakha (including Tiksi Bay, a port of call on the northern shipping route just east of the Lena delta) as well as one on the Verkhoyansk Range; but the key recent work is perhaps *Flora Suntar-Khajata* by B. A. Yurtsev (see below), dealing with the arctalpine flora of the second highest of the above-mentioned ranges (which forms part of the border between Sakha and the Russian Far East). Early records from the north coast appear in *Flora riparia Kolymensis* (1878) by Trautvetter.

Partial work

YURTSEV, B. A., 1968. Флора Сунтар-Хаяата (Северо-Восток Сибирн): лроблемы истории высокогорных ландшафтов Севеор-Востока Сибири (*Flora Suntar-Khajata (Severo-Vostok Sibiri): problemy istorii vysokogornykh landšaftov Severo-Vostoka Sibiri*). 235 pp., 17 illus. Leningrad: 'Nauka'. (Растительность Краинего Севера СССР и ее освоение (Rastitel'nost' Krainego Severa SSSR i ee osvoenie) (ed. B. A. Tikhomirov), vyp. 9.)

Includes a systematic enumeration of the high-mountain vascular flora (301 species), with distribution maps; references and index to scientific names at end.[59]

068

Anadyr and Chukotia

See also **730** (CHARKEVICZ; KOMAROV and KLOBUKOVA-ALISOVA; VOROŠILOV); **734** (KHOKHR-

YAKOV). For the Kamchatka Peninsula see **735**; for Wrangel Island see **058**. Hultén's *Flora of Alaska and neighboring territories* (**110**) also includes records from this area in its species distribution maps. The unit here corresponds approximately to the limits of the Chukot National District (Magadan Oblast') in the Russian Federation. Topographically it is rather broken, with a number of discrete ranges reaching to 1843 m.

With the exception of the recent Magadan Oblast' flora by Khokhryakov (**734**), no separate work fully covering the area is available. Early knowledge was first consolidated by Trautvetter in *Flora terrae Tschuktschorum* (1879) and many scattered lists, particularly for the Chukchi Peninsula, have appeared since including a florula, described below. A general survey of the vegetation over much of the area is given in V. N. VASILEV, 1956. Растительность Анадырского края (*Rastitel'nost' Anadyrskogo kraja*). 218 pp., maps, tables. Moscow: AN SSSR Press.

Partial works

KOŽEVNIKOV, JU. P., 1979. Флора основания Чукотского полуострова (*Flora osnovanija Čukotskogo poluostrova*). 240 pp. Vladivostok/Magadan: 'Depot'.

A 'basic' flora of the Chukchi Peninsula, the most easterly part of the Russian Federation. [Not seen; reference from E. J. Jäger in *Prog. Bot.* **42**: 333, 343 (1980).][60]

KOŽEVNIKOV, JU. P., 1981. Список сосудистых растений Чукотия (Spisok sosudistykh rastenij Čukotija). *Nov. Sist. Vyss. Rast.* **18**: 230–247.

Concise annotated enumeration, comprising a numbered systematic list with for each species three symbols respectively designating a given plant's biogeographical element, floristic province (just two are represented!) and geobotanical/vegetational units. The list is followed by a discussion and analysis of the flora along with references.

Region 07

Nearctic region

This region includes the mainland Arctic and sub-Arctic zones of Alaska, Canada and Greenland, together with the Canadian Arctic Archipelago and the Bering Sea Islands (except for the Aleutian chain).

Until the late nineteenth century, the region, except for Greenland, remained botanically comparatively poorly explored. Serious examination began after World War I, and has intensified in the last 40 or more years. Maguire (1958; see supraregional references) has reviewed the areas visited and the work of the various collectors. Present botanical coverage is now comparatively good.

Floristic information was consolidated firstly by William J. Hooker in his *Flora boreali-americana* (1829–40), by his son Joseph in *Outlines of the distribution of Arctic plants* (1861) and, towards the end of the century, by John Macoun in his *Catalogue of Canadian Plants* (1883–1902). All of these cover the whole of northern North America. More recent compilations include *Circumpolar arctic flora* (1959) by N. Polunin (**050–70**), and *Flora of Canada* (1978–79) by H. J. Scoggan (**120–30**), the latter also inclusive of Alaska.

At territorial level, individual works are by contrast with most of the Eurasian Arctic relatively numerous. The current standards, with one exception, were all published between 1950 and 1980 and in greater or lesser detail cover the Alaskan North Slope, the Canadian Arctic Archipelago, Ungava and the Arctic portions of the mainland Northwest Territories of Canada. Illustrated manuals have been written for the Archipelago and Greenland, both in two or more editions. Works mainly rooted south of the tree-line but covering parts of the Arctic zone include the excellent *Vascular plants of continental Northwest Territories, Canada* (1980) by A. E. Porsild and W. J. Cody (**123**) and *Flora of the Yukon Territory* (1996) by Cody (**122**), both noteworthy for their liberal use of illustrations and species distribution maps, *Géographie floristique du Québec-Labrador* (1974) by C. Rousseau (**133**) covering Ungava, and the Alaskan floras of Hultén and Anderson (**110**). There is also now a flora for the Yukon (1996) by W. J. Cody (**122**).

Bibliographies. General bibliographies as for Superregion 05–07.

Indices. General indices as for Superregion 05–07.

071

Bering Sea Islands

See also **110** (HULTÉN, 1968; WELSH). These include the islands of St. Lawrence and St. Matthew as well as the Pribilofs. The only relatively recent floristic work is that by S. B. Young on St. Lawrence (see below).

YOUNG, S. B., 1971. *The vascular flora of St Lawrence Island with special reference to floristic zonation in the Arctic regions.* Pp. 11–115, 24 figs., incl. map (Contr. Gray Herb., N.S. 201). Cambridge, Mass.

Systematic enumeration, with synonymy, citation of *exsiccatae*, extralimital range, and notes on habitat and the ecological and geographical limits of species along with critical commentary; no separate index. The introductory section provides accounts of physical features, climate, vegetation, history of contact, botanical exploration, and the Bering 'land bridge'. Concluding parts deal with general floristic and biogeographic questions and considerations and include a new attempt at a delimitation of floristic limits on the southern periphery of the Arctic zone.

072

Alaska and Yukon (Arctic and arctalpine zones)

See also **110** (HULTÉN, 1968; WELSH); **122** (CODY).

WIGGINS, I. L. and J. H. THOMAS, 1962. *A flora of the Alaskan Arctic slope.* 425 pp., frontisp., maps. Toronto: University of Toronto Press.

Briefly descriptive flora of vascular plants of that part of Alaska north of the Brooks Range, with keys to all taxa, limited synonymy, detailed accounts of local range (with citations of *exsiccatae*), distribution maps for all species, notes on habitat, special features, etc., and critical taxonomic commentary. The introductory section includes accounts of physical features, climate, soils, permafrost, biological factors, and botanical exploration and research in the area. A gazetteer of localities, glossary, and index to all botanical names conclude the work.

073

Nunavut and Northwest Territories of Canada (Arctic mainland zone)

See **074** (POLUNIN; PORSILD, 1955) and **123** (PORSILD and CODY, 1980). The Arctic zone encompasses the Melville and Boothia Peninsulas (in Nunavut), the greater part of former Keewatin District (all now in Nunavut) and the northern part of Mackenzie District (Great Bear Lake northwards), i.e., the northeastern half of the area covered by the two works of Porsild and Cody (see especially the main map in their 1980 flora).

Key earlier works on this area include *The vascular plants of the Arctic coast of America west of the 100th meridian* by J. Macoun and T. Holm (1921, in *Rep. Canad. Arctic Exped.*, 5 (Botany), A: 1A–24A), and *Materials for a flora of the continental Northwest Territories of Canada* by A. E. Porsild (1943, as *Sargentia* 4).

074

Canadian Arctic Archipelago

This area comprises most of the former Franklin District of the Northwest Territories of Canada, now all in the autonomous Inuit region of Nunavut.

The best flora for field use remains the illustrated manual by Porsild (1957). The other two works here listed are documentary, without keys or many illustrations. The three works were preceded by *A survey of the phytogeography of the Arctic American archipelago with some notes about its exploration* by H. G. Simmons (1913, in *Acta Univ. Lund.*, N.S., afd. II, 9(19)), a work accounting for 204 species.

PORSILD, A. E., 1957. *Illustrated flora of the Canadian Arctic Archipelago.* 209 pp., 70 text-figs., 332 maps (Bull. Natl. Mus. Canada 146). Ottawa. (Reprinted with supplement, 1964, 1973.)

Illustrated manual of vascular plants, with keys to all taxa, essential synonymy, English vernacular

names, generalized indication of formation-zones and local range with individual maps for all species (in a separate section), and notes on habitat, occurrence, frequency, ecology, phenology, etc. The introductory section includes accounts of physical features, climate, soils, habitats, formation-zones, vegetation, and botanical exploration; a glossary and index to all vernacular and botanical names conclude the work. The supplement of 1964 contains revisions of maps 1–332 and additional maps 333–344 as well as a revised index (pp. 205–211) and additional pages 213–218 with figures 71–72. Notes on new collections are also given.[61]

Partial works

POLUNIN, N., 1940. *Botany of the Canadian Eastern Arctic*, I. Flora. 408 pp., 8 text-figs, 2 maps (Bull. Natl. Mus. Canada 92). Ottawa.

Systematic enumeration, without keys, of vascular plants of the eastern Canadian Arctic Archipelago as well as Ungava (extreme northern Québec) and the Melville Peninsula; includes descriptions of new taxa, synonymy (with references but no citations), detailed accounts of local range, *exsiccatae*, general summaries of overall as well as eastern Arctic distribution, and notes on habitat, ecology, biology, etc., along with critical remarks. The introductory section includes notes on geographical and botanical exploration, together with lists of collecting localities, while the main text is followed by a summary (with statistical tables) of the flora and its distribution, a bibliography, and an index to all botanical names.

PORSILD, A. E., 1955. *Vascular plants of the western Canadian Arctic Archipelago.* 266 pp., illus., maps (Bull. Natl. Mus. Canada 135). Ottawa.

Systematic enumeration of vascular plants of the western Canadian Arctic Archipelago and adjacent mainland area, complementing Polunin's work (see above); includes limited synonymy (with references and citations), detailed indication of local range, *exsiccatae*, general summary of overall distribution, taxonomic commentary, and notes on habitat, ecology, biology, and phytogeographic relationships. The introductory sections comprise accounts of physical features, ecology, climate, glaciation, the history of the flora and vegetation, composition of the flora, phytogeographic divisions, and botanical exploration. A bibliography and index to botanical names conclude the work.

075

Ungava (far northern Québec)

See 074 (POLUNIN); 133 (ROUSSEAU). This environmentally vulnerable area, the northern part of the Labrador Peninsula, is not effectively dealt with in standard Québec floras (133) except for Rousseau's *Géographique floristique* and therein only partially. It is politically part of the Inuit region of Nunavut.

GARDNER, G., 1973. *Catalogue analytique des espèces végétales du Québec arctique et subarctique et de quelques autres régions du Canada/Analytic catalogue of plant species from the arctic and subarctic of Quebec and other regions of Canada.* lxii, 142 pp. (French text), lxviii, 234, [2] pp. (English text), illus., maps. Montréal. (Bilingually captioned unpaged maps and photographs between text blocks.)

The main text in each version comprises a systematic checklist of species with localities, data and *exsiccatae* (all collected by the author, encompassing some four decades of field work); this is preceded by descriptions of study sites with indication of representative species. General indices appear at the end of the work. The English text includes non-vascular plants and lichens. Coverage extends south to the Mingan Islands and adjacent north shore of the Gulf of St. Lawrence as well as to Moosoonee (Ontario) and Churchill (Manitoba).

076

Greenland

Area: 2 175 600 km². Vascular flora: 497 species, of which 15 endemic (Böcher *et al.*, 1978). This vast island has long been a Danish possession.

The major earlier floras are the massive, three-part *Conspectus florae groenlandicae* by Johan Lange *et al.* (1880–94, published as *Meddel. Grønland* 3), covering 374 species, and *The flora of Greenland and its origin* by C. H. Ostenfeld (1926, in *Kongel. Danske Vidensk. Selsk., Biol. Meddel.* 6(3)), accounting for 390 species. A brief historical survey of floristic and systematic botanical research appears in the enumeration by

Jørgensen *et al.* (see below), but the most detailed source work is C. Christensen, 1924–26. *Den danske botanisks historie* (in 2 vols.) and its continuations (see 674). Much work on the flora has been published in the topographical serial *Meddeleser om Grønland*, initiated in 1879.

Bibliography
See C. A. JØRGENSEN *et al.* (1958).

BÖCHER, T. W., B. FREDSKILD, K. HOLMEN and K. JAKOBSEN, 1978. *Grønlands flora.* 3rd edn. 326 pp., illus., map. Copenhagen: Haase. (1st edn., 1957.) English edn.: T. W. BÖCHER, K. HOLMEN and K. JAKOBSEN, 1968. *The flora of Greenland*, transl. T. T. Elkington and M. C. Lewis (from the 2nd Danish edition of 1966). 312 pp., 66 text-figs, map. Copenhagen: Haase.

Illustrated manual of vascular plants, with keys to all taxa, limited synonymy, abbreviated indication of local and extralimital range, karyotypes, and notes on habitat, special features, etc. The introductory section includes chapters on earlier published flora accounts, general floristics, and the major plant communities. An illustrated glossary, bibliography, and index to generic and family names conclude the work. [For a systematic list, condensed from this manual, see K. HOLMEN, 1968. *Checklist of the vascular plants of Greenland.* 2nd edn. 40 pp. Copenhagen: Universitets Arktiske Station/Godhavn: Grønlands Botaniske Undersøgelser.]

JØRGENSEN, C. A., T. SØRENSEN and M. WESTERGAARD, 1958. *The flowering plants of Greenland: a taxonomical and cytological survey.* 172 pp. (Kongel. Danske Vidensk. Selsk., Biol. Skr. 9(4)). Copenhagen: Munksgaard.

Systematic enumeration, with references to standard monographic and revisionary treatments, karyological data and sometimes quite detailed critical commentary; no *exsiccatae*. The enumeration proper is followed by a table of species with their chromosome numbers, sources and references (including those especially contributed for the work), summary discussion and acknowledgments, and (pp. 156–172) a bibliography of some 500 items. The introductory part contains on pp. 5–7 a review of botanical exploration.

Superregion 08–09

South Polar regions

The limits of this subdivision are as defined by GREENE and WALTON in their *Checklist* (1975; under **080–90**). It conforms to the southern 'boundary' of woody plant growth, thus incorporating all land below 45° S (excluding the Falkland (Malvinas) Islands, southern South America and southern New Zealand as well as the islands in the New Zealand region from Campbell Island northwards). The Tristan da Cunha group, Gough Island, and the Amsterdam and St. Paul group, formerly also thought to be sub-Antarctic, are now considered by some to be within the southern cool-temperate zone.

Antarctic botany, which in the guise of 'southern zone' botany has had a scope somewhat wider than as here limited, appears to have had its serious beginnings with two voyages by Bougainville. The second voyage, to which Philibert Commerson was attached as naturalist, was a global circumnavigation which, however, would leave him in the southwestern Indian Ocean islands (Superregion 46–49). By the mid-nineteenth century, British contributions had become the most substantial, culminating in the collections and observations of 1839–43 by J. D. Hooker and David Lyall on the voyage of the *Erebus* and *Terror*; these, along with all other available evidence, were used by Hooker for the preparation of the first of his botanical reports, the classic *Flora antarctica* (1843–47; see below). This is still the only comprehensive descriptive account, although for the most part it treats more northerly areas and omits some islands not yet discovered or on which Captain James Ross deemed a landing impossible (the Marion group and the Crozets being among the latter). Hooker would continue his examination of the southern temperate flora in two further voyage reports: *Flora novae-zelandiae* (1852–55) and *Flora tasmaniae* (1855–60).

The remaining islands (including Heard, discovered only in 1853) were explored and documented chiefly by the British *Challenger* expedition in 1873, the Transit of Venus expeditions (including the German *Gazelle*) in 1874–75, voyages conducted in 1882–83

under the aegis of the 'International Polar Year', the German *Valdivia* and *Gauss* expeditions at the turn of the century, and, in 1911–14, the Australasian Antarctic Expedition led by Douglas Mawson. Detailed exploration by temporarily resident students also began in the nineteenth century but did not become general until after World War II when a revival of interest in Antarctic exploration and scientific research occasioned the establishment of base camps (as well as weather stations). In recent years, within the superregion under review, monographic accounts have been published for Macquarie (Taylor, 1955; Selkirk, 1990), Kerguelen (Chastain, 1958), the Marion group (Huntley, 1971), South Georgia (Greene, 1964) and, just outside, the Falklands or Malvinas (Moore, 1968), Fuegia (Moore, 1983) and the Tristan da Cunha group (Wace and Dickson, 1965; Groves, 1981). The Australian islands (Macquarie and the McDonald (Heard) group) were in 1993 written up in *Flora of Australia*. A short account of the miniscule vascular flora of Antarctica itself (with the South Shetlands and South Orkneys) was given by Skottsberg (1954) and extended by Greene and Holtom (1971); and Moss (1988) contributed a monograph of the Antarctic Peninsula.

A new, but brief, consolidated account of the superregion was produced in the form of a checklist by Greene and Greene (1963; revised version 1975). The vascular flora as a whole is now relatively thoroughly known. For the McDonald (Heard) group and the Crozets, however, there are as yet no detailed accounts; and there is a need for a modern comprehensive flora and distribution atlas, as Wace (1965, pp. 254–255; see below under **080–90**) has suggested. Also of interest will be the monitoring of new aliens and of other changes to the vegetation, not to mention more intense studies of individual species.[62]

Progress
Eighteenth- and nineteenth-century exploration is reviewed in the introduction to *Flora antarctica* (Hooker, 1845; see below under **080–90**), and in E. J. GODLEY, 1965. Botany of the southern zone, exploration to 1843. *Tuatara* **13**: 140–181; and *idem*, 1970. Botany of the southern zone: exploration 1847–1891. *Ibid.*, **18**(2): 49–93. No consolidated account of work after 1891 has been seen.

Bibliographies. Bay, 1910; Blake and Atwood, 1942; Frodin, 1964, 1984; Goodale, 1879; Holden and

Wycoff, 1911–14; Jackson, 1881; Pritzel, 1871–77; USDA, 1958.

Indices. BA, 1926; BotA, 1918–26; BC, 1879–1944; BS, 1940– ; CSP, 1800 1900; EB, 1959–98; FB, 1931– ; ICSL, 1901–14; JBJ, 1873–1939; KR, 1971– ; NN, 1879–1943; RŽ, 1954– .

080–90

Superregion in general

The only modern work accounting for the entire superregion is the checklist by GREENE and WALTON (1975; see below). No descriptive flora is available apart from J. D. Hooker's *Flora antarctica* (1844–47) which within the superregion deals only with Kerguelen. For a review of vegetation and phytogeography, see N. M. WACE, 1965. Vascular plants. In J. van Mieghem and P. van Oye (eds.), *Biogeography and ecology of Antarctica*, pp. 201–266. The Hague: Junk. (Monographiae biologicae 15.)

GREENE, S. W. and D. W. H. WALTON, 1975. An annotated checklist of the sub-Antarctic and Antarctic vascular flora. *Polar Rec.* **17**(110): 473–484. (Succeeds S. W. GREENE and D. M. GREENE, 1963. Checklist of the sub-Antarctic and Antarctic vascular flora. *Ibid.*, **11** (73): 411–418.

Tabular list of native vascular plants of the six islands or island groups of the sub-Antarctic floristic zone: Macquarie, McDonald (Heard), Kerguelen, Crozet, Prince Edward (Marion), and South Georgia with presence or absence indicated for each unit; essential synonymy indicated in a separate table. The work also includes descriptive floristic remarks on individual islands and island groups, with notes on aliens and sources of information; a summary of records of vascular plants in the Antarctic floristic zone proper (the South Sandwich, South Orkney, and South Shetland groups together with the Antarctic Peninsula and adjacent islands); and a list of references.

HEMSLEY, W. B., 1885 (1884). Report on the botany of the Bermudas and various other islands of the Atlantic and Southern Oceans, part 2. In W. Thomson and J. Murray (eds.), *Reports on the scientific results of the voyage of HMS* Challenger *during the years 1873–76*, *Botany*, 1(2): 1–299, pls. 14–53. London: HMSO.

This amply dimensioned work incorporates annotated enumerations of non-vascular and vascular plants of the Marion (Prince Edward) Islands (pp. 187–206, pl. 53), the Crozet Islands (pp. 207–210, pl. 40), the Kerguelen Archipelago (pp. 211–243, pl. 40), and the McDonald (Heard) Islands (pp. 245–258, pl. 53), with descriptions of new taxa as well as synonymy, references and citations; also included are general indications of local and extralimital range (with citation of *exsiccatae*), extensive taxonomic commentary, and notes on habitats, special features, biology, etc. (partly compiled from earlier sources), with figures of new and noteworthy plants appended. Each formal treatment is preceded by general notes on physical geography, features of the flora, vegetation formations, and botanical exploration together with tabular analyses of phytogeographical relationships in relation to the island group concerned. The remainder of this second part deals with the South Atlantic Islands (Region 03) along with Amsterdam and St. Paul (**044**).

HOOKER, J. D., 1844–47. *Flora antarctica*. 2 vols. xii, 574 pp., 198 pls., map (The botany of the Antarctic voyage of HM Discovery Ships *Erebus* and *Terror* in the years 1839–1843, I). London: Reeve.

Well-illustrated descriptive flora, in large format, of non-vascular and vascular plants, with synonymy and references; indication of distribution and habitat; extensive taxonomic commentary. The introductory section deals with exploration and phytogeography. The first volume (part 1) covers the Auckland Islands and Campbell (**418**, **419**); the second (part 2) treats Kerguelen, the Falkland (Malvinas) Islands (**389**), and southernmost South America.[63]

SCHENCK, H., 1905. Vergleichende Darstellung der Pflanzengeographie der subantarktischen Inseln insbesondere über Flora und Vegetation von Kerguelen, I–VI. In C. Chun (ed.), *Wissenschaftliche Ergebnisse der deutschen Tiefsee-Expedition auf dem Dampfer 'Valdivia', 1898–1899*, 2(1): 1–178, 33 textfigs., pls. 1–10, map. Jena: Fischer. Supplemented by *idem*, 1906. Die Gefässpflanzen der deutschen Südpolar-Expedition 1901–1903 gesammelt auf der Possession-Insel (Crozet-Gruppe), Kerguelen, Heard-Insel, St Paul und Neu-Amsterdam. In E. von Drygalski (ed.), *Deutsche Südpolar-Expedition 1901–1903*, 8 (Botanik): 97–123, 10 text-figs. Berlin: Reimer.

Schenck's *Valdivia* report comprises individual treatments (to a considerable extent compiled from

outside sources) – with a strong ecological, vegetatiological and phytogeographical bias – of the following
areas: I, 'Kerguelenbezirk', encompassing the
Kerguelen, Marion (Prince Edward), Crozet, and
McDonald (Heard) island groups (pp. 1–82); II,
'Südgeorgien', including also Bouvetøya and the South
Sandwich chain (pp. 82–96); III, 'Falkland-Inseln' (pp.
96–106); IV, 'Feuerland', or Tierra del Fuego (pp.
106–130); V, 'Inseln südlich von Neuseeland', including Campbell Island and the Snares, Antipodes, and
Auckland groups as well as Macquarie Island (pp.
130–161); and VI, 'Antarktisches Polargebiet', comprising various localities on and near the Antarctic (or
Palmer) Peninsula (pp. 161–178). Each section consists
of accounts of botanical exploration, physical geography, climate, general features of the flora and its
'Charakterpflanzen', and ecological adaptations as
given by morphology and anatomy (reflecting the interests of A. F. W. Schimper who had been a member of
this expedition), plant formations, floristics, and phytogeography. Also included (in parts I, II, V and VI) are
tabular lists of vascular and non-vascular plants occurring on individual islands or island groups, with essential synonymy and distributional details (based upon all
available information). Schenck's second report
includes annotated lists of vascular plants collected
during the stay of the *Gauss* in each island group
(Possession Island, Crozets, pp. 99–102; Kerguelen,
pp. 102–119; Heard Island, pp. 119–120).[64]

Region 08

Circum-Antarctic Islands

Included under this heading are the following
islands or island groups: Macquarie, McDonald
(Heard), Kerguelen Archipelago, Crozet (Possession),
Marion (Prince Edward), Bouvetøya, South Georgia,
and South Sandwich. These islands are distinguished
from others in the 'southern oceans' by their remoteness from continents and by their absence of woody
vegetation, a definition narrower than that adopted by
J. D. Hooker but more in keeping with current usage.
The Falkland Islands (Islas Malvinas; see 389) are relatively close to South America, while woody vegetation
is present in Tristan da Cunha, Gough, and the

Amsterdam and St. Paul group, classified by Good
(1974; see **General references**) as 'sub-Antarctic' and
Takhtajan (1986; see **General references**) as part of
his 'Holoantarctic Kingdom'. Additional information
may be found under the supraregional heading.

Bibliographies. General bibliographies as for Superregion 08–09.

Indices. General indices as for Superregion 08–09.

081

Macquarie Island

See also **080–90** (GREENE and WALTON;
SCHENCK, 1905, pp. 130–161); **410** (ALLAN;
CHEESEMAN); **415** (CHEESEMAN). – Area: 120 km².
Vascular flora: 36 native species as of 1985, of which 3,
all grasses, are endemic (PD; see **General references**);
by 1993, 46 were known, of which 41 native (Hnatiuk in
AUSTRALIAN BIOLOGICAL RESOURCES STUDY, 1993;
see below).

Macquarie Island has not unnaturally been the
subject of several florulas, although all date from after
1900. The first critical account of the flora was by the
New Zealand botanist T. F. Cheeseman, published as
The vascular flora of Macquarie Island (1919, in *Scient.
Rep. Australas. Antarctic Expedition 1911–1914*, C,
7(3)). This was followed by the ecologically oriented
survey by B. W. Taylor (1955; see below). In the 1980s
two other papers appeared: G. R. COPSON, 1984. An
annotated atlas of the vascular flora of Macquarie
Island. *ANARE Res. Notes* 18; and R. D. SEPPELT, G.
R. COPSON and M. J. BROWN, 1984. Vascular flora and
vegetation of Macquarie Island. *Tasmanian Nat.* 78:
7–12. For a more general treatment of the island's environment and biology, see P. M. SELKIRK, R. D.
SEPPELT and D. R. SELKIRK, 1990. *Subantarctic
Macquarie Island: environment and biology.* xiv, 285 pp.,
illus. Cambridge: Cambridge University Press.[65]

AUSTRALIAN BIOLOGICAL RESOURCES STUDY,
1993. *Flora of Australia*, 50: *Oceanic Islands* 2. xxvi, 606
pp., 97 figs. (part col. and incl. maps). Canberra:
Australian Government Publishing Service.

Pages 53–62 of this volume comprise an introduction to the Australian sub-Antarctic islands (by

Roger Hnatiuk), encompassing a general description of physical features and vegetation along with concisely annotated lists of species giving their status (unfortunately without indication of previous records where appropriate) and a key to families. The list for Macquarie Island appears on pp. 59–60 and a description (with map) on pp. 53–57. A color topographic photograph appears as fig. 7. All species are further treated in the descriptive part – which forms the bulk of the book and follows the format of other *Flora* volumes – together with those from other islands or groups (Cocos (Keeling), **045**; Christmas, **046**; Ashmore and Cartier, **451**; and Coral Sea Islands, **949**).

TAYLOR, B. W., 1955. *The flora, vegetation and soils of Macquarie Island.* 192 pp., 11 text-figs., 42 pls. (Rep. Austral. Natl. Antarctic Res. Exped., B, 2). Melbourne.

Pages 105–156 comprise an ecologically oriented enumeration, without keys, of known vascular plants; entries include general accounts of local distribution and extensive notes on habitat, life-forms, phenology, floral biology, dispersal mechanisms, and associations. Several species are illustrated. The systematic list is preceded by extensive chapters on physical features, climate, soils, biotic factors, vegetation formations and plant communities.

082

McDonald (Heard) Islands

See also **080–90** (GREENE and WALTON; HEMSLEY, pp. 245–258, pl. 53; SCHENCK, 1905, pp. 1–82 (with Kerguelen); SCHENCK, 1906, pp. 119–120). – Area: Heard I., 700 km². Vascular flora: 12 species, of which 11 native (Hnatiuk in AUSTRALIAN BIOLOGICAL RESOURCES STUDY, 1993); there are no pteridophytes.

The group, an Australian External Territory, comprises Heard, with an actively volcanic central peak rising to 2745 m, and, at 40 km distant, the three small McDonalds. Only some 10–15 percent of the land area on Heard is ice-free. All native species occur on Heard, but only 5 in the McDonalds. The following work contains the first successor to Hemsley's list.

AUSTRALIAN BIOLOGICAL RESOURCES STUDY, 1993. *Flora of Australia,* 50: *Oceanic Islands* 2. xxvi, 606 pp., 97 figs. (part col. and incl. maps). Canberra: Australian Government Publishing Service.

See corresponding entry under **081** for further details. The Heard and McDonald checklist appears on p. 60 and a description of the islands (with map) on pp. 57–59. A color topographic photograph appears as fig. 8.

083

Kerguelen Archipelago

See also **080–90** (GREENE and WALTON; HEMSLEY, pp. 211–244; SCHENCK, 1905, pp. 1–82; SCHENCK, 1906, pp. 102–119). – Area: 7000 km². Vascular flora: 32 native and 6 adventive species (Lourteig and Cour, 1963). The partially glaciated archipelago hosts the headquarters of the French Southern and Antarctic Territory.

The first complete coverage of Kerguelen is embodied in J. D. Hooker's *Flora antarctica* (see **080–90**). Hooker also contributed the first independent treatment, incorporating the Transit of Venus and *Challenger* expeditions' collections (along with records from the related Marion (**085**), Crozet (**084**) and McDonald (**082**) groups) as a chapter of his *Account of the petrological, botanical and zoological collections made in Kerguelen's Land and Rodriguez during the Transit of Venus Expeditions . . . in the years 1874–75* (1879, in *Phil. Trans.* **168**, extra volume). At that time 26 native and one introduced vascular species were recorded. Further data were added by Schenck, following which little was published until the establishment of the French station following World War II.

CHASTAIN, A., 1958. *La flore et la végétation des Îles Kerguelen.* 136 pp., 6 text-figs, 36 pls., 25 tables, 2 maps (Mém. Mus. Natl. Hist. Nat., sér. II/B (Bot.) 11(1)). Paris.

Chapter 3 of this work incorporates an ecologically oriented enumeration (without keys) of known species, with generalized indication of local range and detailed notes on habitat, variability, etc.; each species illustrated. The remainder is devoted to physical features, geology, climate (and wind!), fauna, history of exploration, general features of the flora, phytogeography, vegetation formations and plant communities. A summary list of the flora (pp. 31–33) and a bibliography are also provided.

LOURTEIG, A. and P. COUR, 1963. Essai sur la distribution géographique des plantes vasculaires de l'archipel de Kerguelen. *Publ. CNFRA, Biol.* **3**: 65–78 (incl. 12 maps).

Revision and geographical analysis (with 10 maps) of the known native species (only two strictly endemic); bibliography, p. 69. Two further maps display the work areas of 18 collecting missions from Cook in 1776 to Jolinon in 1959.

084

Crozet (Possession) Islands

See **080–90** (GREENE and WALTON; HEMSLEY, pp. 207–210; SCHENCK, 1905, pp. 1–82 (with Kerguelen); SCHENCK, 1906, pp. 99–102). – Area: 505 km². Vascular flora: 28 species (GREENE and WALTON). The islands are part of the French Southern and Antarctic Territory.

No separate florula covering the group appears to be available. Invasive plants were recently documented in C. CARCAILLET, 1993. Les plantes allochtones envahissantes de l'Archipel Crozet, Océan Austral. *Rev. Écol.* (Terre Vie) **48**(1): 3–20.

085

Marion (Prince Edward) Islands

See also **080–90** (GREENE and WALTON; HEMSLEY, pp. 187–206; SCHENCK, 1905, pp. 1–82 (with Kerguelen)). – Area: 390 km². Vascular flora: 22 native and 13 introduced species on Marion; 21 native and 1 introduced on Prince Edward (Huntley, 1971). Marion rises to an altitude of 1200 m, with the summit permanently under ice. It and the rather smaller Prince Edward Island are dependencies of South Africa.

The standard checklist of Huntley (1971) has been complemented by N. J. M. GREMMEN, 1977. The distribution of alien vascular plants on Marion and Prince Edward Islands. *S. African J. Antarctic Res.* **5**: 25–30.

HUNTLEY, B. J., 1971. Vegetation. In E. M. van Zinderen Bakker, Sr., J. M. Winterbottom and R. A.

Dyer (eds.), *Marion and Prince Edward Islands*, pp. 98–160. Cape Town: Balkema.

Table 3 in this primarily vegetatiological chapter (which forms part of an 'island monograph') comprises a tabular checklist of the vascular flora, based on earlier records and one year's field work by the author; it also includes indication of presence on other groups (Crozet, Kerguelen and South Georgia in particular). Background information on the flora appears on pp. 105–107 and includes historical data; the rest of the chapter deals with vegetation and phytosociology. [Other chapters in this 'island monograph' treat non-vascular plants, animal life, etc.]

086

Bouvetøya (Bouvet Island)

See **080–90** (GREENE and WALTON; SCHENCK, 1905, pp. 82–96 (with South Georgia)). Most ice-free areas on this almost wholly glaciated island, a Norwegian possession, are steep and exposed to avalanches. Recent exploration has yielded, in addition to crustose lichens and algae, four moss species but no other cormophytes. The terrestrial vegetation is described in T. ENGELSKJØN *et al.*, 1987. Botany of Bouvetøya, South Atlantic Ocean, II. The terrestrial vegetation of Bouvetøya. *Polar Research*, N.S., **5**(2): 129–163.[66]

087

South Georgia

See also **080–90** (GREENE and WALTON; SCHENCK, 1905, pp. 82–96). Much of the interior is glaciated. – Area: 3757 km². Vascular flora: 24 native species (PD; see **General references**).

South Georgia was first seriously explored by Hermann Will and others on a German expedition within the International Polar Year programme of 1882–83 (see **Superregion 05–07** for general comments). The collections were described in *Die Phanerogamen von Süd-Georgien* by A. Engler (1886, in *Bot. Jahrb. Syst.* **7**: 281–285, and later in the voyage reports of Neumayer). Greene's flora was founded on more intensive, locally based work after World War II. Naturalized and adventive plants are covered in D. W. H. WALTON and R. I. L. SMITH, 1973. Status of the alien vascular flora of South Georgia. *Bull. British Antarct. Surv.* **36**: 79–97.

GREENE, S. W., 1964. *The vascular flora of South Georgia.* 58 pp., 8 text-figs. (incl. map), 6 pls., 31 text-maps, 1 folding map (Scient. Rep. Brit. Antarctic Surv. 45). London.

Detailed, large-format descriptive flora of native, naturalized, and adventive vascular plants, with keys to all taxa; synonymy, with citations only; detailed distribution given for each species with accompanying gridded maps, with (if appropriate) general indication of extralimital range; taxonomic commentary and notes on habitat, altitudinal distribution, ecology, variability, etc.; glossary and list of references at end. The introductory chapters deal with the history of the island since contact, physical features, geology, climate, botanical exploration and location of collections, vegetation formations and plant communities, status of individual species and records of those introduced, and general comments on the maps. A summary list of species appears on pp. 29–30.

088

South Sandwich Islands

See also **080–90** (GREENE and WALTON; SCHENCK, 1905, pp. 82–96 (with South Georgia)). – Area: 310 km^2. Vascular flora: 1 species (LONGTON and HOLDGATE, 1979; see below). The islands are to a great extent barren, with active vulcanism and extensive glaciation. A sub-Antarctic affinity was demonstrated by Longton and Holdgate on the basis of the bryophyte flora.

LONGTON, R. E. and M. W. HOLDGATE, 1979. *The South Sandwich Islands*, IV: *Botany.* 53 pp., 13 text-figs., 1 pl., tables (Scient. Rep. Brit. Antarctic Surv. 94). Cambridge.

Monographic treatment of the flora and vegetation of this extensively glaciated volcanic island group, covering physical features, climate, geological history, exploration, plant habitats, and treatments of individual islands, concluded with a discussion of different

vegetation types and the origin of the flora; references (pp. 50–51). A table of the flora and its internal (and external) distribution appears on pp. 8–11; *Deschampsia antarctica* Desv. (*D. elegantula* (Steud.) Parodi) is the only vascular species, recorded from just one of the islands (Candlemas).

Region 09

Antarctica

In addition to the Antarctic mainland and the Antarctic Peninsula, the following nearby islands or island groups are included herein: South Orkney, South Shetland, Palmer Archipelago, Peter I, Balleny, and Scott. – Area: South Orkney Islands, 620 km^2; South Shetlands, 4700 km^2; Antarctica, 14 000 000 km^2. Vascular flora: 2 native species (Skottsberg, 1954; see below); naturalization of others has occurred since.

The first collection of a vascular plant in Antarctica was made by an American, James Eights, in 1829 at New South Shetland Island near the Antarctic Peninsula (Graham Land). The presently known southerly limit for such plants is on Neny Island, off the peninsula, at 68° 12′S. With greater summer melt in recent years, however, the two native species are expanding their presence. There is also a considerable non-vascular plant flora in which mosses and lichens are prominent on land whilst larger algae are conspicuous along the littoral.

Bibliographies. General bibliographies as for Superregion 08–09.

Indices. General indices as for Superregion 08–09.

090

Region in general

See also **080–90** (GREENE and WALTON). For further information on the native species, particularly on distribution and performance, see D. M. GREENE and A. HOLTOM, 1971. Studies in *Colobanthus quitensis* (Kunth) Bartl. and *Deschampsia antarctica* Desv., III: Distribution, habitats and performance in the Antarctic botanical zone. *Bull. Brit. Antarctic Surv.* **26**: 1–29. Since 1954, additional vascular plants have become naturalized in the region.

SKOTTSBERG, C. J. F., 1954. Antarctic flowering plants. *Svensk. Bot. Tidskr.* **51**: 330–338, illus., map.

Comprises a descriptive account, with map, of the three vascular plant species (two certainly native) then recorded from the Antarctic mainland along with the Palmer Archipelago, South Shetland Islands, and South Orkney Islands.

Local work: South Orkney Islands

The South Orkneys (and the South Shetlands) are part of the British Antarctic Territory, with Signy home to a British Antarctic Survey research base.

WILLIAMS, J. A., 1972. Studies in *Colobanthus quitensis* (Kunth) Bartl. and *Deschampsia antarctica* Desv., V: Distribution, ecology and performance on Signy Island. *Bull. Brit. Antarctic Surv.* **28**: 11–28.

A detailed study of the native vascular species, with among general conclusions being that their ranges were expanding and abundance increasing as glaciation had receded. Most sites were within 100 m of the shore and no more than 170 m in elevation.

Notes

1 Examples include G. H. M. Lawrence, *Taxonomy of vascular plants* (1951); P. W. Leenhouts, *A guide to the practice of herbarium taxonomy* (1968); A. E. Radford *et al.*, *Vascular plant systematics* (1974); C. Jeffrey, *An introduction to plant taxonomy*, 2nd edn. (1982); V. V. Sivarajan, *Introduction to the principles of plant taxonomy* (1984; 2nd edn., 1991); and D. W. Woodland, *Contemporary plant systematics* (1991; 2nd edn., 1997).

2 P. W. Leenhouts, in his *Guide to the practice of herbarium taxonomy* (1968, Utrecht), and other Dutch botanists considered the original version of this work to be superior to that of Hutchinson (see above); however, due to its rarity, it enjoyed relatively limited usage.

3 A current version of this product may be seen at http://biodiversity.uno.edu/delta/.

4 From 1986 through 1989 new pteridophyte names were also listed in the annual *Kew Index*, now discontinued. For a long time, taxonomic judgments were made in this work with, for example, certain combinations being improperly

omitted as undesirable (Alice F. Tryon, personal communication, 24 January 1986).

5 The index began with a list in *Contr. U.S. Natl. Herb.* **1**: 151–188 (1892) compiled under the direction of the then-librarian of the Department of Agriculture, Josephine Clark, and chief botanist Frederick V. Coville. The cards were introduced in 1894. Compilation continued at Washington until 1903, when increasing duties forced Ms. Clark to arrange for the work to be done elsewhere; it was then assumed by the Gray Herbarium. Non-vascular cryptogams were originally included but dropped after 1927. Inclusion of the infraspecific names dates from about 1945; infrageneric names were added from 1970. The publication of index cards was discontinued in 1983; microfiche was then introduced for new entries. In the 1990s the conversion to electronic format was performed in the Philippines under contract. The index is now entirely virtual; its Web address is at http://herbaria.harvard.edu/. More recently, the *Gray Herbarium index* has joined forces with *Index Kewensis* and the *Australian plant names index* in a new Web-based venture, the *International Plant Names Index* (IPNI). For further information, see B. L. Robinson and Lesley C. Wilcox, 1930. The Gray Herbarium card index. *Science* **71**: 253–256; and Elizabeth A. Shaw, 1971. The Gray Herbarium card index. *Taxon* **20**: 333–336.

6 'Cleanup' of the electronic version continues, and will be reflected in future CD-ROM editions. A second edition of the CD-ROM appeared in 1996.

7 In 1989, a group of botanists under the leadership of the Royal Botanic Gardens, Kew, proposed a 'Species Plantarum Project'. Motivated by increasing world concern with resource management and a perceived need to make botanical information more readily available, these botanists believe that, given an effective organizational structure, international collaboration, adequate finance, and especially the utilization of modern information-handling and communications technology, such an enterprise is possible. The Project is currently established as part of the International Organisation for Plant Information (IOPI), formed in 1991. Reports of activities were initially published in *Taxon* but now appear on the Web at http://iopi.csu.edu.au/iopi/.

8 Until his death in 1841 much of the work was written personally by the senior de Candolle in the old tradition (without extensive collaboration from specialists). Under Alphonse de Candolle, however, the work became, like the contemporary *Flora brasiliensis*, a collaborative enterprise to which most of the leading monographers of the day contributed. With the completion of the *Prodromus*, the decision was taken to proceed with independent monographs, the *Monographiae phanerogamarum*. Detailed surveys of the contents of the *Prodromus* are found in Pritzel's

Thesaurus (see Appendix A) and in vol. 1 of *Taxonomic Literature-2* by F. A. Stafleu and R. S. Cowan (1976). For a modern commentary (also included in the 1966 reprint of vol. 1 of the main work), see F. A. Stafleu, 1966. *The great Prodromus*. Lehre: Cramer. To some observers, however, the work was more encyclopedic than practical; perusal of its pages soon led to a 'sense of despair' (V. H. Heywood, 1980. The impact of Linnaeus on botanical taxonomy – past, present and future. *Veröff. Joachim Jungius-Ges. Wiss. Hamburg* 43: 97–115, esp. p. 105).

9 Adolf Engler early offered to work up Araceae, presenting an outline to Alphonse de Candolle; this became the basis for his family account. See T. S. Ray and S. S. Renner, 1990. Introduction. In A. Engler, *Comparative studies on morphology of Araceae*, II (Englera 12). Berlin. Engler also contributed the Anacardiaceae. The termination of the *Monographiae* may well have influenced him to promote what later became *Das Pflanzenreich*.

10 For a historical account, see H. W. Lack, 1987. Opera magna in Berlin plant taxonomy. *Englera* 7: 253–281; and *idem*, 1990. Opera magna der Berliner Systematischen Botanik. In C. Schnarrenberger and H. Scholz (eds.), *Geschichte der Botanik in Berlin*, pp. 265–296. Berlin: Colloquium. (Wissenschaft und Stadt 15.) The project was launched by the Royal Prussian Academy of Sciences on the occasion, in 1900, of its 200th anniversary. It was financed by the Academy and through a government grant; Hermann Harms was engaged as managing editor, with his salary paid by the Academy (Hermann Sleumer, personal communication); contributors were recruited from both the Botanical Museum and outside. Harms died in 1942 and after World War II the rights to the work passed to what became the Academy of Sciences of the German Democratic Republic; there was, however, little or no effective interest in its continuation. [After German reunification in 1990 this academy underwent a reorganization.]

11 *Species filicum* is the last of its kind on ferns; although now largely of historical value it set precedents in fern taxonomy which were not seriously challenged until the 1930s (through the work of R. C. Ching, E. B. Copeland, and others; B. S. Parris, personal communication).

12 The author himself wrote many precursory papers which appeared in contemporary French botanical journals – though not those of the Muséum National d'Histoire Naturelle.

13 Like Baillon's work, *Genera plantarum* was supported by many precursory studies, large and small. As far as possible, the work was based upon original investigation of specimens and other plant materials, largely those available at Kew. Its classification served as the basis for the arrangement used in many floras, notably the 'Kew series' covering the British Empire (and Commonwealth).

14 This 'opus magnum' was described in H. W. Lack, *Opera magna*. Fifty-seven collaborators completed the phanerogams for the first edition in 12 years, with Engler himself contributing one-sixth of the contents. The original prospectus for the second edition called for 27 *Bände*, with a publication schedule of nine years (1923–31). *Bände* 1–11 were planned to cover non-vascular plants and fungi; *Band* 12 the pteridophytes; *Band* 13, the gymnosperms; *Bände* 14–15, the monocotyledons; *Bände* 16–22, archichlamydean dicotyledons; *Bände* 23–26, metachlamydean dicotyledons; and *Band* 27 was to comprise a *Register*. As the work progressed, however, many *Bände* had to be subdivided; and the numbering of *Band* 28bI (1980) indicates a further revision of the original scheme – at least beyond *Band* 21, Parietales–Opuntiales (published in 1925 and until 1980 the furthest such in the system). As of 1999, 28 individual volumes (*Bände* or parts thereof) covering a miscellany of higher taxa from the Schizophyta through the Angiospermae/Gentianales have appeared. Of these, however, only five date from after World War II (and three of them at least partly based on work accomplished prior to the war), with two since 1959. (Another contribution, on the red algae, was published separately by its author, Harald Kylin, as *Die Gattungen der Rhodophyceen* (1956, Stockholm). Janet Perkins contributed generic revisions of Monimiaceae and Styracaceae; these were published independently in the 1920s but never in the work itself on account of non-completion of related families.) With publication of *Band* 28bI in 1980, the *Pflanzenfamilien* became more international (including a change to English as the language of publication and recruitment of authors from a wider circle). Stylistic changes have also been made. Work on the second edition is again actively in progress; a further part-volume appeared in 1995 and several more are in preparation.

15 Although the Hutchinson system, especially in its 1959 version, is now largely obsolete (its principal intellectual successors being the systems of Takhtajan and Cronquist), it remains worthy of note as one or the other version has served as the basis for the arrangement of several modern floras (notably in Africa). Hutchinson's *Genera* is still 'alive' at Kew in the form of critical publications on individual families (e.g., *Genera of Araceae*, 1997).

16 The project is now under the editorship of Jim Lundqvist, Swedish Museum of Natural History.

17 Revisions of *Ceriops*, *Sonneratia* and *Avicennia* have appeared since publication.

18 For the French version of Hooker's presentation, see J. D. Hooker, 1866. Considérations sur les flores insulaires. *Ann. Sci. Nat., Bot.*, sér. V, 6: 267–299. The English version was reprinted in 1896 (London: Reeve; 36 pp.).

19 The expedition's instructions in fact called for special attention to oceanic islands, including their presumably ancient floras and their fragility as environments.

20 These and other Darwin plants from the *Beagle* voyage have been evaluated in D. M. Porter, 1986. Charles Darwin's vascular plant specimens from the voyage of the *Beagle*. *Bot. J. Linn. Soc.* **93**: 1–172; and *idem*, 1987. Darwin notes on Beagle plants. *Bull. Brit. Mus. (Nat. Hist.)*, Hist. Ser. **14**: 145–233.

21 The collections upon which Hooker's work was based are examined in detail in D. M. Porter, 1980. The vascular plants of Joseph Dalton Hooker's *An enumeration of the plants of the Galapagos . . . Bot. J. Linn. Soc.* **81**: 79–134. For Darwin's plants, see Porter, 1986 (Note 20).

22 For an account of botanical progress and needs, see D. M. Porter, 1990. Taxonomic status and needs. In J. E. Lawesson *et al.* (eds.), *Botanical research and management in Galápagos*, pp. 1–4. St. Louis. (Monogr. Syst. Bot. Missouri Bot. Gard. 32). The remaining papers in this collection are indicative of contemporary interests.

23 The genesis and preparation of this flora is recounted in D. M. Porter, 1968. The flora of the Galápagos Islands. *Ann. Missouri Bot. Gard.* **55**: 173–175.

24 A general assessment of the state of knowledge of Chile's Pacific oceanic islands, with chapters contributed by specialists, is J. C. Castilla (ed.), 1988. *Islas Oceánicas de Chile: estado del conocimiento científico y necesidades de investigación*. Santiago: Ediciones Univ. Católica de Chile.

25 The main work was based in the first instance on collections and a manuscript left by Friedrich Johow.

26 See *ASPT Newsletter* 7(3): 14–15 (July 1993). For a general assessment of the state of knowledge of the islands, see J. C. Castilla, *op. cit.*

27 A new (non-endemic) family, Thyrsopteridaceae, a segregate of the Dicksoniaceae, was here also proposed.

28 Forster's paper initially appeared in *Commentationes societatis regiae scientiarum gottingensis* 9. It was a part of the two Forsters' contributions from the 1772–75 voyage of the *Resolution* and *Adventure* under Captain James Cook.

29 Useful current reviews of the flora may be found in the papers by P. Sunding (pp. 13–40) and C. J. Humphries (pp. 171–199) in D. Bramwell (ed.), 1979. *Plants and islands* (see **009, Islands**). On the 'Flora of Macaronesia' initiative, see D. Bramwell, 1972. Flora of Macaronesia Project. *Taxon* 21: 730–731; and D. Bramwell and D. M. Moore, 1973. Flora of Macaronesia Project: a progress report. In G. Kunkel (ed.), *Proceedings of the I. International Congress pro Flora Macaronesica*, pp. 165–168 (Monographiae biologicae canarienses 4). Las Palmas. A taxonomic database is maintained in Oslo; this has provided support for the successive editions of the *Flora of Macaronesia checklist*.

30 Monod, a well-known French biologist, first visited the islands in 1933.

31 For mid-twentieth century botanical activities, see P. L. Pérez de Paz, 1982. Perspectiva histórica de los últimos 50 años (1932–1982) de la botánica en Canarias. In *Instituto de Estudios Canarios, 50. Aniversario*, 1 (Ciencias): 295–340. Santa Cruz de Tenerife.

32 A German edition has also been published.

33 The history of botanical work is covered in L. A. Grandvaux Barbosa, 1962. Les botanistes dans l'Archipel du Cap-Vert. In A. Fernandes (ed.), *Comptes Rendus de la IV. Réunion Plenière de l'AETFAT*, pp. 77–94. A more recent summary is W. Lobin, 1987. Geschichte der botanischen Erforschungen der Kapverdischen Inseln. *Cour. Forschungsinst. Senckenb.* 95: 23–24.

34 For a description of this project, see M. A. Diniz *et al.*, 1996. Progress of the Cape Verde flora. In L. J. G. van der Maesen *et al.* (eds.). *The biodiversity of African plants*, pp. 783–784. Dordrecht. The bulk of material studied has been from herbaria in Cape Verde, Portugal, Paris, Bonn and Oslo. That such a relatively small flora should have been published in fascicles is mere hubris: vanity rather than utility. The same result could have been achieved in one volume at half or less the cost. Moreover, the project was a bilateral rather than international undertaking (W. Lobin, personal communication).

35 This total has increased only slightly in nearly two centuries. According to Hemsley (1884), following his 1805–06 visit François André Michaux had furnished an estimate of 140–150 native species.

36 A general treatment of the islands is C. Sechin, 1987. *Arquipélago de Fernando de Noronha*. Rio de Janeiro. For a cactus enthusiast's *Reisebericht*, see P. J. Braun, 1990. Arquipélago de Fernando de Noronha (Brasilien). *Kakt. Sukkul.*, 41(11): 254–258.

37 For more historical and topographical information, see W. B. Turrill, 1948. On the flora of St. Helena. *Kew Bull.* 3(3): 358–363. Quentin Cronk has undertaken preparation of a new definitive treatment and Simon Goodenough a field-guide.

38 The author compiled the work in 1975–78 while Agriculture and Forestry Officer in the territory.

39 General works on the islands include B. Lobo, 1919. Conferencia sobre a Ilha da Trindade. *Arq. Mus. Nac. Rio de Janeiro* 22: 107–158, and E. M. Peixoto, 1932. *Ilha da Trindade: memoria historica*. Rio de Janeiro (Publicações do Arquivo Nacional 28). Eyde and Olson, with the aid of material collected by Becker, identified the 'dead trees' and supply an illustration. Suggestions as to the cause of the trees' demise were also advanced.

40 The Tristan group and Gough, sometimes classed as sub-Antarctic, are north of the Antarctic Convergence and, given the presence of a tree, *Phylica arborea*, on all the larger islands, best regarded as cool-temperate oceanic.

41 This expedition arguably was the most important of its kind in the western Indian Ocean before the mid-

twentieth century. A total of 141 papers resulted, all published in eight volumes of the *Transactions of the Linnean Society, Zoology.*

42 S. A. Renvoize, 1979. The origins of Indian Ocean island floras. In D. Bramwell (ed.), *Plants and islands*, pp. 107–129 (see **009**).

43 For a recent general introduction to the islands, see O. Saigal, 1990. *Lakshadweep.* vii, 200 pp., illus. New Delhi: National Book Trust.

44 Some errors occur in the phytogeographical analysis. Among American elements, for example, *Anona muricata* is likely to be an introduction, *Carica papaya* is certainly introduced, and *Dodonaea viscosa* is widely distributed. The real origin of elements has not been satisfactorily analyzed. Of interest is that there are some species not yet recorded on the Indian mainland, including *Guettarda speciosa* and *Pisonia alba* (cf. *P. grandis*).

45 Dennis Adams has completed a new enumeration, accounting for all previous work; this remains in manuscript for the present (C. D. Adams, personal communication; also noted in Fosberg, 1986, under **009**). This arose from a 'flora identification study', inclusive of two months' field work, pursued by Adams in 1982–83 under the aegis of FAO and the main basis for his 1983 report.

46 The history of botanical work in the Maldives to 1964 is related by E. W. Groves in *Atoll Res. Bull.* **116**: 57–59 (1966). It forms part of a general report on Addu.

47 Fosberg has noted in this work that the word 'atoll' is of Maldivian origin.

48 Though sometimes viewed as sub-Antarctic, the native flora is better viewed as cool-temperate oceanic. Support for this view is furnished particularly by the two species with the greatest biomass being also prominent in the Tristan group and Gough (**037, 038**); one of these is the only significant woody plant, *Phylica arborea*. It was this shrub or small tree which contributed to de Labillardière's 1792 sighting of 'forests' on present-day Amsterdam. St. Paul, the smaller and more southerly, by contrast lacks woody vegetation and furthermore has been actively volcanic within historical time.

49 Mariners considered Amsterdam to be very difficult of access; before 1906 the longest stay was nine days, by a French Transit of Venus expedition in 1874–75.

50 This may be compared with the respective figures of 57 and 43 given by S. A. Renvoize, *op. cit.*, p. 116, which reflect what was known by 1912.

51 Baker *et al.* in 1900 had listed 145 native species.

52 Curiously, this work, with its 1826 supplement, has not as such been superseded.

53 Botanical progress may also be recorded in *Arctic Bulletin* (1973– , Washington, D.C.). In the early 1990s, moves towards a new *Flora arctica* were initiated (*Flora of North America Newsletter* **6**: 3 (1992)).

54 According to Lid, however, the French specimens were by 1932 'lost' in the Muséum National d'Histoire Naturelle, Paris.

55 For a general account (based on a survey of 1930; no botany), see N. N. Urvantzev, 1933. *Severnaya Zemlya: a short summary of exploration.* 53 pp., 10 halftones, folding chart. Leningrad: All-Union Arctic Institute, USSR. (Russian title: *Severnaja Zemlja: kratkij očerk issledovanija.*)

56 The author here called for a new comprehensive flora of the Ural. Govoruchin's manual, good in its day, was by the 1960s viewed as obsolete.

57 Annotation originally prepared from notes furnished by the New York Botanical Garden.

58 Annotation prepared with the assistance of M. E. Kirpicznikov, St. Petersburg.

59 Lipschitz no. 2022. Annotation prepared from notes furnished by Mrs. J. Diment, formerly of the Natural History Museum, London.

60 Jäger credited this work as having contributed greatly to an improved understanding of the flora of the then-Soviet Eastern Arctic zone.

61 This work was designated as a 'Basic Flora' for the 1966–73 Flora North America Program.

62 In early 1995, it was reported that with the exposure of more areas of Antarctica during the southern summer, populations of the two native vascular plants have been expanding.

63 Though now out of date and in the present region dealing only with Kerguelen, the work, along with its companions *Flora novae-zelandiae* (1852–55) and *Flora tasmaniae* (1855–60), remains basic to our knowledge of the floristics of the entire cool-temperate southern zone. Fuller evaluations appear in W. B. Turrill, 1963. *Joseph Dalton Hooker.* London; and R. Desmond, 1999. *Sir Joseph Dalton Hooker: traveller and plant collector.* Woodbridge, Suffolk: Antique Collectors' Club.

64 Although the only islands within our circum-Antarctic zone actually visited by the *Valdivia* were Bouvetøya and the Kerguelen Archipelago, it was Schenck's expressed intention to provide a comprehensive survey of the floristics, phytogeography, vegetation and ecology of the whole South Polar zone – much as had Hooker 60 years before.

65 Of the gradually increasing number of recorded native species, Hnatiuk has written: 'These [additional species] are small in size, inconspicuous and occur among relatively dense, short vegetation. Rather than newly arrived taxa they could just as easily be rare taxa that have occurred there for long periods' (in ABRS, 1993, pp. 53ff.).

66 For details of the non-vascular flora, see T. Engelskjøn *et al.*, 1986. *Botany of Bouvetøya, South Atlantic Ocean,* I. *Cryptogamic taxonomy and phytogeography.* 79 pp., illus. (Skr. Norsk Polarinst. 185). Oslo.

Division 1

• • •

North America (north of Mexico)

Compactness being essential, only the leading synonymy and most important references are given, and these briefly.

> A. Gray, preface to *Synoptical flora,* vol. 2, part 1 (1878).

Watson's death will make a big gap in American botany . . . There is no one now to go on with the flora [*Synoptical flora*] and the possibility of our having a North American continental flora seems very remote, a not very creditable state of things for American botanists to contemplate.

> C. S. Sargent to W. T. Thistleton-Dyer, 15 March 1892; quoted from S. B. Sutton, *Charles Sprague Sargent and the Arnold Arboretum,* pp. 130–131 (1970).

A synoptical Flora of North America [on the lines of *Flora Europaea*] is both feasible and desirable at this time.

> S. G. Shetler, *Taxon* 15: 257 (1966).

[FNA is] a new concept of linking modern information systems technology with time-honored means of scientific research and publication to produce a flora – species-based repository of information on plants – *as an electronic data bank and information system.* . . . [In the] 6-year first phase, an intense effort will be mounted to produce the [synoptical] flora.

> Advertisement for FNA, *BioScience* 21: 527–528 (1971).

The Flora North America project was recently revitalized . . . to produce a conventional flora of the vascular plants of North America north of Mexico using *traditional methods* [italics added] . . . It is hoped that the flora project will be completed by 1990.

> Announcement of the 'new' FNA, *Brittonia* 31: 124 (1979).

In the absence of funding from NSF [the U.S. National Science Foundation], the Missouri Botanical Garden cannot continue to function as the organizational center for FNA . . . Center staff except for Jim Zarucchi were laid off June 1 [1999].

> *Flora of North America Newsletter* 13: 2 (1999).

The entire North American continent north of Mexico is here encompassed, save for the tundra zone and Greenland (works referring entirely to these latter areas being covered under Region 07). – Area (exclusive of Greenland): 19 339 000 km^2. Vascular flora: 18 303 species (based on table 3 from 1999 Flora of North America National Science Foundation proposal; at http://www.fna.org/), an increase of over 2000 from the 16 274 redorded in SHETLER and SKOG (1978; see below).

It is difficult in a few paragraphs to present an adequate summary of the background to the present network of standard regional, state and provincial floras now blanketing most of the continent. That network assumes particular importance in the one major part of the northern temperate or Holarctic floristic zone yet without a complete modern comprehensive flora: a glaring gap only partly filled in 1978–79 by *Flora of Canada*, the one substantial work realized from the original Flora North America Program of 1966–73. The current *Flora of North America*, publication of which began in 1993, is still some ways from full realization. Indeed, no continental flora has been completed since the early nineteenth century.

However, this absence of a comprehensive flora should perhaps be set against a long-standing historical trend, evident since the mid-nineteenth century: the gradual increase in specialization at local, state or (less often) regional level, along with increasing emphasis on the solution of systematics problems of ever smaller scale through the use of increasingly diverse and sophisticated methodologies and subdisciplines. Added to this of course is the division of the continent into two large, geographically diverse federal states each comprising a number of unofficial 'regions' (as well as political entities) which among much else have had an effect on professional life.[1]

The overall evolution of floristic botany in the continent and the development of regional activities may be related not only to geography but also to personalities and general social and cultural developments. Its initial development along the Gulf and Atlantic coasts was essentially avocational or paravocational, the work of interested individuals (Mark Catesby, John and William Bartram, Cadwallader and Jane Colden, John Clayton, and Thomas Walter among others) in more or less isolated centers more in contact with colleagues in Europe than with each other. There were also several visitors from Europe, notably Pehr Kalm, charged by Sweden with an official mission in the 1740s, and André Michaux, similarly commissioned by France in the 1780s. Kalm's collections were incorporated by Linnaeus into *Species plantarum* and Michaux eventually compiled the first descriptive flora of the continent, the *Flora boreali-americana* (1803) actually published by his son François-André. Johann Reinhold Forster compiled a 51-page checklist, *Flora Americae septentrionalis* (1771), while in England. Only a few other floras, all local, were published.

The early national period in the United States saw the rise of eastern Pennsylvania as a significant general center with the development of the nursery trade (including the growers John Bartram, Jr., Humphrey Marshall and Bernard M'Mahon), the development of landscaped estates, and the detailed studies of Henry Muhlenberg (Lancaster) and Benjamin Smith Barton (Philadelphia) and, somewhat later, William Barton, William Darlington, Thomas Nuttall and Constabin Rafinesque-Schmaltz (all in Philadelphia) and Lewis von Schweinitz (Bethlehem). Barton, Muhlenberg and Nuttall independently began the task of compiling continent-wide works, although the first-named's project was realized only by his erstwhile assistant, the German Frederick Pursh. These works respectively were *Flora Americae septentrionalis* (1814; reissued 1816); *Catalogus plantarum Americae septentrionalis* (1813; 2nd edn., 1818); and *Genera of North American flowering plants* (1818), the last also enumerating known species (with descriptions of novelties). Works of more local scope were contributed between 1810 and 1830 by Jacob Bigelow, William Barton, William Darlington, Stephen Elliott and John Torrey; and at Lexington, Kentucky, there developed the first center west of the Appalachians where, at Transylvania College (now University), Constabin Rafinesque (1819–26) and, later, Charles Short were on the faculty. Amos Eaton in Troy, New York, published the first of several editions of his school-flora (the last,

still on the Linnaean system, in 1840). As a popularizer, he would be followed by Alphonso Wood.

The period from 1820 until Reconstruction saw the emergence of botany as a distinct profession, during which systematic research on any scale became dominated by a few 'big men', notably John Torrey and – by mid-century – Asa Gray, to whom others by and large began to defer. Ewan (1969) has termed this period the 'Torrey and Gray epoch'. Their contemporaries often acted in a 'field office' capacity, few continuing much work on their own. The 'big men' were based in the Boston area and New York; the earlier centers, especially Philadelphia, waned in importance, any but local work in the former capital largely ceasing by 1841 with Nuttall's return to England, the death of Rafinesque and, concomitantly, the growth to dominance of zoology and paleontology at the Academy of Natural Sciences of Philadelphia.

The Academy and other societies and institutions, including in time the Smithsonian Institution, became the depositories for the considerable collections amassed from about 1830 onwards by systematic state-sponsored geological and natural history surveys as well as the exploring expeditions and surveys of the federal government. It was through these surveys over the following six decades that there came about the great expansion of floristic knowledge of the continent, particularly in the rich West. A large proportion of the new collections, notably from the railway route surveys of the 1850s, were written up by Torrey and Gray with significant contributions from, among others, J. S. Newberry, Elias Durand, W. S. Sullivant and George Engelmann. Torrey and Gray, either alone or in collaboration, also prepared several synthetic works: *Flora of North America* (Torrey and Gray, 1838–43), *Flora of the state of New-York* (Torrey, 1843), and *Manual of botany of the northern United States* (Gray, 1848; 5th edn., 1867). Guided in part by William Hooker's *Flora boreali-americana* (1829–40) on British North America, these works established the descriptive 'manual-flora' style since more or less customary in North America.

With increasing settlement in other parts of the continent and the formation of more tertiary institutions of learning, particularly after the Morrill Act of 1862, the botanical community, largely floristically oriented at first but later widely diversifying, grew greatly, notably in the Midwest and West. Not surprisingly it was the botanists in the San Francisco area, notably Edward L. Greene, who first seriously 'challenged' the northeastern oligarchy, having developed at least a partially independent capability for floristic work and the preparation of local floras. Similar nuclei emerged elsewhere, notably in St. Louis with the Missouri Botanical Garden, at Washington, D.C. with the development of the joint herbarium of the U.S. Department of Agriculture and Smithsonian Institution (where agrostological research was in particular pursued under George Vasey, curator from 1872 to 1893), in New York with the revival of systematics at Columbia University under Nathaniel L. Britton, and in eastern Canada through the work of the Macouns at Ottawa and David Penhallow at Montreal. After 1873 and his retirement, Gray in the 'decades of transition' returned to the ideal of a continental flora, and before his death completed two parts (corresponding to one volume) of *Synoptical flora of North America* and began work on another. Of this latter, two (of four) fascicles were completed by Gray's assistant and curator Sereno Watson (until 1892) and Benjamin Robinson, but after 1897 activity lapsed due to changing personal and institutional interests and circumstances. The indexing work begun by Watson in 1878 has, however, carried on to the present day in the form of the *Gray Herbarium Index*.

The growing decentralization of floristic botany, as well as the development of taxonomy in North America, was strongly influenced by the dominance from 1890 to 1930 by Britton and his 'school' in New York. Indeed, it has been said that the New York Botanical Garden, founded in 1891 largely through his efforts, was established mainly for flora production. Britton's group (which included Leroy Abrams in California) was responsible for several regional works in the genre, some in direct competition with Gray's *Manual*, which between them blanketed the United States and adjacent parts of Canada except for the Great Basin, the Southwest, and parts of Texas and Oklahoma; they also produced a goodly number of state and local floras together with bibliographic and other tools. All became vehicles for the propagation of the so-called 'American Code' of nomenclature, a set of rules largely conceived by Britton who with a precise but rigid mind wanted no part of the 'sensible . . . Kew Rule' and other such nomenclatural subtleties. The 'American Code' was after 1900 adopted by the United States government as it was, perhaps fortuitously, American, but doubtless also as some members of the Britton 'school' were involved at the time in the expanding federal scientific establishment. It was thus that the 'American Code' was used in a large number of the many floras and other contributions on western (and to a lesser extent, southern) states and Mexico, as well as in publications on dendrology, grasses and other topics, prepared by federal or other botanists and published from the 1890s to the 1940s in the *Contributions from the United States National Herbarium*, *Miscellaneous Publications of the U.S. Department of Agriculture*, and other government outlets.[2]

The use of the 'American Code' was of course also extended to Britton's ambitious but never-completed comprehensive project, *North American Flora*, launched in 1905 partly as an answer to Gray's *Synoptical flora* but with a different methodology (involving among other things collective effort, as in *Flora brasiliensis*) and with more extensive geographical limits (it included the whole of Middle America in a gesture smacking more of botanical imperialism than common sense). In many respects, the Britton era was the great age of American descriptive floristics (the use of floras and keys then also forming a part of secondary school curricula), as well as the one exhibiting the greatest number of herbarium starts. During this period additional 'floristic centers' came into being, notably on and near the Pacific coast, in Wyoming and Colorado, and in the Midwest and, in later years, elsewhere.

Although the Britton 'school' had become discredited by the 1920s and the various floras were provocative of much criticism, the wide collective coverage of the latter helped greatly to promote local exploration and collecting in the period of the second transportation revolution, thus laying the groundwork for the later revival of floristic work (linked in part with ecology and 'biosystematics') and the marked proliferation of state floras and related contributions from the 1930s onwards, a development continuing to the present partly under its own momentum. This promotion was also accomplished by the floras from those work-groups (among them Hall and Jepson in California) remaining loyal to the International Rules and broader species and generic limits characteristic of Torrey and Gray; these gained renewed influence after unification of the nomenclatural codes in the 1930s and particularly the findings of early 'biosystematic' work and development of the neo-Darwinian 'biological species concept'.

The interwar period also saw further development of independent capabilities for floristic work and the writing of floras, firstly in eastern Canada and in Texas and later – mainly after World War II – in the Plains, the mountain states, and most notably in the Southeast. These have contributed greatly to a much improved knowledge of the flora, important in local terms, and to the waning of botanical 'colonialism' within North America north of Mexico. The genesis of the popular modern 'documentary' state flora appears to have been in the topographically relatively level Middle West, several of which from the time of Deam's *Flora of Indiana* (1940) included county dot maps, still a commonly used system of floristic mapping in the United States.

By the 1950s, if not before, floristic botany in the continent had become effectively decentralized. The various areas of 'biosystematics', spreading mainly from California and to many possessing a greater 'scientific respectability' with their seemingly more exact approaches to and hopes of 'solving' the 'species problem', attracted great attention in the rapidly expanding graduate schools of the day and, by and large, displaced classical floristic work. Within the relatively diminishing floristic field, 'extralimital' involvement became comparatively limited, relating mainly to Greenland, Alaska, the Canadian North, and parts of the American West, Southwest and South. Some of these, however, remain the least adequately known floristically; interesting discoveries have been made over the last half century and can still be expected. Indeed, much exploration has continued over most of the continent, now strongly influenced by the needs of monographic and revisionary work, 'biosystematic' studies, and ecological and environmental research (including investigation of the status of threatened and endangered species). The areas so named correspond to the three latter of Shetler's five 'states' of plant taxonomy. However, with some exceptions, the involvement by 'national' institutions in North American floristics and flora-writing had become comparatively limited.

Yet changes in the style and range of data presented in floras came slowly, with the evidence and conclusions from 'biosystematics' – including karyotypes, population structure and gene flow, hybridization, apomixis, and pollination and dispersal biology – having little influence until the late 1950s, and then at first mainly in the western United States. Even now, the style of some North American floras would still be quite recognizable to Torrey and Gray. There has,

however, been a greatly increased emphasis on mapping and on distribution, ecological and environmental data in some works, with a corresponding increase in bulk. Significant departures from tradition have been made in British Columbia, however, where both a so-called 'biosystematic' flora (of the Queen Charlotte Islands) and a state checklist based on the methodology of the Flora North America Program have been published. Other examples of nontraditional approaches have appeared patchily elsewhere, but in general little serious thought appears to have been given to the problem of presentation of data.[3]

It was not until the mid-1960s that, under the stimulus of overseas developments in flora-writing such as (notably) the appearance of the first volume of *Flora Europaea* but probably also influenced by a contemporaneous reappraisal of 'biosystematics' as well as contemporary economic optimism, that a mood for synthesis and 'big thinking' again appeared. This led to the initiation in 1966 of the most important and best-documented, but still chimeric, recent development in North American floristic botany: the Flora North America (FNA) Project (later Program).

The genesis, development and subsequent vicissitudes of FNA have been extensively recorded elsewhere. Over a period of more than 30 years, the FNA concept has undergone marked ideological and methodological oscillations, as indicated by the quotations under the divisional heading and reflecting, it would seem, the changing socio-economic and political atmosphere of the day. The project has been involved in controversy from time to time, particularly with respect to the projected information system. It was domestic disagreements, however, which were an immediate cause of its first termination in 1973. A revival on a more modest scale began in 1976 but had faded by 1980. With a deep business recession in progress prospects in the nearer term looked dim; Torrey and Gray's dream seemed as far away as ever.

In 1982, the Flora North America Project was reactivated as a 'private' rather than 'official' undertaking. A secretariat at the Missouri Botanical Garden was created and an extensive network of collaborators organized. After much preliminary work including the establishment of text-processing and database systems and an invited workshop in 1988, the first two of a currently projected 30 volumes appeared in 1993; volume 3 was published in 1997 and vol. 22 was in press by 1999. The *Flora* also has a World Wide Web site.[4]

In the intervening three decades, additional new and revised state, provincial and regional works have been published or are in preparation. There have been some outstanding achievements in individual polities, as in Newfoundland, the Carolinas, Texas, California, Michigan, and Utah, and recently keyed enumerations have been published for Florida and Maine, providing their fullest coverage to date. Species mapping has been pursued in many states and Canadian provinces, with the results incorporated into floras or, as in Pennsylvania and Utah, in separate atlases. Nevertheless, Heiser's 1969 statement that 'many areas of the United States have no manuals, or at least no up-to-date manuals, for the identification of their floras . . .' remains to an extent true.[5] Improvements have been piecemeal and for some states the only available general works are antiquated. Moreover, available manpower for such projects continues to decline.[6]

At regional level, recent projects have included the *Intermountain Flora*, now well advanced, *Flora of Great Plains* (1986), Boivin's *Flora of the Prairie Provinces* (published 1967–81 in installments in *Phytologia*), Budd's *Flora of the Canadian Prairie Provinces* (last revised in 1987), *Vascular plants of continental Northwest Territories* (1980), and Cronquist's revision of Gleason's *Manual of vascular plants of northeastern United States and adjacent Canada* (1991). The last-named was followed by an *Illustrated companion* (1997) edited by Noel Holmgren, based on the figures from Gleason's 1952 *Illustrated flora* but reorganized following the Cronquist system (as adopted for the revised *Manual*) and furnished with many new illustrations. Work has also continued on the *Generic flora of the southeastern United States*, a series of generic accounts begun in the late 1950s. All these followed such major post-World War II undertakings as the eighth edition of *Gray's Manual of Botany* by Fernald, the already-mentioned *Illustrated flora*, *Vascular plants of the Pacific Northwest* by Hitchcock and Cronquist as well as the later volumes of *Illustrated flora of the Pacific States* by Abrams and Ferris. Most were of necessity based at large, well-founded institutional herbaria. In the last decade, however, efforts which might have gone into further regional floras have been directed more towards *Flora of North America*, and will likely continue so for some years.

With respect to states, provinces and territories, new general works have appeared for Arkansas, Alaska, British Columbia, Texas, Utah, Wyoming, Kansas, South Dakota, the Carolinas, and Québec among others, and tangible progress is being made on floras for New York, Virginia, Tennessee, Florida, Louisiana, and Mississippi, to name those where documentation is lacking or markedly outdated. Two progressive field-manuals on the European model by W. Weber together cover Colorado. Projects have also been initiated in New Jersey and Pennsylvania. A considerable number of state checklists have also been published, along with some substantial partial floras (e.g., parts of Florida, northeastern Minnesota, northern Arizona, northern Utah and southern California). Notable gaps yet remain, however; even where the resources are available, professional or official interest or both may be lacking. Some areas may be considered adequately covered for most needs by the regional floras.

As for the woody flora, there is a wide range of more or less modern works now available, with trees being more fully treated than shrubs. A bibliography was published by Little and Honkala in 1976 as a successor to Dayton's list of 1952, and more recently a revised edition of the Forest Service *Check List* was produced by Little to succeed the edition of 1953. All tree species in the United States have now been mapped in some detail in the five-volume *Atlas of United States trees* (1971–78), a work also useful for adjacent parts of Canada and Mexico. However, no successor has yet appeared for the standard descriptive technical manual by Sargent, last revised in 1922; the nearest approach is that by Elias (1980), a more popularized work.[7]

Grasses have been covered most recently by Hitchcock and Chase (1950); a revision is currently underway under the direction of Mary Barkworth (now incorporated into *Flora of North America*). For the pteridophytes a number of regional and many state guides are available, recently listed in detail by Miasek (1977). In 1985 a nationwide semi-popular work by David Lellinger was published by Smithsonian Institution Press, in addition to the 'How to know' field-guide by John Mickel (1979). Manuals on aquatic plants (see 108) have been listed by Stuckey (1975) for the conterminous United States. No bibliography of North American floristic works as detailed as the listing in Blake and Atwood has appeared since completion of that list in 1939, although new works have been recorded in *Index to American Botanical Literature* (if published in serials; however, with transfer of the *Index* to *Brittonia* at the end of 1995, books are now also accounted for), *The Taxonomic Index* (1939–67), and elsewhere. State and regional floras were selectively covered by Gunn (1956; United States only), the present writer (1964, 1984), Shetler (1966), and by Lawyer *et al.* (late 1970s; never published). Popular and semi-popular floras were listed by Blake (1954) as well as in a nine-page leaflet, *Selected guides to the wildflowers of North America* by Elaine R. Shetler (1967, Washington, D.C.: Department of Botany, Smithsonian Institution).

A final noteworthy development of recent decades was the publication of the sumptuous six-volume set (with separate general index), *Wild Flowers of the United States* (1966–73); this was produced, as with many North American floras earlier in the twentieth century, at the New York Botanical Garden. This work, edited by H. W. Rickett, is presently the only 'modern' professionally prepared floristic work collectively covering the conterminous United States, but it is not, and was never intended as, a comprehensive scientific manual-flora.

Progress

The first considerable work on botanical history in the Americas is S. I. Mitchill, 1814. A discourse . . . embracing a concise and comprehensive account of the writings which illustrated the botanical history of North and South America. *Collect. New-York Hist. Soc.* 2: 149–215. The situation as of the mid-nineteenth century was summarized in W. Darlington, 1849. *Memorials of John Bartram and Humphry Marshall.* Philadelphia. His chapter 'Progress of Botany in North America' (pp. 17–33) includes a full coverage of floristic works.[8] In the early twentieth century, after the Civil War and several decades of breakneck industrialization and resource exploitation, came two more surveys: L. M. Underwood, 1907. The progress of our knowledge of the flora of North America. *Popular Sci. Monthly* 70: 497–517, and J. W. Harshberger in his *Phytogeographic survey of North America* (1911, Leipzig), pp. 1–92.

Mid-twentieth century histories include B. Maguire, 1958. Highlights of botanical exploration in the New World. In W. C. Steere (ed.), *Fifty years of botany*, pp. 209–246. New York: McGraw-Hill; L. H. Shinners, 1962. Evolution of the Gray's and Small's manual ranges. *Sida* 1: 1–31; and J. Ewan, 1969. *A short history of botany in the United States*, ix, 174 pp. New York: Hafner. The French contribution is sur-veyed in J.-F. Leroy (ed.), 1957. *Les botanistes françaises en Amérique du Nord avant 1850.* 360 pp., 32 pls., col. frontisp. (Colloq. Int. CNRS 63). Paris. The decade of the 1970s was marked by two additional reviews of North American floristics: in P. H. Raven, 1974. Plant systematics, 1947–1972. *Ann. Missouri Bot. Gard.* 61: 166–178, and in S. G. Shetler, 1979. North America. In I. Hedberg (ed.), *Systematic botany, planned utilization and biosphere conservation*, pp. 47–54. Stockholm: Almqvist & Wiksell.

Surveys published since the original edition of this *Guide* include J. L. Reveal, 1991. Botanical explorations in the American West, 1889–1989: an essay on the last century of a floristic frontier. *Ann. Missouri Bot. Gard.* 78: 65–80; and *idem*, 1992. *Gentle conquest: the botanical discovery of North America with illustrations from the Library of Congress.* 160 pp., illus. Washington, D.C.: Starwood. (Library of Congress Classics Series.) Finally, attention should be drawn to chapter 7, 'Taxonomic botany and floristics' by J. L. Reveal and J. S. Pringle in volume 1 of *Flora of North America* (1993; see below under **100**), as well as Pringle's contributions on Canada (see **Region 12/13**).

Flora North America (1966–73): The following accounts form a representative selection from a considerable literature: S. G. Shetler, 1966. Meeting of Flora of North America Committee. *Taxon* 15: 255–257; *idem*, 1968. Flora North America project. *Ann. Missouri Bot. Gard.* 55: 176–178; *idem*, 1971. Flora North America as an information system. *BioScience* 21: 524, 529–532; R. L. Taylor, 1971. The Flora North America project. *Ibid.*: 521–523. On the termination of the project, see H. Irwin, 1973. Flora North America: austerity casualty? *Ibid.*, **23**: 215; B. MacBryde, 1974. Flora of North America Program suspended. *Biol. Conservation* 6(1): 71; and J. Walsh, 1973. Flora North America: project nipped in the bud. *Science*, N.S. **179**: 778.[9]

A first, abortive revival of the *Flora North America* project was publicized in J. L. Reveal, 1979. Announcement: Vascular plants of North America north of Mexico. *Brittonia* 31: 124 (and other journals). The second, for some time successful, revival began – as already noted – in 1982; progress has been documented in *Flora of North America Newsletter*, on the FNA Web site, and by the published volumes.

General bibliographies. Bay, 1910; Blake and Atwood, 1942; Frodin, 1964, 1984; Goodale, 1879;

Holden and Wycoff, 1911–14; Hultén, 1958; Jackson, 1881; Pritzel, 1871–77; Rehder, 1911; USDA, 1958.

Divisional bibliographies: general floras and related works

In addition to the works listed below, mention should also be made of an unpublished, briefly annotated selection of floristic works and literature on species of special interest to conservationists, viz.: J. I. LAWYER *et al.*, [1979]. *A guide to selected current literature on vascular plant floristics for the contiguous United States, Alaska, Canada, Greenland, and the US Caribbean and Pacific Islands.* [New York.]

BLAKE, S. F., 1954. *Guide to popular floras of the US and Alaska.* 56 pp. (Bibliogr. Bull. USDA 23). Washington, D.C.: U.S. Government Printing Office. [Listing, by states and regions, of popular and semi-popular floristic works, including 'wild-flower books'.]

GUNN, C. R., 1956. An annotated list of the state floras. *Trans. Kentucky Acad. Sci.* **17**: 114–120. Continued as *idem*, 1956. A guide to some recent state floras. *Castanea* **21**: 33–38. [The main list comprises a tersely annotated selection, arranged by states, of 'standard' works. Regional floras are not included.]

MEISEL, M., 1924–29. *A bibliography of American natural history: the pioneer century, 1769–1865.* 3 vols. Brooklyn, N.Y.: Premier Publishing Co. (for the author). (Reprinted 1967, New York: Hafner.) [Includes references to floristic literature.]

SHETLER, E. R., 1966. *Floras of the United States, Canada, and Greenland: a selected bibliography with annotations.* 12 pp. Washington, D.C.: Department of Botany, Smithsonian Institution. (Offset from typescript.) [Concisely annotated list of major national, regional, state, and some local works, with indication of availability and prices. Compiled so as to furnish for the original Flora North America Program a working list of 'standard' floras similar to that drawn up by the *Flora Europaea* project.]

Divisional bibliographies: trees and shrubs

DAYTON, W. A., 1952. *United States tree books: a bibliography of tree identification.* 32 pp. (Bibliogr. Bull. USDA 20). Washington, D.C.: U.S. Government Printing Office. Continued as E. L. LITTLE, JR. and R. H. HONKALA, 1976. *Trees and shrubs of the United States: a bibliography for identification.* ii, 56 pp. (Misc. Publ. USDA 1336). Washington, D.C.. [Selected, classified references on the native and introduced woody flora of the United States and territories, the 1976 work emphasizing more recent literature.]

Divisional bibliographies: distribution maps

PHILLIPS, W. L. and R. L. STUCKEY, 1976. *Index to plant distribution maps in North American periodicals through 1972.* 752 pp. Boston: Hall. [Reproduced from a file of some 29 000 cards.]

Divisional bibliographies: pteridophytes

BLAKE, S. F., 1941. State and local fern floras of the United States. *Amer. Fern J.* **31**: 81–91, 131–143. Continued as *idem*, 1950. State and local fern floras of the United States. Supplement 1. *Ibid.*, **40**: 148–165; and as M. A. MIASEK, 1978. State and local fern floras of the United States of America, supplement 2. *Ibid.*, **68**: 109–118. [Successive classified lists of works through the mid-1970s.][10]

General indices. BA, 1926– ; BotA, 1918–26; BC, 1879–1944; BS, 1940– ; CSP, 1800–1900; EB, 1959–98; FB, 1931– ; ICSL, 1901–14; JBJ, 1873–1939; KR, 1971– ; NN, 1879–1943; RŽ, 1954– .

Divisional indices

TORREY BOTANICAL CLUB, NEW YORK, 1969. *Index to American Botanical Literature, 1886–1966.* 4 vols. Boston: Hall. Continued as *idem*, 1977. *Supplement, 1967–76.* 740 pp. Boston. [Originally published serially in *Bulletin of the Torrey Botanical Club* as well as (for a time) on index cards. In the reprint, taxonomic and other non-author entries appear in vol. 4. The *Index* continued publication in the *Bulletin* through vol. 122 (1995); control then passed directly to the New York Botanical Garden and from 1996 it has appeared in *Brittonia* (vol. 48 onwards) with a more specific emphasis on systematics and related fields.[11] From 1999, the *Index* is offered solely in electronic form.]

AMERICAN SOCIETY OF PLANT TAXONOMISTS, 1939–67. *The Taxonomic Index.* Vols. 1–30. New York (later Cambridge, Mass.: from vol. 20 (1957) published serially in *Brittonia*). (Vols. 1–9 mimeographed.) [Begun on the initiative of W. H. Camp, this index had an existence of 28 years but from the mid-1940s it was reproduced from the appropriate parts of the *Index to American Botanical Literature*. In 1957 it was consolidated with *Brittonia* although retaining separate volumation, and in 1967 discontinued.][12]

Conspectus

100

Division in general

Unless otherwise noted, works described below encompass the entire North American continent north of Mexico. General works at present comprise two floras (neither complete), two recent checklists and one nomenclator. Of those on specific groups, a selection accounting for woody plants, grasses and pteridophytes is furnished here. For boreal North America (Superregion 11–13) as a whole, see **120**. For the conterminous United States as a whole, including *Wild flowers of the United States* by Harold Rickett and collaborators, see **140**.

Keys to families and genera
The complementary keys by Wade T. Batson are both designed for field use. The current version of the eastern key is, however, in typescript too small for easy reading.

BATSON, W. T., 1983. *A guide to the genera of the plants of eastern North America.* 3rd edn. [4], 203 pp. Columbia, S.C.: University of South Carolina Press. [Illustrated keys to families and genera, with general statements of distribution; covers an area from Key West to the Arctic and from the Atlantic to the Plains and southern Texas. – Originally published 1975 as *A guide to the genera of the eastern plants* (2nd

edn., 1977, New York). The current version is a retitled reissue of the latter.]

BATSON, W. T., 1983. *Genera of the western plants.* 209 pp., illus. [Cayce, S.C.]: The author. [Pocket-sized illustrated dichotomous keys, covering 1669 native and 'commonly introduced' genera west of 98° (thus somewhat west of a line from Fort Worth, Texas to Fargo, North Dakota); the individual generic entries, arranged following Engler, include a diagnosis, an estimate of the number of species in the region, and distribution according to U.S. Postal Service state codes; list of poisonous plants and index to generic and vernacular names at end. The introductory part includes directions for use and a glossary.]

Comprehensive floras

BRITTON, N. L. *et al.* (eds.), 1905–57. *North American flora.* Vols. 1–34, *partim.* New York: The New York Botanical Garden. [Many vols. incomplete or not published.] Continued as C. T. ROGERSON and W. R. BUCK (eds.), 1954– . *North American flora,* Series II. Parts 1– . New York.

Briefly descriptive, rather strictly formatted flora of the vascular and non-vascular plants of North and Middle America and the West Indies, with keys to genera and species, type localities, extensive synonymy (with references and citations) and concise indication of local and regional distribution; no taxonomic commentary or ancillary remarks included. Each completed volume is separately indexed. [The original work, with families and higher groups arranged according to the Englerian system, was terminated with some 99 parts over 24 volumes published; its successor, in a similar format, is produced serially at irregular intervals without reference to a taxonomic sequence and with some revisions omitting Middle America and the West Indies.][13]

FLORA OF NORTH AMERICA EDITORIAL COMMITTEE, 1993– . *Flora of North America north of Mexico.* Vols. 1– . Illus., maps. New York/Oxford: Oxford University Press.

A multi-volume, systematically arranged, illustrated descriptive flora of cormophytes. Volume 1 is general, with chapters (by different authors) on physical features, climate, soils, historical and contemporary vegetation, phytogeography, 'floristics' (actually on the history of botanical exploration), weeds, ethno- and economic botany, conservation, and systematics (with chapter 14, by Arthur Cronquist, being a commentary on the 'general system of classification of flowering plants', and chapter 15, by James Reveal, presenting comparative features of different systems). Reveal adds, as appendix 1, a detailed outline of the Cronquist system as used in the work and, as appendix 2, a concordance with other systems. A literature list (for all chapters) and index conclude the volume. Subsequent volumes comprise the flora proper, with all treatments by specialists or knowledgeable authors; features include keys to all taxa, names with places of publication, concise synonymy, indication of distribution, habitats, and altitudinal range, individual distribution maps, critical notes, and comments on particular attributes, biology, etc., along with literature references if appropriate. All generic and familial entries also include references, and these are additionally gathered at the end of each volume along with a complete index. Treatments in each volume are preceded by an explanatory introduction. [Four volumes have been published through early 2000 (vols. 1–2, 1993; vol. 3, 1997; vol. 22, 2000); further volumes are in press or in preparation. Thirty volumes in all are now projected.][14]

GRAY, A. *et al.*, 1878–97. *Synoptical flora of North America*. Vols. 1(1), fasc. 1–2; 1(2); 2(1). New York: Ivison, Blakeman, Taylor & Co. (for 1(2) and 2(1)); American Book Co. (for 1(1)). (Volumes 1(2) and 2(1) reissued with supplementary sections in 1 vol., New York, 1886, the supplements also issued separately; 2nd reissue, 1888, Washington, D.C.: Smithsonian Institution, as *Smithsonian Misc. Collect.* 491. Vol. 1(1) as well as parts after 2(1) never completed.)

Concise descriptive flora of seed plants of North America north of Mexico (arranged according to a slightly modified Bentham and Hooker system); includes synoptical keys to families, genera, and species or groups of species, full synonymy (with references), generalized indication of internal and extralimital range, extensive taxonomic commentary, and notes on phenology, special features, etc.; indices at end of each part. A special supplement to vol. 1(2), containing addenda, corrigenda, and an index, appeared in 1886. [Volume 1(1), planned to cover 'Polypetalae' (but of which only two of a projected four fascicles ever appeared), as published extends from Ranunculaceae through the end of the Disciflorae (Polygalaceae), thus omitting the Calyciflorae; vol. 1(2) from Caprifoliaceae through Compositae; and vol. 2(1) from Goodeniaceae through Plantaginaceae (thus reaching the end of the Gamopetalae). The work as originally conceived was to have had two volumes, each of some 1200 pages. Volume

1 was, in the words of Gray's biographer Hunter Dupree, 'to go over the old ground', i.e., the contents of the original *Flora of North America* as far as completed by Gray with John Torrey, while vol. 2 was to deal with families not so covered. Only part of this plan was realized; when towards 1900 work lapsed, revision of the Leguminosae, though advertised for the third fascicle of vol. 1(1), had not been completed. Thereafter, it and the rest of the Calyciflorae (including the Rosaceae) as well as the Monopetalae, monocotyledons, gymnosperms and vascular cryptogams were simply forgotten. The Polypetalae were, however, accounted for in the first (and only) volume of Sereno Watson's *Bibliographical Index to North American Botany* (1878; see below under **Comprehensive checklists, nomenclators and information systems**).][15]

Comprehensive checklists, nomenclators and information systems

The first North American checklist was *Florae americae septentrionalis* (1771) by Johann Reinhold Forster. No comparable works appeared until after the Civil War, and even those soon became incomplete with the rapid rate of discoveries in the latter half of the nineteenth century. Those published between 1865 and 1914 – apparently to meet demand from private collectors as well as institutions – include *Catalogue of phaenogamous plants of the United States, east of the Mississippi, and of the vascular cryptogamous plants of North America, north of Mexico* (1868; 2nd edn., 1872) by Horace Mann, Jr.; *Catalogue of the phaenogamous and vascular cryptogamous plants of North America* (1885; 2nd edn., 1888) by John Houck Oyster of Kansas; several lists by the Oquawka (Illinois) newspaper proprietor and botanist Harry N. Patterson, particularly *Check-list of North American plants* (1887, with 12 794 taxa including 10 796 species) and *Patterson's numbered check list of North American plants north of Mexico* (1892); and three editions of *Catalogue of North American plants north of Mexico, exclusive of the lower cryptogams* (1898, 1900, 1909–14) by Amos A. Heller. The first edition of the last-named accounted for 14 534 species and non-nominate infraspecific taxa, the second, 16 673, and the third, completed from *Ophioglossum* (Filicinae) only through *Uva-ursi* (Ericaceae) on the Englerian sequence, 15 319 numbered taxa. All were private efforts, their compilation facilitated by publication of Gray's *Synoptical flora* as well as the fully documented but never-completed

Bibliographical Index to North American Botany by Sereno Watson, of which only one volume (Polypetalae, 1878, as *Smithsonian Misc. Collect.* 258) was ever published. The first-ever official United States list is that published in 1982 by the USDA Soil Conservation Service (SCS); this is now a component of the U.S. interdepartmental ITIS information system (see Note 17). The most widely used work of its kind is, however, the *Synonymized checklist* of John T. Kartesz and his associates, first published in 1980 with a second print edition in 1994. The latter has now been supplemented by a CD-ROM-based information system, *Synthesis of the North American flora*, covering publication references (wanting in the printed list) as well as distribution maps and biological data.

KARTESZ, J. T. and R. KARTESZ, 1994. *A synonymized checklist of the vascular flora of the United States, Canada and Greenland*. 2nd edn., 2 vols. (vol. 1, Checklist; vol. 2, Thesaurus). lxi, 622, vii, 816 pp. Portland, Oreg.: Timber Press. (1st edn., 1980, in 1 vol., Chapel Hill, N.C.: University of North Carolina Press.) Accompanied by J. T. KARTESZ (ed.), 1999. *Synthesis of the North American flora*. 1 CD-ROM. Databases, maps. Chapel Hill, N.C.: North Carolina Botanical Garden. (Electronic publication.)

Alphabetically arranged list of species and recognized infraspecific taxa (within three primary groups: pteridophytes, gymnosperms and angiosperms) in vol. 1; alphabetical lexicon of all names in vol. 2, with accepted names in Roman font and synonyms in italics. No basionyms, places of publication, taxonomic commentary or geographical or ecological information are furnished. An extensive introductory part covers the ontology of the work, acknowledgments (including reviewers and cooperating specialists), excluded and 'anomalous' names and (pp. xxxviii–lxi) statistics. A total of 20729 species and 8650 infraspecific taxa are accepted in a compass extending to include Hawaii, Puerto Rico and the Virgin Islands as well as North America north of Mexico, associated islands and Greenland. [The synonymy was selected to account for standard manuals, floras and other works but these are nowhere listed. The printed work is presently the effective standard of its kind for the continent.][16]

SHETLER, S. G. and L. E. SKOG, 1978. *A provisional checklist of species for Flora North America (revised)*. xix, 199 pp. St. Louis. (Monogr. Syst. Bot. Missouri Bot. Gard. 1/FNA Rep. 84.) (Original edn., 1972, Washington, as *FNA Rep.* 64.)

Briefly annotated, computer-generated general checklist of vascular plants of the North American continent (north of Mexico) and Greenland (16274 species in 2350 genera); each one-line species entry includes its accepted name, authority, coded indications of regional distribution as well as status and life-form, and a source of information. Families are indexed on pp. 197–199. An extensive introductory section gives a review of the original Flora North America Program and the genesis of the checklist together with a discussion of the data system used, a consideration of other technical aspects, and an explanation of the various classes of information and their codes. Four appendices respectively deal with references/sources, reviewers, statistics of the flora (with details for each family), and sample data forms. [Effectively superseded.]

SOIL CONSERVATION SERVICE, UNITED STATES DEPARTMENT OF AGRICULTURE, 1982. *National list of scientific plant names*. 2 vols. vi, 416, 438 pp., map (Soil Conservation Service Tech. Publ. 159). Washington, D.C.: U.S. Government Printing Office.

Volume 1 comprises a computer-generated one-line checklist (in three sections: U.S. and Canada, Puerto Rico and Virgin Islands, and Hawaii) of all accepted species and non-nominate infraspecific taxa; each entry includes an identifier code, scientific name, authority(ies), coverage in 'source manuals' (cf. pp. 2–3), habit (cf. p. 4), distribution (cf. p. 5 for map), and family names and numbers. Pp. 7–20 in vol. 1 encompass an index to genera with family equivalents and their numbers. Volume 2 is complementary, covering synonymy (as far as known) and the arguments respecting them (for explanation of codes, see p. 2). Contributors, consultant authorities and reviewers are listed on pp. ii–iv of vol. 1.[17]

Woody plants (including trees)

The North American literature on woody plants (and trees), especially that of a popular and semi-popular nature, is so considerable that of necessity a rigorous selection has had to be made for this *Guide*. State, provincial and territorial tree guides thus have generally not here been accounted for. However, if such works cover all woody plants or, in a few cases, shrubs alone, they are included; indeed, in some polities woody plants may account for a substantial percentage of the vascular flora as, for example, shrubs in California. For more comprehensive lists of references (U.S.A. only), including the many state tree books, the bibliographies by Dayton (1952) and Little and Honkala (1976) should be consulted (under **Divisional bibliographies: trees and shrubs** above).[18]

In the introduction to his 1980 work, Thomas Elias noted that the leading students of the North American woody flora have been Charles S. Sargent, George B. Sudworth, Alfred Rehder and Elbert L. Little. Their work, however, was preceded by that of Humphrey Marshall, the two Michaux and Thomas Nuttall, who between them contributed *Arbustrum americanum* (1785) and the two series of *North American Sylva* (1810–13, 1842–49). Sargent's *Catalogue* (1880) and *Report* (1884), as well as his *Sylva* and *Manual* (see below), and similarly *The geographical distribution of the forest trees of Canada* (1882) by Robert Bell and *Native trees of Canada* (1917) by Boyd R. Morton and Richard G. Lewis, were all produced in the early decades of the forestry and conservation movements; upon them was based much of the later dendrological work of the United States and Canadian forestry services. In the U.S. service, Sudworth, author in 1898 of the first official national checklist (*Check list of the forest trees of the United States*; 2nd edn., 1927), instituted one of Sargent's dreams, a mapping programme of North American trees. Although in publication terms initially abortive (one part alone of his *Forest Atlas* being realized in his lifetime), this programme ultimately attained its principal goal with the completion in 1981 of Little's six-volume work (see below). Maps also began appearing in popular national and regional works, both in Canada and in the United States, and are now a common feature.

Other widely used 'tree books' worthy of note include E. L. LITTLE, JR., 1980. *The Audubon field guide to North American trees: Eastern Region*. 714, 2 pp., text-fig., col. pls., maps; *idem*, 1980. *The Audubon field guide to North American trees: Western Region*. 639, 1 pp., text-fig., col. pls., maps. New York: Knopf/Chanticleer Press. (Borzoi Books.); and R. J. PRESTON, 1989. *North American trees exclusive of Mexico and tropical Florida*. 4th edn. xxvii, 407 pp., illus., maps. Ames, Iowa: Iowa State University Press (simultaneously also in soft cover, Cambridge, Mass.: MIT Press). (1st edn., 1948, Ames; 2nd edn., 1961 (reissued in soft cover, 1966, Cambridge); 3rd edn., 1976 (simultaneously also in soft cover, Cambridge).)

ELIAS, T. S., 1980. *The complete trees of North America: field guide and natural history*. xii, 948 pp., illus., maps. New York: Van Nostrand Reinhold. (Outdoor Life/Nature Book.) (Reprinted 1987, New York: Gramercy.)

Popularly oriented illustrated descriptive guide to native, naturalized and commonly cultivated trees (about 750 species), the entries headed by vernacular and 'standardized common' names; includes pictured-keys, figures (to a large extent reproductions of the Faxon and Gill drawings in Sargent's *Manual*), maps (based on those in Little's *Atlas* and where available including Mexican ranges), and commentary covering distribution, habitats, associates, appearance, biology, wood, uses and other useful information (all preceding the descriptions proper); winter key (pp. 933–938) and complete index at end. The introductory part covers tree names, procedures for identification, organography, trees and wildlife, and North American forest and other biotic regions.

LITTLE, E. L., JR., 1979. *Checklist of United States trees (native and naturalized)*. iv, 375 pp. (Agric. Handb. USDA 541). Washington, D.C.: U.S. Government Printing Office. (1st edn., 1953, Washington, as *Check list of native and naturalized trees of the United States (including Alaska)*).

Alphabetically arranged and critically annotated enumeration of tree species with full synonymy (including references), citations of standard botanical treatments, earlier checklists, and the author's *Atlas of United States trees*, vernacular names, extensive indication of internal and extralimital range, critical commentary, and needs for future study; index to all vernacular names. An extensive introductory section gives details of previous checklists, remarks on nomenclature, species ranges, vernacular names, and naturalized and rare and local species as well as a statistical summary and a list of major references. Eight appendices, including a concise systematic list, are also given. The work includes all Canadian species, but not those of Hawaii or the external U.S. territories (Puerto Rico, Guam, Samoa, etc.).

PETRIDES, G. A., 1988. *A field guide to eastern trees*. xv, 272 pp., illus., [70 pp.] pls. Boston: Houghton Mifflin. (Peterson field guide series 11a.) Complemented by *idem*, 1992. *A field guide to western trees*. xii, 308 pp., illus. (some col.), 50 pp. pls. Boston. (Peterson field guide series 44.)

Artificially arranged, popularly oriented field-guides, with range maps and colored illustrations (by Olivia Petrides) and keys to artificial sections; treatments include essential synonymy, vernacular names and diagnostic and key features, along with habitat, habit and size and, under 'Remarks', notes on properties and uses. Subheadings entitled 'Similar species' bear caveats about possible misidentifications. Keys to trees when leafless, systematic synopses, glossaries and indices appear at the end of each work. [The eastern tree guide is based on *idem*, 1972. *A field guide to trees and shrubs* (see **140**).]

SARGENT, C. S., 1922. *Manual of the trees of North America (exclusive of Mexico)*. 2nd edn. xxvi, 910 pp., 783 text-figs., map. Boston: Houghton. (Reprinted with corrections, 1926, 1933; additional reprint in 2 vols., 1961, New York: Dover. 1st edn., 1905.)

Copiously illustrated, amply descriptive technical 'tree book'; includes keys to all taxa, limited synonymy, vernacular names, fairly detailed indication of internal range, summary of extralimital distribution (where appropriate), and notes on diagnostic features, habitat, wood, bark, ornamental and other uses, etc.; remarks on hybrids; glossary and complete index. The illustrations encompass nearly all included species and are by C. E. Faxon and Mary W. Gill. [The work is essentially a condensed and revised version of *idem*, (1890) 1891–1902. *Silva of North America*. 14 vols., 740 pls. Boston, New York: Houghton Mifflin.]

VAN DERSAL, W. R., 1938. *Native woody plants of the United States: their erosion-control and wildlife values.* ii, 362 pp., 44 pls., 3 maps (Misc. Publ. USDA 303). Washington, D.C.: U. S. Government Printing Office.

Comprises a concise, alphabetically arranged enumeration of woody plants (including trees and shrubs) of the conterminous United States, with limited synonymy, vernacular and 'standardized' plant names, indication of growth-zone within the U.S.A., and extensive notes on habitat, ecology, phenology, biology, fruit types, special features, uses in plantings, etc.; lexicon of vernacular names, with botanical equivalents, and copious bibliography at end. The introductory section covers plant growth-zones, relationship of vegetation to soil conservation, and evaluation of plants with respect to wildlife, planting out, etc.

Woody plants: distribution maps

The first volume of Little's *Atlas* is the functional successor to *The distribution of important forest trees of the United States* (1938) by E. N. Munns.

LITTLE, E. L., JR., 1971–81. *Atlas of United States trees.* 6 vols. 982 maps (Misc. Publ. USDA 1146, 1293, 1314, 1342, 1361, 1410). Washington, D.C.: U.S. Government Printing Office.

This monumental series provides distribution details of all known tree species in the continental United States, with the exception of the 'critical' genus *Crataegus*; the maps also include Canadian and northern Mexican distribution where possible. For Alaska, shrubs are also included. Details of the volumes are as follows: 1. Conifers and important hardwoods (1971); 2. Alaska trees and shrubs (1975); 3. Minor western hardwoods (1976); 4. Minor eastern hardwoods (1977); 5. Florida (1978); 6. Supplement (1981). Each volume is prefaced by technical notes together with commentary on distribution patterns, conservation status, sources of information, etc., and is separately indexed.[19]

Pteridophytes

A continuing popular interest in ferns and fern-allies has led to a host of 'fern floras' covering individual states/provinces and/or larger areas. As there have been in recent decades relatively rapid changes in pteridophyte taxonomy (and consequently nomenclature), an attempt is made to include recent works; some are in any case quite detailed (and may be complementary to treatments of phanerogams).

North American pteridophytes were first monographed by Daniel Cady Eaton in his *Ferns of North America* (1877–80), a luxurious two-volume work. This furnished a solid basis for popularizers, among them Willard Clute who founded the forerunner of the present American Fern Society. In the twentieth century, there have been two critical continent-wide works, a checklist by Maurice Broun (1938) and a field-manual by David Lellinger (1985), as well as a picture-key by John Mickel (1979). Pteridophytes are now also covered in volume 2 of *Flora of North America*.

BROUN, M. (ed.), 1938. *Index to North American ferns.* 217 pp. Orleans, Mass.: The author.

Annotated checklist of native, naturalized, and adventive ferns and fern-allies, with full synonymy (including references), vernacular names, fairly detailed indication of distribution in the continent, and notes on habitat; statistical tables, list of authors and index to all botanical names at end. [The nomenclature and some taxonomic concepts are by now somewhat outdated.]

LELLINGER, D. B., 1985. *A field manual of the ferns and fern-allies of the United States and Canada.* ix, 389 pp., 26 text-figs., 406 col. photographs (by A. M. Evans). Washington, D.C.: Smithsonian Institution Press.

Briefly descriptive systematic manual-flora (with all 406 species and accepted infraspecific taxa illustrated in color), with keys, vernacular names, generalized indication of distribution, critical remarks, and notes on habitat, frequency, status, biology, special features (including hybridization), and cultivation, followed by a closing chapter on hybrid complexes and mechanisms (pp. 313–328, with many examples) and references (pp. 347–357), an alphabetic checklist, an illustrated glossary, and separate indices to vernacular and scientific names. The introductory part encompasses collecting, nomenclature, classification, geography and climate, floristic provinces, ecology and biology, paleohistory, uses, cultivation, and fern societies along with abbreviations and (pp. 42–47) general keys.

MICKEL, J. T., 1979. *How to know the ferns and fern-allies.* 229 pp., illus., maps. Dubuque, Iowa: Brown.

Illustrated field-guide in manual-key format to all species of pteridophytes in North America, including diagnostic illustrations and distribution maps; general systematic conspectus, bibliography of 54 state and regional treatments, glossary, and index at end. Vernacular names are included. An introduction discusses relationships of pteridophytes, organography, life-cycles, cytology, hybridization, spore culture, fern gardening, and preparation of herbarium specimens.

Grasses

In the absence as yet of a complete continental flora, and because of its wide use, botanical importance and continued availability (at a relatively low cost), 'Hitchcock and Chase' is included here. Work towards a successor, now to form volumes 24 and 25 of *Flora of North America* as well as an independently published *Manual of North American grasses*, is currently in progress.

HITCHCOCK, A. S., 1950 (1951). *Manual of the grasses of the United States.* 2nd edn., revised by A. Chase. [2], 1051 pp., 1200 text-figs. (Misc. Publ. USDA 200). Washington, D.C.: U.S. Government Printing Office. (Reprinted 1972 in 2 vols., New York: Dover. 1st edn., 1935.)

Illustrated descriptive manual of grasses occurring in the conterminous United States, with keys to all genera and species, limited synonymy, vernacular names, indication of internal distribution, and notes on habitat, special features, etc.; many distribution maps; complete indices at end.[20]

101

Continental subdivisions and other special areas

As works covering respectively eastern and western North America (or the eastern or western United States) often have been developed in complementary fashion, they are dealt with under **100**.

102

Uplands and highlands

This unit is introduced to cater for the extensive, geologically old Appalachian system in the eastern part of the continent. Its southern portion has a considerable literature, two bibliographies and, from 1989, a manual. There are no comparable works for the Ozark Mountains in the south-central part of the continent or for the Laurentides in Québec.

Appalachian Mountains

Bibliography

WOFFORD, B. E. and P. S. WHITE, 1981. *Systematics and identification of Southern Appalachian phanerogams: an indexed bibliography*. [iv], 69 pp. (U.S. National Park Service, NPS-SER Research Resources Management Report 53). Gatlinburg, Tenn. (distributed by the NPS Southeast Regional Office, Atlanta, Ga.). Complemented by A. M. EVANS, P. S. WHITE and C. PYLE, 1981. *Southern Appalachian pteridophytes: an indexed bibliography, 1833–1980*. 33 pp. (U.S. National Park Service, NPS-SER Research Resources Management Report 44). Gatlinburg. [The first work encompasses 745 titles, largely monographic and revisionary references; an index to available parts of *Generic flora of the southeastern United States* is also included. The material is arranged by families and genera, interspersed

with more general keywords. The second work accounts for 365 titles.]

WOFFORD, B. E., 1989. *Guide to the vascular plants of the Blue Ridge*. 384 pp., 8 figs., 1 map. Athens, Ga.: University of Georgia Press.

Keyed enumeration of vascular plants (2391 native and introduced species and additional lesser taxa); entries, separate from the key, include vernacular names, usual habitat(s), frequency, distribution by states (with 'All' standing for 'throughout'), and phenology; reference to revisions and monographs under genera. The end matter includes excluded taxa, statistics (pp. 327–328), and three indices: vernacular names, genera and lower taxa, and families. In the introductory part are a map of the area and an explanation of the scope and format of the work. [Covers 85 counties from Virginia to Georgia.]

Woody plants

SWANSON, R. E., 1994. *A field guide to the trees and shrubs of the Southern Appalachians*. x, [4], 399 pp., 161 text-figs., map. Baltimore, Md.: Johns Hopkins University Press.

Illustrated manual with keys, vernacular names and *pro forma* dendrological descriptions including distribution and habitat; synoptic list (pp. 369–381), glossary and index to all names at end. The introductory part includes a section on terminology along with summer and winter general keys and 'shortcuts' leading from a range of key distinctive features.

103

Alpine and upper montane zones

In contrast to certain other parts of the world, notably Europe and northern Asia, North America as yet has few serious floras of high-mountain areas. Among them is William Weber's *Rocky Mountain flora*, an outstanding work first produced in 1953 and now in its fifth edition. For a general introduction to the North American alpine environment with chapters on individual mountain systems, see A. H. ZWINGER and B. E. WILLARD, 1972. *Land above the trees: a guide to American alpine tundra*. xv, 487 pp., text-fig. New York: Harper. (Reissued 1986, New York: Perennial Library;

1989, Tucson: University of Arizona Press; 1996, Boulder, Colo.: Johnson Books.)

I. Eastern North America

The potential area is very small and in part rather remote; not surprisingly, little has been published. The largest fragment is in the upper parts of the Presidential Range in the White Mountains of New Hampshire; its flora is arctalpine.

LÖVE, Á. and D. LÖVE, 1966. *Cytotaxonomy of the vascular plants of Mt. Washington.* [iii], 74 pp. (Univ. Colorado Stud., Ser. Biol. 24). Boulder, Colo.

An enumeration of 165 species, with indication of diploid chromosome number(s), a voucher collection, and discussions of taxonomy (where appropriate), cytology and habitat(s); summary and extensive bibliography at end.[21]

II. Rocky Mountains

The manual by William Weber remains among the best of its kind in North America. It served as a model for the author's later manuals for eastern and western Colorado (see **185**). The atlas by R. W. Scott is the first in a major new series.

SCOTT, R. W., 1995 (1997). *The alpine flora of the Rocky Mountains.* Vol. 1: *The Middle Rockies.* ix, 901 pp., illus., maps. Salt Lake City: University of Utah Press.

This first of a projected series of three volumes comprises a descriptive atlas/flora of the middle Rocky Mountains from timberline upwards (53 147 km²) including keys, synonymy, vernacular names, indication of distribution and habitat (but no commentary), and figures and a map for each species; appendices (I–IV) on alpine terminology, a glossary, author abbreviations and biographical accounts, and a list of chromosome numbers; bibliography (pp. 864–870) and indices to botanical and vernacular names at end. The general part (pp. 3–34) features sections on physical features, geology and geomorphology, environments, plant adaptations, floristics (with statistics), taxonomic philosophy, and the plan of the work along with a general key to families. [609 species (700 taxa) are accounted for from altitudes of 2682 to 3293 m upwards, depending upon the prevailing timberline. Species totals are also given for each range (see p. 25).

The work covers all Wyoming, southern Montana, northeastern Utah and eastern and southeastern Idaho. The two further volumes are planned respectively to cover the northern and southern Rockies.][22]

South Central U.S. Rocky Mountains
This area has not yet been covered by Scott (see above).

WEBER, W. A., 1976. *Rocky Mountain flora.* 5th edn. xii, 479 pp., illus. (some col.). Boulder, Colo.: Colorado Associated Universities Press. (1st edn., 1953, under title *Handbook of plants of the Colorado Front Range.*)

Field manual-key to vascular plants, with synonymy, vernacular names, and notes on habitat, occurrence, biology, variation, uses, etc.; includes numerous diagnostic figures as well as a glossary, list of references, and index at the end. An introductory section includes a description of the region and its phytogeographic features as well as an illustrated organography. [Limited to the Colorado Rockies, whose highest peak is Mt. Elbert (4399 m) and most famous is Pike's Peak (14 110 ft.). The author remarks particularly on the affinity of the area with the Altai Mountains in Siberia (**703/VIII**) and notes that for mountain regions in Eurasia works of this type are much more numerous.]

Northern U.S. and Canadian Rocky Mountains
No really representative work is available. Important early studies, made after the opening up of the region by railways, include *Contributions to a catalogue of the flora of the Canadian Rocky Mountains* (1907) by Edith M. Farr, based mainly on her 1904–05 collections, and *Flora of Glacier National Park* by Paul C. Standley (1921). Mention should also be made of *Alpine flora of the Canadian Rocky Mountains* (1907) by Stewardson Brown and Mrs. Charles Schäffer, an illustrated popular guide.

III. Sierra Nevada

In addition to the field-guide by N. F. Weeden, reference may also be made to F. J. SMILEY, 1921. *A report upon the boreal flora of the Sierra Nevada of California.* 423 pp., illus. (Univ. Calif. Publ. Bot. 9). Berkeley.

WEEDEN, N. F., 1986. *A Sierra Nevada flora.* 3rd edn. [4], iv, 406 pp., illus., maps. Berkeley, Calif.: Wilderness Press. (1st edn., 1975; 2nd edn., 1981.)

Pocket field manual-flora of vascular plants; includes keys, vernacular names, diagnostic features, distribution and altitudinal range, and notes on phenology, habitat, uses, and edibility (the work is also intended as a 'survival-guide'); glossary, references (pp.

396–397) and index to all names at end. A brief introduction includes notes on the use of the work, explanation of abbreviations, and general keys. [Inclusive of an
area from the Thousand Lakes Valley (north of Lassen
Peak) south to Walker Pass (Tulare/Kern Counties
border) and descending to *c.* 1060 m on the western fall
and 2450 m on the eastern. It thus encompasses the
black oak (*Quercus kelloggii*), pine, fir and alpine belts.]

104

Ectopotrophic areas

Included here is Clyde Reed's 'green telephone
book' on the flora of serpentine formations in the East.

REED, C. F., 1986. *Floras of the serpentinite formations in eastern North America: with descriptions of
geomorphology and mineralogy of the formations.* vi, 858
pp., 22 text-figs., 30 maps. illus. (Contr. Reed Herb. 30).
Baltimore, Md.: Reed Herbarium.

The largest section (pp. 285–858) of this copiously documented, largely compiled ecofloristic
monograph comprises a systematic enumeration of
non-vascular and vascular plants; the latter are
arranged on the Englerian system and include localities
with citations of *exsiccatae* and published records along
with incidental illustrations. The remainder of the
work covers geology, geomorphology, soils, vegetation
and floristics, mainly area by area including indications
of characteristic and endemic species (pp. 17–264), followed by general considerations including a discussion
of endemism (pp. 265–284); acknowledgments (pp.
285–286). A conventional introduction is lacking, and
there is **no** index.[23]

105

Drylands (steppes and deserts)

Relatively few significant works relating to the
extensive and typologically diverse prairie, steppe and
desert areas of North America have to date been published. P. A. Rydberg's well-known *Flora of the prairies
and plains of central North America* and its successors
appear under **175**. With respect to the great western

desert lands, the leading such is *Flora of the Sonoran
Desert* (1964) by Ira Wiggins; but as the greater part of
that desert lies within Mexico, it appears at **205**. The
Great Basin is now being covered in *Intermountain flora*
(**180/III**). For the Mojave and Chihuahuan Deserts, no
separate works are yet available although underway or
projected.[24] Benson and Darrow's *Trees and shrubs*
appears below under **Woody plants**.

Southwestern deserts (woody plants)

The woody flora of the southwestern deserts has been well
treated by Lyman Benson and Robert Darrow; their manual,
now in its third edition, is listed below. Cacti, not therein
included, have been described in L. BENSON, 1982. *Cacti of
the United States and Canada.* Stanford, Calif.: Stanford
University Press.

BENSON, L. and R. A. DARROW, 1981. *Trees and
shrubs of the southwestern deserts.* 3rd edn. xviii, [2], 416 pp.,
424 illus. (in text), 40 col. pls., 252 maps. Tucson: University
of Arizona Press. (1st edn., 1944, as *A manual of southwestern
desert trees and shrubs*; 2nd edn., 1954, as *The trees and shrubs of
the southwestern deserts.*)

Generously sized illustrated descriptive treatment of
woody plants; includes keys to all taxa, synonymy, vernacular
names, typification, generalized indication of local and extralimital range (with some localities as well as distribution maps
for most species), critical commentary, and usually substantial notes on habitat, phenology, special features, properties,
uses, etc.; complete index at end. The introductory section
includes remarks on climate, vegetation, floristic regions, and
medicinal properties, along with an illustrated organography.
[The current edition features a much enlarged format, and
moreover is somewhat more 'scientific' than its predecessors.]

108

Wetlands

For the two field-guides to coastal wetland plants
by R. W. Tiner, see **109**.

Although no work covers the whole, North
America currently enjoys the most complete coverage
of aquatic and wetland plants among comparable parts
of the world. The manuals of Muenscher and Prescott
encompass the widest area but are largely limited to
more or less open-water aquatic species (hydrophytes).
A greater depth of coverage (encompassing marsh-
dwelling helophytes as well as hydrophytes) as well as
more extensive documentation is employed in the

regional and state manuals here included. No regional coverage is yet in existence for the central and northern mountain United States or for Alaska and Canada, save for parts of the southern fringe of the last-named; for these areas, general floras should be consulted.

Bibliography

STUCKEY, R. L., 1975. A bibliography of manuals and checklists of aquatic vascular plants for regions and states in the conterminous United States. *Sida* 6: 24–29. [Titles only, under three main headings: 'United States', 'Regional', and 'The States'.]

MUENSCHER, W. C., 1944. *Aquatic plants of the United States.* x, 374 pp., 154 figs., 400 maps, 3 tables. Ithaca, N.Y.: Comstock.

Descriptive illustrated manual of vascular plants, with keys to all taxa, vernacular names, and brief notes on habitat, frequency, distribution, etc.; descriptions of genera; dot maps depicting distribution of species by states; references to revisionary treatments; glossary and general index to all names. A brief introductory section defines the work's synusial coverage (hydrophytes) and discusses distribution, biology, dispersal, and seed characteristics (with list of references). The Englerian system is followed except that pteridophytes are placed last.

PRESCOTT, G. W., 1980. *How to know the aquatic plants.* 2nd edn. viii, 158 pp., 229 text-figs. Dubuque, Iowa: Brown. (1st edn., 1969.)

Illustrated field-key to aquatic macrophytes of North America, with small figures accompanied by often extensive morphological, biological and ecological notes; complete index at end (incorporating also an illustrated glossary). The introductory section is pedagogic, but also includes a generic list and 'general references'. [Primarily for student use and limited to hydrophytes and 'obligate' helophytes; encompasses 165 of the 306 North American genera reported as having at least one or more marginally aquatic or marsh species. The second edition has larger, clearer illustrations.]

I. Northeastern and North Central United States (and adjacent parts of Canada)

The long-standard *Manual of aquatic plants* (1940; 2nd edn., 1960) by N. C. Fassett was in 2000 succeeded by a new (and more inclusive) manual by Garrett E. Crow and C. Barre Hellquist. These authors earlier produced a series of eight illustrated extension pamphlets covering New England aquatic plants.

CROW, G. E. and C. B. HELLQUIST, 2000. *Aquatic and wetland plants of Northeastern North America.* 2 vols. Illus. Madison: University of Wisconsin Press. (Based on *A manual of aquatic plants* by N. C. Fassett (1940, New York: McGraw-Hill; 2nd edn. with revision appendix by E. C. Ogden, 1957, Madison: University of Wisconsin Press.))

Descriptive reference flora with keys to all taxa, botanical and vernacular names, synonymy, references, indication of distribution and habitat, commentary, and illustrations (covering over 90 percent of the taxa); complete index in each volume. [1186 taxa in 1139 species are accounted for, an increase of some 40 percent over its predecessor which moreover was more narrowly circumscribed. Vol. 1 covers pteridophytes, gymnosperms and dicotyledons; vol. 2, monocotyledons.]

FASSETT, N. C., 1957. *A manual of aquatic plants.* 2nd edn., with revision appendix by E. C. Ogden. ix, 405 pp., illus. Madison: University of Wisconsin Press. (1st edn., 1940, New York: McGraw-Hill.)

Manual-key to marsh and aquatic vascular and non-vascular macrophytes; includes essential synonymy, vernacular names (given, along with citations of significant monographs and revisions, at generic and family levels), and concise indication of habitat and distribution; complete index at end along with an appendix on uses by wildlife (the latter with a special list of references). The revision appendix in the second edition (pp. 363–384) is cross-referenced to the main text. In the introductory section are directions for use of the work, key source works, and an illustrated artificial general key; for the manual's scope, see the preface. [Covers most of the *Gray's Manual* range, but omitting Newfoundland and not accounting fully for Virginia.]

HELLQUIST, C. B. and G. E. CROW, 1980–85. *Aquatic vascular plants of New England.* Parts 1–8. Illus. (incl. figs., maps) (New Hampshire Agr. Expt. Sta. Bull. 515, 517, 518, 520, 523, 524, 527, 528). Durham, N.H.

Illustrated topographical treatment, with keys, generic descriptions, and species accounts accounting for distribution (with maps), frequency, water quality (including pH and alkalinity), and notes on biology, behavior, formal records of rarity, and some critical remarks; references at the end of each part (or also at the end of each genus). [Intended as contributions towards a manual and based on personal observations and records from New England herbaria. No specific sequence is followed.]

II. Southeastern United States

GODFREY, R. K. and J. W. WOOTEN, 1979. *Aquatic and wetland plants of southeastern United States: monocotyledons.* xii, 712 pp., 397 text-figs., frontisp. Athens, Ga.: University of Georgia Press. Complemented by *idem*, 1981. *Aquatic and wetland plants of southeastern United States: dicotyledons.* x, 933 pp., 399 text-figs., illus. in end-papers. Athens.

Copiously illustrated manual-flora, with keys to all taxa, vernacular names, very limited synonymy, brief indication of regional distribution, some taxonomic commentary, and notes on habitat; glossary and indices to vernacular and scientific names at end. The introductory section includes a consideration of the format adopted – essentially the North American 'manual-flora' style, and very much like that of *Aquatic and wetland plants of southwestern United States* (see below) – as well as remarks on the synusial extent of coverage (here hydrophytes and helophytes), distribution, habitats, environments and land use, and future perspectives, and concludes with an artificial key to families. [Covers North Carolina and Tennessee south to Florida and west to Arkansas and Louisiana.]

III. Great Plains

LINDSTROM, L. E., [1968.] *The aquatic and marsh plants of the Great Plains of North America.* 247 ll. Manhattan, Kans.: Kansas State University. (Ph. D. dissertation; abstracted in *Dissertation Abstracts International* as #68-17519 and distributed on demand by University Microfilms International, Ann Arbor, Mich.)

Manual-flora of flowering plants (603 species), with keys to all taxa, concise descriptions, and indication of range and habitat; no illustrations, maps or index or discussion of critical features or distribution. The introductory part covers the scope of the work, taxonomic concepts and nomenclature, sources of data,

and statistics along with references. [Covers the drier plains north of the 'mesquite belt' in Oklahoma and northernmost Texas and between the Rocky Mountains and the more fertile (and wooded) eastern plains, but centering on Kansas.]

IV. Southwestern United States

CORRELL, D. S. and H. B. CORRELL, 1975. *Aquatic and wetland plants of southwestern United States.* 2 vols. xv, 1777 pp., 785 text-figs., map. Stanford, Calif.: Stanford University Press. [Also published in 1 vol. under the same title but in a somewhat smaller format, 1972, Washington, D.C.: U.S. Government Printing Office (for the Environmental Protection Agency, Department of the Interior).]

Illustrated descriptive manual-flora of vascular plants, with keys to all taxa, occasional synonymy, vernacular names, generalized indication of distribution and habitat, and notes on variability and infraspecific forms; illustrated glossary, list of abbreviations, and index to all vernacular and scientific names at end. The introductory section includes remarks on the background and development of the work, a consideration of what aquatic plants are – a broad definition being taken here, as in the work by Godfrey and Wooten (see above) – and discussion of habitats, special features, distribution, economics and control, significance for wildlife, and pollution problems, followed by a general key to families. [Covers Oklahoma, Texas, Arizona and New Mexico.]

V. Pacific Coast United States (and adjacent parts of Canada)

MASON, H. L., 1957. *A flora of the marshes of California.* [ix], 878 pp., 367 figs. (incl. maps). Berkeley/Los Angeles: University of California Press.

Descriptive flora of marsh and aquatic vascular plants, with keys to all taxa, synonymy and references (no literature citations), vernacular names, critical remarks, and generalized indication of distribution, status and habitat; illustrated glossary and general index to all names at end. The introductory section includes a definition of scope of coverage (hydrophytes and helophytes), and considerations on the nature of habitats (incorporating a suggested classification), marshland geography, species distributions, floristics, vegetation, the phenomenon of rapid change characteristic of marsh and aquatic ecosystems, and stylistic matters, followed by an illustrated general key to families; no separate bibliography included. [A rather methodical and conservative work in a somewhat generous format, but highly regarded and very influential in relation to the current generation of wetland floras.]

STEWARD, A. N., LaR. J. DENNIS and H. M. GILKEY, 1963. *Aquatic plants of the Pacific Northwest.* 2nd edn. ix, 261 pp., 27 pls. Corvallis, Oreg.: Oregon State University Press. (Studies in botany 11. 1st edn., 1960.)

Illustrated manual-flora to aquatic plants of Oregon, Washington, British Columbia and Alaska, with keys to all taxa, limited synonymy, vernacular names, generalized indication of distribution within and without the area covered, and brief indication of habitat (variously within or without the keys); glossary, list of references, and complete general index to names. The introductory section is merely prefatory, and precedes the general key to families. The figures are somewhat crudely executed. [The critical and technical standards set generally for aquatic macrophyte floras by Herbert Mason's work (see above) were here not attained.][25]

109

Oceanic littoral

The Pacific coast has yet to be treated in the same fashion as much of the Atlantic and Gulf coasts in the following (of which the first is the best value).

DUNCAN, W. H. and M. B. DUNCAN, 1987. *The Smithsonian guide to seaside plants of the Gulf and Atlantic coasts from Louisiana to Massachusetts, exclusive of lower peninsular Florida.* 409 pp., illus. (588 col.), map (inside front cover). Washington, D.C.: Smithsonian Institution Press.

Semi-popular, systematically arranged guide to flowering plants (943 species), with keys (pp. 56–71), descriptions, essential synonymy, distribution, and notes on taxonomy (where necessary), frequency, habitat, phenology, and related species; index to all

names at end. The introductory part encompasses physiographic aspects (including descriptions of individual regions), adaptations, cultural and economic notes (including uses), and literature (p. 32) as well as a guide to the use of the work, illustrations of diagnostic features and a glossary. [The northeastern limit is Plum Island, Massachusetts.]

MOUL, E. T., 1973. *Marine flora and fauna of the northeastern U.S.: higher plants of the marine fringe.* iv, 60 pp., illus. (National Marine Fisheries Service (NMFS) Circular 384). Washington, D.C.: National Oceanic and Atmospheric Administration, U.S. Department of Commerce.

Popularly oriented illustrated guide; includes lists summarizing plant distribution, arranged by habitat (beaches, dunes, tidal marshes); references (pp. 53–54) and indices to scientific names and to vernacular names and habitats. The introductory part includes an illustrated glossary. [Part of a series of guides encompassing the coastal biota; the emphasis of this treatment is on southern New England but it may be used along much of the northeastern United States coast.]

TINER, R. W., JR., 1987. *A field guide to coastal wetland plants of the Northeastern United States.* [x], 285 pp., illus., maps. Amherst: University of Massachusetts Press. Complemented by *idem*, 1993. *Field guide to coastal wetland plants of the Southeastern United States.* xiii, 328 pp., illus. Amherst.

Atlas-treatments featuring illustrations with facing text; for each species are given a line drawing along with a description, scientific and vernacular names, indication of distribution, and notes on phenology, habitat, and related species (as appropriate). Suitable study sites (with maps), sources of information, references, glossaries and complete indices appear at the end, while the introductory parts encompass general surveys of coastal wetland ecosystems, guides to identification features, and artificial keys (respectively on pp. 33–80 and 25–38, the latter being illustrated). [The Northeastern guide covers Maine to Maryland with 150 fully treated species and mention of a further 130; the Southeastern, Maryland and Delaware to Florida and along the Gulf coast from there to the Mexican border (the subtropical coasts excepted) with 250 fully treated species and mention of more than 200 others. The arrangement of species in the 1993 volume is more artificial than in the earlier; moreover, its layout is more generous.]

Superregion 11–13

Boreal North America

In this superregion is included that part of the North American continent and associated islands between the 'tree-line' to the north and the boundary of the conterminous United States to the south. For the polar zone (here considered also to include Greenland), see Region 07. It thus accounts for the greater part of both Alaska (Region 11) and Canada (Region 12/13).

The principal all-Canada floristic works in actual fact account for this whole area as well as the contiguous polar zone, but on account of their nominal titles are listed under 120–30. These are in addition to the works covering the entire continent, described under 100.

The history of floristic work and flora-writing is in part the same as that for the continent as a whole, but has certain distinctive features, among them the Russian presence in Alaska until 1867 and British activities in Canada, many of them in relation to the search for the 'Northwest Passage' and the fate of Sir John Franklin as well as Pacific coastal exploration. For the first half of the twentieth century, the general essay of Maguire (1958; see **Progress** under the divisional heading) has a separate section on 'Boreal North America'. Although significant progress has since been made, considerable areas remain only sketchily explored and documented.

Bibliographies. Any coverage is accounted for under the divisional heading.

Indices. Any coverage is accounted for under the divisional heading.

Region 11

Alaska

Area: 1484165 km². Vascular flora (Alaska and Yukon): 1280 species (Hultén, 1941–50; see below). – Comprises the state of Alaska, including the mainland and offshore islands, the Pribilof Islands, and the Aleutian chain. Floristic works dealing only with the Bering Strait Islands or the Arctic Slope are listed respectively under **071** and **072**.

The history of earlier botanical exploration in this region, which until 1867 was Russian territory and thus was treated in Ledebour's *Flora rossica* and other Russian works, has been reviewed in detail by Eric Hultén (1940; see below under **Progress**). The modern critical basis for Alaskan taxonomy was subsequently laid down by him in his *Flora of Alaska and Yukon* (1941–50), the first work entirely to succeed the compiled *Sketch of the flora of Alaska* (1872) by Joseph T. Rothrock. From World War II, activity within the state expanded greatly and the resulting collections form an important additional basis for the definitive modern manual-floras of Hultén (1968) and Anderson/Welsh (1974), both of which also cover the Yukon. Partial floras are available for the Aleutians (Hultén, 1960) as well as some parts of the polar zone, and the trees and shrubs have been covered in an illustrated semi-popular work (Viereck and Little, 1972). Distribution maps have been published in Hultén's manual of 1968, as well as by Elbert Little in the 1970s for the trees as part of his *Atlas of United States trees* (**100**). Knowledge of the Alaskan flora has thus become well integrated at a relatively high level.[26]

Progress

For a very detailed account, see E. HULTÉN, 1940. History of botanical exploration in Alaska and Yukon Territories from the time of their discovery to 1940. *Bot. Notis.* [**93**]: 289–346.

Bibliographies. General and divisional bibliographies as for Division 1.

Indices. General and divisional indices as for Division 1.

110

Region in general

See also **121/I** (TAYLOR). The detailed technical basis for Hultén's manual-flora, incorporating much commentary and details of local distribution as well as documented synonymy omitted from that work, may be found in E. HULTÉN, 1941–50. *Flora of Alaska and Yukon*. 10 parts (Acta Univ. Lund., afd. II, N.S. **37–46** [Kongl. Svenska Fysiogr. Sällskap. Handl. 52–61]). Lund; and *idem*, 1967. Comments on the flora of Alaska and Yukon. *Ark. Bot.*, II, **7**: 1–147.

HULTÉN, E., 1968. *Flora of Alaska and neighboring territories: a manual of the vascular plants*. xxii, 1008 pp., text-figs., maps. Stanford, Calif.: Stanford University Press.

Generously formatted, briefly descriptive illustrated manual-flora of vascular plants, with keys to all taxa, essential synonymy, vernacular names, distribution maps showing local and overall range for most species, taxonomic commentary, and remarks on habitat, special features, etc.; large maps in end-papers; glossary, list of authors, lengthy bibliography, and indices to botanical and vernacular names. An introductory section includes chapters on climate, geology, ecological zones, infraspecific taxa, and botanical exploration. [The area covered by this work encompasses Alaska, Yukon, the Chukotsk district (Russia), the Commander and Aleutian Islands, northwestern British Columbia, and the western part of the Mackenzie District.][27]

WELSH, S. L., 1974. *Anderson's Flora of Alaska and adjacent parts of Canada*. xvi, 724 pp., illus., col. frontisp., end-paper maps. Provo, Utah: Brigham Young University Press.

Descriptive manual of vascular plants, with keys to all taxa, an abundance of good illustrations (all especially drawn for the work), selected vernacular names, generalized indication of local and extralimital range, some taxonomic commentary, and indication of status, habitat, etc.; list of references, glossary and general index at end. No karyotypes or extensive ecological or biological notes are included. An introductory section includes a brief history of Alaskan botanical work and notes on the genesis of the book. [Based on J. P. Anderson's *Flora of Alaska and adjacent parts of Canada* (1959), but so completely revised as to constitute virtually a new work.][28]

Woody plants

VIERECK, L. A. and E. L. LITTLE, JR., 1972. *Alaska trees and shrubs*. vii, 265 pp., 128 text-figs., map (Agric. Handb. USDA 410). Washington, D.C.: U.S. Government Printing Office.

Illustrated, copiously descriptive treatment; includes

keys to all species, vernacular names, indication of local and extralimital range (with distribution maps), some critical remarks, and notes on habitat, biology, special features, properties, uses, etc.; list of references and index to all botanical and vernacular names. The introductory section includes among other topics accounts of vegetation formations, general features of the flora, and phytogeography. [For distribution maps of species, see vol. 2 in the junior author's *Atlas of United States trees* series (**100**).

111

Aleutian Islands

HULTÉN, E., 1960. *Flora of the Aleutian Islands and westernmost Alaska Peninsula, with notes on the flora of the Commander Islands*. 2nd edn. 418 pp., 32 pls., maps (unnumbered pp. 377–413). Weinheim: Cramer. (1st edn., 1937, Stockholm.)

Critical systematic enumeration of vascular plants, with detailed synonymy and references, indication of local range (with distribution maps for all species), and some taxonomic commentary; bibliography and index to generic names. A comprehensive introduction, containing sections on physical features, climate, floristics, vegetation, etc., is also provided.

Region 12/13

Canada

Area: 9 922 387 km². Vascular flora: 3220 native and 880 introduced species (Scoggan 1978–79; see below). – In addition to the Dominion of Canada, this region includes the French territory of St. Pierre and Miquelon (off southern Newfoundland). Certain regional works relating to the western and central provinces appear, however, under **121/I**, while those on the Northwest Territories are given under **123** along with works on the southwestern, forested part of its Mackenzie District and those on Ungava (northernmost Québec) at **075**.

Floristic work in Canada developed slowly, with only 14 francophone and 10 anglophone contributions to and including Michaux's *Flora boreali-americana* (1803). The works of Frederick Pursh (1814) and Thomas Nuttall (1818) also covered what was known of British North American plants, but it was Pursh who, during his residence in Montreal (1815–20), first began serious work towards a Canadian flora. His materials, along with those of Archibald Menzies and David Douglas from the Pacific territories (then also including the present-day U.S. states of Oregon and Washington) and those from several Arctic expeditions including those of John Franklin with John Richardson (1818–22, 1825–27), formed a major basis for a first overall treatment, W. J. Hooker's *Flora boreali-americana* (1829–40).

Completion of the *Flora*, along with the appearance in the United States of the floras and manuals of Amos Eaton, John Torrey and Asa Gray, thus provided an improved basis for local studies. These gradually spread from centers of settlement and especially secondary and higher education (as, for example, in Halifax, Québec City, Montreal, Kingston and Toronto), and over the next half-century some provincial floras and checklists appeared. The most considerable of these was *Flore canadienne* (1862) by Abbé Léon Provancher, covering Canada in its strictest sense: the St. Lawrence Valley and lower Great Lakes west to present-day Windsor, Ontario.

Until the late nineteenth century, however, botanical activities remained limited in scope, as indeed did scientific work in general: the 'intellectual maturing' of the Canadian people was slowed by prevailing economic and social conditions.[29] An important step was, however, taken in 1843 with the formation at Ottawa, the future dominion capital, of the Geological Survey of Canada, so providing for the first time a central focus for the natural sciences. The Survey's Museum (the core of the present National Museums of Canada and in particular the Canadian Museum of Nature) was established in 1860 and in 1882 its herbarium came into being through the engagement of John Macoun as its curator (and Dominion Botanist) and the purchase of his collections. Thus was finally furnished a first opportunity for a Canadian-based national account, quickly realized by Macoun with his *Catalogue of Canadian plants* (1883–1902) covering vascular plants, bryophytes and lichens. Though regrettably not always critical, this account remained standard for almost a century. In the undertaking he was assisted by his son James, who on his own

additionally contributed 18 supplementary papers (1894–1906). The two also collected over 100000 numbers throughout the country, including its Arctic regions.[30]

There was, however, not to be another descriptive provincial or local flora until well after completion of the vascular plants in the *Catalogue*, and even by World War II few such had appeared. Regional manuals for the northern and northeastern United States, which generally also cover much of southern Canada, were, and continue to be, widely used in the eastern provinces; and in southern British Columbia, *A flora of Northwest America* by Thomas Howell, though in coverage not extending north of the 49th parallel, was of value. It was in the latter, however, that the first local manual in English, *A flora of southern British Columbia* (1915) by Joseph K. Henry, appeared. In Québec, Provancher's manual was succeeded in 1931 by *Flore-manuel du province de Québec* by Père Louis-Marie (at Oka Agricultural Station near Montreal) and, in 1935, *Flore laurentienne* by Frère Marie-Victorin (at the University of Montreal). The last-named work, revised in 1974 and again in 1995, remains much-esteemed. The opening up of the Canadian Rockies west of Lethbridge (Alberta) and associated development attracted several collectors and, among other publications, resulted in an early popular work, *Alpine flora of the Canadian Rocky Mountains* (1907) by Stewardson Brown and Mrs. Charles Schäffer (see **103/III**). Extensive collecting in many other parts of the country also continued, as Bassett Maguire has related, and a goodly number of reports and local checklists were published, including first accounts for Manitoba (1909), Saskatchewan (1937) and parts of the Arctic (Polunin, 1940; see **074**).[31]

Since 1945, however, several descriptive provincial and territorial floras as well as checklists have been published. Among the former are floras of the Yukon (covered in the Alaskan floras of Anderson/Welsh and Hultén), the Arctic Archipelago (Porsild, 1957; see **074**), the continental Northwest Territories (Porsild and Cody, 1980), Alberta (Moss, 1959, revised 1983), Manitoba (Scoggan, 1957), the Prairie Provinces (Budd, 1957, with revisions to 1987; Boivin, 1967–81), Québec (Rousseau, 1974, not covering the whole flora and without keys), Nova Scotia (Roland, 1947–49, revised 1966–69), and New Brunswick (Hinds, 1986). The Gaspé Peninsula and Anticosti Island in Québec as well as the never-glaciated Queen

Charlotte Islands in British Columbia have been treated in important partial floras. Many of the afore-mentioned polities, as well as others, also feature modern enumerations or checklists; most recently one has appeared for Ontario, a first for that province (Morton and Venn, 1990). In the early 1990s came the first complete provincial atlas: *Atlas of the vascular plants of the island of Newfoundland* . . . (Rouleau and Lamoreux, 1992).

Canadian botanists early moved into the use of computers as aids in floristic documentation. In British Columbia, a modification of the Flora North America methodology was introduced by Roy L. Taylor and his associates in the 1970s; one result was *Vascular plants of British Columbia: a descriptive resource inventory* (1977).

At national level, two general works appeared in the third quarter of the twentieth century. A successor to Macoun's *Catalogue* with respect to vascular plants appeared in 1978–79 as *Flora of Canada* under the authorship of Homer J. Scoggan at the National Museums of Canada. Somewhat earlier, Bernard Boivin (at Agriculture Canada) had produced, in installments, *Énumération des plantes du Canada* (1966–69).

In a country where forestry is a major player in the economy, an interest in trees not unnaturally has long existed. Highlights include a standard work on tree distribution, *The geographical distribution of the forest trees of Canada* (1882) by Robert Bell, published through the Geological Survey and incorporating the results of his extensive explorations, the many editions of *Native trees of Canada*, a dendrology first published in 1917 under the authorship of Boyd R. Morton and Richard G. Lewis, its successor *Trees in Canada* by J. L. Farrar (1995), and *Identification guide to the trees of Canada* (1989) by Jean Lauriault.

Neither of the recent national accounts, however, contains descriptions; the now-considerable network of provincial works, particularly outside the eastern provinces, was perhaps a consideration. Population distribution, the continuing influence of U.S. regional floras, and realization of a new *Flora of North America* are surely also factors. But, although some projects were not realized (among them Scoggan's plan for a flora of the Atlantic provinces), the level of documentation is in comparison with the 1950s undoubtedly much improved. Only Québec and Labrador are without a complete modern flora or enu-meration.[32]

Progress
In addition to the already-cited review by Maguire, the most useful existing histories of floristic studies in Canada are D. P. PENHALLOW, 1887. A review of Canadian botany from the first settlement of New France to the nineteenth century, I. *Proc. & Trans. Roy. Soc. Canada*, 5/IV: 45–61; *idem*, 1897. A review of Canadian botany from 1800 to 1895. *Ibid.*, II, 3/IV: 3–56; R. B. THOMSON, 1932. A sketch of the past fifty years of Canadian botany. In Royal Society of Canada, *Fifty years retrospect: anniversary volume 1882–1932*, pp. 173–179. [Ottawa.]; and J. S. PRINGLE, 1995. The history of the exploration of the vascular flora of Canada, Saint-Pierre et Miquelon, and Greenland. 3 parts and index. Portr., maps. *Canad. Field-Nat.* 109(3): 291–356, 357–361, 362–382 (combined author index: 378–382).

Bibliographies. General and divisional bibliographies as for Division 1.

Regional bibliographies
Literature to 1895 is also covered in the historical papers by David Penhallow (see **Progress** above) and for 1900 to 1915 in A. H. McKAY, 1901–17. Botanical bibliography of Canada (*later* Bibliography of Canadian botany), 1900(–1915). *Proc. & Trans. Roy. Soc. Canada*, II, 7/IV–12/IV and III, 4/IV–7/IV, *passim*; *Trans. Roy. Soc. Canada*, III, 8/IV–10/IV, *passim*.

ADAMS, J., 1928–30. A bibliography of Canadian plant geography to the end of the year 1920 [I–III]. *Trans. Roy. Canad. Inst.* 16: 293–355; 17: 103–145, 227–265. Continued as *idem*, 1930. A bibliography of Canadian plant geography, 1921–1925 [IV]. *Ibid.*, 17: 267–295; J. ADAMS and M. H. NORWELL, 1932. A bib-liography of Canadian plant geography, 1926–1930 [V]. *Ibid.*, 18: 343–373; *idem*, 1936. A bibliography of Canadian plant geography, 1931–1935 [VI], with an appendix on the flora of Greenland. *Ibid.*, 21: 95–134; H. A. SENN, 1946. A bibliography of Canadian plant geography, VII. Additions, author, geographic and subject indices for the period 1635–1935. *Ibid.*, 26: 9–151; *idem*, 1947. A bibliography of Canadian plant geography, VIII. The period 1936–1940. *Ibid.*, 26: 153–344; and *idem*, 1951. A bibliography of Canadian plant geography, IX. The period 1941–1945. 183 pp. (Publ. Canada Dept. Agric., Res. Br. 863). Ottawa. [5402 references in all, arranged by authors in each installment with geographical and subject/taxonomic indices in parts 7–9; only those parts also annotated. Coverage extends to Greenland and Alaska.]

SOPER, J. H. *et al.*, 1989. *Database of Canadian checklists and bibliographies* (CANLIST). 16 lvs. N.p.: Canadian Botanical Association. (Mimeographed.) [Entries classified by province and partly annotated; bibliographies on p. 2, checklists on pp. 3–16.]

Indices. General and divisional indices as for Division 1.

120–30

Region in general

Boivin's *Énumeration* and Scoggan's *Flora of Canada*, both described below, are the current standard works covering the whole of Canada; neither incorporates descriptions although the latter has keys. Both supplant (for vascular plants) *Catalogue of Canadian plants* (1883–1902) by John Macoun. For a general overview of the flora and its characteristics, see R. L. TAYLOR and R. A. LUDWIG (eds.), 1966. *The evolution of Canada's flora*. viii, 137 pp., illus., tables. Toronto: University of Toronto Press.

BOIVIN, B., 1966–69. *Énumeration des plantes du Canada*. Parts 1–7, index (Provancheria 6). Québec: Université Laval. (Reprinted from *Naturaliste Canad.* **93** (1966): 253–274, 371–437, 583–646, 989–1063; **94** (1967): 131–157, 471–528, 625–655. Index separately published, 1969, Québec. 54 pp.)

Bibliographic enumeration of vascular plants of Canada (together with Alaska, Greenland, and St. Pierre and Miquelon); includes limited synonymy, numerous citations of appropriate floristic and systematic papers, and extensive notes on regional distribution and, at the end, a statistical summary and lists of species limited respectively to Alaska, Greenland, and St. Pierre and Miquelon; separate index to all botanical names.

SCOGGAN, H. J., 1978–79. *Flora of Canada*. 4 parts. xiii, 1711 pp. (Natl. Mus. Nat. Sci. Canada, Publ. Bot. 7). Ottawa.

Concise, non-descriptive flora of native and naturalized vascular plants (4153 species in 934 genera), with keys to all taxa (that for families in part 2), essential synonymy, critical commentary, indication of inter-

nal and overall distribution, biogeographic affinities and ecological preferences, life-form symbols, references to distribution maps, and (for genera) Dalla Torre and Harms numbers. The general index appears in part 4. The whole of part 1 comprises a general survey of the vascular flora, with sections on floral regions, factors affecting plant distribution and distribution patterns (accompanied by biogeographic tables), life-form spectra in different zones in relation to 'bioclimates' (including comparisons with other parts of the Northern Hemisphere), and technical notes, followed by a glossary and bibliography.[33]

Trees

FARRAR, J. L., 1995. *Trees in Canada*. x, 502 pp., illus. (part col.), maps. Markham, Ont.: Fitzhenry & Whiteside/Ottawa: Canadian Forest Service. [Also published as *Trees of the northern United States and Canada*, 1995, Ames, Iowa: Iowa State University Press.]

Dendrological atlas of native and introduced species, with keys; treatments include maps for all native species, illustrations, botanical and vernacular names, and descriptive text covering botanical features, wood, bark, seeds and seedlings, habitat, distribution, and notes on uses, properties, special features, etc.; group and species keys (pp. 409–460, including winter keys); bibliography, authors with biographies, derivations of botanical names, and complete index at end. The general part includes an illustrated organography and introductory notes to tree identification and the use of the book. [A systematic arrangement is here not followed; however, conifers comprise the first of the 12 artificial groups (each with 'thumbnail' marks on the edge of the page). The work is a successor to *Native trees of Canada* (see below).]

HOSIE, R. C., 1979. *Native trees of Canada*. 8th edn. 380 pp., illus., maps (some col.). Hull, Québec: Canadian Government Publishing Centre. (1st edn., 1917, by B. R. Morton and R. G. Lewis.) French edn.: MINISTÈRE DES FORÊTS, CANADA, 1966. *Arbres indigènes du Canada*. 289 pp. Ottawa: Imprimerie de la Reine.

Photographic atlas-manual with descriptive text, the latter giving details of tree form, size, botanical features, properties, uses, habitat, etc.; vernacular names and maps provided for every species. Appendices include a pictorial key to all species, etymology of tree names, historical sketches and a list of references; a complete index concludes the work. The introductory section gives an account of forest regions in Canada.

LAURIAULT, J., 1989. *Identification guide to the trees of Canada*. xvi, 479 pp., illus., maps. Markham, Ont.: Fitzhenry & Whiteside (with the National Museum of Natural Sciences). (Also published in French as *Guide d'identification des arbres du Canada*.)

Artificially arranged illustrated tree guide, the illustrations (usually of twigs and fruit) and maps with text covering vernacular names (English and French), distribution, diagnostic features, and notes on recognition, etymologies, habitats, properties, uses, etc., and related species. The general part covers the genesis of the work, its layout, scope, approaches to identification (with examples), 'what is native', botanical concepts, functions and organs, collection-making, and determination of trees' ages, while at the end are lists of rare trees and of trees by family, a bibliography, and an index to all names.

Pteridophytes

Earlier treatments include Macoun and Burgess, *Canadian Filicineae* (1884) and *The School Fern-flora of Canada* by George Lawson (1889). Cody and Britton in an addendum to their introduction record a trend in more recent years towards narrower species concepts.

CODY, W. J. and D. M. BRITTON, 1989. *Ferns and fern allies of Canada*. iv, 430 pp., 160 text-figs., 159 maps (Publ. Agriculture Canada, Res. Br. 1829). Ottawa: Canadian Government Publishing Centre. (Also in French as *Les fougères et les plantes alliées du Canada*.)

Copiously illustrated descriptive treatment in *pro forma* fashion; includes keys, synonymy, vernacular names, descriptions, karyotypes, habitat, distribution, and commentary along with ample discussions of the genera; glossary, extensive references and index to botanical names at end. The introductory part furnishes background, the plan of the work, discussions of special topics, and a key to genera.

121
Western Canada

Included here are certain floras and other works relating to the Pacific and Rocky Mountain zone (mainly British Columbia and the Yukon) and the 'prairie belt' (covering large portions of Alberta, Saskatchewan and Manitoba). Provincial headings should also be consulted along with, in the conterminous United States, **175** (GREAT PLAINS FLORA ASSOCIATION; RYDBERG), **181** (KIRKWOOD; RYDBERG), and **191** (ELIOT; HITCHCOCK and CRONQUIST; HITCHCOCK *et al.*).

I. Pacific and Cordilleran subregion

Pteridophytes

TAYLOR, T. M. C., 1970. *Pacific Northwest ferns and their allies*. ix, 247 pp., text-figs., maps. Toronto: University of Toronto Press.

Illustrated descriptive treatment with keys to all taxa, synonymy (with references and many literature citations), distribution maps, taxonomic commentary, and notes on habitat, variability, special features, etc.; list of excluded species, list of karyotypes, phytogeographic summary, bibliography, glossary, and index to botanical names at end. The area covered, though centering on British Columbia, extends from Oregon north to Alaska. [Based on the author's *Ferns and fern allies of British Columbia* (see **124** under **Pteridophytes**).]

II. Central Plains subregion (the 'Prairie Belt')

Area: 1 779 000 km^2. Vascular flora: 1974 species and 220 additional infraspecific taxa (Looman and Best, 1987). – The enumeration by Boivin is formally more detailed than *Budd's Flora*, but both works now cover all of Alberta, Saskatchewan and Manitoba. *Budd's Flora* had its origins in Dominion Range Botanist Archibald C. Budd's *Plants of the farming and ranching areas of the Canadian Prairies* (1952). In 1957 this became *Wild plants of the Canadian prairies*; a second edition jointly with K. F. Best followed in 1964.

BOIVIN, B., 1967–81. *Flora of the prairie provinces*. 5 parts (Provancheria 2–5, 12). Québec: Herbier Louis-Marie, Université Laval. (Parts 1–4 reprinted from *Phytologia* **15** (1967): 121–159, 329–446; **16** (1967–68): 1–47, 219–339; **17** (1968): 58–112; **18** (1969): 281–293; **22** (1972): 315–398; **23** (1972): 1–140; **42** (1979): 1–24, 385–414; **43** (1979): 1–106, 223–251. Part 5 (Gramineae) independently published (as *Provancheria* 12); 108 pp.)

Descriptive manual of vascular plants, with keys to all taxa, essential synonymy, English and French vernacular names, generalized indication of local and extralimital range, extensive taxonomic commentary, and notes on habitat, phenology, etc.; index to genera at end of each part. The fifth and last part is devoted to grasses.

LOOMAN, J. and K. F. BEST, 1987. *Budd's Flora of the Canadian prairie provinces.* Revised edn. by J. Waddington. 863 pp., 230 figs. (incl. halftones and map), portr. (Publ. Canada Dept. Agric., Res. Br. 1662). Hull, Quebec: Canadian Government Publishing Centre. (1st edn., 1979.)

Briefly descriptive manual-flora with keys to all taxa, limited synonymy, vernacular names (to family and generic level only), indication of distribution, and brief notes on frequency, status, habitat, and behavior; glossary and indices to scientific and vernacular names at end (along with a note on spelling of vernacular names). The introductory section includes a *curriculum vitae* of Budd, a note on the expansion of the geographical coverage of the work, a description of vegetation formations, pedagogical notes, and the general key to families.

122

Yukon Territory

See also **110** (all works). – Area: 482 681 km² (Cody, 1996; see below). Vascular flora: 1112 species (Cody, 1996). A total of 894 species and 'major' nonnominate infraspecific taxa were recorded by Porsild in 1951.

The recent flora by Cody (1996) succeeds *Botany of southeastern Yukon adjacent to the Canol Road* by A. E. Porsild (1951, published as *Bull. Natl. Mus. Canada* 121 (Biol. Ser. 41)). Ottawa. This included the first general checklist for the Yukon and was itself a successor to various collection reports. Porsild added to this work with *Contributions to the flora of southeastern Yukon Territory* (1966) and *Materials for a flora of the central Yukon Territory* (1975).

CODY, W. J., 1996. *Flora of the Yukon Territory.* xvii, 643 pp., illus., col. end-paper maps. Ottawa: NRC Research Press (National Research Council of Canada). (National Research Council of Canada, Monograph Publishing Program.)

Illustrated descriptive manual of vascular plants with keys to all taxa, vernacular names, limited synonymy, indication of habitat and distribution, and illustrations and maps for each taxon; glossary, bibliography (pp. 617–624) and complete index at end. The general part covers physical features (Yukon features altitudes to 6050 m), geological history (much of area was not glaciated in the Wurmian period), vegetation and floristic statistics, history of botanical work (the main previous surveys being those of Hultén and Porsild, with much of the more recent work being related to government programmes), and a general key to families.

123

Northwest Territories and Nunavut (forested zone)

Area (continental portion): 1 826 400 km². Vascular flora (continental portion): 1113 species (Porsild and Cody, 1980; see below). – Although this unit is set up for works relating to the forested parts of the present Northwest Territories and Nunavut southwest of the 'tree-line', those covering the whole of their continental lands (until 1999 all part of Northwest Territories) are also given here. The 'Arctic' portions – mostly now in Nunavut – are considered under **073** (mainland) and **074** (the Arctic Archipelago). Under the latter is listed Alf Erling Porsild's *Illustrated flora of the Canadian Arctic Archipelago* (1957), a companion work to that by him and William Cody described below.

Porsild and Cody's definitive *Vascular plants of continental Northwest Territories* (1980; see below) supplanted their preliminary *Checklist of the vascular plants of continental Northwest Territories, Canada* (1968), itself the first consolidated account. Hugh Raup's *Botany of southwestern Mackenzie* (1947; see below under **Partial work**), however, remains useful.

PORSILD, A. E. and W. J. CODY, 1980. *Vascular plants of continental Northwest Territories, Canada.* viii, 667 pp., 978 text-figs., 1155 distribution maps. Ottawa: National Museums of Canada (for the National Museum of Natural Sciences).

Copiously illustrated manual-flora of vascular plants, with keys, limited synonymy, vernacular names (not below generic level), indication of internal and extralimital distribution (with dot maps for each species), and notes on habitat and other features (e.g., infraspecific forms); glossary and full general index at end (on green paper). Preceding the formal treatment are *curricula vitarum*, an abstract, and an introductory section covering physical features, climate, biomes, major source works, the history of botanical exploration, and general keys to families. A general bibliography is also included, but references to revisions or other publications are not given in the text. Covers mainland Northwest Territories save for the mainland portion of Franklin District (i.e., the Boothia Peninsula).

Partial work

RAUP, H. M., 1947. *Botany of southwestern Mackenzie.* 275 pp., 16 text-figs., 37 pls. (Sargentia 6). Jamaica Plain, Mass.

Enumeration of vascular plants, without keys; includes synonymy (with references and citations), indication of *exsiccatae* (with localities), distribution maps for most species, critical remarks, and notes on habitat, frequency, associates, special features, etc.; list of references and index to all botanical names. The introductory section gives accounts of physical features, climate, geology, soils, plant communities, agriculture, phytogeography, the origins of the flora, and botanical exploration. The area covered extends north to 64° 30′ N, while the eastern limit follows the Marian and Great Slave Rivers (passing through Great Slave Lake).

124

British Columbia

See also **121/I** (TAYLOR), **191** (HITCHCOCK *et al.*). – Area: 948 600 km². Vascular flora: 2854 native species and infraspecific taxa, of which 2301 native (Douglas, Straley and Meidinger, 1989–94; see below).

A definitive checklist, the first for the province as a whole, is now available. It was developed from the 1977 enumeration of Roy Taylor and Bruce MacBryde – itself based on a never-published 1966 checklist by the late Thomas M. C. Taylor. The new work, rather more critical than its predecessor, effectively supplants the now-obsolete *Flora of the northwest coast* (1915) by Charles Piper and Rolla Kent Beattie (**191**), centered in western Washington State, and *Flora of southern British Columbia* (1915) by Joseph K. Henry and moreover complements *Flora of the Queen Charlotte Islands* (see below under **Partial work: Queen Charlotte Islands**). In addition, since the 1960s detailed revisionary treatments by T. Taylor and other authors have been appearing in 'installments' under the auspices of the British Columbia Museum. In 1998, a new

nomenclator was published by Qian and Klinka (see below).[34]

Bibliography

In addition to the work here cited, an outstanding example in its genre, reference may also be made to *Vascular plants of British Columbia* (see below). Standard and other references are also given in the earlier enumeration by Taylor and MacBryde.

DOUGLAS, G. W., A. ČEŠKA and G. G. RUYLE, 1983. *A floristic bibliography for British Columbia.* iv, 143 pp. (Land Management Report 15). Victoria: Information Services Branch, Ministry of Forests, British Columbia. [A very comprehensive treatment, based on a perceived need for *complete* ecosystem knowledge. Of the seven sections, A (floras) and D (monographs and revisions) are of most interest in the present context.]

Keys to families

RAFIQ, M., G. F. HARCOME and R. T. OGILVIE, 1982. *An illustrated key to dicotyledon families of British Columbia.* 127 pp., illus. Victoria: Ministry of Environment, British Columbia/British Columbia Provincial Museum. Complemented by *idem*, 1982. *An illustrated key to monocotyledon families* . . . . 25 pp., illus. Victoria. [Both works comprise illustrated keys arranged in hierarchical fashion in 20 interconnected charts, followed by family accounts with diagnostic features; glossary, references and index in each volume. Some families are keyed out more than once.]

DOUGLAS, G. W., G. B. STRALEY and D. MEIDINGER, 1989–94. *The vascular plants of British Columbia.* 4 parts (Special Reports Series 1–4). Victoria: Ministry of Forests, British Columbia (distributed by Crown Publications, Victoria).

Keyed enumeration, the entries including synonymy, vernacular names, distribution within and without the province, status and frequency; general and family references, a list of excluded species, and complete index in each fascicle. The introductory section is mainly on methodology and style, but includes a brief background; statistics and a floristic summary appear in part 4 along with a glossary and complete cumulative index.[35]

QIAN, H. and K. KLINKA, 1998. *Plants of British Columbia.* xiv, 534 pp. Vancouver: University of British Columbia Press.

A nomenclator with two major parts of about equal length: (1) lists of accepted names (with codes) and associated synonyms, arranged by major groups (vascular plants (with 2105 native species), bryophytes and lichens), families and genera (pp. 1–255), and (2) an alphabetically arranged list of all names with accepted equivalents (where applicable) or families with major groups (pp. 257–464). These are followed by a lexicon of vernacular names (pp. 465–524) and a list of excluded names (pp. 525–534). The introductory part furnishes the plan and justification for the work, along with principal statistics and a bibliography of main source works.

TAYLOR, R. L. and B. MACBRYDE, 1977. *Vascular plants of British Columbia: a descriptive resource inventory.* xxiv, 752 pp., col. map (Techn. Bull. Univ. Brit. Columbia Bot. Gard. 4). Vancouver: University of British Columbia Press.

An information system-based enumeration of vascular plants (3137 taxa) in a tabular format, with abbreviated indications of karyotypes, habitat, distribution, reproductive biology, life-span, economics, conservation status, etc. Taxa are alphabetically arranged within the major classes, with an index to scientific and vernacular names at the end. Six appendices account for literature references and standard floristic works on British Columbia, taxon–reference links, a lexicon of authorities, and a sample data form, while the colored map depicts biogeoclimatic zones in the province. An introductory section gives technical notes and a description of the information system and its software.[36]

Partial work: Queen Charlotte Islands
These islands, a 'glacial refugium', are covered by J. A. Calder and R. L. Taylor in an uncommon – and outstanding – example of a 'biosystematic' flora.

CALDER, J. A. and R. L. TAYLOR, 1968. *Flora of the Queen Charlotte Islands.* 2 vols. Illus., maps (Monogr. Canad. Dept. Agric., Res. Br. 4). Ottawa: Queen's Printer.

Volume 1 comprises a detailed enumeration of vascular plants, with keys, appropriate synonymy, citations of *exsiccatae* with localities, critical remarks, and extensive notes on habitat, ecology, variation, biology, etc., accompanied by numerous figures and distribution maps; index to all botanical names at end. A fairly comprehensive introductory section is also provided. Volume 2 is a 'cytotaxonomic supplement', containing a synopsis of species with their chromosome numbers where known.

Woody plants
GARMAN, E. H., 1963. *Pocket guide to the trees and shrubs of British Columbia.* 3rd edn. 137 pp., illus. (Special Publ. Brit. Columbia Forest Serv. B-28). Victoria. (1st edn., 1937.)

Pocket manual of trees and shrubs (covering the whole province), with keys to all taxa, essential synonymy, generalized indication of local range (with some details), and notes on habitat, ecology, etc.; map of forest regions; glossary and index to all vernacular and botanical names.

Pteridophytes
In addition to *Pacific Northwest ferns and their allies* (**121/I**), see also T. M. C. TAYLOR, 1963. *The ferns and fern allies of British Columbia.* 2nd edn. (Handb. Brit. Columbia Prov. Mus. 12). Victoria. (1st edn., 1956.)

125

Alberta

See also **121/II** (BOIVIN; LOOMAN and BEST). – Area: 661 185 km². Vascular flora: 1755 species (Moss and Packer, 1983; see below).

Prior to the first edition of *Flora of Alberta* (1959), no floristic works for the province as a whole existed. The eastern prairie region was first fully covered in *Wild plants of the Canadian prairies* (1957) by A. C. Budd (see **121/II**); the Rocky Mountains in *Contributions to a catalogue of the flora of the Canadian Rocky Mountains and the Selkirk Range* (1907) by Edith M. Farr; and the northern forests in survey papers by Hugh Raup beginning in the 1930s.

MOSS, E. H. and J. G. PACKER, 1983. *Flora of Alberta.* 2nd edn. xiii, 687 pp., 2, 1158 maps. Toronto: University of Toronto Press. (1st edn., 1959, by Moss alone.)

Descriptive manual of vascular plants, with keys to all taxa, essential synonymy, vernacular names, generalized indication of local and extralimital range, karyotypes, some taxonomic commentary, and brief notes on habitat, frequency, etc.; individual dot distribution maps (each depicting one or two species), glossary and index to all vernacular and botanical names at end. The introductory part (inclusive of two maps from the *Atlas of Alberta*) covers physical features, geological history, climate, soils, vegetation formations, 'notable collectors and authorities', statistics of the flora and a key to families.

126

Saskatchewan

See also **121/II** (BOIVIN; LOOMAN and BEST). –
Area: 570 113 km². Vascular flora: no data.

No separate keyed descriptive flora is available,
and indeed until 1937 there was no consolidated
account of the provincial flora. In that year *An annotated list of the plants of Saskatchewan* by W. P. Fraser
and R. C. Russell was published, embodying almost 20
years of field work by its senior author (of the
University of Saskatoon) and others; revisions
appeared in 1944 and 1953. A new province-wide treatment by Vernon Harms and associates reportedly is in
preparation.

Breitung's list largely, but not entirely, replaces
W. P. FRASER and R. C. RUSSELL, 1953. *An annotated
list of the plants of Saskatchewan*, revised by R. C.
Russell *et al.* 47 pp., illus. Saskatoon. (1st edn., 1937.)

BREITUNG, A. J., 1957. Annotated catalogue of
the vascular flora of Saskatchewan. *Amer. Midl.
Naturalist* 58: 1–72. Complemented by *idem*, 1959.
Supplement. *Ibid.*, **61**: 510–512.

Concisely annotated systematic checklist, with
synonymy, vernacular names, and indication of localities (with citation of some *exsiccatae* and other authorities), habitat, frequency, etc.; occasional critical
remarks; discussion of doubtful and excluded species
at end. An introductory section gives limits of the flora,
main features of the vegetation, history of collecting,
etc., together with abbreviations.

127

Manitoba

See also **121/II** (BOIVIN; LOOMAN and BEST). –
Area: 540 700 km². Vascular flora: 1417 species and 124
additional infraspecific taxa, both native and introduced (Scoggan, 1957; see below).

The current accounts supplanted two earlier
checklists: *The Phanerogamia and Pteridophyta of
Manitoba* by the Rev. W. A. Burman (1909), published
for a meeting of the British Association for the

Advancement of Science and covering 702 species
south of 52° N, and *Check list of Manitoba flora* by V. W.
Jackson and others (1922, with a revision in 1925).[37]

LOWE, C. W., 1943. *List of the flowering plants,
ferns, club-mosses and liverworts of Manitoba*. 110 pp., 1
halftone, map. Winnipeg: Natural History Society of
Manitoba.

Systematic census of vascular plants, with notes
on local range, habitat, frequency, and phenology;
limited synonymy; vernacular names; list of bryophytes; index to family names. A folding map of vegetation
zones in the province is also provided.

SCOGGAN, H. J., 1957. *Flora of Manitoba*. 619
pp. (Bull. Natl. Mus. Canada 140). Ottawa.

Critical enumeration of vascular plants, with keys
to all taxa, essential synonymy, vernacular names,
detailed indication of internal range with many localities (and authorities), summary of extralimital range,
critical remarks, and notes on habitat, etc.; discussions
of doubtful and excluded taxa; list of references and
index to all botanical names. The introductory section
includes accounts of physical features, climate, vegetation, and affinities of the flora as well as lists of new additions and excluded taxa and a tabular summary of the
flora. [For additional records, see P. M. CATLING and V.
R. BROWNELL, 1987. New and significant plant records
for Manitoba. *Canad. Field-Natur.* **101**: 437–439.]

128

Northern Ontario

For Ontario in general, see **132**. The area circumscribed here is that part of Ontario north of Lake
Superior and the 48th parallel (north of the towns of
Sault Ste. Marie and Cadbury). Encompassing somewhat more than half the province, it is a sparsely populated, largely wilderness region: a sentient 'barrier'
between eastern and western Canada. Separate floristic
coverage is available only in partial treatments, of
which that on the north shore of Lake Superior is
described below.

Partial work

SOPER, J. H., C. E. GARTON and D. R. GIVEN, 1989.
*Flora of the north shore of Lake Superior (vascular plants of the
Ontario portion of the Lake Superior drainage basin)*. 61 pp.

(Syllogeus 63). Ottawa: National Museums of Canada Publishing.

Tabular checklist covering 994 native and 217 introduced species of vascular plants with accepted names, coded indication of habitat(s) (see p. 11 for explanation) and presence/absence in each of nine geographical subdivisions (map, p. 12); statistics (pp. 54–56) and general references (pp. 57–61). The background and plan of the work are covered in the introduction (pp. 5–22) along with a brief account of floristic elements, rare species, and directly cited references. [Based on data and collections going back to extensive surveys in the 1930s by R. C. Hosie, T. M. C. Taylor and others but much added to following completion in 1960 of the Trans-Canada Highway through the region.]

131

Eastern Canada

No overall works are available, mainly as a result of the language division and, for the anglophone community, the presence of the standard manuals for the northern United States and 'adjacent parts of Canada' (see **140–50**). With the 49th parallel their usual northern limit (some also extending to Anticosti Island and Newfoundland), these latter do not in general extend beyond more heavily populated areas.

Bibliography

CATLING, P. M., B. S. BROOKES, Y. M. SKORUPINSKI and S. M. MALETTE, 1986. *Bibliography of vascular plant floristics for New Brunswick, Newfoundland (insular), and Nova Scotia.* 28 pp. (Techn. Bull. Agric. Canada Res. Br. 1986-3E). [Not seen.]

132

Southern Ontario (and Ontario in general)

Area (Ontario): 1 068 582 km². Vascular flora: no data. – Included here are references relating to the province as a whole; for northern Ontario alone, see **128**. Full coverage of southern Ontario has conventionally been provided by the standard manuals for the northeastern United States (**140**).

A checklist of the flora of Ontario: vascular plants (1990) by John K. Morton and J. M. Venn represents the first complete provincial account; regrettably, however, it is a mere name list. It effectively supersedes *The vascular plants of southern Ontario* (1949) by James H. Soper. Earlier floristic treatments, beginning in the mid-nineteenth century, were mostly local checklists or reports of surveys.

Bibliography

HODGINS, J. L., 1978. *A guide to the literature on the herbaceous vascular flora of Ontario.* 2nd edn. 73 pp. Toronto: Botany Press. (1st edn., 1977.) [Classified bibliography, with format (books, papers, maps or 'botanical surveys') as a primary attribute; floras, pp. 1–4. Under 'botanical surveys' is listed much unpublished and semi-published material. Page 69 gives a list of public herbaria, and pp. 70–73 comprise a name list of herbaceous species. Coverage is of post-1930 publications.]

MORTON, J. K. and J. M. VENN, 1990. *A checklist of the flora of Ontario: vascular plants.* x, 218 pp. Waterloo, Ont. (University of Waterloo biology series.)

Alphabetically arranged name list of vascular plants, with accepted names and synonyms in a common sequence (the latter also appearing under their accepted equivalents) and citation of a few references or other sources of information; bibliography (pp. 189–203) and index to family and generic names at end. The introductory part gives the philosophy and plan of the work (with some sharply worded general comments on taxonomy and nomenclature), a list of basic historical and modern reference works (pp. vii–viii), and notes on problem genera and families and on aliens. Varieties are not routinely included for, in the authors' view, an effective understanding more often than not 'must await further floristic and taxonomic studies'.

Woody plants

The following is included here in view of the relative paucity in general of recent identification works for shrubs in eastern North America.

SOPER, J. H. and M. L. HEIMBURGER, 1982. *Shrubs of Ontario.* xxxi, 495 pp., illus., maps. Toronto: Royal Ontario Museum. (A Life Sciences Miscellaneous Publication. Based on the authors' *100 Shrubs of Ontario* (1961).)

Atlas-manual of shrubs and lianas with full-page figures and part-page dot distribution maps along with descriptions, accepted scientific and vernacular names, indication of habitat and distribution, taxonomic notes, and field recognition features; literature, author abbreviations, glossary, and index at end. The introductory part covers early botanists, sequence of species, geography and biogeographic unities (with three maps), rare and endangered species, and (pp. xxi–xxxi) the general keys.

133

Québec (except Ungava)

Area: 1357780 km^2. Vascular flora: 1850 species (Rousseau, 1974; see below). – Included with Québec are Anticosti and Magdalen Islands in the Gulf of St. Lawrence and the neighboring islands in Hudson Bay; for Ungava, however, see **075**. Additional coverage of southern Québec by anglophone manuals appears under **140**.[38]

The St. Lawrence Valley was the center of most of the earliest botanical work in Canada, symbolized by Jacques Philippe Cornuti's *Canadensium plantarum . . . historia* (1635) and a long-unpublished 1708 study by the colonial physician Michel Sarrazin and the Paris botanist Sébastien Vaillant. From the time of Léon Provancher's *Flore canadienne* (1862), however, provincial floras have for the most part been limited to the St. Lawrence basin (west of the 68th parallel, i.e., at Rimouski on the south bank) and adjacent hilly areas including the Laurentide and Notre Dame ranges; the most important work has been *Flore laurentienne* (1935) by Brother Marie-Victorin (Conrad Kirouac). To these Homer Scoggan added a Gaspé flora in 1950 and in 1969 Brother Rolland-Germain published his joint work with Marie-Victorin on Anticosti Island and the adjacent Mingan region (on the north shore of the Gulf of St. Lawrence). The Labrador Peninsula north of the 50th parallel remains but sketchily documented. Camille Rousseau's *Géographie floristique du Québec-Labrador* most nearly approaches complete provincial coverage, but is not strictly a flora.[39]

ROUSSEAU, C., 1974. *Géographie floristique du Québec-Labrador. Distribution des principales espèces vasculaires.* xiii, 799 pp., 1016 area maps. Québec: Les Presses de l'Université Laval. (Travaux et Documents du Centre d'Études Nordiques 7.)

An atlas of the more significant vascular plants of Québec (1020 species, or some 60 percent of the known total), without keys or descriptions; for each up to three separate maps are given along with descriptive text encompassing notes on ecology, biology, geographical range and limits, floristic classification, and history; bibliography (pp. 559–615) and general index at end. The introductory section gives accounts of general geology and physiography, bioclimatic zones, floristic elements and their distribution, and the history of the flora as well as the philosophy and methodology of the work. The maps for technical reasons are grouped into three sections depending on their limits (upper St. Lawrence; southern Québec and the Gaspé Peninsula; Québec as a whole) and do not follow the sequence of the species accounts.

Partial works: southwestern Québec

LOUIS-MARIE, Père (L.-M. Lalonde), 1931. *Flore-manuel de la province de Québec.* 320 pp., over 2100 text-figs. on 90 pls. (Contr. Inst. Agricole Oka 23). Oka, Québec. (Reissued without date, Montréal: Centre de Psychologie et Pédagogie. 2nd edn., 1953; 3rd edn., 1959.)

Well-illustrated students' manual-key of vascular (and some non-vascular) plants; includes limited synonymy, English and French vernacular names, brief notes on local and extralimital range, uses (including the flavoring of gins), pests and diseases, etc., together with indices to family, generic and vernacular names. An extended introductory section includes chapters on descriptive terminology, collecting methods, nomenclature, etc., while appendices include a glossary and accounts of medicinal plants and the essential principles of ecology. [Despite its title, this once much-appreciated 'ouvrage de vulgarisation', with its 'general' and 'special' sections, is essentially limited to the St. Lawrence basin and adjacent areas; moreover, the keys do not run out directly to all species. In some issues, the plates are colored; other issues are pocket-sized.][40]

MARIE-VICTORIN, Frère, 1995. *Flore laurentienne.* 3rd edn., revised by L. Brouillet, S. G. Hay and I. Goulet. xv, 1093 pp., 324 pls. (with 2800 text-figs.), 120 col. pls., 23 maps (1 col.). Montréal: Les Presses de l'Université de Montréal. (Reprinted 1997. 1st edn., 1935; supplement by E. Rouleau, 1947; 2nd edn., 1964, by E. Rouleau, with at least six reprintings.)

Manual-flora of vascular plants with keys to all taxa, essential synonymy, French and English vernacular names, generalized indication of local range with some localities, taxonomic commentary, figures of diagnostic features, and notes on habitat, frequency, occurrence, uses, etc.; glossary, abbreviations of serials and authors, and 46-page index to all botanical and vernacular names at end, along with statistics (pp. 856–857) and the album of color plates with 900 photographs (pp. 863–982). The rather lengthy introductory section (pp. 1–60) includes an account of the history of botany in Québec together with a general survey of the flora and its phytogeography, evolution, and Quaternary history as well as a general artificial key (by J. Rousseau). The work encompasses southern Québec up to 49° N and extends east to the western limits of the Gaspé Peninsula. [Typographically a reproduction of the 1964 edition with marginal additions, followed by a new section of 37 pages with 1140 notes and references and a 6-page appendix with taxa new to the subprovince. The family circumscriptions of 1935 are retained.][41]

Partial works: other parts of Québec

See also **075** (GARDNER).

MARIE-VICTORIN, Frère and Frère ROLLAND-GERMAIN, 1969. *Flore de l'Anticosti–Minganie*. 527 pp., 26 text-figs., portr., map. Montréal: Les Presses de l'Université de Montréal.

Systematic enumeration of vascular plants, without keys, of Anticosti Island and the adjacent mainland to the north; includes synonymy, citation of *exsiccatae* with localities, and taxonomic commentary; bibliography and index to botanical names at end. A lengthy introductory section includes accounts of geography, geology, botanical exploration in the area, etc., while pp. 399–498 are devoted to floristic analyses and phytogeographic considerations.

SCOGGAN, H. J., 1950. *The flora of Bic and the Gaspé Peninsula, Québec*. 399 pp. (Bull. Natl. Mus. Canada 115). Ottawa.

Systematic enumeration of vascular plants; includes keys to all taxa, essential synonymy with some references and citations, vernacular names, localities in some detail along with generalized summaries of local and extralimital range, and brief notes on habitat and life-form; bibliography and index to botanical names at end. The introductory section includes accounts of physical features, climate, soils, life-form spectra, Quaternary and post-Quaternary history, and phytogeography, together with a tabular summary of the flora and a list of Québec plants absent from the Peninsula.

Woody plants

ROULEAU, R. (ed.), 1974. *Petite flore forestière du Québec*. 216 pp., 126 col. illus., map. Québec: Éditeur officiel du Québec (Ministère des Terres et Forêts, P.Q.).

Pocket-sized, empirically organized illustrated atlas of forest trees, shrubs, herbs, ferns and selected bryophytes and lichens; the figures (usually facing the text) include foliage, flower, fruit and bark details, while the text encompasses vernacular names (French-Canadian), dendrological features, distribution and utilization; references (p. 191), glossary and index to all names. The six empirical sections are based on gross features. The introductory part includes a summary of vegetation zones (with map). [Extends north to 52–53°, thus to about the southern boundary of Labrador Territory.]

Pteridophytes

Québecois pteridophytes were separately covered by Marie-Victorin in two works, both now well out of date: *Les Filicinées du Québec* (1923, in *Contr. Inst. Bot. Univ. Montréal* 2) and *Les Lycopodinées du Québec et leurs formes mineures* (1925, in *ibid.*, 3).

134

Labrador

See **133** (ROUSSEAU), **135** (ROULEAU). Labrador Territory has been administratively part of Newfoundland since 1927. – Area: 290 080 km². Vascular flora: no data.

No recent general accounts relating specifically to the territory are available. The earliest flora, based in the first instance on collections forwarded by Herzberg, a Moravian missionary brother in Okak and accounting for 198 species, comprises the first book of *De plantis labradoricis libri tres* (1830) by E. H. F. Meyer.[42] Later works include in particular *List of the plants known to occur on the coast and in the interior of the Labrador Peninsula* (1897) by James Macoun and a plant list by E. B. Delabarre in *Labrador: the country and the people* (1909) edited by W. T. Grenfell. Additional explorations have since been made, but publication has only been in the form of reports, among them a number by Merritt L. Fernald.

Bibliography

See **135** (LAIRD, 1980).

135

Newfoundland

See also **140** (BRITTON and BROWN; FERNALD). – Area: 111 000 km². Vascular flora: 1166 species (Rouleau and Lamoureux, 1992, who record 1197 species from both Newfoundland and St. Pierre and Miquelon). Newfoundland, retained by the United Kingdom after formation of the Canadian Federation in 1867, became part of Canada only in 1949.

The nineteenth-century checklists of Henry Reeks (1873) and Arthur Waghorne (1896–98, also encompassing Labrador but not completed), are long out of date. The extensive field work of Merritt L. Fernald and his associates and successors over the first three-quarters of the twentieth century, including (from 1948) Ernest Rouleau, increased the known number of vascular species by perhaps one-third. The results to 1950 were incorporated into Fernald's edition of *Gray's Manual* (see **140**) and so furnished a basis for Rouleau's first checklist: *Checklist of the vascular plants of the province of Newfoundland* (1956; in *Contr. Inst. Bot. Univ. Montréal* **69**: 41–106); this was revised in 1978. All available data have been accounted for in the 1992 *Atlas*, which also contains a revision of the 1978 checklist.

Bibliography
See also **131** (CATLING *et al.*, 1986).

LAIRD, M., 1980. *Bibliography of the natural history of Newfoundland and Labrador.* lxxi, 376 pp., end-paper maps. London: Academic Press. [Well-annotated catalogue arranged by author; index, pp. 339–376 but no proper subject lexicon.]

ROULEAU, E., 1978. *List of the vascular plants of the province of Newfoundland (Canada).* [vi], 132 pp., illus. (front cover). St. John's: Oxen Pond Botanic Park.

Systematic checklist, with essential synonymy and indication of presence in Labrador, Newfoundland proper, and/or St. Pierre and Miquelon; lexicon of vernacular names with scientific equivalents (pp. 81–96) and index to all names (pp. 97–132). The brief introduction gives the basis for and plan of the work. Arrangement of families is after Takhtajan.

ROULEAU, E. and G. LAMOUREUX, 1992. *Atlas des plantes vasculaires de l'île de Terre-Neuve et des îles de Saint-Pierre-et-Miquelon/Atlas of the vascular plants of the island of Newfoundland and of the islands of Saint-Pierre-et-Miquelon.* 777, [4] pp., 1197 individual distribution maps, portr., maps (end-papers). Lévis, Québec: Fleurbec.

Distribution atlas of individual species maps, each with precisely plotted localities and brief text giving essential synonymy, vernacular names and status; symbols on p. 145; extensive bibliography (pp. 751–758) and complete index at end. The introductory section (bilingual) contains sources for the work, memorials to the senior author, the basis for the work and preparation for publication, the history of local botanical exploration and floristic studies, and lists of

unmapped and cultivated species, followed by a list of mapped species (1197 in all; 8 numbers not represented and 36 in the appendix). A formal checklist (pp. 103–143) precedes the atlas proper.[43]

136
St. Pierre and Miquelon

See also **135** (ROULEAU; ROULEAU and LAMOREUX). – Area: 241 km^2. Vascular flora: 642 species (Rouleau and Lamoreux, 1992 (under **135**)); 627 species and infraspecific entities were recorded by Le Gallo (1954; see below). These islands are French metropolitan territory, and not part of Canada.

The main summary of the flora prior to Le Gallo's enumeration is *Contribution to the flora of the islands of St. Pierre et Miquelon* (1927) by Frère Louis Arsène.

LE GALLO, C., Père, 1954. Les plantes vasculaires des îles Saint-Pierre et Miquelon. *Naturaliste Canad.* **81**: 105–132, 149–164, 181–196, 203–242.

Systematic enumeration, with citation of *exsiccatae* and other sources along with indication of frequency and habitat; for less common or more unusual plants, some notes on taxonomy, uses, distribution, and special features are also furnished. The list proper is preceded (in the first part) by an introduction covering generalities, previous work on the flora, and floristic statistics along with technical notes.

137
Prince Edward Island

See also **138** (ROLAND and SMITH); **140** (all works). – Area: 5656 km^2. Vascular flora: 625 native species (Erskine, 1960).

The two editions of David Erskine's flora were preceded by Blythe Hurst's *Flowering plants and ferns of Prince Edward Island* (1933, in *Trans. Roy. Canad. Inst.* **19**).

ERSKINE, D. S., 1985. *The plants of Prince Edward Island: with new records, nomenclatural changes, and corrections and deletions.* [2nd edn.] xxii, 272 pp., halftones, maps (incl. folding map in inside back cover) (Publ. Canada Dept. Agric., Res. Br. 1798). Ottawa: Queen's Printer. (1st edn., 1960, as Publ. Canada Dept. Agric., Res. Br. 1088.)

Systematic enumeration of native and naturalized vascular plants, with limited synonymy and some vernacular names; detailed indication of local range, with citation of *exsiccatae* and other records; distribution maps given for all species; notes on habitat, frequency, special features, uses, etc.; index to genera and families. The introductory section includes accounts of botanical exploration, physical features, floristics, vegetation associations, and phytogeography, together with a list of references. [The current edition is a reprint of the 1960 work with an added prologue (pp. i–xxii) encompassing a new introduction and all additions and changes, fully documented by *exsiccatae* and other sources of information.]

138

Nova Scotia

See also **140** (all works). – Area: 54566 km². Vascular flora: slightly under 2000 species and non-nominate infraspecific taxa (Roland and Smith 1966–69; see below).

Prior to publication of the first edition of Albert Roland's *Flora of Nova Scotia* in 1944–47, coverage was furnished in an early work from the Nova Scotian Institute of Science: *A catalogue of the flora of Nova Scotia arranged according to Gray's Manual of Botany* (1875–76) by Andrew Lindsay.

Bibliography
See 131 (CATLING *et al.*, 1986).

ROLAND, A. E. and E. C. SMITH, 1966–69. *The flora of Nova Scotia*. Revised edn. 2 parts (as *Proc. Nova Scotian Inst. Sci.* **26**(2): 3–244; **26**(4): 277–743). Halifax. (1st edn., 1944–46 (1947) in *ibid.*, **21**: 97–642.)

Illustrated systematic enumeration of vascular plants; includes diagnostic keys to all taxa, references to key literature on individual plant groups, vernacular names, generalized indication of local and extralimital range (with some citations of *exsiccatae* and other sources), numerous figures of characteristic features

together with dot distribution maps, some taxonomic commentary, and notes on habitat, frequency, biology, etc.; glossary, bibliography and index to all botanical and vernacular names in each part. [The work also accounts for all species on Prince Edward Island.]

139

New Brunswick

See also **140** (all works). – Area: 72481 km². Vascular flora: *c.* 1500 species including introductions (Hinds, 1986; see below).

In 1986, after 101 years, the Rev. James Fowler's *Preliminary list of the plants of New Brunswick* (1885; in *Bull. Nat. Hist. Soc. New Brunswick* 1(4): 8–84 and separately issued) was succeeded by a new manual-flora by Harold Hinds. Fowler's work, however, includes particular consideration of railway and other ballast adventives.

Bibliography
See 131 (CATLING *et al.*, 1986).

HINDS, H. R., 1986. *The flora of New Brunswick.* [ii], xxvi, 460, 121, 85 pp., >1452 figs., 1014 distribution maps. Fredericton, N.B.: Primrose Press.

The textual part of this work is a keyed enumeration of native and naturalized vascular plants, with essential synonymy, vernacular names, local and extralimital distribution, frequency, notes on taxonomy (if appropriate), habitat and dynamics, and cross-references to the figures and distribution maps; cited references, glossary and index at end. The two appendices comprise atlases respectively of figures (pp. A1–A121) and topographic dot maps (pp. B1–B85), each with 12 units per page and in the former depicting all species, subspecies and hybrids. The introductory section covers physical features, geology and earth history, soils, climate, and the history of botanical work along with notes on the use of the book and a general key to families.[44]

Superregion 14–19

Conterminous United States

This superregion comprises all of the 48 conterminous United States of America, 'from sea to shining sea'. Alaska is here designated as Region 11, and Hawaii, being in the Pacific Ocean, is treated as Region 99.

The various 'national' floras, works on special groups (woody plants, pteridophytes, grasses, etc.), checklists, and bibliographies more often than not overtly or covertly cover Canada, St. Pierre and Miquelon, and (in some cases) Greenland as well as Alaska and the conterminous United States and Alaska, and are thus, for convenience, all listed under **100**.

The long, expansive, many-faceted and personality-rich history of botanical exploration and floristic study in the United States is in considerable measure congruent with that of the continent north of Mexico. It has therefore been treated, together with current developments and some indications of the present state of knowledge of the flora, under the divisional heading. It may suffice to note here that in recent decades the largest amounts of new data have come from the Southeast (Region 16), the mountain states (Region 18) and parts of the Pacific states (Region 19).[45]

Bibliographies. Any supraregional works, as well as divisional and general bibliographies, are accounted for under the divisional heading.

Indices. Any supraregional works, as well as divisional and general indices, are accounted for under the divisional heading.

140–90

Superregion in general

No floras covering the conterminous United States as a whole have been published. The classic early works by André Michaux, Henry Muhlenberg, Frederick Pursh and Thomas Nuttall, the great floras begun by John Torrey, Asa Gray and Nathaniel L. Britton – the latter all, alas, never completed – and the recent checklists by Stanwyn G. Shetler and Laurence Skog (1978), the USDA Soil Conservation Service (1982) and John and Rosemary Kartesz (1980; 2nd edn., 1994) all cover at least the whole continent and Greenland, as will the new *Flora of North America* now in preparation at various North American botanical centers. The nearest approach to a 'national' flora, although not scientific in a strict sense, is *Wild flowers of the United States* (1966–73), directed by Harold Rickett (see below).

RICKETT, H. W. and collaborators, 1966–73. *Wild flowers of the United States.* 6 vols. (in 14), illus. (many col.). New York: McGraw-Hill.

Large-scale, lavishly illustrated, popularly oriented guide in six boxed, regionally oriented volumes (each with two or three bound parts) to wild flowers in the conterminous United States; individual volumes include synoptic keys, descriptions, explanations of terms, glossaries, photographs in color, line drawings, and complete indices, but do not arrange the plants in any systematic order. The different volumes represent, respectively, the northeastern states, the southeastern states, Texas, the southwestern states, the northwestern states, and the central mountains and plains states. A complete general index is projected. [Written under professional auspices from an editorial office at the New York Botanical Garden and published with fanfare and gusto under prominent private patronage, this work is a fine monument to a great age of American affluence and to the botanical sensibilities of its public.]

Region 14/15

Northeastern and North Central United States

Area: 666387 km². Vascular flora (including southeastern Canada and Newfoundland): 5523 species, of which some 80 percent native (Fernald, 1950; see below). – This large region comprises the 22 states (and the District of Columbia) stretching from Maine to Virginia and west to Missouri, Iowa and Minnesota. Two 'subregions' may be recognized: the

Northeast forest region (which floristically continues further south along the Appalachians), and the Midwest, comprising basically the lake forest region and the woodland/prairie belt. Both subregions overlap into adjacent Canada. Much of the southern fringe, however, has greater affinities with the Southern forest belt (Region 16); the traditional mutual boundary between the ranges of Northern and Southern manuals is arbitrary, a legacy in part of nineteenth-century commercial circumstances.

The 'Northeast' not unnaturally was the first part of the continental United States to undergo regional consolidation of floristic knowledge. The first contributions were respectively *A manual of botany for the northern states* (1817, with seven further editions to 1840, including increased geographical coverage) by Amos Eaton, *A flora of the northern and middle sections of the United States* (1823–24, not completed) and *Compendium of the flora of the northern and middle states* (1826) by John Torrey, and *Botany of the northern and middle states* (1833; 2nd edn., 1848, with later reissues) by Lewis C. Beck in Albany, New York.[46] In 1845 Alphonso Wood published the first edition of his *Class-book of Botany*, with expansion of coverage in 1848 and again in 1861. In 1848, Asa Gray produced his *Manual of botany of the northern United States*, the first edition of what would become known as *Gray's Manual*; further editions appeared in 1856 (with expansion to Virginia and the Midwest), 1862, 1863, 1867 and 1890.[47] Wood returned to the Northeast alone with *The American botanist and florist* (1870; 2nd edn., 1877). All these greatly facilitated preparation of state and local accounts. Until the mid-twentieth century these took most commonly the form of checklists or enumerations (some with keys) but among them was one outstanding descriptive work: Torrey's *Flora of the state of New-York* (1843).

In the 'Midwest', early studies by Beck and others, including several expedition papers, were consolidated by John L. Riddell in his *Synopsis of the flora of the western states*, published at Cincinnati in 1835.[48] However, Riddell soon after moved to New Orleans; and, with the geographical expansion of *Gray's Manual* and other works westwards consequent to a recognition of the essential similarity of the Midwestern and Northeastern floras, no further independent regional treatments were essayed. Efforts passed into more detailed studies at state and local level, stimulated in part by state surveys; by 1900 all Midwestern states had been catalogued at least once and more substantial enumerations, such as *Metaspermae of the Minnesota Valley* (1892) by pioneer vegetation ecologist Conway MacMillan, had begun to appear, with descriptive state manuals following from about the mid-twentieth century.

The progress of the regional manuals was for much of the modern era marked by rivalries. An early one might be said to be between Beck, Eaton and Torrey, with Eaton to the last remaining with the Linnaean classification system, while Beck was the first to espouse a natural arrangement (in his case that of Lindley), as would also Torrey and Gray in their *Flora of North America*. In the 1840s Gray, at Cambridge (Mass.), prepared his *Manual* in response to Wood's *Class-book*, and these works (along with Wood's *American botanist and florist*) remained in competition for the next generation.[49]

A new rivalry emerged at the close of the nineteenth century with the rapid development of the New York Botanical Garden under its first director, N. L. Britton. Assuming Wood's mantle, Britton produced, firstly, the original edition of the *Illustrated flora* (1896–98, with Addison Brown; 2nd edn., 1913) and then *Manual of the flora of the northern states and Canada* (1901; 2nd edn., 1905; 3rd edn., 1907). Both featured the 'progressive' Englerian system as well as the 'American Code' of nomenclature. By contrast, at Cambridge *Gray's Manual* remained, for many years almost uniquely in the United States, in conformity to the International Code although for the seventh edition (1908) the Englerian arrangement also was adopted.[50] After 1930, with Britton retired, the rivalry diminished as Merritt L. Fernald in Cambridge became the effective 'master', his knowledge enhanced by numerous field trips in the decades before World War II. His definitive edition of *Gray's Manual* appeared in 1950. In 1952 a revision of the *Illustrated flora* was produced by Henry Gleason; the text alone was again revised by Arthur Cronquist and published in 1963 as *Manual of vascular plants of northeastern United States and adjacent Canada* (see below). This became the effective successor to Britton's manual. Cronquist released a new edition of the *Manual* in 1991, with families arranged according to his system.

Both current works are of high quality and the choice is mainly a matter of taxonomic philosophy and personal taste, though a new revision of *Gray's Manual* is much to be desired. A continuing demand, however, existed for Britton's *Illustrated flora* of 1913, with a greater area of coverage and (to at least some) a more satisfying arrangement of text and figures.[51] In 1997, the New York Botanical Garden released a new version of the *Illustrated flora* entitled *An illustrated companion to Gleason and Cronquist's* Manual. This is a thick, one-volume work with full-page plates in part based on the original illustrations; like its predecessors, all species are figured.

With respect to the woody flora, many guides are available, rather the more so to trees alone. The most elegant of the latter is perhaps still Hough's *Handbook* (1907; reissued 1947); but among the more widely used works are the field-guides of Harlow (1942) and Petrides (1972). The former was reissued by Dover in 1957 with minor revisions. Since 1980 they have been joined by Elbert Little's popular *Audubon field guide to North American trees: Eastern Region* and a revision (1988), limited to trees, of Petrides' guide (for both, see **100**). Several fern guides are also available.

At state level, progress has been mixed. In the Midwest, where before World War II individual coverage of the vascular flora had been mainly in the form of enumerations or checklists, publication of the widely acclaimed *Flora of Indiana* (1940) by Charles Deam inaugurated a tradition of more or less detailed treatments, with much attention given to distribution. Such are now also available for Ohio (Braun, 1960, 1967; Fisher, 1988); Illinois (notably *The illustrated flora of Illinois* by Robert Mohlenbrock, 1967–); Michigan (with *Michigan flora* by Edward G. Voss, 1972–96, being a much-esteemed work); northeastern Minnesota (Lakela, 1965); and Missouri (Steyermark, 1963). A number of less expansive works are also available for these and other states, among them Minnesota and Kentucky, each with checklists published in 1992. On the other hand, the critical revision of the Wisconsin flora, begun in 1929, has yet to be consolidated, and no modern checklist has been published.

In the Northeast and mid-Atlantic, fewer modern works are available; the examples in the Midwest have not been widely emulated or have taken other forms. More notable contributions include *Flora of West Virginia* by Perry Daniel Strausbaugh and Earl Core (1952–64; 2nd edn., 1970–77); *Contributions to a flora of New York State* by Robert S. Mitchell and others (1978–, including a 1986 checklist), and *Flora of Vermont* and *Flora of New England*, both by Frank C. Seymour (1969, with a revision of the latter in 1982). In the mid-Atlantic, *Atlas of the flora of Pennsylvania* finally made its appearance in 1979. Since then, its records have become a basis for a floristic information system; a revision of the *Atlas* appeared in 1993 and a state manual is projected. Arthur Massey's *Virginia flora* (1961) gave the 'Old Dominion' its first-ever checklist, and a distribution atlas has also been published.

The Pennsylvania *Atlas* excepted, the use of detailed distribution maps on the Midwest model is not yet that widespread in the Northeast and mid-Atlantic. Moreover there has not been much evidence of any movement towards concise 'critical' manuals or field manual-keys of the kinds common in Europe and, more seriously, greater attention to taxonomic and nomenclatural developments there.

Bibliographies. General and divisional bibliographies as for Division 1.

Indices. General and divisional indices as for Division 1.

140
Region in general

Except as indicated, all works described below cover the whole of the region as outlined. Most of these also cover mainland Canada south of the 49th parallel and the St. Lawrence River, with the older Britton and Brown *Illustrated flora* and Fernald's edition of *Gray's Manual* extending as far as Newfoundland and Anticosti Island. In addition to the regional manuals, there is a goodly number of semi-popular treatments of woody plants (or simply the trees); a selection of these is described under a separate subheading following the general works.

Gleason's 1952 edition of the *Illustrated flora* (itself now superseded) did not for some entirely supplant its predecessor: N. L. BRITTON and A. BROWN, 1913. *An illustrated flora of the northern United States, Canada, and the British possessions*. 2nd edn. 3 vols. 4666 text-figs. New York: Scribners. (Reprinted 1967, New York: Dover. 1st edn., 1898.) This version has a more satisfying arrangement of text and illustrations, in which each species is provided with a discrete figure and parallel text; in addition, it features greater geographical coverage (northeast to Newfoundland and west to the 102nd degree of longitude, or the Kansas/Colorado border).][52]

Keys to families and genera

BATSON, W. T., 1977. *A guide to the genera of the eastern plants.* See **100**.

FERNALD, M. L., 1950. *Gray's manual of botany.* 8th edn. lxiv, 1632 pp., illus. New York: American Book Co. (Reprinted with revisions, 1970; reissued 1987, Portland, Oreg.: Dioscorides Press. 1st edn. by A. Gray, 1848, New York, entitled *A manual of the botany of the northern United States*; 5th edn., 1867; 6th edn., 1890; 7th edn., 1908.)

Briefly descriptive, critical manual-flora of vascular plants with keys to all taxa, essential synonymy, English and French vernacular names, generalized indication of internal and extralimital range, sometimes spicy taxonomic commentary, and notes on habitat, frequency, special features, phenology, etc.; glossary, list of authorities, and indices to all vernacular and botanical names at end. An introductory section includes a detailed statistical table as well as technical and general notes. [Certain 'critical' taxa are here treated in more detail than in Gleason and Cronquist's flora (see below); there is, concomitantly, a tendency towards a narrower delimitation of species.][53]

GLEASON, H. A. (ed.), 1952. *The new Britton and Brown illustrated flora of the northeastern United States and adjacent Canada.* 3 vols. Illus. New York: The New York Botanical Garden. (Reprinted 1958 and subsequently with slight revisions.)

Briefly descriptive atlas-flora of vascular plants, with each plate comprising figures of a number of species; text inclusive of keys to all taxa, limited synonymy, vernacular names, generalized indication of local and extralimital range, taxonomic commentary, and notes on habitat, phenology, variation, special features, etc.; glossary and index to all botanical and vernacular names at end of vol. 3. [Now superseded by the *Illustrated companion* (see below under GLEASON and CRONQUIST).]

GLEASON, H. A. and A. CRONQUIST, 1991. *Manual of vascular plants of northeastern United States and adjacent Canada.* 2nd edn. by A. Cronquist. lxxv, 910 pp., map (front end-papers). New York: The New York Botanical Garden. (Reissued with corrections,

1993, 1995. 1st edn., 1963, New York: van Nostrand.) Complemented by N. HOLMGREN (ed.; assisted by P. K. Holmgren, R. A. Jess, K. M. McCauley and L. Vogel), 1997. *Illustrated companion to Gleason and Cronquist's* Manual: *illustrations of the vascular plants of northeastern United States and adjacent Canada.* [i], xvi, [iii], 937 pp., 827 full-page pls., end-paper maps. New York.

The *Manual* is a briefly descriptive manual-flora of native, naturalized, and adventive vascular plants with keys to all taxa, limited synonymy, vernacular names, concise indication of internal range, and notes on habitat, etc.; complete general index at end. The introductory section includes a glossary, guides to author citations and abbreviations, and general keys (pp. xl–lxxv). The full-page plates in the *Companion* each depict a number of species; all are cross-referenced to the *Manual* and arranged in the same order. An appendix relates its nomenclature to that of vols. 2 and 3 of *Flora of North America*. [The *Manual* covers the northeast and north-central United States and adjacent parts of southern Canada but neither includes southern Missouri nor extends to the Gaspé Peninsula in Québec. It was initially formed from the text of Gleason's 1952 *New Britton and Brown illustrated flora*.][54]

SWINK, F., 1990. *The key to the vascular flora of the northeastern United States and southeastern Canada.* vii, 514, [1, 8], xii pp., illus. Flossmoor, Ill.: Plantsmen's Publications.

A pure key based on 'common field traits', covering some 90 percent of the total Northeastern and North Central regional vascular flora and, with its many 'caveats', written with a sense of humor. Entries for all taxa and vernacular names, including keys to genera and species, are alphabetical after the general key to families (the classical 'variant' forms of family names being used where permitted), while at the end are found a list of localized species, glossary, and brief bibliography.

Guides to woody plants, including trees

Only a selection from the abundant popular and semi-popular literature is given here. Fuller coverage may be had in the various bibliographies listed at the divisional heading, especially that by LITTLE and HONKALA (1976) and, for older literature, DAYTON (1952). There is also a selection in the book of keys by Muenscher (see below). Only the works by Harlow, Muenscher and Petrides remain in print.[55]

BLACKBURN, B., 1952. *Trees and shrubs in eastern North America.* xv, 358 pp., illus. New York: Oxford University Press.

Illustrated manual-key to trees and shrubs, including many cultivated forms; includes notes on regional distribution, habitat, special features, etc., as well as a glossary and index.

BROWN, H. P., 1938. *Trees of northeastern United States, native and naturalized.* Revised and enlarged edn. 490 pp., illus. Boston: Christopher. (Based on *idem*, 1922. *Trees of New York State, native and naturalized.*)

Essentially a dendrological treatment, with large plates accompanied by descriptive text relating to botanical features, habitat, distribution, uses, timber properties, etc.; includes analytical keys to all species together with a systematic conspectus arranged by genera and families; index.

GRAVES, A. H., 1956. *Illustrated guide to trees and shrubs; a handbook of the woody plants of the northeastern United States and adjacent regions.* Revised edn. 271 pp., illus. New York: Harper. (1st edn., 1952, Wallingford, Conn.)

Illustrated descriptive treatment, with essential synonymy, some taxonomic commentary, and notes on habitat, distribution, special features, uses, etc.; includes keys to all species as well as a glossary and index. [Covers only the northeastern states from Delaware and Pennsylvania to Maine.]

HARLOW, W. M., 1942. *Trees of the eastern (and central) United States and Canada.* xiii, 288 pp., 152 text-figs., pls. (some col.). New York: Whittlesey House/McGraw-Hill. (Reprinted with slight revisions, 1957, New York: Dover.)

Illustrated pocket guide to native and naturalized trees, with much emphasis on field features, uses, folklore, etc.; includes keys to all species and indication of regional distribution; index.

HOUGH, R. B., 1907. *Handbook of the trees of the northern states and Canada east of the Rocky Mountains.* x, 470 pp., 498 illus. (incl. maps and halftones). Lowville, N.Y.: The author. (Reprinted 1947, 1950, New York: Macmillan.)

Copiously illustrated descriptive treatment, with numerous fine photographs based on freshly collected material and field characters; includes keys to all species, distribution maps, and notes on habitat, biology, winter recognition features, uses, etc., as well as a glossary and index. [Never excelled.]

KEELER, H., 1903. *Our northern shrubs and how to identify them.* xxx, 521 pp., 35 text-figs., 205 pls. New York: Scribners. (Reissues to 1935; reprinted 1969, New York: Dover, with 18-page appendix of nomenclatural revisions by E. G. Voss and repaged index.)

Descriptive treatment of native and introduced species, with vernacular and scientific names, distribution, *pro forma* treatment of botanical features, and discursive

remarks on likely occurrences, special features, cultivars, phenology, 'plant relations', etc.; glossary, nomenclatural changes (in 1969 version, with a worthwhile prologue by Voss), and indices to all names at end. The introductory part contains a synoptic table of species (pp. xi–xviii).[56]

LI, HUI-LIN, 1972. *Trees of Pennsylvania, the Atlantic States and the Lake States.* x, 276 pp., 724 illus. Philadelphia: University of Pennsylvania Press.

Illustrated treatment of 118 species, with vernacular names, distribution, and notes on habitat, cultivation, etc.; keys for summer and winter identification; glossary and index. Each species is illustrated, with accompanying text.

MUENSCHER, W. C., 1950. *Keys to woody plants.* 6th edn. [iv], 107, [1] pp., 1 pl. Ithaca, N.Y.: Comstock/Cornell University Press. (Many reprintings, most recently in 1990. 1st edn., 1922.)

Comprises indented keys for identification, followed by a synoptic list of species included with authorities, vernacular names and (if appropriate) origin and status; references (pp. 90–93), illustrated glossary and index to all scientific names at end.

PETRIDES, G. A., 1972. *A field guide to trees and shrubs.* 2nd edn. xxxii, 428 pp., illus. (some col.) (Peterson field guide series 11.) Boston: Houghton Mifflin. (1st edn., 1958.)

Pocket manual, with picture-keys to species and notes on diagnostic features, phenology, distribution, biology, etc.; includes also a winter key as well as a glossary and index.

Pteridophytes

A goodly selection is available; we include those by Boughton Cobb, Eugene Ogden, and Edgar Wherry. A recent electronic companion to Ogden's book is R. S. MITCHELL and L. DANAHER, 1999. *Northeastern fern identifier.* 1 CD-ROM. Illus. Albany: New York State Museum.

COBB, B., 1956. *A field guide to the ferns and their related families of Northeastern and Central North America . . .* xviii, 281 pp., illus. (Peterson field guide series 10.) Boston: Houghton Mifflin. (Reprinted.)

Popularly oriented atlas-guide in small format, featuring full-page figures with facing formatted text with information on botanical features, distribution, etc.; the figures include diagnostic pointers. The text proper is preceded by descriptive introductions to families and genera, an introduction to fern biology, and extensive illustrated keys; at end are sections on European pteridophytes, gardening, fern paleo-history (by T. Just), references (pp. 263–267), checklists and complete index. [Limited to the west by the Mississippi.]

OGDEN, E. C., 1981. *Field guide to Northeastern ferns.* 128 pp., illus. (New York State Mus. Bull. 444). Albany, N.Y.

An atlas-treatment of 70 species, with conventional and random-access keys and (pp. 38–114) full-page plates of 60 species accompanied by extended descriptions as well as verncular names and information on distribution and habitat (the 10 species not so treated are extinct, very rare, or 'very similar'); glossary and index at end. References, pp. 1–2.

WHERRY, E. T., 1961. *The fern guide: northeastern and midland United States and adjacent Canada.* 318 pp., illus. Garden City, N.Y.: Doubleday. (Originally published 1937, Lancaster, Pa.: Science Press, as *Guide to eastern ferns*; 2nd edn., 1942.)

Pocket-sized atlas-guide with full-page line drawings and facing text (covering 135 species); the latter includes descriptions, distributions, habitats, and cultivation notes along with scientific and vernacular names. Recognized infraspecific taxa are also accounted for, with justifications, along with hybrids. Authorities, a lexicon of epithets, and indices to scientific and vernacular names are given at the end; while the front matter contains a glossary, list of abbreviations, and discussions of cytology, sporology, 'fern-izing', ferns in the garden, recent literature (p. 35) and a general key (pp. 37–53). [The current version features greater geographical coverage in comparison with its predecessors.]

141

New England States

Area: 107 240 km². Vascular flora: 2882 species, with 660 additional varieties and 24 hybrids; 1995 species wholly native (Seymour, 1982; see below). – New England comprises the states of Maine, New Hampshire, Vermont, Massachusetts, Connecticut, and Rhode Island. Works relating to individual states are listed under separate headings following the main entry.

Plants were touched upon in *New-Englands Rarities Discovered* (1672) by John Josselyn but the first regional florula, with 350 species arranged on the Linnaean system, was *An account of . . . vegetable productions naturally growing in this part of America . . .* (1785) by Manasseh Cutler. The first effective manual was *Florula bostoniensis* by Jacob Bigelow (see **Massachusetts** below). In the mid-nineteenth century, William Oakes of Ipswich, Mass., made extensive travels in the region and accumulated considerable materials towards a flora. Although he assisted Asa Gray with the first edition of the latter's *Manual* (**140**), his own project never materialized – the more so following his death by drowning in 1848, the year Gray's *Manual* was published.

With the organization towards the end of the

century of the New England Botanical Club and its herbarium, however, the basis for a more thoroughly founded general work gradually was developed. The Club itself, almost from its inception, published in their journal *Rhodora* a long series of family accounts as *Preliminary lists of New England plants I–XXXIII* (1899–1936, not completed) with an emphasis on distribution. These and other sources were ultimately utilized by Frank Seymour for his *Flora of New England* (1969; 2nd edn., 1982), the first modern work. A new guidebook for ferns and fern-allies appeared in 1997, and a full manual with illustrations and range maps in 1999. There are also relatively modern manuals for Maine (1998) and Vermont (1969).

Bibliographies

For the works by Day (1899) on New England and Barnhart (1907) on Vermont local floras see Meisel (under **Divisional bibliographies**).

EGLER, F. H., 1948. Regional vegetation literature, I. Connecticut. *Phytologia* 3: 1–26; *idem*, 1949. Regional vegetation literature, II. Rhode Island. *Ibid.*: 49–56; *idem*, 1950. Regional vegetation literature, III. Massachusetts. *Ibid.*: 193–237. [Classified bibliographies, with maps and introductory accounts; floristic literature included. No more published.]

MAGEE, D. W. and H. E. AHLES, 1999. *Flora of the Northeast: a manual of the vascular flora of New England and adjacent New York.* 1264 pp., 995 text-fig., 2433 maps. Amherst: University of Massachusetts Press.

Briefly descriptive flora of vascular plants with keys, species diagnoses, indication of distribution and habitat (including county dot maps for all species), and notes on value to wildlife, properties and uses. [Covers the New England states along with New York east of the Hudson River (including Long Island).]

SEYMOUR, F. C., 1982. *The flora of New England.* 2nd edn. xvii, 611 pp., illus., map (Phytologia, Mem. 5). Plainfield, N.J. (Reissued 1989, Jamaica Plain, Mass., as the 'third printing' with several additional unnumbered pages of corrections and additions by P. M. Brown. 1st edn., 1969, Rutland, Vt.)

Annotated, keyed enumeration of vascular plants, with very limited synonymy, vernacular names, fairly detailed indication of local range, figures depicting critical and diagnostic characters, and notes on phenology, special features, etc.; glossary, index to all vernacular and botanical names, and county map with guide (pp. 610–611 in this edition) at end. The introductory section provides brief accounts of physical features, climate, etc., as well as addenda and corrigenda. [The illustrations in the second edition are at least in part redrawn or changed, and pp. 597–606 comprise further addenda and corrigenda. Otherwise it is an offset-reprint of the original work which, itself printed from reduced typescript, was wearing for the user.]

Distribution maps

ANGELO, R. and D. E. BOUFFORD, 1996 (1997)– . Atlas of the flora of New England. *Rhodora* **98**: 1–79, *passim*; *ibid.*, **100**: 101–233 (1998); *ibid.*, **102**: 1–119 (2000).

Comprises county dot maps accompanied by an enumeration with chromosome numbers, usual habitat(s), essential synonymy, and occasional critical notes; extensive list of references at end. An introduction in the first installment includes notes on the basis and plan of the work. [The first part covers pteridophytes and gymnosperms. The next two respectively encompass Poaceae (1998) and other monocots except Cyperaceae (2000). The *Atlas* is also on the Web at http://www.herbaria.harvard.edu/~rangelo/Neatlas0/Webintro.html.][57]

Woody plants

In addition to the work of Dwelley briefly described here, two others may be noted: L. L. DAME and H. BROOKS, 1972. *Handbook of the trees of New England with ranges throughout the United States and Canada.* 196 pp. New York: Dover. (Reprint of 1901 edn., Boston: Ginn, with table of nomenclatural revisions by E. S. Harrar); and F. L. STEELE and A. R. HOUGHTON, 1968. *Trees and shrubs of New England.* 127 pp., illus. Concord, N.H.: Society for the Protection of New Hampshire Forests.

DWELLEY, M., 1980. *Trees and shrubs of New England.* xii, 275 pp., illus. (part col.). Camden, Maine: Down East Books.

Systematically arranged popular guide with colored illustrations covering all woody plants, including the species of *Salix*; index.

Pteridophytes

Prior to the appearance of the manual described below, New England fern nomenclature was significantly revised in A. F. TRYON, 1978. New England ferns (Filicales). *Rhodora* **80**: 558–569.

TRYON, A. F. and R. C. MORAN, 1997. *The ferns and allied plants of New England.* xv, 325 pp., silhouettes, halftones, maps. Lincoln, Mass.: Center for Biological Conservation, Massachusetts Audubon Society. (Natural history of New England series.)

Descriptive semi-popular atlas covering 92 native

species and nothospecies (the latter not separately indicated) with photographs and maps facing the text, the latter in *pro forma* style and featuring botanical and vernacular names, diagnostic and spore features, regional and extralimital range, chromosome numbers, hybrids (if any), indication of habitat, threats (if any), and more or less extensive taxonomic and general commentary; keys to as well as general and regional distribution maps for all taxa are also furnished. The introductory part comprises an overview of pteridophytes including their spores, hybridity and patterns of distribution as well as a key to genera, while at the end are scanning electron micrographs of spores, remarks on regional geology and climate, a section on horticulture, a glossary, list of references, and index to all names. Genera are alphabetically arranged (without reference to families), and the regional map points are more or less precisely localized.[58]

Maine

Area: 80 587 km². Vascular flora: 2137 species and 714 non-nominate infraspecific taxa (Bean *et al.*, 1966; see below).

The 1998 manual by Haines and Vining is the first substantial statewide floristic account since the beginnings of Maine floristics in 1672 with John Josselyn's florula of Scarborough. The *Flora of Maine* complements a long series of state catalogues and checklists, of which the contemporary version was first produced by Ogden, Steinmetz and Hyland in 1946. This was subsequently three times revised, most recently in 1995. It in turn succeeded *Catalogue of the flowering plants of Maine* (1862) by George L. Goodale and *The Portland catalogue of Maine plants* (1868, by Goodale and J. Blake; 2nd edn., 1892, by M. L. Fernald with a further issue in 1895). The earlier works, published through the Portland (Maine) Society of Natural History, had their origin in Goodale's work on the second Maine Geological Survey of 1861 (which included natural history).[59]

BEAN, R. C., C. D. RICHARD and F. HYLAND, 1966. *Revised check-list of the vascular plants of Maine.* [ii], 71 pp. (Bull. Josselyn Bot. Soc. Maine 8). Orono, Maine (1st edn., 1946, by E. C. Ogden, F. H. Steinmetz and F. Hyland; 3rd edn., 1995 (not seen).)

Tabular systematic checklist of vascular plants, with indication of local range (by counties) in 16 columns; index to genera and families at end. Nomenclature follows the eighth edition (1950) of *Gray's Manual of Botany* (**140**).

HAINES, A. and T. F. VINING, 1998. *The flora of Maine: a manual for identification of native and naturalized vascular plants of Maine.* 846 pp. Bar Harbor, Maine: V. F. Thomas Co.

Keyed enumeration with essential synonymy, vernacular names, and indication of habit, overall range, frequency, preferred habitats, and official and other (including taxonomic) notes; basic bibliography, vegetative key to hydrophyte genera (pp. 780–784), and full index to all names. A preface and brief history of Maine botany, followed by the plan and basis of the work, changes to the flora since the third (1995) edition of the state checklist, glossary, and general keys form the opening sections.

New Hampshire

Area: 23 382 km². Vascular flora: no current data.

The only state checklist to date is that by William F. Flint in *Geology of New Hampshire* (1874) by Charles H. Hitchcock – part of the second state survey established in the late 1860s. The biologically interesting north (including the White Mountains) is, however, thoroughly covered in the florula described below. For pteridophytes, see also E. SCAMMAN, 1947. *Ferns and fern allies of New Hampshire* (Bull. New Hampshire Acad. Sci. 2). Durham, N.H.

PEASE, A. S., 1964. *A flora of northern New Hampshire.* v, 278 pp., 2 maps. Cambridge, Mass.: New England Botanical Club. (Based on *idem*, 1924. *Vascular flora of Coös County, New Hampshire.* Boston.)

Systematic enumeration, with vernacular names, indication of localities with *exsiccatae*, and notes on habitat, frequency, etc.; index at end. There is also a substantial introductory section, along with orientation maps.

Vermont

Area: 24 110 km². Vascular flora: no current data.

The first account of the state flora is *Catalogue of Vermont Plants* (1842) by the already-mentioned William Oakes, published in Zadok Thompson's *History of Vermont*. A second edition appeared in 1853, when Thompson became head of the state Geological Survey. Its successors were *General catalogue of the flora of Vermont* (1882; 2nd edn., 1888) by George H.

Perkins and, in 1900, the first of those sponsored by the Vermont Botanical Club. Revisions of the latter followed in 1915, 1937 and 1969, with only the last featuring descriptions. A further checklist appeared in 1973.

SEYMOUR, F. C., 1969. *The flora of Vermont.* 4th edn. ix, 393 pp., illus. (Vermont Agric. Exp. Sta. Bull. 660). Burlington. (1st edn., 1900, by E. Brainerd, L. R. Jones and W. W. Eggleston.)

Briefly descriptive manual-flora of vascular plants, with keys to all taxa, very limited synonymy, indication of county distribution and/or localities with *exsiccatae*, and notes on habitat, frequency, phenology, etc.; addenda and index to family and generic names. An introductory section provides brief accounts of physical features, climate, and floristics and phytogeography.

VERMONT BOTANICAL AND BIRD CLUB, 1973. *Check list of Vermont plants.* vi, 90 pp. N.p.

Tabular checklist, with county records and sources (those taxa asterisked being considered post-European introductions); indices to genera, families and vernacular names at end.

Massachusetts

Area: 20 342 km². Vascular flora: no current data.

Massachusetts floras begin with *Florula bostoniensis* (1814; 2nd edn., 1824; 3rd edn., 1840) by Jacob Bigelow; this served all New England until the first appearance of *Gray's Manual* in 1848. The initial extensive survey of the flora statewide was connected with the work of the state Geological (and Natural History) Survey organized in 1830 – the first of its kind to include natural history. Its first stage was summarized in the Amherst professor and Survey chief Edward Hitchcock's *Report on the geology, mineralogy, botany, and zoology of Massachusetts* (1833; 2nd edn., 1835), which contains (pp. 599–652 in edn. 2) the only full catalogue of plants of the state ever published. The work of the survey was later divided up; its second stage is represented in *Report on the herbaceous flowering plants of Massachusetts* (1840) by Chester Dewey, resident in the Berkshires, and *Report on the trees and shrubs growing naturally in the forests of Massachusetts* (1846) by George B. Emerson; a second, fancier edition of the latter, with plates by Isaac Sprague, appeared in 1875 and enjoyed several reissues.[60]

Rhode Island

Area: 2743 km². Vascular flora: no current data.

The first state checklist was *Catalogue of plants . . . principally of Rhode Island* (1845) by Stephen T. Olney with three associates. Further lists were published in 1888 (*Plants of Rhode Island* by James Bennett) and 1920 (*The ferns, fern allies and flowering plants of Rhode Island*) by Olney's sponsors, the Providence Franklin Society – to whom floristic survey had from 1836 been delegated by the state survey.

PALMATIER, E. A., 1952. *Flora of Rhode Island.* 75 pp. Kingston: Department of Botany, University of Rhode Island.

Systematic list of vascular plants, with limited synonymy, cross-references to the eighth edition (1950) of *Gray's Manual of Botany*, and vernacular names; includes a list of species properly excluded from the flora of the state.

Connecticut

Area: 12 667 km². Vascular flora: no current data. – The state is also wholly covered in *Flora of the vicinity of New York* (**143**).

There is no separate descriptive flora. Checklists include *Catalogue of the plants of Connecticut* by J. N. Bishop (1885; 3rd edn., 1901); its successors by Graves and Dowhan appear below. All these works have been sponsored by the state Geological and Natural History Survey.[61]

DOWHAN, J. J., 1979. *Preliminary checklist of the vascular flora of Connecticut (growing without cultivation).* x, 176 pp., map. (Report of investigations 8). [Hartford]: State Geological and Natural History Survey of Connecticut.

Briefly annotated systematic checklist; includes essential synonymy, vernacular names and (where necessary) indicators to references (pp. 145–151), and indices to scientific and vernacular names. Two appendices deal with non-persistent members of the flora and their problems. The introductory part furnishes background, indicates where additional work is needed, gives definitions of the floristic units adopted ('ecoregions'), a description of sources and procedures, and discussions of alien species, hybrids and vernacular names. [Less extensively documented than Graves's work.]

GRAVES, C. B. *et al.*, 1910. *Catalogue of the flowering plants and ferns of Connecticut growing without cultivation.* 569 pp. (Connecticut State Geol. Surv. Bull. 14). [Hartford.] Continued as E. B. HARGER *et al.*, 1931. *First supplement . . . Additions to the flora of Connecticut.* 94 pp. (*ibid.*, 48).

Annotated enumeration of vascular plants, with essential synonymy and principal literature citations, vernacular names, localities with indication of *exsiccatae*, and notes on habitat, frequency, phenology, status, biology, medicinal values, uses, etc.; lists of authorities for botanical names as well as of species excluded from the flora; index to all vernacular and botanical names. An introductory section includes accounts of physical features, botanical exploration, and floristics. See also E. H. EAMES, 1931. Further additions to the Connecticut flora. *Rhodora* 33: 167–170.

142

New York State

Area: 123 180 km² (inclusive of Long Island). Vascular flora: 3195 'persisting' (including 2078 native) species (Mitchell, 1997; see below); 3022 were so reported by Mitchell in 1986. – For additional works on the New York City region and Long Island, see also **143**.

The earliest florula is *Plantae coldenghamiae* (1743), Cadwallader Colden's account of the botany of his vast Hudson Valley estate. The first statewide checklist is Jacob Green's *Catalogue of plants indigenous to the state of New York* (1814). The establishment in 1836, however, of a well-founded state natural history survey gave the opportunity for John Torrey to prepare, firstly, a revised *Catalogue of plants of New York State* (1841), a revision of Green's work, and then his classic state flora (1843) – the latter yet not superseded.[62] Later works, all published through the Survey and State Museum, include a further catalogue by Torrey, *Botany of New-York* (1853) with supplements firstly by George W. Clinton (1865–66) and then by Charles H. Peck. Following these were Homer D. House's *Annotated list* (1924), still useful for its citations and references, and Richard S. Mitchell's *Checklist* (1986, 1997), the 1986 edition commemorating the Survey's sesquicentennial. Since 1978, materials towards a new flora have been published as *Contributions to a Flora of New York State*.

The considerable local and regional literature has been surveyed by House. Two key works are noted here: the much-appreciated students' manual for the south-central region by Karl Wiegand and Arthur J. Eames, and Charles A. Zenkert's enumeration for the Buffalo–Niagara 'Frontier' region.

Bibliography

HOUSE, H. D., 1941–42 (1942). *Bibliography of botany of New York State, 1751–1940*. 2 parts. 233 pp. (New York State Mus. Bull. 328/329). Albany. [Exhaustive bibliography, with about 3000 entries (Besterman, in *Biological sciences: a bibliography of bibliographies*, 1971, Totowa, N.J.); statewide works are followed by those for individual counties.]

HOUSE, H. D., 1924. *Annotated list of the ferns and flowering plants of New York State*. 759 pp. (New York State Mus. Bull. 254). Albany.

Systematic enumeration (nomenclaturally following the now-obsolete 'American Code') of known vascular plants; includes full synonymy with references, vernacular names, fairly detailed indication of local ranges (often with citations), and notes on habitat, frequency, special features, etc., as well as occasional descriptive remarks; index to generic and family names at end. An introductory section gives details of previous state floras or lists.

MITCHELL, R. S. and G. C. TUCKER, 1997. *A checklist of New York State plants*. vii, 400 pp. (New York State Mus. Bull. 490). Albany. (Originally published 1986 as R. S. MITCHELL, *A checklist of New York State plants* (New York State Mus. Bull. 458).)

Nomenclatural checklist, with essential synonymy (tied to House's *Annotated list* as well as appropriate Northeastern floras and *Flora Europaea*), vernacular names, status including indication of relative rarity, and occasional notes; annotated appendix of excluded species (pp. 353–357); lexicon of vernacular names (with botanical equivalents) and index to generic and family names at end. The introductory part includes acknowledgments, the plan of the work and conventions along with statistics and major references. Arrangement of flowering plant families follows the Cronquist system.

MITCHELL, R. S. (ed.), 1978– . *Contributions to a flora of New York State*, 1– . Illus., maps. (New York State Mus. Bull. 431, etc., *passim*). Albany.

Comprises a series of illustrated descriptive revisionary accounts of individual families with county

distribution maps in an atlas format, with one page for each species; text includes keys to all taxa and for individual taxa important synonymy, vernacular names, and notes on local and overall range, phytogeographic element(s), habitat, habit, phenology, biology, behavior, special features, uses, properties, etc., and critical remarks on variability, forms, hybridization, and the like. Appendices on associated fungi and insects, a list of references, and an index appear in each fascicle. [As of 1999, 10 parts covering vascular plants had been published, the most recent in 1993.][63]

TORREY, J., 1843. *A flora of the state of New-York*. 2 vols. 161 pls. Albany: State of New-York. (Natural history of New-York, division 2: Botany.)

Detailed descriptive flora of vascular plants in a large format, with synoptic keys to genera and species, vernacular names, full synonymy (with references and citations), generalized indication of local range, taxonomic commentary (with references to related European plants), figures of representative species, and notes on habitat, etymology, uses, etc.; list of probable additions to the flora, statistics, and complete general index in vol. 2. An introductory section (in vol. 1) gives accounts of botanical exploration, floristic regions, and sources for the work, with a list of authorities.[64]

Partial works

The most significant of the many local and regional florulas of 'upstate' New York are the students' manual by Wiegand and Eames for the south-central Finger Lakes district and adjacent areas (centering on Ithaca) and the enumeration by Zenkert for the 'Niagara Frontier' centering on Buffalo.

WIEGAND, K. and A. J. EAMES, 1926. *Flora of the Cayuga Lake basin, New York. Vascular plants*. 491 pp., map (Cornell Univ. Agric. Exp. Sta. Mem. 92). Ithaca.

Concise manual-key to vascular plants in south-central New York and adjacent northern Pennsylvania, with essential synonymy, vernacular names, local and extralimital range, critical remarks, and notes on habitat, frequency, phenology, etc.; map and index. Extensive appendices account for physical features, climate, geological history, soils, floristics and vegetation, botanical exploration, etc.

ZENKERT, C. A., 1934. *The flora of the Niagara Frontier region: ferns and flowering plants of Buffalo, N.Y., and vicinity*. x, 328 pp. (Bull. Buffalo Soc. Nat. Sci. 16). Buffalo, N.Y. For additions and revised checklist, see R. H. ZANDER and G. J. PIERCE, 1979 (1980). *Flora of the Niagara Frontier region: second supplement and checklist*. iii, 110 pp. (*ibid.*, Suppl. 2). Buffalo.

The original work contains on pp. 72–269 an annotated systematic enumeration of native, naturalized and adventive vascular plants (covering 1587 species and 115 additional varieties), with essential synonymy, 'standardized vernacular names', localities and collectors (for less frequently encountered taxa), and notes on habitat, frequency, abundance, and (occasionally) taxonomy; statistics (pp. 270–274), 'ecological areas and plant societies', references (pp. 313–314) and index to vernacular, generic and family names. The substantial introductory part covers geographical and other natural features, human impact, conservation, introduced plants and the plan of the work and conventions; a historical section is also included. The *Second Supplement* encompasses additions and a revised checklist. [Covers an area of *c.* 20 360 km^2 within a radius of 80.5 km (50 mi.) from Buffalo, so including Niagara Falls and part of the Niagara Peninsula in Ontario (**132**).]

143

New Jersey, New York City area, and Long Island

The whole of the area, along with southeastern Pennsylvania as well as all of Connecticut, is covered by Norman Taylor's flora (see below). For New York City and Long Island, see also **142** as well as the subheading below. – Area: New Jersey, 19 417 km^2; Long Island, 3627 km^2. Vascular flora: New Jersey, 3111 taxa (Schmid and Kartesz, 1994; noted below), a notable increase over the 2407 native and naturalized species recorded by Anderson (1979; see below); of these, some 28 percent are not native.[65]

Earlier New Jersey accounts include *Catalogus plantarum in Nova Caesarea repertarum* (1874; revised edn., 1877) by Oliver R. Willis and, soon after, Staten Islander Nathaniel L. Britton's first major floristic essays: *Catalogue of flora found in New Jersey* (1881) and *Catalogue of plants found in New Jersey* (1889, but released in 1890), the latter sponsored by the state Geological Survey.[66] In the early twentieth century New Jersey was covered in Norman Taylor's *Flora of the vicinity of New York* (see above), not entirely superseded by either of the two current statewide works, and, in the 'Pine Barrens' and neighboring areas, Witmer Stone's *The plants of southern New Jersey* (see below).

For Greater New York City and Long Island, separate treatments go back to John Torrey's *Catalogue* (1819, anonymously authored), issued through what is now the New York Academy of Sciences, and its revision in 1840. In 1888 the Torrey Botanical Club published *Preliminary catalogue of plants growing within 100 miles of New York City*. The advent of the New York Botanical Garden and changing nomenclatural philosophy led a decade later to *Preliminary catalogue of Anthophyta and Pteridophyta growing near New York City* (1898) by Britton with Emerson E. Sterns and Justus F. Poggenburg (the 'BSP' of many plant names). Long Island was treated separately in *The flora of Long Island* (1899) by Smith Ely Jelliffe. In 1915 the whole area was encompassed in Norman Taylor's flora and, from the 1930s, for practical purposes, by Gleason's manual-key (for both, see below).[67]

TAYLOR, N., 1915. *Flora of the vicinity of New York.* vi, 683 pp., 9 maps (Mem. New York Bot. Gard. 5). New York.

Enumeration of vascular plants (without vernacular names); includes keys to genera and species, general and altitudinal range (with distribution and habitat details for less common and/or more localized species) and indication of geological formations, edaphic relationships, phenology and other features; statistics (pp. 649–651) and index to all names at end. The introductory part has an extensive consideration of physiographic, climatic and edaphic factors and their relationships with the flora, and plant geography (including relationship with growing season length), along with a list of local floras. [Not entirely superseded by later works although nomenclaturally obsolete.][68]

New Jersey

In addition to the following works, coverage of the state flora may be found on pp. 41–174 in J. A. SCHMID and J. T. KARTESZ, 1994. *New Jersey and Pennsylvania higher plants, II: Desk manual.* 443 pp., tables, graphs. Media, Pa.: Schmid & Company. Coverage of pteridophytes may additionally be had in J. D. MONTGOMERY and D. E. FAIRBROTHERS, 1992. *New Jersey ferns and fern-allies.* 293 pp., illus., maps. New Brunswick, N.J.: Rutgers University Press. This in turn succeeded *The ferns of New Jersey, including the fern allies* (1947) by M. A. Chrysler and J. L. Edwards.

Bibliography

FAIRBROTHERS, D. E., 1964. An annotated bibliography of the floristic publications of New Jersey from 1753 to

1961. *Bull. Torrey Bot. Club* 91: 47–56, 141–151. Continued as *idem*, 1966. An annotated bibliography of the floristic publications of New Jersey from 1962–1965. *Ibid.*, 93: 352–356. [Chronologically arranged by authors; not limited to floras but includes many reports. 660 items in all are included. More substantial titles are annotated, particularly in the 1962–65 supplement.]

ANDERSON, K., 1979. *A check list of the plants of New Jersey.* [2], 54 pp., [1 p. additions]. Mt. Holly, N.J.: The author.

Briefly annotated systematic list, with vernacular names, cross-references to earlier standard works (Britton, Taylor, Stone), and indication (where appropriate) of rarity; references (pp. 32–33) and index to families. The introductory part gives the plan of the work and its scope and coverage; taxonomy as well as arrangement of families after *Gray's Manual*.

HOUGH, M. Y., 1983. *New Jersey wild plants.* [ix], 414 pp., maps. Blairstown, N.J.: The author (distributed by New Jersey Audubon Society).

Concise enumeration, the genera and species all alphabetically arranged; includes vernacular names, cross-references to other works (not, however, to those of Britton, Taylor or Stone), distribution (New Jersey only, with a county dot map for each species), indication of status, frequency, habitat, phenology, and (where appropriate) notes on uses, properties, toxicity, associates, and identification problems; explanation of format (with published sources) and general references (with evaluations) at end. The preface (by D. E. Fairbrothers) gives the background to and reasons for the work; the introduction covers its nomenclatural basis (after Gleason/Cronquist) and plan along with notes on plant collecting and preservation, identification, and available maps. [The dots in the maps are of two sizes, the smaller representing pre-1930 records. The work is oriented towards amateurs.]

Partial work

STONE, W., 1911 (1912). *The plants of southern New Jersey, with especial reference to the flora of the Pine Barrens and the geographic distribution of the species.* 2 vols., pp. 23–829, 5 figs., 129 pls. [Trenton.] (Also published in 1 vol. in *Annual Rep. New Jersey State Mus.* 1910. Reprinted 1973 with added foreword, Boston: Quarterman Publications.)

Includes (pp. 213–779) a systematic enumeration of vascular plants, with keys to all larger genera, synonymy, citation of *exsiccatae* with localities, and extensive notes on local occurrence, habitat, ecology, phenology, etc., together with gazetteer, bibliography and index. An introductory section includes remarks on local herbaria, phytogeography, floristic districts, vegetation, etc. (with a note on the author added in the reprint). [From Burlington and Monmouth Counties southwards.]

New York City area and Long Island

Gleason's 1962 manual covers the entire area here delimited, while Clemants' checklist is limited to a radius of 50 miles from New York City. Research towards a new flora of 'Greater New York City' is proceeding.

CLEMANTS, S. C., 1990. *New York metropolitan flora: preliminary checklist.* [4], 61 pp. (Contr. Brooklyn Bot. Gard. Herb. 1). Brooklyn, N.Y.

Alphabetically arranged checklist for observer use, with indication of presence/absence by state. [Covers a radius of 50 mi. from the 'Big Apple'; intended as a first contribution towards a new metropolitan-area flora, a current Garden project.]

GLEASON, H. A., 1962. *Plants of the vicinity of New York.* 3rd edn. 307 pp., 32 text-figs. New York: Hafner. (1st edn., 1935.)

Semi-popular manual-key to vascular plants in a compact format, with vernacular names and notes on occurrence, habitat and frequency; index to family, generic and vernacular names. An introductory account of basic organography and a glossary are also provided. [Covers the environs of New York City to a radius of some 161 km (100 mi.) together with Long Island.]

144

Pennsylvania

Area: 116 709 km². Vascular flora: 3318 species, of which 2176 native (Rhoads and Klein, 1993; see below). Schmid and Kartesz (1994; noted below) report 3374 taxa, of which some 36 percent are not native.

A *Prodromus Florae Pennsylvaniae* was essayed by Benjamin S. Barton in the early nineteenth century but was never published or apparently even completed.[69]

Progress towards a state flora was later also hindered by an exclusive focus on geology and paleontology on the part of the state Geological Survey, established in 1836.[70] The first statewide account was thus *The botanical class-book and flora of Pennsylvania* (1851; reissue 1852) by Henry R. Noll of Bucknell College (now Bucknell University), a 'grind' not succeeded until publication in 1903 of the Rev. Thomas C. Porter's critical *Flora of Pennsylvania*. Two substantial sets of additions to the latter worthy of note here are those by Harold N. Moldenke (1940) and S. A. Thompson *et al.* (1989).

The need for a successor to Porter's flora was recognized by the early 1930s and, under the leadership of John M. Fogg, Jr. at the University of Pennsylvania, the 'Flora of Pennsylvania Project', a systematic collecting and documentation programme, was initiated. A primary objective was improved knowledge of plant distribution and occurrence and, with the collaboration of Edgar T. Wherry and Herbert A. Wahl (the latter at what is now Pennsylvania State University), efforts accordingly were directed towards what eventually appeared in 1979 as *Atlas of the flora of Pennsylvania*. Since then, the Project (now at the Morris Arboretum of the University of Pennsylvania) has continued its interests in plant distribution, with special attention to methodology and the development of an electronic database. A revised edition of the *Atlas* appeared in 1993 (see below); a descriptive flora remains a goal.[71]

There are many county, local and other floras and checklists, of which *Index Flora Lancastriensis* (1793; supplement, 1799) by Henry Muhlenberg, *Compendium Florae Philadelphicae* (1818) by William P. C. Barton, and the first and second editions of *Flora cestrica* (1826, 1837) by William Darlington, all for parts of southeastern Pennsylvania, are of historical importance. Only the 1953 regional treatment by Otto E. Jennings for the western part of the state is accounted for here. The northern fringe is in part also covered by *Flora of the Cayuga Lake basin* (**142**) and the Philadelphia area by *Flora of the vicinity of New York* (**143**), and in much of the southwest Strausbaugh and Core's West Virginian manual (**148**) is useful.

In addition to the works described below, coverage of the state flora may be found on pp. 175–317 in J. A. Schmid and J. T. Kartesz, 1994. *New Jersey and Pennsylvania higher plants*, II: *Desk manual*. 443 pp., tables, graphs. Media, Pa.: Schmid & Company.

Bibliography

Baron, J. J., 1925. Floral life of western Pennsylvania: a bibliography. *Trillia*, 8: 45–57. [Partly annotated; alphabetically arranged by authors.]

Porter, T. C., 1903. *Flora of Pennsylvania*. Edited with the addition of analytical keys by J. K. Small. xv, 362 pp., map. Boston: Ginn. Complemented by *idem*, 1904. *Catalogue of the Bryophyta . . . and Pteridophyta found in Pennsylvania*. iii, 66 pp. Boston.

Enumeration of seed plants, with keys to all taxa and cross-references to Britton's *Manual* and *Illustrated flora* (see **140**) but no synonymy; entries include vernacular names, detailed indication of local range (by counties) and general summary of extralimital distribution, and brief notes on habitat and miscellaneous features; summary of the flora and indices to vernacular, generic and family names at end. There is no introductory matter save for a brief preface by Small (the work being posthumous). Nomenclature follows the obsolete 'American Code'. [Substantial sets of additions are contained in H. N. Moldenke, 1946. A contribution to the wild and cultivated flora of Pennsylvania. *Amer. Midl. Naturalist* 35: 289–399, and S. A. Thompson, W. E. Baker and M. MacDonald, 1989. *Notes on the distribution of Pennsylvania plants based on specimens in the Carnegie Museum of Natural History herbarium*. 55 pp. (Spec. Publ. Carnegie Mus. Nat. Hist. 14). Pittsburgh, Pa. The first of these works is an annotated catalogue of non-vascular and vascular plants, based on over 4300 collections, while the second, with localities and *exsiccatae*, accounts for many records from the mid-1960s onwards.]

Rhoads, A. and W. McK. Klein, Jr., 1993. *The vascular flora of Pennsylvania: annotated checklist and atlas*. [vi], 636 pp., col. illus., numerous maps (Mem. Amer. Philos. Soc. 207). Philadelphia. (A revision of E. T. Wherry, J. M. Fogg, Jr. and H. A. Wahl, 1979. *Atlas of the flora of Pennsylvania*. xxx, 390 pp. (maps), frontisp. Philadelphia: Morris Arboretum, University of Pennsylvania.)

Systematically arranged (after Cronquist) enumerative atlas with distribution maps, one for each taxon; annotations to the left of each map include scientific and vernacular names, principal synonyms (keyed to major floras and other works), indication of habitat and life-form, while next to the maps are, where applicable, codes relating to wetlands (as used by the

U.S. Fish and Wildlife Service), conservation status, and (if not native) biological status. The collection sites as recorded on the maps are fixed by latitude and longitude, some retrospectively so, and displayed against a uniform gray background. In the introductory part are the basis and plan of the work, the physical and biological setting (including maps related to physical and biological provinces), a list of changes to the flora (pp. 6–8), and statistical summary. No overlays are furnished, in contrast to the 1979 version of the work.[72]

Partial work

JENNINGS, O. E., 1953. *Wild flowers of Western Pennsylvania and the upper Ohio basin.* With watercolors by A. Avinoff. 2 vols. lxxv, 574 pp., maps, 200 col. pls. Pittsburgh: University of Pittsburgh Press.

Volume 1 of this quarto 'flower-book' comprises a comprehensive descriptive flora of seed plants, with keys to all taxa, concise synonymy, vernacular names, rather detailed indication of local range (with many maps), summary of extralimital distribution, some critical remarks, and notes on habitat, ecology, phenology, special features, etc.; index to all botanical and vernacular names. The extensive introductory section includes a bibliography, gazetteer, and glossary together with accounts of botanical exploration, physical features, soils, geology, floristics, and phytogeography. Volume 2 constitutes an atlas of 200 watercolors.

Distribution maps

For *Atlas of the flora of Pennsylvania* and its successor, see RHOADS and KLEIN (1993, above).

145

Delaware (and the Eastern Shore)

Area: Peninsula, 15098 km^2; Delaware, 5023 km^2. Vascular flora: no current data. – See also **146** and **147** for additional coverage of those parts of Maryland and Virginia east of Chesapeake Bay; together with the state of Delaware, they make up the so-called 'Del-Mar-Va' peninsula to which this present unit corresponds.

Only New Castle County in Delaware, which includes Wilmington, was the subject of checklists before the 1940s. Sponsored by local societies, these include *Catalogue of the phaenogamous and filicoid plants of New Castle County* (1844) by John Dolph and its sim-

ilarly titled, more substantial successor (1860) by Edward Tatnall. The latter was superseded only in 1946 by the latter author's nephew Robert Tatnall's *Flora*, the first for Delaware as well as the Peninsula.[73]

TATNALL, R. R., 1946. *Flora of Delaware and the Eastern Shore; an annotated list.* xxvi, 313 pp., 9 pls., map. Wilmington: Society of Natural History of Delaware.

Briefly annotated, concise enumeration of vascular plants, with essential synonymy, vernacular names, more or less detailed indication of local range, some critical remarks, and notes on habitat, frequency, phenology, special features, etc.; tabular summary of the flora, bibliography (pp. 289–290), and index to all botanical and vernacular names. The introductory section includes *inter alia* accounts of botanical exploration, floristics, and phytogeographic regions.

146

Maryland (and District of Columbia)

For the 'Eastern Shore' of Maryland, see also **145** (TATNALL); for the Baltimore area, see also the District of Columbia subheading below. – Area: 25740 km^2 (District of Columbia alone, 164 km^2). Vascular flora: 1437 species listed by Shreve *et al.* for Maryland (1910; see below); now some 2400 or more.[74]

The first account of the Maryland flora is in *Remarks by Mr. James Petiver . . . on some animals, plants, etc. . . . from Maryland* (1698; in *Philosophical Transactions* 20: 393–406). Its basis lay primarily in collections by Hugh Jones.[75] This was followed in the nineteenth century by two florulas for the Baltimore area, respectively by William Aikin (1837, with 1063 species) and Basil Sollers (1881).[76] The first complete state flora was *Plant life of Maryland* (1910) by Forrest Shreve, part of a research programme led by the Johns Hopkins bioclimatologist B. E. Livingston with whom the author was then associated.[77] A revision in the form of a checklist by J. B. S. Norton and R. G. Brown appeared in 1946. Since then, the vascular flora has been covered by three descriptive works, respectively on pteridophytes (Reed, 1953), woody plants (Brown and Brown, 1972), and herbaceous plants (*idem*, 1984).

In the District of Columbia, originally inclusive of most of what is now Arlington County, Virginia, the most substantial of early florulas was *Florae Columbianae prodromus* by John A. Brereton (1830, but published in 1831). It was succeeded in turn by *Flora columbiana* (1876) by George Vasey *et al.*, *Guide to the flora of Washington and vicinity* (1881) by Lester F. Ward (with six supplements to 1901) and then by those described below under **Local works**.

BROWN, R. G. and M. L. BROWN, 1972. *Woody plants of Maryland.* 347 pp., illus. College Park, Md.: University of Maryland. Complemented by M. L. BROWN and R. G. BROWN, 1984. *Herbaceous plants of Maryland.* xlvii, 1127 pp., over 2250 text-figs., 272 col. photographs, col. map. College Park: The Book Center, University of Maryland.

The first-named work (which is inclusive of sub-shrubs) is an atlas-manual with keys, descriptions, synonymy, vernacular names, and notes on distribution, habit, phenology, frequency, uses, etc. The second work, similar in format but with shorter descriptions and commentary, covers 2255 native and introduced species and also features a glossary, illustrated organography, selected references (pp. 1089–1090), and indices to vernacular and scientific names. Its introductory part moreover incorporates accounts of physical features, climate and weather, geology and soils and the features of the flora and main formations (with maps) along with rare species (one not reported since 1698!) as well as statistics, a historical account, and general key (pp. xxxix–xlvii). [The text-figures specify flower color, a useful device.]

NORTON, J. B. S. and R. G. BROWN, 1946. A catalog of the vascular plants of Maryland. *Castanea* 11: 1–50.

A concise enumeration including citations of representative *exsiccatae* and literature records as well as essential synonymy; list of references at end. [Intended as a nomenclatural revision of Shreve's *Plant life*.]

SHREVE, F. *et al.*, 1910. *The plant life of Maryland.* 533 pp., 15 text-figs., 39 pls. (Maryland Weather Serv., Spec. Publ. 3). Baltimore: Johns Hopkins University Press. (Reprinted 1969, Lehre, Germany: Cramer, as *Reprints of US-floras* 5.)

Part VII of this work (pp. 381–497) comprises a systematic enumeration of vascular plants, with very limited synonymy, vernacular names, more or less detailed indication of local range, and notes on habitat, frequency, etc.; no separate index (see general index to work, pp. 507–533). The remaining parts (I–VI) deal with physical features, floristics, ecology, agriculture, forestry, etc. [For a revision of the checklist, see NORTON and BROWN, 1946.]

Pteridophytes

REED, C. F., 1953. *The ferns and fern-allies of Maryland and Delaware, including District of Columbia.* Baltimore: Reed Herbarium.

Illustrated manual with distribution maps; includes keys, vernacular names, descriptions, diagnostic features, and notes covering also growth patterns and care; glossary, references and indices at end. The introductory part includes a synopsis, history of studies and list of treatments, statistics and general notes of distribution and biology; an organography is also furnished.

Local works: District of Columbia and the Baltimore area

The limits of Hitchcock and Standley's flora were based on a radius of 24.15 km (15 mi.) from the Capitol in Washington, D.C., an area now largely built up; those of Hermann extend to Chesapeake Bay, the Pennsylvania border, the east foot of the Blue Ridge and, in the south, the Rappahanock and Rapidan Rivers.

HITCHCOCK, A. S. and P. C. STANDLEY, 1919. *Flora of the District of Columbia and vicinity.* 329 pp., 42 pls. (Contr. U.S. Natl. Herb. 21). Washington, D.C.: Smithsonian Institution.

Manual-key to vascular plants, with essential synonymy, vernacular names, generalized indication of local and extralimital range, and notes on habitat, frequency, phenology, special features, etc.; statistics, glossary and index to vernacular, generic and family names. An introductory section gives background information on the region, its botanical features, and previous investigations. [Partly supplanted by F. J. HERMANN, 1946. *A checklist of plants in the Washington-Baltimore area.* 2nd edn. 134 pp. Washington, D.C.: Conference on District Flora. (1st edn., 1941.) This includes references and citations for new and existing records where not given in Hitchcock and Standley.]

147

Virginia

For the 'Eastern Shore' see also **145** (TATNALL). Arlington County and adjacent areas are also covered by works on the District of Columbia (**146**), and the absence of a modern state flora can be bridged in the south and west by reference to floras of North Carolina (**161**) and West Virginia (**148**). Area: 103230 km². Vascular flora: no current data.

The first account of plants in the 'Old Dominion' is by the Rev. John Banister (1680; published 1688 in the second volume of John Ray's *Historia plantarum*). This was followed by *Flora virginica* (1739–43; 2nd edn., 1762) by the Dutch physician and naturalist Jan Frederik Gronovius.[78] Although long obsolete and not offering effective geographical coverage, *no successor has ever appeared* save for Arthur Massey's 1961 enumeration; indeed, from the mid-nineteenth century until well into the twentieth significant contributions to the flora were largely the work of outsiders. A major essay, intended as a revision of Gronovius's work and drawing on Frederick Pursh's extensive collections of 1806–07, was the Philadelphia botanist Benjamin S. Barton's *Flora virginica* (1 vol., 1812). This was, however, abandoned after the Linnaean class Tetrandria; afterwards, investigations all but ceased for some six decades or more. In the late nineteenth and early twentieth centuries, contributors included Eileen J. Erlanson (with Earl J. Grimes) in the 'Virginia Peninsula' (inclusive of Williamsburg and Newport News), Merritt L. Fernald (partly with Bayard Long and others) in many parts of the southeast, Thomas H. Kearney (cf. **187**) in the Dismal Swamp on the borders of North Carolina, and John Kunkel Small, William Murrill, and A. A. Heller in the southwest (all collecting for Britton in New York).[79]

In the 1920s and 1930s, however, there developed a cultural 'revival' in Virginia (most manifest in the reconstruction of old Williamsburg). In floristics it may be marked by Paul R. Merriman's *Flora of Richmond and vicinity* (1930) and the formation in 1935 under Arthur Massey of the herbarium at the Virginia Polytechnic Institute and State University at Blacksburg. These signaled the effective resumption of resident work, which gathered strength after 1945.[80]

MASSEY, A. B., 1961. *Virginia flora.* 258 pp. (Virginia Agric. Exp. Sta. Tech. Bull. 155). Blacksburg.

Concise systematic enumeration of vascular plants (based in the first instance on the eighth edition of *Gray's Manual* (**140**)), with essential synonymy, abbreviated indication of local range (by counties and physiographic regions), and notation of habitat and frequency; bibliography and index to family and generic names. The introductory section includes a descriptive account of Virginian physiography (with its consequences for the flora) and a list of counties (with index). [For additions, see M. F. JOHNSON, 1970. Additions to the flora of Virginia. *Castanea* **35**: 144–149; and L. J. UTTAL and R. S. MITCHELL, 1970. Amendments to the flora of Virginia, I. *Ibid.*: 293–301. These latter papers both include county localities for specimens studied.]

Distribution maps

HARVILL, A. M., JR. *et al.*, 1992. *Atlas of the Virginia flora,* [edition] III. 144 pp., maps. Burkeville, Va.: Virginia Botanical Associates. (Originally published 1977–81 in 2 vols. as *Atlas of the Virginia flora*, 1: *Pteridophytes through monocotyledons.* Pp. i–iv, 1–59, maps. Farmville, Va., and *Atlas of the Virginia flora*, 2: *Dicotyledons.* Pp. 61–148, maps; reissued in 1 vol., 1986.)

An unannotated distribution atlas for all taxa with county dot maps (pp. 7–131, the families arranged according to the Englerian sequence), followed by remarks (with references) on plant geography and the prehistory of the 'coastal plain' flora (the latter with special attention to 19 species also found in Augusta County near the West Virginia border). Pp. 136–140 cover novelties, waifs and doubtfully established taxa, unfortunately without commentary save for indication of natives in Roman type; an index to genera and synonymous binomials concludes the work. Pp. 1–6 comprise a history of botanical exploration and documentation in the state. [The accounts of physical features, climate, geology, soils, vegetation formations, and plant communities given in the original edition have here been omitted.][81]

Partial work

HARVILL, A. M., JR., 1970. *Spring flora of Virginia.* xxx, 240 pp., illus. (including frontisp.), map (inside front cover). Parsons, W.Va.: McClain Printing Co. (for the author).

Illustrated, briefly descriptive students' manual of seed plants flowering before 1 June, with keys to all taxa, vernacular names, distribution (by county or physiographic province), figures of representative species, and notes on phenology, habitat, occurrence, etc.; index to generic and vernacular names at end. An introductory section covers botanical history, plant classification, use of keys, etc., along with a glossary (with botanical terms also illustrated on inside back cover).

Woody plants

For a popular color-guide, see O. W. GUPTON and F. C. SWOPE, 1981. *Trees and shrubs of Virginia.* [xii], [4], 205 pp., col. illus., map. Charlottesville: University Press of Virginia. [Covers 185 species; no key.][82]

Pteridophytes

MASSEY, A. B., 1960. *The ferns and "fern allies" of Virginia*. 3rd edn. 80 pp., 21 text-figs. (Bull. Virginia Agric. Ext. Serv. 256.) Blacksburg, Va.: VPI Agricultural Extension Service. (First published 1944; 2nd edn., 1958.)

Briefly descriptive treatment with diagnostic figures, synonymy, literature citations, county and physiographic distribution and habitats, and (pp. 68–77) a key to all taxa; literature list and index at end. The general part covers fern morphology, propagation and horticulture, a geographical and county index, statistics, and (pp. 26–32) a checklist with vernacular names. [Encompasses 70 species and 23 additional varieties.]

148

West Virginia

Area: 62709 km^2. Vascular flora: 2200 species (Strausbaugh and Core, 1978; see below).

West Virginia has had since its formation in 1863 a succession of state checklists and floras, the first three by Charles F. Millspaugh: *Preliminary catalogue of the flora of West Virginia* (1892), *Flora of West Virginia* (1896, with L. W. Nuttall) and *The living flora of West Virginia* (1913).[83] These were succeeded by *Catalogue of the vascular plants of West Virginia* (1940) by Earl L. Core and by the two editions of *Flora of West Virginia* (see below). The latter, well founded in the south-central Appalachians, is now effectively useful over much of the inland east-central United States.[84]

Bibliography

CORE, E. L., W. H. GILLESPIE and B. J. GILLESPIE, 1962. *Bibliography of West Virginia plant life*. 46 pp. New York: Scholar's Library. [Encompasses 900 entries (Besterman, in *Biological sciences: a bibliography of bibliographies*, 1971, Totowa, N.J.).]

STRAUSBAUGH, P. D. and E. L. CORE, 1978. *Flora of West Virginia*. 2nd edn., revised by E. L. Core. xl, 1079 pp., illus. Grantsville, W.Va.: Seneca Books. (Originally published in 4 parts, 1970–77, Morgantown, W.Va., as *W. Virginia Univ. Bull.* 70/7-2; 71/12-3; 74/2-1 and 77/12-3. 1st edn., 1952–64.)

Copiously illustrated, briefly descriptive atlas-manual of vascular plants, with keys to all taxa, essential synonymy, vernacular names, citations of relevant papers in headings, indication of local distribution (by counties), critical remarks, and notes on habitat, phenology, variation, special features, etc.; addenda, corrigenda and index to all botanical and vernacular names at end. Text is given on one side, with facing illustrations. An introductory section includes accounts of West Virginian climate, physical features, and vegetation. [This edition, though somewhat expanded and in a slightly larger format, differs little in style from its predecessor.]

149

Kentucky

Area: 103139 km^2. Vascular flora: 3142 species and non-nominate infraspecific taxa (Browne and Athey, 1992; see below). Braun in 1943 accounted for 2803 taxa.

In 1792, what is now Transylvania University was established as a college in Lexington – the first tertiary educational institution west of the Appalachians. Here, in the early nineteenth century, Rafinesque and, after him, Charles W. Short were professors of natural history and *materia medica*. Short, together with Robert Peter and Henry A. Griswold, here produced the first state checklist, *Catalogue of the native phaenogamous plants and ferns of Kentucky* (1833, with supplements to 1840). It was followed by *Catalogue of Kentucky plants* (1853) by Charles H. Spilman. In the same year Peter led a group of scientists advocating a state survey; when organized in 1854, however, no botany was included. Ninety years elapsed before publication of the next state lists, University of Kentucky botanist Frank McFarland's *A catalogue of vascular plants of Kentucky* (1942; in *Castanea* 7: 77–108) and E. Lucy Braun's *Annotated catalog of spermatophytes of Kentucky* (1943).

The lists of McFarland and Braun were succeeded in 1992 by a new catalogue by Edward T. Browne, Jr. and Raymond Athey (see below). The pteridophytes had already been extensively documented by Ray Cranfill (1980). A useful pair of semi-popular works is *Wildflowers and ferns of Kentucky* (1971) and *Trees and shrubs of Kentucky* (1973) by M. E. Wharton and R. W. Barbour (for the latter, see also **Woody plants** below).

Bibliographies

FULLER, M. J., 1979. Field botany in Kentucky: a reference list. *Trans. Kentucky Acad. Sci.* **40**: 43–51. [Unannotated list with 323 entries, including theses (among which are many florulas).]

MEIJER, W., 1970. The flora and vegetation of Kentucky as a field for research and teaching. *Castanea* **35**: 161–176. [Includes a bibliography of Kentucky floristic literature, with classified index.]

BROWNE, E. T., JR. and R. ATHEY, 1992. *Vascular plants of Kentucky: an annotated checklist.* xx, 180 pp., map. Lexington: The University Press of Kentucky.

Systematically arranged enumeration with indication of distribution by coded regions and essential synonymy; excluded species (p. 134), bibliography (pp. 136–145) and complete index but no vernacular names. The introductory part covers historical background and includes as well a summary, county map, physiographic regions, and a guide to the use of the list. An addendum (p. 135) deals with undeterminable records from McFarland's 1942 checklist.[85]

Partial work

GUNN, C. R., 1968. *The floras of Jefferson and seven adjacent counties, Kentucky.* 322 pp. (Ann. Kentucky Soc. Nat. Hist. 2). Louisville.

Annotated flora of the Louisville district in northern Kentucky, with keys and indication of local range. [Not seen; cited by Meijer, 1970.]

Woody plants

WHARTON, M. E. and R. W. BARBOUR, 1973. *Trees and shrubs of Kentucky.* x, 582 pp., illus. (some col.), maps. Lexington: The University Press of Kentucky. (Kentucky nature studies 4.)

An illustrated, essentially popular work organized in three sections: an introduction, an atlas with explanatory text, and systematically arranged 'natural history accounts' (the latter with many attractively written details of habitat, occurrences, uses and other features, especially for the trees and shrubs), followed by an illustrated glossary and general index. The atlas sections have species grouped artificially by gross features (without recourse to keys) and give both vernacular and scientific names. Technical details and background information are furnished in the introductory sections.

Pteridophytes

CRANFILL, R., 1980. *Ferns and fern allies of Kentucky.* [vi], 284 pp., illus., maps. [Frankfort, Ky.]: Kentucky Nature Preserves Commission.

Illustrated documentary treatment, with standardized vernacular names and extensive notes on local and overall distribution, habitats, phenology, karyotypes, biology, toxic (if any) and other properties, along with discussion of races and other taxonomic questions and county dot distribution maps (with source indicated for each); illustrated glossary, references and index at end. The introductory part covers history of studies, geography and ecology, physiographic units with characteristic species, chorology, life history and identification, along with statistics and general keys.

151
Midwest subregion

This category is used for works relating to the 'American Middle West' (the area bounded by Ohio, Michigan, Minnesota and Missouri) but not specifically related to a given state. Included here, for example, is the 'Driftless Area', an area (almost centering on La Crosse, Wis.) straddling southwestern Wisconsin, eastern Iowa, southeastern Minnesota, and a corner of Illinois, and of considerable biological interest as not having been glaciated during the Pleistocene; its flora was documented by Thomas G. Hartley (1966; see below).

The distinctiveness of the Great Lakes region and particularly its shore flora was first recognized by Nuttall during his explorations of 1810–11 and further revealed through the Cass expedition of 1820. Regional floristics for a time likewise enjoyed an identity of its own, particularly through the work of John Leonard Riddell at Cincinnati and William Sullivant at Columbus, both in Ohio, the scientists in the New Harmony colony in Indiana, and Charles Short in Kentucky. Its most important achievement was *Synopsis of the flora of the Western States* (1834) by Riddell, a precursor to a projected descriptive 'Flora of the Western States' and one of the first floristic accounts to espouse a 'natural' system. With the rise of Asa Gray and his 'circle' over the next decade, however, the Midwestern flora in time came to be included in the coverage of Northeastern floras and manuals, and so it has remained.[86]

As regional floristics declined, however, development in the various states led to dispersal of botanical activities. Learned societies and institutions – including, from 1859, the Missouri Botanical Garden – appeared in Chicago, Detroit and Ann Arbor, and

St. Louis, and more foci emerged as the state universities and colleges – notably the 'Big Ten' – gathered strength after the Civil War. Very importantly, state scientific surveys on the Eastern model were established; a number of them included natural history in their brief, either from the beginning or later.[87] Floristic work not unnaturally became very state-centered, resulting in some notable achievements but with overall an uneven record. Moreover, institutions which might have acted as regional foci sometimes oriented their research programmes in other directions, and no organization comparable to the Great Plains Flora Association (GPFA; see **175**) has come into being.[88] The limitation to Illinois of Robert Mohlenbrock's illustrated flora thus could be seen as a lost opportunity. In 1990, however, there appeared the first Northeastern/Midwestern manual-key written from a Midwestern base (see **140**, SWINK).[89]

Included here are two works on the so-called 'Driftless Area' (a territory in Illinois, Iowa, Minnesota and Wisconsin never glaciated during the Wurmian (Wisconsonian) of the Pleistocene) as well as a long-used manual of upper Midwest woody plants.

Partial work

HARTLEY, T. G., 1966. *The flora of the 'Driftless Area'.* 174 pp., map (Stud. Nat. Hist. Iowa Univ. 21(1)). Iowa City, Iowa.

Annotated systematic checklist of vascular plants (1639 species, of which 1344 native), including vernacular names and brief notes on distribution (with indication of counties), occurrence (if less common), usual habitat and, for rarities, occasional citations of *exsiccatae*; statistical summary, list of excluded and doubtful records as well as those based purely on literature, bibliography, and index. An introductory section deals with the plan of the work, earlier studies, geography, topography, geology, climate, and glaciation along with the central question, the relationship of the present flora with Quaternary history. [Pteridophytes were subsequently also covered by Peck (1982; see below).]

Woody plants

ROSENDAHL, C. O., 1955. *Trees and shrubs of the upper Midwest.* 411 pp., illus. Minneapolis: University of Minnesota Press. (Subsequently reissued. Originally published 1928 as *Trees and shrubs of Minnesota* by C. O. Rosendahl and F. K. Butters.)

Copiously illustrated, briefly descriptive treatment with keys to all taxa, synonymy (with references for accepted species), vernacular names, generalized indication of local and extralimital range, and notes on habitat, phenology,

special features, etc.; glossary and index to all botanical and vernacular names. Vegetation regions are described in the introductory section. Covers Iowa, upper Michigan, Minnesota and Wisconsin.

Pteridophytes

PECK, J. H., 1982. *Ferns and fern allies of the Driftless Area of Illinois, Iowa, Minnesota and Wisconsin.* 140 pp., maps (Milwaukee Public Mus. Contr. Biol. Geol. 53). Milwaukee, Wis.

Descriptive treatment of 73 species, 6 additional infraspecific taxa, and 13 nothospecies.

152

Ohio

Area: 106 610 km^2. Vascular flora: 2309 native, naturalized and adventive species (Schaffner, 1932). – The southeastern part of the state, including the upper Ohio Basin, is included in the Western Pennsylvania flora of JENNINGS (**144**).

Plans for the first state survey in the 1830s included natural history (surely advocated by John Riddell, a member of its planning committee) as well as geology, but only its post-Civil War successor (led by John Strong Newberry) produced floristic (and faunistic) work. William A. Kellerman and William C. Werner's *Catalogue of Ohio plants* (1895) appeared in vol. 7 of the survey's 'final reports' series, realizing an account originally commissioned by Newberry from H. C. Beardslee but not published. There were, however, several statewide checklists prior to 1895: Riddell's *Supplementary catalogue of Ohio plants* (1836; in *Western Journal of Medical and Physical Sciences* 9), an extension from his Midwest enumeration (**151**), Newberry's own *Catalogue of the flowering plants and ferns of Ohio* (1860), produced through the state board of agriculture and generally conceded to be the first proper state list, and Beardslee's slender *Catalogue of the plants of Ohio* (1874). Kellerman and Werner's *Catalogue* was soon succeeded by the first author's *The fourth state catalogue of Ohio plants* (1899, with three later supplements).

In 1912 the Ohio Biological Survey was organized at Ohio State University. Through its *Bulletin* and other publications, it has since played an important role in floristic work including, in recent years, documentation relating to threatened and endangered plants. All subsequent statewide checklists (and floras) have appeared under its aegis. Prior to 1950 it released *Catalog of Ohio vascular plants* (1914) and *Revised catalog of Ohio vascular plants* (1932, with supplements to 1939), both by John H. Schaffner; the latter is yet without a successor. For student use, Schaffner produced *Field manual of the flora of Ohio and adjacent territory* (1928). This was succeeded in 1960 by Clara Weishaupt's *Vascular plants of Ohio*, with a third edition in 1971.

The Ohio Flora Project, begun by E. Lucy Braun in 1950 and continuing under the aegis of the Ohio Academy of Science, is responsible for the first descriptive state flora. Four volumes (the first on woody plants) were published between 1961 and 1995, with one more (on dicotyledons) projected.[90]

Bibliographies

MILLER, E. M., 1932. Bibliography of Ohio botany. *Bull. Ohio Biol. Surv.* **27** (= vol. 5(4)): 281–376. [Columbus.] (Also published as *Ohio State Univ. Bull.* 37(5).) [Largely unannotated list arranged by author; coverage from 1755 to 1931.]

ROBERTS, M. L. and R. L. STUCKEY, 1974. *Bibliography of theses and dissertations on Ohio floristics and vegetation.* iv, 92 pp. (Ohio Biological Survey Informative Circular 7). Columbus. [Geographically arranged list with key-letters indicative of emphasis. Under 'Ohio-General' are several statewide taxonomic revisions.]

BRAUN, E. L., 1967. *The Monocotyledoneae: cattails to orchids.* viii, 464 pp., illus., maps (The vascular flora of Ohio 1). Columbus: Ohio State University Press. Continued as T. R. FISHER, 1988. *The Dicotyledoneae of Ohio, 3: Asteraceae.* x, [iv], 280 pp., illus., maps (The vascular flora of Ohio 2(3)). Columbus; and T. S. COOPERRIDER, 1995. *The Dicotyledoneae of Ohio, 2: Linaceae through Campanulaceae.* xxi, 656 pp., illus., maps (The vascular flora of Ohio 2(2)). Columbus.

Briefly descriptive, illustrated atlas-flora; includes keys to all taxa, essential synonymy, vernacular names, generalized indication of local and extralimital range (with county dot distribution maps for all species), and notes on habitat, biology, associates, special features, uses, etc., along with critical commentary (with also references to standard monographic and revisionary treatments); bibliography and indices to subjects as well as to all botanical and vernacular names.

The introductory section gives accounts of physiographic features (emphasizing their impact on vegetation) and of general aspects of the flora and its paleohistory. [In the volumes on dicotyledons references to treatments appear in the bibliographies but are also cross-referenced under genera. One further volume is planned to complete the work. The series also includes *The woody plants of Ohio* (see below).][91]

SCHAFFNER, J. H., 1932. Revised catalog of Ohio vascular plants. *Bull. Ohio Biol. Surv.* **25** (= vol. 5(2)): 87–215, 3 maps. [Columbus.] (Also published as *Ohio State Univ. Bull.* 36(9). Originally published 1914 as *Catalog of Ohio vascular plants*.)

Annotated checklist of vascular plants, with essential synonymy, county records (or more specific data) for less common species, and origin of alien taxa; index to vernacular and generic names. The introductory section covers previous catalogues, the plan of the work, floristic regions and patterns of plant distribution, and conventions. A bibliography is also included. [Based mainly on records in the Ohio State University Herbarium. Nomenclature follows the 'American Code', as used in Britton and Brown's *Illustrated flora*, with synonymy keyed to the seventh edition of *Gray's Manual*. For additions, see *idem*, Additions . . . I–VII. *Ohio J. Sci.* **33**: 288–294 (1933); **34**: 165–174 (1934); **35**: 297–303 (1935); **36**: 195–203 (1936); **37**: 260–265 (1937); **38**: 211–216 (1938); **39**: 232–234 (1939, with C. H. Jones).]

WEISHAUPT, C. G., 1971. *Vascular plants of Ohio.* 3rd edn. iii, 292 pp., [5] pls. Dubuque, Iowa: Kendall/Hunt. (1st edn., 1960, Columbus, Ohio.)

Concise students' manual-key to vascular plants in large format; includes short descriptions of genera and families but not species, vernacular names, indication of status, and miscellaneous notes; glossary (with supporting illustrations) and partial index. An artificial key on vegetative features to woody plants is also included. Synonymy, local distribution, and ecological data are, however, lacking (for these, recourse must be had to the *Revised catalog* and *Vascular flora of Ohio*).

Woody plants

BRAUN, E. L., 1961. *The woody plants of Ohio: trees, shrubs and climbers, native, naturalized and escaped.* viii, 362 pp., text-figs., maps. Columbus: Ohio State University Press. (Reprinted 1969, New York: Hafner; 1989, Columbus: Ohio State University Press.)

Illustrated, briefly descriptive atlas-flora, with keys to all taxa, essential synonymy and citations, vernacular names, indication of local and extralimital range (with dot distribution maps for all species), taxonomic commentary, and notes on occurrence, habitat, uses, properties, special features, etc. (including specific references); glossary, bibliography, list of authors, and indices to subjects and to all botanical and vernacular names. The introductory section includes an illustrated organography as well as an account of Ohio vegetation. [The work is complementary to *Vascular flora of Ohio*, and was the first fruit of the Ohio Flora Project.]

Pteridophytes

VANNORSDALL, H. H., 1956. *Ferns of Ohio.* Wilmington, Ohio: Distributed by Curtis Book Store.

Popularly oriented atlas-guide with descriptive text and 'chat', without keys but incorporating many observations (and special treatment of 'scenic places'); brief glossary and indices at end (including one to places and subjects). A county map appears in the inside covers.

153

Indiana

Area: 93423 km^2. Vascular flora: 2265 native, naturalized and adventive species (Crovello *et al.*, 1983; see below).

Flora of Indiana (1940) by Charles Clemon Deam, the first 'modern' state flora in the Midwest, is by common consent considered a classic and long has served as a model for later floristic writing. It was based on his 73000 collections of Indiana plants as well as other records.[92] Its main predecessors were *Catalogue of plants of Indiana* (1881) by Charles R. Barnes, published through J. M. Coulter's *Botanical Gazette* (then at Crawfordsville) and *A catalogue of the flowering plants and of the ferns and their allies indigenous to Indiana* (1900) by Stanley Coulter, published by the Indiana Geological Survey in its 24th report.[93] To this annotated list he, and later Deam, made many additions through 1935. Recent works include the checklist by Crovello *et al.* (1983) and manual-key by Crankshaw (1989).[94]

Bibliography

See DEAM (1940, pp. 1130–1161); includes 762 titles.

CRANKSHAW, W. B., 1989. *Manual of the seed plants of Indiana.* 278 pp., illus., map. Indianapolis: Indiana Academy of Science. (Monograph 6.)

Manual-key to all taxa with vernacular names and indication of frequency, usual habitats and status; illustrations of plant parts with glossary, bibliography (p. 234) and index to all names at end. The introduction covers the plan, basis and philosophy of the work; this is followed by the key to families.[95]

CROVELLO, T. J., C. A. KELLER and J. T. KARTESZ, 1983. *The vascular plants of Indiana: a computer based checklist.* xxiv, 136 pp. Notre Dame, Ind.: American Midland Naturalist/University of Notre Dame Press.

One-line checklist of vascular plants, with references to the continental checklist of Kartesz and Kartesz (**100**), a taxon code, and indication of status (native or alien). The arrangement is alphabetical within the major vascular plant groups. Pp. 73–124 encompass more extended annotations, including changes to Deam's nomenclature; at end are statistics (by families) and an index to genera and families. The introductory part gives a consideration of the need for a new checklist and the methodology involved, along with general statistics and sources of information. [Derived from an electronic database.]

DEAM, C. C., 1940. *Flora of Indiana.* 1236 pp., 1 pl., 2200+ maps. Indianapolis: Indiana Department of Conservation. (Reprinted 1970, Lehre, Germany: Cramer, as *Reprints of US-floras* 6.)

Detailed but non-descriptive atlas-flora of vascular plants, with keys to all taxa, limited synonymy, citation of standard monographic/revisionary treatments under family and generic headings where appropriate, vernacular names, extensive discussion of local distribution (with references) and summary of extralimital range; and notes on habitat, occurrence, ecology, phenology, uses, etc., as well as critical remarks; dot distribution maps provided for all species; lists of excluded species, collectors, and authorities, a tabular summary of the flora, and a bibliography, gazetteer, and index to all botanical and vernacular names at end. The introductory section includes general accounts of the flora and vegetation of the state.

Woody plants

DEAM, C. C. and T. E. SHAW, 1953. *Trees of Indiana.* 3rd edn. 329 pp., illus., maps (Publ. Indiana Dept. Conservation 13a). Indianapolis. (1st edn., 1911; 2nd edn.,

1932.) Accompanied by C. C. DEAM, 1932. *Shrubs of Indiana.* 2nd edn. 380 pp., illus., maps (*ibid.*, 44). Indianapolis. (1st edn., 1924.)

Atlas-treatments with keys, extensive descriptions, distribution (with county dot maps, each incorporating phenological scales and voucher information), habitat, and economic information. Glossaries and complete indices are also included along with general county maps, forest type maps, and temporal maps of earliest and latest frosts.

154

Illinois

Area: 144677 km². Vascular flora: 3204 species and non-nominate infraspecific taxa (Mohlenbrock, 1986; see below).

Illinois has been well served in the decades since World War II through the indefatigable efforts of a pair of rival flora-writers (and their respective associates), and is now one of the best-documented polities in North America. In this time the total of known vascular species has risen from 2125 to 2853. These greatly improved upon the relatively weak legacy of the nineteenth century, represented only by *Catalogue of Illinois plants* (1857) by Increase A. Lapham from Wisconsin, published by the state agricultural society, and *Catalogue of the phaenogamous and vascular cryptogamous plants of Illinois, native and introduced* (1876) by Oquawka newspaper publisher H. N. Patterson.[96] No catalogue was produced through the state survey which, though organized in 1851, had had a patchy history and was inactive (apart from the state museum and some publications) from 1875 until 1905. One result was decentralization of activity; not surprisingly, 'Chicagoland' has in this respect had the best record. Its latest flora – the seventh since 1872 – is the well-documented fourth edition of *Plants of the Chicago region* by Floyd Swink and Gerould Wilhelm (1994; see below). Other centers include Carbondale in the southernmost tier of counties (notable for the absence of glacial drift as well as the head of the Mississippi Delta); here was produced *Flora of southern Illinois* (1959) by Robert Mohlenbrock, a basis for his later state manual. George N. Jones's works were prepared at Champaign/Urbana, home to the University of Illinois and the Illinois Natural History Survey.

Bibliography

See JONES and FULLER (1955, pp. 537–556).

JONES, G. N., 1963. *Flora of Illinois.* 3rd edn. 401 pp., maps (Monogr. Amer. Midl. Naturalist 7). South Bend, Ind.: University of Notre Dame Press. (1st edn., 1945.)

Manual-key to vascular plants, with essential synonymy, vernacular names, generalized indication of local range (by counties or state regions), and notes on habitat, phenology, and special features; conspectus of work, county map, glossary, and index to all botanical and vernacular names at end. The introductory section includes a general review of the flora and vegetation of the state (with map).

JONES, G. N. and G. D. FULLER, 1955. *Vascular plants of Illinois.* xii, 593 pp., 1375 maps. Urbana: University of Illinois Press. (Subsequently reissued.)

Comprehensive, non-descriptive flora of vascular plants, with detailed synonymy (including references and citations), vernacular names, indication of local and extralimital range (including dot maps for all species), and notes on habitat, phenology, special features, etc.; list of principal collectors, extensive bibliography, and index to all botanical and vernacular names. The introductory section includes a general description of Illinois flora and vegetation. [For supplement, see G. S. WINTERRINGER and R. A. EVERS, 1960. *New records for Illinois vascular plants.* ix, 135 pp. (Illinois State Mus. Sci. Pap. Ser. 11). [Springfield.]]

MOHLENBROCK, R. H., 1986. *Guide to the vascular flora of Illinois.* Revised edn. vii, [4], 507, [1] pp., maps. Carbondale: Southern Illinois University Press. (1st edn., 1975.)

Very concise field-manual of vascular plants in a small format, with somewhat telegraphic keys to all taxa, brief descriptions, vernacular names, and brief indication of habitat, phenology, distribution, etc.; notes on hybrids; glossary, statistics, addenda and indices at end. An introductory section accounts for habitat factors and the natural regions of the state (14 recognized) along with the history of botanical exploration, and in addition includes an artificial key to families.[97]

MOHLENBROCK, R. H., 1967– . *The illustrated flora of Illinois.* [Vols. 1– .] Illus. Carbondale: Southern Illinois University Press.

Detailed descriptive illustrated flora of vascular plants; includes keys to all taxa, limited synonymy (with some literature citations), vernacular names, generalized indication of local and extralimital range (with dot distribution maps for all species), some critical remarks, notes on habitat, occurrence, phenology, ecology and biology, karyotypes, special features, etc.; bibliography, glossary, and index to all botanical and vernacular names in each volume. [As of 1999, 13 volumes had appeared: pteridophytes (one volume), the whole of the monocotyledons including the grasses and sedges (six volumes, with *Carex* the last in 1998), and (so far in six volumes) many dicotyledon families: 'magnolias to pitcher plants' (Magnoliaceae–Sarraceniaceae), 'basswoods to spurges' (Tiliaceae–Euphorbiaceae), 'willows to mustards' (Salicaceae–Cruciferae), 'hollies to loasas' (Aquifoliaceae–Loasaceae), 'nightshades to mistletoe' (Solanaceae–Viscaceae), and 'smartweeds to hazelnuts' (Polygonaceae–Betulaceae). The plan of the work follows the Thorne system, though individual volumes have not been appearing in systematic order.][98]

MYERS, R. M., 1972. *Annotated catalog and index for the Illinois flora.* 64 pp. (Biol. Sci. Ser. Western Illinois Univ. 10). Macomb.

Numbered systematic list of native, naturalized and adventive vascular plants, with indication of status and bibliographic references; statistical table and addenda; no index. Notes on the flora of McDonough County are also appended. The introductory section includes remarks on alien, weedy, and extinct species.

Distribution maps

MOHLENBROCK, R. H. and D. M. LADD, 1978. *Distribution of Illinois vascular plants.* vii, 282 pp., maps. Carbondale: Southern Illinois University Press.

Comprises an atlas of dot distribution maps of the vascular plants of Illinois, with 251 pages (each with 12 maps) accounting for 3001 taxa (species and varieties). The arrangement is alphabetical, and only accepted names as used in the first edition of the senior author's *Guide* (see above) are used. A table of synonymy (relating to other major floras) and a systematic name list appear as appendices.[99]

Partial work

SWINK, F. and G. WILHELM, 1994. *Plants of the Chicago region.* 4th edn. xiv, 921 pp., illus., maps. Indianapolis: Indiana Academy of Science. (1st edn., 1969; 2nd edn., 1974; 3rd edn., 1979, Lisle: The Morton Arboretum.)

Alphabetically arranged, dictionary-style topographical enumeration of vascular plants (2530 species) with all names, scientific and vernacular, in a single sequence;

includes keys, limited synonymy (in relation only to certain key works), vernacular names, references to other floras, detailed account of local distribution with county dot maps for most species, status, taxonomic commentary, and extensive notes on habitat, frequency, associates, etc.; no separate index. The general part includes a definition of the region (from Berrien Co., Michigan through northwestern Indiana and northeastern Illinois to southeastern Wisconsin), an explanation of the plan and scope of the work, mapping philosophy, taxonomic and nomenclatural parameters as well as a glossary, bibliography and keys to families. The appendix on field floristic methodology, a feature of the 1979 edition, is now also included in the general part and retitled 'Floristic Quality Assessment' with a direction towards restoration ecology; it is accompanied by an 'evaluation checklist' of the flora wherein all species have assigned a wetland value and a 'coefficient of conservatism' (the latter on a scale of 1–10 and appearing also in the regular entries). [Now one of the best of its kind in North America if in the current edition somewhat too bulky, this work has a strong field and practical orientation with relatively catholic coverage.]

155

Michigan

See also **151** (ROSENDAHL). – Area: 148 800 km^2. Vascular flora: 2465 seed plant species (Voss, 1972–96; see below). A total of 2365 vascular species were recorded by Beal in 1904–08.

Several checklists, all state-sponsored, appeared in the nineteenth century. As part of the first detailed survey begun under Douglas Houghton in 1837 following statehood, John Wright, as botanist, produced in its second report (1839) *Catalogue of the phaenogamous and filicoid plants,* covering 850 species. Natural history was, however, discontinued the next year and, after Houghton's death in 1845, accrued collections passed to the University of Michigan. A second state survey was launched in 1859 under Alexander Winchell, the university's professor of natural sciences; in 1861 survey assistant Newton H. Winchell published in its first report a new checklist, *Flora of Michigan,* covering the 'Lower Peninsula'. Subsequent accounts were under other auspices, including *Catalogue of the . . . Plants of Michigan* (1881) by Charles F. Wheeler and E. F. Smith, and *Michigan Flora* (1892) by Michigan State College professor William J. Beal with Wheeler. The second edition of that work, by Beal alone, appeared in 1905 (though dated 1904) and would remain nominally current for decades. Beal himself published some additions in 1908 (*Annual Rep. Michigan Acad. Sci.* **10**: 85–89).

From 1918, *Michigan Flora* was supplemented with students' keys by Henry Gleason; these took the form of a European-style manual-key but without the extensive concise annotations usually found in such works. In this period and continuing to 1941 Oliver A. Farwell published numerous additions to Beal's flora. In the 1950s Edward Voss at the University of Michigan began work on his definitive *Michigan flora*; its third and final volume appeared in 1996. A new 'students' manual' by R. K. Rabeler appeared in 1998 as an overt successor to that of Gleason.

Local literature is quite extensive but with published bibliographies and the existence of the series *Michigan plants in print* (see below) no coverage is here essayed.

Bibliographies

DARLINGTON, H. T., 1945. Bibliography. In *idem, Taxonomic and ecologic work on the higher plants of Michigan*, pp. 45–59 (Michigan State Coll. Agr. Expt. Sta. Tech. Bull. 201). East Lansing.

VOSS, E. G., 1962–. Michigan plants in print; guide to literature on the Michigan flora. *Michigan Bot.* **1**–, *passim*. [A running series, organized under three heads (not all of which may appear at any one time) respectively covering (a) physical features and general, (b) separate publications and (c) journal articles.][100]

RABELER, R. K., 1998. *Gleason's plants of Michigan: a field guide.* 398 pp., illus., map. Ann Arbor: Oakleaf Press.

Field manual-key to vascular plants, the leads with vernacular and scientific names, indication of habit, phenology, and usual habitat; glossary and subject and name indices at end. No synonymy or details of distribution are given. The general part furnishes the history, plan and purpose of the work (with particular reference to Gleason and his Michigan connections), pedagogical details, a bibliography, and an illustrated organography as well as the general keys. [Succeeds *Plants of Michigan* (1918; 3rd edn., 1939) by H. A. Gleason.]

VOSS, E. G., 1972–96. *Michigan flora.* 3 vols. Illus. (some col.), 2465 distribution maps. Bloomfield Hills, Mich.: Cranbrook Institute of Science (respec-

tively as their *Bulletins* 55, 59 and 61); Ann Arbor: University of Michigan Herbarium.

Concise annotated manual of seed plants; includes diagnostic keys to all species, limited synonymy, vernacular names, general indication of local range (with dot maps for all species), critical commentary, and notes on habitat, phenology, occurrence, biology, variation, special features, etc.; glossary and index to all botanical and vernacular names. The introductory section (reprinted, with revisions, in vol. 2) includes a historical survey of botanical work together with descriptive remarks on vegetation, phytogeography, and paleohistory of the flora (with specific references). [Vol. 2 published 1985. Ferns and fern-allies were covered separately by Cecil Billington (see below under **Pteridophytes**).][101]

Woody plants

Attention may here be drawn to C. BILLINGTON, 1943. *Shrubs of Michigan.* 249 pp., 161 text-figs., maps. Bloomfield Hills, Mich.: Cranbrook Institute of Science (as their *Bulletin* 20). This is a well-illustrated descriptive work, with keys and county dot distribution maps; glossary, bibliography and index at end. For a more modern account, users might consider *Shrubs of Ontario* (1982; see **132**).

Pteridophytes

BILLINGTON, C., 1952. *Ferns of Michigan.* 240 pp., 78 text-figs., unnumbered figures, 16 halftone pls., maps. Bloomfield Hills, Mich.: Cranbrook Institute of Science (as their *Bulletin* 32).

Illustrated semi-popular descriptive treatment with keys, vernacular names, county dot maps, indication of habitat and overall distribution, and extensive commentary; glossary, literature and index to all names at end. The introductory part covers at length the scientific and popular aspects of pteridophytes with particular reference to the state, along with a list of derivations of scientific names and one of authors of fern names.

156

Wisconsin

See also **151** (HARTLEY, ROSENDAHL). – Area: 141 508 km². Vascular flora: no data.

No modern state flora has been published. The nearest approach is in the long series of *Preliminary reports* begun in 1929 by Norman C. Fassett, with which may be associated his substantial monographs on the grasses and legumes (a volume on pteridophytes was contributed by Rolla Tryon). The *Reports* were preceded by two nineteenth-century checklists (the first including much, if not all, of what is now Minnesota): *Plants of Wisconsin* by Increase A. Lapham (1850; in *Proceedings of the American Association for the Advancement of Science*) and *Catalogue of the phaenogamous and vascular cryptogamous plants of Wisconsin* (1883) by Goodwin D. Swezey.[102] The former was a result of the federally supported 'Northwestern Geological Survey' conducted just prior to statehood by David Owen in 1847–48; the second, in the final report of the third Geological Survey of 1873–79 organized under Lapham. As in Michigan, a set of the 1870s collections went to the state university. A related work is *Vergleichende Flora Wisconsins* (1877; supplements, 1878–79) produced in Vienna by Thomas A. Bruhin, the list therein forming part of a comparative study of the Wisconsin and European floras.

For student use, a staple has been Fassett's *Spring flora of Wisconsin* (1931; 4th edn., 1976). Pteridophytes were documented by Tryon (1953). Earlier local literature was surveyed by Green and Curtis (1955).[103]

Bibliography

GREEN, H. C. and J. T. CURTIS, 1955. *A bibliography of Wisconsin vegetation.* 84 pp. (Milwaukee Public Mus., Publ. Bot. 1). Milwaukee. [Classified by methodology; includes references to local floristic works. A short history of Wisconsin botany is also incorporated.]

FASSETT, N. C. *et al.*, 1929– . Preliminary reports on the flora of Wisconsin, 1– . Illus. *Trans. Wisconsin Acad. Sci.* **24**, etc., *passim.* [Madison.] Complemented by *idem*, 1939. *The leguminous plants of Wisconsin.* xii, 157 pp. Madison: University of Wisconsin Press; and *idem*, 1951. *Grasses of Wisconsin.* 173 pp. Madison.

The *Preliminary reports* comprise a long series of descriptive (sometimes illustrated) revisions of families of Wisconsin vascular plants, each with keys to genera and species, concise synonymy, vernacular names, generalized indication of local range (with commentary and dot distribution maps), extensive critical discussion, and notes on habitat, frequency, karyotypes, biology, special features, etc.; prefatory remarks and lists of references are also included. Most treatments

have been contributed by specialists. The most recent installment is no. 69 (1987). The companion works on legumes and grasses are full descriptive accounts with keys, synonymy, commentary and references.

Partial work

FASSETT, N. C. and O. S. THOMSON, 1975. *Spring flora of Wisconsin.* 4th edn. 416 pp., 564 figs., 4 maps. Madison: University of Wisconsin Press. (1st edn., 1931.)

This well-known, widely used work, the 'doyen' of American 'spring floras', comprises a copiously illustrated, concise students' field-manual in small format of seed plants flowering before 15 June, with complete keys, vernacular names, local distribution, and notes on habitat, special features, etc.; glossary, selected list of references, and index at end. An introductory section includes four maps depicting counties and major geoclimatic and vegetation zones. The fourth edition includes Cyperaceae (omitted from earlier versions), but *Salix* and Juncaceae are not included and the grasses are merely listed (for a fuller treatment of the last-named, see Fassett (1951, noted above under his *Preliminary reports*)).

Pteridophytes

TRYON, R. M., JR. *et al.*, 1953. *The ferns and fern allies of Wisconsin.* 2nd edn. 158 pp. Madison: University of Wisconsin Press. (1st edn., 1940.)

A treatment with keys, similar in style to the author's work for Minnesota (**157**) but here with additional synonymy. [The 1953 edition differs but little from its predecessor, save for nomenclature.]

157

Minnesota

See also **151** (HARTLEY; ROSENDAHL); **175** (GREAT PLAINS FLORA ASSOCIATION). – Area: 206 825 km². Vascular flora: 1618 native and 392 introduced species (Ownbey and Morley, 1991; see below).

Minnesota was constituted as a territory separate from Wisconsin in 1849 and became a state in 1858, several years after publication of Lapham's Wisconsin list. Much botanical exploration had been undertaken in the 1830s and 1840s by Douglas Houghton and Heinrich Beyrich (respectively with Schoolcraft's two expeditions), Jean Nicollet, and under David Owen. A first list of 896 species, *Catalogue of the plants of Minnesota* (1875), was published by Lapham; very likely this was stimulated by the formation in 1872 of the second state survey, which enjoyed a broad brief. The next list, published in the survey's reports, was *Catalogue of the flora of Minnesota* (1884; supplement, 1887) by Warren Upham; covering 1850 species, it remained 'current' for more than a century. The collections of the survey formed the basis of the state herbarium at the University of Minnesota, which assumed responsibility; thus, Conway Macmillan's *Metaspermae of the Minnesota Valley* (see below) also appeared as a survey report. The only other statewide works until 1991 were *A preliminary checklist of the flowering plants, ferns and fern allies of Minnesota* (1946) by John W. Moore and Rolla Tryon, and *Spring flora of Minnesota* by Thomas Morley (see below).

From the regional and local literature, two substantial partial works respectively covering the northeastern triangle and southern third of the state – the main population centers – have been selected.

OWNBEY, G. B. and T. MORLEY, 1991. *Vascular plants of Minnesota: a checklist and atlas*. x, 307 pp., 2 general and 1881 distribution maps. Minneapolis: University of Minnesota.

A checklist with separate atlas section; the list contains essential synonymy, vernacular names, map cross-references and (sometimes) terse notes on taxonomy, biology, etc., and, under genera and families, references to monographs and revisions. The maps, nine to a page, depict county boundaries and rivers but (as in Pennsylvania) localized dots. Two appendices to the list cover excluded names and statistics, and an index to genera concludes the work.

Partial works

LAKELA, O., 1965. *A flora of northeastern Minnesota*. 541 pp., 110 text-figs., maps. Minneapolis: University of Minnesota Press.

Descriptive manual of vascular plants (1179 species); includes keys to all taxa, vernacular names, generalized indication of local range (with *exsiccatae* together with maps for all species), summary of extralimital distribution, figures of representative species, and notes on habitat and phenology; glossary and complete index. The introductory section includes accounts of physical features, geology, soils and vegetation. [The area covered, in St. Louis and Lake Counties, is the triangle between Duluth, Lake Superior, and the Canadian border, and includes Isle Royale (in Lake Superior) and the iron-bearing Mesabi Range.]

MACMILLAN, C., 1892. *The Metaspermae of the Minnesota Valley*. xiii, 826 pp., 2 maps (Minnesota Geol. Surv. Rep., Bot. Ser. 1). Minneapolis.

Detailed enumeration of vascular plants (1165 species), including full synonymy (with references and citations), indication of localities with *exsiccatae* and other sources, summary of extralimital range, and notes on habitat, special features, etc.; bibliography and index to all botanical names. An introductory section gives a delimitation of the area, while several appendices account for physical features, climate, floristics, phytogeography, and statistics of the flora. [The area centers on the Minnesota River and covers approximately the southern third of the state as well as adjacent parts of Iowa and South Dakota.]

MORLEY, T., 1969. *Spring flora of Minnesota*. 2nd edn. 283 pp., 2 maps. Minneapolis: University of Minnesota Press. (Reissued 1974. 1st edn., 1966.)

Descriptive students' manual of seed plants flowering before 7 June, with complete keys, essential synonymy, vernacular names, local range, and notes on habitat; map of vegetation formations, gazetteer, county map, glossary, and index at end. Patterned after Fassett's *Spring flora of Wisconsin* (**156**).

Woody plants

Mention may be made here of the predecessor of C. O. Rosendahl's *Trees and shrubs of the upper Midwest* (1955; see **151**): C. O. ROSENDAHL and F. K. BUTTERS, 1928. *Trees and shrubs of Minnesota*. Minneapolis: University of Minnesota Press. This was in turn based on *Minnesota trees and shrubs* (1912) by F. E. Clements, C. O. Rosendahl and F. K. Butters.

Pteridophytes

TRYON, R. M., JR., 1980. *Ferns of Minnesota*. 2nd edn. 165 pp., 201 text-figs., 81 col. pls., 78 maps. Minneapolis: University of Minnesota Press. (1st edn., 1954.)

Descriptive 'natural history' treatment, with keys, figures and localized dot distribution maps for each species; the text encompasses vernacular names, diagnostic features, local and general distribution, and frequency. The introductory part includes a treatment of physical features, geology, distribution, ferns in the garden, collecting, and use of the handbook, along with a short bibliography.[104]

158

Iowa

See also **151** (HARTLEY; ROSENDAHL); **175** (GREAT PLAINS FLORA ASSOCIATION). – Area: 145509 km². Vascular flora: 1958 species, of which 1516 native (Eilers and Roosa, 1994). Thorne (1954; see Note 105) gave a figure of 1785 species.

Following the early expeditions of Lewis and Clark (1804 and 1806), Long (1819) and Nicollet (1839), more intensive collecting in the state began with the work of C. C. Parry on David Owen's 1847–48 'Northwestern Geological Survey', already noted under Wisconsin (**156**). The first state checklists were *Contributions to the flora of Iowa* (1871), an early work of Charles E. Bessey produced at Ames some years before his final move to Nebraska (**177**), and *Contributions to the flora of Iowa: a catalogue of the phaenogamous plants* (1876) by J. C. Arthur. Then followed Thomas J. Fitzpatrick's *Manual of the flowering plants of Iowa* (1899). However, this last was never completed and to this day the state remains without a descriptive flora.

Only with its third launch in 1892 did the state geological survey include in its brief natural history; their annual reports thenceforth sometimes contained local checklists and other floristic works. The first of the twentieth-century statewide catalogues, *Plants of Iowa* (1907) by Wesley Greene, was, however, published by the state horticultural society; it covers both non-vascular as well as vascular plants (the latter comprising nos. 1506–3090). Greene's work was followed by *The Iowa flora* (1933) by Robert Cratty and, very recently, by *The vascular plants of Iowa* (1994) by Lawrence Eilers and Dean Roosa (see below). Henry Conard (*Plants of Iowa*, 1952) and, later, Richard W. Pohl produced students' keys comparable to those of Gleason in Michigan.

In the 1950s a regional floristic studies programme was instituted by Robert F. Thorne while at the University of Iowa in Iowa City. Eight contributions were prepared; though not all were published they form a key foundation for present and future statewide floristic works.[105]

EILERS, L. J. and D. M. ROOSA, 1994. *The vascular plants of Iowa: an annotated checklist and natural history*. xi, [2], 304, [1] pp., halftones, 7 maps. Iowa City: University of Iowa Press. (A Bur Oak original.)

The general part includes remarks on physical features, climate, geology, history of glaciation, and physiography (for physiographic zones, see map 6), natural regions and their characteristic plants, biogeography with indicator species, and the plan of the checklist along with county abbreviations, a list of excluded taxa (pp. 24–25) and statistics (pp. 25–26). The enumeration proper is alphabetically arranged (with a common ending for all families) and includes essential synonymy, English vernacular names where known,

indication of habitat as well as distribution (sometimes with references to specific counties, regions and dates), abundance, and status. Selected references and indices respectively to taxa and synonyms and to vernacular names conclude the work.

POHL, R. W., 1975. *Keys to Iowa vascular plants*. 198 pp. Dubuque, Iowa: Kendall/Hunt.

Consists entirely of keys to 'common' plants and, in the author's words, as 'most of our land is cultivated and bears an exotic flora', also incorporates most crop and ornamental species, not to mention weeds. No sedges or grasses are included (references being made merely to separate monographs), nor do names have any authorities; the keys, moreover, have no annotations. A glossary and index to families are given at the end, while the introductory section incorporates general keys.

Pteridophytes

COOPERRIDER, T. S., 1958. *The ferns and other pteridophytes of Iowa*. 66 pp., 82 text-figs., 53 maps (Stud. Nat. Hist. Iowa Univ. 20(1)). Iowa City.

Illustrated treatment of 53 species with keys, vernacular names, indication of frequency, distribution and habitat with for rare taxa citations of *exsiccatae* with localities, synonymy, and some commentaries; dot distribution maps for each taxon (pp. 56–62), list of references and complete index at end. The introductory part includes a glossary and illustrations of fern morphology.

159

Missouri

See also **175** (GREAT PLAINS FLORA ASSOCIATION). – Area: 179257 km². Vascular flora: 2438 species and 913 additional infraspecific taxa (Steyermark, 1963).

Like Indiana, Missouri is the beneficiary of one of the best modern comprehensive state manuals, *Flora of Missouri* (1963) by Julian A. Steyermark; its popularity ensured six reissues in 33 years. This grew out of his detailed *Annotated catalogue of the flowering plants of Missouri* (1935, with E. Jesse Palmer as senior author) and *Spring flora of Missouri* (1940; see below). Palmer and Steyermark's work in turn superseded *Flora of Missouri* by Samuel M. Tracy (1886). The various state surveys, the first organized in 1853, were not oriented towards Recent botany; rather, there was an active state horticultural society through which, before 1915, Tracy's catalogue and some other statewide floristic works were published. The St. Louis Academy of Sciences, the Engelmann Botanical Club and, through most of the present century, the Missouri Botanical Garden have also provided foci and outlets. The state Department of Conservation is currently a sponsor of note.

In recent years, a first step towards a new state flora has been taken with publication of *Catalogue of the flora of Missouri* (1990) by George A. Yatskievych and J. Turner. The project was initiated in 1987 with support from the state Department of Conservation, publishers of *Field guide to Missouri ferns* (1982) by J. S. Key (see below). In 1999, the first of two projected volumes of the new flora appeared under the title *Steyermark's Flora of Missouri.*

Bibliography
Good bibliographies for Missouri may be had in Palmer and Steyermark's *Annotated catalogue* (1935, pp. 442–453), noted above, and in Yatskievych, *Steyermark's Flora of Missouri* (1999).

YATSKIEVYCH, G., 1999. *Steyermark's Flora of Missouri.* Vol. 1. xii, 991 pp., 194 pls., halftones, 797 taxon maps, other maps in introduction. St. Louis: Missouri Botanical Garden Press; Jefferson City: Missouri Department of Conservation. (Originally published 1963 as *Flora of Missouri* by J. A. Steyermark; six reissues through 1996.)

Comprehensive illustrated descriptive flora of vascular plants (to encompass two volumes); includes keys to all taxa, essential synonymy, vernacular names, indication of distribution and habitat, county dot maps, and sometimes extensive taxonomic, economic, ecological and other commentary; glossary, bibliography (pp. 929–951) and complete index (with index to families covered on the back end-paper). The general part covers the plan of the work, kinds of information included, a history of floristic work in the state, and an introduction to Missouri geography, geology, climate, soils, vegetation, natural landscape units, plant communities (with principal species), an analysis of the flora (with statistics), conservation, origins of the flora and its affinities, and human influence.

STEYERMARK, J. A., 1963. *Flora of Missouri.* lxxxiii, 1728 pp., 2300 illus., 2400 maps. Ames, Iowa:

Iowa State University Press. (Seventh reprinting, 1996.)

Comprehensive illustrated, non-descriptive atlas-flora and manual of vascular plants, with essential synonymy (including some literature citations), vernacular names, general indication of local and extralimital range (with county dot maps for all species), taxonomic commentary, and extensive notes on habitat, frequency, occurrence, phenology, biology, special features, properties, uses, etc.; tabular summary of the flora, glossary and index to all botanical and vernacular names at end. The introductory section includes a historical sketch as well as accounts of the flora and vegetation in general and lists of species with distributions bordering on Missouri; a general county map is also provided. Encompasses 2438 species and 913 additional infraspecific taxa. [Partly supplanted by *Steyermark's Flora of Missouri* (see above).]

YATSKIEVYCH, G. and J. TURNER, 1990. *Catalogue of the flora of Missouri.* xii, 345 pp. (Monogr. Syst. Bot. Missouri Bot. Gard. 37). St. Louis.

Briefly annotated, alphabetically arranged checklist of vascular plants; includes essential synonymy (tied to Steyermark's flora and other key area works), vernacular names, citations where appropriate (including cross-references to Steyermark), symbols for taxa new to the state, and taxonomic commentary; named hybrids, excluded taxa and references at end (pp. 301–345). The introduction is brief and pragmatic, and includes conventions and explanations of symbols.

Partial work
STEYERMARK, J. A., 1940. *Spring flora of Missouri.* viii, 582 pp., 66 figs., 163 pls., map. St. Louis: Missouri Botanical Garden; Chicago: Field Museum of Natural History. (Reprinted 1954, Columbia, Mo.: Lucas Brothers.)

Illustrated descriptive flora, with keys, vernacular names, indication of distribution, and notes on habit, habitat, etc.; glossary and index.[106]

Pteridophytes
KEY, J. S., 1982. *Field guide to Missouri ferns.* x, 208 pp., illus., map. Jefferson City, Mo.: Missouri Department of Conservation.

An atlas-treatment, with full-page figures (by Paul Nelson) and facing *pro forma*, relatively spare text including botanical features, indication of frequency, and habitat; references, glossary, checklist and complete index at end. A form-key is given on pp. 14–29 following the introductory part.

Region 16

Southeastern United States

Area: *c.* 1 150 000 km^2 (Wood, 1983; see below). Vascular flora: no data. – This region, the heart of 'Dixie', encompasses the nine states from North Carolina to Florida and west to Arkansas and Louisiana, although floristically its Gulf coastal and Atlantic component extends to Virginia as well as east Texas, much of its north is southern Appalachian, and southern Florida is properly Caribbean or West Indian (Superregion 24–29).

The first consolidation of floristic knowledge in the 'Old South' began in the 1840s, spurred, as in the North, by a need for teaching manuals as well as, perhaps, a lack of further progress of work on *Flora of North America*. A first attempt was made by John Darby with *A manual of botany* (1841, Macon) and *Botany of the southern states* (1855, New York, with reissues to 1860 and again in 1866). These were effectively succeeded by Alvan Chapman's *Flora of the southern United States* (1860, New York; later editions 1883, 1897).[107] Chapman's death in 1899, however, created an opening for a putative successor work; with no effective competition, it became, as in the West, an opportunity for 'Lord' Britton and an associate in New York, John Kunkel Small – the latter already becoming known for his field work in the region. The result was Small's *Flora of the southeastern United States* (1903; 2nd edn., 1913), based on examination of over 50 000 specimens as well as literature, and, later, his *Manual* (1933), both still standard works although now all but obsolete. At state level, despite intermittent activities by state surveys and interested individuals, limited means and other factors restricted the production of general works until after World War II. The most substantial of these, Charles Mohr's *Plant life of Alabama*, was published only with U.S. federal support.

In the second half of the twentieth century, Harvard University, through its Arnold Arboretum, re-entered the Southern field with a scheme for a series of floristic projects, including a new manual. The main result has been the definitive *Generic flora of the southeastern United States*, prepared since its inception under the direction of Carroll Wood, Jr. (more recently jointly with Norton Miller at the New York State Museum and Walter Judd at the University of Florida). Until 1990 it appeared in irregular installments in the Arboretum's *Journal*; later it moved to a 'supplementary series' (the *Journal* itself having been suspended) and is now in *Harvard Papers in Botany*. It is now far advanced, with over 140 contributions. But, given the considerable social and economic progress of the South after 1950 and the development there of new or expanded botanical centers, interest in a descriptive regional flora not unnaturally emerged from within; this led to establishment of the *Vascular Plants of the Southeastern United States* project at the University of North Carolina at Chapel Hill under the direction of Albert E. Radford – using as a basis the successful *Manual of the vascular flora of the Carolinas* (1968; see **162**), the first 'modern' Southern flora.[108] Of this regional work, however, only two parts to date have been published. Other key contributions have dealt with pteridophytes as well as trees and other woody plants (see below), the Atlantic and Gulf coastal flora (*The Smithsonian guide to seaside plants of the Gulf and Atlantic coasts . . .* by Wilbur and Marion Duncan (1987; see **109**)), and wetland plants (*Aquatic and wetland plants of southeastern United States*, 1979–81, by Robert K. Godfrey and Jean W. Wooten; see **108/II**).

Nonetheless, this floristically richest part of the conterminous United States remains imperfectly documented. There were in earlier years relatively few resident botanists, and the region was to a greater or lesser extent looked upon by the North as a botanical *Kolonialgebiet* in rather the same way as transalpine Europe saw the Mediterranean Basin (with, inevitably, some reaction: the *Generic flora* has been criticized for 'carpetbagging' if nothing else). Moreover, important collections are, as with many tropical countries, often housed outside the region; and within, resources are relatively decentralized. Much progress has indeed been made, notably in the last half century, but the greatest single desideratum would yet seem to be a new manual-flora to replace Small's all-but-obsolete works (at present the Carolinas manual is a widely used alternative). Beyond this one might look to more statewide inventories or manuals, although over the three decades since 1968 the situation has significantly improved. Among the most notable developments was Wunderlin's Florida manual of 1998 – the first such covering the whole state. Other projects have reached stages or are continuing although, like the regional *Vascular Plants*, some are progressing only slowly.[109] The vascular floras are by North American standards comparatively large, and the number of active workers, many with other commitments, is limited as is also, unfortunately, financial support.[110]

Bibliographies. General and divisional bibliographies as for Division 1.

Indices. General and divisional indices as for Division 1.

160
Region in general

Small's 1913 *Flora* covers the entire region as here defined and, in addition, eastern Texas and Oklahoma; his 1933 *Manual*, however, encompasses only the states or parts thereof **east** of the Mississippi River (a projected companion work for Arkansas, Louisiana and east Texas was projected but never realized). Of the considerable literature on woody plants and pteridophytes, only a selection, given here under separate subheadings, can be described here.

Notable outlets for new floristic information on the region are *Castanea*, the journal of the Southern Appalachian Botanical Club, published since 1936, and *Sida*, established in 1962 and now published from the Botanical Research Institute of Texas in Fort Worth.

Keys to families and genera

BATSON, W. T., 1972. *Genera of the southeastern plants.* iv, 151 pp., 1089 text-figs. Columbia, S.C.: The author. [Keys to all taxa of vascular plants down to the genus, with synonymy, vernacular names, notes on distribution, habitat, etc., and representative figures; glossary, twig key, list of poisonous plants, and index. Revised and expanded 1975 and 1977 as *A guide to the genera of the eastern plants* and reissued 1983 as *A guide to the genera of the plants of eastern North America* (see **100**).]

RADFORD, A. E. *et al.* (eds.); L. S. RADFORD (exec. ed.), 1980– . *Vascular flora of the southeastern United States.* Vols. 1– . Chapel Hill, N.C.: University of North Carolina Press.

A descriptive manual-flora of vascular plants,

without illustrations or maps; includes keys to all taxa, limited synonymy with citations, more or less generalized indication of distribution (by physiographic provinces in the first instance), notes on karyotypes, phenology, habitat, special features, etc., and vernacular names (where known); consolidated index at end of each volume (or part-volume). Generic and other headings include citations of key monographs and revisions, along with notes on waifs or casual adventives not otherwise included. Volume 1 includes in its introductory section a style-manual for the whole flora, with notes on format, geographical and ecological notation, and synonymy as well as abbreviations; while vol. 3(2) features a more extensive geographical chapter, with maps showing physiographic and floristic provinces. [Limited by the Atlantic Ocean, the Gulf of Mexico, the Mason–Dixon line, the Ohio and Mississippi Rivers, the northern boundary of Arkansas, and the western boundaries of that state and Louisiana; presence in all adjacent states is, however, mentioned. Of the five projected volumes, as of 1998 only vol. 1 (Asteraceae, 1981, by A. Cronquist) and 3(2) (Leguminosae, 1990, by D. Isely) have been published; the latter includes an appendix on legume systematics.][111]

SMALL, J. K., 1913. *Flora of the southeastern United States.* 2nd edn. xii, 1394 pp. New York: The author. (1st edn., 1903.)

Briefly descriptive manual-flora of vascular plants, with somewhat awkward keys covering all taxa; limited synonymy; vernacular names; generalized indication of internal and extralimital range; miscellaneous notes on habitat, special features, etc.; indices. [Superseded, except for Louisiana and Arkansas, by the same author's *Manual of the southeastern flora* (next entry).]

SMALL, J. K., 1933. *Manual of the southeastern flora.* xxii, 1554 pp., illus. New York: The author. (Reprinted 1953, Chapel Hill: University of North Carolina Press; 1972, New York: Hafner (in 2 vols.).)

Illustrated, briefly descriptive manual-flora of seed plants; includes keys to all taxa, limited synonymy, vernacular names, generalized indication of internal and extralimital range, and miscellaneous notes; indices to all botanical and vernacular names. [For Pteridophyta east of the Mississippi, see below.]

WOOD, C. E., JR. *et al.*, 1958– . [Generic flora of the southeastern United States] 1– . *J. Arnold Arbor.* **39–71**, *passim* (1958–90; nos. 1–135); *J. Arnold Arbor.*

Suppl. Ser. **1** (1991; nos. 136–140); *Harvard Pap. Bot.,* 1(8)– , *passim* (1995– ; nos. 141–). Complemented by *idem*, 1983. Indexes to papers 1 to 100 published as parts of the *Generic Flora of the Southeastern United States. J. Arnold Arbor.* **64**: 547–563.

An irregular series of family (or tribal) revisions, with one or more in any given part (or 'paper'). Each treatment is written along the lines of *Die natürlichen Pflanzenfamilien*, with family description, discussion of morphological, anatomical, palynological, karyological and other attributes, classification, and keys to and treatments of the genera in the region, with figures of representative species. A rich selection of references is also furnished in each contribution. [As of writing (1999) 157 of a planned 184 treatments had been published, the latest in *Harvard Pap. Bot.* **4**(2): 365–416 (1999). No defined systematic sequence is followed.][112]

Woody plants

The works described below represent a selection from a larger literature, and should generally be useful over the region. However, Florida with its distinctive subtropical tree flora – omitted in two of the three works cited below – deserves special consideration: further references will thus be found under **163**.

COKER, W. C. and H. R. TOTTEN, 1945. *Trees of the southeastern states.* 3rd edn. 419 pp., illus. Chapel Hill, N.C.: University of North Carolina Press. (1st edn., 1934; 2nd edn., 1937.)

Illustrated semi-popular descriptive treatment of native and naturalized trees, with keys to all species, vernacular names, taxonomic commentary, and extensive notes on distribution, habitat, phenology, special features, etc.; remarks on special trees, list of larger shrubs; glossary, bibliography, and index to all names at end. [Inclusive of Virginia but not effectively covering southern Florida.]

DUNCAN, W. H., 1973. *Woody vines of the southeastern United States.* 87 pp., illus., maps. Athens, Ga.: University of Georgia Press. (Originally published 1967 in *Sida* 3(1): 1–79.)

Semi-popular treatment of the rich (for North America) woody vine flora, with keys, descriptive text, dot distribution maps, and illustrations; index.

DUNCAN, W. H. and M. B. DUNCAN, 1988. *Trees of the Southeastern United States.* xi, 322 pp., illus. (incl. 281 col.), maps. Athens, Ga.: University of Georgia Press. (Wormsloe Foundation Publ. 18.)

Descriptive treatment (306 species), with text in relatively plain language, keys, vernacular names, distribution maps (the dots more artistic than accurate), and extensive notes on regional and wider distribution, frequency, habitat, altitudinal range (metric!), seasons, and properties and uses; color illustrations on separate plates, with each depicting 3–4 species photographed mostly from fresh sprays; index to all names at end. The arrangement of species is artificial, in 11 groups (A–K; see general key, p. 22). The introductory section gives the geographical range of the work (with maps in inside covers; the limits are from Delaware through Kentucky to Texas southwards, save peninsular Florida), use of the work, notes on tree structure (with terminology), and a glossary; abbreviations, p. 15.

HARRAR, E. S. and J. G. HARRAR, 1946. *Guide to southern trees.* 709 pp., illus. New York: Whittlesey House/McGraw-Hill. (Reprinted with slight amendments and revised nomenclature, 1962, New York: Dover.)

Copiously illustrated semi-popular descriptive treatment of trees, with simple keys to all taxa, limited synonymy, vernacular names, and extensive remarks on occurrence, habitat, diagnostic features, variation, properties, uses, wood, etc.; glossary, list of references, and complete index. [Covers the entire region, including southern Florida.]

Pteridophytes

SMALL, J. K., 1938. *Ferns of the southeastern states.* 517 pp., illus. (incl. map). New York: The author. (Reprinted 1965, New York: Hafner.)

Fairly copiously descriptive illustrated treatment, with vernacular names, extensive critical and biological observations (with quotations from other workers), and notes on habitat and distribution; table of species of limited range, notes on cultivation, synoptic list with citations (by J. H. Barnhart), authorities, glossary, and general index to all names at end. The introductory part includes a general pictured-key (pp. 17–42) as well as an overview of the fern flora (the Blue Ridge and Caribbean islands being singled out as phytogeographical 'reservoirs'); a map of 'plant provinces' is also given.

WHERRY, E. T., 1964. *The southern fern guide.* 349 pp., illus., map. New York: Doubleday. (Reprinted with corrections and nomenclatural changes, 1972, New York: American Fern Society, New York Chapter.)

Illustrated field-guide to ferns and fern-allies; includes keys, descriptions, limited synonymy, vernacular names, indication of distribution, some critical commentary, and notes on habitat, occurrence, frequency, etc., and an etymological lexicon and indices at end. The introductory part includes chapters on morphology, life-cycles, cytology, classification, biogeography, and cultivation along with a glossary, references (list of fern floras, p. 35) and general keys (pp. 37–55). [A companion to the author's northeastern fern guide (see **140**). The 1972 version features additions and corrections in a table on p. 8.]

161

Tennessee

Area: 107003 km². Vascular flora: 2785 species and additional infraspecific taxa, of which 518 are introduced (Wofford and Kral, 1993, p. 66).

The first state survey was organized in 1831, but in the apparent absence of a botanical brief extensive floristic exploration did not get underway until Augustin Gattinger began his travels after mid-century (although before then the eastern mountains not unnaturally had attracted many outsiders, beginning with the elder Michaux). Gattinger's first general flora was *The flora of Tennessee, with special reference to the flora of Nashville* (1887); this was followed by his *Flora of Tennessee* (see below).[113] By the mid-twentieth century, however, the latter effectively was out of date. Studies towards a new state flora and related works began at the University of Tennessee after World War II under the direction of Aaron J. Sharp, and were initially consolidated in the form of two mimeographed lists: *A preliminary checklist of monocots in Tennessee* (1956) and *A preliminary checklist of dicots in Tennessee* (1960). Both of these included distributional information. Since then, a further 200 or more new state records have been documented. In more recent years a three-part programme has been developed, encompassing a revised checklist (published 1993; see below), a county atlas (1979–80, revised in 1993–97), and a descriptive flora. There are also separate works on woody plants and on ferns.

An idea of the current state of knowledge may be had from E. W. CHESTER (ed.), 1989. The vegetation and flora of Tennessee: proceedings of a symposium . . . *J. Tennessee Acad. Sci.* **64**(3): 57–207.

Bibliography

BATES, V., 1985. The vascular plants of Tennessee: a taxonomic and geographic guide to the floristic literature. *J. Tennessee Acad. Sci.* **60**(3): 66–76. [Includes theses.]

GATTINGER, A., 1901. *The flora of Tennessee and a philosophy of botany.* 296 pp., illus., portr. Nashville: Tennessee Department of Agriculture/Gospel Advocate Publishing Co.

Includes on pp. 1–184 an enumeration of vascular plants, with concise synonymy, very brief indication

of local range (including some specific localities), critical remarks, and notes on habitat, phenology, special features, cultivation (if applicable), uses, etc.; summary tables and index to generic names at end. The introductory section gives remarks on general features of the flora, floristic zones, and lists of plants in particular habitat types. [Now primarily of historical interest. The abbreviation 'OS', not defined in the text, stands for 'over the whole state'.]

WOFFORD, B. E. and R. KRAL, 1993. *Checklist of the vascular plants of Tennessee.* iii, 66 pp. (Sida, Bot. Misc. 10). Fort Worth, Tex.

An alphabetically arranged checklist, with limited synonymy and indication of biological and conservation status where appropriate, and detailed statistical summary at end. The introductory part encompasses short sections on history, goals and listed plants and includes a list of primary references for the state.

Distribution maps

CHESTER, E. W. *et al.*, 1993–97. *Atlas of Tennessee vascular plants.* 2 vols. 118, 242 pp., 2819 taxon maps. Clarksville, Tenn.: Center for Field Biology, Austin Peay State University. (Miscellaneous Publications 9, 13.) (Originally published as B. E. WOFFORD and A. M. EVANS, 1979–80. Atlas of the vascular plants of Tennessee, 1–3 (part 3 by B. E. Wofford alone). *J. Tennessee Acad. Sci.* 54: 32–38, 75–80; 55: 110–114.)

Unannotated atlas of county dot maps for each species and recognized infraspecific taxon (2819 in all; 906 pteridophytes, gymnosperms and monocotyledons in part 1, and 1913 dicotyledons in part 2, with introductions asterisked). Arrangement of species within the major groups follows that of Wofford and Kral (1993; see previous entry). Alphabetical indices (actually nomenclators with synonymy) appear in each part. The introductory section (pp. 3–7 in part 1 and a shorter version in part 2) provides background information, including the history of floristic investigations and the present state of statewide coverage, county and physiographic maps (pp. 6–7 in part 1), statistics (p. 8 in part 1 for non-dicotyledons, pp. 6–8 in part 2 for dicotyledons), a definition of the 'major works' (essentially corresponding to those given here), and comments to the effect that an atlas was superior to a manual-flora for handling a unit as floristically diverse as Tennessee; separate list of references.[114]

Woody plants

SHANKS, R. E., 1952. Checklist of the woody plants of Tennessee. *J. Tennessee Acad. Sci.* 27: 27–50.

Systematic list, with limited synonymy, vernacular names, and indication of local range; index to genera.

Pteridophytes

SHAVER, J. M., 1954. *Ferns of Tennessee, with the fern allies excluded.* xviii, 502 pp., 243 text-figs. Nashville, Tenn.: George Peabody College. (Reprinted 1970, New York: Dover, as *Ferns of the Eastern Central States.*)

Detailed descriptive treatment, with keys, synonymy, vernacular names, illustrations of all species, and extensive documentation of distribution (including *exsiccatae*); index. [One of the most thoroughly documented of all fern floras in North America and, as acknowledged by its 1970 reissue, useful well outside its nominal geographical limits.]

162

The Carolinas

Under this heading are included both North and South Carolina. – Area: 205 520 km² (North Carolina, 126 992 km²; South Carolina, 78 528 km²). Vascular flora: 3360 species (Radford *et al.*, 1968).

Floristic studies in the Carolinas effectively began with *Natural history of Carolina, Florida and the Bahama Islands* by Mark Catesby (1730–47; reissues 1754, 1771). Key additional contributions in the century following were *Flora caroliniana* (1788) by Thomas Walter and *A sketch of the botany of South Carolina and Georgia* (1816–24) by Stephen Elliott. Afterwards, in South Carolina the *Sketch* was not superseded until the mid-twentieth century, while in North Carolina there appeared two successive enumerations: *Catalogue of the indigenous and naturalized plants of the State* or *Botany of North Carolina* (1867) by Moses A. Curtis, a result of the second North Carolina Geological Survey (1852–64), and *Flora of North Carolina* (1899) by Charles W. Hyams.[115]

Work towards a new flora covering both North and South Carolina began seriously after World War II. Key precursors include *Guide to the vascular flora of the Carolinas* (1964) by A. E. Radford, H. E. Ahles and C. R. Bell, and the same authors' *Atlas of the vascular flora of the Carolinas* (1965). In 1968 they were succeeded by the *Manual*, still the best in the Southeast.

Bibliography

HARDIN, J. W. and J. D. SKEAN, JR., 1982. *Guide to the literature on plants of North Carolina.* 2nd edn. 22 pp. Raleigh:

North Carolina Agricultural Extension Service. (1st edn. by J. W. Hardin and C. B. McDonald, 1975, Raleigh.) [Mainly post-1930 references, arranged by author under 14 subject headings and concisely annotated. Floras are grouped in section I, 'Native Wildflowers' (including 'Weeds and other Vascular Plants'). Monographic and revisionary literature is **not** included.]

RADFORD, A. E., H. E. AHLES and C. R. BELL, 1968. *Manual of the vascular flora of the Carolinas.* lxi, 1183 pp., illus., maps. Chapel Hill, N.C.: University of North Carolina Press. (Nine printings to 1992.)

Illustrated descriptive manual of vascular plants; includes keys to all species, essential synonymy, vernacular names, concise indication of local and extralimital range (with dot maps for all species), karyotypes, phenology, figures of diagnostic features along with representative species, and notes on frequency and habitat; list of authors and index to all vernacular and botanical names. The introductory section includes historical remarks, a summary of the flora and a county map.[116]

163

Florida (in general)

Area: 140 798 km². Vascular flora: *c.* 3834 native and naturalized species, of which 2654 native (Wunderlin, 1998; see below). – For works dealing only with the (sub)tropical southern part of the state, see **164**.

The Floridan 'Panhandle' was, following annexation of the territory by the United States in 1819, fairly quickly integrated into the 'South'. Moreover, in the latter half of the nineteenth century the area was home to Alvan Chapman, as already noted a leading authority on the Southeastern flora. Peninsular Florida on the other hand remained almost until the opening of the twentieth century essentially an *Explorationsgebiet*, with floristic knowledge largely embodied in expedition and survey reports by outside workers. An effective state survey was organized in 1907 (following abortive initiatives in the nineteenth century) but did not include natural history in its brief. Attempts to consolidate knowledge of the rich flora remained largely localized until after World War II, although not unnaturally with improved access from the late 1800s the south and the Keys attracted particular attention, with in particular a number of works produced by J. K. Small.

After World War II, Daniel Ward at the University of Florida initiated a statewide checklist. This appeared as D. B. WARD, 1968. *Checklist of the vascular flora of Florida.* Part 1. 72 pp. (Tech. Bull. Univ. Florida Agric. Exp. Sta. 726). Gainesville, Fla. Only the non-dicotyledonous vascular plants were, however, ever covered. Associated with this was his *Contributions to the flora of Florida*, eight parts (2–9) of which appeared *passim* in *Castanea* **28** (1962)–**40** (1975); the introductory first part was deliberately postponed. In 1977 the same author began another series, *Keys to the flora of Florida*; only one installment was published (in *Phytologia* **35**(6)), comprising the plan of the work, a state map, and definition of 19 vegetation associations – effectively what would have been the first part of the *Contributions*.

Given the size of the state and its floristic and other discontinuities, the preparation of regional accounts has proved a more effective way forward.[117] These begin with *A flora of tropical Florida* (1971; see **164**) and were continued with a work from Richard P. Wunderlin (1983) on the central part and one by Andre F. Clewell (1985) on the Panhandle (the last considered by one reviewer to be the best of these three). A similar work for the northeast, an area bounded by Gainesville, St. Augustine and Jacksonville and including much of the St. John's River basin, would have completed state coverage; however, in 1998 Wunderlin published a keyed enumeration for the whole state, *Guide to the vascular plants of Florida* – the first statewide account. This is complemented by an electronic atlas of distribution maps.

Among works on particular groups, we include here *Trees of Florida* by Gil Nelson, *Native trees of Florida* by West and Arnold, *Shrubs of Florida* by Small and *Ferns of Florida* by O. Lakela and R. Long.

Bibliography
WHITTIER, H. O., [1982]. *Florida botanical/ecological bibliography.* [iv], 278, [1] pp. Winter Park, Fla.: Florida Native Plant Society/Florida Conservation Foundation. [Unannotated list, arranged in the first instance by authors; pp. 107–278 encompass 29 mainly popularly oriented subject headings (with general flora in the first). No monographs or revisions are included (for these, Clewell's Panhandle flora, given below, is the best source).]

Keys to families
ZOMLEFER, W. B., 1989. *Flowering plants of Florida: a guide to common families.* xi, 207 pp., 81 text-figs., illus. glossary. Gainesville: The author. [Teaching text, with keys and references covering 73 families. The Thorne system, as published in *Nordic J. Bot.* **3**: 85–117 (1983), is adopted.] An earlier version of this work appeared as *idem*, 1983. *Common Florida angiosperm families*, 1. iii, 107 pp., 41 figs. [Gainesville]: The author, and *idem*, 1986. *Common Florida angiosperm families*, 2. vi, 106 pp., 37 figs. Gainesville: Biological Illustrations.

WUNDERLIN, R. P., 1998. *Guide to the vascular plants of Florida.* ix, 806 pp., illus. (title-page), map (inside front cover). Gainesville: University Press of Florida.

Keyed enumeration of vascular plants with essential synonymy, vernacular names, and indication of habitat, distribution and phenology; no illustrations or critical commentary. In an appendix (pp. 635–662) are a synopsis (p. 663; 3834 species in the flora of which 2654 native and 1180 introduced), glossary, and indices to vernacular and scientific names (with family index in inside back cover and end-paper). The relatively brief introduction relates to the genesis and plan of the work, with references on pp. 5–6. For the associated *Atlas*, see below under **Distribution maps**.

Distribution maps
WUNDERLIN, R. P., B. F. HANSEN and E. L. BRIDGES, 1996. *Atlas of Florida vascular plants.* 1 CD-ROM. Tampa: Institute for Systematic Botany, University of South Florida. (Electronic publication.)[118]

Comprises county maps for each species along with taxonomic, distributional, ecological and status data; includes search and reporting capabilities. [For Windows 3.1, Windows 95 or Macintosh; not for use with Windows NT.]

Partial works
CLEWELL, A. F., 1985. *Guide to the vascular plants of the Florida Panhandle.* 616 pp., illus., map. Tallahassee, Fla.: University Presses of Florida/Florida State University Press.

Manual-key to vascular plants (2359 species in an area of 38 628 km²), with limited synonymy, vernacular names, indication of distribution, some diagnostic figures, and habitat and other notes; synopsis, addenda and indices to vernacular and scientific names (a family index appears in the back end-papers). The introductory part gives the background and scope of the work, methodology, practicalities, descriptions of physical features and habitat and vegetation types, references and an illustrated glossary (general keys to families are given at the beginning of each major plant group; within these latter families and genera are alphabetically arranged, with separate descriptions). [The area covered is northwestern Florida west of the Suwanee River, within which are located Tallahassee, the state capital, and the lower Apalachicola River.]

WUNDERLIN, R. P., 1983. *Guide to the vascular plants of central Florida.* 480 pp. Tampa, Fla.: University Presses of Florida. (A Florida Southern University Book.)

A keyed enumeration of vascular plants (2197 species in 68 749 km², 49 percent of the state's area); entries include essential synonymy, vernacular names, generalized local distribution (with sometimes mention of specific counties), and notes on frequency, habitat and phenology; glossary and indices to vernacular and scientific names (a family index appears in the end-papers). The introductory part covers the plan and scope of the work and includes a statistical synopsis (pp. 6–11) and keys to families (pp. 12ff.) [The work is bounded on the south by 'tropical' Florida (**164**) and by the Suwanee River in the northwest; in all it covers 30 counties within which are found such places as Palm Beach, Sarasota, Tampa/St. Petersburg, Orlando (and Disney World), and Daytona Beach.]

Woody plants

The coverage of Floridan trees in HARRAR and HARRAR (**160**) is nearly as extensive as in West and Arnold's work. For distribution, see also volume 5 (Florida) of LITTLE's *Atlas of United States trees* (**100**). Regional treatments include *Trees of northern Florida* (1962) by H. Kurz and R. K. Godfrey (revised and expanded 1988 by Godfrey as *Trees, shrubs and woody vines of northern Florida and adjacent southern Georgia and Alabama*) and *Trees of central Florida* (1980) by Olga Lakela and R. P. Wunderlin. For southern Florida, see **164**.

SMALL, J. K., 1913. *Shrubs of Florida.* x, 140 pp. New York: The author.

Field-manual to native and naturalized species (in a handy small format), with keys, limited synonymy, local names (where known), generalized indication of distribution with some details, and notes on special features, etc.; index. [Nomenclature follows the now-obsolete 'American Code'. The author's companion *Florida trees* (1913) has been superseded by the following work.]

NELSON, G., 1994. *The trees of Florida.* 338 pp., 83 pls. (line drawings), 153 col. photographs, map. Sarasota, Fla.: Pineapple Press.

Illustrated *pro forma* guide with for each of 342 species vernacular and scientific names as well as statements about form, leaves, flowers, fruit, diagnostic features, and distribution; bibliography (pp. 317–321) and index to all names at end. The general part includes general introductions to all families concerned along with a glossary, but no analytical keys.[119]

WEST, E. and L. ARNOLD, 1956. *The native trees of Florida.* 2nd edn. xx, 218 pp., illus. Gainesville, Fla.: University of Florida Press. (1st edn., 1946.)

Illustrated descriptive atlas-treatment, with keys (mostly to genus only), local names, general and specific indication of distribution, and commentary, including notes on uses; county maps in end-papers; glossary, references and index. Less common or 'borderline' trees are merely noted.

Pteridophytes

Lakela and Long's *Ferns of Florida* is a successor to *Ferns of Florida* by J. K. Small (1931, New York).

LAKELA, O. and R. W. LONG, 1976. *Ferns of Florida.* xiii, 178 pp., 117 text-figs. (incl. photographs). Miami: Banyan Books.

Semi-popular illustrated descriptive treatment (covering 135 species), with keys, indication of distribution, local names, synonymy, commentary, etc.; glossary and index. No *exsiccatae* are cited.

164

Southern Florida

This unit relates particularly to Florida south of 27°, an area characterized by an Antillean tropical and subtropical flora unique within the conterminous United States. It corresponds to the 10 counties from Martin, Glades and Charlotte southwards and is inclusive of Lake Okeechobee. – Area: 31 619 km². Vascular flora: 1647 species and 190 non-nominate infraspecific taxa (Long and Lakela, 1971; see below).

The main predecessors of *Flora of tropical Florida* are *Flora of Miami* and *Flora of the Florida Keys* (1913), both by John Kunkel Small, but Long and Lakela's work is the first for the whole region.

LONG, R. W. and O. LAKELA, 1971. *A flora of tropical Florida.* xvii, 962 pp, text-figs., map, portr. Coral Gables: University of Miami Press. (Reprinted 1976, Miami: Banyan Books.)

Descriptive manual of native, naturalized and commonly cultivated vascular plants of the 10 counties of southern peninsular Florida and the Florida Keys, with limited synonymy, vernacular names, generalized indication of local and extralimital range, figures of representative species, and notes on habitat, phenology, special features, etc.; list of authors, glossary, and index to all botanical and vernacular names. The introductory section gives an account of botanical exploration (pp. 3–9, by J. Ewan) together with chapters on geology, plant communities, and the origin of the flora; included also are statistical tables and a select list of (recent) references. [Covers all of Lake Okeechobee and Martin, Glades and Charlotte Counties southwards.][120]

Woody plants (including trees)

TOMLINSON, P. B., 1980. *The biology of trees native to tropical Florida.* v, 480 pp., illus. [Petersham, Mass.]: The author.

A morphologically and biologically oriented illustrated systematic study of the tree flora, with keys to all species, descriptions, indication of distribution, and discussion; checklist, glossary, artificial key to genera and index at end. The introductory section (with references topically divided) covers the parameters of the work, a discussion of the extent of 'Tropical Florida', physical features and climate, main vegetation types, kinds of distribution, morphological concepts (pp. 29ff., including definitions of 'prolepsis' and 'syllepsis'), phenology, and floral types.

Partial work: Florida Keys

The following florula, encompassing the keys (cays) from Key West to the Dry Tortugas (the home of abandoned Fort Jefferson), is of historical importance in the study of plant demographics.[121]

MILLSPAUGH, C. F., 1907. *Flora of the sand keys of Florida.* Pp. 191–245, maps (Publ. Field Mus. Nat. Hist., Bot. Ser. 2(5)). Chicago.

A series of florulas, one for each cay, with very precise indication of species distribution; also includes notes on individual species dynamics and (pp. 242–243) a summary table of local distribution. [Based on field work carried out in 1904. For a detailed revision, see D. R. STODDART and F. R. FOSBERG, 1981. *Topographic and floristic change, Dry Tortugas, Florida, 1904–1977.* [i], 55, [17] pp., 11 figs. (incl. maps), 11 pls. (Atoll Res. Bull. 253). Washington, D.C.]

165

Georgia

Area: 150946 km². Vascular flora: 3686 species and non-nominate infraspecific taxa (Duncan and Kartesz, 1981); 3049 species (Coile and Jones, 1985; for both see below).

The first important consolidation of knowledge of the flora of Georgia is embodied in Stephen Elliott's *Sketch* (see **162**). This was followed in 1849 by an anonymously authored checklist (actually by John L. Le Conte) in *Statistics of the state of Georgia* by George White, accounting for 2058 species; in the absence of any state natural history survey, it remained for 130 years the last of its kind (with a reprint in 1972). In 1979 the original version of *The Georgia Plant-list* by Nancy Coile and Samuel B. Jones, Jr. appeared; in quick succession it was followed by *Vascular flora of Georgia* (1981) by Wilbur Duncan and John T. Kartesz, the Georgia Botanical Society distribution atlas (1984), and a revision of the *Plant-list* (1985). All of these have laid a good foundation for any descriptive flora. Among partial works the most valuable is the enumeration for the southwestern counties by Robert Thorne.

Pteridophytes have been fully treated twice, most recently in *Field guide to the ferns and other pteridophytes of Georgia* (1986; see below).

Bibliography

See also MELLINGER (1984) below.

VENARD, H., 1969. Bibliography of Georgia botany. *Castanea* **34**: 267–306. [Annotated; arranged by author. Coverage is through the 1950s.][122]

COILE, N. C. and S. B. JONES, JR., 1985. *The Georgia plant list: a list of the Georgia plants in the University of Georgia Herbarium.* 2nd edn. [1], 69 pp. Athens, Ga.: University of Georgia Herbarium. (1st edn., 1979, by S. B. Jones, Jr. and N. C. Coile.)

Alphabetically arranged name list with vernacular equivalents (where known) and some synonymy. The brief introduction covers statistics of the flora and conventions (nomenclature follows the Soil Conservation Service checklist (**100**)). [Based on the University of Georgia Herbarium and accounting for 3049 species.]

DUNCAN, W. H. and J. T. KARTESZ, 1981. *Vascular flora of Georgia; an annotated checklist.* xi, 147 pp., map. Athens, Ga.: University of Georgia Press.

Comprises a briefly annotated systematic checklist of vascular plants (arranged following the Engler and Prantl system), with some synonymy and indications of geographical occurrence (by botanical provinces; see map in frontispiece); index to families at end. The introductory part covers the background and plan of the work, vegetational elements in the state, a brief statistical summary, and literature list. [An outgrowth from materials collected by H. Venard (compiler of a state bibliography; see above).]

Distribution maps

JONES, S. B., JR. and N. C. COILE, 1988. *The distribution of the vascular flora of Georgia.* 230 pp., numerous maps. Athens, Ga.: Department of Botany, University of Georgia.

Preface (nomenclature based on the U.S. Soil Conservation Service national checklist (see **100**)); alphabetically arranged county dot distribution maps (six to a page) with indication of presence/absence of any given taxon; general references on pp. 215–217 and index to genera and families at end.

MELLINGER, M. B. (comp.); H. L. WHIPPLE (ed.), 1984. *Atlas of the vascular flora of Georgia.* Unpaged; 2697 distribution maps (and 1 map in inside front cover). Millidgeville, Ga.: Studio Designs Printing (for the Georgia Botanical Society).

Atlas of county dot maps, one for each species and arranged 12 to a page with indication of scientific and vernacular names; bibliography (including published and unpublished local plant lists), dedication, and indices to vernacular and scientific names (including map references). [Though not stated, arrangement of species follows the Englerian system. A star covering a given map stands for statewide distribution.]

Partial work

THORNE, R. F., 1954. The vascular plants of southwestern Georgia. *Amer. Midl. Naturalist* **52**: 257–327, 2 maps.

Regional 17-county checklist of vascular plants (1747 species, of which 1539 native), based on floristic surveys by the author (as part of dissertation research) in 1946–49; includes essential synonymy, general and more specific indication of distribution, and notes on habitat, frequency, etc., with occasional references (cf. pp. 326–327). A fairly substantial introductory part covers physical features, geology, climate, soils, vegetation, human factors, floristic elements, and botanical history. [Covers some 12 950 km² (5000 mi.²) in two major physiographic zones.]

Pteridophytes

The atlas-guide by Snyder and Bruce succeeds *Ferns of Georgia* (1951) by R. McVaugh and H. Pyron.

SNYDER, L. H., JR. and J. G. BRUCE, 1986. *Field guide to the ferns and other pteridophytes of Georgia.* ix, 220 pp., illus., maps. Athens, Ga.: University of Georgia Press.

Semi-popular atlas-guide with text and facing illustrations (covering 119 species, an increase of 41 over McVaugh and Pyron); each text includes scientific and vernacular names, derivation of scientific names, description, distribution (with county dot map), extralimital range, frequency and habitat; alphabetical checklist of species and hybrids (12 of the latter), glossary and references (pp. 265–266). The introductory section includes an introduction to ferns and 'allied' plants and a discussion of those in Georgia (including two maps) along with a non-dichotomous key to genera (pp. 10–12).

166

Alabama

Area: 131 994 km². Vascular flora: 2476 species of seed plants (Mohr, 1901; see below).

The first state survey was organized in 1848 consequent to the appointment of Michael Tuomey to the geology chair at the University of Alabama; however, only with a reorganization following Reconstruction was a botanical component included under Tuomey's successor Eugene A. Smith. With a view to renewal of collections and other resources destroyed in 1865, formal collaborative arrangements with U.S. federal bodies were also effected. In 1880 the Geological Survey released *A preliminary list of plants growing without cultivation in Alabama* by Mobile physician Charles Mohr, incorporating collections by him as well as by Smith. Mohr continued botanical work under Survey auspices until his death in 1901; in that year his major work, *Plant life of Alabama*, appeared under joint state and federal auspices.[123] In subsequent decades the Alabama Geological Survey published works on forests and the woody flora by Roland W. Harper, a leading regional botanist, student of vegetation and collector from 1900 until the 1940s.

In addition to Mohr's still-definitive work, two recent treatments respectively on woody plants and of pteridophytes are accounted for here.

MOHR, C. T., 1901. *Plant life of Alabama.* 921 pp., 13 pls., incl. map (Contr. U.S. Natl. Herb. 6). Washington, D.C.: U.S. Department of Agriculture. (Also published in an 'Alabama Edition' with added biographical sketch (as pp. i–xii) and portrait, 1901, Montgomery, Ala.: Alabama Geological Survey. Reprinted 1969, Lehre, Germany: Cramer, as *Reprints of US-floras* 2.)

Detailed, non-descriptive flora of non-vascular and vascular plants, without keys; includes synonymy (with references and some citations), vernacular names, fairly extensive indication of local range, summary of extralimital distribution, and critical remarks together with notes on habitat, life-form, frequency, phenology, special features, etc.; catalogue of cultivated plants, statistical summary and index to all vernacular and botanical names and subjects. The extensive introductory section features accounts of physiography, climate, botanical exploration, floristics, phytogeography, vegetation, and agriculture. [Still valuable but now in need of extensive revision (Clark, 1971; see below).]

Woody plants

Clark's account of woody plants partially succeeds R. M. Harper, 1928. *Economic botany of Alabama*, 2: *Catalogue of the trees, shrubs and vines of Alabama*. 357 pp., 66 figs., 23 maps (Monogr. Alabama Geol. Surv. 9). University, Ala.

Clark, K. C., 1971 (1972). The woody plants of Alabama. *Ann. Missouri Bot. Gard.* 58: 99–242, maps.

Systematic enumeration, with keys, relevant synonymy, general and specific indication of distribution and physiographic provinces, and commentary; county dot maps for each species; list of references. An introductory section accounts for the history of Alabama botanical work, sources of information, descriptions of physical features, climate, soils, and natural regions, and discussion of plant distribution and floristics.

Pteridophytes

Dean, B. E., 1969. *Ferns of Alabama*. Revised edn. xxiv, 214 pp., illus. Birmingham: Southern Universities Press. (1st edn., 1964.)

Illustrated semi-popular treatment, with keys, synonymy, local names, and indication of distribution (by counties for less common species); index. A general introduction is also provided.

167

Mississippi

Area: 122 806 km². Vascular flora: no data.

As in Alabama, the state survey from an early date included in its brief natural history and agriculture as well as geology; it, too, was associated with the state university from the latter's establishment in 1850. In 1921 it published the only statewide flora, the still-current but now outdated enumeration by Ephraim Lowe.

In recent years, much new information has become available, particularly through the work of S. B. Jones, Jr., and T. M. Pullen. The former has been reported as having undertaken preparation of a new

Guide to the flora of Mississippi to succeed Lowe's work; towards this end he produced, with collaborators, some statewide revisions. C. T. Bryson has also published many additions and other changes to the flora, chiefly in *Phytologia* and *Sida*.

Bibliography

Stevens, M. P., 1991. *Botanical literature of Mississippi: a taxonomic, geographic and subject guide, 1854–1984*. 50 pp., 2 maps. Jackson, Miss.: Mississippi Museum of Natural Science. [Spiral-bound unannotated list of 327 items arranged by author with nomenclatural, geographical and author indices. The maps depict county boundaries and physiographic regions.]

Jones, S. B., Jr., 1974–76. Mississippi flora, I–VI. *Gulf Res. Rep.* 4(3): 357–379 (1974) (I); 5: 7–22 (1975) (IV); *Castanea* 39: 370–379 (1974) (II); 40: 238–253 (1975) (III); 41: 41–58, 189–212 (1976) (V–VI).

Comprises miscellaneous contributions (including family revisions) towards a new state flora, with keys and county distribution maps. For a related paper on pteridophytes, with keys, indication of distribution by Lowe's physiographic zones, and county maps, see A. M. Evans, 1978. Mississippi flora: a guide to the ferns and fern allies. *Sida* 7: 282–297.

Lowe, E. N., 1921. *Plants of Mississippi. A list of flowering plants and ferns*. 292 pp., map (Mississippi State Geol. Surv. Bull. 17). Jackson.

Enumeration of vascular plants, with synonymy, vernacular names, general indication of local range (with some citations and other details), and brief notes on habitat, phenology, etc.; county map and gazetteer; no index. The introductory section includes accounts of life-zones (Merriam system), ecological types, habitat factors, and floristic and topographic regions in the state. The work was to a large extent based on Mohr's *Plant life of Alabama*. [For substantial additions, see T. M. Pullen *et al.*, 1968. Additions to the vascular flora of Mississippi. *Castanea* 33: 326–334; and L. C. Temple and T. M. Pullen, 1968. A preliminary checklist of the Compositae of Mississippi. *Ibid.*: 106–115, map. – The main paper, including county distributions and based on over 29 000 collections made from 1963 through 1967, features 150 taxa not in Lowe's work.]

168

Arkansas

Area: 135 403 km². Vascular flora: 2510 species (Smith, 1994; see below).

A state survey of relatively liberal scope was instituted in 1857; under its auspices the first checklist, by the noteworthy paleobotanist Leo Lesquereux, was produced as part of his *Botanical and palaeontological report on the Geological State Survey of Arkansas* (1860). A later survey was responsible for its successor, *A list of the plants of Arkansas* (1891) by John C. Branner (the state geologist) and his botanist, Frederick V. Coville; together with *Supplement to the Catalogue of Arkansas Plants* (1926) by John T. Buchholz and E. Jesse Palmer, issued through the St. Louis Academy of Sciences, this annotated catalogue long remained standard.[124] An additional catalogue was produced in 1943 by Delzie Demaree (in *Taxodium* 1(1): 1–88) which, however, had many errors. All have effectively been superseded by Edwin B. Smith's annotated atlas/checklist, now in its second edition (1988); more recently, it has been joined by a set of keys. Smith, however, has indicated that the flora remains 'rather poorly known'.

Bibliography

See also Smith (1988, pp. 438–450).

PECK, J. H. and C. J. PECK, 1988. A bibliographic summary of Arkansas field botany. *Proc. Arkansas Acad. Sci.* **42**: 58–73.

SMITH, E. B., 1988. *An atlas and annotated list of the vascular plants of Arkansas.* 2nd edn. iv, 489 pp., maps. Fayetteville, Ark.: distributed by University of Arkansas Bookstore. (Reproduced from typescript. 1st edn., 1978.)

An alphabetically arranged atlas/checklist comprising county dot maps, one for each species (12 maps per page), with facing annotated text; the latter includes synonymy, vernacular names, and notes on distribution, occurrence, karyotypes (with vouchers), origin of aliens, etc., along with some critical remarks; references to revisions/monographs and floristic papers scattered throughout text. Collecting and other statistics are given at the end along with possible additions, a list of excluded names, an addendum (pp. 429–433) and a bibliography (pp. 438–450). An 'R' on a map indicates a report, for which vouchers have not been seen.

SMITH, E. B., 1994. *Keys to the flora of Arkansas.* xii, 363 pp. Fayetteville, Ark.: University of Arkansas Press.

Comprises analytical keys to 2510 indigenous and over 300 introduced species; in addition there are tables of families arranged alphabetically and in the Cronquist sequence, a glossary, and indices to vernacular and scientific names of genera and families.

Woody plants

A guide by Gary E. Tucker, with keys, was presented as a dissertation at the University of Arkansas, Fayetteville, in 1976. Save for a checklist of taxa (1979), however, it has remained unpublished; the following is an independent undertaking.

HUNTER, C. G., 1989. *Trees, shrubs and vines of Arkansas.* viii, 216 pp., text-fig., 300 col. photographs. Little Rock, Ark.: Ozark Society Foundation.

A semi-popular color photographic atlas with accompanying text, inclusive of vernacular names. The introductory part encompasses Arkansas botanical history and family descriptions, while at the end are a glossary, references and a complete index.

Pteridophytes

TAYLOR, W. C., 1984. *Arkansas ferns and fern-allies.* [vi], 262 pp., illus. Milwaukee, Wis.: Milwaukee Public Museum.

Generously formatted atlas-flora, with each recognized species or hybrid allocated a page of text (with map) and one for a line drawing; text includes scientific and vernacular names, a short description, local and extralimital distribution (with a sample *exsiccata* given if rare or local), habitat, and extensive taxonomic, cytological and other notes, while the maps show distribution by counties. Associated with each map is a list of all *exsiccatae* seen with their place(s) of deposit. Keys to species are given under generic headings, and a glossary, phytogeographic notes and index to all names conclude the work.[125]

169

Louisiana

Area: 115 755 km². Vascular flora: the most recent figure, based on entries in Thomas and Allen (1993–98; see below), is 3249 native and introduced species. A total of 1800 were known by 1900 (as estimated by Cocks; see Note 129).[126]

Although botanical work in the state dates back to the era of French rule and the investigations of Le Page du Pratz (1758), the large size of the flora as presently known, a considerable decentralization of herbarium resources (10 of value within the state alone) and, in the past, a lack of strength therein, have all been impediments to realization of a modern state flora.

There are two significant early accounts: *Flora ludoviciana* (1817) by C. S. Rafinesque (a revision of C. C. Robin's *Flore louisianaise*, a part of the third and final volume (1807) of that author's *Voyage dans l'intérieur de la Louisiane*), and *Catalogus florae ludovicianae* (1852) by J. L. Riddell (a migrant from the Midwest in 1836).[127] In 1864, a formal state survey was organized; however, it initially did not include natural history. In 1871 and 1872, however, Americus Featherman produced two regional floristic checklists in successive survey reports. Operations of the survey were afterwards suspended for 20 years as economic depression set in. Additions to the flora nevertheless continued at a rapid rate, in 1900 causing Reginald Cocks in New Orleans to postpone publication of a projected state checklist.[128] It would be another eight decades before the first modern statewide accounts appeared, made possible by substantial new collections accumulated after World War II. Only the pteridophytes and woody plants were in the meantime fully documented in two works by Clair Brown at Louisiana State University (the former together with Donovan Correll).

Of the current works, the *Atlas* of Thomas and Allen (1993–98) is perhaps the most definitive. None, however, has encompassed *all* collections and other resources.[129]

Bibliographies

In addition to the items given below, attention is here drawn to the annotated references on pp. 130–149 of *Vascular plants of Louisiana* by D. T. MacRoberts (1984; see below).

EWAN, J., 1967 [1968]. A bibliography of Louisiana botany. *Southwestern Louisiana J.* 7: 1–83 (as *Flora of Louisiana* (ed. J. W. Thieret and collaborators), 1). Continued as E. SUNDELL, 1979. *A bibliography of Louisiana botany 1951–1975.* 66 pp. (Tulane Stud. Zool. Bot. 21(1)). New Orleans. [Both works comprise chronologically arranged, annotated lists of titles (433 in the second), with subject and author indices. The first covers all known works through 1950; the second was compiled under Ewan's supervision.][130]

MACROBERTS, D. T., 1984. *The vascular plants of Louisiana: an annotated checklist and bibliography of the vascular plants reported to grow without cultivation in Louisiana.* 165 pp. (Bull. Mus. Life Sciences, Louisiana State Univ., Shreveport 6). Shreveport, La.

Annotated, alphabetically arranged, mainly literature-based checklist; for each species synonymy and sources of information are given, with notes (inclusive of *exsiccatae* where necessary) on questionable or otherwise critical taxa or records given at the end of each family; references (pp. 130–149) and index to generic names at end. The introductory part covers the genesis of and basis for the work, nomenclature, density of coverage (including a map of parishes (i.e., counties) with checklists), a somewhat acerbic history of botanical work in the state including a discussion of the Thieret and Thomas compilations, and (pp. 21–22) statistics.

THIERET, J. W., 1972. *Checklist of the vascular flora of Louisiana, 1: Ferns and fern-allies, gymnosperms and monocotyledons.* 48 pp. (Techn. Bull. Lafayette Nat. Hist. Mus. 2). Lafayette, La. Complemented by R. D. THOMAS and C. M. ALLEN, 1982. *A preliminary checklist of the dicotyledons of Louisiana.* [i], 130 ll. (Contributions of the herbarium of Northeast Louisiana University 3). Monroe, La.

Thieret's work is an unannotated checklist, with synonymy related to that given in major floristic works in or of particular concern to the state. Nomenclature follows the Texas manual of Correll and Johnston (**171**). A brief introductory section gives the plan of and basis for the work (only vouchered records included), and a statistical table. The dicotyledons are covered by Thomas and Allen; their work, like that of Thieret, is an alphabetically arranged name list, with synonymy (in parentheses).

Distribution maps

MACROBERTS, D. T., 1989. *A documented checklist and atlas of the vascular flora of Louisiana, I–III.* 3 parts. Maps. (Bull. Mus. Life Sciences, Louisiana State Univ., Shreveport 7–9). Shreveport, La.

An atlas of parish (county) dot maps (eight to a page) with citations of *exsiccatae* and/or commentary facing; indices to families and genera in each volume. Part I covers all groups except dicotyledons; parts II and III cover dicotyledons. The first part also contains a list of references additional to those in the author's 1984 work. [Based primarily upon literature references.]

THOMAS, R. D. and C. M. ALLEN, 1993–98. *Atlas of the vascular flora of Louisiana.* 3 vols. Maps. Baton Rouge:

Louisiana Department of Wildlife and Fisheries, Natural Heritage Program.

An atlas of parish (county) dot maps (12 per page) with facing text featuring accepted scientific and vernacular names, common synonymy, up to five (or more) validated *exsiccatae* with place of deposit, and status (native or introduced); list of excluded taxa, literature references and index at end. The introductory part (vol. 1) gives the plan and scope of the work (including statistics) along with statewide geological, soil and vegetation maps. [Families within the major groups are alphabetically arranged. Vol. 1, on non-dicotyledonous vascular plants, encompasses 1069 species and infraspecific taxa. Vol. 2 (1996) covers 1005 species of dicotyledons from Acanthaceae to Euphorbiaceae; vol. 3, 1175 species from Fabaceae through Zygophyllaceae. The work was based almost entirely upon holdings in eight larger Louisiana herbaria, surveyed during 1991–92.]

Woody plants
In addition to the works below, the following may be consulted: C. A. BROWN, 1964. *Woody plants of Louisiana, native, naturalized and cultivated: check list.* 16 pp. (Louisiana Forest. Commiss. Bull. 8). Baton Rouge, La. (1st edn., 1957.)

BROWN, C. A., 1945. *Louisiana trees and shrubs.* 262 pp., 147 halftones, 1 col. pl., map (Louisiana Forest. Commiss. Bull. 1). Baton Rouge, La.

Semi-popular illustrated descriptive treatment of trees and shrubs, without keys; includes limited synonymy, vernacular names, generalized indication of local range, critical remarks, and notes on habitat, biology, uses, etc.; glossary, list of references, and complete index at end. An introductory section includes among other topics details (with a map) of tree regions. [For nomenclatural and other revisions see R. D. THOMAS and C. M. ALLEN, 1981. *A checklist of the woody plants of Louisiana.* 59 pp. (Contributions of the herbarium of Northeastern Louisiana University 2). Monroe, La.]

Pteridophytes
Thieret's 1980 work was a successor to *Ferns and fern allies of Louisiana* (1942) by C. A. Brown and D. S. Correll, the first 'fern flora'.

THIERET, J. W., 1980. *Louisiana ferns and fern allies.* vi, 123 pp., 75 extra pls. (Flora of Louisiana 2). Lafayette, La.: Lafayette Natural History Museum/University of Southwestern Louisiana.

Semi-popular illustrated treatment (73 species and hybrids) in two sections; the taxonomic part contains keys, vernacular names, indication of distribution, critical commentary, and notes on habitat, biology, distinguishing features, etc. (but no formal species descriptions), while the atlas comprises county dot maps of each species along with illustrations. The introductory section covers background, sources and the plan and scope of the work, an introduction to

pteridophytes, collecting procedures, rare species, and earlier studies, while at the end (pp. 102ff.) are a synopsis, statistics, glossary, bibliography and general index to all names.

Region 17

Central Plains United States and Texas

In this region are the states of Texas, Oklahoma, Kansas, Nebraska, South Dakota, and North Dakota. The four northern states comprise the Northern Central Plains subregion (**175**), the core territory for three major works which collectively extend west to the Rockies, east to the woodland 'ecotone', and south into Oklahoma, i.e., the total extent of the northern Great Plains.

The Plains region and Texas encompass the cores of two significant floristic areas, respectively the northern Plains and the Texan *llanos* with, in Oklahoma, a transitional belt. The work of botanical circles largely has conformed to this pattern, the result being that there never has been a flora for the plains as a whole.[131] Such has, however, been more easily achieved in the biotically less diverse north, firstly through total or partial inclusion in eastern floras and then, successively, in Per Axel Rydberg's *Flora of the prairies and plains of central North America* (1932) – the last work of that prominent early Western botanist – and the more definitive works of the Great Plains Flora Association (1977, 1986; see **175**).[132] The two latter are accompanied by *Woody plants of the North Central Plains* (1973; also under **175**) by H. A. Stephens. In Texas, an outstanding event has been the publication of *Manual of the vascular plants of Texas* (1970) by Donovan Correll with Marshall C. Johnston, in part distilled from the larger *Flora of Texas* begun in the 1940s by C. L. Lundell with Correll. The *Manual*, the first statewide flora, effectively succeeded Small's *Flora of the southeastern United States* (for east Texas) and Coulter's *Botany of western Texas*, the latter long a standard for areas west of Dallas/Fort Worth. Several editions of an official state checklist have also been published. Individual floras or checklists are also available for the other states, although in Nebraska none has appeared since 1933.

Regional authorities have had to resolve many doubtful or poorly documented species records. In many cases, these have proved to represent misdeterminations, with eastern manuals the most responsible for their perpetration, perpetuation or both. In some cases these have been based on materials in eastern herbaria not also available in Texas and the Plains.

Bibliographies. General and divisional bibliographies as for Division 1.

Indices. General and divisional indices as for Division 1.

170

Region in general

Keys to families and genera
For *A guide to the genera of the plants of eastern North America* (1983) and *Genera of the western plants* (1983), both by W. T. Batson, Jr., see **100**.

171

Texas

Area: 681 244 km^2. Vascular flora: 4862 species (Correll *et al.*, 1970–72); 4834 native species (Hatch *et al.*, 1990). – For eastern Texas trees, see also **160** (HARRAR and HARRAR); for the Panhandle, see also **175** (GREAT PLAINS FLORA ASSOCIATION); for Trans-Pecos Texas, see also **180, 180/II** and **185** (all works).

Texas floristics began early with explorations and surveys and, later, research – some of it controversial – by residents. Knowledge of east Texas, functionally as well as biologically a part of the 'Old South', was from the 1840s incorporated into southeastern floras (**160**). West of Dallas and the 97th meridian of longitude – which passes through the present-day Dallas-Fort Worth Airport – leading early works included *Plantae lindheimerianae* by Asa Gray with George Engelmann (1845–50; a supplement by Joseph W. Blankinship appeared in 1907), *Plantae wrightianae texano-neomexicanae* (1852–53) by Gray alone, and the botany

of William Emory's Mexican boundary surveys by Engelmann (1858) and John Torrey (1859). In 1858 the Texas Geological Survey was organized; under its first director, Benjamin F. Shumard (before 1861), and then again from 1866 under John Glenn as well as during his own directorships Samuel B. Buckley made substantial collections.

Further consolidation of knowledge was effected through the work of Valéry Havard in *Report on the flora of Western and Southern Texas* (1885), published through the U.S. National Museum, and, soon after and on a larger scale, by John Merle Coulter in his *Botany of western Texas* (1891–94).[133] The latter, still featuring synoptic keys, remained standard until *Flora of Texas* by Cyrus L. Lundell and collaborators commenced publication in 1942. It was wholly superseded only in 1970 on publication of the Correll and Johnston *Manual*, a project itself instituted around 1955 when control of *Flora of Texas* passed wholly to the private Texas Research Foundation.

The first complete state checklist was *Catalogue of the flora of Texas* by Victor L. Cory and Harris B. Parks (1937, but issued in 1938), based on the S. M. Tracy Herbarium at Texas A&M University in College Station. This was revised in 1962 by F. W. Gould as *Texas plants: a checklist and ecological summary*, with further revisions in 1969 and 1975 and most recently in 1990 by S. L. Hatch *et al.* as *Checklist of the vascular plants of Texas*. Checklists have also been published at Austin, firstly by Johnston (1990) and lately by Jones *et al.* (1997; for both, see below).

Some partial floras have been published, those in the east for student use. Post-1945 contributions include *Flora of south central Texas* (1947) by R. G. Reeves and D. C. Bain (revised in 1972 by its senior author as *Flora of Central Texas*) and *Spring flora of the Dallas-Fort Worth area* (1958) by Lloyd H. Shinners. The latter has been three times revised, in the 1980s by W. F. Mahler under the title *Shinners' manual of the North Central Texas flora*, and in 1999 by G. M. Diggs, B. M. Lipscomb and R. J. O'Kennon as *Shinners and Mahler's illustrated flora of North Central Texas*. For partial floras of far western (Trans-Pecos) Texas, see **185**.

Bibliographies
For taxonomic purposes, the bibliographies in Hatch *et al.* (1990), Jones *et al.* (1997) and Diggs *et al.* (1999) will likely be found more useful. Those below are given here for the record.

WINKLER, C. H., 1915. *The botany of Texas.* 27 pp., portr. (Bull. Univ. Texas 1915/18). Austin. [Includes an annotated bibliography of 121 titles.]

SMEINS, F. E. and R. B. SHAW, 1978. *Natural vegetation of Texas and adjacent areas, 1675–1975: a bibliography.* 36 pp., map (Texas Agric. Expt. Sta. Publ. MP-1399). College Station, Tex. [458 unannotated, mainly non-taxonomic references (through 1975) arranged by author; classified index by vegetation area (keyed to map on p. 3). Sources used are discussed on pp. 1–2.]

CORRELL, D. S., M. C. JOHNSTON and collaborators, 1970. *Manual of the vascular plants of Texas.* xv, 1881 pp., col. frontisp., 3 maps. Renner, Tex.: Texas Research Foundation (distributed by Stechert-Hafner Service Agency, New York). Supplemented by D. S. CORRELL, 1972. Manual of the vascular plants of Texas: 1. Additions and corrections. *Amer. Midl. Naturalist* 88: 490–496.

Descriptive manual of vascular plants, with keys to all taxa, essential synonymy, vernacular names, generalized indication of local and extralimital range or area of origin, concise critical remarks, and notes on habitat, phenology, special features, etc.; glossary, list of authorities, selected references, and complete index. An introductory section includes descriptions of floristic zones in the state (with maps) and a summary account of the flora. [For updated taxonomy, distribution and nomenclature, see JOHNSTON (1990) below.][134]

HATCH, S. L., K. N. GANDHI and L. E. BROWN, 1990. *Checklist of the vascular plants of Texas.* iv, 158 pp., 4 maps, col. cover (Texas Agric. Expt. Sta. Publ. MP-1655). College Station, Tex. (Originally published 1937 (1938) as *Catalogue of the flora of Texas* by Victor L. Cory and Harris B. Parks; revised by F. W. Gould, 1962, 1969, 1975 as *Texas plants: a checklist and ecological summary.*]

Annotated systematic checklist of vascular plants (in arrangement generally conforming to Correll and Johnston's *Manual*), with accepted and (where known) vernacular names, coded status designations, life-form and longevity, phenology and regional distribution (the last-named in 10 botanical districts as defined in map 1); classified bibliography (pp. 128–140) and index to genera, families and vernacular names at end. Asterisks indicate corrections to authorities, taxonomy and new species and/or records. An introductory section includes accounts of physical features, climate, vegetation and plant communities.

JOHNSTON, M. C., 1990. *The vascular plants of Texas: a list, updating the* Manual of the vascular plants of Texas. [2nd edn.] iii, 107 pp. Austin, Tex.: The author. (1st edn., 1988.)

Annotated checklist, with *Manual* cross-references, added taxa and, where necessary, commentary (covering range extensions, taxonomic and nomenclatural changes, etc.); literature (pp. 86–103) and index to genera at end.

JONES, S. D., J. K. WIPFF and P. M. MONTGOMERY, 1997. *Vascular plants of Texas: a comprehensive checklist including synonymy, bibliography and index.* xii, 404 pp. Austin: University of Texas Press.

A synonymized checklist in a two-column format, the families and genera of angiosperms alphabetically arranged within the 11 classes of the Cronquist system; bibliography (54 pp.) and complete index (84 pp.) at end. The introduction includes a historical survey. [In effect the Austin successor to Johnston (1990; see above) but expanded.]

LUNDELL, C. L. and collaborators, 1942–69. *Flora of Texas.* Vols. 1–3. Illus. Dallas: Southern Methodist University (vol. 3, parts 1–4); Renner: Texas Research Foundation (vols. 1–2 and vol. 3, parts 5–6); all distributed by Stechert-Hafner Service Agency, New York.

This never-completed semi-monographic 'research' flora comprises a series of family revisions for the state, with keys to all subordinate taxa, full synonymy (with references), vernacular names, citation of *exsiccatae* and authorities for localities, generalized summary of internal and extralimital range, extensive taxonomic commentary and notes on habitat, special features, uses, etc.; complete indices at end of each volume. [Originally planned as a 'closed' work in 10 volumes but from 1955 conducted in serial form and suspended after 1969. Families, as in *Flora Malesiana*, not published in systematic sequence although presented in bound volumes.][135]

Partial work: North Central Texas

Included here are the metropolitan areas of Dallas, Fort Worth and Waco. These have long been covered by successive editions of 'Shinners' Manual', which originally appeared in 1958 as *Spring flora of the Dallas-Fort Worth area, Texas* by L. H. Shinners; revisions by William Mahler appeared as *Shinners' Manual of the North Central Texas flora* in 1984 and 1988. The latest version, described below, is effectively a new work.

DIGGS, G. M., B. M. LIPSCOMB and R. J. O'KENNON, 1999. *Shinners and Mahler's Illustrated flora of North Central Texas.* [i], 1626 pp., 2300 figs., 174 col. photographs, maps (Sida, Bot. Misc. 16). Fort Worth: Botanical Research Institute of Texas/Sherman: Austin College.

Copiously illustrated manual of vascular plants (2223 species in an area about the size of Kentucky, of which 1829 native) with keys to all taxa, concise descriptions, vernacular names, derivations of scientific names, distribution, habitat, phenology, essential synonymy, and critical notes; symbols appear for introduced species. Generic entries contain references to monographs, revisions or other sources. Fifteen appendices (pp. 1353–1419) cover classification, phylogeny, Texas endemic species reported in the area, sources of illustrations, Internet (URL) addresses, collecting, conservation and environmental organizations, host plants for Lepidoptera, further reading, actual or potential ornamentals, and a dedication to Benny J. Simpson; these are followed by an illustrated glossary, an extensive bibliography (pp. 1457–1523), complete index, and publication notes.[136] The comprehensive general part covers the plan and use of the work, acknowledgments, and a description of the general features, geology, climate, vegetation, and floristics of the region along with (pp. 63–75) an account of botanical exploration; general keys appear on pp. 77–172. Primary floristic statistics appear in the inside back cover. [A 'resource work' rather than a field-manual.]

Pteridophytes

The number of species in Texas is the second largest for the 48 conterminous United States, after Florida.

CORRELL, D. S., 1956. *Ferns and fern allies of Texas.* xii, 188 pp., 38 pls. (line drawings), map. Renner: Texas Research Foundation. (A reorganized reissue, with added material, of vol. 1(1) (1955) of *Flora of Texas.*)

An amended reissue of the author's *Flora of Texas* treatment (the figures here being integrated with the text) also featuring an added glossary, bibliography and index as well as an introduction including a historical review and coverage of floristic regions, ecology and distribution, morphology, reproduction, horticulture, and uses.

174

Oklahoma

See also **175** (GREAT PLAINS FLORA ASSOCIATION). – Area: 178 503 km². Vascular flora: 2545 species (Taylor and Taylor, 1989; see below).

The first checklist, *Annotated catalog of the ferns and flowering plants of Oklahoma* (1900) by E. E. Bogue, was an agricultural publication issued seven years before statehood and covered some 750 species; it preceded any detailed floristic surveys. This next phase was represented by its successors, all state survey publications: *Plants of Oklahoma* by A. H. Van Vleet (1902), *A preliminary list of the ferns and seed plants of Oklahoma* by Royal E. Jeffs and Elbert L. Little, Jr. (1930), and *The trees and shrubs of Oklahoma* (1913) by C. R. Shannon. The 1930 checklist was complemented by keys by Jeffs (1931) and succeeded in 1952 by that of Theodore Waterfall, who, like Jeffs, produced separate keys. The fourth edition (1969) of these keys (the so-called fifth edition being a mere reprint following Waterfall's death) became the starting-point for the current checklist by R. J. and C. Taylor. Of descriptive works, none exists apart from that of Stemen and Myers; a successor was projected by Waterfall but is as yet unrealized. Mapping, undertaken by the state biological survey and the Bebb Herbarium at the University of Oklahoma, has so far encompassed only the woody plants.

Bibliography

JOHNSON, F. L. and T. H. MILBY, 1989. *Oklahoma botanical literature.* 150 pp. Norman: University of Oklahoma Press (for the Oklahoma Biological Survey). [862 briefly annotated items, alphabetically arranged by author; author and keyword-in context indices. Includes unpublished theses and dissertations.]

STEMEN, T. R. and W. S. MYERS, 1937. *Oklahoma flora.* xxix, 706 pp., 494 text-figs. Oklahoma City: Harlow Publishing Corporation.

Manual of vascular plants, with keys to genera and species, essential synonymy, vernacular names, generalized indication of local range, representative illustrations, and notes on habitat, frequency, life-form, etc.; lists of endangered flowers, hay-fever plants, drug plants, useful water plants, and edible plants at end as well as a glossary and indices to vernacular and botanical names. [Does not include Gramineae, Cyperaceae or Juncaceae; for the first-named, see H. I. FEATHERLY, 1946. *Manual of the grasses of Oklahoma.* 137 pp. (Bull. Oklahoma Agric. Mech. Coll. 43(2)). Stillwater, Okla.]

TAYLOR, R. J. and C. E. S. TAYLOR, 1989. *An annotated list of the ferns, fern allies, gymnosperms and flowering plants of Oklahoma.* vi, 110 pp., 1 fig., map. Durant, Okla.: The authors. [3rd edn., 1994.]

A tabular checklist similar in format to that of Hatch *et al.* (1990) in Texas, with at the end lists of additions since 1969 (the last revision of Waterfall's keys) and excluded species, detailed statistics, literature (pp. 197–202) and index to genera, families and vernacular names. The introduction includes a historical survey of botanical work.

WATERFALL, U. T., 1972. *Keys to the flora of Oklahoma.* 5th edn. 246 pp. Stillwater, Okla.: The author. (1st edn., 1960; 4th edn., 1969.)

Students' manual, with analytical keys to all taxa, a glossary, and generic index. Complementary to *idem*, 1952. *Catalogue of the flora of Oklahoma.* 91 pp. Stillwater: Oklahoma A & M University Research Foundation.

Woody plants

WILLIAMS, J. E., [1973]. *Atlas of the woody plants of Oklahoma.* Unpaged; maps. Norman, Okla.: Oklahoma Biological Survey/Robert Bebb Herbarium [at the University of Oklahoma].

Annotated checklist with county distribution maps, covering 314 species (in what is usually thought of as a 'prairie state', in the words of the author!); the text includes accepted names, synonymy, vernacular names, and occasional commentary. Indices to vernacular and scientific names, excluded species and a bibliography conclude the work.

175

Northern Central Plains subregion

See also **140** (BRITTON and BROWN). Vascular flora: *c.* 3000 species (GREAT PLAINS FLORA ASSOCIATION, 1977; see below). – The heart of the 'Great Plains', this area comprises Kansas, Nebraska, and the two Dakotas; physiographically, however, it extends across eastern Montana, Wyoming and Colorado, northern Oklahoma and the Texas Panhandle, and western Missouri, Iowa and Minnesota, thus ranging from the Rocky Mountains to the woodland 'ecotone' in the east; it also extends into Canada (**121/II**).

The first 'wide' coverage of the northern Plains was in *Catalogue of the phaenogamous and vascular cryptogamous plants collected . . . in Dakota and Montana by*

Dr. Elliot Coues (1878) by J. W. Chickering; this was based on a federal survey of 1873–74 which also included in its brief the Canadian border. This followed earlier surveys of the central Plains (in Nebraska and Kansas as well as the Dakotas, notably that of Lt. G. K. Warren in 1855–57), the advent therein of resident botanists and the first checklists. In individual states, much of the topographic botanical work became the province of the state colleges and universities, addressing more local needs and interests. It fell to outside institutions to integrate knowledge at regional level; similarly, for identification the area was from the 1890s served by western extensions of Gray's and Britton's works (see **140**) and by the floras of the Rocky Mountains (**180/I**). Sponsorship by the New York Botanical Garden and experience from explorations of the Sandhills and Black Hills in the 1890s gave Rydberg a foundation for his 1932 Plains flora (see Region 17), which remains useful for its illustrations. That work has otherwise now been superseded by the publications of the Great Plains Flora Association, organized in 1973 and based at the University of Kansas – its museum (and herbarium) long a notable regional center.

For the woody flora and the pteridophytes, the descriptive works given below are the first of their kind for the region. The woody plants of the northern Plains were, however, earlier documented by D. G. Hoag (mentioned under **179**).

Bibliography

BROOKS, R. E., 1976. *A bibliography of taxonomic literature of the Great Plains flora.* 74 pp. (State Biol. Surv. Kansas, Rpt. 2). Topeka, Kans.

GREAT PLAINS FLORA ASSOCIATION, 1986. *Flora of the Great Plains.* vii, 1392 pp., 2 figs. (maps). Lawrence, Kans.: University Press of Kansas.

Descriptive manual-flora of vascular plants; includes keys to all taxa, references to standard revisions, monographs and other works, essential synonymy (chiefly Rydberg's names), vernacular names, distributional details by and within states, extralimital range, karyotypes, phenology, frequency, habitat, status and (in smaller type) some critical notes; glossary, abbreviations for authorities, and index to all names at end. Arrangement of families follows the Cronquist system. The general part succinctly introduces the region and its floristic and vegetational features (with two maps) and gives the background to the work; abbreviations (pp. 13–14) and the primary keys (pp. 15–37) follow. [The work encompasses, in addition to the subregion, the fringes of Minnesota, Iowa and Missouri, Oklahoma exclusive of its uplands, the Texas Panhandle, northeastern New Mexico, and eastern Colorado, Wyoming and Montana.]

Distribution maps

GREAT PLAINS FLORA ASSOCIATION, 1977. *Atlas of the flora of the Great Plains* (coord. R. L. McGregor; ed. T. M. Barkley). xii, 600 pp., 2217 maps. Ames, Iowa: Iowa State University Press.

Distribution atlas of the Great Plains vascular flora; includes some 3000 species on dot distribution maps, with a list of families, and a county map showing limits of the area covered. [Published as a complement to the *Flora of the Great Plains*.]

Woody plants

STEPHENS, H. A., 1973. *Woody plants of the North Central Plains*. xxx, 530 pp., 2472 figs. on 255 pls., 2+255 maps. Lawrence: University Press of Kansas. (An expansion of *idem*, 1969. *Trees, shrubs and woody vines in Kansas*. 250 pp., illus. (maps). Lawrence.)

Atlas-flora of woody plants of Kansas, Nebraska, and the Dakotas, with a general key to species and separate keys in some 'critical' genera; each species text includes limited synonymy, vernacular name(s), an extensive description, dot maps, summary of extralimital range, and remarks on taxonomy, biology, special features, etc. A glossary, list of references, and complete index are given as well as an appendix on adventive species and doubtful and erroneous records. In the introductory section are remarks on distribution and floristic zones (with maps).

Pteridophytes

PETRIK-OTT, A. J., 1979. *The pteridophytes of Kansas, Nebraska, South Dakota, and North Dakota*. 332 pp., 122 pls. (Nova Hedwigia, Beih. 61). Vaduz, Liechtenstein: Cramer/Gantner.

Lavish illustrated treatment of ferns and fern-allies (65 species), with keys, descriptions, indication of distribution (with maps), synonymy, vernacular names, karyotypes, critical remarks, and indication of habitat, occurrence, frequency, etc.; glossary and complete index.

176

Kansas

See also **175** (all works). – Area: 212 623 km^2. Vascular flora: 2197 species and infraspecific taxa, including 54 hybrids (Brooks, 1986; see below).

Kansas, a territory from 1854, soon became a floristic center for the Plains. J. H. Carruth was appointed as state botanist in 1868, serving until 1892 in connection with the board of agriculture; in 1872 he produced the first checklist, *Catalogue of plants seen in Kansas*, with a revision in 1877 as *Centennial catalogue of the plants of Kansas*. These were followed by *Plants and flowers in Kansas* (1889), *Check list of the plants of Kansas* (1892, covering 1831 vascular plant taxa) and *Catalogue of the flora of Kansas I–II* (1911–12, not completed), all by B. B. Smyth, a curator at the state museum in Topeka (the last with L. C. R. Smyth), and the Kellermans' *Analytical flora of Kansas* (1888) and Ringle and Kenoyer's *Students' botany of eastern Kansas* (1903), both students' manuals. A new development was represented by A. S. Hitchcock's *Flora of Kansas* (1898–99), in which all species were mapped by county, a first of its kind (and, in retrospect, very accurately).

In the mid-twentieth century came F. C. Gates's *Annotated list of the plants of Kansas: ferns and flowering plants* (1940), which like Hitchcock's work featured distribution maps. Gates also surveyed trees, shrubs and vines in his *Woody plants, native and naturalized, in Kansas* (1938). The modern cycle began with the first edition of Barkley's manual (1968); this was followed in 1969 by *Trees, shrubs and woody vines in Kansas* by H. A. Stephens (later expanded; see **175**).

Bibliography

McGREGOR, R. L. and R. E. BROOKS, 1977. *Preliminary bibliography of taxonomic and floristic works on Kansas vascular plants*. 22 pp. (State Biol. Surv. Kansas, Rpt. 13). Lawrence, Kans.

BARKLEY, T. M., 1978. *A manual of the flowering plants of Kansas*. 2nd edn. vi, 402 pp., map. Manhattan, Kans.: Kansas State University Endowment Association. (1st edn., 1968.)

Manual-key to seed plants, with limited synonymy, vernacular names, generalized indication of local range, and notes on habitat, frequency, etc.; county map; glossary and index to family and generic names. A list of principal references appears in a brief introductory section. [Essentially a corrected reprint of the first edition.]

BROOKS, R. E., 1986. *Vascular plants of Kansas: a checklist.* [4], 129 pp., illus. (cover and frontisp.). [Lawrence]: Kansas Biological Survey, University of Kansas. (A revision of R. L. McGregor, R. E. Brooks and L. A. Hauser, 1976. *Checklist of Kansas vascular plants.* 168 pp. (State Biol. Surv. Kansas, Techn. Publ. 2). [Lawrence.])

Alphabetically arranged name list (by families, genera and species), with synonymy, vernacular names and indication of status (including naturalization if applicable); references and index at end. The introductory portion covers previous lists along with technical matters.[137]

177

Nebraska

See also **175** (all works). – Area: 199 274 km². Vascular flora: no data.

Establishment of Nebraska Territory in 1854 was followed by a survey of it and the Dakotas by Lt. G. K. Warren in 1855–57, the plants being worked up by Engelmann; other surveys followed after the Civil War. In 1875, Samuel Aughey at the University of Nebraska compiled the first state checklist, *Catalogue of the flora of Nebraska*, and, in 1880, a study of the woody flora; his efforts towards a state survey, however, did not bear fruit until after 1900. From 1884, botany at the university was developed under Charles E. Bessey and a 'botanical survey' established.[138] Under this aegis were compiled *Catalogue of the plants of Nebraska* by H. J. Webber (1889), a bibliography (by Roscoe Pound, in the 1890s), several family revisions, and a phytogeography (by the young Frederic E. Clements). The Sandhills region (from 98° to 103° W) was explored in 1893 by Rydberg and documented by him for the USDA (and the National Herbarium) in *Flora of the sand hills of Nebraska* (1895). In the twentieth century came *Flora of Nebraska* (1912; 3rd edn., 1923) by Niels Petersen, an enumeration with keys; its successor, still current but covering only flowering plants, was John M. Winter's similar work, given below.

WINTER, J. M., 1936. *An analysis of the flowering plants of Nebraska, with keys* . . . . 203 pp. (Bull. Conservation Dept. Univ. Nebraska 13). Lincoln.

Concise, non-descriptive flora of flowering plants, with diagnostic keys to all taxa, limited synonymy (including literature citations), vernacular names, and notes on local range, habitat, life-form, phenology, critical features, biology, etc.; index to family, generic and vernacular names at end.

178

South Dakota

See also **175** (all works). – Area: 197 475 km². Vascular flora: at least 1500 species (Over, 1932).

Dakota Territory was organized in 1861 but not until 1889 did South and North Dakota become states. More detailed surveys then began, succeeding those by Warren, Hayden, Coues and Custer over the preceding three and a half decades. The first state checklist was *Ferns and flowering plants of South Dakota* by DeAlton Saunders (1899), an agricultural extension publication. It was followed by *Flora of South Dakota* (1932) by William H. Over, produced through the state geological survey and the University of South Dakota at Vermillion. Over had previously (1923) also surveyed the woody flora. The important Black Hills region in the west was first explored in detail by Rydberg in 1892 and, like the Sandhills, documented for the USDA (and the National Herbarium) in his *Flora of the Black Hills of South Dakota* (1896).[139] That work was succeeded by Arthur C. Macintosh's *A botanical survey of the Black Hills of South Dakota* (1931).

VAN BRUGGEN, T., 1985. *The vascular plants of South Dakota.* xxvi, 476 pp., illus., maps. Ames, Iowa: Iowa State University Press. (1st edn., 1976.)

Concise students' manual to vascular plants, with descriptive keys which include essential synonymy, vernacular names, and brief indication of local distribution, habitat, phenology, etc.; selected taxonomic references in family headings; glossary, list of general references, and complete index at end. An introductory section gives details of physical features, vegetation, and floristic elements together with technical notes and statistics. Some adventive and cultivated plants are also included. [For some technical taxonomic details not included in the manual as well as a survey of previous studies, see J. M. WINTER, C. K. WINTER and T. VAN BRUGGEN, 1959. *A checklist of the vascular plants of South Dakota.* 176 pp. Vermillion, S.D.: Department of Botany, State University of South Dakota. (Mimeographed.)]

Partial work: The Black Hills
Although mostly in the southwestern part of the state, the Black Hills extend also into northeast Wyoming (183).

DORN, R. D., 1977. *Flora of the Black Hills.* x, 377 pp., illus., map. N.p.: The author.

A key to all taxa, without diagnoses or annotations save for vernacular names and indication of habitat; unconfirmed species are asterisked. An appendix covers geography, climate, vegetation, and botanical history (the Hills being 'one of the last areas of the West to be explored'); this is followed by references, excluded names, statistics, an illustrated glossary, and a complete index.

179
North Dakota

See also **175** (all works). – Area: 180 180 km². Vascular flora: no data.

As in South Dakota, detailed botanical exploration of North Dakota began relatively late, and a first full checklist of the flora appeared only in 1900 (*A preliminary list of seed-bearing plants of North Dakota* by H. R. Bolley and L. R. Waldron). This was followed by the nomenclaturally peculiar *Enumerantur plantae Dakotae Septentrionalis vasculares* (1915–18) by Joel Lunell and *Flora of North Dakota* (1918) by Herbert F. Bergman, the latter, featuring keys, published as part of a state soil and geological survey. Bergman's work remained current until 1950. The woody plants were described in *Trees and shrubs for the northern plains* (1965) by Donald G. Hoag, a work stressing properties, uses and potential. In more recent years, the state herbarium, at the University of North Dakota, was among the first to introduce computer-based record-keeping.

Bibliography
HEIDEL, B. L. and D. S. ROGERS, 1984. *North Dakota vegetation: a bibliography.* 48 pp. Grand Forks: North Dakota Natural Heritage Program/Institute for Ecological Studies, University of North Dakota. (Mimeographed.) [Unannotated

list, arranged by author; classified cross-references, pp. 43–48, and additions, pp. 49–50.]

STEVENS, O. A., 1963. *Handbook of North Dakota plants.* Revised edn. (third printing). 324 pp., appendix. Fargo, N.D.: North Dakota Institute for Regional Studies. (Originally published 1950.)

Descriptive manual of vascular plants, with illustrated keys to genera and families, essential synonymy, vernacular names, generalized indication of local range (with some county details here and there), and notes on habitat, phenology, frequency, special features, behavior of alien species, etc.; glossary, addenda, and index to all botanical and vernacular names. The introductory section includes chapters on physiography, soils, vegetation, floristics, botanical exploration, and plant names; an illustrated organography is also provided. The 'third printing' has an appendix with new data to 1962.

Region 18
Western United States

Area: 2 305 244 km² (2 223 504 km² without Trans-Pecos Texas). Vascular flora: no data seen. – Included herein are the states of Montana, Idaho, Wyoming, Colorado, New Mexico, Arizona, Utah, and Nevada. The region thus encompasses such subregions as the Rocky Mountains proper, the Southwest desert areas, and the Intermountain Plateau or Great Basin. The subregions also can be taken to include western (Trans-Pecos) Texas, southeastern California, southeastern Oregon, and the Black Hills in South Dakota.

The Western United States, physiographically and floristically very heterogeneous, is not covered by any single manual or other floristic work save for *Rocky Mountain trees* by Richard J. Preston, Jr. (1940; 3rd edn., 1970). Instead, consolidation of the results of the many nineteenth-century expeditions and surveys and subsequent, more detailed floristic research have, above state level, more or less focused around three subregions: the Rocky Mountains (and adjacent plains), the Southwest, and the Intermountain (or Great) Basin.

The first-named (**180/I**) has long been served by the manuals of Coulter and Nelson (1885; 2nd edn.,

1909) and Rydberg (1917; 2nd edn., 1922); both, however, are obsolescent, and the nomenclature in the latter moreover follows the obsolete 'American Code'. A new regional 'doorstop manual' to replace them is badly needed.[140] Modern coverage for the Intermountain Plateau (180/III) is being furnished in 'grand' fashion by the multi-volume *Intermountain flora* (1972–). This was initiated at the New York Botanical Garden by Bassett Maguire, carried forward by Arthur Cronquist and Arthur Holmgren (in continuation of a plan first developed by Marcus E. Jones before his death in 1934), and is now under the editorship of Noel Holmgren. The Southwest (180/II) is, along with 'cactus books' (here not accounted for save a passing mention under **105** of Lyman Benson's 1982 *Cacti of the United States and Canada*), well supplied with regard to its distinctive woody plants but a modern general flora is lacking. Specific 'deserts' therein have, however, been the subject of floristic projects, of which one on the Sonoran Desert appeared in 1964 (see **205**).[141]

In the states, official surveys were for the most part not organized until after 1900; when these came about, much was accomplished through state universities (and their museums where such had been established). Early floristic coverage was mainly in the form of more or less detailed enumerations, several of them published or even also prepared under federal auspices; descriptive manuals by and large did not appear until after the 1930s. With few exceptions, however, all states are now more or less effectively covered; in some, current works represent a second or third 'generation'. Since World War II floras have appeared for Montana (Booth, 1950–66; Dorn, 1984), Idaho (Davis, 1952), Wyoming (Porter, 1962–72; Dorn, 1977, 1988, 1992), Colorado (Harrington, 1964; Weber, 1987, 1990), New Mexico (Martin and Hutchins, 1980–81), Arizona (Kearney and Peebles, 1942, 1960; Tidestrom and Kittel, 1941, with inclusion of New Mexico), and Utah (Welsh *et al.*, 1987, 1993). For Nevada there is a long series of contributions published by the United States National Arboretum (1940–65), one of the last two state floras prepared and published under federal auspices (notably the U.S. Department of Agriculture), and more recently an as yet unpublished manual by John T. Kartesz. A number of partial works, notably Walter B. McDougall's *Seed plants of Northern Arizona* (1973), Arthur H. Holmgren's *Handbook of the vascular plants of northeastern Nevada* (1942), and R. J. Shaw's

Vascular plants of northern Utah (1989), and some state checklists have also been published. Partly because of the region's heterogeneity, few of these works have as yet been widely influential although the most critical would seem to be those by Harrington, Kearney and Peebles, and Welsh *et al.* Some older standards, notably the flora of Utah and Nevada by Ivar Tidestrom (1925, recently reprinted), have yet in nominal terms entirely to be superseded. Parts of the region are also covered by works from outside, such as *Vascular plants of the Pacific Northwest* (see **191**) and *Flora of the Sonoran Desert* (see **205**).

Coverage may thus be considered moderately good and improving, but of variable standard and moreover dependent on collecting density, which in some areas remains relatively low, with yet a potential for new discoveries, biological or geographical.[142] A number of problems remain, however, not the least of which has been, in William Weber's words, a continuing need to 'go east' – travel such as is also required of botanists in many other parts of the world.[143]

Bibliographies. General and divisional bibliographies as for Division 1.

Indices. General and divisional indices as for Division 1.

180
Region in general

The region is, as a whole, botanically too large and diverse to be covered by a single general flora, save for works on North America as a whole (**100**). Larger works on the Western flora have centered on three subregions, considered below as follows: I. The Rocky Mountains and adjacent plains (centering on the states of Montana, Wyoming and Colorado and extending into northern and eastern Idaho, eastern Utah, northeastern Arizona, northern New Mexico and to the Black Hills of South Dakota); II. The Southwest (centering on Arizona and New Mexico and extending into southeastern California and western (Trans-Pecos) Texas along with northern Mexico (**210**; see also **205**); and III. The Intermountain Plateau (containing the 'Great Basin' and centering on the states of Utah and

Nevada with extensions to southeastern Oregon, southern Idaho, southwestern Wyoming and parts of eastern California and northern Arizona. One 'tree book', however, encompasses the whole Western region: *Rocky Mountain trees* by Richard J. Preston, Jr.; and there is also an older students' manual by Aven Nelson, *Spring flora of the Intermountain Region* (1912).

Keys to families and genera

For *Genera of the western plants* (1983) by W. T. Batson, Jr., see the comparable subheading under **100**.

Woody plants (and trees): guides

In addition to the long-standard manual by Preston, reference can be made to J. B. BERRY, 1966. *Western forest trees: a guide to the identification of trees and woods for students, teachers, farmers and woodsmen.* 251 pp. New York: Dover (originally published 1924, Yonkers, N.Y., as *Western forest trees: a guide . . . to accompany farm woodlands*); and M. TRESHOW, S. L. WELSH and G. MOORE, 1970. *Guide to the woody plants of the mountain states.* 178 pp., illus. Provo: Brigham Young University Press. [The latter is a pictured-key, based on its authors' *Guide to the woody plants of Utah.*]

PRESTON, R. J., JR., 1968. *Rocky Mountain trees.* 3rd edn. pp. i–lix, 1–285, lxi–lxxxi, illus., maps. New York: Dover Publications. (1st edn., 1940, Ames, Iowa: Iowa State College Press; 2nd edn., 1947.)

Illustrated, semi-popular descriptive treatment; includes keys to all genera and species, essential synonymy, vernacular names, indication of range by life-zones (with distribution maps for all species), and extensive commentary on diagnostic features, habitat, ecology, biology, wood, silvical characteristics, and related shrubby species; glossary, list of references, and index to all subjects together with botanical and vernacular names. The introductory section gives an account of the region's life-zones (with map), notes on descriptive morphology and use of the work, and checklists of trees for each individual state. [Covers all the Western states or portions thereof encompassed in Region 18. The Dover reprint incorporates corrections and updated nomenclature.][144]

I. Rocky Mountains subregion (with adjacent plains)

See also **103/II**, **181** through **184**, and **188** (all works); **191** (HITCHCOCK *et al.*; HITCHCOCK and CRONQUIST). An estimate of 3000 species for the subregion was given by Aven Nelson in his *An Analytical Key* (1902).

The work by Coulter and Nelson (1909) is included here for its coverage of the full area defined here and its use of the International Code as then adopted; but for practical purposes both it and Rydberg's flora are largely obsolete.[145] For an introductory work with keys covering 355 species, see R. A. NELSON, 1992. *Handbook of Rocky Mountain plants.* 2nd edn. by R. L. Williams. xviii, 444 pp., illus., maps. Denver: Denver Museum of Natural History and Niwot, Colo.: Roberts Rinehart. (1st edn., 1969.)

COULTER, J. M., 1909. *New manual of botany of the central Rocky Mountains (vascular plants).* 2nd edn. by A. Nelson. 646 pp. New York: American Book Co. (1st edn., 1885, under title *Manual of botany . . . of the Rocky Mountain regions.*)

Field-manual of vascular plants (nomenclaturally in accordance with the 1905 International Code); includes concise keys to all taxa, essential synonymy (with references), vernacular names, generalized indication of regional distribution, taxonomic commentary, and notes on habitat, special features, etc.; statistical summary, list of new names and combinations, list of authors' names, and glossary; indices to all vernacular and botanical names. The species concept adopted is broader than in Rydberg's *Flora of the Rocky Mountains and adjacent plains* (see below); the area covered encompasses Colorado, Wyoming, northern New Mexico, northeastern Arizona, eastern Utah, southeastern Idaho, and most of Montana along with the Black Hills in South Dakota.

RYDBERG, P. A., 1922. *Flora of the Rocky Mountains and adjacent plains.* 2nd edn. xii, 1144 pp. New York: The New York Botanical Garden. (Reprinted 1954, 1961, New York: Hafner. 1st edn., 1917.)

Briefly descriptive manual-flora of vascular plants (in accordance with the 'American Code'); includes keys to all taxa, limited synonymy, vernacular names, concise indication of regional range with indication of life-zones (following the Merriam system), and notes on habitat, phenology, etc.; brief sketch of life-zones, glossary, and index to all botanical and vernacular names at end. Encompasses Montana, Idaho, Wyoming, Colorado, and Utah as well as adjacent parts of Canada. The incorporation of an addendum (pp. 1111–1144) is all that distinguishes this edition from its predecessor. [For a very concise field version of this work, see *idem*, 1919. *Key to the Rocky Mountain flora.* 305 pp. New York: The author.]

Woody plants (and trees): guide

See also **180** above.

KIRKWOOD, J. E., 1930. *Northern Rocky Mountain trees and shrubs*. xvii, 340 pp., 87 figs., 35 pls. Stanford, Calif.: Stanford University Press.

A fairly detailed semi-popular descriptive treatment to some 248 species (of which 24 are gymnosperms), with vernacular and scientific names, keys to genera and species, extensive indication of distribution, and notes on first discovery, examples of special interest, and areas where a given species is conspicuous along with photographs of habitats and sprays and figures of technical details; illustrated organography and indices to vernacular and scientific names at end (but no bibliography). The introductory section includes a general description of the region and its salient features and a treatment of the forest vegetation with its relationships to climate, topography, etc. as well as its historical development and biogeography; these are followed by statistics (pp. 9–10) and a general key to families (pp. 12–13). [Covers Yellowstone Park, Idaho and western Montana and parts of southern Alberta and British Columbia.]

II. Southwestern subregion ('the Southwest')

See also **171**, **185**, **186**, **187**, **195** and **197** (all works); **105** (all works) and **205** (SHREVE and WIGGINS), these latter relating specifically to desert areas. General treatments for the southwestern United States are given here, although strictly speaking this subregion more particularly represents Arizona and New Mexico. Subregional works include *Flora of Arizona and New Mexico* by Ivar Tidestrom and Sr. Teresita Kittel (1941), now largely superseded, and works by Little (1950) on the trees and Vines (1960) on woody plants.

TIDESTROM, I. and Sister [M.] T. KITTEL, 1941. *Flora of Arizona and New Mexico*. xxvi, 897 pp., map. Washington, D.C.: Catholic University of America Press.

Concise descriptive manual-flora of vascular plants, with keys to all taxa, limited synonymy, vernacular names, generalized indication of local and extralimital range along with modified Merriam life-zones as appropriate, and notes on habitat, etc.; index to generic and vernacular names at end. The introductory section contains, among other material, an account of early explorers in the area, together with a map.[146]

Bibliography

BOWERS, J. E., 1982. Local floras of the Southwest, 1920–1980: an annotated bibliography. *Great Basin Naturalist* **42**: 105–112. [Annotated list arranged in the first instance by states and then by authors; sources at end. Covers published and unpublished floras and other floristic contributions for Arizona, New Mexico, Utah, Nevada and Colorado.]

Woody plants (and trees): guides

For *Trees and shrubs of the southwestern deserts* by L. Benson and R. A. Darrow, see **105**.

LITTLE, E. L., JR., 1950. *Southwestern trees; a guide to the native species of New Mexico and Arizona*. ii, 109 pp., illus. (Agric. Handb. USDA 9). Washington, D.C.: U.S. Government Printing Office.

Illustrated descriptive popular account, without keys; includes synonymy, vernacular names, local range, some critical remarks, and notes on habitat, ecology, uses, etc.; list of references and index at end.

VINES, R. A., 1960. *Trees, shrubs and woody vines of the Southwest*. 1104 pp., illus. Austin: University of Texas Press.

Copiously illustrated, descriptive systematic treatment of native and naturalized woody plants, with vernacular names, indication of local and extralimital range, and notes on habitat, variation, special features, uses, derivation of names, etc.; glossary, bibliography, and complete indices at end. All species are figured, but keys are wanting. [Covers the wider Southwest from western Texas to southern California.]

III. Intermountain subregion including the Great Basin

See also **182**, **188**, **189** and **193** (all works). Previously without a general work of any kind and still unevenly known botanically, this subregion is now being treated by the fine modern *Intermountain flora*, produced at the New York Botanical Garden under the direction of Arthur Cronquist (until 1992) and Noel Holmgren.

CRONQUIST, A. *et al.*, 1972– . *Intermountain flora: vascular plants of the Intermountain West, U.S.A.* Vols. 1– . Illus., maps. New York: Hafner (vol. 1); Columbia University Press (vol. 6); New York Botanical Garden (vols. 3A, 3B, 4, 5).

Large-scale, illustrated descriptive 'research' flora; includes keys to all taxa, synonymy (with references), vernacular names, generalized indication of local and extralimital range, taxonomic commentary, and notes on habitat, special features, karyotypes, variation, etc.; glossary and comprehensive indices in each volume. The introductory section (vol. 1) includes accounts of physical features, history of the flora, phytogeography, ecology, and an account of botanical exploration. [Encompasses Utah, Nevada except its southernmost part, parts of eastern California, northwest Arizona (from the north rim of the Grand Canyon northwards), south Idaho, bordering parts of Wyoming, and southeast Oregon. Of the six volumes originally projected (later increased to seven), there had appeared at the time of writing vols. 1 (1972), 3A (1997), 3B (1989), 4 (1984), 5 (1994) and 6 (1977), respectively covering pteridophytes and gymnosperms, Rosidae except Fabales, Fabales, Asteridae except Compositae, Asterales, and the monocotyledons, the arrangement following the 1968 version of the senior author's system of families.][147]

HOLMGREN, A. H. and J. L. REVEAL, 1966. *Checklist of the vascular plants of the Intermountain Region.* iv, 160 pp., map (U.S. Forest Serv. Res. Pap. INT-32). Ogden, Utah.

Systematic list of vascular plants, with rather extensive synonymy as well as vernacular names and a map; list of references and complete index. The introductory section gives remarks on the limits of the checklist, physiography of the region, and botanical exploration. [A precursor for *Intermountain flora* (see above).]

Woody plants (and trees): guide

For a popular natural history with good illustrations, see R. M. LANNER, 1984. *Trees of the Great Basin: a natural history.* xvi, 215 pp., illus. (part col.), maps. Reno.

181

Montana

See also **180/I** (all works); **175** (GREAT PLAINS FLORA ASSOCIATION, 1977, 1986); **191** (HITCHCOCK and CRONQUIST, 1955–59, 1973). – Area: 378 009 km². Vascular flora: no data.

A first catalogue of plants of Montana and the Dakotas was published by J. W. Chickering in 1878 (further details under **175**). Its successor, also enumerative, was *Catalogue of the flora of Montana and the Yellowstone National Park* (1900) by Per Axel Rydberg,

the work which began the New York Botanical Garden's now almost century-long documentary association with the Western states. A supplement to that work by Joseph W. Blankinship appeared in 1905. The first descriptive manual was that begun by William E. Booth in 1950 (see below).

Among partial works, a key contribution was *Flora of Glacier National Park, Montana* (1921) by Paul C. Standley; this has been most recently revised as *Checklist of the vascular plants of Glacier National Park, Montana, U.S.A.* (1985) by Peter Lesica.

Bibliography

HABECK, J. R. and E. HARTLEY, 1965. The vegetation of Montana: a bibliography. *Northw. Sci.* **39**(2): 60–72. [Includes coverage of local floristic works.]

BOOTH, W. E., 1950. *Flora of Montana, I: Conifers and monocots.* 232 pp. Bozeman, Mont.: Montana State College Research Foundation. Complemented by W. E. BOOTH and J. C. WRIGHT, 1966. *Flora of Montana, II: Dicotyledons.* 305 pp., illus., maps. Bozeman: Montana State University.

Descriptive manual of seed plants, with keys to all taxa, essential synonymy, vernacular names, general indication of local range (with inclusion of distribution maps in the second volume), and notes on habitat, etc.; figures and photographs of diagnostic features; indices to all botanical and vernacular names in each volume. The introductory section in the first volume includes an illustrated organography. [For a partial successor to the first volume (not inclusive of grasses), see B. E. HAHN, 1973. *Flora of Montana: conifers and monocots.* 143 pp. Bozeman: Montana State University.]

DORN, R. D., 1984. *Vascular plants of Montana.* iv, 276 pp., illus., maps. Cheyenne, Wyo.: Mountain West Publishing.

Manual-key to vascular plants, with briefly descriptive leads inclusive also of vernacular names and notes on habit, distribution (based on the map in the inside front cover) and habitat; figures of critical features; additions, illustrated glossary and general index at end. The introductory section covers the scope of the work, nomenclature (which follows *Flora of the Pacific Northwest* (see **191**)), a chorological map (p. iv), and a general key to families.

182

Idaho

See also **180/I, 180/III** (all works), **191** (all works except Gilkey). – Area: 214 271 km². Vascular flora: no data.

Idaho, a state in which full control came late, was first covered – along with Washington and Oregon – by Thomas J. Howell in his *Flora of Northwest America* (1897–1903). The first state manual was that by Ray Davis (see below); revisions are given in a checklist by D. Fulton (1976). The westernmost part, bordering on the Palouse in Washington, is additionally covered by *Flora of southeastern Washington* (and its predecessors).

Bibliography

DAUBENMIRE, R. F., 1962. Vegetation of the state of Idaho: a bibliography. *Northw. Sci.* **36**: 120–122. [Includes references for local floras.]

DAVIS, R. J. and collaborators, 1952. *Flora of Idaho.* iv, 828 pp. Dubuque, Iowa: Brown. [Later reissues published by Brigham Young University Press, Provo, Utah, with additions and corrections and a total of 836 pp.]

Descriptive manual-flora of vascular plants, without illustrations; includes keys to all taxa, essential synonymy, vernacular names, rather sketchy indication of local range along with summaries of extralimital distribution, and notes on habitat, phenology, etc.; glossary and index to all botanical and vernacular names. The introductory section includes accounts of physical features, vegetation zones, phytogeography, and the origin of the flora (in part by R. F. Daubenmire). No statistics on the flora are, however, presented.

FULTON, D., 1980. *List of scientific and common plant names for Idaho.* iii, 70, 70 pp. [Boise, Idaho]: Soil Conservation Service. [Initial edn., 1976.]

An information system-based tabular list in two complete sequences (respectively by genus and species and by vernacular names), with code, genus, species and infraspecific epithets, vernacular and standardized plant names (SPN), and habit (cf. explanation, pp. ii–iii).

183

Wyoming

See also **180/I, 180/III** (all works); **175** (GREAT PLAINS FLORA ASSOCIATION, 1977, 1986). – Area: 252 171 km². Vascular flora: 2398 species and 709 additional infraspecific taxa (Dorn, 1992; see below).

Locally based botanical work dates to the beginning of Aven Nelson's career at the University of Wyoming at Laramie in the 1890s. His first account, *First report on the flora of Wyoming* (1896), remained for eight decades the only nominally complete work. His successor, Cedric L. Porter, began work towards a descriptive flora, but before its completion his serially published *Flora of Wyoming* (1962–72) ceased publication. In 1977 Robert Dorn brought out a state manual; in 1988 and again in 1992 there appeared a more compact version of this work.[148]

DORN, R. D., 1992. *Vascular plants of Wyoming.* 340 pp., 2 maps, >250 illus. Cheyenne, Wyo.: Mountain West Publishing. (First published 1988.)

Illustrated field manual-key to vascular plants, with leads inclusive of vernacular names and notes on habit, distribution and habitat but not diagnostic/descriptive; figures of critical features; critical notes, late additions, illustrated glossary and index to all names at end. Generic and family headings contain concise references to detailed treatments. The introductory section covers the scope of the work as well as nomenclature and includes a general key to families. [See also *idem*, 1977. *Manual of the vascular plants of Wyoming.* 2 vols. viii, viii, 1498 pp., 27 illus. New York: Garland. This is a partly compiled, somewhat uncritical, briefly descriptive manual-flora with few illustrations.][149]

PORTER, C. L., 1962–72. *A flora of Wyoming.* Parts 1–8 (Wyoming Agric. Exp. Sta. Bull. 402, 404, 418, 434 (parts 1–4); Wyoming Agric. Exp. Sta. Res. J. 14, 20, 64, 65 (parts 5–8)). Laramie, Wyo.

Briefly descriptive flora of vascular plants, with keys to all taxa and accounts of families and genera but species merely enumerated; includes references (where appropriate) to more detailed papers, essential synonymy, vernacular names, generalized indication of local range (with some details), taxonomic commentary, and notes on habitat, frequency, variation, biology, properties, uses, etc.; indices in each part (from the third onwards). The introductory section (part 1) includes accounts of physical features, climate, major floristic elements, and vegetation zones as well as a list of 108 families. [Not completed; covers 45 families through the Fumariaceae on an arrangement following the Englerian system. A precursory series, with keys and many range maps, was *idem*, 1942–61. *Contributions toward a flora of Wyoming*. 34 parts. Laramie: Rocky Mountain Herbarium. (Mimeographed.)]

184

Colorado

See also **180/I** (all works); **175** (GREAT PLAINS FLORA ASSOCIATION, 1977, 1986). For Weber's outstanding *Rocky Mountain flora*, see **103**. – Area: 269 347 km². Vascular flora: 3088 species and additional infraspecific taxa (Weber and Wittmann, 1992).

The first consolidation of floristic knowledge for the state, which established for Colorado a 'leading edge' in Western floristics never really lost since, was accomplished in 1874 by Thomas C. Porter and John M. Coulter in their *Synopsis of the flora of Colorado*. It was produced under the auspices of Frederick V. Hayden's 'Survey of the Territories', and formed a basis for Coulter's later *Flora of the Rocky Mountains and adjacent plains* (1885; see **180/I**).[150] The *Synopsis* itself was succeeded by Rydberg's *Flora of Colorado* (1906), a work which, embodying the narrower taxonomic concepts of the New York 'school', remained current until the first edition of Harrington's manual (1954). The most practical modern works, however, are the two nicely illustrated, complementary *Colorado flora* volumes by William Weber (both revised with Ronald Wittmann, 1996; see below).[151]

With the University of Colorado home to EXIR – a derivative of the pioneering taxonomic information program TAXIR – computerization of floristic inventory began there in the 1970s under Weber. Two successive state checklists, accounted for below, have since been published. The first, *Natural History Inventory* (1976) is, along with its somewhat more detailed British Columbia counterpart (**124**), one of the early representatives in its genre.

HARRINGTON, H. D., 1964. *Manual of the plants*

of Colorado. 2nd edn. 666 pp. Chicago: Swallow Press. (1st edn., 1954, Denver.)

Descriptive manual of vascular plants, with keys to all taxa, limited synonymy, vernacular names, generalized indication of local and extralimital range with details here and there, taxonomic commentary, and notes on habitat, etc.; state map, with counties; glossary, summary of the flora, and index to family, generic and vernacular names. The introductory section includes a summary account of the vegetation and its zonation (by D. Costello). [The second edition differs only slightly from the original.][152]

WEBER, W. A. and R. C. WITTMANN, 1996. *Colorado flora: eastern slope*. Revised edn. xl, [i], 524 pp., text-figs., 64 col. pls. Niwot: University Press of Colorado. Complemented by *idem*, 1996. *Colorado flora: western slope*. xxxvii, [ii], 496 pp., text-figs., 64 col. pls. Niwot. (1st edns. respectively in 1987, Boulder, and 1990, Niwot.)

These complementary works, respectively focused to the east and west of the Continental Divide, are manual-keys to vascular plants (in the European tradition) with diagnostic phytographic text, vernacular names, figures of essential details, concise indication of local range, habitat, and special features along with some critical remarks; glossary and indices to vernacular and generic names are at the end but no list of references. The arrangement of families is alphabetical. The introductory parts contain the plan and thrust of the works along with sections on physical features, topography, floristic zones, plant geography, families treated and statistics of the flora, vernacular names, instructions for collectors, pronunciation of botanical names, abbreviations and codes, and general keys to families. [2300 and 2100 species, respectively, are accounted for.]

WEBER, W. A. and R. C. WITTMANN, 1992. *Catalog of the Colorado flora: a biodiversity baseline*. xi, [ii], 215 pp. Niwot: University Press of Colorado. (Based on W. A. Weber and B. C. Johnston, 1976. *Natural history inventory of Colorado*, I. *Vascular plants, lichens and bryophytes*. ii, 205 pp. Boulder.)

Alphabetically arranged checklist (vascular plants, pp. 1–137), with all doubtful and erroneous records integrated but italicized; entries include references, synonymy, types (if from within the state), status, and (where necessary) brief notes; list of additional synonyms, lists of mosses and lichens, and (pp. 198–215) a bibliography at end. The introductory part includes a treatment of sources for the checklist and its plan.

185

Trans-Pecos Texas

See also **171**, **180/II** (all works). – Area: 81740 km² (12 percent of the total area of Texas, or about the size of Maine). Trans-Pecos Texas includes all that part of the state within the Cordillera west of the Pecos River (including its highest peak, Guadelupe, and Big Bend National Park on the Rio Grande). Also therein is a portion of the Chihuahuan Desert (see **205**).

Along with the rest of Texas as far east as Dallas and the 97th meridian, the area was early documented by John M. Coulter in *Botany of western Texas* (1891–94; see passing reference under **171**). Among twentieth-century works, the most complete separate coverage is in Powell's woody flora (described below) and in O. E. SPERRY, 1938. *A check list of the ferns, gymnosperms and flowering plants of the proposed Big Bend National Park of Texas*. 98 pp. (Bull. Sul Ross State Teachers' Coll. 19(4)). Alpine, Tex. Many mountain ranges remained not or but poorly known until into the 1930s.[153]

Woody plants

POWELL, A. M., 1988. *Trees and shrubs of Trans-Pecos Texas*. 536 pp., 398+3 text-figs. (incl. map), portr.. Big Bend National Park, Tex.: Big Bend Natural History Association.

Semi-popular illustrated systematic treatment of 453 species, with descriptions of families and genera but for species only essential synonymy, detailed local distribution, overall range, and commentary; references (pp. 487–491), glossary and complete index at end. The introductory part includes a description of the region and its physiography, climate and vegetation as well as a key to families.

186

New Mexico

See also **180/II** (all works). – Area: 315471 km². Vascular flora: 3728 species and non-nominate infraspecific taxa (Martin and Hutchins, 1980–81; see below).

The pioneer *Flora of New Mexico* (1915; in *Contr. U.S. Natl. Herb.* **19**) by E. O. Wooton and Paul C.

Standley, consolidating the work of nineteenth-century explorers as well as their own, was supplanted partly by *Flora of Arizona and New Mexico* (1941; see **180/II**) by Ivar Tidestrom and Sister Teresita Kittel, the first state flora with descriptions, and more definitively by the bulkier two-volume *Flora of New Mexico* by Martin and Hutchins, described below. The last-named was a primary result of a long organized programme of state exploration summarized in interim fashion in 1970 by Martin and Castetter's *Checklist* (see below).

Bibliography

STANDLEY, P. C., 1910. A bibliography of New Mexican botany. *Contr. U.S. Natl. Herb.* **13**: 229–246. [Annotated list.]

MARTIN, W. C. and C. R. HUTCHINS, 1980–81. *A flora of New Mexico*. 2 vols. xx, 2592 pp., 463 text-figs., 1300 maps. Vaduz, Liechtenstein: Cramer/Gantner.

Briefly descriptive, well-illustrated flora of vascular plants (3728 entities recognized, many at infraspecific level), with keys to all taxa, limited synonymy, 'standardized vernacular names', generalized indication of internal and extralimital distribution (supplemented by dot distribution maps, each depicting three entities), altitudinal zonation, and notes on habitat, phenology, etc.; citations of pertinent literature for each genus after the last of its species; glossary, list of authorities and abbreviations, bibliography, and general index (all in vol. 2). The introductory section (in vol. 1) encompasses physical features, climate, geology, and vegetation of the state, along with an explanation of the unit distribution maps and family key.[154] [See also W. C. MARTIN and E. F. CASTETTER, 1970. *A checklist of gymnosperms and angiosperms for New Mexico*. 245 pp. Albuquerque, N.M.: The authors (through the Department of Biology, University of New Mexico).]

187

Arizona

See also **180/II**, **180/III**, **205** (all works). – Area: 295121 km². Vascular flora: 3438 species (Kearney and Peebles, 1960); 3370 had been recorded by 1951.

The earliest floristic literature on Arizona dates only from 1848 and until well into the twentieth century botanical work was oriented towards primary exploration and documentation. Consolidation began in the 1930s, during which Thomas H. Kearney and Robert H. Peebles undertook preparation of a state flora and Sr. Mary Teresita Kittel produced a checklist. This, along with a 1941 study of Compositae, became a partial basis for the latter's *Flora of Arizona and New Mexico* (1941, with Ivar Tidestrom; see **180/II**) – the first substantial descriptive flora in the Southwest.[155] Kearney and Peebles' *Flowering plants and ferns of Arizona* appeared the next year and their *Arizona flora* in 1951.[156] All these works are now out of date; research and preliminary contributions towards a new *Vascular plants of Arizona* are presently in progress, with *Journal of the Arizona-Nevada Academy of Science* the outlet for precursory papers.[157]

Among floras of lesser areas, the most substantial is *Seed plants of Northern Arizona* by W. T. McDougall (1973; see below under **Partial work**). This encompasses the Grand Canyon and the San Francisco Peaks.[158]

Bibliographies

BOWERS, J. E., 1981. Local floras of Arizona: an annotated bibliography. *Madroño* **28**: 193–209, map. [Geographically arranged; no author or chronological index. The map depicts the locations and areas covered of the cited floras.]

EWAN, J., 1936. Bibliography of the botany of Arizona. *Amer. Midl. Naturalist* **17**: 430–454. [Coverage to 1934; arrangement chronological with author and geographical indices.]

SCHMUTZ, E. M., 1978. *Classified bibliography on native plants of Arizona*. xii, 160 pp. Tucson: University of Arizona Press. [Comprehensive though unannotated treatment with several topical headings; floras, pp. 34–53, and trees, pp. 130–132.]

KEARNEY, T. H., R. H. PEEBLES and collaborators, 1960. *Arizona flora*. 2nd edn., with supplement by J. T. Howell and E. McClintock. viii, 1085 pp., 45 halftones, map. Berkeley: University of California Press. (Reissued, most recently in 1988. 1st edn., 1951.)

Concise manual of vascular plants, with abbreviated keys to all taxa, very limited synonymy, vernacular names, fairly detailed indication of local range and summary of extralimital distribution, extensive taxonomic commentary, and notes on habitat, frequency, biology, uses, etc.; glossary, list of references, and complete indices at end. An introductory section, with a state map, includes accounts of physical features, geology, soils, climate, vegetation, life-forms, floristic elements, and botanical exploration. The addition of the supplementary pp. 1035–1081 distinguishes the second from the first edition. – Partly succeeds *idem*, 1942. *Flowering plants and ferns of Arizona*. 1069 pp. (Misc. Publ. USDA 423). Washington, D.C.: U.S. Government Printing Office. [Similar but with more detailed keys, synonymy, and accounting of local distribution; lacks vernacular names and extensive commentary.]

LEHR, J. H., 1978. *A catalogue of the flora of Arizona*. vi, 203 pp. Phoenix: Desert Botanical Garden.

Systematic checklist of vascular plants (in arrangement following *Arizona flora*), with essential synonymy (where nomenclature varies from that in the earlier work), vernacular names (where available), and occasional notations relating to taxonomic problems, etc.; references (keyed to main text) and index to genera and families at end (but unfortunately no statistics). The introduction is brief, being mainly a justification for the work.

Partial work

MCDOUGALL, W. B., 1973. *Seed plants of Northern Arizona*. 594 pp., frontisp. Flagstaff, Ariz.: Museum of Northern Arizona.

Concise manual-flora for the five northern counties, with abbreviated keys to all taxa, brief descriptions, scant synonymy, vernacular names, indication of status and of Northern Arizona and overall distribution, a few critical remarks, and notes on habitat, altitudinal zonation, phenology, etc.; glossary and index to vernacular, generic and family names. Based mainly on *Arizona flora* (see above) but written to provide more detail on this botanically important area (centering on the Grand Canyon).

188

Utah

See also **180/II**, **180/III** (all works). – Area: 213 390 km². Vascular flora: 3677 recognized taxa, including 2602 native and 682 introduced species as well as 393 non-nominate infraspecific taxa (Welsh *et al.*, 1993; see below). Albee *et al.* map 2822 species in their atlas.

Nineteenth and early twentieth century work in the state, including the pioneer *Description of the species constituting the botany of the basin of the Great Salt Lake of Utah . . .* (1860) by Elias Durand and, more importantly, the catalogue of Utah and Nevada plants comprising Sereno Watson's *Botany* (1871, in C. J. King's *Report of the geological exploration of the fortieth parallel*), was consolidated by Ivar Tidestrom in his long-standard 1925 *Flora of Utah and Nevada*. These works also encompassed the more than four decades of detailed study of the flora (1880–1923) by Marcus E. Jones. At Provo, Stanley L. Welsh, in association with others, contributed *Guide to common Utah plants* (1964; more definitive reissue, 1965) and, in 1973 with G. Moore, *Utah plants: Tracheophyta*. These were predecessors of *A Utah flora* (see below). The first edition of this last was in 1988 complemented by a distributional atlas published through the Museum of Natural History of the University of Utah in Salt Lake City. Some local floras have also been published, one of which is accounted for here.

Bibliography

CHRISTENSEN, E. M., 1967. *Bibliography of Utah botany and wildland conservation.* 136 pp. (Sci. Bull. Brigham Young Univ., Biol. Ser. 9(1)). Provo, Utah; and *idem*, 1967. *Bibliography . . .* II. *Proc. Utah Acad. Sci.* 44(2): 545–566. [Alphabetically arranged by author, with cross-references; chronological and subject indices; unpublished theses included. Together with the supplement, coverage is provided through 1966.]

TIDESTROM, I., 1925. *Flora of Utah and Nevada.* 665 pp., illus. (Contr. U.S. Natl. Herb. 25). Washington, D.C.: Smithsonian Institution. (Reprinted 1969, Lehre, Germany: Cramer, as *Reprints of US-floras* 3.)

Manual-key to vascular plants, with synonymy and references; vernacular names; generalized indication of local and extralimital range, including life-zones (Merriam system); notes on habitat, etc.; index to all botanical and vernacular names. An introductory section features descriptions of life-zones and plant communities, as well as statistics of the flora (in part by H. L. Shantz and A. W. Sampson).

WELSH, S. L., N. D. ATWOOD, S. GOODRICH and L. C. HIGGINS (eds.), 1993. *A Utah flora.* Revised edn. viii, 986 pp., maps (end-papers). Provo, Utah: Brigham Young University Press. (1st edn., 1987. A successor to S. L. WELSH and G. MOORE, 1973. *Utah plants: Tracheophyta.* iv, 474 pp., illus. Provo.)

Manual-flora of native, naturalized, adventive and commonly cultivated vascular plants, with keys to all species, somewhat extended descriptions, limited synonymy, vernacular names, counties, generalized indication of local and extralimital range (with very occasional *exsiccatae*), altitudinal distribution, karyotypes, critical remarks, numbers of specimens seen, and notes on status, habitats, etc.; references to revisions for some genera; author abbreviations, glossary and index to all names at end. The introductory part covers the work's historical basis, philosophy, plan and methodology along with discussions of nomenclature, vernacular names, the environment, plant geography and human affairs; statistics and a table of abbreviations are also given as well as acknowledgments and general references. Families are by various authors, including the editors. [Based on over 400 000 specimens in Utah and a somewhat lesser total elsewhere. Many generic and other treatments are based on precursory papers in *Great Basin Naturalist*.]

Distribution maps

ALBEE, B. J., L. M. SHULTZ and S. GOODRICH, 1988. *The atlas of the vascular plants of Utah.* xx, 670, [1] pp., 2438 distribution maps, 6 general and thematic maps (Occas. Publ. Utah Museum of Natural History 7). Salt Lake City.

An atlas, with four maps per page depicting for each species locality-based dots on a background of relief, drainage, county lines and scales, and accompanied by annotations incorporating scientific and standardized vernacular names, phenology, habit, status, usual habitat(s), and altitudinal range (in meters); list of subspecies, varieties and plants of very restricted distribution (and not mapped), bibliography (pp. 621–626) and indices at end. The introductory part includes an account of vegetation types (by J. A. MacMahon) and four thematic maps.[159]

Partial work

Shaw's flora is a successor to *Spring flora of the Wasatch Region* by A. V. Garnett (1909, with further editions to 1936) and *Flora of the Central Wasatch Front, Utah* (1980) by L. Arnow *et al.*

SHAW, R. J., 1989. *Vascular plants of northern Utah: an identification manual.* vii, 412 pp., illus. Logan: Utah State University Press.

Very concise flora with keys, vernacular names, representative figures, brief diagnostic descriptions of families and genera, and for species and lesser taxa indication of status, distribution, habitat and phenology; addenda, illustrated glossary, references (p. 373) and complete index at end (but no statistics). [Covers the northern quarter of the state including the Logan, Ogden and Salt Lake City areas, in all nine counties and 32 538 square kilometers.]

Woody plants

TRESHOW, M., S. L. WELSH and G. MOORE, 1963. *Guide to the woody plants of Utah.* 115 pp. Provo: Brigham Young University Press. (Reissued 1964, Boulder, Colo.: Pruett Press.)

Analytical key to all taxa with vernacular and scientific names as well as some halftone illustrations; index to all names at end. The general part includes an illustrated glossary. [Subsequently expanded and reissued in 1970 as *Guide to the woody plants of the mountain states* (see **180**).][160]

189

Nevada

See also **180/II**, **180/III** (all works, especially CRONQUIST *et al.*). – Area: 285 724 km². Vascular flora: over 3000 native and naturalized species (Kartesz, 1987; see below).

The only published flora since 1900, with but sketchy coverage of the state, has been Ivar Tidestrom's *Flora of Utah and Nevada* (1925; see **188**). That work succeeded Sereno Watson's 1871 catalogue of the floras of Utah and Nevada (mentioned under Utah), the subsequent *Catalogue of Nevada flora* by C. L. Anderson (1939), and partial works such as *Early flora of the Truckee Valley* (1894) by Fred H. Hillman. The *Contributions* began as a Works Progress Administration (WPA)-aided joint project of the U.S. National Arboretum with the University of Nevada at Reno; prior to commencement of the series, a preliminary checklist was prepared from Tidestrom's flora and issued in 1939 in mimeographed form (122 pp.). The first part of the larger work appeared in 1940; after 50 parts, it was terminated, incomplete, in 1965.

To this day, the state remains without a modern descriptive flora; a detailed 1987 account by John T. Kartesz so far remains available only 'on demand'. Kartesz claimed in his work that Nevada was botanically among the most poorly known of all the conterminous states, with very ill-consolidated documentation. Indeed, its numerous ranges have been among the last botanical 'frontiers' in the 'lower 48'.[161]

Partial works, both comparatively recent, are available for the northeastern and central-southern areas of the state, the latter encompassing the U.S. federal nuclear weapons test sites.

Bibliography

TUELLER, P. T., J. H. ROBERTSON and B. ZAMORA, 1978. *The vegetation of Nevada – a bibliography.* 28 pp. Reno, Nev.: Agricultural Experimental Station, College of Agriculture, University of Nevada. [Includes local floristic references.]

KARTESZ, J. T., 1987. *A flora of Nevada.* [vii], vii, 1739 ll. (Unpublished Ph.D. dissertation, University of Nevada, Reno; available from University Microfilms International, Ann Arbor, Mich. Abstract in *Dissertation Abstracts* **49**/12B: 5119.)

A copiously annotated, keyed enumeration, without family or generic descriptions (and associated commentary); species entries include diagnostic features, synonymy, vernacular names, county distribution, overall state and extralimital range (with altitudinal spread), usual habitat(s), and taxonomic commentary; references, ll. 1726–1729.

UNITED STATES NATIONAL ARBORETUM, 1940–65. *Contributions toward a flora of Nevada.* 50 parts. Maps. Washington, D.C.: Crops Research Division, Agricultural Research Service, USDA (for the National Arboretum). (Mimeographed.)

Comprises a series of fairly detailed descriptive revisions of seed plant families and other groups, with keys, full synonymy and references, localities with citations of *exsiccatae*, general summary of local and extralimital range (including some distribution maps), taxonomic commentary, and notes on habitat, phenology, special features, etc.; each part separately indexed. Many treatments were contributed by ARS botanist William A. Archer, with others by specialists; for full list, see the state bibliography above.

Partial works

BEATLEY, J. C., 1976. *Vascular plants of the Nevada test site and central-southern Nevada: ecologic and geographic distributions.* viii, 308 pp., 28 figs. (incl. 4 maps, 24 halftones). Washington, D.C.: Energy Research and Development Administration (distributed by U.S. National Technical Information Service, Springfield, Va.). (Original version, 1969.)

Includes a systematic enumeration (pp. 110ff.) of vascular plants, with details of local distribution, altitudes, lifeform, phenology, and associates; no vernacular names. The lengthy general part includes details of physical features, climate, vegetation, floristics (with floristic regions), disturbed vegetation, and threatened and endangered species, with a list of references. [Represents in effect an evaluation of the flora and vegetation in an area long used for nuclear bomb testing and subjected to radiation in varying degrees.]

HOLMGREN, A. H., 1942. *A handbook of the vascular plants of northeastern Nevada.* vi, 214 pp. [Logan], Utah: Utah State Agricultural College and Experiment Station/Grazing Service, U.S. Department of the Interior.

A keyed, annotated systematic enumeration of vascular plants, with 'standardized' vernacular names, indication of overall distribution, and notes on habit, special features, frequency, habitat, etc.; glossary and index at end. The introductory section covers geography, physiography, climate, history of botanical exploration, statistics of the flora, and technical matters; list of references (p. vi). [Compiled under the direction of the Intermountain Herbarium at what is now Utah State University in Logan, this work covers the Elko grazing district (Elko County and northern portions of Eureka and Lander Counties).]

Region 19

Pacific Coast United States

This region in a nominal sense comprises the states of Washington, Oregon and California and thus also encompasses the whole of the Cascade Range and the Sierra Nevada. Three subregions can be recognized: the Pacific Northwest, here formally designated as **191**; Northern and Central California, including the Central Valley; and Southern California with the Channel Islands (**197, 198**). The first-named extends into British Columbia, while the last notionally includes parts of the Southwestern United States (**180/II**) and northwestern Mexico. Parts of eastern California and Oregon fall within the Intermountain Plateau (**180/III**). Area: 829 384 km².

The initial consolidation of floristic knowledge on the Pacific Coast took place during the nineteenth and early twentieth centuries. What are now Oregon and Washington were until 1848 claimed as British territory and so first treated in W. J. Hooker's *Flora boreali-americana* (Region 12–13). They next were the focus of pioneer Oregon botanist Thomas J. Howell's never-completed *A flora of Northwest America* (1897–1903). Washington was then separately treated

in *Flora of the state of Washington* (1906) by Charles V. Piper, published through the U.S. National Herbarium and based on a decade's work (1893–1903) at Washington State College in Pullman (and before that at the University of Washington in Seattle); at rather a later date, Oregon got its first flora, *A manual of the higher plants of Oregon* (1941) by Morton E. Peck of Willamette University in Salem. In California, the crest of primary botanical exploration was reached in the mid- to late 1800s (most notably in connection with the second state geological survey of 1860–73). The bulk of the collections gathered went to the Gray Herbarium, the Torrey herbarium in New York and to public and private herbaria in the San Francisco Bay area (including the California Academy of Sciences, established by Albert Kellogg with five others in 1853, and the University of California at Berkeley, strongly developed under Edward L. Greene, its first professor of botany). These materials, along with those collected by William H. Brewer on the above-mentioned state survey, laid a foundation for *Botany of California* (1876–80) by Brewer, Asa Gray and Sereno Watson.[162] But afterwards, functional control, symbolized in California by the work over 38 years of '49er Kellogg, by the *Botany* itself and the subsequent controversy between Gray and Greene, and by the vigorous activity of Greene's successor Willis Linn Jepson, passed to the region as existing centers were developed and more were established. Many individuals – not least Jepson – developed collections of their own; while over the same period local pedagogical works began to appear, among them *Popular California flora* (1879, with reissues to 1896) and *Analytical key to West Coast botany* (1879, with later editions to 1905), both by Volney Rattan, and *A school flora for the Pacific Coast* (1902) by Jepson. By the first decade of the twentieth century, enough of a foundation existed for Jepson to commence publication of a 'home-grown' critical work, his still-unfinished *A flora of California* (1909–).[163]

After World War I, Leroy Abrams at Stanford University, home to the increasingly substantial Dudley Herbarium, began *Illustrated flora of the Pacific States* as a West Coast analogue of 'Britton and Brown' (**140**). Jepson's *Manual*, which long would remain in print, also appeared in the mid-1920s, and in 1935 Philip Munz published *A manual of southern Californian botany*. The latter would be the first major publication of the new Rancho Santa Ana Botanical Garden, then in Orange County southeast of Los

Angeles. A unified but somewhat sketchy list for the California Channel Islands was produced by Alice Eastwood in 1941.[164]

Subsequent works have now brought knowledge of the vascular plants to a degree equaled only by the Northeast and Midwest, although novelties continue to be found, particularly in California.[165] Abrams' *Illustrated flora* was completed in 1960 by Roxana S. Ferris. State coverage is currently of relatively recent date: Washington in the subregional *Vascular plants of the Pacific Northwest* (1955–65) and *Flora* (1973) by C. Leo Hitchcock and his associates, Oregon in the 1961 revision of Peck's manual, and California in *A California flora* (1959; supplement, 1968) by Munz with David D. Keck. Munz also produced a substantial revision of his Southern California manual in 1974; and in 1993 came the very important revision of the Jepson manual, produced under the editorship of James Hickman. However, no 'doorstop manual' for the entire Pacific Coast region comparable to those for the Northeast, Southeast or other large regions has been published, or perhaps even contemplated; the work of Abrams and Ferris was never 'condensed' as was its eastern model.

The considerable physiographic and habitat diversity in the region has, along with local needs, given rise to several partial floras. The areas in the Great Basin are covered in *Intermountain flora* (**180/III**); the 'Pacific Northwest' in the works already referred to as well as *Pacific Northwest ferns and their allies* (1970; see **121/I**); and the dry southeast in *Trees and shrubs of the southwestern deserts* (3rd edn., 1981; see **180/II**). There are several more or less recent local floras, particularly in the Bay Area, the Seattle–Portland–Eugene axis in the Northwest, parts of Southern California, the Palouse in southeastern Washington and adjacent parts of Idaho (centering on Washington State University in Pullman, Wash.). For the Channel Islands (which can also include Guadalupe (**011**)), Eastwood's list was succeeded in 1985 by one by Gary Wallace (though even this now needs revision), and there are more or less recent enumerations or checklists for individual islands or groups; nevertheless, there is here scope for a 'biological flora'.

So far, no major compendia of distribution maps have been published. As elsewhere in the western states, the county dot style widely used in central and eastern states has been uncommon (and indeed is unsuitable).

A number of books on trees (and shrubs) covering individual states in the region also are available, none of them recent except *Manual of Oregon trees and shrubs* (1988) by W. R. Randall. For the Pacific states as a whole, the most recent and generally useful work is *Illustrated manual of Pacific Coast trees* (1935, with later reissues) by Howard McMinn and Evelyn Maino. Valuable also is *An illustrated manual of California shrubs* (1939; 2nd edn., 1951) by McMinn.[166]

Bibliographies. General and divisional bibliographies as for Division 1.

Indices. General and divisional indices as for Division 1.

190

Region in general

Abrams and Ferris's *Illustrated flora of the Pacific States* covers the whole area and is comparable to Britton and Brown's *Illustrated flora of the northern United States* (**140**).

Keys to families and genera
For *Genera of the western plants* (1983) by W. T. Batson, Jr., see **100**.

ABRAMS, L. and R. S. FERRIS, 1923–60. *Illustrated flora of the Pacific States.* 4 vols. Illus. Stanford: Stanford University Press. (Vol. 1 reprinted with revised nomenclature, 1940; all vols. subsequently reissued.)

Copiously illustrated, briefly descriptive atlas-flora of vascular plants; includes keys to all taxa, rather full synonymy (with references), vernacular names, generalized indication of local, regional and extralimital distribution, including life-zones, and notes on habitat, frequency, phenology, special features, etc.; indices in each volume (complete general index in vol. 4). The introductory section includes among other topics descriptions of physical features, floristic elements, and Merriam life-zones.[167]

Woody plants (and trees): guides
In addition to the following, reference may be made to E. L. LITTLE, *The Audubon field guide to North American trees:*

Western Region (under **100: Woody plants**) and to J. B. BERRY, 1966. *Western forest trees: a guide to the identification of trees and woods for students, teachers, farmers and woodsmen.* 251 pp. New York: Dover. (Originally published 1924, Yonkers, N.Y., as *Western forest trees: a guide ... to accompany farm woodlands.*)

MCMINN, H. E. and E. MAINO, 1935. *An illustrated manual of Pacific Coast trees.* xii, 409 pp., 415 text-figs., 1 col. pl. Berkeley: University of California Press. (Several reissues, the first in 1946.)

Fully keyed, briefly descriptive semi-popular treatment encompassing native, naturalized and many cultivated species; includes synonymy, vernacular names, generalized indication of regional distribution or country of origin, and notes on habitat, phenology, cultivation, special features, etc.; lists of trees recommended for particular uses (by H. W. Shepherd); glossary, list of references, and index to all botanical and vernacular names.

SUDWORTH, G. B., 1908. *Forest trees of the Pacific slope.* 441 pp., 207 illus. (of which 15 halftones), 2 maps. Washington, D.C.: U.S. Government Printing Office (for Forest Service, USDA). (Reprinted with revised nomenclature by E. S. Harrar, 1967, New York: Dover. xv, 455 pp., 207 illus., 15 pls., 2 maps.)

Illustrated, semi-popular descriptive systematic treatment, without keys; includes synonymy, vernacular names, and extensive notes on distribution, habitat, ecology, special features, wood, bark, properties, uses, etc.; complete index. All species are figured. [The now-obsolete 'American Code' of nomenclature was employed in the original work; the Dover edition incorporates necessary nomenclatural changes along with an appreciation of Sudworth by Woodbridge Metcalf.][168]

191
Pacific Northwest subregion

See also **121/I** (TAYLOR). The area as treated here includes, along with Washington, the greater part of Oregon (exclusive of those parts in the 'Great Basin'), northern Idaho, far western Montana, and southern British Columbia.

Following *Flora boreali-americana*, the first general treatment was *A flora of Northwest America* by Thomas Howell (1897–1903). Howell's work was succeeded by state floras for Washington and Oregon, and then by the two major works described below. For the coastal zone, there is also a students' manual. The trees were first separately treated by W. A. Eliot in 1938.

HITCHCOCK, C. L. *et al.*, 1955–65. *Vascular plants of the Pacific Northwest.* 5 vols. Illus. Seattle: University of Washington Press.

Comprehensive, illustrated descriptive 'research' flora of vascular plants; includes keys to all taxa, extensive synonymy and references, vernacular names, relatively detailed indication of local range, general summary of extralimital distribution, extensive taxonomic commentary, well-executed figures of most species, and notes on habitat, phenology, etc.; indices to all botanical and vernacular names in each volume. The introductory section includes a glossary as well as an index to families. The area covered by this work is as indicated above under the heading.[169]

HITCHCOCK, C. L. and A. CRONQUIST, 1973. *Flora of the Pacific Northwest: an illustrated manual.* xix, 730 pp., illus. Seattle: University of Washington Press.

Manual-key to vascular plants, with limited synonymy, vernacular names, concise indication of local range, diagnostic figures, and notes on habitat, phenology, life-form, etc.; complete index at end. [Essentially a condensation, with revisions, of *Vascular plants of the Pacific Northwest.*]

Partial works

The original version of *Handbook of Northwestern plants* succeeded *Flora of the Northwest Coast* (1915) by C. V. Piper and R. K. Beattie, the first regional manual.

GILKEY, H. M. and LaR. J. DENNIS, 1967. *Handbook of Northwestern plants.* [vi], 505 pp., illus. Corvallis, Oreg.: Oregon State University Bookstores. (Originally published by Gilkey alone in 1936 as *Handbook of Northwest flowering plants.*)

Briefly descriptive students' manual of vascular plants, with keys to all taxa, vernacular names, diagnostic features, and notes on local distribution, frequency, habitat, etc.; glossary and complete index. Covers western Oregon and Washington from the Cascade Range to the Pacific coast, and includes the Olympic Peninsula with its 'rain forest'.

Woody plants (including trees): guide

ELIOT, W. A. (assisted by G. B. McLEAN), 1938. *Forest trees of the Pacific Coast.* 565 pp., 249 halftones, map (end-papers). New York: Putnam. (Reissued as a 'revised edn.', 1948.)

Concise, popularly oriented, illustrated descriptive treatment of trees of Alaska, British Columbia, Washington and Oregon, with some keys to species, botanical and vernacular names, very detailed indication of internal and extralimital range, numerous photographs (of both habit and fresh sprays displayed against 1 inch squares), and notes on habitat, frequency, special features, etc.; full synoptic list of trees, with details of distribution (pp. 539–548) and complete index. The 'Acknowledgments' include publications consulted.

192

Washington

Area: 172 929 km². Vascular flora: no data.

The only flora specifically treating Washington is *Flora of the state of Washington* (1906) by C. V. Piper, now obsolete. It was soon after joined by *Elementary flora of the Northwest* (1914) by T. C. Frye and G. R. Rigg, a students' work. Their successors, the works by C. L. Hitchcock (with A. Cronquist and others), have extended coverage to the 'Pacific Northwest' (see **191** above).

Bibliography

DAUBENMIRE, R. F., 1962. Vegetation of the state of Washington: a bibliography. *Northw. Sci.* **36**: 50–54. [Includes local floristic works.]

Partial work

St. John's manual, described below, encompasses the 'Palouse', an important agricultural area centering on Pullman (home to Washington State University) and also taking in the nearby Idaho towns of Moscow and Lewiston. Its first edition was preceded by *The flora of the Palouse region* (1901) and then *Flora of southeastern Washington and of adjacent Idaho* (1914), both by C. V. Piper and R. K. Beattie.

ST. JOHN, H., 1963. *Flora of southeastern Washington and adjacent Idaho*. 3rd edn. xxix, 583 pp., illus. Escondido, Calif.: Outdoor Pictures. (1st edn., 1937, Pullman.)

Students' manual of seed plants, with keys to all taxa, brief descriptions, partial synonymy, vernacular names, local distribution (including Merriam life-zones), some critical commentary, and notes on habitat, frequency, biology, etc.; glossary, lexicon and index at end.

193

Oregon

For the northern part, see also **191**. – Area: 250 078 km². Vascular flora: 3203 native and introduced species (Peck, 1941).

Prior to the manuals by Morton Peck, Oregon was covered by pioneer Oregon botanist Thomas Howell's *A flora of Northwest America*. Additional coverage may be sought in Pacific Northwest floras (**191**) or, for the Great Basin, in *Intermountain flora* (**180/III**). The students' manual by Gilkey and Dennis, which encompasses the Willamette Valley and adjacent areas, is under **191**. In the mid-1990s a project for a new Oregon flora was organized, with its main aims production of a new manual as well as development of an Internet Web site; its headquarters are at Oregon State University.[170]

Bibliography

FRANKLIN, J. F. and N. E. WEST, 1965. Plant communities of Oregon: a bibliography. *Northw. Sci.* **39**: 73–83. [Includes floristic papers.]

PECK, M. E., 1961. *A manual of the higher plants of Oregon*. 936 pp. [Portland]: Binfords and Mort (in cooperation with Oregon State University Press, Corvallis). (1st edn., 1941.)

Descriptive manual of vascular plants, with keys to all taxa, essential synonymy, vernacular names, concise indication of local and extralimital range, and miscellaneous notes; illustrated glossary and index to all botanical and vernacular names at end. The introductory section gives a description of floristic regions in Oregon, with map.

Woody plants

RANDALL, W. R., 1988. *Manual of Oregon trees and shrubs*. 5th edn. 305 pp., illus. Corvallis: Oregon State University Bookstores. (1st edn., 1958.)

A dendrology with *pro forma* text covering habit, botanical features, habitat, distribution and notes on plant relations, uses, special features and potential; references and readings, pp. 256–258. Discussions of family features, shade tolerance, fossil history and U.S. forest regions, a winter twig key, and list of state trees along with an index are also included. The introductory part is pedagogic and includes terminology and (pp. 28–35) a checklist with vernacular names.

195

California (in general)

Area: 406377 km². Vascular flora: 5867 native and naturalized species and 1169 non-nominate infraspecific taxa; of these 4844 species are now considered native with 1416, or 29.3 percent, being endemic (Hickman, 1993; see below).[171] Under this heading are listed works pertaining to the state as a whole. For northern California, see also **196**; for southern California, see also **197**; for the Californian Channel Islands, see **198**. For *Flora of the marshes of California* by Herbert F. Mason, see **108**.

With a large population and, among western states, a long-autonomous botanical tradition, California is now well supplied with floristic literature. Only the principal works can be accounted for here.[172]

Bibliographies

HOWARD, A. Q., 1974. *An annotated reference list to the native plants, weeds, and some of the ornamental plants of California*. 34 pp. Berkeley, Calif: Cooperative Extension, University of California. [Includes coverage of local floras.]

SMITH, J. P., JR., 1985. *California vascular plants: literature on their identification and uses*. [iii], 75 pp. (Humboldt State University Miscellaneous Publication 1). Arcata, Calif. [Twentieth-century publications only, arranged by author with geographical and subject indices.]

TAYLOR, M. S. (comp.), 1983. *California floristic bibliography*. 265 pp. (Fl. Butt.4(2/3)). St. Louis: The author. [A classified, annotated bibliography with keywords arranged in the first instance by regions and counties (with statewide works on pp. 5–13); source bibliographies, pp. 259–261, and county index, p. 265. Unpublished as well as published material is included, theses not excepted. The introduction gives the plan of the work as well as general observations, with acknowledgments on pp. 3–4. – Incorporates *idem*, 1977. *List of regional and local floras and additions of California*. 11 pp. Chico, Calif.: The author; and *idem*, 1982. *California floristics: a preliminary bibliography of unpublished checklists and reports*. 45 pp., map (Fl. Butt. 3(2)). Chico, Calif.]

HICKMAN, J. C. (ed.), 1993. *The Jepson Manual: higher plants of California*. xvii, 1400 pp., illus. Berkeley/Los Angeles: University of California Press. (Succeeds W. L. JEPSON, 1923–25. *A manual of the flowering plants of California*. Pp. 1–24, 24a–24h, 25–1238; 1023 text-figs. Berkeley: Sather Gate

Bookshop and Associated Students' Store; reprinted 1938 with further reissues in 1951 and 1960, Berkeley/Los Angeles: University of California Press.)

Concise, copiously illustrated descriptive desktop flora, with keys, limited synonymy and citations (including links to Munz's flora and its supplement as well as recent literature), vernacular names, indication of distribution within (in coded form) and without the state, as well as habitat, some taxonomic commentary, and horticultural information; revisionary references under genera and complete index at end. Appendices include (1) a summary with statistics, (2) the classification system used (essentially that of Cronquist) with a discussion, and (3) name changes in relation to recent general works (pp. 1323–1363). The introductory part includes acknowledgments, the history of the project, the plan of the work and conventions, pronunciation of names, a glossary, climate zones (with map), geographical subdivisions (with map), paleohistory, and horticultural aspects along with a general key to families (organized in 22 main groups).[173]

JEPSON, W. L., 1909– . *A flora of California*. Vols. 1–4 (in parts). Illus. Berkeley: The author (vols. 1–3; distributed through Associated Students' Store at the University of California and other outlets); Jepson Herbarium and Library, University of California (vol. 4(2)). (Vol. 4(2) edited by L. R. Heckard and R. Ornduff.)

Comprehensive 'research' flora of seed plants; includes keys to genera and species, ample descriptions, full synonymy with references and citations, indication of localities with support of *exsiccatae* and other records, generalized summary of local and extralimital range, extensive taxonomic commentary, photographs and large figures of representative species and/or diagnostic features thereof (480 so far), and brief notes on habitat, special features, etc.; no indices (save for family and generic names in vol. 2, although in vol. 4(2) there appears an index to families treated as of 1979). A historical sketch of Californian botany appears in vol. 2. [The arrangement of families follows the classical Englerian system; vol. 1 (1909–22, completed save for an introduction and indices) covers families from Pinaceae to Fumariaceae (parts 1–7; parts 8(1), pp. 1–32, and 8(2), pp. 579ff., i.e., the index, lacking); vol. 2 (1936, complete) covers Capparidaceae to Cornaceae (in 3 parts); vol. 3 (1939–43, not completed) covers Lennoaceae to Solanaceae (parts 1–2; the introductory pages 1–16 and a final part as well as the index are lacking); and of vol. 4 (begun 1979) only part 2, covering Rubiaceae, has been published. Approximately 20–25 percent of the seed plants are still not covered; furthermore, revision of the earlier parts is now necessary.][174]

MUNZ, P. A. (in collaboration with D. D. KECK), 1959. *A California flora.* 1681 pp., 134 text-figs., 1 col. pl., 7 maps. Berkeley/Los Angeles: University of California Press. Continued as P. A. MUNZ, 1968. *Supplement to 'A California Flora'.* 224 pp. Berkeley/Los Angeles. (Both works subsequently reprinted and bound together as 1 vol.)

Descriptive, somewhat sparsely illustrated manual of vascular plants; includes keys to all taxa, limited synonymy, vernacular names, generalized indication of local range (organized according to community-units) as well as extralimital distribution, taxonomic commentary, representative figures (depicting families and subfamilies), and notes on karyotypes, habitat, special features, variation, etc.; glossary, list of vernacular names with botanical equivalents, and index to all botanical names. The introductory section includes chapters on geology and vegetation, with descriptions of the 29 community-units upon which species distributions are based. [Less 'accessible' than Jepson's manual; one commentator has criticized it for 'obfuscatory monotonous text'.][175]

Woody plants
In addition to the work given below, reference may also be made to *Handbook of the trees of California* (1905) by Alice Eastwood and *The silva of California* (1910), with 85 plates, and *Trees of California* (1909), both by Jepson, as well as to the tree manuals under **190**. A more recent treatment of shrubs is that by P. H. Raven (1966; see **197**).

McMINN, H. E., 1951. *An illustrated manual of California shrubs.* 2nd edn. xi, 663 pp., illus., col. frontisp. Berkeley: University of California Press. (4th printing, 1964. 1st edn., 1939, San Francisco.)

Descriptive manual with numerous figures and photographs, covering 800 species and 200 additional varieties; includes essential synonymy, references, type localities (with *exsiccatae*), distribution and habitats, and sometimes extensive notes covering special features, properties, taxonomy, and SPNs; glossary, lexicon of epithets, proposed nomenclatural changes, bibliography, horticulture (by F. Schumacher) and general index at end. The introductory part covers the scope of the work, sources of material, acknowledgments, and preliminary sections on classification, vegetation and plant formations, terminology, and use of keys as well as general keys to families and genera.

196
Northern California

See also **195** (all works). Briefly accounted for here are floras and manuals covering the greater San Francisco Bay area. Earlier standards included *Catalogue of the plants growing in the vicinity of San Francisco* by H. N. Bolander (1870), *Flora of the vicinity of San Francisco* by H. H. Behr (1888), and *Catalogue of the flowering plants and ferns growing spontaneously in the City of San Francisco* by K. Brandegee (1892, in *Zoe*). Edward L. Greene's *Flora franscicana* (1891–97, not completed) and *Manual of botany of the region of San Francisco Bay* (1894) followed, the latter with analytical keys. More extensive coverage was furnished by Jepson's *A flora of western middle California* (1901; 2nd edn., 1911), a progenitor of his later statewide manual.

Current partial works include *Marin flora* (1949; 2nd edn. with supplement, 1970) by J. T. Howell, *A flora of San Francisco, California* by Howell, P. H. Raven and P. Rubtzoff (1958, in *Wassmann Journal of Biology*), and *A flora of the Santa Cruz Mountains of California* (1961) by J. H. Thomas. The last-named covers the whole of the San Francisco Peninsula.

197
Southern California

See also **181/II, 195** (all works). The first substantial separate florulas in this geographically and botanically distinctive area were *Botany of the Death Valley expedition* by Frederick V. Coville (1893, in *Contr. U.S. Nat. Herb.*), part of the results of an expedition mounted by what would soon afterwards become the U.S. Biological Survey, and *Flora of Los Angeles and vicinity* (1904; revised edns., 1911, 1917) by Leroy Abrams. The latter was prepared at Stanford University as sufficient local resources were not yet available, although by then much collecting had been accomplished by Samuel B. Parish and others. Abrams also revised the woody flora on behalf of the New York Botanical Garden in *A phytogeographic and taxonomic study of the southern California trees and shrubs* (1910).

These works were in 1923 followed by *Flora of Southern California* by Anstreuther Davidson (of the Los Angeles County Museum) and George L. Moxley, in 1935 by Philip Munz's *Manual*, and in 1974 by the latter's very substantial successor.

Among partial works, the San Diego area has recently been covered in *A flora of San Diego County, California* (1986) by R. M. Beauchamp; and Santa Barbara in *Flora of the Santa Barbara area* by C. Smith (1952; revised 1976). Shrubs were separately treated in P. H. RAVEN, 1966. *Native shrubs of Southern California*. 132 pp., illus. (part col.). Berkeley/Los Angeles: University of California Press.

Bibliography

PARISH, S. B., 1909–20. A bibliography of the southern California flora. *Bull. Southern Calif. Acad. Sci.* 8: 71–75; 9: 57–62; 19: 24–79. [Arranged by author.]

MUNZ, P. A., 1974. *A flora of southern California.* 1086 pp., 103 pls., map. Berkeley/Los Angeles: University of California Press. (Based on *idem*, 1935. *A manual of Southern Californian botany*. Claremont, Calif.)

Illustrated manual-flora of vascular plants, with keys to all taxa down to species, concise descriptions, limited synonymy (without references), vernacular names, and brief notes on habitat, occurrence, frequency, local and extralimital range, altitude, karyotypes, and vegetation formations; glossary and indices to all vernacular and botanical names. The short introductory chapter includes sections on geography, geology, vegetation formations, the Californian Channel Islands, phytogeography, and the plan of the work.

198

Californian Channel Islands

See also **197** (all works, especially that of MUNZ). – Area: no information located. Vascular flora: 621 taxa, of which 227 introduced and 137 endemic (Wallace, 1985; see below); these figures are, however, now out of date as Junak *et al.* (1995; see below) have reported for Santa Cruz alone 627 species, of which 157 are considered introduced.

These islands, whose flora is noted for endemism and seemingly unusual biogeographical links, are located in two groups in the Pacific Ocean just off the southern Californian coast. Somewhat related to them floristically is the more distant Mexican island of Guadalupe (**011**). The first overall account – not covering all islands – was *Flora of the Californian Islands* (1890) by Townshend S. Brandegee, published in the Brandegees' private journal, *Zoe*. The next consolidation, encompassing all islands, was Alice Eastwood's imperfectly documented *The islands of Southern California and a list of the recorded plants* (1941, in *Leaflets of Western Botany*). In 1985 a third account, by Gary D. Wallace, was published (see below).

Several accounts for individual islands or groups are now available, some representing a second or higher cycle. In the **southern** group recent contributions include P. H. RAVEN, 1963. A flora of San Clemente Island, California. *Aliso* 5: 289–347; R. F. THORNE, 1967. A flora of Santa Catalina Island, California. *Ibid.*, 6(3): 1–77, figs. 1–28 (incl. map); S. A. JUNAK and J. M. VANDERWIER, 1990. An annotated checklist of the vascular plants of San Nicolas Island, California. In *Fifth Biennial Mugu Lagoon/San Nicolas Island Ecological Research Symposium* (Naval Air Station, Point Mugu, Calif., 1988), pp. 121–145. N.p.; and S. A. JUNAK *et al.*, 1993. *A revised flora of Santa Barbara Island*. 112 pp. Santa Barbara: Santa Barbara Botanic Garden. The **northern** islands are covered in CHANNEL ISLANDS NATIONAL PARK and SANTA BARBARA BOTANIC GARDEN, 1987. *A checklist of vascular plants of Channel Islands National Park*. 16 pp. Tucson: Southwest Parks and Monument[s] Association; to these may be added S. A. JUNAK *et al.*, 1995. *A flora of Santa Cruz Island*. v, 397 pp., text-fig., halftones, maps. Santa Barbara: Santa Barbara Botanic Garden (with the California Native Plant Society).[176]

WALLACE, G. D., 1985. *Vascular plants of the Channel Islands of southern California and Guadalupe Island, Baja California, Mexico*. 136 pp. (Nat. Hist. Mus. Los Angeles County, Contr. Sci. 365). Los Angeles.

The core of this work comprises an alphabetical table of taxa (pp. 2–34) with indication of presence/absence on any or all of nine islands, and whether a given taxon is endemic, native or introduced; where some doubt exists a number is given relating to the 'Literature Cited' (pp. 133–136; now the best available bibliography for the islands). The table is followed by a discussion of floristics and, in appendix I, citation of all *exsiccatae* consulted with locations. The introductory part includes historical background.

Notes

1 The importance of these 'regions', with their concentrations of specialists and key herbaria, was indeed acknowledged by the original FNA Program through provision for seven regional editorial offices in addition to the central unit in Washington.

2 The U.S. National Herbarium, whose *Contributions* were the outlet for many floras from the 1890s through the 1920s, was initially organized in 1868 under the U.S. Department of Agriculture with the Smithsonian Institution's accumulated collections on deposit. Its first curators were Charles C. Parry (1869–72) and George Vasey (1872–93). In 1894 it was re-established at the Institution in its Division of Plants (part of the U.S. National Museum), but remained a joint undertaking with the USDA under Frederick V. Coville, who from 1893 was head curator on behalf of the Smithsonian as well as chief Departmental botanist. The Institution created its own first full-time botanical post in the same year, and in 1896 the collections were physically re-housed. The *Contributions*, first published in 1890, was set up as a consequence of an 1889 Act of Congress providing for exploration and reporting on the biota of then little-known areas of the United States by the Department (then also officially responsible for bird and mammal survey under the well-known biogeographer C. Hart Merriam). From 1902 it became the responsibility of the Institution, developing into a leading journal in the field.

3 The average American manual-flora is bulkier and less well adapted for field use than its European counterpart. This is perhaps as much due to tradition as to factors such as larger areas and floras, reinforced by widespread use of motor transport for botanizing.

4 Progress is documented in the *Flora of North America Newsletter*, published about twice a year or so. In mid-1999, however, the secretariat was discontinued and editorial and technical functions decentralized. The *Newsletter* is now prepared at the Hunt Institute for Botanical Documentation of Carnegie-Mellon University in Pittsburgh, Pennsylvania, which has moreover assumed a greater overall role in the FNA Project. Further details may be found in *Fl. N. Amer. Newsl.* **13** (1999).

5 In J. Ewan, 1969. *A short history of botany in the United States*, p. 113. New York.

6 Noel Holmgren, personal communication, December 1998.

7 Disappointing in some respects, particularly in design and in its absence of references, even a 'further reading' section.

8 Darlington indicated (p. 33) that by the end of the 1840s students of botany were with respect to floras and other texts far better provided for than when he was young.

9 A complete bibliography and classified index to the series of Flora North America Reports (nos. 1–83) may be found in B. R. Rohr *et al.*, 1977. The Flora North America Reports – bibliography and index. *Brittonia* **29**: 419–432.

10 The last of these was based on M. A. Miasek, 1977. *Regional, state, and local fern floras, manuals, checklists, and new fern and fern-ally records for the United States and adjacent Canada since 1950*. 21 pp. New York: Library, New York Botanical Garden (offset from typescript.) [An entirely bibliographic work, not restricted to floras but including listings by state and region.]

11 It is now in effect an expanded version of *The Taxonomic Index*. Coverage was at the same time advanced beyond a former restriction to papers by American authors or appearing in American journals. Books have been added as well as non-American authors and/or journals.

12 From 1996, however, *The Taxonomic Index* has effectively reappeared as a feature of *Brittonia* with the transfer thereunto of *Index of American Botanical Literature*.

13 Bassett Maguire, personal communication. By 1998, 13 parts of series II had appeared, the most recent (not on vascular plants) in 1990.

14 The *Flora* is supported by a substantial database and information system, housed at the Missouri Botanical Garden in St. Louis. The project also produces *Fl. N. Amer. Newsl.*, in practice usually bi-annual; there is also a Web site (http://www.fna.org/). As of writing (1999) the convening editor was Nancy R. Morin and the managing editor was James L. Zarucchi. Publication in CD-ROM format was also planned. The full text is available for search through the FNA Web site. In 1996 the planned volumination was increased from 14 to 30 volumes. Vols. 1–26 will encompass the introduction along with vascular plants, while vols. 27–29 are planned to cover bryophytes. Vol. 30 is to contain a general index. However, continuation of the project was for a time uncertain following loss in mid-1999 of most funding.

15 The deaths of Gray in 1888 and Watson in 1892 removed from the project those most personally involved. Benjamin Robinson lacked the energy and commitment necessary for its completion and, moreover, was more interested in systematic studies, while Merritt L. Fernald, though strongly interested in floristic work, became from the time of his entry into the Gray Herbarium in 1891 almost entirely concerned with the northeastern and mid-Atlantic regions (Harold St. John, personal communication, 1982).

16 The first edition of this work appeared as the second of a series, *Biota of North America*. No other volumes have been seen. Support for the *Synonymized checklist* came in

part from The Nature Conservancy, a private organization in the United States. Extension of coverage to Hawaii, Puerto Rico and the Virgin Islands was made in response to requests from various government agencies, but without geographical distributions the book is all but useless for biogeographical purposes. Kartesz is, however, proceeding with the compilation of geographical information (K. Gandhi, personal communication, September 1997). Additional products have since been announced.

17 The principal compilers of the work, as mentioned in an added loose insert, were W. E. Rice, S. F. Smith and D. C. Wasshausen. While viewed professionally as fairly critical, it is considered to have been less representative of the taxonomic community than the Karteszes' nomenclator (Stephen Edwards, personal communication, 1986). The checklist was subsequently incorporated into the USDA's PLANTS information system and, with revisions, has become available over the Web, both directly and also within the U.S. government's Integrated Taxonomic Information System (ITIS); see http://www.itis.usda.gov/itis/info.html.

18 The effectiveness of existing manuals was examined in J. W. Anderson, 1961. An evaluation of state tree identification guides of the United States. *J. For.* 59: 349–355.

19 The series represents the culmination of more than 20 years' work on the part of its author (who became Forest Service dendrologist in 1942). For a description of the project, see E. L. Little, Jr., 1951. Mapping ranges of the trees of the United States. *Rhodora* 53: 195–203.

20 A 'Basic Flora' in the original Flora North America Program.

21 A full taxonomic and phytogeographic treatment was projected as a separate publication but nothing has materialized.

22 Mountain plants are here called 'orophytes' (after Löve, Löve and Kapoor, 1971, in *Arct. & Alp. Res.* 3: 139–165).

23 The format in this underedited book features too much white space; moreover, herbarium abbreviations are non-standard.

24 The desert research programme of the Carnegie Institution of Washington, organized in 1903 following a proposal by Frederick Coville (author of *Botany of the Death Valley expedition*; see **197**), was from 1928 in the care of Forrest Shreve (at the Desert Laboratory 1909–47). He envisioned floras of the four main desert areas along with ecological and other studies. Only *Vegetation and flora of the Sonoran Desert* by him and Ira Wiggins was, however, actually realized, with the flora (by Wiggins) not completed until well after Shreve's death and the winding-up of the Laboratory. Floras of the other three areas are underway or planned, though under other auspices (the Great Basin at New York

Botanical Garden, as *Intermountain flora*; the Chihuahuan Desert under Marshall C. Johnston and James S. Henrickson; and the Mojave Desert under Henrickson and Barry Prigge).

25 C. D. K. Cook, personal communication, 1979.

26 Very recently, botanists in Alaska have joined with others in North America and Eurasia in a Panarctic Flora Project.

27 The work was a 'Basic Flora' in the original Flora North America Program.

28 Although intended as a field manual, its format renders it too bulky as such.

29 See, for instance, G. Lawson, 1892. On the present state of botany in Canada. *Proc. &. Trans. Roy. Soc. Canada*, I, 9/IV: 17–20; R. Falconer, 1932. The intellectual life of Canada as reflected in its Royal Society. In Royal Society of Canada, *Fifty years retrospect: anniversary volume 1882–1932*, pp. 9–25. [Ottawa.]; and S. E. Zellner, 1987. *Inventing Canada: Early Victorian science and the idea of a transcontinental nation*. Toronto: University of Toronto Press.

30 The work of the Macouns was augmented by the botanical group of the Central Experimental Farms, established in 1886 under William Saunders and especially active in the central prairies and plains. For a recent biography of the elder Macoun, see W. A. Waiser, 1989. *The field naturalist: John Macoun, the Geological Survey, and natural science*. Toronto: University of Toronto Press. William Steere and Howard Crum (in *Bot. Rev.* 43 (1977)) have labeled his treatment of the mosses (with Nils Kindberg) in the *Catalogue* a 'muscological disaster'.

31 B. Maguire, *Highlights of botanical exploration* (see **Progress** under Division 1).

32 In the 1950s, Áskell Löve wrote that, in spite of a long history of botanical work, 'it is a regrettable fact that the Canadian flora, covering an area comparable in size to the whole of Europe, remains so little known botanically that even estimates of the total number of species to be expected in the country vary in thousands rather than in hundreds'. In addition, books comparable to *Flore laurentienne* had 'still to be written for other parts of the country'. Á. Löve, *New Phytol.* 57: 267–269 (1958). The situation is now considerably improved.

33 The lack of differentiation of typefaces in this book is regrettable and wearing for the user. Moreover, parts of this work did not take into account all possible contemporary taxonomic studies, relatively little attention was paid to work outside North America, and publication appears to have been tardy, with all research terminated by 1972 (from review by J. S. Pringle in an undetermined source).

34 An evaluation of floristic literature appears in A. Ceška, 1985. Museum collections and phytogeography. *Occas. Pap. Brit. Columbia Prov. Mus.* 25: 79–91.

35 A two-volume version was also planned.

36 The system used was based on that developed for the original Flora North America Project, with which both authors were associated before its suspension in 1973. The information system created was projected to be continuing, allowing for updating of the inventory and to serve as a basis for a descriptive flora (S. G. Shetler in *Syst. Bot.* **2**: 226 (1977)). According to the authors of the later *Vascular plants of British Columbia*, however, 'several hundred' erroneous citations were introduced based on misidentifications or wrong assumptions. The work had been heavily based on the literature, with rather less emphasis on verification of records.

37 The present northern boundary (60°) was established only in 1912.

38 Some workers consider Québec as including Labrador Territory; the total area is then 1 825 780 km².

39 In addition to the manuals listed here, Marie-Victorin from 1923 to 1931 published several critical family revisions as contributions towards 'un traité complet de la flore de Québec'. This was never realized, in part on account of additional interests and most surely after his death in an automobile accident in 1944. Moreover, such a project may have been at the time simply not feasible. Rousseau's work represents the most extensive subsequent development in the province of Marie-Victorin's interests in phytogeographical documentation.

40 For an obituary of Père Louis-Marie and further information on his *Flore-manuel*, see B. Boivin, 1979. Père Louis-Marie (1896–1978). *Taxon* **28**: 432. The work is now, however, used less than *Flore laurentienne*, at least in university circles (Marcia Waterway, personal communication).

41 When originally published, the work was the first such to include chromosome numbers.

42 Meyer, also a specialist on *Juncus* and a professor at Göttingen University, is best known for his four-volume *Geschichte der Botanik* (1854–57).

43 A vita and bio-bibliography of Rouleau, who early in his career had connections with Fernald, is also included.

44 Hinds has recognized relatively few formal infraspecific taxa, much as have Morton and Venn in Ontario.

45 Cf. J. Reveal, *Botanical explorations in the American West* (see **Progress** under Division 1).

46 For a contemporary review of these and other key works, see Charles W. Short, *Bibliographia botanica* (1835).

47 The Mississippi River was reached by 1856 and the 1890 edition reached well into the Great Plains (there coinciding with the eastern limit of John M. Coulter's Rocky Mountains manual). In 1907 and in 1950 the western limit was set at the Missouri–Red River line, so excluding Kansas, Nebraska, and the Dakotas.

48 This was an impetus to Short's second review, *Sketch of the progress of botany in Western America* (1836).

49 The Gray–Wood 'race' is told in *Asa Gray* by A. Hunter Dupree (1959, pp. 169–173, 202). Wood's work would eventually sell over 800 000 copies in its many versions (Stafleu and Cowan, *Taxonomic Literature-2*, 7: 434 (1988)) and, despite criticism, was in its time and even afterwards much appreciated.

50 This rivalry was a distinguishing feature of the so-called era of 'Bughole Botany' (Reveal, *op. cit.*, pp. 179–183).

51 It was reissued in 1967 by Dover Books, a New York firm then specializing in affordable reprints of 'classics'.

52 The style of the original version reflected its senior author's precise mind and his early education in engineering, and its first edition led, through the French publisher Klincksieck, to the genesis and organization of what became a European classic, *Flore descriptive et illustrée de la France* (1901–06) by Abbé Hippolyte Coste. The two works together served as models for similar illustrated floras in North America, Europe and eastern Asia. See G. G. Aymonin, 1974. La naissance de la 'Flore descriptive et illustrée de la France' de l'Abbé Hippolyte Coste. *Taxon* **23**: 607–611.

53 *Gray's Manual* was a 'Basic Flora' in the original Flora North America Program.

54 The *Manual* was a 'Basic Flora' in the original Flora North America Program. The main changes in the 1991 edition are in the genera *Panicum* and part of *Viola* as well as the adoption of the author's family classification. The *Companion* accounts for subsequent revisions to taxonomy and nomenclature, and moreover is better printed than its 1952 predecessor. The individual figures are new or (for the greater part) have been copied or reused from the *Illustrated flora*. For a floristic summary of the *Manual*, see J. P. Bennett in *Bot. Rev.* **62**: 203–206 (1996); 4285 species (or taxa?) are accounted for. Of the *Companion*, one reviewer has stated that 4403 species are figured.

55 A long-popular older work was *Our native trees and how to identify them* (1900; many reissues to 1931) by Harriet Keeler, to which her book on shrubs of 1903 was a counterpart.

56 The work was written in response to a turn-of-the-century hardy-shrub movement; but, as the reprint publisher's comments suggest, the study of shrubs has suffered neglect. The author also published *Our native trees and how to identify them* (1900), widely popular in the early twentieth century with reissues to 1931.

57 Reference to this long-desired project was made by E. O. Wilson in his *Diversity of life* (1992, Cambridge, Mass.). The use of county-based dots was to 'disguise' the effects of collecting bias towards highways and popular botanical or resort 'spots'.

58 The habit photographs, a particular feature of this book, were made largely by R. L. Coffin of Amherst, Mass., in the 1930s and 1940s.

59 Ogden had plans for a full flora, but owing to his becoming New York State Botanist and moving to Albany these were never realized.

60 The work remains a 'convenient reference' in Massachusetts (S. B. Sutton, *Charles Sprague Sargent and the Arnold Arboretum* (1970, Cambridge, Mass.)).

61 The Survey was founded in 1835 but an expanded brief was introduced only in the last third of the nineteenth century, occasioning the successive lists by Bishop, Graves and Dowhan.

62 The New York State Natural History Survey was well founded with on establishment an appropriation of $104000. It contributed greatly to careers as well as major publications, of which 25 appeared to 1865 and five more from then until 1892. It was the first to have 'departments' (which correspond with those in the present State Museum in Albany).

63 Families covered include Magnoliaceae through Ceratophyllaceae, Ranunculaceae, Berberidaceae through Fumariaceae, Platanaceae through Myricaceae, Betulaceae through Cactaceae, Chenopodiaceae, Amaranthaceae, Portulacaceae through Caryophyllaceae, Polygonaceae and Juncaceae (Cronquist system) as well as gymnosperms and, in the bryophytes, Sphagnaceae. Publication is, however, not sequential but as families, or groups thereof, are completed. The project, guided by a six-man 'Flora Committee', represents the first fruits of many years of preparation and field work by the State Botanist's office, including among others Stanley J. Smith, the present author's earliest mentor in systematic botany.

64 Copies occur with either plain or hand-colored plates.

65 Britton in 1889 had recorded 1919 species and additional varieties.

66 A state survey was first organized in 1835 but only after the Civil War was there an expansion of activities, including development of a state museum in Trenton, the capital.

67 For progress to 1906, see H. H. Rusby in *Torreya*, **6**: 101–111, 133–145 (1906). Study of the metropolitan flora has in recent years become an interest of the Brooklyn Botanic Garden.

68 This book is still in print and available from its original publisher.

69 See F. W. Pennell in *Bartonia* **9**: 17–34 (1946).

70 This elicited strong criticism from William Darlington, author of *Flora cestrica* [a local flora of West Chester, southwest of Philadelphia]. 'The style in which "The Empire State" [New York] has illustrated *every department* of her natural history, is calculated to make a true

Pennsylvanian blush for the contrast exhibited by the authorities of his own state; who, in the first place, meanly restricted the survey of her glorious domain to a *mere geological examination*; and now, when it is done, have not spirit enough to give to the public the benefit, even of *that*!' W. Darlington, 1849. *Memorials of John Bartram and Humphry Marshall*... (Philadelphia), p. 31; emphasis that of the original author.

71 The methodology of the Project was described by Fogg in *Contr. Gray Herb.*, N.S., **175**: 121–132 (1947). For more recent developments, see A. Rhoads and L. Thompson in *Taxon* **40**: 43–49 (1992). The Project is on the Web at http://www.upenn.edu/paflora/.

72 The mapping philosophy adopted was, with its unusual attention to detail, at the time relatively uncommon in North America. Of particular value in the original *Atlas* was the inclusion in each map of physiographic and floristic provinces, river systems, and county names and boundaries. The 1993 edition features smaller maps than in the original with, unfortunately, the elimination of the background information, so destroying much of the effect of the original. The present author believes that the original authors would have been somewhat disappointed.

73 The Delaware state survey, established in 1837, did not include botany.

74 Shreve in 1910 had in fact estimated the Maryland vascular flora as having upwards of 1900 species, suggesting that the state was still not 'fully collected'. Present figures more than bear out his statement.

75 For reviews of early Maryland botany, see J. L. Reveal, 1987. Botanical explorations and discoveries in colonial Maryland: an introduction. *Huntia* **7**: 1–3; and G. F. Frick, 1987. Botanical explorations and discoveries in colonial Maryland, 1688 to 1753. *Ibid.*: 5–59.

76 As in Pennsylvania and Virginia, the early Geological Survey was not concerned with Recent botany.

77 The two were later co-authors of a standard text, *The distribution of vegetation in the United States as related to climatic conditions* (1921, Washington).

78 Gronovius's work was at once the first 'state' flora in North America and the earliest worldwide to adopt Linnaean classification principles (though never binomial nomenclature). His significance in the early career of Linnaeus is discussed in F. A. Stafleu, 1971. *Linnaeus and the Linnaeans* (Regnum Veg. 79). Utrecht. A large proportion of the collections on which the work was based came from John Clayton, a resident of Gloucester County on York River in the Tidewater.

79 Serious development of the sciences came late in Virginia. In the colonial and early federal eras, there were no significant herbaria in the state, even at William and Mary College in Williamsburg (it now has one) or

the University of Virginia in Charlottesville; moreover, key collections such as those of Banister and Clayton were in Europe. The Civil War, waged more fiercely in Virginia than elsewhere, also resulted in losses. The situation was surely not improved by the purely geological orientation of the state survey, organized in 1835. Grimes, at William and Mary College, became the first serious collector in the lowland Peninsula since Clayton.

80 Among active centers in addition to VPI are Longwood College, Farmville (herbarium, 1963), Lynchburg College (herbarium, 1927) and, appropriately, William and Mary College (herbarium, 1969).

81 Unfortunately, the connection between Barton and Pursh was overlooked along with the fact that the original *Flora virginica* of Gronovius and Clayton was never finished.

82 Reference may also be made to F. C. James, 1970. *The woody flora of Virginia*. 215 pp., illus. (maps). Chapel Hill, N.C. (Unpublished Ph.D. dissertation, University of North Carolina; abstract in *Dissertation Abstracts International* 31, B(1): 81.)

83 A state survey did not come into existence until after 1885, and only with the inclusion of a biological brief did they undertake publication of the third of Millspaugh's floras.

84 A further introduction to the flora is in E. L. Core, 1960. *Plant life of West Virginia*. New York. For botanical history, see W. W. Boone, 1965. *A history of botany in West Virginia*. [12], 196 pp., illus. Parsons, W.Va.: McClain Printing Co. (for the author).

85 At least some of these doubtful records may not be traceable on account of the loss of the University of Kentucky Herbarium to fire in 1948.

86 Indeed, the sixth edition of *Gray's Manual* (1890) extended to the 100th meridian and from the seventh adopted the Missouri–Red River line as its western limit. One stimulus – apart from 'empire-building' – was a perceived similarity in the floras of the East and Midwest; another was, as Shinners suggested in 1962, commercial considerations.

87 The early floras were mostly simple checklists; by the end of the century, however, several surveys had published relatively detailed enumerations, indicative of the transition to professionalism in the century's last three decades as well as the support made possible by the new land-grant schools. Designed to be used with the standard Northeastern manuals, they also laid a foundation for the more detailed surveys and floristic works of the twentieth century. Similar works appeared somewhat later in some eastern, southern and western states (some published by, or through, federal agencies); one or more were likely a model for E. D. Merrill's *An enumeration of Philippine flowering plants* (**925**).

88 Among those who did look at the flora in general terms was Merritt L. Fernald at Harvard University.

89 The history of botanical exploration around Lakes Huron, Michigan and Superior is documented in E. G. Voss, 1978. *Botanical beachcombers and explorers: pioneers of the 19th century in the upper Great Lakes*. 100 pp. (Contr. Univ. Michigan Herb. 13). Ann Arbor, Mich.

90 For a fairly detailed historical treatment of Ohio floristics, see R. L. Stuckey, 1984. Early Ohio botanical collections and the development of the State Herbarium. *Ohio J. Sci.* 84: 148–174.

91 The remaining volume on dicotyledons (by John J. Furlow) is planned to cover Saururaceae through Leguminosae; cf. T. S. Cooperrider, 1984. Ohio's herbaria and the Ohio Flora Project. *Ohio J. Sci.* 84: 189–196.

92 The author was a man without formal botanical education. For a recent biography, see R. C. Kriebel, 1987. *Plain ol' Charlie Deam: pioneer Hoosier botanist*. x, 183 pp., illus. West Lafayette, Ind.: Purdue University Press.

93 From its third launch in 1869 the Survey was also active in natural history.

94 Botanical progress to 1917 was reviewed in J. M. Coulter, 1917. A century of botany in Indiana. *Proc. Indiana Acad. Sci.* (1916): 236–260; an unannotated bibliography is included.

95 This work – its manual-key format unusual among North American floristic accounts – came to be seen as idiosyncratic; most copies were withdrawn (G. Yatskievytch, personal communication, 1993). A committee of Indiana botanists has been organized with the aim of revising Deam's flora.

96 Patterson also published some North America-wide checklists, noted in the introduction to Division 1.

97 In format and style, this work resembles the many European manual-keys, but has few peers in North America. Jones's *Flora of Illinois* may, however, have served as a model. It is, however, less satisfactory than that work (D. Nickrent, personal communication, 1999).

98 The work, originally planned also to cover lower cryptogams, algae and higher fungi, has nominally been guided by an advisory committee of five. The adopted version of Thorne's system is embodied in R. F. Thorne, 1968. Synopsis of a putatively phylogenetic classification of the flowering plants. *Aliso* 6: 57–66. At present, it is not likely to be completed (D. Nickrent, personal communication, 1999). To the present author, a work such as this would have been more appropriate if designed to cover the whole Middle West. Moreover, its pseudo-popular presentation, rarely used systematic arrangement, and manner of appearance are, to say the least, disconcerting.

99 Although in this work specific reference is made to the atlas supplement of Winterringer and Evans, no mention is made of the original work by Jones and Fuller. As with Java (**918**), Illinois floristics has had its share of 'controversy'.

100 As of 1996, the latest installment was in **31**(2) (March 1992).

101 *Michigan Flora* is regarded by many as of exceptional quality. For part 2 alone 100000 specimens were seen; these have constituted the sole basis for records and distribution.

102 Lapham, author of several early state (and local) checklists in the Midwest, was for most of his career a Milwaukee businessman. He recognized, however, the great importance of North American forest resources as capital for development; in 1867 he wrote a report for the state on excessive forest destruction. See J. Perlin, *A forest journey*, pp. 354–355 (1989).

103 A state flora is reportedly projected by Hugh Iltis and Thomas Cochrane at the University of Wisconsin.

104 Relatively few changes were made for the second edition. The author notes that only much more work would warrant a serious technical treatment.

105 For reviews of floristic studies in Iowa to 1975, see C. L. Gilly, 1947. The flora of Iowa: a progress report based on past contributions. *Proc. Iowa Acad. Sci.* **54**: 99–106; R. F. Thorne, 1954. Present status of our knowledge of the vascular plant flora of Iowa. *Ibid.*, **61**: 177–183; and L. J. Eilers, 1975. History of studies on the Iowa flora. *Ibid.*, **82**: 59–64. The current annotated checklist by Eilers and Roosa is based on a statewide database; see L. J. Eilers, 1979. BIOBANK, a computerized data storage and processing system for the vascular flora of Iowa. *Ibid.*, **86**: 15–18.

106 A more concise format, as in Fassett's *Spring flora of Wisconsin*, might have been desirable. It was, however, the first descriptive flora in the state and one of few then available in its region.

107 All editions of this work were distributed or published by the same firm (Ivison, Phinney) as *Gray's Manual* from its third edition (1862) onwards; later, the rights passed to the American Book Company.

108 The *Vascular Plants* project is described in A. E. Radford et al., 1967. *Contributor's guide for the Vascular Plants of the Southeastern United States.* Chapel Hill, N.C.; and idem, 1974. *Vascular plant systematics*, pp. 501–521. New York.

109 There has been a distinct loss of momentum in the *Vascular Plants* project (David Boufford, personal communication); moreover, during the 1990s greater attention surely has been given to *Flora of North America*.

110 A 'stocktake' symposium was held in Fort Worth in April 1998. Its papers are in *Sida, Bot. Misc.* 18 (2000).

111 At present Jimmy Massey is editorial chair. Of projected volumes, vol. 2 was to encompass pteridophytes, gymnosperms and the first three Cronquist subclasses of dicotyledons, and vol. 3(1) the Rosidae other than Leguminosae. The first volume was reviewed by T. M. Barkley in *Brittonia* 32: 103–104 (1980) with the comment that it resembles Correll and Johnson's *Manual of the vascular plants of Texas* (**171**); to the present writer there is also a resemblance to *North American flora* (**100**). In a world ever more oriented to the visual its presentation seems quite dull. For a recent discussion of the project, see *Fl. N. Amer. Newsl.* **11**: 22–33 (1997).

112 Full details of contents are available at the New York State Museum Web site at http://www.nysm.nysed.gov/biogenflora.html. The cumulative index of 1983 accounts for progress during the first 25 years of the project and has four parts: a chronological list, authors and titles, and indices by genera and families. The series was originally begun as part of a project to produce a new regional manual-flora, guided by the belief that compilation of a critical foundation would be desirable in what was then a relatively lesser-known part of North America. The *Journal of the Arnold Arboretum, Supplementary Series* was established to continue publication of the *Generic flora* following suspension (and subsequent termination) of its parent. One number appeared in 1991, containing among other treatments a portion of the Gramineae. Though announced, no second number was ever published; subsequent contributions have instead appeared in *Harvard Papers in Botany*, in a two-column format slightly larger than the single-column layout previously used. The *Generic flora* remains an exemplary work, with even greater breadth and thoroughness in later numbers.

113 For a biography of Gattinger, see Oakes, 1934. *A brief sketch of the life and works of Augustin Gattinger.* Nashville. Tragically, his collection, along with the rest of the University of Tennessee Herbarium, was in 1934 destroyed by fire at Knoxville; duplicates survive in other institutions.

114 The second edition of the *Atlas* reflects moves in recent years towards development of an electronic database comparable to that in Pennsylvania. Archives are housed at the University of Tennessee Herbarium, Knoxville. The Vanderbilt University Herbarium (VDB), also a major source for the *Atlas*, has now been for the most part transferred to the Botanical Research Institute of Texas (BRIT).

115 For more on Curtis, see E. and D. S. Berkeley, 1986. *A Yankee botanist in the Carolinas.* Berlin.

116 The *Manual* was a 'Basic Flora' in the original Flora North America Program. As of 1996 a revision was in preparation (D. Porter, personal communication).

117 Ward himself (*Florida Scientist* **49**: 265 (1986)) believed a statewide descriptive flora to be 'still an indeterminate number of years in the future'.

118 Also on the Web at http://www.usf.edu/~isb/projects/hb-atlas.html.

119 Strangely, no reference is made in the work to that of West and Arnold.

120 The work was a 'Basic Flora' in the original Flora North America Program.

121 Millspaugh's hopes that his flora would constitute a 'base line' for the future were recognized by Robert MacArthur and Edward O. Wilson; the work featured in a discussion of island dynamics in their *Theory of island biogeography* (1967, Princeton).

122 This bibliography was a precursor to a larger work pursued by the author to 1953 but only realized in 1981 with publication of Duncan and Kartesz's *Vascular flora of Georgia*.

123 K. C. Clark in his *Woody plants of Alabama* has written that on collecting trips Mohr traveled largely on horseback, so conforming with social etiquette in the countryside. For further information, see also L. J. Davenport, 1979. Charles Mohr and *Plant life of Alabama*. *Sida* **8**(1): 1–13.

124 This was Coville's first major work, prior to his move to the USDA as its Chief Botanist and, soon afterwards, Honorary Curator (1893–1937) of the U.S. National Herbarium.

125 Regarded as 'one of the best' of its kind by A. R. Smith, *Taxon* **35**: 208 (1986).

126 Thieret (1972) and Thomas (1982; for both, see main text) respectively covered 799 non-dicotyledonous species and 1848 species and non-nominate infraspecific taxa of dicotyledons, while MacRoberts (1984; see main text) admitted 2952 species as confirmed and 390 as questionable. MacRoberts (1984, p. 9) believed the estimate of upwards of 4000 or more species given by Brown in *Wildflowers of Louisiana and adjoining states* (1972) and cited in the previous edition of the *Guide* to be too high.

127 Riddell's catalogue was intended as a precursor to a larger work which, however, was 'suppressed' before publication on advice from Asa Gray. This, along with the War between the States, surely had a serious effect on progress in floristics.

128 Cocks was subsequently a professor at Tulane University in New Orleans.

129 For early Louisiana botanical history, see R. S. Cocks, 1900. Historical sketch of the botany of Louisiana. *Proc. Louisiana Soc. Naturalists* (1897–99): 69–74. (Reprinted in R. L. Stuckey (ed.), 1978. *Development of botany in selected regions of North America before 1900*. New York: Arno.)

130 Unfortunately, many errors exist in the original work (Joseph Ewan, personal communication, 1987).

131 Differing histories of nineteenth-century botanical exploration represent another factor. Only in Texas was a formal state survey effectively organized and mounted before 1895. Elsewhere in the plains, following the work of the mid-century federal exploring expeditions, state colleges and universities – with here and there participation by the USDA and its bureaux (including in this period the U.S. National Herbarium) – played a major part in botanical topography. In the mountain and Pacific states, botanical topography similarly was largely the province of state universities, supported by contributions from federal agencies (including the early Bureau of Biological Survey) and private individuals. State survey bureaux were for the most part not organized until the twentieth century and then were largely concerned with the earth sciences, specialization having overtaken the more general natural-scientific compass established in many eastern states. Only in California did a general state survey operate, and then only until 1874. Nevertheless, biological surveys were in time established in a number of states.

132 Rydberg had in the 1890s served as a USDA field agent under Frederick Coville, exploring the Sandhills region (Nebraska) and the Black Hills (South Dakota).

133 This work was published by the USDA in *Contributions from the U.S. National Herbarium* with a grant-in-aid from funds then recently appropriated by Congress towards furtherance of surveys.

134 The *Manual* was a 'Basic Flora' in the first Flora North America Program.

135 Vols. 1–2 originally were planned to contain general chapters (including a history of botanical work in the state), a key to families, checklist of the flora, and distribution maps; vols. 3–10 would encompass family revisions. From 1955 the first two volumes were also allocated to revisions, with the first part of vol. 1 covering pteridophytes (later reissued with revisions as *Ferns and fern allies of Texas*).

136 The work is the first in a series of 'Illustrated Texas Floras' projected to appear under the joint sponsorship of the Botanical Research Institute of Texas and Austin College.

137 In the 1976 version of this work the authors noted that a number of Kansas records had proved to be based on misidentifications. Sometimes, the necessary evidence was only to be had in 'Eastern' herbaria.

138 For a biography of Bessey and his 'new botany', see R. A. Overfield, 1993. *Science with practice: Charles E. Bessey and the maturing of American botany*. Ames, Iowa.

139 Both Rydberg's survey reports were financed under the same line item as Coulter's *Botany of western Texas*.

140 Plans for such have been made at the Rocky Mountain Herbarium in Laramie, Wyo. This, established in 1894 by Aven Nelson, is among major regional collections one of the two oldest, the other being the herbarium at the University of Colorado at Boulder. Nelson was, however, the first resident teacher of systematic and floristic botany between California and Kansas.

141 A work comparable to *Intermountain flora* has been projected for the Southwest by James Reveal.

142 Recent field work in Wyoming, for example, has yielded many new distributional records and some taxa new to science (R. L. Hartman at 41st American Institute of Biological Sciences Meeting, Richmond, Va., 1990).

143 On Western botanical history, see further J. and N. D. Ewan, 1972. Botanical explorations in the Intermountain Region. In A. Cronquist *et al.*, *Intermountain flora*, 1: 40–76 (see **180/I**); and *idem*, 1981. *Biographical dictionary of Rocky Mountain naturalists*. xvi, 253 pp. (Regnum Veg. 107). Utrecht/Antwerp.

144 The original edition was well produced and physically is entirely satisfactory for use even after more than 50 years.

145 A new flora is projected; see note by Ronald Hartman in *Fl. N. Amer. Newsl.* 6: 19–20.

146 The work is an early example of a book reproduced by offset from typescript.

147 The work was a 'Basic Flora' in the original Flora North America Program, and fills 'probably one of the largest gaps in the floristic coverage in the United States' (from *Scientific Publications of the New York Botanical Garden* (1992): 3). Only vol. 2 remains to be published.

148 A revised checklist by Hartman is also available, but only in photocopied form from Laramie. He has noted (*loc. cit.*) that since publication of Dorn's manual in 1977, 217 taxa (196 species) have been added to the flora, a reminder that goodly parts of North America have until recently remained, and often still remain, imperfectly known. Many portions of Wyoming had been significantly undercollected until the 1970s, when additional funds and opportunities became available. Field work in the northeast and north-central parts, including the Big Horn valley, has so far yielded substantial increases in local records, new state records and a few novelties as collecting density increased to 4.2 specimens/mi.^{2} (*c.* 1.5/km^2) – a level comparable with Costa Rica and Peninsular Malaysia. Limestone country was said to be particularly rewarding.

149 The original manual was prepared under contract between 1971 and 1974 and suffered in publication. Unduly bulky, it was reproduced by offset lithography from typescript on thick paper with over-generous margins, and so justly criticized by W. A. Weber (*Syst. Bot.* 2: 230–231 (1977)) who suggested that the manual

otherwise could, without loss of content, have been compressed to a single volume of 631 pp. These points have been partly addressed in the 1988 and 1992 versions.

150 Hayden's undertaking was one of a number of *ad hoc*, competing U.S. government surveys. In 1879 they were consolidated through formation of the U.S. Geological Survey under Clarence J. King; Hayden took a position at the University of Pennsylvania. Federal-level responsibility for floristic studies in the West (and elsewhere), after more than two generations of sponsorship by government expeditions and surveys, then passed to, among others, the Department of Agriculture (with at the time responsibility for the herbarium collections of the Smithsonian Institution as well as its own). Soon after, and also in the Department of Agriculture, the Bureau of Biological Survey was organized with as a primary brief work on vertebrate animals. As already noted, state surveys in the West did not effectively come into being until after 1900 – at a time when two were still territories – and even today vast areas remain under federal control. The Biological Survey passed to the Interior Department in the 1930s before its abolition. Revived in 1993 as the National Biological Service, it soon came under the wing of the Geological Survey, thus restoring the pre-1879 relationship.

151 A successor to Harrington's manual was projected by Dieter Wilken and W. Kelley. Weber's recent works have in their introductions included pungent remarks on the prevalence of short-term thinking, the influence of 'fads' and other factors which have all contributed to what he sees as a worsening of floristic knowledge in the United States in recent decades. They also feature a partial return to Rydberg's generic concepts, in contrast to those of Coulter and Harrington which have been more representative of the North American 'mainstream'.

152 The *Manual* was a 'Basic Flora' in the original Flora North America Program.

153 For appropriate comments, see J. A. Steyermark and J. A. Moore, 1933. Flora of western Texas. *Ann. Missouri Bot. Gard.* 20: 791–806.

154 The contents of this over-generously produced work could have been spatially considerably compressed without loss of content, and, with the use of less substantial paper, contained in less bulky volumes. It also is surprising that the authors had to go outside North America for publication.

155 The checklist (1937) was Kittel's master's thesis at Catholic University of America in Washington, D.C. The revision of the Compositae of Arizona and New Mexico formed her doctoral dissertation, also at CUA. See A. O. Tucker *et al.*, 1989. History of the LCU Herbarium, 1895–1986. *Taxon* 38: 196–203.

156 For preliminary information, see Kearney and Peebles in *J. Washington Acad. Sci.* **29**: 474–492 (1939). The 1942 work was the last state flora to be published under federal auspices. Kearney was for many years with the USDA, like C. V. Piper (see **192**) from 1903.

157 Information in *Fl. N. Amer. Newsl.* **6**: 4 (1992). Principal investigators include T. F. Daniel and D. J. Pinkava.

158 The San Francisco Peaks are of interest as a 'type locality' for the biologist (and head of the U.S. Bureau of Biological Survey from its formation to 1910) Charles Hart Merriam's 'life-zone' biotic classification, widely used in natural history writing including several western North American floras.

159 Unfortunately, no plastic overlays have been furnished as has been so in some contemporary European works.

160 The 1964 edition is the one more likely to be encountered.

161 These ranges, along with other parts of the state, have been the focus of collecting by Arnold Tiehm and others through the 1980s. In the words of Reveal (1991; see **Progress** under Division 1), 'Tiehm . . . may be considered the last to have explored on the botanical frontier in the continental United States' (p. 71).

162 This survey initially was well founded and could afford to operate on the scale and scope of that in New York. It was, however, abruptly terminated by the state legislature in 1873; accordingly, some of the results – including the botany – had to be privately published. For the purpose, the survey leader, J. D. Whitney (after whom is named the state's highest mountain), raised money from 'a few citizens of San Francisco' – among them Leland Stanford – enabling their appearance in a quarto format comparable to the New York survey volumes. *Botany of California* thus became the most substantial western state flora to appear until well into the twentieth century, a felicitous development given the rich and relatively strongly endemic plant life.

163 John H. Thomas (Note 166) considered 1880 to represent the end of the period of domination by eastern U.S. institutions, although their influence would linger on in academic links (and rivalries) and through federal bureaux such as the U.S. Biological Survey, whose first major expedition was to Death Valley in 1890–91.

164 Some of this enterprise was surely stimulated by what became a West Coast version of the Harvard–New York rivalry over codes of nomenclature and other matters. Jepson followed the Harvard school, while Abrams was a Brittonian. See J. Ewan, 1987. Roots of the California Botanical Society. *Madroño* **34**: 1–17.

165 California is among the top three parts of North America north of Mexico currently yielding taxa new to science (R. L. Hartman in *Fl. N. Amer. Newsl.* **6**: 20 (1992)).

166 For further details on botanical progress in the Pacific states, see J. Ewan, 1955. San Francisco as a mecca for nineteenth century naturalists. In California Academy of Sciences, *A century of progress in the natural sciences, 1853–1953*, pp. 1–63. San Francisco; and J. H. Thomas, 1969. *Botanical explorations in Washington, Oregon, California and adjacent regions.* Pittsburgh. (Preprint of *Huntia* **3**(1): 5–62; reissued in serial, 1979.)

167 The 'American Code' nomenclature in the original edition of vol. 1 was in its reprint altered to conform to the International Code.

168 Metcalf notes that the work was part of a projected but never-realized regional series.

169 The work was a 'Basic Flora' in the original Flora North America Program of 1967–73.

170 Some US$2 000 000 in all was being sought for the 10-year project.

171 See also R. Ornduff, 1974. *An introduction to California plant life.* Berkeley, Los Angeles.

172 For a recent survey of progress and prospects in California floristics (including some historical material), see 'The future of California floristics and systematics: research, education, conservation.' *Madroño* **42**(2): 93–306 (1995). It represents the proceedings of a symposium at the Jepson Herbarium, UCB, held in 1994 following publication of the new *Jepson Manual*. Barbara Ertter's comment on p. 120 is worthy of note here: 'To become a reality, the Jepson Manual Project depended largely on grass-roots funding from non-academic sources within California, with mainstream NSF funding only at the tail end'.

173 The new *Jepson Manual* was something of a best-seller; its first run, of 7000 copies, sold out in only three months. Nearly 2000 people were involved in one way or another with its preparation and publication.

174 With the development of the SMASCH floristic information system, completion of the work in its original form is not likely.

175 *A California Flora* was a 'Basic Flora' in the original Flora North America Program of 1967–73.

176 Junak's flora of Santa Cruz is the most substantial of its kind for the Channel Islands since that of Millspaugh and Nuttall on Santa Catalina.

Division 2

•••

Middle America

No country of equal area presents a richer or more varied vegetation than Mexico.

> J. D. Hooker, 'Commentary'; in Godman and Salvin, *Biologia centrali-americana, Botany*, vol. 1: LXII (1888).

What I would venture to suggest is a work in 8vo, without plates, scientific yet intelligible to any man of ordinary education; and, the country that I particularly have in view is the British West Indian Islands, so rich in useful vegetable products. I have reason to know that a very able botanist, Dr Griesbach [*sic*], is only deterred from publishing this Flora, by the fact that such works are not remunerative to the author . . . A sum of £300 would be required.

> W. J. Hooker to the Colonial Office, 14 May 1857; quoted from Thistleton-Dyer, Botanical survey of the Empire, in *Bull. Misc. Inform.* (Kew) **1905**: 12 (1906).

In tropical America . . . the flora has been studied from isolated centers with little regard for the species accepted at other centers, but with the assumption that each area is floristically distinct. Correlation through monographic work, covering a genus throughout its range, will reduce the species that have been multiplied unnecessarily.

> P. C. Standley in *Flora of the Panama Canal Zone* (1928); quoted from Prance in *Ann. Missouri Bot. Gard.* **64** (1978).

Adverso ha sido el destino de las obras que sobre la flora de nuestro país se han escrito.

> E. Beltrán, 'Prologo'; in C. Conzatti, *Flora taxonómica mexicana*, vol. 1: viii (1946).

This division encompasses Mexico, the Central American countries from Guatemala to Panama, and the West Indies from Cuba and the Bahamas east and south around the Caribbean arc to Trinidad and the southern Netherlands Antilles (Curaçao, Aruba and Bonaire). Apart from the seven geopolitical regions here delimited, physiographic headings have been established for the Sonoran Desert, for which a significant flora is available, and for wetlands, bibliographic coverage for which has appeared in recent years along with some floras.

Introductory remarks on the history and progress of floristic botany in Middle America are here given under two superregional headings: 21–23, Mexico and Central America, and 24–29, The West Indies. This separation arises from their largely mutually distinctive patterns of exploration, botanical 'development' and floristic writing, which moreover have involved to a considerable extent different sets of personalities – a situation which has also existed, and continues to exist, within each superregion. This renders the creation of new overall floras potentially difficult if not impossible.

It was, however, just this challenge which served as a stimulus towards formation of the supranational Organization for Flora Neotropica. Its main undertaking, *Flora Neotropica*, was initiated in 1964 and encompasses the greater part of Divisions 2 and 3 (see **300** for primary listing). For descriptions of this project, see B. MAGUIRE, 1973. The organization for 'Flora Neotropica'. *Nature and Resources* 9(3): 18–20, and A. GENTRY, 1979. Flora Neotropica news. *Taxon* **28**: 647–649. There is also a Flora Neotropica Web site (http://www.nybg.org/bsci/ofn/).

General bibliographies. Bay, 1910; Blake and Atwood, 1942; Frodin, 1964, 1984; Goodale, 1879; Holden and Wycoff, 1911–14; Jackson, 1881; Pritzel, 1871–77; Rehder, 1911; Sachet and Fosberg, 1955,

1971; USDA, 1958. See also under the super-regions.

Divisional bibliography

AGOSTINI, G., 1974. Taxonomic bibliography for the neotropical flora. *Acta Bot. Venez.* **9**: 253–285. [Covers the six families from Acanthaceae through Anacardiaceae in alphabetical order. Not continued.]

General indices. BA, 1926– ; BotA, 1918–26; BC, 1879–1944; BS, 1940– ; CSP, 1800–1900; EB, 1959–98; FB, 1931– ; IBBT, 1963–69; ICSL, 1901–14; JBJ, 1873–1939; KR, 1971– ; NN, 1879–1943; RŽ, 1954– . See also under the superregions.

Divisional indices

The first two are also listed under **Division 1**.

TORREY BOTANICAL CLUB, NEW YORK, 1969. *Index to American Botanical Literature, 1886–1966.* 4 vols. Boston: Hall. Continued as *idem*, 1977. *Supplement, 1967–76.* 740 pp. Boston. [Originally published serially in *Bulletin of the Torrey Botanical Club* as well as (for a time) on index cards. In the reprint, taxonomic and other non-author entries appear in vol. 4. The *Index* continued publication in the *Bulletin* through vol. 122 (1995); control then passed directly to the New York Botanical Garden and from 1996 it has appeared in *Brittonia* (vol. 48 onwards) with a more specific emphasis on systematics and related fields. From 1999, the *Index* is offered solely in electronic form.]

AMERICAN SOCIETY OF PLANT TAXONOMISTS, 1939–67. *The Taxonomic Index.* Vols. 1–30. New York (later Cambridge, Mass.: from vol. 20 (1957) published serially in *Brittonia*). (Vols. 1–9 mimeographed.) [Begun on the initiative of W. H. Camp, this index had an existence of 28 years but from the mid-1940s it was reproduced from the appropriate parts of the *Index to American Botanical Literature*. In 1957 it was consolidated with *Brittonia* although retaining separate volumation, and in 1967 discontinued. (From 1996, however, it was effectively resumed with the conversion of the *Index to American Botanical Literature* (see above) into a primarily taxonomic index.)]

FORERO, E. (coord.), 1978– . *Boletín botánico latinoamericano.* Nos. 1– . Bogotá, etc. (Offset-printed.) [Primarily a newsletter, but includes news of floras and related projects as well as a selection of other literature references. 40 numbers published through 1999.]

Conspectus

200	Division in general
201	Tropical Middle America
205	Drylands
208	Wetlands
	Mexico
	Central America

Superregion 21–23: Mexico and Central America
210–30 Superregion in general
Region 21/22 Mexico

210	Region in general
211	Baja California
212	Sonora
213	Sinaloa
214	Chihuahua
215	Coahuila
216	Nuevo León and Tamaulipas
217	San Luis Potosí
218	Zacatecas
219	Durango
221	'Nueva Galicia' (including Jalisco, Nayarit, Colima and Aguascalientes)
222	'El Bajío' (Guanajuato and Querétaro with northern Michoacán)
223	Michoacán
224	Guerrero
225	'Central Highlands' (including the Valle de México)
226	Veracruz
227	Oaxaca
228	Chiapas
229	Yucatán Peninsula (including Campeche, Yucatán and Quintana Roo) and Tabasco

Region 23 Central America

230	Region in general
231	Guatemala
232	Belize
233	El Salvador
234	Honduras
235	Nicaragua
236	Costa Rica
237	Panama
238	San Andrés and Providencia Islands
239	Swan Islands

Superregion 24–29: The West Indies
240–90 Superregion in general
Region 24 Bahama Archipelago (including the Bahamas and Turks and Caicos Islands)

200

Division in general

See **100** (BRITTON, *North American flora*).

201

Tropical Middle America

The major works covering the whole of the Neotropics are described under **301/I** (tropical South America).

Keys to and treatises on families and genera

MAAS, P. J. M. and L. Y. TH. WESTRA, 1998. *Familias de plantas neotropicales.* See **301/I**.

PITTIER, H., 1939. *Clave analítica de las familias de plantas superiores de la América tropical.* 4th edn. See **301/I**.

ORGANIZATION FOR FLORA NEOTROPICA, 1967– . *Flora Neotropica: a series of monographs.* See **301/I**.

Pteridophytes

TRYON, R. M. and A. F. TRYON, 1982. *Ferns and allied plants: with special reference to tropical America.* See **301/I**.

205

Drylands

Of the extensive areas in Middle America which can be considered as deserts, only the Sonoran Desert, which extends across southeastern California, southwestern Arizona, the central (and largest) part of the Baja California peninsula, and most of the northern and central parts of the Mexican state of Sonora, has well-consolidated floras.[1] The portion within the United States is additionally covered by the floras and related works on the Southwest in general (**105**, **180/II**), Arizona (**187**), and California (**195**); however, no similar coverage, apart from what is recorded in general works (**210**), exists for the Mexican portions except in the work described below and in *Flora of Baja California* (**211**).

A complement to Wiggins' *Flora* is R. M. TURNER, J. E. BOWERS and T. L. BURGESS, 1995. *Sonoran Desert plants: an ecological atlas.* xvi, [ii], 504 pp., 85 halftones, maps. Tucson, Ariz.: University of Arizona Press. (Originally published 1972 as J. R. HASTINGS, R. M. TURNER and D. K. WARREN, 1972. *An atlas of some plant distributions in the Sonoran Desert.* Tucson.) This furnishes detailed range maps of 339 species occurring below 1300 m altitude, with altitudinal scatter diagrams, extensive text and some halftone illustrations.

WIGGINS, I. L., 1964. Flora of the Sonoran desert. In F. Shreve and I. L. Wiggins (eds.), *Vegetation and flora of the Sonoran desert.* 2 vols. 1740 pp., 37 pls., 27 maps. Stanford: Stanford University Press.

The greater part of this work comprises a descriptive flora of vascular plants of those parts of Baja California and Sonora, together with corresponding parts of the states of California and Arizona, which comprise the Sonoran Desert; included are keys to all taxa, concise synonymy, references and some citations (to significant revisions and monographs), fairly detailed indication of local and extralimital range, taxonomic commentary, and notes on habitat, special features, etc., and an index to all botanical names. The introductory section of the Flora incorporates a history of botanical exploration in the area as well as itineraries of the two authors. The Flora proper is preceded by an extensive illustrated descriptive account of the vegetation by F. Shreve (originally published in 1951 as *Publ. Carnegie Inst. Wash.* 591).

208

Wetlands

Bibliography

HURLBERT, S. H. and A. VILLALOBOS-FIGUEROA (eds.), 1982. *Aquatic biota of Mexico, Central America and the West Indies.* xv, 529 pp. San Diego: San Diego State University (for the editors). [Tracheophyta by Antonio Lot, pp. 33–42; classified listing by topics, including floras, but titles not annotated.]

Mexico

LOT, A., A. NOVELO and P. RAMÍREZ, 1986. *Angiospermas acuáticas mexicanas,* 1. 60 pp. Mexico City: Instituto de Biológia, UNAM. (Listados florísticos de México V.)

Checklist with *exsiccatae* arranged by state and territory (covering 111 species). The introductory part gives the rationale and scope of the work as well as highlights of collections and areas examined. [Not yet completed. A relatively strict definition of the synusia has been adopted.]

Central America

See also K. B. ARMITAGE and N. C. FASSETT, 1971. Aquatic plants of El Salvador. *Arch. Hydrobiol.* **62**: 234–255.

GÓMEZ P., L. D., 1984. *Las plantas acuáticas y anfibias de Costa Rica y Centroamérica,* 1: *Liliopsida.* 430, [1] pp., 126 figs. (incl. halftones). San José: Editorial Universidad Estatal a Distancia (EUNED).

Systematic account (including sea-grasses) with keys, diagnostic features, generally short descriptions of species (those of families and orders being longer), references for accepted names, limited synonymy, representative illustrations, and brief notes on local and extralimital distribution, habitat, diagnostic and special features, and taxonomy; references (at end of each family) and indices to genera and higher taxa (at end). No vernacular names are included. The introductory part contains a brief account of hydroseres, a classified table of habitats and descriptions of each system. [A second volume was planned to cover dicotyledons.]

Superregion 21–23

Mexico and Central America

This superregion comprises the continental portion of Middle America, together with closely associated islands. It thus extends from the northern boundaries of Mexico to the southern boundary of Panama. For the nearby islands of the eastern Pacific Ocean, including Guadalupe, Rocas Alijos, the Revillagigedo group, Clipperton and Cocos, see Region 01.

The southern portion of the superregion, from the Mexican state of Chiapas southwards, is estimated as having 18000–20000 species of vascular plants (L. O. Williams in GENTRY, 1978). It is thus likely that the whole area, with its extremes of relief and climate, has at least 30000 species. In spite of a long history of botanical exploration, particularly in Mexico, knowledge of the different parts of the superregion is very uneven, and no more than one-third is covered by floras published since 1940 (a proportion which will increase substantially on completion of *Flora Mesoamericana*, which will partially supplant the still-current but long-outdated *Biologia centrali-americana: Botany*). An additional proportion is covered by incomplete or other older works.

In Mexico, the gathering of botanical information dates back to the 'Aztec Herbal' and the work of Francisco Hernández, through which some pre-Columbian traditions were preserved; unfortunately, what survives of Hernández's writings was published only in 1651. Serious field work began at the end of the eighteenth century when, as part of the Spanish king Charles III's botanical programme, a 17-year 'Real Expedición Botánica a Nueva España', ultimately ranging from Nicaragua and Guatemala through Mexico to California and Vancouver Island and into Cuba and Puerto Rico, was organized and conducted from 1787 to 1804 by Martín de Sessé y Lacasta with a number of associates, chief among them the Spanish-Mexican José M. Mociño. Most of the scientific value of this work was, however, lost through lack of publication; only during a national cultural and scientific revival beginning after the overthrow of Maximilian, and mostly coinciding with the long presidency of Porfirio Diáz, did a renewed interest in the work manifest itself and two preliminary manuscripts left in Mexico appear in print. Evaluation of the names in these works with regard to those now used as well as the original collection began in 1930 at the instigation of Paul C. Standley in Chicago, with Rogers McVaugh continuing the work from 1956; until recently, however, the results have only been published in haphazard fashion. The 2000 unpublished plates, briefly made available to Augustin-Pyramus de Candolle in Geneva

and there copied and later partly distributed, were afterwards not seen again until the late 1970s (they are now housed at the Hunt Institute for Botanical Documentation).[2]

Since the Real Expedición, field work in Mexico – now recognized as a major world center of plant diversity – has been more or less continuous but until recent decades rather unevenly distributed. The lowland tropics and the southern states of Oaxaca and Chiapas in particular have remained imperfectly known, with few useful logistical bases.[3]

During the nineteenth century, collecting was largely dominated by outside field workers but since then very much more work has been undertaken by Mexicans. Significant foci of recent activity have included Baja California, the Northwest and Northeast, Nueva Galicia, the Mexican highlands, Veracruz, the Yucatán Peninsula and (partly in connection with work in northwestern Central America, floristically closely related) the southern states of Oaxaca and Chiapas. Major summaries have included *Biologia centrali-americana: Botany* by William Botting Hemsley (1879–88), *Trees and shrubs of Mexico* by Paul Standley (1920–26) and *Flora taxonómica mexicana* by Cassiano Conzatti (1939–47, incomplete but now being continued). In the third quarter or so of the twentieth century, activities became somewhat decentralized, with a number of 'state' projects published or in progress (notably *Flora de Veracruz*), but the flora was considered too insufficiently known for a new general work. The national herbarium (at the Instituto de Biología, UNAM, Mexico City) also undertook publication of a number of state, territorial and other lists and began to build up a national database. The major publication of the era, though, was *A selected guide to the literature on the flowering plants of Mexico* (1964) by Ida Langman. In 1980, a substantial manifesto towards a national flora was published in *Macpalxochitl* and in 1983 the 'Consejo Nacional de la Flora de México' was organized. The vascular flora was then estimated at some 20000 species. The first part of a national checklist was published in 1992, and the initial installments (all on pteridophytes) of *Flora de México* appeared in 1993.

Floristic work in Central America, except in central Panama and at scattered points along the coasts (including British-influenced Mosquitia in eastern Nicaragua) and here and there in the interior (notably by Martín Sessé and José Mociño in 1795–99 and by

Anders Ørsted in Costa Rica and Nicaragua in 1846–48), began seriously only in the latter third of the nineteenth century. In Costa Rica, notable work was done by Henri Pittier and his associates as a coffee and settlement boom followed completion of the Atlantic railway from San José and the Museo Nacional, the first in the region, was established (1887). In Guatemala, much work was done by W. Shannon, John Donnell Smith, Hans von Türckheim and William Kellerman, partly in connection with the so-called Inter-continental Railway Commission. In Panama (independent from 1903) Pittier, Standley and others were very active as infrastructure was developed during the canal construction era and afterwards.[4]

A Nicaraguan botanist, Miguel Ramírez Goyena, returned to his country on a contract to write *Flora nicaraguënse* (1909–11), the first of its kind in Central America. Until the late 1920s, however, collecting remained relatively localized. An additional logistical center, and focus for detailed floristic and biological investigations, was provided by the establishment in 1923 of the Barro Colorado Island research station (now the Smithsonian Tropical Research Institute or STRI).

From 1921 until the 1950s nearly all of Central America was dominated by Paul C. Standley's collecting and publications, including several floras and checklists of which the first was *Flora of the Panama Canal Zone* (1928). In that year Standley moved from the Smithsonian Institution, base for his work on the Mexican and Panamanian floras, to the Field Museum of Natural History at Chicago, and renewed its role, begun with Charles Millspaugh's interest in the flora of the Yucatán Peninsula, as a center for Central American botanical studies.[5] Later, he was partly responsible for the establishment, at the 'Esculea Agrícola Panamericana' in Honduras, of the first major Central American herbarium outside Costa Rica.[6]

The Field Museum was about 1930 joined by the Missouri Botanical Garden, which began a long-term programme of floristic studies in Panama in association with the then-Canal Zone Experiment Gardens at Summit. After issuing a series of contributions in the late 1930s, Ralph Woodson and Robert E. Schery began publication of *Flora of Panama* in 1943.

Continuing activity since 1940 by a large number of local and overseas botanists, prominent among the latter those from the Field Museum and the Missouri Botanical Garden, has advanced studies to a point where most countries, with the exception of Nicaragua and patches elsewhere, are now considered more or less moderately well collected, with an overall average somewhat exceeding Malesia.[7] The successors to Standley have also been largely responsible for three modern national floras, some new local works, and a host of other contributions.

The most recent development in Central American floristic botany, and one of a significance comparable to *Biologia centrali-americana*, is the *Flora Mesoamericana* project. Under plans published in 1981, the projected work would be a descriptive flora (along the lines of *Flora Europaea*, but with some variations such as selected specimen citations) written in Spanish and encompassing seven volumes. Some 16 000 vascular plant species are to be included within the projected compass of this work, which extends from the southern boundary of Panama north to, in Mexico, the Chiapas–Oaxaca and Tabasco–Veracruz state boundaries (thus including the whole of the state of Chiapas and the Yucatán Peninsula). Preparation of the new flora would be supported by programmes of field work focused on inadequately known areas. Among other advantages the new work would cut through a host of synonyms, the result of a long past preoccupation with single polities – as noted already by a number of writers, including Standley himself (as quoted above in the divisional epigraphs). Sponsors include the Missouri Botanical Garden, the Natural History Museum in London, and the Universidad Nacional Autónoma de México (UNAM). Two volumes have now (1999) been published.[8]

Progress
For early and recent history, see F. VERDOORN, 1945. *Plants and plant science in Latin America*. Waltham, Mass., especially the 'Historical sketch' by F. W. Pennell (pp. 35–48); B. MAGUIRE, 1958. Highlights of botanical exploration in the New World. In W. C. Steere (ed.), *Fifty years of botany*, pp. 209–246. New York; S. G. SHETLER, 1979. North America. In I. Hedberg (ed.), *Systematic botany, plant utilization and biosphere conservation*, pp. 47–54. Stockholm; and V. M. TOLEDO, 1985. *A critical evaluation of the floristic knowledge in Latin America and the Caribbean*. 78, 17 pp. Washington, D.C.: The Nature Conservancy. (Mimeographed.) Further remarks appear in the general review of tropical floristics by Prance (1978). Central America was well summarized in A. H.

GENTRY, 1978. Floristic knowledge and needs in Pacific tropical America. *Brittonia* **30**: 134–153, and Toledo's above-cited review has a particular emphasis on Mexico.

On the *Flora Mesoamericana* project, see M. SOUSA SÁNCHEZ, 1981. Flora Mesoamericana. *Bol. Bot. Latinoamer.* **8**: 11; also ANON., 1981. Flora Mesoamericana. *Syst. Bot.* **5**: 447.

Bibliographies. Bay, 1910; Blake and Atwood, 1942; Frodin, 1964, 1984; Goodale, 1879; Holden and Wycoff, 1911–14; Jackson, 1881; Pritzel, 1871–77; Rehder, 1911; Sachet and Fosberg, 1955, 1971; USDA, 1958. See also the divisional bibliography by AGOSTINI, 1974 (full reference under Division 3).

Regional bibliography

HAMPSHIRE, R. J. and D. A. SUTTON, 1988. *Flora Mesoamericana: a preliminary bibliography of the Mesoamerican flora.* [vi], 193, [1] pp., map (end-paper). London: British Museum (Natural History). [5237 annotated references arranged by authors, with indices on other keys (family, region, subject). Produced from a computerized database built up for the *Flora Mesoamericana* project.]

Indices. BA, 1926– ; BotA, 1918–26; BC, 1879–1944; BS, 1940– ; CSP, 1800–1900; EB, 1959–98; FB, 1931– ; IBBT, 1963–69; ICSL, 1901–14; JBJ, 1873–1939; KR, 1971– ; NN, 1879–1943; RŽ, 1954– . See also the divisional indices (full references under Division 1).

210–30

Superregion in general

No work has yet appeared to supersede Hemsley's comprehensive enumeration of vascular plants for the monumental *Biologia centrali-americana*, now over a century old but still a standard reference for all of Mexico and Central America, just as Bentham's *Flora australiensis* long has so remained for that continent. A gauge as to the increase in botanical knowledge is given by the fact that Hemsley's work, which admittedly was to a large extent compiled, covers perhaps only one-third of the presently known and estimated number of vascular plants. Its projected successor,

Flora Mesoamericana, will cover only a portion of the area covered by the earlier work, where perhaps the need is greatest: the countries and states from the Yucatán Peninsula and the Isthmus of Tehuantepec eastwards and southwards, with their patchwork of state and national floras (or lack of them) of which only three are more or less current (with none of them in Spanish, the language of the new work).

DAVIDSE, G., M. SOUSA SÁNCHEZ et al. (gen. eds.) [FLORA MESOAMERICANA COMMITTEE], 1994– . *Flora Mesoamericana.* Vols. 1, 6. Maps (end-papers). Mexico City: Universidad Nacional Autónoma de México; St. Louis: Missouri Botanical Garden; London: The Natural History Museum.

Concise flora of vascular plants (in Spanish) with keys, limited synonymy, typification, descriptions, distribution with (sometimes) selected *exsiccatae*, and occasional brief notes; bibliography and index to all scientific names at end of each volume. Genera and families are described along with standard references (all fully cited in the bibliographies). The general parts contain lists of contributors, a preface (with statistics), acknowledgments, an introduction, conventions, abbreviations and signs, and keys to families. [The work covers Mexico from Chiapas, Tabasco and the Yucatán Peninsula south through Central America to Panama. Vol. 1 (1995) covers ferns and fern-allies (1382 species) and vol. 6 (1994), monocotyledons except for Scitamineae and Orchidales (1891 species). Vol. 4, the next planned for publication, will cover dicotyledons from Cucurbitaceae through Polemoniaceae (1957 species.) The Flora Mesoamericana Web site is at http://www.mobot.org/ MOBOT/fm/.][9]

HEMSLEY, W. B., 1879–88. *Biologia centrali-americana: or, contributions to the knowledge of the fauna and flora of Mexico and Central America* (eds. F. D. Godman and O. Salvin), *Botany.* 5 vols. (1–4, text; 5, atlas). 111 pls. (incl. map; some col.). London: R. H. Porter; Dulau (for the editors).

Systematic regional enumeration of vascular plants (exclusive of Baja California) with descriptions of new taxa; includes synonymy, references, citations of significant revisions and monographic works under family and genus headings, detailed indication of internal distribution with citations of *exsiccatae*, general summary of extralimital range, and critical commentary; complete indices to botanical names in each volume (general index in vol. 4). An introductory section (vol. 1) includes chapters on phytogeography and the composition of the flora, while an appendix by J. D. Hooker (vol. 4: 116–332) includes accounts of geography, vegetation formations, botanical exploration, the high-mountain flora, distribution of the more prominent families, and florulas of certain offshore islands, along with a bibliography and a summary and statistical analysis of the flora. A total of 12233 species is covered in this work.[10]

Region 21/22

Mexico

For Guadalupe, Rocas Alijos and the Revillagigedo Islands, see respectively **011**, **012** and **013**. – Area: 1972546 km^2, comprising the entire Federal Republic of Mexico, including Baja California and the Yucatán Peninsula. Vascular flora: 26285 species (Espejo Serna and López Ferrari, 1992; see below); other estimates are 20000 (Lot and Toledo, 1980; see below) and (phanerogams only) *c.* 22000 (Rzedowski, 1991). A total of 1075 ferns have been recorded (Palacios in Magaño A., 1992; see **229**).[11]

The history of floristic work in Mexico to the last decade of the nineteenth century has been summarized in N. LEÓN, 1895. *Biblioteca botánico-mexicana*. 372 pp. Mexico City (especially pp. 297–368). Some of the subsequent activity has been covered in the previously cited reviews of VERDOORN, MAGUIRE and SHETLER (see **Superregion 21–23, Progress**) and in P. DÁVILA A. and M. GERMÁN R., 1991. *Herbario Nacional de México*. 122 pp., illus. Mexico City; but no relatively comprehensive account of recent history and current work has been seen. The principal results of the 'Real Expedición Botánica a Nueva España' led by Martín Sessé y Lacasta and José M. Mociño (1787–1803) long remained in manuscript, seeing publication only late in the nineteenth century: *Plantae novae hispaniae* (1887–90; reissued 1893) and *Flora mexicana* (1891–97; reissued 1894).[12] By this time William B. Hemsley had already 'codified' all available floristic information on Mexico and Central America as part of *Biologia centrali-americana* (see Superregion 21–23) and Cassiano Conzatti, Nicolás León and Manuel Urbina had begun to develop an autochthonous floristic literature. As already noted, León contributed the *Biblioteca*, Urbina

(at the Museo Nacional) a *Catálogo de plantas Mexicanas (fanerógamas)* (1897), and Conzatti some didactic works including, with Lucius Smith, the first part of a *Flora sinóptica Mexicana* (1895–97).[13] This last covered 1573 species in 35 'sympetalous' families; though the part was revised in 1910 with 2505 species in the same families, the work as a whole was never completed. Conzatti also produced the first of a projected three-volume *Los géneros vegetales mexicanos* (1903–05), styled after Bentham and Hooker's *Genera plantarum*.[14] At a regional level, a significant contribution was *Pteridografía del sur de México* by José N. Rovirosa (1909; reprinted 1976). However, the growing size of the known flora and the very small and in part widely scattered active group of local botanists impeded more effective coverage, and the 1911 revolution and subsequent instability resulted in further disruption.[15]

Foreign botanical activities have been considerable, with the result that as late as the 1980s more collections were deposited outside Mexico than in (Toledo, 1985; see **Division 2, Progress**). The work of the many visitors before the 1880s was summarized by Hemsley, with those from North America later partly accounted for by Paul Standley in his *Trees and shrubs of Mexico* (1920–26), still a standard work. The latter, though, gave a new stimulus to local activities; and late in life Conzatti turned again to his manual, publishing in 1938–43 (with revision and extension in 1946–47) the early parts of his projected nine-volume-plus *Flora taxonómica mexicana*, covering pteridophytes, gymnosperms and monocotyledons.[16] At regional level, North American influence remained prominent; some older projects were completed (Yucatán Peninsula, by Standley in 1930 concluding work begun at the Field Museum under Charles Millspaugh) and new ones begun (Sonoran Desert, by Forrest Shreve at the Desert Laboratory in Tucson; the Mayan and other areas, by Howard S. Gentry and Cyrus L. Lundell under sponsorship from the Carnegie Institution of Washington). The western islands (see also **011–013**) were documented by Alice Eastwood and others through the California Academy of Sciences.

Since World War II and the subsequent expansion of higher education and research centers, floristic work has become somewhat decentralized, with a number of active groups covering 'work areas' logistically of manageable size. These centers have since 1960 built up 'state' and 'regional' herbaria complementary to the now large national (UNAM) herbarium; the largest of them now are in Jalisco and Veracruz.[17] Additional state and regional floras began to appear: the bibliographically complicated *Flora del Estado de México* (Martínez *et al.*, 1953–81), a fern flora of Chihuahua (Knobloch and Correll, 1962), *Flora of the Sonoran Desert* (Wiggins, 1964, the realization of Shreve's project; see **205**) *Flora novo-galiciana* (McVaugh and others, 1974–), *Flora del Bajío* (Rzedowski and Rzedowski, 1991–), *Flora de Veracruz* (INIREB, 1978–), *Flora of Baja California* (Wiggins, 1979), and *Flora fanerogámica del Valle de México* (Rzedowski and Rzedowski, 1979–90). Since 1980 the UNAM herbarium has produced several state and other checklists; a further state flora, the *Flora of Chiapas*, has commenced publication; and pteridophyte floras have appeared for Chiapas (within the *Flora de Chiapas* series) and Oaxaca. The available parts of Conzatti's manual were also reissued in one volume, with additional families, in 1988. One students' manual is available: *La flora del Valle de México* (Sánchez Sánchez, 1969). Miscellaneous local florulas and other works also exist.

Nonetheless, since 1900 at national level no nominally complete accounts have been made, although the extremely detailed bibliography by Ida Langman (1964; see below) as well as those by Jones and by Riba and Butanda have prepared the bibliographic ground. Stanwyn Shetler in 1979 wrote that 'as yet no serious effort to write a Flora of Mexico is underway' although the *Flora of Veracruz* programme had been visualized as a precursor to such a project. In 1983, however, a dedicated body, the 'Consejo Nacional de la Flora de México', was organized and incorporated with the goal of promoting and producing a national *Flora de México*; in 1993 its first fascicles were published.[18] Monocotyledons are separately being documented in the form of an enumeration, *Las monocotiledóneas mexicanas* (1992–) by A. Espejo Serna and A. R. López-Ferrari.

Bibliographies. General and divisional bibliographies as for Divison 2.

Regional bibliographies

JONES, G. N., 1966. *An annotated bibliography of Mexican ferns.* xxxiii, 297 pp. Urbana: University of Illinois Press. [Includes geographical, botanical, subject and author cross-references as well as indices.]

LANGMAN, I. K. 1964. *A selected guide to the literature on the flowering plants of Mexico.* 1015 pp. Philadelphia: University of Pennsylvania Press. (A Morris Arboretum Monograph.) [Arranged in the first instance by author and partly annotated; detailed subject index. Exhaustive; among the best national botanical bibliographies in existence.][19]

LEÓN, N., 1895. *Biblioteca botánico-mexicana.* 372 pp. Mexico City. [805 annotated titles arranged by author. Pp. 297–368 comprise a historical review of Mexican botany.]

Indices. General and divisional indices as for Division 2.

210

Region in general

See also **210–30** (HEMSLEY). Only three other general works for Mexico as a whole exist: Conzatti's lately revived but still-incomplete *Flora taxonómica mexicana*, the recently initiated *Flora de México*, and Standley's classic *Trees and shrubs of Mexico*. The last-named has been three times reprinted and remains much used in spite of its age, obsolete nomenclature, and relatively limited basis (for tropical lowlands it has been supplemented by a manual by Pennington and Sarukhán).

Families and genera

ESPEJO SERNA, A. and A. R. LÓPEZ-FERRARI, 1990. *Clave artificial para géneros y familias de monocotiledóneas mexicanas.* 64 pp. Mexico City: Consejo Nacional de la Flora de México/Universidad Autonóma Metropolitana Iztapalapa. [Artificial keys to monocotyledon families and genera. The work complements their *Monocotiledóneas mexicanas* (see below).]

CONZATTI, C., 1903. *Los géneros vegetales mexicanos.* Vol. 1. 449 pp. Mexico City: Secretaría del Fomento, México. [Generic flora with synoptic keys, descriptions of genera and lists of known Mexican species. – Originally projected to cover 1900 genera in 172 families over three volumes but not completed. Vol. 1 covers 70 families of Thalamiflorae, Disciflorae and Calyciflorae following the Bentham and Hooker system (i.e., Ranunculaceae through Cornaceae and including Garryaceae). The orchids, intended for vol. 3, appeared as *Taxonomía de la Orquideas Mexicanas* (1904, in

Mem. Soc. Cient. 'Antonio Alzate' **21**(5–8, 9–12): 249–272, 273–341), and the pteridophytes as *Las criptogamas vasculares de México* (1907, in *ibid.*, **25**(2, 3): 59–106, 107–176 but also issued independently: 72 pp., 60 pls.). A preliminary treatment of Compositae, in the form of keys to tribes and genera, appeared as *Sinanteraceas* (1931, in *Mem. Rev. Acad. Nac. Cienc. 'Antonio Alzate'* **53**(3): 65–88 with a separate reissue in 1934).]

CONSEJO NACIONAL DE LA FLORA DE MÉXICO, 1993– . *Flora de México*. In fascicles. Illus. Mexico City.

Descriptive flora of vascular plants with keys, synonymy, types, indication of distribution, vernacular names, uses, and taxonomic and ecological notes along with representative *exsiccatae*; references and index at end of each fascicle. [As of 1999 fascicles 1–5 of vol. 6 and 1–2 of vol. 7 had been published; the first-named cover pteridophyte families and the latter respectively Caricaceae and Hernandiaceae.]

CONZATTI, C., 1988. *Flora taxonómica mexicana*. 3rd edn. Vol. 1. xxix, 1064 pp. Mexico City: Consejo Nacional de Ciencia y Tecnológia (CONACyT)/ Instituto Politécnico Nacional y Centro Nacional de Enseñanza Técnica e Industrial. (1st edn., 1939–43, Oaxaca, corresponding to vol. 1 of the 1946–47 edition; 2nd edn. (vols. 1–2), 1946–47, Mexico City: Sociedad Mexicana de Historia Natural.)

Briefly descriptive flora with keys to all taxa; includes synonymy with some references and citations, vernacular names, and fairly detailed indication of local range (with some citations of *exsiccatae*); indices to all botanical and vernacular names in each part. The introductory section includes a brief summary of botanical exploration in Mexico, a glossary, and a selected bibliography. [The two published volumes of the second edition cover pteridophytes and all the monocotyledons (families 1–51). Several further volumes were projected to cover the gymnosperms and dicotyledons. The 1988 edition encompasses *in addition* gymnosperms (families 52–57) and monochlamydeous dicotyledons from Balanophoraceae through 'Amentiferae' and 'Centrospermae' to the Podostemonaceae, including Euphorbiaceae (families 58–99). Two further volumes to complete the work were planned.][20]

ESPEJO SERNA, A. and A. R. LÓPEZ-FERRARI, 1992– . *Las monocotiledóneas mexicanas; una sinopsis florística*, 1. *Lista de referencia*. Parts 1– . Illus. Mexico City: Consejo Nacional de la Flora de México/ Universidad Autonóma Metropolitana Iztapalapa.

Alphabetically arranged enumeration with synonymy, type collection(s), and distribution by states; index to all names in each part at end. Families and subordinate taxa are arranged in alphabetical order. [Family and generic entries are unannotated though family headings feature illustrations. About 4600 species in all will be covered, some 17–18 percent of the total number of Mexican vascular plants. As of 1999 five parts had been published covering families from Agavaceae through Cannaceae (1–3), Dioscoreaceae through Nolinaceae (6) and the second of two parts for Orchidaceae (8).]

Woody plants (including trees)

For the lowland tropical areas in the country, Standley's work has been supplemented by that of Pennington and Sarukhán (1968).

PENNINGTON, T. D. and J. SARUKHÁN, 1998. *Árboles tropicales de México: manual para la identificación de las principales especies*. 521 pp., illus., maps. Mexico City: Universidad Nacional Autónoma de México/Fondo de Cultura Económica. (1st edn., 1968, Mexico City: Instituto Nacional de Investigaciones Forestales.)

Illustrated dendrological manual of 190 more important lowland trees, the figures for each species facing the text; the latter contains botanical and dendrological details along with vernacular names, distribution, habitat and uses. The general part covers climate and vegetation along with the main key and an illustrated organography, while at the end are a list of specimens seen, a bibliography and a full index. [About 40 species have been added to those in the original edition.]

STANDLEY, P. C., 1920–26. *Trees and shrubs of Mexico*. 5 parts. 1721 pp. (Contr. U.S. Natl. Herb. 23). Washington, D.C. (Parts 1–3, 5 reprinted 1961 as Smithsonian Institution Publication 4461; complete work reprinted in 2 vols., 1967, 1969, Smithsonian Institution Press. Subsequently reprinted by Koeltz.)

Briefly descriptive flora of native, naturalized and widely cultivated woody vascular plants, with keys to all taxa; includes synonymy, references, citations (of relevant monographs and revisions under each genus), vernacular names, concise indication of local and extralimital range, more or less extensive notes on properties, uses, cultivation, local significance, etc., and some critical remarks; complete indices in each part, with a general index to families and genera at end of work. An introductory section includes a short general account of early botanical exploration and research in Mexico, with special attention to Hernández, Sessé and Mociño, etc. [This work, to a large extent compiled - principally from literature sources and the large Mexican collections of the United States National Herbarium – and regrettably also with nomenclature following the then official (but now obsolete) American Code, was mostly written by Standley alone, but several family treatments were contributed by specialists.][21]

211

Baja California

See also **205** (WIGGINS). – Area: 143 396 km^2, encompassing the political units of Baja California Norte and Baja California Sur. Vascular flora: 2705 species and 253 additional non-nominate infraspecific taxa, of which 427 endemic (Toledo, 1985; see **Division 2, Progress**).

Ira Wiggins' flora represents the first complete work on the peninsula, the fruit of some 50 years of study. The Cape region, in the extreme south, was first documented by Thomas S. Brandegee in 1891 and, a century later, by Lee W. Lenz.

WIGGINS, I. L., 1980. *Flora of Baja California.* [xiii], 1025 pp., 970 text-figs., 3 tables, 4 maps. Stanford: Stanford University Press.

Copiously illustrated descriptive manual-key to all taxa of native, naturalized and adventive vascular plants, with nearly 1000 figures; includes synonymy with references, localities, indication of local and extralimital range, critical taxonomic commentary, and notes on phenology, habitat, special features, etc.; index. An introductory section treats physical features, geology, soils, climate, characteristics of the flora, plant communities, endemism and the history of botanical exploration.[22]

Partial work: Cape region

Lenz's catalogue succeeds *Flora of the Cape Region of Baja California* by T. S. Brandegee (1891–92, in *Proc. Calif. Acad. Sci.*, II, 2: 108–182, 218–227) with additions in *Zoe* 4: 398–408 (1894), altogether inclusive of 795 vascular species.

LENZ, L. W., 1992. *An annotated catalogue of the plants of the Cape Region, Baja California Sur, Mexico.* xii, 114, [1] pp., cover illus., map. Claremont, Calif.: The Cape Press (distributed by Rancho Santa Ana Botanic Garden).

Annotated enumeration of vascular plants (1053 species and additional infraspecific taxa), with references, locality reports and citations of *exsiccatae*, and (sometimes quoted) notes; lexicon of vernacular names (with sources) and index to genera and families at end. Citations of key revisions or other papers are also included. The brief intro-

duction furnishes background, remarks on the region and its limits, and statistics, along with a table of abbreviations.

212

Sonora

See **205** (WIGGINS) for coverage of those parts of the state falling within the Sonoran Desert (amounting to about three-fifths of the area). – Area: 182 052 km^2. Vascular flora: *c.* 4000 phanerogam species (Rzedowski, 1993).

No general works specifically relating to the state are available. In the far south around Río Mayo (at 26°30′ to 28°30′ N, south of the Sonoran Desert zone), the following may be found useful: P. S. MARTIN *et al.* (eds.), 1998. *Gentry's Río Mayo plants: the tropical deciduous forest and environs of northwest Mexico.* xvi, 558 pp., illus. (halftones), maps (also in inside back cover pocket). Tucson: University of Arizona Press. (Southwest Center series.) This contains in its part 4 (pp. 167–522) an annotated enumeration of some 2825 species and infraspecific taxa of vascular plants inclusive of records subsequent to the original *Río Mayo plants* (1942) by H. S. Gentry.

213

Sinaloa

Area: 58 328 km^2. Vascular flora: 2128 phanerogam species known by 1929 (González Ortega).

Two earlier works covering the state are *Catálogo sistemático de la plantas de Sinaloa* by Jesús González Ortega (1929, in *Bol. Pro-Cult. Reg.* 1(1, 2–4, 7–12)) and *Contributions to the flora of Sinaloa* by Laurence Riley (see below). The former covered 2128 phanerogam species; the latter was never completed.

VEGA AVIÑA, R., G. A. BOJÓRQUEZ B. and F. HERNÁNDEZ ALVÁREZ, 1989. *Flora de Sinaloa.* 49 pp., map. Culiacán, Sinaloa: Universidad Autónoma de Sinaloa, Coordinación General de Investigación y Postgrado.

Name list with numbers of collections represented in the 'González Ortega' herbarium (EACS) at the University; summary of novelties and new records, discussion, conclusions, and list of literature at end. [Very much a 'working document'.]²³

RILEY, L. A. M., 1923–24. Contributions to the flora of Sinaloa. *Bull. Misc. Inform.* (Kew) **1923**: 103–115, 163–175, 333–346, 388–401; **1924**: 206–222.

Enumeration; includes citations of *exsiccatae*, vernacular names, and a few critical remarks. Covers only Polypetalae (Bentham and Hooker system).

214

Chihuahua

Area: 244 938 km². Vascular flora: Chihuahuan Desert, 3500 species (Toledo, 1985; see **Division 2, Progress**).

The sole general work on this state deals only with the pteridophytes (see below). Much of the area should, however, be covered by the projected Chihuahuan Desert flora of Marshall C. Johnston and James Henrickson. In northern parts Texas floras (**171**) will be found useful.

Pteridophytes

KNOBLOCH, I. W. and D. S. CORRELL, 1962. *Ferns and fern allies of Chihuahua, Mexico.* 198 pp., 57 pls. Renner, Texas: Texas Research Foundation (distributed by Stechert-Hafner Service Agency, New York).

Detailed illustrated descriptive treatment with keys to all taxa; includes complete synonymy, citations of *exsiccatae*, taxonomic commentary, and notes on habitat and special features; gazetteer, glossary, and index at end.

215

Coahuila

Area: 149 982 km². Vascular flora: no data.

No separate general works for this state are available. However, for the northern part, general works on Texas (**171**) may be found useful.

216

Nuevo León and Tamaulipas

Area: 144 308 km² (Nuevo León, 64 924 km²; Tamaulipas, 79 384 km²). Vascular flora: *c.* 5000 phanerogams (Rzedowski, 1993). For Nuevo León alone, an estimate of 2400 species has been furnished by Alanís (1993). A total of 60 ferns and fern allies were recorded by R. Aguirre-Claverán (1983) in a thesis, *Contribución al conocimiento de la pteridoflora del estado de Nuevo León, México.*²⁴

No separate general floras or enumerations for either of these states are available. However, as with Coahuila, partial coverage may be obtained from general floras and checklists for Texas (**171**). Nuevo León, in addition to being the site of Mexico's second largest city, Monterrey, is of particular phytogeographic importance due to its effective backbone, the Sierra Madre Oriental; the flora of its higher elevations is notably related to that of the Appalachian system in the eastern United States. Background information on the state is furnished in G. J. ALANÍS FLORES, G. CANO Y CANO and M. ROVALO MERINO, 1996. *Vegetación y flora de Nuevo León: una guia botánica-ecológica.* 251 pp., illus. (part col.). Monterrey, N.L.: The authors.

ALANÍS FLORES, G. J., 1993. *Tipos de vegetación y lista florística de plantas vasculares registradas en Nuevo León.* 50 pp. Linares, N.L.: Facultad de Ciencias Forestales, UANL. (Publicacion especial.)

Includes a floristic checklist. [Not seen.]

217

San Luis Potosí

Area: 63 068 km². Vascular flora: no data.

No separate general works for this state are available.

218

Zacatecas

Area: 73 252 km^2. Vascular flora: no data.

No separate general flora or enumeration is available. The more southerly parts within the Sierra Madre Occidental are, however, covered by *Flora novo-galiciana* (**221**).

219

Durango

See also **221** (McVaugh and McVaugh *et al.*). – Area: 123 181 km^2. Vascular flora: *c.* 2500 species (Toledo, 1985; see **Division 2, Progress**), but now some 4000 phanerogams are known (Rzedowski, 1993). A total of 3630 species and infraspecific taxa were documented by González *et al.* (1991).

A statewide checklist was published in 1991 as part of the National Herbarium's *Listados florísticos*. The southernmost part is also encompassed by *Flora novo-galiciana* (**221**), and research towards a state flora is in progress under S. González E. (a co-author of the *Listado*) at the Instituto Politécnico Nacional, Mexico City.

GONZÁLEZ ELIZONDO, M., S. GONZÁLEZ E. and Y. HERRERA ARRIETA, 1991. *Flora de Durango.* [2], 167 pp. Mexico City: Instituto de Biológia, UNAM. (Listados florísticos de México IX.)

Checklist, with citation of *exsiccatae* for each specific and non-nominate infraspecific taxon; no indication of local distribution. The introduction also covers the background and plan of the work.

221

'Nueva Galicia' (including Jalisco, Nayarit, Colima and Aguascalientes)

Area: 118 477 km^2. Vascular flora: 7500 species (Toledo, 1985; see **Division 2, Progress**); there is also an estimate of 7000 phanerogams (Rzedowski, 1993). The former Spanish colonial province of this name centered on the present city of Guadalajara and included the modern states of Jalisco, Colima and Aguascalientes as well as parts of adjoining Nayarit (wholly included herein for convenience), Zacatecas (**218**), Durango (**219**), Guanjuato (**222**) and Michoacán (**223**). *Flora novo-galiciana* essentially covers the whole of this area.

In 1974, the first fascicle of *Flora novo-galiciana*, directed by Rogers McVaugh and long in preparation, was published. This followed precursory accounts on botanical history and the vegetation (published in the same serial as the flora itself; see *Contr. Univ. Michigan Herb.* **9**: 1–123 (1966), 205–357 (1972)). From 1983, however, the flora began to appear as independently published volumes; as of 1999 seven had been issued.

McVAUGH, R., 1974. *Flora novo-galiciana: Fagaceae.* Illus. (Contr. Univ. Michigan Herb. 12(1), no. 3). Ann Arbor. Continued as R. McVAUGH *et al.* (ed. W. R. Anderson), 1983– . *Flora novo-galiciana.* Vols. 5, 12–17. Illus. Ann Arbor: University of Michigan Press (vols. 5, 12, 14, 16); University of Michigan Herbarium (vols. 13, 15, 17).

Large-scale, copiously illustrated serial 'research' flora, with keys; includes rather long descriptions, full synonymy with references and citations, detailed indication of localities with *exsiccatae*, extralimital distribution, vernacular names, critical remarks, and notes on habitat, altitude, phenology, etc.; index. At this writing (1999), apart from the Fagaceae in the first series, seven parts (of a planned 17) have appeared covering pteridophytes and gymnosperms (in vol. 17), monocotyledons (vols. 13–16), and dicotyledons (vol. 5 on Leguminosae; vol. 12 on Compositae).

Partial work: Tres Marias Islands
These islands are part of Nayarit but lie some ways off in the Gulf of California and biologically are somewhat distinctive.

EASTWOOD, A., 1929. A list of plants recorded from the Tres Marias Islands, Mexico. *Proc. Calif. Acad. Sci.*, IV, 18: 442–468.

Enumeration of vascular plants (324 species); includes citations of *exsiccatae*, localities, and critical and other notes. An account of botanical exploration and list of species first described from the group precede the systematic part.

222

'El Bajío' (Guanajuato and Querétaro with northern Michoacán)

Area: 41 940 km^2 (Guanajuato, 30 491 km^2; Querétaro, 11 449 km^2). Vascular flora: *c.* 4000 species (Rzedowski, 1993). The recently initiated *Flora del Bajío* covers the area, while parts of Guanajuato are moreover also encompassed by *Flora novo-galiciana* (**221**).

RZEDOWSKI, J. and G. C. DE RZEDOWSKI (eds.), 1991– . *Flora del Bajío y de regiones adyacentes*. Fasc. 1– . Pátzcuaro, Michoacán: Instituto de Ecología, A.C., Centro Regional del Bajío.

Documentary descriptive flora comprising family revisions; each treatment includes keys to genera and species, synonymy, references, typification, vernacular names, localities with *exsiccatae*, internal and extralimital distribution, altitudinal range, indication of phenology and habitat(s), distribution maps and illustrations, and index. [78 families (or parts thereof) had appeared by 1999; a list of all those published appears in the inside back cover of each successive number. The so-called 'Fascículos Complementarios' contain independent contributions and where appropriate are here separately accounted for.]

Partial work: Querétaro

ARGÜELLES, E., R. FERNÁNDEZ and S. ZAMUDIO, 1991. *Listado florístico preliminar del Estado de Querétaro*. 155 pp., map (Flora del Bajío y de regiones adyacentes, Fascículo Complementario II). Pátzcuaro, Michoacán: Instituto de Ecología, A.C., Centro Regional del Bajío.

Pages 11–154 comprise a checklist with *exsiccatae*, collectors, localities and their places of deposit; explanation of collector abbreviations at beginning (pp. 11–18) and main references at end. The introductory part covers past and current

botanical exploration along with notes on topography, climate, vegetation and methodology. [2334 species accounted for in an area of 11 270 km^2, one of the smallest of the Mexican states.]

Pteridophytes

DÍAZ BARRIGA, H. and M. PALACIOS-RIOS, 1992. *Listado preliminar de especies pteridofitas del los Estados de Guanajuato, Michoacán y Querétaro*. 57 pp. (Flora del Bajío y de regiones adyacentes, Fascículo Complementario III). Pátzcuaro, Michoacán: Instituto de Ecología, A.C., Centro Regional del Bajío.

Includes (pp. 35–54) a species checklist with *exsiccatae*, collectors and locality codes (defined on pp. 15–34). The treatment is preceded by an analysis (with statistics) and phytogeography. [Covers 300 species with 172 new records for at least one of the three states.]

223

Michoacán

Area: 59 928 km^2. Vascular flora: 214 species of pteridophytes; no data for phanerogams. Parts of the state are encompassed by *Flora novo-galiciana* (**221**) and *Flora del Bajío* (**222**), with the preliminary treatment of pteridophytes in the latter covering the whole. All known seed plants have been documented in *Listado florístico del estado de Michoacán* (see below).

RODRÍGUEZ JIMÉNEZ, L. DEL S. and J. E. GARDUÑO, 1995–96. *Listado florístico del estado de Michoacán*, I–V. 5 vols. (Flora del Bajío y de regiones adyacentes, Fascículos Complementarios VI, VII, X, XII, XV). Pátzcuaro, Michoacán: Instituto de Ecología, A.C., Centro Regional del Bajío.

A checklist of seed plants with citation of *exsiccatae* (through 1992), produced from a database; entries include municipalities (but not precise localities), dates of collection and herbaria where deposited. Presentation of synonymy is very limited and there is no commentary. Each part contains explanatory matter and a list of collectors as well as two maps respectively of physiographical units and municipalities. [Gymnosperms in part I; angiosperm families generally alphabetically arranged save that part II is wholly on Compositae and part IV is limited to Fagaceae, Gramineae, Krameriaceae and Leguminosae. No statistical summary is presented.]

224

Guerrero

Area: 64281 km². Vascular flora: 355 pteridophytes; no data for phanerogams. – A serial state flora was initiated in 1989.

PÉREZ, N. D. *et al.* (eds.), 1989– . *Flora de Guerrero*. Fasc. 1– . Mexico City: Facultad de Ciencias, UNAM.

Documentary descriptive flora comprising family revisions; each treatment includes keys to genera and species, synonymy, references, typification, vernacular names, localities with *exsiccatae*, internal and extralimital distribution, altitudinal range, indication of phenology and habitat(s), distribution maps and illustrations; index at end. No introductory fascicle(s) have yet been published. [As of 1999 eight fascicles, by various authors, had been published, covering selected families and genera.][25]

225

'Central Highlands' (including the Valle de México)

Area: 86515 km² (México, 11449 km²; Morelos, 4950 km²; Puebla, 33902 km²; Tlaxcala, 4016 km²; Hidalgo, 20813 km²). – This subregion is here defined as encompassing the Distrito Federal and the states of México, Morelos, Puebla, Tlaxcala and Hidalgo. The physiographically defined Valle de México (see subheading below) centers on Mexico City and includes, along with the Federal District, parts of the states of Mexico, Hidalgo and Tlaxcala.

The first separate flora covering any part of the Mexican heartland was *Flora excursoria en el Valle Central de México* by Carlos F. Reiche (1926; reissued 1963 and subsequently at least to 1981), covering 1500 species. It was complemented in 1969 by *Flora del Valle de México* by O. Sánchez S. and then by the Rzedowskis' *Flora fanerogamica* (1979–90), both described below.

Valle de México (in general)

Area: 7500 km². Vascular flora: 2071 phanerogams, of which 1910 native (J. RZEDOWSKI and G. C. DE RZEDOWSKI, 1989.

Sinopsis numérica de la flora fanerogámica del Valle de México. *Acta Bot. Méx.* 8: 15–30). – A general background work, published in relation to a larger study of the water system of the Valley, is J. RZEDOWSKI, 1975. Flora y vegetación en la cuenca del Valle de México. In Anonymous, *Memoria de las obras del sistema del drenaje profundo del Distrito Federal*, 1: 79–134. Mexico City.

RZEDOWSKI, J. and G. C. DE RZEDOWSKI (eds.), 1979–90. *Flora fanerogámica del Valle de México*. 3 vols. Illus., maps. Mexico City: Continental (vol. 1); Instituto de Ecológia/Esculea Nacional de Ciencias Biológicos, Instituto Politécnico Nacional (vol. 2, 1985); Instituto de Ecológia, Pátzcuaro, Michoacán (vol. 3).

Descriptive manual-flora of seed plants, with keys to all taxa, occurrence, habitat, altitudinal range, special features, some taxonomic commentary, and references to revisionary treatments of families and genera; index at end of each volume. The introductory part (vol. 1) includes sections on botanical exploration, geography, geology, hydrology, climate, floristics and vegetation, uses, and conservation conditions; literature (pp. 57–60).

SÁNCHEZ SÁNCHEZ, O., 1969. *La flora del Valle de México*. viii, 519 pp., 368 text-figs. Mexico City: Herrero. (Subsequently reprinted; 6th edn., 1980.)

Illustrated students' manual of seed plants (1120 species), with keys to all taxa, vernacular names, local range, and notes on habitat, phenology, etc.; glossary, list of references, and full index.

Estado de México

MARTÍNEZ, M., E. MATUDA *et al.*, 1953–81. [*Familias de la flora del Estado de México.*] Published in fascicles. Illus. Toluca (later Metapec): Dirección de Agricultura y Ganaderia (later, in succession, Dirección de Recursos Naturales and Conjunto CODAGEM (for Dirección de Recursos Naturales)), Edo. de México [here referred to as Series A]. Complemented by M. MARTÍNEZ, 1956–58. *La flora del Estado de México*. 5 fasc. Illus. Toluca [here referred to as Series B]. (All parts from 1953 to 1972 in both series reissued 1979 as M. MARTÍNEZ and E. MATUDA, *Flora del Estado de México*. 3 vols. Illus. Mexico City. (Biblioteca Enciclopédica del Edo. de México 74, 75, 76.))

Semi-popular illustrated flora of vascular plants; each family treatment includes keys to genera, limited synonymy, and local distribution. Fascicles are separately indexed. ['Series A' as designated here comprises single family (or generic) fascicles, mainly dealing with larger families; of these, 24 were published, the last in 1981. 'Series B', the so-called *Flora*, is in fact complementary; its five fascicles deal with a large number of families having but few local representatives. Until 1962 progress on 'Series A' was relatively rapid but afterwards only six fascicles were published (the last three, however, respectively cover Orchidaceae, Gramineae and Leguminosae, the treatment of the last-named being Eizi Matuda's final publication). The 1979 reissue includes in its first volume a new introduction by I. P. Lujan on physical features, climate and vegetation along with biographical sketches of the original authors.][26]

Hidalgo

A bibliography has been published: M. A. VILLAVICENCIO and B. E. PÉREZ ESCANDON, 1994. *Literatura básica sobre Flora de Hidalgo*. 55 pp. Pachuca, Hgo.: Universidad Autónoma de Hidalgo. [Alphabetical by authors; not annotated.]

Puebla

Much effort in recent years has gone into documentation of the flora of the Tehuacán-Cuicatlán drainage, which straddles the border with Oaxaca and comprises 2703 vascular species in an area of a little over 10 000 km^2. The area is of archeological importance and from 1979 an intensive collecting programme has taken place under the direction of the Instituto de Biología, UNAM. Two floristic works have been published or initiated: *Flora del Valle de Tehuacán-Cuicatlán* (1993, as *Listados florísticos de México* X) and a series of revisions, *Flora del Valle de Tehuacán-Cuicatlán* (1993–), both under the direction of Patricia Dávila A. Of the latter, 14 fascicles had appeared by 1999; publication is continuing.

226

Veracruz

Area: 71 699 km^2. Vascular flora: *c.* 8000 species (Toledo, 1985; see **Division 2, Progress**); includes 573 pteridophytes.

Veracruz has had a long history of botanical exploration and by the 1960s the literature had become large and scattered. No attempts had been made to undertake a state flora or checklist. However, from the mid-1960s the 'Flora de Veracruz' project, a detailed biological survey by Arturo Gómez-Pompa and his associates at the present Instituto de Ecológia, Xalapa (from 1975 to 1988 the Instituto Nacional de Investigaciones sobre Recursos Bióticos) and the University of California, Riverside, has been in progress. Collaborators at other institutions have also been involved. Early work was devoted to an extensive collecting programme, literature search, computer database development (the first successful such effort in the world), tracking down materials in other her-

baria, and development of activities in ecology, resource development and land-use planning. Publication of a descriptive *Flora de Veracruz* commenced in 1978. An important precursor, which includes – as part of a floristic analysis of the Misantla area (*c.* 50 km north of Jalapa, near the coast) – a partial treatment of the families and genera of woody plants in Veracruz, is A. GÓMEZ-POMPA, 1966. *Estudios botánicos en la region de Misantla, Veracruz*. xvi, 173 pp. Mexico City.[27]

GÓMEZ-POMPA, A. and V. SOSA (eds.), 1978– . *Flora de Veracruz*. Published in fascicles. Illus. Xalapa: Instituto Nacional de Investigaciones sobre Recursos Bióticos (INIREB); Instituto de Ecología, A.C. (from 1988; from 1990 also in collaboration with the University of California, Riverside).

Serially published illustrated descriptive 'research' flora; each family fascicle includes keys to all taxa, synonymy with references and citations (especially of significant monographs, revisions, and floras), typification, localities with altitudes and citations of *exsiccatae* and altitudes, generalized distribution, some critical notes, vernacular names, and many remarks on habitat, vegetation types, phenology, special features, etc., and an index. Each species is individually mapped. The emphasis is very much towards distribution and ecology. Fascicles are issued without regard to systematic sequence. [As of 1999, 102 fascicles had appeared; a list of those published appears in the back of each successive issue. A computer-generated state checklist has also been circulated but not formally published.]

227

Oaxaca

Area: 93 952 km^2. Vascular flora: 9000 phanerogams (Rzedowski, 1993) with an additional 690 pteridophytes (Mickel and Beitel, 1998; see below). A figure of 8000 species was earlier recorded (Toledo, 1985; see **Division 2, Progress**).

Apart from the recent pteridoflora by John Mickel and Joseph Beitel, no separate general works for this state are available. A *Flore de Oaxaca* was projected by the Instituto de Biología, UNAM, and one fascicle (*Guía de autores*) published in 1991, but the project was not continued. There is also a state bibliography, possibly derived in large part from that of Langman (see **Region 21/22**), as well as a recent checklist for the south-central, Pacific coastal Tehuantepec district, considered to be about the best-collected part of the state.

Bibliography

ACEVES DE LA MORA, J. L., 1988. *Compilación bibliografica sobre flora Oaxaqueña.* 141 pp., 1 col. pl. México: Instituto Politécnico Nacional. [Arranged by authors with a subject and geographical index; annotated. Coverage through 1979, but references after 1964 are few.]

Partial work

TORRES COLÍN, R. *et al.*, 1997. *Flora del distrito de Tehuantepec, Oaxaca.* 68 pp., map. Mexico City: Instituto de Biológia, UNAM. (Listados florísticos de México XVI.)

Checklist, with *exsiccatae* for each specific and non-nominate infraspecific taxon (all collectors being coded; see pp. 9–11 for explanation); no indication of local distribution. The introduction covers the background and plan of the work, and includes short notes on six principal collectors including two of the four compilers of this list. Arrangement of families is alphabetical within major groups. [1720 species in an area of 6600 km^2.]

Pteridophytes

MICKEL, J. T. and J. M. BEITEL, 1988. *Pteridophyte flora of Oaxaca, Mexico.* [ii], 568 pp., 129 text-figs., 2 maps, 2 portrs. (Mem. New York Bot. Gard. 46). New York.

Documentary descriptive treatment (690 species) with keys to all taxa, synonymy (with references), typification, *exsiccatae*, generalized Mexican (by state) and extra-Mexican distribution, some taxonomic commentary, and indication of habitat, phenology, distinctive features, etc.; references to revisions and other relevant work in generic headings; complete index at end. The introductory section covers the background to and plan of the work, progress in collecting, and phytogeography, along with references.[28]

228

Chiapas

Area: 74 211 km^2. Vascular flora: 8248 species, of which 609 are pteridophytes (Breedlove, 1981, 1986; see below).

The *Flora of Chiapas*, initiated under the direction of Dennis Breedlove at the California Academy of Sciences, began publication in 1981. This has begun to fill the gap caused by the lack of existing general floras or enumerations for this most southeasterly of the states of Mexico. The state is also being covered by *Flora Mesoamericana* (**210–30**). In addition, Breedlove published a checklist in 1986 as part of the National Herbarium's *Listados florísticos*. The Chiapas flora is furthermore closely related to that of Guatemala; thus, *Flora of Guatemala* (**231**) will be found useful. For an introduction to plant communities and the vegetation, see F. MIRANDA, 1952–53. *La vegetación de Chiapas.* 2 vols. Illus. Tuxtla Gutiérrez, Chiapas.[29]

BREEDLOVE, D. E., 1986. *Flora de Chiapas.* v, 246 pp. Mexico City: Instituto de Biológia, UNAM. (Listados florísticos de México IV.)

Checklist, with *exsiccatae* for each specific and non-nominate infraspecific taxon (more notable collectors are coded and where numbers appear without a code or name they are the author's; see pp. iii–iv for explanation); no indication of local distribution. The introduction also covers the background and plan of the work. Arrangement of families is alphabetical within major groups. [Several families have been contributed or advised on by specialists.]

BREEDLOVE, D. E. [and collaborators], 1981– . *Flora of Chiapas.* Parts 1– . Illus., maps. San Francisco: California Academy of Sciences.

Submonographic 'research' flora of vascular plants; includes keys, descriptions, synonymy (selective where recent revisions are available), typifications, selected *exsiccatae*, taxonomic commentary, citations of illustrations, and notes on distribution, altitudinal range, habitat, etc.; generic headings with extensive taxonomic and biological commentary as well as key literature; addenda and index at end of each part. The introduction (in part 1) covers physical features, vegetation formations (with characteristic plants and associations), history of botanical work (pp. 24–27), the scope and plan of the work, and acknowledgments; statistics for each family given in inside front cover. [As of 1999 five parts had been published including the introduction, pteridophytes, Malvaceae; Compositae-Heliantheae and Acanthaceae. For the flowering plants the scope of families was to follow a version of the Thorne system.]

229

Yucatán Peninsula (including Campeche, Yucatán and Quintana Roo) and Tabasco

Included here are the three states of the Yucatán Peninsula (Campeche, Yucatán, and Quintana Roo) as well as Tabasco. – Area: 164943 km^2 (Yucatán Peninsula, 139426 km^2; Tabasco, 25267 km^2). Vascular flora: Yucatán Peninsula, *c.* 2100 species (Toledo, 1985; see **Division 2, Progress**); Quintana Roo, 1257 species and non-nominate subspecies (Sousa Sánchez *et al.*, 1983; see below). For Tabasco 2147 species have been recorded (Rzedowski, 1993).

The greater part of the area is covered by the now-outdated *Flora of Yucatán* by Standley (see below), itself based partly on earlier but never-completed research by C. F. Millspaugh and others including some family revisions. Parts are, however, contiguous with the northern province of Petén in Guatemala so that the *Flora of Guatemala* (**231**) will also be found useful. In addition, field work and research towards a new general flora was initiated in the late 1970s by the Instituto de Ecológia (formerly INIREB) in Veracruz (**226**) in association with the Field Museum of Natural History in Chicago; a recent major contribution towards the project is *Manual para la identificación de la Compositae de la Península de Yucatán y Tabasco* (1989) by J. L. Villaseñor. Parts of the Peninsula also lie within the scope of the *Flora Mesoamericana* project.[30]

A key additional series is that represented by the 21 contributions comprising Carnegie Institution of Washington, 1935–40. *Botany of the Maya Area.* 2 vols. Washington (Publ. Carnegie Inst. Wash. 461, 522). These arose from botanical investigations carried out in relation to archeological studies in the Peninsula. Within are several family revisions as well as additions to the flora by Standley.

Standley, P. C., 1930. *Flora of Yucatán.* Pp. 157–492 (Publ. Field Columbian Mus., Bot. Ser. 3(3)). Chicago.

Briefly descriptive systematic enumeration of fungi, bryophytes and vascular plants of the three states and territories of the Yucatán Peninsula, without keys;

includes limited synonymy (with references), vernacular names, general indication of local range, taxonomic commentary and notes on habitat, frequency, special features, properties and uses; index to families, genera and vernacular names. The introductory section includes accounts of physical features, climate, vegetation, botanical exploration, and vernacular nomenclature together with a floristic analysis and list of references.

Tabasco

In addition to the work given below, reference may also be made to C. L. Lundell, 1942. *Flora of eastern Tabasco and adjacent Mexican areas.* Pp. 5–74 (Contr. Univ. Michigan Herb. 8). Ann Arbor. For an illustrated guide to ferns (without keys), see M. A. Magaño Alejandro, 1992. *Helechos de Tabasco.* 273 pp., col. photographs. Villahermosa, Tabasco: Centro de Investigación de Ciencias, UJAT.

Cowan, C. P., 1983. *Flora de Tabasco.* 123 pp. Mexico City: Instituto de Biología, UNAM. (Listados florísticos de México I.)

Checklist of vascular plants with *exsiccatae* for each specific and non-nominate infraspecific taxon (more notable collectors or parties are coded; see pp. 9–12 for explanation); bibliography, pp. 122–123. The introduction also covers the background and plan of the work. Arrangement of families is alphabetical within major groups. [Several families were advised on or contributed by specialists.]

Campeche

No separate floras or checklists are available.

Yucatán

No separate floras or checklists are available.

Quintana Roo

Area: 50212 km^2. Vascular flora: see main heading.

Until 1980 the state (a territory until well into the second half of the twentieth century) was in its own right botanically poorly documented. Increasing development, particularly on and near the coast, called for a response; this has included the checklist given below as well as a more popularly oriented work emphasizing trees and other more conspicuous plants: O. Téllez Valdés and M. Sousa Sánchez, 1982. *Imagenes de la flora quintanarroense: proyecto de investigación.* 224 pp., illus. (some col.). Puerto Morelos, Q.R.: Centro de Investigaciones de Quintana Roo.

The island of Cozumel, now an important tourist destination, has also been separately documented; see O. Téllez Valdés and E. F. Cabrera Cano, 1987. *Florula de la isla de Cozumel, Q.R.* Mexico City: Instituto de Biología, UNAM. (Listados florísticos de México VI.)

SOUSA SÁNCHEZ, M. and E. F. CABRERA CANO, 1983. *Flora de Quintana Roo*. [i], 100 pp. Mexico City: Instituto de Biología, UNAM. (Listados florísticos de México II.)

Checklist with *exsiccatae* for each specific and non-nominate infraspecific taxon (more notable collectors are coded; see introductory section for explanation). The introduction covers the background and plan of the work, sources, and (p. 14) statistics; bibliography, pp. 16–18. Arrangement of families follows that of Engler/Diels. [Several families were advised on or contributed by specialists.]

Region 23

Central America

As circumscribed here, this region includes the countries from Guatemala and Belize to Panama inclusive, along with the Swan Islands and the San Andrés and Providencia archipelago in the nearby parts of the Caribbean. For Cocos Island, a dependency of Costa Rica, see **015**.

The history and progress of botanical work in the region has been well reviewed in VERDOORN (1945) and MAGUIRE (1958) and, for the late twentieth century, the status of knowledge and contemporary activities by GENTRY (1978) and in CAMPBELL and HAMMOND (1989) (for all references, see **Superregion 21–23, Progress**). For additional discussion, see W. G. D'ARCY, 1977. Endangered landscapes in Panama and Central America: the threat to plant species. In G. T. Prance and T. S. Elias (eds.), *Extinction is forever*, pp. 89–104. New York. Current belief is that present and future efforts should be directed towards adequate collecting in threatened areas, lands poorly explored in the past (particularly Nicaragua), and for more specialized studies as well as to the preparation of florulas for areas of special interest, enlargement of the growing literature in Spanish, and the achievement of a new regional synthesis (the latter now underway as the *Flora Mesoamericana* project, publication of which began in 1994).

It is only in the twentieth century that nationally oriented floristic documentation has seriously moved forward. Early collections, nearly all by non-residents, were sporadic and documentation haphazard, although as the nineteenth century progressed contributions became more substantial. Significant works of the time include *Flora of the isthmus of Panama* (1852–54) by

Berthold Seemann, published shortly before completion of the Panama Railroad, and *L'Amérique centrale* (1863) by Anders Ørsted, the latter embodying results of exploration in Costa Rica and Nicaragua. Further work by Seemann in southern Central America was frustrated by his death in Nicaragua in 1871. The major European undertaking in consolidation is represented by the botanical part of *Biologia centrali-americana*, compiled by W. B. Hemsley (1879–88; see **210–30**). In the closing decades of the nineteenth century, North American activities increased as interest grew in further trans-isthmian transportation developments. Their first major botanical expression was in the eight-volume *Enumeratio plantarum guatemalensium necnon salvadorensium hondurensium nicaraguensium costaricensium* (1889–1907) by John Donnell Smith, a private publication covering 3736 species or some 36 percent more than in Hemsley's work.[31]

The present level of knowledge in relation to major works varies considerably from country to country. At opposite ends of the region, Guatemala and Panama are each covered by nominally complete multi-volume 'research' floras produced over many years. Moreover, in the latter Standley's *Flora of the Panama Canal Zone* (1928) and Thomas Croat's *Flora of Barro Colorado Island* (1979) are also available, the latter written for a small, intensively studied area of special scientific interest. The larger *Flora of Panama* was, however, begun on a collection base now understood to have been less than adequate; early parts are more often than not quite out of date. For Costa Rica, Standley's partly compiled *Flora of Costa Rica* is gradually being supplanted by *Flora costaricensis* (Burger, 1971–), a critical work comparable with the *Flora of Guatemala* and *Flora of Panama*. Unfortunately, no such works are yet available for the remaining countries. Ramírez Goyena's *Flora nicaraguense* (1909–11) is obsolete, while Honduras and El Salvador are covered simply by checklists in addition to a miscellany of florulas, such as Standley's *Flora of the Lancetilla Valley, Honduras* (1931) and (particularly in El Salvador) other works of more limited scope. Belize is wholly covered by the *Flora of Guatemala* as well as provided with separate enumerations (Spellman and Dwyer, 1975–81; Standley and Record, 1936). The *Flora of Guatemala* can also be used in El Salvador. San Andrés and Providencia in the Atlantic are covered by miscellaneous reports (Toro, 1929–30; Proctor, 1950) and a popular work (González *et al.*, 1995).

On the whole, Central America is now, for the tropics, comparatively well collected – far ahead of tropical South America and, according to Prance (1978; see **General references**), nearly half again as thoroughly covered as Malesia (save for Peninsular Malaysia and Java). However, apart from differences between units, variations also exist with respect to given life-forms: the tree flora is evidently better known in Malesia than in Central (and South) America, whereas in certain parts of Malesia the non-woody flora has received relatively little attention. Only comparatively recently have there begun to appear publications on the tree flora in Mexico and Central America more or less comparable to those long available in Malesia. One of the first was Allen's *Rain forests of Golfo Dulce* (1956); others (all partial or selective) have followed for humid lowland Mexico, Honduras, Nicaragua, Costa Rica, and Panama. The rich pteridophyte flora has been, or is being, documented in Guatemala, Costa Rica and Panama in works respectively by Robert Stolze and David Lellinger.

The vascular flora estimates given for each polity are based on those provided by Gentry (1978, p. 153; see Superregion 21–23).

Bibliographies. General and divisional bibliographies as for Division 2 or Superregion 21–23.

Indices. General and divisional indices as for Division 2 or Superregion 21–23.

230
Region in general

HEMSLEY, W. B. *Biologia centrali-americana: Botany.* See **210–30**.

DAVIDSE, G., M. SOUSA SÁNCHEZ *et al.* (gen. eds.) [FLORA MESOAMERICANA COMMITTEE]. *Flora Mesoamericana.* See **210–30**.

231
Guatemala

Area: 108888 km^2. Vascular flora: about 8000 species (Gentry, 1978); 7749 were accounted for in *Flora of Guatemala* and its satellites, including 605 ferns and fern-allies.

The first organized floristic coverage of Guatemala and its neighbors was a *Flora de Guatemala* by J. M. Mociño, based on collections made in 1795–99. Like its Mexican counterparts from the Real Expedición Botánica (1787–1803), it long remained unpublished.[32] It was only through *Florula guatimalensis* (1840), a 'portfolio of curiosities' by the Bologna botanist Antonio Bertoloni, that the Guatemalan flora was first seriously introduced to science. Then followed *Biologia centrali-americana* (**210–30**) and Donnell Smith's *Enumeratio* (see regional heading), the latter incorporating the collections of Ohio botanist William Kellerman as well as his own and those of his contemporary Hans von Türckheim. In the 1930s, as part of a Field Museum botanical programme in Central America, Paul Standley initiated the *Flora of Guatemala* project. Completed in 1977 after 31 years in publication, this comprehensive work (along with its complements on orchids and pteridophytes, respectively by Oakes Ames and Donovan S. Correll and by Robert Stolze) is considered to cover most of the presently known flora. Grounded in a good representation of collections (aided also by the contemporary Maya programme of the Carnegie Institution of Washington), it still remains, after 3–5 decades, relatively complete.

The recent *Catálogo de árboles de Guatemala* (1982) by J. M. Aguilar is primarily a lexicon of vernacular names; it moreover lacks a botanical index.

STANDLEY, P. C. *et al.*, 1946–77. *Flora of Guatemala.* 13 parts. Illus. (Fieldiana, Bot. 24). Chicago. Complemented by O. AMES and D. S. CORRELL, 1953–54. *Orchids of Guatemala.* 2 parts. 739 pp., 200 text-figs. (*ibid.*, 26); D. S. CORRELL, 1965. Supplement to *Orchids of Guatemala and British Honduras.* Illus. (*ibid.*, 31(7): 177–221); and R. G. STOLZE, 1976–83. *Ferns and fern allies of Guatemala*, I–III. 3 parts. 80 text-figs. (*ibid.*, 39; N.S. 6, 12). (Treatment of the orchids reprinted 1985, New York: Dover, as *Orchids of Guatemala and Belize.*)

The main work comprises a copiously illustrated (in later parts) descriptive flora of native, naturalized and commonly cultivated seed plants of Guatemala and Belize, with keys to genera and species; also included are appropriate synonymy, references and citations (with indication of appropriate revisions and monographs under family and generic headings), typifications, vernacular names, detailed indication of local range (with citation of *exsiccatae*), general summary of extralimital distribution, taxonomic commentary, and often extensive notes on habitat, occurrence, frequency, special features, and ethnobotany (properties, uses, cultivation, etc.); index to genera in each part (with part 13 comprising a comprehensive index). Representative illustrations appear in all except parts 3–6 (the earliest published). The Englerian sequence of families is for the most part followed. The associated treatments of orchids and pteridophytes are similar in format to the main work and, like it, include keys to all taxa as well as representative illustrations.[33]

232

Belize

See also **229** (CARNEGIE INSTITUTION OF WASHINGTON), **231** (all works). Area: 22965 km². Vascular flora: 2500–3000 species (Gentry, 1978); 3200 species (cf. Spellman and Dwyer).

Belize (formerly British Honduras), in which serious collecting began only in 1907, is thought by some now to be relatively well covered; sampling, however, has been uneven with parts of the country still not well known (e.g., the Maya Mountains) and/or difficult of access. The checklists of Spellman, Dwyer and Davidse (1975–81) have effectively supplanted that in Standley and Record's *Forests and flora of British Honduras* (1936), with for the known dicotyledons alone a 127 percent increase. The country was moreover completely covered in *Flora of Guatemala*.[34]

SPELLMAN, D. L., J. D. DWYER and G. DAVIDSE, 1975. A list of the Monocotyledoneae of Belize including a historical introduction to plant collecting in Belize. *Rhodora* **77**: 105–140, illus. Complemented by J. D. DWYER and D. L. SPELLMAN, 1981. A list of the Dicotyledoneae of Belize. *Ibid.*, **83**: 161–236.

Alphabetically arranged nomenclatural checklist; includes citations of representative *exsiccatae* and occasional notes but lacks literature citations (except cross-references to *Flora of Guatemala* in family headings), notes on distribution, or indication of habitat. The introductory section treats physical features, climate, vegetation, and the history of botanical exploration. [The checklists, slanted towards the work of 16 recent collectors, cover 721 species of monocotyledons and 2500 species of dicotyledons, or 90 percent of the vascular flora.]

STANDLEY, P. C. and S. J. RECORD, 1936. *The forests and flora of British Honduras*. 432 pp., 16 pls. (Publ. Field Mus. Nat. Hist., Bot. Ser. 12). Chicago.

Partially annotated enumeration of vascular plants, with keys and descriptions relating largely to the woody species; includes limited synonymy (with some references), some citations of *exsiccatae* along with general indication of local and extralimital range, vernacular names, and notes on habitat, frequency, special features, properties, uses, etc.; index to genera and to vernacular names. The introductory section includes general remarks on geography, geology, soils, climate, agriculture, forest formations, principal timber species, other forest products, vernacular nomenclature, and relationships of the flora, together with a bibliography.

Partial work: Belize Cays

For physical and other background information, see D. R. STODDART *et al.*, 1982. *Cays of the barrier reef and lagoon*. 76 pp., illus., maps (Atoll Res. Bull. 256). Washington, D.C.

FOSBERG, F. R. *et al.*, 1982. *Plants of the Belize Cays*. 77 pp. (Atoll Res. Bull. 258). Washington, D.C.

Systematic checklist of vascular plants (182 species including four sea-grasses), with English vernacular names and localities supported by *exsiccatae*, literature and sight records but without habitat data; references, gazetteer and index to species at end.

233

El Salvador

Area: 20865 km². Vascular flora: about 2500 species (Gentry, 1978).

Botanical exploration before the twentieth century was sporadic; the only general floristic coverage prior to the 1920s is embodied in *Biologia centrali-americana* and Donnell Smith's *Enumeratio* (see Region 23). With the *Lista preliminar* of Standley and Calderón (1925; 2nd edn., 1941) now largely of historical interest, the only really useful general work, which contains many Salvadorean records, is the *Flora of Guatemala* (**231**). There is also a photographic collection, *Flora salvadoreña* (1926–32; reissued 1975–78) by Félix Choussy.[35] In 1981 a floristic project, now appearing as *Listado básico de la Flora Salvadoreña*, was organized. Various other works on the flora exist; the pteridophytes are now fairly well documented. Natural vegetation has been destroyed or greatly altered over most of the country; only a few pockets of old forest remain.

Families and genera

LAGOS, J. A., 1983. *Compendio de botánica systematica*. 2nd edn. 318 pp., illus. San Salvador: Ministerio de Educación. [Students' text, including accounts of families.]

BERENDSOHN, W. G. (ed.), 1989– . *Listado básico de la Flora Salvadoreña*. In fascicles (Cuscatlania 1(1–)). San Salvador.

A nomenclator with citations; accepted names in bold face, names of erroneously reported taxa in Roman, and synonymy in italics, with introduced species asterisked. Commentaries are provided for families. *Exsiccatae* are also given (not, however, in the initial fascicles). A list of cited references including taxonomic papers and floras appears at the end of each part. [Circumscription of families is after Dahlgren; numbering is after the 12th edition of Engler's *Syllabus*. As of 1999 nine fascicles had been published (including one each on bryophytes and pteridophytes), the most recent in 1995.]

CALDERÓN, S. and P. C. STANDLEY, 1941 (1944). *Flora salvadoreña. Lista preliminar de las plantas de El Salvador*. 2nd edn. 450 pp., 2 portrs. San Salvador: Imprensa Nacional. (1st edn. incorporating first supplement, 1925, as *Lista preliminar de las plantas de El Salvador* by P. C. Standley and S. Calderón; second supplement, 1927.)

Pages 1–378 of this work comprise an annotated enumeration (with two supplementary sections) of known fungi, bryophytes and vascular plants, with brief general indication of local range or country of origin, vernacular names, and notes on habitat, uses, etc.; index (pp. 293–344) to all botanical and vernacular names. An introductory section includes an account of local botanical exploration. [The remainder of the so-called second edition comprises unrelated posthumous papers by its senior author. The *Lista* itself is simply a reissue of the original, along with the two supplementary sections, intended to make them again available.][36]

Pteridophytes

The known pteridoflora has increased markedly since a 1930 account by W. R. Maxon and Standley, and the pteridophyte section (the first part) of the *Listado básico* (see above) should now be considered standard. Seiler's treatment is best used together with *Ferns and fern allies of Guatemala* (**231**).

LÖTSCHERT, W., 1953. Ferns of the republic of El Salvador. *Ceiba* **4**(1): 241–250.

Systematic checklist (174 species), with *exsiccatae* and field data; references, pp. 249–250. [Included here mainly for its introductory part.]

SEILER, R., 1980. *Una guía taxonómica para helechos de El Salvador*. 58 pp., 5 figs. (incl. 1 halftone). San Salvador: Dirección de Publicaciones, Ministerio de Educación (for Museo de Historia Natural de El Salvador).

Systematic checklist of ferns and fern-allies (279 species) as part of a students' manual; includes keys to families and (in Polypodiaceae sensu lato) to genera. See also *idem*, 1982. Contribuciones a la pteridológia centroamericana: Enumeratio filicum sancti-salvatoris. *Brenesia* **19/20**: 381–391.

234

Honduras

Area: 112 087 km². Vascular flora: about 5000 species (Gentry, 1978).

The only national coverage prior to 1920 was provided by *Biologia centrali-americana* and Donnell Smith's *Enumeratio* (see Region 23). The country initially was investigated by the 'Real Expedición Botánica a Nueva España' in the 1790s; and in 1919 there appeared *Apuntes de la flora de Honduras* by Eusebio Fallas. The advent of 'El Pulpo' and its development of the banana industry on the north coast furnished opportunities for further exploration which resulted in further florulas and other accounts; these are given below under **Partial works**.[37] In 1943 the herbarium of the Escuela Agrícola Panamericana at Tegucigalpa came into being, and in 1975 its curator Antonio Molina R. published his unannotated *Enumeración*, still standard although based largely on the herbarium's holdings. In western parts, the *Flora of Guatemala* (231) will be found useful.

The country is considered still to be rather unevenly collected, with a basis insufficient for a modern flora.[38] However, a catalogue is now in process of publication (Nelson *et al.*, 1996); initial results suggest that, at least with respect to pteridophytes, the Molina catalogue is now seriously out of date.[39]

MOLINA R., A., 1975. *Enumeración de las plantas de Honduras.* 118 pp. (Ceiba 19(1)). Tegucigalpa.

Unannotated systematic list of vascular plants, with indication if a given species is merely cultivated. A brief introduction outlines the present status of botanical exploration and knowledge.

NELSON SUTHERLAND, C., R. GAMMARA GAMMARA and J. FERNÁNDEZ CASAS, 1996. *Hondurensis plantarum vascularium catalogus: Pteridophyta.* 143 pp., 4 halftones (3 portrs.) (Fontqueria 43). Madrid.

Enumeration of 651 species of ferns and fernallies (in 109 genera) with accepted names, authorities and places of publication, typification, citations in literature, and (where known) vernacular names; literature list, portraits of authors and index to all scientific names at end. The introduction includes statistics and a list of families and genera (with an arrangement and circumscription much as in Crabbe *et al.* (see **General references**)). [Continuation of the series to cover seed plants is projected.]

Partial works

The most substantial of these focus on the northern Atlantic coast.

STANDLEY, P. C., 1931. *Flora of the Lancetilla Valley, Honduras.* 418 pp., 68 pls. (Publ. Field Mus. Nat. Hist., Bot. Ser. 10). Chicago.

Annotated enumeration of vascular plants, without keys; includes miscellaneous descriptive notes and vernacular names; index. A list of non-vascular plants is also included. [The Lancetilla valley lies near Tela on the Atlantic coast.]

YUNCKER, T. G., 1938. *A contribution to the flora of Honduras.* 18 pls. (*Publ. Field Mus. Nat. Hist.*, Bot. Ser. **17**(4): 283–407). Chicago.

Enumeration of collections made by the author in 1934 and 1936 from various points along a route from Tegucigalpa (the capital) to San Pedro Sula and the Lancetilla basin around Tela. Includes additions to Standley's florula (see above).

YUNCKER, T. G., 1940. *Flora of the Aguan Valley and the coastal regions near La Ceiba, Honduras.* 4 pls. (*ibid.*, **9**(4): 245–346). Chicago.

Similar in form to preceding; the area covered is about 75 km east of Tela.

Woody plants (including trees): guides

PACQUET, J., 1981. *Manual de dendrología de algunas especies de Honduras.* 232 pp. Québec: Cohdefor. [Not seen.]

RECORD, S. J., 1927. Trees of Honduras. *Trop. Woods* **10**: 10–47. Complemented by P. C. STANDLEY, 1930. A second list of the trees of Honduras. *Ibid.*, **21**: 9–41.

Briefly annotated systematic lists of known tree species, with indication of vernacular names, uses, etc.; indices (vernacular names only). For further additions, see P. C. STANDLEY, 1934. Additions . . . *Ibid.*, **37**: 27–39.

235

Nicaragua

Area: 148005 km². Vascular flora: about 5000 species (Gentry, 1978).

The flora of Nicaragua until recent years was very poorly collected and documented. It was initially investigated in 1797 by the 'Real Expedición Botánica a Nueva España'. Prior to 1910 the only available coverage was embodied in *Biologia centrali-americana* and Donnell Smith's *Enumeratio* (see Region 23). In 1903 a repatriate, Miguel Ramírez Goyena, began work towards his *Flora nicaragüense* (1909–11); this remains the only general flora.[40] Realization of a modern work is seen as one of the great desiderata in Central American floristic botany, particularly as the area lies on a floristic transition zone. A *Flora of Nicaragua* project was initiated in 1977 and is now very advanced (W. D. Stevens, personal communication). From 1981 forest surveys were made; one result is *Árboles de Nicaragua* (see below).

A complementary work to the two following is J. B. SALAS ESTRADA, 1966. *Lista de la Flora nicaraguensis con especimenes en la Herbario de la Esculea Nacional de Agricultura y Ganaderia, Managua.* 60 pp. Managua. (Mimeographed.) In the southeast, works on Costa Rica (236) will be found useful.

RAMÍREZ GOYENA, M., 1909–11. *Flora nicaragüense.* 2 vols. 1064 pp. Managua.

Pages 123–1064 of this work comprise a briefly descriptive flora of vascular plants and charophytes, with synoptic keys to families, genera and groups of species; also includes vernacular names, an appendix on medicinal plants and indices to generic, family and vernacular names. The general part of the work is pedagogic and includes an extensive glossary.

SEYMOUR, F. C. (comp.), 1980. *A check list of the vascular plants of Nicaragua.* x, 314 pp. (Phytologia, Mem. 1). Plainfield, N.J.

Systematic checklist of vascular plants; includes citations of key floristic works (relating to Central America) and indication of *exsiccatae* together with departments and places of deposit. [Based largely upon collections made by the author and others from 1968 to 1976 and a number of precursory papers, mainly in *Phytologia* from 1973 onwards.]

Woody plants (including trees): guide

Some 2000 species are considered by Salas (1993) to be trees; only the most important are fully treated in his dendrology.

SALAS ESTRADA, J. B., 1993. *Árboles de Nicaragua.* 388 pp., 117 col. pls., 9 maps. Managua: Instituto Nicaragüense de Recursos Naturales y del Ambiente (IRENA).

Atlas of 117 species with colored plates and facing text with botanical descriptions, indication of distribution and habitat, and notes; alphabetically arranged list of tree species with scientific and vernacular names, lexicon of vernacular names with scientific equivalents, illustrated glossary, list of literature, and complete index at end. The introductory part includes generalities on tree and shrub distribution and floral regions along with (pp. 33–79) a survey (with maps) of vegetation and its classification and its resolution into floral regions.

236

Costa Rica

Area: 50 899 km². Vascular flora: about 8000 species (Gentry, 1978: 153), with other estimates to 10 000–12 000 including 1500 orchids (PD; see **General references**). For Cocos Island, see **015**. For a general introduction to the biota of Costa Rica, see D. H. JANZEN, 1983. *Costa Rican natural history.* Chicago; for the flora, see also M. B. MONTIEL LONGHI, 1991. *Introducción a la flora de Costa Rica.* 2nd edn. San José, C.R.: Universidad de Costa Rica. (1st edn., 1980.)

In addition to *Biologia centrali-americana* and Donnell Smith's *Enumeratio* (see Region 23), earlier floristic treatments of Costa Rica include *Lista de la plantas encontradas hasta ahora en Costa Rica . . . extractada de a "Biologia Centrali-Americana"* (1887) by Anastásio Alfaro G., and *Primitiae florae costaricensis* (1891–1901, not completed) by Théophile Durand and Henri Pittier, the latter incorporating the results of a national survey carried out during a late nineteenth century 'boom'.[41] In the 1930s, as part of the Central American botanical programme of the Field Museum of Natural History in Chicago, Standley compiled his *Flora of Costa Rica* (1937–39) which to a considerable extent remains a standard work. Subsequent collections rather quickly rendered it out of date as the great richness of the flora was further revealed. Research towards its more critical and phytographic successor, *Flora costaricensis*, began at the Field Museum in the late 1960s under the direction of William Burger, with publication commencing in 1971; by 1999, 10 parts had been published. The tree flora is partly covered in three works. Pteridophytes are being studied by David B. Lellinger at the Smithsonian Institution, and the first part (of two planned) of a pteridoflora (covering also Panama and the Colombian Chocó) appeared in 1989. Mention should also be made of *The vascular flora of the La Selva Biological Station*, appearing in installments in the journal *Selbyana* from 1986, and related treatments in other journals.

BURGER, W. (ed.), 1971– . *Flora costaricensis.* In parts. Illus. (Fieldiana, Bot. 35, 40; N.S. 4– , *passim*). Chicago.

Detailed descriptive flora of seed plants; includes keys to genera and species, full synonymy (with references) and citations, summary of local and extra-limital range (with *exsiccatae* where relatively few collections are available), often extended taxonomic commentary, and notes on habitat, phenology, variability, biology, etc.; index to botanical names and subjects and list of references in each part. [Arrangement of families follows the Engler/Diels system of 1936; as of 1999, 10 parts had been published, respectively covering Gramineae (family 15), Orchidaceae in part (family 39), Casuarinaceae–Piperaceae (40–41), Chloranthaceae–Urticaceae (42–53), Podostemaceae–Caryophyllaceae (54–70), Lauraceae—Hernandiaceae (80–81), Krameriaceae–Erythroxylaceae (97–101, 101a, 102), Euphorbiaceae (113), Acanthaceae–Plantaginaceae (200–201) and Rubiaceae (202).][42]

STANDLEY, P. C., 1937–39. *Flora of Costa Rica.* 4 parts, index. 1616 pp., map (Publ. Field Mus. Nat. Hist., Bot. Ser. 18). Chicago.

Largely compiled descriptive enumeration of seed plants (covering also Cocos Island), arranged following the Englerian system with keys in selected larger genera and families; includes concise synonymy (with some references), vernacular names, generalized indication of local and extralimital distribution (with citations of some *exsiccatae*), critical commentary, and notes on special features, properties, uses, etc.; index to all botanical and vernacular names at end. An introductory section includes accounts of physical features, climate, vegetation formations, relationships of the flora, botanical exploration, and regional geography, together with a bibliography. [The work, now very out of date, is gradually being supplanted by the more critical *Flora costaricensis* (see above). A slightly amended Spanish edition, covering only the Cycadaceae through the Araceae, was also published: *idem*, 1937–40. *Flora de Costa Rica*, 1(1–4) (Mus. Nac. Costa Rica, Ser. Bot. 1). San José.]

Woody plants (including trees): guides

The tree flora contains some 1400 species (Zamora, 1989). The works of Holdridge and Poveda and of Zamora are in coverage complementary, but the latter is as yet far from complete.

ALLEN, P. H., 1956. *The rain forests of Golfo Dulce*. xi, 417 pp., 22 text-figs., 34 pls. Gainesville: University of Florida Press. (Reprinted with new preface, 1977, Stanford: Stanford University Press.)

Illustrated manual of forest trees, with artificial keys, vernacular names, synonymy, local range, and notes on taxonomy, habitat, biology, timber, uses, etc.; glossary. All families, genera, species and vernacular names are *alphabetically arranged* in the main text, with appropriate cross-referencing. The 1977 reprint contains a new preface by P. H. Raven. [Golfo Dulce is on the Pacific coast, in southeastern Puntarenas Province.][43]

HOLDRIDGE, L. R. and L. J. POVEDA A., 1975. *Árboles de Costa Rica*. Vol. 1. 646 pp., 527 text-figs. (largely half-tones). San José: Centro Científico Tropical.

Artificially arranged treatment of trees in atlas form, with illustrations accompanied by descriptive text; the latter includes details of habit, leaves, flowers, and fruit (with essential features in bold face) together with notes on habitat, distribution, uses, properties, special features, and taxonomy. Artificial keys for identification (using, as far as possible, vegetative features) and indices are also provided. The scope of the work and its style are discussed in an introductory section, where the main key to the artificial groups is given. The present volume treats palms, other monocotyledonous trees, and dicotyledonous trees with lobed or compound leaves. [Not so far completed. Trees with simple leaves are covered in the following work.]

ZAMORA, N., 1989. *Flora arborescente de Costa Rica*, 1.

Especies de hojas simples. 262 pp., illus. Cartago: Editorial Tecnológica de Costa Rica.

Atlas-manual of illustrations with facing text, the latter *pro forma* with vernacular names, botanical details, indication of distribution, and notes on recognition, uses, special features, etc.; glossary, bibliography and index at end. The general part includes statistics as well as an overall key. [Covers 100 of the 1024 known species with simple leaves.]

Pteridophytes

LELLINGER, D. B., 1989. *The ferns and fern-allies of Costa Rica, Panama and the Chocó*, 1. 364 pp., illus., maps (Pteridologia 2A). Washington: American Fern Society.

Systematic enumeration with diagnostic keys and illustrations, descriptions of and key references for genera, synonymy (limited to basionymy), typification (with *exsiccatae*), indication of distribution and habitat, illustrations of details for most species, and occasional commentary; no index in part 1. The introductory part (with maps) covers the basis and plan of the work along with basic references, a list of families, genera and infrageneric taxa, and general key to families. [Coverage is thought to be about 90 percent complete. As so far published the work covers fern-allies and the 'lower' and 'middle' ferns from Ophioglossaceae through Dicksoniaceae (562 species) mostly following the arrangement of Crabbe *et al.* (see **General references**); a second part will complete the work, with over 600 species more.][44]

237

Panama

Area: 78046 km^2. Vascular flora: 9000–10000 species (W. G. D'Arcy in *Flora of Panama*, Introduction, p. vii (1981; see below).[45] Until 1903 the country, part of the old Spanish viceroyalty of New Granada, was under Colombian rule.

The first separate work was *Flora of the isthmus of Panama* by Berthold Seemann (1852–54), published as the second part of his *Botany of the voyage of H.M.S. 'Herald'* (on which he had been naturalist). From then until the 1920s, plant hunting, though considerable – after 1900 much of it by staff and associates of the U.S. National Herbarium, including Paul Standley and Henri Pittier – was limited by and large to comparatively few areas, with particular emphasis on the Isthmus proper. A further early stimulus was the establishment in the 1920s of the Barro Colorado Island Laboratory (now the Smithsonian Tropical Research

Institute). The Missouri Botanical Garden entered the field late in that decade, beginning a collecting programme in association with the Canal Zone Experiment Gardens at Summit. Inspired by Standley's *Flora of the Panama Canal Zone* (1928) and other works, one of the Garden's botanists, Robert Woodson, in 1935 expanded this into a *Flora of Panama* project. After some preliminary contributions, the *Flora* itself was commenced in 1943 and formally 'completed' in 1981. Indices followed in 1987.

It generally is now realized that the *Flora of Panama* was premature, even though at its commencement a substantial number of collections were available. An early remark by Standley doubting that 'they have half the total number of species in the published flora' would prove to be very close to the mark.[46] In the last three decades the known vascular flora of Panama has proved to be far larger than was commonly believed; it will likely increase still further as detailed exploration continues. The species-richness of the poorly known wet forest zone facing the Caribbean, including Bocas del Toro Province, Comisaria de San Blas, and much of the Darién district as well as the numerous 'low' mountains was not suspected; in addition, many records represent range extensions from northern South America (and notably northwestern Colombia) as well as Costa Rica.

Panama has also been the beneficiary of some good local floras as well as a tree manual, and these are accounted for below. Moreover, the first part of a pteridoflora by David Lellinger is now available (see **236**).

Useful general surveys of the flora include in addition to the 1987 introduction by William D'Arcy in part 1(1) of *Flora of Panama* (see below): R. L. DRESSLER, 1972. Terrestrial plants of Panama. In M. L. Jones (ed.), *The Panamic Biota*, pp. 179–185 (Bull. Biol. Sci. Washington, 2). Washington; and W. G. D'ARCY and M. D. CORREA A., 1985. *The botany and natural history of Panama/La botánica e historia natural de Panamá*. xxi, 455 pp., illus. (Monogr. Syst. Bot. Missouri Bot. Gard. 10). St. Louis.

Bibliography

D'ARCY, W. G., 1987. Bibliography relating to *Flora of Panama* database. In *idem, Flora of Panama*, 1(1): xxiii–xxvii.

WOODSON, R. E., JR., R. W. SCHERY and collaborators, 1943–81. *Flora of Panama*. Parts 2–9 (in 41 issues). Illus. (Ann. Missouri Bot. Gard. 30–67, *passim*). St. Louis. (From 1981 distributed in separate issues or as complete sets by Allen Press, Lawrence, Kans.) Continued as W. G. D'ARCY, 1987. *Flora of Panama: checklist and index* (Flora of Panama, part 1). 2 parts (Monogr. Syst. Bot. Missouri Bot. Gard. 17, 18). St. Louis.

Parts 2–9 comprise a detailed descriptive flora of seed plants, covering some 6200 species; included are keys to genera and species, extensive synonymy with references and citations (including indication of appropriate revisions and monographs under family and generic headings; for later documentation see the bibliography in part 1), typification, vernacular names, general indication of local range (with citations of *exsiccatae*) and extralimital distribution, critical commentary, and notes on habitat, special features, uses, etc.; selected illustrations; and an index to taxa at end of each part (i.e., each issue; after 1965 family treatments within issues are separately indexed). Part 1 comprises two sections (labeled parts): the first contains a fixed-format systematic checklist with indication of distribution, altitudinal range, and habit, and the second a complete index to scientific names and place of publication (whether in the *Flora* or more recent papers). It also includes (pp. vii–xxx) a 'definitive' introduction, including a project history, statistics of the flora and changes in our knowledge of it, an almanac, and an index to families and list of authors. [Some installments include supplementary family treatments, but much of the earlier work is now outdated. Numbering of families (and the division of the work as well as the 1987 checklist) essentially follows the 1936 Engler/Diels sequence.][47]

Partial works

In addition to the two key works on the Canal Area, a relatively recent florula is also available for San José Island in the Gulf of Panama. For Barro Colorado Island, Croat's work entirely supersedes *The flora of Barro Colorado Island* by P. C. Standley (1933, as *Contr. Arnold Arbor.* 5).

CROAT, T. B., 1979. *Flora of Barro Colorado Island.* 960 pp., illus. Stanford: Stanford University Press.

Descriptive flora of this 14.8 km^2 island in Gatun Lake in the 'Canal Area' (covering 1369 species); includes keys to all taxa (with separate key to woody plants in sterile condition), references and citations, synonymy, indication of local and extralimital range, critical remarks, and often extensive notes on habitat, phenology, and biology (including pollination and dispersal mechanisms); index at end. An introductory section deals with physical features, climate, geology, soils, life-forms, floristics, and phytogeography as well as general and botanical history of the area.

JOHNSTON, I. M., 1949. *The botany of San José Island* (Gulf of Panama). ii, 306 pp., 17 pls. (Sargentia 8). Jamaica Plain, Mass.

This 'biological florula' (somewhat comparable with that by Thomas Croat, cited above) includes a systematic enumeration of 627 indigenous and alien species of vascular plants, with very detailed descriptions and notes written in semi-popular style; supporting the plant descriptions are English vernacular names, citation of *exsiccatae* as well as some literature (mainly the Canal Zone flora of Standley, cited below), localities, critical commentary, and remarks on overall distribution, habitat, and biology; index to vernacular and scientific names at end. The introductory section contains background historical, physical and biotic information, along with descriptions of vegetation types and synusiae and special biological features, beach drift, and leaf fall and renewal; keys to tree and shrub species and to orchids are also provided. [Based in the first instance upon field work by the author in 1943–46.]

STANDLEY, P. C., 1928. *Flora of the Panama Canal Zone.* x, 416 pp., 7 text-figs., 67 pls. (Contr. U.S. Natl. Herb. 27). Washington, D.C. (Reprinted 1968, Lehre, Germany: Cramer.)

Manual-key to seed plants, with essential synonymy, local range, vernacular names, and notes on special features, uses, etc.; extensive introductory section treating physiography, floristics, land uses, and botanical exploration; index at end. [This work covers in a compact fashion some 2000 species and possesses a good general key to families (actually based on that of Franz Thonner; see **General keys** under **000**). It is, however, now rather incomplete; moreover, plant names follow the obsolete American Code of nomenclature.]

Woody plants (including trees): guide

HOLDRIDGE, L. R., 1970. *Manual dendrológico para 1000 especies arbóreas en la República de Panamá.* xi, 325 pp., 9 text-figs. (Inventariación y demonstraciones forestales: Panamá). Panamá: Food and Agriculture Organization, United Nations.

Artificially arranged tree-manual for foresters and ecologists, similar in style to his *Árboles de Costa Rica* (**236**) but without illustrations; species entries include vernacular names as well as habit, botanical, dendrological, and other data. The introductory section gives the plan and scope of the work, a selected bibliography, an illustrated glossary, and the general analytical keys; indices to scientific and vernacular names appear at the end (pp. 293ff.).

Pteridophytes

LELLINGER, D. B., 1989. *The ferns and fern-allies of Costa Rica, Panama and the Chocó,* 1. 364 pp., illus.

(Pteridologia 2A). Washington: American Fern Society. – See **236** for description.

238

San Andrés and Providencia Islands

Area: 44 km^2 (San Andrés and Providencia). These two hilly islands, an *intendencia* of Colombia, lie east of Nicaragua in the Caribbean Sea. Included here also are the scattered islets and cays in the south and southeast near San Andrés (Courtown Cays and Cayos de Albuquerque) and to the northeast towards Jamaica (North Cay and Southwest Cay on Serrana Bank; Middle Cay and East Cay on Serranilla Bank; Bajo Nuevo; Roncador Cay). The flora, a subset of that in the Yucatán Peninsula, is noteworthy among those of the Antilles for a relatively high diversity per unit area.

No modern separate technical account is available. A popular work, oriented towards uses and featuring numerous color photographs with parallel text in Spanish and English, is F. A. GONZÁLEZ G., J. NELSON DÍAZ and P. LOWY CERÓN, 1995. *Flora ilustrada de San Andrés y Providencia/An illustrated flora of San Andrés and Providencia.* 280 pp., illus. (part col.). [Bogotá]: Servicio Nacional de Aprendizaje/ Universidad Nacional de Colombia. Additional information is available in E. BARRIGA-BONILLA *et al.*, 1969. *La isla de San Andrés.* Bogotá: Instituto de Ciencias Naturales.[48]

PROCTOR, G. R., 1950. Results of the Catherwood-Chaplin West Indies expedition, 1948. I: Plants of Cayo Largo (Cuba), San Andrés and Providencia. *Proc. Acad. Nat. Sci. Philadelphia* **102**: 27–42, 1 pl., 2 maps.

The second part of this paper comprises an enumeration of pteridophytes of San Andrés and Providencia, with citations of *exsiccatae*, habit and habitat notes, and critical commentary, followed by a general discussion; no list of references. The plant list is preceded by a physical description of the islands and remarks on their current vegetation.[49]

TORO, R. A., 1929. Una contribución a nuestro conocimiento de la flora silvestre y cultivada de San Andrés. *Revista Soc. Colomb. Ci. Nat.* **18**: 201–207.

Complemented by *idem*, 1930. Una contribución a nuestro conocimiento de la flora de San Andrés y Providencia. *Ibid.*, **19**: 56–58.

Unannotated systematic lists of flowering plants (respectively 96 and 40 species), with inclusion of some English vernacular names. Most records appear to relate to San Andrés.

239

Swan Islands

Area: 2.5 km^2. These two small islands lie in the Caribbean to the northeast of Honduras at 84° W, and now are part of that country. No floristic accounts are available so far as is known.

Superregion 24–29

The West Indies

As delimited here, this includes all the islands from the Bahamas and Cuba through the Greater and Lesser Antilles to Curaçao, Aruba and Bonaire. Barbados, Tobago, Trinidad, Margarita and Coche and their associated islets are also included. For Bermuda, see **Region 03**.

Due to their economic importance in the seventeenth and eighteenth centuries as well as their relative accessibility, the West Indies have among tropical lands one of the longest histories of botanical exploration, rivaled only by southern India, Sri Lanka, and parts of the East Indies. The earliest serious work was done in 1687 by Hans Sloane in Jamaica and elsewhere and then, over several years from 1689, by Charles Plumier in Hispaniola and Martinique. Others, including Mark Catesby in the Bahamas and Nikolaus von Jacquin generally in the Caribbean, followed in the eighteenth century. Linnaeus was thus able to account for (and so transfer to modern nomenclature) a great many West Indian species. His consolidation was followed by the field work and publications of Patrick Browne (*Civil and natural history of Jamaica*, 1756), Jacquin

(*Enumeratio stirpium plantarum*, 1760, and *Selectarum stirpium americanarum historia*, 1763) and Olof Swartz (*Prodromus*, 1788, and *Flora Indiae occidentalis*, 1797–1806). In 1808–28 François de Tussac, forced to leave Haiti after a long residence, published his four-volume color-plate *Flore des Antilles*, a selection of more conspicuous or economically important species.

Important as this legacy is, however, these efforts – perhaps inevitably – were somewhat uncoordinated and incidental. They have left considerable difficulties with respect to typification, nomenclature and geography, a situation not always – in view of a prevailing emphasis on local floristics – fully resolved by later scholarship.[50] An attempt at consolidation of a miscellany of local efforts in the British-controlled islands (many in economic difficulties from the 1830s onwards) was essayed from afar by August Grisebach, firstly in his *Systematische Untersuchungen . . .* (1857), covering the Lesser Antilles, and then in his landmark *Flora of the British West Indian islands* (1859–64, in W. J. Hooker's 'Kew Series'). Activity also emerged in Cuba, mainly through the efforts of Ramón de la Sagra, part-author of its first complete flora. After 1850 many collectors, both resident and visiting and including both amateurs and professionals, were active in the superregion, particularly in the Greater Antilles. Intensive collections in Cuba (Charles Wright), Hispaniola (Henrik Eggers), Puerto Rico (Domingo Bello y Espinosa, Leopold Krug, and Paul Sintenis) and the Virgin Islands (Eggers and others) were made at this time, and in the French Lesser Antilles Charles Bélanger, Père Antoine Duss and Otto Hahn were all active. In Havana Manuel Gómez de la Maza y Jiménez was in charge of the botanical garden earlier run by Sagra.

In the last third of the century, systematic and floristic studies became more sophisticated and, with an increasing dependence on large collections – difficult to maintain in the humid tropics – began to be organized around a limited number of foci outside the West Indies. These latter, directed by professionals, included Berlin (Leopold Krug and Ignatz Urban); New York and Chicago (respectively Nathaniel Britton and Charles Millspaugh); and Kingston (Jamaica) and London (William Fawcett and Alfred Rendle, the latter at the Natural History Museum). The young Albert S. Hitchcock, later a leading student of grasses at Washington, D.C., also contributed floristic work, particularly in the Cayman Islands and Jamaica (and, late in his career, *Manual of the grasses of the West Indies*);

the Utrecht botanist Isaac Boldingh did likewise in the hitherto somewhat-neglected Dutch islands and from 1865 Père Antoine Duss had been active in the French possessions. In Trinidad, the government botanist George S. Jenman (also noted for his work in British Guiana) paid considerable attention to pteridophytes; his work eventually was completed by his successor Henry C. Hart as *The ferns and fern allies of the British West Indies and Guiana* (1898–1909). They, and the collectors and local workers with whom they were associated, were responsible for the greater part of botanical work in the West Indies over the some 40 or so years from 1879; to them also are owed a not inconsiderable number of the standard floras of the superregion. Urban also attempted to bring some synthesis of knowledge through his privately financed, nine-volume *Symbolae antillanae* (1898–1928), which included many family revisions as well as some island floras (notably of Hispaniola and Puerto Rico, where complete accounts of vascular plants had been lacking), along with a bibliography, historical account and a cyclopedia of collectors.

Local work, however, continued or developed in a number of centers, among them Havana, Santiago de Cuba, Santo Domingo, Kingston, Basse-Terre and Port-of-Spain. In some, a serious start was made towards production of a flora, and this trend has continued, with several works appearing in the last seven decades. Many local herbaria, notably in Cuba, Hispaniola, Puerto Rico, Jamaica, Guadeloupe, Barbados and Trinidad, are now in existence, some of long standing and with substantial resources. Additional 'standard' floras, of which the most considerable are *Flora de Cuba* by Brother Léon (Joseph Sauget-Barbier) and Brother Alain (Alain Henri Liogier) and *Flora of the Lesser Antilles: Leeward and Windward Islands* by Richard Howard and his collaborators (1974–89), have been published or are in progress. The flora, estimated as having from 12 000 to 15 000 vascular species, has become comparatively well known and, although novelties continue to be found here and there, it 'cannot be regarded as . . . needing immediate study or a massive collecting program' (Howard, 1977). Yet, as with Central America, further scope for consolidation exists which, among other advantages, would overcome a legacy of fragmented study and artificial taxa, particularly in the Greater Antilles. Monographic studies (within the framework of *Flora Neotropica* or otherwise) form one direction;

but as the superregion is now perhaps about as well collected and documented as much of the Mediterranean Basin, it would also not be unreasonable to suggest the compilation of a work similar to the *Med-Checklist* (601/I). This would, among other undoubted benefits, draw attention to problems requiring solution.[51]

Progress

For early and recent history, see I. URBAN, 1902. Notae biographicae peregrinatorum Indiae occidentalis botanicorum. In *idem* (ed.), *Symbolae antillanae*, 3: 14–158. Berlin; F. VERDOORN, 1945. *Plants and plant science in Latin America*. Waltham, Mass. (especially the 'Historical Sketch' by F. W. Pennell, pp. 35–48); B. MAGUIRE, 1958. Highlights of botanical exploration in the New World. In W. C. Steere (ed.), *Fifty years of botany*, pp. 209–246. New York; and G. R. PROCTOR, 1961. Our knowledge of the flora of the West Indies. In Ninth International Botanical Congress, *Recent advances in botany*, 1: 929–932. Toronto.

Other papers, in part dealing with the current situation, are R. A. HOWARD, 1977. Conservation and the endangered species of plants in the Caribbean Islands. In G. T. Prance and T. S. Elias (eds.), *Extinction is forever*, pp. 105–114. New York; *idem*, 1979. Flora of the West Indies. In K. Larsen and L. B. Holm-Nielsen (eds.), *Tropical botany*, pp. 239–250. London; the contributions respectively by Rene P. C. López *et al.*, T. Zanoni, J. L. Vivaldi and R. A. Howard on Cuba, Hispaniola, Puerto Rico and the Lesser Antilles in D. G. CAMPBELL and H. D. HAMMOND (eds.), *Floristic inventory of tropical countries* (under **General references**), pp. 315–349; and in the more general reviews of PRANCE (1978; see **General references**) and S. G. SHETLER (1979; see **Superregion 21–23**, introduction, p. 000).

Bibliographies. Bay, 1910; Blake and Atwood, 1942; Frodin, 1964, 1984; Goodale, 1879; Holden and Wycoff, 1911–14; Jackson, 1881; Pritzel, 1871–77; Rehder, 1911; Sachet and Fosberg, 1955, 1971; USDA, 1958. See also the incomplete divisional bibliography by Agostini.

Supraregional bibliography

URBAN, I., 1898. Bibliographia indiae occidentalis botanica. In *idem* (ed.), *Symbolae antillanae* 1: 3–195. Berlin; supplemented in *idem*, 1900–04. Continuatio I–III. *Ibid.*, 2: 1–7; 3: 1–13; 5: 1–16.

(Reprinted collectively in 1 vol., 1964, Amsterdam: Asher.)

Indices. BA, 1926– ; BotA, 1918–26; BC, 1879–1944; BS, 1940– ; CSP, 1800–1900; EB, 1959–98; FB, 1931– ; IBBT, 1963–69; ICSL, 1901–14; JBJ, 1873–1939; KR, 1971– ; NN, 1879–1943; RŽ, 1954– .

Supraregional index

ZANONI, T. A., 1986– . Bibliografía botánica del Caribe, I– . *Moscosoa* **4**, *passim*. [Concisely annotated, alphabetically arranged lists of recent papers and other works from 1984 onwards. Four parts appeared through 1993, the latest in *Moscosoa* **7**.]

240–90

Superregion in general

The only comprehensive works are Grisebach's *Flora of the British West Indian islands* (1859–64) and the more detailed *Symbolae antillanae* edited by Urban (1898–1928), the former a part of the 'Kew Series' of British colonial floras begun in the late 1850s by W. J. Hooker. Although by now out of date, these works serve as a foundation for all subsequent floras. For some islands, they have yet wholly to be superseded; in particular, the *Symbolae* remains useful for its family treatments. A full revision of the Gramineae is found in A. S. HITCHCOCK, 1936. *Manual of the grasses of the West Indies.* 439 pp. (Misc. Publ. USDA 243). Washington, D.C. With respect to pteridophytes, no overall account has appeared since *Histoire des fougères et des Lycopodiacées des Antilles* by Antoine Fée (1866, as part 11 of his *Mémoires sur la famille des fougères*).

GRISEBACH, A. H. R., 1859–64. *Flora of the British West Indian islands.* xvi, 789 pp. London: Reeve. (Reprinted 1963, Weinheim, Germany: Cramer.)

Concise, briefly descriptive flora of vascular plants of those islands (Cayman Islands, Jamaica, British Virgin Islands, the Bahamas, Trinidad and Tobago, and the greater part of the Leeward and Windward Islands) then under British rule; includes partly synoptic keys to genera and groups of species, limited synonymy (with some citations), indication of distribution by island or island group, brief summary of extralimital range, vernacular names, critical commentary, and notes on phenology, uses, etc.; indices to all botanical and vernacular names at end. An introductory section includes a note on geographical limits of the flora as well as chapters on collectors, sources of information, etc.[52]

URBAN, I. (ed.), 1898–1928. *Symbolae antillanae.* 9 vols., portr. Berlin: Borntraeger. (Reprinted 1964, Amsterdam: Asher.)

This monumental serial work – in addition to incorporating complete or partial floras of a number of West Indian islands (e.g., Hispaniola) – is largely devoted to critical regional revisions of many families (in part by specialists) along with reports of various collecting trips (in later years especially those of Erik Ekman). The revisions include keys, descriptions, extensive synonymy (with references and citations), detailed indication of West Indian range (with citation of *exsiccatae*), and critical taxonomic commentary together with miscellaneous notes. [Vol. 4 comprises *Flora portoricensis* and was dedicated to Leopold Krug; vol. 6 included a chapter on additions to the Jamaican flora, with a historical account especially of the period after 1860; vol. 8 is *Flora domingensis*, much aided by Ekman's first collecting trip in Haiti; and a considerable part of vol. 9 covers Ekman's Cuban collections. For a general index to the work as well as a review (by Richard A. Howard) of the author's career and the foundations for the *Symbolae*, see E. CARROLL and S. SUTTON (compil.), 1965. *A cumulative index to the nine volumes of the 'Symbolae Antillanae' edited by I. Urban.* 272 pp. Jamaica Plain: Arnold Arboretum of Harvard University.][53]

Pteridophytes

The most recent separate work covering any considerable part of the West Indies as a whole is G. S. JENMAN, 1898–1909. *The ferns and fern allies of the British West Indies and Guiana.* viii, ix, 407 pp. (Issued in installments in *Bull. Misc. Inform.* (Trinidad).) Trinidad. For the anglophone islands it has been superseded by more recent pteridofloras or general floras of vascular plants.

Region 24

Bahama Archipelago (including the Bahamas and Turks and Caicos Islands)

Area: 11 836 km² (Bahamas, 11 406 km²; Turks and Caicos, 430 km²). Vascular flora: 1350 species (Correll and Correll, 1982; see below). – This comprises the entire chain of some 35 islands and 700 cays from Grand Bahama, Great Abaco and Bimini southeast to the Caicos and Turks, thus encompassing the two polities of the Bahamas and the Turks and Caicos Islands. No elevation in these relatively low islands, about half of them extra-tropical, exceeds 63 m.

Exploration of the Bahamas began early in the eighteenth century and by the present time they are considered reasonably well known. A *Provisional list of the plants of the Bahama Islands* (1889) by John Gardiner and Lewis J. K. Brace – the latter an islands native – summarized available knowledge and provided a foundation for an extensive series of surveys in the last decade of the nineteenth and first two decades of the twentieth century. These were largely the work of N. L. Britton and his associates, among them Charles Millspaugh in Chicago and Stewardson Brown in Philadelphia. The resulting collections and observations formed a principal basis for Britton and Millspaugh's long-standard *Bahama flora* (1920). With such a flora completed, however, botanical interest in the islands then declined; it would not resume until well after World War II and at new locations.

The Fairchild Tropical Garden in Coral Gables near Miami became particularly active from the 1960s, working in association with the Arnold Arboretum of Harvard University. From the former extensive collecting was carried out – notably by the Corrells – with the goal of a new descriptive flora and related works. Additions and corrections to the known flora were published from 1973 onwards in *Rhodora*, *Sida* and *Phytologia*. In 1982, the primary aim was realized with the appearance of *Flora of the Bahama Archipelago* by Donovan S. and Helen B. Correll.[54]

For a concise review of floristics and major plant communities, see D. S. CORRELL, 1979. The Bahama archipelago and its plant communities. *Taxon* **28**: 35–40.

Bibliographies. General, divisional and supraregional bibliographies as for Division 2 and Superregion 24–29.

Regional bibliography

GILLIS, W. T., R. BYRNE and W. HARRISON, 1975. *Bibliography of the natural history of the Bahama Islands*. [ii], 123 pp., map (Atoll Res. Bull. 191). Washington, D.C. [Botany, pp. 17–29.]

Indices. General and divisional indices as for Division 2 or Superregion 24–29.

240

Region in general

The work described below is a successor to N. L. BRITTON and C. F. MILLSPAUGH, 1920. *The Bahama flora*. viii, 695 pp. New York: The authors. (Reprinted 1963, Hafner.)

CORRELL, D. S. and H. B. CORRELL, 1982. *Flora of the Bahama Archipelago (including the Turks and Caicos Islands)*. [50], 1692 pp., 689 text-figs., 1 col. pl., 2 maps (1 in end-papers). Lehre: Cramer. (Reprinted 1996, Vaduz: Gantner.)

Well-illustrated descriptive manual-flora, with bracketed keys to all taxa, publication references, limited synonymy, indication of local (coded, with reference to the end-papers map) and extralimital distribution, one-line notes on habitat and phenology, and (sometimes) brief taxonomic commentary; table of vouchers for each species in any one or more of 11 botanico-geographic units, with explanation (pp. 1554–1604); authors, itineraries, illustrated glossary, bibliography and complete index at end. The introductory part includes a preface (with historical notes and basis for the present work as well as its plan and scope) and sections on physical features, geology, origin of the flora, plant communities, natural resources, agriculture, and economic aspects (including plant products, bush medicine, etc.); statistics (p. [29]) and general key (pp. [31]–[50]). [The unaltered reissue is on slightly thinner paper.]

Woody plants (including trees): guide

PATTERSON, J. and G. STEVENSON, 1977. *Native trees of the Bahamas.* 128 pp., illus. (part col.). Hope Town, Abaco (Bahamas): Patterson.

Artificially arranged popular account of native, naturalized and commonly cultivated trees, with the species accounts inclusive of vernacular and botanical names, local distribution and extralimital range, illustrations, and features enabling distinction from related species; appendices on tree geography, very rare species, figures of very common shrubs and herbs, a list of families, and indices to all names at end. An introductory section gives background information as well as a preliminary artificial key (p. 7).

Region 25

The Greater Antilles (except the Bahamas)

As delimited here, this region comprises the islands from Cuba and the Cayman Islands eastward to Puerto Rico and the Virgin Islands. The last-named group, although often considered as belonging to the Lesser Antilles, is included here for bio-historical reasons as well as convenience.

Politically and culturally diverse, the region has had a long history of botanical activity and publication and by now is relatively well known. Within it, Havana was before 1900 perhaps the most prominent center (along with Kingston in Jamaica). However, while much had, as indicated above in the introduction to the West Indies, become known from seventeenth- and eighteenth-century collections and publications, serious work in the mountainous interiors of the islands remained for Wright, Eggers, Sintenis, and others in the nineteenth century, followed in the early twentieth century by Harris, Fawcett, Britton, Shafer, Wilson, and Hitchcock and culminating with Ekman's extensive botanizing up to 1931. Much of this was in the 1850s and 1860s synthesized by August Grisebach at Göttingen and from 1898 to 1925 on a grand scale by Ignatz Urban in his *Symbolae antillanae*, which included separate floras for Puerto Rico (1903–11) and Hispaniola (1920–21, a first), as well as many family revisions covering the whole of the West Indies. Urban's Puerto Rico treatment became a basis for

Britton and Percy Wilson's *Botany of Porto Rico and the Virgin Islands* (1923–30), among North American programmes the largest single contribution. Contemporaneously with the work of Urban, a detailed *Flora of Jamaica* was initiated by William Fawcett, for some years director of the botanic gardens in the island, in associaton with Alfred Rendle at the Natural History Museum in London, and five volumes of a projected seven appeared between 1910 and 1936. The latter, however, remained incomplete despite some further work after World War II at the Museum.

The deaths in the early 1930s of Britton, Urban and Ekman coincided with a new, and more permanent, devolution of activity. Serious contributions to knowledge of the Greater Antilles flora began again to come from local centers, particularly in the three decades after World War II. Until the close of the 1950s, a significant focus in Cuba was the Museo de Historia Natural of the Colegio de La Salle, to which Brothers Léon and Alain were attached and with whom Brother Marie-Victorin from Montreal collected.[55] Léon and Alain's major work was *Flora de Cuba* (1949–62; supplement by Alain alone, 1969). After reorganization and reorientation following the advent of the Castro government, botanical work in Cuba has continued, partly in collaboration with botanists in Hungary and the former German Democratic Republic, and in the 1980s a new flora project was mounted; publication commenced in 1992. In Jamaica, the postwar era saw development of the Institute of Jamaica and the University of the West Indies, each of which became 'home' to major works: George Proctor's *Preliminary checklist of Jamaican pteridophytes* (1953), *Flora of the Cayman Islands* (1984) and *Ferns of Jamaica* (1985) at the former, and Dennis Adams's *Flowering plants of Jamaica* (1972) at the latter. Leaving Cuba after the change of government, Brother Alain (as Alain H. Liogier) shifted his botanical activities to Hispaniola and Puerto Rico. Becoming associated firstly with the New York Botanical Garden and later with the University of Puerto Rico, he has published several floristic works on these two islands (including descriptive floras, both completed but without the monocotyledons). In Hispaniola, the work of Ekman and Urban was continued by Rafael Moscoso and J. de J. Jiménez; their work was summarized respectively in *Catalogus florae domingensis* (1943) and its *Suplemento 1* (1967). Another flora of the island has been begun in Santo Domingo. In Puerto Rico, the presence of the U.S.

Forest Service's tropical forest research center (established in 1940) provided a considerable stimulus to new research on the tree flora there and in the Virgin Islands; the principal contribution has been the two large volumes by Elbert Little, Frank Wadsworth and R. O. Woodbury (1964–74).[56]

All islands are now covered by more or less adequate floras or enumerations. Adams's *Flowering plants of Jamaica* is moreover by general consent considered to be one of the best of its kind among tropical floras. Recent coverage of the pteridophytes, from a relatively early date often treated apart from the seed plants, is also available almost throughout the region. In addition, a *Flora of the Greater Antilles* has been initiated at the New York Botanical Garden under the direction there of William R. Buck.[57]

Bibliographies. General, divisional and supraregional bibliographies as for Division 2 and Superregion 24–29.

Indices. General and divisional indices as for Division 2 or Superregion 24–29.

251

Cuba

Area: 114 524 km². Vascular flora: about 6000 species (Alain H. Liogier in Prance, 1978; see **General references**); 5785 species of seed plants in 1296 genera accounted for by León and Alain, 1946–62 and 1969).

The first Cuban flora by Balthasar Boldó and José Estévez, based on materials collected on the 'Real Comisión de Guantanamo' of Conde Mopox y Jaruco (1796–1802) and illustrated by José Guio, was, like the work of the botanical expedition to New Spain which also visited the island, never published in its time and effectively lost to science.[58] Nevertheless, from a relatively early date local resources, based on the Havana Botanic Garden, were developed with the island enjoying some stability for several decades under continuing Spanish rule. The main nineteenth-century works include those composed after 1830 by Achille Richard with the Cuban naturalist Ramón de la Sagra (and, for cryptogams, Camille Montagne), the Spanish version (the only one completed) appearing in volumes 9–12

(1845–55) of Sagra's *Historia fisica*, and, in 1866, *Catalogus plantarum cubensium exhibens collectionem wrightianum* by August Grisebach, based on the extensive collections of Charles Wright in the 1860s and long a 'benchmark'. The *Catalogus* was revised and expanded in Cuba in 1868–73 by Francisco Sauvalle and by then accounted for 3350 vascular species.

The last third of the nineteenth century was evidently less favorable to botanical work; Sauvalle's successor Manuel Gómez de la Maza published only a *Flora habanera* (1897) and a *Catálogo de las plantas cubanas* (1896; in *Progreso Médico* 8). The 1898 revolution and subsequent political changes led to a period of several decades during which, as the historian Conde has noted, collecting and taxonomic studies were dominated by the activities of visiting North Americans. Gómez de la Maza and Juan T. Roig y Mesa did, however, produce in the latter period an illustrated introductory *Flora de Cuba* (1914) for pedagogical use.

The present standard work by Brothers León and Alain (1946–62; supplement, 1969), the result of sustained locally based work from 1918, covers only seed plants. Pteridophytes have been accounted for separately in an enumeration by Duek (1971). After 1959, several smaller collections were amalgamated into two central herbaria (one dating back to 1860). A new, collaborative *Flora de Cuba* project was also organized; to this end, a number of papers have appeared, mostly in Cuban, Hungarian and eastern German journals. *Flora de la República de Cuba* itself commenced publication in 1992, initially in the Spanish journal *Fontqueria* but from 1998 as a separate series.[59]

For a general review of the flora and vegetation, see A. BORHIDI, 1991. *Phytogeography and vegetation ecology of Cuba*. Budapest: Akademiai Kiadó.

Bibliographies

MANITZ, H., 1999. *Bibliography of the flora of Cuba*. 1008 pp. (Regnum Veg. 136). Königstein/Ts., Germany: Koeltz. [11 600 entries, arranged in the first instance by author; subject and taxonomic indices.]

SAMKOVA, H. and V. SAMEK, 1967. *Bibliografia botánica cubana (téorica y aplicada) con enfasis en la silvicultura*. 36 pp. (Acad. Ci. Cuba, Ser. Biol. 1). Havana. [Literature from 1900 through 1967.]

LEIVA, A. *et al.*, 1992. *Flora de la República de Cuba*. In parts. Illus., maps (Fontqueria 34–35). Madrid. Continued as A. LEIVA SÁNCHEZ *et al.*,

1998– . *Flora de la República de Cuba*, serie A: *Plantas vasculares.* In fascicles. Illus., maps. Königstein/Ts., Germany: Koeltz.

Descriptive 'research' flora with keys, synonymy, references, typification, indication of distribution (with maps), citation of *exsiccatae*, illustrations, and notes on habitat, frequency, phenology, etc. Family and generic descriptions are accompanied by concise taxonomic and general commentary, while for families relevant references are listed at the end of each account. [Families published in *Fontqueria* include Loranthaceae (34) and Eremolepidaceae, Clethraceae and Ericaceae (35); those so far in the independent series include Araceae, Aristolochiaceae, Bombacaceae, Droseraceae, Linaceae and Mimosaceae.][60]

LEÓN, Hermano and Hermano ALAIN, 1946–62. *Flora de Cuba.* 5 vols. 794 text-figs., portr., maps (vols. 1–4 as *Contr. Ocas. Mus. Hist. Nat. Colegio 'La Salle'* 8, 10, 13, 16). Havana (vols. 1–4); Río Piedras, Puerto Rico: Editorial de Univ. de Puerto Rico (vol. 5). (Reissued in 3 vols., 1974, Königstein/Ts., Germany: Koeltz.) (Vol. 1 by León alone; vol. 5 by Alain alone.) Continued by ALAIN (as A. H. LIOGIER), 1969. *Flora de Cuba: suplemento.* 150 pp. Caracas: the author.

Copiously illustrated, concisely descriptive flora of seed plants; includes keys to all taxa, limited synonymy, abbreviated indication of local and extralimital range, vernacular names, critical commentary, and brief notes on habitat, properties, uses, etc. Bibliography, summaries, and indices to botanical and vernacular names are provided at the end of each volume. An introductory section (vol.1) incorporates a glossary along with accounts of phytogeography and the history of floristic botany in Cuba.

Woody plants (including trees): guide

BISSE, J., 1988. *Árboles de Cuba.* Revised edn. xvi, 384 pp., 55 text-figs. Havana: Editorial Científico-Técnica. (Original edn., 1981.)

Illustrated manual of trees with scientific and vernacular names, limited synonymy, keys, concise descriptions, and indication of internal and extralimital distribution and habitat; brief list of references and indices to all names at end. The work is preceded by a preface, introduction, and general sections on scientific names, vegetation and soils, climate, floristics and forest types with characteristic species along with the general key to genera (pp. 29–50). Arrangement of families is alphabetical, but not that of genera. [A compact work but limited to the more common or widespread species.]

Pteridophytes

DUEK, J. J., 1971. Lista de las especies cubanas de Lycopodiophyta, Psilotophyta, Equisetophyta, y Polypodiophyta (Pteridophyta). *Adansonia*, II, 11: 559–578, 717–731.

Systematic list, with synonymy, references, vernacular names, typification, indication of distribution in the Greater Antilles and in Florida, and karyotypes; bibliography.

252

Cayman Islands

See also **253** (FAWCETT and RENDLE). – Area: 259 km². Vascular flora: 601 native, naturalized and commonly cultivated species, of which 102 are introductions (Proctor, 1984). This latter-day Caribbean 'tax shelter' encompasses the islands of Grand Cayman, Little Cayman, and Cayman Brac together with several islets, all of low elevation.

George Proctor's descriptive flora replaces *List of plants collected in the Bahamas, Jamaica and Grand Cayman* by A. S. Hitchcock (1893, in *Annual Rep. Missouri Bot. Gard.* 4: 47–179, 4 pls.).

PROCTOR, G. R., 1984. *Flora of the Cayman Islands.* xii, 834 pp., 256 text-figs., 4 pls. (in 3), 25 diagrams, 3 maps (Kew Bull. Addit. Ser. 11). London: HMSO.

Illustrated, briefly descriptive flora of vascular plants; includes essential synonymy, vernacular names, keys, localities with *exsiccatae* and indication of overall distribution, and notes on local habitat and special features; glossary, lexicon of vernacular names with equivalents, and index to all botanical names at end. The introductory part is fairly comprehensive, with sections on physical features, geology, environment, vegetation and plant communities, floristics and phytogeography along with the history of collecting and the basis for the present project.

253

Jamaica

Area: 11 425 km². Vascular flora: 3003 phanerogams (including naturalized species) and 579 pteridophytes (Proctor, 1985; see below).

Among the earliest floristic accounts in the West Indies is *Catalogus plantarum insulae Jamaica* (1696) by Hans Sloane, based on his field trip of 1687. Following this came *Natural history of Jamaica* (1756) by Patrick Browne, *Plantarum jamaicensium pugillus* and *Flora jamaicensis* (1759), both by Linnaeus, and the never-completed *Flora of Jamaica* (1837–50) by James Macfadyen. Grisebach's *Flora of the British West Indian islands* encompassed the island. From the late nineteenth century there was increased exploration of the intricate flora and a local resource of collections and literature was built up at the Hope Botanic Gardens, particularly through the efforts of two successive superintendents, William Fawcett and William Harris, as well as visitors from North America. With earlier records, this became the basis for the *Flora of Jamaica* at the British Museum (Natural History), of which five volumes appeared before 1939. Revival of that project was essayed some years after World War II, but although William T. Stearn and others published some precursory papers and area revisions no new *Flora* volumes materialized. From 1949 another herbarium was developed with establishment of the University of the West Indies; incorporating the Hope collections in 1952, this became the base for C. Dennis Adams's *Flowering plants of Jamaica* (1972) which ably bridges the incomplete *Flora*. For pteridophytes, George Proctor at the Institute of Jamaica prepared two works, the latter the definitive *Ferns of Jamaica* (1985).

ADAMS, C. D., 1972. *Flowering plants of Jamaica*. 848 pp. Mona, Jamaica: University of the West Indies (distributed by Maclehose, Glasgow).

Concise manual-flora of native, naturalized and commonly cultivated flowering plants; includes keys to genera and species, essential synonymy, vernacular names (where known), brief indication of local and extralimital range (with representative *exsiccatae*), some taxonomic commentary, notes on habitat, frequency, phenology, altitude, etc., and indices to all botanical and vernacular names. The introductory section incorporates a summary of the composition of the flora as well as the plan of the work and its sources, while at the end are found a general key to monocotyledons, references to keys for dicotyledons, and a list of relevant references. For a substantial suite of additions, see G. R. PROCTOR, 1982. More additions to the flora of Jamaica. *J. Arnold Arbor.* **63**: 199–315.

FAWCETT, W. and A. B. RENDLE, 1910–36. *Flora of Jamaica*. Vols. 1, 3–5, 7. Illus. London: British Museum (Natural History). (Vol. 1 reprinted in enlarged format, 1963, Kingston, as *Orchids of Jamaica*.)

Detailed descriptive flora of native and naturalized seed plants (arranged according to Rendle's modified Englerian system); includes keys to all taxa, complete synonymy (with references and citations), vernacular names, fairly detailed indication of local range (with *exsiccatae*) and general summary of extralimital distribution, taxonomic commentary, notes on habitat, properties, uses, etc., and an index to vernacular and botanical names in each volume. Coverage is complete apart from the monocotyledons (except Orchidaceae) and the 'early' metachlamydeous dicotyledons, respectively intended for vols. 2 and 6. For supplement to the published volumes, see G. R. PROCTOR, 1967. *Additions to the Flora of Jamaica*. 84 pp., 35 halftones. Kingston: Institute of Jamaica.

Pteridophytes

Proctor's 1985 pteridoflora, with 579 species, succeeds *idem*, 1953. *A preliminary checklist of Jamaican pteridophytes*. 89 pp., 3 pls., 2 maps. (Bull. Inst. Jamaica, Sci. Ser. 5). Kingston.

PROCTOR, G. R., 1985. *Ferns of Jamaica*. viii, 631 pp., 135 text-figs., maps. London: British Museum (Natural History).

Briefly descriptive treatment with keys to taxa below families, full synonymy, typification, generalized local and extralimital distribution (with citation of *exsiccatae* for uncommon species as well as dot maps), altitudinal range, selective critical commentaries, and habitat and other notes; references (pp. 600–602), glossary and index to scientific names. No vernacular names or internal references to revisions and monographs are included. The fairly brief introductory part includes historical background, floristics and phytogeography with statistics, and taxonomy (including a conspectus, pp. 12–13).

254

Hispaniola

Area: 76 190 km^2 (Haiti, 27 749 km^2; Dominican Republic, 48 441 km^2). Vascular flora: about 5000 species of seed plants and 500–550 species of pteridophytes (Zanoni in Campbell and Hammond (see **General references**)). – The island comprises two states: Haiti and the Dominican Republic (Santo

Domingo). Published floras, however, have generally covered both.[61]

The most recent complete work is *Catalogus florae domingensis* by Rafael Moscoso, with additions in 1963–67 by José de Jésus Jiménez Almonte. Earlier exploration and floristic studies to and including the first expedition of Erik Ekman (1917) were reviewed by Urban in 1923 (*Symbolae antillanae*, 9: 1–54) along with vegetation and plant geography. This followed his *Flora domingensis*, the first flora of the island. Progress to 1959 was examined by Jiménez.[62] From the 1970s he, with others, was involved in preparation of a new descriptive flora under the direction of Alain H. Liogier. From 1982 through 1996, eight (of an expected nine) parts were published, covering all but three families of the dicotyledons.[63]

Bibliography

ZANONI, T. A., C. R. LONG and G. McKIERNAN, 1984– . An annotated bibliography of the flora and vegetation of Hispaniola (Dominican Republic and Haiti)/Bibliografia de la flora y de la vegetación de la isla Española, I– . *Moscosoa* 3: 1–61; 4: 39–48 (1986); 5: 340–348 (1989); 6: 242–253 (1990); 7: 242–253 (1993). [Concisely annotated lists of titles in alphabetical order. Prepared in relation to a projected *Flora vascular de la isla Española*.]

LIOGIER, A. H., 1982– . *La flora de la Española.* Parts 1– . Illus. (Serie científica, Universidad Central del Este 12– , *passim*.) San Pedro de Macorís, D.R. (Part 1 also published as *idem*, 1981. *Flora of Hispaniola*. Part 1. 218 pp., 78 text-figs. (Phytologia, Mem. 3). Plainfield, N.J.)

Descriptive manual-flora, with keys, concise synonymy, vernacular names, generalized indication of local and extralimital range (with some citations of *exsiccatae* for, among others, less common species), representative figures, and notes on habitat, habit, special features, etc.; indices at end of each part. A brief introduction, including a sequence of families, is furnished in part 1. [Eight parts have been published as of writing (1999), encompassing (1) Cyrillaceae–Begoniaceae (families 122–149), (2) (1983) Proteaceae–Connaraceae exclusive of Loranthaceae and Cactaceae (families 54–100 except 56 and 69), (3) (1985) Leguminosae–Meliaceae (families 101–113), (4) (1986) Malpighiaceae–Lythraceae (families 114–121, 150–151), (5) (1989), Myrtaceae–Apocynaceae (families 152–177 except 155), (6) (1994) Asclepiadaceae–Scrophulariaceae (families 178–186), (7) (1995)

Bignoniaceae–Campanulaceae (families 187–200), and (8) (1996) Compositae (201) and Casuarinaceae–Viscaceae (43–53, 56–56b). The classical Englerian system has been followed.][64]

MOSCOSO, R. M., 1943. *Catalogus florae domingensis*, I: *Spermatophyta*. xlviii, 732 pp., 3 pls., 2 maps. New York: University of Santo Domingo.

Systematic enumeration of seed plants of Hispaniola; includes synonymy and references, vernacular names, detailed indication of local range (by botanical districts) and general summary of extralimital distribution, brief descriptive notes for some taxa and indices to botanical and to Haitian and Dominican vernacular names. The introductory section includes a general description of Hispaniola, including remarks on its botanical districts (with maps), a summary account of botanical exploration, and a bibliography. An intended second part, to have contained pteridophytes, was not published. For additions and corrections, see J. DE J. JIMÉNEZ, 1967. *Catalogus florae domingensis: suplemento 1.* 278 pp., 3 pls. Santiago de los Caballeros, D.R.: Association para el Desarrollo. (Originally published 1963–67 in six installments in *Arch. Bot. Biogeogr. Ital.* 39: 81–132; 40: 54–149; 41: 47–87; 42: 46–97, 107–128; 43: 1–18, pls. 1–3.)

URBAN, I., 1920–21. Flora domingensis. In *idem* (ed.), *Symbolae antillanae*, 8: 1–860. Leipzig: Borntraeger. (Reprinted 1964, Amsterdam: Asher.) Complemented by *idem*, 1925. Pteridophyta domingensis. Pp. 273–397 (text), 544–568 (index) in *ibid.*, 9.

Systematic enumerations respectively of seed plants and pteridophytes of Hispaniola, with full synonymy, references and citations, vernacular names, detailed indication of local range (with citation of *exsiccatae*), general summary of extralimital distribution, and brief notes on phenology, special features, etc.; critical remarks; indices to all vernacular and botanical names.

I. Haiti

BARKER, H. D. and W. S. DARDEAU, 1930. *Flore d'Haiti*. viii, 456 pp. Port-au-Prince: Départment de l'Agriculture. (Ensemble des ouvrages universitaires 7.)

Briefly descriptive generic flora of seed plants; includes keys to families and genera, lists of species, vernacular names, indication of some localities, occasional notes on uses, etc., and a glossary and index to vernacular and botanical names. The introductory section includes a synopsis of families and a brief list of references.[65]

II. Dominican Republic

All works except those by Schiffino and Liogier on trees (described below) cover Hispaniola as a whole (see main entry above).

Woody plants (including trees): guides

LIOGIER, A. H., 1978 (1979). *Árboles dominicanos*. [viii], 220 pp., illus. (Publ. Rama Botánica, Comisión de Biología, Academia de Ciencias de la Republica Dominicana 3). Santo Domingo.

Atlas-style treatment of 100 selected native and introduced species, each with a two-page spread comprising a figure and text covering dendrological features, vernacular names, and notes on uses, potential value, etc.

SCHIFFINO, J., 1945–47. *Riqueza forestal dominicana*. 3 vols. Illus. Ciudad Trujillo (Santo Domingo): Montalvo. (Earlier editions in 1 vol., 1927, 1936.)

Atlas of native and introduced tree species with vernacular and trade names, descriptions, and commentaries including timber features; index to all names at end of each volume (no cumulative index). The introduction includes a description of the plan of the work along with a religious invocation.

255

Mona and Monito Islands

Area: 50 km². Vascular flora: 417 species (Woodbury *et al.*, 1977). Mona is situated midway between Hispaniola and Puerto Rico, but is politically part of the latter. An earlier list appeared in *The vegetation of Mona Island* (1915; in *Ann. Missouri Bot. Gard.* **2**: 33–58) by N. L. Britton.

WOODBURY, R. O., L. F. MARTORELL and J. G. GARCÍA-TUDURÍ, 1977. *The flora of Mona and Monito Islands, P.R..* 60 pp., 3 figs. (incl. map) (Bull. Agric. Exp. Sta. Puerto Rico (Río Piedras) 252). Río Piedras.

Systematic enumeration of vascular plants, with appropriate synonymy, references to taxonomic studies, indication of status, localities with observations and/or *exsiccatae*, and notes on overall range,

special features, etc.; general references (31 items) and index to all botanical names at end. The introductory part covers physical features, environment, vegetation, plant geography, rare species and an account of botanical work.

256

Puerto Rico

Area: 8794 km² (inclusive of associated islands). Vascular flora: 3039 native and naturalized vascular plants (Liogier and Martorell, 1982). Puerto Rico is here considered to encompass the main island together with Vieques and Culebra. For Mona and Monito, see 255.

The two main floristic works of the nineteenth century, *Apuntes para la flora de Puerto-Rico* (1881–83) by Domingo Bello y Espinosa, in Mayagüez from 1848 to 1876, and *Estudios sobre la flora de Puerto-Rico* (1883–88) by the native Puerto Rican Augustin Stahl, were essentially one-man efforts by long-time residents working with limited resources; neither was in any measure complete or reliable and Stahl's work did not proceed beyond the Candollean groups of Polypetalae and Gamopetalae.[66] The many new collections of the last quarter of the century, particularly from Baron von Eggers and Paul Sintenis (the latter sponsored by Leopold Krug, later closely associated with Ignatz Urban) and the absence of a complete flora prompted Urban in later years to compile *Flora portoricensis* (1903–11) as volume 4 of his *Symbolae antillanae*.[67] At the same time, large-scale North American collecting commenced and within a comparatively short time Urban's work was supplanted by Britton and Wilson's *Botany of Porto Rico and the Virgin Islands* (1923–30). This, with two supplements by Alain Liogier (1965 and 1967), has remained standard for some seven decades.

A return to locally based work became possible with the development after 1939 of internally situated resources. A revised checklist was published in 1982 and a new descriptive work, Liogier's *Descriptive flora of Puerto Rico and adjacent islands*, commenced publication in 1985. For trees, Britton and Wilson's work has been entirely supplanted by the two detailed, well-illustrated books by Elbert Little and co-workers (1964, 1974). These, as well as Britton and Wilson's flora, also cover the Virgin Islands (**257**); Liogier's do not.[68]

BRITTON, N. L. and P. WILSON, 1923–30. Botany of Porto Rico and the Virgin Islands. In New York Academy of Sciences, *Scientific survey of Porto Rico*, 5–6. 626, 663 pp. New York.

Briefly descriptive, relatively detailed but concise flora of vascular plants; includes keys to all taxa, limited synonymy (with references), general indication of local and extralimital range, vernacular names, taxonomic commentary, and notes on habit, etc.; annotated bibliography and general index to all vernacular and botanical names at end of work. The introductory section includes general descriptive remarks on the islands. Mona Island, between Puerto Rico and Hispaniola, is also encompassed by this work. For revisions, see ALAIN H. LIOGIER, 1965. Nomenclatural changes and additions to Britton and Wilson's *Flora of Porto Rico and the Virgin Islands*. *Rhodora* **67**: 315–361; and *idem*, 1967. Further changes and additions . . . *Ibid.*, **69**: 372–376.

LIOGIER, ALAIN H. (as HENRI ALAIN LIOGIER), 1985–97. *Descriptive flora of Puerto Rico and adjacent islands: Spermatophyta*. 5 vols. Illus. Río Piedras (vol. 5: San Juan), P.R.: Editorial de la Universidad de Puerto Rico.

Manual-flora of seed plants; includes keys to genera and species, vernacular names, internal and extralimital distribution, altitudinal range, status, and notes on habitat, etc., along with some critical remarks; representative illustrations; indices to vernacular and scientific names at end of each volume. A short introduction, with list of families (pp. 9–11), appears in volume 1. [Of the 181 seed plant families in the flora, 143 are covered as of writing (1999). Vol. 1 encompasses Casuarinaceae–Connaraceae; vol. 2 (1988), Leguminosae–Anacardiaceae; vol. 3 (1994), Cyrillaceae–Myrtaceae; vol. 4 (1995), Melastomataceae–Lentibulariaceae; and vol. 5, Acanthaceae–Compositae. The arrangement follows the classical Englerian system.][69]

LIOGIER, ALAIN H. (as HENRI ALAIN LIOGIER) and L. F. MARTORELL, 1982. *Flora of Puerto Rico and adjacent islands: a systematic synopsis*. [viii], 342 pp. Río Piedras, P.R.: Editorial de la Universidad de Puerto Rico.

Systematic enumeration and nomenclator with essential synonymy, vernacular names, status (including autochthony), and references to Britton and Wilson's flora and other works (bibliography, pp. 253–267); addenda and corrigenda and indices to vernacular and scientific names at end. The introductory part, in Spanish and English, gives the plan of the work along with some background.

Woody plants (including trees): guides

LITTLE, E. L., JR. and F. H. WADSWORTH, 1964. *Common trees of Puerto Rico and the Virgin Islands*. x, 548 pp., 250 text-figs., 4 maps (Agric. Handb. USDA 249). Washington, D.C.: U.S. Government Printing Office. [Privately reprinted after 1989.] Continued as E. L. LITTLE, JR., R. O. WOODBURY and F. H. WADSWORTH, 1974. *Trees of Puerto Rico and the Virgin Islands: second volume*. xiv. 1024 pp., illus., 4 maps (Agric. Handb. USDA 449). Washington. Spanish edns. (vol. 1 with expanded text and many colored illustrations): E. L. LITTLE, JR., F. H. WADSWORTH and J. MARRERO, 1967. *Árboles comunes de Puerto Rico y las Islas Vírgenes*. xxxix, 827 pp., 247 pls. (some col.), 4 maps. Río Piedras, P.R.: Editorial de la Universidad de Puerto Rico; and E. L. LITTLE, JR., R. O. WOODBURY and F. H. WADSWORTH, 1988. *Árboles de Puerto Rico y las Islas Vírgenes. Segundo volumen*. xvi, 1177 pp., illus., 4 maps (*ibid.*, 449–S). Washington.

These two works comprise a copiously illustrated descriptive account in atlas form of all trees in Puerto Rico and the Virgin Islands (with the less common species in the second volume); they include keys to all taxa (including artificial direct keys), outline systematic lists of species, limited synonymy, generalized indication of local range (supplemented by individual distribution maps in the first volume), vernacular names, a few critical remarks, extensive notes on natural history, phenology, timber properties, uses, etc., and indices to vernacular and botanical names (complete for the whole work in the second volume). In the introductory section are remarks on forests in the area as well as on earlier studies of the forest flora. The second volume – similar in format to the first – covers 500 species of less common trees, 460 of them in full detail with illustrations.[70]

Pteridophytes

PROCTOR, G. R., 1989. *Ferns of Puerto Rico and the Virgin Islands*. [ii], 389 pp., 105 pls., 5 maps (Mem. New York Bot. Gard. 53). New York.

Descriptive pteridoflora with keys, illustrations, synonymy, typification, indication of local and extralimital distribution, and notes on habitat, variation, and taxonomy; bibliography, list of Puerto Rican municipalities, list of Virgin Islands pteridophytes with distribution, alphabetic list of Puerto Rican pteridophytes with collection numbers of the author, glossary (with definitions also in Spanish), index to all names, and author's profile at end. The introductory part includes a history of pteridology in Puerto Rico along with biogeographical and taxonomic notes and a conspectus of the major groups (but no general analytical key). [Covers 408 taxa of ferns and fern-allies, including 376 named species and nothospecies.]

257

Virgin Islands

Area: 343 km². Vascular flora: estimated at about 1500 species. The islands are divided between the United States (St. Thomas, St. Croix and St. John) and Great Britain (Tortola, Virgin Gorda and Anegada). For general coverage of the group see **260** (STEHLÉ; VÉLEZ) and **256** (BRITTON and WILSON; LITTLE *et al.*).

The first (and most recent) separate overall flora is *Flora of St. Croix and Virgin Islands* (1879) by the Dane Baron von Eggers, a leading nineteenth-century collector. For the American Virgin Islands the latest separate treatment is that by Britton (see below). In recent years, however, much new field work has been accomplished and published results include additions for St. Croix, a substantial flora of St. John (1996), and smaller florulas for each of the three British Virgin Islands. A new overall flora would be desirable, particularly as the islands are omitted from the most recent Puerto Rican floras (though not the pteridoflora of Proctor).

Bibliography
See **American Virgin Islands** (Britton, 1918).

American Virgin Islands
BRITTON, N. L., 1918. The flora of the American Virgin Islands. *Mem. Brooklyn Bot. Gard.* 1: 19–118.

Systematic enumeration of non-vascular and vascular plants; includes synonymy, brief indication of local range, and notes on habitat, special features, etc. The introductory section incorporates a general description of the islands, an annotated list of significant botanical literature, and an account of botanical exploration; at the end a table of statistics of the flora and a list of endemic species are given. The work encompasses native, naturalized and commonly cultivated species. For changes relating to St. Croix, see F. R. FOSBERG, 1976. Revisions in the flora of St. Croix, U.S. Virgin Islands. *Rhodora* 78: 79–119.

ACEVEDO RODRÍGUEZ, P. and collaborators, 1996. *Flora of St. John, U.S. Virgin Islands.* [3], 581 pp., 242 pls., maps (end-papers) (Mem. New York Bot. Gard. 78). New York.

Illustrated descriptive flora of vascular plants (747 species of which 642 native) with keys to all taxa, vernacular names, indication of distribution and habitat with selected *exsiccatae* but no critical commentary; illustration vouchers, botanical glossary and complete index at end. The introduction covers physical features, geology, climate, vegetation, the nature and affinities of the flora, and as well botanical history, history of the project and its scope and materials, contributors, and a selected bibliography.[71]

British Virgin Islands
Until publication of William D'Arcy's checklist, Tortola, the largest of the British Virgin Islands, was the most poorly known.

D'ARCY, W. G., 1967. Annotated checklist of the dicotyledons of Tortola, Virgin Islands. *Rhodora* 69: 385–450, illus. (map).

Systematic checklist, with localities, *exsiccatae*, and literature references together with notes on habitat, biology, etc. Floral statistics and floristic affinities as well as background information are treated in an introductory section.

D'ARCY, W. G., 1971. *The island of Anegada and its flora.* 21 pp., illus. (Atoll Res. Bull. 139). Washington, D.C.

Systematic checklist, with localities, *exsiccatae*, and literature references as well as notes on habitat and biology; includes critical taxonomic remarks and coverage of cultivated species. The island is low and relatively flat (with 198 native vascular species; see Howard and Kellogg under **262**).

LITTLE, E. L., JR., R. O. WOODBURY and F. H. WADSWORTH, 1976. *Flora of Virgin Gorda (British Virgin Islands).* 36 pp., illus. (incl. map) (U.S. Forest Serv. Res. Pap. ITF-21). Río Piedras, P.R.

Annotated systematic checklist, with limited synonymy and one-line notes on distribution, habitat, etc.; a few literature references. An introductory section provides background information together with remarks on rare and endemic species and available collections, and a list of references appears at the end. A statistical table is also given.

258

Cay Sal Bank

Vascular flora (Cay Sal only): 105 species. – This bank, almost 4000 km² in area, lies between Cuba, Florida and the Bahamas and has several small cays at its margins. Although Bahamian territory, the bank and its cays are physically separate from the rest of that group.

Although some botanical investigations have been made, the cays are still considered little known (Fosberg, 1986; see **002**). Prior to a visit by W. T. Gillis in 1976, the only collections had been made in 1907 by Percy Wilson. For a general botanical account of Cay Sal proper (without, however, a full plant list), see W. T. GILLIS, 1976. Flora and vegetation of Cay Sal. *Bahamas Naturalist* 2(1): 36–41.

259

Jamaican Cays and Navassa

The Jamaican Cays are located south and south-east of Jamaica, with their chief groups being the Pedro Cays and Morant Cays. These islands, nowhere more than 4 m and mostly 2.5 m or less high, are Jamaican territory. For convenience, Navassa Island (a United States possession west of Haiti) is also here accounted for.

Prior to publication of the checklist by Stoddart and Fosberg (see below), no systematic treatment was available for the Jamaican Cays. They had been, however, briefly described along with notes on their flora and vegetation in C. B. LEWIS, 1947. A trip to the Morant and Pedro Cays. *Nat. Hist. Notes* (Jamaica) 3: 105–108. No checklist for Navassa has been seen.

STODDART, D. R. and F. R. FOSBERG, 1991. *Plants of the Jamaican Cays.* 24 pp. (Atoll Res. Bull. 352). Washington, D.C.

Annotated systematic checklist, with citations of *exsiccatae* and documentary references. [Includes both the Morant and Pedro Cays.]

Region 26/27

The Lesser Antilles (Leeward and Windward Islands)

Within this region are the Windward and Leeward Islands, from Sombrero and Anguilla in the north to Grenada in the south. The 'vanishing' island of Aves, somewhat west of the main island arc at 15°

40′N, 63° 30′W, is also included. For the Virgin Islands, see **257**. The horstian islands along the coast of South America, sometimes also considered as part of the Lesser Antilles, make up Region 28.

The 140-year-old *Systematische Untersuchungen über die Vegetation der Karaiben* by August Grisebach (1857; in *Abh. Königl. Ges. Wiss. Göttingen* 7: 151–286) has now been superseded by the definitive, relatively detailed *Flora of the Lesser Antilles* by Richard Howard and collaborators in six volumes (1974–89). Other floristic accounts exist for herbaceous plants (Vélez, 1957) and trees (Beard, 1944; Stehlé and Stehlé, 1947) but these too are now effectively superseded by the new *Flora*. Of the individual islands, only the 2½ Dutch and 4½ French islands as well as Dominica have had substantial separate coverage. The French islands have most recently been covered by J. Fournet in *Flore illustrée des phanérogames de Guadéloupe et de Martinique* (1978), and the Dutch by A. Stoffers in his still-incomplete *Flora of the Netherlands Antilles* (1962–, also encompassing Curaçao, Aruba and Bonaire in Region 28). The *Flora of Dominica* begun by Walter Hodge in 1954 was completed in 1990, and older coverage is available for St. Bartholomew (Questel, 1941), the Grenadines (Howard, 1952), St. Vincent (Royal Botanic Gardens, Kew, 1893) and Anguilla (Boldingh, 1909).

Like the Greater Antilles, the region is botanically relatively well known and the new regional flora will help place species distribution in a proper perspective.

Bibliographies. General, divisional and supraregional bibliographies as for Division 2 and Superregion 24–29.

Indices. General and divisional indices as for Division 2 or Superregion 24–29.

260

Region in general

Bibliography
See Beard (1944), Stehlé and Stehlé (1947) and Vélez (1957).

HOWARD, R. A. (ed.), 1974–89. *Flora of the Lesser Antilles: Leeward and Windward Islands*. 6 parts.

Illus., maps. Jamaica Plain, Mass.: Arnold Arboretum of Harvard University.

Moderately concise illustrated descriptive flora of vascular plants; includes keys, fairly detailed synonymy (with references and citations), indication of types, local range (with some *exsiccatae*) and extralimital distribution, taxonomic commentary, and notes on habit, phenology, special features, etc.; indices in each part. An introductory section in the first part (on Orchidaceae) deals with the background of the work with an indication of the area covered, and includes a chapter on phytogeography and floristics (by R. A. Howard); there is also a brief introduction in the first of the three dicotyledon parts. [Two full volumes on pteridophytes and Orchidaceae were contributed respectively by G. R. Proctor and L. A. Garay.]

Woody plants

BEARD, J. S., 1944. Provisional list of trees and shrubs of the Lesser Antilles. *Caribbean Forest.* 5: 48–67. Complemented by H. STEHLÉ and M. STEHLÉ, 1947. Liste complementaire des arbres et arbustes des Petites Antilles. *Ibid.*, 8: 91–123.

Tabular lists of trees and shrubs recorded from the Lesser Antilles, with presence or absence indicated for each island (the smaller Leeward Islands being grouped for this purpose). The introductory sections include summary tables and lists of references. The lists include the Virgin Islands but not Barbados.

Herbaceous flowering plants

VÉLEZ, I., 1957. *Herbaceous angiosperms of the Lesser Antilles.* [iv], 121 pp., 51 illus. San Juan, P.R.: Inter-American University of Puerto Rico.

Mainly devoted to an illustrated descriptive account of herbaceous and subwoody vegetation but includes a tabular list of known herbaceous and subwoody flowering plants of the region, arranged alphabetically by families with indication (including authorities) of presence or absence for each island or island group from the Virgin Islands to Grenada. Extralimital distribution in the Greater Antilles is also indicated. A bibliography concludes the work.

261

Sombrero

See **257** (D'ARCY, 1971) for general remarks on this island (administratively part of Anguilla). – Area:

2 km² or less. Vascular flora: no data, but limited. No specific checklists appear to be available.

262

Anguilla (and Dog Island)

See also **265** (BOLDINGH). – Area: 91 km². Vascular flora: 443 seed plants, of which 122 introduced (Howard and Kellogg, 1987; see below). For this one-time 'mouse that roared', two checklists are available.

BOLDINGH, I., 1909. A contribution to the knowledge of the flora of Anguilla (BWI). *Recueil Trav. Bot. Néerl.* 6: 1–36.

Systematic enumeration of seed plants, based mainly upon collections of the author but also incorporating older records; includes synonymy (with references), local range with citation of *exsiccatae*, general indication of extralimital range, and a short list of references.

HOWARD, R. A. and E. A. KELLOGG, 1987. Contribution to a flora of Anguilla and adjacent islets. *J. Arnold Arbor.* 68: 105–131, illus.

This otherwise geographical-vegetatiological account includes in an appendix a concise checklist of seed plants, with observations and/or representative *exsiccatae*.

263

St. Martin/Sint Maarten

See **265** (BOLDINGH); **271** (FOURNET); **281** (BOLDINGH, 1913; STOFFERS). – Area: 88 km² (Dutch part, 34 km²; French part, 54 km²). Vascular flora: 433 species, of which 392 native (Howard and Kellogg; see **262**). For this half Dutch, half French 'tourist haven' (the Dutch half being part of the Netherlands Antilles), no separate flora is available.

264

St. Bartholomew (St. Barthélemy)

See also **271** (FOURNET). – Area: 21 km². Vascular flora: 389 species of which 336 native (Howard and Kellogg, 1987; see **262**).

The island was initially documented by Johan E. Wikström in *Öfversigt af ön Sanct Barthelemi's flora* (1826); 301 vascular species were listed. Two further works appeared in the early 1940s, that by Questel (described below) and, almost simultaneously, J. MONACHINO, 1940–41. A check list of the spermatophytes of St Bartholomew. *Caribbean Forest.* 2: 24–66.

QUESTEL, A., 1941. *The flora of the island of St Bartholomew and its origin*. vii, 244 pp., 1 pl., 2 maps. Basse-Terre, Guadeloupe: Imprimerie Catholique. (French edn. published simultaneously as *La flore de l'île de St-Barthélemy et son origin*.)

Systematic enumeration of seed plants, with occasional descriptive notes, limited synonymy (with some references and citations), vernacular names, general indication of local and extralimital range (with citation of the author's *exsiccatae*), and notes on habitat, special features, etc., appendices on economic and medicinal uses and indices to families and genera at end. The introductory section includes a historical review of botanical work on the island along with general comments on the flora and vegetation and on phytogeographical relationships.

265

St. Eustatius and Saba (Netherlands Antilles)

Area: St. Eustatius, 21 km²; Saba, 13 km². Vascular flora: no data. – As in the southern Netherlands Antilles, detailed botanical exploration began only in the late nineteenth century. Boldingh's 1909 account was the first (and so far only) separate flora. The islands are also covered in his general Dutch

Antillean key of 1913 and by Stoffers' so far incomplete flora (for full details, see **281**).

BOLDINGH, I., 1909. The flora of St Eustatius, Saba and St Martin. In *idem, Flora of the Dutch West Indian islands*, 1. xii, 321 pp., 3 maps. Leiden: Brill.

Systematic enumeration of vascular plants (including also Anguilla and St. Croix), with occasional descriptive notes; includes limited synonymy (with references), vernacular names, rather detailed indication of local range with citations of *exsiccatae* and summary of extralimital range, and separate indices to vernacular and botanical names. The remainder of this work comprises appendices on collectors, physical features, soils, vegetation, phytogeography, etc., together with a bibliography.

BOLDINGH, I., 1913. *Flora voor de Nederlandsch West-Indische eilanden*. See **281**.

STOFFERS, A. L., 1962– . *Flora of the Netherlands Antilles*. See **281**.

266

St. Christopher (St. Kitts) and Nevis

Area: Nevis, 93 km²; St. Christopher, 168 km². Vascular flora: no data.

For an annotated treatment of pteridophytes, see H. E. BOX and A. H. G. ALSTON, 1937. Pteridophyta of St. Kitts. *J. Bot.* **75**: 241–258. No account of phanerogams is available.

267

Barbuda

Area: 160 km². Vascular flora: 261 species, of which 229 native (Howard and Kellogg, 1987; see **262**). No separate floras or lists are available. The island politically is part of Antigua (**268**).

268

Antigua (with Redonda)

Area: Antigua, 279 km²; Redonda, 1.3 km². Vascular flora: 724 angiosperms (PD; see **General references**).

No separate complete floras or lists are available for Antigua save for A. H. G. ALSTON and H. E. BOX, 1935. Pteridophyta of Antigua. *J. Bot.* **73**: 33–40, an annotated enumeration (only of ferns). A full checklist by Box remains in manuscript at the Natural History Museum, London, with copies in the West Indies. For Redonda, see R. A. HOWARD, 1962. Botanical and other observations on Redonda, the West Indies. *J. Arnold Arbor.* **43**: 51–66.

269

Montserrat

Area: 104 km². Vascular flora: no data. – No separate complete floras or lists are yet available. [In the latter part of the 1990s, the vegetation of Montserrat was largely laid waste by volcanic activity.]

271

Guadeloupe and Marie-Galante

Area: 1759 km² (Guadeloupe, 1603 km²; Marie-Galante, 155 km²; Îles des Saintes, 1 km²). Vascular flora (French Antilles): 1750–1850 native species (Fournet, 1978; see below). Works covering the French Antilles in general are listed under this heading.

Save for the first, floras of the group all have been local undertakings. The outside work was Olof Swartz's successor Johan E. Wikström's *Öfversigt af ön Guadeloupe's flora* (1827). The next did not appear until near the end of the century: *Contribution à la flore de Guadeloupe* (1892) by Hippolyte Mazé.

Key works since 1892 include Père A. DUSS, 1897. *Flore phanérogamique des Antilles françaises (Guadeloupe et Martinique).* xxvii, 656 pp. (Ann. Inst. Bot.-Géol. Colon. Univ. Marseille 3). Marseille (reprinted with revised index in 2 parts, 1972, Fort-de-France, Martinique: Société de Distribution et de Culture); and H. STEHLÉ, M. STEHLÉ and Père L. QUENTIN, 1937–49. Catalogue des phanérogames et fougères. In their *Flore de la Guadeloupe et dépendances (et de la Martinique)*, 2(1–3). Illus., maps. Basse-Terre, Guadeloupe: Imprimerie Catholique (part 1); Paris: Lechevalier (parts 2–3). The latter work was never completed.

FOURNET, J., 1978. *Flore illustrée des phanérogames de Guadeloupe et de Martinique.* 1654 pp., 745 text-figs. Paris: Institut National de la Recherche Agronomique (INRA).

Copiously illustrated, briefly descriptive manual-flora of seed plants (1668 native, 149 naturalized); includes keys to all taxa, limited synonymy (with many references to Duss's flora), vernacular names, local distribution, some critical remarks, notes on habitat, frequency, origin (if introduced), altitudinal range, phenology, and indices to family, generic and vernacular names. The sequence follows the Emberger system as used in *Flore de France* by Guinochet and Vilmorin (**651**).

272

Dominica

Area: 772 km². Vascular flora: 1226 species (Hodge, 1954 and Nicolson et al., 1990; see below).

The second part of *Flora of Dominica* unfortunately was published in a format larger than that of the first part. Nevertheless, apart from the volume by Fournet on the French islands the two parts comprise the most substantial account for any island in the Lesser Antilles, the more appropriate as floristically Dominica is the most diverse of any in the chain. Pteridophytes were earlier treated by Karel Domin in *The Pteridophyta of the island of Dominica* (1929).

HODGE, W. H., 1954. Flora of Dominica (BWI), part 1. *Lloydia* **17**: 1–238, 112 figs. Completed as D. H. NICOLSON et al., 1990. *Flora of Dominica*. Part 2. iii, 274 pp. (Smithsonian Contr. Bot. 77). Washington, D.C.

Part 1, covering vascular plants except dicotyledons, comprises a systematic enumeration of vascular plants, with keys to all taxa, extensive synonymy (with references), vernacular names, citation of *exsiccatae* with localities and general indication of extralimital range, figures of representative species, brief taxonomic commentary, and notes on habitat, phenology, special features, uses, etc.; index to botanical and vernacular names at end. The introductory section includes a general description of the island and its physical features together with accounts of botanical exploration, plant communities, agriculture, etc. (with separate bibliography). Part 2, covering 844 dicotyledons, is similar in style but includes short, diagnostic descriptions, compact synonymy, and no illustrations; moreover the families and genera are alphabetically arranged. The introductory part adds to that of Hodge and includes an account of vegetation formations and notes on floristics; at the end are an extensive bibliography and index to all names.[72]

273

Martinique

Area: 984 km^2. Vascular flora: separate figures not located.

No separate systematic accounts have been published apart from that of CHAUVIN (1977–85; see below), the island historically having been treated together with the Guadeloupe group (**271**). For a general introduction to flora and vegetation, see C. T. KIMBER, 1988. *Martinique revisited: the changing plant geographies of a West Indian island.* xx, 458 pp. College Station, Tex.: Texas A&M University Press.

CHAUVIN, G., 1977–85. *Étude illustrée des familles de plantes à fleurs de la Martinique.* 4 parts. Illus. Fort-de-France: CDDP de Fort-de-France (later Centre Régional de Documentation Pédagogique Antilles Guyane (CRDP)). (Cahiers Documentaires Education Enseignement 16, 18, . . . 21.) (Mimeographed.)

An illustrated school-flora of native and introduced plants with emphasis on families, genera and more common species; includes descriptions and biological and other notes. Part 4 contains a glossary (with figures) and index to scientific names in the four parts. The introductory chapter contains a somewhat unorthodox general key.

FOURNET, J., 1978. *Flore illustrée des phanérogames de Guadeloupe et de Martinique.* See **271**.

Woody plants
For a students' introduction (with keys) to the principal trees of Martinique, see J. POUPON and G. CHAUVIN, 1983. *Les arbres de la Martinique.* 256 pp., illus. N.p.: Office National des Forêts, Direction Régionale pour la Martinique.

274

St. Lucia

Area: 616 km^2. Vascular flora: no data located. – No separate complete floras or lists have been published; however, *Plants in danger* (PD; see **General references**) indicates that an unpublished list by C. Dennis Adams is held at the Natural History Museum, London.

275

St. Vincent

Area: 345 km^2. Vascular flora: no data.

The island was the site of one of the earliest tropical botanic gardens, formed in the 1760s. Its existence is reflected – after a refoundation in the late nineteenth century – in the only separate floristic account (which also covers the northern Grenadines; see **276**): ROYAL BOTANIC GARDENS, KEW, 1893. Flora of St Vincent and adjacent islets. *Bull. Misc. Inform.* (Kew) 1893: 231–296.[73]

276

The Grenadines

See also **275** (ROYAL BOTANIC GARDENS, KEW). – Area: 78 km^2. Vascular flora: no data. The Grenadines form a chain of some 600 small islands from Bequia (south of St. Vincent) through Mustique to Carriacou and Ronde (north of Grenada). The northern islands are part of St. Vincent; the southern, part of Grenada.

HOWARD, R. A., 1952. *The vegetation of the Grenadines, Windward Islands, BWI*. 132 pp., 12 pls. (Contr. Gray Herb., N.S. 174). Cambridge, Mass.

Includes (pp. 71–125) a list of vascular plants recorded for these islands, with vernacular names, citations of *exsiccatae* with localities, and brief notes on habitat, special features, uses, etc.; discussion of the origin and relationships of the flora. The general part includes accounts of the physical features of the islands, geology, botanical exploration, agriculture, etc., as well as a detailed descriptive treatment of the vegetation.

277
Grenada

Area: 311 km². Vascular flora: no data. – No separate complete floras or lists have been published.

278
Barbados

Area: 430 km². Vascular flora: *c.* 700 native species (PD; see **General references**). There are also many introduced and cultivated plants.

Revision of the current standard flora of this island by its senior author was reportedly in progress at the end of the 1970s (HOWARD, 1979; see **Superregion 24–29**, introduction, p. 284); as of 1999, however, nothing had been published. The 1965 edition succeeded several earlier works, among them *The natural history of Barbados* (1750) by the Rev. Griffith Hughes, *Flora barbadensis* (1830) by James D. Maycock (a key work), and *The history of Barbados* (1848) by Robert H. Schomburgk; the last-named encompassed 898 native and cultivated phanerogams.

GOODING, E. G. B., A. R. LOVELESS and G. R. PROCTOR, 1965. *Flora of Barbados*. xvi, 486 pp., 27 text-figs., frontisp., map. London: HMSO.

Briefly descriptive flora of native, naturalized, adventive and commonly cultivated vascular plants; includes keys to all taxa, synonymy (with references and citations), vernacular names, local range (with citation of representative *exsiccatae*) and general indication of extralimital distribution, and notes on habitat, properties, uses, special features, frequency, etc., followed by a list of extinct species and doubtful records, a lexicon of vernacular names with botanical equivalents, and an index to all botanical names. The introductory section includes a select bibliography and general key to families.

279
Aves Island

The minute number of vascular plants on this shifting and reputedly shrinking sandbank, a territory of Venezuela, has not attracted botanical writers. For general coverage apart from that in Zuloaga's paper, reference may be made to W. BROWNELL and C. GUZMÁN, 1974. Ecológia de la Isla de Aves con especial referencia a los peces. *Mem. Soc. Ci. Nat. 'La Salle'* 34 (whole no. 98): 91–168. This contains a complete bibliography and research review.

ZULOAGA, G., 1955. The Isla de Aves story. *Geogr. Rev.* (New York) 45: 172–180.

In this short general account, the author records the presence of two vascular plants, both of them pantropic littorals: *Portulaca oleracea* and *Sesuvium portulacastrum*.

Region 28

The Lesser Antilles (southern chain)

This comprises the horstian chain of islands running from Aruba in the west to La Blanquilla in the east, passing through Curaçao, Bonaire, Las Aves, Los Roques, and La Orchila. The first three of these constitute the major part of the Netherlands Antilles, while the remainder are part of Venezuela. For islands on the continental shelf, see Region 29.

Botanical exploration in the region, though of comparatively long duration, has been haphazard. In the three largest islands, serious efforts by Dutch botanists got underway only in 1884 when a mixed expedition including the Leiden botanist Willem Suringar visited the whole group (along with Surinam and the Dutch Leeward Islands) for several months. Later collecting and publications have been largely under the aegis of Utrecht University, culminating in the floras of Isaac Boldingh produced before World War I and in Antonius Stoffers' current *Flora of the Netherlands Antilles*, for which preparatory work began in the 1950s. This latter, however, is still to be completed. A substantial independent contribution was also made by Brother Arnoldo. For the Venezuelan islands, floristic coverage is patchy, with evidently no work on La Blanquilla in existence.

Bibliographies. General, divisional and supraregional bibliographies as for Division 2 and Superregion 24–29.

Indices. General and divisional indices as for Division 2 or Superregion 24–29.

281

Curaçao and Bonaire (Netherlands Antilles) with Aruba

Area: 925 km² (Aruba, 193 km²; Bonaire, 288 km²; Curaçao, 444 km²). Vascular flora: no current data. The Netherlands Antilles now consists of Curaçao and Bonaire together with St. Eustatius (Statia) and Saba (**265**) and the southern half of St. Martin (**263**). Aruba in 1986 became autonomous in association with the Netherlands. For convenience, however, works relating to the whole of these possessions are listed here.

The modern flora by Antonius Stoffers is as yet incomplete and thus has not entirely displaced Isaac Boldingh's treatments of 1913 and 1914, the first such for the islands in their own right.[74]

ARNOLDO, M., Br., 1964. *Zakflora: wat in het wild groeit en bloeit op Curaçao, Aruba en Bonaire*. 2nd edn. 232 pp., 3 pp. of text-figs., 68 pls. (Uitgaven

Natuurw. Werkgroep Ned. Antillen 16). Curaçao. (1st edn., 1954.)

Pocket-sized, semi-popular manual of seed plants (excluding Gramineae and Cyperaceae), with keys to all taxa, vernacular names, general indication of local range, figures and plates of representative species, and summary in English; indices to botanical and vernacular names at end. The introductory section includes a glossary.

BOLDINGH, I., 1913. *Flora voor de Nederlandsch West-Indische eilanden*. xx, 450 pp. Amsterdam: Koloniaal-Institut.

Descriptive manual of vascular plants of all the Netherlands Antilles; includes keys to all taxa, vernacular names, brief general indication of range by islands (but without extralimital distribution), and indices to botanical and vernacular names. The introductory section includes a glossary.

BOLDINGH, I., 1914. The flora of Curaçao, Aruba, and Bonaire. In *idem*, *Flora of the Dutch West Indian islands*, 2. xiv, 197 pp., 9 pls., map. Leiden: Brill.

Systematic enumeration of vascular plants; includes limited synonymy, references, vernacular names, occasional descriptive notes, fairly detailed indication of local range (with citations of *exsiccatae*), general summary of extralimital distribution, and indices to botanical and vernacular names at end. The remainder of the work comprises appendices on collectors, physical features, soils, vegetation, phytogeography, etc., together with a bibliography. Some notes on the plants of Margarita (**292**) are also included.

STOFFERS, A. L., 1962– . *Flora of the Netherlands Antilles*, 1(1–2); 2(1–3); 3(1–3) (Uitgaven Natuurwet. Studiekring Suriname Ned. Antillen 25–113, *passim*). Utrecht.

This conservatively styled work comprises a critical descriptive flora of native, naturalized and commonly cultivated vascular plants of all the Netherlands West Indian islands, with keys to all taxa within respective families, full synonymy (with references), vernacular names, detailed indication of local range (by island in first instance) with citations of *exsiccatae*, general summary of extralimital distribution, and brief notes on habitat, special features, etc. [At this writing (1999) a substantial proportion of each of the three projected volumes has been published, although none is yet complete; the most recent part seen, published as *Uitgave* 113, appeared in 1984. The sequence of the work is approximately that of the Pulle system of 1952 (as regards the angiosperms); present coverage includes the pteridophytes (treated by K. U. Kramer), part of the monocotyledons (to date only Typhaceae, Cyperaceae and Gramineae) and most dicotyledons in Series I–XII (some families by specialists), in all amounting to perhaps two-thirds of the total vascular flora. Notable gaps include the remaining monocotyledons, the Solanales sensu latiss., and Asterales.][75]

283

Territorio Colón (Las Aves, Los Roques and La Orchila)

Area: no information. Vascular flora: no data.

Only local checklists, respectively covering Los Roques/La Orchila and Las Aves, are available. Also useful is R. C. MURPHY, 1952. Bird islands of Venezuela. *Geogr. Rev.* (New York) **42**: 551–561.

ARISTEGUIETA, L., 1956. Florula de la region. In Sociedad de Ciencias Naturales "La Salle" (ed.), *El archipelago de Los Roques y La Orchila*, pp. 47–67, 4 pls. (some col.). Caracas.

Enumeration of non-vascular and vascular plants; includes vernacular names, local and extralimital range, and notes on habitat, frequency, etc. There is also a general description of the vegetation.

GINÉS, Hermano and G. YEPEZ T., 1960. *Aspectos de la naturaleza de las islas Las Aves, Venezuela.* 53 pp. (Trab. Soc. Ci. Nat. "La Salle" 20). Caracas.

Although concerned mainly with bird life, this work includes (pp. 12–20) an account of the vegetation and a short list of vascular plants observed, with vernacular names.

285

La Blanquilla (and associated islets)

Area: no information. Vascular flora: no data. – No separate floras or lists appear to be available for these low islands.

Region 29

West Indian continental shelf islands

Included here are those islands which lie on the South American continental shelf south of the Windward Islands but which traditionally have been considered part of the West Indies. The largest island is Trinidad, along with Tobago noteworthy for a long association with the other anglophone West Indian islands in regard to botanical work as otherwise in matters cultural and political. The remaining islands, all Venezuelan, include La Tortuga, the Nueva Esparta group (Margarita, Coche and Cubagua), and Los Testigos.

Of these islands, Trinidad and Tobago are after two or more centuries of activity botanically relatively well explored. The flora is currently being documented in *Flora of Trinidad and Tobago* (1928–), now all but completed in spite of a spasmodic history of progress. A companion to this work is Richard Schultes's *Orchids of Trinidad and Tobago* (1960). Knowledge of the Venezuelan islands is rather more uneven and deficient, but in the last two decades Margarita, by far the largest, has been the beneficiary of a new manual by Hoyos (1985). More exploration is, however, needed (A. L. Stoffers, personal communication).

Bibliographies. General, divisional and supraregional bibliographies as for Division 2 and Superregion 24–29.

Indices. General and divisional indices as for Division 2 or Superregion 24–29.

290

Region in general

GENTRY, H. S., 1948. *Land plants collected by the Allan Hancock Atlantic Expedition of 1939.* 46 pp., 2 pls. (Allan Hancock Atlantic Expedition Report 6). Los Angeles: University of Southern California Press.

Contains reports on plants collected by the Expedition's botanist Francis H. Elmore as well as others from Tortuga, Cubagua and Trinidad and Tobago as well as the Netherlands Antilles and among the Atlantic shore islands of Panama.

291

La Tortuga

Area: 220 km². Vascular flora: 69 non-vascular and vascular plants (Ernst, 1876; see below). The island is Venezuelan territory.

ERNST, A., 1876. Florula chelonesiaca; or, a list of plants collected in January, 1874, in the island Tortuga, Venezuela. *J. Bot.* **14**: 176–179.

Enumeration of lichens and vascular plants, based on the author's collections; includes brief indication of local range, vernacular names, and notes on habitat, frequency, special features, etc. A short introduction includes remarks on physical features of the island and the composition of its flora.

GENTRY, H. S., 1948. La Tortuga. In *idem*, *Land plants collected by the Allan Hancock Atlantic Expedition of 1939* (Allan Hancock Atlantic Expedition Report 6). Los Angeles.

Enumeration of plants collected (12 species), with localities, *exsiccatae*, general distribution and brief notes.

292

Nueva Esparta (Margarita, Coche and Cubagua)

See also **290** (GENTRY). – Area: 1149 km². Vascular flora: Cubagua, 15 species (Gentry, 1948); Margarita, 1143 species (Hoyos, 1985). The three islands comprise the Venezuelan state of Nueva Esparta, but floristic coverage is only individual.

Hoyos's flora of Margarita succeeds J. R. JOHNSTON, 1909. Flora of the islands of Margarita and Coche, Venezuela. *Proc. Boston Soc. Nat. Hist.* **34**: 163–312, pls. 23–30 (incl. 2 maps). (Also published as *Contr. Gray Herb.*, N.S. 37.)[76]

GENTRY, H. S., 1948. Cubagua. In *idem*, *Land plants collected by the Allan Hancock Atlantic Expedition of 1939* (Allan Hancock Atlantic Expedition Report 6), pp. 9–11. Los Angeles.

Enumeration of plants collected, with localities, *exsiccatae*, distribution, and brief notes. [Claimed by its author to be the first of its kind for the island.]

HOYOS, J., 1985. *Flora de la Isla Margarita, Venezuela.* 929 pp., 482 photographs, 2 maps (Monogr. Soc. Ci. Nat. "La Salle" 34). Caracas.

Descriptive manual-flora of vascular plants, with keys to genera and species, synonymy, citations, distribution (with *exsiccatae*, mainly of recent collections), and notes on habitat, etc.; glossary, bibliography (pp. 901–908) and complete index at end. The introductory part encompasses the aim and scope of the work, physical features, climate, rainfall, phytogeography and vegetation (including a land-use map), areas of special botanical interest, and the history of botanical work; statistics (pp. 90–128).

293

Los Testigos

Area: no information seen. Vascular flora: no data. – No separate floras or lists appear to be available for these low islands (which are Venezuelan territory).

295

Trinidad

See also **240–90** (GRISEBACH). – Area: 4828 km². Vascular flora: 2281 flowering plant species (PD; see **General references**). General works on the state of Trinidad and Tobago appear under this heading.

The only significant floristic works prior to the initiation of *Flora of Trinidad and Tobago* in 1928 were *Outline of the flora of Trinidad* by Hermann Crüger (1858; also in *Trinidad* by L. A. A. de Verteuil) and *The useful and ornamental plants of Trinidad and Tobago* (1927; 2nd edn., 1928) by W. G. Freeman and R. O. Williams. The orchids, written by Richard E. Schultes and published in the *Flora* in 1967, were also covered in that author's *Orchids of Trinidad and Tobago* (1960). Pteridophytes have been separately treated by Arthur Hombersley (1933) and John T. Mickel (1985; see below) but an account was also planned for the *Flora*.

WILLIAMS, R. O. *et al.*, 1928– . *Flora of Trinidad and Tobago*, 1(1–8, index); 2(1–11, index); 3(1–5). [Port-of- Spain], Trinidad: Department of Agriculture (later Ministry of Agriculture, Lands, and Fisheries and its successors, most recently the Ministry of Food Production, Marine Exploitation, Forestry and the Environment).

Briefly descriptive flora of flowering plants; includes keys to families, genera and species, synonymy with citations of pertinent literature, vernacular names, indication of localities with *exsiccatae* and authorities, generalized summary of local and extralimital range, some critical remarks, and notes on habitat, properties, uses, etc.; index to all vernacular and botanical names (vols. 1–2 only; vol. 3 has yet to be completed). No introductory part exists. [Arrangement of families follows the Bentham and Hooker system. As of 1999 coverage encompasses all of the dicotyledons and five and a half of the seven classes of monocotyledons. The remaining classes, Calycineae (including Palmae) and Glumaceae (Gramineae) are completed or in preparation. Continuation of the work to cover pteridophytes is projected.][77]

Woody plants (including trees): guides

DE FREITAS, K. and J. PACQUET, 1980. *Manual of dendrology*, 4: *Inventory of the indigenous forests of Trinidad and Tobago*. 181 pp. Port-of-Spain: Wildlife Conservation Committee, Ministry of Agriculture, Trinidad.

Encompasses descriptions and photographs of bole, bark and foliage of mostly indigenous trees of potential economic value. [Not seen; description from *Plants in danger*, p. 358.]

MARSHALL, R. C., 1934. *Trees of Trinidad and Tobago*. ii, 101, viii pp., 20 pls. Port-of-Spain: Government Printing Office.

Briefly descriptive account of native and commonly cultivated trees; includes synoptic keys and vernacular names, general indication of local range, and notes on properties, uses, timber characteristics, etc.; illustrations of representative species; index to vernacular and botanical names.

Pteridophytes

MICKEL, J. T., 1985. *Trinidad pteridophytes*. 62 pp., 49 pls. New York: The New York Botanical Garden. (Mimeographed.)

For student use; comprises keys to all species and plates with diagnostic figures. The introductory section encompasses use of the book, an index to plates, and (pp. 7–24) authorities, distribution, habitats and abundance in Trinidad for each species.

296

Tobago

See **295** (all works). The offshore islet of Little Tobago is included here. Area: 300 km².

Notes

1 A flora of the Chihuahuan Desert region, by Marshall C. Johnston with James Henrickson, is also in preparation.

2 A recent review and analysis of the expedition is contained in Sociedad Estatal Quinto Centenário/Real Jardín Botánico, Spain, 1987. *La Real Expedición Botánica a Nueva España, 1787–1803*. Madrid. See also H. W. Rickett, 1947. The Royal Botanical Expedition to New Spain. *Chron. Bot.* 11: 1–86, pls. 44–52; and R. McVaugh, 1977. Botanical results of the Sessé and Mociño Expedition (1787–1803), I. Summary of excursions and travels. *Contr. Univ. Michigan Herb.* 11(3): 97–195. A CD-ROM of the plates held at the Hunt Institute has also been released.

3 See A. Gómez-Pompa, 1973. The thrust of present and future research in the lowland tropics of Mexico. *Ann. Missouri Bot. Gard.* 60: 169–173.

4 Donnell Smith reported in 1888 upon plants collected during the Railway Commission's exploratory work, which in turn led him to his *Enumeratio*. The railway itself was humorously described by O. Henry (William Sydney Porter) in his *Cabbages and Kings* (1904, New York).

5 For a compendium of essays on this 'Linnaeus americanus neotropicus', see L. O. Williams (ed.), 1963. *Homage to Standley*. 115 pp., illus. Chicago.

6 There are now 16 or more herbaria in the region, which have in recent years acted as a marked spur to locally based field work, vitally necessary in the face of widespread vegetation degradation and destruction. See A. H. Gentry, 1978. Floristic knowledge and needs in Pacific tropical America. *Brittonia* 30: 134–153.

7 G. T. Prance, 1978, Floristic inventory of the tropics . . ., p. 673 and errata sheet, i–ii.

8 I thank C. J. Humphries of the British Museum (Natural History), one of the four project coordinators, for information on this project.

9 A complementary database is maintained at the Missouri Botanical Garden.

10 The work was largely based on collections then available in Hemsley's base, the Herbarium of the Royal Botanic Gardens, Kew. This limitation was made on account of potential problems with acceptance of determinations at face value on specimens in other leading herbaria, given constraints on the time available for the project. The two editors initially conceived the work as a whole in 1876; the first volume appeared in 1879 and the last in 1915. Of its 63 volumes, 51 were on zoology, six on archeology, five on botany, and one was a general introduction. Quaritch acquired all the stock on completion of publication, issuing a separate prospectus in 1918 (47 pp., map).

11 J. Rzedowski, 1991. Diversidad y origenes de la flora fanerogámica de México. *Acta Bot. Méx.* **14**: 3–21; also *idem*, 1993. Diversity and origins of the phanerogamic flora of Mexico. In T. P. Ramamoorthy *et al.* (eds.), *Biological diversity of Mexico: origins and distribution*, pp. 129–144. Oxford. The current estimates represent at least a doubling from the known total in 1888. Pteridophytes number 1075 species.

12 This was an initiative of the Sociedad Mexicana de Historia Natural (established in 1868). The work also appeared in installments in their journal, *La Naturaleza*.

13 Urbina's work was based on the herbarium of the Museo Nacional, first established in 1825. This and two others were amalgamated in 1915 under Conzatti as part of the Instituto de Biología General y Médico and in 1929 were transferred to the Universidad Nacional Autónoma de México (UNAM). They now form the core of the present Herbario Nacional in its Instituto de Biología (established in 1930).

14 This work is comparable to *Handleiding tot de kennis der flora van Nederlandsch Indië* by J. G. Boerlage (1890–1900; see **910–30**). The vascular cryptogams, never published in the main work, appeared separately as *Las criptogamas vasculares de México* (1907).

15 One never-realized project was a *Flora de los alrededores de México* proposed in the 1870s by the Sociedad Mexicana de Historia Natural. See E. Trabulse, 1992. *José María Velasco: un paisaje de la ciencia en México*. Toluca. The earliest work of a comparable scope is that by Reiche (1926; see **225**).

16 Conzatti's project was first described in S. F. Blake, 1931. A flora of Mexico. *Science*, N.S. **74**: 458–459. It was never completed by the author; but, as part of recent national flora initiatives, plans have been put in hand for a complete issue of this work in three volumes. The first, published in 1988, encompasses the monochlamydeous dicotyledons as well as all the families in the 1946–47 issue.

17 From only 60000 Mexican plants as late as 1956, by 1990 UNAM had expanded to 600000 (P. D. Dávila Aranda and Ma. T. Germán Ramírez, 1991. *Herbario Nacional de México*, Mexico City: Instituto de Biología, Universidad Nacional Autónoma de México).

18 Contemporary efforts towards a national flora have been described by A. Lot and V. M. Toledo, 1980. Hacia una Flora de México: vamos por buen camino. *Macpalxochitl* **88–89** [app.]: 12, 19 pp.

19 Unfortunately, the paper used was less than satisfactory and by the end of the twentieth century had noticeably deteriorated.

20 Conzatti's autochthonous, monothetic system, developed from late in the nineteenth century, is with its several 'progressions' conceptually similar to that of Engler; for a layout, see p. xxxv of the 1946–47 edition. For a commemorative booklet, see A. Espejo Serna and A. R. López-Ferrari, 1990. *Homenaje a Cassiano Conzatti, 1989*. Mexico City: Instituto de Biología, UNAM. The projected continuation of the 1988 edition was planned to cover respectively the dichlamydeous dicotyledons (Moringaceae–Koeberliniaceae in the 'Dialypetalae', or families 100–199, and Orobanchaceae–Compositae in the 'Gamopetalae', or families 200–247). The latter volume would thus be inclusive of the author's 'Coroliferas', representing all that was published of his *Flora sinóptica Mexicana* (1895–97 and 1910 editions).

21 The *Trees and shrubs* may be seen as not only related to Standley's work on the flora of his home state, New Mexico, but also a means of consolidating knowledge derived from the many surveys of the flora of Mexico by U.S. botanists and naturalists. It was also related to U.S. National Herbarium head curator F. V. Coville's programme of producing floras of lesser-known areas in North America.

22 As of 1981 the work was already considered incomplete (Sabine Bladt, personal communication).

23 The 'Flora de Sinaloa' represents one stage in a longer-term project, 'Contribución al estudio taxonómico de la flora de Sinaloa', initiated in 1984. For a progress report, see R. Vega A., 1990. Flora de Sinaloa. In *XI. Congreso Mexicano de Botánica* (Oaxtepec, Morelos): *Programas y Resumenes*, p. 366.

24 Claverán's thesis is cited in Díaz Barriga and Palacios-Rios (1992; see **222**).

25 A supplementary series, *Estudios florísticos en Guerrero*, features floristico-vegetatiological accounts of selected localities.

26 The revisions were part of the results of a state botanical survey. Related works by Martínez include a lexicon of vernacular and scientific names (1956) and a treatise on medicinal plants (1958), both also included in the 1979 reissue.

27 The literature on the *Flora de Veracruz* programme is considerable, with its data-processing aspects not unnaturally having been of particular interest. See, for instance, A. Gómez-Pompa and L. I. Nevling, 1973. The use of electronic data-processing methods in the *Flora of Veracruz* program. *Contr. Gray Herb.*, N.S., **203**: 49–64, and A. Gómez-Pompa *et al.*, 1984. Flora of Veracruz: progress and prospects. In R. Allkin and F. A. Bisby (eds.), *Databases in systematics*, pp. 165–174 (Systematics Association Spec. Vol. 26). London: Academic Press.

28 Alan Smith's pteridoflora of Chiapas (in the *Flora of Chiapas*) and the present work account for some 84 percent of the Mexican pteridophyte flora (*c.* 1000 species). If the flora of Knobloch and Johnston for Chihuahua be also included this percentage would be somewhat higher.

29 Miranda's work also includes particulars on botanical progress in the state.

30 See also a resumé of the *Flora Mesoamericana* project as part of the introduction to Superregion 21–23, p. 261.

31 This latter work was in part intended to accompany sets of specimens.

32 Mociño's flora, arranged on the Linnaean system, was finally published as I. M. Mociño (eds. J. Fernández-Casas, M. A. Puig-Samper and F. J. Sánchez García), 1993. *Guatimalensis prima flora.* 140 pp. (Fontqueria 37). Madrid. See also J. L. Maldonado Palo (ed.), 1996. «Flora de Guatemala» *de José Mociño.* 363 pp., col. pls. Madrid: Doce Calles/Consejo Superior de Investigaciones Científicas.

33 To a great extent the flora proper is the work of J. A. Steyermark and L. O. Williams as well as others then at the Field Museum; few 'outside' collaborators were involved except in the parts on pteridophytes.

34 According to W. T. Thistleton-Dyer, *Botanical survey of the Empire* (see **General references**), coverage of Belize in the 'Kew floras' was projected but by 1905 it was thought still to be too little known.

35 I thank Peter Bernhardt, at the time of writing of the original edition a graduate student at the University of Melbourne (and now at St. Louis University, St. Louis, U.S.A.), for information on Salvadorean botany. See also P. Bernhardt and E. A. Montalvo, 1978. Selected collecting sites in El Salvador, I. Private property. *Bull. Torrey Bot. Club* **105**: 9–13.

36 On seeing this second edition, Standley, in no way involved with its preparation, is reported to have exclaimed, 'Heaven help us!' See Williams, *Homage to Standley*, p. 19.

37 'El Pulpo' was local slang for the U.S.-based United Fruit Company (now United Brands, a part of the U.S. firm Chiquita).

38 See also C. Nelson S. in D. G. Campbell and H. D. Hammond (eds.), 1989. *Floristic inventory . . .*, pp. 290–294.

39 Some 651 species of pteridophytes were recorded by Nelson as opposed to 339 in Molina, a 92 percent increase.

40 The geographical coverage of this work has been disputed (cf. L. O. Williams in Gentry 1978; see Superregion 21–23, p. 262), but it has now been shown that all its localities are Nicaraguan. See W. D. Stevens, 1987. On the identity and recognition of the genus *Pochota* Ramírez Goyena (Bombacaceae). *Taxon* **36**: 458–464.

41 The work of Alfaro, an early publication of the Museo Nacional de Costa Rica, covered 3386 species of which but 1218 were surely recorded for the country.

42 A single set of covers will encompass one or more families, but unlike the *Flora of Guatemala* they will not form parts of a single volume of *Fieldiana, Bot.*

43 Designed for the non-specialist, this work, containing much original field information, ranks with *Indigenous trees of Hawaii* by J. F. Rock (**990**), *Wayside trees of Malaya* by E. J. H. Corner (**911**), and *Ceylon trees* by T. B. Worthington (**829**) as among the best introductory works to tropical trees in a given area.

44 The synonymy reportedly should be handled with care.

45 The *Flora* accounts only for 5314 seed plant species (and 283 additional infraspecific taxa). By 1987 the total as accounted for in the *Checklist* and *Index* by William D'Arcy (vols. 1(1) and 1(2) of the *Flora*) stood at 7345. There are in addition 800 or more pteridophytes.

46 In a letter to Conrad V. Morton in 1945; see *Homage to Standley*, p. 20. By 1987 the known species of seed plants had grown by 38.2 percent over the actual coverage in the *Flora*. However, D'Arcy's 1987 index goes far towards bringing earlier parts up to date.

47 The long and somewhat tortuous history of the *Flora of Panama* project has been well documented in the introduction in part 1 as well as in W. H. Lewis, 1968. The *Flora of Panama*. *Ann. Missouri Bot. Gard.* **55**: 171–173. A critical bibliographic analysis of the first phase of the project along with its precursors may be found in A. Robyns, 1965. Index to the 'Contributions toward a Flora of Panama' and to the 'Flora of Panama' through March 1965. *Ibid.*, **52**: 234–247. In the first two decades of the project, publication of families was systematic; but in later years families appeared when ready, with elements from any one or more of the originally designated formal parts present in a given issue (the practice of the Garden's *Annals* having been to give over one or more of the four annual issues entirely to the Flora). Early issues of the Flora carried dual pagination, but this practice was discontinued in 1965. Worthy also of note is the evolution of this work from a largely 'in-house' project utilizing little more than the Garden's own resources to a semi-monographic collaborative enterprise

based upon a multiplicity of sources. It demonstrates the likely practical limits of such large-scale floristic enterprises if they are not to 'drag on forever' (unless there be non-conventional circumstances) – as well as the importance of long-term institutional commitments, so dependent on financial and societal stability as well as progressive management. In the work also is an admission that a flora can never truly be completed: this contrasts with the outlook of the *Flora of British India* and other works (and personalities) of its era.

48 The *Flora ilustrada* is part of 'Proyecto Flora del Archipélago de San Andrés y Providencia', a joint undertaking of the Servicio Nacional de Aprendizaje and the Universidad Nacional de Colombia under the leadership of F. A. González.

49 A planned contribution on flowering plants appears not to have materialized.

50 See R. A. Howard, 1975. Modern problems of the years 1492–1800 in the Lesser Antilles. *Ann. Missouri Bot. Gard.* **62**: 368–379. For Olof Swartz, see also W. T. Stearn, 1980. Swartz's contributions to West Indian botany. *Taxon* **29**: 1–13.

51 In 1992, such an initiative was indeed taken with organization of a *Flora of the Greater Antilles* programme at the New York Botanical Garden.

52 For further information and documents on the *Flora*, see Thistleton-Dyer, *Botanical survey of the Empire* (see **General references**), pp. 11–13; and W. T. Stearn, 1965. Grisebach's *Flora of the British West Indian islands*: a biographical and bibliographical introduction. *J. Arnold Arbor.* **46**: 243–285.

53 Most of the *Symbolae* was based on the then-private Krug-Urban herbarium, later given to Germany and in 1943 largely destroyed with the rest of the Berlin-Dahlem Herbarium. Duplicates and fragments are widely dispersed, but as for other parts of the world taxonomic work has been seriously hindered.

54 Symposia on Bahamian botany were also organized in 1985 and 1988 (cf. *Taxon* **38**: 259 (1989)).

55 Marie-Victorin's *Itineraires botaniques* are particularly illustrative.

56 During and shortly after World War II the Forest Service center also promoted study of the tree flora of the Lesser Antilles.

57 The *Flora of the Greater Antilles* project publishes a newsletter; among other items it carries bibliographic additions.

58 This Cuban flora was finally published as *Cubensis prima flora* (1990, as *Fontqueria* 29) under the editorship of J. Fernández-Casas, M. A. Puig-Samper and F. J. Sánchez García.

59 The earlier history of botanical exploration was covered by Brother León in *Las exploraciones botánicas de Cuba* (1918, Havana) and again in *Flora de Cuba* (see below), as well as by J. Alvarez Conde, *Historia de la botánica en Cuba* (1958). Recent progress, particularly after 1959, has been reviewed in R. A. Howard, 1977. Current work on the flora of Cuba – a commentary. *Taxon* **26**: 417–423, and by R. P. Capote López, R. Berazaín Iturralde and A. Leiva Sánchez in Campbell and Hammond (1989; see **General references**), pp. 315–335.

60 Unfortunately, the format adopted by Koeltz is more wasteful of space than was that in *Fontqueria*. The project is currently under the sponsorship of four institutions: two in Cuba and two in Germany.

61 J. de J. Jiménez in 1961 credited the island with 5747 non-cultivated species of seed plants, which appears now to be too high.

62 See 'A new catalogue of the Dominican flora'. In Ninth International Botanical Congress, *Recent advances in botany*, 1: 932–936 (1961). Toronto.

63 For additional information on the flora, see T. Zanoni, 'Hispaniola', in Campbell and Hammond (1989; see **General references**), pp. 336–340.

64 A ninth and final volume is projected to cover family 69, Cactaceae, and family 155, Melastomataceae. The author has indicated that he has no plans to undertake the monocotyledons and gymnosperms (families 1–42); they would be left to 'others better prepared' (from a preface in vol. 7).

65 Illustrations were also prepared but for financial reasons could not be included.

66 Stahl's work was reissued by Carlos Chardón in 3 vols. in 1936 with as a basis an original manuscript.

67 Krug, a businessman, had been from 1857 to 1876 resident in the Puerto Rican city of Mayagüez. His support also made possible much of Urban's work on the flora of the West Indies.

68 The history of botanical work in Puerto Rico is treated in pp. 664–674 of Urban's flora (**240–90**) and pp. 4–21 in L. F. Martorell, A. H. Liogier and R. O. Woodbury, 1981. *Catálogo de la nombres vulgares y científicos de la plantas de Puerto Rico*. 231 pp. (Estación Experimental Agricola, U.P.R. Mayagüez, Bol. 263). Río Piedras, P.R. Additional information appears in J. L. Vivaldi, 'Puerto Rico', in Campbell and Hammond (1989; see **General references**), pp. 341–346.

69 As with *La flora de la Española*, the author has indicated that he has no plans to undertake the monocotyledons and gymnosperms (families 1–42); they would be left to others (from the foreword in vol. 5).

70 The Spanish edition of vol. 2 was, contrary to the colophon, not for sale through the U.S. Superintendent of Documents but made available through the Institute of Tropical Forestry in Puerto Rico (letter from E. L. Little, Jr. dated 14 February 1989 accompanying copies of the book).

71 This autochthonous project arose from ecological studies made in the 1980s. The collaboration of R. O. Woodbury followed from preparation of a checklist commissioned by the U.S. National Park Service (which controls much of the island).

72 Production of vol. 2 was part of the programme of the Bredin-Archbold-Smithsonian Biological Survey, instituted in the 1960s.

73 This checklist was published as a result of collecting by H. H. Smith with the support of F. duC. Godman (cf. Superregion 21–23) and appeared shortly after revival of the former botanical garden (first established in 1765).

74 Nicholas Jacquin collected in the islands in 1755–57, with the results appearing in his *Selectarum stirpium americanarum historia* (1763); but not unnaturally with a greater national interest in present-day Indonesia and the relatively small flora in their Antillean possessions Dutch

biologists prior to the 1880s paid them scant attention. See M. J. Sirks, *Indisch natuuronderzoek* (1915, Amsterdam), especially chapter 13.

75 I thank Dr. Stoffers for advice regarding this work.

76 Hoyos notes that the collections of Johnston and his party are not available in Venezuela.

77 Substantial contributions were made by Ernest E. Cheesman from the 1930s through the 1950s, and up through 1967 work continued in fits and starts, with much assistance from botanists in Jamaica and elsewhere. After a lapse of several years, work was resumed in the mid-1970s and the latest parts, all edited by David Philcox, appeared in 1977–92 (with parts 3 and 4 of vol. 3 in one fascicle). Of the outstanding families and other groups, the grasses are in preparation while work on palms and pteridophytes has been completed. [I thank Messrs. C. D. Adams and D. Philcox for advice respecting this work.]

Division 3

• • •

South America

Gross ist der Einfluss gewesen, welchen die von Humboldt und Bonpland zurückgebrachten botanischen Materialen auf die Ausbuildung der Systematik und die umfassendere Kenntniss der Gestalten im Pflanzenreiche ausgeübt haben.

> von Martius, 1860, *Akademische Denkreden* (1866); quoted in Stearn, *Humboldt, Bonpland, Kunth and tropical American botany*: 6 (1968).

At the foot of Chimborazo, with the zoning of its vegetation before his eyes, Humboldt drafted his *Essai sur la Géographie des Plantes* (1807), which established plant geography as a discipline.

> Stearn, *ibid.*: 116.

Botanically South America is the least explored area of the world. This survey . . . has shown that many large areas and important habitats are still uncollected, in spite of the long history of collection and collectors in South America, and a long tradition of local botanists in some of the countries.

> Prance in I. Hedberg (ed.), *Systematic botany, plant utilization and biosphere conservation*: 55 (1979).

This division comprises the entire continent from the Panamanian border southwards to Tierra del Fuego, together with the Falkland (Malvinas) Islands. Associated islands in the eastern Pacific Ocean come within Region 01, while those in the South Atlantic form part of Region 03. The chain of islands off the Caribbean coast, which encompasses Trinidad (with Tobago), the islands of the Venezuelan state of Nueva Esparta, and thence westwards to Aruba – once, with the nearby mainland, known as the 'Spanish Main' – is here included with the West Indies (as Regions 28 and 29). Over and above the eight politico-geographic regions adopted here, four major physiographic units, each with a distinct history of floristic studies, are recognized: Amazonia, Guayana and Patagonia (all in **301**), and the Andinoalpine zone (**303**).

Although 'many investigators have taken part both in early times and [in] our own in the exploration . . . of tropical South America, the complex and enormous flora of no part of the area can [yet] be considered sufficiently known to prepare a descriptive treatise which could satisfy the standards of contemporary demand'. Thus wrote Bassett Maguire, a leading twentieth-century botanical explorer in South America, in 1958. While much new work has been accomplished over most of the continent in the intervening four decades, his words are still true. Home to the largest vascular flora of any continent or island aggregation – more than 90 000 species, or a quarter of the world total – South America remains botanically the least explored and most imperfectly documented part of the world, and, as Ghillean Prance has emphasized, its basic taxonomic inventory is far from complete. Recent writers have continued to indicate that the level of collecting in the infratropical zone remains well behind many other parts of the tropics, a contrast to the 1850s when, according to Francis Pennell, South America was thought botanically to be among the better-known parts of the world (apart from western

Europe and eastern North America). Although the continent as a whole is more or less well into Colin Symington's 'second cycle' of exploration, interesting new discoveries are constantly being made, and in many remote areas will continue while opportunities remain.

The sheer size of the flora, especially in the infra-tropical zone, has also impeded documentation. The preparation of even partial floras has become a daunting task beyond the resources of individuals and even most institutions. At a national level, nominally complete floras exist only for Chile, Uruguay and Brazil and (to a considerable extent) Peru and Surinam, the three last prepared outside their respective countries. National flora projects are currently underway for Chile, Colombia, Ecuador, Venezuela, and the countries and territories of the Guianas, but all are long-term projects, some of them being wholly or partially externally based. In Venezuela the national flora programme is still continuing after more than 50 years (admittedly with a major change of direction in the 1950s). Coverage of more limited areas is very patchy, and few practical works for field use are available. Moreover, herbaria are often not well stocked or supported, hampering local taxonomic research and flora-writing.

The present level of knowledge and documentation of the flora should be viewed not only in terms of its size and accessibility but also in relation to political, social and cultural factors. These include an antagonistic attitude to nature, superstition, official suspicion (notably in the era of Portuguese and Spanish colonial rule), chronic political instability in many countries, economic conditions (exacerbated by foreign debt in the 1980s), decades of scientific imperialism (in the nineteenth and twentieth centuries), the social position of the sciences, and, as an expression of all these factors, the generally slow, limited growth of local resources, human and material, for floristic and systematic studies. These include national herbaria with adequate, well-named collections and stocks of literature – so necessary for serious locally based work, as Enrique Forero has stressed. In many countries, progress has depended upon the interest and enthusiasm of isolated individuals or groups, often in the face of official indifference or more pressing demands upon limited national resources. Continuity has been a serious problem and, until recently, effective institutionalization uncommon.

All these factors were major obstacles not only to botany but also in general to scientific progress in colo-nial South America. While in some reigns (notably those of Charles III and IV in Spain and under the Marquis de Pombal in Portugal) floristic exploration and research in the Spanish viceroyalties and in Brazil were more or less encouraged, the publication of results was but imperfectly so and by the early nineteenth century became all but impossible with war, post-colonial impoverishment, and even looting of collections. Thus was lost, or greatly lessened, the impact of much serious activity, notably the general and biological surveys of the second half of the eighteenth century. These included, on Spain's part, the 1754 *Comisión Botánica* (part of a border survey delimiting Spanish and Portuguese possessions), the three *Reales Expediciones Botánicas* of Charles III which were extended surveys of, respectively, Peru, Chile and Ecuador (by H. Ruiz and J. Pavón, 1777–88 with a continuation by Juan Tafalla and Juan Manzanilla through 1815), New Granada (by J. C. Mutis, 1783–1808) and New Spain (by M. Sessé and J. M. Mociño, 1787–1803), the *Real Comisión de Guantanamo* in Cuba (by Conde Mopox y Jaruco, 1796–1802), and the Malaspina world voyage (1789–94) which numbered among its objectives a survey of the Philippines but also examined the viceroyalty of La Plata and parts of the Pacific coast of South America, including Ecuador.[1] For the Portuguese the outstanding enterprise was the *viagem filosófica* of 1783–92 to the Amazon Basin led by the Coimbra scholar Alexandre Rodrigues Ferreira.

Other key early contributions to South (and Central) American botany came from non-Iberians with fortunate opportunities: Marggraf in Brazil (1637–54) when much of its *Nordéste* was under Dutch rule; Charles Marie de la Condamine and Joseph de Jussieu in Amazonia, Peru and Ecuador in the 1730s and 1740s; Pehr Loefling in Venezuela in 1754–56 (in charge of the above-mentioned *Comisión Botánica*); Fusée Aublet in French Guiana in 1762–64 (whose 1775 work, *Histoire des plantes de la Guiane Françoise*, is not only an outstanding work but also is notable for its many generic names and specific epithets based on the local vernacular); many visitors to Rio de Janeiro (who were, however, usually not allowed inland travel); world voyagers like Commerson, Banks, Solander, the Forsters, Sparrman and others in Fuegia in the late eighteenth century; and, most importantly, the famous expedition of Alexander von Humboldt and Aimé Bonpland in northwestern South America, Cuba and Mexico in 1799–1804.[2] Humboldt and Bonpland's expedition gained wide popular acceptance, exercised considerable indirect political influence, and set the pattern for future scientific exploration of the continent (in much the same way as did Thomas Horsfield and Stamford Raffles in the Malesian region). Published with relative celerity, their results created a sensation in revealing the wealth of plant and animal life to be found.[3]

Six of the 30 volumes of *Voyage de Humboldt et Bonpland* were botanical. The most important of these, the *Nova genera et species plantarum* (1816–25), was largely written by Carl Kunth on a modified Jussieuan (natural) system. It is basic to all later neotropical botany as being, in the words of William T. Stearn, 'essentially a pioneer flora of northern South America and of Central America'.[4] Possessing many of the now-standard principles and features of modern floras, it was intellectually well ahead of most of its contemporaries. Its German author had for the purpose moved to Paris, the Muséum d'Histoire Naturelle then being the leading center for systematics. In it, 3600 species were described for the first time, a staggering four-fifths of its total coverage – so effectively eclipsing much of the Spanish work.

The efforts of Humboldt and Bonpland were, after the Napoleonic Wars, emulated in politically more stable Brazil by the expedition of Spix and C. F. P. von Martius in 1817–20, which included among its contributions the first serious biological reconnaissance of Amazonia. The botanical results, written by von Martius as *Nova genera et species plantarum* (1824–32) in emulation of Kunth's work, would be a major foundation for its author's later *grand projet*, *Flora brasiliensis* (1840–1906). This latter was able in time to draw on the results of many more expeditions, Brazil, under its emperors Pedro I and II, being for years a fashionable destination. As conditions stabilized in other countries, they were in turn visited by numerous, sometimes competitive, organized expeditions. Patagonia and Fuegia continued to be visited by world voyages through the 1840s and witnessed a renewed rush of activity at the end of the century as Antarctic research expanded and the sheep industry developed. However, following the mid-century expeditions of the Schomburgks, Alfred Russel Wallace and Richard Spruce the continent came to be considered in a primary sense relatively well known, and general exploratory interest shifted to Africa, East and Southeast Asia, and finally New Guinea. Prosecution of floristic work in South America

became a more specialized undertaking, with increasing contributions by resident botanists; nevertheless, significant areas – notably the Guayana region between Brazil and the north coast – remained geographically little known and until the latter part of the twentieth century attracted substantial general expeditions.

Local progress in the nineteenth century – including preparation of floras – was, unfortunately, impeded by often unsettled political conditions, adverse economic situations, and inimical patterns of social and cultural development. This was compounded by, in most cases, an absence of formal institutional ties analogous to those characterizing floristic botany as it developed in Africa and much of the Old World (with the accompanying stimulus towards development of local centers) and a corresponding decline in involvement in South American botany by many metropolitan centers. All these contributed to a pervasive problem: lack of continuity. What was accomplished after 1815, at first through general expeditions, then through more specialized efforts, remained until well into the twentieth century largely the work of overseas explorers. With the exception of the southern 'triangle', significant resident contributions have been a feature only of the last five to six decades or so.

The visitors from Europe and, from late in the nineteenth century, North America, came under various types of patronage or for commercial reasons or settlement and were aided by scientists at diverse European (and, later, North American) centers. A large proportion of these visitors were from Central Europe and Scandinavia, where some official or private support at metropolitan centers was often available. A significant influence in much of Europe was surely the *Flora brasiliensis* project (aided by the move in 1877 of its middle editor, August Eichler, to Berlin University and its Botanical Museum, and, at his death, continuation of work there under Ignatz Urban). In Sweden, an important development was the establishment, about 1898, of the Regnell Foundation in the Swedish Natural History Museum at Stockholm (Regnell himself having been a long-time resident in Brazil in the nineteeenth century). Contributions from North Americans, beginning under private auspices in Bolivia in the 1880s and at the turn of the century in Colombia, rapidly rose after 1920, supported mainly by two programmes: an inter-institutional effort, oriented towards northern South America and based at the New York Botanical Garden under Henry A. Gleason, and one on Peruvian biota organized by the Field Museum in Chicago. They were joined by Francis Pennell from the Academy of Natural Sciences in Philadelphia (who collected extensively at higher altitudes).

Serious development of local research in infratropical South America began in Brazil in the late nineteenth century, stimulated by the palm specialist João Barbosa-Rodrigues at Rio de Janeiro who also was the first to turn the old Jardim Botânico there into a serious scientific institution. In the north the establishment of the Museu Goeldi at Belém do Pará provided (until its suspension in 1915 as the regional economy collapsed) a strong impetus to more detailed Amazon exploration; contributors included Emílio Goeldi, Jacques Huber and (later) Adolfo Ducke. At São Paulo, riding on a coffee boom, Alberto Löfgren, Alfred Usteri and (later) Federico C. Hoehne were active, the last-named doing much to build up the Jardim Botânico in the south of the city.

In Peru, local interest was stimulated by the work of Antonio Raimondi at Lima and after 1900 renewed by August Weberbauer, who settled in the country some years after his initial vegetation and floristic studies. A secondary center later developed at Cuzco around César Vargas and Fortunato L. Herrera. In Venezuela and Colombia, however, the work of Adolf Ernst in the former and Mutis, Caldas and José Triana in the latter did not immediately lead to effective continuity; only after World War I were the respective national herbaria established, in Venezuela with Henri Pittier playing a prominent role. In Ecuador, locally based research remained until after World War II entirely in the hands of interested individuals such as Jameson, Luis Sodiro and (later) Acosta-Solís. Progress in Bolivia similarly revolved around individuals; even in the mid-twentieth century, a sense of continuity was still evidently so lacking that Martin Cárdenas, on retirement from scientific work, sent his private herbarium to Tucumán in Argentina! In the Guianas, local development was scant and sporadic until formation of the universities in Georgetown and Paramaribo and the ORSTOM center in Cayenne.

Botanical development in subtropical and temperate South America was generally more rapid, except in poor and long-isolated Paraguay, and most of the region has been more or less well floristically treated. Argentina in particular developed very rapidly from the late nineteenth century to World War II and beyond, but the very diversity of centers and technical considerations there have guided major efforts towards regional and provincial floras. Argentine influence has also spread to Bolivia and Paraguay as well as Uruguay; surprisingly, the last-named country, though relatively well collected, lacks a modern manual. Chile, long provided for by single-handed works and with fewer resources, has of late moved towards international cooperation in mounting a new definitive flora. In Paraguay, the old Swiss connection embodied by Moises Bertoni and Emil Hassler has been reactivated in recent years.

The abundant results have been embodied in floras, enumerations and often poorly coordinated, widely scattered 'contributions' of varying quality. Along with a widespread tendency towards specimen description, this dispersion of efforts has had, as in some other parts of the tropics, the effect of creating unnecessary synonymy, a problem further aggravated by national and even continental boundaries. An outstanding exception was *Flora brasiliensis*, that splendid example of international official and professional collaborative effort which, along with the Candollean *Prodromus*, dominated mid- and late nineteenth century European systematics; and it is only in recent decades, with the establishment of a variety of bilateral and multilateral links aimed at the promotion of floristic work and flora-writing, that something like a return to this tradition has taken place, albeit in most cases with, so to speak, the shoe on the other foot. For the long period of overseas scientific domination, through which vast collections including some 80 percent of the type specimens left the continent, had inevitably brought in its wake various forms of retaliation and restrictions as locally based science began serious development.

A further complicating factor, also with effects upon the level and nature of documentation, has been the distribution of resources within the continent. The bulk yet lies within its south-central part, from Rio de Janeiro to São Paulo, Tucumán, Córdoba, and 'Gran Buenos Aires'. Elsewhere they have been more scattered and often less well founded, with certain countries virtually without herbaria of any kind. Recent decades, however, have seen expansion in Venezuela, Colombia, Peru and Amazonia, with the national herbaria in the two first-named serving as centers for significant flora projects.

Thus, much work in South American botany has been of a disordered and to some extent *ad hoc* nature.

Large-scale organized projects have been few (and often externally administered) and transnational collaboration and integration uncommon. Until recently, decentralization, largely country-centered, has been the rule, and the historical background pertaining to the various principal floristic works has therefore been recounted in the individual introductions to Divisions 31–39. Further details can be sought in the country reports collected in the references below as well as in individual papers. Until recent years, however, botanical progress in South (and Middle) America has been less easy to evaluate than for Africa and the Asia-Pacific tropics. In 1978, however, *Boletín Botánico Latinoamericano* was established, giving the Americas an outlet comparable to *Flora Malesiana Bulletin* or the *AETFAT Index* (although pan-American botanical congresses have a somewhat longer history than those related to Africa or the East).

In recent decades, significant developments towards integration have taken place, involving not only South America but also Mexico, Central America, and the West Indies. The most important of these has been the establishment of *Flora Neotropica*, an unstructured, irregularly appearing series of monographs covering the whole of the Neotropics. This work is under the control of the international Organization for Flora Neotropica (OFN), formed in 1964 under the auspices of the UNESCO Humid Tropics Research Programme (1956–65) and since 1976 designated by UNESCO as a Category B organization. Since publication began in 1967, more than 75 monographs (of families or parts thereof in all groups of plants) have been published or are in press. OFN has also been active in promotion of botanical exploration, study and conservation in Middle and South America, but only recently has begun moves into ancillary documentation of the kind long a feature of AETFAT and the Flora Malesiana Foundation, including areas such as bibliography, biohistory, topography and general studies. Indeed, *Flora Neotropica* arguably more closely resembles *Das Pflanzenreich* than it does *Flora Malesiana* or the several African regional floras. Other attempts at coordination have included the already-mentioned *Boletín Botánico Latinoamericano*, currently issued from Bogotá, and *Taxonomic bibliography for the neotropical flora* by G. Agostini (1974), of which only a small proportion was ever published. Latin American botanical congresses were instituted before World War II, but are not limited to systematics.

Current larger flora projects include *Flora of Suriname* (1932–86, in later years expanded to cover the three Guianas) and its successor, *Flora of the Guianas* (1985–); *Flora de Venezuela* (1964–); *Flora de Colombia* (1983–); *Flora of Ecuador* (1973–), representing a continuation of the long Swedish South American tradition; *Flora of Peru* (1936–), begun by the Field Museum in Chicago and reorganized in the 1970s under multi-institutional sponsorship including the Universidad National Mayor de San Marcos at Lima; the *Flora de Paraguay* (1983–), renewing, as already noted, the Swiss connection; the recently established *Flora de Chile* (1995–), based at the Universidad de Concepción; and the regional floras in Argentina, representing a coordinated project developed in the 1950s after the termination of the elephantine *Genera et species plantarum argentinarum* (1943–56), and their successor *Flora fanerogámica argentina* (1994–). The contemporary *Flora brasílica* (1940–68), begun by Hoehne on the anniversary of its predecessor, never enjoyed the necessary resources and was discontinued far from completion; but a number of state floras have been published or are in progress, with the most significant being *Flora ilustrada catarinense* (1965–). Partial floras are also being produced in Venezuela, Colombia and Chile, and some one-volume florulas of limited areas (e.g., Ávila near Caracas, Río Palenque Science Center in Ecuador, and the environs of Buenos Aires) and generic synopses (e.g., for Chile and Venezuela) have also been published. There is also a modern flora of the Falkland (Malvinas) Islands, and a flora of Fuegia appeared in 1983. Last, but not least, there is *Botany of the Guayana Highland* (1953–89), embodying the results of the many expeditions conducted from the late 1920s to the author Arthur Conan Doyle's 'Lost World' and covering a region botanically one of the most important in South America. This will in the coming years be supplemented by *Flora of the Venezuelan Guayana*, covering the major part of the area; its first volumes appeared in 1995. Another current project is aimed towards a new flora of the high Andes in succession to *Chloris andina*; a first fruit is *Páramos* (1999; see **303**).

Older floras of significance include, apart from *Flora brasiliensis* (1840–1906), *Symbolae ad floram argentinarum* (1879) by Grisebach, *Flora chilena* (1845–54) by Gay and the incomplete *Flora de Chile* (1896–1911) by Reiche, *Plantae hasslerianae* (1898–1907) by Chodat and collaborators on Paraguay, *Flora uruguaya* (1898–1911) by Arechavaleta and W. Herter's works on Uruguay, *Prodromus florae novo-granatensis* (1862–73) by Triana and Planchon on Colombia (terminated far from completion), the largely compiled *Initia florae venezuelensis* (1926–28) by R. Knuth, and the early works by Schomburgk and Pulle in the Guianas. Of more local works, mention may be made of *Flora fluminensis* and its *Icones* (1829–31) by Frei Velloso, *Synopsis plantarum aequatoriensium* (1865) by Jameson, *Flora der Umgebung der Stadt Sao Paulo* (1911) by Usteri, and some state and provincial works in the southern 'triangle'. Weddell's incomplete *Chloris andina* (1855–61) partially covered the Andes, while the 1899 Princeton University Expedition to Patagonia together with the contemporary field work by the Swedish botanists Per Dusén and Carl Skottsberg there and in the Falkland Islands resulted in a major compilatory *Flora patagonica* (1903–14) by George Macloskie and collaborators (including Dusén). This took in the whole area south of about the 40th parallel; earlier, Fuegia had been treated in the second volume of *Flora antarctica* (1845–47) by J. D. Hooker.

Most of the nineteenth-century works more or less represent the primary stage of modern documentation for their respective areas, comparable to *Flora of tropical Africa* (**501**) or *Flora indiae batavae* (**910–30**), while the works of recent decades are representative of the secondary stage as outlined by Symington and may be compared with, for example, *Flora Malesiana* (**910–30**) or *Flora reipublicae popularis sinicae* (**860–80**) although, as already noted, most of the infratropical zone is still very sketchily covered. They remain for the most part without, or with but incomplete, successors. The most poorly documented countries are Bolivia (without any overall work save an uncritical name list and, more recently, a tree flora), Paraguay (though improvement is on the way with *Flora del Paraguay*), and (to a lesser extent) Guyana (now being covered in *Flora of the Guianas*) and Ecuador (with no overall work prior to the initiation of *Flora of Ecuador*).

Of historical interest but benefiting from modern revision is the multi-volume *Flora de la Real Expedición Botánica del Nuevo Reyno de Granada*, a latter-day version of the results of José C. Mutis's 'tribe of botanical adventurers' and containing reproductions of its numerous fine colored figures. Publication began in 1954 and at present is well advanced.

A few words about future flora-writing in South America may be apropos. While production of *Flora Neotropica* and other large-scale national and state/departmental works should be continued as they represent significant contributions to floristic knowledge, consideration should also be given to the preparation of more 'genera', synopses and local florulas. As Alwyn Gentry, Ghillean Prance and others have suggested, these will reach a wider local audience and for teaching purposes are more useful.[5] Such works may be the best tools for raising cultural and environmental awareness and hence an effective interest in conservation, so necessary in the face of extensive destruction in the name of 'development' in Amazonia and elsewhere, however 'undesirable' they may be from a more idealistic botanist's point of view. That a number of such works, all well illustrated, have appeared within the last few years is heartening; they serve to show clearly how rich and diverse is the tropical South American flora even over small areas. They continue a trend also developing elsewhere, as at Barro Colorado Island in Panama (**237**). More of them are needed. Looking beyond floristic information, efforts should also be made at increasing the collection of biological and experimental data, already a well-established activity in several other parts of the tropics as well as in parts of South America and, as Lorenzo Parodi early emphasized, of considerable value to critical flora-writing.

Progress

The most complete introduction to the state of botanical knowledge in South America, with a historical sketch (by F. W. Pennell) and many country or territorial surveys, remains F. VERDOORN, 1945. *Plants and plant science in Latin America.* xxxvii, 383 pp., illus., maps. Waltham, Mass.: Chronica Botanica. (Chronica Botanica Plant Science Books 16.) This has been substantially augmented by several country and regional chapters in CAMPBELL and HAMMOND, *Floristic inventory of tropical countries* (1989; see **General references**). Shorter, discursive reports include H. A. GLEASON, 1927. The botanical investigation of northern South America, a tri-institutional project: its aims and needs. *J. New York Bot. Gard.* 28: 261–265; *idem*, 1932. The progress of botanical exploration in tropical South America. *Bull. Torrey Botanical Club* 59: 21–28; B. MAGUIRE, 1958. Highlights of botanical exploration in the New World. In W. C. Steere (ed.), *Fifty years of botany*, pp. 209–246. New York; and G. T. PRANCE, 1979. South America. In I. Hedberg (ed.), *Systematic botany, plant utilization and biosphere conservation*, pp. 55–70. Stockholm. The infratropical zone was also reviewed in Prance's (1978) general survey of floristic progress in the tropics (see **General references**), and the Pacific wet belt in A. H. GENTRY, 1978. Floristic needs in Pacific Tropical America. *Brittonia* **30**: 134–153.[6] The role of national herbaria in promotion of locally based floristic and taxonomic study has been outlined in E. FORERO, 1975. La importancia de los herbarios nacionales de America Latina para las investigaciones botánicas modernas. *Taxon* **24**: 133–138.

Flora Neotropica: Descriptions of this project may be found in B. MAGUIRE, 1973. The organization for 'Flora Neotropica'. *Nature and Resources* **9**(3): 18–20, and in F. R. FOSBERG, 1985. Origin and early history of the Flora Neotropica project. *Flora Neotropica*, Monogr. 2, Suppl.: 25–27. Progress reports have appeared as 'Flora Neotropica News' in *Taxon* from volume 28 (1979) onwards into the 1980s as well as in *Boletín Botánico Latinoamericano*.[7]

General bibliographies. Bay, 1910; Blake and Atwood, 1942; Frodin, 1964, 1984; Goodale, 1879; Holden and Wycoff, 1911–14; Jackson, 1881; Pritzel, 1871–77; Rehder, 1911; USDA, 1958.

Divisional bibliographies

In addition to the works cited below, there are several national or area bibliographies; these are listed under their appropriate headings.

AGOSTINI, G., 1974. Taxonomic bibliography for the neotropical flora. *Acta Bot. Venez.* **9**: 253–285. [Six families, Acanthaceae–Anacardiaceae, in alphabetical order. Not continued.]

ANGELY, J., 1980. *The South America botanical bibliography.* 5 vols. 1897 ll. São Paulo: 'Phyton'. (Coleção Saint-Hilaire 9.) [Annotated, analytical bibliography of 4733 titles (some duplicated!) from 1753 through 1977, extensively indexed and arranged alphabetically within any one (or more) of seven regions: 1, Southern Brazil (with Paraguay and Uruguay); 2, Northeastern Brazil; 3, Amazonian Brazil and the Guianas; 4, Argentina; 5, Chile; 6, Andean South America (Venezuela to Peru); 7, Bolivia. Only the contents of Region 1 have been published. Titles are heavily annotated or analyzed.]

General indices. BA, 1926– ; BotA, 1918–26; BC, 1879–1944; BS, 1940– ; CSP, 1800–1900; EB, 1959–98; FB, 1931– ; IBBT, 1963–69; ICSL, 1901–14; JBJ,

1873–1939; KR, 1971– ; NN, 1879–1943; RŽ, 1954– . See also under the regions.

Divisional indices

AMERICAN SOCIETY OF PLANT TAXONOMISTS, 1939–67. *The Taxonomic Index.* Vols. 1–30. New York (later Cambridge, Mass.: from vol. 20 (1957) published serially in *Brittonia*). (Vols. 1–9 mimeographed.) [Begun on the initiative of W. H. Camp, this index had an existence of 28 years but from the mid-1940s it was reproduced from the appropriate parts of thc *Index to American Botanical Literature.* In 1957 it was consolidated with *Brittonia* although retaining separate volumation, and in 1967 discontinued.]

ANONYMOUS (SOCIEDAD ARGENTINA DE BOTÁNICA), 1945–86. Bibliografía botánica para [la] América Latina. *Bol. Soc. Arg. Bot.* **1**(1)–**24**(3/4), *passim.* [A running index to botanical literature, published in every number of the journal (four issues per volume, but appearing over every 18 months to two years). Continued through 1986.]

FORERO, E., (coord.), 1978– . *Boletín Botánico Latinoamericano.* Nos. 1– . Bogotá, etc. (Offsetprinted.) [Primarily a newsletter, but includes news of floras and related projects as well as a selection of other literature references. Forty numbers published through 1999.]

TORREY BOTANICAL CLUB, NEW YORK, 1969. *Index to American Botanical Literature, 1886–1966.* 4 vols. Boston: Hall. Continued as *idem*, 1977. *Supplement, 1967–76.* 740 pp. Boston. [Originally published serially in *Bulletin of the Torrey Botanical Club* as well as (for a time) on index cards. In the reprint, taxonomic and other non-author entries appear in vol. 4. The *Index* continued publication in the *Bulletin* through vol. 122 (1995); control then passed directly to the New York Botanical Garden and from 1996 it has appeared in *Brittonia* (vol. 48 onwards) with a more specific emphasis on systematics and related fields. From 1999, the *Index* is offered solely in electronic form.]

Conspectus

300

Division in general

No floristic works cover the entire continent. For *Flora Neotropica* and the Tryons' *Ferns and allied plants*, see **301/I**.

301

Major continental subdivisions

Under this heading are included works on tropical South America (**I**), Amazonia (**Ia**), Guayana (**Ib**), Patagonia (**III**), and Fuegia (**IIIa**). For the high Andes, see **303**.

I. Tropical South America

Keys to and treatises on families and genera

GENTRY, A. H., 1993. *A field guide to the families and genera of woody plants of Northwest South America (Colombia, Ecuador, Peru) with supplementary notes on herbaceous taxa.* xxiii, 895 pp., illus. Washington, D.C.: Conservation International. (Reissued in a slightly smaller format, 1996, Chicago: University of Chicago Press.) [Illustrated identification manual with an artificial general descriptive, non-dichotomous key followed by alphabetically arranged family accounts with keys to genera; indices to vernacular and scientific names at end.][8]

MAAS, P. J. M. and L. Y. TH. WESTRA, 1998. *Familias de plantas neotropicales.* vii, 315, [1] pp., 88 pls. (drawings). Vaduz: Koeltz (apud Gantner). English version: *idem*, 1992. *Neotropical plant families.* vii, 289 pp. Vaduz (reissued 1998). [Descriptions of families (without keys); included are represented genera (with statistics of species in Neotropics), distribution, pollination/dispersal types, uses, and selected literature (including references to key flora accounts; codes, p. 4). The introductory section (pp. 1–7) includes general references, among them major floras (pp. 6–7) and keys to families.]

PITTIER, H., 1939. *Clave analítica de las familias de plantas superiores de la América tropical.* 4th edn. vii, 94 pp. Caracas: 'La Nación'. (1st edn., 1917, as *Clave analítica de las familias de plantas fanerógamas de Venezuela y partes adyacentes de la América tropical*; 2nd edn., 1926, and 3rd edn., 1937, as above.) [Includes a glossary. Based upon F. Thonner's *Anleitung zum Bestimmen der Familien der Blütenpflanzen (Phanerogamen)* (1st edn., 1895; 2nd edn., 1917: see **000**, p. 95).][9]

ORGANIZATION FOR FLORA NEOTROPICA, 1967– . *Flora Neotropica: a series of monographs.* Monogrs. 1– . Illus., maps. New York: Hafner (later New York Botanical Garden) (for the Organization for Flora Neotropica).

Comprises an open-ended series of detailed monographic or submonographic treatments of families and infrafamilial taxa of fungi, non-vascular and vascular plants within the tropics of the Americas; each treatment includes descriptions, analytical keys, synonymy with references and citations, geographically arranged localities with *exsiccatae*, summaries of internal and extralimital range, taxonomic commentary, notes on habitat, special features, etc., along with illustrations and distribution maps of a wide selection of species, and complete indices to all names and to *exsiccatae* in each fascicle as well as introductory sections on morphology, biology, taxonomy, etc., of the group(s) as a whole with synoptic lists of taxa. By 1999/2000, 77 monographs had been published (some with supplements), covering 6388 species of flowering plants (roughly 7 percent of the estimated total of 90 000) and 47 species of gymnosperms.[10]

Pteridophytes

TRYON, R. M. and A. F. TRYON, 1982. *Ferns and allied plants: with special reference to tropical America.* xii, [ii], 857 pp., 127 figs. (incl. photographs and dot maps). New York: Springer.

Generously formatted generic (and family) flora of ferns and 'fern-allies'; within each genus appears a fairly detailed description, synonymy, taxonomic commentary, notes on ecology, geography, sporology, cytology, and special features, a listing of tropical American species (with keys), illustrations and distribution maps of selected species, and basic literature. Family headings include keys to genera, and literature relating to tribes appears under tribal headings; full index (pp. 835–857). The introductory section includes a general review of pteridophyte systematics, a systematic synopsis, and considerations of biogeography, sporology, cytology, and karyology along with references and a key to families. The figures are more comprehensive than the collation suggests; each has several components.

Ia. Amazonia

No general works for the Amazon Basin as a whole are available. Estimates of its extent vary from 5 to 6 million square kilometers, of which 3.7 million are in Brazil (somewhat less than the 'Amazônia Leal' of Brazil, which goes somewhat beyond natural boundaries); the remainder is in Venezuela, Colombia, Ecuador, Peru and Bolivia. Four bibliographies, respectively for Brazil, Colombia, Peru and Venezuela, have been published.

The largest single set of contributions to the Amazonian flora is *Plantes nouvelles ou peu connues de la région amazonienne* by Adolfo Ducke (11 parts, 1915–40, the first 10 in *Arquivos do Jardim Botânico de Rio de Janeiro*, the last in *Arquivos do Servicio Florestal*).

Bibliographies

Fuller details on these Amazonian bibliographies appear under their respective national headings.

ANONYMOUS, 1963–70. *Amazônia: bibliografia.* 2 vols. Rio de Janeiro: Instituto Brasileiro de Bibliografia e Documentação/Instituto Nacional de Pesquisas da Amazônia, Conselho Nacional de Pesquisas. See **350/I**.

DOMINGUEZ, C. (comp.), 1985. *Bibliografía de la Amazonia colombiana y areas fronterizas amazonicas.* ix, 226 pp. Bogotá: DAINCO/Corporación Araracuara/Colciencias. See **321**.

ENCARNACIÓN, F., R. SPICHIGER and J.-M. MASCHERPA, 1982. *Bibliografía selectiva de las familias y de los géneros de fanerogamas: primera contribución al estudio de la flora y de la vegetación de la Amazonia peruana.* 197 pp. (Boissiera 34). Geneva. See **330**.

WEIBEZAHN, F. H. and B. E. JANSSEN-WEIBEZAHN, 1990. *El Territorio Federal Amazonas, Venezuela: una bibliografía/The Amazonas Federal Territory, Venezuela: a bibliography.* See **316**.

Ib. Guayana

Guayana is here defined as that part of northern South America associated with the crystalline Guayana Shield, an ancient formation extending from just east of the Andes (under the Sierra de la Macarena in Colombia) to French Guiana and centering in the spectacular table mountains (tepuis) of southern and eastern Venezuela south and west of the Orinoco, the highest of which is Cerro de la Neblina (3015 m). The whole Guayana region contains perhaps 8000 species of higher plants, 75 percent of them endemic; of these, the Highland flora numbers some 2000, with 90–95 percent endemism.

In addition to the comprehensive series of revisions of the flora of this remarkable region by B. Maguire and collaborators published between 1953 and 1989, there have also appeared florulas for some individual tepuis: the Tafelberg (Surinam), Mt. Roraima (Guyana/Venezuela), Cerro Jáua (Venezuela), Auyántepui (Venezuela), and Cerro Duida (Venezuela). Additional coverage is provided in

J. A. STEYERMARK *et al.*, *Contributions to the flora of Venezuela* (**315**), and a manual-flora for Venezuelan Guayana commenced publication in 1995. The best general review of the physical features, flora, vegetation, and endemicity of the Highland is now in the first volume of that new manual, *Flora of the Venezuelan Guayana* (**316**).[11]

MAGUIRE, B. and collaborators, 1953–89. *The botany of the Guayana Highland*, I–XIII. Illus., map (Mem. New York Bot. Gard. 8–51, *passim*). New York. Complemented by W. R. BUCK, 1990. Indices to "The Botany of the Guayana Highland". *Mem. New York Bot. Gard.* **64**: 45–122.

A series of papers incorporating critical descriptive revisions of selected vascular plant families in Guayana (68 in all, some in two or more installments); each treatment includes keys to species (in large genera), synonymy and references, localities with *exsiccatae*, taxonomic commentary, notes on habitat, special features, etc., and figures of representative species. The general introductory section (part I) gives a historical review of botanical exploration in the Guayana Highland as well as overall remarks on the geography of the region and its flora. [A promised 'introductory part' formally concluding the work was never published; however, there are full cumulative indices.]

III. Patagonia

Area: not known. Vascular flora: 2200 species (Macloskie). – This unit refers to geographical Patagonia (and Fuegia), i.e., all of South America from approximately 40° S to Cape Horn. In the Argentine this refers to the territory south of Río Negro and Río Limay, while in Chile it includes the island of Chiloé and all territory southwards.

For additional coverage, see respectively **380** and **390** for general Argentine and Chilean works, and **388** for Argentine Patagonia (CORREA, *Flora patagónica*). Exploration and research to 1891 are dealt with in some detail by E. J. Godley (see **Superregion 08–09, Progress**).

Bibliography

PÉREZ-MOREAU, R. A., 1965. *Bibliografía geobotanica patagónica.* 110 pp. (Publ. Inst. Nac. Hielo Continental Patagonica 8). Buenos Aires. [Includes systematic, phytogeographical, and geobotanical references; no subject index.]

MACLOSKIE, G. *et al.*, 1903–06. Botany (including 'Flora Patagonica'). 9 parts. xxii, 982 pp., 106 text-figs., 31 pls. (some col.; incl. map). In W. B. Scott (ed.), *Reports of the Princeton University expeditions to Patagonia, 1896–1899*, 8 (divisions 1–2). Princeton, N.J.: Princeton University; Stuttgart: Schweizerbart. Continued as G. MACLOSKIE and P. DUSÉN, 1914 (1915). Botany: supplement (Revision of 'Flora patagonica'). 307 pp., 4 pls. *Ibid.* (division 3). Princeton/Stuttgart.

This luxuriously formatted work has in its nine parts a variety of mainly floristic contributions of which much the largest is part 5, *Flora patagonica*, by G. Macloskie (pp. 139–906). That comprises a partly compiled, concisely descriptive flora in four 'sections' with keys to genera and species, limited synonymy (with some references), a few vernacular names, generalized indication of internal and extralimital range with denotion in some cases of authorities but almost no *exsiccatae*, taxonomic commentary, and notes on uses, special features, etc.; brief introduction at beginning of work. Part 6 (pp. 907–920) comprises an analytical key to the 113 families of seed plants covered in the *Flora*. The concluding three parts (7–9) respectively constitute a historical account (with bibliography), a gazetteer, and a general review of the flora and its possible origin; these are followed by additions, corrections and errata (pp. 961–964) and an index to vernacular, family and generic names (pp. 965–982). Parts 1–3 deal with vegetation (by P. Dusén), hepatics and mosses, while part 4 (pp. 127–138) treats pteridophytes (the determinations by L. M. Underwood). The supplement, dated 1914 but actually published early in 1915, is devoted to revisions of the contents of parts 4 and 5 in the main work, based partly on further field work in Patagonia to 1903 (especially by P. Dusén). [In this work, the approximate northern limit of Patagonia is taken as 40° S.][12]

IIIa. Fuegia

Area: 48 000 km². Vascular flora: 545 species (417 native).

MOORE, D. M., 1983. *Flora of Tierra del Fuego*. ix, 396 pp., 283 text-figs. and photographic reproductions, 42 maps. Oswestry, Salop.: Anthony Nelson; St. Louis, Mo.: Missouri Botanical Garden.

Briefly descriptive flora of native, adventive and naturalized vascular plants; includes keys to all taxa, synonymy with references and concise citations, localities, general indication of extralimital range (with special reference to Argentina and Chile), numerous illustrations, dot distribution maps, and some critical commentary. Key monographs and revisions are cited where appropriate. The introductory section covers physical features, climate, soils, botanical exploration (with a summary of collectors), floristics and biogeography, and vegetation and prehistory (reckoned at 16 000 years since end of the last ice cover), followed by the plan of the work and a general key to families; at the end of the work (pp. 354ff.) are a glossary, references, lexica of Fuegian, Spanish and English vernacular names, abbreviations, and index to scientific names.

302

'Old' upland and montane regions: Guayana Highland

See **301/Ib** and **316**.

303

'Alpine' regions: The Andes

Since Weddell's classic but never-completed *Chloris andina* (1855–61), no comprehensive work for the high-altitude flora of the Andes has been published. The northern *páramo* areas, ranging from Costa Rica to Venezuela and northern Peru, were first covered anew as a whole only in 1999. The more southerly *puna* has had no overall modern coverage. There are several works for more limited areas, both old and more recent.

WEDDELL, H. A., 1855–61. *Chloris andina. Essai d'une flore de la région alpine des cordillères de l'Amérique du Sud*. In F. L. De Laporte de Castelnau (ed.), *Expédition dans les parties centrales de l'Amérique du Sud . . . exécutée par ordre du gouvernement français pendant les années 1843 à 1847, sous la direction du comte Francis de Castelnau*, part 6: Botanique. Vols. 1–2. 90 pls. Paris: Bertrand. (Reprint of whole work in Germany, 1922; botanical part reprinted 1971 in 1 vol., Lehre, Germany: Cramer.)

Detailed descriptive flora of seed plants; includes synonymy and references, *exsiccatae* with localities (not confined to Weddell's own collections and extending from Venezuela to Chile), overall distribution, extensive taxonomic commentary, figures of representative species, and generic indices. [Not completed; vol. 1 encompasses Compositae, vol. 2, Calyceraceae through Ranunculaceae (44 families in all). Until the publication of *Alpine flora of New Guinea* by Pieter van Royen (**903**), this remained the most ambitious of extra-European 'alpine floras'; personal difficulties precluded the work's completion.][13]

I. Northern Andes (Venezuela, Colombia, Ecuador and northern Peru)

Encompasses all the areas of *páramo* in north-western South America.

LUTEYN, J. *et al.*, 1999. *Páramos: a checklist of plant diversity, geographical distribution, and botanical literature*. xv, 278 pp., 20 halftone pls., 6 col. pls., maps (end-papers) (Mem. New York Bot. Gard. 84). New York.

Comprehensive introduction to *páramo* geography, climate, rainfall, soils, paleohistory, vegetation, biological adaptations and growth forms, floristics and phytogeography, fauna, and direct and indirect human influence; checklists of 1298 lichenized fungi and bryophytes and (pp. 74–135) 3399 vascular plants, the taxon entries in the latter with indication of country and altitudinal range and bibliographic sources; gazetteer (pp. 137–225) and general geographical and botanical bibliographies (annotated by codes) followed by a list of genera and families with species numbers as appendix 1. [Covers all known areas of *páramo* from 11° N to 8° S, thus ranging from Costa Rica and Venezuela to northern Peru. Builds on earlier studies by J. Cuatrecasas – from whose archive many of the photographs were taken – and L. E. Ruiz-Terán.]

Partial works

JØRGENSEN, P. M. and C. ULLOA ULLOA, 1994. *Seed plants of the High Andes of Ecuador – a checklist*. 443 pp., 22 text-figs. Aarhus/Quito. (AAU Reports 34.)

Enumeration of plants recorded from 2400 m or more in altitude with indication of habit and life-forms, distribu-

tion, altitudinal range, and (in smaller type) *exsiccatae* with provinces; some vernacular names and, under genera and families, references to standard treatments; appendix (with climate diagrams) and index to vernacular names, genera and families at end. An extensive introductory part furnishes coverage of vegetation physiognomy, floristics and biogeography as well as general references.

ULLOA ULLOA, C. and P. M. JØRGENSEN, 1993. *Árboles y arbustos de los Andes del Ecuador*. xx, 264 pp. Aarhus. (AAU Reports 30.)

Alphabetically arranged generic flora with keys, descriptions, Spanish vernacular names, species lists, and notes; references to key literature throughout and indices at end to all generic and family names as well as to vernacular names. An appendix depicts, in tabular form, altitudinal ranges of all genera scaled by increasing minimum altitude. The introductory part covers botanical history, geography, climate, human modifications, the scope and plan of the work, vegetation formations and history, and floristics and biogeography.

VARESCHI, V., 1970. *Flora de los páramos de Venezuela*. 429 pp., 126 figs. (some col.). Mérida: Ediciones del Rectorado, Universidad de los Andes.

Illustrated field-manual of the alpine flora, comprising a keyed, partly descriptive enumeration of vascular (pp. 89–408) and non-vascular plants, with brief notes on distribution, altitudinal zonation, habitat, frequency, and special features as well as vernacular names (in Spanish, German and English); list of references (pp. 409–411) and indices to vernacular and to family and generic names at end. The introductory section deals with physical features, soils, aspect effects, geology, etc., of the region and its special character, factors influencing the flora, and life-forms (biotypes), followed by a primary key (pp 45–55).

II. Central and southern Andes (central Peru southwards)

Covers the Andes from central Peru to Fuegia and encompasses all areas of the *puna* formation.

Partial works
In addition to the following, reference may usefully also be made to A. HOFFMANN *et al.*, 1998. *Plantas altoandinas en la flore silvestre de Chile*. 281 pp., col. illus. Santiago: Editiones 'Fundación Claudio Gay'. The work is similar in format to the senior author's popular guides for central and southern Chile (see **390**). A history of botanical work and list of references are also included; unfortunately, a map is wanting.

DIMITRI, M. J., 1974. *Pequeña flora ilustrada de los parques nacionales andino-patagónicos*. 122 pp., illus. (Anales Parques Nacionales 13). Buenos Aires.

Illustrated introductory guide to the more characteristic or frequently encountered vascular plants in six national parks (south of the 38th parallel and including Nahuel Huapi), organized in the first instance by life-form; the descriptive text accompanying the figures includes Spanish vernacular names, botanical details, localities, and ecology. The introduction includes sections on geography, basic botany, plant life-forms, floristics, and human influences, along with artificial keys, and a glossary, index to all names, and references (p. 122) appear at the end. [An earlier version of this work appeared in mimeographed form in 1960. A projected *Flora andino-patagónica* by the same author has so far not been published.]

FRIES, R. E., 1905. *Zur Kenntnis der alpinen Flora im nördlichen Argentinien*. 205 pp., 2 figs., 9 pls., map (Nova Acta Regiae Soc. Sci. Upsal., IV, 1(1)). Uppsala.

Systematic enumeration of vascular plants from the alpine regions of the provinces of Jujuy, Salta (northern part) and Los Andes (in part), based on 1901–02 field work by the author; includes synonymy, references, localities with *exsiccatae*, and summary of overall range; list of references and index at end. The introduction includes sections on climate, plant formations, and phytogeography. Much critical commentary on distinguishing features, variability, growth forms, etc., of individual taxa is also included.

305
Drylands

Extensive areas of steppe and desert occur in South America but at present there are few if any substantial recent floristic works on these formations. A recent general review of floristics and vegetation of the Atacama and the *lomas* formations along the Pacific coast of Chile and Peru provides a useful introduction to this driest part of the continent: P. W. RUNDEL *et al.*, 1991. The phytogeography and ecology of the coastal Atacama and Peruvian deserts. *Aliso* 13: 1–49, illus., maps. The authors consider that some 1000 species are represented (including two endemic or nearly endemic families), with greater richness in Peru.

308
Wetlands

Only one general work on the aquatic plants of South America (with special reference to Brazil) has been published; although described below, it is, however, not strictly a flora. There are four others of national or more limited scope: A. L. CABRERA and H. A. FABRIS, 1948. *Plantas acuáticas de la provincia de Buenos Aires.* 131 pp., 55 pls. (Publicaciones Técnicas, Dirección Agropecuaria (Buenos Aires) 5(2)). La Plata; F. KAHN, B. LEÓN and K. R. YOUNG (eds.), 1993. *Las plantas vasculares en las aguas continentales del Perú.* 357 pp., 67 figs. (partly halftones). Lima: Instituto Francés de Estudios Andinos (as *Travaux de l'Institut Français d'Études Andines* 75); A. LOMBARDO, 1970. *Las plantas acuáticas y las plantas florales.* 293, [1] pp., text-fig., 19 col. pls. Montevideo (aquatic plants, pp. 13–64); and J. VELÁSQUEZ, 1994. *Plantas acuáticas vasculares de Venezuela.* 992 pp., illus., maps. Caracas: Universidad Central de Venezuela. Local specialists also consider some of the aquatic floras of the United States (**108**) to be useful in South America. Two bibliographies relating respectively to the extratropical and infratropical zones have also been published (Hurlbert *et al.*, 1977, 1981; see below). A popular, pocket-sized, non-taxonomic introduction to aquatic plant vegetation and diversity (including mangal) is A. L. CABRERA, 1964. *Las plantas acuáticas.* 93 pp., illus. Buenos Aires: Editorial Universitaria de Buenos Aires.

Bibliographies

HURLBERT, S. H. (ed.), 1977. *Aquatic biota of southern South America / Biota acuática de Sudamerica austral.* xiv, 342 pp. San Diego: San Diego State University (for the editor). [Vascular plants by Nuncia M. Tur, pp. 37–45.]

HURLBERT, S. H., G. RODRÍGUEZ and N. D. DOS SANTOS (eds.), 1981. *Aquatic biota of tropical South America / Biota acuática de Sudamerica tropical.* 2 parts. San Diego. [Vascular plants by G. Agostini, pp. 50–68. Floras and other works are given therein under separate headings.]

HOEHNE, F. C., 1948 (1955). *Plantas aquáticas.* 168 pp., 81 pls. (some col.), frontisp.. São Paulo: Instituto de Botânica. (Publicação da Serie D.)

Conceived by its author as a 'rapidíssimo trabalho', this is a discursive systematic account of non-vascular and vascular hydrophytes in the style of

Kerner von Marilaun's *Pflanzenleben* or of Engler's *Pflanzenwelt Afrikas* (**500**), without keys but with notes on occurrence, special features, local references, etc., and excellent illustrations; index to scientific names at end. The introductory section covers ecology and sub-divisions of aquatic plants (with summaries of classifications by Warming and Clements) as well as their special characteristics, geographical distribution, and uses.

309

Oceanic littoral (including the *restingas* of Brazil)

In addition to mangrove, this unit encompasses the vegetation on coastal sands known in eastern Brazil as *restinga* (a term not yet taken up in other parts of the tropics and subtropics). The most detailed floristic treatment available, *Flora ecológica de restingas do sudeste do Brasil* by F. Segadas-Vianna and others, is listed under **350/IV**. A general study is L. D. DE LACERDA, R. CERQUEIRA and B. TURCQ (eds.), 1984. *Restingas: origem, estrutura, processos.* Niterói: Universidade Federal Fluminense/CEUFF.

Region 31

Venezuela and the Guianas

This region consists of French Guiana, Surinam, Guyana (formerly British Guiana), and Venezuela. The islands to the north are all treated as Regions 28 and 29.

All countries have had a long history of floristic exploration, beginning seriously in the eighteenth century. However, after 200 years or more they remain imperfectly known due to the size of the flora, the diverse vegetation, the relatively high percentage of endemism – a striking feature related to the presence of the Guayana Shield under much of the region – along with the comparatively limited accessibility of many areas (for example, some of the mesa-like summits, or tepuis, in Venezuelan Guayana can be reached only by helicopter) and a paucity of research resources.

Early contributors to Venezuelan botanical knowledge included Linnaeus's student Pehr Löfling (1754–56) and, from the end of the century, Alexander von Humboldt, Aimé Bonpland, Jean-Jules Linden, Funck and Schlim, Herman Karsten, August Fendler, Richard Spruce, and Johann Moritz all collected extensively, with Spruce mostly in what is now Territorio Federal Amazonas. Late in the nineteenth century, the Danish botanist and ecologist Eugen Warming made collections in the north, and Adolfo Ernst – the first resident professional in Caracas – collected on the Caribbean islands (see Regions 28 and 29) as well as elsewhere. The modern era is generally agreed to have begun with the arrival of Henri Pittier in 1913. In 1917 he published a first general key, *Clave analítica de las familias de plantas fanerógamas de Venezuela*, and in 1921 founded the first local herbarium (now the National Herbarium in Caracas). Collecting has gradually passed into the hands of resident botanists, although many outside workers have continued to prosecute field work, particularly in the Guayana region (Territorio Federal Amazonas and Estado de Bolívar).

Serious exploration of Venezuelan Guayana began only in the 1920s. It was initiated by the New York Botanical Garden as part of a coordinated inter-institutional research programme on the flora of northern South America begun under H. A. Gleason at the end of World War I. Some of the early expeditions were mounted in collaboration with the American Museum of Natural History. During World War II, work was extended to the mountains of Guyana and the Tafelberg and other mountains in Surinam under the sponsorship of the United States Foreign Economic Administration.

After the war, exploration was resumed by the Garden under the direction of Bassett Maguire. Maguire conducted several expeditions to various tepuis and other summits from 1947 to 1966, and in 1953 was the discoverer of the Neblina massif, rising to nearly 3000 m on the Brazilian border. He was joined in the 1950s by Julian Steyermark, who had begun his own explorations during the war and would continue so until 1985. The last two decades have been most notable for the UNESCO Man and Biosphere (MAB) project in the San Carlos de Río Negro area (visited by Spruce some 130 years before), Otto Huber's detailed explorations of several tepuis and other peaks, and the

1984–85 multidisciplinary expedition to the Neblina massif under the direction of Charles Brewer-Carías. In 1995, publication of a long-awaited definitive flora of the region, *Flora of the Venezuelan Guayana*, commenced.

Exploration and floristic study has also been carried out in other parts of Venezuela as interests and needs have dictated; for some areas the flora is now considered relatively well collected.[14] This belief probably contributed to a decision to launch a definitive *Flora de Venezuela*, publication of which began in 1964; as of the mid-1990s it was still incomplete, although several large families had been published. With a weak economy and limited local resources, a successful conclusion to this work would appear distant; at the same time, some already-published parts are certainly out of date.

In the Guianas, distinct polities from the early modern period to the present, botanical progress has been fragmented, sometimes with nomenclatural consequences. Until recent years, despite visits by scores of collectors, few have visited more than one polity. At present, parts of Guyana and the interior of French Guiana are perhaps the least well known. The latter is currently the object of a flora project.

All three countries were visited by several collectors in the eighteenth and early nineteenth centuries. French Guiana, and to a lesser extent Surinam, were the first to be explored in some detail. The former was made known particularly through Fusée Aublet's 1762–64 collections, published in 1775 as *Histoire des Plantes de la Guyane Françoise* – a work in which many large, well-known South American genera were described for the first time. Hard on Aublet's heels came *Plantae surinamensis* (1775), Linnaeus's last work, based on collections by F. Allamand and D. Rolander. These works stimulated much further collecting up to the mid-nineteenth century, some of it by missionaries and other residents, and two additional floras were published: *Primitiae florae essequeboensis* (1818) by G. F. W. Meyer, developing on work by Moravian missionaries, and *Stirpes surinamenses selectae* (1850) by F. A. W. Miquel. From Guyana, however, significant collections were not forthcoming until Robert and Richard Schomburgk's visits in 1835–44. These latter were written up – with specialist assistance – by the latter in *Versuch einer Fauna und Flora von Britisch-Guiana* (1848). In the latter part of the century, collecting in the Guianas declined except in Guyana, where G. S.

Jenman, Government Botanist from 1879 to 1902, obtained some 7000 numbers and E. F. im Thurn and, later, McConnell and Quelch, undertook the first ascents of Mt. Roraima – the first of the Guayana tepuis to be botanically explored. Im Thurn also established a Forestry Service (1887).

Activities in the Guianas after 1900 at first developed most strongly in Surinam. From Utrecht University, F. A. F. C. Went and his student August A. Pulle made their initial visits which laid a basis for a future comprehensive research programme in the Dutch territory which, with modifications, has continued to the present time. The work of these botanists and their predecessors and contemporaries was first consolidated in Pulle's *Enumeration* of 1906. From 1904, their studies were reinforced by the establishment of a Forestry Service, which soon collected extensively and vigorously. By the late 1920s, collections were thought to be sufficiently numerous to go ahead with a 'research' flora. The first part of the *Flora of Suriname* actually to appear was Posthumus's *Ferns of Surinam* (1928); in 1932, the main work, edited by Pulle, began publication. In Guyana, following a hiatus left after the death of Jenman and aided by Noel Y. Sandwith, botanist for tropical America at Kew, the Government Botanist and Forest Service after World War I renewed serious study of the flora and vegetation. Their work was supplemented by the New York Botanical Garden floristic survey programme for northern South America, beginning in 1921 with a trip by Henry A. Gleason and shortly afterwards engaging J. S. de la Cruz as a contract collector. Staff and associates from that institution have continued from time to time to collect there and elsewhere in the Guianas, most recently in particular Scott Mori (a co-author of *Guide to the vascular plants of central French Guiana*; see **312, Partial works**). Other visitors included E. H. Graham, author of *Flora of the Kartabo region* (1934), and the noted forest ecologist Paul Richards. French Guiana was comparatively neglected until after World War II; however, with the advent firstly of a Forestry Service and, in 1960, a research center at Cayenne under the metropolitan Office de la Recherche Scientifique et Technique d'Outre-Mer (ORSTOM) came considerable opportunities for new collecting and field research.[15]

In 1983, floristic efforts in the three polities were pooled and, although far advanced, the *Flora of Surinam* was wound up in favor of a new project, the *Flora of the Guianas*. This latter is presently a novagonal undertaking

(Utrecht, Paris, New York, Kew, Berlin, the Smithsonian, the University of Guyana at Georgetown, the University of Surinam in Paramaribo, and ORSTOM-Cayenne); its editorial center is in Utrecht. By 1998, a goodly number of fascicles of the new flora had been published. Completion was in the mid-1980s estimated to require 20–30 years. Associated specialized publications have already appeared or are planned, and precursory papers in various serials have been collected and distributed in decades as *Studies on the flora of the Guianas*; 90 such had been issued through 2000. A related programme is 'Biological Diversity of the Guianas', set up between the University of Guyana and the Smithsonian Institution; it maintains a taxonomic database among other activities.

Recent publications on the Venezuelan flora include, apart from the already-mentioned *Flora de Venezuela, Botany of the Guayana Highland* (see **301/Ib**), Steyermark's five-part *Contributions to the flora of Venezuela* (1951–57, 1966), and two now very incomplete checklists by Knuth (1926–28) and Pittier (1945–47). Analytical keys and a treatment of the woody flora, the latter by Aristeguieta (1973), are also available. Some partial floras of extra-Guayanan Venezuela have also been published: the excellent *Flora del Ávila* (Steyermark and O. Huber, 1978), the first part of *Gehölzflora der Anden von Mérida* (H. Huber, 1977), *Flora de los páramos de Venezuela* (Vareschi, 1970; see **303/I**), *Flora de la Isla Margarita* (Hoyos, 1985; see **292**), and *Flora y vegetación del estado Táchira* (Bono, 1996). The entire area south of the Orinoco is being covered in *Flora of the Venezuelan Guayana* (1995–). Several additional floras are in progress or planned (Huber *et al.*, 1998; see below).

The major current floras of the Guianas are *Flora of the Guianas* (1985–) and its predecessor, *Flora of Surinam* (1932–86). There have also been two editions of a checklist published under the auspices of the Biological Diversity of the Guianas Program. For French Guiana, an enumeration, *Inventaire taxonomique des plantes de la Guyane française*, began publication in 1990; seven parts have so far appeared. This has now been complemented by *Guide to the vascular plants of central French Guiana* (vol. 1, 1997). Older though still-current works of more limited scope include the already-mentioned *Versuch* of Richard Schomburgk (1848), *Enumeration of the vascular plants known from Surinam* (Pulle, 1906), the largely compiled *Flore de la Guyane française* (Lemée, 1952–56),

Flora of the Kartabo region (Graham, 1934) and *Contributions to the botany of Guiana* (Maguire *et al.*, 1966). The already-mentioned *Botany of the Guayana Highland* as well as *Flora of the Venezuelan Guayana* (1995– ; see **316**) may also usefully be consulted, especially for Guyana.

Timber has long been a not insignificant part of Guianan economies and, as already noted, forest services have been long established. Forest floras include *Bomenboek voor Suriname* (Lindeman and Mennega, 1963), based on extensive collecting and other work by the Surinam Forestry Service but covering only larger forest trees; *Essences forestières de Guyane* (Bena, 1960), similar in scope but somewhat less thoroughly founded and effectively covering only the coastal fringe; and *Checklist of the indigenous woody plants of British Guiana* (Fanshawe, 1949; revised and reissued in 1988 by the Netherlands-based TROPENBOS Foundation).

Progress
Venezuelan history is reviewed in the introduction to *Manual de las plantas usuales de Venezuela y su suplemento* by Pittier (under **315**: Woody plants). A recent comprehensive review is O. HUBER *et al.*, 1998. *Estado actual del conocimiento de la flora en Venezuela*. 153 pp., maps. Caracas: Fundación Instituto Botánico de Venezuela. Early work in Surinam was reviewed by Pulle in his *Enumeration* (**313**) and again as A. A. PULLE, 1939. Explorações botânicas de Surinam. In *Anais I. Reunião Sul-Americana de Botânica* (Rio de Janeiro, 1938), 1: 239–257. Other substantial reviews of floristic work include O. HUBER and D. FRAME, 1989. Venezuela, and J. C. LINDEMAN and S. A. MORI, 1989. The Guianas. Pp. 362–374 and 375–390 respectively in *Floristic inventory of tropical countries* (see **General references**).

Flora of the Guianas: The last-named undertaking was first outlined by its coordinator, A. R. A. ('Ara') Görts-van Rijn, in *Taxon* 33: 371–372 (1984). Reports of meetings and notices related to that project appear in *Flora of the Guianas Newsletter* of which 12 numbers had appeared by 1999. Two editions of a general checklist have also been published, respectively in 1992 and 1997.[16]

Bibliographies. General and divisional bibliographies as for Division 3.

Indices. General and divisional indices as for Division 3.

311

The Guianas (in general)

See also **313** (more recent parts of *Flora of Surinam*; POSTHUMUS for pteridophytes). – Area: 470 950 km². Vascular flora: at least 8000 species.

Until the launch in 1985 of *Flora of the Guianas*, the fern flora of Posthumus and more recent parts of the *Flora of Surinam* were the only fully inclusive works. A not inconsiderable number of families in the first-named have now been published. An additional relatively recent document, also described below, is Bassett Maguire's *Contributions to the botany of Guiana* (1966). A full checklist is additionally available.

BOGGAN, J. *et al.*, 1997. *Checklist of the plants of the Guianas (Guyana, Surinam, French Guiana)*. 2nd edn. v, 238 pp., 2 maps. Washington, D.C.: Biological Diversity of the Guianas Program, Department of Botany, Smithsonian Institution. (Publ. 30. 1st edn., 1992.)

Nomenclator with indication of presence or absence in each of the three polities, separate synonymy, and (under family headings) key references (including *Flora of the Guianas* and other major works) all arranged alphabetically by families (within major taxa); addresses of contributors and generic and family indices at end. The introductory part (by V. A. Funk) includes summaries (with maps) of geology, vegetation types, current research and extension programmes, and an explanation of the work along with general references. [Covers 9200 species of green plants from charophytes through bryophytes to tracheophytes. There are, however, no statistical tables.]

GÖRTS-VAN RIJN, A. R. A. (ed.), 1985– . *Flora of the Guianas*. In fascicles. Illus. (part col.). Königstein/Ts., Germany: Koeltz (from 1995 through the Royal Botanic Gardens, Kew).

Documentary 'research' flora (organized into five series corresponding to the fungi and major plant groups, with Series A devoted to phanerogams and B to pteridophytes); each fascicle, devoted to a given family or subfamilial taxon, includes keys, synonymy with references and citations, typification, generalized indication of distribution, taxonomic commentary, notes on habitat, frequency, and special features (including wood anatomy), representative illustra-

tions, and (at end) vernacular names, uses and other ethnobotanical information, lists of *exsiccatae*, and index. Key literature is listed at the beginning of each family or subfamilial taxon. [As of 1999, 21 fascicles of seed plants and three of pteridophytes had been published, each covering one or more families. Within the angiosperms the delimitation of families follows the Cronquist system.]

Partial work

MAGUIRE, B. and collaborators, 1966. Contributions to the botany of Guiana, I–IV. *Mem. New York Bot. Gard.* **15**: 50–128.

Comprises a series of revisionary treatments of families and genera, with descriptions of new taxa, synonymy, localities with *exsiccatae*, critical notes, and figures of representative species. Also included are some records of plants from adjacent parts of Brazil, including Amapá.

312

French Guiana

For additional coverage, see **311** (GÖRTS-VAN RIJN; MAGUIRE); **313** (POSTHUMUS; PULLE *et al.*). – Area: 91 000 km². Vascular flora: *c.* 5000 species (Cremers in PD; see **General references**).

The *Inventaire* is now well advanced, while in 1997 there appeared the first of the two planned volumes of *Guide to the vascular plants of central French Guiana*, a major product of a detailed floristico-ecological study around the central interior upland base of Saül – an area scarcely covered in Lemée's flora. Also of use may be G. CREMERS, 1988. *Liste des espèces de phanérogames et de ptéridophytes de Guyane française d'apres d'herbier de Centre ORSTOM de Cayenne*. Cayenne: ORSTOM.[17]

CREMERS, G., M. HOFF and collaborators, 1990– . *Inventaire taxonomique des plantes de la Guyane française*, I– . Illus., maps. Paris: Muséum National d'Histoire Naturelle. (Secrétariat de la Faune et de la Flore, Inventaires de Faune et de Flore, 54 (part I); Collection Patrimoines Naturelles, 7, *passim* (parts II–).)

A checklist of native, introduced and cultivated species, with each part organized in several *Listes*; the first covers accepted taxa with life-form, synonymy, generalized distribution, habitat, and a dot map while the second is a lexicon of synonyms. The remaining lists cover species according to vegetation and to biological classes, collectors, verifiers, a gazetteer with coordinates and number of known specimens for each locality, habitats with species numbers, and a bibliography of key works. [318 species of pteridophytes and all of the monocotyledons (about 1225 species) have been published in parts I (1990) to IV (1994). Parts V (1995), VI (1997) and VII (1998) respectively cover dicotyledons from Acanthaceae to Bixaceae, Bombacaceae to Combretaceae, and Connaraceae to Fabaceae.][18]

LEMÉE, A., 1952–56. *Flore de la Guyane française.* 4 vols. 3 frontisps. Paris: Lechevalier.

A largely compiled, briefly descriptive flora of vascular plants, without keys; includes limited synonymy (with references), general indication of local range, miscellaneous notes, and indices to botanical names in each volume. A bibliography is provided in vol. 2 and the introductory section in vol. 1 includes a key to families. Volume 4 encompasses addenda to the main work as well as an annotated, separately indexed flora of useful plants. [Only the coastal districts up to about 40 km inland are effectively covered; however, some species from Amapá (Brazil) and Surinam are also included.][19]

Partial work: central French Guiana

MORI, S. A. *et al.*, 1997. *Guide to the vascular plants of central French Guiana. Part 1: Pteridophytes, gymnosperms and monocotyledons.* 422 pp., 72 col. pls., 168 full-page text-figs. (incl. maps) (Mem. New York Bot. Gard. 76(1)). New York.

Lavishly illustrated descriptive manual with keys, very limited synonymy, French names (where known), indication of phenology, and brief notes on biology and habitat but no specific localities or *exsiccatae*; illustrated glossary, acknowledgments, bibliography, synopses of families, and indices to all botanical names and illustrations at end. The general part is largely devoted to distinguishing features in aid of identification (with figures); the rest is on the plan of the work. [The area covered is roughly 1402.5 km[2] centering on the inland base of Saül and is largely at 200–400 m (occasionally reaching over 500 m with some cloud-forest, e.g., in the Monts La Fumée.][20]

Woody plants (including trees): guide

BÉNA, P., 1960. *Essences forestières de Guyane.* vii, 488 pp., text-figs., 10 pls., 4 maps. Paris: Imprimerie Nationale.

Illustrated foresters' manual of the more important trees, without keys; includes synonymy, vernacular names, general indication of local range, notes on habitat, field char-

acteristics, phenology, timber properties, uses, etc., and index to botanical and vernacular names. The introductory section incorporates a bibliography. [As with Lemée's flora (see above), this work limits itself largely to the coastal districts.]

313

Surinam

See also **311** (GÖRTS-VAN RIJN; MAGUIRE). – Area: 165 942 km[2]. Vascular flora: 4500 species (Lindeman in PD; see **General references**).

More recent parts of the *Flora of Surinam* in addition cover adjacent French Guiana and Guyana. At its conclusion in the mid-1980s, the *Flora* was about 80 percent complete, with 137 of the 173 known families revised and many pre-World War II treatments furnished with addenda and corrigenda. A summary of *Flora* families appears in *Floristic inventory of tropical countries* (see **General references**), p. 386. To these may be added Marga Werkhoven's enthusiast work, *Orchideeën van Suriname* (1986), a family never published in the *Flora*.

Of other works on the flora, the first part of an introductory guide by August Pulle, *Zakflora voor Suriname*, appeared in 1911. This, however, did not proceed beyond a rather technical key to families and genera. A more recent, illustrated introduction to the flora is *Fa joe kan tak' mi no moi / Inleiding in de flora en vegetatie van Suriname* by J. G. Wessels Boer, W. H. A. Hekking and J. P. Schulz (1976).

Bibliography

TEUNISSEN, P. A. and M. C. M. WERKHOVEN, 1979. *Biological bibliography of Suriname.* Paramaribo: I.O.L. [Not seen.]

PULLE, A. A., 1906. *An enumeration of the vascular plants known from Surinam.* 8, 555 pp., 17 pls., map. Leiden: Brill.

Systematic enumeration of known vascular plants (2101 species), with descriptions of new taxa; includes limited synonymy (with references), vernacular names, localities with *exsiccatae*, summary of extralimital range, short critical notes, and (at end) summary comments on the flora and its phytogeographical relationships, a bibliography, and indices to vernacular and botanical names. The introductory section includes an account of botanical exploration.

PULLE, A. A., J. LANJOUW, A. L. STOFFERS and J. C. LINDEMAN (eds.), 1932–86. *Flora of Suriname.* Vols. 1–6. Illus., map. Amsterdam: Foundation 'van Eedenfonds'/Koloniaal Institut (1932–57, vols. 1–4 in part. Also published as Koloniaal Institut te Amsterdam, Afd. Handelsmuseum, Mededeelingen 11); Leiden: Brill, for Foundation 'van Eedenfonds' (1964–86 as *Flora of Surinam*, vols. 5–6 in part and continuations of vols. 1–3). (Early parts or portions thereof in each of vols. 1–4, except the first (and only) installments of vol. 4(2), reprinted 1966–86 with additional indices, Leiden: Brill.) Complemented by J. LANJOUW, 1935. Additions to Pulle's *Flora of Surinam. Recueil Trav. Bot. Néerl.* **32**: 215–261. For ferns, see O. POSTHUMUS, 1928. *The ferns of Surinam and of French and British Guiana.* [ii], 196 pp. Pasoeroean, Java: The author (also issued 1928, Utrecht, as *Flora of Surinam*, vol. 1, Supplement).

A detailed 'research' flora of vascular plants and bryophytes; includes keys to genera and species, full synonymy, references and citations (including listing of standard monographs and revisions under family and generic headings), vernacular names, localities with *exsiccatae*, general indication of extralimital range, taxonomic commentary, miscellaneous notes; index to botanical names at end of each part. Posthumus's now largely obsolete treatment of the ferns (and fern-allies) is briefly descriptive, with keys to all taxa, synonymy (with references), localities with *exsiccatae*, extralimital range, notes on features of the fern flora, phytogeographic relationships, and index.[21]

Pteridophytes

POSTHUMUS's *Ferns of Surinam* (see above) succeeded F. L. Splitgerber's *Enumeratio filicum et lycopodiacearum quas in Surinamo legit F. L. Splitgerber* (1840, Leiden; in *Tijdschr. Natuurl. Gesch. Physiol.* 7).

KRAMER, K. U., 1978. *The pteridophytes of Surinam: an enumeration with keys of the ferns and fern-allies.* 198 pp. (Uitgaven Natuurw. Studiekring Suriname Ned. Antillen 93). Utrecht.

Keyed enumeration with synonymy, references, selected *exsiccatae* (and number of available collections), general indication of local and extralimital distribution, taxonomic commentary, and notes on habitat, frequency, etc., and (at end) an index. A special polyclave to genera is also included, and recent monographs and revisions are cited in the text. [Designed in part for field use.]

Woody plants (including trees): guide

LINDEMAN, J. C. and A. M. W. MENNEGA, 1963. *Bomenboek voor Suriname.* 312 pp., illus. (1 col.) (Meded. Bot. Mus. Herb. Rijks Univ. Utrecht 200). Paramaribo: 's Lands Bosbeheer Suriname; Utrecht.

Copiously illustrated foresters' manual of the more important trees of Surinam; includes two general keys to species (based respectively on foliar, twig and bark attributes and on wood structure), essential synonymy, vernacular names, local range, notes on uses, special features and timber properties, bibliography, and indices to vernacular and botanical names.

The introductory section includes an illustrated glossary and explanation of botanical and wood-anatomical terms. General summaries of the work in Spanish and English are also provided.

314

Guyana

Area: 214 970 km^2. Vascular flora: no data. – For alternative coverage of what was formerly British Guiana, see **311** (all works); **313** (PULLE *et al.*); **316** (STEYERMARK, BERRY and HOLST).

The checklist by Fanshawe received only limited circulation when originally issued, but was revised and republished in 1988 by the Dutch TROPENBOS Foundation. Some current floristic research is conducted under the Biological Diversity of the Guianas Program of the University of Guyana and the Smithsonian Institution.

SCHOMBURGK, M. R., 1848. Versuch einer Fauna und Flora von Britisch-Guiana. In *idem, Reisen in Britisch-Guiana in den Jahren 1840–44*, 3: 563–1260. Leipzig: Weber.

Includes (pp. 787–1212) four systematic lists (based mainly upon the author's collections) of non-vascular and vascular plants with limited synonymy, local range, notes on phenology and habitat, and index to all taxa (pp. 1226–1260). The four lists are based on habitat (Coastal region; Primary forest; Sandstone region; Savanna). The introductory section includes general remarks on physical features, vegetation, and phytogeographical relationships.

Partial work

Of the few florulas in Guyana, the most significant is E. H. Graham's flora of the Kartabo region in the north-central part, west of the Essequibo and southwest of Georgetown. Its keys are particularly useful.

GRAHAM, E. H., 1934. Flora of the Kartabo region, British Guiana. *Ann. Carnegie Mus.* **22**: 17–292, 2 text-figs. (maps), pls. 3–18.

Descriptive flora of vascular plants, with complete keys, synonymy, localities with *exsiccatae*, extralimital range, miscellaneous notes, and index; list of references included. An introductory section deals with general natural history of the area and with botanical exploration.

Woody plants (including trees): guide

The checklist by Mennega, Tammens-de Rooij and Jansen-Jacobs (1988) was based on and is a successor to *Check-list of the indigenous woody plants of British Guiana* (1949) by D. B. Fanshawe.

MENNEGA, E. A., W. C. M. TAMMENS-DE ROOIJ and M. J. JANSEN-JACOBS (eds.), 1988. *Check-list of woody plants of Guyana*. 281 pp. (Tropenbos Technical Series 2). Ede: Tropenbos Foundation.

Comprises an alphabetically arranged checklist, with Arawak, English and other vernacular names and notes on habitat, frequency, distinguishing features, etc.; lexicon of vernacular names, with botanical equivalents. An introductory section includes technical notes and a list of references.

315
Venezuela (in general)

See also **301/II** (all works); **303** (VARESCHI). – Area: 912 050 km². Vascular flora: 20 000–35 000 species (Steyermark, 1977). Only general Venezuelan works are listed here; the various state and local manuals which have appeared in recent years are listed at **316** through **319**.[22]

Knuth's somewhat uncritical *Initia florae venezuelensis*, the first separate national flora or enumeration, is now way out of date, as is Pittier's *Catálogo*; however, a *Nuevo Catálogo de la Flora de Venezuela* under the editorship of Giovannina Orsini is in prospect. The *Flora de Venezuela* is still far from completion, although treatments of some large and important families are available. Worthy also of note, though not strictly a flora, is L. SCHNEE, 1973. *Plantas comunes de Venezuela*. 2nd edn. 822 pp. Maracay: Instituto de Botánica Agricola, Univ. Central de Venezuela. (1st edn., 1960.)

Bibliography

A useful bibliography is found in *Estado actual del conocimiento de la flora en Venezuela* (see **Region 31, Progress**).

Keys to families and genera

BADILLO, V. M. and L. SCHNEE, 1965. *Clave de las familias de plantas superiores de Venezuela*. 4th edn. 255 pp. (Revista Fac. Agron. (Maracay) 9). Maracay, Venezuela. (1st edn., 1951.) [Analytical keys, based on Franz Thonner's *Anleitung zum Bestimmen der Familien der Blütenpflanzen* (1917; see **000**, p. 95, with descriptions of families; glossary and index.]

PITTIER, H., 1939. *Genera plantarum venezuelensium. Clave analítica de los géneros de plantas hasta hoy conocidos en Venezuela*. 354 pp. Caracas: Tipografia Americana. [Analytical keys to all Venezuelan genera of vascular plants, with index. Based in part on the author's *Clave analítica de las familias de plantas fanerógamas de Venezuela y partes adyacentes de la América tropical* (1917; see **301/I**).]

RICARDI S., M. H., 1988 (1989). *Familias de monocotiledóneas Venezolanas*. 119 pp., 42 pls. Mérida: Universidad de Los Andes. (Ciencias Básicas 3008.) Complemented by *idem*, 1992. *Familias de dicotiledóneas Venezolanas*. 2 vols. 170, 192 pp., illus. Mérida: Centro Jardín Botánico. [Students' manuals with keys and detailed ordinal and family descriptions, their circumscriptions following the 1980 version of the Takhtajan system; texts include world and Venezuelan distribution, genera present in Venezuela, taxonomic commentary, and figures of representative species. Principal references and an index appear at the end of each volume. – The volumes on dicotyledons represent an amplification of the author's *Sinopsis de las familias de espermatófitos leñosos venezolanos* (1977).]

KNUTH, R., 1926–28. *Initia florae venezuelensis*. 768 pp., map (Repert. Spec. Nov. Regni Veg., Beih. 43). Berlin: Fedde.

A largely compiled systematic enumeration of known vascular plants, with synonymy, references, vernacular names, and geographically arranged *exsiccatae* with localities; cyclopedia of collectors (in German); index to genera at end. For additions, see K. SUESSENGUTH and R. BEYERLE, 1939. Ergänzungen zu den 'Initia florae venezuelensis' von R. Knuth. *Bot. Arch.* **39**: 373–381.

LASSER, T., Z. LUCES DE FEBRES, J. A. STEYERMARK and G. MORILLO (eds.), 1964– . *Flora de Venezuela*. Vols. 1– . Illus. Caracas: Instituto Botánico, Ministerio de Agricultura y Cria (MAC), Venezuela (from 1981 Fund. Ed. Amb., INPARQUES; currently Fundación Instituto Botánico de Venezuela).

Illustrated 'research' flora of vascular plants; includes keys to genera and lower taxa, synonymy (with references) and citations, vernacular names, geographically arranged *exsiccatae* with localities, general indication of extralimital range, taxonomic commentary, notes on habitat, properties, uses, etc., and indices to vernacular and botanical names in each volume. [Of the 15 volumes planned, each with one or more parts, by 1998 1(1–2), 2(2), 3(1), 4(1), 4(2), 5(1), 8(1–4), 9(1–3), 10(1–2), 11(2), 12(1) and 15 had been published (the most recent in 1992). The 30 flowering plant families now available include, among others, the Begoniaceae, Bromeliaceae, Cactaceae, Cucurbitaceae, Orchidaceae, Piperaceae, Melastomataceae, Rubiaceae and Compositae. Additional families are scheduled for publication. A rough systematic order has been followed for the main work.][23]

PITTIER, H. *et al.*, 1945–47. *Catálogo de la flora venezolana.* 2 vols. 423, 577 pp. (Third Inter-American Conference on Agriculture, Caracas. Cuadernos Verdes: Serie nacional 20, 52). Caracas: Vargas.

Systematic enumeration of vascular plants, with keys to genera (and some species), vernacular names, citation of *exsiccatae* with localities, taxonomic commentary, notes on habitat, etc., and index to genera in each volume; general index to vernacular names in vol. 2.

Partial works

Apart from contributions on the Guayana Highland (301/Ib), the most important recent 'supplementary' work is that by Steyermark and others described below. It usefully bridges the enumerations of Knuth and Pittier and the modern *Flora de Venezuela*.

STEYERMARK, J. A. and collaborators, 1951–57. *Contributions to the flora of Venezuela.* 4 parts. vii, 1225 pp., 151 text-figs. (Fieldiana, Bot. 28). Chicago. Continued as *idem*, 1966. Contribuciones a la flora de Venezuela, 5. *Acta Bot. Venez.* 1(3/4): 9–256.

Systematically arranged descriptions of and critical notes on new or little-known taxa of non-vascular and vascular plants from the Guayana and other parts of Venezuela; includes a few species keys, synonymy (with references), *exsiccatae* with localities, and miscellaneous notes; general index at end. An introductory section provides an account of botanical exploration, with special reference to the author's itineraries of the 1940s and 1950s. Part 5 includes an index to the first four parts along with four papers respectively on the floras of Roraima, Duida, Ptari-tepui, and Cerro Turumiquire.

Woody plants (including trees): guides

A revision of the first of these works is in progress.

ARISTEGUIETA, L., 1973. *Familias y géneros de los árboles de Venezuela.* 845 pp. (Edición especial, Instituto Botánico). Caracas: Instituto Botánico, MAC. (A revised and expanded version of *idem*, 1944. *Clave y descripción de las familias de los árboles de Venezuela.* Caracas.)

A 'generic flora' of woody plants, with alphabetically arranged family accounts; includes artificial keys to families and genera, descriptions, vernacular names, and notes on number of species, representatives in Venezuela, uses, etc. Key cross-references are also given. [Designed in part as a students' manual.]

PITTIER, H., 1970. *Manual de las plantas usuales de Venezuela y su suplemento.* 2nd edn., revised by E. Mendoza. xxii, 620 pp., illus. Caracas: Fundación 'Eugenio Mendoza'. (Reprinted 1978. Original edn., 1926; Suplemento, 1939.)

Included here for a list (in the *Suplemento*) of known woody plants, with vernacular names.

Pteridophytes

There is also a treatment of pteridophytes of Falcón State by H. van der Werff and A. R. Smith in *Opera Bot.* **33** (1980; see **319**). The group was completely covered by V. Vareschi in *Flora de Venezuela*, 1 (1969; see above) but this has now become out of date; by the mid-1980s, 1059 species were known, 430 of them added since 1969 (Smith, 1985).

SMITH, A. R., 1985. *Pteridophytes of Venezuela, an annotated list.* vii, 254 pp. Berkeley, Calif.: The author. (Mimeographed.)

Introduction, with basic references; checklist with cross-references to Vareschi's *Flora* treatment where applicable, synonymy, altitudinal range, habitat, distribution, and representative *exsiccatae*; generic headings with standard revisionary and monographic references; excluded species. Page 254 contains 'recent additions'.

316

Venezuelan Guayana

See also **301/Ib** (all works). Comprises Territorio Federal Amazonas, Delta Amacuro, and Bolívar, or 45 percent of the area of Venezuela. Some *tepui* florulas were published by Steyermark in the 1960s and 1970s, preparatory to a more general work. This latter is now being realized as *Flora of the Venezuelan Guayana*; it was initiated by Steyermark and continues under the direction of Paul Berry and Bruce Holst. Publication began in 1995, with in all nine volumes planned. Some precursory notes have appeared as J. A. STEYERMARK *et al.*, 1984–89. Flora of the Venezuelan Guayana, I–VII. *Ann. Missouri Bot. Gard.* **71–76**, *passim*.

Bibliography

WEIBEZAHN, F. H. and B. E. JANSSEN-WEIBEZAHN, 1990. *El Territorio Federal Amazonas, Venezuela: una bibliografía/The Amazonas Federal Territory, Venezuela: a bibliography.* 294 pp., map (Scientia Guianae 1). Caracas. [In both Spanish and English; encompasses 1468 entries on all aspects of natural sciences; arranged by author with keywords and indexed by keywords. Also includes a description of the region (pp. 9–18) and source bibliographies (pp. 19–20).]

STEYERMARK, J. A., P. E. BERRY and B. K. HOLST (gen. eds.), 1995– . *Flora of the Venezuelan Guayana.* Vols. 1– . Illus., maps. St. Louis, Mo.: Missouri Botanical Garden; Portland, Oreg.: Timber Press.

Volumes 2 onwards comprise an alphabetically arranged illustrated descriptive flora of vascular plants with keys, synonymy, references and concise accounts encompassing diagnostic features, habitat, and local and extralimital distribution (but no vernacular names); index to all botanical names at end of each volume. Volume 1 is a collectively written general introduction with chapters on physical features, climate and soils, vegetation, floristics and biogeography, history of botanical exploration, and conservation; there is also a list of families, a general key, and a core bibliography. [As of 1999 five volumes had appeared, comprising the general introduction along with treatments of pteridophytes, gymnosperms and flowering plant families from Acanthaceae through Lentibulariaceae.][24]

317

Orinoco Basin and Llanos

Comprises the plain between the Andes and Coastal Ranges and the Orinoco, incorporating the states of Apure, Barinas, Portuguesa, Cojedes, Guárico, Anzoátegui and Monangas. Only an introductory work has been published; however, additional floras are in progress or planned.

RAMIA, M., 1974. *Plantas de las Sabanas Llaneras.* 287 pp., illus. Caracas: Editores 'Monte Ávila'.

Pocket semi-popular illustrated flora, covering more common or conspicuous species; includes brief descriptions, vernacular names, habitat, and other notes, with (at end) a lexicon (vernacular names with botanical equivalents, pp. 241–250) and index to botanical names.

318

The Andes and Coastal Ranges

Comprises the mountain ranges from the Colombian border to the Paria Peninsula; includes Táchira, Mérida, Trujillo, Lara, Yaracuy, Carabobo, Aragua, Miranda, and Sucre together with the Distrito Federal. Florulas are available for the Mérida Andes and for the Ávila range north of Caracas; the latter is especially recommended. A flora of Táchira has also been published.

BONO, G., 1996. *Flora y vegetación del estado Táchira, Venezuela.* 952 pp., text-fig., 93 col. pls. (photographs), maps (Monografie di Museo Regionale di Scienze Naturali, Torino 20). Turin.

The main part of this work (pp. 340–831) comprises an alphabetically arranged enumeration respectively of pteridophytes and seed plants with principal synonyms, distribution with localities and *exsiccatae* (without citation of places of deposit), and indication of habitat and habit; a detailed statistical survey is also given. The general part is detailed with expositions of physical features, climate, rainfall (and its distribution), phytogeography, endemism (with geographically centered lists), and vegetation (including floristic lists for several 'stations' at different altitudes along with a review of secondary formations). The photographic album, a bibliography (pp. 927–934) and index to the enumeration (pp. 935–949) conclude the work. [4076 species were recorded for an area of 11 117 km^2. The Táchira Depression, within the state, is biogeographically significant in being the northeastern limit for many widely ranging Andean species.]

HUBER, H., 1977. Gehölzflora der Anden von Mérida, I. *Mitt. Bot. Staatssamml. München* **13**: 1–127.

Keyed enumeration of woody plants of Mérida and parts of adjacent states; species entries include synonymy, references, *exsiccatae* with localities, critical notes, and remarks on habitat, occurrence, etc., along with typification and cross-references to monographs and revisions. A short introductory section is provided. Families and genera are alphabetically arranged, with coverage from Acanthaceae through Asteraceae. [Not continued.][25]

STEYERMARK, J. A. and O. HUBER, 1978. *Flora del Ávila*. 971 pp., 308 pls., photographs (some col.), folding map. Caracas: Sociedad Venezolana de Ciencias Naturales/Ministerio del Ambiente y de los Recursos Naturales Renovables, Venezuela.

Copiously illustrated, alphabetically arranged enumeration of vascular plants (1892 species); includes descriptive keys, synonymy, vernacular names, local range, notes on habitat, frequency, altitude, special features, properties, etc., and (in appendices) guides to determination by special features and an illustrated glossary along with a bibliography (pp. 923–933) and complete index. An introductory section gives details of physical features (the area reaches over 2500 m), climate, vegetation and phytogeography, and introduced species, with a floristic summary; this is followed by an illustrated key to families. [A supplement, by W. Meier, is planned.]

319

Lake Maracaibo region

This area comprises the states of Zulia and Falcón. A flora of Falcón, which includes the isolated, moist Cerro Santa Ana and Sierra de San Luis, was projected in the 1970s but the only result to date has been a revision of pteridophytes: H. VAN DER WERFF and A. R. SMITH, 1980. *Pteridophytes of the state of Falcón, Venezuela*. 34 pp. (Opera Bot. 56). Copenhagen. This covers 216 species and features keys, limited synonymy, habitats, distribution, and notes. A flora of Zulia is currently planned.

Region 32

Colombia and Ecuador

Included here are mainland Colombia and Ecuador, together with their immediately adjacent islands. Until the middle of the nineteenth century they constituted one country, New Granada, which comprised the greater part of the historical viceroyalty. The Pacific islands of Malpelo and the Galápagos are respectively designated as **016** and **017**, while the Caribbean islands of San Andrés and Providencía constitute **238**.

The botanical history of Colombia dates back to the latter part of the eighteenth century with the explorations of Nicholas J. von Jacquin (1755–59, particularly on the Caribbean coast), Pehr Löfling on the Orinoco and, later, José Celestino Mutis and his associates, among them Francisco José de Caldas and Juan A. Céspedes. Mutis was active in the country from 1760 but under royal patronage from 31 March 1783 became head of the 'Real Expedición Botánica del Nuevo Reyno de Granada' and remained so until his death in 1808 (the expedition officially continuing until 1816). He was thus in Colombia when the next notable expedition arrived, that of Humboldt and Bonpland in 1801–03. Yet, though numerous specimens were obtained and more than 6000 drawings made by 39 different artists, the major flora planned did not begin publication until 1954.[26]

A 'Museo de Ciencias Naturales' was organized in Bogotá in 1823 by Francisco Zea, an Expedición member, following a directive of Simon Bolívar; in 1827 it was incorporated into the then-Universidad Central in Bogotá and in 1855 José Triana organized its herbarium in the wake of his work on the *Comisión Corográfica*, mentioned below. A university reorganization in 1867 brought into existence an 'Escuela de Ciencias Naturales'. Little was there accomplished, however, until well into the twentieth century. With Triana mostly in Europe and renewed political instability botanists such as Santiago Cortés labored mainly on their own, as did also those at the Instituto de La Salle in Bogotá and its museum under Brother Apolinar María. From early in the twentieth century Medellín began its development as a botanical center in its own right through the work of Andrés Posada Arango, Joaquín A. Uribe, Lorenzo Uribe-Uribe, and Brother Daniel; from there also came Enrique Pérez Arbeláez, later author of *Plantas útiles de Colombia* (1936). The Uribes were responsible for the first regional flora in the country, *Flora de Antioquia* (1940).

The major contributions to knowledge of the Colombian flora came thus from abroad for more than a century following independence. More than twenty well-known collectors, nearly all European, were active throughout the remainder of the nineteenth century and beyond, some for extended periods. Their collections went for the most part to European herbaria; only those of Triana, botanist to *La Comisión Corográfica* (or the 'Segunda Expedición Botánica', organized in May 1849 under Agustín Codazzi and active until 1867), remained to any extent in the country. The Europeans were joined by North Americans from the 1850s, though their numbers remained few until after 1900. Among these latter were I. F. Holton in the 1850s and H. H. Smith in 1898–1901; Smith was especially active around the Sierra Nevada de Santa Marta.

The organization of the inter-institutional floristic programme for northern South America by Henry Gleason, already described under Region 31, led to a substantial increase in collecting; in turn this acted as a stimulus to Pérez Arbeláez and other Colombians. Notable contributors of the 1920s and 1930s included Ellsworth P. Killip, Francis Pennell and Albert C. Smith. It was Killip who began a long association between the country and the Smithsonian Institution. The botanists of the World War II-era U.S. Foreign Economic Administration and related agencies would during the 'Cinchona Expeditions' of the 1940s add over 20 000 more collections, and in the postwar years Richard E. Schultes (with Hernando García-Barriga and other Colombians) was active in the then still poorly collected Oriente, obtaining some 10 000 numbers.

The modern history of systematic botany in the country is generally agreed to have begun with the formation in 1928 of the Herbario Nacional Colombiano as a distinct entity under the direction initially of Pérez Arbeláez, recently returned from studies in Europe. In 1936 the developing local botanical community was bolstered by the definitive arrival from Spain of José Cuatrecasas; he would remain for 15 years. Cuatrecasas, Armando Dugand and others including García-Barriga, Jesús M. Idrobo, and Fr. Uribe-Uribe

(who would initiate the working-up and publication of the *Flora de la Real Expedición Botánica*) made substantial contributions over the next quarter-century, and their successors have continued this tradition. The Herbario Nacional, from 1936 a part of a new Departemento de Botánica (and in 1940 expanded into an Instituto de Ciencias Naturales), incorporated all its collections into an electronic database in the 1970s – one of the earliest worldwide to do so. In North America, the large collections accumulated in the half-century before 1950 served as a major basis for several family revisions, written as contributions towards a possible flora.

At the present time, no complete treatment of the Colombian flora has been written. The major accounts are the incomplete *Prodromus florae novo-granatensis* (Triana and Planchon, 1862–73) and the *Flora de la Real Expedición Botánica* (1954–), that long-delayed cultural symbol. A *Flora de Colombia* by Cortés did not progress beyond a first volume (1897); substantial materials including some 2000 plates remained unpublished. A widely used introductory work has been Pérez Arbeláez's *Plantas útiles de Colombia* (1936; 3rd edn., 1956). Killip projected a flora during his term at the Smithsonian, but although several family treatments appeared from 1931 to 1960 (including some by Cuatrecasas) this plan was not continued and North American contributions subsequently became more oriented towards *Flora Neotropica*. The tradition of national family revisions was, however, revived with the establishment of *Flora de Colombia* (1983–).

The contemporary growth of the botanical profession has been accompanied by a trend towards decentralization and the development of 'regional' herbaria. Associated with this has been progress on a number of partial floras. The most important of these is the unfinished *Catálogo ilustrado de las plantas de Cundinamarca* (1966–79), a project of the Herbario Nacional Colombiano.[27]

While much collecting is currently in progress, the size and diversity of the flora are such that many more man-years will be required before knowledge is even sketchily adequate for more than limited areas. The Amazonian zone has been rather poorly studied, while in other zones are many neglected areas.

Mainland Ecuador is also still relatively imperfectly known and unevenly collected, and until recent years its resident botanical community has been very small. Like Colombia, it has a long history of botanical exploration, beginning with the equatorial 'Misión Geodésica' of La Condamine and Joseph de Jussieu in 1735–45 (from which in 1743 La Condamine made his epic trip down the Amazon). They were followed at the end of that century by Humboldt and Bonpland in 1802–03 and, at different periods from 1799 to 1808, by Juan Tafalla and his associates from the 'Real Expedición Botánica al Virreinato del Perú'. Tafalla contributed a *Flora huayaquilensis* covering 625 species, but this was never published in his lifetime. Later nine-teenth-century visitors included Cuming, Spruce, Pearce, Isern y Batlló, André, and Eggers (of whom Spruce and Isern traveled in the Oriente, which remained relatively little collected until after World War I). Notable resident botanists in this and the fol-lowing century included the Scot William Jameson (from 1826 to 1870) and the Italian Jesuit Luigi (Luis) Sodiro (from 1870 to 1908), both for many years on the staff of the College (later University) of Quito, and, in the mid-twentieth century, Misael Acosta-Solís. It was Jameson who contributed the first local flora, *Synopsis plantarum aequatoriensium* (1865) which deals mainly with the altiplano, especially in the north around Quito. An early plan by Sodiro to complete this work gave way to research on vascular cryptogams (including a sub-stantial fern flora), Araceae, Piperaceae and other groups.

German and Swedish contributions became significant in the mid-twentieth century, the former largely before World War II. Germans included H. J. F. Schimpff, Ludwig Diels, and the Schultze-Rhonhofs; Diels's *Beiträge* remains a standard work. Swedish interest began in the 1930s but only after World War II was there mounted the first of several 'Regnellian Expeditions' from the Stockholm Natural History Museum, led by Erik Asplund. The later ones, and a number of others organized from Göteborg University, were conducted by Gunnar Harling, Benkt Sparre and their students and associates. The Swedish efforts were reinforced by Danes from 1968, when Aarhus University and the Universidad Católica in Quito orga-nized a joint floristic programme.

North Americans were active mainly during World War II, when the previously mentioned 'Cinchona Expeditions' brought Wendell Camp and Julian Steyermark to the country, and from the 1960s, beginning with the work of Calaway Dodson. Much recent work has been organized through the Missouri Botanical Garden. An expedition from Oxford University did significant ecological work in the Oriente in 1960.

The rapid accumulation of collections at Stockholm, then claimed as being the largest holding of Ecuadorean plants in the world, along with new hold-ings at Göteborg and Aarhus, led to a decision in 1966 to go ahead with a definitive *Flora of Ecuador*. Publication began in 1973 and is currently under the editorship of Harling and Lennart Andersson; by 1999, 63 fascicles had been published. However, as with Panama, the stock of collections, even with the substan-tial increases of recent decades, is even now not consid-ered at all adequate, with the result that early parts of this work are, or may soon become, significantly incom-plete. Ecuador, in relation to its area, is now regarded as having the richest flora in all South America.

In the last 20 years or so several local and zonal floras have been published, covering the lowland Oriente, the mountains above 2400 m, and some sites west of the Andes. More are needed.[28]

Progress
Botanical progress generally in northwestern South America as of the mid-1980s was surveyed in A. H. GENTRY, 1989. Northwestern South America. In *Floristic inventory of tropical countries* (see **General references**), pp. 392–400. For Colombia alone, mid-twentieth century reviews of work include F. A. BARKLEY and V. G. GUTIERREZ, 1948. Colectores de plantas en Colombia: tal como estan representades en los herbarios del pais. *Revista Fac. Nac. Agron., Medellín* 8: 85–107; and R. E. SCHULTES, 1951. La riqueza de la flora Colombiana. *Revista Acad. Col.* 8: 230–242. The Mutis expedition is reviewed in the first volume of *Flora de la Real Expedición Botánica* (1954; see below). The situation as of the mid-1980s was twice examined by Enrique Forero: E. FORERO, 1985. Estado actual de la investigación y la docencia en botánica en Colombia. In E. Forero (ed.), *Memoria de la Reunión de Botánicos de los países miembros del Convenio "Andrés Bello"*, pp. 24–41. Bogotá: SECAB/FUN/Botánica (Ciencia y Tecnología 6); and *idem*, 1989. Colombia. In *Floristic inventory of tropical countries* (see **General references**), pp. 353–361. For Ecuador, a very full survey of collectors is M. ACOSTA-SOLÍS, 1968. Naturalistas y viajeros científicos que han contribuido al conocimiento floristico fitogeográfico del Ecuador. *Contr. Inst. Ecuat. Ci. Nat.* 65: 1–138. Work in the Oriente was surveyed by S. S. Renner in

S. Lægaard and F. Borschenius (eds.), 1990. *Nordic botanical research in the Andes and Western Amazonia*, pp. 8–9. Aarhus. (AAU Reports 25.)[29]

Bibliographies. General and divisional bibliographies as for Division 3.

Indices. General and divisional indices as for Division 3.

321

Colombia (in general)

Area: 1141748 km². Vascular flora: 35000–55000 species.[30] – The unit as delimited here encompasses present-day mainland Colombia and nearby offshore islands. Until 1861 the country was known as the United States of New Granada, taking its name from the former viceroyalty which through 1811 included Venezuela (**315**) and from 1819 to 1830, as 'Gran Colombia', included Ecuador (**329**). Colombia also encompasses the islands of San Andrés and Providencia in the Caribbean (**238**) and Malpelo Island in the Pacific (**016**); until 1903, it also controlled Panama (**237**).

No complete modern floras or enumerations of the vascular plants were until recent years available for what is botanically one of the richest countries in the world. The only nominally general works at present are the incomplete and now long-outdated *Prodromus florae novo-granatensis* of José Triana and Jules-Émile Planchon (1862–73) and, as far as published, the family treatments in the sumptuously illustrated *Flora de la Real Expedición Botánica del Nuevo Reyno de Granada* (1954–). There are also several country-wide revisions, separately published from 1931 to the 1960s in a variety of serials but in no inconsiderable part prepared at the U.S. National Herbarium under the stimulus of Ellsworth P. Killip; one author termed these an unofficial *'Flora of Colombia'*.[31] These have recently been joined by *Flora de Colombia* (1983–), a project of the Instituto de Ciencias Naturales in Bogotá. Further coverage is also developing as *Flora Neotropica* (**301/I**) continues. For smaller areas, a still-limited number of departmental floras have also been published or are underway, and, with the existence and development of a number of 'regional' herbaria – especially at Medellín

and Cali, but now in most larger centers – this trend may continue.

As introductions to the flora, the following have long been widely used: E. Pérez Arbeláez, 1956. *Plantas útiles de Colombia*. 3rd edn. 832 pp., 752 text-figs., 47 pls. (2 col.). Bogotá: 'Colombiana' (1st edn., 1936); and J. Cuatrecasas, 1958. Aspectos de la vegetación natural de Colombia. *Revista Acad. Col.* **10**: 221–268. A more recent survey is P. Pinto Escobar, 1993. *Vegetación y Flora de Colombia*. [vi], 72 pp., illus., maps. Bogotá: Fundación Segunda Expedición Botánica/Fondo Nacional Universitario.

Bibliographies

Estrada, J., J. Fuertes and J. M. Cardiel (eds.), 1991. *Bibliografía botánica para Colombia*, 1. xvii, 400 pp. Bogotá: Agencia Española de Cooperación Internacional (AECI-ICI), Oficina Técnica de Cooperación en Colombia. [Detailed listing of mainly systematic works with 6844 unannotated entries on pp. 1–214; author and taxonomic cross-referential indices. In major floristic works all families are indexed separately. Coverage extends well beyond national boundaries.][32]

Dominguez, C. (comp.), 1985. *Bibliografía de la Amazonia colombiana y areas fronterizas amazonicas*. ix, 226 pp. Bogotá: DAINCO/Corporación Araracuara/Colciencias. (Contribución a la Segunda Expedición Botánica.) [4092 unannotated numbered entries arranged by author, with locations where known; indices by plant and animal taxa, indigenous *poblaciones*, geography, and themes (botany, p. 214).]

[Mutis, J. C.], 1954– . *Flora de la Real Expedición Botánica del Nuevo Reyno de Granada*. Vols. 1– . Illus. (some col.). Madrid: 'Cultura Hispánica' (for Instituto de Cultura Hispánica, Madrid, and Instituto Colombiano de Cultura Hispánica, Bogotá).

A 'Great Flower Book', this emerald represents the realization of the long-unpublished results of the extensive botanical surveys conducted by J. C. Mutis and his pupils, associates and assistants – the 'tribe of botanical adventurers' of their English contemporary, J. E. Smith – in the late eighteenth and early nineteenth centuries. Family treatments, some revised by specialists, feature extensive descriptions, keys, synonymy (with references), citations, *exsiccatae* (both from Mutis's time and more recent), geographical range, taxonomic commentary, miscellaneous notes, and indices. Of special interest are the reproductions in color of the many beautiful paintings by Mutis's artists. The work, initially appearing slowly but since 1982 with greater speed, was projected to encompass 51 volumes. As of 1999 vols. 1, 2, 3(1–2), 4(1–2), 5(1), 7, 8, 9(3), 10, 13, 20(2), 23, 25(2), 27, 30–32, 36, 41, 44, 45(1) and 47 have appeared. Volume 1 includes a history of the 'Real Expedición Botánica' as well as other introductory matter.

PINTO-ESCOBAR, P. and P. M. RUÍZ (eds.), 1983– . *Flora de Colombia*. Monografia 1– . Illus., maps. Bogotá: Universidad Nacional de Colombia (for Instituto de Ciencias Naturales/Museo de Historia Natural).

An open-ended series of 'research' flora-monographs; each treatment includes keys, full descriptions, synonymy (with references) and citations, typification, etymology of scientific names, vernacular names, localities with *exsiccatae*, summary of distribution (Colombia only), habitat and altitudinal range, taxonomic commentary, illustrations of most species (including photographs, unfortunately not well reproduced), and (at end) key literature, lists of *exsiccatae* and indices. For genera and subgenera their total distribution is given along with commentary and the number of species in Colombia. The taxonomic treatments are preceded by general sections on morphology, anatomy, biology, cytology, phylogeny, distribution, and ethnobotany (properties, uses, etc.). [As of 1999, 17 fascicles had appeared.][33]

TRIANA, J. and J.-É. PLANCHON, 1862–67. *Prodromus florae novo-granatensis*. Vols. 1–2. 4 pls. Paris: Masson. (Reprinted with changed pagination from *Ann. Sci. Nat., Bot.*, IV, 17–20 (1862); V, 1–5, 7 (1863–67), *passim*.) Continued as *idem*, 1872–73. Prodromus florae novo-granatensis. *Ann. Sci. Nat., Bot.*, V, **14** (1872): 286–325; **15** (1872): 352–382; **16** (1872): 361–382; **17** (1873): 111–194.

Systematic enumeration of fungi, non-vascular (except algae) and vascular plants, without indices; includes descriptions of new taxa, synonymy (with references), vernacular names, *exsiccatae* with localities, taxonomic commentary, and notes on habitat, etc. The treatment of vascular plants generally follows the Candollean system as far as the Papayaceae (= Caricaceae), although some large families are omitted. The second volume is entirely devoted to non-seed plants and fungi (lower vascular plants, pp. 275–396).

Woody plants (including trees): guide

DEL VALLE A., J. I., 1972. *Introducción a la dendrología de Colombia*. 351 pp., illus. Medellín: Centro de

Publicaciones, Universidad Nacional de Colombia. (Mimeographed.)

An introductory student handbook featuring family accounts with representative species, with some keys (not to families and genera) and many figures, notes on field and wood characters, and internal and extralimital distribution; references (pp. 300–307), an illustrated lexicon to botanical features, a key to fruits, and index to vernacular and botanical names (followed by figures of buttress types) at end.

Pteridophytes

Ferns and fern-allies have been covered in *Flora de la Real Expedición Botánica*, in volume 2 of the *Prodromus* and, in part, *Flora de Colombia* but the following is the first full local treatment to generic level.

MURILLO-PULIDO, M. T. and M. A. HARKER-USECHE, 1990. *Helechos y plantas afines de Colombia*. 323 pp., 148 text-figs. Bogotá: Academia Colombiana de Ciencias Exactas, Fisicas y Naturales. (Colección Jorge Alvarez Lleras 2.)

An illustrated generic flora with keys throughout and, for each genus, indication of type species, synonymy, Spanish vernacular names, etymology, a description, indication of distribution, and list of species known from the country; bibliography and indices to scientific and vernacular names and to illustrations at end. The introductory part is relatively brief and to the point.

322

Colombia: northern departments

Included here are Guajira, Magdalena, Bolívar, Atlántico and Córdoba. The large, isolated massif of Sierra Nevada da Santa Marta, in Magdalena, is botanically still imperfectly known.

ROMERO CASTAÑEDA, R., 1965. *Flora del Centro de Bolívar*. Vol. 1. Illus. Bogotá: Instituto de Ciencias Naturales, Universidad Nacional de Colombia.

Illustrated descriptive treatment of relatively common species of vascular plants; includes synonymy, vernacular names, local range, notes on uses, and (at end) a glossary, key to families, and general index. [Not completed; only the first volume ever published.]

ROMERO CASTAÑEDA, R., 1966–71. *Plantas del Magdalena*. Fasc. 1–2. Illus. Bogotá: Instituto de Ciencias Naturales, Universidad Nacional de Colombia.

Illustrated descriptive treatment, with synonymy, vernacular names, local range, and notes on wood and other botanical features; glossary. [Not completed; by 1999 only two fascicles (1966: Zygophyllaceae; 1971: Flora de la Isla de Salamanca) had appeared.]

323

Colombia: Antioquia, Caldas and Quindió

Included here are the departments of Antioquia (with its capital city of Medellín), Caldas and Quindió.

URIBE, J. A. (ed. Pe. L. URIBE-URIBE), 1940 (1941 on cover). *Flora de Antioquia*. 383 pp., 25 pls. Medellín: Imprensa Departemental.

Systematic enumeration of more commonly encountered non-vascular and vascular plants inclusive of some keys, limited synonymy, vernacular names, and notes on special features, properties, uses, and cultivation; list of references and indices to generic, family and vernacular names at end. The plan and basis of the work are given in a brief introductory section.

Woody plants
An illustrated account (without keys) of more common trees in Antioquia may be found in L. S. ESPINAL T., 1986. *Árboles de Antioquia*. 251 pp., illus. Medellín: Universidad Nacional de Colombia, Seccional Medellín.

324

Colombia: Chocó and El Valle de Cauca

Comprises the departments of Chocó (area: 47 205 km^2) and El Valle de Cauca, the latter including the city of Cali. Chocó, one of the wettest and apparently botanically richest areas in the Neotropics, was until recent decades very poorly known. A checklist by Enrique Forero and Alwyn Gentry appeared in 1989 using as a primary basis the more than 14 000 collection numbers made by them and others between 1973 and 1983. The pteridophytes are being revised by David

Lellinger; the first of two volumes of a work extending from Chocó to Costa Rica also appeared in 1989. For El Valle, with a longer history of exploration, José Cuatrecasas (who in the 1940s had collected extensively in the department) was said by Gentry (1978; see **General references**) to have been at work on a checklist; with his passing such is not now an immediate prospect.[34]

FORERO, E. and A. H. GENTRY, 1989. *Lista anotada de las plantas del Departamento del Chocó, Colombia*. 142 pp. Bogotá: Instituto de Ciencias Naturales – Museo de Historia Natural, Universidad Nacional de Colombia. (Biblioteca José Jeronimo Triana 10.)

Concise, alphabetically arranged checklist with names, localities and *exsiccatae*. The introductory part (pp. 1–22) covers geography, climate, vegetation, collecting, the plan of the work, and main references and includes a list of the authors' papers on the area. [Accounts for 3866 species.]

Pteridophytes
LELLINGER, D. B., 1989. *The ferns and fern-allies of Costa Rica, Panama and the Chocó*, 1. 364 pp., illus. (Pteridologia 2A). Washington: American Fern Society.

Systematic enumeration with diagnostic keys and illustrations, descriptions of and key references for genera, synonymy (limited to basionymy), typification (with *exsiccatae*), indication of distribution and habitat, illustrations of details for most species, and occasional commentary; no index in part 1. The introductory part (with maps) covers the basis and plan of the work along with basic references, a list of families, genera and infrageneric taxa, and general key to families. [As so far published the work covers fern-allies and the 'lower' and 'middle' ferns from Ophioglossaceae through Dicksoniaceae (562 species) mostly following the arrangement of Crabbe *et al.* (see **General references**); a second part will complete the work.]

325

Colombia: northeastern highlands

Included here are the departments of Norte de Santander, Santander and part of Boyacá. No separate floras are available, but works on nearby Mérida in Venezuela (**318**) may be of some use. A flora of

Santander, by F. Llanos and E. Rentería A., was reportedly in preparation in the 1970s.[35]

326

Colombia: Central Valley

Included here are the departments of Cundinamarca, Tolima and Huila. Within the first is Bogotá, the national capital, and its 'special district'. For Cundinamarca, the *Catálogo ilustrado* of Polidoro Pinto-Escobar and his associates (not completed) is the most detailed and comprehensive of present departmental floras.

PINTO-ESCOBAR, P. *et al.* (eds.), 1966–79. *Catálogo ilustrado de las plantas de Cundinamarca.* Parts 1–7. Illus. Bogotá: Imprensa Nacional.

Illustrated enumeration of vascular plants; includes descriptive keys to genera, extensive synonymy, vernacular names, local range (with maps), and notes on habitat, etc., with each fascicle (which may contain one or more families) separately indexed. [Seven parts were published; no preset family sequence has been followed, although some fascicles contain treatments of groups of related families.]

327

Colombia: Cauca and Nariño

Included here are the departments of Cauca and Nariño. Active collecting has been in progress in both departments but no floras exist or are yet in prospect.

328

Colombia: Oriente

This comprises all Colombia east of the Andes – more than one-half the country. Included here are the departments of Arauca, Boyacá (in part), Meta, Vichada, Vaupés, Caquetá, Putumayo and Amazonas. A considerable part of the area lies on the Guayana crystalline shield, which terminates to the west in the Sierra de la Macarena (in Meta), botanically still poorly known and now seriously threatened by human disturbance.

ROA TORRES, A., 1973. *Descripción dendrológica de algunas especies forestales de Caquetá.* 120 pp., illus. Bogotá: Ministerio de Agricultura, División Forestal (for Instituto de Desarrollo de los Recursos Naturales Renovables (INDERENA), Colombia).

Dendrological treatment, with unpaged plates, of principal timber species.

329

Ecuador

Area: 284000 km². Vascular flora: a figure of 10000 species was suggested by Gentry (1978; see **General references**) but a more current estimate is 19000–21000.[36] Only the mainland is considered here; for the Galápagos, see **017**. The alpine flora is also covered under **303/II** (ULLOA ULLOA and JØRGENSEN). The eastern boundary follows current convention.

The *Flora of Ecuador*, initiated in 1966 with publication beginning in 1973, represents the first attempt at a large-scale flora of a small but botanically very rich country. Associated with this project is a database, presently located at Aarhus University in Denmark. Partial works include the old, never-completed *Sinopsis* of Jameson (reissued about 1940) and four recent florulas of sites of special interest on the Pacific side of the country. For trees, the northwesterly Esmeraldas Province has been treated in a partial dendroflora by E. L. Little and R. G. Dixon; while F. Latorre (at Universidad Central) and I. Padilla have contributed a checklist for the rich fern flora.

A useful introduction to the flora and vegetation (with a list of references) is L. DIELS, 1937. *Beiträge zur Kenntnis der Vegetation und Flora von Ecuador.* 190 pp., 2 text-figs., map (Biblioth. Bot. 29 (heft 116)). Stuttgart; in Spanish: *idem*, 1938. *Contribuciones al conocimiento de la vegetación . . .* Transl. R. Espinosa. 364 pp., map. Quito. More recent is M. ACOSTA-SOLÍS, 1968. *Divisiones fitogeográficas y formaciones geobotánicas del Ecuador.* 307 pp., 29 pls., 1 map. Quito: Casa de la Cultura Ecuatoriana.

Families and genera

CERÓN MATÍNEZ, C. E., 1993. *Manual de botánica ecuatoriana*. 191 pp., illus. Quito: The author. [Students' manual with illustrated family accounts and lists of useful species along with treatments of botanical survey techniques, collecting methods, terminology, systems of classification, etc. Not useful for individual genera.]

NARANJO, P., 1981–83. *Indice de la flora del Ecuador*. Vols. 1–2. Illus. Quito: Casa de la Cultura Ecuatoriana. [Vol. 1: introduction, history of botany (pp. 13–19), physical features, biogeography, and lists of algae, fungi, bryophytes and pteridophytes; vol. 2: 'early dicotyledons' from Casuarinaceae through Leguminosae (Cronquist system). References for each species listed are given, and families and genera are alphabetically indexed. Not completed.]

VALVERDE BADILLO, F. DE M., 1974. *Claves taxonómicas de órdines, familias y géneros de dicotyledoneas*. 1. parte. 385, [1] pp. Guayaquil: Publicaciones, Univ. de Guayaquil. [Keys to families and genera, with representative species (including their distribution) listed; descriptions of families; some vernacular names; references (pp. 363–364) and index at end. Not completed; covers 'Apetalae' and 'Polypetalae'.]

HARLING, G., B. SPARRE *et al*. (eds.), 1973– . *Flora of Ecuador*. In fascicles. Lund, Sweden (fasc. 1–4, 1973–75 (as *Opera Bot.*, B)); Stockholm: Swedish Research Councils (fasc. 5–27, 1976–86); Copenhagen: Nordic Journal of Botany (fasc. 28–48, 1987–93); Copenhagen: Council for Nordic Publications in Botany (fasc. 49– , 1994–).

A 'research' flora of mainland Ecuador, comprising a series of revisions of vascular plant families (and lesser taxa); each treatment includes detailed descriptions, keys, synonymy (with references) and citations, localities with *exsiccatae*, summary of local and extralimital distribution, taxonomic commentary, notes on habitat, special features, etc., and (at end) index. [Parts are appearing as revisions are ready, although family numbers have been assigned according to the modified Englerian sequence of the 12th edition of the *Syllabus*. As of 1999, 63 fascicles had been published.][37]

JØRGENSEN, P. M. and S. LÉON-YÁNEZ (eds.), 1999. *Catalogue of the vascular plants of Ecuador/ Catálogo de las plantas vasculares del Ecuador*. viii, 1181 pp., col. end-paper maps (Monogr. Syst. Bot. Missouri Bot. Gard. 75). St Louis.

Concise enumeration (16 807 species of which 15 306 native with 4173 endemic) with for each taxon key synonyms, references, literature citations or up to two voucher collections, life-forms, status, provinces, altitudinal ranges, and geographical classes; new names

and combinations, general bibliography, and indices to synonyms and to families and genera at end. The general part (in both English and Spanish, with parallel text) covers geography, climate, vegetation, characteristic species, the history of collecting (pp. 25–41), the format of the catalogue, and an extended analysis of the flora (over various parameters or combinations thereof) with (pp. 103–104) a summary. [Among seed plant families, the Orchidaceae are the most speciose at 3013, and there are 1298 known or possible ferns and fern-allies.]

Partial works

The vegetation classification referred to in Dodson's florulas is based on the Holdridge life-zone system. These works were written partly to demonstrate the relationship of a given flora to its typology in that scheme. Two of the florulas were published; a third, covering a site in the 'Tropical Dry Forest' zone, was announced for 1987 but evidently not published: *Flora of Capeira and the Guayaquil region* by C. Dodson and A. H. Gentry.

DODSON, C. H. and A. H. GENTRY, 1978. *Flora of the Río Palenque Science Center, Los Rios Province, Ecuador*. xxx, 628 pp., 22 halftones (incl. maps), 278 pls., col. frontisp., maps (end-papers) (Selbyana 4(1–6)). Sarasota, Fla.: Marie Selby Botanical Gardens.

Atlas-flora (in octavo format) of vascular plants (1112 species), with keys to genera and species and including short species descriptions with notes on ecology, biology, phenology and distribution; index to all taxa. Appendices include a list of specimens cited, acknowledgments and a list of rare species. An introductory section gives a general description of the area, plant-geographic patterns, major features of the flora, statistics, and vegetation formations with characteristic species. [Covers an area of some 167 ha, north of Quevedo in the central western lowlands (00° 35′ S, 79° 25′ W) classified as 'Tropical Wet Forest'.]

DODSON, C., A. H. GENTRY and F. DE M. VALVERDE BADILLO, 1984. *Flora de Jauneche*. xxix, 512 pp., 184 pls., 14 figs. (halftones, maps) (Selbyana 8). Sarasota, Fla.: Marie Selby Botanical Gardens/Banco Central del Ecuador. (Spanish edn., 1985, Quito, as *La flora de Jauneche*.)

Atlas-flora of vascular plants (728 species), similar in style to *Flora of Río Palenque Science Center*. [Based on a 130 ha site at the Pedro Franco Davila Biological Station in Los Rios Province in the western lowlands classified as 'Tropical Moist Forest', with an annual rainfall of 1865 mm and a five-month dry season.]

JAMESON, W., 1865. *Synopsis plantarum aequatoriensium . . . viribus medicatis et usibus oeconomicis plurimarum adjectis*. Vols. 1–3. Quito: 'del Pueblo' (Joannis Paulus Sanz). (Reprinted 1938, Quito: Universidad Central.)

Descriptive treatment of vascular plants, without keys;

includes localities (with some citations) and (at end of each family) extensive notes on properties, uses, etc. (including medicinal values) of the species concerned; indices to family and generic names in vols. 1–2. Purposely limited to the plants of the altiplano; botanical text in Latin, commentary in Spanish. Not completed; vols. 1–2 cover Ranunculaceae through Labiatae and vol. 3 (unfinished) Verbenaceae through *Plantago* (Candollean system).

RENNER, S. S., H. BALSLEV and L. B. HOLM-NIELSEN, 1990. *Flowering plants of Amazonian Ecuador – a checklist.* vi, 241 pp. (Rep. Bot. Inst. Aarhus Univ. 24). Risskov.

An alphabetic checklist with entries inclusive of provinces, habit, *exsiccatae* and miscellaneous notes. Family headings include references to key literature.

VALVERDE BADILLO, F. DE M., G. R. DE TAZAN and C. G. RIZZO, 1979. *Cubierta vegetal de la Península de Santa Elena.* 236 pp., 18 halftones, map (Publicaciones de la Facultad de Ciencias Naturales, Universidad de Guayaquil 2). Guayaquil.

Includes (pp. 79–132) a descriptive checklist of 295 species with synonymy and local distribution. [The area – west of Guayaquil – is distinguished by very low rainfall, in parts as little as 100 mm/annum.]

VALVERDE BADILLO, F. DE M., 1966–69. Fanerogamas de la Zona de Guayaquil. 4 parts. *Ciencia y Naturaleza* (Quito) **9**: 13–52; **10**: 25–56; **11**: 20–53; **12**: 31–62. (First two parts also published as *idem*, 1966. Flora de Guayaquil y sus alrededores, [I]. *Naturaleza Ecuatoriana* **1**(1): 23–67.)

An illustrated descriptive florula of flowering plants with keys to genera (but not species). The introductory part includes background (including collectors), an account of the vegetation, and a key to families. Part 1 covers monocotyledons; parts 2–4, dicotyledons. [The *Naturaleza Ecuatoriana* version is in a larger format.]

Woody plants (including trees): guides

For *Árboles y arbustos de los Andes del Ecuador* by C. ULLOA ULLOA and P. M. JØRGENSEN, see **303/II**. There are no tree guides for the country as a whole.

LITTLE, E. L., JR. and R. G. DIXON, 1969. *Árboles comunes de la Provincia de Esmeraldas.* 536 pp., illus., map. Rome: FAO/Programa de las Naciones Unidas para del Desarrollo. (Project FAO/SF: 76/ECU/13.)

Illustrated systematic dendrology of a selection of forest trees (230 species out of 500 or more); includes copious descriptions along with vernacular names, synonymy (if any), voucher specimens, notes on general and local distribution, wood, uses, ornamental value, etc., and (at end) index to all names, along with their derivations. The introductory section covers forest types, wood characteristics and properties, sources of figures, and regional chorology with characteristic species; references (pp. 10–11); general key.

NEILL, D. A. and W. A. PALACIOS, 1989. *Árboles de la Amazonia ecuatoriana: lista preliminar de especies.* 120 pp., 2 maps. Quito: Dirección Nacional Forestal, Ecuador.

Alphabetically arranged, annotated checklist (1009 species) with botanical and vernacular names, indication of habitat and distribution, and selected *exsiccatae*; no index. The general part includes an introduction to the region and its tree flora, with some statistics (p. 10) and a suggested possible total of some 3000 species (these forests being among the richest in the world). A list of proposed new species (not here published) appears on pp. 13–14, references on pp. 15–16, and terminology and abbreviations on p. 17.

Pteridophytes

The work of Latorre and Padilla, covering about 600 species, was preceded by *Cryptogamae vasculares quitenses* by Luis Sodiro (1892–94, in *Anales de la Universidad de Quito*), a descriptive work with keys covering the whole country but with emphasis on the central provinces.

LATORRE A., F. and I. PADILLA C., 1973–74 (1974–75). Lista preliminar de helechos del Ecuador. *Ciencia y Naturaleza* (Quito) **14**: 21–57; **15**: 55–59.

Systematic enumeration of ferns (Ophioglossaceae through *Polypodium* in part 1; *Cuspidaria* to *Drymoglossum* in Polypodiaceae, Grammitidaceae, and the heterosporous orders in part 2); includes synonymy, references and citations, and summary of local distribution (by provinces, with altitudes). The introductory section includes a synoptic conspectus of genera, while part 2 contains addenda, a bibliography and an index.

Region 33

Peru

Area: 1 290 000 km^2. Vascular flora: perhaps over 20 000 species (Gentry, 1980, in Macbride, *Flora of Peru* [N.S., fasc. 1]); however, Iltis has suggested a figure of 30 000.[38] A total of 17 143 seed plants were recorded as of the early 1990s (Brako and Zarucchi, 1993; see below). The region corresponds to the presently recognized limits of Peru. Parts of northern Loreto, south of the Colombian border, have been claimed by Ecuador.

The country has had a long history of botanical work, but remains comparatively poorly known due to its geography, floristic diversity, and relatively small resident botanical community. Early knowledge of the flora is based largely upon the collections and observations of the 'Real Expedición Botánica al Virreinato del Perú' of 1778–1815, in which Hipolito Ruíz and José Pavón participated from 1778 to 1788 and Joseph Dombey through 1784. The main results were embodied in the never-completed *Flora peruviana, et chilensis* (1794, 1798–1802).[39] Following the political and other changes of the early nineteenth century, significant but sometimes not large contributions were made by several foreign explorers and a few residents: Poeppig, Mathews, Ball, Weddell, Cuming, Spruce and Antonio Raimondi, with Kuntze following towards the end of the century. From 1901 until 1940 August Weberbauer was active in Peru, finally settling in Lima. His vegetatiological monograph on the Andean zone appeared in 1911, with a revision in Spanish in 1945. Other key collectors – mostly active after World War I – included Macbride, Klug, Erik Asplund, Carlos Schunke, Killip and A. C. Smith, Ll. Williams, Thomas H. Goodspeed, and, at Cuzco, F. L. Herrera (from 1912 a professor at Cuzco University) and C. Vargas; they would continue after World War II or be succeeded by others.

A decision by the Field Museum of Natural History in Chicago around 1920 to mount an extensive survey of Peruvian biota brought J. Francis Macbride to the staff and laid the groundwork for a major flora. After some years of preparation, including field trips (the first in 1922) and comprehensive type photography in Europe (which would, owing to events, become of incalculable value), publication of the *Flora of Peru* began in 1936; by 1971, when the last contribution in the original series appeared, it was over three-quarters completed. Until Macbride's departure from the Museum in the 1950s the *Flora* was very largely run 'in-house'; later contributions had a more diversified authorship, although Macbride himself continued to study lots of specimens (loaned to him in California) and produce contributions up to 1962. The project became dormant in the latter part of the 1960s (although publication of the original series continued to 1971); however, in 1975 it was reactivated as a cooperative venture of the Field Museum, the Missouri Botanical Garden, the Universidad Nacional de Amazonia Peruana in Iquitos and the Universidad Nacional Mayor de San Marcos in Lima. The revived project, initially planned to take six years or perhaps longer, was aimed at covering the remaining seed plant families (including such important groups as the Guttiferae, Cactaceae and Compositae) as well as the remaining lower vascular plants not treated by Rolla Tryon. There was also to be an emphasis on new collections, particularly from the botanically poorly known eastern (Amazonian) part of the country.

Though the *Flora* truly is a monumental effort – among the standard floras of South America only *Flora brasiliensis* is larger – it is, inevitably, deficient in its coverage, particularly of the large Amazonian region, and in addition much of the work generally is now regarded as less than critical as well as obsolescent. No plans exist, however, for revision of parts in the original series, and of the second series relatively few fascicles have been published. Fifteen families, as well as part of the Compositae, remain outstanding. Given that situation, along with the availability of a database management system, the Missouri Botanical Garden undertook preparation of an enumeration; this appeared in 1993 as *Catalogue of the flowering plants and gymnosperms of Peru* by Lois Brako and James L. Zarucchi. It accounts for more than 17000 species, almost half as many again as in the *Flora*.

Pteridophytes were also taken in hand by the *Flora of Peru*, but only one part (Tryon, 1964) appeared during the lifetime of the programme's original phase. Realization of the full work by Rolla Tryon with Robert Stolze and others has been part of the project's second phase; its six parts were published by the Field Museum in 1989–94.

Local coverage is scattered and rather limited. There are a number of collection reports and minor florulas, some centered on Lima and Cuzco. The most substantial of these, *Sinopsis del Flora del Cuzco* (1941) by Fortunado L. Herrera, is listed below. There are also several revisionary papers published (in *Boissiera* and *Candollea*) as part of the inventory and ecological studies of the Arboretum 'Jenaro Herrera' on the Ucayali southwest of Iquitos, in Requena Province (Loreto), but these have not yet been consolidated into a florula.[40]

There is no full coverage of woody plants. The only 'tree book' for long was *Woods of northeastern Peru* (1936) by Llewelyn Williams. It has since been joined by *Los árboles del Arborétum 'Jenaro Herrera'* (1989–90), a well-illustrated dendrology.

Progress

A local review of collectors, institutions, etc., appears in F. L. HERRERA, 1937. Exploraciones botánicas en el

Perú. *Revista Mus. Nac.* (Lima) **6**: 296–358. The situation to the 1980s is considered in A. H. GENTRY, 1989. Northwestern South America. In *Floristic inventory of tropical countries* (see **General references**), pp. 392–400. The historical commentary on the *Flora of Peru* presented here is based upon personal knowledge as well as a survey by Gentry in the first fascicle of its 'new series' (see below).

Bibliographies. General and divisional bibliographies as for Division 3.

Regional bibliographies

There is a list of 686 references through the 1940s in the work of Weberbauer (1945; see above). The *Catalogue* (see below) is also supported by a substantial reference list. Both of these are arranged by author.

ENCARNACIÓN, F., R. SPICHIGER and J.-M. MASCHERPA, 1982. *Bibliografía selectiva de las familias y de los géneros de fanerogamas: primera contribución al estudio de la flora y de la vegetación de la Amazonia peruana.* 197 pp. (Boissiera 34). Geneva. [Two-part bibliography of 1616 items derived from a database; main part with entries alphabetized by author, followed by a systematically arranged index. Families, with included genera, listed at end.]

Indices. General and divisional indices as for Division 3.

330

Region in general

In addition to the comprehensive floristic coverage provided by the still-incomplete *Flora of Peru*, good introductions to Peruvian plant life are found in A. WEBERBAUER, 1911. *Die Pflanzenwelt der peruanischen Anden in ihren Grundzügen dargestellt.* xxi, 355 pp., 63 text-figs., 40 pls., 2 maps (Die Vegetation der Erde 12). Leipzig: Engelmann, and its revision in Spanish: *idem*, 1945. *El mundo vegetal de los Andes peruanos.* xix, 776 pp., 63 text-figs., 40 pls., folding map. Lima: Estación Experimental de La Molina. There is also an introductory survey in English by the same author in the first volume (1936) of the *Flora of Peru.*

Families and genera

FERREYRA, R., 1979. *Sinopsis de la flora peruana: gymnospermas y monocotiledoneas.* 60 pp., illus. Lima. Continued as *idem*, 1979. *Flora del Perú: dicotiledoneas (Sinopsis de la flora peruana).* [1], 191, [1] pp. Lima. (Mimeographed.) [Keys to and accounts of families along with keys to genera; lists of vernacular names with botanical equivalents; indices to all names. The first number includes a short history of botanical work in Peru, while both the first and second contain a brief bibliography. – The emphasis is on families; not all genera are accounted for.]

MOSTACERO LEÓN, J. and F. MEJÍA COICO, 1993. *Taxonomía de fanerogamas peruanas.* 602 pp., illus., col. pls. [Trujillo.] [Illustrated students' text to families, genera and selected species; also includes an illustrated glossary, index to vernacular, generic and family names, and bibliography of principal references (pp. 601–602).]

BRAKO, L. and J. L. ZARUCCHI, 1993. *Catalogue of the flowering plants and gymnosperms of Peru/Catálogo de las angiospermas y gimnospermas del Perú.* xl, 1286 pp., maps (end-papers) (Monogr. Syst. Bot. Missouri Bot. Gard. 45). St. Louis, Mo.: Missouri Botanical Garden Press.

Concise enumeration of seed plants with synonymy, literature references (keyed to main bibliography, pp. 1197–1251), indication of habitat, habit and distribution (by departments), brief notes, and one or two *exsiccatae*. Families and genera are arranged alphabetically with an initial separation into gymnosperms and angiosperms, and synonymous names are intercalated with those of accepted taxa. The introductory part covers the basis and plan of the work, with conventions and abbreviations, contributors and reviewers of individual families, statistics, families published in *Flora of Peru*, and an introduction to the Peruvian flora (by A. H. Gentry). The main bibliography and two appendices, one with comprehensive statistics to generic level, follow the body of the work.

MACBRIDE, J. F. *et al.*, 1936–71. *Flora of Peru.* 10 parts in 24 nos. Illus. (in part), map (Publ. Field Mus. Nat. Hist., Bot. Ser. 13(1/1–3, 2/1–3, 3/1–3, 3A/1–2, 4/1–2, 5/1–2, 5A/1–3, 5B/1–3, 5C/1, 6/1–2)). Chicago. (Part 3A/1, containing the Leguminosae, reprinted, Chicago.) Continued as J. F. MACBRIDE and collaborators, 1980– . *Flora of Peru* [N.S.]. Published in fascicles. Illus. (Fieldiana, Bot., N.S. 5, *passim*). Accompanied by C. SCHWEINFURTH, 1958–61. *Orchids of Peru.* viii, 1026 pp., 194 text-figs., portr. (Fieldiana, Bot. 30(1–4, index)). Chicago; and *idem*,

1970. *First supplement to the* Orchids of Peru. 80 pp. (*ibid.*, 33).

The main work is a comprehensive (but in its first phase extensively compiled) descriptive flora of seed plants (save for Orchidaceae), the families arranged generally according to the Englerian system; included are keys to genera and species, full synonymy (with references), citations (including standard monographs and revisions under family and generic headings), vernacular names, localities with *exsiccatae*, summary of extralimital range, notes on habitat, uses, etc., extensive and sometimes pointed taxonomic commentary, and (at end of each part) an index to all botanical names. The introductory section (part 1) includes a concise account of vegetation and phytogeography by A. Weberbauer. The revision of Orchidaceae by Schweinfurth at the Botanical Museum of Harvard University is in most respects similar in style to the main work. In the 'new series', issues follow no established order. The objectives have been to cover the remaining seed plant families (including important ones like the Guttiferae, Cactaceae and Compositae) as well as the lower vascular plants (see below under **Pteridophytes**). The style resembles that of the original series but descriptive details are in smaller type and the commentary is more detached. As of 1998 seven fascicles, comprising Gentry's index to families (see below) and various parts of the Asteraceae, had been published, the most recent in 1995. [Most treatments after 1960 in the original work, and all of those in the second series, have been by specialists. For indices to the original series, see 'The *Flora of Peru*: a conspectus' by A. H. Gentry on pp. 1–11 in the first fascicle (1980) of the new series as well as D. C. DALY, 1980 (1981). Families of spermatophytes included in the *Flora of Peru*. *Brittonia* 32: 548–550. There is also an index in the Brako and Zarucchi *Catalogue* (1993; described above).][41]

Partial works

HERRERA, F. L., 1941. *Sinopsis de la flora del Cuzco*, I. *Parte sistemática*. 528, [1] pp. Lima. Complemented by C. VARGAS, 1941. *Addenda á* Sinopsis de la Flora del Cuzco *de F. L. Herrera, 1941*. 36 pp. (Revista Univ. (Cuzco) 94). Cuzco.

The main work is a systematic enumeration (on the Englerian system) of fungi, lichenized fungi, and non-vascular and vascular plants (2157 species) of Cuzco department; entries include vernacular names, localities with *exsiccatae*, extralimital distribution, notes (sometimes) on habit, uses, etc. (especially on crop plants), and (in appendices) an account of new taxa collected by the author, a lexicon of vernacular names with equivalents, and index to genera.

MCDANIEL, S., 1996. *Guía de la flora de Iquitos*. iii, 225 pp., 23 text-figs. Mississippi State [*sic*], Mississippi: Institute for Botanical Exploration.

A key to families and genera with family descriptions, summaries and references to standard works (keyed to the general bibliography, pp. 5–6; abbreviations, p. 7). Numbers of species are indicated following the generic names in the keys. An index to all family and generic names appears at the end. [Statistics, p. 212; the author gives figures of 157 families, 1089 genera and 4758 species.]

VÁSQUEZ MARTÍNEZ, R. (eds. A. RUDAS LLERAS and C. M. TAYLOR), 1997. *Flórula de las Reservas Biológicas de Iquitos, Perú*. xii, 1046 pp., 4 figs. (maps), 117 pls. (Monogr. Syst. Bot. Missouri Bot. Gard. 63). St. Louis, Mo.: Missouri Botanical Garden Press.

A concise but hefty descriptive flora of vascular plants with keys, indication of habitat, location by reserve, and occasional critical notes; little or no citation of synonymy (the nomenclature largely following the 1993 *Catalogue*). The general part encompasses geography, geology, climate, soils, hydrography, features of the vegetation, life-forms (with statistics), and the composition of the flora (also with statistics), while at the end are the plates (pp. 861–977), a glossary, a bibliography, a lexicon of vernacular names, and a complete botanical index. [Accounts for 2740 species in an area of just over 4670 ha, being the combined compass of Allpahuayo-Mishana, Explornapo Camp, and Explorama Lodge. The illustrations are of representative species (one for each genus).]

Woody plants (including trees): guides

The work of Llewelyn Williams in the Amazon region has been reinforced by contributions from the Arboretum 'Jenaro Herrera'. In addition to many family revisions, published in *Candollea* from 1980 through 1990, a reserve 'tree flora' has been published and is accounted for here.

SPICHIGER, R. *et al.*, 1989–90. *Los árboles del Arborétum Jenaro Herrera*. 2 vols. 359, 565 pp., 402 text-figs., map (Boissiera 43, 44). Geneva.

An illustrated descriptive treatment with keys, vernacular names, indication of distribution in general and within the Arboretum, and a representative *exsiccata*. Standard references appear at the end of each family, and indices to all names are given at the end of each part (that in the second part covering both, along with indices to *exsiccatae* and their collectors). The introduction in the first part gives brief accounts of climate, soil, vegetation, the arboretum, and the plan and scope of the work. [The arrangement of the 54 families covered follows the Englerian system with, however, the palms last (vol. 1: Moraceae to Leguminosae; vol. 2: Linaceae to Palmae).]

WILLIAMS, LL., 1936. *Woods of northeastern Peru.* 587 pp., 18 text-figs., 2 maps (Publ. Field Mus. Nat. Hist., Bot. Ser. 15). Chicago.

Systematic dendrological account of forest tree species and their woods; includes vernacular names, localities with the author's *exsiccatae*, summary of distribution, and notes on botanical and wood-anatomical features, properties, uses, etc., with tables of anatomical characters, lexicon of vernacular names with botanical equivalents, list of references, and index at end. An introductory section accounts for the explorations of the author, vegetation formations, climate, physical features, characteristic species, forest products, etc. [Encompasses parts of Loreto, San Martin and Amazonas departments, treating an area stretching from Moyobamba into the Amazon Basin towards Iquitos.]

Pteridophytes

The current *Pteridophyta of Peru* was preceded by R. M. TRYON, 1964. *The ferns of Peru.* 253 pp., 196 text-figs., 46 maps (Contr. Gray Herb., N.S. 194). Cambridge, Mass. This incomplete work covered 187 species in Polypodiaceae sensu lato from Dennstaedtieae through Oleandreae (about one-quarter of the then-estimated total).

TRYON, R. M., R. G. STOLZE *et al.*, 1989–94. *Pteridophyta of Peru.* 6 parts. Illus. (Fieldiana, Bot., N.S. 20–34, *passim*). Chicago.

Descriptive flora with keys to genera and species, detailed synonymy (with references), types, localities with *exsiccatae*, summary of local and extralimital range, taxonomic commentary, and notes on habitat, special features, etc.; addenda, floristic and ecological summary (with two maps), and full cumulative index in part 6 (the five earlier parts also separately indexed). Some new taxa are described. [Covers 1060 species in 28 families, with the six most species-rich departments being all mainly montane.][42]

Region 34

Bolivia

Area: 1 098 581 km^2. Vascular flora: *c.* 18 000 species (Solomon in Campbell and Hammond, 1989; see **General references**). The region corresponds to the political limits of Bolivia.

Floristic botany has hitherto developed only to a limited extent and, as James Solomon has written, with only some 90 000 unit collections of vascular plants, its 'collecting index,' about 0.08/km^2, is at present among the world's lowest. The country is quite unevenly explored, with many parts still uncollected.

The earliest recorded collections are by the Czech Thaddeus (Tadeo) Haenke, made following his departure from the Spanish 'Malaspina' expedition in 1794. He arrived in the country in the following year and later settled in Cochabamba, remaining there until his death in 1817. His collections, including those from outside Bolivia, were written up by Karl Presl in his *Reliquiae Haenkeanae* (1825–35). He was followed by Alcide d'Orbigny (1826 33), Joseph Pentland (1826–51), Hugh Cuming, Claudio Gay, Hugo A. Weddell (1845–48, entering as a member of the Castelnau expedition), Gilbert Mandon (1848–53 or 1856–61), Richard Pearce, Juan Isern y Batlló (mid-1860s), and, from Argentina, Paul Lorentz (1870–76) and Georg Hieronymus (1874–83). More or less sustained activity, however, began only in the late nineteenth century as economic activities, particularly tin-mining, expanded and railways and other communications were constructed. During this period and continuing to the 1940s, Bolivia was visited by several outside expeditions and individuals; among them were Henry Rusby, Pierre Jay, Otto Kuntze, Sir William Martin Conway, Robert E. Fries (near the Argentine border), Robert S. Williams, Otto Buchtien, Karl Fiebrig-Gertz, Theodor Herzog and, after 1915, José Steinbach, the Vavilov Institute, St. Petersburg (for germplasm), Erik Asplund, the Mulford Expedition (Rusby, Cárdenas and White), Carl Troll, the Ladew Expedition (with George Tate), Boris Krukoff, Allan Beetle and Walter Eyerdam, and Edward K. Balls. Bolivians active during this period included Miguel Bang, Belisario Díaz de Romero A. and Pedro Rafael Peña de Flores in the 'Rusby era' and Martin Cárdenas H. from then until his death in 1973.

After World War II, and especially following the popular revolution of 1952, collecting declined sharply. Until the late 1970s, it was the mainly pre-war collections of the above-mentioned visitors and residents which constituted the main basis for knowledge of the flora. Most of these were held outside the country, in Argentina or in northern metropolitan countries; only in 1978 was the forerunner of the present Herbario Nacional de Bolivia established (following renewal of that at Cochabamba in 1976). In the last two decades many new collections have been made over much of the country.

The reports of the nineteenth and early twentieth century collections were by and large written up in isolation from one another, a situation with parallels in other (at least initially) 'open-field' parts of the tropics, such as Borneo and New Guinea. Indeed, Funk and Mori speak of 'four independent projects' in the early twentieth century (in addition to the results of the New York botanists).[43] The only consolidated treatments of the flora following Weddell's *Chloris andina* (**303**) have been the vegetatiological account of Herzog (1923) and the checklist by Foster (1958), the latter one of only a few publications from an abortive Bolivian flora project at the Gray Herbarium. No sufficient basis yet exists for a definitive national flora and the only current general coverage is thus to be had in the revisions of *Flora Neotropica*. Some local floras are, however, available, and plants of economic interest were treated by Cárdenas (1969). In addition, two important recent contributions are *Guía de árboles de Bolivia* (1993; see below), initiated by David N. Smith and written under the editorship of Timothy Killeen, Emilia García E. and Stephan G. Beck, and *Gramineas de Bolivia* (1998) by Stephen A. Renvoize with Ana Antón and Beck.[44]

Progress
The period to 1923 is covered in T. HERZOG, 1923. *Die Pflanzenwelt der bolivischen Anden und ihres östlichen Vorlandes* (see below under **340**), pp. 1–4. Leipzig. A useful review of the state of knowledge through the 1980s is presented in J. C. SOLOMON, 1989. Northwestern South America. In *Floristic inventory of tropical countries* (see **General references**), pp. 455–463. Collectors have been surveyed by M. CÁRDENAS, 1952. Exploradores botánicos de Bolivia. *Revista Agric.* (Cochabamba) **10**(7): 26–45, and in cyclopedic form by V. A. FUNK and S. A. MORI, 1989. *A bibliography of plant collectors in Bolivia.* 20 pp. (Smithsonian Contr. Bot. 70). Washington, D.C..

Bibliographies. General and divisional bibliographies as for Division 3.

Regional [national] bibliography
ALIAGA vda. DE VIZCARRA, I., 1978. *Bibliografía boliviana de recursos vegetales.* 14 pp. La Paz: Academia Nacional de Ciencias de Bolivia. [Classified alphabetical listing, with index to authors; sources. Oriented towards applications.]

Indices. General and divisional indices as for Division 3.

340
Region in general

Apart from Foster's uncritical compiled checklist, no general flora or enumeration of Bolivian vascular plants is available. The best introduction to the flora, by far, is the following: T. HERZOG, 1923. *Die Pflanzenwelt der bolivischen Anden und ihres östlichen Vorlandes.* viii, 258 pp., 25 text-figs., 3 maps (Die Vegetation der Erde 15). Leipzig: Engelmann. For a partial list of series of 'contributions' upon which much of present floristic knowledge is based, see below under **Partial works**. For economic plants, see M. CÁRDENAS, 1969. *Manual de plantas económicas de Bolivia.* 421 pp., illus. Cochabamba: 'Icthus'.

FOSTER, R. C. (comp.), 1958. *Catalogue of the ferns and flowering plants of Bolivia.* 223 pp. (Contr. Gray Herb., N.S. 184). Cambridge, Mass.

Unannotated, uncritical systematic list of vascular plants, without index; includes references to place of publication and limited synonymy.

Partial works
All the following works are lists of collections with descriptions of novelties and other notes, usually in a systematic arrangement. In some cases, family treatments are by specialists, notably in Herzog's report. For further bibliographic and other details, see Blake and Atwood (1942), pp. 237–238 (under **General bibliographies**). There are also some local florulas: Santa Cruz, by R. Peña de Flores (*Flora crucena*, 1901; 2nd edn., 1944; 3rd edn., 1976), La Paz, by B. Díaz de Romero (*Florula pacensis*, 1919–20); Potosí, by M. Cárdenas (*Plantae potosinae: catálogo*, 1932), Cochabamba, by Hno. Adolfo María (A. Jiménez) (*Nomina de las plantas recolectadas en el valle de Cochabamba*, 1962–66; *Flora de Cochabamba*, 1984) and, on the Amazon region, by N. Kempff (*Flora amazonica boliviana*, 1976).

BUCHTIEN, O., 1910. *Contribuciones á la flora del Bolivia*, I. 197 pp. La Paz. [Localities, without *exsiccatae*. Not continued.]

HERZOG, T., 1913–22. *Die von Dr. Herzog auf seiner zweiten Reise durch Bolivien in den Jahren 1910 und 1911 gesammelten Pflanzen*, I–VI (Meded. Rijks-Herb. Leiden 19, 27, 29, 33, 40, 46). Leiden. Continued as *idem*, 1945. Plantae a Th. Herzogio in itinere eius boliviensi altero annis 1910 et 1911 collectae, VII. *Blumea* 5: 641–685; J. T. KOSTER, 1948. Plants collected by Th. Herzog on his second Bolivian journey, [VIII]. *Ibid.*, 6: 266–273; and C. EPLING, 1950. Plants collected by Th. Herzog on his second Bolivian journey, 1910–11, [IX]. *Ibid.*: 355–357. [Many specialist treatments incorporated in the nine parts of this series.]

KUNTZE, O., 1898. *Revisio generum plantarum*, 3(2): 1–384. Leipzig. [Report of his Bolivian collections.]

RUSBY, H. H., 1893–96. On the collections of Mr Miguel Bang in Bolivia, I–III. *Mem. Torrey Bot. Club* 3(3): 1–67; **4**: 203–274; **6**: 1–130. Continued as *idem*, 1907. An enumeration of the plants collected in Bolivia by Miguel Bang, with descriptions of new genera and species. Part IV. *Bull. New York Bot. Gard.* **4**: 309–470. [Systematic treatment, lacking the grasses.]

RUSBY, H. H., 1910–12. New species from Bolivia collected by R. S. Williams. *Bull. New York Bot. Gard.* **6**: 487–517; **8**: 89–135. [Lacks pteridophytes, grasses and orchids.]

RUSBY, H. H., 1927. Descriptions of new genera and species of plants collected on the Mulford Biological Expedition of the Amazon Valley, 1921–2. *Mem. New York Bot. Gard.* **7**: 205–387, figs. 1–8. [Descriptions of new taxa and miscellanous notes.]

RUSBY, H. H., E. G. BRITTON and N. L. BRITTON, 1888–1902. An enumeration of the plants collected by Dr. H. H. Rusby in South America, 1885–1886, I–XXXII. *Bull. Torrey Bot. Club* **15–29**, *passim*. [Covers non-vascular and vascular plants; most collections from Bolivia.]

Woody plants (including trees): guide

KILLEEN, T. J., E. GARCÍA E. and S. G. BECK (eds.), 1993. *Guía de árboles de Bolivia*. vi, 958 pp., 126 text-figs., unnumbered fig., maps (part col.). [La Paz]: Herbario Nacional de Bolivia; [St. Louis, Mo.]: Missouri Botanical Garden.

Illustrated descriptive tree flora with keys, limited synonymy, indication of distribution, representative *exsiccatae*, habitat, and special features; families and genera also described, with key literature. A general bibliography, departmental maps, an illustrated glossary, lexica to vernacular names, and index to all scientific and vernacular names follow. The introductory part encompasses the background to the work, a chapter on vegetation (with map), the system of families adopted (along with statistics), and the general keys (pp. 31–73).

Region 35/36

Brazil

Area: 8 511 965 km^2. Vascular flora: over 55 000 species (Prance, 1979, p. 55; see **General references**). – The region corresponds to the political limits of mainland Brazil. The Atlantic islands of Fernando Noronha (with Rocas) and Trindade (with Martin Vaz) are respectively at **033** and **036**.

Brazil, the country with the largest vascular flora on earth, has also had a long and illustrious botanical tradition. The period up to the completion of *Flora brasiliensis* was thoroughly reviewed in that work in 1906 by Ignatz Urban, its final editor-in-chief. Important early contributors include Georg Marggraf, associated with the short-lived seventeenth-century Dutch colony in the northeast, his contemporary Frei Cristóvão in Maranhão, and, from 1786, Frei Velloso in Rio de Janeiro – by then the Brazilian capital. Rio de Janeiro and other coastal ports were also visited for short periods by foreign naturalists on circumnavigations and survey voyages, including Philibert Commerson and Banks and Solander, but movement was restricted. The Amazon Basin was the main objective of the *viagem filosófica* of Alexandro Rodrigues Ferreira in 1783–92 but botanically it achieved little. The modern history of Brazilian botany thus only follows the arrival of the Portuguese court under Dom João VI in 1807–08 and the opening of the country to commerce.

Under the Empire the arts and sciences in general flourished. In this more progressive era a 'veritable parade' of European visitors arrived. These included Friedrich Sellow, Ludwig Riedel (who would remain for most of the rest of his life, becoming in 1842 the first director of the Museu Nacional) and Georg Langsdorff, Prince Maximilian zu Wied-Neuwied, Auguste de Saint-Hilaire, Giuseppe Raddi, P. W. Lund, James Bowie, Allan Cunningham, William Burchell, Johann Mikan and his pupil Johann Pohl and, most influentially, Karl von Martius along with the zoologist Johann von Spix in 1817–20. Naturalists on post-Napoleonic exploring voyages included Adalbert von Chamisso, Johann Eschscholtz, Charles Gaudichaud, George Lay and Alexander Collie (with Beechey), and Charles Darwin. Most concentrated on the south and central parts of the country but Darwin also explored in Bahia and Spix and Martius carried out in 1819–20 the first real biological survey of the Amazon.

Foreign collectors continued to be active through most of the Brazilian Empire through most of the remainder of the nineteenth century; indeed, with a relatively stable government it became one of the most popular areas in the world for biological exploration.

Notable botanically active naturalists of mid-century included the Dane Peter W. Lund (from the 1830s, later settling at Lagoa Santa in Minas Gerais), Eduard Poeppig (1832, following his departure from Peru), George Gardner (1836–40), focusing on northern Minas Gerais including the sensitive diamond-mining districts, and Alfred Russel Wallace (1848–52) and Richard Spruce (1849–55), forever associated with the Amazon Basin. Later in the nineteenth century and up to 1914, visiting and resident collectors included Anders Regnell, Fritz Müller (in Santa Catarina, where many Germans settled after 1848), Auguste Glaziou (under contract to the Brazilian government), the pioneer ecologist Eugen Warming (especially around Lagoa Santa), the palm specialist João Barbosa Rodrigues, and Spencer Moore (in Mato Grosso). In the early twentieth century they were followed by Carl Lindman and Gustav Malme, Ernst Ule (including Amazonia), Per Dusén (especially in Paraná), Jacob Huber (in Amazonia), the young Adolfo Ducke and Frederico C. Hoehne (the former in Amazonia, the latter with the Rondon frontier geographical commission and the Roosevelt-Rondon expedition), and Philipp von Luetzelburg (with the Inspectoria Federal de Obras contra as Sécas, particularly in Bahia and the Northeast).

The first national institutions, both established in 1808 in Rio de Janeiro, were the National Library, repository of Vellozo's manuscripts and until 1842 also responsible for natural history collections, and the Jardim Botânico. The latter, however, functioned largely as an imperial vegetable and botanic garden until late in the century. Later foundations included, in addition to the already-mentioned Museu Nacional, the Museu Paraense in Belém, established by Emílio Goeldi in 1895, and, in São Paulo, the Museu Paulista and the Instituto de Botânica (the latter organized as a distinct entity in 1917 and later strongly developed by Hoehne). In Rio de Janeiro, the Jardim Botânico was from 1890 – after the fall of the monarchy – elevated into a scientific institution under João Barbosa Rodrigues. By then, however, a very large proportion of collections from Brazil were only represented abroad, as elsewhere an all too common obstacle for local students.

The only comprehensive work, *Flora brasiliensis*, was begun in 1840 under von Martius and completed, save for a few families, in 1906. It is the largest single work of its kind, covering 22767 species of vascular plants and Musci. With the great increase in known taxa in most families since their publication, however, the work is now outdated. A successor, *Flora brasílica*, was initiated in 1940 by Hoehne following consolidation of his Instituto de Botânica but only a very small proportion of the flora was covered before an institutional shift in goals led to termination of that project.[45]

In the twentieth century, the local botanical profession has steadily grown and is now the largest in South America. It has made considerable contributions to floristic knowledge in all parts of the country, sometimes in conjunction with overseas collaborators. Considerable collecting programmes have been prosecuted in Amazonia (from Huber and Ducke onwards, with a renewal taking place during and after World War II including the foundation in 1945 of the Instituto Agronomico do Norte in Belém and, in Manaus, postwar plans for an international tropical research institute), the Nordéste and Bahia (through the Comissão das Obras contra as Sécas do Nordéste and, at Pernambuco, by Artura Dardano de Andrade-Lima), Mato Grosso (initially through the Comissão Rondón), the Planalto (especially from about 1930 with accelerated development under Presidents Getulio Vargas and Juscelino Kubitschek), and in the southern states.[46] There was also a small continuing flow of visiting botanists. To these and more recent developments, some of the latter in the form of joint operations with North American, British or other foreign institutions, should be added the independent publications of João Angely (mainly on southern Brazil), the work of A. Curt Brade and his Herbarium Bradeanum in Rio de Janeiro, and, in the 1970s and 1980s, the national government's 'Programa Flora', an extensive collecting, inventorization and educational programme.[47]

The existence of 'Programa Flora' and similar initiatives may be seen as attempts to cope with and counter a gradual tendency towards decentralization, inevitable in a large, populous country. Nearly 50 herbaria were on record in 1977 and 85 by 1990. Floristic efforts have also become more oriented to state and local level, resulting in the appearance in recent decades of several state (Rio Grande do Sul, Santa Catarina, Paraná, São Paulo and Goiás) and local floras. Those for the two southernmost states and Goiás are serial works (comparable with the 'tropical' floras prepared in Europe), with one, *Flora ilustrada catarinense*, now far advanced (and arguably the best in south-central South America). In two states these floras are associated with graduate schools. The others are largely compiled, and their author had made a beginning on a similar work for Minas Gerais. For more restricted areas, a number of florulas, some of them now fairly old, are available for city environs (e.g., São Paulo), discrete areas (e.g., the Ilha da Santa Catarina), points of interest (such as the Itatiaia massif, the Serra dos Orgãos or the Chapada Diamantina, the last-named partly covered by the recent *Florula of Mucugê* and the *Flora of the Pico das Almas*) or on particular vegetation formations (such as the east coast *restingas*, those in Rio de Janeiro being the subject of the *Flora ecológica de restingas* edited by Fernando Segadas-Vianna). Further coverage of the Chapada Diamantina, with its unusual *campo rupestre* flora, is underway, while in Amazonia a florula for the Ducke Reserve near Manaus appeared in 1999. A *Flora of Bahia* and a new *Flora of São Paulo* are also projected.[48]

At the present time, the Brazilian flora remains imperfectly known, with collections and documentation as yet inadequate. Poorly known areas, of which several are under threat, abound; these have been noted by Prance (1979), Daly and Prance (1989) and Mori (1989). In general, they are found in the central, western, northern and northeastern parts of the country; in the better-known south they are more localized.

Progress

No single general study of the history of botany in Brazil has been seen; available information remains widely scattered among Brazilian and extra-Brazilian works. Studies prior to the arrival of Piso and Marcgrave in 1637, the conventionally accepted starting-date for serious natural history and botanical studies in the country, are treated in F. C. HOEHNE, 1937. *História da botânica e agricultura do Brasil do século XVI*. São Paulo: Companhia Editôra Nacional. There is also a short history in A. LÖFGREN, 1914. Breve historico das explorações botanicos no Brasil. *Cháceras e Quintais* 10(5): 350–360.

The principal, more or less exhaustive anthologies of collectors are respectively I. URBAN, 1906. Vitae itineraque collectorum botanicorum. In G. F. P. von Martius *et al.* (eds.), *Flora brasiliensis*, 1(1): 1–152. Munich; F. C. HOEHNE, 1941. Notas biobibliográficas de naturalistas botânicos. In *idem* (with M. Kuhlmann and O. Handro), *O jardim botânico de São Paulo*, pp. 19–246, illus. São Paulo: Departamento de Botânica,

Secretaria da Agricultura, Indústria e Comércio [do Estado]; and N. PAPAVERO, 1971. *Essays on the history of neotropical dipterology, with special reference to São Paulo*. 2 vols. São Paulo: Museu de Zoologia, Universidade de São Paulo. Amazonian collectors are accounted for in G. T. PRANCE, 1971. An index of plant collectors in Brazilian Amazonia. *Acta Amazonica* 1(1): 25–65. Information on collectors and authors not covered in these works is scattered and, although some area reviews (e.g., Paraná, Santa Catarina) exist, reference should be made to general contributions on South America (see **Division 3**, introduction, p. 314).

Parts of Brazil were reviewed in the 1980s for Campbell and Hammond, *Floristic inventory of tropical countries* (under **General references**): D. C. DALY and G. T. PRANCE, 1989. Brazilian Amazon (pp. 401–426), and S. A. MORI, 1989. Eastern, extra-Amazonian Brazil (pp. 427–454).

From 1957 to 1973, a more or less regular botanical bibliographic index was produced (Anonymous, 1957–73; see below).

Bibliographies. General and divisional bibliographies as for Division 3.

Regional bibliographies
In addition to the botanical bibliographies described below, reference may be made to R. B. DE MORAES, 1958. *Bibliographia brasiliana*. 2 vols. Amsterdam/Rio de Janeiro: Colibris. This accounts for books by Brazilian authors from 1504 to 1900 including those published abroad before independence. An older, historically oriented but very incomplete botanical bibliography is A. J. DE SAMPAIO, 1924–28. Bibliographia botânica, relativa á flora brasileira, com inclusão dos trabalhos indispensaveis aos estudos botânicos no Brasil. *Bol. Mus. Nac. Rio de Janeiro* 1 (1924): 111–125, 225–245; **2**(3) (1926): 35–61; **2**(5) (1926):19–38; **3**(1) (1927): 37–45; **4**(3) (1928): 97–119.

BENEVIDES DE ABREU, C. L., N. M. FERREIRA DA SILVA *et al.*, 1972–79. Bibliografia de botânica, I–VI. *Rodriguésia* 27(39, Anexo)–31(51, Anexo), *passim*. [Taxonomic bibliography, to date covering dicotyledonous families from A through N. Published as supplementary parts to successive volumes of the journal.]

Indices. General and divisional indices as for Division 3.

Regional indices

Anonymous, [1957]–73 . *Bibliografia brasileira de botânica*, 1–7 (in 6). Rio de Janeiro: Instituto Brasileiro de Bibliografia e Documentação. [1: 1950–55; 2: 1956–58; 3: 1959–60; 4: 1961–69; 5: 1970; 6–7: 1971–73 .][49] [Unannotated lists.]

350

Region in general

The only comprehensive floras of Brazil are *Flora brasiliensis* (1840–1906) and the incomplete *Flora brasílica* (1940–68). Much of the country is, however, covered by the serially published, still far from complete *Flora Neotropica* (1967–; see **301/I**). A further development was the launching by the federal government in 1975 of 'Programa Flora', a floristic inventorization programme including five regional projects (Teixeira, 1984; see Note 47); this has in particular added considerably to knowledge of the Amazon region. In addition, there are a number of sets of keys to families, one of which also covers genera; four of these are listed below.

Keys to families and genera
Several sets of keys for student and general use have appeared in Brazil during the twentieth century. Some, including that by Joly, were derivatives of Thonner's *Analytical key to the natural orders of flowering plants* (1917; see **000**, p. 95), as discussed in the 1981 version of that work by R. Geesink *et al.* Joly has also named as Thonner derivatives two other keys (not seen): *Chave para a determinação das famílias das plantas Pteridophytas, Gimnospermas e Angiospermas brasileiras ou exóticas encontradas no Brasil* by P. de Tarso Alvim (1943, 1950) and a work by Rawitscher and Rachid-Edwards (1956). The first work described below is the most substantial since that of Löfgren (1917). For dicotyledons, it succeeds in particular L. J. Barroso, 1946. *Chaves para a determinação de gêneros indígenas e exóticas das dicotiledôneas do Brasil.* 2nd edn. 272 pp. Rio de Janeiro: Serviço de Documentação do Ministério da Agricultura.

Barroso, G. M. *et al.*, 1978–86. *Sistemática de angiospermas do Brasil.* Vols. 1–3. 1068 text-figs. São Paulo: Editora da Universidade de São Paulo/Livros Técnicos e Científicos Editora (vol. 1); Viçosa/Belo Horizonte, Minas Gerais: Imprensa Universitária, Universidade Federal de Viçosa. [Descriptions of orders and families (with figures of technical characters) and keys to genera. Derivations of names of

genera appear at intervals and, for each order, key references for it and all its families are furnished. The introductory section in vol. 1 includes a history of systematic botany. – Arrangement is after the 1968 Cronquist system; subclasses Magnoliidae, Hamamelidae, Caryophyllidae and Dilleniidae in vol. 1, Rosidae in vol. 2 and Asteridae in vol. 3. A final volume, to cover monocotyledons, has not so far been published.][50]

Barroso, L. J., 1946. Chaves para a determinação de gêneros indígenas e exóticas das monocotiledôneas do Brasil. *Rodriguésia* 9(20): 55–77. [Keys to genera in 25 families from Alismaceae through Zingiberaceae; synonymous genera listed, pp. 66–74 and synoptic list of accepted genera by families, pp. 74–77. Not complete; 12 families omitted including for example Cyperaceae, Gramineae and Orchidaceae.]

Goldberg, A. and L. B. Smith, 1975. *Chave para das famílias espermatofíticas do Brasil*. 204 pp., 69 pls. In R. Reitz (ed.), *Flora ilustrada catarinense*, 1. parte, *hors série*. Itajaí, Santa Catarina. [Illustrated key to families, with glossary and references; the plates include explanations.] See also *Chave para das famílias de pteridófitas da Região Sul do Brasil. Ibid.*, 1. parte, *hors série*. Itajaí.

Joly, A. B., 1977. *Botânica. Chaves de identifição das famílias de plantas vasculares que ocorrem no Brasil, baseadas em chaves de Franz Thonner*. 3rd edn. 159 pp. São Paulo: Companhia Editôra Nacional. (Provisional edn., 1968; 1st edn., 1970; 2nd edn., 1975.) [Analytical keys.][51]

Löfgren, A., 1917. *Manual das familias naturaes phanerogamas: con chaves dichotomicas das familias e dos gêneros brasileiros*. xviii, 611 pp. Rio de Janeiro: Imprensa Nacional. [Keys to families and genera, with descriptions of the latter; also includes some vernacular names and notes on uses.]

Hoehne, F. C. *et al.*, 1940–68. *Flora brasílica*. Fasc. 1–12. Illus. (some col.). São Paulo: Instituto de Botânica.

A comprehensive, illustrated documentary descriptive flora of vascular plants, projected as a successor to *Flora brasiliensis* but in a somewhat smaller format; each treatment includes keys to genera and species, synonymy (with references) and citations, vernacular names, localities with *exsiccatae* and general summary of extralimital distribution, taxonomic commentary, notes on habitat, special features, biology, etc., and (in some parts) indices to botanical names. [Not completed. The 12 fascicles published before termination of the project comprise vols. 2(2), 9(2), 12(1, 2, 6, 7), 15(2), 25(2, 3, 4), 41(1) and 48. Vol. 12 treats Orchidaceae, a specialty of Hoehne.]

Martius, K. F. P. von, A. W. Eichler and I. Urban (eds.), 1840–1906. *Flora brasiliensis*. 15 vols. (in 40). 20 733 pp. (i.e., columns), 3811 pls., 2 maps.

Munich: Fleischer (later Vienna; Leipzig: Hiersemann). (Pp. 317–672 and pls. 76–133 of vol. 3(4) reprinted 1920, Leipzig: Hiersemann; whole work (in reduced format), 1966–67, Lehre, Germany: Cramer.)

A large-scale, comprehensive, generally critical documentary descriptive flora of vascular plants and Musci of Brazil and adjacent lands (22767 species in all); includes synoptic or analytical keys to genera and species (or groups of species), full synonymy (with references) and citations (including appropriate revisions and monographs), localities with *exsiccatae* or authorities, general indication of internal and extralimital range, taxonomic commentary, notes on habitat, phenology, properties, uses, phytogeography, etc., figures of representative species (some executed by nature-printing, a widely practiced graphic art form in the mid-nineteenth century) on separate plates, and index to all botanical names at end of each partial volume. The general introduction to the work in volume 1(1), partly by von Martius and partly by Urban, incorporates (pp. 1–152) a cyclopedia of collectors and collaborators (the latter numbering about 60), a chronologically ordered list of all 130 fascicles, floristic statistics, and a synopsis of and general index to families together with the 59 celebrated 'Tabulae physiognomicae' illustrating various aspects of the Brazilian landscape and vegetation.[52]

Woody plants (including trees)

As with vascular plants in general, Brazil has more species of trees than any other country, with a total in the thousands. While several dendrological and census publications exist, few cover the whole country and none is 'accessible'. Since 1992, however, Harri Lorenzi has been producing an attractive series of well-illustrated, all-color dendrological atlases worthy of note here: H. LORENZI, 1992– . *Árvores brasileiras: manual de identificação e cultivo de plantas arbóreas nativas do Brasil*. Vols. 1–3. Nova Odessa, São Paulo: Instituto Plantarum de Estudios da Flora. (Vol. 1 revised 1998; vol. 2, 1998; vol. 3, 1999.) Families and genera in each volume are alphabetically arranged.

I. Amazonia

Incorporating the 'Naiades' of Martius, Brazilian Amazonia is taken here to correspond to that recognized by J. M. Pires in 1972 (Daly and Prance, 1989, p. 402; see **Progress** above): an area of some 3700000 km² or anywhere from 62 to 74 percent of all Amazonia. This differs from the larger 'Amazônia Legal', defined by the Brazilian government as a politico-economic area of 4975527 km². – Vascular flora: *c.* 30000 species (Prance in Goodland and Irwin, 1975, pp. 101–111).

Probably because of its great metaphysical attraction, Brazilian Amazonia has had a 'long and illustrious history of botanical collecting' resulting in what, for comparable parts of the tropics, is a relatively good and well-distributed stock, both in Brazil and abroad. The early work, particularly of Ferreira, Martius and Spruce, was followed by a great increase in mainly domestically based activity during the first economic 'boom' of the late nineteenth and early twentieth centuries. This included establishment of a regional herbarium at Belém which from 1895 to 1915 served as a base for expeditions and local research, particularly by Jacques Huber and Adolfo Ducke. Many publications on the flora also appeared. In the 1920s and 1930s, however, collecting decreased considerably, save for the work of Ducke and Boris Krukoff and their associates, and would not revive until the 1940s.[53]

The modern development of Brazilian Amazonia may be said to have begun with initiatives around 1940 by the federal government under then-president Getulio Vargas. A second 'rubber boom' during World War II, the 'Batalha da Boracha', and associated economic aid gave rise to new infrastructure. Among the many developments was the foundation at Belém, in 1945, of the Instituto Agronômico do Norte (now the Centro de Pesquisa Agropecuária do Trópico Úmido). This was followed in 1954 by a more general scientific research center for the Amazon at Manaus, the Instituto Nacional de Pesquisas da Amazônia (INPA). These, along with the revived Museu Paraense (originally established by Emílio Goeldi in the late nineteenth century), have since become the leading botanical centers in the region. In 1971 INPA established the serial *Acta Amazonica*, which continues.

Land conversion, major projects, conservation and environmental concerns, and a recognition that most of the region, along with many other parts of Brazil, was still imperfectly known botanically led to organization of a special collecting and inventorization programme of the Brazilian National Research Council (CNPq) in 1975, the 'Programa Flora'. Its Amazonian component, Projeto Flora Amazônica (PFA), was initiated in 1976 and by 1986 had conducted 30 expeditions along with associated database development, technical work and other activities including graduate education. Nearly 50000 collections of plants and fungi were obtained through 1984, of which some 2 percent have been new to science.[54]

The PFA collections are, however, only a part of the more than 200000 thought to have been collected since 1953. While collecting should, and will, continue, more attention should be given to florulas for limited areas, especially around Manaus, Belém and other places where educational programmes are active. Such a project is now in place for the INPA Reserva Ducke near Manaus; already, intensive collecting has added considerably to the known flora, with many novelties as well as significant range extensions.

With relatively few floras or florulas yet available, much use must be made of separate revisions, monographs, collection reports and miscellaneous papers. Published work, however, may become rapidly 'dated' given the still-imperfect level of botanical knowledge.[55] Nevertheless, recent family and infrafamilial treatments, notably those appearing in *Flora Neotropica*, are gradually contributing to a future synthesis.

On noteworthy trees and other plants see P. LE COINTE, 1947. *Árvores e plantas uteis da Amazonia brasileira*. 506 pp. São Paulo. Contributions from 'Projeto Flora Amazônica' have appeared in *Acta Amazonica* 14(1/2), Suplemento (1984) and subsequent issues as well as in *Flora Neotropica* and elsewhere.[56]

Bibliography

ANONYMOUS, 1963–70. *Amazônia: bibliografia*. 2 vols. Rio de Janeiro: Instituto Brasileiro de Bibliografia e Documentação/Instituto Nacional de Pesquisas da Amazônia, Conselho Nacional de Pesquisas. (Reprinted 1975, Königstein/Ts., Germany: Koeltz.) [An unannotated general bibliography covering the years 1614–1962, arranged in the first instance following the UDC (see pp. xxvii–xxix) and then by author; botany, pp. 417–473. An author index appears at end. The second volume is a supplement, covering additional items as well as extending coverage back to 1601 and forward through 1970; botany, pp. 226–249.]

II. The Planalto

This includes all of Central Brazil from Bahia to Rondônia and southwards to Mato Grosso and Minas Gerais, with an extension into eastern Bolivia (340). In many parts it overlaps or interdigitates with Amazonia and other regions, a phenomenon also visible in the vegetation. – Vascular flora: *c.* 10000–15000 species.

The region incorporates the woody and herbaceous *campos* (or woodlands, tree and shrub savanna, and open savanna) and associated gallery forests characteristic of the central highlands – the major part of Martius's 'Oreades', with its rich *flora geral* – and the ranges of the Serra do Cipó in the east. A substantial amount of collecting has been accomplished in the past 175 years and, with the influx of inhabitants, the establishment of tertiary institutions beginning around 1930 and especially the occupation of the new national capital in 1960 and the foundation of the Universidade de Brasília, much more detailed systematic, ecological and biological research has been initiated. A substantial literature is now available, including many species lists associated with vegetation studies; however, no modern consolidated floras or enumerations have yet been published. Useful alternatives are given below under **Partial works**. Convenient introductions to the vegetation include G. EITEN, 1972. The cerrado vegetation of Brazil. *Bot. Rev.* 38: 201–341, and R. J. A. GOODLAND and M. G. FERRI, 1979. *Ecologia do cerrado*. Illus. Belo Horizonte: Itatiaia. Mention may also be made of the four volumes of 'Simposiões sobre o Cerrado' published between 1963 and 1977 under the editorship of M. G. Ferri; these contain some floristic papers.

Partial works

For *Florula of Mucugê* (1986) by R. M. Harley and N. A. Simmons and *Flora of the Pico das Almas* (1995) edited by B. Stannard, see 359. For *A flora de Matto Grosso* (1916) by A. J. de Sampaio and *Prodromus florae matogrossensis* (1998) by Balthasar Dubs, see 363.

FERRI, M. G., 1969. *Plantas do Brasil: espécies do cerrado/Plants of Brazil: species of the cerrado*. 239 pp., 100 illus. São Paulo: Blücher.

An atlas-flora (in Portuguese, with captions also in English) of selected *cerrado* species, with full-page figures and facing descriptive text with scientific and vernacular names; glossary (pp. 227–239). The introduction is bilingual.

HOEHNE, F. C. and J. G. KUHLMANN (comp.), 1951. *Índice bibliográfico e numérico das plantas colhidas pela Comissão Rondón*. 400 pp. São Paulo: Instituto de Botânica.

Pages 110–400 of this work comprise a systematic enumeration of plants (mainly of the southwestern part of the Planalto, in Mato Grosso and Rondônia), based in the first instance on the collections of the Comissão Rondón but incorporating other published records. Entries include synonymy, citations, taxonomic commentary, citations of *exsiccatae* with localities, and field observations, and (at end) an index to families.

WARMING, E. (ed.), 1867–94. *Symbolae ad floram Brasiliae centralis cognoscendam*. Particulae i–xl (in 4 series). 1239 pp., [15 pp. prefaces]. Copenhagen. (Reprinted from *Vidensk. Meddel. Dansk. Naturhist. Foren. Kjøbenhavn* (1867–93, *passim*.) Incorporates H. KIÆRSKOU, 1893. *Enumeratio myrtacearum brasiliensium*. 200 pp., 24 pls. (Symbolae ad floram Brasiliae centralis cognoscendam. Particula xxxix). Copenhagen: J. Gjellerup; Christiana: Cammermeyer.

A series of family revisions, the majority by specialists; each consists of a systematic enumeration of taxa, with nomenclature, descriptions of novelties, localities with *exsiccatae*, distribution, phenology, biology and taxonomic commentary. The work was based in the first instance upon Warming's collections from central Minas Gerais in 1863–66 but also included those of P. W. Lund (1820s and 1830s) and others from Brazil represented in the Botanical Museum in Copenhagen.

Woody plants

RIZZINI, C. T., 1971. Árvores e arbustos do cerrado. *Rodriguésia* 26(38): 63–77. [Alphabetically arranged nomenclator of species with their families; 653 accounted for. Floristic statistics and lists of 'vicariant species' paired between wet forest and *cerrado* are appended.]

III. 'O Nordéste'

This encompasses northeastern Brazil from the limits of Amazonia in Maranhão to Bahia. It includes the vast dry *sertão*, an area of low and uncertain rainfall. – Vascular flora: no data.

The major plant cover is the thorny *caatinga* – a subset of Martius's 'Hamadryades'. The eastern part of the Atlantic coast, however, features an attenuated northwards extension of the humid *mata atlantica*, and scattered through the interior are the *brejos*, pockets of closed forest at higher elevations. The latter contain many phytogeographical 'outliers', notably from the southeast.

In spite of many gaps, the subregion is considered botanically moderately well known. Nevertheless, in the words of Adolfo Ducke, 'até o presente, a botânica nordestina era produto quase exclusivo de trabalhos de campo feitos por coletores itinerantes, com identificação científica posterior do material reunido'. Though written from Fortaleza, in his adopted state of Ceará, they are equally valid elsewhere. No concerted

effort towards production of complete floras or enumerations for the whole or any part of the subregion has so far been attempted although, as Ducke has indicated, documentation of various kinds has long existed. The earliest dates from 1648 with Willem Piso and Georg Marggraf's *Historia naturalis Brasiliae*; in the twentieth century it has been augmented by substantial contributions from Philipp von Luetzelburg, Bento Pickel, Lyman B. Smith and, after World War II, Artura Dárdano de Andrade-Lima as well as Ducke. Revisions of Gramineae and Leguminosae for Bahia have been published, and a flora of that state is projected.[57]

IV. The East Coast

Includes Atlantic coastal Brazil from Bahia southwards. – Vascular flora: over 15000 species. The vicinity of Rio de Janeiro, including the Serra do Mar, shows exceptionally high diversity.

For Fernando Segadas-Vianna's flora of the southeastern *restingas*, see below. No consolidated accounts yet exist for other major vegetation formations, including the much-reduced *mata atlantica*. As elsewhere, recent work and flora-writing has focused on circumscribed sites, such as Macaé do Cima near Nova Friburgo (see **365**).

SEGADAS-VIANNA, F. (ed.), 1965–78. *Flora ecológica de restingas do sudeste do Brasil*, II: *Flora*. Fasc. 2–23. Illus. Rio de Janeiro: Museu Nacional/ Universidade Federal do Rio de Janeiro.

Illustrated descriptive flora; each fascicle incorporates a single family, with keys, synonymy (including references), vernacular names, local and extralimital range, and notes on habitat, ecology, etc. A general introduction appears in fasc. 2. [Not completed; fasc. 1 moreover was not published. No set sequence of families was followed.][58]

351

Rondônia

Area: 238379 km^2. Vascular flora: no data. – This state is now, along with western Mato Grosso, the heart of the 'Polonoroeste', wherein much settlement has taken place since the 1960s. It was formerly part of Mato Grosso; when first separated, it was known as Guaporé.

Under the 'Projeto Flora Amazônica' a lot of new collecting was done, but more is needed. The only organized floristic account, based mainly on field work in 1908–23, remains *Indice bibliográfico numérico das plantas colhidas pela Comissão Rondón* by F. C. Hoehne and J. G. Kuhlmann (1951; see **350/II**). The state is also covered in *A flora do Matto Grosso* by A. J. de Sampaio (1916; see **363**).

352

Acre

Area: 153 698 km^2. Vascular flora: no data.

No floras or florulas are available for this state, part of Bolivia before 1903. Under 'Projeto Flora Amazônica' and other programmes, considerable collecting has been carried out since the 1960s. Works relating to eastern Peru (**330**) may be useful for identification.

353

Amazonas

Area: 1 567 954 km^2. Vascular flora: no data.

No floras or florulas have yet been published; however, work is in hand towards a florula of Reserva Ducke near Manaus. The majority of collections in this still-undercollected large state have come from the environs of Manaus, the capital and veritable 'center' of the Amazon, and along the rivers and roads radiating from it. An additional 'node' of collections exists for the Humaitá region in the south near the Rondônia boundary.

Partial work

An important addition to the literature is the long-awaited *Flora da Reserva Ducke*, covering an area of Amazonian forest of 100 km^2 north of Manaus.

RIBEIRO, J. E. L. DA S. *et al.*, 1999. *Flora da Reserva Ducke*. xvi, 800 pp., illus. (mostly col.). Manaus: Instituto Nacional de Pesquisas da Amazônia (in association with the Department for International Development, United Kingdom).

Copiously illustrated guide to vascular plants (*c.* 2200 species) with artificial multi-access keys (based principally upon more or less evident habit and vegetative features; pp. 94–95 comprise a 'guia rápido' of 18 characters each with up to six color-coded states linked to marks on relevant page edges), descriptive accounts of families each with one or more keyed pages of genera and species (some with sidebars covering additional species, particular morphological and other features, or uses) with photographs of key features of habit, bark, blaze, twig, leaf (fresh and dry), flower and fruit as well as vernacular names, distribution, frequency and habitat; alphabetically arranged checklist of families, genera and species (pp. 751–770); index to all names and corrigenda at end. The general part – following the table of contents – features an introduction to humid tropical forest, the phenomenon of diversity, patterns of geographical distribution, the Ducke Reserve and its habitats, the background and plan of the floristic documentation programme (of which the guide is part), participants, and an introduction to principal literature and the use of the book (pp. 14–19), taxonomy and angiosperm classification (pp. 20–23), and life-form statistics and an illustrated glossary (pp. 24–84). Pp. 85–93 encompass a nested-box key to 'guilds', in effect a modern folk-key.

354

Roraima Territory

Area: 225 017 km^2. Vascular flora: no data.

No territorial flora has been published. Recent studies have focused on limited areas; an example is the Ilha de Maracá, a group of islands in the Rio Uraricuera in the north-central part of the Territory. This was an objective of the 1987–88 Maracá Rainforest Project, a joint undertaking of Brazilian and British botanists. The first contribution towards a florula is G. P. LEWIS and P. E. OWEN, 1989. *Legumes of the Ilha de Maracá*. xvi, 95 pp., 8 col. pls., 7 figs., 2 maps. Kew: Royal Botanic Gardens.

355

Amapá Territory

Area: 142359 km². Vascular flora: no data.

No flora or florula has yet been published, though a local herbarium has been established as part of the Museu Ângelo Moreira da Costa Lima in Macapá.

356

Pará

Area: 1246833 km². Vascular flora: no data. Within the state are the Zona Bragantina (30000 km²), a focus of colonization in the early twentieth century, the area of the Projeto Grand Carajás through the center and into Maranhão, and, along the Amapá boundary in the north, the Jari integrated development. All of these have undergone extensive forest conversion.

No separate general flora or targeted florulas have yet been published. The Museu Goeldi and the former Instituto Agronomico do Norte (presently the Centro de Pesquisa Agropecuária do Trópico Úmido, or CPATU), with between them the largest collections stock in Amazonia, would provide a good foundation for any future projects.

357

Maranhão, Piauí and Tocantins

Area: 858151 km² (Maranhão, 329556 km²; Piauí, 251273 km²; Tocantins, 277322 km²). Vascular flora: no data. – This subregion, in which extensive *carnauba* palm forests occur, is intermediate between Amazonia and the Nordéste, and also contains the northern limits of the *campos*. Tocantins has only relatively recently been separated from Goiás (**361**).

Tocantins continues to be encompassed by *Flora do Estado de Goiás* (now *Flora dos Estados de Goiás e Tocantins*), but while the work is progressing it is still far from completion. Maranhão and Piauí are botani-

cally comparatively poorly known despite early beginnings to exploration, and no separate floras or enumerations for either are available. The 53 illustrated trees and other plants recorded in the early seventeenth century manuscript *História dos animais e árvores do Maranhão* by Frei Cristóvão de Lisboa were published in 1968 (ii, 187 pp. Curitiba: Universidade Federal do Paraná).

358

Ceará, Rio Grande do Norte, Paraíba, Pernambuco, Alagoas and Sergipe

Area (including Fernando Noronha): 404812 km² (Ceará, 145694 km²; Rio Grande do Norte, 53167 km²; Paraíba, 53958 km²; Pernambuco, 101023 km²; Alagoas, 29107 km²; Sergipe, 21863 km²). Vascular flora: no data. – This unit, the heart of 'O Nordeste', comprises the states of Ceará, Rio Grande do Norte, Paraíba, Pernambuco, Alagoas and Sergipe. For Fernando Noronha and Rocas, northeast of the tip of the mainland and since 1988 politically part of Pernambuco, see **033**.

The relatively extensive record of collecting and publication so far contains no recognizably floristic work. For additional comments, see above under **350/III**.

359

Bahia

Area: 566979 km². Vascular flora: 5000–6000 presently known; perhaps as many as 10000 (Harley and Mayo, 1980).

This large state is still in many parts botanically not well known, although since the 1960s much collecting has been carried out and some research centers have been established. Among the latter is the Centro de Pesquisas do Cacau (CEPEC) station west of Ilhéus in the coastal forest zone. Many new taxa and records have been found, in both 'new' and 'old' areas. No separate floras are available, although there is a long history of exploration and much scattered documentation exists.

A joint Kew–Bahian working group has been active in floristic exploration for more than 20 years. The first results were embodied in a mimeographed annotated systematic checklist of 1596 species: R. M. HARLEY and S. J. MAYO, 1980. *Towards a checklist of the flora of Bahia*. 250 pp., 4 maps, illus. (on cover). Kew: Royal Botanic Gardens. This has been followed by two florulas (see below) and by a keyed enumeration of Leguminosae (741 species, of which 698 native): G. P. LEWIS, 1987. *Legumes of Bahia*. 370 pp., illus. Kew. A full flora is now in preparation.

Partial works: Chapada Diamantina

Additional florulas in progress include *Flórula de Serra de Chapada e do Pai Inácio, Bahia* and *Checklist of Catolés, Bahia*.

HARLEY, R. M. and N. A. SIMMONS, 1986. *Florula of Mucugê*. xii, 228 pp., 2 maps. Kew: Royal Botanic Gardens.

Briefly descriptive preliminary flora of the Chapada Diamantina National Park (including the Serra do Sincorá) in the central highlands; includes accepted names, *exsiccatae* and habitats but no keys. Based mainly on post-1974 collections by Brazilian and British botanists and encompassing 670 species.

STANNARD, B. (ed.), 1995. *Flora of Pico das Almas, Chapada Diamantina – Bahia, Brazil*. xxiv, 853 pp., 51 textfigs. (incl. maps), 8 col. pls.

Manual-flora with keys, limited synonymy, descriptions, indication of habitat, localities with *exsiccatae* (no more than two per taxon) and commentary; enumeration of bryophytes, list of contributors, list of collectors on Pico das Almas from 1974 onwards, list of taxa first described from area, glossary and index to all botanical names at end. The introductory part (in both English and Portuguese) covers background, physical features, geology, climate, botanical history, vegetation formations and characteristic species, floristics, biogeography, and plant uses along with a bibliography and the plan of the work. [1044 species are accounted for here in an area of about 20 km^2 ranging from *c.* 500 to 1958 m.]

361

Goiás

The limits of the state adopted here are those since separation of Tocantins (see **357**). – Area: 340166 km^2 (since separation of Tocantins; previously 617488

km^2). Vascular flora (including Tocantins): 9605 species (Rizzo).

RIZZO, J. (ed.), 1981– . *Flora do Estado de Goiás* (later *Flora dos Estados de Goiás e Tocantins*): *Coleção Rizzo*. In fascicles. Illus. Goiânia: Universidade Federal de Goiás.

Documentary descriptive flora; each family account includes keys, synonymy (with references), citations, localities with *exsiccatae* (mostly in-house), distribution (with dot maps), notes on habitat, phenology, rarity, uses and potential, special features, and (at end) conclusions, references and index. An introductory fascicle outlines the background and reasons for the flora and includes accounts of physical features, geology, climate and hydrography, vegetation formations, collecting strategy, and revision methodology along with general references. [As of 1999, 23 fascicles covering vascular plants had been published, the latest in 1998. The change of title was effective with fasc. 15.][59]

362

Distrito Federal (Brasília)

Area: 5794 km^2. Vascular flora: no data.

The Federal Territory is now relatively well collected, and around 1990 a flora project was initiated under the direction of Taciana Cavalcanti and Carolyn Proença, Universidade de Brasília. Floras of Goiás (**361**) are useful in the absence of a more specific work.

363

Mato Grosso and Mato Grosso do Sul

Area: 1258893 km^2 (Mato Grosso, 901421 km^2; Mato Grosso do Sul, 357472 km^2). Vascular flora: *c.* 10000 species. The mutual boundary between Mato Grosso and Mato Grosso do Sul, separated in 1979, crosses the northern part of one of the area's outstanding features, the vast marshy Pantanal Matogrossense on the Bolivian and Paraguayan borders.

Serious botanical work in these states began relatively late. The first century or more of botanical work was recorded in many different outlets in addition to *Flora brasiliensis*. Among the few large-scale accounts is one of the results of an 1890s British expedition led by William Evans: *The phanerogamic botany of the Matto Grosso expedition, 1891–92* by S. LeM. Moore (1895; in *Trans. Linn. Soc. Bot.*, II, **4**: 265–516, pls. 21–39, map). The expeditions of Gustav Malme, another of the few notable early collectors, in the 1890s and 1900s did not, on the other hand, lead to a single systematically arranged report. It was not until after the work of the Comissão Rondón in the early twentieth century that interest grew in a consolidated work. The first attempt was *A flora de Matto Grosso* (1916; see below) by A. J. de Sampaio; this was followed 35 years later by *Índice bibliográfico e numérico das plantas colhidas pela Comissão Rondón* by F. C. Hoehne and J. G. Kuhlmann (1951; see **Partial works** under 350/II). Much collecting and related field work, including the joint Universidade de Brasília/Royal Society of London expeditions in the latter half of the 1960s (mainly in the northeast of Mato Grosso), has, however, since been accomplished, rendering both works quite incomplete. They have now been succeeded by *Prodromus florae matogrossensis* by B. Dubs (1998).

DUBS, B., 1998. *Prodromus florae matogrossensis*. [6], 444 pp., map, portr. Kusnacht: Betrona. (The botany of Mato Grosso, B 3.)

This work, covering the two states of the old Mato Grosso, is organized in two main sections: I, the checklist proper, and II, a catalogue of types described from the state. The checklist (pp. 1–306) includes synonymy, references and citations, and localities with *exsiccatae* but no keys or commentary. Family headings include acknowledgments, some taxonomic notes and key family and generic references. Part II includes taxonomic reductions where relevant. [The author acknowledges this as the first overall work on the states since the 1951 *Índice bibliográfico*. Some of it, inevitably, shows evidence of compilation, and moreover there is no statistical section.]

SAMPAIO, A. J. DE, 1916. *A flora de Matto Grosso. Memória em homenagem aos trabalhos botânicos da Commissão Rondón*. 125 pp., 11 maps (*Arq. Mus. Nac. Rio de Janeiro* **19**: 1–125). Rio de Janeiro.

Part 2 (pp. 26–120) of this three-part work includes an alphabetical checklist (arranged within the major groups of non-vascular plants, pteridophytes, gymnosperms, dicotyledons and monocotyledons) inclusive of scientific names, localities and sources, accounting not only for the records of the Rondon surveys of 1908–15 but also for relevant literature. Part 1 of the work is historical and part 3 bibliographic; the 10 maps are of botanical itineraries.

364

Minas Gerais

See also **350/II** (WARMING). – Area: 586624 km². Vascular flora: over 11156 species (from estimate given by João Angely in *Plants in danger* (see **General references**)).

Botanical activities relating to Minas Gerais are dispersed among several widely scattered research and teaching centers both inside and outside the state, none of them evidently with collections sufficient for a critical state flora project. A number of florulas of limited areas are, however, at present underway.

Partial works

Current locally based projects include *Flora de Serra do Cipó* and *Flora de Grão Mogol*, both in the Serra de Espinhaço. The former is represented by a series of family revisions (in *Bol. Bot. Univ. São Paulo* **9–17**, *passim*) as well as a preliminary checklist: A. M. GIULIETTI, J. R. PIRANI, M. MEGURO and M. G. WANDERLEY, 1987. Flora de Serra do Cipó, Minas Gerais: Caracterização e lista das espécies. *Bol. Bot. Univ. São Paulo* **9**: 1–151. For an earlier checklist for the Triângulo Mineiro in westernmost Minas Gerais, see R. J. A. GOODLAND, 1970. Plants of the cerrado vegetation of Brasil. *Phytologia* **20**: 57–70.

365

Rio de Janeiro and Espírito Santo

Area: 89386 km² (Rio de Janeiro, 43653 km²; Espírito Santo, 45733 km²). Vascular flora: no data. This unit encompasses the states of Rio de Janeiro (inclusive of the former capital territory of Guanabara, united with Rio de Janeiro in 1975) and Espírito Santo.

I. Espírito Santo

This state, lying between Rio de Janeiro and southeastern Bahia, has been undercollected and hitherto has been relatively poorly documented. No general or significant partial floras or enumerations are yet available; however, some family treatments have been published for the CVRD reserve at Linhares, the largest remaining tract of *mata atlantica* in the state.

II. Rio de Janeiro (with former Guanabara)

No separate works of a general character on the vascular plants of Rio de Janeiro State, the center of the *mata atlantica* or coastal forest belt – Martius's 'Dryades' – have appeared since the release of most of Frei Velloso's pioneer *Florae fluminensis* and its *Icones* in 1829–31 (the rest of the text was published only in 1881). That work, despite its imperfections, is symbolically important, coming soon after Pedro I's declaration of independence in 1822. In the absence of any true successors (all that is available being miscellaneous family revisions) and because there exists an interpretation of its names in modern nomenclatural terms it is included here as a standard work.

Some partial floras are available for areas of special botanical interest: the *restingas* in sandy areas along the coast (Segadas-Vianna, 1965–78; see **350/IV**), mountain forest (Rizzini, 1954, Serra dos Orgãos; Lima and Guedes-Bruni, 1994–96, Macaé de Cima) and the highest mountain range in southeastern Brazil, the Itatiaia massif (Dusén, 1905, 1909–10; Brade, 1956). These are selected for inclusion here from an extensive local literature. There was also one contribution towards a Guanabara flora: G. M. BARROSO *et al.*, 1974. *Flora de Guanabara. Familia Dioscoreaceae.* 248 pp., 213 text-figs., tables (Sellowia 23). Many family revisions covering all or part of the area have in recent decades appeared in *Arquivos do Jardim Botânico do Rio de Janeiro, Rodriguésia* and elsewhere.[60]

VELLOSO, J. M. DA C., Frei, 1825 (1829). *Florae fluminensis.* Pp. 1–352. Rio de Janeiro: Imprensa Nacional. (Reissued and completed 1881, Rio de Janeiro, as *Arq. Mus. Nac. Rio de Janeiro* 5: i–xii, i*–xii*, 1–461.) Accompanied by *idem*, 1827 (1831).

Florae fluminensis icones. 11 vols. 1639 pls. Rio de Janeiro: Bibliotheca in Urbe Fluminensi [now the National Library of Brazil]. (Lithographed and printed by Senefelder, Paris; indices in vol. 1 (pp. 1–14, 1–21) reprinted separately, 1929, Berlin: Junk.)

The text of this controversial work – strongly criticized by Karl von Martius in the years after its publication – comprises a briefly descriptive flora of vascular (and a few non-vascular) plants, without synonymy or detailed geographical distributions but including information on phenology and habitats and cross-references to plates; no separate index. Arrangement of genera in both text and atlas follows the Linnaean system, but after the foreword in vol. 1 of the latter are two indices: an *index alphabeticus* (14 pp.) and an *index methodicus* (21 pp.), the second being a conspectus of the work according to the then-new Candollean 'natural' system. [The taxonomic part in the original printing of the text terminated on p. 329 with the composite *Sabbata romana* (Syngenesia, i.e., the 19th of the 24 Linnaean classes, corresponding to the end of vol. 8 of the *Icones*); its conclusion, with additional prefatory notes by L. Netto, appeared as pp. 329–461 in the 1881 reissue. Many genera (and species) described in the work have been reduced to synonymy but some taxa have remained hard to interpret.][61]

Partial works

The Itatiaia range, the highest in southern Brazil, is in the southwestern part of Rio de Janeiro State, while Guanabara, the former federal district, encompassed the city and its immediate surroundings. The Serra dos Orgãos and Serra de Macaé are located in the central part of the state, north and northeast of the former capital. These and other ranges contain the heart of what is left of the *mata atlantica*.[62]

BRADE, A. C., 1956. A flora do Parque Nacional do Itatiaia. *Bol. Minist. Agric., Serv. Florest.* 5: 1–85, figs. 1–14, 8 maps.

Includes annotated lists with references and limited synonymy, emphasizing endemic or otherwise noteworthy species.

DUSÉN, P., 1903. *Sur la flore de la Serra do Itatiaya au Brésil.* 119 pp. (Arq. Mus. Nac. Rio de Janeiro 13). Rio de Janeiro. Continued as *idem*, 1909–10. Beiträge zur Flora des Itatiaia. *Ark. Bot.* 8(7): 1–26, illus.; 9(5): 1–50, illus. (The latter two papers were translated into Portuguese and reissued in 1955 as *Contribuições para a flora do Itatiaia.* 91 pp., illus., map (Parque Nacional do Itatiaia, Bol. 4). Rio de Janeiro.) [The 1903 work is an annotated enumeration of bryophytes and vascular plants with descriptions of novelties, collection records of the writer and others, overall distribution (if applicable) and commentary; no index. The introduction includes details of the range and prior botanical exploration. Additions and corrections as well as notes on the vegetation appear in the 1909–10 papers.]

LIMA, M. P. MORIM DE and R. R. GUEDES-BRUNI (coords.), 1994–96. *Reserva Ecológica de Macaé de Cima (Nova Friburgo, RJ): aspectos florísticos das espécies vasculares.* Vols. 1–2. Illus., map (vol. 1). Rio de Janeiro: Jardim Botânico do Rio de Janeiro.

Descriptive flora of vascular plants with keys, vernacular names, phenology, habitat, overall distribution, commentary, and *exsiccatae* (all or most of collections made for the project); no index. The general part (vol. 1) encompasses a chapter on geography, vegetation, floristics and taxonomy of the Reserve (by the coordinators; a checklist of species ultimately to be covered is included) and three general keys respectively for woody and herbaceous plants and for climbers. [The first two volumes cover 462 species in 57 families; a summary appears in each volume. The Reserve, 7200 ha (72 km²) in extent, ranges in altitude from 880 to 1720 m and features humid forest; in all some 1011 vascular species have been recorded.]

RIZZINI, C. T., 1954. Flora organensis. *Arq. Jard. Bot. Rio de Janeiro* **13**: 117–243. (Reprinted separately.)

Partly compiled systematic enumeration of bryophytes and vascular plants, with limited synonymy and indication of occurrences.

366

São Paulo

Area: 248256 km². Vascular flora: *c.* 7200–7300 species (cf. Angely, 1969–71, below).

In addition to the enumeration by Angely, there is also a descriptive state flora (in Portuguese), abandoned after covering only eight 'sympetalous' families: A. LÖFGREN and G. EDWALL, 1897–1905. *Flora paulista*, I–IV. 4 vols. (Boletim da Commisão Geographica e Geologica de São Paulo 12–15). São Paulo. Reference may also be made to the incomplete results of a 1901 Austrian expedition: R. VON WETTSTEIN and V. SCHIFFNER, 1908–31. *Ergebnisse der botanischen Expedition der kaiserlichen Akademie der Wissenschaften nach Südbrasilien 1901.* Bde. 1–2 (in 3 parts). vi, 454 pp., 23 figs. (incl. map), 41 pls.; 358 pp., 15 figs., 24 pls. (Akad. Wiss. Wien, Math.-Naturwiss. Kl., Denkschr. 79(1–2), 83). Vienna. Among local floras, two (one now

rather old) are listed below. A new state flora is now in preparation.

ANGELY, J., 1969–71. *Flora analítica e fitogeográfica do Estado de São Paulo.* 6 vols. 1901 maps. São Paulo: 'Phyton'.

Uncritical, largely compiled enumeration of vascular plants (7251 native, naturalized, adventive and commonly cultivated species), without keys; includes synonymy, references, vernacular names, superficial summaries of local and extralimital range (with many distribution maps), miscellaneous notes on habitat, occurrence, karyotypes, etc., and (in each volume) an index to botanical names (with a general index in vol. 6). An introductory section (vol. 1) includes a plan and synopsis of the work, family statistics, and a general comparison of the state's flora with that of areas elsewhere; it also gives a reconsideration of the publication dates of Velloso's *Florae fluminensis* (365/I).

Partial works

Mention should also be made here of a series of family revisions of the 'Fontes do Ipiranga' reserve in metropolitan São Paulo; these have appeared in *Hoehnea*.

KUHLMANN, M. and E. KÜHN, 1947. *A flora do distrito de Ibiti.* 221 pp., 94 illus. (some col.). São Paulo: Instituto de Botânica.

Annotated enumeration of non-vascular (except algae) and vascular plants of the Ibiti district (about 100 km north of São Paulo, near the Minas Gerais border); includes vernacular names, localities with *exsiccatae*, ecological and phenological data, and (at end) glossary, bibliography and indices. The latter part of the work deals with general natural history.

USTERI, A., 1911. *Flora der Umgebung der Stadt São Paulo.* 271 pp., 72 text-figs., 1 col. pl., map. Jena: Fischer.

Concise manual of vascular plants occurring in the vicinity of São Paulo (now largely built up); includes keys, localities, notes on habitat, and index. An extensive, pedagogically oriented introductory section contains remarks on vegetation and phytogeography.

Woody plants (including trees)

With respect to trees, as yet there is no successor to E. NAVARRO DE ANDRADE and O. VECCHI, 1916. *Les bois indigènes de São Paulo.* v, 376 pp., illus. São Paulo. Though well illustrated, it covers only principal species.

367

Paraná

Area: 199 324 km². Vascular flora: 6395 phanerogams (Angely, 1978; see below).

In this state is found the heart of the great Paraná pine forest zone, part of Martius's 'Napaeae'. An impression of this vegetation, now much decimated, may be had from F. C. HOEHNE, 1930. *Araucarilandia.* Illus. São Paulo. João Angely's *Flora analítica* was succeeded not by a critical state flora but by his *Flora descritiva,* a further sacrifice. The only current alternative is reference to works on Santa Catarina (**368**).

Bibliography

ANGELY, J., 1964. *Bibliografia vegetal do Paraná.* 304 pp. Curitiba: 'Instituto Paranaense do Botânica'.

ANGELY, J., 1965. *Flora analítica do Paraná: nomenclator species plantarum paranaënsium.* xii, 728 pp., map. São Paulo: 'Instituto Paranaense do Botânica'.

Uncritical, largely compiled enumeration of native and naturalized mosses and vascular plants (5287 native and introduced species); includes synonymy (with references), status (endemic or otherwise), miscellaneous notes, and (at end) an index to genera and families. An introductory section includes various statistical tables as well as historical and polemical remarks.[63]

ANGELY, J., 1978. *Flora descritiva do Paraná.* 10 vols. xxxv, 3483 ll. (Coleção Saint-Hilaire). São Paulo: 'Phyton'. (Reproduced from typescript.)

Compiled, annotated enumeration of native, naturalized and adventive phanerogams (6395 species), arranged according to Engler-12; species entries include synonymy (with references), (sometimes) a brief description, vernacular names, local range (without *exsiccatae*), and Brazilian and extralimital distribution, while family entries include (largely older) references, description, karyotypes, commentary on distribution, (sometimes) infrafamilial taxa, and notes on species of special interest, medicinal or otherwise. An index to all genera and families appears in volume 10. Background information appears on pp. xviii–xxxv in volume 1. [A regrettable second edition of his *Flora analítica,* wastefully produced on A4 leaves.]

368

Santa Catarina

Area: 95 318 km². Vascular flora: 4709 species (R. Reitz, unpublished; quoted in Campbell and Hammond, 1989, p. 444 (see **General references**)).

The progress of botanical work in the state up to and through the foundation (in 1942) of the *Flora ilustrada catarinense*'s main resource, the Herbário 'Barbosa Rodrigues' at Itajaí, is related in R. Reitz, 1949. História da botânica catarinense. *Anais botânicos [Sellowia]* 1: 23–110. This was planned for incorporation into part IV of the *Flora*; the first section, 'Plano de coleção', appeared in 1965. For the general keys to pteridophyte and phanerogam families published within part I of the *Flora*, see **350**.

REITZ, R., Pe. and A. REIS (eds.), 1965– . *Flora ilustrada catarinense.* I. parte, *As plantas.* In fascicles. Illus., maps. Itajaí: Herbário 'Barbosa Rodrigues'.

Comprehensive illustrated descriptive flora of vascular plants; each family includes keys to genera and species, full synonymy, references, vernacular names, citation of *exsiccatae* with localities, general indication of local and extralimital range (including distribution maps), extensive taxonomic commentary, and notes on habitat, ecology, phenology, etc.; index to botanical names and list of references in each fascicle. Many treatments are by specialists. [As of 1998, 153 families had been published (covering 3380 species or *c.* 70 percent of the total), with 67 still to appear. The work has set a high standard for floristic writing in southeastern Brazil. Of the four other projected series or 'partes', II (*As zonas fitogeográficas*), III (*As associações vegetais*), IV (*História*) and V (*Mapa fitogeográfico*), only the fifth and the first part of the fourth have appeared.][64]

Partial works: Ilha de Santa Catarina

This island, the site of the state capital, is a large mainland fragment parallel to the east coast. In addition to the *Florula,* reference may be made to A. BRESOLIN, 1979 (1981). *Flora da restinga da Ilha de Santa Catarina.* 54 pp., 9 halftones in text, 2 maps (Insula 10). Florianópolis. [Essentially on vegetation, but includes references (p. 54).]

SOUZA SOBRINHO, R. J. DE and A. BRESOLIN (eds.), 1970– . *Florula da Ilha de Santa Catarina.* [Fasc. 1– .] Illus. Florianópolis: Universidade Federal de Santa Catarina, Divisão de Botânica, etc.

Similar in style to *Flora ilustrada catarinense*; published in fascicles. As of 1979 at least 16 fascicles had appeared.

369

Rio Grande do Sul

Area: 280674 km². Vascular flora: *c.* 4500 species of seed plants (estimate by B. Rambo in *Pesquisas* 1: 8 (1957)). – The prevailing *campinas* vegetation of the central and southern parts of the state is closely related to that of Uruguay (375).

Locally based botanical work began seriously only in the twentieth century; a first text in plant systematics by Ir. Augusto appeared in 1930. Together with Ir. Edésio and Karl Emrich, Augusto then began a state flora, one complete volume of which appeared in 1946. Some additional families were published by João Rodrigues de Mattos (1965, 1983). In 1938 A. R. H. Schultz produced the first edition of *Introdução ao estudo da botânica sistemática* (2nd edn., 1943; 3rd edn. in 2 vols., 1963) and in 1955 initiated *Flora ilustrada do Rio Grande do Sul*, a critical work which has continued to the present. An additional considerable contribution is represented by the many family treatments published in *Pesquisas, Botânica* by Fr. Balduíno Rambo, based on the substantial herbarium of his Jesuit foundation, the 'Instituto Anchietano de Pesquisas' now in São Leopoldo near Porto Alegre. Pteridophytes were rather fully treated by Rambo's associate Fr. A. Sehnem in five installments of *Uma coleção de Pteridófitos do Rio Grande do Sul* (I in *Sellowia* 7 (1956): 299–326; II–III in *Pesquisas* 2 (1958): 223–229 and 3 (1959): 495–576; IV–V in *Pesquisas, Bot.* 10 (1960): 1–44 and 13 (1961): 1–42, illus.). A good dendrology by Pᵉ. R. Reitz, R. M. Klein and A. Reis appeared in 1983.[65]

Bibliography

QUINTAS, A. TAVARES, 1967. Bibliografia botânica Sul-Riograndense. *Anais do XV. Congresso da Sociedade Botânica do Brasil* (Pôrto Alegre, 1964), pp. 175–195. Pôrto Alegre. [Unannotated list arranged by author. Some key works not accounted for.]

AUGUSTO, Irmão [AUGUST DUFLOT], 1946. *Flora do Rio Grande do Sul, Brasil*, 1. 639, vii pp., 341

text-figs. Pôrto Alegre: Imprensa Oficial (published under gubernatorial patronage).

Copiously illustrated manual of seed plants, with keys to all taxa. Each family treatment contains two sections: an analytical key to genera and species and a systematic catalogue of all taxa known from the state with localities and collectors thereof; however, no synonymy, references, critical commentary, or habitat and biological data are provided. The work is prefaced by a short introduction and a general key to all families, and concludes with a complete index. [Not completed; covers only Clethraceae through Compositae (Englerian system). A continuation for families from Anacardiaceae through Vitaceae may be found in J. R. DE MATTOS, 1965. *Flora do Rio Grande do Sul*, 7. 110 pp., 34 text-figs. São Paulo. A compilation of Myrtaceae appears as *idem*, 1983. *Myrtaceae do Rio Grande do Sul*. 721 pp. Porto Alegre.][66]

SCHULTZ, A. R., M. H. HOMRICH and L. R. M. BATISTA (coords.), 1955– . *Flora ilustrada do Rio Grande do Sul*. Fasc. 1– . Illus., maps (Bol. Inst. Cent. de Biociências (Porto Alegre) 3, *passim*). Porto Alegre.

Large-scale illustrated serial flora; each fascicle encompasses lengthy descriptions, full synonymy (with references), citations, vernacular names, localities with *exsiccatae*, dot distribution maps, extralimital range, taxonomic commentary, notes on habitat, phenology, etc., representative illustrations, family summaries, and indices. [As of 1999, 23 fascicles had been published.]

Partial work: Porto Alegre

TEODORO LUIS, Irmão, 1960. *Flora analítica de Pôrto Alegre*. iv, (unpaged) pp., illus. Canoas: Instituto Geobiológico "La Salle".

Very concise pocket-sized field checklist-key of higher plants of the Pôrto Alegre metropolitan area; includes keys to genera and species as well as representative figures but no synonymy, vernacular names, distributional data, or ecological notes. Each entry is hierarchically numbered.

Woody plants (including trees): guide

REITZ, R., Pᵉ., R. M. KLEIN and A. REIS, 1983. *Projeto madeira do Rio Grande do Sul*. 525 pp., 138 pls. (partly halftones), maps (Sellowia 34/35). Itajaí.

Illustrated dendrology of principal trees, arranged by vernacular names; species entries include vernacular and scientific names (with families), botanical and field descriptions, phenology, local and extralimital distribution (with dot distribution maps), frequency, biology, and wood properties and uses. The main part is preceded by an alphabetical list of all known trees and large shrubs, with distribution by eight phytogeographic units.

Region 37

Paraguay and Uruguay

These two countries are grouped together merely for convenience. Uruguay is, however, rather better known botanically than Paraguay; the latter, along with Bolivia, is the least-known country in South America.

Botanical work in Paraguay began in the nineteenth century but has been very sporadic, with only some 30 or so collectors on record and but few contributions by residents. Leading collectors and writers include Mosé Bertoni, Balansa, Emil Hassler (with Rojas and Chodat), Swedish Regnellian explorers such as Dusén and Malme, and (in part indirectly, through revisionary studies) a number of Argentine botanists, mainly in the period between the War of the Triple Alliance (1864–70), which led to the country's independence in 1871, and the Chaco War (1932–35). The largest contributor was Hassler, a physician who had followed Mosé Bertoni from Switzerland and whose collections over many years provided the main basis for the Geneva University professor Robert Chodat's *Plantae hasslerianae* (1898–1907) – for long the one work approaching a general flora. After World War I, contributions declined, and until recent years remained few in number.

Developments in forestry in the eastern region in the 1970s occasioned publication of a dendrology by Juan A. López Villalba (1979; new version with Elbert L. Little *et al.*, 1987), and this period also saw renewed support from Switzerland, particularly through the Conservatoire et Jardin Botaniques in Geneva, repository for Hassler's collections. This has led to additional botanical exploration, preparation of a tree manual (initially by Lucien Bernardi), and the organization of the *Flora del Paraguay* project; the latter is run jointly with the Missouri Botanical Garden. Publication of the *Flora* began in 1983; by 1999, 27 fascicles had appeared. The work by Bernardi on the tree flora resulted in two contributions (1984–85); in 1989 E.

Ortega Torres, L. Stutz de Ortega and R. Spichiger produced a suite of 90 trees as an addition to the López and Little dendrologies. Serious development of local herbaria also has begun. An outcome of the recent studies has been recognition of the La Plata drainage (including the Paraná and Paraguai Rivers) as a distinct floristic province.[67]

A rather different situation exists in Uruguay, where there are substantial local collections and a small resident botanical community. In the latter part of the eighteenth and first half of the nineteenth centuries, many foreign botanists made visits, among them Philibert Commerson, Auguste F. C. P. de St. Hilaire, Friedrich Sellow, and John ('James') Tweedie. St. Hilare's *Flora Brasiliae meridionalis* (1835), written with Adrien de Jussieu and Jacques Cambessèdes, is basic to the floristic knowledge of the country as well as present-day southern Brazil. Locally based work began to develop as prosperity increased in the latter part of the nineteenth century; a first contribution was *Enumeratio plantarum sponte nascentium agro Montevidensi* (1873) by E. Gibert. In 1890 the Museo Nacional de Historia Natural in Montevideo was established, and the first full descriptive flora, *Flora uruguaya* by José Arechavaleta (1898–1911, not completed) was begun. Contributions also began to be made by Argentine botanists. In the period from the mid-1920s until World War II the controversial German botanist Wilhelm (Guillermo) Herter was resident for a considerable time and made many contributions, among them *Enumeratio plantarum vascularium . . .* (1930), *Flora ilustrada del Uruguay* (1939–57, not completed), and *Flora del Uruguay* (1949–56, not completed). None of these, however, is a descriptive work: the first is a checklist, the second an iconography, and the third an annotated enumeration with keys.

In the 1950s a new descriptive *Flora del Uruguay* was begun by the Museo Nacional de Historia Natural in succession to that by Arechavaleta. Progress was, however, minimal, only four fascicles being published (the last in 1972). This and the present low level of collecting have perhaps been due to changing interests as well as national regression, the latter further exacerbated in the 1970s by political strife. This is an unfortunate state of affairs as although the flora is in general relatively well known, intensive collecting is needed in many areas and the present level of documentation is unsatisfactory. The only larger floristic works to be completed since 1960 are both by Attilo Lombardo: the second edition of his *Flora arbórea y arborescente del Uruguay* (1964) and *Flora montevidensis* (1982–84).

Progress
A brief history of botanical work in Paraguay, with special reference to Swiss contributions, appears in *Flora del Paraguay: Guia para los autores* (1983). With respect to Uruguay, the life and activities of W. G. Herter, a long-time student of the flora, have been reviewed by H. M. BURDET, 1978. L'oeuvre et les tribulations du botaniste Herter: une étude biographique et bibliographique germano-uruguayenne. *Candollea* 33: 107–134. Herter himself in his *Enumeración* reviewed botanical work to 1930, while collectors before 1900 were memorialized by Arechavaleta in his *Flora uruguaya*.

Bibliographies. General and divisional bibliographies as for Division 3.

Indices. General and divisional indices as for Division 3.

371

Paraguay

Area: 406750 km². Vascular flora: 7000–8000 species.

In addition to the cumulated series of family revisions based on Hassler's collections (see below), reference should also be made to the following introductory work, which contains general surveys of various families as they occur in Paraguay: R. CHODAT and W. VISCHER, 1916–26(–27). La végétation du Paraguay, I–XIV. *Bull. Soc. Bot. Genève*, II, **8–18**, *passim.* (Separately reprinted, Geneva: 509, 49 pp.; 2nd reprint, 1977, Lehre: Cramer.)

Bibliography
BERTONI, B. S., J.-M. MASCHERPA and R. SPICHIGER, 1982. Datos bibliográficos para el estudio de la vegetación y de la flora del Paraguay. *Candollea* 37: 277–313. [Introduction (in French and Spanish); includes classified and author index. All parts of *Plantae hasslerianae* are accounted for.]

CHODAT, R., 1898–1907. *Plantae hasslerianae, soit énumération des plantes récoltées au Paraguay par le dr. Émile Hassler d'Aarau (Suisse) de 1885 à 1902 et déterminées par le prof. dr. R. Chodat avec l'aide de plusieurs collaborateurs*, I–II. 2 vols. Geneva: Romet (for Institut de Botanique, Université de Genève). [Reprinted from *Bull. Herb. Boissier*, I, **6**: app. 1 (1898), **7**: app. 1 (1899); *idem*, II, **1–5, 7** (1901–07), *passim.* Continued as E. HASSLER, 1917. *Addenda ad Plantas hasslerianas.* 20 pp. Geneva.]

Polyglot enumeration of vascular plants (not in systematic order), based mainly upon the collections of Hassler made in Paraguay from 1885 to 1902; includes synonymy (with references) and a few citations, localities with *exsiccatae*, brief summary of extralimital range, taxonomic commentary, notes on distribution, special features, etc. (in part by Hassler), and addenda and corrigenda. A general index to families and genera and (p. 713) full bibliographic details are furnished at the end of part II, which also includes a short account of Paraguayan floristic regions. Treatment of some families contributed by specialists. [For additional collations of the original work, see BLAKE and ATWOOD, 1942, p. 252, and STAFLEU and COWAN, 1976–88, 2: 98–99 under **General references** as well as in the above-cited 1982 bibliography of Bertoni, Mascherpa and Spichiger. Additional papers on Hassler plants are listed in *Flora del Paraguay: Guia para los autores* by R. Spichiger and J.-M. Mascherpa (see Note 68).]

SPICHIGER, R. *et al.* (eds.), 1986– . *Flora del Paraguay.* In fascicles. Illus. Geneva: Conservatoire et Jardin Botaniques; St. Louis, Mo.: Missouri Botanical Garden.

An illustrated 'research' flora of vascular plants, issued in fascicles each devoted to a family (or major part thereof) with four suggested arrangements (Bentham and Hooker, Engler, Hutchinson and Cronquist); treatments include keys, descriptions, synonymy (with references), citations, localities with *exsiccatae*, summary of local and extralimital distribution, taxonomic commentary, notes on phenology, ecology, biology, etc., and (at end) an index to scientific names. Generic headings include a description, statistics, and brief notes, on uses, special features, etc., common to the species treated; major headings include key literature. Most, if not all, species are figured. The *Guia para los autores* includes a historical review as well as instructions for contributors, resources at the Conservatoire and a complete index to contributions on Hassler's collections. [Published and unpublished families are listed in the inside covers of each fascicle. Thirty fascicles had appeared by 1999.][68]

Woody plants (including trees): guides
The eastern region of the country includes its only 'tropical rain forests'. The main dendrological works represent parallel undertakings; combined, they still do not offer complete coverage.

BERNARDI, L., 1984–85. *Contribución a la dendrología paraguayana.* 2 parts. 343, 295 pp., illus. (Boissiera 35, 37). Geneva.

Comprises illustrated family revisions, with significant literature, keys, descriptions, synonymy, vernacular names, localities with *exsiccatae*, local and extralimital distribution, taxonomic commentary, 'imagery', and notes on habitat, properties and uses; list of all *exsiccatae* and index (only in part 2). [A precursor to a projected tree manual; covers Apocynaceae, Bombacaceae, Euphorbiaceae, Flacourtiaceae, Leguminosae (in part 1) and Meliaceae, Moraceae, Myrsinaceae, Myrtaceae, Rubiaceae and Vochysiaceae (in part 2).]

LÓPEZ VILLALBA, J. A., E. L. LITTLE, JR. *et al.*, 1987. *Árboles comunes del Paraguay / Ñande yvyra mata kuera.* v, 425 pp., illus. N.p.: Cuerpo de Paz (EE. UU.). (Colección e Intercambio de Información.) (Based on J. A. LÓPEZ VILLALBA, 1979. *Árboles de la región oriental del Paraguay.* 277 pp. Asunción.)

Dendrological atlas of 156 species with illustrations and extended descriptions of habit, botanical features, habitat, distribution, wood attributes, uses, potential, etc. [Covers both eastern and western Paraguay.]

ORTEGA TORRES, E., L. STUTZ DE ORTEGA and R. SPICHIGER, 1989. *Noventa especies forestales de Paraguay.* 218, xiv pp., illus. (part col.). (Flora del Paraguay, série especial 3). Geneva.

An illustrated treatment of 90 species of secondary importance with habit silhouettes and illustrations of trunks and sprays of foliage along with botanical and wood descriptions and habitat and dendrological notes; vernacular names and etymologies are also included. An artificial key is also furnished, along with indices to botanical and vernacular names. [Designed to complement the work of López and Little (see above).]

375

Uruguay

Area: 186 925 km². Vascular flora: 3000–3500 species (2998 in Herter, 1930–37).

In spite of the number of (mostly incomplete) works available, there is no descriptive floristic treatment as yet for certain families or groups thereof, notably in the 'Sympetalae'. As comparatively little has been published to date in the Museo Nacional's *Flora del Uruguay* series, modern coverage is still dominated by the works (two of them left unfinished at his death) of the German-Uruguayan botanist Wilhelm (Guillermo) Herter. In the last decade, the literature has been expanded by a flora of Montevideo and vicinity by Attilo Lombardo (1982–84).

ARECHAVALETA, J., 1898–1911. *Flora uruguaya.* Vols. 1–3; vol. 4, parts 1–3. Illus. (Anales Mus. Nac. Montevideo 3, 5–7). Montevideo.

Illustrated descriptive flora of seed plants; includes some synoptic keys to genera and species, full synonymy (with references), citations, vernacular names, general indication of local range, critical commentary, and notes on habitat, phenology, uses, etc.; index to botanical names in each volume (except the last). The introductory sections of volumes 2–3 include sketches of botanists associated with the Uruguayan flora. Not completed (owing to the death of the author in 1912); extends from Ranunculaceae through *Cuscuta* (Convolvulaceae) on the Bentham and Hooker system.

HERTER, W., 1930. *Enumeración de las plantas vasculares que crecen espontáneamente en la República oriental del Uruguay agregando las plantas adventicias, las principales plantas cultivadas, los nombres vulgares, la repartición en la República y los números de las collecciones Gibert y Herter.* 191 pp., 18 pls. (some col.), col. map. Montevideo: The author [with official support]. (Estudios botánicos en la región Uruguaya 4; Florula uruguayensis II.) [Full title also in Latin.] Continued as *idem*, 1935–37. Additamenta ad floram uruguayensem, I–III. *Revista Sudamer. Bot.* **2**: 111–128, 2 figs., 1 col. pl.; **3**: 146–178, 1 col. pl.; **4**: 179–232, 1 col. pl.

Systematic checklist of native, adventive, and commonly cultivated vascular plants (2998 species) with vernacular names, very limited synonymy, and concise indication of local range by provinces (with selected *exsiccatae*); list of collections by Gibert and Herter; indices to vernacular and botanical names at end. An introductory section includes a historical account of botanical exploration in Uruguay. Additions are embodied in three supplements.

HERTER, W., 1939–43. *Flora ilustrada del Uruguay.* Fasc. 1–5 (comprising vol. 1). pp. i–xvi, 1–13, pls. 1–256 (figs. 1–1024). Montevideo, Berlin, Kraków:

The author. Continued as *idem*, 1952–57. *Flora ilustrada del Uruguay*. Fasc. 6–13. pp. xvii–xxxii, pls. 257–352, 373–600 (figs. 1025–2320). Basel (later Hamburg): The author. [Fasc. 1 also published as *Repert. Spec. Nov. Regni Veg., Beih.* 118(1).]

An atlas of 580 plates of the vascular flora of Uruguay, with four figures to a plate; the drawing of each species is accompanied by vernacular and botanical names together with a concise indication of local range. Not completed; extends from the lower vascular plants through the Cactaceae (Englerian system). The figures are continuously numbered; the gap in the numbering of the plates is the result of an error. Only fascicles 1–5 (vol. 1) possess a general index.

HERTER, W., 1949–56. *Flora del Uruguay*. Vol. 1. 10 fascicles. 280 pp. Montevideo (later Berne, Basel, Hamburg): The author. (Mimeographed.)

Enumeration of vascular plants, complementing the author's *Flora ilustrada del Uruguay*; includes synonymy (with references), vernacular names, *exsiccatae*, brief general summaries of local and extralimital range, and complete index (at end of fasc. 10). An introductory section (fasc. 1) includes a short list of references. Not completed; encompasses Pteridophyta, Gymnospermae, and Monocotyledoneae (Englerian system).

MUSEO NACIONAL DE HISTORIA NATURAL, URUGUAY, 1958–72. *Flora del Uruguay*. Fasc. 1–4. Illus. Montevideo.

Detailed 'research' flora of vascular plants; each treatment (of a family or group of families) includes synonymy, local range, miscellaneous notes, full-page illustrations, and separate index. As of 1999, only four fascicles had appeared, the latest in 1972; groups covered include the pteridophytes and miscellaneous families of dicotyledons. Some treatments are by specialists.

Partial work: Montevideo

LOMBARDO, A., 1982–84. *Flora montevidensis*. 3 vols. Illus. Montevideo: Intendencia Municipal de Montevideo.

Briefly descriptive, well-illustrated flora of native and other vascular plants of Montevideo and vicinity (664 km²); includes keys to all taxa, synonymy (with references) and citations, vernacular names, local and regional distribution, phenology, origin (if not native), occurrence and habitat, some taxonomic commentary, and (at end of vol. 3) a glossary, references, and general index to vernacular names (scientific names are indexed only for each volume). Volume 1 includes the pteridophytes and 'polypetalous' families', vol. 2 the 'gamopetalous' families, and vol. 3, the monocotyledons.

Woody plants (including trees): guide

LOMBARDO, A., 1964. *Flora arbórea y arborescente del Uruguay, con clave para determinar las especies.* 2nd edn. 151 pp., 223 text-figs. Montevideo: Concejo Departemental de Montevideo. (1st edn., 1946.)

Illustrated manual of trees and shrubs, with keys to all taxa, limited synonymy, a few citations, vernacular names, summary of local range, taxonomic commentary, and notes on habitat, uses, etc.; indices to vernacular and botanical names. A glossary is included in the introductory section.

Region 38

Argentina

Area: 2 777 815 km². Vascular flora: 9260 species (Zuloaga and Morrone, 1994, 1996–99; see below). This region corresponds to the political limits of the Argentine Republic, with the addition of the Falkland Islands (Islas Malvinas). Like Chile, it includes a portion of Patagonia (and Fuegia).

Botany in the Argentine, though with a less extended history than some other parts of South America, developed strongly after independence in 1810 and even more so following a revolution in 1852. In the second half of the nineteenth century came extensive settlement from Europe and a corresponding development of a scientific community and institutions, not only in Buenos Aires but also in other, widely dispersed centers. This process continued through the apogee of the country's economy in the first half of the twentieth century and into the 1950s. Although a number of neglected areas are on record and some key localities require more intensive work, given such a tradition it is probably true to say along with Toledo (1985, p. 34; see **Superregion 21–23, Progress**) that 'Argentina is, botanically, the best developed country in the [Latin American and Caribbean] region'. Yet, as in the United States, no modern national flora has been completed; and its somewhat fragmented systematic botanical structure in a country of almost Roman transport geography is in large part the legacy of several eminent individuals as well as organized development of research groups. Indeed, three of the more important institutions began life as 'casas particulares'.

Visitors in the eighteenth century included Mylam, Commerson, Pernetty, Banks and Solander, the Forsters, and the Malaspina Expedition (including its botanists Haenke and Neé). After independence they were followed by Friedrich Sellow, Aimé Bonpland (1817–58, from 1820 in northern Argentina), John ('James') Tweedie (1825–62), Charles Darwin (1832–34), Alcide d'Orbigny, Charles Gaudichaud, Joseph Hooker, the U.S. Exploring Expedition, J. S. C. Dumont d'Urville, and others. Local activities, initiated by the Jesuits before their expulsion in the 1760s, were given a new impetus after independence by the foundation, in 1812, of the first museum in Buenos Aires.[69] As elsewhere in most of South America, however, political and other difficulties prevented much national scientific development before mid-century; Bonpland in particular was a victim of very difficult circumstances. Only in remote Fuegia and the Malvinas (which in 1833 passed to British control) did collecting become sufficiently far advanced to prompt preparation of a full flora; this was accomplished by Hooker in the second volume of his *Flora antarctica* (1845–47).

The period after the fall of the tyranny of Juan Manuel Rosás – whose rise was recorded by Darwin – was one of intellectual liberation and a broadly based educational movement.[70] Floristic and systematic work, however, developed more rapidly at Córdoba than in Buenos Aires. A prominent German naturalist, Germán Burmeister, settled in the inland city in 1857; he then became director of the National Museum in Buenos Aires in 1862 and, in 1869, dean of a new science faculty at the Universidad de Córdoba. From 1870 to 1915 (and beyond) its botany was directed by successive Germans who also collected extensively. These included Paul Lorentz (1870–74, afterwards residing in Uruguay), Georg Hieronymus (1873–83, afterwards at Berlin as an associate of Eichler, Engler and Urban), and Fritz (Federico) Kurtz (1884–1915). The collections of Lorentz and Hieronymus were the primary resource for August Grisebach's *Plantae lorentzianae* (1874) and, most importantly, his yet-to-be-superseded general account, *Symbolae ad floram argentinarum* (1879). However, the *Symbolae* was largely limited to northern Argentina, the south being still rather little explored save for, as already noted, Fuegia and the Malvinas (or Falklands). In 1879, Lorentz (with Gustav Niederlein) accompanied a major expedition into the deserts of Patagonia; the results appeared as *Expedición al Río Negro*, II: *Botánica* (1881). The two

later explored the subtropical province of Misiones and other parts of the northeast from 1883 to 1888. Kurtz worked extensively in Mendoza and other western provinces.

Buenos Aires (along with satellites such as La Plata, founded in 1882, and San Isidro) did not, by contrast, emerge as a significant botanical center until the 1880s. This development resulted from more stable government, an economic 'boom' in that decade, and the advent of Eduardo Holmberg and Carlo Spegazzini.[71] Holmberg remained in the capital, eventually as a professor at the university, while Spegazzini built up the field at the new Universidad (and Museo) de La Plata. Both traveled widely, but the latter would contribute more floristic works (including several on Patagonia in the 1890s and, in 1905, the first volume of a Buenos Aires flora). The recession of 1890 prompted by the first Baring bank collapse was temporary; by the second decade of the twentieth century the two men, along with younger colleagues such as the medical botanist Juan Domínguez, Holmberg's student Cristóbal Hicken and the Belgian Lucien Hauman, had effectively established Buenos Aires (with neighboring La Plata) as a leading botanical center in South America.

Foreign visiting botanists continued to be active, the most prominent around the turn of the century being the Swedes Robert E. Fries (in the northern Andes) and Per Dusén and Carl Skottsberg in Patagonia and Fuegia. The North Americans J. B. Hatcher and O. A. Peterson also were active in both Argentine and Chilean Patagonia. All their collections, with earlier records, were largely accounted for by G. Macloskie and others in the John Pierpont Morgan-sponsored *Flora patagonica* (1903–06, 1914; see **301/III**), Dusén being a co-author of the 1914 supplement. Skottsberg also worked in the Malvinas, producing a flora in 1913.

An apogee was reached in the years before World War I, landmarks being the foundation by Hickén of his private research center, the 'Darwinion' (1911), the effective establishment of a botany department at the National Museum (1912), and, at La Plata, Spegazzini's 'going private' (1912). In this period, the chemist and autodidact Miguel Lillo in the northwestern city of Tucumán also began to make significant contributions, including two on the tree flora (1910–17). At the Museo National, Hauman believed the time was now right for a nationwide flora, and accordingly began (with Germainne Vanderveken and Luis H. Irigoyen) his *Catalogue des phanérogames de l'Argentine* (1917–23). However, Hauman returned to Belgium well before finishing the work, and his successor at the Museo, Alberto Castellanos, could not or would not complete it; moreover, as Cabrera has indicated, conditions in the interwar years were less directly favorable to systematics.

The foundation of the 'Miguel Lillo' and 'Spegazzini' institutes, the appointment of Arturo Burkart as director of the 'Darwinion' following the death of Hicken and a reorganization, and the advent of Lorenzo Parodi to the curatorship of vascular plants at the Museo de La Plata in the mid-1930s (in addition to his professorship in Buenos Aires, where he succeeded Hauman and remained until 1966) along with improved circumstances and new research approaches gave rise to a strong renewal, which was to last through the 1950s. To these was added, in 1945, the Instituto de Botánica of the Ministry of Agriculture (from 1956 the Instituto de Botánica Agrícola of the Instituto Nacional de Tecnología Agropecuaria (INTA) at Castelar). In the 1940s Angel L. Cabrera succeeded Parodi at La Plata and Armando Hunziker replaced Curt Hosseus (Kurtz's successor) at Córdoba. Additional research centers would develop from the 1950s as more universities and institutes were founded; the largest of these is at Corrientes, in the northeast.

The situation of 'Miguel Lillo' enabled it to attract able foreign as well as local botanists to its staff, and it was here that the idea of a nationwide flora was seriously conceived or revived. Under the direction of Horatio Descole this began to be realized in the massive *Genera et species plantarum argentinarum* (5 vols., 1943–56), published during the years of the first Perón regime. This was, however, discontinued while still far from completion, and soon after economic circumstances forced a reduction of the institute's staff and activities. The general flora plan then gave way to one, promoted by INTA, which called for a network of 'regional' floras (akin to those in South Asia sponsored by the Botanical Survey of India). Research towards these would be supported at Castelar or at other suitably oriented centers. The INTA programme has continued to the present, and eight floras, all in multiple parts, have been completed, are in process of publication or are in preparation. These cover Buenos Aires Province (completed), Patagonia, Jujuy, Entre Rios, and Central Argentina, the Chaco, Corrientes and La Pampa (in preparation). For the Falklands (Malvinas), Skottsberg's 1913 flora was succeeded in 1968 by David Moore's modern treatment. This has been complemented for Fuegia by the same author's attractive *Flora of Tierra del Fuego* (1983; see **301/IIIa**). In addition to these, there is a miscellany of older provincial and local works.

Nevertheless, coverage outside of Patagonia and the pampas is inadequate, and within the last decade the INTA programme appears to have lost some momentum. However, a *Flora fanerogámica argentina* project has been organized with a secretariat at Córdoba; its first fascicles appeared in 1994. It is complemented by *Catálogo de las plantas vasculares de la República Argentina* by Zuloaga and Morrone (1994, 1996–99), an effective successor to the unfinished work of Hauman, Vanderveken and Irigoyen.

Progress

The state of affairs in Argentine botany towards the end of the nineteenth century was reviewed in E. L. HOLMBERG, 1898. La flora de la República Argentina. In Comisión directiva de Censo, República Argentina (ed.), *Segundo censo de la República Argentina, mayo 10 de 1895*, 1: 383–474. Buenos Aires. Later reports include W. B. TURRILL, 1920. Botanical exploration in Chile and Argentina. *Bull. Misc. Inform.* (Kew) **1920**: 57–66, 223–224; C. M. HICKEN, 1923. *Los estudios botánicos.* 167 pp. (Evolución de las ciencias en la República Argentina VII). Buenos Aires: Coni (Papeles del cincuentenario de la Sociedad científica argentina (1872–1922)); A. CASTELLANOS, 1945. Las exploraciones botánicas en la epoca de la independencia, 1810–1853. *Holmbergia* **4**(8): 3–14; L. R. PARODI, 1961. Ciento cinquentia años de botánica en la República Argentina. *Bol. Soc. Arg. Bot.* **9**: 1–68; A. L. CABRERA, 1972. Estado actual del conocimiento de la flora Argentina. In Sociedad Botánica de México (ed.), *Memórias de symposia del primero congreso latino-americano y mexicano de botánica*, pp. 183–197. Mexico City; and *idem*, 1977. *Evolución de las ciencias en la República Argentina, 1923–1972*, VI: *Botánica*. Buenos Aires: Sociedad Científica Argentina.

For separate reviews on Patagonia and Fuegia, see the introduction to *Flora antarctica* by J. D. Hooker and two papers in *Tuatara* by E. J. Godley (both under **Superregion 08–09**) as well as A. SORIANO, 1948. Las exploraciones botánicas en la Patagonia Argentina. *Ci. Invest.* **4**: 443–453.

Bibliographies. General and divisional bibliographies as for Division 3.

Regional bibliography
The bibliography of Castellanos and R. A. Pérez-Moreau (1941) succeeds earlier compilations by Kurtz (1913–15), Hauman and Castellanos (1922–27) and Hicken (1927–30).

CASTELLANOS, A. and R. A. PÉREZ-MOREAU, 1941. Contribución á la bibliografía botánica argentina, I–II. *Lilloa* **6**: 5–161; **7**: 5–549, [iv], map. [Classified lists of floristic works published through 1937 appear in part II.]

Indices. General and divisional indices as for Division 3.

380

Region in general

In addition to the general works described below, the regional floras being published under the auspices of the Centro de Investigaciones de Recursos Naturales of Instituto Nacional de Técnologia Agropecuaria (INTA) are intended collectively to comprise a modern general flora of the Argentine. Published as part of INTA's *Colección científica*, they are successors to the *Genera et species plantarum argentinarum* project, terminated in 1956. Those available or in progress cover Buenos Aires (vol. 4), Entre Rios (vol. 6), Patagonia (vol. 8), and Jujuy (vol. 13). Several fascicles of a *Flora chaqueña* have also been published and work is said to be in progress on others.[72] In the 1990s, two new national works commenced publication.[73]

Keys to families and genera
BOELCKE, O., 1992. *Plantas vasculares de la Argentina, nativas y exóticas.* 2nd edn. x, 334 pp. Buenos Aires: Hemisferio Sur. (Originally published 1981; reissued with supplement, 1986). Accompanied by O. BOELCKE and A. VIZINIS, 1986–93. *Plantas vasculares de la Argentina, nativas y exóticas. Illustraciones,* I–IV. Illus. Buenos Aires. [The main text is a pedagogical work on vascular plant families with references, keys and descriptions together with notes on significant species; general references (pp. 274–288) and index to all names. The companion atlas covers all families of vascular plants.]

HAUMAN, L., 1984. Los géneros de fanerogamas de Argentina: claves para su identificación. *Bol. Soc. Arg. Bot.* **23**: 1–384, portrs. [Prologue by A. T. Hunziker; keys to genera with footnoted references and other notes with each family signed by its reviewer; dubiously present and excluded families (pp. 351–352) and epilogue with statistics (1538 genera of seed plants native, 211 introduced); author and other abbreviations (pp. 357–359) and index to all names at end.]

DESCOLE, H. *et al.*, 1943–56. *Genera et species plantarum argentinarum.* Vols. 1–5 (in 7). Illus. (some col.), maps. Buenos Aires: Kraft.

A massively formatted, illustrated documentary descriptive flora of seed plants; includes keys to genera and species, full synonymy (with references and citations), local range with *exsiccatae*, summary of extralimital range (including distribution maps), miscellaneous notes, and (at end of each family) indices to botanical names. The families revised are as follows: 1, Zygophyllaceae, Cactaceae, Euphorbiaceae; 2, Asclepiadaceae, Valerianaceae; 3, Centrolepidaceae, Mayacaceae, Xyridaceae, Eriocaulaceae, Bromeliaceae; 4 (in 2 parts), Cyperaceae; 5 (in 2 parts), Scrophulariaceae. [Following the withdrawal of official support after the fall of the first Peronist government and consequent suspension of this work, further family treatments appeared in *Revista del Museo de La Plata, Sección Botánica* as a series, 'Flora Argentina'.]

PRO FLORA (CONICET, REPÚBLICA ARGENTINA), 1994– . *Flora fanerogámica argentina.* Fasc. 1– . [Córdoba.]

Comprises descriptive accounts of families or parts thereof with keys to all taxa, accepted names, derivations, synonymy, vernacular names, references to illustrations, representative *exsiccatae*, indication of overall and national distribution, habitat and altitudinal range, chromosome numbers, and commentary where appropriate; index to all names at end of each fascicle. The family and generic accounts include key bibliographic references. [A list of families (the numbering of which broadly follows that of Dalla Torre and Harms) appears in the inside covers of each fascicle; parts published are progressively indicated. Through 1999, 63 fascicles had been published.]

ZULOAGA, F. O. and O. MORRONE (eds.), 1996–99. *Catálogo de las plantas vasculares de la República Argentina.* 2 vols. (in 3). Maps (end-papers). (Monogr. Syst. Bot. Missouri Bot. Gard. 60, 74). St. Louis: Missouri Botanical Garden Press. Accompanied by F. O. ZULOAGA *et al.*, 1994. *Catálogo de la familia*

Poaceae en la República Argentina. 178 pp. (*Ibid.*, 47). St. Louis.

Alphabetically arranged (within major taxa), annotated enumeration of vascular plants with synonymy, literature citations, representative *exsiccatae*, and internal and external distribution; synonymized and doubtful taxa, master bibliography, and standard list of accepted and synonymous family and generic names apropos to the volume at end. The general part (in both Spanish and English) furnishes the plan and scope of the work; this is followed by acknowledgments, the list of authors, and detailed statistics. [Vol. 1 covers 354 pteridophyte, 21 gymnosperm and 1138 non-glumaceous monocotyledon species; vol. 2 covers 6896 species of dicotyledons; Poaceae, 1151 species.]

Partial work

Grisebach's *Symbolae* was preceded by that author's *Plantae lorentzianae* (1874).

GRISEBACH, A. H. R., 1879. Symbolae ad floram argentinarum. *Abh. Königl. Ges. Wiss. Göttingen* 24: 3–345. (Reprinted separately, 1879, Göttingen; 346 pp.)

Systematic enumeration of vascular plants, without keys; includes descriptions of new taxa, limited synonymy (with references and a few citations), vernacular names, abbreviated summary of local range with some *exsiccatae*, general indication of extralimital range, short critical remarks here and there, and (p. 346) an index to families. The introductory section includes a phytogeographical discussion. [The emphasis of this work is on the plants of northern Argentina, ranging from the provinces of Catamarca, Córdoba, Santa Fé, and Entre Rios to the north and northwest.]

Woody plants (including trees): guides

No treatments are available apart from Latzina's outdated checklist or detailed dendrologies covering a restricted range of commercial species. The standard dendrology remains that of Tortorelli (1956).

DIRECCIÓN DE INVESTIGACIONES FORESTALES, REPÚBLICA ARGENTINA, 1961. *Árboles forestales argentinos*. 2nd edn. Unpaged in 1 vol., illus. (part col.), maps. Buenos Aires: Ministerio de Agricultura y Ganaderia (distributed by Librart). (1st edn., 1956.)

A 48 cm high dendrological atlas, arranged by vernacular names, comprising colored plates (with habitat figures, botanical details, wood sections and range maps) and facing descriptive text and commentary.

LATZINA, E., 1937. Index de la flora dendrológica argentina. *Lilloa* 1: 95–211, 14 pls.

Comprises a briefly annotated systematic checklist of woody plants (839 species), with indication of localities, prov-

inces, habit, vital statistics, special notes, and plates of selected tree species. The introductory section contains a synopsis and indices to vernacular names as well as to families and genera, and a good bibliography appears at the end (pp. 198–211). Special emphasis is given to Tucumán Province.

TORTORELLI, L. A., 1956. *Maderas y bosques argentinas*. xxvii, 910 pp., 104 figs. (mostly halftones), 111 pls. (micrographs), 22 tables, map, col. frontisp. Buenos Aires: Acme.

The third of the six sections of this book (pp. 211–629) comprises a dendrological treatment of 111 leading tree species (with mention here and there of additional species); each entry includes trade, scientific and vernacular names, distribution, frequency and volume, botanical and biological features, and features of the wood: aesthetics, structure, histology, properties, applications and exploitation. Other sections include general treatments of Argentine forests and woods, a wood key, and a bibliography (pp. 871–885) and indices.

Pteridophytes

CAPURRO, R. H., 1938. Catálogo de las pteridófitas argentinas. In P. Campos Porto *et al*. (eds.), *Anais da Primeira Reunião Sul-Americana de Botânica*, 2: 69–210. Rio de Janeiro: Jardim Botânico.

Systematic enumeration of pteridophytes (arranged according to Christensen's system, as in *Index filicum*), covering 359 species and 90 additional infraspecific taxa; includes synonymy and extensive citations (indicated as having descriptions or mere observations) all cross-referenced to the *Index*; references (pp. 196–210); no separate index. Species without full verification of status are given in smaller type.

381

'Mesopotamia'

Comprises the northeastern provinces of Misiones, Corrientes and Entre Rios, as well as (for convenience) the province of Santa Fé just to the west of the Rio Paraná. No recent general works are available for the area or any of the political units, except for the new flora of Entre Rios (see below) begun by Antonio Burkart and part of the INTA programme. However, since 1979 Burkart's flora has been joined by another INTA project, a flora of Corrientes, currently in preparation under the direction of Antonio Krápovicas.

BURKART, A. *et al*. (eds.), 1969– . *Flora ilustrada de Entre Rios*. Parts 2, 3, 5, 6. Illus., maps (Colecc. Ci. INTA 6). Buenos Aires: Librart.

Illustrated descriptive flora of vascular plants; includes keys to all taxa, synonymy (with references) and citations, vernacular names, generalized local range with selected *exsiccatae*, summary of extralimital distribution, taxonomic commentary, and notes on habitat, special features, uses, etc.; glossary and index to all botanical and vernacular names at end of each part. [Four parts of a projected six have been published.]

Woody plants (including trees): guide

JOZAMI, J. M. and J. DE D. MUÑOZ, 1982. *Árboles y arbustos indigenas de la Província de Entre Rios.* 407 pp., illus. Santa Fé: 'Ipnays'.

Systematically arranged students' manual for woody plants; includes extensive treatment of each species, including a description, vernacular name(s), properties and uses, and chemical characteristics; references for each family; general references (pp. 388–396) and extensive synoptic index/table of contents at end of the work. The introductory section, mainly pedagogical, includes a history of botanical work in the province as well as topographical matter and a discussion of the forests.

382

'El Chaco'

Includes the provinces of Santiago del Estero and Formosa as well as the Chaco Territory. For long the only available provincial treatment was a rather uncritical checklist of Santiago del Estero by Antenor Alvarez (described below; part of the province is also covered by E. R. de la Sota's treatment of northwestern pteridophytes listed under **383**). Work is, however, in progress on a new *Flora chaqueña* as part of the INTA series of regional floras, and to this end some fascicles have been published. A complete account, however, is still well into the future.

DIGILIO, A. P. L., 1971–74. *Notas preliminares para la flora chaqueña (Formosa, Chaco y Santiago del Estero).* Fasc. 1–7. Castelar, Buenos Aires: Centro de Investigaciones de Recursos Naturales, INTA. Continued as R. L. PÉREZ-MOREAU, 1994– . *Flora chaqueña.* Fasc. 8– . Illus. Castelar, Buenos Aires.

Descriptive 'research' flora with keys, synonymy, vernacular names, localities with *exsiccatae*, and notes on distribution, habitat, and variability; index at end of

each fascicle. Family entries include lists of key references. [Through 1999, 11 fascicles in all had appeared, the most recent in 1994. A list of families with indication of those published appears on the outside and inside back covers.]

Partial works

Two early works are A. ALVÁREZ, 1919. *Flora y fauna de la Provincia de Santiago del Estero.* 176 pp., illus. Santiago del Estero: sine publ.; and A. GANCEDO, 1916. *Flora arbórea del territorio nacional del Chaco y proyecto de ley.* 237 pp., illus., map. Buenos Aires: Talleres gráficos de la penitenciaria nacional. The former includes a poorly annotated checklist of the flora, while the latter features photographs of nearly every tree species accounted for.

383

Northwestern Argentina

Area: 380000 km^2 (Jujuy, Salta, Catamarca, Tucumán and parts of Santiago del Estero (**382**) and La Rioja (**384**)). Vascular flora: Jujuy, 3500 species (PD; see **General references**); no data for other provinces. – Includes the provinces of Los Andes, Catamarca, Salta, Jujuy and Tucumán; the well-known Instituto Botánico 'Miguel Lillo' is located in the last-named.

Save for recent contributions on pteridophytes and trees, no general work on this partly subtropical, partly Andine region has been published. Since 1977, however, partial coverage has begun to be realized in modern provincial floras. Those for Jujuy (in the INTA series) and Tucumán have commenced publication, but as of early 1990 neither was yet complete. For separate accounts of the alpine flora, see **303/II** (FRIES).

Woody plants (including trees): guide

LEGNAME, P. R., 1982. *Árboles indigenas del Noroeste Argentino.* 226, [1] pp., 120 pls. (Opera Lilloana 34). Tucumán.

Illustrated descriptive account, with artificial keys (pp. 6–19); text and atlas in separate sections. References, pp. 98–99; indices, pp. 100–106. Covers Jujuy, Tucumán, Salta, Catamarca and, in addition, adjacent Santiago del Estero (**382**). [A revision and expansion of the author's 1966 tree book for Tucumán (see below) and designed to be used with it (some illustrations in that work are, however, not reproduced here).]

Pteridophytes

Sota, E. R. de la, 1972–77. Sinopsis de las pteridófitas del noroeste de Argentina, I–IV. *Darwiniana* (Buenos Aires) **17**: 11–103; **18**: 173–263; **20**: 225–232; **21**: 120–138. Illus.

A systematic enumeration (171 species and 15 infraspecific taxa) with keys; includes synonymy (with references) and citations, type localities, overall and regional distribution, *exsiccatae*, taxonomic notes, and references to revisions and other treatments. New taxa are illustrated. The introductory section (part I) deals with floristic formations of Argentina, regional background and problems, and the rationale for the work (as a precursor to more definitive treatments in the INTA programme), while part IV is devoted to a review of the pteridophyte flora, including its nomenclature, ecology, biogeography and utilization, along with an index (pp. 130–138). Covers Jujuy, Salta, Catamarca, Tucumán and parts of Santiago del Estero (**382**) and La Rioja (**384**).

I. Jujuy

Cabrera, A. L. (coord.), 1977– . *Flora de la provincia de Jujuy*. Parts 2, 8–10. Illus., maps (Colecc. Ci. INTA 13). Buenos Aires: Librart.

Illustrated semi-monographic flora of vascular plants; includes keys to all taxa, synonymy (with references) and citations, localities with selected *exsiccatae* along with summary of local and extralimital distribution and altitudinal zonation, taxonomic commentary, and notes on habitat, special features, uses, etc.; citations of revisionary treatments under family and generic headings; index to all scientific names at end of each volume. As of 1999 four of the 10 projected parts, covering pteridophytes, the families Clethraceae through Solanaceae (part 8), Verbenaceae through Calyceraceae (part 9) and the Compositae (part 10), had been published; arrangement of angiosperms follows the Englerian system. Part 2 (pteridophytes) includes a general introduction to the ferns and fern-allies as well as a short introduction to the work as a whole.

II. Tucumán

Meyer, T., M. Villa Carenzo and P. R. Legname, 1977. *Flora ilustrada de la provincia de Tucumán*. Fasc. 1. 305 pp., 79 text-figs. Tucumán: Ministerio de Cultura y Educación de la Nacion, Fundación 'Miguel Lillo'.

Semi-monographic descriptive flora; includes full-page plates illustrating most species (one or more to a plate) and text including synonymy (with references), citations, vernacular names, representative *exsiccatae*, summary of Argentine and overall distribution, taxonomic notes, and remarks on origin or occurrence, habitat, etc.; references to family or generic revisions given under appropriate headings; index to all scientific and vernacular names at end. The introductory section is very brief and merely explanatory. This first of several projected fascicles covers 12 mainly 'sympetalous' families, although no specific systematic sequence is being followed. [Apparently not continued.]

Woody plants (including trees): guide

Digilio, A. P. L. and P. R. Legname, 1966. *Los árboles indigenas le la provincia de Tucumán*. xxvii, 214, 29 pp., illus., maps (Opera Lilloana 15). Tucumán.

Descriptive account with keys, limited synonymy (including some citations), vernacular names, local and extralimital range, illustrations, and notes on habitat, uses, etc.; list of references and indices at end.

384

West central Argentina

Includes the provinces of La Rioja, San Juan and Mendoza. Save for the new work described below (along with coverage of the pteridophytes of part of La Rioja by de la Sota; see **383**), no general floras or enumerations for the subregion or its parts appear to be available.

Partial work (San Juan)

Kiesling, R. (coord.), 1994– . *Flora de San Juan, República Argentina*. Vol. 1 (eds. R. Kiesling, M. Múlgera and E. A. Ulibarri). [8], 348, [2] pp., illus. Buenos Aires: Vázquez Mazzini Editores.

Illustrated descriptive handbook-flora with keys, key literature, vernacular names, distribution and habitat (with selected *exsiccatae*), and overall range along with some taxonomic and other notes; summary of new combinations and records (pp. 347–348); index to all names at end. The introduction includes background to the work, major sources, abbreviations and other conventions, a list of collaborators, and a history of exploration and of research towards the current work. [Encompasses pteridophytes, gymnosperms (all Ephedraceae), and dicotyledons from Salicaceae through Leguminosae, essentially following the Englerian system. The section on Chenopodiaceae (pp. 86–117) is, not surprisingly, noteworthy. The work is a new member of the INTA provincial floras programme.]

385

Central Argentina

This encompasses the provinces of Córdoba and San Luis. Although the Museo Botánico has been a center for taxonomic studies since its formation under Grisebach, Hieronymus and Kurtz in the second half of the nineteenth century, no general floras or enumerations for the subregion or either of the provinces are available save for Seckt's *Flora cordobensis* (1929–30; see below). A new INTA flora, *Flora del Centro de Argentina*, was around the 1970s being elaborated at Córdoba; parts are now appearing under other auspices.

Bibliography

SPARN, E., 1938. *Bibliografía botánica de Córdoba*. 108 pp., 2 pls. (Revista Mus. Prov. Ci. Nat. Córdoba 3). Córdoba. [Arranged by author but includes a systematic index. The plates depict botanists closely associated with the province.]

MUSEO BOTÁNICO, FACULDAD DE CIENCIAS EXACTAS, FÍSICAS Y NATURALES, UNIVERSIDAD NACIONAL DE CÓRDOBA, 1994– . *Pródromo de la flora fanerogámica de Argentina Central*. No. 1. Illus. Córdoba.

Detailed 'research' accounts with keys, synonymy, references, citations, descriptions, localities with *exsiccatae*, distribution, and (where necessary) commentary; indices at end. No vernacular names or typifications are given. Key references appear in the introductory paragraphs, but not in the text. [Part 1 covers Asteraceae–Vernonieae and –Eupatorieae.]

SECKT, H., 1929–30. *Flora cordobensis*. 632 pp., 22 pls. (Revista Univ. Nac. Córdoba 16, 17). Córdoba.

Manual-key and enumeration of native, naturalized and cultivated seed plants of Córdoba Province; includes limited synonymy, vernacular names, general indication of local range, notes on habitat, phenology, etc., and (at end) a lengthy glossary and index to vernacular, family and generic names. Species are enumerated under their respective genera but no keys to them are supplied.

386

Buenos Aires

Corresponds with the limits of the province, thus including the greater part of the pampas. Relatively considerable collections and a fair basis of literature were keys to the rapid appearance of Angel Cabrera's provincial flora, the first in the INTA programme to be initiated and completed as well as being the first complete Argentine provincial flora. Earlier floras and enumerations include *Enumeración sistemática de las plantas recogidas en Buenos Aires y sus alrededores* by Carlos Bettfreund with the assistance of G. Hieronymus and G. Niederlein (1890; 2nd edn., 1898), *Chloris platensis argentina* by C. M. Hicken (1910), covering 1261 species, and *Flora bonariensis* by Hans Seckt (1918), a key to families and genera with lists of species, and two manuals by C. Spegazzini: *Flórula de la ciudad de La Plata y su partido* (1901) and the never-completed *Flora de la provincia de Buenos Aires* (1905).

Bibliography

CABRERA, A. L. and M. FERRARIO, 1970. *Bibliografía botánica de la provincia de Buenos Aires: plantas vasculares*. 96 pp. La Plata: Comisión de Investigaciones Científicas, Provincia de Buenos Aires. [Annotated alphabetical listing, followed by classified indices with categories inclusive of floras, tree books and medicinal and poisonous plants.]

CABRERA, A. L. (ed.), 1963–70. *Flora de la provincia de Buenos Aires*. 6 parts. Illus. (Colecc. Ci. INTA 4). Buenos Aires: Librart.

Illustrated descriptive flora of native, adventive and commonly cultivated vascular plants; includes keys to all taxa, synonymy (with references) and citations, vernacular names, general indication of local range (with representative *exsiccatae*) and brief summary of extralimital distribution, some taxonomic commentary, miscellaneous notes, and indices to all botanical and vernacular names (in each part). A general introduction to the work appears in part 6.

Partial work: Buenos Aires metropolitan region

CABRERA, A. L. and E. M. ZARDINI, 1978. *Manual de la flora de los alrededores de Buenos Aires*. 755 pp., illus. Buenos Aires: Acme. (Originally published 1953 as *La flora des alrededores de Buenos Aires*, by Cabrera alone.)

Concise illustrated manual of vascular plants; includes keys, synonymy, vernacular names, and notes on local range, habitats and phenology; glossary and indices to all names at end. An introductory section includes general information and a primer on descriptive terminology and floristics.

387

La Pampa

No general floras or enumerations appear to be available for this south-central province. Preparation has been said, however, to be in progress on a provincial flora (under the direction of G. Covas) as part of the INTA regional floras programme.

388

Patagonian Argentina

For works relating to Patagonia as a whole (including the Chilean portions as well as all of Tierra del Fuego), see **301/III** and **/IIIa**. The Argentine part here includes the provinces and territories of Río Negro, Neuquén, Chubut, Santa Cruz, and Tierra del Fuego. For the Malvinas, see **389**.

The basic work on the region is *Flora patagonica* (MACLOSKIE *et al.*, 1903–06, 1914; **301/III**); this in turn succeeded Hooker's *Flora antarctica*. Correa's modern, similarly titled work, one of the INTA regional floras, is limited to Argentine territory.

CORREA, M. N., 1969–99. *Flora Patagónica*. Parts 1, 2, 3, 4a, 4b, 5, 6, 7. Illus. (Colecc. Ci. INTA 8). Buenos Aires: Librart.

Illustrated descriptive flora of vascular plants, with keys to all taxa, synonymy, references and citations, vernacular names, citation of *exsiccatae* and general indication of local and extralimital range, critical remarks, and notes on habitat, etc.; indices to all botanical and vernacular names in each part.

Encompasses Argentine Patagonia and the Islas Malvinas/Falkland Islands. [Completed in 1999.]

389

Falkland Islands (Islas Malvinas)

See also **388** (CORREA). – Area: 12 172 km^2. Vascular flora: 256 species, of which 163 native (Moore, 1968; see below). The islands are administered by the United Kingdom but have long been claimed by Argentina.

David Moore's definitive flora effectively summarizes an extensive earlier literature, beginning with Gaudichaud's first report of 1825 (in *Annales de Sciences Naturales* **5**: 89–110) and including Skottsberg's *Botanical survey of the Falkland Islands* (1913, in *Kongl. Svenska Vetenskapsakad. Handl.* **50**(3)). However, the following is of particular value for its illustrations: E. F. VALLENTIN and E. M. COTTON, 1921. *Illustrations of the flowering plants and ferns of the Falkland Islands*. xii, [65] pp., 64 col. pls. London: Reeve.

MOORE, D. M., 1968. *The vascular flora of the Falkland Islands*. 202 pp., 24 text-figs. (incl. maps), 6 pls., 2 folding maps (Scient. Rep. Brit. Antarctic Surv. 60). London.

Critical descriptive flora of vascular plants; includes keys to genera and species, appropriate synonymy (with references and citations), vernacular names, localities with *exsiccatae*, summary of local and extralimital range, taxonomic commentary, diagnostic figures, and notes on habitat, special features, biology, etc.; glossary, bibliography, and indices to botanical and vernacular names at end. The introductory section includes descriptions of physical features, climate, and vegetation formations as well as an account of botanical exploration and analyses of floristics and phytogeography.

Region 39

Chile

Area: 751656 km^2. Vascular flora: 5215 species (Marticorena and Quezada, 1985, see below).[74] – The region corresponds to mainland Chile. For the country's Pacific insular possessions, see respectively **018** (Desventuradas Islands), **019** (the Robinson Crusoe/Juan Fernández group) and **988** (Rapa Nui/Easter Island).

Important early observations (first published in Italy as *Saggio sulla storia naturale del Chili* in 1782, with a precursor in 1776 and a second edition in 1810) were made by the Chilean Jesuit Juan Ignacio Molina before the expulsion of his order in 1768 by the Spanish king Carlos III. Serious botanical collecting in the country began, however, with the 1782–83 visit of Hipólito Ruiz, José Pavón and Joseph Dombey of the 'Real Expedición Botánica' to Peru and Chile. Their results appeared primarily in the never-completed *Flora peruviana, et chilensis* (1798–1802, 1954–59). The expeditions 'rounding the Horn' or traversing the Straits of Magellan naturally herborized in Chile; among their collectors were Banks and Solander, the Forsters, Haenke and Neé, Gaudichaud, Darwin, J. D. Hooker, and the U.S. Exploring Expedition. Early nineteenth-century resident collectors included Hugh Cuming (1819–31), Carlo Bertero (1828–31) and Claude Gay (1828–32, 1834–42); Gay's contribution was by far the most considerable. The second half of the nineteenth century was dominated by the work of Rudolf Philippi, in Chile from 1851 to 1897; he was effectively responsible, along with his son and successor Federico, for the establishment of the present national herbarium in Santiago. Other botanists from late in the nineteenth century onwards have included, besides Karl Reiche (resident from 1890 to 1911), Friedrich Johow (notable particularly for studies of Chile's Pacific islands), Per Dusén, Carl Skottsberg, Ivan M. Johnston (active especially in the north), Gualterio Looser, and Carlos Muñoz Pizarro.

Bertero's main work was *Lista de las plantas que han sido observadas en Chile por el Dr. Bertero en 1828* (1829, in *Mercurio Chileno* **12–16**, *passim*; an English edition appeared in 1831–33 as *List of the plants of Chile*, in *American Journal of Science* **19–23**, *passim*). His early death prevented any further contributions. Claude Gay's two long sojourns in Chile, devoted to natural history in general, resulted in a large illustrated encyclopedic work, now a classic of Chileana: *Historia física y política de Chile*. Its eight-volume botanical part contains the first complete flora. This was followed by the *Catalogus* of F. Philippi (1881) and the incomplete flora of Reiche (1896–1911). In the second half of the twentieth century the most important contributions through 1990 have been the one-volume *Sinopsis de la flora chilena* (1959; 2nd edn., 1966) by Muñoz Pizarro, *Catálogo de la flora vascular de Chile* (1985) by C. Marticorena and M. Quezada, and *Flora arbórea de Chile* (1983) by R. Rodríguez, M. Quezada and O. Matthei.

Among the few partial and local floras, a manual for the Santiago district was produced by L. E. Navas-Bustamante (1973–79). Elsewhere, the best coverage is in the south. The Fuegian flora has been revised by several authors, firstly by Joseph D. Hooker in his *Flora antarctica* (vol. 2, 1845–47) and latterly by David M. Moore in *Flora of Tierra del Fuego* (1983; see **301/IIIa**). Patagonia as a whole was treated firstly by Hooker and then by George Macloskie and others in their *Flora patagonica* (1903–06, 1914; see **301/III**).

Chile remains in need of much detailed botanical exploration as well as improved documentation. General collecting is extensive but uneven; local institutions, though of long standing, have an inadequate representation of the flora. Moreover, suggestions have been made that many genera are in Chile still imperfectly known.[75] In contemporary terms, only a beginning, albeit significant, was made by Muñoz Pizarro before his untimely death in 1976. His projected five-volume 'generic' flora, then in preparation, has now given way to a new, definitive *Flora de Chile*, based at the University of Concepción under the care of Clodomiro Marticorena and Roberto Rodríguez. Organized on a collaborative basis and with some outside assistance, this project began in 1987. Its first volume appeared in 1995.

Progress

For early history, see C. E. PORTER, 1910. Bosquejo histórico, desarollo y estado actual de los estudios sobre antropología, flora y fauna chilenas. *Anales Soc. Cient. Argent.* **70**: 267ff. (separately reprinted: 45 pp.), based on an address to an Inter-American Scientific Congress; and W. B. TURRILL, 1920. Botanical exploration in Chile and Argentina. *Bull. Misc. Inform.* (Kew) 1920: 57–66, 223–224. A historical survey is also given on pp. 1–27 of Reiche's *Flora*, and a chronologically arranged cyclopedia of collectors and botanists appears in vol. 1 (pp. 1–62) of the current *Flora de Chile*.[76] News of the latter project appears in *Flora of Chile Project Newsletter* (July 1991–).

For separate reviews on Patagonia and Fuegia, see the introduction to *Flora antarctica* by J. D. Hooker as well as two papers by E. J. Godley in *Tuatara* (1965, 1970; both under **Superregion 08–09**).

Bibliographies. General and divisional bibliographies as for Division 3.

Regional bibliography

Sinopsis de la flora chilena by Muñoz Pizarro (see **390** below) also features an extensive bibliography.

MARTICORENA, C., 1992. *Bibliografía botánica taxonómica de Chile.* 587 pp. (Monogr. Syst. Bot. Missouri Bot. Gard. 41). St. Louis. [6100 references arranged by author, with taxonomic, geographical and other subject indices; includes numerous monographs and revisions as well as floristic works. References relating to Chile's Pacific insular possessions are also included.]

Indices. General and divisional indices as for Division 3.

390

Region in general

The only complete descriptive flora to date is the one by Gay (1845–54); that by Reiche was never finished and the new *Flora de Chile* has only recently commenced publication. The best modern introduction to the flora, with a valuable bibliography, is Carlos Muñoz Pizarro's *Sinopsis* (1966). For additional coverage of Patagonia (in Chile from Valdivia southwards, including Chiloé) and Fuegia, see **301/III** and **/IIIa**. Good popular works include, in addition to those on the tree flora (see below) and one on alpines (see **303**), two field-guides by Adriana Hoffmann: A. HOFFMAN

J., 1995. *Flora silvestre de Chile: zona central.* 3rd edn. 255 pp., illus. (mostly col.). [Santiago]: Fundación 'Claudio Gay' (originally published 1980; 2nd edn., 1989); and *idem*, 1982. *Flora silvestre de Chile: zona austral.* 258 pp., illus. (mostly col.). [Santiago.]

Keys to families and genera

MUÑOZ PIZARRO, C., 1966. *Sinopsis de la flora chilena (claves para la identificación de familias y generos).* 2nd edn. 500 pp., 248 pls. (5 col.). Santiago: Ediciones de la Universidad de Chile. (1st edn., 1959.) [Illustrated, partly pedagogical introduction to the vascular flora; includes keys to and descriptions of families and keys to genera, limited synonymy, citations, indication of more important species in each family, vernacular names, full-page figures of representative species (particularly trees), a glossary and extensive bibliography, and indices to all vernacular and scientific names. An appendix includes a separate key to all known tree species. (The work also encompasses the Robinson Crusoe/Juan Fernández group, the Desventuradas, and Rapa Nui/Easter Island.)]

GAY, C., 1845–54. *Historia física y política de Chile. Botánica (Flora chilena).* 8 vols., atlas. 103 col. pls., map. Paris: The author; Santiago de Chile: Museo de Historia Natural de Chile.

Descriptive flora, without keys, of vascular plants (vols. 1–6, inclusive of Characeae), the botanical text in Latin and commentary in Spanish; includes limited synonymy, references, vernacular names, some *exsiccatae*, generalized local and extralimital range, taxonomic commentary, notes on phenology, special features, etc., and a lexicon of vernacular names with botanical equivalents. A general index to all botanical names appears in vol. 8 following the treatments (in vols. 7–8) of other plants and fungi. The introductory section (vol. 1) includes a survey of the Chilean flora and its major features.

MARTICORENA, C. and M. QUEZADA M., 1985. Catálogo de la flora vascular de Chile. *Gayana, Bot.* 42(1–2): 1–157.

Alphabetically arranged name list of vascular plants (including infraspecific taxa) with authorities, in four major sections: pteridophytes, gymnosperms, monocotyledons and dicotyledons. Introduced species are indicated by an asterisk. No synonymy is given. The main list is followed by a long statistical section (pp. 97ff.), a 'resumen' (p. 114), indices to genera and author abbreviations, special notes (pp. 151–152) on sources and unusual problems, and (relatively few) references

(p. 155). [A precursor to the critical *Flora de Chile*, at present in preparation at the University of Concepción.]

MARTICORENA, C. and R. RODRÍGUEZ R., 1995– . *Flora de Chile*. Vols. 1– . Illus. (some col.), maps. Concepción: Universidad de Concepción.

Descriptive flora of native and introduced vascular plants with keys, basionymy and synonymy, citations of illustrations, vernacular names, indication of distribution, altitudinal range and habitat, some commentary, figures of diagnostic features, and dot distribution maps; indices to all scientific and vernacular names at end of each volume. Volume 1 also contains chapters on the chronology of botanical exploration, plant geography, and the prehistory and evolution of the flora, each with selected references. [Covers at present pteridophytes and gymnosperms.][77]

PHILIPPI, F., 1881. *Catalogus plantarum vascularium chilensium adhuc descriptarum*. viii, 377 pp. Santiago: Ediciones de la Universidad de Chile. (Reprinted from *Anales Univ. Chile* 75 (1881).)

Systematic list (5358 species) of vascular plants and Characeae, with synonymy, references, and citations (the last chiefly to Gay's *Flora*); index to genera and families. The introductory section includes a cyclopedia of collectors and others associated with the Chilean flora.

REICHE, C. F., (1894–)1896–1911. *Flora de Chile*. Vols. 1–6(1). Santiago: Ediciones de la Universidad de Chile. (Reprinted in part from *Anales Univ. Chile* 88 (1894)-116 (1905), *passim*.)

Detailed descriptive flora of seed plants; includes keys to genera and species, full synonymy (without references), citations of key source literature (Gay, Philippi, and others), generalized account of local range, and extensive taxonomic commentary; synopses of contents and indices to botanical names at end of each volume. [Not completed; extends from Ranunculaceae through Chenopodiaceae (except for Cactaceae) on the Candollean system. For a survey of additions (covering about 500 species mostly described after 1918), see G. LOOSER, 1938. Catálogo de plantas vasculares nuevas de Chile. *Revista Univ.* (Santiago) 23: 215–275.]

Partial works

A treatment of the Atacama Desert, *Flora atacamensis*, was published by R. A. Philippi in 1860. The only two such of significance to appear since World War II (Fuegia excepted; see 301/IIIa) are described below.

JOHOW, F., 1948. Flora de la plantas vasculares de Zapallar. *Revista Chilena Hist. Nat.* 49/50: 1–566.

Descriptive flora without keys; entries include references to standard (Gay, Reiche) and other works, limited synonymy, local distribution and habitat, and (in some cases) notes on biology, special features, etc.; index. The final part of the work (pp. 365–513) is a treatment of cultivated plants. [Zapallar is by the coast in southwestern Aconcagua Province (32° 30' to 32° 38' S, north of Valparaíso). Accounts of some taxa are missing on account of the author's death prior to completion of the work.]

NAVAS-BUSTAMANTE, L. E., 1973–79. *Flora de la cuenca de Santiago de Chile*. 3 vols. Illus. (some col.), map. Santiago: Ediciones de la Universidad de Chile.

Students' manual-flora of vascular plants (covering 882 native and introduced species); includes keys, brief descriptions, synonymy, references and key citations, vernacular names, general indication of local and Chilean range, notes on habitat preferences, properties, uses, etc., and (at end of each volume) statistics, lists of references, glossary, indices to all names, and technical figures with captions on facing pages followed by the color plates. An introductory section (vol. 1) includes a map of the area covered (the environs of Santiago de Chile), author abbreviations, floristic elements and kinds of vegetation, useful study areas, and technical and students' notes; the map and abbreviations are repeated in the remaining volumes.

Woody plants (including trees): guides

In recent years, forestry development and an increasing public interest in trees and shrubs have spawned a number of contemporary works on the woody flora. That by Rodríguez, Matthei and Quesada (1983) is a worthy counterpart to J. C. Salmon's 1980 book on the trees of New Zealand (410) in view of the many similarities between the floras.

DONOSO ZEGERS, C., 1978. *Dendrología: árboles y arbustos chilenos*. 2nd edn. 142 pp., illus. (Faculdad de Ciencias Forestales, Universidad de Chile, Manual 2). Santiago.

A formatted, systematically arranged, illustrated dendrological treatment; under each species are vernacular names and headings for local and (if applicable) extra-Chilean distribution, habitat, frequency, ecology, specific features, and importance (properties, uses, general economics), along with small halftone photographs of fertile branchlets. References and indices to vernacular and scientific names are given at the end.

DONOSO ZEGERS, C., 1981. *Árboles nativos de Chile: guía de reconocimiento*. 116 pp., illus. (all col.), maps. Valdivia: Alborada.

A 12×15cm pocket guide, arranged by vernacular names with descriptive text (in Spanish and English), distribution maps, and many photographs of habit and details. The introduction contains necessary aids for the book's use.

DONOSO ZEGERS, C. and L. R. LANDRUM, [1975]. *Manual de identificación de especies leñosas del bosque humedo de Chile.* 168 pp., illus. N.p.: Corporación National Forestal.

An artificially arranged, illustrated dendrological treatment of the moist and wet forest zones (rainfall of 1000–4000 mm/annum). The body of the work is preceded by an introduction, a review of organography and terminology, a glossary, a fruit key, and a general key; it concludes with a systematic list of species. [The work was quickly produced and contains many printing errors. It is similar in format to W. D. Francis's *Australian rain-forest trees* (420–50).]

RODRÍGUEZ R., R., O. MATTHEI S. and M. QUEZADA M., 1983. *Flora arbórea de Chile.* 408 pp., illus. (part col.). Concepción: Universidad de Concepción. (Biblioteca de recursos renovables y norenovables de Chile 1.)

Definitive, large-format dendrological atlas (123 species) with many color paintings and other figures; the alphabetically arranged species treatments (unfortunately without indication of their family) each include vernacular names, synonymy, a full description, distribution (with *exsiccatae*), taxonomic commentary, biological and ecological notes, and an indication of properties, cultivation, and other uses. The extensive introductory section includes historical background, the plan and basis for the work, phytogeographical considerations, a table of distribution covering all species, characteristic trees for each major area (mainland and insular), and a general key to species; at the end are an unillustrated glossary, a gazetteer (with coordinates), an extensive bibliography, and indices to vernacular and scientific names.

Pteridophytes

An early separate work on the pteridophyte flora is *Enumeratio plantarum vascularium cryptogamarum chilensium* (1858) by J. W. Sturm. Three relatively recent works are described below.

DUEK, J. J. and R. RODRÍGUEZ R., 1972. Lista preliminar de las especies de Pteridophyta en Chile continental y insular. *Bol. Soc. Biol. Concepción* **45**: 129–174.

Nomenclatural checklist (193 specific and infraspecific taxa); includes for each species the accepted name, a basionym (where appropriate) and synonymy (with references), and typification. Bibliography, pp. 158–161; index to names, pp. 162–173.

LOOSER, G., 1961–68. Los pteridófitos o helechos de Chile (excepto Isla de Pascua), I–IV. *Revista Univ.* (Santiago) **46**: 213–262; **47**: 17–31; **50/51**: 75–93; **53**: 27–39.

Detailed enumeration, with illustrations, of vascular cryptogams; includes keys to species within genera, etymology, synonymy, indication of habitat, general and Chilean distribution, *exsiccatae*, taxonomic commentary, quotations from original sources, 'folklore', and other notes, with for each family a review of ecology and references at the end of that family.

LÜER, H. G., 1984. *Helechos de Chile.* 245 pp., 43 text-figs. Santiago. (Monografias anexas a los Anales de la Universidad de Chile 1.)

Descriptive treatment with keys, synonymy, derivations, vernacular names, and notes on distribution, habitat, and uses; key literature cited at end of each family or subfamily. The introductory parts cover pteridophyte basics, their uses and cultivation, and (pp. 27–29) a history of their study in Chile, while at the end are short biographies, a list of general literature, an unillustrated glossary, and indices to scientific and vernacular (Mapuche) names. [150 species covered, or some 3 percent of the Chilean vascular flora. The Pacific Island possessions are included.]

Notes

1 A concise review of the work of these expeditions is furnished by S. Castroviejo, 1989. Spanish floristic exploration in America: past and present. In L. B. Holm-Nielsen, I. Nielsen and H. Balslev (eds.), *Tropical forests: botanical dynamics, speciation and diversity,* pp. 347–353. London: Academic Press.

2 For a review of Aublet's generic names, see J. L. Zarucchi, 1984. The treatment of Aublet's generic names by his contemporaries and by present-day taxonomists. *J. Arnold Arbor.* **65**: 215–242.

3 It is thus forever to be regretted that a botanical work truly to complement those of Humboldt, Bonpland and Carl Kunth did not emerge from Horsfield's explorations in Java (918), *Plantae javanicae rariores* (1838–52) notwithstanding.

4 For further particulars on this work, see W. T. Stearn, 1968. The plan of the 'Nova Genera'. In *idem* (ed.), *Humboldt, Bonpland, Kunth and tropical American botany: a miscellany on the 'Nova Genera et Species Plantarum',* pp. 10–21 (quotation from p. 12).

5 The educational value of skillfully illustrated florulas of limited areas has been forcefully argued in two reviews, respectively by Rupert Barneby and Thomas Croat, of *Flora del Ávila* (318) in *Brittonia* 31: 496–497 (1979). This contrasts strongly with the views of Cornelius A. Backer regarding illustrations in his *Flora of Java* (918), although in some of his other works on Java the Dutch author did make liberal use of them. Barneby noted that comparable works were in preparation for other parts of Middle and South America, thus filling long-standing gaps.

6 The lag in the knowledge of the tropical South American flora as compared with that of Africa, strongly pointed out by Prance, was earlier emphasized by André Aubreville in a state-of-knowledge review of sub-Saharan Africa (*Adansonia,* sér. 2, **4**: 4–7 (1964)).

7 There is also a Web site: http://www.nybg.org./bsci/ofn/.

8 The author on p. 1 expressed himself as having no time for niceties about phylogenetic 'purity'.

9 Pittier's work was in turn adapted by Standley for his *Trees and shrubs of Mexico* and *Flora of the Panama Canal Zone*.

10 The Organization for Flora Neotropica (OFN) is a UNESCO Category B non-governmental body, comprising at present 130 members from 30 countries. Control is vested in a 'staff committee' of four, while the *Flora* itself is under the management of an editorial committee of six. OFN also promotes conservation of the neotropical flora as well as botanical education. For a recent review, see E. Forero and S. Mori, 1995. The Organization for Flora Neotropica. *Brittonia* 47: 379–393. A Web site has since been established (see Note 7).

11 Earlier general reviews, all with extensive lists of references, include: B. Maguire, 1970. On the flora of the Guayana Highland. *Biotropica* 2: 85–100; *idem*, 1979. Guayana, region of the Roraima sandstone formation. In K. Larsen and L. B. Holm-Nielsen (eds.), *Tropical botany*, pp. 223–238; J. A. Steyermark, 1979. Flora of the Guayana Highland: endemicity of the generic flora of the summits of the Venezuelan tepuís. *Taxon* 28: 45–54; and O. Huber, 1988. Guayana Highlands versus Guayana Lowlands, a reappraisal. *Taxon* 37: 595–614.

12 Though based in the first instance upon the expedition collections of John B. Hatcher and O. A. Peterson, with some additions from a collection of B. Brown, Macloskie used the opportunity to produce the first, and so far only, work of its kind: a synthesis flora of the whole of Patagonia, pulling together the scattered literature of the eighteenth and nineteenth centuries and incorporating the results of contemporary exploration by other (mainly Swedish) workers. The sumptuous appearance of the work, produced with the support of the J. Pierpont Morgan Publication Fund, is nevertheless misleading: the type is of a size appropriate for the 'Large Print' edition of the *New York Times* whilst the text is surprisingly poorly documented. It could by the standards of the late twentieth century be seen as something of a 'white elephant' – though not the only one in southern South America.

13 At the Muséum National d'Histoire Naturelle, the abolition of Adrien de Jussieu's chair following his death in 1853 led to a departmental transfer for Weddell, who had been an *aide naturaliste* to Jussieu. Problems in his new position, as well as family concerns, brought about a departure from the Muséum in 1857. See J. Léandri, 1967. La fin de la dynastie des Jussieu et l'éclipse d'une chaire au Muséum (1853 à 1873). *Adansonia*, II, 7: 443–450. With respect to the expedition reports as a whole, they were in all published in seven parts with 15 volumes from 1850 to 1861, and have been much in demand for artistic as well as scientific reasons.

14 On the other hand, Steyermark claimed in 1974 that less than 2 percent of the country was botanically adequately known.

15 ORSTOM is now officially the Institut Français de la Recherche Scientifique et Technique d'Outre-Mer (IFRSTOM). However, 'ORSTOM' – easier to pronounce – remains in wide use.

16 An encyclopedia of collectors (begun at Utrecht by Florence E. Vermuelen and being continued by Renske C. Ek) is also planned (the part on Guyana being published in 1990 as no. 1 in *Flora of the Guianas, Supplementary Series*).

17 The French Guiana entry was kindly reviewed by Christian Feuillet in 1993.

18 Underlying the work is an information system dubbed AUBLET; for description, see M. Hoff *et al.*, 1990. La banque de donnés «AUBLET» de l'herbier du Centre ORSTOM de Cayenne (CAY). *Bull. Jard. Bot. Nat. Belgique* 59: 171–178. The *Inventaire* should be used with caution: families not reviewed by a specialist may contain errors, as in Araliaceae (part V, personal observation).

19 This work drew extensively on *Flora of Surinam* but still contains numerous misidentifications (C. Feuillet, personal communication).

20 The *Guide* is part of a long-term floristic-ecological study of some 50000 km^2 of central upland French Guiana under the aegis of ORSTOM-Cayenne and the New York Botanical Garden, the latter represented by S. A. Mori. It was written in relation to U.S. National Research Council guidelines regarding intensive study of selected sites in aid of ecology and evolution. Its design was made 'simple' so as not to overlap with the larger-scale *Flora of the Guianas* and *Flora Neotropica*. Mori (with Carol Gracie) has also used the occasion to provide a detailed analysis of the intellectual, technical and economic aspects of the *Guide* project (*Brittonia*, in press).

21 The bibliographic history of *Flora of Suriname* is complicated. Of the original four volumes as projected (each with two parts), only the first part and a portion of the second in each volume (save for vol. 1, where part 2 had yet to be commenced) were published prior to interruption of the work in the early 1940s due to the closure of Utrecht University under the German occupation. Between 1945 and 1957 the first installment of volume 1(2) and the second of 3(2) appeared. On resumption of the project in the 1960s, unpublished phanerogam families were assigned to a new volume 5, with bryophytes going to volume 6. Volume 1(2) was finally 'closed' in 1968 with a tranche of additions and corrections to volume 1(1). In 1976, volume 2(2) was similarly terminated but by

re-publication in a complete form, with more than 300 new pages comprising additions and corrections to volume 2(1). A similar procedure was followed for volume 3(2), completed and published by 1986. However, by the 1980s the work had outlived its warrant and in 1983 a decision was made to terminate it in favor of *Flora of the Guianas* (see **311** above). Thus, volume 4(2) – not reprinted after World War II and never continued – remains unfinished; while volumes 5 and 6 likewise were not completed (of vol. 5(1), three and of 6(1), two fascicles were published). The Orchidaceae were never published in the *Flora*; for a national revision see M. C. M. Werkhoven, 1986. *Orchideeën van Suriname/Orchids of Suriname.* 256 pp., illus. (part col.). Paramaribo: Vaco.

22 J. A. Steyermark, 1977. Future outlook for threatened and endangered species in Venezuela. In G. T. Prance and T. S. Elias (eds.), *Extinction is forever*, pp. 128–135. New York.

23 Complete, though separately published, revisions also exist for the Araceae, Arecaceae, Cactaceae, Droseraceae, Lauraceae, Verbenaceae and some pteridophyte families. Altogether, some 25 percent of the total vascular flora has been documented, rising to 32 percent if coverage in *Flora Neotropica* is also accounted for (Huber *et al.*, 1998, pp. 50–51; see **Region 31, Progress**). For the Verbenaceae, independently published but in a format very similar to that of the main work, see S. López-Palacios, 1977. *Flora de Venezuela: Verbenaceae.* 654 pp., 146 text-figs. Mérida: Facultad de Farmacia/Consejo de Publicaciones, Universidad de los Andes.

24 Would that more floras have good introductory volumes! Fortunately, there are now several good modern examples, among them *Flora of North America* (**100**) and *Flora of Victoria* (**431**).

25 This work should be used with caution as errors exist, for example in Araliaceae where good revisionary studies were not then available.

26 A recent retrospective on the expedition, with reproductions of many of the plates, is Real Jardín Botánico, Madrid, 1992. *Mutis and the Royal Botanical Expedition of the Nuevo Reyno de Granada.* 2 vols. Illus. (part col.), maps. Bogotá/Barcelona: Villegas Editores/Lunwerg Editores. See also L. Uribe Uribe, 1953. La Expedición Botánica del Nuevo Reyno de Granada: su obra y sus pintores. *Revista Acad. Col.* 9(33–34): 1–13.

27 This was wound up following a scientific review of the institute in 1978; in its place was developed the broader *Flora de Colombia* programme, with from 1982 the inclusion therein of the *Flora de la Real Expedición Botánica* project. These were in 1983 formalized as the 'Programa Segunda Expedición Botánica', 200 years after its namesake. See *Flora de Colombia*, 1, Introduction (1983), and P. Pinto-Escobar, *Vegetación y flora de Colombia* (1993).

28 For Colombian botanical history, see E. Pérez Arbaláez, 1971. Las ciencias botánicas en Colombia. In A. Bateman *et al.* (eds.), *Apuntes para la historia de la ciencia en Colombia* (Coloquio sobre Historia de la Ciencia en Colombia, Rionegro, 1970), 1: 101–161. Bogotá: Fondo Colombiano de Investigaciones Científicas 'Francisco José de Caldas' (Documentación e historia de la ciencia en Colombia); and V. M. Patiño, 1985. *Historia de la botánica y de las ciencias afines en Colombia.* 255 pp., portr. Bogotá: Lerner. (Academia Colombiana de Historia, Historia Extensa de Colombia, 16.) A historical introduction is also furnished by Polidoro Pinto-Escobar in his *Vegetación y flora de Colombia* (1993; see **321**).

29 This last work was part of a 1989 conference, the third in a series begun in 1984. All these have been published in *AAU Reports*.

30 The figure of 35000 species is from *Global biodiversity* (see **General references**); 45000, from Prance (1978; see **General references**); 55000, from J. Hernández-Camacho *et al.*, 1992. Estado de la biodiversidad en Colombia. In G. Halffter (comp.), *La diversidad biologica de Iberoamérica*, I: 39–236. Xalapa, México: Instituto de Ecología, A.C.

31 For list, see p. 15 of D. G. Frodin, 1964. *Guide to standard floras of the world (arranged geographically).*

32 The work is derived from a database, 'COLIFLOR', accessed with an application written in FOXBASE+, a D-base class database program for personal computers.

33 From fasc. 14 (1995) the *Flora de Colombia* programme has had three sponsors: the Instituto de Ciencias Naturales (Colombia) as principal; the Instituto Colombiano de Cultura Hispanica (Colombia), and the Real Jardín Botánico, Madrid (Spain). The editorial board is drawn from all three. Nevertheless, at present rates of publication many years more are needed for substantial advancement.

34 According to Enrique Forero and Alwyn Gentry (personal communication, 1981) the mountains of Chocó are nearly always shrouded in cloud, and exceptionally heavy rainfalls (some over 10000 mm/annum) have been reported. Western Valle (including the Bajo Calima area), on the Pacific, is also quite wet. For extensive discussion of floristic knowledge and progress in Chocó and El Valle, see A. H. Gentry, 1978. Floristic knowledge and needs in Pacific tropical America. *Brittonia* 30: 134–153. On the projected Chocó flora, see E. Forero, 1982. La flora y la vegetación del Chocó y sus relaciones fito-geográficas. *Colombia Geográfica* 10(1): 77–90.

35 See E. Rentería A., 1979. Contribución al estudio de la flora de Santander del Sur de Colombia. *Actualidades Biológicas* **621**: 70–79.

36 See P. M. Jørgensen *et al.*, 1995. A floristic analysis of the high Andes of Ecuador. In S. P. Churchill *et al.* (eds.),

Biodiversity and conservation of Neotropical montane forests, pp. 221–237. New York: The New York Botanical Garden.

37 Progress on the *Flora of Ecuador* to 1986 was summarized by G. Harling in *Reports Bot. Inst. Univ. Aarhus* 15: 9–10 (1986). Preliminary work was started in 1968; in that year the Botanical Institute of Aarhus University became a collaborator along with Göteborg (Harling) and Stockholm (Sparre until 1986) and mounted its first expedition. Further expeditions followed, with reports following. The Catholic University of Quito joined in 1977, and by 1986, 50 of 230 families had been published, with 65 under study. Some 2160 species had appeared or were ready out of an estimated total vascular flora of 20000. A want of collaborators for some large families was reported. An information system was also developed during the 1980s (see P. Frost-Olsen and L. B. Holm-Nielsen, 1986. A brief introduction to the AAU-Flora of Ecuador information system. *Reports Bot. Inst. Univ. Aarhus* 14).

38 H. H. Iltis, 1988. Serendipity in the exploration of biodiversity. In E. O. Wilson and F. M. Peter (eds.), *Bio Diversity*, pp. 98–105. Washington, D.C.: National Academy Press.

39 There are several publications on the Expedición, of which a useful recent retrospective is Sociedad Estatal Quinto Centenário/Real Jardín Botánico, [1988]. *La Expedición Botánica al Virreinato del Perú (1777–1788)*. 2 vols. Barcelona/Madrid: Lunwerg Editores.

40 These latter studies are part of programmes first organized in 1976 and renewed in 1983 in collaboration with the Conservatoire et Jardin Botaniques, Geneva (Switzerland).

41 Upon publication of the last part in the original series, the *Flora* (inclusive of the Orchidaceae) was about 80 percent completed, covering 11789 of an expected 'final' coverage of 13937 species in all save 15 seed plant families. Macbride's pithy style of commentary was in the manner of his teacher, M. L. Fernald (cf. the latter's edition of *Gray's Manual of Botany* (140)). Where families were contributed by specialists, however, preparation inevitably was sometimes protracted. The longest time taken was surely for the Orchidaceae. From as early as 1923 orchid specimens were loaned to Harvard Botanical Museum for study by Charles Schweinfurth for *Orchids of Peru*; some remained there for 38 years!

42 The two principal authors were responsible for most of the work. The original part 1 of 1954 was produced by arrangement with T. K. Just, then chief curator of botany at the Field Museum, and the later, complete work was founded on a continuation of the Harvard–Field agreement. A similar agreement had also covered *Orchids of Peru*.

43 These were respectively the responsibility of R. E. Fries, O. Buchtien, Th. Herzog and J. R. Perkins.

44 A second edition of the *Guía de Árboles* is planned.

45 The Instituto's then-director was A. R. Teixeira, who would from 1975 also lead 'Programa Flora'.

46 The UNESCO-supported international tropical institute project at Manaus was aborted in 1948 but in 1954 the Instituto Nacional de Pesquisas da Amazônia (INPA) came into being.

47 On the 'Programa Flora', see A. R. Teixeira, 1984. O "Programa Flora" do Brasil – história e situação atual. *Acta Amazonica* 14(1–2), Suppl.: 31–47. Under the 'Programa Flora' four of the five planned regional programmes, or 'projetos', were implemented. Several 'nuclei' were established: two in Amazonia, one in the Planalto (at the Universidade de Brasília), seven in the Nordéste (from Bahia to Piauí), and one in the southeast (in São Paulo). In addition to collecting and related efforts, over 400000 herbarium records were entered into a master database. Unfortunately, the total programme was not realized.

48 It has been argued that, where no state floras exist, florulas of selected small areas of special interest represent the most valuable and practicable kind of contribution of floristic treatment. Although not in the eyes of some botanists academically 'correct', they are more or less realizable within the constraints of the prevailing project-grant method of funding and provide a good introduction to a flora, as has the florula of Río Palenque in Ecuador (329). Coverage of the whole vascular flora of Bahia would, for example, be on the order of that of *Flora of tropical East Africa*. Such a direction is now being pursued in several parts of Middle and South America as well as elsewhere in the tropics.

49 As of 1979 seven numbers had been published, with coverage through 1973. From vol. 4 an A4 format was adopted.

50 This set of keys to dicotyledon genera is widely used in Brazil. Unfortunately, publication of the final volume is not in prospect (D. Zappi, personal communication, 1996).

51 As already noted, this key was one of the Brazilian derivatives of Thonner's *Anleitung zum Bestimmen der Familien der Blütenpflanzen (Phanerogamen)* (1895; 2nd edn., 1917: see 000, p. 95). As related by Joly for the 1981 version of that work (by R. Geesink *et al.*), its original source was a manuscript derived from the *Anleitung* and used before 1939 at the present Universidade de Viçosa (Minas Gerais).

52 This greatest of all floras stands as a monument to nineteenth-century botanical scholarship as, in their different ways, do *Flora SSSR* and *Flora Europaea* for the twentieth century. It is one of the first of modern large-scale floras, and one of the earliest involving international

specialist collaboration and the services of 'in-house' writers and, as 'managers', successively Stephan Endlicher (until 1849) and Eduard Fenzl (through 1879). Sponsors included the kings of Bavaria (including Ludwig II), the Austro-Hungarian emperor, Franz Josef I, and the last emperor of Brazil, Dom Pedro II.

53 The Belém herbarium was, and remains, part of the Museu Paraense 'Emílio Goeldi'. Support for the museum declined sharply in the wake of economic collapse in Amazonia as well as Huber's death and the herbarium was essentially closed from 1915 to 1955. Renewal of official interest in the region then led to the museum's rehabilitation. The quotation on collecting is from Prance and Daly in Campbell and Hammond (1989, p. 414; see **General references**).

54 PFA came to an end in 1987.

55 Cf. B. W. Nelson *et al.*, 1990. Endemism centres, refugia and botanical collection density in Brazilian Amazonia. *Nature, London* 345: 714–716; P. H. Williams *et al.*, 1996. Promise and problems in applying quantitative complementary areas for representing the diversity of some Neotropical plants (families Dichapetalaceae, Lecythidaceae, Caryocaraceae, Chrysobalanaceae and Proteaceae). *Biol. J. Linn. Soc.* 58: 125–157.

56 On floristic education in Brazil, see G. T. Prance, 1975. Botanical training in Amazonia. *AIBS Education Review* 4: 1–4. Projeto Flora Amazonica has been discussed in G. T. Prance, B. W. Nelson, M. F. Da Silva and D. C. Daly, 1984. Projeto Flora Amazônica: eight years of binational botanical expeditions. *Acta Amazonica* 14(1–2): 5–29. For more on the practicality of florulas, see Note 5.

57 For Ducke's comments, see A. Ducke, 1959. Estudos botânicos no Ceará. *Anais de Academia Brasileira de Ciências* 31(2): 211–308. (Facsimile with new index, 1979, Mossoró: Escola Superior de Agricultura. 130 pp. (Coleção Mossoroense XC).) Knowledge of the *brejos* has been summarized by Andrade-Lima in D. de Andrade-Lima, 1981. Present-day forest refuges in northeastern Brazil. In G. T. Prance (ed.), *Biological diversification in the tropics*, pp. 245–251. New York: Columbia University Press. Andrade-Lima also contributed to knowledge of Marggraf's work with identifications of surviving collections; see D. de Andrade-Lima *et al.*, *Bot. Tidsskr.* 71: 121–160 (1971). [No complete evaluation of the eight natural history 'books' of the *Historia* exists; von Martius's *Versuch eines Commentars über die Pflanzen in den Werken von Marcgrav und Piso über Brasilien* (1853, Munich) never advanced beyond its first part, on cryptogams.] Luetzelburg's main work is P. von Luetzelburg, 1925–26. *Estudo botanico do nordéste*. 3 vols. (with pls. and map), atlas (Ministerio da Viação e Obras Publicas, Inspectoria Federal de Obras Contra as Seccas, Publicação 57). Rio de Janeiro.

58 The *Flora* was part of a larger project, 'Ecological Survey of the Vegetation of the States of Guanabara and Rio de Janeiro'. A list of all project publications appears at the beginning of each fascicle.

59 An additional series on non-vascular plants is also in progress, and there has moreover been one *volume especial* entitled *Goiás de Saint-Hilaire e de hoje* (1996), a 'retro-travelogue' with itineraries and maps.

60 These include, for example, E. F. Guimarães and J. R. Miguel, 1985. Flora do Estado do Rio de Janeiro – Família Trigoniaceae. *Rodriguésia* 37(63): 57–72.

61 The history of *Florae fluminensis* – including its author's intellectual isolation, its working-up for publication by Frei Antonio da Arrabida (the director of the National Library) after the author's death, and its difficult passage to the light of day (the abdication of Pedro I in 1831 being all but catastrophic) – has been examined in T. Borgmeier, 1937. A historia da 'Florae fluminensis' de Frei Velloso. *Rodriguésia* 3(9): 77–96, 2 pls. See also C. Stellfeld, 1952. *Os dois Vellozo: biografas de Frei José Mariano da Conceição Vellozo e padre doutor Joaquim Vellozo de Miranda*. 266 pp., portrs. Rio de Janeiro: Sousa. Martius's review is in *Flora* 20(2), Beibl. 1–4: 9–13 (1837). A nomenclatural revision was produced by A. J. de Sampaio and O. Peckolt in 1943 (*Arq. Mus. Nac. Rio de Janeiro* 37: 331–394) and a revised index by M. Cruz in 1946 (17 pp.; Rio de Janeiro) but evaluation of taxa continues, an example being the Bignoniaceae (see A. H. Gentry in *Taxon* 24: 337–344 (1975)).

62 The need for a renewed and improved botanical knowledge of the *mata atlantica* was underscored by the creation in the 1980s of a *Programa Mata Atlantica*, based at the Jardim Botânico de Rio de Janeiro. The florula of the Macaé de Cima represents the results of its first major undertaking in floristics and phytosociology.

63 The author has also published several precursory works, too numerous to list here.

64 The project was for a time dormant following the death of Padre Reitz in 1990.

65 The citations and descriptions below were partly based on information supplied by Lyman B. Smith, Washington, and the Library, Universidade Federal do Rio Grande do Sul, Porto Alegre.

66 Brother Augusto's work represents a consolidation and expansion of several precursory papers of the author and Br. Edésio (partly with K. Emrich), viz.: Irmão Augusto and Irmão Edésio, 1941–42. Flora do Rio Grande do Sul: plantas catalogadas neste Estado até hoje, 1820–1940. *Revista Agron.* (Rio Grande do Sul) 5: 561–570, 639–645, 731–737; 6: 79–81, 101–103, 205–208, 309–312, 380–382, 417–420, 497–500, 635–640 (reprinted in 2 parts without date; 62 pp., 21 text-figs., 7 pls.), and continued independently as *idem*, 1943. *Flora do Rio Grande do Sul: Famílias*

Solanáceas e Labiadas. 43 pp. Porto Alegre: Klein; *idem,* 1943. *Flora do Rio Grande do Sul: Famílias das Escrofulariáceas.* 32 pp.; *idem,* 1944. *Flora do Rio Grande do Sul: Famílias Boragináceas . . . Teofrastáceas.* 82 pp.; and *idem,* 1944. *Flora do Rio Grande do Sul: Famílias Myrsináceas . . . Compostas.* 120 pp. See also *idem,* 1943. *Flora do Rio Grande do Sul: quadro geral das famílias.* 32 pp. Porto Alegre: Klein.

67 Part of the Swiss contribution was under the terms of a 1978 agreement between the Swiss Technical Cooperation Foundation (COTESU) and the City of Geneva (for the Conservatoire et Jardin Botaniques). This was with particular reference to preparation of a new tree manual.

68 There is also a guide to authors: R. Spichiger and J.-M. Mascherpa, 1983. *Guia para los autores.* 50 pp. (Flora del Paraguay, série especial 1). St. Louis. This includes additional references to papers on Hassler's collections.

69 The museum's creator, Bernardino Rivadavia, was also responsible for bringing to the country Aimé Bonpland, out of sorts in France after Waterloo and contemplating emigration.

70 In 1872, the first scientific society, the Sociedad Científica Argentina, was formed. Its 50th and 100th anniversaries were both commemorated by historical proceedings.

71 On the Argentine economy, see, for example, T. Duncan and J. Fogarty, 1984. *Australia and Argentina: on parallel paths.* Melbourne.

72 See note by L. B. Smith in *Taxon* **24**: 580 (1975).

73 I wish to acknowledge the assistance of Troels Myndel Pedersen, a collaborator of the Botanical Museum and Herbarium, Copenhagen, and resident of Corrientes Province, Argentina, with the choices for Region 38 as adopted for the original edition of this work. Advice has also been received from Elsa M. Zardini, formerly of the Museo de La Plata and currently at Missouri Botanical Garden, St. Louis, U.S.A.

74 Other estimates suggest a total of around 7000 species.

75 For example *Adesmia, Oxalis, Senecio* and *Calceolaria*; see T. Stuessy and C. Marticorena, 1988. The flora of Chile project (abstract). *Amer. J. Bot.* **75**(6, part 2): 209–210.

76 News of the *Flora de Chile* project appears in *Flora of Chile Project Newsletter* (July 1991–).

77 Supporting data, including specimens seen, are maintained at the Universidad de Concepción, Chile.

Division 4

• • •

Australasia and islands of the southwest Indian Ocean (Malagassia)

I always think that some of the other governments might have followed the example of yours. Out of the £250 I get, I have to pay £100 down to Reeve, . . . [and] I have much to pay in carriage of specimens from the continent, in postage and various minor expenses attending in the work, so that on the whole I scarcely clear £125 per volume, which is very poor pay for a 12-month hard work, after being nearly 40 years in the trade.

> Bentham to Mueller on *Flora australiensis*, 24 November 1864; quoted from Daley, The history of *Flora australiensis*, IV. *Victorian Naturalist* **44**: 153 (1927).

It will afford me very sincere pleasure to see a beginning made [to a flora of Mauritius] during my residence here, as has been the case in regard to the two last colonies [Jamaica and Victoria] over which I have presided.

> Sir Henry Barkly, Governor of Mauritius, to the Royal Society of Mauritius, January 1864; quoted from Thistleton-Dyer, Botanical survey of the Empire, in *Bull. Misc. Inform.* (Kew) **1905**: 36 (1906). [Barkly, earlier a Governor of Victoria and supporter there of Mueller, was later to accomplish the same in Cape Colony, his next charge.]

Within this division are grouped some fragments of ancient Gondwanaland not readily placed elsewhere: New Zealand and surrounding islands (Region (Superregion) 41); the Australian continent with Tasmania (Superregion 42–45); and 'Malagassia' for the islands of the southwest Indian Ocean, including Madagascar (Superregion 46–49).[1] New Caledonia and dependencies, although with some claim for inclusion here, have been placed in Division 9 (as Region 94) for practical and bio-historical reasons. Other Indian Ocean islands, including Christmas Island, are in Division 0.

A unifying feature of this otherwise fairly heterogeneous assemblage of islands and island-continents is the presence of peculiar floras with high endemism, odd life-forms and many relicts. The special features of Australasia are well known; but of Madagascar Commerson already in 1770 wrote that 'it seems that nature has retired to this spot, as to a special sanctuary, to work on forms other than those she has used elsewhere; at every step one comes upon the most unusual and most marvellous shapes'. On the other hand, the bio-histories of New Zealand, Australia and Malagassia have followed quite different courses and are thus treated here under the respective superregions.

General bibliographies. Bay, 1910; Blake and Atwood, 1942; Frodin, 1964, 1984; Goodale, 1879; Holden and Wycoff, 1911–14; Jackson, 1881; Pritzel, 1871–77; Rehder, 1911; Sachet and Fosberg, 1955, 1971; USDA, 1958. For other bibliographies, see under the regions and superregions.

General indices. BA, 1926– ; BotA, 1918–26; BC, 1879–1944; BS, 1940– ; CSP, 1800–1900; EB, 1959–98; FB, 1931– ; IBBT, 1963–69; ICSL, 1901–14; JBJ, 1873–1939; KR, 1971– ; NN, 1879–1943; RŽ, 1954– . For other indices, see under the regions and superregions.

Conspectus

403

'Alpine' regions

The flora of these zones, best expressed in Australia, Tasmania and New Zealand, is of the highest botanical interest but until the 1970s no significant separate accounts were available. Two useful accounts have now appeared, respectively for New Zealand and Australia; the latter, at once popular and scientific, is particularly noteworthy. Floristic studies have also been made of the high mountains of Madagascar and Réunion, but no proper florulas have been published.

I. New Zealand Alps

In addition to the work described below, the following may also be consulted: J. T. SALMON, 1992. *A field guide to the alpine plants of New Zealand*. 3rd edn. 333 pp., 460 col. photographs, maps (end-papers). Auckland: Godwit Press. (1st edn., 1968, Wellington; 2nd edn., 1985.)

MARK, A. F. and N. M. ADAMS, 1995. *New Zealand alpine plants*. 2nd edn. 269, [iii] pp., 10 text-figs., 118 col. pls., frontisp., maps (end-papers). Auckland: Godwit Press. (1st edn., 1973, Wellington; revised reprintings, 1979, 1986.)

Semi-popular, systematically arranged illustrated guide to vascular plants occurring above the tree-line (varying from 900 m to 1450 m from south to north); the plates (some only partly in color), with four to five plants each, are faced by text containing short descriptions, indication of distribution, altitudinal range, and notes on habitat, phenology and special features (but no vernacular names). Also given is the source of the voucher collection for the respective illustration. Five introductory chapters encompass the definition of the area and coverage of its physical features, environment, general features of the flora and its elements, and the vegetation (with illustrations); a glossary and index conclude the work.

II. Australian Alps

COSTIN, A. B., M. GRAY, C. J. TOTTERDELL and D. J. WIMBUSH, 1979. *Kosciusko alpine flora*. 408 pp., illus. (incl. 351 col. pls.), tables, maps. Melbourne: Commonwealth Scientific and Industrial Research Organisation, Australia; Sydney: Collins.

Lavishly illustrated, semi-technical ecologically oriented field-guide to the alpine flora of the Kosciusko 'Primitive Area' (from 36° 20' to 36° 35' S) from 1830 m upwards; includes a systematic treatment with notes for the user (pp. 109–110), keys, scientific names with references, literature citations, descriptions, vernacular names, local and extralimital distribution, a few critical remarks, and notes on habitat, frequency, occurrence, biology, special features, etc.; bibliography (pp. 381–389), glossary, and complete index at end. The introductory section includes considerations of physical features, earth history, the delimitation of the alpine region, human impact and land use, and general features of the plant life: elements, life-forms, communities and vegetation formations, etc. [Although at 100 km² the actual area covered by this field-guide is small, it will be useful over much of the Australian alpine country, a scattering of areas in southern New South Wales, Victoria and Tasmania.]

HARRIS, T. Y., 1970. *Alpine plants of Australia*. xiii, 193 pp., 305 text-figs., halftones, col. pls. Sydney: Angus & Robertson.

Descriptive systematic enumeration of vascular plants (except *Eucalyptus*) with vernacular names, natural history notes, flower color, altitudinal zones and distribution, and figures of representative species; glossary, literature list and index to all names at end. The introductory part covers alpine habitats, plant adaptations, distribution, and communities (alpine, subalpine and montane). [Eucalypts are dealt with under plant communities rather than in the enumeration proper.]

405

Drylands

JESSOP, J. (ed.), 1981. *Flora of central Australia*. See **420–50**.

408

Wetlands

Several treatments of aquatic and marshland plants have been published in New Zealand and Australia in relation to weed control as well as the growth of environmental and wildlife concerns. No similar works are yet available for Madagascar and other southwestern Indian Ocean islands.

I. New Zealand

JOHNSON, P. N. and P. BROOKE, 1989. *Wetland plants in New Zealand*. vii, 319 pp., text-fig., halftones, 9 col. pls., map (DSIR Information Series 167). Wellington: DSIR Publishing.

Atlas-manual of plates with facing text, the latter encompassing scientific and vernacular names, descriptions, distribution and habitat, and notes with synoptic keys; literature, glossary and indices to botanical and vernacular names at end. The introductory chapters encompass habitats and vegetation types in some detail; conventions and abbreviations are on p. 22. [Covers Characeae as well as vascular plants, with in all 689 species of which 531 are described in detail.]

II. Australia (with Tasmania)

Although emphasizing the state of Victoria (**431**), the 1973 *Aquatic plants of Australia* by Helen Aston accounts for aquatic macrophytes over the whole of Australia and Tasmania.

ASTON, H. I., 1973. *Aquatic plants of Australia.* xv, 368 pp., 138 text-figs. (incl. 4 maps), 81 distribution maps, 2 location maps, 2 end-paper maps. Melbourne: Melbourne University Press. (Supplementary leaflet, 1978.)

Detailed descriptive flora of herbaceous non-vascular and vascular aquatic macrophytes; includes keys to all taxa, synonymy, vernacular names, indication of distribution (and origin, if appropriate), and notes on habitat, biology, variability, phenology, special features, etc., and some critical remarks. Distribution maps are provided only for species occurring in Victoria. The introductory section includes notes for the user and a glossary as well as a discussion of geography, climate, etc., related to aquatic plants and a consideration of the scope of coverage (mainly hydrophytes), while at the end are appendices on the menacing water-hyacinth (*Eichhornia crassipes*) and on sea-grass species. [Covers 222 species (109 in Victoria).]

SAINTY, G. R. and S. W. L. JACOBS, 1981. *Waterplants of New South Wales.* 550 pp., col. illus. [Sydney]: Water Resources Commission of New South Wales.

Copiously illustrated, practically oriented atlas-flora of helophytes and hydrophytes; the text includes scientific names (without synonymy or authorities), descriptions, distribution within and without the state (with maps showing internal range), and notes on habitat, biology and economic significance; glossary, organographic lexicon, references (pp. 530–531) and complete index at end. The introductory section includes the main keys and directions for the book's

use, while an assortment of appendices includes chapters on rice-fields, seeds, algae, aquatic fauna, bacteria, eutrophication and nutrient load, and management and control. Arrangement of species is generally alphabetical by families, but five 'mangroves' are segregated. [Designed as an extension of, and to be used together with, Aston's flora (see above).]

III. Malagassia

Bibliography

DAVIES, B. and F. GASSE (eds.), 1988. *African wetlands and shallow water bodies/Zones humides et lacs peu profonds d'Afrique. Bibliography/Bibliographie.* xxiii, 502 pp., maps. Bondy: Éditions ORSTOM. (Collection Travaux et Documents 211.) – See **508**. Madagascar comprises region 9 of this work (pp. 469–485).

409

Oceans, islands, reefs and the littoral

Included here is the Great Barrier Reef of Australia and its islands. For the Coral Sea Islands Territory, see **949**.

Great Barrier Reef
Until 1991, no formal flora of any considerable part of the Great Barrier Reef islands had been published, although several florulas of individual islands, some in *Atoll Research Bulletin*, were available. An attractive, well-illustrated popular guide to more conspicuous plants is A. B. and J. W. CRIBB, 1985. *Plant life of the Great Barrier Reef and adjacent shores.* xviii, 294 pp., col. illus. St. Lucia: University of Queensland Press.

FOSBERG, F. R. and D. R. STODDART, 1991. *Plants of the Reef Islands of the northern Great Barrier Reef.* 82 pp. (Atoll Res. Bull. 348). Washington.

Annotated systematic checklist, with citations of key literature and *exsiccatae*.

Western Australia
The best (and only) introduction to the coastal flora is E. RIPPEY and B. ROWLAND, 1995. *Plants of the Perth coast and islands.* xi, 292 pp., illus. (part col.), maps, end-paper pls. Perth (Nedlands): University of Western Australia Press. [A

semi-popular work, with in part 2 an atlas with facing text covering 120 or so species; sea-grasses covered in part 3. The introductory part (part 1) is extensive, including general remarks on physical and biological features, communities and coastal dynamics.]

New Zealand (Aotearoa)

No account has yet appeared in succession to L. B. Moore and N. M. Adams, 1963. *Plants of the New Zealand coast*, 113 pp., illus. Auckland: Paul's Book Arcade. [This small illustrated work with simple text, originally published in installments in the late 1950s for early secondary-level scholars, covers selected algae as well as higher plants. Some endemic species of offshore islands are also included.]

Region (Superregion) 41

New Zealand (Aotearoa) and surrounding islands

This region includes New Zealand proper (North, South, and Stewart Islands, and other offshore entities) and Lord Howe, Norfolk, the Kermadec and Chatham Islands, and the 'sub-Antarctic' islands to the south and southwest (except Macquarie).

Early New Zealand botanical exploration was accomplished through expeditions from Britain and France beginning with Banks and Solander in the *Endeavour*; until the 1820s, however, almost all collecting was confined to the coasts. This was reflected in the first attempt at a flora, *Essai d'une Flore de la Nouvelle-Zélande* (1832) by Achille Richard, part of the results of the *Astrolabe* expedition of Dumont d'Urville but accounting for earlier records. A total of 380 species was accounted for in this work. Inland field trips began with the visits of the Cunningham brothers, Charles Fraser, Bidwill (1839, penetrating the mountains of the North Islands), and Dieffenbach (the last active in 1839–41 as botanist to the New Zealand Company and first to visit the Chatham Islands). In 1839 Allan Cunningham produced, in several parts, *Florae insularum Novae-Zelandiae praecursor*; this accounted for 394 phanerogams and 294 cryptogams. In 1840–42 M. E. Raoul, a French naval surgeon, collected extensively, especially in South Island; his *Choix des plantes de la Nouvelle-Zélande* appeared in 1846 and accounted for 950 species, over 500 of them flowering plants. Norfolk Island was explored from the early 1800s, beginning with Ferdinand Bauer, while the 'sub-Antarctic' islands of Campbell Island and the Auckland group were first studied extensively by J. D. Hooker in 1841 while on the *Erebus* and *Terror* voyage. Other collectors visited the Chathams, the Kermadecs and Lord Howe Island; but the latter two groups, along with the Snares and the Antipodes, apparently remained poorly known until late in the nineteenth century. With the publication by Hooker of his *Flora antarctica* (vol. 1, 1843–45) and *Flora novae-zelandiae* (1851–53), the primary phase of botanical survey in the region was largely completed.

The signing of the Treaty of Waitangi in 1840 ushered in a detailed 'secondary' phase of botanical exploration and 'alpha' taxonomy, one largely dominated by British settlers beginning as early as 1834 with the arrival of William Colenso and, some years later, Andrew Sinclair. Aided by the appearance of Hooker's *Handbook of the New Zealand flora* (1864–67), one of the 'Kew Series', this movement, accompanied by much additional collecting and the initial formation of local herbaria, led firstly to Thomas Kirk's never-completed *Students' flora of New Zealand and the outlying islands* (1899) and then to T. F. Cheeseman's *Manual of the New Zealand flora* (1906; 2nd edn., 1925), the first complete (and still much valued) 'indigenous' work. Kirk had earlier produced *The forest flora of New Zealand* (1889), now something of a collector's item. Detailed exploration was also carried out in most of the surrounding islands and island groups, including Norfolk and Lord Howe, with the main cycles of activity taking place early this century and from World War II and the so-called 'Cape Expeditions' onwards.[2] The first major monograph devoted to the southern islands was *The subantarctic islands of New Zealand* by C. Chilton (1909), with the botany by Cheeseman. The current standard flora, which has a considerable 'biosystematic' content with great stress on variability and hybridism, is H. H. Allan's *Flora of New Zealand* (1961–), for both native and introduced vascular plants now complete save for the Gramineae. This work extends its limits to include Macquarie Island but omits Norfolk and Lord Howe Islands. Good separate coverage is also available for the woody flora, with several publications, some lavishly illustrated, appearing in succession to Kirk's 1889 flora.[3]

With the considerable number of local checklists and other treatments, the native vascular flora is now about the best known in the Southern Hemisphere. A biological flora and a series of distribution maps have been in course of publication in *New Zealand Journal of Botany*. For many years, however, inclusion in a 'flora' of the now numerous adventive and naturalized plants, today a conspicuous part of the landscape of settled areas, was frowned upon. A first essay, *A handbook of the naturalized flora of New Zealand* (1940), was authored by Allan in a Department of Scientific and Industrial Research bulletin, some 12 years after organization of its Botany Division. Only within the last 15 years has a more definitive treatment appeared, forming volumes 3 and 4 of the *Flora of New Zealand*; additions were published in 1995.

Coverage of the outer islands has further improved since 1980, and save for the Chatham group all are now covered by more or less modern treatments. For Lord Howe and Norfolk these are found in volume 49 of *Flora of Australia*. Of the New Zealand islands, good modern treatments are available for the Kermadecs (Sykes, 1977), the Snares (Fineran, 1969), the Antipodes (Godley, 1989), the Aucklands (Johnson and Campbell, 1975), and Campbell Island (Sorensen, 1951, with additions to 1975). The Chatham Islands have seemingly been neglected: no descriptive flora has appeared since 1864 (albeit with additions to 1902).

Progress

For historical information, see pp. 17–30 in L. COCKAYNE, 1967. *New Zealand plants and their story*. 4th edn., revised by E. J. Godley. Wellington; and the introductory section (pp. xv–xl) of the general flora by Cheeseman (1925). The latter is the more detailed. An additional source is R. GLENN, 1950. *The botanical explorers of New Zealand*. 176 pp., illus. Wellington.

Nomenclatural and taxonomic changes, including additions to the flora, have been consistently recorded since publication of the first volume of *Flora of New Zealand* in 1961. A cumulation was issued in 1987: H. E. CONNOR and E. EDGAR, 1987. Name changes in the indigenous New Zealand flora, 1960 1986 and Nomina nova IV, 1983–1986. *New Zealand J. Bot.* **25**: 115–170. (Also separately issued in spiral-bound form without change of pagination.)

Bibliographies. General bibliographies as for Division 4.

Regional bibliographies

ALLAN, H. H. *et al.*, 1961–88. Annals of New Zealand botany. In H. H. Allan *et al.* (eds.), *Flora of New Zealand*, vols. 1–4 (see below). [Chronologically arranged lists of references relating to New Zealand systematic botany, including subject indices in vols. 2–4.]

DRUCE, A. P., 1981. Index to areas for which check-lists of vascular plants have been compiled: supplement 1. *Bulletin of the Wellington Botanical Society* **41**: 73–77.

Indices. General indices as for Division 4.

Regional index

STEENIS, C. G. G. J. VAN *et al.* (eds.), 1948– . *Flora Malesiana Bulletin*, 1– . Leiden: Flora Malesiana Foundation. (Vols. 1–9 mimeographed.) [At present issued bi-annually, this bulletin contains in one of its two annual issues a substantial bibliographic section also rather thoroughly covering New Zealand floristic literature. A fuller description is given under Superregion 91–93.]

410
Region in general

Although the handbook-flora by Allan *et al.* is the latest available for the New Zealand region, no section on grasses has yet been published and for this and other reasons Cheeseman's manual remains in use. In addition, the first volume of *Flora of New Zealand* is now itself outdated. Recent general introductions to the flora are *Forest vines to snow tussocks: the story of New Zealand plants* (1988, Wellington: Victoria University Press) by J. W. Dawson and *Flowering plants of New Zealand* (1990, Christchurch: DSIR Botany) by C. J. Webb, P. Johnson and W. Sykes.

ALLAN, H. H., 1961. *Flora of New Zealand*, 1: *Indigenous Tracheophyta*. liv, 1085 pp., 40 text-figs., 4 maps (end-papers). Wellington: Government Printer. Continued as L. B. MOORE and E. EDGAR, 1971. *Flora of New Zealand*, 2: *Indigenous Tracheophyta*. xl, 354 pp., 43 text-figs., 4 maps (end-papers). Wellington; A. J. HEALY and E. EDGAR, 1980. *Flora of New Zealand*, 3: *Adventive cyperaceous, petalous and spathaceous monocotyledons.* xlii, 220 pp., 31 text-figs., 4 maps (end-papers). Wellington; and C. J. WEBB, W. R. SYKES and P. J. GARNOCK-JONES, 1988. *Flora of New Zealand*, 4: *Naturalised pteridophytes, gymnosperms, dicotyledons.* lxviii, 1365 pp., 122 text-figs., 16 col. pls., maps (end-papers). Christchurch: Botany Division, DSIR, New Zealand.

Volumes 1 and 2 comprise a concise, well-documented descriptive flora of native vascular plants (except the Gramineae), with keys to all taxa; synonymy, with references; Maori and English vernacular names; indication of type localities; generalized indication of local (and extralimital) range, with more details for less common taxa; extensive notes on taxonomy, phenology, ecology, heteroblasty, hybridism, etc.; extensive bibliographies and, at the end, diagnoses of novelties, glossaries, addenda, and indices to all botanical and other names in each volume. In vol. 2 there is also a list of known chromosome numbers. The introductory sections in each volume include *inter alia* the bibliographies, lists of authors, a short account of floristic regions, and synopses of families. A special feature of this work is the presence of extended biosystematic discussions on individual genera and species (with suggestions for future research). Volumes 3 and 4 comprise a treatment, in similar style and format, of the now-considerable and wonderous naturalized and adventive flora (representing a change from past practice in New Zealand, as noted in the extensive introduction); additions to the bibliographic annals in volumes 1 and 2 are also included. [For additions to vols. 3 and 4, see C. J. WEBB *et al.*, 1995. Checklist of dicotyledons, gymnosperms, and pteridophytes naturalized or casual in New Zealand: additional records 1988–1993. *New Zealand J. Bot.* 33: 151–182.]

CHEESEMAN, T. F., 1925. *Manual of the New Zealand flora.* 2nd edn. xliv, 1163 pp. Wellington: Government Printer. (1st edn., 1906.)

Descriptive manual of native vascular plants, with keys to all taxa; synonymy and references; generalized indication of local (and extralimital) range, with some citations for less common or interesting species; vernacular names; brief notes on habitat, etc., sometimes given; list of naturalized plants; glossary and indices to all botanical and Maori names. The introductory section gives an account of botanical exploration. [Associated with this work is an excellent selective atlas, with plates by Matilda Smith (of Kew) and descriptive text: T. F. CHEESEMAN and W. B. HEMSLEY, 1914.

Illustrations of the New Zealand flora. Vols. 1–2. 250 (251) pls. Wellington: Government Printer.]

PARSONS, M., P. DOUGLASS and B. H. MACMILLAN (comp.), 1998. *Current names list for wild gymnosperms, dicotyledons and monocotyledons (except grasses) in New Zealand as used in herbarium CHR.* Version 1 (to 31 December 1995). 206 pp. New Zealand: Manaaki Whenua Press (Landcare Research New Zealand).

Nomenclatural checklist of native, naturalized and adventive ('casual') seed plants with principal synonymy, references to *Flora of New Zealand* and subsequent revisions, distribution, Maori and English vernacular names, status, and brief additional notes (usually referring to additional undescribed species or other taxa); index to all names at end. The general part encompasses the plan and basis of the work as well as conventions and abbreviations. [The work is in the first instance based on the four published volumes of *Flora of New Zealand.* References to revisions of grasses are furnished in the introduction. Arrangement of the families basically follows the Dahlgren system of 1983.]

Local works

Included here are separate florulas for two significant offshore entities, the Three Kings Islands northwest of North Island and Stewart Island just south of South Island. Both have species (and genera) not presently native on the two main islands.

OLIVER, W. R. B., 1948. The flora of the Three Kings Islands. *Rec. Auckland Inst. Mus.* 3: 211–238, pls. 31–34.

Introduction with brief description of the group and past and current botanical exploration; enumeration with previous records and vouchers, synonymy, localities and notes; introduced species; list of literature. Accounts for 178 species of vascular plants, of which 168 are also in New Zealand.[4]

WILSON, H. D., 1982. *Stewart Island plants.* 528 pp., 814 figs. (part col.). Christchurch: Field Guide Publications.

Field-guide to vascular plants (580 species) and the more conspicuous species of bryophytes, lichens and fungi, with vernacular names, brief descriptions, indication of local and extralimital range, and notes on habit, ecology and biology; glossary, acknowledgments and index to all names at end. The introductory section includes a (modern) history of the island (area: 1720 km^2), the scope of the work, notes on nomenclature, and keys (pp. 14–36).

Woody plants (including trees)

J. T. Salmon's *The native trees of New Zealand* (1980) was the functional successor to Kirk's *Forest flora of New Zealand*

(1889). For its condensed version, see *idem*, 1986. *A field guide to the native trees of New Zealand.* vi, 228 pp., col. pls. Auckland (Birkenhead): Reed Methuen. [Limited to North, South and Stewart Islands; accounts for name changes since 1980.]

COCKAYNE, L. and E. P. TURNER, 1967. *The trees of New Zealand.* 5th edn. 182 pp., 126 figs. (mostly halftones). Wellington: Government Printer. (1st edn., 1928.)

Small atlas-guide to forest trees, comprising photographs of freshly collected material accompanied by descriptions, indication of distribution, Maori vernacular names, and miscellaneous notes; genera alphabetically arranged. Appendices cover timber species, seedlings, hybridism, borderline species occasionally becoming treelike, and monocotyledon trees; a general introduction to forest vegetation is also given. [Not significantly revised since 1939, and effectively supplanted by the smaller version of Salmon's dendroflora (see below).]

EAGLE, A., 1975. *Eagle's trees and shrubs of New Zealand in colour.* 311 pp., 228 col. illus. Auckland: Collins. Complemented by *idem*, 1982. *Eagle's trees and shrubs of New Zealand: second series.* 382 pp., 405 col. illus. Auckland.

Semi-scientific color-plate atlas of trees and shrubs; each volume incorporates separate text containing phytographic and other notes on each species, along with references (p. 305 in the first volume, pp. 375–376 in the second), glossary and index. The second volume contains short biographies of many botanists and others after whom plants have been named. Much biological and other information incorporated in this work is otherwise not easily accessible.

POOLE, A. L. and N. M. ADAMS, 1963. *Trees and shrubs of New Zealand.* 250 pp, illus., map. Wellington: Government Printer.

Briefly descriptive, semi-popular illustrated manual of trees and shrubs; includes simple keys to genera and species, Maori and English vernacular names, concise indication of distribution, habitat, and diagnostic features, occasional additional notes; glossary and complete index at end. The introductory section includes chapters on vegetation, forest lands (and their utilization and management), and biotic factors. No synonymy is given, and all illustrations are based on drawings.

SALMON, J. T., 1980. *The native trees of New Zealand.* 384 pp., over 1500 col. photographs, 2 maps. Wellington: Reed.

Semi-scientific atlas of trees accompanied by descriptive text with a strong 'natural history' orientation; includes vernacular and scientific names and notes on habit, distribution and habitats but no keys or synonymy. An introductory section covers basic botany, the place of trees in human life, and the kinds and functions of forest along with a history of vegetation and human interference; at the end of the work are a gazetteer, glossary, references and index to vernacular names (scientific names are indexed on p. 49).

Pteridophytes

With ferns and allied plants something of a national trademark, there have not unnaturally been several treatments since the nineteenth century. Brownsey and Smith-Dodsworth's current *New Zealand ferns and allied plants* succeeds M. E. CROOKES, 1963. *New Zealand ferns.* 6th edn. Christchurch: Whitcombe & Tombs; and E. HEATH and R. J. CHINNOCK, 1974. *Ferns and fern allies of New Zealand.* Wellington: Reed. A nineteenth-century predecessor was *The ferns and fern allies of New Zealand* (1882) by G. M. Thompson.

BROWNSEY, P. and J. C. SMITH-DODSWORTH, 1989. *New Zealand ferns and allied plants.* viii, 168 pp., 198 text-figs., 36 col. pls., 2 maps. Auckland: David Bateman.

Semi-popular manual with vernacular and scientific names, keys, descriptions, indication of distribution and habitat, critical commentary, and notes on cultivation; 36 color plates each with six illustrations in center of work; halftones and line drawings in 198 text-figures. A glossary, bibliography, locality map and index appear at the end. The general part includes an illustrated organography but no floristic summary. Covers 215 species (194 of ferns proper) in 66 genera and 25 families; 88 are endemic and 22 naturalized. – Taxonomy and nomenclature are based on P. BROWNSEY, D. R. GIVEN and J. D. LOVIS, 1985. A revised classification of New Zealand pteridophytes, with a synonymic checklist of species. *New Zealand J. Bot.* **23**: 431–489. [An enumeration with nomenclature, citations and, for all taxa included, a detailed review of classification. Adventives are asterisked.]

411

Lord Howe Island group

Area: 13 km². Vascular flora: 379 species and additional infraspecific taxa, of which 219 native (74 endemic) (Rodd and Pickard, 1983).

The first collections were made only in 1854, and a first checklist published by Charles Moore in 1870 as part of an official enquiry following a murder case. A more substantial account, by Ferdinand von Mueller, followed in 1875 as *Index omnium Insulae Howeanae plantarum* by Mueller (in the ninth volume of his *Fragmenta*). Further revisions by W. B. Hemsley and J. H. Maiden followed respectively in 1896 and 1898. The affinities of the flora with New Zealand naturally attracted interest from that country, and in 1917 the Auckland botanist W. R. B. Oliver made a visit which resulted in a substantial and long-standard account,

The vegetation and flora of Lord Howe Island (1917, in *Trans. & Proc. New Zealand Inst.* **49**: 94–161). Interest then largely lapsed until the 1960s and the growth of conservation concerns.

These concerns led the Island Board to commission a new biological survey, and this was undertaken by the Australian Museum in Sydney in the early 1970s. The results, which include a revised checklist by J. Pickard and A. N. Rodd, appeared as H. F. RECHER and S. S. CLARK (eds.), 1974. *Environmental survey of Lord Howe Island*. viii, 86 pp., illus. (some col.), maps, graphs. Sydney: New South Wales Government Printer. This work remains a worthwhile general introduction to the biota and environment; its botanical part along with subsequent contributions and collections forms a basis for the modern account of the flora by Peter Green in volume 49 of *Flora of Australia* (see below).

Bibliography

References are also given in AUSTRALIAN BIOLOGICAL RESOURCES STUDY (1994; see below).

PICKARD, J., 1973. An annotated botanical bibliography of Lord Howe Island. *Contr. New South Wales Natl. Herb.* **4**(7): 470–491.

AUSTRALIAN BIOLOGICAL RESOURCES STUDY, 1994. *Flora of Australia*, 49: *Oceanic Islands* 1. xxiii, 681 pp., 107 figs. (part col. and incl. maps). Canberra: Australian Government Publishing Service.

Pages 13–26 comprise an introduction (by Peter Green) to the Lord Howe group (with map); this encompasses background information, vegetation formations, collection history, fauna, key references, and (pp. 18–24) an alphabetically arranged species list. This is followed by a general key to families covering both Lord Howe and Norfolk island groups. All species are further treated by Green in the descriptive part – which follows the format of other *Flora* volumes (cf. **420–50**) – together with those from the Norfolk Island group (**412**).

RODD, A. N. and J. PICKARD, 1983. Census of the vascular flora of Lord Howe Island. *Cunninghamia* **1**: 267–280.

Systematic checklist, with synonymy given only where names have changed since publication of the Oliver checklist (see above). Covers native and naturalized species; endemics are marked with a '+'. [Accompanies and follows a vegetation survey by J. Pickard in *ibid.*: 133–265.]

412

Norfolk Island group

Area: 39 km^2 (Norfolk, 36 km^2). Vascular flora: 445 species, of which 171 native with 47 endemic (P. S. Green in Australian Biological Resources Study, 1994; see below). – In addition to Norfolk proper, the territory, an autonomous dependency of Australia, includes the nearby Philip and Nepean Islands.

The Norfolk Island group was from settlement until 1914 under the jurisdiction of New South Wales or Tasmania; control was then assumed by the Commonwealth of Australia. There are four earlier floras or enumerations: *Prodromus florae norfolkicae* by Stephan Endlicher (1833), based on the artist Ferdinand Bauer's collections, *The flora of Norfolk Island*, I (1904, in *Proc. Linn. Soc. New South Wales* **28**: 692–785) by J. H. Maiden, *A revised list of the Norfolk Island flora* by R. M. Laing (1915, in *Trans. & Proc. New Zealand Inst.* **47**: 1–39), and J. S. Turner, C. N. Smithers and R. D. Hoogland, *The conservation of Norfolk Island* (1968). All represent different cycles of interest, the last resulting from concerns over threats to the flora which became manifest in the mid-1960s. The plants are now well documented in a treatment by Peter Green in volume 49 of *Flora of Australia* (see below).

AUSTRALIAN BIOLOGICAL RESOURCES STUDY, 1994. *Flora of Australia*, 49: *Oceanic Islands* 1. xxiii, 681 pp., 107 figs. (part col. and incl. maps). Canberra: Australian Government Publishing Service.

Pages 2–13 comprise an introduction (by Peter Green) to the Norfolk group (with map); this encompasses background information, vegetation formations, collection history, fauna, key references, and (pp. 7–13) an alphabetically arranged species list. This is followed (pp. 27–42) by a general key to families covering both Lord Howe and Norfolk island groups. All species are further treated by Green in the descriptive part – which follows the format of other *Flora* volumes (cf. **420–50**) – together with those from the Lord Howe Island group (**411**).

413

Kermadec Islands

Area: 38 km². Vascular flora: 195 species, of which 113 native (Sykes, 1977). – The main account of the flora prior to that of Sykes is *The vegetation of the Kermadec Islands* by W. R. B. Oliver (1910, in *Trans. & Proc. New Zealand Inst.* 42: 118–175, pls. 12–23, map).

SYKES, W. R., 1977. *Kermadec Islands flora: an annotated check list.* 216 pp., 48 halftones, 9 tables (New Zealand Div. Sci. Indust. Res. Bull. 219). Wellington.

Enumeration of native, naturalized and adventive vascular plants, bryophytes and lichens; includes commonly used synonyms, vernacular names, status, local distribution in the group, and occurrence, habitat preferences and associates, and location of collections; table of local distribution (pp. 187–191), list of references, and complete index at end. The general part covers physical features, climate, introduced biota, human activities, vegetation and phytogeography, history of botanical work, and statistics of the flora. [Musci and Lichenes also are fully treated, but Hepaticae and Anthocerotae are merely listed in an appendix.]

414

Chatham Islands

Area: 1235 km². Vascular flora: *c.* 300 species (PD; see **General references**). – It remains regrettable that no more recent flora or checklist is available for this interesting and remote group. The Mueller work and the Buchanan list, both well over 100 years old, are largely of historical interest, and subsequent information is scattered.

MUELLER, F., 1864. *The vegetation of the Chatham-Islands.* 86 pp., 7 pls. Melbourne: Victorian Government Printer.

Descriptive flora of vascular plants (87 species), with synonymy, references and citations; vernacular names; extensive notes on habitat, phenology, biology and taxonomy; generalized indication of local (and extralimital) range. The quaintly written introduction includes remarks on the general features of the flora as well as a panegyric on the need for field studies. – For a revised list accounting for 205 species with vernacular names along with descriptions of new or little-known taxa, see J. BUCHANAN, 1875. On the flowering plants and ferns of the Chatham Islands. *Trans. & Proc. New Zealand Inst.* 7: 333–341, pls. 12–15. For further additions in an otherwise mainly vegetatiological account, see L. COCKAYNE, 1902. A short account of the plant-covering of Chatham Island. *Ibid.*, 34: 243–325.

415

Islands to the south of New Zealand (in general)

See also 080–90 (SCHENCK). For works relating specifically to Macquarie Island, see 081. Much recent work on the flora and vegetation of these islands, which (except in part for Campbell) are not truly sub-Antarctic, has been accomplished in the last half-century so that Cheeseman's account, though still fundamental, is now for practical purposes superseded.

CHEESEMAN, T. F., 1909. On the systematic botany of the islands to the south of New Zealand. In C. Chilton (ed.), *The subantarctic islands of New Zealand*, 2: 389–471. Wellington: Government Printer (in association with the Philosophical Institute of Canterbury).

Systematic enumeration of vascular plants (except Gramineae); includes synonymy, references and literature citations, taxonomic commentary, extensive remarks on habitat, life-form, biology and local distribution, and general indication of wider range (in the first instance by island or island group). A concluding section gives a tabular summary of geographical ranges, with columns for the individual island units, and a summary of the overall affinities of the flora and its presumed historical origins. The introductory part includes an account of botanical exploration. – For a revision of the Gramineae, see V. D. ZOTOV, 1965. Grasses of the subantarctic islands of the New Zealand region. *Rec. Domin. Mus.* 5: 101–146, 4 text-figs., 4 pls.

416

Snares Islands

See also **415** (CHEESEMAN; ZOTOV). – No figures for area or number of vascular plants have been seen.

FINERAN, B. A., 1969. The flora of the Snares Islands, New Zealand. *Trans. Roy. Soc. New Zealand, Bot.* **3**: 237–270, 5 pls.

Enumeration of known non-vascular and vascular plants; includes synonymy (with references and citations), *exsiccatae* and literature records, and extensive remarks on ecology and biology. The concluding section includes a phytogeographic and floristic summary, together with a table of extralimital distribution and a list of references.

417

Antipodes and Bounty Islands

See also **415** (CHEESEMAN; ZOTOV). – Area: 21 km² (Antipodes group); 1.3 km² (Bounty group). Vascular flora: 69 native and 3 introduced species, all from Antipodes (Godley, 1989; see below). The land area of Antipodes Island and its immediate satellites is quite limited, while the Bounty Islands are mere rocks, without vascular plants. The woody vascular flora of Antipodes features no trees and but one shrub, *Coprosma rugosa*.

Eric Godley's enumeration for the Antipodes was the first of its kind to appear since *On the botany of Antipodes Island* by T. Kirk (1891, in *Trans. & Proc. New Zealand Inst.* **23**: 436–441). This accounted for 55 vascular plants.

GODLEY, E. J., 1989. The flora of Antipodes Island. *New Zealand J. Bot.* **27**: 531–563, 3 figs. (incl. map).

Detailed, alphabetically arranged enumeration of vascular and non-vascular plants and fungi, with accepted names, synonymy as used in other standard works, citations of *exsiccatae*, and sometimes extensive documentation and commentary on local range, habitat, biology, etc.; excluded species, phenological patterns, phytogeographical elements, history of the flora and its dispersal and establishment, botanical

provinces and districts, statistics, acknowledgments and extensive list of references at end. The opening paragraphs cover the history of botanical work, inclusive of the activities of individual collectors (the first visit being in 1890).

418

Auckland Islands

See also **415** (CHEESEMAN; ZOTOV). – Area: 625 km². Vascular flora: 257 native and 'alien' species (Johnson and Campbell, 1975; see below); native species: 187 (Godley, 1989; see **417**).

JOHNSON, P. N. and D. J. CAMPBELL, 1975. Vascular plants of the Auckland Islands. *New Zealand J. Bot.* **13**: 665–720, 5 figs. (incl. map).

Annotated systematic enumeration of vascular plants; includes synonymy, localities, critical commentary, and notes on status, occurrence, special features, etc. The systematic account is followed by a discussion of species of limited occurrence and uncertain status, a list of adventives and species to be excluded from the flora, an analysis of past and possible future history of the flora in relation to introduced animals and their possible increase or elimination, and a list of references. A brief general part accounts for past and current work in the island group and the bases for knowledge.[5] – For additions, see C. D. MEURK, 1982. Supplementary notes on plant distributions of the subantarctic Auckland Islands. *New Zealand J. Bot.* **20**: 373–380.

419

Campbell Island

See also **415** (CHEESEMAN; ZOTOV). – Area: 113 km². Vascular flora: 308 species, of which 223 native (PD from unpublished figures of Colin Meurk and David Given; see **General references**); 225 native species recorded by 1989 (Godley; see **417**).

A renewal of field work on this near sub-Antarctic island began in the 1960s, leading to preparation of a new flora in the mid-1970s (Meurk, 1975; see below). By the late 1980s this was far advanced. The recent exploration has greatly increased the known vascular flora in comparison with that recorded by Sorensen. For a general description and natural history of the island, see A. M. BAILEY and J. H. SORENSEN, 1962. *Subantarctic Campbell Island*. (6), 305 pp., illus. (Proc. Denver Mus. Nat. Hist. 10). Denver, Colo.

SORENSEN, J. H., 1951. Botanical investigations on Campbell Island, II. An annotated list of the vascular plants. *New Zealand Div. Sci. Indust. Res., Cape Exped. Sci. Bull.* 7: 25–38.

Briefly annotated systematic list of native and introduced vascular plants collected between 1941 and 1947; for each species are given notes on localities, occurrence, frequency, habitat, and general range within the island, but no extralimital distribution is indicated. Section A of the list covers native species: section B, the introductions. – For additions, see E. J. GODLEY, 1969. Additions and corrections to the flora of the Auckland and Campbell Islands. *New Zealand J. Bot.* 7: 336–348, 2 figs.; and C. D. MEURK, 1975. Contributions to the flora and plant ecology of Campbell Island. *Ibid.*, 13: 721–742.[6]

Superregion 42–45

Australia

Area: 7 691 891 km^2. Vascular flora: 20 000 species (A. Orchard through R. Makinson, personal communication, November 1995). Hnatiuk (1990; see below) listed 17 590 species, of which 15 638 were native and 1952 naturalized.[7] Included here are the Australian continent and Tasmania together with immediately adjacent islands, the Torres Strait Islands, and the Ashmore and Cartier Islands in the Timor Sea. For Lord Howe and Norfolk Islands, see Region 41; for the Coral Sea Islands (Territory), see Region 96; for the Cocos (Keeling) group and Christmas Island, see Region 04; and for Macquarie Island and the McDonald (Heard) group, see Region 08. The boundary with Papuasia (Region 93) is formed by the seabed boundary between Australia and Papua New Guinea through the Torres Strait and north of the Great North East Channel, thus here incorporating (as **439**) all islands except those immediately off the Papuan coast.

The flora of this smallest but botanically most interesting of continents has excited botanists since the landfall of Joseph Banks and Daniel Solander at Botany Bay and their voyage along the east coast in Lt. James Cook's *Endeavour* in 1770. Through the subsequent field work and systematic studies of J. B. de La Billiardière, Robert Brown and other workers, it also contributed greatly to the development of 'natural' classifications of angiosperms in the first half of the nineteenth century and their ouster of the Linnaean sexual system. After 1810, the year of publication of Brown's epochal (but financially unsuccessful) *Prodromus florae Novae Hollandiae*, locally based exploration began in earnest and from then on widened more or less steadily from the main centers of settlement, partly through expeditions and partly through the activities of amateur and professional collectors. From 1847, there developed a residential research botanical capability with the settlement of Ferdinand (later Baron Ferdinand von) Mueller, who for close to half a century would dominate Australian botany in much the same way as did his near-contemporary Asa Gray in North America. With the establishment of the Kew colonial floras scheme at the end of the 1850s, his writings and collections, as well as those of nearly all others, were drawn upon by George Bentham – albeit not without some resistance on the part of Mueller who had entertained a similar ambition – for his *Flora australensis* (1863–78), a work which remains the longest one-man flora in existence and by general consent is one of the best.

Nevertheless, even the finest flora is out of date from the day of its publication and Bentham's 8125 species of vascular plants are now considered to represent only some two-fifths of the probable total. More than once were calls heard for a supplementary volume or volumes, but the continuing flow of novelties and other problems reduced these to a receding dream of overloaded state botanical officers; in addition, materials said to have been accumulated by Mueller for such a work disappeared after his death in 1896 and have never been found. At Bentham's behest, however, Mueller did produce his *Systematic census of Australian plants, with chronologic, literary and geographic annotations* (1882; 2nd edn., 1889) of which only the part on vascular plants was ever completed, in its second edition covering 8839 species. A century would pass before a successor appeared – with more than twice as many species.[8]

With the passing of Mueller and the onset of the 'cultural cringe' came decades of taxonomic twilight and fragmentation, pierced only here and there by a few rays of energy. Political federation in 1901 did not bring unity but rather the opposite: efforts went into state herbaria and in the 50 years preceding World War II most of the state and territory floras and enumerations standard by mid-century (and here and there still current) were published. At first largely based on *Flora australiensis* with the addition of later discoveries, significant original research began to feature in these works starting in the 1920s with Black's *Flora of South Australia*. This array, however, gave in some quarters a misleading impression that the continent was relatively well documented. Blake and Atwood in the introduction to their *Geographical guide to floras of the world* (1942, p. 10) noted that 'the flora as a whole can be regarded as more satisfactorily covered by published works than that of any equally extensive division of the earth's surface except Europe'. That this was far from the truth, and that *Flora australiensis* in particular was becoming obsolete, began seriously to be perceived only after World War II, although as early as 1907 the energetic Joseph H. Maiden, then director of the Botanic Gardens at Sydney, had urged the writing of a new work.

Proponents of state floras continued, nevertheless, ascendant for some time and until mid-century or even later Australian systematic botany remained a knobbed wheel without a hub. Not until 1959 did Stanley T. Blake of the Queensland Herbarium make the first formal modern proposal for a new general flora. Years of discussion, both ideological and technical, and fruitless proposals and negotiations then followed. A prime mover during this period was Nancy T. Burbidge at CSIRO, who in 1963 contributed her *Dictionary of Australian plant genera*. Only in 1973, in the face of steadily mounting local demands for such a work as well as external pressure, did preliminary investigations towards a new flora along with compilation of source materials definitively begin with support from the Australian Academy of Science and a private benefactor.[9] Sponsorship was shortly afterwards assumed by the Australian Biological Resources Study (ABRS) of the federal government, which in 1978–79 became formally established as the 'Bureau of Flora and Fauna' with among its briefs the preparation of a national flora along lines similar to those suggested in 1959. The *Flora of Australia* project was thus assured. Writing of the *Flora* commenced in 1980 and the first of

the projected 48 volumes on vascular plants, containing general chapters, appeared at the XIII International Botanical Congress in Sydney in 1981. Later volumes have been numbered with reference to the sequence of the then-current version of the Cronquist system.[10] Though initially seen as being 'based largely on current knowledge' (George, 1981, p. 10), the *Flora* has, while remaining relatively concise, developed into a semi-monographic critically based work thanks to continuing official support, through a grants scheme administered and directed by ABRS, for associated revisionary studies.[11]

The obstacles, though, remain great: the already-mentioned size of the flora, with its many species still to be described, the amount of basic taxonomic work to be done, current standards of documentation, and the still limited number of specialists in various isolated centers with mainly regional collections will all make full realization a daunting task.[12] The success of *Flora of central Australia* (1981) edited by John Jessop with contributions from over 50 collaborators and covering the central third of the continent, has, however, augured well for the *Flora of Australia*, and its subsequent progress has been steady if slower than anticipated.[13] Recently, two important associated works were published: the *Census of Australian vascular plants* (Hnatiuk, 1990) and the *Australian plant name index* (Chapman, 1991; for both, see below), the latter fulfilling one of the original objectives of 1973. There have also been plans for a comprehensive taxonomic bibliography.

Botanical work has been active in all states (and territories), particularly in recent decades, and the Australian north, much of it difficult of access, is becoming botanically better explored although much remains to be done. Except in the southeast and in patches elsewhere, however, knowledge of the continent is still considered inadequate and documentation of much of it thin. Other problems have included changes of scientific names at state boundaries (although with the *Flora of Australia* and the recent *Census* such cases should be diminishing) and as yet some dispersal and fragmentation of research efforts.

Progress

For early history see F. M. BAILEY, 1891. A concise history of Australian botany. *Proc. Roy. Soc. Queensland* 8(2): xvii–xli, xlv–xlvii; also the several papers by J. H. Maiden dating from 1907 to 1921 and cited by Blake and Atwood (1942; see **General references**) as well as below under individual states. These have been complemented in more recent years by additional works, notably D. J. and S. G. M. CARR (eds.), 1981. *People and plants in Australia.* xxi, 416 pp., illus. Sydney; and P. S. SHORT (ed.), 1990. *History of systematic botany in Australasia.* v, 326 pp., illus., maps. Melbourne.

Developments relating to the current *Flora of Australia* and the establishment of the present Biodiversity Group of Environment Australia (previously the Bureau of Fauna and Flora) were described in: S. T. BLAKE, 1960. A new flora of Australia. *Austral. J. Sci.* **23**: 173–176; N. T. BURBIDGE, 1974. Progress towards a new flora of Australia. *Annual Rep. Div. Plant Industry, CSIRO, Australia* (1973): 31–34; W. D. L. RIDE, 1978. Towards a national biological survey. *Search* **9**: 73–82; DEPARTMENT OF SCIENCE, AUSTRALIA, 1979. *Australian Biological Resources Study 1973–8* (Commonwealth Parliamentary Paper 354/1978). Canberra; and ANONYMOUS, 1979. Surveying Australia's plants and animals. *Ecos* **21**: 28–31. *Biologue*, the presently bi-annual newsletter of the Australian Biological Resources Study (a programme of the Biodiversity Group), has since reported on progress with *Flora of Australia* and related Commonwealth programmes. A retrospective account appears in the revised edition (1999) of vol. 1 of *Flora of Australia*.

General bibliographies. Bay, 1910; Blake and Atwood, 1942; Frodin, 1964, 1984; Goodale, 1879; Holden and Wycoff, 1911–14; Jackson, 1881; Pritzel, 1871–77; Rehder, 1911; Sachet and Fosberg, 1955, 1971; USDA, 1958.

Supraregional bibliographies
In addition to the following work, Australian floristic literature is also listed in C. G. G. J. VAN STEENIS, 1955. Annotated selected bibliography. In *idem* (ed.), *Flora Malesiana*, I, 5: i–cxliv. Groningen: Noordhoff. There is also an annotated bibliography in the revised edition (1999) of vol. 1 of *Flora of Australia* (see below).

MAKINSON, R., 1995. *Plant identification: a basic bibliography.* 40 pp. Canberra: Australian National Botanic Gardens. [Classified by subject areas, with synopsis on pp. 2–3; a good overall introduction to the literature. Many listings are for 'special groups' largely not catered for here. All titles are concisely annotated.]

General indices. BA, 1926– ; BotA, 1918–26; BC, 1879–1944; BS, 1940– ; CSP, 1800–1900; EB, 1959–98; FB, 1931– ; IBBT, 1963–69; ICSL, 1901–14; JBJ, 1873–1939; KR, 1971– ; NN, 1879–1943; RŽ, 1954– .

Supraregional indices
FERGUSON, I. K., 1970–72. *Index to Australasian taxonomic literature for 1968–70.* 3 nos. (Regnum Veg. 66, 75, 83). Utrecht. [Annual index to floristic, systematic and bio-historical literature of Australasia, Oceania and Papuasia. Subsumed by *Kew Record*.]

STEENIS, C. G. G. J. VAN *et al.* (eds.), 1948– . *Flora Malesiana Bulletin*, 1– . Leiden: Flora Malesiana Foundation. (Vols. 1–9 mimeographed.) [At present issued bi-annually, this bulletin contains in one of its two annual issues a substantial bibliographic section also rather thoroughly covering Australian floristic literature. A fuller description is given under Superregion 91–93.]

420–50
Superregion in general

Since publication began to coincide with the XIII International Botanical Congress in Sydney in 1981, the *Flora of Australia* gradually has been complementing or superseding existing literature, including *Flora australiensis* as well as the various state and territory floras published in the intervening century. The latter have furnished but partial coverage, being largely centered in the southeastern 'boomerang'. For much of the Australian west and north Bentham's work has until recently remained a primary resource. The year 1981 also saw the appearance of the first modern flora of the arid center, *Flora of central Australia*, prepared under the auspices of the Australian Systematic Botany Society and edited by J. P. Jessop. Since then, two major landmarks have been the *Australian plant name index* and the *Census of Australian vascular plants* (for both, see below). Good recent coverage of pteridophytes has been furnished by D. L. Jones and S. C. Clemensha in their semi-popular *Australian ferns and fern allies* (2nd edn., 1981; see below under **Pteridophytes**). An alternative, lavishly illustrated, alphabetically arranged guide (not described in full here) to much of the Australian flora is W. R. ELLIOT and D. L. JONES,

1980– . *Encyclopaedia of Australian plants (suitable for cultivation)*. Vols. 1– . Melbourne: Lothian (to date, seven volumes have appeared). For Australian vegetation, the most recent work is R. H. GROVES, 1994. *Australian vegetation*. 2nd edn. Cambridge: Cambridge University Press. (1st edn., 1974.)

Families and genera: dictionaries and keys

BURBIDGE, N. T., 1963. *Dictionary of Australian plant genera (gymnosperms and angiosperms)*. xviii, 345 pp., 2 maps. Sydney: Angus & Robertson. [Annotated enumeration of seed plant genera known from Australia, Tasmania, and Lord Howe and Norfolk Islands, including synonymy, citations, and indication of distribution. Now out of date.]

MORLEY, B. and H. R. TÖLKEN (eds.), 1983. *Flowering plants of Australia*. 416 pp., 230 figs. (part col.), maps. Adelaide: Rigby. [Semi-popular systematic guide to families, with keys to genera, descriptions, distributions (with maps), statistics and references, and notes on uses and special features; complete index at end. Patterned after *Flowering plants of the world* by V. H. Heywood (1978; see **000**). The work is of particular value for its conspecti of Myrtaceae and Proteaceae, two families prominent in the flora.]

Nomenclatural index

CHAPMAN, A. D., 1991. *Australian plant name index*. 4 vols. xxi, 3055 pp. Canberra: Australian Government Publishing Service. (ABRS Australian Flora and Fauna Series 12–15.) [Covers 62 000 names of genera and all subordinate taxa. All four volumes are arranged alphabetically by genus downwards; the last also contains synoptic indices. 99.8 percent of the names have been checked against their sources.][14]

BENTHAM, G., 1863–78. *Flora australiensis*. 7 vols. London: Reeve. (Reprinted 1967, Amsterdam, Asher; Brook, Ashford, Kent, England: Reeve.)

Concise descriptive flora of vascular plants; includes keys to genera and species, brief indication of synonymy (with references and citations), detailed indication of internal range (by state and territory) with *exsiccatae* and summary of extralimital distribution, concise, informative critical commentary, and notes on habitat, special features, etc. based on available information; index to all botanical names in each volume (cumulative index lacking). In the introductory section (vol. 1) are chapters on botanical exploration and sources of information, while the concluding postscript (vol. 7) consists of general remarks on the flora and its composition.[15]

AUSTRALIAN BIOLOGICAL RESOURCES STUDY (A. GEORGE *et al.*, exec. eds.), 1981– . *Flora of Australia*. Vols. 1– . Illus. (part col.), maps. Canberra: Australian Government Publishing Service.

Briefly descriptive flora of vascular plants; includes keys to all taxa (with a general key to families in volume 1), synonymy with references and key citations, typification, generalized indication of range (including small maps), external distribution (where appropriate), selected *exsiccatae* with localities, concise taxonomic commentary, and representative illustrations; references to *Flora australiensis* and standard modern treatments under genera and families; abbreviations and indices at end of each volume. Treatments may include description of novelties. The plan of the work and the arrangement of families follows Cronquist (and is set out in the front end-papers of each volume, the back end-papers containing an index to families), but actual publication is not sequential. The revised edition of volume 1 (1999) contains several background chapters including a history of systematic botany in Australia, the genesis of the project, and various aspects of the flora as well as a bibliography, glossary and key to flowering plant families.[16]

HNATIUK, R. J., 1990. *Census of Australian vascular plants*. xvi, 650 pp. Canberra: Australian Government Publishing Service. (ABRS Australian Flora and Fauna Series 11.)

Tabular checklist in landscape format, with indication of presence or absence in each of 97 recording units in seven major units (see map, p. xiv) or for the unit in general if internal distribution not more precisely known; index to genera at end. Arrangement of families, genera and species is alphabetical; significant synonyms are indexed along with accepted taxa and cross-referenced. The recording units are based on those existing in states and territories and do not follow a uniform scheme. The introduction contains an explanation of the work and a list of assessors as well as statistics (which include comparisons with the work's direct predecessors, the two editions of Mueller's *Systematic census of Australian plants* (1882, 1889); the number of species has about doubled in the intervening century). Some 15 638 native and 1952 naturalized species for a total of 17 590 are accounted for, with non-nominate subspecies numbering a further 861, non-nominate varieties 1462, and additional forms 82.

Subcontinental and partial works

Included here are *Beiträge zur Flora und Pflanzengeographie Australiens* by Karel Domin as well as *Flora of central Australia* edited by J. P. Jessop. The first-named work, though with a strong emphasis on Queensland, was the only substantial continent-wide systematic work to appear in the first half of the twentieth century.

The region is too large and diverse to have been covered by a specific flora. The works treated below, however, cover differing aspects of eastern Australian plant life, that by Domin being of particular importance for Queensland.

DOMIN, K., (1914) 1915–29(–1930). *Beiträge zur Flora und Pflanzengeographie Australiens, I. Systematische Bearbeitung des eigenen sowie auch fremden besonders des von Frau Amalie Dietrich in Queensland (1863–73) und von Dr Clement in Nordwest-Australien gesammelten Materiales mit teilweiser Berücksichtigung der gesamten Flora Australiens.* Abt. 1–3 (in 12 parts). 1317 pp., 207 text-figs., 38 pls. (some col.) (Biblioth. Bot. 20 (Heft 85: Lfg. 1–4) (1913–15): 1–554 (1–554 of whole work), figs. 1–117, pls. 1–18; 22 (Heft 89: Lfg. 1–8) (1921–29): 1–763 (555–1317 of whole work), figs. 118–207, pls. 19–38). Stuttgart: Schweizerbart.

Systematic enumeration of vascular plants (covering all families represented in the flora), based in large part on the extensive collections (mainly from Queensland) made by the author in 1909–10 but also including records of others, notably A. Dietrich and E. Clement; includes descriptions of new taxa, full synonymy (with references and citations), localities with citation of *exsiccatae*, summary of distribution, and often extensive taxonomic and phytogeographical commentary as well as miscellaneous notes; complete index to botanical names at end (pp. 1245–1317). [Not completed as originally planned. The *Beiträge* collectively comprise a work of particular importance for Queensland botany, the most important following the publication of F. M. Bailey's *Queensland flora* and *Comprehensive catalogue*.][17]

JESSOP, J. P. (ed.), 1981. *Flora of central Australia.* xxxiv, 537 pp., 648 text-figs., 18 col. photographs, maps (endpapers, insert). Sydney: Reed/Australian Systematic Botany Society.

Briefly descriptive manual-flora; includes keys, references in family and generic headings to pertinent literature, limited synonymy (without references or citations), generalized indication of internal and (in parentheses) external distribution, some taxonomic commentary, and notes on habitat, special features, etc.; complete index at end. The introductory section covers the plan of the work, sources, and sections on the nature of the flora (much of it in geological terms relatively recent), vegetation (by J. S. Beard, pp. xxi–xxv) and history of botanical exploration, followed by a glossary and list of abbreviations; key references, pp. xxv–xxvi. [The work, while covering the whole of nominal Central Australia (**443**) as well as large parts of other units, is not a flora of arid Australia as a whole; the Nullarbor Plain, for example, is largely excluded.]

Woody plants (including trees): guides

The extensive total tree flora (in part imperfectly known) and its disparate nature have precluded overall guides of the kind common in North America, Europe, northeast Asia, New Zealand and even southern Africa. Indeed, woody plants constitute a substantial proportion of the total Australian flora. The initial version of *Forest trees of Australia* (1957), published under the Commonwealth, was a slender work with not many more than 100 species, the majority eucalypts. It followed another Commonwealth publication on trees, originally prepared in Queensland but first produced with the aid of a special Executive Council fund for meritorious works: *Australian rain-forest trees* by W. D. Francis (1929; see **430**). The latter work has now been succeeded for tropical northeastern Australia by the three-volume *Australian tropical rainforest trees* and for subtropical and temperate rain forest trees by *Rainforest trees of mainland south-eastern Australia* (for both, see **430**). State tree guides have had a longer history, mainly in South Australia, Victoria and New South Wales.[18]

BOLAND, D. J. *et al.*, 1984. *Forest trees of Australia.* 4th edn. xvi, 687 pp., illus. (incl. 60 col. pls.), maps. Melbourne: Commonwealth Scientific and Industrial Research Organisation/Nelson. (1st edn., 1957.)

Dendrological atlas of more important trees (223 species, including 137 eucalypts); treatments include illustrations of habit, bark, foliage and other key features, distribution map, vernacular names, notes on climate, habitat, soils, associates, special features, history of use, and related species and (in smaller type) extensive descriptions of botanical characteristics, bark, wood, etc.; special introductions to some major genera (e.g., *Acacia, Eucalyptus*) and key to *Eucalyptus*; abbreviations, glossary, list of references (pp. 662–664) and complete index at end. For further coverage of *Eucalyptus* see M. I. H. BROOKER and D. A. KLEINIG, 1983–94. *Field guide to eucalypts.* 3 vols. Melbourne: Inkata Press; for rain forest trees, see **430**.

Pteridophytes

JONES, D. L. and S. C. CLEMENSHA, 1981. *Australian ferns and fern-allies.* 2nd edn. 232 pp., 297 text-figs., 6 l. pls., 60 col. photographs. Terrey Hills, N.S.W.: Reed. (1st edn., 1976.)

Well-illustrated, semi-popular descriptive treatment, without keys, of pteridophytes of the Australian continent and Tasmania (358 species in 108 genera, up from 312 and 101 respectively in the original edition, with yet more species to be described); includes remarks on distinguishing and other special features, vernacular names, indication of geographical range, and notes on cultivation, with extensive commentaries on genera and their distribution included;

glossary, list of references (pp. 227–228) and complete index at end. Four appendices encompass naturalized species, an alphabetical synopsis, a systematic lexicon (Holttum scheme), and synonymy, while the general part provides an introduction to ferns and their cultivation. Arrangement of genera in the main text is alphabetical within the Filicinae. The work, despite its popular orientation, was for some two decades the only scientifically based review of the group for Australia, but is now partly supplanted by the account in vol. 48 of *Flora of Australia*. – For keys see H. T. CLIFFORD and J. CONSTANTINE, 1980. *Ferns, fern allies and conifers of Australia*. xvii, 150 pp., 24 figs., tables. Brisbane: University of Queensland Press.

Region 42

Tasmania

Area: 68 281 km^2 (exclusive of Macquarie Island). Vascular flora: 2197 species, of which 1627 native and 570 naturalized (R. Makinson, personal communication, 1995). A total of 94 of these are pteridophytes. – This region is considered to include the main island of Tasmania, adjacent small islands, and the Bass Strait islands of King and Flinders with their islets. Macquarie Island, administratively part of the state of Tasmania, is, however, botanically sub-Antarctic (see **081** in Region 08).

The primary stage in the exploration of the peculiar Tasmanian flora, which began in the 1790s with the visit of the French ships *La Recherche* and *L'Esperance*, may be said essentially to have ended with the publication of J. D. Hooker's splendid *Flora tasmaniae* (1855–59), the last part of his tripartite *Botany of the Antarctic Voyage of 'Erebus' and 'Terror'*. In addition to the flora proper, botanical exploration in the then–colony – and in Australia generally – up to that time was reviewed and important and original biogeographical observations were made. Resident botanical work was, however, already in progress, with much of it in time to find its way into Hooker's pages. But *Flora tasmaniae* was not unnaturally too bulky for field use; local demand called for a more convenient handbook. In 1878 – the year *Flora australiensis* was completed – there appeared a somewhat cramped field-manual, *A handbook of the plants of Tasmania* by the Rev. William W. Spicer, an Anglican clergyman resident in the island from 1874 to 1878. Spicer's manual was in time succeeded by Leonard Rodway's *Tasmanian flora* (1903), also now long out of date but still the most recent single-volume manual.[19]

The decades since World War II have witnessed the publication of the current standard work, *The student's flora of Tasmania* by Winifred Curtis (1956–94; 2nd edn., 1975, not completed), which in the fashion of Allan's *Flora of New Zealand* (**410**) highlights taxonomic problems, and a magnificent modern 'Great Flower Book' by the same author, the six-volume *Endemic flora of Tasmania* (1965–78) with over 150 paintings of peculiar and colorful species by Margaret Stones. A reorganization of the state herbarium took place in the mid-1970s, which with the appearance of most of the *Student's flora* as well as the *Endemic flora* stimulated more detailed study of the flora. A state census appeared in 1989 (with further editions in 1995 and 1999). The state is at present considered moderately well collected and documented, although less so in the southwest, an area largely remote and difficult of access.[20]

Bibliographies. General and supraregional bibliographies as for Superregion 42–45.

Indices. General and supraregional indices as for Superregion 42–45.

420

Region in general

The recent flora by Curtis and Morris – not really a students' manual in the sense used in this *Guide*, but a conventional descriptive flora – was completed in 1994. With five volumes, however, it remains a 'desk flora'. The *Endemic flora of Tasmania* is another key source of information and illustrations. Those wanting a one-volume manual may consult M. CAMERON (ed.), 1981. *Guide to flowers and plants of Tasmania*. Launceston: Launceston Field Naturalists' Club (with Reed); or the now very dated L. RODWAY, 1903. *The Tasmanian flora*. xix, 320 pp., 50 pls. Hobart: Government Printer. All main works also cover Macquarie Island (**081**).

BUCHANAN, A. M. (ed.), 1999. *A census of the vascular plants of Tasmania and index to* The Student's Flora of Tasmania. iv, 99 pp. Hobart: Tasmanian Museum and Art Gallery. (Tasmanian Herbarium Occ. Publ. 6.) (1st edn. by A. M. Buchanan, A. McGeary-Brown and A. E. Orchard, 1989, Hobart, as *A census of the vascular plants of Tasmania*; 2nd edn. (under present title), 1995.)

Alphabetically arranged checklist, the taxa arranged by families; entries include authorities, places of publication, status in the flora, and a reference to the *Student's flora*; some synonymy is incorporated; lexicon of genera and their families at end. Pp. ii–iv encompass the introduction and a list of key references; pp. 89–91 comprise a checklist for Macquarie Island (081).

CURTIS, W. M., 1956–94. *The student's flora of Tasmania.* Parts 1–3, 4A, 4B. xlvii, 661 pp., 138 figs. (parts 1–3); 138 pp., illus. (some col.) (part 4A); xxvii, 459 pp., illus. (part 4B). Hobart: Government Printer (parts 1–3, 4A); St. David's Park Publishing (part 4B). Partly revised as W. M. CURTIS and D. I. MORRIS, 1975. *The student's flora of Tasmania.* 2nd edn. Part 1.1, 240 pp., 60 text-figs. Hobart (reissued 1993, Hobart: St. David's Park Publishing).

Illustrated descriptive flora of native and naturalized seed plants, with keys to all taxa; includes limited synonymy, vernacular names, generalized indication of local range, brief taxonomic commentary, and a few notes on other aspects; glossaries (in parts 1 and 4B) and indices to all names in each part (but no cumulative index). Parts 1–3 cover the gymnosperms and dicotyledons (the latter arranged according to the Bentham and Hooker system); part 4A, the Orchidaceae (with many colored illustrations); and part 4B, the remaining monocotyledons (arranged according to Cronquist). The first part of the second edition is essentially a slightly expanded version of its predecessor, covering gymnosperms and dicotyledons from Ranunculaceae through Myrtaceae.[21]

Partial work

STONES, M. and W. M. CURTIS, 1965–78. *The endemic flora of Tasmania.* 6 vols. 478 pp., 155 col. pls., maps (endpapers). London: Ariel.

Not a flora in the strict sense, but a sumptuous, large-scale 'Great Flower Book', containing 155 plates by the first author depicting the more colorful and characteristic endemic plants of the state, each with accompanying descrip-

tive text by the second author furnishing details of form, technical details, biology, ecology, and special features; general index and systematic lexicon in volume 6 (the arrangement of plates follows no system). The introductory section (in vol. 1) includes an account of Tasmanian vegetation by the second author. Notes on cultivation have been added by the work's sponsor, Lord Talbot de Malahide.

Region 43

Eastern Australia

Within this region are found the states of Victoria, New South Wales and Queensland, the Australian Capital Territory (here treated as part of New South Wales), and the Torres Strait Islands south of the Australia/Papua New Guinea seabed boundary. Under **430** are described *Beiträge zur Flora und Pflanzengeographie Australiens* by K. Domin and *Australian rain-forest trees* by W. D. Francis; despite their titles, these works are limited almost entirely to eastern Australia. Both have provided valuable supplementation for state manuals, particularly in Queensland, but now are more of historical rather than practical value. For rain forest trees in tropical Australia both works are superseded by *Australian tropical rainforest trees* (1994) by Bernard P. M. ('Bernie') Hyland and Trevor Whiffin, also described below.

As is well known, botanical exploration in eastern Australia began with the landfall of the *Endeavour* at Botany Bay, now enveloped within metropolitan Sydney. Early collecting took place mostly near the coast, with a substantial contribution coming from Robert Brown in the *Investigator*, but from approximately the end of the Napoleonic Wars serious interior exploration began, first in New South Wales proper but later extending to what are now Victoria and Queensland. Most collections, however, went overseas until the establishment of the first major local herbarium by Mueller at Melbourne in 1853.

While J. D. Hooker's *Flora tasmaniae* is nominally the first 'state' flora in Australia, the first for the mainland, and the first to be locally produced, is Ferdinand Mueller's *Plants indigenous to the colony of Victoria* (1860–65). Though only part 1 (Thalamiflorae) was ever published, it was a primary result of his extensive botanical surveys of the then-new colony in the 1850s with in particular the support of then-Lt. Governor Charles Joseph Latrobe and his successors, Charles Hotham and Henry Barkly. The 'iron horse', paddle steamer and Cobb and Co. followed, and with such inland transportation networks and the spread of settlement, floristic exploration became for the greater part the province of resident botanists. Support for natural sciences research also largely passed to colonial governments, although some explorations, such as that of Augustus Gregory in northern Australia (1855–56), were at least partly financed from outside.[22] A major stimulus was provided by the completion in 1878 of *Flora australiensis*, in the form firstly of Mueller's *Systematic census of Australian plants* (1882; 2nd edn., 1889) and secondly of several more state floras, now mostly – though in part only recently – superseded: *Handbook of the plants of Tasmania* (1878; already mentioned under **42**), *Synopsis of the Queensland flora* (1883, with supplements to 1890) by Frederick Manson Bailey, *Key to the system of Victorian plants* (1885–88) by Mueller, *Flora of extra-tropical South Australia* (1890; mentioned under **44**), *Handbook of the flora of New South Wales* (1893) by Charles Moore with Ernst Betche, and finally Bailey's *The Queensland flora* (1899–1902) and Rodway's *The Tasmanian flora* (1903; mentioned under **420**).

After publication of the last-named, the only state-level manuals to appear in the decades before World War II were J. M. Black's *Flora of South Australia* (1922–29, mentioned under **445**) and A. J. Ewart's *Flora of Victoria* (1931). Some state checklists also appeared or were revised; of particular note was Bailey's *Comprehensive catalogue of Queensland plants* (1913). Other significant floristic undertakings of the decades before and after World War I were Joseph H. Maiden's *Forest flora of New South Wales* (1902–25), one of a number of large-scale works by this energetic author, and the Czech botanist Karel Domin's *Beiträge zur Flora und Pflanzengeographie Australiens* (1915–29), the results of his collecting trip of 1909–10, mostly in eastern Australia. Some one-volume 'tree books' also made their debut, among them *Trees of New South Wales* (Anderson, 1932; 4th edn., 1968) and *Australian rain-forest trees* (Francis, 1929; 2nd edn., 1951; 3rd edn., 1970). Both these works now are well out of date.

The first of a new generation of state floras began with the initiation of a *Flora of New South Wales* in 1961. Rather than being a self-contained work, however, it was a submonographic series effectively intended to cover southeastern Australia. After publication of several families (including an account of the Gramineae), re-evaluation of user needs along with the advent of *Flora of Australia* caused the series to be wound up in 1984. The National Herbarium of New South Wales then directed its efforts towards a synoptic manual-style *Flora of New South Wales*; this appeared in 1990–93 in four volumes under the editorship of Gwen Harden. Other recent but now inevitably aging works include *Flora of the Sydney region* (Beadle, Evans and Carolin, 1963; 2nd edn., 1972; 4th edn., 1994); *Flora of the Australian Capital Territory* (Burbidge and Gray, 1970); *Students' flora of North Eastern New South Wales* (Beadle *et al.*, 1971–87; 2nd edn. of vol. 1, 1989); *Plants of Western New South Wales* (Cunningham *et al.*, 1981, recently reissued); and *Kosciusko alpine flora* (Costin *et al.*, 1979; see **403**). The Sydney region is also served by two good recent popular field-guides.

In Victoria, Ewart's flora was between 1962 and 1972 partly supplanted by James H. Willis's *Handbook to plants in Victoria*; in style this represented a partial return to that of Mueller's *Key* though in two volumes rather than one. More recently a new *Flora of Victoria* (Foreman, Walsh and Entwisle, 1993–99) has made its appearance in four volumes to much acclaim; and, for the Melbourne region, there is also *Flora of Melbourne* (1991) issued by the Maroondah chapter of the Society for Growing Australian Plants. In Queensland, multi-volume handbooks have been projected for each of four subregions in place of a successor to *Queensland flora*. The first, covering the four southeastern districts, has appeared in three volumes as *Flora of southeastern Queensland* (Stanley and Ross, 1983–89). Outside eastern Australia, a very heavily revised fourth edition of Black's South Australian manual appeared as *Flora of South Australia* in four volumes in 1986 (in connection with the state's sesquicentennial); and in Western Australia the first good regional floras were the two-volume *Flora of the Perth region* (Marchant and Wheeler, 1987) and *Flora of the Kimberley region* (Wheeler, 1992) – the latter for a botanically important but previously little-documented area. The two latter works complement the longer-established but idiosyncratic *How to know Western Australian wildflowers* begun by W. E. Blackall in the 1940s and continued by B. J. Grieve, with the second edition completed in 1998. The 'Top End' has followed with *Flora of the Darwin region* (1992–), four volumes of which are projected.

Woody plants and other 'special groups' have continued to draw attention. *Aquatic plants of Australia* by Helen Aston appeared in 1973 (see **408**) and there are also some guides to the littoral flora (**409**). The woody flora received further coverage in two somewhat complementary works: *Native trees and shrubs of South-eastern Australia* by Leon Costermans (1981, with reissues) and *Rainforest trees of mainland south-eastern Australia* (1989) by A. G. Floyd. Temperate pteridophytes were treated in *Ferns and allied plants of Victoria, Tasmania and South Australia* by Betty Duncan and Golda Isaac (1986; 2nd edn., 1994) in response to another 'fern craze'.

The present state of floristic knowledge and documentation is thus variable, ranging from moderately good in the south to poor or fair in the north, particularly in the monsoon- and rain forest areas. The formal scientific floras have in recent years been joined by many popular or semi-popular guides covering a wide variety of areas and habitats.

Bibliographies. General and supraregional bibliographies as for Superregion 42–45.

Indices. General and supraregional indices as for Superregion 42–45.

430
Region in general

All the works under this heading concern the woody flora.

Woody plants (including trees): guides
See also **420** (BOLAND *et al.*, *Forest trees of Australia*; BROOKER and KLEINIG, *Field guide to eucalypts*).

No single work covers the whole of the woody flora of eastern Australia, even the trees alone. Its elucidation has come but gradually, with new species (and genera) being described even in recent years, notably in the north. *Australian rain-forest trees* by William D. Francis (see below), long the only available work for the rain forest formations, is now mainly of historic interest. Its main successors (for both, see below) are *Australian tropical rainforest trees* by B. P. M. Hyland and T. Whiffin, itself a successor to B. P. M. HYLAND, 1971. *A key to the common rainforest trees between Townsville and Cooktown based on leaf and bark features.* 103+

pp., illus.; 93 punch cards in separate packet. Brisbane: Department of Forestry, Queensland; and *Rainforest trees of mainland south-eastern Australia* (1989) by A. G. Floyd. This latter was based on A. G. FLOYD *et al.*, 1960–84. *New South Wales rainforest trees.* 12 parts, index (parts 1–5 also in second editions, 1979–81). Illus. (Forestry Comm. New South Wales Res. Note 3–49 and *hors série, passim*). [Sydney.] Other works include A. G. FLOYD, 1977. *Key to major rainforest trees in N.S.W.* 2nd edn. 20 pp., illus. (Forestry Comm. New South Wales Res. Note 27). [Sydney]; and J. B. WILLIAMS *et al.*, 1984. *Trees and shrubs in rainforests of New South Wales and southern Queensland.* 142 pp., illus. Armidale: University of New England. For southeastern Australia more generally, a major gap was filled with publication of *Native trees and shrubs of South-eastern Australia* (1981) by Leon Costermans (described below). Enthusiasts should also consider K. J. SIMPFENDORFER, 1992. *An introduction to trees for South-eastern Australia.* Revised edn. xiii, 376 pp., illus., maps. Melbourne: Inkata Press.

COSTERMANS, L. F., 1981. *Native trees and shrubs of South-eastern Australia.* viii, 422 pp., illus. (part col.), maps. Adelaide: Rigby. (Subsequent edns., 1983, 1985, 1986 with minor revisions.)

Popularly oriented, copiously illustrated guide to *c.* 900 species, organized in six chapters of which chapters 4, 'Regional guide-lists', and 5, 'Description of species', comprise the keys and systematic treatment. Eight lists make up chapter 4, each corresponding to a biotic or physiographic zone; each list is illustrated with diagnostic notes and a reference to a fuller treatment in chapter 5 (pp. 141–293) but standard keys are lacking. Chapter 5 is laid out in atlas format, the colored illustrations (four to a page) faced by descriptions with vernacular and scientific names, preferred habitats and substrates, distributional and biological notes, and indication of similar taxa along with diagnostic figures and small distribution maps. An extension to chapter 5 (pp. 295–379) covers *Acacia* sensu lato and *Eucalyptus*. Comments on major families and genera comprise chapter 6; this is followed by an illustrated glossary, explanation of abbreviations, and a complete index. The remaining chapters relate to the form and philosophy of the work, land and vegetation, and places of special interest.

FLOYD, A. G., 1989. *Rainforest trees of mainland south-eastern Australia.* xii, 420 pp., illus. (some col.), map. Sydney: Inkata Press.

Illustrated, alphabetically arranged dendrological manual (covering 385 species) with keys (on pp. 23–55 preceding species accounts), botanical and vernacular names, descriptions, and notes on habitat, distribution, timber and uses, and regeneration; localities of reserves and index to all botanical and vernacular names at end. The general part covers the history and extent of rain forest, habitat and biotic factors, forest form and structure, special features and synusiae, classification (by structure and physiognomy), and selected references, followed by (pp. 17–22) an illustrated glossary. [Covers most species in southeastern Queensland, New South Wales and Victoria.]

FRANCIS, W. D., 1970. *Australian rain-forest trees.* 3rd edn., revised by G. Chippendale. xvi, 468 pp., 270 text-figs., frontisp., 4 maps. Canberra: Australian Government Publishing Service. (Subsequently reissued. 1st edn., 1929, Brisbane; 2nd edn., 1951, Sydney.)

Descriptive treatment of the more important rain forest tree species in eastern Australia, especially in north-eastern New South Wales and southeastern Queensland; keys to extra-tropical species; limited synonymy, with occasional references and citations; vernacular names; generalized indication of distribution, with some localities; notes on special features, wood, etc.; index to all botanical and vernacular names. The introductory section includes subsections on physical features, vegetation, forest structure, plant morphology, etc. The so-called third edition differs from the second only with respect to nomenclature.

HYLAND, B. P. M. and T. WHIFFIN, 1994. *Australian tropical rainforest trees.* 3 vols. with three 5.25 in. diskettes. Illus. Melbourne: CSIRO Publications. Succeeded by B. P. M. HYLAND, T. WHIFFIN, D. C. CHRISTOPHEL, B. GRAY, R. W. ELICK and A. J. FORD, 1999. *Australian tropical rain forest trees and shrubs.* 1 CD-ROM, manual of 95 pp. Collingwood, Vic.: CSIRO Publishing.

The 1994 work is a manual of rain forest trees, mainly for northeastern Queensland, in both print and electronic form. Volume 1 (303 pp.) comprises directions for use of the electronic materials, including the interactive key, and also includes a list of all 1056 taxa documented (many of them as yet not formally described) and details of characters scored for the matrix upon which the key is based. Volume 2 (564 pp.) contains concise taxonomic descriptions and details of distribution, habitat, etc. Volume 3 (260 pp., by D. C. Christophel and B. P. M. Hyland) comprises a leaf atlas covering all taxa, with details of size, shape and venation (reminiscent of the 'nature-printed' leaves in *Flora brasiliensis*). The 1999 edition – with most content released only in electronic form – includes other woody plants as well as data from Northern Territory and Western Australia, in all accounting for 1733 species. A range of color photographs is also included. The printed manual encompasses an introduction, glossary, list of species, and bibliography.[23]

Pteridophytes

Eastern Australia is easily the richest part of the continent for ferns and fern-allies. Recent separate works include, in addition to that by Jones and Clemensha for the whole of Australia (**420–50**), Andrews' *Ferns of Queensland* (1990; see **434**) and the following covering South Australia, Victoria and Tasmania.

DUNCAN, B. and G. ISAAC, 1994. *Ferns and allied plants of Victoria, Tasmania and South Australia (with distribution maps for the Victorian species).* 2nd edn. 269 pp., 8 pp. col. pls., 122 figs., 252 photographs, 104 distribution maps. Melbourne: Melbourne University Press (in association with Monash University). (1st edn., 1986.)

Semi-popular illustrated guide in 22 chapters with keys to all taxa, scientific and vernacular names (with derivations of the former), descriptions, diagnostic field characters, indication of distribution and discussion of habitat and occurrence, distribution maps (for Victoria), and taxonomic remarks (mainly at generic and family level, but also in footnotes to species); bibliography, glossary, authors of plant names, and complete index at end. Chapter 22 (by C. J. Goudley and R. L. Hill) is on propagation and cultivation. The introductory section (chapter 1) includes a fold-out tabular key to genera. [Pp. 259–269 of the reset, soft-cover reissue comprise additions and corrections to the main text (not accounted for in the index) as well as to the bibliography, glossary, and list of authors of plant names.]

431

Victoria

Area: 227416 km². Vascular flora: 4100 native and 1035 non-native species (*Flora of Victoria*, 4: 1033 (1999)); 118 of these are pteridophytes (Duncan and Isaac, 1994). Ross (1996) listed 4160 species and infraspecific taxa, with 3105 native.

Though not in a stricter sense a descriptive work, the only presently useful smaller-format flora (and a 'base line' until completion of the new *Flora of Victoria*) is James H. ('Jim') Willis's *Handbook*. Alfred Ewart's *Flora of Victoria* (1931) is now essentially of historical interest; it in turn succeeded *Plants indigenous to the colony of Victoria* (1862–65, not completed) and *Key to the system of Victorian plants* (1886–88), both by Ferdinand von Mueller with the latter serving as a model for Willis's *Handbook*. Mueller's works were based on his systematic botanical surveys of the then-colony in the 1850s (with further, shorter tours through to the mid-1880s). There are also two checklists (Churchill and de Corona, 1972; Ross, 1996); the latter is a current successor to *Census of the plants of Victoria* by the Field Naturalists' Club of Victoria (1923; 2nd edn., 1928). A new *Flora of Victoria*, comparable to *Plants indigenous to the colony of Victoria*, was completed in 1999, and a good modern *Flora of Melbourne* has recently also appeared.[24]

CHURCHILL, D. M. and A. DE CORONA, 1972. *The distribution of Victorian plants.* [130] pp., 4 figs. (incl. maps). Melbourne: The authors (distributed by National Herbarium of Victoria).

Tabular census of vascular plants of the state, with letters designating presence/absence of species in each of 23 squares; genera alphabetically arranged. The introductory section gives an explanation of the work, and at the end (pp. 122ff.) is given a table of authorities (in the main part designated only by numbers).

FOREMAN, D. B., N. G. WALSH and T. J. ENTWISLE, 1993–99. *Flora of Victoria.* 4 vols. Illus. (some col.), maps (part col.). Melbourne: Inkata Press.

The first volume ('Introduction', 1993) of this large-scale four-volume series is a comprehensive 'tome preliminaire' with chapters (by various authors) on prehistory, geology and geomorphology, climate, botanical exploration, natural regions and vegetation, soils, aboriginal use of plants, the pyric interaction, rare and threatened plants, and the impact of introduced flora, all fully indexed. The remaining volumes comprise the flora proper with keys, descriptions, limited synonymy, vernacular names, indication of distribution (with for Victoria both Churchill/Corona and current codes, the latter relating to the 'Natural Regions of Victoria' map in the inside covers of each volume), habitat, more or less extensive critical notes, figures, and for all taxa distribution maps; illustrated glossary, bibliography and complete index at end of each volume. A short introduction and a general key to included families precede the main texts. [Vol. 2 (1994) covers ferns and allied plants, gymnosperms and monocotyledons; vol. 3 (1996), dicotyledons from Winteraceae through Myrtaceae. Vol. 4 (Cornaceae–Asteraceae, additions and corrections, statistics and a cumulative index) appeared in 1999. Arrangement of the families is after Cronquist. Statistics for families appear in vol. 4, pp. 1030–1033; also given therein are additions and corrections to vols. 2 and 3 and a cumulative index to the work.][25]

ROSS, J. H., 1996. *A census of the vascular plants of Victoria.* 5th edn. iii, 230 pp. Melbourne: National Herbarium of Victoria. (1st edn. by S. J. Forbes *et al.*, 1984; 2nd edn., 1988; 3rd edn., 1990; 4th edn., 1993.)

Alphabetically arranged nomenclator with each entry on one line; for accepted names (in bold face) authorities and the place of publication are given. A

species included in Willis's *Handbook* is indicated with a 'W' and naturalized species are given an asterisk. Two alphabetical arrangements are presented: by families (pp. 1–94) and by genera and species (pp. 95–198).[26]

WILLIS, J. H., 1970–72. *A handbook to plants in Victoria.* 2 vols. Melbourne: Melbourne University Press. (1st edn. of vol. 1, 1962.)

Concise diagnostic manual-key to vascular plants, with synonymy and references, citations of illustrations, vernacular names, general indication of local and extralimital range, and brief notes on taxonomy, habitat, phenology, special features, etc.; addenda and indices to all botanical and vernacular names in each volume. In vol. 2 Victorian distribution is expressed on the basis of a grid system; there are also a few minor changes in format. [The second edition of vol. 1 is a reprint of the first edition with a supplement. A further supplement appeared independently in 1988.]

Partial work: Greater Melbourne

SOCIETY FOR GROWING AUSTRALIAN PLANTS MAROONDAH, INC., 1991. *Flora of Melbourne: a guide to the indigenous plants of the Greater Melbourne area.* xi, 360 pp., illus. (incl. 68 col. photographs), col. maps (end-papers and facing p. 20). South Melbourne, Vic.: Hyland House.

An 'illustrated flora' covering nearly 1000 species, the figures bordering the text down the outer margin of each page; each *pro forma* entry (alphabetically arranged within major plant groups, the monocotyledons being artificially subdivided) covers vernacular and scientific names, plant size and form, habitat, foliage, flowers, localities, overall distribution, requirements (including degree of insolation) and propagation, essential synonymy, and special features. Notes are also given for some (but not all) genera; there are, however, no keys. There is an extensive general part with a description of plant communities in chapter 1 and soils in chapter 2, while five appendices (a table of properties for every species, a list of localities, local nurseries, a glossary and a list of references) and a complete general index conclude the work.

Woody plants (including trees): guides

The most recent separate work is L. F. COSTERMANS, 1994. *Trees of Victoria and adjoining areas.* 5th edn. xii, 164 pp. Frankston, Vic.: Costermans Publishing. (1st edn., 1966, as *Trees of Victoria: an illustrated guide*; 4th edn., 1981.) This field-guide covers some 250 species; its small format follows that of an earlier popular work (limited to *Eucalyptus*) by Reuben T. Patton: *Know your own trees* (1942; revised 1961).

432

New South Wales (mainland in general, with the Australian Capital and Jervis Bay Territories)

Area: 803 059 km^2 (New South Wales mainland, 800 628 km^2; Australian Capital Territory, 2358 km^2; Jervis Bay Territory, 73 km^2). Vascular flora (New South Wales): 5930 species, of which 4677 native and 1253 naturalized (R. O. Makinson, personal communication, 1995); 170 of these are pteridophytes. – For Lord Howe Island, see **411**. The unit as circumscribed here includes mainland New South Wales, the Australian Capital Territory, and the Jervis Bay Territory.

In contrast to the situation in Victoria, nineteenth-century botanical exploration in New South Wales proceeded in fits and starts. Until 1850 or so most collections were deposited in Northern Hemisphere collections. Charles Moore, director of the Botanic Gardens in Sydney from 1848 to 1895, was primarily a horticulturalist and made little effort to develop a herbarium; after 1853, much of his material as well as that of his contemporaries went to Mueller in Melbourne. An important collection, including many Cunningham specimens, was lost in 1879 with the burning of the Garden Palace exhibition building in Sydney. In 1881, however, Moore appointed a Gardens' collector, Ernst Betche; later the two collaborated on the first state manual, *Handbook of the flora of New South Wales* (1893), with Betche perhaps the more effective contributor. Although largely intended as a students' manual and not comprehensive, the work was not to be entirely superseded for a century. The first serious institutional herbarium in the colony was that developed after 1880 at the then-Technological Museum (later the Museum of Applied Arts and Sciences) in Sydney; but not until its creator Joseph Henry Maiden's appointment in 1896 to the Botanic Gardens directorship, in succession to Moore, was a state herbarium (the National Herbarium of New South Wales) effectively established and an enlarged collecting programme instituted.

Maiden published vigorously until the end of his tenure in 1924. His main floristic contributions were *The forest flora of New South Wales* (1902–25), a stately but somewhat disorganized work (now something of a collector's item) and, with Betche, *A census of New South Wales plants* (1916); the latter would remain current for 65 years. After Maiden's time there was a long 'slump', with no new major floristic initiatives until after 1950 save for Robert H. Anderson's *Trees of New South Wales*, first published in 1932 with three further editions to 1968. Anderson also developed the state system of botanical districts which, with modifications, remains in use; and in 1961 he initiated the serial *Flora of New South Wales* as a semi-monographic 'research' work comparable to the *Flora of Texas*. For more than two decades it served as a vehicle for revisionary studies, latterly expanded to southeastern Australia as a whole. It was wound up in the mid-1980s, and the National Herbarium then gave attention to preparation of a state manual. This appeared in 1990–93 as *Flora of New South Wales* (see below). A new census to replace that of Maiden and Betche had already appeared in 1981 on the occasion of the XIII International Botanical Congress in Sydney; this is, however, now in need of replacement.[27]

For the needs of field work and identification in the main centers of population (including the Australian Capital Territory), the sizeable flora has also been partly digested in a range of regional manual-floras; these are given under **433**. All are post-World War II developments. Costermans' *Native trees and shrubs of South-eastern Australia* and Floyd's *Rainforest trees of mainland south-eastern Australia* appear under **430**. For *Waterplants of New South Wales*, see **408**.

Bibliography

PICKARD, J., 1972. Annotated bibliography of floristic lists of New South Wales. *Contr. New South Wales Natl. Herb.* 4(5): 291–317. [Briefly annotated list, including references to unpublished material. Important for the local literature.]

HARDEN, G., 1990–93. *Flora of New South Wales.* 4 vols. Illus., col. pls. Sydney: University of New South Wales Press.

A briefly descriptive 'illustrated flora', the figures arranged vertically to the left of the text; includes keys, concise descriptions, vernacular names, essential synonymy, distribution (by the Anderson system of botanical regions, shown on maps near the front of each volume), extralimital range and (if appropriate) origin, habitats, and season; occasional critical commentary.

Generic entries include standard revisionary references (those for each volume all gathered at its beginning). The first three volumes include an illustrated glossary and index to all botanical and vernacular names, while the fourth volume in addition concludes with addenda and corrigenda and cumulative indices to families and to all names. An introductory section in vol. 1 gives the plan of the work and how to use it, botanical essentials, a list of protected plants, natural regions and vegetation, a list of contributors (repeated in the remaining volumes, amended as appropriate), and abbreviations and symbols (also repeated in subsequent volumes). General keys to families precede each major plant group. Their arrangement in the text follows Dahlgren with vol. 1 covering pteridophytes, gymnosperms, and dicotyledons from Annonaceae through Amygdalaceae (Rosaceae in part); vol. 2, Proteaceae through Leguminosae; vol. 3, Balsaminaceae through Lamiaceae, and vol. 4, monocotyledons. [Covers 6363 native and introduced species in 1616 genera and 255 families. A number of taxa remain unnamed.][28]

JACOBS, S. W. L. and J. PICKARD, 1981. *Plants of New South Wales.* 226 pp., maps (end-papers). Sydney: Government Printer (for the National Herbarium of New South Wales).

Alphabetically arranged (by major groups and families) tabular checklist of native and naturalized seed plants, with indication of presence or absence (based on collections or other reliable information) for each of 13 mainland botanical districts and Lord Howe Island; index to genera at end. Unsubstantiated published records are included for convenience, and naturalized species are indicated by a dagger (†). The districts are those of Anderson in *Trees of New South Wales* except for subdivision of his western and far western districts. [Distributional information has now been superseded by that in *Flora of New South Wales* (see above).][29]

NATIONAL HERBARIUM OF NEW SOUTH WALES, 1961–84. *Flora of New South Wales* (by various botanists). Published in fascicles. Sydney: National Herbarium of New South Wales. (Until 1975 published as *Contr. New South Wales Natl. Herb., Flora Series.*)

Serially published semi-monographic 'research' flora; includes keys to genera and species, lengthy descriptions, full synonymy (with references and citations), vernacular names, citation of *exsiccatae* with generalized indication of local range, summary of extra-limital distribution, often extensive taxonomic com-

mentary, and notes on ecology, special features, karyotypes, etc.; indices in each fascicle. The introductory fascicle (1961) includes a general description of the different parts of the state, remarks on botanical history, and a map with 'botanical districts' (essentially those of Anderson as used in *Trees of New South Wales*). Although pre-numbered according to the Englerian system, families were published when ready. For the Orchidaceae, originally published separately but with its 1969 reissue included as part of this series, see H. M. R. RUPP, 1943. *The orchids of New South Wales.* 152 pp. Sydney: Government Printer. (Reprinted 1969, Sydney, with supplement by D. J. McGillivray.)

Woody plants (including trees): guides

ANDERSON, R. H., 1968. *The trees of New South Wales.* 4th edn. xxvi, 510 pp., illus., 4 col. pls., map. Sydney: Government Printer. (1st edn., 1932; 3rd edn., 1957.)

Descriptive manual of all native, naturalized and commonly cultivated tree species in the state, arranged in four main sections corresponding to natural regions with an additional section for introduced trees; includes limited synonymy (without references), vernacular names, general indication of range, and notes on field attributes, habitat, uses, etc.; general analytical keys to *Acacia*, *Eucalyptus* and all other trees (the last in two versions); glossary and complete index at end. The introductory section has several parts dealing with trees in general and their ecology.[30]

MAIDEN, J. H., 1902–25. *The forest flora of New South Wales.* Vols. 1–8. 295 pls. Sydney: Government Printer.

Large-scale descriptive botanical and dendrological atlas of trees, with full-page plates and copious accompanying texts; the latter include ample descriptions, full synonymy (with references and citations), aboriginal and English vernacular names, localities, and notes on habitat, special features, wood, properties, uses, etc.; complete index in each volume (no cumulative index). No particular system of classification has been followed. [Much of the text was taken from Bentham's *Flora australiensis.*][31]

433

New South Wales: regions (including the Australian Capital Territory)

The 13 botanical/geographical districts commonly recognized in mainland New South Wales (inclusive of the Australian Capital and Jervis Bay Territories) are here expressed as six regions. All have separate works for all or part of their area; where coverage is wanting, *Flora of New South Wales* (**432**) or works on neighboring regions or states can to a greater or lesser extent be utilized. There are many other works of more restricted scope; Pickard (1972; see **432**) covers earlier floristic lists while some later ones have appeared in the serial *Cunninghamia*.

I. Central and south coast, and central tablelands

In addition to *Flora of the Sydney region* two other floristic guides are suggested: A. FAIRLEY and P. MOORE, 1989. *Native plants of the Sydney district*. 432 pp., illus. Kenthurst, N.S.W.: Kangaroo Press/Society for Growing Australian Plants; and L. ROBINSON, 1991. *Field guide to the native plants of Sydney*. 448 pp., illus., maps. Kenthurst: Kangaroo Press. The first is strongly photographic, while the second features an abundance of line drawings; neither, however, has many keys.

BEADLE, N. C. W., O. D. EVANS and R. C. CAROLIN (with M. D. TINDALE), 1994. *Flora of the Sydney region*. 4th edn. by R. C. Carolin and M. D. Tindale. 840 pp., illus., 16 col. pls., map. Chatswood, N.S.W.: Reed. (Originally published 1963, Armidale, N.S.W., as *Handbook of the vascular plants of the Sydney district and Blue Mountains*; 2nd edn. (with current title), 1972, Sydney; 3rd edn., 1982 (reprinted 1986).)

Concise manual-key to the native and naturalized vascular plants of the Sydney region (comprising the central coast and adjacent tablelands, including the 'Blue Mountains' west to somewhat beyond Lithgow and, from north to south, Newcastle to Nowra); includes diagnoses, limited synonymy, vernacular names, notes on variability and hybridization, and brief indication of habitat, substrate, local range, etc.; list of authorities and complete index at end. Critical botanical details are depicted in the figures, while the photographs contain a limited representation of the district's rich flora. An illustrated glossary appears in the brief introductory section along with technical notes. [The latest edition features space-saving bracketed keys.]

II. Southern tablelands (including the ACT)

BURBIDGE, N. T. and M. GRAY, 1970. *Flora of the Australian Capital Territory*. vii, 447 pp., 409 text-figs., map. Canberra: Australian National University Press.

Illustrated manual-flora of seed plants; includes keys to all taxa, references to standard monographs or revisions in family and generic headings (but no synonymy), vernacular names, general indication of local and extralimital range, and notes on habitat, special features, etc., as well as taxonomic commentary; glossary, list of references, and complete index at end. An introductory section includes, *inter alia*, an account of the vegetation of the area. [The work is of use generally in the Southern tablelands.]

III. Northern tablelands, north coast and northern western slopes (including the New England district)

The *Students' flora* was preceded by M. GRAY, 1961. A list of vascular plants occurring in the New England Tablelands, NSW, with notes on distribution. *Contr. New South Wales Natl. Herb.* 3(1): 1–82, map. This systematic checklist (with limited synonymy, *exsiccatae* and ecological notes) covers a more limited area than does the *Students' flora*.

BEADLE, N. C. W. *et al.*, 1971–87. *Students' flora of North Eastern New South Wales*. 6 parts, index. vii, 1177 pp. (+ unnumbered pp. of index), 492 text-figs., map. Armidale: The authors (distributed by Department of Botany, University of New England). (Part 1 also in a 2nd edn., 1989, revised by N. Prakash, G. J. White and J. B. Williams. vi, 116 pp.)

Manual-key to vascular plants; includes extensive diagnoses, limited synonymy, vernacular names, and brief indication of local and extralimital distribution, habitat, phenology, etc., along with analytical keys to genera and to families. The illustrations mostly show critical details. Lists of references and full botanical indices are provided in each part, and part 1 includes an illustrated glossary. The work largely covers the region as defined in the heading, with the addition of part of the northern western plains (around Moree). For flowering plants, a modified Bentham and Hooker system has been followed.

IV–VI. Other regions

These respectively comprise (**IV**) the central and southern western slopes, (**V**) the Riverina, and (**VI**) the western and far western districts. Much of the area is now covered by *Plants of western New South Wales* (1981; see below). The state floras of Victoria (**431**) and South Australia (**445**) may also be useful.

CUNNINGHAM, G. M. *et al.*, 1981. *Plants of western New South Wales*. 766 pp., over 1400 col. photographs, 67 text-figs., maps (end-papers). Sydney: Government Printer. (Reprinted 1992, Sydney: Inkata Press.)

Atlas-flora of vascular plants (2027 native and introduced species and infraspecific taxa), arranged according to the Englerian system and without keys but with nearly all species illustrated in color; the text includes description, vernacular name(s), local (and extralimital) distribution, habitat, phenology, and more or less extensive notes on frequency, biology and dispersal, behavior under grazing, weediness (if appropriate), etc.; addendum, references (pp. 731–733), glossary and complete index at end. The introductory section includes a foreword (by Neville Wran, then-state premier) and sections on climate, geology, geomorphology, soils, vegetation and (pp. 23–25) the plan and basis of the work. [Covers all territory west of the 500 mm isohyet, about half of the state. Some 46 percent of the species are in four families: Chenopodiaceae, Compositae, Gramineae, and Leguminosae sensu lato.]

434

Queensland (in general)

Area: 1730148 km^2 (exclusive of Torres Strait Islands; see **439**). Vascular flora: 8696 species, of which 7535 native and 1161 naturalized (R. O. Makinson, personal communication, 1995); 350 of these are pteridophytes. The Queensland Herbarium (1997; see below) listed 9271 (8106 native and 1165 established)

species; this may be compared with the respective figures of 4566 (4259 and 307) recorded by Bailey in 1913. The state thus encompasses some 40 percent of the Australian flora. – For convenience five separate units have been established for the four handbook regions (**435** through **438**) and the Torres Strait Islands (**439**). Other works of importance for the Queensland flora are given under **420–50** (DOMIN) and **430** (FLOYD; FRANCIS; HYLAND and WHIFFIN). For the Great Barrier Reef, see **409**.

The first flora of Queensland was *A synopsis of the Queensland flora* (1883) by Frederick Manson Bailey, produced two years after he became Colonial Botanist and organized the Queensland Herbarium. This work, with concise descriptions but no keys, covered all plants and plant-like organisms. Three supplements appeared between 1886 and 1890, Bailey also bringing out in the latter year a systematic checklist, *Catalogue of the indigenous and naturalised plants of Queensland*. 'Austrobaileya' also produced two books on ferns – the latter of which would remain current for a century – as well as several other scientific and popular works. His final major works were the well-known and long-standard *Queensland flora* (1899–1902, index 1905) and *Comprehensive catalogue of Queensland plants* (1913), neither entirely superseded though both for practical purposes now obsolete.

Through much of the twentieth century, any successor to the *Queensland flora* seemed too great a task given the continuing increase in the known vascular flora and relatively scant resources at the state herbarium. In the 1970s, however, improved prospects led to a decision to initiate work towards a new state flora. Because of the size and varying levels of knowledge of the flora as well as geography and user distribution, the project was, however, divided into four regional handbooks. One of these, *Flora of southeastern Queensland*, appeared between 1983 and 1989 (see **435**). At the present time, the only modern statewide works are an official checklist, *Queensland plants* (1997; see below), and K. A. W. WILLIAMS, 1979–87. *Native plants of Queensland*. 3 vols. North Ipswich, Qld.: The author. The latter is a copiously illustrated color atlas with accompanying text but no keys.[32]

Keys to families and genera (flowering plants)

CLIFFORD, H. T. and G. LUDLOW, 1978. *Keys to the families and genera of Queensland flowering plants*. xiv, 202 pp., 8 pls., 1 text-fig. Brisbane: University of Queensland

Press. [Analytical keys to all Queensland genera and families (both native and naturalized), with family diagnoses; glossary and index. Not easy to use without experience.]

BAILEY, F. M., 1899–1902. *The Queensland flora.* 6 parts. 2015 pp., 88 pls. Brisbane: Government Printer; and *idem*, 1905. *The Queensland flora: general index*. 66 pp. Brisbane.

Descriptive flora of vascular plants; includes keys to genera and species, limited synonymy (with references), vernacular names, generalized indication of local and extralimital range with localities and *exsiccatae* for some species, taxonomic commentary, and notes on special features, wood, uses, etc.; indices to all botanical and vernacular names in each part (the General Index is limited to botanical names). The bulk of this work is essentially a plagiarization of the relevant parts of Bentham's *Flora australiensis*, although subsequently discovered taxa, records and other data were incorporated as far as possible. [Now largely of historical interest.]

BAILEY, F. M. [1913]. *Comprehensive catalogue of Queensland plants, both indigenous and naturalized.* 879 pp., 976 text-figs., 7 halftones, 16 col. pls. Brisbane: Government Printer.

Copiously illustrated systematic list of non-vascular and vascular plants; includes limited synonymy with occasional references, aboriginal and English vernacular names, and short notes relating to uses, poisonous properties, etc.; addenda and indices to vernacular, generic and family names. [The figures, largely by Cyril T. White (later Government Botanist), resulted from a suggestion made by Sir William MacGregor, who at that time was state governor. The work remains listed here for these illustrations.]

QUEENSLAND HERBARIUM (ed. R. J. F. HENDERSON), 1997. *Queensland plants: names and distribution*. 286 pp., 2 maps, col. pl. on cover. Brisbane: Queensland Herbarium, Queensland Department of Environment and Heritage. (1st edn., 1994, as *Queensland vascular plants: names and distribution*.)

Alphabetical checklist of accepted taxa with distribution by state districts (including the number of available records) and more generally without the state (based on regions as depicted in the back end-paper map); no synonymy. Symbols for status are also given, naturalized taxa featuring an asterisk. Unidentified or undescribed species are furnished with *exsiccatae*. The introductory part includes explanations of conventions and abbreviations as well as a list of assessors and a brief statistical summary.

Pteridophytes

Until publication of *Ferns of Queensland* in 1990 the state continued to be served by two Bailey 'classics': *Handbook to the ferns of Queensland* (1874) and *Lithograms of the ferns of Queensland* (1892).

ANDREWS, S. B., 1990. *Ferns of Queensland*. [1], xx, 427 pp., illus. Brisbane: Department of Primary Industries, Queensland.

An illustrated descriptive flora with keys to all taxa, vernacular names, synonymy, references and citations, and indication of habitat and distribution; includes misapplied names. The many illustrations are largely full-page. Appendices 1 and 2 cover respectively new taxa and additions to the original work through 1988 (the latter coordinated by Les Pedley); these are followed by a list of families and genera, a lexicon of genera with their families, and complete index. The introductory part includes a glossary, with illustrations, and key to families, but there is no real survey of the fern flora and no figure for species save that Queensland encompasses over 80 percent of the pteridophyte flora of Australia (from the preface by Bob Johnson). [According to the 1997 census, there are presently 377 native and 9 naturalized pteridophytes in the state.][33]

435

Southeastern Queensland

Area: 184 600 km^2 (Stanley and Ross). Vascular flora: about 3600 species. – Comprises the four southeastern pastoral districts (Moreton, Darling Downs, Burnett and Wide Bay), thus including the Brisbane and Maryborough coastal belt, the Darling Downs and the Bunya Mountains.

The biologically important Fraser Island is off the northeastern corner of Wide Bay district. For its woody flora, see M. PODBERSCEK, 1993. *Field guide to rainforest trees, shrubs, and climbers of Fraser Island, using vegetative characters*. 114 pp. Brisbane: Department of Primary Industries.

STANLEY, T. D. and E. M. ROSS, 1983–89. *Flora of southeastern Queensland*. 3 vols. Illus. (DPI Misc. Publ. 81-020, 84-007, 88-001). Brisbane: Queensland Department of Primary Industries.

Manual-flora of seed plants, with representative illustrations; includes basionyms and other essential synonymy, usually concise diagnostic descriptions, local range (and, if not native, a plant's origin), and notes on habitat, phenology, uses, weediness (if appropriate) and other properties, and (sometimes) pathology and special features; separate indices at end of each volume to botanical and vernacular names. The introductory section (vol. 1) includes the background to the work, its scope, and sections on climate, physical features, soil, vegetation formations, and plant collecting and use of keys, followed by a glossary, abbreviations and symbols, and a general key to dicotyledons (based on that of Clifford and Ludlow). Volume 3 features cumulative indices to all names.

436

Central coastal Queensland

Included here are the Port Curtis, Leichhardt and South Kennedy pastoral districts, representing east central Queensland (centered on Gladstone, Rockhampton and Mackay). In addition to the work listed below, a selection of 240 species, with illustrations and maps, may be found in *Plants of Central Queensland* (1993) by E. Anderson.[34]

Partial work

BATIANOFF, G. N. and H. A. DILLEWAARD, 1988. *Port Curtis flora and early botanists*. (5), 121 pp. [Gladstone, Qld.]: Society for Growing Australian Plants, Gladstone Branch (in association with the Queensland Herbarium, Brisbane).

Pages 64–120 comprise an alphabetically arranged tabular checklist of native and introduced vascular plants of the Port Curtis district, with indication of distribution, frequency, life-forms, and special features (all conventions explained on p. 64). The general part incorporates biographies as well as a description of the area and an overview of the flora.

437

Western Queensland

Area: 776365 km². Vascular flora: 2499 species and 149 additional infraspecific taxa. – Includes the pastoral districts of Maranoa, Warrego, Gregory

South, Gregory North, and Mitchell, covering the 'Channel Country' and all the inland south of 21° S (a line not far south of Cloncurry).

NELDNER, V. J., 1992. *Vascular plants of Western Queensland*. iv, 171 pp., 18 halftones, 23 text-figs. (maps) (Queensland Bot. Bull., [N.S.] 11.). Brisbane: Queensland Herbarium.

Pages 58–138 comprise (as tables 5.1, 5.2 and 5.3, respectively for pteridophytes, gymnosperms and angiosperms) alphabetically arranged tabular checklists of species (by families) with presence or absence indicated for one or more of 18 vegetation types (as described in the first part of the work) along with an 'information qualifier', indication of normal life-form, up to three vernacular names, and illustration references. The general part includes sections on physical features, climate, geology, soils and the use of the work along with a general floristic summary (pp. 7–8) and accounts (with distribution maps) of the general and floristic features of each of the 18 recognized vegetational units; following the checklist are a lexicon of vernacular names with botanical equivalents, a summary of 'recent' name changes, rare and threatened plants, a list of genera with their families, and acknowledgments and a general bibliography.

438

North Queensland

Included here are the North Kennedy, Burke and Cooktown pastoral districts, comprising all the state north of about 21° S. Major features are the northern tropical forests, the Bellenden Ker Range, the Atherton Tableland, Cape York Peninsula and the 'Gulf country'. The most important separate publication is now *Australian tropical rainforest trees* by B. P. M. Hyland and T. Whiffin (1994; see **430**). For *Plant life of the Great Barrier Reef and adjacent shores* by A. B. and J. W. Cribb (1985; see **409**).[35]

439

Torres Strait Islands

While much additional collecting has been done in recent decades in this botanically important island group, no contemporary checklist has yet appeared; that of Bailey (1898) is now quite incomplete and moreover does not cover all islands. The group straddles the traditional boundary between the Oriental and Australian floral kingdoms (as recognized by Ludwig Diels and many others) but is most likely only part of a broad zone of floristic transition involving substantial parts of northern Australia and southern New Guinea. The area as recognized here is wholly included within *Flora Malesiana* (**910–30**).[36]

BAILEY, F. M., 1898. A few words about the flora of the islands of Torres Strait and the mainland about Somerset. *Rep. Meetings Australas. Assoc. Advancem. Sci.* **7**: 423–447.

The main feature is a briefly annotated list of plants from Thursday Island (the main administrative center); this includes English common names and miscellaneous notes on special features, phenology, habitat, behavior, and uses. The descriptive introduction includes desultory observations about the vegetation and flora and potential uses. [Based on collections made by the author in the dry season of 1897 along with earlier records. All observations pertain to islands south of 10° 30′, i.e., south of the Strait proper.]

Region 44

Middle Australia

This region encompasses the state of South Australia and the Northern Territory, now a self-governing federal territory. From 1863 to 1911 South Australia encompassed the whole region. Within the biogeographically diverse Northern Territory, two major areas are designated: 'Central Australia', the southern dry region centering on Alice Springs with less than 300 mm rainfall/annum, and 'The Top End',

the northern infratropical region centering on Darwin (and including Arnhem Land).

The coastal areas in north and south alike were explored botanically at the beginning of the nineteenth century by Robert Brown and (in the south alone) Jean-Baptiste Leschenault de la Tour, but not until after the establishment of South Australia in 1836 and the foundation of Adelaide did interior botanical exploration commence. From his arrival in 1847, substantial contributions were made by Ferdinand Mueller, both through his own collecting and through reports on those made by others, contributions which continued even after his departure for Melbourne in 1852. In the latter part of the century, public demand and the organization of higher education manifested themselves in the first 'state flora', the very concise *Handbook of the flora of extra-tropical South Australia* by Ralph Tate (1890). Tate also published two censuses of the extra-tropical flora (1880 and 1889, the latter supplemented in 1895) and some expedition botanical reports. Tate's flora was in the 1920s succeeded by J. M. Black's classic *Flora of South Australia* (1922–29; 2nd edn., 1943–57 and supplement, 1965; 3rd edn. (not completed), 1978; 4th edn., 1986), published originally as a one-volume handbook but in its present four volumes more now a reference work.[37]

In addition to present-day South Australia, Tate's two checklists also covered the present Northern Territory south of the Tropic of Capricorn. The first separate overall account did not appear until after the 1911 transfer of the Territory from South Australia to the Commonwealth. This was *Flora of the Northern Territory* (Ewart and Davies, 1917), a largely compiled, partially keyed checklist produced at the National Herbarium of Victoria for the Commonwealth Government. Now very incomplete and of little practical value, it was supplemented for Arnhem Land by R. L. Specht and the ethnologist Charles P. Mountford's report on the botanical results of the 1948 American-Australian expedition (1958) and concise checklists by George Chippendale (1972) and Clyde Dunlop (1987). A *Flora of the Darwin region* has also commenced publication.

Extra-tropical 'Central Australia' was in 1959 furnished with its first separate checklist, also by Chippendale. This has been subsumed in the Territory-wide checklists and in *Flora of central Australia*, edited by J. P. Jessop (1981; see **420–50**) which covers all of arid interior Australia.

With the establishment in 1954 of the State Herbarium of South Australia and, later, two herbaria in the Northern Territory (the latter now administratively consolidated), professional collecting has increased and expanded to more areas, but much of the region remains poorly known. Some recent floristic surveys have been published.[38]

Bibliographies. General and supraregional bibliographies as for Superregion 42–45.

Indices. General and supraregional indices as for Superregion 42–45.

441
Northern Territory (in general)

Area: 1 349 129 km^2. Vascular flora: 3555 species, of which 3293 native and 262 naturalized (R. O. Makinson, personal communication). Partial works, respectively for the 'Top End' and Central Australia, appear below under **442** and **443**. *Flora of central Australia* is under **420–50**.

The Northern Territory was from 1863 to 1911 administered by South Australia; since then it has been under the Commonwealth. Early exploration was before 1850 primarily coastal and afterwards in the form of overland expeditions, notably those of Gregory (1856), Forrest (1879), Tietkens (1889) and Horn (1894). A substantial survey, with as its main botanical collector G. F. Hill, was mounted in 1911 shortly after transfer of the administration. His collections formed a principal basis for the first full account, *Flora of the Northern Territory* (1917) by Alfred Ewart and Olive Davies with the aid of J. H. Maiden and others. Though now obsolete it provides some information not available in more recent works. The next significant work was the botanical volume of the American-Australian scientific expedition to Arnhem Land (Specht and Mountford, 1958; see **442**). With establishment of herbaria in the Territory, firstly in Alice Springs and later (1967) in the Darwin area, locally based collecting increased and new checklists were published, the first for the whole polity in 1972 (with a successor in 1987).

Families and genera: keys and distribution maps

CLIFFORD, H. T. and I. D. COWIE, 1992. *Northern Territory flowering plants: a key to families.* [2], 92 pp. (Northern Territory Bot. Bull. 12). Darwin. [Notes on use of keys and conventions and list of families and orders (after Cronquist); key (pp. 8–54) and descriptions of families (pp. 55–83); glossary and index at end. – Key leads are short and checking of a plant against a family description is recommended.]

DUNLOP, C. R. and D. M. J. S. BOWMAN, 1986. *Atlas of the vascular plant genera of the Northern Territory.* 116 pp., numerous maps. Canberra: Australian Government Publishing Service. (ABRS Australian Flora and Fauna Series 6.) [A series of grid-based dot maps (nine to a page) with one for each genus (save in *Eucalyptus* where infrageneric taxa have been independently mapped), representing collections and records to 1984; index to all generic and family names at end with, respectively, statistics for species per genus and genera per family. The map grid is by half-degree squares, with a leading objective being more precise biogeographic regionalization.]

CHIPPENDALE, G. M., 1972. Check list of Northern Territory plants. *Proc. Linn. Soc. New South Wales* **96**: 207–267, 2 tables, map.

Systematic list of vascular plants (both native and naturalized), without synonymy or references; indication of distribution according to four main divisions (based on established pastoral districts); addenda. The introductory section has accounts of floristics and vegetation, statistics of the flora, and a list of references. [Superseded by Dunlop (1987; see following entry) but retained here for its introductory accounts.]

DUNLOP, C. R., 1987. *Checklist of vascular plants of the Northern Territory.* v, 87 pp., map (Conservation Commission of the Northern Territory, Technical Report 26). Darwin.

Alphabetically arranged list, without synonymy, references or notes; includes indication of distribution according to five main divisions (based on those of Chippendale (1972) but with his 'Central Australia' subdivided at 23° S) and an index to genera (by family numbers) at end. The brief introductory section includes contributors, basic references on the Territory, and a map of the divisions. Family circumscriptions are mainly after Cronquist (1981; see **000**).

EWART, A. J. and O. B. DAVIES, 1917. *The flora of the Northern Territory.* viii, 387 pp., 14 tables, 27 pls., map. Melbourne: [Published through the Government Botanist of Victoria on authority of the Commonwealth Minister for Home and Territories; printed by McCarron, Bird].

Systematic enumeration of known vascular plants; includes descriptions of new taxa, synoptic keys to genera and species, limited synonymy, vernacular names, localities with citations of *exsiccatae*, and notes on habitat, ecology, phenology, properties and uses; appendices with detailed treatments of Cyperaceae, *Acacia* and *Eucalyptus* (the last along with other Myrtaceae); special lists of plants arranged by properties and uses; general index to all botanical and vernacular names. – Coverage in this work of Arnhem Land and the southern part of the Territory is very imperfect, and localization of collection records is often vague.[39]

Distribution maps

For Dunlop and Bowman (1986) see above under **Families and genera**.

442

'The Top End'

See also **441** (all works). 'The Top End', the local term for the northern part of the Territory, has been variously delimited but essentially covers all the area with a monsoonal distribution of rainfall. It includes all the monsoon vine forest or 'rain forest' lands.

With no contemporary overall works, the best introduction to date is J. Brock's *Native plants of Northern Australia* (first published as *Top End native plants*), described below. Reference can also be made to G. WIGHTMAN and M. ANDREWS, 1989. *Plants of Northern Territory monsoon vine forests*. Vol. 1. 163 pp., illus. Darwin: Conservation Commission of the Northern Territory. Among other works, that by Specht and Mountford is now relatively out of date but no successor has been published. *Flora of the Darwin region* will represent a valuable addition when complete.

BROCK, J., 1993. *Native plants of Northern Australia*. xii, 355 pp., *c.* 700 col. photographs, 26 line drawings, maps. Chatswood, N.S.W.: Reed. (Originally published 1988, Winnellie, Darwin, N.T.: The author, as *Top End native plants*.)

Semi-popular atlas predominantly of woody plants, each of the 450 selected species being furnished with one or more color photographs, a dot or solid-area distribution map, and a description with notes on phenology, special features, habitat, distribution, uses and cultivation; references, illustrated glossary, and index to all names at end. The general part features notes on the use of the book, vegetation and communities (in part by collaborators), and a reference table listing all included species with their life-form, seasons and known habitats. [Covers the 'Top End' south to about 16° S. Also useful in the Kimberley district (Western Australia) and the Gulf country (Queensland). Arrangement of species is alphabetical by genus.]

Partial works: Darwin region, Arnhem Land
The 'Darwin region' encompasses an area of the mainland inland to Pine Creek as well as Bathurst and Melville Islands. Arnhem Land lies to its east, beginning with Kakadu National Park.

COUSINS, S. N., 1989. *Checklist of vascular plants of the Darwin Region, Northern Territory, Australia*. 54 pp., map (Northern Territory Bot. Bull. 8). Darwin.

Unannotated systematic name list based on holdings in the herbaria at Darwin and Alice Springs; introduced species asterisked.

DUNLOP, C. R., G. J. LEACH and I. D. COWIE, 1995. *Flora of the Darwin Region*. Vol. 2. vi, 261 pp., 80 text-figs., map (Northern Territory Bot. Bull. 20). Darwin.

Descriptive manual-flora with keys, English vernacular names, and notes on season, local and extralimital distribution, habitat and uses but few critical notes; glossary, list of literature and index to all names at end. The brief introduction includes conventions for use of the work. The arrangement of families follows Cronquist. [Covers Leguminosae (as three families) through Euphorbiaceae and including Myrtaceae. The work is planned to encompass four volumes, the first containing an expanded introduction as well as pteridophytes, gymnosperms and dicotyledons from Annonaceae through Surianaceae.]

SPECHT, R. L. and C. P. MOUNTFORD (eds.), 1958. Botany and plant ecology. xv, 522 pp., illus., maps. In American-Australian Scientific Expedition to Arnhem Land, *Records of the American-Australian scientific expedition to Arnhem Land*, vol. 3. Melbourne: Melbourne University Press.

Includes among other contributions separate annotated enumerations of the vascular cryptogams (pp. 171–184) by M. D. Tindale and the seed plants (pp. 185–317) by R. L. Specht and C. P. Mountford, based mainly on the expedition's collections of 1948. Each paper includes descriptions of new taxa, synonymy (with references and citations), citation of some *exsiccatae* and generalized indication of local and extralimital range, limited taxonomic commentary, and notes on habitat, life-form, special features, etc., and representative illustrations. The first article in the volume includes an account of previous botanical exploration in Arnhem Land; others cover climate, geology, floristics, phytogeography, vegetation and ecology, and ethnobotany. All botanical names in the work are included in its general index.[40]

Distribution maps

LIDDLE, D. T., J. RUSSELL-SMITH, J. BROCK, G. J. LEACH and G. T. CONNORS, 1994. *Atlas of the vascular rainforest plants of the Northern Territory.* xxii, 164 pp., illus., numerous maps. Canberra: Australian Biological Resources Study. (Flora of Australia, Suppl. Ser. 3.)

Alphabetically arranged distribution maps (six to a page) for 585 native and 19 naturalized taxa north of about 18° 40′ S, with for each a distinction between survey site data and herbarium records, botanical names, families, and codes for life-form, number of records, endemism or otherwise, reservation status, and conservation status (with explanation of codes and conventions on pp. vii–xvii). Appendices 1 and 2 respectively encompass reference specimen and literature sources and extralimital distribution in tabular fashion; a literature list and index follow. The introduction covers, in addition to codes and conventions, methodology, floristics, and history of the flora. [Based on 58 783 site records and 21 187 herbarium vouchers, the latter mostly dating from after World War II.]

443

'Central Australia'

See also **441** (all works). The illustrated *Flora of central Australia* covers the whole of the arid interior of the continent and has therefore been listed under **420–50**. In Chippendale's 1959 list (expanded in 1972 to cover the whole Northern Territory and succeeded in 1987 by Dunlop's list) the 'Central Australia' district of the Northern Territory is bounded on the north by the 20th parallel or the 12-inch rainfall isohyet (for Dunlop's list it was divided at the 23rd parallel into northern and southern sectors).

CHIPPENDALE, G. M., 1959. Check list of Central Australian plants. *Trans. Roy. Soc. South Australia* 82: 321–358.

Unannotated systematic list of vascular plants reported from the southern part of the Northern Territory (south of the 12-inch isohyet); includes list of

references. A discussion of previous botanical work in the area is also incorporated. [For supplements, see *idem*, 1960–63. Contributions to the flora of Central Australia, 1–3. *Ibid.*, 83: 199–203; 84: 99–103; 86: 7–9. All records were later incorporated into the author's checklist for the whole of the Northern Territory (1972; see **441**).]

445

South Australia

Until 1911 South Australia was also responsible for the administration of the Northern Territory (**441**). – Area: 983 482 km². Vascular flora: 4354 species and non-nominate infraspecific taxa (Jessop, 1989); of these 2748 species are native and 927 naturalized (R. O. Makinson, personal communication). A total of 50 of these are pteridophytes.

Black's *Flora of South Australia*, widely used without as well as within the state, has long been a standard reference in south-central and southeastern Australia. A fourth edition appeared in 1986. Three editions of a state checklist have also been published, the latest in 1989. Current dendrological coverage is in Boomsma (1981).

JESSOP, J. P. and H. R. TOELKEN (eds.), 1986. *Flora of South Australia.* 4th edn. 4 vols. 2248 pp., 1009 text-figs., 64 col. pls., portr., maps (end-papers). Adelaide: South Australian Government Printing Division (for Flora and Fauna Handbooks Committee). (1st edn., 1922–29, by J. M. Black; 2nd edn. with supplement, 1943–65, by J. M. Black, Hj. Eichler *et al.* (partly reprinted); 3rd edn., 1978, by J. P. Jessop (not completed; only vol. 1 published).)

Concise, well-illustrated descriptive manual-flora of native, naturalized and adventive vascular plants; includes in each volume a glossary, keys to all taxa, synonymy (with references and citations, including illustrations), vernacular names, local distribution (by botanical districts, as shown in end-paper maps), generalized indication of extralimital range, taxonomic commentary, and notes on ecology, life-forms, phenology, etc., and complete index (with references and cumulative index to all names in vol. 4). The introductory section (vol. 1) encompasses notes on the plan and use of the work, collecting, abbreviations, a glossary (on pink paper), summary of new names and combinations, and family key (pp. 65–77). No synopsis of families is given, but the sequence followed, as in previous editions, is that of Engler and Gilg (here, however, with monocotyledons in vol. 4).

JESSOP, J. P. (ed.), 1989. *A list of the vascular plants of South Australia.* 3rd edn. 163 pp., map (J. Adelaide Bot. Gard. 12(1)). Adelaide. (1st edn., 1983, 2nd edn., 1984; both independently published.)

Systematically arranged nomenclatural list (as in the *Flora,* its sequence a modification of Englcr and Gilg), with synonymy and indication of distribution by botanical districts (13 recognized; see map in inside back cover); index to genera and families at end. Naturalized taxa are asterisked. The concise introduction lists family reviewers in addition to an explanation of the list.

Partial works: Adelaide area
Two more or less popular guides treat the flora of Adelaide and vicinity. These are *It's blue with five petals – wildflowers of the Adelaide region* (1988) by A. Prescott and *Plants of the Adelaide Plains and Hills* (1990) by G. R. M. Dashorst and J. P. Jessop.

Woody plants
Until 1972 the only tree flora was *The forest flora of South Australia* (1882) by J. E. Brown, a 'prestige' work of generous dimensions.

BOOMSMA, C. D., 1981. *Native trees of South Australia.* 2nd edn. [8], 288 pp., illus., maps (some col.) (South Australian Woods and Forests Dept. Bull. 19). [Adelaide.] (1st edn., 1972.)

Dendrological atlas of 126 tree and 'mallee' species; includes for each illustrations and distribution maps along with descriptive text covering trade names, botanical features, distribution, habitat, overall form, timber, properties and uses, etc.; glossary and complete index at end. In the introductory section are general remarks on botany, distribution, phenology, timber properties, pollen and honey production along with an artificial general key and a list of references.

Region 45

Western Australia

This region corresponds to the state of Western Australia (along with the Ashmore and Cartier Islands, a Commonwealth territory) in the Timor Sea. – Area: 2 529 878 km² (of which the Ashmore and Cartier Islands account for 3 km²). Vascular flora: 8316 species of which 7463 native and 853 naturalized (R. O. Makinson, personal communication). A total of 65 of these are pteridophytes.

In the history of Australian floristic botany, Western Australia was something of a 'late-bloomer', with two significant cycles of internal growth related to general economic movements in the state. The desolate appearance of much of the coast, as reported by William Dampier and other explorers, deterred early settlement; not until 1829 was the Swan River colony founded. The earliest botanical collections were made by Dampier, Archibald Menzies, Jacques J. H. de Labilliardière, J. B. L. T. Leschenault de la Tour, and (particularly) Robert Brown; the last explored coasts in the tropical north as well as in the south and west. The collections of Brown caused great general interest and in the remainder of the period to 1850 several collectors visited the Swan River and other parts of the southwest, and their results were written up by, among others, John Lindley (especially his *A sketch of the vegetation of the Swan River Colony,* 1839–40), Carl Meisner, Johann G. C. Lehmann (*Plantae preissianae,* 1844–48), Nikolai Turczaninow and George Bentham. Among the best of these collectors were James Drummond, who first went out in 1829 as agriculturalist and curator of a botanical garden on the Swan River, and Ludwig Preiss (in 1838–42).

The initial hopes raised by the Swan River colony were not realized and settlement remained limited until the 1890s. The remainder of the century was characterized by the criss-crossings of several exploring expeditions, the plants of which were largely written up by Ferdinand von Mueller (who participated in one of them). Mueller also furnished botanical and related advice to the colonial authorities. Late in the period the tropical Kimberley district began more seriously to be explored with, in 1885, the discovery of gold; this was in subsequent years followed by like finds elsewhere, most spectacularly at Kalgoorlie in 1892–93. Development and settlement were thereafter rapid and in the 25 years before World War I serious resident botanical activity experienced a renewal. The transition may suitably be marked by the first colonial checklist (covering extra-tropical areas), prepared by Mueller in 1895 and published in the 1896 edition of the colonial yearbook. There were also, as in the early years of the colony, some notable visits by overseas botanists, particularly Ludwig Diels with Ernst Pritzel (1900–01), Eric Mjöberg (1910–13), Carl H. Ostenfield (1914), and Karel Domin. That by Diels and Pritzel in the southwestern region resulted in the period's most notable floristic contribution: *Fragmenta phytographiae Australiae occidentalis* (1904–05), a work basic to all later floristic and phytogeographic studies in that area.

More local contributions appeared after 1900, one of the first being a revision of Mueller's checklist for the extra-tropical regions by the then-Government Botanist, A. Morrison, for the *Western Australian Year Book for 1900–01* (1903).[41] Successors to this list, expanded to cover the state, are *Enumeratio plantarum Australiae occidentalis* by C. A. Gardner (1931) and two of more recent date, described below: the *Descriptive catalogue* of Beard (1970; incomplete and in part uncritical), and the *Census* of Green (1981; 2nd edn., 1985, with supplements to 1988); the latter is the current 'official' work (at least in print).

No state flora, however, exists complete (save for the virtual FloraBase, described below); Gardner's *Flora of Western Australia* (1952) did not progress beyond the Gramineae although materials for other families exist. Until 1928 no state herbarium existed (apart from a small number of collections in the Western Australian Museum) and only in recent decades, consequent to extensive professional and other collecting throughout the state since the 1950s, has sufficient material accumulated to make feasible the preparation of a modern state flora. As in Queensland, however, subdivision of the state was considered necessary; present plans 'contemplate' eventual compilation of five regional handbooks, respectively covering the greater Perth area, the southwest, the south coast, the Eremaea, and the Kimberley district. The first and the last of these appeared respectively in 1987 and 1992.

The first three of these working units together more or less comprise the temperate extra-Eremaean zone, until recently the only area with a regional hand-book, *How to know Western Australian wildflowers* (Blackall and Grieve, 1954–75; 2nd edn., 1981–98), aimed at non-specialists and students and covering some 5800 species. The only one of its kind (thankfully!) in Australasia, it consists mainly of analytical pictured-keys. In 1987 it was joined by the two-volume official *Flora of the Perth region* (described below). The Kimberley district is now covered by *Flora of the Kimberley region* (also described below) as well as a checklist. Other parts of the state still have only scattered local documentation. Much new material, precursory to the proposed state floras, has been published in *Nuytsia* and elsewhere.

Nevertheless, with an area nearly four times the size of Texas – one-third of a continent – much is yet to be learned about the state's flora. It is thus not surprising, perhaps, that in the 1980s serious interest emerged in developing a floristic database. With the advent of the World Wide Web and intranet technology, various formerly disparate elements including the *Census* and the *Descriptive catalogue* as well as interactive keys, illustrations and maps have been combined and edited into a system known as 'FloraBase'. One of the first 'virtual' floras in the world, it was formally 'published' in late 1998. Enhancement of the system continues.[42]

Bibliographies. General and supraregional bibliographies as for Superregion 42–45.

Indices. General and supraregional indices as for Superregion 42–45.

450

Region in general

Apart from Bentham's *Flora australiensis*, still the only work with nominally statewide coverage, there is one current checklist which attempts to account for all known Western Australian vascular plants. Gardner's detailed *Flora of Western Australia* has never gone beyond the Gramineae. With much new information coming in from all parts of the state, including some interesting discoveries, and a large amount of basic taxonomic work to be done, a modern state flora is some way off and, as in Queensland, a more regional orientation in floristic documentation has been adopted. Two substantial regional works have now been published, respectively for the Kimberley district (**452**) and the Perth region (under **455**).

BEARD, J. S., 1970. *A descriptive catalogue of West Australian plants.* xiv, 142 pp., 16 col. pls., map. Perth: King's Park Board/Society for Growing Australian Plants. (1st edn., 1965.)

Systematic list of seed plants, with very brief indication of distribution and terse notes on habitat, life-form, flower color, etc.; index to generic names. The brief introductory section incorporates remarks on vegetation regions, especially in the southwestern part of the state. [The work is retained here for its biological and ecological information. It omits at least the Cyperaceae and Gramineae, and is considered not critical.][43]

GARDNER, C. A., 1952. *Flora of Western Australia.* Vol. 1(1). xii, 400 pp., 104 pls. (1 col.), map. Perth: Government Printer.

Copiously illustrated descriptive flora (covering only Gramineae); includes keys to genera and species, synonymy (with references), vernacular names, generalized indication of local distribution, and notes on habitat, uses, special features, etc.; index to botanical names. [Not continued.]

GREEN, J. W., 1985. *Census of the vascular plants of Western Australia.* 2nd edn. 312 pp. South Perth: Department of Agriculture/Western Australian Herbarium. (1st edn., 1981.) Also *idem*, 1988. *Census of the vascular plants of Western Australia, Cumulative Supplement* 7. 40 pp. South Perth.

The main part of the work comprises two single-line nomenclatural checklists of identical content: one arranged systematically by families (mainly following the Engler and Gilg sequence of 1912, with monocot families delimited after Dahlgren and Clifford, 1982) and one alphabetically by genus and species. Accepted taxa and synonyms are intermingled, with places of publication given for the former, and asterisks indicate naturalized species. An index to genera and families appears at the end. The introductory section includes statistical tables and abbreviations of publications. The supplement was published on 30 November 1988.[44]

451

Ashmore Reef and Cartier Island

Area: 3 km². Vascular flora: 27 species, of which 23 native and 4 naturalized (Kenneally in Australian Biological Resources Study, 1993; see below). – These reefs and low islands in the Timor Sea, initially British but in 1931 ceded to Australia, officially constitute an Australian External Territory. Ashmore Reef (an Australian national nature reserve) features three small, low, partly vegetated islands and several sandbanks supporting sea-grass meadows; while Cartier is an unvegetated sand cay in a platform reef with the sea-grass *Thalassia hemprichii* the only vascular plant. Also within the territory are Scott and Seringapatam Reefs southwest of Ashmore and Hibernia Reef to the north (south of Roti I. near Timor; see **919**).

AUSTRALIAN BIOLOGICAL RESOURCES STUDY, 1993. *Flora of Australia*, 50: *Oceanic Islands* 2. xxvi, 606 pp., 97 figs. (part col. and incl. maps). Canberra: Australian Government Publishing Service.

Pages 43–47 comprise an introduction to Ashmore Reef and Cartier Island (by Kevin F. Kenneally), encompassing background information, vegetation formations and characteristic species on the four islands of the group, together with key references, a key to families and a concisely annotated list of species with indication of their status. All species are further treated in the descriptive part – which follows the format of other *Flora* volumes – together with those from other islands or groups (Cocos (Keeling), **045**; Christmas Island, **046**; Macquarie, **081**; the McDonald (Heard) group, **082**; and Coral Sea Islands, **949**).

PIKE, G. D. and G. J. LEACH, 1997. *Handbook of the vascular plants of Ashmore and Cartier Islands.* iv, 156 pp., illus., maps. Darwin: Parks and Wildlife Commission of the Northern Territory/Parks Australia.

Popular illustrated descriptive treatment with for each species general and botanical descriptions, distribution and ecology, notes on other features and uses, and English, Indonesian, Rotinese and Bajauan vernacular names. Flotsam is covered on pp. 109–140, and a glossary, list of references, and index to all names and subjects follow. The introductory part (pp. 1–10) covers physical features, climate, soils, vegetation, fauna, human activities and conservation.

452

Northeastern zone (Kimberley district)

Area: *c.* 300 000 km². Vascular flora: 2085 native and introduced species, of which *c.* 290 are endemic or poorly known (Wheeler *et al.*, 1992; see below). The unit here corresponds to the Northern Botanical District of Western Australia as depicted on the maps in *Flora of the Kimberley region*.

The works described below represent the main organized basis for botanical knowledge of this partly highland region, which lies within the tropical monsoon zone of northern Australia. In the fiordlike coastal belt, many important collections were made by William Dampier, Robert Brown, Allan Cunningham and other botanists on early exploring expeditions. Inland, key contributions were those of the Grey expedition (1837–38), the Forrest expedition (1879), W. V. Fitzgerald (1905–06) and Charles Gardner (1921). The last two undertakings respectively yielded *The Botany of the Kimberleys, North-West Australia*, by W. V. Fitzgerald (1918, in *J. & Proc. Roy. Soc. Western Australia* 3: 102–224), and *Botanical Notes, Kimberley Division of Western Australia* by C. A. Gardner (1923, Perth, as *Bull. Forests Dept. Western Australia* 32), long the standard references. Since the 1960s, however, much new collecting has been accomplished in the area and a better basis for a new floristic account has become available (K. F. Kenneally, personal communication). In 1988 a large-scale Kimberleys expedition was mounted as part of the bicentennial celebrations of the Linnean Society of London.

KENNEALLY, K. F., 1989. *Checklist of vascular plants of the Kimberley, Western Australia*. 108 pp., illus., map. Nedlands, W.A.: Western Australian Naturalists' Club (as their Handbook 14).

Includes systematic and alphabetical lists, with a list of literature and a family and generic index.

WHEELER, J. R. (ed.) with B. L. RYE, B. L. KOCH and A. J. G. WILSON, 1992. *Flora of the Kimberley region*. xxii, 1327 pp., 354 text-figs., map (end-papers). Como, W.A.: Department of Conservation and Land Management (with the Western Australian Herbarium).

Manual-flora of vascular plants with keys, vernacular names, descriptions, notes on habitat and dis-

tribution, seasons, and here and there critical remarks; an unillustrated glossary and complete index are at the end. Generic entries include principal references. The relatively brief introductory part covers limits of the work and geography, botanical history, climate, botanical districts, vegetation, and floristics (with statistics) along with (pp. xvi–xviii) the plan of the work and (pp. 1–23) the general key to families.

Partial work

PETERHAM, R. J. and B. KOK, 1983. *Plants of the Kimberley region of Western Australia*. xii, 556 pp., illus. (col.). Nedlands, W.A.: University of Western Australia Press.

Atlas-flora of 242 selected species, organized in three main sections (I: grasses and other herbs; II: woody shrubs and shrubby trees; III: trees); the illustrations are accompanied by organized text with vernacular names, botanical details, distribution, ecology, properties and uses, and (where appropriate) management information related to grazing; photographic techniques (appendix II), references (appendix III), glossary and index at end. The introductory section covers the physical and biological background, settlement history, and present land use, along with notes on the use of the work. [Focuses on the Kunnunura area in the Ord River basin, East Kimberleys.]

453

The Eremaea

This vast, dry and partly desertic region, which includes such areas as the Kalgoorlie goldfields, the Pilbara and the Western Australian part of the Nullarbor Plain, is not covered by any complete flora. Some assistance can be had from the works on the southwestern zone, whose coverage may extend more or less into the southerly parts of the Eremaea, and to *Flora of central Australia* (see **420–50**). The work described below offers only a selection of the more significant species.

Partial work

MITCHELL, A. A. and D. G. WILCOX, 1994. *Arid shrubland plants of Western Australia*. 2nd edn. x, 478 pp., col. illus. (incl. maps). Nedlands, W.A.: University of Western Australian Press (with Department of Agriculture, Western Australia). (1st edn., 1988.)

Atlas-flora of some 200 selected species, organized in four main sections (I: annual herbs, short-lived perennials, ferns and vines; II: annual and perennial grasses; III: shrubs less than 3 m high; IV, trees and shrubs more than 3 m high); the photographs and maps are accompanied by organized text with English vernacular names, botanical details, distribution, ecology, forage value, indicator value, and (sometimes) special features. The introductory part has sections on the use of the work as well as on rangeland management and site rehabilitation, while at the end are an etymology of generic names and indices to botanical and vernacular names. [Covers an area of some 600 000 km².]

455

Southwestern zone

Home to one of the more peculiar floras of the world, rich in endemics and moreover very floriferous, southwestern Western Australia is presently covered by two substantial floristic works: *How to know Western Australian wildflowers* (Blackall and Grieve, 1981–98), covering some 5800 species, and *Flora of the Perth region* (Marchant *et al.*, 1987; see below). To these may be added the still-important *Fragmenta phytographiae Australiae occidentalis* of Diels and Pritzel (1904–05).

BLACKALL, W. E. and B. J. GRIEVE, 1981–98. *How to know Western Australian wildflowers.* 2nd edn. restructured and revised by B. J. Grieve. 5 vols. (parts 1, 2, 3A, 3B, 4). Illus., col. pls. Perth (Nedlands): University of Western Australia Press. (1st edn. in 4 parts, 1954–75; parts 1–3 reprinted 1974 in 1 vol. with omission of col. pls.; parts 1–2 alone later similarly reprinted.)

Subtitled *A key to the flora of the extratropical regions of Western Australia*, this individualistic work is more than a mere 'wild-flower book': it comprises a rather detailed, copiously illustrated analytical key to the vascular plants of temperate Western Australia south of the 26th parallel, accompanied by vernacular names and concise indication of distribution. Glossaries, lists of references, user aids, and complete indices are also included in each part. The color plates depict, inevitably, only a small range of the vast variety of the region's wild flowers. With some variations, the arrangement of families broadly follows the Englerian sequence. [Part 4 of the original edition and all parts of the second edition were published in a larger, more elegant format than earlier parts. The reprint of parts 1–3 (later only parts 1–2) is compact and adaptable for field use; it incorporates merely the main illustrated

key, a glossary, and botanical indices. The revised edition encompasses some 5800 species, with 433 *Acacia* and 77 *Drosera* – of sundews more than the rest of the world combined.][45]

DIELS, L. and E. PRITZEL, 1904–05. Fragmenta phytographiae Australiae occidentalis. *Bot. Jahrb. Syst.* **35**: 55–662, 70 text-figs., map.

Systematic enumeration of vascular plants, based primarily on collections made by the two authors in 1900–01; includes descriptions of new taxa, some keys to species, synonymy (with references and citations), relatively detailed indication of local range with listing of *exsiccatae*, and extensive taxonomic commentary; separate index to all botanical names. Most commentary is in German, but the botanical text is in Latin.[46]

Partial works: Perth region

The area encompassed by *Flora of the Perth region* (about 10 500 km²) extends from Perth to Yanchep, Armadale, Pinjarra and Bunbury and includes the coastal plain and western scarp. For *Plants of the Perth coast and islands* (1995) by E. Rippey and B. Rowland, see **409**.

MARCHANT, N. G. *et al.* (eds. N. G. MARCHANT and J. R. WHEELER), 1987. *Flora of the Perth region.* 2 vols. 1080 pp. 317 text-figs., maps (end-papers). Perth: Western Australian Herbarium.

Manual-flora of vascular plants (2057 species); includes references to revisions and monographs in family and generic headings, keys to all taxa, descriptions, limited synonymy, vernacular names, distribution, occasional taxonomic commentary, selective illustrations, and notes on habitat, occurrence, phenology, etc.; glossary (on blue paper) and index to all names at end of work.

Superregion 46–49

Islands of the southwestern Indian Ocean (Malagassia)

Within this superregion are incorporated Madagascar (together with the Comoros and the low islands west of the Red Island); the Mascarene high islands; the Seychelles group (sensu stricto); and Aldabra with all other low coralline and sandy islands and atolls to the east and southeast towards the Mascarenes and north to the Seychelles. Floristically the strongest links are with Africa but there are many groups absent, or not well developed, on that continent, some with connections with Greater Malesia/Oceania and northern (and eastern) South America, and several others peculiar to the superregion.

Although the Dutch were early entrants into this botanically curious superregion, progress in floristic botany has been largely dominated by French workers, particularly in the eighteenth and twentieth centuries. The first important 'modern' collector, who worked extensively in Mauritius but also visited Madagascar, was Philibert Commerson, active from 1768 to 1773. In the nineteenth century, much collecting was done both by visiting workers and by residents, notably Wenceslaus Bojer (in the Mascarenes) and the missionary Richard Baron (35 years in Madagascar, and responsible also for the first general enumeration of its flora, *Compendium des plantes malgaches*, published 1901–07 in Tananarive and accounting for some 4100 species). The Mascarenes and the Seychelles had already been explored in some detail by 1877, the year of publication of the standard 'Kew Series' colonial flora by J. G. Baker, but serious professional field work in Madagascar developed only relatively late, following assumption in 1895 of full French control. Over the next 40 years, the many French expeditions ('missions'), particularly those of Perrier de la Bâthie and Humbert, together with the work of Baron and the botanical contributions in Grandidier's lavish *Histoire physique, naturelle et politique de Madagascar* laid the foundation for the large-scale serial flora begun in 1936. Since World War II, considerable additional field work has been done throughout the superregion by both resident and visiting workers, and two large flora projects are currently in progress. The smaller islands have not been neglected; Aldabra and adjacent islands were the subject of intensive field work carried out under the auspices of the Royal Society of London from the end of the 1960s through the 1970s and most, if not all, islands and atolls were visited by the International Indian Ocean and other oceanological expeditions from the early 1960s. Overall, the flora is considered 'moderately' well known, but, like Hawaii, much of it is very fragile and the severe pressures to which it has, and is being, subjected are causes for grave concern.

Progress

Most reports, as well as historical accounts, are incorporated with those for the African mainland (**Division 5**).

For Madagascar and the Comoros, the accounts of Humbert, Keraudren-Aymonin, and Schatz *et al.* may be consulted (see **Region 46, Progress**). Humbert's review contains a select bibliography covering the whole superregion. For the Mascarenes and Seychelles, the introductions to Baker's flora and to subsequent accounts should be consulted. Progress since 1960 has from time to time been reviewed in reports published in the proceedings of the AETFAT congresses as well as in some overall Afro-Malagassian surveys, notably J. LÉONARD, 1965. Statistiques des progrès accomplis en 10 ans dans la connaissance de la flore phanérogamique africaine et malgache (1953–1962). *Webbia* 19: 869–875, map, and P. MORAT and P. P. LOWRY II, 1997. Floristic richness in the Africa-Madagascar region: a brief history and perspective. *Adansonia*, III, **19**: 101–115. Léonard's map was revised by F. N. Hepper and presented at the Ninth AETFAT Congress (Las Palmas, 1978), with publication in its proceedings: *Taxonomic aspects of African economic botany* (ed. G. Kunkel; 1979, Las Palmas) and more recently in the tropical African checklist of Lebrun and Stork (see **501**). For another summary of current knowledge, see S. A. RENVOIZE, 1979. The origins of Indian Ocean island floras. In D. Bramwell (ed.), *Plants and islands*, pp. 107–129. London: Academic Press.

General bibliographies. Bay, 1910; Blake and Atwood, 1942; Frodin, 1964, 1984; Goodale, 1879; Holden and Wycoff, 1911–14; Jackson, 1881; Pritzel, 1871–77; Rehder, 1911; Sachet and Fosberg, 1955, 1971; USDA, 1958. For bibliographies on particular regions, see appropriate headings.

Supraregional bibliography

LÉONARD, J., 1965. Liste des flores africaines et malgaches récentes. *Webbia* 19: 865–867. [List of principal floristic works, arranged by countries. Nearly all are described or referenced in this *Guide*.]

General indices. BA, 1926– ; BotA, 1918–26; BC, 1879–1944; BS, 1940– ; CSP, 1800–1900; EB, 1959–98; FB, 1931– ; IBBT, 1963–69; ICSL, 1901–14; JBJ, 1873–1939; KR, 1971– ; NN, 1879–1943; RŽ, 1954– .

Supraregional indices

ASSOCIATION POUR L'ÉTUDE TAXONOMIQUE DE LA FLORE D'AFRIQUE TROPICALE, 1952–86. *AETFAT Index*, 1– . Brussels. [Extensive coverage of

African and southwest Indian Ocean islands regional and national floristic literature; includes lists of papers and books, with country index. Published annually. In addition, the periodic symposium proceedings of AETFAT, which include country progress reports with lists of references, should be consulted.]

Region 46

Madagascar and associated islands

Area (Madagascar): 594 180 km². Vascular flora (Madagascar): 10 000–12 000 species (PD; see **General references**). In addition to the large island of Madagascar (Malagasy), this region also includes the Comoros and the small low islands of Juan de Nova, Bassas da India, Europa, and the Gloriosos.

Botanical exploration of Madagascar began in the early seventeenth century with the visit of Flacourt, but it was not until the early nineteenth century that significant collections began to be made in the interior. The spread of the Imerina kingdom over most of the island as the century progressed facilitated primary exploration, with the largest single contribution being by the resident British Congregational (London Missionary Society) missionary Richard Baron. His collections were written up by J. G. Baker and others in the periodical literature, mainly *Journal of the Linnean Society, Botany*; and from 1881 he built up what became his *Compendium des plantes malgaches* (definitive version in 23 installments, 1901–06, in *Revue de Madagascar*, 3(10)–8(10)).[47] At nearly the same time, but without access to Baron's work, there appeared *Catalogus plantarum madagascariensium* (1905–07, in five fascicles) and *Filices madagascarienses* (1906), both by the Czech politician and natural scientist Jan Palacký. French interests were represented by the botanical volumes (1886–1903, by Henri Baillon and Emmanuel Drake del Castillo) of Alfred Grandidier's *Histoire physique, naturelle et politique de Madagascar*, a multi-volume privately sponsored encyclopedic treatise comparable to *Biologia centrali-americana* but of even broader scope. Unfortunately, only one volume of the text, a descriptive flora (not completed), was ever

published (in 1902); the remaining five volumes are wholly of plates. The pioneer era may be said to have closed with the appearance of the three foregoing compilations along with the deaths of Drake (1904) and Baron (1907).

The imposition and consolidation of French rule following the Battle of Antananarivo in 1895 brought renewed exploration by both visitors and residents. From early in the century until World War II major contributors included the explorer and naturalist Henri Perrier de la Bâthie and, from Paris, Henri Humbert whose initial contribution, *La destruction d'une flore insulaire par le feu* (1927), first publicized the fragility of the native flora (and later would play a large part in the establishment of the present system of natural reserves). In 1930 Perrier organized production of a successor to Baron's *Compendium*, the *Catalogue des plantes de Madagascar* (Academie Malgache, 1930–40, in 29 fascicles). Owing to World War II, however, it was not completed; afterwards, all efforts were directed towards Humbert's 'grande flore', *Flore de Madagascar et des Comores*, commenced in 1936.

Humbert had become a professor in the Muséum d'Histoire Naturelle in Paris and head of its Laboratoire de Phanérogamie in 1933. In that capacity, he was able to put into place his new Malagasy flora (alongside *Flore générale de l'Indochine*, a major project of his predecessor Henri Lecomte). Until relatively recent years, this work has dominated progress in floristic botany in Madagascar. By the mid-1970s it was far advanced with three-quarters of the 189 planned fascicles having been published; Perrier himself had, before his death in 1958, contributed 71. Publication, however, then sharply slowed and, with additional discoveries in the flora, not only was the estimate of coverage by 1996 fixed at just 60 percent but parts published before 1960 were considered to be out of date, sometimes badly so.[48]

There are also many local accounts, based on field trips in various parts of Madagascar; these may be located in the cited bibliographies. Precursory revisions and other contributions have been published in *Adansonia* (for a time in the 1980s and 1990s known as *Bulletin du Muséum National d'Histoire Naturelle (Paris)*, IV, B (*Adansonia*)) and in other outlets. An introduction to the forest flora (Capuron, 1957) and a school-work (Cabanis *et al.*, 1969–70) are also available, the latter also designed for use in the Mascarenes. From the 1980s, contributions by Malagasy as well as by other non-French botanists have increased substantially, with a more diversified range of publications including, for example, *Palms of Madagascar* (1995) by J. Dransfield and H. Beentje.[49]

Separate florulas are also available for the Gloriosos, Europa, Juan de Nova, and the Comoros, although for the last-named the only available work, *Flora und Fauna der Comoren* (Voeltzkow, 1917) is long out of date.

The region as a whole is considered moderately, if variably, well explored but somewhat more unevenly documented. As already noted, many parts of Humbert's flora are now quite dated, and certain major families remain outstanding. Indeed, for some of the latter even basic research is all but wanting. A progressive though expedient recent development – following in particular the precedent set by Luis Carrisso for Angola (**522**) – is the creation of an electronic *Conspectus*, with access available through the World Wide Web; integrated with this would be conservation data, imagery and a geographical information system (GIS). In addition, preparation of a revision and expansion of Capuron's 1957 *Introduction à l'étude de la flore forestière de Madagascar* into a field-key to all genera and families is underway.

Progress
Earlier history has been reviewed in H. HUMBERT, 1962. Histoire de l'exploration botanique à Madagascar. In A. Fernandes (ed.), *Comptes rendus de la IV. réunion plénière de l'AETFAT* (Lisbonne et Coimbre, 1960), pp. 127–144. Lisbon. For progress to 1974, see M. KERAUDREN-AYMONIN, 1976. Progrès accompli dans la connaissance de la *Flore de Madagascar et des Comores*. *Boissiera* **24**: 635–639. A well-documented review through the mid-1980s is L. J. DORR, L. C. BARNETT and A. RAKOTOZAFY, 1989. Madagascar. In D. G. Campbell and H. D. Hammond (eds.), *Floristic inventory of tropical countries* (see **General references**), pp. 236–250. Current international work, with particular reference to the *Conspectus* database, is described in G. E. SCHATZ *et al.*, 1996. Conspectus of the vascular plants of Madagascar: a taxonomic and conservation electronic database. In L. J. G. van der Maesen *et al.* (eds.), *The biodiversity of African plants*, pp. 10–17. Dordrecht: Kluwer.[50]

Bibliographies. General and supraregional bibliographies as for Superregion 46–49.

Regional bibliography
GRANDIDIER, G. *et al.*, 1905–57. *Bibliographie de Madagascar*. 3 vols. (in 4). Paris (vols. 1–2); Tananarive (vol. 3). [Includes 23 003 references. A further cumulation for 1956–63 and subsequent annual continuations were published in Antananarivo.]

TAYLOR, M. S. (comp.), 1991. *Flora and ethnobotany of Madagascar: contributions towards a taxonomic bibliography*. iv, 377 pp. St. Louis: Missouri Botanical Garden. [This classified work, by the author's admission a 'very rough draft', includes references through 1989.]

Indices. General and supraregional indices as for Superregion 46–49.

460

Region in general

All general works on Madagascar, by far the major land mass in the region, are placed under this heading. Separate works on the Comoros and the low islands appear under **465** to **469** inclusive.

As a 'tome préliminaire' for the *Flore de Madagascar et des Comores* is still wanting, a useful general introduction to the Malagasy flora is J. KOECHLIN, J.-L. GUILLAUMET and P. MORAT, 1974. *Flore et végétation de Madagascar*. xvi, 688 pp., illus., maps. Lehre, Germany: Cramer. (Flora et vegetatio mundi 5; reissued 1997.) Attention is also called to the following school-work: Y. CABANIS, L. CABANIS and F. CHABOUIS, 1969–70. *Végétaux et groupements végétaux de Madagascar et des Mascareignes*. 4 vols. 260 pls. Tananarive. Key precursory works relating to the *Flore* are ACADÉMIE MALGACHE, 1931–35. *Catalogue des plantes de Madagascar*. 23 fascicles. Tananarive; and C. CHRISTENSEN, 1932. *The Pteridophyta of Madagascar*. xv, 253 pp., 80 pls. (Dansk Bot. Ark. 7). Copenhagen. Both these latter works, the first largely written by Perrier de la Bâthie, contain enumerations of families or of larger taxa, with synonymy, references, indication of *exsiccatae* with localities, general summaries of internal and extralimital range, and miscellaneous notes; that by Christensen also contains partial keys and illustrations, as well as a floristic and phytogeographical analysis.

Keys to genera and families (woody plants)

CAPURON, R., 1957. *Introduction à l'étude de la flore forestière de Madagascar*. iv, 125 pp., illus. Tananarive: Inspection Générale des Eaux et Forêts. (Mimeographed.) [Analytical keys to families and genera of arborescent plants with family descriptions, accompanied by special commentaries (pp. 97–104), glossary and index.][51]

HUMBERT, H. *et al.* (eds.), 1936– . *Flore de Madagascar et des Comores*. In fascicles. Illus. Tananarive: Gouvernement Général de Madagascar; Paris: Laboratoire de Phanérogamie, Muséum National d'Histoire Naturelle.

Comprehensive illustrated 'research' flora of vascular plants, with keys to genera and species, lengthy descriptions, full synonymy (with references), detailed indication of local range (with citation of *exsiccatae*) and general summary of extralimital range (if applicable!), taxonomic commentary and notes on habitat, special features, etc.; index to botanical names in each fascicle. No introductory section has yet been published. [Families have been pre-numbered following the Englerian system but fascicles have been appearing only as family accounts were ready. Some fascicles contain more than one family. The work was some two-thirds or so complete as of 1998, with the latest fascicle, on Labiatae, appearing in that year; many earlier parts are, however, more or less outdated – some markedly so – by virtue of the results of continuing botanical exploration in the more than 60 years since commencement of the work along with concomitant taxonomic studies and revisions.][52]

465

Comoro Islands

Area: 2238 km^2. Vascular flora: 416 native species (out of 935) were recorded by Voeltzkow (1917) but this is certainly too low; more recently a figure of 1500 has been proposed (Lowry, Pascal and Labat, *Adansonia*, III, **21**: 67–73 (1999), especially p. 68). – Grande-Comore, Moheli and Anjouan now form the Republic of the Comoros, while the most easterly island, Mayotte, is French territory.

Comoro plants are treated in *Flore de Madagascar et des Comores* (**460**) as far as published but

the only separate survey remains that of Voeltzkow. Subsequent to publication of that work relatively few new collections were made; in the 1990s, however, renewed floristic surveys, at least on Mayotte, have shown that Voeltzkow's account is now quite out of date, with several new species as well as new records now to hand.

VOELTZKOW, A., 1917. Flora und Fauna der Comoren. In *idem* (ed.), *Reise in Ostafrika in den Jahren 1903–1905 mit Mitteln der Hermann und Elise geb. Heckmann-Wentzel-Stiftung ausgeführt. Wissenschaftliche Ergebnisse*, 3: 429–480. Stuttgart: Schweizerbart.

Pages 430–454 of this work, prepared with the assistance of G. Schellenberg, contain a checklist of non-vascular and vascular plants, with localities, citations of *exsiccatae*, and indication of extralimital distribution; this is preceded by an account of prior exploration of the islands and some general remarks on the flora as a whole. [Now considered to be relatively incomplete.]

466

Glorioso Islands

See also **490** (HEMSLEY, p. 145). – Vascular flora: 48 species (Battistini and Cremers, 1972).

BATTISTINI, R. and G. CREMERS, 1972. Geomorphology and vegetation of Îles Glorieuses. 10 pp., illus., maps (Atoll Res. Bull. 159). Washington.

In this general survey are included lists of plants collected in 1971 from Grand Glorioso, Lily Island (Île du Lys) and other islets, with new records and changes since the appearance of Hemsley's 1919 list. [Fifteen records given by Hemsley were not reconfirmed in 1971.]

467

Juan de Nova Island

The island is small and low, with mostly widespread species.

BOSSER, J., 1952. Notes sur la végétation des îles Europa et Juan de Nova. See **469**.

468

Bassas da India Island

The island is small and low, with mostly widespread species. No separate floristic lists appear to be available (but see **469** below).

469

Europa Island

The island is small and low, with for the most part only widespread species.

BOSSER, J., 1952. Notes sur la végétation des îles Europa et Juan de Nova. *Naturaliste Malg.* **4**: 41–42, illus.

Includes short lists of plants recorded from Europa and Juan de Nova as well as brief remarks on their vegetation. See also R. CAPURON, 1966. Rapport succinct sur la végétation et la flore de l'île Europa. *Mém. Mus. Natl. Hist. Nat.*, sér. II/A (Zool.) **41**: 19–21.

Region 47

The Mascarenes

The Mascarenes are here considered to include the islands of Réunion, Mauritius and Rodrigues together with their associated islets. For the several small low islands to the north, certain of which are dependencies of Réunion and Mauritius, see Region 49.

The standard floras of the group have long been *Flora of Mauritius and the Seychelles* (Baker, 1877) and *Flore de l'Île de Réunion* (Jacob de Cordemoy, 1895), both very out of date though recently reprinted. A new cooperative work to replace them, *Flore des Mascareignes* (Bosser *et al.*, 1976–), is in progress. Account has had to be taken of additional introduced plants (a task accomplished for Mauritius through 1937

by R. E. Vaughan), as well as of the current status of the fragile native floras (on which research is continuing).

Bibliographies. General and supraregional bibliographies as for Superregion 46–49. See also Region 48 (Peters and Lionnet, 1973).

Regional bibliography

LORENCE, D. H. and R. E. VAUGHAN, 1992. *Annotated bibliography of Mascarene plant life, including the useful and ornamental plants of the region, covering the period 1609–1990.* [vi], 274 pp., illus. Lawai, Kauai, Hawaii: National Tropical Botanical Garden. [About 1600 entries, briefly annotated and arranged by author with a keyworded subject index. The latter includes all taxa to generic level. There is also a lexicon of cited places of publication.]

Indices. General and supraregional indices as for Superregion 46–49.

470

Region in general

For Baker's *Flora of Mauritius and the Seychelles*, see **472**. Successors to this and other current but largely outdated works on the region include a checklist of the pteridophytes by Tardieu-Blot (1960) as well as *Flore des Mascareignes*, the latter published under joint British, French and Mauritian auspices.

BOSSER, J. *et al.* (eds.), 1976– . *Flore des Mascareignes (La Réunion, Maurice, Rodrigues)*. In fascicles. Illus. Mauritius: Sugar Industry Research Institute; Kew: Royal Botanic Gardens; Paris: ORSTOM.

Descriptive 'research' flora of vascular plants; includes keys, extensive synonymy (with citations and references), local range with *exsiccatae* and extralimital distribution, taxonomic commentary, illustrations of representative species, and notes on habitat, status, occurrence, etc.; index in each fascicle. Naturalized and commonly cultivated plants are also included. Families have been pre-numbered following the modified Bentham and Hooker system as used at Kew but fascicles are appearing only as family accounts are ready; more than one, however, may appear at any given time as in the *Flore de Madagascar et des Comores* (**460**). By 1995, 17 fascicles (the *Glossaire* and families 31–50, Ranunculaceae–Theaceae; 51–63, Malvaceae–Oxalidaceae (incl. Averrhoaceae); 64–68, Balsaminaceae–Burseraceae; 80, Leguminosae; 90–106, Rhizophoraceae–Araliaceae; 107–108 and 108bis, Caprifoliaceae, Rubiaceae and Valerianaceae; 109, Compositae; 110–111, Goodeniaceae–Campanulaceae; 111, Campanulaceae (Supplement) and 112–120, Ericaceae–Salvadoraceae; 136–148, Myoporaceae–Hydnoraceae; 153–160, Lauraceae–Euphorbiaceae; 161–169 and 169bis, Urticaceae–Ceratophyllaceae; 171–176, Zingiberaceae (incl. Costaceae)–Bromeliaceae; 177–188, Iridaceae–Juncaceae; 189, Palmae; 191–201, Typhaceae–Eriocaulaceae) had been published.[53]

Pteridophytes

Pteridophytes comprise a significant percentage of the flora of the Mascarenes. New surveys in each of the main islands have been made within the last three decades in succession to earlier work by Christensen and Tardieu-Blot. These include: F. BADRÉ and T. CADET, 1978. The pteridophytes of Réunion Island. *Fern Gaz.* **11**: 349–365; D. H. LORENCE, 1976. The pteridophytes of Rodrigues Island. *Bot. J. Linn. Soc.* **72**: 269–283, map; and *idem*, 1978. The pteridophytes of Mauritius Island (Indian Ocean): ecology and distribution. *Bot. J. Linn. Soc.* **76**: 207–247, map.

TARDIEU-BLOT, M. L., 1960. Les fougères des Mascareignes et des Seychelles. *Notul. Syst.* (Paris) **16**: 151–201.

Systematic enumeration of known ferns and fern-allies in the whole Mascarenes/Seychelles region, with synonymy, citations of *exsiccatae* with localities, and index.

471

Réunion

Area: 2510 km². Vascular flora: 550 indigenous seed plants and 220 indigenous pteridophytes (PD; see **General references**). A total of 1156 species were accounted for by Jacob de Cordemoy (1895), the first full flora of the island.

CORDEMOY, E. JACOB DE, 1895. *Flore de l'Île de la Réunion.* xxvii, 574 pp. Paris: Klincksieck (altered by label to Jacques Chevalier). (The consolidated issue; originally published at least in part, 1891 onwards, St.

Denis, Réunion. Reprinted 1972, Lehre, Germany: Cramer.)

Briefly descriptive flora of non-vascular and vascular plants (except fungi), with more detail accorded to new taxa; includes keys to genera and species (or groups of species), limited synonymy (with references), vernacular names, generalized indication of local range, and notes on habitat, phenology, properties, uses, etc.; index to vernacular, generic and family names at end. In the introduction are chapters on geography, physical features, climate, and the general aspects of the flora, together with statistical tables and an account of botanical exploration.

472

Mauritius

Area: 1865 km². Vascular flora: 800–900 species, including 186 ferns (PD; see **General references**). A total of 869 species for Mauritius, Rodrigues, Seychelles and dependencies were accounted for by Baker (1877).

The first consolidation of floristic knowledge for Mauritius and its dependencies was *Hortus Mauritianus* (1837) by Wenceslaus Bojer, then curator of the Botanical Gardens (at Pamplemousses). While including cultivated as well as native species, the two were mutually distinguished by the author. It was followed by Baker's well-known flora. The latter is now gradually being supplanted by *Flore des Mascareignes* (**470**).

BAKER, J. G., 1877. *Flora of Mauritius and the Seychelles: a description of the flowering plants and ferns of those islands.* 19, 1, 557 pp. London: Reeve. (Reprinted 1970, Lehre, Germany: Cramer.)

Briefly descriptive flora of vascular plants, with keys to genera and species; limited synonymy, with references; vernacular names; citations of *exsiccatae*, with localities, and general indication of extralimital range (where applicable); notes on habitat, etc., and taxonomic commentary; indices to all vernacular and botanical names. The introductory section incorporates general descriptive remarks on the islands covered and notes on the main features of their floras, an account of botanical exploration, a synopsis of orders and families, and Bentham's *Outlines of Botany*.

– For additions, see H. H. Johnston, 1895. Additions to the flora of Mauritius as recorded in Baker's 'Flora of Mauritius and the Seychelles'. *Trans. Proc. Bot. Soc. Edinburgh* **20**: 391–407; also R. E. Vaughan, 1937. Contributions to the flora of Mauritius, I. An account of the naturalized flowering plants recorded from Mauritius since the publication of Baker's *Flora of Mauritius and the Seychelles. J. Linn. Soc., Bot.* **51**: 285–308.[54]

473

Rodrigues (Rodriguez)

See also **472** (Baker). – Area: 104 km². Vascular flora: 145 indigenous species, of which 40 endemic (PD; see **General references**). Balfour (1879) recorded 322 species, with already 108 seed plants viewed as having been introduced by 1874. Cadet (1971) recorded 375+ species, including over 100 introductions since Balfour's time.

No full separate account of Rodrigues vascular plants has been published since 1879. However, the highly vulnerable endemic flora, some species of which are already now lost, is well documented and figured in W. Strahm, 1989. *Plant red data book for Rodrigues*. viii, 241 pp., illus., maps. Königstein/Ts., Germany: Koeltz (for IUCN/World Conservation Union).

Balfour, I. B., 1879. Flowering plants and ferns. In J. D. Hooker and A. Günther (eds.), *An account of the petrological, botanical and zoological collections made in Kerguelen's Land and Rodriguez during the Transit of Venus expedition carried out by order of Her Majesty's Government in the years 1874–75*, [II]: *Collections from Rodriguez*, pp. 302–387, pls. 19–36 (Philos. Trans. 168: extra vol.). London. (Reprinted as pp. 1–86 in I. B. Balfour *et al.*, *Botany* [of Rodriguez]. 118 pp. London [for private circulation].)

Systematic enumeration of vascular plants, with descriptions of new taxa discovered; includes limited synonymy (with references), generalized indication of local and extralimital range (with mention of some localities), taxonomic commentary, and notes on habitat, uses, etc. A general part includes remarks on overall features of the flora and vegetation (already in 1874 very disturbed) and its history, together with a floristic analysis. All records were accounted for by

Baker (**472**). For additions (introduced species), see T. Cadet, 1971. Flore d'Île Rodrigues: espèces spontanées introduites depuis Balfour (1874). *Bull. Mauritius Inst.* **7**: 1–12.

Region 48

The Seychelles

Area: 212 km² ('high' islands, except Curieuse). Vascular flora: Robertson (1989) gives a figure of 250 indigenous flowering plant species but Friedmann (1994) only 200 with a further *c.* 80 pteridophytes; there are no native gymnosperms. Summerhayes in 1931 recorded a total of 480 flowering plants (including introductions). – The Seychelles group, home of the 'Coco de Mer' (the palm *Lodoicea maldivica*), is here defined as comprising the five granitic 'high' islands together with Frégate, Curieuse and other low islands and islets on the Seychelles Bank (including Bird and Denis on its northern edge). Other islands, cays and atolls of the Seychelles, including Platte and the Aldabra group, are assigned to Region 49.

The only complete descriptive treatment remains Baker's *Flora of Mauritius and the Seychelles* (1877; see **472**); however, the dicotyledons have now been fully rewritten by Francis Friedmann for *Flore des Seychelles* (1994; see **480** below). Other accounts drawing upon botanical work of the nineteenth and early twentieth centuries include the enumerations of Christensen for the ferns (1912) and Summerhayes for flowering plants (1931), both mentioned below under **480**. The pteridophytes were again revised by Tardieu-Blot in 1960 (see **470**).

Beginning with field work by Charles Jeffrey early in the 1960s, there has been considerable interest in improving knowledge of the native flora of the granitic islands, given their great botanical importance and perceived threats. Several floristic surveys have been made, but only in 1989 did a new enumeration make its appearance: S. A. Robertson's *Flowering plants of Seychelles* (see **480** below). Though covering both native and introduced species, it was designed as an interim work, pending completion of Friedmann's more definitive flora. An atlas of the endemic flora, with paintings by Rosemary Wise, has also been published.

Bibliographies. General and supraregional bibliographies as for Superregion 46–49.

Regional bibliography

PETERS, A. J. and J. G. LIONNET, 1973. *Central western Indian Ocean bibliography.* 322 pp., 3 maps (Atoll Res. Bull. 165). Washington. [Centering on the Seychelles and arranged by authors with subject indices, this substantial bibliography (not limited to botany) covers the area from 2° to 11° S and 45° to 75° E, thus encompassing island groups from Aldabra and the Seychelles through Agalega and Coëtivy to the Chagos Archipelago (including Diego Garcia).]

Indices. General and supraregional indices as for Superregion 46–49.

480

Region in general

See also **470** (TARDIEU-BLOT); **472** (BAKER). Both Robertson's enumeration and Friedmann's flora encompass the territory of the Seychelles, so including all the low islands to the southwest (**496** to **499**) as well as Coëtivy (**494**) and Platte (**495**). Separate florulas of Cousin and Frégate appear in SACHET *et al.* (see **490**). Tardieu-Blot's *Les fougères des Mascareignes et des Seychelles* (1960; see **470**) remains current for pteridophytes. All these add to and largely supplant V. S. SUMMERHAYES, 1931. An enumeration of the angiosperms of the Seychelles Archipelago. *Trans. Linn. Soc. London*, II, Zool. 19(2): 261–299, table; and C. CHRISTENSEN, 1912. On the ferns of the Seychelles and the Aldabra group. *Trans. Linn. Soc. London*, II, Bot. 7(19): 409–425, pl. 45.

FRIEDMANN, F., 1994. *Flore des Seychelles: Dicotylédones.* 663 pp., 209 text-figs. Paris: Éditions ORSTOM. (Collection Didactiques.)

Well-illustrated descriptive manual-flora with keys, synonymy, references and citations, vernacular names, and notes on distribution (sometimes fairly detailed, especially for endemic species), habitat, morphology, biology, and (if relevant) origin; indices to all vernacular and botanical names at end. An introduction (with selected references), abbreviations and conventions, and an illustrated glossary comprise the general part along with (pp. 55–64) the main key to families.

ROBERTSON, S. A., 1989. *Flowering plants of Seychelles.* xvi, 327 pp., 212 text-figs., 2 maps. Kew: Royal Botanic Gardens.

Descriptive enumeration of seed plants (including native, naturalized, adventive, cultivated and ornamental species: 1139 in all); includes very limited synonymy, vernacular names, localities with literature references and *exsiccatae*, and notes on uses and special features; addenda and indices at end. The introductory section gives the plan of and basis for the work, sources and other references, localities and collectors.[55]

Outer northern islands

STODDART, D. R. and F. R. FOSBERG, 1981. *Bird and Denis Islands, Seychelles.* iii, 50 pp., 4 figs. (incl. maps), 12 halftone pls. (Atoll Res. Bull. 252). Washington.

Topographic accounts of two reef islands on the northern edge of the Seychelles Bank; each includes an annotated systematic checklist of vascular plants with *exsiccatae* and/or published records; references, pp. 45–50. Detailed maps of each island with distributions of substrates, vegetation and key species are also presented.

Region 49

Aldabra and other low islands north and east of Madagascar

The islands or island groups in this region include the Cargados Carajos group, Tromelin, Agalega, Platte, Coëtivy, the Amirante group, Alphonse, Providence (with St. Pierre and Cerf), Farquhar and Aldabra (with the Assumptions, the Cosmoledos and Astove). All are low coralline or sandy islands or atolls, of which Aldabra, a slightly raised atoll, is much the largest.

For decades, the only general flora was the patchy and in part very sketchy 1919 study by W. B. Hemsley, cited below; it paid particular attention to Aldabra but also included lists for a number of other islands and groups. In the wake of the IIOE cruises and the Aldabra research programme of the 1960s and early 1970s, most of the units here recognized were subjected to renewed studies, including floristic surveys, and most now have modern florulas or at least checklists. There have also been some comparative floristic analyses, e.g., Renvoize (1975, with later additions and revisions). Of particular note was the publication, in 1980, of *Flora of Aldabra and neighbouring islands* by F. Raymond Fosberg and Stephen Renvoize (under **499**).

Progress

For a brief review of botanical progress, see D. R. STODDART and F. R. FOSBERG, 1984. Vegetation and floristics of western Indian Ocean coral islands. In D. R. Stoddart (ed.), *Biogeography and ecology of the Seychelles Islands*, pp. 221–238. The Hague: Junk. A more detailed survey would be desirable.

Bibliographies. General and supraregional bibliographies as for Superregion 46–49.

Regional bibliography

PETERS, A. J. and J. G. LIONNET, 1973. *Central western Indian Ocean bibliography*. See **Region 48**.

Indices. General and supraregional indices as for Superregion 46–49.

490

Region in general

See also **470** (TARDIEU-BLOT); **472** (BAKER); **480** (FRIEDMANN; ROBERTSON). Reference should be made also to *Island bibliographies* by Sachet and Fosberg (see **General bibliographies**) as well as to the Peters and Lionnet bibliography (**Region 48**) for details of the considerable but scattered literature on these islands. – Overall vascular flora: 228 species or more (Renvoize, 1975, pp. 134–135).

The sketchy omnibus work of Hemsley, long the effective authority on the region, has been largely supplanted by more recent and detailed island accounts, either separately published or included in the collections below. It remains, however, basic to all floristic studies in the region.

GWYNNE, M. D. and D. WOOD, 1969. *Plants collected on islands in the western Indian Ocean during a cruise of the MFRV 'Manihine', September–October 1967*. 15 pp. (Atoll Res. Bull. 134). Washington.

Incorporates a two-dimensional tabular list of plants, with species given on one coordinate and islands or island groups on the other; list of references included. Accounts for the vascular plants, with citations of representative *exsiccatae* where appropriate, are in this collective way presented for Assumption, Astove, Coëtivy, Cosmoledo and Farquhar as well as Daros, Desroches and Remire in the Amirante group.

HEMSLEY, W. B., 1919. Flora of Aldabra: with notes on the flora of the neighbouring islands. *Bull. Misc. Inform.* (Kew) **1919**: 108–153, 451–452, map.

Annotated list of vascular plants of Aldabra, with descriptions of some new species; includes limited synonymy, principal citations, *exsiccatae* with localities and general summary of extralimital range, and occasional taxonomic commentary. The general part includes sections on principal features of the island, climate, vegetation, endemic species, phytogeography and botanical exploration. The main Aldabra list (now superseded by the illustrated flora of FOSBERG and RENVOIZE; see **499**) is followed by separate accounts, partly discursive and all derived from literature or unpublished sources, dealing with the known plants of Assumption, Astove, Cosmoledo (the first three being in the Aldabra group), Farquhar, Providence, St. Pierre, Gloriosa, Alphonse and the Amirante group, Coëtivy, Agalega and the Cargados Carajos, as well as the Chagos Archipelago (**043**). A few novelties apropos to this section of the paper are described on pp. 140–142.

RENVOIZE, S. A., 1975. A floristic analysis of the western Indian Ocean coral islands. *Kew Bull.* **30**: 133–152.

Includes (as table 3, pp. 146–152) a tabular systematic checklist, without keys or synonymy but with indication of presence or absence, of the vascular plants of 14 island units scattered over the whole region as here defined. This list is included as an adjunct to a semi-quantitative floristic and phytogeographic analysis of these islands, chosen on the basis of the availability of more or less adequate modern collections. [Since publication of this survey, a greater or lesser number of additions have been made for a number of the given island units.][56]

SACHET, M.-H. *et al.* (eds.), 1983. *Floristics and ecology of western Indian Ocean islands.* 253 pp. (Atoll Res. Bull. 273). Washington.

A sequel to the collection of Stoddart (1970); includes several checklists for previously never- or long-undocumented islands as well as additional data for others (see **491** through **499** for individual papers).

STODDART, D. R. (ed.), 1970. *Coral islands of the western Indian Ocean.* x, 224 pp. (Atoll Res. Bull. 136). Washington.

A collection of papers by various authors on geographical, physical, botanical, zoological and other aspects of many coral atolls and sand cays in the southwest Indian Ocean; several deal in whole or in part with floristics and vegetation and include lists of the non-vascular and vascular plants of individual islands or island groups with citations of *exsiccatae* and some general notes on the flora (see **491** through **499** for individual papers).

491

Cargados Carajos group

See also **490** (HEMSLEY, p. 147). – Area: *c.* 4 km². Vascular flora: 41 species including native, weedy and cultivated plants (Staub and Guého, 1968; see below). The Cargados Carajos (or St. Brandon Islands) comprise 22 low coralline islets with few trees; they are at present part of Mauritius.

STAUB, F. and J. GUÉHO, 1968. The Cargados Carajos shoals or St Brandon: resources, avifauna and vegetation. *Proc. Roy. Soc. Arts Sci. Mauritius* 3(1): 7–46, pls. 1–5 (incl. maps).

Includes (pp. 25–40) a descriptive account of vegetation and flora, with a census of the known plants.

492

Tromelin

Area: 1 km². Vascular flora: 6 species (Staub, 1970). Tromelin is a small, isolated coral islet with scrub and herbaceous vegetation, presently under French administration.

STAUB, F., 1970. Geography and ecology of Tromelin Island. In D. R. Stoddart (ed.), *Coral islands of the western Indian Ocean*, pp. 197–210, 10 halftones (Atoll Res. Bull. 136). Washington.

Includes general remarks on the vegetation with notes on the vascular plants observed.

493

Agalega

See **490** (HEMSLEY, p. 140; RENVOIZE) for checklists. – Area: 21 km². Vascular flora: Hemsley recorded 6 species, but Bojer in 1835 – the first and, for the next 100 years, the only botanical visitor – already many more; a realistic figure is 60. Guého and Staub (1983; see below) gave a figure of 43 as truly indigenous; there are now also a great many introductions.[57] Agalega, a pair of relatively well-wooded low islands, is a dependency of Mauritius. For further background, see A. S. CHEKE and J. C. LAWREY, 1983. Biological history of Agalega, with special reference to birds and other land vertebrates. *Atoll Res. Bull.* 273: 65–108.

FOSBERG, F. R., M.-H. SACHET and D. R. STODDART, 1983. List of the recorded vascular flora of Agalega. In M.-H. Sachet *et al.* (eds.), *Floristics and ecology of western Indian Ocean islands*, pp. 109–142 (Atoll Res. Bull. 273). Washington.

Concise annotated systematic enumeration, documented by literature records and/or *exsiccatae* and indicating status and location on either (or both) of the two islands; references at end. An introductory section covers the history of botanical work as well as general features.

GUÉHO, J. and F. STAUB, 1983. Observations botaniques et ornithologiques à l'atoll d'Agaléga. *Proc. Roy. Soc. Arts Sci. Mauritius* 4(4): 15–110, pls. 1–9 (part col.).

Incorporates (pp. 35–70) an annotated enumeration of vascular plants with vernacular names and natural history notes. This is followed by remarks on cultivated and ornamental plants, an examination of biogeography and dispersal, a list of material seen, and literature cited. The introductory part includes a history of botanical work (with a reissue of an 1835 list of Wenceslaus Bojer) and a review of vegetation formations with significant species. No index is included.

494

Coëtivy

See also **490** (GWYNNE and WOOD; HEMSLEY, p. 146) for other checklists. – Area: 9.2 km². Vascular flora: 65 species of which 49 are native (Gwynne and Wood). The island is a dependency of the Seychelles.

ROBERTSON, S. A. and F. R. FOSBERG, 1983. List of plants collected on Coëtivy Island, Seychelles. In M.-H. Sachet *et al.* (eds.), *Floristics and ecology of western Indian Ocean islands*, pp. 143–156 (Atoll Res. Bull. 273). Washington.

Annotated systematic enumeration, documented by *exsiccatae* and/or literature records (the letter 'R' indicates a plant's inclusion in Renvoize's 1975 list (see **490**)). Accounts for collections made since the time of Gwynne and Wood, especially that of Robertson in 1980. [The Cyperaceae are here much more fully accounted for than in earlier lists.]

495

Platte

Area: 0.65 km². Vascular flora: 56 native and established species. – This sand cay, though close to the Seychelles 'high' islands and politically part of that country, is not on the Seychelles Bank. Most of it is presently a coconut plantation with a central airstrip. The first collections ever were made in 1980.

ROBERTSON, S. A. and F. R. FOSBERG, 1983. List of plants collected on Platte Island, Seychelles. In M.-H. Sachet *et al.* (eds.), *Floristics and ecology of western Indian Ocean islands*, pp. 157–164, map (Atoll Res. Bull. 273). Washington.

Annotated systematic enumeration of vascular plants with citation of *exsiccatae* and brief notes. The map depicts current vegetation and land use.

496

Amirante group (including Desroches and Alphonse)

See also **490** (GWYNNE and WOOD, for Desroches, Daros and Remire; HEMSLEY, p. 145, for these islands as a whole). – Area: no data. Vascular flora: 97 species of which 72 are native (Gwynne and Wood). The main group comprises several small islets on a submarine bank, e.g., the African Banks, Daros, Poivre, Remire and Desnoeufs; Desroches Atoll is just to the east and the Alphonse atolls lie to the south.

There is no single list for the group. Desroches, Remire and African Banks were covered in *Coral islands of the western Indian Ocean* (1970; see also **490**), while Alphonse and Remire were listed in *Floristics and ecology of western Indian Ocean islands* (1983; see also **490**). The latter work also includes treatments of Marie-Louise and Desnoeufs by J. R. Wilson, each with references (respectively pp. 185–202 and 203–222).

Partial work: Alphonse

A sand cay of 172 ha on an atoll. Oddly, in the following checklist no reference therein is made to Hemsley, although his work has a specific list for this islet.

ROBERTSON, I. A. D., S. A. ROBERTSON and F. R. FOSBERG, 1983. List of plants collected on Alphonse Island, Amirantes. In M.-H. Sachet *et al.* (eds.), *Floristics and ecology of western Indian Ocean islands*, pp. 177–184 (Atoll Res. Bull. 273). Washington.

Concise annotated enumeration, documented by *exsiccatae* and/or sight records and references; no indication of status. [Based on a 1979 visit by the senior author.]

Partial works: Desroches, Remire and African Banks

FOSBERG, F. R. and S. A. RENVOIZE, 1970. Plants of Desroches Atoll. In D. R. Stoddart (ed.), *Coral islands of the western Indian Ocean*, pp. 167–170 (Atoll Res. Bull. 136). Washington; *idem*, Plants of Remire (Eagle) Island, Amirantes. *Ibid.*, pp. 183–186; and *idem*, Plants of African Banks (Îles Africaines). *Ibid.*, pp. 193–194.

Annotated lists of non-vascular and vascular plants recorded, with indication of *exsiccatae* and notes on distribution, ecology, dispersal, etc.

Partial work: Poivre

ROBERTSON, S. A. and F. R. FOSBERG, 1983. List of plants of Poivre Island, Amirantes. In M.-H. Sachet *et al.*

(ed.), *Floristics and ecology of western Indian Ocean islands*, pp. 165–176 (Atoll Res. Bull. 273). Washington.

Concise annotated enumeration with *exsiccatae* and/or sight records (as well as references to a list published in a 1969 land evaluation study); status of species not indicated.

497

Farquhar group

See **490** (GWYNNE and WOOD; HEMSLEY, pp. 143–145) for other checklists. – Area: 7.5 km² (Farquhar). Vascular flora: Providence, 14 species; St. Pierre, 24 species; Farquhar, 44 species.[58] The unit as here circumscribed includes Providence Atoll (including Providence and Cerf) and St. Pierre as well as the slightly removed Farquhar Island together with associated reefs and cays.

FOSBERG, F. R. and S. A. RENVOIZE, 1970. Plants of Farquhar Island. In D. R. Stoddart (ed.), *Coral islands of the western Indian Ocean*, pp. 27–33 (Atoll Res. Bull. 136). Washington.

Enumeration of non-vascular and vascular plants, with citations of *exsiccatae* and notes on distribution, ecology, dispersal, etc.; references.

498

Rajaswaree Reef

No reports have been seen concerning plants from this isolated reef, 170 km southeast from Farquhar Island.

499

Aldabra group

Area: 175 km² (Aldabra, 155 km²). Vascular flora: 274 species and varieties, of which 185 indigenous and 43 endemic (Fosberg and Renvoize, 1980, p. 6). As here delimited, this unit includes the large raised-coral island of Aldabra and the smaller, nearby raised atolls of Assumption, Cosmoledo and Astove.

The group has lately been treated in *Flora of Aldabra and neighbouring islands* by Fosberg and Renvoize (1980); this supplants Hemsley's lists (**490**) and scattered other sources. For convenience, however, modern separate accounts for Cosmoledo and Astove also are accounted for here. All the islands, especially Aldabra, received intense study in the late 1960s and 1970s in relation to conservation efforts, beginning with a Royal Society expedition in 1967–68. *The Flora of Aldabra and neighbouring islands* was preceded by G. E. WICKENS, 1974 (1975). *A field guide to the flora of Aldabra.* xlii, 99 pp. London: The Royal Society.

Bibliography

STODDART, D. R., 1997. *Bibliography of Aldabra and nearby atolls.* 3rd edn. 153 pp. Victoria, Mahé, Seychelles: Seychelles Islands Foundation. (1st edn., 1967; 2nd edn., 1986 with reissue in 1987.) [Entries (nearly 1200) arranged by author, with cross-referential taxonomic and subject index at end. Outside the Aldabra group, coverage extends to the Glorioso Islands (**466**) and the Farquhar group (**497**).]

FOSBERG, F. R. and S. A. RENVOIZE, 1980. *The flora of Aldabra and neighbouring islands.* [vi], 358 pp., 55 figs., 2 maps (Kew Bull. Addit. Ser. 7). London: HMSO.

Descriptive flora mainly of vascular plants, with keys to all taxa, synonymy, indication of local distribution, vernacular names, some taxonomic commentary, and notes on habitat, occurrence, phenology, origin, dispersal and special features; list of mosses (by C. C. Townsend); indices to all names. The introductory section of this work deals with physical features and geography, climate and other habitat factors, vegetation, and botanical exploration; statistics, references (pp. 7–8), a glossary, and a general key (pp. 17–31) are also included. No *exsiccatae* are cited (available upon separate application).

Partial works: Astove and Cosmoledo

Included here are separate works on Astove and Cosmoledo. Other treatments, accounting also for Assumption, may be found under **490** (GWYNNE and WOOD, 1969; HEMSLEY, pp. 140–143).

FOSBERG, F. R. and S. A. RENVOIZE, 1970. Plants of Cosmoledo Atoll. In D. R. Stoddart (ed.), *Coral islands of the western Indian Ocean*, pp. 57–65 (Atoll Res. Bull. 136).

Washington; and *idem*, 1970. Plants of Astove Island. *Ibid.*, pp. 101–111.

    Annotated enumerations of non-vascular and vascular plants, with indication of *exsiccatae* and/or other data sources and notes on distribution, ecology, dispersal, etc.

Notes

1 The term 'Malagassia' is here, as in the previous edition, introduced as a uninomial for the southwest Indian Ocean islands, a geographically discrete and floristically interrelated group. It is derived from the word 'Malagasy' (of or for the people of Madagascar) and the ending is related to the region's archipelagic composition.

2 The collections of this epoch largely made their way to herbaria in the major cities. These, the first such establishments in New Zealand, were (as in Argentina or the United States) parts of museums developed partly as 'by-products' of geological, natural history and ethnographic surveys, rather than, as in parts of Australia or southern Africa, developed in association with botanical gardens. The Department of Scientific and Industrial Research herbarium outside Christchurch came into being only in the 1920s following calls for a properly founded national unit in support of agricultural and biological research. It is now part of a corporatized body known as Manaaki Wenhua/Landcare Research.

3 Mention should also be made here of the considerable contributions of Leonard Cockayne, author of several botanical survey reports as well as popular works (e.g., *New Zealand plants and their story* (1910; see **Region 41, Progress**) and *The trees of New Zealand* (1928, with E. Phillips Turner; see **410, Woody plants**), both revised several times through the 1960s) and two editions of *The vegetation of New Zealand* (1921, 1928), the latter published in Engler's series *Die Vegetation der Erde* and long a standard reference.

4 An endemic genus described here, *Plectomirtha*, has since been reduced to *Pennantia* (Icacinaceae), a genus also in Australia and New Zealand; *P. baylisiana* is moreover synonymous with the Norfolk Island species *Pennantia endlicheri*.

5 The paper's biogeographical analysis is exceptionally interesting.

6 Another version of Sorensen's checklist, with some revisions, is found in J. L. Gressitt, 1964. *Insects of Campbell Island*, pp. 1–20 (Pacific Insects Monogr. 7). Honolulu. The so-called 'Cape Expeditions' were a euphemism for military-related activities in the southern islands during World War II (Colin Meurk, personal communication).

7 Since 1980 revisions of Australian vascular plant genera have generally resulted in increases in numbers of taxa of from 10 to 50 percent with an average of 25–30 percent (A. Orchard through R. Makinson, personal communication, November 1995). In very few have numbers been reduced.

8 On the life of Mueller, the best account remains C. Daley, 1924. Baron Sir Ferdinand von Mueller, K.C.M.G., M.D., F.R.S., botanist, explorer and geographer. *Victorian Hist. Mag.* **10**: 23–75. The relationship between him and the mid-century 'Kewites' is described there and in Daley's history of *Flora australiensis* (Daley, 1927–28; see Note 15) as well as by Bentham in B. Daydon Jackson, 1906. *George Bentham*. London: Dent. For a definitive edition of Mueller's correspondence (in progress), see R. W. Home *et al.* (eds.), 1998– . *Regardfully yours*. Vols. 1– . Bern: Peter Lang. (Life and letters of Ferdinand von Mueller.) Among his many extra-botanical activities (more so later in his career), Mueller was one of the leaders in an Australian inter-colonial movement for scientific independence during the latter part of the nineteenth century.

9 The major part of the source materials became what is now the *Australian plant name index*.

10 For more on the organization of the *Flora of Australia*, see Report of 1979 meeting of Committee of Heads of Australian Herbaria. *Australian Systematic Botany Society Newsletter* **21** (December 1979): 7–9, and Report from Bureau of Flora and Fauna, Department of Science (ABRS). *Ibid.*, **23** (June 1980): 5–6, as well as personal knowledge. The plan adopted for the vascular plants was for a series of revisionary treatments in 48 bound volumes, with two additional volumes for the floras of outlying island territories. The original target for completion was 15 years (but now is estimated to be in about 2010 at the earliest or, more realistically, 2015). Other groups of plants and similar organisms were to occupy volumes 51 and beyond. Mosses were originally due in 1997, hepatics perhaps by 2010, lichenized fungi by about the same time (vols. 54 and 55 have already appeared), and other fungi by 2050. A second, much enlarged edition of volume 1 appeared in 1999. The work is currently under the general editorship of A. Orchard. The *ASBS Newsletter*, begun in 1973, has been chronicling many of the developments and discussions relating to the *Flora of Australia*. (I thank R. Makinson, Canberra, for information with respect to current targets.)

11 Written by taxonomists throughout Australia and elsewhere, these have appeared in *Australian Journal of Systematic Botany* or its predecessors, in the journals of the state herbaria, or elsewhere.

12 Until the 1980s, it could be said that the fragmentation criticized by Bentham more than a century ago – a factor

contributing to his assuming sole responsibility for the execution of *Flora australiensis* – continued to be seen as a serious incubus to a national flora. There was even expressed a wish that a 'Bentham redivivus', perhaps from outside Australia, should 'take charge'. The existence of a separate Commonwealth bureau – much as had been envisaged by Stanley Blake – has, however, provided a worthy 'seat' for the *Flora of Australia* (and similar faunistic projects); in return they have promoted national self-confidence.

13 The very success of the *Flora* has to a degree contributed to this apparent lesser rate of progress. A by-product was an increased demand for new state and local floras which fell within the responsibility of collaborating state botanists. These are nevertheless seen as complementary and help to maintain a social and political constituency for plant taxonomy in Australia (R. Makinson, personal communication). The new works now cover most of the more populous areas of the country.

14 Publication of the *Index* had originally commenced in 1980 on microfiche, but this was soon discontinued – most likely on account of technological changes in information transmission, paradoxically contributing in the nearer term to a return to paper. It is being maintained and updated by the Centre for Plant Biodiversity Research (representing an amalgamation of the botanical research activities of the Australian National Botanic Garden (ANBG) and the Commonwealth Scientific and Industrial Research Organisation (CSIRO)). The interface is, however, 'barely useable' and updates reportedly indifferent (H. Cohn, personal communication, 1999). Projected improvements will encompass both the interface and content; the latter will eventually incorporate distributional information for accepted taxa and as well link known synonyms. The *Index* is now also available through the Web-based International Plant Names Index (http://www.ipni.org/).

15 The work forms part of the 'Kew Series' of British colonial floras initiated by W. J. Hooker. The 'assistance' of F. von Mueller, indicated on the title-pages, included the loan of Australian holdings in the National Herbarium of Victoria, Melbourne, and much correspondence; however, actual collaboration was out of the question, as, according to Bentham, 'four months were required for an answer to the smallest query' though, as already indicated, there was rather more to it than that. For more on *Flora australiensis*, see Thistleton-Dyer (1906, under **General references**) as well as: G. Daley, 1927–28. The history of *Flora australiensis*, I–VIII. *Victorian Naturalist* 44(3–10): 63–74, 91–100, 127–138, 153–164, 183–187, 213–221, 248–256, 271–278; and F. A. Stafleu, 1967. The *Flora australiensis*. *Taxon* 16: 538–542.

16 For vascular plants, as of 1999 published volumes encompassed 1 (Introduction; 2nd edn., 1999), 3 (Hamamelidales–Casuarinales), 4 (Caryophyllales in part), 8 (Lecythidales–Batales), 12 (Mimosaceae and Caesalpiniaceae), 16 (Proteales in part), 17B (Proteales in part), 18 (Podostemales, Haloragales and Myrtales except Myrtaceae), 19 (Myrtales: *Eucalyptus* and *Angophora*), 22 (Rhizophorales–Celastrales), 25 (Sapindales in part), 28 (Gentianales), 29 (Solanales: Solanaceae), 35 (Campanulales: Brunoniaceae and Goodeniaceae), 45 and 46 (Hydatellales–Liliales), 48 (gymnosperms and pteridophytes), and 49 and 50 (Oceanic Islands I–II). Two further volumes (54 and 55, on lichens) have also been published.

17 Most standard bibliographies have lacked complete bibliographic information on this work, even in concise form. See D. J. McGillivray, *Contr. New South Wales Natl. Herb.* 4(6): 366–368 (1973) for a full review.

18 *Australian rain-forest trees* was actually prepared under Queensland auspices, its author being in fact on the staff of the Queensland Herbarium. The state remained responsible for all official research on its rain forest trees until 1971, when the Atherton branch of the Commonwealth Forestry and Timber Bureau was established; indeed, the predecessor of *Australian tropical rainforest trees* appeared as a Queensland Forestry Department publication. In 1975 control of the Atherton station passed to the Commonwealth Scientific and Industrial Research Organisation (CSIRO), with which it has remained to the present as its Tropical Forest Research Centre. Both polities nevertheless must be credited with support for the field, herbarium and other research which has led in recent decades to the vast increase in our knowledge and public perception of the trees of the northeast of the continent.

19 Rodway, a dental surgeon who migrated to Tasmania in 1880, was from 1896 to 1932 Honorary Government Botanist.

20 For the early history of Tasmanian botany, see J. H. Maiden, 1910. Records of Tasmanian botanists. *Pap. & Proc. Roy. Soc. Tasmania* 1909: 9–29.

21 This work was first conceived by Winifred Curtis about 1945 with the mycologist Hugh D. Gordon (then at the University of Tasmania but from 1947 in New Zealand) when Rodway's manual became unavailable. Subsequent progress was necessarily slow given the authors' maintenance of critical standards; indeed, much of the work was accomplished only following the formal retirements of both Curtis (1967) and Dennis Morris (1985). Plans exist for a revised version of the whole work entitled *A flora of Tasmania* and arranged according to the Cronquist system.

22 A major stimulus was the search for Ludwig Leichhardt, who disappeared in 1848.

23 The diskettes associated with the 1994 edition included

the interactive keys, and represented the first step towards a full floristic information system. The CD-ROM of the 1999 edition, with interactive keys and searchable formatted text, is a second step. The future will also include object-oriented, query-based interactive access.

24 Major contributions to field work and documentation in the state following Mueller's pioneering activities were made by members of the Field Naturalists' Club of Victoria, founded in 1880 and publishers of *Victorian Naturalist*. A survey of the history of botanical work in the state forms chapter 5 of the first volume of *Flora of Victoria*. Reference can also be made to J. H. Maiden, 1908. Records of Victorian botanists. *Victorian Naturalist* **25**: 101–117.

25 The *Flora of Victoria* is, according to Schmid (1997; see references to the General introduction), one of the best modern works of its kind worldwide.

26 The authors of the first edition noted that, while Willis covered 91 percent of the species known by 1984, only 79 percent of the names adopted were still current.

27 For the history of botany in New South Wales to the 1900s, see J. H. Maiden, 1908. Records of Australian botanists: (a) General; (b) New South Wales. *J. & Proc. Roy. Soc. New South Wales* **42**: 60–132.

28 One reviewer thought the book 'splashy' in appearance, in contrast to the hitherto 'dull look' of most floras (R. Schmid, *Taxon* **41**: 627 (1992)), but it is better termed 'striking' or 'elegant'. Much as now do many magazines, the use of rules as well as shading to separate text elements are contributing features. Such a presentation was, however, foreshadowed in the 1981 *Plants of New South Wales*.

29 A grid system for recording was considered but dropped as in many areas collections were too sparse to be meaningful.

30 The 1957 edition was one of the first botanical books bought by the then 17-year-old author of this book, shortly after its publication.

31 By the 1920s, this and Maiden's other 'grand' works were not viewed kindly by the governments of the day; on his death in 1925 they were discontinued.

32 Queensland is larger than Alaska or the southeastern United States (respectively **Regions 11** and **16**), and the centers of population (and users of a flora) are moreover widely spaced although the largest concentration is in the southeast (**435**). This surely influenced the decision in the 1970s to go ahead with regional floras. More recently, however, the future of that strategy has become less certain; with the development and implementation of a full herbarium database (HERBRECS), another project begun in the 1970s, attention has shifted to the production of regional checklists derived from it and other sources. At the same time, there was scope for a new statewide check-list in succession to that of Bailey; its first edition appeared in 1994. – For early Queensland botanical history, see J. H. Maiden, 1910. Records of Queensland botanists. *Report of the Twelfth Meeting of the Australasian Association for the Advancement of Science* (Brisbane, 1909), pp. 373–384. The flora project is considered more fully in L. Pedley, 1978. A new flora of Queensland. *Queensland Naturalist* **22**: 8–12. HERBRECS as well as the recent changes in the Queensland Herbarium's approach to floristic documentation are discussed in *Port Curtis District Flora* (**436**) as well as R. W. Johnson, 1991. HERBRECS: the Queensland Herbarium records system, its development and use. *Taxon* **40**: 285–300.

33 The original project ran from 1973 to 1978 but final preparation and publication of this book was drawn out over more than 10 years, in goodly part due to production problems (Trevor Stanley, personal communication).

34 Another work, *Plants of central Queensland* by S. and A. Pearson, was announced for 1989 by the Society for Growing Australian Plants but has not been published.

35 A regional enumeration, *An annotated checklist of the vascular flora of Cape York Peninsula* by J. Clarkson, has been announced but not so far published.

36 For a review of the 'Torres Strait problem', with contributions from several disciplines, see D. Walker (ed.), 1972. *Bridge and barrier.* xviii, 437 pp. (Australian National University, Department of Biogeography and Geomorphology, Publ. BG-3). Canberra.

37 Black's flora was, and continues to be, published as part of a series of state handbooks on flora and fauna. Established in 1921 and produced under the guidance of an advisory committee, these handbooks have no counterparts elsewhere in Australia save for the publications of the federal Bureau of Fauna and Flora, established in the 1970s. See Rutherford Robertson, *Australian Systematic Botany Soc. Newsletter* **53**: 18–19 (1987) and the book there reviewed, *Ideas and Endeavours – The natural sciences in South Australia* (1986) by C. R. Twidale *et al.* (eds.; botany by E. Robertson).

38 Early South Australian (and Northern Territory) botanical history is recorded in J. H. Maiden, 1908. A century of botanical endeavour in South Australia. *Report of the Eleventh Meeting of the Australasian Association for the Advancement of Science* (Adelaide, 1907), pp. 158–199. For more on Black, see W. R. Barker (comp.), 1992. Memories of J. M. Black. *Australian Systematic Botany Soc. Newsletter* **70**: 2–13.

39 These defects were pointed out in a review in *J. Bot.* **57**: 69–71 (1919).

40 The botanical volume forms one of a series of four on various aspects of the expedition, published between 1956 and 1964.

41 Also published separately in M. A. C. Fraser (ed.), 1903.

Notes on the natural history, etc., of Western Australia. 250 pp., illus., maps. Perth.

42 On the antecedents and development of the Western Australian Herbarium, see J. W. Green, 1990. History of early Western Australian herbaria. In P. S. Short (ed.), *History of systematic botany in Australasia*, pp. 23–27. [Melbourne]: Australian Systematic Botany Society. For early history of Western Australian botany, see J. H. Maiden, 1909. Records of Western Australian botanists. *J. Western Australia Nat. Hist. Soc.* 2(6): 5–27. For an introduction to FloraBase, see [A. Chapman], 'About FloraBase' on the World Wide Web at http://florabase.calm.wa.gov.au/about.html/.

43 For critique, see J. H. Willis in *Muelleria* 1: 240–241 (1967). The catalogue was in 1998 being incorporated and much enhanced as part of FloraBase.

44 The *Census* has since become a primary component of FloraBase.

45 With publication of part 2 of the second edition in 1998 – a year after Grieve's death – the revision of what has been, in the words of one colleague, an 'infuriating but indispensable' work was completed. Grieve had been associated with the *Wildflowers* since 1948 when Blackall's papers were received. The original model was *Flore complète portative de la France et de la Suisse* by Gaston Bonnier and Georges de Layens (**651**).

46 The work is in many ways a western counterpart of Domin's *Beiträge zur Flora und Pflanzengeographie Australiens* (see **430**); in the absence of a modern state flora, it forms a valuable and useful supplement to Bentham's *Flora australiensis* (R. D. Royce, personal communication).

47 For further particulars on the *Compendium*, see L. J. Dorr, 1987. Rev. Richard Baron's *Compendium des plantes malgaches. Taxon* 36: 39–46. Serious typographical errors mar the last eight installments, a problem ascribed by Dorr to the death in 1904 of Emmanuel Drake del Castillo who had helped with proofs.

48 The 1996 estimate of completeness of *Flore de Madagascar et des Comores* is taken from Morat and Lowry, 1997 (see **Superregion 46–49, Progress**).

49 This elegant new study of Malagasy palms provides a good example of how incomplete earlier parts of the *Flore de Madagascar et des Comores* have become (the palms having been revised by Jumelle in 1945). A similar situation obtains with caudiciform and other succulents, where much new information is embodied in the two volumes of Werner Rauh's *Succulents and xerophytic plants of Madagascar* (1995–98), and orchids, well documented and lavishly illustrated in *Orchids of Madagascar* by D. DuPuy *et al.* (1999). Such books are indeed what

was earlier called for in these lines: they would more effectively serve the cause of the peculiar indigenous flora, to a great extent under threat of serious degradation or extinction.

50 French activities to mid-century are additionally reviewed in H. Humbert and J. Léandri, 1954. Cinquante ans de recherches botaniques en Madagascar. *Bull. Trimestriel, Acad. Malgache* (1954): 33–43. For a cyclopedia of collectors, see L. J. Dorr, 1997. *Plant collectors in Madagascar and the Comoro Islands*. xlvi, 524 pp., illus. Kew: Royal Botanic Gardens.

51 This work is currently being revised and expanded into a key to all genera and families in Madagascar.

52 Progress to 1974 was reviewed by Keraudren-Aymonin and to 1994 by Schatz *et al.* (for both references, see **Progress**). By 1974, 153 families with some 5900 species had been published. In the subsequent 20 years a further 23, with 330 species, appeared; this includes one revised account (Iridaceae). Labiatae appeared in 1998; with that issue 47 families, in all covering some 3000 species, still awaited treatment. Among these are large groups such as Leguminosae, Rubiaceae, the greater part of the Euphorbiaceae (including *Croton*, with some 200 species expected), and the rest of the Acanthaceae. Schatz *et al.* moreover suggested that if the useful life of flora accounts is about 40 years, then revision of the Iridaceae represents just a fraction of what would be desirable. Palmae and Orchidaceae have been revised as separate monographs.

53 Among key outstanding groups are the pteridophytes, Loganiales, Acanthaceae, Orchidaceae, Pandanaceae, Cyperaceae and Gramineae.

54 A projected second contribution from Vaughan, covering newly found native species, did not materialize. The original work was one of the 'Kew Series' of British colonial floras initiated in the 1850s by W. J. Hooker; it covers Mauritius, Rodrigues, the Seychelles proper, Aldabra, and all the other smaller islets and atolls in the region then under British administration.

55 The work was completed in 1982 but afterwards only slightly revised, thus not accounting for some more recent studies. It was intended as an 'interim' document, pending completion of Friedmann's more definitive flora. Indigenous species have not clearly been indicated, nor is there any form of distinction between species of 'high' islands and those of atolls and other 'low' islands.

56 See S. A. Renvoize, 1979. The origins of Indian Ocean island floras. In D. Bramwell (ed.), *Plants and islands*, pp. 107–129. London: Academic Press.

57 J. Procter in Renvoize, 1979, p. 117.

58 Renvoize, 1979, p. 117.

I can only, in conclusion, express the hope that this somewhat monumental and, at any rate, laborious work [*Flora of tropical Africa*], may be found, as I believe certainly it will be, of real service to the material development of the resources of our African possessions. At the moment it perhaps is more appreciated in France and Germany than by our own countrymen.

> Portion of letter from W. T. Thistleton-Dyer to R. L. Antrobus, Colonial Office, 8 December 1905; quoted from Thistleton-Dyer, Botanical survey of the Empire, in *Bull. Misc. Inform.* (Kew) **1905**: 33 (1906).

[Even with the present state of knowledge] . . . there is already a strong tendency [to experimental research], perhaps regrettable, because those engaged in it prefer the laboratory to the field.

> E. B. Worthington, *Science in the development of Africa*, p. 157 (1958).

The *Flora of Tropical Africa* and its twentieth century successors . . . are valuable tools for the developing nations and are therefore being supported by them. It is encouraging to see the publication of national floras by these independent countries, involving indigenous taxonomic research.

> F. N. Hepper in I. Hedberg (ed.), *Systematic botany, plant utilization and biosphere conservation*, p. 43 (1979).

. . . The Cape has a Mediterranean climate of winter rains, in which the Bantu summer rain crops do not grow. By 1652, the year the Dutch arrived at Cape Town with their winter-rain crops of Near Eastern origin, the Xhosa had still not spread beyond the Fish River.

> J. Diamond, *Guns, germs and steel*, p. 397 (1997).

. . . The Limpopo . . . forms a much larger barrier heuristically, than it does empirically. . . . The majority of taxonomic studies are geographically restricted, with the British, French and Belgians working on the tropical African floras, while the South Africans, Germans, and to a large extent the Scandinavians work on the southern African flora. Where studies have been pan-African, they have illustrated rather the integrity of the African flora . . . The separation, therefore, into 'tropical' and 'southern' African floras obscures the real phytochorology of Africa.

> H. P. Linder in H.-D. Ihlenfeldt (ed.), *Proceedings of the Twelfth Plenary Meeting of AETFAT* (in *Mitt. Inst. Allg. Bot. Hamburg* **23**), p. 779 (1990).

Une Flore est un instrument de formation initiale et continue, ce qui justife la recherche d'une certaine qualité pédagogique.

> J. Mathez in *Bot. Chron.* **10**: 107 (1991).

Division 5

• • •

Africa

Africa here includes the entire continent (bounded on the northeast by the Suez Canal) together with immediately adjacent islands, Socotra (with 'Abd-el-Kuri), and the Benin or Guinea Islands (Bioko (Fernando Póo) to Pagalu (Annobon)). All Mediterranean islands are, however, incorporated in Region 62. Madeira, the Canaries and the Cape Verdes are part of Macaronesia (Region 02); Ascension, St. Helena, Tristan da Cunha and Gough are all in Region 03; and the Marion (Prince Edward) group is in Region 08. Islands in the southwest Indian Ocean, including Madagascar, form Superregion 46–49 in Division 4.

With the largest land area ($30\,319\,000$ km^2) of our 10 divisions, Africa is geologically ancient but presently supports a tropical flora less rich than that in the Americas or the Indo-Pacific region, a phenomenon historically associated with more erratic rainfall and a greater extent and severity of dry periods.[1] Complementary to this is a more limited potential for tall evergreen forest with only the Atlantic belt from Cameroon to Angola, including the Mayombe, being relatively rich with many localized taxa.[2] Woodland areas also are less rich than their counterparts in South America, although in both cases the zones south of the equator are more diverse floristically than those to the north. Also relatively rich are northeast tropical Africa, with high broken mountains and diverse habitats, the central highlands from eastern Zaïre eastwards, and parts of the eastern coastal region. Tropical Africa is estimated to have 30 000 species of vascular plants as opposed to at least 35 000 (more likely 40 000) for the Asia-Pacific tropics and about 90 000 in the Americas. However, to tropical Africa must be added the estimated 22 000 species in southern Africa, probably the richest region of its kind in the world (and especially so in the Cape Peninsula and southwestern Cape Province), and the perhaps 10 000 species of temperate northern Africa, with in the Atlas Mountains a key center of endemism and others elsewhere. Moreover, for tropical and southern Africa together generic endemism is over 58 percent (Brenan, 1979).

Much of the history of African botany has been told in a series of contributions edited by Abílio Fernandes (1962). Further details on early botanists have been added by Nigel Hepper (1979). The successive proceedings of the meetings of the Association pour l'Étude Taxonomique de la Flore de l'Afrique Tropicale, further described below, and the association's *Bulletin*, account for contemporary developments. While numerous contributions were made from the mid-seventeenth to the mid-nineteenth centuries, they were at first directed largely to the northern fringe and to southern Africa. Indeed, tropical Africa yielded less information to Linnaeus in 1753 than either the Neotropics or the tropical Asia-Pacific region. The first major contribution from within the tropics was made by the French botanist Michel Adanson, resident in Senegal from 1749 to 1754; his collections were in part accounted for by Linnaeus in his various publications and moreover formed a major basis for his theoretical work *Familles des plantes* (1763). More extensive inland botanical exploration began in the temperate fringes and in tropical west and northeast Africa in the late eighteenth century (e.g., Isert and Thonning; see Hepper, 1979) but spread more widely only in the nineteenth century as the rest of the continent became crisscrossed by numerous expeditions. The development by William J. Hooker of his 'Kew floras' scheme was another stimulus to botanical exploration; for Africa it enabled the initiation in 1864 of work on a *Flora of tropical Africa* and the 1877 revival of *Flora capensis*, both as Kew projects. It was, however, the imposition of colonial rule and development of (largely rural) resources, along with the gradual development (at first mainly in northern and southern Africa, but after 1900, with some stimulus from Germany, also in the infratropical belt) of professional cadres with associated institutions including herbaria, gardens and laboratories, which led to a very large increase in collections and related botanical knowledge. The process was not uniform, with the result that *Flora of tropical Africa*, while by present standards largely obsolete, when written covered some regions, particularly West Africa, better than others.

The early years of the present century witnessed other attempts at comprehensive 'primary' coverage. Adolf Engler directed a large portion of his resources at Berlin into energetic study of the African flora, beginning more or less with his own study of the montane flora (1892) and a first account of the plants of German East Africa (1895). Numerous revisions and other papers were produced by Engler and his staff and associates, many in the series 'Beiträge zur Flora von Afrika' (published in the periodical *Botanische Jahrbücher*, with 54 parts from 1891 to 1929) and in *Monographien afrikanischer Pflanzen-Familien und -Gattungen* (1898–1904), of which eight numbers were published. These provided a basis for a first 'grand account', *Die Pflanzenwelt Afrikas* (1908–25), a detailed, discursive account by Engler treating the whole continent (not since emulated and itself never completed, lapsing with its author's death in 1930). Complementing these was Franz Thonner's *Die Blütenpflanzen Afrikas* (1908; published in English in 1915 as *The flowering plants of Africa*), a work ever since much esteemed. The *Monographien* were relatively soon succeeded by the more comprehensive series, *Das Pflanzenreich*. Commendably, German efforts were more synthetic than floristic, with less attention merely to the writing of colonial floras.

Other metropolitan powers became involved in African botany in the decades preceding World War I as they acquired territories in the 'Scramble'. However, with more limited or less well-organized resources and in most cases faced with a need for primary botanical exploration and inventory, their contributions were, perhaps inevitably, more limited or fragmented. Much emphasis was given in some areas to applied botany, and several accounts of useful plants were published. In Belgium, Léopold II, becoming flush with funds, allocated considerable resources for scientific studies in his Congo State (now the Democratic Republic of Congo), including the establishment in 1898 of a Musée du Congo at Tervuren (near Brussels) and support for an aggressive programme of botanical exploration. This latter provided a basis for several large works, both statewide and regional; among them was Théophile and Hélène Durand's *Sylloge florae congolanae* (1909), still the only nominally complete floristic work for Zaïre. Théophile Durand in the same year surveyed botanical progress in the colony.[3] French contributions developed more slowly, not making a real mark until well after World War I; such key works as *Flore forestière de la Côte d'Ivoire* by André Aubréville and the never-completed *Flore vivante de l'Afrique occidentale française* by Auguste Chevalier appeared only in the late 1930s. Likewise, Portugal only slowly developed a programme of botanical exploration and research in Africa, and, in the same way, *Conspectus florae angolensis* began publication relatively late. Nevertheless, rivalries arose which have left their mark on the synonymy of many plant species; much taxonomic work since World War II has involved integration of the published taxa of the different national-imperial cadres.

Along with more proactive colonial policies and the concomitant development of local needs, the period after 1919 became notable for a greatly increased number of floras, checklists, and other floristic works of more limited scope and a turn away from more spectacular, all-embracing comprehensive accounts. Postwar inflationary pressures were one factor, but as significant were limitations on the ability of any one institution to handle the growing pool of information as well as an increased emphasis on individual regions. Both *Flora of tropical Africa* and *Flora capensis* were all but completed (the former admittedly still lacking Gramineae), and any immediate justification for German contributions had disappeared with the loss of their African possessions. The near-withdrawal of Germany was, however, partially offset by increased French, Belgian, Portuguese and Italian activities, although German activity in what is now Namibia revived from the mid-1930s.

The most important development was a 'new movement' at Kew in 1924–25 under then-director Arthur Hill which sought, as part of a revised approach to general colonial development which emerged in Britain after World War I, to provide more focused floras and related works. This led to a 'division' of tropical Africa into four regions, the northeast apart: (1) west tropical Africa; (2) Central Africa; (3) tropical East Africa; and (4) south tropical Africa. The most immediate result was the initiation of the *Flora of west tropical Africa*, which began to appear in 1927, but the same principles guided the planning for *Flora of tropical East Africa*, begun after World War II, and the later *Flora Zambesiaca*. The regional approach very likely also created a place for major contributions from other metropolitan countries. For Belgian Central Africa this was taken up as early as 1926 when the Brussels botanist Walter Robyns formulated a plan for a Congo flora (publication of which, however, did not begin until 1948 as *Flore du Congo Belge et du Ruanda-Urundi*). The 'division', with modifications, has persisted to the present, much influenced by the relative success of *Flora of west tropical Africa*. Practical interests continued, with the interwar period most notable for the

appearance of the first serious dendrofloras. These latter were prepared and published largely through the efforts of the Imperial Forestry Institute at Oxford and sister organizations in other metropolitan countries.

The more or less comprehensive regional flora – termed by Colin Symington 'the next evolutionary step' – became a more general goal during the reassessments of work programmes forced upon metropolitan institutions by World War II and its aftermath, one which also was associated with changes in planned development strategies. A greater role was foreseen for resident workers and institutions, especially in the infratropical zone with which most externally directed 'development' was concerned. French botanists already had regional flora programmes underway for North Africa and Madagascar, and South Africa was to follow suit in the 1950s. The only African regions without such a project were equatorial Africa and the Sudan. An ostensibly complete flora – a 'third-generation work' – was published in the 1950s in the latter, while in the former, no large-scale projects were begun until after independence of its constituent territories. Gabon and Cameroon entered into cooperative agreements for full-scale floras of the regional genre, but not Chad, Congo or the Central African Republic; coverage in this part of Central Africa has thus become distinctly uneven. In southern Central Africa, the prior existence of *Conspectus florae angolensis*, prepared in Portugal, and the Southwest Africa flora project at Munich were recognized by the organizers of *Flora Zambesiaca* who thereby limited the latter to the central and eastern parts of the region.

Since World War II, then, African floristic botany has to a large extent been dominated by the ponderous progress of the great institutional regional floras. Their styles vary considerably as do the resources available for their production; political and economic events have inevitably affected their progress. Lying behind all is the size of the respective flora to be covered. Most of these works, however, are appearing at such a rate as to require many decades for their completion – and that may be not soon enough in the face of rapidly growing conservation and land-use problems to which attention has, repeatedly, been drawn. Most of the regional floras have been run as semi-monographic outlets, in some respects an unhappy compromise between the Berlin tradition of comprehensive systematic studies espoused by Engler and practical needs at regional or national level. The *Flora of west tropical Africa*

(FWTA) remains the only African regional flora with a concise manual-type format – in large part possible because of the relatively extensive and strong cumulation of knowledge existing in the region.

Recalling the original *Flora of tropical Africa* (FTA) and the large-scale works it inspired, interest also revived in treatises covering the whole of tropical Africa, particularly after FTA was definitively abandoned (without plans for revision). In 1957, a Belgian botanist, Jean Léonard, formally suggested (and has subsequently reiterated) that a pan-African work was a legitimate ideal; and in the same year Rodolfo E. G. Pichi-Sermolli proposed a 'Genera plantarum africanarum' (*Mem. Soc. Brot.* **13**: 113–157). Pichi-Sermolli's suggestion was not further pursued; but at Geneva, the Conservatoire et Jardin Botaniques under Jacques Miège, an Africa specialist, adopted and developed a more modest approach. This in effect was a return to the plan adopted by Théophile Durand and Hans Schinz in the 1890s for their *Conspectus florae Africae* but took advantage of improved information technology. The French botanist and Sahel specialist Jean-Pierre Lebrun, together with Miège's associate Adélaïde Stork, joined this project, and in 1991 the first part (of four) of *Énumération des plantes à fleurs d'Afrique tropicale* appeared under the joint authorship of Lebrun and Stork. The final part appeared in 1997, and a summary paper followed in 1998.

Many works of various kinds have also been produced at subregional or more local level, varying with needs, interests and available resources, some emanating from metropolitan institutions but others from the growing number of African botanical institutions and herbaria, some of which have had a checkered history.[4] The 'national' works which have appeared in recent decades largely have taken the form of checklists; keyed descriptive floras have been written only for a few – Egypt, Libya, Algeria, Senegal, Ethiopia, Nigeria, upland Kenya, Swaziland and Mozambique among them. Both resident and overseas botanists have been involved. Some 'dendrofloras' have also been produced, but relatively few local florulas for areas of special interest. This apparent oversight was deplored by André Aubréville who, in turn, was criticized for his stand by 'opponents' of local florulas. More of the latter are surely needed – but their style, contents and format should be carefully considered in relation to their potential audience. In Africa this is by no means uniform and, in many parts, remains relatively small.

An important post-World War II development in African botany – a part of the contemporary general movement towards international cooperation described by Brenan (1953) – was the formation in 1950 of the 'Association pour l'Étude Taxonomique de la Flore de l'Afrique Tropicale' (AETFAT).[5] In addition to a *Bulletin*, produced since 1951, AETFAT has held congresses every three or four years, each followed by published proceedings. These latter give a good idea of ongoing developments, and have often included 'progress reports'. The proceedings of the fourth congress, published in 1962, gave an extensive series of historical reviews which collectively constitute the best available. From 1952 to 1986 (with interruptions) AETFAT produced an annual index to new taxa as well as literature (a function now partly absorbed into *Kew Record*; additions to the AETFAT library in Brussels – now with more than 11000 items – continue in the *Bulletin*). The activities of AETFAT also encompass the southwest Indian Ocean islands and to a considerable extent southern Africa as well. Those interested in North Africa – not part of AETFAT's warrant – have from 1975 been served by the 'Organization for the Phyto-Taxonomic Investigation of the Mediterranean Area' (OPTIMA); this produces a semi-annual newsletter and distributes collected separates as 'Cahiers OPTIMA'.

The current pattern of botanical progress involves an increasing amount of more specialized systematic and biological research and inventory as the general floristic picture has become better known, with a corresponding shift of emphasis in collecting to more critical, detailed work (though this, coupled with continuing 'primary' exploration, has continued to yield on average one new species daily since 1953, albeit on a somewhat declining curve). Associated with this has been the development of a pattern of cooperation between African institutions, where now there are more African botanists, and the metropolitan institutions in the Northern Hemisphere. This has included the appointment in some cases of 'liaison officers', including specialist collectors associated with flora projects. The shifts of emphases have tended to favor the local institution, most clearly demonstrated where research into difficult-to-preserve plant parts or plant groups has been concerned. However, with the semi-monographic nature of most current African regional floras, the more critical approaches now demanded have also reduced quantitative taxonomic productivity (in numbers of species yearly). Nevertheless, the regional floras have already contributed to the creation of a sound taxonomic foundation in African botany and have stimulated much new research and the development of local cadres. In turn, 'national' checklists and floras have begun to make their appearance – some metropolitan, some local (with one of the former, *Flore du Cameroun*, now published therein).

Jean Léonard, who over four decades surveyed the progress of African botany and moreover has contributed a map showing the degree of African floristic exploration (south of the Tropic of Cancer), in 1975 summarized future needs as follows (*Boissiera* 24a: 19):

- Il convient donc d'intensifier la campagne d'exploration floristique de l'Afrique principalement en des régions jusqu'ici peu explorées;
- Il importe de poursuivre, et si possible d'intensifer, la publication des Flores régionales (de préférence du type de Flora of West Tropical Africa) se rapportant à de vastes contrées (grands pays ou groupe de pays);
- Il devient urgent d'envisager la préparation d'une Flora africana afin d'inaugurer la période synthétique de l'étude de la flore africaine.

While the third of these may remain more conjectural (but now, happily, has been partly filled by the already-mentioned *Énumération* by Lebrun and Stork), the first two are in accord with what is the greatest need: adequate documentation in the face of environmental destruction, for effective conservation recommendations and action plans, and for education. The last also calls for the production of more local field floras *in situ* – the third of Symington's evolutionary steps – and related scholastic works, a direction already taken long ago by John Corner in his *Wayside trees of Malaya* (911) and, more recently, cogently argued by Alwyn Gentry for the Neotropics.

The 1980s saw continuation of major regional flora projects such as *Flora of tropical East Africa* (now projected for completion in 2006), *Flora Zambesiaca*, *Flora of Southern Africa* and *Flore d'Afrique Centrale*, and national floras such as *Flore de Cameroun*, *Flore de Gabon*, *Flora of Libya*, *Conspectus florae angolensis*, *Flora de Moçambique*, *Flore du Rwanda* (completed in 1988) and *Flore de la Tunisie* (completed in 1981). A *Flore analytique du Togo* appeared in 1984, Laurent Aké Assi published his multi-faceted *Flore de la Côte d'Ivoire* in

the same year, and two new national floras, the *Flora of Ethiopia* (now *Flora of Ethiopia and Eritrea*) and *Flora of Somalia*, commenced publication respectively in 1988 and 1994. Lebrun continued his series of enumerations for countries in the Sahel and southern Saharan zones, publishing additions for Niger in 1983, an account for Mali in 1986 and one for Burkina Faso in 1991. The *Flore descriptive des Monts Nimba* commenced by Jacques-Georges Adam in the 1960s was completed in 1983, three years after his death; with P. Jaeger, he also prepared a treatment for the nearby Loma Mountains in Sierra Leone (1980–81). These geologically old West African massifs are the highest west of Cameroon Mountain and are of considerable biogeographic importance. However, nearly all bigger projects have remained dependent on outside assistance, in spite of some steps such as printing (the *Flore de Cameroun* is presently printed in that country) and organization and direction (such as the *Flora of Ethiopia and Eritrea*).

Changing interests, available resources, and, in some cases, the sheer numbers of species to be covered are now causing re-evaluation of projects. The latter, along with a perceived lack of resources, temporarily forced abandonment of the *Flora of Southern Africa* project in the early 1990s. There has in some quarters developed a belief that descriptive works are more suited to restricted areas (in view of being 'doable' in a short time) and that larger areas (or species numbers, as in tropical or South Africa) are best coped with in the form of enumerations or checklists derived from databases, also easier to 'do' (a situation which in fact faced Elmer D. Merrill in the Philippines after 1917, resulting, instead of a flora, in his famous *Enumeration*). Political, economic, social and other circumstances have changed; not the least of these is a seemingly lesser sense at higher levels of the kind of long-term commitment and engagement detailed botanical exploration and major flora projects require.

Progress

Many 'progress reports' on African floras and floristics have been published. They are here grouped thematically for convenience, but considerable overlap in content exists. Material relating to North Africa alone appears under Region 59.

State-of-knowledge maps: The first such to appear was J. LÉONARD, 1965. Carte du degré d'exploration floristique de l'Afrique au sud du Sahara. *Webbia* 19: 907–914, map. A revision, by F. N. Hepper, appears in

G. KUNKEL (ed.), 1979. *Taxonomic aspects of African economic botany.* Las Palmas. Further revisions have appeared in one or another of the reports cited below as well as in Lebrun and Stork's *Énumération des plantes à fleurs d'Afrique tropicale* (see **501**).

Africa in general: An early pair of accounts is M. DRAR, 1963. Flora of Africa north of the Sahara. In UNESCO, *A review of the natural resources of the African continent,* pp. 249–261. Paris (Natural resources research 1); and J. KOECHLIN, 1963. Flora of Africa south of the Sahara. In *ibid.,* pp. 263–275. More extensive general surveys, including some historical background, are contained in the detailed zonal state-of-knowledge botanical reports by various specialists in M. CROSBY (convenor), 1978. Systematic studies in Africa. *Ann. Missouri Bot. Gard.* **65**: 367–589. Other reports to 1990 include J. LEBRUN, 1960. Sur la richesse de la flore de divers territoires africains. *Bull. Acad. Roy. Soc. Outre-Mer, Bruxelles,* N.S. **6**: 669–690; J.-P. LEBRUN, 1976. Richesses spécifiques de la flore vasculaire des divers pays ou régions d'Afrique. *Candollea* **31**: 11–15; F. N. HEPPER, 1979. The present stage of botanical exploration in Africa. In I. Hedberg (ed.), *Systematic botany, plant utilization and biosphere conservation,* pp. 41–46, map. Stockholm; J.-P. LEBRUN, 1982. *Introduction à la flore d'Afrique: faits et chiffres.* 90 pp., illus., maps. Maisons-Alfort, France: Division de l'Enseignement, IEMVPT; and D. CAMPBELL, 1990. Rates of botanical exploration in Asia and Latin America; similarities and dissimilarities with Africa. In H.-D. Ihlenfeldt (ed.), *Proceedings of the Twelfth Plenary Meeting of AETFAT,* pp. 155–167 (Mitt. Inst. Allg. Bot. Hamburg 23). The situation by the mid-1990s was reviewed in P. MORAT and P. P. LOWRY II, 1997. Floristic richness in the Africa-Madagascar region: a brief history and perspective. *Adansonia,* III, **19**: 101–115.

Tropical and southern Africa: Many detailed country-by-country historical and current progress reviews have been published in successive proceedings of AETFAT congresses.[6] Particular mention is, however, here made of the following: Histoire de l'exploration botanique de l'Afrique au sud du Sahara. In A. FERNANDES (ed.), 1962. *Comptes rendus de la IV. réunion plenière de l'AETFAT* (Lisbonne, Coïmbre, 1960), pp. 45–248, illus. Lisbon: Junta de Investigações do Ultramar; Progress in the preparation of African floras. In H. MERXMÜLLER (ed.), 1971. *Proceedings of the VII. plenary meeting of the AETFAT* (München,

1970); pp. 13–90. Munich (in *Mitt. Bot. Staatssamml. München* 10); Progrès accomplis dans l'étude de la flore et de la végétation d'Afrique. In J. MIÈGE and A. STORK (eds.), 1976. *Comptes rendus de la VIII. réunion plenière de l'AETFAT* (Genève, 1974); pp. 519–639 (in *Boissiera* **24b**); and 'Flora reports' in *Proceedings of the X. plenary meeting of the AETFAT* (Pretoria, 1982), pp. 1015–1023 (in *Bothalia* **14**(3/4)). Useful information also appears in the (more or less) annual *AETFAT Bulletin.*

Other reports relating to tropical and southern Africa include A. AUBRÉVILLE, 1964. État actuel de la connaissance de la flore phanérogamique tropicale africaine et malgache. *Adansonia,* II, **4**: 47; J. P. M. BRENAN, 1979. The flora and vegetation of tropical Africa today and tomorrow. In K. Larsen and L. B. Holm-Nielsen (eds.), *Tropical botany,* pp. 49–58; and J.-P. LEBRUN and A. L. STORK, 1998. Analyse structurelle de la flore des Angiospermes d'Afrique tropicale. *Candollea* **53**: 365–385. Pertinent comments may also be found in the pantropical reviews of Symington (1943) and Prance (1978), both cited in Part I of this *Guide,* and Campbell and Hammond (1989; see **General references**).[7]

General bibliographies. Bay, 1910; Blake and Atwood, 1942; Frodin, 1964, 1984; Goodale, 1879; Holden and Wycoff, 1911–14; Jackson, 1881; Pritzel, 1871–77; Rehder, 1911; Sachet and Fosberg, 1955, 1971; USDA, 1958. For bibliographies on particular regions, see the appropriate headings.

Divisional bibliographies

Lebrun and Stork's *Énumération des plantes à fleurs* (1991–97; see **501**) includes for each family representative bibliographic references.

FERNANDES, A., 1982. *Bibliografia mais relevante sobre botânica pura e aplicada referente aos países africanos de expressão portuguesa.* 288, [3] pp. Lisbon: Academia das Ciências de Lisboa. (Publicações do II. Centenário de Academia das Ciências de Lisboa.) [Unannotated lists of contributions on pure and applied botany, arranged firstly by countries (Cape Verde, Guinea-Bissau, São Tomé and Príncipe, Angola and Mozambique) and major areas of interest and then by author; analysis of the literature for each country but no locality or subject indices. There is also a history of Portuguese institutions and other bodies concerned with overseas territories and development.]

LÉONARD, J., 1965. Liste des flores africaines et malgaches récentes. *Webbia* 19: 865–867. [List of principal recent floristic works, arranged by countries. Nearly all of them are also included in this *Guide*.]

General indices. BA, 1926– ; BotA, 1918–26; BC, 1879–1944; BS, 1940– ; CSP, 1800–1900; EB, 1959–98; FB, 1931– ; IBBT, 1963–69; ICSL, 1901–14; JBJ, 1873–1939; KR, 1971– ; NN, 1879–1943; RŽ, 1954– .

Divisional indices

ASSOCIATION POUR L'ÉTUDE TAXONOMIQUE DE LA FLORE DE L'AFRIQUE TROPICALE, 1952–86. *AETFAT Index*, 1– . Brussels. [Annual bibliography recording new African taxa as well as floristic and systematic literature, with country index. Its functions have been effectively absorbed by *Kew Record*. However, publication of accessions to the AETFAT library (in Belgium) continues in *AETFAT Bulletin*; these lists are indexed.]

BAMPS, P., 1994. *Répertoire des familles des phanérogames traitées dans les principales Flores de l'Afrique tropicale*. 4th edn. 24, [1] pp. Meise, Belgium. [Tabular summary with families alphabetically arranged and coverage (or not) indicated for several major works.]

LEBRUN, J.-P. and A. L. STORK, 1977. *Index des cartes de répartition: plantes vasculaires d'Afrique (1935–1976)/Index of distribution maps: vascular plants of Africa (1935–1976)*. x, 138 pp. Geneva: Conservatoire et Jardin Botaniques. Continued as A. L. STORK and J.-P. LEBRUN, 1981. *Index des cartes de répartition des plantes vasculaires d'Afrique: complément 1935–1976 [et] supplément 1977–1981 avec addendum A–Z*. [iii], 98 pp., portr. (Études Bot. IEMVPT 8). Maisons-Alfort, France; and *idem*, 1988. *Index des cartes de répartition des plantes vasculaires d'Afrique: complement 1935–1981 [et] supplément 1982–1985*. lxvii, 128 pp., illus., maps (*ibid.*, 13). [Alphabetical species lists, with citation of maps; bibliographies included.]

Conspectus

500

Division in general

For tropical Africa, see **501**; for floras of 'old' mountains, see **502**; for the high-mountain and 'alpine' flora, see **503**; for the Sahara and other parts of dry northern Africa, see **505**. The comprehensive survey by Engler, though not a flora in the strict sense and never completed, remains the most recent continent-wide floristic treatment; in addition, Thonner's keys are still much valued.

Keys to families and genera

THONNER, F., 1908. *Die Blütenpflanzen Afrikas*. xvi, 672 pp., 150 pls. (on 75 ll.), map. Berlin: Friedländer. Supplemented by *idem*, 1913. *Nachtrage und Verbesserungen* ... 88 pp. Berlin. English edn.: *idem*, 1915. *The flowering plants of Africa: an analytical key to the genera of African phanerogams*. xvi, 647 pp., 150 pls. (on 75 ll.), map. London: Dulau (reprinted 1963, Weinheim, Germany: Cramer). [Analytical keys to families and genera of seed plants in Africa, with synonymy, distribution notes, and remarks on cultivation, other uses, etc.; statistical tables; index. An introductory section includes a bibliography of principal works on the African flora.]

DURAND, T. and H. SCHINZ, 1895–98. *Conspectus florae Africae*. Vols. 1(2), 5. Brussels: Jardin Botanique de l'État; Berlin: Friedländer; Paris: Klincksieck.

Enumeration of vascular plants (the flowering plants arranged according to the Bentham and Hooker system) with synonymy, references, and indication of local and overall distribution. [Not completed; covers only Ranunculaceae–Frankeniaceae, monocotyledons and gymnosperms. In part superseded by Lebrun and Stork's *Énumération* (see **501** below).]

ENGLER, A., 1908–25. *Die Pflanzenwelt Afrikas insbesondere seiner tropischen Gebiete. Grundzüge der Pflanzenverbreitung in Afrika und die Charakterpflanzen Afrikas*. Parts 1–3, 5(1). Illus., pls., maps (Die Vegetation der Erde 9). Leipzig: Engelmann. (Reprinted 1976, Lehre: Cramer.)

Parts 2 and 3 of this work ('Charakterpflanzen Afrikas') contain a comprehensive, discursive, illustrated systematic account of the African flora, with emphasis on the families, genera, and more significant or characteristic species and their role in the makeup of the vegetation; included also are detailed accounts of local and regional distribution as well as, in each volume, complete indices. [Not completed; extends from the Pteridophyta to the end of the Archichlamydeae, i.e., the Cornaceae (Englerian system). Part 1 is entirely devoted to a general account of the vegetation and phytogeography of the continent, while part 5(1) provides supplementary material on vegetation in tropical Africa resulting from research subsequent to 1910. Part 4, to have covered the Sympetalae and the non-vascular plants, was never published.]

Distribution maps

JARDIN BOTANIQUE NATIONAL DE BELGIQUE, 1969– . *Distributiones plantarum africanarum.* Fasc. 1– . Loose maps in portfolios. Brussels (later Meise).

Comprises a series of small folios of dot maps of individual species showing their distribution throughout the African continent, with indices. Each fascicle, containing some 30 maps, is devoted to one or more genera. [As of 1999, 43 fascicles, with 1459 maps, had been published.][8]

Pteridophytes

The only work nominally covering the whole of Africa is *Filices africanae* (1868) by Maximilian Kuhn. No recent survey comparable to that of Tryon and Tryon for the Americas has been published.

501

Tropical Africa

Area: just over 20000000 km^2 (Brenan, 1979; see above under **Progress**). Vascular flora: *c.* 30000 species; 26274 seed plants (exclusive of Namibia and Botswana) accounted for by Lebrun and Stork (1998; see **Division 5, Progress**)

What became the comprehensive *Flora of tropical Africa* was initially conceived by W. J. Hooker as a flora of 'Upper Guinea', or present-day West Africa. By 1863, however, with the important new collections coming in from such explorers as Mann, Burton, Speke, Grant, Kirk, Livingstone and Welwitsch as well as such other sources of information as Richard's *Tentamen florae abyssinicae*, expansion of the work to cover all tropical Africa began to be considered. A further recommendation in this direction came from

David Livingstone, the African explorer, in 1864 upon his return to Great Britain from the Zambesi Basin. The enlarged proposal was accordingly adopted as a 'Kew flora'; work began that same year under the direction of Daniel Oliver, with four volumes planned.

This attempt at coverage of more than half a continent – an area as large as North and Middle America *combined* but then little known geographically, let alone botanically – was, even given the optimism of the mid-nineteenth century, an act of hubris. The need for additional volumes soon became clear, and ultimately 10 in all, in 12 physical parts, would be realized. Publication of the first three volumes, of which the first appeared in 1868, proceeded uneventfully. In the mid-1870s, however, Oliver discontinued work in disgust because of loss of access to a key collection and the project was accordingly interrupted. Only in 1891, after a nudge from Lord Salisbury, was it revived – and then as 'extra-official' due to prior Kew commitments to *Flora capensis* as well as other duties. The results, prepared by various botanists, began to appear in 1897 and by World War I had covered all spermatophytes save the Gramineae. Accounts for a considerable proportion of that family were afterwards published in installments (comprising volumes 9 and part of 10) up to 1937; completion was, however, not realized due to termination of the work in the wake of World War II and a shift of corporate interest, already apparent in the 1920s, to the preparation (or revision) of regional floras.

Flora of tropical Africa was bold but arguably premature. For many of the areas encompassed by it decades would pass before even a sketchily adequate basis for botanical knowledge was available. It also inevitably became out of date for 'Upper Guinea', arguably one of its better-known regions; at Kew this was used as partial justification to move to *Flora of west tropical Africa*. Nevertheless, in spite of its now very imperfect coverage, it furnished an effective platform for later taxonomic and floristic work and indeed for some regions has yet to be wholly superseded.

A revised 'FTA' therefore became a key goal of AETFAT following its organization after World War II. As realization of such a project was inevitably a distant prospect, the Conservatoire et Jardin Botaniques, Geneva, began developing a plan for a comprehensive checklist as part of a research programme in tropical African botany. This was initiated in the 1960s under the directorship of Jacques Miège and in the 1970s joined by Adélaïde Stork. This was viewed as the most useful potential contribution given the institution's outstanding library and developing information systems capability (cf. *Med-Checklist* (**601**)). The scheme was further strengthened by the addition of the French botanist Jean-Pierre Lebrun, who later became senior author.

The result was *Énumération des plantes à fleurs d'Afrique tropicale* by Lebrun and Stork, whose four volumes appeared between 1991 and 1997. It furnishes a link among the various regional and national floras completed or underway and, in some cases, breaks ground not covered since FTA or not at all. With its valuable bibliographic references, it moreover is among the kinds of African works most useful for a non-specialist's bookshelf.

LEBRUN, J.-P. and A. STORK, 1991–97. *Énumération des plantes à fleurs d'Afrique tropicale*, I–IV. 4 vols. Maps (Publ. Conservatoire Jard. Bot. *hors série* 7). Geneva.

Annotated checklist of flowering plants; includes all accepted taxa with authorities and sources (abbreviated according to conventions on p. 30 in part I) with their distribution. Relevant monographs, revisions and other taxonomic literature where available are listed under each family and generic heading. The sequence of families partially follows that of the *Flora of west tropical Africa*, but genera and species are listed alphabetically. The introductory section contains a preface (by Rodolphe Spichiger), historical background to the project and (with a francophone emphasis) tropical African botany in general, and chapters on progress, statistics, and richness of floras, as well as a bibliography of 'standard' sources used for the project. Addenda, statistics and indices to genera and families conclude each part. [Part I covers Annonaceae through Euphorbiaceae and Pandaceae; part II (1992) extends the work to Apiaceae; part III (1994) covers monocotyledons; part IV, the remaining dicotyledons.]

OLIVER, D. *et al.*, 1868–77. *Flora of tropical Africa*. Vols. 1–3. London: Reeve. Continued as W. T. THISTLETON-DYER *et al.* (eds.), 1897–1937. *Flora of tropical Africa*. Vols. 4–9, 10(1). London (later Ashford, Kent): Reeve. (Vols. 1–8 partly reprinted, n.d., Ashford.)

Comprehensive, relatively concise descriptive flora of seed plants of infratropical Africa, with keys to genera and species, synonymy (with references), fairly detailed indication of distribution (including citations of *exsiccatae*), grouped into six broad regions (Upper Guinea, North Central, Nile Land, Lower Guinea, South Central, and Mozambique; not, however, corresponding to the regions adopted in this *Guide*), and notes on habitat, special features, etc., where known; indices to all taxa in each volume (no general index). A synopsis of families, together with Bentham's *Outlines of Botany*, appears in volume 1 and a definition of the six regions used is given in volume 7. [Not completed; lacks part of the Gramineae.][9]

Woody plants (including trees): guides

Included here are several dendrological works of the latter half of the twentieth century relating to all or to major parts of tropical Africa. Those of Talifer (1990) and Vivien and Fauré (1985) related principally to closed forest, while those of Aubréville (1950) and Geerling (1988) cover the northern woodland zones. For the southern woodland zone, the only supranational work is that of Coates Palgrave and Coates Palgrave (1957). [For other works, see under the regions or individual polities.]

LETOUZEY, R., 1986. *Manual of forest botany: tropical Africa*. Transl. R. Huggett. 3 parts in 2 vols. 194, 451 pp., text-fig. Nogent-sur-Marne: Centre Technique Forestier Tropical; Yaoundé: Institute de la Recherche Agronomique, Cameroon. (French edns., 1969, 1982.)

A students' manual of which vol. 1 encompasses basic botany and organography and vol. 2 illustrated family treatments with discursive text.

*(a) Guineo-Congolian equatorial closed forest (*forêt dense*)*

TALIFER, Y., 1990. *La forêt dense d'Afrique centrale: identification pratique des principaux arbres*. 2 vols. 1271 pp., illus., folding map (vol. 2). Wageningen: Centre Technique de Coopération Agricole et Rurale (C.T.A.) (for Agence de Cooperation Culturelle et Technique, France, and C.T.A.).

Volume 1 comprises eight artificial analytical keys, each based on a different set of lead characters from habitat through fruit, and 24 'quick keys' each covering a more or less limited range of species; these are followed by a lexicon of vernacular names (pp. 333–426) and glossary. Volume 2 is the dendrology proper, with a general key to families and genera followed by systematically arranged 'fiches' for each species (sequence, pp. 497–506); each 'fiche' has text with facing illustrations, the text embodying botanical and dendrological descriptions, vernacular and botanical names (with derivations of the latter), indication of distribution, and notes. Generic and family headings incorporate additional keys. The main text is concluded by an index to botanical names (pp. 1251–1268) and bibliography (pp. 1269–1271). [Covers Guineo-Congolian closed forests (*Le Massif forestier Guinéen Equatorial*) from Sierra Leone to the African Great Lakes and from roughly between 10°–4° N and 7°–4° S, an area of some 4800 km east–west and 800–340 km north–south from the Benin Gulf eastwards. There is unfortunately no statistical section; moreover the text is reproduced from typescript and the photographs often small and murky. The work nevertheless covers the largest area of any African tropical dendroflora.]

VIVIEN, J. and J. J. FAURÉ, 1985. *Arbres des forêts denses d'Afrique centrale*. viii, 565 pp., illus. (part col.), maps. Paris: Agence de Cooperation Culturelle et Technique.

Practically oriented dendrological atlas of trees of commercial interest of 0.6m diameter or more; the text includes scientific, commercial and local names, description of botanical and technical features, and distribution (with special reference to Cameroon and including maps); indices to generic, specific and vernacular names and references (p. 565) at end. The introductory section includes a map of vegetation formations.

(b) Sudano-Guinean woodlands

AUBRÉVILLE, A., 1950. *Flore forestière soudano-guinéenne [AOF–Caméroun–AEF]*. 523 pp., 115 pls., 40 maps. Paris: Société d'Éditions Géographiques, Maritimes, et Coloniales. (Reprinted 1975, Nogent-sur-Marne: Centre Technique de Forestier Tropical.)

Copiously illustrated forest flora of the forest, woodland, and tree savanna zones of west (and central) Africa; includes some keys to species, limited synonymy, vernacular names, local and extralimital distribution (with range maps and some *exsiccatae*), and extensive notes on habitat, biology, special features, etc.; complete indices at end. An introductory section includes remarks on physical features, climate, features of the forest flora, and taxonomic problems, together with species lists relating to a given special feature, property, use, etc.

GEERLING, C., 1988. *Guide de terrain des ligneux sahéliens et soudano-guinéens*. 2nd edn. [iv], 340 pp., 92 pls., map (Agricultural University of Wageningen Papers 87-4). Wageningen: PUDOC. (Originally published 1982, Wageningen, as *Ligneux sahéliens et soudano-guinéens*.)

Illustrated guide to woody plants of the Sahel and Sudano-Guinean zones of West Africa (map, p. 2); species entries fairly formal, with subheads for habit, leaves, leaf venation, flowers and fruit as well as for habitat and ecological notes, taxonomic commentaries (where appropriate), distribution, and representative *exsiccatae*. Scientific names are accompanied by concise references, essential synonymy, and necessary cross-references to *Flora of west tropical Africa*. The arrangement is alphabetical, by families and genera, and most, if not all, species are figured. The systematic treatment is preceded by an illustrated glossary as well as general artificial keys and 'character groups' with special features, and a general index to all scientific names is given at the end.

MAYDELL, H.-J. VON, 1983. *Arbres et arbustes du Sahel: leurs charactéristiques et leurs utilisations*. See **505**.

(c) Zambesian woodlands

COATES PALGRAVE, O. H. and K. COATES PALGRAVE, 1957. *Trees of Central Africa*. See **520**.

502
'Old' upland and montane regions

See also **503** (ENGLER) and individual regions and units. The relative geological age of most of Africa means there are a substantial number of 'old' mountains scattered across the continent. So far, there have been few attempts to write detailed floras although there are now a considerable number of 'floristic studies' as well as vegetatiological accounts. Of particular interest are, however, the Loma and Nimba massifs in West Africa for which separate florulas are fortunately available; these are described below. Over 2000 species are known from the Nimba and 1576 species and infraspecific taxa from the Loma; many of these are endemic, with some of them relict, and there are two endemic families.

ADAM, J.-G., 1971–83. *Flore descriptive des monts Nimba*. 6 parts. 2181 pp., 1057 pls. (parts 1–4 in *Mém. Mus. Natl. Hist. Nat.*, sér. II/B, Bot. **20, 22, 24, 25**; parts 5–6 independently published). Paris: Muséum National d'Histoire Naturelle (parts 1–4, 1971–75); Centre National de la Recherche Scientifique (parts 5–6, 1981–83).

The second section of this work (pp. 145ff.) comprises a detailed illustrated descriptive flora (1989 species and additional infraspecific taxa) of this isolated, biogeographically important mountain massif (which falls largely within Liberia and the Republic of Guinea, with Mt. Nimba in the latter); keys are lacking but the text includes limited synonymy, vernacular names, *exsiccatae* with localities, summary of extralimital range, notes on habitat, phenology, etc. Indices to all species in each part are provided, with a cumulative index in part 6 (along with a supplement to parts 1–4). The lengthy general section (1) includes an introduction to the region, the basis for the work, and chapters (with special reference to the Liberian portion of the massif) on climate, geology, and vegetation (principal zones 500–1200 m and over 1200 m, with a maximum altitude of 1924 m).

JAEGER, P. and J.-G. ADAM, 1980–81. *Recensement des végétaux vasculaires des Monts Loma (Sierra Leone) et des pays de piedmont.* 2 parts. 301, 397 pp., 92 text-figs., 55 halftones, map (Boissiera 32, 33). Geneva.

Illustrated systematic treatment of the Loma Mountains (northeast Sierra Leone), a Nimba Mountains outlier rising to 1947 m (Mt. Bintumane); comprises an enumeration with limited synonymy, vernacular names (where known), source of record(s) (including *exsiccatae* with localities), generalized distribution in area, habitat, biological and other notes; references and index at end of part 2 (pp. 355–397) following the general chapters on physical features, climate, habitat factors, floristics and phytogeography.

503

'Alpine' regions

The tropical-high mountain flora is well documented in the accounts by Engler and by Hedberg; the latter is less broad in scope but is more critical and detailed, and includes keys. Although its 'alpine' component is small (278 species, according to Hedberg (1968, in *Acta Phytogeog. Suec.* 54)), endemism reaches 81 percent. The North African mountains were surveyed in *Le peuplement végétal des hautes montagnes de l'Afrique du Nord* (1957) by P. Quézel.

ENGLER, A., 1892. *Über die Hochgebirgsflora des tropischen Afrika.* 461 pp. (Abh. Königl. Akad. Wiss. Berlin, Phys.-Math. Cl. (1891), 2). Berlin. (Reprinted 1975, Königstein/Ts., Germany: Koeltz.)

This mainly phytogeographical work incorporates an enumeration of known vascular plants, with synonymy, references, citations of *exsiccatae* with localities and altitudes, phytogeographic categories, and overall distribution; index. The main enumeration is preceded by area lists of plants, with a review and discussion for each, and by an account of botanical exploration, with references. [Covers the mountains from Ethiopia through to Zimbabwe (Rhodesia) and Mozambique, the West African (Cameroon) mountains, the islands in the Gulf of Guinea, and the mountains of Angola.]

HEDBERG, O., 1957. *Afroalpine vascular plants: a taxonomic revision.* 411 pp., 52 text-figs., 12 pls., tables (Symb. Bot. Upsal. 15(1)). Uppsala.

The major part of this work (pp. 21–253) is an illustrated critical enumeration of vascular plants of the alpine zone of central and east Africa, with descriptions of new taxa, keys to species (but not genera or families), full synonymy (with references), detailed indication of local range with citations of *exsiccatae* or other records, and notes and references on ecology, karyotypes, etc. An appendix (pp. 254–371) incorporates extensive taxonomic commentaries on individual species and genera and notes on vicariant taxa, and a bibliography and index to all botanical names conclude the work. The introductory section includes notes on prior botanical exploration, taxonomic concepts, and general notes on Afroalpine vegetation and ecology.

505

Drylands (including the Sahara and Sahel)

There are three major dryland zones in Africa: (1) northern Africa, comprising the vast area of the Sahara and, to its south, the Sahel (with an axis ranging from 17° N in the west to 14° N in the east and passing through Timbuktu, with two recognized vegetation types); (2) northeast Africa (the Ogaden and peripheral areas); and (3) southwest Africa (the Namib and peripheral areas). Only the Sahara has had a complete, discrete flora (Ozenda, 1977); the guide for the Sahel by von Maydell (1983, 1986) is limited to woody plants (as is also that of Geerling; see **580**). The preparation of distribution maps of selected species is, however, currently in progress under the direction of J.-P. Lebrun and has also been essayed by Frankenberg and Klaus (for both, see below). For a general survey of arid steppe floristics and vegetation (on pp. 7–153), see H. N. LE HOUÉROU, 1995. *Bioclimatologie et biogéographie des steppes arides du Nord de l'Afrique: diversité biologique, développement durable et désertisation.* Paris: Centre International de Hautes Études Agronomiques Méditerranéennes. (Options méditerranéennes, sér. B, Études et recherches 10.)

The Sahara

OZENDA, P., 1991. *Flore et végétation du Sahara.* 3rd edn. 662 pp., 186, 60 text-figs. (incl. maps), 18 pls., 1 map (at end). Paris: Centre National de la Recherche Scientifique. (1st edn., 1958, as *Flore du Sahara septentrional et central*; 2nd edn., 1977, as *Flore du Sahara.*)

Part II of this work is an illustrated manual-key to vascular plants, incorporating limited synonymy, generalized indication of local and extralimital range, and brief notes on habitat, special features, etc. Marginal notations indicate references to the first supplementary section, the *Compléments* (part III, pp. 465–589) added for the 1977 edition, while the concluding part IV encompasses a glossary, notes on collecting and preservation, an extensive bibliography, and an index to generic and family names. The introductory part I includes an extensive treatment of floristics, geobotany, and phytogeography of the region (also supplemented in part III with considerations of karyotypes and transitions in the flora); a detailed table of contents is in both French and English. No vernacular names are included. [The limits of this edition and its predecessor encompass the whole of the Sahara from the Atlantic to 20° E longitude (just east of the Tibesti massif) and, southwards, from the Maghreb to the overlap with the Sahelian zone (about 18° N, or the approximate latitude of Noukachott in Mauritania), with some parts to 20° N – this being also the approximate northern limit of *Flora of west tropical Africa* (**580**). The third edition is a corrected reissue of its predecessor with the addition of a 40-page *second supplement* covering all parts of the work.]

The Sahel

MAYDELL, H.-J. VON, 1983. *Arbres et arbustes du Sahel: leurs charactéristiques et leurs utilisations.* 531 pp., illus. (part col.) (Schriftenreihe GTZ 147). Eschborn: Deutsche Gesellschaft für Technische Zusammenarbeit. English edn.: *idem*, 1986. *Trees and shrubs of the Sahel: their characteristics and uses.* 1, iii, 525 pp., illus. (part col.) (Schriftenreihe GTZ 196). Eschborn. (Both editions reissued 1990, Weikersheim: Josef Margraf.)

A practical color-guide to woody plants (pp. 87–385 in the French edition, 98–407 in the English);

entries include illustrations of plants and plant parts, with accompanying text giving names, limited synonymy, phytography, generalized indication of distribution, habitats, uses and properties, propagation procedures, and references (bibliography, pp. 517–531 in the French edition, 505–525 in the English). The general section (through p. 85 and 96 respectively) covers botanical parameters, ecology and geography, propagation and culture, protection, and all forms of direct and indirect utilization. Appendices contain lists of scientific names, families with genera represented, lexica for local languages, samples of *pro formae* for data collection as used in the preparation of the book, a key to spiny species (appendix 5), a lexicon of botanical terms, a table of habitat requirements (pp. 477–487, French edition only), seed weights, seed photographs, and notes on plant protection. [The English edition incorporates substantial revisions, but some of the appendices in the French version have been omitted and in addition it is typographically less attractive.]

Distribution maps

The only thematic sets of distribution maps available are for the Sahara and adjacent parts of northern Africa. Reference must otherwise be made to the *Index des cartes de répartition.*

FRANKENBERG, P. and D. KLAUS, 1980. *Atlas der Pflanzenwelt des Nordafrikanischen Trockenraumes: Computerkarten wesentlicher Pflanzenarten und Pflanzenfamilien.* xix pp., 474 maps on 237 ll., folding map (Arb. Geog. Inst. Univ. Bonn, A, 133). Bonn.

A series of plotter-generated distribution maps of selected families as well as individual species based on a grid of 80 × 80 km; arrangement follows the 12th edition of the Engler *Syllabus.* The introductory section puts the problem and covers sources, method of production, mapping points, and selection of species and families (evidently governed by availability of source maps). Contents are listed on pp. vii–xix. [Covers Africa north of 16° N (about the latitude of St. Louis on the Senegal/Mauritania border and from somewhat south of Timbuktu in Mali).]

LEBRUN, J.-P., 1977–79. *Éléments pour un atlas des plantes vasculaires de l'Afrique sèche*, 1–2. 50 + 40 maps (with overlays) (Études Bot. IEMVPT 4, 6). Maisons-Alfort, Val-de-Marne: Institut d'Élevage et de Médecine Vétérinaire des Pays Tropicaux.

Comprises a series of distribution maps with accompanying descriptive text treating representative species of the drier parts of northern Africa, including the Sahel and the Sahara, with extensions to encompass southwestern Asia and nearby Macaronesian islands. The text contains synonymy (with references), brief descriptions of the plant concerned, habitat preferences, place in the ecosystem, and overall distribution, with some citations of *exsiccatae* and literature sources. No specified order is followed, but lists of the species covered appear at the beginning of each number.

506

Rivers

While rheophyte diversity in Africa is relatively low compared with the Americas and the Asia-Pacific region, no systematic work of continental scope on riverine flora has yet been published. The only available account, now long-dated, is on riverine vegetation: O. DEUERLING, 1910. *Die Pflanzenbarren der afrikanischen Flüsse.* 253 pp., pls., maps. Munich: Ackermann. A manual with special reference to West Africa was in preparation as of 1998.[10]

508

Wetlands

No systematic work of continental scope on aquatic macrophytes has been published. Weedy, aggressive species have been treated and illustrated in H. WILD, 1961. Harmful aquatic plants in Africa and Madagascar. *Kirkia* 2: 1–66.

Bibliography

DAVIES, B. R. and F. GASSE (eds.), 1988. *African wetlands and shallow water bodies/Zones humides et lacs peu profonds d'Afrique: bibliography/bibliographie.* xxiii, 502 pp., maps. Bondy: Éditions ORSTOM. (Collection Travaux et Documents 211.) [In 10 parts (each by a different coordinator or coordinators) covering respectively Africa in general, eight mainland regions and Madagascar (see map, p. ix). Within each part the bibliographic entries are further classified according to schemes depicted in a map or maps following its introduction, listed therein by author and furnished with a taxonomic cross-referential index. A general index to authors, however, is wanting. – Bibliographic entries are not limited to those exclusively on wetland environments or biota; nevertheless, there are some significant botanical omissions.]

Sahara, Sahel and Sudano-Guinean regions

DURAND, J. R. and C. LÉVÊQUE (eds.), 1980–81. *Flore et faune aquatiques de l'Afrique sahelo-soudanienne.* 2 vols. xii, 873 pp., illus. Paris: ORSTOM. (Collection Initiations-Documentations Techniques 44–45.)

A multi-authored, mostly zoological work (in aim comparable to *Ward and Whipple's Fresh-water Biology* in North America); vascular macrophytes by Aline Raynal-Roques in vol. 1 (pp. 63–152). Raynal's treatment comprises a key to genera and an illustrated generic-level systematic enumeration (in the manner of *Die natürlichen Pflanzenfamilien*), with descriptions of genera and notes on individual species as well as their habitats, morphology, biology, distribution, and properties and uses; references (p. 152). An introductory section covers general aspects of morphology and biology as well as methods of collection and study. A map of the Sahelo-Sudanian region as covered in the work appears as part of the general introduction in vol. 1, and a *general* index to genera and higher ranks of all organisms treated in the work is given in vol. 2.[11]

East Africa

Peter's work relates to mainland Tanzania but may be useful throughout the region.

PETER, A., 1928. *Wasserpflanzen und Sumpfgewächse in Deutsch-Ostafrika.* 129 pp., 21 text-figs., 19 pls. (part col.) (Abh. Ges. Wiss. Göttingen, Math.-Phys. Kl., N.F. 13(2)). Berlin: Weidmann.

Illustrated descriptive treatment (with keys) relating to the author's 1913 collections, followed by a systematic enumeration of all known species with local and extralimital distributions. The plates depict representative species and habitats. The introductory part (pp. 3–38) is almost entirely devoted to a consideration of wetland environments and associations.

South Central Africa

G. E. Gibbs Russell's modern treatment for Zimbabwe has unfortunately never formally been published. It includes distribution information for the countries of the Zambesi Basin in addition to its nominal limits.

GIBBS RUSSELL, G. E., 1974. *The vascular aquatic plants of Rhodesia.* xvii, 563 ll., 220 text-figs., graphs, maps. (Diss., University of Georgia, Athens, Georgia, U.S.A.; available in Xerox from University Microfilms International, Ann Arbor, Mich.)

Illustrated descriptive treatment, with artificial keys to genera and, within genera, to species; also includes synonymy with dates and citations, localities with *exsiccatae*, summaries of local, Zambesi-region and extralimital distribution, phenology, and habitat information; no index. References, ll. 530–538, followed by lists of species organized by habitat-type and life-form (a list for Lake Kariba is also included). The introductory part (ll. 1–32) is geographical and vegetatiological, with analyses of the wetland flora.

Southern Africa

Bibliography

GIBBS RUSSELL, G. E., 1975. Taxonomic bibliography of vascular aquatic plants in Southern Africa. *J. Limnol. Soc. South Africa* 1(1): 53–65. [638 references, arranged by author; taxonomic index (families and genera). Individual family treatments in major floras included.]

Region 51

Southern Africa

Area: 1 271 227 km^2 (South Africa, Lesotho and Swaziland). Vascular flora: 23 000 species (H. P. Linder in PD; see **General references**). Arnold and de Wet (1993; see below) accounted for 23 686 taxa (in 21 397 species) of vascular plants in southern Africa including South Africa, Namibia, Botswana, Swaziland and Lesotho. – Included here are all parts of the Republic of South Africa as well as Lesotho and Swaziland. For Namibia and Botswana, see **521** and **524** respectively.

The region is home to the rich and peculiar 'Cape' flora, horticulturally highly esteemed and collectively distinct enough to merit 'kingdom' status in the world-schemes of some phytogeographers, beginning with Diels. Interest in the flora began to develop

from the time of the first Dutch settlements at the Cape in 1652 and records in the Sloane Herbarium show that some 20 people had collected in the area by 1753, the year of publication of Linnaeus's *Species plantarum*. For many, these collections were made merely in the course of stopovers; serious botanical work dates only from the latter part of the eighteenth century although prior to then some consolidated works were produced, notably *Rariorum africanum plantarum* (1738–39) by Johannes Burman. Thereafter, progress was rapid, with the first significant flora, *Flora capensis* by Carl Thunberg, appearing in 1807–13. From 1815 field work increased greatly and by mid-century enough collections were available to justify preparation of a general flora.

A first contemporary synthesis was *Genera of South African plants* (1838), William H. Harvey's first major contribution to southern African botany (a second edition appeared in 1868). Species accounts included *Enumeratio plantarum Africae australis extratropicae* (1834–37) by C. F. Ecklon and C. Zeyher and *Catalogus plantarum . . . Africae australioris* (1837) by J. F. Drège, both vehicles for their many collections. Among the German collectors' sponsors was Otto Wilhelm Sonder in Hamburg. In 1857 Harvey and Sonder joined forces to initiate *Flora capensis*; its first volume appeared in 1859 and covered Cape Colony, Kaffraria and Natal, the three largely English-dominated colonial entities of the period in southern Africa. However, after publication of the third volume in 1865 and Harvey's death the following year, the project lapsed. Increasing demand from the South African colonies, caught up in northwards territorial and economic expansion and with a developing scientific commmunity led by an interested Cape Governor, Henry Barkly, caused *Flora capensis* to be revived in 1877. Under the direction of Kew Gardens, the work, now taking in the whole of southern Africa to the 'great grey-green, greasy' Limpopo River, was all but completed between 1896 and 1925 (a final installment appeared in 1933). Many other works of lesser scope appeared during this time; among these were fern floras beginning with *Synopsis filicum Africae australis* by C. Pappe and R. W. Rawson (1858) and ending with *Ferns of South Africa* by T. R. Sim (1892; 2nd edn., 1915).

The decades between 1877 and 1925 also saw the formation of a number of the major southern African herbaria, botanical gardens, and other institutions. What was long known as the Botanical Research

Institute – since 1989 a principal component of the National Botanical Institute (NBI) of South Africa – was first established in Pretoria in 1903 as part of a new Department of Agriculture for Transvaal; it acquired a Union-wide remit from 1911 and some years later a botanical survey was formally established. Since then the Institute and its main herbarium have been responsible for botanical work at a national level. Important contributions have also been made by the large regional herbaria in Durban and Cape Town (both now also part of NBI, the latter since 1989) and (more locally) by the smaller, mostly university, herbaria. The southern African region is now considered botanically moderately well known, with the Southern and Eastern Cape and patches elsewhere fairly well known and some remote areas still poorly collected.

In the second half of the twentieth century, *Flora capensis* as well as Thomas Sim's *Ferns of South Africa* (1915) became all but obsolete. Recognizing this, the then-Botanical Research Institute at Pretoria initiated in 1957 a project for a new general *Flora of Southern Africa* in 33 volumes. The first volume appeared in 1963; however, progress to date has been slow on account of the large size of the currently known vascular flora and the limited available professional resources. By the 1990s no more than 10 percent of the known species had been revised and published. The pteridophytes were more fortunate: Sim's fern manual was ultimately succeeded by two checklists (respectively in 1969 and 1985) and two descriptive works (respectively by Jacobsen and Burrows in 1983 and 1990) as well as a *Flora* treatment (1986).

Another important NBI contribution is *The genera of Southern African flowering plants* (Phillips, 1926; revised in 1951 and again by Dyer in 1975–76), heir to a tradition of 'generic floras' first established by Harvey. Phillips' *Genera* (and its successor) have long been regarded as models of taxonomic work of their kind and for several decades have been indispensable to southern African botanists (along with, for vegetation and floristics, J. P. H. Acocks' *Veld types of South Africa* of which three editions have appeared, the last in 1988). To these works have been added a classified bibliography (Bullock, 1978) and three editions of a computerized checklist of bryophyte and vascular plant taxa (Gibbs Russell, 1984, 1985–87; Arnold and de Wet, 1993).[12]

To these purely scientific works must be added a large range of more or less semi-popular treatments of different components of the southern African flora. Premier among these is *The flora of South Africa* by Rudolf Marloth (1913–32), a copiously illustrated, discursive work patterned upon Engler's *Die Pflanzenwelt Afrikas* and a forerunner of the many lavish monographs of different groups of 'Cape' plants published in the last six decades. The tree flora is especially well catered for, with four general works to suit different tastes and functions; the most comprehensive is *Trees of southern Africa* by Palmer and Pitman (1972–73).

Coverage at the more local level, however, is still patchy, and what exists is usually in the form of checklists. Substantial descriptive manual-floras have been written only for the Cape Peninsula and Swaziland, both treating over 2000 species each, and a concise manual-key (not completed) for the Transvaal. Significant checklists have been published for parts of the Southern and Eastern Cape, Lesotho, KwaZulu/Natal, Swaziland, the Kruger National Park, and Griqualand West, covering from 1500 to nearly 5000 species each. Some more or less semi-popular works on the woody plants (or trees alone) have been written for the Southern Cape, Natal, the Kruger National Park, the Witwatersrand, and Orange Free State where particular interests have manifested themselves. Many of these works have been produced privately or through local institutions. There are signs that interest in local works is increasing (A. Nicholas, personal communication).

Progress
The period to 1951 is covered in two surveys by Phillips: E. P. PHILLIPS, 1930. A brief historical sketch of the development of botanical science in South Africa. *South African J. Sci.* **27**: 39–80; and a 1951 supplement by the same author in *South African Biol. Soc. Pamphlet* **15**: 19–35. A historical survey (and literature review) are also given by J. Hutchinson in *A botanist in southern Africa* (see **510** below).[13]

Bibliographies. General and divisional bibliographies as for Division 5.

Regional bibliographies
Several bibliographies have been published, the first in 1951 within the second edition of *The genera of South African flowering plants* as indicated below. This was revised and expanded by the Kew botanist (and assistant for southern Africa) Arthur A. Bullock in 1978.

Current literature is processed and maintained within the PRECIS system of the National Botanical Institute of South Africa; one hard copy version has been published (MEYER *et al.*, 1997; see below).

BULLOCK, A. A. (ed. O. A. LEISTNER), 1978. *Bibliography of South African botany (up to 1951)*. i, iii, 194 pp. Pretoria: South African Government Printer (for the Botanical Research Institute, Department of Agricultural Technical Services, Republic of South Africa; part of *Flora of Southern Africa*). [Detailed coverage, organized by authors with a taxonomic/subject cross-referential index. The work succeeds the bibliography on pp. 857–905 in the second edition of *The genera of South African flowering plants* by E. P. Phillips (1951, as *Mem. Bot. Surv. S. Africa* 25; for the third and current edition of that work, see DYER under **510** below).]

FOURIE, D. M. C., 1990. *Guide to publications on the southern African flora*. iii, 43 pp. Cape Town: National Botanical Institute. [An unannotated listing, in the first instance arranged by subjects; short-title author index at end.]

KERKHAM, A. S. (comp.), 1988. *Southern African botanical literature, 1600–1988 (SABLIT)*. 256 pp., illus. (part col.). Cape Town: South African Library. (Grey bibliographies 16.) [820 bibliographic entries with library locations, some also furnished with summaries of contents; index to authors, taxa and subjects at end. A potted history of botany in southern Africa is also given (pp. 13–30, with references). The entries are essentially bibliographic in style, sometimes also with summaries of contents, and include library locations. – The work is oriented towards more general users, with an emphasis on illustration and connoisseurship. – A revision of W. TYRELL-GLYNN and M. R. LEVYNS, 1963. *Flora africana: South African botanical books, 1600–1963*. [viii], 77 pp., pls. Cape Town: South African Public Library.]

MEYER, N. L., M. MÖSSMER and G. F. SMITH (eds.), 1997. *Taxonomic literature of Southern African plants*. v, 164 pp. (Strelitzia 5). Pretoria: National Botanical Institute, South Africa. [Classified bibliography (to 31 March 1996) of taxonomic literature for bryophytes and vascular plants, the titles annotated by three-letter codes for content and arranged alphabetically by families and genera. Individual compilers are acknowledged, and for some taxa critiques are given. The bibliography proper is followed by an index to genera and families.]

Indices. General and divisional indices as for Division 5.

510
Region in general

The classical *Flora capensis* begun by Harvey and Sonder in the mid-nineteenth century is now to a considerable extent obsolete, with the three early volumes (1859–65) chiefly of historical interest. Publication of a successor work, the *Flora of Southern Africa*, has been underway since 1963. However, by 1990–91 only a relatively small proportion of the flora had been covered; a decision was then taken to discontinue the programme (though in the end this was rescinded, and modifications introduced). Partly filling this gap since 1926 have been successive editions of *The genera of Southern African flowering plants*, originally written by E. P. Phillips. Marloth's semi-popular *Flora of South Africa* also remains useful for much information not readily available in the more 'scientific' works. The implementation of PRECIS greatly facilitated the establishment and maintenance of checklists; such have been available since 1984, with changes to the current standard (ARNOLD and DE WET, 1993; see below) published annually in *Bothalia*. Several books are devoted to the trees, and one recent work covers pteridophytes. For floristics and vegetation, good background works are J. P. H. ACOCKS, 1988. *Veld types of South Africa*. 3rd edn., ed. O. A. Leistner (with B. A. Momberg). x, 146 pp., 104 halftones, 5 col. maps, separate col. wall map (Mem. Bot. Surv. S. Africa 57). Pretoria (1st edn., 1952); and B. J. HUNTLEY (ed.), 1994. *Botanical diversity in Southern Africa*. 412 pp., illus. (Strelitzia 1). Pretoria. Also of interest is J. HUTCHINSON, 1946. *A botanist in Southern Africa*. xii, 686 pp., illus., maps. London: Gawthorn.

Keys to and surveys of families and genera

DYER, R. A., 1975–76. *The genera of Southern African flowering plants*. 2 vols. viii, 1040 pp., map. Pretoria: South African Government Printer (for the Botanical Research Institute, Department of Agricultural Technical Services, Republic of South Africa; part of *Flora of Southern Africa*). (Originally published as E. P. PHILLIPS, 1926. *The genera of South African flowering plants*. Pretoria; 2nd edn., 1951 (respectively as Mem. Bot. Surv. S. Africa 10, 25).) [Concise descriptive generic flora of seed plants, with analytical keys; includes synonymy (with references), citations of key floras or revisionary studies (for both genera and families), generalized indication of local and overall distribution of genera, and 1–2 lines of commentary on number of species, preferred habitat, and a few critical remarks; separate complete indices in each volume (for consolidated list, see Dyer, above) with addenda. An introductory section includes technical notes, statistics (nearly 2200 genera in the third edition as against 1645 recorded in 1926), and miscellaneous remarks. Volume 2 was prepared with the assistance of Ms. A. Amelia Obermayer-Mauve and others. The work, like its predecessors, covers the Republic of South Africa (with associated states), Namibia (Southwest Africa), Lesotho and Swaziland, but the bibliography of the earlier editions has here been omitted.][14]

DYER, R. A., 1977. *Flora of Southern Africa: key to families and index to the genera of Southern African flowering plants.* 60 pp. [Pretoria]: Botanical Research Institute, Department of Agricultural Technical Services, Republic of South Africa. [Based on his *Genera of Southern African flowering plants* (see above).]

RILEY, H. P., 1963. *Families of flowering plants of southern Africa.* xviii, 269 pp., 144 col. photographs, 3 maps. Lexington, Ky.: University of Kentucky Press. [Discursive introductory survey, with accounts of each family accompanied by (in most cases) a photograph of a representative species. Arrangement follows the 1959 Hutchinson system. No keys are provided.]

ARNOLD, T. H. and B. C. DE WET, 1993. *Plants of southern Africa: names and distribution.* 825 pp., portr. (Mem. Bot. Surv. S. Africa 62). Pretoria. (Originally published as G. E. GIBBS RUSSELL AND STAFF OF THE NATIONAL HERBARIUM, 1984. *List of species of Southern African plants* (Mem. Bot. Surv. S. Africa 48). Pretoria; 2nd edn. in 2 parts, 1985–87 (*ibid.*, 51, 56).)[15]

Systematically arranged, partly tabular checklist of bryophytes and vascular plants with assigned code, synonymy, references to standard literature, and distribution in eight defined geographical units; excluded taxa and index to all generic names at end. Synonyms are cross-referenced within the list proper, and 'standard literature' is further listed under family and generic headings along with names of reviewers. [Accounts as of 1992 for 23 686 taxa of vascular plants (in 21 397 species) in southern Africa including Namibia, Botswana, Swaziland and Lesotho as well as South Africa. Arrangement of pteridophytes follows that of Schelpe in *Flora Zambesiaca*, while seed plants

follow in the first instance the Englerian system as used in Dyer's *Genera* (with, however, different arrangements for monocotyledons and the genera of grasses).

DYER, R. A. *et al.* (eds.), 1963– . *Flora of Southern Africa.* Vols. 1– . Illus. Pretoria: South African Government Printer (for the Botanical Research Institute, Department of Agricultural Technical Services (later Department of Agriculture and Water Supply)). Accompanied by E. A. SCHELPE and N. C. ANTHONY, 1986. *Flora of Southern Africa: Pteridophyta.* xv, 292 pp., illus., maps. Pretoria.

Large-scale descriptive 'research' flora, with family treatments mainly by specialists and staff botanists; in each family (or portion thereof) are keys to genera and species, full synonymy (with references), literature citations, detailed indication of distribution with citations of *exsiccatae* and other records, summary of overall range, extensive taxonomic commentary, notes on habitat, phenology, special features, etc.; and figures of representative species and/or diagnostic attributes; indices to all botanical names appear at the end of each volume. Volume 1 includes a brief introductory section, with map. [Designed to be used together with *The genera of Southern African flowering plants* (see above). Volumes have been numbered and families arranged according to a modified Englerian sequence; these (or their component fascicles) have, however, been issued only as accounts are ready. A single volume may contain any range of higher taxa.][16]

HARVEY, W. H. and O. W. SONDER, 1859 (1860)–65. *Flora capensis.* Vols. 1–3. Dublin: Hodges, Smith. Continued as W. T. THISTLETON-DYER and A. W. HILL (eds.), 1896–1933. *Flora capensis.* Vols. 4–7. London (later Ashford, Kent): Reeve. (Vols. 1–3 reprinted 1894, London; some volumes reprinted, n.d., Ashford; vols. 4 and 5 reprinted 1973, Lehre, Germany: Cramer.)

Descriptive flora of seed plants; includes analytical keys to genera and species (the latter sometimes represented only by synoptic devices, particularly in vols. 1–3), synonymy (with references), citation of *exsiccatae* with localities (in a more detailed fashion in vols. 4–7), generalized indication of overall distribution, taxonomic commentary, and notes on habitat, etc.; indices to all botanical names in each volume (no general index). [Vols. 1–3 encompass only the Cape Colony, Kaffraria, and Natal (then under British control); later volumes include the whole of southern Africa north to the Tropic of Capricorn and the Limpopo River. By

1925 it was concluded (save for the Cycadaceae sensu lato, published as a supplement in 1933); no revision, however, of the already-obsolescent early volumes was undertaken. The volumes published to 1925 accounted for 11 705 species – now only some one-half of the currently estimated total.][17]

MARLOTH, R., 1913–32. *The flora of South Africa, with synoptical tables of the genera of the higher plants.* 4 vols. (in 6). Illus., pls. (some col.). Cape Town: Darter.

Copiously illustrated, semi-popular descriptive account of the higher plants of southern Africa north to the Tropic of Capricorn and the Limpopo River (between Transvaal and Zimbabwe); includes vernacular names and keys to families and genera together with a consideration of the more important species as well as extensive remarks on their biology, adaptations, and special features. Cross-references to *Flora capensis* and indices are also incorporated. [Although not a flora in a strict sense, this scientifically soundly based work is valuable for its excellent illustrations and biological and other information not readily available elsewhere. In overall style it bears comparison with Engler's *Die Pflanzenwelt Afrikas*; it may also have been inspired by Otto Warburg's three-volume *Pflanzenwelt* (1913–22). Its influence on later works on the flora of southern Africa and its characteristic groups has been considerable.]

Woody plants (including trees)

The study of trees in southern Africa has attracted exceptional interest and for a comparatively small active population there is a wide and varied range of 'tree books' now available. All known trees (at present 1073 species and infraspecific taxa) have been given a three-digit number (with additions since 1972 being intercalary) in the systematically arranged, pocket-sized, semi-official *Nasionale lys van inheemse bome/National list of indigenous trees* (3rd edn., 1987, Pretoria: Dendrologiese Stiftung) by F. von Breitenbach. These numbers are used in the books by Breitenbach and Breitenbach (1992), Palgrave, Palmer, and Palmer and Pitman, all described below.

BREITENBACH, F. VON, 1965. *The indigenous trees of southern Africa.* 5 vols. 345 text-figs. Pretoria: Government Printer (for the Department of Forestry, Republic of South Africa). (Mimeographed.)

Illustrated systematic descriptive account of the trees and larger shrubs of southern Africa, with keys to all taxa, limited synonymy, fairly detailed summary of internal range, vernacular names (in Afrikaans, English, and Bantu lan-

guages), and extensive notes on habitat, ecology, karyotypes, special features, pests and diseases, timber, various uses, and forestry practice; complete index in vol. 1 together with a key to families and bibliography.[18]

BREITENBACH, F. VON and J. VON BREITENBACH, 1992. *Tree atlas of Southern Africa/Boomatlas van Suider-Afrika.* Vol. 1. Illus., maps. Pretoria: Dendrological Foundation.

An illustrated dendrological atlas with descriptive text, halftone photographs and gridded distribution maps with records for each square where a taxon is known; text parallel in English and Afrikaans and inclusive of vernacular names, botanical features, distribution, ecology, taxonomy, conservation, cultivation and derivation of names. References to literature at end of families. Index at end. The general part gives the plan of the work and major references as well as a list of subscribers. [Only gymnosperms published to date.]

COATES PALGRAVE, K. (in association with R. B. DRUMMOND; ed. E. J. MOLL), 1983. *Trees of southern Africa.* 2nd edn. 959 pp., 375 pls. (314 col.), numerous text-figs., maps. Cape Town: Struik. (Slightly revised reissue with extra pages 16a–16l, 1988. 1st edn., 1977.)

Copiously illustrated, amply descriptive semi-popular treatment of trees (including tree-ferns), with keys to all taxa, important synonymy, southern African and Zimbabwean tree numbers, vernacular and standardized names (Afrikaans and English), general indication of distribution, with maps for all species, and notes on habitat, occurrence, phenology, properties and uses, and potential (with stress on arboriculture). The numerous fine illustrations include color plates, some photographic, and line drawings depicting technical details. An introductory section outlines the style and scope of the work and gives conservation rules, a list of southern African herbaria, and a general key to families, while an illustrated glossary, selected bibliography, and index conclude the work. In addition to South Africa, Lesotho and Swaziland, the book extends to include Namibia, Botswana, Zimbabwe, and Mozambique south of the Zambesi.[19]

PALMER, E., 1977. *A field guide to the trees of southern Africa.* 352 pp., 395 text-figs., 32 col. pls., map (Collins Field Guide Series). London: Collins.

Concise well-illustrated field manual-key to the more commonly encountered tree species (some 800 of the total of 1000) of southern Africa; includes vernacular names (Afrikaans, English, and Bantu languages), southern African tree numbers (full list, pp. 304–320), indication of distribution and habitat, and notes on properties, uses, potential, special (and diagnostic) features, etc.; select bibliography (pp. 321–322) and indices to botanical (including synonymous), Afrikaans, English and Bantu- (mainly Zulu-)language names. Illustrations are numerous but compact, mainly stressing diagnostic technical details. In the introductory section is an illustrated glossary and vegetation-zone map as well as technical remarks and aids for the tyro. Based on the author's larger *Trees of southern Africa* (see next entry), this book is a remarkable attempt at compression; almost no other current 'tree book' packs so many species into a limited compass while maintaining high standards.

PALMER, E. and N. PITMAN, 1972–73. *Trees of southern Africa*. 3 vols. 2000 halftones, 900 text-figs., 24 col. pls., map. Cape Town: Balkema.

Popularly oriented, discursive systematic and biological treatment of the native trees (and some large shrubs) of southern Africa south of the Kunene and Limpopo Rivers; includes non-dichotomous keys to genera and species, fairly detailed synonymy (without references), vernacular names (Afrikaans and English), and extensive commentary on taxonomic questions as well as habitat, life-form, phenology, biology, variability, special features, and local and extralimital distribution; indices to all botanical and vernacular names in each volume (general bibliography and indices in vol. 3). The introductory section (vol. 1) incorporates chapters on botanical exploration and research 'personalities', uses, properties, and special features of trees, and vegetation zones. Volume 1 also includes a chronological list of references and the *National list of indigenous trees* (see above under the subheading).

Pteridophytes
Ferns and fern-allies have been covered by E. A. Schelpe with N. C. Anthony in *Flora of southern Africa* (see above). There are also, however, two semi-popular works produced in succession to the long-standard *Ferns of South Africa* by T. R. Sim (1892; 2nd edn., 1915). Associated checklists include E. A. SCHELPE, 1969. A revised checklist of the Pteridophyta of Southern Africa. *J. S. African Bot.* **35**: 127–140; and N. C. ANTHONY and E. A. SCHELPE, 1985. A checklist of the pteridophytes of the 'Flora of Southern Africa' region. *Bothalia* **15**(3–4): 541–544. This last accounts for 250 indigenous and naturalized species.

BURROWS, J. E., 1990. *Southern African ferns and fern allies*. xiii, 359, [5] pp., illus., maps, 56 col. pls. Sandton: Frandsen.

Illustrated descriptive semi-popular atlas of 'Southern African' pteridophytes (inclusive also of Namibia, Botswana, Zimbabwe, and Mozambique south of the Zambezi) with keys, synonymy, types, derivations of names, indication of distribution, and notes on habitat, biology, taxonomy and variability as well as on similar species; list of references (pp. 340–342), derivations of generic names, illustrated glossary, and indices to English and scientific names at end. The general part covers the plan of the work as well as an outline of pteridophyte habitats (with illustrations and lists of representative species). [The text is here more precisely categorized than in Jacobsen's work; on the other hand, there is no comparative ecological table.]

JACOBSEN, W. B. G., 1983. *Ferns and fern allies of Southern Africa*. [18], 542 pp., 373 halftones, 186 distribution maps, 2 text-maps, end-paper maps. Durban: Butterworths.

Illustrated descriptive semi-popular atlas-flora (arranged in accordance with E. A. Schelpe's treatment of pteridophytes in *Flora Zambesiaca*); text includes synonymy (with references and citations), distribution (including maps), and discussions of habitat, biology, phenology and special features. Keys are also provided. The lengthy general part furnishes information on distribution, ecology, life-forms, associations (at a fairly substantial level), and phytogeography as well as the background to the work and acknowledgments; there are also sections on collecting, herbaria, taxonomy, and history of investigations and a glossary of terms. Two appendices show respectively the relationships of species (and habitat) distributions to rainfall and altitude; they are followed by references (pp. 521–524) and indices to subjects as well as names. [Covers South Africa, Botswana, Lesotho, Namibia, Swaziland, southern Mozambique and Zimbabwe.]

511

Western Cape Province

Area: 129 370 km^2. Vascular flora: 8579 species (in the wholly included Southern Cape botanical province; Bond and Goldblatt, 1984, under **Partial works**). Within this area is the historic territory of Cape Colony. For Griqualand West and Bechuanaland, which lie north of the Orange River, see **519**.

No separate floras of the former Cape Province or its successors have been published. Apart from several works of relatively limited geographical scope, the most comprehensive works are, firstly, *Plants of the Cape flora* by Pauline Bond and Peter Goldblatt (1984) and, for woody plants, Sim's forest flora (1907). The former covers the famously rich Southwestern Cape floral region.

Partial works: Southern Cape
Bond and Goldblatt's work has succeeded *Check-list of flowering plants of the divisions of George, Knysna, Humansdorp and Uniondale* by H. G. Fourcade (1941; in *Mem. Bot. Surv. S. Africa*, **20**). The work by Breitenbach (under **Woody plants**) provides a useful introduction to the forest flora of the important temperate moist forest belt near the coast.

BOND, P. and P. GOLDBLATT, 1984. *Plants of the Cape flora: a descriptive catalogue*. xi, 455 pp., 13 illus. (12 col.), end-paper maps (J. S. African Bot., Suppl. 13). Kirstenbosch: National Botanic Gardens of South Africa.

Alphabetically arranged enumeration of seed plants; each entry includes correct name, English and Afrikaans vernacular names, diagnostic features, habitat, phenology, and local and extralimital distribution. Generic entries include the total number of species and approximate range. The general part gives the background to the work and its plan and basis along with concise discussions of physical features, climate, landscape and soils, vegetation, and general botany including an analysis of the flora and its diversity. [Covers the Southern Cape botanical province, an area of 90 000 km^2 with 8504 species of seed plants (and an additional 75 species of pteridophytes).][20]

Partial works: Cape Peninsula

See also BOND and GOLDBLATT (1984).

ADAMSON, R. S. and T. M. SALTER (eds.), 1950. *Flora of the Cape Peninsula.* xix, 889 pp., maps. Cape Town: Juta.

Descriptive manual-flora of vascular plants of the Cape Peninsula from Cape Town southwards; includes keys, limited synonymy, vernacular names, indication of distribution (with many localities), taxonomic commentary, and notes on habitat, phenology, special features, etc., and complete indices. An introductory section gives accounts of physical features, vegetation, and human pressures as well as technical notes and a glossary. The Cape Peninsula, which includes Table Mountain, has perhaps the richest flora for an area of equivalent size anywhere in the world; this manual treats no fewer than 2622 species.[21]

Partial works: interior areas and Namaqualand

No significant separate florulas or checklists evidently exist for any of these areas, which together include the Great Karroo and Namaqualand as well as the dry central and western highlands south of the Orange River.

Woody plants

For a condensed field version of *Southern Cape forests and trees,* see F. VON BREITENBACH, 1985. *Southern Cape tree guide.* iv, 114 pp., illus. (Department of Environment Affairs, Forestry Branch Pamphlet 360). Pretoria.

BREITENBACH, F. VON, 1974. *Southern Cape forests and trees.* [iv], 328 pp., 2 pp. (loose addenda), text-figs., halftones, maps. Pretoria: Government Printer. (Simultaneously published in Afrikaans as *Suid-Kaapse Bosse en Bome.*)

Semi-popular, atlas-style illustrated botanical and dendrological treatment to 100 trees of the temperate moist forest of the Southern Cape (from Mossel Bay through George and Knysna to Humansdorp and formerly known as the 'Knysna forests'), with artificial leaf-key (pp. 87–162) and additional notes on tree species not illustrated or fully treated. The text includes descriptions of habit, leaves,

flowers, fruits, bark and wood along with associates, distribution and relationships and vernacular names. The systematic part and key (the latter on blue paper) are preceded by an extended introductory section on the forests (with maps) and their structure, composition, associated fauna and history. An index to all names in all three sections appears at the end of the work.

SIM, T. R., 1907. *The forests and forest flora of the colony of the Cape of Good Hope.* vii, 361 pp., 161 pls., map. Aberdeen, Scotland: Taylor and Henderson.

The bulk (pp. 99–343) of this quarto work comprises an illustrated descriptive account of the trees and shrubs of the Cape Province south of the Orange River; included are partially synoptic keys to all taxa together with synonymy (and references), fairly detailed accounts of local range, vernacular names, and notes on habitat, timber properties, history of utilization, etc., and a complete index. The lengthy general part is mostly devoted to a discussion of the forests in the region and their composition, ecology, economics, utilization and conservation.

512

Eastern Cape Province

Area: 169 600 km^2. – Within this area is the former territory of Kaffraria as used in *Flora capensis* as well as by Sim in his *Sketch and check-list* (1894). Also included here are the former 'states' of Ciskei and Transkei. Phytogeographically the flora is part of F. White's 'Tongaland-Pondoland' regional mosaic, with the Drakensberg considered to be Afromontane.[22]

Partial works

MARTIN, A. R. H. and A. R. A. NOEL, 1960. *The flora of Albany and Bathurst.* xxiv, 128 pp., map. Grahamstown: Department of Botany, Rhodes University.

Annotated checklist of vascular plants (2390 species), with vernacular names and notes on habitat, occurrence, variability and phenology, preceded by general notes and including an index. The introductory section comprises a survey of the vegetation. [Covers an area centering on Grahamstown, between Port Elizabeth and East London.]

SCHÖNLAND, S., 1919. *Phanerogamic flora of the divisions of Uitenhage and Port Elizabeth.* 118 pp., map (Mem. Bot. Surv. S. Africa 1). Pretoria. [Afrikaans edn., 1920, as *Verhandeling over de fanerogamiese flora van de afdelingen Uitenhage en Port Elizabeth.*]

Briefly annotated enumeration of seed plants (2312 species), with indication of status, localities, flowering times, occurrence in the Cape Peninsula and/or Natal, and some critical notes; statistical summary. An introductory section covers geography, geology, climate, composition of the flora, vegetation formations, and phytogeography with also history of collecting.[23]

SIM, T. R., 1894. *Sketch and check-list of the flora of Kaffraria.* 92 pp. Cape Town: Argus Printing and Publishing (for King William's Town Natural History Society).

Systematic list of vascular plants (2449 species), with vernacular names and essential synonymy; no index. An introduction includes accounts of physical and biological features of the region, botanical exploration, vegetation formations, and aspects of utilization. Encompasses what is now the Eastern Cape, encompassing Port Elizabeth and East London along the southeastern coast. [Now well out-of-date; partly superseded by the checklists of Martin and Noel and of Schönland (see above).]

513

Lesotho

Area: 30460 km². Vascular flora: 1591 species (partly from Jacot Guillarmod, 1971). This state, under direct British rule before independence, was formerly known as Basutoland.

JACOT GUILLARMOD, A., 1971. *Flora of Lesotho (Basutoland).* 474 pp., maps. Lehre: Cramer.

Systematic enumeration of non-vascular and vascular plants, without keys or descriptions; includes limited synonymy, detailed indication of local range (with *exsiccatae* and other records), and summary of extralimital distribution; lexicon of Sotho names with equivalents; lists of plant uses; bibliography; and index to generic and family names. The introductory section includes accounts of the history, physical features, land utilization, and vegetation of Lesotho as well as a statistical analysis, a cyclopedia of collectors, and a gazetteer.

514

KwaZulu/Natal

Area: 92180 km². Vascular flora: 4818 seed plant species (Ross, 1972); further work by 1980 had increased this to 5000 (J. H. Ross, personal communi-

cation). The province is by southern African standards comparatively well supplied with provincial-level floristic works. Until 1972 the standard work was *An introduction to the flora of Natal and Zululand* (1921) by J. W. Bews.

ROSS, J. H., 1972 (1973). *The flora of Natal.* 418 pp., 5 text-figs. (incl. 2 maps) (Mem. Bot. Surv. S. Africa 39). Pretoria.

Comprises a systematic enumeration of seed plants, with diagnoses of and keys to families and genera but having species only listed; includes essential synonymy, brief indication of local range with citation of representative, mostly recent *exsiccatae*, and taxonomic commentary; index to generic and family names at end. The introductory section includes a glossary and list of references.

WOOD, J. M. (and M. S. EVANS), 1898–1912. *Natal plants.* Vols. 1–6. 600 pls. Durban: Natal Government and Durban Botanic Society. (Reprinted 1970 in 1 vol., Lehre, Germany: Cramer.)

Comprises an atlas of seed plants, with plates and accompanying descriptive text; the latter includes references to *Flora capensis*, generalized indication of local range (with some citations of *exsiccatae*), taxonomic commentary, and notes on habitat, uses, etc.; index to all taxa in each volume (no general index). No systematic order is followed in this work, which was discontinued before completion. [M. S. Evans was joint author of vol. 1.]

Woody plants

The works of Moll and Pooley effectively succeed J. S. HENKEL, 1934. *A field book of woody plants of Natal and Zululand.* xii, 252 pp., 2 pls. Pietermaritzburg: Natal University Development Fund Committee.

MOLL, E. J., 1992. *Trees of Natal.* 2nd edn. xx, 542 pp., illus., maps, diagrams. Cape Town: Eco-lab Trust Fund, University of Cape Town. (1st edn., 1981; revised and expanded from *idem*, [1967]. *Forest trees of Natal.* 180 pp., illus., maps. [Pietermaritzburg].)

Concise field-guide to over 800 species (numbered according to the National Tree List (see **510**) but artificially arranged in 10 major form-groups), with keys to all taxa (main key, pp. 1–2 and in inside back cover); entries include vernacular (Afrikaans, English and Zulu) and scientific names and indication of heights, phenology, habitat and diagnostic features, along with distribution maps (with symbols) and drawings (usually of leaves). Introduced trees, along with some shrubs and climbers, are included. A brief introduction includes a guide to the use of the book along with vegetational features, and a glossary, references and complete index conclude the work.

POOLEY, E., 1993. *The complete field guide to trees of Natal, Zululand and Transkei.* 512 pp., illus., col. photographs, maps (incl. end-papers). Durban: Natal Flora Publications Trust.

Semi-popular field-guide to 780 species (inclusive of some shrubs) comprising text with facing photographs; features artificial keys, descriptions, botanical and vernacular names, indication of distribution and habitat, small diagnostic figures, and notes on special features and arboriculture. References (p. 494) and a complete index conclude the work.

515

Northern Provinces (Gauteng, Mpumalanga, Northern and North West)

Area: 336 650 km^2. Vascular flora: 5700 species (Retief and Herman, 1997; see below). – Comprises North West, Gauteng, Northern and Mpumalanga Provinces, thus including the whole of former Transvaal along with the northeastern part of former Cape Province (encompassing Mafikeng). For separate works on Kruger National Park, see **517**.

All previous general floras were in 1997 supplanted by *Plants of the northern provinces of South Africa* by Retief and Herman (see below). This concise work succeeds *A first check-list of the flowering plants and ferns of the Transvaal and Swaziland* by Burtt-Davy and Pott-Leendertz (1912; in *Ann. Transvaal Mus.* 3: 119–182), its supplement (1920; in *ibid.*, 6: 119–135), and *A manual of the flowering plants and ferns of the Transvaal with Swaziland, South Africa* by Burtt-Davy (1926–32, not completed).[24] Miscellaneous local checklists and other works exist, the most substantial being a 'tree book' for the Witwatersrand (in Gauteng), the great gold-mining district within which Johannesburg is centered.[25]

RETIEF, E. and P. P. J. HERMAN, 1997. *Plants of the northern provinces of South Africa: keys and diagnostic characters.* vii, 681 pp., end-paper maps (Strelitzia 6). Pretoria.

Concise manual of vascular plants with keys, diagnostic descriptions (with key characters italicized), limited synonymy, a representative *exsiccata* for each

taxon, indication of phenology and habitat, and distribution codes with presence or absence in five 'subregions' (keyed to the maps). Generic headings include references to standard treatments in floras or other works as well as National Botanical Institute standard taxon numbers. The general part covers the background to the work, geography, climate, soils, general features of vegetation and flora, descriptions of the subregions (with included quarter-degree grids in table 1 on p. 4), plan of the work, abbreviations, and key to main groups (with all subsequent keys in main sections or following families and genera). Within the four main vascular groups arrangement of families, genera and species is alphabetical. A general bibliography (pp. 652–653), an outline of the Cronquist system, a glossary (illustrations on pp. 670–671), and an index to all family and generic names conclude the work. [Covers 5700 species – nearly 25 percent of the southern African flora including Botswana and Namibia – in an area of 325 000 km^2 inclusive of North West, Gauteng, Northern and Mpumalanga Provinces (corresponding to former Transvaal Province and the northeastern part of former Cape Province). Unfortunately there is no statistical table.]

Partial works

LOWREY, T. K. and S. WRIGHT (eds.), 1987. *The flora of the Witwatersrand*, 1: *The Monocotyledoneae.* xi, 365 pp., illus. Johannesburg: Witwatersrand University Press.[26]

Illustrated manual-enumeration of native and introduced species with diagnostic analytical keys; species accounts feature additional key features, habitat, English vernacular names, biology and phenology; glossary and index at end. A general part covers physiography, geology and vegetation as well as use of the book. [Covers an area of 435 km^2 in Gauteng Province.]

TREE SOCIETY OF SOUTHERN AFRICA, 1969. *Trees and shrubs of the Witwatersrand.* 2nd edn., illus. by B. Jeppe. xxi, 309 pp., 126 text-figs. (some col.). Johannesburg: Witwatersrand University Press.

Semi-popular illustrated guide, with pictured-keys to species, field descriptions, vernacular names, local range, and notes on habitats, special features, etc.; glossary, references and complete index. The introductory section includes an indication of the area covered as well as propagation notes.

516

Swaziland

Area: 17363 km². Vascular flora: 2715 native and 110 naturalized species (Kemp, 1983, *Flora checklist*). Prior to independence, this state was under direct British rule.

In recent years, it has been supplied with a descriptive flora, checklists and a 'tree book', which supersede the outdated treatments of BURTT-DAVY (**515**).

COMPTON, R. H., 1966. *An annotated check list of the flora of Swaziland.* iii, 191 pp., 3 maps (J. S. African Bot., Suppl. 6). Kirstenbosch/Mbabane.

Annotated list (pp. 25–79) of vascular plants, with notes on local range, habitat, vegetation formation class, life-form, etc. An extensive appendix incorporates descriptive notes on individual plant families, as represented in Swaziland, a lexicon of vernacular names, and an index to family and generic names. The introductory section contains descriptions of vegetation formations and a list of symbols.

COMPTON, R. H., 1976. *The flora of Swaziland.* 684 pp. (J. S. African Bot., Suppl. 11). Kirstenbosch/Mbabane.

Concise descriptive manual-flora of seed plants with keys to genera and species, Swazi vernacular names, indication of distribution, with localities (list, pp. 7–8), representative *exsiccatae*, and altitudinal zonation (by three veld belts), and casual notes on habitat, phenology, etc.; appendix (pp. 675ff.) of name changes from the checklist of 1966; index to family and generic names. An introductory section, with dedication to King Sobhuza II, includes an account of botanical exploration (beginning with Ernest Edward Galpin in 1886), the background to the book, and technical notes. No family key is provided (see that of Dyer under **510**). For additions, see E. S. KEMP, 1981. *Additions and name changes for* The Flora of Swaziland. 74 pp. Lobamba: Swaziland National Trust Commission.[27]

KEMP, E. S., 1983. *A flora checklist for Swaziland.* i, 5, 5, 101 pp. (Occasional Papers, Swaziland National Trust Commission 2). Lobamba.

Unannotated systematic checklist (pp. 1–77, covering 2715 native and 110 naturalized species), followed by an appendix of additions to the flora since 1981 (with indication of grid squares and *exsiccatae*) and an index to genera and families, a map (p. 99), and references (p. 101). The introductory section gives the rationale and basis for the work (in part an update of Compton's 1966 checklist).

Woody plants

For an introductory work on the tree flora, see E. S. KEMP, 1983. *Trees of Swaziland.* vi, 59 pp. (Occasional Papers, Swaziland National Trust Commission 3). Lobamba.

517

Kruger National Park

See also **515** (RETIEF and HERMAN). – Area: *c.* 19200 km². This great wildlife reserve, created in 1926 and one of the largest in Africa, is located in Northern and Mpumalanga Provinces east of the Drakensberg divide from the Swazi border to the Limpopo, with Mozambique on the eastern boundary. Apart from a checklist of the flora (van der Schijff, 1969) there is a fine work on the trees (van Wyk, 1972–74) partly supplanting Codd's sketchy treatment of 1951.

SCHIJFF, H. F. VAN DER, 1969. *A check-list of the vascular plants of the Kruger National Park.* [ii], 100 pp. (Publ. Univ. Pretoria 53). Pretoria.

Includes an annotated checklist of 1838 vascular plant species, with limited synonymy, localities with citations of *exsiccatae*, and indication of habitat, etc. The introductory section covers plant communities and their significance in relation to park management, biogeography and the affinities of the flora, new records, statistics, and references.

Woody plants

Van Wyk's 'desktop' *Trees of the Kruger National Park* is accompanied by a compact field-guide covering 380 species: P. VAN WYK, 1994. *Field guide to the trees of the Kruger National Park.* 3rd edn. 272 pp., line drawings, numerous col. photographs Cape Town: Struik (with National Parks Board of Trustees, Pretoria). (1st edn., 1984; 2nd edn., 1990.) Both works were preceded by E. W. CODD, 1951. *Trees and shrubs of the Kruger National Park.* 192 pp., 65 text-figs., 6 pls., map (Mem. Bot. Surv. S. Africa 26). Pretoria.

WYK, P. VAN, 1972–74. *Trees of the Kruger National Park.* 2 vols. xxii, 597 pp., 200 col. pls., maps. Cape Town: Purnell.

Copiously illustrated semi-popular descriptive treatment of trees with keys, vernacular names, many distribution maps as well as indication of localities, and notes on habitat, biology, special features, etc.; indices to all names at end.

518
Free State (Vrystaat)

Area: 129 480 km². Vascular flora: no data. – The Free State has had a relatively small botanical literature. The field-guide to woody plants by Johann Venter, now revised, was its first general floristic work as well as one of the few in South Africa also in Afrikaans.

Woody plants

VENTER, H. J. T. and A. M. JOUBERT, 1984. *Klimplante, bome en struike van die Oranje-Vrystaat/Climbers, trees and shrubs of the Orange Free State.* 356 pp., illus. Bloemfontein: de Villiers. (1st edn. by Venter alone, 1976, as *Bome en struike van die Oranje-Vrystaat.*)

Atlas-style, artificially arranged field-guide to native and naturalized climbers and erect woody plants 1m or more in height; includes small technical figures and distribution maps for each species along with bilingual text treating diagnostic features, habit and habitat and Afrikaans and English vernacular names. An illustrated glossary and (pp. 335–336) list of references are provided at the end, and the introductory section includes a general key (additional keys appear before each form-group of species).

519
Northern Cape Province

Area: 361 800 km². – This area was part of the former Cape Province, but now no longer includes the area around and to the west of Mafikeng (presently in North West Province; see **515**). It encompasses such territories as western Bechuanaland, Great Bushman Land and Griqualand West. The last-named incorporates the diamond-mining center (and provincial capital) of Kimberley, while in northwestern Bechuanaland is the remote Kalahari Gamsbok National Park. The mostly thinly populated and botanically still imperfectly known province is not covered

by any general works apart from those for South Africa as a whole. Only one local florula is available (Wilman, 1946; see below).

Partial work: Griqualand West

WILMAN, M., 1946. *Preliminary check list of the flowering plants and ferns of Griqualand West (South Africa).* vii, 381 pp., map. Cambridge, England: Deighton Bell; Kimberley: Alexander McGregor Memorial Museum.

Systematic enumeration of vascular plants of the Kimberley area and other parts of Griqualand West, without keys; includes cross-references to *Flora capensis*, vernacular names, local range (with citation of *exsiccatae*), and notes on habitat, ecology, phenology, uses, etc.; glossary and indices.

Region 52
South Central Africa

Area (inclusive of the Caprivi Strip but exclusive of Angola and the rest of Namibia, i.e., that of *Flora Zambesiaca*): 2 643 861 km²; inclusive of those states the area is 4 714 000 km². Vascular flora: *c.* 9860–9900 species (excluding Angola and Namibia; perhaps 11 000 for the whole region). For the *Flora Zambesiaca* area Morat and Lowry (1997; see **500**, **Progress**) give a figure of 8000 species. South Central Africa here comprises the Caprivi Strip, Botswana, Namibia (former South West Africa), Zimbabwe, Zambia, Malawi, Angola and Mozambique. Four serial floras, three of them still in progress, collectively cover the area: *Prodromus einer Flora von Südwestafrika*, *Conspectus florae angolensis*, *Flora de Moçambique* and *Flora Zambesiaca*.

Scattered collecting along the coasts and in the interior was accomplished in the seventeenth and eighteenth centuries, notably in Angola and Mozambique after 1780 following the inception of the Portuguese *Viagens Philosophicas*. In the mid-nineteenth century the Zambezi Basin was more fully explored by David Livingstone while notable field work was done by Friedrich Welwitsch in the western part of the region. Only after 1870, however, did serious botanical collecting begin with penetration by more European expeditions and subsequent imposition of colonial rule. The first overall checklist (for the newly established British territories) appeared in 1898, and collections of the day were also worked into *Flora of tropical Africa*. In 1903 a German team under the leadership of H. Baum crossed Africa from the Kunene to the Zambesi. Its botanical results, written up at Berlin by Otto Warburg for Baum's *Kunene-Sambesi-Expedition* (1903), long remained the only other transterritorial floristic account before *Flora Zambesiaca*. In this period, also, began the association of the British Museum (Natural History) with the region through its production of several expedition botanical reports. Greater links with Kew and the Imperial Forestry Institute, Oxford, developed from the 1920s onwards; the latter contributed considerably to the preparation of accounts of the woody flora. There was also at this time a significant increase in Portuguese activities in their colonies, including systematic geographical and biological surveys or *missões*. The first regional herbarium was begun at present-day Harare in 1909, and remains the largest (and most active) in this part of Africa.

Following World War II and the abandonment of *Flora of tropical Africa*, there was an intensification of development activities amongst all the responsible powers. In the 1950s, following the creation of AETFAT and, significantly, the Federation of Rhodesia and Nyasaland, exploration of the possibility of a new regional flora began; the memorandum of understanding that created *Flora Zambesiaca* was signed between Britain and Portugal in 1954. With *Conspectus florae angolensis* having commenced publication in 1937 and with Namibia accounted for by research work towards the *Prodromus einer Flora von Südwestafrika*, the organizers of the new flora limited its scope to the central and eastern polities, i.e., the Zambesian subregion. The first part of a projected 10 volumes (later increased to 12) appeared in 1960.

Although now quite far advanced after some 40 years, production of the *Flora* has been at times relatively slow. External events, particularly in the 1970s, are only partially responsible. A major limiting factor has been a continuing adherence to publication according to a predetermined family sequence (in the tradition of the earlier Kew floras) rather than in family fascicles (as in *Flora of tropical East Africa*) or in volumes of randomly revised families (as in *Flora Malesiana*) – while at the same time depending upon contributions from a large number of specialists. This has been overcome by partly 'bypassing' certain volumes (allocated entirely to a single family) and dividing some of them into parts; and with an improved environment the work is now proceeding more quickly, more than half of the species at present being in print. A similar technical problem arose with *Conspectus florae angolensis*, but after a hiatus of several years the sponsors and editors moved to a fascicular format. In 1969 *Flora de Moçambique*, like the *Conspectus* sponsored by the Junta de Investigações do Ultramar (now the Instituto de Investigação Científica Tropical) in Portugal, commenced publication and, save for the years 1974–79 and 1983–90, is continuing.

The only large-scale national work so far completed is the Namibian *Prodromus* (1966–72), published following 15 years of preparation. It represents an object lesson to flora-writers in careful planning, extensive preparation, thorough distillation, and rapid publication (but not so as to break one's pocket). The work is in many ways reminiscent of Merrill's *Enumeration of Philippine flowering plants*, but here improved by the addition of keys and short descriptions.

With *Flora Zambesiaca* in progress, no general floras for Malawi, Zambia, Zimbabwe and Botswana have been published. The sole national-level coverage is in checklists – available only for Botswana, Malawi and Zimbabwe – but there are also some florulas and other accounts of more restricted scope, and the tree flora has by now been relatively well treated (except for Angola and Namibia), both locally and (in southern parts, including Namibia) by southern African works (**510**). Among tree books, *Trees of central Africa* (Coates Palgrave and Coates Palgrave, 1957) is particularly handsome, with colored paintings and descriptive text.

Botanical exploration to date has been considerable but somewhat patchy, and extensive remote areas are still poorly known.

Progress

South Central Africa was not specifically reviewed in *Floristic inventory of tropical countries* (Campbell and Hammond, 1989; see **General references**). Some reports have appeared in AETFAT proceedings and elsewhere, among them respective accounts for anglophone and lusophone areas by Launert and Mendes in *Bothalia* 14(3/4): 1015–1023 (1983).

Bibliographies. General and divisional bibliographies as for Division 5.

Regional bibliography

POPE, G. V. and R. K. BRUMMITT, 1991. *A bibliography for Flora Zambesiaca*. 26 pp. Kew: Royal Botanic Gardens. [Unannotated list, arranged by countries with a chronological index. Covers the more important floristic literature but does not account for monographs, revisions, or ecological, phytogeographical and other botanical literature without species lists.]

Indices. General and divisional indices as for Division 5.

520

Region in general

See also **501** (LEBRUN and STORK; OLIVER *et al.*). *Flora Zambesiaca* does not cover Angola and, in Namibia, only the Caprivi Strip (**523**).

EXELL, A. W., E. LAUNERT and G. V. POPE (eds., for FLORA ZAMBESIACA EDITORIAL BOARD), 1960– . *Flora Zambesiaca*. Vols. 1–12 (in progress). Illus. (some col.), maps. London: Crown Agents (for the Flora Zambesiaca Managing Committee, through 1978); Natural History Museum Publications (through 1992); Kew: Flora Zambesiaca Managing Committee (from 1993); Lisbon: Instituto de Investigação Científica Tropical. Accompanied by E. A. SCHELPE, 1970. *Flora Zambesiaca, Supplement: Pteridophyta*. 254 pp., map. London.

Illustrated, detailed 'research' flora of vascular plants, with keys to all taxa; includes full synonymy (with references and citations), symbolic summary of local range, with indication of *exsiccatae*, and generalized summary of overall distribution, taxonomic commentary, and notes on habitat, special features, etc.; colored frontispiece and indices to all botanical names in each volume. Volume 1 includes a general introduction in Portuguese and English, a historical account of botanical exploration, a glossary, and a list of principal references, along with a general key to families in English (Portuguese version in vol. 2). Covers the entire region except Angola and Namibia. [In progress. As of the end of 1999 the Pteridophyta and volumes 1, 2, 3(1), 4, 5(1–2), 6(1), 7(1, 2 and 4), 8(1–3), 9(1, 2, 4 and 6), 10(1–3), 11(1–2) and 12(4) had been published. Coverage of the vascular flora has increased to between

50 and 60 percent (of a total which has increased by over a third since inception of the project). Yet to appear are, among others, the remainder of the numerous Fabaceae (vol. 3), Rubiaceae (vol. 5), Asteraceae (vol. 6), Asclepiadaceae (vol. 7(3)), Poaceae (vol. 10) and the Euphorbiaceae.][28]

Partial work

The following is now very much out of date but covers some taxa not so far encompassed in *Flora Zambesiaca*.

BURKILL, I. H., 1898. List of the known plants occurring in British Central Africa, Nyasaland, and the British territory north of the Zambesi. In H. H. Johnston (ed.), *British Central Africa*, 2nd edn., App. II, pp. 233–284, 284a–284l. London: Methuen.

List of vascular and non-vascular plants recorded from British Central African territories (corresponding to Zimbabwe, Zambia and Malawi, with emphasis on the last two named); includes limited synonymy and local distribution (with citation of some *exsiccatae*), but no index. A brief introductory section includes a short history of botanical exploration.

Trees

The southern part of the region (Botswana, Zimbabwe and southern Mozambique) is also included in *Trees of southern Africa* by K. Coates Palgrave (see **510**).

COATES PALGRAVE, O. H. and K. COATES PALGRAVE, 1957. *Trees of Central Africa*. xxviii, 466 pp., 110 col. pls., 267 halftones, col. frontisp., map. [Salisbury]: National Publications Trust, Rhodesia and Nyasaland.

Semi-popular, richly illustrated 'coffee table' guide to the more common and/or conspicuous trees of present-day Zimbabwe, Zambia and Malawi; presented in atlas format with paintings (by the senior author) and relatively detailed descriptive text (by the junior author) as well as notes on habitat, biology, special features, uses, folklore, etc.; index.

521

Namibia

Area: 824293 km² (inclusive of the Caprivi Strip). Vascular flora: 3159 species (Merxmüller in PD; see **General references**). The Caprivi Strip, the only part of Namibia included in *Flora Zambesiaca*, is here also a distinct unit (**523**). The rest of Namibia is, in addition to the recent flora by Merxmüller and collaborators, variously covered by works on southern Africa (see **510**). Some of these reach only the Tropic of Capricorn, which passes not far south of Windhoek, while others – generally the more recent, including Dyer's edition of *Genera of southern African flowering plants* and the tree book of Palgrave (1983) – extend right up to the Angolan boundary.

Serious exploration of Namibia began in the era of German rule (1884–1914); it is best symbolized by the work of the botanist and forester Kurt Dinter, author of the first organized account of the flora: *Index der aus Deutsch-Südwest-Afrika bis zum Jahre 1917 bekannt gewordenen Pflanzenarten I–XXVII* (1917–28, in *Repert. Spec. Nov. Regni Veg.* **15–25**). Another leading early collector was Hans Schinz, botanist to an 1884–87 exploring expedition led by A. E. Lüderitz. After 1918, however, scientific activity declined sharply and in botany did not effectively revive until the mid-1930s with increased interest on the part of southern African botanists. Contacts with Germany were resumed, and in that decade Berlin botanists undertook preliminary studies towards a flora. Following the close of World War II, more residents became active and, following an extended visit in 1952 by the German botanist Heinrich Walter and his wife, the present Namibian Herbarium was established. Shortly afterwards, an association with the Bayerische Botanische Staatssammlung in Munich came into being. This institute had, firstly under Karl Suessenguth and (from 1955) Herman Merxmüller (who visited Namibia in 1957), initiated work towards a flora of the then-territory, with preliminary papers appearing since 1950. The Munich project, first conceived as a full flora, was completed in 1972 as *Prodromus einer Flora von Südwestafrika*; this remains the standard work. There have been many other contributions in the last half-century, a number of them in the local journal *Dinteria*.[29]

Bibliography

GIESS, W., 1989. *Bibliography of South West African botany*. 236 pp. (Wissenschaftliche Forschung in Südwestafrika 18). Windhoek: SWA Wissenschaftlichen Gesellschaft. [3158 items; introductory account and historical survey in Afrikaans, English and German.]

MERXMÜLLER, H. (ed.), 1966–72. *Prodromus einer Flora von Südwestafrika*. 35 fascicles. 2188 pp., maps. Lehre, Germany: Cramer.

Enumeration of vascular plants with analytical keys to genera and descriptive–diagnostic keys to species, full relevant synonymy with references, types (or country from where first described) and citations, localities with *exsiccatae* and other sources, concise summary of local distribution, taxonomic commentary, and notes on special features; no individual indices. A map and index to families appear on the covers of each part. Fascicle 35 includes a brief introduction to the work, a list of family names, a map of districts, a general key to families (pp. 9–60), and complete index (pp. 61–175). For additions, see H. ROESSLER and H. MERXMÜLLER, 1976. Nachträge zum *Prodromus einer Flora von Südwestafrika. Mitt. Bot. Staatssamml. München* **12**: 361–373.

522

Angola

Area (country): 1 245 790 km^2; exclusive of Cabinda: 1 238 520 km^2. Vascular flora (exclusive of Cabinda): *c.* 5000 species (Lebrun and Stork 1991; see **501**), a figure surely too low.

Portuguese contact with what is now Angola began in 1483, but significant botanical exploration did not begin until late in the eighteenth century, following establishment of the Real Museu e Jardim Botânico da Ajuda in Lisbon. Outstanding explorers of the nineteenth century included Christian Smith and Friedrich Welwitsch. A set of the latter's collections in London became the basis for the first large floristic account for Angola, *Catalogue of African Plants collected by Dr. Friedrich Welwitsch in 1853–1861* (6 parts in 2 vols., 1896–1901), edited by staff at the then-British Museum (Natural History). Collections were also made in this period by other European visitors, including Heinrich Wawra with the exploring voyage of the 'Carolina'. Collections from the latter resulted in *Sertum Benguellense* (1860) by Wawra and J. Peyritsch. In 1903, the German Kunene-Zambesi expedition crossed the southern part of the country; its plants were written up by Otto Warburg in *Kunene-Sambesi-Expedition* by H. Baum (1903). For a half-century from 1900, an important further basis was provided by the collections of John Gossweiler, a Swiss attached to the provincial agricultural service; like those of Welwitsch, a set was worked up at the British Museum (Natural History) in London and partly published as supplements to the

Journal of Botany as *Mr. John Gossweiler's plants from Angola and Portuguese Congo* (1926–36).

A renewal of Portuguese metropolitan efforts began in 1927 with field work by Coimbra University botanists Luis Carrisso and Francisco Mendonça. These and British activities were afterwards combined under the general direction initially of Carrisso and *Conspectus florae angolensis* begun.[30] The two decades or so following World War II were years of considerable botanical activity in Angola, in part related to development activities, and until the early 1970s the *Conspectus* made good progress.[31] It was not unnaturally accounted for by the organizers of *Flora Zambesiaca* and accordingly their areas do not overlap. In the 1960s, however, rebellion in Angola largely ended further collecting, and since independence the almost unbroken civil war there and political and structural changes in Portugal have adversely affected the *Conspectus*. Despite a change to a fascicular format little has been published since 1975 (although this has included the pteridophytes, which appeared in 1977). As of 1996 more than half of the vascular flora had been accounted for but earlier parts are now more or less out of date.[32]

Bibliography

FERNANDES, A., 1982. See **Divisional bibliographies** above. [Angola, pp. 131–210; 603 titles in pure and 297 in applied botany.]

CARRISSO, L. *et al.*, 1937–70. *Conspectus florae angolensis.* Vols. 1–4(1). Illus. Coimbra: Instituto Botanico de Coimbra (vol. 1(1)); Lisbon: Junta de Investigações Colonais (later Junta de Investigações do Ultramar; Junta de Investigações Científicas do Ultramar) (vols. 1(2)–4, 1951–70). Continued as A. FERNANDES *et al.* (eds.), 1977– . *Conspectus florae angolensis.* In fascicles. Lisbon: Junta de Investigações Científicas do Ultramar (later Instituto de Investigação Científica Tropical). Accompanied by E. A. SCHELPE, 1977. *Conspectus florae angolensis: Pteridophyta.* Lisbon.

Systematic documentary enumeration of vascular plants; includes descriptions of new taxa, keys to genera and species, extensive synonymy with references and principal citations, *exsiccatae* with local range and general summary of extralimital distribution, some taxonomic commentary, notes on habitat, phenology, special features, etc., and figures of representative species (from vol. 3 onwards more abundantly supplied); index to all taxa in each volume (or fascicle). A

general introduction in both Portuguese and English appears in vol. 1. [Vols. 1–3 and 4(1) cover families from 1. Ranunculaceae through 94. Alangiaceae (save for no. 70). Three fascicles under the revised publication arrangements had appeared through 1998, covering pteridophytes (1977), Crassulaceae (no. 70, 1982) and Bignoniaceae (no. 122, 1993). Thus coverage as yet omits monocotyledons and most 'Gamopetalae'.][33]

523

Caprivi Strip

This narrow strip of northern Namibia separates Botswana from Angola and Zambia and ends to the east at a very short common boundary between Botswana and Zambia. It is the only part of Namibia included in *Flora Zambesiaca* (**520**). For other coverage, see **521**.

524

Botswana

See also **510** (all works except Dyer, although some do not cover the infratropical zone). – Area: 582096 km². Vascular flora: size not known (PD; see **General references**).

As Bechuanaland Protectorate, Botswana became a British territory in 1885; independence was achieved in 1966. Only since then have local resources for floristic work been developed. Much of the country is botanically still poorly known. The first general checklists of the flora appeared independently from one another in 1986 but neither can be considered entirely critical. Prior to then the plants had been documented in expedition reports such as *List of plants collected in Ngamiland and the northern part of the Kalahari Desert* (1909, in *Bull. Misc. Inform.* (Kew): 89–146) by N. E. Brown, *Sertum kalahariense* (1935, in *Ann. Transv. Mus.* 16) by C. E. B. Bremekamp and A. Obermeyer, and the results of tours in 1931 and 1937 by I. B. Pole Evans (1948, in *Bot. Surv. S. Africa Mem.* 21), in ecological surveys, or in regional floras. The woody flora has enjoyed relatively more documentation, beginning with O. B. Miller's first enumeration of 1948.

BARNES, J. D. and L. M. TURTON (eds.), 1986. *A list of the flowering plants of Botswana in the herbaria at the National Museum, Sebele, and University of Botswana*. 52 pp. Gaborone: Botswana Society/ National Museum and Art Gallery.

A catalogue of holdings in three herbaria in Gaborone and vicinity.

GRIGNON, I. and P. JOHNSEN, 1986. *Towards a check-list of the vascular plants of Botswana*. 91 pp. Aarhus: Zoological Laboratory, University of Aarhus.

Nomenclatural list with sources; bibliography (pp. 79–84) and index at end.

Woody plants

O. B. Miller's work (see below) was derived from *idem*, 1948. *Bechuanaland Protectorate (provisional list)*. 77 pp., 12 pls., map. Oxford: Imperial Forestry Institute (Check-lists of the forest trees and shrubs of the British Empire 6) and *idem*, 1949. *Additions and corrections to December, 1948*. 16 pp. Ladybrand.

MILLER, O. B., 1952. The woody plants of the Bechuanaland Protectorate. *J. S. African Bot.* **18**: 1–100, map. (Separately reprinted, Kirstenbosch; corrigenda in *ibid.*, **19**: 177–182 (1953).)

Systematic enumeration with briefly descriptive notes; includes synonymy, vernacular names, generalized indication of local range with some *exsiccatae*, taxonomic commentary, and notes on phenology, wood properties, etc.; glossary, lexicon of vernacular names, and index to generic and family names at end. The introductory section includes lists of principal collectors and descriptive remarks on the physical features and vegetation formations of the country.

525

Zimbabwe

Area: 391 090 km². Vascular flora: 5428 species (Gibbs Russell in Lebrun and Stork, 1991; see **501**). Initially organized in the 1880s under the British South Africa Company, the country (after 1923 a British colony) was from then until 1965 known as Southern Rhodesia and until 1980 as Rhodesia.

Rather than produce a 'national' flora, much locally based effort in the decades since World War II has gone into contributions to *Flora Zambesiaca*, a work conceived as part of the externally inspired scientific development of the former Federation of Rhodesia and

Nyasaland and with a project 'center' in the country. Until then, most works on Zimbabwe were expedition and survey reports; among the earlier are *A contribution to the botany of Southern Rhodesia* by Lilian S. Gibbs (1906; in *J. Linn. Soc., Bot.* **37**), *A contribution to our knowledge of the flora of Gazaland* by A. B. Rendle and others (1911; in *ibid.*, **40**) covering collections by Swynnerton, and *Beiträge zur Kenntnis der Flora on Süd-Rhodesia* I–X by the Swedish botanists Tycho Norlindh and Hennig Weimarck (1932–58; in *Bot. Notis.*) and based on a 1930–31 expedition. The only country-wide account remains that of EYLES (1916). There are, however, several works on the woody flora, and the aquatic macrophytes were revised by GIBBS RUSSELL (see **507**).

EYLES, F., 1916. A record of plants collected in Southern Rhodesia. *Trans. Roy. Soc. South Africa* **5**(4): 273–564.

Systematic enumeration of non-vascular and vascular plants (2397 species), with limited synonymy (including references and citations), *exsiccatae* with localities, and index to all botanical names. A list of basic references on the flora appears in the introduction.

Woody plants

There is a considerable literature on trees and other woody plants in Zimbabwe dating from early in the twentieth century, in addition to the good semi-popular regional tree works of Coates Palgrave and Coates Palgrave (1957; see **520**) and Coates Palgrave (1983; see **510**). The current Zimbabwean list by Drummond (1975; see below) succeeded *A check list of the trees of Southern Rhodesia* by A. S. Boughey (1964; in *J. S. African Bot.* **30**: 151–176). This has been complemented by a set of keys by Coates Palgrave (1996; see below). Other works include B. GOLDSMITH and D. T. CARTER, 1981. *Indigenous timbers of Zimbabwe*. x. 406 pp., illus. (Zimbabwe Bulletin of Forestry Research 9). [Harare]: Forestry Commission; A. A. PARDY, 1951–56. Notes on indigenous trees and shrubs of Southern Rhodesia. *Rhodesian Agric. J.* **48**(3)–**53**(6), *passim*, with halftones (reprinted under separate cover, Salisbury: Government Printer); and E. C. STEEDMAN, 1933. *A description of some trees, shrubs and lianes of Southern Rhodesia*. xix, 191 pp., 92 halftone pls. [Bulawayo]: The author. [The works of Steedman and Parey are illustrated descriptive treatments of selected species, with vernacular names and notes on uses. That of Goldsmith and Carter is a dendrology covering 350 species.]

COATES PALGRAVE, M., 1996. *Key to the trees of Zimbabwe*. xii, 365 pp., illus. Harare: The author.

A field-guide comprising analytical keys; these are

organized around six primary habitat types (each on different-colored paper) with recent references and index at end. The introductory part features an illustrated glossary.

DRUMMOND, R. B., 1975. A list of trees, shrubs, and woody climbers indigenous or naturalized in Rhodesia. *Kirkia* 10(1): 229–285, map.

Systematic checklist of woody plants (1172 species), including standard numbers, limited synonymy, habit, abbreviated indication of local and extralimital range, and representative *exsiccatae* (deposited in Harare); references to standard floras, revisions, etc., provided; index to family and generic names at end. The map gives distribution units as used in the work.

Pteridophytes

BURROWS, J. E. and S. M. BURROWS, 1983. A checklist of the pteridophytes of Zimbabwe. *J. S. African Bot.* 49(3): 193–212.

Tabular systematic list with synonymy, indication of external and local distribution, a representative *exsiccata* (where possible, deposited in Harare), and habitat; list of references and index to generic and family names at end. [The arrangement of families and genera adopted is that of Jermy *et al.* (1975) which differs radically from that used by Schelpe for *Flora Zambesiaca*.]

526

Zambia

Area: 752 613 km². Vascular flora: 4600 species (Lebrun in PD; see **General references**), a figure probably too low. First organized in the 1880s under the British South Africa Company, the country, known as Northern Rhodesia, became a protectorate in 1924 and achieved independence in 1964. From 1953 to 1963 it was part of the Federation of Rhodesia and Nyasaland.

Except for the woody plants, no separate general works are available, save an introduction to the flora with keys to families prepared by D. M. Nath Nair for use at the University of Zambia. Several expedition and survey reports exist; an important early account was *Botanische Untersuchungen* by R. E. Fries (1914–16; 1921) covering the plants of a 1911–12 Swedish Central African expedition. The woody flora was more thoroughly studied consequent to establishment of an association with the then-Imperial Forestry Institute (IFI) in Oxford in 1927 and, two years later,

the creation of a forest service; a first typed checklist from IFI appeared in 1934. In 1950 the Institute and the government agreed to a more definitive list as part of the former's established series of imperial checklists; two years later, however, that concept gave way to a descriptive work similar in style to *Flora of west tropical Africa* and the *Forest flora of Northern Rhodesia* thus came into being.

Keys to families and genera

NATH NAIR, D. M., 1967. *A numbered dichotomous key to the selected families of Zambian flowering plants.* 62 pp. Lusaka; and *idem*, 1967. *The selected families of Zambian flowering plants.* 154 pp., illus. Lusaka. (Both mimeographed.) [Respectively keys to and descriptions of flowering plant families for student use; with pedagogical chapters. The former covers 118 families; the latter, 30; the latter also features a bibliography of key works (pp. 133–134).]

Woody plants

White's *Forest flora* was partially updated by D. B. FANSHAWE, 1973. *Checklist of the woody plants of Zambia showing their distribution.* iv, 48 pp., map (Zambia Forest Res. Bull. 22). Lusaka.

WHITE, F. (assisted by A. ANGUS), 1962. *Forest flora of Northern Rhodesia.* 482 pp., 72 text-figs., map. Oxford: Oxford University Press.

Briefly descriptive account of trees and shrubs, with keys to genera and species; limited synonymy, with citations of principal works; vernacular names; indication of *exsiccatae*, with fairly detailed summary of local range; taxonomic commentary and notes on habitat, associates, special features, uses, etc.; indices to vernacular, generic and family names. A bibliography appears in the introductory section. [Covers some 1400 species, 250 of them introduced.][34]

Pteridophytes

KORNÁS, J., 1979. *Distribution and biology of the pteridophytes in Zambia.* 207 pp., 83 figs. (mostly maps), 6 pls. Warsaw/Kraków: Pánstwowe Wydawnictwo Naukowe.

Systematically arranged record and analysis of the biology and local distribution of Zambian pteridophytes (146 species), based on extensive field work by the author; includes *exsiccatae*, critical remarks (with revisions to Schelpe's *Flora Zambesiaca* treatment), and notes on chorology, habitat, phenology, and other aspects of biology, with altitudinal zonation and distribution maps given for each species. A general introductory section is also provided. [One of the most thorough current treatments of African pteridophytes.]

527

Malawi

Area: 118 484 km^2 (in all); 94 081 km^2 (on land). Vascular flora: *c.* 3600 species (Lebrun in PD; see **General references**). First penetrated by David Livingstone in 1859, the country, initially known as Nyasaland, became a protectorate in 1891 and a British colony in 1907; independence was achieved in 1964. From 1953 to 1963 it was part of the Federation of Rhodesia and Nyasaland.

No complete account of vascular plants is available. Various checklists and other works covering parts of the flora exist, the most substantial being that for woody plants by Burtt-Davy and Hoyle (1958; see below). As in Zambia, this was originally based on an association with what is now the Oxford Forestry Institute (OFI) in Oxford.[35] Lists complementary to that of Burtt-Davy and Hoyle are available for grasses and other herbaceous plants: B. BINNS, 1968. *A first check-list of the herbaceous flora of Malawi.* 113 pp., map. Zomba: Government Printer; and R. G. JACKSON and P. O. WIEHE, 1958. *An annotated check-list of Nyasaland grasses.* 75 pp., illus. Zomba.[36] A significant additional contribution, based on collections by Leonard J. Brass, is J. P. M. BRENAN *et al.*, 1953–54. Plants of the Vernay Nyasaland expedition (1946), I–III. *Mem. New York Bot. Gard.* 8: 191–256, 409–510; 9: 1–132.

Woody plants

BURTT-DAVY, J. and A. C. HOYLE, 1958. *Check list of the trees and shrubs of the Nyasaland Protectorate.* 2nd edn., revised by P. Topham. 137 pp., tables. Zomba: Government Printer. (1st edn., 1936, Oxford, as *Check-lists of the forest trees and shrubs of the British Empire*, 2: *Nyasaland Protectorate.*)

List of trees and shrubs, with limited synonymy, vernacular names, and (in tabular form) brief notes on habitat, life-form, etc.; lexicon of vernacular names, with botanical equivalents; index to genera. The introductory section includes an account of the forest types of the region, with indication of important species.[37]

528

Mozambique

Area: 789 800 km^2. Vascular flora: 5500 species (Lebrun in PD; see **General references**), a figure probably too low. – Much of the country north of the Zambezi River was also covered in *Die Pflanzenwelt Ost-Afrikas* by Engler (see **531**). For the tree flora south of that river, see also COATES PALGRAVE (1983; under **510**).

Prior to initiation of *Flora Zambesiaca* and, later, *Flora de Moçambique*, the only overall works on the flora were the dendrologies by Thomas R. Sim and provincial botanist António F. Gomes e Sousa, noted below, and a herbarium catalogue by the latter author: *Flora de Moçambique: lista de algumas plantas clasificadas do herbário da Direcção de Agricultura* (1933; in *Bol. Agr. Pec. Moçambique* 1932). The first significant geographical expedition was the *viagem filosófica* of Manoel Galvão da Sylva in 1784–88, but few botanical collections were made. In the nineteenth and early twentieth centuries, numerous mostly non-Portuguese expeditions and individuals collected principally in the north (Quelimane), around Beira (a later railhead for Zimbabwe), the Inhambane district, and in the so-called Delagoa Bay region centering on Maputo (but extending into Swaziland and former Transvaal). These included Sim's tours, executed in 1908. Only in the second quarter of the twentieth century did Portuguese efforts increase; particularly notable were the work from 1930 of the colonial agronomist António de F. Gomes e Sousa and, from 1942, the *Missões Botânicas* organized from Portugal under the then-Junta de Investigações Colonais. As had Sim before him, Gomes e Sousa paid particular attention to the tree flora, his work resulting in two major publications.

The *Missões*, of which eight in all were mounted through 1971, comprised the first serious botanical survey of the country. The collections from these and other postwar sources laid a foundation deemed sufficient for a definitive *Flora de Moçambique*. Based on the Anglo-Portuguese *Flora Zambesiaca*, this was initiated in 1969; publication has continued to the present but is not yet complete. Partial works are few; a noteworthy contribution for the Maputo area was *A natural history of Inhaca Island* (1958) edited by W. Macnae and M. Kalk.

From the 1960s, the country was dominated by rebellion and civil war and their aftermath; these have long disrupted field work. Many parts of the country remain imperfectly known.[38]

Bibliography

FERNANDES, A., 1982. See **Divisional bibliographies** above. [Mozambique, pp. 211–277; 546 titles in pure and 214 in applied botany.]

FERNANDES, A. and E. J. MENDES (eds.), 1969– . *Flora de Moçambique.* In fascicles. Illus. Lisbon: Junta de Investigações Científicas do Ultramar (now Instituto de Investigação Científica Tropical). Accompanied by E. A. SCHELPE and M. A. DINIZ, 1979. *Flora de Moçambique: Pteridophyta.* 257 pp., 15 pls. Lisbon.

Copiously illustrated, serially issued flora of vascular plants; each family fascicle includes keys to genera and families, full synonymy (with references and citations), vernacular names, localities with *exsiccatae*, general summary of extralimital distribution, taxonomic commentary, notes on habitat, special features, occurrence, etc., and complete botanical index. [In progress. 'Readjustment' resulted in suspension for five years in the 1970s, but by the mid-1980s the Pteridophyta (1979, by E. A. Schelpe and M. A. Diniz) and 64 seed plant families had appeared, covering some 55 percent of the vascular flora. Further fascicles have appeared since. Numbering follows a predetermined sequence but publication is not in systematic order.][39]

Woody plants

Gomes e Sousa's 1966–67 *Dendrologia de Moçambique* was preceded by an earlier, incomplete work in large format by the same author: A. DE F. GOMES E SOUSA, 1948–49, 1958–60. *Dendrologia de Moçambique.* Parts 1–2, 4–5. Illus., maps. Lourenço Marques: Imprensa Nacional de Moçambique (for Junta de Comércio externo da Província de Moçambique). (Part 1 also published in English.) This in turn followed *Forest flora and forest resources of Portuguese East Africa* (1909) by T. R. Sim.

GOMES E SOUSA, A. DE F., 1966–67. *Dendrologia de Moçambique: estudo geral.* 2 parts. 817 pp., 229 pls., text-figs., frontisp., 2 col. maps (Mem. Inst. Invest. Agron. Moçambique, Centr. Doc. Agrar. 1). Lourenço Marques [Maputo]: Imprensa Nacional de Moçambique.

Illustrated *in extenso* descriptive treatment of trees and shrubs; includes keys (only to genera), full synonymy, vernacular names, fairly detailed treatment of local range with indication of formation-types, cursory summation of extralimital distribution, taxonomic commentary, and notes on habitat, phenology, wood structure, etc.; index only to illustrations. The introductory section includes general geographical notes and accounts of floristic zones, vegetation formations, and early and recent botanical exploration, with two accompanying maps.

Region 53

East tropical Africa

Area: 1 813 148 km². Vascular flora: at least 12 000 species (Knox and Vanden Berghe, 1998; see **531**). The region as circumscribed here includes Tanzania (with the offshore islands of Zanzibar and Pemba), Uganda, Rwanda, Burundi and Kenya.

Because of Arab domination as well as distance from trade routes, early European contacts were limited until well into the nineteenth century. There thus exist few collections prior to about 1860, with these mainly originating from Zanzibar and the area of Mombasa. The first major inland expeditions were those of British explorers Speke and Grant and Burton and Speke in the 1860s and David Livingstone in the 1870s; afterwards contacts, aided greatly by the opening of the Suez Canal in 1869, increased rapidly and many other expeditions were mounted. With these, serious collecting began in the region, continuing under colonial rule and afterwards until the greater part of the area, except for more remote parts of Kenya and Tanzania, is now considered well to moderately well known. Particularly energetic collecting programmes were prosecuted in present-day Tanzania by the Germans, who also established in 1902 in Amani a forerunner of the present East African Herbarium (now at Nairobi). This latter is today one of the largest in Africa and has been an important contributor to floristic work.

By the late 1940s collections had reached a level officially deemed sufficient for initiation of a regional successor to the abandoned *Flora of tropical Africa*. The serial *Flora of tropical East Africa* was therefore organized in 1949 and afterwards incorporated into the programmes of the East African Community.[40] The first installment appeared in 1952 and ever since the work's ponderous progress has dominated regional floristic botany. By 1974, it had covered about one-third of the vascular plants, and by 1999 some 68 percent or so; formal completion is now planned for 2006. New discoveries, however, continue.

Rwanda and Burundi, although geographically part of East Africa and at one time German territory, were mandated to Belgium after World War I and thus are covered in the present *Flore d'Afrique centrale* (**560**) and in other works prepared by Belgian botanists (particularly Georges Troupin). They are thus not accounted for in *Flora of tropical East Africa*. Burundi is said to remain botanically somewhat underexplored.

Modern descriptive floras of subregional scope are few. The best are for upland Kenya (Agnew, 1974; revision currently in progress) and Rwanda (Troupin, 1971, 1978–87). Other coverage is very sketchy and some of it much outdated, such as the *Verzeichnis* in Engler's *Pflanzenwelt Ost-Afrikas* of 1895 which remains the latest complete 'separate' work for Tanzania. Better coverage exists for the woody plants, for which manuals, some lavishly illustrated, have been produced for each of the three former East African Community countries. The 'Afroalpine' flora has been covered in O. Hedberg's *Afroalpine vascular plants* (1957) as well as in the earlier, more general review of the African mountain flora by Engler (1892; for both works, see **503**).

In recent years renewed attention has been given to the 'islands' of rain forest in the eastern part of East Africa, especially in the 'Zanzibar-Inhambane Regional Mosaic' near the coast, which despite their relatively small extent are of great biogeographical interest. Some major ecological and biogeographic studies have been presented but as yet no flora, which would be a considerable aid to a more secure future for these forests.

Progress

Recent accounts include R. POLHILL, 1989. East Africa (Kenya, Tanzania, Uganda). In D. G. Campbell and H. D. Hammond (eds.), *Floristic inventory of tropical countries* (under **General references**), pp. 217–231; and J. C. LOVETT, 1989. Tanzania. *ibid.*, pp. 232–235. The latter relates particularly to the Tanzanian forest flora. Earlier history is recorded in J. B. GILLETT, 1962. *History of the botanical exploration of the areas of "The Flora of Tropical East Africa" (Uganda, Kenya, Tanganyika, and Zanzibar)* (pp. 205–229 in A. Fernandes (ed.), *Histoire de l'exploration botanique de l'Afrique au sud de Sahara*; see **Division 5, Progress**).

Bibliographies. General and divisional bibliographies as for Division 5.

Indices. General and divisional indices as for Division 5.

530
Region in general

See also **501** (LEBRUN and STORK; OLIVER *et al.*); **503** (ENGLER; HEDBERG). The comprehensive *Flora of tropical East Africa* does not include Rwanda and Burundi.

Accounts of families

KOKWARO, J. O., 1994. *Flowering plant families of East Africa: an introduction to plant taxonomy.* [8], 292 pp., illus. Nairobi: East African Educational Publishers. [An illustrated introduction to East African seed plants with descriptions of selected families and keys to their genera (as well as some species, e.g., in *Eucalyptus*). The general part covers botanical exploration, the principles of taxonomy, the origins of botanical names, plant collecting, and botanical terminology; selected references appear on p. 280.]

TURRILL, W. B. *et al.* (eds.), 1952– . *Flora of tropical East Africa*. In fascicles. Illus., maps. London: Crown Agents (from 1980, Rotterdam: Balkema).

Comprehensive, detailed documentary 'research' flora of vascular plants; each family fascicle includes keys to genera, species and infraspecific taxa, full synonymy, with references and citations, typification, abbreviated designation of internal floristic districts, summary of internal range, with citation of (mainly representative) *exsiccatae*, and generalized indication of extralimital distribution, taxonomic commentary (often lengthy), and notes on habitat, special features, associates, etc., and complete botanical index. A general introduction to the work (1952) and a glossary appear in special fascicles. [Publication of fascicles follows no set sequence, but tables for arrangement according to any one of three common systematic schemes (Bentham and Hooker, Engler, and the first version of Hutchinson) are provided. Prepared mainly at the Kew Herbarium, the progress of this work has generally been steady; as of 1990, 6425 species, some 61 percent of the projected total, had been documented, and by the end of 1997, 7456 species (in 178 parts), or 67.8 percent.][41]

Pteridophytes

The publication of *Pteridophytes of Tropical East Africa* (see below) 'represents an attempt to fill the gap in our knowledge of the Ferns and Fern-allies of East Africa until . . . publication . . . in the *Flora of Tropical East Africa*'.[42] No similar regional summary is available.

JOHNS, R. J., 1991. *Pteridophytes of Tropical East Africa: a preliminary check-list of the species.* 131 pp., 120 text-figs., maps. Kew: Royal Botanic Gardens.

Annotated illustrated enumeration (covering some 510 species), without keys; entries include synonymy and indication of distribution along with (sometimes) notes on habitat and altitudinal range along with representative *exsiccatae*. Following the main list under each genus may be one or two further unannotated lists of names; those from the so-called 'List 2' are unverified names used by Peter (1929; see 531 below) and from 'List 3' other, as yet unconfirmed literature records. A list of references, abbreviations to be used in a definitive *Flora* account, a guide to key works on ferns in Africa, and an index to genera follow. The general part is explanatory, including a section on the East African distribution recording system, and also (pp. 6–7) furnishes a synoptic list of families and genera. [The illustrations in this 'working document' are mostly derived from other publications.]

531

Tanzania

Area: 939 652 km² (country); 937 011 km² (mainland, inclusive of water areas). Vascular flora (country, including Zanzibar and Pemba): 9105 species (Knox and Vanden Berghe, 1998; in *Journal of East African Natural History* 85: 65–79).[43] – Mainland Tanzania, first seriously explored from the 1850s, was initially colonized as German East Africa and from 1921 until 1963 administered as Tanganyika Territory under a mandate to the United Kingdom from the League of Nations. From the 1950s to the 1970s, many government activities, including botanical services, were delegated to the East African Community.

No separate general coverage is available apart from that presented below. A flora for the biogeographically significant montane forest regions, particularly those of the eastern arc, was proposed in 1984 but so far has not been realized.[44]

ENGLER, A. *et al.*, 1895. *Die Pflanzenwelt Ost-Afrikas und der Nachbargebiete*, Teil C: *Verzeichniss der*

bis jetzt aus Ost-Afrika bekannt gewordenen Pflanzen. [ii], ii, 433, 40 pp., 45 pls. Berlin: Reimer. (Deutsch-Ost-Afrika V.)

Systematic enumeration of vascular and non-vascular plants; includes descriptions of new taxa and diagnoses of genera, synonymy with references and principal citations, vernacular names (in several languages), symbolic indication of local range with some *exsiccatae* and summary of African distribution, and notes on habitat, properties, uses, etc.; indices to all taxa and to vernacular names. The introductory part includes a subdivision of Africa (into 39 phytogeographic regions) and a concise account of botanical exploration in former German East Africa and neighboring areas. The work covers the mainland part of Tanzania, Rwanda, Burundi, northern Mozambique, southern Uganda and southern Kenya.

PETER, A., 1929 (1930)–38. *Flora von Deutsch-Ostafrika.* Lfg. 1–3. 135 pls. (Repert. Spec. Nov. Regni Veg., Beih. 40(1): 540, 142 pp., 91 pls.; 40(2): 272, 36 pp., 44 pls.). Berlin: Fedde.

A rather uncritical keyed enumeration, arranged according to the Englerian system and including descriptions of new taxa (in separately paged appendices), synonymy with references and citations, localities with *exsiccatae* and summary of general distribution, and notes on phenology, abundance, uses, etc.; no indices. The introductory section includes a listing of Peter's itineraries of 1914–19 and 1925–26. [Incomplete; issues 1 and 2 cover pteridophytes, gymnosperms, and mono-cotyledons (Typhaceae–Cyperaceae), while issue 3 covers 'early' dicotyledons (Casuarinaceae–Basellaceae). Based almost entirely on the author's own as well as other German collections; now largely of historical interest save where not yet superseded by *Flora of tropical East Africa*.][45]

Woody plants

BRENAN, J. P. M. and P. J. GREENWAY, 1949. *Checklists of the forest trees and shrubs of the British Empire*, 5: *Tanganyika Territory*, II. xviii, 653, [ii] pp. Oxford: Imperial Forestry Institute.

Concise manual-checklist of woody plants; includes keys (only to species or groups thereof), limited synonymy, vernacular names, abbreviated indication of local range with representative *exsiccatae*, and notes on habitat, phenology, associates, uses, etc.; references to key literature (under family headings); index to genera. (Part I of this work (1940) comprises botanical–vernacular and vernacular–botanical lexica, compiled by F. B. Hora.)

Pteridophytes

Apart from the work of Peter (see above) and Johns (see **530**), pteridophytes have been documented in L. S. GILL and L. B. MWASUMBI, 1976. An annotated checklist of the pteridophytes of Tanzania. *Nova Hedwigia* **27**: 931–952. This last comprises a systematic list of accepted taxa with authors and for each a voucher specimen and locality; references, p. 952.

532

Zanzibar and Pemba

Area: 2642 km^2 (Zanzibar, 1658 km^2; Pemba, 984 km^2). Vascular flora: no data. These islands, the seat of a sultanate formerly in control of much of the East African mainland, were from 1890 administered as a British protectorate; in 1964 they merged with Tanganyika to form Tanzania. From the 1950s the area was part of the so-called 'East African Community'.

The most general work on the flora (in spite of its title) is R. O. WILLIAMS, 1949. *The useful and ornamental plants in Zanzibar and Pemba.* ix, 497 pp., illus., 1 col. pl. Zanzibar: [Protectorate Government]. This is a descriptive account, without keys, of a wide cross-section of the flora including the more common native and introduced plants. For more recent coverage of Pemba, see L. KOENDERS, 1992. *Flora of Pemba Island: a checklist of plant species.* 104 pp., illus. N.p.: Wildlife Conservation Society of Tanzania. (Publication 2.)

533

Kenya

Area: 582644 km^2 (including inland waters). Vascular flora: 6302 species (Knox and Vanden Berghe, 1998; see **531**). – Kenya, formerly a part of the Sultan of Zanzibar's domains, was from 1885 recognized as a British 'sphere of influence' and in 1888 came under the administration of the British East Africa Company. From 1895 it was ruled as the British East Africa Protectorate and, from 1920 until 1963, as Kenya Colony (with a small portion on the coast, legally under lease from the Sultan, remaining under protectorate status). From the 1950s to the 1970s it was the center of the East African Community.

The southern part of the present state is also covered by *Die Pflanzenwelt Ost-Afrikas* (ENGLER *et al.*; see **531**). This and the upland zone are the most completely treated; by contrast, coverage of the east and north is sketchy apart from revisions in *Flora of tropical East Africa*. No separate general checklist or other account is available, save for the woody plants, first enumerated in 1922. The Agnews' *Upland Kenya wildflowers*, first published in 1974, reappeared in a new edition in 1994. This work lacks grasses, sedges and Juncaceae; these, however, have been covered respectively in K. M. IBRAHIM and C. H. S. KABUYE, 1987. *An illustrated manual of Kenya grasses.* 765 pp., illus. Rome: FAO; and R. W. HAINES and K. A. LYE, 1983. *The sedges and rushes of East Africa: a flora of the families Juncaceae and Cyperaceae in East Africa.* 404 pp., illus. Nairobi: East African Natural History Society.

Partial work

AGNEW, A. D. Q. and S. AGNEW, 1994. *Upland Kenya wildflowers: a flora of the ferns and herbaceous flowering plants of upland Kenya.* 2nd edn. [4], 374 pp., 175 pls., 5 other figs., map. Nairobi: East African Natural History Society. (Original edn., 1974, Oxford: Oxford University Press; based in turn on mimeographed checklists of 1966 and 1968 entitled *Flora of upland Kenya*.)

Concise, well-illustrated manual-flora of (largely) non-woody plants (except Gramineae, Cyperaceae and Juncaceae; pteridophytes by R. Faden) of 'upland' (southwestern) Kenya (map, p. 7); includes brief descriptions, complete keys, limited synonymy, symbolic indication of local range, and short notes on habitat, altitude, frequency, etc.; glossary, index to unnamed taxa with citation of *exsiccatae*, all illustrations, and index to all botanical names at end. The introduction gives the background and scope of the work, key references, acknowledgments, and ecogeographic zones. Just over 3000 species are accounted for; about one-third are illustrated.[46]

Woody plants

Beentje's *Kenya trees, shrubs and lianas* succeeds the much-appreciated *Kenya trees and shrubs* (1961) by I. R. Dale and P. J. Greenway.[47]

BEENTJE, H., 1994. *Kenya trees, shrubs and lianas.* ix, 722 pp., illus. (incl. end-papers), maps. Nairobi: National Museums of Kenya.

Concise descriptive manual (covering some 1850 species) with keys, limited synonymy, conservation status, distribution (with map for each taxon), altitudinal range, phenology, habitat, frequency, vegetation associations, vernacular names, indication of uses and properties, and occasional notes; name changes and additions (pp. 648–650), sources of illustrations, indices to local names, and taxonomic indices at end. The introductory part encompasses background, acknowledgments, a tribute to Dale and Greenway, use of the book, an introduction to vegetation types, and general keys (pp. 12–38).

GILLETT, J. B. and P. G. MCDONALD, 1970. *A numbered check-list of trees, shrubs and noteworthy lianes indigenous to Kenya.* iii, 67 pp. Nairobi: Government Printer.

Systematic checklist of woody plants, with each species numbered and provided with known trade and vernacular equivalents; indices. The introduction includes an explanation of the numbering system. [Now out of date.]

534

Uganda

Area: 236 578 km² (including waters). Vascular flora: 4318 species (Knox and Vanden Berghe, 1998; see **531**). – A land with four kingdoms, external rights were acquired by the British East Africa Company in 1888. The kingdom of Buganda, within which are situated the major centers of Kampala and Entebbe, became in 1893 the British Uganda Protectorate; by 1903 the three other kingdoms had also come under British administration. The Rudolf district in the northeast was transferred to Kenya in 1926, and independence was achieved in 1962. Many government services, including botanical matters, were administered as part of the 'East African Community' from the 1950s through the 1970s.

The southern part of Uganda (near Victoria Nyanza) is also covered by *Die Pflanzenwelt Ost-Afrikas* (ENGLER *et al.*; see **531**) but apart from works on the trees no separate general flora or checklist is available beyond *List of the plants occurring in the Uganda Protectorate* (1902) by C. H. Wright (pp. 329–351 in H. H. Johnston, *The Uganda Protectorate*). Local botanical work was from the mid-twentieth century further stimulated by the formation of Makerere University.

Woody plants (including trees)

EGGELING, W. J. and I. R. DALE, 1951. *The indigenous trees of the Uganda Protectorate.* 2nd edn. xxx, 491 pp., 94 text-figs., 55 halftones, 20 col. pls., frontisp., map. Entebbe: Government Printer. (1st edn., 1936.)

Briefly descriptive manual of trees (in layout similar to *Kenya trees and shrubs*); includes keys to all taxa, limited synonymy, vernacular names, local range with representative *exsiccatae*, taxonomic commentary, and notes on habitat, special features, timber properties and other uses, etc.; glossary and indices to botanical, trade and vernacular names at end. The introductory section includes a list of principal references.

HAMILTON, A., [1981]. *A field guide to Uganda forest trees.* 279 pp., illus. N.p. [Printed by Makerere University Printery.]

Artificially arranged, illustrated descriptive treatment of 447 numbered species (with additional less common or prominent ones accounted for in smaller type within the text); includes keys (pp. 16–72), vernacular names, essential synonymy, diagnostic characters, distribution, and habitat; glossary and indices to all names at end. The arrangement of species is by vegetative characteristics. An introductory section gives an explanation of the book's approach (less 'botanical' than Eggeling and Dale), instructions for use, and a bibliography (p. 11).[48]

535

Rwanda

Area: 26 338 km². Vascular flora: 2150 species (Troupin, 1971; see below); a figure of 2288 is given by Morat and Lowry (1997; see **Division 5, Progress**). – A Tutsi kingdom from the middle of the second millennium C.E., the country was from 1899 until 1962 ruled firstly as part of German East Africa and, after World War I, under mandate to Belgium (from 1946 as part of the territory of Ruanda-Urundi).

Biological exploration of Rwanda began with the trans-African expedition of Graf von Götzen in 1894; some of his results were incorporated into Engler's *Pflanzenwelt Ost-Afrikas* (1895; see **531**). More intensive botanical collecting was undertaken from 1962 onwards by Georges Troupin and others, with many new records (but few novelties); new ligneous plants alone amounted to 350 species. The recent works of Troupin, produced as part of a Rwandan-Belgian development programme, have effectively supplanted all other treatments and greatly improved knowledge of the flora. The only other work useful for Rwanda is the unfinished *Flore d'Afrique centrale* (**560**). A lavishly illustrated general introduction to Rwandan natural history, arranged by biomes with emphasis on characteristic higher plants and the herpetofauna, is E. FISCHER and H. HINKEL, 1992. *Natur Ruandas/ La nature du Rwanda*. 452 pp., illus. (350 col.), maps. Mainz: Ministerium des Innern und für Sport, Rheinland-Pfalz (Germany). (Materialien zur Partnerschaft Rheinland-Pfalz/Ruanda 1992/1.)

Troupin's definitive flora was preceded by what remains useful as a school-flora: G. TROUPIN, 1971. *Syllabus de la flore du Rwanda (Spermatophytes)*. viii, 6, 76, 10, 24, 20, 31, 340, 5, 16 pp., illus. (Ann. Mus. Roy. Afrique Centr., Série en-8°, Sci. Econ. 7). Butare, Rwanda (as *Publ. Inst. Natl. Rech. Sci. Rwanda* 8); Tervuren, Belgium. This illustrated treatment, with keys, covers about 500 species in all families and the more important genera.

TROUPIN, G., 1978–87. *Flore du Rwanda: spermatophytes*. 4 vols., illus. maps (Ann. Mus. Roy. Afrique Centr., Série-en-8°, Sci. Econ. 9, 13, 15, 16). Butare (as *Publ. Inst. Natl. Rech. Sci. Rwanda* 18); Tervuren, Belgium.

Keyed enumeration of seed plants; includes limited synonymy, citation of key references, French and Kinyarwanda vernacular names, illustrations of representative members of each included family, indication of local range (by districts, with representative *exsiccatae*), overall distribution, taxonomic remarks, and notes on occurrence, biology, related species possibly to be found, etc.; complete indices at end of each part. The introductory section (vol. 1) includes historical information, technical details, a list of major references, an extensive illustrated glossary, and a key to families. Family arrangement follows the 12th edition of Engler's *Syllabus*.

Woody plants

TROUPIN, G. (collab. D. BRIDSON), 1982. *Flore des plantes ligneuses du Rwanda*. xi, 747 pp., 245 text-figs., map (Ann. Mus. Roy. Afrique Centr., Série-en-8°, Sci. Econ. 12). Tervuren.

Concise, copiously illustrated descriptive flora covering about 700 species (including some significant non-woody plants), with keys, limited synonymy (with citations), vernacular names, selected *exsiccatae*, and brief indication of altitudinal range, occurrence, habitat, frequency (where reasonably known), and local range (by *préfectures* or districts); no external distribution; supplementary plates and indices to Kinyarwanda and scientific names. The introductory section includes the plan of the work (p. 2), acknowledgments, references (p. 5), an illustrated glossary, explanation of abbreviations and symbols, and keys to families.

Pteridophytes

PICHI-SERMOLLI, R. E. G., 1983–85. A contribution to the knowledge of the Pteridophyta of Rwanda, Burundi and Kivu (Zaïre), I–II. *Bull. Jard. Bot. Natl. Belg.* 53: 177–284; 55: 123–206.

Critical account with synonymy, references, localities with *exsiccatae*, generalized indication of distribution, habitat, habit and altitudinal range, taxonomic notes, and maps; index to all names at end of the second part. The introductory section (in part I) incorporates background information including data on collectors. [Based mainly on material in Belgium, Florence and Nairobi.]

536

Burundi

Area: 27 834 km². Vascular flora: 2500 species (PD; see **General references**). Burundi came under German occupation in 1890; from then until 1962 it was ruled firstly as part of German East Africa and, after World War I, under mandate to Belgium (from 1946 as part of the territory of Ruanda-Urundi).

Botanically the country is less well explored than Rwanda. Limited local coverage has been provided by J. Lewalle in recent years but otherwise reference must be made to *Pflanzenwelt Ost-Afrikas* (ENGLER *et al.*; see **531**) and the serial *Flore d'Afrique centrale* (**560**). For pteridophytes, see **535** (PICHI-SERMOLLI, 1983–85).

Partial work

LEWALLE, J. (comp.), 1970. *Liste floristique et repartition altitudinale de la flore du Burundi occidental*. 84 pp., maps, figs. Bujumbura: Université officielle de Bujumbura. (Mimeographed.)

Includes a tabular list of vascular plants, with indication of life-form, chorology, and altitudinal zonation; based on collections made by the author in western Burundi.

Woody plants

For a partial treatment of trees, see J. LEWALLE, 1971. *Arbres du Burundi*, sér. 1 (essences autochtones). Fasc. 1 (with G. Gilbert). 61 pp. Bujumbura.

Region 54

Northeast tropical Africa

Area: 1935916 km². Vascular flora: *c.* 7000 species (Moggi); 6323 spermatophytes were recorded by Cufodontis, and Lebrun and Stork (1991) give an estimate of 5500 species. This region, the 'Horn of Africa', encompasses Somalia, Djibouti (former French Somaliland), Eritrea and Ethiopia and, east of Cape Guardafui, the islands of Socotra, 'Abd-al-Kuri and the Brothers (politically part of Yemen).

Although little was known of the flora of the region until the latter part of the eighteenth century due to a paucity of contacts, collections and knowledge, particularly from the Ethiopian highlands, thereafter increased fairly rapidly, giving botanists their first real idea of the African tropical-montane flora. Early contributions by Robert Brown, the zoologist Eduard Rüppel (from Frankfurt, leading the first Senckenberg Museum expedition), Wilhelm Schimper (whose plants were distributed by Hochstetter and Steudel), and a French scientific survey under Théophile Lefebvre were cumulated by Achille Richard into his notable *Tentamen florae abyssinicae* (1847–51). Although regarded by its author as only a preliminary exercise, this work has long remained definitive.[49] A further substantial checklist, compiled by Georg Schweinfurth with Paul Ascherson, appeared in the former's *Beitrag zur Flora Aethiopiens*, 1 (1867). In the following decades, Schweinfurth was to carry out extensive field work in the Eritrean lowlands. As exploration, colonial penetration and development proceeded, collecting spread into Djibouti, the Ogaden, Somalia (where material was first gathered only in 1842) and Socotra (the last-named being of particular biological interest). In Abyssinia, the modernization of the country initiated by Menelik, including the foundation of the present capital, Addis Ababa, brought in more collectors. Among them were a number of Italians – particularly after 1936 when the country came under Italian rule. By the 1940s it was considered that there was sufficient material to form a basis for a regional work.

This led to not one but *two* projects. One was in Austria, where Georg Cufodontis – himself a collector in Ethiopia – elaborated his *Enumeratio plantarum Aethiopiae.* This appeared from 1953 to 1972 in 26 systematically arranged parts (later issued in book form). The other was in Italy where, because of colonial interests, the 'Erbario coloniale' (later 'Erbario tropicale') within the state herbarium at Florence University began work on the comprehensive *Adumbratio florae aethopicae.* The first fascicle of this work, which features species distribution maps, appeared in 1953; by January 1980, 32 fascicles had been published, covering some 5 percent of the vascular flora (with several fascicles devoted to ferns, the founding editor's specialty). No further contributions have been produced since.

Other general works are few. They include an illustrated key to families by Burger (1967; see **548**) as well as *Afroalpine vascular plants* (HEDBERG; see **503**) and *Flora of tropical Africa* (**501**).

Coverage of individual territories long remained patchy, with most works dating from before 1940. The Italian presence in the region resulted in floras of Eritrea (Pirotta, 1903–07, not completed) and Somalia (Chiovenda, 1929–36), the latter also accounting for some collections from then-British Somaliland. The only significant official British contribution was a compiled checklist by Glover (1946) for the whole of present-day Somalia. For Socotra (then a part of the territory of Aden) important extra-official contributions were made by Balfour (1888, 1903). The few French contributions subsequent to Richard's *Tentamen* relate mainly to Djibouti but only in 1989 was a modern checklist for that country published.

Recent collecting throughout the region has rendered Cufodontis's work noticeably incomplete. Beginning in 1980, work was undertaken towards a new *Flora of Ethiopia* with centers in Addis Ababa (Ethiopia), Asmara (Eritrea), Kew and Uppsala and support from the four concerned countries. The first of the eight projected volumes of what is now the *Flora of Ethiopia and Eritrea* appeared in 1989. A *Flora of Somalia*, initiated in 1988 as a collaborative undertaking of the Somalian National Range Agency and Uppsala University, commenced publication in 1993; three of its four planned volumes are now available. With respect to woody plants, the older works by Fiori (1909–12) for Eritrea, Senni (1935) for Somalia, and Breitenbach (1963) for Ethiopia have recently been complemented by *Forests and forest trees of Northeast Tropical Africa* (1992) by Ib Friis; a practical work of wider scope, however, would also be welcome.

Progress

Progress to 1958 was reviewed by G. CUFODONTIS, 1958. Über den gegenwärtigen Stand der botanischen Erforschung Aethiopiens. *Mem. Soc. Brot.* **13**: 65–68. For a chronology of collectors intercalated with historical events, see *idem*, 1962. A preliminary contribution to the knowledge of the botanical exploration of Northeastern Tropical Africa. In A. Fernandes (ed.), *Comptes rendus de la IV. réunion plenière de l'AETFAT* (Lisbonne, Coïmbre, 1960), pp. 233–248, illus. Lisbon: Junta de Investigações do Ultramar. A relatively detailed map and further commentary appears in G. MOGGI, 1976. Aperçu sur la connaissance floristique d'Ethiopie et de Somalie en vue d'une nouvelle édition de la carte du degré d'exploration floristique de l'Afrique. *Boissiera* **24**: 593–596, maps.

Bibliographies. General and divisional bibliographies as for Division 5.

Indices. General and divisional indices as for Division 5.

540
Region in general

See also **501** (LEBRUN and STORK; OLIVER *et al.*); **503** (ENGLER; HEDBERG). For works concerned largely with Ethiopia alone, see **548**.

CUFODONTIS, G., 1953–72. *Enumeratio plantarum Aethopiae: Spermatophyta.* 26 parts (Bull. Jard. Bot. État 23–42, *passim*, Suppl.: i–xxvi, 1–1657). Brussels. (Reprinted in 2 vols., 1975, Brussels.)

Systematic enumeration of seed plants of Ethiopia, Eritrea, Somalia and Djibouti; includes synonymy, references, principal citations, vernacular names, and partly symbolic indication of local and extralimital distribution (with some *exsiccatae*); general index (in the book version). An introductory section (in German) in part 1 includes a map, bibliography, and list of abbreviations; a revised map appeared in part 15 (1964). [Originally issued as a series of supplements to successive volumes of *Bulletin du Jardin Botanique de l'État* (later *Bulletin du Jardin Botanique National de Belgique*); after completion reissued in book form. New collecting and subsequent research has rendered the

work out of date; it is being succeeded by new floras of Ethiopia and Eritrea and of Somalia (see **541** and **548** below).][50]

PICHI-SERMOLLI, R. E. G. *et al.*, 1953–78. *Adumbratio florae aethopicae.* Parts 1–32. Illus., maps. *Webbia* 9–33, *passim*. (Also issued separately, Florence).

A series of critical revisions of the vascular plants of the region, including, in addition to Ethiopia, Eritrea, Djibouti, Socotra and 'Abd-el Kuri; each family fascicle is enumerative with synonymy, references and citations, vernacular names, detailed account of local distribution with *exsiccatae* and some distribution maps, taxonomic commentary, and notes on habitat, associates, etc. The area covered is delineated in the introductory section (part 1). [No systematic sequence of publication has been observed (a plan of the work with indication of parts published appears in the covers of the reissues). 20 parts in all have been on pteridophytes. The last part seen is no. 32 (in *Webbia* **33**(1) (1978)). Each part usually covers one family.][51]

Woody plants (including trees)

FRIIS, I., 1992. *Forests and forest trees of Northeast Tropical Africa.* iv, 396 pp., 22 figs. (incl. tables and maps), 153 distribution maps (Kew Bull. Addit. Ser. 15). London: HMSO.

Includes a treatment of over 250 species of trees (of at least 8 m in height) from ecological, geographical and chorological points of view, without keys or descriptions; entries include references (only the year of publication), synonymy, citations, and a *pro forma* consideration of taxonomy, diagnostic features, diaspore types, habitats, regional and extralimital distribution, and chorological class(es); distribution maps, literature, appendix (on eight accepted phytochoria), and index to all botanical names at end. The general part of the work includes descriptions of physical features, climate, vegetation, forest types, and the history of botanical work.[52] [For a summary of this and other work by the author (in Danish and English), see *idem*, 1991. *Skovtræsfloraen i det Nordøstlige Tropiske Afrika (Etiopien, Djibouti og Somalia): en sammenfattende redegørelse for tidligere studier (The forest tree flora of northeast tropical Africa (Ethiopia, Djibouti and Somalia): a conspectus of previous studies)*. 108 pp., illus. Copenhagen: Botanical Museum.]

Pteridophytes

Reference should also be made to R. E. G. Pichi-Sermolli's contributions in *Adumbratio florae aethopicae* (see above), and to *idem*, 1977. Novitates pteridologicae aethopiae. *Webbia* **32**(1): 51–68.

CUFODONTIS, G., 1952. Enumeratio plantarum aethiopiae, III: Pteridophyta. *Phyton* (Horn) 4(1–3): 176–193.

A systematic enumeration in the same style as that on seed plants (see above).

541

Somalia

Area: 686 803 km². Vascular flora: 3015 species (Kuchar in Lebrun and Stork, 1991); some 600 are endemic. The present country was formed in 1960 from the union of former British and Italian Somaliland, both under outside rule from the latter part of the nineteenth century (and from 1941 to 1949 entirely under British administration).

The only nineteenth-century account is *Faune et flore des pays Çomalis (Afrique orientale)* by G. Révoil (1882), covering 144 species. For the twentieth century, two checklists of limited scope (Glover, 1947; Kuchar, 1986) are available, reinforced by four major collection reports (Chiovenda, 1916, 1929–36). A definitive flora commenced publication in 1993. The country is also accounted for in the general regional works described under **540**.

GLOVER, P. E., 1947. *A provisional check-list of British and Italian Somaliland trees, shrubs and herbs including the reserved areas adjacent to Abyssinia.* xxviii, 446 pp., 13 pls., map. London: Crown Agents.

List of recorded vascular plants, with limited synonymy and brief descriptive notes; lexicon of vernacular names, with botanical equivalents; bibliographical index to families and index to genera; general bibliography. The introductory section includes a description of physical features and the vegetation.

KUCHAR, P., 1986. *The plants of Somalia: an overview and checklist.* 335 pp. [Mogadishu]: Central Rangelands Development Project, National Range Agency, Somalia/Louis Berger International. (Crop Technical Report Series 16.) (Mimeographed.)

Annotated enumeration with vernacular names; includes also a good brief history of botanical work in the country with particular reference to the works of Chiovenda, Glover and Cufodontis.

THULIN, M. (ed.), 1993– . *Flora of Somalia.* Vols. 1– . Illus. (text-figs.), map. Kew: Royal Botanic Gardens.

Descriptive manual of native and commonly naturalized vascular plants with dates of publication of accepted names and synonymy, typifications, vernacular names, distribution within and without the country, representative *exsiccatae*, and critical notes; indices to vernacular and scientific names at end of each volume. The introduction is limited to background and the plan and scope of the work. [A total of four volumes are planned; vol. 1 (1993), vol. 2 (1999) and vol. 4 (1995) have to date been published, the first respectively covering pteridophytes, gymnosperms and Annonaceae–Fabaceae, the second, Tiliaceae–Apiaceae, and the fourth, monocotyledons. The arrangement of dicotyledons follows the original version of the Hutchinson system, as in *Flora of Ethiopia and Eritrea*, while the monocotyledons follow the Dahlgren system.][53]

Partial works: former Italian Somaliland
The following represent the principal floristic works of the Italian colonial era.

CHIOVENDA, E., 1916. *Resultati scientifici della Missione Stefanini-Paoli nella Somalia Italiana*, 1: *Le collezioni botaniche.* 241 pp., 6 figs. (incl. map), 24 halftones. Florence: Museo de Erbario Coloniale, R. Istituto Studi Superiori di Firenze.

Systematically arranged enumeration of non-vascular and vascular plants of the 1913–14 Stefanini-Paoli expedition; includes descriptions of novelties, localities with *exsiccatae*, distribution, sometimes copious commentary, and (at end) an appendix with records from other surveys and an index to vernacular names. The introductory part largely comprises a chronology of botanical exploration. [Records in this book, mostly from northwest of Mogadishu towards the Ogaden, are *not* repeated in the author's *Flora somala*.]

CHIOVENDA, E., 1929. *Flora somala*, I. xvi, 436 pp., 50 pls. (on 26 ll.), map. Rome: Sindicato italiano arti grafiche (for Ministero delle Colonie). Continued as *idem*, 1932. *Flora somala*, II. xvi, 482 pp., 247 text-figs. (Lav. Ist. Bot. Reale Univ. Modena 3). Modena: Reale Orto Botanico; and *idem*, 1936. *Flora somala*, III. Raccolte somale dei G. Pollacci, L. Maffei, R. Ciferri e N. Puccioni fatte negli anni 1934 e 1935. *Atti Ist. Bot. 'Giovanni Briosi' (R. Univ. Pavia)*, IV, 7: 117–160, pls. 1–12.

These three works each comprise self-contained annotated enumerations of individual collections of non-vascular and vascular plants, with descriptions of new taxa, synonymy, literature citations, vernacular names, *exsiccatae* with localities, and indices. An introductory section in the first contribution includes chapters on physical features, climate, vegetation, phytogeography, and floristic relationships. No cumulative index has been published. [The 1929 volume is based largely on collections made by G. Stefanini and G. Paoli both in former British and Italian Somaliland and in the Ogaden of eastern Ethiopia, the others on later collections.][54]

Woody plants: former Italian Somaliland

SENNI, L., 1935 (A. Fas. 13). *Gli alberi e le formazioni legnosae della Somalia*. 305 pp., 89 text-figs. (partly halftones). Florence: Istituto Agricolo Coloniale Italiano.

Systematic enumeration of woody plants with vernacular names, references to illustrations, distribution, habit and habitat, and (for some species) descriptions and/or notes on uses; list of literature (pp. 265–271), lexicon of vernacular names, and index to family and generic names at end. The introductory sections cover environment, natural regions and botanical districts, statistics, and forest resources. [Accounts for 759 species, of which 124 are trees and another 129, treelets.]

544

Djibouti

Area: 23 310 km^2. Vascular flora: 641 species, of which 13 are pteridophytes (Lebrun *et al.*, 1989; see below). A figure of 671 appears in Morat and Lowry (1997; see **Division 5, Progress**). – This small state, a French interest from 1862 and formally organized in 1882–84 as French Somaliland, was reconstituted in 1967 as the Territory of the Afars and Issas and achieved independence in 1977. Collecting began in the mid-nineteenth century; however, interior penetration was slow, with effective military control achieved only in the 1930s, and until the 1970s a basis sufficient for a flora was thought to be wanting.

A cursory introduction to the flora was presented by Auguste Chevalier in his *La Somalie française: sa flore et ses productions végétales* (1938; in *Rev. Int. Bot. Appl. Agric. Trop.* **19**: 663–687, illus.). This had been prompted in particular by work in 1937–38 by the geologist and naturalist Edgar Aubert de la Rüe, notably on Mt. Goudah with its relict forest. A more substantial contribution was *Contributo alla conescenza della flora del Territorio Francese degli Afar e degli Issa* by R. Bavazzano (1972; in *Webbia* **26**: 267–384, map); this covers 534 species based in particular on the collections of Col. E. Chedeville, resident from 1953 to 1964. In 1989 these works were succeeded by a new enumeration, described below; a descriptive flora followed in 1994.

LEBRUN, J.-P., J. AUDRU and J. CÉSAR, 1989. *Catalogue des plantes vasculaires de la République de Djibouti*. [ii], 277, [7] pp., illus. (Études et syntheses IEMVT 34). Maisons-Alfort, France.

Annotated enumeration of native, naturalized, adventive and commonly cultivated vascular plants with synonymy, references, key citations, localities and elevations with *exsiccatae*, overall distribution, habitat, and (where necessary) some taxonomic commentary; additions (pp. 275–276), statistics, and indices to families and genera at end. The introductory part contains a history of botanical exploration, an overview of the flora, and (pp. 21–27) a bibliography.

AUDRU, J., J. CÉSAR and J.-P. LEBRUN, 1994. *Les plantes vasculaires de la République de Djibouti: flore illustrée*. 2 vols. (in 3) 366, 968 pp., illus. (incl. 301 col. photographs), maps. Djibouti: CIRAD, Département d'Élevage et de Médecine Vétérinaire.

Volume 1 comprises an illustrated keyed enumeration covering 783 native, introduced and commonly cultivated species, the taxonomic entries including indication of habit and habitat along with brief diagnostic features and explanations of the accompanying technical figures; list of references (pp. 299–308) and indices to families and to all botanical names. The introductory part furnishes a justification for and the plan of the work, an account of the vegetation with colored photographs, a map and a diagram of relationships, plant classification and nomenclature, and a glossary; general keys are on pp. 41–52. Volume 2, the complementary and more encyclopedic part of the work, features taxonomic entries with derivations of accepted names, synonymy, vernacular names, distribution, diagnostic features, halftone illustrations, and sometimes many descriptive and other notes compiled without much digestion; index to genera at end.[55]

545

Socotra and 'Abd-al-Kuri Islands

Area: 3885 km^2. Vascular flora: 680 species (Radcliffe-Smith in J.-P. Lebrun, 1976; see **Division 5, Progress**). A part of the South Arabian Mahra (Mahri) sultanate of Qishn and Socotra, the islands came under

British rule with its inclusion in the Aden Protectorate; after 1967 Mahra was taken over by the Democratic Republic of South Yemen and, in 1990, it passed into the united state of Yemen (**782**). They are being covered in *Flora of the Arabian Peninsula* (**780**).

The islands are of considerable biological and biogeographical interest, with a number of relics and evolutionary oddities. The contributions by Isaac B. Balfour (1888, 1903; see below) remain fundamental. Important additional works on the flora, the latter with many fine figures, are F. VIERHAPPER, 1907. Kenntnis der Flora Südarabiens und der Inseln Sokotra, Semha und Abd el Kuri. *Akad. Wiss. Wien, Math.-Naturwiss. Kl., Denkschr.* **71**: 321–490, and ROYAL BOTANIC GARDENS, KEW, 1971. New or noteworthy species from Socotra and Abd-al-Kuri. *Hooker's Icon. Pl.,* V, 7(4): pls. 3673–3700, map. An account of the vegetation, along with a species list, appears in G. B. POPOV, 1957. The vegetation of Socotra. *J. Linn. Soc., Bot.* **55**: 706–720. Relatively little was published after 1971 but improved political stability in recent years has facilitated renewed collecting efforts.

Bibliography
See **782** (GABALI, 1993).

BALFOUR, I. B., 1888. *Botany of Socotra.* lxxv, 446 pp., 100 pls., map (Trans. Roy. Soc. Edinburgh 31). Edinburgh.

Amply descriptive flora of non-vascular and vascular plants, without keys; includes synonymy (with references and citations, including references to standard treatments under generic headings), vernacular names, localities with *exsiccatae*, summary of extralimital range where applicable, extensive taxonomic commentary, and notes on habitat, special features, etc.; figures of representative species; indices to all botanical and vernacular names. The introductory section includes accounts of physical features, history and geography, animal life, botanical exploration, major features of the flora and vegetation, endemic species, floristics and phytogeography.

BALFOUR, I. B., 1903. Botany of Sokotra and Abd-el-Kuri. Angiospermae [and] Pteridophyta. In H. O. Forbes (ed.), *The natural history of Sokotra and Abd-el-Kuri,* pp. 447–542, illus., pls. 26A–26B (col.). Liverpool: The Free Public Museums; London: Porter. (Liverpool Public Museums Special Bulletin.)

Comprises revised systematic enumerations of lower vascular and flowering plants, supplementary to *Botany of Socotra* (see above) and incorporating the results of the 1898–99 expedition of Forbes and W. R. Ogilvie-Grant; includes some descriptions of new taxa, localities with *exsiccatae* and field data, taxonomic commentaries, and indication of status (separate indices are, however, wanting). [Other chapters in this definitive natural history account for the non-vascular plants as well as animal groups.]

546

Eritrea

Area: 117 600 km^2. Vascular flora: no data. After 109 years, firstly as an Italian colony and then under British and (after 1952) Ethiopian rule, Eritrea recovered independence in 1991. The country is being covered in the current *Flora of Ethiopia and Eritrea* (**548**); this will in time replace the now-antiquated works of Fiori and Pirotta.

PIROTTA, R., 1903–07. *Flora della colonia Eritrea.* Parts 1–3, pp. 1–464, pls. 1–12 (Annuario Reale Ist. Bot. Roma 8) Rome.

Non-systematic enumeration of vascular plants; includes partial keys, descriptions of new taxa, synonymy (with references and citations), *exsiccatae* with localities, taxonomic commentary, and notes on habitat, diagnostic features, etc. The introductory section includes an account of botanical exploration. [Not completed.][56]

Woody plants
FIORI, A., 1909–12. *Boschi e piante legnose dell'Eritrea.* 428 pp., 177 text-figs. (some halftones). Florence: Istituto Agricolo Coloniale Italiano.

A copiously illustrated work in three sections: (I) a general survey of forests, deforestation, protection, and reforestation; (II) a descriptive treatment of arborescent vegetation (with illustrations), vegetation zones, and habitat factors, with references; and (III) a systematic descriptive flora of woody plants, with keys, synonymy, references, citations (especially to Pirotta's flora), vernacular names, indication of localities with collectors as well as generalized summary of local and extralimital range and altitudinal zonation, habitats, and some critical commentary. These three parts are followed by addenda, a general key to genera (pp. 378–398), a table of specific gravities, a lexicon of vernacular names with scientific equivalents, references (pp. 401–405), and an index to scientific names. For more important tree species further notes on properties, wood, uses, etc., are given.

548

Ethiopia

Area: 1 023 050 km². Vascular flora (including Somalia and Eritrea): 5500 species (Lebrun, 1976; see **Division 5, Progress**). Morat and Lowry (1997; see **Division 5, Progress**) propose a figure of *c.* 6500 (including Eritrea).

The history of botanical exploration and the significance of the Ethiopian flora have been summarized by Hedberg (1983).[57] The present *Flora of Ethiopia and Eritrea* project came into being in 1980 after 13 years of planning and fund-raising; publication began in 1989 with three full volumes and one part-volume completed by 1998. Some recent studies on the Ethiopian flora feature in I. HEDBERG (ed.), 1986. *Research on the Ethiopian flora*. 212 pp. (Symb. Bot. Upsal. 26(2)). Uppsala.

Keys to families

BURGER, W., 1967. *Families of flowering plants in Ethiopia*. 236 pp., 74 pls. (Oklahoma Agric. Exp. Sta. Bull. 45). Stillwater: Oklahoma State University Press. [Comprises keys to and descriptions of families of seed plants of Ethiopia and adjacent regions, with references and representative illustrations; glossary, bibliography, and index at end.][58]

HEDBERG, I., S. EDWARDS and MESFIN TADESSE (eds.), 1989– . *Flora of Ethiopia*. Vols. 2(2), 3, 6, 7. Illus. Addis Ababa, Asmara and Uppsala: National Herbarium of Ethiopia; University of Asmara; University of Uppsala. (From vol. 2(2) retitled *Flora of Ethiopia and Eritrea*.)

Illustrated documentary descriptive flora; includes keys, synonymy with dates, type citations, selected *exsiccatae*, partly symbolic indication of local and extralimital range, altitudinal amplitude, brief taxonomic commentary where necessary, and notes on habitat and uncertain records; glossary, complete indices and lexica to vernacular names at end (no vernacular names are included in the text proper). The introductory portion covers the background and scope

of the work, notes for its use, and abbreviations and conventions. [Arrangement of families largely follows the 1926–34 version of the Hutchinson system. As of 1999, volumes 2(2) (Canellaceae–Euphorbiaceae, 1995), 3 (Pittosporaceae–Araliaceae, 1989), 6 (Hydrocharitaceae–Arecaceae, 1997) and 7 (Poaceae, 1995) had been published.][59]

Woody plants (including trees)

BREITENBACH, F. VON, 1963. *The indigenous trees of Ethiopia*. 2nd edn. 306 pp., 129 text-figs. Addis Ababa: Ethiopian Forestry Association (Mimeographed.) (1st edn., 1960.)

Briefly descriptive, illustrated treatment of the more common trees of Ethiopia, with keys to families; index. The introductory section includes notes on physical features and vegetation formations together with a list of references. [Based partly on EGGELING and DALE's *Indigenous trees of the Uganda Protectorate* (**534**).]

SOUANE THIRAKUL, [1994]. *Manual of dendrology for the south, south-east and south-west of Ethiopia*. iv, 478 pp., 50 text-figs., numerous halftones. [Addis Ababa]: Woody Biomass Inventory and Strategic Planning Project, Ministry of Natural Resources Development and Environmental Protection, Ethiopia/Canadian International Development Agency.

Illustrated dendrology of 144 species comprising usually muddy halftones of leaves, bark and other features with facing *pro forma* text; the latter includes vernacular names, limited synonymy, diagnostic features, indication of distribution and habitat, and descriptions of habit, bark, leaves, flowers, fruit and wood; list of references, glossary, alphabetical list of families and species covered (pp. 467–469) and index to all vernacular and botanical species names. The general part includes an extensive illustrated organography along with (pp. 53–63) keys to all species.[60]

Region 55

Sudan (middle Nile Basin)

Area: 2 505 815 km². Vascular flora: Andrews (1950–56; see below) treats 3137 flowering plant species. Other estimates are 3500 (Wickens) and 3200 (Friis in PD; see **General references**). This region corresponds to the Sudan Republic, the largest country in Africa and wholly within the tropics. From the end of the nineteenth century until 1956 the country was under a British and Egyptian condominium government and known as Anglo-Egyptian Sudan.

Serious botanical exploration dates from the mid-nineteenth century with the expeditions of Schweinfurth and of Speke and Grant, but with for long a nebulous political status, tardy development of general administrative control, and a paucity of resident botanists subsequent floristic progress was erratic. The results of the early expeditions were cumulated into a substantial checklist for the whole Nile Basin compiled by Georg Schweinfurth with Paul Ascherson in the former's *Beitrag zur Flora Aethiopiens*, 1 (1867). A first separate enumeration appeared in 1906: *Catalogue of Sudan flowering plants* by A. F. Broun, then Government Botanist. This was revised by Broun with his successor R. E. Massey as *Flora of the Sudan* (1929), and further revised and expanded by F. Andrews as *Flowering plants of the Anglo-Egyptian Sudan* (1950–56; additions by G. Wickens, 1969). Early collections have also been accounted for in *Flora of tropical Africa*.

Floristic work, with additional collections, has continued since independence – within the country chiefly at Khartoum University – and already by 1970 Andrews' flora was out of date (Ekhlas a-Bari, personal communication). A new tree manual appeared in 1990. Much of the country and especially the long-troubled and remote south, however, remains poorly collected and documented; the white areas on Léonard's map of 1965 cover more than two-thirds of the country. For southern and southeastern areas, the standard works on northeast and east tropical Africa (Regions 53 and 54) are regarded as more useful than that of Andrews. Among them is the new *Flora of Ethiopia and Eritrea* (**548**).

Of the several partial floras and field reports, significant recent contributions, described below, include *Flora of Jebel Marra* (Wickens, 1976), a monograph on one of the great isolated Saharan massifs (in the westerly Darfur district and until 1920–21 not visited by a botanist), and a 1990 field-guide to the woody plants of the Imatong massif, the most northerly extension of the East African highlands (and similarly not first explored until 1929). A fuller account of the Imatong and neighboring border massifs began to appear in 1998.

Progress
Collectors in Sudan have been accounted for by Cufodontis (1962; see **Region 54, Progress**). Recent

progress has been limited for social and political reasons.

Bibliographies. General and divisional bibliographies as for Division 5.

Indices. General and divisional indices as for Division 5.

550
Region in general

See also **501** (Lebrun and Stork; Oliver *et al.*). The Andrews flora (1950–56) is most useful only in the central and northern parts of the Sudan, i.e., the Khartoum district and Nubia. Floras of neighboring areas should be used for the south and west.

Andrews, F. W., 1950–56. *Flowering plants of the Anglo-Egyptian Sudan*. 3 vols. Illus., map. Arbroath, Scotland: Buncle. (Vol. 3 entitled *Flowering plants of the Sudan*.)

Briefly descriptive illustrated flora of seed plants; includes keys to genera and some species, limited synonymy, generalized indication of internal range (with some specific localities), and miscellaneous notes; index to all botanical names in each volume (no general index). The introductory section includes an illustrated glossary-organography. [For a supplement to this now very incomplete work, see G. E. Wickens, 1969. Some additions and corrections to F. W. Andrews' *Flowering plants of the Sudan*. 49 pp. (Sudan Forests Bull., N.S. 14). Khartoum. Pteridophytes have been covered in a complementary enumeration: K. N. G. Macleay, 1953. The ferns and fern-allies of the Sudan. *Sudan Notes & Rec.* **34**: 286–298.]

Partial works
Two major ranges of mountains respectively dominate western and southern Sudan. That in the west, the phytogeographically significant, over 3000 m high isolated Sahelian massif of Jebel Marra, was the subject of a recent detailed monograph by Gerald Wickens, thus complementing Pierre Quézel's study of the Tibesti massif in northern Chad (**593**) and moreover supplementing the sketchy coverage in Andrews' flora. In the south are the Imatong and other mountains on the Ugandan border, the most northerly extension of the East African highlands; the highest peak is Mt. Kluyeti

(3181 m). These latter have been the subject of an account by Friis and Vollesen (1998).

FRIIS, I. and K. VOLLESEN, 1998. *Flora of the Sudan–Uganda border area east of the Nile*, I: *Catalogue of vascular plants, first part*. Pp. 1–389, figs. 1–3 (maps) (Biol. Skr. 51(1)). Copenhagen: Munksgaard (for Det Kongelige Danske Videnskabernes Selskab).

Annotated enumeration of vascular plants with limited synonymy, literature citations (for key to abbreviations, see pp. 26–29), detailed indication of distribution with localities, *exsiccatae*, and places of deposit; summary of habitat and overall distribution; general references at end. The general part (pp. 9–24) includes (1) an introduction to the border mountains area (with map; the greater part of the area is in Sudan) with descriptions of the massifs and the defined study area (a large part of the former East Equatoria province in Sudan and a small part of Acholi province in Uganda), its phytogeographic interest, and the basis and plan of the work; and (2) a history of general and botanical exploration (the latter first seriously initiated in 1929). [The first part as described here covers pteridophytes, gymnosperms and dicotyledons from Annonaceae through Asteraceae (the last-named arranged primarily in accordance with the original 1926–34 version of the Hutchinson system); the second part will cover the 'tubiflorous' dicotyledons, the monocotyledons and in addition include a list of collectors, a gazetteer of localities, and vegetatiological and phytogeographical analyses.]

WICKENS, G. E., 1976. *The flora of Jebel Marra (Sudan Republic) and its geographical affinities*. [ix], 368 pp., 34 text-figs., 8 pls., 208 distribution maps, frontisp. (area map), overlays (in pocket) (Kew Bull. Addit. Ser. 5). London: HMSO.

Includes (pp. 83–191) a systematic checklist of 982 species (966 vascular), without keys or descriptions; includes synonymy, references, citations of *exsiccatae*, and critical and ecological notes. This is followed (pp. 233–337) by a long series of species distribution maps and, in turn, an extensive list of references as well as indices. The introductory part includes a survey of plant geography (pp. 35–59) as well as vegetation and its history.

Woody plants (including trees)

EL AMIN, H. MD., 1990. *Trees and shrubs of the Sudan*. vii, 484 pp., 153 text-figs. Exeter: Ithaca Press.

Descriptive manual with some keys to species; includes national distribution, habitat, phenology, diagnostic or representative illustrations, occasional commentary, and index to all scientific names. [Based on a 1980 doctoral dissertation by the author; after his death edited for publication by Ekhlas Abdel Bari (Khartoum) with the omission of synonymy, literature citations, extra-Sudanian distribution, notes on uses, and the bibliography.][61]

SOMMERLATTE, H. and M. SOMMERLATTE, 1990. *A field guide to the trees and shrubs of the Imatong Mountains, southern Sudan*. ix, 373 pp., illus., maps. Nairobi: GTZ.

Section 2 of this work comprises a descriptive atlas of woody plants covering 138 species with full-page figures and facing text; the latter includes scientific and vernacular names, botanical details, and notes on uses and properties and on related or less common species (where appropriate). Artificial keys and a glossary are also provided. Species are arranged according to the vegetation association (six in all, more or less altitudinally zoned) in which they are most commonly found. Section 1 covers vegetation ecology and phytogeography, while at the end are a list of references (pp. 365–366) and indices to vernacular and scientific names.[62]

SOUANE THIRAKUL, 1984. *Manual of dendrology of Bahr el Ghazal and Central Regions*. [Place of publication and publisher not known.] [Not seen; cited from, and presumably similar to, the author's *Manual of dendrology for the south, south-east and south-west of Ethiopia* (1994; see **548**).]

Region 56

Democratic Republic of Congo (formerly Zaïre)

Area: 2 345 409 km². Vascular flora: 11 000 species (Léonard in Lebrun and Stork, 1991; see **501**); however, a figure of 9300 seed plants is suggested by Morat and Lowry (1997; see **Division 5, Progress**). The region as delimited here corresponds to the present Democratic Republic of Congo (until 1908 Congo Free State and from then until 1960, Belgian Congo; from the 1970s until 1998 it was known as Zaïre). For Rwanda and Burundi, formerly administered (as Ruanda-Urundi) by Belgium but geographically part of East Africa, see respectively **535** and **536**.

Serious botanical exploration in Congo began only in the wake of the great expeditions (including those of David Livingstone and Henry Stanley) of the nineteenth century and, from 1884, the establishment of the so-called 'Congo Free State'. The first to collect extensively was Georg Schweinfurth in 1870–71, while Belgians began serious work after 1884. In 1895 the Jardin Botanique de l'État in Brussels assumed responsibility for the growing official collections and Émile De Wildeman took charge of their curation. From then until the 1930s and beyond De Wildeman was a prolific contributor on the Congolese flora; his output included several major mission reports (including among others those of Émile Laurent, the Compagnie du Kasai, J. Bequaert, and Franz Thonner – the last-named the author of *Die Blütenpflanzen Afrikas*, cited here under **500**) and preliminary regional floras. Among the latter were *Études sur la Flore du Katanga* (1902–13) and *Études de Systématique et de Géographie Botaniques sur la Flore du Bas- et du Moyen-Congo* (1903–12). After World War I there appeared his *Plantae Bequaertianae* (1921–32) and *Contribution à l'Étude de la flore du Katanga* (1921; supplements, 1927–32). Some coverage of Zaïre was also furnished by *Flora of tropical Africa* (**501**).[63]

The rapidly increasing knowledge of the flora called for summaries, and before 1914 three appeared in relatively quick succession: *Études sur la flore de l'État Indépendant du Congo* (1896) by Théophile Durand and the Swiss botanist Hans Schinz, published through the Académie Royale de Belgique and covering some 1100 species of vascular plants; *Census plantarum congolensium* (1900, in *Actes du 1er Congrès International de Botanique*, pp. 277–340) by De Wildeman and Durand, covering 2062 vascular species; and, a year after transfer of the 'Free State' from Léopold II to the Belgian government, *Sylloge florae congolanae* (1909) by Durand and Hélène Durand, covering 3546 phanerogams. The value of the *Sylloge* was formally recognized by the Académie with the award for 1907–08 of its 'Prix Émile Laurent' (for Congo-related botanical studies). Continuing exploration inevitably soon rendered the *Sylloge* out of date; however, with the aftermath of World War I and changing interests, it never was replaced.[64]

In 1926 Walter Robyns, then early in his career, argued that the number of collections had become sufficient to warrant preparation of a large-scale descriptive flora of the Belgian territories – even though, as De Wildeman wrote in 1928, many areas were not or but imperfectly known. Robyns was also inspired by the development by Kew of its regional floras programme (and the appearance of the first volume of *Flora of west tropical Africa*) in succession to *Flora of tropical Africa*. Opportunities for additional field work were opened following the formation in 1933 of the 'Institut National pour l'Étude Agronomique du Congo' (INEAC) – part of a general interwar movement towards increased metropolitan European capital and social investment in dependent territories – and the

organization in 1938 of a local herbarium at their research station in Yangambi (on the Congo below modern Kisangani). Preparation of works on forest trees and other groups of actual or potential economic interest was also undertaken; for trees, this resulted in area manuals by Vermoesen (1923), Delevoy (1929) and Lebrun (1935). In 1942 the *Flore du Congo Belge et du Ruanda-Urundi* project was established; publication commenced in 1948 and by 1990 about half the estimated seed plant flora had been documented. Production has continued in spite of changes in sponsorship, name and format. It came to be regarded as a 'regional' flora like those in the Kew plan but, in contrast to *Flora of west tropical Africa*, adopted a more fully descriptive format (a development subsequently followed for most other regions of tropical Africa by the various 'metropolitan' botanical centers).[65]

Botanical exploration in Congo has continued and some local floristic publication has been effected in the second half of the twentieth century, notably through the universities at Kinshasa, Kisangani and Lubumbashi. Most organized effort has, however, gone into *Flore d'Afrique centrale* (so renamed in 1971). Expert belief is that completion of this work should have priority (Kendrick, 1989, p. 213; see **Progress** below). Among the few partial floras and enumerations are Robyns' *Flore des spermatophytes du Parc National Albert* (1948–55), covering what is now the Virunga Volcanoes National Park, and Georges Troupin's incomplete *Flora des spermatophytes du Parc National de la Garamba* (1956) for the important wildlife reserve along the northern border. Another postwar flora, now no longer recommended, is the three-volume mimeographed *Flore du Kwango* (1948) by Fr. M. Renier; however, L. Pauwels in *Plantes vasculaires des environs de Kinshasa* (1982 and subsequent releases) has noted it as the work most often utilized by students and naturalists in the western part of the country. Another recent partial work is *Catalogue informatisé des plantes vasculaires des Sous-Régions de Kisangani et de la Tshopo* (1983) by J. Lejoly, S. Lisowski and M. Ndejele. The extensive collections and related studies of Lisowski, F. Malaisse and J. J. Symoens on the southeastern Shaba plateau (centering on Lubumbashi) have, however, yet to be organized into a regional florula; here, *Flora Zambesiaca* – now relatively far advanced – may be effectively utilized. Some guides to forest trees have been published, the most recent being those of Gillardin (1959, never completed) and Pieters (1977).

Progress

Belgian reviews include: T. DURAND, 1909. Les explorations botaniques au Congo Belge et leurs résultats. *Bull. Acad. Roy. Sci. Belgique, Cl. Sci.,* sér. IV, **1909**(12): 1347–1374; É. DE WILDEMAN, 1928. La flore du Congo Belge. Zones explorées – Zones à étudier. *Congo* **9**(1): 459–476; W. ROBYNS, 1946–47. Statistiques de nos connaissances sur les spermatophytes du Congo Belge et du Ruanda-Urundi. *Bull. Jard. Bot. État* **18**: 133–144; and J. LÉONARD, 1967. *Stand und Aufgabe der belgischen Afrikabotanik.* 10 pp. (Sonderbeilage, Afrika Heute **17**/67). Information on Congo is also contained in K. KENDRICK, 1989. Equatorial Africa. In D. G. Campbell and H. D. Hammond (eds.), *Floristic inventory of tropical countries* (under **General references**), pp. 203–216.

Bibliographies. General and divisional bibliographies as for Division 5.

Indices. General and divisional indices as for Division 5.

560
Region in general

See also **501** (LEBRUN and STORK; OLIVER *et al.*); **503** (ENGLER; HEDBERG). For Rwanda, see **535**; Burundi, **536**. Although *Flore d'Afrique centrale* has yet to be completed, the change to publication in fascicles has allowed revision of some early parts as circumstances permit. The 3539 species treated to 1970 were thought to represent about 35 percent of the seed plants; this resulted in hints that the work would not be finished before 2020 (Léonard in Merxmüller, 1971; see **Division 5, Progress**). With publication of the first part of Orchidaceae by 1984, total coverage had passed 50 percent (Kendrick, 1989, p. 212; see **Progress** above). Subsequent progress has been slower; however, by 1996 Morat and Lowry (1997; see **Division 5, Progress**) reported the *Flore* – after nearly 50 years – as being 67 percent 'published or accessible'.

Keys to families

ROBYNS, W., 1958. *Flore du Congo belge et du Ruanda-Urundi: tableau analytique des familles.* 67 pp. Brussels:

Institut National pour l'Étude Agronomique du Congo. [Analytical keys.]

DURAND, T. and H. DURAND, 1909. *Sylloge florae congolanae (Phanerogamae).* iii, 716 pp. Brussels: A. de Boeck (for the Ministère des Colonies). (Also published as *Bull. Jard. Bot. État* 2.)

Systematic enumeration (in French) of known seed plants; includes full synonymy, references and citations, vernacular names, and localities with *exsiccatae*; floristic statistics and indices to all vernacular and botanical names. The introductory section includes a list of collectors. [Rwanda and Burundi, then part of German East Africa (**Region 53**), are not included.]

ROBYNS, W. *et al.* (eds.), 1948–63. *Flore du Congo Belge et du Ruanda-Urundi* (from 1960 *Flore du Congo, du Rwanda et du Burundi*). Vols. 1–7, 8(1), 9–10. Text-figs., pl. (some col.). Brussels: Institut National pour l'Étude Agronomique du Congo. Continued as JARDIN BOTANIQUE NATIONAL DE BELGIQUE, 1967– . *Flore du Congo, du Rwanda et du Burundi* (from 1971 *Flore d'Afrique centrale (Zaïre-Rwanda-Burundi)*). Published in fascicles. Illus. Brussels.

Illustrated descriptive 'research' vascular flora of the present-day Democratic Republic of Congo together with Rwanda (**535**) and Burundi (**536**); includes keys to all taxa up to generic rank, full synonymy with references, literature citations, vernacular names, fairly detailed summary of local range with *exsiccatae*, generalized extralimital distribution, taxonomic commentary, notes on habitat, biology, special features, etc., and illustrations of representative species; indices in each volume (or fascicle) to vernacular and botanical names. The early volumes followed a modified Englerian sequence, and cover the gymnosperms and about one-third of the known angiosperms; the later fascicles, although organized in two series (Ptéridophytes and Spermatophytes) and each containing one family, do not follow a set sequence.[66]

Woody plants (including forest trees)

See also **501** (LETOUZEY, 1982, 1986; TALIFER, 1990; VIVIEN and FAURÉ, 1985). Forests are regarded by Pieters (1977; see below) as covering about 1 million km², or 42 percent of the country. Timber has been for many decades a significant contributor to the Congolese economy and, not surprisingly, several dendrological works have been published, not only for the state as a whole but also for individual regions. There is, however, no complete 'forest flora'; the manual by Gillardin

(1959) evidently was never completed. Pieters' dendrology, the most recent available, covers but 112 species.

GILLARDIN, J., 1959. *Les essences forestières du Congo belge et du Ruanda-Urundi.* 378 pp, illus., map. Bruxelles: Ministère du Congo Belge et du Ruanda-Urundi.

Foresters' checklist/manual of trees and important shrubs, vines and herbs, without keys; includes limited synonymy, vernacular names, fairly detailed indication of local range, notes on habitat, special features, etc.; indices to vernacular, trade and botanical names, the latter being cross-referenced to the *Flore du Congo* (see above). [Not completed; extends up to the Leeaceae (modified Englerian system), not including Monocotyledoneae.]

PIETERS, A., 1977. *Essences forestières du Zaïre.* [iv], 349 pp., 44 halftones. Ghent: Onderzoekscentrum voor Bosbouw, Bosbedrijfsfoering en Bospolitiek, Faculteit Landbouwwetenschappen, Univ. Gent.

Illustrated dendrological treatment of principal trees (112 species), without keys; includes scientific and trade names, distribution, and notes on habitat, followed by a forest regions map (p. 322), references (pp. 323–324), and indices. The introductory part is devoted to an exhaustive treatment of organography and terminology.

I. Central Congo

Includes the so-called 'Cuvette Centrale' along with Kisangani (Stanleyville) and environs and the research center of Yangambi.

LEJOLY, J., S. LISOWSKI and M. NDEJELE, 1988. *Catalogue des plantes vasculaires des Sous-Régions de Kisangani et de la Tshopo (Haut-Zaïre).* 3rd edn. [v], 122 pp. Brussels. (Trav. Lab. Bot. Syst. Ecol., Univ. Libre Bruxelles.) (1st edn. as J. LEJOLY and S. LISOWSKI, 1978. *Plantes vasculaires des Sous-Régions de Kisangani et de la Tshopo (Haut-Zaïre).* Kisangani; 2nd edn. by current authors, 1983, as *Catalogue informatisé des plantes vasculaires des Sous-Régions de Kisangani et de la Tshopo (Haut-Zaïre).*)

Alphabetically arranged checklist of 2608 vascular plants (including 94 pteridophytes, pp. 117–122) with indication of principal synonyms, chorological class, and coded indication of life-form and status; brief introduction on p. 1. No map is presented.

Forest trees

PACQUET, J., [Date not known]. *Manuel de dendrologie de la Cuvette Centrale du Zaïre.* [Not seen; information from Souane Thirakul, *Manual of dendrology* (see 577).]

VERMOESEN, C., 1931. *Manuel des essences forestières de la région équatoriale et du Mayombe.* Revised edn. xii, 282 pp., illus. (some col.). Brussels: Ministère des Colonies. (Original edn., 1923, Brussels, as *Les essences forestières du Congo Belge (région équatoriale et Mayombe).*)

Pocket descriptive manual of principal forest trees and other woody elements of the moist high forest and seasonal monsoon forest/woodland zones, with vernacular names, indication of distribution, and notes on habitat, special features, uses, etc.; indices.

II. Eastern and southeastern Congo

Encompasses the Kivu or Great Lakes provinces along with Shaba. The Albert (now Virunga Volcanoes) National Park is along the Rwandan frontier. For Kivu pteridophytes, see 535 (PICHI-SERMOLLI, 1983–85).

ROBYNS, W. *et al.*, 1948–55. *Flore des spermatophytes du Parc National Albert.* 3 vols. Text-figs., pls. (some col.), maps. Brussels: Institut des Parcs Nationaux du Congo Belge.

Illustrated descriptive treatment of seed plants; includes keys, synonymy, vernacular names, detailed local range (including *exsiccatae*), summary of extralimital distribution, taxonomic commentary, and notes on habitat, biology, etc.; full indices. The introductory section (in vol. 1) includes accounts of physical features, vegetation, phytogeography, and botanical exploration.

Forest trees

DELEVOY, G., 1929. Les essences forestières du Katanga. In *idem, La question forestière du Katanga*, vol. 2. xv, 525, 15 pp., illus. Brussels.

Enumeration of woody plants, alphabetically arranged by genera; includes synonymy, citations, vernacular names, descriptions (sometimes with extended notes), indication of distribution and habitat, occasional taxonomic commentary, and uses and potential; table of utilization (arranged by families) with indication of importance to forestry, indices to all botanical and vernacular names, and added addenda and errata. [The total work comprises three volumes; the others respectively are on forest vegetation and on utilization and development.]

LEBRUN, J., 1935. *Les essences forestières des régions montagneuses du Congo oriental.* 263 pp., 28 text-figs., 18 pls. (Série scientifique 1). Brussels: Institut National pour l'Étude Agronomique du Congo.

Foresters' manual of principal trees and other woody plants in eastern Congo; includes keys, references, vernacular names, *exsiccatae*, summary of distribution, and notes on habitat, special features, properties, uses, etc.; indices. An account of floristic zones and a bibliography are also featured.

III. Northern Congo

Much of this area, towards Sudan, comprises woodland and savanna. The Garamba National Park is along the Sudanese border.

TROUPIN, G., 1956. *Flore des spermatophytes du Parc National de la Garamba.* Vol. 1. 349 pp., illus. (some col.), map. Brussels: Institut des Parcs Nationaux du Congo Belge.

Descriptive treatment of seed plants, without keys; includes synonymy, vernacular names, localities with *exsiccatae*, summary of extralimital range, and brief notes on habitat, biology, etc.; index. [Not completed; encompasses the Gymnospermae and Monocotyledoneae.]

IV. Western Congo

Includes the Mayombe hill belt along with Kinshasa and environs.

PAUWELS, L., 1982. *Plantes vasculaires des environs de Kinshasa.* xi, 118, [3] pp., 2 maps. Bruxelles: The author. [Subsequent versions produced through 1992.]

Alphabetically arranged checklist, with abbreviated indication of habit(s) and essential synonymy; index to genera (with references by number and family) at end. An introduction covers the scope and plan of the work and sources utilized, with a brief list of references (pp. v–vi). [Covers 2433 species in an area of 22 863 km² in Bas-Congo as well as the Congo River island of Mbamu in Congo-Brazzaville (572).[67]]

Woody plants (including forest trees)

PAUWELS, L., 1993. *Nzayilu N'ti: guide des arbres et arbustes de la région de Kinshasa-Brazzaville.* 495 pp., 20 text-figs. (incl. maps), 241 pls. Meise: Jardin Botanique National de Belgique.

Students' manual of trees and shrubs, the taxa all alphabetically arranged; includes general keys to all taxa (pp. 27–85), concise descriptions, indication of habitat and distribution, and occasional notes on related species followed by the plates; glossary and indices to French, Kongo, and generic and family names. The introductory part covers the plan and scope of the work and includes brief passages on physical extent, climate, phytogeography, vegetation (with lists of characteristic species by formation), and classification; key references, pp. 22–23. [Accounts for some 800 species in an area of 34 741 km², of which 317 are fully described and keyed out; to these latter have been added 100 commonly cultivated ones.]

Region 57

Equatorial Central Africa and Chad

Area (from *c.* 16° N southwards, i.e., without northern Chad): *c.* 2 600 000 km². Vascular flora: here estimated at between 12 000 and 15 000 species. This somewhat heterogeneous region includes the states of Congo (Brazzaville), Gabon, Central African Republic, Chad, Cameroon, Equatorial Guinea, the territory of Cabinda (part of Angola), and the islands in the Gulf of Guinea (comprising the state of São Tomé and Príncipe and the insular possessions of Equatorial Guinea (Pagalu or Annobon and Bioko)). Of particular botanical interest are the islands, the Mt. Cameroon massif, and the Mayombe upland formation inland from the coast.[68]

Some early collections were made along the coast and notably in the easily accessible Benin Islands in the eighteenth and early nineteenth centuries, and Cameroon Mountain was ascended by the mid-nineteenth century. Systematic botanical exploration, however, began only in the wake of European penetration and colonial rule in the latter part of that century. Collections increased rapidly in succeeding decades, notably in the German Cameroons (where, at Victoria – now Limbé – a 'botanical station' was also established), Ubangi-Shari (the Central African Republic), where Auguste Chevalier made extensive collections in 1902–04, and Gabon, where Georges Le Testu prospected for plants in the 1920s and 1930s. Much was also contributed by Johannes Mildbraed as a member of the 1910–11 'second African expedition' of the Duke of Mecklenburg, Adolf Friedrich, in Congo (Kinshasa), Congo (Brazzaville), Cameroon and (afterwards) in the Benin Islands (579). Nevertheless, in a region with some of the richest floras in tropical Africa a botanical basis barely sufficient even for 'national' floras was not attained until well after World War II.

Earlier records were incorporated into *Flora of tropical Africa* (**501**) and a miscellany of expedition and collection reports. Prominent among the latter are works by Chevalier (1913; see **575**), Mildbraed (1922; see **570**), and Pellegrin (1924–38; see **573**). In addition, western Cameroon, mandated to Britain after World War I, was included (along with Bioko or Fernando Po) within the two editions of *Flora of west tropical Africa*. In contrast to most other parts of tropical Africa, however, the post-1945 period did not witness the launch of a 'regional' flora. This has led to very mixed floristic progress, reflecting the geographical and political diversity of the region. Much of it – particularly in the Congo and Central African Republics – by 1960 was still botanically poorly known (and to a goodly extent so remains); a comprehensive account may thus not have been thought practicable if not politically impossible.[69]

In place of any regional treatise, there have appeared some 'national' accounts of varying scope. The most elaborate are the serially published *Flore du Cameroun* and *Flore du Gabon*, both begun in the 1960s and by 1996 covering respectively one-quarter and one-third of their vascular flora. For other areas, only checklists are available, although serial descriptive floras were begun for the Benin Islands. Congo (Brazzaville) was covered by Bernard Descoings (1961) and later by Sita and Moutasmboté (1988), mainland Equatorial Guinea (Río Muni) by Emilio Guinea López (1947), the present Central African Republic by Charles Tisserant (1950) and later by Boulvert (1977), southern Chad by Jean-Pierre Lebrun *et al.* (1972), and the Benin (Guinea) Islands by Arthur W. Exell (1944–56, 1973). The last-named are additionally covered in serial floras by Escarré (for Fernando Póo or Bioko, 1968–69, not continued apart from a series on the pteridophytes by G. Benl) and the Jardim e Museu Agrícola do Ultramar (now part of IICT) in Lisbon (for São Tomé and Príncipe, 1972–). The high-mountain flora in the Cameroons was reviewed by Adolf Engler (1892; see **503**) and the whole of Mt. Cameroon in a checklist by Cable and Cheek (1998; see **578**).

Without a general flora in progress or contemplated the current situation cannot be considered satisfactory. Two large-scale projects continue to run side by side in seemingly wasteful fashion amidst a patchwork of checklists of variable quality and scope.

Progress
See particularly K. KENDRICK, 1989. Equatorial Africa. In D. G. Campbell & H. D. Hammond (eds.), *Floristic inventory of tropical countries* (under **General references**), pp. 203–216.

Bibliographies. General and divisional bibliographies as for Division 5.

Indices. General and divisional indices as for Division 5.

570

Region in general

See also **501** (LEBRUN and STORK; OLIVER *et al.*); **580** (AUBRÉVILLE; CHEVALIER; TARDIEU-BLOT). The only substantial supranational treatment (not, however, a systematic flora) is J. MILDBRAED, 1922. *Wissenschaftliche Ergebnisse des Zweiten Deutschen Zentral-Afrika-Expedition 1910–11, 2: Botanik.* 202 pp., 90 pls. Leipzig: Klinkhardt & Biermann (for Hamburgische Wissenschaftliche Stiftung); this includes floristic lists for different expedition stations on the mainland as well as Bioko and Pagalu, visited by Mildbraed after the close of the expedition.

Woody plants (including forest trees)
Both the following works are of general application in wet tropical west and west-central Africa, not merely in Region 57.

TALIFER, Y., 1990. *La forêt dense d'Afrique centrale: identification pratique des principaux arbres.* See **501**.

VIVIEN, J. and J. J. FAURÉ, 1985. *Arbres des forêts denses d'Afrique centrale.* See **501**.

571

Cabinda

See **522** (CARRISSO *et al.*); **560** (ROBYNS *et al.*). – Area: 7270 km². Vascular flora: no data. This petroleum-producing district politically is part of Angola.

572

Congo Republic

Area: 341 945 km². Vascular flora: *c.* 4350 species (Sita and Moutsamboté, 1988; see below); a figure of 6000 is suggested by Morat and Lowry (1997; see **Division 5, Progress**). The country has also been known as Congo (Brazzaville), and formerly was 'Middle Congo'.

Brazzaville, the capital, was before 1958 the main administrative and cultural center for this and other territories constituting French Equatorial Africa. Here, too, was located the Institut d'Études Centrafricaines of ORSTOM, first established in 1947 and home to the principal regional herbarium. Organized collecting in the former territory, however, began relatively late, and a first, relatively limited checklist appeared only in 1961. Resident collectors have been few, although among their ranks is J. de Brazza, the country's effective 'founder'. The vascular flora alone is even now regarded as poorly known. Works on neighboring countries, especially Gabon (**573**), Cameroon (**577**) and Congo (Kinshasa) (**560**), should be consulted. A general flora was projected in the 1970s but to date only the grasses have been treated.[70]

DESCOINGS, B., 1961. *Inventaire des plantes vasculaires de la République du Congo, déposées dans l'herbier de l'Institut d'Études Centre-Africaines à Brazzaville.* [2], 63 pp. Brazzaville: ORSTOM. (Mimeographed.)

Unannotated checklist of vascular plants in the then-ORSTOM herbarium in Brazzaville, without synonymy or indication of local range.

SITA, P. and J.-M. MOUTASMBOTÉ, 1988. *Catalogue des plantes vasculaires du Congo.* 195 pp. Brazzaville: ORSTOM.

[Not seen; reference from Lebrun and Stork 1991–97; see **501**).]

573

Gabon

Area: 267 667 km². Vascular flora: *c.* 6000 species (Floret, 1976; see Note 72). Breteler (1989; see below),

however, claims a total of *c.* 8000 species, while a figure of 7000 is suggested by Morat and Lowry (1997; see **Division 5, Progress**). Whatever the best figure, the richness of the flora is related to a high and relatively regular equatorial rainfall, particularly along the coast and in the nearby Mayombe hills.

An early work was F. JARDIN, *Aperçu sur la flore du Gabon* (1891; in *Bull. Soc. Linn. Normandie*, IV, **4**), based on field work of 1845–48. Much the most important predecessor of the *Flore* is, however, F. PELLEGRIN, 1924–38. *La flore du Mayombe d'après les récoltes de M. Georges Le Testu.* 3 parts. Illus. (Mém. Soc. Linn. Normandie 26(2) (1924); N.S., Bot. 1(3, 4) (1928–38)). Caen.[71] This deals in particular with the southwestern and central uplands inland from the coast. Pellegrin later contributed a revision of Leguminosae (1948; in *Mém. Inst. Études Centrafr.* **1**). The northeast was the object of studies initiated in the 1960s by Nicholas Hallé and Annick Le Thomas and continued by Annette Hladik and Jacques Florence through 1980; six successive floristic lists were published. Substantial contributions have also been made within the last three decades by botanists from Wageningen Agricultural University. Nevertheless, with a broken terrain, a small infrastructure, and difficulties of access botanical exploration has been limited and uneven; the flora consequently remains imperfectly known. The elaborate *Flore du Gabon* project, after over 35 years, is still some way from 'completion' with only a third of the probable total vascular flora accounted for.

Progress has been reviewed in F. J. BRETELER, 1989. Gabon. In D. G. Campbell and H. D. Hammond (eds.), *Floristic inventory of tropical countries* (under **General references**), pp. 198–202, as well as by Floret (1976; see Note 72).

AUBRÉVILLE, A. *et al.* (eds.), 1961– . *Flore du Gabon.* Fasc. 1– . Illus. Paris: Muséum National d'Histoire Naturelle, Laboratorie de Phanérogamie.

Copiously descriptive, illustrated serial 'research' flora of vascular plants; includes in each family keys to genera and species, synonymy with references and citations, vernacular names, *exsiccatae* with localities and summary of overall range, taxonomic commentary, notes on habitat, special features, etc., and indices to all botanical and vernacular names. [As of 1999, 35 fascicles had appeared, covering some 35 percent of the vascular flora; each fascicle contains revisions of one or more families (with fasc. 8 comprising pteridophytes, by

M. L. Tardieu-Blot). Fascicles are numbered consecutively, with no formal sequence of families followed in publication.][72]

Woody plants (including trees)

The diverse equatorial tree flora and the corresponding importance of forestry to the national economy in the twentieth century has been reflected in the existence of successive dendrological manuals. Earlier works in this genre include A. CHEVALIER, 1917. *La forêt et les bois du Gabon.* vii, 468 pp., illus. Paris, and H. HEITZ, 1943. *La forêt du Gabon.* 292 pp., illus. Paris. Gabon is also covered by TALIFER and by VIVIEN and FAURÉ (**501**).

SAINT AUBIN, G. DE, 1963. *La forêt du Gabon.* 208 pp., pls., maps (Publ. Centre Technique de Forestier Tropical 21). Nogent-sur-Marne.

Illustrated field-manual of the more important forest trees, without keys; vernacular and trade names; generalized indication of local range; notes on habitat, timber properties, sylvical characteristics, etc.; indices to botanical and other names. The introductory section includes accounts of physical features, climate, vegetation, and forest composition (with maps).

574

Mbini (mainland Equatorial Guinea)

Area: 28 051 km[2] (country); 26 016 km[2] (Mbini). Vascular flora: according to PD (see **General references**) probably 'rich', while a figure of 4076 species is given by Morat and Lowry (1997; see **Division 5, Progress**). Only the mainland portion of Equatorial Guinea, Mbini (formerly Río Muni), is accounted for here. Bioko (Fernando Po) and Pagalu (Annobon) are included with other Benin Gulf islands (**579**).

Collectors active in Mbini are listed by Aedo *et al.* (1999). The major contributors have been Tessmann (before World War I), Guinea in 1945 (as part of an agricultural and natural resources survey), and, after 1986, various local and Madrid botanists including the compilers of the *Bases documentales*. The only major floristic work prior to the *Bases documentales* has been the partly compiled checklist in Guinea's *Ensayo geobotánico* (1946). A full flora is projected.

AEDO, C., Mª. T. TELLERÍA and M. VELAYOS, 1999. *Bases documentales para la flora de Guinea*

Ecuatorial: plantas vasculares y hongos. 414 pp., col. illus., halftones. Madrid: Real Jardín Botánico/CSIC, Spain.

The main part (pp. 23–380) is a census of records (55 398 in all) arranged alphabetically (within major groups) by families and scientific names; entries include both collection citations and literature records (the latter referenced to the main bibliography, pp. 381–396). The census proper is followed by indices to fungi (all names) and vascular plants (families and genera). No vernacular names, habitat data or taxonomic notes are included. The general part includes an appreciation of the work of Emilio Guinea López, a brief survey of recent field work (starting in 1986), a gazetteer with coordinates, a list of known collectors with numbers of specimens collected, and the plan of the work. [Covers Bioko and Pagalu as well as Mbini. Not to be regarded as a 'critical' enumeration; many records are effectively duplicates or lack modern determinations. Coverage is, however, thorough. The section on fungi has a wider geographical circumscription, from Nigeria to former Zaïre.][73]

GUINEA LÓPEZ, E., 1946. *Ensayo geobotánico de la Guinea continental española.* 389 pp., illus., maps. Madrid: Dirección General de Marruecos y Colonias.

The fourth and last section of this work (pp. 219–388) comprises an enumeration of the known vascular plants of Mbini (Río Muni), with limited synonymy, concise indication of local and extralimital range (with some citations and range maps), representative illustrations, and a bibliography; no index. Sections 1–3 deal with general aspects of tropical wet forest, its parameters and limiting factors, and vegetation classification; also included is a summary of the author's 1945 investigations (with itinerary map).

575

Central African Republic

Area: 628 780 km[2]. Vascular flora: >3600 species (Lebrun and Stork, 1991; see **501**); almost certainly this is an underestimate, although on the whole the climate is more seasonal than in western Central Africa and woodland and savanna formations consequently are extensive. – Formerly known as Ubangi-Shari and, for a time, the Central African Empire.

The flora remains poorly known and the checklists by Tisserant and Boulvert are respectively out of date or uncritical. Among earlier compendia, a key contribution (based on collections made in 1902–04 during a primary exploratory expedition) is A. CHEVALIER, 1913. *Études sur la flore de l'Afrique Centrale française (bassins de l'Oubangui et du Chari)*, 1. *Énumération des plantes récoltées*. xii, 451 pp., frontisp. Paris: Challamel. A useful more contemporary survey is Y. BOULVERT, 1995. *Documents phytogéographiques sur les savannes centrafricaines*. 140 pp., maps. Paris: Éditions ORSTOM (Collection Études et Thèses); this comprises a series of detailed distribution maps at 1:5000000 of selected species based to a great extent on field observations. Its taxonomy, however, is not necessarily up to date.

BOULVERT, Y., 1977. *Catalogue de la flore de Centrafrique: Écologie sommaire – Distribution (texte provisoire)*. 3 vols. (in 4). 114, 84, 94, 89 pp., maps. Bangui: Centre ORSTOM. (Mimeographed.)

A compiled checklist comprising names with usually brief annotations indicative of habitat and local distribution as well as some vernacular names. Cultivated plants are also included. Taxa are all alphabetically arranged within each of the three volumes, which refer respectively to 'forêt dense et galeries', 'strate herbacée des savannes' and 'strate ligneuse des savannes'. The second volume is further divided into two parts covering (1) dicotyledons and (2) monocotyledons, pteridophytes and bryophytes. [Not critical and intended by its author very much as a working document.][74]

TISSERANT, C., Père, 1950. *Catalogue de la flore de l'Oubangi-Chari*. 166 pp., map (Mém. Inst. Études Centrafr. 2). Brazzaville. [? Reprinted 1965.]

Briefly annotated enumeration of seed plants, with vernacular names, generalized indication of local range, taxonomic commentary, and notes on habitat, local ecology, biology, uses, etc.; no index. The introductory section includes a discussion of the value of vernacular names.[75]

Forest trees

In addition to the following, reference may be made to D. NORMAND, 1965. *Identification des arbres et des bois des principales essences forestières de la République Centrafricaine*. Nogent-sur-Marne: Centre Technique de Forestier Tropical. Of the two accounts given below, that of Guigonis was based on that of Chevalier but is less informative. For more recent dendrological works, see TALIFER (1990) as well as VIVIEN and FAURÉ (1985; both under **501**).

CHEVALIER, A., 1951. Catalogue des arbres vivant dans la forêt dense et les galeries en Afrique centrale (Bassin de l'Oubangui, de la Hte Sangha et du Haut Chari). *Rev. Int. Bot. Appl. Agric. Trop.* **31**: 489–504, 605–623.

Alphabetically arranged enumeration with vernacular names and notes on habit, height and girth, distribution and biology, along with an autobiographical introduction; no separate index.

GUIGONIS, G., 1970. *Liste des arbres et arbustes vivant dans la forêt dense et les galéries de la République Centrafricaine*. 30 pp. N.p. (Mimeographed.)

An alphabetically arranged bare list of scientific names with authors and their vernacular and trade equivalents; covers 645 species (but not necessarily all those in Chevalier).

576

Chad Republic

Area: 1284640 km^2 (684290 km^2 exclusive of the large northern prefecture of Borkou-Ennedi-Tibesti). Vascular flora: *c.* 1600 species (1516 in the southern part, *fide* Lebrun and Stork, 1991 (see **501**)). That part of Chad north of the 16th parallel, largely high Saharan plateaux culminating in the large Tibesti massif (and for practical purposes corresponding to Borkou-Ennedi-Tibesti), is here also treated under **593**.

Chad, including its botanically somewhat better-known southern part, lies outside the wet tropics and has thus a relatively small flora. Accumulated knowledge was summarized in an enumeration by Jean-Pierre Lebrun (1972). Prior to that treatment, one of the few floristic accounts was F. PELLEGRIN, 1911–12. Collections botaniques rapportées par la Mission Tilho de la région Niger-Tchad. *Bull. Mus. Natl. Hist. Nat.* (Paris) **17**: 459–466, 566–571; **18**: 46–50.

LEBRUN, J.-P. *et al.*, 1972. *Catalogue des plantes vasculaires du Tchad méridional*. 289 pp., 18 maps (Études Bot. IEMVPT 1). Maisons-Alfort, Val-de-Marne: Institut d'Élevage et de Médecine Vétérinaire des Pays Tropicaux.

Annotated enumeration of vascular plants of the Chad Republic south of the 16th parallel, thus more or less excluding the Saharan zone; includes limited synonymy, indication of localities, and notes on habitat, special features, etc.; bibliography, statistics and indices. Appendices cover floristics, phytogeography, useful plants, and other topics, while an account of botanical exploration appears in an introductory section. For additions, see J.-P. LEBRUN and A. GASTON in *Adansonia*, II, **15**: 381–390 (1976) and *Publ. Cairo Univ. Herb.* **7/8**: 109–114 (1977).

577

Cameroon

Area (present republic): 475 439 km². Vascular flora: >8000 species (Le Thomas in Lebrun and Stork 1991; see **501**). – Cameroon is here taken to comprise that part of the former German territory mandated to France after World War I. The southern part of the British mandate, including the Bamenda Plateau and Cameroon Mountain (now also part of Cameroon), forms **578**. As in Gabon, the zone near the coast has a rich flora associated with high and relatively regular rainfall.

Like the *Flore du Gabon*, the *Flore du Cameroun* is a large-scale work which is still a long way from 'completion'. However, a cyclopedia of collectors (as part of the *Flore*) and a phytogeographical summary, *Étude phytogéographique du Cameroun* (1968), both by René Letouzey, may also be consulted. The only earlier studies of the Cameroon flora (apart from the Cameroon Mountain area, here covered under **578**) took the form of expedition and field reports along with works on forest trees and other plants of economic or ethnobiological interest. Nevertheless, the overall foundation for floristic study in Cameroon is presently perhaps better than in most of the rest of Central Africa, with a comparatively high 'density index' of collections based initially upon German work before World War I and, later, the activities of Johannes Mildbraed and René Letouzey among many others. A national herbarium, at Yaoundé, was established in the 1960s.

AUBRÉVILLE, A. *et al.* (eds.), 1963– . *Flore du Cameroun*. Fasc. 1– . Illus. Paris: Muséum National d'Histoire Naturelle, Laboratoire de Phanérogamie/ Yaoundé: Ministère de l'Enseignement Supérieur et de la Recherche Scientifique (MESRES).

Detailed illustrated 'research' flora of vascular plants; each family treatment includes keys to genera and species, synonymy with references and citations, vernacular names, local range with *exsiccatae* and summary of overall distribution, taxonomic commentary, notes on habitat, ecology, uses, etc., and indices to all botanical and vernacular names. [As of 1998, 34 fascicles had appeared, covering some 25 percent of the vascular flora with each containing revisions of one (or more) families. Fasc. 1, containing a phytogeographical summary, and fasc. 7, comprising a cyclopedia of collectors and a history of botanical exploration, are by R. Letouzey; fasc. 8, on pteridophytes, is by M. L. Tardieu-Blot. Families have been published serially as ready, without regard to any systematic scheme.]

Woody plants (including trees)

As in Gabon, forestry plays a not insignificant role in the national economy and a number of dendrologies have as a result been published. The first substantial treatment was *Étude sur la forêt et les bois du Cameroun sous mandat français* (1930) by Louis Hédin. The dendrologies described below are complemented by those of TALIFER (1990) and VIVIEN and FAURÉ (1985; for both, see **501**).[76]

SOUANE THIRAKUL, 1985. *Manual of dendrology: Cameroon*. 640 pp., 59 text-figs., numerous halftones. Québec: Groupe Poulin Thériault (for the National Centre for Forestry Development, Cameroun/Canadian International Development Agency).

Loose-leaf dendrological atlas comprising often muddy halftones of habit, bark, leaves, slash and (sometimes) fruits with facing *pro forma* text; the latter includes vernacular names, limited synonymy, diagnostic features, indication of distribution and habitat, and descriptions of habit, bark, leaves, flowers, fruit and wood; list of references, glossary, and indices to genera and families and to vernacular names. Arrangement is alphabetical by families and genera. The general part encompasses tree morphology, botanical organography (with illustrations), and (pp. 71–94) a key to all species. [Covers trees of moist evergreen and semi-deciduous forests in central and southern Cameroon.][77]

SOUANE THIRAKUL, 1990. *Manuel de dendrologie des savanes boisées, République du Cameroun*. Illus. [Québec.]

Similar in style to the preceding work, but oriented towards the woodlands of northern Cameroon.

578

Bamenda Plateau and Cameroon Mountain

These formerly British-administered parts of Cameroon have since 1960 been reunited with the former Trust Territory to form the Republic of Cameroon (577). Special attention is called to it here as an area of great botanical interest. Included here is the highest peak in all West Equatorial Africa, Cameroon Mountain (4070 m), first ascended by Gustav Mann with Richard F. Burton in 1861. Also within this area is the long-established botanical garden of Limbé (Victoria).

Mann's (and other) plants from Cameroon Mountain (and the Benin Islands) were first written up by J. D. Hooker in *On the plants of the temperate regions of the Cameroons Mountains and islands in the Bight of Benin* (1864; in *J. Linn. Soc., Bot.* 7: 171–240). For subsequent coverage of the mountain flora, see 503 (ENGLER). Apart from general coverage in floras and other works on Cameroon, West Africa (580) and Nigeria (581), more recent treatments of the Cameroon massif were for long not available. Since the mid-1980s, however, renewed botanical exploration has been in progress and the first of a projected group of checklists appeared in 1998.

CABLE, S. and M. CHEEK (comp. and ed.), 1998. *The plants of Mount Cameroon: a conservation checklist.* lxxix, 198 pp., 5 maps, portr. Kew: Royal Botanic Gardens.

The main part comprises an alphabetically arranged enumeration of vascular plants with principal synonyms and citations, *exsiccatae*, and indication of ecological and chorological classes, overall distribution, usual habitat, habit, altitudinal range, and a conservation 'star' rating; no index. The general part includes sections on physical features, geology, geomorphology, climate and rainfall, soils, vegetation, and phytogeography along with statistics, a bibliography, and (pp. xxvii–lxxvi) a 'Red Data Book' with general and specific evaluations; these are followed by the plan of the work (pp. lxxvii–lxxix).[78]

579
Benin (Gulf of Guinea) Islands

Area: Bioko (Fernando Po), 2017 km^2; Príncipe, 110 km^2; São Tomé, 854 km^2; Pagalu (Annobon), 17 km^2. Angiosperm flora: Bioko, 1105 species; Príncipe, 314; São Tomé, 601; Pagalu, 208 (Exell, 1973); total, 1666 native taxa of which 176 are now considered to be endemic, a lesser number than in the past (Figueiredo, 1994). – The Benin Islands comprise the four main islands in the Gulf of Guinea southwest of Cameroon. For further coverage of Bioko, nearest the mainland and connected with it at times during the Pleistocene, see 580 (HUTCHINSON and DALZIEL). Bioko and Pagalu are part of Equatorial Guinea; São Tomé and Príncipe form an independent state.[79] A close floristic affinity is shown with 578, particularly with respect to the montane plants.

Although agricultural exploitation of the islands began under Portuguese rule early in the sixteenth century, little was collected until after 1800 (following Bioko's cession to Spain) and the first general account of the flora is that of Hooker published in 1864 (see 578) – unfortunately after loss of much of the natural vegetation at lower elevations. In the late nineteenth and early twentieth centuries, substantial collecting was carried out by and on behalf of Júlio Henriques at Coimbra; his main contributions were *Flora de S. Thomé* (1886) and *Catalogo da flora da Ilha de S. Thomé* (1892; both in *Bol. Soc. Brot.*) and, finally, *A ilha de S. Tomé sob o ponto de vista histórico-natural e agrícola* (1917). In 1911 Johannes Mildbraed visited the then-Spanish islands, listing his collections in his Central Africa expedition report (see 570). The most useful works remain the mid-century lists by Arthur Exell (a visitor to all the islands in 1932–33) and others, which include phytogeographical remarks. Reference should also be made to ENGLER (503). The serial flora of São Tomé and Príncipe begun in the 1970s has not been continued, but a new enumeration of pteridophytes appeared in 1998. In the two islands of Equatorial Guinea, renewed botanical exploration has taken place in recent years under a cooperative programme with Madrid instituted in 1985 and several new contributions published, notably in *Fontqueria*. The two islands are also included in *Bases documentales* (1999; see below).[80]

Bibliographies

FERNANDES, A., 1982. See **Divisional bibliographies** above. [São Tomé e Príncipe, pp. 109–129; 120 titles in pure and 93 in applied botany.]

FIGUEIREDO, E., 1994. Contribution towards a botanical literature for the islands of the Gulf of Guinea. *Fontqueria* 39: 1–8. [Unannotated and alphabetically arranged by author; no taxonomic or subject index. Revisions and floras accounting for endemics indicated by a '+' sign.]

EXELL, A. W. *et al.* 1944. *Catalogue of the vascular plants of São Tomé (with Príncipe and Annobon).* xi, 428 pp., 26 text-figs., 3 maps. London: British Museum (Natural History). Continued in *idem*,1956. *Supplement . . .* v, 58 pp., 3 text-figs. London.

Critical systematic enumeration of vascular plants, without keys; includes descriptions of new taxa, rather extensive synonymy with references and citations, vernacular names, *exsiccatae* with localities, summaries of local and extralimital range, taxonomic commentary, miscellaneous other notes, and index to all botanical names. The introductory section includes chapters on physical features, geology, climate, history and geography, and botanical exploration, together with remarks on the affinities of the flora and on phytogeography.

EXELL, A. W., 1973. Angiosperms of the islands of the Gulf of Guinea (Fernando Póo, Príncipe, S. Tomé, and Annobon). *Bull. Brit. Mus. (Nat. Hist.), Bot.* **4**(8): 325–411.

Comprises a concise, systematically arranged tabular checklist of flowering plants of the Benin Islands (including Bioko or Fernando Po), with numbers designating post-1944 bibliographic references and letters for each of the four islands as appropriate; index to genera and families at end. The introductory section gives a brief account of the main features of the combined flora and a phytogeographical analysis together with a numbered bibliography of relevant works published since the 1944 enumeration (see above).

Bioko and Pagalu (Equatorial Guinea)

AEDO, C., Mª. T. TELLERÍA and M. VELAYOS, 1999. *Bases documentales para la flora de Guinea Ecuatorial: plantas vasculares y hongos.* See **574**.

ESCARRÉ, A., 1968–70. *Aportaciones al conocimiento de la flora de Fernando Póo.* Fasc. 1–3 (Acta Phytotax. Barcinon. 2, 3, 5). Barcelona (fasc. 3 with T. Reinares). Complemented by G. BENL, 1978–91. *The Pteridophyta of Fernando Po* (later *The Pteridophyta of Bioko (Fernando Po)*), I–V. 31, 34, 46, 69, 109 pp. (Acta Bot. Barcinon. 31–33, 38, 40).

Semi-revisionary serial flora; each fascicle includes keys, substantial descriptions, synonymy, citations, *exsiccatae* with localities, taxonomic commentary, notes on habitat, special features, etc., and indices. The treatment of pteridophytes is similar in scope. [The pteridophytes were completed, the phanerogams not.]

São Tomé and Príncipe

FIGUEIREDO, E., 1998. The pteridophytes of São Tomé and Príncipe (Gulf of Guinea). *Bull. Nat. Hist. Mus. (Bot.)* 28(1): 41–66, 2 maps.

Enumeration (153 taxa) with names, references, synonymy, citations (but no types), localities with *exsiccatae*, and generalized indication of distribution along with conservation status; doubtful records (p. 64) and general references and full index (pp. 64–65). The introductory part encompasses floristics, statistics, and the somewhat episodic collecting history along with the plan of the work and general remarks on the conservation status of the flora.

JARDIM E MUSEU AGRÍCOLA DO ULTRAMAR, PORTUGAL, 1972–82. *Flora de São Tomé e Príncipe.* Fasc. 1–7. Lisbon: Ministerio do Ultramar (later Ministerio da Cooperação).

A serial 'research' flora; each fascicle includes keys, descriptions, synonymy, citations, vernacular names, localities with *exsiccatae*, extralimital distribution, some special notes, and separate index. All treatments are by specialists or staff botanists, the latter mainly M. C. Liberato and J. do Espirito Santo. [Not continued.]

Region 58

West tropical Africa

Area: *c.* 3 150 000 km². Vascular flora: 7343 species (Lebrun and Stork, 1991); 7072 species were recorded in the second edition of *Flora of west tropical Africa.* West tropical Africa here ranges from the eastern boundaries of Nigeria and the Niger Republic westward to the Atlantic at Senegal, with the northern limit approximately defined by the limits of the Sahara. This latter at present ranges from 19° to 16° N west to east, so loosely corresponding with the 18° N limit of coverage of the current edition of the standard regional work, *Flora of west tropical Africa.*[81] More northerly countries or areas thereof, including western Sahara, are here treated as part of northern Africa (**Region 59**).

'West Africa' is botanically about the best-known part of tropical Africa, and is considered relatively well if unevenly explored. Early contacts in the seventeenth and eighteenth centuries were extensive and much botanical material was collected around the many coastal 'castles' (forts) and elsewhere by both visitors and residents. Though not a flora, an outstanding contribution of the era was *Familles des plantes* (1763) by Michel Adanson, based partly on its author's residence in Senegal in the 1750s. Extensive inland exploration and travel began in the late eighteenth century;

representative floras of this period include *Genera plantarum guineensium* (1804) by Adam Afzelius, *Flore d'Oware et de Benin en Afrique* (1804–07) by A. M. F. J. Palisot de Beauvois, *Florae Senegambiae tentamen* (1830–33) by Antoine Guillemin, Samuel Perrottet and Achille Richard, and finally *Niger Flora* (1849) by William J. Hooker. By 1860 Hooker and others believed that a basis sufficient for a general flora had been attained. This, with its warrant considerably extended, became *Flora of tropical Africa* (1868–1937). With the French in time following the British, colonial penetration and control hastened collecting and floristic study, but not until after World War I and consequent policy changes did local herbaria begin to be established. The largest of these today is the Forest Herbarium at Ibadan (Nigeria). The considerable time lag, however, has meant that substantial West African collections are held only outside the region, above all in Europe.

The gradual obsolescence as well as unwieldiness of the *Flora of tropical Africa*, along with the development at Kew under Arthur Hill (director from 1922) of a regional approach to African floras, gave rise to the concise *Flora of west tropical Africa* (Hutchinson and Dalziel, 1927–36; 2nd edn., 1953–72).[82] This much-esteemed work remains one of the most concise but informative of tropical floras.

The existence of such a 'staple' flora, however, tended for some time to deflect publication of 'national' floras and (to a lesser extent) checklists and 'tree books'. The only current national floras are those for Senegal (Berhaut, 1967, 1971– , the illustrated work as yet not complete); Guinea-Bissau (*Flora da Guiné Portuguesa* (later *Flora de Guiné-Bissau*), 1971– , but now dormant); Togo (Brunel *et al.*, 1984 – the best in this class), and Nigeria (*Flora of Nigeria*, 1970– , a never-completed continuation of *Nigerian trees* – itself revised in 1989). An introductory manual covering the major francophone centers is Roberty's *Petite flore de l'Ouest-Africain* (1954). Checklists of varying scope and quality exist for the Ivory Coast, Liberia, Sierra Leone, Guinea-Bissau, Mauritania, Gambia, Senegal, Burkina Faso, Mali, and Niger (the last four sponsored by the French cooperation agency Institut d'Élevage et de Médecine Vétérinaire des Pays Tropicaux (IEMVPT); see below).

The principal regional 'tree book' remains André Aubréville's *Flore forestière soudano-guinéenne* (1950), reprinted in the 1970s. This has been complemented by an illustrated field-guide by C. Geerling (1982; 2nd edn., 1988) and, for the Sahel in the north, by the picture-guide by H.-J. von Maydell (1983; see **505**). Dendrofloras are also available for Nigeria, the Ivory Coast, Liberia (two), Sierra Leone, and Ghana, some of them illustrated; the most recent is, as already noted, the revision of Keay's *Nigerian trees* (1989). For pteridophytes, treatments in both French (Tardieu-Blot, 1953; includes also **Region 57**) and English (Alston, 1959, as part of *Flora of west tropical Africa*) are available. Descriptive floras have also been written by Jacques-Georges Adam for the Nimba massif in Liberia and Guinea (1971–83) and, with P. Jaeger, for the Loma Mountains in Sierra Leone (1980–81); both massifs are phytogeographically significant with a number of endemic and relict species (see **502**).

Because of language needs – as well as, perhaps, *Kulturpolitik* – non-anglophone states (and other areas) have not unnaturally been better supplied with state or local floras. To these may be added the several contributions of the botanical branch (formed in the 1960s) of the French cooperative research organization, Institut d'Élevage et de Médecine Vétérinaire des Pays Tropicaux. These formed part of a programme of work covering most of northern dryland tropical Africa.[83]

Among lesser-known areas in West Africa is the 'Guinea corner', especially Liberia where the flora is the richest for the region. Also worth further attention are the southern border areas of Ivory Coast/Ghana and southeastern Nigeria (with adjacent parts of Cameroon), the latter an area of high rainfall. A few countries, notably Benin (Dahomey) and Guinea, lack any suitable checklist; and for Liberia and Sierra Leone the available offerings are poor. From a floristic point of view, the most valuable new contributions would be on Liberia and adjacent areas, where endemism is high, and in Benin, where extensive species transition reportedly occurs. Elsewhere, local endemism is limited, with distribution being constrained mainly by bioclimate.[84]

Progress

The best recent account is F. N. HEPPER, 1989. West Africa. In D. G. Campbell and H. D. Hammond (eds.), *Floristic inventory of tropical countries* (under **General references**), pp. 189–197, map. See also *idem*, 1958. Progress accomplished in the study of the flora of West Tropical Africa. *Mem. Soc. Brot.* **13**: 23–24.

Bibliographies. General and divisional bibliographies as for Division 5.

Indices. General and divisional indices as for Division 5.

580

Region in general

See also **501** (LEBRUN and STORK; OLIVER *et al.*). In the first edition of *Flora of west tropical Africa*, coverage extended into northern Africa as far as the Tropic of Cancer, thus including a substantial part of the Sahara.

HUTCHINSON, J. and J. M. DALZIEL, 1953–72. *Flora of west tropical Africa.* 2nd edn., revised by R. W. J. Keay and F. N. Hepper. 3 vols. 462 text-figs., map. London: Crown Agents (distributed by HMSO). (1st edn. in 2 vols., 1927–36, London.) Complemented by A. H. G. ALSTON, 1959. *The ferns and fern allies of West Tropical Africa.* 89 pp. London.

Manual-flora of vascular plants; includes keys to all taxa, concise synonymy, principal citations, representative *exsiccatae*, summaries of local and extralimital distribution, taxonomic commentary, and brief notes on habitat, frequency, special features, etc.; index to all botanical names in each volume (cumulative index in vol. 3). An introductory section (vol. 1) includes an account of botanical exploration in the region along with a glossary and bibliography.[85]

Partial works

CHEVALIER, A., 1938. *Flore vivante de l'Afrique occidentale française.* Vol. 1. 360 pp., illus. Paris: Muséum National d'Histoire Naturelle.

Copiously descriptive flora of seed plants; includes keys, synonymy with references and citations, local and extralimital range, and notes on phenology, life-form, biology, etc.; indices to generic and family names at end. Along with francophone territories in west tropical Africa, this work extends to the Sahara and the northern part of francophone equatorial Africa. [Discontinued by the author after World War II.][86]

ROBERTY, G., 1954. *Petite flore de l'Ouest-Africain.* 441 pp. Paris: Larose (for ORSTOM).

Concise manual of seed plants (particularly for schools); includes keys, limited synonymy, literature citations, local range, and notes on habitat, frequency, cultivation, etc.; lexicon of vernacular names; index. Only the more important species are described and keyed out. Covers Burkina Faso, the Ivory Coast, Guinea, Mali (in part), and Senegal, i.e., the region south of 16°N and west of 0° longitude.

Woody plants

AUBRÉVILLE, A., 1950. *Flore forestière soudano-guinéenne [AOF-Caméroun-AEF].* 523 pp., 115 pls., 40 maps. Paris: Société d'Éditions Géographiques, Maritimes, et Coloniales. (Reprinted 1975, Nogent-sur-Marne: Centre Technique de Forestier Tropical.)

Copiously illustrated forest flora of the forest, woodland, and tree savanna zones of West (and Central) Africa; includes some keys to species, limited synonymy, vernacular names, local and extralimital distribution (with range maps and some *exsiccatae*), and extensive notes on habitat, biology, special features, etc.; complete indices at end. An introductory section includes remarks on physical features, climate, features of the forest flora, and taxonomic problems, together with species lists relating to a given special feature, property, use, etc.

GEERLING, C., 1988. *Guide de terrain des ligneux sahéliens et soudano-guinéens.* 2nd edn. [iv], 340 pp., 92 pls., map (Agricultural University of Wageningen Papers 87-4). Wageningen. PUDOC. (Originally published 1982, Wageningen, as *Ligneux sahéliens et soudano-guinéens.*)

Illustrated guide to woody plants of the Sahel and Sudano-Guinean zones of West Africa (map, p. 2); species entries fairly formal, with subheads for habit, leaves, leaf venation, flowers and fruit as well as for habitat and ecological notes, taxonomic commentaries (where appropriate), distribution, and representative *exsiccatae*. Scientific names are accompanied by concise references, essential synonymy, and necessary cross-references to *Flora of west tropical Africa.* The arrangement is alphabetical, by families and genera, and most, if not all, species are figured. The systematic treatment is preceded by an illustrated glossary as well as general artificial keys and 'character groups' with special features, and a general index to all scientific names is given at the end.

Pteridophytes

See also ALSTON (under HUTCHINSON and DALZIEL, above). Alston's work does not, however, extend to equatorial central Africa.

TARDIEU-BLOT, M. L., 1953. *Les ptéridophytes de l'Afrique intertropicale française.* 241 pp., illus. (Mém. Inst. Franç. Afrique Noire 28). Dakar.

Descriptive treatment; includes keys, full synonymy and references, *exsiccatae* with localities, summary of overall distribution, and notes on habitat, etc.; bibliography and index at end. An introductory section includes a list of principal collectors and remarks on floristic regions and fern ecology. Covers former French territories in West and Central Africa, Cameroon Mountain, and islands in the Gulf of Guinea.

581

Nigeria

Area: 923 850 km². Vascular flora: 4715 species (Morat and Lowry, 1997; see **Division 5, Progress**). A figure of 4614 species was listed by Lebrun and Stork, 1991 (see **501**).

Stanfield's *Flora of Nigeria* project, initiated as a continuation of *Nigerian trees*, has not progressed beyond three fascicles. A notable feature in the *Flora* was the use of polyclave tables in place of the traditional Lamarckian analytical keys. This was part of attempts to make the *Flora* easier to use, following a lead set by *Nigerian trees*.

STANFIELD, D. P. and J. LOWE (eds.), 1970–89. *Flora of Nigeria*. Published in fascicles. Illus. Ibadan: Ibadan University Press (fasc. 1–2, 1970–74; 2nd edn. of fasc. 1, 1989). Continued as S. A. GHAZANFAR, 1991. *Flora of Nigeria: Caryophyllales*. 39 pp., illus. (Monogr. Syst. Bot. Missouri Bot. Gard. 34). St. Louis, Mo.

Briefly descriptive, semi-technical account; includes polyclave keys to all taxa, limited synonymy, principal citations, vernacular names, generalized indication of local and extralimital range (with representative *exsiccatae*), and notes on habitat, biology, special features, uses, etc.; general index to family and generic names in vol. 2. An introductory section contains an account of botanical exploration as well as an illustrated glossary. The Caryophyllales features Lamarckian analytical keys, as well as types (but no literature citations) and taxonomic commentary but lacks an index. [Only two fascicles of the original series (Gramineae and Cyperaceae respectively, with a new edition of the former, by J. Lowe, in 1989) appeared through 1990. The work by S. A. Ghazanfar was an independent undertaking.]

Woody plants

KEAY, R. W. J., 1989. *Trees of Nigeria*. x, 476 pp., 165 text-figs. Oxford: Oxford University Press. (Originally published in 2 vols., 1960–64, Lagos: Nigerian Government Printer, as *Nigerian trees* by R. W. J. Keay, C. Onochie and D. P. Stanfield.)

Descriptive and diagnostic illustrated flora, with keys to genera and species, essential synonymy, cross-references to *Flora of west tropical Africa*, vernacular names, summary of overall distribution, representative illustrations, and notes on habitat, uses and special features; lexica for Edo, Hausa, Igbo, Yoruba and other local vernacular names, and index to botanical names at end. The introductory section is brief; it is followed by an illustrated glossary and key to families. [Not all trees are fully described; smaller or rarer ones are merely diagnosed.]

Pteridophytes

GBILE, Z. O., 1981. Dichotomous key to the Nigerian species of ferns and fern-allies. *Nigerian J. Forest.* 11(1–2): 33–48.

A bracketed analytical key to 190 species, the botanical names without authors; glossary, list of families, acknowledgments and brief bibliography at end. [Coverage of species was in the first instance based on Nigerian records in Alston's *Ferns and fern allies of West Tropical Africa* (see **580**).]

582

Benin (Dahomey) and Togo

These two states encompass the so-called 'Dahomey Gap', a biogeographical discontinuity partly associated with climatic patterns.[87] A step towards their individual coverage was taken with the 1984 publication of *Flore analytique du Togo* edited by J. F. Brunel, P. Hiepko and H. Scholz. For Benin, a first checklist was produced by S. de Souza in 1987.

Togo

Area: 112 622 km². Vascular flora: 2423 seed plant species (Akpagana and Guelly, 1994; see below); this represents a significant increase over the 2195 recorded by Brunel *et al.* (1984; see below) but is consistent with the prediction of A. G. Johnson in the preface to the latter work.

Togo was initially energetically explored following proclamation of German rule; a first list of the flora was made in A. ENGLER, 1911. Mittel-Guinea: Togo. In *idem, Die Pflanzenwelt Afrikas*, 1: 777–811 (see **500**). After World War I, the division of the original territory (the western fifth falling to British rule and eventually uniting with Ghana), and the advent of a French mandate (later trusteeship, with independence in 1960), collections increased more gradually. In 1972 a programme of collaboration was established between Benin University in Lomé and the Botanischer Garten und Botanisches Museum in Berlin; a large-scale collecting programme was begun, thus forming a good basis for a principal goal: *Flore analytique du Togo*, published in 1984. A summary of subsequent additions appeared in 1994. There is also a detailed treatment of grasses and Cyperaceae: H. SCHOLZ and U. SCHOLZ, 1983. *Flore descriptive des Cyperacées et Graminées du Togo.* 360 pp. Vaduz: Cramer/Gantner. (Phanerogamarum monographiae 15.)[88]

BRUNEL, J. F., P. HIEPKO and H. SCHOLZ (eds.), 1984. *Flore analytique du Togo: phanérogames.* 751 pp., map (Englera 4). Berlin. (Also published through GTZ, Eschborn, and distributed by TZ-Verlagsgesellschaft, Rossdorf.)

A manual-enumeration of seed plants with more or less descriptive keys; species entries with citations of *exsiccatae* (grouped according to five vegetational districts), illustrations and references, and family headings with identification of authors and citations of monographs and revisions; main bibliography (pp. 690–720) and index to all botanical names at end. The introductory section treats physical features and vegetation, with characteristic species and other features (including rainfall) for each district as defined, and collectors since 1970. [For a summary of additions (233 species) with indication of districts and vouchers, see K. AKPAGANA and K. A. GUELLY, 1994. Espèces d'angiospermes nouvelles pour la flore du Togo. *Acta Bot. Gall.* **141**: 781–787.][89]

Benin

Area: 112 622 km^2. Vascular flora: *c.* 2000 species (H. Ern in PD; see **General references**). A figure of 1105 is given by Morat and Lowry (1997; see **Division 5, Progress**).

The country has received relatively little attention from collectors and remains among the lesser known in West Africa. De Souza's 1987 checklist is the first of its kind, but lacks any statistics for the flora.

DE SOUZA, S., 1987. *Flore du Bénin*, 1: *Catalogue des plantes du Bénin.* [4], 87 pp. Cotonou: distributed through Université National de Bénin.

Alphabetically arranged name list of native and introduced bryophytes and vascular plants; brief list of references, p. 87. [Volumes 2 (1987) and 3 (1988) of this flora, both by the same author, respectively comprise an illustrated descriptive treatment of mangrove and strand plants and a lexicon of local vernacular names.]

583
Ghana

Area: 238 537 km^2. Vascular flora: 3725 species (Morat and Lowry, 1987; see **Division 5, Progress**). A figure of 3600 species had earlier been proposed (J. B. Hall in Lebrun and Stork, 1991; see **501**). Present-day Ghana, before 1957 the Gold Coast, includes a part of the former German colony of Togo.

Irvine's work on the woody flora, the only recent extensive systematic work on Ghana, lacks keys and moreover does not truly represent a revision of his *Plants of the Gold Coast* (1930), an ethnobotanically oriented work but in effect the first flora of the country. A complementary work relating to species distribution, with many individual maps, is J. B. HALL and M. D. SWAINE, 1981. *Distribution and ecology of vascular plants in a tropical rain forest: forest vegetation in Ghana.* 383 pp., maps. The Hague: Junk. (Geobotany 1.)

Woody plants

HAWTHORNE, W., 1990. *Field guide to the forest trees of Ghana.* 276 pp., col. photographs, many full-page text-figs. Chatham Marine, Kent: Natural Resources Institute (for Overseas Development Administration, United Kingdom).

Sections 3 and 4 of this work (with yellow pages) comprise respectively a '100 main species key' (with leads not always dichotomous) and glossary and indices to scientific names. The remainder of the work comprises introductory matter including nomenclature, use of keys, and the main (pp. 11–23) and family and group (pp. 24–225) keys, all with facing figures of key features and ultimate leads descriptive including as necessary habitat data and footnotes. [The photographs are too small for satisfactory use.]

IRVINE, F. R., 1961. *Woody plants of Ghana.* xcv, 868 pp., 142 text-figs., 34 pls. (some col.), frontisp. Oxford: Oxford University Press.

Briefly descriptive forest flora, without keys; includes limited synonymy, vernacular names, generalized local and extralimital range, with representative *exsiccatae*, and extensive notes on habitat, field features, properties, uses, etc.;

lexicon of vernacular names, bibliography, illustrated glossary, and index to generic and family names at end. The introductory section includes lists of plants arranged by uses, with a tabular key.

584

Ivory Coast

Area: 322463 km². Vascular flora: 3720 species, including 143 pteridophytes (Aké Assi, 1984; see below).

The major recent floras are those by André Aubréville (1959, on forest trees) and Laurent Aké Assi (1984). Aké Assi earlier contributed a large precursory paper, *Contribution à l'étude floristique de la Côte d'Ivoire et des territoires limitrophes* (1964).

AKÉ ASSI, L., 1984. *Flore de la Côte d'Ivoire: étude descriptive et biogéographique, avec quelques notes ethnobotaniques.* 3 vols. (in 6). 1206 ll., 86 ll. pls., illus. Abidjan. (Thèse du doctorat, Université d'Abidjan.)

This copious work (in A4 format) comprises three 'tomes' (in six volumes) of what may best be described as *matériaux*: systematic analysis, catalogue and biogeographic considerations, of which the catalogue is the second (physically the fifth) volume of the work. This latter is an alphabetically arranged, concisely annotated enumeration of native and well-established introduced vascular plants, without keys but including indication of life-form and chorological classes (the latter both within Africa and worldwide). The systematic analysis (tome I, or volumes 1–4) is a lengthy illustrated family-by-family analysis (angiosperms only) of the kind which might be used as a basis for a teaching manual or other introductory work for local use; included here is much on the habitat, biology and medicinal and other local uses of individual species not included in the enumeration proper, as well as an introduction to the whole work. Floristics are dealt with as tome III (physically the sixth volume), along with the references (pp. 1157–1200) and index to families (pp. 1201–1206) which conclude the work.

Woody plants

AUBRÉVILLE, A., 1959. *La flore forestière de la Côte d'Ivoire.* 2nd edn. 3 vols. 351 pls., maps (Publ. Centre Technique Forestier Tropical 15). Nogent-sur-Marne. (1st edn., 1936, Paris.)

Generously formatted, briefly descriptive forest flora; includes analytical keys to genera and species and tabular keys to families, limited synonymy, vernacular names, generalized indication of local range, with citation of representative *exsiccatae*, taxonomic commentary, notes on habitat, biology, etc., and indices to generic and family names (the latter only in vol. 3). The introductory section includes a short bibliography and a discussion of the forest zones of the Ivory Coast. [In practice an awkward work to use.]

585

'Greater Guinea'

Included here are Liberia, Sierra Leone, Guinea Republic, and Guinea-Bissau (former Portuguese Guinea).

These polities together form a somewhat natural unit within which are to be found much the greater part of the biotically significant Nimba and Loma massifs (see **502**). Aided by high and fairly consistent rainfall (especially in Liberia), the flora is the richest in West Africa yet remains somewhat imperfectly known. Beyond coverage in general works on West Africa (**580**) and those on the Loma and Nimba massifs (**502**) there is only a miscellany of local works.

I. Liberia

Area: 111370 km². Vascular flora: *c.* 2200 species (Morat and Lowry, 1997; see **Division 5, Progress**). – For Adam's flora of the Nimba massif, see **502**. The country has the highest and most consistent rainfall in West Africa, giving rise to a relatively rich flora (much of it now under threat). However, there never has been a consistent history of botanical exploration and with the mountainous interior and difficulties of travel as noted by Winifred Harley (1955; see below) it remains imperfectly known, in spite of much input in the 1950s and early 1960s by development agencies and contracted persons. In recent years civil disturbances have limited further exploration.

In addition to the outdated enumeration of Max Dinklage, miscellaneous works are available for forest trees and pteridophytes. Prior to the *Verzeichnis* Otto Stapf had contributed an account of the flora in *Liberia*

(1906, 2 vols.) by H. Johnston, covering 700–800 species arranged by families.

Bibliography

See also M. S. BRISCOE, 1963. *Bibliography of the scientific literature of Liberia*. Washington: College of Medicine, Howard University.

THIRGOOD, J. V., 1964. *An annotated bibliography of Liberian forestry and botany*. 9 pp. Monrovia: College of Forestry, University of Liberia. [Introduction; chronologically arranged list with annotations; no indices.]

DINKLAGE, M. (ed. J. MILDBRAED), 1937. Verzeichnis der Flora von Liberia. *Repert. Spec. Nov. Regni Veg.*, **41**: 235–271.

Unannotated list of known vascular plants, with very limited synonymy and no index; includes *exsiccatae* with localities. An account of botanical exploration appears in the introductory section. [Now very incomplete.]

Woody plants

KUNKEL, G., 1965. *The trees of Liberia*. 270 pp., text-figs., 80 halftones. Munich: Bayerischen Landwirtschafts-Verlag. (German Forestry Mission to Liberia, Report 3.)

Copiously illustrated account of the more important trees; includes tabular keys, synonymy, citations, vernacular names, notes on habitat, and indication of related species of lesser importance; lexica of vernacular and trade names; index. A glossary and an account of forest types appear in the introductory section. [Based on the author's experience with the German Forestry Mission in 1961–63.][90]

VOORHOEVE, A. G., 1965. *Liberian high forest trees*. 416 pp., 72 text-figs., 32 halftones. Wageningen: PUDOC. (Agricultural Research Reports, 652. Reissued 1979.)

Critical, well-illustrated account of 75 more important rain forest trees; includes keys, extensive synonymy, literature citations, indication of local and extralimital range, vernacular and trade names, extensive critical remarks and copious notes on habitat, field attributes, utilization potential, etc., with mention of related species of lesser importance; glossary and general index. An introductory section includes accounts of physical features, climate, vegetation, forest types, and botanical exploration in Liberia. [Based primarily on the author's experience at the University of Liberia from 1960 to 1963.][91]

Pteridophytes

See also G. KUNKEL, 1962. Über die Farne Liberias (West-Afrika), ihre Vorkommen und ihre Verbreitung. *Ber. Schweiz. Bot. Ges.* **72**: 21–66. This accounts for 132 species. Additions are in *idem*, 1963. Neue Farne für die Flora Liberias. *Ibid.*, **73**: 1–17.

HARLEY, W. J., 1955. The ferns of Liberia. *Contr. Gray Herb.*, N.S., **177**: 58–101, 1 pl., map.

Systematic enumeration of ferns; includes synonymy, literature citations, critical remarks, and notes on habitat, frequency, etc., followed by a list of references and index. A fairly lengthy introductory section covers physical features of Liberia, geography and travel problems, notes on prior collecting work, and discussion of phytogeography and vegetation with special reference to ferns. A phytogeographical map is included, as well as separate index.[92]

II. Sierra Leone

Area: 71 740 km². Vascular flora: 2480 native and introduced species were listed in Gledhill [1962]. Morat and Lowry (1997; see **Division 5, Progress**) give a figure of 1685; this is in turn based on Good (1974; see **General references**).

Sierra Leone has had a long history of botanical collecting, but as in Liberia useful publications on the flora are few; forest trees are rather more fully treated. Gledhill's list lacks any documentation. For Jaeger and Adam's flora of the Loma massif, see **502**.

GLEDHILL, D., [1962]. *Check list of the flowering plants of Sierra Leone*. 38 pp. Sierra Leone: University College.

Unannotated simple list of native and naturalized flowering plants, based upon the *Flora of west tropical Africa*; no index.

LANE-POOLE, C. E., 1916. *The trees, shrubs, herbs, and climbers of Sierra Leone*. 159 pp. Freetown.

Comprises alphabetically arranged descriptive notes covering botanical details and uses, the entries including vouchers or other records; index to families, a synoptic list, and lexicon of English and Creole names with botanical equivalents. An appendix features an extract from Afzelius's report of 1795 to the Sierra Leone Company on vegetable crops.

Woody plants

SAVILL, P. S. and J. E. D. FOX, 1971. *Trees of Sierra Leone*. 316 pp., illus., 1 col. map. Freetown: Forestry Division, Sierra Leone. (Mimeographed.)

Descriptive manual of forest trees, with keys, local range, and notes on frequency, phenology, timber properties, uses, etc.; lexicon of vernacular names and index. An introductory section includes accounts of physical features, geology, climate, vegetation formations, and forest management.

III. Guinea Republic

Area: 245855 km². Vascular flora: *c.* 3000 species (Morat and Lowry, 1997; see **Division 5, Progress**).

No proper flora of Guinea has been published. The first separate floristic work, strongly oriented towards useful plants, is H. POBÉGUIN, 1906. *Essai sur la flore de la Guinée française.* 392 pp., 80 pls., map. Paris. Conditions for some two or more decades after independence in 1960 were not favorable to botanical work. For Adam's flora on the Nimba massif (in the southeast), biotically the most important part of the country, see **502**.

CARRIÈRE, M., 1994. *Plantes de Guinée à l'usage des éleveurs et des vétérinaires.* 235, [2] pp., 130 text-figs. Maisons-Alfort: Centre de Coopération Internationale en Recherche Agronomique por le Développement, Départment d'Élevage et de Médecine Vétérinaire (CIRAD-EMVT), France.

Alphabetically arranged illustrated enumeration of 371 species with limited synonymy, vernacular names, diagnostic features, information on uses (directly indicated or through references), and indication of distribution, habitat, and botanical zones; list of species (with numbers) collected by the author, names of animal ailments in different languages, classified lists of plants by categories of use, literature (pp. 209–211), and lexicon of vernacular names with designation of their language (pp. 213–235). The introductory part includes a brief history of botanical work along with objectives, sources of information, the plan of the work, and directions for use. [Not a complete flora, but the only organized floristic work for the country since 1906.]

IV. Guinea-Bissau

For the Cape Verde Islands (now a separate polity), see **025**. – Area: 36125 km². Vascular flora: *c.* 1000 species (D'Orey in Lebrun, 1976; see **Division 5, Progress**).

Formerly known as Portuguese Guinea, this state, the smallest in West Africa, was after the eighteenth century the only remnant of a larger Portuguese territory, 'Senegâmbia'. A considerable but scattered floristic literature has built up since the appearance of the first organized account, *Subsídos para o conhecimento da flora de Guiné portuguesa* (1930) by António de Figueiredo Gomes e Sousa. The most significant works – described below – are a series of contributions from the Centro de Botânica of the Junta de Investigações do Ultramar (Sousa, 1946–63) and a serial flora by the Jardim e Museu Agrícola do Ultramar (1971–83), of which, however, only seven fascicles have appeared.[93]

Bibliography

FERNANDES, A., 1982. See **Divisional bibliographies** above. [Guinea-Bissau, pp. 83–107; 133 titles in pure and 141 in applied botany.]

JARDIM E MUSEU AGRÍCOLA DO ULTRAMAR, PORTUGAL, 1971–83. *Flora da Guiné Portuguesa* (from fasc. 6 *Flora da Guiné-Bissau*). Fasc. 1–7. Lisbon: Ministerio do Ultramar (later Ministerio da Cooperação).

A serial 'research' flora; each fascicle, comprising one family or a part thereof, contains keys, synonymy (with references and citations), vernacular names, *exsiccatae* with localities, and notes on overall range, habitat, life-form, phenology, etc.; indices at end. All treatments are by specialists or staff botanists.[94]

SOUSA, E. C. PEREIRA DE, 1946–57. Contribuições para o conhecimento da flora da Guiné Portuguesa, I–VIII. *Anais Junta Invest. Colon.* **1**: 45–152; **3**(3), fasc. 2: 5–85; **4**(3), fasc. 1: 1–63; **5**(5): 1–64; **6**(3): 1–62; and *Anais Junta Invest. Ultram.* **7**(2): 1–78; **11**(4), fasc. 2: 1–38; **12**(3): 1–27. Continued as *idem*, 1960. *Contribuições para o conhecimento da flora da Guiné Portuguesa*, IX. 101 pp. (Estudos, Ensaios e Documentos, Junta Invest. Ultram. 77). Lisbon; and *idem*, 1963. *Contribuições para o conhecimento da flora da Guiné Portuguesa*, X. 76 pp. (Mem. Junta Invest. Ultram., II, 46). Lisbon.

Annotated enumeration of seed plants of Guinea-Bissau in the form of a series of 10 contributions; the individual accounts include for each taxon synonymy, references and citations of principal works, *exsiccatae* studied, summaries of local and extralimital range, and notes on habitat, life-form, phenology, etc., along with an index. Part 10 includes a collation (p. 48) and a complete general index (pp. 49–76); this latter should be consulted in any use of this work.[95]

586

'Senegambia'

Includes the states of Senegal and The Gambia.

I. Senegal

Area: 196 722 km². Vascular flora: 2086 species (Morat and Lowry, 1997; see **Division 5, Progress**). This is close to the figure of 2100 species earlier given by Lebrun (in Lebrun and Stork, 1991; see **501**).

Senegal, with a history of botanical activity stretching back to the eighteenth century and the pioneer work of Michel Adanson, is botanically one of the best-known (and documented) countries in West Africa. This was aided by the establishment, in 1942, of the herbarium of the Institut Français d'Afrique Noire (later the Institut Fondamental d'Afrique Noire), covering the whole of French West Africa.

BERHAUT, J., Père, 1967. *Flore du Sénégal.* 2nd edn. [viii], 485 pp., text-figs., col. pls. Dakar: Clairafrique. (1st edn., 1954.)

Artificially arranged, analytical manual-key to seed plants; includes vernacular names, brief indication of local range (with some *exsiccatae*), and figures of representative species; glossary and list of families covered; indices to all botanical and vernacular names.

BERHAUT, J., Père (continued by C. VANDEN BERGHEN), 1971– . *Flore illustrée du Sénégal.* Vols. 1–6, 9. Illus. Dakar: Ministère du Développement Rural (later Ministère de la Protection de la Nature), Direction des Eaux et Forêts (distributed by Clairafrique, Dakar, and Maisonneuve, Moulins-lès-Metz, France).

Spaciously planned atlas-flora of flowering plants, with amply descriptive text; includes synonymy, vernacular names, localities with *exsiccatae*, and notes on habitat, occurrence, properties, uses, etc.; bibliography and indices to botanical names, vernacular names, and uses and properties. Of the 11 volumes projected, seven have appeared as of 1998 (but none since vol. 9, 1988); these cover dicotyledon families in alphabetical order from Acanthaceae through Nymphaeaceae and (in vol. 9, by C. vanden Berghen) about half of the monocotyledons.[96]

LEBRUN, J.-P., 1973. *Énumeration des plantes vasculaires du Sénégal.* 209 pp., 6 pls., map (Études Bot. IEMVPT 2). Maisons-Alfort, Val-de-Marne: Institut d'Élevage et de Médecine Vétérinaire des Pays Tropicaux.

Briefly annotated, alphabetically arranged (by genera) checklist of vascular plants; includes family cross-references, synonymy with references and citations, distribution and some critical commentary (in footnotes); bibliography of Senegal botany, 1930–72. An introductory section (with map) includes an account of botanical exploration and a survey of floristic regions. Additions to Berhaut's manual of 1967 (see above) are specially indicated.

II. The Gambia

Area: 10 689 km². Vascular flora: 966 species (Jones, 1991; see below); a figure of 974 is furnished by Morat and Lowry (1997; see **Division 5, Progress**).

For an introductory account to the biota, see E. EDBERG, 1982. *A naturalist's guide to The Gambia.* 96 pp., illus., maps. St. Anne: J. G. Sanders. The Williams checklist is more fully documented than that of Jarvis but is now rather incomplete. That of Jones has no documentation (its basis being explained in the introduction).

JARVIS, A. E. C., 1980. *A checklist of Gambian plants.* 29 pp. Banjul, The Gambia: Book Production and Material Resources Unit. (Mimeographed.)

Unannotated checklist of 530 species, with local vernacular names. Sources (not including Williams!) listed in the introduction on p. 1.

JONES, M., 1991. *A checklist of Gambian plants.* ii, 46 pp., map. N.p.: The author (Gambia College).

Systematic name list of 966 species (arranged following *Flora of west tropical Africa*) with vernacular names; references (pp. 5–6) and index (pp. 34–46). [Of the recorded species, some 205 were trees and 110 shrubs.]

WILLIAMS, F. N., 1907. Florula gambica. *Bull. Herb. Boissier,* II, 7: 81–96, 193–208, 369–386.

Annotated enumeration of flowering plants (285 species), with descriptions of new or little-known species, some literature citations, localities with *exsiccatae*, taxonomic remarks, and notes on distribution, uses, etc.; index. An introduction includes a survey of collectors and local botanical history.

587

Mauritania

Area (whole country): 1 030 700 km². Vascular flora (including **596**, Western Sahara): 853 species (Lebrun, 1998; see below). The northern (Saharan) part is also accounted for under **595** and in floras of the Sahara (**505**). The inclusion of Western Sahara by Lebrun reflects political claims.

Jacques-Georges Adam was the first to produce consolidated works on Mauritania, notably the check-list in *Itinéraires botaniques en Afrique occidentale* (1962, initially in *J. Agric. Trop. Bot. Appl.* **9**: 97–200, 297–416, but also separately published). Earlier work was largely in the form of collection reports and ecological and applied studies along with contributions arising from resource reconnaissance. In 1984 X. Jaouen produced a key for identification of plants in the area of the capital, Noukachott (*c.* 18° N, about the limit of *Flora of west tropical Africa* coverage). The 1991 manual was considered by its authors as merely a first step towards a more definitive work; however, geographically it is not critical with both it and Adam's work having been based to a large extent on unconfirmed assumptions of presence. Lebrun's 1998 enumeration presents a significantly smaller, but better-documented, flora.[97]

BARRY, J. P. and J. C. CELLES, [1991]. *Flore de Mauritaine.* 2 vols. xlviii, 550 pp., pp. A1–A4; illus., 4 maps. Noukachott/Sophia-Antipolis: Institut Supérieur Scientifique de Noukachott/Université de Nice.

Comprises analytical keys to vascular plants (dicotyledons in vol. 1, other groups in vol. 2), the individual families alphabetically arranged and copiously illustrated, along with an unannotated checklist (with authors), an unillustrated glossary, and index to all botanical names at end. The key leads include indication of habitats. Some family accounts include extensive pedagogical aids; however, there are no vernacular names or distribution maps. The extensive introductory part (pp. i–xlviii) encompasses physical features, climate, biogeography (with maps), a chronologically arranged list of references, an illustrated basic organography, nomenclature, sources of illustrations, and non-botanical references along with (pp. xxvii–xlviii) the general keys. [Covers 1400 species, but geographi-

cally not critical; coverage based largely on that of Adam.]

LEBRUN, J.-P., 1998. *Catalogue des plantes vasculaires de la Mauritanie et du Sahara occidental.* 322 pp., 18 halftones, 1 pl., 45 maps (Boissiera 55). Geneva.

Systematic enumeration with essential synonymy, some citations of revisions and taxonomic studies, specimens seen, literature records, overall distribution, and critical notes; statistics (p. 312) and index to genera and families at end. Accepted native species are continuously numbered (853 in all, in 443 genera; 19 percent in Poaceae). The general part encompasses the background, plan and basis of the work, and includes a historical account of botanical work through 1962 (with a list of collectors and an extensive bibliography) as well as a floristic and phytogeographical survey.

588

Burkina Faso and Mali

The northern part of Mali, lying within the Sahara, is here treated under **595**. J.-P. Lebrun's team, as elsewhere in the West African drylands, have provided the first separate floristic accounts for the two countries included here.

I. Burkina Faso

Area: 274 122 km². Vascular flora: 1203 species (Lebrun *et al.* 1991; see below).

LEBRUN, J.-P. *et al.*, 1991. *Catalogue des plantes vasculaires du Burkina Faso.* 341 pp. (incl. 3 pp. of facsimiles of early herbarium labels) (Études et synthèses de l'IEMVPT 40). Maisons-Alfort, Val-de-Marne, France: Institut d'Élevage et de Médecine Vétérinaire des Pays Tropicaux.

Annotated enumeration of vascular plants, with synonymy, references and literature citations, indication of *exsiccatae*, extralimital range, and notes on habitat, soils, palatability for stock, and other special features; detailed bibliography (pp. 309–326), statistics (p. 327) and indices to families and genera. The introduction contains a useful history of botanical work along with its primary bases.[98]

II. Mali

Area: 1 240 000 km². Vascular flora: 1741 species (Lebrun in Lebrun and Stork, 1991; see **501**).

In addition to the enumeration by Boudet and Lebrun (see below), there is also a dendrological work: J. PARKAN, 1974. *Dendrologie forestière: République du Mali*. This, however, is not cited either by Boudet and Lebrun or by Maydell (see **505**) in their bibliographies.

BOUDET, G. and J.-P. LEBRUN, 1986. *Catalogue des plantes vasculaires du Mali*. 480 pp. (Études et synthèses de l'IEMVPT 16). Maisons-Alfort, Val-de-Marne, France: Institut d'Élevage et de Médecine Vétérinaire des Pays Tropicaux.

Systematic enumeration of native alien and cultivated vascular plants (1739 species); includes synonymy (with references), localities with *exsiccatae*, summary of total distribution, and habitat; indices to genera and families at end. The introductory section includes a historical account and an alphabetical list of collectors. A list of 316 references (pp. 435–461) is also furnished.

589

Niger Republic

Area: 1 267 000 km². Vascular flora: 1178 species (Lebrun in Lebrun and Stork, 1991; see **501**). The northern part of Niger, lying within the Sahara, is treated under **593**.

Although botanical exploration goes back to the time of Mungo Park, the country was covered for the first time in its own right by J.-P. Lebrun (with B. Peyre de Fabrègues) only in 1976 in the first of what has become an important suite of enumerations for northern West and Central Africa.

PEYRE DE FABRÈGUES, B. and J.-P. LEBRUN, 1976. *Catalogue des plantes vasculaires du Niger*. 433 pp., map (Études Bot. IEMVPT 3). Maisons-Alfort, Val-de-Marne: Institut d'Élevage et de Médecine Vétérinaire des Pays Tropicaux.

Annotated systematic enumeration of vascular plants; includes synonymy, references to principal works, localities and general range, with addition of *exsiccatae* for less common species (those not in *Flora of west tropical Africa* being specially indicated), some critical remarks, and notes on alimentary and other uses; list of references and indices to family and generic names. An introductory section includes technical details as well as an account of botanical exploration (beginning with Mungo Park) and an introduction to the vegetation and phytogeography, with definition of three major chorological zones and indication of the practical limits of cultivation (about 15° N) on a map. For additions, see E. BOUDOURESQUE, S. KAGHAN and J.-P. LEBRUN, 1978. Premier supplément au 'Catalogue . . .'. *Adansonia*, II, **18**: 377–390; and J.-P. LEBRUN *et al.*, 1983. Second supplément au 'Catalogue des plantes vasculaires du Niger'. *Bull. Soc. Bot. France* **130** (Lettres Bot.): 249–256. [74 additional species for a total of 1178, a number of the new records being aliens.][99]

Region 59

North Africa and Egypt

Area: 6 018 683 km² (exclusive of northern parts of Chad, Niger, Mali and Mauritania). Inclusion of these takes the region to over 8 000 000 km². Vascular flora: no data. – This region includes Egypt, Libya, Tunisia, Algeria and Morocco and the whole of the Western Sahara territory (presently controlled by Morocco) along with the northern parts of Chad, Niger, Mali and Mauritania. It thus encompasses the 'Maghreb' which borders the Mediterranean Sea from Libya to Morocco as well as most of the Sahara. On the east it is bounded by the Red Sea and the Suez Canal; the southern boundary is the Sudanian border (along the 22nd parallel), turning south (at Jebel Uweinat) to the 16th parallel of latitude (the northern limit of Lebrun's catalogue of southern Chad plants; **576**), thence westwards along the approximate southern limit of the Sahara (*c.* 16–18° N and inclusive of the Aïr massif in northern Niger) to the Atlantic Ocean at Noukachott (Mauritania).[100]

Sustained botanical exploration in the region dates from the latter part of the eighteenth century and has continued to the present, although in some areas activity has lessened since decolonization (and, in Algeria, the advent of civil strife). From the nineteenth century there has been a resident botanical community with herbaria and other institutions. However, for bio-historical reasons as well as the sheer size of the region, no comprehensive work has been achieved (a gap, however, being in the late twentieth century bridged to a considerable extent by *Med-Checklist*). The only 'regional' work is *Flore de l'Afrique du Nord* by René Maire, never completed by its author and without plans for continuation apart from publication of available manuscript.

Maire's *Flore* was preceded principally by the pioneering *Flora atlantica* (1798–1800) of René L. Desfontaines and the detailed, but not completed, *Compendium florae atlanticae* (1881–87) of Ernest Cosson. As precursors towards his projected *Flore*, Maire – a professor of botany at Algiers University for 38 years until his death in 1949 – himself from 1918 wrote more than 3100 separately numbered notes in his long-running *Contributions à l'étude de la flore de l'Afrique du Nord*. These were published in the main North African natural history journals. In 1978, J.-P. Lebrun and A. Stork published a classified index (see **Regional bibliographies** below). Around this time, preparation of a successor work, *Med-Checklist*, was also begun in accordance with a resolution of the European Science Research Councils.[101] As of 1998 three of the projected six volumes of that work had been published.

Egypt, long viewed as part of the 'Oriental' floristic sphere, was in the latter part of the nineteenth century included in *Flora orientalis*. It was also among the first countries in Africa to develop a local floristic literature. Such factors likely induced Maire to limit his *Flore* to the territories west of Egypt. Southwards, Maire took his southern boundary as corresponding to those of Libya, Algeria and Morocco; at that time, the 25th parallel was the northern limit of *Flora of west tropical Africa*. Further work has shown, however, that overall the Sahara is floristically more closely allied to the Mediterranean Basin than to the tropics.[102]

Botanical knowledge for individual countries is uneven, with the best-known areas evidently being Morocco and northern Algeria and the Nile Valley and Delta of Egypt and from these east to the Sinai and west along the coast to Cyrenacia (Libya). Coverage of the Sahara is sketchy, most attention having been given to the various mountain massifs and oases. Most countries and other units recognized here possess floristic works of one or another kind. In addition, the Sahara west of the Nile is largely covered by Ozenda's *Flore du Sahara*, now in its third edition (1991; see **505**).

The most practical works are Täckholm's *Students' flora of Egypt* (2nd edn., 1974) and Quézel and Santa's *Nouvelle flore de l'Algérie* (1962–63). The *Flore analytique et synoptique de la Tunisie* by Cuénod *et al.* (1954) was, after 25 years, completed by one of its junior authors, Mme. G. Pottier-Alapetite, as *Flore de la Tunisie* (1979–81). These two latter publications have succeeded the now-obsolete manual-key for Algeria and Tunisia by Battandier and Trabut (1902). In early 1999 the first volume (of three) of a new manual for Egypt by L. Boulos was published. Among enumerations and checklists, the most detailed (and still widely used) work is *Catalogue des plantes du Maroc* (Jahandiez *et al.*, 1931–41, with later additions). This, however, needs considerable revision or replacement; the rather elaborate use of infraspecific categories in this work is moreover now regarded by some as old-fashioned.[103]

A number of partial floras and enumerations, both old and new, exist for areas of special interest. Noteworthy are Quézel's monograph on the Tibesti massif and that of H. Gillet and G. Carvalho on the Ennedi, both of them 'monadnocks' in northern Chad, the flora of R. Maire on the Hoggar and Ajjer ranges in Saharan Algeria, and the florulas of Nègre for the Marrakesh district (in a wide sense) and Sauvage for the oak-woodlands of the Atlas and Rif, both in Morocco. On the other hand, no modern 'tree books' are available for any except limited areas. The only work of broader scope in this genre is *Flore forestière de l'Algérie* by Lapie and Maige (1915; see **598**).

Progress

Reports for each North African country, with lists of recent literature, have been given in V. H. HEYWOOD (coord.), 1975. Données disponibles et lacunes de la connaissance floristique des pays méditeranéens. In Centre National de la Recherche Scientifique (France), *La flore du bassin méditerranéen: essai de systematique synthétique*, pp. 15–142 (Colloques internationaux CNRS 235). Paris. Additions to the country reports are contained in S. CASTROVIEJO, 1979. Synthèse des progrès dans le domaine de la recherche floristique et littérature sur la flore de la région méditerranéenne. *Webbia* **34**: 117–131.[104] Later progress reports, usually for individual projects or countries, have appeared in *OPTIMA Newsletter* as well as in OPTIMA symposium proceedings (see **601/I**).

Bibliographies. General and divisional bibliographies as for Division 5.

Regional bibliographies

For country bibliographies, see the above-cited report by Heywood.

LEBRUN, J.-P. and A. L. STORK, 1978. *Index général des 'Contributions à l'étude de la flore de l'Afrique du Nord'* [par R. Maire]. 365 pp. (Études Bot. IEMVPT 5). Maisons-Alfort, Val-de-Marne: Institut d'Élevage et de Médecine Vétérinaire des Pays Tropicaux. [Bio-bibliography of R. Maire and complete cross-index to 3697 papers on North African botany written from 1918 through 1949 as background for *Flore de l'Afrique du Nord*, Maire having been by far the largest contributor. With the inclusion of a systematic table of species and their citations in the various parts (with references) the work becomes a kind of floristic checklist for the region.]

Indices. General and divisional indices as for Division 5.

590

Region in general

René Maire's legacy for posterity included manuscript sufficient for at least 20 volumes of his monumental *Flore de l'Afrique du Nord*, encompassing lower vascular plants, gymnosperms, monocotyledons, and dicotyledons through a part of Leguminosae (Englerian system). As of 1998, 16 volumes had been published, with a nearly 10-year hiatus in the 1960s and 1970s due to technical and legal problems and, more recently, no issues since 1987. No plans exist for continuation of the work beyond publication of available manuscript, itself now increasingly out of date. The considerable gap thus left is, however, partly bridged by the recent publication of an index to Maire's numerous but scattered *Contributions à l'étude de la flore de l'Afrique du Nord* (Lebrun and Stork, 1978; see **Regional bibliographies** above). In any case, neither of these works covers Egypt and, moreover, *Med-Checklist* (1981– ; see **601/I**) will in time furnish a new 'base line'.

A more general introductory survey on the North African flora is P. QUÉZEL, 1978. Analysis of the

flora of Mediterranean and Saharan Africa. *Ann. Missouri Bot. Gard.* **65**(2): 479–534.

MAIRE, R. *et al.* (ultimate ed.: P. QUÉZEL), 1952–87. *Flore de l'Afrique du Nord*. Vols. 1–16. Illus., map. Paris: Lechevalier. (Encyclopédie Biologique 33–73, *passim*.)

Comprehensive illustrated descriptive flora of vascular plants; each volume includes keys to genera and species, synonymy (with references and citations), detailed account of local range with *exsiccatae*, summary of extralimital distribution, some taxonomic commentary, notes on habitat, life-form, karyotypes, special features, etc., and index to botanical names at end. [Of the 22 volumes projected (corresponding to the extent of the original manuscript), 16 had been published by 1998. All but one, however, date from before 1980, with vol. 16 appearing in 1987. Vols. 1–15 cover families from the pteridophytes through the Rosaceae (Englerian system), while vol. 16 covers Leguminosae in part (Mimosoideae, Caesalpinoideae, and Sophoreae and Genisteae in the Papilionoideae). For an alphabetical index to vols. 1–16, see M. CARAZO-MONTIJANO and C. FERNÁNDEZ-LÓPEZ, 1990. *Index de la "Flore de l'Afrique du Nord" de Maire (1–16)*. 63 pp., map. Jaén: Facultad de Ciencias Experimentales (distributed by Libraria Agricola, Madrid).

591

Egypt

See also **770–90** (BOISSIER). – Area: 1 000 250 km². Vascular flora: 2094 native and naturalized species and 153 additional infraspecific taxa (Boulos, 1995; see below). The Sinai Peninsula is considered geographically to be part of Asia and so has been designated as **776**.

There have been more floras of Egypt than for any other African country, spanning more than 400 years. The earliest is *De plantis Aegypti* (1592; further issues in 1640 and 1735) by Prospero Alpino. Soon after the Linnaean reforms, Pehr Forskål's *Flora aegyptiaco-arabica* (1775) appeared in Copenhagen. The 'French era', symbolized by Napoleon's great expedition of 1798–99, is reflected in the successors to Forskål's work: *Mémoires botaniques* (1813) and *Flore d'Égypte* (1824) by Alire R. Delile. The first German contribution was *Flora von Aegypten und Arabien* (1834) by Johann Fresenius; from then until World War I German influence would predominate through such works as *Beitrag zur Flora Aethiopiens* (1867) by Georg Schweinfurth and *Illustration de la flore d'Égypte* (1887–89; in *Mém. Inst. Egypt.* 2) by Schweinfurth and Paul Ascherson. The latter, a thorough compilation, remains the most important of the older accounts. It was supplemented in 1901 by *Contributions à la flore d'Égypte* (1901) by E. Sickenberger; in addition to vascular plants this featured sections (by various authors) on fungi, lichens, bryophytes and Characeae. In 1912 there appeared *Manual flora of Egypt* by Reno Muschler; despite its faults it has yet entirely to be superseded. Keys were later added by M. T. Hefnawi (1922). The last work to appear before *Flora of Egypt* was *Bestimmungstabellen zur Flora von Aegypten* (1929) by A. Ibrahim Ramis, a notable Cairo university professor of surgery.[105]

Of more contemporary works, the most useful and satisfactory is Täckholm's *Students' flora of Egypt* (2nd edn., 1974, accompanied by a book of supplementary notes co-authored with Loutfy Boulos). These have been complemented by a modern checklist (Boulos, 1995). Täckholm and Drar's *Flora of Egypt*, published in book form between 1941 and 1969 but not then completed, has been continued as irregular fascicles since 1980 under M. Nabil El Hadidi but is making only slow progress. Some 80 percent or more of the vascular flora has, however, been the subject of revisionary studies (bibliography in El Hadidi and Fayed, 1995), and in early 1999 the first volume (of three) of Boulos's *Flora of Egypt* was published.

Bibliographies

SHERBORN, C. D., 1915. *Bibliography of scientific and technical literature relating to Egypt, 1800–1900.* ii, 155 pp. Cairo. [Unannotated; arranged by authors with subject index.]

TÄCKHOLM, V., 1932. Bibliographical notes to the flora of Egypt. In *Festskrift till Verner Söderberg*, pp. 193–208, [1–2]. Stockholm (reprint). [Classified and well annotated, with connecting commentary; effectively also a history of floristic botany in Egypt.]

BOULOS, L., 1999– . *Flora of Egypt.* Vols. 1– . Illus. (part col.), map. Cairo: Al Hadara.

Concise illustrated descriptive flora with bracketed keys to all taxa, limited synonymy, local distribution (in codes, tied to map of phytogeographical regions, p. xii), overall distribution and indication of

habitat(s), and some critical notes; color section (pp. 377–400 in vol. 1, with 96 photographs) and index to all botanical names. The general part in vol. 1 includes foreword, preface, acknowledgments, the map, a list of endemic taxa in the volume, abbreviations, and (p. xv) principal floras. No family key, vernacular names or statistical section are presented. [Vol. 1 covers 719 pteridophytes, gymnosperms and dicotyledons from Salicaceae through Oxalidaceae; vols. 2–3 will cover the remaining dicotyledons as well as monocotyledons. The sequence of families is that of the 12th edition of the Engler *Syllabus*, as also used in Täckholm's *Students' flora*. Naturalized as well as native species are included.]

BOULOS, L., 1995. *Flora of Egypt checklist.* 283 pp. Cairo: Al Hadara.

Systematically arranged checklist (2094 indigenous species) with authors, places of publication, synonymy, life-form and distribution codes; lexicon of author abbreviations, lists of basic floras (p. 231), supporting recent references (pp. 233–236), and index to all botanical names at end. The introductory part includes background and some statistics, along with a list of endemic taxa (61 in all).

EL-HADIDI, M. N. and A.-A. FAYED, 1995. *Materials for* Excursion flora of Egypt. [1–3], a–e, i–x, 1–233 pp., map (Taeckholmia 15). Cairo.

Annotated enumeration ('Working List') of vascular plants with essential synonymy, references, and indication of distribution and habitat; bibliography (pp. 224–233; unpublished theses are bulleted) but no index. The introduction includes a brief history along with general and regional floristics as well as the plan of and basis for the *Materials*. [Accounts for 2076 native and established species, 124 more than in the 1974 *Students' flora*.][106]

MONTASIR, A. H. and M. HASSIB, 1956. *Illustrated manual flora of Egypt.* Vol. 1. 615 pp., 81 text-figs. [Cairo]: 'Misr'.

Descriptive manual of vascular plants; includes keys to all taxa, essential synonymy, abbreviated indication of local distribution, and notes on habitat, abundance, etc.; figures of representative species; index to all botanical names. [Offered as a successor to Muschler's *Manual flora of Egypt*, but encompasses only the Dicotyledoneae.]

MUSCHLER, R., 1912. *A manual flora of Egypt.* 2 vols. xii, 1342 pp. Berlin: Friedländer. (Reprinted 1970, Lehre, Germany: Cramer.)

Briefly descriptive manual of vascular plants; includes keys to genera and species (but not families), synonymy with references and citations, vernacular names, concise but relatively detailed indication of local range, summary of extralimital distribution, and notes on life-form, phenology, etc.; lexicon of vernacular names with botanical equivalents; index to all botanical names. Appendices include a glossary, a list of cultivated plants, and a lengthy tabular summary of local and regional distribution of all species in relation to Egypt, the Mediterranean Basin, and elsewhere. [For additions and emendations, see N. D. SIMPSON, 1930. *Some supplementary records to Muschler's* Manual flora of Egypt. 59 pp. (Bull. Techn. Sci. Serv., Minist. Agric. (Egypt) 93). Cairo; for a key to families, see M. T. HEFNAWI, 1922. *Key to the flora of Egypt*. Cairo: Ministry of Agriculture, Egypt.][107]

TÄCKHOLM, V., 1974. *Students' flora of Egypt*. 2nd edn. 888 pp., 292 pls., 16 col. pls., map. Cairo: Cairo University Press. (1st edn., 1956.) Accompanied by V. TÄCKHOLM and L. BOULOS, 1974. *Supplementary notes to* Students' flora of Egypt, edn. 2. 136 pp., 16 pls. (Publ. Cairo Univ. Herb. 5). Cairo.

The *Students' flora* is a concise manual of vascular plants, including keys to genera and species, a synoptic table of families, essential synonymy, concise indication of local range, and partly abbreviated notes on habitat, frequency, special features, etc., followed by illustrations of representative taxa, a list of Arab names in script and in transliteration with botanical equivalents, and an index to generic and family names at end. The introductory section includes among other topics remarks on plant collecting. The *Supplementary notes*, arguably a 'critical supplement', contain a more extensive treatment of matters, including taxonomic problems, not readily accommodated in the main work.

TÄCKHOLM, V., G. TÄCKHOLM and M. DRAR, 1941–69. *Flora of Egypt*. Vols. 1–4 (Bull. Fac. Sci. (Cairo Univ.) 17, 28, 30, 36). Cairo: Cairo (formerly Fouad I) University. (Vols. 2–4 by V. Täckholm and M. Drar alone.) Continued as M. NABIL EL-HADIDI *et al.* (eds.), 1980– . *Flora of Egypt*. Published in fascicles (as *Taeckholmia*, Additional Series). Königstein/Ts., Germany: Koeltz (fasc. 1); Cairo: Herbarium, University of Cairo (fasc. 2–).

Documentary descriptive flora of vascular plants; includes keys to all taxa, extensive synonymy (with references and citations), vernacular names, detailed indication of local range with *exsiccatae*, summary of extralimital distribution, and quite lengthy notes on habitat, ecology, biological features, properties and uses, local 'lore' historical and modern, and (in each volume) a complete index. The continuation comprises family revisions with keys, descriptions, synonymy, references and citations, localities with *exsiccatae*, overall distribution, notes, illustrations, and maps; wild and cultivated species are included. Each fascicle has an extensive bibliography along with an index. [Original work not completed; coverage runs from the Pteridophyta through the Gymnospermae and Monocotyledoneae and the beginning (in reverse order) of the Dicotyledoneae (Casuarinaceae to Piperaceae). The Candollean system as used in Boissier's *Flora orientalis* was the basis for the arrangement of families, and with publication of volume 4 some 30 percent of the Egyptian vascular flora had been accounted for. From 1980 fascicular publication was introduced, but subsequent progress has been slow.][108]

592

Libya

Area: 1759 540 km^2 (coastal, non-desertic portion: *c.* 93 000 km^2, some 5.2 percent). Vascular flora: *c.* 1600 species (1440 in coastal region) (PD; see **General references**).

The first Libyan flora is *Flora libycae specimen* (1824) by Domenico Viviani. Later in the nineteenth century, a principal contribution was the plant checklist by Paul Ascherson in *Kufra* (1881) by the explorer Gerhard Rohlfs, covering 775 species. More organized exploration followed in the next three decades, leading to a first definitive work, *Florae libycae prodromus* (1910; see below) by Ernest Durand and Gustave Barratte. The era of Italian rule is reflected in the works of Renato Pampanini (1914, 1930, 1936, 1938; see below) and in *Flora e vegetazione del Fezzán e della regione di Gat* by Roberto Corti (1942). A renewal of floristic studies, in part an extension of activities in Egypt, followed the revolution of 1969; since then a bibliography, two floras and a checklist have appeared or are in progress. Plans also exist for a one-volume manual based on the current *Flora of Libya*. Pending completion of that work, however, the floras of Durand and

Barratte and of Pampanini remain current. Much of the Saharan region in Libya is also covered in *Flore du Sahara* by OZENDA (see **505**).

Bibliography

BOULOS, L., 1972. Our present knowledge on the flora and vegetation of Libya: bibliography. *Webbia* **26**: 365–400. (Erbario Tropicale di Firenze, Pubbl. 23.) [Unannotated list of 791 entries arranged by authors, preceded by a brief review of general features, rainfall, botanical resources and history; no indices.]

ALI, S., A. A. EL-GADI and S. M. H. JAFRI (eds.), 1976–89. *Flora of Libya*. Parts 1–145; Gymnosperms; Pteridophytes. Illus. Tripoli: Botany Department, Al-Faateh University of Tripoli/Arab Development Institute (distributed by Koeltz, Königstein/Ts., Germany).

Illustrated, serially published full-scale flora; each fascicle includes descriptions of taxa, keys, synonymy with references and literature citations, indication of local distribution by grid references (with some citations of *exsiccatae*), extralimital range, critical commentary, notes on phenology, a grid-map of Libya and index to taxa. An introduction to the project appears in fascicle 1, with the comment that a related single-volume work is also planned. [No systematic sequence is followed. Only the angiosperm fascicles were serially numbered; all those published are listed in fasc. 145, in numerical order. The gymnosperms (1986) and pteridophytes (1989) appeared *hors série*. The general layout of the work resembles that of *Flora of Pakistan* (**793**) with which the first editor was associated. The *exsiccatae* are indicative of much new collecting since 1969, but the work has little information on habitat, ecology, biology or variability.]

BOULOS, L., 1977. A check-list of the Libyan flora, 1. *Publ. Cairo Univ. Herb.* **7/8**: 115–141. Continued as *idem*, 1979. A check-list of the Libyan flora, 2–3. *Candollea* **34**: 21–48, 307–332. (Compositae by C. Jeffrey; corrections to that treatment in *ibid.*, **35**: 565–567 (1980).)

A systematic name list, including basionyms and selected synonymy with literature references. The first part covers pteridophytes, gymnosperms and monocotyledons (Typhaceae–Orchidaceae), while part 2 covers 'early' dicotyledons (Salicaceae–Neuradaceae). Part 3 includes a treatment of the Compositae. In general, the same Englerian sequence as was followed in Pampanini's *Prodromo della flora cirenaica* (see below) is

employed for this list. Part 1 is prefaced by a brief general review of studies on the Libyan flora. [Discontinued by the author with the advent of the *Flora of Libya* project.]

KEITH, H. G., 1392 A.H. [1973]. *A preliminary check list of Libyan flora*. 2 vols. 1047 pp., illus. N.p.: Published for Government of Libya. (Cover entitled *Libyan flora*.)

Volume 2 of this work (pp. 173–1028) comprises a generically arranged alphabetical checklist (with indication of families) of vascular plants, with synonymy, indication of status (and origin), vernacular names, local distribution, citation of some *exsiccatae*, notes on life-form, and miscellaneous casually presented notes; photographs of characteristic plants; list of references (through 1964). In Vol. 1 is a relatively brief introduction, with the plan of the work as well as miscellaneous observations on climate, vegetation and history (dated 1965), followed by a lengthy list of vernacular names with botanical equivalents and derivations.[109]

Partial works

The three works listed here comprised the main coverage for Libya before 1952.

DURAND, E. and G. BARRATTE, 1910. *Florae libycae prodromus: ou Catalogue raisonné des plantes de Tripolitaine*. cxxvii, 330 pp., 20 pls., map. Geneva: Romet.

Systematic enumeration of non-vascular and vascular plants; includes descriptions of new taxa, synonymy (with references and citations), detailed account of local range with indication of *exsiccatae*, general summary of extralimital range, taxonomic commentary, notes on phenology, life-form, etc., addenda, and index to generic and family names. The introductory section includes chapters on physical features and botanical exploration, a guide to collectors (by P. Ascherson), phytogeography (including synoptic tables, pp. xxvii–lxx), endemic species, and geology together with a list of cultivated plants and a bibliography. [The work covers the Egyptian coastal district of Mamara (west of the Nile Delta) together with the Libyan districts of Cyrenacia, Tripolitania and Fezzan. Collaborators included Ascherson, W. Barbey and R. Muschler.][110]

PAMPANINI, R., 1914. *Plantae tripolitanae ab auctore anno 1913 ab lectae et Repertorium florae vascularis tripolitanae*. xiv, 334 pp., 1 text-fig., 9 pls., map. Florence: 'Pellas' (for Società Italiana per lo Studio della Libia).

Systematic enumeration of non-vascular and vascular plants, based mainly upon collections by the author in 1913; includes synonymy and references, localities with citation of *exsiccatae*, summary of overall range, bibliography, and index to generic and family names.

PAMPANINI, R., 1931. *Prodromo della flora cirenaica.* xxxviii, 577 pp., 2 text-figs., 6 pls. Forlì: 'Valbonesi' (for Ministero delle Colonie).

Systematic enumeration of non-vascular and vascular plants; includes descriptions of new or little-known taxa, synonymy (with references), localities with citation of *exsiccatae*, miscellaneous notes, a gazetteer, list of references, and complete index. An introductory section includes among other material an account of botanical exploration, notes on collectors, and tables of additions to the flora after 1910. For supplements, see *idem*, 1936. Aggiunte e correzione al 'Prodromo della flora cirenaica'. *Arch. Bot.* (Forlì) **12**: 17–53; also *idem*, 1938. Aggiunte al 'Prodromo . . .' delle mie raccolte in Cirenacia negli anni 1933/1934. *Rendiconti Seminarii Fac. Sci. Reale Univ. Cagliari* **8**: 53–79.

593

Northern Chad and Niger

See also **589** (PEYRE DE FABRÈGUES and LEBRUN); **505** (OZENDA). Area: northern Chad (prefecture of Borkou-Enedi-Tibesti), 600 350 km²; northern Niger, no information. This otherwise almost entirely Saharan expanse is notable for the isolated Tibesti and Ennedi massifs in far northern Chad and the Aïr massif in northern Niger. Along with Jebel Marra (Sudan, **550**) and the Ahaggar and Ajjer (southern Algeria, **594**) these are of considerable phytogeographical interest as lofty 'islands' between the mountains of tropical Africa and the Maghreb.

Partial works: Tibesti and Ennedi massifs (Chad)
The highest peak in the sizeable Tibesti massif (Emi Koussi) of northwest Chad rises to 3415 m; three others reach over 3000 m. The somewhat lower Ennedi massif, an area of 43 000 km² (the size of Switzerland), is in northeast Chad near the Sudan border (and Jebel Marra). The studies cited below fill a void not bridged by other floras (except perhaps for Ozenda's *Flore du Sahara*, see **505**). The Ennedi massif is additionally covered, with a revised checklist, in H. GILLET, 1968. *La peuplement végétal du massif de l'Ennedi (Tchad).* 206 pp., 21 text-figs., 33 halftone pls., 2 maps (Mém. Mus. Natl. Hist. Nat., sér. II/B, Bot. 17). Paris.

GILLET, H. and G. CARVALHO, 1960. Catalogue raisonné et commenté des plantes de l'Ennedi (Tchad septentrional). *J. Agric. Trop. Bot. Appl.* **7**: 49–96 (with map), 193–240, 317–378 (with 6 pls.).

Annotated enumeration of vascular plants (410 species), incorporating localities with *exsiccatae*, summary of overall distribution, and more or less extensive commentary on habit, status, habitat, phenology and biology, local range, and systematics. The first installment contains a brief introduction and in the third installment are a lexicon of vernacular names and complete index.[111]

MAIRE, R. and T. MONOD, 1950. *Études sur la flore et la végétation du Tibesti.* 140, [1] pp., 6 pls. (Mém. Inst. Franç. Afrique Noire 8). Paris: Larose.

Contains (pp. 17–57) an enumeration of 40 non-vascular and 356 vascular plants, with chorological data, localities and citations of *exsiccatae*. The remainder of the work encompasses considerations of physical features, climate, biogeography, and ecology along with a lexicon of Teda names (by T. Monod) and the botanical itinerary (pp. 125–140). References, pp. 9–14. [Partly supplanted by the following work.]

QUÉZEL, P., 1958. *Mission botanique au Tibesti.* 357 pp., pls., maps (Mém. Inst. Rech. Saharien. Univ. Alger 4). Algiers.

Includes separate enumerations of non-vascular plants, pteridophytes and seed plants, with descriptions of novelties, general indication of local and extralimital range, and notes on habitat, frequency, etc.; statistical summary and index. An introductory section includes accounts of physical features, climate, etc., while the final chapters cover floristics, vegetation classification, and phytogeography.

Partial work: Aïr massif (Niger)
The Aïr massif in northwest Niger is the highest mountain system in that country. In addition to the work cited below, reference may be made to A. PITOT, 1950. Contribution à l'étude de l'Aïr: contribution à l'étude de la flore. *Mém. Inst. Franç. Afrique Noire* **10**: 31–81.

BRUNEAU DE MIRÉ, P. and H. GILLET, 1956. Contribution à l'étude de la flore du massif de l'Aïr. *J. Agric. Trop. Bot. Appl.* **3**: 221–247, 422–438, 701–740, 857–886 (with map).

Systematic enumeration of vascular plants (412 species); includes localities (with some *exsiccatae*), local occurrence and habitat, and summary of extralimital distribution. The final installment treats the environment, vegetation and altitudinal zonation along with biogeography (in contrast to Tibesti and the Hoggar, no Mediterranean elements are present); bibliography, pp. 884–886.

594

Algeria (Saharan zone)

For coverage of this vast area of *c.* 2 000 000 km², see **505** (OZENDA) and **598** (QUÉZEL and SANTA). In the southeastern quarter are the massifs of Ahaggar (Hoggar), whose highest peak (Tahat) attains 2918 m, and the less lofty Ajjer, where Mt. Azao reaches 2158 m. Together with the Tibesti massif in northern Chad the Ahaggar has been described as a botanical 'stepping-stone' between the Mediterranean and the mountains of tropical Africa; its affinities, however, lie with the former (and were even more pronounced during the Pleistocene).

Partial works

MAIRE, R., 1933–40. *Études sur la flore et la végétation du Sahara central*. 433 pp., 25 text-figs., 36 pls., 1 col. pl., 2 maps (Mém. Soc. Hist. Nat. Afrique Nord 3). Algiers.

Of the three parts of this work, part 1 comprises a detailed critical enumeration of non-vascular and vascular plants of upland southern Algeria (including the Ahaggar (Hoggar) and Ajjer massifs, with their relict Mediterranean floristic elements); this includes descriptions of novelties, synonymy, localities with *exsiccatae*, indication of overall distribution, altitudinal range and habitat, notes on uses and properties along with critical commentary, and (after part 2) an index. Part 2 comprises a Tamachek botanical vocabulary and lexicon, a supplement to part 1 (pp. 264–267) and the index. Part 3 (1940) covers vegetation and (pp. 407–433) further additions and corrections to the earlier parts. The Introduction includes a history of botanical exploration as well as a bibliography.

QUÉZEL, P., 1954. *Contribution à l'étude de la flore et de la végétation du Hoggar*. 164 pp., 10 pls. (Monogr. Rég. Inst. Rech. Saharien. Univ. Alger 2). Algiers.

Part I of this work comprises a floristic account with accepted names and for each notes on habitats, frequency, altitudinal range, localities, and (sometimes) botanical notes; a few novelties are described and discussed. About 300 vascular species were recorded, with general floristic discussion (pp. 45–52). Part II is on vegetation, while part III furnishes conclusions (the Mediterranean element is slight and temporally relatively late). The main parts of the work are preceded by the field itinerary and notes on climate and rainfall. [Based primarily on field work of 1950–53.][112]

595

Northern Mali and Mauritania (Saharan zone)

See also **505** (OZENDA), **587** (BARRY and CELLES; LEBRUN) and **588** (BOUDET and LEBRUN).

Partial work

MONOD, T., 1939. Phanérogames. In *idem* (ed.), *Contributions à l'étude du Sahara occidental*, 2: 53–211, pls. 1–24 (Publications du Comité d'études historiques et scientifiques de l'Afrique occidentale française B 5). Paris: Larose.

Includes in part 1 an annotated list of charophytes and vascular plants collected with critical notes and indication of overall distribution within and beyond the area. The remainder of the report (parts 2 and 3) describes itineraries followed (with a map of the region studied) and gives general remarks on floristic elements, divisions, etc., along with a bibliography. [Based largely upon field work by the author in 1934–35 and part of a group of geographical, ethnographic, botanical and zoological studies.]

596

Western Sahara

Area: 266 000 km^2. Vascular flora: 330 species (Lebrun in PD; see **General references**); *c.* 600 species if Saharan Mauritania is included (Quézel in PD). – This former Spanish territory, previously known as Spanish West Africa, is politically disputed but through the 1990s has been effectively under the control of Morocco.

Published literature on Western Sahara is limited. In addition to the works described here, reference may also be made to E. GUINEA LÓPEZ, 1945. *La vegetación leñosa y los pastos del Sahara español*. 152 pp., illus., maps (Publication *hors série*). Madrid: Instituto Forestal de Investigaciones y Experiencias.

GUINEA LÓPEZ, E., 1948. Catálogo razonado de las plantas del Sahara español. *Anales Jard. Bot. Madrid* 8: 357–442, pls.

Systematic enumeration of seed plants, including limited synonymy, vernacular names, indication of *exsiccatae* with localities, generalized summary of overall range, taxonomic commentary, and notes on habitat, properties, uses, etc.; photographs of representative vegetation formations; bibliography and author's itinerary; no separate index. [Not wholly critical; some records based on extrapolations from neighboring lands.]

LEBRUN, J.-P., 1998. *Catalogue des plantes vasculaires de la Mauritanie et du Sahara occidental*. See **587**.

597

Tunisia

Area: 164 148 km^2. Vascular flora: 2100–2200 species (PD; see **General references**).

The belated completion of the *Flore de Tunisie* begun by A. Cuénod in 1954 has freed Tunisia from dependence on the now-elderly *Flore analytique et synoptique de l'Algérie et de la Tunisie* by Battandier and Trabut. There has, however, been as such no successor to the enumeration of Bonnet and Barratte, part of the results of a French government scientific survey in the 1890s. For practical use the Algerian flora of Quézel and Santa (**598**) can also be recommended, there being considerable overlap between the floras of Tunisia and Algeria.

BONNET, E. and G. BARRATTE, 1896. *Catalogue raisonné des plantes vasculaires de la Tunisie.* xlix, 519 pp. Paris: Imprimerie Nationale.

Systematic enumeration of vascular plants, including descriptions of new or little-known taxa, synonymy (with references and citations), fairly detailed indication of local range, summary of extralimital distribution, taxonomic commentary, and notes on habitat, life-form, phenology, associates, etc.; index to generic and family names. The introductory section includes chapters on physical features, geology, climate, botanical exploration, cultivated plants, vegetation and phytogeography.

CUÉNOD, A., G. POTTIER-ALAPETITE and A. LABBÉ, 1954. *Flore analytique et synoptique de la Tunisie.* Vol. 1. 39, 287 pp., 65+ text-figs., 2 maps. Tunis: Office de l'Expérimentation et de la Vulgarisation Agricoles, Tunisie. Continued as G. POTTIER-ALAPETITE, 1979–81. *Flore de la Tunisie: Angiospermes – Dicotylédones.* 2 parts. xix, xiv, 1190 pp., illus., maps. Tunis: Imprimerie Officielle/Ministère de l'Enseignement Supérieur et de la Recherche Scientifique et Ministère de l'Agriculture.

Illustrated (for the dicotyledons, more generously so), briefly descriptive flora of vascular plants; includes keys to all taxa, limited synonymy, French vernacular names, fairly detailed indication of local range, summary of extralimital distribution, and notes on habitat, life-form, phenology, frequency, etc.; figures of diagnostic features; indices to all botanical names. The introductory section (vol. 1) includes a brief historical review and a lengthy glossary. Arrangement of the families follows the Englerian system, with volume 1 covering pteridophytes, gymnosperms and monocotyledons. [Substantial contributions towards the latter two parts were made by Md. Abdelhamid Nabli following the death in 1971 of Mme. Pottier-Alapetite.]

598

Algeria (general and Maghrebian zone)

Area (whole country): 2 381 745 km^2. Vascular flora (whole country): 3150 species (Le Houérou, 1975 in PD; see **General references**); 3139 species (Quézel and Santa; see below).

The well-illustrated *Nouvelle flore* of Quézel and Santa is now three decades old, but no supplement or successor work has yet been published. This work in turn supplanted *Flore d'Alger* (1884, not completed), *Flore de l'Algérie* (1888–95; supplement, 1910) and *Flore analytique et synoptique de l'Algérie et de la Tunisie* (1902 (1904)), all by J. A. Battandier and L. Trabut.[113] In the first quarter-century of French rule, Ernest Cosson with Jean Baptiste Bory de St.-Vincent and Michel C. Durieu de Maisonneuve wrote up the plants of the official surveys in *Exploration scientifique de l'Algérie, Botanique* (1846–69), laying a foundation for his own *Compendium* (see **Region 59**) and later floras. The local literature is considerable.

QUÉZEL, P. and S. SANTA, 1962–63. *Nouvelle flore de l'Algérie et des regions désertiques meridionales.* 2 vols. 1180 pp., numerous text-figs., pls. 1–112, A–J, 2 maps. Paris: Centre Nationale de la Recherche Scientifique.

Briefly descriptive illustrated manual-flora of vascular plants; includes keys to all taxa, limited synonymy, concise but relatively detailed indication of local range, summary of extralimital distribution (by phytogeographic zones), French and local vernacular names, and notes on infraspecific delimitation, habitat, frequency, associates, etc.; index to all botanical names (vol. 2). The introductory section contains aids to the use of the work, with maps. [Covers all of Algeria except the extreme south.]

Woody plants

Lapie, G. and A. Maige, [1915]. *Flore forestière de l'Algérie.* viii, 359 pp., 881 text-figs., 1 halftone, map. Paris: Orlhac.

Popularly oriented manual of woody plants; includes artificial analytical illustrated keys to species (along with keys to woods and to winter features), generalized indication of local range, French and local vernacular names, and discursive descriptive notes on habitat, biology, special features, uses, etc. (with illustrations); glossary and indices to botanical and vernacular names. In addition to Algeria, the work also includes the more common woody plants of Tunisia, Morocco and southern France. A total of 482 species are accounted for.

599

Morocco

Area: 447 000 km² (exclusive of Western Sahara). Vascular flora: 3700 species and 500 additional infraspecific taxa (Fennane and Mathez); 800 species apparently endemic. Included here are the Spanish enclaves of Ceuta and Melilla, but not Western Sahara.

The flora, the richest and most varied in the Maghreb, includes in addition to its marked endemism many 'archaic' elements. A considerable literature has accumulated consequent to the commencement after 1900 of relatively detailed exploration and, 21 years later, the establishment at Rabat of the 'Institut Scientifique Chérifien' (now the Institut Scientifique). The latter furnished resources for the most important work on the flora, the still-standard *Catalogue* of Jahandiez and Maire (1931–33; supplement, 1941). This succeeded Battandier and Trabut's *Flore de l'Algérie et catalogue des plantes du Maroc*, 1 (1888–90, 1895), covering flowering plants (with a listing of Moroccan plants collected by John Ball) as well as Ball's earlier *Spicilegium florae maroccanae* (1877–78; in *J. Linn. Soc., Bot.* 16). Additions to the *Catalogue* appeared through 1956; in this period also a definitive flora was commenced, but soon was passed over in favor of a more practical approach. Continuing research and field work as well as changing concepts of taxa have rendered these works out of date. Two significant partial floras are also in existence, covering respectively the Marrakech Basin and the oak-forest zone, and there is one general 'tree book'.

Since 1961, attempts have been made towards elaboration of a *Flore pratique du Maroc* along the lines of *Nouvelle flore de l'Algérie* (**598**) which was becoming much used in teaching in Morocco. Towards this end various treatments were written and circulated and, under a five-year programme begun in 1981, a trial sample was published by M. Fennane *et al.* in *Naturalia Monspesulana* 50 (1986). The first volume of the *Flore pratique* appeared in late 1999, covering pteridophytes, gymnosperms, and dicotyledons from Lauraceae through Neuradaceae. Another recent development envisages detailed examination of floristic links between Andalucía (Spain) and the northerly Rif Mountains, encompassing two tectonic systems with a common history.[114]

Bibliography

Annual bibliographic cumulations on natural sciences in Morocco appeared in *Bulletin de la Société des Sciences Naturelles et Physiques du Maroc* through vol. 47 (1967). These were followed by biennial lists for botany in vol. 49 (1969): 227–233; vol. 51 (1971): 247–257 (by J. P. Peltier); and vol. 53 (1973): 247–251 (also by J. P. Peltier).

Gattefossé, J. and É. Jahandiez, 1922. Essai de bibliographique botanique marocaine. *Bull. Soc. Sci. Nat. (Phys.) Maroc* 2(3–4): 71–86. [Unannotated, chronologically arranged treatment of 188 titles with an introduction.]

Jahandiez, É. and R. Maire, 1931–33. *Catalogue des plantes du Maroc (spermatophytes et ptéridophytes).* 3 vols., pp. i–lvii, 1–913, map. Algiers: 'Minerva' (distributed by Lechevalier, Paris). Continued as L. Emberger and R. Maire, 1941. *Catalogue des plantes du Maroc*, 4 (*Supplément*)... Pp. lix–lxxv, 915–1181 (Mém. Soc. Sci. Nat. Maroc, *hors série*). Algiers.

Systematic enumeration of vascular plants, including relevant synonymy (with references), abbreviated but detailed indication of local range, citation of representative *exsiccatae*, general summary of extralimital distribution, critical commentary, and notes on habitat, phenology, associates, etc.; index to all botanical names in vol. 3. A lengthy bibliography appears in the introductory section. The treatment of infraspecific taxa is relatively detailed, but this has provided a basis towards improved understanding of variation and speciation in this region. [For additions, see C. Sauvage and J. Vindt, 1949–56. Notes botaniques marocaines: mise à jour du *Catalogue des plantes du Maroc*. 4 parts. *Bull. Soc. Sci. Nat. Maroc* 29: 131–162; 32: 27–51; 34: 217–234; 36: 185–222.]

SAUVAGE, C. and J. VINDT, 1952–54. *Flore du Maroc: analytique, descriptive et illustrée.* Parts 1–2. 102 text-figs. (Trav. Inst. Sci. Chérifien, Sér. Bot. 1, 3). Rabat.

Illustrated descriptive flora, including keys to genera and species, synonymy (with references and citations), detailed account of local range, taxonomic commentary, and notes on habitat, phenology, habit, etc.; indices in each part (part 2 also includes a botanical gazetteer). [Not completed; only a small number of families were published (Ericaceae–Boraginaceae, following the sequence of vol. 3 of Jahandiez and Maire's *Catalogue*). The work was projected as a continuation for Morocco of the interrupted *Flore de l'Afrique du Nord* (**590**).][115]

Partial works

NÈGRE, R., 1962–63. *Petite flore des régions arides du Maroc occidental.* 2 vols. 982 pp., 150 pls. (2 col.), 2 maps. Paris: Centre National de la Recherche Scientifique.

Illustrated descriptive flora of vascular plants of the Marrakech Basin north of the Grand Atlas, including keys to all taxa, limited synonymy, French and local vernacular names, general indication of local and extralimital range, taxonomic commentary, notes on habitat, life-form, associates, special features, etc., and a glossary, gazetteer and bibliography; general index to all vernacular and botanical names at end.

SAUVAGE, C., 1961. *Flore des subéraires marocaines: catalogue des cryptogames vasculaires et des phanérogames.* xvi, 252 pp., map (Trav. Inst. Sci. Chérifien, Sér. Bot. 22). Rabat.

Systematic enumeration of vascular plants of the oak-forest zone, with symbolic indication of habitat, associates, life-form and local range; more or less extensive taxonomic commentary and notes on variation and occurrence; indices to family and generic and to French vernacular names.

Woody plants

For additional coverage, see **598** (LAPIE and MAIGE).

EMBERGER, L., 1938. *Les arbres du Maroc et comment les reconnaître.* 317 pp. Paris: Larose.

Semi-popular field-guide to trees and shrubs, including keys, descriptions, local range, French and local vernacular names, and miscellaneous notes on habitat, associates, etc.; index to genera and French vernacular names. Shrubs are dealt with in less detail, although included in the keys.

Notes

1 P. W. Richards, 1973. Africa, the 'odd man out'. In B. S. Meggers *et al.* (ed.), *Tropical forest ecosystems in Africa and South America: a comparative review*, pp. 21–26. Washington.

2 The Mayombe is further discussed in Note 68.

3 T. Durand, 1909. Les explorations botaniques au Congo belge et leurs résultats. *Bull. Acad. Roy. Sci. Belgique, Cl. Sci.*, sér. IV, **1909**(12): 1347–1374.

4 The Makerere University herbarium in Uganda was hidden in an attic during the country's civil troubles of the 1970s and 1980s (Campbell, 1990, p. 158). Others, sadly, have been entirely destroyed through war, fire or neglect.

5 See J. P. M. Brenan, 1953. International co-operation in African botany. *Nature* 172: 987–989. For a history of AETFAT, see P. Bamps, 1993. Fourty [*sic*] years activities of AETFAT. *Webbia* 48: 735–742; also in French as *idem*, 1995. Quarante années des activités de l'AETFAT. *AETFAT Bulletin* 42: 25–31.

6 The first congress was held in 1951 at Brussels, the second at Oxford in 1953, and since then at intervals of three or four years; for each, a set of proceedings has been published. Congresses since 1980 have been at Las Palmas, Canary Islands (1978); Pretoria (1982); St. Louis (1985); Hamburg (1988); Zomba, Malawi (1991); Wageningen, the Netherlands (1994); and Harare, Zimbabwe (1997). The *AETFAT Index* was begun in 1952 but ceased publication in 1986; during this period it was used as a basis for statistical analyses of progress on the floras of tropical and southern Africa and the south-western Indian Ocean islands, compiled for many years by J. Léonard and published in various of the proceedings. The *AETFAT Bulletin*, however, continues.

7 Mention may also be made of general reviews of the sciences in Africa. The leading such for the later colonial period, both with chapters on biology, are E. B. Worthington, 1938. *Science in Africa: a review of scientific research relating to tropical and southern Africa.* 764 pp. London; and *idem*, 1958. *Science in the development of Africa: a review of the contribution of physical and biological knowledge south of the Sahara.* xix, 462 pp., illus., maps. [Paris/London.] A brief summary account, written after decolonization but receptive to pure as well as applied research, is R. W. J. Keay, 1965. The natural sciences in Africa. *African Affairs* (1965), Suppl.: 50–54. A marked emphasis on applied biology, however, pervades E. S. Ayensu and J. Marton-Lefèvre (eds., for ICSU/UNESCO), 1981. *International biosciences networks/African biosciences network: proceedings of the*

symposium on the state of biology in Africa (Accra, Ghana, April 1981). Washington. Nevertheless, one contributor therein noted (p. 63) that 'it does not seem that the continent has a single natural history museum worthy of its vast treasures'.

8 A review of progress to 1982 appeared in 'Flora Reports' in *Bothalia* 14: 1015–1023 (1983). Usually two fascicles per year, each with some 30 maps, appeared through 1988 but since then only three have been published, the latest in 1994. The methodology of this work may be seen as old-fashioned; not only should electronic versions in addition to, or in place of, the print series be introduced but GIS approaches also adopted. Electronic versions (either on-line or issued as a CD-ROM) could also potentially have provision for textual commentary, representative illustrations and supporting background maps.

9 The *Flora* was part of the 'Kew Series' of British colonial floras initiated by W. J. Hooker in the 1850s. It is moreover one of a group of 'primary' or 'parent' floras antecedent to later regional floras. Vol. 10(1), the last part to appear, enjoyed only a limited distribution.

10 See G. Ameka *et al.*, 1996. Rheophytes of Ghana. In L. J. G. van der Maesen *et al.* (eds.), *The biodiversity of African plants*, pp. 780–782. Dordrecht: Kluwer.

11 The work was a response to the consequences of the long Sahelian drought of the 1970s.

12 On advantages of generic floras and synopses, see T. Just, 1953. Generic synopses and modern taxonomy: introductory essay. *Chron. Bot.* 14(3): 103–114. Among others, the generic floras of southern Africa, with a long history, are considered. The computerized checklists, however, have unfortunately conveyed the misleading impression that the flora is by now well known.

13 See also *Lantern* 13(1) (September 1963) as well as C. Lighton, *Cape floral kingdom* (1973), for additional information. For individual collectors and botanists, the most complete source is M. Gunn and L. E. Codd, 1981. *Botanical exploration of Southern Africa*. Cape Town: Balkema (for the Botanical Research Institute, Pretoria). The history of the former Botanical Research Institute prior to its union with the National Botanical Gardens in Kirstenbosch and assumption of its present name is reviewed in D. M. C. Fourie, 1998. The history of the Botanical Research Institute 1903–1989. *Bothalia* 28(2): 271–297. There is also a chapter on botany (by R. A. Dyer) in A. C. Brown (ed.), 1977. *A history of scientific endeavour in South Africa*. Cape Town.

14 A revision of this work in one volume by O. A. Leistner, entitled *Seed plants of southern Africa: families and genera*, appeared in 2000.

15 The foreword to the 1985–87 edition (by B. de Winter) indicates that previous estimates of the size of the vascu-

lar flora were low by 3000 or more species. The role of the computer and the PRECIS database from which the work is derived were also discussed. Further developments were outlined in the introduction to the 1993 edition.

16 The *Flora of Southern Africa* project was initiated in 1953 and sanctioned in 1955, but without provision for extra staff; completion in 40 years was foreseen (W. Marais, 1958. The proposed *Flora of Southern Africa*. *Mem. Soc. Brot.* 13: 51–52). Separate series were established for non-vascular plants, pteridophytes and seed plants. As of 1993, its general editorship was under the charge of O. A. Leistner. By December 1996 the work was 13 percent complete (Morat and Lowry, 1997; see **Division 5, Progess**). Parts in both the non-vascular and seed plant series have continued to appear; the most recent among the latter – after a break of four years – covers Iridaceae–Ixioideae (1999). Of the 33 volumes planned for seed plants, to date (1999) vols. 1, 4(2), 5(3), 7(2, fasc. 1–2), 8(2), 10(1), 13, 14, 16(1–2 and 3, fasc. 6), 18(3), 19(3, fasc. 1), 21(1), 22, 26, 27(4), 28(4), 31(1, fasc. 2) and 33(7, fasc. 2) have been published. A parallel series, *Flora of South Africa Contributions*, began publication in *Bothalia* in 1993; it continues, with 11 contributions by 1998 each devoted to a family or genus and following the *Flora* format.

17 Although *Flora capensis* became part of the 'Kew Series' of British colonial floras upon publication of its first volume, it did not until 1877 become an official charge for Kew. The remaining volumes were prepared largely by herbarium botanists at that institution with some contributions by workers in southern Africa. Colonial (and dominion) subventions totalling £3250 were received towards its preparation and publication. For a short historical review, see W. T. Thistleton-Dyer, 1925. Flora capensis. *Bull. Misc. Inform.* (Kew), **1925**: 289–293.

18 This largely compiled work was regarded as an 'interim' publication, but its more definitive replacement, the *Tree Atlas*, began to appear only in 1992. It has attracted some criticism as supposedly not entirely reliable; moreover, its circulation was limited due to withdrawal of many copies by the distributor following a copyright dispute.

19 The contribution of R. B. Drummond (cf. **525**) is represented in the northward extension of the geographical range of this book.

20 The work was written to fill a gap resulting from the age of *Flora capensis* and the limited coverage of *Flora of Southern Africa*. It was partly based on J. S. Beard's *A descriptive catalogue of West Australian plants* (**450**), the 1970 revision of which was the responsibility of the first author. A revised edition, under the leadership of J. Manning, was in preparation as of writing (1999).

21 The extraordinary richness of the Cape Peninsula flora in relation to its area, perhaps the highest such in the world, is singled out in R. d'O. Good, 1974. *The geography of the flowering plants* (see **General references**), p. 224. The *Flora* is currently under revision by P. Linder and T. Joinder-Smith.

22 See I. Friis, 1998. Frank White and the development of African chorology. In C. R. Huxley, J. M. Lock and D. F. Cutler (eds.), *Chorology, taxonomy and ecology of the floras of Africa and Madagascar*, pp. 25–51. Kew: Royal Botanic Gardens.

23 This first publication of the Botanical Survey of South Africa (now the Botanical Research Institute) was intended to be a contribution to a 'regional floras' programme.

24 Only parts 1 and 2 were published. Parts 3 and 4, which would have concluded the *Manual*, were in galley proof by 1941, but owing to the author's death in 1940 and the disruption of World War II publication was not effected. The proofs are deposited in the archives of the Royal Botanic Gardens, Kew (Stafleu and Cowan, *Taxonomic Literature-2*, 1: 603).

25 The herbarium of the Transvaal Museum, upon which the checklist of 1912–20 was based, has long been part of the present National Botanical Institute.

26 Work towards this flora was initiated by Hamish B. Gilliland in 1952 with publication of a students' key for monocotyledons. There appear to exist no plans for its extension to the dicotyledons or other vascular plants.

27 The main work, although designed as a practical manual-flora, regrettably is somewhat too bulky due to the format imposed by the journal.

28 Publication of *Flora Zambesiaca*, begun in the late 1950s as a joint project of the Natural History Museum (London) and the Centro de Botânica of the present Instituto de Investigação Científica Tropical (Lisbon) and also involving the Royal Botanic Gardens, Kew, and the National Herbarium of Zimbabwe at Harare, was seriously delayed for a decade or more after the mid-1960s due to political and other difficulties; by then only volumes 1, 2, 3(1), 10(1) and the Pteridophyta, covering 2228 species, had appeared. Volume 4, however, appeared in 1978 and since 1980 a sustained renewal has taken place; as of the end of 1997, 4983 species, or some 50 percent of the whole, had been published with a further 15 percent ready in typescript (Gerald Pope, personal communication). Nevertheless, in the early 1990s the British Museum (Natural History), now The Natural History Museum, reduced its interest in *Flora Zambesiaca* following retirement of its editorial board member (E. Launert); it is no longer part of its research programme. The IICT Centro de Botânica in Lisbon has maintained its commitment while at Kew completion of the work by 2006 is seen as a priority. The original estimate of the time required for the project (H. Wild, 1958. The *Flora Zambesiaca*. *Mem. Soc. Brot.* **13**: 53–55) was given as 20 years, but this was based on an estimate of *c.* 6000 species.

29 The history of botanical work to 1961 was reviewed in A. D. J. Meeuse, 1962. A short history of botanical exploration in the territory of South West Africa. *Comptes rendus de la IV. réunion plenière de l'AETFAT* (ed. A. Fernandes), pp. 123–126. Lisbon. A short historical account also appears in the bibliography by Giess.

30 Following Carrisso's death in 1937, most of the first 1½ volumes were written by Mendonça with Arthur Exell at the British Museum (Natural History).

31 Responsibility in Portugal for the *Conspectus* was after World War II assumed by the Junta das Missões Geográficas e de Investigações Colonais (now the Instituto de Investigação Científica Tropical), originally created in 1936. Its Centro de Botânica came into being in 1948.

32 The history of botanical work to 1961 was reviewed in F. A. Mendonça, 1962. Botanical collectors in Angola. *Comptes rendus de la IV. réunion plenière de l'AETFAT* (ed. A. Fernandes), pp. 111–121, 12 pls. Lisbon. See also A. Fernandes, 1958. Progrès dans l'étude de la flore de l'Angola. *Mem. Soc. Brot.* **13**: 33–46.

33 The *Conspectus* was by 1996 considered to be 55 percent complete (Morat and Lowry, 1997; see **Division 5, Progress**). Although interest in the project continues in some quarters with several additional fascicles having appeared in the 1990s, its future remains uncertain (Gerald Pope, personal communication, 1997). A portion of vol. 1(2) was distributed in 1939 to delegates at an international congress of tropical agriculture; this was for nomenclatural purposes disowned by the authors.

34 A description of the *Forest flora* project appears in F. White, 1958. Forest Flora of Northern Rhodesia. *Mem. Soc. Brot.* **13**: 47–49.

35 The association with OFI also gave rise to a standard work on forest vegetation, *The evergreen forests of Malawi* (1970) by J. D. Chapman and F. White.

36 The Binns checklist is now very incomplete (R. K. Brummitt, personal communication).

37 A revised, expanded version by F. White of this work was in 1997 being edited for publication (R. K. Brummitt, personal communication).

38 Particulars of Mozambican botanical progress are recounted in A. Cavaco, 1958. Aperçu de l'état actual de nos connaissances sur la flore du Moçambique. *Mem. Soc. Brot.* **13**: 69–77, and in F. A. Mendonça, 1962. Botanical collectors in Mozambique. *Comptes rendus de la IV. réunion plenière de l'AETFAT* (ed. A. Fernandes),

pp. 145–152. Lisbon. Another account is A. Fernandes, 1972. Progressos no estudo da flora da Moçambique. *Mem. Acad. Ci. Lisboa, Cl. Ci.* **16**: 141–175.

39 Prosecution of the work remains among the research objectives of the IICT Centro de Botânica. For the most part, accounts are adapted from *Flora Zambesiaca*. The most recent fascicle appeared in 1993.

40 The East African Community was an ambitious British-organized administrative scheme for Kenya, Uganda and Tanzania involving transport networks, utilities, research services, and many other government activities. It continued to operate for some years after independence, but collapsed in 1977 resulting for the *Flora* in an interruption of publication. 'Privatization', including a change of publisher, enabled resumption in 1980 (R. M. Polhill in *Bothalia* **14**: 1015–1023 (1983)).

41 Progress reports have been presented by E. Milne-Redhead in *Mem. Soc. Brot.* **13**: 57–59 (1958) and by R. M. Polhill in *Bothalia* **14**(3–4): 1020–1021 (1983) and in *Modern Systematic Studies in African Botany*, pp. 709–710 (1988). The 1990 figure is taken from a report by Polhill in *Fl. Males. Bull. Special Vol.* **1**: 11–20 (1990), while that for 1997 was kindly communicated by Christopher Whitehouse (Kew). Preparation of the *Flora* was from the 1950s aided by the contribution of six British Colonial Office-supported posts to the Royal Botanic Gardens, Kew.

42 From a publication flyer of 1991. [There should be more such works, produced and distributed at relatively modest prices (here for £9.20 surface postpaid).]

43 See also I. Hedberg and O. Hedberg (eds.), 1968. *Conservation of vegetation in Africa south of the Sahara.* 320 pp. (Acta Phytogeogr. Suec. 54). Stockholm.

44 J. C. Lovett and R. M. Polhill, 1984. [Montane flora of Tanzania: a feasibility study.] Kew. (Mimeographed.); cited in Lovett, 1989 (see **Region 53, Progress**).

45 The author being an ardent Germanophile, collections made by British and other workers were singularly left out of account. The work is regarded by Brenan (in Brenan and Greenway; see above) as 'useful in providing a partial catalogue of Peter's very copious collections made in the [Tanganyika] Territory'. The first part actually appeared in 1930.

46 The 1994 edition accounts for about 5 percent more species than that of 1974.

47 Dale and Greenway's work, with its black cover, became over the course of 30 years a 'Tree Bible'. Details appear in the original edition of this *Guide*.

48 The work, originally completed in 1971, was finally published with the support of a Norwegian cooperative aid programme.

49 A reprint of the *Tentamen* was produced in Uppsala in 1982 for the *Flora of Ethiopia and Eritrea* project.

50 For details on the *Adumbratio*, see F. G. Meyer in *Taxon* **22**: 655–657 (1973).

51 The 'Erbario tropicale' initially had planned a full flora with descriptions, but as no funds for additional staff and expenses were forthcoming efforts were then directed into the more modest, irregular *Adumbratio* (Drar, 1963; see **Division 5, Progress**).

52 A splendid monograph!

53 The project has been an undertaking of the Department of Systematic Botany, University of Uppsala in collaboration with the National Herbarium of Somalia, Mogadishu (sadly, now destroyed), with support from the Swedish Agency for Research Cooperation with Developing Countries; currently it is part of a more general programme, 'Biodiversity of NE African Arid Regions'. Field work was resumed on a local basis in 1995. The two volumes published by 1996 together encompass some 46 percent of the vascular flora (Morat and Lowry, 1997; see **Division 5, Progress**).

54 The paper of parts II and III has now become badly deteriorated, an ironic symbol.

55 Perhaps the weightiest of African floras per species, this is a work more of prestige than of real utility.

56 According to Blake and Atwood (1942, p. 33; see **General bibliographies**) the manuscript of a fourth and final part (by E. Chiovenda) was lost by the printing establishment to whom it had been sent.

57 O. Hedberg. 1983. Ethiopian flora project. *Bothalia* **14**: 571–574.

58 This work was a contribution of the 'Oklahoma-Ethiopia Agricultural Group'. The country is a major center of diversity for durum (hard) wheats (used for pasta).

59 For descriptions of the *Flora of Ethiopia and Eritrea* project see, in addition to Hedberg (1983; above), I. Hedberg in *Symbol. Bot. Upsal.* **26**(2): 17–18 (1986), Mesfin Tadesse in *Taxon* **40**: 267–272 (1991) and I. Hedberg in *The biodiversity of African plants*, pp. 802–804 (1996, Dordrecht).

60 Reputedly an unsatisfactory work, with little original content.

61 The omissions may have been necessary for reasons of space. Publication was made possible by the Netherlands through an award to the FAO programme, 'Fuelwood Development for Energy in the Sudan'.

62 The work was written in particular for students at the University of Juba, *c.* 190 km to the northwest.

63 These works were published through King Léopold II's 'Musée du Congo' in Tervuren near Brussels (now the Musée Royale d'Afrique Centrale), established in 1898.

64 Belgian activities through the end of the 'Free State' are reviewed in T. Durand, 1909 (see Note 3).

65 Belgian botanists under Robyns and Jean Léonard also played a leading role in the creation of the present

'umbrella' organization for the coordination of African floristic and systematic botany, the 'Association pour l'Étude Taxonomique de la Flore de l'Afrique Tropicale' (AETFAT), which came into being about 1950.

66 Progress to 1958 was described in G. Gilbert, 1958. La *Flore du Congo Belge et du Ruanda-Urundi. Mem. Soc. Brot.* **13**: 87–92. For a 50-year statistical survey, see J. Léonard, 1994. Statistiques des spermatophytes de la *Flore d'Afrique Centrale* de 1940 à 1990. *Bull. Jard. Bot. Natl. Belg.* **63**: 181–194; here, it is shown that in recent years reductions of names have more than kept pace with novelties, with the overall total for seed plant species falling from 9705 to 9377. The change in the 1960s to publication in fascicles was made in the interests of flexibility. Control of the work is vested in a managing editor. It was noted that 4636 species had been covered through 1990.

67 The author in his *Nzayilu N'ti* (1993; see main text) claims, however, that the work is not formally published.

68 The Mayombe is an area of high rainfall and optimally forested uplands, here and there to over 1000 m, in the western part of Central Africa west of the central Zaïrean basin and facing the Atlantic. Its largest part is in Gabon, from south of the Ogowé River and extending southwards into the Congo Republic. The name originates from the coastal Gabonese town of Mayombe. A good map is to hand in J.-M. Moutsamboté, 1990. Dynamique de reconstitution forestière au Congo: cas du Mayombe congolais. In H.-D. Ihlenfeldt (ed.), *Proceedings of the Twelfth Plenary Meeting of AETFAT*, pp. 225–232 (Mitt. Inst. Allg. Bot. Hamburg 23). Hamburg.

69 The publication of Lebrun and Stork's *Énumération* (**501**) will to an extent bridge this gap.

70 The flora proposal was reported by G. Bocquet in *Boissiera* **24**: 581 (1976).

71 Le Testu was one of two main collectors in Gabon before World War II.

72 For progress on the flora to 1974, see J. Floret, 1976. Flore du Gabon. *Boissiera* **24**: 575–580.

73 The illustrations include many photographs from Guinea's 1945 survey as well as reproductions of the paintings in his 1946 book.

74 I thank David Harris, Oxford, for a loan of a copy of this work.

75 The author was a brother of Eugène Cardinal Tisserant, dean of the Sacred College of Cardinals in the reign of Pope John XXIII. The *Catalogue* was followed by a series of papers in *Notulae Systematicae* (Paris) comprising revisions entitled *Matériaux pour la flore de l'Oubangui-Chari*, in part with R. Sillans.

76 The release in the same year of two independently produced Cameroon dendrologies provokes some disbelief, but reflects an inherent problem with bilateral aid programmes. Souane Thirakul's work was supported by the Canada–Cameroon Cooperation Program, while that of Vivien and Fauré enjoyed French support.

77 In addition to the poorly reproduced halftones, this work is also marred by misprints. With current paper and printing technology, it is possible to combine durability with illustration quality as is shown in the field-manuals accompanying *Biodiversity assessment* (1996, London: HMSO).

78 The Cape Debundscha area to the southwest of Mt. Cameroon features rainfalls of 10–15 m per annum, among the three highest such in the world.

79 Bioko was, however, Portuguese from 1494 to the 1780s; it then passed to Britain which held it until 1843. Spanish rule began in 1844.

80 The history of botanical work to 1961 was reviewed in A. W. Exell, 1962. Botanical collecting on the islands of the Gulf of Guinea. *Comptes rendus de la IV. réunion plenière de l'AETFAT* (ed. A. Fernandes), pp. 95–102. Lisbon.

81 For discussion of the limits of the Sahara, see H. Walter, 1973. *Die Vegetation der Erde*, 1: 609–610, map. Jena.

82 From 1909 Hutchinson had been 'assistant for tropical Africa' in the Kew Herbarium.

83 The IEMVPT botanical programme was reviewed in J.-P. Lebrun, 1971. Les activités botaniques de l'Institut d'Élevage et de Médecine Vétérinaire des Pays Tropicaux. *Mitt. Bot. Staatssamml. München* **10**: 86–90.

84 Richards, 1973; see Note 1.

85 For an account of the history of this work, see R. W. J. Keay, 1998. Third best taxonomy. *The Linnean* **14**(3): 48–49. Not mentioned therein, however, were the reasons for the shift southwards of the northern boundary between the first and second editions. This was not only for biogeographical reasons (as noted in the *Flora* itself) but also on account of the poor representation of material from the Sahara in British herbaria (Drar, 1963; see **Division 5, Progress**). Additional notes on the second edition appear in Hepper, 1958 (see **Region 58, Progress**).

86 Noted by its author in *Rev. Bot. Appl. Agric. Trop.* **30**: 264–265 (1950).

87 Richards, 1973; see Note 1.

88 For a summary of work to 1993, see K. Akpagana and P. Bouchet, 1994. État actuel des connaissances sur la flore et la végétation du Togo. *Acta Bot. Gall.* **141**: 367–372.

89 For details of this flora project, see *Bothalia* **14**(3/4): 1015–1023. 1984.

90 An example of the kind of, now rather too common, 'results' of 'development contract' research, of which another example is Whitmore's guide to Solomon Islands trees (**938**).

91 Represents, within its chosen compass, a thorough piece

of work, beside which Kunkel's contemporaneously published book looks like a piker. Both works, however, are more for the office, taking little heed of the user in the field.

92 Thirgood in his bibliography records the existence of an undated 114-page *Handbook of Liberian Ferns* by the same author, produced at Ganta Mission, Liberia.

93 As with *Flora de Cabo Verde* and given a relatively small flora, it would have been more beneficial if *Flora da Guiné-Bissau* had emerged as a one-volume illustrated manual rather than yet another 'flora-series'. As endemism is low, however, reference can be made to the several works on nearby Senegal (586).

94 Continuation of this flora seems doubtful; it is not among current projects of the Instituto de Investigação Científica Tropical in Lisbon.

95 The somewhat peculiar collation reflects nomenclatural and administrative changes with regard to the publications of the issuing body.

96 Père Berhaut died in 1977 but only in 1983 was work resumed under the direction of Constant vanden Berghen. Volume 9 (1988), his first (and so far only) contribution, has more copious descriptions and supporting matter (but fewer *exsiccatae*) in a reorganized, more economical page format.

97 Lebrun in his *Catalogue* notes that for historical reasons Mauritanian collections from before 1962 were widely dispersed, making retrospective study difficult.

98 The largest families in Burkina Faso are Gramineae, Cyperaceae and Leguminosae. The bulk of collecting has been accomplished since World War II; before then only six collectors had been active, starting with August Chevalier in 1898–99.

99 Only nine of the 1045 species covered in the main work are not in the current edition of *Flora of west tropical Africa*; striking features are the almost total lack of endemism and orchid species.

100 This is approximately in agreement with the suggestions of P. Bruneau de Miré, 1960. Le 18ème parallèle: constitue-t-il une limite floristique en Afrique Occidentale? *J. Agr. Trop. Bot. Appl.* 3: 439–442.

101 European Science Foundation, *Report 1977*, pp. 81–82; *Report 1978*, pp. 95–97. In the latter reference, the plan and scope of the projected checklist are described.

102 For the second edition of the *Flora of west tropical Africa* a 'retreat' was made to 18° N (cf. Note 85).

103 While the seemingly excessive recognition of infraspecific taxa in this work (and others by Maire) may be criticized, Maire's approach at least had the merit of drawing formal attention to the 'polymorphism' of many species (C. Sauvage, 1975. L'état actual de nos connaissances sur la flore du Maroc. In Centre National de la Recherche Scientifique, France (ed.), *La flore du bassin*

méditerranéen, pp.131–139). With the application of less 'traditional' classificatory parameters (ecological, phytochemical, molecular, etc.), these apparently polymorphic species may in fact represent two or more discrete taxa, some likely deserving of specific rank – a view supported by the work of F. Ehrendorfer *et al.* on the anthemid composite genus *Anacyclus* (*Taxon* 26: 387–394 (1977)) and further reinforced in phylogenetic approaches to classification.

104 An outcome of the CNRS meeting (at Montpellier, 1974) was the formation in 1975 of the Organization for the Phyto-taxonomical Investigation of the Mediterranean Area (OPTIMA), inclusive of North Africa. Its semi-annual *Newsletter* gives some indication of ongoing botanical work in the region (also from time to time reported in AETFAT congress proceedings).

105 Owing to the author's sudden death in 1928 a planned English version as well as illustrations were not realized.

106 This work evidently does not account for Sinai Peninsula records listed by Danin but unrepresented in Egyptian herbaria (W. Greuter in *OPTIMA Newsl.* 31: 12 (1997)). Those species listed here but not in that of Boulos are mainly naturalized weeds. The *Materials* were intended as a precursor to a *Flora Aegyptiaca* (*Taeckholmia* 17: 16 (1998)).

107 Muschler's *Manual flora* has been considered to be 'presumably as unreliable as his other publications' (Blake and Atwood, 1942 (see **General bibliographies**), p. 9). Earlier, Täckholm in her *Bibliographical Notes* (p. 194) had noted that 'it was certainly a great pity for Egyptian science that this Flora was allowed to appear, because any reference made from this book, with its mixture of correct and false identifications, must be considered worthless'. For this and even more for bogus descriptive work published in the *Botanische Jahrbücher* and elsewhere, Reno Muschler was dismissed from the Berlin Botanical Museum in 1913; a lawsuit against him was, however, disallowed on the grounds of psychological instability (Stafleu and Cowan, *Taxonomic literature-2*, 3: 674). Muschler later became a successful novelist.

108 V. Täckholm in Merxmüller 1971, pp. 17–18 (see **Division 5, Introduction**). In 1970, Täckholm thought that a further 20 years would be needed to complete the *Flora*. However, by the time of her death in 1978 (Drar had died in 1964) nothing had been published since 1969 and its continuation seemed uncertain. The considerable amount of non-floristic matter had moreover attracted some criticism. A reorganization enabled resumption of production; the revised plan appeared in 1980 as *An outline of the planned 'Flora of Egypt'* by M. Nabil el-Hadidi, the new editor. By 1999 six installments had been published (1, 1980; 2, 1983; 3, 1988; 4, 1991; 5, 1994; 6, 1998).

109 The author, formerly in the British colonial service with a term in present-day Sabah (**917**), was in 1955–60 FAO forestry advisor in Libya. The *Check list*, evidently delayed in publication, is based to a large extent on his observations and collections as well as on government herbarium holdings and thus is not fully retrospective.

110 The work, sponsored by Barbey, was principally a vehicle for the results of field work by Paul Taubert in 1887 but in the end encompassed many other collectors.

111 The massif was then under military administration.

112 Of the latter, the Hoggar receives a peak of rainfall in the summer, this varying from year to year and in longer cycles; 1950–53 were relatively wet. *Cupressus* cf. *dupreziana* was located, but dead; there were two species of *Ficus* recorded.

113 Botanical progress during the first hundred years of French rule in Algeria was summarized by R. Maire in *Les progrès des connaissances botaniques en Algérie depuis 1830* (Paris, 1931).

114 With respect to the *Flore pratique du Maroc*, see J. Mathez, 1988. Le programme franco-marocain de rédaction de la *Flore Pratique du Maroc. Lagascalia* **15**, Suppl.: 143–153; and *idem*, 1991. Les perspectives d'une nouvelle Flore du Maroc. *Bot. Chron.* **10**: 105–116. The flora remains imperfectly explored and understood (M. Fennane, personal communication, 1998). On the Andalucia/Rif project, see B. Valdés, 1991. Andalucia and the Rif: floristic links and a common Flora. *Ibid.*: 117–124.

115 Work on a third part, covering Lamiaceae, was commenced but not continued. In the early 1960s, the *Flore* was abandoned in favor of what is now the projected *Flore pratique du Maroc*.

Division 6

•••

Europe

Flora URSS is thus completed. We remember all our colleagues, many of them long dead, who contributed to its achievement.

We have done what we could. We welcome the young botanists and wish them success.

Fecimus quod potuimus. Vivant sequentes.

E. G. Bobrov, *Nature* **205**: 1049 (1965).

It is the hope of the editors [of *Flora Europaea*] that by wrestling with these problems they have, at least to some extent, made it unnecessary for the next generation to do so again, and have thus enabled them to devote more time to plants.

D. A. Webb, *Taxon* **27**: 14 (1978).

The geographical limits of Europe adopted here are essentially the same as those adopted for *Flora Europaea*, with the omission of the Azores, the Arctic islands and the tundra zone (respectively included in Regions 02, 05 and 06). The Caucasus, sometimes considered to be part of Europe, is treated as Region 74 within Division 7 (northern, central and southwestern Asia). Outside of the eight groups of polities here delimited, separate units have been allocated for a number of major physiographic entities such as the Mediterranean, the Alps and Carpathians, and the Ural.

Organized floristic study and the writing of floras as they are today internationally known began in Renaissance Europe, but had antecedents in ancient Greece and in China and Japan.[1] Significant advances in methodology came through the work of such scholars as Charles de l'Écluse (Clusius), Johann and Caspar Bauhin, John Ray, Albrecht von Haller, Carl Linnaeus, Jean-Baptiste de Lamarck, Augustin-Pyramus de Candolle, William J. and Joseph D. Hooker, Karl Philip von Martius, George Bentham, Nathaniel L. Britton and Gustav Hegi and, in the modern era, the *Flora SSSR* work-group, the Flora Europaea Organisation, the *Flora of Australia* project, and the original, abortive Flora North America Program. The most significant developments before the 'modern era' took place in Europe in the late eighteenth and early to mid-nineteenth centuries, influencing floras there and in other parts of the world. By 1860, European (and most other) floras had assumed their more or less 'modern' forms which since have changed little except in details. Twentieth-century contributions have been at higher levels of integration: project management and application of information technology.

The formative period of floristic botany in Europe may be taken as lasting from 1500 to 1623. In the latter year there was published in Basel the first general systematic survey, Caspar Bauhin's *Pinax theatri botanici*, which encompassed ancient, Renaissance and early modern writers. During this period, floristics – then largely synonymous with botany as a whole – gradually emerged from herbalism, particularly between 1530 and 1550. A significant mover was Luca Ghini, professor of botany and *materia medica* at Pisa University, through his technical innovations in the 1540s of the herbarium and organized botanical garden. Through his students and others, notably Conrad Gesner, national floristic exploration began in Italy, Central Europe, France, the Low Countries, Spain and England and in succeeding centuries gradually spread through the rest of the continent and beyond. Among the more notable traveler-writers – on account of his explorations of the Iberian Peninsula and, later, the Pannonian Basin – was Charles de l'Écluse. The first half of the seventeenth century, dominated by the Thirty Years' War and, in England, by the Civil War, was not conducive to much beyond local efforts; nevertheless, what was then known was incorporated by Bauhin, whose *Pinax* would form a new point of reference: a first European as well as world enumeration.

With more settled conditions after 1650, however, came renewal. The travels of John Ray furnished the experience necessary for his more or less pan-European checklists, the first of their kind: *Catalogue stirpium in exteris regionibus a nobis observatarum* (1673, in his *Travels through the Low Countries*) and *Stirpium Europaearum extra Britannias nascentium sylloge* (1694); while he and Joseph P. de Tournefort – the latter also a traveler, mainly in the eastern Mediterranean – contributed significantly to classification. In the eighteenth century a further impetus was furnished by Linnaeus, who traveled extensively in Lapland, other parts of Sweden and elsewhere, and his students, most notably Pehr Löfling who explored much of Spain in the 1750s prior to his ill-fated trip to South America. Their output includes many florulas and larger floras as well as Linnaeus's own *Species plantarum* – like Bauhin's *Pinax* arguably a European as well as a world flora. *Species plantarum* was later adapted for Europe by Jean (Jan) Gilibert and published as *Systema plantarum Europae* (1785–87, especially vols. 3 and 4).

The French Revolution and the Napoleonic Wars, like the Thirty Years' War, caused prolonged disruption over much of the continent but served to direct new attention to Greece and the Aegean as well as northwest Africa and Egypt. After 1815 exploration became more systematic, with the progressive penetration of more of the Alps, remoter parts of the Iberian Peninsula, European Russia, and particularly southeastern Europe. The last-named long remained a politically and socially unstable region of regressive development, though Greek independence in 1830 furnished an important opportunity for further exploration. In the late nineteenth and early twentieth centuries detailed attention was directed towards the floras of some lesser-known mountainous areas such as the Carpathians, the Balkan highlands and the Alpes-Maritimes.[2]

Prior to the mid-eighteenth century, the study of floristic botany in Europe was largely the province of scholars and other professionals; among the more notable were Conrad Gesner, L'Écluse, Ray, Joseph P. de Tournefort, Albrecht Haller, and of course Linnaeus. With the spread of the Linnaean reforms, however, floristic botany became, particularly in countries with a growing middle class, a respectable 'leisure' activity, aided by comparatively simple texts. This greatly increased the demand for 'local' floras and resulted in the development of amateur/professional 'national' botanical circles. In the hundred years from Waterloo to the outbreak of World War I, extensive and detailed floristic exploration spread all through Europe, albeit more thinly in many parts of the south and distant east where amateurs were few and 'expeditions' correspondingly more important, as was then also the case in most of the tropics, Australia, North America, and China and Japan.

As the nineteenth century progressed, the growing professional and semi-professional community in many countries, particularly from around 1830, shifted their interests to a large extent to Macaronesia, North Africa, the Orient, and further afield for scientific and other reasons. In addition, there was a great growth in other areas of botany than floristics or taxonomy, and upon these were developed many of the great university 'schools' (many of which, however, retained strong interests in systematics despite the seemingly disparaging remarks of Julius Sachs). An increasing share of the conduct of floristic botany in Europe thus came into the hands of dedicated amateurs, schoolmasters and clergy, who for the latter part of the century and the first decades of that following were more or less to dominate the field, carrying on local (or more distant) exploration and observation and more or less blanketing the map with floras and other works. Many of them were to contribute major 'national', provincial political and physiographic, and even all-European works, especially following the Franco-Prussian War.

The years from 1870 to 1914 (and somewhat beyond into the 1920s) represented the zenith of this development (and, as Stafleu has noted, of descriptive floristic and taxonomic botany in general). Especially characteristic of this era were the large-scale 'national' or supranational floras and other works – some very detailed – compiled by (usually) one or two (sometimes more) authors. Although varying considerably in detail (as well as quality), these corresponded in scope to many modern tropical 'research' floras, with detailed descriptions, elaborate synonymies and infraspecific treatments, more or less extensive chorological information, and taxonomic and other commentary. Examples included *Synopsis der mitteleuropäischen Flora* by Paul Ascherson and Carl Graebner, *Conspectus florae graecae* by Halácsy, *Flore de France* by Rouy and Foucauld, the *Cambridge British Flora* by Moss, *De flora van Nederland* by Heukels, *Prodrome de la flore belge* by de Wildeman and Théophile Durand, *Prodromus florae hispanicae* by Willkomm and Lange, and *Flore des Alpes-Maritimes* by Burnat and Briquet. The most ambitious project, differing from the others by its greater engagement of specialists, was Hegi's *Illustrierte Flora von Mitteleuropa*, almost the last to be begun before the outbreak of World War I. While some of these works were the product in whole or in part of national herbaria, others, probably the majority, were sponsored privately (like Burnat's *Flore des Alpes-Maritimes*) or through non-governmental institutions. Certain works, like those for Spain, Greece and parts of the Balkans, were in effect 'colonial' floras, although in Spain the local botanists of the period, particularly Miguel Colmeiro, Mariano del Amo y Mora and Blas Lázaro e Ibiza, also produced major compilations. A similar development, mostly sponsored through state bodies, began in the 1890s in the Russian Empire, of which the leading contribution for Europe was *Flora evropejskoj Rossii* by B. A. Fedtschenko and A. Flerow (1908–10). This, along with contributions for extra-European areas, was later to provide valuable source material for *Flora SSSR*.

The era also witnessed renewed attempts at pan-European works. These included *Sylloge florae Europaeae* (1854–55; supplement, 1865) and *Conspectus florae europaeae* (1878–82, with supplements to 1890) by Carl Nyman, *Plantae Europeae* (1890–1903, not completed) by Karl Richter, the bizarre *Flora Europae terra-rumque adjacentium* (1883–91) and *Novus conspectus florae Europae* (1910), both by Michel Gandoger, and, for the traveler, *Exkursionsflora von Europa* (1901; supplement, 1918) by Franz Thonner. Nyman's *Conspectus* remained of lasting worth, and in the mid-twentieth century it became a major source for *Flora Europaea*.

The travel revolution of the mid-nineteenth century and the demands of an interested public for works in 'national' languages resulted in another significant development: the 'excursion' flora. In Europe, these have taken the form of analytical keys incorporating diagnoses and related information about any given plant but lacking separate descriptions. Evidently inspired in the first instance by the analytical keys in *Flore française* of Lamarck and de Candolle (1805), this generally condensed, purely practical type of work spread through Europe in train, so to speak, with Thomas Cook, although in some countries with large floras such manuals were sometimes abridged. The greatest number were produced in western, northern and central Europe. While in general only state-wide works are accounted for in this *Guide*, many others were also written for internal administrative units or for areas of special scientific or other (e.g., touristic) interest, such as the Alps. The manual-key 'excursion flora' was to migrate to other parts of the world and its forms sometimes used for floras of a higher order, but as a style it has but rarely been adopted in North America. In some manuals, however, descriptions and keys have continued to be discrete; examples include George Bentham's *Handbook of the British flora* (first published in 1858) and Clive Stace's *New flora of the British Isles* (1991; revised edn. 1997). In Europe, these works have been more or less frequently reissued and/or revised and occasionally entirely rewritten, but by and large they do not represent 'creative' work; for that reason they have by some been dubbed 'routine' floras.

The 1914–18 war and the changes brought by the rise of ecology, genetics and cytology as well as the relative decline of taxonomy and floristics within botany and general shifts in fashion brought the European floristic 'boom' to a close, although production of a few of the larger works lingered on afterwards for varying lengths of time, and one new project, *Flora polska* (with associated atlas) began publication in 1919. In the newly organized Soviet Union, however, after the dislocations caused by the war and the revolution, large-scale floristic work revived strongly during the interwar period. At the end of the 1920s, the complete reorganization of social, cultural and scientific life and the formulation of national priorities led in 1931, following preliminary planning, to the initiation of *Flora SSSR*. Publication commenced in 1934. As the new All-Union flora began to appear, major republican and other 'regional' floras were commenced, continued or revised. Contemporaneously in Germany, the 'second edition' of Hegi's *Illustrierte Flora*, eventually even more detailed than its predecessor, was begun; its preparation and publication have continued off and on to the present.[3]

Almost all activity ceased or was redirected during World War II, although Rechinger's key *Flora aegaea* appeared in 1943 and heroic efforts were made to keep *Flora SSSR* going. Professional botanists who were preparing tropical floras became involved in assignments involving European plants, and it gradually became perceived how little was really known when a more comprehensive view was taken of the European flora. In the preceding decade or two, the needs of the developing subdisciplines of 'biosystematics' had also begun to have a 'recycling' effect with demands for more and better overall taxonomic accounts. The war more or less blew open the tight 'national' schools of floristic botany which had been largely dominant from the late eighteenth and early nineteenth centuries. Thinking was forced into new directions, which included consideration of the possibilities of collaborative floras covering large areas, like the tropical in-house 'colonial' floras of the large western European or North American herbaria or *Flora SSSR* (with the latter exerting a very strong influence, whatever the merits or demerits of the Komarovian species concept).

The successes of the Third Reich in the early part of World War II, along with the visions of its *Führer*, expanded German horizons, and in 1943 Werner Rothmaler – whose work in Portugal before the war doubtless had enlarged his own *Weltanschauung* – conceived a scheme for a collective *Flora Europaea* which would encompass not only the continent but also the Ural, the Caucasus, the Levant, North Africa, and Macaronesia. This was in its original form made abortive by the ultimate outcome of the war and the disintegration of the Central European botanical profession along with the destruction of herbaria in Berlin and elsewhere. Nevertheless, a 'seed' had been sown, although 10 years would pass before 'germination'.

After the war, there was a renewal in the preparation of floras as well as in other systematic studies,

surely in part connected with social and cultural reconstruction. Major projects, some including distribution maps, were initiated in Scandinavia, Hungary, Romania, and (later) Serbia, Bulgaria, Switzerland and France, while others, such as *Flora SSSR*, Hegi's *Illustrierte Flora*, and *Flora polska* were continued. For Great Britain, Ireland and associated islands, the first wholly new flora since the nineteenth century appeared in 1952. There was also a new round of new or revised 'routine' manuals, partly promoted by the expanding secondary and higher education market and in later years by the increasing scientific and public interest in ecology and the environment.

For a large number of professional botanists, who now once more played a significant role in European floristic botany, this, however, was not enough. Inspired by Rothmaler's proposals, which after 1945 became more widely circulated, lobbying for further integration continued and in 1954 there was mounted at the Eighth International Botanical Congress at Paris a session, 'Progress of Work on the European Flora', which reportedly closed on a somber, inconclusive note apart from sonorous words about the great desirability of a general Flora Europaea. Like the challenge of *Flora SSSR* thrust upon the botanists in Soviet-era Leningrad, there was a feeling of doubt and fear in the face of such a project; it was still spoken of as a nebulous future ideal.

However, while big meetings may pass resolutions and appoint committees, it is often small, relatively homogeneous collectives of cadres which may actually initiate and push through major projects. The present *Flora Europaea* is said to have been born when a group of botanists, mainly British and, moreover, all or most of them once students at Cambridge University, met at a café opposite the Sorbonne after the inconclusive Congress session and 'continued the discussion with the aid of appropriate refreshment'. This was in July 1954. By the end of 1955 the relatively radical basic principles had been evolved by this group and the first formal meeting of the Editorial Committee was held in January 1956 at Leicester. Twenty-one years later, at the final Flora Europaea Conference in Cambridge, the work was declared completed. A total of 187 authors had been involved from all over Europe and overseas, and upwards of £200000 expended on a project which dominated European botany for nearly a generation. Yet, in contrast to too many tropical floras, *Flora Europaea*, although taking twice as long as originally projected, was completed within one set of working lifetimes. The first volume was published in 1964, and the fifth and final volume appeared in 1980. A revision of at least part of the work was commenced at Reading University during the 1980s; volume 1 appeared in 1993. The Flora Europaea Trust, however, has since determined not to proceed *per se* with the remaining volumes but to support further development of an electronically based information system (from which printed volumes could be derived), an objective in fact in place since the late 1970s as discussed below.

Although some controversy has surrounded *Flora Europaea*, great benefits, both scientific and practical but too numerous to detail herein, have accrued from this vast undertaking. Already in 1962 Gérard Aymonin had hinted that more emphasis should be placed on southern Europe and the Mediterranean; his argument was only reinforced as the *Flora* project brought out the deficencies in floristic knowledge over most parts of that area. Greater interest also developed in floristic mapping, and from experience in Britain and Ireland, Scandinavia and Finland, and Central Europe, the *Atlas florae europaeae* project was mounted in the 1960s and has since then progressed gradually, with at this writing 12 parts published (though with still a long way to go). As expositions of flora-methodology, the 'Green Books' (1958, 1960) published by the Flora Europaea Organisation also became very influential.[4]

In an age when future planning and more or less rigid 'project development' have become all but necessities, various ideas have been put forward as to the directions European floristic botany should pursue. Some authors, in recommending additional research, have suggested new directions including studies in groups of key scientific and practical importance, but also stressed the need for improvements in the quality and meaning of, as well as access to, old and new information. The need for keeping in motion the synthetic temper engendered by *Flora Europaea*, including the mapping of species, the increase in and promotion of integrated 'biosystematic' and 'ecosystematic' studies, investigations into the status of and changes to the flora, and the maintenance of control over the flow of information, has likewise been stressed. Others suggest a greater selectivity in priorities, with an increased emphasis on assistance to, and cooperation with, botanists concerned with the plant life of the Mediterranean and the tropics who all too frequently lack adequate support and may be too few to cope

effectively with rich, threatened and often poorly known floras. The style and content of floras and manuals are said also to require review, with the additional point that in view of changes in the flora much of the stock of existing works must be viewed as largely obsolete in what they cover as well as in their method of treatment of the material. The determinism and 'finality' inherent in much floristic writing before 1914, including the detailed large-scale works, are, or should be, philosophically and psychologically past.

In recent years, one formal review has been made: by the 'Ad Hoc Group on Biological Recording, Systematics and Taxonomy' of the European Science Research Councils (ESRC), constituted in 1975 and acting with support from the European Science Foundation (ESF) through its 'Additional Activity in Taxonomy'. This group made its first reports in 1977. In botany, four priorities for European floristics were recognized: (1) a European floristic information system (as a partial means of carrying on the work of *Flora Europaea*); (2) coordination in 'biosystematic' research and studies of critical groups; (3) extension of the work of the Threatened Plants Committee of IUCN; and (4) a flora of the Mediterranean Basin. Subsequent work has focused on amateur workers, the education of systematists and the extent of involvement of European botanists with tropical floras.

Partly because of the wide agreement amongst the European botanical community on the need for more investigation of the flora of southern Europe and the Mediterranean, significant progress was already being made with the fourth of the four priorities. Existing knowledge, including literature surveys, and needs with respect to the Mediterranean flora were surveyed at a conference at Montpellier in 1974, and in 1975 the 'Organization for the Phyto-Taxonomic Investigation of the Mediterranean Area' (OPTIMA) came into being. Currently based in Berlin, the OPTIMA Secretariat publishes a semi-annual newsletter along the lines of the *Flora Malesiana Bulletin* and circulates collections of separates. It early became accepted that realization of a descriptive flora for the whole of the Mediterranean Basin was in the nearer term not practicable and attention turned to the preparation of a critical enumeration. This became the *Med-Checklist*; the first formal meeting of its steering committee took place in December 1978. By 1989 three handsomely produced volumes of a projected total of six had been published (although none since).

Discussions also took place with respect to the European floristic information system concept, and in 1979 a proposal to the ESF was approved. Work on this (as the European Floristic, Taxonomic and Biosystematic Information System or European Documentation System, shortened here as ESFEDS), an amalgam of the first and second priorities of the ESRC Taxonomy Group, was formally initiated late in 1981. Later in the decade, however, the Group was disbanded and the ESFEDS initiative expired without corporatization. The computer files of the ESFEDS are held at the Royal Botanic Garden Edinburgh, and have been made available on the World Wide Web.[5]

Thus the trend towards interaction and cooperation amongst botanists in Europe, which began after World War II and was greatly accentuated by *Flora Europaea*, is likely to, and should, continue in so far as existing political, ideological and financial frameworks allow. Proper assessment of priorities within generally limited means will be a key feature, associated with greater formalization of supranational cooperation. Progress is likely to be relatively steady and unspectacular, with perhaps the most considerable developments in the southern fringes and in parts of adjacent continents. There also remains the hope that the current Euro+Med PlantBase initiative, seen as a successor to *Flora Europaea* as well as a lasting and effective information system, will one day be fully realized.[6]

Progress

No comprehensive or comparative historical accounts have been seen, but a sketchy general survey is given by V. H. HEYWOOD, 1978. European floristics: past, present and future. In H. E. Street (ed.), *Essays in plant taxonomy*, pp. 275–289. London. British contributions to overall study of the European flora are reviewed by W. T. STEARN, 1975. History of the British contribution to the study of the European flora. In S. M. Walters and C. J. King (eds.), *Floristic and taxonomic studies in Europe*, pp. 1–17 (BSBI Conf. Rep. 15). Abingdon, England. Many national histories are also available. Other reviews of European floristic botany include G. AYMONIN, 1962. Où en sont les flores européennes? *Adansonia*, II, **2**: 159–171; EUROPEAN SCIENCE FOUNDATION, 1977. *ESF Annual Report 1977*, pp. 78–82 and Annex 2. Strasbourg; and B. JONSELL, 1979. The present stage of botanical exploration: Europe. In I. Hedberg (ed.), *Systematic botany, plant utilization and biosphere conservation*, pp. 34–40. Stockholm.

Flora Europaea: The genesis, development and execution of this landmark have been described in many papers. The mid-twentieth century setting and contemporary German proposals are given in W. ROTHMALER, 1944. Aufforderung zur Mitarbeit an einer Flora von Europa. *Repert. Spec. Nov. Regni Veg. [Fedde]* **53**: 254–270. Developments before and after the Paris Botanical Congress are traced in H. GAMS, 1954. Flores européennes modernes: résultats acquis, objectifs à atteindre. In Huitième Congrès International de Botanique, *Rapports et communications avant le Congrès*, sects. 2, 4, 5 & 6, p. 101. Paris; and D. H. VALENTINE, 1954. Progress of work on the European flora. *Ibid.*, pp. 87–92. Early developments in the *Flora Europaea* project are surveyed by V. H. HEYWOOD, 1957. A proposed flora of Europe. *Taxon* **6**: 33–42; and *idem*, 1961. The *Flora Europaea* project: a report of progress. In Ninth International Botanical Congress, *Recent advances in botany*, 1: 941–944. Toronto.[7] A 'mid-level' view was afforded by D. H. VALENTINE, 1971. Floristics in Europe. *BioScience* **21**: 512–514. 'Valedictory' papers include ANONYMOUS, 1977. Floras: [a] brief history of the *Flora Europaea* project. *Australian Syst. Bot. Soc. Newsletter* **13**: 9–12; and D. A. WEBB, 1978. *Flora Europaea* – a retrospect. *Taxon* **27**: 3–14. The publication in 1993 of volume 1 of the second edition occasioned another personal view: S. M. WALTERS, 1995. The taxonomy of European vascular plants: a review of the past half-century and the influence of the *Flora Europaea* project. *Biol. Reviews* **70**: 361–374.[8] For the two sets of country-by-country reports associated with *Flora Europaea*, presented respectively at the second and seventh Flora Europaea symposia, and the Mediterranean area country reports of 1975, see below under **Divisional bibliographies**.

Other projects: The ESFEDS was described in some detail in V. H. HEYWOOD and L. N. DERRICK, 1984. The European Taxonomic, Floristic and Biosystematic Information System. *Norrlinia* **2**: 33–54; a further summary is in V. H. HEYWOOD, 1989. *Flora Europaea* and the European Documentation System. In N. R. Morin *et al.* (eds.), *Floristics for the 21st century* (see **General references**), pp. 8–10. Walters (1995; see above) has reviewed this and the *Atlas florae europaeae* as well as other initiatives.

General bibliographies. Bay, 1910; Blake and Atwood, 1942 (with the omission of Germany, Austria, eastern Europe and the European part of the former

Soviet Union); Frodin, 1964, 1984; Goodale, 1879; Holden and Wycoff, 1911–14; Jackson, 1881; Pritzel, 1871–77; Rehder, 1911; Sachet and Fosberg, 1955, 1971; USDA, 1958. For national and regional bibliographies, see the appropriate headings.

Divisional bibliographies

See also **601** (CENTRE NATIONAL DE LA RECHERCHE SCIENTIFIQUE, FRANCE). Indications of works regarded as 'standard' floras are given by Lawalrée (see below) and, without annotations, in Hamann and Wagenitz (1977; see Region 64) as well as in each of the volumes of *Flora Europaea* (and its style manuals, the 'Green Books').

HEYWOOD, V. H. (ed.), 1963. A survey of taxonomic and floristic research in Europe since 1945. In V. H. Heywood and R. E. G. Pichi-Sermolli (eds.), *Proceedings of the II. Flora Europaea Symposium* (Florence, 1961), pp. 95–562 (Webbia 18). Florence. [Contains more or less complete bibliographies for every European country up to 1960 or 1961. For the period 1961–71, see Heywood, 1974–75 below.]

HEYWOOD, V. H. (coord.), 1974–75. Floristic reports, 1961–71. In *VII. Flora Europaea Symposium* (Coimbra, 1971). 2 parts. 834 pp. (Mem. Soc. Brot. 24). Coimbra. [An extension to the reports in the preceding work; covers all countries except Iceland, Norway, Poland, Portugal, Turkey-in-Europe, the USSR, and Yugoslavia. For Spain see *Boissiera* **19**: 23–60 (1971).]

LAWALRÉE, A., 1960. Indication des principaux ouvrages de floristique européenne (plantes vasculaires). *Natura Mosana* **13**(2–3): 29–68. [A list, uneven in places, of 'standard' floras with relatively complete bibliographic data. Some works on the adjacent regions of North Africa, the Caucasus, and Macaronesia are also included. Now out of date.]

General indices. BA, 1926– ; BotA, 1918–26; BC, 1879–1944; BS, 1940– ; CSP, 1800–1900; EB, 1959–98; FB, 1931– ; ICSL, 1901–14; JBJ, 1873–1939; KR, 1971– ; NN, 1879–1943; RŽ, 1954– .

Divisional indices

The question of the continued flow of information in European floristic botany consequent to the completion of the *Flora Europaea* project remains, as already noted, a matter of concern among the botanical profession in Europe. Without a dedicated mechanism the best sources are *Kew Record of Taxonomic Literature*, symposia, and informal contacts.

BRUMMITT, R. K. *et al.* (comp.), 1966–71. *Index to European taxonomic literature for 1965, 1966, 1967, 1968, 1969* (respectively in *Regnum Veg.* **45, 53, 61, 70, 80**). Utrecht; and R. K. BRUMMITT and D. H. KENT (comp.), 1977. *Index to European taxonomic literature for 1970.* Kew: Bentham-Moxon Trustees. [Classified bibliography, with a separate section for floras and related works. From 1971 incorporated into *Kew Record of Taxonomic Literature*.][9]

KENT, D. H., 1954–70. Abstracts from literature. In *Bot. Soc. British Isles, Proc.* **1–17**. [Summary accounts of new European floristic and taxonomic literature. From 1971 – coincident with the initiation of *Kew Record of Taxonomic Literature* – continued with a different emphasis as *BSBI Abstracts* (see **Region 66**) but references to significant European floristic literature are therein still furnished.]

Conspectus

---

612   Spain (in general)
613   Galicia, Asturias and Cantabria
614   Basque Lands (Vascongadas) and Navarra
615   Catalonia (Catalunya)
616   Valencia and Alicante (País Valencià)
617   Central Spain
618   Andalucía (with Gibraltar), Almería and Murcia
619   Balearic Islands
Region 62   Italian peninsula and associated islands
620   Region in general
621   Corsica
622   Sardinia
624   Lipari (Aeolian) Islands and Ustica
625   Sicily
626   Pantelleria and Isole Pelagie
627   Malta
628   San Marino
629   Istra, the Soča valley and Trieste
      (former Austrian Küstenland)
Region 63   Southeastern Europe
630   Region in general (including former Yugoslavia)
631   Croatia (Hrvatska)
632   Bosnia-Hercegovina (with former Sanjak Novipazar)
633   Yugoslavia (Serbia and Črna gora)
      I.   Vojvodina
      II.   Serbia
      III. Kosovo i Metonija
      IV.   Montenegro (Črna gora)
634   Former Yugoslav Republic of Macedonia (FYROM)
635   Albania
636   Greece (mainland and western islands)
637   Aegean Islands
      I.   Crete (with Karpathos)
      II.   Southern Sporodhes (Dodecanese)
      III. Other islands
638   Turkey-in-Europe
639   Bulgaria
Region 64   Central Europe
640   Region in general
      I.   'Mitteleuropa'
      II.   'Carpatho-Pannonia'
641   Romania
642   Hungary
643   Slovenia
644   Austria
645   Czech Republic (and former Czechoslovakia)
      I.   Czechoslovakia (1918–92)
      II.   Czcch lands (Čcské zeme)
646   Slovakia
647   Poland
648   Germany
649   Switzerland (with Liechtenstein)
Region 65   Western Europe (mainland)

651   France (including Monaco)
655   Channel Islands (Îles Normandes)
656   Belgium
657   Luxembourg
658   The Netherlands
Region 66   Western European archipelago
      (British Isles and Ireland)
660   Region in general
662   Wales
663   Scotland
664   Isle of Man
665   Ireland
666   Outer Hebrides
667   Orkney Islands
668   Shetland Islands
669   Rockall Island
Region 67   Northwestern Europe (Scandinavia, Finland and
      the Baltic States)
670   Region in general
      I.   Scandinavia and Finland
      II.   Baltic States
671   The Faeroes
672   Iceland
673   Denmark
674   Norway
675   Sweden
676   Finland
677   Estonia
678   Latvia
679   Lithuania

**Region (Superregion) 68/69   CIS-in-Europe (and works
covering the former USSR)**
680   Region in general
      I.   'All-Union' (Vsesojuz) works
      II.   Belarus', Moldova, Ukraine and Russia-in-
            Europe (CIS-in-Europe)
681   Kaliningrad Oblast' (Russia)
684   Belarus' (White Russia)
685   Greater St. Petersburg region (Russia)
686   Karelian Autonomous Republic (Russia)
687   Kola Peninsula
688   Northern Russia
689   Northern Middle Russia
691   Middle Volga Basin
692   Middle Russia
693   Moldova (Moldavia)
694   Ukraine (in general)
695   Southwestern Ukraine (including Ruthenia)
696   Crimea (Ukraine)
697   Lower Don region (Russia)
698   Southeastern Russia (and associated republics)
699   Eastern Russia (and associated republics)

600

Division in general

Area (including associated islands): *c.* 10 507 000 km². Vascular flora: 13 650 species and non-nominate subspecies accounted for in *Flora Europaea*, of which 542 are in so-called 'critical' genera (*Alchemilla*, *Hieracium*, *Rubus*, *Sorbus* and *Taraxacum*). Nyman listed 11 111 such taxa (exclusive of those in the above 'critical' genera) and Thonner gave a figure of 9876 seed plant species. The second edition of vol. 1 of *Flora Europaea* would suggest an increase of 10 percent over the number of species in the original work to be in order.

Figures for the size of the vascular flora in a given polity are taken where possible from recent floras. Otherwise, they come from tables provided by Webb in his retrospective review (see **Progress** under **Division 6** above). The larger figure in each of his pairs of figures is inclusive of naturalized and adventive species.

The conspectus of Nyman (with its supplements) has been superseded but is retained here in the absence of any modern equivalent. Also described are Thonner's *Exkursionsflora von Europa* (1901–18), reprinted in the 1980s, and *Atlas florae europaeae* of Jalas *et al.* (1972–).

Keys to families and genera

MARTENSEN, H. O. and W. PROBST, 1990. *Farn- und Samenpflanzen in Europa: mit Bestimmungsschlüsseln bis zu den Gattungen.* x, 525 pp., 51 text-figs., 233 illustrated identification tables, 21 'overview' illus. Stuttgart: Fischer. [An introduction to the morphological and other features of each major plant group (classes or, in the angiosperms, subclasses) with illustrated tabular keys to families and conventional analytical keys to genera. Generic entries include distribution and status, species numbers in Europe, and an indication of whether it is hardy (G), of economic value (K), and/or requiring protection (Z). The general part is largely pedagogical; at its end are references to sources and indices to terms and to generic (and higher) names.]

MOORE, D. M., 1982. *Flora Europaea: check-list and chromosome index.* x, 423 pp. Cambridge: Cambridge University Press.

Compiled list of all species, each with (where available) a single diploid chromosome number supported by a source country and one reference; full bibliography and index to genera and families at end of work. [As far as possible, the karyotypes included are based on verifiable material; limitations of space, however, forbade inclusion of a range of counts for any given taxon.]

NYMAN, C. F., 1878–84. *Conspectus florae europaeae* (incl. *Supplementum I*). 1046 pp. Örebrö, Sweden: Bohlin. Continued as *idem*, 1889–90. *Conspectus florae europaeae: Supplementum II.* 404 pp. Örebrö.

Concise, partly compiled enumeration of the phanerogamic flora of Europe, with synoptic keys down to groups of species; includes the more important synonymy, generalized indication of distribution, brief taxonomic commentaries and miscellaneous notes, with index at end. [For an 'unofficial' supplement, see E. ROTH, 1886 (1885). *Additamenta ad 'Conspectum Florae Europaeae'.* 46 pp. Berlin.][10]

THONNER, F., 1901. *Exkursionsflora von Europa.* x, 50, 356 pp. Berlin: Friedländer; and *idem*, 1918. *Exkursionsflora von Europa. Nachträge und Verbesserungen.* 55 pp. Berlin. (Reprinted in 1 vol. with added foreword, 1980, Leiden: Rijksherbarium.) French edn.: *idem*, 1903. *Flore analytique de l'Europe.* vi, 322 pp. Paris: Baillière.

Compact diagnostic and descriptive manual-key to the seed plant genera of Europe, with indication of geographical ranges of those more limited in range along with some synonymy and including German vernacular names. The introductory section includes a definition of geographical limits of the work along with its scope (which encompasses native, naturalized, adventive and/or extensively cultivated taxa), family and generic circumscriptions (after *Die natürlichen Pflanzenfamilien*), the method of key construction, and sources, while at the end are a glossary, a list of author abbreviations, and a full index. [In the *Nachträge* (pp. 52–53) Thonner gave figures for the total European flora: 9876 species of seed plants (9740 native) in 1264 genera (1164 native) and 143 families (31 native). The German and French editions are almost identical, save that no French version of the supplement appears to have been published.][11]

TUTIN, T. G. *et al.* (eds.), 1964–80. *Flora Europaea.* 5 vols. Maps. Cambridge: Cambridge University Press. Accompanied by G. HALLIDAY and M. BEADLE, 1983. *Flora Europaea: consolidated index.* 210 pp. Cambridge. Partly revised as T. G. TUTIN *et al.* (eds.; assisted by J. R. Akeroyd and M. E. Newton),

1993. *Flora Europaea.* 2nd edn. Vol. 1 (Psilotaceae–Platanaceae). xlvi, 581 pp. Cambridge.

Briefly descriptive, synthetic flora of native, naturalized and commonly cultivated vascular plants of Europe; includes keys to all taxa down to subspecies, references for accepted names, limited synonymy, concise indication of distribution (both in general and by country, with some extra-European ranges also given where relevant), indication of habitat, karyotypes (where known), and, for particular genera and families, critical taxonomic commentary; complete index in each volume (with cumulative index to genera and families in vol. 5). The introductory sections (in each volume) include a discussion of the scope and aims of the *Flora*, lists of basic and standard national and regional floras, citations of serials and other important literature, and a glossary of authors of plant names. The *Consolidated index* is comprehensive, with all accepted names also furnished with their numbers as used in the *Flora* proper. The second edition of volume 1 adds 350 taxa, the majority new to science, and includes revised appendices.[12]

Distribution maps

JALAS, J., J. SUOMINEN, R. LAMPINEN and A. KURTTO (eds.), 1972– . *Atlas florae europaeae.* Parts 1– . Maps [i–iv], 1– . Helsinki: Akateeminen Kirjakauppa (Academic Bookstore). (Parts 1–7 reissued in 3 vols., 1987, Cambridge: Cambridge University Press, with new preface by D. A. Webb.)

An annotated atlas of distribution maps of the European flora, based on some 4400 fifty-kilometer squares. For any given species or recognized infraspecific taxon, there is for each square an indication of occurrence, dates of records (before 1900, between 1900 and 1939, and since 1939), and status; the accompanying text gives the correct botanical name (usually that adopted for *Flora Europaea*), references to other published maps or relevant distribution reports, and occasional taxonomic and nomenclatural commentary (with references). The introductory section (part 1) gives remarks on the history of the project and the methodology pursued, as well as a list of collaborators. The sequence of families generally follows *Flora Europaea* as also do circumscriptions of species and botanical names, any exceptions being duly noted. [As of 1999, 12 parts (the latest in 1999) have been published, comprising maps 1–3270 with coverage corresponding to vol. 1 of the *Flora* proper (pteridophytes, gymnosperms, and 'early' dicotyledons through Platanaceae).]

Woody plants (including trees)

For the manuals by Humphries *et al.*, Press, Mitchell and Wilkinson, and Sutton, see **601/II**.

'Tree books' have a long history in Europe, particularly north of the Mediterranean, and moreover it was uniquely in the continent that a tradition of comprehensive documentation of known temperate woody plants became established, encompassing cultivated as well as native species. The earliest supranational work is *Dendrologia Europeae mediae* (1763) by J. J. Ott, while major landmarks of the first two-thirds or so of the nineteenth century are *Arboretum et fruticetum britannicum* by John Claudius Loudon (1835–38; supplement, 1850; abridged as *An encyclopaedia of trees and shrubs*, 1842, with reissues to 1883) and *Dendrologie* by Karl H. E. Koch (1869–73). National and more local works also began to appear in some numbers in the last third of the century and beyond, but for some time most remained relatively technical.

A great impetus to dendrology was given by the many discoveries in the latter part of the nineteenth century of trees and shrubs in eastern Asia new to science. This became manifest in such works as *Handbuch der Laubholzkunde* (1889–93) by Leopold Dippel, *Traité des arbres et arbrisseaux* by Pierre Mouillefert (1891–98), *Trees and shrubs of Great Britain and Ireland* by Henry J. Elwes and Augustine Henry (1906–13), *Illustrierte Handbuch der Laubholzkunde* by Camillo Schneider (1906–12), and Kew curator W. J. Bean's *Trees and shrubs hardy in the British Isles* (1914; 6th edn., 1936). Alfred Rehder continued the German tradition in North America with *Manual of cultivated trees and shrubs* (1927; 2nd edn., 1940), still a standard work.

After World War II, S. Ja. Sokolov undertook the six-volume *Trees and shrubs of the USSR* (1949–62; supplement, 1965), and soon after there appeared Gerd Krüssmann's *Handbuch der Laubgehölze* (1960–62; 2nd edn., 1976–78) and *Handbuch der Nadelgehölze* (1970–71; 2nd edn., 1983). Krüssmann's works, both also available in English, represent the nearest contemporary approach to a universal temperate woody flora. Two further editions of Bean's *Trees and shrubs* were also published, the most recent (the eighth) in 1970–80 with a supplement in 1988. There was also a great growth in popular and semi-popular guides; four have been selected for description below: *Die Bäume Europas* (1968; 2nd edn., 1979) by Krüssmann, *Trees and bushes of Europe* by Oleg Polunin with Barbara Evrard (1976), *The Hamlyn guide to trees of Britain and Europe* by C. J. Humphries, J. R. Press and D. A. Sutton (first published in 1981), and *Field guide to the trees of Britain and Europe* by D. A. Sutton (1990). The two middle titles have been translated into a number of European languages. A fifth title [not seen] is M. GOLDSTEIN, G. SIMONETTI and M. WATSCHINGER, 1987. *Guida al riconoscimento degli alberi d'Europa.* Milan: Arnoldo Mondadori. In comparison with North American guides, however, it has in Europe been more usual to include a substantial number of cultivated species.

In addition to the titles given below, others appear

respectively under **601/I** (southern Europe and the Mediterranean) and **601/II** (central and northern Europe). Many regional and national 'tree books' and woody floras are also available, and a number of these have been accounted for in the *Guide*. No overall listing *per se* has, however, been published.

HUMPHRIES, C. J., J. R. PRESS and D. A. SUTTON, 1992. *The Hamlyn guide to trees of Britain and Europe*. 320 pp., 1000+ col. illus. London: Hamlyn. (Hamlyn Guides.) (Original edn., 1981. French edn., 1982, as *Tous les arbres d'Europe*. 320 pp., illus. Bordas; editions in other languages also available.)

A field-guide in atlas format with illustrations and facing text; the latter features scientific and vernacular names and running accounts of individual species including botanical details, phenology, distribution (in general and by two-letter codes as in *Flora Europaea*), dates of introduction (if relevant), uses, special features, and notes on cultivars, similar species, etc.; key references (p. 310 but not revised since 1981), glossary and index to all names at end. Illustrations include habits as well as foliage sprays with flowers and fruit. The introductory part covers organography, systematics and taxonomy, and hybrids and is followed (pp. 14–33) by a general key to genera and species. [Accounts for about 400 species; arrangement follows *Flora Europaea*.]

KRÜSSMANN, G., 1979. *Die Bäume Europas*. 2. Aufl. 172 pp., 405 figs. (on 88 pls., 8 col.), 127 maps. Berlin: Parey. (1st edn., 1968.)

Profusely illustrated, popularly oriented pocket-sized manual of native, naturalized and commonly cultivated trees of Europe (154 angiospermous, 59 gymnospermous species), with tabular pictured-keys to species, limited synonymy, vernacular names (in German, French, Italian and English), generalized indication of range (with maps), brief notes on key distinguishing features, phenology, habitat, etc., and indices. The introductory section includes a terminological table and notes on use of the work. A list of key references is also included. [This second edition accounts more fully for Mediterranean trees than its predecessor. The photographs, though numerous, are regrettably murky.]

POLUNIN, O., 1976. *Trees and bushes of Europe*. xvi, 208 pp., col. illus. (in part by B. Everard). London: Oxford University Press. [Also in French as *Arbres et arbustes d'Europe*; in German as *Bäume und Sträucher Europas*; in Italian as *Guida agli alberi e arbusti d'Europa*; in Spanish as *Árboles y arbustos de Europa*.]

Copiously illustrated, systematically arranged popular color atlas-guide to more commonly encountered woody plants 2 m in height or over, with photographs and/or drawings accompanied by descriptive and distributional notes and vernacular names in principal western European languages; related, less common species are briefly mentioned. An introductory section and index are also included, but no bibliography or reading list.

SUTTON, D. A., 1990. *Field guide to the trees of Britain and Europe*. 192 pp., illus., map. London: Kingfisher.

A copiously illustrated color guide accounting in all for 430 species; the paintings are accompanied by descriptive and distributional notes encompassing one species per page with pointers to others. The species are color-coded by 'class' accessed by a somewhat awkwardly arranged opening key. A short reading list and sources of further information conclude the work.

Pteridophytes

All ferns and fern-allies are also covered in vol. 1 of the second edition of *Flora Europaea* (see above). For *Die Farnpflanzen Zentraleuropas* by K. Rasbach, H. Rasbach and O. Wilmanns, see **640**.

DERRICK, L. N., A. C. JERMY and A. M. PAUL, 1987. *Checklist of European pteridophytes*. xx, 94 pp. (Sommerfeltia 6). Oslo.

Alphabetically arranged checklist with synonymy, places of publication, and distribution within and without Europe; index to all families and genera at end (pp. 69–94). The introductory part encompasses taxonomic and nomenclatural notes (pp. viii–xiii), an explanation of the work with conventions, acknowledgments, a synoptic list of families and genera, a list of geographical codes, and key sources. [Based on the ESFEDS database.]

GAMS, H., 1995. *Kleine Kryptogamenflora*, IV: *Die Moos- und Farnpflanzen Europas*. 6th edn., revised by W. Frey, J.-P. Frahm, E. Fischer and W. Lobin. xi, 426 pp., 149 pls. Stuttgart/Jena: Fischer.

Manual-key to bryophytes and pteridophytes (the latter by Fischer and Lobin on pp. 319–391 with 29 plates) featuring diagnostic leads accounting for status (showing an evident decline for many species since 1945!), essential synonymy, and overall distribution but no vernacular names. A number of additional species are listed but not keyed out (being naturalized or adventive or known only from Macaronesia or other bordering areas or not generally recognized 'critical' taxa). At the end of the work are key references, sources of figures, a glossary, and (pp. 401–426) an index to all botanical names.

601

Major continental subdivisions

Included here are works for the Mediterranean Basin and for central and northern Europe.

I. Mediterranean Basin

The Mediterranean Basin, with its distinctive flora, occupies parts of three continents but as the largest part of its land area lies in Europe it is accounted for here. The vascular plant flora has been estimated at 24000 ± 600 species (59.1 percent endemic) with a further 5700 ± 700 additional infraspecific taxa.[13]

Until the appearance of the first volume of *Med-Checklist* in 1984, no overall floras or enumerations had been published. Interest in the region among European botanists had, however, noticeably increased from the 1960s and with it came attempts at a greater coordination of activities. A general conference at Montpellier in June 1974, sponsored by the French Centre National de la Recherche Scientifique, led to the formation of the Organization for the Phyto-Taxonomic Investigation of the Mediterranean Area (OPTIMA), a body resembling the Flora Malesiana Foundation (see Superregion 91–93) and Organization for Flora Neotropica (see Division 3) but differing from both in structure and activities. Its secretariat is currently in Berlin, at the Botanical Garden and Museum. Publications include a *Newsletter* (1975–), currently (1999) produced in Madrid, and a series of recirculated reprints, *Cahiers OPTIMA*.[14]

One of the major objectives of OPTIMA has been the preparation of a checklist of the Mediterranean flora. This approach was considered to be more feasible than any form of more comprehensive work given available and likely resources. An organization for the projected *Med-Checklist* was accordingly created, with three centers: Berlin, the Institut de Botanique, USTL, Montpellier, and the Conservatoire et Jardin Botaniques in Geneva. An electronic database was developed, a basic bibliography (see below) produced in 1979 and a preliminary 'beta' version of the pteridophytes appeared in 1981. Three years later the first definitive volume of the *Med-Checklist* was published; plans presently call for six volumes, of which by 1998 three had appeared, covering 44.5 percent of the vascular flora.[15]

Bibliographies/Progress reports

In addition to the works considered below, there has also been circulated a *Liste codée des flores de base* (1979, Geneva, as *Med-Check List Document* 7), with 66 references (most also in this *Guide*). It has been reproduced in the *Med-Checklist* volumes as appendix 1.

HEYWOOD, V. H. (coord.), 1975. Données disponibles et lacunes de la connaissance floristique des pays méditerrannéens. In Centre National de la Recherche Scientifique (France), *La flore du bassin méditerranéen: essai de systématique synthétique* (Montpellier, 1974), pp. 15–142 (Colloques internationaux du CNRS 235). Paris. [Country-by-country surveys, with lists of references. For extensions of these reports up to 1977 (as contributed to the second OPTIMA congress in Florence), see S. CASTROVIEJO, 1979. Synthèse de progrès dans le domaine de la recherche floristique et littérature sur la flore de la région méditerranéenne. *Webbia* **34**: 117–131.][16]

GREUTER, W., H. M. BURDET and G. LONG (eds.), 1984– . *Med-Checklist*. Vols. 1, 3 and 4. Maps (in inside covers). Geneva: Conservatoire et Jardin Botaniques; Berlin-Dahlem: Sécretariat Med-Checklist, Botanischer Garten und Botanisches Museum.

Tabular checklist of vascular plants, the pages in landscape format at 90°; each entry includes a code, synonymy, status (with an 'E' indicating the plant as endemic), and presence or absence in one or more of 27 geographical units (with extensive supplementary notes). Family limits follow the modified Englerian system as used in *Flora Europaea*, but within the major groups of pteridophytes, gymnosperms, dicotyledons and monocotyledons the families, genera and species are for the most part alphabetically arranged. A loose plastic leaf serves as a guide to symbols and abbreviations. [Three of the planned six volumes have been published (the most recent in 1989), covering the non-flowering plants as well as dicotyledons from Acanthaceae through Rhamnaceae (except Compositae). Those on Compositae (volume 2), the remaining dicotyledons (volume 5) and the monocotyledons (volume 6) remain to be realized. The treatment of the many infraspecific taxa is quite thorough.][17]

Distribution maps

DAGET, P. (ed.), 1980. *Atlas d'aréologie périméditerranéenne*. Fasc. 1. 412 maps on 206 pp. (Naturalia Monspeliensia, *hors série*). Montpellier.

Distributional atlas of selected species, with two maps on each page; synopses of contents arranged according to the Emberger system (pp. 9–16) and alphabetically (pp. 18–22). A bibliography is given on pp. 5–8, following the plan of the work. [No additional fascicles have been published.]

Woody plants

The following was the first modern illustrated guide to woody plants of the Mediterranean area as a whole.

GÖTZ, E., 1975. *Die Gehölze der Mittelmeerländer: ein Bestimmungsbuch nach Blattmerkmalen.* 114 pp., 577 text-figs., map. Stuttgart: Ulmer.

Comprises an illustrated artificial key to woody plants of the Mediterranean Basin, based on vegetative features, especially leaves, with German vernacular names and countries of occurrence indicated for each species. Indices as well as a short list of references for individual countries are also included.

II. Central and northern Europe

Under this heading are given Friedrich Hermann's *Flora von Nord- und Mitteleuropa* (1956) and (for trees) a selection of recent field-guides.

HERMANN, F., 1956. *Flora von Nord- und Mitteleuropa.* xi, 1134 pp. Stuttgart: Fischer. (Originally published 1912, Leipzig, as *Flora von Deutschland und Fennoskandinavien sowie von Island und Spitzbergen.*)

Concise manual-key to vascular plants, with limited synonymy, relatively extensive general indication of internal distribution (including altitudes), summary of extralimital range or source area, and symbolic designation of life-form, etc.; index to all botanical names at end. The introductory section contains among other matters a synopsis of orders and families. [Does not include the Iberian Peninsula, western France, Great Britain and Ireland, most of Italy, southeastern Europe, or the European part of the USSR.][18]

Woody plants
In addition to the title described below, the following, accounting for some 600 native and introduced species, is also available: A. MITCHELL and J. WILKINSON, 1982. *The trees of Britain and northern Europe.* 288 pp., illus. London: Collins. (French edn., 1991, as *Les arbres de France et d'Europe occidentale.* 272 pp., illus. Arthaud.)

MITCHELL, A., 1978. *A field guide to the trees of Britain and northern Europe.* 2nd edn. 415 pp., 640 text-figs., 40 col. pls. London: Collins. (1st edn., 1974.)

Illustrated descriptive semi-popular treatment; includes keys to all species, critical notes, vernacular names, and extensive commentary on special features, biology, phenology, properties, uses, occurrence, notable individual examples, etc. and concluding index. Many naturalized or cultivated species are included, either as full entries or as notations. Forestry aspects are also discussed. [Most useful in Great Britain and Ireland and from northern France through the north European plain to Scandinavia; not, however, an efficiently organized work.]

PRESS, J. R., 1992. *Bob Press's field guide to the trees of Britain and Europe.* 247 pp., text-figs., col. photographs London: New Holland.

A field-guide in atlas format, similar to the preceding in arrangement and style; it includes, however, appendices on field equipment, conservation, useful organizations (without, however, mention of the International Dendrology Society) and arboreta, a list of references (all in English), and indices. The introductory part is similar to that in *The Hamlyn guide to trees of Britain and Europe* (see **600**) but additionally features a general consideration of the distribution and past history of the tree flora as well as a leaf key. [The work is of most value in continental and insular northern Europe.]

602

'Old' upland and montane regions

See also **603** for other works. The focus here is on geologically ancient mountain systems, notably the Ural and the mountains of Fennoscandia, Britain and Ireland.

I. The Ural and the mountains of northern Fennoscandia

See also **Region 67** (most works); **710** (KRYLOV); **699** (KUČEROV); **064** (IGOSHINA). The mountains of both northern Fennoscandia and the Ural are geologically ancient, with the latter being also, so to speak, a 'state of mind': more than a long chain of mountains, the Ural are the traditional dividing line between Europe and Asia.

The higher parts of the Ural and Fennoscandian mountain systems are characterized by a flora which, like Mt. Washington in North America, has been dubbed 'arctalpine', notably by Russian and other CIS authors. Works limited to these mainly treeless areas are considered under **Region 06**; of particular note are those by H. Roweck (1981) and Örjan Nilsson (1986) for Fennoscandia (**061**) and Kapitolina N. Igoshina (1966, 1969) for the Ural (**064**). The Ural as a whole has by some authors – among them the agrostologist Nikolai Tzvelev – been accorded independence as a floristic unit (TZVELEV, 1976; see **680** under KOMAROV); such a step, however, has not been essayed in more general schemes such as, for example, that of Takhtajan.[19]

A summary work on the Ural appeared a year following Sergei Korshinsky's *Tentamen florae Rossiae orientalis* (1898; see **699**): *Die Flora des Urals, gouvernment Perm, Ufa und Orenburg* by Bolesław Hryniewiecki (in *Sitzungsber. Naturf. Ges. Dorpat* **12**(1): 90–124 (1899)). It was succeeded by *Konspekt flory Urala v predelakh Permskoj gubernii* (*Florae uralensis in finibus provinciae Permensis conspectus*) (1912) by P. V. Siuzev (Ssüsew), an annotated checklist and floristic analysis covering an area of some 332 000 km^2 in the present *oblasti* of Perm' and Ekaterinburg. Current coverage of the Ural is embodied in the works described below: the outdated manual of Govoruchin (1937), covering all the range; two partial works: Vakar (1964) for more common plants of the central and southern Ural and Mamaev (1965) for woody plants; and the manual for the Central Ural by Gorchakovskij (1994). All these latter have keys and illustrations.

GOVORUCHIN, V. S., 1937. Флора Урала/*Flora Urala: flora medio, boreali et polari uralensis*. 536 pp., 164 text-figs., col. pl., portr., folding map. Sverdlovsk [Ekaterinburg]: Oblast' Publishing Service.

Concise manual of vascular plants (1574 species) of the entire Ural (apart from the steppe region in the south), with keys to all taxa, limited synonymy, vernacular names, and brief notes on distribution, habitat, phenology, etc.; indices to generic, family and vernacular names at end. The introductory section gives accounts of floristic regions and botanical exploration as well as aids for the use of the work.

Partial work

GORCHAKOVSKIJ, P. L. *et al.*, 1994. Определитель сосудистых растений Среднего Урала (*Opredelitel' sosudistykh rastenij Srednego Urala*). 526 pp., 39 text-figs., map. Moscow: 'Nauka'.

Manual-key to vascular plants, the ultimate leads with diagnostic descriptions and notes on habitat, height and phenology; indices to Russian and botanical names at end. The brief introductory section includes background, abbreviations, and a summary account of floristic districts (with map, p. 6). The figures are of the species most characteristic of the Ural. [Accounts for 2000 species in Perm' and Ekaterinburg *oblasti*. Ekaterinburg is near the southern limit of the flora area, Perm' in the central-west.]

Herbaceous plants

VAKAR, B. A., 1964. Определитель растений Урала (*Opredelitel' rastenij Urala*). 2nd edn. 416 pp., 96 text-figs. Sverdlovsk [Ekaterinburg]: Middle Urals Book Publishing Service. (1st edn., 1961.)

Popularly oriented field-guide to the vascular flora of the southern Ural; includes Russian binomial and vernacular names, diagnostic keys, brief general indication of local range, notes on habitat, uses, special features, etc., and figures of representative species, with indices to genera and families and to Russian binomial names at the end. The introductory section is pedagogical. Covers some 640 herbaceous and sub-woody species which comprise those more likely to be encountered. For trees and shrubs, the complementary work by Mamaev described below should be consulted. [Description in part based on information supplied by M. E. Kirpicznikov, St. Petersburg.]

Woody plants

MAMAEV, S. A., 1965. Определитель деревьев и кустарников Урала (*Opredelitel' derev'ev i kustarnikov Urala*). 116 pp., 35 text-figs., map (Trudy Inst. Biol. Akad. Nauk SSSR, Ural'sk. Fil. 41). Sverdlovsk [Ekaterinburg].

Illustrated manual of the spontaneous and commonly cultivated trees and shrubs of (mainly) the central and southern Ural (194 species); includes diagnostic keys to all species, Russian binomial and vernacular names, general indication of local range or country of origin, and notes on habitat, along with a list of references, aids for the use of the work, and an index to botanical names. The keys are wholly artificial, being arranged in six primary tables (as in Krüssman's *Die Bäume Europas*; see **600**).

II. Mountains of Britain and Ireland

WILLIAMS, F. N., 1908–10. The high alpine flora of Britain. *Ann. Scottish Nat. Hist.* (1908): 163–169, 242–251; (1909): 30–36, 108–114, 164–167, 229–234; (1910): 34–39.

Annotated enumeration of 164 species with ranges to 1000 m or more in elevation; includes localities and altitudinal ranges. Covers three mountains in Wales, two in Ireland and 64 in Scotland.

603

Alpine and upper montane zones

The greater part of the alpine and high-mountain floras of Europe are rooted on geologically younger formations, many of them calcareous. For those of the north (including the Ural) and in Britain and Ireland, see **602**. These floras have long been of special interest to many people, and it thus seems useful here to list a range of key works representative of an extensive (and fragmented) literature. For convenience, they have been grouped under several subheadings, viz.: I. Iberian Peninsula; II. The Pyrenees (including Andorra); III. Alpes-Maritimes; IV. The Alps; V. The Apennines (with Corsica and Sicily); VI. The Sudetens; VII. The Carpathians; and VIII. The Dinaric and Balkan Alps and Mt. Olympus.[20]

Only one serious all-European 'alpinarium', *Icones florae alpinae plantarum* by L. Marret, L. Capitaine and R. Farrer (described below), has ever been essayed. After a vigorous start, it fell victim to the unsettled conditions following World War I as well as the death in 1920 of one of its authors (Reginald Farrer) and was never completed. Some more popularly oriented works also exist, though only two date from after World War I: T. OHBA, 1973. (*Yōrropa kosan shokubutsu*)/*Alpine plants of Europe*. 184 pp., 298 col. photographic pls. Tokyo: Gakken (in Japanese); and C. GREY-WILSON and M. BLAMEY, 1995. *Alpine flowers of Britain and Europe*. 2nd edn. 384 pp., col. pls. London: HarperCollins (1st edn., 1979).[21] That by Ohba is a 'coffee-table' photographic album covering 289 species, descriptive text on pp. 145–161 being followed by a discussion of vegetation formations (pp. 162–181) and index to plates and text (pp. 182–183).

MARRET, L., L. CAPITAINE and R. FARRER (with Á. VON DEGEN, A. VON HAYEK and C. H. OSTENFIELD), 1911–24. *Icones florae alpinae plantarum*. Fasc. 1–. Pls., maps. Paris: Société d'Édition des Sciences Naturelles (later Lechevalier).

An atlas with plates, maps and descriptive text; the heliotype plates include photographs of herbarium specimens as well as text-figures while the short text mainly relates to distribution and cultivation. The numbering system used is not immediately evident.

[Not completed; only three of the projected 10 annual series (each with 100 plates and 90 maps in five fascicles) were published, with the second being interrupted by World War I. In all, 255 plates and 55 maps appeared.]

I. Iberian Peninsula (Sierra Nevada, Picos de Europa)

The first modern scientific treatment of the Sierra Nevada flora appeared in 1975. The range reaches a maximum elevation of 3481 m. Picos de Europa in the Cantabrian range, reaching 2598 m, were covered in 1988 and 1995.

MOLERO-MESA, J. and F. PÉREZ-RAYA, 1987. *La flora de Sierra Nevada: avance sobre el catalogo florístico nevadense*. 397 pp., 11 text-figs., 64 col. photographs on 8 pls., 2 maps. Granada: Universidad de Granada.

Enumeration of vascular plants (1935 species and subspecies), with essential synonymy, localities (including grid references to 10 km UTM squares) supported by *exsiccatae* or literature references, habitats, and symbolic indication of vegetation type(s), chorology and frequency; bibliography (pp. 371–387) and index at end. The introductory part accounts for the background of the work, a general description of the area (including associations and chorological units), history of botanical exploration, and (pp. 31–33) the plan. The color photographs illustrate representative species but are not cross-referenced to or from the text. [Covers the whole mountain system, not merely the tundra zone as in Prieto's work.]

NAVA, H. S. and M. Á. FERNÁNDEZ CASADO, 1995. *Flora de Alta Montaña*. xvii, 265 pp., text-fig., 300 col. illus. Madrid: ICONA. (Picos de Europa, Parque Nacional.)

Semi-popular illustrated manual with keys to all taxa, descriptions, accepted scientific and vernacular names, phenology, status and chorological class, usual habitat, frequency and vulnerability; illustrated glossary and complete index at end. The brief introduction covers the plan of the work and includes an outline of geobotanical zones with characteristic species and a general key to families. All the colored illustrations are grouped on pp. 117–243. [Covers the zone from the tree-line at 1500–1600 m upwards. For a more technical treatment, see H. S. NAVA, 1988. *Flora y vegetación orófila de los Picos de Europa.* 243 pp. (Ruizia 6). Madrid.]

PRIETO FERNÁNDEZ, P., 1975. *Flora de la tundra de Sierra Nevada.* 236 pp., 31 pls. (part col.), halftones, text-fig., folding map. Granada: Universidad de Granada.

Illustrated monographic guide to vascular plants occurring over 2500 m, with general key (pp. 129–146) and annotated descriptions including vernacular names and distribution and habitat details (pp. 149–183); glossary, bibliography and indices at end. The general part deals with geography, environment (including comparisons of the Sierra Nevada with other parts of the arctic and alpine world), soils and substrates, vegetation, plant adaptations, biogeography and endemism, fauna and human activities.

II. The Pyrenees (including Andorra)

The highest peak in this chain of mountains between France and Spain is Aneto, west of Andorra (3404 m).

The first modern general flora, Henri Gaussen's serial *Catalogue-flore des Pyrénées,* was all but complete at the time of its author's death in 1981; as of writing, however, publication has not been resumed. The *Catalogue* has recently been joined by *La grande flore illustrée des Pyrénées* by M. Saule. Significant earlier works include *Catalogue des plantes vasculaires des Pyrénées principales* (1857) by Johan E. Zetterstedt, *Flore des Pyrénées* (2 vols., 1859) by Xavier Philippe, and *Flora pyrenaea* by Pietro Bubani (edited by O. Penzig in 4 vols., 1897–1902). All are, however, now very outdated or untrustworthy. Of the several floras of more limited areas, reference is made here to the account of Andorran plants by Losa and Montserrat (1951; see below).[22]

GAUSSEN, H. *et al.* (comp.), 1953–81. *Catalogue-flore des Pyrénées.* In installments. Illus., map (in first installment and in no. 333, 1961). *Monde Pl.* **48** (nos. 293–297)–**76** (nos. 408–410), *passim.*

Systematic enumeration of vascular plants of the Pyrenees chain, with synonymy, notes on habitat, and detailed symbolic indication of local range arranged by districts (indicated in the map). [Not completed; when suspended coverage extended to pteridophytes, gymnosperms, monocotyledons, and dicotyledons through *Centaurea* in the Compositae (following the classical Englerian sequence of families).][23]

SAULE, M., 1991 (1992). *La grande flore illustrée des Pyrénées.* 765 pp., 330 pls. (figs.), 12 col. pls., col. photographs, map. Toulouse: Milau/Torbes: Randonnées pyrénéennes.

Atlas-flora of vascular plants, with each plate of figures and facing text covering 4–6 species; includes descriptions, French standard names (based on the botanical names), vernacular names (French, Occitan, Catalan, Basque, Aragonais), indication of distribution and habitat, overall range, and notes on special features and uses; list of protected species, an album of color photographs arranged as a floral calendar (but cross-referenced to the figures in the main text), an illustrated lexicon, index to French and Latin names (the latter only to genera and above), references (p. 762) and acknowledgments at end. The introductory part includes notes for users, a detailed area map, a systematic treatment of vegetation types as well as zonation in space and altitude, and an illustrated general key (in the manner of Bonnier (**651**) or Blackall and Grieve (**458**)). [The order of the plants in the main text is artificial, following the key, although members of a given family appear together.][24]

Andorra

Area: 465 km². Vascular flora: 985 species. – Andorra, as a 'Catalan land', is encompassed in recent floras of Catalan-speaking regions (see **615**). Reference may also be made to J. BOUCHARD, 1981. *Primer herbari de la Flora d'Andorra.* 180 pp., illus. (incl. text-figs., halftones). [Perpignan]: Institut d'Estudis Andorrans, Centre de Perpinyà. [This accounts for 985 species, covering nearly all those in Losa and Montserrat's work (see below) 'que ens va adjudar molt', and 100 mainly synanthropic additions.]

LOSA, M. and P. MONTSERRAT, 1951 (1950). *Aportación al conocimiento de la flora de Andorra.* 184 pp., illus.; 6 pls. (on 3), maps (Monogr. Inst. Estud. Pirenaicos 53 (Botánica 6)). Zaragoza. (Also published as *Actas del I. Congreso Internacional del Pirineo del Instituto de Estudios Pirenaicos* 1. San Sebastian.)

Systematic enumeration of vascular plants, bryophytes and fungi, with limited synonymy, indication of local range, and notes on habitat; index at end. An introductory section deals with local geography, geology, botanical exploration, major features of the flora, vegetation associations, and floristics.

III. Alpes-Maritimes

See also FENAROLI (under **IV. The Alps**) as well as general floras of Italy (**620**) and France (**651**). The long-unfinished *Flore des Alpes-Maritimes* was recently completed in two volumes of *Matériaux* compiled by A. Charpin and R. Salanon at the Conservatoire et Jardin Botaniques in Geneva, now home to Émile Burnat's herbarium.

BURNAT, É., J. BRIQUET and F. G. CAVILLIER, 1894–1931. *Flore des Alpes-Maritimes*. Vols. 1–7. Geneva, Basel, Lyon: Georg (vols. 1–6); Geneva: Conservatoire et Jardin Botaniques (vol. 7).

Copiously annotated, critical enumeration of vascular plants, with synonymy, indication of local range (including *exsiccatae* and other published records), and extensive taxonomic commentary; indices. [Not completed; extends from Ranunculaceae through Compositae–Cynarioideae (Candollean system). A related series of precursory revisions, containing material neither published in the *Flore* nor subsequently accounted for in the 1985–88 *Matériaux* (see below), is É. BURNAT (ed.), 1891–1916. *Matériaux pour servir à l'histoire de la flore des Alpes-Maritimes*. Geveva, Basel: Georg. (This contains, for example, full treatments of the Labiatae (by J. Briquet) and ferns (by H. Christ) not available elsewhere.)][25]

CHARPIN, A. and R. SALANON, 1985–88. *Matériaux pour la Flore des Alpes Maritimes*, I–II. 258, vii, 339 pp., folding map (Boissiera 36, 41). Geneva.

This work serves as a complement to Burnat's *Flore* and *Matériaux*. All species in the Alpes-Maritimes are enumerated, whether covered in the *Flore*, or only by *exsiccatae* or other revisions (including the 1891–1916 *Matériaux*), or not at all. Any hitherto undocumented *exsiccatae* are cited in full. Part II concludes with an account of field trips from 1871 to 1914 and a list, with dates, of specialists who studied and identified Alpes-Maritimes material in the Burnat Herbarium; these are followed by an index to genera in the whole work. [The arrangement of families here, in contrast to the original *Flore*, follows the modified Englerian system of *Flora Europaea*.]

IV. The Alps

Of the considerable floristic literature on the Alps – beginning with Conrad Gesner's observations on Mt. Pilatus near Lucerne and quickening in the eighteenth century with the works of Johann J. Scheuchzer and Albrecht von Haller and, in 1756, Linnaeus's *Flora alpina* – three only appear to be sufficiently comprehensive in coverage to justify full entries here. The two first are the five-volume German-language *Atlas der Alpenflora* (1881–84; 2nd edn., 1896–97 and 1899; French version, 1899) published by the Deutscher und Österreichischer Alpenverein and the fine one-volume manual by Fenaroli, *Flora delle Alpi* (most recently revised in 1971 but with antecedents going back to 1902). The third is *Unsere Alpenflora* by Elias Landolt, now considerably expanded and available in German (1992), English (1989) and French (1986) versions. Other current guides to Alpine plants, of which there is a considerable range, take the form of more or less popular and selective pocket-sized pictorial treatments with accompanying concise descriptive and ecological notes and thus strictly speaking fall outside the scope of the *Guide*. Some, however, are very well known and have had a wide influence in other parts of the world; a range of titles, cited in brief, is therefore presented following the description of the three main works.[26]

There are several natural histories of the Alps, but the following is particularly recommended: C. FAVARGER and P.-A. ROBERT, 1995. *Flore et végétation des Alpes*. 3rd edn. 2 vols. 76 figs., 6 col. pls. Lausanne: Delachaux et Niestlé. (1st edn., 1956–58, Neuchâtel. German version: *idem*, 1958–59. *Alpenflora*. 2 vols. Illus. Bern.) Included are reviews of physical features, geology, climate, floristics, plant biology, vegetation formations, and associations along with narrative accounts of principal families. Volume 1 focuses on the alpine zone (*c.* 2000–2900 m) and vol. 2, the subalpine zone (*c.* 1300–2000 m). Less detailed, but also useful (focusing particularly on associations), is H. REISIGL and R. KELLER, 1987. *Alpenpflanzen im Lebensraum*. 149 pp., 291 figs. (part col., incl. maps). Stuttgart: Fischer.

DEUTSCHER UND ÖSTERREICHISCHER ALPENVEREIN, 1896–97. *Atlas der Alpenflora*. 2nd edn. (ed. E. Palla). 5 vols. Pp. i–xv, 500 col. pls. (by A. Hartinger). Graz: D. und Ö. Alpenverein (also distributed by Lindauer, Munich). Accompanied by K. W. VON DALLA TORRE, 1899. *Die Alpenflora der österreichischen Alpenländer, Südbaierns und der Schweiz*. xvi, 270, [1] pp. Munich. French edn.: CLUB ALPEN ALLEMAND ET AUTRICHIEN, 1899. *Atlas de la flore alpine* (ed. H. Correvon). 6 vols. (vols. 1–5: atlas of 500 col. pls.; vol. 6: vii, 193 pp.). Geneva/Basle: Georg. (Reissued 1901, Paris.) (1st [bilingual] edn. in 5 vols., 1881–84, Salzburg, as A. Hartinger and K. W. von Dalla Torre, *Atlas der Alpenflora/Atlas du Alpenflora*.)

The text volume in both German- and French-language versions of this work comprises an analytical manual-key for field use (with on pp. xii–xvi of the introductory section a reference list of key works on Alpine floristics). The atlas consists of full-page chromolithographic plates, each of a single species (with an index in vol. 5). The plates in the German edition were arranged on the Englerian system while in the French version the Candollean system (as used in the 1881–84 edition) was followed (the plates, however, bearing dual numbering).[27]

FENAROLI, L., 1971. *Flora delle Alpi*. 2nd revised edn. xi, 429 pp., 262 text-figs., 61 halftones, 44 col. pls., 2 maps (in end-papers). Milan: Martello. (1st revised edn., 1955. Originally published 1902, Milan, as O. PENZIG, *Flora delle Alpi illustrata*; 2nd edn., 1915; 3rd edn., 1932, as L. FENAROLI, *Flora delle Alpi e degli altri monti d'Italia*.)

Copiously illustrated, comprehensive manual-key to the vascular plants of the high country (above 2000 m) throughout the Alps as well as of the mountains of the Italian region (including Corsica, Sardinia and Sicily); includes vernacular names (in several European languages), limited synonymy, generalized indication of internal and extralimital range, altitudinal zonation, and fairly detailed habitat and phytosociological notes together with some remarks on properties, uses, etc.; indices to vernacular and botanical names as well as associations. The introductory section, partly pedagogical, includes general remarks on physical, climatic, edaphic and biotic factors as well as on vegetation associations and seres.

LANDOLT, E., 1992. *Unsere Alpenflora*. 6. Aufl. 318 pp., 52 text-figs. (incl. maps), 480 col. photographs on 120 pls. [Zürich-Zollikon]: Schweizer Alpen-Club (in Germany distributed by Fischer, Stuttgart). (3. Aufl., 1964; 5. Aufl., 1984.) English version: E. LANDOLT, 1989. *Our alpine flora*. Transl. and revised by K. M. Urbanska. 303 pp., text-figs., col. pls. Zürich: Swiss Alpine Club Publications. (Earlier edn., 1963.) French version: E. LANDOLT and D. AESCHIMANN, 1986. *Notre flore alpine*. 333 pp., 52 text-figs. (incl. maps), 480 col. photographs on 120 pls. Zürich. (1st edn., 1963.) Italian version: E. LANDOLT, 1962. *La*

nostra flora alpina. Transl. G. Kauffman. 256 pp., 25 text-figs., 72 col. pls. Zürich.

Copiously illustrated, briefly descriptive pocket-manual of vascular plants known from 1500 m and above, with keys, vernacular names, very concise indication of distribution, frequency, altitudinal range, chorological class, phenology, legal status, taxonomic commentary, and coded 'indicative values' covering several habitat factors; references and indices at end. An extensive introductory part features chapters on the origin of the flora, its distribution and relationships, habitat factors, plant relations (i.e., autecology), and vegetation formations and communities; these are followed by general keys and a consideration of conventions and abbreviations. [The present edition of this work is considerably expanded over its predecessors, but remains compact. Current French and English versions are based on the fifth German edition, the most recent with major revisions.]

Popular guides

Alpine plant guides have appeared since the advent in the mid-nineteenth century of large-scale tourism. The following is merely a selection.

CORREVON, H. (with P. ROBERT), 1909. *La flore alpine*. 440 pp., 100 col. pls. Geneva: Atar. (2nd edn., 1917; further reissues to 1951.) English version: *idem*, 1912. *The alpine flora*. Transl. and revised by E. W. Clayforth. 436 pp., 102 pls. (100 col.). London: Methuen. [Comprises color plates followed by species descriptions.]

HEGI, G., H. MERXMÜLLER and H. REISIGL, [1977]. *Alpenflora*. 25th edn., revised by H. Reisigl. 194 pp., 1, 40 pls. (32 col.), 1, 48 maps. Berlin: Parey. (1st edn., 1905, Munich, by G. Hegi and G. Dunzinger.) [Pocket atlas-flora; comprises plates with accompanying text and distribution maps but no keys.]

PILÁT, A., 1973. *Atlas alpínek*. 506 pp., 629 halftones, 40 col. photographs Prague: 'Academia'/ČSAV. [A photographic atlas (in Czech) of 629 species, accompanied by alphabetically arranged descriptive accounts.]

RASETTI, F., 1996. *I fiori delle Alpi*. 2nd edn., revised by W. Rossi. 222 pp., 12 text-figs., 568 col. photographs Rome: Accademia Nazionale dei Lincei/Selcom. (1st edn., 1980.) [Photographic atlas with integrated descriptive text but no keys or vernacular names; annotated bibliography and index to all names at end. The extensive introductory part covers botanical classification, scientific and vernacular names, plant parts, life-forms, and floristics, vegetation and the environment along with a discussion of photographic methods.]

RAUH, W., 1951–53. *Alpenpflanzen*. 4 vols. Illus. (some

col.). Heidelberg: Winter. (Winters Naturwissenschaftliche Taschenbücher 15, 16, 21, 22.) (Later editions of vols. 1–2 published by Quelle & Meyer, Heidelberg. Original version (in 2 vols.) published as L. KLEIN, *Alpenblumen* and limited to the western Alps.) [Small-format atlas-flora, with extensive introductory sections on environment, phytogeography, life-forms, ecology, biology, environment, etc. spread across the four volumes; no keys. Vol. 1 covers Ranunculaceae–Rosaceae; vol. 2, Onagraceae–Gentianaceae; vol. 3, Labiatae–Compositae; and vol. 4, Rubiaceae–Campanulaceae, monocotyledons, gymnosperms and pteridophytes. There are in each volume additional species listed supplementary to the main treatments. All volumes are indexed, with that in vol. 4 cumulative.]

SCHRÖTER, L. and C. SCHRÖTER, 1963. *Taschenflora des Alpenwanderers.* 29th edn. by W. Lüdi. 26 pls. (24 col.). Zürich: Schumann. (1st edn., 1889; 25th edn., 1940.) [Pocket field-atlas with short descriptive notes in German, French and English; no keys.][28]

V. The Apennines (and mountains of Corsica and Sicily)

See FENAROLI (under **IV. The Alps**) for complete coverage of the high-mountain areas of the Italian peninsula and islands to the west. The highest peak is Mt. Corno northeast of Rome (2914 m).

VI. The Sudetens

See also PILÁT (under **IV. The Alps**) for other coverage. Popular guides include J. FABISZEWSKI, 1971. *Rośliny Sudetów: atlas.* 154 pp., 60 col. pls. Warsaw: Państwowe Zakłady Wydawnictwo Szkolnych (PZWS); and A. ZLATNÍK and A. KAVINOVÁ, 1966. *Květiny a hory.* 104 pp., 96 col. pls. Prague: Státní Pedagogické Nakladatelství.

KRUBER, P., n.d. *Exkursionsflora für das Riesen- und Isergebirge.* vii, 345 pp., 42 text-figs., 6 col. pls. Warmbrünn (Cieplice Śląskie Zdroj): Lepelt.

Manual-flora with synoptic keys, brief descriptions, German names, and notes on phenology, habit, habitat, frequency, special features, etc.; indices to family, generic and local names. The introductory section includes keys to classes and to families based on the Linnaean sexual system. [No date of publication is given for this work, but it appears to have been in the 1920s or early 1930s.]

ŠOUREK, J., 1969 (1970). *Květena Krkonoš.* 452 pp., 60 halftones, col. pls. Prague: 'Academia'.

Annotated enumeration of vascular plants (1149 species and infraspecific taxa) with Czech vernacular names and detailed indication of distribution and habitats; list of neoendemics (p. 410), statistical summary by chorological classes (p. 411), textual summaries in Polish and German, and references (pp. 420–426; the work by Kruber is, however, not listed); indices to all vernacular and botanical names at end. The very detailed general part covers physical features, floristics, vegetation, conservation and notes on the arrangement of families (the Novák system is adopted here) as well as on the history of botanical work in the range.

VII. The Carpathians

The Carpathians form perhaps the longest mountain range in Europe and pass through the Czech Republic, Slovakia, Poland, Ukraine and Romania. No floras deal with the chain as a whole, but there are several partial floras. Of these, the best, or at least most comprehensive, modern work is the Ukrainian Botanical Institute's Визначник рослин Українських Карпат/*Vyznačnyk roslyn Ukrajins'kykh Karpat* (1976; see **695**) on the Ukrainian Carpathians and its northern approaches south of the Dniestr. One of that work's editors, V. J. Čopyk, also produced a detailed enumeration of its high-mountain flora (1976; see below); this includes an extensive bibliography useful for the whole of the Carpathians. Plant geography was earlier surveyed in F. PAX, 1898–1908. *Grundzüge der Pflanzenverbreitung in den Karpathen,* I–II. 2 vols. Illus., maps. Leipzig: Engelmann. (Die Vegetation der Erde 2, 10.)

The key works are here grouped under several subheadings.

Bibliography

See also **642** (GOMBOCZ). Pax's *Grundzüge* (see above) contains geographically classified lists of references through 1908 in its first (pp. 26–63) and second (pp. 273–281) volumes.

DOSTÁL, J., 1931. *Bibliografie prací Československý botaniků o území Karpatském.* 35 pp. Praha. (Publ. Geobot. Unie Karpatské, sekce Československá 1.) [365 works and three sets of *exsiccatae* cited. General and regionally oriented titles in four sections, of which the Tatra comprise section 3. Unannotated.]

Central Carpathians

Includes the Tatry (Tatra Range), within which is Mt. Gerlachovski-Sztit, the highest point in the Carpathians (2655 m). The first flora to appear was *Flora carpatorum principalium* (1814) by Göran Wahlenberg.

PAWŁOWSKI, B., 1956. *Flora Tatr/Flora tatrorum*. Vol. 1. 672 pp., 130 text-figs. Warsaw: Państwowe Wydawnictwo Naukowe.

Descriptive flora (in Polish) of vascular plants, with keys to all taxa, synonymy, vernacular names, local range and distributional remarks, and notes on variation, habitat, phenology, etc.; indices. An account of vegetation formations in the area appears in the introduction. [Not completed; covers families from the Pteridophyta through the end of the Choripetalae (Englerian system).][29]

SAGORSKI, E. and G. SCHNEIDER, 1891. *Flora der Centralkarpathen/Flora carpatorum centralium*. 2 vols. 2 pls. Leipzig: Kummer.

Volume 2 consists of a concise manual of vascular plants in the central Carpathians (especially the Tatry), with keys, limited synonymy, local range, and notes on habitat, phenology, etc.; index at end. Volume 1 incorporates a general introduction, with accounts of geography, geology, climate, vegetation formations, and botanical exploration; bibliography, p. 112.

Ukrainian Carpathians

See also 695 (all works). Much floristic and geobotanical work has been carried out in the Ukrainian Carpathians since World War II, and the area is now relatively well documented. Strictly speaking, however, only the following is limited to the high-mountain flora; but its extensive bibliography is of value over a far greater area, including effectively the whole of the Carpathians.

ČOPYK, V. J., 1976. Високогірна флора Українських Карпат/*Vysokogirna flora Ukrajins'kykh Karpat*. 267 pp., 12 figs. Kiev (Kyjiv): 'Naukova dumka'.

Systematic enumeration of the high-mountain vascular flora (475 species), with synonymy, local and extralimital distribution, karyotypes, some critical remarks, and notes on habit, ecology, phytosociological groups, etc.; references to *Flora SSSR* or other major works given under species headings. Pp. 167ff. comprise a detailed floristic/phytogeographic analysis, with emphasis also given to the biology and the development of the flora, while pp. 235–246 feature a rather detailed bibliography. In the introductory section are notes on physical features, climate, and floristic zones, with coverage extended to account for most of the Carpathian chain and its geographical and botanical subdivisions from Poland and former Czechoslovakia to Romania. A complete index concludes the work.

Romanian Carpathians (Carpații Moldavi, Carpații Meridionali)

The Moldavian Carpathians divide Transylvania from Moldavia, and extend southwards to a point east of Brașov (Kronstadt). The Southern Carpathians, also known as the Transylvanian Alps, lie on an east–west axis, dividing Transylvania from the plains of Wallachia, and link with the eastern Carpathians east of Brașov. Three portions of the Southern Carpathians rise to over 2500 m: the Bucegi (Butschesch) massif southwest of Brașov (reaching 2507 m at Omu Peak); Retezat (Retyezat), in the western part of the range (reaching 2506 m); and Negoi, at 2543 m the highest peak in the range and the second highest in the whole of the Carpathians. To the west of Retezat, at the extreme west of the range towards the Danubian 'Iron Gates', are Țarcu, Godeanu and Cernei, a group of three peaks with a maximum elevation of 2292 m. The Masivul Postăvarul (Schulergebirge), treated by Fink, and the Ciucaș range studied by Ciucă and Beldie are, like the Bucegi, near Brașov.

The only general manual is *Flora alpină și montană* (1959) by Ana M. Paucă and Stefana Roman. For the Bucegi massif, a manual was produced by Beldie (1972). Both of these are described below. An older illustrated field-guide is K. UNGAR, 1913. *Die Alpenflora der Südkarpathen*. 92 pp., 24 pls. Hermannstadt (Sibiu). Local checklists, compiled in conjunction with geobotanical studies, are, however, available for many peaks or ranges (though not Negoi). Among these are A. BELDIE, 1967. *Flora și vegetația Munților Bucegi*. 578 pp., 68 halftones, tables, map. Bucharest: Academia RSR; N. BOȘCAIU, 1971. *Flora și vegetația Munților Țarcu, Godeanu și Cernei*. 494 pp., 39 figs., tables, maps. Bucharest; M. CIUCĂ and A. BELDIE, 1989. *Flora Munților Ciucaș*. 193 pp., 269 text-figs. Bucharest; H. G. FINK, 1975. Flora des Schulergebirges (Südostkarpaten). *Linzer Biol. Beitr.* 7(2): 121–223; and E. J. NYÁRÁDY, 1958. *Flora și vegetația Munților Retezat*. 195 pp., 42 text-figs., 1 folding pl., tables. Bucharest.

BELDIE, A., 1972. *Plantele din Munții Bucegi: determinator*. 409 pp., 62 text-figs. Bucharest: Academia RSR Press.

Manual-key to vascular plants, the leads with limited synonymy, generally concise but relatively detailed indication of local distribution (and sometimes altitudinal range), preferred habitat, life-form, habit, phenology, chorological and phytosociological classes, frequency, etc., as well as illustrations of critical features; index to genera at end. In the introduction are a zonation table (500–2500 m), glossary, and explanation of abbreviations and signs. [Based on the 1967 monograph of the author (see above), where full references are given.]

PAUCĂ, A. M. and S. ROMAN, 1959. *Flora alpină si montană (îndrumător botanic)*. 306, 1 pp., 365 figs. Bucharest: 'Ştiinţifica'.

Well-illustrated manual of vascular plants with concise keys to families and genera (pp. 43–89) and descriptive entries (pp. 90–289); the latter include a short diagnosis, vernacular names, habit symbols, phenology, and (for some taxa) habitat; indices to families and genera and to Romanian names at end. Species keys are provided only for the larger genera such as

Ranunculus. The introductory section includes an organography, glossary (pp. 21–28), notes on the use of the work, and conventions along with a history of botanical work in the mountains, a discussion of altitudinal zonation, and a synopsis of families and genera. [Covers the whole of the Romanian mountains, encompassing about 800 species.]

VIII. The Dinaric and Balkan Alps and Mt. Olympus

Until recently, there has been little documentation of the flora of the high mountains of Bulgaria, Greece, and immediately adjacent areas. The highest peaks here are Botev in the Stara Planina of central Bulgaria (2376 m), Musala in the western Rhodope Mountains just to the south (2925 m), and Pirin (2914 m) in southwestern Bulgaria. Nearby Mt. Olympus (2911 m) in Greece is the highest in its so-called Balkan Range. The two last-named have now been covered in some detail: Olympus by Strid in the work described below and in *Mountain flora of Greece* (**636**), and Pirin in a monograph by Kitanov and Kitanov, also here described. There is in addition a popular guide to Bulgarian alpines by Stojanov and Kitanov (1966; under **Popular guide** below) and a manual of the Planina Vitoša (a range reaching 2286 m) south of Sofia: B. KITANOV and I. PENEV, 1963. Флора на Витоша (*Flora na Vitoša*). 514 pp., 14 text-figs., 80 pls. Sofia: 'Nauka i Izkustvo'.

KITANOV, B. and G. KITANOV, 1990. Флора на Пирин (*Flora na Pirin*). 478, [2] pp., 14 text-figs., 80 pls. Sofia: 'Nauka i Izkustvo'.

Manual of vascular plants, the keys and descriptions separate; includes Bulgarian standard names, habit symbols, concise diagnoses, altitudinal zonation, phenology, height, habitat(s), and references to the figures (all in one section after the main text); illustrated organography and detailed table of contents at end. The introduction (dated 1986) includes highlights as well as notes on the use of the book. Bulgarian names precede the binomials in the keys and text. [Covers some 1700 species in 547 genera in 111 families (a notably rich flora); family limits are as in *Flora Europaea*. The work includes some species new for Bulgaria; there is, however, neither coverage of infraspecific taxa nor an index.][30]

STRID, A., 1980. *Wild flowers of Mount Olympus*. 380 pp., 106 col. pls., 2 maps. Kifissia: Goulandris Natural History Museum.

The latter part (pp. 223–353) of this iconographic book is a floristic treatment, with keys, brief descriptions, distribution, chorological information (not in all cases), notes on uses and properties, and (in more detail) discussions of ecology and biology as well as taxonomic and biogeographical notes; index to all scientific names at end. The extensive introduction includes considerations of geography, geology, vegetation, roads and the built environment, and biogeography. The individual plates are grouped by habitat.

Popular guide

STOJANOV, N. and B. KITANOV, 1966. Високопланинските растения в България (*Visokoplaninskite rastenija v Bǎlgarija*)/ *Hochgebirgspflanzen Bulgariens*. 150 pp., 70 col. pls. Sofia: 'Nauka i Izkustvo'.

Popularly oriented descriptive enumeration of and field-guide to 300 high-mountain species, with vernacular names (Bulgarian, Russian, German, French and English) as well as local and wider distribution; indices to all names after the text. The introductory section contains sources, explanatory matter and a list of protected plants. The plates follow the text proper.[31]

605

Drylands

Though extensive steppes (prairies) occur from Romania, Hungary and Poland eastwards, no overall flora for this formation has been written.

606

Rivers

The river flora and vegetation of Europe has received particular attention in recent years, notably in the studies of S. M. Haslam and others. Haslam's *River plants of Western Europe* (1987, Cambridge), is, however, not a floristic work.

608

Wetlands

In contrast to North America, relatively few significant works have been published on European aquatic and marsh plants, although the 1936 treatment by Hugo Glück in the original edition of Adolf Pascher's *Die Süßwasser-Flora Mitteleuropas* is one of the earliest of its kind. A greatly expanded version of that work was published in 1980–81. Keys for Portugal, Great Britain and Ireland, and Central European Russia have also been seen.

Northern and Northwestern Europe

HASLAM, S. M., C. SINKER and P. WOLSELEY, 1975. British water plants. *Field Studies* 4: 243–351, illus. (incl. 28 pls.). (Reprinted separately, London.)

This work is in two parts: an illustrated key (with glossary), and a systematic enumeration of aquatic macrophytes with abbreviated but fairly detailed indication of habitat, substrates, habit, vernacular names, and geographical distribution. Illustrative plates are provided. In the introduction are given technical notes and tips for collecting as well as a brief indication of the scope of the work, while at the end is a complete index.[32]

MOESLUND, B. *et al.* (ed. N. THYSSEN), 1990. *Danske vandplanter*. 192 pp., 67 pls. Copenhagen: Miljøstyrelsen. (Miljønyt 2.)

Atlas of aquatic macrophytes (mostly vascular plants and Characeae) with descriptive text and (pp. 12–26) keys; references and indices to botanical and vernacular names at end.

PRESTON, C. D. and J. M. CROFT, 1997. *Aquatic plants in Britain and Ireland*. 365 pp., illus., maps. Great Horkeley, Colchester, Essex: Harley Books.

Introduction with rationale for book and sources of records; illustrated descriptive atlas with extensive taxonomic, geographical and biological commentary (including dynamics as well as extralimital range) as well as dot distribution maps in the style of the *Atlas of the British flora*; conservation status, gazetteer, bibliography and index to vernacular and scientific names. The work encompasses introduced as well a truly native species. [The authors consider the aquatic macroflora relatively rich by European standards; however, no statistics are given in the book.]

Central Europe

CASPAR, S. J. and H.-D. KRAUSCH, 1980–81. Pteridophyta und Anthophyta, I–II. In A. Pascher, *Süßwasserflora von Mitteleuropa*, 2. Aufl. (eds. H. Ettl *et al.*), vols. 23–24. 2 vols. 109 + 119 pls. (with 2733 figs.), maps (end-papers). Stuttgart: Fischer. (A revision of H. GLÜCK, 1936. Pteridophyten und Phanerogamen. In A. Pascher (ed.), *Die Süßwasser-Flora Mitteleuropas*, 15. 486 pp., 258 figs. Jena: Fischer.)

Illustrated descriptive manual-flora of hydro- and helophytic vascular plants; includes keys to all taxa, synonymy, indication of regional occurrence and overall distribution, citations of published distribution maps, karyotypes, phenology, and detailed discussion of habitat, preferences, associates, and special ecological and biological features. An index is in the second part. The work is opened by a concise introduction, with a general key and definition of ecological and phytosociological classes. [Pteridophytes and monocotyledons covered in part I; dicotyledons in part II.]

Iberian Peninsula

VASCONCELLOS, J. DE CARVALHO E, 1970. *Plantas (angiospérmicas) aquáticas, anfibias e riberinhas*. 253 pp., 36 pls. Lisbon: Secretaria de Estado da Agricultura, D.-G. dos Serviços Florestais e Aquícolas.

Manual-key to hydrophytes, helophytes and rheophytes with descriptions of families and species entries featuring vernacular names and indication of habitat and distribution; indices and a full table of contents (in effect a synoptic list) at end. Pp. 9–10 comprise a brief introduction.

Russia-in Europe and neighboring states

The work of Ryczin (1946) is probably a considerably expanded version of JU. V. RYCZIN and P. B. SERGEEVA, 1939. Водная и прибрежная флора (*Vodnaja i pribrežnaja flora*). 184 pp. Moscow: Učpedgiz. This was in turn preceded by B. A. FEDTSCHENKO and A. F. FLEROW, 1913. Водная флора Европейской России (*Vodnaja flora Evropejskoj Rossii*). 3rd edn. 65 pp., 49 figs. Moscow: Sabašnikovykh. (1st edn., 1897, Moscow; 2nd edn., 1900, both as Водяные растения Средней России (*Vodjanye rastenija Srednej Rossii*).)[33]

RYCZIN, JU. V., 1948 (1946). Флора гигрофитов (*Flora gigrofitov*). 448 pp., 164 text-figs., 14 pls. Moscow: 'Sovetskaya Nauka'.

Artificially organized manual-key to aquatic macrophytes, the arrangement following vegetative features and the leads to species fairly copiously descriptive; entries also contain indication of family, synonymy, Russian binomial and vernacular names, distribution, habit and life-class, phenology, habitat, and miscellaneous (including some taxonomic) remarks along with notes on varieties and related species. Some species are keyed out more than once. The main key is followed by several miscellaneous sections: pp. 322–324 and 325–326 respectively cover *Cuscuta* and charophytes (and include invalidly published descriptions of new taxa);

pp. 327–400, expanded treatments of individual species introduced in the main key; pp. 401–409, symbols, abbreviations and authorities; pp. 410–412, references; pp. 415–428, 14 plates of diagnostic features; and pp. 429–448, lexical indices. The introductory part gives the work's background together with sections on ecology, biology and structure, including pedagogics (pp. 44–48) and collecting technique (pp. 48–52); there are also notes on the systematics of particular families. [Covers Central European Russia; about 530 species in 65 families.]

609

Oceanic littoral

There are currently few guides to littoral plants in Europe in spite of the extent and significance of its coastlines.

CLAUSTRES, G. and C. LEMOINE, 1980. *Connaître et reconnaître la flore et la végétation des côtes Manche-Atlantique.* 332 pp., text-fig., col. pls. Rennes: Ouest-France.

A well-illustrated popular guide encompassing some 600 species and covering cultivated as well as native plants; analytical keys wanting. References, p. 323.

Region 61

Iberian Peninsula

Area: 596 510 km². Vascular flora: *c.* 8000 species (S. Castroviejo, personal communication, 1995). – In addition to Spain, Portugal and Gibraltar, this region also includes the Balearic Islands. In addition to the national floras, linguistic and geographical diversity has occasioned several regional floras of note; these are individually accounted for.

Serious floristic study of the Iberian Peninsula began in response to a critique by Linnaeus, reiterated in the second edition of his *Bibliotheca botanica* (1751): 'Hispanicae Florae nullae nobis innotuerunt, adeoque plantae istae rarissime, in locis Hispaniae fertilissimus, minus detectae sunt. Dolendum est, quod in locis

Europae cultioribus, tanta existat nostro tempore barbaries botanices!' Linnaeus's student Pehr Löfling was given exceptional facilities to travel in Spain from 1751 to 1754 and in 1755, in Madrid, the Real Jardín Botánico was established, moving to its present site in 1781. Löfling's work resulted in his posthumous *Iter hispanicum* (1758); and, under that great patron of the arts and sciences, Carlos III, the first full account was published: *Flora española* (1762–84) by José Quer y Martínez and Casimiro Gómez Ortega. Many additions were made by Antonio J. Cavanilles in *Icones et descriptiones plantarum quae in Hispania crescunt* (1791–1801). In Portugal, the key early works are both by Felix A. Brotero: *Flora lusitanica* (1804, covering 1900 species) and an atlas, *Phytographiae Lusitaniae selectior* (1816–27). War, disorder, the loss of the Americas, political changes and decline then ended the botanical 'golden age' in Iberia, and through the rest of the nineteenth century and into the twentieth floristic work by residents by and large became more oriented to proximal needs. It was outsiders who by and large would make the effective advances.

In Spain, the multi-volume national or peninsula-wide works of this period were largely compilations: the descriptive/phytographic *Flora fanerogámica de la península ibérica* (1871–73) by Granadian-based Mariano del Amo y Mora, the topographic but uncritical *Enumeración y revisión de las plantas de la Península Hispano-Lusitana é islas Baleares* (1885–90) by the Gallegan Miguel Colmeiro at the Real Jardín Botánico, and, also at Madrid, *Botánica descriptiva* (1896; revisions in 1906–07 and 1920–21) by Blas Lázaro é Ibiza. The major outside contributions included *Iter hispaniense* (1838) and *Otia hispanica* (1853) by Philip Barker Webb, *Voyage botanique dans la Midi d'Espagne* (1839–45) by Edmond Boissier, *Icones et descriptiones plantarum novarum* (1852–62) and *Illustrationes florae hispaniae* (1881–92) by Moritz Willkomm, and *Descriptio iconibus illustrata plantarum novarum* (1864–66) and *Diagnoses plantarum peninsulae Ibericae novarum* (1878–81) by Johan Lange. These were brought together by Willkomm and Lange in their *Prodromus florae hispanicae* (1861–80; supplement, 1893) – one of the best floras of the day in all Europe. In Portugal after Brotero, the only flora was the 'luxe' (and by 1909 already rare) *Flore portugaise* (1809–40), the fruit of a tour by two Germans: Johann, Graf von Hoffmannsegg and Heinrich F. Link (sadly, their projected *Flora lusitanica* was never published and in 1943 its manuscript was destroyed). The

Balearics were as a group first treated in 1827 by the Frenchman Jacques Cambessèdes.

In the late nineteenth century, linguistic and cultural revivals developed in different parts of the Peninsula. At the same time there were new botanical initiatives which resulted in a number of important works, some still current.[34] Among these were *Manual da flora portuguesa* (1909–14; 2nd edn., 1947) by Gonçalo Sampaio at Porto and *A flora de Portugal* (1913; 2nd edn., 1939) by António X. Pereira Coutinho at Lisbon, *Flora descriptiva é ilustrada de Galicia* (1905–09) by Baltasar Merino y Román, and especially *Flora de Catalunya* (1913–37) by Juan Cadevall i Diars and his associates. The last-named, modeled on Abbé Coste's illustrated flora of France (**651**), has in turn influenced some current Spanish works. For the Balearics, the Montpellier connection was renewed with *Flora balearica* (1921–23) by Herman Knoche, an independently wealthy *norteamericano* and student of Charles Flahault.[35] Works on the woody flora emerged slightly earlier; among them were *Flora forestal española* (1875; revised edn., 1883–90) by Máximo Laguna y Villanueva and Pedro de Ávila y Zumarán and *Esbôço de uma flora lenhosa portuguesa* (1887; 2nd edn., 1936) by Pereira Coutinho.

Few new works were produced in the mid-twentieth century. For Spain Willkomm and Lange remained standard though increasingly out of date. A selective students' manual, *Flora analítica de España* by Arturo Caballero y Segares (then at the Real Jardín Botánico), appeared in 1940. In Portugal, revisions of both standard manuals appeared but, not surprisingly, there was relatively more interest in the flora of the African territories.[36] From time to time there were movements towards a successor to the *Prodromus*; that most strongly prosecuted was one led by Pio Font i Quer in the 1950s. A further revival has developed, however, from the 1970s, doubtless influenced by *Flora Europaea* of which Font i Quer had been a strong supporter.[37] Among its features are a new Portuguese flora and new national and regional floras and manuals in Spain, all described below.

Of modern undertakings, the two largest have been mounted at the Real Jardín Botánico within the framework of a more general biological programme: a new flora of the Iberian Peninsula and, in succession to the work of Colmeiro, detailed documentation of ancillary data including plant distribution and chorology, vernacular names, and chromosome numbers. The first, *Flora iberica*, began publication in 1986; six of a planned 21 volumes have now appeared. Results from the second project are appearing by groups of publications as *Archivos de* Flora iberica; seven installments had appeared by 1998. Both projects would have had, as Heywood has noted, to have taken account of parallel but long-distinct floristic traditions, respectively of Spanish and of foreign (mainly northern European) botanists working in Spain: a problem not unlike that faced by writers of many (if not most) tropical floras. Spain also remains, along with parts of the Italian region and southeastern Europe, one of the less well-documented areas in the continent; some parts furthermore are still botanically underexplored, much as in the West Indies, India, and parts of West Africa or Australia. Nevertheless, in recent decades improvement has been rapid, and particularly in Spain there has been a great increase in collection resources.[38]

Bibliographies. General and divisional bibliographies as for Division 6. Very thorough coverage through 1959 is in particular provided by Blake, 1961.

Indices. General and divisional indices as for Division 6.

Regional index

PAJARÓN SOTOMAYOR, S. (ed.), 1989– . Bibliografía botánica ibérica, 1989 [*et passim*]. *Botanica complutensis* **15**–, *passim*. [Annual compilation covering all plants and plant-like organisms, arranged by author with subject keywords. As of 1999 eight installments, the most recent in vol. 22, had been published.]

610

Region in general

Flora iberica is accompanied by a series of volumes of references to topographical data, extracted from runs of relevant botanical journals and arranged by botanical names: M. VELAYOS, F. CASTILLA and R. GAMARRA, 1991– . *Corología ibérica*, I– . Madrid: Real Jardín Botánico/ICyT. (Archivos de Flora Iberica 2, *passim*.)

CASTROVIEJO, S. *et al.* (eds.), 1986– . *Flora iberica: plantas vasculares de la Península Ibérica e Islas*

Baleares. Vols. 1– . Illus. Madrid: Real Jardín Botánico (distributed by Servicio de Publicaciones, Consejo Superior de Investigaciones Científicas).

Detailed descriptive flora with keys, synonymy (with particular reference to 'standard' works), types, vernacular names, local range (by geographical units, relating to the end-paper maps), critical commentary where appropriate, representative illustrations, and notes on overall distribution, habitat, phenology, etc.; appendices (author abbreviations, book and journal abbreviations, and derivation of scientific names) and indices to plates and to scientific and vernacular names. The introductory section in volume 1 includes the plan of the work and its background, along with abbreviations, geographic divisions and codes, and signs (pp. xxiv–xxvi) and a general key to families. [In progress; as of 1999 vols. 1–6, 7(1) and 8 had been published. Arrangement of flowering plants follows the Stebbins system (see Heywood, 1978. *Flowering plants of the world*, pp. 12–15). Some 21 volumes are currently projected, to cover a probable total of 8000 species. An outline of the planned arrangement along with a list of published and unpublished families may be found in the back end-pages of each successive volume.][39]

Woody plants

For trees and shrubs, see also P. GALÁN, R. GAMARRA and J. A. GARCÍA, 1998. *Árboles y arbustos de la Península Ibérica e Islas Baleares*. Madrid: Jaguar. For trees alone, see also A. M. ROMO, 1997. *Árboles de la Península Ibérica y Baleares*. Madrid: Planeta.

LÓPEZ GONZÁLEZ, G., 1982. *La guía de INCAFO de las árboles y arbustos de la Península Ibérica*. 866 pp., text-figs., col. photographs Madrid: INCAFO. (Guias verdes de INCAFO 4.) (Several reissues, the fifth in 1995.)

Well-illustrated descriptive account of native, naturalized and frequently grown introduced species, with scientific and vernacular names (the latter in Spanish, Catalan, Balearic Catalan, Valencian, Portuguese, Galician and Basque), distribution, status, small marginal figures of habit and technical details, and notes on habitat, paleohistory, properties, uses, derivation of scientific names, etc.; comprehensive index at end. The introductory section gives the plan and scope of the work, a guide to its use, extensive discussions of the flora and of woody vegetation formations, and (pp. 45–90) an artificial general key.

Pteridophytes

Ferns and fern-allies are also included in vol. 1 of *Flora iberica*.

SALVO TIERRA, A. E., 1990. *Guia de helechos da la*

Peninsula Iberica. 377 pp., 256 illus. (some col.), maps. Madrid: Pirámide.

Field-guide to ferns and allied plants with diagnostic keys, illustrations (including photographs), distribution maps (within the Peninsula and Balearics as well as overall, the latter as small insets), and detailed descriptions including derivation of names, overall and Iberian distribution, habitat, uses and properties (if any), critical notes, and conservation status if appropriate; reading list (pp. 363–366); complete index at end. A substantial general part (written with assistance from specialists) covers all botanical aspects including biogeography (map, p. 73); an explanation of symbols is on p. 85.

SALVO TIERRA, A. E., B. CABEZUDO and L. ESPAÑA, 1984. Atlas de la pteridoflora ibérica y balear. *Acta Bot. Malacitana* 9: 105–128.

An analysis of distribution, with individual gridded dot maps, of 122 species; includes (p. 106) collecting density indices. [Based on *Atlas florae europaeae* with post-1972 data.]

611

Portugal

For the Azores, see **012**. – Area: 91 631 km². Vascular flora: 2750–2950 species, of which 2400–2600 native (Webb, 1978). Pinto da Silva (in Heywood, 1975; see **601/I**) suggested a total of 3000 native species.

Franco's flora, still incomplete, has yet entirely to supersede the two older manuals. Not accounted for here are several regional and local floras. The country is also covered in *Prodromus florae hispanicae* by Willkomm and Lange (see **612**).

FRANCO, J. M. A. DO AMARAL (partly with M. DA LUZ DA ROCHA AFONSO), 1971–98. *Nova flora de Portugal (continente e Açores)*. Vols. 1–2, 3(1–2). Maps (part. col.). Lisbon: The author (printed by Sociedade Astoria) (vols. 1–2); Escolar Editora (vol. 3(1–2)). (Maria da Luz da Rocha Afonso co-author for vol. 3.)

Concise descriptive account of vascular plants (encompassing mainland Portugal and the Azores); includes citations for standard treatments of particular genera under their headings, keys to all taxa, limited synonymy, and notes on habitats and local range; list of authors, guide to references, and index to botanical names at end. [The work is an adaptation of *Flora Europaea*. The families in the first volume correspond to those in volumes 1–2 of that work, those in the second to volumes 3–4, and in the last (to be completed in three parts), to volume 5. Volume 3(1) covers Alismataceae–Iridaceae and 3(2), Poaceae; 3(3) will cover Palmae, Araceae, Lemnaceae, Sparganiaceae, Zingiberaceae and Orchidaceae.]

COUTINHO, A. X. PEREIRA, 1939. *Flora de Portugal (plantas vasculares)*. 2nd edn., revised by R. T. Palhinha. ii, 938 pp. Lisbon: Bertrand. (Reprinted 1973, Lehre: Cramer, as *Historia naturalis classica* 98.) (1st edn., 1913, as *A flora de Portugal*.)

Manual-key to native and naturalized vascular plants, with limited synonymy (no references), vernacular names, short descriptions, concise indication of local range (by provinces), and brief notes on phenology and ecology; glossary and indices to vernacular and botanical names at end. The introductory section includes a list of authors' names.

SAMPAIO, G., 1946 (1947). *Flora portuguesa*. 2nd edn., revised by A. Pires de Lima. xliii, 792 pp., 850 text-figs., 13 pls. Porto: Imprensa Moderno. (1st edn. (not completed), n.d. [1909–14] as *Manual da flora portugueza*; 3rd (facsimile) edn., 1988; 4th (facsimile) edn., 1990, Lisbon: Instituto Nacional de Investigação Científica.) Accompanied by *idem*, 1949 (1950). *Iconografia selecta da flora portuguesa*. 150 pls. Lisbon.

The *Flora* is an illustrated manual-key to native and naturalized vascular plants including vernacular names, brief general indication of local range, and phenological data; list of authors' names, glossary and indices to vernacular and botanical names at end. No synonymy is included, and aliens and widely cultivated species are merely noted in smaller print. The *Iconografia* comprises full-size plates without descriptions and, for the most part, details.[40]

Woody plants

The most recent manual is A. X. PEREIRA COUTINHO, 1936. *Esbôço de uma flora lenhosa Portuguesa*. 2nd edn. 368, [3] pp., 50 text-figs. [Lisbon.] (Publicações da Direcção Geral dos Serviços Florestais e Aqüícolas 3(1).) (1st edn., 1887.) This textbook includes keys, descriptions and supporting data, with some genera (e.g., *Quercus*, *Ulex*, *Cytisus*) covered in considerable detail; family keys, pp. 18–27, and glossary and indices, pp. 287–327.

Pteridophytes

In addition to the following, reference may also be made to SALVO TIERRA *et al.* (**610**) and to J. M. A. DO AMARAL FRANCO and M. DA LUZ DA ROCHA AFONSO, 1982. *Distribuição de pteridófitos e gimnospérmos em Portugal*. 327 pp.,

illus., maps (with overlays). Lisbon. (Colecção «Parques Naturais».)

VASCONCELLOS, J. DE CARVALHO E, 1968. *Pteridófitas de Portugal continental e ilhas adjacentes*. 188 pp., illus. (incl. 44 pls.). Lisbon: Fundação Calouste Gulbenkian.

Manual-key with botanical and vernacular names, indication of distribution and habitat, and occasional notes; literature (pp. 115–118) and indices to all names at end of text. Diagnostic figures in the text along with full-page plates (at the end) are furnished. [Covers mainland Portugal, the Azores and the Madeira Islands. A preface by Abílio Fernandes acknowledges the work as a contribution to the *Nova flora de Portugal* project.]

612
Spain (in general)

Area: 504 879 km². Vascular flora: 5150–5200 species, of which 4750–4900 native (Webb); more recent figures suggest 7500 or higher (S. Castroviejo, personal communication). – Only works pertaining to Peninsular Spain as a whole are given under this heading. The important separate works on Galicia, Catalonia, Andalucía and the Balearic Islands are treated respectively under **613**, **615**, **618** and **619**. Gibraltar appears under **618**. For Spanish settlements in Morocco (Ceuta, Melilla, etc.), see **599**.

The long-standard major national and regional works are, particularly over the last decade, being succeeded by a new 'generation'. *Flora iberica*, when completed, will in particular fill a major gap.

Bibliography
Literature to 1885 is summarized in the prefatory part of the first volume (1885) of *Enumeración y revisión de las plantas de la Península* . . . by Miguel Colmeiro (see introduction to **Region 61**). In the twentieth century, significant retrospective contributions were made by E. F. Galiano with B. Valdés in *Boissiera* **19**: 23–60 (1971) and by Galiano alone (in HEYWOOD, 1975; see **601/I**). These were continued by Galiano and Valdés in *Mem. Soc. Brot.* **24**: 377–394 (1974) and *Lagascalia* **4**: 239–258 (1974), **7**: 83–120 (1977), and **9**: 3–28 (1979). For literature from 1989 onwards, see PAJARÓN under **Region 61, Regional index**.

GARCÍA ROLLÁN, M., 1996–97. *Atlas clasificatorio de la flora de España peninsular y balear*. 2 vols. Illus. (text-figs., col. photographs). Madrid:

Mundi-Prensa/Ministerio de Agricultura, Pesca y Alimentación (MAPA). (A revision of *idem*, 1981–83. *Claves de la flora de España (Península y Baleares)*. 2 vols. 675, 764 pp., illus. Madrid: Mundi-Prensa. (2nd edn. of vol. 2, 1985.))

Copiously illustrated manual-key to vascular plants (5250 species), adapted from *Flora Europaea*; includes vernacular names, notes on distribution and habitat (diagnostic features being given in larger type), and, in each volume, indices to families and to all binomials and vernacular names. The introductory section gives the background to and plan of the work and general keys to families. The arrangement of families and genera is alphabetical within each of the four major groups. The numerous color photographs are small but very well reproduced.

GUINEA LÓPEZ, E. and A. CEBALLOS JIMÉNEZ, 1974. *Elenco de la flora vascular española (Península y Baleares)*. ii, 403 pp. Madrid: Instituto Nacional de Conservación de la Naturaleza (ICONA).

Systematic enumeration of vascular plants, with synonymy, citations, karyotypes (where known), habitat, altitudes, habit, and symbolic indication of local and overall distribution, but no critical remarks; addenda and indices to genera and families at end. A short introductory section contains a list of works consulted. Based in part on *Flora Europaea*.

LÁZARO É IBIZA, B., 1920–21. *Botánica descriptiva: compendio de la flora española*. 3rd edn. 3 vols. 1000 text-figs., map. Madrid. (1st edn., 1896.)

Briefly descriptive flora of non-vascular and vascular plants, with synoptic keys to groups of species and to sections, genera and families and incorporating vernacular names, brief indication of local range, floral diagrams and figures of representative species, and notes on habitat, phenology and special features; appendix on vegetation and phytogeography and indices to all vernacular and botanical names. The introductory section (vol. 1) includes a glossary and several alternative general keys, as well as a historical account of botanical work in Spain. [Designed mainly as a students' manual; largely compiled from various sources including Willkomm and Lange's *Prodromus* (see below).]

SMYTHIES, B. E., 1984–86. *Flora of Spain and the Balearic Islands: checklist of vascular plants*. xxxiii, 882 pp., 2 maps (Englera 3; in three parts). Berlin.

Alphabetically arranged enumeration (within each of the major vascular taxa), without keys; species entries include essential synonymy, references, citations of 'standard' works, status, and distribution by compass directions, along with extralimital range. The introductory section gives details of the work as well as a discussion of Spanish geography (including the former kingdoms and modern provinces); it also commemorates the 100th anniversary of completion of the *Prodromus* (next entry), acknowledging it and *Flora Europaea* as primary sources.

WILLKOMM, M. and J. LANGE, 1861–80. *Prodromus florae hispanicae*. 3 vols. Stuttgart: Schweizerbart. Continued as M. WILLKOMM, 1893. *Supplementum prodromi florae hispanicae 1862–93*. ix, 370 pp. Stuttgart. (Both works reprinted 1972, Stuttgart.)

Concise, briefly descriptive flora (in Latin) of native, naturalized and commonly cultivated plants of both Spain and Portugal (in spite of the title, though the Balearic Islands are omitted); includes non-dichotomous analytical keys to genera (and, in the larger genera, synoptic keys to species), full synonymy (with references), vernacular names, detailed account of internal range with *exsiccatae* and localities, general summary of extralimital distribution, taxonomic commentary, and partly abbreviated notes on habitat, life-form, phenology, etc.; index to genera in vols. 1–2, addenda and corrigenda, and complete indices to all botanical and known vernacular names in vol. 3. The introductory section includes a synopsis of families as well as lists of abbreviations and authors' names. The *Supplementum* comprises further addenda and corrigenda.[41]

Partial work

CABALLERO, A., 1940. *Flora analítica de España*. xiv, 617 pp., 268 text-figs. Madrid: Sociedad Anónima Española de Traductores y Autores. (Reprinted 1983, Königstein/Ts., Germany: Koeltz.)

Concise students' manual-key to some 3000 more common or significant native, naturalized and cultivated plants; comprises descriptive but unannotated keys to all taxa (with figures of representative species) followed by an illustrated glossary, list of generic synonyms, and lastly an alphabetical index-enumeration of accepted names, those of species accounting for vernacular names, habitat, and altitudinal range and furnished with figures of representative species.

Woody plants

The book by Ruiz has less than a quarter of the species covered in its 1883–90 predecessor, *Flora forestal española*.

For more complete coverage, *La guía de INCAFO* (**610** above) may be consulted.[42] Trees alone (including introduced species) are covered in A. LÓPEZ LILLO and J. M. SÁNCHEZ DE LORENZO CÁCERES, 1999. *Árboles en España.* 643 pp., illus. Madrid: Mundi-Prensa.

RUIZ DE LA TORRE, J. and L. CEBALLOS, 1971. *Árboles y arbustos de la España Peninsular.* xxii, 512 pp., 133 pls. Madrid: Instituto Forestal de Investigaciones y Experiencias/E.T.S. de Ingenieros de Montes. (Later edition by Ruiz alone, 1979. See also RUIZ DE LA TORRE, J., 1984. *Árboles y arbustos de España.* Madrid: Salvat.)

Dendrological atlas-manual of trees and treelike shrubs, with extensive descriptions of species, Spanish, Catalan, French, English and German vernacular names, and copious notes on ecology, phenology, distribution, diseases, and forestry aspects, and full-page drawings; illustrated glossary, periodical abbreviations, and indices at end. The introductory section includes a synoptic list (encompassing 125 species), technical notes, and general and artificial keys to all species (pp. 1–35). An extensive bibliography appears on pp. 505–512.

613

Galicia, Asturias and Cantabria

See also **612** (all works). The only significant regional flora is that for Galicia by Merino, which with its supplement remains a 'standard' work within the Iberian Peninsula. For Asturias, *Mis contribuciones al conocimiento de la flora de Asturias* (1982) by M. Laínz may be consulted.

MERINO Y ROMÁN, B., 1905–09. *Flora descriptiva é ilustrada de Galicia.* 3 vols. Illus. Santiago de Compostela: The author. (Reprinted 1980, La Coruña: 'La Voz de Galicia' (in their collection «Biblioteca Gallega»).) For additions (collectively reprinted from *Brotéria*, Sér. Bot. **10–14**, *passim* (1912–16)), see *idem*, 1917. *Adiciones a la flora de Galicia.* [iii], 176 pp., 10 pls. Braga.

Descriptive flora of vascular plants (1854 species); includes partly synoptic keys to all taxa, vernacular names, fairly detailed indication of local range, figures of representative species, and notes on habitat, phenology, etc.; addenda and corrigenda, remarks on floristics and phytogeography, and indices to all botanical and vernacular names in vol. 3. The introductory section includes an illustrated glossary. [The reissue includes a new foreword and a short biography with an appreciation of the author's scientific work.][43]

Woody plants

Galicia is covered in Peninsular and Spanish works, but for a Gallegan-language tree book, see F. J. SILVA PANDO and A. RIGUEIRO RODRÍGUEZ, [1992]. *Guía das árboles é bosques de Galicia.* 294 pp., 106 col. illus. on 64 pp., folding map, unpaged index and table of contents. Vigo: Galaxia. [Field-guide with descriptions of native and introduced trees along with family and artificial keys and gridded distribution maps.]

Pteridophytes

For an atlas-guide to ferns and fern-allies (in Gallegan), see I. BARRERA MARTÍNEZ, 1980. *Os fentos de Galicia.* 124 pp., illus. (Cuadernos da área de ciencias biolóxicas 1). Coruna.

614

Basque Lands (Vascongadas) and Navarra

See also **612** (all works). Woody plants are additionally covered in I. AIZPURU, P. CATALÁN and F. GARÍN, 1996. *Guía de los árboles y arbustos de Euskal Herria.* 2nd edn. 477 pp., illus., maps (part col.). Vitoria-Gasteiz: Servicio Central de Publicaciones del Gobierno Vasco. (1st edn., 1990.) Aseginolaza's work (see below) does not cover Navarra.[44]

ASEGINOLAZA IPARRAGIRRE, C. *et al.*, 1984–85. *Catálogo florístico de Álava, Vizcaya y Guipúzcoa/ Araba, Bizkaia eta Gipuzkoako landare katalogoa.* xxii, 1149 pp., figs., maps (one folding). Vitoria-Gasteiz: Gobierno Vasco, Viceconsejera de Medio Ambiente.

Annotated atlas-catalogue of vascular plants with gridded maps for all species; includes essential synonymy, standardized vernacular names (Basque and Spanish), some taxonomic commentary, and details of distribution, altitudinal range, frequency code(s), habitat and (if any) special features; bibliography (pp. 1121–1138) and index to genera at end. The map grid is 10×10 km (UTM-coordinates), but points, based on *exsiccatae* and literature records alike (and depicted respectively as solid and open circles), are calculated to the nearest 1 km. The introductory section (in part bilingual) includes details on the background of the work and its execution and sections on physical features, climate, vegetation and previous botanical work; conventions follow (pp. 20–23).

615

Catalonia (Catalunya)

See also **612** (all works). The area covered by *Flora dels països Catalans* is about 70 000 km^2, but this also accounts for Perpinyà (France), Andorra, Franja de Ponent (part of Aragón), País Valencià and the Balearic Islands as well as Catalonia proper.

Earlier floras include *Catálogo metódico de plantas observadas en Cataluña* (1846) by Miguel Colmeiro and *Introducción a la flora de Cataluña* (1864; 2nd edn., 1877) by Antonio C. Costa y Cuxart. Also of importance is *Catalogue raisonné de la flore des Pyrénées-Orientales* (1898) by G. Gautier. The modern works are both associated with major Catalan revivals and have been promoted by the Institut d'Estudis Catalans.[45]

BOLÒS I CAPDEVILA, O. DE et al., 1985– . *Atlas corològic de la flora vascular dels Països Catalans*. Vols. 1– . Maps 1– , plastic overlays. Barcelona: Institut d'Estudis Catalans (distributed by ORCA). Complemented by O. DE BOLÒS I CAPDEVILA, 1998. *Atlas corològic de la flora vascular dels Països Catalans: primera compilació general*. 2 parts. 1102 pp., over 4400 maps. Barcelona.

The *Atlas* proper features full-page gridded distribution maps, with occurrence by squares depicted according to presence/absence; sources of data are given overleaf (for explanation in both Catalan and English, see the prefatory pages in each volume). The overlays supplied in vol. 1 relate distribution to selected parameters including altitudinal zones, population, climate and rainfall, and, for 'fidelity', phytosociological units. Maps are cross-referenced to *Flora Europaea* and *Flora dels Països Catalans* (the latter by number). The *Compilació general* covers the entire vascular flora, including maps not so far published in the main work; the genera and species are alphabetically arranged. Four maps appear on each page, and there is no text. [As of late 1999 nine volumes of the main work, encompassing maps 1–2106, had been published. No systematic order has been followed, but the first five volumes are cumulatively indexed in vol. 5. Volume 1 is loose-leaf; the remainder paperbound. Maps in the *Compilació general* numbered above 1519 are 'advance versions' which when published in the *Atlas* will be more complete.]

BOLÒS I CAPDEVILA, O. DE and J. VIGO, 1984–95. *Flora dels Països Catalans*. Vols. 1–3. Illus., maps. Barcelona: Barcino.

Illustrated atlas-flora in manual-key form; includes diagnostic descriptions, essential synonymy, vernacular names, detailed indication of local distribution, habitat(s) and association(s), altitudinal range and frequency (grouped by Catalonia proper, Valencia and the Balearics), life-form, habit and phenology along with some critical commentary; individual figures and maps of local and extralimital range (the latter only in Europe and adjacent areas), lexicon of adjectives and other essential words (with Latin, German and English equivalents) and index to genera and families in each volume. The introductory section (in vol. 1) includes substantial chapters on geography and geobotany (including a gazetteer and coverage of bioclimates, chorological classes, associations and vegetation formations) and history of floristic work; principal works (p. 108) and local floras (pp. 108–110), conventions and abbreviations (pp. 111–114) and a general key to families follow. [All vascular plants are now accounted for except monocotyledons, to occupy the fourth and final volume.][46]

BOLÒS I CAPDEVILA, O. DE, J. VIGO, R. M. MASALLES and J. M. NINOT, 1990. *Flora manual dels Països Catalans*. 1247 pp., text-figs. Barcelona: Pòrtic. (Conèixer la Natura 9.)

Concise illustrated manual-flora (covering 3580 species and infraspecific taxa) with keys to all taxa, brief descriptions, essential synonymy, life-forms, habit, phenology and notes (in part abbreviated) on status, local distribution, habitat, plant-sociological associations, and chorology (including overall range-classes); numerous diagnostic figures; glossary, necessary nomenclatural changes and index to all scientific names at end. The introductory part covers the background, purpose and scope of the work, phytogeography (including chorological provinces) in general and in the Catalan lands, vegetation formations, and plant associations and their subdivisions along with abbreviations and a general key to families. No vernacular names or taxonomic commentaries are included. [A second edition appeared in 1993.]

CADEVALL I DIARS, J., A. SALLENT I GOTÉS and P. FONT I QUER, 1913–37. *Flora de Catalunya*. 6 vols. Illus., 6 col. pls. Barcelona: Institut de Ciencias.

Illustrated descriptive atlas-flora of vascular plants; includes keys to all taxa, limited synonymy, vernacular names, fairly detailed indication of local range, figures of each species, notes on habitat, phenology, etc., and in each volume complete indices; illustrated glossary and general index in volume 6. The introductory section (vol. 1) includes accounts of botanical exploration in the province and definitions of floristic zones.

Woody plants

MASCLANS, F., 1988. *Guia per a conèixer els arbres*. 8th edn. 254 pp., 16 halftone pls., 35 full-page or text-figs., 110 + 1 maps. Barcelona: Montblanc/Centre Excursionista de Catalunya. (1st edn., 1963; 2nd through 5th edns., 1973–82.) Accompanied by *idem*, 1988. *Guia per a conèixier els arbusts i les lianes*. 6th edn. 268 pp., 53 full-page or text-figs., map. Barcelona. (1st edn., 1963.)

Semi-popular field-guides respectively to trees and to shrubs and lianas, relying in the first instance on vernacular names with scientific names subordinate; includes keys to all taxa, vernacular names, descriptions, figures, distribution maps, and indication of local and extralimital range, phenology, habit, and vegetation type(s); index to all names at end. The introductory parts include treatments of distribution, vegetation, and organography (with illustrations) along with the general keys and notes on how to identify plants.

616

Valencia and Alicante (País Valencià)

See also **612** (all works). País Valencià is also covered by *Flora dels Països Catalans* and *Flora manual dels Països Catalans* (**615**) as well as the two works described below.

MATEO SANZ, G. and M. B. CRESPO VILLALBA, 1998. *Manual para la determinación de la flora Valenciana*. 495 pp. Valencia: The authors. (Monografías de flora Montibérica 3.) (Based on the authors' *Claves para la flora valenciana* (1990, Valencia) and *Flora abreviada de la Comunidad Valenciana* (1995, Alicante).)

Comprises analytical keys (in Spanish) with for individual taxa essential synonymy, vernacular names, distribution (by province: Alicante, Castellón and Valencia), habitat, chorological and life-form classes, height, phenology and frequency (for explanation, see pp. 7–14); illustrated glossary and index to generic and family names at end.

MATEO SANZ, G. and R. FIGUEROLA LAMATA, 1987. *Flora analítica de la provincia de Valencia*. 384, [2] pp., 4 maps. Valencia: Alfons el Magnanim/Institució Valenciana d'Estudis i Investigació.

Concise keyed enumeration of vascular plants (covering 2033 species and 102 non-nominate subspecies); includes essential synonymy (with references), habitat, critical notes where appropriate, and indication of habit, life-form, phenology, chorological class and frequency; statistics (pp. 366–367), new combinations and taxa, references (pp. 372–377) and index to genera and families at end. The introductory section gives a justification for the work, a history of botanical investigations, and chapters on bioclimate, chorology, life-forms, taxonomic philosophy and technical aspects of the work.

617

Central Spain (Aragón, Rioja, Castilla y León, Castilla-La Mancha, Extremadura and Madrid)

See **612** (all works). Several regional and provincial floras are also available, although some still current are very old. For the Madrid region in particular, see J. RUIZ DE LA TORRE, 1982. *Aproximación al catálogo de plantas vasculares de Madrid*. Madrid: Comunidad de Madrid, Consejeria de Agricultura; and S. RIVAS-MARTÍNEZ *et al.*, 1981. *Flora matritensis*, 1 (Pteridophyta). *Lazaroa* 3: 25–61.

618

Andalucía (with Gibraltar), Almería and Murcia

See also **612** (all works). For Andalucía as a whole, reference may additionally be made to the following 'provisional' (and uncritical) work: C. FERNÁNDEZ LÓPEZ et al., 1991. *Flora de Andalucía: catálogo bibliográfico de las plantas vasculares.* 99 pp., map. Jaén: Facultad de Ciencias Experimentales, Jaén.

Western Andalucía and Gibraltar

In addition to its inclusion in *Flora vascular de Andalucía Occidental*, Gibraltar is covered in successive local floras. Of these, the most recent are L. L. LINARES, 1983. A checklist of the Gibraltar flora. *Alectoris* 5: 24–39; and its supplement: *idem*, 1990. Amendments to the checklist of the Gibraltar flora. *Ibid.*, 7: 77–82. Both works are based on a collection of slide diapositives made by the author and A. Harper and include accepted scientific and vernacular names, status and phenology.

VALDÉS, B., S. TALAVERA and E. FERNÁNDEZ-GALIANO, 1987. *Flora vascular de Andalucía Occidental.* 3 vols. Illus., maps. Barcelona: Ketres.

Illustrated descriptive atlas-flora with individual distribution maps, two species to a page; includes keys to all taxa, synonymy (with references), local (by 'natural' districts) and extralimital distribution, critical commentary where necessary, karyotypes, phenology, and notes on habitats and status; index in each volume (with general index to families and genera at end of vol. 3). Distributions are depicted on the maps by shading. The introductory part (in vol. 1) deals with the plan of the work, major physiographic units (as used on the maps), chorology (with maps of Spain and Europe), the sequence of families (flowering plants follow Cronquist), and a general key to families.[47]

Almería

SAGREDO, R., 1987. *Flora de Almería: plantas vasculares de la provincia.* xxiii, 552, [7] pp., illus., col. pls. Almería: Instituto de Estudios Almerienses.

Illustrated quarto atlas (in the style of Britton and Brown's *Illustrated flora of the northern United States, Canada, and the British possessions* but with a somewhat more fluid arrangement of figures and text); includes synonymy, descriptions, localities (of the author and from other sources, including *exsiccatae*), overall range, phenology, and (sometimes) notes; illustrated glossary and indices to vernacular and scientific names at end. The introductory part includes a history of botanical work in the region. [Covers 2658 taxa in 149 families, all sequentially numbered; an additional 208 cultivated species, designated by Roman numerals, are intercalated.]

Murcia

SÁNCHEZ GÓMEZ, P. et al., 1996. *Flora de Murcia: claves de identificación de plantas vasculares.* 378 pp., 3 maps, 2 pls. Murcia: DM.

Keyed enumeration of vascular plants, the species entries featuring vernacular names, indication of distribution and habitat, and codes for bioclimate, humidity and frequency; new taxa and combinations (p. 347), glossary, plates and index to genera and families at end. The introduction covers geography, bioclimatology (with maps), biogeography and chorology (with map), and vegetation, with the general keys following (pp. 39–52).

619
Balearic Islands

See also **612** (all works except Willkomm and Lange); **615** (*Flora dels països Catalans*; *Flora manual dels països Catalans*). – Area: 5014 km². Vascular flora: 1450–1550 species, of which 1250–1450 native (Webb). Ibiza and Formantera together form a subgroup known as the Pityusic Islands.

No modern one-volume manual has appeared to succeed *Flora de las Islas Baleares* (1879–81) by Francisco Barceló y Combis or the multi-volume *Flora balearica* (1921–23) by Herman Knoche. The illustrated *Flora de Mallorca* is limited to that island. Important earlier accounts include *Enumeratio plantarum quas in insulis Balearibus collegit J. Cambessèdes* (1827) by Jacques Cambessèdes, encompassing 1320 native species, and *Catalogue raisonné des plantes vasculaires des îles Baléares* (1880) by P. Marès and G. Vigineix.

BARCELÓ Y COMBIS, F., 1879–81. *Flora de las Islas Baleares.* xlviii, 645 pp. Palma de Mallorca: Gelabert.

Briefly descriptive flora of the vascular (and some non-vascular) plants (in Spanish): includes polychotomous keys to taxa above species level, limited synonymy, vernacular names, indication of range by island and locality (with citations), and notes on habitat, phenology, uses, etc.; supplement (pp. 563–592) and indices to generic, family and vernacular names. The introductory section includes descriptions of the geography, geology, climate, and history of botanical work in the islands, as well as a list of authors. [The work incorporates all records from the checklist of Marès and Vigineix.]

DUVIGNEAUD, J., 1979. *Catalogue provisoire de la flore des Baléares.* 2nd edn. 43 pp. Liège: Département de Botanique, Université de Liège. (Mimeographed. 1st edn., 1974.) (Also published as *Société pour l'Échange des Plantes Vasculaires de l'Europe Occidentale et du Bassin Méditerranéen*, Fasc. 17, Supplément.)

Systematic checklist of vascular plants, with indication by island based on available records and the author's collections and observations over several years; addenda and corrigenda (pp. 42–44); no index.

KNOCHE, H., 1921–23. *Flora balearica*. 4 vols. 47 pls., frontisp., tables, maps. Montpellier: Roumégouis et Déhan. (Reprinted 1974, Königstein/Ts., Germany: Koeltz.)

A detailed floristic and phytosociological treatise, but lacking keys. Volumes 1–2 contain an enumeration of non-vascular and vascular plants (covering all of the latter), with complete synonymy, references, fairly detailed indication of local range, general summary of extralimital distribution, and extensive notes on habitat, life-form, phenology, etc.; index. An introductory section (in vol. 1) includes a list of authors and tables of signs and abbreviations, as well as a gazetteer. Volume 3 is wholly devoted to floristics, phytogeography, and phytosociological analyses; vol. 4 contains plates of illustrations and a map.[48]

Mallorca

BONAFÈ BARCELÓ, F., 1977–80. *Flora de Mallorca*. 4 vols. Illus., maps. Palma de Mallorca: Moll.

Atlas-flora of vascular plants (in Mallorcan), arranged according to the Englerian system; contains photographs of all species with accompanying descriptive text, synonymy, vernacular names (Mallorcan, Catalan, Spanish), extralimital distribution, and indication of local occurrence, habitat and frequency; index at end of each volume (general index in volume 4). The introductory section contains a list of references (pp. xi–xviii) and an illustrated glossary, while in volume 4 are three appendices: an index to authors' names, a tabular list of vascular plants found throughout the Balearics (Ibiza and Formantera together here termed the Pityusic Islands), and a lexicon of vernacular names with their scientific equivalents.

Region 62

Italian peninsula and associated islands

Area: 310 120 km^2 (region); 251 447 km^2 (mainland Italy). Vascular flora: 6190 species in the whole of Region 62 and some adjacent areas (Zangheri); 5250–5350 species in the mainland, of which 4750–4900 native (Webb). Pignatti (1982) accounts for 5599 species. – Within this region are the Italian mainland (including San Marino), Corsica, Sardinia, the Lipari Islands, Sicily, Malta, and Pantelleria (with the Pelagic Islands). Most, if not all, general works on Italy cover the whole of the region; however, as a fair literature exists for each of the main islands or island groups, they are here separately accounted for.

Floristic exploration and documentation after four centuries are in general more advanced than in the Iberian Peninsula, but until recent years have remained behind most of northern Europe. Reflecting the social and political situation, floristic study and publication remained essentially provincial or regional until well into the nineteenth century, although a slim nomenclator, *Florae italicae prodromus* by Antonio Turra, was prepared by 1768 and appeared in Vicenza in 1780. The first serious general flora was Antonio Bertoloni's 10-volume *Flora italica* (1833–57) which, though already accounting for 4309 species and in its time well regarded and influential, was anachronistic in its retention of the Linnaean system.[49] More in accord with contemporary ideas of systematics was the *Flora italiana* (1848–96, not completed) of Filippo Parlatore (founder of the present Museo Botanico in Florence) and his successor Théodore Caruel. The Mezzogiorno responded with *Compendio della flora italica* (1868–89) by Vincenzo de Cesati and others, a work first issued in installments as part of *L'Italia*, a national encyclopedia, but also published separately. The need for a compact flora for field use was filled by Giovanni Arcangeli's *Compendio* (1882; 2nd edn., 1894), in physical terms still not superseded as its successors are either selective (Baroni) or better suited to four wheels (Zangheri).

The *Flora analitica d'Italia* (1895–1908) of Adriano Fiori and Giulio Paoletti established a new tradition, continued by all its successors with Sandro Pignatti adding maps to what is now the current standard, his *Flora d'Italia* (1983). The second edition of Fiori's work, the *Nuova flora* (1923–29; illustrations 1933), was unfortunately distinguished by an unusually broad species concept and has been considered as of lesser quality than the original edition. For student and field use, Eugenio Baroni's *Guida* (1907; 4th edn., 1969, with several reissues), first written for central Italy but with the second edition of 1932 expanded to the whole peninsula and the islands from Malta to Corsica, has long been current but not now substantively revised for more than 60 years; a similar work for northern Italy by Silvia Zenari, *Flora escursionistica* (1957), has likewise never been reworked. As for checklists, that by Raffaele Ciferri and Valerio Giacomini, *Nomenclator florae italicae* (1950–54), was never completed; however, at a later date the database for *Flora d'Italia*, developed by the Pignatti group, was used to produce a nomenclator, *Check-list of the flora of Italy with codified plant names for computer use* (1980).

Several provincial and insular floras have also been published, some only the latest in a long succession of works. A distinct gradient in coverage from north to south has now become manifest, with a varied picture in the islands (Corsica and Malta being perhaps the best documented; by contrast, no separate flora of Sicily has appeared since Lojacono-Pojero's *Flora sicula*). Mention should also be made of the works of Luigi Fenaroli: *Alberi: dendroflora italica* (1976, with G. Gambi; see below) and *Flore delle Alpi* (1955, 1971; see **603/IV**), the latter perhaps the most complete of Alpine floras, and *Alberi e arbusti in Italia* by M. Ferrari and D. Medici (1996; see below).[50]

Bibliographies. General and divisional bibliographies as for Division 6.

Regional bibliography
Very thorough coverage through 1959 is available in Blake, 1961 (see **General bibliographies**). A *Bibliografia floristica italiana* (1969) and first *Supplemento* (1973) by Guido Moggi have been circulated but not formally published.

SACCARDO, P. A., 1895–1901. *La botanica in Italia*. 2 parts. 236 + xv, 172 pp. (Mem. Reale Ist. Veneto Sci. 25(4), 26(6)). Venice. (Reprint in 1 vol., n.d., Bologna: Forni, as no. 1 in their series *Biblioteca botanica*.) [Includes an annotated bio-bibliography by

authors (pp. 11–180 and 7–127), the references in short-title form.]

Indices. General and divisional indices as for Division 6.

620
Region in general

Works dealing with Italy and associated lands and islands are treated here. Most, if not all, include in addition to present-day Italy the Nizzardo (the eastern French Riviera from Nice to Ventimiglia), Ticino (Switzerland), eastern Friuli (**629**; now in Slovenia), the Istrian peninsula (**629**; the greater part now in Croatia), Corsica (**621**) and Malta (**627**). Baroni's *Guida* is, however, limited to the mainland. For Fenaroli's *Flora delle Alpi*, see **603/IV**.

Pignatti's *Flora d'Italia* is the functional successor to A. FIORI, 1923–29. *Nuova flora analitica d'Italia*. 2 vols. 22 text-figs. Florence: Ricci (reprinted 1969, 1974, 1981 in somewhat reduced format, Bologna: Edagricole), its atlas: A. FIORI and G. PAOLETTI, 1933. *Iconographia florae italicae, ossia flora italiana illustrata*. 3rd edn. vii, 549 pp., 4419 text-figs. Florence (reprinted 1970, 1974, 1981 in somewhat reduced format, Bologna), and A. FIORI, 1943. *Flora italica cryptogama*, 5: *Pteridophyta*. 601 pp. Florence: Società Botanica Italiana.

ARCANGELI, G., 1894. *Compendio della flora italiana*. 2nd edn. xix, 836 pp. Turin: Loescher. (1st edn., 1882.)

Concise manual of vascular plants, with non-dichotomous keys to families and genera along with synoptic keys to groups of species; also includes limited synonymy, general indication of internal and extralimital range (the latter very abbreviated), and notes on habitat, life-form, phenology, etc.; index to all botanical names at end. The work covers the Italian peninsula, Istria, Corsica, Sardinia, Sicily and Pantelleria, accounting for 4932 species.

BARONI, E., 1969. *Guida botanica d'Italia*. 4th edn., revised and corr. by S. Baroni Zanetti. xxxii, 545 pp., 16 col. pls. (with 144 figs.). Bologna: Cappelli. (Several reissues, at least to 1982. 1st edn., 1907, Rocca S. Casciano; 2nd edn., 1932, Bologna (reissued 1935);

3rd edn., 1955, Rocca S. Casciano (reissued 1963, Milan).)

Field manual-key to selected native, naturalized and commonly cultivated vascular plants (largely based on Fiori's *Nuova flora analitica*), with very limited synonymy, concise summary of distribution in the Italian region, and indication of habitat, life-form, phenology, etc.; colored figures of representative species; glossary and index to illustrations as well as to family and generic names. The brief introductory section is pedagogical. [3446 species in all accounted for; many uncommon and rare taxa omitted.]

PIGNATTI, S. (with C. ANZALDI, F. DUSA, P. NIMIS and L. PASSERINI), 1980. *Check-list of the flora of Italy with codified plant names for computer use*. 256 pp. Rome: Consiglio Nazionale delle Richerche. (Serie di Contributi a Congressi AQ/5/13.)

A nomenclator in two parts: an alphabetical list of all taxa with numerical 'genspec' codes and (from p. 117) a list of genera and species arranged systematically by families, with codes (their generic component being based on those of Dalla Torre and Harms). The introductory part appears in both Italian and English. Some 6000 species are accounted for, but without synonymy (a future version to include such was projected). [Representative of the database for *Flora d'Italia* (see below). Covers Corsica, Ticino and the Maltese Islands as well as Italy.]

PIGNATTI, S., 1982. *Flora d'Italia*. 3 vols. 2302 pp., *c.* 8400 text-figs., maps. Bologna: Edagricole.

Copiously illustrated, briefly descriptive flora of vascular plants; includes keys to all taxa, limited synonymy, vernacular names, distribution (with dot maps of varying detail), and notes on habitat, altitudinal range, frequency, habit, phenology, and chorology along with critical remarks including evolutionary and ecological dynamics; diagnoses for families and genera with derivations of names and references to revisions and other taxonomic works; sources of figures and indices to names at end. The introductory section deals with the plan of the work, past botanical activity, a definition of regional chorological units (20 in all), and the role of data processing; these sections are followed by standard references (pp. 19–21), a limited list of technical terms, protected plants, abbreviations and a general key (pp. 27–36). [Covers 5499 species in mainland Italy, the Nizzardo, Ticino, the Istrian peninsula and parts of Austria and Slovenia along with all associated islands south to Malta, including Corsica, Sardinia and Sicily.][51]

ZÁNGHERI, P. (assisted by A. J. B. BRILLI-CATTARINI), 1976. *Flora italica (Pteridophyta–Spermatophyta)*. 2 vols. xxii, 1157 pp. (text); xxii pp., 7750 figs. on 210 pls. Padua: A. Milani (CEDAM).

A concise manual of vascular plants, with separate text and atlas volumes. Two sections comprise the text volume: a general part covering life-forms, altitudinal zones, technical notes, symbols and abbreviations, and a glossary; and a special part comprising a manual-key to native, naturalized and commonly cultivated vascular plants, with essential synonymy, vernacular names, karyotypes, indication of habit and life-form, phenology, altitudinal range, and internal and extralimital distribution given in 1–2 lines for each; bibliography of key works, addenda and corrigenda, and complete indices at end. The accompanying atlas volume contains analytical figures for identification, with a separate skeleton index. Encompasses 8452 terminal taxa in 6190 species, of which 5692 species are keyed out and the remainder merely referred to in the text. The whole of the Italian region as defined here is included, along with Ticino, the Nizzardo and the Istrian peninsula. [A functional, if bulkier, successor to Arcangeli's manual.]

Woody plants

For more compact field works see L. FENAROLI, 1974. *Gli alberi d'Italia*. 320 pp., illus. Milan: Giunti-Martello (1st edn., 1967, Aldo Martello); and E. FERIOLI, 1987. *Atlante degli alberi d'Italia*. 231 pp., illus. (mostly col., by G. Pozzi), portr., maps. [Milan]: Mondadori. The latter covers 103 species, some not native. An older work is *Compendio della flora forestale italiana* (1885) by Antonio Borzì.

FENAROLI, L. and G. GAMBI, 1976. *Alberi: dendroflora italica*. 717 pp., illus. (part col.). Trent: Museo Tridentino di Scienze Naturali. (Additional issues through 1989.)

A well-illustrated, semi-popular dendrology of native and key non-native species, with references for each species, synonymy, local and other European vernacular names, distribution (with maps), and habitat and other notes, along with critical notes including discussions of alternative classifications; general references and indices to all names at end. The introductory section covers the scope of the work and includes a discussion of phytosociological units.[52]

FERRARI, M. and D. MEDICI, 1996. *Alberi e arbusti in Italia: manuale di riconoscimento*. xvi, 967 pp., illus. (part col.). Bologna: Edagricole.

A semi-popular but detailed atlas (in landscape format) of *c.* 350 species of native and common introduced trees, shrubs and woody climbers, with text facing artwork and photographs; individual species accounts in a *pro forma* style with botanical and vernacular names, description, indication of distribution (and origin, if appropriate), taxonomic and cultural commentary, and notes on actual or potential pests and diseases as well as appearance, cultivation and tolerances. Extra illustrations are furnished for problematic species such as *Juniperus communis*, and the notes moreover may contain references to additional species not otherwise fully accounted for. This main part (chapter 2) is preceded by a general part (chapter 1) including (pp. 1–40) an illustrated organography and (pp. 41–42) an introduction to classification; it is followed (pp. 856ff.) by two further chapters on the development of botany, collecting and herbaria, landscaping and garden design, ecology, and plant protection. A glossary, bibliography and index to all names conclude the work.

621

Corsica

Area: 8723 km². Vascular flora: 2524 native taxa with 2092 species (Gamisans and Jeanmonod, 1993; see below). Webb (1978) estimated 2250–2400 species (2150–2250 native). Apart from the two general floras described below (neither providing complete coverage, although the continuation of Briquet's work is attempting to close this lacuna), Corsica is covered by all standard floras and related works on Italy (**620**) as well as France (**651**), and known species are all accounted for by Gamisans and Jeanmonod. The high-mountain zone is additionally covered by Fenaroli's *Flora delle Alpi* (**603/IV**).

Prior to the work of John Briquet and René de Litardière, the most substantial treatment of the flora was *Catalogue des plantes vasculaires indigènes ou généralement cultivées en Corse* (1872) by L. J. A. de Marsilly.

Bibliography
References to 1955 are listed in volume 1 of Briquet and de Litardière's *Prodrome*. For additional references see Gamisans and Jeanmonod (1993; below).

GAMISANS, J., 1980. Bibliographie botanique corse, 1955–1979. *Candollea* 35: 211–221. [233 references on vascular and non-vascular plants, fungi and lichens, arranged by author.]

BOUCHARD, J., 1977 (1978). *Flore pratique de la Corse.* 3rd edn. 405 pp., 50 pls. (incl. 2 maps) (Bull. Soc. Sci. Hist. Nat. Corse, numéro spécial 7). Bastia: Société des Sciences Historiques et Naturelles de la Corse. (1st edn., 1968.)

Concise manual-key to vascular plants (1833 species); includes limited synonymy, French and local vernacular names, general indication of local range (with some citations), figures and photographs of representative species, and brief, partly symbolic notes on habitat, phenology, life-form, frequency, uses, etc.; list of endemic and phytogeographically interesting taxa, lexicon of Corsican local names, and index to generic and family names and their French equivalents at end. A brief introductory section includes a glossary, abbreviations of authors' names, and a general key to families.

BRIQUET, J. and R. DE LITARDIÈRE, 1910–55. *Prodrome de la flore corse.* Vols. 1–3. Geneva: Georg (vols. 1, 2, part 1), Paris: Lechevalier (vols. 2, part 2, 3). Continued as D. JEANMONOD *et al.*, 1987– . *Compléments au* Prodrome de la flore corse. Published in fascicles. Illus., maps, microfiches. Geneva: Conservatoire et Jardin Botaniques.

The *Prodrome* is a detailed, critical systematic enumeration of vascular plants and includes full synonymy (with references and citations), detailed indication of local range with citation of *exsiccatae* and general summary of extralimital distribution, extensive taxonomic commentary, and notes on habitat, life-form, phenology, ecology, etc.; index to generic and family names in each volume. The introductory section (vol. 1) includes a bibliographic part as well as notes on the plan of the work and records of field trips. The *Compléments* comprise detailed revisionary treatments, including keys (not supplied in the original work), synonymy and references, substantial descriptions, local distribution (with full list of *exsiccatae*) and chorological position (including documented maps for all species), sometimes extensive taxonomic commentary, and notes on habitat, local and extralimital distribution, phytosociological units, and special features; references and an index to all scientific names appear at end of each fascicle. An introductory fascicle, published as *Annexe* 1, gives the background to the project, the plan of the work, conventions and abbreviations, and contents of maps and map overlays (provided for use with all the species maps). The inside covers include an index to all families with their listings here and in the original work as published. In addition to new family treatments, supplements to families in the

Prodrome have been projected. [The original work extends from the Pteridophyta up through the Solanaceae (Englerian system). Of the *Compléments*, still in progress, nine fascicles had been issued through 1998, the most recent, Asteraceae I, in that year.][53]

GAMISANS, J. and D. JEANMONOD, 1993. *Catalogue des plantes vasculaires de la Corse*. 2nd edn. 258 pp., illus., map (Compléments au Prodrome de la Flore Corse, Annexe 3). Geneva. (Originally published as J. GAMISANS, 1985. *Catalogue des plantes vasculaires de la Corse*. 231 pp. Ajaccio: Parc Naturel Régional de la Corse.)

Alphabetically arranged, annotated checklist with synonymy, indication of biogeographic class and status, habitat codes (in capital letters if more than five localities known), and where necessary documented commentary including standardized and other references; frequency in right-hand column of text. A master bibliography and indices to family and generic names are given at the end of the work, while the general part includes necessary preliminaries to the catalogue along with statistics on the flora and sections on its origin, history and affinities and on vegetation formations.

Pteridophytes

BADRÉ, F., R. DESCHARTRES and J. GAMISANS, 1986. Les ptéridophytes de la Corse. *Bull. Mus. Natl. Hist. Nat.*, IV/B (Adansonia) 8: 423–461, illus., maps.

Systematically arranged, ecologically oriented enumeration, with extended notes on overall distribution, local range and habitat, and frequency; list of excluded species; bibliography (pp. 458–461). The introductory part treats plant geography, physical and geological features, climate, and discussion of habitats with characteristic species.

622

Sardinia

See also **620** (all works). – Area: 24 090 km². Vascular flora: 1900–2100 species, of which 1900–2000 native (Webb). Pignatti (1982; see **620**) gives a figure of 2089 species.

Though additional literature on Sardinia has become available in the last three decades, nothing has yet appeared effectively to succeed the works of Moris

and Barbey. A systematically arranged collection of recent records, with citations of *exsiccatae*, critical notes, and (pp. 63–70) references appears as A. D. ATZEI and V. PICCI, 1973. Note sulla nuova entità della flora Sarda non indicate in '*Nuova flora analitica d'Italia*' di A. Fiori per la Sardegna. *Arch. Bot. Biogeogr. Ital.* **49**(1–2): 1–70. At a semi-popular level, a well-illustrated 'guidebook', strongly oriented towards plants as representative of one (or more) vegetation formations, is M. CHIAPPINI, 1985. *Guida alla flora pratica della Sardegna*. 460 pp., pls., col. illus., col. map. Sassari: Delfino.[54]

BARBEY, W., 1884 (1885). *Florae sardoae compendium. Catalogue raisonné des végétaux observés dans l'île de Sardaigne*. 263 pp., 7 pls., portr. Lausanne: Bridel.

Systematic enumeration of non-vascular and vascular plants, with descriptions of new taxa (being mainly a supplement to Moris's *Flora sardoa*); includes information on synonymy and geographical distribution (limited to data not found in the earlier work) as well as some notes on habitat, etc.; addenda (by P. Ascherson and E. Levier) and index to generic and family names at end. The introductory section includes an account of the phytogeography of Sardinia in relation to neighboring lands as well as a list of endemic species, while on pp. 123–169 is an account of the island journey undertaken in 1858 by Georg Schweinfurth, the later-renowned African explorer.[55]

COSSU, A., 1968. *Flora pratica sarda illustrata*. 365 pp., 22 pls. Sassari: Gallizzi.

Alphabetically arranged enumeration (by genera) of the more frequently encountered native, naturalized and cultivated cellular and vascular plants; includes vernacular names (in Sardinian dialect as well as in Italian), general indication of local range, and notes on habitat, life-form, cultivation, properties, uses, etc.; glossaries of vernacular names with botanical equivalents as well as of cultivars; short bibliography (pp. 263–264). The plates, of selected species, are reproduced from Moris's *Flora sardoa*.

MORIS, G. G., 1837–59. *Flora sardoa*. 3 vols., atlas. 114 pls., map. Turin: Piedmont Royal Printery. Continued as U. MARTELLI, 1896–1904. *Monocotyledones sardoae*. Fasc. 1–3. viii, 152 pp., 10 pls. Florence: Niccolai; Rocca S. Casciano: Cappelli.

Descriptive flora (in Latin) of the spontaneous and commonly cultivated seed plants of Sardinia; includes synoptic keys to groups of species, full synonymy (with references and citations), vernacular names, indication of local range, taxonomic commentary, and notes on habitat, life-form, phenology, uses, etc.; indices in each volume to generic and family names. The introductory section includes an account of the physical features of the island and its climate, as well as a synopsis of families. [Moris's work covers the dicotyledons and gymnosperms. That of Martelli was not completed; it covers only Orchidaceae, Iridaceae, Amaryllidaceae, Dioscoreaceae, and part of the Liliaceae sensu lato.][56]

Woody plants

The following pair of atlases provides attractive coverage of the island's native trees, shrubs, subshrubs and lianas.

CAMARDA, I. and F. VALSECCHI, 1983. *Alberi e arbusti spontanei della Sardegna*. 477 pp., illus. (part col.), maps. Sassari: Gallizzi. Complemented by *idem*, 1990. *Piccoli arbusti, liane e suffrutici spontanei della Sardegna*. 349 pp., illus. (part col.), maps. Sassari: Delfino.

Alberi e arbusti is a dendrological atlas of the larger woody plants inclusive of text, line drawings, colored figures and distribution maps; the first-named features descriptions, local (both Italian and Sardegnan) and foreign vernacular names, and notes on distribution, floristics and phytogeography, life-form, phenology, ecology, uses and properties, wood and silviculture. A glossary, author abbreviations, bibliography (pp. 461–463) and indices follow. The introductory part covers nomenclature, botany, ecology and vegetation, silviculture, wood and uses in general. *Piccoli arbusti, liane e suffrutici* is similar in format and detail but covers smaller shrubs, subshrubs and lianas.

624

Lipari (Aeolian) Islands and Ustica

The nine Lipari or Aeolian Islands are north of Sicily, and the more isolated Ustica to the north–north-west. Area: no data. Vascular flora (Lipari Islands): 499 species (Lojacono-Pojero, 1878; see below), a figure now considered too low.

Lipari Islands

No modern work has appeared accounting for the archipelago as a whole, though individual islands have been the subject of detailed floristico-vegetatiological studies (for references, see Raimondo (1988; Note 57)).

LOJACONO-POJERO, M., 1878. *Le isole Eolie e la loro vegetazione, con enumerazione delle piante spontanee vascolari*. 140 pp. Palermo: Lorsnaider. (Also published as *Atti Soc. Acclim. Agric. Sicilia* 17: 177–328.)

Includes (pp. 85–140) an enumeration of vascular plants, with synonymy and citations (chiefly of regional literature), phenology, habitat, frequency, and more or less detailed indication of local range. The lengthy introductory section gives accounts of the geography, main features and vegetation of the group and descriptions of the individual islands. [The author's own collections were a major basis for the work.]

Ustica

Prior to the work of Ronsisvalle (1973) the flora of this somewhat isolated island (55 km north of Palermo) had been documented in an unannotated list by N. Borzì in LUDWIG SALVATOR, ARCHDUKE OF AUSTRIA, 1898. *Ustica*. xii, 132 pp., illus., 58 pls., maps. Prague: Mercy.

RONSISVALLE, G. A., 1973. Flora e vegetazione dell'isola di Ustica. In Società Italiana di Biogeographia, *Il popolamento animale et vegetale delle Isole Circumsiciliane*, pp. 1–63, illus., maps (Lav. Soc. Ital. Biogeogr., N.S. 3). Forlì.

Includes a checklist with indication of distribution and habitat, phytogeographical classes, life-forms, essential synonymy and citations in major floras; this is followed by reviews of biological spectra, floristic composition, endemism, and vegetation formations with a bibliography on pp. 61–63. The introductory part encompasses geography, geomorphology, history, and the structure and relationships of the flora (with explanations of the life-form codes and chorological abbreviations as used in the list). Accounts for 528 species (555 taxa in all), with 114 reported for the first time, in an area of 8.65 km^2.

625

Sicily

See also **620** (all works). – Area: 25708 km^2. Vascular flora: 2350–2600 species, of which 2250–2450 native (Webb); Raimondo (1988; see Note 57) gives figures of 2424 species and 2650 taxa. The alpine zone is also covered by Fenaroli's *Flora delle Alpi* (**603/IV**). No descriptive work, however, has succeeded the luxurious *Flora sicula*, the last of a number of separate Sicilian floras beginning with *Catalogus plantarum sicularum* (1692), *Syllabus plantarum Siciliae* (1694) and *Panphyton siculum* (1713), all by Fr. Francesco Cupani.[57]

LOJACONO-POJERO, M., 1886–1909. *Flora sicula o descrizione delle piante vascolari spontanee o indigenate in Sicilia*. 3 vols. (in 5). 101 pls. Palermo: L. Pedone/Lauriel di Carlo Clausen.

Descriptive flora (partly in Latin) of vascular plants; includes synoptic keys to genera and groups of species, full synonymy (with references and citations), detailed indication of local range, extensive taxonomic commentary, notes on phenology, etc., and index to all botanical names in each volume; addenda and emendata at end of work (vol. 3, pp. 412–447). The introductory section (vol. 1) includes material on the characters of Sicilian vegetation, the affinities of its flora, and on phytogeography. The plates depict representative species.[58]

Systematic enumeration of vascular plants (with non-vascular plants listed in an appendix); each entry gives authorities along with localities, frequency, abundance and habitat as observed by the author along with his predecessors and codes for life-form and habit in the left margin. The introductory part gives a general account of the island and its physical features, climate, geology and flora, while at the end are a floristic summary and a treatment of the vegetation as well as (p. 243) references.

SOMMIER, S., 1922. *Flora dell'isola di Pantelleria*. 110 pp., portr. Florence: Ricci (for Reale Istituto Botanico di Firenze).

Systematic enumeration of non-vascular (except algae) and vascular plants, with descriptions of new or less well-known taxa, synonymy, references and citations, taxonomic commentary, and notes on habitat, local occurrence, frequency, etc.; no index.[59]

626

Pantelleria and Isole Pelagie (Lampedusa, Linosa and Lampione)

See also **620** (all works). – Area: Pantelleria, 88 km²; Lampedusa, 20.2 km²; Linosa, 5.4 km²; Lampione, 0.8 km². Vascular flora: Pantelleria, 569 taxa (Di Martino, 1963; see below); Lampedusa, Linosa and Lampione, respectively 482, 310 and 30 taxa (Di Martino, 1960, see below). The composition of the flora has also varied over time; some recorded taxa have become extinct (or did not establish) while others are more recent records (Bartolo *et al.*, 1988).

Pantelleria

Stefano Sommier's 1922 work, published posthumously from an apparently incomplete manuscript left by its author on his death in that year, was a revision of part of the work of 1906–08 listed below under **Isole Pelagie** (Sommier, 1908). An unannotated name list (pp. 22–39) forms part of S. BRULLO, A. DI MARTINO and C. MARCENÒ, 1977. *La vegetazione di Pantelleria*. 110 pp., 32 text-figs. (incl. maps). Catania: Istituto di Botanica, Università di Catania. (Also in *Pubblicazioni dell'Istituto di Botanica dell'Università di Catania*, 1977, and secondarily distributed as *Lavori dell'Istituto ed Orto Botanico dell'Università di Palermo* 1(4) (1980).)

DI MARTINO, A., 1963. Flora e vegetazione dell'Isola di Pantelleria. *Lav. Ist. Bot. Giard. Colon. Palermo* **19**: 87–243, 44 halftones, 1 map.

Isole Pelagie

DI MARTINO, A., 1960 (1961). Flora e vegetazione. In E. Zavattari *et al.*, *Biogeografia delle Isola Pelagie*, pp. 163–261, pls. 21–28 (Atti Accad. Naz. Lincei, Rendiconti Cl. Sci. Fis. Mat. Nat., VIII, 11). Rome.

Includes separate reports for Lampedusa (pp. 171–210), Linosa (pp. 211–239) and Lampione (pp. 240–246) along with comparative lists (pp. 247–259) and references (p. 261). Each account includes a checklist of vascular plants with previous reports, distribution, frequency, chorological class, and (in left margin) life-form and habit codes, followed by a name list of non-vascular and other plants, a floral spectrum, statistics, a descriptive account of vegetation formations with relevés, and summary of past work. [Comprises the main botanical contribution in a general natural scientific survey of the islands comprising the whole volume (463 pp.). For Lampedusa, see also P. MINISSALE and G. SPAMPINATO, 1987. Segnalazioni di piante nuove per la flora di Lampedusa. *Inform. Bot. Ital.* **19**: 136–143; and G. BARTOLO, S. BRULLO, P. MINISSALE and G. SPAMPINATO, 1988. Flora e vegetazione dell'isola di Lampedusa. *Boll. Accad. Gioenia Sci. Nat. Catania*, III, **21**(334): 119–255, maps (also issued in *Pubblicazioni dell'Istituto di Botanica dell'Università di Catania*, 1988). Minissale and Spampinato add 62 new records, with notes, and Bartolo *et al.* include (pp. 134–159) a systematic list with chorological classes and status (with indication of taxa newly recorded since 1985, those not seen in recent years, and doubtful records with sources).]

SOMMIER, S., 1908. *Le isole Pelagie: Lampedusa, Linosa, Lampione e la loro flora con un elenco completo delle piante di Pantelleria*. iv, 344, [1] pp. Florence. (Originally published 1906–08 as appendices to *Bollettino del R. Orto Botanico (e Giardino Coloniale) di Palermo* 5(1–2)–7(1–3).)

Includes for each island enumerations of non-vascular and vascular plants along with a comparative table of occurrences for these islands, Pantelleria and Malta and a list of species from Pantelleria not found in the Pelagic Islands. There are also sections on physical features, geology, history, botanical exploration, plant geography and general features of the flora as well as statistics and (pp. 13–29) a well-annotated bibliography.

627

Malta

Some works listed under **620** also cover these islands. – Area: 316 km². Vascular flora: 900 native species (Lanfranco in PD; see **General references**). Earlier floras include *Florae melitensis thesaurus* (1827–31) by S. Zerafa, covering 489 species, *Flora melitensis* (1853) by G. C. Grech Delicata, and the relatively detailed *Flora melitensis nova* (1915) by S. Sommier and A. C. Gatto.

BORG, J., 1927. *Descriptive flora of the Maltese Islands, including the ferns and flowering plants.* 846 pp. Malta: Government Printing Office. (Reprinted 1976, Königstein/Ts., Germany: Koeltz.)

Briefly descriptive flora of vascular plants; includes synoptic keys to families, some synonymy, vernacular names (Maltese and English), general indication of local and extralimital range, taxonomic commentary, and notes on habitat, phenology, ecology, etc.; index to all botanical names. The introduction includes accounts of Maltese geography, geology, fossil records, climate, general floristics, phytogeography and botanical exploration.

HASLAM, S. M., P. D. SELL and P. A. WOLSELEY, 1977. *A flora of the Maltese islands.* lxxi, 560 pp., 29 figs., 66 pls. Msida, Malta: Malta University Press.

Concise manual-flora of vascular plants; includes keys, very limited synonymy (where accepted name differs from that in Borg's flora), Maltese and English vernacular names, karyotypes, local and overall distribution (with citations of earlier authorities), and notes on habitat, phenology, uses and taxonomy; illustrations (on plates) of representative plant species; illustrated glossary, list of key works, and complete index at end. The introductory section gives notes on the genesis of the work, the history of floristic studies in Malta, physical features, climate, soils, geology, and biotic

influences including man, as well as descriptions of ecological areas and plant communities, a phytogeographical analysis, and technical notes, along with separate chapters on medicinal plants (by H. Micallef) and fruit tree growing (by J. Borg).

628

San Marino

Area: 61 km². Vascular flora: 721 species (Pampanini, 1930).

PAMPANINI, R., 1930. *Flora della Repubblica di San Marino.* 228 pp. San Marino: 'Arti Grafiche Sammarinesi' (a spese del Pubblico Erario).

Systematic enumeration (pp. 35–218) covering all plants and similar organisms, with synonymy, citations, localities (sometimes with dates of collection) and habitat; index to genera and families at end. The introductory part is largely devoted to the history of botanical work with notes on various personalities; there is also a list of references.[60]

629

Istra (Istria), Soča (Isonzo) valley and Trieste (Trst) (former Austrian Küstenland)

See also **631**, **640**, **643** and **644** for other coverage, the latter three particularly for the alpine zone (not encompassed in Eduard Pospichal's flora, the only separate standard work). – The Istrian peninsula and the rest of the former Austrian province of Küstenland, including the city of Trieste and the Brioni Islands along with the Soča (Isonzo) valley to the north, are now divided among Croatia, Italy and Slovenia, but for convenience and in view of a strong autochthonous identity are treated here as discrete. Southern Istria (also omitted from Pospichal's work) is separately treated by Josef Freyn in his *Die Flora von Süd-Istrien* (1877–81; see below). Additional coverage of the Trieste area may be had in *Flora di Trieste e de' suoi dintorni* (1896–97) by Carlo de

Marchesetti and the very detailed *Atlante corologico delle piante vascolari nel Friuli-Venezia Giulia* (1991) by Livio Poldini.

Bibliography

DE MARCHESETTI, C., 1895. Bibliografia botanica ossia catalogo delle publicazioni intorno alla flora del Litorale Austriaco. *Atti Mus. Civico Storia Nat. Trieste* 9: 129–210. (Reprinted separately; 82 pp.) Continued as *idem*, 1931. Aggiunte alla bibliografia botanica della Venezia Giulia. *Ibid.*, 11: 217–356. [1738 items in the two works, covering all branches of botany.]

POSPICHAL, E., 1897–99. *Flora des österreichischen Küstenlandes.* 2 vols. xliii, 574, 946 pp., 25 pls., map. Vienna/Leipzig: Deuticke.

Partially keyed descriptive flora of vascular plants with synonymy, localities and other annotations; index. The introductory section covers physical features and general aspects of the flora. [Does not cover the alpine zone or the southern part of the peninsula. The latter is encompassed in *Die Flora von Süd-Istrien* (1879) by J. F. Freyn, and the former is best covered in the contributions by A. Cohrs (1953–54 and 1961; for all, see below).]

Partial works

COHRS, A. 1953–54. Beiträge zur Flora des nordadriatischen Küstenlandes mit besonderer Berücksichtigung der Görzer Umgebung. *Repert. Spec. Nov. Regni Veg.* [Fedde], 56: 66–143; and *idem*, 1963. Beiträge zur Flora des nordadriatischen Küstenlandes mit besonderer Berücksichtigung von Friaul, den Julischen und Karnischen Alpen. *Ibid.*, 68: 12–80.

Systematic enumerations with commentary and any necessary descriptions.

FREYN, J. F., 1877–81. Die Flora von Süd-Istrien [mit Nachtrag]. 284 pp. Vienna. (Reprinted from *Verh. K. K. Zool.-Bot. Ges. Wien* 27: 241–490; 31: 359–392.)

Systematic enumeration with commentary and descriptions of novelties.

Region 63

Southeastern Europe

Area: 517 122 km². Vascular flora: 7000–8000 species south and east of a line along the Sava and Danube Rivers, passing through Belgrade (Stevanović and Vasić, 1995, p. 185; see Note 68).[61] – This region comprises the countries of the Balkan Peninsula (excluding Romania and Slovenia), Crete, and the western and central Aegean Islands. The eastern limit against Asia is as accepted for *Flora Europaea*. For Slovenia see **643**.

The history and progress of botany in this region has been as balkanized as its politics. Distinct traditions have largely prevailed in Bulgaria, Greece, and different parts of the former six-state Yugoslavia, with the most effective unifying forces emanating mainly from outside. The standard of documentation, along with knowledge of the flora, is best in the north and parts of the center. The remainder, encompassing Albania, Greece, FYROM (Former Yugoslav Republic of Macedonia), and parts of Yugoslavia (Montenegro and Serbia), was botanically long the most underexplored and documented part of Europe, and, along with Bosnia-Herzegovina in an earlier period, a phytological *Kolonialgebiet* for students and other visitors from central and northern Europe. Prominent among these latter have been Austrians and others from lands once part of Austria-Hungary; in addition Bulgarian botanists have been active in FYROM and Thrace. Local institutions, including herbaria and museums, were established only during and after the second half of the nineteenth century. The leading externally based work remains that which covers the region as a whole: *Prodromus florae peninsulae balcanicae* (1924–33) by August von Hayek. It has been fundamental to all later floristic study, and for some parts provides the only overall coverage.[62]

Floristic botany in the former Yugoslavia largely revolved around its component states, all with distinct patterns and traditions in development of floras and related botanical literature. No 'All-Union' flora ever was completed although in the 1960s two such works were begun (*Analitička flora Jugoslavije*, a descriptive work with keys, and *Catalogus florae Jugoslaviae*, an enumeration).[63] Neither was there – perhaps inevitably – a Federation-wide woody flora, though in its early years the Belgrade botanist Nedeljko Košanin proposed a *Dendrologija Jugoslavije*. National floras, manuals or checklists of varying formats exist for Slovenia (now **643**), Croatia, Bosnia-Herzegovina, Montenegro, Serbia, and former Yugoslav Macedonia and there are some key partial works such as *Flora velebitica* as well as (in the north) students' manuals. By necessity or preference, however, in some areas floras of neighboring countries are used.

Until 1988, Albania was covered only by manuals, particularly *Flora ekskursioniste e Shqipërisë* (1983) by M. Demiri and *Dendroflora e Shqipërisë* (1966) by I. Mitrushi. From 1988 a more definitive *Flore e Shqipërisë* began its appearance under the editorship initially of K. Paparisto, M. Demiri, I. Mitrushi and X. Qosja; three volumes have now been published.

From the time of the 'Greek Revival' (with *Flora graeca* its botanical manifesto), Greece and the Aegean Islands have attracted the largest number of foreign botanists. The richness and intricacies of the plant life, the still-developing local resources, and a past marked by, as in Spain, separated threads of floristic work makes preparation of any definitive critical flora of Greece (such as the current *Flora hellenica*) a complex project. Notable earlier works, still widely used, include *Conspectus florae graecae* (1900–08, 1912) by Eugen von Halácsy and *Flora aegaea* (1943; supplement, 1949) by Karl-Heinz Rechinger.

Several works cover Bulgaria; among them *Flora na narodna republika Bălgarija* (1963–), initially directed by Daki Jordanov and of which 10 volumes have appeared to date, and the manuals *Flora na Bălgarija* (1966–67) by Nikolai Stojanov (with others) and *Ekskurzionna flora na Bălgarija* (1960) by S. T. Vălev *et al.* The last-named was, like Caballero's manual in Spain (**612**), an abridgment for student use. A revised field-manual by Dimitàr Delipavlov appeared in 1983.

Turkey-in-Europe has recent coverage in Peter Davis's *Flora of Turkey* (**771**) – for which a new supplement is currently in preparation – as well as more particularly in a 1966 checklist by David Webb.

In general, the standard of documentation is at a higher level in the northwest than elsewhere; there it has been influenced directly or indirectly by Central European circles. In Greece, major planks are Halácsy's and Rechinger's floras, to which may be added Strid's work on the mountains.

Bibliographies. General and divisional bibliographies as for Division 6.

Indices. General and divisional indices as for Division 6.

630
Region in general (including former Yugoslavia)

Hayek's otherwise comprehensive *Prodromus* is limited on the northwest by the Sava River, thus skirting Zagreb and excluding the Istrian peninsula, Slovenia, Slavonia (northeastern Croatia) and Vojvodina; from Belgrade eastwards the northern limit follows the Danube.

Keys to genera and families
HORVATIĆ, S., 1954. *Ilustrirani bilinar. Priručnik za određivanje porodica i rodova višega bilja*. 767 pp., 174 text-figs. Zagreb: Institut za Botaniku Sveučilišta.

Illustrated students' handbook to genera and families, with analytical keys; also includes descriptions, diagnostic features, critical remarks, and notes on distribution, special attributes, etc.

HAYEK, A. VON, 1924–33. *Prodromus florac peninsulae balcanicae*. 3 vols. (vol. 3 ed. F. Markgraf). Map (Repert. Spec. Nov. Regni Veg., Beih. 30). Berlin. (Reprinted 1970–71, Königstein/Ts., Germany: Koeltz.)

Briefly descriptive flora of vascular plants; includes analytical keys to families and genera and synoptic devices to species or groups thereof, extensive synonymy (with abbreviated references), concise symbolic indications and notes on local range, life-form, habitat, etc.; complete indices in each volume. The introductory section includes lists of symbols and abbreviations together with a methodological exposition concerning the work.[64]

Partial works: former Yugoslavia (1918–91)
Only general works dealing with the whole of former Yugoslavia (all of which are in Serbo-Croat) are listed here. Those covering individual states or combinations thereof, including present Yugoslavia, will be found under **631** to **634**.

DOMAC, R., 1950. *Flora: za određivanje i poznavanje bilja*. 552 pp., 23 text-figs. Zagreb: Institut za Botaniku Sveučilišta.

Concise students' manual-key (covering 3350 species); includes limited synonymy, abbreviated indication of internal range (by national units or 'Ras.' for 'throughout'), an illustrated glossary and index to generic names. The brief introduction contains on p. 7 a list of significant floristic references. [Covers the whole of former Yugoslavia.]

HORVATIĆ, S. and I. TRINAJSTIĆ (eds.), 1967–86. *Analitička flora Jugoslavije/ Flora analytica Jugoslaviae.* Vols. 1(1–7), 2(1–4). Maps. Zagreb: Institut za Botaniku Sveučilišta (but most parts from 'Sveučilišna Naklada Liber').

Descriptive flora of vascular plants, with keys to all taxa, some synonymy, detailed indication of local range and summary of extralimital distribution, indication of habitat, symbolic notes on life-form, karyotypes, phenology, etc., and extensive infraspecific treatments; no indices. The introductory section contains chapters on physical features, geography, vegetation formations and plant communities (with references), and a summary of administrative subdivisions. [Prior to the senior editor's death in 1975 only the pteridophytes, gymnosperms, and Ranunculaceae through Juglandaceae (together with family keys to dicotyledons) had officially been published (as vol. 1(1–3), 1967–74). Between 1978 and 1986 the remainder of volume 1 (parts 4–7) and parts 1–4 of volume 2, respectively covering Phytolaccaceae through Plumbaginaceae and Paeoniaceae through (a part of) Resedaceae, appeared. Nine supplements, all edited by Trinajstić and largely containing precursory material, accompanied the main series, appearing from 1973 to 1986.][65]

MAYER, E. *et al.*, 1964–73. *Catalogus florae Jugoslaviae.* Vols. 1(1–2), 2(1). Map. Ljubljana: Academia Scientiarum RP Socialistica Foedrativae Jugoslaviae/ Academia Scientiarum et Artium Slovenica.

Comprehensive enumeration of non-vascular and vascular plants; includes full synonymy, references, concise indication of local range and karyotypes, lists of references, and indices to generic names. No keys are provided. [Not continued; covers only Pteridophyta and Gymnospermae (vol. 1 (1–2)) and Musci (vol. 2(1)).]

Woody plants

JOVANOVIĆ, B., 1967. *Dendrologija s osnovima fitocenologije.* iv, 576 pp., illus., maps. Belgrade: 'Naučna knjiga' (for Belgrade University).

A textbook (in Serbo-Croat with Roman script) featuring principally a descriptive dendrology (pp. 65–446). For major species (whatever their origin) the text features vernacular names, full descriptions, maps (for native species), illustrations, silvicultural notes, and references at the end of genera, while for minor species (e.g., in *Cytisus*) references are wanting and descriptions are much briefer although some taxa may be illustrated. The introductory sections cover the scope of dendrology, the parts of trees and shrubs, chorology, synecology, taxonomy, and Humboldtian/Rübelian vegetation typology, while pp. 447–543 cover phytocenology including ecological 'schools', general principles, habitat factors and syntaxonomy, European vegetation regions and those of former Yugoslavia (the latter with 19 recognized syntaxonomic classes (pp. 490–491) followed by descriptions of

each). An extensive bibliography (pp. 545–556) and complete index conclude the work.

ŠILIĆ, Č., 1973. *Atlas drveća i grmlja.* 218 pp., 1150 text-figs., 166 col. photographs. Sarajevo: Zavod za Izdavanje Udžbenika. (Reissued 1983, Sarajevo: 'Svjetlost'.)

Well-illustrated semi-popular atlas-guide (in Serbo-Croat) to native trees and shrubs; each species normally features a one-page treatment, with photographs and figures accompanied by text with botanical descriptions, vernacular names (local and in English, French, German, Italian and Russian), distribution (in varying detail), and notes on habitat, ecology and biology, uses and horticultural and arboricultural value as well as on allied species (not always from southeastern Europe) with, where appropriate, references; literature cited (pp. 181–182), glossary, author abbreviations, and indices to all names at end. [For a companion work on forest herbs, see *idem*, 1983. *Šumske zeljaste biljke.* 2nd edn. 272 pp. Sarajevo.]

631

Croatia (Hrvatska)

Area: 56 538 km^2. Vascular flora: no data seen. – Includes the territories of Slavonia, Croatia proper and Dalmatia (before 1918 an Austrian territory). Only the latter two are covered by Hayek's *Prodromus*. For the Istrian peninsula see also **629**.

The main nineteenth-century flora of Croatia is *Flora croatica* (1869) by Joseph C. Schlosser [von Klekowski] and Ljudevit F. Vukotinović. From this Schlosser derived *Bilinar* or *Flora excursoria Croatiae* (1876, Zagreb). Both cover the three subdivisions of modern Croatia (without Istria; see **629**). Earlier syntheses pertained to individual territories: the classical *Flora dalmatica* (3 vols., 1842–51; supplements 1 and 2, 1872–78) by Roberto de Visiani, summarizing some three centuries of study; Schlosser and Vukotinović's *Syllabus florae Croaticae* (1857) for Croatia proper; and *Die bisher bekannten Pflanzen Slavoniens* (1866) by Stephan Schulzer von Müggenburg *et al.* At the outset of the twentieth century Dragutin Hirc began *Revizija Hrvatske flore*, publishing it from 1903 in installments in the Yugoslav Academy journal; after the Leguminosae (1912), however, the work was discontinued. In the unitary Yugoslavia of the 1920s and 1930s Croatia did not exist as a distinct polity; only some contributions on the coastal regions distinguished these interwar years. Among these are Árpad von Degen's *Flora velebitca*, covering the near-coastal limestone Velebit Mountains (between Rijeka and Knin in Kvarneria) and described below, and *Pregled flore Hrv. Primorja* (1930) by Ljudevit Rossi, unfortunately not completed due to the death in that year of the author. The main works from World War II until after independence in 1991 were successive manuals by Radovan Domac; there was no documentary flora apart from the unfinished *Analitička flora Jugoslavije* (see **630**). Works on the forest flora have also been published; these include *Šumsko drveće i grmlje* (1890) by J. Ettinger and *Dendrologija* (1946) by M. Anic.

Since independence a new floristic programme ('*Flora Hrvatske*') has been seen as a contribution to national consciousness. Apart from creation of a Web site (as part of a biodiversity information system) a first step has been publication of a revised checklist in installments as supplements to *Natura croatica* (Nikolić, 1994, 1997). A mapping programme has also commenced.

Bibliography
Substantial bibliographic sections are a feature of the current *Index florae croaticae* (see below). For pre-1918 works on Croatia proper and Slavonia (but not Dalmatia), see also **642** (Gombocz).

DOMAC, R., 1994. *Flora Hrvatske: priručnik za određivanje bilja.* Zagreb.
Concise manual-key to vascular plants. [Succeeds Domac, 1973. Not seen.]

DOMAC, R., 1973. *Mala flora Hrvatske i susjednik područja.* 543 pp. Zagreb: 'Skolska Knjiga'. (A corrected reissue of *idem*, 1967. *Ekskurzijska flora Hrvatske i susjednik područja.* 543 pp. Zagreb.)
Concise manual-key to vascular plants; includes limited synonymy, concise indication of local range, occasional taxonomic commentary, and notes on habitat, etc.; index to generic and family names at end. A list of source works is also included.

NIKOLIĆ, T., 1994–97. *Flora croatica: Popis flore hrvatske / Index florae croaticae / The flora of Croatia: check list.* Parts 1–2. 116, 232 pp. Zagreb: Hrvatski prirodoslovni musej. (Natura Croatica 3: Suppl. 2; 7: Suppl. 1.)
Annotated checklist with accepted botanical names and references (but without numeric codes for taxa), selected synonymy (without references), vernac-ular names, and indication of distribution, taxonomic status, and IUCN threat categories (for explanations, see beginning of text); extensive classified bibliographies (pp. 97–110 in part 1, 103 pp. in part 2) and indices to family and generic names at end of each part. [At present (1998) complete for pteridophytes, gymnosperms, and dicotyledons from Magnoliidae through part of the Asteridae (Gentianales, Oleales and Solanales).][66]

Partial work
DEGEN, Á. VON, 1936–38. *Flora velebitica.* 4 vols. Illus. Budapest: Akadémiai Kiadó / Ungarischen Akademie der Wissenschaften.
Very detailed, critical enumeration (in German) of the vascular and higher non-vascular plants of the Velebit Range, without keys; includes full synonymy, citations of *exsiccatae*, taxonomic commentary, and copious supporting notes; full bibliography and complete index in vol. 3. The introductory section deals extensively with physical features, geography, vegetation, forestry, agriculture, botanical exploration, and other topics. Volume 4 is a condensed Hungarian version of the work.

632

Bosnia-Hercegovina (with former Sanjak Novipazar)

Area: 51 129 km². Vascular flora: no data. – The present state of Bosnia-Hercegovina encompasses what were before 1878 territories of the Ottoman Empire and from then until 1914 were under Austrian administration (and, from 1908, rule). These lands were reconstituted after 1945 as part of federal Yugoslavia and became independent in 1992 (with two cantons from 1995). The adjacent Sanjak Novipazar, an Ottoman administrative unit occupied by Austria from 1878 to 1908–09 and attached to Bosnia-Hercegovina, is now divided between Serbia and Montenegro.

Flora Bosnae et Herzegovinae, begun by Günther Beck in 1903, was completed only in 1983. Thus was concluded a cycle begun with the publication at Cluj of *Catalogus cormophytorum et anthophytorum Serbiae, Bosniae, Hercegovinae, Montis Scodri, Albaniae hucusque cognitorum* (1877) by Paul Ascherson and August Kanitz – coinciding with the First Balkan War – and passing through Beck's *Flora von Südbosnien und der angrenzenden Herzegovina* (1886–98). The main work was materially aided by the existence of the present national museum in Sarajevo, formed in 1888, and its herbarium. There is also a dendrology, *Pregled dendroflore Bosne i Hercegovinae* (1959) by P. Fukarek. For *Atlas drveća i grmlja* (Šilić, 1973), see **630, Woody plants**.

Beck von Mannagetta und Lerchenau, G., 1903–23. *Flora Bosne, Hercegovine i Novopazarskog sandžaka*, 1–2, 1[bis]. 484, 26 pp., illus. Sarajevo. (Reprinted in fascicles from *Glasn. Zemaljsk. Muz. Bosni Hercegovini* **16–35**, *passim*.) Continued as G. Beck-Mannagetta, 1927. *Flora Bosnae, Hercegovinae et regionis Novipazar*, 3. viii, 487 pp. (Posebna Izd. Srpska (Kral.) Akad. 63). Belgrade; completed as G. Beck and K. Malý, 1950 (1951)–83. *Flora Bosnae et Hercegovinae*, 4(1–4). Revised and ed. by Ž. Bjelčić *et al.* (parts 2–4). 73, 111, 83, 188 pp. Sarajevo. (Part 1 published as Posebna Izd. Biol. Inst. Sarajevo 1; parts 2–4 as Posebna Izd. Zemaljsk. Muz. Bosne Hercegovine, Prir. Odjel. 2–4.) German edn.: G. Beck von Mannagetta und Lerchenau, 1904–16. *Flora von Bosnien, der Herzegowina und des Sandzaks Novipazar*, 1, 2(1–3). Pp. 1–261. Vienna: Gerold. (Reprinted in fascicles from *Wiss. Mitt. Bosnien-Herzegowina* **9–13**, *passim*.)

Detailed systematic enumeration (in Roman script) of vascular plants of Bosnia, Hercegovina, and the former district of Novipazar, with descriptions of new taxa; includes full synonymy, references and citations, indication of *exsiccatae* with localities, taxonomic commentary, and notes on habitat, phenology, etc. (in Latin or Serbo-Croat, depending upon the part); indices to generic names (vols. 3 and 4 only). [The reprint of volumes 1 and 2 from *Glasnik Zemaljskog Muzeja u Bosni i Hercegovini* is continuously paginated save for the section on Pteridophyta, which I have here designated as vol. 1[bis]. The German edition of this work, containing many additions, was terminated after the Ranunculaceae (corresponding to p. 225 of the Serbo-Croat version).][67]

633

Yugoslavia (Serbia and Črna gora)

Area (Federation): 102 173 km². Vascular flora (Federation): 3905 species and 277 infraspecific taxa (Stevanović *et al.* in Stevanović and Vasić, 1995).[68] – Included here are the republics of Serbia (with the former autonomous regions of Vojvodina and Kosovo) and Montenegro (Črna gora). The former Ottoman Sanjak Novipazar, for a time administered as part of Bosnia-Hercegovina (**632**), is now divided between the two constituent states.

No flora exists for the Federation as a whole (apart from the incomplete undertakings described under **630**). A checklist has been prepared by V. Stevanović, M. Niketić and D. Lakušić but remains unpublished.

Bibliography
See **642** (Gombocz) for coverage of Vojvodina through 1918.

I. Vojvodina

See **640/II** (Jávorka and Csapody); **642** (Jávorka). Situated north of Belgrade, this Pannonian territory was before 1918 part of the Kingdom of Hungary. It includes the western Banat. For some time after World War II it was an 'autonomous province' of Serbia. It is encompassed in both editions of *Flora SR Srbije/Flora Srbije*. A provincial flora was reported to have been in preparation under the direction of B. Butorac in Novi Sad.[69]

Partial work
Prior to publication of *Flora Deliblatske peščare*, the flora of the Deliblat sandhill zone in the southeast of the province had attracted much interest.

Gajić, M. (ed.), 1983. *Flora Deliblatske peščare* (*The flora of the Deliblato Sand*). 474, [2] pp., 103 pls. (part col.), additional unnumbered halftone and col. pls. Novi Sad/Pančevo: Prirodno-matematički fakultet, OOUR Institut za biologiju/Šumsko-industrijski Kombinat »Pančevo«/ OOUR Specijalni prirodni rezervat »Deliblatski pesak«.

A copiously illustrated manual-flora covering vascular plants, bryophytes, lichens and fungi; includes summaries of special features including forest history and silviculture. Covers about one-third of the total provincial flora. [In extreme southeast part, east of Pančevo on the Danube.]

II. Serbia

Area (entire republic): 88 361 km^2. Vascular flora (entire republic): 3272 species and 390 additional infraspecific taxa (Stevanović *et al.* in Stevanović and Vasić, 1995).[70] Although plant records from what is now Serbia had filtered into the literature from the time of Clusius, floristic studies were effectively initiated by the Croatian Josip Pančić following his arrival in Belgrade in 1846. Between then and 1884, he increased the total of known species in the then-small principality from 134 (as known to Grisebach in *Spicilegium florae Rumelicae et Bythinicae*, 1843–45) to 2422. Among his works prior to *Flora kneževine Srbije* (described below) were a first enumeration, *Verzeichniss der in Serbien wild-wachsenden Phanerogamen* (1856), a 'primer', *Flora agri belgradensis* (1865, with further editions to 1892), and a dendroflora (*Sumsko drveće i šiblje u Srbiji*, 1871). The principality was also covered by Ascherson and Kanitz in their *Catalogus* of 1877 (mentioned under **632**), and in Niš – then not part of Serbia – another local flora, *Flora okoline Niša* by Sava Petrović, appeared in 1882 (with a supplement in 1885). Their successors Lujo Adamović and Nedeljko Košanin (and others) largely confined themselves to further exploration, additions, and critical notes, while from 1934 to 1966 even less was accomplished. A 'Committee for the Exploration of the Flora [later Flora and Vegetation] of Serbia' was then set up by the Serbian Academy and the *Flora SR Srbije* project initiated under Mladen Josifović. Publication began in 1970 and was completed by 1977 with a further supplement in 1986. In 1992, following exhaustion of stock, a second edition began to appear.[71]

JOSIFOVIĆ, M. (coord.), 1970–77, 1986. Флора СР Србије (*Flora SR Srbije*)/*Flore de la Republique Socialiste de Serbie*. 10 vols (vol. 10 edited by M. R. Sarić and N. Diklić). Illus. Belgrade: Srpska Akademija Nauka i Umetnosti. Partly succeeded by M. R. SARIĆ (ed.), 1992. Флора Србије (*Flora Srbije*). 2nd edn. Vol. 1. Illus., maps. Belgrade.

Large-scale, illustrated descriptive documentary flora of native, naturalized and commonly cultivated vascular plants of Vojvodina, Kosovo i Metonija, and Serbia proper (including its part of the former Sandžak

Novipazar); includes keys to all taxa, full synonymy (with references), vernacular names, citation of *exsiccatae* with localities and fairly detailed indication of local range, and notes on habitat, phenology, ecology, uses, etc.; indices to all botanical and vernacular names in each volume (general index to families and genera in vol. 9). The introductory section (vol. 1) includes an extensive historical account (with references) and an illustrated organography. Volume 9 (1977) also contains a supplement (*dodenik*) to vols. 1–8, and vol. 10 (1986) an extensive bibliography as well as a second supplement. The second edition is similar in style to its predecessor.[72]

PANČIĆ, J., 1874. Флора кнежевине Србије (*Flora kneževine Srbije*)/*Flora principatus Serbiae*. xxxii, 798, [2] pp. Belgrade: Royal Serbian Printer. Complemented by *idem*, 1884. *Additamenta* .... 254, [2] pp. Belgrade. (Whole work reprinted in 1 vol., 1976, Belgrade, as *Posebna Izd. Srpska Akad.* 492.)

Concise manual-key (arranged following the Candollean system) of the vascular plants of Serbia proper (excluding Novipazar and Kosovo i Metonija); includes generalized indication of local range, notes on life-form, phenology, etc., and an index to family and generic names. The introductory section gives a general account of features of the flora, an account of botanical work in the area including the author's own investigations, and (pp. 1–102) an analytical key to genera. The 1884 supplement includes an account of additional exploration, a revision of the general key (pp. 20–102), additions (pp. 103–249), and index; while the 1976 reissue includes a 13-page commemorative appendix by M. Janković and N. Diklić.[73]

III. Kosovo i Metonija (Kosova, Stara Srbija)

This territory, inhabited mostly by Albanians, was until 1989 an 'autonomous province' of Serbia; it is currently an international protectorate. There are no separate works of significance, but, like Vojvodina, it is encompassed in both editions of *Flora Srbije*.

IV. Montenegro (Črna gora)

Area: 13 820 km^2. Vascular flora: 2920 species and 216 additional infraspecific taxa (Stevanović *et al.* in

Stevanović and Vasić, 1995).[74] Pulević and Karaman (1996; see Note 75) suggest a total of over 3100 species. Rohlena (1942; see below) recorded 2623 species for the former kingdom in its less extensive pre-1913 limits. This is perhaps the highest number of species per unit area for a European country. – Long more or less independent, Montenegro acquired part of the Sandžak Novipazar upon its division in 1913. That area is covered by part of *Flora Bosne* (Beck, 1903–83; see **632**). In 1918, however, the state was absorbed into what became Yugoslavia and under unitary rule from Belgrade lost its separate identity until World War II. Reconstituted in 1945, it became a republic in the Yugoslav Federation and therein remained following reorganization of the latter in 1992.

Early explorers included Roberto de Visiani and Muzio Tommasini (Mucio Tomazini), but only in the 1870s did serious examination of the flora begin. The first separate account is *Elenchus plantarum vascularium quas aestatae a. 1873 in Črna gora, legit dr. Jos. Pančić* (1875) by Josip Pančić, covering 1298 species. In 1900–06 the Czech botanist Josef Rohlena undertook six expeditions which formed the primary basis for his later *Beiträge* and *Conspectus*. Locally based work became possible with the postwar formation of a museum in Titograd (now Podgorica). Additions to the flora have since continued to flow. A supplement to Rohlena's *Conspectus* was ready by 1979 but has never been published.[75]

Bibliography

PULEVIĆ, V., 1980. *Bibliografija o flori i vegetaciji Črne gore/The bibliography of flora and vegetation of Montenegro.* 235 pp. (Črnog. Akad. Nauk. Umjetn., Odj. Prir. Nauk., Bibliografije 1). Titograd (Podgorica). Continued as *idem*, 1987 (1988). Dopuna bibliografiji o flori i vegetaciji Črne gore. *Glasn. Republ. Zavoda Zaštitu Prir.-Prirodnjačke Muz., Titogradu* 18: 5–95. [The main work encompasses 1055 annotated, alphabetically arranged (chronologically by authors) numbered entries with coverage through 1978; author and systematic indices and lexicon of serial abbreviations at end. A further 428 entries through 1987 appear in the supplement.]

ROHLENA, J., 1942. *Conspectus florae montenegrinae.* 506 pp. (Preslia 20/21). Prague.

Systematic enumeration (in Latin) of vascular plants, with limited synonymy, citation of *exsiccatae* with localities, and brief indication of habitat; index to generic and family names at end. [The work covers 'old

Montenegro', i.e., the former kingdom within its pre-1913 limits, and was based on several *Beiträge* published in journals from 1902 to 1924.][76]

634

Former Yugoslav Republic of Macedonia (FYROM)

Area: 25 713 km^2. Vascular flora: no data seen. – The Former Yugoslav Republic of Macedonia (FYROM), part of Serbia from 1913 (and Yugoslavia from 1918, when it became known as South Serbia), was constituted in its present form only in 1945; at the same time its language, related to Bulgarian, became formally recognized. In 1992 it achieved independence.

The territory, which botanically remains imperfectly known yet is of some importance, is the subject of several collection reports but until relatively recently no separate flora or enumeration. Early contributions include *Beitrag zur Flora von Serbien, Macedonien, Bulgarien und Thessalien* (6 parts, 1891–98) by Eduard Formánek and *Beiträge zur Flora von Macedonien und Altserbien* (1904) by Lujo Adamovič. Bornmüller's *Beiträge*, based partly on his collections of 1917–18, followed in the 1920s along with several papers by Nedeljko Košanin. There is also a small dendroflora, *Pregled na dendroflorata na Makedonija* (1967) by Hans Em, long a professor of forestry in Skopje. Resident florists remain very few; the most active worker has been Kiril Micevski (Mitsevski), also of Skopje, who in 1985 began publication of the country's first flora (described below).

Bibliographies

MICEVSKI (MITSEVSKI), K., 1956. Bibliographie der Flora und der Vegetation Mazedoniens. *Godišen Zborn. Prir.-Mat. Fak. Univ. Skopje* 7: 1–32; and *idem*, 1960. Bibliographie der Flora und der Vegetation Mazedoniens. *Posebni Izd. Filoz. Fak. Univ. Skopje* 8: 99–118. [Not seen.]

TODOROVSKI, A. Đ., 1975. Bibliografija no raboti od prirodnite i nivnite primeneti nauki (voglavno botanicki i drugo) na SR Makedonija. 2 parts. Bitola. [Based largely on the author's library; 852 titles or so on natural sciences and related subjects but without those already in Micevski's two bibliographies or the author's own *Botanička bibliografija na Bitolsko* (1967). In Serbo-Croat and in Roman script.]

BORNMÜLLER, J., 1925–28. Beiträge zur Flora Mazedoniens, I–III. 18 pls. *Bot. Jahrb. Syst.* **59**: 293–504; **60**, Beibl. 136: 1–125; **61**, Beibl. 140: 1–196.

Systematic enumeration of vascular plants, with synonymy, localities with citation of *exsiccatae* and other records, and extensive critical commentary; index to families in the third part. The introductory section includes a gazetteer and details of the author's itineraries. [The work accounts for other collections in addition to those of the author. According to Micevski (see below), it is for the country its most valuable precursory compendium.]

MICEVSKI (MITSEVSKI), K., 1985, 1993, 1995. Флора на СР Македонија (*Flora na SR Makedonija*)/ *The flora of SR Macedonia*. Vol. 1(1–3). Pp. 1–772. Skopje: Makedonska Akademija na Naukite i Umetnostite. (Vol. 1(2) and subsequent installments entitled *Flora na Republika Makedonija/ The flora of the Republic of Macedonia*.)

Descriptive flora with keys, significant synonymy, and indication of frequency, habitat and distribution with localities and authorities but little or no direct citation of *exsiccatae*; indices to genera and families in each part. The introductory section (in Macedonian and English) in vol. 1(1) gives general particulars and a summary of botanical exploration along with general keys to families. [Some 20 percent of the flora has so far been covered, including among other groups pteridophytes, gymnosperms, Ranunculaceae, Urticaceae, Fagaceae, Betulaceae, Caryophyllaceae, Amaranthaceae, Polygonaceae, Capparaceae, Cruciferae and Salicaceae.]

Woody plants

EM, H., 1967. *Pregled na dendroflorata na Makedonija* [*Survey of Macedonian dendroflora*]. 125 pp., illus., 34 halftones, 15 maps. Skopje.

A textbook (in Roman script), the dendrology proper on pp. 51–87 (covering 302 species and 70 infraspecific taxa) with botanical and vernacular names (the latter also in Serbo-Croat), descriptions, indication of distribution and altitudinal range, uses, growth and dimensions, and sociological affinities; no keys. The species accounts are preceded by a treatment of the physical features, climate, vegetation and formations in the republic, along with (pp. 15–25) a summary list of species, with statistics. The maps are a selection to show characteristic distribution patterns, and may depict more than one species.

635

Albania

Area: 28 748 km². Vascular flora: 2965 species as of 1988 (Akeroyd in *Global biodiversity*; see **General references**). About 3200 species are treated in Demiri's manual, while Webb proposed a figure of 3150–3300 species (3100–3300 native). – The country became separate from the Ottoman Empire in 1912–13; its boundaries then were as shown in Hayek's *Prodromus*. In 1923 the present limits became fixed.

Albania was until well into the twentieth century an 'expedition area', and botanically it remains among the least well-known countries in Europe. A first consolidation was the 1877 *Catalogus cormophytorum et anthophytorum* by Ascherson and Kanitz (referred to under **633**); further contributions were made by R. von Wettstein in *Beiträge zur Flora Albaniens* (1892) and in his *Zweiter Beitrag zur Kenntnis der Flora von Albanien* (1924). These were followed by *Pflanzen aus Albanien 1928* by F. Markgraf (1931; in *Akad. Wiss. Wien, Math.-Naturwiss. Kl., Denkschr.* **102**: 317–360, 1 pl.).

Manuals and other works in Albanian appeared only after World War II; the first was *Drurët dhe shkurret e Shqipërisë* (1955) by I. Mitrushi. All are notable for relatively limited use of synonymy and other means of independent taxonomic documentary control. On the other hand, the figures in the new atlas-flora, *Flora e Shqipërisë*, are all original and effectively depict the plants in question.

DEMIRI, M., 1983. *Flora ekskursioniste e Shqipërisë*. 986, 1 pp., 2484 numbered and some unnumbered text-figs., col. photograph (covers). Tirana: 'Shtëpia Botuese e Librit Shkollor'.

Manual-key to vascular plants in two sections (text and atlas); the text (pp. 25–504) includes diagnoses, essential synonymy (based on *Flora Europaea*), standardized vernacular names, and notes on habit, phenology, habitat, etc., while the atlas (pp. 507–920) contains all the numbered figures, six to a page; indices to all vernacular and botanical names; errata (p. 986). The introductory section covers conventions and the use of the work. [An outgrowth of *Flora ë Tiranës* to which the author was a contributor.]

PAPARISTO, K., M. DEMIRI, I. MITRUSHI, X. QOSJA, J. VANGJELI, E. BALZA and B. RUCI (eds.),

1988– . *Flora e Shqipërisë/Flore de l'Albanie*. Vols. 1– . Illus. Tirana: Qendra e Kërkimeve Biologijke (vols. 2–3: Instituti i Kërkimeve Biologjike), Akademia e Shkencave.

Descriptive atlas-flora (in landscape format), the figures appearing in parallel with the text; includes keys to all taxa, botanical and vernacular names, limited synonymy (without references), and indication of local distribution, habitats, phenology, and some notes on special features; indices to all names at end of each volume. Pp. 403–439 in vol. 1 comprise an illustrated glossary with figures for each term. The general part (vol. 1) covers the origins of the flora, climate, geology, land use, and vegetation with principal zones, features and indicator species along with a brief list of principal literature, the order of families (arrangement is that of *Flora Europaea*), and the main key to the families (pp. 37–55). [Vol. 1 covers pteridophytes, gymnosperms and dicotyledons through Platanaceae; vol. 2 (1992), Rosaceae through Umbelliferae; vol. 3 (1996), Pyrolaceae through Campanulaceae. Paparisto was senior editor for vol. 1; Qosja for vols. 2 and 3.][77]

Partial work: Tirana district

PAPARISTO, K. *et al.*, 1961 (1962). *Flora e Tiranës*. 521 pp., 216 text-figs., 2 maps (end-papers). Tirana: Univ. Shtetëvor Tiranës. Accompanied by *idem*, 1965. *Flora e Tiranës (Ikonografia)*. 515 pp., 1300 text-figs. Tirana.

The text volume comprises a students' manual-key to vascular plants of the Tirana district, with limited synonymy, vernacular names and notes on habitat, occurrences, phenology, etc.; illustrated glossary and lists of plants classified by special features, uses, properties, etc., in appendices; indices to botanical and to vernacular names at end. The atlas volume contains half-page figures, with cross-references to the text.

Woody plants

For additional descriptive details and commentary, see I. MITRUSHI, 1955. *Drurët dhe shkurret e Shqipërisë*. 604, vii pp., 289 text-figs., 9 maps (1 col.). Tirana: Univ. Shtetëvor Tiranës.

MITRUSHI, I., 1966. *Dendroflora e Shqipërisë*. 520 pp., 617 text-figs. Tirana: Univ. Shtetëvor Tiranës.

Illustrated manual-key to woody and subwoody plants; includes brief diagnoses, very limited synonymy, vernacular names, and notes on local range, habitat, phenology, status, origin, etc.; illustrated organography/glossary and indices to vernacular and botanical names at end. The introductory section includes general keys to families and genera.

636

Greece (mainland and western islands)

For the Aegean Islands (including Crete), see **637**. Reference should also be made to **630** (HAYEK). – Area (of country): 131957 km^2. Vascular flora (of country): *c.* 4900 species as of 1989 (Akeroyd in *Global biodiversity*; see **General references**). The *Flora hellenica* project gives an estimate of 4700.[78]

Greece has one of the richest floras in Europe, if not by unit area. Following the contributions of the classical and early modern eras, a first, inevitably only partial, synthesis was achieved by Sibthorp, Smith, Lindley and Ferdinand Bauer, firstly in *Florae graecae prodromus* (1806–13) and then in that epitome of illustrated 'great national flower-books', the 10-volume *Flora graeca* (1806–40). The nineteenth century after 1810 was a time of substantial exploration, both by visitors (among them Dumont d'Urville, Bory de Saint Vincent, Grisebach, Fiedler and Boissier) and by native and foreign residents (including Heldrech and Orphanides). However, apart from some partial floras such as Michele Pieri's florulas of Kerkira (1814, 1824) and *Nouvelle flore du Péloponnèse et des Cyclades* (1838) by Bory de Saint Vincent, the checklist (Uebersicht) by K. G. Fiedler in his *Reise durch alle Thiele des Königreichs Griechenland* (1840–41), and *Enumeratio chloridis Hellenicae* (1866) by Orphanides, no consolidation was forthcoming until *Flora orientalis* (1867–88) by Boissier. The next major work, the *Conspectus* by Eugen von Halácsy (1900–08, 1912) is historically of the greatest importance; it takes into account the results from a peak of exploration in the 30 or so years following the end of the 1877–78 Balkans War and the addition of Thessaly to Greece. Further national expansion followed in 1912–14 (Epirus, southern Macedonia and Crete) and 1920 (Thrace), with exploration continuing at varying intensities; nevertheless, the publication of Hayek's *Prodromus* (1924–33) arguably marks the end of the post-Napoleonic investigative cycle. In the 1930s two largely compiled manuals in Greek were commenced (with one brought to completion in 1949).

World War II brought renewed attention from outside, with a major product being K.-H. Rechinger's

Flora aegaea (1943; see **637**). This afterwards led to a new round of detailed investigations as the intricacies of the flora and its potential relative to advances in botanical understanding became more apparent; a new general flora was, however, seen as premature. Nevertheless, over some three decades numerous contributions (including partial floras) were published, many relating to the Aegean Islands.

A new phase of consolidation began in the 1980s, firstly in the two volumes of Arne Strid and Kit Tan's montane flora (1985–91), furnishing full coverage of areas over 1800 m, and secondly with the commencement in 1988 of the *Flora hellenica* project, also an undertaking of Strid's team. In 1997, the first of the projected 10 volumes of the definitive *Flora* was published.

In addition to the works described here, notice may be taken of D. S. CAVADAS (CAVVADAS), 1957–64. Εικονογραφημενον βοτανικον-φυτολογικον λεξικον (*Eikonografēmenon botanikon-fytologikon lexikon*)/[*Illustrated botanical-phytological dictionary*]. 9 vols. Athens. For illustrations, nothing has surpassed the 966 plates of Ferdinand Bauer in *Flora graeca*.[79]

Bibliographies

Economidou's bibliography, being in the first instance classified, continues to be useful consequent to publication of that of Strid.

ECONOMIDOU, E., 1976. Bibliographie botanique sur la Grèce (plantes vasculaires – végétation). In S. Dafis and E. Landolt (eds.), *Vegetation und Flora von Greichenland*, 2: 190–242 (Veröff. Geobot. Inst. ETH, Stiftung Rübel, Zürich 56). Zürich. [A classified, unannotated listing; headings include Greece in general, its regions, and major taxonomic groups. Accompanies *Zur Vegetation und Flora von Griechenland: Ergebnisse der 15. Internationalen Pflanzengeographischen Exkursion (IPE) durch Griechenland 1971.*]

STRID, A., 1996. *Flora Hellenica bibliography: a critical survey of floristic, taxonomic and phytogeographical literature relevant to the vascular plants of Greece, 1753–1994.* x, [2], 508 pp., map (Fragm. Flor. Geobot., Suppl. 4). Krakow. [10 241 entries, alphabetically arranged by author with taxonomic (to generic level) and geographical indices.]

CAVADAS (CAVVADAS), D. S., 1938. Η χλωρις τες Ελλαδος. Εικονογραφημενη και μετα δικοτομικων κλειδων. Τομος Α. Πτεριδοφυτα, Γυμνοσπερμα και Μονοκοτυλα (*Ē khlōris tēs Ellados. Eikonografēmenē kai meta dikhotmikōn kleidōn. Tomos A. Pteridophyta, Gymnosperma kai Monoko-*

tula). xxii, 384 pp., 274 text-figs. Thessaloniki: The author.

Illustrated, synoptically keyed manual (covering pteridophytes, gymnosperms and Gramineae) with descriptions of species and higher taxa, synonymy, vernacular names, indication of distribution and habitat, and commentaries; indices at end respectively to genera and families, non-Greek vernacular names, and Greek names. The introductory section (pp. 1–101) includes a glossary, author abbreviations, pedagogy, and an illustrated key to major groups and orders. [Not completed. Some cultivated plants are also included.]

DIAPOULIS, K. A., 1939–49. Ελληνικη χλωρις (*Ellēnikē khlōris*)/*Synopsis florae graecae*. Tomos A–B (in 3 vols.). Illus. Athens.

Manual-key to the vascular plants of Greece (chiefly the central and southern parts); includes vernacular names, brief general indication of local range, numerous small text-figures, and indices to generic, family and vernacular names. The introductory section (vol. A) includes an illustrated organography/glossary, a Latin–Greek lexicon of terms, and lists of authors and references. [The work was to a large extent compiled from other sources and lacks a critical basis.]

HALÁCSY, E. VON, 1900–08. *Conspectus florae graecae*. 3 vols. and supplement. Leipzig: Engelmann. (Reprinted 1968, Lehre, Germany: Cramer.) Continued as *idem*, 1912. Supplementum secundum *Conspectus florae graecae*. *Magyar Bot. Lapok* 11: 114–202.

Briefly descriptive flora (in Latin) of vascular plants; includes synoptic keys to genera and species (or groups thereof), full synonymy (with references and citations), indication of *exsiccatae* with localities, critical taxonomic remarks, and notes on habitat, phenology, life-form, etc.; indices to all botanical names in each volume. [Parts of northern and eastern Greece and the Aegean Islands are not included; nevertheless, the work stands almost without peer and has retained its value in spite of the lapse of time.]

STRID, A. and K. TAN (eds.), 1997– . *Flora hellenica*. Vols. 1– . Frontisp., maps. Königstein/Ts., Germany: Koeltz.

Documentary flora with keys to all taxa, synonymy, references, types, sometimes lengthy descriptions, indication of distribution and habitat and altitudinal range, extralimital range, and status; literature cited (pp. 375–392), maps 1–722 (six to a page), and index to all names at end of work. No vernacular names are included. The opening section in vol. 1

covers the plan of the work, conventions and abbreviations, physical features and geography (with maps), geology and soils, climate and environment, vegetation formations, phytogeography and endemism, and history of botanical work, followed by selected key references (pp. xxxiv–xxxv) and acknowledgments.[80]

Partial works

Included here are recent works on Kerkira (Corfu), the Peloponnesus, and mountain regions.

BORKOWSKY, O., 1994. *Übersicht der Flora von Korfu/ Floristic investigations of Corfu*. [1], 202 pp., map. Brunswick. (Braunschweiger Geobotanische Arbeiten.)

Pages 35–156 comprise an enumeration of vascular plants (1410 species) of Kerkira (Corfu) with essential synonymy and localities with recorded habitats and associations (and including vouchers; for abbreviations, see p. 34); new records (some 500 in all) are asterisked. A bibliography, lists of alien, cultivated and adventive species, and species on nearby islands possibly to be found in Kerkira, a full tabular species list (pp. 174–200) and index to collectors follow. The general part includes a history of botanical work along with summaries of geography, geology, climates, and plant geography, sociology, and vegetation formations with indicator species.

STRASSER, W., 1999. *Plants of the Peloponnese*. [2], 350 pp., >1900 illus., map. Ruggell, Liechtenstein: Gantner. German edn.: *idem*, 1997. *Pflanzen des Peloponnes*. [2], 321 pp., >1800 illus., map. Vaduz: Gantner.

Compact field-guide, the plants artificially arranged by form and flower color in 10 categories; each page features text and figures for a number of species, the former inclusive of diagnoses and accepted names. Separate keys are provided for *Trifolium* and yellow-flowered Compositae. A list of families with English or German vernacular names, an index to all generic names and below, and a list of references conclude the work. The general part covers the philosophy and plan of the work along with abbreviations, symbols, and a brief glossary. [Some 2300 or more species are accounted for in the English version, which incorporates revisions not in its German predecessor.]

STRID, A. and K. TAN (eds.), 1985–91. *Mountain flora of Greece*. 2 vols. Illus. (incl. maps). Cambridge: Cambridge University Press (vol. 1); Edinburgh: Edinburgh University Press (vol. 2).

Briefly descriptive flora; includes keys to genera and species (with family key on pp. 1–10), concise synonymy with contexts, local distribution in greater or lesser detail with altitudinal compass, citation of *exsiccatae* and discussion, generalized overall range, critical and sometimes quite lengthy taxonomic commentary, and notes on habit, phenology, karyotypes and habitat; indices to scientific and to geographical (including mountain) names at end. Generic and family

headings are descriptive, with literature citations. The introductory section includes conventions and an account of botanical exploration, along with discussions of geography, topography and geology and a list of mountains (map, p. xvii) as well as brief remarks on phytogeographical elements and endemism. Volume 2 includes a detailed bibliography (pp. 885–918) and index to all scientific names. [The work recognizes 'full' and 'associate' taxa; the former are plants known from at least 1800 m and above, the latter, those in open habitats with ranges down to 1500 m.][81]

Woody plants

In addition to the following work, a useful popular guide (but without keys and featuring in some cases outdated nomenclature) is G. SFIKAS, 1978. *Self-propagating trees and shrubs of Greece*. 213 pp., col. illus. Athens/Thessaloniki: Eustathiadē. ('Living Greece' series; reissued 1991.) Reference may also be made to D. VOLIOTIS and N. ATHANASIADIS, 1971. Δενδρα κη Θαμνι (*Dendra kē thamni*). 294 pp. Thessaloniki. [Not seen.]

BORATYŃSKI, A., K. BROWICZ and J. ZIELIŃSKI, 1992. *Chorology of trees and shrubs in Greece*. 2nd edn. 286 pp., 270 maps. Poznán/Kórnik: Soros. (1st edn., 1990, Kórnik: PWN/Polish Academy of Sciences, Institute of Dendrology.)

An alphabetically arranged atlas of distribution maps, one species to a page; for each taxon a dot map is furnished along with a brief description, essential synonymy, indication of overall range-types (with country codes but no chorological classes), and discussion of more local distribution, altitudinal range, and habitats with literature list and a table of contents at end. There is also a brief introduction. [The work is based primarily on herbarium and field records and only secondarily on floristic literature. Some 70 species were added for the second, current edition of this work.]

637

Aegean Islands (including Crete, Kithira, the Cyclades, the Sporodhes and Evvia (Euboea))

See also **636** (all works); **770–90** (BOISSIER). – Area: Crete, 8336 km^2; Evvia (Euboea), 4167 km^2; no data for other islands. Vascular flora: Crete, 1586 native species with a further 104 of doubtful status and 92 certainly introduced (Barclay, 1986); former Dodecanese, 1186 species (Ciferri, 1944).

The Aegean Islands, in addition to their undoubted charms, are a great natural laboratory for the study of plant speciation and evolution and in the half-century since Rechinger's *Flora aegaea* many floristic and systematic papers and other works have been published. Apart from those accounted for here, extensive contributions have come from K.-H. Rechinger, Werner Greuter and his associates in Geneva and Berlin, and Hans Runemark and his associates (including Arne Strid) in Lund. Greek authors have also contributed, particularly in the last two decades. The smaller islands were of particular interest to the Lund group; see H. RUNEMARK *et al.*, 1960–80. Studies in the Aegean flora, 1–23. *Bot. Notis.* **113–133** and *Opera Bot.*, A, **13** (1967), *passim.*

RECHINGER, K.-H., 1943. *Flora aegaea.* xx, 924 pp., 25 pls., 3 maps (Akad. Wiss. Wien, Math.-Naturwiss. Kl., Denkschr. 105/1). Vienna. Continued as *idem*, 1949. Florae aegaeae supplementum. *Phyton* (Horn) **1**: 194–228. (Both works reprinted in 1 vol., 1973, Königstein/Ts., Germany: Koeltz.)

Critical systematic enumeration (text in German, keys in Latin) of the vascular plants of the foreshores of the Aegean Sea and the Aegean Islands together with Crete; includes analytical keys to species within genera, descriptions of new taxa, full synonymy (with references and citations), detailed indication of local range with citation of *exsiccatae* and other records, and notes on habitat, special features, etc.; annotated list of collectors; bibliography and index to all botanical names at end. The introductory section includes accounts of physical features, climate, etc., together with a gazetteer. The supplement is similar in style. [Fundamental for all subsequent work.]

I. Crete (with Karpathos)

Crete has had a substantially longer post-Renaissance botanical history than most other parts of modern Greece. Control by Venice until 1669, with moreover an active interest on the part of their governors, facilitated contacts with Italian and other botanists and thus laid a foundation for knowledge of this largest and highest of the Greek islands. From the imposition of Ottoman rule until the nineteenth century, however, botanical contacts were much reduced. A first synthesis was essayed by F. W. Sieber in *Herbarium florae creticae* (1820). The first half of the twentieth century featured, in addition to coverage in Halácsy's *Conspectus* (**636**) and Rechinger's *Flora aegaea* (see above), the uncritical

Flora cretica of Michel Gandoger (1916), written following some double-edged trips to the island, and *Neue Beiträge zur Flora von Kreta* (1943) by Rechinger (an extension of *Flora aegaea*). No further consolidated works appeared until *Prodromus florae creticae* by Schönefelder and *Crete: checklist of the vascular plants* by Sir Coville Barclay (1986, as *Englera* **6**: i–xiii, 1–138). This was in turn a basis for the two current manuals (Turland, Chilton and Press, 1993, with supplement by Turland and Chilton, 1997; Jahn and Schönfelder, 1995). The four main mountain massifs were a particular focus of Jacques Zaffran's *Contributions* (1990): 25 percent of the island is over 800 m with three of the massifs topping out at more than 2000 m. These have continued to yield novelties and new records, as Chilton and Turland have recorded in their supplement.[82]

JAHN, R. and P. SCHÖNFELDER, 1995. *Exkursionsflora für Kreta.* 446 pp., 6 text-figs. (incl. maps), 101 col. photographs, map (end-papers). Stuttgart: Ulmer. (Based on P. SCHÖNFELDER (ed.), *Prodromus florae creticae.* iv, 316 pp., 54 figs.)

Manual-key to vascular plants with annotated leads including local range, chorological class, altitude, habitat, habit, life-form, phenology, status and essential synonymy; references and index to family and generic names at end. The introductory part is concise but fairly detailed, with sections on physical features, climate, vegetation formations (the reversion in recent decades of much formerly cultivated land being of particular note) and floristics (including statistics) as well as (pp. 28–32) the plan of the work and explanation of all abbreviations and codes.

TURLAND, N. J., L. CHILTON and J. R. PRESS, 1993. *Flora of the Cretan area: annotated checklist and atlas.* xii, 439 pp., text-figs, 1738 distribution maps. London: HMSO (for the Natural History Museum). (Reissued with corrections, 1995.) Complemented by L. CHILTON and N. J. TURLAND, 1997. *Flora of Crete: a supplement.* 172 pp., 47 pp. maps. Hunstanton, Norfolk: Marengo Publications.

Annotated enumeration and distribution atlas of native and naturalized vascular plants of Crete and the Karpathos Islands (in the Dodecanese), the families alphabetically arranged and preceding the species maps; entries include key references and citations, distribution (as abbreviations; see p. 27 for explanation), map reference (if applicable), habitat, altitudinal range, frequency and (sometimes) localities in Crete and

Karpathos, status and (as necessary) commentary; atlas (pp. 196–416), appendices, references and index to genera at end. The introductory section includes descriptions of physical features, vegetation formations and land-use history as well as the plan of the work, its background, and (p. xii) statistics of the flora. The 1997 supplement comprises a checklist covering 1107 species either newly recorded or with revised data and 282 new or revised maps (some 15 percent of the total).[83]

ZAFFRAN, J., 1990. *Contributions à la flore et à la végétation de la Crète.* 615 pp., halftones, 10 text-figs. (incl. maps). Aix-en-Provence: Université de Provence Aix-Marseille 1.

Of the three main parts of this monograph, the second (pp. 51–383) comprises an annotated enumeration of 1061 vascular species, each with place of publication, frequency, *Med-Checklist* number, record(s) by geographic unit (for explanation, see map in pl. 10 (after p. 52)), chorological unit, and more or less extensive commentary; index to genera (at very end of work). The remaining parts cover (1) the history of botanical work, physical features, climate and rainfall, cloudiness, soils, and geological history, and (2) vegetation associations, their composition and relationships, and floristic features including endemics.

II. Southern Sporodhes (Dodecanese)

The Southern Sporodhes or Dodecanese – of which Rhodos is the largest – were for the most part an Italian possession from 1912 to 1947 following centuries of Ottoman rule. Though politically now all Greek, they are botanically part of Southwest Asia. The nearby Karpathos group is, however, floristically associated with Crete. Early collectors included Forskål, Sibthorp and particularly Aucher-Eloy (in 1832).

In addition to the works below, the Southern Sporodhes are covered in *Flora of Turkey* (771).

CARLSTRÖM, A., 1987. *A survey on the flora and phytogeography of Rodhos, Simi, Tilos and the Marmaris Peninsula (SE Greece, SW Turkey).* 302, [22] pp., 1604 distribution maps. Lund: Department of Systematic Botany, University of Lund.

The bulk of this work comprises a checklist (pp. 43–135) with distribution maps (pp. 137–290) and list of references (pp. 292–302). The checklist features essential synonymy (with some references), distribution (by islands), citation of *exsiccatae* where necessary, indication of habitat, chorological type, substrate and vegetation formation(s), and taxonomic commentary as appropriate. The introductory section covers climate, physical features, geology (including the role of ultramafic rock, mainly on the Marmaris Peninsula), vegetation, a review of past botanical work, and an extensive floristic analysis. [A contribution to the Aegean programme of Runemark (see above).]

CIFERRI, R., 1944. *Flora e vegetazione delle Isole italianae dell'Egeo.* 200 pp. (Atti Ist. Bot. Univ. Pavia, Lab. Crittogam., V, Suppl. A). Pavia. (Mimeographed.)

Pages 21–136 comprise a systematic enumeration of vascular plants (1186 species), with references and key citations of literature along with documented localities on the 14 islands or island groups (including Rhodos but not Ikaria or Samos). The remainder of the work covers vegetation, chorology, and floristic plant geography as well as phytosociological classification, with references (pp. 183–192) and indices at the end, while in the introductory section is an account of the area and of previous botanical work in each of the islands or island groups.

III. Other islands

Individual syntheses are quite scattered; the most substantial are those by Rechinger (1961) for Evvia (Euboea) and Greuter and Rechinger (1967) for Kithira, both described below.

GREUTER, W. and K.-H. RECHINGER, 1967. *Flora der Insel Kythera.* 206 pp., 4 full-page illus. (Boissiera 13). Geneva.

Concise, partly descriptive account of vascular plants of Kithira (Kythera), with synonymy, indication of local range (with *exsiccatae*), taxonomic commentary, and notes on habitat, special features, etc.; phytogeographical considerations (in appendix); index to new names and combinations and generic index at end. The introductory section covers botanical exploration and previous literature. [The work, ostensibly an account treating Kithira and Antikithira, was also designed to serve as a first contribution to nomenclatural 'modernization' of the Greek vascular flora.]

RECHINGER, K.-H., 1961. Die Flora von Euboea. *Bot. Jahrb. Syst.* **80**: 294–465, pls. 4–10, 3 maps.

Consolidated report of collections made in the Euboea subsequent to publication of *Flora aegaea* (see under main heading); includes synonymy, localities with citations of *exsiccatae*, taxonomic commentary, and habitat and other data.

638

Turkey-in-Europe

Area: 23764 km². Vascular flora: 2100–2250 species, of which 2000–2100 native (Webb). For general works and bibliographies on Turkey, including *Flora of Turkey*, see **771**.

The flora of the Istanbul area has additionally been covered in K. H. RECHINGER, 1938. *Enumeratio florae Constantinopolitanae. Aufzählung der nach dem Erscheinen von Boissiers* Flora orientalis *aus der Umgebung von Konstantinopel bekannt gewordenen Farn- und Blütenpflanzen.* 73 pp. (Repert. Spec. Nov. Regni Veg., Beih. 98). Berlin.

WEBB, D. A., 1966. *The flora of European Turkey.* 100 pp., 2 maps (Proc. Roy. Irish Acad. 65–B(1)). Dublin.

Systematic enumeration of vascular plants (2006 species), with limited synonymy and concise indication of local range (with some localities); no index. The introductory section includes remarks on major floristic elements, while a gazetteer, references, and desiderata for further exploration conclude the work.

639

Bulgaria

Area: 110912 km². Vascular flora: 3867 species as of 1992 (Kožuharov, *Opredelitel*; see below). Akeroyd in *Global biodiversity* (see **General references**) gave a total of 3505, while Webb indicated one of 3550–3750 species, of which 3500–3650 were native. – Included here are Bulgaria proper, Rumelia and parts of historical Macedonia and Thrace.

A first contribution to the Bulgarian flora is represented in *Spicilegium florae rumelicae et bithynicae* (1843–45) by August Grisebach, written when the country was still under Ottoman rule. This was based to a large extent on collections by Hungarians as well as on his own travels in 1839, and in Bulgaria focused particularly on the central Stara Planina mountain system. Other early explorers included Victor Janka, also from Hungary (1871–72), and Josip Pančić from Serbia (1882–83). After independence in 1878 (and acquisition of Rumelia in 1885), locally based botanical work developed rapidly. The first national flora was *Flora bulgarica* by the Czech botanist Josef Velenovský (1891; supplement, 1898 and further addenda to 1910). Another Czech contributor of this period was Eduard Formánek with his two *Beiträge* of 1892 and 1898. A herbarium was also established within the Natural History Museum. In 1913, Stephan Petkov (Petkoff) at Sofia University published a retrospective bibliography. As the twentieth century progressed, a national cadre of systematists developed, particularly under the Belarusian Nikolai Stojanov. His first overall contribution was a checklist (*Spisăk na rastenijata, koito se srještat v Bălgarija*, 1921), an admittedly 'rough' work accounting for 3571 species. This was followed by a manual, *Flora na Bălgarija* (1924–25). Both were written together with a contemporary, the Bulgarian forester, ecologist and dendrologist Boris Stefanov. The *Flora* passed through three more editions, the last in 1966–67 shortly before Stojanov's death. Stefanov also contributed a work on trees, *Dendrologija* (1934; revised edn., 1953) and, with Daki Jordanov, *Topographische Flora von Bulgarien* (1931, in *Bot. Jahrb. Syst.* **64**: 388–536). Bulgarian botanists have also contributed to the floristics of FYROM and Thrace beyond present national frontiers.[84]

Following World War II, the Bulgarian Academy of Sciences initiated work towards a definitive national flora; publication began in 1963. Since then, 10 of the projected 11 volumes have appeared, the most recent in 1995. A contemporary review concluded that the flora was quite rich but to some degree imperfectly known; the southern mountains were in particular thought to merit further exploration.[85]

Bibliographies

A long series of national bibliographies covers all general and systematic literature on Bulgarian plants through 1978 (save for 1913–17).

PETKOV, S., 1913. Библіографія Болгарской флоры (*Bibliografija Bolgarskoj flory*). 62 pp. St. Petersburg. (Reprinted from *Russkij Botaničeskij Žurnal* **1911**(7–8):

201–262.) [264 annotated entries covering all aspects of botany from 1794 to 1912, arranged by author within two major areas, non-vascular and vascular plants. In Russian.]

STOJANOV, N., 1928. Литература върху флората на България за последните десет години 1918–1927 (Literatura vărkhu florata na Bălgarija za poslednite decet godini 1918–1927)/Die im letzten Jahrzehnte (1918–1927) erschienene Literatur über die bulgarische Flora. *Izv. Tsarsk. Prirodonaučni Inst.* (Sofija) **1**: 182–189. Continued as *idem*, 1939. Литература върху флората на България за последните единадесет години (1928–1938 год.) (Literatura vărkhu florata na Bălgarija za poslednite edinadecet godini (1928–1938 god.))/Die in den letzten elf Jahren (1928–1938) erschienene Literatur über die Flora Bulgariens. *Izv. Tsarsk. Prirodonaučni Inst.* (Sofija) **12**: 209–230; N. STOJANOV and B. KITANOV, 1950. Литература върху флора и растителната география на България за десетилетието 1939–1948 год. (Literatura vărkhu flora i rastitelnata geografija na Bălgarija za decetiletieto 1939–1948 god.)/Literatur über die Flora und Pflanzengeographie Bulgariens im Jahrzehnte 1939–1948. *Izv. Bot. Inst.* (Sofia) **1**: 480–506; B. KITANOV, 1960. Литература върху висшата флора и растителната география на България 1949–1958 (*Literatura vărkhu visšata flora i rastitelnata geografija na Bălgarija 1949–1958*)/*Literatur über die Flora und Pflanzengeographie Bulgariens 1949–1958*. 76 pp. Sofia: Tsentralna biblioteka, BAN; *idem*, 1975 (1976). Литература върху висшата флора и растителната география на България (*Literatura vărkhu visšata flora i rastitelnata geografija na Bălgarija*) 1959–1968. 270 pp. Sofia; and *idem*, 1984. Литература върху висшата флора и растителната география на България (*Literatura vărkhu visšata flora i rastitelnata geografija na Bălgarija*) 1969–1978. 240 pp. Sofia. [Entries in all installments alphabetically arranged (from year to year beginning with the 1960 compilation) and annotated, sometimes in detail; systematic, title and author indices with, from the 1960 compilation, separation of Cyrillic- and Roman-script titles.]

DELIPAVLOV, D. (ed.), 1983. Определител на растения в България (*Opredelitel na rastenija v Bălgarija*). 431 pp., 571 text-figs., map (back end-papers). Sofia: Zemizdat. (Revised edn. [not seen], 1992.)

Concise manual-key to native, naturalized and commonly cultivated vascular plants (3036 species in 1983), the families arranged alphabetically; includes limited synonymy, Bulgarian nomenclatural equivalents, and partly abbreviated or coded notes on distribution, habit, phenology, and habitat along with many small diagnostic figures; indices to families and genera. [A functional successor to the manual by Vălev *et al.* (see below), but to a considerable extent taxonomically based on *Flora na Bălgarija* by Stojanov *et al.* (see below).]

JORDANOV, D. *et al.* (eds.), 1963– . Флора на народна република България (*Flora na narodna republika Bălgarija*)/*Flora reipublicae popularis bulgaricae* (from vol. 10: Флора на република България/*Flora reipublicae bulgaricae*). Vols. 1– . Illus., map. Sofia: Bălgarska Akademija na Naukite.

Comprehensive descriptive flora of vascular plants (along the lines of *Flora SSSR*); includes keys to all taxa, fairly copious descriptions, full synonymy (with references), Bulgarian and vernacular nomenclature, detailed exposition of local range and general summary of extralimital distribution, some taxonomic commentary, and notes on habitat, phenology, ecology, etc.; indices to all plant names in each volume. The introductory section (in vol. 1) contains chapters on botanical exploration and research in Bulgaria (this also in English), an illustrated organography, a glossary, and a list of authors (the latter three items also appearing in succeeding volumes). [Of a probable total of 11 volumes, 10, the last in 1995, had been published covering all families in the traditional Englerian sequence from the pteridophytes through Rosaceae (vols. 1–5), Leguminosae (vol. 6), Oxalidaceae through Araliaceae (vol. 7), Umbelliferae through Cuscutaceae (vol. 8), families through Solanaceae (including Labiatae) as well as Rubiaceae (vol. 9), and Scrophulariaceae through Valerianaceae (vol. 10).]

KOŽUHAROV, S. (coord.), 1992. Определител на висшите растения в България (*Opredelitel na visšite rastenija v Bălgarija*). Sofia: 'Nauka i Izkustvo'.

A manual-key in landscape format, the entries for species and subspecies including vernacular and scientific names, habitat, distribution, phenology and habitat; family index with generic names and addenda (novelties, new combinations and new records) at end. The general part covers symbols and geographical abbreviations (p. 6), the use of the book (pp. 9–10), abbreviations, an organography with illustrations (on 15 plates), contributors with families covered (p. 35), and family keys (pp. 36–68). [Covers 3867 species and 330 additional subspecies; accounts for taxonomic and nomenclatural changes associated with *Flora Europaea* and other leading contemporary works. There is, however, no bibliography.]

KOŽUHAROV, S., N. ANDREEV and D. PEEV, 1980. Конспект на висшите растения в България (*Konspekt na visšite rastenija v Bălgarija*). Sofia.

A revised checklist reflecting the progress of *Flora na narodna republika Bălgarija*. [Not seen.]

STOJANOV, N., B. STEFANOV and B. KITANOV, 1966–67. Флора на България (*Flora na Bălgarija*)/*Flora bulgarica*. 4th edn. 2 vols. 1326 pp., 1549 text-figs. Sofia: 'Nauka i Izkustvo'. (1st edn., 1924–25; 2nd edn., 1933; 3rd edn., 1948, all by Stojanov and Stefanov alone.)

Illustrated manual-key to vascular plants, with limited synonymy, vernacular and transliterated names, fairly detailed indication of local range, and notes on habitat, phenology, ecology, etc.; indices to all botanical and other names (in vol. 2). The introductory section is pedagogic, while a list of authors' names and an account of organography appear in the appendices. [The manual edited by Delipavlov (see above) is a shortened 'field' version of this now-aging work.]

VĂLEV, S., I. GANČEV and V. VELČEV, 1960. Екскурзионна флора на България (*Ekskurzionna flora na Bălgarija*). 735 pp., illus. Sofia: 'Narodna Prosveta'.

Illustrated, somewhat selective 'rucksack-sized' field manual-key to native, naturalized and commonly cultivated vascular plants (covering *c.* 2250 species); includes diagnoses, very limited synonymy, vernacular and transliterated names, indication of local range (or country of origin), and concise notes on life-form, habitat, ecology, vertical zonation, special features, etc.; extra key to cultivated grasses (including varieties of *Triticum durum* and *T. vulgare*), lists of flowering times, ranges, vertical zonation, etc. of individual species and indices to generic, family and vernacular names at end. The introductory section has a chapter on descriptive organography as well as remarks on the use of the work. Abbreviations and conventions are given in the inside covers. [Intended mainly for student use; tends to omit rare or local species.][86]

Woody plants

Various dendrologies as well as manuals for woody plant identification have appeared in the twentieth century, most recently D. GRAMATIKOV, 1992. Определител на дървета и храсти в България (*Opredelitel' na dărveta i khrasti v Bălgarija*). 2nd edn. Plovdiv. An older dendrological work [not seen] is B. STEFANOV and A. GANČEV, 1958. *Dendrologija*. 2nd edn. 651 pp., illus. Sofia: Sel'skostopanska literatura. (1st edn., 1934, by Stefanov alone.)

CHERNJAVSKI, P., S. NEDJALKOV, L. PLOSCHAKOVA and I. DIMITROV, 1959. Дървета и храсти в горите на България (*Dărveta i khrasti v gorite na Bălgarija*)/*Bäume*

und Sträucher in den Wäldern Bulgariens. 399 pp., 13 col. pls., 427 text-figs. (incl. halftones). Sofia: D'rzhavno izdatelstvo za selskostopanska literatura (for Nauchnoizsledovatelski institut za gorata i gorskoto stopanstvo).

Well-illustrated dendrology in quarto format (covering native and a few commonly grown non-native trees and shrubs, the latter including, among others, *Quercus rubra*), with keys to species and illustrations of habitat, twigs, bark, wood, leaf variation, flower, fruit, etc.; text entries include synonymy, description, botanical and technical data (encompassing distribution, habitat, phenology, biology, populations, growth, wood properties, uses, etc.); Russian and German summaries (pp. 390–391), literature (pp. 392–394) and indices to botanical and vernacular names at end. The foreword and introduction are brief, the latter including remarks on taxonomy and classification. [The authors offer the work as a contribution to a more sophisticated forest practice, related to increasing forest utilization. Only species of greater importance for forestry and park and street use are accounted for here.]

Region 64

Central Europe

Area: 1 273 616 km². – This region encompasses Romania, Hungary, Slovenia, Austria, the Czech Republic, Slovakia, Poland, Germany, Liechtenstein and Switzerland. Included thus are most of the territories of the pre-1918 German Reich as well as the greater part of the former Austro-Hungarian Dual Monarchy. It incorporates botanical 'Mitteleuropa' and reflects historical usage as a region 'lying between Emden and Geneva on the west and the Masurian Lakes and Transylvania on the east'.[87]

Central Europe is by and large one of the floristically best-known and documented parts of Europe (and indeed the world), with four to five centuries of continuous study, the longest north of the Alps.[88] It is densely covered by manuals and other works of all kinds, only the most important of which can be listed here. The region as a whole was not neglected: already in the nineteenth century the works of Ludwig Reichenbach (*Flora germanica excursoria*, 1830–33, and *Icones florae germanicae et helveticae*, 1837–1914) and Wilhelm Koch (*Synopsis florae germanicae et helveticae*, 1837, with further editions to 1903) encompassed most of its western, mainly German-speaking part, while in Austria-Hungary a first overall work was *Enumeratio plantarum phanerogamicarum imperiis austriaci universi* by J. C. Maly (1848; supplement by A. Neilreich, 1861). Neilreich shortly afterwards published two further works: *Aufzählung der in Ungarn und Slavonien bisher beobachten Gefässpflanzen* (1866; Nachtrage, 1870) and a supplement (1867) to Koch's work: *Diagnosen der in Ungarn und Slavonien bisher beobachteten Gefässpflanzen welche in Koch's Synopsis nicht enthalten sind*. In the twentieth century the dominant feature has been Gustav Hegi's ponderous but critical *Illustrierte Flora von Mitteleuropa*. Its influence has spread far and wide, with sets now widely dispersed through the botanical world; its bulk, however, has given rise to questions about the role and contents of floras.[89] The 'second edition' has yet to be 'completed', while a 'third edition' was initiated in 1966 as a revision of early parts of its predecessor. The work is accompanied by a checklist (1967; revised 1973) by Friderich Ehrendorfer and associates, as well as a regional bibliography. No comparable work is yet available for the rest of Central Europe; this includes most of the Carpatho-Pannonian floristic region, for which the best modern reference is *Ikonographie der Flora des südöstlichen Mitteleuropas* (1979; Hungarian version, 1975) by S. Jávorka and V. Csapody, a reissue of their *Iconographia florae hungaricae* (1929–34). In the 1980s Josef Dostál and some colleagues undertook an enumeration of the plants of this same area (see Note 95); a first draft was ready by 1986.

With respect to individual countries, all have more or less recent descriptive floras or detailed enumerations. Associated with these are field manual-keys, available for all countries and mostly recent. The most important of the modern contributions are *Flora der Schweiz und angrenzender Gebiete* (1967–73) by Hans Hess, Elias Landolt and Willy Hirzel, *Flora Republicii Socialiste România* (1952–76) directed by Traian Săvulescu, and *A Magyar flóra és vegetácio rendszertaninövenyföldrajzi kezikönyve* (1964–80) by Reszö Soó, the last a critical chorological–ecological synopsis of the Hungarian flora. *Flora polska* (1919–80; 2nd edn., 1985–) and *Atlas flory polskiej i ziem ósciennych* (1930–) should also be mentioned. The most useful book of illustrations is, however, the already mentioned iconography of Jávorka and Csapody. Of the concise manuals, those by Werner Rothmaler published under the title *Exkursionsflora von Deutschland* (or, for a time,

Exkursionsflora für die Gebiete der DDR und der BRD) enjoy wide popularity.

Nevertheless, in a mature field the majority of works are now 'routine', and likely to continue so. Significant changes are likely to relate as much to social, cultural and political as to scientific developments. Much current activity relates to mapping and vegetatiological studies with a view to conservation and land stewardship. This includes the introduction of floristic and biodiversity databases and information systems, now a general requirement.[90]

Bibliographies. General and divisional bibliographies as for Division 6 (except Blake, 1961, which covers only Switzerland).

Regional bibliography

HAMANN, U. and G. WAGENITZ, 1977. *Bibliographie zur Flora von Mitteleuropa*. 2nd edn., 374 pp. Berlin: Parey. (1st edn., 1970, Munich.) [Provides detailed coverage of floristic and other botanical works, mainly for 'Mitteleuropa' but also includes a listing of 'standard' floras from other parts of the Holarctic zone. Designed as a companion to Hegi's *Illustrierte Flora*. In the revised edition, additions appear as a supplement, pp. 329–374.]

Indices. General and divisional indices as for Division 6.

640

Region in general

The two main subregions given below are the Germanic 'Mitteleuropa' in the northwest and Carpatho-Pannonia in the southeast. For the Alps, see also **603/IV**; for the Carpathians, see also **603/VII**.

I. 'Mitteleuropa'

Included here are German-language works relating to 'Mitteleuropa', corresponding to western Central Europe. This botanical 'circle' developed as a comprehensive working unit long before World War I; it encompassed the pre-1918 German Reich,

Switzerland, Liechtenstein, and 'historical' Austria (i.e., the former Kingdom of Austria in a narrower sense, including the present state, Alto Adige (South Tyrol), the Czech lands of Bohemia and Moravia (the latter with Moravian Silesia), most of modern Slovenia, and Küstenland (encompassing Istria, Trieste and Venezia Giulia)). In effect, it is (or was) the region with the greatest concentration of German-speaking peoples. The works listed below are, however, of value over a far wider area, especially Hegi's *Illustrierte Flora von Mitteleuropa* – about the most comprehensive floristic work ever produced.[91]

Two other key floristic works relating to 'Mitteleuropa' are *Vergleichende Chorologie der zentraleuropäischen Flora* (Meusel *et al.*; see **001**) and R. Lindacher (ed.), 1995. *PHANART – Datenbank der Gefässpflanzen Mitteleuropas: Erklarung der Kennzahlen, Aufbau und Inhalt*. 436 pp., map (Veröff. Geobot. Inst. Rübel, Zürich 125). Zürich. The latter contains a tabular list of all species in the subregion with vernacular names as well as numerous *Zeigerwerte* (indicator-values) relating to environmental and biological parameters such as status, distribution, chorology, habitat factors, phytosociology, biology, eco-anatomy, etc.; references to sources appear on pp. 42–43.[92]

Ehrendorfer, F. (ed.), 1973. *Liste der Gefässpflanzen Mitteleuropas*. 2nd edn., xii, 318 pp., 2 figs., map. Stuttgart: Fischer. (1st edn., 1967, Graz.)

Checklist (with genera arranged alphabetically) of the native and naturalized vascular plants of 'Mitteleuropa' (delimited as above, but omitting much of eastern Germany); includes accepted names of species and infraspecific taxa, standardized acronyms of names with numbers, symbolic indication of citations and of 'critical taxa', and geographical range within the region. The introductory section gives a list of abbreviations and of 'standard' floras utilized.[93]

Hegi, G. (ed.), 1906–31. *Illustrierte Flora von Mittel-Europa* (later *Mitteleuropa*). 7 Bde. (in 13 vols.). 280 pls. (mostly col.), 3434 + 1273 figs. (incl. halftones, maps). Munich: Lehmann; Vienna: Pichler. (Bd. 6(2) reprinted 1954, Munich: Hanser; Bde. 4(3), 5(1), 5(2), 5(3) and 5(4) reprinted with additions and corrections, 1964–66, Munich; pp. 580–1386 of Bd. 6(2), covering part of Compositae, reprinted with new supplement as Bd. 6(4) of the second edition, 1987, Berlin: Parey. Addenda and emendata of 1964–66 collectively reissued in 1 vol. as H. Merxmüller (comp.), 1968.

Nachträge, Berichtigungen und Ergänzungen zu den unveränderten Nachdrucken der Bände 4(3) und 5(1) bis 5(4) mit Verzeichnissen der lateinischen und deutschen Pfanzennamen. 168 pp. Munich: Hanser.) – Largely succeeded by G. Hegi *et al.*, 1935– . *Illustrierte Flora von Mitteleuropa*. 2nd edn., revised and ed. by K. Suessenguth *et al.* Bde. 1–6, *partim* (planned to occupy 7 Bde. in 18 vols.). Illus. (incl. col. pls., halftones, maps). Munich: Lehmann (Bde. 1–2, 1935–39); Hanser (Bde. 3(1), 3(2) in part, 3(3), 4(1), 4(2A), 6(1), 6(2) in part, 6(3) in part, 1957–74); Berlin: Parey (remainder of Bde. 3(2) and 6(3) and parts of 4(2B) and 6(2) including Lfg. A of the last-named, 1978–94); Blackwell (now Parey; remainder of Bd. 4(2B) and rest of work, 1994–). (In progress. Bde. 4(2C) and 6(2) not yet completed; Bd. 7 yet to be published. Bd. 1 reprinted 1965, Munich. Bde. 4(3) and 5(1)–5(4) are reprints, with supplements, from the first edition. For Bd. 6(4), see above.) – Further continued as G. Hegi *et al.*, 1966– . *Illustrierte Flora von Mitteleuropa*. 3. Aufl., revised and ed. by H.-J. Conert, E. J. Jäger, W. Schultze-Motel, J. W. Kadereit, G. Wagenitz and H. E. Weber. Bde. 1–4, *partim*. Illus. (incl. col. pls., halftones, maps). Munich: Hanser (Bd. 2(1) in part, 1966–69); Berlin: Parey (Bde. 1(1), 1(2) and part of 1(3), 1979–92; remainder of Bd. 2(1), 1977–80; Bd. 3(1), 1981; Bd. 4(1), 1986); Blackwell (now Parey; remainder of Bd. 1(3), 1994–97, and Bd. 4(2A), 1995). (In progress. Bd. 1(3) now complete in 10 fascicles; Bd. 2(2), 2(3) and 2(4) still in preparation. Bd. 3(1) and 4(1) are reprints, with some new matter, of the corresponding volumes in the Zweite Auflage, and Bd. 4(2A) is a complete revision of its predecessor.)

This work, perhaps the most compendious and copiously illustrated of all modern floras, covers in great detail the vascular plants of 'Mitteleuropa' (as defined above in the unit heading); includes keys to and lengthy descriptions of all taxa along with principal synonymy and references, vernacular names (in German, English, French and Italian), detailed exposition of local and regional ranges (with many distribution maps), generalized account of extralimital range, more or less detailed critical discussion, and extensive notes on habitat, phenology, biology, ecology, karyotypes (in later editions), morphology, anatomy, palynology, chemistry, etymology of names, properties, uses and (where appropriate) cultivation as related to families, genera and individual species; complete indices to botanical names in each volume; general index in Bd. 7 (available only for original edition) incorporating all botanical and vernacular names in the entire work (along with a key to families, guides to special categories of plants, a synopsis down to genera, a list of botanical authors, and an extensive glossary). The introductory section (in Bd. 1) includes a list of abbreviations and an additional glossary. The sequence of families follows the Englerian system; Bde. 1–2 cover pteridophytes, gymnosperms and monocotyledons and Bde. 3–6, dicotyledons. [This is the standard detailed flora for Central Europe and was one of the five 'Basic Floras' designated as a primary source for the preparation of *Flora Europaea*. A small part of the 'Zweite Auflage' is yet to be completed; as of 1998 the index volume (Bd. 7) was unpublished and Bd. 6(2) remains incomplete, with part of Valerianaceae and the Dipsacaceae, Rubiaceae and Campanulaceae outstanding. Furthermore, some of the other volumes (Bd. 4(3), the four subdivisions of Bd. 5, and Bd. 6(4)) are simply reprints (with supplements) of the corresponding sections of the first edition. The 'Dritte Auflage' initially was conceived as a six-volume revision of Bde. 1–2 of the second edition, published before World War II; by 1998 Bde. 1(1), 1(2), all of 1(3), and 2(1) had appeared, with 2(2) and 2(3) yet to come. However, in 1981 a reprint, with supplement, of Bd. 3(1) was issued and in 1995 a completely revised Bd. 4(2A) made its appearance. Revisions for this edition of 4(3), mostly on legumes, and 5(4), focusing on Lamiaceae, are now planned. Publication and distribution were from July 1975 in the hands of Verlag Paul Parey, Berlin and from 1994 with Blackwell Wissenschafts-Verlag (now Parey), Berlin. Accompanying this work is an extensive source bibliography by U. Hamann and G. Wagenitz (1970, 2nd edn., 1977; see **Regional bibliography** above).][94]

Distribution maps

For *Vergleichende Chorologie der zentraleuropäischen Flora* by H. Meusel *et al.* (1965–92, Jena), see **001** (Eurasia).

Woody plants

FITSCHEN, J., 1994. *Gehölzflora*. 10. Aufl., revised and ed. by F. H. Meyer *et al.* Various paginations with numerous text-figs.; maps (back end-papers). Heidelberg: Quelle & Meyer. (1st edn., 1920; 5th edn., 1959; 6th edn., 1977; 8th edn., 1987.)

Concise pocket manual-key for identification of native, naturalized and commonly cultivated trees and larger shrubs in 'Mitteleuropa'. Section B incorporates artificial general keys (140 pp.) and section C the systematically arranged (after Takhtajan for flowering plants) manual proper, with entries in the latter including limited synonymy, vernacular names, diagnostic figures, and indication of area-class (and more limited range if necessary), life-form, phenology, habitat, frost tolerance and toxicity. Section A (65 pp.) covers organography, terminology, classification, taxonomy and nomenclature, plant geography and ecological formations, hardiness, autecology and poisonous species along with an explanation of codes used in the text; and section D includes a list of key literature and subject and name indices. Abbreviations and symbols now appear in the inside front covers (as well as on a loose card insert), and the maps in the inside back covers.

Pteridophytes

No systematically arranged regional work comparable to *Scandinavian ferns* by B. ØLLGAARD and K. TIND (see **670**) has yet been published. For the pteridophyte section of *Kleine Kryptogamenflora*, see **600**. Mention should be made, however, of K. RASBACH, H. RASBACH and O. WILMANNS, 1976. *Die Farnpflanzen Zentraleuropas*. 2nd edn., 304 pp., 154 pls. (halftones). Stuttgart: Fischer (1st edn., 1968), an ecologically arranged illustrated 'natural history' covering 95 species.

II. 'Carpatho-Pannonia'

Most of 'Carpatho-Pannonia' was in the nineteenth and early twentieth centuries part of the Austro-Hungarian Empire and more particularly the Kingdom of Hungary as established under the terms of the 'Ausgleich' of 1867. Prior to 1918, the main overall work was *Enumeratio plantarum phanerogamarum imperii austriaci universi* by Joseph C. Malý (1848; *Nachträge* by A. Neilreich, 1861). This was succeeded by Neilreich's *Aufzählung der in Ungarn und Slavonien bisher beobachteten Gefässpflanzen* (1866; *Nachträge und Verbesserungen*, 1870). Neilreich also expanded the coverage of Wilhelm Koch's *Synopsis florae germanicae et helveticae* in his *Diagnosen der in Ungarn und Slavonien bisher beobachteten Gefässpflanzen welche in Koch's* Synopsis *nicht enthalten sind* (1867).[95]

JÁVORKA, S., 1925. *Magyar flóra / Flora hungarica*. cii, 1307 pp., 13 pls., map. Budapest: 'Studium'.

Manual-key to vascular plants; includes limited synonymy, vernacular names, concise indication of local range, brief summaries of overall distribution outside the Hungarian Basin, critical commentary, and notes on habitat, phenology, etc.; indices to botanical and vernacular names. The introductory section includes general chapters on biology, ecology, organography (with glossary), use of keys, etc. Although for Hungary proper this work is now superseded by Soó and Kárpáti's *Magyar Flóra*, it is included here for its coverage of most of Carpatho-Pannonia. It also provides the original textual basis for Jávorka and Csapody's *Ikonographie* (see next entry).

JÁVORKA, S. and V. CSAPODY, 1979. *Ikonographie der Flora des südöstlichen Mitteleuropas/ Iconographia florae partis austro-orientalis Europae centralis.* Revised edn. 704, 80 pp., 576 monotone and 40 col. pls. (with 4090 figs.). Stuttgart: Fischer. Hungarian edn.: *idem*, 1975. *Közep-Európa délkeleti részének flórája képekben.* Budapest: Akadémiai Kiadó. (Originally published 1929–34, Budapest, as *A Magyar flóra képekben/Iconographia florae hungaricae*.)

A systematically organized atlas of the vascular plants of those parts of southeastern Central Europe corresponding to 'historical' Hungary; each lithographed plate includes separate figures for a number of species, while the captions include Hungarian vernacular and Latin names as well as concise notes (in Hungarian) on distribution. The 80 pages added in the 1979 German reissue comprise a version of the captions in that language (translated and augmented by Sz. Priszter). The introductory section in both versions of the reissue includes a new introduction with bibliographic notes (by Sz. Priszter), taxonomic changes (by R. Soó), a trilingual glossary, and a gazetteer; several indices conclude the work. All figures are the work of Vera Csapody. [The area of coverage here espoused, corresponding to that of Jávorka's *Magyar flora* (see above), is a considerable asset, the more so as its geographical coverage is distinct from Hegi's *Illustrierte Flora* (see above).]

641

Romania

Area: 237 500 km². Vascular flora: 3550–3750 species, of which 3300–3400 native (Webb). The latest survey by Popescu and Sanda (1998) accounts for 3630 species, somewhat less than the 3759 numbered species in Ciocîrlan (1988). – The northwestern region of Transylvania (Siebenbürgen) along with the Banat in

the central west (some of the latter being now part of Vojvodina) were before 1918 parts of the Kingdom of Hungary. They are thus, in addition to standard Romanian floras, covered in older Hungarian works including those cited under **640/II**.

The first substantial work on Romania to appear after national independence was *Prodromul florei Române* (1879–83) by Dimitrie Brandza. This covered Moldavia and Walachia (Old Romania). It was followed by *Enumeraţia plantelor din România* (1880) and *Conspectul florei României* (1898; supplement, 1909), both by Dimitrie Grecescu. Both works appeared in Bucharest. The Transylvanian botanical circle also contributed from an early date: August Kanitz, at Cluj (Koloszvár) University from 1872, produced *Plantas Romaniae hucusque cognitas* (1879–81). This latter built upon an already long history of Transylvanian floristic studies which included the major documentary works of Johann C. G. Baumgarten, *Enumeratio stirpium Magno Transsilvanie principatui* (1816–46) and J. Ferdinand Schur, *Enumeratio plantarum Transsilvaniae* (1866) as well as a manual by Michael Fuss, *Flora Transsilvanica excursoria* (1866). A Hungarian-language work appeared in 1887: *Erdély edényes flórájának helyesbitett foglalata* (*Enumeratio florae transsilvanicae vasculosae critica*) by Lajos (Ludwig) Simonkai. Later botanists in Cluj included Alexandru Borza, Erasmus Nyárády, Emil Pop and Iuliu (Gyula) Prodan, while in southern Transylvania another active contributor was K. Ungar (*Die Alpenflora der Südkarpathen*, 1913, and *Die Flora Siebenbürgens*, 1925, both published in Sibiu (Hermannstadt)). The latter, the last German-language manual for Transylvania, was written as a successor to that of Fuss. The first of Prodan's floristic publications appeared in 1914; in 1923, three years after the transfer of Transylvania (along with Bukovina and Bessarabia) to Romania, he produced the two-volume *Flora pentru determinarea şi descrierea plantelor ce cresc în România*, the major interwar national manual (an expanded second edition followed in 1939). Five years later, Prodan derived from it *Flora mică ilustrată a României*, a school-flora and field-guide which went through several editions (the last, edited by Alexandra Buia and titled *Flora mică ilustrată a Republici populare Romîne*, in 1961 with a reissue in 1966).[96] The others (particularly Nyárády, whose publications spanned 63 years) would contribute, some of them substantially, to *Flora R. P. R.*

In addition to Transylvania and parts of the Banat, Romania after World War I expanded to include also Bukovina (now in Ukraine, **694**) and Bessarabia (after World War II, however, incorporated into the Soviet Union and now mostly in Moldova, **693**, with the coastal districts in Ukraine, **694**). For the last-named, the Bucharest botanist Traian Săvulescu (mainly with Tscharna Rayss) wrote several contributions between 1924 and 1934, notably *Materiale pentru flora Basarabiei*.

All earlier general works on the Romanian flora have been effectively superseded by *Flora R. P. R.* (Săvulescu, 1952–76), *Flora României* (Beldie, 1977–79) and *Flora ilustrată a României* (Ciocîrlan, 1988). *Flora R. P. R.* has been accepted as having a high standard and so forms an effective base line for further floristic studies in central and southeastern Europe and elsewhere. For pedagogical and field use, Buia's work was succeeded by I. TARNAVSCHI and M. ANDREI, 1971. *Determinator de plante superioare*. 443 pp., illus. Bucharest: 'Didactica şi pedagogica'. In 1998 a new conspectus was published by the Bucharest Botanical Garden, with a revised bibliography. It effectively succeeds the 1947 *Conspectus* of Borza (kept here, however, for its detailed bibliography).[97]

Bibliographies

See also **642** (GOMBOCZ, 1936, 1939) for coverage of Transylvania through 1918, as well as **603/VII** (PAX, 1898, 1908). A classified list of works also appears in the second part of Borza's *Conspectus* (1947) and a bibliography in that of Popescu and Sanda (1998; for both, see below). These should be consulted for the numerous partial and local floras.

BORZA, A. and E. POP, 1921–47. *Bibliographia botanica Romaniae*. Parts 1–38 (Bul. Grăd. Muz. Bot. Univ. Cluj 1–27, *passim*). Cluj. [Unannotated periodic series of alphabetically arranged additions to the literature.]

ŢOPA, E., 1935. Bibliografia manualelor româneşti de botanică. *Revista Pedagog.* 5(3): 231–268. [Not seen.]

BELDIE, A., 1977–79. *Flora României: determinator ilustrat al plantelor vasculare*. 2 vols. 1439 text-figs. Bucharest: Academia RSR.

Illustrated field manual-key to vascular plants (3350 species); includes essential synonymy, abbreviated indication of internal and extralimital distribution, and terse notes on habitat, life-form, habit, phenology, ecology, etc.; index to generic and family names at end. An introductory section gives technical details, main sources, and statistics, followed by two keys respectively to families and woody plants. The taxonomic arrangement is that of *Flora R. P. R.*

BORZA, A., 1947. *Conspectus florae Romaniae regionumque affinium*. viii, 360 pp., 3 col. pls. Cluj: 'Cartea Romaneasca'.

Systematic checklist of the vascular plants of Romania with concise indication of internal range (by regions); addenda and corrigenda, classified list of references, and index to generic names at end. Nomenclature follows Mansfeld's *Verzeichnis* (1940; see **648**) where applicable. [Originally intended to accompany the series *Flora Romaniae Exsiccata*. The labels for the *exsiccatae* include synonymy as well as habitat and other information. Now largely superseded by Popescu and Sanda (1998; see below) but remains useful for its bibliography.]

CIOCÎRLAN, V., 1988. *Flora ilustrată a României*. 2 vols. Illus. Bucharest: 'Ceres'. (Vol. 1 reprinted 1992, Bucharest: 'Ştiinţa'.)

Manual-key to vascular plants with essential synonymy, vernacular names, and indication of distribution, habitat, phenology, life-form, habit-class, ecology and biology, karyotype(s) and chorological class; index to families and genera at end of vol. 2. The work is preceded by a discussion of species treatment, vegetation and altitudinal zonation, ecology, biosystematics, properties and chorology along with abbreviations, an illustrated organography and notes on the use of the keys. The illustrations are grouped on plates and not dispersed through the text. [Plants are arranged according to Ehrendorfer's system as presented in the 1978 edition of 'Strasburger's Textbook'. Vol. 1 covers non-angiosperms and angiosperms through Cistaceae; vol. 2, Cruciferae to Lemnaceae.]

POPESCU, A. and V. SANDA, 1998. *Conspectul florei cormofitelor spontane din România*. 336 pp. Bucharest: Editura Universităţii Bucureşti. (Acta Botanica Horti Bucurestiensis, ex-ser.)

Enumeration of vascular plants (3630 numbered species) with limited synonymy, chromosome numbers, frequency, distribution, and codes or abbreviations for habitat factors ('geoelemente'), life-form ('bioforme'), status, and chorological and phytosociological affinities; bibliography (pp. 311–322) and index to generic and family names at end. An explanation of abbreviations and codes appears on p. 4. [Nothospecies and infraspecific taxa are included but not numbered. The work largely succeeds Borza's *Conspectus*.]

SĂVULESCU, T. *et al.* (eds.); NYÁRÁDY, E. J. *et al.* (comp.), 1952–76. *Flora Republicii Populare Romîne/ Flora reipublicae popularis romanicae* (from vol. 11 *Flora*

Republicii Socialiste România/ Flora reipublicae socialisticae romaniae). 13 vols. 1486 pls. Bucharest: Academia RPR (from vol. 11: Academia RSR).

Comprehensive descriptive documentary flora of vascular plants, including native, naturalized and commonly cultivated plants, with keys to genera, species and infraspecific taxa; includes extensive synonymy, references and citations (including *Flora Romaniae Exsiccata*, a series of distributions), vernacular names in several languages, fairly detailed exposition of local range and summary of extralimital distribution, critical commentary, and extensive notes on habitat, lifeforms, phenology, phytosociology, etc.; index to all botanical names in each volume. The introductory section in vol. 1 includes a historical account of Romanian botanical exploration and floristic and taxonomic research. In vol. 13 are a synopsis of the flora, general keys to families, addenda and corrigenda to vols. 1–12, notes on useful plants, endemics, rare species, etc., author abbreviations, and a general index to the whole work.

Woody plants

A number of dendrologies and manuals have been published since 1945. These include, in addition to that of Beldie described below, INSTITUTUL DE CERCETĂRI FORESTIERE, 1950. *Manual pentru determinarea plantelor lemnoase din R. P. R.* 238 pp., 53 pls. Bucharest; E. G. NEGULESCU and A. SĂVULESCU, 1965. *Dendrologie.* 2nd edn., 511 pp., 335 textfigs., folding map. Bucharest (1st edn., 1957); and V. STANESCU, 1979. *Dendrologie.* 470 pp., illus. Bucharest: 'Didactica'.

BELDIE, A., 1953. *Plantele lemnoase din R. P. R.* 464 pp., 79 pls., 15 figs. Bucharest: 'Agro-Silvica'.

Descriptive treatment of woody plants; includes keys, vernacular names, indication of distribution and/or origin, and illustrations of representative features. Indices and an alphabetical glossary are also incorporated.

642

Hungary

Area: 93 036 km^2. Vascular flora: 2550–2700 species, of which 2250–2450 native (Webb). Priszter (1985; see below) lists 2268 species. – Prior to 1918, the Hungarian Kingdom, as a part of the Austro-Hungarian Empire (or, after 1867, the Dual Monarchy),

encompassed Hungary proper, Slovakia, Ruthenia, the Banat, Transylvania (Siebenbürgen), Slavonia, Croatia and, in Austria, Bürgenland. These areas (except Croatia) are wholly covered by the complementary works of Jávorka and Jávorka and Csapody, including the 1979 reissue of the latter, *Ikonographie der Flora des südöstlichen Mitteleuropas* (for all, see **640/II**). The floras described here are limited to the territory of post-World War II Hungary.

The Hungarian lands were first treated by Carolus Clusius in *Rariorum aliquot stirpium . . . historia* and *Stirpium nomenclator pannonicus* (1583–84). The next significant work is *Descriptiones et icones plantarum rariorum Hungariae* (1802–12) by Adam von Waldstein and Pál Kitaibel. Then followed *Flora comitatus Pestiensis* (1825–26; 2nd edn., 1840) by Joseph Sadler of Budapest University and *Aufzählung der in Ungarn und Slavonien bisher beobachteten Gefässpflanzen* (1866; *Nachträge und Verbesserungen*, 1870) by August Neilreich in Vienna. A contemporary account for the Austrian Empire as a whole was *Enumeratio plantarum phanerogamarum imperii austriaci universi* by J. C. Malý (1848; *Nachträge* by Neilreich, 1861). After the *Ausgleich* of 1867, new developments became possible. Beyond Budapest, a Hungarian botanical school developed from 1872 in Kolozsvár (Cluj, now in Romania) under August Kanitz (a student of Neilreich), and a first manual in that language, with a key to genera, appeared under its auspices in 1882: *Növényhatározó a Dráva, Alsó-Duna és Kápátok övezte Magyarföldön* by Lajos (Ludwig) Simonkai.[98] All these efforts, however, were at best provisional. With full independence after World War I a new round of flora-writing took place under the leadership of Sándor Jávorka in Budapest, resulting in *Magyar flora* (1924–25) – the first significant national manual – and, with the artist Vera Csapody, *Iconographia florae hungaricae* (1929–34); the latter became a cultural as well as scientific landmark (for both, see **640/II**).[99]

Bibliography

GOMBOCZ, E., 1936. *A Magyar tudományos irodalom bibliográfiája 1901–1925,* IV: *Növenytán.* 440 pp. Budapest: Országos könyforgalmi ès bibliográfiai központ/Bureau Central Bibliographique des Bibliothèques Publiques de Hongrie. Complemented by *idem*, 1939. *A Magyar növénytani irodalom bibliográfiája 1578–1900/ Bibliographie der ungarischen botanischen Literatur 1578–1900.* 360 pp. Budapest: Az országos Magyar Természettudományi Múzeum Növenytára. [Both works cover all areas of botany; the entries are classified (relatively finely), with author indices and a scheme at end of the volumes. All Hungarian titles are also rendered in German. Items up to 1918 are drawn from the whole of the former Kingdom of Hungary.]

Soó, R. (ed.), 1964–80. *A Magyar flóra és vegetáció rendszertani-növényföldrajzi kézikönyve/Synopsis systematico-geobotanica florae vegetationsque Hungariae.* Vols. 1–6. Budapest: Akadémiai Kiadó. Complemented by S. PRISZTER, 1985. *A Magyar flóra és vegetáció rendszertani-növényföldrajzi kézikönyve*, vol. 7: *Kiegészítések és mutatók az I–VI. kötethez.* 683 pp. Budapest.

Critical systematic-geobotanical enumeration and synopsis of vascular plants (and bryophytes) of Hungary, without keys (except to subspecific taxa); includes full relevant synonymy (with references and citations), detailed account of local range, general summary of extra-Hungarian distribution, taxonomic commentary, and notes on karyotypes, habitat, life-form, phenology, associates, properties, uses, etc.; indices in each volume to genera and to authors. The introductory section (in vol. 1) includes general chapters on taxonomy, geobotany, and the history of the flora, as well as a detailed phytosociological classification and an account of floristic regions. An appendix with addenda and corrigenda to vols. 1–2 appears in vol. 3 together with a key to plant societies. Volume 6 covers additions and corrections to vols. 1–5, together with a list of officinal drugs and an overall conspectus of the flora together with its geobotanical associations. Volume 7 encompasses addenda and corrigenda (pp. 27–84), collections of *exsiccatae* (pp. 85–117), a catalogue of the Bryophyta and vascular plants with standard numbers, essential synonymy, and biological and legal status (pp. 119–229), indices to these three sections, and complete indices to the series (pp. 255–621).[100]

Soó, R., S. JÁVORKA *et al.*, 1951. *A Magyar növényvilág kézikönyve.* 2 vols. xlvi, 1120 pp., 170 text-figs., map. Budapest: Akadémiai Kiadó.

Descriptive manual of native, naturalized and commonly cultivated non-vascular and vascular plants; includes keys to all taxa and citations of standard monographic and revisionary treatments along with synonymy, references, vernacular names, fairly detailed account of local range, generalized summary of extra-limital distribution, and notes on habitat, phenology, habit, associations, karyology, properties, uses, etc.; lists of authors, medicinal plants, and errata; indices to

botanical and vernacular names. The introductory section includes an explanation of the work, an extensive bibliography, and accounts of plant associations and floristic regions. Illustrated glossaries are given at the beginning of each major plant group. [Now partly superseded by the senior author's *Synopsis* (preceding entry) and the following work.]

Soó, R. and Z. KÁRPÁTI, 1968. Magyar flóra: harasztok [Pteridophyta] – virágos növények [Anthophyta]. In T. Hortobágyi (ed.), *Növényhatározó,* II. 4th edn. 846 pp., 1990 text-figs. Budapest: Tankönyvkiadó. (1st edn., 1952.)

Illustrated manual-key to native, naturalized and commonly cultivated vascular plants; includes abbreviated summary of local range, vernacular names, and notes on habitat, life-forms, associations, etc., and index to family, generic and vernacular names. The introductory section includes summaries of phytosociological units and floristic regions, technical notes, an illustrated glossary, lists of plants with particular features or properties, and a short reference list. [Based on preceding work. For additions, see R. Soó, 1978 (1979). Kiegészítések és javítások Soó-KÁRPÁTI: Magyar Flórá jához az újabb kutatások eredményei alapján/Ergänzungen und Verbesserungen zu „Magyar Flora" von Soó und KÁRPÁTI auf Grund der Ergebnisse neuerer Forschungen. *Bot. Közlem.* **65**: 149–164.]

Woody plants

CSAPODY, I., V. CSAPODY and F. ROTT, 1966. *Erdei fák és cserjék.* 327 pp., illus., col. pl. Budapest: Országos Erdészeti Föigazgatóság.

An illustrated manual of woody plants. [Not seen.]

643

Slovenia

See also **630** (HORVATIĆ, 1954, 1967–86; MAYER); **640/I** (HEGI). For those parts of extra-alpine Slovenia formerly in the Austrian province of Küstenland, see also **629**. – Area: 20 251 km². Vascular flora: almost 3000 species (Martinčić and Sušnik, 1984; see below).

Until 1918 all Slovenia was part of the 'crown lands' of Austria, comprising the whole or parts of the provinces of Küstenland, Kärnten (Carinthia), Krain

(Carniola) and Steiermark (Styria). Krain and the Slovene-speaking parts of Kärnten and Steiermark joined the new South Slav state after World War I (but without internal political recognition) and in 1945 the present republic was organized as a Yugoslav federal state. Between the world wars Küstenland, including Venezia Giulia and Istria, was part of Italy and so included in Italian floras (**620**); postwar works in that country have continued this tradition. Since 1945, however, most of Küstenland has been divided between Slovenia and Croatia, with only small parts, including Trieste, remaining in Italy. In 1992 Slovenia became independent.

During Austrian rule, Slovenia and the Istrian peninsula were, beginning with J. A. Scopoli and his *Flora carniolica* (1760; 2nd edn., 1771–72) and B. Hacquet in *Plantae alpinae carniolicae* (1782), early well explored and documented.[101] In 1844 for Carniola (Krain) there appeared at Ljubljana an annotated checklist, *Übersicht der Flora Krains* by A. Fleischmann. Other parts of the Slovene lands were covered in floras of Steiermark, Kärnten and Küstenland. From 1897 these lands, like the rest of Austria, were served by Fritsch's *Excursionsflora* (3rd edn., 1922). During World War II there appeared a Slovene-language manual, *Ključ za določanje cvetnic in praprotnic* by Angela Piskernik[ova] (1941, Ljubljana; 2nd edn., 1951). Its successor was *Mala flora Slovenije* (see below). A critical enumeration, also described below, was published by Ernest Mayer in 1952; and a 'chromosome atlas', *Cytotaxonomical atlas of the Slovenian flora* by Áskell and Doris Löve (1974, Lehre, Germany) is also available. In 1995 a revised checklist by Trpin and Vreš was published consequent to the new Austrian manual of Adler, Oswald and Fischer (see **644**).

MARTINČIČ, A. and F. SUŠNIK (eds.), 1984. *Mala flora Slovenije*. 2nd edn., 793 pp., 435 text-figs., maps (end-papers). Ljubljana: Državna založba Slovenije. (1st edn., 1969, Ljubljana; 3rd edn., 1999, not seen.)

Concise manual-key to vascular plants; includes very limited synonymy, vernacular names, and notes on local range (with reference to floristic districts or, if less common, more specific areas), habitat, occurrence, associates, phenology, etc.; indices to vernacular and botanical names of families and genera. The introductory section includes a chapter on descriptive morphology, a table of phytosociological units, and a key to families.

MAYER, E., 1952. *Seznam praprotnic in cvetnic Slovenskega ozemlja/Verzeichnis der Farn- und Blütenpflanzen des slowenischen Gebietes*. 427 pp. (Razpr. Slovensk. Akad., IV, 5; Inst. Biol. 3). Ljubljana.

Systematic enumeration of the vascular plants of Slovenia, with extensive synonymy (no references) and concise general indication of local range; full bibliography and index to generic and family names at end.

TRPIN, D. and B. VRES, 1995. *Register flore Slovenije: praprotnice in cvetnice*. 143 pp., 8 pls., 3.5 in. diskette. Ljubljana: Znanstvenoraziskovalni center SAZU. (Zbirka ZRC 7.)

Introduction (in both Slovene and English) with explanation and plan of work; symbols (p. 13) and list of principal references (p. 14); numbered tabular checklist with computer codes, accepted names and 'aggregates', more important synonyms, Slovene vernacular names, and sources (pp. 15–96); Slovene–Latin lexicon (pp. 97–136); detailed references for sources (pp. 137–139), abstract (p. 140), and portfolio of color photographs. [Accounts for 3216 taxa (and an additional 61 plants known purely in cultivation). 216 new or overlooked plants have been added to the flora since publication of *Mala flora Slovenije* in 1984; 25 taxa considered to be extinct and 13 records dropped as incorrect.]

Woody plants

Three works (one a textbook) have been published: R. ERKER, 1957. *Opiz gazdnoga drevja in grmovja* (Dendrografija). Ljubljana; *idem*, 1962. *Dendrologija*. 173 pp. Ljubljana; and E. MAYER, 1958. *Pregled spontane dendroflore Slovenije*. Ljubljana. (Gozdarski vestnik 6–7.)

644

Austria

Area: 83853 km². Vascular flora: 3300–3450 species, of which 2900–3100 native (Webb). Adler, Oswald and Fischer (1994; see below) account for 3300 species and infraspecific taxa.

Before 1918, the Austrian Kingdom in a narrower sense (the 'Erzherzogsthum', exclusive of Dalmatia, Bukovina and Galicia) included, in addition to most of modern Austria, the Istrian peninsula, present-day Slovenia, Trent and South Tyrol (Trentino-Alto

Adige), and the Czech lands of Bohemia and Moravia (with Moravian Silesia). It is this area which is covered in Fritsch's *Exkursionsflora* as well as within works on 'Mitteleuropa' (**640/I**). Janchen's enumeration by contrast deals only with modern Austria (inclusive of the province of Bürgenland, a seat of the Esterházy family and before 1918 in the Kingdom of Hungary). For works covering the Alpine region, see also **603/IV**.

Austria, like Hungary, was first floristically documented by Carolus Clusius in 1583. The first Austrian floras are those by N. J. Jacquin, *Enumeratio stirpium plerarumque quae sponte crescunt in agro Vindobonensi, montisque confinibus* (1762) and H. J. N. Crantz, *Stirpes austriacae* (1762–67). Later Jacquin produced *Florae austriacae . . . icones* (5 vols., 1773–78), a response to *Flora danica*. These were followed by the first excursion flora, *Östreichs Flora* (1794; 2nd edn., 1814) by J. A. Schultes, a work later recognized as permeated with errors. Then followed *Synopsis plantarum in Austria . . .* (1797) by N. T. Host, *Flora des Oesterreichischen Kaiserthumes* (1816–22, with 200 color plates) by Leopold Trattinick, and *Flora austriaca* (1827–31), also by Host. An important and critical step forward came with *Flora von Niederösterreich* (1859) by the jurist and judge August Neilreich. This was over the following half-century joined by various new provincial floras, some quite substantial. National excursion floras not unnaturally continued to appear, however; in 1854 Schultes's flora was succeeded by *Botanisches Excursionsbuch für die deutsch-österreichischen Krönlander und das angrenzende Gebiet* by Gustav Loriser, with further editions to 1883. Its successors in turn were the well-known *Exkursionsflora für Österreich* by Graz professor Karl Fritsch, which first appeared in 1897 (3rd edn., 1922, as *Exkursionsflora für Österreich und die ehemals österreichischen Nachbargebiete*) and, currently, *Exkursionsflora von Österreich* by Wolfgang Adler, Karl Oswald and Raimund Fischer (1994; see below). A further key step, the first of its kind in the Republic, was Erwin Janchen's *Catalogus* (1956–60; also described below). For the school and enthusiast market, Moritz Willkomm while at Prague produced *Schulflora von Österreich* (1888; 2nd edn., 1892); this was succeeded by a long-used work, *Schulflora für Österreich* by Anton Heimerl (1903; 2nd edn., 1912, reissued 1923).

Other developments of recent years have included new provincial floras, mapping initiatives, and preparation (at Vienna) of a so-called 'critical flora', *Flora von Österreich*.[102]

Bibliographies

In addition to the following, there is an extensive bibliography (pp. 1050–1066) in the new *Exkursionsflora* of Adler *et al.* (see below).

EHRENDORFER, F. *et al.*, 1974. Fortschritte der Gefässpflanzensystematik, Floristik und Vegetationskunde in Österreich, 1961–1971. *Verh. Zool.-Bot. Ges. Wien* 114: 63–143. [A continuation of the Janchen bibliography.]

JANCHEN, E., 1956. [Bibliography.] In *Catalogus florae austriae* (see below), I: 1–50. Vienna. [Alphabetical by authors. Continued in the four *Ergänzungshefte* to 1967.]

ADLER, W., K. OSWALD and R. FISCHER, 1994. *Exkursionsflora von Österreich*. 1180 pp., 510+ text-figs. (incl. maps), maps in end-papers. Stuttgart/Vienna: Ulmer.

Concise manual-key to native, naturalized and adventive vascular plants with for each taxon diagnostic and other key features, vernacular names, essential synonymy, and indication of habit, life-form, phenology, habitat(s), altitudinal zonation, frequency, distribution (by Ländern) with in some cases further detail, chorology including main area(s) of distribution beyond Austria, properties and uses, toxicity (if relevant), legal status, and critical and other notes; there are also a few diagnostic figures. The substantial general part includes an introduction covering the plan of the work, organography (pp. 38–91), systematics, taxonomy and nomenclature, ecological morphology including life-forms and biology, chorology and areology, regional floristics (including chorological classes and endemism), natural areas, aut- and synecology, sociology, vegetation classification, geology, climate, zonation, conservation, and the history of floristic botany, along with notes on the use of the keys and the layout of the work (pp. 164–171; the boxes on pp. 170–171 are essential), collecting, herbaria and useful addresses (pp. 172–178), the system of classification adopted (pp. 179–184), and the general key (pp. 185–232); a considerable literature list, glossary, subject index, table of derivation of epithets, and a general index to all names conclude the work.[103]

JANCHEN, E., 1956–60. Pteridophyten und Anthophyten (Farne- und Blütenpflanzen). 4 parts. In K. Hafler and F. Knoll (eds.), *Catalogus florae austriae* I. 999 pp. Vienna: Österreichische Akademie der Wissenschaften (distributed by Springer, Vienna/New

York). Continued as *idem*, 1963–67. *Ergänzungshefte* I–IV. 4 parts. 128, 83, 84, 221 pp. Vienna.

Detailed systematic enumeration of native, naturalized and commonly cultivated vascular plants; includes extensive synonymy (with references and citations), vernacular names, symbolic indication of local range, taxonomic commentary, and notes on habitat, ecology, etc.; addenda (pp. 881–974); indices to vernacular, generic and family names (complete index to all parts of the work in *Ergänzungsheft* IV). Numerous references to relevant revisions and monographs of individual genera and families as well as synopses of infrafamilial classifications are also included. The introductory section includes a very full bibliography (separately cited above).

Woody plants
There are two older works: *Die Bäume und Sträucher des Waldes* (3 vols., 1889–99) by G. Hempel and K. Wilhelm and *Kleiner Bilder-Atlas zur Forstbotanik* (1907) by Wilhelm. These both cover the 'Erzherzogsthum', the Austrian crown lands without Dalmatia, Bukovina and Galicia. For everyday identification reference should be made to Fitschen's *Gehölzflora* (see 640/I).

645

Czech Republic (and former Czechoslovakia)

Area (Czech Republic): 78 864 km². Vascular flora (Czechoslovakia): 3000–3150 species, of which 2600–2750 native (Webb). – Included here are modern works relating to former Czechoslovakia as well as the Czech lands alone (Čechy or Bohemia, Morava or Moravia and Moravian Silesia). For Slovakia, see **646**.

Until 1918 what is now the Czech Republic comprised three of the 15 'crown lands' of Austria. Between 1918 and 1939 the republic of Czechoslovakia consisted of the Czech lands, Slovakia and Ruthenia (the last-named also known as Carpatho-Ukraine and now part of Ukraine (**694**; for more specific works, see also **603/VII**, Carpathians, and **695**, Southwestern Ukraine)). From 1968 to 1992 Czechoslovakia was a federal union of two states, each now independent.

I. Czechoslovakia (1918–92)

Bibliography
In addition to the 1952 monograph by Futák and Domin, from 1923 through 1961 there ran in *Preslia* (3–33, *passim*) a periodic listing, *Bibliografie československých botaniků*, initially edited by Jan Vilhelm, then by K. Cejp and lastly by Vladimír Skalický. With some breaks, this covered literature through 1958. From 1967 coverage was resumed in *Bibliographia botanica Čechoslovaca*, produced as annual to triennial separate issues under the editorship of Z. Neuhäuslová-Novotná and published under the auspices of the Botanický ústav ČSAV and the Institute of Botany, Průhonice. Through 1994, 16 installments had been published covering the years from 1959 to 1990.[104]

FUTÁK, J. and K. DOMIN, 1960. *Bibliografia k flóre ČSR do r. 1952*. 883 pp. Bratislava: Vydavateľstvo Slovenskej akadémie vied. [20000 entries, arranged by author with subject, taxonomic and regional indices; addenda and tables of contents (the latter in Czech and German) at end. Some entries are annotated.]

DOSTÁL, J., 1958. *Klíč k úplné květeně ČSR*. 2nd edn., 982 pp.; xviii, 408 text-figs. Prague: Československé akademie věd. (1st edn., 1953.)

Illustrated, briefly descriptive manual of vascular plants; includes keys to all taxa, limited synonymy, vernacular names, numerous diagnostic figures, and symbolic indication of distribution and altitude, habitat, life-form, phenology, etc.; list of authors and indices to all botanical and vernacular names at end. The introductory section includes a glossary. [Essentially a condensation and revision of the author's *Květena ČSR*.]

DOSTÁL, J., 1982. *Seznam cévnatých rostlin květeny Československe*. vii, 408 pp. Prague-Troja: Pražská botanická zahrada.

Nomenclator of vascular plants, with synonymy and vernacular names. [Partially succeeds K. DOMIN, 1935. *Plantarum Čechoslovakiae enumeratio*. 363 pp. (Preslia 13–15). Prague. This latter is a systematic census with references and literature citations, the latter not a feature of Dostál's work.]

DOSTÁL, J., 1989. *Nová květena ČSSR*. 1548 pp., 336 pls. Praha: 'Academia'. Slovak version: J. DOSTÁL and M. ČERVENKA, 1991–92. *Veľký kľúč na určovanie vyšších rastlín*. 2 vols. 1576 pp., illus. Bratislava: Slovenské pedagogické nakladateľstvo. (Both works based on *idem*, 1949–50. *Květena ČSR*. 64, 2269 pp.; xviii, 711 text-figs. (Sbírka příruček

Československé botanické společnosti, sv. 2). Prague: Přírodovedecké nakladatelství.)

The Czech-language version is a well-illustrated descriptive flora (in A4 format) of vascular plants with keys to all taxa, concise synonymy, Czech vernacular names, synonymy, phenology, karyotype(s), local distribution in the Czech lands, Moravia and Slovakia, external range, and notes on life-form, floristic classes (including distinctive features, if any), coenotypes (in italics), habitat, phenology and life-form; list of corrections and indices to all Czech and Latin names at end. The general part includes a preface (by S. Hejný), introduction, acknowledgments, signs and abbreviations (pp. 15–16), phytocoenological units (pp. 17–34), bibliography (pp. 34–35) and general key to families (pp. 36–46). In the Slovak-language version there is also an illustrated glossary (pp. 28–56) and contrasts between Czech and Slovak terms (pp. 57–58); Slovak as well as Czech vernacular names are moreover incorporated throughout. Indices to botanical, Czech and Slovak names appear at the end of each volume. The floristic-regional maps of the Czech lands and Slovakia appear respectively in the front and back end-papers.[105]

II. Czech lands (České zeme)

See also **I. Czechoslovakia** above (all works). In addition to those in Czech, German-language floras and checklists include EHRENDORFER and HEGI (**640**) and MANSFELD (**648**) along with the old *Exkursionsflora für Österreich und die ehemals österreichischen Nachbargebiete* (3rd edn., 1922) by Karl Fritsch (cf. **644**).

The earliest floras in what is now the Czech Republic were in Latin: *Flora boëmica inchoata* (1793–94) by F. W. Schmidt, an incomplete work with only 400 species, and *Clavis analytica in floram Bohemiae phanerogamarum* (1824) by Vincenz Franz Kosteletzky. The first flora in Czech was *Flora Čechica/Kvetena Česka* (1819) by the brothers Jan S. and Karl B. Presl. These were followed by checklists and documentary works such as *Seznam rostlin květeny České* (1852) by Filip M. Opiz, *Prodromus der Flora von Böhmen* by Ladislav J. Čelakovský (1867–81; Czech version: *Prodromus květeny České*, 1868–83), *Flora von Mähren und österreichischen Schlesien* by Adolf Oborny (1882–86), and *Květena Moravy a Rak. Slezska* by Eduard Formánek (1896–97). The first compact

manual was Čelakovský's *Analytická květena česka* (1879; 3rd edn., 1897, as *Analytická květena Čech, Moravy a Rak. Slezska*). These gave way in the early twentieth century to the four-volume *Názorná květena zemí koruny české* (1900–04) and another manual, *Klíc k úplné květeně zemí koruny české* (1912), both by Frantisek Polívka. A second edition of the latter, revised by Karel Domin and Josef Podpěra and expanded to cover Slovakia (**646**) and Ruthenia (**695**), appeared in 1928 as *Klíc k úplné květeně republiky Československé*; and *Plantarum Čechoslovakiae enumeratio* (noted above) was published in 1935.

HEJNÝ, S. and B. SLAVÍK (eds.), 1988– . *Květena České Socialistické Republiky*, 1– (from vol. 2, 1990: *Květena České Republiky*). Illus. Prague: 'Academia'.

Detailed illustrated descriptive flora in a large format with keys to all taxa, synonymy with references, vernacular names, citations of *exsiccata*-series (not limited to those from the Czech Republic), indication of karyotypes and phenology, and notes on ecology, phytosociology, distribution (including presence/absence in 99 botanical districts and localities for rarer plants) and extralimital range (with references to published distribution maps); indices to vernacular and botanical names at end of each volume. Genus entries include references to relevant literature. An extensive introductory section (pp. 1–164, with 156–162 being a summary in English) covers the history of floristic and chorological studies (by Vladimir Skalický and others), history and composition of the vegetation and plant cover and its individual formations, classification of phytosociological units (141 in all), and a discussion of phytogeography and chorology with definition of 'types' and grid-maps of representative species (at the highest level of affinity, most of the Czech Republic is Central European, but in the southeast towards Břeclav becomes Pannonian), along with a literature list, taxonomic categories, conventions and terminology, summary of the arrangement of families (the Takhtajan system is adopted for flowering plants), and lexicon. The inside covers include (in front) a map of the botanical districts and (in back) descriptions thereof. [Eight volumes in all were originally projected. Vols. 1–5 (1988–97) cover families through the Rosales, Fabales, and Droseraceae through Dipsacaceae (nos. 1–128 out of 179). Vol. 6 is to include many 'Tubiflorae', vol. 7 will cover Asterales, and a remaining volume (or volumes) are to account for monocotyledons.][106]

646

Slovakia

See also **645** (works on former Czechoslovakia). For Hungarian-language works, see **640/II** (Jávorka; Jávorka and Csapody). – Area: 48 667 km². Vascular flora: *c.* 3000 species (Bertová in Marhold, 1988; see Note 107).

The main early floras of Slovakia are *Flora posoniensis* by Stephan (István) Lumnitzer (1791) and a work of the same title by Stephan Endlicher (1830), both relating to the Bratislava (Posonium) district, and *Flora Carpathorum principalium* (1813) by the Swede Göran Wahlenberg (cf. **603/VII**). These were followed by the first general account, *Května Slovenska* (1853) by Gustav Reuss. This latter, moreover, was the first in Slovak, a fruit of the systematization of that language in the early nineteenth century. Reuss's flora covered 2000 species and in addition furnished a basis for descriptive botanical terminology. Not unnaturally the country was also covered in *Descriptiones et icones plantarum rariorum Hungariae* (1802–12) by Waldstein and Kitaibel and in *Aufzählung der in Ungarn und Slavonien bisher beobachten Gefässpflanzen* by Nielreich (1866–70) as well as in later Hungarian floras. After 1918 Czech-language works (**645**) were extended to cover Slovakia. Autonomy in 1939, though, gave rise to *Flóra Slovenskej republiky* by J. M. Novacký (1943); and, consequent to the formation of the Slovak Academy of Sciences and establishment of its botanical research group, preparation of *Flóra Slovenska* commenced in 1954.[107]

DostÁl, J. and M. Červenka, 1991–92. *Vel'ký kl'uč na určovanie vyšších rastlín.* See **645/I**.

FutÁk, J., L. Bertová and K. Goliašová (eds.), 1966– . *Flóra Slovenska.* Vols. 1– . Illus. Bratislava: Vydavatel'stvo Slovenskej akadémie vied (later 'veda').

Comprehensive documentary descriptive flora (vol. 2 onward) of vascular plants, with keys to all taxa (from vol. 5(1) also in English); limited synonymy, with references; vernacular names; detailed exposition of local range, including citations; detailed taxonomic commentary and notes on habitat, ecology, biology, karyotypes, etc.; numerous figures and distributional maps (covering most species); indices to botanical and vernacular names. Volume 1 comprises a comprehensive illustrated Slovak botanical glossary and lexicon, compiled by Josef Dostál and Jan Futák. [Not completed, though well advanced; by 1999 (although with a gap from 1966 to 1982) vols. 1–3, the four parts of vol. 4 (1983–88), vol. 5(1) (1993), and 5(2) (1997) had been published, covering pteridophytes, gymnosperms and dicotyledons from Magnoliales through Scrophulariales and Plantaginales (following the Novák/Futák system).][108]

Marhold, K. and F. HindÁk (eds.), 1998. *Zoznam nižších a vyšších rastlín Slovenska/Checklist of non-vascular and vascular plants of Slovakia.* 687, [1] pp. Bratislava: 'Veda' (Vydavatel'stvo Slovenskej akadémie vied).

Tabular checklist of Cyanobacteria, fungi, non-charophytic and charophytic algae, and non-vascular and vascular plants, arranged in the first instance by major groups and then alphabetically (vascular plants, pp. 333–687); for vascular plants entries include accepted names and synonyms, authors, taxonomic or geographical status (if doubtful), protected status (shown by a § if applicable), threat categories, endemicity, alien status (if applicable), Slovak vernacular names, and whether or not a given taxon is known exclusively in cultivation. An explanation in Slovak and English precedes the list proper (with similar introductions being furnished for the other major groups in the work), and references (main sources) appear on pp. 338–339.

647

Poland

Area: 312 683 km². Vascular flora: 2350–2600 species, of which 2250–2450 native (Webb). – Those areas of modern Poland which formed part of Prussia and, later, the Second German Reich are covered by all German-language floras and related works relating to 'Mitteleuropa', as well as by some pre-1939 works on Germany. Sections of *Flora polska* and *Atlas flory polskiej* published before World War II include large parts of present-day Belarus' and Ukraine and a portion of Lithuania but omit Pomerania, Silesia, and part of East Prussia. For additional coverage of the Sudetens and the Carpathians see respectively **603/VI** and **603/VII**.

Botanical studies in Poland date back to the six-teenth century, with centers in Kraków (where Anton Schneeberger's nomenclator, *Catalogus stirpium qua-rundam*, appeared in 1557) and, later, Warsaw and Wilno (Vilnius; see **679**). The first local flora (for the vicinity of Warsaw, by Martin Bernhard) appeared in 1652. Further development took place during the edu-cationally enlightened Poniatowski reign (1766–95), with botanical terminology becoming systematized from 1777. In the absence of a state between 1795 and 1918, however, documentation of the Polish flora as a whole developed but slowly. Key nineteenth-century works include *Flora polonica phanerogama* (1847–48) by Jakub Waga in Warsaw and *Flora polonicae prodromus* (1873) by Jozef Rostafiński in Kraków.[109] Professional activity and literature developed most strongly in the latter as well as in Galicia as a whole (see also **694**) where political conditions, particularly after 1867, were more favorable than elsewhere.[110] Successors of Rostafiński at Kraków included, in turn, Marian Raciborski and Władysław Szafer, founders of *Flora polska* (1919–80, 1995). Szafer, together with Stanisław Kulczyński (founder of *Atlas flory polskiej*) and Bogumił Pawłowski, also compiled the first edition (1924) of the now long-standard manual, *Rośliny Polskie*. In 1995, a new checklist was published.[111]

Bibliographies

See also MANSFELD (**648**).

NOWAK, M., 1991. *Bibliografia flory Polskiej: rośliny naczyniowe za lata 1971–1980/ Bibliography of Polish Flora – Vascular plants for the years 1971–1980*. 272 pp. Kraków: PAN, Instytut Botaniki im. 'W. Szafera'. (Bibliografie bota-niczne 5.) [2235 unannotated numbered titles, arranged by author with subject index.]

SZYMKIEWICZ, D., 1925. *Bibljografja flory Polskiej*. 159 pp. (Prace Monogr. Komis. Fizjogr. 2). Kraków: Polska Akademji Umiejętności. [Entries alphabetical by author within major subject areas (see table of contents, p. 159) and also featuring a geographical index. Covers literature from 1753 to 1923 in all plant groups but not guaranteed com-plete.]

MIREK, Z., H. PIĘKOŚ-MIREK, A. ZAJĄC and M. ZAJĄC, 1995. *Vascular plants of Poland: a checklist* (*Krztzcyna lista roślin naczyniowych Polski*). 303 pp., 5.25 in. magnetic diskette in pocket. Kraków: W. Szafer Institute of Botany, Polish Academy of Sciences. (Polish Botanical Studies, Guidebook Series 15.)

Alphabetically arranged nomenclator with accepted names in bold type and all synonymy interca-lated; includes vernacular names and indication of status. Notes on individual taxa (569 in all) are flagged in the text and arranged on pp. 217–267 following the list proper; they are followed by a bibliography and a lexicon of authors' names. The general part (in both Polish and English) furnishes the rationale and basis for the work, taxonomic philosophy, and application of Latin and Polish names. [A necessary complement to *Flora polska*, *Flora Polski* and *Rosliny Polskie*.]

RACIBORSKI, M., W. SZAFER, B. PAWŁOWSKI, A. JASIEWICZ and Z. MIREK (eds.), 1919–95. *Flora polska*. 15 vols. (vol. 15 as *Flora polski*). Illus. Kraków: Polska Akademji Umiejętności (1919–47, vols. 1–6); Warsaw (later Warsaw/Kraków): Państwowe Wydawnictwo Naukowe (PWN) (1955–80, vols. 7–14); Kraków: Instityt Botaniki, Polska Akademia Nauk (1995, vol. 15). Partial successor: A. JASIEWICZ (ed.), 1985–92. *Flora Polski: rośliny naczyniowe*. Wydanie 2. Vols. 3–5. Illus. Warsaw/Kraków: PWN (vols. 4–5, 1985–87); Kraków: Instityt Botaniki, Polska Akademia Nauk (vol. 3, 1992). Associated atlas: S. KULCZYŃSKI, and J. MĄDALSKI (eds.), 1930– . *Atlas flory pols-kiej/Florae polonicae iconographia* (from 1954 *Atlas flory polskiej i ziem ościennych/Florae polonicae terra-rumque adiacientium iconographia*). Vols. 1– (published in fascicles). Pls. Kraków: Polska Akademji Umiejętności (1930–36); Warsaw: Polska Akademia Nauk (later PWN) (1954–).

Copiously descriptive documentary flora of vas-cular plants, arranged following the Englerian system; includes keys to genera and species, synonymy and ver-nacular names, fairly detailed indication of local range in Poland and adjacent areas, general summary of extralimital distribution, comments on taxonomic problems, figures of diagnostic features, and notes on habitat, phenology, special features, etc.; indices to all botanical names in each volume.[112] The second edition, commenced in 1985, has aimed at a revision of volumes 1–5; as of 1995 vol. 3 (1992; Betulaceae–Caryophyllaceae), vol. 4 (1985; Aristolochiaceae–Resedaceae) and vol. 5 (1987; *Rosa*), with references to family and generic revisions and monographs, had been published. Volume 15 is a cumulative index, also inclu-sive of notes on the history and authorship of the work. The *Atlas* is in a folio format with large figures and cross-referenced descriptive commentary in Polish and Latin. It was planned to occupy 23 volumes in the same sequence as the *Flora*.[113]

SZAFER, W., S. KULCZYŃSKI and B. PAWŁOWSKI, 1988. *Rosliny Polskie*. 6th edn. xxxi, 1019 pp.; vii, 2187 text-figs., map (in 2 vols.). Warsaw: PWN. (1st edn. in 1 vol., 1953.)

Concise illustrated manual-key to native, naturalized and commonly cultivated vascular plants, with brief diagnoses, limited synonymy, vernacular names, indication of local range, and partly symbolic notes on habitat, life-form, phenology, etc.; indices to botanical and vernacular names. An introductory section, partly pedagogic, includes a list of authors and an illustrated glossary/organography. In addition to modern (post-1945) Poland, this work covers adjacent parts of neighboring countries. [The so-called third edition contains an additional preface by Pawłowski, dated 1969; later issues are merely reprints. The work as a whole has not been revised since 1953.]

Woody plants

BIAŁOBOK, S. and K. BROWICZ (eds.), 1963–81. *Atlas rozmieszczenia drzew i krzewów w Polsce/Atlas of distribution of trees and shrubs in Poland*. 32 fascicles. Maps. Warsaw: PWN (for Zakład Dendrologii i Arboretum Kórnicke PAN).

A distributional atlas, with large-scale dot maps (on a grid) of ranges within modern Poland for 162 individual species of trees and shrubs accompanied by explanatory remarks (in Polish, Russian and English) with much emphasis on distribution, habitats and occurrence and summary maps of extralimital distribution; many references provided. No systematic sequence is followed, and *Rubus* is omitted. A species list for the whole series, with cross-references, appears in fascicle 32.

KOŚCIELNY, S. and B. SĘKOWSKI, 1971 (1972). *Drzew i krzewy: klucze do oznaczania*. 535 pp., 179 text-figs., 65 pls. Warsaw: Państwowe Wydawnictwo Rolnicze i Leśne (PWRL).

Illustrated manual-key to native, naturalized and cultivated trees and shrubs, with limited synonymy, vernacular names, indication of distribution and/or origin, and notes on height, form, habitat (or other situation), special features, etc.; indices to vernacular and botanical names. The introductory section includes an organography and illustrated artificial general keys. [The work includes substantial coverage of introduced and cultivated species.]

648

Germany

Area: 357868 km². Vascular flora: 2674–2705 species, exclusive of an additional 936 apomicts and 109–114 nothospecies (Wisskirchen and Haeupler, 1998; see below). Webb had suggested 3000–3150 species, of which 2600–2750 were native. – Between 1945 and 1990 Germany was divided into two states: the German Democratic Republic (DDR) and the Federal Republic of Germany (BRD). For the Alpine region, see also **603/IV**. Only floras relating to Germany proper are described here.

'Germanorum diligentia in perquirendis plantis spontaneis, prae aliis nationibus, maxime laudanda est.' So wrote Linnaeus in the first edition of his *Bibliotheca botanica* (1736). The density of floristic coverage is perhaps greater than anywhere else on earth, a phenomenon approached only by Scandinavia and Great Britain. 'National' floras are particularly numerous, a likely consequence of the many centers of activity which developed from the late sixteenth century in a long-decentralized nation. From the mid-1800s, several works attempted to cover not only Germany but 'Mitteleuropa' (**640/I**), variously ranging as far as East Prussia, the Carpatho-Ukraine, Transylvania and Slovenia: there being a considerable market, perhaps strongest before 1914 but continuing to the present (and since 1989 with renewed potential).

Early accounts, beginning with Johannes Thal's 1588 *Sylva hercynia*, were largely 'local'; over the next two or more centuries accounts of urban environs, princely states and areas of botanical interest proliferated. The first general checklist, which also introduced Linnaean nomenclature, was an anonymous account of 1764. More substantial were *Vollständinges systematisches Verzeichniss aller Gewächse Teutschlandes* (1782) and *Synopsis plantarum Germaniae* (1792–93), both by Gerhard A. Honckeny, and *Tentamen florae germanicae* (1788–1800; 2nd edn., 1827, not completed) by Albrecht W. Roth, all arranged on the Linnaean system. Notable post-Napoleonic developments include *Enumeratio plantarum Germaniae Helvetiaeque* (1826) by Ernst G. von Steudel and Christian F. Hochstetter, *Flora germanica excursoria* (1830–33) by Ludwig Reichenbach, *Synopsis florae germanicae et helveticae* (1837; 3rd edn., 1857; German edns., 1838, 1846–48,

1892–1903) and *Taschenbuch der Deutschen und Schweizer Flora* (1844; 8th edn., 1878), both by Wilhelm D. J. Koch, and two large-scale, multi-volume, illustrated works: *Icones florae germanicae et helveticae* (1837–1914) begun by Reichenbach, and *Flora von Deutschland* (1840–73; 5th edn., 1880–88) by Diedrich F. L. von Schlechtendal, Christian Langethal and Ernst Schenk as editors (5th edn. by Ernst H. Hallier).

In the first half of the twentieth century the main developments – apart from the *Synopsis* of Ascherson and Graebner and the *Illustrierte Flora* of Hegi, noted below – were the initiation of detailed floristic mapping in the 1930s (the results appearing in 1988 and 1996 respectively for western and eastern Germany) and the publication of *Verzeichnis der Farn- und Blütenpflanzen des Deutschen Reiches* edited by Rudolf Mansfeld (1940 (1941), as *Ber. Deutsch. Bot. Ges.* 58a). This latter work covers a larger area than the modern federation, and remained definitive until the 1970s.

Of current manuals, Garcke's work first appeared in 1849 as *Flora von Nord- und Mittel-Deutschland* and that of Schmeil and Fitschen initially in 1903. The editions of Erich Oberdorfer's *Pflanzensoziologische Exkursionsflora* and Werner Rothmaler's *Exkursionsflora* date from after World War II. These have lately been reinforced by the publication of a new *Standardliste* to succeed Mansfeld's *Verzeichnis*; this, however, is arranged by genera rather than families.

For 'Mitteleuropa' (**640/I**), Reichenbach's *Icones* was followed by the extremely detailed *Synopsis der mitteleuropäischen Flora* (1896–1939, not completed) by Paul Ascherson and Paul Graebner and the still-current, renowned *Illustrierte Flora von Mitteleuropa* by Gustav Hegi, whose first edition began publication in 1906. Other essays in expanded coverage included *Flora von Deutschland und Fennoskandinavien* by F. Hermann (1912, with a revised edition in 1956 as *Flora von Nord- und Mitteleuropa*; see **601/II**) and *Exkursionsflora von Europa* by Franz Thonner (1901, 1918; see **600**).

In addition to the standard floras described below, much current floristic information appears in *Floristische Rundbriefe* (1967–), published in Göttingen. There is also much interest in mapping (and lately development of geographic/floristic information systems), essential to environmental affairs.[114]

Bibliographies

BERGMEIER, E. (comp.), 1994. *Bestimmungshilfen zur Flora Deutschlands: eine kommentierte bibliographische Über-*

sicht. 420 pp. Göttingen. (Floristische Rundbriefe, Beih. 4.) [Alphabetically arranged (by families, genera and species, with limited synonymy) critical taxonomic summaries with references to revisionary and other literature (full bibliographic details in the *Literaturverzeichnis*, pp. 335–414); index to genera and families at end. The references relate not only to Germany but to other parts of Europe as well as elsewhere.]

MANSFELD, R. (ed.), 1940 (1941). *Verzeichnis der Farn- und Blütenpflanzen des Deutschen Reiches.* [See below for full entry. Remains useful for its detailed, classified bibliography (pp. 283–308) which includes a regional section.]

SUKOPP, H., 1960. Übersicht über die in der Zeit von 1945 bis 1959 erscheinen Gefässpflanzenflora Deutschlands. *Willdenowia* 2: 563–583. [An analytical survey of post-World War II floristic work; includes (pp. 565–569) an annotated bibliography of national, regional and local floras.]

GARCKE, A., 1972. *Illustrierte Flora von Deutschland und angrenzende Gebiete.* 23rd edn., revised and ed. by K. von Weihe. xx, 1607 pp., 460 text-figs., 5 pls. Berlin: Parey. (1st edn., 1849, Berlin, entitled *Flora von Nord- und Mittel-Deutschland.*)

Briefly descriptive manual-flora of vascular plants; includes keys to all taxa, limited synonymy, vernacular names, brief summary of local and extralimital range, frequency and occurrence, remarks on variation and taxonomy, symbolic indication of habitat, life-form, phenology, karyotype, special features, etc., and representative figures; illustrated glossary and index to all botanical and vernacular names. The introductory section includes aids to the use of the work, a synopsis, and key floristic references.[115] While no longer effectively current, this work continues in use.]

MANSFELD, R. (ed.), 1940 (1941). *Verzeichnis der Farn- und Blütenpflanzen des Deutschen Reiches.* 323 pp. (Ber. Deutsch. Bot. Ges. 58a). Jena.

Systematic census of the vascular plants of the Greater German Reich of 1940 (including Austria, the Czech lands and parts of Poland and France), with limited synonymy, vernacular names of genera, abbreviated indication of local range, classified bibliography (pp. 283–308) and indices to vernacular and generic names and botanical synonyms. [Now well out of date. Its successors include the *Standardliste* of Wisskirchen and Haeupler (see below) as well as the *Bestimmungshilfen* of Bergmeier (under **Bibliographies** above).]

OBERDORFER, E., 1994. *Pflanzensoziologische Exkursionsflora.* 7. Aufl. (with T. Müller). 1050 pp., 58 text-figs. Stuttgart: Ulmer. (Originally published 1949 as *Pflanzensoziologische Exkursionsflora für Süddeutschland;*

2. Aufl., 1960, and 3. Aufl., 1969 as *Pflanzensoziologische Exkursionsflora für Süddeutschland und die angrenzenden Gebiete*.)

A field-manual to vascular plants with concise keys and separate *non-descriptive* taxonomic entries including vernacular and botanical names, essential synonymy, habitats, substrates, phytosociological class(es), generalized distribution with elevations, chorological and ecological types, life-forms and karyotypes; indices at end. An Englerian arrangement of orders has been followed. [From the 4. Aufl. the work has covered the whole of Germany along with the Vosges and the Swiss Alps. The 7. Aufl. accounts for 3320 taxa and 'bis' numbers. It is considered good for infraspecific taxa and 'Kleinarten'.]

ROTHMALER, W., 1996. *Exkursionsflora von Deutschland*, II. *Gefässpflanzen*. 16. Aufl., ed. R. Schubert. 640 pp., 601 text-figs., map. Jena: Fischer. (1st edn., 1958, Berlin; 15th edn., 1990; reissue of latter, 1994, Jena.) Complemented by *idem*, 1991. *Exkursionsflora von Deutschland*, III. *Atlas der Gefässpflanzen*. 8. Aufl., eds. E. Jäger, K. Werner and R. Schubert. 750 pp., 2800 text-figs. Berlin (reissued 1994, Jena. 1st edn., 1959, Berlin); and *idem*, 1990. *Exkursionsflora von Deutschland*, IV. *Gefässpflanzen: Kritischer Band*. 8. Aufl., eds. R. Schubert and W. Vent. 812 pp., 743 text-figs. Berlin (reissued 1994, Jena. 1st edn., 1963, Berlin). (Entire work preceded by *idem*, 1952. *Exkursionsflora*. Berlin, with six reissues to 1956.)

Teil II, the flora proper, comprises a manual-key to vascular plants, with limited synonymy, vernacular names, general indication of local range, small text-figures depicting diagnostic features, brief notes on habitat, phenology and association-type, and a combined index to generic, family and vernacular names. Its introductory section includes an illustrated glossary and chapters on the biology and distribution of plants. Treatment of 'critical groups' is relatively generalized. Teil III (the atlas) consists entirely of page-sized figures of nearly all recognized species and species aggregates of vascular plants. Teil IV, the 'critical supplement', is actually an expanded manual of the flora, cross-referenced to Teil II, but with more detailed treatment and discussion (including diagnostic figures) of subspecies, aggregates, microspecies, varieties, hybrids, etc., with indication of habit, karyotypes, habitat, phenology, distribution, and chorological and phytosociological affinities, and including medicinal notes and vernacular names along with limited synonymy; list of more

important literature (now somewhat dated), author abbreviations, and complete index at end. An introductory section features sections on organography (with illustrations), plant geography, and sociological divisions (pp. 11–45); general key to genera, etc. (pp. 46–73). [Of Teil II, the 12. Auflage is the latest with significant revisions.][116]

SCHMEIL, O. and J. FITSCHEN, 1993. *Flora von Deutschland und angrenzender Länder*. 89. Aufl., revised by K. Senghas and S. Seybold. x, 802, [ii] pp., 1241 text-figs., 2 maps (end-papers). Heidelberg: Quelle & Meyer. (1st edn., 1903; 88th edn., 1988, as *Flora von Deutschland und seinen angrenzenden Gebieten*.)

Concise illustrated manual-key, with limited synonymy, vernacular names, and partly symbolic indication of distribution, habitat, life-form, special features, etc.; list of protected species, abbreviations of authorities, and index. The introductory section includes an illustrated organography, a discussion of geographical limits, and sections on taxonomy, nomenclature, floristics and current and historical plant geography, conservation and protection of plants, collecting, and use of keys along with (pp. 59–153) general keys to families (and genera) as well as explanations of abbreviations. [The work is considered good for beginners on account of its extensive introductory or general part.][117]

WISSKIRCHEN, R. and H. HAEUPLER, 1998. *Standardliste der Farn- und Blütenpflanzen Deutschlands (mit Chromosomenatlas von Focke Albers)*. 765 pp. Stuttgart: Ulmer (in association with the Bundesamt für Naturschutz). (Die Farn- und Blütenpflanzen Deutschlands (ed. H. Haeupler), I. Based on H. HAEUPLER (ed., for Zentralstelle für die floristische Kartierung der Bundesrepublik Deutschland), 1993. *Standardliste der Farn- und Blütenpflanzen der Bundesrepublik Deutschland* (*vorläufige Fassung*). 478 pp. (Floristische Rundbriefe, Beih. 3). Bochum.)

An alphabetically arranged synonymized checklist of some 4000 taxa of vascular plants; each entry includes accepted botanical and German vernacular names, basionyms, synonyms (the widely used ones in red type), places of publication, floristic status, and one-letter citations of major national and European floras (along with standard monographs in generic headings). Taxonomic and nomenclatural notes appear where appropriate in red type, while problematic or critical groups (such as *Cystopteris*) are highlighted in gray. Synonyms are interspersed with accepted names.

The *Chromosomenatlas* (pp. 553–616) is systematically arranged. Pp. 617–644 comprise an extensive source bibliography and pp. 645–765 an index to all botanical names. The introductory section covers the historical development of checklists as well as abbreviations, conventions, and the plan and methodology of the work (with a separate bibliography). [The work nomenclaturally is a successor to Mansfeld's *Verzeichnis* (see above). Parts were reviewed by specialists.][118]

Distribution maps

HAEUPLER, H. and P. SCHÖNFELDER, 1988. *Atlas der Farn- und Blütenpflanzen der Bundesrepublik Deutschland*. 768 pp., 96 col. photographs, 2490 maps, 8 plastic overlays in pocket. Stuttgart: Ulmer. (2. Aufl., 1989.) Complemented by D. BENKERT, F. FUKAREK and H. KORSCH (eds.), 1996. *Verbreitungsatlas der Farn- und Blütenpflanzen Ostdeutschlands*. 615 pp., maps (mostly col.), plastic overlays in back pocket. Jena: Fischer.

The main part of the first-named work comprises gridded distribution maps of individual species, four to a page. The color photographs depict rare or endangered species. The introductory general part includes an introduction to geobotany (by H. Ellenberg), the work's methodology, organization and sources (pp. 25–36), and commentaries on specific taxa (pp. 38–72). The complementary work for the eastern states, with 1998 individual maps, is similar in format and organization. It is notable for its detailed source bibliography (pp. 557–589).[119]

Woody plants

The current standard manual is *Gehölzflora* by J. FITSCHEN (see **640/I**). Of older works, the most substantial is the three-volume *Die Bäume und Sträucher des Waldes in botanischer und forstwirthschaftlicher Beziehung* (1898–99) by G. Hempel and K. Wilhelm.

649

Switzerland (with Liechtenstein)

See also **603/IV** (all works); **651** (BONNIER and DOUIN). – Area: 41 228 km^2. Vascular flora: 3000–3150 species, of which 2600–2750 native (Webb). *Le nouveau Binz* (1994; see below), not entirely limited to the Federation, accounts for 3040 species and 182 additional subspecies.

With the work of Conrad Gesner in the mid-sixteenth century Switzerland may claim to be the birthplace in the 'Western world' of floristic botany; however, the first substantive national floras did not appear for another two centuries. These are *Enumeratio methodica stirpium Helvetiae indigenarum* (1742; addenda, 1759) and *Historia stirpium indigenarum Helvetiae inchoata* (1768; *Nomenclator*, 1769), all by Albrecht von Haller. Among numerous nineteenth-century works are the seven-volume *Flora helvetica* (1828–33) and the more compact *Synopsis florae helveticae* (1836), both by Jean Gaudin, two works entitled *Die Flora von Schweiz* respectively by Johannes Hegetschweiler (1838–40) and Alexander Moritzi (1844), and, after 1860, field-manuals: *Exkursionsflora für die Schweiz* (1866–67; 9th edn., 1901; French versions, 1886 and 1898; English version, 1889) by August Gremli and *Flore analytique de la Suisse* (1870; 7th edn., 1893) by Paul Morthier. The still-valued *Flora der Schweiz* by Hans Schinz and Robert Keller (later also with Albert Thellung) first appeared in 1900, with its fourth (and last) German edition in 1923. Alpine floras (**603/IV**) made their debut in the mid-nineteenth century.[120]

Current general and excursion floras extend into neighboring countries, sometimes, as in the *Flora der Schweiz* of Hess, Landolt and Hirzel, well beyond Swiss frontiers. A table of accepted nomenclature with full synonymy (accounting for over 10 000 names for some 3250 taxa) as well as vernacular names is presented in D. AESCHIMANN and C. HEITZ, 1996. *Index synonymique de la flore de Suisse et territoires limitrophes/Synonymie-Index der Schweizer Flora und des angrenzenden Gebiete/Indice sinonimico della flora della Svizzera e territori limitrofi*. li, 317 pp. Geneva. For a general introduction to the flora and vegetation, see A. BECHERER, 1972. *Führer durch die Flora der Schweiz mit Berücksichtigung der Grenzgebiete*. 207 pp., illus. Basel: Schwabe.[121]

Bibliography

In addition to Fischer's bibliographies, annual lists of new literature were published in *Berichte der schweizerischen botanischen Gesellschaft* from 1891 through 1930. In 1932 Alfred Becherer instituted for the same journal (later renamed *Botanica Helvetica*) biennial 'Fortschritte' covering new floristic information (with mention in later years of some key floristic works). These were continued by him until the 1974–75 biennium (1977) and then, with an increased bibliographic content, by Hans P. Fuchs and Christian Heitz for a further three biennia. Becherer's *Führer* (1972; see above) also contains (pp. 165–178) an extensive bibliography.[122]

FISCHER, E., 1901. *Flora helvetica, 1530–1900.* xviii, 341 pp. Bern. Continued as *idem*, 1922. *Nachtrage.* ix, 40 pp. (Original work also issued as *Bibliographie der schweizerischen Landeskunde*, IV, 5.) [Classified firstly by centuries, the nineteenth being subdivided into major plant groups and subject areas; author and subject index at end.]

BINZ, A., 1990. *Schul- und Exkursionsflora für die Schweiz mit Berücksichtigung der Grenzgebiete.* 19. Aufl., revised by C. Heitz. 624 pp., 860 text-figs. Basel: Schwabe. (1st edn., 1920; 18th edn., 1986.) French version: D. AESCHIMANN and H. M. BURDET, 1994. *Flore de la Suisse et des territoires limitrophes: le nouveau Binz.* 2nd edn., lxxi, 603 pp., 408 text-figs., 2 maps and geographical cross-section. Neuchâtel: Griffon. (1st edn., 1989. Succeeds A. BINZ and É. THOMMEN, *Flore de la Suisse* (1941, Neuchâtel; 4th edn., 1976).)

Illustrated students' manuals of vascular plants with keys to all taxa, essential synonymy, German (French) vernacular names, numerous small figures of diagnostic features, brief, partly symbolic notes on distribution, altitudinal range, occurrence, habitat, phenology, etc., and cross-references to the figures in the atlas by THOMMEN (see below); lists of toxic plants and indices to German (French) and generic and family names at end. The introductory sections have illustrated glossaries, pedagogies, systematic conspecti and lists of references. [The 1986 and 1990 editions of the German version featured all new figures (by M. Rieder), part of an increased emphasis on illustrations as compared with earlier editions; there is also an index/glossary to morphological terms. The general keys are in two parts: directly to families and via the 24 Linnaean classes. The French version was entirely rewritten for the 1989 edition, making it a new work; for 1994 the figures were grouped in a block (pp. lv–lxxi).]

HESS, H. E., E. LANDOLT and R. HIRZEL, 1976–80. *Flora der Schweiz und angrenzender Gebiete.* 2. Aufl. 3 vols. Illus., maps (some col.), diagrams. Basel: Birkhäuser. (1st edn., 1967–73.) Complemented by H. E. HESS, E. LANDOLT, R. HIRZEL and M. BALTISBERGER, 1998. *Bestimmungsschlüssel zur Flora der Schweiz und angrenzender Gebiete.* 4. Aufl. [5], 659 pp., *c.* 1500 text-figs. Basel. (1st edn., 1976; 3rd edn., 1991.)

The landscape-formatted *Flora* is a bulky but copiously illustrated descriptive treatment of native, naturalized and commonly cultivated vascular plants,

with keys to all taxa, complete synonymy (without references), fairly detailed account of local and extra-Swiss range, taxonomic commentary, notes on habitat, ecology, karyotypes, variability, etc., and indices to all taxa in each volume. The introductory section includes chapters on the Tertiary and Quaternary history of the flora, vegetation formations, phytosociology, floristics, and phytogeography. The *Bestimmungsschlüssel*, based on the larger work, is a 'sideways'-printed illustrated manual-key with notes on occurrence and small marginal analytical figures along with a glossary (pp. 526–533) and indices; no synonymy is included.

LAUBER, K. and G. WAGNER, 1998. *Flora helvetica.* 1613, [3] pp., 2 portrs., col. illus., maps. Bern: Haupt. Accompanied by *idem*, 1998. *Bestimmungsschlüssel zur* Flora Helvetica. 267 pp., illus. Bern. (1st edn. of both works, 1996. *Flora helvetica* also issued as a CD-ROM, 1997.)

The main work is a lavishly illustrated lay-guide to native, naturalized and commonly cultivated vascular plants with usually four species on each page spread, text and distribution maps to the left and photographs to the right; in the text are scientific and vernacular names, concise descriptions with diagnostic features in bold face, and other notes (relating to phenology, habitat, ecological factors (*Zeigerwerte*) following Landolt's system (see Note 121), local distribution, frequency, chorology, karyotypes, properties, essential synonymy, etc.) and the species number in Thommen's *Taschenatlas* (1993; see below); indices to all vernacular (German, French and Italian) and scientific names at end. The general part gives in detail the plan of the work along with physical features, general notes on the flora, and main sources. The *Bestimmungsschlüssel*, handy for field use, comprises the keys and indices and also features an illustrated organography/glossary. [Some 3000 species have been illustrated in this superb handbook.][123]

SCHINZ, H. and R. KELLER, 1923. *Flora der Schweiz*, I: *Exkursionsflora.* 4th edn., revised by H. Schinz with A. Thellung. xxxvi, 792 pp., 172 text-figs. Zürich: Raustein. (1st edn., 1900; 3rd edn., 1909.) Associated work: *idem*, 1914. *Flora der Schweiz*, II: *Kritische Flora.* 3rd edn., revised by H. Schinz with A. Thellung. xviii, 582 pp. Zürich. (1st edn., 1900, together with the main work.) French edn.: *idem*, 1909 (1908). *Flore de la Suisse*, I: *Flore d'excursion.* Transl. E. Wilczek and H. Schinz (from 3rd German edn.). xxii, 690 pp., 128 text-figs. Lausanne: Rouge.

Teil I, the *Exkursionsflora* proper, comprises a general manual-key to native, naturalized and commonly cultivated vascular plant species (and some subspecies), with limited synonymy, vernacular names, symbolic indication of distribution, and notes on taxonomy, habitat, phenology, etc., and indices to all vernacular and botanical names. The introductory section includes a glossary as well as other aids to the use of the manual. Teil II, the *Kritische Flora*, includes detailed treatments of subspecies, varieties, 'microspecies', hybrids, etc., together with extra-Swiss distribution of all taxa listed in both parts; a list of adventive plants is also given.

THOMMEN, É. and A. BECHERER, 1993. *Taschenatlas der schweizer Flora mit Berücksichtigung der ausländischen Nachbarschaft/ Atlas de poche de la flore suisse comprenant les régions étrangères limitrophes*. 7. Aufl./6ème ed., revised by A. Antonietti. xviii, 352 pp., over 3000 text-figs. Basel: Birkhäuser. (1st edns., 1945.)

Small pocket atlas of vascular plants, with limited synonymy, vernacular names, a small figure (or figures) of each species, and indices to genera and families as well as to vernacular names. An appendix gives an account of species found around the borders of Switzerland, especially in the vicinity of Geneva. [The German and French versions, formerly separate, are now published as one.]

Woody plants

Kienli's manual is listed here on account of its inclusion as a 'key reference' for users of the Binz flora (see above).

KIENLI, W., 1948. *Die Gehölze der schweizerischen Flora und des schweizerischen Obstbaues*. xxviii, 404 pp., 199 text-figs. Münsigen: Fischer.

Popularly oriented pocket-manual encompassing native trees, shrubs, shrublets and climbers as well as all cultivated fruit- and some introduced ornamental and useful trees, organized as illustrations with facing descriptive text; index to German and scientific names at end. The introductory part encompasses generalities along with tree growth, light, water relations and other topics as well as (pp. xxiv–xxvii) an illustrated non-dichotomous general key. Other generalities, along with many quotations from local authors, are scattered throughout the main text (all separately indexed, pp. 402–404). Nearly all species are illustrated.[124]

Distribution maps

WELTEN, M. and R. SUTTER, 1982. *Verbreitungsatlas der Farn- und Blütenpflanzen der Schweiz/ Atlas de distribution des ptéridophytes et des phanérogames de la Suisse/ Atlante della distribuzione delle pteridofite e fanerogame della Svizzera*. 2 vols. 716, 698 pp., maps, plastic overlays (in vol. 2). Basel: Birkhäuser.

Atlas of 2572 distribution maps, based on a survey in 1966–79 together with specimen and literature records (but without a specific cutoff date or dates apart from the survey period); index to all names, with map numbers, at end of vol. 2. The 600 mapping units (outlined, with statistics, on pp. 76–86 in vol. 1) are delimited on a 'natural' basis. The trilingual introductory section (vol. 1) gives details of the project, sources of data, the organization of the maps, and special notes on selected maps along with summary maps A–R (reproduced also as plastic overlays) covering mapping schemes (and comparisons of them), cantonal boundaries, physiography, vegetation types, land use, etc.[125]

Liechtenstein

Area: 160 km². Vascular flora: over 1400 native taxa (Seitter, 1977).

SEITTER, H., 1977. *Die Flora des Fürstentums Liechtenstein*. 573 pp., [20] col. pls., illus. Vaduz: Botanisch-Zoologische Gesellschaft Liechtenstein-Sargans-Werdenberg.

Systematic enumeration of vascular plants, with vernacular names, distribution, chorological classes, habitat, and critical commentary; botanists active (pp. 551–553), references (pp. 555–557) and indices to vernacular and generic names. The introduction is fairly detailed, with extensive environmental background and discussion of the flora and its affinities and associations.

Region 65

Western Europe (mainland)

This region consists of the Channel Islands, France, Belgium, Luxembourg and the Netherlands. Like Central Europe, it by and large is a well-collected and well-documented region, densely covered by manuals and other works, although in France many provincial works are comparatively old and until recently approaches to floristics have been less sophisticated (though *Flore du Massif Armoricain*, covering the spine of Brittany, is a good modern work).

France has a number of general floras, none, however, in a single volume save for those by Fournier and Bonnier and de Layens (both recently reissued). The best flora remains Abbé Coste's renowned *Flore descriptive et illustrée de la France* (1901–06), a work directly inspired by a North American counterpart, the *Illustrated flora* by Britton and Brown (**140**). A supplementary fourth volume was published in seven installments from 1972 to 1990. Other floras include those by Gaston Bonnier (the *Flore complète portative* . . . (1909) with Georges de Layens and the 13-volume *Flore complète illustrée* . . . (1911–35) with Robert Douin, the latter featuring 721 colored plates (with eight more in its 1990 reissue)), *Quatre flores de France* (1934–40) by Paul Fournier (reissued in 1977 in two volumes and in 1990 in one volume but not revised), the 14 volume *Flore de France* (1893–1913) by Georges Rouy, Julien Foucaud, Edmond G. Camus and Nicholas Boulay, featuring in the fashion of its time detailed treatments of infraspecific forms, and, most recently, the five-volume *Flore de France* (1973–84) by Marcel Guinochet and Roger de Vilmorin, a descriptive manual with an autochthonous arrangement. A good one-volume manual – in particular a critical revision of Fournier's *Quatre flores* – should be feasible.[126]

The Low Countries and Luxembourg are furnished with a wide range of floras. Both Belgium and the Netherlands have had a number of large-scale documentary floras, of which the current offerings – neither of them completed – respectively are *Flore générale de Belgique* by Lawalrée (1952–66, 1993–) and *Flora neerlandica* of the Royal Netherlands Botanical Society (1952– , not completed). In addition, there are many manual-keys, of which the leaders, both of them illustrated, are *Nouvelle Flore de la Belgique/Flora van België* by de Langhe *et al.* (latest French version, 1992; latest in Dutch, 1998) and *Flora van Nederland* by Heukels and van Ooststroom (latest version, 1996). The Belgian works conventionally include Luxembourg, since 1875 without a separate flora.

Bibliographies. General and divisional bibliographies as for Division 6. Very thorough coverage through 1959 is in particular provided by Blake, 1961.

Indices. General and divisional indices as for Division 6.

651
France (including Monaco)

Area: 549 619 km^2. Vascular flora (exclusive of Corsica): 5000–5150 species, of which 4300–4450 native (Webb). – The region of Alsace-Lorraine has long also been included within botanical 'Mitteleuropa' and so is covered by a number of German-language works listed under **640** and **648**. Northern France is also covered in *Nouvelle flore de la Belgique* by DE LANGHE *et al.* (**656**). The areas bordering on Switzerland (**649**), especially Geneva, are covered in some Swiss works. The Îles Normandes (Channel Islands), though part and parcel also of general works on the British Isles and Ireland (**660**), are at **655**. For the Pyrenees (and Andorra), see **603/II**; for the Alpes-Maritimes, see **603/III**; for the Alps, see **603/IV** (all works, but especially FENAROLI, *Flora delle alpi*). Mediterranean regions are also covered by *Med-Checklist* (see **601**).

Landmarks in earlier French botanical literature include *Flore françoise* (1778) by Jean Baptiste de Lamarck, notable for its effective introduction of analytical keys for identification, its equally notable successor *Flore française* (1805; supplement, 1815) by Lamarck and Augustin-Pyramus de Candolle, probably the first recognizably modern manual-flora, and *Flore de France* (1848–56) by Jean Charles Grenier and Dominique Godron. The last-named, in the eyes of Daydon Jackson (in *Guide to the literature of botany*, 1881, p. 274) was 'an admirable work, indispensable to every student of European plants', but did not cover the Nizzardo or Upper Savoy, lands acquired by France only in 1860. Other early works of interest include *Plantes de la France* (4 vols., 1808–09) by Jean Henri Jaume St. Hilaire, illustrated in the manner of *Flora danica*, *Synopsis plantarum in flora Gallica descriptarum* (1806) by Lamarck and de Candolle, a condensation in one volume of their *Flore française* but with diagnoses and indications of distribution and habit, and the original version, published in 1894, of the *Flore complète portative* of Bonnier and Layens (itself based on their *Nouvelle flore du Nord de la France et de la Belgique* of the 1880s).

The 'Belle Époque' in France as elsewhere was the era of greatest activity in research and writing of national floras. The major works of Gaston Bonnier, Hippolyte Coste and Georges Rouy were all written or at least commenced during this period; that of Fournier followed in the interwar era. By the 1930s, however, it came to be held in some quarters – notably by Paul Jovet – that a new national flora was needed. An initial presentation in 1934 with Marcel Senay met, however, with a negative response; at local as well as at national level, the existing works were thought to be adequate. Antipathy was widespread, and Jovet was led thus to compare France unfavorably with other countries. Only in 1958 did the Centre National de la Recherche Scientifique (CNRS) show more interest; Jovet then responded with a proposal for a biological flora as well as a review of other works. A 'Centre de Floristique' was accordingly organized at the Muséum National d'Histoire Naturelle. It came to be seen that the varying quality and wide scattering of national floristic data would preclude the early appearance of a 'critical' flora, and interest shifted to 'difficult' taxa. Collaborators were found or came forward, and the *Flore de France* came to be planned around them without reference to any established sequence. Only one volume, however, ever appeared (by H. D. Schotsman on Callitrichaceae, 1967) although work on additional families had commenced. In the 1970s, perhaps disenchanted with slow progress, CNRS turned to another project: the *Flore de France* of Marcel Guinochet and Roger de Vilmorin (1973–84; see below). At the same time, Jovet and his associates redirected their activities towards a revision of Coste's *Flore* in the form of supplements; seven of these, encompassing 875 pages, have now been published (1972–90). Current nomenclature and distributional information was summarized in 1993 by Michel Kerguélen, one of Jovet's collaborators.[127]

Bibliographies

Additional sources may be found in the list on pp. 117–120 of Ferrari (see below).

DILLEMANN, G., 1939–48. Bibliographie des flores régionales de la France. *Monde Pl.* **40**(237)–**43**(250/251), *passim*. [Unannotated, geographically arranged lists.]

FERRARI, J.-P., 1994. *Bibliographie des flores françaises*. 121 pp., map (outside back cover). Marseilles: Jardin Botanique 'E. M. Heckel'. (Delectus Seminum quae Botanicus Massiliae Hortus 81: Supplément. Based on JARDINS BOTANIQUES, VILLE DE MARSEILLE, D.G.S.T.-D.E.E.V., 1985. *Bibliographie générale des flores regionales et françaises*. 57 pp. [Marseilles.]) [Entries arranged by authors

but each work keyed to the classified index; bibliographic sources, pp. 117–120.]

BONNIER, G. and R. DOUIN, 1911–35. *Flore complète illustrée en couleurs de France, Suisse et Belgique (comprenant la plupart des plantes d'Europe)*. 13 vols. 721 col. pls. (La végétation de la France, Suisse et Belgique II). Paris, Brussels, Neuchâtel (vols. 1–8); Paris, Brussels (vol. 9); Paris (vols. 10–13): Librarie Générale d'Enseignement (distributed by Orlhac, Paris; Lebègue (later Office de Publicité), Brussels; and Delachaux & Niestlé, Neuchâtel). (Reissued with revisions and eight additional color plates in 5 vols., 1990, Paris: Belin, as *La grande flore en couleurs de Gaston Bonnier: France, Suisse, Belgique et pays voisins*.) Accompanied by G. BONNIER and G. DE LAYENS, [1909]. *Flore complète portative de la France et de la Suisse (comprenant aussi toutes les espèces de Belgique)*. xxvii, 426 pp., 5338 text-figs., 2 maps. Paris: Librarie Générale d'Enseignement. (La végétation de la France, Suisse et Belgique I.) (Many reissues, that of 1986 the first by Belin as *Flore complète portative de la France, de la Suisse et de la Belgique*; again reissued in 1995. Based on *idem*, 1894. *Tableaux synoptiques des plantes vasculaires de la Flore de la France*. xxvii, 412 pp., map. Paris: Dupont.)

The first 12 volumes of the *Flore complète illustrée* comprise a folio atlas of colored plates, with accompanying descriptive text; this latter includes synonymy, vernacular names, local and extralimital range, and notes on habitat, properties, uses, etc. There are also a general commentary on characteristics and relationships of families and genera and indices to all plant names. Volume 13 (Tableau génerale) is devoted to a general index to the entire work as well as addenda and corrigenda. The *Flore complète portative*, in small octavo format, comprises tabular illustrated analytical keys to all vascular plant species, with vernacular names and brief indication of distribution and frequency, and includes an index. [In the Belin reissue of the *Flore complète illustrée* all plates, including the eight new ones (nos. 722–729), were gathered into the first two volumes and the text, including that for the new plates (pp. 1392–1401) and incorporating revised nomenclature, into volumes 3–4. Revised indices respectively to plates and to French, German and botanical names (by R. Palese and D. Aeschimann) comprise the unnumbered fifth volume. The additional plants covered are mainly Corsican.][128]

COSTE, H., *l'abbé*, 1901–06. *Flore descriptive et illustrée de la France, de la Corse, et des contrées limitrophes.* 3 vols. 4354 text-figs., col. map. Paris: Klincksieck. (Reissued 1937, 1950, 1983, Blanchard.) Continued as *idem*, 1972–90. *Flore descriptive et illustrée de la France, de la Corse, et des contrées limitrophes. Suppléments 1–7*, by P. Jovet, R. de Vilmorin, M. Kerguélen and A. Elizabeth. Pp. i–xviii, 1–744, illus. Paris: Blanchard. (Some supplements reprinted.) For revised nomenclature, see M. BALAYER and L. NAPOLI, 1992. *Flore de l'abbé Coste: nomenclature actualisée sur Flora Europaea.* 194 pp. (Ginebre 9). Perpignan.

Illustrated descriptive account of the vascular plants of France; includes keys to all taxa, synonymy, vernacular names, generalized indication of local range, and concise notes on overall distribution, habitat, phenology, etc.; indices to all botanical and vernacular names in vol. 3. The introductory section in volume 1 contains an illustrated glossary and a survey of the general features of the flora together with an essay (by Charles Flahault) on French floristics and vegetation. [Supplements 6 and 7 incorporate revisions to supplements 1–5. The alphabetically arranged index by Balayer and Napoli presents all Coste names with their equivalents in *Flora Europaea*, 'la seule référence valide et reconnue partout'.][129]

FOURNIER, P., 1990. *Les quatre flores de France, Corse comprise.* 3rd revised edn. 1104 pp., 4216 text-figs. Paris: Lechevalier. (Original edn., 1934–40; revised edn., 1961; reissued 1977 (in 2 vols.).)

Manual-key to vascular plants, with brief diagnoses, limited synonymy, vernacular names, and symbolic indication of floristic zone(s), local range, frequency, special features, etc., followed by a general index to all botanical and vernacular names. An introductory section gives a glossary, translations of the commonest specific epithets, and a general key to families. The illustrations are placed at the bottom of the text on each page.[130]

GUINOCHET, M. and R. DE VILMORIN, 1973–84. *Flore de France.* 5 parts. 1879 pp., 296 pls. Paris: Centre National de la Recherche Scientifique (distributed by Doin).

Manual-key to vascular plants of mainland France and Corsica (4565 numbered species); includes very limited synonymy, brief indication of local and extralimital range, figures of critical details, and notes on habitat, life-forms, ecology, phytosociological affinities, phenology, special features, etc.; figures of

critical details, and indices (in each fascicle); glossary, additions and corrections, indices to all names in each volume, and (pp. 1867–1879 in part 5) a cumulative index to families and genera. The introductory section encompasses theoretical considerations, the plan of the work (with signs and abbreviations) and a phytosociological key and synopsis. The work is complete; 172 families are covered following the Emberger system from pteridophytes, gymnosperms, and Santalaceae through Orchidaceae (parts 1–3) followed by Parietales, Rhoedales and Synanthcralcs (in part 4) and Rosales through Myrtales (in part 5).[131]

KERGUÉLEN, M., 1993. *Index synonymique de la flore de France.* xxxviii, 196, [1] pp. Paris: Muséum National d'Histoire Naturelle, Sécretariat de la Faune et de la Flore. (Collection Patrimoines Naturels, Série Patrimoine Scientifique 8.)

Alphabetically arranged list of genera, species and infraspecific taxa including synonymy as represented in five key works (Bonnier, Coste, Fournier, Guinochet and Vilmorin, and *Flora Europaea*) or, if new to the flora, the *départements* concerned; present status also accounted for. For genera, references to a revision or other major work cited in the bibliography (pp. xix–xxvii) are given. The introductory section covers the plan of the work, codes (p. ix), a list of families, new combinations or names (pp. xii–xvi), main floras, conventions and examples of entries.

ROUY, G. *et al.*, 1893–1913. *Flore de France ou description des plantes qui croissent spontanément en France, en Corse et en Alsace-Lorraine.* 14 vols. Asnières, Rochefort: The authors; Paris: Deyrolle. (Also published as supplements to *Annales de la Société des Sciences Naturelles de la Charente-Inférieure* (La Rochelle).) Complemented by *idem*, 1927. *Conspectus de la flore de France.* xvi, 319 pp., portr. Paris: Lechevalier.

The *Flore* is a comprehensive, detailed descriptive account of vascular plants with keys to all taxa, full synonymy (with references and citations), indication of localities with records, detailed summary of local and extralimital range, extensive taxonomic commentary, notes on habitat, and indices to all botanical names in each volume (as well as a general index to family and generic names in vol. 14). The introductory section includes a bibliography. The *Conspectus* is a summary catalogue of vascular taxa as accepted in the main work. [Rouy's *Flore* is now more of historical than contemporary interest, but remains valuable for its documentation. Its species concept is rather broad, but concomitant with this is a very detailed treatment of infraspecific taxa. Collaborators included J. Foucaud (vols. 1–3), E. G. Camus (vols. 6–7), and N. Boulay (vol. 6).]

Distribution maps

DUPONT, P., 1990. *Atlas partial de la Flore de France*. 442 pp., illus. (including 628 distribution maps). Paris: Muséum National d'Histoire Naturelle, Secrétariat de la Faune et de la Flore. (Collection Patrimoines Naturels, Série Patrimoine Génétique 3.)

An atlas of gridded distribution maps (two to a page, each with 1551 cells of 20 km²; legend on p. 97), the commentaries (which precede the maps) including notes on habitats, distribution, frequency and occurrence; index at end. The introductory part includes a map of species density, acknowledgments and references, and pp. 437–440 describe the inventory project (representative of more than 20 years of effort). Covers 645 species and subspecies from the French mainland, Corsica and Andorra.[132]

Woody plants

The work by Rameau *et al.* also includes herbaceous plants of forest areas.

MATHIEU, A., 1897. *Flore forestière*. 4th edn. xxxii, 705 pp. Paris/Nancy. (1st edn., 1858.)

Descriptive treatment of woody plants (including lianas) with keys to all species, indication of habitat, distribution and phenology, and extensive commentary on habit, ecology, soil preferences, rooting, ontology, and properties of the wood and other parts; a winter key, table of densities and an index are also supplied. The introductory part is mostly a glossary; there is also a chapter on wood structure (with a view towards wood identification).

RAMEAU, J. C., D. MANSION and G. DUMÉ, 1989–93. *Flore forestière française: guide écologique illustré*. 2 vols. (1: *Plaines et collines*, 1785 pp.; 2: *Montagnes*, 2421 pp.). Illus., maps. Paris: Institut pour le Développement Forestier.

Atlas-flora with keys; each taxon entry includes a figure, standardized French and other vernacular names (in French, German and English), biological and diagnostic features, distribution (with map showing frequency), chorological classes, 'autecology', phytosociological data (including classed formations), properties and uses. Within the individual volumes woody and herbaceous plants are separated, both being preceded by cryptogamic groups (bryophytes and pteridophytes). References, indices and appendices conclude each volume (with appendix 3 accounting for protected species). [A third volume, to cover southeastern France and Corsica, is projected.]

ROL, R. and M. JACAMON, 1963–69. *Flore des arbres, arbustes, et arbrisseaux*. 4 parts, index. Illus. (covers col.). Paris: Agricole (Maison Rustique).

An illustrated series of atlases of woody plants (including climbers), the botanical and dendrological descriptions accompanied by extensive notes and photographs of leaves, buds, flowering parts, fruits, habit, etc. The primary arrangement of species is geographical, with the individual parts of the work respectively covering (1) 'plains and hills', (2) 'mountains', (3) 'Mediterranean region', and (4) 'introduced species'.[133]

Pteridophytes

The first separate fern flora in France was *Les fougères de France* (1893) by C. de Rey-Pailhade. This was later partly supplanted by a students' textbook, *Pteridophytes (fougères et plantes alliées)* (1954) by M. L. Tardieu-Blot.

BADRÉ, F. and R. DESCHARTRES, 1979. Les ptéridophytes de la France: liste commentée des espèces (taxonomie, cytologie, écologie et répartition générale). *Candollea* 34: 379–457.

Annotated checklist with key synonymy and references, karyotypes and ploidy levels, indication of habitat and distribution in more or less detail, and commentary where necessary; genera are additionally furnished with references to revisions. Known hybrids are also listed. Bibliography, pp. 449–457; index to genera at end of journal volume (but not within the paper).

PRELLI, R., 1990. *Guide des fougères et plantes alliées*. 232 pp., illus. (partly halftones). Paris: Lechevalier. (1st edn., 1985.)

A handy-sized illustrated work (covering 114 native species in the French mainland and Corsica) with key to families, descriptions, and notes on distribution (both within and without France), habitats, vegetation formations, altitudinal range and special features and (for larger genera such as *Dryopteris*) keys to species; list of literature and index at end. The introduction includes generalities on organization and biology, classification, ecology, distribution, history, reproductive biology, karyology, hybridization, and vegetation reproduction along with notes on uses.

655

Channel Islands (Îles Normandes)

See also **660** (all works). – Area: 194 km². Vascular flora: about 1800 taxa, a considerable percentage alien (McClintock, 1975; see below). The five main islands are Jersey, Guernsey, Alderney, Sark and Herm.

No current flora for the Channel Islands as a whole is available, but fair coverage is assured by comparatively recent works on the two largest islands. These succeed two long-used standards, respectively *A flora of the island of Jersey* (1903) by L. V. Lester-Garland and *Flora of Guernsey and the lesser Channel Islands* (1901) by E. D. Marquand.

Partial works

LE SUEUR, F., 1984. *Flora of Jersey*. 243 pp., illus., 18 halftones, 17 col. pls., 600 distribution maps. [St. Helier, Jersey]: Société Jersiaise.

Annotated distribution atlas of native, naturalized, adventive and commonly cultivated vascular plants (over 1500 species), with gridded dot maps, vernacular names (in Jersey-Norman-French and English), and associated commentary; general maps of the Channel Islands (including physical features and land use), bibliography (p. 36), and indices to botanical and vernacular names at end. The introductory section covers physical features, climate, geology, archeology, habitats, wildlife, and man's influence, as well as history of botanical work (pp. xxxiii–xlii) and the problems of data collection and organization.

MCCLINTOCK, D., 1975. *The wild flowers of Guernsey with notes on the frequencies of all species recorded for the Channel Islands*. 288 pp., illus. (line drawings). London: Collins; and *idem*, 1987. *Supplement to* The Wild Flowers of Guernsey. 53 pp. St. Peter Port, Guernsey: Société Guernesiaise.

Systematic enumeration (1340 taxa), with some synonymy, vernacular names (in Guernsey-Norman-French and English), and notes on occurrence (including also Jersey, Sark, Aldernley and Herm), behavior, previous records (where available), etc.; list of botanical references (p. 276), appendix (including general references), and complete index at end. The introductory section covers general features of the flora, with particular emphasis on introduced species and their progress, and includes notes on the history of botanical work. The 1987 supplement adds 70 species and hybrids, not counting microspecies in 'critical' genera.

656

Belgium

See also **648** (SCHMEIL and FITSCHEN, 1988); **651** (BONNIER *et al.*); **658** (HEIMANS and THIJSSE). – Area: 30 519 km². Vascular flora (including Luxembourg): 1900–2100 species, of which 1600–1800 native (Webb); a

figure of 1300 has also been given (PD; see **General references**).

The three-volume 'Belle Époque'-era *Prodrome de la flore belge* (1898–1907) by Émile de Wildeman and Théophile Durand, covering non-vascular and vascular plants alike, long was – and in part remains – the principal documentary flora. A successor, *Flore générale de Belgique*, was begun in 1950 under the editorship of Walter Robyns; though it remains to be completed, publication resumed in 1993 after a 27-year hiatus. Standard manuals prior to 1967 included *Manuel de la flore de Belgique* by François Crépin (1860; 5th edn., 1884, with subsequent 'editions' merely being reissues) and its successor (*nieuwe bewerking*) *Nouveau manuel de la flore de Belgique et des régions limitrophes* by Jules Goffart (1935; 3rd edn., 1945). In the Flemish lands, a long-used work was *Geillustreerde Flora van België* (1892; 5th edn., 1930; last issue, 1959) by J. MacLeod and G. Staes.

In 1967 a new initiative of seven botanists, sponsored by Willem Mullenders, gave rise to *Flore de la Belgique, du Nord de la France et des régions voisines*; heavy demand soon led to the present *Nouvelle flore*, first published in 1973 and in Dutch in 1983. Both have enjoyed successive revisions.[134]

In addition to the two principal floras, note should be taken of a recent introductory students' work covering 1365 species: B. BASTIN, J. R. DE SLOOVER, C. EVRARD and P. MOENS, 1988. *Flore de la Belgique*. 3rd edn. Louvain-le-Neuve.[135]

Bibliographies

HAUMAN, L. and S. BALLE, 1934. *Catalogue des ptéridophytes et des phanérogames de la flore belge*. 126 pp. (Bull. Soc. Roy. Bot. Belgique 66, Suppl.). Gembloux. [Includes a bibliography of relevant literature from 1903 to 1934.]

TOURNAY, R. *et al.*, 1947–86. Bibliographie de l'histoire naturelle en Belgique, B: Botanique. *Natura Mosana* 1–39, Suppl. B, *passim* (also continuously paged from 1 through 1055). [More or less annual bibliographical index to floristic botany and related subjects in Belgium, Luxembourg, and adjacent parts of France, Germany and the Netherlands; entries classified by subject area and briefly annotated. There is no coverage for 1958–60.]

LAMBINON, J. *et al.*, 1992. *Nouvelle flore de la Belgique, du Grand-Duché de Luxembourg, du nord de la France et des régions voisines (ptéridophytes et spermatophytes)*. 4th edn. cxx, 1092 pp., 1490 text-figs., map. Meise, Brabant: Éditions du Patrimoine, Jardin

Botanique National de Belgique. (1st edn., 1973; 2nd edn., 1978; 3rd edn., 1983. Based on J. E. DE LANGHE *et al.*, 1967. *Flore de la Belgique, du nord de la France et de régions voisines.* Liège: Desoer.) Dutch edn.: J. LAMBINON *et al.*, 1998. *Flora van België, het Groothertogdom Luxemburg, Noord-Frankrijk en de aangrenzende gebieden (pteridofyten en spermatofyten).* 3rd edn. cxxiii, 1091 pp., text-figs., map. Meise: Patrimonium, Nationale Plantentuin van België. (1st edn., 1983; 2nd edn., 1988.)

Illustrated manual-key to vascular plants, the keys and species entries separated; the latter include limited synonymy, vernacular names (French, Flemish and German), figures of essential details, indication of local and extralimital range, habit, life-form, phenology and status, critical remarks (to a large extent by specialists), and notes on habitat, special features, etc.; list of new combinations and illustrated glossary; indices to all botanical and vernacular names. The introductory part includes sections on the geographical limits of the work and its plan as well as notes on life-forms, nomenclature, and phytogeographical units, a brief bibliography (pp. xxv–xxvii), and the general key to families (pp. xxxi–lxxi). Also incorporated (pp. lxxii–cviii) is a key to woody plants based on vegetative features. Families are delimited and ordered after Takhtajan and Cronquist. [The work also extends its coverage to the southern Netherlands (i.e., its southern phytogeographic district) and a portion of Germany. Pagination cited here is based on the 1992 French edition.]

LAWALRÉE, A., 1950. Ptéridophytes. In W. Robyns (ed.), *Flore générale de Belgique* III. iv, 194 pp., illus., maps. Brussels: Jardin Botanique de l'État. Complemented by *idem* [and collaborators], 1952–66, 1993– . Spermatophytes. *Ibid.*, IV, 1–4, 5(1), 5(2). Illus., map. Brussels.

Copiously descriptive illustrated flora of pteridophytes (sér. III) and spermatophytes (sér. IV); includes keys to genera, species and infraspecific taxa, and extensive documentation with full synonymy, references and citations, detailed indication of local range with sources, summary of extralimital distribution, critical commentary, and extensive notes on karyotypes, habitat, ecology, biology, phenology, etc.; distribution maps and illustrations of representative species; index to all botanical names at end of each volume. [Not completed; among the seed plants only gymnosperms and (following the Englerian sequence) dicotyledons

from Salicaceae to Thymelaeaceae and (in 5(2)) through Apiaceae have been treated. The other series (I and II) were designated for non-vascular plants.][136]

Distribution maps

VAN ROMPAEY, E., L. DELVOSALLE and collaborators, 1978–79. *Atlas de la flore belge et luxembourgeoise: ptéridophytes et spermatophytes/ Atlas van de Belgische en Luxemburgse flora: pteridofyten en spermatofyten.* 2nd edn., 2 vols. (text, 116 pp., 19 maps; atlas, 1543 maps in 287 pp.). Brussels: Jardin Botanique National de Belgique. (1st edn., 1972.)

Comprises distribution maps of most Belgian vascular plant species, save those of extreme ubiquity, with indication of era of collection and of presence or absence; usually only one species depicted per map. Six maps appear on a page and the grid used is composed of 4 km squares. For the second edition a volume of explanatory text, *Commentaries* (1978) by Van Rompaey and Delvosalle, was added.

657

Luxembourg

Area: 2586 km². Vascular flora: *c.* 1200 native and naturalized species (Reichling in PD; see **Gereral references**). – The Grand Duchy is at present effectively covered by floras of adjacent states, especially Belgium. No separate works have been published since 1900; prior to then there had appeared *Flore luxembourgoise* (1836) by F. A. Tinant, *Prodrome de la flore du Grand-Duché de Luxembourg* (1873–97) by J. P. J. Koltz, and the descriptive *Flore du Grand-Duché de Luxembourg* (1875) by J. H. G. Krombach. Since 1955, much information has been documented by Léopold Reichling in *Bulletin de la Société des Naturalistes Luxembourgeois.*[137]

658

The Netherlands

See also **647** (SCHMEIL and FITSCHEN), and, for the southern part of the country, **656** (LAMBINON *et al.*). – Area: 41 160 km². Vascular flora: 1750–1900 species, of which 1400–1600 native (Webb); 1483 taxa treated in the *Atlas* and 1448 in the *Standaardlijst* of 1990 (see below).

The first floras of the Netherlands, both manuals in the Linnaean style, were *Flora belgica* (1767) and *Flora VII provinciarum Belgii foederati indigena* (1781) by David de Gorter. These were followed by the illustrated encyclopedic *Flora batava* (1800–1934) by Jan Kops and several associates and successors, ultimately encompassing 28 volumes. More practical alternatives as successors to Gorter's floras were *Flora Belgii septentrionalis* (1825–40, in both Latin and Dutch; supplements, 1827–32) by Hermann C. van Hall and, later, *De flora van Nederland* (1859–62; 2nd edn., 1872–74) by Cornelius A. J. A. Oudemans. The growth of interest in floristic documentation was signaled by the formation of what is now the Koniklijke Nederlandse Botanische Vereeniging (KNBV) and its sponsorship of *Prodromus florae batavae* (1st edn., 1850–53, by R. B. van den Bosch; 2nd edn., 1893 (vol. 2, bryophytes) and 1901–16 (vol. 1, angiosperms, by Laurens Vuyck; gymnosperms and pteridophytes were not published)). New manuals for educational use – notably those by Hendrik Heukels and Eli Heimans – began to appear in the later decades of the nineteenth century; and fairly soon after 1900 Heukels published his three-volume, well-illustrated *De flora van Nederland* (1909–11; see below). This latter remained for several decades a definitive work (and in its genre will not now entirely be superseded).

After World War II, the KNBV and the Netherlands flora section of the then-Rijksherbarium at Leiden, under the direction of Theodorus Weevers, began work on a successor, *Flora neerlandica*. Interest in it, however, waned after the mid-1960s although some parts appeared as late as the mid-1970s. It has now been discontinued. Efforts were then directed towards documentation of the Dutch flora in atlas form, resulting in a fine three-volume *Atlas of the Netherlands flora* (1980–89; see below) edited by Jacob Mennema (vols. 1–2) and Ruud van der Meijden with others (vol. 3). An 'ecological flora', one of the few of its kind worldwide, was subsequently produced: E. J. WEEDA, 1985–94. *Nederlandse oecologische flora*. 5 vols. Illus. (mostly col.). Amsterdam: IVN. This features numerous habit- and habitat-photographs of the plants along with extensive text, but has no keys. The momentum generated by these topographic works led to the formation in 1988 of FLORON (Floristisch Onderzoek Nederland), a 'foundation' (based at what in 1999 became the National Herbarium of the Netherlands, Leiden Branch) charged with coordination of floristic

observation and recording in the Netherlands. This body is presently responsible for FLORBASE, a national database with records from 1975 onwards.

The two main current identification manuals, *Geïllustreerde Schoolflora voor Nederland* (from 1956 *Flora van Nederland*) and *Geïllustreerde Flora van Nederland* (now *Geïllustreerde Flora van Nederland, België en Luxemburg*), were initiated respectively by Heukels in 1900 and Heimans with Johannes P. Thijsse in 1899. Heukels' work, however, grew out of his earlier, unillustrated *Schoolflora van Nederland*, first published in 1883 (and continued through 18 everlarger editions until 1932, when it was succeeded by the more concise but similarly unillustrated *Beknopte schoolflora voor Nederland* (noted below)). Relatively frequent revisions have characterized both works; their latest versions both date from the 1990s. A CD-ROM version of Heukels' flora, with additional features including many more illustrations, has also been released.

The 'shorter' field flora, successor to Heukels' *Schoolflora van Nederland* and current over the middle third of the twentieth century, is H. HEUKELS, *Beknopte school- en excursieflora voor Nederland*. Groningen. (1st edn. with W. H. Wachter, 1932, as *Beknopte schoolflora voor Nederland*; 12th and last edn., 1968, by S. J. van Oost   strom.)[138]

Bibliography

BOERLAGE, J. G. and P. P. C. HOEK, 1888 (1975). *Bibliografie van de flora en fauna van Nederland*. ix, 150 pp. Amsterdam: Bakhuys & Meesters. [A classified, concisely annotated bibliography. In this reissue the original text (reproduced) is complemented by new author and geographical indices commissioned by the publishers.]

HEIMANS, E. and J. P. THIJSSE, 1994. *Geïllustreerde flora van Nederland, België en Luxemburg*. 23rd edn., revised by J. Mennema *et al*. viii, 1080 pp., 6000+ text-figs., map. Antwerp: Gulden Engel. (1st edn., 1899, Amsterdam; 21st edn., 1965; 22nd edn., 1983.)

Copiously illustrated manual-key in 'landscape' format to spontaneous and commonly cultivated vascular plants, with limited synonymy, vernacular names, notes on habitat and on local and extralimital range, and symbolic indication of life-form and phenology; notes on scientific names, derivations, habitats, a key to phytosociological units, and indices to all names at end.

The introductory section includes instructions on the use of the work and the general family keys (also fully illustrated). [Coverage was first extended into Flanders in 1935 and from 1994 to all Belgium and Luxembourg as well as adjacent parts of France and Germany to a distance of 50 km. Some material from former editions, including coverage of non-vascular plants, has been omitted but its characteristic 'landscape' format remains.]

HEUKELS, H., 1909–11. *De flora van Nederland*. 3 vols. 2047 text-figs., map. Leiden: Brill.

Illustrated descriptive flora with keys, synonymy, vernacular names, detailed indication of local range, indication of habitat, and commentary; complete indices. [Outdated but not yet superseded by a comparable work, as *Flora neerlandica* (q.v.) has lapsed. It was inspired by Coste's *Flore descriptive et illustrée de la France* (**651**).]

HEUKELS, H., 1996. *Flora van Nederland*. 22nd edn., revised by R. van der Meijden, E. J. Weeda, W. J. Holverda and P. H. Hovenkamp. 676 pp., illus. Groningen: Wolters-Noordhoff. (1st edn., 1900, entitled *Geïllustreerde Schoolflora voor Nederland*; 11th edn., 1934; 13th edn., 1949; with present title from the 14th edn., 1956; present format from the 20th edn., 1983; 21st edn., 1990.) See also *idem*, 1998. *Interactieve Flora van Nederland*. 1 CD-ROM. Beek: Natuur en Techniek.

Illustrated manual-key to vascular plants in a double-column format; includes limited synonymy, vernacular names, brief notes on habitat, distribution and frequency along with symbolic indication of life-form, phenology and status, and references to key taxonomic works; glossary, guide to abbreviations, tables of name changes, and indices to all botanical and vernacular names at end. Each taxon also carries a four-digit number keyed to a standard-list of the vascular flora (VAN DER MEIJDEN *et al.*, 1990; see below). For grasses a separate key to vegetative features is furnished. The general section includes chapters on systematics (by D. J. Kornet, R. Geesink and E. J. Weeda) and the history of botanical work in the Netherlands, plant geography including floristic districts, geobotany and phytosociological 'taxa' (by E. J. Weeda), tables of orders and of ecological units, and the general key to families and genera. The *Interactieve Flora van Nederland* – an electronic version of the 22nd edition – features maps and color illustrations not in the print work as well as standard and multi-entry keys.

KONINKLIJKE NEDERLANDSE BOTANISCHE VERENIGING (T. WEEVERS *et al.*), 1948– . *Flora neerlandica/Flora van Nederland*. Vols. 1– . Illus. Amsterdam.

Documentary descriptive flora in the form of a series of revisions; each treatment includes keys to genera and species, extensive synonymy, references and citations, detailed indication of local range, taxonomic commentary, and notes on ecology, biology, etc., and figures of diagnostic or critical attributes. An index to all taxa concludes each fascicle. [By 1988, 12 fascicles had been published, comprising vol. 1(1–6) and vol. 4(1–2, 6, 9, 10a, 10b). No systematic sequence has been followed in publication but the arrangement of families follows the Englerian system. Presently available are the pteridophytes, gymnosperms, monocotyledons, and a few families (and genera) of the dicotyledons (among them the 'critical' genus *Taraxacum*).][139]

MEIJDEN, R. VAN DER, L. VAN DUUREN, E. J. WEEDA and C. L. PLATE, 1990. Standaardlijst van de Nederlandse flora 1990. *Gorteria* **17**(5): 75–126. (Previous editions in 1971, 1976 and 1983; additions and corrections from time to time in subsequent issues of *Gorteria*.)

Alphabetically arranged tabular checklist with taxon code numbers, accepted names, and codes for frequency in 1940 and 1990 (related to the number of squares in which present in the respective years), old and new ecological groupings, and Red Data Book categories. An explanation of the list, and criteria for inclusion, appear on pp. 76–90, and name changes on pp. 121–126 with standard references on p. 127. Some 1448 taxa are accounted for, with a 'starting-point' of 1825 (corresponding to van Hall's *Flora Belgii septentrionalis*).

Distribution maps

Maps are also given in the CD-ROM *Interactieve Flora van Nederland* (see above). These are based on the records in FLORBASE.

MENNEMA, J. *et al.* (eds.), 1980–89. *Atlas van de Nederlandse flora/Atlas of the Netherlands flora*. 3 vols. (vols. 1–2, ed. J. Mennema; vol. 3, ed. R. van der Meijden *et al.*). Maps, plastic overlays. Amsterdam: Kosmos (vol. 1); Utrecht: Bohn, Schltema & Holkema (vol. 2); Leiden: Rijksherbarium (vol. 3). English edn. of vol. 1 (subtitled *Extinct and very rare species*): 266 pp., 333 maps. The Hague: Junk (distributed by Kluwer).

A distributional atlas of the native vascular flora (accounting for 1458 taxa); comprises gridded maps with indication in green of presence before and in black for presence after 1950 along with associated descriptive text (in Dutch, with, in volume 1 up to p. 41, a summary in English) covering the pattern of records, first reports, current status, distribution elsewhere, and probable reasons (if appropriate) for decline. Cross-references to *Flora Europaea* are also given. Volume 1 in addition includes an extensive introductory section (featuring among other topics a history of geographical and mapping investigations and a review of earlier floristic literature), a list of more general literature, and at its end a comprehensive source bibliography; volume 2 a chapter (by E. J. Weeda) on changes in occurrence of vascular plants in the flora. In volume 3 there is less species commentary but additional general commentary. All volumes are indexed. [The decline of much of the Dutch vascular flora in recent decades is starkly illustrated in this fine work, a model of its kind.][140]

Region 66

Western European archipelago (British Isles and Ireland)

Area: 313707 km² (Great Britain, 229856 km²; England, 130346 km²). Vascular flora: 3354 species and microspecies, 2297 native and 1057 alien (Kent, 1992; see below). In Great Britain 2350–2600 species, of which 1700–1850 native (Webb). – Comprises the Western European archipelago encompassing Great Britain, Ireland, the Isle of Man, the Hebrides, the Orkney and Shetland groups, and all associated smaller islands including isolated Rockall in the North Atlantic. The Channel Islands, although politically part of the United Kingdom, are geographically closer to France (see **655**).

Floristic work in the Western European Islands has a history of over 450 years, beginning in England with William Turner in 1538–48 and somewhat later in Scotland and Ireland; by now the vascular plants are by world standards supremely well documented, rivaled perhaps only by Central Europe (Region 64), the Low Countries (Region 65 in part), and Scandinavia and Finland (Region 67 in part). There is a vast number of general and local floras and related works, and the recording of new information is highly organized through such bodies as the Biological Records Centre, the Botanical Society of the British Isles and other national and regional groups. Only a small selection from the literature can be recorded here.

Already in 1736 Linnaeus, in his *Bibliotheca botanica*, said of John Ray's *Synopsis methodica stirpium britannicarum* (1690; 2nd edn., 1696; 3rd edn., 1724): 'Flora haec, ultima, non habet parum'. He himself introduced his classification system and new nomenclature to the British flora in a 1759 dissertation, *Flora anglia*. The Linnaean nomenclatural reform, added knowledge and a desire for more information soon made new works necessary, and several descriptive floras, large and small, followed: *Flora britannica* (1759 and 1760) by John Hill (a reissue of Ray's work but arranged according to the Linnaean system), *Flora anglica* (1762; 2nd edn., 1778, 1798) by William Hudson (the first truly Linnaean flora of Britain), *Flora britannica* (1800–04), *Compendium florae britannicae* (1800; 4th edn., 1828), *The English flora* (1824–36) and *Compendium of the English flora* (1829; 2nd edn., 1836), all by J. E. Smith, *British flora* (1812) by R. J. Thornton, and *British flora* (1830) by W. J. Hooker. Also very popular was *Botanical arrangement* by William Withering, which passed through 14 editions from 1776 to 1877. These served until – and even after – the advent of 'natural' systems, the philosophy of which in Britain was adopted rather later than in France. The most ambitious work of the period, an illustrated encyclopedia in the tradition of *Flora batava* or *Flora danica*, was *English Botany* by James Sowerby and J. E. Smith (1790–1814 and supplements to 1866; 2nd edn., 1832–46).

Change was, however, inevitable as the advantages of 'natural' systems became apparent. The first flora so arranged was *Natural arrangement of British plants* (1821) by S. F. Gray with J. E. Gray. Others followed in the next two decades, including *Synopsis of the British flora* (1829; 3rd edn., 1841) by John Lindley and the fifth edition of *British flora* (1842) by Hooker with G. A. Walker Arnott. Interest in checklists remained, however, and in 1829 *A catalogue of British plants* by J. S. Henslow appeared in Cambridge. Successors have been produced down to the present, including in the recent past *List of British Vascular Plants* (1958) by J. E. Dandy.

The mid-nineteenth century saw publication of a new generation of descriptive manuals, three of which, with revisions, continued to be used well into the twentieth century: *Manual of British botany* (1843) by C. C. Babington (with its 10th and last edition, by A. J.

Wilmott, in 1922), *Handbook of the British flora* (1858) by George Bentham (revised by J. D. Hooker in 1887 and finally in 1924 by A. B. Rendle, with reissues to 1954); and *The student's flora of the British Islands* by J. D. Hooker (1870; 3rd edn., 1884, with reissues to 1937). The works of Bentham and Hooker were long (and continue to be) admired for their method and conciseness, while that of Babington is useful for infraspecific taxa. There was also a new, third edition of *English Botany*, by J. T. Boswell Syme (1863–86, with supplement in 1892); it, too, was arranged following a 'natural' system.

By the mid-twentieth century, accumulating knowledge and advances in the philosophy and methodology of systematics made an entirely new flora necessary. The result was *Flora of the British Isles* by A. R. Clapham, T. G. Tutin and E. F. Warburg (1952; revised 1962, 1987); this long remained the leading modern work. 'CTW' was later joined by the authors' *Excursion flora of the British Isles* (1959; revised 1968, 1981). In 1991 came a new entry, *New flora of the British Isles* by Clive Stace; with a second edition in 1997 this has now become the 'standard' manual. It, too, now has an 'excursion' stablemate.

With respect to collections of illustrations, many exist beginning with, as already mentioned, *English Botany*. The second edition (1865, in two volumes) of Bentham's *Handbook* initially appeared as *Illustrated Handbook of the British Flora*, with no fewer than 1295 figures by the noted Victorian botanical artist, Walter Fitch – a first in its genre.[141] Of more contemporary sets, the best known is Stella Ross-Craig's *Drawings of British Plants* (1948–74); another series, by Sybil Roles (1959–64), accompanied *Flora of the British Isles*, and a third, by R. W. Butcher, was 'descended' from the Fitch illustrations done originally for Bentham's *Illustrated Handbook* but later released separately as the first compact one-volume atlas. This latter genre has itself become a standard; one mid-twentieth century example which became famous (though in a format impractical for excursions) was *The concise British flora in colour* (1965) by William Keble Martin (see below).

All these works have furnished a wider range of choice in comparison with the situation in 1950. A further bonus has been the initiation of a new formal 'critical flora', *Flora of Great Britain and Ireland* by Peter Sell and Gina Murrell. The first of five projected volumes made its appearance in 1996. That for so long such a work was lacking when compared with other

parts of Europe may be seen as a product of interests and perceptions: not only was much of the botanical profession during the age of elaborate pure phytography preoccupied with overseas floras and related research, but possibly also because *English Botany*, in addition to its illustrations, had come to be regarded as a definitive descriptive 'regional' flora. Joseph Hooker had in mind a detailed critical work, but the only enterprise contemporary to such undertakings as Rouy and Foucaud's *Flore de France* or, for that matter, Hegi's *Illustrierte Flora* was the *Cambridge British Flora* by C. E. Moss (1914–20). Of this projected 10-volume illustrated account, only two volumes were realized.[142]

Topographical botany, by contrast, has enjoyed a strong development. This was in the nineteenth and early twentieth centuries most notably expressed in the work of Hewett Cottrell Watson and George C. Druce in Britain and David Moore (with A. G. More) and R. L. Praeger in Ireland. After World War II an interest in distribution and mapping again developed, coupled with a growing concern about changes to the flora (and fauna) of the Western European Islands. One significant outcome was *Atlas of the British flora* (1962; supplement, 1968), the first to be produced using electromechanically controlled tabulating and plotting technology as well as to adopt a gridded presentation of distribution. It is now regarded as one of the classics in its genre, and has inspired similar efforts elsewhere.

The county and vice-county floras in the Western European Islands are far too numerous to account for here. I include, however, separate floras for Wales (*Welsh ferns*, 6th edn., 1978, and *Flowering plants of Wales*, 1983), Ireland (*An Irish flora*, 1943; 6th edn., 1977; 7th edn., 1996), and the more discrete islands and archipelagoes (Man, the Outer Hebrides, Orkney and Shetland). The Channel Islands appear at 655.[143]

Bibliographies. General and divisional bibliographies as for Division 6. Very thorough coverage through 1959 is in particular provided by Blake, 1961.

Regional bibliographies

Sims, R. W., P. Freeman and D. L. Hawksworth, 1988. *Key works to the fauna and flora of the British Isles and north-western Europe*. 5th edn. 328 pp. Oxford: Oxford University Press (for the Systematics Association). (4th edn., 1978; first published 1953 as *Bibliography of key works for the identification of the British fauna and flora*.) [Coverage from Ireland to the Arctic Circle, the Franco/Swiss border, the former intra-German frontier, Denmark and Sweden; content considerably increased with respect to the 4th edn. Plants, pp. 271–309.][144]

SIMPSON, N. D., 1960. *A bibliographical index of the British flora*. xix, 429 pp. Bournemouth: The author. [This most exhaustive of botanical bibliographies has 35 000 entries (Besterman in *Biological sciences: a bibliography of bibliographies*, 1971, Totowa, N.J.).]

Indices. General and divisional indices as for Division 6.

Regional indices

BOTANICAL SOCIETY OF THE BRITISH ISLES, 1971– . *BSBI Abstracts*, 1– , comp. D. H. Kent. London. [Classified taxonomic, biosystematic and floristic abstracts relating to British and Irish vascular plants. Issued more or less annually; part 28 published 1998.]

660
Region in general

The titles described below represent only a selection from an extensive literature. Widely used also have been *The concise British flora in colour* by W. Keble Martin (1965, with several reprintings; last revised 1982 by D. H. Kent as *The new concise British flora*) and various other guides such as *Collins new generation guide to the wildflowers of Britain and Northern Europe* (1996) by Alistair Fitter.[145]

BUTCHER, R. W., 1961. *A new illustrated British flora*. 2 vols. viii, 1016; viii, 1080 pp. London: Leonard Hill [Books]. (Based upon W. H. FITCH, W. G. SMITH *et al.*, *Illustrations of the British flora* (1880, London, Reeve; revisions and reissues to 1946); and R. W. BUTCHER and F. E. STRUDWICK, *Further illustrations of the British flora* (1930, Ashford, Kent: Reeve; reissued 1944, 1946).)

Atlas-flora of vascular plants, with each species occupying one page and comprising a full-page figure with modest descriptive text; the latter includes scientific and vernacular names, diagnostic features, some notes on the habit, habitat, occurrence and phenology of the plant concerned, and origin of the material used for the drawing (most are from life). Synonymy appears only where the name used differs from that in *Flora of the British Isles* (see below). Short generic descriptions are also included. The introductory portion includes a section on 'descriptive botany' akin to that incorporated by Bentham in his colonial floras as well as the plan of the work (pp. 29–30) and an extensive artificial key, while at the end of vol. 2 is a complete general index. The arrangement of the work follows Dandy's *List* (see below under KENT).[146]

CLAPHAM, A. R., T. G. TUTIN and D. M. MOORE, 1987. *Flora of the British Isles*. 3rd edn. xxvii, [i], 688 pp., 82 text-figs. Cambridge: Cambridge University Press. (1st edn., 1952, by A. R. Clapham, T. G. Tutin and E. F. Warburg; 2nd edn., 1962.) Accompanied by S. J. ROLES, 1957–64. *Flora of the British Isles: Illustrations*. 4 vols. 1910 figs. Cambridge.

The *Flora* is a descriptive manual of native, naturalized and commonly cultivated species and subspecies of vascular plants; included are keys to all taxa, limited synonymy, vernacular names, frequency, generalized indication of local and extralimital range, figures of critical characteristics, taxonomic commentary, notes on habitat, floristic classes, soil preferences and special features, and symbolic indication of phenology, karyotypes, life-form(s) and number of vice-counties in which present. A glossary and index to all botanical and vernacular names follow. The introductory section includes a synopsis of families. [References to specialist treatments are, however, few.][147]

KENT, D. H., 1992. *List of vascular plants of the British Isles*. xvi, 384 pp. London: Botanical Society of the British Isles. Continued as *idem*, 1996 (1997). *Supplement to* List of vascular plants of the British Isles. 36 pp. London: The author (for BSBI).

Introduction (with indication that the arrangement of the flowering plants follows the Cronquist system), conventions, census, excluded species, acknowledgments and (pp. xiv–xv) bibliography; nomenclatural checklist of native and alien species and infraspecific taxa with authorities, essential synonymy and indication of status; index to all generic and family names (and to those of species in *Carex*, *Hieracium*, *Rubus* and *Taraxacum*). [Succeeds J. E. DANDY, *List of British vascular plants*, xvi, 176 pp. London: British Museum (Natural History).][148]

ROSS-CRAIG, S., 1948–74. *Drawings of British plants*. 31 parts, index. Pls. (index, 39 pp.). London: Bell.

This work comprises octavo-sized, systematically arranged (on a modified Bentham and Hooker system) plates of all native and naturalized species of British vascular plants (except Cyperaceae, Poaceae and pteridophytes, available in semi-popular form elsewhere). [The *Drawings* are considered to be among the best such set essayed in the twentieth century. The artist's figures of Cyperaceae and Poaceae are available elsewhere in semi-popular form.][149]

SELL, P. and G. MURRELL, 1996– . *Flora of Great Britain and Ireland.* Vol. 5. Illus., maps. Cambridge: Cambridge University Press.

Detailed descriptive flora of native and introduced species with keys to all taxa, selective synonymy, vernacular names, phenology, chromosome numbers, and commentary including status, distribution, origin (if introduced), habitats, frequency, and illustrations of key characters. Taxonomic remarks are usually confined to generic notes which also include references to key literature. The introductory part of vol. 5 covers the historical background, the scope and contents of the work, geographical conventions (with map), principal sources, a conspectus of major groups and families, and a key to monocotyledons; an appendix of new taxa and combinations, a glossary, and a complete index conclude the volume. [Five volumes in all are planned; that now published encompasses all the monocotyledons.][150]

STACE, C. A., 1997. *New flora of the British Isles.* 2nd edn., xxvii, [ii], 1130 pp., illus. (incl. halftones), end-paper maps. Cambridge: Cambridge University Press. (1st edn., 1991.)

Tersely descriptive manual-flora of vascular plants (accounting for over 4500 taxa including native, naturalized, adventive/recurrent casual and crop plants) with keys, essential synonymy (covering a few select works), English vernacular names, concise statement of distribution in the United Kingdom and Ireland and elsewhere, origin (if applicable), status, habitat, probable date of introduction (if alien), indication of rarity (in margins), limited taxonomic commentary and figures of selected taxa; illustrated glossary and complete index at end. The general part covers the rationale of the work, its scope, arrangement of families (the flowering plants following Cronquist), criteria for synonymy, and other parameters and also includes a bibliography (pp. xxiii–xxiv) and synopsis of families; unfortunately, there is no analysis of the flora.[151]

STACE, C. A., 1999. *Field flora of the British Isles.*

xvi, 736 pp., illus., maps (end-papers). Cambridge: Cambridge University Press.

Manual-key to native and established alien vascular plants, the leads with diagnostic features, status, habitat, generalized distribution, and vernacular and botanical names; brief family and generic descriptions; illustrated glossary and index to vernacular as well as generic and family names at end. Use of the book as well as signs and abbreviations are covered on pp. xv–xvi, and the introduction is purely practical. [Based wholly upon the 1997 edition of the author's *New flora*, including the arrangement of families. Illustrations, taken from the larger work, are limited to critical characters necessary for identification. Apomictic microspecies are accounted for except in *Rubus*, *Taraxacum* and *Hieracium*.]

Distribution maps

An 'Atlas 2000' project has been launched as a successor to the 1962–68 *Atlas*. The original work was last reissued, with a supplementary section, in 1990.

PERRING, F. H. and S. M. WALTERS (eds.), 1962. *Atlas of the British flora.* xxiv, 432 pp., numerous dot maps. London: Nelson (for the Botanical Society of the British Isles). (Reissued 1976, 1982, East Ardsley, Yorkshire: EP Publishing; 1990 (with added bibliography and index to maps of 1962–89 by C. D. Preston), Oundle via Peterborough: Botanical Society of the British Isles.) Complemented by F. H. PERRING (ed., assisted by P. D. SELL), 1968. *Critical supplement to the* Atlas of the British flora. vii, 159 pp., numerous dot maps. London. (Reissued 1978, East Ardsley.)

These two works together comprise a distribution atlas of native and naturalized vascular plant species of the Western European Islands, with the 'critical groups' mapped in the supplement. Mapping is by presence or absence in a square, with distinction according to the era of collection (this sometimes, however, not made) and occasional other qualifications. Four maps appear on a page. The base map is founded on the British National Grid (with an extension to Ireland), subdivided into 10×10 km squares. The introductory section provides an explanation of the work, while at the end is an index of names. The *Critical supplement* treats in more detail such genera as *Sorbus*, *Alchemilla*, *Taraxacum* and *Hieracium* and is separately indexed.[152]

Pteridophytes

JERMY, A. C. and J. CAMUS, 1991. *The illustrated field guide to ferns and allied plants of the British Isles.* x, 208 pp., illus. London: Natural History Museum.

Descriptive atlas-format field-guide treatment of native and significant alien species; includes keys to all taxa (largely at the beginning but in some cases within the text), English vernacular names, distinguishing features of each species (including contrasts with similar-looking plants), distribution, notes on conservation and legal status, and illustrations (with habit in silhouette); common aliens (pp. 186–191) and indices to botanical and vernacular names at end. The introductory part is morphological and includes a concise glossary and brief bibliography.

PAGE, C. N., 1997. *The ferns of Britain and Ireland.* 2nd edn., 540 pp., illus., maps. Cambridge: Cambridge University Press. (1st edn., 1981.)

A biological 'atlas-flora' (covering 100 species and nothospecies) with photographic illustrations, distribution maps, and associated text; the last-named gives botanical and vernacular names and covers preliminary recognition, identification, possible forms of confusion, technical confirmation, and extensive 'field notes' (sometimes quite extensive) on variation, habitats, occurrence, karyotypes, etc.; references and index at end. Phenological charts are also included showing inactive, active and spore-bearing periods. The general part encompasses fern organography, a glossary, a synoptic list of taxa, chart- and multi-access keys to main groups, Watsonian and grid-based botanical subdivisions, altitudinal distribution (with graphs), and (pp. 28–46) a consideration of environmental factors (with several maps, the last indicating degrees of sulfur dioxide (SO_2) pollution).[153]

662

Wales

Area: 20761 km^2. Vascular flora: 1606 species (Hyde and Wade). – Ellis's *Flowering plants of Wales* succeeds *Welsh flowering plants* by H. A. Hyde and A. E. Wade (1934; 2nd edn., 1957).

ELLIS, R. G., 1983. *Flowering plants of Wales.* ix, 338 pp., illus., maps (incl. folding map and plastic grid map). Caerddydd (Cardiff): Amgueddfa Cenedlaethol Cymru (National Museum of Wales). Complemented by G. HUTCHINSON and B. A. THOMAS, 1996. *Welsh ferns.* 7th edn. 265 pp., text-figs., maps. Caerddydd: Amgueddfeydd ac Orielau Cenedlaethol Cymru/ National Museums and Galleries of Wales. (1st edn., 1940, by H. A. Hyde and A. E. Wade; 6th edn., 1978, by S. G. Harrison.)

Flowering plants of Wales is a concise annotated enumeration, with text and map sections; the text includes vernacular names (Welsh and English), indication of vice-county (Watsonian) and 10 km^2 square

distribution (Watsonian system) and source(s), status, origin (if introduced), habitat and phenology, while the 1028 grid-maps are arranged 12 to a page; list of sources used (with numbers as employed in main text), summary of chorological elements of the flora (16 in all), and index to all names and geographical elements at end. The introductory section encompasses the history of botanical collecting (with portraits), special features of the flora (including given vice-counties), statistics (p. 9), physical features and geology, climate and bioclimatic zones, and floristics; references (p. 32); abbreviations and conventions (p. 40). *Welsh ferns* comprises an illustrated semi-popular guide with detailed descriptions, botanical and Welsh and English vernacular names, common synonymy, distribution maps, dates of first records, and indication of Welsh and wider distribution and habitat, and chromosome numbers; bibliography, glossary and index to all names at end. Its general part covers life history, chromosomes, hybridization, identification, the collections of the National Museums of Wales, geography, and a synopsis of classification and general key to all taxa.

663

Scotland

Area: 78749 km^2. Vascular flora: 1451 species (Blake, 1961 (see **General bibliographies**:), p. 18).

The most recent separate Scottish flora remains W. J. HOOKER, 1821. *Flora scotica, or a description of Scottish plants arranged both according to the artificial and natural methods.* 2 vols. London: Hurst-Robinson. There are, however, several county and regional floras (those for the Outer Hebrides, Orkney and Shetland being covered respectively under **666, 667** and **668**).

664

Isle of Man

Area: 588 km^2. Vascular flora: 721 species (Paton, 1933); no figure in Allen. – Allen's flora succeeds *A list of flowering plants, ferns and horsetails of the Isle of Man* by C. I. Paton (1933, Liverpool, as *North West. Nat.* 8,

Suppl.; reprinted 1934 in *Proc. Isle of Man Nat. Hist. Antiq. Soc.*, N.S. **3**: 547–619).

ALLEN, D. E., 1984 (1986). *Flora of the Isle of Man*. xiv, 250 pp., 5 figs. (maps), frontisp. (map). Douglas, Isle of Man: Manx Museum.

Systematic enumeration of vascular plants, with vernacular names, distribution (with localities for less common to rare species and including vouchers where appropriate), and notes on frequency, occurrence and dynamics (many plants being introductions of one or another kind); possible additions, references to works on non-vascular plants (p. 224), main bibliography (pp. 225–228), acknowledgments, and index to all names at end. The introductory section covers physical features, geology, climate and rainfall, land use, human influences, and history and composition of the flora along with habitats, associations and botanical (floristic) districts and a history of botanical endeavor; plan of work (pp. 45–49).

665

Ireland

Included here are both the Irish Republic and the United Kingdom province of Northern Ireland. – Area: 83 851 km^2 (Irish Republic, 68 895 km^2). Vascular flora (whole of Ireland): 1350–1450 species, of which 1000–1150 native (Webb). Scannell and Synott list *c.* 1000 species.

The first Irish flora was *Synopsis stirpium hibernicarum* (1726) by Caleb Threlkeld; this focused on the vicinity of Dublin. The whole island was first encompassed in *Flora hibernica* (1836) by J. T. Mackay; this long remained the only complete flora. For field use, however, *The Irish flora* by Katherine Baily (1833; reissued 1845) was preferred. In 1866 appeared *A contribution towards a* Cybele Hibernica by D. Moore and A. G. More; this was succeeded by *Irish topographical botany* by R. L. Praeger (1901, as *Proc. Roy. Irish Acad.*, III, 7). It and several supplements were later distilled into that author's key work, *A botanist in Ireland* (1934); this was noted by S. F. Blake (1961, p. 281) as 'an admirable work, without exact equivalent in the floristic literature of other countries'.[154] It has been in turn supplemented by *Sources for the* Census Catalogue of the Flora of Ireland (1989) by M. J. P. Scannell and D. M.

Synnott.[155] In 1931 came a new *Students' illustrated Irish flora* by John Adams, and in 1943 the first edition of the now long-standard *An Irish flora* by D. A. Webb. The *Census catalogue* (1972; 2nd edn., 1987) was preceded by a hand-list, first published in 1910 with three further editions to 1943.

SCANNELL, M. J. P. and D. M. SYNOTT, 1987. *Clár de phlandaí na hÉireann / Census catalogue of the flora of Ireland*. 2nd edn., xxvii, 171 pp., 3 text-figs., 2 maps (one folding). Dublin: Stationery Office. (1st edn., 1972.)

Systematic list of native and established alien vascular plants, with relevant synonymy, corresponding English and Irish-Gaelic vernacular names as known, and indication of internal range (by vice-counties); indices to generic and family and to vernacular names. The introductory part gives brief accounts of Irish geography, authorities for key plant groups, vernacular names (with authorities), floristic districts, officially protected plants, noxious plants, acknowledgments, and an explanation of the text (with symbols); bibliography (pp. xxiii–xxv).

WEBB, D. A., J. PARNELL and D. DOOGUE, 1996. *An Irish flora*. 7th edn. xxxiv, 337 pp., text-figs. Dundalk: W. Tempest (Dundalgan Press). (1st edn., 1943; 6th edn., 1977.)

Concise manual-flora of native, naturalized and commonly cultivated vascular plants with keys to all taxa, limited synonymy, Irish (Gaelic) and English vernacular names, floral phenology, generalized indication of local range, and notes on habitat, special features, etc.; figures of critical features; illustrated glossary, lexicon of Latin epithets, and indices to Irish and to family, generic and English vernacular names at end. The introductory section includes notes on the use of the book and its plan, plant sites of interest, a bibliography of key works, and a general key to families.[156]

Trees

NELSON, E. C. and W. F. WELSH, 1993. *Trees of Ireland*. 247 pp., illus. (part col.). Dublin: Lilliput Press.

A 'coffee-table' but scholarly book accounting for 33 native species, comprising a color atlas with descriptive text (including botany, ecology, history, wood structure and properties, uses, cultivation and propagation, folklore and outstanding examples), English and Irish vernacular names, and indication of distribution; full index at end. The general part includes histories of forests as well as of botanical and dendrological studies in Ireland.[157]

666

Outer Hebrides

Area: 2898 km² (inclusive of St. Kilda). Vascular flora: over 700 species. – The work of Pankhurst and Mullin succeeds the controversial *Preliminary flora of the Outer Hebrides* by J. W. Heslop-Harrison (1941, in *Proc. Univ. Durham Philos. Soc.* **10**: 228–273, 7 pls. (incl. maps)).[158]

PANKHURST, R. J. and J. M. MULLIN, 1991. *Flora of the Outer Hebrides*. [vi], 171 pp. 12 figs. incl. frontisp. (map). London: Natural History Museum Publications. (Reissued 1994, London: HMSO.)

Systematic enumeration of vascular plants (arrangement following *Flora of the British Isles*); includes keys (only to *Taraxacum* and *Hieracium*), essential synonymy, vernacular names, status, habitat and distributional and critical notes (the numbers referring to zones as shown in the map); appendix (herbaria, collectors, definition of the 29 map zones, and gazetteer), list of special sites, classified bibliographic lexicon, general bibliography and index at end. The introductory part covers physical features, climate, vegetation, floristics and vegetational history, botanical exploration, and 'plant lore' (including notes on Scottish Gaelic vernacular names) along with the plan of the work. [Parts contributed by collaborators.]

667

Orkney Islands

Area: 976 km². Vascular flora: no data. Bullard's checklists partially succeed M. SPENCE, 1914. *Flora orcadensis*. xcv, 148 pp., 2 portrs., 2 maps. Kirkwall, Orkney.

BULLARD, E. R., 1985. Flowering plants and ferns. In R. J. Berry, *The natural history of Orkney*, pp. 250–261. London: Collins. (New naturalist library.) Based on *idem*, 1972. *Orkney – a checklist of vascular plants and ferns*. Stromness, Orkney: Rendall.

Annotated systematic checklist of native, naturalized, adventive and casual vascular plants, with limited synonymy, vernacular names, frequency and (if uncommon or rare) localities. [General material, including the history of natural history studies in the islands, is covered in other chapters.]

668

Shetland Islands

Area: 1433 km² (inclusive of Fair Isle and Foula). Vascular flora: 827 named taxa, of which 383 native including microspecies (Scott and Palmer, 1987). – The 1987 flora of Scott and Palmer was preceded by W. SCOTT and R. C. PALMER, 1980. Flowering plants and ferns. In R. J. Berry and J. L. Johnston, *The natural history of Shetland*, pp. 282–303. London: Collins. (New naturalist library.) Both are successors to G. C. DRUCE, 1921 (1922). Flora zetlandica. *Bot. Soc. Exch. Club British Isles Rep.* **6**: 457–546.

SCOTT, W. and R. C. PALMER, 1987. *The flowering plants and ferns of the Shetland Islands*. 468 pp., 16 col. pls., 26 maps, col. frontisp. Lerwick, Shetland: The Shetland Times.

Systematic enumeration, including limited synonymy, vernacular names, first reports, distribution (with additional detail for less common or rare species, sometimes with several *exsiccatae*), status, frequency, overall distribution (including Orkney and the Faeroes), and critical notes; casuals in smaller type; bibliography and complete index at end. The extensive general part encompasses a general description of the islands and their geology, climate (the average temperature being low and cloudiness high), and habitats (with characteristic plants) as well as vegetation dynamics and changes, including their relationship to land use, sites of special interest, the wider relationships of the flora (especially with Orkney and the Faeroes), the history of botanical work, and explanatory notes.[159]

669

Rockall Island

Only algae and lichens have been recorded for this desolate North Atlantic rock, a British possession about 250 km northwest of Ireland.

Region 67

Northwestern Europe (Scandinavia, Finland and the Baltic States)

Area: 1431871 km^2 (Scandinavia with Finland, 1257871 km^2; Baltic States, 174000 km^2). – This region consists of Iceland, the Faeroes, Denmark, Norway, Sweden and Finland (**670/I**) and the Baltic States of Estonia, Latvia and Lithuania (**670/II**). However, floras and other works dealing exclusively with Lapland or the Arctic islands to the north are not described here but rather in **Regions 05** and **06**. Some works given here also cover East Fennoscandia, which corresponds to the Russian territories of Karelia and the Kola Peninsula.

Like other parts of central and northern Europe, the region has enjoyed a long history of floristic study, in the eighteenth and nineteenth centuries dominated by the work of Carl Linnaeus and his students as well as by *Flora danica* and in the twentieth strongly influenced by the documentary efforts of Nils Hylander at Uppsala and Eric Hultén at Stockholm. With a wide range of works at all levels, it is today one of the best-documented areas in the world, a status favored also by the comparatively small vascular flora.

The leading current descriptive regional flora in Scandinavia is Hylander's *Nordisk kärlväxtflora* (1953–66), alas, not completed. Associated with this critical work is the same author's regional checklist, *Förteckning över Nordens växter*, I: *Kärlväxter* (1955; supplement, 1959), actually a revision of a work first published under other authorship in 1907. Both works drew support from his *Nomenklatorische und systematische Studien über Nordische Gefässpflanzen* (1945). Hylander's flora was moreover a successor to *Florae Scandinaviae prodromus* (1779; 2nd edn., 1795) by A. J. Retzius in Lund, *Handbok i Skandinaviens flora* by Carl J. Hartman (1820; 12th and last edn., 1889) and *Skandinaviens flora* by Otto Rudolph Holmberg (1922–31).[160] A somewhat more popularly oriented pan-Scandinavian work is *Nordens flora* by Carl A. M. Lindman (originally published as *Bilder ur Nordens flora* in 1904–05 and most recently revised in 1974–75).

The classic *Flora danica* (1761–1883), too, is a work of Nordic scope, a view particularly held by its last editor (Johann Lange). A new *Flora nordica* (to appear in English), directed by Bengt Jonsell, is in preparation under the sponsorship of the Bergius Foundation in Stockholm. The first of its five planned volumes appeared in early 2000, covering vascular plants through Polygonaceae ordered as in *Flora Europaea*.

Floristic work in the Baltic States was for political and social reasons more fragmented. Before 1918, most of the area now in the two northern states – the former Baltic-German lands – had been treated as a unit several times, lastly by Johannes Klinge at the university center of Dorpat (Tartu). These works, some also extending to Ingria and thus encompassing St. Petersburg (**685**), are mentioned under Estonia (**677**). With the formation of Estonia and Latvia, emphasis passed to national-language floras and manuals. For present-day Lithuania, no separate flora existed at independence; indeed, its western part (Samogitia) in particular had remained botanically little studied until the twentieth century. Subsequent materials were largely in Lithuanian although Polish authors were responsible for a national enumeration in 1933 and a florula of the Vilnius region (Polish from 1919 to 1939) in 1957–59. In the 1980s, however, a joint project towards an overall enumeration (in Russian and English) was organized by the botanical working-groups in the three national academies; the first volume appeared in 1994 and the second in 1996.

The ranges of Northwestern European species are well mapped in *Atlas över växternas utbredning i Norden* (2nd edn., 1971) by Eric Hultén, a work now expanded in scope as *Atlas of North European vascular plants north of the Tropic of Cancer* (1986) by Hultén and Magnus Fries (see **001**).

At state and territory level, more or less recent editions of manual-keys are available for all polities. A special feature of some Scandinavian works is the presentation of family keys inclusive of the 24 Linnaean classes. Several countries moreover have large-scale documentary and illustrated floras, of which that by Hjalmar Hjelt for Finland is perhaps the most detailed and *Flora danica* (1761–1883), which also covers Norway and Iceland, culturally the best known. Local floras and manuals are numerous; the only one accounted for here, however, is *Skånes flora* by A. Hennig Weimarck (1963; reissued 1977) which covers Denmark as well as the floristically distinctive – and formerly Danish – southernmost province of Sweden.[161]

Bibliographies. General and divisional bibliographies as for Division 6. Very thorough coverage through 1959 is in particular provided by Blake, 1961 (not, however, for the Baltic States). Selected coverage of the Baltic States may be had in M. E. Kirpicznikov, 1969. Краткий обзор важнейших флор и определителей, изданных в СССР за 50 лет, II/Kratkij obzor važnejšikh flor i opredelitelej, izdannykh v SSSR za 50 let, II. *Bot. Žurn. SSSR* 54: 121–136.

Indices. General and divisional indices as for Division 6.

670

Region in general

I. Scandinavia and Finland

In addition to the more technical works here presented, reference may be made to C. A. M. LINDMAN, 1974–75. *Nordens flora.* 2nd edn., revised by M. Fries. 10 vols. 663 pls. Stockholm: Wahlström and Widstrand (1st edn. 1964. Originally published 1904–05 as *Bilder ur Nordens flora;* 3rd edn. 1921–26); and B. MOSSBERG, L. STENBERG and S. ERICSSON, 1992. *Den nordiska floran.* 696 pp., col. illus., maps. [Stockholm]: Wahlström and Widstrand. From 1992 local editions of *The illustrated flora of Britain and northern Europe* by C. Grey-Wilson and M. Blamey have also been available.[162]

HYLANDER, N., 1955. *Förteckning över Nordens växter,* I: *Kärlväxter/List of the plants of NW Europe,* I: *Vascular plants.* 4th edn. x, 175 pp. Lund: Gleerup. (1st edn., 1907, Lund: Lunds Botaniska Förening, as *Förteckning över Skandinaviens växter.*)

Systematic list of the vascular plants of Northwestern Europe (including also East Fennoscandia, Iceland and the Faeroes), with symbolic indication of distribution and/or origin; table of protected plants and indices to genera and to synonymous names at end. The explanation of signs and abbreviations appears in both Swedish and English. [For additions, corrections

and deletions, see *idem,* 1959. Tillägg och rattelser . . . *Bot. Not.* 112: 90–100.]

HYLANDER, N., 1953–66. *Nordisk kärlväxtflora.* Vols. 1–2. Illus., maps. Stockholm: Almqvist & Wiksell.

Comprehensive critical descriptive flora of the vascular plants of Northwestern Europe, with geographical limits as for the author's *Förteckning över Nordens växter;* incorporates keys to genera, species and infraspecific taxa, concise synonymy (with dates), vernacular names, detailed exposition of distribution within the region along with summaries of extralimital range, taxonomic commentary, figures of diagnostic and critical features, and notes on karyotypes, habitat, phenology, life-form, etc.; short lists of references but no indices (except to genera in inside covers). [Not completed; covers families from the Pteridophyta up through the Polygonaceae (Englerian system). This important work was one of the five 'Basic Floras' constituting a major source for *Flora Europaea.*][163]

Distribution maps

HULTÉN, E., 1971. *Atlas över växternas utbredning i Norden/Atlas of the distribution of vascular plants in Northwestern Europe.* 2nd edn., 56, 531 pp., 1984 distribution maps on 496 pls. Stockholm: Generalstabens Litografiska Anstalt (distributed by Almqvist & Wiksell). (1st edn., 1950.)

Distribution atlas of vascular plants based upon Scandinavia and Finland, with individual colored maps for each species (four to a page); includes limited synonymy, Swedish vernacular names, and phenological data (for some species); indices to vernacular and botanical names. The general part preceding the plates includes sections on general questions of plant distribution, historical phytogeography, physical features of the region, geology, climate, etc. (with special extra maps). Apart from Scandinavia and Finland, the maps also include East Fennoscandia (a part of the Russian Federation) and the Baltic republics. [Now supplanted by E. HULTÉN and M. FRIES (eds.), *Atlas of North European vascular plants north of the Tropic of Cancer* (1986; see **001**, **Eurasia**).]

Pteridophytes

ØLLGAARD, B. (illustrated by K. TIND), 1993. *Scandinavian ferns.* 317 pp., 103 text-figs. (incl. 4 maps), 114 col. pls. Copenhagen: Rhodos.

Illustrated descriptive atlas in a large format (the text and plates in separate blocks) with in the text figures of key features, keys to species where needed, scientific and vernacular names (the latter in Danish and English) with etymology, occasional synonymy, and cursively written notes on variation and relationships, distribution (with references to maps in other publications), ecology, uses and hybrids. The atlas proper, a glossary, a substantial literature list (pp. 301–311) and full index conclude the work. The introductory part is extensive, with sections on background, climate, geography, the plan of the work, pteridophytes in general including polyploidy, chromosomes, apogamy and taxonomy, and a key to families and genera. [Covers Denmark, Norway and Sweden. The lack of distribution maps is something of a loss.]

II. Baltic States

LAASIMER, L., V. KUUSK, L. TABAKA, A. LEKAVIČIUS and R. JANKEVIČIENĖ, (eds.), 1993– . *Flora of the Baltic countries: compendium of vascular plants*/Флора Балтийских Республик: Сводка сосудистых растений. Vols. 1– . Maps. Tartu: Estonian Academy of Sciences, Institute of Zoology and Botany.

A bilingual enumeration of vascular plants, with relevant synonymy including references and regional literature citations, vernacular names (Estonian, Latvian, Lithuanian and Russian), indication of distribution including mention of particular geobotanical districts where necessary (with explanation in figure 1 and on pp. 32–90 in vol. 1), indication of status, frequency, habit, height, phenology, site types (pp. 118–125 in vol. 1 for explanation) and soil preferences, biology and notes on special features including variability, hybrids, cultivation, changes in range, and generic comments; index to scientific names at end. The general chapters (in both English and Russian) cover the plan and principles of the work, physical features including climate and soils, geobotanical units (31 recognized), nature conservation and protected species, site types with indicator species, synecological taxa, abbreviations, and relevant references. No list of background floristic literature is here presented, nor references for specific genera and families. The first volume covers pteridophytes, gymnosperms and dicotyledons following the revised Englerian system through Droseraceae; the second (1996), Crassulaceae through Dipsacaceae.[164]

671

The Faeroes

See also **672** (OSTENFIELD and GRÖNTVED). – Area: 1399 km². Vascular flora: 427 species, including naturalized and synanthropic introductions (Bloch, 1980); of these *c.* 300 are truly native.

In addition to the works described below, the following illustrated guide (covering 164 species) may also be useful: D. BLOCH, 1980. *Farøflora.* 156 pp., 136 col. figs. Tórshavn: Føroya Fróðskaparfelag (simultaneously published in English (153 pp.) as *Faroese flowers*). The works of Rasmussen and Hansen were preceded by *Færøernes Flora* (1870–71, in *Botanisk Tidsskrift*) by Emil Rostrup, based on an initial botanical survey by the Botanical Society of Copenhagen in the summer of 1867, and the three-part wide-ranging documentary *Botany of the Færöes* (1901–08) coordinated by Eugen Warming (with 'Flora of the Færöes' by C. H. Ostenfield in part 1, 1901).[165]

HANSEN, K., 1966. *Vascular plants in the Faeroes: horizontal and vertical distribution.* 141 pp., maps (Dansk Bot. Ark. 23(3)). Copenhagen.

Atlas-account of vascular plant records, with mapped treatments of 363 species (based on field work in 1960–61 and earlier records) including indication of vertical distribution (three life-zones); records of adventives and index. An introductory section describes physical features, geological and vegetation history, climate, and biotic influences, along with botanical exploration.

RASMUSSEN, R., 1952. *Føroya flora.* 2nd edn., xxviii, 232 pp., 108 pls. Tórshavn: Skúlabókagrunnur Løgtingsins. (Reprinted 1970. 1st edn., 1936.)

Illustrated school- and excursion-manual (in Faeroese) of vascular plants with keys to all taxa, essential synonymy, vernacular names, concise indication of local range, and notes on habitat, life-form, etc.; illustrated glossary/ organography and index to vernacular and botanical names at end. The introductory section is pedagogic. The last two plates depict critical features of *Hieracium* microspecies, reused from *Botany of the Færöes.*

672

Iceland

Area: 102 819 km^2 (of which *c.* 83 percent lacks vegetation). Vascular flora: 500–600 species, of which 450–550 native (Webb); PD (see **General references**) gives a figure of 470.

Floristic studies began in 1752 as part of a five-year general survey, with J. G. Koenig taking part. This was followed by visits from Banks, W. J. Hooker and others over the following century and a half but the island remained relatively little known until after 1900. Nevertheless, local publications began to appear from 1830 with *Íslenzk Grasafræði*, an excursion-flora by O. Hjaltalín. *The botany of Iceland*, begun in 1912, encompasses the results of several botanical-topographical surveys; its contribution on vascular plants (Gröntved, 1942) remains the major 'documentary' flora.

In addition to it and the field manual-keys listed below, a useful recent semi-popular photo-atlas for visitors, covering 438 species, is H. Kristinsson, 1987. *A guide to the flowering plants and ferns of Iceland.* 311, [i] pp., text-fig., col. illus., maps. Reykjavik: Örn og Örlygur Publ. House. Kristinsson notes that some disagreement remains with respect to the taxonomy of the Icelandic flora, firstly over the status of aliens and, secondly, over species limits. Attention should also be drawn to another flora [not seen]: Á. Bjarnason, 1983. *Íslenzk flóra.* Reykjavik: Iðunn.

Gröntved, J., 1942. The Pteridophyta and Spermatophyta of Iceland. In J. Gröntved, O. Paulsen, and T. Sørensen (eds.), *The botany of Iceland*, 4(1). 427 pp., 177 text-figs. (incl. maps). (Copenhagen), London: Munksgaard.

Comprehensive, critical enumeration of vascular plants (including full accounts of *Taraxacum* and *Hieracium*); extensive synonymy, with references and citations; vernacular names; detailed accounts of local distribution (with numerous maps), with distribution maps, and general summaries of extralimital range; taxonomic commentary and notes on ecology, life-form, phenology, etc.; bibliography and indices to vernacular and botanical names. The introductory section includes accounts of physical features, climate, soils, vegetation, phytogeography, and botanical exploration and research.[166]

Löve, Á., 1977. *Íslenzk ferðaflóra.* 2nd edn., 429 pp., 576 text-figs., 25 pls. (some col.), map. Reykjavik: Almenna Bókafélagið. English edn.: *idem*, 1983. *Flora of Iceland.* 403 pp., 516 text-figs., 16 col. pls. Reykjavik. (1st Icelandic edn., 1970. A successor to *idem*, 1945. *Íslenzkar jurtir.* 291 pp., illus. Copenhagen: Munksgaard.)

Illustrated manual-flora of vascular plants; includes keys to all taxa, very limited synonymy, vernacular names, symbolic indication of local range, and notes on karyotypes, ecology, life-form, phenology, etc.; indices to vernacular and botanical names. The introductory section includes a glossary, notes on the use of the work, a general key and a list of protected plants. The English edition includes, and separately indexes, vernacular names in that language.

Ostenfield, C. H. and J. Gröntved, 1934. *The flora of Iceland and the Færoes.* cciv, 195 pp., 2 maps. Copenhagen: Levin & Munksgaard.

Manual-key (in English) to vascular plants; includes limited synonymy, vernacular names, generalized indication of local range, and notes on habitat, life-form, phenology, etc.; glossary and indices to family and generic names as well as to Icelandic and Faeroese vernacular names. The introductory section includes a short bibliography. [Originally written for visitors.]

Stefánsson, S., 1948. *Flóra Íslands.* 3rd edn., revised by S. Steindórsson. lviii, 407 pp., 253 text-figs., map. Aðalumboð: Norðri (for Íslenzka Náttúrufræðifélag). (1st edn., 1901, Copenhagen; 2nd edn., 1924.)

Illustrated descriptive manual of vascular plants, with relatively limited treatment of 'microspecies'; includes limited synonymy, vernacular names, generalized indication of local range, and notes on habitat, phenology, life-form, etc.; illustrated glossary and map of districts; indices to all botanical and vernacular names at end.

673

Denmark

See also **675**, **Partial works** (Weimarck). The part of southern Denmark under German rule from 1866 to 1918 is also covered by all works listed under **640**. – Area: 42 936 km^2 (exclusive of the Faeroes). Vascular flora: 1750–1900 species, of which 1400–1600 native (Webb); later work suggests that only 1000 species may truly be native (PD; see **General references**). There are numerous additional 'microspecies' of *Rubus*, *Hieracium* and *Taraxacum*.

A primary basis for subsequent works on the Danish (and Norwegian) flora, as well as on other parts of Scandinavia and some nearby German states, is the royally sponsored *Flora danica* (1762–1883), begun by George Oeder and said to be the first comprehensive national flora project.[167] Manuals and handbooks began to appear in the nineteenth century; the key work was *Haandbog i den Danske flora* (1851; 4th edn., 1886–88 with supplement in 1897) by Johan Lange. Lange also published a nomenclator for *Flora danica* (1887).[168] The first editions of Emil Rostrup's and Christen Raunkiær's field-keys appeared respectively in 1860 and 1890; these and others have since covered that niche. From 1904 detailed mapping of the flora was undertaken under sponsorship of the Botanical Society of Denmark and initially led by Morten Mortensen and Carl Ostenfeld; the results were published in a long series of papers from 1931 to 1980 and in 1989 summarized (in English) by P. Vestergaard and K. Hansen. A new series of illustrations appeared in 1956–60 under the direction of Olaf Hagerup. – Mention may also be made of *Danmarks vilde planter* (1958–59) by M. S. Christiansen and H. Anthon, an illustrated work (produced in 20 installments) in the tradition of *Flora danica* but with more copious text (by K. Faegri, E. Hultén and T. W. Böcher).

Bibliography

CHRISTENSEN, C., 1924–26. *Den danske botaniske historie med tilhørende bibliografi.* 2 vols. 884, 680 pp. Copenhagen. [Incorporates all botanical literature to 1880; supersedes E. Warming, 1880–81. Den danske botaniske litteratur fra de aeldste tider til 1880. *Bot. Tidsskr.* 12: 42–131, 158–247.] For titles after 1880, see *idem*, 1913. *Den danske botaniske litteratur 1880–1911.* 279 pp. Copenhagen; *idem*, 1940. *Den danske botaniske litteratur 1912–1939.* 350 pp. Copenhagen; and A. HANSEN, 1963. *Den danske botaniske litteratur 1940–1959.* 318 pp. (Dansk Bot. Ark. 21). Copenhagen.

HAGERUP, O. and V. PETERSSON, 1956–60. *Botanisk atlas: Danmarks dækfrøede planter, I–II.* 2 vols. Pls. Copenhagen: Munksgaard. English edn.: *idem*, 1959. *A botanical atlas*, I: *Angiosperms.* Transl. H. Gilbert-Carter. xvi, 550 pp., [515] pls. Copenhagen.

Atlas of illustrations of the flowering plants of Denmark, often with details (in some cases, only details are furnished) but without text except for some family or specific descriptions and botanical and Danish (or English) vernacular names. Essentially all native and some adventive species are included, but microspecies (for example in *Hieracium* and *Taraxacum*) are not treated in detail. Volume 1 covers angiosperms; vol. 2, gymnosperms, ferns, fern-allies, and genera of mosses.[169]

HANSEN, K. (ed.), 1981. *Dansk feltflora.* 2nd edn., 757 pp., illus. Copenhagen: Gyldendal.

Manual-key to vascular plants, with limited synonymy, vernacular names, diagnostic figures, and abbreviated or symbolic indication of distribution, habitat, life-form class, habit, phenology, etc.; winter keys (pp. 724–740) and indices to vernacular and generic names. The introductory part covers the plan of the work, the system followed (flowering plants after Cronquist), organography and life-forms (with illustrations), principles of taxonomy, plant geography and floristics, abbreviations and symbols, and (pp. 45–69) a general key to families.

RAUNKIÆR, C., 1950. *Dansk ekskursions-flora.* 7th edn., revised by K. Wiinstedt. xxxi, 380 pp. Copenhagen: Gyldendal (Nordisk Forlag). (1st edn., 1890.)

Manual-key to vascular plants, with limited synonymy, concise indication of local range, vernacular names, and symbolic notes on ecology, life-form, phenology, etc.; glossary and index to vernacular, generic and family names at end. The work also includes an explanation of the author's well-known life-form system.

ROSTRUP, E. and C. A. JØRGENSEN, 1973. *Den danske flora.* 20th edn., revised by A. Hansen. 664 pp., 141 text-figs., map (in pocket). Copenhagen: Gyldendal. (1st edn., 1860, as *Vijledning i den danske flora.*)

Illustrated manual-flora of vascular plants, with keys to all taxa, limited synonymy, vernacular names, concise indication of local range, and notes on habitat, phenology, etc.; index to vernacular, generic and family names. The introductory section includes an illustrated glossary/organography and an explanation of 'botanical districts'. [In this work vernacular and scientific names together are given in all key leads and taxon headings.]

Distribution maps

VESTERGAARD, P. and K. HANSEN (eds.), 1989. *Distribution of vascular plants in Denmark.* 163 pp., 30 text-figs., 563 maps (Opera Bot. 96). Copenhagen.

A summary (in English), with distribution maps of selected species, of the results of the topographical-botanical survey of Denmark (originally instituted in 1904 and published in the form of 1342 maps with associated commentary in 44 installments between 1931 and 1980, mainly in *Botanisk Tidsskrift*). The maps in this anthology are designed to demonstrate a range of biogeographic phenomena, and follow eight analytical/synthetic chapters on different ecosystems. At the end of the work (pp. 145–157) is a systematic checklist of the native and naturalized vascular flora, arranged following *Flora Europaea* and featuring cross-references to both earlier and present maps.

674

Norway

For Lapland, see also **061**. For Jan Mayen and Svalbard (Spitsbergen and Bear Island), see also **051** to **053**. – Area: 323895 km². Vascular flora: 1700–1850 species, of which 1400–1600 native (Webb). Blake (1959; see **General references**) gave a figure of 2341 native and naturalized species of which 477 were 'microspecies' of *Hieracium* and *Taraxacum*.

The Norwegian flora was extensively covered in *Flora danica* during the eighteenth and nineteenth centuries. The first major national work was the three-volume *Norges Flora* (1861–76; supplement, 1877) by Mathias and Axel Blytt. From it the younger Blytt and Ove Dahl derived *Haandbog i Norges flora* (1902–06; reissued 1926), a work with a particular concern for infraspecific taxa. More concise were *Norsk flora* (1891; 10th edn., 1957) by O. A. Hoffstad and *Norsk skoleflora* (1873; 19th edn., 1951) originally by Henrik Sørensen.

Standard manuals from the mid-twentieth century have been those by Nordhagen (1940) and Lid (1944; 5th edn., 1985; 6th edn., 1994).

A mapping project was initiated by Jens Holmboe at Bergen early in the twentieth century and continued there by Knut Fægri and his successors. Two volumes of results have been published: K. FÆGRI, 1960. *Maps of distribution of Norwegian vascular plants*, I: *Coast plants*. 134 pp., 54 maps (Universitet i Bergen Skrifter 26). Oslo: Oslo University Press; and O. GJÆREVOLL, 1990. *Idem*, II: *Alpine plants*. 126 pp., 37 maps. Trondheim: Tapir.

Bibliographies

For earlier literature, see 673 (CHRISTENSEN).

KLEPPA, P., 1955. Norske floraer/Norwegian floras. *Blyttia* **13**: 113–117. [Unannotated list, alphabetically arranged by authors; comprises mostly general works.]

KLEPPA, P., 1973. *Norsk botanisk bibliografi 1814–1964*. [10], 234 pp. (Universitetsbiblioteket i Oslo, Skrifter 2). Oslo: Universitetsforlaget; and *idem*, 1979. *Norsk botanisk bibliografi 1964–1975*. Oslo. [Subject-based.]

LID, J. and D. T. LID, 1994. *Norsk flora*. 6th edn., revised by R. Elven. lxxiii, 1014 pp., numerous text-figs., map. Oslo: Norske Samlaget. (Originally published 1944 as *Norsk flora*; 3rd edn., 1963, as *Norsk og svensk flora*; 5th edn. as *Norsk, svensk, finsk flora*.)

Illustrated manual of the vascular plants of Norway; includes keys to all taxa, concise descriptions, limited synonymy, vernacular names, relatively detailed indication of local range (by regions; see map, p. 931), summary of world distribution, and notes on karyotypes, habitat, phenology, etc.; etymological and author dictionaries and indices to all names at end. The general part includes a summary of the basis and plan of the work, an illustrated glossary, definitions of ecological and geographical terms, and the general keys.[170]

NORDHAGEN, R., 1940. *Norsk flora*. xxiii, 766 pp. Oslo: Aschehoug (W. Nygaard). Accompanied by *idem*, 1970. *Norsk flora: Illustrasjonsbind*, I. xxxvi pp., pls. 1–638 (incl. 772 text-figs.). Oslo.

Descriptive manual of vascular plants, with keys to all taxa, limited synonymy, brief generalized indication of local range, vernacular names, and notes on ecology, phenology, etc.; indices to vernacular and botanical names. The introductory section includes a glossary as well as a general key to families. The accompanying atlas, with figures by Miranda Bødtker, comprises three parts of which the first two originally appeared in 1944–48; it covers lower vascular plants, gymnosperms, monocotyledons, and Salicaceae through Fumariaceae. [The planned second volume of the atlas has never materialized.]

675

Sweden

For Lapland, see also **061**. – Area: 449790 km². Vascular flora: 1900–2000 species, of which 1600–1800 native (Webb). Lindman accounted for 2104 native and naturalized species including 206 'microspecies' of *Hieracium* and *Taraxacum*.

The first Swedish flora was *Flora suecica* (1745; 2nd edn., 1755) by Carl Linnaeus. Both editions benefited from his geographical and natural resource surveys from 1732 to 1749. It was succeeded by Göran Wahlenberg's *Flora suecica* (1824–26; 2nd edn., 1831–32). Manuals in Swedish for professional and school use began to appear from the end of the eighteenth century in different parts of the country, beginning with *Utkast till en svensk flora* (1792; 3rd edn., 1816) by Samuel Liljeblad. Later works of note include *Utkast till svenska växternas naturhistoria* (1867–68) by Carl Nyman, which included first records for each species, *Svensk flora för skolor* by T .O. B. N. Krok and S. Almqvist (first published in 1883 and, as *Svensk flora*, still current), and *Svensk fanerogamflora* (1918; 2nd edn., 1926) by C. A. M. Lindman.[171] The Krok flora itself was based on Carl Hartman's *Handbok i Skandinaviens flora* (noted under the regional heading).

The example of *Flora danica* was in Sweden followed early in the nineteenth century by the 11-volume *Svensk botanik* (1802–43) begun by J. W. Palmstruch and C. W. Venus (with text initially by Conrad Quensel) and sponsored by the Royal Swedish Academy of Sciences. The 774 plates were in their greater part later used in what is now *Nordens flora* (**670**).

Many partial and local works have been published. I include here only the outstanding *Skånes flora* of 1963 by Hennig Weimarck, covering Sweden's floristically distinctive southernmost province (and, in addition, all Danish plants).[172]

Bibliography

Annual supplements to the Krok bibliography were published in *Svensk Botanisk Tidskrift* through 1976.

KROK, T. O. B. N., 1925. *Bibliotheca botanica suecana*. xvi, 799 pp., portr. Uppsala/Stockholm. [Bio-bibliographic, with some 10 000 titles from 2000 authors through 1918 but lacking a subject index (never completed). Titles of prolific contributors are arranged not chronologically but by subject.]

KROK, T. O. B. N. and S. ALMQVIST, 1984. *Svensk flora: fanerogamer och ormbunkväxter*. 26th edn., revised by L. and B. Jonsell. 570 pp., text-figs., map. Uppsala: Esselte Studium. (Reprinted 1985, 1986. 1st edn., 1883; 25th edn., 1960.)

Illustrated manual of vascular plants, with keys to all taxa, concise indication of local range, vernacular names, and symbolic notes on life-form, phenology, etc; lexicon of Latin terms; indices to vernacular, generic and family names as well as to synonymous names. The introductory section includes a glossary, a synopsis and Linnaean system-based general key to families, and a key to plant groups with small or difficult flowers based on vegetative attributes. [With the 26th edition the work became less of a pure 'school-flora' than hitherto.]

LINDMAN, C. A. M. 1926. *Svensk fanerogam-flora*. 2nd edn. x, 644 pp., 329 text-figs. Stockholm: Norstedt and Söners. (1st edn., 1918.)

Manual of vascular plants, with keys to all taxa (including a general key to families and genera based on the Linnaean method), limited synonymy, vernacular names, concise indication of local and extralimital range, and notes on ecology, life-form, phenology, etc.; indices to vernacular, generic and family names. The introductory section includes a glossary and, in a bow to tradition, a key to Linnaean classes.

Partial works: Scania

This province of 11 283 km^2, with for Sweden a somewhat distinctive flora, was Danish territory until 1648.

WEIMARCK, H., 1963. *Skånes flora*. xxiv, 720 pp., 1 col. pl., maps (end-papers). Lund: Corona. Complemented by H. and G. WEIMARCK, 1985. *Atlas över Skånes flora*. 640 pp. Stockholm: Förlagstjänsten.

The *Flora* is a descriptive manual of vascular plants, with keys, limited synonymy, vernacular names, fairly detailed indication of local range, summary of overall range in southern Sweden and Denmark, and notes on karyotypes, ecology, phenology, etc.; complete indices. A glossary is also included. The *Atlas* comprises distribution maps with a substantial general introduction (and summary in English, pp. 130–141).[173]

676

Finland

Area: 337 032 km^2 (including the Åland Islands, an autonomous region). Vascular flora: 1450–1550 species, of which 1250–1450 native (Webb); Hämet-Ahti *et al.* in *Retkeilykasvio* (see below) suggest the latter figure to be about 1100. Some works described below (Hjelt; Hiitonen, 1933, 1934) also cover the Russian regions of Karelia (**686**) and the Kola Peninsula (**687**).

The earliest flora covering Finnish territory is *Catalogus plantarum tam in excultis quam in cultis locis prope Aboam superiori aestati nasci observatarum* (1673; 2nd edn., 1683) by Elias Tillandz; the second edition features Swedish as well as Latin text. No floras in Finnish appeared until the latter part of the nineteenth century following codification of the literary language; the first was *Lyhykäinen kasvioppi ja kasvio* (1877; 2nd edn., as *Suomen kasvio*, 1906) by Aukusti Mela. At the same time, however, Hjalmar Hjelt began what is the primary basis for all modern floras, the detailed *Conspectus florae fennicae* (1888–1926), with geographical information in Latin and taxonomic and other notes in Swedish. Hjelt also published *Kännedomen om växternas utbredning i Finland* (1891, as *Acta Soc. Fauna Fl. Fenn.* 5(2)), a historical review including initial records in Finland for all species.[174] The national language became dominant after independence and post-1920 social reform; in botany this gave rise to *Suomen kasvio* (1933) by Ilmari Hiitonen, in its class still current.

There are also some school- and excursion-floras in both Finnish and Swedish; these include, in Finnish, *Flora fennica* (1860; 2nd edn., 1866) by E. Lönnrot and, in Swedish, *Finlands kärlväxter* by O. Alcenius (1863; last revised 1930). After independence there appeared *Koulu- ja retkeilykasvio* by Hiitonen and A. Poijärvi (1932, with nine editions to 1958), a work considered 'standard' in the mid-twentieth century. It has now been succeeded by *Retkeilykasvio* (1984) by Leena Hämet-Ahti *et al.* A recent almost-comprehensive photographic atlas is *Suomen kasvit* (1992) by N. Lounamaa.

A computerized floristic database, with over 1 000 000 records from all sources by the mid-1980s, is presently maintained at the Botanical Museum, University of Helsinki. There is, however, as yet no national distribution atlas; the nearest comparable work is the three-volume *Suuri kasvikirja* (1958–80) edited by Jaakko Jalas (see below).

Bibliographies

SAELÁN, T., 1916. *Finlands botaniska litteratur till och med år 1900.* xi, 633 pp. (Acta Soc. Fauna Fl. Fenn. 43). Helsinki. [Annotated list, alphabetically by author with publications chronologically arranged.] Continued in R. COLLANDER, V. ERKAMO and P. LEHTONEN, 1973. *Bibliographia botanica Fenniae 1901–1950.* 646, [1] pp. (Acta Soc. Fauna Fl. Fenn. 81). Helsinki. [Briefly annotated list, arranged by authors; subject index, pp. 599–631.]

ALCENIUS, O., 1953. *Finlands kärlväxter: de vilt växande och allmännast odlade.* 12th edn., revised by Å. Nordström. 428 pp. Helsinki: Söderström. (1st edn., 1863.)

The standard Swedish-language manual of the vascular plants of Finland; includes keys to all taxa (among them a Linnaean general key), limited synonymy, vernacular names, symbolized indication of local range, and notes on habitat, phenology, life-form, etc.; index to generic names. [Apparently last revised in 1930 and afterwards merely reprinted.]

HÄMET-AHTI, L., J. SUOMINEN, T. ULVINEN and P. UOTILA (eds.), 1998. *Retkeilykasvio.* 4th edn. 656 pp., illus., maps, family index in rear endpapers. Helsinki: [Suomen] Luonnontieteellinen Keskusmuseo, Kasvimuseo (Finnish Museum of Natural History, Botanical Museum). (1st edn., 1984, Helsinki: Suomen Luonnonsuojelum Tuki).

Manual-flora of vascular plants (arrangement after Takhtajan); includes English summary (p. 4), essential synonymy, vernacular names (Finnish and Swedish), descriptions, internal distribution according to standard botanical districts (map, p. 12), individual range maps, karyotypes, critical remarks, illustrations of key features (where necessary), and partly symbolic notes on phenology, frequency, habitat, associations, status (and level of protection where appropriate), uses and properties, etc.; glossary and indices to vernacular and to generic and family names. The general part gives the plan of the work, an introduction to taxonomy and collecting, a discussion of plant geography and botanical districts, and key to families. No extralimital distributions are furnished.

HIITONEN, I., 1933. *Suomen kasvio.* 771 pp., 437 text-figs., map. Helsinki: Kustannusosakeyhtiö 'Otava'.

Illustrated descriptive manual of vascular plants, with keys to all taxa (synopsis and general key organized on the Linnaean method), limited synonymy, vernacular names, concise indication of local range, and notes on ecology, life-form, etc.; indices to vernacular and to generic and family names. The introductory section includes a glossary together with the general key. [Partly based upon Hjelt's *Conspectus florae fennicae* (see below).]

HIITONEN, I., 1934. *Suomen putkilokasvit.* 158 pp., map. Helsinki: Kustannusosakeyhtiö 'Otava'.

Systematic list of the vascular plants of Finland and adjacent parts of Russia, paralleling the author's *Suomen kasvio*; includes limited synonymy, symbolic indication of local range, gazetteer, and index to generic names; a map, with abbreviations of botanical districts, is also incorporated.

HJELT, H., 1888–1926. *Conspectus florae fennicae.* 9 parts in 2 vols. Illus. (Acta Soc. Fauna Fl. Fenn. 5(1–3), 21(1), 30(1), 35(1), 41(1), 51(1), 54). Helsinki.

Detailed critical enumeration of vascular plants (in the form of a series of family revisions for East Fennoscandia, including Finland), with synonymy, detailed indication of internal range including localities, taxonomic commentary, and other notes; bibliography and index in each part. Volume 1 (1888–95, in three parts) treats vascular plants other than dicotyledons; vol. 2 (1902–26, in six parts), the dicotyledons. [Symbols are explained in *idem*, 1888. *Notae 'Conspectus florae fennicae'.* 20, [4] pp. Helsinki (also published as a supplement, or 'Bihang', to the first part of the *Conspectus*).]

JALAS, J., 1958–80. *Suuri kasvikirja.* 3 vols. Illus., maps. Helsinki: 'Otava'.

A documentary work without keys or descriptions, comprising a detailed enumeration with extensive local distribution records and indication of extralimital range. [A functional continuation of Hjelt's *Conspectus*.]

KURTTO, A. and T. LAHTI, 1987. *Suomen putkilokasvien luettelo (Checklist of the vascular plants of Finland).* [6], 163 pp. (Pamphlet of Botanical Museum, University of Helsinki 11). Helsinki.

Systematically arranged nomenclator with symbolic indication of biological and legal status; does not include distributions.

LOUNAMAA, N., 1992. *Suomen kasvit.* 340 pp., col. photographs Forssa: Forssan Kustanuus Oy.

Photographic atlas of 1223 spontaneous taxa with coded documentation (explanation in the introduction, in Finnish, Swedish and English) and four illustrations to a page; index. [Does not account for microspecies of *Alchemilla*, *Taraxacum* and *Hieracium*.]

Woody plants

HÄMET-AHTI, L. *et al.*, 1989. *Suomen puu- ja pensaskasvio/Finsk träd och boskflora/Woody flora of Finland/Gehölzflora Finlands.* 290 pp., illus. Helsinki: Dendrologian Seura. (Publications of the Finnish Dendrological Society 5.)

Descriptive manual (covering 425 native and cultivated species and forms), with keys, indication of distribution and occurrence (with maps), symbolic notes including ecological and other preferences, other notes as necessary, and illustrations of critical details; indices to Swedish, Finnish and scientific names at end following an illustrated glossary. The introductory matter is relatively extensive, with several sections including references (summaries in Swedish, English and German are included).

677

Estonia

Area: 45 100 km^2. Vascular flora (including also **678** and **679**): 1600–1750 species, of which 1400–1600 native (Webb). – From 1721 to 1918 the entire territory of the modern state was part of the Russian Empire and from 1940 until 1991 it was a Soviet republic. The northern part, including Tallinn, constituted the former duchy of Estland; the remainder, including Tartu, was part of the former duchy of Livonia (see **678**).

Tartu (Dorpat or Jurjew) was long home to the only university in the northern Baltic lands, and remains the national botanical center.[175] Here, Johannes Klinge prepared two of the collective floras published before 1918: *Flora von Est-, Liv- und Curland* (1882) and *Schulflora für Est-, Liv- und Curland und die angrenzenden Gouvernements* (1885). These covered the three provinces corresponding to most of modern Estonia and Latvia. Klinge's flora was, however, the last in a long Baltic-German sequence beginning with *Versuch einer Naturgeschichte von Livland* (1778; 2nd edn., 1792) by Jacob B. Fischer and passing through *Botanisches Taschenbuch für Liv-, Cur- und Ehstland* (1803) by David H. von Grindel, *Flora der deutschen Ostseeprovinzen Esth-, Liv- und Curland* by Johan G. Fleischer and Emanuel Lindemann (1839; 2nd edn., by Alexander Bunge, 1853) and *Beschreibung der phanerogamischen Gewächse Esth-, Liv- und Curlands* (1852) by F. J. Weidemann and E. Weber. Klinge also produced a manual of woody plants, *Die Holzgewächse von Est-, Liv- und Curland* (1883). The two decades before World War I and the years between the world wars were ones of transition, including changes of research interests at Tartu University consequent to successive political movements. The only new manuals were *Eesti taimestik koolidele* (1922, with reissues to 1929) by G. Vilberg and *Kodumaa taimestik* (1943) by L. Enari and others. After World War II more Estonian-language floras began to appear, the first being a manual, *Taimemääraja: Eesti NSV-s sagedamini esinevaid kõrgemaid eos- ja õistaimi* (1948) by K. Eichwald and others. Research for *Eesti NSV floora* was also commenced at this time and concluded in 1968.[176]

Eesti NSV Teaduste Akadeemia, Zoologia ja Botaanika Instituut, 1953–84. *Eesti NSV floora/Флора Эстонской ССР.* 11 vols. Illus., maps. Tallinn: Eesti Riiklik Kirjastus ('Valgus'). (Revised editions of vols. 1–2 were published respectively in 1960 and 1962.)

Illustrated documentary descriptive flora of vascular plants (arranged according to the modified ranalean system of Grossheim); includes keys to genera, species and infraspecific taxa, full synonymy (with references), Estonian and Russian vernacular (and binomial) names, generalized indication of local and extralimital range (with numerous dot maps), taxonomic commentary, and notes on habitat, phenology, uses, special features, etc.; summary in Russian and complete indices in each volume.[177]

Kask, M. and A. Vaga (eds.), 1966. *Eesti taimede määraja.* 1188 pp., 1149 text-figs. Tallinn: 'Valgus'.

Concise descriptive manual of vascular plants, with keys to all taxa, limited synonymy, Estonian and Russian vernacular (and binomial) names, brief indication of local range, figures of diagnostic features, and notes on habitat, phenology, life-form, associates, etc.; indices to all botanical and other names at end.

Kask, M. et al., 1972. *Taimede välimääraja.* 526 pp., 662 text-figs., 5 pls. Tallinn: 'Valgus'.

Concise illustrated manual-key to vascular plants; includes brief diagnoses, limited synonymy, vernacular names, and partly symbolized indication of habitat, life-form class, habit, frequency and phenology; list of protected plants and indices to all botanical and Estonian names at end. Naturalized and widely cultivated plants are also included. The introductory section has a synopsis of families and genera and an illustrated glossary. [In part condensed from *Eesti taimede määraja.*]

678

Latvia

Area: 63700 km^2. Vascular flora: 1650 species of which 1205 indigenous (Tabaka *et al.*, 1988; see below). – Modern Latvia is made up of three parts with differing political histories. South (and west) of the Daugava (Dvina) River is the former Curonia (Courland or Kurland), while in the southeast (Latgale Province) is former Polish Livonia. Both of these were part of Poland-Lithuania until passing to Russia in the partitions of 1772–95. The remainder, along with the southern part of present-day Estonia, comprised the duchy of Livonia, organized by the Teutonic Knights and from the mid-sixteenth century variously under Polish and Swedish rule; it was absorbed into Russia in 1721. Independence was achieved in 1918, but from 1940 to 1991 the country was a Soviet republic.

Apart from the floras of the northern Baltic provinces mentioned under Estonia (**677**), the main pre-1918 work was *Flora von Polnisch-Livland* and its *Nachtrag* (1895–96) by Eduard Lehmann, a friend of Johannes Klinge at Dorpat (Tartu), then in Livonia. Rather earlier there had been a 'fancy' work, by Ernst W. Drümpelmann: *Flora livonica* (1809–10), of which 10 fascicles were published. Without a national herbarium at independence, however, no flora of modern Latvia was (or perhaps could be) produced for some time. In the years before 1939 the most notable works to appear were a students' manual-key by J. Bickis (Bitzkis), *Latvijas augu noteicējs* (1920; 2nd through 4th edns., 1923, 1927 and 1935), and a well-illustrated manual of trees and shrubs, *Koku un krūmu noteicējs* (1925) by Kārlis Starcs. Starcs also published a series of floristic contributions, *Latvijas pavasaraugu noteicējs* (1924–27). A final edition of *Latvijas augu noteicējs*, revised by A. Rasiņš, appeared in 1946. Preparation of *Latvijas PSR flora*, along with a new field-manual, began afterwards, and these and a new manual of woody plants appeared in the 1950s.[178]

Bibliography

No separate botanical bibliography for Latvia has been published. The most useful survey is in L. V. Tabaka, 1974. Основные этапы развития флористических исследований в Латвии (Osnovnye etapy razvitija floristiceskikh issledovanij v Latvii). In Akademija nauk

Latvijskoj SSR, Institut Biologii, Флора и растительность Латвийской ССР: Приморская низменность (*Flora i rastitel'nost' Latvijskoj SSR: Primorskja nizmennost'*), pp. 7–21. Riga: 'Zinātne'. [A semi-chronological unannotated bibliography, the entries mostly arranged by author but grouped in three periods: 1778–1919, 1920–41 and 1945–73.]

GALENIEKS, P. (ed.), 1953–59. *Latvijas PSR flora.* 4 vols. Illus. (some col.). Riga: Latvijas Valsts Izdevniecība.

Illustrated descriptive flora (in Latvian) of vascular plants, with keys to all taxa; includes relatively limited synonymy, references, Latvian and Russian vernacular names (and Russian binomials), concise indication of local and extralimital distribution, and notes on habitat, life-form, phenology, etc.; summaries in Russian as well as complete indices at end of each volume. [Now out of date.]

PĒTERSONE, A. and K. BIRKMANE, 1980. *Latvijas PSR augu noteicējs.* 2nd edn., 590 pp., 189 text-figs. Riga: 'Zvaigzne'. (1st edn., 1958.)

Manual-key to native, naturalized and commonly cultivated vascular plants (arranged following Takhtajan); includes essential synonymy, vernacular names, diagnostic figures, and concise, partly symbolic notes on local distribution, habitat, frequency, life-form, habit, phenology, etc.; systematic synopsis, illustrated glossary, list of authors, and indices to all botanical and other names at end. The general part includes a key to families.

TABAKA, L. V., Ģ. GAVRILOVA and I. FATARE, 1988. Флора сосудистых растений Латвийской ССР (*Flora sosudistykh rastenij Latvijskoj SSR*). 194, [ii] pp. Riga: 'Zinātne'.

Annotated systematic enumeration of known vascular plants, with indication of habitat(s), status and frequency; additions are indicated by special signs (explanation on pp. 15–16). A short introductory section (partly also in English) gives the basis for and scope of the work as well as an analysis of the flora; there is also a bibliography (69 titles).

Distribution maps

FATARE, I. (ed.), 1978– . Хорология флоры Латвийской ССР. Перспективные для охраны вилы растений (*Khorologija flory Latvijskoj SSR. Perspektivnye dlja okrany vidy rastenij*). Parts 1– . Maps. Riga: 'Zinātne'.

Chorological treatise with distribution maps of native, naturalized and adventive vascular plants; each species or infraspecific taxon entry in the text of each part comprises general notes on range and habitat, specific references for each square of 70.7 km² (related to a grid plan on the cover) with localities (and vouchers) and literature source(s), more general references not assignable to a specific square, and a discussion. The text is followed by a reference list and the maps. Abbreviations and symbols are on p. 6 in part 1. No systematic order is followed, but in each part entries are alphabetical by genus. [As of 1998 four parts had been published, the last seen in 1986. The maps have been contributed by several authors.]

Woody plants

The most recent guide to woody plants [not seen] is A. MAURIŅŠ, M. MORKONS and A. ZVIRGDIS, 1958. *Latvijas PSR koki un krūmi.* 303 pp., illus. Riga: LPSR Zin. Akad. Izdeb. There is also a dendrology, *Dendrologia of Latvia* (1986) by R. Cinovskis, M. Bitse, Dž. Knape and D. Smite.

STARCS, K., 1925. *Koku un krūmu noteicējs.* lxx, [1], 444 pp., 395 pls. with 2065 text-figs. Riga: Mežu Departamenta Izdevums.

Illustrated manual of trees and shrubs; includes keys, limited synonymy, vernacular names (Latvian, German and Russian), descriptions, and indication of distribution (or origin, if appropriate) and habitat; indices to scientific and Latvian names at end. The general part encompasses an illustrated organography/glossary, author abbreviations, list of literature consulted (pp. xxxi–xxxiii) and the general key (illustrated) to families (and genera).

679

Lithuania

Area: 65 200 km². Vascular flora: 1072 species listed in Hyrniewiecki (1933). – The final partition of Poland-Lithuania in 1795 resulted in the absorption of present-day Lithuania into the Russian Empire. From 1918 the country was independent, but the Vilnius region passed to Poland in 1920 (until 1923 as part of the 'autonomous' territory of 'Central Lithuania'). The state assumed its present boundaries after the outbreak of World War II and from 1940 to 1991 was a Soviet republic.

The first flora, which covered the Grand Duchy (and thus also present-day Belarus'), was *Flora Lithuanica inchoata* (Grodno, 1781; Vilnius, 1782) by Jean (Jan) Gilibert, then at Vilnius (Wilno) University and effective founder of the 'Vilna School' of botany (1781–1832). Then followed the synopses by Stanisław

B. and Józef Jundziłł; the former was responsible for *Opisanie roślin litewskich* (1791; 2nd edn., 1811), covering the Grand Duchy, the latter for its expanded successor, *Opisanie roślin w Litwie, na Wołyniu, Podolu i Ukrainie dziko rosnących, iako i oswoionych* (1830). The latter accounted for 2305 species in present-day Lithuania, Belarus' and Ukraine (cf. **684, 694**). Both works followed the Linnaean arrangement. In this same period there also appeared *Naturhistorische Skizze von Litthauen, Volhynien und Podolien* by Eduard Eichwald (1830), also on 'Greater Lithuania'; its botanical part was by Stanisław B. Gorski. With the closure of the university by the Russians in 1832, local activity became considerably reduced.[179]

After World War I, the areas which came under Polish rule were until 1939 covered by that country's floristic works, and the remainder in the third part (*Materjały do flory Żmudzi i Litwy spółczesnej/ Prodromus florae Samogitiae necnon Lithuaniae contemporalis*) of B. Hyrniewiecki, 1933. *Zarys flory Litwy/Tentamen florae Lithuaniae.* xvi, 368 pp., 61 textfigs., 2 maps. (Arch. Nauk Biol. Towarz. Nauk. Warszawsk. 4). Warsaw. The Vilnius region was in this period studied by J. Mowszowicz, but only after World War II did his floristic contributions appear as *Przegląd flory Wileńskiej/Conspectus florae vilnensis* (1957–59, in *Prace Łódźkie Towarzystwo Naukowe*, III, **47, 51** and **59**).

Major floristic works in Lithuanian began to appear only in the 1930s. Konstatin Regel'is (Regel) organized botanical education at Kaunas University from its establishment in 1922 and from 1931 to 1942 published seven installments (458 pp.) of his *Lietuvos floros šaltiniai* (*Fontes florae Lituanae*), a *mélange* of floristic, historical and bibliographical materials but fully indexed in part 7 (pp. 440–445). In 1934 there appeared, also at Kaunas, a manual: *Vadovas Lietuvos augalams pažinti* by J. Dagys, J. Kuprevičius and A. Minkevičius. After World War II they were succeeded by the current offerings, notably *Lietuvos TSR flora* edited by Marie P. Natkevičaitė, a student of Regel'is.[180]

Bibliography

Regel'is' *Lietuvos floros šaltiniai*, noted above, also contains bibliographies.

Šapiraitė, S. (ed.), 1971. *Lietuvos botanikos bibliografija 1800–1965.* 528 pp. Vilnius: Lietuvos TSR Mokslu Akademijos, Centrinė Biblioteka. Continued as

S. Norkunienė, 1984. *Lietuvos botanika, 1966–1970: literaturos rodykle.* 534 pp. Vilnius. [The main work was 'scientifically' vetted by K. Brundza; it contains 4051 entries in all areas of botany, most of them concisely annotated. The main part is classified, with locations of holdings, and is followed by author and subject indices. The supplement has not been seen.]

Natkevičaitė-Ivanauskienė, M. P. (ed.), 1959– . *Lietuvos TSR flora.* Vols. 1– . Illus. (some col.), maps. Vilnius: Valstybinė Politinės ir Mokslinės Literatūros Leidykla (later 'Mokslas').

Illustrated descriptive flora of vascular plants; includes keys to all taxa, full synonymy with references and citations, Lithuanian and Russian vernacular (and Russian binomial) names, rather detailed indication of local range, summary of extralimital distribution, and notes on habitat, phenology, life-form, associates, uses, etc., as well as some critical commentary; Russian- and German-language summaries as well as complete indices at end of each volume. The introductory section (in vol. 1) provides a historical account of botanical work in Lithuania. [The fifth of the projected six volumes appeared in 1976, completing coverage of the flora save for Compositae; this last volume, however, has yet to be published.]

Snarskis, P., 1968. *Vadovas Lietuvos augalams pažinti.* Revised edn. 502 pp., illus. Vilnius: 'Mintis'. (1st edn., 1954, under title *Vadovas Lietuvos TSR augalams pažinti.*)

Manual-key to vascular plants; includes concise diagnoses, very limited synonymy, vernacular names, indication of local range, brief notes on habitat, phenology, etc., and figures of some more common species; indices to generic, family, and Russian and Lithuanian vernacular names at end. [Not seen; description based on information supplied by A. O. Chater, formerly of London.]

Woody plants

A recent guide [not seen] is L. Janushkevičius and R. A. Budriunas, 1987. [*The trees and bushes growing in Lithuania.*] Vilnius. There is also a *Dendrologia*, published in 1973 (Vilnius: 'Mintis').

Region (Superregion) 68/69

CIS-in-Europe (and works covering the former USSR)

Area: not given here. Vascular flora (former Soviet Union as a whole): 21770 species with an additional 500 subspecies (Czerepanov, 1995; see below). Some 17520 were accounted for in *Flora SSSR*, with additions of 3070 collectively from 1934 to 1971 (Czerepanov, 1973; see below), 873 to 1981 (in Czerepanov, *Sosudistye rastenija SSSR*) and 307 to the early 1990s. For figures for the European part of the former Soviet Union alone, see below under **680**. – This largest of European regions comprises Belarus', Ukraine, Moldova and the European part of the Russian Federation, the last-named from its western borders to Asia (with limits approximately as defined for *Flora Europaea*). The Arctic tundra zone and the islands of the Arctic Ocean are, however, excluded; for these, see **Regions 05** and **06**. – A lexicon of Russian words used in the titles of floras in the CIS is given in Table 1.

The many-sided, sometimes distinctive and compelling history of botanical exploration and floristic recording and analysis of the Russian Empire and its successors has been recorded in various ways by many authors. Early contributors mostly worked out of St. Petersburg (and in particular the original botanical garden of the Russian Academy of Sciences, on Vasil'evskij Island); among them were Johann G. Siegesbeck, Johann Ammann, Stefan P. Krascheninnikov, David de Gorter and the Siberian explorer Johann G. Gmelin.[181] In the latter part of the eighteenth century, Empress Catherine II was an important promoter of exploration and research; during her reign and those of her son and grandson, respectively Paul I and Alexander I, activities at St. Petersburg were strengthened and additional foci of activity developed, particularly at Moscow, Tartu (Dorpat or Jurjew, now in Estonia) and Kharkiv (Kharkov, now in Ukraine). Among active botanists of the time were Samuel G. Gmelin and Peter S. von Pallas (both noted explorers), Gregor Sobolewsky, C. Friedrich Stephan and George F. Hoffmann (at Moscow), and Joseph Liboschitz, Carl Trinius and Carl Ledebour (at Tartu).

Table 1. *Lexicon of Russian words used in the titles of floras in the CIS*[a]

AR	Autonomous or Associated Republic
ASSR (now AR)	Autonomous Soviet Socialist Republic
čast'	portion
Dal'nyj Vostok (Dal'nego Vostoka)	the [Russian] Far East
derevo (derevja)	tree (trees)
dikorastuščij	wild, spontaneous
flora	flora *or* a Flora
izdanie	edition, publication
južnyj	southern
Kavkaz	the Caucasus
konspekt	conspectus or enumeration (without extensive descriptions; may or may not have keys)
kustarnik	shrub
lesnoj	forested
oblast'	province (or region)
okrestnost'	environs, vicinity
opredelitel'	'manual-key': a work with descriptive keys and concise indications of habit, range, phenology, etc. Often (but in the present author's opinion clumsily) translated as 'the keys', 'determinator', or 'determination'
osennij	autumnal
ostrov	island
ozero	lake
polosa	zone, belt
poluostrov	peninsula
Predbajkal' ja	the Cis-Baikal region in Eastern Siberia, west of the lake and including Bratsk and Irkutsk
Prijenisej	the Jenisei (Yenisei) River basin
RSFSR	[former] Russian Soviet Federative Socialist Republic (now the Russian Federation). The largest of the 15 former union republics, with 89 discrete provinces, territories and autonomous republics
ranne–letnij	early summer
rastenie (rastenij)	plant (or plants)
SSR	[former] Soviet Socialist Republic. One of the 14 former 'national' republics, all now independent states
SSSR	[former] Union of Soviet Socialist Republics
severnyj	northern
sosudistyj (sosudistykh)	vascular

Table 1. (*cont.*)

spisok	list (or 'elenchus'): a *concise* annotated enumeration, but partly synonymous with *konspekt*
srednij	middle, central; see also *tsentral'nyj*
tsentra	center
tsentral'nyj	central
vesennij	spring, vernal
vostočnyj	eastern
vysokogornyj	alpine, high-mountain
vysšij (vysšikh)	higher (of higher plants). Now used as equivalent to 'vascular' when referring to plants
Zabajkal'ja	the Trans-Baikal region, east of the lake; includes most of Buryat-Mongolia as well as Chita Oblast'
zapadnyj	western

Notes:

[a]Nominative singular forms are used, with other forms given in parentheses where significantly different.

It was from the three northern centers that more detailed botanical exploration and the compilation of floras largely began. Effective local coverage was, in European Russia, first attained in the Baltic provinces (now Estonia and Latvia; **677, 678**) and northwest Russia (including the St. Petersburg region or Ingria), followed by Middle Russia and, from the early nineteenth century, Ukraine and Belarus'. The first general flora, the sumptuous *Flora rossica* by Pallas, appeared in 1784–88(–1815) but unfortunately never was completed, funds for its publication having been discontinued. Its successor was Ledebour's more modest but complete (and highly regarded) *Flora rossica* (1842–53).

St. Petersburg began its real rise to prominence as a botanical center only in the 1820s. In 1823, two years before the Decembrists' revolt, the Imperial Botanical Garden, hitherto most noted for its services in medicine, was reorganized as a scientific institution though remaining under the Crown. In the same year, Carl Trinius became a member of the Academy of Sciences and took effective charge of its botanical collections, and in 1835 organized the Botanical Museum as a separate Academy unit. Together the two institutions, along with the school of botany at St. Petersburg University organized in 1830, were to contribute greatly to exploration and the increase of botanical knowledge. This included production of many floras,

floristic reports and related works for Russian territories in Europe, Siberia and Central Asia as well as of Alaska and other lands (notably Japan and the 'Chinese Border') through mid-century and in the three or four decades beyond.

Leading botanists at the St. Petersburg institutes included Friedrich E. L. Fischer, Carl A. Meyer (who directed both Garden and Museum from 1850 to 1855), Franz J. Ruprecht, Ernst von Trautvetter (founder of *Acta Horti Petropolitani*), Eduard A. Regel (a prolific writer and long editor of *Gartenflora*, a leading horticultural journal), Ferdinand von Herder, and Carl J. Maximowicz (author of *Primitiae florae amurensis*). Later, as the Caucasus, Central Asia and the Far East were penetrated, transport and communications improved, additional universities and other institutes were established, their work was carried on by their many (and increasingly more dispersed) successors.[182] In both periods, a considerable number of floras and related works were produced. With the gradual development of more detailed approaches to floristics and systematics, the Garden and Museum alike from the 1890s organized systematic programmes for floristic exploration and flora preparation. This very probably reflected initiatives of Sergei I. Korshinsky who held senior posts at both institutes from 1892 to 1900.[183] Aided by increases in staff and the impact of government survey programmes under the Transmigration Bureau (which yielded vast collections in the first two decades of the twentieth century) and other development schemes initiated by the imperial minister for finance Sergius de Witte, these objectives were to a considerable extent realized and intellectual horizons correspondingly widened. Several, in part overlapping, flora projects were begun; for Russia west of the Ural this resulted in the preparation at the Museum of Korshinsky's own *Tentamen florae Rossiae orientalis* (1898) and, at the Garden, of *Flora Evropejskoj Rossii* (1908–10) by Boris A. Fedtschenko and Alexander F. Flerow.[184] Works on special groups also appeared: woody plants were revised in, for example, *Russkaja dendrologija* (1870–82; 2nd edn., 1883–89) by E. Regel and aquatic plants in *Vodjanye rastenija Srednej Rossii* (1897; 2nd edn., 1900; 3rd edn., 1913, as *Vodnaja flora Evropejskoj Rossii*) by Fedtschenko and Flerow.

There was also significant activity in floristics at other centers, notably Moscow (Wassily Zinger in the 1880s and, later, Dimtri P. Syreistschikov), Tartu, then known as Dorpat and later Jurjew (Carl Ledebour,

Alexander von Bunge, Johannes Klinge and Nicolai I. Kuznetsow through to 1917; see also **677**), Kiev (Wilibald S. von Besser in 1834–41, Trautvetter in 1838–59 and Ivan F. Schmal'hausen in 1879–94; see also **694**), Kharkov (Nicolai S. Turczaninow in 1847–64 after earlier periods in Irkutsk and Khabarovsk, and Vassili M. Czernjaev), Kazan' (from 1836, with V. I. Taliev there prior to 1900; see also **691**), and Odessa (Eduard von Lindemann, author of an early checklist of then-European Russia, *Index plantarum quas in variis Rossiae provinciis hucusque invenit et observavit* (1860), and student of the South Ukraine flora). At Novaja Alexandrija (Pulawy) in Russian Poland Petr F. Majevski and Nikolaj V. Zinger, authors of popular identification manuals, were active before and after 1900. All these centers were at universities or polytechnics which, apart from Moscow, were established at various times during the century. Outside Europe early centers included, in the Caucasus, Tbilisi (the Caucasian Museum, established in 1865) and, in Siberia, Tomsk (from 1885, where Porphyry N. Krylov founded the first Siberian herbarium).[185]

Contemporary general floras and related works pertaining to Russia-in-Europe from before 1900 and prepared outside Moscow or St. Petersburg included *Sbornik svedenij o flore Srednej Rossii* (1886) by Zinger, *Flora Srednej Rossii* (1892; 9th edn., 1964) by Majevski – the latter long the 'standard' manual-key for the Russian heartland – and the still well-regarded *Flora srednej i južnoj Rossii, Kryma, i Severnogo Kavkaza* (1895–97) by Schmal'hausen, covering Ukraine, Crimea, the Lower Don region, and Ciscaucasia.[186] The years to 1914 brought a geographically more extensive but also more concise successor to Schmal'hausen's manual, *Opredelitel' vysšikh rastenij Evropejskoj časti Rossii* (1907) by Taliev, now in Kharkov. This went on to nine editions by 1941 (with the sixth also in Ukrainian). In 1913 there appeared *Opredelitel' rastenij lesnoj polosy Severo-Vostoka Evropejskoj časti Rossii* by A. A. Snyatkov, Grigorii Širjaev and Ivan A. Perfil'ev, the first of its kind for northern Russia and a development from several preliminary studies working outwards from Kazan' and Vologda. Some works also appeared on the Baltic and Lithuanian provinces (**677** to **679**); and in Finland (**676**) Hjalmar Hjelt was documenting Finland proper, Karelia and the Kola Peninsula for his monumental *Conspectus florae fennicae* (1888–1926).

By 1917, therefore, a regional flora tradition had become well established, apparently in lieu of a new

comprehensive flora as a successor to Ledebour's work. Such a project may have seemed too daunting or that, given a rapidly increasing known flora, studies were better manageable or more functional on the basis of major regions. Only Fedtschenko attempted an all-imperial checklist, *Spisok Russkikh'' rastenij/ Enumeratio plantarum Imperii Rossici*, of which but a single installment, covering 868 species, was published (1914; as Suppl. 2 to *Izvest. Imp. Bot. Sada* 14). For the extra-European parts of the empire, an approach featuring extensive documentation characterized the work of Nikolai Kuznetsov and Vladimir I. Lipsky in the Caucasus and Fedtschenko and others in Middle Asia, Siberia and the Far East. Similar detailed coverage at a local level was, however, comparatively limited outside of Finland (and Karelia), northwest and central Russia, the Baltic provinces, and parts of Lithuania, Belarus', Poland and Ukraine.

After the two revolutions of 1917, the pattern of floristic work and publication for the next 10 years or so remained largely as in earlier decades save for the period of disruption and civil war up to 1922 (which had more effect on the Imperial Botanic Garden). Some pre-war flora projects were continued, and under the new Soviet regime strong support for biological surveys resumed.[187] A new regional flora, *Flora Rossiae austro-orientalis*, written by several authors under the editorship of first Fedtschenko and then Boris K. Shishkin, was begun in 1927 and completed in 1936. This flora was noteworthy: it was at once the last of the 'primary' regional floras of European Russia and a forerunner of the collective method of preparation which characterized *Flora SSSR*, which would heavily involve both principals (and so delay publication of the last volume).

The conception of *Flora SSSR* was influenced by the thorough reorganization of science conducted under Communist Party direction beginning from the late 1920s, and it featured as a primary project of the new Botanical Institute (formed from a merger of the Botanic Garden and Museum in 1931) in the first Five Year Plan. Initiated in 1929 and formally begun two years later, the project dominated Soviet floristic botany for more than three decades, with in all the involvement of 92 collaborators. Its origin, progress and completion have been described by several writers. Contributing greatly to its ultimate success was the authority wielded by Vladimir L. Komarov, from 1902 to 1930 at the Botanic Garden, then foundation head of the Institute's Department of Higher Plants and finally (1936–45) president of the Soviet Academy of Sciences, and his associate Shishkin, a foundation member in Higher Plants and for 11 years (1938–49) director of the Botanical Institute.

The concentration of efforts upon *Flora SSSR* did not, however, entirely disrupt the regional flora tradition. In particular, with the development under Soviet rule of the educational and research network, notably outside the Russian Federation, more botanical centers came into being (among others Minsk, Baku and Tashkent may be mentioned). These contributed greatly to the large number of regional, republican and local floras which, promoted by the success of *Flora SSSR*, were produced after World War II. Already before the war there had appeared *Flora Severnogo Kraja* (1934–36) by Ivan A. Perfil'ev and, in Ukraine, the early volumes of *Flora URSR* (1937–65) begun initially by Alexander Fomin and, after his death in 1935, continued under the direction of Eugen Bordzilovsky. Afterwards, regional floras and manuals were written for the Kola Peninsula, Karelia, the St. Petersburg region, the Baltic republics (see **677** to **679**), Moldavia, Belarus', the Crimea, and parts of northern, middle and eastern Russia (including among the latter a new manual for the Moscow region). The Arctic zone was also the subject of a critical regional flora, begun in 1960 under the editorship of Alexander I. Tolmatchev. Tolmatchev, from 1922 a student of the Far North, was later responsible for *Flora Severo-Vostoka Evropejskoj časti SSSR* (1974–77), a successor to Perfil'ev's work.

Coverage by regional works is, however, not uniform and there are some areas for which no recent works in this genre are available. Moreover, such as exist vary considerably in scope from manual-keys to full-scale documentary works similar to those for many other European countries. The most notable lacuna was a successor to *Flora Evropejskoj Rossii*, both in itself and as a necessary basis for new revisions or replacements of standard manuals. This, and a desire to introduce new taxonomic principles and methods of data presentation, led Andrei A. Fëdorov and others at St. Petersburg to undertake *Flora Evropejskoj časti SSSR*. Publication began in 1974 and by 1995 volumes 1–6 (to 1987), vol. 8 (1989) and vol. 7 (1994) of a planned 11 had been published. The standards set by *Flora Europaea* have been used as guidelines.

Other recent works of wide scope have included a volume of additions and corrections to *Flora SSSR* by

S. K. Czerepanov (1973) and *Zlaki SSSR*, a revision of the grasses by N. N. Tzelev (1976; English edn., 1983). Czerepanov produced, successively, Russian (1981) and English (1995) editions of a new checklist for the Soviet Union, *Sosudistye rastenija SSSR*; and a revised *Flora rossica* is projected.

The woody plants are covered in several works of varying scope, the most comprehensive being S. Ja. Sokolov's *Derevja i kustarniki SSSR* (1949–62; supplement, 1965). To these has now been added a series of maps, *Arealy derev'ev i kustarnikov SSSR*, prepared under Sokolov's direction and issued in three parts from 1977 to 1986. Aquatic plants (see **608**) were last revised in *Flora gigrofitov* (1948) by Ju. V. Ryczin.

The political, social and economic changes beginning in the late 1980s and including the formation of the Commonwealth of Independent States in December 1991 have naturally affected institutions and flora projects.[188] Among current undertakings, completion of *Flora Evropejskoj časti SSSR* was still outstanding; this has fortunately been continued (with vol. 7 appearing in 1994) and will be completed under the title *Flora Vostočnoj Evropy* (an English version has also commenced publication). In Moscow, there have been two further editions (1992 and 1995) of a manual for Middle Russia first published in 1983 as *Opredelitel' vysšikh rastenij srednej polosy SSSR* and now entitled *Opredelitel' sosudistykh rastenij tsentra Evropejskoj Rossii*. In addition, a revision of *Sosudistye rastenija SSSR* appeared in 1995 in the United Kingdom under the title *Vascular plants of Russia and adjacent states*, and progress has continued with regional and national floras in other parts of the CIS.[189]

Progress

The corpus of literature on the history of botany in the Russian Empire and its successors is considerable, and only a small selection most useful in the floristic context can be given here. Two *vademeca* are: V. C. ASMOUS, 1947. Fontes historiae botanicae rossicae. Pp. 87–118 (Chron. Bot. **11**(2)). Waltham, Mass.; and S. JU. LIPSCHITZ, 1947. Sistematika, floristika i geografija rastenij. In L. P. Breslavets *et al.* (eds.), Очерки по истории русской ботаники (*Očerki po istorii russkoj botaniki*), pp. 9–114. Moscow: Moscow Society of Naturalists Press. A number of histories for particular regions, institutions, societies or other special categories have also been published; those documented by Asmous are classified in his subject index. For English-speaking readers, the best introductory account (with a bibliography including many Russian references) is S. G. Shetler, 1967. *The Komarov Botanical Institute: 250 years of Russian research*. Washington: Smithsonian Institution Press. Systematic botany and floras are also briefly covered in E. Ashby, 1947. Survey of botany in the Soviet Union. *Rep. Twenty-fifth Meeting, Austral. New Zeal. Assoc. Adv. Sci.* (Adelaide, 1946), pp. 245–266. Mid-twentieth century summaries in Russian on floristics and systematics include E. G. Bobrov, 1963. Sostojanie i perspktivy izučenia otečestvennoj flory. *Bot. Žurn. SSSR* **48**: 1729–1740; A. L. Takhtajan, A. I. Tolmatchev and An. A. Fëdorov, 1965. Izučenie flory SSSR, dostiženija i perspektivy. *Ibid.*, **50**: 1365–1373 (see also A. I. Tolmatchev, 1966. *J. Gen. Biol.* (USSR) **27**: 411–422); and An. A. Fëdorov, 1971. Floristics in the USSR. *BioScience* (USA) **21**: 514–521.

Flora SSSR: The secondary references on this work are numerous, and only a selection is essayed here. In Russian: E. G. Bobrov, 1965. 'Flora SSSR', rabota nad nej i znacenie etogo izdanija. *Bot. Žurn. SSSR* **50**: 1374–1383; I. A. Linczevski, 1966. 'Flora SSSR' (notula bibliographica). *Nov. Sist. Vysš. Rast.* (St. Petersburg) **3**: 316–330; and M. E. Kirpicznikov, 1967. 'Flora SSSR' – krupnejšee dostiženie sovetskikh sistematikov. *Bot. Žurn. SSSR* **52**: 1503–1530 (with extensive bibliography). In English: V. H. Heywood and E. G. Bobrov, 1965. Preparation of 'Flora URSS'. *Nature* **205**: 1046–1049 (abridged and transl. from Bobrov, 1965); and M. E. Kirpicznikov, 1969. The Flora of the USSR. *Taxon* **18**: 685–708 (abridged and transl. from *idem*, 1967; includes bibliography).

European part of the CIS: A useful review of general works, with critical remarks, floristic statistics, details of editions and many supporting references, is M. E. Kirpicznikov, 1968. Краткий обзор важнейших флор и определитель, изданных в СССР за 50 лет, I/Kratkij obzor važnejšikh flor i opredelitel', izdannykh v SSSR za 50 let, I. *Bot. Žurn. SSSR* **53**: 845–855.

Bibliographies. General and divisional bibliographies as for Division 6 (except Blake, 1961).

Regional bibliographies
Other bibliographic sources are given in Lebedev (1956; see below).

Aleksandrova, K. V. (comp.), 1975. Ботаника: основные отечественные библиографические источники и словари 1917–1974 (*Botanika: osnovnye otečestvennye bibliografičeskie istočniki i slovari 1917–1974*). 70 pp. Leningrad: Biblioteka AN SSSR (for the XII International Botanical Congress). [308 bibliographic works along with 59 dictionaries and lexica.]

Lebedev, D. V., 1956. Введение в ботаническую литературу СССР (*Vvedenie v botaničeskuju literaturu SSSR*). 382 pp. Moscow/Leningrad: AN SSSR Press. [Selective, briefly annotated guide to Russian systematic, floristic and geobotanical literature, arranged by subject and region; indices. 3000 entries.]

Lipschitz, S. Ju., 1975. Литературные источники по флоре СССР (*Liternaturnye istočniki po flore SSSR*)/*Florae URSS fontes*. 232 pp. Leningrad: 'Nauka'. [Bibliography of 2368 items, arranged by author within five major geographical divisions. A short section on 'school-floras' is also included.]

Trautvetter, E. R., 1880. *Florae Rossicae fontes*. 382 pp. (Acta Horti. Petrop. 7(1)). St. Petersburg. [1650 items with annotations; coverage through 1878.]

Indices. General and divisional indices as for Division 6.

680
Region in general

The works described here fall into two general categories, respectively for those covering the territory of the former USSR (designated below as 'All-Union' works) and those limited to the European part of the Commonwealth of Independent States (Belarus', Moldova, Ukraine and the European part of Russia). Each category includes some key works on the woody flora (including 'tree books').

I. 'All-Union' (Vsesojuz, 1917–91) works

Czerepanov, S. K., 1995. *Vascular plants of Russia and adjacent states (the former USSR)*. x, 516 pp., map. Cambridge: Cambridge University Press.

(Revised and enlarged from *idem*, 1981. Сосудистые растения СССР (*Sosudistye rastenija SSSR*)/*Plantae vasculares URSS*. 509 pp., colophon. Leningrad: 'Nauka'.)

An alphabetically arranged checklist and thesaurus of the native, naturalized, and cultivated but spreading vascular flora of the former Soviet Union (21 770 species in 1945 genera in 216 families, as well as 594 nothospecies, 203 species possibly to be found within the range, and 533 commonly cultivated plants); accepted taxa appear with indication of distribution (given as one or more of six regions as depicted by the map in the frontispiece), while synonyms are depicted with the equivalent accepted names. Pp. ix–x include a list of major floristic works published from 1981 in the CIS and Baltic States, while at the end (pp. 501–516) is an index to all family and generic names. Family and generic entries are unannotated and moreover are without references even to selected literature. Cultivated species are marked with an asterisk.[190]

KOMAROV, V. L., B. K. SHISHKIN *et al.* (eds.); E. G. BOBROV *et al.* (comp.), 1933–64. Флора СССР(*Flora SSSR*)/*Flora URSS*. 30 vols., index. Illus., maps. Moscow/Leningrad: AN SSSR Press (index: 'Nauka'). (Vol. 11 reprinted 1945, Moscow/Leningrad; vols. 1–13 reprinted 1964, Lehre, Germany: Cramer.) English edn. (1): *idem*, 1963– . *Flora of the USSR*. Transl. N. Landau and P. Lavoott (vols. 1–21, 24) and K. Bakaya (vols. 22, 25, 27); sci. eds. N. Landau, J. Lorch and U. Plitman (vols. 1–21, 24) and S. G. Shetler and G. N. Fet (vols. 22, 25, 27). Vols. 1–21, 24 (1963–77), 22 (1997), 25 (1999), 27 (1998). Jerusalem: Israel Program for Scientific Translations (vols. 1–21, 24); New Delhi: Amerind/Plymouth, England and Enfield, N.H.: Science Press (vol. 22, 25, 27; all in association with the Smithsonian Institution, Washington, D.C.). (The remaining volumes are expected. Vols. 1–21 and 24 variously reprinted 1985–87, Königstein/Ts., Germany: Koeltz.) English edn. (2): *idem*, 1990– . *Flora of the USSR*. Transl. Doon Scientific Translations Co.; sci. ed. G. Panigrahi. Vols. 22–23, 25–30, *passim*. Dehra Dun: Bishen Singh Mahendra Pal Singh/Königstein/ Ts., Germany: Koeltz. (Vols. 22–23, 25–28 and 30 published and 29 in preparation by early 2000.) Complemented by N. N. TZVELEV, 1976. Злаки СССР (*Zlaki SSSR*). 788 pp., 9 text-figs., 16 pls. Leningrad: 'Nauka'. (English edn.: *idem*, 1983. *Grasses of the Soviet Union*. Transl. B. R. Sharma. xvi, 1196 pp., 16 pls., 10 figs. Rotterdam: Balkema/New Delhi: Amerind.) A list

of additions to the *Flora* through 1971 is found in S. K. CZEREPANOV, 1973. Свод дополнений и изменений к 'Флоре СССР' (*Svod dopolnenij i izmenenij k 'Flore SSSR'*) (тт. I–XXX)/*Additamenta et corrigenda ad 'Floram URSS'* (*vols. I–XXX*). 667 pp. Leningrad: 'Nauka'.[191]

Flora SSSR is a comprehensive descriptive treatment of the vascular plants of the former Soviet Union, including keys to all taxa, extensive synonymy (with references and citations), vernacular and Russian binomial names, fairly detailed accounts of distribution within and outside the former Union, taxonomic commentary, and notes on habitat, life-form, phenology, associates, paleobotany, uses, etc., and references to published illustrations; indices to all botanical and other names in each volume (and general index to complete work in a separate, concluding volume). Volume 1 includes a brief historical account and an exposition (by Komarov) of the taxonomic principles employed, while the index volume includes maps showing floristic regions (the basis for individual distribution accounts). [*Zlaki SSSR* is a revision of volume 2 of the *Flora*, while *Svod dopolnenij i izmenenij k 'Flore SSSR'* is an alphabetically arranged list of additions and changes to *Flora SSSR* from 1934 to 1971 with synonymy, references, citations, and indication of types along with an extensive introduction (in English and Russian) and a bibliography (pp. 20–23).][192]

Woody plants (including trees)

The major contribution in this class remains the six-volume woody flora edited by Sokolov, which includes an extensive range of cultivated as well as spontaneous species. There is in addition a distributional atlas in three folios. A selection of students' keys and a popular guide are also given here.

Sokolov's work built upon a long tradition of dendrological research and publication in the Russian Empire and its successors. The earliest major work is the six-volume *Russkaja dendrologija* (1870–82) by Eduard Regel, covering 471 species; a revision (not completed) appeared in 1883–89. Many texts and practical manuals have appeared since; only that of Lapin (1966) is fully described here. For a synopsis (without keys) covering some 3000 native and introduced species, see A. A. KAČALOV, 1970 (1969). Деревья и кустарники: справочник (*Derev'ja i kustarniki: spravočnik*). 407 pp. Moscow: 'Lesnaja promyšlennost''; for a students' manual comprising keys to 187 more common and significant species, see S. I. VANIN, 1967. Определитель деревьев и кустарников (*Opredelitel' derev'ev i kustarnikov*). 2nd edn., 236 pp., illus. Moscow: 'Lesnaja promyšlennost''. (1st edn., 1956.)

LAPIN, P. I. (ed.); N. A. BORODINA *et al.* (comp.), 1966. Деревья и кустарники: СССР (*Derevja i kustarniki SSSR*). 637 pp., illus. (some col.), maps. Moscow: 'Mysl'.

Semi-popular illustrated manual to native and principal introduced woody plants of the then-USSR, divided for the respective groups into two major sections. The first section (pp. 210–540) includes descriptions of individual native species as well as notes on their internal and extralimital range, altitudinal zonation, special features, uses, potential, etc.; there are also short associated notes on closely related but less common species. The second section (pp. 543–606) provides a similar treatment for exotic species. The introductory chapters encompass general remarks, a definition of terms (pp. 7–24), a map of geobotanical regions, and a series of polyclaves based on regions and morphological characters (pp. 30–209), while at the end is a list of references (pp. 609–611) and an index to Russian names.

SOKOLOV, S. JA., 1949–62. Деревя и кустарники СССР (*Derevja i kustarniki SSSR*). 6 vols. Illus., maps. Moscow/Leningrad: AN SSSR Press. Accompanied by S. JA. SOKOLOV and O. A. SVJAZEVA, 1965. География древесных растений СССР (*Geografija drevesnykh rastenij SSSR*). 265 pp., map. Moscow/Leningrad: 'Nauka'. (Derevja i kustarniki SSSR 7.)

The main work is a comprehensive descriptive account of the native, naturalized, and more or less commonly cultivated trees and shrubs of the former USSR; included are keys to all taxa, extensive synonymy (without references), Russian binomial and vernacular names, fairly detailed indication of internal distribution (with some maps), summary of extra-Union distribution or country or region of origin, notes on habitat, phenology, biology, dendrological features, cultivation, hybrids, uses, etc., and indices to all botanical and other names in each volume (with a general index in vol. 6). The supplementary volume of 1965 is devoted to statistics, lists, and a summary and general analysis of the woody flora (with an extensive bibliography).

SOKOLOV, S. JA. *et al.* (eds.), 1977–86. Ареалы деревьев и кустарников СССР (*Arealy derev'ev i kustarnikov SSSR*)/*Areographia arborum fruticumque URSS*. 3 vols. (each with portfolio of maps). Leningrad: 'Nauka'.

The textual parts of this distributional atlas comprise analyses of the distribution of each species of tree, shrub and woody vine mapped (covering distribution, habitat, associates, ecology and biology and including specific references) while the distributions themselves are presented on loose maps in portfolios. Each textual part includes a bibliography and Russian and Latin name indices but there is no general index. The volumes respectively cover, on the Englerian system, families from Taxaceae to Aristolochiaceae, Polygonaceae to Rosaceae, and Leguminosae to Caprifoliaceae.[193]

II. Belarus', Moldova, Ukraine and Russia-in-Europe (CIS-in-Europe)

Vascular flora: 4450–4600 species, of which 4100–4300 native (Webb). The limits adopted here correspond to those used for 'Russia' in *Flora Europaea*, save for the transfer of the Baltic States to Region 67.

Flora Evropejskoj Rossii (1908–10) by B. A. Fedtschenko and A. F. Flerow, an illustrated work covering 3542 species, and other manuals – especially those of Schmal'hausen (1895–97) and Stankov and Taliev (1957) – are all well out of date given the lapse of time as well as taxonomic and nomenclatural changes in the wake of *Flora Europaea*. A new, detailed *Flora Evropejskoj časti SSSR* was begun in 1974 and is now some three-quarters or so completed, and an English translation has commenced publication.

FĚDOROV, AN. A., N. N. TZVELEV and D. GEL'TMAN (eds.), 1974– . Флора европейской части СССР (*Flora evropejskoj časti SSSR*)/*Flora partis europaeae URSS* (from vol. 9 Флора восточной Европы (*Flora vostočnoj Evropy*)/*Flora europae orientalis*). Vols. 1– . Illus., map, portrs. Leningrad: 'Nauka'. English edn.: AN. A. FĚDOROV *et al.*, 1999– . *Flora of Russia: the European part and bordering regions*. Vols. 1– . Rotterdam: Balkema.

Partially descriptive flora of vascular plants, with keys to all taxa, diagnoses (of families and genera), full synonymy (with references and principal citations), indication of internal range (by geobotanical districts) and extralimital distribution, localities, karyotypes, taxonomic commentary, and notes on habitat, special features, etc. Diagnoses of subspecies are provided, and indices to all Latin and Russian names appear at the end of each volume. An introductory section (in volume 1) includes historical and philosophical chapters, a synopsis of orders and families, the plan and scope of the work, a map of floristic districts, a bibliography of key works (pp. 22–27) and general key to families. [By 1999 volumes 1–9 (of a planned 11) had been published, covering Pteridophyta, Gymnospermae, all the Monocotyledoneae and various blocks of Dicotyledoneae. Although families have been pre-numbered with respect to the 1969 version of the Takhtajan system, actual publication, as in *Conspectus florae asiae mediae* (**750**) has taken place as families or blocks of them are ready. Of the English version, as of 1999 the first two volumes had been published; two more were expected by 2000.][194]

STANKOV, S. S. and V. I. TALIEV, 1957. Определитель высших растений Еворлейской части СССР (*Opredelitel' vysšikh rastenij Evropejskoj časti SSSR*). 2nd edn., 740 pp., 645 text-figs. Moscow: 'Sovetskaja Nauka'. (1st edn., 1949 (1948). Based on V. I. TALIEV, Определитель высших растений Европейской части России (*Opredelitel' vysšikh rastenij Evropejskoj časti Rossii*): 1st edn., 1907, 9th edn., 1941.)

Manual-key to the vascular plants of the European part of the former USSR, with limited synonymy, vernacular and binomial Russian names, very concise indication of internal distribution, figures of representative species, and notes on habitat, life-form, phenology, etc.; indices to all botanical and other names at end. The introductory section includes an illustrated organography/glossary, a synopsis of the system adopted, and a list of references. [4473 species were included in the 1948 edition, and 5090 in its successor. The original version covered a smaller area (omitting the Carpathians, Crimea and far northern regions).][195]

Woody plants (including trees)

A goodly number of works exist for different parts of eastern Europe; those before 1956 are documented by Lebedev (see **Region 68/69, Regional bibliographies**). No recent dendroflora, however, covers the whole area; reference should be made to 'All-Union' works (see above).

(Sostav, rasprostranenie po rajonam i khozjajstvennoe značenie flory Kaliningradskoj oblasti). *Trudy Bot. Inst. AN SSSR*, III (Geobotanika), **10**: 225–329.

Includes a concise systematic enumeration of vascular plants, with notes on local distribution (keyed to the map on p. 239) and habitat, life-form, habit, phenology, special features, potential, etc.; list of non-vascular plants; bibliography. The introduction includes a map and itinerary details, while the general index to the volume (pp. 373–392) incorporates all Latin names.[196]

STEFFEN, H., 1940. *Flora von Ostpreussen*. iii, 319 pp. Königsberg: Gräfe und Unzer.

Concise field-manual of vascular plants, with analytical keys, brief descriptions, German vernacular names, and partly symbolized notes on habitat, life-form, phenology, and extralimital range; local distribution usually not specifically indicated. An introductory section includes remarks on plant geography and vegetation.

Woody plants

BITSE, M. A., D. A. KNAPE, G. G. KUCHENEVA *et al.* (comp.); R. E. TSINOVSKIS (otv. red.), 1983. *Konspekt dendroflory Kaliningradskoj oblasti*. 161, [1] pp., 15 text-figs. (maps). Riga: 'Zinatne' (for AN Latv. SSR, Bot. Sad.).

Checklist of woody plants with detailed indication of distribution and sizes encountered (explanation, p. 5; districts and sites on map, p. 6). A discussion of forest areas with notable features (pp. 12–19) precedes the checklist proper, and a gazetteer, literature list and table of contents (but no index) appear at the end. [Builds on earlier surveys including those in former East Prussia back to 1922.]

681

Kaliningrad Oblast' (Russia)

Vascular flora: *c.* 1200 species (Pobedimova, 1955). – Kaliningrad Oblast' represents the northern half of the former German province of East Prussia, and includes Kaliningrad (Königsberg or Królewiec), a Russian naval base. The remainder of East Prussia (Mazuria) is part of Poland (**647**). The area is also covered by some works listed under **640/I** ('Mitteleuropa') and **648** (Germany).

POBEDIMOVA, E. G., 1955 (1956). Состав, распространение по районам и хозяйственное значение флоры Каилининградской области

684

Belarus' (White Russia)

Area: 208000 km². Vascular flora: 1841 species (Shishkin *et al.*, 1948–59; see below). The country west of a line running some way west of Minsk (and including Grodno and Brest-Litovsk) was from 1921 to 1939 part of Poland (**647**) and so is covered by Polish works of the period. Its present limits date from 1945. Save for some nine months in 1918–19, the country has been independent only since 1991; earlier, it was variously under the rule of Lithuania, Poland-Lithuania, Russia or the Soviet Union (and between World Wars I and II divided between Poland and the Soviet Union).

Early coverage of Belarus', or historical 'White Ruthenia', is furnished by works covering 'greater Lithuania', among them *Flora Lithuanica inchoata* by Jan Gilibert (at Grodno from 1775 to 1781), the works of the Jundziłłs (see **679**), and *Naturhistorische Skizze von Litthauen, Volhynien und Podolien in geognostisch-mineralogischer, botanischer und zoologischer Hinsicht* (1830) by Eduard Eichwald. The last, in its 'Botanische Bemerkungen', included a list of 1961 vascular plant species.[197] From 1831 until late in the century, with no universities and few other centers of learning, botanical studies developed only slowly and major floristic publications were few. Eduard E. Lindemann's *Prodromus florarum Tschernigovianae, Mokilovianae, Minskianae nec non Grodnovianae* (1850) is the main landmark. In the 1890s came further consolidation: Jozef K. Paczoski's *Flora Poles'ja i prileža\u0161\u010dikh mestnostej* (1897–1900) in the south and, indirectly, *Flora von Polnisch-Livland* (1895–96; cf. **678**) by Eduard Lehmann for the north. More intensive studies commenced after organization of the Belarusian (later Soviet) republic in 1918. A central herbarium was established in 1923 in Minsk, previously without any systematics resources, and in 1948 the definitive *Flora BSSR* commenced publication. The first standard manual dates from 1967.[198]

Bibliography

See also *Flora Belorussii* (1978) by N. V. Kozlovskaya (Note 198).

OSIPCHIK, L. A. and V. S. GEL'TMAN, 1970. Советская литература по флоре и растительности Белоруссии. Библиография 1919–1968 (*Sovetskaja literatura po flore i rastitel'nosti Belorussii. Bibliografia 1919–1968*). 433 pp. Minsk. [3800 entries, arranged according to a relatively detailed classification with author and geographical indices. Continued in annual supplements at least through 1975.]

SHISHKIN, B. K. and N. A. DOROŽKIN (eds.), 1948 (1949)–59. Флора БССР (*Flora BSSR*). 5 vols. Illus. Moscow: Sel'khozgiz (vol. 1); Minsk: AN BSSR Press (vols. 2–5).

Illustrated descriptive flora of vascular plants; includes keys to all taxa, brief indication of synonymy (with references), vernacular and binomial Russian names; generalized summary of local and extralimital range, and partly symbolized notes on habitat, life-form, phenology, etc.; complete indices in each volume (with general index to family and generic names as well

as a list of principal references in vol. 5). The introductory section in vol. 1 contains a historical account of floristic work in the area. [Shishkin was general editor for vols. 1, 4 and 5; Dorožkin for vols. 2 and 3.]

SHISHKIN, B. K., M. P. TOMIN and M. N. GONČARIK (eds.), 1967. Определитель растений Белоруссии (*Opredelitel' rastenij Belorussii*). 871 pp., 325 figs., maps. Minsk: 'Vyšejšaja škola'.

Manual-key to vascular plants; includes limited synonymy, scientific and standardized Russian (and also, for genera, Belarusian) vernacular names, concise indication of local range (by botanical districts), and notes on habitat, life-form, phenology, etc.; indices to all botanical and other names at end. The introductory section includes an organography/glossary. Figures of representative species appear throughout the text, and distribution maps of the botanical districts and a range of species are given on pp. 790–800, before the index.

685

Greater St. Petersburg region (Russia)

See also **678** (HIITONEN, 1933, 1934). Vascular flora: no data. – Part of this area, which includes the former Ingria, was for a time administered separately from Middle Russia; at present, it corresponds to the area covered by St. Petersburgskaja, Pskovskaja and Novgorodskaja *oblasti*. It is outside the area covered by floras of Middle Russia (**692**).

The considerable literature, effectively originating with *Flora ingrica* (1761) by the Dutch botanist David de Gorter (using notes left by Stephan Krascheninnikow) and passing through several successive handbooks, has been digested in the manual by Minjaev, Orlova and Shmidt described below. Mention may be made here, however, of the most recent of the earlier floras: B. K. SHISHKIN, 1955–65. Флора Ленинградской области (*Flora Leningradskoj oblasti*). 4 vols. Leningrad; and N. A. MINJAEV, V. M. SHMIDT and M. V. SOKOLOV, 1970. Конспект флоры Псковской области (*Konspekt flory Pskovskoj oblasti*). Leningrad.

MINJAEV, N. A., N. I. ORLOVA and V. M. SHMIDT, 1981. Определител' высших растений

Северо-Запада Европейской части РСФСР (Ленинградская, Псковская и Новгородская области) (*Opredelitel' vysšikh rastenij Severo-Zapada Evropejskoj časti RSFSR (Leningradskaja, Pskovskaja i Novgorodskaja oblasti)*). 376 pp., map. Leningrad: Leningrad State University [Press].

Manual-key to vascular plants, with limited synonymy, vernacular names, and indication of local range, habitat, phenology, habit, biology, properties, uses, etc.; explanation of floristic regions (with map) and indices at end. An introductory section includes sources, consideration of floristic regions, and a general key to families.

686

Karelian Autonomous Republic (Russia)

See also **676** (HIITONEN, 1933, 1934); **687** (RAMENSKAYA, 1983; RAMENSKAYA and ANDREEVA, 1982); **688** (PERFIL'EV). – Vascular flora (including the Kola Peninsula): 1317 species, of which 1052 native (Ramenskaya, 1983; see **687**).

Plants flowering before 15 June are also covered in V. N. CHERNOV, 1955. Весенняя флора Карело-Финской ССР: определитель весенних растений (*Vesennjaja flora Karelo-Finskoj SSR: opredelitel' vesennikh rastenij*). 154 pp., 90 text-figs. Petrozavodsk. [Includes glossary and bibliography.]

RAMENSKAYA, M. L., 1960. Определител' высших растений Карелии (*Opredelitel' vysšikh rastenij Karelii*). 485 pp., 94 text-figs. (incl. 2 maps), 12 col. pls. Petrozavodsk: Karel'skoj ASSR Government Publishing Service.

Manual-key to vascular plants, with limited synonymy, generalized indication of local range, vernacular and Russian binomial names, and notes on habitat, phenology, biology, etc.; figures of diagnostic features; index to all botanical and other names. The introductory section includes a brief account of vegetation associations and floristic regions.

687

Kola Peninsula

See also **676** (HIITONEN, 1933, 1934); **688** (PERFIL'EV). The area corresponds approximately to that of Murmanskaja Oblast'. The Arctic zone, which includes the isolated mountain of Khibiny Gora, has also been designated as **062**.

In addition to the works described below, note should also be taken of the following which includes a revised checklist for both the Kola Peninsula and Karelia (and list of references): M. L. RAMENSKAYA, 1983. Анализ флоры Мурманской области и Карелии (*Analiz flory Murmanskoj oblasti i Karelii*). 213, [i] pp., maps. Leningrad: 'Nauka'.

GORODKOV, B. N. and A. J. POJARKOVA (eds.), 1953–66. Флора Мурманской области (*Flora Murmanskoj oblasti*). 5 vols. Illus., maps. Moscow/Leningrad: AN SSSR Press (later 'Nauka').

Illustrated descriptive flora of vascular plants; includes keys to all taxa, synonymy (with references and citations), vernacular and Russian binomial names, generalized indication of local and extralimital range (with dot maps for each species), taxonomic commentary, and notes on habitat, phenology, associates, etc.; addenda and corrigenda (vol. 5); indices in each volume to botanical and other names. The introductory section (vol. 1) includes a short list of important floristic references.

RAMENSKAYA, M. L. and V. N. ANDREEVA, 1982. Определитель высших растений Мурманской области и Карелии (*Opredelitel' vysšikh rastenij Murmanskoj oblasti i Karelii*). 432 pp., 82 text-figs., folding map. Leningrad: 'Nauka'.

A manual-key to vascular plants based on the *Flora* of 1953–66 and the Karelian manual of 1960; includes essential synonymy and indication of habit, height and phenology in the leads followed by notes on habitat, distribution and frequency as well as (where necessary) on identification and taxonomy; lexicon of name changes, synopsis and indices at end. The introductory part includes a description of the area and its features (cf. map on p. 5) and a general key to families (pp. 8–26).

688

Northern Russia

Vascular flora (including **686** and **687**): 1300–1450 species, of which 1250–1450 native (Webb). This area represents a combination of Arkhangel'skaja Oblast' of Russia (except for the tundra zone; see **063**) and the Komi Autonomous Republic. For Vologda Oblast', see **689**.

Bibliography (Komi Republic)

BOLOTOVA, V. M., 1957. Флора и растительность Коми АССР (1814–1955 гг.): библиографический указатель (*Flora i rastitel'nost' Komi ASSR (1814–1955 gg): bibliografičeskij ukazatel'.*) 179 pp. Moscow/Syktyvkar. [534 entries.]

PERFIL'EV, I. A., 1934–36. Флора Северного Края (*Flora Severnogo Kraja*). Vyp. 1, 2–3 (in 2 vols.). 160, 407 pp.; 111 text-figs. (in vol. 2). Archangel: Severnoj Kraj Press.

Partially illustrated manual-key to vascular plants; includes essential synonymy, Russian vernacular and binomial names, fairly copious diagnoses, and concise notes on local range, habitat, life-form and phenology; list of references and indices to Russian and botanical names of genera and families at end of volume 2. No extralimital distribution is indicated. The second volume also includes a list of new taxa discovered in the area from 1500 onwards. Geographical coverage extends from the islands in the Arctic Ocean to Nizhni Novgorod and Ivanovsk *oblasti* in the south and from Karelia to the Ural. [To a considerable extent superseded by Tolmatchev's 1974–77 flora, described below.][199]

SNYATKOV, A. A., G. I. ŠIRJAEV and I. A. PERFIL'EV, 1922. Определитель растений лесной полосы Северо-Востока Европейской части России (*Opredelitel' rastenij lesnoj polosy Severo-Vostoka Evropejskoj časti Rossii*). 2nd edn., revised and ed. by I. A. Perfil'ev. 215 pp., illus. Vologda: State Press, Vologda Oblast' Branch. (1st edn., 1913.)

Manual-key to vascular plants of the forest region of northern and northeastern Russia; includes limited synonymy, vernacular and Russian binomial names, concise indication of local range, and notes on habitat, life-form, phenology, etc.; indices to generic, family and Russian names at end. The introductory section is mostly pedagogical. [Although superseded for the whole of **688** by later works, it has been included here for the record and for its coverage of **689** and the northern half of **699**, i.e., down to the southern limits of the zone of potential continuous forest vegetation.]

TOLMATCHEV, A. I. (ed.), 1974–77. Флора Северо-Востока Европейской части СССР (*Flora Severo-Vostoka Evropejskoj časti SSSR*) / *Flora regionis boreali-orientalis territoriae europaeae URSS.* 4 vols. Maps. Leningrad: 'Nauka'.

Non-descriptive flora of vascular plants; includes analytical keys, full synonymy (with references and relevant citations), Russian binomial names, indication of localities (with citations) and summary of internal and extralimital distribution, and notes on phenology, special features, etc. At the end of each volume are dot maps for all species and subspecies as well as indices for all Komi, Russian and Latin names; there is, however, no consolidated index. [Covers the entire area as defined above, with the addition of the tundra zone to the north (**063**); expands on the editor's *Opredelitel' vysšikh rastenij Komi ASSR* (1962; see below).]

Komi Republic

TOLMATCHEV, A. I. (ed.), 1962. Определитель высших растений Коми АССР (*Opredelitel' vysšikh rastenij Komi ASSR*). Moscow/Leningrad: AN SSSR Press.

Manual-key to vascular plants; includes essential synonymy, Russian binomials and Russian and Komi vernacular names, habit, seasons and notes on local range, habitat, phenology, etc.; indices to all botanical and other names. A brief introduction, with keys to families, precedes the main text.

689

Northern Middle Russia

See **688** (PERFIL'EV; SNYATKOV *et al.*). The area delimited here centers on the present-day Vologda Oblast' but additionally includes the northeastern part of Kostroma Oblast' and the northern lobe of Kirov Oblast' (both formerly included in Vologda Oblast'). Although wholly included within the northern forest zone accounted for in the earlier subregional floras, it is *not* encompassed by Tolmatchev's recent *Flora Severo-Vostoka Evropejskoj časti SSSR*. An early but unfinished work was *Catalogue des plantes croissant dans les Gouvernments de Vologda et d'Archangel: Monopétales et Apétales* by N. Ivanitzky (1894, in installments in *Le monde des plantes* 3–4, *passim*). In the early twentieth century, Perfil'ev, author of *Flora Severnogo Kraja* (see 688), also focused on Vologda.

691

Middle Volga Basin

This area, centering on the course of the Volga from Kazan' to Samara, encompasses from north to south Marij El, the Chuvash Republic, Tatarstan west of the Zaj and Vyatka Rivers, and the Ul'yanov and Samara *oblasti*. Included are the cities of Kazan' (a botanical center from 1805), Ul'yanovsk and Samara (Kujbysev), all on the Volga. It lies between Middle Russia (692), Southeastern Russia (698) and Eastern Russia (699) and falls mostly within the mixed forest-and-steppe (prairie) vegetation formation.[200]

There are students' manuals respectively centered on Ul'yanovsk and Samara, and a manual and bibliography for Tatarstan. Additional coverage may be sought in works covering adjacent units as indicated above.[201]

Bibliography

VLASOV, I. V., A. S. KAZANTSEVA, V. V. TUGANAEV and A. KH. KHUSAINOVA, 1971. Флора и растительность Татарской АССР. Указатель литературы XVIII в.-1967 г. (*Flora i rastitel'nost' Tatarskoj ASSR. Ukazatel' literatury XVIII v.-1967 g*). 120 pp. Kazan': Kazan' University. [1425 scientific and popular entries, in a systematic arrangement with indices.]

Partial works

BLAGOVESHCHENSKIJ, V. V. *et al.*, 1984. Определитель растений Среднего Поволжья (*Opredelitel' rastenij Srednego Povolž'ja*). 392 pp. Leningrad: 'Nauka'.

Manual-key to vascular plants (1366 species) in Ul'yanov Oblast', with limited synonymy, Russian names, local distribution, and brief indication of habit, phenology, habit, etc.; indices. Cross-references to Majevski's manual (692) are given.

MARKOV, M. V., 1979. Определитель растений Татарской АССР (*Opredelitel' rastenij Tatarskoj ASSR*). 371 pp., 37 text-figs., map (inside front cover), glossary figures (other end-papers). Kazan': Izd. Kazansk. Univ.

Illustrated manual-key to native, naturalized and commonly cultivated vascular plants with indication of habit, phenology, habitat, distribution by districts (map in inside front cover), and properties and uses; indices to botanical and Russian names of families and genera at end. The general part includes a preface, a brief chapter on the botanical-geographical districts (nine recognized), and notes on how to use the work along with abbreviations (p. 7); these are followed by the key to families. The glossary, with illustrations, is wholly within the inside covers. [The author notes in his preface that Majevski's manual (692) did not cover all of the republic, only the portion west of the Volga.]

TEREKHOV, A. F., 1969. Определитель весенних и осенних растений Среднего Поволжья и Заволжья (*Opredelitel' vesennikh i osennikh rastenij Srednego Polvolž'ja i Zavolž'ja*). 3rd edn. 464 pp., 100 text-figs., 4 pls. Kujbyšev (= Samara): Kniz. Izd. (1st edn., 1939; 2nd edn., 1948. Based on a combination of the author's *Opredelitel' vesennikh rastenij Srednego Polvolz'ja* (1930, Moscow/Samara; 2nd edn., 1932) and *Opredelitel' ossenikh rastenij Srednego Polvolz'ja* (1931, Samara).)

Illustrated students' manual-key to the spring and autumn flora (about 900 species) of the central Volga Basin and areas eastward, with brief notes on phenology, habitat, local range, habit and life-form, etc.; glossary and index. [The second and third editions are evidently mere reissues of the 1939 original, without revision.]

692

Middle Russia

Vascular flora: *c.* 2500 species (Gubanov *et al.*, 1981; see Gubanov *et al.*, 1995, below). This area – the Russian heartland – is bounded on the west and south by Belarus' and Ukraine, on the north and east by the Volga Valley (see also 691) and the Volga-Don Canal, and on the northwest by the Greater St. Petersburg region (685). Southeastern Russia (698), including the Saratov region, forms the remainder of the boundary.

The literature on Middle Russia is extensive. In addition to more than 130 years of general manuals (of which the recent leaders have been those of Petr Majevski, with nine editions (10 if the reissue of the seventh edition is counted) from 1892 to 1964, and of I. A. Gubanov *et al.* from 1981), several of its individual administrative units are covered by separate works. Of these latter, only the greater Moscow region is accounted for here. A historically important work is

Флора средней и южной России, Крима и Северного Кавказа (*Flora srednej i južnoj Rossii, Krima i Severnogo Kavkaza*) by I. F. Schmal'hausen (1895–97). Research current in the 1980s was summarized in a long series of contributions in V. N. Тікномікоv (ed.), 1984. *Sostojanie i perspektivy issledovanija flory srednej polosy Evropejskoj časti SSSR.* 90 pp. Moscow: Moscow Society of Naturalists.

Bibliography

The work listed below pertains effectively to the whole of Middle Russia. Other bibliographies are also available for some individual administrative units; details are given by LEBEDEV (see **Region 68/69, Regional bibliographies**) and in Путеводитель для биологов по библиографическим изданиям (*Putevoditel' dlja biologov po bibliografičeskim izdanijam*) (1978; see Appendix A).

TIKHOMIROV, V. N., I. A. GUBANOV, I. M. KALINIČENKO and R. A. LOZAR', and 1998. Флора средней России: аннотированная библиография (*Flora srednej Rossii: annotirovannaja bibliografija*). 199, [i] pp. Moscow: «Russkij Universitet». [Concisely annotated bibliography of 3627 items; author and regional indices at end, with summaries in English and German. Covers literature from the eighteenth century through 1997 (with some 1998 works also featured). The bibliography proper is preceded by a historical review, with for each region bibliohistoric surveys (these are alphabetically arranged). Works pertaining to the former Soviet Union, Russia, and Russia-in-Europe in general are also accounted for. – A thorough piece of work.]

GUBANOV, I. A., K. V. KISELEVA, V. S. NOVIKOV and V. N. TIKHOMIROV, 1995. Определитель сосудистых растений центра Европейской России (*Opredelitel' sosudistykh rastenij tsentra Evropejskoj Rossii*). Revised edn. 558, [2] pp., 335 text-figs. Moscow: 'Argus'. (Originally published 1981, Moscow: 'Prosveščenie', as Определитель высших растений средней пологы Европейской части СССР (*Opredelitel' vysšikh rastenij srednej polosy Evropejskoj časti SSSR*) under the authorship of I. A. Gubanov, V. S. Novikov and V. N. Tikhomirov; 1st edn. by current authorship and under present title, 1991 (1992), Moscow: Moscow State University Press.)

Manual-key to vascular plants, with essential synonymy, Russian nomenclature, selective illustrations, and indication of habit, season, and general indication of local distribution; index to family and generic names. Mention is made of some species not included in the keys. The brief introduction, inclusive of the circumscription of the area covered, is followed by the key to families (pp. 6–49).[202]

MAJEVSKI, P. F., 1964. Флора средней полосы Европейской части СССР (*Flora srednej polosy Evropejskoj časti SSSR*). 9th edn., revised and ed. by B. K. Shishkin. 880 pp., 325 text-figs. Leningrad: 'Kolos'. (1st edn., 1892, Moscow, as Флора средней России (*Flora srednej Rossii*).)

Manual-key to vascular plants, with limited synonymy, Russian vernacular and binomial names, concise indication of local range, and partly symbolized notes on habitat, phenology, life-form, associates, uses, etc.; small text-figures of diagnostic features; list of authors and complete indices to botanical and vernacular names. [Succeeded for practical use by *Opredelitel' vysšikh rastenij srednej polosy SSSR* (1981) and *Opredelitel' sosudistykh rastenij tsentra Evropejskoj Rossii* (1992, 1995; see above).]

Partial work: Moscow region

In addition to the standard manuals for Middle Russia, key works are: D. P. SYREISHCHIKOV, 1906–14. Иллюстрированная флора Московской губернии (*Illjustrirovannaja flora Moskovskoj gubernii*). 4 parts. Illus. Moscow: Lakhtin and Syreishchikov; and V. N. VOROSHILOV, A. K. SKVORTSOV and V. N. TIKHOMIROV, 1966. Определитель растечий Московской области (*Opredelitel' rastenij Moskovskoj oblasti*). 367 pp., 313 text-figs., 1 map (in end-papers). Moscow: 'Nauka'. [The latter work was a precursor of the 1981 *Opredelitel' vysšikh rastenij srednej polosy Evropejskoj časti SSSR* (see above).]

693

Moldova (Moldavia)

Area: 33 700 km^2. Vascular flora: 1650 species (Geideman, 1954; see below). Also known as Bessarabia, Moldova east of the Prut fell to Russia as a consequence of settlements with the Ottoman Empire in the nineteenth century. Between the two world wars, however, the territory was part of Romania. From 1945 until 1991, shorn of the Gagauz region bordering the Black Sea (transferred to Ukraine) but with the addition of the former 'New Russian' strip known as Transdnestria (and its capital, Tiraspol), it was a Soviet republic. In 1990, the country adopted Roman orthography.

The Moldovan flora was first accounted for by Besser in his *Enumeratio* (see **694**) but afterwards more specifically in *Catalogue des plantes qui croissent naturellement en Bessarabie et aux environs d'Odessa* (1842; in *Zapiski Imp. O-va Sel'skogo Khozjastva Južnoj Rossii* 1: 134–177). This was followed by *Očerk flory Khersonskoj gubernii* (1872) and *Flora Chersonensis* (1881–82) by Eduard Lindemann. These latter works represented a tradition of botanical exploration and study based in particular in Odessa and elsewhere in 'New Russia' (southern Ukraine) which continued until World War I. Afterwards, Romanians, notably Traian Săvulescu and Tscharna Rayss (see **641**), became the chief contributors to floristic knowledge. The Moldavian Academy of Sciences herbarium, with which was connected Tatiana Geideman, author of the standard national flora, was first organized in 1948.

In addition to Geideman's manuals, Romanian works can be used for identification.

Bibliography

L'VINA, I. N. and G. E. KUŠNIRENKO, 1970. Флора и растительность Молдавии: библиография 1715–1965 гг. (*Flora i rastitel'nost' Moldavii: bibliografija 1715–1965 gg.*). 206 pp. Kishinev. [1698 mostly unannotated entries, arranged into broad subject areas with author indices at end. Brief introduction.]

GEIDEMAN, T. S., 1986. Определитель ысших растений Молдавской ССР (*Opredelitel' vysšikh rastenij Moldavskoj SSR*). 3rd edn. 637 pp., 124 text-figs. (incl. map). Kishinev: 'Štiintsa'. (1st edn., 1954; 2nd edn., 1975.)

Illustrated manual-key to native, introduced and cultivated (the last-named asterisked) vascular plants; includes very limited synonymy, short diagnoses, vernacular names (in Russian and Moldovan), and partly symbolized notes on local distribution, habitat, lifeform, phenology, special features, etc.; bibliography, glossary, and indices to all Russian, Moldovan and Latin names at end. The 10 floristic regions to which local distribution is related are defined on the map.

Woody plants

Two works [neither seen] have appeared since 1945, respectively a detailed descriptive account and a concise identification manual (the latter covering 140 species): V. N. ANDREJEV, 1957–68. Деревья и кустарники Молдавии (*Derevja i kustarniki Moldavii*). 3 vols. (vol. 3 without author's name but edited by T. S. Geideman). Illus. Moscow: AN SSSR Press (vol. 1); Kishinev: 'Kartja Moldovenjaske' (vols. 2–3); and T. S. GEIDEMAN, 1965. Краткий определитель дикорастущих деревьев и кустарников Молдавской ССР (*Kratkij opredelitel' dikorastuščikh derev'ev i kustarnikov Moldavskoj SSR*). 101 pp., illus. Kishinev: 'Kartja Moldovenjaske'.

694

Ukraine (in general)

For Southwestern Ukraine see also **695**. – Area: 603 700 km². Vascular flora: 4997 species covered by Prokudin (1987) and 3669 in *Flora URSR* (for both, see below; the latter does not altogether account for the present-day national territory). Webb's corresponding unit (inclusive of **693** and part of **684** but *not* **696**) encompassed 3100–3250 species, of which 2900–3100 were native. Much of western Ukraine in particular eastern Galicia (Rus or Red Ruthenia), the Carpatho-Ukraine, and Bukovina – was before 1918 part of the Austro-Hungarian Empire. Between 1919 and 1939 eastern Galicia and most of adjacent Wolhynia was part of Poland, while Bukovina was under Romanian rule and the Carpatho-Ukraine – a component of 'historical Hungary' – formed part of Czechoslovakia and then, from 1939 until shortly after World War II, was Hungarian. Since 1954 Ukraine has also encompassed the Crimea (**696**).

The considerable eighteenth- and nineteenth-century literature on central and eastern Ukraine, among it *Enumeratio plantarum hucusque in Volhynia, Podolia, gub. Kiioviensi, Bessarabia cis-tyracia et circa Odessam collectarum* (1822) by Wilibald S. von Besser and *Flora Ukrainy* (1869, posthumously) by Besser's associate Anton Andrzejowski, along with several floras of more restricted scope, was digested by Ivan F. Schmal'hausen in the late nineteenth century, firstly in *Flora jugo-zapadnoj Rossii* (1886) and then in his two-volume, briefly descriptive *Flora srednej i južnoj Rossii, Kryma i Severnogo Kavkaza* (1895–97; with a list of references). Austrian western Ukraine was initially treated by Besser in *Primitiae florae Galiciae austriacae utriusque* (1809) and later by Joseph Knapp in *Die bisher bekannten Pflanzen Galiziens und der Bukowina* (1872) and Hugo Zapałowicz in *Krytyczny przegląd roślinności Galicyi/Conspectus florae Galiciae criticus* (1904–14).

The Ukrainian Academy of Sciences came into being in 1918 and organized its Botanical Institute in 1921. Alexander V. Fomin, long active in the Caucasus, began a *Flora Ukrajiny* (in Ukrainian) in 1926, but it was not continued beyond its first fascicle due to policy changes as well as the author's increasing Academy commitments. In the 1930s two directions towards a national flora were pursued. The first was *Flora USRR: vyznačnyk kvitkovykh ta vyščykh sporovykh roslyn USRR* (1935), a manual edited by N. Ljaskivs'ky, E. M. Lavrenko and P. O. Opperman. Three volumes were planned but only the first, on pteridophytes, gymnosperms and monocotyledons, was published. The second became *Flora URSR*. Fomin was foundation editor but after his death in 1935 he was succeeded by Evgeny Bordzilowski (until 1949), Mikhail Kotov and others. *Flora USRR* would later be realized, albeit in a more condensed form, as M. V. Klokov, *Vyznačnyk roslyn URSR* (1950); a second edition appeared in 1965 as *Vyznačnyk roslyn Ukrajiny* under the principal editorship of D. K. Zerov and a Russian-language successor, *Opredelitel' vysšikh rastenij Ukrainy*, was published in 1987 (see below). A new checklist appeared in 1999 (see Note 204).[203]

Bibliographies

In addition to the two works by M. G. Mikhajlova, A. I. Barbarycz produced annualized lists irregularly from 1956 to 1967 in *Ukrajinskij botaničnyj žurnal*, N.S. 13–24, covering the years 1954–64.

BOURDEILLE, COMTE DE MONTRÉSOR, 1892–1900. Les sources de la flore des provinces qui entrent dans la composition de l'arrondissement scolaire de Kieff. *Byull. Mosk. Obšč. Ispytat. Prirod.* (*Bull. Soc. Nat. Moscou*), N.S. **6**: 322–381; **7**: 420–496; **14**: 485–505. (First two installments reprinted 1894, 136 pp.; third installment, 23 pp., 1900 (including 2 pp. of corrigenda).) [624 works in all, arranged by authors and well annotated with many citations of plant records. Locations of copies seen are also given (with explanation on p. 2 of reprint).]

MIKHAJLOVA (MICHAILOFF), M. G., 1938. Флора та рослинність УРСР: бібліографія (*Flora ta roslynnist' URSR: bibliografija*). 63 pp. Kiev (Kyjiv): AN URSR. Continued as *idem*, 1941. Флора та рослинність УРСР: бібліографія. Допов. 1 (*Flora ta roslynnist' URSR: bibliografija. Dopov.* 1). 41 pp. Kiev. [Altogether almost 2600 entries of books and journal articles, 1747–1938; arrangement by authors (those in Cyrillic scripts first). No indices.]

BORDZILOWSKI, E. I. *et al.* (eds.), 1938–65. *Flora URSR/Flora RSS Ucr.* 12 vols. Illus., 2 maps. Kiev (Kyjiv): AN URSR Press (vols. 1–11); 'Naukova dumka' (vol. 12). (A first edition of vol. 1 by A. V. Fomin as general editor, dated 1936, appeared in 1937.) Poaceae (Gramineae) revised as YU. N. PROKUDIN *et al.*, 1977. Злаки Украины (*Zlaki Ukrainy*). 518 pp., 74 figs., 165 maps. Kiev: 'Naukova dumka'.

The *Flora* is a comprehensive illustrated descriptive treatment (in Ukrainian) of vascular plants; it includes keys to all taxa (with a general key to families in vol. 3), synonymy with references and citations (together with illustration references), Ukrainian binomial and vernacular names, extensive indication of Ukrainian and Union-wide distribution with citation of some *exsiccatae*, summary of range outside the then-USSR, taxonomic commentary, and notes on habitat, life-form, phenology, associates, etc.; Latin diagnoses of new taxa and complete indices to botanical and other names at end of each volume. The introductory section (vol. 1) includes an account, with maps, of the floristic regions of Ukraine. No comprehensive index was published. The 1977 work by Prokudin on grasses succeeds the by-then out-of-date account in the *Flora*.

PROKUDIN, YU. N., 1987. Определитель высших растений Украины (*Opredelitel' vysšikh rastenij Ukrainy*). 546, [2] pp. 570 text-figs. Kiev: 'Naukova dumka'.

Manual-key (in Russian) to native, naturalized and commonly cultivated vascular plants; includes essential synonymy, Russian and Ukrainian binomials, habit and dimensions, seasons, and notes on habitat, distribution, and uses and properties along with figures of selected species; list of references (p. 471), individual credits (p. 545) and complete indices to Russian, Ukrainian and scientific names. The introductory part includes an analysis of the flora and its features as well as a discussion of the arrangement (the Takhtajan system having been adopted).[204]

Woody plants

The still-current major manual (in Russian) by Lypa is a more concise version of the same author's Ukrainian-language Дендрофлора УРСР (*Dendroflora URSR*) (1939), of which only one part, on gymnosperms, was ever published. A good, relatively recent illustrated Russian-language popular guide for western Ukraine, covering native and cultivated species, is T. M. and M. M. BRODOVICH, 1979. Деревья и кустарники Запада УССР: Атлас (*Derev'ja i kustarniki Zapada UkSSR: Atlas*). 251 pp., 116 pls. (part col.). L'vov:

'Višča Škola' (for L'vov State University). (First published in Ukrainian, 1973, as Атлас дерев та кущів заходу України (*Atlas derev ta kuščiv zakhodu Ukrajiny*).)

Lypa, A. L., 1955–57. Определитель деревьев и кустарников (дикорастущих и культивируемых) в Ук.ССР (*Opredelitel' derev'ev i kustarnikov (dikorastuščikh i kul'tiviruemykh) v Uk. SSR*). 2 vols. Illus. Kiev: Kiev State University Press.

Descriptive account of wild and cultivated Ukrainian trees and shrubs; includes keys, synonymy, Russian binomial equivalents and vernacular names, illustrations, and extensive notes on local and extralimital distribution, habitat, phenology, variation, cultivation, etc.; indices to all botanical and other names at end. [Not seen; description based on information supplied by A. O. Chater, formerly of London.]

695

Southwestern Ukraine (including Ruthenia)

See also **603/VII** and **694** (all works). – Area (Carpathians): 37000 km². Vascular flora (including native, naturalized, adventive and commonly cultivated plants): 2012 species (Čopyk *et al.*, 1977; see below). Southwestern Ukraine is here defined as that part of western Ukraine southwest of the upper Dniester and Prut Rivers; it encompasses the Ukrainian Carpathians and includes Bukovina and much of 'Red' Ruthenia. The part lying south of the mountain divide is the area once called Carpathian Ruthenia or Carpatho-Ukraine but now comprising Zakarpats'ka Oblast'. Until 1918 this latter was part of Hungary, while the remainder was part of Austrian Galicia.

As with western Ukraine in general there is a substantial floristic literature, reflecting the presence of two long-established centers of learning (L'viv, just without the area, and Chernivtsi, within) and a third, more recent, at the Ruthenian 'capital', Uzhhorod.[205]

Bibliography

For bibliographical coverage in addition to that in Popov (1949) and other works listed here, see **641** (Borza and Pop), **642** (Gombocz), **647** (Szymkiewicz) and possibly also **645** (Futak and Domin); the works of Bourdeille and Mikhijlova given under **694** do not cover this area.

Čopyk, V. J., M. J. Kotov and V. V. Protopopova (eds.), 1977. Визначник рослин Українських Карпат (*Vyznačnyk roslyn Ukrajins'kykh Karpat*). 435 pp., 338 text-figs., map. Kyjiv (Kiev): 'Naukova dumka'.

Manual-key to native, naturalized, adventive and commonly cultivated vascular plants with species-leads incorporating limited synonymy and Ukrainian-binomial names along with indication of habitat, phenology, occurrence (by floristic districts), habit, and some localities and taxonomic notes; complete indices at end. The introductory section gives the background to the work, technical details, a description (with map) of the 10 floristic districts in the area, and a general key to families.

Partial works: Ruthenia (Carpatho-Ukraine)

Separate works limited to Ruthenia or Carpatho-Ukraine (in Ukrainian except for that by Popov) include S. S. Fodor, 1974. Флора Закарпаття (*Flora Zakarpattja*). 208 pp. L'viv (L'vov): 'Vyšča Škola'; P. D. Jarošenko, 1947. Короткий визначник рослин Закарпаття (*Korotkyj vyznačnyk roslyn Zakarpattja*). 99 pp. Uzhhorod: Uzhhorod State University; and the chapter 'Конспект флоры Закарпатской области (Konspekt flory Zakarpatskoj oblasti)' in M. G. Popov, 1949. Очерк растительности и флоры Карпат (*Očerk rastitel'nosti i flory Karpat*), pp. 176–300 (Mater. Pozn. Fauny Fl. SSSR, n.s. 13 [Otd. Bot. 5]). Moscow.[206]

696

Crimea (Ukraine)

See also **694** (all works, though not early volumes of *Flora URSR*). – Area: 25775 km². Vascular flora: 2250–2400 species, of which 2150–2250 native (Webb). Wulff's flora accounts for 2277.

The Crimea, a Tatar land first incorporated into the Russian Empire in 1783 but since 1954 part of Ukraine, is, due to its geographical situation, mountainous character, and (for the country) distinctive vegetation, characterized by considerable local endemism. An early major work was *Flora taurico-caucasica* (1808–19) by Friedrich August Marschall von Bieberstein; in subsequent floras it usually was similarly combined with the Caucasus or 'Southern Russia' (Ukraine), an example being the unfinished *Opredelitel' rastenij Kavkaza i Kryma* (1909–14) by A. V. Fomin (later a leading early light for *Flora URSR*) and Yu. Voronov. The establishment of the Nikita Botanic Garden in Yalta in 1812 by Christian von Steven for practical as well as horticultural purposes – it played a key role in the development of viticulture as well as sericulture in the Crimea – later provided a local focus for floristic studies. In the 1920s the Garden became the sponsor of *Flora Kryma*.[207]

RUBTSOV, N. I. (ed.), 1972. Определитель высших растений Крыма (*Opredelitel' vysšikh rastenij Kryma*). 550 pp., 504 text-figs. Leningrad: 'Nauka'.

Copiously illustrated manual-key to vascular plants; includes concise diagnoses, limited synonymy, Russian vernacular and binomial names, local range, short notes (partly in symbolic form) on habitat, life-form, phenology, ecology and frequency, and indices to all botanical and other names. An introductory section gives aids for the use of the work.

WULFF, E. V. *et al.*, 1927–69. Флора Крыма (*Flora Kryma*)/*Flora taurica*. 3 vols. Portr., maps. Yalta: Nikitskij Botaničeskij Sad (1(1–3)); Moscow/ Leningrad: Otiz-Sel'khozgiz (1(4), 2(1)); Moscow: Sel'khozgiz (2(2), 3(1)), 'Sovetskaja Nauka' (2(3)), 'Kolos' (3(2)); Yalta: Nikitskij Botaničeskij Sad (3(3)). Continued as L. A. PRIVALOVA and YU. N. PROKUDIN (with S. S. STANKOV and N. I. RUBTSOV, eds.), 1959. Дополнения к первому тому «Флоры Крыма» (*Dopolnenija k pervomu tomu «Flory Kryma»*)/*Addenda et corrigenda ad Vol. 1 'Florae tauricae'*. 127 pp. (Trudy Gosud. Nikitsk. Bot. Sada 31). Yalta.

Comprehensive critical enumeration of vascular plants; includes diagnostic keys to all taxa and full descriptions of novelties along with extensive synonymy and references, literature citations, detailed indication of local range (with many *exsiccatae* and other records), generalized statement of extralimital distribution, taxonomic commentary, and notes on habitat, associates, etc.; tables of contents and literature lists in each part (but only vol. 3 fully indexed). The introductory section (in vol. 1(1)) contains a historical account of botanical work in the Crimea and a brief description of its floristic districts (mapped in vol. 1(4)).

697

Lower Don region (Russia)

Lying between the eastern border of Ukraine, the North Caucasian lowlands, and the lower Volga Basin, this area centers on Rostov Oblast' but also includes a part of Stavropol Krai, the southern part of Volgograd Oblast' (south of the Volga-Don Canal), and the western lobe of Kalmytsia. The eastern limit is here taken as the western boundary of *Flora jugo-vostoka Evropejskoj časti SSSR*, i.e., along the Jergeni Rise at 44° E.

The only regional work actually covering the area is the old *Flora srednej i južnoj Rossii, Kryma i Severnogo Kavkaza* by I. F. Schmal'hausen (see **694**).

Bibliography

NOVOPOKROVSKY, I., 1938. Литература по ботанической географии Ростовской области, Краснодарского края, Орджоникидзевского края и Дагестана (Literatura po botaničeskoj geografii Rostovskoj oblasti, Krasnodarskogo kraja, Ordžonikidzevskogo kraja i Dagestana). *Trudy Rostovsk. Obl. Biol. Obšč.* **2**: 5–45. [442 titles.]

Woody plants

PERESELENKOVA, L. M., 1950. Деревья и кустарники Ростовской области (*Derevja i kustarniki Rostovskoj oblasti*). 107 pp., illus., map. Rostov-na-Donu: Rostov Oblast' Press.

Semi-technical illustrated descriptive account of 105 native and commonly cultivated woody species and lesser taxa; includes notes on habitat, local range, properties, uses and potential, and an artificial general key to all species but no index. The introductory section gives aids to the use of the work, basic organography, and an account of the area and its vegetation (with map).

698

Southeastern Russia (and associated republics)

Vascular flora (including parts of **697** and **699**): 2300–2500 species, of which 2200–2450 native (Webb). Fedtschenko and Shishkin's flora encompasses 2146 species. – This unevenly known subregion comprises the lower Volga Basin from the Caspian Sea at 45° N, upstream through Volgograd to a point north of Saratov and ranges eastwards into extreme western Kazakhstan; the eastern limit follows the Ural River, part of the traditionally recognized boundary between Europe and Asia.

The flora by Fedtschenko and Shishkin is comprehensive, forming a basis for later manuals, all so far more local in scope (see also **699**).

Bibliography

Byčkova, E. Kн., 1950. Флора и растительность Юго-Востока Эвропейской части СССР: за 1772–1948 гг. *(Flora i rastitel'nost' jugo-vostoka Evropejskoj časti SSSR*, [1]: *za 1772–1948 gg.*). 202 pp. Saratov: Saratov State University. (Bibliografija Saratovskoj oblasti 2.) Continued as R. M. Akčurina, 1963. Флора и растительность Юго-Востока Эвропейской части СССР: 1949–1960 *(Flora i rastitel'nost' jugo-vostoka Evropejskoj časti SSSR*, 2: *1949–1960*). 219 pp. Saratov. (Bibliografija Saratovskoj oblasti 6.) [4250 systematically arranged entries in the two volumes, the latter encompassing a broader range of subdisciplines than the former; indexed.]

Fedtschenko, B. A. and B. K. Shishkin, 1927–38. Флора Юго-Востока Эвропейской части СССР *(Flora jugo-vostoka Evropejskoj časti SSSR)/ Flora Rossiae austro-orientalis* (later *Flora partis austro-orientalis Europensis URSS*). 3 vols. (in 6 parts), index. Illus., maps. Leningrad: Glavnyj Botaničeskij Sad (vols. 1–2, comprising parts 1–5); Moscow/ Leningrad: AN SSSR Press (vol. 3, comprising part 6, and index). (Parts 1–5 also published as *Trudy Glavn. Bot. Sada SSSR* 40(1–3) and 43(1–2). Index entitled *Alfabitnyj ukazatel' nazvanij rastenij tomov 1–6*.)

Large-scale descriptive flora of vascular plants; includes keys to all taxa, limited synonymy, Russian binomial and vernacular names, detailed account of local range (by botanical regions; see maps in parts 1 and 2) with *exsiccatae* as well as some range maps, taxonomic commentary, figures of representative species, and notes on habitat, associates, etc.; complete index (in separate volume). [In addition to the area as circumscribed in **698**, this work also includes much of Eastern Russia as delimited in **699**, with a northern limit at the Kama and Bjeloja Rivers and an eastern limit in the southern Ural.]

699

Eastern Russia (and associated republics)

Eastern Russia here represents the area between the Volga and Ural Rivers, the Ural Mountains, and the southern boundary of the Komi Autonomous Republic, thus including among other areas the whole of Ekaterinburg and Perm' *oblasti*, most of Kirov Oblast' (except its northern lobe), Udmurtia, the eastern part of Tatarstan, Bashkortostan, and a large part of Orenburg Oblast'. The circumscription of this area has been based on the geographical limits adopted and clearly mapped by Sergei Korshinsky in his *Tentamen florae Rossiae orientalis*, except that the middle Volga Basin from Kazan' to Samara, including areas west of the river encompassed by Majevski's flora of Middle Russia (**692**), comprise **691**. The northern and southern parts (separately considered below) are respectively also covered by other, more recent works, noted under the relevant subheadings. For the Ural Mountains, see also **602**.

In addition to the titles given below, a never-completed work of some consequence – on account of its introductory essay on taxonomic philosophy as well as floristic and vegetation zones – is S. I. Korshinsky, 1892. Флора востока Эвропейской России *(Flora vostoka Evropejskoj Rossii*). Vyp. 1. 227 pp., 3 pls. Tomsk.[208]

Bibliography

Skal'naja, G. D., 1971. Растительный мир Кировской области: указатель литературы. 76 pp. Kirov. [For Kirov Oblast'; not seen. Lipschitz (1975) gives the number of entries as 522.]

Korshinsky, S. I., 1898. *Tentamen florae Rossiae orientalis*. xix, 566 pp., 2 maps (Zap. Imp. Akad. Nauk, Fiz.-Mat. Otd., VIII, 7(1)). St. Petersburg.

Systematic enumeration (in Latin) of vascular plants, without keys; includes synonymy (with references), detailed indication of local range with citations of *exsiccatae* and generalized summary of extralimital range, taxonomic commentary, and notes on habitat, etc.; index to all botanical names at end. The introductory section includes a description of the region (with maps), an account of botanical districts, and an extensive literature survey together with an index to collectors. [Now superseded over much of its geographical range.]

Partial works: non-forest zone

This zone, except for the Ural (**602**), extends north to about 54–55° N and in its greatest part lies within the mixed forest-and-steppe vegetation formation. In addition to Korshinsky's *Tentamen*, the area is almost entirely covered by Fedtschenko and Shishkin's *Flora jugo-vostoka Evropejskoj časti SSSR* (**698**). The most useful local work is the manual for Bashkortostan by Kučerov and Muldasev (1988–89). [For floras of the Volga Basin, see **691**.]

KUČEROV, E. V. and A. A. MULDASEV (eds.), 1988–89. Определитель высших растений Башкирской АССР (*Opredelitel' vysšikh rastenij Baškirskoj ASSR*). 2 vols. Illus., maps. Moscow: 'Nauka'. (Based on A. I. BARBARICZ *et al.* (comp.), B. K. SHISHKIN and V. I. GRUBOV (eds.), 1966. Определитель растений Башкирской АССР (*Opredelitel' rastenij Baškirskoj ASSR*). Moscow.)

Manual-key to vascular plants (in two unnumbered volumes); includes essential synonymy, Russian binomials, habit, life-form, season(s), and notes on phenology, habitat, distribution and (in some cases) taxonomy; representative figures and distribution maps; authorships and indices in each part to Russian binomials and generic names and to Latin generic and family names at end (cumulative in vol. 2). The relatively brief introduction contains information about signs and abbreviations, a discussion of floristics and vegetation, and (pp. 14–38 in the first volume) family keys.[209]

Partial works: forest zone

The Ural (**602**) excepted, this zone has its southern limit between 54 and 55° N, corresponding to the approximate limit of potentially continuous forest. Besides Korshinsky's *Tentamen*, it is entirely, or almost so, covered in *Opredelitel' rastenij lesnoj polosy severo-vostoka Evropejskoj časti Rossii* of Snyatkov *et al.* (1922; see **688**). Modern manuals, however, have followed current administrative boundaries; those included here are anonymous author (1975) for the Kirov Oblast' (comprising in large part the former Vyatka *gubernija*) and Efimova (1972) for Udmurtia.

ANONYMOUS, 1975. Определитель растений Кировской области (*Opredelitel' rastenij Kirovskoj oblasti*). Vol. 2. Kirov.

Manual-key to higher plants. [Not seen; cited from *Prog. Bot.* **40**: 415, 427 (1978).]

EFIMOVA, T. P., 1972 (1971). Определитель растений Удмуртии (*Opredelitel' rastenij Udmurtii*). 224 pp., illus., 8 col. pls. Iževsk: 'Udmurtija'.

Concise field manual-key to vascular plants, with brief diagnoses, Russian nomenclature, and partly symbolic notes on habitat, life-form, phenology, local and extralimital range, and special features; indices at end. The introductory section includes an illustrated glossary and notes on the use of the work. A general key to families is also provided.

Notes

1 The works of Theophrastus give a fair impression of the flora of classical Greece and Asia Minor, but the earliest specifically floristic treatise is the fourth-century *Nan fang ts'ao mu chuang*, covering southern China (of which an English translation by H.-L. Li was published in

1979). The earliest regional flora in Europe is Johannes Thal's *Sylva hercynia* (1588); in Britain, *Catalogus plantarum circa Cantabrigiam nascentium* (1660) by John Ray.

2 For pre-Linnaean exploration see W. T. Stearn, 1958. Botanical exploration to the time of Linnaeus. *Proc. Linn. Soc. London* **169**: 173–196.

3 The early volumes of the 2. Auflage have been, or are to be, succeeded by a completely revised 3. Auflage. This commenced publication in 1966.

4 The 'Green Books', first seen by the author in 1963, directly influenced the development of this *Guide*.

5 The ESFEDS was conceived and developed at Reading University under the direction of Vernon Heywood and David Moore as an extension of *Flora Europaea*. Its data were to be organized in seven linked tables in a relational DBMS (using Oracle software), with a videotext interface design conforming to the 'Viewdata' standards in use by the 'Prestel' public data network then under development in the United Kingdom. The latter was chosen for 'ease of use by potential consumers' (V. Heywood, *in litt.* 27 April 1982). By the end of the project in 1985, only the enumerative component of the database had been compiled, along with endemism information and karyotypes. No descriptions were included, nor had any work been begun on the bibliographical and informational elements of the programme or in new areas such as chorology, phytosociology, biological data, uses or conservation. The database was later converted from Oracle to Advanced Revelation and dBASE formats. No printed version is generally available, save for pteridophytes; a trial covering ferns and fern-allies appeared in 1987, at the end of the original project (see Derrick *et al.* under **600**, **Pteridophytes**). It was, however, not utilized directly in the preparation of the second edition of vol. 1 of *Flora Europaea* (S. L. Jury, 1991. Some recent computer-based developments in plant taxonomy. *Bot. J. Linn. Soc.* **106**: 121–128). The whole enterprise consumed a total of £400 000 expended over five and a half years. A lack of a long-term commitment as well as effective advice on the part of ESF and failure to secure a permanent institutional 'home' were significant factors in its collapse (J. Edmondson, personal communication). It has now been succeeded by the Euro+Med PlantBase initiative (S. L. Jury in *Linnean Society Annual Report 1998*, p. 17 (1999)). This programme, after much discussion and several meetings over four years, was funded by the European Union late in 1999 (*Linnean* **16**(1): 3 (2000)).

6 Indeed, for such works print and electronic media are becoming inextricably linked. New projects (and even the continuation of some existing ones) are now seen as incomplete without provision for both. Such a 'dual' approach is a key feature of the Euro+Med PlantBase initiative.

7 Heywood, as committee secretary, published further progress reports in *Nature* **191**: 446–449 (1961); *Rev. Roum. Biol. (Bot.)* **10**: 119–121 (1965); *Boissiera* **19**: 17–20, 345–347 (1971); and *Bol. Soc. Brot.*, II, **47**(Suppl.): 25–27 (1973).

8 Progress on the *Flora*, regional floristic reports, proposals for satellite or later projects, and scientific contributions were the compass of eight Flora Europaea symposia held in different parts of the continent from 1959 to 1977: [1] V. H. Heywood (ed.), 1960. Problems of taxonomy and distribution in the European flora. *Feddes Repert.* **63**(2): 105–228; [2] V. H. Heywood and R. E. G. Pichi-Sermolli, 1963. Proceedings of the Second Flora Europaea Symposium. *Webbia* **18**: 1–562 (including 'Floristic Reports', pp. 95–562); [3] Anonymous, 1965. Volume dédié au Symposium «Flora Europaea». *Rev. Roum. Biol. (Bot.)* **10**(1–2): 1–181; [4] Auctores varii, 1966. Fourth Flora Europaea Symposium. *Bot. Tidsskr.* **61**: 325–334; [5] E. Fernández-Galiano, 1969. *V. Simposio de Flora Europaea*. 350, [2] pp. Seville: Universidad de Sevilla; [6] J. Miège and W. Greuter (eds.), 1971. Actes du VIᵉ Symposium de Flora Europaea. *Boissiera* **19**: 1–350; [7] [A. Fernandes, ed.], 1973–74 (1974). Proceedings of the VII. Flora Europaea Symposium. *Bol. Soc. Brot.*, II, **47** (Suppl.): 1–412, [2] pp. (with 'Floristic Reports' in *Mem. Soc. Brot.* **24**(1–2) (1974–75)); and [8] Anonymous, 1977. *Flora Europaea Final Symposium*. 17 pp. Cambridge: Cambridge University Press.

9 For retrospective commentary, see R. K. Brummitt, 1973. A survey of the *Index to European taxonomic literature 1965–1970*. *Bol. Soc. Brot.*, II, **47** (Suppl.): 41–55.

10 The work was an important source for the *Flora Europaea* project; and it was said that in making estimates of the number of species in the European flora, a useful rule of thumb was 'Nyman plus 20%'. D. A. Webb, 1978. *Flora Europaea* – a retrospect. *Taxon* **27**: 3–14.

11 As with many of its author's other works, the *Exkursionsflora* became shrouded in obscurity; this may be the reason for its omission in vol. 2 of Blake's *Geographical guide* (1961). Nevertheless, according to R. Geesink and R. van der Meijden in their foreword to the 1980 reprint, it possesses considerable intrinsic merit.

12 The best retrospective account of *Flora Europaea* is by Webb, *loc. cit.* The original work covers 11 948 species and 2102 additional subspecies (in all, 13 650 taxa) in 1544 genera and 203 families, arranged on a modified Englerian system (*Syllabus*, 12th edn.) with some amendments. Grants from the former Science Research Council in the United Kingdom, for the original project its main backer, totaled in excess of £150 000; there was also aid from other countries as well as private sources. The revised vol. 1 was financed largely from the Flora Europaea trust fund (Linnean Society of London), but no definitive plans exist for revisions of the remaining volumes; the editorial committee in 1993 placed itself 'in hibernation'. For progress reports, see J. R. Akeroyd and S. L. Jury, 1991. Updating "Flora Europaea". *Bot. Chron.* **10**: 49–54; and J. Edmondson, 1993. News from the Flora Europaea Secretariat. *Taxon* **42**: 730–732.

13 See W. Greuter, 1991. Botanical diversity, endemism, rarity, and extinction in the Mediterranean area: an analysis based on the published volumes of Med-Checklist. *Bot. Chron.* **10**: 63–79.

14 In addition, congresses have been held in different Mediterranean botanical centers (1975, Heraklion (Crete); 1977, Florence; 1979, Madrid; 1983, Palermo; 1986, Istanbul; 1989, Delphi; 1993, Borovetz (Bulgaria); 1995, Seville). Each has been followed by published proceedings which usually include floristic progress reports.

15 Early support for OPTIMA and the *Med-Checklist* came from the then-recently organized European Science Foundation. See European Science Foundation, *Report* 1977: 78–82; *idem*, 1978: 95–97. Strasbourg. The 1978 report contains a description of the plan and scope of the projected checklist.

16 This latter report was a product of the OPTIMA 'Commission pour l'encouragement et la coordination de l'exploration floristique, la préparation des flores et de travaux monographiques'.

17 Some 44.5 percent of the total flora has been covered in the three published volumes. Of the rest, Compositae account for 17 percent, the remaining dicotyledons, 19 percent and monocotyledons, 19.5 percent (Greuter, *loc. cit.*). Lack of funds has delayed publication of the remaining volumes; however, typescript for vol. 5 has been completed while work on vol. 2 (Compositae) is currently in progress. No start has yet been made on vol. 6. [Part of the database is now available on the International Organization for Plant Information (IOPI) Web site.]

18 For review, see *Taxon* **7**: 86 (1958). The nomenclature used in this work was not entirely in conformity with the International Code of Botanical Nomenclature as prevailing at the time of publication.

19 A. Takhtajan, 1986. *Floristic regions of the world*. Berkeley/Los Angeles: University of California Press.

20 The alpine areas of Europe are mapped in the inside covers of Ohba's work.

21 Grey-Wilson and Blamey's work is the first anglophone successor to H. S. Thompson, [1911]. *Alpine plants of Europe*. vii, 287 pp., 64 col. pls., map. London: Routledge.

22 Pyrenean floristic history is reviewed by C. Dendaletche in Saule (1991; see main text). Regarding Bubani's flora, opinions differ. It was credited by Saule (*loc. cit.*) with being 'la première vrai travail d'ensemble sur la flore de la chaîne pyrénéenne'; on the other hand, Blake in vol. 2 of his *Geographical guide* (1961) was critical of its idiosyncratic nomenclature though believing the critical commentaries to be valuable. Its records are oriented towards the Iberian side of the range.

23 An eventual reissue of the *Catalogue-flore* in one volume was originally contemplated. The author himself financed its publication so it did not become a charge on the Société de Biogéographie, publishers of *Le monde des plantes.*

24 Some nomenclature appears to be not quite current (e.g., *Cytisus sessilifolius* for *Cytisophyllum sessilifolium*).

25 Preparation of this work was from 1912 entirely entrusted by Burnat to his co-authors, and at his death in 1920 his substantial herbarium passed to the Conservatoire, enabling continuation of the work by its staff. Unfortunately, progress was interrupted by the death of Briquet in 1931 and not effectively resumed for half a century.

26 Notice should also be taken of H. Pitschmann and A. Reisigl, 1965. *Flora der Südalpen*. Stuttgart: Fischer; and L. Bouvier, 1882. *Flore des Alpes*. viii, 812 pp. Paris: Didot. (1st edn., 1878.) The former is inclusive of the Italian Trentino-Alto Adige region, while the latter covered Switzerland and Savoy. With respect to a modern pan-Alpine flora 'de Nice à Vienne' – long a major *desiratum* – a Swiss proposal was funded in 1991; the projected *Flora alpina* would cover up to 5500 species in the form of a database and portable manual (D. Aeschimann, 1991. Projêt pour une Flore des Alpes. *Rev. Valdôtaine Hist. Nat.* **45**: 155–158).

27 The plates were termed by Blake in his *Geographical guide* (1961, p. 559) as 'fair to excellent . . . [but] without dissections'. They appear to have been issued unbound, in book-like boxes. Some copies were subsequently bound, but with small plate margins the result is not very satisfactory. An earlier version of dalla Torre's introductory survey and keys appears as 'Anleitung zur Beobachtung und zum Bestimmen der Alpenpflanzen', comprising pp. 117–434 of volume 2 of the Deutscher und Österreichischer Alpenverein's *Anleitung zu wissenschaftlichen Beobachtungen auf Alpenreisen* (1882, Wien) but also separately published.

28 This work inspired *The mountain flora of Java* (**903**).

29 The author died suddenly in 1971, while on Mt. Olympus in Greece.

30 Reviewed in *OPTIMA Newsl.*, **25–29** (1991): (26)–(27). The species concept adopted is less narrow than that in *Flora na NR Bălgarija*; however, such a 'synthetic' approach is thought by the reviewer to obscure significant details in an area of key floristic interest. The mountain has long been a Bulgarian botanical 'Eldorado'.

31 The work is comparable to Hegi's *Alpenflora* or Schröter and Schröter's *Taschenflora des Alpenwanderers* (**603/IV**).

32 Considered to be of a good standard (C. D. K. Cook, personal communication).

33 Of the works cited under this subheading, only that of Ryczin (1948) has been seen by the author. All, however, are listed in Lipschitz (1975; see **Region 68/69**).

34 In Portugal the work of Júlio Henriques at Coimbra was of lasting importance. He was a founder of the Sociedade Broteriana and in 1880 purchased the Willkomm Herbarium for Coimbra University (A. Fernandes, 1986; see Note 38 below).

35 The work was Knoche's dissertation at the Université de Montpellier. Later, he purchased Engler's reprint collection for Stanford University.

36 African studies, for instance, long accounted for a major part of the systematics research programme at Coimbra University. Pereira Coutinho and Sampaio were both at 'city' universities (then polytechnics).

37 *Flora Europaea* has since its advent been widely used to identify Spanish vascular plants (G. López González, 1991; see Note 38 below).

38 Historical studies through 1940 include M. Colmeiro, 1858. *La botánica y los botánicos de la península hispanolusitana*. Madrid; A. de Brito, 1883. *Historia da botanica em Portugal*. Lisbon; and F. Bellot Rodríguez, 1940. Notas bibliográficas sobre la botánica portuguesa. *Anales R. Acad. Farm. Madrid*, II, **6**: 217–277. There is also a historical review in M. Willkomm, 1896. *Grundzüge der Pflanzenverbreitung auf der iberischen Halbinsel* (Die Vegetation der Erde 1). Leipzig. Accounts from after 1940 include J. Ma. Muñoz Medina, 1969. *Historia del desarrollo de la botánica en España*. 34 pp. Pamplona; those respectively for Portugal and Spain by A. R. Pinto da Silva (pp. 19–28) and E. F. Galiano (pp. 29–39, with map) in *La flore du bassin méditerranéen* (1975; see **601/I**, **Bibliographies/Progress reports**); and A. Fernandes, 1986. História da botânica em Portugal até finais do Seculo XIX. In Academia das Ciências de Lisboa, *História e desenvolvimento da ciência em Portugal*, 2: 851–916. Lisbon (Publicações do II. Centenário da Academia das Ciências de Lisboa). Recent developments are covered in G. López González, 1991. The progress of "Flora Iberica" and other Spanish floras. *Bot. Chron.* **10**: 91–104. The pertinent comments by Heywood may be found in V. H. Heywood, 1978. European floristics: past, present and future. In H. E. Street (ed.), *Essays in plant taxonomy*, pp. 275–289. London: Academic Press.

39 For an extensive review following publication of volumes 3 and 4, see F. Sales and I. C. Hedge in *Taxon* **44**(3): 470–472 (1995). In their view, 'common' material, including the general key to families and lists of abbreviations, was unnecessarily repeated from volume to volume, and perhaps too many formal infraspecific taxa were accepted; nevertheless, the work would become one of Europe's 'Basic Floras'. Other complications include insufficient knowledge of floristically related areas (in part due to a weak representation of extra-Iberian material in Peninsular herbaria) as well as imperfectly understood and/or undercollected genera (G. López González, 1991; see Note 38 above). *Flora iberica* is nevertheless among the most important current floristic enterprises in Europe.

40 The 1988 facsimile edition was produced in commemoration of the 50th anniversary of the author's death as an initiative of the Botanical Institute of Porto University. The 1990 issue has a new prefatory note. These two issues surely reflect a demand for an illustrated one-volume manual, for which in Portugal there are no alternatives.

41 Continuation of the *Prodromus* was sometimes in doubt. Contemporary sales in 1866 were, at only some 100 copies, considered low enough that without at least a 50 percent increase its continuation might be at risk; Willkomm therefore appealed for more subscriptions (*Flora* **49**: 159–160 (1866)). Even afterwards, some years might pass before another installment appeared, prompting comment (*Bull. Soc. Bot. France* **20**, Bibl.: 213 (1873)); at that time, however, relative remoteness may also have played a part as for four years (1868–72) Willkomm was professor of botany at the University of Dorpat (Tartu) in present-day Estonia (**677**). Health considerations subsequently – and perhaps fortunately! – necessitated a return to Central Europe; in 1874, following an extended field trip in Spain and the Balearics, he took up an appointment to the Charles University in Prague, remaining there for the rest of his life. His Mediterranean herbarium, on which the *Flora* was to a large extent based, was sold in 1880 to Coimbra University at the initiative of Júlio Henriques, then professor of botany (A. Fernandes in *Anuário Soc. Brot.* **43**: 15–44 (1977)).

42 The literature on woody plants in Spain is now quite substantial. For lists, see http://www.guidverde.com/arboles/bibliografia.html.

43 A reprint of the 1917 supplement and a further volume summarizing the additions and changes to the flora incorporated in the many subsequent papers as cited by Blake (1961, p. 502) and since would be welcome.

44 A similar account for that province has been projected.

45 Botanical progress in the Catalan lands to the mid-1980s is summarized in J. Vigo, 1985. Els estudis florístics i fitocenològics als Països Catalans: situació actual. *Butlletí Inst. Catal. Nat. Hist.* **50**: 241–247, map.

46 Vol. 4 was in press at the time of writing (1999).

47 An account of the project appears in B. Valdés, 1988. Realización de la *Flora Vascular de Andalucía Occidental. Lagascalia* **15**(extra): 67–73.

48 Prepared within a relatively short time, this work is mainly useful for data. Bòlos in *Flora dels Països Catalans*, 1: 107 referred to its taxonomy as 'excessivament sintètica'.

49 The earlier *Flora italiana* (1818–24) by the Pisa botanist Gaetano Savi is botanophilic: a 'choix des plus belles fleurs' from Italian gardens. For a modern appreciation of Bertoloni's *Flora italica*, see V. Giacomini, 1970. Un monument de la littérature floristique italienne: la 'Flora Italica' de Ant. Bertoloni (1833–56). In P. Smit and R. J. Ch. V. ter Lange (eds.), *Essays in biohistory*, pp. 1–13 (Regnum Veg. 71). Utrecht.

50 For historical treatments of floristic botany in Italy, see P. A. Saccardo, 1895–1901 (under **Regional bibliography**) and *idem*, 1909. *Cronologia della flora italiana*. xxxvii, 390 pp. Padua (also as A. Fiori, *Flora analitica d'Italia* 5). [Both works subsequently reissued, the second by Edagricole, Bologna, with new preface.] In 1909 Augusto Béguinot proposed a permanent topographical committee for floristic recording; see A. Béguinot *et al.*, 1909. *Lo stato attuale delle conoscenze sulla vegetazione dell'Italia* . . . 107 pp. Padua. In 1968, a working group for floristics was organized by the Società Botanica Italiana and a variety of activities initiated. A summary of the state of knowledge to the mid-1970s (with map depicting levels of exploration to 1969) was furnished by G. Moggi (pp. 53–63) in *La flore du bassin méditerranéen* (see **601/I, Bibliographies/Progress reports**). The most recent overview of the state of knowledge is in R. Venanzoni, 1988. Le flore d'Italia. In F. Pedrotti (ed.), *100 anni di richerce botaniche in Italia*, pp. 533–538 (Pubblicazione in occasione del centenario della Società Botanica Italiana 2). Florence. Nine further papers (among others) in this commemorative volume deal respectively with the different regions of mainland Italy, Sardinia and Sicily and with forest botany and pteridology.

51 Reviewed by W. Greuter in *OPTIMA Newsl.* **14–16**: 53–54 (1983); the only significant criticisms related to nomenclature, particularly in families for which *Flora Europaea* was not then available. The author, called a philosopher as well as a scientist in the stricter sense, believed that his work 'does not pretend to be closer to the truth than its predecessors, but . . . attempts better to correspond to the present-day needs of research'. It was written in five years (1969–74) but then required eight before publication on account of new technology, industrial action, finances, publishing changes, etc.

52 The work has something of the flavor of the nineteenth-century Michaux/Nuttall *Sylva of North America*.

53 Work on the original *Prodrome* was suspended with the death of René de Litardière in 1957. Some notes on families from Scrophulariaceae through Ambrosiaceae (Compositae in part), intended for volume 4, were shortly afterwards published as an enumeration (with bibliography): J. Bouchard, 1961. *Materiaux pour un géographie botanique de la Corse*. 92 pp. N.p. Not until Gilbert Bocquet assumed the directorship of the Conservatoire et Jardin Botaniques in 1979 did active interest in completing Briquet's project resume. A trust, *Projet Flore Corse*, was established to assume sponsorship and a scientific and editorial committee organized.

54 For a survey of floristic work and knowledge to 1987, see B. Corrias, 1988. L'esplorazione floristica della Sardegna negli ultimi 100 anni. In F. Pedrotti (ed.), *op. cit.* (Note 50), pp. 667–679.

55 The work is in general something of a miscellany. Blake (1961 (see **General bibliographies**), p. 387) considered it not very reliable.

56 Moris himself did not attempt the monocotyledons on account of his duties in later life as a political administrator (Stafleu and Cowan, *Taxonomic Literature-2*, 3: 586).

57 For summaries of current knowledge of the flora of the Sicilian region (including the Lipari Islands, Ustica, Pantelleria and the Pelagic Islands), see F. M. Raimondo, 1988. Stato delle conoscenze floristiche della Sicilia al 1987. In F. Pedrotti (ed.), *op. cit.*, pp. 637–665 (with map), and A. Di Martino and F. M. Raimondo, 1979. Biological and chorological survey of the Sicilian flora. *Webbia* 34(1): 309–335.

58 The additions and corrections in vol. 3 are reprinted from *Malpighia* 20: 37–48, 95–119, 180–218 and 290–300 (1906). Vol. 1(1) initially was published in parts in *Giornale del Comizio Agrario di Palermo* and reissued complete in 1889.

59 Stefano Sommier was additionally a considerable contributor to knowledge of the flora of Malta as well as other Italian islands, and moreover collected in Siberia and the Caucasus. He first visited the Pelagic Islands and Pantelleria in 1873 with the Conti di Lampedusa.

60 Acknowledged by Pampanini as a significant contributor was Giovanni B. de Gaspari, a collaborator from 1912 but killed in battle in 1916.

61 W. B. Turrill in his *Plant life of the Balkan Peninsula* (1929) gave a total for the Balkans of 6753 species. Some 5000–5150 species were reckoned by Webb to occur in former Yugoslavia, of which 4750–4900 were native.

62 Calls for a 'new Hayek' have been heard in recent decades as interest developed in floristic mapping of the Balkans. See V. I. Velčev and S. I. Kozuharov, 1981. *Kartirovanie flory Balkanskogo poluostrova/Mapping the flora of the Balkan Peninsula*. 247 pp., tables, maps. Sofia: Bulgarian Academy of Sciences.

63 The two all-Yugoslavia works were founded through the efforts of an inter-academy committee on flora and fauna headquartered in Ljubljana. For the plants, three works were planned: the *Catalogus*, for which Ernest Mayer took responsibility, the *Analitička flora* initially under Stjepan Horvatić in Zagreb, and in addition a more definitive documentary and descriptive flora (never realized).

64 The work was one of the five 'Basic Floras' constituting a primary source for the preparation of *Flora Europaea*. An English version of the introduction is known.

65 Unofficial issues of three parts (vols. 1(3), 1(4) and 2(1)) were printed and distributed by others after Horvatić became seriously ill in 1974. They were disowned on ethical and other grounds by the Yugoslav botanical community at a 1977 congress (A.-Ž. Lovrić, *OPTIMA Newsl.* **6**: 38–40 (1978)). Doubts were then also expressed about formal continuation of the project, at least in the near term (E. Jäger, *Fortschr. Bot.* **40**: 413ff. (1978)).

66 For a description of the mapping project (*Karte rasprostranjenosti*) associated with *Flora Hrvatske*, see T. Nikolić, D. Bukovec, J. Šopf and S. D. Jelaska, 1998. *Kartiranje flore Hrvatske: mogućnosti i standardi*. [2], 62 pp., maps, overlays. Zagreb: Hrvatski prirodoslovni musej. (Natura Croatica 7: Suppl. 1.) It has been reported that some four-fifths of available floristic data are over 50 years old.

67 The survival of this work to its completion – fortunately before the Sixth Balkan War of 1992–95 – is remarkable. Publication of the third volume was facilitated by Nedeljko Košanin of Belgrade University; and from 1958 completion of the fourth was largely in the hands of Željka Bjelčić, with contributions from others including Ernest Mayer, Pavel Fukarek and Čedomil Šilić. A bibliographic guide to the separate parts of vols. 1–2 and l[bis] appears in vol. 3, pp. vii–viii, and a full collation of the whole work (save vol. 4, parts 3 and 4) in Stafleu and Cowan, *Taxonomic literature-2*, 1: 158–159 (1976; see **General references**).

68 V. Stevanović and V. Vasić (eds.), 1995. *Biodiverzitet Jugoslavije*. viii, 586 pp., tables, maps. Belgrade: Biološki Fakultet/Ecolibri, especially the chapter "Diverzitet vaskularne flore Jugoslavije: sa pregledom vrsta od međunarodnog značna" by V. Stevanović, S. Jovanović, D. Lakušić and M. Niketić, a survey of the flora of the present Federation.

69 See K. Marhold, 1990. *List of taxonomical projects concerning the Carpathian and Pannonian Flora*. Bratislava.

70 V. Stevanović and V. Vasić, *loc. cit.* (Note 68).

71 For Serbian botanical history, see M. M. Janković, 1988. Od Pančićeve *Flore Kneževine Srbije* do Akademijine

Flore SR Srbije. Glasn. Inst. Bot. Baste Univ. Beograd **22**: 1–25; O. Vasić and N. Diklić, 1996. Flora of Serbia: a review of phytotaxonomic studies published during the period 1981–1993. *Bocconea* **5**: 45–53; and M. R. Sarić, N. Diklić and O. Vasić, 1998. Committee for Exploration of the Flora and Vegetation of Serbia: three decades of work. In I. Tsekos and M. Moustakas, *Progress in botanical research: proceedings of the First Balkan Botanical Congress*, pp. 141–148. Dordrecht: Kluwer.

72 *Flora SR Srbije* is said to have been to an extent compiled, with some authorities regarding it as uncritical (E. Jäger, *Fortschr. Bot.* **40**: 413ff. (1978)). Production of the second edition, interrupted by the Balkan war of 1992–95, was subsequently again taken in hand (Vladimir Stevanović, personal communication, December 1996). As of 1998 the texts of vols. 2 and 3 were ready (Sarić, Diklić and Vasić, 1998).

73 Entitled 'Pogovor uz "Flory Kneževine Srbije"', this appendix incorporates a history of botanical work in Serbia. The authors note that a remark by Grisebach in his *Spicilegium florae rumelicae et bithynicae* to the effect that Serbia then was very poorly known may have presented a challenge to the young Pančić. By the 1880s this argument had been almost single-handedly refuted. Comparisons with similarly isolated but dedicated contemporaries such as Ferdinand von Mueller in Australia, Alvin Chapman in the southeastern United States and John Macoun in Canada are not out of place.

74 V. Stevanović and V. Vasić, *loc. cit.* (Note 68).

75 For the history of botanical work in Montenegro before World War II, see A. Baldacci, 1924. Le esplorazioni botaniche nel Montenegro, primo periodo (1827–1841). *Mem. R. Accad. Sci. (Bologna)*, *Cl. Sci. Fis.*, VIII, **1**: 27–33; *idem*, 1926. Le esplorazioni botaniche nel Montenegro, secondo periodo (1841–1878). *Ibid.*, **3**: 13 pp. (as separate); and *idem*, 1932. Le esplorazioni botaniche nel Montenegro, terzo periodo (1878–1930). *Ibid.*, **9**: 18 pp. (as separate). Post-1945 accounts include V. Pulević, 1970. Istorijski pregled florističkih i vegetacijskih istraživanja u Črnoj Gori. *Glasn. Republ. Zavoda Zaštitu Prir. Prirodnjačk. Zbirke, Titograd* **3**: 109–123; E. Mayer, 1982. Primitiae florae montenegrinae. *Ibid.*, **15**: 27–48 (in German, with particular reference to the work of Tommasini and Visiani), and V. Pulević and V. Karaman, 1996. Additional literature data on the flora of Črna gora (Montenegro). *Bocconea* **5**: 53–54.

76 The format of this well-regarded work resembles Mayer's 1952 enumeration of the Slovenian flora (**643**). A planned supplement, *Dodatak Rohleninom* Conspectus Florae Montenegrinae by V. Blečić, R. Lakušić and V. Pulević, was ready by 1979 but has never been published (V. Pulević and V. Karaman, *loc. cit.*). Reference to it, however, exists in the literature (V. Stevanović and V.

Vasić, *loc. cit.*). It would have had records of over 500 additional species and 200 additional infraspecific taxa.

77 I thank John Akeroyd for information respecting this work. It has also been reviewed in *OPTIMA Newsl.* **32**: (7)–(8) (1997). A copy of vol. 1 was seen in the British Library.

78 Exclusive of Crete and the East Aegean, a figure of 4100–4250 species (of which 3950–4100 native) was given by Webb. For Greece as a whole, Rechinger suggested a figure of *c.* 5500 species and subspecies (in PD; see **General references**).

79 For historical surveys and assessment of floristic progress in this botanically still imperfectly known country, see K.-H. Rechinger, 1968. Bericht über die botanische Erforschung von Griechenland. *Webbia* **18**: 234–259; W. Greuter, 1975. Floristic studies in Greece. In S. M. Walters and C. J. King (eds.), *Floristic and taxonomic studies in Europe*, pp. 18–37, 4 maps. Abingdon, Berks. (BSBI Conf. Rep. 15); W. Greuter and D. Phitos, 1975. Greece and the Greek Islands: a report on the available floristic information and on current floristic and phytotaxonomic research. In *La flore du bassin méditerranéen* (see **601/I, Bibliographies/Progress reports**), pp. 67–89; and W. Greuter, 1998. The early botanical exploration of Greece. In I. Tsekos and M. Moustakas (eds.), 1998. *Progress in botanical research: proceedings of the 1st Balkan Botanical Congress*, pp. 9–20. Dordrecht: Kluwer.

80 The *Flora hellenica* project was described by Strid in *Bot. Chron.* **10**: 81–94. The secretariat is at the Botanical Laboratory, Copenhagen University; other centers are at Patras, Berlin, Edinburgh and Lund. It is supported by a database of over 330000 records. The current project is successor to a proposal first made in 1970 at the Sixth Flora Europaea Symposium; see V. H. Heywood, 1971. *The Flora of Greece project.* 11 pp. Kifissia: Goulandris Botanical Museum. – The text is unfortunately characterized by too much 'white space' and an absence of illustrations.

81 The presently imperfect understanding of many taxa is clearly recognized in this work; an example is the treatment of *Alyssum montanum*, dubbed 'provisional' (p. 286).

82 For an account of Cretan botanical history additional to those in the cited floras and manuals, see H. W. Lack, 1996. Die frühe botanische Erforschung der Insel Kreta. *Ann. Naturh. Mus. Wien* **98B**, Suppl.: 183–236. This makes particular reference to the contributions made in the era of Venetian rule.

83 Sir Colville Barclay indicated in his foreword to *Flora of the Cretan area* that such consolidated works greatly increase access to information which, while 'invaluable', is 'scattered in publications which have a limited circulation and are generally inaccessible to amateur botanists'.

84 A significant stimulus to the study of Bulgarian natural history in the post-independence period was a personal interest on the part of Ferdinand of Bulgaria (Prince from 1889, Tsar from 1908 to 1918). A natural history museum was established in 1889 as one of six 'Royal Institutions' and Ferdinand himself made many collecting trips. The tradition was continued by his successor, Boris III, until World War II. See E. Dimitrova, 1993. A little known Bulgarian biological connection surveyed at the Linnean Library collection. *Linnean* **10**(3): 24–29.

85 For reviews of the progress of botanical exploration and floristic work in Bulgaria, see B. Stefanoff, 1930. Historische Übersicht der Untersuchungen über die Flora Bulgariens. *Mitt. Naturw. Inst. Sofia* **3**: 61–112; E. G. Bobrov, 1974. Ботанические исследования в Болгарии (Botaničeskie issledovanija v Bolgarii). *Bot. Žurn. SSSR* **55**: 1046–1050; S. Stanev, 1982. (*Belezhiti bălgarski botanitsi.*) Sofia; S. I. Kožuharov and A. V. Petrova, 1988. The flora of the PR Bulgaria: state, prospects, development. In K. Marhold (ed.), *Karpatskaja flora*, pp. 50–57. Bratislava; and E. Ančev, 1996. Phytotaxonomic and phytogeographical studies in Bulgaria during the last decade (1983–1993). *Bocconea* **5**: 33–44. The first volume of *Flora na NR Bălgarija* also contains a historical survey, in both Bulgarian and English. Stefanov believed the flora to be completely studied, a claim disputed by Kožuharov and Petrova as too extreme. Bobrov considered the early volumes of *Flora na NR Bălgarija* to be stylistically too much like *Flora SSSR*.

86 Description partly based on information supplied by Arthur O. Chater, formerly of the Natural History Museum, London.

87 D. A. Webb, 1978. *Flora Europaea* – a retrospect. *Taxon* **27**: 3–14, esp. p. 13.

88 For additional background, see W. T. Stearn, 1975. History of the British contribution to the study of the European flora. In S. M. Walters and C. J. King (eds.), *Floristic and taxonomic studies in Europe*, pp. 1–17, map (BSBI Conf. Rep. 15). Abingdon, Berks. The eastern 'half' of the region – Carpatho-Pannonia – has not been documented to the same degree of thoroughness as the western; see J. Dostál *et al.*, A survey of the present knowledge of the Carpatho-Pannonian flora. In Marhold, *Karpatskaja flora*, pp. 26–30.

89 V. H. Heywood, 1970. The new Hegi Compositae [review]. *Taxon* **19**: 937–938.

90 Projects current in the late 1980s in the southeastern part of the region were enumerated by K. Marhold, 1990. *List of taxonomical projects concerning the Carpathian and Pannonian flora.* 79, [1] pp. Bratislava. A related series of progress reports, contributed to a Comecon workshop in Smolenice (Slovakia) in 1988, appeared as *idem* (ed.), 1988. Карпатсекая флора/ *Die Flora der Karpaten/ Carpathian Flora.* 148 pp. Bratislava.

91 For further discussion, see R. Schmid, *Taxon* **36**: 777 (1987). The political changes of 1989–90 have furthered the process of reintegration of the Central European botanical 'circle' begun in the 1980s; however, some countries have faced constraints on research activities.

92 An additional exposition of indicator-values is to be found in H. Ellenberg *et al.*, 1992. *Zeigerwerte von Pflanzen in Mitteleuropa.* 258 pp. (Scripta Geobotanica 18). Göttingen. A tabular list of species is also included.

93 The overall form of this work resembles to some extent that of Ferdinand von Mueller's *Second systematic census of Australian plants* (**420–50**). A new edition under the leadership of Walter Gütermann (at Vienna) – managing editor for the 1973 version – is in prospect.

94 Schmid, 1987 (Note 50).

95 In view of past uneven floristic treatment, interest has been expressed in a new overall flora of 'Carpatho-Pannonia'. Already in 1978, at the 2nd OPTIMA congress, Oleg Polunin had called for such a work. Josef Dostál took up this challenge and to this end a 'circle' was organized by the late 1980s (Dostál in Marhold, *op. cit.*, pp. 26–30). Dostál himself had by 1986 prepared a preliminary 'Enumeratio plantarum cormophytarum florae Pannonicae-Carpaticae'. However, this initiative became dormant due to the death of one key collaborator, Michaela Šourková (Jan Kirchner, personal communication, 1996).

96 Beldie in his *Flora Romîniei* indicates that the text of *Flora mică ilustrată* was probably not substantially revised after 1944. A Hungarian edition was also published (not seen; information from Alexandru Borza, Cluj-Napoca, 1965).

97 For the early history of botany in Romania, see E. Pop, 1931. Contributii la istoria botanici Romanesţi/Beitrag zur Geschichte der Botanik in Rumänien. *Bul. Grăd. Muz. Bot. Univ. Cluj* **10**: 185–196. There are also historical notes (also in Russian) in the first volume of *Flora R. P. R.*

98 Like several other eastern and southeastern European tongues, Hungarian as a literary and scientific language was still in development during the nineteenth century.

99 For the early history of botany in Hungary and adjacent lands, see A. Kanitz, 1863. *Geschichte der Botanik in Ungarn.* Hannover/Pest (revised 1864–65 as Versuch einer Geschichte der ungarischen Botanik. *Linnaea* **33**: 401–664 (subsequently separately reprinted)); and E. Gombocz, 1936. *A Magyar botanika története/ Geschichte der ungarischen Botanik.* 636 pp. Budapest.

100 The final pages of vol. 7 comprise a bibliography of Soó's publications.

101 Scopoli was from 1754 to 1767 a physician in Idria (*c.* 40 km west of Ljubljana); F. A. Stafleu, *Linnaeus and the Linnaeans*, pp. 193–194 (1971).

102 A concise history of floristic studies in Austria appears on pp. 157–163 in Adler *et al.*, *Exkursionsflora von Österreich* (see main entries).

103 A considerable proportion of the manual represents the work of its 'managing editor', Manfred Fischer.

104 The installments appeared in 1967, 1968, 1969, 1970, 1972, 1974, 1976, 1978, 1980, 1982, 1984, 1986, 1989, 1990, 1991–92, and 1994.

105 The author was an advocate of relatively narrow generic concepts; in the *Nová květena*, for example, the segregates *Agaloma*, *Chamaesyce* and *Tithymalus* are adopted in place of *Euphorbia* sensu lato and *Corothamnus* as well as *Chamaecytisus* are separated from *Cytisus*. These limits are also adopted in *Veľký kľúč*.

106 For further discussion, see B. Slavík, 1988. Flora der tschechischen sozialistischen Republik – vom Plan zur Realisierung. In K. Marhold (*op. cit.*), pp. 108–117. A number of models were considered, one being *Flora der Schweiz und angrenzender Gebiete* by Hess *et al.* (**649**).

107 For long, 'all that mattered in Bratislava . . . was the dominant Hungarian element, or perhaps to some extent the Austro-German one; the substratum of the Slovak peasantry had no standing or relevance whatsoever'. C. Magris, *Danube* (1989, New York), p. 220. Magris also notes, however, that on a more intellectual level there were arguments over language: should Czech be adopted, or Slovak developed? In systematic botany at least, the latter course ultimately was followed, resulting in textbooks, terminologies, checklists and *Flóra Slovenska*. For another account of Slovak botanical history, see L. Bertová, 1988. Die flora der Slowakei: die Geschichte ihrer Entstehung und gegenwärtiger Stand. In Marhold (*op. cit.*), pp. 14–20.

108 Volumes on non-vascular plants and fungi have also been included in the publication plan; vol. 10(1), on eurysiphalean fungi, appeared in 1995. The terminological volume, with few peers, is especially noteworthy. The initial appearance of the *Flóra* was reviewed by Á. Löve in *Taxon* **16**: 133–135 (1967).

109 As Rostafiński wrote in the latter work (p. 81), 'man spricht öfters vom Königreich Polen als von einer "terra incognita" in botanischer Hinsicht'.

110 Particular stimuli were university renewal, the establishment in 1872 of the *Academia Umiejętności* at Kraków, and the development of Lwów (L'viv, now in Ukraine) as a viceregal capital. See N. Davies, *God's Playground* (1982, New York), 2: 150, 155.

111 Historical accounts include B. Pawłowski, 1928. *Roswój florystyki i systematyki roślin w Polsce w latach 1875–1925.*

Kraków; B. Hryniewiecki, 1931. *Précis de l'histoire de la botanique en Pologne.* 45 pp. Warsaw; *idem*, 1948. *Roswój botaniki w Polsce.* 52 pp. Kraków (as *Historia nauki polskiej, mon.* 8); J. Kornaś and J. Zwizycki, 1959. *Botany in Poland.* 35 pp. Kraków; and W. Szafer, 1970. Outline of the history of botany in Poland from the Middle Ages to the year 1918. In P. Smit and R. J. Ch. V. ter Lange (eds.), *Essays in biohistory*, pp. 381–388 (Regnum Veg. 71). Utrecht.

112 The work was only completed after much delay in finishing the Compositae. A particular problem was finding interested specialists for such 'critical' genera as *Taraxacum* and *Hieracium* (Krzysztof Rostánski, personal communication, 1975).

113 As of writing at least 35 fascicles in vols. 1–7, 9, 11 and 17 have been published, but only vols. 1, 3, 5 and 9 have been completed. The plan of the work as well as an indication of fascicles published appears in the covers of each new fascicle. A given volume may contain anywhere from two to seven fascicles. The editorial committee as of 1990 comprised K. Browicz, A. Jasiewicz, J. Mądalski and K. Zarzycki.

114 Recent summaries of floristic work and mapping include: W. Ahlmer and E. Bergmeier, 1991. Floristische Erhebungen in der Bundesrepublik Deutschland – Übersicht und Ausblick. *Natur und Landschaft* **66**(9): 423–425; and E. Bergmeier (ed.), 1992. *Grundlagen und Methoden floristischen Kartierungen in Deutschland.* Göttingen (as *Floristische Rundbriefe, Beih.* 2). The latter contains critiques of current German and Central European floras.

115 Though remaining in use in some quarters, this work is no longer effectively current. It is, for example, not among the standard cross-references adopted for the *Standardliste* of Weisskirchen and Haeupler (1998, q.v.).

116 Both Teil II and IV are extremely compact works, representing much distillation from a vast body of information; they have been, and deserve to be, an example for others of their kind.

117 The most important of the many past 'editions' are those with substantial revisions: the 2nd (1905), 4th (1908), 7th (1910, the first under the current publisher), 10th (1912), 32nd (1923), 37th (1927), 64th (1954), 70th (1960), 81st (1967), 87th (1982), and the current (1993). See also *Taxon* **40**: 182 (1991) as well as p. vii of the 1993 edition. The '67/68. Aufl.' published in 1958 in the former German Democratic Republic and cited in the last edition of the *Guide* is now omitted.

118 The *Standardliste* is the first of a projected trilogy. Its two planned companions are respectively an atlas of figures and a distribution atlas, the latter in succession to those of 1988/89 and 1996 (see **Distribution maps**). A related work is H. Haeupler and J. Paeger, 1989.

Checkliste der Farn- und Samenpflanzen der Bundes-republik Deutschland. Teil 1: *Pteridophyta.* 2. Aufl. viii, 52 pp., map. Bochum: Ruhr-Universität, Spezielle Botanik, Arbeitsgruppe Geobotanik. This 'pilot work' includes details by state (in western and southern Germany), physiographic provinces, ecological classes, karyotypes, life-forms, and distributional and chorological classes.

119 The Haeupler and Schönfelder *Atlas* was an outgrowth of a more general project, 'Mapping the flora of Central Europe', to which is also related the checklist by Ehrendorfer *et al.* (**640/I**). That of Benkert *et al.* was a culmination of work begun by Hermann Meusel and Wilhelm Troll in the 1930s.

120 Early history is surveyed in T. A. Bruhin, 1863–64. *Übersicht der Geschichte und Literatur der Schweizer-Floren.* 2 parts. 30, 40 pp. (Jahresbericht über die Erziehungsanstalt des Benediktiner-Stiftes Maria-Einsiedeln im Studiensjahre 1862/63, 1863/64). Einsiedeln. The period from 1890 to 1990 is examined in a collection of papers on the various Swiss botanical centers entitled 'Hundert Jahre Schweizerische Botanische Gesellschaft' with an introduction by Heinrich Zöller (in *Bot. Helv.* **100**(3): 269–395 (1991)).

121 Attention should also be drawn to *Ökologische Zeigerwerte zur Schweizer Flora* (1977, in *Veröff. Geobot. Inst. ETH, Stiftung Rübel, Zürich* **64**) by Elias Landolt. This has been drawn upon by flora and manual writers, notably Lauber and Wagner in their *Flora helvetica.* For all species in the Swiss flora it gives codes for selected ecological factors and related biological features.

122 The *Fortschritte* were suspended following the 52nd number (1983). Following ratification of the Convention on Biological Diversity, however, a new body, the 'Centre du Réseau Suisse de Floristique', came into being. As part of its functions, it assumed responsibility for the *Fortschritte.* In 1995 the 53rd number appeared under the title *Documenta Floristicae Helveticae.*

123 A French edition, translated by E. Gfeller, was planned for 1999.

124 Kienli's work has a touch of the 'rustic' not unlike Harlow's *Trees of the eastern (and central) United States and Canada* (see **Guides to woody plants** under **140**).

125 Summary map C, for example, compares the mapping system of this atlas with the 50 km² grid of *Atlas florae europaeae.*

126 The size of the French flora falls between that of Texas (**171**) and California (**195**), both with highly regarded single-volume manual-floras. Guinochet and Vilmorin's work is marred by too much 'white space', whereas Bonnier and Layens, writing originally for northern France, took a product too far (which after over a century of reissues has also lost some of its clarity).

127 For French botanical history and progress, see A. Davy de Virville, 1954. *Histoire de la botanique en France.* Paris; and R. de Vilmorin, 1975. Données disponibles et lacunes de la connaissance floristique de la France méditerranéenne continentale et de la Corse. In CNRS, *La flore du bassin méditerranéen* (**601/I**), pp. 41–52. A short account also exists in the introduction to Kerguélen's 1993 checklist. Current activities of the Sécretariat de la Faune et de la Flore are summarized on pp. 437–440 of *Atlas partial de la Flore de France* by P. Dupont (see main text under **Distribution maps**).

128 Bonnier died in 1922 while volume 6 was in press and the work was completed by Robert Douin. In 1985, Éditions Belin acquired all rights to the publications of the Librarie Générale d'Enseignement (itself founded by Bonnier). For the reissue of the *Flore complète illustrée* considerable efforts were made to retrieve the original artwork, by then widely dispersed. The reproductions, though brighter than in the original impression, may be to some less satisfactory. That the work was still highly valued, however, was indicated by the exhaustion in four months or so of the first printing of the reissued plates, released in spring 1990. A second printing appeared in the autumn of that year. The *Flore complète portative* remains in print, now over 105 years without substantive revision; this has borne out the prediction of a leading contemporary florist that '[elle] est certainement destiné à un grand et légitime succès à cause de la simplicité du plan . . . et de l'attrait des nombreuses figures qu'elle contient' (Saint-Lager, 1894. *Les nouvelles flores de France: étude bibliographique.* 31 pp. Paris). Without practice, however, it is not that easy to use. For a commentary on the work of Bonnier, in his day a notable 'vulgarisateur', see A. Davy de Virville, 1970. L'oeuvre scientifique de Gaston Bonnier. In P. Smit and R. J. Ch. V. ter Lange (eds.), *Essays in biohistory,* pp. 1–13 (Regnum Veg. 71). Utrecht.

129 Coste's flora is by general consent considered to be a classic among European floras. For the *Flora Europaea* project, it was designated as a 'Basic Flora' of primary importance. Like several others of its genre, its plan follows that of Britton and Brown's *Illustrated flora of the northern [United] States* (see **140**). Completion of the first edition of that work in 1898 stimulated the Paris publisher Klincksieck to realize a similar semi-popular work for France. Connections in the French botanical community put him into contact with Coste in 1899, who agreed to the undertaking. The value of Coste's flora is largely due to the author's insistence on forming descriptions from fresh plants (or specimens) and to Klincksieck's choice of artists. It is also notable for its application of geobotanical data. Flahault's preface is also of great historical importance; many principles of what

became the subdiscipline of phytosociology were there outlined. Coste's voucher collection is now preserved at the present Université de Montpellier (MPU) where Flahault was then professor. See G. G. Aymonin, 1974. La naissance de la 'Flore descriptive et illustrée de la France' de l'Abbé Hippolyte Coste. *Taxon* 23: 607–611.

130 The 1977 issue allocated text and illustrations to separate volumes, with the latter considerably enlarged. For the 1990 issue the illustrations again were reduced but placed at the foot of the text; what resulted was a somewhat elongated format comparable to *The plant-book* (D. J. Mabberley, 1987, 1997). The contents, however, are those of 1935–40.

131 Emberger's system is embodied in its author's textbook *Traité de Botanique (Systématique)*, II: *Les végétaux vasculaires* (Paris, 1960). With a basis in a pleiophyletic origin of angiosperms and containing evolutionary lines based on 'stachyospory' and 'phyllospory', it has not been taken up outside francophone circles; however, in certain respects it foreshadows current thought that the dicotyledons, in contrast to the monocotyledons, are not monophyletic. For France, it was in its time revolutionary in doing away with the divisions of Apetalae, Dialypetalae and Gamopetalae inherited from A.-L. de Jussieu; in its place, five main lines were proposed, with monocots embedded among dicotyledons. – On another note, Mangenot in his introduction to the work noted that this *Flore de France* would not have the 'belle unité' of works such as *Flora der Schweiz* by Hess, Landolt and Hirzel (**649**) or the more local *Flore du Massif Armoricain* by des Abbayes *et al.* as it involved many different collaborators and thus a diversity of styles and personalities.

132 The work commemorates the centenary of the first proposal for systematic mapping of the flora, made in 1889 by G. Rouy.

133 In style the work recalls *Handbook of the trees of the northern states and Canada* by R. B. Hough (**140**). A single-volume issue also exists.

134 Support for the work continued, with the run of 5000 copies of the 1973 edition sold out in less than five years and that of the second edition in four years.

135 Belgian botanical history has been documented in F. Crépin, 1878. *Guide du botaniste en Belgique.* Brussels/Paris (also a topographical work); A. Gravis, 1914. *La botanique en Belgique de 1830 à 1905.* Brussels; and W. Robyns, 1951. Une flore générale de Belgique. *Bull. Jard. Bot. État* 21: 1–10. The last-named covers national floristic works from 1876, and was written in support of the *Flore générale de Belgique* project.

136 In a discussion of floristic undertakings in Europe and elsewhere, Gérard Aymonin likened the work to the several tropical floras then recently initiated at his institution (the Laboratoire de Phanérogamie of the Muséum National d'Histoire Naturelle, Paris). See G. G. Aymonin, 1962. Où en sont les flores européennes? *Adansonia*, II, **2**: 159–171.

137 Luxembourgois botanical history is covered in F. L. Lefart, 1950. Contribution à l'histoire botanique du Luxembourg. *Bull. Mens. Soc. Nat. Luxemb.* **54**: 31–160, 9 pls.

138 For botanical history, see M. J. Sirks, 1935. *Botany in the Netherlands.* 140 pp. Leiden. On FLORON, see 'Vijf jaar FLORON', in *Gorteria* 21(4/5): 133–188 (1995).

139 Rather little time was available after the early 1960s for preparation of *Flora neerlandica* (J. Mennema, 1979. The Rijksherbarium and its contribution to the research on the Netherlands and European flora. *Blumea* 25: 115–119). The work was formally discontinued in the early 1990s.

140 Mennema in *Blumea* 25: 118 (1979) noted that the *Atlas* was prepared at the Netherlands and European Flora section of the Rijksherbarium, Leiden. The lesser amount of specific text in the third volume was the result of an editorial decision by van der Meijden, who succeeded Mennema in the mid-1980s. This change elicited some criticism (Aljos Farjon, personal communication).

141 This was reissued in 1873. From the fourth edition of 1878, however, the figures were separated and issued as *Illustrations of the British flora* by W. H. Fitch *et al.* (1880 (1879), with several revisions and reissues). A companion volume, *Further illustrations of the British flora*, first appeared in 1930 (for further details of both, see main text under R. W. BUTCHER). Altogether 1800 species were illustrated.

142 Preparation and production of a new 'critical flora' was discussed between 1973 and 1985 by a group of interested botanists but then abandoned, with the result that one of the participants, Clive Stace, proceeded with his *New flora*. Peter Sell, with the collaboration of Gina Murrell, then undertook the project beginning in 1987.

143 For histories, see among others J. R. Green, 1914. *A history of botany in the United Kingdom from the earliest times to the end of the 19th century.* London; W. T. Stearn, 1975. History of the British contribution to the study of the European flora. In S. M. Walters and C. J. King (eds.), *Floristic and taxonomic studies in Europe*, pp. 1–17, map. Abingdon, Berks (BSBI Conf. Rep. 15); and D. E. Allen, 1986. *The Botanists.* Winchester: St. Paul's Bibliographies. The latter in particular examines the Botanical Society of the British Isles and its antecedents. The *Cambridge British Flora* and its author are discussed by Allen, pp. 97–103. Ideological and political opposition and, ultimately, financial defeat ended the project; Moss spent his later years at Witwatersrand University. Reviews of the work were mixed, with a better reception in Central Europe.

144 This work is rather more important for groups such as the invertebrates (including arthropods), fungi (which occupy 70 pages!) and non-vascular cryptogams where the literature on the whole is less consolidated. Vertebrates and vascular plants correspondingly have more compact coverage.

145 Keble Martin's work was in its time something of a phenomenon. In his foreword to the 1982 revision, HRH the Duke of Edinburgh noted that the total run of the original work (with reprintings) was 558000 copies. The plates, however, are now considered out of date (Douglas Kent, personal communication).

146 The figures in *Further illustrations* were all by Strudwick. They were reused for the *New illustrated flora* along with some 1500 newly drawn by various artists in succession to those of Fitch and Smith which were presumably too small.

147 The 1987 edition contained numerous mistakes, and generally has been deemed less satisfactory than its predecessor (Douglas Kent, personal communication). It has now fallen out of common use.

148 An associated work is D. H. Kent, 1967. *Index to botanical monographs*. xi, 163 pp. London.

149 The artist particularly acknowledged *British Sedges* (1968) by A. C. Jermy and T. G. Tutin and *Grasses* (1954; 3rd edn., 1984) by C. E. Hubbard; the figures in the latter are her work.

150 The ensemble of *Flora of Great Britain and Ireland* and Stace's *New flora* represents an ideal division of function with respect to floristic works. In contrast to *Flora iberica*, however, no searchable CD-ROM or on-line version of *Flora of Great Britain and Ireland* has so far been released.

151 All descriptions were newly written for the original edition of this work; however, at significant leads some keys may be too limiting. Nearly all the species added for the second edition, were recently established aliens. It would have been helpful to have had the native taxa typographically more clearly indicated. Cambridge University Press acquired the rights to this work prior to its initial publication when the original publishers, Edward Arnold, disposed of their biological sciences list (Maria Murphy, personal communication, 1993).

152 The *Atlas* was the first of its kind to be produced with the aid of mechanized means of data processing and plotting. For further details (with illustrations), see S. M. Walters, 1954. The distribution maps scheme. *Proc. Bot. Soc. British Isles* 1: 121–130, and Allen, *The Botanists*, pp. 154–158. The tabulating machine and card-sorter used for the project are now in the Computer Museum, Boston, Mass., U.S.A.

153 This work is reportedly under revision.

154 It is a botanists' equivalent of the 1930s-era U.S. Works Progress Administration state guidebooks.

155 On the relative merits of earlier Irish floras, see E. C. Nelson and J. Parnell in *Taxon* **41**: 35–42 (1992). Mackay's work was botanically the better; that of Baily was regarded as a 'grind' – as Alphonso Wood's works were viewed by some in North America. However, Mackay sold for 16/-, while Baily's more compact work went for 5/-.

156 Keys were rewritten and units metricated for the sixth edition.

157 Nelson and Welsh's historical account is sobering. Ireland was well forested until 1600; by 1698 conservation laws were needed and by 1800 only 2 percent of the land was under forest.

158 On *Preliminary flora of the Outer Hebrides*, see K. Sabbagh, 1999. *A rum affair: how botany's 'Piltdown Man' was unmasked.* ix, 224 pp., illus. London: Allen Lane/Penguin.

159 A very good work (Peter Sell, personal communication). A mild chauvinism emerges in the preface: of earlier floras, one was 'written by an English botanist after only two visits'. More seriously, the authors indicate that a major influence in the Shetlands landscape – as in the Falkland (Malvinas) Islands (**389**), to which they have been likened and where the average annual temperature is only 1°C lower – is the presence of 300000 sheep. A supplement to Scott and Palmer's flora is projected.

160 The Hartman handbook was expanded to cover all Scandinavia (including Finland) only for the last edition, a development maintained by Holmberg. Neither work, however, was finished.

161 For a Nordic 'state-of-knowledge' collection, dedicated to B. Jonsell, see U.-M. Hultgård, K. Martinsson and R. Moberg, 1996. *The Nordic flora – towards the twenty-first century.* 363 pp., illus. (Symb. Bot. Upsal. 31(3)). Uppsala. An early survey of knowledge in the Baltic States is in K. Winkler, 1877. Literatur- und Pflanzenverzeichnis der Flora Baltica. Dorpat (in *Archiv für die Naturkunde Liv-, Est- und Kurland*, II, 7: 387–490).

162 There is also a 'chromosome atlas': Á. and D. Löve, 1948. *Chromosome numbers of northern plant species.* 131 pp. Reykjavik.

163 The flora was very much a personal undertaking of its author. On a visit to Uppsala University in 1970 some months after Hylander's death, the writer was advised by Rolf Santesson that its continuation was in doubt. Some manuscript for the remaining volumes exists, however, and has been utilized for *Flora nordica* (Bengt Jonsell, personal communication).

164 The impetus for this work arose in the 1970s with a recognition that the national floras were becoming out of date. A distribution atlas is also planned.

165 *Botany of the Færöes* also includes a history of botanical investigation by Warming. Serious topographical botany began in 1895; by 1901, 285 vascular species had been recorded.

166 The *Botany of Iceland* was an occasional series of research contributions initiated in 1912 by L. Kolderup Rosenvinge and E. Warming and continued until 1949. Five volumes, two not completed, were published.

167 For a historical and bibliographical account of this work, see G. Buchheim in *Huntia* 3: 161–178.

168 Lange also co-wrote *Prodromus florae hispanicae* with Moritz Willkomm (**612**) and was chief author of the first major Greenland flora, *Conspectus florae groenlandicae* (**076**).

169 A review by Á. Löve (*Rhodora* **59**: 269–272 (1957)) of the Danish edition has pointed comments with respect to North America: the quality of the illustrations in the *Atlas*, when compared with the relative lack of comparable atlases west of the Atlantic, was 'perhaps . . . a part of the secret affecting the considerably more widespread interest in botany on the eastern side of the ocean'.

170 The maps of the fifth edition have been omitted from the current version, which features more critical taxonomic details including infraspecific taxa and nothospecies.

171 An Italian counterpart of Nyman's *Utkast* is *Cronologia della flora italiana* (1909) by Pier A. Saccardo.

172 Skåne in fact was once Danish territory, but fell to Sweden in the seventeenth century.

173 The mapping programme for the *Atlas* ran from 1938 through 1972.

174 A comparable work to Nyman's *Utkast* in Sweden and Saccardo's *Cronologia* in Italy.

175 The University of Dorpat was founded in 1632 and the herbarium in 1802; the latter antedates all others in the former Russian Empire except those of Vilnius University (before 1831), Moscow University and the Academy and Imperial Botanical Garden in St. Petersburg (with the two latter not effectively active until after 1820). Carl Ledebour was professor of botany there in 1811–36. Among his Germanic successors was Moritz Willkomm, the noted student of the Iberian flora, in 1868–73. In 1893 the university was Russified, being renamed the University of Jurjew. In the early twentieth century, Nikolai Kuznetsov, a noted plant geographer and specialist on the Caucasus, as well as the Polish botanist Bolesław Hyrniewiecki, were on the staff. The university was for a short time also home to the meteorologist and proponent of continental drift, Alfred Wegener. Its historical importance in floristic botany, particularly in the early nineteenth century, partly derives – like the 'Vilna (Vilnius) School' in former Poland-Lithuania – from its relatively unique situation. The comparative proximity of St. Petersburg doubtless ensured continuing relations as its botanical institutions developed.

176 For histories of botany in Estonia, see L. Laasimer (ed.), 1969. *Plant taxonomy, geography and ecology in the Estonian S.S.R.* 112 pp., illus. Tallinn: 'Valgus'; and *idem*, 1975. *Some aspects of botanical research in the Estonian S.S.R.* 238 pp., illus., maps. Tartu. An earlier survey is that in Winkler (1877; see Note 161).

177 For an introduction to the Grossheim system, see A. A. Grossheim, 1945. K voprosu o grafičeskom izobrazenii sistemy tsvetkovykh rastenij. *Sovetsk. Bot.* **13**(3): 3–27.

178 Some comments on Latvian botanical history are available in V. Mühlenbach, 1985. Reflections of an oldtimer on the flora of Latvia. *Phytologia* **58**: 305–323, and in L. Tabaka, 1974. Основные этапы развития флористичецких исследований в Латвии (Principal periods in the development of floristic research in Latvia) (see **Bibliography**).

179 S. B. Jundziłł was a student of Georg Forster, who taught at the university from 1784 to 1787. On the 'Vilna School', see W. Grębecka, 1993. The Vilna school of botany. In A. and B. Zemanek (eds.), *Studies on the history of botanical gardens and arboreta in Poland*, pp. 59–76 (Polish Botanical Studies, Guidebook Series 9). Kráków. With two centers (the other being in Kremenets, in western Ukraine near Rovno), the 'School' in half a century accomplished the initial floristic documentation of the whole of the *kresy* or borderlands, from the Baltic to the Black Sea – an area geographically comparable to the American Middle West.

180 Some historical information is available in Hyrniewiecki, *Tentamen*.

181 On the Academy garden, see V. L. Nekrasov, 1945. K istorii Botaničesogo sada Akademii Nauk. *Sovetsk. Bot.* **13**(2): 13–37.

182 Many of the botanists active in Russia in the late nineteenth and early twentieth centuries were students of Andrei N. Beketov at St. Petersburg University, one of the best-regarded botanists of his day. Among them was Vladimir L. Komarov.

183 Korshinsky has been additionally known as a Russian Hugo de Vries. See D. P. Todes, 1989. *Darwin without Malthus: the struggle for existence in Russian evolutionary thought.* Oxford: Oxford University Press. [This work as a whole is an examination of a distinctive scientific outlook: one which rejected, or was strongly critical, of pure Malthusianism. See also A. Vucinich, 1989. *Darwin in Russian thought.* Berkeley, Calif.]

184 An 'exchange of views' took place with respect to *Flora Evropejskoj Rossii* during and after its publication; references are given in Lipschitz (1975, p. 102).

185 B. K. Shishkin, the largest contributor to *Flora SSSR*, was a pupil of Krylov, graduating from Tomsk University in 1911.

186 Majevski also wrote spring and autumn floras, both first published in 1886 and revised many times to the 1960s.

187 Fedtschenko and Flerow's *Flora Sibiri i Dal'nego Vostoka*, however, ceased publication in 1924, freeing the authors for other projects.

188 The Komarov Botanical Institute of the Russian Academy of Sciences, the home of many floras, has not been immune to the changes of the 1980s and 1990s. Not only have recurrent funds been reduced, but the effects of years of capital underinvestment have become better known. The main Institute building, built in 1913 and already considered 'outmoded' by S. G. Shetler in the 1960s, was by the early 1990s in poor condition and perhaps beyond economic renovation (C. Jeffrey, 1991. The condition of the Leningrad Herbarium. *Taxon* **40**: 459–460; see also V. I. Grubov and L. A. Sergienko, 1988. [On the condition of herbaria in the USSR.] *Bot. Žurn. SSSR* **73**: 1507–1511). It is, however, 'listed' and subsequent efforts have been devoted to structural conservation and renovation of basic services. Similar problems have been reported at libraries. Regrettably, the main library of the Academy in St. Petersburg, where in 1975 some research for the first edition of this book was carried out, was damaged by fire in the 1980s.

189 This so far does not appear to include the second edition of *Flora Kavkaza*, of which nothing has appeared since 1967.

190 The major geographical regions in this work are those of *Flora SSSR* but without the Arctic being separate; in addition, the western boundary of the Russian Far East region has been revised.

191 On the bizarre phenomenon of parallel translations of the last eight volumes of the *Flora*, with full bibliographic details and commentary, see R. Schmid, 1998. A bibliographically Kafkaesque situation. *Taxon* **47**: 788–790. The original translation project had been financed with US PL-480 surplus currency funds until those for Israel were exhausted. Subsequent arrangements were made through Amerind but scientific editing was substantially delayed, resulting in a 20-year hiatus. In the meantime, Koeltz independently organized a translation and editing programme for the remaining volumes. Both series are considered by Schmid to be competent, with the Koeltz format clearer and that of Amerind/Science Press better translated and edited.

192 *Flora SSSR* was one of the five 'Basic Floras' considered fundamental to the preparation of *Flora Europaea*. A considerable historico-bibliographic and interpretative literature has grown up around this work, which con-

tains some 22 000 pages of text and 1250 plates and may be regarded as a *tour de force* comparable to *Flora brasiliensis*. Fëdorov (1971; see **Region 68/69, Progress**) has stated that the reviews of Linczewski (1966) and Kirpicznikov (1967) give the course of events regarding *Flora SSSR* 'most accurately' [and dispassionately]. Kirpicznikov's review additionally furnishes statistics as well as a consideration of competing taxonomic philosophies in relation to treatments in the *Flora*, including the advantages and disadvantages of the Komarovian method. Bobrov (1965), even in its edited English version, is more personal, as might be expected from one associated with the project 'for the duration'. The remarks of Shetler (1967) in his study of the Komarov Institute are detached but detailed and well informed.

193 The work is a counterpart, albeit of broader scope, to the atlas of United States trees by Little (**100**).

194 For a description of the project, whose philosophy was strongly influenced by *Flora Europaea*, see An. A. Fëdorov, 1971. Floristics in the USSR. *BioScience* (USA) **21**: 514–521. The concept of polytypic species, not recognized in *Flora SSSR*, was accepted. Vol. 7 (1994) was at the publication office when the Commonwealth of Independent States was formed; as a result, the consequent political changes are not reflected in the text. This took effect only with vol. 9 (vol. 8 having appeared before vol. 7). Magnoliales through Passiflorales (families 27–64), Malvales through Rosales (families 79–90) and Myrtales through Umbellales to Elaeagnales (families 94–126) have yet to be published. Large plant groups thus wanting include the Centrospermae, Rosales and Umbellales.

195 Any future revision and production of this work would do well to follow the example of the new *Jepson Manual* (**195**). Interestingly, both editions, like the Californian work, sold out fast following publication, according to Fëdorov in his introduction to *Flora evropejskoj časti SSSR* (1974).

196 This account is part of a collection of seven papers on the results of a 1949–51 survey of the region edited by I. V. Lapin and E. P. Matvejeva. Among areas covered in other papers are plant biology and dynamics, forage plant physiology, vegetation, ornamental plants, and forests.

197 This attracted a critique by Wilibald Besser (1832, in *Flora* **15**, Beibl. 2) and a reply by Eichwald (1833, in *ibid.*, **16**, Beibl. 2).

198 For a historical treatment, see N. V. Kozlovskaja, 1978. *Flora Belorussii, zakonomernosti ee formirovanija, naučnye osnovy ispol'zovanija i okhrany*. 128 pp. Minsk. There are also historical accounts in the bibliography by Osipchik and Gel'tman (under **684, Bibliography**) as well as in the first volume of *Flora BSSR*.

199 Description originally prepared from notes and facsimile pages supplied by M. E. Kirpicznikov, St. Petersburg, and the late E. Hultén, Stockholm, and modified following examination of copies in Geneva and St. Petersburg.

200 Kazan' is on or near the southern boundary of continuous forest, and Samara on or near the northern limit of continuous steppe (or prairie). See L. S. Berg, *Natural regions of the USSR* (1950, New York), p. 351 (map 13).

201 For botanical history, see M. V. Markov, 1980. *Botanika v Kazanskom universitete za 175 let.* 101 pp., 2 portrs. Kazan': Kazan University Press.

202 The area covered is given on p. 4; it goes as far as Smolensk, Kursk (part), Vladimir, Nizhni Novgorod, Mordovia, Chuvash, Yaroslavl, and Kostroma, thus omitting the northwest, Vologda, and the central Volga although furnishing partial coverage of these areas. With less fully descriptive keys than in Majevski's work, it more resembles Robert Dorn's key for Wyoming (**183**) and thus is handier for field use, especially the 1995 edition (20 as opposed to 23 cm). The authors have consulted the new *Flora evropejskoj časti SSSR* (Flora of the European part of the USSR) as published through 1994. As the print run of 1983 copies for the first edition was thought 'miserly', 10000 were produced for its successor. The 1981 version, by contrast, had a run of 100000 copies.

203 For history, see J. Paczoski, 1896. Przyczynek do historiji badań flory krajowej. *Pamiętnik Fizyograficzny* **14**: 145–151; K. M. Sytnik, 1993. Akademija nauk i botanika. *Ukrajins'kyj Bot. Žurn.* **50**(6): 5–12; and *idem*, 1995. Akademična botanika v Ukrajiny: narodžennja pershi kroki, pershi imeni. *Ibid.*, **52**(1): 5–18. [The last paper covers mainly the period from 1918 to 1930.]

204 The original version of this manual was some 10 years in preparation; however, the typescript was lost while with the publishing house, necessitating an all-out rewrite within a year. It is thus regarded as not of 'last resort' (I. Yeremko, personal communication 1996). Reference should now also be made to S. L. Mosyakin and M. M. Fedoronchuk, 1999. *Vascular plants of the Ukraine; a nomenclatural checklist.* 345 pp. Kiev.

205 To the north of the range, in the foothills southwest of Chernivtsi, the geneticist, crop scientist and plant hunter Nikolai I. Vavilov in August 1940 made his last field trip.

206 For a more recent but non-enumerative general account, see S. M. Stojko and L. O. Tasenkewitsch, 1991. Pflanzengeographische Stellung und Schutz von Flora und Vegetation der Ukrainschen Karpaten. *Verh. Zool.-Bot. Ges. Österreich* **128**: 165–177. These authors indicate that the 2012 species of Čopyk's 1977 flora represent 44 percent of the vascular plants of Ukraine; the phytocoenoses correspondingly are intricate.

207 The Garden is also near to the site of the historic Three Power Talks of 1945.

208 Like de Candolle's *Regni vegetabilis systema naturale* of 1818–21 and Hooker and Thomson's *Flora indica* of 1855, this precursor of the *Tentamen* was a first attempt towards an objective later otherwise realized.

209 The 1966 version of this work had its genesis in the evacuation of the Ukrainian Academy of Sciences Botanical Institute to the republic's capital, Ufa, during World War II as well as an interest on the part of one of its senior scientists, M. I. Kotov (see **694**). Kotov, along with fellow Kievian botanists Barbaricz and Lypa, was part of its collective authorship.

Division 7

• • •

Northern, central and southwestern (extra-monsoonal) Asia

On a publié des catalogues des flores locales, mais tous [les] riches matériaux étaient épars, sans liaison entr'eux, souvent difficiles à consulter; il était [donc] indispensable de les reunir, de les comparer, de les relier ensemble, et c'est le travail que j'aborde aujourd'hui.

Boissier, *Flora orientalis*, 1: i (1867).

The methodological basis of the work on the *Flora URSS* was the morphological–geographical method, the concept of the race as an actual biogeographical unit. That [this method] was generally adopted . . . is by no means fortuitous. Those who live and work in the USSR . . . cannot but think in terms of geography . . . This ideological trend and the progressive character of the *Flora URSS* are frequently ignored, and are possibly not sufficiently well understood by foreign botanists.

E. G. Bobrov, *Nature* 205: 1048 (1965).

For me, the genesis of any Flora is always an interesting topic . . . The genesis of [the *Flora of Turkey*, *Flora Iranica* and *Flora of Pakistan* are] particularly so in that all of them owe their inception and development to an individual – not to a government, or an institute or an advisory committee.

I. C. Hedge in *Plant life of South Asia* (eds. S. I. Ali and A. Ghaffar), p. 32 (1991).

Generally speaking, Division 7 comprises that half of the Asiatic continent which is for the most part boreal, cool-temperate, semi-arid or arid; in other words, it is largely beyond the influence of the summer monsoon. The sub-Arctic and Arctic zones are excluded; for these, see Regions 05 and 06. The western limits correspond to those of Division 6 (Europe), thus taking in the Caucasus (as Region 74), while the eastern and southern limits border for their entire length on Division 8 (southern, eastern and southeastern Asia). Within the division, nine geopolitical regions have been delimited; those constituting the Asiatic part of the USSR are additionally grouped as Superregion 71–75, while those in southwestern Asia (except the Caucasus) are grouped as Superregion 77–79. That part of central Asia which is outside the USSR remains separate as Region (Superregion) 76. Additional units have been allocated for key physiographic areas, at present the various high-mountain chains along the southern border such as the Karakoram, Pamirs and Altai-Sayan system.

The history of floristic work and current programmes in flora-writing in extra-monsoonal Asia cannot be readily reviewed as a whole, and is therefore accounted for in the introductions to the superregions. It may be noted here, however, that a significant field of development in the floristics of the semi-continent has been in studies of the genesis, spread and present distribution of the high-mountain flora of northern and central Asia, as evidenced by the considerable body of published work now available which includes a number of floras worthy of inclusion in this *Guide*. Also worthy of mention, however, is a real or potential obstacle to the successful early completion of some current floras: the sheer size – anywhere from 1500 to 2000 species – of the legume genus *Astragalus*, which in extra-monsoonal Asia attains its maximum development.

General bibliographies. Bay, 1910; Frodin, 1964; Goodale, 1879; Holden and Wycoff, 1911–14; Hultén, 1958; Jackson, 1881; Pritzel, 1871–77; Rehder, 1911; USDA, 1958. See also superregional headings.

General indices. BA, 1926– ; BotA, 1918–26; BC, 1879–1944; BS, 1940– ; CSP, 1800–1900; EB, 1959–98; FB, 1931– ; ICSL, 1901–14; JBJ, 1873–1939; KR, 1971– ; NN, 1879–1943; RŽ 1954– .

Conspectus

700

Division in general

For comprehensive floras, see supraregional headings.

702

'Old' mountains: The Ural

See **602** (all works).

703

'Alpine' regions of extra-monsoonal Asia

The works accounted for within this unit all relate to the high-mountain or 'alpine' flora of various mountain ranges or systems in extra-monsoonal Asia from the Caucasus to the mountains of northeast Siberia. Among these is one work on the forest flora. Only one treatment, described below, provides something approaching an *overall* treatment of part of the high-mountain flora. All others selected relate only to a single range or system of ranges, and are listed under specific subheadings arranged roughly from west to east. The overall output of floras and related works on the high-mountain flora of extra-monsoonal Asia has been considerable, greatly exceeding even North America (**103**) though not of course Europe (**603**). The greater part of this is the work of Russian and Soviet botanists. – For useful background material on the progress and range of high-mountain floristic, chorological and geobotanical studies in the former Soviet Union, see V. N. SUKACHEV (ed.), 1960. Материалы по изучению флоры и растительности высокогорий (*Materialy po izučeniju flory i rastitel'nosti vysokogorij*).* 304 pp., illus., maps (I. Vsesojuznoe soveščanie [Proceedings] po problemam izučenija i osvoenija flory i rastitel'nosti vysokogorij (Leningrad, 1958)). Moscow/ Leningrad: AN SSSR Press (transl. into English as *idem*, 1965. *Studies on the flora and vegetation of high-mountain areas.* viii, 293 pp., illus., maps. Jerusalem: Israel Program for Scientific Translations); and A. I. TOLMATCHEV (ed.), 1974. Растительный мир высокогорий и его освоение (*Rastitel'nyj mir vysokogorij i ego osvoenie*). 339 pp., illus. Leningrad: 'Nauka'.[1] Alpine and high-mountain areas in western and northern China may be covered in F. KONTA, 1996. *Shina tenzan no shokubutsu.* 228 pp., col. illus., col. map. Osaka: Tosobo.

TOLMATCHEV, A. I. (ed.), 1974. Эндемичные высокогорные растения Северной Азии: атпас (*Endemičnye vysokogornye rastenija Severnoj Azii: atlas*). 335 pp., 228 maps. Novosibirsk: 'Nauka'.

An atlas of distribution maps of 228 mostly herbaceous high-mountain endemic species depicting their spread throughout northern Asia, with a separate section of signed commentaries (by a sizeable number of contributors) containing notes on chorology and taxonomy, karyotypes, and references; bibliography (pp. 95–98). A brief preface is given on pp. 3–4.[2]

---

*For a lexicon of Russian words used in titles see Table 1, p. 618.

I. High Caucasus and Transcaucasus

Comprises the Bolshoi (High) Kavkaz (including 5633 m Mt. El'bruz) and the Malyi Kavkaz (including 5165 m Mt. Ararat, now in Turkey, and 4095 m Mt. Aragats). No successor to the key and enumeration of Medwedew (1915–19) has been published.

MEDWEDEW, JA. S., 1915–19(–18). Список высокогорных растений Кавказа с пособием для их определенния (*Spisok vysokogornykh rastenij Kavakza s posobiem dlja ikh opredelenija*). [2], 592, xxiv pp. (Trudy Tiflissk. Bot. Sad 18). Tiflis (Tbilisi). (Растительность Кавказа, I. Область бысокогорной растительности Кавказа (*Rastitel'nost' Kavakza, I. Oblast' vysokogornoj rastitel'nosti Kavkaza*), Приложение (*Priloženije*) [Appendices], 1–2.)

The *Spisok*, the first of the two 'appendices', comprises a manual-key to vascular plants at altitudes from 1830 m upwards with indication of distribution according to six main geographical units (and their subdivisions) as well as habit. This is followed by an index to genera (pp. 581–586) and list of references (pp. 587–592). The second 'appendix' (*Kritiko-sistematičeskie primečanija k spisku vysokogornykh rastenij Kavakza*, pp. i–xxiv) comprises 55 critical memoranda relating to taxonomic problems signaled in the manual-key. The substantial general part (*Obščaja čast'*; 108 pp. with three chapters) is devoted mainly to phytogeography (the third chapter containing tables of plant classes and vegetatiological units, the latter with rainfall parameters). [No further contributions in this series were published. Some territory covered in the work is now in present-day Turkey.]

II. Elburz Ranges (Iran)

Several contributions, or sets thereof, have been published. Besides that described below, there are works by V. Lipsky (1894), F. A. Buhse (1899), J. Bornmüller (1904–08), and A. Gilli (1941); details appear in regional bibliographies (see **Superregion 77–79**).

BORNMÜLLER, J. and E. GAUBA, 1935–42. Florulae keredjensis fundamentia (Plantae Gaubaeanae iranicae), 1–4 et suppl. 2. 44 pls. *Repert. Spec. Nov. Regni Veg.* **39**: 73–124, 370–372; **41**: 297–344; **47**: 52–80, 131–137; **50**: 365–376; **51**: 33–48, 84–112, 209–239. (Also designated as *Repertorium Europaeum et Mediterraneum*, Bd. 4–6, *passim*, with alternative pagination.)

Enumeration with authors, principal floristic citations, localities with *exsiccatae*, and critical notes; summary of work, p. 233, and full indices to genera and to new species and varieties, pp. 233–239 in final installment. [Parts 1–4 (1935–39) comprise the main enumeration, while supplement 2 (1941–42) is a suite of additions. Karaj (Keredj), northwest of Tehran and below the Elburz, was the site of an agricultural research institute which served as Gauba's base.]

III. Hindu Kush

No separate florulas have been seen.

IV. The Karakoram

For other works, see **798**. This highest of ranges in Central Asia forms, with the Pamir (see below), the celebrated 'Roof of the World'; large parts are under permanent ice and snow and its highest point, K2 (Mt. Chogoru or Godwin-Austen) reaches 8611 m. No work has yet appeared to succeed fully the relatively comprehensive treatment by Pampanini (1930–34), although the range has been visited by a number of expeditions in succeeding decades. Hartmann's papers represent important supplements to Pampanini's work as well as that by Kitamura (see **798**), and that of Dickoré a complete revision of monocotyledons.

DICKORÉ, W. B., 1995. *Flora Karakorumensis*, I: *Angiospermae, Monocotyledoneae*. x, [viii], 298 pp., unnumbered pp. distribution maps, 9 text-figs., 3 col. pls. (Stapfia 39). Linz.

Detailed documentary enumeration of 282 species with keys, synonymy, references to illustrations, localities with *exsiccatae* and literature citations, indication of regional and overall distribution, altitudinal range, frequency, habitat and associates, species in neighboring areas, and critical notes; survey of vegetation, phytogeography, history, biology and biosystematics, list of references (pp. 255–268), indices to all botanical names and to families, synoptic list (pp. 293–297), and full-page distribution maps of selected species at end. The general part covers the plan and scope of the work, conventions, sources, background, geography and topography, climate, and general features of the flora but a general key is wanting. [Volumes on dicotyledons are in preparation.]

HARTMANN, H., 1966. Beiträge zur Kenntnis der Flora des Karakorum. *Bot. Jahrb. Syst.* 85: 259–409, 8 text-figs. (incl. map).

Incorporates a systematic enumeration of 340 vascular plant species collected in a range of 3000–5000 m (the upper limit being the average summer snow line) in the central Karakorum (in the vicinity of the Biafo Glacier; map, p. 263), with indication of habitat and local, regional and general distribution as well as taxonomic commentary; index to families (p. 404) and references (pp. 406–409). The introductory part includes a review of past work along with a consideration of physical features, climate, vegetation, plant physiognomy, habitations and localities and the author's own itinerary. For additions, see *idem*, 1984. Neue und wenig bekannte Blütenpflanzen aus Ladakh mit einem Nachtrag zur Flora des Karakorum. *Candollea* 39: 507–537, 13 halftones.

PAMPANINI, R., 1930. La flora del Caracorùm. In Spedizione Italiana De Filippi nell' Himàlaia, Caracorùm e Turchestàn Cinese (ed.), *Relazione scientifiche della Spedizione italiana De Filippi nell'Himàlaia, Caracorùm e Turchestàn Cinese, 1913–14*, II: *Geologici e geografici* (dir. G. Dainelli), 10: 1–290, 32 text-figs. (incl. maps), pls. 1–7 (incl. 1 map). Bologna: Zanichelli. Continued as *idem*, 1934. Aggiunte alla flora del Caracorùm. In *ibid.*, 11: 143–178, pls. 7–10. Bologna.

Critical systematic enumeration of the vascular plants, bryophytes and fungi of the Karakoram north of the upper Indus River from Gilgit southeastwards to Lake Pancong, with descriptions of new taxa; includes full synonymy, references, citations, indication of *exsiccatae* and other records with localities, some taxonomic commentary, and notes on habitat, special features, etc.; index to genera at end. The introductory section includes chapters on the history of botanical work in the area (with provision of individual route maps), the itinerary of the De Filippi expedition, and a list of maximum altitudes for various species, while following the enumeration proper is a full bibliography. All known records have been incorporated, so this work is retrospective and thus fundamental to all floristic studies in the Karakoram. The 1934 supplement covers additions arising from a return expedition by Dainelli

in 1930, and a corresponding revision of phytogeographical tables.

V. Kunlun Shan

No separate florulas are available for any part of this remote northern bordering range of the Tibetan Plateau, whose highest peak is Ulugh Muztagh or Mu-tzu T'a-ko (7723 m) in the central part of the range.

VI. The Pamir

Like the Karakoram, the Pamir forms part of the so-called 'Roof of the World'. Its highest summit is the former Pik Kommunizma (Mt. Communism; 7495 m). The central plateau lies mainly to its south within the Tajik Republic, with extensions into northeastern Afghanistan and extreme western Xinjiang. The early *Flora pamira* (1903), its five supplements (1904–14) and the *Opredelitel' pamirskikh rastenij* or manual (1907) by Olga A. Fedtschenko, all published in *Trudy Imperatorskago S.-Peterburgskago Botaničeskago Sada*, were supplanted in 1963 by Ikonnikov's 'alpine flora', which remains current.

IKONNIKOV, S. S., 1963. Определитель растений Памира (*Opredelitel' rastenij Pamira*). 281 pp., 31 pls., map (Trudy Bot. Inst. Akad. Nauk Tadžiksk. SSR 20). Dushanbe.

Manual-key to vascular plants of the Pamir, with limited synonymy, Russian and local vernacular names, fairly detailed general indication of local range, brief summary of extralimital distribution, taxonomic commentary, and notes on habitat, ecology and phenology; list of references and indices to all botanical and other plant names. The introductory section incorporates chapters on physical features, climate, floristics, phytogeography, and botanical exploration.

VII. Tian Shan

This extensive complex of mountains, with several more or less parallel and overlapping ranges of different geological ages and with altitudes ranging up to 7439 m (Pik Pobedy), runs north of the Syr Darya from just east of Tashkent across Middle Asia (largely within Krygyzstan to the south of Bishkek) and into western Xinjiang in China, therein (with a maximum altitude of 5445 m in Mt. Pokota near Urumqi) separating the Great Tarim Basin to its south from Dzungaria to its north. The latest independent treatment, with keys and descriptions, is the following: B. A. FEDTSCHENKO, 1904–05. Флора западного Тянь-Шаня (*Flora zapadnogo Tjan-Šanja*), [pts. I–II]. (Trudy Imp. S.-Peterburgsk. Bot. Sada **23**: 249–532; **24**: 155–260). St. Petersburg. However, by far the greater part of the range outside China is now encompassed by the more recent *Flora Kirghizskoj SSR* (**755**) and other relevant floras of Central Asia, as well as *Plantae asiae centralis* (GRUBOV; see **760**).

VIII. Altai-Sayan mountain system

This group of ranges in southern Siberia consists of three major divisions, each here treated separately: the Altai Mountains, the Western Sayans, and the Eastern Sayans. Separate works of relatively recent vintage are available for the latter two ranges, along with the forest flora of I. Koropachinsky (1975) and the 'alpine flora' by Leonid Malyschev (1968). The latter work is comparable to *Rocky Mountain flora* by William A. Weber (**103/II**).

MALYSCHEV, L. I., 1968. Определитель высокогорных растений Южной Сибири (*Opredelitel' vysokogornykh rastenij južnoj Sibiri*). 284 pp., 37 text-figs., map. Leningrad: 'Nauka'.

Manual-key to vascular plants (840 species) of the alpine regions of southern Siberia (inclusive of the Altai and the Sayans), with limited synonymy, concise indication of local range, Russian binomial and other vernacular names, and notes on habitat, life-form, phenology, etc.; list of references and indices to all botanical and other names at end.

Altai

This range, with its highest peak (Belukha) reaching 4506 m, arises in the extreme southeastern part of Western Siberia and extends far into Xinjiang (Sinkiang) in China, where it forms the boundary between that territory and the Republic of Mongolia. To its south lies the Dzungarian Basin. It has the distinction of being among the earliest 'alpine' areas in extra-monsoonal Asia to be botanically explored; but until the appearance of Revushkin's monograph in 1988 no specific

treatment of the mountain flora (said to exhibit affinities with that of the central Rocky Mountains in Colorado, U.S.A.; see **103/II**), had been published since before 1840. Key predecessors include *Flora altaica* by K. F. Ledebour (1821–34) and the seven-volume *Flora Altaja i Tomskoj gubernii* (1901–14) by the Tomsk professor Porphyry Krylov. Modern works for regional identification include Krylov's monumental *Flora Zapadnoj Sibiri* (**710/I**), the current *Flora Sibiri* begun in 1988 (**710**) and Malyschev's manual-key for the southern Siberian mountains (see above).

REVUSHKIN, A. S., 1988. Высокогорная флора Алтая (*Vysokogornaja flora Altaja*). 318, [1] pp., 28 text-figs. (incl. maps). Tomsk: Tomsk University Press.

Includes (pp. 26–134) an enumeration of 996 vascular plants with limited synonymy, essential literature citations, habitat and local distribution, indication of frequency, and taxonomic commentary. Families and genera are arranged following *Flora SSSR* but species appear alphabetically. The introductory part covers the background in exploration, topography, climate, soils and vegetation, while following the formal account are analytical chapters on the taxonomic spectrum, altitudinal zonation, relationship with climate, ecology and chorology as well as on the 'structure' of the mountain flora (in 22 selected sites; maps, pp. 182 and 200ff.) and its typology (following the *konkretnaja flora* concept); summary (pp. 247–250), bibliography, index to genera (pp. 261–264) and table of presence/absence of every species for the 22 study sites (pp. 266–309) at end.

Western Sayans

The Western Sayans form the boundary between Tuva and Krasnoyarsk Krai, and are somewhat lower than their eastern and western neighbors.

KRASNOBOROV, I. M., 1976. Высокогорная флора Западного Саяна (*Vysokogornaja flora Zapadnogo Sajana*). 379 pp., 32 figs., 96 maps, tables. Novosibirsk: 'Nauka'.

Concise keyed enumeration of high-mountain vascular plants, with synonymy, references, local distribution (with dot maps for all species), and notes on habitat, special features, potential, etc.; tabular floristic summary with for each species karyotypes, range type, soil preferences and record of presence/absence; bibliography (pp. 289–298) and complete indices at end. An extensive introductory section accounts for botanical exploration (with portraits and itineraries), physical features, climate, vegetation formations with *Charakterpflanzen*, and statement and review of species concepts.

Eastern Sayans

Situated along the southern fringe of Cisbaikalia, with a maximum altitude of 2922 m, the Eastern Sayans arise to the south of Krasnoyarsk and run southeastwards, following the boundaries of Tuva and Mongolian People's Republic until petering out to the south of Lake Baikal.

MALYSCHEV, L. I., 1965. Высокогорная флора восточного Саяна (*Vysokogornaja flora vostočnogo Sajana*)/*Flora alpina montium Sajanensium orientalium*. 368 pp., illus., maps. Moscow/Leningrad: 'Nauka'.

A primarily geobotanical and biogeographic work but including a briefly descriptive flora of 540 species with diagnostic keys, synonymy, citations, Russian binomials, fairly detailed indication of local range (with some dot maps), and notes on habitat, special features, ecology, etc.; list of references (pp. 353–362) and indices at end of work.

Woody plants

KOROPAČINSKIJ, I. JU., 1975. Дендрофлора Алтайско-Саянской области (*Dendroflora Altajsko-Sajanskoj gornoj oblasti*). 290 pp., 141 illus. (incl. halftones, maps). Novosibirsk: 'Nauka'.

A descriptive treatment of the woody plants of the southern Siberian mountains (including the whole of Tuva), with keys, synonymy and references, regional distribution (including dot maps for each species), and extensive remarks on external features and geographical patterns as well as uses and properties; this is followed (pp. 226ff.) by a floristic analysis of the region covered, characteristic species in each 'type', and history and dynamics of the flora and vegetation. A table of altitudinal zones, bibliography, and index to botanical names appear at the end. In the introductory section are illustrated accounts of physical features, climate, vegetation and some altitudinal and aspect zonation, vegetation zones and formations, forest composition with size ranges, and statistics of the woody flora. [The work covers substantial parts of southern Siberia and adjacent Middle Asia and Mongolia, and is an expansion of an earlier work by the author on Tuva (1966; see **724**).]

IX. The Stanovoy Uplands

This area, northeast of Lake Baikal, generally lies just south of the line of the Baikal–Amur Magistral (BAM) railway. A useful introduction is the analysis by Malyschev (1972; see below).

MALYSCHEV, L. I. (ed.), 1972. Высокогорная флора Станового Нагорья: состав, особенности и генезис (*Vysokogornaja flora Stanovogo Nagor'ja: sostav, osobennosti i genezis*). 270 pp., 7 figs. (incl. maps), tables, 318 distribution maps. Novosibirsk: 'Nauka'.

Pages 36–149 of this work encompass a systematic enumeration, without keys (by the editor with several collaborators), of more than 600 species of vascular plants in a range of *c.* 1200–2200 m (the latter the apparent limit of growth: the range reaches a maximum

of 2999 m); entries include essential synonymy, citations, local geographical and altitudinal ranges (with on pp. 206–259 dot distribution maps for many species) and sometimes extensive geographical and ecological notes. The introductory parts include a history of botanical work, the physical geography and other features of the habitat, vegetation formations, and floristic units, along with a review of new taxa and records, while the enumeration proper is followed (pp. 150ff.) by the analytical part, with sections on the composition of the flora and its origins (55 percent of the species have arctic affinities, 22 percent with western and central Europe), relics, endemism, and a discussion of the forms of evidence employed; references (pp. 260–265) and indices to families and genera at end.

X. Mongolian Alps

For other works see **761**. The Changaja are situated in the central-western part of Mongolia, west of Ulan Bator.

BJAZROV, L. G., E. GAN'BOLD, I. A. GUBANOV and N. ÜLZIJCHUTAG, 1989. *Flora Khangaja/ Khangajn urgamlyn ajmag*. 192 pp., illus. Leningrad: 'Nauka'. (Biologičeskie resursy i prirodnye uslovija MNR 33.)

Enumeration of vascular plants with names, essential synonymy, distribution, habitat, and selected *exsiccatae*; statistics (pp. 157–159, with table), references (pp. 162–164) and index to all botanical names at end. The general part includes a section on floristic regions (with map, p. 14; six main units are recognized) along with others on physical features, climate, rainfall and vegetation. [1468 vascular plants are accounted for over an area the size of France.]

705

Drylands

Steppes (or prairies) and deserts make up a large proportion of the area of Division 7, and in addition to these biomes in their own right, there not unnaturally has also been an interest in the intergradation of their floras with those of the forests. But, because of relatively late formal scientific development – the first modern universities in all Division 7 being the American College (later University) in Beirut, founded in 1865, and Tomsk University, established in 1885 – and pressures or sensibilities for working up floras for politically delimited areas before any concerning physiographic and/or vegetational units few, if any, specific works of suitable scope yet exist.

Siberia

PESCHKOVA, G. A., 1972. Степная флора Байкальской Сибири (*Stepnaja flora Bajkal'skoj Sibiri*). 207 pp., 9 text-figs. (in 10), 23 tables. Moscow: 'Nauka'.

Incorporates (pp. 38–105) an enumeration of 710 species in the Lake Baikal region of Eastern Siberia with steppe ('prairie') affinities; entries include references and indication of frequency and habitat as well as some commentary. The introductory part furnishes a historical review as well as discussions of physical features, vegetation, and floristic regions along with the author's investigations, while the enumeration proper is followed by an analysis with coverage of the various floristic elements represented in the steppe-related flora (seven in all, but with endemism of only 3.7 percent) as well as paleogeography, vegetation history and the influence of man; references (pp. 186–201) and indices to families and genera.

China and Mongolia (including the Gobi)

Norlindh's never-completed flora has been complemented by a descriptive work prepared under the editorship of Y.-H. Liu. For alternative coverage, see **760, 761** and **763.** Pteridophytes are, however, found only in Norlindh's account.

LIU YING-HSIN (LIU YING-XIN) (ed.), 1985–92. (*Chung-kuo sha mo chih wu chih*)/ *Flora in desertis reipublicae populorum sinarum.* 3 vols. Text-figs., maps. Beijing: K'o hsüeh chu pan she (Academia Sinica Press).

Illustrated descriptive flora of seed plants with Latin and Chinese names, synonymy, references, literature citations, and indication of distribution and habitat (with also for some species critical commentary);

descriptions of new taxa and indices to all names at end of each volume. [Covers Xinjiang, western Gansu and Inner Mongolia, thus encompassing the Takla Makan, Dzungaria and the Gobi. Vol. 1 encompasses gymnosperms, monocotyledons and dicotyledons from Salicaceae through Menispermaceae; vol. 2 (1987), Papaveraceae through Umbelliferae; vol. 3, Primulaceae through Compositae.]

NORLINDH, T., 1949. *Flora of the Mongolian steppe and desert areas.* Fasc. 1. 155 pp., ix text-figs., 16 pls., maps (Reports from the scientific expedition to the northwestern provinces of China under the leadership of Dr Sven Hedin 11 (Botany, 4)). Stockholm.

Systematic enumeration of vascular plants, with keys to species within genera, full synonymy, references and citations, detailed indication of local range (with several dot maps), summary of overall distribution, biological and horticultural notes, and extensive critical commentary. An introductory section includes an account of prior botanical exploration in the region and the work of the expedition. [Not completed; covers pteridophytes, gymnosperms, and monocotyledons from Typhaceae through Gramineae on the Englerian system.]

Southwestern Asia

In addition to the coverage provided by current and forthcoming national and regional floras, reference may also be made to the following study by Jean Léonard, given below. For the Thar (Indian) Desert, see **805.**

LÉONARD, J., 1981–89. *Contribution à l'étude de la flore et de la végétation des deserts d'Iran.* Fasc. 1–9. Illus., maps. Meise: Jardin Botanique National de Belgique.

Fasc. 1–7 comprise an introduction, bibliography (1, pp. 15–17) and systematic enumeration of 509 species, with keys, essential synonymy, localities with *exsiccatae* and field data, and geographical distribution; separate indices to all names in each fascicle with in fasc. 7 (pp. 119–125) a cumulative index to families and genera. Fasc. 8–9 are phytogeographical, with an exposition of the author's chorological system in fasc. 8 and comparison with other schemes in fasc. 9. [Based primarily on the results of a 1972 expedition.]

707

Lake Baikal and environs

Among the outstanding natural features of northern Asia is Lake Baikal. This unique and spectacular body of fresh water, the deepest in the world, is furthermore surrounded by high mountains. Its littoral flora was first documented by M. Popov and V. Busik (1966; see below). Further additions may be found in the contributions featured in L. I. MALYSCHEV and G. A. PESCHKOVA (eds.), 1978. Флора Прибайкалья (*Flora Pribajkal'ja*). 320 pp., 6 tables. Novosibirsk: 'Nauka'.[3] For other coverage see **725, 726** and **727** (all works, but especially *Flora tsentral'noj Sibiri* by MALYSCHEV and PESCHKOVA, 1979).

POPOV, M. G. and V. V. BUSIK, 1966. Конспект флоры побережий озера Байкал (*Konspekt flory poberežij ozera Bajkal*)/*Conspectus florae litorum laci Baical*. 214 pp., 15 text-figs. (incl. maps). Leningrad: 'Nauka'.

Concise enumeration of the vascular plants along the foreshores of Lake Baikal; includes essential synonymy, citation of *exsiccatae*, and indication of localities, followed by diagnoses of new taxa and a bibliography; no index. The introductory section incorporates chapters on the physical features of the lake and its environs, floral regions, and botanical exploration past and present.

Superregion 71–75

CIS-in-Asia

The whole of this superregion corresponds to the Asiatic part of the former Soviet Union, including Siberia (Region 71/72), the Russian Far East (Region 73), the Caucasus (Region 74), and Middle Asia (Region 75). All 'standard' regional and other floras for this area are here accounted for, except for those few pertaining entirely to the sub-Arctic and Arctic zone (Regions 05 and 06). The remainder of Central Asia constitutes Region 76.

A pioneer work, in part preceding the advent of binomial nomenclature, was *Flora sibirica* by Johann G. Gmelin (1747–69). This was followed by the first regional floras: Marschall von Bieberstein's *Flora taurico-caucasica* (1808–19) and the already-mentioned *Flora altaica* by Karl Ledebour and others (1829–34). The next work to cover the whole of Asiatic Russia was Ledebour's *Flora rossica* (1842–53), extending to include Alaska (**110**) but written when botanical exploration was not particularly far advanced. Additional regional floras also appeared in the mid-nineteenth century; among these were *Flora baicalensi-dahurica* (Turczaninow, 1842–56), *Primitiae florae amurensis* (Maximowicz, 1859), and the unfinished *Flora caucasi* by Franz Ruprecht (1869; in *Mém. Acad. Sci. St.-Pétersb.*, VII, **15**(2)), each respectively a source work for Regions 71 through 74 (Region 75, Middle Asia, was then mostly not yet part of the Russian Empire).

After these mid-century developments, however, further efforts at comprehensive coverage did not take place until the 1890s. A general rise in development in Russia then began which had a positive effect on botanical (and floristic) research. This manifested itself in the appointment of additional staff to the two large botanical institutions in St. Petersburg, the Imperial Botanic Garden and the Botanical Museum of the Academy of Sciences. With greater opportunities, both institutions launched major projects on the Russian flora. The Botanic Garden instituted a somewhat mutable programme of critical regional works, in Asia covering (a) Siberia, under Boris A. Fedtschenko and Aleksandr F. Flerow (*Illjustrirovannyj opredelitel' rastenij Sibiri*, 1909, not completed save for the first two parts, on pteridophytes and gymnosperms); (b) the Caucasus, under Nikolai I. Kuznetsov (at what is now Tartu University in Estonia) and others (*Flora caucasica critica*, 1901–18); (c) Middle Asia, under V. I. Lipsky (*Materialy dlja flory Srednej Azii*, I–III, 1900–09 (covering 480 species), and *Flora Srednej Azii*, 1902–05, of which only the 'background' volume was ever published), and then under Olga A. and B. A. Fedtschenko (*Conspectus florae Turkestanicae*, 1906–16 in six parts and covering 4145 species, and *Perečen' rastenij Turkestana i Kirgizskogo kraja*, 1924 (in *Trudy Glavn. Bot. Sada* **38**(1), afterwards discontinued));[4] and lastly (d) Asiatic Russia in general under B. A. Fedtschenko (*Flora asiatskoj Rossii*, 18 fascicles in two series, 1912–24). The Museum, with lesser resources, established a series entitled *Flora Sibiri i Dal'nego*

Vostoka (1913–31, in six installments), produced under the direction of I. P. Borodin and Dimitri I. Litwinow. As a precursor to the Museum series, Litwinow prepared a historical review (1908) and a floristic bibliography (1909). At the same time, one-volume manuals for identification were promoted for 'on-the-spot' use. Lipsky wrote *Flora Kavkaza* (1899; supplement, 1902) covering 4430 species, while B. A. Fedtschenko contributed *Rastitel'nost' Turkestana* (1915), an illustrated work covering 5031 species. Ja. Medwedew contributed a separate manual for high-elevation Caucasian species (1915–19; see **703/I**). Much new collecting was also accomplished in the two decades before World War I as topographical and resource surveys were carried out, activities resumed in the 1920s with the return of internal order.

Gradual imposition of central control over the sciences as in all other areas of life was under Soviet rule inevitable. Widespread administrative changes and critical project review began from the mid-1920s. Transfer of the Botanic Garden to the Academy of Sciences took place in 1925, and in 1931 came the merger of the Botanical Museum with the Botanic Garden to form what is now the Komarov Botanical Institute. Special mention of wasteful duplication of effort was made in the critique for this rationalization, notably the parallel flora projects on Asiatic Russia; by 1931 both had been discontinued, neither covering more than 10 percent of the possible flora. In 1929, as part of the programme for the first Five Year Plan (1929–34), planning for an All-Union flora was initiated; the discredited projects were terminated and in 1931–32, concomitant with the merger, work began on *Flora SSSR*, ultimately completed in 1964. The two earlier projects, however, provided a useful basis for the new flora. For woody plants, the *Flora* was later complemented by Sokolov's *Derev'ja i kustarniki SSSR* (1949–62).

The *Flora SSSR* achieved a tremendous task: effective consolidation in modern terms of floristic knowledge for most of northern Asia. The project, built upon large new collections resulting from systematic biological surveys carried out in many parts of the Soviet Union as well as other available resources, has provided a major basis for subsequent regional, republican, territorial and more local floras, enumerations and manuals in the Asiatic parts of the Union. Many of the later floras were prepared as part of planned scientific development and the survey and inventory of

natural and other resources in a given area, at first under guidance from Leningrad (and Moscow) and later from (or together with) republican or regional institutions. This continues to the present, with such works as the Tashkent-based *Opredelitel' rastenij Srednej Azii* (*Conspectus florae Asiae mediae*) (1968–93) and *Sosudistye rastenija sovetskogo Dal'nego Vostoka* (1985–), a project of the Far East Centre of the present Russian Academy of Sciences at Vladivostok. A botanical research group at the Academy of Sciences Siberian Branch in Novosibirsk has become active in recent decades, particularly in high-mountain floristic and vegetation studies as well as studies of Lake Baikal and its vicinity. This group was also responsible for *Flora tsentral'noj Sibiri* (1979) and may now be considered a leading contributor in floristic research in Siberia. In 1981 the group, in conjunction with another at Tomsk University, launched a major new initiative, a complete *Flora Sibiri*; this was completed in 14 volumes in 1999. It effectively succeeds all other Siberian floras, including the good, but aging, *Flora zapadnoj Sibiri*. With respect to the now-independent states of the Caucasus and Middle Asia, *Opredelitel' rastenij Srednej Azii* was, as already noted, completed in 1993 but the second edition of *Flora Kavkaza* appears to have been abandoned unfinished. *Flora Tadžikiskoj SSR*, the last of the multi-volume republican floras in Middle Asia, was completed in 1991; however, *Flora Armenii* and the second edition of *Flora Gruzii* remain unfinished.

Progress
For history, see I. P. BORODIN, 1908. *Kollektory i kollektsii po flore Sibiri*. 245 pp. (Trudy Bot. Muz. Imp. Akad. Nauk 4). St. Petersburg; G. V. KRYLOV and N. G. SALATOVA, 1969. *Istorija botaničeskikh i lesnykh issledovanij v Sibiri i na Dal'nem Vostoke*. Novosibirsk; and L. I. MALYSCHEV, 1979. Razvitie botaniki v Sibiri v Sovetskij period. *Bot. Žurn. SSSR* 64: 112–120. Other accounts are cited under **Region 68/69**. Early work in the Caucasus is reviewed by V. I. LIPSKY in his *Flora Kavkaza* (1899; see **Region 74**), and in Middle Asia likewise by Lipsky in his pioneering work *Flora Srednej Azii* [*Flora Asiae mediae*] (1903, pp. 249–337; see **Region 75**).

General bibliographies. Bay, 1910; Frodin, 1964; Goodale, 1879; Holden and Wycoff, 1911–14; Hultén, 1958; Jackson, 1881; Pritzel, 1871–77; Rehder, 1911; USDA, 1958.

Supraregional bibliographies

The two following works both have separate sections for the major parts of former Soviet Asia: the Caucasus, Siberia and the Far East, and Middle Asia.

LEBEDEV, D. V., 1956. Введение в ботаническую литературу СССР (*Vvedenie v botaničeskuju literaturu SSSR*). 382 pp. Moscow/Leningrad: AN SSSR. [Described in **Region 68/69** under **Regional bibliographies**.]

LIPSCHITZ, S. JU., 1975. Литературные источники ио флоре СССР (*Literaturnye istočniki po flore SSSR*). 232 pp., table. Leningrad: 'Nauka'. [Described in **Region 68/69** under **Regional bibliographies**.]

General indices. BA, 1926; BotA, 1918–26; BC, 1879–1939; BS, 1940– ; CSP, 1800–1900; EB, 1959–98; FB, 1931– ; ICSL, 1901–14; JBJ, 1873–1939; KR, 1971– ; NN, 1879–1943; RŽ, 1954– .

710–50

Superregion in general

All works formally cited below are fully described under **680**. They succeed I. P. BORODIN and D. I. LITWINOW (eds.), 1913–31. Флора Сибири и Дальнего Востока (*Flora Sibiri i Dal'nego Vostoka*). Fasc. 1–6 (in 9 parts). St. Petersburg: Botaničeskij Muzej, (I)AN; and B. A. FEDTSCHENKO and A. F. FLEROW (eds.), 1912–24. Флора Азиатской России (*Flora aziatskoj Rossii*). Fasc. 1–15; N.S., fasc. 1–3 (the latter as *Flora Rossii*, ser. 1). St. Petersburg: Glavnyj (Imperatorskij) botaničeskij sad. Both these series comprised revisions of various vascular plant and bryophyte families. The first-named covered only Siberia and the Far East, without Middle Asia or the Caucasus. Although never completed and covering no more than 10–15 percent of the total flora, both projects have provided a basis for subsequent floristic studies as well as *Flora SSSR*.

CZEREPANOV, S. K., 1995. *Vascular plants of Russia and adjacent states (the former USSR)*. x, 516 pp., map. Cambridge: Cambridge University Press. [Based on *idem*, 1981. Сосудистые растения СССР (*Sosudistye rastenija SSSR*)/*Plantae vasculares URSS*. 509 pp. Leningrad: 'Nauka'. Both works are described under **680**.]

KOMAROV, V. L. *et al.* (eds.), 1934–64. Флора СССР (*Flora SSSR*)/*Flora URSS*. 30 vols., index. Illus., maps. Moscow/Leningrad: AN SSSR Press (later 'Nauka'). [Described under **680**.]

SOKOLOV, S. JA. (ed.), 1949–62. Деревья и кустарники СССР (*Derev'ja i kustarniki SSSR*). 6 vols., supplement. Illus., maps. Moscow/Leningrad: AN SSSR Press. [Described under **680**.]

SOKOLOV, S. JA. *et al.*, 1977–86. Ареалы деревьев и кустарников СССР (*Arealy derev'ev i kustarnikov SSSR*). 3 vols. (with portfolios). Leningrad: 'Nauka'. [Described under **680**.]

Region 71/72

Siberia

Area: 9 724 200 km^2. Vascular flora: *c.* 4200 species and additional infraspecific taxa (Krasnoborov *et al.*, 1988–99; see below).

Siberia's vast extent and varied geography, with several centers of development, led inevitably to different courses of exploration. Much of Western Siberia is low-lying and swampy, with its southern reaches (partly now in Kazakhstan) characterized by progressively increasing dryness. There were relatively few natural features to attract explorers apart from the river systems of the Ob' and Irtysh and especially the Altai and other hilly areas in the southeast. Not unnaturally, much collecting was conducted along the trans-Siberian *Trakt*, the earlier Russian counterpart of the nineteenth-century 'National Road' in the U.S.A. and to many a 'road of tears'. By contrast, Central and Eastern Siberia are to a large extent topographically rough, with extensive hilly and mountainous areas and a less integrated river network. Here, too, the *Trakt* as well as the river systems, notably the Yenisei and Lena, and Lake Baikal were the focus of much early botanical exploration. Work was, however, facilitated by the existence of the relatively long-established towns of Krasnoyarsk and especially Irkutsk. Parts of the Arctic coast were also relatively early targets of botanical exploration by both Russians and foreigners as the Arctic Ocean was surveyed and coastal stations were established. Botanical knowledge through the mid-eighteenth century was summarized by J. G. and S. G.

Gmelin in their *Flora sibirica* (1747–69); this included data from the decade-long Siberian expedition (1733–43) conducted by the elder Gmelin on behalf of the Russian Academy of Sciences. This flora remained a standard work until publication of Ledebour's *Flora rossica* a century later. The explorations of Pallas under Catherine II also yielded additions; these were incorporated into his *Flora rossica* and other works.

The Altai was the setting for the first nineteenth-century regional flora, *Flora altaica* (1829–33) by K. F. Ledebour, K. A. Meyer and A. A. Bunge. This was followed in the next decade by the more easterly *Flora baicalensi-dahurica* (1842–56) of Nikolai I. Turczaninow, a civil servant and botanist resident in Irkutsk from 1828 to 1835 and then (as provincial governor) in Krasnoyarsk from 1835 to 1845.[5] This furnished a strong basis for the future in Eastern Siberia. It did not, however, effectively cover the Yenisei Basin and as a consequence early botanical documentation in Central Siberia was piecemeal. Few other significant advances took place until the last quarter of the nineteenth century. In 1888 *Plantae vasculares jeniseensis* by Nils J. W. Scheutz appeared in Stockholm and in the 1890s Stefano Sommier published *Flora dell'Ob Inferiore*, while a first documentation of the flora of Sakha (Yakutia) appeared in *Reisen und Forschungen in Jakutischen Gebiet Ostsiberiens in den Jahren 1861–71* by G. Maydell (1893–96). Perhaps the most significant development, however, was the formation of a botanical school at Tomsk University in 1885. There, its founder Porphyry N. Krylov and, in two periods from 1911 to 1931, his student and associate Boris K. Shishkin, promoted more intensive botanical exploration. In later years this work was continued by Viktor Reverdatto and Lydia P. Sergievskaya; their field of work expanded to cover just about the whole of Siberia, there being until after World War II no other significant regional botanical center.[6] Their work was aided by increasing settlement, development of communications in the wake of the construction of the Trans-Siberian Railway in the 1890s, and subsequent agricultural and resource surveys before and after 1917.[7]

The first significant modern flora in Western Siberia was Krylov's seven-part *Flora Altaja i Tomskoj gubernii* (1901–14). A quarter-century later this was succeeded by the first great Soviet-era regional flora, *Flora zapadnoj Sibiri* (11 vols., 1927–49 and supplement, 1961–64). Begun by Krylov and after his death continued by Shishkin and Sergievskaja, it is based to a

large extent on critical studies, and has long been regarded as a model of its kind.[8] New floras in Central and Eastern Siberia remained few, although Boris A. Fedtschenko commenced *Jakutskaja flora* in 1907. Indeed, in the early twentieth century interest had shifted to documentation of Siberia as a whole through two semi-monographic works, both based in St. Petersburg: *Flora Asiatskoj Rossii* (1913–24) and *Flora Sibiri i Dal'nego Vostoka* (1913–31). Neither of these was, however, completed in the wake of post-World War I administrative and policy changes.

Following the 1917 revolution, Lenin created a number of new universities in what in 1924 became the Soviet Union. One of these was at Irkutsk; there, a local students' manual was produced by V. F. Diagilev in 1938. Elsewhere in Central and Eastern Siberia, however, floristic work was largely the province of Tomsk or Leningrad botanists (the latter including Fedtschenko and Komarov as well as A. I. Tolmatchev); between them they were responsible for *Flora Zabajkal'ja* (1929–), *Flora Jakutii* (1930, not completed), *Flora tsentral'noj časti Vostočnogo Tajmyra* (1932–35; see **066**), and *Konspekt Prijenisejskoj flory* (1937; revived in 1960 as *Flora Krasnojarskogo kraja* and completed in 1983). In Western Siberia the main event was publication of *Flora Urala* (1937) by V. S. Govoruchin, covering the whole of that mountain system.

The post-World War II years saw a further expansion of activities. Among significant developments was the establishment of the Siberian research center of the then-Soviet Academy of Sciences at Novosibirsk; this included both a botanical garden and a research institute. This latter became the most active in floristic research and publication. From the 1950s it was involved in surveys of the many mountain systems as well as hitherto little-collected areas like the Tuva Republic (incorporated into the USSR in 1945), the Stanovoy Uplands, and the alignment of the new Baikal–Amur Railway (begun in 1975) and new geographical discoveries such as the Putorany Plateau (on the divide between the Evenk and Taimyr national areas in Central Siberia). A series of floras and dendrological works were produced by I. Ju. Koropachinsky, I. M. Krasnoborov and L. I. Malyschev for these remote areas in the 1950s and 1960s, followed by *Flora Putorana* in 1976 (see **066**) and *Flora učastka raionov osvoenija BAM* in 1983 for the western part of the new railway. In 1973 the Novosibirsk group brought out

Opredelitel' rastenij Novosibirskoj oblasti; this was followed by other identification manuals, respectively covering Sakha (1978, by A. I. Tolmatchev), southern Krasnoyarsk (1979, by M. I. Begljanova and others), and Tuva (1984, by Krasnoborov). In 1979 the two-volume *Flora Tsentral'noj Sibiri* of Malyschev and G. A. Peschkova appeared; this centers on Lake Baikal. This was, however, merely an 'overture' to what is the group's main achievement, the multi-volume *Flora Sibiri*. This was begun in the mid-1980s as a joint project with Tomsk University under the editorship of Krasnoborov; the first of 14 volumes appeared in 1988 and the last in 1999.[9]

Elsewhere, floras were published in continuation of earlier projects or wherever a strong interest was manifest. A revision of Turczaninow's flora had been undertaken by Mikhail G. Popov while at Irkutsk University; but his *Flora Srednej Sibiri* (1957–59), covering over 2000 species from southern Krasnoyarsk to Lake Baikal, appeared only after his death (in 1955) and lacks analytical keys. Popov was also partly responsible for a flora of the foreshores of Lake Baikal (1965, with V. V. Busik). L. M. Čerepnin, in Krasnoyarsk, produced six volumes of *Flora južnoj časti Krasnojarskogo kraja* (1957–67), which became a basis for the already-mentioned 1979 identification manual. The Tomsk group completed *Flora Krasnojarskogo kraja* in 1983 and have, as already noted, contributed to *Flora Sibiri*.[10] Additional manuals were produced for the Ural, the most recent being *Opredelitel' sosudistykh rastenij Srednego Urala* (1994; see **602/II**); these enjoyed support from the Urals Centre of the Academy of Sciences (in Ekaterinburg).

A more or less 'full' coverage of Siberia has thus been achieved, although considerable areas remain difficult of access and have been but little collected. Some documentary floras, notably *Flora zabajkal'ja*, remain unfinished. Nevertheless, *Flora Sibiri*, when complete, will represent as significant a consolidation of knowledge as did its predecessors.

Bibliographies. General and supraregional bibliographies as for Superregion 71–75.

Regional bibliographies

For Eastern Siberia, see also Superregion 86–88 (MERRILL and WALKER).

AKSENOVA, N. N., 1965. Флора западной Сибири: библиографический указатель (*Flora*

zapadnoj Sibiri: bibliografičeskij ukazatel'). 105 pp. Tomsk: Tomsk State University Press. [An unannotated list, the titles arranged in the first instance by subject with author index; covers Western Siberia.]

KOSOVANOV, V. P., 1923–30. Библиография Приенисейского края (*Bibliografija Prijenisejskogo kraja*)/ *Bibliographie des Jenisseier Gebietes*. Vols. 2–3. Krasnoyarsk: Jenisejsk. Gub. Ekon. Soveščanija (vol. 2); Sibir. Kraev. Izd-vo (vol. 3). Continued as ANONYMOUS, 1963. Библиография Красноярского края (1924–1960 гг.). 2 vols. Krasnoyarsk. [Both works cover the Yenisei Basin. Pp. 107–122 in vol. 2 of the original work cover botany.]

LITWINOW, D. I., 1909. Библиография флоры Сибири (*Bibliografija flory Sibiri*). ix, 458 pp. (Trudy Bot. Muz. Imp. Akad. Nauk 5). St. Petersburg. [Over 1200 annotated entries arranged by author; gazetteer, taxonomic, regional and subject indices.]

Indices. General indices as for Superregion 71–75.

710

Region in general

KRASNOBOROV, I. M. *et al.*, 1987–99. Флора Сибири (*Flora Sibiri*). Vols. [1]–14. Illus., maps. Novosibirsk: 'Nauka'.

Descriptive 'research' flora with keys to all taxa, synonymy, references and citations, Russian nomenclature, detailed indication of distribution (but without *exsiccatae*), chorological class(es), external range, and habitats, representative illustrations, and dot maps; indices to all names in each volume. The introductory part (in vol. 1) furnishes background and gives the philosophy and plan of the work and its coverage and sources; conventions (pp. 10–11), acknowledgments and general key to families follow before the flora proper. Volume 14 comprises keys to families, taxonomic and chorological supplements, and cumulative indices. [The family arrangement is basically that of Engler. Of the 14 planned volumes, all had been published by 1999.][11]

Woody plants

KOROPAČINSKIJ, I. JU., 1983. Древесные растения Сибири (*Drevesnye rastenija Sibiri*). 382 pp., 253 illus. (incl. halftones), maps. Novosibirsk: 'Nauka'.

Illustrated descriptive flora of woody plants (and sub-woody forest herbs), with an emphasis on their ecology and geography and including keys; individual accounts include limited synonymy, sometimes plentiful citations, distribution (with most species individually mapped), karyotypes, habitats, growth limits, and some critical remarks. The introductory part covers sources and the plan of the work along with sections on geography, climate, frost duration, botanical areas, and regional accounts as well as a key to families, while following the species accounts is an ecological-geographical analysis including considerations of chorological classification, life-forms and typological and regional differences; references (pp. 369–377) and index to scientific names at end.

I. Western Siberia

Area: 2 500 000 km². Vascular flora: 2805 species (Krylov *et al.*, 1927–64). – This region includes most of the territory of the Ob'–Irtysh river system and its tributaries; it extends from the Ural Mountains in the west to the boundary of Krasnoyarsk Krai in the east, and from the Altai Mountains and the Kazakhstan boundary in the south to the 'tree-line' in the north (for the Arctic zone, see **064**). Its primary political/administrative subdivisions include Kurgan, Tyumen', Omsk, Tomsk, Novosibirsk and Kemerovo *oblasti* along with Altai Territory.

The long-standard *Flora zapadnoj Sibiri* (see below) was based on P. N. KRYLOV, 1901–14. Флора Алтая и Томской губернии (*Flora Altaja i Tomskoj gubernii*). 7 parts. Tomsk.

KRYLOV, P. N. *et al.*, 1927–64. Флора Западной Сибири (*Flora zapadnoj Sibiri*)/ *Flora sibiriae occidentalis*. 12 vols. Tomsk: Russkoe botaničeskoe obščestvo, Tomskoe otdelenie (and successors) (vols. 1–11, 1927–49); Tomsk State University Press (vol. 12, 1961–64). (Vol. 1 reprinted 1955, Tomsk; vol. 5 reprinted 1958.)

Comprehensive, large-scale descriptive flora of vascular plants of Western Siberia (and parts of northern Kazakhstan); includes keys to all taxa, synonymy with references and citations, rather detailed indication of local and regional distribution (with summary of extralimital range), Russian binomials and vernacular names, taxonomic commentary, and notes on habitat, special features, etc.; indices to all botanical and other names in each volume (complete general index in volume 12). The introductory section (vol. 1) includes chapters on administrative subdivisions (the old *guber-*

nija system of Czarist Russia has been used in this work) and botanical districts as well as an artificial general key to families. Volume 12 is supplementary. [The limits of the *Flora* range from longitude 59° to 90° E and from latitude 49° to 73° N.][12]

II. Central and Eastern Siberia

Area: *c.* 7 224 200 km². – As here delimited, this part of Siberia extends from the western limits of the Yenisei Basin (approximately 84° to 88° E) east to the eastern border of Chita Oblast' and from the southern boundary of the Russian Federation and the Far Eastern region (**Region 73**) north to the 'tree-line' (the Arctic and sub-Arctic zones form part of **Regions 05 and 06**). It thus includes Krasnoyarsk Krai (and Khakassk Autonomous Oblast'), Irkutsk and Chita *oblasti*, the Buryat and Tuva Republics, and Sakha (Yakutia). For separate works on the southern alpine zone, see **703/VIII** and **703/IX**.

Popov's *Flora srednej Sibiri* covers the entire southern half of Central and Eastern Siberia, while the less geographically extensive *Flora tsentral'noj Sibiri* by Malyschev and Peschkova centers on Lake Baikal. Both are now largely superseded by *Flora Sibiri*.

MALYSCHEV, L. I. and G. A. PESCHKOVA, 1979. Флора центральной Сибири (*Flora tsentral'noj Sibiri*)/ *Flora Sibiriae centralis*. 2 vols. 1048 pp., 1 + 1284 maps, frontisp. Novosibirsk: 'Nauka'.

Briefly descriptive flora of vascular plants, with keys to all taxa, limited synonymy, Russian nomenclatural equivalents, karyotypes, some critical remarks, and indication of habitat, occurrence and distribution (individual dot maps, covering 55 percent of the species, grouped together at end of each volume) along with special features; complete general indices (volume 2). The introductory section (volume 1) gives information on the background of the work, earlier floras, definition of the limits of the area covered (p. 8) and of its subdivisions, and principal references (p. 10).[13]

POPOV, M. G., 1957–59. Флора Средней Сибири (*Flora srednej Sibiri*). 2 vols. 918 pp., 104 text-figs., 3 maps, 2 portrs. Moscow/Leningrad: AN SSSR Press.

Briefly descriptive flora of vascular plants (covering *c.* 2000 species) with synoptic keys to genera and groups of species (but no analytical keys), limited synonymy (with citations), vernacular and binomial

Russian names, generalized indication of local and extralimital range, taxonomic commentary, notes on habitat, etc., and figures of representative species; general indices to all botanical and other names in vol. 2. The introductory section includes a list of principal references, a synopsis of families, and a survey of local floristics (with maps). [Covers an area of some two million square kilometers of Siberia from the Upper Yenisei Basin through Cis- and Transbaikalia to Dahuria and from the Sajan Mountains on the southern frontier to approximately 61° N.][14]

Partial work

IVANOVA, M. M. and A. A. ČEPURNOV, 1983. Флора западного участка районов освоения БАМ (*Flora zapadnogo učastka raionov osvoenija BAM*). 222, [1] pp., maps (2 general, 770 individual). Novosibirsk: 'Nauka'.

Includes an enumeration of vascular plants (1352 species), with scientific and Russian names, habitats, phenology, and detailed distribution by five defined areas (as depicted in map 2 on p. 8, with addition of specific localities for less common or localized species; 232 species are *vse raïony*, i.e., in all areas) including cross-references to the individual distribution maps where these are provided; no index (the order is as in Czerepanov's 1981 checklist; see **680**). The introductory part furnishes the basis of the work (little collecting in the area prior to the 1960s, with much more from 1975 when railway construction was initiated), while on pp. 118–125 is an analysis and discussion of the characteristics of the flora. [Covers the line from Ust'-Kut (its western end) to Chara in the Yablonoy ranges, including the northern end of Lake Baikal.]

711

Yekaterinburg (Sverdlovsk)

No separate works are available apart from local checklists and florulas, including those on the Ural (**602**). The region is sometimes considered part of CIS-in-Europe (**680**).

712

Chelyabinsk and Kurgan

No separate works are available for this industrial region apart from local checklists and florulas.

713

Tyumen'

No separate works are available apart from local checklists and florulas. The region contains the old trading center of Tobol'sk.

714

Khanty-Mansai and Yamal-Nenets Republics

No separate works are available for these two autonomous republics apart from local checklists and florulas.

715

Omsk

No separate works are available apart from local checklists and florulas.

716

Novosibirsk

Area: 178 746 km². Vascular flora: no data.

KOROLEVA, A. S., I. M. KRASNOBOROV and E. F. PEN'KOVSKAJA, 1973. Определитель растений Новосибирской области (*Opredelitel' rastenij*

Novosibirskoj oblasti). 368 pp., 281 text-figs. Novosibirsk: 'Nauka'.

Illustrated manual-key to vascular plants; includes diagnoses, very limited synonymy, Russian binomials and vernacular names, and brief notes on habitat, phenology and special features (but without specific indication of local range); complete indices at end. The introductory section includes definitions of morphological terms.

717

Tomsk

Although here inclusive of Tomsk Oblast' alone, floras from the time of Krylov onwards have taken in larger tracts and therefore appear under other headings.

718

Altai Krai

See **Altai** under **703/VIII, Altai-Sayan mountain system**. No separate works have been published since 1914.

719

Kemerovo

This area is inclusive of the Kuznetsk coal basin. The only separate work relates to the woody flora.

Woody plants

KHLONOV, JU. P., 1979. Деревья и кустарники юго-восточной части Западной Сибири (Кузнетское нагорье, Салаир, Кузнетская котловина) (*Derev'ja i kustarniki jugo-vostočnoj časti Zapadnoj Sibiri (Kuznetskoe nagor'e, Salair, Kuznetskaja kotlovina)*). 126 pp., 24 pls., 50 maps. Novosibirsk: 'Nauka'.

Dendrological treatment, with keys to species (pp. 35–42), Russian and botanical nomenclature, and maps and photographs of diagnostic features. Pp. 4–5 contain the master map, with definitions of all sites sampled. The treatment proper is preceded by considerations of physical and climatic features, plant geography and characteristic species, vegetation types and their distribution, and patterns of individual species distributions. References (pp. 119–122) and index to botanical and Russian names appear at the end.

721

Krasnoyarsk Krai (in general)

Area (whole territory, inclusive of Khakassia, Evenk and Taimyr): 2 404 000 km². – For more coverage of the southern part of this vast territory, which includes by far the greater part of the Yenisei River basin, see **722**.

The general flora begun by Reverdatto in 1937 was finally completed in 1980 with publication of the Compositae; a revision of the first installment appeared in 1983.

Bibliography
For KOSOVANOV and its successor, see **Region 71/72**.

REVERDATTO, V. V. *et al.* (eds.), 1960–83. Флора Красноярского края (*Flora Krasnojarskogo kraja*). Vyp. 1–10 (14 parts in 11 installments). Illus., maps. Tomsk: Tomsk University Press (fasc. 1 (1983), 2 (1964), 5(2) (1971), 5(3), 5(4), 6 (1960), 7/8, 9(2), 10 (1980)); Novosibirsk: 'Nauka' (fasc. 3/9 (1965), 4/5(1) (1967)).

A detailed systematic treatment of the vascular plants of Krasnoyarsk (before 1935, Prijenisejsk) Krai, with descriptions of species, diagnostic keys to genera and species, Russian binomials and vernacular names, synonymy, citations of pertinent literature, generalized indication of local range, northern limits, records and *exsiccatae*, illustrations of representative species, and miscellaneous notes; each part fully indexed. [Arrangement of families follows the Englerian system. Fasc. 1 includes a key to families as well as pteridophytes, gymnosperms, and monocotyledons from Typhaceae through Hydrocharitaceae; fasc. 2–4, the remaining monocotyledons; fasc. 5(1) and 5(2), Salicaceae and Betulaceae through Amaranthaceae; fasc. 5(3), Portulacaceae through Menispermaceae; fasc. 5(4), Papaveraceae through Rosaceae; fasc. 6,

Leguminosae; fasc. 7/8, Geraniaceae through Cornaceae and Pyrolaceae through Boraginaceae; fasc. 9, Labiatae; fasc. 9(2), Solanaceae, Scrophulariaceae and Campanulaceae; and fasc. 10, Compositae. Fasc. 1 is a revision (without, however, its coverage of exploration and sources) of V. V. REVERDATTO and L. P. SERGIEVSKAJA, 1937. Конспект Приенисейской флоры (*Konspekt Prijenisejskoj flory*) / *Conspectus florae jenisseensis*. Fasc. 1. 46 pp. Tomsk: Biologičeskij Naučno-Issledovatel'skij Institut, Tomsk State University.][15]

(part 3 (1961) from 'Book Publishing Service', part 4 (1963) as *Uchen. Zap. Krasnojarsk. Pedagog. Inst.* 24/4).

Documentary, non-descriptive manual of vascular plants (1828 species) with diagnostic keys to all taxa, synonymy with references and citations, Russian vernacular names and binomials, localities with *exsiccatae*, and many notes on habitat, occurrence, properties, uses, habit, phenology, etc.; area maps in fascicles 1 and 6 and general indices to all vernacular, Russian binomial and Latin names in fascicle 6 (fasc. 5 separately indexed). A general key to families appears in fascicles 1 and 5. [The 58th parallel (just north of the lower Angara) forms the northern limit of this flora.][16]

722

Krasnoyarsk Krai (southern part)

See also **721** (REVERDATTO *et al.*). – Area: not known. Vascular flora: 1848 species (Čerepnin, 1957–67), 1800 species (Begljanova *et al.*, 1979). This unit covers the whole of the territory north to the limit of the Upper Yenisei floristic region (*c.* 59°–62°N, north of the Angara River), and centers on Krasnoyarsk. Included also is the Khakassk Autonomous Oblast'. Most centers of population and scientific activity in the territory are situated within these limits. Two standard works are available: a documentary flora (Čerepnin, 1957–67) and a manual-key (Begljanova *et al.*, 1979).

BEGLJANOVA, M. I. *et al.*, 1979. Определитель растений юга Красноярского края (*Opredelitel' rastenij juga Krasnojarskogo kraja*). 669 pp., 324 text-figs. Novosibirsk: 'Nauka'.

Illustrated manual-key to vascular plants, with abbreviated indication of habit, phenology, habitat, distribution, properties, uses and potential, etc., Russian nomenclatural equivalents, and limited synonymy; general indices to all names at end. The introductory section contains an illustrated glossary as well as a general key to families. [Distribution units within the southern Krasnojarsk Krai are as used in Čerepnin's flora (see below).]

ČEREPNIN, L. M., 1957–67. Флора южной части Красноярского края (*Flora južnoj časti Krasnojarskogo kraja*). 6 parts. 122 text-figs., map. Krasnoyarsk: Krasnojarsk State Pedagogical Institute

723

Krasnoyarsk Krai (northern part) (including Evenk and Taimyr Republics)

Stretching from the northern limit of the Upper Yenisei floristic region north through the Central Siberian Plateau and on to the limits of tree growth, this part of Krasnoyarsk includes the middle Yenisei, Tungusia and Putorania. Within this expanse are the territories of Evenk and the southern part of Taimyr including the town of Noril'sk. For the sub-Arctic north, including the Taimyr Peninsula, and the arctalpine Putorany Uplands (on the Evenk–Taimyr border), see **066**; for Severnaja Zemlja, see **056**.

The only consolidated coverage is in the territory-wide flora of REVERDATTO *et al.* (**721**). Most of the area is north of the northern limits of Popov's *Flora srednej Sibiri* (**710/II**).

724

Tuva Republic

See also **703/IX** (all works). – Area: 172000 km². Vascular flora: 1782 species (Krasnoborov, 1984; see below); 1326 had been accounted for by Sobolevskaja in 1953. At one time loosely associated with the Manchu Empire and later known as 'Tannu Tuva', this territory was after World War II incorporated into the Soviet Union; from 1991 it has been part of the Russian Federation.

KRASNOBOROV, I. M. (ed.), 1984. Определитель растений Тувинской АССР (*Opredelitel' rastenij Tuvinskoj ASSR*). 335 pp., illus., map. Novosibirsk: 'Nauka'.

Manual-key to vascular plants, with keys, very limited synonymy, botanical and Russian names, distribution by districts, figures of selected species, and additional notes on habitat, altitudinal range, habit and frequency; indices at end. The general part includes a historical review and definition of the 10 districts adopted (with map) as well as an illustrated organographic treatment, an introduction to the use of the book, and a general key to families. Nomenclature follows CZEREPANOV (1981; see **680**).

SOBOLEVSKAJA, K. A., 1953. Конспект флоры Тувы (*Konspekt flory Tuvy*). 245 pp., portr., map. Novosibirsk: AN SSSR Press (West Siberian Branch).

Systematic enumeration of vascular plants, with limited synonymy and citations of relevant literature, Russian binomials and vernacular names, fairly discursive indication of local range (with some citations) but no extralimital distribution, and some taxonomic commentary and notes on uses; indices to all Russian and botanical names. The prefatory section contains a short account of botanical exploration in Tuva. [Now out of date; covers 1326 species, rather fewer than currently known.][17]

Woody plants

KOROPAČINSKIJ, I. JU. and A. V. SKVORTSOVA, 1966. Деревья и кустарники Тувинской АССР (*Derev'ja i kustarniki Tuvinskoj ASSR*). 184 pp., 76 text-figs., map. Novosibirsk: 'Nauka'.

Illustrated forest flora with analytical keys, synonymy, general indication of local and extralimital range (with local distribution maps), and extensive notes on habitat, biology, special features, properties, uses, etc.; index to botanical names. The introductory section includes chapters on floristics, vegetation, ecology, and botanical exploration.

725

Eastern Central Siberia (in general)

See **710/II** (all works). – Area: 1 569 000 km². Vascular flora: 2311 native and naturalized species (Malyschev and Peschkova, 1979; see **710/II**). This part of Central Siberia, which includes areas geographically usually thought of as part of Eastern Siberia, here comprises Irkutsk and Chita *oblasti* together with the Buryat Mongol Republic, with Lake Baikal in the center (for other coverage of the lake and its margins, see **707**).

726

Cisbaikalia (Predbajkal'ja)

This area is here considered to correspond for the most part with the limits of Irkutsk Oblast' (including the Ust-Ordyn Buryat National District). Full modern coverage is available in MALYSCHEV and PESCHKOVA (1979; see **710/II**). There is also an earlier 'spring flora' (Djagilev, 1938; see below).

Partial work

DJAGILEV, V. F., 1938. Определитель весенних и ранне-летних растений Предбайкалья (*Opredelitel' vesennikh i ranne-letnikh rastenij Predbajkal'ja*). 223 pp., illus. Irkutsk: Oblast' Publishing Service.

Illustrated, semi-popular students' manual-key to the spring and early summer vascular plants of Irkutsk Oblast' and the Lake Baikal littoral, with Russian binomial and vernacular names and notes on habitat, phenology, local distribution, uses, biological features, etc.; list of references and index to Russian binomial names. The introductory section includes chapters on collecting methods and terminology as well as on general features of the vegetation. Arrangement of families is on the Englerian system.[18]

727

Transbaikalia (Zabajkal'ja) (including Buryat and Chita)

This includes the region formerly known as Dahuria. As delimited here, Transbaikalia corresponds approximately to the area of Buryatia (the Buryat Mongol Republic) and Chita Oblast' (including the Aga Buryat district), with the eastern limit formed by a line from the Silka–Amur river junction north to the Olyomka River (120°–122° E). The western limit is largely formed by Lake Baikal, and in the north are the Stanovoy Uplands. See also **725** (MALYSCHEV and PESCHKOVA).

FEDTSCHENKO, B. A. *et al.* (eds.), 1929– . Флора Забайкалья (*Flora Zabajkalja*)/ *Flora transbaicalica*. Fasc. 1–8. Pp. 1–842, figs. 1–372, maps. Leningrad: Geografičeskoe obščestvo (fasc. 1–2, 1929–31); Moscow/Leningrad: AN SSSR Press (fasc. 3–6, 1937–54); Leningrad: 'Nauka' (fasc. 7, 1975; fasc. 8, 1980). (Published on behalf of the Vsesoyuznoe Botaničeskoe Obščestvo, St. Petersburg.) Partly superseded by L. P. SERGIEVSKAJA *et al.*, 1966–72. Флора Забайкалья (*Flora Zabajkalja*). 2nd edn. Vyp. 1–4. Portr., map. Tomsk: Tomsk University Press.

The original work – in manual-key format – comprises a briefly descriptive account of vascular plants, with limited synonymy, Russian binomial and vernacular names, fairly detailed indication of local range (by subdivisions as delimited on an area map, supplied in fascicles 1, 6 and 7), and notes on habitat, phenology, special features, biology, etc., and representative illustrations; no indices. Fascicle 1 includes a consideration of floristic districts as adopted for the work. The revised edition of 1966–72 – in the format of a briefly descriptive flora – includes synonymy, references, citations of key literature, detailed indication of local range (including localities and authorities), extralimital distribution, some taxonomic commentary, and notes on habitat, but no illustrations. The introductory section (in fascicle 1) includes an account of floristic districts as used in the work (with map) and a concise encyclopedia of collectors in Transbaikalia. [Original edition at present (1999) complete from pteridophytes through the Convolvulaceae and Cuscutaceae (Englerian system). The revised edition covers pteri-

dophytes, gymnosperms and monocotyledons, corresponding to the first two fascicles of the original work.][19]

728

Sakha (Yakutia)

Area: 3 103 200 km². Vascular flora: 1523 species (Karavaev, 1958; see below). – The zone north of the 'tree-line' has also been designated as **067**. For parts of the Aldan and Kolyma floristic districts not in Sakha, see **Region 73**.

Apart from *Flora SSSR* and associated works, no effective separate coverage of Sakha (the former Yakut Republic) is available except in the works given below. The incomplete, expansively formatted flora of Petrov (1930) is complemented by a checklist (Karavaev, 1958) and a manual-key (Tolmatchev, 1974). These three principal works were preceded by Boris Fedtschenko's *Jakutskaja Flora* (1907; not completed), a consolidation of information gathered from several nineteenth-century expeditions and surveys, and V. L. Komarov's *Vvedenie v izučhenie rastitel'nosti Jakutii* (1926), with a checklist covering 1190 species.

KARAVAEV, M. N., 1958. Конспект флоры Якутии (*Konspekt flory Jakutii*). 190 pp., 12 text-figs. Moscow/Leningrad: AN SSSR Press.

Systematic enumeration of vascular plants, with limited synonymy, concise indication of local range, and brief notes on habitat, special features, etc.; bibliography but no indices. The introductory section includes chapters on the major features of the flora, floristic regions, and past and present botanical exploration.

PETROV, V. A., 1930. Флора Якутии (*Flora Jakutii*). Vol. 1. xii, 221 pp., 75 text-figs. (incl. maps). Leningrad: AN SSSR Press.

Descriptive flora of vascular plants, including keys to all taxa, full synonymy with references, Russian nomenclature, citation of *exsiccatae* with indication of local range (including some dot maps), general summary of extralimital range, and notes on habitat, etc. (these latter in Latin); index to all botanical names at end. [Not completed; covers pteridophytes, gymnosperms, and monocotyledons from Typhaceae through Poaceae (Gramineae) on the Englerian system.]

TOLMATCHEV, A. I. (ed.), 1974. Определитель высших растений Якутии (*Opredelitel' vysšikh rastenij Jakutii*). 543 pp., 70 pls. Novosibirsk: 'Nauka'.

Concise illustrated manual-key to vascular plants of this vast territory, with brief diagnoses, Russian and Yakut vernacular names, and short notes on habitat and internal range. A very brief introduction is given and there are indices to Russian and Sakhan family and generic names as well as to all botanical names. [A helpful feature of this work is a clear indication in range summaries for individual species of their presence in the tundra or sub-Arctic zone.]

Region 73

Russian Far East

Area: 3 591 800 km². Vascular flora: *c.* 4000 species (Charkevicz *et al.*, 1985– , 1: 5; see below). A total of 3215 native and established species were recorded for the forested zone by Voroshilov (1985; see below). – This region encompasses all the areas of the Russian Federation associated with the north Pacific Ocean (except for Anadyr-Chukotia which is wholly in the sub-Arctic zone; see 068) together with the Amur Basin west to the Silka River junction. Principal political/administrative units include Primorsk Krai, Khabarovsk Krai, Amur Oblast', the Jewish Autonomous Region, Magadan Oblast', the Chukotsk National District, Kamchatka Oblast' (including the Commander Islands), the Koryak National District, and Sakhalin Oblast' (including the Kurile Islands). For convenience, however, this region has been subdivided into seven geographical units (731 to 737) corresponding to the 'districts' defined by Vladimir N. Voroshilov in his *Flora sovetskogo Dal'nego Vostoka* (1966).

Early botanical exploration was sporadic and localized. The Kamchatka Peninsula, discovered in the late seventeenth century, furnished an initial focus; for it, a natural history sketch, *Opisanie zemli Kamčatka* by Stephan Krascheninnikov, was published in 1755. However, the first regional works appeared only in the 1850s. These included *Florula ochotensis phaenogamia* by Ernst R. Trautvetter and Carl A. Meyer (1856); *Florula ajanensis* by Eduard Regel and S. H. Tiling and the more substantial *Primitiae florae amurensis* by Carl Maximowicz (1859); and *Tentamen florae Ussuriensis* by Regel (1861). In 1860, Russia secured formal control of the lower Amur and the Primorsk area from the Chinese Empire; in the next three decades several papers and monographs appeared for these and other parts of the Far East. Effective settlement – and, with it, more detailed surveys and collecting work – came only after 1890 with the construction of the Trans-Siberian and Chinese Eastern Railways and increased settlement. A particularly important figure in this phase was Vladimir Komarov, whose field work in the region laid a basis for his floras of Manchuria (1901–07; see 860/I), the Kamchatka Peninsula (prepared around 1910 but not published until the late 1920s), and the Russian Far East generally (*Malyj opredelitel' rastenij Dal'nevostočnogo kraja*, 1925, and *Opredelitel' rastenij Dal'nevostočnogo kraja*, 1931–32, both with Evgenija Klobukova-Alisova). He also compiled a bibliography on the flora and vegetation of the region (1928; a continuation was published by others in 1973). The works of the 1920s were associated with the contemporaneous formation of the Far East Branch of the then-Soviet Academy of Sciences. Further collecting, practical needs, wartime evacuation and the influence of *Flora SSSR* resulted in a new cycle of works for the region (or parts thereof) appearing after World War II; these began with a young person's guide, *Opredelitel' vesennikh rastenij Primor'ja* (1949) by D. P. Vorobiev. If account is taken also of the Japanese contributions on Sakhalin and the Kuriles as well as the critical works of Eric Hultén on Kamchatka (1927–30) and elsewhere, floristic coverage is now quite substantial although still sketchy for some remote areas. Through 1982 much of this was the work of Vorobiev. Afterwards, reassessment along with a felt need for a new definitive account induced a new generation at the Far Eastern Centre to launch *Sosudistye rastenija Sovetskogo Dal'nego Vostoka*; eight volumes have appeared as of this writing (1999). Good coverage is also available for the diverse woody flora.

Bibliographies. General and supraregional bibliographies as for Superregion 71–75; see also Superregion 86–88 (MERRILL and WALKER).

Regional bibliographies

KOMAROV, V. L., 1928. Библиография к флоре и описанию растительности Дальнего

Востока (*Bibliografija k flore i opisaniju rastitel'nosti Dal'nego Vostoka*)/*Bibliography of the flora and of the vegetation of the Far East.* 279 pp. (Zap. Juzno-Ussurijsk. Otd. Gosud. Russk. Geogr. Obsc., vyp. 2). Vladivostok. Continued as P. G. GOROVOJ and M. S. SOPOVA (eds.), 1973. Флора, растительность и растительные ресурсы Дальнего Востока. Указатель литературы (1928–1969 гг.) (*Flora, rastitel'nost' i rastitel'nye resursy Dal'nego Vostoka. Ukazatel' literatury (1928–1969 gg.*))/*Flora, vegetation and plant resources of the Far East. Bibliography (1928–1969).* 552 pp. Vladivostok. [The first work is a comprehensive list of 1202 items with subject and author indices. The second is an extension, arranged in the first instance by subject headings.]

MERRILL, E. D. and E. H. WALKER, 1938. *A bibliography of Eastern Asiatic botany.* xlii, 719 pp., maps. Jamaica Plain, Mass.; Arnold Arboretum of Harvard University. Continued as E. H. WALKER, 1960. *A bibliography of Eastern Asiatic botany: Supplement 1.* xl, 552 pp. Washington: American Institute of Biological Sciences. [Region 73 is fully included in both works and Region 71/72 partly so. For full description, see Superregion 86–88.]

Indices. General indices as for Division 7.

730

Region in general

Until 1982 the only overall work including keys remained the manual by Komarov and Klobukova-Alisova (1931–32). The flora by Voroshilov (1966) was merely an enumeration, with only partial keys. A separate manual-key is now available for Primorja and Priamurja (**731, 732**), the areas with the largest potential audience, and in 1985 publication of a new 'research' flora, *Sosudistye rastenija Sovetskogo Dal'nego Vostoka*, commenced. Manuals have also been published for Sakhalin and Magadan Oblast', and the important woody flora is now also relatively well documented.[20]

CHARKEVICZ (KHARKEVICH), S. S. (ed.), 1985– . Сосудистые растения Советского Дальнего Востока (*Sosudistye rastenija Sovetskogo Dal'nego Vostoka*)/*Plantae vasculares orientis extremi sovietici.*

Vols. 1– . Illus., maps (in text and in end-papers). Leningrad (from vol. 5, St. Petersburg): 'Nauka'.

A full 'research' flora (10 volumes in all projected), with keys, botanical and Russian nomenclature, descriptions, synonymy (including references and citations), internal range (by districts or part-districts, with for less common or recently described taxa some localities and *exsiccatae*) and extralimital distribution, habitat, phenology, habit, karyotypes, individual distribution maps (not, however, for all species), and selective illustrations; indices to all names in each volume. The delimitation of families follows the 1980 version of the Takhtajan system, but as in *Flora evropejskoj časti SSSR* (see **680**) publication is not in a fixed sequence. The area covered encompasses Primorya, Priamurya, Ochotia, Anadyr and Chukotia, the Kamchatka Peninsula, Sakhalin, the Commander and Kurile Islands, and Wrangel Island, i.e., the Far East in a broad sense. [As of 1999 eight of a planned 10 volumes had been published, the most recent in 1998; all the pteridophytes, gymnosperms and most 'early' families are accounted for, including the Leguminosae. Among families still outstanding are the Rosaceae, Caryophyllaceae and Orchidaceae.]

KOMAROV, V. L. and E. N. KLOBUKOVA-ALISOVA, 1931–32. Определитель растений Дальневосточного края (*Opredelitel' rastenij Dal'nevostočnogo kraja*)/*Key for the plants of the Far Eastern Region of the USSR.* 2 parts. x, 1175 pp., 330 pls., col. frontisp. Leningrad: AN SSSR Press.

Concise illustrated manual-key to vascular plants, with limited synonymy, Russian vernacular and binomial names, brief indication of internal range, and short notes on special features of particular genera; separate keys to genera of aquatic and woody plants; abbreviations; indices to all Russian and botanical names. [Encompasses the entire Russian Far East together with the Kamchatka Peninsula and Sakhalin, but now largely superseded. The illustrations remain useful.]

VOROSHILOV, V. N., 1966. Флора Советского Дальнего Востока (*Flora sovetskogo Dal'nego Vostoka*). 477 pp. Moscow: 'Nauka'.

A concise enumeration of vascular plants of the entire region, with keys only to species within genera; includes synonymy (with references), Russian binomial and vernacular names, fairly detailed indication of local range (by districts), and some taxonomic commentary; bibliography and indices to family, generic and vernacular names. The introductory section includes statistical details together with an outline of the features of the flora. [Does not include the sub-Arctic Chukotsk National District.]

VOROSHILOV, V. N., 1982. Определитель растений Советского Дальнего Востока (*Opredelitel' rastenij Sovetskogo Dal'nego Vostoka*). 672 pp., map. Moscow: 'Nauka'.

Manual-key to vascular plants (2700 native and 400 adventive species), with botanical and Russian nomenclature, limited synonymy, indication of distribution and habitat, and some critical commentary including treatments where appropriate of 'microspecies'; index to botanical names at end. The general part (pp. 4–10) encompasses taxonomic and other developments, phytogeographical notes and definition of floristic units. [Complements and updates the author's 1966 enumeration.]

VOROSHILOV, V. N., 1985. Список сосутистых растений Советского Дальнего Востока (Spisok sosudistykh rastenij Sovetskogo Dal'nego Vostoka). In A. K. Skvortsov (ed.), Флористические исследования в расных раюнах СССР (*Florističeskie issledovanija v rasnykh rajonakh SSSR*), pp. 139–200. Moscow: 'Nauka'.

Concise checklist of vascular plants (3215 species), with numbers indicating presence in any or all of nine floristic/phytogeographic units (as follows: (1) Primorye; (2) western Amur; (3) southern Amur; (4) eastern Amur; (5) Ochotia; (6) Kamchatka; (7) the Commander Islands; (8) Sakhalin; (9) the Kuriles). [This scheme is modified from that proposed by the author in 1966, and in addition differs somewhat from that in the current *Sosudistye rastenija Sovetskogo Dal'nego Vostoka* (a work also accounting for the sub-Arctic lands).]

Woody plants

None of the following works can be considered 'complete in itself'.

AGEENKO, A. S. *et al.*, 1982. Древесная флора Дальнего Востока (*Drevesnaja flora Dal'nego Vostoka*)/ *Woody flora of the Far East*. 224 pp., 25 text-figs. (incl. map), 31 photographs (part col.). Moscow: 'Lesnaja Promyšlennost'.

A semi-popular 'natural history' with on pp. 56–182 discursive treatments of individual species including Russian and botanical nomenclature. The general part covers the forest flora and vegetation as a whole, forest types and forest regions, while following the species accounts are sections on parks, reserves and institutions as well as introduced species; classification of uses and properties (pp. 213–215), bibliography (pp. 216–218) and index to Russian names.[21]

USENKO, N. V., 1969. Деревья, кустарники и лианы Дальнего Востока (*Derev'ja, kustarniki, i liany Dal'nego Vostoka*). 416 pp., 58 pls., maps (end-papers). Khabarovsk: Khabarovsk Book Publishing Service.

Dendrological flora covering 498 species of woody plants, with descriptions, illustrations, regional distribution, notes on special features, and silvicultural data; ranges of 32 leading species shown on end-paper maps, references (p. 384) and index to Russian names with Latin equivalents. No keys are provided, and less common species are treated in small type.

VOROBIEV, D. P., 1968. Дикорастущие деревья и кустарники Дальнего Востока: определитель (*Dikorastuščije derev'ja i kustarniki Dal'nego Vostoka: opredelitel'*). 277 pp., 40 text-figs. Leningrad: 'Nauka'.

Descriptive manual treating 475 species of indigenous woody plants of the Russian Far East, with keys to genera and species, general indication of internal and extralimital range, some taxonomic commentary, and Russian vernacular and binomial names; lists of references and indices to all Russian and botanical names. No synonymy is included.

731

Primorja

Area: 168 200 km². Vascular flora: no data. – This unit corresponds with Primorsk Krai, including the Vladivostok district. A general manual and a 'tree book', both relatively recent, are available specifically for this area and Priamurja (**732**). There is also a florula for the environs of Vladivostok.

VOROBIEV, D. P. *et al.*, 1966. Определитель растений Приморя и Приамуря (*Opredelitel' rastenij Primorja i Priamurja*). 491 pp., 192 text-figs. Moscow: 'Nauka'.

Concise manual-key to the vascular plants of Primorja (and the lower Amur Basin, including the Khabarovsk district), with limited synonymy, Russian nomenclature, brief indication of habitat and local range, and occasional taxonomic remarks; indices to all botanical and other names at end. The introductory section includes statistical details as well as a short chapter on floristics. There is also a special key to genera of woody plants.

Local work: Vladivostok

VOROBIEV, D. P., 1982. Определитель сосудистых растений окрестность Владивостока (*Opredelitel' sosudistykh rastenij okrestnost' Vladivostoka*). 253 pp., 50 text-figs. Leningrad: 'Nauka'.

Manual-key to vascular plants (1184 species), with indication of habitat and occurrence; references (p. 221) and indices at end. [Accounts for some 61 percent of all species in Primorya.]

Woody plants

VOROBIEV, D. P., 1958. Определитель деревьев и кустарников Приморья и Приамурья (*Opredelitel' derev'ev i kustarnikov Primorja i Priamurja*). 184 pp., text-figs., 50 pls. Blagoveščensk: Amur Book Publishing Service.

Illustrated artificial analytical manual-key to woody plants (353 species) for field use, with separate keys based on different field and other characters; systematic synopsis; no index.

732

Priamurja

Comprises the Amur Basin (including the Khabarovsk district) as far upstream as the Šilka River junction at the boundary of Chita Oblast' (Region 71/72), thus inclusive also of Amur Oblast' and the Jewish Autonomous Region. The area is covered by the same manuals as Primorja (see **731** above) but not by any separate works. A few recent florulas of study areas exist, including *Flora i rastitel'nost' khrebta Tukuringra* edited by I. A. Gubanov (1981, Moscow). This interior mountain range divides the upper and lower Zeya basins in central Amur Oblast' and is crossed by the BAM railway.

733

Sakhalin

Area: 74066 km². Vascular flora: 1173 species (Vorobiev *et al.*, 1974; see below). – The southern half of this island, known in Japanese as Karafuto, was from 1905 to 1945 under that country's administration.

General floras are available in both Russian (Vorobiev *et al.*, 1974) and Japanese (Sugawara, 1937, 1937–40). Also accounted for here are two partial works by Japanese workers and a 'tree book' by Tolmatchev.

The first accounts, both in expedition reports, are by the explorer-naturalist F. B. Schmidt: *Flora sachalinensis* (1868), forming pp. 79–227 in his *Reisen im Amur-lande und auf der Insel Sachalin*, and *Sakhalinskaja flora* (1874). The next, relatively elaborate (but never-completed) account was K. MIYABE and Y. KUDO, 1930–34. *Flora of Hokkaido and Saghalien*. Parts 1–4. Pp. 1–528 (J. Fac. Agric. Hokkaido Univ. 26). Sapporo. This covers families from the pteridophytes through Polygonaceae on the Englerian system.

SUGAWARA, S., 1937. *Karafuto-no shokubutsu* [*Plants of Saghalien*]. [iv], 490 pp., map. Toyohara (Sakhalin).

Pages 50–138 comprise a systematic enumeration, without keys, of vascular plants, including synonymy and references, generalized indication of local range throughout the island, vernacular and transliterated scientific names, and indices to all botanical and Japanese names. The introductory section has a chronological table of local botanical research and exploration as well as a descriptive historical account, while extensive appendices provide descriptive remarks on woody, useful and naturalized plants (separately indexed).

SUGAWARA, S., 1937–40. *Karafuto shokubutsu zushi* [*Illustrated flora of Saghalien*]. 4 vols. 42 text-figs., 892 pls., 2 maps. Tokyo: The author. (Reprinted 1975.)

Generously scaled atlas-flora of vascular plants, without keys; each species plate is accompanied by text containing vernacular and transliterated scientific names, a short description, synonymy (with references and citations), citations of *exsiccatae* and literature records, generalized indication of local and extralimital distribution, and notes on habitat, special features, biology, etc.; indices to all names in each volume. The introductory section (vol. 1) includes an account (with itinerary maps) of botanical research and exploration in the island as well as a bibliography.[22]

VOROBIEV, D. P. *et al.* (coord. A. I. TOLMATCHEV), 1974. Определитель высших растений Сахалина и Курильских островов (*Opredelitel' vysšikh rastenij Sakhalina i Kuril'skikh ostrovov*)/*Key for the vascular plants of Sakhalin and Kurile Islands*. 372 pp., 64 pls. Leningrad: 'Nauka'.

Concise illustrated manual-key to vascular plants with diagnoses, limited synonymy, vernacular and transliterated scientific names, and indications of local range, habitat and endemicity; some critical remarks also included. The introductory section gives the plan of and basis for the work, while at the end are a separate key to genera with woody species, a list of literature (p. 362), and complete indices. No information on physical features, climate, vegetation, biogeography or floristics is presented.

Partial works

KUDO, Y., 1923. A contribution to our knowledge of the flora of northern Saghalien. *J. Fac. Agric. Hokkaido Univ.* 12(1): 1–68, pls. 1–12.

Enumeration of vascular plants (based mainly on collections by the author), with synonymy, relevant citations, notation of *exsiccatae* with localities, some taxonomic commentary, and remarks on habitat, etc.; index. An introductory section (in English) includes descriptive remarks on the vegetation.

MIYABE, K. and T. MIYAKE, 1915. [*Flora of Saghalin.*] 26, 648, 19, 10 pp.; 13 pls. Toyohara: Saghalin Government.

Concise, keyed descriptive flora (in Japanese) of the vascular plants of southern Sakhalin, with limited synonymy, essential citations, generalized indication of local and extralimital distribution, and brief taxonomic and other notes; complete indices.

Woody plants

TOLMATCHEV, A. I., 1956. Деревья, кустарники и деревьянистые лианы острова Сахалина: краткий определитель (*Derev'ja, kustarniki, i derev'janistye liany ostrova Sakhalina: kratkij opredelitel'*). 171 pp., 83 halftones. Moscow/Leningrad: AN SSSR Press.

Illustrated semi-popular manual-key to woody plants, with Russian and botanical nomenclature, general indication of local range, and more or less extensive remarks on habitat, ecology, phenology, biology, etc.; no index.[23]

734

Ochotia

Vascular flora (Magadan Oblast'): 1675 species (Khokhryakov, 1985; see below). – Includes Magadan Oblast' (exclusive of Chukotia) and the northern (Ochotian) part of Khabarovsk Krai. For Chukotia see also **068**.

KHOKHRYAKOV, A. P., 1985. Флора Магаданской области (*Flora Magadanskoj oblasti*). 396, [4] pp., 150 figs. (incl. halftones). Moscow: 'Nauka'.

Manual-key to vascular plants; species entries include botanical and Russian nomenclature, limited synonymy, diagnoses, figures of critical features where necessary, indication of local distribution (by 12 districts as defined on p. 8) and habitat, taxonomic notes, segregates and infraspecific taxa (the treatment of the two latter sometimes relatively detailed), and figures of critical features; diagnoses of new taxa (pp. 347–348), references (pp. 349–358, with emphasis on more recent works) and indices to botanical and Russian names at end. The general part includes coverage of the plan and scope of the work, a discussion of 'good' and 'cryptic' species, and the use of ecological criteria in identification.

735

Kamchatka Peninsula

Includes the Peninsula and the region immediately to the north of it, so encompassing Kamchatka Oblast' and the Koryak National District. – Area: 273 000 km^2 (Peninsula proper). Vascular flora: 1168 species (Belaja *et al.*, 1981; see below).

In spite of the similarity in publication dates of the two general floras, Hultén's work, as well as being more critical, is the more recent by 15 years. Both are, however, for practical purposes superseded by the manual by Belaja *et al.* (1981) which, moreover, includes many more species.

BELAJA, G. A. *et al.*, 1981. Определитель сосудистых растений Камчатской области (*Opredelitel' sosudistykh rastenij Kamčatskoj oblasti*). 411 pp., 103 text-figs., 4 tables. Moscow: 'Nauka'.

Manual-key to vascular plants, with botanical and Russian nomenclature, indication of distribution and habitat, and occasional critical notes, followed by an analysis of the flora (including statistics, p. 374, and endemism); bibliography (pp. 380–381) and indices to all names at end. The general part covers physical features of the region, history of work on the vascular flora, floristic regions and indicator species, and use of the work (along with an illustrated organography).

HULTÉN, E., 1927–30. *Flora of Kamchatka and adjacent islands.* 4 vols. Illus., maps (Kongl. Svenska Vetenskapsakad. Handl., III, 5(1–2); III, 8(1–2)). Stockholm.

Detailed descriptive flora of vascular plants, without keys; includes extensive synonymy with references and literature citations, notation of *exsiccatae* with indication of local range (including distribution maps), general summary of extralimital range, taxonomic commentary, and extensive notes on habitat, ecology, etc.; index to all botanical names in vol. 4. The introductory section (vol. 1) incorporates accounts of prior botanical exploration and the author's Kamchatka itinerary of 1920–22 together with descriptions of the various plant communities and formations. In addition to the Peninsula, this work also covers the Commander Islands and the northern Kuriles.

KOMAROV, V. I., 1927–30. Флора полуострова Камчатки (*Flora poluostrova Kamčatki*) / *Flora peninsulae Kamtschatka*. 3 vols. Illus., pls., maps. Leningrad: AN SSSR. (Reprinted 1951 in 2 vols., Moscow/Leningrad: AN SSSR, as vols. 7–8 of V. L. KOMAROV, *Opera selecta*.)

Copiously descriptive flora of vascular plants, with keys to all taxa, synonymy (with references), some Russian names, citation of *exsiccatae* with localities, general summary of extralimital range, taxonomic commentary, and notes on habitat, occurrence, special features, etc. (habitat and occurrences also given in Latin); indices to botanical names in each volume. The introductory section includes an account of botanical exploration, while in vol. 3 is found a summary discussion (in very defective English) of the general features of the flora and vegetation.[24]

736

Commander Islands

See also **110** (HULTÉN); **111** (HULTÉN); **735** (HULTÉN; BELAJA *et al.*). – Area: 1855 km². Vascular flora: 345 species (Vasil'ev, 1957).

Vasil'ev's work succeeds E. R. Trautvetter, *Plantas quasdam in insulis Praeffectoriis nuper lectas lustravit* (1886), with 132 species, and B. A. Fedtschenko, *Flore des îles du Commandeur* (1906), with 252 species.

VASIL'EV, V. N., 1957. Флора и папеогеография Командорских островов (*Flora i paleogeografija Komandorskikh ostrovov*). 260 pp. Moscow/Leningrad: AN SSSR Press.

Briefly descriptive flora of vascular plants, with keys to all taxa, full synonymy (with references and literature citations), localities with *exsiccatae*, general summary of extralimital range (including distribution classes), 'standardized' Russian names, taxonomic commentary, and notes on habitat, life-form, etc.; detailed floristic, phytogeographic and historical analysis (pp. 201–236); annotated bibliography (pp. 237–242) and index to all botanical and other names. The general part contains sections on physical features, climate, floristics, vegetation, and (pp. 6–8) botanical exploration and research.

737

Kurile Islands

See also **733** (VOROBIEV *et al.*); **735** (HULTÉN). – Area: 15 600 km². Vascular flora: 1098 species (Vorobiev *et al.*; see **733**). A total of 317 species were reported by Miyabe in 1890 and 994 by Vorobiev in 1956.

Russian surveying ships made the first territorial claims to these often fog-bound islands. The southern group, however, was from the 1870s to 1905 under Japanese administration, and in 1905 all the islands came under Japanese control. From 1945 they have been part of the Russian Federation (although Japan continues to claim the southernmost four). Miyabe's work incorporates all then-available records, including those in a list prepared by Maximowicz as well as his collections from a tour in the southern islands in 1884.[25] For the many papers of Misao Tatewaki, in the interwar years active throughout the Kuriles, general and supraregional bibliographies should be consulted along with the regional bibliographies (see **Region 73, Bibliographies**).

MIYABE, K., 1890. The flora of the Kurile Islands. *Mem. Boston Soc. Nat. Hist.* **4**(7): 203–275, map.

Systematic enumeration of known vascular plants, without keys or descriptions (except for new taxa); includes synonymy (with references), general indication of local range, taxonomic commentary, and notes on habitat and special features; no index. The introductory section has chapters on physical features and floristics together with a statistical summary. [Now well out of date.]

VOROBIEV, D. P., 1956. Материалы к флоре Курильских островов (*Materialy k flore Kuril'skikh ostrovov*). 79 pp. (Trudy Dal'nevostočnogo Fil. 'Komarova' AN SSSR, ser. Bot. 3). Vladivostok.

Pages 6–78 comprise a systematic enumeration of known vascular plants of the archipelago, accounting for all earlier reports and giving an indication of local range (by islands), Russian names, and habitat; bibliography (pp. 78–79). The introductory section gives an account of botanical exploration and research in the area and statistics of the flora.

Region 74

The Caucasus

Area: 422 500 km². Vascular flora: over 6200 species at present known. Some 5767 were recorded in the last complete flora (Grossheim, 1928–34; see **740** below).[26] – This region is here considered to include the republics of Armenia, Georgia and Azerbaijan along with, in the Russian Federation, Ciscaucasia (the north slope) and most of the region lying between the Sea of Azov to the west and the Caspian Sea. The northern boundary is approximately equivalent to that adopted for *Flora Europaea,* i.e., along the Kuma River in the east from its mouth as far as Priumsk, then on a line running north of Stavropol' as far as the Jeja River, thence along this to its mouth in the Sea of Azov northwest of Novorossiysk.[27]

With a history of botanical exploration dating from the beginning of the eighteenth century and a 'first flora', Marschall von Bieberstein's *Flora taurico-caucasica* (1808–19) contemporary with such works as Pursh's *Flora Americae septentrionalis,* the rich, varied Caucasian flora, containing relict Eurasian elements and including some unusual warm-temperate – in Russian parlance, 'subtropical' – pockets, is now relatively well known; an extensive range of floras and other works is available. The work of outside bodies and individuals in exploration, research and publication has from the formation of the Georgian State Museum in Tbilisi in 1865 been added to by regional and national institutes, universities and gardens. Yet since Bieberstein's time only one full descriptive flora, the original edition of Alexandr Grossheim's *Flora*

Kavkaza (1928–34), has fully covered the vascular plants. An enlarged second edition of this work, begun in 1939 and featuring maps for all species, lapsed after its seventh volume (1967) without completing the sympetalous families after Scrophulariaceae (on the Englerian system). Other nominally complete works were *Flora Kavkaza* (1899; supplement, 1902) by Vladimir I. Lipsky, including an annotated enumeration of 4430 species as well as a historical account, cyclopedia of collectors, and bibliography, and Grossheim's *Opredelitel' rastenij Kavkaza* (1949). *Flora caucasica critica* (1901–16), the very detailed series by Nikolai Kuznetsov and his associates and students at Jurjew (Dorpat; now Tartu, Estonia), was abandoned in the wake of World War I, as was also *Opredelitel' rastenij Kavkaza i Kryma* (1907–19) by Aleksandr V. Fomin and Yuri N. Voronov (Woronow).[28]

All these works followed almost a century of more or less continuous study following *Flora taurico-caucasica.* The first significant contribution is *Verzeichniss der Pflanzen* (1831) by Carl A. Meyer, its coverage of 1965 species for long furnishing an important source of information. This was followed by the Tbilisi-based *Opyt russko-kavkazskoj flory i primenenii k sel'skomy khozjajstvu i domašnemu bytu* (1858–59) by Alexander P. Owerin and N. P. Sitovskij. Though left incomplete, it covered 2016 species from Ranunculaceae through Umbelliferae on the Candollean system, reflecting the rapid increase of knowledge in a generation. In the second half of the century, contributors included Gustav Radde, upon whose collections Trautvetter based several papers (1873–87), and Émile Levier, who afterwards prepared a number of papers with Stefano Sommier (1892–1900, the largest being *Enumeratio plantarum anno 1890 in Caucaso lectarum* (1900, in *Acta Hort. Petrop.* **16**)). W. Barbey, by the 1880s proprietor of the Boissier Herbarium in Geneva, sponsored explorations by Nikolai Albov in Colchidia, resulting in a first flora of this important 'subtropical enclave' (*Prodromus florae colchicae,* 1895, in *Trudy Tiflis. Bot. Sada* 1, prilož. 1), covering 1500 species. All these works enabled elaboration of floristic and phytogeographical syntheses: Radde in *Grundzüge der Pflanzenverbreitung in den Kaukasusländern von der unteren Volga über den Manytsch-Scheider bis zur Scheitelfläche Hocharmeniens* (1899, as *Vegetation der Erde* 3) and, after completing the first full modern flora, Grossheim in his *Analiz flory Kavkaza* (1936, Baku).

Of particular elements in the flora, the woody plants early attracted attention; the German dendrologist Karl Koch visited the region in the mid-nineteenth century and soon after separate treatises began their appearance. Notable early syntheses were the first (1883) and third (1919) editions of *Derev'ja i kustarniki Kavkaza* by Jakob S. Medwedew. These have now given way to the six-volume *Dendroflora Kavkaza* (1959–86; see below), encompassing native and introduced species. Medwedew also prepared a detailed survey and manual of the high-mountain flora as part of his unfinished *Rastitel'nost' Kavkaza* (1915–19; see **703/I**).

Among the national and other larger floras are two complete works on the Georgian Republic, both in Georgian (*Flora Gruzii*, 1941–52, 2nd edn., 1971– ; *Opredelitel' rastenij Gruzii*, 1964–69), one on Azerbaijan (*Flora Azerbajdžana*, 1950–61), and Armen Takhtajan's *Flora Armenii* (1954– ; nine volumes now published with a further three expected). For the north slope, or Ciscaucasia, there is a concise checklist, *Spisok rastenij Severnogo Kavkaza i Dagestana*, by Alexandr Flerow (1938), and a three-volume manual, *Flora Severnogo Kavkaza* (1978–80), by Anatol Galushko; these succeeded an earlier, incomplete *Flora ciscaucasica* (1894) by Lipsky. At district and republican level, more or less recent complete floras are available for Krasnodarsk Krai, Dagestan, western Georgia, the Black Sea republics of Abkhazia and Adzharia, the environs of Tbilisi and Jerevan, the Talysh Basin in southeastern Azerbaijan, and the Apsheron Peninsula near the oil city of Baku. A number of dendrologies and other guides to the woody flora have also been published, in national languages as well as in Russian.[29]

Progress

The only separate study seen is V. I. LIPSKY, 1899. Historija botaničeskogo issledovanija Kavkaza. In *idem*, *Flora Kavkaza*. St. Petersburg. There is, however, also a survey of botanical exploration in Radde's already-mentioned *Grundzüge der Pflanzenverbreitung in den Kaukasusländern*.

Bibliographies. General and supraregional bibliographies as for Superregion 71–75. See also Superregion 77–79 (FIELD).

Regional bibliography

The following may be consulted in addition to Lipsky's bibliography: A. A. GROSSHEIM, 1948.

Растительный покров Кавказа (*Rastitel'nyj pokrov Kavakza*). 268 pp. (Mater. Pozn. Fauny Fl. SSSR, Mosk. Obšč. Ispytat. Prirod., N.S., Otd. Bot. 4). Moscow. [Pp. 207–238 comprise a bibliography of 1013 titles on flora and vegetation.]

LIPSKY, V. I., 1899. Литературьа по флоре Кавказа (Literatura po flore Kavkaza). In *idem*, Флоре Кавказа (*Flora Kavkaza*), pp. 1–116. St. Petersburg (also as *Trudy Tiflissk. Bot. Sada* 4: 1–116). [396 titles with annotations by the compiler; essential for any study of floristic work in the nineteenth century and before. Unpublished material is considered in a following chapter.]

Indices. General indices as for Superregion 71–75.

740

Region in general

The second edition of Grossheim's *Flora Kavkaza*, begun in 1939, has yet to be completed although after seven volumes it was far advanced. The first edition of that work as well as Lipsky's enumerative *Flora Kavkaza* (1899–1902) thus remain the only works providing nominally complete regional coverage. For the woody plants, *Dendroflora Kavkaza* was completed in 1986. All of these are in Russian.[30]

GROSSHEIM, A. A., 1949. Определитель растений Кавказа (*Opredelitel' rastenij Kavkaza*). 747 pp., map. Moscow: 'Sovetskaja Nauka'.

Concise manual-key to vascular plants of the whole Caucasus region, with limited synonymy, vernacular and binomial Russian names, brief indication of local and extralimital range, and notes on habitat, phenology, life-form, etc.; special keys to families represented only by cultivated species; indices to botanical and Russian names of families and genera. The brief introductory section includes a map with floristic regions.

GROSSHEIM, A. A. *et al.* (eds.), 1939–67. Флора Кавказа (*Flora Kavkaza*). 2nd edn. Vols. 1–7. Illus., maps. Baku: AN Azerbajdžanskoj SSR (vols. 1–3); Moscow/Leningrad: AN SSSR Press (later Leningrad: 'Nauka') (vols. 4–7). (1st edn., 1928–34, Tiflis (later Baku).)

Comprehensive descriptive 'research' flora of

vascular plants, with keys to all taxa, full synonymy, references and citations, Russian and local vernacular names (but no Russian 'standardized' binomials), detailed indication of local range, with distribution maps for every species, taxonomic commentary, and notes on habitat, life-form, special features, etc.; diagnoses of new taxa and indices to generic names in Russian as well as all scientific names at end of each volume. An introductory section (vol. 1) includes a map of floral regions (repeated in later volumes). [Not completed; seven volumes have been published, covering families through the Scrophulariaceae (on a modified Englerian system).]

Woody plants

For scientific if not practical purposes, *Dendroflora Kavkaza* supersedes Ja. S. Medwedew, 1919. Деревья и кустарники Кавказа (*Derev'ja i kustarniki Kavkaza*). 3rd edn. iv, 485 pp. Tiflis.

Gulisashvili, V. Z. *et al.*, 1959–86. Дендрофлора Кавказа (*Dendroflora Kavkaza*). 6 vols. Illus. Tbilisi: AN Gruzinskoj SSR Press.

Copiously illustrated descriptive dendrological treatment of the woody plants of the Caucasus (1829 native and introduced species), with partial keys, limited synonymy, references, 'standardized' Russian binomials and Russian and local vernacular names, general indication of local and extralimital distribution of native species, with distribution maps, origin of exotic species, and notes on habitat, special features, ecology, biology, adaptation, silvicultural aspects, management, etc.; lists of references and complete indices in each volume.

741

Ciscaucasia (Predkavkaz) in general

Area: 236400 km² (exclusive of southern Rostov Oblast' and western Kalmytsia). Vascular flora: *c.* 3900 species including introductions (Galushko, 1978–80). – This region consists of the north slope and foothills of the main Caucasus range, extending north to the mutual boundary of Asia with Europe as defined in this *Guide* (see p. 675); on the south is the southern boundary of the Russian Federation. It includes the Krasnodar and Stavropol' Kraiy in the north and west, the Chechen, Ingush and Dagestan Republics in the

east, and the Kabardian and North Ossetian Republics along the southern mountain crest. The checklist of Flerow (1938) and the two manuals of Galushko (1967, 1978–80) more or less cover the area as a whole; some substantial partial works are also available, here listed under **742, 743** and **744**. There are two earlier treatments, neither completed: *Opyt russko-kavkazskoj flory i primenenii k sel'skomy khozjajstvu i domašnemu bytu* (1858–59) by A. P. Overin and N. P. Sitovskij and *Flora ciscaucasica* (1894) by Lipsky. The latter was soon followed by Ivan F. Schmal'hauscn's more inclusive *Flora srednej i južnoj Rossii, Kryma i Severnogo Kavkaza* (1895–97), a standard for the following eight decades.

Bibliography

Novopokrovsky, I., 1938. Литература по ботанической географии Ростовской области, Краснодарского края, Орджоникидзевского края и Дагестана (Literatura po botaničeskoj geografii Rostovskoj oblasti, Krasnodarskogo kraja, Ordžonikidzevskogo kraja i Dagestana). *Trudy Rostovsk. Obl. Biol. Obšč.* **2**: 5–45. [442 titles.]

Flerow, A. F., 1938. Список растений Северного Кавказа и Дагестана (*Spisok rastenij Severnogo Kavkaza i Dagestana*)/*Elenchus plantarum in Caucaso septentrionale nec non Daghestana sponte crescentium.* 693 pp., portr. Rostov-on-Don: Rostovizdat.

Concise systematic enumeration of vascular plants, with essential synonymy, detailed indication of local and altitudinal range, and brief notes on taxonomy, habitat, special features, etc.; addenda, extensive bibliography (755 titles, pp. 623–690), and index to orders and families at end.

Galushko, A. I., 1978–80. Флора Северного Кавказа (*Flora Severnogo Kavkaza*). 3 vols. Illus., maps. Rostov-on-Don: Rostov University Press.

Field manual-key to vascular plants (covering some 3900 species), with limited synonymy, Russian botanical nomenclature, and abbreviated indication of internal distribution, altitudinal limits, habitat and occurrence; indices only to scientific names (with an afterword, explanation of abbreviations, and complete general index in vol. 3). The brief introductory section contains a general key to families (repeated in each volume) as well as brief remarks on the background for the work (but without mention of other standard references on the area).[31]

Woody plants

GALUSHKO, A. I. (ed.), 1967. Деревья и кустарники Северного Кавказа (*Derev'ja i kustarniki Severnogo Kavkaza*). 536 pp., 118 text-figs. Nal'čik, Kabardian Republic.

Illustrated flora of woody plants, with moderately long descriptions, Russian names, relatively detailed indication of local range (for native species) or region of origin and places of cultivation (for introduced species), and notes on habitat, special preferences, biological forms, cultivation, uses and potential; bibliography and indices to all Russian and Latin plant names at end. The introductory section gives sources for the work, its plan, and a general key to families. [Encompasses the entire North Caucasian Slope, including Dagestan, and extends into parts of Rostov Oblast'.]

742

Western Ciscaucasia

Corresponds to Krasnodarsk Krai (including Adygeyskaya Territory). Kosenko's manual covers most if not all of the area; a portion is also covered by *Flora Europaea* whose southern boundary follows the Kuban River. A predecessor work was *Opredelitel' rastenij ravnin i predgorij Kuban i častju Černomor'ja* (1924) by Pavel I. Misczenko and Nathalie A. Desjatova-Shostenko.

KOSENKO, I. S., 1970. Определитель высших растений Северо-Западного Кавказа и Предкавказа (*Opredelitel' vysšikh rastenij Severo-Zapadnogo Kavkaza i Predkavkaza*). 613 pp., 9 text-figs. Moskva: 'Kolos'.

Concise manual-key to all vascular plants of this area (3150 species), with diagnoses, essential synonymy, Russian names, and brief (partly symbolic) indication of local range, habitat, phenology, life-form, uses and potential; complete indices at end. The introductory section includes an illustrated glossary. [An expansion of the author's *Opredelitel' glavnejšikh dikorastuščikh kormovykh zlakov i bubovykh Krasnodarskogo kraja* (1949), a grass and legume manual.]

743

Central Ciscaucasia

Corresponds to Stavropol' Krai, the Karachaievo-Cherkessia Territory, and the republics of Kabardino-Balkaria and North Ossetia (Vladikavkaz). No recent separate floras or related works are available. An early work on Stavropol' is *Stavropol'skaja flora / Florula stavropolensis* (1880, published 1881) by A. Normann, accounting for 700 species.

744

Eastern Ciscaucasia

See also **698** (FEDTSCHENKO and SHISHKIN, 1927–38). Eastern Ciscaucasia here encompasses the republics of Ingushetia, Chechenya and Dagestan. Separate works exist only for the last-named. – *Flora Europaea* covers the northern lowland half of Dagestan (south to the Terek River north of Makhačkala).

Dagestan
Area: 38 210 km². Vascular flora: 1250 species (L'vov; see below). The northern part is low-lying, while the southern is very mountainous and broken. The work of Lepëkhina includes an extensive bibliography.

L'VOV (LJVOV), P. L., 1960. Определитель растений Дагестана (*Opredelitel' rastenij Dagestana*). 422 pp., illus. Makhačkala: Dagestan State University.

Manual-key to 1250 species. [Not seen; cited from Lipschitz, 1975, p. 131 (no. 1448). For additions, see *idem*, 1971. К флоре Дагестана (K flore Dagestana)/ Additamenta ad floram Daghestaniae. *Novost. Sist. Vysš. Rast.* 8: 284–289.]

LEPËKHINA, A. A., 1971. Определитель деревьев и кустарников Дагестана (*Opredelitel' derev'ev i kustarnikov Dagestana*). 243 pp., 31 text-figs., map. Makhačkala: Dagestanskoe Učebno-Pedagogičeskogo Press.

Descriptive manual of native and introduced woody plants (409 species and additional infraspecific taxa, of which 278 native) with keys to all taxa, diagnoses, Russian names, and notes on habitat, local range, phenology, habit, special features, uses, potential, etc.; short glossary, list of literature (pp. 223–226) and complete indices at end. The introductory section includes among other topics an account (with map) of dendrofloristic areas and a general key to families.

746

Georgia (Gruzia) (including Abkhazia and Ajaria)

Area: 69 700 km² (inclusive of South Ossetia, Ajaria and Abkhazia). Vascular flora: no data.

Two works, both in Georgian and described below, cover the republic and associated territories: a multi-volume flora (currently under revision) and a manual-key. There are also a number of florulas of more limited areas; those for Ajaria and Abkhazia, on the Black Sea, are in Russian as is also a more concise treatment (Kolakovsky, 1961) on the Colchis region in general.

BOTANIČESKIJ INSTITUT, AKADEMIA NAUK GRUZINSKOJ SSR, 1964–69. Определитель растений Грузии (*Opredelitel' rastenij Gruzii*). 2 vols. 2 maps. Tbilisi: 'Metsniereba'. (Also a title-page in Georgian.)

Concise manual-key to vascular plants (in Georgian), based mainly on *Flora Gruzii*; includes limited synonymy, nomenclatural equivalents, local vernacular names, concise indication of local range, and notes on habitat, life-form, phenology, etc.; indices to all botanical and other names in each volume (combined index to families in vol. 2). A map of the region with floristic districts is furnished in the introductory section.

MAKASHVILI, A. K., D. I. SOSNOVSKIJ and A. L. KHARADZE (eds.), 1941–52. Флора Грузии (*Flora Gruzii*)/*Flora Georgiae*. 8 vols. Illus. Tbilisi: AN Gruzinskoj SSR Press. Partially succeeded by N. N. KETSKHOVELI et al. (gen. eds.), 1971– . Флора Грузии (*Flora Gruzii*). 2nd edn. Vols. 1– . Illus., maps. Tbilisi: 'Metsniereba'. (Both editions also with title-pages in Georgian.)

A detailed descriptive flora (in Georgian) of vascular plants, with botanical and Georgian nomenclature, keys to genera and species, significant synonymy with references and literature citations, vernacular names, concise generalized indication of local and extralimital range, figures of representative species, and notes on habitat, phenology, special features, etc.; indices to all botanical and other names in each volume (no cumulative index available). The second edition is similar in format and style to its predecessor but additionally features species distribution maps (in later volumes integrated with the illustrations in the text). [As of 1998, 11 volumes of the second edition, covering pteridophytes, gymnosperms, and angiosperms through Labiatae, had been published, the last in 1987; this corresponds to vols. 3–6 and some three-fifths of vol. 7 of the first edition. There thus is no revised coverage of monocotyledons or the 'final' dicotyledon families including Compositae.][32]

Partial works: Black Sea region (Colchis)

Florulas for the Abkhazian Republic (centering on Sukhumi) and the Ajarian Republic (centering on Batumi) as well as for 'subtropical' Colchidia in general are listed here. There is also a dendroflora: A. V. VASIL'EV, 1955–59. Флора деревьев и кустарников субтропиков Западной Грузии (*Flora derev'ev i kustarnikov subtropikov Zapadnoj Gruzii*). 4 vols. (in 5). 91 illus. Sukhumi.

DIMITRIEVA, A. A., 1990. Определитель растений Аджарии (*Opredelitel' rastenij Adžarii*). 2nd edn. 2 vols. 328, 280 pp., 28 text-pls. Tbilisi: 'Metsniereba'. (1st edn. in 1 vol., 1959, Tbilisi.)

Brief introduction, abbreviations and key to families (vol. 1); concise manual-key with essential synonymy, Russian nomenclature, and brief notes on habitat, local range, etc.; indices to Russian and botanical generic names in each volume along with plates (no consolidated index). [The Grossheim system of classification is followed.][33]

KOLAKOVSKY, A. A., 1980–86. Флора Абхазии (*Flora Abkhazii*). 2nd edn. 4 vols. Illus. Tbilisi: 'Metsniereba'. (1st edn., 1938–49, Sukhumi: Institut dlja Abkhazskoj Kulturu, AN SSSR (later AN Gruzinskoj SSR).)

Descriptive flora (in Russian) of vascular plants (1978 species) with keys, synonymy, references, Russian and Georgian nomenclature, local distribution with altitudinal range, notes on habitat, phenology, karyotypes, phytochoria, etc., and indices in each volume (with cumulative indices to all names in vol. 4). Generic headings are commentative rather than descriptive. [Within the major groups of vascular plants the arrangement of families and genera is alphabetical.][34]

KOLAKOVSKY, A. A., 1961. Растительный мир Колхиды (*Rastitel'nyj mir Kolkhidy*). 460 pp., 231 illus., map (Mater. Pozn. Fauny Fl. SSSR, Mosk. Obšč. Ispytat. Prirod., N.S., Otd. Bot. 10). Moscow.

Pages 121–459 constitute a very concise keyed and extensively illustrated manual of vascular plants of subtropical Colchidia, with brief notes on ecology, phenology and local range; glossary.

Partial work: Tbilisi and vicinity

There have been successive treatments for the Georgian capital and its environs. An early work was *Očerk Tiflisskoj flory, c opisaniem Ljutikovykh, ej prinadležaščikh* (1853) by A. N. Beketov, with keys to genera. Successors included *Opredelitel' rastenij okrestnostej Tiflisa* (1920) by D. I. Sosnovskij and A. A. Grossheim and the unfinished *Flora Tiflisa* (part 1, 1925; in *Trudy Muzej Gruzii* 3) by Grossheim, Sosnovskij and B. K. Shishkin. There is also a series *Materialy dlja flory Tiflisa i ego bližajšikh okrestnostej* (1914 onwards) by O. Mitskevich and M. Neprintseva.

MAKASHVILI, A. K., 1952–53. Флора окрестностей Тбилиси (*Flora okrestnostej Tbilisi*). 2 vols. Illus. Tbilisi: Tbilisi State University. (Also a title-page in Georgian.)

Descriptive manual-flora (in Georgian) of the environs of Tbilisi (formerly Tiflis), encompassing 700 species. [Not seen; cited by both Lebedev and Lipschitz (see **Superregion 71–75, Bibliographies**).]

Woody plants

Two standard works, not seen but both listed by Lipschitz (1975; under **Superregion 71–75, Bibliographies**) are: V. I. MIRZASHVILI, 1947–48. Дендрология (*Dendrologija*). 2 parts. 840 pp., illus. Tbilisi: Tbilisi State University Press; and A. D. GOGOLADZE, 1964. Определитель деревьев и кустарников (*Opredelitel' derev'ev i kustarnikov*). 261 pp. Tbilisi: 'Tsodna'. (In Georgian.) More popular is A. K. MAKASHVILI, 1960. Наши деревья и кустарники (*Naši derev'ja i kustarniki*). 115 pp. Tbilisi: 'Nakaduni'. (In Georgian.)

748

Armenia

Area: 29 800 km². Vascular flora: *c.* 3200 species.

In addition to the as yet incomplete multi-volume flora begun by Armen Takhtajan in 1954, a concise manual for woody plants is also available (Sosnovskij and Makhatadze, 1950) and a florula for the Yerevan district (Takhtajan and An. Fëdorov, 1972). Both these latter are in Russian.

TAKHTAJAN, A. L. (ed.), 1954– . Флора Армении (*Flora Armenii*). Vols. 1– . Illus., map. Jerevan: AN Armjanskoj SSR Press (vols. 1–8); Havlickuv Brod, Czech Republic: Koeltz (vol. 9).

Copiously illustrated flora of spontaneous and commonly cultivated vascular plants, with descriptive keys to all taxa, full synonymy, references, 'standardized' Russian binomials and Armenian vernacular names, concise indication of local and extralimital distribution, taxonomic commentary, and notes on habitat, special features, biology, cultivation, etc.; indices to all vernacular and scientific names in each volume. [As of 1999 nine of the 12 or 13 volumes projected have appeared, the most recent in 1995; coverage currently runs through the greater part of the Compositae following an early version of the author's well-known system of flowering plants. Remaining to be published are two tribes of the Compositae and the monocotyledons.]

Partial work: Yerevan and vicinity

The current flora of the Armenian capital and its environs supplants two earlier works by the same authors: *Flora Jerevana: opredelitel' rastenij okrestnostej Jerevana* and *Atlas risunkov k 'Flore Jerevana'* (both published 1945, Yerevan).

TAKHTAJAN, A. L. and AN. A. FËDOROV, 1972. Флора Еревана: определитель дисорастущих растений Араратской котловины (*Flora Jerevana: opredelitel' dikorastuščikh rastenij Araratskoj kotloviny*). 394 pp., 118 text-figs. Leningrad: 'Nauka'.

Illustrated manual-key to vascular plants (1452 species) of the Ararat Basin around Yerevan, with limited synonymy, vernacular names, and notes on habitat, phenology, local distribution, etc.; index.

Woody plants

In addition to the Russian-language manual by Sosnovskij and Makhatadze, there is also a manual in Armenian (not seen; listed by Lipschitz, 1975): A. G. ARARATJAN and JA. I. MYLKIDZHANJAN, 1951. Деревья и кустарники Армении: определитель по облиственным побегам (*Derev'ja i kustarniki Armenii: opredelitel' po oblistvennym pobegam*). 114 pp. Yerevan: AN Armjanskoj SSR.

SOSNOVSKIJ, D. I. and L. B. MAKHATADZE, 1950. Краткий определитель деревьев и кустарников Армянской ССР (*Kratkij opredelitel' derev'ev i kustarnikov Armjanskoj SSR*). 103 pp. Jerevan: AN Armjanskoj SSR Press.

Systematic manual-key to woody plants (in Russian, with Armenian summary on pp. 76–94); includes Russian and Armenian names, short diagnoses, and notes on habit, distribution, special features, uses, potential, etc.; artificial key to genera and complete indices.

Pteridophytes

GABRIËLAN, Ė. C. and W. GREUTER, 1984. A revised catalogue of the Pteridophyta of the Armenian SSR. *Willdenowia* 14: 145–158.

A critical enumeration (38 species) with synonymy and references, some vernacular names, localities with *exsiccatae* and depositories, taxonomic notes, and indication of distribution, altitudinal range, frequency, karyotypes, habitat and associates; references (pp. 157–158). [Offered as a revision of the *Flora* account, with the same arrangement of taxa.]

749

Azerbaijan

See also **790** (RECHINGER). – Area (including Nagorno-Karabakh and Nakhičevan): 86 600 km². Vascular flora: 3359 species (Karjagin, 1950–61; see below). A portion of Azerbaijan Republic, especially the Talysh lowland and parts of the Kara-Araks Basin near the Caspian Sea, is encompassed within the limits of *Flora Iranica*.

Botanically distinctive areas in Azerbaijan include the somewhat insular, oleiferous Apsheron Peninsula and the 'subtropical' Talysh Basin. The first study of the latter was *Die Fauna und Flora des südwestlichen Caspi-Gebietes* (1886) by Gustav Radde; this includes a list of 1600 plants. The basin was later the subject of Alexandr Grossheim's first flora in the republic, *Flora talysha* (1926). Grossheim later wrote the only general flora in Azeri (in Roman script, standard in the country before 1940), *Azerbajçan floras/ Flora Azerbajdžana* (1934–35, in 3 vols.). After World War II that work was succeeded by the current Russian-language flora. Three volumes of a standard dendrology have also been published; this is in Azeri with modified Cyrillic script.

KARJAGIN, I. I. *et al.* (eds.), 1950–61. Флора Азербайджана (*Flora Azerbajdžana*). 8 vols. 368 pls., maps. Baku: AN Azerbajdžanskoj SSR Press.

Detailed descriptive flora of vascular plants, with keys to all taxa, extensive synonymy, references, 'standardized' Russian binomials and local vernacular names, generalized indication of local and extralimital distribution, taxonomic commentary, and notes on habitat, special features, phenology, etc.; figures of representative species; complete indices in each volume (cumulative generic index in vol. 8). An introductory section (vol. 1) has a list of principal references for the area as well as a map of local floristic regions.

Partial work: Baku and the Apsheron Peninsula
Baku, the Azeri capital, and the Apsheron Peninsula were earlier the subject of *Plantae bakuenses bruhnsii* (1867) by L. Gruner, *Flora Apsherona i ju.-v. Šivanskoj stepi* (1928) by P. V.

Švan-Gurijskij, and *Opredelitel' rastenij Apšerona* (1931), compiled by I. I. Karjagin and others.

KARJAGIN, I. I., 1953. Флора Апшерона (*Flora Apšerona*). 439 pp., illus. Baku: AN Azerbajdžanskoj SSR Press.

Manual-key to vascular plants (729 species) of the Apsheron Peninsula (on the Caspian Sea, Baku being situated on its southern side). [Not seen; cited by both Lebedev and Lipschitz (see **Superregion 71–75, Bibliographies**).]

Woody plants
AKHUDOV, G. F., U. M. AGHAMIROV and A. R. ÆLIJEV, 1961–70. *Azerbaijanyn aghaj ve kollary*/Деревья и кустарники Азербайджана (*Derev'ja i kustarniki Azerbajdžana*). Vols. 1–3. Illus. Baku: 'Elm'.

Illustrated dendrological treatment (in Azeri) with keys, descriptions, botanical and Russian nomenclature, Azeri vernacular names, distribution, ecology, origin (if applicable), and commentary; indices to all names at end of each volume as well as detailed tables of contents. Species of particular importance feature extensive coverage of properties, uses, agronomy, silviculture, protection, etc. Towards the end of vol. 2 is a treatment of floristic regions, with three maps respectively covering Azerbaijan, the former Soviet Union and Eurasia; there is also a 10-page bibliography. Volume 3 includes further references (pp. 320–321). Volumes 1 (1961) and 2 (1964) cover Ginkgoaceae through Platanaceae; vol. 3 (1970), Rosaceae through Rhamnaceae. [A projected fourth and final volume was not published.]

Region 75

Middle Asia

Area: 3 996 500 km². Vascular flora: 8096 species (Vvedensky *et al.*, 1968–93; see **750** below). Some 5031 species had been recorded by the second decade of the twentieth century (Fedtschenko, 1915, noted below). – Comprises the republics of Kazakhstan, Krygystan (Kirghizstan), Uzbekistan (including the Kara-Kalpak Republic), Tajikistan and Turkmenistan. The limits of the region correspond to political boundaries except in the extreme west of Kazakhstan, where the boundary (along the Ural River from Orsk downstream) is that shared with *Flora Europaea*. A small portion of Kazakhstan thus is nominally excluded from Region 75.

As in the Far East, more intensive botanical exploration began only in the latter half of the nineteenth century with the spread of Russian influence into what then was known as Turkestan. This followed occasional expeditions such as that of Alexander Lehmann in 1839–42, whose plants were written up by Alexander Bunge in 1852. The later work included surveys of high-mountain areas such as the Pamir and Tien Shan as well as the lowlands and lower mountain chains. Much was accomplished between the Alexis P. Fedtschenko expedition of the 1870s (partly written up by Eduard Regel as *Turkestanskaja flora*) and 1917. Among notable contributors were the central botanical institutions in St. Petersburg through several expeditions and surveys and four floristic works: (1) Lipsky's *Materialy dlja flory Srednej Azii* (1901–09, not completed); (2) Olga A. and Boris A. Fedtschenko's Candollean-arranged *Conspectus florae Turkestanicae* (1906–16, covering 4141 dicotyledons to Monochlamydeae; German version (1905–13, 1923, in *Beih. Bot. Centralbl.*, II, **18–40**, *passim*, not continued past Monotropaceae)); (3) *Conspectus florae Turkestanicae et Kirghisicae* by B. A. Fedtschenko (1924, in *Acta Horti Petropolitani* **38**(1)), covering 609 pteridophytes, gymnosperms, and monocotyledons to Eriocaulaceae on the Englerian system; and (4) a first identification manual, *Rastitel'nost' Turkestana* (1915) by B. A. Fedtschenko. This last-named was until very recently the only nominally complete account. A *Flora srednej Azii* by Lipsky (1902–05) did not proceed beyond its first volume (a historical survey, cyclopedia and annotated bibliography).

After the 1917 revolution and the establishment in the following year of a university at Tashkent (Toshkent), the capital of Russian Turkestan, widespread collecting was resumed. Much effort focused on the foothills of the mountain ranges from Ashkabad and Dushanbe (Stalinabad) to Almaty (Alma-Ata, Verny) but also in the plains and lower hills. As already noted, B. A. Fedtschenko undertook a revised *Conspectus* in 1924, but only a single installment was published (but due to the use of a different systematic arrangement not overlapping the coverage of its predecessor). The trend thereafter was – in addition to contributions to the All-Union *Flora SSSR* – towards works relating to the new polities which came into being in the 1920s and 1930s. Among the first of these were *Flora tsentral'nogo Kazakhstana* (1928–38) by Nikolai V. Pavlov, and *Flora turkmenii* (begun in 1932, with Fedtschenko one of its authors). In 1937 there appeared the first volume of *Flora Tadžikistana*, a treatment of the legumes by Nikolai Goncharov. With

World War II came a further renewal of efforts, including institutional development (the Toshkent Branch of the Soviet Academy of Sciences being the first in the region to gain 'independence', in 1943) and in the 1950s and 1960s, as production of *Flora SSSR* progressed, multi-volume floras were achieved for all the republics (that for Tajikistan not, however, being completed until 1991, shortly before its independence). Local works were also written for autonomous areas or for centers of demand, including the Kara-Kalpak Republic and Gorno-Badakshan (in Tajikistan; see also **703/VI**) and the environs of Ashkabad, Dushanbe, Khudzand (Leninabad), and Toshkent, in the last-named initially in the 1920s.

At present, the network of coverage may be said to be extensive though of uneven quality and completeness. The main development in recent years has been the production of a concise comprehensive regional manual-flora, *Opredelitel' rastenij Srednej Azii* (*Conspectus florae Asiae mediae*) (1968–93). This, a project of the Botanical Institute of the Uzbek Academy of Sciences, is now complete in 10 volumes. A number of works on woody plants, all relatively technical and some quite detailed, are also available; but more local treatments are few. The whole publication record ostensibly is impressive, given a region larger than Europe (west of the CIS) and some two-thirds the size of Australia; however, Andrei Fëdorov in his review of Soviet floristic studies (1971; see **Region 68/69, Progress**), considered the standard of some work to be less than satisfactory.

Progress

The only study seen has been V. I. LIPSKY, 1903. Историа ботаническаго изследования Средней Азии (Istoria botaničeskago izsledovanija [*sic*] Srednej Azii). In *idem*, Флора Средией Азии (*Flora Srednej Azii*), 1(2). St. Petersburg (also as *Trudy Tiflissk. Bot. Sad* 7(2)).[35]

Bibliographies. General and supraregional bibliographies as for Superregion 71–75; see also Superregion 77–79 (FIELD).

Regional bibliography

There are additionally separate bibliographies for Kazakhstan (to 1965), Tajikistan (to 1941) and Uzbekistan (1917–52); these are cited under their respective national headings.

LIPSKY, V. I., 1902. Литература по флоре Средней Азии (Literatura po flore Srednej Azii). In *idem*, Флора Средней Азии (*Flora Srednej Azii*), 1(1). St. Petersburg (also as *Trudy Tiflissk. Bot. Sad* 7(1)). [333 items, each with detailed summaries by the compiler. Essential for any study of pre-1900 floristic work.]

Indices. General indices as for Superregion 71–75.

nomenclature; included are synonymy, references, key citations, type localities (where appropriate), distribution and altitudinal limits, habitat and frequency along with a list of references (p. 21).[36]

Woody plants

For a regional guide to woody plants, see A. U. USMANOV and G. S. KOSTELOVA, 1974. Деревя и кустарники Средней Азии (*Derevja i kustarniki Srednej Azii*). 151 pp. ?Tashkent. [Not seen; cited from *Koeltz Catalog* 886: 15 (2000).]

750

Region in general

The already-mentioned early works by Lipsky and the Fedtschenkos have been entirely succeeded by the 10-volume *Conspectus florae Asiae mediae* (see below).

VVEDENSKY, A. I. *et al.* (gen. eds.), S. S. KOVALEVSKAJA *et al.* (vol. eds.), 1968–93. Определитель растений Средней Азии (*Opredelitel' rastenij Srednej Azii*)/*Conspectus florae Asiae mediae*. 10 vols. Illus. Tashkent: 'FAN'.

Concisely formatted manual of native and introduced vascular plants (8096 species) of Middle Asia; includes descriptive keys to all taxa, full synonymy, references and citations, Russian binomial as well as local vernacular names, generalized indication of internal distribution, occasional taxonomic commentary, and brief notes on habitat, life-form, phenology, special features, etc.; indices to all botanical and other names in each volume (but none cumulative). A map of the area covered, with subdivisions and their explanation, appears on pp. 624–625 in vol. 10. [Based principally on the herbaria of the Tashkent State University and the Uzbek Academy of Sciences, the most substantial of their kind in Middle Asia. The traditional Englerian arrangement of families is followed, the monocotyledons preceding dicotyledons within the angiosperms.]

Pteridophytes

BOBROV, A. E., 1984. Конспект папрортников Средней Азии и Казахстан (Konspekt paporotnikov Srednej Azii i Kazakhstan)/*Conspectus filicarum Asiae Mediae et Kazakhstaniae*. *Novost. Sist. Vyss. Rast.* 21: 5–21.

A review and enumeration of ferns as represented in the first volume of the *Opredelitel'* (see above) with revised

751

Kazakhstan

Area: 2 717 300 km^2. Vascular flora: 5631 species listed in *Flora Kazakhstana*; a more recent figure is 4759 species (*Flora Tadzhikskoj SSR*, 10: 6; see **756**).

This largest of Middle Asian republics, with vast stretches of steppe and desert in addition to its mountainous southeast, has an illustrated manual and a 'tree book' as well as a multi-volume general flora, all in Russian. A precursor to the last-named was, as already noted, Pavlov's 1928–38 flora of Central Kazakhstan, its imprints reflecting an era of rapid political and social changes. After World War II Pavlov produced a general sketch, *Rastitel'noe syr'e Kazakhstana* (1946). In 1991 Kazakhstan became independent and in 1999 the first volume of a new national flora appeared.

Bibliography

PAVLOV, N. V., 1940. Литературные источники по флоре и растительности Казахстана (*Literaturnyje istočniki po flore i rastitel'nosti Kazakhstana*). 182 pp. (Trudy Kazakhstansk. Fil. Akad. Nauk SSSR 19). Moscow/Leningrad. Complemented by G. A. DEMEŠEVA, 1971. Ботаническая литература Казахстана 1937–1965 гг. (*Botaničeskaja literatura Kazakhstana 1937–1965*). 425 pp. Alma-Ata. [2377 works on higher plants to January 1937 in Pavlov's work; 3000 from 1937 to 1965.]

Keys to families and genera

The following represents the first volume of a new national flora.

BAJTENOV, M. S., 1999. Флора Казахстана, 1: Иллюстрированный определитель семейств и родов (*Flora Kazakhstana*, 1: *Illjustrirovannyj opredelitel'semejstv i rodov*). 396, [2] pp., 173 pls. Almaty: 'Hylym'. [Illustrated key to 1100 genera in 150 families of vascular plants.]

INSTITUT BOTANIKI, AKADEMIJA NAUK KAZAKHSKAJA SSR, 1969–72. Иллюстрированный определитель растений Казахстана (*Illjustrirovannyj opredelitel' rastenij Kazakhstana*). 2 vols. Illus., map. Alma-Ata: 'Nauka', AN Kazakhskoj SSR.

Copiously illustrated, succinct manual-key to the vascular plants of Kazakhstan, with limited synonymy, Russian binomial and local vernacular names, brief indication of local range, and notes on habitat, life-form, phenology, etc.; general index (in vol. 2). A brief introduction to the use of the work appears in vol. 1.

PAVLOV, N. V. (ed.), 1956–66. Флора Казахстана (*Flora Kazakhstana*). 9 vols. Illus., maps. Alma-Ata: AN Kazakhsk. SSR Press (later 'Nauka', AN Kazakhskoj SSR).

Briefly descriptive flora of vascular plants, with keys to all taxa; full synonymy, with references and citations; Russian binomial and local vernacular names; detailed indication of local range and generalized summary of extralimital distribution; some taxonomic commentary; notes on habitat, life-form, phenology, uses, etc.; diagnoses of novelties; indices to all botanical and other names in each volume (cumulative index to families and genera in vol. 9). The introductory section (vol. 1) contains an illustrated organography and a map with floristic regions. [A successor work has been initiated; see BAJTENOV above.]

Woody plants

MUŠEGJAN, A. M., 1962–66. Деревья и кустарники Казахстана: дикорастущие и интродуцированные (*Derev'ja i kustarniki Kazakhstana: dikorastuščie i introdutsirovannye*). 2 vols. 20 pls. Alma-Ata: Kazsel'khozgiz (vol. 1); 'Kainar' (vol. 2).

The main part (section III) of this work comprises a descriptive account of the native and exotic trees and shrubs of Kazakhstan (739 species), with analytical keys, synonymy, principal citations, Kazakh and Russian names, general indication of local and extralimital range (as well as countries of origin), and notes on occurrence, habitat, cultivation, properties, uses, etc., both in the republic as well as (where relevant) elsewhere in the former USSR; extensive bibliographies and full indices in both volumes. The introductory sections (I and II) include a short account of sources for the work and a scheme for dendrofloristic regions, while two concluding sections (IV and V) deal respectively with the origin and distribution of the native and introduced woody species. [Based in part on Sokolov's *Derev'ja i kustarniki SSSR*. Not seen; description prepared from details supplied by M. E. Kirpicznikov, St. Petersburg.]

752

Turkmenistan

Area: 488000 km^2. Vascular flora: 2800 species (Nikitin and Gel'dikhanov, 1988; see below); 2607 in *Flora Turkmenii*. Some 2200 apparently truly native (*Flora Tadžikskoj SSR*, 10: 6; see **756**). – With regard to the southern (Iranian/Afghan) fringe, including the Kopetdagh mountains, see also **790** (RECHINGER).

Begun in 1932 – some 50 years after the start of serious botanical exploration of the present-day republic and before *Flora SSSR* itself began publication – *Flora Turkmenii* was the first of the great mid-century floras of Middle Asia. The three parts (1980–85) of the illustrated manual-key written under the direction of P. C. Čopanov and S. K. Czerepanov correspond to the first three volumes of the larger work. Both are in turn complemented by a compact manual initiated by Vasilii V. Nikitin, earlier the author of a florula for the Ashkabad district. For a general introduction to the flora and vegetation, as well as a historical review of biological exploration, see V. FET and K. I. ATAMURADOV (eds.), 1994. *Biogeography and ecology of Turkmenistan*. viii, 650 pp., illus. Dordrecht: Kluwer. (Monographiae biologicae 72.)[37]

ČOPANOV, P. C. *et al.*, 1978. Определитель хвощеобразных, папоротникообразных, голосемянных и однодольных растений Туркменистана (*Opredelitel' khvoščeobraznykh, paporotnikoobraznykh, golosemjannykh i odnodol'nykh rastenij Turkmenistana*). 330 pp., 33 text-figs. Ashkabad: 'Ylym'. Continued as S. K. CZEREPANOV (ed.), 1980–85. Определитель растений Туркменистана (*Opredelitel' rastenij Turkmenistana*). Parts 2–3. 136, 204 pp., illus. Ashkabad.

The first part is a manual-key to lower vascular plants, gymnosperms, and monocotyledons (519 species), with annotated key-leads, limited synonymy, and vernacular names; index. Parts 2–3, in similar style, respectively cover Salicaceae through Chenopodiaceae (the latter accounting for two-thirds of the contents!) and Amaranthaceae through Resedaceae. Each volume corresponds in coverage to its sister volume in *Flora Turkmenii*. [Not completed.][38]

FEDTSCHENKO, B. A., M. G. POPOV and B. K. SHISHKIN (eds.), 1932–60. Флора Туркмении (*Flora Turkmenii*). 7 vols. Illus., map. Leningrad: AN SSSR (vol. 1); Askhabad: Turkmensk. Gosizdat (vol. 2); Turkmensk. fil. AN SSSR (later AN Turkmensk. SSR) Press (vols. 3–7).

Full-scale descriptive flora of vascular plants, with keys to all taxa; includes limited synonymy, localities with citations of *exsiccatae*, generalized indication of internal and extralimital range, and notes on habitat, life-form, phenology, etc.; indices to botanical names in each volume (except vols. 1–2). Volume 1 contains a brief preface (with map of floristic regions) and a general discussion of the significant botanical features of all vascular families in Turkmenia. [Vols. 1–3 complemented and updated by *Opredelitel rastenij Turkmenistana* (see above).]

NIKITIN, V. V. and A. M. GEL'DIKHANOV, 1988. Определитель растений Туркменистана (*Opredelitel' rastenij Turkmenistana*). 680 pp., 170 text-figs., map (inside front cover). Leningrad: 'Nauka'/'Ylym'.

Illustrated manual-key to vascular plants, the leads with brief diagnoses and the species entries giving essential synonymy, Russian and Turkmen names, habit, phenology, habitat and local distribution (with all abbreviations explained in the inside back cover); indices to botanical and Russian names (of families and genera) and to Turkmen names (where known). A brief introduction furnishes particulars of geography, climate, and botanical history as well as the basis of the work, its scope and philosophy, and the subdivisions of the country and their description (accompanied by the map); key to families, pp. 9–21.[39]

Local work: Ashkabad

NIKITIN, V. V., 1965. Иллюстрированный определитель растений окрестностей Ашкабада (*Illjustrirovannyj opredelitel' rastenij okrestnostej Aškabada*). 458 pp., 116 text-figs. Moscow/Leningrad: 'Nauka'.

Illustrated manual-key to vascular plants (1525 species, inclusive of weeds) of the vicinity of Ashkabad, with limited synonymy and indication of distribution, occurrences and habitat; indices. [A predecessor of the author's national manual (see above).]

753

Uzbekistan

Area: 449 480 km^2 (including Kara-Kalpakia). Vascular flora: 4147 species (*Flora Uzbekistana*), of which 3663 native (*Flora Tadžikskoj SSR*, 10: 6; see **756**).

The multi-volume republic flora, promoted as an aid to agriculture and land economy and recipient in 1968 of a state prize, has been supplemented for the Kara-Kalpak ASSR (by the Aral Sea) by two editions of a separate work (see **754**). These have been in recent years complemented by a students' manual in Uzbek, *Üzbekiston Üsimliklari Aniḳlagichi* (see below).[40]

Bibliography

DEVJATKINA, A. V., 1966. Растительный и животный мир Узбекистана: библиографический указатель (1917–52 гг.) (*Rastitel'nyj i životnyj mir Uzbekistana: bibliografičeskij ukazatel' (1917–52)*). 468 pp. Tashkent: 'FAN' (AN Uzbekskoj SSR). [Not seen.]

SCHREDER, R. R., E. P. KOROVIN and A. I. VVEDENSKY (eds.), 1941–62. Флора Узбекистана (*Flora Uzbekistana*)/*Flora Uzbekistanica*. 6 vols. Illus. Tashkent: Uzbekskij filial AN SSSR (later AN Uzbekskoj SSR) Press.

Full-scale descriptive flora of vascular plants; includes keys to all taxa, full synonymy (with references), Russian nomenclature, some Uzbek vernacular names, general indication of local and extralimital range, taxonomic commentary, and notes on habitat, life-form, phenology, uses, etc.; diagnoses of new taxa and indices to all botanical and other names in each volume (cumulative indices in vol. 6). The introduction (vol. 1) contains an illustrated organography, a short history of botanical exploration and research, and an account of Uzbek botanical regions. [Vol. 1, 1941; vols. 2–6, 1953–62.]

Partial works

A first illustrated manual for the Toshkent (Tashkent) region, *Opredelitel' rastenij okrestnostej Taškenta* (1923–24) by Alexei I. Vvedensky and others (and directed by Mikhail G. Popov), ceased publication after the appearance of two parts. It covered 221 species in families through part of the Cruciferae. It was succeeded by *Opredelitel' rastenij Taškentskogo oazisa* (two parts, 1938–41) under the general direction of A. M. Lapin (the manual proper, directed by Vvedensky, being in the first part). A comparable introductory work in Uzbek (see below) appeared in 1988. For identification of woody plants around Toshkent, see L. I. NAZARENKO, 1973. Определитель деревьев и кустарников Ташкентской области (*Opredelitel' derev'ev i kustarnikov Taškentskoj oblasti*). 169 pp., illus. Tashkent.

KHAMIDOV, A., M. M. NABIEV and T. ADILOV (ODILOV), 1987 (1988). *Üzbekiston Üsimliklari Aniķlagichi/* Иллюстрированный определитель растений Узбекистана (*Illustrirovannyj opredelitel' rastenij Uzbekistana*). 328 pp., 68 + 8 pls., 16 col. pls. Toshkent: 'Üķituvči'.

Illustrated manual-key (in Uzbek) to 1168 species of vascular plants, with keys, vernacular names, and annotations; illustrated glossary and index at end.

754

Kara-Kalpak Republic

See also **753** (SCHREDER *et al.*, 1941–62). – Area: 164900 km^2. Vascular flora: 979 species were recorded by Korovina *et al.* (1982–83), an increase of 105 from the 874 recorded by Bondarenko in 1964; the former, however, also covers an adjacent district of Uzbekistan proper. From 1936 Kara-Kalpakia, on the southern side of the now-shrinking Aral Sea and in the delta of the Amu Darya (Oxus), has been politically associated with Uzbekistan. Floristically it is rich in Chenopodiaceae but only two ferns and two horsetails are known. The current manual succeeds O. N. BONDARENKO, 1964. Определитель высщих растений Каракалпакий (*Opredeliel' vysšikh rastenij Karakalpakij*). 303 pp., 94 text-figs. Tashkent: 'Nauka' (Uzbekskoj SSR).

KOROVINA, O. N. *et al.*, 1982–83. Иллюстрированный определитель высших растений Каракалпакий и Хорезма (*Illjustrirovannyi opredelitel' vysšikh rastenij Karakalpakii i Khorezma*). 2 vols. Illus. Tashkent: 'FAN'.

Illustrated manual-key to vascular plants (979 species) of the Kara-Kalpak Autonomous Republic and the adjacent (and enclosed) Chorezm Oblast' of Uzbekistan; includes botanical and Russian nomenclature, synonymy, relatively extensive citations, local distribution, habitat, occurrence, and phenology but no indication of overall distribution and with but relatively brief diagnoses in the leads; indices in vol. 2.[41]

755

Krygyzstan (Kirghizia)

Area: 198652 km^2. Vascular flora: 3276 species (*Flora Tadžikskoj SSR*, 10: 6; see **756**); 3576 in *Flora Kirgizskoj SSR*. For other treatments of the Tian Shan, a large proportion of which lies within Krygyzstan, see **703/VII** (FEDTSCHENKO) and **760** (GRUBOV).

SHISHKIN, B. K. and A. I. VVEDENSKY (eds.), 1950–62. Флора Киргизской ССР: определитель растний Киргизской ССР (*Flora Kirgizskoj SSR: opredelitel' rastenij Kirgizskoj SSR*). 11 vols. Illus. Frunze [Bishkek]: Kirgizskij filial AN SSSR (later AN Kirgizskoj SSR) Press. Continued as Y. V. VYKHOTSEV, 1967–70. Флора Киргизской ССР: Дополнение (*Flora Kirgizskoj SSR: Dopolnenie*). Vyp. 1–2. Illus. Frunze: 'Ilim'.

Briefly descriptive flora of vascular plants, with keys to all taxa; includes limited synonymy (without references), Russian botanical binomials and local vernacular names, generalized indication of local and extralimital distribution, and notes on habitat, life-form, phenology, uses, etc.; diagnoses of new taxa; indices to all botanical and other names in each volume (general index in volume 11). In volume 1 there is an illustrated organography along with a general key to families, while volume 2 includes a general introduction to the flora and its phytogeographic relationships. The supplement of 1967–70 includes corrections and additions in all families as well as some revised keys.

Woody plants

GAN, P. A. *et al.*,1959–61. Деревья и кустарники Киргизииа (*Derev'ja i kustarniki Kirghizii*). 2 vols. Illus. Frunze [Bishkek]: AN Kirgizskoj SSR Press. (Alternative title-page and summary in the Kirghiz language.)

A rather detailed descriptive dendrological treatment with keys and illustrations; there is also in vol. 1 (covering gymnosperms) a substantial general part covering tree growth, development, and vegetation, along with (pp. 111–113) an extensive bibliography. [P. A. Gan is simply the first-named of eight compilers, whose shares are acknowledged on p. 4. Volume 2 not seen.]

756

Tajikistan

Area: 143071 km^2. Vascular flora: 4513 species (*Flora Tadžikskoj SSR*, 10: 6; see below). – For the Pamirs, see also **703/VI** (IKONNIKOV).

Flora Tadžikskoj SSR, covering some of the highest country and greatest altitudinal ranges in the former Soviet Union, was completed in 1991 after 34 years, the last amongst its peers. Thus was brought to fruition a project first launched in the 1930s as *Flora Tadžikistana* but of which only vol. 5 (1937; on Leguminosae), edited by V. L. Komarov and N. F. Goncharov, ever appeared. It was relaunched in the 1950s, with a revision of Leguminosae allocated to vols. 5 and 6. Local florulas are also available respectively for the Dushanbe (Stalinabad) and Khudzand (Leninabad) districts (Grigorev, 1953; Komarov, 1967) and for the highland district of Badakshan (Ikonnikov, 1979).

Bibliography

MARGOLINA, D. L., 1941. Флора и растительность Таджикистана: библиография (*Flora i rastitel'nost' Tadžikistana: bibliografija*)/ *Flore et végétation du Tadjikistan: bibliographie*. 346 pp., portrs. Moscow/Leningrad: AN SSSR Press. [1413 entries with annotations.]

OVCHINNIKOV, P. N. *et al.* (eds.), 1957–91. *Flora Tadžikskoj SSR*. 10 vols. Moscow/Leningrad: AN SSSR Press (later Leningrad: 'Nauka').

Full-scale semi-revisionary descriptive flora of vascular plants, with keys to all taxa, detailed synonymy, references, localities with citations of *exsiccatae* and general indication of local and extralimital distribution, notes on habitat, life-form, phenology, special features, uses, etc., and some taxonomic commentary; detailed lists of references and complete indices at end of each volume (with a general index in vol. 10). The introductory section (vol. 1) contains an account of botanical districts (with map) and a historical survey.[42]

Partial works

GRIGOR'ËV, JU. S., 1953. Определитель растений окрестностей Сталинабада (*Opredelitel' rastenij okrestnostej Stalinabada*). 299 pp., 56 text-figs. Moscow/Leningrad: AN SSSR Press.

Concise manual-key to vascular plants of the Dushanbe district, with Russian binomials and vernacular names, occasional taxonomic notes, and brief indication of local range, life-form, habitat, phenology, special features, uses, etc.; indices at end.

IKONNIKOV, S. S., 1979. Определитель высших растений Бадакшана (*Opredelitel' vysšikh rastenij Badakšana*). 400 pp., 25 pls. Leningrad: 'Nauka'.

Manual-key to vascular plants (1567 species) with limited synonymy, Russian nomenclature, and partly symbolic indications of habit, phenology, habitat, local range and altitudinal zonation; addenda, references (pp. 372–374) and index at end. The introductory section includes background information and a brief conspectus of the vegetation as well as of the history of floristic work in the area. [Badakshan, a 'Roof of the World' with altitudes ranging from 1700 to 7000 m, centers on the town of Khorog and is bordered to the west by the Amu Darya.]

KOMAROV, B. M., 1967. Определитель растений Северного Таджикистана (*Opredelitel' rastenij Severnogo Tadžikistana*). 495 pp., illus. Dushanbe: 'Donis'.

Manual-key to the vascular plants in the northern 'lobe' of Tajikistan (centering on Khudzand and the area south of Toshkent), with diagnoses, limited synonymy, Russian binomial and local vernacular names, local distribution, and notes on habitat, life-form, phenology, cultivation, uses, etc.; indices.

Region (Superregion) 76

Central Asia

Area: *c.* 6500000 km^2 (6355900 km^2 without western Gansu). Vascular flora: no data. This region, the greatest part of which was once known as the 'Chinese Border', essentially comprises all the territory between Soviet Asia to the north and west, 'China proper' to the south and east, and the Indian subcontinent to the south – with the exception of northeast China (Manchuria), the non-Mongolian parts of which are for practical reasons referred to Region 86. Save for the Republic of Mongolia, the entire region is governed as part of the People's Republic of China. The Chinese portion includes Xizang (Sitsang or Tibet), Qinghai (Chinghai), Xinjiang (Sinkiang or 'Chinese Turkestan', i.e., the Tarim Basin and Dzungaria), Nei Mongol (Inner Mongolia), and the western part of Gansu (Kansu).

Botanical exploration in the region prior to about 1920 was already extensive but, save for its northern parts, relatively sporadic. Much early work was carried out by scientific parties attached to Russian exploring expeditions; their botanical collections are now largely in St. Petersburg. Contributions were also made by British parties (mainly in Xizang) and others. Difficulties of travel, extremes of climate, and sparse (and sometimes hostile) population along with an absence of scientific centers limited detailed collecting, and published work largely took the form of 'reports' and 'contributions'. Few consolidated floristic treatments appeared; notable among them are the works (some not completed) of Carl Maximowicz. The region was moreover excluded from *Index florae sinensis*. From 1920, several expeditions from Europe and the United States, as well as China and the new Soviet Union, penetrated different parts of the region and more sizeable collections were made, although, as in earlier years, few major floristic works resulted. Among the latter was the never-completed *Flora of the Mongolian steppe and desert areas* by Tycho Norlindh (1949; see **705**). Information on the vegetation was consolidated by Chi-wu Wang (1961; see below).

Since World War II, however, consolidation of political control, the formation of scientific establishments in Mongolia, Nei Mongol, Xinjiang, Qinghai and Xizang, and a felt need for better documentation has resulted in the appearance of modern floras and enumerations for nearly all parts of the region. In addition to *Flora reipublicae popularis sinicae*, the most important such work is the serially published *Rastenija Tsentral'noj Azii* led by Valery Grubov. Based primarily on collections available in St. Petersburg, the first of a projected 15 parts appeared in 1963; as of 1999, 11 fascicles, one in two parts, have been published. Recent unit floras include, for Mongolia, two enumerations (the first in 1955, the second in 1996) and a manual (*Opredelitel' sosudistykh rastenij Mongolii* (1982)), and, in China, *Flora intramongolica* by Hiang-chian Fu and others (1977–85; second edition commenced 1989), and *Flora in desertis reipublicae populorum sinarum* (1985–). In previously poorly documented Xizang, Chinese botanists have carried out a floristic and vegetation survey, working from a scientific center in Qinghai Province. A first publication on central Xizangian vegetation appeared in 1967 and a second survey, on the flora and fauna of the A-li region in the west, was issued in 1979. In the 1980s, two general

works, written under the leadership of Cheng-i Wu, followed: *Enumeratio xizangensis* and *Flora xizangica*, the latter in five volumes. For Qinghai itself, a flora began publication in 1996.

Despite such activity, however, the region, along with the Arabian Peninsula, remains in general one of the least-known areas of its size in Asia. Moreover, much of the recent work is not easily accessible, and until recent years political circumstances have impeded consolidation of the floristic information based on collections now in widely different parts of the world. The English edition of *Flora reipublicae popularis sinicae*, now underway, should further aid the process of synthesis begun with the Chinese original as well as *Rastenija Tsentral'noj Azii*.

A useful general introduction to the flora and vegetation is C. W. WANG, 1961. *The forests of China: with a survey of grassland and desert vegetation.* xiv, 313 pp., 78 text-figs., map (Maria Moors Cabot Foundation, Harvard University, Spec. Publ. 5). Petersham, Mass.

Progress

Early and recent activities are usually treated as part of Chinese botanical history (see Superregion 86–88). The most useful sources are the works by Bretschneider and Cox. Expeditions from British India are reviewed in South Asian sources (see Superregion 81–84).

General bibliographies. Bay, 1910; Frodin, 1964; Goodale, 1879; Holden and Wycoff, 1911–14; Hultén, 1958; Jackson, 1881; Pritzel, 1871–77; Rehder, 1911; USDA, 1958.

Regional bibliographies

LEBEDEV, D. V., 1963. Материалы к библиографии по флоре и растительности Центральной Азии (Materialy k bibliografii po flore i rastitel'nosti Tsentral'noj Azii). In V. I. Grubov (ed.), Растения Центральной Азии (*Rastenija Tsentral'noj Azii*), fasc. 1: 99–166. Moscow. [Unannotated, alphabetically arranged list. For supplement, see fasc. 6 (1971) of the same work.]

MERRILL, E. D. and E. H. WALKER, 1938. *A bibliography of Eastern Asiatic botany.* xlii, 719 pp., maps. Jamaica Plain, Mass.: Arnold Arboretum of Harvard University. Continued as E. H. WALKER, 1960. *A bibliography of Eastern Asiatic botany: Supplement 1.* xl,

552 pp. Washington: American Institute of Biological Sciences. [Save for Xinjiang (Sinkiang), Region 76 is fully accounted for in both works. For full description, see Superregion 86–88.]

General indices. BA, 1921– ; BotA, 1918–26; BC, 1879–1939; BS, 1940– ; CSP, 1800–1900; F.B, 1959–98; FB, 1931– ; ICSL, 1901–14; JBJ, 1873–1939; KR, 1971– ; NN, 1879–1943; RŽ, 1954– .

760
Region in general

See also general works on China (**860–80**: all references except Forbes and Hemsley, with coverage in at least some apt to be sketchy). Virtually the whole of the region is now being covered in the critical *Rastenija Tsentral'noj Azii* produced at the Komarov Botanical Institute in St. Petersburg under the direction of Valery Grubov.

GRUBOV, V. I. (ed.), 1963– . Растения Центральной Азии (*Rastenija Tsentral'noj Azii*)/*Plantae asiae centralis*. Vyp. 1– . Illus., maps. Moscow/Leningrad: AN SSSR Press (later Leningrad, now St. Petersburg: 'Nauka'). English edn.: *idem*, 1965. *Plants of Central Asia*. Fasc. 1 (all published). Jerusalem: Israel Program for Scientific Translations. New version, 1999– . Fasc. 1– . Plymouth, England/Enfield, N.H.: Science Publishers.

Detailed regional conspectus of vascular plants, with diagnostic keys to genera, species and infraspecific taxa: includes descriptions of novelties, extensive synonymy with references and citations of relevant literature, detailed indication of internal distribution with citation of *exsiccatae* and inclusion of some range maps, general summary of extralimital range, taxonomic commentary, and notes on habitat, special features, etc.; index to all botanical names in each fascicle. The first fascicle includes an extensive account on floristics and geobotanical regions, along with a large regional bibliography (the latter supplemented in fasc. 6). [Based upon the Central Asian collections at the Komarov Botanical Institute, the work is expected to encompass 20 fascicles; as of 1999, 11 had been published (with vyp. 8, Fabaceae, still wanting *Astragalus*). Fascicles may contain one or more families; if the latter

be the case a pre-numbered systematic sequence is followed (although this is independent of the order of publication). Some treatments are by specialists. Of the long-dormant English edition, a new version was initiated in 1999.]

761
Mongolia

Area: 1 565 000 km². Vascular flora: 2823 species, nothospecies and infraspecific taxa (Gubanov, 1996; see below).[43] Formerly known as Outer Mongolia and before World War I a territory of the Qing (Manchu) Empire.

Early collections in Mongolia were all by Russians, including Gmelin, Pallas, Turczaninow and Bunge. Przewalski and Potanin added significantly to knowledge in the late nineteenth century; these precipitated compilation by Maximowicz of *Enumeratio plantarum hucusque in Mongolia nec non adjacente parte Turkestaniae sinensis lectarum* (1889), of which only one volume ever was published. Systematic survey and investigations on any scale began only after 1921, when the country allied itself with the Soviet Union and the advent of organized scientific and technical assistance; in that decade N. V. Pavlov (Kazakhstan) and P. A. Baranov from St. Petersburg were active in the field. Much of their work had to do with vegetation and land use and improvement studies. In the years after World War II, much work was coordinated through the Mongolian Commission of the Soviet Academy of Sciences (in whose *Trudy* Grubov's *Konspekt* of 1955 appeared), and research stations set up. A further round of field studies began in 1970 under the aegis of the Soviet-Mongolian Integrated Biological Expedition, and additions were made to the *Konspekt* in 1972. In 1982 the first manual-key, also by Grubov, was published. Continuing field work and additions to the flora, however, necessitated a new *Konspekt* by 1996. There have also been two partial manuals (Žamsran *et al.*, 1972; Sančir *et al.*, 1985).

GRUBOV, V. I., 1955. Конспект флоры Монгольской Народной Республики (*Konspekt flory Mongol'skoj Narodnoj Respubliki*). 308 pp., 1 col. pl., 4 portrs., 3 maps (Trudy Mongol'sk. Komiss. Akad. Nauk SSSR 67). Moscow: AN SSSR Press.

Systematic enumeration of vascular plants, without keys, synonymy or references; species headings include vernacular names, a few critical remarks, and notes on habitat, biology, special features, etc. Indices to family and generic and to vernacular names are also provided. An introductory section incorporates chapters on local botanical exploration and research, geobotanical and floristic regions, and statistics of the flora together with a glossary, bibliography, and list of collectors. [Additions and corrections are contained in *idem*, 1972. Дополнения и исправления к 'Конспекту флоры Монгольской Народной Республики' (Dopolnenija i ispravlenija k 'Konspektu flory Mongol'skoj Narodnoy Respubliki'). *Nov. Sist. Vyss. Rast.* **9**: 270–298.][44]

GRUBOV, V. I., 1982. Определитель сосудистых растений Монголии (с атласом) (*Opredelitel' sosudistykh rastenij Mongolii (s atlasom)*)/ *Key to the vascular plants of Mongolia (with atlas)*. 442, [2] pp., 144 pls., map. Leningrad: 'Nauka'.

Manual-key to vascular plants (2239 species), with botanical and standardized Mongolian names, limited synonymy, figures of representative species (in a separate block of pages and covering some one-third of the total), and concise indication of distribution and habitat; references (pp. 414–415), name changes from the *Konspekt*, and indices to botanical and Mongolian names. The general part contains a history of the Mongolian floristic project and an account of the present work, with acknowledgment of the participants.[45]

GUBANOV, I. A., 1996. Конспект флоры внешней Монголии (сосудистые растения) (*Konspekt florz vnesnej Mongolii (sosudistye rastenija)*)/ *Conspectus of flora of Outer Mongolia (vascular plants)*. [ii], 136 pp., map. Moscow: «Valang» (for D. P. Syreishchikov Herbarium, Moscow State University).

Concise enumeration of vascular plants (2823 species and infraspecific taxa) with synonymy, references, essential citations, and indication of distribution (in 16 botanical regions; cf. map, p. 8); summary (pp. 110–112, also in English); detailed bibliography (pp. 113–131) and index to generic and family names at end. [In the introduction (also summarized in English) the author acknowledges the considerable additions to the flora (marked with a plus sign) as well as extensions of range as a result of new exploration from 1971 to 1991.]

Partial and local works

ŽAMSRAN, C., N. ÖLSI-KHUTAG and Č. SANČIR, 1972. *Ulaanbataar orčmyn urgamal tanikh bičig*. 294 pp. Ulaan Bataar. [Local flora; in Mongolian. Not seen; cited from *Prog. Bot.* **35**: 309, 320 (1973) as well as from Gubanov's *Konspekt* (see above).]

SANČIR, Č. *et al.*, 1985. *Urgamal tanikh bičig* [*Opredelitel' rastenij*]. 336 pp. Ulaan Bataar. [Students' flora; in Mongolian. Not seen; cited from Gubanov's *Konspekt* (see above).]

763

Nei Mongol (Inner Mongolia)

See also **705** (NORLINDH); **761** (both works); **860/I** (KITAGAWA; KOMAROV; NODA). – Area (Inner Mongolia Autonomous Region): 1 177 500 km^2. Vascular flora: 2167 species (*Flora intramongolica*, 1st edn.; see below). As delimited here, Nei Mongol (Inner Mongolia) encompasses the greater part of the former Chinese provinces of Ningxia, Suiyuan and Chahar, part of the former district of Jehol, and western Manchuria; this reflects the political boundaries in effect since 1979.[46]

The *Flora intramongolica* project was started about 1957. Following the interruption resulting from the Cultural Revolution of 1966–70, a 'flora group' was organized in 1976 and more formally constituted in 1981. The project was completed in 1984, and a revision is well advanced.

COMMISSIO REDACTORUM FLORAE INTRAMONGOLICAE (eds. MA YU-CHAN *et al.*), 1987–98. *Flora intramongolica*. 2nd edn. 5 vols. Illus. Huhhot: Department of Biology, University of Inner Mongolia. (1st edn., 1977–85, as SECTIO EDITORUM ET SCRIPTORUM FLORAE INTRAMONGOLICAE (later COMMISSIO REDACTORUM FLORAE INTRAMONGOLICAE) (eds. FU HIANG-CHIAN *et al.*), *Flora intramongolica*. 8 vols. (vol. 3 entitled *Flora chinae intramongolicae*). Illus. Huhhot.)

Copiously illustrated, briefly descriptive flora of vascular plants; includes keys, botanical and Chinese nomenclature, synonymy (with references and key citations), distribution, and notes on taxonomy, habitat, etc.; indices in each volume. A brief introduction to each volume is also provided. Volume 1 contains prefaces (in both Chinese and English), coverage (in Chinese only) of botanical exploration (pp. 11–64, with itinerary maps), physical features, climate, soils, vegetation, floristics, chorology and biogeography (with a map of floristic districts), accounts of pteridophytes and gymnosperms, and (pp. 282–408) cumulative indices to all names.

764

Xinjiang Weiwuer (Sinkiang Uighur) Autonomous Region

Area: 1646800 km². Vascular flora: 3537 species (Anonymous, 1976; see below); *c.* 4000 (Yang *et al.*, 1992–). – Apart from the recent works listed below, the most important accounts are Maximowicz's incomplete *Enumeratio* (see **761**) covering Mongolia and eastern Xinjiang ('Chinese Turkestan') and *Flora in desertis reipublicae populorum sinarum* (see **705**). Reference otherwise must be made to general works on China (**860–80**, except Forbes and Hemsley) or to *Rastenija Tsentral'noj Azii* by GRUBOV (**760**). There is also a nomenclator: ANONYMOUS, 1976. (*A list of Xinjiang plants.*) 72 pp.

COLLEGIUM AGRICULTURAE AUG.-1 XINJIANG-ENSE, 1982–85. (*Hsin-chiang wu chien so piao*)/ *Claves plantarum xinjiangiensium.* Vols. 1–3. Illus. Urumchi: Xinjiang People's Press/Editio Popularis Xinjiangensis.

Keyed enumeration of vascular plants; entries include Chinese nomenclature, synonymy (with references and key citations), distribution, altitudinal range, habitat, and karyotypes; illustrations of representative species and complete indices in each volume. The first volume includes an illustrated organography/glossary. [Not yet completed. Vol. 1 covers pteridophytes, gymnosperms and monocotyledons; vol. 2, dicotyledons from Salicaceae through Rosaceae; and vol. 3, Fabaceae through Cornaceae, all following the 1936 Engler/Diels system; 2300 species so far are accounted for. A further two volumes were planned.]

COMMISSIO REDACTORUM FLORAE XINJIANG-ENSIS (eds. YANG CHANG-YOU *et al.*), 1992– . *Xin jiang zhi wu zhi*/*Flora xinjiangensis.* Vols. 1– . Illus. [Urumqi]: Xin jiang ke ji wei sheng chu ban she (Xinjiang Science, Technology and Hygiene Publ. House).

Descriptive flora of vascular plants with keys to all taxa, references and citations, synonymy, indication of distribution and habitat, properties and uses, and commentary; index to Chinese names, systematic list of Chinese and Urumqi names of taxa, and index to all botanical names in each volume. A synoptic list of taxa prefaces each volume. [As of 1999 three volumes (of six planned), the second in two parts, had been published. Vol. 1 encompasses pteridophytes, gymnosperms, and dicotyledons from Salicaceae through Polygonaceae; vol. 2 (1994–95), Chenopodiaceae through Circaeaster-aceae and Berberidaceae through Rosaceae, and vol. 6 (1996) all the monocotyledons. A modification of the 1936 Engler/Diels *Syllabus* system is followed.]

765

Gansu (Kansu): western part

Area (Gansu as a whole): 366500 km². – Western Gansu, as here defined, includes all that part of the province lying northwest of the Huang Ho (with Lanzhou (Lanchow) City lying just outside on the south bank of the river). To the north are the deserts and steppes of Nei Mongol (**763**); to the south, the highland province of Qinghai (**767**).

No provincial flora has been published. Alternative coverage is available in *Flora in desertis reipublicae populorum sinarum* (**705**) and, for part of the area, in Carl Maximowicz's incomplete *Flora tangutica*, which centers on northeastern Qinghai (see below). Reference may also be made to general works on China (**860–80**, except Forbes and Hemsley) or to *Rastenija Tsentral'noj Azii* by GRUBOV (**760**).

767

Qinghai (Tsinghai)

See also **768** (HEMSLEY). – Area: 721000 km². Vascular flora: no data. The Tangut (Amdo) district, on which *Flora tangutica* was focused, lies around the great lake, Koko Nor, while the area around Xining

(Hsi-ning), the present provincial capital, was in the late nineteenth century a part of Gansu. Qinghai has also been administered as a district of Xizang (**768**).

Among the earliest 'Western' collectors in Qinghai were the Russians Przewalski and Potanin, who had also been active in Xinjiang and Mongolia. Their collections and data went into the first flora for any part of the modern province, *Flora tangutica* (1889) by Carl Maximowicz; this work, however, was unfortunately never completed. The next account covering Qinghai was *The flora of Tibet or High Asia* (1902; see **768**), based on collections from the Younghusband expedition. Plants obtained from the province by Joseph F. Rock in 1925 were accounted for by REHDER and WILSON, and by REHDER and KOBUSKI (see **860–80** under **Partial works**; general works on China can also be found therein). After 1949, exploration was renewed, and in 1996 the first of four volumes of a provincial flora, *Flora qinghaica*, made its appearance.

EDITORIAL COMMITTEE OF FLORA QINGHAICA (eds. LIU SHANG-WU *et al.*), 1996– . *Flora qinghaica*. Vols. 1– . Illus. Xining: Qinghai People's Publishing House.

Illustrated descriptive flora with keys, accepted names, Chinese equivalents, limited synonymy, places of publication, principal literature citations, and indication of distribution, altitudinal range, habitat, phenology, etc.; addenda with new taxa and indices to all Chinese and botanical names at end of each volume. [As of 1998 vols. 1–3 had appeared; these cover pteridophytes, gymnosperms, and all dicotyledons arranged following the 12th edition of Engler's *Syllabus*. Monocotyledons are expected in vol. 4.]

Woody plants

ANONYMOUS, 1987. (*Ligneous flora of Qinghai*.) 669 pp., illus. [Accounts for 521 native and introduced taxa, with keys, descriptions and line drawings. Not seen; information from Chen, 1993, p. 72.]

768

Xizang (Sitsang; Tibet)

Area: 1 221 600 km². Vascular flora: 5766 species (including those in the northern two-thirds of Arunachal Pradesh, an area claimed by China).

Until publication of *Xizang tzu-chih-ch'ü* (an enumeration) in 1981 and *Flora xizangica* in 1983–87 no general coverage of Xizang was available. Two long-standard earlier papers were the botanical results of the Younghusband expedition, published as *The flora of Tibet or High Asia* by W. B. Hemsley with H. H. W. Pearson (1902; in *J. Linn. Soc., Bot.* 35: 124–265), and *A sketch of the geography and botany of Tibet, being materials for a flora of that country* by F. Kingdon-Ward (1935; in *ibid.*, 50: 239–265). The first-named paper also encompassed Qinghai. After 1951, China instituted a programme of geographical and biological exploration of both Xizang and Qinghai. The two works cited below represent the main contributions in floristic botany for Xizang resulting from this programme. Other information may be found in general works on China (**860–80**).

The Chinese works also cover a major portion of the Indian state of Arunachal Pradesh (**847**) over which China has had territorial claims.

QINGHAI-XIZANG PLATEAU COMPLEX EXPEDITION, ACADEMIA SINICA, 1980 (1981). (*Xizang tzu-chih-ch'ü*)/ *An enumeration of the plants of Xizang (Tibet)*. [viii], 463 pp. (including covers), map (on front cover). [Beijing.]

Compiled enumeration of vascular plants, without keys or bibliography; species entries encompass botanical and Chinese nomenclature, distribution (by county), altitudinal range, and general indication of extralimital range.

QINGHAI-XIZANG PLATEAU COMPLEX EXPEDITION, ACADEMIA SINICA (Wu Cheng-i and collaborators), 1983–87. *Flora xizangica*. 5 vols. Illus. Beijing: K'o hsüeh chu pan she (Academia Sinica Press).

Briefly descriptive flora of vascular plants; includes keys to all taxa, essential synonymy, Chinese and botanical nomenclature, localities, distribution and altitudinal range, representative illustrations, notes on taxonomy, habitat, etc., references at end of each family; indices to all names in each volume. Volume 1 contains a foreword (by Wu) and an account of the plan of the work, along with a loose map.[47]

Superregion 77–79

Southwestern Asia

This large region includes most of the traditional 'Orient', now expanded and called 'the Middle East'. As delimited here it extends from the Bosphorus and Dardanelles and from the Suez Canal eastward to the long escarpment bordering the Indus Basin, and from the CIS border in the north to the Indian Ocean, including the whole of the Arabian Peninsula. Infratropical Arabia apart, this substantially corresponds to the area in Asia covered by Edmond Boissier's comprehensive *Flora orientalis* (1867–88). – Area: 7 636 000 km^2.

Prominent names in the early botanical exploration of the superregion were Leonhard Rauwolff (whose collections of 1573–75 first saw publication only in 1755 in *Flora orientalis* by Johan F. Gronovius), Joseph P. de Tournefort (1700–02), Fredric Hasselquist, and Pehr Forsskål (author (posthumously) of *Flora aegyptiaco-arabica*, 1775, and *Icones rerum naturalium*, 1776). The work of these men and their more numerous followers in the nineteenth century – including William Griffith in Afghanistan – was effectively consolidated by Edmond Boissier. A model of good floristic exposition, comparable to other works of the Candollean school as well as the Kew colonial floras, Boissier's *Flora* ever since has represented a foundation for regional floristic research and flora-writing. The Candollean classification system as used therein was moreover adopted in some subsequent floras.

So well satisfied were user needs by *Flora orientalis* that until the mid-twentieth century there was little apparent demand for more up-to-date works. Modern institutions of higher education were few; among the earliest was the American College (later University) in Beirut where in 1875 the Rev. George E. Post established what is now the Post Herbarium. This formed a primary basis for the first of a new cycle of local works, *Flora of Syria, Palestine and Sinai* (1896); a second edition, by J. E. Dinsmore, appeared in 1932–33. In Cyprus, the advent of British control led to more intensive local work, including research by the Norwegian botanist Jens Holmboe which resulted in his *Studies on the vegetation of Cyprus* (1914). The inclusion of southern Yemen in the Raj led to a flora of Aden (1914–16) by the Bombay botanist Ethelbert Blatter. Botanical work also followed in present-day western and northern Pakistan as the Raj in India was consolidated. With respect to Asia Minor and Iran, Joseph Bornmüller, curator of the then Weimar-situated Haussknecht Herbarium, contributed considerably to further knowledge in numerous papers (partly with others, among them Erwin Gauba) both before and after World War I.

The advent of francophone tertiary studies in Beirut in the late nineteenth century opened the way to an enduring tradition of French-language floristic works in the Lebanon and Syria. The first initiative was that of Fr. Louis Bouloumoy, who organized a herbarium and undertook floristic studies, largely completing them by 1914; his *Flore du Liban et de la Syrie* eventually appeared in 1930, its combination of text and atlas setting a precedent for later works along the Eastern Mediterranean.

Extensive Jewish settlement in Palestine and, in 1918, the imposition of the British mandate also presented new opportunities, one of them being the formation of Hebrew University in Jerusalem; under the foundation professorship of Otto Warburg it became a leading botanical center. A manual for identification was early seen as necessary; the result was *The plants of Palestine: an analytical key* by Alexander Eig, Michael Zohary and Naomi Feinbrun (1931), shortly before publication of Dinsmore's edition of Post's flora. Local botanical work was similarly initiated in Iraq, beginning effectively with the arrival in 1933 of Evan R. Guest in the Directorate of Agriculture and Forestry at Baghdad. Coverage of the Arabian Peninsula – of which only the extra-tropical part had been covered by Boissier – also improved through Blatter's somewhat sketchy *Flora arabica* (1919–23, 1936) and the critical, gap-filling *Flora des tropisches Arabien* by Oscar Schwartz (1939). A miscellany of more local floristic reports and other works was also produced for such scattered areas as the North-West Frontier, Baluchistan, Transjordan, the Sinai, and parts of Turkey. In the last-named country, the first local flora in Turkish (*Ankaranın Flora* by Kurt Krause, 1929, 1937) was, as in Palestine, associated with educational reform and the advent of tertiary-level studies. Nevertheless, because these developments were relatively modest in their reach (and not unnaturally disrupted by World

War II), it was still possible by 1951 to say, as did George Lawrence in his *Taxonomy of vascular plants*, that 'most of the literature . . . is very old', if not obsolete.[48]

Lawrence was, however, writing as a new cycle was beginning. He did not, or was not in a position to, perceive the great burst of new floristic activity which would take place almost throughout the superregion, both by outside botanists (mainly from Europe) but also by nationals, either in their own countries or in others. This was in a sense a continuation of developments which had, as already noted, been initiated before World War II: the growth of tertiary institutions, the foundation of more local herbaria, and the development of a regionally based botanical profession. Widespread flora-writing, however, was limited by the constraints of local resources and prevailing fashions. By contrast, much collecting had been accomplished between the wars, and this activity accelerated after 1945. From the 1950s and notably in the 1960s, as botanical interest revived and, with dedicated leadership, a favorable climate for support of science, and the means available resulting from petroleum extraction revenues or other sources (such as U.S. PL-480 surplus currency funds), new floras were published or begun for every country in the superregion. Leading works of the era included the *Flora of Turkey*, *Flora of Cyprus*, *Nouvelle flore du Liban et de la Syrie*, *Flora palaestina*, *Flora of Iraq*, *Flora iranica*, *Flora of Pakistan*, and (from 1980) *Conspectus florae orientalis*. More recently these have been joined by *Flora of the Arabian Peninsula*, like *Flora of Turkey* a project of the Royal Botanic Garden Edinburgh (with the collaboration of the Royal Botanic Gardens, Kew). A dendrological distribution atlas was also compiled at Kórnik (Poland). Some assistance to flora writers concerned with the superregion has been given by the publication of *Flora SSSR* and regional floras in the Caucasus and Middle Asia.

Many of the new floras, however, effectively remain little more than *prodromi* or preliminary surveys covering more or less sketchily known areas. They indeed represent an advance over what was known to Boissier, but over a century our expectations have also increased. The level of coverage is about that of much of Middle America, South Asia, or some parts of Africa. More detailed documentation, including the results of biosystematic studies, has started to develop only within the last three decades. Other problems are continuing political turbulence and the apparent lack of any but informal coordination between major flora projects. The choice of season is also extremely important in planning collecting trips.[49] The 'umbrella' organization, OPTIMA, has, however, since its formation in 1975 served to some extent as a force for collective progress. It has 'rescued' one flora project accounted for here and was the inspiration for another (*Conspectus florae orientalis*). Focused symposia, of which the first was held in 1970, have also brought together specialists with the contributions presented later being published as proceedings (see Note 51).

At the present time, almost the whole of the 'Boissier area' has benefited from the new standard works. As Hedge has also indicated, this represents a great advance on 1939 when, with the composition of *Flora orientalis* and other great nineteenth-century floras in spirit not yet too remote, floristic work was thought to have 'run its course'. Yet, as we have seen, the contemporary works are by no means final; they are there to promote further, more detailed studies including florulas and checklists of local scope and as tools for conservation, management and rehabilitation. It is also to be hoped that the *Flora of the Arabian Peninsula* and – at some time – the *Flora of Iraq* can be completed, and that accounts of *Astragalus* – richly represented in the region – will be forthcoming for *Flora iranica* and *Conspectus florae orientalis*.

Progress

Early history was surveyed by Boissier in volume 1 of *Flora orientalis* (1867).[50] Recent progress in the countries on and near the Mediterranean (Region 77), with lists of references, has been surveyed in separate articles in V. H. HEYWOOD (coord.), 1975. Données disponibles et lacunes de la connaissance floristique des pays méditerranéens. In Centre National de la Recherche Scientifique (France), *La flore du bassin méditerranéen: essai de systématique synthétique*, pp. 15–142 (Colloques internationaux du CNRS 235). Paris (see also **590, 601**). These have been continued in the irregular *OPTIMA Newsletter* and the proceedings of OPTIMA conferences. The various collections of proceedings of conferences on Southwest and Central Asia also include reports on floristic progress.[51]

General bibliographies. Bay, 1910; Frodin, 1964, 1984; Goodale, 1879; Holden and Wycoff, 1911–14; Hultén, 1958; Jackson, 1881; Pritzel, 1871–77; Rehder, 1911; USDA, 1958.

Supraregional bibliographies

In addition to the titles given below, the volumes of *Conspectus florae orientalis* (C. C. Heyn *et al.*) encompass a list of key literature, with 'standard floras' along with 'selected regional floras and papers' and other floristic and vegetatiological literature. There is also a list of 'selected standard floras' on pp. xii–xiv in the first volume of *Flora palaestina* (eds. M. Zohary and N. Feinbrun-Dothan; see **775** below).

YUDKISS, H. (comp.); D. HELLER (ed.), 1987. *Bibliography of botanical research of the Middle Eastern area*. 439 pp. (Boissiera 39). Geneva. [Arranged by author, with coverage to 1982; includes chronological, geographical and subject (including systematic) indices. Covers Egypt and Turkey to the Arabian Peninsula, Afghanistan and Iran.]

FIELD, H. (comp.), 1953–64. *Bibliography on Southwestern Asia*. 7 vols. Coral Gables, Fla.: University of Miami Press; and H. FIELD and E. M. LAIRD, 1969–72. *Bibliography of Southwestern Asia: Supplements I–VIII*. Coconut Grove, Fla.: Field Research Projects. [Encompasses the Caucasus, Egypt, and Middle Asia as well as the whole of southwestern Asia as here delimited; coverage in the original work extends up through 1959. Cumulative botanical subject indices have been provided by R. C. Foster.]

General indices. BA, 1926– ; BotA, 1918–26; BC, 1879–1944; BS, 1940– ; CSP, 1800–1900; EB, 1959–98; FB, 1931– ; ICSL, 1901–14; JBJ, 1873–1939; KR, 1971– ; NN, 1879–1943; RŽ 1954– .

Supraregional indices

Limited coverage is provided in the currently irregular *OPTIMA Newsletter* (OPTIMA Secretariat, Madrid).

770–90

Superregion in general

The whole of the superregion was long dominated by the comprehensive *Flora orientalis* of Edmond Boissier, written in the style of the contemporaneous 'Kew floras' then being prepared for different parts of the British Empire. Since 1980 it has been complemented by *Conspectus florae orientalis*, a taxonomically validated enumeration covering a somewhat lesser area.

BOISSIER, E., 1867–88. *Flora orientalis, sive enumeratio plantarum in Oriente a Graecia et Aegypto ad Indiae fines hucusque observatarum*. 5 vols. and supplement. 6 pls., portr. Basel, Geneva: Georg. (Vol. 1 reprinted 1936, Geneva, Herbier Boissier. Whole work reprinted 1963–64, Amsterdam: Asher; 2nd reprint, 1984.)

Descriptive flora (in Latin) of vascular plants, with synoptic keys to genera and species; includes essential synonymy, references and citations, vernacular names, generalized indication of internal and extralimital range (with citation of some *exsiccatae* and localities), taxonomic commentary, and miscellaneous observations; indices to all botanical names in each volume (cumulative index to family, generic and vernacular names in vol. 5). The introductory section in volume 1 incorporates chapters on botanical exploration, floristic regions, limits of the *Flora*, and philosophical and methodological considerations. Volume 6, the *Supplementum*, edited by Boissier's associate Robert Buser following the author's death in 1885, comprises additions and corrections to the five volumes of the main work.[52]

ZOHARY, M., C. C. HEYN and D. HELLER, 1980–94. *Conspectus florae orientalis: an annotated catalogue of the flora of the Middle East*. Fascicles 1–9 [of 10]. Maps. Jerusalem: Israel Academy of Sciences and Humanities.

An annotated systematic enumeration featuring accepted names, basionyms (where appropriate), common synonyms, references, and Boissier and other citations along with regional distribution (by defined plant-geographical units as shown in map 1) and general chorotypes (following the scheme in map 2; both maps appear in each fascicle); addenda, corrigenda, new names and combinations, and index to family and generic names at end of each fascicle. [Covers countries from Turkey (Asia Minor) and the east Aegean islands of Greece along with Egypt to Iran and southward to the foot of the Arabian Peninsula (corresponding to **Regions 77, 78** and parts of **59** and **79**). The arrangement and limits of families are based on the 12th edition of the Engler *Syllabus*, but in order of publication the blocks of families as presented do not conform to that sequence. The final fascicle, not yet published, will supplement the first nine and as well contain Hydrangeaceae (no. 77a) and *Astragalus* (Leguminosae).][53]

Woody plants: distribution maps

BROWICZ, K., assisted by J. ZIELIŃSKI (from vol. 4 as co-author), 1982–94, 1996. *Chorology of trees and shrubs in south-west Asia and adjacent regions.* 10 vols., supplement. Maps. Poznán: Pánstwowe Wydawnictwo Naukowe/Polish Scientific Publishers (vols. 1–8); Sorus (vol. 9); Bogucki (vol. 10 and Supplement) (all for the Institute of Dendrology, Polish Academy of Sciences).

A chorological atlas of 575 dot distribution maps of individual woody plant species, each with separate text containing details of distribution, altitudinal range, habitat, associates, and phytogeographic relations; bibliography and indices in each part (the former represented in vols. 2 *et seq.* as supplements to that in vol. 1) and general index in vol. 10. The choice of species is arbitrary, depending on 'availability of sufficient data', and the area covered ranges from 19° to 78° E and 23° 30′ to 47° 20′ N, thus reaching into parts of Europe and Middle Asia.[54]

Region 77

The Levant

Also known loosely as 'The Orient', the region here called the Levant consists of Turkey-in-Asia (Asia Minor), the adjacent islands of the eastern Aegean, Cyprus, Syria, Lebanon, Israel and the 'West Bank' (in Palestine), the Sinai Peninsula, Jordan and Iraq. Within this region is the subregion known as the 'Holy Land'. The overall limits correspond almost precisely with those of the Asiatic possessions of the former Ottoman Empire (except for the Hejaz, now part of Saudi Arabia (Region 78)). – Area: 1 573 000 km². Vascular flora: no data.

The first subregional or unit manual-flora within the Levant as here defined was the first edition of George Post's *Flora of Syria, Palestine and Sinai* (1896); an earlier work by the author in Arabic, *Flora of Syria* (1884), was not completed. Post's flora was followed by a period of collecting and miscellaneous publications, including studies of Biblical plants. His work and, slightly later, that of Fr. Louis Bouloumoy found a parallel further south in the activities of the British-based Palestine Exploration Fund; in 1884 the results of a natural history survey, oriented in particular towards Transjordanian Palestine, appeared as *The fauna and flora of Palestine* under the editorship of Henry B. Tristram (vascular plants, pp. 205–455). This work was followed by in particular the studies of Aaron Aaronshon, whose checklists, however, appeared posthumously. A second cycle of floras began about 1930

with the appearance (posthumously) of Bouloumoy's *Flore du Liban et de la Syrie*, the second edition of Post's manual, and a Hebrew University manual for Palestine (the last with successive editions to 1991). Only well after World War II, however, did modern floras for Turkey and Iraq begin publication; at this writing the *Flora of Iraq* remains unfinished. For Syria and Lebanon, additional floras were published by Joseph-Marie Thiébaut (1936–53) and Paul Mouterde (1966–83), the latter with atlases and based on four decades' residence in Lebanon. For Israel, the West Bank, Palestine and Jordan, *Flora palaestina* (Zohary and Feinbrun-Dothan, 1966–86) provides fairly detailed coverage. The Sinai Peninsula is covered by floras of Egypt (**591**) as well as by Post's floras; there is also a recent separate checklist by M. N. el-Hadidi *et al.* (1991), successor to those of Danin, Zohary and others. A modern flora of Cyprus was realized by Desmond Meikle in 1977–85 after prolonged preliminary work.

The degree to which floras have been produced locally is related to the penetration of modern higher education, the availability of collections and related research resources, and cultural perceptions. With the exception of Beirut (and Cairo), no significant resources were developed in the region until well into the twentieth century. Thus several major floras were prepared and produced elsewhere. At the present time, however, resources are more numerous and better distributed, with locally based works more readily possible; this will, for example, be a feature of the planned new supplement to *Flora of Turkey*. Botanical knowledge has thus become by southwestern Asian standards relatively far advanced, save perhaps for less accessible desert or mountainous areas such as the eastern part of Turkey and eastern Syria and Jordan.[55]

Bibliographies. General and supraregional bibliographies as for Division 7 and Superregion 77–79.

Indices. General indices as for Superregion 77–79.

770

Region in general

See **770–90** (*Conspectus florae orientalis*; *Flora orientalis*). For areas bordering on the Mediterranean, see also **601/I** (GREUTER *et al.*, *Med-Checklist*).

771

Turkey-in-Asia (Asia Minor)

Area: 755 688 km². Vascular flora (whole country): 8800 species including non-natives (Rechinger, 1991; see reference under *Flora iranica*). *Flora of Turkey* accounts for 8579 species, with 30.9 percent endemism (GB; see **General references**). – General works for the whole of modern Turkey are given under this heading. For the European part alone, see **638**.

Flora orientalis was the first, and long remained the only, flora for Asia Minor. It summarized the many prior contributions of its author and others, including those of Petr Tchihatcheff in the botanical part (1866) of his *Asie mineure*. Only after political and educational reforms consequent to the formation of modern Turkey did local initiatives increase. A first bibliography was published by Kurt Krause, then in Ankara, in 1927–31 (*Die botanische Literatur über die Türkei*; in *Repert. Spec. Nov. Regni Veg.* **24** and **28**). Krause subsequently published two editions of *Ankaranın Floru* (1934, 1937; in Turkish). Turkish-language manuals of larger scope appeared after World War II, firstly the incomplete and unreliable *Türkiye bitkileri* (1952) by H. Ahmet Birand, an associate of Krause at Ankara, and later a much larger version by Kâmil Karamanodlu (1974– ; not so far completed), like its predecessor based on resources in Ankara. The *Flora of Turkey* project was also initiated after 1945; publication began in 1964 and was completed by 1988. A new supplement is planned.

Bibliographies

The 1993 bibliography of Demiriz effectively supersedes earlier works. All, including those given in the previous edition of this *Guide*, are listed by him in his introduction. Mention may also be made of the bibliographic resources in *Flora of Turkey*, in particular in volume 10.

DAVIS, P. H. and J. R. EDMONDSON, 1979. Flora of Turkey: a floristic bibliography. *Notes Roy. Bot. Gard. Edinburgh* **37**: 273–283. [Literature base for *Flora of Turkey* project including precursory papers.]

DEMİRİZ, H., [1993]. *An annotated bibliography of Turkish flora and vegetation/Türkiye flora ve vejetasyonu bibliyografyasi.* xvii, 670 pp., map. [Ankara]: Temel Bilimler Araştırma Grubu (Basic Sciences Research Committee), TÜBİTAK (The Scientific and Technical Research Council of Turkey). [Introduction, with conventions and notes on previous bibliographies; alphabetical list of some 5000 numbered titles with brief annotations including geographical focus, if any, and language of publication; classified indices at end covering taxa, geography, institutions and personalia.]

DAVIS, P. H. (ed.), 1965–88. *Flora of Turkey and the east Aegean islands.* 10 vols. Illus., maps. Edinburgh: Edinburgh University Press.

Comprehensive, briefly descriptive flora of vascular plants; includes keys to all taxa, extensive synonymy with references and principal citations, vernacular names, localities with indication of *exsiccatae*, and general summary of internal and extralimital range (together with a number of distributional maps), notes on habitat, phenology, associates, etc., and taxonomic commentary; indices to all botanical names in each volume. The introductory section (vol. 1) provides chapters on geography, climate, floristics, and phytogeography as well as a list of references. Volume 10 contains a supplement and general index.[56]

KARAMANOĞLU, K., 1974. *Türkiye bitkileri.* I. cilt: *Pteridophyta, Gymnospermae, Dicotyledoneae.* 1277 pp. (T. C. Ankara Üniv., Eczaclik Fakültesi, Yayınları, sayı 32). Ankara.

Annotated systematic enumeration (in Turkish) of vascular plants; includes places of publication of accepted names, synonymy, localities with *exsiccatae* and published reports, overall distribution, and habit but little if any commentary; index to all botanical names (pp. 1153–1277). The introduction includes a historical review, a list of authority abbreviations and explanation of symbols. [Covers in the dicotyledons the 'archichlamydean' families, ending with Umbelliferae. No continuation has yet been published.]

YAKAR-TAN *et al.*, 1982–83, 1993– . *Türkiye florasi atlasi*, fasc. 1– . Illus. Istanbul: Fen Fakültesi Basımevi.

Atlas of illustrations with facing text inclusive of description, phenology, distribution and altitudinal range, notes and lists of literature; fascicles not indexed. [As of 1995 seven numbers in four parts had been published, covering only a minute proportion of the flora.]

Partial work: Ankara and vicinity

KRAUSE, K. (assisted by H. AHMET BIRAND), 1937. *Ankaranın floru.* 2. baski (revised edn.). 207 pp., 12 halftone pls. (T. C. Yüksek Ziraat Enstitüsü, Çalışmalarından, sayı 2). Ankara. (1st edn., 1934.)

An enumeration (in Turkish) of *c.* 800 species of vascular plants of the vicinity of Ankara, with vernacular names (only for families), diagnostic descriptions (in Turkish), localities (with authorities but no *exsiccatae*), distribution in Anatolia and elsewhere, and indication of habitat (the latter also in German); index at end. An introductory section, in Turkish and German, gives a history of collecting in Turkey and the background to the work, a description of the area and its climate, soil, and vegetation formations with characteristic species (the ruderal flora being of special importance), phenology, and phytogeographical relationships.

Woody plants

A Turkish-language work is F. YALTIRIK, 1988. *Dendrologi ders Kitabi*, II: *Angiospermae* 1. Istanbul: Orman Fakültesi, Istanbul University.

Pteridophytes

PARRIS, B. S. and C. R. FRASER-JENKINS, 1980. A provisional checklist of Turkish Pteridophyta. *Notes Roy. Bot. Gard. Edinburgh* 38: 273–281.

A concise tabular checklist, with supplementary annotations giving details of new records or keys to relevant references (fully listed in pp. 278–279); one new taxon proposed. [Intended as a revision of the treatment in the first volume of the *Flora of Turkey*.]

772

Cyprus

Area: 9254 km^2 (33 percent being in 'Northern Cyprus'). Vascular flora: 1682 native and naturalized species (cf. Meikle, 1977–85; see below).

The works by Chapman and Holmboe and the recent flora by Meikle are to some extent interrelated. The last-named represents a splendid addition to botanical literature on the Middle East. Holmboe's work succeeded the floristic enumeration (pp. 150–392) in *Die Insel Cypern* (1865) by Franz Unger and Theodor Kotschy.

HOLMBOE, J., 1914. *Studies on the vegetation of Cyprus.* 344 pp., 143 text-figs., 7 pls. (Bergens Mus. Skrifter, N.S. 1(2)). Bergen.

Incorporates a systematic enumeration of known vascular plants (based partly upon the author's collections of 1905), with synonymy and references, indication of *exsiccatae*, with localities, brief descriptive notes with emphasis on special features, and extra details of new or interesting plants; list of references; index to generic names. The introductory section includes accounts of the physical features and geography of the island as well as a history of botanical exploration and the author's itinerary, while at the end of the work is a discussion of vegetation formations and plant communities along with an account of historical plant geography.

MEIKLE, R. D., 1977–85. *Flora of Cyprus.* 2 vols. xii, 832 pp., 52 pls., map. Kew: Bentham-Moxon Trustees (Royal Botanic Gardens).

Concise descriptive flora of vascular plants; includes keys to genera and species (no key to families), full synonymy with references, citations of principal regional works, typifications, local range with altitudes and citations of (selected) *exsiccatae*, generalized indication of overall distribution, critical remarks, and notes on habitat, special features, etc.; appendices with abbreviations, list of collectors, list of new taxa and names, and complete index at end. A precursory part covers physical features, climate, botanical subdivisions (with map and remarks on *Charakterpflanzen*; eight units recognized), and history of botanical exploration (mainly subsequent to that of Holmboe), and technical details. Volume 2 includes three appendices covering the dates of activity of cited collectors, new taxa, names and combinations, and a key to families, as well as an index to both volumes.[57]

OSORIO-TAFALL, B. F. and G. M. SERAPHIM, 1973. *List of the vascular plants of Cyprus.* v, 137 pp. Nicosia: Ministry of Agriculture and Natural Resources, Cyprus.

Systematic name list, with synonymy and indication of endemics and introduced species as well as those of uncertain status; summary of the flora with statistics (1810 species in 668 genera); index to families and genera. A brief introduction is also given, outlining the basis for the work.

Woody plants

CHAPMAN, E. F., 1949. *Cyprus trees and shrubs.* 88 pp., illus. Nicosia: Government Printer. (Reprinted 1967.)

Briefly descriptive, somewhat non-technical treatment of Cyprus woody plants; includes keys to all taxa, Greek vernacular names, indication of local range with citation of some *exsiccatae*, and notes on habitat, phenology, local uses, and wood properties; indices to all botanical and vernacular names. An illustrated glossary is included together with a general key to genera.

773

'Holy Land' (in general)

Vascular flora: *c.* 3500 species (from a 1907 estimate by A. Aaronsohn in Oppenheimer, 1930, p. 9; see 775), now surely too low. – This is here defined as a purely geographical subregion, extending approximately from the Turkish border to the Gulf of Suez and from the Mediterranean Sea to the Syrian-Jordanian desert ('Arabia deserta'). It overlaps parts of 774, 775 and 776.

The two editions of Post's manual followed *Icones plantarum Syriae rariorum* (1791–1812) by J. J. H. de Labilliardière and *Catalogue des plantes* by E. Cosson and L. Kralik (1854). There is also a four-century-long tradition of Biblical and Judaic floristics.

POST, G. E., 1932–33. *Flora of Syria, Palestine and Sinai.* 2nd edn., revised by J. E. Dinsmore. 2 vols. 774 text-figs., map. Beirut: American University of Beirut. (1st edn., 1896.)

Briefly descriptive flora of vascular plants; includes analytical keys to families, genera, and species in larger genera as well as synoptic keys to species, essential synonymy with principal citations (including the first three of the four volumes of Immanuel Löw's *Die Flora der Juden*, 1926–34), indication of internal range by districts, with some citations and localities, Arabic and English vernacular names, and notes on habitat, life-form and phenology; complete indices to all botanical and vernacular names at end of vol. 2. The introductory section gives details of sources for the work as well as a glossary and bibliography. [Still basic though long out of date, this work covers a region extending from the Taurus Mountains in the north to Ras Muhammad (near Sharm-el-Sheikh) in the south and from the Mediterranean to the Syrian-Jordanian desert. The arrangement of families in this work follows the Candollean system as used in *Flora orientalis*.]

774

Syria and Lebanon

See also 773 (POST). – Area: 195 632 km² (Syria, 185 180 km²; Lebanon, 10 452 km²). Vascular flora: Syria, 3459 species and subspecies (Al-Hakim, 1986); Lebanon, 2000 species (GB). Mouterde planned for 3600 species (*Nouvelle flore*, 1: ix). Partly for historical reasons (both were French mandates after World War I), these two countries have always been grouped together for floristic purposes and are accordingly treated here as a single unit.

The first flora in the modern era (apart from that of Post) was *Flore du Liban et de la Syrie* by Fr. Louis Bouloumoy (1930, in 2 vols.), prepared mostly before 1914 but not published until after financial aid was made available in 1925. It was relatively soon after succeeded by *Flore libano-syrienne* (1936–53, in 3 vols.) by Joseph-Marie Thiébaut, a French official and amateur botanist active in Lebanon until 1934. Preparation of the current standard flora was undertaken by its author from 1931; to this end he went through all material in the Post Herbarium as well as that of Bouloumoy and others now housed at Lebanon University.

MOUTERDE, P., Fr., 1966–83(–84). *Nouvelle flore du Liban et de la Syrie.* 3 vols. (each in 2 parts, respectively text and atlas). Pls. Beirut: Dar el-Machreq (Imprimerie Catholique) (distributed by Orientale; vol. 3 by Orientale and by OPTIMA Secretariat, Berlin-Dahlem).

Briefly descriptive flora of vascular plants; includes keys to all taxa, limited synonymy, fairly detailed internal distribution with localities and citations, brief summary of extralimital range, extensive taxonomic commentary and notes on habitat, phenology, etc.; indices to all botanical names. Each text volume is accompanied by an atlas with full-page plates of nearly all species. The introductory section (in vol. 1) covers floristic zones, general features of the flora, life-form categories, and the history of botanical exploration; a gazetteer of localities and bibliography are also provided.[58] [The *Nouvelle flore* was preceded by a mimeographed checklist: *idem*, 1965. *Flore du Liban et de la Syrie.* i, 101 pp. Beirut: The author.]

Syria

AL-HAKIM, W., 1986. *Enumeratio plantarum Syriae*. 77 pp. (Naturalia Monspeliensia 51). Montpellier.

A nomenclatural checklist of vascular plants in tabular form (based on Mouterde's *Nouvelle flore*) produced in two sequences (alphabetically by genus and species and following the Hutchinson system), with each entry including a code number, scientific name and authority; family and generic lists (pp. 1–5), plan of the work and references cited.

775

Palestine sensu lato (including Israel, Palestine, the 'West Bank' and Jordan)

See also **773** (POST). – Area: 116 218 km² (Israel, 20 770 km²; Gaza, 363 km²; West Bank, 5879 km²; Jordan, 89 206 km²). Vascular flora: 2470 species (Zohary and Feinbrun-Dothan, 1966–86); in Israel alone: 2294 species (GB); in Jordan alone: 2200 species (GB). For floristic purposes, these polities have often been considered as one unit and are accordingly treated together. They correspond to the former British mandate of Palestine (or Cis- and Transjordanian Palestine).

A separate floristic literature began to develop from late in the nineteenth century, soon also based on the work of resident collectors including the agricultural botanist Aaron Aaronsohn. Aaronshon's results appeared in *Reliquiae aaronsohnianae* by H. R. Oppenheimer (2 vols., 1930–40, respectively on Trans- and Cisjordanian Palestine). The establishment of the Hebrew University herbarium in the 1920s enabled the preparation of local works; the first was the Hebrew-language *The plants of Palestine: an analytical key* (1931) by Alexander Eig, Michael Zohary and Naomi Feinbrun-Dothan, a predecessor of the current standard manual (see below under **Israel**).

ZOHARY, M. and N. FEINBRUN-DOTHAN, 1966–86. *Flora palaestina*. 4 vols. (each in 2 parts). Pls. Jerusalem: Israel Academy of Sciences and Humanities.

Briefly descriptive flora (in English) of vascular plants, with keys to all taxa, synonymy with references and principal citations, generalized indication of local and extralimital distribution, taxonomic commentary, and notes on habitat, phenology, special features, ecology, uses, etc.; maps of botanical districts; diagnoses of new taxa and index to all botanical names in each volume (with in vol. 4 a list of Biblical names (pp. 416–417) and general index to families and genera). As in Mouterde's *Nouvelle flore*, each text volume is accompanied by an atlas with illustrations of most species captioned with botanical and Hebrew vernacular names. [The work, like the *Analytical flora* (see below), covers former Cis- and Transjordanian Palestine, with an extension into the eastern desert.][59]

Israel

In addition to the *Analytical flora* (see below), a useful semi-popular illustrated guide, with photographs by D. Darom, is U. PLITMANN *et al.*, 1982. *Pictorial flora of Israel*. 338 pp., illus. (part col.). Givatayim: Massada. Both works are in Hebrew, with in the *Pictorial flora* the preface being also in English. The current edition of the *Analytical flora* is the fifth in a series begun in 1931.

FEINBRUN-DOTHAN, N. and A. DANIN, 1991. (*Analytical flora of Eretz-Israel*.) 1040 pp., illus., maps. Jerusalem: CANA. (Originally published 1931 as *The plants of Palestine: an analytical key* by A. Eig, M. Zohary and N. Feinbrun; revised 1948, 1960 as *Analytical flora of Palestine* by A. Eig *et al.* and 1976 as *A new analytical flora of Israel* by M. Zohary.)

Manual-key to vascular plants with scientific names and Hebrew equivalents, generalized indication of local range, phenology, habitat, etc., small figures, and computer-generated maps, the latter showing presence or absence in one or more of 31 units; synonymic lexicon (pp. 977–1014), colored illustrations, and indices to vernacular and scientific names. The general part contains an explanation of the work, a map of distributional units, phytogeographic provinces (map, p. 10), references (p. 21), a glossary and organography, and the general key (pp. 58–79). The 31 map units encompass Israel, the West Bank, Palestine, western Jordan and the Golan Heights.

Jordan

AL-EISAWI, D. M., 1982. List of Jordan vascular plants. *Mitt. Bot. Staatssamml. München* **18**: 79–182.

An alphabetically arranged nomenclatural list of vascular plants, with most useful synonyms. The brief introduction gives the basis of the work (the first such for the present state of Jordan), which reflects considerable new collecting accomplished in the decade or so before publication. [Written as a precursor to a descriptive flora in preparation by the author.]

776

Sinai Peninsula

See also **591** (all works); **773** (POST). – Area: 58 714 km². Vascular flora: 984 species (El-Hadidi *et al.*, 1991; see below). Danin *et al.* (1985; see below) recorded 886 species. The peninsula is politically part of Egypt but geographically belongs to Asia; the Suez Canal is treated here as a continental boundary.

A first flora was *Florula sinaica* by Joseph Decaisne (1854), covering 233 species. Further collecting raised the total by 1912 to 800 species. The next general checklist was *Die phytogeographische Gliederung der Flora der Halbinsel Sinai* (1935) by M. Zohary (1935; in *Beih. Bot. Centralbl.*, B, **52**: 549–621), accounting for 942 species.[60]

DANIN, A., A. SHMIDA and A. LISTON, 1985. Contributions to the flora of Sinai, III. Checklist of the species collected and recorded by the Jerusalem team 1967–1982. *Willdenowia* **15**: 255–322, 4 figs. (incl. map).

Designed primarily as a complement to *Flora palaestina* (see **775**) and Täckholm's *Students' flora* (see **591**), this work is an annotated enumeration of vascular plants collected as part of an extended, comprehensive Sinai flora and vegetation survey (accounting for 886 species); entries include distribution by subdivision(s), habitat(s), frequency level (in both qualitative and quantitative terms), and references to illustrations. The list proper is followed by a discussion and summary of the flora (40 species are recorded for the first time).

EL-HADIDI, M. N. *et al.*, 1991. *Annotated list of the flora of Sinai*, 1–10. 100 pp., map (Taeckholmia 12). Cairo.

Annotated enumeration in 10 parts, each by different authors and featuring authorities, places of publication, geographical codes, and taxonomic notes with indication of some *exsiccatae*. A relatively brief introduction appears in part 1, and appropriate literature references in each part.

778

Iraq

For the northern highland region, see also **790** (*Flora iranica*). – Area: 438 446 km². Vascular flora: 2937 species (Al-Khayat in PD; see **General references**).

The area of present-day Iraq, a state formed only in the 1920s, was until well into the twentieth century a land without separate floras or checklists. Establishment of a botanical unit – with a herbarium – in the then-directorate of agriculture was effected in 1933 with the appointment of Evan Guest, who remained there until the 1950s. An early visitor was Michael Zohary, author of the first national checklist, *The flora of Iraq and its phytogeographical subdivision* (1950). This latter was largely succeeded in 1954 by the still-current checklist by Ali al-Rawi. Karl-Heinz Rechinger's *Lowland flora of Iraq* was prepared as a result of a sojourn by its author at present-day Baghdad University in 1956–57, with assistance from Guest and al-Rawi at the Ministry. Soon afterwards, the *Flora of Iraq* project was initiated as a joint undertaking of the Iraqi Government and the Royal Botanic Gardens, Kew. Until its indefinite suspension in the 1980s, this definitive work was published in the manner of *Flora Zambesiaca*, in volumes or half-volumes as ready, with families following a predetermined systematic sequence.

Bibliography

In addition to the reference below, there is also a selected bibliography in *Flora of Iraq*, 1: 184–206 (1966).

GUEST, E. and R. A. BLAKELOCK, 1954. Bibliography of 'Iraq. *Kew Bull.* **9**: 243–249. [A list of titles arranged by author, without annotations.][61]

AL-RAWI, A., 1964. *Wild plants of Iraq, with their distribution*. 232, 18 pp., 2 maps (Dir. Gen. Agric. Res. (Iraq), Tech. Bull. 14). Baghdad: Government Press of Iraq.

Briefly annotated systematic list of vascular plants of Iraq; includes concise indication of local range (arranged by districts) and indices to generic and family names. The introductory section incorporates accounts of physical features, climate, vegetation, floristic zones, and progress in botanical exploration; a list of references appears at the end of the work.

GUEST, E., C. C. TOWNSEND and A. AL-RAWI (eds.), 1966– . *Flora of Iraq.* Vols. 1– . Illus., maps. Baghdad: Ministry of Agriculture (later Ministry of Agriculture and Agrarian Reform), Republic of Iraq.

Comprehensive, detailed but concise descriptive flora of vascular plants; includes keys to all taxa, extensive synonymy with references and relevant citations, vernacular names, local range with citation of *exsiccatae* and summary of extralimital distribution, taxonomic commentary, and notes on habitat, karyology, phenology, ecology, uses, special features, etc.; indices in each volume. The introductory sections are spread over both vols. 1 and 2. In the former are those on physical features, climate, vegetation, floristic zones, local names and languages, and the history of botanical exploration along with a gazetteer, a glossary, and a bibliography; in the latter appear notes on systems of classification, the projected sequence of families (the flowering plants following the 1959 version of the Hutchinson system), and general keys. A summary in Arabic appears in vol. 1. [Not yet completed. As of 1999, of the nine volumes planned vols. 1 (Introduction, 1966), 2 (lower vascular plants, gymnosperms, and Magnoliaceae–Rosaceae, 1966), 3 (Leguminosae, 1974), 4(1) (Cornaceae–Rubiaceae, 1980), 4(2) (Bignoniaceae–Resedaceae, 1980), 8 (monocotyledons except Gramineae, 1985) and 9 (Gramineae, 1968) had been published. A considerable proportion of 'herbaceous' dicotyledons thus remains to be documented.][62]

RECHINGER, K. H., 1964. *Flora of lowland Iraq.* 746 pp. Weinheim: Cramer.

Briefly descriptive manual-flora of vascular plants; includes keys to all taxa, limited synonymy, fairly detailed indication of local range, occasional taxonomic commentary, and notes on habitat, etc.; index to all botanical names. The introductory part incorporates notes on floristic zones, a list of frequently cited collectors, glossary, addenda, and a list of references.[63]

Partial work: Baghdad district

For the vicinity of the Iraqi capital, reference may be made to the following concise illustrated students' manual (never completed): A. D. Q. AGNEW, 1962. *Flora of the Baghdad district*, 1. *Monocotyledons.* ii, 170 pp. (Bulletin of the College of Science, Baghdad 6, Supplement). Baghdad.

Region 78

Arabian Peninsula

Area: *c.* 3 313 000 km². Vascular flora: *c.* 3400 species (A. G. Miller and T. A. Cope in *Flora of the Arabian Peninsula and Socotra*; see below). Works dealing with Saudi Arabia alone are accounted for under **781**. – As here delimited, the Arabian Peninsula includes all the land (and associated islands) south of the southern boundaries of Jordan and Iraq. The bulk of the area is occupied by Saudi Arabia, with the remainder comprising the mainland part of the Yemen Republic (including Aden), Oman (including Muscat), the United Arab Emirates (formerly the Trucial States), Qatar, Bahrain and Kuwait. For Socotra (politically part of Yemen), see **545**.

General coverage north of the Tropic of Cancer has hitherto been provided by *Flora orientalis* and the partly compiled *Flora arabica* of Ethelbert Blatter (1919–23, 1936) while the critical *Flora des tropischen Arabien* of Oskar Schwartz (1939) was available southwards. These works in turn built upon a foundation developed since publication of *Flora Aegyptiaco-Arabica* (1775) by Pehr Forsskål and passing through *Plantes de la Arabie heureuse* (1834) by Joseph Decaisne. Their coverage of the Peninsula was, however, patchy; many areas had botanically not or been but scarcely studied. With the great increase in collections in recent decades, beginning in particular with those of Violet Dickson in the northeast and more recently in previously little-collected areas such as Qatar and Oman as well as 'Arabia Felix' north of Yemen, these works can now be seen to be seriously out of date. Following some precursory work (with a bibliography in 1982), a new general peninsular flora was initiated at Edinburgh in the mid-1980s under the direction of A. G. Miller (with its first volume being published in 1996); for Saudi Arabia, a students' manual, *Flora of Saudi Arabia*, has also been published (Migahid and Hammouda, 1974; 3rd edn. by Migahid alone, 1988–90). Since 1982, many precursory papers and regional family studies, of which the bibliography by Miller *et al.* was the first, have appeared in a series *Studies in the flora of Arabia*, published in various journals.

Of works covering more limited areas, the older

Flora of Aden (1914–16) by Blatter and *The wild flowers of Kuwait and Bahrain* by Dickson (1955) have been joined by several modern floras, floristic reports and checklists, most recently *Flora of eastern Saudi Arabia* (1990) by James Mandaville, *Annotated catalogue of the vascular plants of Oman* (1992) by Shahina Ghazanfar, and *Handbook of the Yemen flora* (1997) by J. R. I. Wood. Very useful also is *An illustrated guide to the flowers of Saudi Arabia* (1985; 2nd edn., 1999) by S. A. Collenette, with its numerous photographs.

Progress

The earlier history of work in the Peninsula is covered by E. BLATTER, 1933. *The botanical exploration of Arabia.* 51 pp. Calcutta. [Reprinted from *Rec. Bot. Surv. India* 8(5): 451 [bis]–501 [bis], Delhi, as an addition to his *Flora arabica* (see below).] Useful summaries are also to be had in *Flora of eastern Saudi Arabia* by James Mandaville (1990; see **781**) and in the first volume of *Flora of the Arabian Peninsula and Socotra* (1996; see below).

Bibliographies. General and supraregional bibliographies as for Division 7 and Superregion 77–79.

Regional bibliographies

MILLER, A. G., I. C. HEDGE and R. A. KING (comp.), 1982. Studies in the flora of Arabia, I: A botanical bibliography of the Arabian Peninsula. *Notes Roy. Bot. Gard. Edinburgh* **40**: 43–61. [Briefly annotated list by authors arranged under three primary headings (I: general; II: cryptogams; III: references on adjacent countries) and covering floristic, taxonomic, phytogeographic, ecological and travel works.]

RAHIM, M. A., 1979. *Biology of the Arabian Peninsula: a bibliographic study from 1557–1978.* xxiv, 180 pp. (Saudi Biol. Soc. Publ. 3). N.p. [Not seen.]

Indices. General and supraregional indices as for Superregion 77–79.

780

Region in general

For the extra-tropical zone, see also **770–90** (*Flora orientalis*).

BLATTER, E., 1919–23, 1936. *Flora arabica.* Fasc. 1–5. ii, 519, xlix pp. (Rec. Bot. Surv. India 8(1–4, 6)). Calcutta (later Delhi).

Systematic enumeration, without keys, of the seed plants of the Arabian Peninsula (including parts of the Sinai Peninsula, Negev, etc., north to 31°N), with synonymy, references, and fairly extensive literature citations; vernacular names; generalized indication of internal range (based on four large quadrants), with localities and indication of *exsiccatae*; summary of extralimital distribution; general index to all botanical names. Also part of this volume (as fasc. 5) is Blatter's historical account of botanical exploration in Arabia (see above under **Progress**).

MILLER, A. G. and T. COPE, 1996– . *Flora of the Arabian Peninsula and Socotra.* Vols. 1– . Illus., maps. Edinburgh: Edinburgh University Press (in association with Royal Botanic Garden Edinburgh, and Royal Botanic Gardens, Kew).

Concisely descriptive flora with keys, synonymy, literature citations, indication of distribution, altitudinal range and habitat, and taxonomic and other notes with *exsiccatae* where necessary; individual dot distribution maps for all taxa and indices at end. An extended general part (vol. 1) features a selected bibliography (pp. xiii–xvi) and sections on physical features, geology, climate, vegetation (including representative transects), floristics (with statistics) and phytogeography (chorology), conservation, and the history of botanical exploration, with the plan of the work on pp. 30–31. [Family arrangement follows the 12th edition of the Engler *Syllabus*. Vol. 1 covers pteridophytes (56 species), gymnosperms, and angiosperms from Casuarinaceae through Rosaceae and Neuradaceae.]

Partial work: infratropical Arabia

The Arabian Peninsula south of the Tropic of Cancer was excluded from Boissier's *Flora orientalis*. It includes southern Saudi Arabia, Yemen, most of Oman and the southern portion of the United Arab Emirates.

SCHWARTZ, O., 1939. *Flora des tropischen Arabien.* 393 pp. (Mitt. Inst. Allg. Bot. Hamburg 10). Hamburg.

Systematic 'research' enumeration of vascular plants of the Arabian Peninsula south of the Tropic of Cancer, without keys; includes descriptions of new taxa, extensive synonymy with references and citations, detailed indication of local range with citation of *exsiccatae* and other records, and general summary of extralimital distribution; short list of references; index to all botanical names. The introductory section includes remarks on phytogeographical zones and on botanical exploration in the region. [Most complete for Yemen and western Saudi Arabia from Mecca southwards.]

781

Saudi Arabia

Area: 2 401 554 km². Vascular flora: 1759 species (GB; see **General references**).

In addition to the formal floras, the following – covering 2250 species – is very useful for its numerous individual photographs: S. A. COLLENETTE, 1999. *Wildflowers of Saudi Arabia*. 832 pp., 2400 col. pls. Riyadh: National Commission for Wildlife Conservation and Development, Saudi Arabia. (1st edn., 1985, London.)

Bibliography
   BATANOUNY, K. H., 1978 (A.H. 1356). *Natural history of Saudi Arabia: a bibliography*. xii, 113, 8 pp. (Publ. Biol. King Abdulaziz Univ. 1). Jeddah. [Botany, pp. 21–31. Lists 77 references, in part on non-vascular plants.]

   MIGAHID, A. M., 1988–90 (A.H. 1369–70). *Flora of Saudi Arabia*. 3rd edn. 3 vols. 593 col. pls. Riyadh: University Libraries, King Saud University. (1st edn., 1974, with M. A. Hammouda; 2nd edn., 1978.)

   Atlas-flora of vascular plants, without keys; includes more or less brief descriptions, indication of phytogeographical unit(s) (keyed to map in vol. 1) and habitat, and mainly colored photographs; lexica for Arab vernacular names and indices in each volume and (vol. 3) glossary (in Arabic). The introductory sections in vol. 1 encompass the scope of the work, additions (especially for this edition), sequence of families (continued in vols. 2 and 3), phytogeographical regions (with map), systematic synopsis, and diagnoses for all orders and families (covering the whole work). [With the second edition, the format of the work was expanded from the manual-style of the first and the extensive range of color illustrations introduced; the larger format is retained here. There are now 115 families covered vs. 94 in the original version.]

Partial work: eastern Saudi Arabia
   MANDAVILLE, J. P., 1990. *The flora of eastern Saudi Arabia*. x, 482 pp., maps, graphs, 267 col. photographs

(Studies in the flora of Saudi Arabia 1). London: Kegan Paul International (in association with the National Commission for Wildlife Conservation and Development, Saudi Arabia).

Briefly descriptive flora, with keys, essential synonymy, vernacular names, localities with citations of *exsiccatae* and other references (keyed to gazetteer, pp. 423–434), and, for some species, commentary related to taxonomic and other questions; glossary, gazetteer, bibliography (with separate foreword), Arabic index and glossary (oriented right to left), Romanized index to vernacular names, and general index at end. The introductory part covers botanical history, topography, climate, vegetation and plant communities, and notes on the work and its scope and presentation along with statistics (p. 38) and a key to families (pp. 38–44). [Covers 565 spontaneously occurring vascular species in an area of 605 000 km² (see map 2.1), a pestle-shaped area covering most of eastern and southern Saudi Arabia (including two-thirds of the Rub' al-Khali).]

782

Yemen

Area (including Socotra and 'Abd-al-Kuri): 477 530 km². Vascular flora: former South Yemen, 1417 species (GB), North Yemen, 974 (GB; see **General references**). – The Yemen Arab Republic (North Yemen) and the People's Democratic Republic of Yemen (South Yemen, including Aden, Hadhramaut and Socotra) merged in 1990. South Yemen was until 1967 a British territory, firstly under India and then, after World War II, the Colonial Office. For Socotra and 'Abd-al-Kuri, see **545** (they are, however, also covered in *Flora of the Arabian Peninsula and Socotra*; see **780**).

Yemen is noteworthy for succulent plants, many of quite local occurrence. Before 1900, different parts of the country were variously covered in Forsskål's *Flora Aegypto-Arabica*, by Decaisne in *Plantes de l'Arabie heureuse* (1841), by Thomas Anderson in *Florula adenensis* (1860), and by the French explorer Albert Deflers *Voyage au Yemen* (1889; plants, pp. 107–222). The next consolidations were by Blatter (for Aden; see below) and Schwartz (see **780**), but almost 60 years ensued before the appearance of their (partial) successor: *A handbook of the Yemen flora* by J. R. I. Wood (1997; see below). Yet overall botanical knowledge remains imperfect, with the south in particular once called one of the least-known parts of the world.

An introduction to the flora, with an account of botanical exploration, appears in F. N. HEPPER, 1977. Outline of the vegetation of the Yemen Arab Republic. *Publ. Cairo Univ. Herb.* 7/8: 307–322.

Bibliography

GABALI, S. A., 1993. *Yemen: a botanical bibliography.* 41 pp. Aden: University of Aden. [Annotated list of 275 titles, arranged by author; no indices.]

GABALI, S. A., 1995. *Plant life in Yemen; a general survey and preliminary checklist of the flowering plant species.* 85 pp. (Roman), [2 pp. maps], 21 pp. (Arabic). Aden: Aden University Press. (Publication of University of Aden, Reference Book Series 4.)

An alphabetically arranged nomenclator with indication of presence in one or more of 12 numbered regions (tied to map 1 in the middle of the work). The general part, in Arabic, occupies the last (or first) 21 pages; it includes a history of botanical exploration. [The work covers the Yemeni mainland.]

Partial works

Blatter's flora draws on the critical studies by Anderson embodied in his *Florula adenensis* (1860). Wood's flora is largely limited to former North Yemen.

BLATTER, E., 1914–16. *Flora of Aden.* 3 parts. 418, xix pp., pls., maps (Rec. Bot. Surv. India 7). Calcutta.

Briefly descriptive flora; includes keys, synonymy, indication of local and extralimital range (with citation of *exsiccatae*), vernacular names, and index. An introductory section incorporates remarks on physical features, vegetation, and botanical exploration.

WOOD, J. R. I., 1997. *A handbook of the Yemen flora.* 434 pp., 35 text-figs., 40 col. pls., 32 maps. Kew: Royal Botanic Gardens.

A concise manual of vascular plants with keys to all taxa, essential synonymy, brief descriptions, diagnostic features, and notes on distribution, altitudinal range, habitat, critical features and possible dynamics; Arabic plant name lexicon, list of poisonous plants, gazetteer, annotated 'select bibliography' (pp. 427–428) and index to genera and families at end. The general part includes a botanical history. [All previous records and literature have been as far as possible accounted for.][64]

784

Oman

Area: 271950 km^2. Vascular flora: 1174 species (Ghazanfar, 1992; see below). – Until the 1970s, few, if any, collections had been made beyond the environs of the capital, Muscat. Since then, two collections of papers reporting on exploration of the biota and a fine color-book, *Plants of Dhofar*, have been published through the Sultanate. A full checklist, by Shahina Ghazanfar, appeared in 1992. Additional contributions to the flora have appeared in the main series of *Journal of Oman Studies* and elsewhere. Ghazanfar has indicated that while the flora has become relatively well known, more exploration is still needed, especially following good rains.

GHAZANFAR, S., 1992. *An annotated catalogue of the vascular plants of Oman and their vernacular names.* 153 pp., 2 maps (Scripta Botanica Belgica 2). Meise: National Botanic Gardens of Belgium.

An alphabetically arranged, annotated enumeration (accounting for 1174 native and naturalized vascular plants) with partial keys (though not in Cyperaceae or Poaceae), diagnostic features for each species, limited synonymy, vernacular names, selected *exsiccatae* with localities, distribution (by administrative units), and notes on frequency, habitat and occurrence; references (p. 143) and indices to genera and to vernacular names at end. The general part encompasses collecting history, geography, climate, general features of the flora and main vegetation formations, and the plan of the work.

Southern region

In addition to the following, reference can usefully be made to A. G. MILLER and M. MORRIS, 1988. *Plants of Dhofar.* xxvii, 361 pp., col. illus., 1 chart. [Muscat]: Office of the Adviser for Conservation of the Environment, Diwan of Royal Court, Oman. This is a relatively detailed color-plate atlas of 190 selected species, with substantial descriptive text.

SHAW-READE, S. N. *et al.* (eds.), 1980. *The scientific results of the Oman Flora and Fauna Survey 1977 (Dhofar).* 400 pp., illus. (text-fig., halftone and col. photographs), folding maps (Journal of Oman Studies, Special Report 2). [Muscat]: Office of Government Adviser for Conservation of Environment, Oman.

Includes (pp. 59–86) a paper by Alan Radcliffe-Smith on the vegetation of Dhofar which incorporates a systematic checklist of all plants collected in 1976 and 1977.

Northern region

MANDAVILLE, J. P., JR., [1977]. Plants. In D. L. Harrison *et al.* (eds.), *Scientific results of the Oman flora and fauna survey 1975*, pp. 229–267, pls. 1–24 (some col.) (Journal of Oman Studies, Special Report [1]). [Muscat]: Ministry of Education and Culture, Oman.

Includes (pp. 253–259) a checklist of plants collected in 1972 and 1975 in northern Oman, with determinations by D. Hillcoat; references, p. 267. [The work, a narrative account of the vegetation and plant life, emphasizes montane areas scarcely explored in the past, including the 3000 m high, juniper-clad Jabal al-Akhdar and the slightly less lofty Jabal Aswad.]

785

United Arab Emirates

Area: 75150 km². Vascular flora: 380+ species (Western, 1989; see below). – This federation encompasses the seven polities of Abu Dhabi, Ajman, Dubai, Fujairah, Ras al Khaimah, Sharjah (including Kalba) and Umm al Qaiwain, formerly known as the 'Trucial States'. Of these, much the largest is Abu Dhabi.

Few accounts of the flora are currently available. In addition to the checklist included in the work of Western (1989; see below), a general introduction may be had in D. F. VESEY-FITZGERALD, 1957. The vegetation of central and eastern Arabia. *J. Ecol.* **45**: 779–798. The Al-Ain area in eastern Abu Dhabi, on the Oman border, is covered in *Ecology and flora of Al Ain region* (1985) by A. A. El-Ghonemy.

WESTERN, R. A., 1989. *The flora of the United Arab Emirates: an introduction.* [vii], 188 pp., 1 p. portr., col. photographs, maps. N.p. (Abu Dhabi): United Arab Emirates University.

An album of photographs, color-coded range maps and descriptive text covering some 250 species, the latter including details of distribution and habitat, uses, and some observations. An appendix (pp. 171–181) contains a checklist of all known species with indication of frequency, distribution, and a representative voucher. The general part covers physical features, climate, water resources, soils, human influence, vegetation and its adaptations, associations, the plan of the work, a short glossary, and maps with localities and vegetation associations. A list of references and index to botanical names conclude the work. [The color coding of the maps relates to frequency.]

786

Qatar

Area: 11437 km². Vascular flora: 301 species (Batanouny, 1981; see below). – The first collections in this state were made in 1971, but those of Obeid and Boulos in 1975–77 were the first effectively published.[65]

ARAB ORGANISATION FOR AGRICULTURAL DEVELOPMENT, 1983. *Wild plants of Qatar.* xiv, 161 pp., illus. (part. col.) on 28 pls., 1 map. [Khartoum]: Arab Organisation for Agricultural Development (for Ministry of Industry and Agriculture, Qatar).

Incorporates a briefly descriptive flora (by H. M. El-Amin, covering 297 species), with essential synonymy, Arabic names in transliteration, local distribution, and indication of habitat, phenology, properties and uses; list of domestic and ornamental cultivated trees and shrubs, references (p. 151), and indices to botanical and Arabic names. [The work was based primarily upon three months' field work in 1982.][66]

BATANOUNY, K. H., 1981. *Ecology and flora of Qatar.* xiv, 246 pp., 11 text-figs. (including table and maps), 124 col. pls., col. folding map. Oxford: Alden Press (on behalf of the Centre for Scientific and Applied Research, University of Qatar).

Briefly descriptive flora of seed plants (301 species) in part II of the work; includes keys to genera and species, synonymy (with references and citations), generalized indication of local distribution, and habitat and phenology. Families and genera are also described, and most species are illustrated. The flora is followed by references (pp. 203–204), abbreviations, an addendum, Arabic glossary, lexicon (Latin–vernacular names), and index to all botanical names. Part I encompasses physical features, climate, topography, soils, geology, vegetation and plant communities, and human activities and their impact, with separate references (p. 51).

BOULOS, L., 1978. Materials for a flora of Qatar. *Webbia* **32**(2): 369–396, 10 figs., map.

Enumeration, with families alphabetically arranged; includes citation of *exsiccatae* and, where appropriate, critical commentary; plants of special interest illustrated. The 260 species treated are based on the collections of Obeid and Boulos. A map appears in the introduction.

787

Bahrain

Area: 688 km². Vascular flora: *c.* 175 species (Good in Dickson, 1955; see below). Much detailed work has been accomplished by the Bahrain Natural History Society in the 1970s and 1980s. A useful field-guide, with photographs and descriptions of 133 species, is D. C. PHILLIPS, 1988. *Wild Flowers of Bahrain.* ii, 207 pp., illus. Manama, Bahrain: The author.

DICKSON, V., 1955. *The wild flowers of Kuwait and Bahrain.* 144 pp, 7 pls. (some col.), text-figs., 6 maps. London: Allen & Unwin.

Semi-popular work; includes lists of the Kuwait flora (by Violet Dickson) and that of Bahrain (by Ronald d'O. Good), each incorporating vernacular names, local range, and short descriptive notes (the latter only in the Kuwait list). Remarks on land vegetation, marine algae, and plant collectors are also included.[67]

VIRGO, K. J., 1980. An introduction to the vegetation of Bahrain. In T. J. Hallam (ed.), *Wildlife in Bahrain; the annual reports of the Bahrain Natural History Society for 1978–79,* pp. 65–109. N.p.

This treatment comprises three sections: (1) ecology, habitats, and main associations; (2) species descriptions; and (3) a checklist. Profile diagrams and site plant lists appear in the first section; in the second section the species descriptions emphasize field characteristics, frequence, phenology, sites from which collected, and uses (if known). The checklist in the third section is, as the species descriptions, alphabetical, and is followed (p. 109) by references. For additions, see D. A. BELLAMY, 1984. Additional flowering plants of Bahrain. In M. Hill and T. Nightingale (eds.), *Wildlife in Bahrain: third biennial report of the Bahrain Natural History Society,* pp. 90–96. Bahrain: B.N.H.S./Caltex.

789

Kuwait

Area: 17819 km² (Kuwait proper), 5700 km² (Neutral Zone). Vascular flora: 236 species (GB; see **General references**).

The two-volume *Flora of Kuwait* by H. S. Daoud and Ali al-Rawi oddly contains no reference to Dickson's *Wild flowers of Kuwait and Bahrain* (1955; see 787 for description). The Kuwait list in that book drew upon *On the flora of Kuweit* by B. L. Burtt and P. Lewis (1949–54; in *Kew Bull.* 4: 273–308, 7: 333–352 (1952), and 9: 377–410), although this never was continued past Rutaceae. Additional floristic data, particularly on distribution, appears in B. S. MIDDLEDITCH and AMER A. AMER, 1991. *Kuwaiti plants: distribution, traditional medicine, phytochemistry, pharmacology and economic value.* xiii, 322 pp., 173 text-figs. (maps). Amsterdam: Elsevier. (Studies in plant science 2.)

DAOUD, H. S. and A. AL-RAWI, 1985–87. *Flora of Kuwait.* Vol. 1: pp. i–xi, 1–224, 14 text-figs., 247 col. photographs London: Kegan Paul International; vol. 2: pp. 225–455, 66 text-figs., col. photographs 249–338. Oxford: Alden Press (for Faculty of Science, University of Kuwait). Complemented by L. BOULOS, 1988. *The weed flora of Kuwait.* xi, 175 pp., illus. Kuwait: Kuwait University.

The main work is an illustrated descriptive flora, with keys, synonymy (including references and citations), phenology, local and extralimital range, indication of major phytochoria, and notes on habitat; indices to families and to all botanical names at end in both volumes. Families and genera are also described and incorporate citations of standard monographs and revisions. The introductory part encompasses chapters on geography, topography, climate, and vegetation and ecosystems as well as a general key to families, but no word about background, methodology or design. The second volume additionally contains an organographic iconography, a glossary, a list of authorities, and a vernacular–Latin lexicon, along with a folding map. The weed flora by Boulos adds a number of species not accounted for in the main work.[68]

Region 79

Iran, Afghanistan and Pakistan ('Iranian Highland')

Area: *c.* 2750000 km². Vascular flora: over 10000 species (Rechinger, 1991; see note after *Flora iranica*). – This region comprises the largely 'highland' area from northeastern Iraq through Iran and Afghanistan to the upland western and northern parts of Pakistan, finding its eastern limit at the base of the Indus escarpment and in the Hindu Kush and Pamirs; it thus includes the Pakistani tribal areas of the North-West Frontier, Swat, Dir and Chitral. For convenience, however, Hazara and Azad Kashmir (east of the Indus) and northern Kashmir (Gilgit, Diamir, Baltistan and Ladakh) are also included here although their floras respectively are essentially of Himalayan (Region 84) and Central Asiatic (Region 76) affinities. The region thus includes all of present-day Pakistan except for the lowland areas along the Indus (**811, 812**). The Iranian Highland additionally extends into Azerbaijan (**749**) and the southern fringe of Middle Asia including the Kopetdagh mountains (**Region 75**).

The whole region, with the exception of Hazara, Azad Kashmir, Gilgit, Baltistan and Ladakh (**796** to **799**) and the addition of the above-mentioned parts of Middle Asia, is covered by the modern comprehensive *Flora iranica*, the first general flora for a century and one accounting for the extensive collecting and other floristic and systematic work accomplished since World War II – much of it by its founder and long-time general editor, Karl-Heinz Rechinger. For the three individual countries within the region, however, few practically useful floras are yet available. In Iran, there was for long only Ahmad Parsa's voluminous but largely compiled *Flore de l'Iran* (1943–52), subsequently added to and of which a new edition was begun (as *Flora of Iran*) in 1978. Later Iranian works include Ghahreman's *Flore de l'Iran en couleurs naturelles*, a photographic collection, and the Farsi-language *Flora of Iran*, begun in 1989. Parsa's flora nominally also covered Afghanistan, a country otherwise provided with only two general works, neither of them, properly speaking, a complete flora: *Flora of Afghanistan* (Kitamura, 1960; supplement, 1966) and *Symbolae afghanicae* (Køie and Rechinger, 1954–65). For Pakistan, however, there is the serial *Flora of Pakistan* (1970–), now far advanced, and *An annotated catalogue of the vascular plants of West Pakistan and Kashmir* (Stewart, 1972). In Pakistan there are also several floras and checklists – both old and modern – for individual highland provinces and territories: Baluchistan (Burkill, 1909), the North-West Frontier and associated tribal districts (several works), and 'Pothohar', a geographical area centering on Rawalpindi (Bhopal and Chaudhri, 1977–78). A number of miscellaneous works

exist for northern Kashmir, the most important being *La flora del Caracorùm* (Pampanini, 1930, 1934; see **703**).

The only significant dendrological works are *Forests, trees and shrubs of Iran* (1976) by H. Sabeti and *Trees of Pakistan* (1995) by S. R. Baquar. In the eastern part of the region, Brandis's *Indian trees* can also be used.

Although in all countries collecting was extensive in the three decades or so from the end of World War II, coverage is still uneven, with Baluchistan sensu lato perhaps the least known. Northern Pakistan and northern and western Iran, where a number of scientific centers and tertiary institutions have come into existence within the last 60–70 years, have been the most thoroughly studied. The documentary emphasis, however, has been on the regional and national floras, with at a more local level relatively few organized contributions.

Progress
No fully retrospective regional account has been published; however, a useful survey of progress since 1939 is in K.-H. RECHINGER, 1989. Fifty years of botanical research in the *Flora iranica* area. In K. Tan (ed.), *Plant taxonomy, phytogeography and related subjects: the Davis and Hedge Festschrift*, pp. 301–349. Edinburgh. Separate retrospective historical accounts are available respectively for Iran and Pakistan: B. A. FEDTSCHENKO, 1945. [The investigators of the flora of Iran.] *Bot. Žurn. SSSR* **30**: 31–43; and R. R. STEWART, 1967. Plant collectors in West Pakistan and Kashmir. *Pakistan J. Forest.* **17**: 337–363. Some subsequent developments in Pakistan are described in MD. N. CHAUDHRI, 1977. The Pakistan herbarium. *Pakistan Syst.*, **1**(2): 100–105. Also of interest is R. R. STEWART, 1982. Missionaries and clergymen as botanists in India and Pakistan. *Taxon* **31**: 57–64.

Bibliographies. General and supraregional bibliographies as for Division 7 and Superregion 77–79.

Indices. General indices as for Division 7 and Superregion 77–79.

790

Region in general

Dominating current activities is the critical *Flora iranica*, produced in fascicles under the direction of Karl-Heinz Rechinger at Vienna. The first number appeared in 1963 and, after three and a half decades of steady progress, is at this writing (1999) about 90 percent complete. Hitherto, the region has been covered only by *Flora orientalis* (**770–90**), now long out of date, as well as by a motley variety of 'national' and local floras, either of local or of outside origin.

RECHINGER, K.-H. (ed.), 1963– . *Flora iranica*. Fasc. 1– . Illus., map (on covers). Graz: Akademische Druck- und Verlagsanstalt.

Comprehensive but concise descriptive flora of vascular plants; each fascicle – covering a family or part thereof – includes keys to genera and species, important synonymy with references and principal citations, detailed indication of internal range, with citations of *exsiccatae* and other records, summary of extralimital distribution, critical commentary, notes on habitat, biology, special features, etc., and index to all botanical names. No systematic sequence is followed in publication, but tables of five possible systems are given with every fascicle, with families published specially designated. Except for the Karakoram, the work encompasses the entire region (for definition see under the regional heading). Some 173 fascicles have now been published, covering 8645 species (through fasc. 170, thus not accounting for Ranunculaceae, Chenopodiaceae, Cyperaceae, and the some 1200 species of *Astragalus* in the Leguminosae); only three 'units' now remain. Volumes on phytogeography and topography (the latter including a gazetteer and itineraries) are planned.[69]

791

Iran

Area: 1648000 km^2. Vascular flora: *c.* 6550 species (cf. GB; see **General references**).

No separate floras of Iran were published until after the advent of the Pahlavi dynasty. Ahmad Parsa was the first professor of botany in Tehran in the modern era, from 1933; in addition to his teaching he also helped to establish a natural history museum with a herbarium as well as to compile (in French) his *Flore de l'Iran* (1943–59). Also active in Iran during the years before World War II was Erwin Gauba, a collaborator of Bornmüller and in service with the Higher School of Agriculture and Rural Craft (1935–42).

In the last year of the Pahlavi regime, a new edition of Parsa's flora, now in English, was commenced along with a series of photographic albums by Ahmed Ghahreman. Publication of both was continued under the Islamic Republic, although no volumes of Parsa's flora have appeared since 1986. In 1989, a 'serious' flora in Farsi by Mostafa Assadi and others began publication, and in 1990–95 Ghahreman brought out a four-volume elementary flora.[70] Some families of an additional flora of Iran by S. Mobayen appeared in 1996 in *Tehran University Publications* **228** (as *Flora of Iran*, 4).

Bibliographies

BURGESS, R. L., A. MOKHTARZADEH and L. CORNWALLIS, n.d. [1966]. *A preliminary bibliography of the natural history of Iran*. 220, 140 pp. (Pahlavi University, College of Arts and Sciences, Sci. Bull. 1). Shiraz. (Reproduced by offset from typescript.) [1719 serially numbered references, grouped in the first instance by main subject areas with indices by author, geographical areas and dates of publication; 10 percent of references annotated. Includes 339 references in systematics, floristics and vegetation. The first section is in English; the second, in Farsi.]

FREY, W. and H.-J. MEYER, 1971. Botanische Literatur über den Iran. *Bot. Jahrb. Syst.* **91**: 348–382. [Unannotated list arranged firstly by five classes (general works, cryptogams, phanerogams, geobotany and ecology, and 'verschiednes') and then by author.]

GHAHREMAN, A., 1978[A.H. 1356]–1997[A.H. 1376]. *Flore de l'Iran en couleurs naturelles*. Vols. 1–16. Pls. 1–2000. Tehran: Société Nationale pour le Conservation des Resources Naturelles et de l'Environment Humain (fasc. 1–2); Research Organization of Agriculture and Natural Resources, Ministry of Agriculture, Iran (fasc. 3–11); Research Institute of Forests and Rangelands, Ministry of Jahad-e-Sazandegi [Reconstruction] (fasc. 12–16): all in collaboration with the University of Tehran.[71]

A portfolio-atlas of colored photographs of native and introduced species, with text in Farsi and

French on the reverse side; the latter gives botanical details, information on distribution, habit and phenology, and the provenance of the photograph. [Numbering is continuous but the selection of plates follows no systematic sequence. Each fascicle contains 125 numbered plates, the genera therein arranged alphabetically.]

GHAHREMAN, A., 1990–95. *Plant systematics: cormophytes of Iran.* 4 vols. Illus. Tehran: Iran University Press.

An illustrated Farsi-language textbook, with keys to families and genera.

ORGANIZATION OF THE FLORA OF IRAN, 1367 A.H. [1989]– . *Flora of Iran.* In numbered parts (one or more to a fascicle). Illus. (text-figs.), map (front cover). [Tehran]: Research Organization of Agriculture and Natural Resources, Ministry of Agriculture and Rural Development, Iran (nos. 1–5); Research Institute of Forests and Rangelands, Ministry of Jahad-e-Sazandegi [Reconstruction] (nos. 6–).

Large-scale illustrated descriptive flora (in Farsi); includes keys, synonymy with references (no literature citations, except in family headings), indication of distribution, some commentary, and indices in each fascicle. As of 1999, 27 numbers had appeared; no set sequence has been adopted (a list of numbers published appears in the inside front cover of each successive fascicle).

PARSA, A., 1943–52. *Flore de l'Iran.* 5 vols. and supplement (in 6 vols.). Illus. Tehran: Muséum d'Histoire Naturelle. Continued as *idem*, 1959. *Flore de l'Iran*, 7(1–2) (Publications of the University of Tehran 504). Tehran; *idem*, 1960. *Flore de l'Iran*, 8 (*ibid.*, 613); *idem*, 1966. *Flore de l'Iran*, 9 (*ibid.*, 613/9); and *idem*, 1981. *Flore de l'Iran*, 10 (*ibid.*, 1777). Tehran. Revised edn.: *idem*, 1978–86. *Flora of Iran.* Vols. 1–2. Illus. (some col.). Tehran: Ministry of Science and Higher Education (vol. 1); Ministry of Culture and Higher Education, Iran (vol. 2). (Vols. 1–2 respectively issued as Publ. Natl. Sci. Res. Council, Iran 21 and 22.)

Largely compiled, briefly descriptive flora (in French) of vascular plants of Iran and Afghanistan; includes synoptic keys to genera and groups of species, synonymy with references and principal citations, detailed local range with citation of *exsiccatae*, general summary of extralimital distribution, Farsi vernacular names, and notes on phenology, life-form, etc.; indices to vernacular and botanical names in each volume. The introductory section (in vol. 1) contains a general key to families as well as chapters on physical features and botanical exploration, a gazetteer, a glossary, and a list of references. Further supplements to this work include (as vol. 7, 1979, in two parts) a complete general index and the rarely seen volumes 8 and 9 (1960, 1966) with additions and changes to volumes 1 and 2. Volume 10 contains additions and corrections to vol. 3 of the original work, in English as well as in French and, as the series editor notes, the fifth supplement in all. The revised edition, in English, was planned for 12 volumes; by 1999 only two had appeared (although work on others was said in 1985 to have been in progress).[72]

Woody plants

SABETI, H., 1976 (A.F. 2535). *Forests, trees and shrubs of Iran.* 810 pp. (Farsi text), 64 pp. (English text), illus. (some col.), portr., maps. Tehran: National Agriculture and Natural Resources Research Organization, Iran. (A revision of *idem*, 1966. [*Native and exotic trees and shrubs of Iran.*] xii, 430, [ii] pp., illus. Tehran.)

This generously dimensioned work, in Farsi with a summary section in English, is a descriptive atlas, without keys, of 986 species of woody plants in a dictionary arrangement (by scientific names, including synonyms). For accepted species, the text includes synonymy, vernacular names (in Persian, English, French and German), botanical description, and commentary, and, for native species, photographs (partly in color) of habit, technical details, etc. and distribution maps with summary (the latter also in English). Persian names are separately indexed. An introductory section deals with floristic regions and phytogeography. The English section, separately paged at the 'back', contains the same introductory matter with, in addition, an index of families, indices to English, French and German vernacular names, and a selected bibliography (pp. 57–62).[73]

Pteridophytes

No pteridophytes have yet been published in *Flora iranica*.

WENDELBO, P., 1976. An annotated checklist of the ferns of Iran. *Iran J. Bot.* 1: 11–17, illus., maps.

Annotated enumeration of ferns, with necessary critical notes, illustrations and maps; references (p. 17).[74]

792

Afghanistan

See also **791** (PARSA). – Area: 636 267 km². Vascular flora: *c.* 3500 species (GB; see **General references**).

The two works described below are the most comprehensive available, although neither provides complete coverage of the flora. That by Kitamura is essentially an amplified expedition report, while the series by Mogens Køie and Karl-Heinz Rechinger is a partial precursor to *Flora iranica* (which provides complete coverage of the country).

Bibliography

BRECKLE, S. W. *et al.*, 1969. Botanical literature of Afghanistan. *Notes Roy. Bot. Gard. Edinburgh* **29**: 357–371. Continued as *idem*, 1975. Botanical literature of Afghanistan: supplement I. *Ibid.*, **33**: 503–521. [Unannotated lists arranged alphabetically by author, supplemented by lists of selected references for adjacent countries and maps and map series.]

KITAMURA, S., 1960. Flora of Afghanistan. In Committee of the Kyoto University Scientific Expedition to the Karakoram and Hindukush (ed.), *Results of the Kyoto University scientific expedition to the Karakoram and Hindukush, 1955*, 2. ix, 486 pp., 105 text-figs., col. frontisp., 2 maps. Kyoto. Continued as *idem*, 1966. Additions and corrections to 'Flora of Afghanistan'. *Ibid.*, 8: 67–154.

Systematic enumeration of seed plants; includes synonymy with references and citations, vernacular names, localities with *exsiccatae* and general summary of local and extralimital distribution; list of references; index to botanical names. An introductory section includes chapters on prior botanical exploration and research, the Kyoto expedition itinerary, and definition of floristic and phytogeographic zones. The supplement of 1966 is based on later field work in the region.

KØIE, M. and K. H. RECHINGER, 1954–65. *Symbolae afghanicae.* 6 parts. Illus. (Kongel. Danske Vidensk. Selsk. (Biol. Skr.) 8(1), etc.). Copenhagen.

Comprises a series of critical enumerations of Afghan representatives in various families of vascular plants, with keys for some genera and descriptions of novelties, limited synonymy, indication of *exsiccatae* with localities, and extensive taxonomic commentary; indices in each part. [Not in the strict sense a flora, but with its wide range of family treatments complementary to Kitamura's work.]

Partial work: Wakhan

PODLECH, D. and O. ANDERS, 1977. Florula des Wakhans (Nordost-Afghanistan). *Mitt. Bot. Staatssamml. München* 13: 361–502, 3 maps.

An enumeration of vascular plants (including but one fern), with limited synonymy, citations, and *exsiccatae* with localities; endemic species are asterisked. References at end (pp. 368–370). The introduction contains a geographical description of the area and an account of botanical collecting.[75]

793

Pakistan (in general)

Area (inclusive of Jammu and Kashmir): 803941 km². Vascular flora: *c.* 5000 species (extrapolated from GB; see **General references**). – Under this heading are included bibliographies and floristic works pertaining to Pakistan as a whole or to substantial portions thereof not corresponding to a given province or territory. These latter appear as separate units: **794**, Baluchistan (with Quetta); **795**, North-West Frontier Province (except Hazara); **796**, Hazara; **797**, Azad Kashmir; **798**, Gilgit Agency and Baltistan; **799**, Ladakh; **811**, Sind; **812**, (West) Punjab (with Bahawalpur). For Indian-occupied Jammu and southern Kashmir, east of the so-called 'cease-fire line', see **841**. Ladakh is, in part, also under Indian administration. The former East Pakistan separated in 1971 as Bangladesh (**835**).

Pakistan lies abreast of the Indus escarpment zone and the deep dry upper Indus valley of Kohistan, a major biogeographic discontinuity used in the nineteenth century as the mutual boundary between *Flora orientalis* and *Flora of British India*, and in the twentieth as the eastern limit of *Flora iranica*. Internal coverage prior to 1947 amounted to little more than a smattering of scattered florulas and enumerations (except in the Punjab), and few centers for detailed floristic work existed.[76] Knowledge of the mountainous regions developed only in piecemeal fashion, often in conjunction with pacification, border demarcation or other security missions.

Since independence, much new botanical exploration has been accomplished and some local floras produced. From 1967 (with publication beginning in 1970), available information has been consolidated in two nationwide works: the *Flora of Pakistan* begun by Eugene Nasir and Syed I. Ali (1970– , initially as *Flora of West Pakistan*), published in chronologically numbered fascicles in the manner of *Flora iranica*, and the *Annotated catalogue* (1972) by Ralph R. Stewart. Much of the northern hill and mountain country centering on Rawalpindi and Islamabad and extending to the northern and western frontiers has additionally been treated in *Flora of Pothohar and the adjoining areas* by F. G. Bhopal and Md. N. Chaudhri (1977–78). However, it has been suggested that many areas, particularly in the mountainous northern districts, are in need of further collecting and study.[77]

Bibliographies

Selected references, with a classified 'Resumé', also appear in Ali (1978; see Note 77).

KAZMI, S. M. A., 1970–77. *Bibliography on the botany of West Pakistan and Kashmir and adjacent regions*, I–V. Coconut Grove, Fla.: Field Research Projects. [The 'adjacent regions' include nearly all of Afghanistan and parts of Iran, Middle Asia, China and India.]

STEWART, R. R., 1956. A bibliography of the flowering plants of West Pakistan and Kashmir. *Biologia* (Lahore) **2**: 221–230. [Somewhat selective.]

NASIR, E., S. I. ALI, Y. J. NASIR and M. QAISER (eds.), 1970– . *Flora of West Pakistan* (later *Flora of Pakistan*). Fasc. 1– . Illus., maps. Karachi: The editors (distributed from Department of Botany, University of Karachi).

Detailed descriptive flora of seed plants, published in family fascicles; each fascicle contains keys to genera and species, synonymy, references and citations, localities with indication of representative *exsiccatae* (referred to a grid system), general indication of overall range (internally and externally), notes on habitat, frequency, phenology, special features, uses, etc., and index to all botanical names. As of 1998, 197 fascicles had been published (the latest in 1995) with, however, some families or parts thereof still due. In part produced with the aid of surplus United States PL-480 agricultural funds, the work has been appearing fairly steadily; however, a systematic order has not been followed and no preset scheme or alternative schemes have been presented. For some families, interim mimeographed reports under Ali's editorship also exist.

STEWART, R. R., 1972. *An annotated catalogue of the vascular plants of West Pakistan and Kashmir*. xviii, 1028 pp., portr., map. Karachi (as part of NASIR, ALI, NASIR and QAISER, *Flora of West Pakistan*; see previous entry).

Systematic enumeration of vascular plants known from Pakistan and the whole of Kashmir; includes essential synonymy, references and citations, localities with indication of selected *exsiccatae*, critical remarks, and notes on habitat, local range, etc.; list of authorities, index of collaborators and index to all botanical names. An introductory section incorporates a general description of the region together with accounts of floristics, vegetation, botanical exploration and analysis, and future research needs. [The work is now considered to be in need of revision, with perhaps 10 percent of known Pakistan species yet to be included.][78]

Partial work: Pothohar

BHOPAL, F. G. and MD. N. CHAUDHRI, 1977–78. Flora of Pothohar and the adjoining areas, I–II. *Pakistan Syst.* 1(1): 38–128; (2): 1–98, illus. (some col. in part II).

Concise descriptive floristic treatment; includes keys to all taxa, limited synonymy with references and citations (usually major floristic works or revisions), local distribution, overall range, and some critical notes but no habitat or biological data. [Only the Centrospermae as well as Casuarinaceae through Polygonaceae (Englerian system) have been published. The work, centering on Rawalpindi (and Islamabad), covers the Margalla Hills (northern Punjab), Hazara, and Azad Kashmir as well as adjoining parts of North-West Frontier Province (from Peshawar to Chitral), Gilgit Agency, and Baltistan – an area nowhere, however, defined in the text.][79]

Woody plants

BAQUR, S. R., 1995. *Trees of Pakistan: their natural history, characteristics and utilization*. [14], viii, 634 pp., text-figs., col. photographs, maps. Karachi: Royal Book Co.

The taxonomic part of this work (pp. 177–465) comprises an alphabetically arranged descriptive treatment with limited synonymy, vernacular names (Urdu, Punjabi, Sindhi and English), indication of distribution, habitat and phenology, and remarks on properties and special features; references (pp. 466–483), a glossary, abbreviations (p. 498), a list of figures, an index to vernacular names, a synoptic list of species with chromosome numbers arranged by families, an index to botanical names, and illustrations (pp. 569–633) conclude the work. The general part includes forest history, a justification for the work, and an introduction to the botany, physiology, management, utilization, propagation and environmental role of trees.[80]

794

Baluchistan (with Quetta)

See also **793** (all works). – Area: 347 190 km². This Pakistani province is made up of Quetta and the former Baluchistan Agency.

BURKILL, I. H., 1909. *A working list of the flowering plants of Baluchistan*. 136 pp. Calcutta: Superintendent of Government Printing, India. (Reprinted after 1950, Baluchistan Forestry Department.)

Systematic enumeration of known seed plants, with limited synonymy, localities with citations of *exsiccatae* or other authorities, vernacular names, and notes on economics, uses, etc.; index. An introductory section includes an account of local botanical exploration. [For additions, see E. BLATTER, F. HALLBERG and C. C. McCANN, 1919–20. Contributions towards a flora of Baluchistan. *J. Indian Bot.* **1**: 56–61, 84–91, 128–138, 169–178, 226–236, 263–270, 344–352.]

795

North-West Frontier (except Hazara)

See also **793** (all works). – Area: North-West Frontier, 74 521 km²; tribal territories, 27 219 km². Hazara appears here as **796**, while **795** comprises the remainder of the present North-West Frontier Province from Waziristan and Dera Ismail Khan north through Malakand and Mardan to Swat, Kalam, Dir and Chitral. [The last four districts named have been centrally administered as 'tribal trust territories'.]

No general flora for this area is available. For convenience the various local florulas have been grouped under two subheadings: the North-West Frontier proper (Waziristan to Malakand and Mardan) and the tribal territories (Swat to Chitral). Floristic literature before 1920 largely took the form of expedition and survey reports such as *The botany of the Chitral Relief Expedition* (Duthie, 1898), published by the Botanical Survey of India and described below.

Local works: North-West Frontier

BLATTER, E. and J. FERNANDEZ, 1933–34. Flora of Waziristan, I–V. *J. Bombay Nat. Hist. Soc.* **36**: 665–687, 950–977, pls. 1–3, map; **37**: 150–171, 391–424, 604–619, pl. 48.

Concise enumeration of higher plants; includes descriptions of new taxa, essential synonymy with references and citations of key literature, vernacular names, localities with *exsiccatae* and summary of overall (extralimital) range, and brief notes on habitat, phenology, etc.; short list of references. A prefatory section includes remarks on botanical exploration and a definition of the limits of the area. The few plates depict a range of vegetation formations including notable *Charakterpflanzen*. [Waziristan is in the southern part of the province, from the Baluchi border north to about 33° 30′ N; its northern limit is thus south of the Peshawar district.]

QURAISHI, M. A. and S. A. KHAN, 1965–72. Flora of Peshawar district and Khyber agency, 1–2. *Pakistan J. Forest.* **15**: 364–393, 41 figs.; **17** (1967): 203–254; **22**: 153–219, 323–383. (First two installments reissued 1971 as *An illustrated flora of Peshawar district and Khyber Agency*, 1. 212 pp., 119 text-figs., map. Peshawar: Pakistan Forest Institute.)

Copiously illustrated, briefly descriptive flora of seed plants, with keys, limited synonymy, general indication of local and extralimital range, and brief notes on habitat, etc.; no indices. An introduction in part 1 discusses physical features and geography of the area, which centers on Peshawar and extends westward to include Khyber Pass, the famous western gateway to South Asia. [Not completed; extends from Ranunculaceae through Crassulaceae on the Bentham and Hooker system (the reissue, however, covering only the Thalamiflorae and Disciflorae, ending with Moringaceae (Sapindales).]

Local works: Swat, Kalam, Dir and Chitral

DUTHIE, J. F., 1898. The botany of the Chitral relief expedition. *Rec. Bot. Surv. India* **1**: 139–181, map.

Systematic enumeration of vascular plants and bryophytes collected and/or observed by the writer; includes descriptions of some new taxa, citation of *exsiccatae* with localities, and brief commentary; no index. An introductory section gives a description of physical and botanical features of the territory as well as a map of the expedition's routes.

STEWART, R. R., 1967. Checklist of the plants of Swat State, NW Pakistan. *Pakistan J. Forest.* **17**: 457–528.

Systematic list of vascular plants, with limited synonymy, local range (including some *exsiccatae*), and habitat notes; no index. The introductory section gives a general description of physical and botanical features of the region, a list of collectors, and statistical tables.

796

Hazara

See **793** (all works, but especially BHOPAL and CHAUDHRI); **812** (BAMBER; PARKER; STEWART). This part of the North-West Frontier Province lies *east* of the Indus and centers on Abbotabad, north of Rawalpindi.

797

Azad ('Free') Kashmir

See **793** (all works, but especially BHOPAL and CHAUDHRI). – Area: *c.* 84 000 km². Under Pakistani administration but with an autonomous regional government at Muzaffarabad, Azad Kashmir includes the Mirpur, Poonch (Punch) and Muzaffarabad districts of Kashmir. For the remainder of Pakistani-administered Kashmir, including Gilgit, Diamir, Baltistan and Ladakh, see **798** and **799**.

798

Gilgit, Diamir and Baltistan

See also **703/IV** (PAMPANINI); **760** (GRUBOV *et al.*); **793** (all works); **810–40** (*Flora of India*). This area, at present under direct rule from Islamabad, corresponds to the part of northern Kashmir north of the upper Indus River and north and west of the 'cease-fire line' between Pakistan and India. Dominating the area is the greater part of the Karakoram (**703/IV**), much of it under permanent ice and snow and rising to a height of 8611 m at K2 (Mt. Godwin Austen). Most of the flora is of Central Asiatic affinity; however, *Flora iranica* extends as far as Gilgit. The above cross-referenced works furnish the best coverage, but some further details can be found in Kitamura's enumeration (1964), the only other 'separate' florula.

KITAMURA, S., 1964. Flowering plants of West Pakistan. In Committee of the Kyoto University

Scientific Expedition to the Karakoram and Hindukush (ed.), *Results of the Kyoto University scientific expedition to the Karakoram and Hindukush, 1955*, 3: 1–161, figs. 1–60. Kyoto.

Partial enumeration of the seed plants in the mountainous areas of Gilgit and Baltistan; includes descriptions of novelties, synonymy with references and literature citations, localities with *exsiccatae* and general summary of overall range, and notes on habitat, etc.; no separate index. An introductory section gives accounts of prior botanical exploration, the itinerary of the Kyoto expedition, and phytogeographical relationships.

799

Ladakh

See also **703** (PAMPANINI); **760** (GRUBOV *et al.*); **793** (all works); **810–40** (*Flora of India*). This area, centering on Leh and including the Ladakh Range and the eastern Karakoram, is part of northern Kashmir *north* of the Indus but falls abreast of the 'cease-fire line' and thus is divided between Pakistani and Indian administration, with, furthermore, the northeastern Ladakh Plateau (Aksai Chin) effectively now under Chinese control (cf. **768**).

More than in Gilgit, the flora is of Central Asian (and Tibetan) affinity. Apart from coverage in general works (but not *Flora iranica*), two separate floras are available (Stewart, 1916; Kachroo *et al.*, 1977).

KACHROO, P., B. L. SAPRU and U. DHAR, 1977. *Flora of Ladakh: an ecological and taxonomical appraisal.* x, 172 pp., figs. 1–6, 4–24, tables (incl. maps). Dehra Dun: Bishen Singh Mahendra Pal Singh.

Contains an annotated systematic enumeration of seed plants, including diagnoses, synonymy, references, localities with representative *exsiccatae*, and notes on habitat, phenology, etc.; references (p. 166) and index to all botanical names at end. An extended introductory section gives generalities on physical features, climate, biota, people and land use, the economy of the area and planned development, and finally on vegetation, floristics and phytogeography, and life-forms.

STEWART, R. R., 1916. The flora of Ladak, western Tibet, I–II. *Bull. Torrey Bot. Club* **43**:

571–590, 625–650. (Reprinted separately 1973, Dehra Dun: Bishen Singh Mahendra Pal Singh.)

Includes a systematic list of vascular plants (based mainly on collections by the author), with occasional synonymy, localities with *exsiccatae* and published records, taxonomic commentary, and miscellaneous observations; list of references. An introductory section covers geography, climate, geology, and botanical exploration as well as details of the author's itinerary. The area covered falls wholly within eastern Kashmir, west of the Chinese boundary. [All records are now incorporated in the author's *Annotated catalogue of the vascular plants of West Pakistan and Kashmir* (1972; see **793**), and the work moreover has been at least partly superseded by Kachroo *et al.*'s flora (see above).]

Notes

1 The first-named of these collections also represents the work of the first in a series of specialist meetings on high-mountain flora and vegetation, organized under the auspices of the then-All-Union Botanical Society of the USSR and its affiliated bodies and held at approximately three-yearly intervals in various centers in the former Union. The papers (*Tezisy dokladov*) of the fifth (Baku, 1971) and sixth (Stavropol', 1974) meetings have in addition been seen.

2 Preparation of this atlas was in the hands of a working committee directed by Tolmatchev. It may be regarded as a tool in support of his broader studies in Northern Hemisphere chorology and biogeography.

3 In spite of its title, *Flora Pribajkal'ja* is not properly a flora but rather a collection of floristic and other contributions by various authors, including checklists (E. E. Jäger in *Prog. Bot.* **42**: 333 (1980)).

4 The *Conspectus* also appeared in German in installments in *Beih. Bot. Centralbl.* **18**/II–**31**/II (1905–13, *passim*), covering the first four parts of the Russian version before being terminated with the outbreak of World War I.

5 Turczaninow was aided in gathering material for his flora by a grant of special leave in 1830–34. For an appreciation, see L. Bernardi, 1967. Une lettre d'Irkoutsk pour M. de Candolle. *Musées Genève* **79**: 7–13.

6 For a history (with bibliography) of the Krylov Herbarium, see A. V. Polozhij, 1986. Гербарий имени П. Н. Крылова в Томском Университете (к 100-летию со времени основания)(*Gerbarij imeni P. N. Krylova v Tomskom Universitete (k 100-letiyu so vremeni osnovanija)*). 86, [2] pp., illus. Tomsk: Tomsk State University Press.

7 The present 'capital' of Siberia, Novosibirsk, was a product of the railway.

8 S. G. Shetler, *Komarov Botanical Institute* (1967), p. 74.

9 As in Canada, a relative concentration of resources in a few locations has perhaps facilitated realization of such a major work.

10 An account of the Tomsk University herbarium is contained in Polozhij (*op. cit.*).

11 Publication of an English translation commenced in 2000 (Plymouth, England/Enfield, N.H.: Science Publishers).

12 B. K. Shishkin and L. P. Sergievskaja were largely responsible for completing this work. The latter was also editor of the two-part supplementary volume 12, which in addition to the amendments to vols. 1–11 and the general index contains an account of botanical work in the region from 1927 onwards. For a general and bibliographic review of the work, see S. J. Lipschitz in *Bot. Žurn. SSSR* **51**: 1016–1018 (1966).

13 For a progress report (with bibliography), see L. I. Malyschev, 1978. Svodka 'Flora tsentral'noj Sibiri' i rabota nad nej. *Bot. Žurn. SSSR* **63**: 1358–1363. A precursory work, containing floristic lists but not as such a flora, is L. I. Malyschev and G. A. Peschkova, 1978. *Flora Pribajkalja*. 320 pp., 6 tables. Novosibirsk: 'Nauka'.

14 The lack of analytical keys in this work was a source of some criticism; moreover, its coverage of the eastern part of its area is weaker than to the west. These deficiencies prompted preparation of *Flora tsentral'noj Sibiri* in the 1970s.

15 The whole work was a long-term project of the Krylov Herbarium at Tomsk University, in more recent years under the leadership of A. V. Polozhij. In the 1960s, some fascicles were for publication paired under a single cover.

16 This flora evidently was written as a source work for use in middle and senior secondary schools.

17 Description in part prepared from notes supplied by M. E. Kirpicznikov, St. Petersburg. The press run for this flora was, for the USSR, small: only 300 copies.

18 Description prepared in part from notes supplied by M. E. Kirpicznikov, St. Petersburg. No 'alpines' are included in this work.

19 The progress of the original work has been intermittent due perhaps to its [semi-official] status and the availability and interest of particular individuals as editors. The project has been a charge of the All-Union Botanical Society (VBO) of the USSR (from 1991 the Russian Botanical Society). The second edition, a project of the herbarium at Tomsk University and the Tomsk Branch of the Botanical Society, was not continued, perhaps – at least in part – due to the death of the senior author in

1970. [Descriptions of both editions prepared with the assistance of information supplied in 1975 by M. E. Kirpicznikov and V. I. Grubov, St. Petersburg. The latter has from fascicle 7 acted as editor of the work.]

20 For a useful review of current knowledge and research trends, including comments on floras and floristic regions as well as a helpful bibliography, see T. S. Elias, 1986. The distribution and phytogeographic relations of woody plants of the Soviet Far East. *Aliso* **11**: 335–354.

21 This work is functionally and in scope comparable to the natural histories of trees in North America written in the 1950s by Donald C. Peattie.

22 This well-produced and documented work was originally published in a limited edition (500 sets) which evidently was never much circulated beyond Japan.

23 Functionally this work is comparable to *Trees and shrubs of northern New England* by Steele and Houghton (**141**) but features a stronger scientific orientation.

24 Eric Hultén (personal communication, 1970) indicated that Komarov's flora was written some time after 1908–09, when its author traveled extensively in Kamchatka, but almost certainly well before 1917. Unfortunately, the manuscript was lost following submission to the publishing house and did not turn up for nearly 10 years. Following its rediscovery in the mid-1920s, however, no attempt at revision was evidently made by Komarov despite a copy of all Hultén's notes having been made available. When finally published the work consequently was already some 15 or more years out of date.

25 Shortly afterwards Miyabe moved to Harvard University for advanced studies with Asa Gray and Sereno Watson. This accounts for publication of his Kuriles flora in *Memoirs of the Boston Society of Natural History*.

26 In his *Analiz flory Kavkaza* (referred to below) Grossheim suggested an eventual total of 6000 or more species, with 1158 endemic.

27 This is further south than the boundary adopted by Grossheim; from west to east that extended along the Manyč through Manyč Lake, then on the Stavropol' and Dagestan line to the mouth of the Kuma.

28 The rapid development of locally based floristic work in the Caucasus in the decades before 1920 was aided by the early emergence of Tbilisi as a regional scientific center, along with Tomsk in Siberia one of the very few in the Russian Empire beyond the European heartland. This is surely related to contemporary cultural and social progress. Indeed, Tbilisi has been called 'the most sophisticated, elegant and educated of former Soviet cities' (Andrew Higgins, *Independent*, 27 August 1993).

29 Andrei Fëdorov, however, criticized some of these works for excessive imitation of *Flora SSSR*; to him, among the large Caucasian works only the *Flora Armenii* had the most suitable design for a regional flora, with [rightly, in the present author's view] lesser emphasis on descriptions and more on geographical distribution and critical commentary where local knowledge and information were important.

30 A new conspectus, by Yuri Menitsky, is currently in preparation at St. Petersburg.

31 The preparation and publication of this work sparked controversy, akin to if perhaps at least in print more subdued than that of Backer and Koorders in Java (**918**). See A. I. Galushko, 1975. *Rukovodstvo dlja avtorov 'Flory Severnogo Kavkaz'*. 22 pp. Rostov-on-Don, Stavropol'; M. E. Kirpicznikov, 1977. Tak li nužno pisat' rukovodstva [Is this how manuals should be written]? *Bot. Žurn. SSSR* **62**: 136–138; and A. I. Galushko, 1979. Po povodu retsenzii M. E. Kirpicznikova 'Tak li nužno pisat' rukovodstva?' *Ibid.*, **64**: 1354–1359.

32 The second edition follows Grossheim's system. For an introduction, see A. A. Grossheim, 1945. K voprosu o graficeskom izobrazenii sistemy tsvetkovykh rastenij. *Sovetsk Bot.* **13**(3): 3–27.

33 Some 1600 species were covered in the first edition.

34 The format is reminiscent of a North American manual and, if produced in the United States or Britain, the work could fairly readily be fitted into one volume.

35 A literary treatment of the work of Russian naturalists in Middle and Central Asia may be found in Vladimir Nabokov's *The Gift* (1952).

36 The author in his introductory remarks indicates that he did not consider the original 1968 account to be sufficiently critical.

37 Chapters 6 through 11 of this work cover the flora, with chapter 10 in particular (pp. 173–186) devoted to the woody flora of the southern mountains – the only surviving such modern link between West and East Eurasia south of the boreal forests but now much reduced through human activities. Chapter 1 of this work, by Fet, comprises a brief historical review of natural history and biotic resource exploration in the republic, serious survey of which began only in the 1880s.

38 Of part 1, reviewed by V. Botchantzev in *Bot. Žurn. SSSR* **64**: 277–278 (1979), only 500 copies were printed. The run for part 2 was 1000, and for part 3, 700.

39 A concise and by then-prevailing standards well-presented and informative work; better paper and a stiff cover would make it even more suitable for the field.

40 One of the compilers of the 1988 manual, M. M. Nabiev, was also the author of an Uzbek botanical glossary and dictionary (1969).

41 The authors note that in the two decades from completion of Bondarenko's work the number of available collections had more than doubled.

42 In contrast to the other regional and republican floras in Middle Asia (save some of their early volumes), *Flora*

Tadžikskoj SSR was from the beginning a St. Petersburg-based project.

43 Some 489 species were known to Maximowicz by 1889, 1875 to Grubov in 1955, 2008 to Grubov in 1972, and 2239 to Grubov in 1982.

44 Further changes have been published in a series of papers published in *Botaničeskij Žurnal SSSR* from volume 56 (1971) onwards; by 1978 seven such contributions had appeared.

45 An English edition was released in 2000 (Plymouth, England/Enfield, N.H.: Science Publishers).

46 The boundaries did, however, fluctuate considerably for 30 years after 1949. All of Ningxia (including the lands of the Mongol Alashan League and the Chinese Hui) was reabsorbed by Gansu in 1954, with only Ningxia Hui, or Cis-Ningxia, reverting to separate status in 1958. In 1969, the four northern leagues of Nei Mongol were returned to the three reconstituted provinces of North-East China and the Alashan League in Trans-Ningxia transferred to Cis-Ningxia, Ningxia retaining its 'provincial' status. This is reflected in vols. 2–4 of *Flora intramongolica*. In 1979 the Alashan as well as the northern leagues were restored to Nei Mongol; the revised boundaries have been followed in vols. 1 and 5–8 of the *Flora* and all volumes of its second edition.

47 The species concept in this work is relatively narrow.

48 With the advent of an Arabic edition of this book Lawrence's comment would have widened its impact.

49 P. Davis and I. C. Hedge in *Candollea* **30**: 332–335 (1975).

50 On Edmond Boissier, see H. M. Burdet (ed.), 1985. *Edmond Boissier, botaniste genevois 1810–1885–1985* (Publ. Conservatoire Jard. Bot., Sér. Doc. 17). Geneva.

51 For OPTIMA congresses, see **601/I**. Collections from the 'Plant Life in Southwest (and Central) Asia' symposia include P. H. Davis, P. C. Harper and I. C. Hedge (eds.), 1971. *Plant life of South-west Asia*. Edinburgh: Botanical Society of Edinburgh; I. C. Hedge (ed.), 1986. *Symposium: plant life of South-west Asia* (Proc. Roy. Soc. Edinburgh, B, 89). Edinburgh; T. Engel, W. Frey and H. Kürschner (eds.), 1991. *Contributiones selectae ad floram et vegetationem Orientis*. Berlin: Cramer/Borntraeger (Flora et vegetatio mundi 9); and M. Öztürk, Ö. Seçmen and G. Görk (eds.), 1996. *Plant life in Southwest and Central Asia*. 2 vols. Izmir: Ege University Press. Two other collections, both *Festschriften*, are K. Tan (ed.), 1989. *Plant taxonomy, phytogeography and related subjects: the Davis and Hedge Festschrift*. Edinburgh; and S. I. Ali and A. Ghaffar (eds.), 1991. *Plant life of South Asia*. Karachi. All these variously include papers on floristic progress. A fairly personal but pragmatic view of developments was that of R. D. Meikle, 'Co-ordination of floristic work and how we might improve the situation' (in Davis *et al.*, pp. 313–331). Hedge has reviewed floristic progress in 'The genesis and

results of some SW Asiatic floras' (in Ali and Ghaffar, pp. 29–38) and 'Quo vadimus: after the Floras what then?' (in Engel *et al.*, pp. 311–318). In the latter he calls for better coordination of work with the republics of Middle Asia. Karl-Heinz Rechinger and others have reported on specific flora and other projects; references are noted where relevant.

52 The close association of Boissier with the 'Kewites' is exemplified by the 'gentleman's agreement' with J. D. Hooker by which the western limits of the *Flora of British India* would be matched along a natural physical and botanical boundary with the eastern limits of *Flora orientalis*. The overall geographical limits of the latter also recall the Alexandrine Empire and the sphere of Hellenistic civilization (the extent of which was then being rediscovered by travelers and archeologists). The southern limit, however, was arbitrarily set at the Tropic of Cancer (which almost bisects the Arabian Peninsula and runs not far south of Aswan). It would be another half-century before any overall coverage of the infratropical part of the Peninsula became available.

53 The *Conspectus* attempts to present a modern 'overview' accounting for recent floristic and other taxonomic work (and facilitating among other things a better grasp of large genera such as *Cousinia* in the Compositae). In contrast to the *Med-Checklist* (see **601/I**), this Hebrew University-based work emphasizes distribution and chorology rather than critical taxonomy and nomenclature.

54 The authors indicate in their introduction to the work (vol. 1) that its compilation became a possibility only after 1960 with the initiation of several critical national and regional floras.

55 Large areas are, however, currently inaccessible due to civil and other conflicts.

56 For a report on the *Flora of Turkey*, see P. H. Davis and I. C. Hedge, 1975. The *Flora of Turkey*: past, present and future. *Candollea* **30**: 331–351 (also in, or as, *Istfea* 1).

57 This is one of the best of 'traditional' floras to have appeared in the last two or three decades. In format and style it may be seen as a continuation of the classical 'Kew Series' begun in the mid-nineteenth century, striking a compromise between practicality and documentation.

58 Although volume 3 was thought to be virtually ready for press on the author's death in 1972, publication became protracted not only due to the severe political turmoil and onset of civil war in the Lebanon in 1975 but also by a recognition by Mouterde's botanical editors, André Charpin and Werner Greuter, that considerable revision was necessary. It was thus decided to bring out the text part of that volume in four fascicles; the first appeared in 1978.

59 *Flora palaestina* embodied contrary philosophical forces: Michael Zohary, the initial senior author, was representative

of the 'expediency' school, while Naomi Feinbrun espoused a more 'scholarly' approach (Aaron Liston, *in litt.* 20 July 1986). The work could in some respects be seen as an expansion of N. Feinbrun, M. Zohary and R. Koppel, 1949–58. *Iconographia florae terrae Israëlis* (later *Flora of the land of Israel: iconography*). 3 parts. 150 pls. (15 col.). Jerusalem: Palestine Journal of Botany (parts 1–2); Weizmann Science Press (part 3).

60 Botanical progress is more fully surveyed in K. H. Batanouny, 1985. Botanical exploration of Sinai. *Qatar Univ. Sci. Bull.* **5**: 187–211.

61 Along with some localities not shown on standard maps, this bibliography forms the eighth installment in a series of nine in *Kew Bulletin*, **3–9**, *passim* (1948–54) entitled 'The Rustam Herbarium, 'Iraq'. Further bibliographic references appear in other installments of this series.

62 The introductory sections in vol. 1 are outstanding. The unusual choice of Mark 2 of the Hutchinson system was made for domestic reasons (and over the better judgment of its compilers; Jan Gillett, personal communication, 1991). A non-technical description of the *Flora* appears in C. C. Townsend, 1980. The 'Flora of Iraq' project. *Ur* **2**: 19–23. See also M. M. Harley and J. Gillett, 1992. Evan Rhuvon Guest (1902–1992). *Taxon* **41**: 616–617.

63 The work, written at the invitation of Baghdad University, was intended as a 'students' flora'. At the price asked, however, it would have been largely out of reach of its potential clientele – a not-uncommon problem in other parts of the world.

64 This work was announced in the 1980s but various circumstances long frustrated publication.

65 The publication of two or three independent floristic undertakings relatively so close together is a reflection of the vagaries of development desires, sponsorship and programming.

66 *Kew Record* credits El-Amin as author.

67 A singularly miscellaneous book, in which only the Bahrain checklist remains somewhat definitive.

68 Stronger editing and design would have improved this work; a small flora such as Kuwait possesses could easily be accommodated in one volume. Vol. 1 represents the compass of Daoud's contributions.

69 For a progress report, see K.-H. Rechinger, 1991. Report on *Flora iranica*. In S. I. Ali and A. Ghaffar (eds.), *Plant life of South Asia*, pp. 39–46. Karachi. As of 1999, only the pteridophytes, Rubiaceae and the numerous species of *Astragalus* (Leguminosae) – the last-named currently under study by Dietrich Podlech (personal communication, 1996) – were yet to appear; however, in the latter part of that year the first of four installments of *Astragalus* appeared as 'Leguminosae III: *Astragalus*, 1'.

70 Some remarks on the development of botany in Iran appear in A. Ghahreman, 1989. Problems of the flora of Iran. *J. Sci. Univ. Tehran* **18**: 29–33.

71 The first part is also dated in Persian chronology (2537, corresponding to 1978 C.E.); parts 2–16, in Islamic chronology (1358 through 1376, corresponding approximately to 1980 to 1997 C.E.).

72 Volume 10 of the original work was reported to exist as early as the mid-1970s; see *Acta Ecologia Iranica* **1**: 85 (1976). Of this onion of a flora, a review and bibliographic data have been presented in J. M. Lamond, 1978. Notes on Parsa's 'Flore de l'Iran', volumes 8 and 9. *Notes Roy. Bot. Gard. Edinburgh* **35**: 349–364. Lamond commented that (to her) the work was 'badly written and ill-presented', with numerous errors and questionable typification practices. The second edition as so far published covers pteridophytes, gymnosperms, and dicotyledons from Ranunculaceae through Fumariaceae (on a modified Bentham and Hooker system).

73 Invalidly published descriptions of two new species (in English) are also included.

74 The author prepared this account as older works were out of date, and that 'little interest has been shown in the ferns of Iran'; in addition, he considered the fern flora to be undercollected.

75 The Wakhan was effectively opened to climbers and others only from the 1960s; it had hitherto been among the least-known parts of the country.

76 The only herbaria at independence were a small one at Lahore University and a larger one at Gordon College, Rawalpindi; the latter now forms the core of the Pakistan National Herbarium. The Blatter Herbarium, the base for the work of Ethelbert Blatter and his associates in Sind (formerly part of the Bombay Presidency) and towards the frontier, is in Mumbai (Bombay).

77 See M. N. Chaudhri, 1977 (1978). The Pakistan herbarium. *Pakistan Syst.* **1**(2): 100–105; S. I. Ali, 1978. The flora of Pakistan: some general and analytical remarks. *Notes Roy. Bot. Gard. Edinburgh* **36**: 427–439; and R. R. Stewart, 1982. *History and exploration of plants in Pakistan and adjoining areas*. 186 pp. (Flora of Pakistan, Suppl.). Islamabad: Pakistan Agricultural Research Council. Further historical contributions may be found in chapters by Stewart and others in Ali and Ghaffar (1991; see **Superregion 77–79, Progress**) as well as in Burkill and other works on India (see **Superregion 81–84**).

78 M. N. Chaudhri, 1977 (1978).

79 I thank Mr. Abid beg Mirza, formerly of the Wau Ecology Institute, Wau, Papua New Guinea, for assistance with the definition of 'Pothohar'.

80 This tree book is well documented but would have greatly benefited from rigorous editing!

Division 8

•••

Southern, eastern and southeastern (monsoonal) Asia

In these matters (of style and phraseology) my Flora of the British Islands has been followed; the style there adopted having been suggested by the requirements of the Professors of Botany in the Scotch universities and approved by them, seemed to me to be equally applicable to a more extended [work]. . . . It is as a hand-book to what is already known, and a pioneer to more complete works, that the present is put forward.

> J. D. Hooker, Preface to *Flora of British India* (1872).

The time for the preparation of a complete Local Flora of the Lower Provinces [Bengal] has not yet come.

> D. Prain, *Bengal plants*, 1: 5 (1903).

It should be remembered that outside the small herbarium and botanical library in Hong Kong herbaria and botanical libraries were non-existent in China previous to 1916.

> H.-L. Li, *Proc. Linn. Soc. London* **156**: 39 (1944).

Beyond the Great Wall [of China] there is not enough rainfall to support agriculture; below it there is.

> R. M. Nixon, *The Real War* (1980).

This division essentially consists of that half of the Asiatic continent under the influence of the summer monsoon. Its western and northern limits, which impinge upon Division 7, originate at the Arabian Sea coast west of Karachi and run along the western edge of the Indus Basin and then along the upper course of the Indus River and the 'cease-fire line' to the Kashmiri–Chinese border. From there it runs along the Himalayan frontier of China as far as the Indian–Myanmar border and from there northward along the Tibet–Sichuan provincial boundary (west of the 'Tibetan Marches') and the eastern limit of Qinghai (Chinghai) Province; thence along the Huang Ho through Gansu (passing by the city of Lanzhou) and the Ningxia Hui Autonomous Region to the boundary of the Inner Mongolian Autonomous Region; and thence eastward approximately along the Great Wall, the south-eastern boundary of the Region, and the Chinese–Mongolian and Chinese–Russian frontiers to the Sea of Japan southwest of Vladivostok. Also included in Division 8 are Korea, Japan, the Ryukyus, the Bonin and Volcano Islands, Parece Vela, the Daito Islands, Taiwan, the South China Sea islands (Paracels and Spratlys), and the Coco, Andaman and Nicobar Islands. However, Peninsular Malaysia is referred to West Malesia (Region 91) in Division 9, and the Laccadive and Maldive Islands are referred to Region 04 in Division 0. Nine politico-geographic regions have been delimited, additionally grouped into four superregions (South Asia, 81–84; Japan and Korea, 85; China 'proper' and Manchuria, 86–88; and Southeastern Asia, 89). These latter by and large also represent the bio-historical foci of the hemi-continent, and it is thus appropriate to present background surveys under their respective headings.

General bibliographies. Bay, 1910; Frodin, 1964, 1984; Goodale, 1879; Holden and Wycoff, 1911–14; Jackson, 1881; Pritzel, 1871–77; Rehder, 1911; Sachet and Fosberg, 1955, 1971; USDA, 1958.

General indices. BA, 1926– ; BotA, 1918–26; BC, 1879–1944; BS, 1940– ; CSP, 1800–1900; EB, 1959–98; FB, 1931– ; IBBT, 1963–69; ICSL, 1901–14; JBJ, 1873–1939; KR, 1971– ; NN, 1879–1943; RŽ 1954– .

Conspectus

800

Division in general

No flora or related work covers the whole of the division, its size and diversity alone all but dictating that several foci for research have developed. The geographically most expansive floras are *Flora of British India* (Superregion 81–84) and *Flora republicae popularis sinicae* (Superregion 86–88 along with Region 76), whereto reference should be made.

802

'Old' mountain systems

South Indian Hills (Western Ghats)

Fischer's work on the Anamalai Hills is included as they are not covered in Fyson's flora. For the Palnis, a new flora is being realized by Fr. K. M. Matthew; a volume of illustrations appeared in 1996 with a supplement in 1998 and text volumes in 1999. Complementary to these is the tree species distribution atlas of Ramseh and Pascal (1997), a work which in matters of presentation has broken new ground.

FISCHER, C. E. C., 1921. *A survey of the flora of the Annamalai Hills in the Coimbatore District, Madras Presidency.* 218 pp., 5 pls. on 3, folding map (Rec. Bot. Surv. India 9). Howrah. (Reprinted 1978, Dehra Dun: Bishen Singh Mahendra Pal Singh.)

Systematic checklist (1798 vascular plants along with some non-vascular plants and fungi) with brief annotations featuring *exsiccatae* (mainly those of the author), phenology, altitudes, and miscellaneous notes including some critical commentary; appendix, addendum and index to scientific names at end. The general part includes an account of the range and its vegetation, notes on naturalized species, and a synopsis and explanation of abbreviations.

FYSON, P. F., 1932. *The flora of the South Indian hill stations: Ootacamund, Coonor, Kotagiri, Kodaicanal, Yercaud, and the country round.* 2 vols. xxix, 697 pp.; atlas of 611 pls. Madras (now Chennai): Government Press. (Originally published 1915–20 in 3 vols. as *The flora of the Nilghiri and Pulney hill-tops*; this edition reprinted 1974, Dehra Dun: Bishen Singh Mahendra Pal Singh.)

Copiously illustrated, semi-popular descriptive flora of the seed plants of the high country on the Nilghiris, Palnis and Shevaroys above about 4000 ft. (1220 m); includes limited synonymy, English common names, fairly detailed indication of local range, general summary of extralimital distribution, and notes on habitat, phenology, biology, special features, etc.; complete index at end.[1]

MATTHEW, K. M., 1996. *Illustrations on the flora of the Palni Hills, South India*. xlvi, 979 pp., 950 pls. Tiruchirapalli: Rapinat Herbarium, St. Joseph's College. Complemented by *idem*, 1998. *Supplement*. Pp. (pls.) 951–1223, pp. 1225–1266. Tiruchirapalli; and *idem*, 1999. *The flora of the Palni Hills, South India*. 3 vols. xcvi, 1880 pp., 4 maps. Tiruchirapalli.

The atlas volumes comprise a systematically arranged atlas of plates (1171 species in 660 genera over both volumes) covering native and introduced seed plants, the captions with Tamil (in Tamil script) and botanical names and voucher specimens; cumulative index, pp. 1225–1266. The characteristically polemical introductory part (in the main volume) includes the plan and programme for the work along with a statistical summary. The text volumes comprise a detailed descriptive account with keys, synonymy, citations, vernacular names, representative *exsiccatae*, and commentary together with (in vol. 1) an extended general introduction and (in vol. 3) several appendices and complete indices. [2478 species in all have been recorded for the Palnis in an area of 2068 km², a considerable percentage of them also known from Tamilnadu Carnatic (**828**).]

RAMESH, B. R. and J.-P. PASCAL, 1997. *Atlas of endemics of the Western Ghats (India): distribution of tree species in the evergreen and semi-evergreen forests*. iv, 403 pp., 352 pls. (with maps), 6 additional figs. (in part maps), CD-ROM. Pondicherry: Institut Français de Pondichéry. (Publ. Dépt. Écol. (Pondichéry) 38.)

Comprises 352 distribution maps with accurately plotted records against a topographic base; standard insets depict for each species its preferred forest strata, time and duration of the prevailing dry season, and altitudinal distribution respectively along the eastern and western falls of the Ghats (for explanation of the insets, see pp. 13–15); index to all species at end. Arrangement of the maps is alphabetical, in the first instance by families. The introductory part (pp. 3–21) encompasses the justification, plan, methodology and 'perspectives' (i.e., possible uses) of the work along with major features of climate and vegetation (including definition of the 19 forest types to which the species distributions have been related and their potential areas) and a bibliography; it is followed by four appendices (a lexicon of synonyms, herbaria consulted, specimens seen, and literature sources – all 'standard floras' – for records). The CD-ROM (by C. Noguier and R. Datta) encompasses the contents of the printed work (with color versions of the maps) as well as additional photographic images of landscapes, vegetation and individual (but not all) species.[2]

803

'Alpine' regions

Substantial portions of the hemi-continent are at comparatively high elevations, but apart from the Himalaya – delimited here in more or less political terms as Region 84, as most floras within it are so bounded and/or cover much more than the alpine and upper-montane zones – and extra-Chinese East Asia, with as in Europe several semi-popular guides, there has hitherto been little specific literature falling within the scope of this *Guide*. A recent major contribution, however, is the two-volume flora (1993–94) of the Heng Duan Shan mountain system running through western Sichuan and northwestern Yunnan and including among others the Lixiang range and Gongga Shan (Minya Konka).

I. The Himalaya

For other works, including the good illustrated popular guide *Flowers of the Himalaya* (1984; supplement, 1988) by Oleg Polunin and Adam Stainton, see Region 84.

OHBA, H., 1988. The Alpine flora of the Nepal Himalayas: an introductory note. In H. Ohba and S. B. Malla (eds.), *The Himalayan plants*, 1: 19–46 (Bull. Univ. Tokyo Mus. 31). Tokyo.

Includes a survey and name list of the flora, precursory to 'compilation of a modern and comprehensive manual of the Himalayan alpine flora'. Some 1227 species of 'alpine' seed plants are now known from 4000 m or more in elevation, a large percentage being autochthonous.

RAU, M. A., 1975. *High altitude flowering plants of West Himalaya*. x, 234 pp., 25 figs., 5 photographs. Howrah, West Bengal: Botanical Survey of India.

Concise systematic account (encompassing over 1000 species); includes descriptions of and general notes on families, keys to genera, and lists of species in each genus with localities (arranged by districts) and altitudinal zonation and occasional synonymy; addenda, glossary, references (pp. 207–210), and index to botanical names. The introductory section covers physical features, climate, insolation, the extent of the alpine flora, habitats and general features of the flora, noteworthy plants, the origin of the Himalaya, and phytographic history. [Covers the mountain tracts of Uttar Pradesh (now Uttaranchal) and Himachal Pradesh in northwest India above 3300–3600 m.]

II. Japan

Semi-popular alpine guides have been published since early in the twentieth century. The first, in two volumes, was *Pocket atlas of the Alpine plants of Japan* (1906–07) by M. Miyoshi and J. Makino. This was followed by the first edition of *Kôzan shokubutsu zu-i* (*Pocket-book of alpine plants*) (1933; revised 1937) by Hisayoshi Takeda.[3] A one-volume manual, *Nihon kôzan shokubutsu zukan* (*Illustrated manual of alpine plants of Japan*), by the same author followed in 1950. Subsequent editions of the *Illustrations* and the *Illustrated manual* followed respectively in 1957 and 1961, the latter with K. Tanabe. At this time, Takeda also produced a new two-volume work, *Genshoku Nihon kôzan shokubutsu zukan* (*Alpine flora of Japan in color*) (1959–62) as nos. 12 and 28 in the flora and fauna series produced by the Osaka publisher Hoikusha. This last-named has been succeeded by Shimizu (1982–83).[4]

SHIMIZU, T., 1982–83. [*The new alpine flora of Japan in color.*] 2 vols. 39 text-figs., 24 halftones, 144 col. pls. (Flora and fauna of Japan 59, 60). Osaka: Hoikusha.

Copiously illustrated manual-flora of 'alpine' vascular plants (484 species) with keys, brief descriptions, synonymy, vernacular names and transliterations, indication of distribution with localities, some critical comments, and notes on habitat, frequency, etc.; full keys in English, karyological tables, notes on underground parts and seed structure, and indices to vernacular and scientific names at end of each volume. A historical review is given on pp. 249–251 of the first volume, with a chronology of key works.

TAKEDA, H. and K. TANABE, 1961. *Nihon kôzan shokubutsu zukan/Illustrated manual of alpine plants of Japan.* Revised edn. 15, x, 347 pp., illus. (some col.). Tokyo: Hokuryukan. (1st edn. by H. Takeda alone, 1950.)

Pocket-manual of alpine plants. [Not seen; cited from *National Union Catalog 1963–67*, 52: 499. Earlier editions listed in Merrill and Walker, *A bibliography of Eastern Asiatic botany, Supplement*, p. 347 (cited under Superregion 86–88).]

III. Korea

PAK MAN-KYU (BOKU, MANKYU), 1942. [*A list of Korean alpine plants.*] *J. Chosen Nat. Hist. Soc.* **33**: 1–12.

A checklist with distributions of 383 species on 10 higher mountains. [Not seen.]

IV. China (Taiwan)

An early flora of alpine and high-mountain areas in China is *Flora montana Formosae* (1908) by Bunzo Hayata. This has been succeeded by two manuals, published respectively in 1929 and 1975–78. Associated with the latter is YING SHAO-SUN, 1975. A list of alpine plants of Taiwan. *Quart. J. Chinese Forest.* **8**(3): 89–122; **8**(4): 123–151.

ITÔ, T., 1929. *Taiwan kôzan shokubutsu zusetsu* [*Flora alpina Formosana*]. 100, 30, 8, 12, 16, 11 pp., 200 text-figs., 14 pls., 2 folded col. pls.

An illustrated handbook. [Not seen; cited from Merrill and Walker, *A bibliography of Eastern Asiatic botany, Supplement*, p. 140 (cited under Superregion 86–88). There it was noted that the work includes habitat data, altitudinal distribution and a list of mountains with elevations.]

YING SHAO-SUN, 1975–78. *Tai-wan kao shan chih wu tsai se tu chien* (*Alpine plants of Taiwan in color*). 2 vols. Illus. (text-figs., col. pls.). Taipei: Kuo li Tai-wan ta hsueh sen lin hsi. (Tai-wan chih wu tsai se tu chien (Colored illustrations of plants of Taiwan) 1.)

Illustrated manual with Chinese and botanical names, synonymy inclusive of references and citations, keys in some genera, and descriptions with distributional and ecological details in both Chinese and English; brief bibliography and consolidated indices to botanical and Chinese names in vol. 2.

V. China (mainland)

An outstanding recent contribution is the massive flora for the Heng Duan range (inclusive of Gongga Shan or Minya Konka along with many other areas of interest to horticulture) on the eastern side of the Qinghai-Xizang (Tibet) Plateau.

ZHONG-GUO KE XUE YUAN QING ZHAN GAO YUAN ZONG HE KE XUE KAO CA DUI (COMPREHENSIVE

SCIENTIFIC EXPEDITION TO THE QINGHAI-XIZANG PLATEAU, CHINESE ACADEMY OF SCIENCES), 1993–94. (*Heng Duan Shan qu wei guan zhi wu*)/ *Vascular plants of the Heng Duan Mountains.* 2 vols. xiv, xii, 2608, cxxii, cxxiv pp. Beijing: Ke xue chu ban she (Science Press). (Series of the Scientific Expedition to Heng Duan Mountains, Qinghai-Xizang Plateau.)

Descriptive flora of vascular plants (8559 species including 597 pteridophytes and 63 gymnosperms) with botanical and standardized Chinese names, synonymy, references and citations, localities with *exsiccatae*, distribution and habitat, and commentary; indices in Chinese (by character strokes) and to botanical names at end of each volume (not, however, cumulative in vol. 2). [Arrangement of families follows the Engler/Diels system but with monocotyledons following dicotyledons. Vol. 1 covers pteridophytes, gymnosperms, and dicotyledons from Saururaceae through Cornaceae sensu lato; vol. 2, Diapensiaceae through Orchidaceae. Encompasses parts of Xizang, Sichuan and Yunnan including many classically important plantsmen's 'findspots' but based in the first instance on surveys made from 1981 onwards.][5]

804

Ectopotrophic areas

Included here is an extensive study by Xu Zhaoran of the flora of limestone hills, a widespread geological phenomenon in southern China and Southeast Asia.

Limestone hills

XU ZHAO-RAN, 1993. A species list of limestone plants in China. In *idem* (ed.), *Studies of the limestone forests in China: floristics, ecology, conservation and taxonomy*, pp. 155–258 (Guihaia, Addit. 4). Yanshan, Guilin: Guangxi Institute of Botany/Botanical Society of Guangxi.

A checklist with for each terminal taxon one herbarium voucher as well as botanical and Chinese standard names; many unnamed or undescribed species included. Covers 3976 species, 11 subspecies and 276 varieties or 4263 terminal taxa of vascular plants, based largely on collections made since 1981 and deposited at Zhongshan University (SYS) in Guangzhou. Most records are from Yunnan, Guangxi and Guizhou; a small portion are from Guangdong, Hunan, Hubei and Sichuan. Distribution details are not included.

[The rest of the monograph includes an introduction to the flora by the editor (pp. 5–54) and studies of rare and threatened species.]

805

Drylands

There are no extensive steppes and only one major desert in Division 8. The last-named is the Indian (Thar) Desert, covering much of the western part of Rajasthan in India.

Indian (Thar) Desert

The Thar Desert has been documented in two successive works, the first being the enumeration by Blatter and Hallberg (1918–21, *passim*; in *Journal of the Bombay Natural History Society* **26–27**) and the second the descriptive flora by Bhandari (see below), first published in 1978 and revised in 1990. The work by Blatter and Hallberg covers the former princely states of Jodhpur and Jaisalmer and includes an extensive discussion of the vegetation. Bhandari's flora reflects the establishment in 1972 at Jodhpur of the Arid Zone Circle of the Botanical Survey of India (BSI-AZC).

BHANDARI, M. M., 1990. *Flora of the Indian Desert.* 2nd edn. viii, 459 pp., 136 text-figs., 114 col. photographs. Jodhpur: MPS Reproductions. (1st edn., 1978, Jodhpur: Scientific Publishers.)

Briefly descriptive illustrated flora of seed plants (682 species over 20000 km^2, of which 619 species are considered native) with keys to all taxa, synonymy (with references), citations, vernacular names, local range (with representative *exsiccatae*), overall distribution, critical remarks, and notes on habitat, frequency, occurrence, phenology, uses, etc.; list of references and indices to all vernacular and scientific names at end. An introductory section includes descriptions of physical features, climate, soils, vegetation, floristics (with statistics), plant geography and biotic factors along with a review of past botanical endeavor and the foundations for the present work.

808

Wetlands

Major works have so far been published for Japan and China as well as for the countries of South Asia.

South Asia (including India) and Myanmar (Burma)

For India the definitive flora of Cook (1996) has made earlier works, including the two editions of Biswas and Calder (1937, 1955) and the handbook of Subramanyam (1962), obsolete except over 1000 m in the Himalaya where there is no alternative coverage. In Myanmar, however, there has been no successor to Biswas and Calder (1937; revised 1955). Mention should also be made of a popular work for Sri Lanka, aimed primarily at visitors, amateurs and aquarists and covering more commonly encountered species: W. H. and W. V. DE THABREW, 1983. *Water plants of Sri Lanka*. 126 pp., illus. (part col.). New Malden, Surrey: Suhada Press. (Heritage of Lanka Series 1.)

BISWAS, K. and C. C. CALDER, 1955. *Handbook of common water and marsh plants of India and Burma*. 2nd edn., revised by K. Biswas. xvi, 216 pp. (incl. 32 pls.), 6 halftones (Health Bulletin (India) no. 24). Delhi: Manager of Publications, Government of India. (Reprinted 1984, Dehra Dun: Bishen Singh Mahendra Pal Singh. 1st edn., 1937.)

Amply descriptive account of the more common aquatic macrophytes (encompassing 178 vascular plant species) with accompanying figures; includes keys (to genera and species only), indication of habitat and distribution, and special features, with plates (pp. 179–216) and index to scientific names at end. A glossary is also included. The introductory section deals with limnological processes, general features of aquatic vegetation, periodicity, control methods, and surveys of coastal halophytes and mangroves as well as other topics.

COOK, C. D. K., 1996. *Aquatic and wetland plants of India*. [v], 385 pp., 374 text-figs. Oxford: Oxford University Press.

Descriptive flora with keys to all taxa, essential synonymy, indication of local and extralimital distribution and habitat, life- and growth-form classes (with explanation of conventions and abbreviations on pp. 6–7), notes on biology, phenology, and related but excluded taxa; generic and family descriptions supported by key literature citations; index to all botanical names at end. The introductory part covers the scope and plan of the book and also features a key to growth forms, a summary of excluded species, and (pp. 13–21) general keys. [Covers India south of the Himalaya, there being a considerable floristic break corresponding to the hills north of the plains of the Indus, Sutlej, Ganges and Brahmaputra; an altitudinal limit of 1000 m was therein set. The work also covers Sri Lanka and most of Myanmar (Burma) as well as the Andaman Islands and the Laccadives.][6]

KHAN, M. S. and M. HALIM, 1987. *Aquatic angiosperms of Bangladesh*. 120 pp., 26 pls. Dhaka: Bangladesh National Herbarium.

Illustrated descriptive manual (covering 123 species) with keys to all taxa, synonymy, references, and indication of distribution and habitat; bibliography and index at end. There is a brief, one-page introduction. [About 100 species are illustrated.]

NASKAR, K. R., 1990. *Aquatic and semi-aquatic plants of the lower Ganga Delta*. 408 pp., 16 col. photographs Delhi: Daya Publishing House.

Briefly descriptive account of 327 aquatic and wetland angiosperms with keys, synonymy, references and citations, vernacular names, localities, overall distribution, habitats, phenology, and uses and properties; extensive list of references and index to all names at end. The extensive general part features an introduction to all aspects of aquatic macrophytes, vegetation, and interaction with humans with respect to weediness or economic importance; this is followed by a tabular list with principal features (pp. 60–83) and general keys (pp. 84–97).

SUBRAMANYAM, K., 1962. *Aquatic angiosperms*. viii, 190 pp., 63 text-figs., col. frontisp. (Botanical Monographs, Council of Scientific and Industrial Research, India 3). New Delhi.

Illustrated descriptive account of more common flowering hydrophytes, with keys, synonymy and references, literature citations, generalized indication of range, habitat, phenology, etc., and notes on specialized morphological features. Illustrations separate from text; references (pp. 177–180), list of karyotypes, and general index at end. The introduction includes a general key to families along with a consideration of degree of coverage.

Japan

KADONO, Y., 1994. *Aquatic plants of Japan*. viii, 179 pp., col. illus., maps. Tokyo: Bun-ichi Sogo Shuppan.

Illustrated descriptive account with Japanese and scientific names and notes on distribution, habitat and similar species along with dot distribution maps. The generic and family headings contain references to key literature listed in the bibliography (pp. 156–160). A general key appears on pp. 161–171, followed by indices to Japanese and scientific names. The general chapter contains sections on aquatic plant morphology, habitats, and conservation. [Covers Japan proper and the Ryukyus.]

China

In addition to the work described below, there are many treatments for provinces or special regions. The latter are listed in Chen *et al.* (1993; see **Supraregional bibliographies** under **Superregion 86–88**).

ANONYMOUS, 1983. (*Illustrated handbook of aquatic vascular plants of China*.) 683 pp., illus. [Not seen; cited from Chen *et al.* (1993, p. 71; see **Superregion 86–88, Supraregional bibliographies**). Treats 317 species in 145 genera.]

YEN [YAN] SU-ZHU, 1983. *Chung-kuo shui sheng kao teng chih wu tu shuo* (*Higher water plants of China*). v, 335 pp., illus. Beijing: Ko hsüeh chu pan she (Science Press). [Not seen; cited from Cook, *Aquatic plant book* (1990, p. 7; see **008**) as well as Kadono, *Aquatic plants of Japan* (see above), p. 160.]

Superregion 81–84

South Asia (Indian subcontinent)

Vascular flora: perhaps 20000 species (based on current estimates of 17000 for India by the Botanical Survey of India). – The limits of South Asia (or the Indian subcontinent) are here considered to be the western edge of the Indus Basin (the mutual boundary between Boissier's *Flora orientalis* and Hooker's *Flora of British India*), the upper course of the Indus River in northern Kashmir; the Chinese border in the Himalaya (in the east as recognized by India); and the Burmese frontier in the east. It thus includes the

southern and eastern parts of Pakistan, almost all of India, Nepal, Sikkim, Bhutan, Bangladesh, and Sri Lanka (Ceylon).

Until the arrival of the Dutch in South Asia at the end of the sixteenth century, a sketchy knowledge of the flora of the subcontinent and Sri Lanka had been built up through the activities of local and foreign authorities, but without formal consolidation. Increased awareness abroad came with the work of Garcia da Orta (Garcia ab Horto, Garcia del Huerto) in the Portuguese colony of Goa in the mid-sixteenth century, the publication there, in 1563, of his *Coloquios dos simples*, and its broader dissemination after 1567 by Clusius in five successive Latin editions.

It is all but certain that the circumstances of the publication of the Latin editions of the *Coloquios*, and Clusius's eventual residence in Leiden, contributed greatly to Dutch interest in Asian natural history. With their effective economic control by the seventeenth century of the western part of the Peninsula and Sri Lanka, the means were to hand for serious studies of the flora. These were taken advantage of by Hendrik Rheede tot Draakestein and his assistants in the latter part of the seventeenth century. Their *Hortus malabaricus* (1678–1703), based on freshly gathered material as well as centuries-old manuscripts and other local botanical knowledge, effectively laid a foundation for *all* future botanical work in the superregion. These activities early spread to Sri Lanka through one of the assistants, Paul Hermann (who, like Clusius, eventually moved to Leiden). His collections were, however, published or written up by others, most notably in *Museum zeylanicum* (1717, 1726) and in Linnaeus's *Flora zeylanica* (1747). The last major Dutch contribution for the subcontinent was *Flora indica* (1768), a 'digest' by Nicolaus Burmann covering all the tropical East as well as southern Africa.

The Seven Years' War of 1755–63 and especially the victory by the British over the French at the Battle of Plassey in 1757 were preludes to a new cycle of botanical activities, especially in Bengal (where Calcutta became the English East India Company's regional seat of government and administration) and in the south. The most notable manifestation was in the work of the 'United Brotherhood' of Moravian pastors and their associates in southern India and Sri Lanka. One of its 'leaders' was Pastor Benjamin Heyne, whose plants were written up by A. W. Roth in Germany and published as *Novae plantarum species praesertim Indiae orientalis* (1821). Another was the East India Company physician–botanist William Roxburgh who after service in Madras moved on to Calcutta in 1793 as Superintendent of the then recently founded Botanic Garden, remaining until 1814. He was responsible for two major works: *Plants of the coast of Coromandel* (1795–1819, assisted by Sir Joseph Banks), and, most importantly, his posthumous *Flora indica* (1820–24, ed. Nathaniel Wallich, not completed; 2nd edn., 1832, by William Carey). The latter – with a reissue in 1874 – was to serve as the only complete flora of the subcontinent until full realization of *Flora of British India* in 1897. Roxburgh's flora was soon followed by *Prodromus florae peninsulae Indiae orientalis* (1834) by Robert Wight, another Company physician, and George Walker-Arnott in Edinburgh; this, however, was never completed.

The gradual extension of the Raj, its reorganization in the mid-nineteenth century, and the development of infrastructure was accompanied by a rapid and widespread expansion of botanical work, especially into Baluchistan, the North-West Frontier, the Himalaya, Kashmir, Assam, and other mountainous interior country as well as many parts of the Peninsula. Additional study centers came into existence and a great number of floristic contributions, large and small, were published in response to increasing demand. The most ambitious of these prior to 1870 was a three-volume *Handbook of the Indian flora* (1864–69) by Heber Drury, working initially in Trivandrum; although largely compiled (and limited to the subcontinent south of the Himalaya) it was the first such to be arranged on a natural system (but cross-referenced throughout to the sexual system).[7] A first attempt by Joseph D. Hooker and Thomas Thomson towards a modern, fully critical *Flora indica* was, however, abandoned after its first volume (1855) on account of official duties, a format yet too ambitious, and – significantly – the dissolution of the East India Company in 1858 following the Indian Mutiny.

Following establishment of formal government, however, botanical resources were reorganized and increased, making realization of a comprehensive, if more compact, Indian flora more feasible. The project was moreover adopted as a 'Kew flora' in 1863. With this basis, Hooker, initially with Thomson but eventually also with C. B. Clarke and other specialists, was able to summarize most available information in his *Flora of British India* (1872–97). Though regarded by its chief author as 'a hurried sweeping up of nearly a century of undigested materials, and . . . in no sense a flora like Bentham's [*Flora australiensis*]',[8] with over 11 000 species it belongs, like its peer, among the greatest floras ever written. It would be nearly a century before a necessary successor commenced publication; and, for historical purposes, it never will be superseded. Coverage by the *Flora* also extended to Burma, the Straits Settlements, the Malay States, and Ceylon. In the last-named, a separately governed British colony from 1818, it gathered together the work of, among others, the successive Peradeniya botanists Henry Moon, George Gardner, George Thwaites and Henry Trimen. With the *Flora* to hand, Trimen (and, after 1895, Hooker himself) realized a long-standing Peradeniya goal: a descriptive flora for the island. Accompanied by an atlas of 100 mostly locally drawn plates and in its day regarded as a model of its kind, this appeared as the five-volume *Handbook of the flora of Ceylon* (1893–1900; supplement, 1931, by A. H. G. Alston).

Forest botany also began serious development after 1860. Notable appointments to the expanding forestry services were J. L. Stewart, R. H. Beddome, Dietrich Brandis, Sulpiz Kurz and (somewhat later) J. S. Gamble. Early results were incorporated into regional forest floras for the south (1869–73, by Beddome), the northwest (1874, by Stewart and Brandis), and the then-British part of Myanmar (1877, by Kurz; see **895**). These and other contributions formed the basis for Brandis's final work, *Indian trees* (1906). Beddome also contributed a large-scale *Ferns of British India* (1866–68; supplement, 1876) and the still-current *Handbook*, noted below.

In 1890, the Botanical Survey of India was established on the initiative of George King, then Superintendent of the Royal Botanic Garden, Calcutta, and William T. Thistleton-Dyer, director of the Royal Botanic Gardens, Kew. Two goals were envisioned: coordination of the work of government botanical services in Bengal (at Sibpur), Madras (at Coimbatore), Bombay (at Pune) and the Northwest Provinces (at Saharanpur), and the examination of various remote and little-known areas (including, for example, northern Myanmar and the North-west Frontier in present-day Pakistan) to which British rule had then but recently been extended. Contributions were produced and for the most part published in the Survey's *Records* (initiated in 1893); annual reports also were issued.

Assistance was moreover given to the several regional floras, projects intended to follow and to complement Hooker's *Flora* and largely realized in the decades to 1930.[9] At the same time, there developed in the Indian Forest Service a parallel movement for the publication of additional forest floras and checklists; this was promoted by Brandis and more particularly by James S. Gamble and Robert S. Hole at Dehra Dun. Other important floristic contributions came from Fr. Ethelbert Blatter and his circle at St. Xavier's College, Bombay (among whom after World War I was Fr. Hermengild Santapau, a later director of the Botanical Survey).

It was the forest botanists (especially at Dehra Dun) and the Blatter circle who, along with interested individuals at some other tertiary institutions (including, for example, Ralph R. Stewart at Gordon College, Rawalpindi, S. R. Kashyap in Lahore, and H. Mooney in Bihar), were largely responsible for pursuing organized floristic work and education after the era of the regional floras. This followed a reorientation of the Botanical Survey in the decade to 1910 and, in that stagnant but ominous interwar period so well characterized by the novelist E. M. Forster and other writers, its virtual abolition through suspension of the directorship.[10] The Dehra Dun forest research center was ironically a beneficiary of this process through its acquisition in 1908 of the Saharanpur herbarium, a basis for Duthie's flora of the Upper Gangetic Plains.[11] The universities and institutes established or further developed after 1900 did not on the whole contribute much to floristic work; rather they followed trends in Europe and North America and focused to a great extent on 'general botany', including morphology, anatomy, embryology, cytology, genetics, paleobotany, ecology, cryptogams, and (notably) fungi as well as all kinds of applied botany.[12]

After the political changes of 1947–50, a recognition of the increasing obsolescence of available floristic works and the need for a renewal of floristic and systematic botany was gradually perceived in the new nations of South Asia. In India, the Botanical Survey was revived in 1947 with reinstatement of the directorship and, more importantly, reorganized in 1954 as a statutory body. In addition to the already-affiliated Industrial Section of the Indian Museum in Calcutta, it came in time to encompass the Central National Herbarium, seven new or revived regional centers and (from 1963) the Botanic Garden at Sibpur.[13] A new enumeration of Indian plants by Sunil K. Mukerjee was begun in 1959 in the BSI *Bulletin* but lapsed in 1962. A realization that many parts of the country were undercollected led in the 1960s to the institution of grants for field studies at district level. In Pakistan, with few systematics resources at independence, recent progress has been quite evident. Much effort went into establishment of a national herbarium and projection of a national flora. Ralph Stewart's checklist was published in 1972 and the *Flora of West Pakistan* (now *Flora of Pakistan*; see 793), begun in 1970, is now far advanced, with over 195 parts now available. In Sri Lanka, a Ceylon Flora project was organized in the late 1960s under sponsorship from Sri Lankan institutions and the Smithsonian Institution; publication of the results as *A revised handbook to the flora of Ceylon* began in 1973. This was, however, abandoned after two installments and a new series of volumes under the same title instituted in 1980, with 12 published by 1998. A *Flora of Bangladesh* began publication the year after its nation's independence and has continued to the present.

All these new productions, however, have not replaced the basic superregional treatises: Beddome's *Handbook to the ferns of British India, Ceylon, and the Malay Peninsula* (1883; 2nd edn., 1892), the already-mentioned *Indian trees* by Brandis, and, most importantly, *Flora of British India* itself (apart from revisions of certain families). Additions and corrections to the last-named were published by Calder *et al.* (1926), Razi (1959) and Nayar and Ramamurthy (1976); a consolidation of published records appeared in *Name changes in flowering plants of India and adjacent regions* (1987) by S. S. R. Bennet. The grasses (except for the Bambuseae) were wholly revised by N. L. Bor (1960). For pteridophytes, a 'companion' to Beddome's *Handbook* by Nayar and Kaur appeared in 1972 (with a reissue in 1974) and a revised checklist by R. D. Dixit was published in 1984. In spite of their nomenclature, based on the now-obsolete 'Kew Rule' and other nineteenth-century practices, the old national and regional works continue to be widely used and have all been reprinted within recent decades.

Since World War II, a considerable amount of new field work has been carried out throughout the superregion, much of it under the Botanical Survey of India; however, relatively little specialist collecting has been done and large forest trees reportedly have often been neglected. Notable field projects have been mounted in Sri Lanka, the Himalaya, Karnataka, Tamil Nadu, northeastern India, and Pakistan among others. But 'many densely vegetated areas are yet to be explored', and more effort in this and in the writing of floras and related works is required. The considerable collections now to hand will serve as a partial basis for a much-needed new round of subregional, state, and district as well as national floras to supplant the many works published in the last half-century of the Raj; more details appear under the regional headings.

Through 1977 the work of the Botanical Survey had been directed towards eventual production of new state or regional floras to replace those of the early part of the century as well as to fill gaps; less emphasis was given towards individual districts or groups thereof.[14] In that year a general symposium was held at Sibpur for assessment of progress and needs. This led to introduction of a more comprehensive programme of floristic documentation as well as greater coordination of efforts inside and outside the Survey. Four 'Flora of India' series were established: (1) a national flora, to appear serially in fascicles; (2) state flora 'analyses' (enumerations); (3) district (or local) floras, with keys, descriptions, etc.; and (4) works on non-vascular cryptogams and special topics as well as monographs.[15] Works in all series have since been published; the regional and local floras are, however, of uneven quality and, very importantly, the national flora was making but slow progress.[16]

A further review was accordingly held in 1984, with two key decisions being made respecting all-Indian works. A national checklist, to cover some 17 000 species, was initiated at Howrah; two parts have now been published, covering monocotyledons and Asteraceae (the pteridophytes having been previously covered in similar fashion by Dixit). At the same time the style of *Flora of India* was changed from semi-monographic fascicles to a concise flora in systematically arranged volumes of 400–600 species. From about 1986, family treatments were assigned to BSI staff in Howrah and the regional circles as well as elsewhere, with any precursory treatments to be published in series 4 of 'Flora of India'. In 1993 there appeared the first three volumes of the new *Flora of India*, appropriately commemorating the 200th anniversary of the initiation under Roxburgh of systematic botanical research in Calcutta. By 1998 seven volumes of a projected 25–30 had appeared, one of them comprising a general introduction.[17] At district level, an 'All-India

Co-ordinated Project on District Flora' was initiated in 1982 as a collaborative undertaking of the Survey and tertiary institutions with an ultimate objective of covering some 100 districts in various parts of the country as well as strengthening plant systematics. A substantial number of these district floras – with varying critical standards – have by the time of writing been published. Monography for its own sake, however, continues to be undervalued.[18]

The long period over which the present stock of standard floras in South Asia has appeared has also been one of many boundary changes, especially within India as reorganization of internal boundaries took place according to linguistic, ethnic and other criteria. The units adopted here represent a compromise between the political mosaic of the last decade of the Raj (circa 1937) and the modern states, provinces and territories in the successor nations.[19]

Progress

For early and recent history (largely referring to India), see particularly H. SANTAPAU, 1958. *History of botanical researches in India, Burma and Ceylon*, 2: *Systematic botany of angiosperms*. vi, 77 pp. Bangalore (for the Indian Botanical Society); P. MAHESHWARI and R. N. KAPIL, 1963. *Fifty years of science in India: progress of botany*. viii, 178 pp., illus., map. Calcutta; I. H. BURKILL, 1965. *Chapters on the history of botany in India*. xi, 245 pp., maps. Delhi; and R. DESMOND, 1992. *The European discovery of the Indian flora*. xii, 355 pp., halftones, 32 col. pls. Oxford: Oxford University Press (with the Royal Botanic Gardens, Kew). Briefer accounts (but all inclusive of the twentieth century) include S. P. AGHARKAR, 1938. Progress of botany during the last twenty-five years. In B. Prasad (ed.), *Progress of science in India*, pp. 742–767. Calcutta; K. BISWAS, 1943. Systematic and taxonomic studies on the flora of India and Burma. In *Proceedings 30th Indian Science Congress (Calcutta)*, 2: 101–152, figs. 1–11. Calcutta; H. SANTAPAU, 1962. The present state of taxonomy and floristics in India after independence. *Bull. Bot. Surv. India* 4: 209–216; and M. A. RAU, 1994. Plant exploration in India and floras. In B. M. Johri (ed.), *Botany in India: history and progress*, 1: 17–41. Lebanon, N.H.: Science Publishers. Other surveys of progress are found in: P. LEGRIS, 1974. Vegetation and floristic composition of humid tropical continental Asia. In UNESCO, *Natural resources of humid tropical Asia*, pp. 217–238 (Natural resources research 12).

Paris; the general survey of tropical floristics by Prance (1978; see **General references**); and country surveys by K. Kendrick for India (pp. 133–140) and Sri Lanka (pp. 141–145) in *Floristic inventory of Tropical Countries* (Campbell and Hammond, 1989; see **General references**).[20] Pteridophyte studies have been reviewed in C. R. FRASER-JENKINS, 1984. An introduction to fern genera of the Indian subcontinent. *Bull. British Mus. (Nat. Hist.), Bot.* **12**(2): 37–56.

Hooker's Flora indica *(with T. Thomson) and* Flora of British India*:* Of particular interest here is R. DESMOND, 1993. Sir Joseph Hooker and India. *Linnean* **9**(1): 27–49, illus. Passing references to these works are also made in *idem*, 1999. *Sir Joseph Dalton Hooker: traveller and plant collector.* 286 pp., illus., maps. Woodbridge, Suffolk: Antique Collectors' Club/Royal Botanic Gardens, Kew.

Botanical Survey of India: Early history and the period following its reorganization are dealt with in H. SANTAPAU, 1964. The Botanical Survey of India. *Sci. Cult.* **30**: 2–11. The state of knowledge by the mid-1970s was surveyed extensively although unevenly in the many contributions in BOTANICAL SURVEY OF INDIA, 1977. *All-India symposium on floristic studies in India: present status and future strategies* (Howrah, 1977): *Abstracts.* 52 pp. Howrah, West Bengal; and *idem*, 1977 (1979). *All-India symposium . . . Proceedings.* iv, 336, 6 pp. (Bull. Bot. Surv. India 19). Howrah. A book-length account of the Survey is R. K. BASAK, 1982. *Botanical Survey of India.* 300 pp. Howrah. The Survey also produces annual reports as well as publicity booklets from time to time.

Bibliographies. General bibliographies as for Division 8.

Supraregional bibliographies
In addition to the following, notice should be taken of M. P. NAYAR, 1984– . *Key works to the taxonomy of flowering plants of India.* Vols. 1–5. Howrah: Botanical Survey of India. This annotated work is systematically arranged by families and genera. It was not continued past Polygonaceae; however, the resulting gap is partly covered in the second section of the bibliography by SANTAPAU (see below).

BLATTER, E., 1909. A bibliography of the botany of British India and Ceylon. *J. Bombay Nat. Hist. Soc.* **20**: lxxix–clxxvi. [Classified, unannotated list of 1508 titles with generic and subject index. The titles are arranged in the first instance under general and regional headings and then by author. For continuation, see SANTAPAU below.]

NARAYANASWAMI, V., 1961–65. Indian botany, 1–2. In *A bibliography of Indology*, 2(1–2). Calcutta: National Library of India. [5770 numbered titles through 1958, arranged by author and briefly annotated. A projected third part, containing classified indices by subjects and regions, has not been seen.]

NAYAR, M. P. and S. G. GIRI, 1988. *Key works of floristics of India.* Vol. 1. Illus. (part col.). Howrah. [Geographically arranged classified listing; individual titles unannotated but each section is preceded by a floristic and geographical summary with notes on local species, reserves, etc. The first volume covers India as a whole and states and territories from A through Madhya Pradesh; addenda, pp. 348–350.]

SANTAPAU, H., 1952. Contributions to the bibliography of Indian botany. *J. Bombay Nat. Hist. Soc.* **50**: 520–548; **51**: 205–259. [In two main parts; general works and regional floras in the first (**50**: 521–540); works on specific families in the second. Titles are unannotated, and any relating specifically to Pakistan, Bangladesh, Sri Lanka and Myanmar are omitted.]

STEWART, R. R., 1956. A bibliography of the flowering plants of West Pakistan and Kashmir. *Biologia* (Lahore) 2(2): 221–230. [Unannotated list of the most important literature, arranged by authors. Among sources, special mention is made of the bibliographies of BLATTER and SANTAPAU.]

Indices. General indices as for Division 8.

Supraregional indices
STEENIS, C. G. G. J. VAN *et al.* (eds.), 1948– . *Flora Malesiana Bulletin*, 1– . Leiden: Flora Malesiana Foundation (later Rijksherbarium and Rijksherbarium/Hortus Botanicus). (Earlier issues mimeographed.) [Variously issued annually or bi-annually, this contains a substantial bibliographic section also very thoroughly covering South Asian literature. A fuller description is given under Superregion 91–93.]

810–40

Superregion in general (including all-India works)

Hooker's *Flora of British India* remains the leading work on the superregion, although time has rendered it incomplete and keys and nomenclature are now obsolete. A modern treatment of non-bambusoid grasses (Bor, 1960) has succeeded that in volume 7 of the original work and from time to time revisions of other families as well as individual genera have appeared independently or in journals. The first of several suites of additions and corrections appeared in 1926, with all recently consolidated in one volume (Naithani, 1990). Two attempts at an enumeration of the Indian flora have been undertaken, firstly by S. K. Mukerjee in *Enumeration of Indian flowering plants* (1959–62, in *Bull. Bot. Surv. India* 1: 138–141; 2: 99–107, 293–297; 3: 99–101, 351–355; 4: 39–47) and then by the Botanical Survey in the 1980s. The former, documented in some detail, was left incomplete after covering only Ranunculaceae through Annonaceae (on the Bentham and Hooker system); of the latter, to date only monocotyledons and Asteraceae have so far been published. General manuals are also available for the ferns (Beddome, 1892, with supplement in 1972) and trees (Brandis, 1906). To these classics have been added a dictionary of genera for India (Santapau and Henry, 1973) and a forest botany manual (Bor, 1953).

Prior to *Flora of British India*, Hooker had commenced, in collaboration with Thomas Thomson, a *Flora indica*. Its first volume (1855) – all that was ever published – contains a rather complete (and still useful) general introduction to the Indian flora, along with a classical philosophical chapter on the principles and practice of taxonomy. *Flora indica* was itself preceded by a work of the same title by William Roxburgh (1820, ed. N. Wallich, not completed; 1832, ed. S. Carey; reissued 1874 in 1 vol. with additional preface by C. B. Clarke). Although arranged on the Linnaean system, this work accounted for most of the more frequently encountered plants of the subcontinent and as such has had no successors.

At the end of 1978 a serial, *Fascicles of the flora of India*, commenced publication; it was issued in fascicles like the *Flora of Pakistan* and *Flora of Bangladesh*.[21]

As already noted, however, in 1984 it was de-emphasized in favor of a volume-based *Flora of India*; given a customarily long lag in publication, however, new fascicles have continued to appear. The first three volumes in the new format appeared in 1993. Floras for Sri Lanka and Bhutan are in progress or nearly complete, and a critical enumeration for Nepal has also been published as a prelude to a more definitive flora.

Dictionary

SANTAPAU, H. and A. N. HENRY (assisted by B. ROY and P. BASU), 1973. *A dictionary of the flowering plants in India*. vii, 198 pp. New Delhi: Council of Scientific and Industrial Research, India. (Reprinted 1994.) [Annotated alphabetical checklist of 2900 generic names (both accepted and synonymous) incorporating their families, essential references, brief descriptions, range, approximate number of species overall and in India, and any outstanding forms; index to vernacular and other common names.]

BOTANICAL SURVEY OF INDIA, 1978–96. *Fascicles of flora of India*. Fasc. 1–22. Illus. Howrah, West Bengal: Botanical Survey of India.

A serially published documentary descriptive flora; within each fascicle (comprising one or more families) are keys to all taxa, synonymy (with references and citations), typification, 6–9 line descriptions, vernacular names, phenology, general distribution and altitudinal zonation, ecology, pollen, karyotypes, and brief commentary (including notes on uses, properties, etc.) along with illustrations of representative species; index to all botanical names at end. [Families have been published when ready, without regard to sequence. Treatments of species are more detailed than in *Flora of India*, particularly in installments published after 1984 when the *Fascicles* became effectively a series of national monographs.][22]

HOOKER, J. D. *et al.*, 1872–97. *Flora of British India*. 7 vols. London: Reeve. (Reprinted in covers entitled 'Flora of India' at various dates to 1961, Brook nr. Ashford, Kent: Reeve; 1973, 1978, Dehra Dun: Bishen Singh Mahendra Pal Singh. A Peking reprint of about 1956 is also known.) Supplemented by N. L. BOR, 1960. *The grasses of India, Burma, Ceylon and Pakistan, excluding Bambuseae*. xviii, 767 pp., illus. Oxford: Pergamon. (Reprinted with additions and corrections, 1973, Königstein/Ts., Germany: Koeltz.); S. S. R. BENNET, 1987. *Name changes in flowering plants of India and adjacent regions*. xvi, 772 pp. Dehra Dun: Triseas Publishers; and H. B. NAITHANI, 1990. *Flowering plants of India, Nepal and Bhutan (not recorded in Sir J. D. Hooker's* Flora of British India*).* viii, 711 pp. Dehra Dun: Surya.

The *Flora* is a concise, briefly descriptive account of seed plants with modified synoptic keys to genera and species, synonymy (with references and citation of key literature), generalized indication of internal range (with some localities and *exsiccatae*), summary of extralimital distribution, taxonomic commentary, and notes on habitat, special features, etc.; indices at end of each volume (with a cumulative index to botanical names in vol. 7). Treatments of some families are by specialists. – Bor's *Grasses* is a successor, with descriptions, analytical keys, distributions and notes, to the corresponding account in volume 7 of the *Flora*; Bennet's *Name changes* summarizes nomenclatural or taxonomic changes since 1897 (and for non-bambusoid grasses, 1960) for 5175 taxa (about a third of the Indian total); and Naithani's *Flowering plants* is a systematically concordant enumeration with synonymy, references and indication of states and territories where reported, followed by late additions (pp. 517–528, 695–711) and a complete index (a statistical summary is, however, wanting). [The *Flora* encompasses the entire superregion as here delimited along with Myanmar, peninsular Thailand (at the isthmus of Kra), and the then-Straits Settlements (as well as other parts of the Malay Peninsula). Coverage of Myanmar is, however, reasonably adequate only for its southern part; for southern Thailand and the Malay Peninsula coverage is rather sketchy and moreover has – save for historical purposes – long been succeeded by other works. The preface to vol. 1 is very brief; for a more complete introduction to the flora, as well as a folding map and an explanation of the geographical divisions used, the earlier *Flora indica* by Hooker and T. Thomson should be consulted.][23]

RAO, R. R. *et al.*, 1988. *Florae indicae enumeratio: Asteraceae.* viii, 119 pp. Calcutta: BSI; and S. KARTHIKEYAN, S. K. JAIN, M. P. NAYAR and M. SANJAPPA, 1989. *Florae indicae enumeratio: Monocotyledoneae.* [viii], 435 pp. Calcutta. (Both works in Flora of India, series 4.)

Alphabetically arranged checklists with references, citations of key works, synonymy and distribution; indices to all names at end. [The monocotyledon list covers 4081 species and that on Asteraceae, 1052 (mostly specific) taxa. A few novelties are concisely described, for example in *Bambusa*. Issue of the remaining dicotyledons is projected.][24]

SHARMA, B. D. *et al.*, 1993– . *Flora of India*. Vols. 1– . Illus., col. pls. Calcutta: Botanical Survey of India.

Concisely descriptive flora with keys, limited synonymy, selected vernacular names, references and key citations, and indication of phenology, geographical status, local and extralimital range, occasional notes, figures of selected species, and under family and generic headings key monographic and revisionary literature; indices to all botanical and cited vernacular names at end of each volume. [As of 1998 seven volumes had appeared, the general introductory volume in 1998, vols. 1–3 in 1993, 12–13 in 1995, and 4 in 1998; coverage presently extends from Ranunculaceae (1) through Dichapetalaceae (44) and the Asteraceae (88) in the Bentham and Hooker sequence. Volume 5 appeared in 1999 and additional volumes are in preparation.][25]

Woody plants

BOR, N. L., 1953. *Manual of Indian forest botany.* xv, 441 pp., 31 pls. Bombay: Oxford University Press (India).

Elementary descriptive treatment (designed as a students' manual) of the more important trees, shrubs and other woody plants of India; includes keys to families, genera and (in larger genera) species, limited synonymy (without references or citations), and notes on local range, habitat, life-form, special botanical features, wood properties and uses; index to all botanical names.

BRANDIS, D., 1906. *Indian trees.* xxxii, 767 pp., 201 text-figs. London: Archibald, Constable. (Reprinted 1907, 1908, 1911, London; 1971, 1978, Dehra Dun: Bishen Singh Mahendra Pal Singh.)

Descriptive treatment of trees, shrubs, woody climbers, bamboos and palms of the former Indian Empire, Myanmar and Sri Lanka; includes synonymy (with references and citations), local and English vernacular names, generalized indication of internal and extralimital distribution, taxonomic commentary, and notes on habitat, phenology, special features, properties and uses; addenda and index to all botanical and vernacular names at end. Less important and lesser-known species are featured in smaller type. The introductory section includes chapters on forest districts and forest vegetation, sources for the work, and special botanical features, together with a synopsis of families.

Pteridophytes

Beddome's *Handbook*, with revisions, has been a standard on the subcontinent for more than a century; a 'companion' was published in 1972. The value of this work was acknowledged by Hooker who accordingly omitted pteridophytes from *Flora of British India* and *Handbook of the flora of Ceylon*.

Beddome's earlier, larger-format *pteridologiae* – *Ferns of South India* (1863), *Ferns of British India* (1866–68) and *Supplement* (1876) – also continue to be appreciated, as recognized by the *Nomenclatural guide* of S. Chandra and S. Kaur (1987). Complementary to Beddome's works were *A review of the ferns of Northern India* by C. B. Clarke (1880, in *Trans. Linn. Soc.*, II, Bot. 1) and, somewhat later, *The ferns of Northwestern India* by C. W. Hope (1899–1904, in *J. Bombay Nat. Hist. Soc.* 12–15). That a new handbook is now required, however, was highlighted by the census of Dixit (1984; see below); therein are listed over 1000 species, more than twice Beddome's 466 for a larger area.[26]

In addition to the works cited below, note should be taken of R. D. DIXIT and J. N. VOHRA, 1984. *A dictionary of the pteridophytes of India*. 48 pp., illus. Howrah: BSI (Flora of India, series 4), and C. R. FRASER-JENKINS, 1984. An introduction to fern genera of the Indian subcontinent. *Bull. British Mus. (Nat. Hist.), Bot.* 12(2): 37–76.

BEDDOME, R. H., 1892. *Handbook to the ferns of British India, Ceylon and the Malay Peninsula*. 2nd edn., with supplement. xiv, 500, 110 pp., 299 text-figs., frontisp. Calcutta: Thacker, Spink. (Reprinted 1969, Delhi: Today and Tomorrow's.) (1st edn., 1883.)

Briefly descriptive, illustrated fern flora with synoptic keys to genera and species, citations of major sources, generalized indication of internal range and elevations, more or less extensive taxonomic commentary, and index to all botanical names. The supplementary section in the 1892 edition comprises additions and alterations to the otherwise unchanged 1883 text. For further additions and changes as well as revised nomenclature, see: B. K. NAYAR and S. KAUR, 1972. *Companion to R. H. Beddome's 'Handbook to the ferns of British India, Ceylon, and the Malay Peninsula'*. 196, vii pp. New Delhi: Pama Primlane. (Reissued in larger format and re-paged with additional introductory matter, 1974, New Delhi: Pama Primlane, and Königstein/Ts., Germany: Koeltz. iv, 244 pp.); and S. CHANDRA and S. KAUR, 1987. *A nomenclatural guide to R. H. Beddome's* Ferns of South India *and* Ferns of British India. x, 139 pp. New Delhi: Today and Tomorrow's.

DIXIT, R. D., 1984. *A census of the Indian pteridophytes*. [iv], 6, iv, 177 pp. Howrah: BSI. (Flora of India, series 4.)

Systematic enumeration, with synonymy, references, citations, Indian and extra-Indian distribution, and occasional notes; index to genera at end. Arrangement of families and genera is after Pichi-Sermolli, and delimitation of genera after R. C. Ching.[27]

Region 81

Indus Basin and valley (Pakistan); northwest and central India

Within this region are the Pakistani provinces of Sind and [West] Punjab along with the Indian states (and territories) of [East] Punjab, Haryana, Chandigarh, Delhi, Uttar Pradesh (with the exception of the Himalayan hill and mountain tracts), Rajasthan, and Madhya Pradesh (the last-named inclusive of the former Central India Agency as well as parts of the Central Provinces as constituted before 1947). Generally speaking, it corresponds to 'North-West and Central India' of the Raj, except that all the hill and mountain tracts in the Himalaya and beyond the Indus have been excluded (for these, see Regions 79 and 84).

Botanical exploration in the region has been fairly extensive but varying in thoroughness; lesser-known areas include the arid lands centering on the Indian desert and parts of the central plateau and other hilly upland tracts. Early general accounts include *A catalogue of the plants of the Punjab and Sindh* (1869) by J. E. T. Aitchison and *Punjab Plants* (1869) by J. L. Stewart (reprinted 1977). Stewart later undertook, and Dietrich Brandis completed, *Forest flora of North-West and Central India* (1874). The 'regional floras' movement of the early twentieth century yielded *Plants of the Punjab* (1916) by C. J. Bamber and *Flora of the upper Gangetic Plain* (1903–22, unfinished) by J. L. Duthie at Saharanpur (to which were added, decades later, a supplement as well as the grasses by M. B. Raizada and S. K. Jain and the pteridophytes by M. N. Chowdhury). Two editions of a forest flora of the Punjab were produced around the same time by R. N. Parker, and in the Central Provinces handy-sized forest manuals were produced by H. H. Haines and D. O. Witt. Afterwards, few state or provincial floras or forest manuals were published until the 1960s.

A revival of floristic work began gradually in the 1950s, in India under the auspices of the reorganized Botanical Survey of India which established its Northern Circle (BSI-NC) at Dehra Dun and Central Circle (BSI-CC) at Allahabad, and in Pakistan at

Gordon College in Rawalpindi and a new National Herbarium in Islamabad. Work in India was further aided by grants for district-level research. Since then floras of varying quality have been, in addition to Pakistan in general (**793**: *Flora of (West) Pakistan*, commenced in 1969), published for parts of West Punjab (Bhopal and Chaudhuri, 1977; see **793**), the (East) Punjab plains (Nair, 1978), Delhi (Maheshwari, 1963–66), Uttar Pradesh (the already-mentioned additions to Duthie's flora), western Rajasthan (Puri, 1964), and the Indian Desert (Bhandari, 1978, with a revision in 1990; see **805**). Several district floras have also been published, both through BSI and independently. BSI-CC paid particular attention to Rajasthan and Madhya Pradesh, for most of which no regional floristic coverage existed. New state floras for both have now been completed or are in progress (respectively by Shetty and Singh, 1988–93 and Verma *et al.*, 1993–).

Bibliographies. General and supraregional bibliographies as for Division 8 and Superregion 81–84.

Indices. General and supraregional indices as for Division 8 and Superregion 81–84.

810

Region in general

See also **793** (all works); **810–40** (all works).

Woody plants

STEWART, J. L. and D. BRANDIS, 1874. *The forest flora of North-West and Central India.* 2 vols. xxxi, 608 pp.; atlas of 70 pls. (by W. Fitch). London: Allen. (Reprinted 1972, Dehra Dun: Bishen Singh Mahendra Pal Singh.)

Briefly descriptive treatment of trees and shrubs with accompanying atlas; includes non-dichotomous keys to genera and species, synonymy (with references), vernacular names, generalized indication of local and extralimital range, taxonomic commentary, and extensive notes on habitat, life-form, phenology, associates, timber, uses, etc.; indices to all botanical, English and vernacular names at end. The introductory section includes a general description of the woody flora of the region and its composition, a short glossary, and a list of references.

811

Sind (Pakistan)

See also **793** (all works); **822** (COOKE; TALBOT). – Area: 140 914 km². Vascular flora: no data. Prior to 1947 Sind formed part of the Bombay Presidency. Karachi and Hyderabad are the main population centers, and for these and the Indus Delta between them local florulas are available in addition to the provincial flora by Sabnis (1923–24). Earlier, Sind had been documented in *A catalogue of the plants of the Punjab and Sindh* (1869; reprinted 1882) by J. E. T. Aitchison.

SABNIS, T. S., 1923–24. The flora of Sind. *J. Indian Bot. Soc.* **3**: 151–153, 178–180, 204–206, 227–232, 277–284; **4**: 25–27, 50–70, 101–115, 134–148.

Systematic enumeration of seed plants; includes references, detailed indication of localities with *exsiccatae* and other sources, and general summary of overall range; list of references (in introductory section). [Forms an areal supplement to Cooke's *Flora of the Presidency of Bombay* (**822**), which covers the province.]

Partial works

BLATTER, E., C. C. McCANN and T. S. SABNIS, 1929. *The flora of the Indus Delta.* vi, 172 pp., 140 text-figs, 50 half-tones. Bombay: Indian Botanical Society.

This mainly ecological and phytogeographical work includes (pp. 1–38) a systematic enumeration of vascular plants with synonymy, localities and citations.

JAFRI, S. M. H., 1966. *The flora of Karachi (coastal West Pakistan).* iv, 375 pp., 345 text-figs., 3 pls. Karachi: The Book Corporation.

Students' manual-flora of vascular plants; includes keys, descriptions, illustrations, fairly extensive synonymy, relatively detailed indication of local range, summary of extralimital distribution, taxonomic commentary, and notes on habitat, uses, etc.; list of references and index to all botanical names at end.

812

(West) Punjab and Islamabad (Pakistan)

See also **793** (all works); **813** (all works). – Area: 206 241 km² (of which Islamabad Federal Territory accounts for 907 km²). Following Partition, the large princely state of Bahawalpur was incorporated into West Punjab; at a later date, the Federal Territory was excised on establishment of the new national capital near Rawalpindi. Floras pertaining to the whole of the Punjab and adjacent tracts published before 1947 are described under **813**.

District and local works

Among the several population centers, florulas are available for the Lahore and Rawalpindi districts, both centers of higher education.

KASHYAP, S. R. and A. C. JOSHI, 1936. *Lahore district flora.* 285 pp., 218 text-figs. Lahore: University of the Punjab. (Reprinted 1980.)

Illustrated, briefly descriptive students' manual-flora; includes keys to genera and species, vernacular names, summary of local range, and notes on habitat, phenology, etc.; glossary and index to generic, family and vernacular names at end.

STEWART, R. R., [1958]. *The flora of Rawalpindi district, West Pakistan.* x, 163 pp. Rawalpindi. (Originally published 1957–58 in installments in *Pakistan Journal of Forestry.*)

Systematic enumeration of vascular plants; includes synonymy, vernacular names, summary of local range (with some localities), taxonomic commentary, and notes on habitat, biology, special features, cultivation, uses, etc.; index to family and generic names at end. [For additions and corrections, see *idem*, 1961. Additions and corrections to the flora of Rawalpindi district, West Pakistan. *Pakistan J. Forest.* 11: 51–63, 222–295.]

813

Former Punjab

Area: Punjab, 300 939 km²; Delhi, 1497 km². Vascular flora: no data. – The region was simultane-

ously first documented by J. E. T. Aitchison and J. L. Stewart, the former in *A catalogue of the plants of the Punjab and Sindh* (1869; reprinted 1882), the latter in *Punjab Plants* (1869; reprinted 1977). Stewart's work emphasized useful plants, a reflection of his profession as a forestry official (he was also co-author with Dietrich Brandis of the text of *Forest flora of North-West and Central India*; see **810**). Its coverage ranged well beyond Punjab to cover Garhwal and Kumaon in the Himalaya and to Jammu, Kashmir, and the North-West Frontier, and in its introductory chapters is a discussion of the flora and vegetation and its phytogeographical relationships. Trees and shrubs were further documented in *A catalogue of the trees and shrubs of the Punjab* (1901) by E. M. Coventry.

BAMBER, C. J., 1916. *Plants of the Punjab.* iii, 652, xxviii pp., 6 pls. Lahore: Superintendent of Government Printing, Punjab. (Reprinted 1976, Dehra Dun: Bishen Singh Mahendra Pal Singh.) Complemented by T. S. SABNIS, 1940–41. A contribution to the flora of the Punjab plains and the associated hill regions. *J. Bombay Nat. Hist. Soc.* 42: 124–149, 342–379, 533–586. (Reprinted 1986.)

Plants of the Punjab is a descriptive manual of seed plants (the nomenclature mostly after *Flora of British India*) with artificial 'synoptic' keys to genera and groups of species, Urdu and English vernacular names, generalized indication of local range (including some specific records), and occasional notes on uses, special features, etc.; complete index to botanical and vernacular names at end. A brief glossary is incorporated in the introductory section. [Encompasses Jammu, Kashmir, and the North-West Frontier Province (of Pakistan) as well as Bahawalpur and the East and West Punjab. The work was designed with non-specialists in mind and does not follow a usual flora format; arrangement of the plants follows folk conventions. Grasses are not included.]

Woody plants

PARKER, R. N., 1924. *A forest flora of the Punjab with Hazara and Delhi.* 2nd edn. 591 pp. Lahore: Superintendent of Government Printing, Punjab. (Reprinted as '3rd edn.', 1956; 2nd reprint, 1973, Dehra Dun: Bishen Singh Mahendra Pal Singh. 1st edn., 1918.)

Briefly descriptive manual of trees and shrubs, with keys to genera and species, limited synonymy, references, vernacular names, generalized indication of local range and elevation, taxonomic commentary, and notes on habitat, phenology, special features, uses, etc.; complete index at end.

An introductory section has a glossary and an account of the general features of the flora. [For additions, corrections and commentary, see R. R. STEWART, 1957. Parker's *Forest Flora*. *Pakistan J. Forest*. 7: 8–19.]²⁸

814

(East) Punjab, Haryana, Chandigarh and Delhi (India)

Area: Punjab, 50 362 km²; Haryana, 44 222 km²; Chandigarh Territory, 114 km²; Delhi, 1497 km². Vascular flora: Punjab plains, 807 native and naturalized species (Nair, 1978; see below); Punjab, 1879 native, naturalized and cultivated species and 43 additional infraspecific taxa (Sharma, 1990); Delhi, 531 native and naturalized species (Maheshwari, 1963). No figures for Chandigarh or Haryana have been seen. – The Himalayan hill and mountain tracts formerly part of Punjab, including Shimla, are now Himachal Pradesh (**842**) and thus not covered here. The still-standard pre-Partition general works on the Punjab are described under **813**.

The manual by Nair (1978) is the only modern work purporting to cover most of the present area but according to Sharma (1990) it is incomplete. Sharma's nomenclator is a first attempt at documentation of the present state of Punjab. He has also contributed papers on Chandigarh in *Research Bulletin of the Panjab University*. No separate flora or list for Haryana has been published.

NAIR, N. C., 1978. *Flora of the Punjab Plains*. xx, 326 pp. (Rec. Bot. Surv. India 21(1)). Howrah, West Bengal.

Briefly descriptive manual-flora; includes keys to all taxa, synonymy (with references and citations), vernacular names, local range with representative *exsiccatae*, some taxonomic remarks, and notes on phenology, habitats, associates, frequency, uses, etc.; sources (for naturalized species); indices to botanical and vernacular names. An introductory section deals with physical features, geology, soils, climate, earlier botanical work and sources for the present flora, and floristics and vegetation; principal references are also listed. [Covers an area from 28° 50′ N to 32° 50′ N and 73° 30′ E to 78° E, i.e., from Delhi northwards.]

Partial work (except Delhi)

SHARMA, M., 1990. *Punjab plants: check-list*. [iii], 115 pp. Dehra Dun: Bishen Singh Mahendra Pal Singh.

Systematically arranged nomenclator of vascular plants with essential synonymy; bibliography (pp. 104–111) and index to families. The introduction includes a floristic summary. [Based largely on the author's own collections and regional herbarium resources. The author's data show recruitment of exotics as continuing. Cultivated useful and ornamental species, not separately indicated in the list, are numerous.]

Partial works: Delhi

Delhi lies between the Punjab plains and the Upper Ganges Plain; as the Red Fort and city center are on the right bank of the Jamuna the territory is included here with the former. The state was created in 1912 following transfer of the central government from Calcutta.

MAHESHWARI, J. K., 1963. *The flora of Delhi*. viii, 447 pp. New Delhi: Council for Scientific and Industrial Research, India. Complemented by *idem*, 1966. *Illustrations of the flora of Delhi*. xx, 282 pp., 278 text-figs. New Delhi.

The *Flora* is a briefly descriptive account of seed plants featuring keys, synonymy (with references and citations), vernacular names, general summary of local range with some *exsiccatae*, vernacular names, and notes on habitat, phenology, uses, special features, etc., and bibliography and complete index. Its introductory part covers physical features, climate, soils, earlier botanical work, floristics and vegetation. The *Illustrations* comprise an atlas of full-page figures preceded by an introduction to illustrated works on Indian flora (pp. vii–viii) and nomenclatural corrections to the text volume; index at end.

815

Upper Gangetic Plain (including Uttar Pradesh)

Area: *c.* 480 000 km² (excluding the hill districts in **843** above 700 m); Uttar Pradesh alone (including the hill districts in **843**), 294 413 km². Vascular flora: no data. – The Himalayan hill and mountain tracts of Uttar Pradesh (or Uttaranchal), including the Dehra Dun district, are here removed to **843**.²⁹

The standard flora of Duthie (1903–29) covers the lowland and colline parts of modern Uttar Pradesh and, in addition, the hilly parts of present-day northern Madhya Pradesh to the south (**816**) and the dry lands of eastern Rajasthan (**817**). It was designed as a subregional flora for the former United Provinces (and Oudh), extending to the former princely states of Gwalior and the Central India Agency (now part of Madhya Pradesh (**819**) but from 1948 to 1956 the separate states of Bhopal, Madhya Bharat and Vindhya Pradesh, or **816**) and the eastern part of the former Rajputana Agency of princely states (present-day Rajasthan). The Narmada River forms the approximate southern limit for the work. As left by its author, it lacked an index as well as the grasses and pteridophytes; these were added respectively by Jain (1952), Raizada (1961–83) and Chowdhury (1973). A revised checklist was published by M. A. Rau in 1968, and a consolidated supplement by Raizada (1976). There is also a forest flora of several northerly and southerly peripheral areas by P. C. Kanjilal (1933). Floristic work in the region has been led in recent years by the Forest Research Institute, Dehra Dun, as well as the Botanical Survey of India with its Northern Circle (BSI-NC), established there in 1956, and Central Circle (BSI-CC, established in Allahabad in 1962.[30]

Duthie, J. F., 1903–29. *Flora of the upper Gangetic Plain and of the adjacent Siwalik and sub-Himalayan tracts.* 3 vols. Map. Calcutta: Indian Government Printer. (Reprinted in 2 vols., 1960, Calcutta: Botanical Survey of India.) For index, see S. K. Jain, 1952. *Index to* Flora of the upper Gangetic Plain and of the adjacent Siwalik and sub-Himalayan tracts. 150 pp. Delhi: Council for Scientific and Industrial Research [of India]. For Poaceae, see M. B. Raizada *et al.*, 1957–66. Grasses of upper Gangetic Plain, I–[III]. *Indian For. Rec.*, n.s., Bot. 4(7): (iv), 171–277, 12 pls. [Panicoideae, I]; 5(3) (1964): 149–226, 10 pls. [Panicoideae, II]; *Indian Forester* 92: 637–642 [Poöideae]. For supplement, see M. B. Raizada, 1976. *Supplement to Duthie's 'Flora of the upper Gangetic Plain and of the adjacent Siwalik and sub-Himalayan tracts'.* vii, 355 pp. Dehra Dun: Bishen Singh Mahendra Pal Singh. Pteridophytes are covered in N. P. Chowdhury, 1973. *The pteridophyte flora of the upper Gangetic Plain.* vii, 91 pp., 10 pls. New Delhi: Navayug Traders.

The *Flora* is a briefly descriptive manual-flora of seed plants (less Poaceae) with keys to genera and species, synonymy (with references and citations), generalized indication of internal range, taxonomic commentary, and notes on phenology, biology, uses, etc; no index (this separately compiled and published by S. K. Jain). The substantial supplement by Raizada is in the same style and sequence as the original work though in a slightly larger format. The first two installments on Poaceae are descriptive; the third is a checklist.[31]

Rau, M. A., 1968. *Flora of the Upper Gangetic Plain and of the adjacent Siwalik and sub-Himalayan tracts: check list.* 87 pp. (Bull. Bot. Surv. India, Suppl. 2). Calcutta.

Systematic checklist of vascular plants (except Gramineae) with brief introduction; incorporates additions (275 species), nomenclatural changes, and revised synonymy.

Woody plants

Kanjilal, P. C., 1933. *A forest flora of Pilibhit, Oudh, Gorakhpur and Bundelkhand.* li, 427 pp. Allahabad: Supt. of Printing, United Provinces. (Reprinted 1982.)

Descriptive manual-flora with essential synonymy, vernacular names, distribution (without *exsiccatae*), indication of habitat, phenology, uses and status, and occasional diagnostic notes; index to all scientific names. Keys are provided in larger families and genera. The general part covers the scope of the work and main sources along with physical features, climate, forest vegetation with main formations and characteristic species; these are followed by a synopsis of families and glossary. [Covers several forest divisions in the Upper Gangetic Plain; the first three are near the Nepal border, the fourth in the southern hills (initially after 1947 in Vindhya Pradesh but since 1956 in Madhya Pradesh (**819**)).]

District and local works (Uttar Pradesh, without Uttarkhand)

Recent coverage exists for Allahabad district (surrounding the state capital) and Gorakhpur district in the northeast. A flora of Moradabad district in the northwestern plains has been a project of BSI-NC (Dehra Dun), and other district floras have been in preparation or are planned at BSI-CC (Allahabad).

Misra, B. K. and B. K. Verma, 1992. *Flora of Allahabad District.* 527 pp. Dehra Dun: Bishen Singh Mahendra Pal Singh.

Briefly descriptive flora of flowering plants (713 species in 7254 km^2) with keys, synonymy, references and citations, vernacular names, occurrence and situation, phenology, one or more representative *exsiccatae*, and a few taxonomic notes; bibliography (pp. 475–478) and indices to vernacular, English and botanical names. The general part encompasses physical features, climate, vegetation and forests, synusiae, biotic factors, cultivated and useful plants, a floristic analysis, and the key to families along with the methodology of the work and conventions. [Allahabad is the state capital; the district is representative of the plains flora.]

SHARMA, A. K. and J. S. DHAKRE, 1995. *Flora of Agra District*. 356 pp., map. Calcutta: Botanical Survey of India. (Flora of India, series 3.)

Briefly descriptive flora of flowering plants (609 species in 4836 km^2) with keys, synonymy, references, citations, vernacular names, selected localities with *exsiccatae* (mostly those of the authors), and indication of habitats, phenology and behavior; some cultivated taxa listed. The general part covers geography, geology, soils, climate, rainfall, botanical exploration and the present investigations, the plan of the work, vegetation and floristics (with statistics), abbreviations, and (pp. 18–29) a key to families, while a bibliography (pp. 332–338) and indices to botanical and vernacular names conclude the work.

SRIVASTAVA, T. N., 1976. *Flora Gorakhpurensis*. vii, [iv], 411 pp., unpaged index, 40 pls. (incl. 2 maps). New Delhi: Today and Tomorrow's.

Descriptive flora with keys and (pp. 385–389) references. [Gorakhpur district is between the *terai* on the Nepalese border and the Ghagra River to the south.]

816

Former Madhya Bharat, Bhopal State and Vindhya Pradesh

See **815** (all works), **819** (VERMA *et al.*). Madhya Bharat and Vindhya Pradesh were originally formed in 1950 from Gwalior and the princely states comprising the erstwhile Central India Agency (including those of Bundelkhand, covered by Kanjilal in his forest flora described under **815**). They were, however, in 1956 taken over – along with the former state of Bhopal – by Madhya Pradesh (**819**) which then chose the city of Bhopal as a state capital.

817

Rajasthan (in general)

See **815** (all works). – Area: 342 214 km^2. Vascular flora: 1911 indigenous and naturalized seed plants (Shetty and Singh, 1988–93; see below). The unit, before 1947 the Rajputana Agency of princely states, covers general works as well as those pertaining to that part of the state from the Aravalli Hills eastwards and including Jaipur and Ajmer.

A key botanical feature in an otherwise dry state is the isolated patch of evergreen forest around the summit of Mt. Abu (1724 m) in the geologically old Aravallis. Its flora was first treated by Miss McAdam in *A list of trees and plants of Mt. Abu* (1890) and later by R. N. Sutaria in *Journal of the University of Bombay*, N.S. **9B**(5): 64–68 (1941) and M. B. Raizada in *Indian Forester* **80**: 207–215 (1954). Further studies were published in the 1960s by S. K. Jain.[32]

SHETTY, B. V. and V. SINGH (eds.), 1988–93. *Flora of Rajasthan*. 3 vols. 1246 pp., illus. (part. col.). Calcutta: BSI. (Flora of India, series 2.)

Briefly descriptive flora (with families arranged according to Bentham and Hooker but the genera and species alphabetical); includes keys to genera, species and infraspecific taxa, synonymy (with references and citations), vernacular names, selected *exsiccatae* with localities, distribution, and notes on phenology, frequency and occurrence, and (at end of each genus) doubtful and cultivated species. The introductory section (vol. 1) treats prior exploration and studies, geography and physiography, geology, soils, climate, main features of the vegetation and characteristic species (with 16 color photographs of formations), weeds, cultivated plants and other exotics, and the basis for and plan and methodology of the work, and includes a general key to families. There are also 20 halftone plates at the end of the volume. Volume 3 includes in addition to the last third of the flora proper a floristic analysis (pp. 1140–1143), general bibliography, and complete index (to pages only, without volume numbers) as well as further illustrations (12 color plates and 37 full-page text-figures). [Vol. 1 (pp. i–vii, 1–451) covers Ranunculaceae through Compositae; vol. 2 (1991; pp. 453–860) covers Campanulaceae through Commelinaceae; vol. 3 (pp. 861–1246) covers Palmae through Gramineae along with the bibliography and index. The lone gymnosperm (*Ephedra ciliata*) is found in vol. 2 between dicotyledons and monocotyledons. Some 274 cultivated plants are also listed but not keyed.]

District and other partial works

All of the following cover parts of eastern Rajasthan, encompassed – but in parts only sketchily – by *Flora of the upper Gangetic Plain* (see **815**). For western Rajasthan see **818**.

SHARMA, S. and B. TIAGI, 1979. *Flora of North-East Rajasthan*. xx, 540, 4 pp. New Delhi/Ludhiana: Kalyani.

Briefly descriptive flora (627 species) with keys. [Covers Jaipur, Sawai Madhopur and Alwar districts.]

SHETTY, B. V. and R. P. PANDEY, 1983. *Flora of Tonk District, Rajasthan*. [vi], 253 pp., map. Howrah: BSI. (Flora of India, series 3.)

Briefly descriptive flora with keys. [Covers the present Tonk district in the northeastern part of the state.]

SINGH, V., 1983. *Flora of Banswara Rajasthan*. [ix], 312 pp., map. Howrah: BSI. (Flora of India, series 3.)

Briefly descriptive flora with keys; includes references (pp. 289–291). [Covers Banswara district in the southeastern corner of the state, an area physiographically related to Gujarat and Madhya Pradesh.]

818

Western Rajasthan

This area, which includes the Thar (or Indian) Desert (see also **805**), constitutes all that part of the state west of the Aravalli Hills. The eastern boundary lies approximately along a line from the Gulf of Cambay through and along the Aravallis as suggested by Blatter and Hallberg (see **805**).

Floristically the region is predominantly Saharo-Sindhian rather than Indo-Malayan. It is not covered in any regional flora published before 1947; existing knowledge, including previous work on the desert by Blatter and Hallberg (1918–21; see **805**), was first consolidated by Puri *et al.* (1964). Puri's partly compiled account has since been supplanted by other works, notably a flora of the desert quarter by Bhandari (1990; see **805**) as well as a state flora (**817**).

PURI, G. S. *et al.*, 1964. *Flora of Rajasthan (west of the Aravallis)*. 159 pp., 1 text-fig., 1 pl. (Rec. Bot. Surv. India 19(1)). Calcutta.

Systematic enumeration of seed plants recorded from the western part of Rajasthan (floristically the poorest in India); includes keys to all taxa, synonymy (with references and citations of more important literature), local range with citations of *exsiccatae* and other sources, and notes on habitat, special features, etc.; list of references and index to generic and family names at end. The introductory section provides chapters on physical features, geography, other biota, vegetation, and prior botanical exploration and research.[33]

Partial works: Indian Desert

See **805** for all works, including M. M. Bhandari's *Flora of the Indian Desert*.

819

Madhya Pradesh and adjacent areas

Area: 463452 km². Vascular flora: *c.* 2400 species (Verma *et al.*, 1993; see below). – Modern Madhya Pradesh, the largest state in India, takes its name from but no longer corresponds to the former Central Provinces of the Raj (first organized in 1903). In boundary changes consequent to implementation of the language policy of the central government and dating mostly from late 1956, Berar, Nagpur and neighboring Marathi-speaking areas were transferred to the then-state of Bombay (now Maharashtra), Madhya Bharat, Vindhya Pradesh and Bhopal (corresponding to the princely states of Gwalior and the former Central India Agency) were annexed, and other boundary adjustments were made. Rather earlier, in 1905, five of the Chota Nagpur states, all of which had been in the Presidency of Bengal, were transferred to Central Provinces; these, however, became part of a new Eastern States Agency in 1933 before reverting to Madhya Pradesh in 1948. Unit **819** as circumscribed here corresponds to the modern state; for convenience, however, **816** is retained for remarks on Madhya Bharat, Vindhya Pradesh and Bhopal while earlier works on Chota Nagpur appear under Bihar (**831**). More recent references on Marathi-speaking areas, including Berar, are treated with Maharashtra (**822**).

Botanical progress in this landlocked, somewhat isolated upland area was relatively limited until formation of the Central Circle of the Botanical Survey of India (BSI-CC) at Allahabad in 1962. No separate flora for the Central Provinces and adjacent areas was apparently ever undertaken before 1947, nor was it included in any others. Indeed, many districts were very poorly collected, and largely remained so until after independence. The relatively extensive forests did attract some interest, however, and the woody plants and larger forest herbs were covered in various lists as well as in two manuals respectively by H. H. Haines (1916) and D. O. Witt (1916; for both, see below).

The advent of the BSI-CC led to the first serious state surveys and with them more intensive collecting and study. An early development was a series of seven contributions by G. Panigrahi and collaborators (1965–67, with two further numbers in the 1980s); many other floristic and taxonomic studies including family revisions have followed. Since 1977, preparation of a state flora has been undertaken by staff at BSI-CC; a first volume appeared in 1993 and a second in 1997. Some district floras have also been compiled, both by Circle staff and by non-BSI authors; to date, however, fewer than half of the 45 districts have been covered. Among larger families not yet covered in the state flora, Poaceae have been revised in G. P. ROY, 1984. *Grasses of Madhya Pradesh*. 180 pp., 19 pls. (incl. 3 folding). Howrah: BSI (Flora of India, series 4), and Cyperaceae in D. M. VERMA and A. CHANDRA, 1981. Cyperaceae of Madhya Pradesh. *Rec. Bot. Surv. India* **21**(2): 221–275.[34]

PANIGRAHI, G. *et al.*, 1965–85. Contributions to the botany of Madhya Pradesh, I–IX. *Bull. Bot. Surv. India* **8** (1966): 117–125 (I); *Proc. Natl. Acad. Sci. India*, B (Biol. Sci.) **35** (1965): 87–98, 99–109 (II–III); *ibid.*, **36** (1966): 553–564 (IV); *ibid.*, **37** (1967): 77–104 (V); *Bull. Bot. Surv. India* **9** (1967): 262–267 (VI); *ibid.*, **9** (1967(1968)): 268–276 (VII); *J. Econ. Taxon. Bot.* **4** (1983): 421–434 (VIII); *ibid.*, **7** (1985): 77–93 (IX). (Parts I–III (Dilleniaceae through Convolvulaceae) by G. Panigrahi alone; part IV (Euphorbiaceae and Urticaceae) by G. Panigrahi and R. Prasad; part V (Leguminosae) by G. Panigrahi and A. N. Singh; part VI (Compositae through Sapotaceae) by Ram Lal and G. Panigrahi; part VII (Poaceae) by U. Shukla and G. Panigrahi; part VIII (Hydrocharitaceae through Eriocaulaceae) by J. Lal and A. Kumar; part IX (15 assorted dicotyledon families not previously covered, among them the Lamiaceae) by A. Kumar and J. Lal.)

Systematic accounts of various flowering plant families with descriptions of novelties, synonymy, references, local range, and miscellaneous notes. [Intended as additions to existing knowledge of the flora of the state and as precursors to a state flora; family sequences follow the Bentham and Hooker system.]

VERMA, D. M. *et al.* (eds.), 1993– . *Flora of Madhya Pradesh*. Vols. 1– . Illus. (part col.), maps (end-papers). Calcutta: BSI. (Flora of India, series 2.)

Concise descriptive flora of vascular plants with keys, synonymy, references, key citations, indication of localities, habitat and phenology, occasional *exsiccatae* and other notes, and full-page figures of selected species; indices to botanical and vernacular names at end. The general part covers the background to and plan of the work as well as aspects of geography, climate, rainfall, geology, soils, land use, the people, vegetation and forest types (forests covering in theory some 31.5 percent of the total area), aquatic macrophytes, botanical history and floristic works, and useful and medicinal plants; bibliography (pp. 24–30). Keys to families appear at the beginning of major groups (pteridophytes, angiosperms, gymnosperms). [Vol. 1 encompasses Ranunculaceae through Plumbaginaceae (essentially following the Bentham and Hooker system); vol. 2 (1997), Primulaceae through Ceratophyllaceae.][35]

District and local works

For the Nagpur and Berar forest circles, now in Maharashtra, see **822**. A list of the 45 districts established in the state as of 1994, with map, appears in *Floral elements of Madhya Pradesh* by A. K. Tripathi, B. K. Shukla and V. Mudgal (1994, New Delhi: Ashish). Consequent to establishment of the district flora programme in the 1960s, several have been documented, while other district floras are in progress or await release.

MUKHERJEE, A. K., 1984. *Flora of Pachmarhi and Bori Reserves*. vi, 407 pp., map. Howrah: BSI. (Flora of India, series 3.)

Briefly descriptive flora of 778 species with keys, synonymy, distribution, and other details; index. A relatively brief introduction includes notes on prior exploration and sources for the flora but no list of literature. [The reserves, in the eastern Satpura Ranges between Betul, Chhindwara and Hoshangabad districts in the south-central part of the state, include its highest peaks; Pachmarhi, home to a 'hill station', reaches 1351 m. The area has thus considerable botanical significance.]

OOMMACHAN, M., 1977. *The flora of Bhopal*. xi, 475 pp., 15 pls., tables, 2 maps. Bhopal: Jain Brothers.

Descriptive flora of seed plants (836 species in 2763 km²); includes keys to taxa from generic level downwards, synonymy and references, literature citations, local and overall distribution (including some specimen citations), vernacular names, some critical remarks (in footnotes), and notes on variability, habitat, occurrence, phenology, etc.; lists of plants in particular situations, references, lexicon, and index to families and genera at end. The introductory section covers physical and biotic features, vegetation and forests, human influence, plant life-forms, and floristic statistics. [Bhopal is the state capital.]

OOMMACHAN, M. and J. L. SHRIVASTAVA, 1996. *Flora of Jabalpur*. [ix], 354 pp., 128 text-figs., 2 maps. Jodhpur: Scientific Publishers.

Concise descriptive flora of seed plants (933 species, all angiosperms, in an area of 10 164 km^2) with keys, synonymy, phenology, vernacular names, indication of distribution, vouchers, and some critical notes; bibliography (pp. 321–323) and complete indices at end. The general part furnishes a description of the district and its vegetation and forests, with floristic analysis (pp. 14–16) but no overall key to families. [Jabalpur is in the east of the state.]

PANIGRAHI, G. and S. K. MURTI, 1989. *Flora of Bilaspur District.* Vol. 1. 396 pp., 17 text-figs. Howrah: BSI. (Flora of India, series 3.)

Descriptive flora (852 species in 19 755 km^2) with keys and a marked emphasis on references and citations as well as practical information. An extensive general part covers past botanical work and sources (the area had scarcely been surveyed before 1954). [Covers Ranunculaceae through Convolvulaceae; a second volume will complete the work. Bilaspur is in the central-eastern part of the state.]

ROY, G. P., B. K. SHUKLA and B. DATT, 1992. *Flora of Madhya Pradesh (Chhatarpur and Damoh).* xxxi, [i], 639 pp. New Delhi: Ashish Publ. House.

Briefly descriptive flora of seed plants (881 species in 15 991 km^2) with keys, synonymy and references, phenology, distribution, frequency, occurrence and habit; literature (pp. 568–574) and an index (not properly alphabetized) to scientific names. There is an extensive general part including past botanical work and sources, statistics and a key to families. [Chhatarpur district is one of those formed from Bundelkhand (**816**); it was previously covered in Duthie's flora of the Upper Gangetic Plain (**815**). Damoh, to its south, was in the former Central Provinces.]

SAMVATSAR, S., 1996. *The flora of western tribal Madhya Pradesh (India).* xxi, 441 pp., 59 text-figs., 32 col. photographs, 3 maps. Jodhpur: Scientific Publishers.

Concise descriptive flora (1159 seed plants in 14 931 km^2) with keys, synonymy, references and citations, representative *exsiccatae* (all made by the author), and indication of distribution and habitat; bibliography (pp. 399–403) and index to all botanical names. The somewhat disorganized general part includes a glossary and key to families as well as the scope of the work and reviews of physical features, climate, ethnography, vegetation with characteristic species, and floristic and life-form analyses. [Covers the districts of Dhar and Jhabua in westernmost Madhya Pradesh; the Narmada River forms their southern boundary.]

SINGH, V. P. and V. S. KHARE, 1996. *Flora of Ujjain District.* vii, 540 pp., 43 pls. Delhi: Periodical Export Book Agency.

A work of several chapters; chapter 11 comprises a concise descriptive flora of flowering plants (631 species) with keys, synonymy, references, citations, vernacular names, voucher collections, figures or photographs of selected species, and notes on phenology, distribution, habitat, uses,

and other features. Chapters 1–10 comprise accounts of geography, climate, soils, history, prior botanical exploration, vegetation, floristics, floral biology and the foundation for the work, while chapters 12–14 comprise a general (essentially vegetatiological) discussion, summary and bibliography. Chapter 15 contains addenda; this is followed by two appendices (the first statistical, the second a basic systematic checklist) and index to families. [Ujjain, formerly part of the state of Gwalior (**816**), is in the west of Madhya Pradesh just northeast of Dhar.]

VERMA, D. M., P. C. PANT and M. I. HANFI, 1985. *Flora of Raipur, Durg and Rajnandgaon.* xx, 524, [i] pp., 64 pls., map. Howrah: BSI. (Flora of India, series 3.)

Briefly descriptive flora (1032 species in *c.* 45 000 km^2) with keys, synonymy, and distributional and other data; extensive introductory part. [The three districts, in part former princely states of the Eastern States Agency, are in the northern part of the southern 'panhandle' of Madhya Pradesh, south of Bilaspur.]

Woody plants

The work of Witt was preceded by *A list of trees, shrubs and economic herbs in Northern Forest Circle, Central Provinces* (1906) by R. S. Hole.

HAINES, H. H., 1916. *Descriptive list of trees, shrubs, and economic herbs of the Southern Circle, Central Provinces.* xxviii, 384 pp. Allahabad: Pioneer Press. (Reprinted, with revisions, from a series in *Indian Forester* **38–40** (1912–14).) Complemented by D. O. WITT, 1916. *Descriptive list of trees, shrubs, climbers and economic herbs of the Northern and Berar Forest Circles, Central Provinces.* xiii, 79, 247, viii, xviii, ii, xvi, ii, ii pp. Allahabad.

Descriptive manual-floras of native, naturalized and commonly cultivated woody plants (and herbs of economic significance). Haines's work includes non-dichotomous keys to genera and species (including also an artificial key at the end of the work), while that of Witt features analytical keys; both cover essential synonymy, vernacular names, generalized indication of local range, and notes on habitat, phenology, properties, uses, etc., and include a glossary and index to vernacular and generic names. The introductory sections include a description of the respective regions along with sections on geology and soils (and, in Witt's work, accounts of forest formations and a recognition of botanical subregions) as well as synopses of families.

Region 82

Peninsular India and Sri Lanka

This region comprises all of India south of the boundaries of Madhya Pradesh and Andhra Pradesh, together with Sri Lanka (Ceylon). Included are all of the former Madras Presidency, the southern part of the Bombay Presidency, the Nagpur and Berar districts of the former Central Provinces, and some former princely states including the Nizam's Dominions (Hyderabad-Deccan) and Mysore.

With a long history of botanical exploration, beginning in the sixteenth century, peninsular India and Sri Lanka may currently be considered to be floristically relatively well known. The basic stock of subregional floras is, however, mostly 60–100 or more years old; within it are *Flora of the Presidency of Bombay* by Cooke (1901–09), *Flora of the Presidency of Madras* by Gamble and Fischer (1915–38), and *Handbook of the flora of Ceylon* by Trimen and Hooker (1893–1900, with a supplement by Alston, 1931). Coverage of many areas by these works is moreover sketchy, and for the former Nizam's Dominions no separate flora was ever published. Revisions of these have not (or only recently) materialized or have taken other forms. On the other hand, since the mid-twentieth century active collecting projects have taken place in Karnataka, the Peninsula proper, Sri Lanka, and elsewhere, thereby laying the basis for new district and state floras. The largest amount of recent work has come from Karnataka, in the interior uplands (where active systematics centers have existed at universities in Mysore and Bangalore), Tamil Nadu (from the school of Fr. Koyapillil M. Matthew), and Sri Lanka (as part of the Ceylon Flora project). Several publications have resulted, of which the most considerable are Fr. Matthew's *Flora of Tamilnadu Carnatic* series and *Revised handbook to the flora of Ceylon*, the latter yet to be completed. The Botanical Survey of India, with regional centers at Pune (BSI-WC) and Coimbatore (BSI-SC), has also been active in collecting, research and publication; their largest contribution to date has been a state flora of Tamil Nadu but several district floras have also been produced, particularly in hitherto poorly documented Andhra Pradesh and Kerala as well

as parts of Maharashtra. Some regional, state and district forest flora-manuals have also been published. Nevertheless, more exploration (before it is too late in many areas) and documentation, particularly from the eastern part of the region, is greatly needed in terms of currently accepted standards in floristics.

Bibliographies. General and supraregional bibliographies as for Division 8 and Superregion 81–84.

Regional bibliography
KARTHIKEYAN, S., M. P. NAYAR and R. S. RAGHAVAN, 1981. *An annotated bibliography of taxonomic botany of peninsular India 1959–1978*. 201 pp. Howrah: BSI. [Classified, with sections on floristic and systematic works, the latter arranged by families.]

Indices. General and supraregional indices as for Division 8 and Superregion 81–84.

820

Region in general

See also **810–40** (all works). For *Flora of the South Indian hill stations* (1932) by P. Fyson, see **802**. Among nineteenth-century works subsequent to those of Wight, *Ferns of south India* (1863; 2nd edn., 1873; supplement, 1876) by R. H. Beddome remains important.

Woody plants
BEDDOME, R. H., 1869–73. *Flora sylvatica for South India*. 2 vols. 330 + 28 pls. Madras (now Chennai): Gantz (for the Madras Government). (Reprinted 1978, Dehra Dun: Bishen Singh Mahendra Pal Singh.)

Comprises two main sections: (1) a 'Foresters' Manual of Botany for Southern India' (pp. i–ccxxxviii), including Bentham's *Outlines of Botany* as written for colonial floras, a synopsis and index of families and genera, and a descriptive treatment of genera and species (with keys to genera) and extensive notes on botanical attributes, wood, properties, uses, etc.; and (2) an iconography of 330 plates (not in systematic order) with accompanying descriptive text, featuring every species treated in the first volume; in this text are given formal details of synonymy, local distribution, and occurrence. The additional 28 plates in vol. 2 feature morphological analyses of representatives of a number of different South Indian genera. A synoptic index to plates appears in vol. 1, which, along with the 'Foresters' Manual', follows the Bentham and Hooker system.

PASCAL, J.-P. and B. R. RAMESH, 1987. *A field key to the trees and lianas of the evergreen forests of the Western Ghats (India)*. iii, 236 pp., 143 pls. (Trav. Sect. Sci. Tech. Inst. Franç. Pondichéry 23). Pondicherry.

Analytical keys accompanied by rubbed leaf drawings (some of the details being, however, less clear than desirable); index.

SOMASUNDARAM, T. R., SRI, 1967. *A handbook on the identification and description of trees, shrubs and some important herbs of the forests of the southern states for the use of the Southern Forest Rangers' College, Coimbatore.* iv, 563, 9 pp. Delhi: Manager of Publications, Government of India.

Concise field-manual of native and principal introduced woody plants (as well as the larger and/or more important herbs), with keys to nearly all taxa, concise descriptions, limited synonymy and key references, vernacular names, generalized indication of local range, remarks on size, habitat, phenology, special features, properties and uses; errata, glossary and complete indices at end. The introductory section includes a general key to families.

821

Gujarat (with Daman, Diu, Dadra and Nagar Haveli)

See also **823** for Daman, Diu, Dadra and Nagar Haveli. – Area: Gujarat, 195 984 km^2; Daman, 57 km^2; Diu, 40 km^2; Dadra and Nagar Haveli, 490 km^2. Vascular flora: 1800 species in 791 genera, of which 150 species are cultivated (Shah, 1978; see below).[36] The state of Gujarat was separated from Bombay in 1960, culminating a series of political changes initiated after 1947. It encompasses those parts of the old Bombay Presidency north of the formerly Portuguese Union Territory of Dadra and Nagar Haveli as well as that Territory (incorporated into India in 1954) and moreover incorporates the Western India and Baroda and Gujarat States Agencies of princely states (parts of which were from 1948 to 1956 constituted as the states of Kutch and Saurashtra). Also included here are the former Portuguese enclaves of Daman and Diu (which remain administratively combined with Goa as a Union Territory).

Prior to publication of a state flora by Gopalkrishna Shah in 1978, coverage of this floristically diverse polity was furnished in haphazard fashion in the Bombay Presidency floras of Cooke and Talbot (**822**) as well as in some partial and local works (of which the largest is *Flora of Saurashtra* (1962, 1988)). The flora and the checklist by R. S. Raghavan (1981) take account of extensive collections assembled after 1955 which have added many taxa for the state. Woody plants are in addition covered statewide by a pocket-sized manual by R. I. Patel.

RAGHAVAN, R. S. *et al.*, 1981. *A checklist of the plants of Gujarat.* 127 pp. (Rec. Bot. Surv. India 21(2)). Howrah.

Systematic checklist, with limited synonymy (including references and citations, the latter mainly to key floristic works with special reference to Cooke (**822**)); bibliography (pp. 111–127). The introductory part furnishes background and an analysis of the flora. [Additions for the state and to Cooke's flora are indicated respectively by one (*) or two (**) asterisks.]

SHAH, G. L., 1978. *Flora of Gujarat State.* 2 vols. vii, 1074 pp., maps. Vallabh Vidyanagar (Kheda or Kaira district, Gujarat): Sardar Patel University.

Briefly descriptive flora of seed plants; includes keys, synonymy (with references and citations), vernacular names, distribution, and indication of habitat, phenology, frequency, etc.; references (pp. 883–904) and indices to vernacular and botanical names in volume 2. The introductory part (vol. 1) gives the plan of the undertaking along with reviews of physical features, geology, soils, climate, vegetation, forests and their composition, phenology, biotic factors and man's influence, and useful and ornamental plants but lacks any review of past botanical work; these are followed by a key to families.

District and local works

The political history of what is now Gujarat is paralleled in the existence of several regional, district and other local florulas. The principal ones are given below; of these the most substantial covers Saurashtra which comprises the south-central part of Gujarat north of the former Portuguese island of Diu and includes the famous Gir Forest.

BLATTER, E., 1908–09. On the flora of Cutch, [I–II]. *J. Bombay Nat. Hist. Soc.* **18**: 756–777; **19**: 157–176.

Systematic enumeration of seed plants of this arid, salty, seasonally marshy area near the Sind (Pakistan) border, with generalized indication of local and extralimital range and brief notes on habitat, phenology and frequency; appendices on the features of the flora, vegetation, cultivated plants, and species biology are also included.

CHAVAN, A. R. and G. M. OZA, 1966. *The flora of Pavagadh (Gujarat State, India)*. vii, 296 pp., 2 pls. (Botanical Memoirs 1). Baroda: University of Baroda.

Briefly descriptive, keyed flora of seed plants of Pavagadh Hill (a monadnock near Baroda); includes synonymy (with references), vernacular names, indication of local range (with some *exsiccatae*) and notes on habitat and phenology; list of references and index to all botanical names. [Intended partially as a students' manual.]

SANTAPAU, H., 1962. *The flora of Saurashtra*: Part 1. x, 270 pp. Rajkot: Saurashtra Research Society. Continued as P. V. BOLE and J. M. PATHAK, 1988. *Flora of Saurashtra*: Parts 2–3. Pp. 1–302, 303–553. Howrah: BSI. (Flora of India, series 2.) Associated work: H. SANTAPAU and K. P. JANARDHANAN, 1966 (1967). *The flora of Saurashtra: check list*. 58 pp. (Bull. Bot. Surv. India, Suppl. 1). Calcutta.

The *Flora* is a briefly descriptive account with keys, synonymy, vernacular names, internal range (with indication of many localities), and notes on special features, phenology, uses, etc.; indices at end of vol. 3 (but only to vols. 2–3) as well as an appendix to vol. 1 (pp. 487–508) and bibliography (pp. 509–513). The *Check list* is unannotated. [Arrangement of the *Flora* follows the Bentham and Hooker system; vol. 1 covers Ranunculaceae–Rubiaceae and vols. 2–3 Compositae through Gramineae.]

SAXTON, W. T. and L. J. SEDGWICK, 1918. Plants of northern Gujarat. *Rec. Bot. Surv. India* 6: 207–323, i–xiii, map.

Annotated systematic checklist of seed plants (614 species); includes citations of major works, vernacular names, localities, critical remarks, and notes on uses, occurrence, behavior, etc.; index. An introductory section covers physical features, climate, soil, features of the vegetation, floristics and ecology, and statistics; it also notes the inadequacy of coverage of the area by Cooke's Bombay Presidency flora (**822**). In a short appendix are notes on the cryptogamic flora.

Woody plants

PATEL, R. I., 1971. *Forest flora of Gujarat State*. ii, 381 pp. Baroda: Forest Department, Gujarat State (distributed by Gujarat Government Press, Ahmadabad).

Briefly descriptive flora of woody plants and larger herbs; includes analytical keys to all taxa, synonymy with citations of key literature, vernacular and English names, general indication of local range, and notes on habitat, lifeform, phenology, etc.; indices to botanical, vernacular and English names at end. The introductory section includes a description of the region along with remarks on its climate, geology, soils, vegetation, and the history of botanical work. [Not seen; account prepared from notes supplied by the book's author.]

822

Maharashtra (Bombay)

Area: 307 762 km^2. Vascular flora: no data. – This unit is here defined as all that part of the former Bombay Presidency south of the Territory of Dadra and Nagar Haveli (in **821**). For Goa, see **823**; for Kurg (Coorg), see **826** (with Karnataka). In 1956, Bombay assumed control of all of what is now Gujarat (**821**) as well as former Hyderabad (**824**) and Marathi-speaking districts of Madhya Pradesh (**819**) (including Nagpur, the former capital of the Central Provinces, as well as the rest of Berar). At the same time, Kannada-speaking districts in the south were ceded to Mysore (**826**; renamed Karnataka in 1973). In 1960, Gujarat was for linguistic reasons separated from Bombay which then was renamed Maharashtra. Floristic works dealing with the whole of the old Presidency for convenience appear under this heading.

Botanical studies in Bombay and formerly dependent areas are embodied in the now-elderly floras of Cooke (1901–06) and Talbot (1909–11), the supplements to the former by Blatter and McCann (1926–35) and some partial and local works. Among the last-named should be mentioned A. K. NAIRNE, [1894]. *The flowering plants of Western India*. xlvii, 401 pp. London: Allen (reprinted 1976), an introductory work covering Bombay and Gujarat.

The state is at present botanically a charge of the Western Circle of the Botanical Survey of India (BSI-WC), established in Pune (Poona) in 1955.[37] Of particular interest have been the exploration and documentation of lesser-known areas; among these are the districts formerly in Madhya Pradesh and the Nizam's Dominions. Several of their floras have now been published, and the first volume of a state flora, covering monocotyledons, appeared in 1996.

Bibliography

See also pp. 637–747 in *Flora of Maharashtra State: monocotyledons* for an extensive list arranged by authors.

KAMAT, N. D., 1978. *Botanists and botanical researches in Maharashtra, 1951–1975*. vi, [ii], 310 pp. Aurangabad: Vimal Prakashnan. [Covers 2415 papers in all areas of botany along with 217 theses and dissertations; author, geographical and subject indices included.]

COOKE, T., 1901–09. *The flora of the Presidency of Bombay*. 2 vols. (in 3). ix, 2, 645, 1083 pp. London: Taylor and Francis. (Reprinted 1958, 1967 in 3 vols., Calcutta: Botanical Survey of India.) For additions and corrections, see E. BLATTER and C. C. McCANN, 1926–35. Revision of the flora of the Bombay Presidency, I–XXVII. Illus. *J. Bombay Nat. Hist. Soc.* **31–38**, *passim*; and S. KARTHIKEYAN, M. P. NAYAR and R. S. RAGHAVAN, 1981. A catalogue of species added to Cooke's *Flora of the Presidency of Bombay* during 1908–1978. *Rec. Bot. Surv. India* **21**(2): 153–205.

The *Flora* is a descriptive account of vascular plants (covering North Kanara in Karnataka, the greater part of Maharashtra, Goa Union Territory, Gujarat and Sind Province in Pakistan) with keys to genera and species, synonymy (with references and citations), vernacular names, indication of local range (with citation of some *exsiccatae*), general summary of extralimital distribution, taxonomic commentary, and notes on phenology, uses, etc.; complete index at end of vol. 2. The *Revisions* by Blatter and McCann are in enumerative form and account for all families except Poaceae, following in sequence the original version of the Hutchinson system. The cumulative *Catalogue* by Karthikeyan *et al.* accounts for five genera, 715 species and 113 non-nominate infraspecific taxa.[38]

SHARMA, B. D., S. KARTHIKEYAN and N. P. SINGH (eds.), 1996. *Flora of Maharashtra State: monocotyledons*. ix, 794 pp., text-figs., 4 col. pls. Calcutta: BSI. (Flora of India, series 2.)

Concisely descriptive flora with keys, synonymy, references, citations, Marathi vernacular names (where recorded), references to illustrations, and indication of distribution (by districts), habitat, occurrence and phenology; bibliography (pp. 637–747) and indices to all botanical and vernacular names. No general part is included; a brief foreword features comparisons between the present work and that of Cooke. [A total of three or four volumes for the whole *Flora* was foreseen, with the companion volumes on dicotyledons said to be 'in press'.][39]

District and local works

Except around Mumbai as well as Pune and other 'hill stations' there was relatively little detailed floristic work until the mid-twentieth century. In the 1960s the Botanical Survey of India as well as universities with specified grants-in-aid initiated the study of district floras. Several have now been published while others exist as theses. These may be seen as complementary to, and a partial basis for, the state flora. Current coverage is well scattered through the state.

ALMEIDA, S. M., 1990. *The flora of Savantvadi*. 2 vols. 2 maps, col. pls. Jodhpur: Scientific Publishers.

Briefly descriptive flora of vascular plants (1685 species in 1336.2 km^2) with keys, synonymy, references and citations, vernacular names, localities, selected *exsiccatae*, phenology, and indication of habitat; bibliography (pp. 191–192) and indices to all botanical and vernacular names. The general part encompasses the history of botanical work as well as reviews of physical attributes (a particular feature is relatively high though strongly seasonal rainfall at 4305 mm/annum in Savantvadi town, rising to 7443 mm inland) and vegetation (a satisfactory floristic analysis is, however, wanting); it also contains a key to families. [Savantvadi *taluk* (subdistrict) is the southernmost in Sindhudurg, adjacent to Goa. Records of Dalgado and Vartak (see **823**) are incorporated, as are also those of Kulkarni (see below). Of particular interest is the marked increase in known species from the 778 recorded in Kulkarni's flora.]

DESHPANDE, S., B. D. SHARMA and M. P. NAYAR, 1993. *Flora of Mahabaleshwar and adjoinings, Maharashtra*. 2 vols. 776 pp., 8 col. pls., 4 maps. Calcutta: BSI. (Flora of India, series 3.)

Briefly descriptive flora of flowering plants (1398 species – of which 1201 spontaneous – in 10 492 km^2) with keys, synonymy, references, citations, vernacular names (where recorded), references to illustrations, and indication of habitat, occurrence and phenology; bibliography (pp. 723–732) and indices to all botanical and vernacular names. The general part includes introductory material on the area and its floristics and vegetation along with a key to families. [In Satara district (south of Pune) on the Ghats, with mountains to 1436 m; includes two hill stations.]

KAMBLE, S. Y. and S. G. PRADHAN, 1988. *Flora of Akola District, Maharashtra*. [xii], xix, 320 pp., 4 pls., map. Howrah: Botanical Survey of India. (Flora of India, series 3.)

Briefly descriptive flora of vascular plants (651 species in *c.* 10 606 km^2) with keys, synonymy, references, citations, vernacular names, phenology, indication of distribution, and references to illustrations; bibliography (pp. 292–295) and index to scientific names. The general part includes abbreviations, reasons for the work and its plan and basis, and reviews of physical features, climate and rainfall, forests, vegetation types, plants of economic value, rare and endangered taxa, and a floristic analysis along with a general key to families. [Akola district, in the north-central part of the state, is one of the Berar lands formerly in what is now Madhya Pradesh (see **819**).]

KARTHIKEYAN, S. and A. KUMAR, 1993. *Flora of Yavatmal District*. 344 pp., 4 col. pls., 2 maps. Calcutta: Botanical Survey of India. (Flora of India, series 3.)

Briefly descriptive flora of seed plants (577 species)

with keys, synonymy, references, citations, vernacular names (where recorded), references to illustrations, and indication of distribution, habitat, occurrence and phenology; bibliography (pp. 315–319) and indices to all botanical and vernacular names. The general part includes introductory material on the district and its botanical history, floristics and vegetation along with a key to families. [In the east of the state, partly bordered on the south by Andhra Pradesh. Prior to its inclusion in Maharashtra the district was part of Berar in the former Central Provinces (Madhya Pradesh; see **819**).]

KOTHARI, M. J. and S. MOORTHY, 1993. *Flora of Raigad District*. 581 pp., 12 halftone pls., 3 maps. Calcutta: Botanical Survey of India. (Flora of India, series 3.)

Briefly descriptive flora of vascular plants (1248 species) with keys, synonymy, references, citations, vernacular names (where recorded) and indication of habitat, occurrence and phenology; bibliography (pp. 521–525) and indices to all botanical and vernacular names. The general part includes introductory material on the district and its floristics and vegetation along with a key to families. [On west coast southeast of Mumbai, bordered to the east by Pune (Poona) district.]

KULKARNI, B. G., 1988. *Flora of Sindhudurg*. xx, 605 pp., 10 halftones, map. Howrah: Botanical Survey of India. (Flora of India, series 3.)

Briefly descriptive flora (1123 species) with keys, synonymy, references and citations, indication of distribution, frequency and habitat, phenology, references to illustrations and (where appropriate) taxonomic commentary; index at end. The general part includes historical notes and reviews of physical features, geology, soils, climate, wildlife, vegetation formations and species of economic interest as well as a floristic analysis; there is also a key to families. [Separated from Ratnagiri district in 1981, Sindhudurg is the southernmost in Maharashtra, adjacent to Goa (**823**) and North Kanara in Karnataka (**826**); the potential natural forest vegetation represents a transition from moist deciduous to semi-evergreen and evergreen states. Already very incomplete for Savantvadi, the southernmost of the seven *taluks* or subdistricts in Sindhudurg; see the work of ALMEIDA above.][40]

LAKSHMINARASIMHAN, P. and B. D. SHARMA, 1991. *Flora of Nasik District*. [vi], 644 pp., 52 text-figs., 12 pls. (4 col.), 4 maps. Calcutta: Botanical Survey of India. (Flora of India, series 3.)

Briefly descriptive flora of vascular plants (958 species or 35 percent of the flora of Maharashtra in 15 582 km^2) with keys, synonymy, references and citations, vernacular names, references to illustrations, indication of phenology, and notes on habitat, associates, places found, etc.; bibliography (pp. 599–602) and indices to botanical and vernacular names; addenda and corrigenda (pp. 643–644). The general part includes historical notes as well as reviews of physical features, vegetation, floristics, and useful plants and a key to

families. [Nasik district is in the northwest of Maharashtra near Surat.]

NAIK, V. N., 1979. *The flora of Osmanabad*. xiv, 466 pp. [1] p. map. Aurangabad: Venus.

Briefly descriptive flora (804 species of seed plants in 8894 km^2 on the Deccan at an average elevation of 600 m) with keys, synonymy, references, citations, vernacular names, indication of habitat, occurrence and phenology, representative *exsiccatae* with localities, and occasional taxonomic commentary; bibliography (pp. 422–427) and index to all scientific names. The general part covers physical features, soils, vegetation, phenology, botanical history, useful, medicinal and cultivated plants, and a floristic analysis. [Osmanabad, in south-central Maharashtra, was one of five Marathwada districts formerly in the Nizam's Dominions (**824**).]

SANTAPAU, H., 1967. *The flora of Khandala on the western Ghats of India*. 3rd edn. xxv, 372 pp., map (Rec. Bot. Surv. India 16(1)). Calcutta. (1st edn., 1953.)

Systematic enumeration of seed plants; includes synonymy and references, vernacular names, some citations of *exsiccatae*, taxonomic commentary, and notes on occurrence, habitat, phenology, special features, etc.; list of references and complete index. [Khandala is a 'hill station' near Mumbai (Bombay).]

UGEMUGE, N. R., 1986. *Flora of Nagpur District*. [x], 497 pp., [48] pp. of pls. with 192 figs. and map. Nagpur: U. Rekha for Shree Prakashan.

Briefly descriptive flora of seed plants; includes keys, vernacular names, synonymy, references and citations, and notes on distribution (with representative *exsiccatae*), habitat, occurrence, phenology, etc.; references (pp. 446–452) and indices to vernacular and scientific names along with (at end) illustrations of representative species. The general part includes reviews of physical features, geology, soils, minerals, climate, vegetation, and previous botanical work. [Nagpur district, home to the administrative center of the pre-1947 Central Provinces (Madhya Pradesh; see **819**), is in the east of the state.]

Woody plants

Talbot's definitive *Forest flora* was preceded by his *Systematic list of the trees, shrubs and woody-climbers of the Bombay Presidency* (1894) and *The trees, shrubs and woody-climbers of the Bombay Presidency* (1902). Both appeared in an octavo format handier for the field. Also included here is Witt's checklist for the Berar forest circle (within the former Central Provinces when published but now in Maharashtra); for an expanded version of that work, see **819** (Witt, 1916). A modern forest flora of the contemporary state has yet to be produced.

TALBOT, W. A., 1909–11. *Forest flora of the Bombay Presidency and Sind*. 2 vols. 541 text-figs. Poona: Bombay

Presidency Government. (Reprinted 1975 with corrections by M. B. Raizada, Dehra Dun: Bishen Singh Mahendra Pal Singh.)

Generously dimensioned, illustrated descriptive forest flora; includes keys to genera and species, extensive synonymy, references and citations, vernacular names, generalized indication of local and extralimital range, and notes on habitat, phenology, special features, wood details, properties, uses, etc.; index to all botanical and vernacular names in each volume. The introductory section incorporates descriptive treatments of the vegetation and the forest lands of the Presidency.

PATEL, R. I., 1968. *Forest flora of Melghat.* xlviii, 380 pp. Dehra Dun: Bishen Singh Mahendra Pal Singh.

Keyed descriptive manual-flora of woody plants, with limited synonymy, indication of local range, and notes on occurrence, habitat, special features, etc.; index. [The Melghat area is part of the Berar circle, formerly in Madhya Pradesh.]

WITT, D. O., 1908. *List of trees, shrubs and climbers and other plants of economic importance found in the Berar Forest Circle of the Central Provinces.* vii, 103 pp. Nagpur: Government Press [Central Provinces].

Systematically arranged descriptive enumeration (333 species) with cross-references to key works (*Flora of British India, Flora of Bombay Presidency, Indian trees*), vernacular names, and (sometimes) extensive notes on habitat, phenology, uses, taxonomy and identity, frequency, wood, etc., with forest divisions indicated only for less common or localized species; brief glossary and indices to vernacular and scientific names at end. [Covers seven forest divisions.]

823

Goa

For Daman, Diu, Dadra and Nagar Haveli see also **821**. – Area: 3806 km². Vascular flora: 1115 angiosperms and 27 pteridophytes in the two Union Territories of Goa, Daman and Diu and Dadra and Nagar Haveli (Rao, 1985–86; see below). Goa (also known as Gomantak), Portuguese for over 450 years and the setting for Garcia da Orta's mid-sixteenth century *Colloquios dos simples* as well as work around 1780 by the explorer Manual Galvão da Silva, was in 1961 incorporated into India along with Daman and Diu. Together they are presently a Union Territory. Savantvadi, a former princely state just north of Goa and for a time also under Portuguese control, is now

part of the Sindhudurg district of Maharashtra (**822**).

Rolla Rao's flora of 1985–86, which incorporates the results of new collecting dating to 1962, supersedes *Flora de Goa e Savantvadi* (1898) by D. G. Dalgado (issued in commemoration of the 400th anniversary of Portuguese settlement), *Catálogo botânico das plantas de Goa e terras vizinhas* by J. C. d'E. Souza (1944; in *Bol. Inst. Vasco da Gama* **60**: 54–186; **61**: 69–79) and *Enumeration of plants from Gomantak, India, with a note on botanical excursions around Castle Rock* (1966) by V. D. Vartak. The last, in part a compilation from literature but accounting for 1512 species, covers a somewhat larger area including neighboring parts of Savantvadi (**822**) as well as Castle Rock, Anmode and Karwar in Karnataka (**826**).

RAO, R. S., 1985–86. *Flora of Goa, Diu, Daman, Dadra and Nagar Haveli.* 2 vols. 544 pp., illus., 4 maps. Howrah: BSI. (Flora of India, series 2.)

Keyed enumeration of flowering plants; includes synonymy (with references and citations; key to codes on p. ix in vol. 1), vernacular names (Konkani and Portuguese), local distribution (with representative *exsiccatae*), phenology, frequency, references to illustrations, and notes on properties, potential, etc.; index in volume 2. The introductory section in volume 1 covers previous studies, the basis for and plan of the work, geography and physical features, climate, vegetation, useful and potentially useful plants, a floristic discussion and analysis, and (pp. xxxix–xli) concluding remarks. [All relevant information from Dalgado's *Flora* was absorbed.][41]

Woody plants

NAITHANI, H. B., K. C. SAHNI and S. S. R. BENNET, 1997. *Forest flora of Goa.* 666 pp., 132 pls., map. Dehra Dun: International Book Distributors.

Illustrated descriptive manual of trees, shrubs, woody climbers and bamboos (578 species, of which 496 native) with keys, synonymy, references, citations, vernacular names, distribution with *exsiccatae*, and indication of habitat, phenology, timber and other uses, etc.; figures of representative species throughout text; bibliography (pp. 618–622) and indices to vernacular, Portuguese, trade (English) and botanical names at end. The general part includes material on physical features, climate, geology, forest areas, vegetation types, main timbers and other forest products, wildlife, and history, demography and tourism with botanical history on pp. 23–25. There is no general key to families.

824

Former Hyderabad (Deccan), or Nizam's Dominions

Before 1947 the Nizam's Dominions, with their capital at Hyderabad in the Deccan, constituted the largest princely state in relations with Britain and the Indian Empire. They acceded to India in 1949 and in the following year were reorganized as Hyderabad. In 1956, however, the state was dismembered and on linguistic grounds its districts dispersed amongst Bombay (subsequently Maharashtra), Mysore (later Karnataka) and Andhra Pradesh, the last-named receiving the cities of Hyderabad and Nizamabad. For pre-1956 botanical works, however, Hyderabad has here been retained.

The former Dominions, though spreading over a considerable part of the Deccan (and once extended to include Berar, now in Maharashtra), were never botanically well documented. They were largely bypassed in the general expansion of the field in the late nineteenth and early twentieth centuries and moreover neither encompassed in a regional flora nor furnished with a separate general work (save for the selective forest floras of E. A. Partridge and M. S. Khan listed below). Some contributions towards a flora were published by M. Sayeedhuddin from 1935 through 1941 in *Journal of the Asiatic Society of Bengal* and *Journal of the Bombay Natural History Society*. Since 1956, parts of the Dominions have been covered by floras of individual districts.

Woody plants

KHAN, M. S., 1953. *Forest flora of Hyderabad State*. ix, 364 pp., illus. Hyderabad.

Descriptive manual covering 557 species, with keys to families and notes on distribution. [Not seen; cited by Nayar and Giri (1988; see **Superregion 81–84, Supraregional bibliographies**).]

PARTRIDGE, E. A., 1911. *Forest flora of HH the Nizam's Dominions (Hyderabad-Deccan)*. 2, 3, 60, viii, 433, iii, xi, xi pp. Hyderabad: Pillai.

Descriptive field-manual of more representative woody plants; includes key literature citations, vernacular names, and copious notes on local distribution, frequency, localities, timber, other products, uses, special features, and phenology (but not much on habitats); analytical keys to

genera; indices to all English, local and scientific names. The introductory section includes a descriptive organography (separately indexed) and a synoptic key to families.

825

Andhra Pradesh

Area: 276814 km^2 (exclusive of Yanam). Vascular flora: 2351 species (Pullaiah *et al.*, 1995; see below). – The greater part of modern Andhra Pradesh was before 1947 (save for a single princely state, Banganapalle) part of the old Madras Presidency. A renewal of the pre-1947 movement towards linguistically based states resulted in separation of the latter's Telugu-speaking areas in 1953 to form Andhra; in 1956 the Telugu districts of Hyderabad (**824**) were added. Also included here for convenience is Yanam, a former French enclave now part of Pondicherry Union Territory (see **828**).

Botanical exploration, especially in less accessible areas, has been uneven, with many gaps remaining, and apart from *Flora of the Presidency of Madras* by Gamble and Fischer (see **828**), no general coverage was available until the mid-1990s. Key early works include the large-scale color-plate *Plants of the coast of Coromandel* (1795–1819) by William Roxburgh and *Flora andhrica* (1859) by Walter Elliot. The latter, however, was left unfinished after its first volume, a lexicon of vernacular names; a second part would have contained a systematic treatment. It relates particularly to the vicinity of Vishakhapatnam (Vizagapatnam) in the northeastern part of the state and also covers the Northern Circars, areas previously examined in the late eighteenth century by Roxburgh. A successor *Flora andhrica* was submitted by S. K. Wagh as a thesis in Bombay in 1960 but never published. A renewal of botanical exploration began in the 1950s, with work towards the new state flora commencing in the 1980s.[42]

PULLAIAH, T., E. CHENNAIAH and P. SURYA PRAKASH BABU, 1995–98. *Flora of Andhra Pradesh State (India)*. 4 vols. 2071 pp., illus. Jodhpur: Scientific Publishers.

Briefly descriptive flora of seed plants (vols. 1–2, dicotyledons; vol. 3, monocotyledons) with keys, synonymy, references and citations, vernacular names, indication of distribution, habitat and phenology, and selected *exsiccatae*; bibliography (vol. 1, pp. 439–456) and index to botanical names at end of each volume (with a consolidated index in vol. 3). The general part (vol. 1) includes reviews of physical features, geology, climate, soils, vegetation and characteristic species, forests, botanical exploration and progress of knowledge, and a key to families. Volume 4 comprises full-page illustrations of selected species, with separate index.[43]

District and local works

Except in the Vizagapatnam area and Northern Circars – where Roxburgh and Heyne had been active – relatively little detailed floristic work was done prior to the mid-twentieth century. In the 1960s the Botanical Survey of India and universities with the aid of grants initiated the study of district floras as contributions towards a statewide account. A goodly number have now been published while others exist as theses. Current coverage is geographically well distributed and fairly representative of different habitats.

ELLIS, J. L., 1987–90. *Flora of Nallamalais*. 2 vols. 490 pp., 6 pls. (incl. map). Howrah: BSI. (Flora of India, series 3.)

Briefly descriptive flora of vascular plants with keys, documentation and some critical notes; bibliography (pp. 456–460) and index to botanical names at end. The general part contains brief reviews of physical features and vegetation along with a key to families. [A hilly, hitherto very poorly collected area of forest reserves in the south-central part of the state, mainly in Prakasam and Kurnool districts. Its highest peak reaches 917 m.][44]

LAKSHMINARAYANA, K., P. VENKANNA and T. PULLAIAH, 1997. *Flora of Krishna District, Andhra Pradesh, India*. New Delhi: MD Publications.

Concisely descriptive flora of seed plants with keys, synonymy, references, citations, vernacular names, representative localities with *exsiccatae*, and indication of phenology, distribution, habitat, uses, special features, etc.; bibliography and index to families at end. The general part is extensive, covering all aspects, and incorporates a key to families. [The district is on the central coast, between West Godavari to the northeast and Guntur to the southwest (separated by the Krishna River); most of it is low-lying.]

PULLAIAH, T. and B. RAVI PRASAD RAO, 1995. *Flora of Nizamabad (Andhra Pradesh, India)*. 452 pp., 30 text-figs., 2 maps. Dehra Dun: Bishen Singh Mahendra Pal Singh.

Briefly descriptive flora of vascular plants (708 species in 7969 km^2) with keys, synonymy, references and citations, localities with *exsiccatae*, and notes on distribution, habitat, frequency and phenology; bibliography (pp. 403–420) and index to all botanical names at end. The general part covers physical features, history, geology, soils, climate, land use, past botanical work and project methodology, vegetation and characteristic species, and a floristic analysis along with (pp. 29–41) a general key to families. [Formerly in Hyderabad (Deccan) (**824**) and situated northwest of Hyderabad; Adilabad is to the northeast and the state of Maharashtra to the northwest.]

PULLAIAH, T. and N. YESODA, 1989. *Flora of Anantapur District, Andhra Pradesh*. [iv], 325 pp. Dehra Dun: Bishen Singh Mahendra Pal Singh.

Descriptive flora of vascular plants (707 species over 19130 km^2); includes keys, brief synonymy, references and citations, vernacular names, and notes on occurrence, habitat, seasons, uses, and representative *exsiccatae*; references and indices to vernacular and botanical names at end. The general part covers environmental features, vegetation and details of the flora along with a general key to families (but no district map). [In the southwest part of the state; adjoins Chitradurg and Bellary districts in Karnataka.]

PULLAIAH, T., P. V. PRASANNA and G. OBULESU, 1992. *Flora of Adilabad District*. x, 284 pp. Delhi: CBS Publishers and Distributors.

Briefly descriptive flora (covering an area of 16128 km^2) with keys, synonymy with references and citations, representative *exsiccatae*, indication of frequency, and index to scientific names at end. The general part encompasses physical features, vegetation, forests (an important feature), characteristic plants, and previous and current botanical work along with a floristic analysis and key to families. [The district, in the northwest of the state, was before 1956 part of Hyderabad (Deccan) (**824**) and is northeast of Nizamabad district.]

RAO, R. S. and S. H. SREERAMALU, 1986. *Flora of Srikakulam District, Andhra Pradesh, India*. vi, 640 pp., 4 text-figs., 4 pls., 4 maps. Howrah: BSI. (Flora of India, series 3.)

Descriptive flora similar to that for West Godavari (see below); contains an addendum after the bibliography. [In extreme northeast of the state, bordered by Orissa State on the north and Vishakhapatnam district to the southwest; includes part of the Circars.]

RAO, R. S., P. VENKANNA and T. APPI REDDY, 1986. *Flora of West Godavari District, Andhra Pradesh, India*. [iv], ix, 520, [1] pp., text-fig., 4 pls., 4 folding maps. Meerut: Indian Botanical Society. (Flora of India series.)

Briefly descriptive flora (809 species in an area of 7795 km^2); includes keys to genera and lower taxa, synonymy with references and citations, vernacular names, localities, frequency, habitat, uses and, where appropriate, taxonomic commentary; literature (pp. 447–489) and indices at end along with a key to families. The general part includes historical notes and the background to the work, methodology, and reviews of physical features, geology, climate, vegetation and synusiae, biota, human activities, species of special interest, and a floristic analysis. [In Godavari River basin in east-central part of the state.]

VENKATA RAJU, R. R. and T. PULLAIAH, 1995. *Flora of Kurnool (Andhra Pradesh)*. iv, [4], 595 pp., 29 text-figs. (incl. 2 maps). Dehra Dun: Bishen Singh Mahendra Pal Singh.

Briefly descriptive flora of vascular plants (836 taxa in 18 799 km²) with keys, synonymy, references and citations, localities with *exsiccatae*, indication of distribution, frequency, phenology and habitat, and critical notes; bibliography (pp. 551–558) and index to all botanical names at end. The general part covers situation, physical features, geology, soils and climate, vegetation and characteristic species, features of the flora, useful plants, statistics, and new records along with (pp. 36–46) a key to families. [Kurnool is in the central-west of the state adjacent to the Karnataka border.]

826

Karnataka

Area: 192 204 km². Vascular flora: at least 3400 species (Saldanha, 1984; see below). Sharma *et al.* (1984; see below) list 3924 taxa inclusive of cultigens. – The former princely state of Mysore, centering on Bangalore and Mysore and also including the Hassan district in the northwest, was included in *Flora of the Presidency of Madras* by Gamble and Fischer (**828**). It was first enlarged in 1953 by the transfer of Bellary district from Madras and in 1956 transformed to encompass all predominately Kannada-speaking districts in northwestern Madras (i.e., South Kanara), Bombay (**822**) and Hyderabad (**824**). At the same time the linguistically distinct state of Kurg (or Coorg, in the Western Ghats) was incorporated. In 1973 the enlarged state acquired its present name.

The imperfect coverage by the Presidency flora as well as the works on the old state of Mysore by Lal Bagh superintendent J. Cameron (described below under **Woody plants**) and B. A. Razi of Mysore University have now been supplemented by two statewide treatments (one still in progress) and several district floras, with additional works in preparation. Nevertheless, the south remains better known than the north.[45]

SALDANHA, C. J. (assisted by S. R. RAMESH *et al.*), 1984–96. *Flora of Karnataka*. Vols. 1–2. Illus., col. pls. New Delhi: Oxford and IBH Publishing; Rotterdam: Balkema (vol. 1); IBH Publishing.

Descriptive flora with keys to all taxa, vernacular names, synonymy with references and citations, indication of habitat, phenology, local distribution and overall range, representative *exsiccatae* and illustrations and taxonomic commentary; index to all scientific names at end of volume. The general part furnishes a justification for the project, its geographical scope, sections on physical features, geology, soils, bioclimate, vegetation classification and description of formations with characteristic species, a history of floristic work, a general key to flowering plants (pp. 25–36), and a list of families covered. [In progress. Vol. 1 covers Magnoliaceae through Leguminosae (families 1–64); vol. 2 (by Saldanha alone), Podostemonaceae to Apiaceae (families 65–112).]

SHARMA, B. D. *et al.*, 1984. *Flora of Karnataka: analysis*. xi, 394 pp., 4 pls., map. Howrah: BSI. (Flora of India, series 2.)

Concise enumeration of seed plants (the gymnosperms coming between monocotyledons and dicotyledons) with synonymy, main citations, and occurrence by districts (those with records based only on literature asterisked); list of literature (pp. 349–352) and index to generic names at end. The introductory section covers geography, soils, climate, past work on the flora, vegetation and characteristic component species, endemic and threatened plants, a plan of the work and an analysis of the flora.

District and other local works

In addition to the works described below, the following systematic checklist, derived largely from *Flora of the Presidency of Madras* (**828**), covers the former princely state of Mysore: B. A. RAZI, 1950. A list of Mysore plants. *J. Mysore Univ.*, B, 7(4): 39–81. As elsewhere in India, there has been since the 1960s an interest in the production of district floras complementary to statewide works. Such floras at present cover, in the north, South Kanara (Arora *et al.*), the Deccan districts of the east including Bellary (Singh), and, in the south and west, Chikmagalur (Yoganarasimhan *et al.*), Coorg (Keshavamurthy and Yoganarasimhan), Hassan (Saldanha and Nicolson), Bangalore (Ramaswamy and Razi), and Mysore (Raghavendra Rao).

ARORA, R. K., B. M. WADHWA and M. B. RAIZADA, 1981. *The botany of South Kanara District*. [iv], 64 pp., 1 fig. (Indian J. For., Add. Ser. 3). Dehra Dun: Bishen Singh Mahendra Pal Singh.

Systematic enumeration (593 species) with localities, *exsiccatae* (all of the senior author), and indication of habitat; references (p. 64). The introduction surveys previous work, general features of the area and of its flora, and characteristic or rare or otherwise special plants. [An 'interim' work.]

KESHAVAMURTHY, K. R. and S. N. YOGANARASI-
MHAN, 1990. *Flora of Coorg.* 711 pp. Bangalore: Vimshat
Publishers.

Descriptive flora, with keys. [Not seen; cited from
Ramesh and Pascal (1997; see **802**).]

RAGHAVENDRA RAO, R., 1981. *A synoptic flora of
Mysore District.* xii, 674 pp., folding map. New Delhi: Today
and Tomorrow's. (International Bioscience Series 7.)

Systematic enumeration of native, naturalized and
other introduced species (1601 in all) including diagnostic
features, synonymy, vernacular names, distribution (with
representative *exsiccatae*), phenology, some critical remarks,
and notes on uses, etc.; references and indices. The enumeration is preceded by a general key to all taxa (pp. 27–382).
[Arrangement of families follows Cronquist.]

RAMASWAMY, S. V. and B. A. RAZI, 1973. *Flora of
Bangalore district.* 1, 739 pp., map. Mysore: Prasaranga,
University of Mysore.

Descriptive flora of higher plants, with keys, synonymy, references and citations, localities, and notes on habitat,
phenology, frequency, etc.; indices. An introductory section
covers sources for the work, geography, geology, vegetation,
history of botanical exploration, and agriculture and includes
a general key to families and a statistical synopsis of the flora.

SALDANHA, C. J. and D. H. NICOLSON, 1976. *Flora of
Hassan District, Karnataka, India.* viii, 915 pp., 132 text-figs.,
19 col. pls., map. New Delhi: Amerind Publishing Co. (for the
Smithsonian Institution and U.S. National Science
Foundation, Washington). (Reprinted with addenda, 1978.)

Concise descriptive flora of vascular plants (without
vernacular names); includes keys to all taxa, synonymy, references, indication of local range (with selected *exsiccatae*),
overall distribution, critical remarks, representative illustrations, and notes on phenology, occurrence, habitat, uses,
desiderata for future work, etc.; complete index. The introductory part gives descriptions of physical features, climate,
vegetation, and man's influence along with economic botany,
botanical history and the origin and progress of the flora
project, and a general key to families.[46]

SINGH, N. P., 1988. *Flora of eastern Karnataka.* 2 vols.
xiv, 794 pp., illus. (incl. some halftones). Delhi: Mittal.

Briefly descriptive but well-annotated flora of native,
introduced and commonly cultivated vascular plants (1421
taxa, with 138 on the basis of literature only, over an area of
92 353 km^2); includes keys to all taxa, synonymy, references
and citations (including illustrations), taxonomic and
nomenclatural commentary, phenology, distribution by districts with localities, notes on habitat, occurrence, uses,
special features, etc., and figures of representative species but
no vernacular names; list of references (pp. 726–729), supplementary name changes, and complete index at end. The
general part furnishes a justification for the work, prior
exploration and botanical studies, and sections on physical

features, geology and soils, climate, human influences, vegetation with formations and characteristic species, synusiae, an
analysis of the flora and its life-forms, lists of endemics and
threatened taxa and of plants of economic interest, new
records, and conventions. [The eight districts covered (Bidar,
Gulbarga, Raicher, Bijapur, Bellary, Chitradurg, Tumkur and
Kolar) encompass 48 percent of the total area of the state.][47]

YOGANARASIMHAN, S. N., K. SUBRAMANYAM and B.
A. RAZI, 1981. *Flora of Chikmagalur district, Karnataka,
India.* Dehra Dun: International Book Distributors.

Briefly descriptive flora; includes keys to genera and
species, synonymy (with references and citations), local
and overall distribution (with selected *exsiccatae*), phenology,
and notes on special features, etc.

Woody plants

CAMERON, J., 1894. *The forest trees of Mysore and
Coorg.* 3rd edn. viii, 334, xxxvi pp., maps. Bangalore: Mysore
Government Central Press. (Reprinted 1978, Dehra Dun:
Bishen Singh Mahendra Pal Singh. 1st edn., 1880.)

Systematic descriptive treatment of more common
native and introduced woody plants, with emphasis on trees;
includes vernacular names, citations of key literature and
figures, local range, and notes on uses, special features, silviculture, etc.; indices to all English and local vernacular as well
as scientific names. Appendices include classified tables of
uses, potential, special features, etc. [Not a complete woody
flora.]

827

Kerala

Area: 38 864 km^2 (except Mahé). Vascular flora:
no data, but Rama Rao (1914) already recorded 3535
species of seed plants for the former state of
Travancore. It is thus one of the diversity 'hot spots' in
India. – Kerala was organized in 1956 from
Travancore-Cochin (itself a 1949 merger of two
princely states) and adjoining western parts of the old
Madras Presidency; it encompasses predominately
Malayalam-speaking areas along the old Malabar
Coast. Included here also is Mahé, a former French
enclave now part of the Union Territory of
Pondicherry (see **828**).

Kerala's botanical history effectively begins in
the seventeenth century when Hendrik van Rheede,
governor of Cochin (then recently acquired by the
Dutch East India Company), caused *Hortus indicus*

malabaricus to be compiled, the first Western-style systematic botanical work in Asia (1678–93). It remains to the present the only 'general' account for the state. Several analyses, or 'interpretations', of this famous 12-volume work have been essayed, the latest (based in part on new field studies) being *An interpretation of van Rheede's* Hortus malabaricus by D. H. Nicolson, C. R. Suresh and K. S. Manilal (1988).[48] Subsequent studies of the flora, although beginning early in the nineteenth century with collections by Buchanan, were sporadic. However, the state of Travancore soon became relatively well documented through the works of Thomas F. Bourdillon (1908), its first conservator of forests, and his successor Rao Saahib M. Rama Rao (1914). The whole of the present state was covered in *Flora of the Presidency of Madras* (**828**).

From the mid-twentieth century there was a renewal of botanical exploration with the organization of the Botanical Survey of India circles at Pune in then-Bombay (BSI-WC) and Coimbatore in Tamil Nadu (BSI-SC) as well as the state Forest Research Institute, Calicut University, and the Tropical Botanical Garden. As elsewhere, from the early 1960s an emphasis was placed on district floras ahead of preparation of a state flora. Kerala was considered to be relatively underexplored with many poorly known species worthy of further investigation.[49] The rich forest flora in particular had been largely neglected since the time of Bourdillon.[50] Several florulas have now been published, some for administrative units, others for forest divisions; more await publication. At state level, the Poaceae have been revised in P. V. SREEKUMAR and V. J. NAIR, 1991. *Flora of Kerala – Grasses.* Calcutta: Botanical Survey of India.

District and local works

Kerala also has now a considerable range of district floras, most of them the result of directed support since the 1960s. Available accounts cover, from south to north, former Travancore (Rama Rao, 1914), Thiruvanathapuram (Mohanan and Henry, 1994), Calicut and Malappuram (Manilal and Sivarajan, 1982), Palghat and Kozhikode (Vajravelu, 1990), and Cannanore (Ramachandran and Nair, 1988). There are also floras for smaller areas: the forests of Thrissur (Sasidharan and Sivarajan, 1996), Palghat (Subramanian *et al.*, 1987), Thenmala (Subramanian, 1995), and 'Silent Valley' southwest of the Nilagiri Hills (Manilal *et al.*, 1988). For the Anamalais, see **802**. Research for a flora of Idukki in the botanically important central eastern uplands or 'High Range' is in progress.

MANILAL, K. S. and V. V. SIVARAJAN, 1982. *Flora of Calicut.* [ii], 2, 387 pp. Dehra Dun: Bishen Singh Mahendra Pal Singh.

Briefly descriptive manual of flowering plants (983 species in an area of about 600 km^2) with keys, synonymy, references, literature citations, indication of distribution, habitat and phenology, representative *exsiccatae*, and (where appropriate) uses along with some taxonomic commentary; indices to families and other scientific names at end. The general part covers physical features, climate, prior studies of the flora, the plan of the work, conventions, vegetation and characteristic species, synusiae, cultivated and introduced plants, and a floristic analysis; these are followed by the general key to families.

MANILAL, K. S. *et al.*, 1988. *Flora of Silent Valley: tropical rain forests of India.* xl, 398 pp., 9 text-figs. (incl. map), 9 tables. Calicut: The senior author.

Briefly descriptive flora without keys; includes synonymy with references and citations, distribution and occurrence, phenology, and representative *exsiccatae*; index to all scientific names at end. The general part covers physical features, geology, climate, origin of the name 'Silent Valley', the fauna, the biological and environmental importance of the basin, the forest and its composition, and special features of the flora (including several lists); references (pp. xxxix–xl). [A natural area in Malappuram district of great biological importance, with a high rainfall. See also VAJRAVELU below.]

MOHANAN, M. and A. N. HENRY, 1994. *Flora of Thiruvananthapuram.* [8], 621, [4] pp., some text-figs., map. Calcutta: BSI. (Flora of India, series 3.)

Briefly descriptive flora (covering 1270 species of seed plants in 2192 km^2) with keys, synonymy, references and citations, vernacular names, phenology, frequency, and local distribution; list of literature (pp. 566–576) and indices to botanical and vernacular (Malayalam) names. The general part encompasses geography, geology, climate and soils, an analysis of the flora, vegetation formations and characteristic species, human influences and prospects for conservation, special observations based on the work, and a key to families. [The district center of the same name is also known as Trivandrum.]

RAMA RAO, M., 1914. *Flowering plants of Travancore.* xiv, 495 pp. Trivandrum [= Thiruvananthapuram], Kerala: Travancore Government Press. (Reprinted 1976, Dehra Dun: Bishen Singh Mahendra Pal Singh.)

Systematic enumeration of seed plants (3535 species); includes Malayalam and English vernacular names, references to illustrations, and variously extended notes on habitat, distribution, uses and properties; indices to vernacular, generic and family names. A brief introductory section supplies the background to the work. [Partly superseded by MOHANAN and HENRY, 1994.]

RAMACHANDRAN, V. S. and V. J. NAIR, 1988. *Flora of Cannanore.* iii, 599 pp. Howrah: BSI. (Flora of India, series 3.)

A briefly descriptive flora (1132 species of seed plants) with keys, limited synonymy, references and literature citations, vernacular names, distribution (with some localities), habitat and phenology, and occasional taxonomic commentary; bibliography (pp. 557–564) and index to scientific names at end. The introductory section reviews past and current work along with physical features, geology, climate, soils, vegetation formations and characteristic species, the composition of the flora (with statistics) and plants of special interest.

SASIDHARAN, N. and V. V. SIVARAJAN, 1996. *Flowering plants of Thrissur forests.* 579 pp., 48 text-figs., 19 col. photographs on 3 pls., portr., map. Jodhpur: Scientific Publishers.

Briefly descriptive flora of flowering plants (1225 species in 1041 km²) with keys, synonymy, references and chief citations, frequency, habitat and phenology, wider distribution, occasional commentary, and selected *exsiccatae* with localities of collection; bibliography (pp. 529–531) and index to all botanical names. The general part furnishes background on botany in Kerala and on the Thrissur forests and reviews of physical features, geology, climate (but no rainfall data), soils, and (in some detail) vegetation formations and characteristic species, and phytogeography and endemic plants; there is also a floristic analysis and a general key to families. [Most of the area is below 1200 m.]

SUBRAMANIAN, K. N., 1995. *Flora of Thenmala (and its environs).* liii, 516 pp., halftones, map. Dehra Dun: International Book Distributors.

Briefly descriptive flora of seed plants (875 species) with keys, synonymy, references and principal citations, vernacular names, phenology, representative *exsiccatae* with localities, and notes on uses, medicinal properties, etc.; bibliography and indices to scientific and Malayalam names. The general part includes reviews of physical features, geology, climate, rainfall, forest types with characteristic species, altitudinal zonation, stratification, and plantations as well as a floristic analysis, considerations of phytogeography and biological spectra (life-forms), special features of the flora, and the key to families. [The area is part of the old Shencottah forest division (between Shencottah and Quilon) in former Travancore; it faces serious threats to its biota.][51]

SUBRAMANIAN, K. N., N. VENKATASUBRIMANIAN and V. K. NALLASWAMY, 1987. *Flora of Palghat.* 149 pp., 23 text-figs., map. Dehra Dun: Bishen Singh Mahendra Pal Singh.

Annotated systematic checklist of seed plants with limited synonymy, diagnostic features, citations of *exsiccatae*, and uses; literature (pp. 147–149) but no index. [Arrangement follows the Bentham and Hooker system. The work covers the Palghat forest division in the Palghat Gap between Kerala and central Tamil Nadu, straddling the Palghat and Kozhikode districts and partly south of Silent Valley; it is part of a more general programme on the forest flora of Kerala.]

VAJRAVELU, E., 1990. *Flora of Palghat District including Silent Valley National Park, Kerala.* [viii], 646 pp., 14 pls., 2 maps. Calcutta: BSI. (Flora of India, series 3.)

Briefly descriptive flora (1355 species of seed plants in 4400 km²) with keys, synonymy, references and citations, phenology, and local distribution; list of references and index to all botanical names at end. The general part covers geography, climate (parts of the district are seasonally very wet), soil, botanical history, vegetation, synusiae, plants of special interest or of economic importance, statistics, and the plan of the work; main key to families, pp. 33–42.

Woody plants

In addition to the classic account by Bourdillon (see below), reference can be made to K. BALASUBRAMANYAN, K. SWARUPANANDAN and N. SASIDHARAN, 1985. *A field key to the identification of indigenous arborescent species of Kerala forests.* Peechi: Kerala Forest Research Institute; and N. SASIDHARAN, 1987. *Forest trees of Kerala: a checklist with an index to important exotics.* Peechi. Readers should also consult the field-key by PASCAL and RAMESH and the manual by SOMASUNDARAM (for both, see **820, Woody plants**).

BOURDILLON, T. F., 1908. *Forest trees of Travancore.* xxxii, 456 pp., pls. Trivandrum [= Thiruvananthapuram], Kerala: Travancore Government Press. (Reprinted 1976, Dehra Dun: Bishen Singh Mahendra Pal Singh.)

Briefly descriptive flora of trees and other woody plants (582 species) with keys, vernacular names, localities, habitats, and notes on wood structure and properties, the timber trade, and forest protection where appropriate; indices to vernacular and botanical names at end. The general part includes a glossary and a synopsis of families but no general key.

Pteridophytes

NAYAR, B. K. and K. K. GEEVARGHESE, 1993. *Fern flora of Malabar.* 424 pp., 179 text-figs. (incl. map). New Delhi: Indus Publishing.

Copiously descriptive treatment with keys, synonymy, references and citations, notes on distribution and habitat, and a few taxonomic remarks but no *exsiccatae*; concise bibliography and index to all botanical names at end. The general part includes an account of earlier studies as well as sections on physical features, climate, vegetation, and fern ecology and biology. [Covers northern Kerala from Kannur (Cannanore) and Wynad to Palakkad, or somewhat under half the area of the state.]

828

Tamil Nadu

Area: 130 069 km² (exclusive of Pondicherry and Karikal). Vascular flora: *c.* 5540 species and additional infraspecific taxa (Nair *et al.*, 1983–89; see below). – Included here are works covering the whole of the former Madras Presidency, including the well-known *Flora of the Presidency of Madras* (1915–38) by J. S. Gamble and C. E. C. Fischer. This historic polity of the Raj after 1947 absorbed some contiguous princely states but then in accordance with the language policy underwent successive territorial excisions, notably in 1953 with the formation of Andhra (now Andhra Pradesh; see **825**) and the expansion of Mysore (now Karnataka; see **826**) and in 1956 with the formation of Kerala (**827**). In 1969 what remained of Madras, the largely Tamil-speaking districts, acquired its present name. Also included here are the formerly French enclaves of Karikal and Pondicherry, politically part of the Union Territory of Pondicherry.[52]

Madras was in the late eighteenth and early nineteenth centuries the center of activities of the botanical 'United Brotherhood', with the missionary (and student of Linnaeus) Johann G. Koenig being from 1768 the first resident botanist. He was followed most notably by Johan Rottler, Benjamin Heyne, Roxburgh, Bernhard Schmid and, finally, Robert Wight who began his 35-year stay in 1819. Including its botanically important hill districts, the Presidency continued as a primary focus for another century and more, marked by the establishment of a local (and regional) herbarium in 1853 (initially at Udaghamandalam (Ootacamund) but afterwards in Coimbatore) and, later, another at Madras Museum. A botanical survey was established in 1890 under M. A. Lawson in connection with the new Botanical Survey of India, while R. H. Beddome, J. S. Gamble and C. E. C. Fischer were in turn associated with the forestry services. In the first third of the twentieth century, the Presidency became the subject of the most critical of the five key regional floras of British India, the already-mentioned – and still much-consulted – *Flora of the Presidency of Madras*. Contemporaneously, the 'hill stations' of the Nilagiri, Palni and Shevaroy hills were worked up by P. F. Fyson in two floras (1915–20, 1932; see **802**). By independence

the future Tamil Nadu had become one of the best-known tracts in India.

Key contributors since 1947 have been the Botanical Survey of India, which assumed direct control of the Madras Herbarium in 1955 and established there its Southern Circle (BSI-SC), and Fr. K. M. Matthew at St. Joseph's College in Tiruchirapalli and what had been the Sacred Heart College in Shembanganur in the Palni Hills. The BSI-SC has published an enumeration of the state flora, while Fr. Matthew has continued in the tradition of earlier writers with his series on the flora of the Tamilnadu Carnatic (covering approximately the central third of the state and including the Shevaroys). Both his works, particularly the recent *Excursion flora*, will be generally useful in the Tamil Nadu lowlands and lower hills.[53]

GAMBLE, J. S. and C. E. C. FISCHER, 1915–38. *Flora of the Presidency of Madras.* 3 vols. 2017 pp., folding map. London: Adland. (Reprinted 1957, 1967, Calcutta: Botanical Survey of India.)

Concise descriptive manual-flora of seed plants of the former Presidency of Madras together with neighboring princely states (including Travancore, Cochin and Mysore), with diagnostic keys to all taxa; the enumerative species headings incorporate synonymy (with references and citations), vernacular names, fairly detailed general summaries of local range, and notes on habitat, special botanical features, uses, etc. In the introductory part are a glossary and a description of the general features of the flora, while in volume 3 are complete indices to vernacular and botanical names.[54] For additions, see S. KARTHIKEYAN and B. D. SHARMA, 1983. A catalogue of species added to the Gamble's *Flora of the Presidency of Madras. J. Bombay Nat. Hist. Soc.* **80**: 63–79.

NAIR, N. C., A. N. HENRY, G. R. KUMARI, V. CHITHRA and N. P. BALAKRISHNAN, 1983–89. *Flora of Tamil Nadu, I: Analysis.* 3 vols. Coimbatore: Botanical Survey of India, Southern Circle.

Systematic enumeration of seed plants (*c.* 5540 taxa) with synonymy (including references and citations), diagnosis, and distribution by districts; index to families in each volume (but no general index in vol. 3). The introductory part includes sections on geography, vegetation (including specialized formations), forests, reserves, man's impact, ethnobotany, crops, and plant-based industries along with botanical history before and after 1955. In vol. 3 is a bibliography (pp. 152–171). [The family arrangement basically follows the

Bentham and Hooker system. Vol. 1 covers Ranunculaceae through Sambucaceae; vol. 2, Rubiaceae through Ceratophyllaceae; vol. 3, monocotyledons and gymnosperms.]

Partial works: Tamilnadu Carnatic

The Tamilnadu Carnatic as covered by Fr. Matthew and his team contains 2037 species in an area of 27794 km² in the central-eastern part of the state.

MATTHEW, K. M., Fr., 1983. *The flora of the Tamilnadu Carnatic.* 3 vols. lxxxiv, 2154, [ii] pp., 113 pls., 2 maps. Tiruchirapalli: Rapinat Herbarium. (Flora of the Tamilnadu Carnatic 3.) Accompanied by *idem*, 1982. *Illustrations on the flora of the Tamilnadu Carnatic.* 980 pls. (pls. 961–980 halftones). Tiruchirapalli. (Flora of the Tamilnadu Carnatic 2.); *idem*, 1988. *Further illustrations on the flora of the Tamilnadu Carnatic.* 834 pls. Tiruchirapalli. (Flora of the Tamilnadu Carnatic 4.); and *idem*, 1981. *Materials for a flora of the Tamilnadu Carnatic.* x, 469, [i] pp., 2 maps, 20 pls., 1 plan. Tiruchirapalli. (Flora of the Tamilnadu Carnatic 1.)

The main work is a documentary descriptive flora of seed plants with keys to all taxa, synonymy (with references and key citations), Tamil names (in local and Roman script), local and overall distribution, altitudinal range (where appropriate), some critical notes, and indication of habitat, frequency, phenology, etc. The introductory part gives the basis and plan of the work and its conventions along with a treatment of biogeography and general key to families (pp. xliii–lxxxiv), while at the end are author abbreviations, several appendices, a gazetteer and complete indices to the flora along with its two companion works (save of course for the second volume of *Illustrations*). The *Materials* furnish an extensive general treatment of the region and its vegetation and main botanical features along with a systematically arranged list of all *exsiccatae*, while the two volumes of *Illustrations* furnish an iconography of full-page figures of nearly all species (with a complete index in the second volume).[55]

MATTHEW, K. M., Fr., 1991. *An excursion flora of central Tamilnadu, India.* xlv, 647 pp., 5 pp. figs., 2 folding maps. New Delhi: Oxford and IBH Publishing Co. (Reprinted 1995 with 8 col. pls.) Tamil version (in Tamil script): *idem*, 1991. (*Maiya Thamizhaga KalaVagai Thavaraviyai.*) Transl. S. John Britto. liii, [v], 940 pp., 4 pls., 2 folding maps. Tiruchirapalli: Rapinat Herbarium, St. Joseph's College.

Concise, briefly descriptive manual of flowering plants with keys, essential synonymy, indication of phenology, and notes on habitat, frequency, and local and extralimital distribution (with altitudinal range where appropriate); index to families and genera at end. A relatively brief introduction reviews the basis of the Tamilnadu Carnatic flora-series and furnishes conventions, signs and abbreviations along with a key to families. The Tamil version includes additional vernacular/botanical lexica (pp. 905–940).

Partial and local works: other areas

In Tamil Nadu there evidently has not been a developed programme for district floras, perhaps because as a whole the polity has botanically been better collected than many other parts of India. The only works thus included here are for Coimbatore, long a teaching center, and Courtallum, a 'classical' botanical locality near the Kerala border. For the western mountains, see also **802**.

CHANDRABOSE, M. and N. C. NAIR, 1988. *Flora of Coimbatore.* xxviii, 398 pp., 6 text-figs., map. Dehra Dun: Bishen Singh Mahendra Pal Singh.

Briefly descriptive flora of native and introduced seed plants (850 species and non-nominate infraspecific taxa); includes keys, synonymy (with references and citations), local range (with selected *exsiccatae*), phenology, occasional taxonomic commentary, and notes on uses, etc.; references and complete index at end. The introductory part includes notes on physical features, climate, soils, botanical exploration and the plan of and basis for the work. [In the western part of state; site of some tertiary institutions.]

NAIR, K. K. N. and M. P. NAYAR, 1986–87. *Flora of Courtallum.* 2 vols. 442 pp., 11 text-figs., map, 4 pp. pls. Howrah: BSI. (Flora of India, series 3.)

Briefly descriptive flora of flowering plants with keys, synonymy, full references and citations, representative *exsiccatae*, indication of phenology, distribution, frequency, occurrence and habitat, and figures of selected species; bibliography and complete index in vol. 2. The general part covers the locality and its importance, botanical history, and reviews of geography, geology, climate, rainfall, soils, vegetation and characteristic species, synusiae and phytogeography; a floristic analysis with statistics and the key to families follow. [Courtallum, 7 km from Tenkasi at the northern end of the Tirunelvelli Hills, though an area small in extent (14 × 7 km), is a 'classical locality' of nineteenth-century collectors, especially Robert Wight. Families are arranged following the Cronquist system.]

829

Sri Lanka (Ceylon)

Area: 65610 km². Vascular flora: 2900 native flowering plant species (PD; see **General references**); 314 native pteridophytes (Sledge, 1982; see below).

After more than three centuries of formal investigation, the flora of Sri Lanka, also known as

Taprobane or Serendib, is tolerably well known, with a relatively considerable stock of floristic works. Summaries of the flora in the period of Dutch rule included *Museum zeylanicum* (1717) by Paul Hermann (published posthumously by William Sherard) and *Flora zeylanica* (1747) by Linnaeus, and *Thesaurus zeylanicus* (1737) by Johannes Burman; these were based on different Hermann collections of 1672–77 respectively now in the Natural History Museum, London and in the Institut de France in Paris. These were succeeded under the British by *Catalogue of the indigenous and exotic plants growing in Ceylon* (1824) by Alexander Moon, *Enumeratio plantarum Zeylaniae* (1864) by G. H. K. Thwaites, and *Handbook to the flora of Ceylon* (1893–1900; see below) by Henry Trimen, the last of which supplied a solid foundation for the phytogeographical studies of J. H. Willis in the early twentieth century. Two checklists were also published: *A systematic catalogue of the flowering plants and ferns . . .* (1885) by Trimen and *A revised catalogue* by Willis (1911). Between the wars, Arthur H. G. Alston, active at Peradeniya Botanic Garden from 1925 to 1929 and later at the Natural History Museum, prepared a supplement to the *Handbook* (1931) and, in response to the needs of higher education, a students' manual for the Kandy/Peradeniya area (1938).[56]

After independence, however, a perception grew that Henry Trimen's *Handbook* was becoming not only obsolete but also scarce and fragile.[57] In 1959 Bartholomeusz A. Abeywickrama of the University of Ceylon published a revised checklist, a precursor to a possible new flora. The initiation in the late 1960s of a Smithsonian Institution cooperative research programme, including projects in plant ecology, then led Abeywickrama to propose inclusion of a revision of the *Handbook*. With the aid of 'surplus funds' available under United States Public Law 480, this undertaking was launched in 1968 by Abeywickrama and F. Raymond Fosberg; much new collecting, some of it by visiting specialists, and other investigations were accordingly carried out. The first fascicle of the *Revised handbook* was published in 1973, but progress was delayed and by 1979 only one more had been produced. Alternative arrangements became necessary, and in 1980 a new series of volumes made its debut under the editorship of Fosberg and Meliyasena D. Dassanayake. Meanwhile, many precursory papers had begun to appear in *Ceylon Journal of Science* and elsewhere. By 1991 seven volumes had been published, but termination of the U.S. support programme and changing interests then resulted in a transfer of the project to Kew with support from the then-Overseas Development Administration of the United Kingdom. Under the new arrangements, six further volumes were produced through 1999 and the work is now close to completion.[58]

Bibliographies

ALWIS, N. A., 1978. *Bibliography of scientific publications relating to Sri Lanka*. v, 244 pp. Colombo: Social Science Research Centre, National Science Council of Sri Lanka. (Mimeographed.) [Botany, pp. 121–150; 304 main and cross-references.]

GOONETILEKE, H. A. I., 1970–76. *A bibliography of Ceylon*. 3 vols. Zug, Switzerland: Inter-Documentation Co. (Bibliotheca asiatica 5 (vols. 1–2), 14 (vol. 3).) [Botany and horticulture, 222 items; vegetation, forests and soils, 180 items.]

ABEYWICKRAMA, B. A., 1959. A provisional check list of the flowering plants of Ceylon. *Ceylon J. Sci., Biol. Sci.* 2(2): 119–240.

Systematic census of flowering plants, with limited synonymy and cross-references to Trimen and Hooker's *Handbook* (see below); index to families. Partly supersedes J. C. and M. WILLIS, 1911. *A revised catalogue of the flowering plants and ferns of Ceylon, native and introduced*. 188 pp. Colombo.

DASSANAYAKE, M. D., F. R. FOSBERG and W. D. CLAYTON (eds.), 1980– . *A revised handbook to the flora of Ceylon*. [2nd edn.] Vols. 1– . Illus., maps. New Delhi: Amerind/Oxford and IBH; Rotterdam: Balkema. (Vols. 1–7 also published for the Smithsonian Institution and the National Science Foundation, Washington, D.C.) (1st edn. (not completed), 1973–77, Peradeniya.) Complemented by R. PETHIYAGODA and R. RODRIGO (comp.), 1993. *A provisional index to* A revised handbook to the flora of Ceylon. [iii], 251 pp. Colombo: The Wildlife Heritage Trust of Sri Lanka.

The *Revised handbook* is an amply descriptive flora of higher plants with keys to genera and species, synonymy (with references and citations, in some cases rather extensive), vernacular names, indication of distribution (and origin) together with *exsiccatae* and localities, taxonomic commentary, and notes on habitat, vegetation type, frequency, phenology, elevations etc.; no indices. Some species distribution maps and illustrations have been published, but not consistently. A brief, partly historical preface precedes the systematic accounts (covering given families or parts thereof) in each volume. The *Provisional index* covers the first seven volumes. [The work is intended as a successor to Trimen's *Handbook* (see next entry) and somewhat resembles it in style. As of 1999, 13 volumes had been published, with the fourteenth and final volume scheduled for 2000. A full cumulative index is planned.][59]

TRIMEN, H. and J. D. HOOKER, 1893–1900. *Handbook of the flora of Ceylon*. 5 vols., atlas of 100 col. pls. London: Dulau. (Reprinted 1974, Dehra Dun: Bishen Singh Mahendra Pal Singh.) Continued as A. H. G. ALSTON, 1931. *Handbook of the flora of Ceylon*, 6: *Supplement*. vi, 350 pp. London.

Briefly descriptive flora of vascular plants, with keys to genera and species, synonymy, references and pertinent citations, vernacular names (English, Sinhalese and Tamil), fairly detailed indication of local range (with some *exsiccatae*), generalized summary of extralimital distribution, taxonomic commentary, and notes on habitat, uses, etc. Volume 1 incorporates a general sketch of the physical features and climate of the island, while vol. 5 contains a synopsis and general key to orders and families, accounts of the vegetation and of the history of botanical exploration, and complete indices. Volume 6 comprises additions and corrections to the original work. The fine atlas (in larger format) contains 100 hand-colored plates of characteristic species.[60]

Partial work: Kandy/Peradeniya area

The following work by Alston is useful as a students' flora in the central region.

ALSTON, A. H. G., 1938. *The Kandy flora*. xvii, 109 pp., 404 text-figs. Colombo: Ceylon Government Press.

Keyed, briefly descriptive illustrated manual-flora of seed plants of the Kandy/Peradeniya district, with limited synonymy, vernacular names, and notes on phenology, life-form, etc.; complete index at end. The introductory section includes a description of the general features of the flora and vegetation, while an appendix contains a glossary of terms as well as a 'List of grasses, rushes, and flowerless plants'.

Woody plants

ABEYESUNDERE, L. A. J. and R. A. DE ROSAYRO, 1939. *Draft of first descriptive check-list for Ceylon*. 115 pp. (Checklists of the forest trees and shrubs of the British Empire 4). Oxford: Imperial Forestry Institute.

Enumeration of trees and shrubs recorded from Ceylon, with synonymy, citations of pertinent literature, vernacular names, generalized indication of local range, and notes on habitat, phenology, special botanical and dendrological features, and uses; list of references and indices to generic and vernacular names. An appendix lists records of exotic Coniferae.

ASHTON, M. *et al.*, 1997. *A field guide to the common trees and shrubs of Sri Lanka*. vii, 432 pp., numerous text-figs., 16 col. pls., maps. Colombo: WHT Publications.

Field-guide to 704 native and common introduced species in 95 families featuring keys (to families only, pp. 43–50) and formatted species accounts with closely integrated line drawings; the text encompasses scientific and vernacular names, cross-references to Trimen and Dassanayake *et al.*, status (and place of origin if required), maximum height, habit, botanical description, indication of habitats, and notes on uses. Family descriptions are also furnished. The guide proper is preceded by an introduction to Sri Lanka, its vegetation, and history of land use along with an illustrated organography and ecography as well as (pp. 33–42) a glossary, while the final sections are on plant uses (including landscape architecture, a genre with a history in the country of perhaps two millennia) and medicinal plants; indices to all names conclude the work.

WORTHINGTON, T. B., 1959. *Ceylon trees*. 429 pls., map. Colombo: Colombo Apothecaries' Co.

Illustrated atlas of photographic plates with accompanying text, depicting the most commonly encountered native and exotic trees of Ceylon; the text includes descriptive notes on habitat, local range, life-form, biology, distinguishing features, wood, properties, uses, etc., as well as limited synonymy (with cross-references to Trimen and Hooker's *Handbook*), vernacular names, and extralimital range; index to all vernacular, generic and family names. The introductory section includes addenda, corrigenda, a glossary, and a list of references.[61]

Pteridophytes

Until publication of the following contributions, no separate work had ever been produced; recourse was customarily had to Beddome's *Handbook* (**810–40**). A new treatment for the *Revised handbook* (see above) is in preparation.

ABEYWICKRAMA, B. A., 1978. *A check list of pteridophytes of Sri Lanka*. 16 pp. (Sri Lanka MAB Natl. Committee, Publ. 3). Colombo: National Science Council of Sri Lanka.

Systematic checklist, with synonymy and numbered literature references (cf. pp. 14–15); index to genera. [At least partly superseded by the following work.]

SLEDGE, W. A., 1982. An annotated check-list of the Pteridophyta of Ceylon. *Bot. J. Linn. Soc.* **84**: 1–30.

Nomenclatural checklist (314 native and 18 naturalized species), with indication of status (an asterisk for alien taxa) and, where appropriate, corresponding page references to and names in Beddome's *Handbook* or other works; necessary commentary is given in 76 end-notes preceding the general references (pp. 28–30). The preliminary matter includes a historical review, phytogeographic considerations, status lists and necessary new combinations. [The work summarizes more than 30 years of studies (including field research in 1950–51), mostly published as separate papers.]

Region 83

Eastern India and Bangladesh

Area: no data. Vascular flora: no data, but of all India its northeast has the richest flora, with some 7000 flowering plants in 121 828 km^2 (Arunachal Pradesh, Assam, Meghalaya, Tripura, Mizoram, Manipur and Nagaland).[62] – Included herein are all the states and territories from Bihar and Orissa eastward to the Burmese frontier, as well as Bangladesh. However, the hill and mountain zones of the Himalaya, including the Darjiling district, are excluded (for these, see **Region 84**).

From 1861 to 1905, the entire area was under two administrative units: the great Presidency of Bengal, stretching from Bihar and Orissa to East Bengal (with before 1861 an even larger area), and Assam (separated from Bengal in 1874). It was the former territory which was covered by David Prain in his *Bengal plants* (1903). Associated with these were the princely states of eastern India and various tribal territories. In 1905 and 1912 two different partitions – the first very controversial – of the British-ruled territories resulted in three provinces: Bihar and Orissa, Bengal, and Assam. In 1933 the Eastern States Agency was organized and three years later Orissa was separated from Bihar. Subsequent changes have resulted in the formation of 10 distinct polities (exclusive of Arunachal Pradesh), of which one, Bangladesh, is an independent state.

From the time of Roxburgh to the early twentieth century, substantial botanical work was done in the region and a number of basic floras written, notably *Bengal plants* (Prain, 1903), *Botany of Bihar and Orissa* (Haines, 1921–25, with supplement by Mooney, 1950),

and the incomplete *Flora of Assam* (U. N. Kanjilal *et al.*, 1934–40). Comparatively few floras of substance have appeared since 1950, but exploration in more remote areas has continued and some source materials, recently joined by a few district floras in better-known areas such as Meghalaya and West Bengal, have been published. To these must be added *Flora of Bangladesh*, begun in 1972 as a serial work but appearing less rapidly than the *Flora of Pakistan*. Significant contributions of the 1980s were *Flora of Tripura* by D. B. Deb and *Forest Flora of Meghalaya* by Haridasan and Rao. The *Flowering plants of eastern India* (Mitra, 1958), a students' manual, ceased to appear after one volume and moreover suffers from flawed and tortuous keys. Generally speaking, therefore, much work has been done in the region, both before and after 1950, but by modern standards available botanical knowledge cannot be considered well consolidated.

Bibliographies. General and supraregional bibliographies as for Division 8 and Superregion 81–84.

Indices. General and supraregional indices as for Division 8 and Superregion 81–84.

830

Region in general

See also **810–40** (all works). The whole region (with adjacent parts of the Himalaya) is nominally covered in J. N. Mitra's manual but of this largely compiled work the only volume published has been that on the monocotyledons.

MITRA, J. N., 1958. *Flowering plants of eastern India*. Vol. 1: *Monocotyledons*. xx, 389 pp., map. Calcutta: World Press (Private).

Manual-key to flowering plants with brief diagnoses, limited synonymy, vernacular names, and generalized indication of internal range (with some localities mentioned); glossary and index to all botanical names at end. The introductory section includes a synopsis of orders and families. [Covers the entire region from Bihar and Orissa to Assam; however, the rather tortured layout of the keys (which are in part non-analytical) may render this book somewhat difficult to use. No further volumes were published.]

831
Bihar (including Chota Nagpur)

See also 833 (PRAIN). – Area: 173 876 km². Vascular flora: no data. The present state of Bihar was initially organized in 1936 from division of the former province of Bihar and Orissa, itself formed in 1912 from the second partition of Bengal (and Assam). Several neighboring princely states were, however, grouped administratively in 1933 as the Eastern States Agency; some of these acceded to Bihar after 1947. The Sambalpur district, part of the Chota Nagpur group of forest districts and from 1905 to 1912 wholly associated with Bengal, is, however, now in Madhya Pradesh.[63]

Botanical work in Bihar began with the Bengal survey by Buchanan-Hamilton after 1800, but it was the holy mountain of Parasnath (close to the Great Trunk Road) and other parts of the hilly south (including the districts of Chota Nagpur) which attracted the most attention, beginning with a visit by J. D. Hooker in 1848. A first list was produced by T. Anderson in 1863 (*On the flora of Bihar and the mountains of Parasnath*; in *J. Asiat. Soc. Bengal*, II, **32**: 189–218). Other works through the early twentieth century came from C. B. Clarke (on Parasnath), J. S. Gamble, H. H. Haines, and J. J. Wood.

The 'regional' flora of Haines (1921–25) and its supplement by Mooney (1950) both deal with the former province along with neighboring princely states. Recent work, however, suggests that some areas of the state are insufficiently collected, and exploration has accordingly continued.

HAINES, H. H., 1921–25. *The botany of Bihar and Orissa*. 2 vols. x, 199, 1350 pp., 2 maps. London: West, Newman. (Reprinted 1961 in 3 vols., Calcutta, Botanical Survey of India.) Complemented by H. MOONEY, 1950. *Supplement to* The botany of Bihar and Orissa. iii, 294 pp. Ranchi: Catholic Press.

Briefly descriptive flora of vascular plants, with keys to genera and species, synonymy, vernacular names, fairly detailed indication of local range with citation of *exsiccatae* and general summary of extralimital distribution, taxonomic commentary, and notes on habitat, uses, etc.; complete index to botanical and vernacular names. The introductory section includes chapters on physical features, climate, vegetation,

ecology, and botanical exploration. The *Supplement* includes numerous additions and corrections along with accounts of the physical features and floristics of Bihar and Orissa.

District and local works

A few new district floras have appeared to complement that of Wood published nearly a century ago. Nevertheless, improved knowledge of the uplands is still needed.

SINGH, M. P., 1986. *Flora of Patna: dicotyledons*. iii, 343 pp., map, 2 figs., 32 pls. New Delhi: International Books and Periodicals Supply Service.

Briefly descriptive flora (674 species in 80 km²) with keys, vernacular names, and the author's *exsiccatae*; no index. The general part contains reviews of physical and biological features along with a floristic analysis and key to families.

VARMA, S. K., 1981. *Flora of Bhagalpur: dicotyledons*. xii, 414 pp., 33 pls., 2 maps. New Delhi: Today and Tomorrow's.

General features and previous work, plan and methodology, analysis of the flora, new records (pp. 29–30) and general key; descriptive treatment with keys, synonymy, vernacular names, 'field notes', author's voucher, indication of phenology, and occasional commentary; bibliography (pp. 368–378) and appendix but no index.

WOOD, J. J., 1902. *Plants of Chutia Nagpur including Jaspur and Surguja*. 170 pp., folding map, 2 folding tables (Rec. Bot. Surv. India 2(1)). Calcutta. (Reprinted 1977, Delhi: Periodical Export Book Agency/Dehra Dun: International Book Distributors.)

Concise systematic enumeration (pp. 77–159, covering 1424 vascular plant species and a few from other groups) with brief notes and vernacular names; index to Santali names at end. The enumeration is preceded by an alphabetical lexicon of all botanical and local vernacular names with their respective equivalents and, additionally for the taxa, occasional comments on economic significance. [The area of coverage extends from Santal Parganas in the east through the Hazaribagh and Ranchi plateaux and Singhbhum as far as 82° E, thus also including Surguja and part of Raigarh in Madhya Pradesh (**819**).]

Woody plants

HAINES, H. H., 1910. *Forest flora of Chota Nagpur*. ii, 634, xxxvii pp., map. Calcutta: Superintendent of Government Printing. (Reprinted 1974, Dehra Dun: Bishen Singh Mahendra Pal Singh.)

Keyed, briefly descriptive manual-flora of trees, shrubs, lianas, vines and more important herbs of this then-extensively forested upland area, with vernacular names, indication of local range, and notes on habitat, special features, etc.; indices. [In southern Bihar, with western portions including Surguja in Madhya Pradesh (**819**). Remains a key reference for the southern uplands.]

832

Orissa

See also **831** (HAINES); **833** (PRAIN). – Area: 155 707 km^2. Vascular flora: 2561 spontaneous species (after Saxena and Brahmam, 1994–96; see below). Part of the Presidency of Bengal until 1904, the greater portion of the then-British districts of modern Orissa were included with Bihar in 1912 as the province of Bihar and Orissa. They became a separate province in 1936 and at the same time the mainly Oriya-speaking districts of Koraput and Ganjam (formerly in the Madras Presidency) were incorporated. After 1947 the state of Orissa attained its present extent through accession of neighboring princely lands formerly in the Eastern States Agency.

Earlier studies of Orissa were largely a function of activities in Bihar; as a distinct field of study it attracted separate attention mainly after World War II. The Mooney supplement to Haines's flora (see **831**), while focusing on Orissa, did not effectively account for its present southern part; likewise, Gamble and Fischer's flora of the Madras Presidency (**828**) contained but few references to the districts now in Orissa. As in adjacent Madhya Pradesh (**819**) and Andhra Pradesh (**825**), many hilly areas remained under-collected.

The reorganization of the Botanical Survey of India in the 1950s brought about renewed interest in Orissa as it did elsewhere. Much new collecting – some in connection with wildlife studies – was done in the ensuing three decades and beyond, furnishing a better basis for a state flora. Such a work (see below) finally appeared in 1994–96, though not under BSI auspices; a major gap in second-level coverage was thus closed.

SAXENA, H. O. and M. BRAHMAM, 1994–96. *The flora of Orissa.* 4 vols. 2918 pp., 75 text-figs., map (folded in inside back cover of vol. 1). Bhubaneswar: Regional Research Laboratory, Council of Scientific and Industrial Research, India/Orissa Forest Development Corporation, Plantation (A) Division.

Descriptive flora (2727 species of which 166 commonly cultivated) with keys to all taxa, common synonymy, references and citations, vernacular names, local and extralimital distribution and habitat (with representative *exsiccatae* for less common species), indication of frequency, and notes on phenology, taxonomy, uses, etc.; illustrations of lesser-known species; excluded records; indices to generic names in each volume with complete cumulative index in vol. 4. The general part in vol. 1 includes a historical survey of the province, its physical and environmental features, vegetation, botanical history and current work, and the plan of the work; general key to families, pp. xxxvii–lxiv. Lists of endemic and threatened plants are also included. [Arrangement of families follows the Bentham and Hooker system, but their limits are after Hutchinson.]

District and local work
The only work of consequence is one on Similipahar in the Mayurkhanj district; this may be considered preparatory to the same authors' state flora.

SAXENA, H. O. and M. BRAHMAM, 1989. *The flora of Similpahar (Similipal), Orissa.* x, 231 pp., folding map. Bhubaneswar: Regional Research Laboratory, Council of Scientific and Industrial Research, India.

Systematic enumeration of a portion of the Mayurbhanj district (1076 species, of which 64 cultivated), with emphasis on a planned national park; entries include essential synonymy, citations, localities with *exsiccatae*, vernacular names, and notes on phenology, occurrence, habitat, etc.; indices to all vernacular and scientific names at end. The general part covers physical features, vegetation formations, ethnography, potential medicinal plants (pp. 10–21) and timber species (pp. 21–53) as well as the plan of the work and conventions.[64]

833

Old Bengal Region

This here refers to the whole of the old Presidency of Bengal as constituted up to 1904 and for which Prain's *Bengal plants* (1903), still in part current, was written. It then included most of modern Bihar and Orissa (separated in 1912) as well as the state of West Bengal and the now-independent nation of Bangladesh. Associated with it were Tripura and the realms of what later became the Eastern States Agency (**819**). Prain's flora, however, does not cover the Darjiling district (**845**, with Sikkim) or Sylhet (now part of Bangladesh but then in Assam).

Serious botanical work began with the establishment of the Asiatic Society of Bengal in 1784 and, soon after, the Botanic Garden. Key works of the ensuing century or so include *Hortus Bengalensis* of Roxburgh (1814), *Calcutta Flora* by J. W. Masters (1839–40, not completed), *Hortus suburbanus Calcuttensis* by J. O. Voigt (1845), *Indigenous plants of Bengal* by J. Long (1859), a list by George King of principal species in volume 20 of Hunter's *Statistical account of Bengal* (1877), and the works of David Prain including his *Bengal plants* (see below), *Flora of the Sundribuns* (1903), and *The vegetation of the districts of Hughli-Howrah and the 24-Pergunnahs* (1905). The two last-named summarized current knowledge of the important Lower Ganges Delta and its extensive mangals as well as greater Calcutta. Considerable attention was also paid to the exploration of hilly and mountainous parts including Chittagong and Sylhet (now in Bangladesh), Darjiling, the Khasi Hills (later in Assam and now in Meghalaya), and Parasnath and Chota Nagpur (now in Bihar). After 1900 or so, however, botanical documentation generally was considered adequate and little new work was done until the 1950s save that prosecuted by the forestry services.

Bibliography

BASAK, R. K., 1973 (1976). The bibliography on the flora and vegetation of Bengal with an introductory note. *Bull. Bot. Surv. India* 15: 22–38. [Quite comprehensive and inclusive of North Bengal (845) as well as West Bengal and Bangladesh. The main list is in three parts, preceded by a fairly detailed historical review.]

PRAIN, D., 1903. *Bengal plants.* 2 vols. 1319 pp., map. Calcutta: Government of Bengal. (Reprinted 1963 in 2 vols., Calcutta: Botanical Survey of India.)

Descriptive manual of vascular plants, with keys to genera and species, an artificial general key to genera, limited synonymy, principal literature citations, vernacular names, generalized indication of local range (with some localities), and notes on taxonomy, habitat, occurrence, diagnostic features, life-form, properties, uses, etc.; complete general index. Extensive descriptions are provided only for genera and families; those for species are brief. An introductory section includes a synopsis of families and a general description of Bengal (in the pre-1905 sense) with notes on botanical subdivisions and floristics. [The work includes all the area defined under the heading save for, as noted, the Darjiling and Sylhet districts.]

834

West Bengal (India)

See also 833 (PRAIN) and (for Darjiling district) 845. – Area: 87 853 km². Vascular flora: no data. Modern West Bengal includes the former princely state of Cooch Behar, part of the district of Purnea (formerly in Bihar), and the Himalayan foothill districts of Jaipalguri and Siliguri as well as, along the Hooghly, the erstwhile Danish settlement of Serampore and the French enclave of Chandernagore. It also includes the former Sikkimese district of Darjiling, for geographical reasons here combined with Sikkim (845).

The story of botanical exploration and documentation is until 1947 that of old Bengal, with the last significant floristic works appearing before 1910. Independence and social changes, however, brought about a renewed awareness of the importance of natural resources and by the 1960s the state came to be viewed as imperfectly known (Basak, 1973; see 833, **Bibliography**). As part of the renewal of the Botanical Survey of India a strategy was developed for new district surveys and the first, for Howrah across from Calcutta – previously documented in 1905 by Prain as noted under 833 – was completed in 1968. Most districts have now been examined and their florulas compiled in the form of theses at Calcutta and other universities in the state. Some of these have been published, generally well after completion of the original research: Howrah (1979), Murshidabad (1984), the former 24–Parganas in part (1993) and Bankura (1994).[65] In 1997 the first volume of a state flora appeared; this also encompasses Darjiling district, excluded from *Bengal plants*.

BOTANICAL SURVEY OF INDIA, 1997– . *Flora of West Bengal.* Vols. 1– . Illus. (part col.), maps (endpapers). Calcutta: Botanical Survey of India. (Flora of India, series 2.)

Briefly descriptive flora of seed plants with keys, synonymy, references, literature citations, vernacular names (where known), local range (by district), and indication of phenology and habitat; indices at end. The extensive general part in vol. 1 covers physical features, geology, soil, climate, botanical history, vegetation and floristic types (with lists), and uses (including medicine) and continues with a list of references (pp. 67–78) and general key to families (pp. 79–111).

The color plates depict 'highlights'. [Vol. 1 covers Ranunculaceae through Moringaceae following the Bentham and Hooker system. Three more volumes will complete the work.]

District and local works

For works on Darjiling district, see **845**. Some pre-1947 local floras are noted under **833**.

BENNET, S. S. R., 1979. *Flora of Howrah district*. viii, 406 pp. New Delhi: Periodical Export Book Agency; Dehra Dun: International Book Distributors.

Descriptive flora (encompassing 605 species) with keys, synonymy and references, literature citations (key works), brief indication of local range along with the author's *exsiccatae*, and notes on habitat, frequency, occurrence, phenology, uses, special features, etc.; index to scientific names. In the introductory section are brief surveys of physical and biotic features, vegetation (nearly all anthropogenic), land use, and earlier botanical work in the area along with floristic statistics and a general key to families.

GUHA BAKSHI, D. N., 1984. *Flora of Murshidabad District, West Bengal*. xvi, 440 pp., 12 figs. (incl. 4 maps). Jodhpur: Scientific Publishers.

Briefly descriptive flora of native, naturalized and commonly cultivated flowering plants (636 species), with keys, synonymy (including references and citations), vernacular names, local range (with representative *exsiccatae*), overall distribution, indication of phenology, habitat and frequency, and notes on uses, etc.; index to all names. The introductory part encompasses physical features, climate, soil, vegetation (with characteristic or common species), previous work in the area, and the scope of the present study along with statistics (pp. 31–33). [Murshidabad is about in the middle of the state, north of Calcutta.]

NASKAR, K., 1993. *Plant wealth of the Lower Ganga Delta*. 2 vols. xxv, 810 pp., 41 text-figs., 25 col. pls., 2 maps. Delhi: Daya Publishing House.

Briefly descriptive flora of flowering plants in the former 24–Parganas district (1175 species over 14 136 km²) with synonymy, citations, vernacular names, indication of distribution and habitat, phenology, selected *exsiccatae* and notes on uses, properties and special features; selected literature (pp. 747–753), indices to all botanical and recorded Bengali vernacular names, and a classified index to recorded plant uses. The substantial introductory part (vol. 1) encompasses geography, climate, land use, wildlife, past botanical work and current research, vegetation, floristics, and a key to families. [This work covers a large proportion of the area encompassed in Prain's *Vegetation of the districts of Hughli-Howrah and the 24-Pergunnahs* (noted under **833**). Calcutta was separated from 24-Parganas in 1903 and in 1988 the remainder divided respectively into South and North 24-Parganas.]

SANYAL, M. N., 1994. *Flora of Bankura District (West Bengal)*. v, 555 pp., map. Dehra Dun: Bishen Singh Mahendra Pal Singh.

Descriptive flora of seed plants (938 species in 6871 km²), the native and introduced taxa not separately indicated; includes keys, synonymy, citations, Bengali names, selected *exsiccatae*, and notes on habitat, frequency and phenology but no extralimital ranges. The introductory section includes sections on physical features, vegetation and floristics, and previous work as well as a general key to families; at the end are an appendix of recently recorded species and indices to botanical and vernacular names. [Situated *c.* 150 km west-northwest of Calcutta.]

Woody plants

[SHEBBEARE, E. O.], 1957. *Trees of the Duars and Terai*. ii, 114, vii pp., illus. Alipore: Superintendent of Government Printing, West Bengal (for the Directorate of Forests).

Illustrated descriptive account with vernacular names and distinguishing features highlighted; index to all binomials at end. The brief general part (by V. S. Rao) precedes a short foreword by the author. [Covers the lower hills and adjacent plains in Buxa, Kalimpong, Kurseong and Jalpaiguri Divisions. Only trees with a trunk of 6 in. (*c.* 15 cm) or more in diameter are included; 'exotics' are asterisked.]

835

Bangladesh

See also **833** (PRAIN). Area: 143 998 km². Vascular flora: no data. – Formerly part of Bengal, Bangladesh was between 1947 and 1971 part of Pakistan; since then it has been independent. It includes the former Assamese district of Sylhet but not the former princely state of Tripura (**836**) which after 1947 acceded to India.

Key early works on Bangladesh, which among other features encompasses some two-thirds of the Sundarbuns, are noted under **833**; further details appear in the bibliography by Basak (1973; see there). Relatively few floristic works appeared from 1900 until independence; among these was *List of plants of Chittagong Collectorate and Hill Tracts* (1925) by the forester R. L. Heinig.

A serial documentary work, *Flora of Bangladesh*, commenced publication in 1972; it has yet to be completed. It has given a renewed impetus to botanical exploration. Among recent works of more restricted scope is M. K. ALAM, 1988. *Annotated check list of the woody flora of Sylhet forests.* 153 pp. (Bull. For. Res. Inst. Chittagong, Pl. Tax. Ser. 5). Chittagong.

KHAN, M. S. and A. M. HUQ (eds.), 1972– . *Flora of Bangladesh.* Fasc. 1– . Illus.; map in each fascicle. Dacca: Agricultural Research Council, Bangladesh (for the Bangladesh National Herbarium). (Parts 1–3 reprinted 1975.)

Documentary descriptive flora of higher plants, published in fascicles with one or more families; each family treatment encompasses keys to genera and species, synonymy (with references and citations), vernacular (including English) names, detailed indication of local distribution in 19 (now 64) districts (with representative *exsiccatae*), summary of extralimital distribution, notes on origin (if naturalized), habitat, occurrence, karyotypes, special features, etc., and a list of references consulted and index. [62 families in 51 fascicles have so far been published, the most recent in 1996; no systematic order has been followed.]

836

Tripura (India)

See also **833** (PRAIN). – Area: 10477 km². Vascular flora: 1545 species (Deb, 1981–83; see below). After 1947 this state, once known as Tipperah or Hill Tipperah and previously part of the East India States Agency, opted to remain with India. Until the mid-1950s it was botanically scarcely known, having been visited by but two explorers (Deb, 1963; see **Bibliography** below). In 1956–61 and again in 1962, however, considerable collections were made for the Botanical Survey by D. B. Deb and an initial bibliographical paper published in 1963. Deb subsequently produced a full flora (described below).

Bibliography

DEB, D. B., 1963. Bibliographical review on the botanical studies in Tripura. *Bull. Bot. Surv. India* **5**: 49–58. [Includes extensive background information, including previous botanical and forestry work.]

DEB, D. B., 1981–83. *The flora of Tripura State.* 2 vols. 30 figs. (part halftones). New Delhi: Today and Tomorrow's. (International Bioscience Series 9–10.)

Briefly descriptive flora (with family arrangement after Hutchinson); includes keys, synonymy (with references and many citations), vernacular names, local range with selected *exsiccatae*, overall distribution, phenology, and indication of habitat and occurrence; index to all scientific names in each volume (no consolidated index). The introductory part (volume 1) covers the plan, basis and scope of the work along with physical features, climate, soil, vegetation, phytogeography, forest types (with characteristic species), the fossil record, history of botanical work (including forestry and plant introduction), and a summary of the flora with statistics; bibliography (pp. 29–30).

837

Old Assam Region (Assam, Meghalaya and Mizoram)

Area: 122099 km² (Assam, 78523 km²; Meghalaya, 22489 km²; Mizoram, 21087 km²). Vascular flora: no data. – Encompassed here are the states of Assam and Meghalaya (the latter created in 1972) and the former Lushai or Mizo Hills district of the first-named, now the Union Territory of Mizoram (its status likewise granted in 1972). For Arunachal Pradesh (formerly the North-East Frontier Tracts of Assam, but separated in 1954 as a centrally adminstered agency), see **847**. Before 1947 the old province of Assam also included Nagaland (**839**) and, south of the Khasi Hills, the former Sylhet district (now part of Bangladesh; see **834**).

The Assam Region was not covered by a separate flora until the launch of the *Flora of Assam* project by P. C. Kanjilal (apart from a checklist of woody plants, *Descriptive List of Trees and Shrubs of the Eastern Circle* (1925), by the same author). The *Flora*, begun in 1934, unfortunately was never completed (of the monocotyledons only the grasses were ever published) and its earlier volumes have, not unnaturally, a bias towards woody plants reflecting the authors' professional interests. The remaining monocotyledons were covered in Mitra's 1958 manual for eastern India (**830**). The botanical importance of the Assam Region was recognized with the formation at Shillong in 1956 of the Eastern Circle of the Botanical Survey of India (BSI-EC); this assumed control of the existing Assamese government herbarium.[66]

KANJILAL, U. N. *et al.*, 1934–40. *Flora of Assam.* Vols. 1–5 (in 6). Shillong: Assam Government.

Briefly descriptive flora of seed plants (with in earlier volumes a bias towards woody species); includes keys to genera and species, vernacular names, relatively detailed indication of local range, and notes on phenology, etc.; indices to all botanical and vernacular names in each volume. The introductory section (vol. 1) includes descriptive accounts of physical features, climate, ecology, etc., a glossary, a synopsis of families (following the Bentham and Hooker system), and a list of references. [Not completed; lacks any families of monocotyledons except for the Poaceae (by N. L. Bor) in volume 5. For contributions on non-graminoid monocotyledons, see A. S. RAO and D. M. VERMA, 1972–79. Materials towards a monocot flora of Assam, I–V. *Bull. Bot. Surv. India* **12**: 139–142; **14**: 114–143 (1975); **15**: 189–203 (1976); **16**: 1–20 (1977); **18**: 1–48.]

I. Assam

No state flora has been published in succession to that of Kanjilal *et al.* (see above). A useful recent local contribution is M. ISLAM, 1990. *The flora of Majuli.* [i], ii, [ii], [6], 436 pp. Dehra Dun: Bishen Singh Mahendra Pal Singh. It covers 692 flowering plants recorded from the Brahmaputra island of Majuli (1080 km²) located in Jorhat district.

II. Meghalaya

District and local works

BALAKRISHNAN, N. P., 1981–83. *Flora of Jowai and vicinity, Meghalaya.* 2 vols. 19 text-figs., halftones, 2 maps. Howrah: BSI. (Flora of India, series 3.)

Briefly descriptive flora, with keys, vernacular names, local and altitudinal range (with selected *exsiccatae*), phenology, and notes on habitat, etc.; index (vol. 2). The introductory part covers physical and biological features, human impacts and botanical history, and includes references (pp. 35–37) and a key to families. [Jowai is in the eastern part of the state.]

Woody plants

HARIDASAN, K. and R. R. RAO, 1985–87. *Forest flora of Meghalaya.* 2 vols. Illus. (incl. 7 halftone pls.). Dehra Dun: Bishen Singh Mahendra Pal Singh.

Briefly descriptive flora of woody plants (1151

species), with keys to all taxa, synonymy (with references and citations), vernacular names, local range (with selected *exsiccatae*), overall distribution, representative figures, phenology, and notes on habitat, associates, etc.; lists of important monocotyledons and gymnosperms in vol. 2 along with a bibliography (pp. 870–879) and complete indices to botanical and vernacular names. The introductory part in vol. 1 deals with geography (at present five districts in the state), physical features, geology, climate, soil, forest types, human impacts, conservation matters (including list of rare and/or threatened species), history of botanical studies, publications, and the plan of and basis for the work.

Pteridophytes

BAISHYA, A. K. and R. R. RAO, 1982. *Ferns and fern allies of Meghalaya State, India.* x, 162 pp., 36 pls. Jodhpur: Scientific Publishers.

Descriptive flora (nearly 250 species), with keys, synonymy (including references), citations, distribution, and notes on habitat, frequency, etc.; bibliography (pp. 149–151) and complete index at end. The introductory part includes a map of the state, collecting localities, physical and biological background, basis for the work, and (pp. 17–20) a general key to families.

III. Mizoram

The area of present-day Mizoram was delimited on organization of the Lushai Hills district of Assam in 1891.

FISCHER, C. E. C., 1938. The flora of the Lushai Hills. *Rec. Bot. Surv. India* **12**(2): 75–161.

Systematic enumeration of one moss and 1359 vascular plants with vernacular names, localities with *exsiccatae*, and indication of phenology. [Not claimed to be complete.]

838

Manipur

See also **837** (KANJILAL *et al.*). – Area: 22356 km². Vascular flora: no data. A former princely state, Manipur acceded to India in 1948 but between 1956 and 1972 was accorded Union Territory status.

Botanical exploration and documentation of Manipur has been spasmodic. A first checklist, *Plants of Kohima and Muneypore (India)* (1887) was produced by C. B. Clarke from collections made during border-lands exploration and surveys in the late nineteenth century (the latter in part related to the contemporary extension of British control in Burma). In the late 1950s, the reorganized Botanical Survey through Debendra B. Deb conducted a new field programme; this was a primary basis for his enumeration of flowering plants in 1961–62, the most recent floristic account available. For 2000, the first volume of a new state flora has been announced.

DEB, D. B., 1961. Monocotyledonous plants of Manipur Territory. *Bull. Bot. Surv. India* 3: 115–138; and *idem*, 1962. Dicotyledonous plants of Manipur Territory. *Ibid.*: 253–350.

Enumeration of known flowering plants, with brief diagnoses, synonymy (with references), and generalized indication of local range (including citations of *exsiccatae*). The introductory sections of both parts contain accounts of physical features, climate, and botanical exploration and long lists of references, together with statistics of the flora.

839

Nagaland

See also **837** (KANJILAL *et al.*). – Area: 16 527 km². Vascular flora: no data. The present state of Nagaland, previously part of Assam of which it became a district in 1881, was first organized as a separate administrative unit in 1957 (absorbing part of the North-East Frontier Agency) and granted statehood in 1963. Although some lists of collections from this area (including that by C. B. Clarke mentioned under Manipur) have been published, botanical penetration of this mountainous land on the Burmese border has been relatively slow and to date no separate flora or checklist has been published, except for the pteridophytes. For 2000 two volumes of a new state flora – respectively on Orchidaceae and a part of the dicotyledons – have been announced.

Pteridophytes

JAMIR, N. S. and R. R. RAO, 1988. *The ferns of Nagaland.* [v], 426 pp., illus., map. Dehra Dun: Bishen Singh Mahendra Pal Singh.

Descriptive manual (280 species) with keys, synonymy,

references and citations, phenology, Indian and overall distribution, indication of frequency and habitat, and localities with *exsiccatae*; bibliography and index to all botanical names at end. The general part includes accounts of physical features, geology, climate, vegetation and forests, the pteridophyte flora, past botanical work, the present study, and statistics along with a key to families.

Region 84

The Himalaya: *terai* (foothill) and mountain zones

Area: *c.* 490 000 km². Vascular flora: no precise figures seen, but the Himalaya are a 'hot spot' for plant biodiversity (cf. Barthlott *et al.*, 1996 under **General references**). Kendrick (in Campbell and Hammond, 1989, see **General references**) estimated 3000 flowering plants for the Western Himalaya and 4000 for the Eastern, with high endemism. – This region extends along the range from the upper Indus River (in northern Kashmir) to the North-East Frontier Agency north of Assam, passing through Kashmir (with Jammu), the Punjabi Himalaya, the hill and mountain tracts of Uttar Pradesh, Nepal, Sikkim and northern Bengal, and Bhutan. Its southern limit is along the base of the foothills; to the north it is bounded by China.

Botanical exploration in the Himalaya, a unit of great phytogeographic importance and now listed as a world diversity 'hot spot', dates from early in the nineteenth century; until at least the 1950s, however, collecting was comparatively patchy, with much emphasis on a small number of areas. Large tracts, including Nepal, Bhutan and what is now Arunachal Pradesh, remained for a variety of reasons poorly known, documented mainly by a hodgepodge of technically uneven works of relatively local scope. The mountainous terrain and multiplicity of polities has in any case hindered attempts at overall floristic syntheses of the region. In the last three to four decades, however, several field trips and systematic/floristic analyses by Japanese botanists under the overall direction of Hiroshi Hara (and his successors at Tokyo University) have been carried out. With a special interest in floristic and phytogeographic links between Japan, East Asia and the Himalaya, this team have compiled several synthetic works of varying scope on the east-central zone (eastern Nepal to Bhutan). In the late 1960s Hara's group combined with botanists at the then-British Museum (Natural History), home to major early and modern collections from Nepal, to put together the first modern systematic account for that country. Much collecting has reportedly also been carried out in the Northwest Himalaya, Sikkim and Arunachal Pradesh under the Botanical Survey of India and other agencies.

The results of the Japanese and British groups mentioned above are embodied in some outstanding modern floras: *Flora of Eastern Himalaya*, of which three 'reports' (1966–75) have appeared, and *Enumeration of the flowering plants of Nepal*, in three volumes (1978–82; see **844**). The pteridophytes have been revised for the Northwest Himalaya by Dhir (1980). District and state floras and other works have also appeared for all polities except Arunachal Pradesh. There is also a recent 'Alpenflora' for the Northwest Himalaya (Rau, 1975; see **803/I**). A desirable future step, with the material now available, would be the compilation of a critical synthetic enumeration for the whole region, which would serve to eliminate, as in Mexico and Central America, much superfluous synonymy.[67]

Bibliographies. General and supraregional bibliographies as for Division 8 and Superregion 81–84.

Regional bibliographies

OHBA, H. and S. AKIYAMA, 1992. A bibliography of literature of Himalayan plants (taxonomy and flora in 1980–1990). *Newsl. Nat. Hist.* **9**: 14–34, map (with an addendum of 2 pp. in offprints). [Alphabetically arranged with annotations relating to novelties or other taxonomic changes.]

THOTHATHRI, K. and A. R. DAS, 1984. *Bibliography on the botany of the Eastern Himalayas.* Vol. 1. [iv], 72 pp. Howrah: Botanical Survey of India. [Organized into five main sections: general, Bhutan, Nepal, Sikkim and Tibet; arrangement within each by author. Volume 1 covers literature to 1972. No subsequent volumes have been seen.]

Indices. General and supraregional indices as for Division 8 and Superregion 81–84.

840

Region in general

See also **810–40** (all works). Apart from the general floras for South Asia, no single work covers the whole of the Himalaya and its southern foothills. The Eastern Himalaya is covered by the critical series described below; this is now complemented by the irregular serial *The Himalayan Plants*, edited by H. Ohba and (for the first two volumes) S. B. Malla (1988– , Tokyo).[68] For the Western Himalaya (from Kashmir and Ladakh east to the Nepal–Sikkim border), a work of considerable utility is O. POLUNIN and J. D. A. STAINTON, 1984. *Flowers of the Himalaya* (Oxford: Oxford University Press; supplement, 1988, Delhi, by Stainton alone; paperback reissue, 1997: Delhi). This includes concise descriptions, indication of distribution and habitat, and 'synoptic' keys.

HARA, H. (ed.), 1966. *The flora of Eastern Himalaya.* 744 pp., 68 text-figs, 40 pls. (some col.). Tokyo: University of Tokyo Press. Continued as *idem*, 1971. *The flora of Eastern Himalaya: second report.* 393 pp., 24 pls. (some col.) (Bull. Univ. Tokyo Mus. 2). Tokyo; and H. OHASHI (ed.), 1975. *The flora of Eastern Himalaya: third report.* xv, 458 pp., 33 pls., other illus., maps (Bull. Univ. Tokyo Mus. 8). Tokyo.

The main work comprises a systematic enumeration, without keys, of the non-vascular (except algae) and vascular plants of eastern Nepal, Sikkim and northern Bengal; it includes rather extensive synonymy (with references and citations), detailed indication of local ranges, general summaries of extralimital distribution, some taxonomic commentary, and notes on habitat, uses, etc.; gazetteer and index to all botanical names. The introductory section includes chapters on karyology and phytogeography, together with a selected list of references. The *Second report* contains additional records based upon further collecting in Nepal, Sikkim and Bhutan as well as three revisionary and cytological studies; there is also an additional list of references. The *Third report* is based on the fifth expedition in 1972 and includes an additional list of vascular plants (pp. 13–205) as well as a list of bryophytes, a number of special studies including revisions, an additional bibliography, and indices to localities and to botanical names. [These well-executed volumes have been based on the results of at least five successive Tokyo University-sponsored expeditions, but also account for older literature.]

Pteridophytes

DHIR, K. K., 1980. *Ferns of North-Western Himalayas.* 158 pp., 20 text-figs. (incl. map). Vaduz: Cramer/Gantner. (Bibliotheca pteridologica 1.)

Systematic enumeration (*c.* 264 species), with synonymy and citations, distribution (with *exsiccatae*), altitudinal range, and indication of frequency; references (pp. 135–146) and index. The introductory part accounts for geography and physical features, climate, soils, vegetation (including zones, 'guilds' and characteristic species), and species biology. [Covers Jammu and Kashmir, Himachal Pradesh and the hill districts of Uttar Pradesh or Uttarkhand.]

KHULLAR, S. P., 1994. *An illustrated fern flora of West Himalaya.* Vol. 1. xl, 506 pp., 168 pls., 3 maps. Dehra Dun: International Book Distributors.

Descriptive flora with synonymy, references and citations, detailed local distribution along with overall range, indication of habitat and altitudinal spread, and short critical notes; bibliography (pp. 459–496) and index to all scientific names at end. The introductory part includes a synoptic list, a summary of the classification adopted, and a general key to families. [This first of two volumes covers 28 families from Botrychiaceae through Aspleniaceae and 173 species and 9 *Asplenium* nothospecies in Jammu and Kashmir, Himachal Pradesh and the hill tracts of Uttar Pradesh. The total fern flora is some 325 species.]

841

Kashmir (with Jammu)

See also **793** (all works); **803** (all works), **812** (BAMBER; PARKER). – Area (whole of Jammu and Kashmir): 101 283 km². Vascular flora (whole of Kashmir): 2928 species (Singh and Kachroo, 1977; see below). As circumscribed here, Kashmir (with Jammu) is bounded on the north and northwest by the upper course of the Indus River and on the west by the 'cease-fire line' which forms the *de facto* boundary between Pakistan and India (Jammu and Kashmir being claimed by both countries). The remaining areas of Kashmir not here included – among them Gilgit, Baltistan and Ladakh – are treated with Region 79 except for Aksai Chin, here referred to Region 76.

Botanical exploration of Kashmir began early in the nineteenth century, with one of the best known of the early collectors being the Frenchman Victor Jacquemont, in 1831. Activity has continued ever since, especially in the early and mid-twentieth century, and there have been numerous floristic and popular publications on what is primarily a Eurasian flora. Jammu, at somewhat lower elevation, has had relatively less activity, with few published accounts.

No general flora of the state has been published, though still-useful works on wild flowers (by Fr. E. Blatter) and on the woody flora (by W. J. Lambert) appeared between the wars; the latter was revised by Javeid (1978–79). There are, however, two recent florulas respectively for Jammu and the vicinity of Srinagar (the latter in the Vale of Kashmir).[69]

Woody plants

LAMBERT, W. J., 1933. *List of trees and shrubs for Kashmir and Jammu forest circles.* ii, 40 pp. (Forest Bull. India 80). Calcutta.

Annotated list of known trees and shrubs, with concise indication of localities (including observations by the author or literature citations); index to generic names. Supplements Parker's *Forest flora of the Punjab* (1924; see **812**), which covers part of this area.

I. Kashmir

Partial works

BLATTER, E., [1927–29]. *Beautiful flowers of Kashmir.* 2 vols. 2 col. frontisps., 62 pls. (mostly col.). London: Staples and Staples. (Reprinted 1984.)

Popular descriptive treatment, with keys only to species; includes notes on diagnostic features, local occurrence, habitat, altitudinal zonation, phenology and general distribution, some English names, and indices to scientific and English names in both volumes. A glossary is also included. Only a selection of the more conspicuous species is treated but the work features a lavish array of watercolors. Nomenclature follows *Flora simlensis* (**842**).

Woody plants

JAVEID, G. N., 1978–79. Forest flora of Kashmir: a check list, I–II. *Indian Forester* **104**: 772–779; **105**: 148–170.

Part I of this work is introductory, covering physical features, climate, past explorations, historical phytogeography, and forest types and composition along with a summary. Part II comprises the checklist (with localities) and a list of references; 375 tree and shrub species are accounted for. [A revision of Lambert's checklist.]

SINGH, G. and P. KACHROO, 1976 (1977). *Forest flora of Srinagar and plants of neighbourhood.* x, 278 pp., 11 pls., tables, map. Dehra Dun: Bishen Singh Mahendra Pal Singh.

Part of this work comprises a systematic enumeration of the vascular plants of the Srinagar district (661 species), inclusive of synonymy (with references), vernacular names, local distribution, phenology, occasional critical remarks, and indication of habitat and altitudinal zonation; bibliography (pp. 233–238), extensive appendix on ethnobotany, and index to all names. The lengthy preliminary section includes descriptions of physical and biotic features, history of botanical work, floristic statistics and units, an account of the outstanding features of the flora in different localities, with lists (special emphasis being given to the Srinagar woodland), and a description of the vegetation and floristics of the Srinagar area; a list of new plant records and species concludes the section.

II. Jammu

SHARMA, B. M. and P. KACHROO, 1981–83. *Flora of Jammu and plants of the neighbourhood.* vii, 413 pp. Dehra Dun: Bishen Singh Mahendra Pal Singh. Accompanied by *idem*, 1983. *Illustrations to the flora of Jammu and plants of neighbourhood.* 303 pls. Dehra Dun.

Briefly descriptive manual-flora of seed plants; includes keys to species, synonymy (with references and citations of principal works), and notes on habitat, overall and altitudinal range, phenology, etc.; index to scientific names. The introductory part comprises sections on physical features, climate, vegetation, phytogeography and floristics, and an appendix (pp. 361–373) comprises a plant list of alpine Poonch Kashmir (over 2000 m) with summary table. The accompanying atlas comprises full-page figures. [Species known only from outside Jammu are merely enumerated.]

842

Himachal Pradesh

See also **810** (STEWART and BRANDIS); **812** (BAMBER; PARKER). – Area: 55 673 km². Vascular flora: no data. Prior to 1947 this area consisted of the Punjab Hill States, Chamba, and the Punjabi hill district of Kangra. Himachal Pradesh was initially formed from the princely states and in 1966 the Kangra district was incorporated. From 1956 to 1971 the state was a Union Territory. Its present capital is the renowned 'hill station' of Shimla (Simla), in the time of the Raj a summer seat of Indian governments.

Botanical exploration began early in the nineteenth century, facilitated by the establishment of Shimla in 1822; the area was over the next century and beyond extensively collected by visitors and residents. Some local accounts were written, of which the most important (and still useful) was *Flora simlensis* (1902; 2nd edn., 1921) by Henry Collett (see below).

Establishment of the BSI Northern Circle in Dehra Dun in 1955 provided an additional stimulus to detailed exploration. A 'state flora analysis' appeared in three volumes in 1984 as a project of that Circle, the first consolidated account – albeit not a descriptive flora – to be published. Some partial works have also appeared.

CHOWDHERY, H. J. and B. M. WADHWA, 1984. *Flora of Himachal Pradesh: analysis.* 3 vols. 860 pp., map. Howrah: BSI. (Flora of India, series 2.)

Systematic enumeration of flowering plants; includes limited synonymy (with references), citations, localities with *exsiccatae* or other sources, and overall distribution; index (in volume 3). The introductory section covers vegetation, forests, parks, land economy and the history of botanical work. [Now out of date.]

District and local works

ASWAL, B. S. and B. N. MEHROTRA, 1994. *Flora of Lahaul-Spiti.* iii, 761 pp., 27 pls. (part col. and incl. 1 map). Dehra Dun: Bishen Singh Mahendra Pal Singh.

Briefly descriptive flora of 985 species of seed plants with keys, synonymy, literature citations, indication of distribution and habitat, selected *exsiccatae*, phenology and notes; lexicon of abbreviations, bibliography and index at end. The ample introduction includes a key to families. [Comprises an area of 12 210 km² in the north-central and northeast parts of the state, partly bordering on Kashmir.]

COLLETT, H., 1921. *Flora simlensis.* 2nd edn. lxviii, 652 pp., 200 text-figs., map. Calcutta: Thacker, Spink. (Reprinted 1971, Dehra Dun: Bishen Singh Mahendra Pal Singh. 1st edn., 1902.)

Briefly descriptive illustrated semi-popular manual-flora of seed plants (about 1300 species) of the Shimla district; includes keys to genera and species, English names, local and extralimital range, and notes on habitat, phenology, biology, special features, etc.; index to all botanical and English names at end. An introductory section (by the Kew botanists W. B. Hemsley and W. T. Thistleton-Dyer) includes accounts of botanical exploration, vegetation, phytogeography, and notes on particular life-form classes along with a synopsis of families, a glossary, and a list of references.[70]

DHALIWAL, D. S. and M. SHARMA, *Flora of Kulu District (Himachal Pradesh)*. l, 744 pp., 33 col. photographs. Dehra Dun: Bishen Singh Mahendra Pal Singh.

Descriptive flora of 930 species with keys to all taxa, synonymy, references, citations of key works, indication of distribution, overall range, habitat, frequency and phenology, representative *exsiccatae*, and some critical notes; principal references (pp. 713–716) and index to all botanical names at end. The relatively detailed general part includes a floristic analysis and phytogeographical account (pp. 55–73) and general key to families (pp. 79–93). [The area, of 5503 km², ranges in elevation from 1100 to 4300 m and centers on the Kullu Valley.]

NAIR, N. C., 1977. *Flora of Bashahar Himalayas*. xxxii, 360 pp., 24 pls. Hissar, Haryana: International Bioscience Publishers.

Descriptive enumeration of seed plants (1629 species); includes synonymy, references, literature citations, brief diagnoses, localities with *exsiccatae* (mainly from Dehra Dun Forest Herbarium), indication of phenology, and other notes; index to all scientific names at end. The introductory section covers physical and biotic features, previous work on the area, and floristic statistics along with a list of references (pp. xxx–xxxi). [Covers the former princely state of Bashahar (now divided between Kinnaur and Mahasu (= Shimla) districts), an area of some 10 000 km² in the east and southeast of the state bordered by Xizang (China) to the north and the Himalayan tracts of Uttar Pradesh to the east.]

843

Uttar Pradesh: Himalayan tracts (Uttaranchal)

Area: no data. Vascular flora: Garhwal Division, 2996 seed plant species (Gupta, 1989; see below). – This mountainous area, from the early nineteenth century part of the Northwestern Province (and, later, the United Provinces), comprises two divisions: Garhwal and Kumaon. Notable 'hill stations' therein include Mussorie (in Garhwal) and Naini Tal (in Kumaon), while Dehra Dun, long established as a forestry center and presently also home to the Northern Circle of the Botanical Survey (BSI-NC), is in the Garhwal *terai*. The area largely falls outside the

scope of *Flora of the upper Gangetic Plain* (815), Dehra Dun being at about its northern limit.

Although the modern botanical history of the Uttar Pradesh Himalayan tracts is as extended as in many other parts of India, and with Dehra Dun moreover home to good herbaria and resident botanists since the late nineteenth century, a broken terrain and fragmented past administrative history have resulted in a variety of local florulas but no general work. The most useful handbooks and other accounts are listed below under subheadings representing respectively Garhwal and Kumaon Divisions. Not surprisingly, woody plants are quite well documented.

I. Garhwal Division

Encompasses the Chakrata-Dehra Dun valley and foothill area bordering the plains as well as Tehri Garhwal (a former princely state), Garhwal proper, Chamoli and Jaunsar. Also within the area is the 'hill station' of Mussorie.

GUPTA, R. K., 1989. *The living Himalayas*, 2: *Aspects of plant explorations and phytogeography*. [iii], clxiv, 512 pp., illus., 2 foldouts. New Delhi: Today and Tomorrow's.

The main part of this work comprises a detailed enumeration of seed plants (2996 species in 1091 genera, with an additional 161 cultigens) with synonymy, references and citations, localities with *exsiccatae*, generalized indication of distribution and habitat, altitudinal range, associates (where known), phenology, uses and properties, and some critical notes; bibliography (pp. 506–512) but no index. The general part (to a great part derived from literature) covers physical features, climate, floristics, vegetation, forest types with characteristic species, phytogeography, and floristic patterns (a high endemism being especially noteworthy). [Covers Dehra Dun, Garhwal, Jaunsar, Tehri Garhwal and the 'hill station' of Mussorie.]

Partial works (except pteridophytes)

The works by Babu and Kanjilal for the Dehra Dun area are more or less complementary. The latter extends its coverage into Saharanpur in the lowlands (815). A recent work (not described here) for part of Tehri Garhwal is *Plant diversity in the Tehri Dam submersible area* by B. P. Uniyal *et al.* (1995, Calcutta: BSI). A new flora of Garhwal has been announced for 2000.

BABU, C. R., 1977. *Herbaceous flora of Dehra Dun*. vii,

721 pp., map. Delhi: Publications and Information Directorate, CSIR, India.

Descriptive florula of herbaceous and subwoody seed plants (1230 species); includes keys to all taxa, synonymy, references and citations, indication of localities with selected *exsiccatae* and notes on origin (for naturalized and other alien plants), habitat, phenology, etc., as well as critical remarks; index to all botanical names at end. An introductory section includes accounts of physical features, geology, soils, climate, vegetation and biotic factors, and previous botanical exploration along with a floristic and phytogeographic analysis (with statistical table). No vernacular names are included. [Complements Kanjilal's long-standard manual for woody plants (see below under **Woody plants**).]

NAITHANI, B. D., 1984–85. *Flora of Chamoli*. 2 vols. Illus. (halftones), maps. Howrah: BSI. (Flora of India, series 3.)

Briefly descriptive flora (2022 seed plants including cultigens), with keys, synonymy (including references and citations), vernacular names, local range (including some *exsiccatae*), phenology, and other notes; addenda and indices in volume 2. The introductory part covers physical features, vegetation (with significant species), reserves, phytogeography, history of botanical work, and sources, along with an analysis with statistics (pp. xiv–xvii) and key to families.

RAIZADA, M. B. and H. O. SAXENA, 1978 (1979). *Flora of Mussorie*. Vol. 1. lvi, 645 pp. Dehra Dun: Bishen Singh Mahendra Pal Singh.

Briefly descriptive flora of native, naturalized, adventive and commonly cultivated seed plants (1219 species), with keys to all taxa, synonymy with references, literature citations, local range including indication of *exsiccatae* seen as well as overall distribution, vernacular names, and notes on habitat, phenology, properties and uses, etc.; indices to both scientific and vernacular names at end of volume. The introductory section covers physical and biotic features, floristics and vegetation, phytogeography, plants particular to given habitats, adventive and naturalized species, and history, followed by a more detailed consideration of forest vegetation and a general key. A comparison of the area with Shimla and vicinity, with 1326 species, is also presented. [This first of two planned volumes covers Ranunculaceae through Labiatae (Bentham and Hooker system). The second volume has not been published.]

Woody plants

KANJILAL, U. N., 1928. *Forest flora of the Chakrata, Dehra Dun and Saharanpur forest divisions, United Provinces.* 3rd edn., revised by B. L. Gupta. xxiii, 558 pp. Calcutta: Central Publications Branch, Government of India. (1st edn., 1901, as *Forest flora of the School Circle, NWP*; 2nd edn., 1911, as *Forest flora of the Siwalik and Jaunsar divisions of the United Provinces of Agra and Oudh*.)

Keyed, briefly descriptive manual-flora of woody plants, with limited synonymy, vernacular names, indication of local range, and notes on habitat, special features, etc.; complete index at end. A glossary is also included. [In addition to the foothill zone around Dehra Dun and northwestwards, coverage of this work also extends to Saharanpur in the adjacent plains (**815**).][71]

Pteridophytes

BIR, S. S. *et al.*, 1982. *Pteridophytic flora of Garhwal Himalaya*. 83 pp. Dehra Dun: Jugal Kishore.

Annotated enumeration (149 species) with synonymy, references and citations, and taxonomic, geographical and ecological notes; summary, bibliography and index at end. [Covers Dehra Dun, Mussorie, Chakrata and adjoining hills.]

DHIR, K. K. and A. SOOD, 1981. *Fern-flora of Mussorie Hills*. 99 pp., 85 figs. (on additional pages). Vaduz: Cramer/Gantner. (Bibliotheca pteridologica 2.)

Descriptive flora, with keys, indication of distribution and altitudinal range; includes references.

II. Kumaon Division

Kumaon Division encompasses Naini Tal (including the eponymous 'hill station'), Almora and Pithorgarh. In addition to the works described below, mention may be made of *Flora of Corbett National Park* by P. C. Pant (1986, Howrah: BSI).

Partial works (except pteridophytes)

GUPTA, R. K., 1968. *Flora nainitalensis: a handbook of the flowering plants of Naini Tal*. xxix, 489 pp., 40 pls. New Delhi: Navayug Traders.

Keyed, briefly descriptive manual-flora of seed plants; includes synonymy and citations, indication of local and extralimital range, taxonomic commentary, and notes on habitat, phenology, uses, etc.; glossary, lexica of vernacular names, and botanical index at end. The introductory section covers botanical exploration, general features of the area and its vegetation, floristics, and statistics of the flora. [Resembles in style and format Collett's *Flora simlensis* (**842**).]

STRACHEY, R., 1906. *Catalogue of the plants of Kumaon and of the adjacent parts of Garhwal and Tibet.* Revised by J. F. Duthie. 269 pp. London: Reeve. (Reprinted 1974, Dehra Dun: Bishen Singh Mahendra Pal Singh. Original version in Atkinson's *Gazetteer of the Himalayan districts of the NW Provinces and Oudh* (1882).)

Annotated list of known vascular plants, bryophytes and lichens recorded from the given area, with tabular indication of habitat, diagnostic features, phenology, local and extralimital range (with some localities), and elevations; index to generic and family names at end. An introductory section includes floral statistics and a list of references.

Woody plants

OSMASTON, A. E., 1927. *A forest flora of Kumaon.* xxxiv, 605 pp., map. Allahabad: Superintendent of Government Press, United Provinces. (Reprinted 1978, Dehra Dun: Bishen Singh Mahendra Pal Singh.)

Briefly descriptive flora of woody plants of the Kumaon district, with keys, essential synonymy, vernacular names, local range, and notes on taxonomy, habitat, phenology, etc.; glossary, list of references, and complete indices. An introductory section covers physical features, climate and vegetation of the area.

Pteridophytes

KHULLAR, S. P. *et al.*, 1991. *Ferns of Nainital.* vi, 234 pp., map. Dehra Dun: Bishen Singh Mahendra Pal Singh.

Descriptive manual (143 species in 208.5 km^2) with keys, synonymy, indication of distribution and habitat, localities with *exsiccatae*, and short notes; bibliography and index.

844
Nepal

Area: 145 391 km^2. Vascular flora: *c.* 7000 species.

The rich flora of this Himalayan kingdom remained relatively little explored from about 1830 until the 1960s, although important collections had been made early in the nineteenth century. For long the only standard work was *Prodromus florae nepalensis* by David Don (1825), although all available records were later incorporated into *Flora of British India* (**810–40**). The present greater ease of access to Nepal has resulted in a great deal of collecting, including visiting British and Japanese expeditions, and substantial materials are now available. Several precursory and other publications have been published in addition to the main work by H. Hara and others, *An enumeration of the flowering plants of Nepal* (1978–82).

In addition to the floristic works accounted for here, a useful, attractively laid out general introduction (by the co-author, with Oleg Polunin, of *Flowers of the Himalaya*, noted under **840**) is J. D. A. STAINTON, 1972. *Forests of Nepal.* xvi, 181 pp., 32 col. pls., 5 maps. London: Murray.

Bibliographies

DOBREMEZ, J. F., F. VIGNY and L. H. J. WILLIAMS, 1972. *Bibliographie du Népal*, 3: *Sciences naturelles.* Part 2, Botanique. 126 pp., 4 pls., 9 maps. Paris: Centre National de la Recherche Scientifique. [732 references, arranged by author with detailed subject index; notes on herbaria, botanical gardens, and collectors.]

RAJBHANDARI, K. S., 1994. *A bibliography of the plant science of Nepal.* [11], 247 pp. Kathmandu: R. L. Rajbhandari. [Introduction and sources; titles alphabetically arranged by author and partly annotated, with subject, geographical and taxonomic indices at end; addendum, pp. 245–247. Includes theses and dissertations. Coverage runs through 1993.][72]

Keys to families and genera

PANDE, P. R. *et al.* 1967–68. *Keys to the dicot genera in Nepal.* 2 parts. Kathmandu: HM Government of Nepal. Complemented by DEPARTMENT OF MEDICINAL PLANTS, NEPAL, 1981. *Keys to the pteridophytes, gymnosperms and monocotyledonous genera of Nepal.* [7], 90, (x), [i] pp. Kathmandu. [Together these works comprise analytical keys to the families and genera of Nepalese vascular plants. Sources are given in the introductory sections, along with lists of references.]

HARA, H., W. T. STEARN, A. O. CHATER and L. H. J. WILLIAMS, 1978–82. *An enumeration of the flowering plants of Nepal.* 3 vols. (vol. 3 with assistance from S. Y. Sutton). Illus., maps. London: British Museum (Natural History).

Systematic enumeration of seed plants, with synonymy, detailed references and citations, local distribution and altitudinal zonation (with representative *exsiccatae*) and overall range outside Nepal, critical remarks (sometimes extensive), and miscellaneous notes; list of new names and index to family and generic names. Some keys are included, as well as citation of key literature under family and generic headings, but no descriptions are given. The introductory section includes sections on the genesis of the work, the history of botanical work in Nepal, geography and physical features, vegetation (with map), horticultural contributions, and technical notes; lists of depository herbaria, collectors (by S. Y. Sutton), and route maps of recent field trips are also included. [See also KOBA *et al.*, below.]

KOBA, H., S. AKIYAMA, Y. ENDO and H. OHBA (preface by H. Ohba), 1994. *Name list of the flowering plants and gymnosperms of Nepal.* [vii], 569 pp. (Univ. Tokyo Mus., Material Reports 32). Tokyo.

Alphabetically arranged nomenclator with all accepted names and synonyms, cross-referenced to the *Enumeration* (see above). Additions since the appearance of that work, with their dates of publication, are listed on pp. 557–569. [An iconographic database of specimens was projected.]

MALLA, S. B. *et al.* (eds.), 1976. *Catalogue of Nepalese vascular plants.* ix, 211, 12, 40 pp. (Bull. Dept. Medicinal Pl., Nepal 7). Kathmandu: Ministry of Forests, Nepal.

Unannotated name list of vascular plants, with synonymy, covering 3453 species; based on herbarium holdings in Kathmandu. An introductory section gives the basis for the work, collecting history (with map of routes and areas), and a brief account of the vegetation with a synoptic table. Selected references and an index to genera and families are also provided.[73]

Pteridophytes

IWATSUKI, K., 1988. An enumeration of the pteridophytes of Nepal. In H. Ohba and S. B. Malla (eds.), *The Himalayan plants*, 1: 231–339 (Bull. Univ. Tokyo Mus. 31). Tokyo.

Introduction with historical background and an account of modern work along with the scope of the present study and conventions and general key to families; keyed enumeration of ferns and fern-allies including synonymy, references, generalized indication of distribution and habitat, extralimital range, representative *exsiccatae*, and occasional critical notes; references to revisions and monographs under families; general references (p. 339) but no separate index.[74]

845

Sikkim and northern West Bengal (Darjiling district)

Area (Sikkim): 7299 km². – Sikkim became a state of India in 1975 following a long period of autonomy or semi-autonomy under monarchical rule. Neither it nor Darjiling district (until 1850 part of Sikkim) is covered in Prain's *Bengal plants* (833).

The botanical exploration of Sikkim and Darjiling district began with British penetration into the area in the mid-nineteenth century, best symbolized by the work of J. D. Hooker there in 1849–50. Later in the century, as economic development proceeded in

Darjiling, forestry work was initiated and J. S. Gamble collected actively in the 1870s. The main result was his *List of trees, shrubs and large climbers found in the Darjeeling district, Bengal* (1878, Calcutta, 2nd edn., 1898). Further collecting was carried out in the early twentieth century by William W. Smith and John McQ. Cowan; in 1929 the latter produced (with A. M. Cowan) a revision of Gamble's list. Botanical survey was renewed after independence; in 1979, following its major review of activities, the Botanical Survey of India opened its Sikkim Himalaya Circle (BSI-SHC) in Gangtok.

The best available modern work, centering more or less in this area, is *Flora of Eastern Himalaya* by H. Hara and others (see **840**). Also listed here is the incomplete descriptive flora by Biswas (1966) which covers most of the present area. Pteridophytes were revised by Mehra and Bir (1964). Miscellaneous florulas of boroughs and other small areas also exist. For the woody plants, the Cowan list was partially supplanted by the *Provisional checklist* of Grierson and Long (1980; see **846**). Darjiling district is also covered in *Flora of West Bengal* (1997– ; see **834**). A brief bibliography for Sikkim is also available.

Bibliography

See also **Region 84 (Bibliographies)**.

MATTHEW, K. M., 1970. A bibliography of the botany of Sikkim. *Bull. Bot. Soc. Bengal* **24**: 57–59. [Short, not exhaustive list.]

BISWAS, K., 1966. *Plants of Darjeeling and the Sikkim Himalayas.* Vol. 1. 8, vii, 540, xxviii pp., *c.* 100 halftones, 47 col. pls. Alipore: West Bengal Government Press.

This generously proportioned work comprises a copiously illustrated descriptive treatment, without keys but including synonymy, vernacular names, local range, notes on habitat, special features, etc.; index at end. [Not completed; extends from Ranunculaceae through Ericaceae (Bentham and Hooker system).]

HAJRA, P. K. and D. M. VERMA (eds.), 1996. *Flora of Sikkim.* Vol. 1 (Monocotyledons). vii, 336 pp., 4 col. pls. Calcutta: Botanical Survey of India. (Flora of India, series 2.).

Tersely descriptive flora without keys; includes references, synonymy, key citations, occasional vernacular names, phenology, and local distribution and altitudinal range; index to all botanical names at end. The general part includes an account of the vegetation and (pp. 15–22) principal literature but neither a floristic analysis nor a general key to families. [Treatments contributed by various authors.]

Partial and other works

Sikkim is also covered in the *Provisional checklist* of Grierson and Long (1980) for Bhutan (see **836**).

MUKHERJEE, A., 1988. *The flowering plants of Darjiling*. xiv, [i], 288 pp., folding map. Delhi/Lucknow: Atma Ram.

Briefly descriptive flora with keys, synonymy, and indication of phenology and overall distribution; list of adventive, cultivated and ornamental plants; corrigenda and index. The general part includes a description of the area, notes on past botanical work (including the role of the Lloyd Garden in Darjiling), a bibliography (pp. 12–18) and the main key.

Woody plants

For the lower hills, see also **834** (SHEBBEARE). The Cowans' work nomenclaturally has been supplanted by GRIERSON and LONG (1980; see **846**).

COWAN, A. M. and J. M. COWAN, 1929. *The trees of Northern Bengal, including shrubs, woody climbers, bamboos, palms and tree ferns*. 178 pp. Calcutta: Government of Bengal.

Systematic enumeration, without keys, of the woody plants of northern Bengal (and Sikkim); includes limited synonymy, vernacular names, general indication of local range, altitudinal distribution, and notes on taxonomy, habitat, life-form, phenology, associates, wood, uses, etc.; index to all botanical and vernacular names.

Pteridophytes

MEHRA, P. N. and S. S. BIR, 1964. Pteridophytic flora of Darjeeling and Sikkim Himalayas. *Res. Bull. Panjab Univ.* (N.S.) **15**(1–2): 69–181, illus., map.

Systematic enumeration with synonymy, references, distributions and commentary; floristic summary (pp. 177–180) and references. The substantial general part covers physical features, climate, vegetation, and features of the flora. [362 species accounted for.]

846

Bhutan

See also **840** (HARA *et al.*). – Area: 41 000 km². Vascular flora: 5468 seed plant species (GB; see **General references**).

As with Nepal, some collecting trips were made into this feudal state in the nineteenth century and the first half of the twentieth, but restrictions on access delayed any more detailed botanical surveys until the mid-twentieth century. In 1963–65 the Botanical Survey of India made extensive collections, and they have been followed by Japanese, British and other botanists. The results of the Indian survey were published in 1973, those of the Japanese in their series on the Himalayan flora (see **840**), and the initial results of the Edinburgh-based Flora of Bhutan project (begun in 1975) appeared in 1980. The definitive *Flora of Bhutan* commenced publication in 1983 and is now far advanced.

GRIERSON, A. J. C., D. G. LONG and H. J. NOLTIE, 1983– . *Flora of Bhutan: including a record of plants from Sikkim*. Vols. 1– . Illus. (part col.), maps. Edinburgh: Royal Botanic Garden (for the U.K. Overseas Development Administration and the Royal Government of Bhutan).

Briefly descriptive illustrated flora of seed plants; includes concise synonymy, vernacular names, distribution and altitudinal range, origin of introduced species, and notes on habitat, occurrence, uses and properties, special features, etc.; indices at end of each volume. The introductory part covers physical features, climate, vegetation, phytogeography, conservation, and horticultural resources and introductions as well as botanical exploration and the basis for and plan of the work; also furnished is a bibliography (pp. 37–43 in vol. 1 with continuations in succeeding volumes). [In progress; as of 2000 volumes 1(1), 1(2), 1(3), 2(1), 2(2), 3(1) and 3(2) had been published (with the most recent, 3(2), appearing in 2000). These cover pteridophytes, gymnosperms, a major part of the dicotyledons (through Lamiaceae), and monocotyledons except Orchidaceae (planned for vol. 3(3)). The remaining 'higher' dicotyledons are to appear in vol. 2(3).][75]

SUBRAMANYAM, K. (ed.), 1973. *Materials for the flora of Bhutan*. xii, 278 pp. (Rec. Bot. Surv. India 20(2)). Calcutta.

Annotated enumeration of vascular plants, without keys; includes principal synonymy (with references) and citations, short diagnoses (of key features), and localities (with indication of *exsiccatae*); a few other notes are given as required. An appendix gives a list of more abundant and important medicinal plants, and a selected bibliography and index to all botanical names conclude the work. The introductory section gives a summary of places visited, previous botanical exploration, and general remarks on physical features, climate, and vegetation. [Based principally on collections of 1963–65 made by the Botanical Survey of India.]

Woody plants

GRIERSON, A. J. C. and D. G. LONG (comp.), 1980. *A provisional checklist of the trees and major shrubs (excluding woody climbers) of Bhutan and Sikkim.* 51, 2 pp. Edinburgh: Royal Botanic Garden. (Mimeographed.)

Interim nomenclatural checklist, arranged according to the Englerian system. [Partially succeeds the treatment by Cowan and Cowan for northern Bengal and Sikkim (845).]

847

Arunachal Pradesh

See also **768** (all works), **837** (KANJILAL *et al.*). – Area: 83 578 km². Vascular flora: 4055 species (Hajra *et al.*, 1996; this, however, seems too low). For the most part also claimed by China, Arunachal Pradesh was before 1954 a part of Assam known as the North-East Frontier Tracts. It then came under central rule as the North-East Frontier Agency. In 1972 it was granted Union Territory status under its present name and in 1987, statehood. One of the last parts of India formally to be explored and surveyed, its 12 districts – created from the original five in 1980 and 1993 – are highly mountainous and for the most part difficult of access. In addition, visits from outside have been controlled.

Botanical collecting has been sporadic and, except for the work of William Griffith, J. L. Lister, I. H. Burkill, N. L. Bor, U. N. Kanjilal and Frank Kingdon Ward, the area remained very imperfectly known until after World War II.[76] Among the few florulas of this period are *Botany of the Abor Expedition* (Burkill, 1924–25; see below) and *The flora of the Aka Hills* by K. P. Biswas (1941; in *Indian For. Rec.*, N.S., Bot. 3(1): 1–62). Since 1955, considerable additional collections have been made and floristic studies are in progress. Towards this end, the Botanical Survey in 1977 established at New Itanagar a field station (BSI-APFS) responsible to the Central National Herbarium in Howrah. By 1996 a first synthesis, *Materials for the flora of Arunachal Pradesh*, had begun publication. For areas claimed by China available records have been incorporated into the current Xizang flora (**768**).[77]

HAJRA, P. K., D. M. VERMA and G. S. GIRI (eds.), 1996– . *Materials for the flora of Arunachal Pradesh.* Vols. 1– . Illus. (part col.), maps. Calcutta: Botanical Survey of India. (Flora of India, series 2.)

Tersely descriptive flora of flowering plants, without keys; includes full synonymy, references and literature citations, vernacular names, and indication of distribution (by the historical districts), habitat, altitudinal range, phenology, uses and properties; references and indices to all botanical and vernacular names at end of each volume. The general part in vol. 1 features sections on geography, geology, physiography, soil, climate, peoples (26 tribes being formally recognized), biota and sanctuaries, botanical exploration (pp. 11–13), vegetation (with map) and component species, special floristic elements (pp. 29–34), useful plants (including a list of commercial timber species), a summary of gymnosperms (pp. 34–37), statistics (pp. 37–39) and plan of the work, and language abbreviations. [Vol. 1 covers Ranunculaceae through Dipsacaceae, following the Bentham and Hooker sequence.][78]

Partial works

Burkill's work is set within the former Siang district south of the McMahon line, while that of Chauhan *et al.* is in the present Changlang and Lohit districts in the east, towards the Myanmar frontier.

BURKILL, I. H., 1924–25. *The botany of the Abor Expedition.* 2 parts. 420 pp., 10 pls., map (Rec. Bot. Surv. India 10). Calcutta.

The first part of this work presents an account of the author's journey along with narrative sections on earlier work in the area, geography, climate, soils, general phytogeography, plant biology and ecology, human influences, forest types, synusiae, etc.; then follows a list (pp. 44–64) of seed plants with altitudinal zones and distribution and afterwards a very detailed analysis covering every species, a phytogeographical analysis, a discussion of the role of humidity and moisture, a consideration of species of apparently restricted range, and an account of paleohistory and the genesis of the flora. The enumeration proper (pp. 203–412) encompasses fungi, lichens, bryophytes and vascular plants with field notes, distribution, and taxonomic commentary; this is followed by a summary (pp. 413–418) but no index.[79]

CHAUHAN, A. S. (lichens by K. P. Singh; bryophytes by D. K. Singh), 1996. *A contribution to the flora of Namdapha, Arunachal Pradesh.* viii, 422 pp., 48 unnumbered pls. (col. photographs). Calcutta: Botanical Survey of India.

Concisely descriptive enumeration of lichens, Hepaticae and Anthocerotae, and vascular plants (1119 species in some 2500 km² ranging from 200 to 4571 m in elevation, of which 977 are vascular) including synonymy, references, citations, vernacular names, phenology, *exsiccatae* with localities, and overall distribution; list of ethnobotanically important species with parts as used by tribespeople (pp. 357–367), medicinal plants (pp. 367–372), wild relatives of cultivated plants (p. 373, including *Camellia caudata*, a wild tea), endemic species (pp. 374–377), records of particular interest (pp. 378–380), fauna (pp. 380–383), threats and conservation (pp. 383–386), and bibliography and index to all botanical names. The general part gives the background to the work as well as notes on physical features, geology, soil, climate and rainfall, vegetation, composition of the flora (with analyses and conclusion that the proposed Biosphere Reserve is a 'hot spot'). [The authors acknowledge that their survey is incomplete, with further studies of the flora at higher altitudes required.]

Region (Superregion) 85

Japan, Korea and associated islands

Included under this heading are the main islands of Japan proper, the Ryukyu Islands, the Nanpō Shotō (south of Honshu), the Ōgasawara (Bonin) and Kazan (Volcano) Islands, the Daitō (Borodino) group, and Korea. For Parece Vela (Okino-tori-shima), see **964**.

The interaction of Japanese and Western botany began early in the seventeenth century with the establishment of the Dutch East India Company in Deshima off Nagasaki. For more than two centuries, this channel remained virtually the only means by which knowledge of the Japanese flora flowed out to other parts of the world, particularly the West; through this sieve also came Western approaches to floristic study. Notable European observers and collectors, all associated with the Company, included Engelbert Kaempfer in 1690–92, Carl Thunberg in 1773 and Philipp F. von Siebold in 1826–29 and again in the 1850s. All afterwards prepared floristic accounts (the last jointly with Joseph G. Zuccarini), respectively *Amoenitatum exoticarum politico-physico-medicarum fasciculi V* (1712), *Flora japonica* (1784) and *Flora japonica* (1835–70, completed by Friedrich Miquel). The final purely European work is *Prolusio florae japonicae* (1865–67) by Miquel, prepared after the full resources of the Dutch

state herbarium had been opened.[80] Edo Japan also contributed plant accounts, beginning with *materia medica* but by the closing years of the seventeenth century branching into the purely botanical. The most notable pre-1800 accounts are *Ka-fu* [*Manual of flowering plants*] by Atsunobu (Tokushin) Kaibara (1698; reprinted 1844), *(Zôko-)chikin-shô* [*An illustrated treatise on plants, flowers and herbs*] by Ihei Itô (1710–19; supplement, 1733), and *Ka-i* [*A collection of flowering plants*] by Mitsufusa (Mochifusa) Shimada (1759; reprinted 1765; French edn. by P. L. Savatier, 1873, as *Livres Kwa-wi*).[81] Later Edo works showed increasing European influence; the most notable are the great iconographic *Honzō zufu* (1828; enlarged versions in 1884 and 1920–21 with eventually over 2000 *ukiyo-e*, or multiple wood-block, prints in 95 volumes) by Tsunemasa Iwasaki, an acquaintance of Siebold, and *Somoku-zusetsu* (1856; 2nd edn., 1874; 3rd edn., 1907–12) by Yokusai Iinuma, first written in 1832 in 30 volumes. In the later editions of both works attempts were made to identify the plants in 'modern' terms through separate indices, but even in the original edition of *Somoku-zusetsu* some Latin and Dutch names appear on the plates along with the Japanese.[82]

After the 'opening' of Japan in 1853 and prior to the Meiji restoration of 1868, collecting was done by Keiské Itô (a pupil of Siebold and associate of Miquel), Motoyoshi Ono and Yosiwo Tanaka (the last-named associated with Paul Savatier) and, from Europe and North America, by S. Wells Williams and James Morrow (with Commodore Perry's squadron), Charles Wright (on the U.S. North Pacific Exploring Expedition), John Gould Veitch, Robert Fortune, Carl Maximowicz (in the 1860s), and Richard Oldham as well as by Savatier (in the country from 1866 to 1871 and 1873 to 1876). The last-named also collaborated with Adrien Franchet (in Paris) on *Enumeratio plantarum in Japonia sponte crescentium* (1873–79). This fundamental early synthesis of Japanese and Western knowledge of the flora of Japan covers 2487 vascular plants and has been a basis for all later research and writing. The *Enumeratio* was complemented in particular by *Diagnoses plantarum novarum Japoniae et Mandshuriae* (1866–77) and *Diagnoses plantarum novarum Asiaticarum* (1877–93), both by Maximowicz, and, somewhat later, *Forest flora of Japan* (1894) by the U.S. dendrologist Charles Sprague Sargent.

In 1877, botanical teaching and research began under Cornell-educated Ryōkichi Yatabe at the newly established Tokyo Imperial University. The Botanical Society of Japan followed in 1882, and in 1887 the *Botanical Magazine* was established. This marked the emergence of the modern Japanese botanical profession along with the scientific community as a whole. Some collecting trips continued to be made by Western botanists, among them Charles Maries, Sargent (in 1892), Ernest Henry Wilson and especially Urbain Faurie (from 1873 to 1914, including Taiwan) but by 1900 most detailed botanical exploration was being carried out by Japanese scientists and plantsmen. The first European-style checklist, *Nippon shokubutsu mei-i* by Yatabe with Jinzô Matsumura (also of Tokyo University), appeared in 1884 and passed through nine editions, the last in 1915–16. This was followed in 1904–12 by a more formal checklist by Matsumura alone, *Index plantarum japonicarum*, functionally a successor to the work of Franchet and Savatier, and in 1906 by a first manual, *Nihon shokubutsu-hen* by Yatabe. The iconographic tradition of previous centuries continued in works by Yatabe and, independently, Tomitarô Makino (a largely self-educated follower of Maximowicz).

Over the next three decades, further investigations and consolidations of earlier work enabled Makino and others to prepare a number of manuals. These included the large unillustrated *Nihon shokubutsu sôran* (*Flora of Japan*) (1925; 2nd edn., 1931) by Makino and Kwanji Nemoto, with the latter adding a substantial supplement, *Nihon shokubutsu sôran-hoi*, in 1936; the smaller illustrated *Nihon shokubutsu zukan* (1925) by Makino alone; and the abridged *Manual of the flora of Nippon* (1927) by Makino with Kôichi Tanaka. Only the first-named extended coverage to southern Sakhalin, the Kuriles and Taiwan, then also Japanese territory. A concise illustrated manual-key by M. Murakoshi, *Dai shokubutsu zukan*, covering 4339 species, appeared in 1928 (with a second edition in 1932). In 1939 Masaji Honda published the first edition of his *Nihon shokubutsu mei-i* (*Nomina plantarum japonicarum*) and in 1940 Makino brought out what is probably his best-known work, *Nippon shokubutsu zukan* (*An illustrated flora of Nippon, with the cultivated and naturalized plants*). This last is an atlas-flora along the lines of Britton and Brown's *Illustrated flora* (140) but without keys; it includes some non-vascular plants and fungi as well as most vascular plants. With additions in 1951 and a substantial supplement in 1955 (aided by H. Hara and T. Tuyama), it enjoyed at least annual reissues until a few years after the author's death in 1957; afterwards it was

entirely revised by Hara, Tuyama and F. Maekawa, reappearing in 1961 (see below). The absence of a work on woody plants prompted Takenoshin Nakai to begin *Trees and shrubs indigenous in Japan proper* of which, however, only one volume (of two or three projected) appeared, in 1922 (with a revision in 1927). Complete coverage was first achieved only in the 1950s (by S. Kitamura and S. Okamoto; see below). In 1938, Nakai also began *Nippon shokubutsu-shi* (*Nova flora japonica*), a series of monographic studies with contributions by specialists, but after his death in 1952 this largely lapsed; afterwards, under Hara and his successors the 'Tokyo school' widened its interests to East Asian monography and the flora of the Himalaya.

The Japanese flora thus had become comparatively well known by World War II, and today, with a new suite of works (several of them from the 'Kyoto school' first developed under Genichi Koidzumi), is covered at a standard corresponding to much of Europe or North America. Current major works all date from after 1945 and one's choice is mainly a matter of taste. The 'technical' floras are those by Ohwi (*Nihon shokubutsu*, 1953–57, with an English edition as *Flora of Japan*, 1965, and second Japanese edition, 1982–83, as *Shin Nihon shokubutsu-ki*), a North American-style manual-flora with keys, and Makino's executors (*Makino's new illustrated flora of Japan*, 1961), a keyless atlas-flora like its predecessor but larger in format. These have from 1993 been joined by a new definitive *Flora of Japan* in English; this also covers the Ryukyu and Daito Islands, the Izu and Nanpo Islands, and the Bonin and Kazan groups along with the southern Kuriles (737). Of the semi-popular illustrated works, the best are those in the 'Flora and fauna of Japan' series of the Hoikusha publishing house in Osaka, written under the direction of Siro Kitamura (1954–64; revision of the woody plants, 1971–79). An unfortunately incomplete detailed nomenclator by Hiroshi Hara (*Enumeratio spermatophytarum japonicarum*, 1948–54) attempted to provide a critical basis for modern studies. A large range of regional and local florulas is also available for the islands of Japan proper, as also are dendrological, pteridological and other works of more restricted scope.

In other parts of the region now under Japanese rule, the level of documentation has since 1950 improved markedly. Some works on the Bonin (Ogasawara) and Volcano (Kazan) Islands have been published since their return to Japan in 1969, beginning with *The nature in the Bonin Islands* whose senior author, Takasi Tuyama, had been before World War II a significant contributor to knowledge of the islands' floristics. No 'scientific' flora, however, has yet been published. A checklist and a field-guide have also appeared. With respect to the Ryukyu Islands, in earlier years almost without separate works, two large floras have been published in recent years: Sumihiko Hatusima's *Flora of the Ryukyu Islands* (1971; 2nd edn., 1975) and Egbert H. Walker's *Flora of Okinawa and the southern Ryukyu Islands* (1976). These complement Genkei Masamune's critical enumeration of 1951–64, the first modern complete account of the vascular flora. Some 'tree books' have also been published. Hatusima's work also covers the Daitō Islands, for which there is apparently only one separate account, a 1941 enumeration by Masamune. All these areas are encompassed by the new *Flora of Japan*.

Botanical exploration in Korea was before 1905 very sporadic. The first general flora was Ivan Palibin's *Conspectus florae Koreae* (1898–1901), an outcome of an era of Russian penetration into northeast Asia with which is also associated the work of Vladimir Komarov in Manchuria and elsewhere. At the same time the country was covered in *Index florae sinensis* (1886–1905). From then onwards, ending only with World War II, most collecting and flora-writing was done by Japanese, with notable contributions being made by Nakai in his *Flora koreana* (1909–11) and, after his appointment as government botanist in 1913, the detailed *Flora sylvatica koreana* (1915–39). A major Korean contribution of mid-century is the two-volume flora by Tai-hyun Chŏng (1956–57; revamped 1965), but this is regarded as less than satisfactory and a new critical flora is much needed. Chŏng's work has been complemented by other treatments. The country evidently is less well explored botanically than Japan and not presently as well documented, in either the north or the south. An illustrated guide to woody plants has also been published.

Progress

A useful although somewhat sketchy review of history and current knowledge, mainly for Japan, is T. KOYAMA, 1961. Available knowledge on the flora of Japan and its neighboring regions. In *Ninth International Botanical Congress, Recent advances in botany*, 1: 937–940. Toronto.[83] The introduction in the English version of Ohwi's *Flora of Japan* should also be consulted. No similar reports on progress for other

areas have been seen; reference should be made to the floras concerned.[84]

Bibliographies. General bibliographies as for Division 8.

Regional bibliographies

MERRILL, E. D. and E. H. WALKER, 1938. *A bibliography of Eastern Asiatic botany*. Jamaica Plain, Mass.; and E. H. WALKER, 1960. *A bibliography of Eastern Asiatic botany: Supplement*. Washington. [Arranged by author, with detailed subject indices. For fuller description, see **Superregion 86–88**.][85]

Indices. General indices as for Division 8.

Regional indices

STEENIS, C. G. G. J. VAN *et al.* (eds.), 1948– . *Flora Malesiana Bulletin*, 1– . Leiden: Flora Malesiana Foundation (later Rijksherbarium). (Mimeographed.) [At present issued bi-annually; the autumn installment contains a substantial bibliographic section which rather thoroughly covers ongoing new literature on Southeast Asia. For fuller description, see **Superregion 91–93**.]

851

Japan

Area: 369968 km². Vascular flora: 4022 species (Ohwi, 1965). – Japan proper is here defined as encompassing the main Japanese islands of Hokkaido (Yezo), Honshu (Hondo), Shikoku, and Kyushu, along with a number of nearby small islands and groups including Osumi Gunto (Yakushima and Tanegashima), Goto Retto, Oki Gunto, Sado, and Izu Shichito; this corresponds to the limits of Ohwi's standard modern flora of Japan.

Floristic works have been outlined under the regional heading. For a general introduction to the flora and vegetation, see M. NUMATA, 1974. *The flora and vegetation of Japan*. Tokyo: Kodansha.

Bibliographies

Kanai's 1994 bibliography is, along with that of Langman for Mexico (**210**), one of the most exhaustive available.

KANAI, H. (comp.), 1994. *List of literatures related to plant taxonomy and phytogeography of Japan.* 2 vols. Kamakura: Abac-sha Co. Ltd. [Covers publications from 1887 (when the *Botanical Magazine* was founded in Tokyo) through 1993. Vol. 1 comprises the bibliography proper, the arrangement by author (in Romanized order) and with concise annotations including taxa (if relevant); each of the 61 182 items carries a unique number based on the year of publication. Vol. 2 contains subject, taxonomic, regional, personal, and Japanese vernacular name (in *katakana*) indices. Explanations in both Japanese and English. – A successor to *idem*, 1975–79. *List of literature related to plant taxonomy and phytogeography of Japan.* In issues. Tokyo (serially covering the years 1971 through 1976); and *idem*, 1985. *List and index of literatures related to plant taxonomy and phytogeography of Japan published in 1973 through 1982.* xii, 602 pp. Tokyo.]

Dictionary

HONDA, M., 1957. *Nihon shokubutsu mei-i/Nomina plantarum japonicarum.* 2nd edn. [4], 389, 126 pp. Tokyo: Koseisya-Koseikaku. (1st edn., 1939.) [Systematic list of vascular plants, with standard Japanese nomenclatural equivalents in *katakana* script. A list of necessary new combinations, with required documentation, and indices are also provided.]

HARA, H., 1948–54. *Enumeratio spermatophytarum japonicarum.* Vols. 1–3. Tokyo: Iwanami Shoten. (Reprinted in 1 vol., 1972, Königstein/Ts., Germany: Koeltz.)

Critically compiled, systematic enumeration of native, naturalized and commonly cultivated seed plants of Japan and adjacent islands; includes detailed synonymy, references and pertinent citations, Japanese nomenclature, and general indication of local range; index to genera at end. Text in Japanese except for Latin-language references and botanical names. [Not completed; covers only the Metachlamydeae and part of the Archichlamydeae (Geraniaceae through Cornaceae) in the Englerian system.]

IWATSUKI, K., T. YAMAZAKI, D. E. BOUFFORD and H. OHBA, 1993– . *Flora of Japan.* Vols. 1, 2c, 3a, 3b. Maps (end-papers). Tokyo: Kodansha.

Detailed descriptive flora of native, naturalized and introduced vascular plants with keys to all taxa, synonymy, references, Japanese vernacular names, indication of chromosome numbers, summary of distribution within and without Japan and of habitat, taxonomic notes and references to illustrations; references to key literature under family and generic headings; indices to all scientific names at end of each volume or part-volume. Volume 1 contains a foreword by the editors and a preface by the first-named editor, but no 'general part'. [Four volumes, some with part-volumes, are planned, respectively to cover pteridophytes and gymnosperms, archichlamydeous dicotyledons, metachlamydeous dicotyledons, and monocotyledons. To date (1999) vols. 1 (1995), 2c (1999), 3a (1993) and 3b (1995) have been published.][86]

KITAMURA, S. *et al.*, 1957–64. *Coloured illustrations of herbaceous plants of Japan.* 3 vols. Text-figs., 198 col. pls. Osaka: Hoikusha. (Flora and fauna of Japan 15, 16, 17.) (Revised edn. of the first volume (Sympetalae), 1958; all vols. subsequently reprinted more or less annually.) Complemented by S. KITAMURA and G. MURATA, 1971–79. *Coloured illustrations of woody plants of Japan.* Revised edn. 2 vols. 438 text-figs., 144 col. pls. Osaka. (Flora and fauna of Japan 49, 50. Originally published as S. KITAMURA and S. OKAMOTO, 1959. *Coloured illustrations of trees and shrubs of Japan.* viii, 306 pp., 72 text-figs., 68 col. pls. (Flora and fauna of Japan 21; subsequently reprinted)); and M. TAGAWA, 1959. *Coloured illustrations of the Pteridophyta of Japan.* iv, 270 pp., 8 text-figs., 72 col. pls. Osaka. (Flora and fauna of Japan 24; subsequently reprinted.)

These complementary works together constitute an outstanding semi-popular series of copiously illustrated manuals covering all (or most) of the vascular plants in Japan; with slight variations they include keys to all taxa, limited synonymy, standard Japanese names, vernacular names in English, French and German (for genera), general summaries of local and extralimital distribution, and notes on habitat, phenology, karyotypes, special features, etc. Each volume has a list of references and complete indices. The three volumes on herbaceous plants respectively cover Sympetalae, Choripetalae and Monocotyledoneae. The volume on pteridophytes also contains an annotated enumeration of all taxa. Collaborators have included G. Murata, M. Hori and T. Koyama.

MAKINO, T., 1961. *Makino's new illustrated flora of Japan.* 2nd edn., revised under supervision of F. Maekawa, H. Hara and T. Tuyama. vii, 12, 1060, 77 pp., 3894 text-figs., 4 col. pls., portr. Tokyo: Hokuryukan. (1st edn., 1940, as *Illustrated flora of Nippon, with cultivated and naturalized plants.* Both editions reissued several times, the first sometimes with revisions and in 1955 with a substantial supplement.)

Illustrated descriptive atlas-flora of most native and commonly cultivated vascular and the more important non-vascular plants, without keys (in all 3617 vascular plants); includes scientific names, standardized and vernacular Japanese nomenclature, general summaries of local range, and notes on habitat, special features, uses, etc.; etymological and terminological glossaries, a lexicon of genera, and indices to botanical and other names at end. This work, modeled to some extent on Coste's *Flore de France* (**651**) or the earlier versions of Britton and Brown's *Illustrated flora of the northern United States* (**140–50**), has for many years been the standard traditional-style flora of Japan, with roots in classical Japanese (and Chinese) botanical works but written in a manner compatible with international conventions.[87]

OHWI, J. and M. KITAGAWA, 1983. *Shin Nihon shokubutsu-ki kenkaken (New flora of Japan)*. 1716 pp., 20 text-figs., 32 halftones, [30 pp.] pls. Tokyo: Shibundo. Complemented by T. NAKAIKE, 1982. (*New flora of Japan: Pteridophytes*.) [iv], 808, [i] pp., 849 halftones. Tokyo. (Both works originally published 1953–57 as *Nihon shokubutsu-shi* by Ohwi alone; 2nd edn., 1965.) English version: J. OHWI, 1965. *Flora of Japan* (eds. F. G. Meyer and E. H. Walker). vii, 1067 pp., 17 text-figs., 16 pls., maps (partly in end-papers). Washington: Smithsonian Institution. (Reissued 1984.)

The main work is a manual-flora of seed plants, with keys to all taxa, limited synonymy, Japanese nomenclature, generalized indication of internal and external range, phenology, and notes on habitat, special features, etc.; lexicon of authorities and indices to all Japanese and scientific names. The introductory part includes a brief history of botanical work, floristic zones, illustrated glossary, and (pp. 36–50) general keys to families. The 32 plates include 27 photographs of types of species described by Ohwi. [In the English version of 1965, which covers all vascular plants, the Japanese names are in the Roman alphabet.] The volume on pteridophytes is a lavishly illustrated atlas-flora, with text incorporating concise descriptions, synonymy, Japanese and scientific nomenclature, and notes on distribution, habitats, etc. as well as critical remarks but no keys; illustrations are nearly all of herbarium specimens. The treatment of hybrids (pp. 724–787) is followed by indices to all names.

Distribution maps

In addition to the work of Horikawa, reference may also be made to H. HARA and H. KANAI, 1958–59. *Distribution maps of flowering plants in Japan*. 200 maps. Tokyo.

HORIKAWA, Y., 1972–76. *Atlas of the Japanese flora. An introduction to the plant sociology of East Asia*. Vols. 1–2. 862 maps. Tokyo: Gakken.

Systematically arranged folio atlas (in English) of distribution maps of the flora of Japan and neighboring regions, based on records from grid squares 20 × 20 km and incorporating associated descriptive text; the latter incorporates principal synonymy and notes on life-form, phenology and uses. A general index is given at the end of each volume. Each species map also has 'satellite' maps depicting altitudinal zonation. The introduction to the work incorporates a concise sketch of the vegetation of Japan. A further three volumes were projected. [Citation and description based partly on a review by C. G. G. J. van Steenis in *Fl. Males. Bull.* 27: 2214–2216 (1974).]

Woody plants

For an introduction to trees, shrubs and woody lianas, the best reference is *Coloured illustrations of woody plants of Japan* by Kitamura and Murata (see above under KITAMURA *et al.*, 1957–64). The following is, however, useful for dendrological details and in addition has fine color plates as well as parallel Japanese and English text.

JAPAN FOREST TECHNICAL ASSOCIATION (S. KURATA, comp.), 1964–76. (*Genshoku Nihon ringy o jumoku zukan*)/ *Illustrated important forest trees of Japan*. 5 vols. Illus., maps. Tokyo: Chikyu Shuppan. (1st edn., 1964.)

Dendrological atlas of significant tree species, with colored illustrations, descriptive text (in Japanese and English), and distribution maps; the latter includes a botanical description and remarks on wood, properties, uses, etc. No systematic sequence is followed. A general index is given in vol. 5.[88]

Pteridophytes

With some 500 species the fern flora is relatively rich. Besides those below, the works of Murata (1959; under KITAMURA *et al.*, 1957–64) and Nakaike (1982; under OHWI and KITAGAWA, 1983) given above may be consulted.

NAKAIKE, T., 1975. *Enumeratio pteridophytarum japonicum*. xiii, 375 pp. Tokyo: University of Tokyo Press.

Systematic enumeration, with synonymy, references, known citations, Japanese nomenclature, and generalized indication of overall distribution; indices to Japanese and generic names at end. The introductory part includes a discussion of the system adopted (pp. xi–xii) – one newly proposed here – and the scope of the work (geographically it is inclusive of Nansei-shoto). [For pteridophytes the work is complementary to *Enumeratio spermatophytarum japonicarum* (see above).]

KURATA, S. and T. NAKAIKE (eds.), 1979–97. *Nihon no shida shokubutsu zukan/ Illustrations of pteridophytes of Japan*. 8 vols. Illus., col. frontisp., maps, folding maps in inside back covers. Tokyo: University of Tokyo Press.

An iconography in large format with plates, spore photographs and distribution maps, each volume covering some 100 species; following each map are one or more pages of records, arranged by prefectures (north to south) and further organized by named squares and quarter-squares related to latitude and longitude (keyed to the folding map furnished in each volume). Tables of contents (with provenance and sources of photographs), sporographic appendices, and indices are given at the end of each volume. Encompasses Japan proper, the Ryukyus, and the Ogasawara and Kazan groups.

852

Nanpo-shoto

See also **851** (IWATSUKI *et al.*). – This archipelago, lying between the Izu-shichito just off central Honshu (**851**) and the Bonin Islands (**853**), includes Hachijo-jima, Aoga-shima, Urania, Sumisu-jima, Tori-shima, and Sofu-gan (Lot's Wife). No general floras or enumerations specifically covering these islands appear to be available. Aoga-shima has been treated in three papers by M. Mizushima in *Misc. Rept. Res. Inst. Nat. Resources* (Japan) **38**: 106–126 (1955); **41/42**: 76–80 (1956); **45**: 648 (1957).

853

Ogasawara-shoto (Bonin Is.) and Kazan-retto (Volcano Is.)

See also **851** (IWATSUKI *et al.*). – Area: Ogasawara-shoto, 73 km²; Kazan-retto, 27 km². Vascular flora: Ogasawara-shoto, 483 species of which 369 native (152 endemic) and 114 introduced or commonly cultivated (Kobayashi, 1978; see below); Kazan-retto, 257 flowering plants of which 118 on Minami-iwo-jima (PD after Okutomi, 1982; see below).

The biogeographical affinities of this distinctive, just extra-tropical pair of island groups have been by some authorities thought to be with Oceania; but in the 1930s Takahide Hosokawa showed that the vascular flora as a whole was more closely related to that of mainland Asia, Nansei-shoto and Japan although

Oceanian elements are indeed present. Hosokawa's view (upheld by M. van Balgooy), is maintained here with the inclusion of these groups in Region 85.[89]

Botanical exploration of the two groups began relatively late, following their occupation (1862) and annexation (1879) by Japan. Extensive pioneer studies were made by Ernest H. Wilson, Takenoshin Nakai and Takasi Tuyama before World War II, but no consolidated flora or enumeration was published for another three decades. From 1945 until 1968 the islands were under U.S. occupation and only in 1970 did there appear the first new natural history work, *The nature in the Bonin Islands*. Additional works have since been published, including two editions of an enumeration (Kobayashi, 1978; Kobayashi and Ono, 1987) and a field-guide (Toyoda, 1981). The woody plants have been additionally reviewed in E. H. WILSON, 1919. The Bonin Islands and their ligneous vegetation. *J. Arnold Arbor.* **1**: 97–115. Pteridophytes have been separately accounted for in H. OHBA, 1971. A taxonomic study on Pteridophyta of the Bonin and Volcano Islands. *Sci. Rep. Tohoku Univ.* **36**: 75–127.

KOBAYASHI, S. and M. ONO, 1987. *A revised list of vascular plants indigenous and introduced to the Bonin (Ogasawara) and the Volcano (Kazan) Islands*. vii, 55 pp., 3 maps (Ogasawara Research 13). Tokyo: Ogasawara Research Committee, Tokyo Metropolitan University. (First published as S. KOBAYASHI, 1978. *A list of the vascular plants occurring in the Ogasawara (Bonin) Islands*. 33 pp., 3 maps (Ogasawara Research 1).)

An annotated systematic checklist (arranged following the 11th edition of Engler's *Syllabus*) with Romanized Japanese names, notes on habit, habitat and associates, indication of presence in the various islands, and (if applicable) external distribution or origin. A tabular synopsis of taxa with distribution by island follows on pp. 39–55. No synonymy or specimen citations are generally furnished. The brief general part includes maps of the islands and a list of key references.[90]

Ogasawara-shoto

The three groups of islands are all of volcanic origin, but are no longer active. – In addition to the works given below, reference may be made to the following well-illustrated general natural history (with vegetation and flora in vol. 1, pp. 109–141): T. TUYAMA and S. ASAMI, 1969–70. *The nature in the Bonin Islands*. 2 vols. Illus. (mostly col.), maps. Tokyo: Hirokawa Shoten. Endemism and dispersal have been reviewed in M. ONO, 1991. The flora of the Bonin

(Ogasawara) Islands: endemism and dispersal modes. *Aliso* 13: 95–105.

NAKAI, T., 1930. [Plants of the Bonin Islands.] *Bull. Biogeogr. Soc. Japan* 1: 249–278, pls. 16–18.

Comprises an unannotated systematic list of vascular plants, with botanical names and their Japanese equivalents, together with a detailed phytogeographical analysis of the flora and discussion of distribution patterns. The illustrations depict a selection of endemic species. The significant disjunction between the flora of these islands and the Marianas (subsequently known as 'Hosokawa's Line') is pointed out.

TOYODA, T., 1981. *Ogasawara shokubutsu zufu* [*Flora of Bonin Islands*]. 396 pp., illus., maps. Kamakura: Aboc-sha.

Illustrated, artificially arranged field-guide to 236 native and naturalized species, with a focus on the endemics; includes for some taxa detailed distributions (with small maps) for Chichi-jima and Haha-jima (the two largest of the island groups) along with photographs. Pp. 301–378 comprise general chapters on floristics, vegetation, endemic species, disturbance (p. 252 features a picture of a goat and p. 276 one of eroded soil), and the history of botanical work, while in the front are 'scene-setting' color photographs of vegetation and characteristic plants. A bibliography and indices to Japanese and scientific names are also given. [A full species list is unfortunately lacking.][91]

TUYAMA, T., 1935–39. Plantae boninenses novae vel criticae, I–XII. *Bot. Mag.* (Tokyo) 49 (1935): 367–374, 445–452, 505–512, pls. 1–17; 50: 25–32, 129–134, 374–379, 425–430, pls. 18–36; 51: 22–24, 125–132, pls. 37–43; 52: 463–467, 567–572, 6 pls.; 53 (1939): 1–7, 3 pls.

Descriptions of, and notes upon, new or little-known species of vascular plants, with citations of *exsiccatae* and remarks on taxonomy, habitat, special features, etc. Illustrations are furnished for the more interesting taxa.

Kazan-retto

The three small islands of this group are all volcanic, with Minami-iwo-jima (San Augustino) reaching 918 m and Kita-iwo-jima, 792 m. Tectonically they are younger than the more northerly Ogasawara group. Minami-iwo-jima is difficult of access and remains biotically undisturbed, while the less lofty Iwo-jima – made famous during World War II – was completely devastated in 1945. The natural flora overall is very similar to that of their northern neighbors.

OKUTOMI, K. (ed.), 1982. *Conservation reports of the Minami-Iwojima Wilderness Area*. 403 pp. Tokyo: Nature Conservation Bureau, Environmental Agency of Japan.

[Plants on pp. 61–143. – Not seen; cited from *Plants in danger* (see **General references**) as well as Kobayashi and Ono (1987; see above).]

854

Oagari-jima (Daito or Borodino Islands)

See also 851 (IWATSUKI *et al.*); 855 (HATUSIMA; MASAMUNE). – Area: no information. Vascular flora: no data. One separate account is available for these limestone outliers of Nansei-shoto in thc Philippine Sea.

MASAMUNE, G. and M. YANAGIHARA, 1941. (On the flora of Oagari-jima (Daito Islands), Ryukyu Archipelago), I–III. *Trans. Nat. Hist. Soc. Taiwan* 31: 237–250, 268–274, 317–330.

Parts II–III of this work contain an enumeration (in Japanese) of all known vascular plants from this island group, with botanical and equivalent Japanese names and miscellaneous notes on local range, habitat, etc.

855

Nansei-shoto (Ryukyu Islands)

See also 851 (IWATSUKI *et al.*). – Area: 2196 km^2 (Okinawa, 1176 km^2). Vascular flora: over 2000 native and introduced in the southern districts (Walker, 1976). The group is here considered to include the four sets of islands (Tokara-gunto, Amami-gunto, Okinawa-gunto and Sakishima-gunto) between the Osumi Islands just off Kyushu (**851**) and Taiwan (**886**). From 1945 to 1972, the two southern groups were under United States administration.

The first modern separate treatment of Nansei-shoto was *Tentamen florae lutchuensis* (1899, not completed) by Tokutaro Itô and Jinzô Matsumura. Through the first half of the twentieth century the islands were included in metropolitan Japanese floras, manuals and enumerations, including those by Matsumura, Makino and Nemoto. Serious destruction of resources accompanied World War II, but afterwards, with outside assistance, a tradition of more particular study and documentation was reintroduced and several floras and related treatments published. At present, two overall works are available: a descriptive flora in Japanese (Hatusima, 1971) and an enumeration in English (Masamune, 1951–64), both also covering the more remote Daito Islands (**854**). These are accompanied by an illustrated 'tree book' (Walker, 1954). For the southern groups, a substantial descriptive flora in English (Walker, 1976) is available.[92]

HATUSIMA, S., 1971. [*Flora of the Ryukyu Islands.*] x, 940 pp., 18 text-figs. (incl. 6 maps), 30 pls. (some col.), portrs. Okinawa: Okinawa Association for Biology Education. (Subsequently revised as *idem*, 1975. *Ryukyu shokubutsu-shi* [*Flora of the Ryukyus*]. 16, 27, 1002 (actually 1023) pp., 27 half-tone pls., 62 col. photographs, 6 maps. Okinawa.)

Descriptive flora of vascular plants, with keys to all taxa, synonymy (with dates), standardized Japanese and local vernacular names, general summary of local and extralimital range, taxonomic commentary, and notes on habitat, special features, uses, etc.; indices to all botanical and other names at end. An introductory section includes chapters on physical features, geography, geology, climate, special features of the flora, history of botanical exploration and research, and an illustrated organography. [The 1975 revision (not seen) is somewhat expanded and more copiously illustrated than that described here.]

MASAMUNE, G., 1951–64. Enumeratio tracheophytarum Ryukyu insularum. 10 parts. *Sci. Rep. Kanazawa Univ., Biol.* **1** (1951): 33–54, 167–199; **2** (1953): 87–114; **2A** (1954): 59–117; **3** (1955): 101–182, 253–338, figs. 1–7; **4** (1956): 45–134, 201–280, figs. 8–23; **5** (1957): 85–121, figs. 24–31; **9** (1964): 119–156, 1 col. pl.

Systematic enumeration (in English) of vascular plants, with synonymy, references and citations, generalized indication of local and extralimital range, taxonomic commentary, and notes on habitat, special features, etc.; figures of plants of special interest; index to families and genera in part 10. Synonymy and literature citations have been relatively thoroughly worked out in this compilation, the last of several by this author. [The final part, on Orchidaceae, was published significantly later than its predecessors.]

Partial works: southern islands

Egbert H. Walker's large flora, covering the areas under postwar United States administration, was preceded by three accounts for Okinawa: a checklist by Soichiro Sakaguchi, *Okinawa skokubutsu somokuroku* (1924), *Flora of Okinawa* (1952) by S. Sonohara, S. Tanada and T. Amano (all native Okinawans), and [*Flora of Okinawa*] (1958; 2nd edn., 1967) by Sumihiko Hatusima and T. Amano. The last was later developed by Hatusima into a full flora of the Ryukyus (see above); in 1976 and 1994, however, Hatusima produced two editions of a nomenclator (see below), initially to be used alongside Walker's work. This covers Okinawa-gunto and Sakishima-gunto.

HATUSIMA, S. and T. AMANO, 1994. *Flora of the Ryukyus, south of Amami Island.* 2nd edn. by S. Hatusima. v, 393 pp., map. [N.p.]: Biological Society of Okinawa. (Flora and fauna in Okinawa 2.)

A systematically arranged nomenclator with scientific and Japanese names and distribution of the taxa accepted; encompasses native (shown in ordinary face), naturalized and introduced plants including those commonly cultivated. A summary of the work appears on pp. iii–iv; at the end are indices to all generic and family names and to Japanese names. [The work also encompasses Oagari-jima (**855**).]

WALKER, E. H., 1976. *Flora of Okinawa and the southern Ryukyu Islands.* 1159 pp., 209 text-figs. (incl. 1 col. pl.). Washington: Smithsonian Institution Press.

Descriptive manual of vascular plants, with keys to all taxa, synonymy and key references, standard Japanese (in transliteration, with their derivation) and local Ryukyu names, fairly detailed indication of local distribution (with mention of representative *exsiccatae* arranged by district), critical commentary, and notes on habitat, occurrence, behavior, associates, etc.; glossary, author abbreviations, cyclopedia of collectors, selective bibliography (pp. 1099–1104), and complete indices at end. The illustrations are largely of woody plants. An introductory section accounts for the genesis of the work and gives technical notes as well as a key to families. [This conservatively planned work is similar in style and format to the English edition of Ohwi's *Flora of Japan* (**851**). It was an outgrowth of the 1952 *Flora of Okinawa* by Sonohara *et al.* (referred to under the subheading).][93]

Woody plants

WALKER, E. H., 1954. *Important trees of the Ryukyu Islands.* v, 350 pp., 209 text-figs., frontisp., 2 maps. Okinawa: United States Civil Administration.

Illustrated descriptive treatment of the more significant tree species, with keys to all taxa (including an artificial general key), Okinawan and standard Japanese names, generalized indication of local range, and notes on habitat, wood structure, uses, etc.; references and indices to all botanical and other names at end.

857

Islands in Korea Strait (Ullung, Tsushima group and Cheju)

The Tsushima group, between northwestern Kyushu and Pusan in South Korea, is Japanese, while Cheju Do (Quelpaert), in the western entrance to the Strait, and Ullyn (Ullung) Do (Dagelet), in the Sea of Japan, are part of the Republic of Korea. For phytogeographic relationships, see T. NAKAI, 1928. The floras of Tsushima and Quelpaert as related to those of Japan and Korea. In Third Pan-Pacific Science Congress, *Proceedings*, 1: 893–911. Tokyo.

Local works

NAKAI, T., 1914. (*Flora of Saishu and Kwan (Quelpaert) Islands.*) 156, 35 pp. Seoul: Chosen Government. Accompanied by *idem*, 1914. Enumeratio specierum filicum in insula Quelpaert adhunc lectarum. *Bot. Mag.* (Tokyo) 28: 65–104.

The latter work includes some keys along with synonymy, records and commentary. [The former has not been seen.]

NAKAI, T., 1919. (*Report on the vegetation of the island Ooryongto or Dagelet Island, Corea, February, 1818* [sic].) [vii], 87, [1] pp., illus. (incl. 2 folding pls.), map. Seoul: Chosen Government.

Includes checklist (pp. 13–27), notes on species of special interest (pp. 33–43), and key to woody plants (pp. 58–73).

NAKAJIMA, K., 1942. Preliminary report on the flora of the Tushima Islands, I–VIII. *Bot. Mag.* (Tokyo) 56: 193–198, 245–248, 286–292, 344–349, 413–419, 462–468, 515–521, 610–614.

Systematic checklist, with annotations, including Japanese nomenclature, and localities with *exsiccatae*; covers 1130 species and additional infraspecific taxa.

858

Korea

Area: 222273 km^2 (North Korea, 122098 km^2; Demilitarized Zone, 1262 km^2; South Korea, 98913 km^2). Vascular flora: 2898 species (Y. N. Lee in PD). Nakai (1952) listed 3176 species and Chŏng (1970),

3406 species and non-nominate infraspecific taxa. Korea here includes both the Democratic People's Republic of Korea (North) and the Republic of Korea (South), the latter extending to the Cheju Do (Quelpaert) group in Korea Strait (see also 857).

Floras and related works have been published in a number of languages. The first was *Conspectus florae koreanae* (1899–1901) by Ivan V. Palibin, covering 644 species.[94] The peninsula was in this period also covered by *Index florae sinensis* (see 860–80). Following imposition of Japanese control, Takenoshin Nakai contributed *Flora koreana* (1909–11; see below), the standard modern basis for subsequent floristic treatments, *Chôson shokubutsu* (1914), a manual (of which only one volume was published), and the large-scale *Flora sylvatica koreana* (1915–39). Tamezō Mori published an enumeration (1922; see below), covering 2904 species, and T. Kawamoto (T'ae hyŏn Chŏng) a forest manual, *Illustrated book of Korean forest plants* (1934). The emphasis among these publications on the woody flora reflected a marked concern over the apparently poor state of the forests as they existed early in the century. Nakai's final work, produced after what was, for him, a 'stupid war', was his *Synoptical sketch* (1952), a very concise checklist and summary of the vascular flora. In 1943 Chŏng published the first volume of a new manual, *Chosen sinrin shokubutsu zusetsu*, covering woody plants.

By 1950, substantial progress had been made in botanical exploration and documentation and the basic composition of the flora had become known. Nevertheless, it has been said that the level of knowledge was significantly below that in Japan and the basis for floras correspondingly weaker.[95] Chŏng's works – fully issued or reissued by 1957 – were moreover not highly regarded.[96] The problems of an inadequate foundation had been disrupted by World War II and effective partition and were certainly compounded by the Korean War of 1950–53.[97] Nevertheless, much has been accomplished, not only to improve knowledge of the flora but also better to relate it to those of neighboring countries. There is now a not inconsiderable literature in Korean and other journals; furthermore, several illustrated manuals have been published at different times since the 1960s. A good critical modern flora is nevertheless much needed.[98]

In addition to the floras, manuals and other works accounted for below, note should be taken of two other recent publications: W. T. LEE, 1996. *Lineamenta florae*

Koreae. Seoul: Academy Publishing; and *idem*, 1996. (*Coloured standard illustrations of Korean plants.*) Seoul. There is also a six-volume illustrated descriptive atlas-flora published in North Korea: PEOPLE'S DEMOCRATIC REPUBLIC OF KOREA, 1976. *Choson sikmul chi* (*Flora coreana*). 6 vols. Illus. Pyongyang: Edittio [*sic*] Scientiarum; and *idem*, 1979. (*Flora coreana: appendix.*) Pyongyang.

Bibliography

A substantial list may also be found in Chŏng (1986, pp. 146–190; see Note 98).

GOODE, A. M., SR., 1956. An annotated bibliography of the flora of Korea. *Trans. Kentucky Acad. Sci.* **17**: 1–32. (Provisional edn., 1955, Washington, as separate work.) [Includes 305 analytical entries, along with a brief introduction. A strong agricultural and applied element is present.]

Dictionaries

LEE, C. N. and H. S. AHN, 1963. *Nomina plantarum koreanum.* 353, 172 pp. Seoul: Beom-Hak-Sa. [Not seen.]

MORI, T., 1922. *Chôson shokubutsu meii/ An enumeration of plants hitherto known from Corea.* 10, 3, 3, vii, vii, 372, 174 pp., 5 pls. Seoul: Chosen Government. [Systematic list of all vascular plants, with equivalents in Japanese, Korean and Chinese; tabular summary of the flora and list of references; complete indices.]

TO, PONG SŎP and IM, NOK CHAE, 1955. *Chosŏn singmul myŏng chip/*Сборник названий растений Кореи (*Sbornik nazvanij rastenij Korei*)/*Nomina plantarum koreanum.* 10, 364 pp. Pyongyang: Academy of Sciences, People's Democratic Republic of Korea. [Systematic list of vascular plants, with Korean equivalents, covering all families, genera and species; complete indices.]

CHŎNG, T'AE HYŎN [CHUNG TAI HYUN], 1957. *Han'guk singmul togam* [*Korean flora*], 1: [*Trees and shrubs*]. 2nd edn. [7], 13, 7, 507, 82 pp., 1013 text-figs., frontisp., map. Seoul: Synji Sa. (1st edn., 1943 (in Japanese), 683 pp., as *Chôson shinrin shokubutsu zusetsu* under the author's Japanese name (Taigen Kawamoto).) Complemented by *idem*, 1956. *Han'guk singmul togam* [*Korean flora*], 2: *Ha kwŏn ch'o bon pu* [*Herbaceous plants*]. [2], 2, 3, 11, 1025, 129 pp., 2050 text-figs., 2 pls., map. Seoul. (Reprinted 1957, 1962, Seoul.)

A pair of works comprising an atlas-flora of all Korean vascular plants, without keys but including large figures with accompanying descriptive text and (in vol. 1 only) dot distribution maps. The text accounts (in Korean) feature occasional synonymy, Korean and Japanese scientific and vernacular equivalents (with transliterations in the Roman alphabet), general indication of provincial and extra-Korean range, and notes on habitat, phenology, special features, uses, etc.; indices to all botanical and other names at end. The layout of vol. 1 features three species (with figures) and one range map for every two pages; in vol. 2, there are two species (with figures) per page. [The work was to a large extent based on those of Nakai (see below).]

CHŎNG, T'AE HYŎN [CHUNG TAI HYUN], 1965. *Tracheophyta.* 1824 pp., 340 pls. (some col.) with 3051 figs., 12 plain pls. (in glossary), 2 col. pls., 2 col. maps (frontisp.), portr. Seoul: Samhwa (for the Ministry of Education, Republic of Korea). Continued as *idem*, 1970. *Tracheophyta: appendix.* 232 pp., 40 pls. with 355 figs. Seoul. (Illustrated encyclopedia of fauna and flora of Korea 5, 5a.)

Atlas-flora of vascular plants, with text and figures divided, the latter on separate plates; the text includes descriptions, Korean and Romanized Japanese nomenclatural equivalents, distributional data, altitudinal zonation (where necessary), and notes on habitat, origin, occurrence, uses, etc., while the plates each have nine small figures (these partly in color). The introductory section begins with an explanatory section and continues with glossary-tables, a list of references (p. 26), and synoptic tables of contents (in Korean and in Latin), while at the end of the work are separate indices to all scientific names and their Korean and Romanized Japanese equivalents. The supplement of 1970 includes additional species in the same format, but with all illustrations in monochrome; a statistical table is also included. [Nomenclature largely follows the 1952 checklist by Nakai (see below).]

LEE, TCHANG BOK, 1979 (1980). *Illustrated flora of Korea.* 990, 2 pp., text-figs., 2 col. pls., col. signature. Seoul: Hyangmunsa. (Reprinted 1985, 1989.)

Amply dimensioned, systematically arranged atlas-flora of vascular plants (3160 mainly specific entities), with botanical and Korean nomenclature, descriptions, and small figures (four to a page, in the style of Makino's flora of Japan (851)). The several appendices comprise tables of illustrations of botanical features, a glossary of derivations of botanical names, an explanation of author abbreviations, Latin–Korean, Korean–Japanese and Japanese–Korean nomenclatural lexica, and indices to specific and infraspecific taxa in Korean and Latin. A short introduction and two synoptic tables of contents (respectively in Korean and in

Latin) precede the main text. [The work is very well produced, but contains virtually no information on ecology, distribution, etc., as usually provided in floras.]

LEE, YEONG NO, 1996. (*Flora of Korea.*) 1237, [1] pp., col. illus. Seoul: Kyohak Publishing.

Lavishly illustrated descriptive treatment of 3637 native, naturalized and commonly cultivated seed plant species with essential synonymy, vernacular names in Korean, Chinese, English and (most commonly) Romanized Japanese, and indication of phenology, distribution and habitat. A detailed table of contents precedes the work; at the end are an illustrated glossary, a summary of new species and infraspecific taxa and new names, complete indices to botanical and Korean names, and (pp. 1234–1236) a bibliography.

NAKAI, T., 1909–11. *Flora koreana.* 2 vols. Illus. (J. Coll. Sci. Imp. Univ. Tokyo **26**: 1–304, i–ii, pls. 1–15; **31**: 1–573, pls. 1–20). Tokyo.

Systematic enumeration (in Latin) of known vascular plants; includes keys to genera and species, descriptions of new taxa (with illustrations), detailed synonymy (with references and citations), and localities with records and representative *exsiccatae* as well as generalized indication of extralimital range; addenda and complete index to botanical names in vol. 2. [Partly supplanted by the following work.]

NAKAI, T., 1952. A synoptical sketch of Korean flora. *Bull. Natl. Sci. Mus. Tokyo* **31**: 1–152.

Systematic checklist (in English), following an arrangement proposed therein (but not explained) by the author; places of publication given only for genera and families. A narrative summary of the flora is also included. [An unrelated scheme of Loranthales, with a very narrow concept of families therein, is also included.]

Partial work

LEE, YEONG NO, 1976. [Flowering plants.] 895 pp., 186 col. pls. (with 889 photographs) (Illustrated encyclopedia of fauna and flora of Korea 18). Seoul: Samhwa (for the Ministry of Education, Republic of Korea).

Copiously illustrated large-scale semi-popular guide to wild flowers and flowering trees of Korea (covering in all 889 species), with four main sections corresponding to the seasons (each with a separate table of contents); within each section species are described, annotated and accompanied by photographs, and a general summary of each season's special features is given. A synopsis in English (pp. 779–826) covering all species, a bibliography (pp. 827–830) and complete

indices (Latin, Korean, Japanese and Chinese) conclude the work. The introductory sections (in both Korean and English) give the basis of the work, its plan, and an introduction to climate, floristics, life-forms, and phenological patterns; table of families (pp. 25–29). No species is covered in more than one of the four main sections. Some species keys are provided, but none for genera and families.

Woody plants

In addition to the following, reference may be made to T. W. KIM, 1994. (*The woody plants of Korea in color.*) 643, [1] pp., col. illus. Seoul: Kyohak Publishing. This attractive publication accounts for 513 native and introduced species with vernacular names in Korean, English, Chinese and Japanese; there are, however, no keys.

FOREST EXPERIMENTAL STATION, KOREA [T. B. LEE, comp.], 1973. *Illustrated woody plants of Korea.* Revised edn. iv, 237, 15, 10 pp., 755 text-figs. Seoul. (Original edn., 1966. Additional editions in 1987 and 1992.)

Atlas-manual of woody plants, with small figures accompanied by descriptive text and scientific names with Korean transliterations; keys to all species (pp. 190–237); complete indices to Latin and other names at end. [1987 and 1992 editions not seen.]

NAKAI, T., 1915–39. *Flora sylvatica koreana.* Parts 1–22. Pls. Seoul: Forest Experiment Station, Chosen Government. (Reprinted in 10 vols. including index, 1976, Tokyo: Tosho Kankokai.)

Large-scale atlas (in Latin and Japanese) of woody plants organized around revisions of families (not, however, arranged in systematic order), with engraved plates accompanied by copious descriptions (in Latin and Japanese); species accounts include synonymy, references and citations, Japanese and Korean names, generalized indication of local and extralimital distribution, and taxonomic commentary. Preceding each family treatment is a conspectus of all known species (both woody and non-woody) recorded for Korea, with references, literature citations, vernacular names, and notes on distribution.[99]

Pteridophytes

PARK, MAN KYU, 1975. *Pteridophyta.* 549 pp., 43 text-figs., 60 pls. with 258 figs. (some col.), 7 pls. (some col.) incl. map, portr. (Illustrated encyclopedia of fauna and flora of Korea 16). Seoul: Samhwa (for Ministry of Education, Republic of Korea).

The text of this semi-monographic treatment (admitting 272 species) is in two sections: (a) a descriptive portion, including indication of distribution, some taxonomic commentary, and notes on habitat, etc., along with critical figures and keys to all taxa; and (b) an appendix with a synoptic list, a key to rhizome sections, a floristic summary and statistical table, an account of life-forms, distributional and chorological tables, and (in sect. 6) formal nomenclature, with synonymy, references, citations, life-form, Romanized Japanese names, and Korean equivalents. The introductory section recounts the plan of the work and the history of fern studies concerning Korea, followed by a synoptic table of contents, while at the end are the list of references (pp. 529–531) and general indices to Japanese, Korean and Latin names. [The work is in part a companion to the treatment of flowering plants by Yeong No Lee (see above).]

Superregion 86–88

China (except Chinese central Asia)

Area (of country): 9 597 000 km². Vascular flora: 27 283 species in 3339 genera (*Flora of Zhejiang*, general volume; see **872**). At the beginning of the twentieth century, 8271 seed plant species were recorded by Hemsley in *Index florae sinensis* (a work not inclusive of Region 76). By 1930, W. Y. Chun (in a report to the Fifth International Botanical Congress), estimated there to be some 18 000 species; and in 1993 Keng *et al.* in their *Orders and families of seed plants of China* (see below) reported *c.* 26 766 seed plant species in 2871 genera. Even higher figures come from Bartholomew *et al.* in their 1979 trip report (see **Progress** below) and *Plants in danger* (1986; see **General references**), which suggest a range of from 30 000 to 32 000.[100] Of this number, tropical and subtropical regions are said to harbor half, with 7000 in Yunnan alone – a province altogether with some 14 000 species and thus one of the six 'hottest' plant biodiversity spots worldwide. – This superregion corresponds approximately to the area encompassed by the 18 Qing provinces of 'China proper' together with the three provinces of the Dongbei (Northeast China or Chinese Manchuria), i.e., that part of the People's Republic of China bounded on the south by Myanmar and Vietnam, on the west by Xizang (Tibet) and Qinghai Province, on the north by Nei Monggol (Inner Mongolia) and the eastern part of the Russian Federation, and on the east by Korea and the North and South China Seas – with the inclusion of Taiwan, Hainan, the Pratas group, and

Table 2. *Chinese provincial names*

Conventional or 'Post Office' (as formerly used)	Pinyin (current Chinese usage)
Anhwei	Anhui
Chekiang	Zhejiang
Fukien	Fujian
Hainan (Tao)	Hainan (Dao)
Heilungkiang	Heilongjiang
Honan	Henan
Hong Kong	Xianggang
Hopeh (or Hopei)	Hebei
Hunan	Hunan
Hupeh	Hubei
Inner Mongolian A.R.	Neimenggu A.R.
Kansu	Gansu
Kiangsi	Jiangxi
Kiangsu	Jiangsu
Kirin	Jilin
Kwangsi Chuang A.R.	Guangxi Zhuang A.R.
Kwangtung	Guangdong
Kweichow	Guizhou
Liaoning	Liaoning
Ningsia Hui A.R.	Ningxia Huizu A.R.
Shansi	Shanxi
Shantung	Shandong
Shensi	Shaanxi
Sinkiang Uighur A.R.	Xinjiang Weiwuer A.R.
Szechwan	Sichuan
Tibet/Sitsang A.R.	Xizang A.R.
Tsinghai	Qinghai
Yunnan	Yunnan

Notes:
No Pinyin equivalent is to be used for Macao or Taiwan. Not shown here are the South China Sea Islands.
A.R., Autonomous Region.

the Xisha (Paracel) and Nansha (Spratly) Archipelagoes. In effect, it is the China of the summer monsoon; excluded is the dry interior, the former 'Chinese Border'. These latter, with Mongolia, comprise Region 76. For a list of Chinese provincial names in 'conventional' Romanized form with current Pinyin equivalents, see Table 2.

From the mid-nineteenth century when its richness and diversity became generally known, the vascular flora of China has been of great interest to the botanical world. An added feature was the existence of a large body of 'traditional' phytographic literature, with many copiously illustrated works dating back in some cases for nearly two millennia. However, botany in China as a modern profession began only after the revolution of 1911, and thus the publication of two major national works, *Chung kuo chung tz'u chih wu t'u chih (Flora reipublicae popularis sinicae)* (1959–), now far advanced, and *Iconographia cormophytorum sinicorum* (5 vols., 1972–76; supplements 1–2, 1982–83), are testimonials to the remarkable if at times interrupted progress of floristic botany in China since early in the century, when local resources for such work were virtually non-existent.

The development of botany in China was treated by Chun (1931; see **Progress** below) as having had by 1930 three phases. To these must now be added a fourth (and possibly fifth) covering the period since 1949. The first of Chun's periods is that of traditional 'encyclopedic' research and the writing of herbals, dictionaries and other compilations – the *pen ts'ao* ('woody and herbaceous plants') class of literature. Such works began to appear about 30 B.C.E. but, until the time of *Nan fang ts'ao mu chuang* (an account of useful plants of southern China and Vietnam first written in 304 C.E. by Chi Han), have not survived in complete form.[101] More general natural histories and *materia medica* include the tenth-century *Cheng lei pen ts'ao*, a classified work, and the famous *Pen ts'ao kang mu* or 'General guide to materia medica' by Li Shih-chen (1578, but actually first published in 1590 by Li's son and afterwards many times revised, with the best edition being that of 1883–85).[102] The outstanding 'pre-modern' classified botanical work is the 40-volume illustrated *Chih wu ming shih t'u k'ao* (1848) by Wu Ch'i-chün (reprinted in 1880 and in 1912–22 indexed by the U.S. Department of Agriculture). Many of these works remain of value; the task of systematization in modern terms of the material involved is, however, very great and remains unfinished, the work of Emil Bretschneider being but a beginning.[103]

The second phase was that of collecting and research by foreigners. This concurred with late developments in the first phase and in the twentieth century overlapped with the third phase. Until the 1840s the Qing (Ch'ing) rulers kept tight controls on foreign contacts; outsiders were limited geographically to Canton (Guangzhou) and the Portuguese colony of Macao, and it is thus from these places that most early collections from China originate. Through 1771 and beyond these

remained a trickle; in that year only some 260 species were listed by J. R. Forster in his *Flora sinensis*, published in Pehr Osbeck's *A voyage to China and the East Indies*. After 1793 and the Macaulay mission to the Qing court in Beijing, an increasing number of collectors and recorders ensured that a considerable rise in the number of known species would take place in the early nineteenth century, particularly in the two decades before the First Opium War.

The first of the so-called 'unequal treaties' which followed that war included cession of the island of Hong Kong, more treaty ports, and more freedom of movement. From a botanical point of view this created important opportunities for plant introduction. The collections of Champion, R. B. Hinds, Charles Wright, Robert Fortune, and others from Hong Kong enabled George Bentham to write the first florula of any part of China, *Flora hong-kongiensis* (1861; supplement, 1872). Collections were also made in other treaty ports and in accessible parts of the eastern interior, notably by Fortune and, afterwards, H. F. Hance.

A second war in 1860 was followed by further treaties, in which most of the interior became open to outside travel and the customs and some other services were contracted to Western countries, seen at that time as more worthy of trust. From then onwards came the great surge in botanical exploration in China; this also extended to Taiwan, the Ryukyus and Korea although the interior of the first was hardly penetrable and contact with the last was very limited. In the three decades or so before the end of the nineteenth century, collecting was largely the province of the French, Russians and British. Roman Catholic missionaries, among them Armand David, who first made the biological riches of the mountainous west known to the outside world, and Pierre Delavay, the first of the renowned Yunnan collectors, were most prominent among the French (with in addition Odon Debeaux in the east); the Russians, active mainly in the north, included Grigorii N. Potanin, Nicolai M. Przewalski, Carl J. Maximowicz, and (at a later date) Bretschneider and the young Vladimir L. Komarov; while among the British (and Irish) were the customs official Augustine Henry and (at Hong Kong) the gardens superintendent C. Ford. For garden introductions, however, it was relatively a less active period.[104]

The turn of the century was fittingly marked by *Index florae sinensis* (1886–1905; supplement, 1911), initiated by Charles N. Forbes and sponsored in part by the Linnean Society of London. This first modern catalogue of Chinese seed plants, covering (to 1905) over 8200 species, included the mainland, Taiwan, Hong Kong, Korea and the Ryukyu Islands, and has provided, like its primary compiler's earlier *magnum opus*, the botanical volumes of *Biologia centrali-americana*, a key foundation for future research. Other important contributions came from Adrien Franchet, the leading French writer on East Asian floristics, and Maximowicz; however, apart from Franchet's *Plantae davidianae* none reached completion.

From 1900 began the last phase of 'Western-dominated' activity. The first two decades comprise the so-called 'Wilson era', named after the British plant hunter Ernest Henry Wilson. It was dominated by collectors sponsored by arboricultural and gardening interests, particularly in Britain and the United States. There, the new century was opening to a marked change in gardening fashion, one favorable to temperate plants and naturally receptive to the unfolding richness of the Chinese flora. Notable collectors of the period to about 1940 included, in addition to Wilson, Reginald Farrer, William Purdom, George Forrest, Frank Kingdon Ward, Filippo Silvestri, Camillo Schneider, Heinrich Handel-Mazzetti, Joseph Henry Rock and Harald (Harry) Smith. Western China and adjacent parts of Xizang were the main focus of activities, although other parts, notably western Hubei and adjacent parts of Sichuan, were also visited in the search for 'hardy' plants. Although the main emphasis was on seeds and other material for gardens, most collectors obtained large quantities of herbarium material which was energetically studied in the botanical centers of Europe, Japan and the United States. As in the latter part of the preceding century, large, autarchic 'prestige' reports, sometimes with outside sponsorship, were prepared for particular collections – perhaps, let it be said, without too much regard for their coordination. The best is *Plantae wilsonianae* (1911–17) in which, like *Index florae sinensis*, a serious effort at consolidation was attempted. Others include *Die Flora von Central-China* (1900–01) from Berlin, *Plantae chinenses forrestianae* (1911–30) compiled at Edinburgh, *Symbolae sinicae* (1929–37) from Vienna, and *Plantae sinenses* (1924–47) from Uppsala. Large contributions were also made from Paris, where after the death of Franchet those most active included Achille Finet, Henri Lecomte and François Gagnepain, all with a special interest in Indochina, and Le Mans, where Joseph-Henri Léveillé

worked up, in an uncritical fashion, many collections from western China, Guizhou and parts of the east where French missionaries continued to be active.[105]

Conflict with Japan flared in 1931 with the latter's seizure of the Northeast and from then on conditions for foreign collectors became more difficult. Publication of the results of Handel-Mazzetti's and Smith's expeditions continued, but with World War II, the subsequent change of government in China, and changes in scientific fashion, such a pattern of research and publication largely came to an end. Consolidation, above all, was the next step. A first, perhaps ill-considered, attempt was the short-lived *Flora of China* project of the Arnold Arboretum, initiated in the early 1950s; this resulted in a limited amount of published work (Malvaceae and Compositae by Shiu-ying Hu, the latter independently published in Taiwan). On the other hand, the need for national and provincial floras would early be recognized in the new People's Republic of China; to this end, consolidation of research resources would take place and, in the Academy of Sciences, a flora committee would be organized.

Any serious efforts at a 'Western-style' synthesis would, however, be difficult on account of the great scattering of collections and other resources over a century or more. Already by 1930 some 21 centers outside China had been identified by Merrill as having at one or another time had an interest in Chinese floristics. In many cases, little of the material in their possession has been duplicated by specimens in Chinese herbaria (with much of that in Berlin destroyed in 1943). The location outside China of so many of the types of Chinese plants – nearly 18 000 by 1930, and far more than has been the case with North America – has created a serious and lasting problem for Chinese floristic botanists. Chun, a correspondent of (and sometime co-author with) Merrill and later co-founder of *Flora reipublicae popularis sinicae*, made an early plea for assistance in overcoming this unfortunate hiatus. This plea was repeated by a new generation as foreign contacts were resumed in the 1970s (Bartholomew *et al.*, 1979) and has been followed by more concrete action as the *Flora* has progressed and its English version initiated.

The progress of the 'Wilson era' and its aftermath were also to see the development of Chun's third phase of Chinese botany, namely the entry of Chinese into the profession and their emergence as collectors and researchers, accompanied by the formation of herbaria, libraries, laboratories, botanical gardens and field stations. This had been strongly influenced by the great interest of foreign botanists in China, but was given impetus in the cultural reforms which followed the 1911 revolution which ended the Qing dynasty. The 'first generation' of students went abroad for further education as that decade progressed; among them was Sungshu Chien, with Chun the eventual founder of *Flora reipublicae popularis sinicae*. From the 1920s – an early landmark being Hsin-hsuan Chung's *A catalogue of the trees of China* (1924), covering 4840 species – floristic publications by Chinese authors began to appear in considerable numbers. At the same time a strong tradition of biological survey developed which over a period of three decades yielded some hundreds of thousands of collections for herbaria; as in North America in the nineteenth century the often arduous surveys launched the careers of many subsequently prominent botanists. The survey collections were determined by Chun, Hsien-hsu Hu, Ren-chan Ching, and other botanists at the emerging Chinese herbaria, as well as by Handel-Mazzetti (in Vienna), William Wright Smith (at Edinburgh), Diels (at Berlin) and others outside China. They did not, however, themselves become the subject of 'prestige' works as had those from earlier periods. It was during this time that the real richness and diversity of the floras of Yunnan and some other parts of southern and southeastern China – including the island of Hainan, a particular interest of Chun – was more fully recognized; there also developed an improved understanding of the floristic relationships among different parts of the country and abroad.[106]

Much emphasis was also given to education and institution-building. National floristic works mainly took the form of students' manuals and dictionaries while those few at provincial level were generally checklists. Some 'showpiece' items, such as Liou's *Atlas*, were also published; but more important were several works on woody plants such as *Chinese economic trees* by Chun (1922), Chung's already-mentioned catalogue of trees and shrubs, *Forest botany of China* by S. C. Lee (1935; supplement, 1973), and *Chung kuo shu mu fên lei hsüeh* (*Illustrated manual of Chinese trees and shrubs*) by Y. Ch'en (1937; revised 1957). A large number of works outside our scope were also produced. For large descriptive floras, however, collections, facilities and experience very likely were, even more than in India at the same period, quite inadequate; such undertakings were for a later generation.[107]

In the Northeast, floristic work progressed rapidly after 1931 and the foundation was laid for a number of works. This was aided in particular by the existence of *Flora Mančžurii* (1901–07) by V. L. Komarov, one of several other works on northeast Asia by the energetic Russian. The most important work was *Lineamenta florae mandshuricae* by Masao Kitagawa (1939; revised 1979). In Taiwan, floristic work also was aggressively prosecuted, notable discoveries flowing as the interior mountains were opened up. A suite of major floristic works quickly followed, notably the 10-volume *Icones plantarum formosanarum* by Bunzo Hayata (1911–21; supplement by Yoshimatsu Yamamoto, 1925–32) and *Formosan trees* (1917; 2nd edn., 1936) by Ryôzo Kanehira. By the 1940s basic knowledge of these areas had reached a level comparable with that of Hainan, Guangdong and parts of northern and eastern China.

Closely associated with the growth of the Chinese profession were a number of botanists from the United States. They became active in China in the wake of the visits to East Asia by Sargent, the work by Wilson for Sargent's Arnold Arboretum and, from the 1900s, the establishment of schools and colleges by Christian organizations. Among the best-known were Merrill, acting as a 'consultant' to Canton Christian College (later Lingnan University); Lingnan faculty G. Wiedmann Groff, Carl O. Levine and Floyd A. McClure and, later, Egbert Walker; Albert N. Steward at Nanking University; and Franklin P. Metcalf (firstly at Lingnan and then Fukien Christian University). Apart from teaching, these 'botanical missionaries' collected more or less extensively, established herbaria and living collections, and began or contributed to floristic projects including some of the floras and enumerations documented here. In the temperate north, considerable assistance on woody plants was received from Alfred Rehder, part-author of *Plantae wilsonianae*, and others at the Arnold Arboretum.

The 1937–45 war with Japan was marked by great disruption of scientific activities but, because of the move of many institutions to western China, led to an improved knowledge of the flora and contributed to the establishment of regional herbaria in Shaanxi, Sichuan and Yunnan. Only a short time was available after the war for reorganization before the 1949 revolution brought in its train profound structural changes, ending – it may be said – Chun's third phase of Chinese botanical history.

The formation of the People's Republic initiated an era of close relations with the Soviet Union and the adoption of many Russian models. Among them was the formation of an all-embracing, centrally responsible Academy of Sciences (AS), successor to the Academia Sinica which accompanied the ousted Nationalist government in its flight to Taiwan. As already indicated, a number of formerly separate herbaria were amalgamated at this time. Among the new bodies which resulted was the AS Institute of Botany in Beijing, formed in 1950. The South China Institute of Botany (first established in 1928) and certain other botanical institutions and university herbaria in the provinces similarly took their present form at this time. Among incorporated herbaria were those of Christian colleges and universities such as Lingnan University (Guangzhou), Fujian Christian University, and Université l'Aurore (Shanghai). In 1952 the Beijing Academy established *Acta Phytotaxonomica Sinica* as a successor to various pre-revolutionary serials; for a quarter-century (including eight years of suspension in 1966–74), it was effectively the sole outlet for floristic contributions in China. Since the late 1970s, however, other journals have come into being.[108]

Through most of the 1950s there was much interaction between Chinese and Soviet (and other eastern European) scientists. Very likely a desire to emulate *Flora SSSR* and the network of regional Soviet floras led to the establishment of the *Flora reipublicae popularis sinicae* project at Beijing and Guangzhou and, later, its provincial counterparts, most similar to their North Asian models. An added impetus was furnished by the events of 1957–58 (including the 'Great Leap Forward'). A *Flora illustrata plantarum primarum sinicarum* was begun, along with the project eventually published in the 1970s and early 1980s as *Iconographia cormophytorum sinicorum* (in seven volumes). The first parts of a regional flora for Northeast China commenced publication and, some years later, the first of the four volumes of *Flora hainanica*, Chun's *magnum opus* (he had already produced a local flora for Guangzhou (Canton) in 1956). By 1966, however, only three volumes of the national flora had been issued.

The next six years were dominated by the Cultural Revolution, with disruption to almost all scientific work. A semblance of normal conditions began to return in 1972, and since the mid-1970s publication of floras and related works has been rapid. Some 270 botanists have contributed to the national flora (Ma and Liu, 1998; see

Progress below); many families have been, and are being, revised through teamwork at a number of locations. Since 1989, there has also been in place a joint project of the Beijing Institute of Botany and the Missouri Botanical Garden (along with other centers) for an English-language version of the national flora; this has included provision for 'liaison officers' in herbaria with large East Asian holdings. As of 1998 four volumes of text and one of illustrations had been published. In addition, floras have appeared, or will likely appear, for all subregions, provinces and major autonomous areas; of these there has been a veritable flood in the last decade and a half, too numerous to mention here (although special mention should be made of *Flora yunnanica* and *Flora xizangica*, the one on an area renowned for diversity and the other on a land hardly before documented).[109] A similar comprehensive work was also projected for woody plants, but the resulting *Sylva sinica*, though far advanced, is as yet incomplete. On the 'fringes', the first edition of a *Flora of Taiwan* has been published (with a second currently in progress) and work has been underway for some years on a Hong Kong flora.[110]

At the present time, floristic botany in China might be considered to be still in its fourth (post-1949) phase, but with an increased general emphasis on science and technology, the dominance of the national flora project, and renewed interaction with botanists in other parts of the world, the period from the mid-1970s should perhaps be seen as a fifth phase. It has been one during which, as suggested in the original edition of this *Guide*, floristic documentation has expanded rapidly, reaching a level comparable to the West Indies, parts of Africa and Central America, and the Commonwealth of Independent States.

Progress

Developments before 1950: For foreign involvement, see E. BRETSCHNEIDER, 1898. *History of European botanical discoveries in China*. 2 vols. St. Petersburg (renowned for its thoroughness); E. H. M. COX, 1945. *Plant hunting in China*. London; and L. A. LAUENER (ed. D. K. Ferguson), 1996. *The introduction of Chinese plants into Europe*. Amsterdam. The period from about 1915 to 1940 or so, during which Chinese entered the profession on a large scale, is well treated, firstly in separate papers by Chun and Merrill in FIFTH INTERNATIONAL BOTANICAL CONGRESS, 1931. *Proceedings of the Fifth International Botanical Congress* (Cambridge, 1930), pp. 513–536 (part of Symposium on the flora of China) and later by H. H. Hu, 1938. Recent progress in botanical exploration in China. *J. Roy. Hort. Soc.* **63**: 381–389, figs. 99–106, and Li Hui-Lin, 1944. Botanical exploration in China during the last twenty-five years. *Proc. Linn. Soc. London* **156**: 25–44. Progress relating to specific families is outlined in Y. Tsiang, 1950. The development of plant taxonomy in China during the past thirty years. *Quart. J. Taiwan Mus.* **3**: 1–11.

Developments since 1950: S. H. Gould (ed.), 1961. *Sciences in Communist China.* Washington (American Association for the Advancement of Science, Publ. 68); Shiu-Ying Hu, 1975. The tour of a botanist in China. *Arnoldia* **35**: 265–295; B. Bartholomew *et al.*, 1979. Phytotaxonomy in the People's Republic of China. *Brittonia* **31**: 1–25; Ching Ren-Chang, 1979. (Twenty years of Chinese systematic botany.) *Acta Phytotax. Sin.* **17**: 1–6; Hsu Ping-Sheng, 1987. Progress in plant taxonomy in the past two decades. *J. Wuhan Bot. Res.* **5**(1): 77–92; and Ma Jin-shuang and Liu Quan-ru, 1998. The present situation and prospects of plant taxonomy in China. *Taxon* **47**: 67–74.

Flora of China (Flora reipublicae popularis sinicae). At least two papers relate especially to this project: Yü Te-Tsun, 1979. Special report: status of the *Flora of China. Syst. Bot.* **4**(3): 257–260; and I. C. Hedge, 1979. The 'Flora of China'. *Notes Roy. Bot. Gard. Edinburgh* **37**: 467–468. Yü described its organization and progress, while Hedge in particular called attention to defects such as a narrow species concept and too much of an adherence to Komarovian methodology and style. Hsu (1987) and Ma and Liu (1998; for both, see above) have reported on subsequent progress, while developments with the English edition (of which publication commenced in 1994) have been reported in *Flora of China Newsletter*, issued from the Missouri Botanical Garden.

Bibliographies. General bibliographies as for Division 8.

Supraregional bibliographies
See also Ma Jin-Shuang, 1989. Introduction to materials for phytotaxonomy in the People's Republic of China. *Taxon* **38**: 617–620, and Ma and Liu, 1998 (above, under **Progress**).

Chen Sing-Chi (Chen Hsin-Chi), Li Jiao-Lan, Zhu Xiang-Yun and Zhang Zhi-Yun, 1993. *Bibliography of Chinese systematic botany (1949–1990)/ Zhongguo chiwu xitongxue wenxian yaolan, 1949–1990/ Chung kuo chih wu hsi tung hsüeh wen hsien yao lan, 1949–1990.* [10], 810 pp. Guangdong: Guangdong Science and Technology Press. [Arrangement by author, with five appendices (including details of *Flora reipublicae popularis sinicae* and the regional floras) and three indices (general, regional and systematic). A most valuable complement to the works of Merrill and Walker. The regional floras in appendix 2 are not, however, also cited in the main author–title list.]

Liu Ju-Ch'iang, 1930. Important bibliography on the taxonomy of Chinese plants. *Bull. Peking Soc. Nat. Hist.* **4**(3): 17–32. [Key floristic and systematic works.]

Merrill, E. D. and E. H. Walker, 1938. *A bibliography of Eastern Asiatic botany.* xlii, 719 pp., maps. Jamaica Plain, Mass.: Arnold Arboretum of Harvard University. Continued as E. H. Walker, 1960. *A bibliography of Eastern Asiatic botany: Supplement 1.* xl, 552 pp. Washington: American Institute of Biological Sciences. [Detailed coverage of floristic and geobotanical literature, with regional and subject indices for China proper as well as for Korea and Japan (Region 85), the 'Chinese Border' (Region 76), and Central and Eastern Siberia and the Russian Far East (Regions 72, 73). The original work has 35000 entries, and goes up through about 1936–37; the supplement takes coverage through about 1958.]

Wang Zong-xun (Wang Tsung-hsun) (ed., for the Chinese Botanical Society/Chung-kuo chih wu hsüeh hui), 1983. (*Bibliography of Chinese botany.*) 3 vols. xxx, 1793 pp. Beijing: Ke xue chu ban she (Science Press). Continued as *idem*, 1995. (*Bibliography of Chinese botany.*) Vol. 4. 1463 pp. Beijing. [The original bibliography, on pp. 1–1657 in vols. 1–3, is arranged by authors and dates and covers all areas of botany from 1857 through 1981; this is followed by a chronological list of older Chinese works (pp. 1658–1667), a classified index (pp. 1668–1791; plant taxonomy, pp. 1711–1721; English synopsis, pp. 1668–1670), and addenda (p. 1793). The arrangement of authors in the main bibliography is 'Romanized', enabling a single sequence for entries irrespective of script. The introductory part includes a list of serials cited as well as (p. xviii) sources. Vol. 4 is similarly organized, with classified index on pp. 1363–1463 (the taxonomic portion from p. 1402).]

Indices. General indices as for Division 8.

Supraregional index

STEENIS, C. G. G. J. VAN *et al.* (eds.), 1948– . *Flora Malesiana Bulletin*, 1– . Leiden: Flora Malesiana Foundation (later Rijksherbarium). (Mimeographed.) [At present issued bi-annually; the autumn installment contains a substantial bibliographic section which rather thoroughly covers ongoing new literature on Southeast Asia. For fuller description, see **Superregion 91–93.**]

860–80

Superregion in general

Apart from the general works accounted for here, which appear under four subheadings (Comprehensive works, Abridged general works, Woody plants, and Pteridophytes), a representative selection of botanical reports and related literature relating to the vast collections made in central and western China from the 1870s to the 1940s – including the still often-consulted *Plantae wilsonianae* – is given under a further subheading (**Partial works**). For details of the immense literature, however, reference should be made to the standard bibliographies (enriched now by a new retrospective for the 1949–90 period).

The publication of *Flora reipublicae popularis sinicae*, begun in 1959 and projected to comprise 80 volumes, was greatly accelerated from the mid-1970s. When completed, it will provide the first modern general account of the Chinese vascular flora, succeeding *Index florae sinensis* by Forbes and Hemsley (1886–1905; supplement, 1911), long a standard reference but now accounting for little more than a quarter of the known vascular flora (and omitting pteridophytes as well as the so-called 'Chinese Border' territories).

Abridged manuals are also available. Of these, by far the most comprehensive is *Iconographia cormophytorum sinicorum*, complete in seven volumes (1972–82). This has effectively superseded two earlier works: *Chung kuo chih wu t'u chien* by Chia and Chia (1937; revised 1955), similar to the *Iconographia* but much more selective, and *Chung kuo chung tz'u chih wu fen lei hsüeh* by Mien Chêng (1954–59), a never-completed three-volume descriptive account with partial keys. For the vast woody flora, *Sylva sinica* (1985–), a work begun by the prominent forest botanist and dendrologist W. C. Cheng, will if completed supersede an aging checklist and two less than satisfactory descriptive works; by 1999 three volumes (of an expected four) had been published. Other works of interest are the unfinished *Flora illustrata plantarum primarum sinicarum*, with useful introductory accounts to pteridophytes, grasses and legumes, two standard dictionaries and two sets of keys to families.

For readers not literate in Chinese, the latest work is *Flora of China*, the English edition of *Flora reipublicae popularis sinicae*, of which the first of 25 projected volumes appeared in 1994 with illustrations following from 1998. Older works include the classic expedition reports as well as the now-obsolete Forbes and Hemsley *Index* and Lee's *Forest botany of China*. All Chinese-language works, however, employ Roman script and Arabic numerals at least for botanical names and, where necessary, references, citations and certain descriptive elements. An excellent general survey of vegetation and flora may be found in WANG CHI-WU, 1961. *The forests of China: with a survey of grassland and desert vegetation.* xiv, 313 pp., 78 text-figs., map (Maria Moors Cabot Foundation, Harvard Univ., Spec. Publ. 5). Petersham, Mass.

Dictionaries

HOW FOON-CHEW [HOU K'UAN-CHAO], 1982. *Chung kuo chung tz'u chih wu ko shu tz'u tien (A dictionary of the families and genera of Chinese seed plants).* 2nd edn, revised by Wu Te-lin, Ko Wan-cheung, Chen Te-chou *et al.* v, 632 pp. Beijing: (Science Press; for the South China Institute of Botany, Guangzhou). (1st edn., 1958.) [Includes all families and genera as well as some species, accompanied by formal Chinese names and, for each taxon, descriptions and brief notes (with citations of pertinent revisions and monographs and, under families, lists of genera). Definitions of descriptive botanical terms are also incorporated. Cross-indices conclude the work.]

K'UNG CH'ING-LAI (and 12 collaborators), 1918. *Chih wu hsüeh ta tz'u tien* [*Botanical nomenclature: a complete dictionary of botanical terms*]. 1726 pp., illus. Shanghai: Shang Wu Yin Shu Kuan (Commercial Press). (5th edn., 1923; reprinted 1933.) [Illustrated dictionary of vascular plants, with short descriptive text, some synonymy, and generalized indication of internal range; cross-indices to botanical, common European, and Japanese nomenclatural equivalents. Botanical terms are also defined, so altogether the work is something of a Chinese 'Willis'. – Not seen by the author; description and citation prepared with the aid of notes supplied by Joseph Needham, Cambridge.]

Keys to families and genera

The Institute of Botany's *Claves* succeeds HU HSIEN-HSU *et al.*, 1954–55. Chung kuo chih wu k'o shu chien so piao [Claves familiarum generumque plantarum sinicarum]. *Acta Phytotax. Sin.* **2**: 173–586.

INSTITUTE OF BOTANY, ACADEMIA SINICA, 1979. *Claves familiarum generumque cormophytorum sinicorum.* iv, 733 pp., illus. Beijing: Ke xue chu ban she (Academia Sinica Press). [Analytical keys, together with a systematic synopsis of families and genera and an index. Based in the first instance on Engler's *Syllabus der Pflanzenfamilien* and other standard works.]

KENG YI-LI and KENG PAI-CHIEH, 1958. *Chung kuo chung tz'u chih wu fên ko chien so piao.* [3rd edn.] ii, 108 pp. Beijing. (1st edn., 1948; 2nd edn., 1951); and KENG YI-LI *et al.*, 1988. (*Illustrations and keys to the families of seed plants in China.*) 541 pp., illus. Beijing. [This 'key to the families of phanerogams in China' also includes a synopsis of the families recorded. The 1988 edition incorporates illustrations. – Not seen by the author; cited from Chen *et al.* (1993; see above under **Supraregional bibliographies**).]

KENG, HSUAN, HONG DE-YUAN and CHEN CHIA-JUI, 1993. *Orders and families of seed plants of China.* xx, 444 pp., 163 text-figs. Singapore: World Scientific. [A synoptic students' text, with descriptions of orders, a description and representative illustration for each family and keys to genera (or groups of genera); index to genera and families at end. Appendix I is an outline of the Cronquist system of 1988, and in appendix II is a key to families. – Based on *Claves familiarum generumque cormophytorum sinicorum* (1979; see above); circumscription of taxa largely follows the Engler/Diels system of the 1936 *Syllabus* (as used in *Flora reipublicae popularis sinicae*).]

Comprehensive works

Included here are the two major accounts of the Chinese flora, *Flora reipublicae popularis sinicae/Flora of China* and *Index florae sinensis.*

FORBES, F. B. and W. B. HEMSLEY, (1886–) 1888–1905. *Index florae sinensis. An enumeration of all plants known from China proper, Formosa, Hainan, Corea, the Luchu Archipelago and the island of Hongkong, together with their distribution and synonymy.* 3 vols. 14 pls., map (*J. Linn. Soc., Bot.* **23** (1886–88): 1–521, pls. 1–14, map; **26** (1889–1902): 1–592; **36** (1903–05): 1–686). London. (Originally issued both as journal and under separate cover; only the latter carries the title *Index florae sinensis.* Reprinted in 3 vols., 1980, Königstein/Ts., Germany: Koeltz.) Continued as S. T. DUNN, 1911. A supplementary list of Chinese flowering plants. *J. Linn. Soc., Bot.* **39**: 411–581.

The main work is a systematic enumeration (in English) of seed plants recorded from the territories indicated in the title; includes descriptions of new taxa, full synonymy with references and citations, fairly detailed indication of internal distribution, brief summary of extralimital range, and miscellaneous taxonomic and other notes. The third volume includes at its end some remarks on the history of botanical exploration in the area, a 'first supplement' (by M. Smith) for the period 1886–1904, and a complete index to botanical names. The *Supplementary list*, actually the 'second supplement', contains an alphabetically arranged list of species under their respective genera.[111]

INSTITUTE OF BOTANY, ACADEMIA SINICA, 1959– . *Chung kuo chung tz'u chih wu t'u chih/Flora reipublicae popularis sinicae.* Vols. 1–80, *passim.* Illus. Beijing: Ke xue chu ban she (Academia Sinica Press). English edn.: WU CHENG-I and P. H. RAVEN (for the Flora of China Editorial Committee), 1994– . *Flora of China.* Beijing: Science Press [Ke xue chu ban she]/St. Louis, Mo.: Missouri Botanical Garden.

A multi-volume documentary descriptive flora of vascular plants; in each family treatment (encompassing one or more volumes or part-volumes or in a part thereof) are keys to genera and species, full synonymy (with references and citations), Chinese and scientific names, generalized indication of local and extralimital distribution, and notes on habitat, special features, uses, etc. Indices to all Chinese and Latin names appear at the end of each volume or part-volume. Volume 1 has not yet been published. – The English edition began publication with its vol. 17; as of writing (1999) five volumes (4, 15–18) of text and two (16–17) of illustrations have appeared. The text volumes differ from the original in their larger, double-column format; moreover, any Latinized text in the Chinese edition (notably taxonomic documentation) is omitted. [Volumes (or part-volumes) in the Chinese edition – with 125 in all expected – have been issued only as family or infrafamilial treatments have become ready. As in its model, the *Flora SSSR*, the arrangement of families follows the Englerian system (in particular the 11th edition of the *Syllabus* (1936) by Diels). For a partial index to all botanical names (with on pp. 620–655 an index by genera to volumes and authors), see C.-Y. MA *et al.* (comp.), 1997. *Indices nomenum* [sic] *latinorum (1959–1992).* 697 pp. Beijing.][112]

Abridged general works

Here are listed *Iconographia cormophytorum sinicorum* and two students' manuals. These were as of writing being joined by the 13-volume quarto *Higher plants of China* edited by Fu Likuo, Chan Tanqing, Lang Kaiyung and Hong Tao (1999– , Qingdao); with keys, descriptions, figures and distribution maps this will cover some 17 000 species of cormophytes.

CHENG MIEN, 1954–59. *Chung kuo chung tz'u chih wu fên lei hsüeh*. Vols. 1–2 (in 3). Illus. [Shanghai]: Hsin Ya Shu Tien (New Asia Book Co.).

This 'manual of Chinese seed plants' is a concise, illustrated systematic account (on the Englerian system) designed for students, with analytical keys to families and genera and synoptic keys for species or groups thereof, limited synonymy, diagnostic descriptions, Chinese transliterations of scientific names, notes on distribution, habitat, special features, etc., and many small figures of representative species. Descriptions of genera and families are also given. Appendices include a synopsis of families and a glossary, and (presumably) indices to names in Chinese and Latin. [Never completed (the third and final volume remained unpublished) and now effectively supplanted by *Iconographia cormophytorum sinicorum* (see below).][113]

CHIA TSU-CHANG (KIA, TCHOU-TSANG) and CHIA TSU-SHAN (KIA, TCHOU-SHAN), 1955. *Chung kuo chih wu t'u chien*. Revised edn. 4, 4, 16, 1529, 62, 70 pp., 2602 text-figs. Beijing: Chung-hua shu chü. (Reprinted 1958. 1st edn., 1937.)

This 'atlas of Chinese plants' is an illustrated descriptive atlas covering a selected range of non-vascular and vascular plants; it features analytical keys, limited synonymy, Chinese transliterations of scientific names, generalized indication of internal distribution, and notes on habitat, life-form, special features, properties, uses, etc., with at the end indices to Latin and Chinese botanical names as well as some European vernacular names. Line drawings (6 × 4 cm) are provided for each species included. [Now largely superseded by the more comprehensive *Iconographia cormophytorum sinicorum* (see below).][114]

INSTITUTE OF BOTANY, ACADEMIA SINICA, 1972–83. *Iconographia cormophytorum sinicorum*. 7 vols. (1–5 and Supplements I–II). 8374 figs. Beijing: Ke xue chu ban she (Academia Sinica Press).

Semi-technical atlas-flora (in simplified ideography) of a substantial selection of Chinese vascular plants and bryophytes (some 10 000 species in all), with accompanying descriptive text; the latter includes for each species a botanical description, Chinese transliteration of the scientific name, and general notes on internal distribution, occurrence, habitat, properties, uses, special features, etc., but omits botanical synonymy and references. The figures, of almost uniformly excellent quality, contain habit and botanical details of the plants depicted. Volume 1 also includes an illustrated glossary and organography, while all volumes contain analytical keys to genera and families (and, in many cases, species) as well as complete indices (with an abridged general index to the work proper in vol. 5). The two supplementary volumes of 1982–83 comprise important additions and corrections with many new and substitute keys. A modified Englerian system is followed with respect to seed plants. [The *Iconographia* in format resembles Coste's *Flore de France* (see **651**) but usually depicts two species per page. Coverage seems to be more complete for northern and eastern provinces than for other regions of China, and full treatment was also related to a given taxon's presumed importance (additional species may appear only in the keys).][115]

Partial works

As noted under the main heading, there is here included a selection of the more important botanical reports on the many collections from (mainly) central and western China made by visiting explorers from 1870 to the 1930s. The background of these field trips, some of long duration, has been ably presented in the works by Bretschneider, Cox and Lauener noted under **Superregion 86–88, Progress**. Cox's work is especially recommended as an introduction to the subject. Together, the reports provide a useful though scattered coverage of the rich flora of a part of China in the past somewhat sketchily covered by the various general works.

DIELS, L., 1900–01. Die Flora von Central-China. *Bot. Jahrb. Syst.* **29**: 169–659, figs. 1–5, pls. 2–5. Continued in *idem*, 1905. Beiträge zur Flora des Tsin Ling Shan und andere Zusätze zur Flora von Central-China. *Ibid.*, **36**, Beibl. 82: 1–138. Chinese version of original work: *idem*, 1932–34. [The flora of Central China] (transl. Tong Koe-yang). *J. Coll. Sci. Sun Yat-sen Univ.* **3** (1932): 684–723; **4** (1932–33): 85–132, 221–281, 427–479, 619–671; **5** (1933–34): 267–316, 455–506; **6** (1934): 25–60, 277–316.

The original work comprises a concise systematic enumeration of records, based upon collections received at Berlin in 1899–1900 (mainly from Rosthorn, Bock, Giraldi and Niederlein) as well as literature sources, originating chiefly from western Sichuan to western Hubei, north to southern Shaanxi, and south to Yunnan and Guizhou (see pp. 176–180 for explanation of the author's four geographical districts), accompanied by a phytogeographic summary (pp. 635–652). Descriptions of novelties are included along with distribution, some *exsiccatae*, citations of sources, and critical notes; a classified list of Chinese names of useful plants and an index to genera conclude the work. The *Nachträge*, similar in style, accounts for additional collections and records. [An important milestone.]

FRANCHET, A., (1883–)1884–88. *Plantae davidianae ex Sinarum imperio.* 2 vols. 44 pls. (Nouv. Arch. Mus. Hist. Nat. Paris, II, **5** (1883): 153–272, pls. 7–16; **6** (1883): 1–126, pls. 11–18; **7** (1884): 55–200, pls. 6–14; **8** (1885): 183–254, pls. 2–10; **10** (1888): 33–198, pls. 10–17). Paris. (Reprinted separately, Paris: Masson; 2nd reprint (in 1 vol. in a smaller format), 1970, Lehre, Germany: Cramer.)

Enumeration (in two parts) of collections made by Père Armand David, the renowned pioneer explorer of western China and the 'Tibetan Marches', from 1866 to 1874 in many parts of China and Mongolia (notably during his second trip of 1868–70 in western Sichuan and adjacent areas). [The work is of great historical importance as it gives one of the first outlines of the overall floristics and phytogeography of China as well as the 'borderlands' in central and northern Asia.]

FRANCHET, A., 1889–90. *Plantae delavayanae, sive enumeratio plantarum quas in provincia chinensi Yun-nan collegit J.-M. Delavay.* Fasc. 1–3. Pp. 1–240, pls. 1–48. Paris: Klincksieck.

Systematic enumeration, with descriptions of new taxa, of the collections of Père Jean-Marie Delavay from northwest Yunnan. [Not completed; extends from Ranunculaceae through Saxifragaceae (Candollean system).]

HANDEL-MAZZETTI, H. (ed.), 1929–37. *Symbolae sinicae. Botanische Ergebnisse der Expedition der Akademie der Wissenschaften in Wien nach Südwest-China, 1914–18.* 7 vols. (in 11 parts). Illus. Vienna: Springer.

Systematic enumeration (with descriptions of new taxa) of non-vascular (vols. 1–5) and vascular plants, based mainly on the collections of the writer-editor and Camillo Schneider made in central and western China between 1914 and 1918, largely in northern Yunnan, southwestern Szechwan, and the adjacent 'Tibetan Marches'. Treatments of many groups contributed by specialists.

LAUENER, L. A. *et al.*, 1961–88. Catalogue of the names published by Hector Léveillé, I–XX. *Notes Roy. Bot. Gard. Edinburgh* **23–45**, *passim.*

Systematic evaluation of all taxa from eastern Asia published by Abbé Léveillé, including synonymy, reductions, and literature references and citations; in some cases indication of holotypes, paratypes and syntypes is made. Arrangement of families follows the Bentham and Hooker system; while the 'traditional' anglicized spelling of provinces has been maintained throughout. [A revision and

amplification of A. REHDER, 1929–37. Notes on the ligneous plants described by H. Léveillé from eastern China. [14 parts.] *J. Arnold Arbor.* **10–18**, *passim.*]

MAXIMOWICZ, C. J., 1890–92. Plantae chinenses Potaninianae nec non Piaszekianae. *Trudy Imp. S.-Peterburgsk. Bot. Sada* **11**: 1–112.

Systematic enumeration of higher plants from the collections of Grigorii N. Potanin and others, mainly from the subregion of 'northwestern China' (Shaanxi and Gansu) and adjacent parts of Mongolia.

PAX, F., 1922. Aufzählung der von Dr Limpricht in Ostasien gesammelten Pflanzen. In W. Limpricht (ed.), *Botanische Reisen in den Hochgebirgen Chinas und Ost-Tibets*, pp. 298–515, illus. (Repert. Spec. Nov. Regni Veg., Beih. 12). Berlin: Fedde.

Comprises a systematic enumeration, with descriptions of new taxa, of Limpricht's collections from western Szechwan and the adjacent 'Tibetan Marches' made in 1913–14. Treatments of several groups by specialists.

REHDER, A., 1929–37. See under LAUENER *et al.* above.

REHDER, A. and E. H. WILSON, 1928–32. Enumeration of the ligneous plants collected by J. F. Rock on the Arnold Arboretum expedition to northwestern China and northeastern Tibet. *J. Arnold Arbor.* **9**: 4–27, 37–125, pls. 12–23; **13**: 385–409. Complemented by A. REHDER and C. E. KOBUSKI, 1933. An enumeration of the herbaceous plants collected by J. F. Rock for the Arnold Arboretum. *Ibid.*, **14**: 1–52.

Systematic enumerations of Joseph Rock's vascular plants from northwestern Yunnan, western Szechwan, Chinghai Province, and southern Kansu; includes descriptions of new taxa, citations of *exsiccatae* and descriptive commentary.

ROYAL BOTANIC GARDEN, EDINBURGH, 1911–12. Plantae chinenses forrestianae. *Notes Roy. Bot. Gard. Edinburgh* **5**: 65–148, 161–308.

Systematic enumeration, with descriptions of new taxa, of collections from George Forrest's first Yunnan expedition (1904–07).[116]

SARGENT, C. S. (ed.), 1911–17. *Plantae wilsonianae.* 3 vols. (in 9 parts) (Publ. Arnold Arbor. 4). Jamaica Plain, Mass. (Reprinted 1988 by Dioscorides Press as Biosystematics, floristic and phylogeny series 3.)

An elaborately produced systematic enumeration of the woody plants collected in particular by E. H. Wilson during the years 1907–09 and 1910, principally in western Hupeh, eastern and western Szechwan, and Kiangsi; includes several regional revisions of genera, descriptions of new taxa, extensive synonymy (with references and citations), localities with *exsiccatae*, and keys. [Treatments of many groups contributed by specialists; much of the text written by Alfred Rehder.]

SMITH, H. *et al.*, 1924–47. Plantae sinenses a Dre H. Smith annis 1921–2 lectae [1924 et 1934]. *Acta Horti Gothob.* **1** (1924): 1–187; **2** (1926): 83–121, 143–184, 285–328; **3** (1927): 1–10, 65–71, 151–155; **5** (1930): 1–54; **6** (1931): 67–78; **8** (1933): 77–81, 127–146; **9** (1934): 67–145, 167–183; **12** (1938): 203–359; **13** (1940): 37–235; **15** (1944): 1–30, figs. 1–195, pls. 1–6; **17** (1947): 113–164.

Enumeration, with descriptions of new taxa, of the author's collections from western and northwestern Szechwan and the adjacent 'Tibetan Marches' as well as from Shansi. [Not completed; contributions, mostly by specialists, not in systematic sequence. Like *Plantae wilsonianae*, the work contains some important synthetic systematic accounts of particular taxa.]

Woody plants

The first modern systematic checklist of tree species is CHUNG HSIN-HSUAN, 1924. *A catalogue of the trees and shrubs of China*. 271 pp. (Mem. Sci. Soc. China 1(1)). Shanghai. This is a compiled work with synonymy and indication of distribution by provinces with information to 1920. Though not in its genre superseded, it is now effectively out of date. The principal modern manual used in China (though not comprehensive and now also outdated) has been that by Chen Yung (1957; see below); a successor, *Sylva sinica*, began to appear in 1985 under the editorship of Wen-chun Cheng. For English-literate readers Lee's work of 1935 remains useful; his 1973 supplement, however, lacks illustrations and moreover is uncritical.

ZHENG WAN-JUN (CHENG WAN-CHUN) (chief ed.), 1985– . *Sylva sinica / Zhongguo shumu zhi* [*Flora of woody plants in China*]. Vols. 1– . Illus. Beijing: (Forestry Press).

Illustrated descriptive manual of native and introduced species, with keys to all taxa, botanical and formal Chinese names, essential synonymy, and notes on distribution, altitudinal range, habitat, etc.; indices to Chinese and botanical names. Volume 1 contains introductory chapters, including illustrated accounts of forest formations with their characteristic species (arranged generally from north to south and east to west), and (pp. 96–136) a general key to families. [Not yet completed; by 1999 three of a projected four volumes, with pages 1–3969 of main text and figures 1–2060, had been published, covering pteridophytes, gymnosperms, and dicotyledons ranging from the 'Magnoliidae' and 'Hamamelideae' (including the important Laurales and Fagales) through the Rosales, Fabales, Saxifragales, Cornales, Umbellales, Caprifoliales, Malvales, Myrtales, Ericales, Euphorbiales, Urticales and Rhamnales among others. The remaining dicotyledons (including Sapindales) as well as woody monocotyledons (including Poales) should appear in volume 4.]

CH'EN YUNG, 1957. *Chung kuo shu mu fēn lei hsüeh* [*Illustrated manual of Chinese trees and shrubs*]. 2nd edn. 2, 2, 6, 14, 2, xxxxiv, 6, 106, 13, 1191, 4, 64, 67, 6, 18, 13 pp. (main

text); 11, 60 pp. (appendix); 1159 text-figs. Shanghai: Science and Technology Press. (1st edn., 1937, Shanghai: Agricultural Association of China.)

A briefly descriptive, extensively illustrated but largely compiled account of about 2100 selected native and introduced species, with keys to all taxa, limited synonymy, Chinese transliterations of scientific names, generalized indication of internal range, and notes on habitat, special features, uses, etc.; glossary, bibliography, summary of the work, and indices to all Latin, Chinese and English names. Introductory chapters include a list of sources, an illustrated organography, detailed synopses of families and genera, and a general key to families. [Essentially a reprint of the 1937 edition with subsequent additions and corrections covered in the appendix.]

LEE SHUN-CHING, 1935. *Forest botany of China*. xlvii, 991 pp., 272 text-figs. Shanghai: Commercial Press. Continued in *idem*, 1973. *Forest botany of China: supplement*. xi, 477 pp. Taipei: Chinese Forestry Association.

The main work is an illustrated descriptive treatment (in English) of the more common trees (and some shrubs) of China, with keys to families and genera, limited synonymy, generalized indication of local range and altitudinal zonation, and notes on habitat, special features, etc.; family summaries, glossary, and index to botanical names at end. An introductory section includes a synoptic list of orders and families. In general, this work is more useful for the northern and eastern parts of old China; coverage of other areas seems weak. The 1973 supplement accounts for 4073 species and additional varieties (encompassing all taxa in the original work) but mainly consists of additions and corrections (including some new keys). It accounts, however, for new information only through 1963 and moreover is unillustrated, poorly documented and uncritical.[117]

Pteridophytes

For a recent overall review of this group in China, see CHING REN-CHANG, 1978. The Chinese fern families and genera: systematic arrangement and historical origin. *Acta Phytotax. Sin.* **16**(3): 1–19; **16**(4): 16–37. An earlier account of genera is FU SHU-HSIA, 1954. (*Genera of Chinese ferns.*) viii, 203 pp., 102 text-figs. Peking. [Not seen; reference from Koeltz Book List No. 205 as well as Chen *et al.* (1993; see above under **Supraregional bibliographies**).] Some accounts of ferns and fern-allies in *Flora reipublicae popularis sinicae* remain outstanding.

FU SHU-HSIA, 1957. *Flora illustrata plantarum primarium sinicarum: Pteridophyta*. 6, ii, 280 pp., 346 text-figs. Beijing: Ke xue chu ban she (Academia Sinica Press).

This work comprises an illustrated descriptive manual of more important or widespread pteridophytes in China, covering 437 species in 130 genera with limited synonymy, keys, internal and extralimital distribution, and habitat and other notes; index at end. (Other volumes in this series, not accounted for here, cover grasses and legumes.)

Region 86

North China (including Northeast China)

As delimited here, North China includes the northeastern provinces of Heilongjiang (Heilungkiang), Jilin (Kirin) and Liaoning (long the core of Manchuria, and later the North-East Province), the northern provinces of Shandong (Shantung), Hebei (Hopeh), Henan (Honan) and Shanxi (Shansi), and the northwestern provinces and territories of Shaanxi (Shensi), Gansu (Kansu) (eastern part) and Ningxia Hui Autonomous Region. For convenience, the former territory of Jehol is here considered part of Hebei. The western part of Gansu, beyond the Huang Ho and Tao Ho, is floristically part of Central Asia and therefore is treated as part of Region 76.

A considerable amount of exploration was accomplished in North China by non-Chinese botanists in the nineteenth and early twentieth centuries, with later extensions to Manchuria and northwestern China. Additional extensive work was carried out during the biological surveys mounted after 1920, and from 1931 in Manchuria collecting was assiduously pursued by the Japanese. Regional botanical centers have been established at Mukden (Shenyang) and Wukung (near Xian) for the Northeast and Northwest respectively, and at both floras have been compiled in recent years.

At present, no separate flora covers the whole region, which for botanical purposes has in any case been treated as three subregions. There is, however, an overall flora of woody plants (1984), supplanting the incomplete account of 1923–26 by Alfred Rehder. It is, however, relatively well covered in general floras of China, most thoroughly of course in *Flora reipublicae popularis sinicae*, and in subregional and, latterly, provincial works which initially were more numerous in the Northeast than elsewhere.

Bibliographies. General and supraregional bibliographies as for Division 8 and Superregion 86–88.

Indices. General and supraregional indices as for Division 8 and Superregion 86–88.

860

Region in general

See also **860–80** (all general works). While no flora encompasses the whole of the region as here defined, most is covered by partial works centering on one of three subregions: Northeast China, North China and Northwest China. Each of these is here allocated a subheading. Northeast China corresponds almost entirely to the former Manchuria (now Heilongjiang, Jilin, Liaoning, and northern Inner Mongolia); North China centers on Beijing and includes the provinces of Hebei (with Jehol), Shandong, Henan and Shanxi; while Northwest China here includes Shaanxi, southern Gansu, and Ningxia Hui Autonomous Region.

Keys to families
LIU JU-CH'IANG, 1934. *Systematic botany of the flowering families in North China.* 2nd edn. xvi, 218 pp., 300 text-figs. Peiping (Beijing): Vetch (The French Bookstore). (1st edn., 1931.) [Students' handbook, with keys to and descriptions of families as well as illustrations of representative species, vernacular names, taxonomic commentary, and notes on cultivation, uses, etc.; glossary and descriptive organography.]

Woody plants
HUA BEI SHU MU ZHI BIAN XIE ZU BIAN (WOODY FLORA OF NORTH CHINA WORKING GROUP), 1984. *Hua bei shu mu zhi (Woody flora of North China).* [4], 3, 743 pp., 733 text-figs. Beijing: Zhong guo lin ye chu ban she (Forestry Publishing House of China).

Illustrated descriptive woody flora (encompassing native and introduced species); includes keys to all taxa, formal Chinese and vernacular names, limited synonymy, indication of distribution and (sometimes) altitudinal range, notes on habitat, etc., and indices to all names. The introductory part (pp. 1–22) includes the general key to the 89 families covered in the work.

REHDER, A., 1923–26. Enumeration of the ligneous plants of northern China, I–III. *J. Arnold Arbor.* **4**: 117–192; **5**: 137–224; **7**: 151–227, 1 pl.

Systematic enumeration of the known woody plants of North China, with full synonymy (including references and citations), notation of *exsiccatae* with localities, general indication of local and extralimital distribution, taxonomic commentary, and miscellaneous other notes; no separate index. [Not completed; covers families from Ginkgoaceae through Sapindaceae (Englerian system). The work, whose limits include Northwest as well as North China, was based primarily on collections held in what are now the Harvard University Herbaria.]

I. Northeast China (former Manchuria)

This area at present encompasses three provinces: Heilongjiang (**861**), Jilin (**862**) and Liaoning (**863**). Many works cited here also include the western hills and mountains, now mostly in Nei Monggol (**763**). From the late nineteenth century until 1979 political and administrative changes were frequent. Northern Manchuria, including most of modern Heilongjiang, was for a time at the turn of the twentieth century a Russian concession granted in relation to the construction and operation of the 'Chinese Eastern Railway' from Eastern Siberia to Vladivostok (the railway, however, passed to the Japanese in 1905, one of the spoils of the Russo-Japanese War). From 1931 to 1945 all Manchuria, together with the border territory of Jehol, comprised the Japanese state of Manchukuo. After 1949 it became a single North-East Province. In the following decades the western hill zone (including the Great Khingan Mountains) was attached to Nei Monggol (Inner Mongolia) and the remainder divided into three provinces. Political arrangements have remained thus, except for a period of reintegration of this zone into the provinces which ended in 1979.

Five different floras are available and, given their diversity of languages, all are accounted for here. Three are copiously illustrated, and only the work in English lacks keys. The two works of Liu (Liou), the first (1955) on the woody flora and the second (1958– , not completed) on the herbaceous plants, are complementary and so combined in a single entry despite differences in style. Kitagawa in 1979 recorded the vascular flora at just over 2700 species. The latest offering, a second edition of *Claves plantarum Chinae boreali-orientalis*, appeared in 1995 and records 3103 species.[118]

Kitagawa, M., 1979. *Neo-lineamenta florae manshuricae.* viii, 716 pp., 12 pls., map. Vaduz, Liechtenstein: Cramer/Gantner (Flora et vegetatio mundi 6). (A revision of *idem, Lineamenta florae manshuricae* (1939, Hsinking; in *Reports of the Institute for Scientific Research, Manchukuo* 3, App. 1(1)).)

Systematic enumeration (in English) of spontaneously occurring vascular plants (2708 species) with synonymy, references, Japanese names in transliteration, and indication of local and extralimital distribution; list of principal literature and index to genera at end. The introductory section includes notes on the preparation of the work and accounts of the vegetation, floristics and phytogeography of the subregion. The few illustrations depict species of special interest.

Komarov, V. L., 1901–07. *Flora Mančžurii.* 3 vols. Illus. (Trudy Imp. St. Peterburgsk. Bot. Sada 20, 22, 25). St. Petersburg. (Reprinted 1949–50, Moscow/Leningrad: AN SSSR Press, as vols. 3–5 of *V.L. Komarov: Opera Selecta.*)

Amply descriptive flora (in Russian and Latin) of vascular plants, without keys (except to species in the larger genera); includes full synonymy, references and citations, indication of *exsiccatae* with localities, general summary of local range, taxonomic commentary, and extensive notes on habitat, life-form, special features, etc.; indices to all botanical names in each volume. An introductory section (vol. 1) incorporates a discussion of species concepts as well as accounts of floristic regions and botanical exploration.

Liu Shen-o (Liou Tchen-ngo), [1955 (1959)]. *Tung-pei mu pen chih wu t'u shih* (*Illustrated flora of ligneous plants of North-East China*/Иллюстрированная флора деревьев и кустарников С.–В. Китая). [ii], ii, 568 pp., 26 text-figs., 175 pls., 2 maps. Mukden: Institutum Silviculturae et Pedologiae, Academia Sinica. Complemented by Liu Shen-o (ed.), 1958– . *Tung-pei ts'ao pen chih wu shih* (*Flora plantarum herbacearum Chinae boreali-orientalis*). Fasc. 1– . Illus. Beijing: Ke xue chu ban she (Academia Sinica Press).[119]

Extensively illustrated descriptive accounts respectively of the woody and herbaceous vascular plants of the northeastern subregion; both include keys to all taxa, formal Chinese and vernacular names, synonymy with references and citations (these last more copious in the second work), indication of local and extralimital range, critical remarks, and notes on habitat, special features, etc.; lists of references and indices to all names. Latin diagnoses of new taxa in the second work appear in appendices; while in the introductory section of the first work there successively appear an account of general features of the flora, a floristic analysis, and an illustrated glossary. The descriptions in the woody flora are more 'dendrological' in style, with separate short paragraphs for each feature. [The *Tung-pei ts'ao pen chih wu shih* was by the early 1980s some two-thirds complete, but of the 12 projected parts only 1–2 (1958–59), 3–7 (1975–81) and 11 (1976) have been published. These cover pteridophytes, dicotyledons from the 'Amentiferae' through the Lamiales, and Cyperaceae.][120]

Fu Pei-yun (ed.), 1995. *Dong bei zhi wu jian su biao/Claves plantarum Chinae boreali-orientalis.* 2nd edn. vii, 1007 pp., 456 pls. [Beijing]: Ke xue chu ban she (Science Press). (1st edn., 1959, under Liu Shen-o (Liou Tchen-ngo).)

Well-illustrated manual-key to vascular plants, with essential synonymy, references, Chinese and botanical names, and indication of distribution and habitat; complete indices at end. A complete table of contents and general keys as well as separate descriptions of families with keys to genera are also provided. [The family sequence is a modification of that in the Engler/Diels *Syllabus* of 1936; genera are alphabetically arranged.]

Noda, M., 1971. *Flora of the North-East Province (Manchuria) of China.* 10, 17, 3, 1613 pp., 237 pls., text-figs., portr., maps. Tokyo: Kazama Bookshop.

Illustrated, briefly descriptive flora (in Japanese, with portions in English) of non-vascular (chiefly algae) and vascular plants, with keys to genera, species, and infraspecific taxa; also includes synonymy, references and citations, Japanese transliterations of scientific names, local names, generalized indication of local and extralimital distribution, and miscellaneous notes; complete indices. The introductory section (in Japanese and English) includes accounts of Manchurian floristics and vegetation formations as well as botanical exploration and research (with a separate list of references). The limits of this work are those of Chinese Manchuria (Manchukuo) before World War II.[121]

II. North China

See also **860, Woody plants**. This area is traditional 'north China', from Beijing to Shandong and the western borders of Shaanxi.

Liu Shen-o (Liou Tchen-ngo) (ed.), 1931–36. *Flore illustrée du nord de la Chine: Hopei*

(Chih-li) et ses provinces voisines. Fasc. 1–4. Pls. Peking: Académie Nationale de Peiping.

Large-scale atlas (in Chinese and French) of flowering plants, with descriptive text including synonymy, general indication of local range, taxonomic commentary, and keys to genera and species for each family. [Interrupted by the outbreak of war with Japan in 1937 and never resumed; covers only four families.]

III. Northwest China

This area encompasses much of the upland 'loess region' of Gansu, Shaanxi and Ningsia (north of Xian and west of the Huang Ho) as well as the large Qinling (Tsinling) range in the southern part. The two main works, *Flora tsinlingensis* (1974–85) and *Shen Kan Ning p'ang ti chih wu shih* (1957), are somewhat complementary, but together do not cover the whole area. In the last decade or so, two new works have commenced publication: *Flora loess-plateaus sinicae*, an expanded complement to the 'Shen Kan Ning' manual, and *Flora sinensis in area Tan-Yang*, a work covering southern Ningxia and adjacent Gansu as far west as Lanzhou.

Fu Kun-tsun *et al.* (eds.) (for North Shaanxi Construction Commission (vol. 5 only) and Northwest Institute of Botany), 1989– . *Huang tu gao yuan zhi wu zhi/Flora Loess-Plateaus sinicae.* Vols. 2, 5. Illus. [N.p.]: Ke xue ji shu wen xian chu ban she (Scientific and Technical Documents Publishing House) (vol. 5); Zhong guo lin ye chu ban she (Forestry Publishing House of China) (vol. 2). [Vol. 1 was announced for 2000.]

Illustrated descriptive flora with keys to all taxa, botanical and Chinese names, synonymy with references and citations, indication of phenology, distribution and habitat, and notes of varying extent; diagnoses of new taxa and complete indices at end of each volume. [In progress. Vol. 2 (1992) covers Crassulaceae through Leguminosae including Saxifragaceae and Rosaceae; vol. 5 (1989), Phrymataceae through Compositae including Caprifoliaceae, Cucurbitaceae and Campanulaceae. A modification of the 1936 Engler/Diels *Syllabus* family sequence is followed.]

Le Tien-yu (Lo Tien-yu) and Hsu Wei-ying, 1957. *Shen Kan Ning p'ang ti chih wu shih* (*Flora of Shaanxi-Gansu-Ningxia Basin*). 2, 274 pp., 266 illus. (incl. maps). Beijing: Ke xue chu ban she (Academia Sinica Press).

A briefly descriptive manual-flora with keys to all genera and species, Chinese nomenclature, essential synonymy, indication of local range, and notes on special features, uses, etc.; photographic illustrations of pressed specimens; indices to all names at end. An introductory section includes chapters on geography, climate, and vegetation (with map), notes on particular woody plants, a synoptic table of contents (pp. 26–49), and a key to families (pp. 50–57).[122]

NORTHWEST INSTITUTE OF BOTANY, ACADEMIA SINICA, 1974–85. *Flora tsinlingensis*. Vol. 1, parts 1–5; vol. 2. Illus. Beijing: Ke xue chu ban she (Academia Sinica Press).

Illustrated descriptive manual-flora of vascular plants; includes analytical keys to all taxa, Chinese nomenclature, extensive synonymy with references and citations, vernacular names, indication of local range and altitudinal zonation as well as extralimital distribution, and notes on special features, properties, uses, etc. An appendix (in each part) includes Latin diagnoses of new taxa, and complete indices to Chinese and Latin plant names are also furnished. The five parts of vol. 1 encompass seed plants, the Englerian system being followed; vol. 2 covers pteridophytes. [A third volume covers bryophytes, beyond our scope.]

861

Heilongjiang (Heilungkiang)

Area: 463600 km^2. Vascular flora: no data seen.

Prior to the start of publication of *Flora Heilongjiangensis*, the province was covered only in works on Northeast China (**860/I**). A woody flora has also appeared.

CHOU YI-LIANG (gen. ed.), 1985– . *Flora Heilongjiangensis*. Vols. 1, 4, 5, 6, 9, 11. [N.p.]: Dong bei lin ye da xue chu ban she (North-eastern Forestry University Press).

Documentary descriptive flora of bryophytes and vascular plants; includes keys, botanical and Chinese names and synonymy with references and citations, distribution, and notes on occurrence, properties, uses, etc.; indices in each volume. [In progress, with 12 volumes projected. Of those published, only vols. 4–5 (1992), 6 (1998), 9 (1999) and 11 (1993) cover angiosperms (from the 'Amentiferae' through the Centrospermae to Magnoliales and Papaverales and part of the monocotyledons including Cyperaceae and Orchidaceae). Vol. 1 (1985) covers bryophytes.][123]

Woody plants

CHOU YI-LIANG, TUNG SHIH-LIN and NIE SHAO-CHUAN, 1986. *Ligneous flora of Heilongjiang*. 585 pp., 171 text-figs., 16 halftone pls., map. [Harbin]: Heilongjiang Science Press.

A descriptive manual of trees, shrubs and woody climbers (332 species and 124 additional infraspecific taxa) with keys, synonymy, references, indication of phenology, distribution, habitat and uses, and Chinese names; complete indices at end. Forest and floristic zones are considered in the general part which also features a general key to families. [The poorly reproduced plates depict characteristic vegetation types.]

862

Jilin (Kirin)

Area: 187000 km^2. Vascular flora: no data seen. – The province is at present covered only in works on Northeast China (**860/I**).[124]

863

Liaoning

Area: 151000 km^2. Vascular flora: *c.* 2200 species (Li, 1988–92).

Prior to publication of *Flora liaoningica*, the province was covered only in works on Northeast China (**860/I**). There is also a local flora of the Dalian (Darien) district on the southern Liaotung Peninsula.

LI SHU-XIN (ed.), 1988–92. *Liao ning zhi wu zhi/ Flora liaoningica*. 2 vols. Illus. [Shenyang]: Liao ning ke xue ji su chu ban she (Liaoning Science and Technology Press).

Illustrated descriptive manual of vascular plants (*c.* 2200 species) with keys to all taxa, synonymy with references, and indication of phenology, distribution, habitat, uses, etc.; addenda with diagnoses of new taxa and indices to all names in each volume (no cumulative index). The general part (vol. 1) covers geography, climate, floristics and protected plants along with a general key to families.

Partial work: Dalian district

ANONYMOUS, 1982. (*Flora of Dalian district.*) 3 vols. Illus.

Illustrated descriptive flora of vascular plants. [Not seen; cited from Chen *et al.* (1993; see **Superregion 86–88, Supraregional bibliographies**).]

864
Hebei (Hopei, Hopeh, Chih-li)

Area: 230 800 km^2 (including Beijing Urban Region, 16 800 km^2, and Tianjin Urban Region, 11 000 km^2). Vascular flora: 2724 species of cormophytes (Bai Yu-hua *et al.*, 1986–91; see below). – The limits of Hebei, known under the Qing dynasty as Chih-li, have varied within the last century or so. As presently delimited, it includes the greater part of the former Jehol; on the other hand, the Beijing and Tianjin regions are now autonomous.

Until 1986, the only general works on the province proper (and Beijing) had been those by Abbé Léveillé (1917*a,b*), now obsolete. In that year *Flora hebeiensis* commenced publication; its three volumes had appeared by 1991, covering all cormophytes. There is also a set of prefectural checklists grouped together in HEBEI PROVINCIAL PLANT ASSOCIATION, 1992. *Wild plants in Hebei Province*. [x], 402 pp. [N.p.]: Publishing House of Hebei University. The more common trees are additionally covered in the following: CHOU HANG-FAN, 1934. *Ho-pei hsi chien shu mu t'u shuo*. xii, 370 pp., 43 text-figs. (Handb. Peking Soc. Nat. Hist. 4). Beijing (also published in English as *The familiar trees of Hopei*). At the local level, floras exist for Beijing and the Jehol districts; these are accounted for below under individual subheadings. For Tianjin, however, no work seems to have supplanted that of Odon Debeaux, published in Bordeaux in 1879.

BAI YU-HUA *et al.*, 1986–91. *He bei zhi wu zhi/ Flora hebeiensis*. 3 vols. Illus. N.p.: He bei ke xue ji shu chu ban she (Hebei Science and Technology Publishing House).

Illustrated manual-flora of bryophytes and vascular plants with keys to all taxa, synonymy, Chinese and botanical names, publication references, indication of phenology, habitat and distribution, and notes; complete indices in each volume with a cumulation in vol. 3. A relatively brief general part in vol. 1 includes sections on physical features, climate and vegetation but no floristic summary. [Vol. 1 covers bryophytes, pteridophytes, gymnosperms and 'early' dicotyledons; vol. 2, remaining dicotyledons except Compositae; vol. 3, Compositae and the monocotyledons.]

LÉVEILLÉ, H., *l'abbé*, 1917*a*. Catalogue des plantes de Pékin et du Tché-Li. *Bull. Acad. Int. Géogr. Bot.* **27**: 70–87; and *idem*, 1917*b*. *Flore de Pékin et du Tché-Li*. 5th edn. 122 pp. [Le Mans]: The author. Continued as *idem*, 1917*c*. *Flore de Pékin et de Chang Hai et des provinces du Tché-Li et du Kiang-Sou*. 2nd edn. 112 pp. [Le Mans]: The author. (The latter two issued in limited, holographed editions from author's MSS.)

The first-named is an unannotated list of vascular plants, the families alphabetically arranged; a phytogeographic summary follows. The second is similar in style and content, but also features keys to families. The third contains keys to genera as well as additions and corrections.

Partial works: Beijing Urban Region

In addition to the work described below, reference may be made to the following field-guide: FACULTY OF BIOLOGY, BEIJING NORMAL UNIVERSITY, 1993. *Bei jing zhi wu jian suo biao* (*Claves plantarum pekinensis*). 3rd edn. [v], 6, 3, 468, [1] pp. Beijing: Bei jing chu ban she (Beijing Press). (1st edn., 1978; 2nd edn., 1981.) This work is a manual-key to more than 2000 vascular species; in the current edition additions and emendations are arranged on new pages following the original text and index.

HE SHI-YUAN *et al.* (FACULTY OF BIOLOGY, BEIJING NORMAL UNIVERSITY), 1992 (1993). *Bei jing zhi wu zhi* (*Flora of Beijing*). Revised edn. 2 vols. 1797 text-figs. Beijing: Bei jing chu ban she (Beijing Press). (1st edn. in 3 vols., 1962–64, 1975, under anonymous authorship. 2nd edn. in 2 vols., 1984–87, by He Shi-yuan, Xing Qi-hua, Yin Zu-tang and Jiang Xian-fu.)

Concise manual of vascular plants; includes family keys and keys to all other taxa, brief descriptions, Chinese and botanical names, place of publication, and notes on distribution, occurrence, etc.; indices to Chinese and botanical names and addenda in vol. 2 (the latter on pp. 1477–1510 following the indices).

Partial works: Tianjin Urban Region

No work has so far supplanted the now-obsolete *Florule de Tien-tsin* by Odon Debeaux (1879, as fasc. 4 of his *Contributions à la flore de la Chine*).

Partial works: former Jehol Territory

This tract was first carved out from the northern part of Hebei and parts of adjacent Manchuria and Inner Mongolia after 1911. Between 1931 and 1945 it came under Japanese control and was incorporated into Manchukuo. After 1949 it was broken up once more, the greater portion being reunited with Hebei.

NAKAI, T., M. HONDA, Y. SATAKE and M. KITAGAWA, 1936. *Index florae jeholensis.* [ii], 108 pp., 3 pls. (Report of the first scientific expedition to Manchukuo, sect. IV, 4). Tokyo.

Systematic list (in Japanese) of vascular plants (924 species) with Japanese standard names as well as localities, citations of *exsiccatae* and literature records, and descriptions of new or little-known taxa; index. A brief introduction (in English) precedes the main text.

Woody plants

SUN LI-YUAN and REN XIAN-WEI (eds.), 1997. *Dendrologia hebeiensis.* [7], 3, 616 pp., 651 text-figs. Beijing: Zhong guo lin ye chu ban she (China Forestry Publishing House).

Well-illustrated descriptive manual-flora with keys, names with places of publication, descriptions, and indication of distribution, phenology, habitat, other features and (if any) cultivars; list of threatened species, winter key with diagnostic figures (pp. 550–584), brief bibliography, indices and list of compilers (their names also Romanized) at end. The general part accounts for features of the woody flora and vegetation and includes (pp. 5–18) a key to families. [Covers also Beijing and Tianjin; accounts for 625 native and cultivated species and a further 150 infraspecific taxa. Nomenclature in some cases appears to be out of date.]

865

Shandong (Shantung)

Area: 153 300 km². Vascular flora: no data seen. – Until publication of the first volume of *Flora of Shandong* in 1992, no separate accounts for the province as a whole were available. The only consolidated works were for long Theodor Loesener's florula of the Qingdao (Tsingtao) district (see below) and *Florule du Tché-fou* by Odon Debeaux (1876, as fasc. 3 of his *Contributions à la flore de la Chine*). For other coverage, see **860/II**.

CHEN HAN-PIN, CHENG I-CHIN and LI FA-TSENG, 1992–97. *Shan dong zhi wu zhi (Flora of*

Shandong). 2 vols. Illus. (773 figs., 2 maps in vol. 1; 1241 figs. in vol. 2). [Qingdao]: Qing dao chu ban she (Qingdao Press).

Illustrated manual-flora of vascular plants with keys to all taxa, accepted Chinese and scientific names, indication of phenology, habitat and distribution, and notes; indices to all names at end of each volume (not consolidated, and those in vol. 2 are entirely without page references). [Vol. 1 covers pteridophytes, gymnosperms, monocotyledons, and dicotyledons from Saururaceae through Caryophyllaceae; vol. 2, Nymphaeaceae through Compositae.]

Woody plants

[WOODY FLORA OF SHANDONG WORKING GROUP], 1984. *Shan dong shu mu zhi (Woody flora of Shandong).* [5], 30, 956 pp., numerous full-page figs. Jinan: Shan dong ke xue ji shu chu ban she (Shandong Science and Technology Publishing House).

Illustrated descriptive account with keys to all taxa, formal Chinese and vernacular names, limited synonymy, and concise notes covering local distribution, habitat, special features, and economics; treatments of vegetation types and floristic features and indices to all names at end. Pp. 1–9 feature a general key to families.

Partial work: Qingdao (Tsingtao) district

Qingdao, situated on the southern side of the Shandong peninsula, was from the 1890s until World War I part of the German 'concession' of Kiautschou.

LOESENER, T., 1919. Prodromus florae tsingtauensis: Die Pflanzenwelt des Kiautschou-Gebietes. *Beih. Bot. Centralbl.* 37(2): 1–206, pls. 1–10.

Systematic enumeration of all known vascular and non-vascular plants, with descriptions of new taxa, synonymy, citations of *exsiccatae* with localities, etc.; no keys. Certain family treatments were contributed by specialists.

866

Henan (Honan)

Area: 167 000 km². Vascular flora: c. 3900 species (Chang *et al.* in *He nan zhi wu zhi*, vol. 2, preface). – For other coverage, see **860/II** (all works) as well as, for the southern part, Steward's manual-flora for the lower Yangtze Basin (**870/I**).

The first provincial flora, described below, commenced publication in 1981. There is also a florula for

Song Shan (*Song shan zhi wu zhi* by Ye Yong-zhong and Wu Shun-qing, 1993).

CHANG ZHE-XIN *et al.* (for FLORA OF HENAN COMMITTEE), 1981–98. *He nan zhi wu zhi*. Vols. 1–4. Illus. (figs. 1–2951). [N.p.]: Henan Science and Technology Press.

Illustrated manual-flora of vascular plants with keys to genera, species and infraspecific taxa, limited synonymy, botanical and Chinese names, indication of phenology, habitats and distribution, and notes; family keys in vol. 1 (pp. 548–581) and indices to all names in each volume. [In progress. Vol. 1 covers pteridophytes, gymnosperms and dicotyledons from Saururaceae through Lauraceae; vol. 2 (1988), Papaveraceae through Vitaceae; vol. 3 (1997), Tiliaceae through Compositae (Asteraceae). One or two further volumes will complete the work.]

867

Shanxi (Shansi)

See **860/II** (all works). – Area: 157 100 km^2. Vascular flora: no data seen.

The first of a projected six volumes of a provincial flora, the first of its kind, appeared in 1992. There is also a recent local flora for the Taiyuan district, in the center of the province and inclusive of the provincial capital of the same name.

LIU TIAN-WEI (ed.) (for FLORA OF SHANXI EDITORIAL COMMITTEE), 1992– . *Shan xi zhi wu zhi* (*Flora Shanxiensis*). Vols. 1– . Illus., map. [N.p.]: Zhong guo ke xue ju shu chu ban she (China Science and Technology Publishing House).

Illustrated descriptive flora of vascular plants with keys to all taxa, Chinese and botanical names, and indication of phenology, distribution and habitat along with brief notes; indices to all names at end. The general part includes a historical survey and accounts of physical features, climate, floristics and vegetation; the plan and scope of the work appear in an introduction preceding the table of contents. [Vols. 1–2, all so far published, cover pteridophytes, gymnosperms, and dicotyledons from Chloranthaceae through Geraniaceae (inclusive of Rosaceae and Fabaceae). Arrangement of families follows a modified version of the Engler/Diels system as used in the 1936 *Syllabus*.]

Partial work: Taiyuan district
Flora Taiyuanica covers the surroundings of the provincial capital.

LIU TIAN-WEI (for EDITORIAL COMMITTEE OF FLORA TAIYUANICA), 1990–92. *Tai yuan zhi wu zhi* (*Flora Taiyuanica*). Vols. 1–2. Illus. (some col.), map. [N.p.]: Xue shu shu kan chu ban she (Academic Publishing House) (vol. 1).

Well-illustrated concise descriptive flora of vascular plants with keys to all taxa, Chinese and botanical names, synonymy with references and citations, indication of phenology, distribution and habitat, and brief notes; complete indices in each volume. The general part in vol. 1 includes a brief introduction and statistics of the flora. [Vols. 1–2 cover pteridophytes, gymnosperms and dicotyledons; a third volume, on monocotyledons, is expected.]

868

Shaanxi (Shensi)

See **860/III** (all works). – Area: 195 800 km^2. Vascular flora: no data seen.

There is reportedly a general work by Tsi-an Peh (1959), but no details were available at the time of the original writing of this entry in 1979. The only recent province-wide work is on woody plants; there is still no modern general flora.[125]

Woody plants
NIU CHUN-SHAN and ZHU BIAN, 1990. *Shaan xi shu mu zhi/Dendrologia schensiensis*. [5], 12, 12, 1261 pp., 1163 text-figs. Beijing: Zhong guo lin ye chu ban she (Chinese Forestry Press).

Illustrated descriptive flora of woody plants (1224 native and introduced species in 101 families); includes keys, accepted botanical and Chinese names and synonymy with references, distribution, altitudinal range, and notes encompassing phenology, habitat, etc., followed by a treatment of vegetation and forest types (including map) with physical and environmental characteristics and representative species; indices to scientific and Chinese names at end. A general key to families follows the table of contents.

869

Ningxia (Ninghsia) Hui Autonomous Region and southern Gansu (Kansu)

See also **860/III** (all works). For Ningxia, reference may also be made to **705** and to *Flora intramongolica* (**763**). – Area: Ningxia, 170 000 km²; Gansu, 530 000 km². Vascular flora: no data seen.

As delimited here, the unit encompasses the eastern and southern parts of Gansu (east of Lanzhou) and the Ningxia Hui Autonomous Region (separated from Gansu in 1949). The drier western parts of Gansu, reaching towards Xinjiang, comprise **765**; their flora is more like that of Central Asia.

For Ningxia, a first provincial flora appeared between 1986 and 1988, while for Gansu no provincial flora has been published.[126] The more mountainous parts in the south and southeast are partially covered in some of the botanical reports listed under **860–80** (**Partial works**), especially those by Maximowicz and by Rehder and Wilson; more recently they have been encompassed by the flora of the great Chinese western mountain chain, *Heng Duan Shan qu wei guan zhi wa* (see **803/IV**). There are also some partial works.[127]

MA DE-ZI and LIU HUI-LAN, 1986–88. *Flora Ningxiaensis.* 2 vols. Illus. Yinchuan: Ningxia People's Press.

Descriptive manual-flora with keys, botanical names, Chinese nomenclature, and indication of distribution, phenology, habitat and other features along with critical notes as needed; new taxa (pp. 521–523 in vol. 2) and indices to botanical and Chinese names in each volume (no cumulative index). [Vol. 2, the second of two volumes, covers Tamaricaceae through Asteraceae and all monocotyledons.]

Partial works

The Liu-p'an Shan and He-lan Shan run south to north through the central part of the Region.

DI WEI-ZHONG (ed.) [for COMPREHENSIVE SURVEY OF MT. HELAN, NINGXIA HUI AUTONOMOUS REGION], 1986 (1987). *He Lan Shan wei guan zhi wu (Plantae vasculares helanshanicae).* [iii], 3, 4, 378, [1] pp., 53 pls., map. [N.p.]: Xi bei da xue chu ban she (Northwest University Press).

Keyed enumeration of vascular plants (690 species)

with botanical and Chinese nomenclature and indication of distribution and habitat; descriptions of novelties and indices to all names at end. The general part furnishes general descriptions of the area and its flora and an account of the vegetation.

FENG XIAN-KUI et al., 1979. *Ning hsia Liu p'an shan Ho lan shan mu pen chih wu t'u chien.* [vi], 211 pp., 266 halftones on 60 pls., 28 text-figs., 16 halftones in text, 1 unnumbered fig. [N.p.]: (Ningxia People's Press).

Illustrated manual-flora of woody plants with keys; concise commentary; indices. [Covers 243 species and 36 varieties in 95 genera. Partly supplanted by Di, 1986.]

Region 87

Central China

This region comprises the provinces of Jiangsu (Kiangsu), Zhejiang (Chekiang), Anhui (Anhwei), Jiangxi (Kiangsi), Hunan, Hubei (Hupeh) and Sichuan (Szechwan). What are here called the 'Tibetan Marches' are separately accounted for, although the area, at one time a separate province known as Sikang, is now divided between Xizang (Sitsang or Tibet) and Sichuan.

Save for the treaty ports and their hinterland in the east, botanical penetration of the region began only after 1860. Much of the outside interest was directed towards the mountainous western subregion, then hardly known. A large number of the collections were described in imposing reports, and in the early twentieth century much material for gardens was also obtained. Later work, undertaken largely by Chinese botanists, was directed at general biological survey with among the aims comprehensive floristic knowledge and eventual flora-writing. From the 1920s regional botanical centers came into being in both subregions: eastern (notably in Nanjing, Hangzhou and Wuhan) and western (in Chengdu).

The eastern subregion has for the most part been treated in *Vascular plants of the lower Yangtze* (Steward, 1958), an English-language work based on long field experience on the part of the author and his students. Since 1976, however, Chinese-language provincial floras for Jiangsu, Zhejiang, Anhui and Hubei have been completed in whole or in part. By contrast, information on the vascular flora of the western subregion remained imperfectly consolidated until recent years (save for some key areas such as Mt. Omei which had attracted special attention). The only provincial flora was for long Abbé Léveillé's uncritical compilation for Sichuan (1918); this was based mainly on the French missionary-priests' collections of the late nineteenth and early twentieth centuries. In 1981, under the leadership of W. P. Fang, *Flora sichuanica* began publication; 11 volumes have now appeared. Elsewhere, an enumeration for Hunan appeared in 1987 and a flora of Jiangxi commenced publication in 1993. The mountainous 'Tibetan Marches' in the far west have also been covered in the two-volume *Heng Duan Shan qu wei guan zhi wu* (see **803/IV**).

The least well-documented provinces have hitherto been Hunan and Jiangxi. An enumeration for the former was, however, published in 1987 and a flora for the latter commenced publication in 1993.

Bibliographies. General and supraregional bibliographies as for Division 8 and Superregion 86–88.

Indices. General and supraregional indices as for Division 8 and Superregion 86–88.

870

Region in general

See also **860–80** (all general works). Two subregions are for convenience here recognized: east Central China, from I-ch'ang on the Yangtze River eastwards, and west Central China, from mountainous western Hupeh and Hunan through Sichuan to the Tibetan Marches. Much of the eastern subregion is covered by Steward's 1958 manual, while coverage of the western subregion is very scattered although fairly abundant. With regard to the latter, a selection of the more important botanical reports (some of which also extend to Hunan and Jiangxi) is given under **860–80** (**Partial works**).

I. East Central China

Keys to genera and families

HSU PING-SHENG (SUI BIN-SHEN), 1957. (*Keys to families and genera of seed plants.*) 188 pp. [Covers 175 families and 1105 genera, native and introduced, in Jiangsu, Anhui,

Jiangxi and Zhejiang Provinces. Not seen; from Chen *et al.* (1993, p. 206).]

STEWARD, A. N., 1958. *Manual of vascular plants of the lower Yangtze Valley, China.* vii, [6], 621 pp., 510 text-figs., 2 maps. Corvallis: Oregon State College. (Reprinted 1986, Cambridge, Mass.: Arnold Arboretum of Harvard University.)

Illustrated descriptive flora, with keys to all taxa, synonymy with references, Chinese and English vernacular names, fairly detailed indication of local distribution, summary of extralimital range, and miscellaneous notes on habitat, special features, etc.; glossary (in English, with Chinese equivalents); index to botanical and English vernacular names. The introductory section contains remarks on physical features, geography and vegetation together with a list of references. [The geographical limits of this work encompass Jiangsu, Zhejiang, Anhui, the northern parts of Hunan and Jiangxi, eastern Hubei, and southern Henan. Some 1959 species are accounted for.]

ZHANG MEI-ZHEN, LAI MIN-ZHOU *et al.*, 1993. *Hua dong wu sheng yi shi zhi wu ming lu/ (A list of plants in East China).* [vi], 13, 491 pp. Shanghai: Shang hai ke xue pu ji chu ban she (Shanghai Scientific and Technical Publishers).

Nomenclator for lichens, bryophytes and vascular plants with botanical and Chinese nomenclature and indication of habitats; bibliography (pp. 443–444) and indices to families and genera and to Chinese names at end. The general part includes a description of the area and accounts of the flora, vegetation and phytogeography. [Accounts for 6640 species in all; also covers parts of South China (**Region 88**).]

II. West Central China

See also **860–80** (**Partial works**) as well as **875** through **878** and **882**.

Partial works
Also of interest is S. Y. Hu, 1980. The *Metasequoia* flora and its phytogeographic significance. *J. Arnold Arbor.* **61**: 41–94. This contains a checklist of collections made in the 'Metasequoia Area' of western Hubei and adjacent eastern Sichuan (with map, p. 45).

WANG WEN-TSAI and LI ZHEN-YU (eds.), 1995. *Keys to the vascular plants of the Wuling Mountains.* vii, 626 pp., 2 maps. [Beijing]: Ke xue chu ban she (Science Press).

Concise manual-key to native, introduced and commonly cultivated vascular plants with botanical and Chinese names, essential synonymy, and indication of distribution and habitat; descriptions of new taxa and index to botanical names at end. The background and scope of the work are given in a brief introduction. [3807 species are accounted for in an area covering northwest Hunan, northeast Guizhou, southwest Hubei and extreme southeast Sichuan.]

871

Jiangsu (Kiangsu) and Shanghai

See also **870/I** (STEWARD). – Area: Jiangsu, 102200 km^2; Shanghai Municipality, 5800 km^2. Vascular flora: no data seen.

With respect to the works by Abbé Léveillé described below, the nomenclatural reviews of LAUENER and REHDER (cited under **860–80, Partial works**) should be consulted. His long-obsolete lists have been succeeded by a descriptive provincial flora (Jiangsu Institute of Botany, 1977–82). With two major centers of botanical activity respectively in Nanjing and Shanghai, the flora is now relatively well collected and documented.

JIANGSU INSTITUTE OF BOTANY, 1977–82. *Jiang su zhi wu zhi* [*Flora of Jiangsu*]. 2 vols. [iii], 502 pp., 742 figs. (incl. 12 col. pls.), 23 pls. (in glossary); 1010 pp., figs. 743–2269. Nanjing: Jiang su ke xue ji shu chu ban she (Jiangsu Science and Technology Press).

Illustrated descriptive flora of vascular plants with keys to all taxa, Chinese and botanical nomenclature, very limited synonymy, indication of local distribution and altitudinal range, and notes on habitat, occurrence, uses and properties, etc.; summary of new taxa and general indices at end of each volume (with an illustrated organography/glossary on pp. 423–463 and short list of references on p. 464 in vol. 1). The introductory section (pp. 1–6 in vol. 1) features accounts of physical features, climate, vegetation, land use, etc. [Arrangement of families follows the Engler/Diels system of the 1936 *Syllabus*. Vol. 1 covers pteridophytes, gymnosperms and monocotyledons; vol. 2, all the dicotyledons.]

LÉVEILLÉ, H., *l'abbé*, 1916. Catalogus plantarum provinciae chinensis Kiang-Sou hucusque cognitarum. *Mem. Real Acad. Ci. Barcelona*, III, **12**: 543–565.

(Separately reprinted, 25 pp.) Complemented by *idem*, 1917. *Flores de Chang-Hai et du Kiang-Sou*. 5th edn. 132 pp. [Le Mans, France]: The author. (Lithographed MS.)

The *Catalogus* is an unannotated list of vascular plants, with descriptions of some new taxa; the families are alphabetically arranged. The *Flores* comprise keys to families and genera. Further additions appear in the author's *Flore de Pékin et de Chang Hai* (**864**).

Partial works: Shanghai Urban Region and South Jiangsu

Shanghai, the largest city in China and currently autonomous, was formerly also home to a substantial foreign resident population. There is, not unsurprisingly, an accumulation of local works, among which those of French writers predominate in the period from 1860 to 1920. The first treatment of the flora was *Florule de Shang-Haï* (1875) by Odon Debeaux, published as the second of his four *Contributions à la flore de la Chine*. In later years the French and their associates (including, from 1935 to 1952, O. William Borrell) continued to be active, working in particular from the former Université de l'Aurore and Musée Heude (Xujiawei or Zhikawei). A number of manuscripts were compiled; from these has been derived the florula by Borrell (1996). Many additions to the literature were also made by Chinese authors, notably Pei Chien *et al.* (1958, 1959) and Hsu Ping-shang (1959); a florula by the latter, however, remains unpublished.

BORRELL, O. W., 1996. *Flora of the Shanghai Area*. Vols. 1–2 (in 1). 130 pp., illus. (part col.), 2 maps. Bulleen, Vic., Australia: The author.

Illustrated descriptive account with limited synonymy, English and Chinese names, and indication of origin (if appropriate), distribution and habitat; indices to all names at end. A foreword has been contributed by Hsu Ping-sheng. [Covers only pteridophytes and gymnosperms; publication of angiosperms planned.][128]

HSU PING-SHENG (SUI BIN-SHEN), 1959. *Enumeratio plantarum civitatis Shanghai*. vi, 138 pp., 80 text-figs. Shanghai.

Briefly annotated systematic list, with limited synonymy and symbolic indication of abundance and local range; illustrations of representative species; indices. A general section includes a summary account of the flora (1450 species and 269 additional varieties, native and introduced).

PEI CHIEN *et al.*, 1958. (*Keys to the seed plants of South Jiangsu.*) 311 pp.; and *idem*, 1959. (*A manual of seed plants in South Jiangsu.*) 881 pp.

The first of these works covers 1351 taxa; the second, 1340 taxa with also two appendices covering economic uses and phenology. [Neither work seen; details from Chen *et al.* (1993, p. 361).]

872

Zhejiang (Chekiang)

See also **870/I** (STEWARD). – Area: 101 800 km². Vascular flora: 3884 species (from *Zhe jiang zhi wu zhi*; see below).

The first provincial account, interrupted by the Japanese invasion of 1937 and never resumed, is the English-language *Enumeration of vascular plants from Chekiang* by W. C. Cheng, S. S. Chien and C. Pei (1933–36; in *Contr. Biol. Lab. Chin. Assoc. Advancem. Sci., Sect. Bot.* **8** (1933): 298–307; **9** (1933–34): 58–91, 240–304, pls. 4–6, 23–28; **10** (1936): 93–155, pls. 13–18). Its four parts cover seed plants from Ginkgoaceae through Rosaceae. Its successor is the eight-volume *Zhe jiang zhi wu zhi* (1989–93; see below). In contrast to most Chinese provincial floras, this latter has a substantial 'general' volume.

[EDITORIAL BOARD OF FLORA OF ZHEJIANG], 1989–93(–94). *Zhe jiang zhi wu zhi* (*Flora of Zhejiang*). General volume and vols. 1–7. Illus. (part col.). Hangzhou: Zhe jiang ke xue ji shu chu ban she (Zhejiang Science and Technology Press).

Large-scale illustrated descriptive flora of native, naturalized and commonly cultivated vascular plants with keys, limited synonymy, and indication of phenology, local and extralimital distribution, habitats and uses; indices to all names in each volume. The general volume (published in 1994) includes statistics and general features of the fossil and Recent flora as well as a history of botanical work and extended treatments of principal economic and ornamental species; references appear on pp. 331–343 and a list of plants first described from the province on pp. 305–330. [Arrangement of the families follows a modified Englerian system; pteridophytes and gymnosperms are in vol. 1 and all monocotyledons in vol. 7.]

Woody plants

No recent forest flora has been published. The only account available is a systematic checklist: LIN KANG, 1936. Enumeration of woody plants in Chekiang Province, China. *J. Forest.* (Tsinan) **6**: 1–32. [Not seen; cited from Walker (1960; see **Superregion 86–88, Supraregional bibliographies**).]

873

Anhui (Anhwei)

See also **870/I** (STEWARD). – Area: 139 900 km². Vascular flora: *c.* 3100 species (*An hui zhi wu zhi*; see below).

A first provincial flora commenced publication in 1986 and was completed in 1992. The only substantial earlier works have been a series on the woody plants by Rehder and Wilson and a treatment of the plants of Huang Shan by Chen Pang-chieh (1965). A checklist of woody plants appeared in 1983. The whole province was also covered by Steward.

AN HUI ZHI XIE ZUO ZU BIAN [FLORA OF ANHUI COMMITTEE], 1986–92. *An hui zhi wu zhi* (*Flora of Anhui*). 5 vols. 3103 text-figs. Beijing: Zhong guo chan wang chu ban she (Chinese Prospect Publishing House) (vols. 1–3); [Hefei]: An hui ke xue ji shu chu ban she (Anhui Science and Technology Press) (vols. 4–5).

Illustrated descriptive flora of vascular plants with keys, vernacular and standardized names, essential synonymy, indication of distribution and habitat, and brief notes; diagnoses of new taxa and indices to all names at end of each volume (no cumulative index). [Volume 1 covers pteridophytes and gymnosperms; vols. 2–4, all dicotyledons; vol. 5, monocotyledons.]

Local work: Huang Shan

The Huang Shan (mountains) are in the southern part of Anhui. They are covered in two works: CHEN PANG-CHIEH *et al.*, 1965. *Observationes ad floram Hwangshanicum.* [iv], 2, 335 pp., text-fig., maps. Shanghai; and HA JIN-QI and LIEM SHI-WEN (eds.), 1996. *Huang-shan zhi wu* (*Plants of Huang Shan*). [7], 6, 601 pp., illus. (part col.), map. [Shanghai]: Fu dan da xue chu ban she. The former includes a systematic checklist, the latter is a descriptive atlas accounting for 1483 species of seed plants and 131 of pteridophytes.

Woody plants

No descriptive manual has been published. Of other works, there is in addition to the 1927 account of Rehder and Wilson (see below) a more recent annotated checklist of 1983 with indication of altitudinal range and habitat but no keys.

LI SHU-CHUN *et al.* (comp.); WU CHENG-HE (ed.), 1983. *Anhui muben zhiwu* [*Woody flora of Anhui*]. 293 pp. Hefei: Anhui Forestry Research Institute.

Systematic enumeration of 1237 species of woody seed plants (in 109 families and 342 genera), the arrangement following the Hutchinson system; included are short diagnoses along with botanical and Chinese names, distribution, altitudinal range, habitat, and notes on special features; separate indexes at end to Chinese and botanical names. [Not limited to native species; introductions and cultivars are also accounted for. No keys are furnished in what is a rucksack-sized book.]

REHDER, A. and E. H. WILSON, 1927. An enumeration of the ligneous plants of Anhwei. *J. Arnold Arbor.* 8: 87–129, 150–199, 238–240.

Systematic enumeration of woody plants, based mainly on collections made from 1922 through 1925; includes relevant synonymy, references and citations, and localities with *exsiccatae*; some taxonomic commentary.

874

Jiangxi (Kiangsi)

Area: 164800 km². Vascular flora: no data seen. – The northern part of this province is covered by STEWARD (see **870/I**). Elsewhere, only scattered coverage was for long available; the most numerous references can be found in the botanical reports on the collections of Handel-Mazzetti and Wilson (respectively by HANDEL-MAZZETTI and SARGENT; see **860–80, Partial works**). A first essay at a provincial flora appeared in 1960; that begun in 1993 is more definitive.

[EDITORIAL COMMITTEE FOR FLORA OF JIANGXI], 1993– . *Jiang xi zhi wu zhi* (*Flora of Jiangxi*). Vols. 1– . Illus. Nanchang: Jiang xi ke xue chu ban she (Jiangxi Science and Technology Press).

Illustrated, amply descriptive flora with keys, synonymy with references and citations, vernacular and standard names, indication of distribution and habitat, and notes; diagnoses of new taxa and indices to all botanical and local names at end of each volume. [Vol. 1, all published as of 1999, covers pteridophytes and gymnosperms; it also includes a concise general introduction with sections on floristics and vegetation and, in the preface, the background to the work. A total of five volumes is projected.]

875

Hunan

Area: 210500 km². Vascular flora: 4324 species (Qi *et al.*, 1987; see below). – The northeastern part of the province is also covered by STEWARD (see **870/I**).

The obscure and now outdated catalogue of Chou Hang-fan, *Hu-nan chiao-yü tsa-chih* (1924; in *Special Number Educ. Misc. Hunan* 4(7)), was succeeded in 1987 by a new enumeration, a first step towards a provincial flora.[129] Otherwise, little save scattered papers has been available. Chou's list was notable for its references to traditional Chinese botanical literature.

QI CHENG-JING, SUN XI-RU and LIN SHI-RONG, 1987. *Hu nan zhi wu ming lu* (*The list of Hunan flora*). [i], 2, 11, 466 pp. [N.p.]: Hu nan ke xue ji shu chu ban she (Hunan Science and Technology Publishing House).

A briefly annotated systematic enumeration of vascular plants (the genera and species alphabetically arranged), with localities; index to genera and families at end.

876

Hubei (Hupeh)

Area: 187500 km². Vascular flora: 5650 species of seed plants (Zheng Zhong, 1993, prelim. p. 1; see below).

Publication of a provincial flora began in 1978, but as of the present writing only two volumes have appeared. There are two other provincial works of more limited scope: *Keys to Hubei plants* (1976) and *Hubei plants complete* (1993). Coverage of the generally low eastern part of this province hitherto has been furnished by Steward's flora of the lower Yangtze Basin (**870/I**), but information on the hilly western part was scarcely consolidated. For that area, the most useful sources have been the botanical reports of SARGENT (on Wilson's collections) and DIELS (for both, see **860–80, Partial works**). There are also separate keys for the Wuhan district.

ANONYMOUS, 1976. (*Keys to Hubei plants.*) 621 pp. [Not seen; reference from Chen *et al.* (1993, p. 67).]

[INSTITUTE OF BOTANY, PROVINCE OF HUBEI], 1978–80. *Flora hupehensis*. Parts 1–2. Illus. Wuhan.

Copiously illustrated descriptive manual-flora of seed plants, with keys to all taxa, synonymy, Chinese transcriptions of botanical names, indication of local range, and notes on habitat, status, ecological preferences, etc.; indices to botanical names in Chinese and Latin. All species treated are illustrated by small figures in boxes. Arrangement of the families is after Wettstein; part 1 covers Gymnospermae and Saururaceae through Lauraceae and part 2, Papaveraceae through Sabiaceae. [Four parts in all were projected but nothing else has to date appeared.]

ZHENG ZHONG, 1993. *Hu bei zhi wu da quan/Hubei plants complete*. [iii], 8, 677 pp. [Wuhan]: Wu han za xue chu ban she (Wuhan University Press).

Annotated systematic enumeration of seed plants with Chinese and botanical names and indication of distribution, habitat, altitudinal range (if relevant) and special features; index to families and genera at end. A brief introduction precedes the table of contents.

877

Sichuan (Szechwan)

Area (present-day province): 569 000 km². Vascular flora: no data seen. – Part of modern Sichuan is in what is here termed the 'Tibetan Marches' (878), an area at one time under separate administration.

Botanical exploration of this large and rather diverse province, the central part of which comprises the Chengdu Plain, effectively began with the activities of Père Armand David in the 1860s. He was followed by other French missionary-priests as well as explorers from other countries. The French data and collections were synthesized by Abbé Léveillé in his *Catalogue illustré* (1918; for evaluations by LAUENER and REHDER, see **860–80**). Those of others largely went into the various botanical reports listed under **860–80** (which – especially *Plantae wilsonianae* – remain useful for those not literate in Chinese). Much work was done with the aim of finding new plants for European and North American horticulture, an activity which continues. Active exploration by Chinese botanists began in the 1920s, and institutions followed in the 1930s. A

relatively copious, albeit very scattered, corpus of literature was built up; among the contributors was W. P. Fang, working from Chengdu and later chief editor of what would be one of the two largest provincial floras: *Si chuan zhi wu zhi*. Research towards this began in the mid-twentieth century but only in 1981 did the first volume appear. By the late 1990s, 14 volumes or part-volumes had been published, each covering, as accounts have been completed, a single family or assorted families.

FANG, W. P. *et al.*, 1981– . *Si chuan zhi wu zhi/ Flora sichuanica*. Vols. 1– . Illus. Chengdu: Sichuan People's Press (from vol. 3: Sichuan Science and Technology Press).

Large-scale illustrated descriptive flora; includes keys to all taxa, formal Chinese and vernacular names, synonymy with references and citations, indication of distribution, altitudinal zonation and habitat, and other notes; indices at end of each volume. No systematic order is followed in the publication of families although a group of related families may appear together and some volumes are limited to a single family or other group (vol. 5, when completed, will encompass Poaceae (except Bambuseae), while vol. 6 is devoted to pteridophytes and vol. 12 is entirely on Bambuseae). [Twenty volumes in all are projected; of these, vols. 1–4, 5(2) and 6–14 had appeared as of writing in 1999 with vol. 15 also announced.][130]

LÉVEILLÉ, H., *l'abbé*, 1918. *Catalogue illustré et alphabétique des plantes du Seu-tchouen*. [2], 221 pp., 66 pls. (some col.). Le Mans: The author. (Limited, holographed edition of 10 copies from author's MS.)

Systematic enumeration of vascular plants, with limited synonymy and references; no index. A tabular summary of the flora and some phytogeographical remarks are appended. [Now very incomplete and moreover as uncritical as his other floras (see **881**, Yunnan; **882**, Guizhou).]

878

'Tibetan Marches'

See also **768**; **803/IV**; **877**. – This mountainous but horticulturally important source area, formerly consisting of petty Tibetan states and for a time in the mid-twentieth century comprising a separate province of Sikang or Changdu (Chamdo), is now divided between Tibet (Xizang) and Sichuan, with the present eastern boundary falling along the upper Yangtze. Its biogeographical links are, however, with monsoon Asia. As defined here it extends from the Ta-hsüeh Shan (in which is situated Gongga Shan or Minya Konka) west to approximately 96° E or the former western boundary, with Qinghai (Tsinghai, Chinghai) to the north and Yunnan, Myanmar and northeastern India to the south (the southern limit being the so-called McMahon line).

Much early collecting in the 'Marches' was, as in Sichuan proper, accomplished by French missionary-priests beginning with Père Armand David; in the twentieth century they were joined by many other botanists from Europe and the United States. From this work there grew up a considerable but scattered literature; of this, perhaps the most useful are the general reports of FRANCHET (*Plantae davidianae*), HANDEL-MAZZETTI, PAX, REHDER and WILSON and SMITH *et al.* (for all see **860–80, Partial works**). More recently, a major new flora for western China, inclusive of the whole of the 'Marches', has appeared: *Heng Duan Shan qu wei guan zhi wu* (WANG WEN-TSAI (ed.), 1993–94; see **803/IV**). In the 'Marches' this concise keyed enumeration covers Changdu district in Xizang and Aba, Garzê and Liangshan districts in Sichuan.

Region 88

South China

As here circumscribed, South China includes the provinces of Yunnan, Guizhou (Kweichow), Guangdong (Kwangtung) and Fujian (Fukien), the Guangxi (Kwangsi) Autonomous Region, the territories of Hong Kong and Macao (here included with Guangdong), the large islands of Taiwan (with its satellites) and Hainan, and the low island groups of the Pratas, Paracels (Hsi-sha) and Spratlys (Nan-sha) in the South China Sea.

Extensive botanical exploration in the modern sense began only after 1860. Foreign collectors had previously been restricted closely to Guangzhou (Canton), with the addition after 1842 of Hong Kong and the environs of the additional treaty ports made available. Activities, however, remained relatively local-

ized until well after 1900, although some individual contributions were considerable. As in west-central China at the same period, diseases (including malaria), wild animals, and other travel difficulties all presented problems. Nevertheless, much South China material is recorded in *Index florae sinensis*, and some pioneer provincial floras were written which long remained current. These include *Flora hong-kongiensis* (Bentham, 1861), *Flora of Kwangtung and Hong Kong* (Dunn and Tutcher, 1912), *Flore de Kouy-Tchéou* and *Catalogue des plantes de Yun-nan* (Léveillé, 1914–15, 1915–17), and a number of works on Taiwan culminating in *Icones plantarum formosanarum* (Hayata, 1911–21; supplement, 1925–32). The importance of mountainous northwestern Yunnan as a source of good garden plants for Europe and North America soon became recognized and in the 30 years or so after 1900 it was a major objective for foreign and, later, Chinese collectors. Their results appeared in the great botanical reports (**860–80**) and in countless other monographs and papers. Less attention was, however, paid to areas not likely to yield hardy plants.

With the establishment after 1920 of Chinese biological surveys there took place systematic collecting through most of South China and Hainan, and the warm-temperate, subtropical and tropical flora became much better known. Until well after the 1949 revolution, however, few synthetic floristic works appeared apart from checklists and a never-completed *Flora of Fukien* (Metcalf, 1942). Since 1956, floras have been published for the environs of Guangzhou (1956, 1957), Hainan (*Flora hainanica*, 1964–77), Yunnan (*Flora yunnanica*, 1977–) and Taiwan (*Flora of Taiwan*, 1975–78; 2nd edn., 1993–); and since 1980 floras of Guizhou, Fujian, Guangdong and Guangxi have begun publication. Preparation of a new florula for Hong Kong was also in progress. Floristic accounts also exist for the Paracels and Spratlys. The woody flora is best documented in Yunnan (Hsi nan lin hsueh yuan, 1988–91) and Taiwan (Liu, 1960–61; Li, 1963).

The earliest botanical centers were in Hong Kong (from 1864), Taiwan (from 1904), and Guangzhou (from 1907, with the Institute of Botany for South China following in 1927). Other centers, notably in Kunming and Guilin, were established in the 1930s. These, and the larger size of the flora, have in southern China – in contrast to the north and east-central parts, where for long the few separate works were subregional – early resulted in establishment of the province as a working unit for floristic studies and flora-writing, with Hainan moreover seen as a distinct entity. Taiwan has been effectively an area apart for most of the twentieth century. There have thus been few if any works of regional or subregional scope.

Bibliographies. General and supraregional bibliographies as for Division 8 and Superregion 86–88.

Indices. General and supraregional indices as for Division 8 and Superregion 86–88.

881

Yunnan

Area: 436200 km². Vascular flora: 14000 species of spermatophytes (*Index florae yunnanensis*; see below). – Yunnan comprises one of the six 'hottest spots' in the world for plant diversity (Barthlott *et al.*, 1996; see **General references**); its known vascular flora is larger than that of all Europe.

It was only in the twentieth century that the great richness of the Yunnan flora was revealed. Early work, particularly by Père Jean-Marie Delavay in the west and Augustine Henry in the south, was followed after 1900 by active collecting by other European travelers and residents, particularly in the west and northwest as its 'garden potential' became known. Areas considered merely of botanical interest were less frequented. The results were reported upon in widely scattered reports and papers, sometimes not specifically related to the province. The most important are accounted for under **860–80**. The collections and data of the French missionary-priests and brethren are the main basis for what long was the only flora, the compiled *Catalogue* (1915–17) by Abbé Léveillé (see below). This uncritical and now largely obsolete work has been evaluated along with other contributions in a series of papers by Lauener and others (see **860–80**).

From the 1920s, a great deal of systematic field work was carried out by Chinese botanists in Yunnan, yielding tens of thousands of collections; over the years, these have been covered in revisionary studies in *Acta Phytotaxonomica Sinica* and its predecessors as well as elsewhere. A provincial botanical center was opened in Kunming in 1930. The first volume of a modern provincial flora appeared in 1977, incorporating as far as possible all previous botanical work; publication of further volumes has continued to the present. There is also a complete enumeration in two volumes as well as a dendrological atlas.

While the northwest may be that part best known to temperate botany and horticulture, the southwest is of particular note for its tropical plant life. Not surprisingly, it too has attracted writers, mostly Chinese and Russian (e.g., Wu Cheng-i, 1965 (noted below), and the Xishuangbanna checklists of 1984 and 1996). A prime advocate for the southwest for many years was H. T. Tsai, one of the notable explorers of the 1930s.

Bibliography
Kun Ming Chih Wu Yuan, 1984. *Chung-kuo ko hsüeh yuan Kun-ming chih wu yen chiu so chu tso lun men mu lu A list of the works and monographs of Kunming Institute of Botany, Academia Sinica), 1958–1983.* 86 pp., illus., (Kunming: Institute of Botany, Scientific Archives. [Bilingual unannotated classified bibliography. A brief institutional history is incorporated. A revision (in two volumes) appeared in 1986.]

Léveillé, H., *l'abbé*, 1915–17. *Catalogue des plantes du Yun-Nan avec renvoi aux diagnoses originales, observations et descriptions d'espèces nouvelles.* 299 pp., 69 text-figs. Le Mans: The author. (Limited, holographed edition from author's MS.)

Systematic list of vascular plants, with descriptions of new taxa as well as references; no keys and no index. [Now obsolete and very incomplete.][131]

Wu Cheng-i *et al.* (eds.) [for Kunming Institute of Botany, Academia Sinica], 1977– . *Flora yunnanica.* Vols. 1– . Illus. Beijing: Ke xue chu ban she (Academia Sinica Press).

Illustrated descriptive flora, presented as a series of provincial family revisions; each treatment includes keys to all taxa, synonymy, references and abbreviated citations, Chinese transliterations and local names, generalized indication of distribution, sometimes lengthy critical commentary, and notes on habitat, occurrence, etc.; tables of uses and complete indices at end of each volume. [Volumes are appearing as family treatments are ready, in the same manner as *A revised handbook to the flora of Ceylon* (**829**) or *Conspectus florae asiae mediae* (**750**). Eight of a projected 20 volumes have now been published, the most recent in 1997; an index to all families published may be found at the end of each successive volume.]

Wu Cheng-i (ed.) [for Kunming Institute of Botany, Academia Sinica], 1984. *Yun-nan chung tzu chih wu ming lu/ Index florae yunnanensis.* 2 vols. 2259 pp., map. Kunming: Yun nan jen min chu ban she (Yunnan People's Publishing House).

Annotated enumeration of seed plants (14000 species), with botanical and Chinese nomenclature, synonymy, references and citations (the latter arranged chronologically), floristic districts (see the introductory part for explanation), localities with *exsiccatae*, altitudinal range, and where appropriate brief commentaries; indices at end of each volume covering all scientific names, with a complete character-stroke index to Chinese names in vol. 2. The floristic districts are mapped in vol. 1. [Vol. 1 covers gymnosperms and the Dicotyledoneae–Dialypetalae; vol. 2, Dicotyledoneae–Sympetalae and Monocotyledoneae.][132]

Partial work: Kunming region

(Yunnan University Botany Section), 1981. (*Catalogue of flowering plants of the Kunming region.*) xii, 189, 7 pp. Kunming: University of Yunnan.

An unannotated systematic checklist, with index to families.

Partial works: southern districts

In addition to the Xishuangbanna checklist described below, note may be taken of Wu Cheng-i and Li Hsi-wen (eds.), 1965. *Yün-nan jo ti ya jo tai chih wu ch'ü hsi yen chiu pao kao* (*Reports on studies of the plants of tropical and subtropical regions of Yunnan, I*). 146 pp., 38 pls. Beijing: Ke xue chu ban she (Academia Sinica Press). This is a series of contributions on the flora of tropical and subtropical southern Yunnan with descriptions of new taxa, indication of new records, and commentary. It is a continuation from three papers by the senior author published in volumes 6 and 7 of *Acta Phytotaxonomica Sinica* (1959–60).

Li Yan-hui (ed.) [for Xizhuangbanna Tropical Botanical Garden and Department of Ethnobotany, Kunming Institute of Botany], 1996. *Xi shuan ban na gao den zhi wu ming lu* (*List of plants in Xishuangbanna*). 2nd edn. [iv], 7, 16, 702, [1] pp., map. [Kunming.] (1st edn., 1984.)

A systematic checklist (4669 taxa in 1697 genera in 282 families) with formal Chinese nomenclature and brief annotations (including English, Thai, Hani, Jinuo and Yao vernacular names where known); addenda (p. 625) and indices to families and genera in Chinese and Latin and to English vernacular names (pp. 626–702). [Covers the Xishuangbanna Thai autonomous district (prefecture) in southwestern Yunnan, south of Simao.]

Woody plants

Hsi nan lin hsueh yuan (Southwestern Forestry College) and Yun nan sheng lin yeh ting pien chu (Forestry Department, Yunnan Province), 1988–91. *Yun-nan shu mu tu chih/ Iconographia arbororum* [sic!] *yunnanicorum.* 3 vols. 1612 pls. Kunming: Yun nan ke ji chu ban she (Yunnan Science and Technology Press).

Descriptive account with full-page illustrations of Yunnan trees; includes Chinese and scientific names, phytography, and notes on distribution, altitudinal range, economics, etc.; each volume fully indexed (no cumulative index). Species not illustrated are merely enumerated, without descriptions. Some 2000 or more species are covered, usually one or two to a plate. [The arrangement of families is modified Englerian; numbering of families is not continuous. Some illustrations have been reused from other works.]

882

Guizhou (Kweichow)

Area: 174000 km². Vascular flora: no data seen.

Before *Flora guizhouensis* commenced publication, the only overall work was the obsolete, uncritical and rare (only 20 copies having been published) enumeration by Abbé Léveillé (1914–15). It has been evaluated, along with his other works, in a series of papers by Lauener and others (see **860–80, Partial works**).

Gui zhou zhi wu zhi bian ji wei yuan hui (Flora of Guizhou Editorial Committee), 1982– . *Gui zhou zhi wu zhi/ Flora guizhouensis.* Vols. 1– . Illus. Guiyang: Gui zhou jen min chu ban she.

Illustrated documentary flora of seed plants; includes keys, formal Chinese nomenclature, synonymy with (in later volumes) references and citations, more or less ample descriptions, indication of distribution, altitudinal range and habitat, and other notes; indices at end of each volume. [By 1999 nine of a projected 10 volumes had been published. As in both *Flora sichuanica* and *Flora yunnanica*, no systematic sequence has been followed (although gymnosperms are all in vol. 1, Poaceae in vol. 5, Rosaceae and Fabaceae in vol. 7, and Asteraceae fill the whole of vol. 9).]

Léveillé, H., *l'abbé*, 1914–15. *Flore de Kouy-Tchéou.* 532 pp., portr. Le Mans: The author. (Limited, holographed edition from author's MS.)

Systematic enumeration of vascular plants, with keys to all taxa and brief descriptions of novelties;

citations of *exsiccatae*, with localities; index to genera at end. [Based largely on the collections of resident French missionary-priests active in the province in the late nineteenth and early twentieth centuries.]

883

Guangxi (Kwangsi) Zhuang Autonomous Region

Area: 220 400 km². Vascular flora: no data seen. – Since 1949 this province, home to large 'minority' populations, has been recognized as an 'Autonomous Region'.

With relatively little to offer 'plant hunters', botanical exploration in Guangxi began seriously only in the 1920s; a botanical center was established in 1935. New plants or new records continue to be found, and the Guilin center now publishes its own journal, *Guihaia*. An enumeration (not completed) was produced by Wang Chen-ju (1940–42); this was apparently concluded and re-published in 1955. A provincial flora commenced publication in 1991. Both works follow the 1926–34 version of the Hutchinson system.

Limestone hills are a prominent feature in parts of the Region; a selection of their characteristic plants appears in ZHONG JI-XIN, 1982. (*Illustrated limestone mountain plants of Guangxi.*) 337 pp. Nanning.

LI SHU-GUANG and LIANG CHOU-FENG (eds.) (for GUANGXI INSTITUTE OF BOTANY), 1991– . *Guang xi zhi wu zhi/Flora of Guangxi.* Vols. 1– . Illus. [N.p.]: Guang xi ke xue ji shu chu ban she (Guangxi Science and Technology Press).

Concise descriptive flora; includes Chinese names, accepted botanical names and essential synonymy with references and citations, brief descriptions, indication of distribution and altitudinal range, and other notes. Indices to Chinese and botanical names are given at the end of each volume. [Not yet completed; vol. 1 covers gymnosperms and dicotyledons to the Myrtaceae, accounting for 1380 native and introduced species and in angiosperms arranged following the 1926–34 version of the Hutchinson system. Five volumes are projected.]

WANG CHEN-JU (WANG YEN-CHIEH) *et al.*, 1940–42. [An enumeration of seed plants collected in

Kwangsi.] Parts 1–9. *Kwangsi Agric.* **1** (1940): 68–77, 403–415; **2** (1941): 134–141, 223–229, 285–294, 371–384, 468–472; **3** (1942): 57–60, 121–124.

Systematic enumeration of seed plants, with synonymy, botanical names with Chinese equivalents, localities with *exsiccatae*, and field data; no taxonomic commentary or indices. A brief introduction appears in part 1. [Not completed; covers the Gymnospermae and families from Ranunculaceae through Rhizophoraceae in parts 1–8 (part 9 has not been seen). A revised (and presumably completed) version of this work, entitled *Catalogue of the plants of Kwangsi*, was apparently published in 1955 but no details have been available.]

884

Guangdong (Kwangtung), Xianggang (Hong Kong) and Macao (Macau)

Area (Guangdong): 197 400 km² (exclusive of Hainan, now a separate province). Vascular flora: no data seen.

The now-incomplete general flora by Dunn and Tutcher (for Hong Kong still the most recent keyed manual) is presently being superseded by a new provincial flora. A separate florula and field-manual have also been published for Guangzhou (Canton) and its environs, and for Hong Kong there is a relatively recent checklist. The minute territory of Macao is covered only in a 1933 catalogue (reissued in 1984) by A. C. de Sá Nogueira. Mountain areas are documented, with a list of 3656 species, in ANONYMOUS, 1990. (*Floristic notes on the mountain area of Guangdong.*) 150 pp. [Not seen; reference from Chen *et al.* (1993; see **Super-region 86–88, Supraregional bibliographies**).]

CHEN FENG-HWAI and WU TE-LIN (eds.), 1987– . *Guang dong zhi wu zhi (Flora of Guangdong).* Vols. 1– . Illus. [Guangzhou]: Guang dong ke ji chu ban she (Guangdong Science and Technology Press).

Illustrated descriptive flora of seed plants with keys, synonymy (without references), Chinese standard and vernacular names, indication of distribution and habitat, external range, and notes; indices to Chinese and vernacular names at end of each volume. [As with *Flora yunnanica*, no systematic sequence has been followed; volumes are published upon completion of an appropriate array of family accounts. By 1999 three of a projected five volumes had been published.]

DUNN, S. T. and W. TUTCHER, 1912. *Flora of Kwangtung and Hong Kong.* 370 pp., map (Bull. Misc. Inform. (Kew), Add. Ser. 10). London.

Concise manual-key to vascular plants, with diagnoses, synonymy and references, fairly detailed general indication of local range, and notes on habitat, phenology, special features, etc.; gazetteer and index to family and generic names. The introductory section includes remarks on physical features, characteristics of the flora, history of botanical investigation, and desiderata. [Hainan records are not included.]

Partial works: Guangzhou (Canton) and district

For a condensed version in pocket-size format for field use of *Guang zhou zhi wu zhi*, see HOW FOON-CHEW (HOU K'WAN-CHAO) *et al.*, 1957. *Guang zhou zhi wu chien so piao.* [i], 226 pp. Guangzhou: Ke xue chu ban she (Science Press).

HOW FOON-CHEW (HOU K'WAN-CHAO) (ed.), 1956. *Guang zhou zhi wu zhi.* 8, 953 pp., 415 text-figs., 4 pls. Beijing: Ke xue chu ban she (Academia Sinica Press). (Reprinted 1959; unofficially also reprinted in Hong Kong or Taiwan, n.d.)

Descriptive manual-flora of vascular plants of Canton and its environs, with keys, formal Chinese nomenclature and vernacular names, synonymy, local range (with some localities), taxonomic remarks, and notes on habitat, etc.; complete indices at end. The work also includes chapters on the basics of taxonomy, collecting, etc. for students as well as lexica and remarks on local floristics and vegetation.

Xianggang (Hong Kong)

Apart from the manual by Dunn and Tutcher cited above, two works are available relating specifically to the former British colony (that by Bentham, however, does not cover the New Territories, not leased from China until 1898). A number of attractive semi-popular illustrated works on different aspects of the local flora have also appeared in the last two or so decades, but their scope does not permit inclusion here. Some 2815 species and varieties are listed in the current *Check list* (1993).[133]

BENTHAM, G., 1861. *Flora hong-kongiensis.* 20, li, 482 pp., map. London: Reeve. Complemented by H. F. HANCE, 1872. *Supplement to the 'Flora hong-kongiensis'.* 59 pp. London.

Descriptive flora of vascular plants; includes keys, limited synonymy, localities with *exsiccatae*, taxonomic commentary, and miscellaneous other notes; glossary and index at end. Introductory chapters include remarks on botanical exploration, floristics and phytogeography as well as *Outlines of Botany*.[134]

HONG KONG HERBARIUM, 1993. *Check list of Hong Kong plants.* 6th edn. 159 pp. (Hong Kong Agriculture and Fisheries Dept., Bulletin 1). Hong Kong: Hong Kong Herbarium. (1st edn., 1962; 2nd edn., 1965; 3rd edn., 1966, all mimeographed; 4th edn., 1974 (1975); 5th edn., 1978.)

Unannotated systematic checklist of vascular plants with Chinese and English vernacular names and limited synonymy; indices to all names at end. The bilingual general part furnishes concise treatments of physical features and the flora and the history of the Hong Kong Herbarium; a bibliography appears on pp. 18–20.

Macao (Macau, Aomin)

Only one general flora has appeared in the twentieth century: A. C. DE SÁ NOGUEIRA, 1933. *Catálogo descritivó de 380 especies botânicas da Colóni de Macau.* 138 pp. Macao. (Reprinted 1984, Macao, by Imprensa Nacional/Serviços Florestais e Agrícolas de Macau. 181 pp.) For introductions to plant life on the islands of Taipa and Coloane, see A. J. E. ESTÁCIO, 1978. *Flora da Ilha de Taipa: monografia e carta temática.* 31 pp., illus., folding map. Macao; and *idem*, 1982. *Flora da Ilha de Coloane.* 49 pp., illus. Macao.

885

Fujian (Fukien)

Area: 123 100 km². Vascular flora: no data seen.

Two botanical centers were in existence from the 1920s, respectively in Xiamen (Amoy) and Fuzhou (Foochow). This facilitated penetration of the interior hills, which came rather later than the coastal districts where nineteenth-century collecting had been concentrated. Only one fascicle of the flora prepared by Franklin P. Metcalf at Fukien Christian University, extending through Fagaceae, was ever published. The herbarium of Amoy University, begun by H. H. Chung, is a major basis for the current provincial flora (1985–96).

Although the provincial flora is now complete, the older *Flora of Fukien* remains useful for its general section, applicable to southeastern China as a whole: F. P. METCALF, 1942. *Flora of Fukien and floristic notes on southeastern China.* Fasc. 1. xiv, [3], 82 pp., 2 maps. Canton: Lingnan University. This covers floristics, vegetation, and the history of botanical investigations.

[FLORA OF FUJIAN EDITORIAL COMMITTEE], 1985–96. *Fu jian zhi wu zhi / Flora Fujianica.* Vols. 1

(2nd edn., 1991), 2–6. Illus. [Xiamen]: Fu jian ke xue ji su chu ban she (Fujian Science and Technology Press). (1st edn. of vol. 1, 1982.)

Descriptive flora of vascular plants; includes keys to genera and species (no family key), relatively extensive phytography, Chinese and scientific names, synonymy, references and citations, and notes on occurrence, habitat, uses, distribution, etc.; diagnoses of new taxa and full indices in each volume. [Now complete; although following a systematic sequence, the volumes appeared only when all the included families were ready.][135]

886

Taiwan (and the Pescadores)

Area: 36 000 km². Vascular flora: slightly over 4000 species and additional infraspecific taxa, including over 600 pteridophytes and 28 gymnosperms (Hsieh *et al.*, 1994, in Editorial Committee of the Flora of Taiwan, vol. 1; see below). Traditionally considered as Chinese territory and at one time part of Fujian Province, Taiwan – also commonly known as Formosa, its Portuguese name – was under Japanese administration from 1895 to 1945. From 1949 it has been the seat of the (Nationalist) Republic of China.

The first separate checklist is that of Augustine Henry, *A list of plants from Formosa*; this appeared in the *Transactions of the Asiatic Society of Japan* in 1896 and incorporated records of his many collections made while resident during two years prior to the Sino-Japanese War. Subsequent to their entry, botanical exploration was vigorously pursued by the Japanese and the flora, including that of the hitherto almost unexplored mountains, was documented in several substantial publications, of which the largest was Bunzo Hayata's *Icones plantarum formosanarum* (1911–21). This work, with its supplement by Y. Yamamoto (1925–32), remains useful for its illustrations. Among the others were nomenclators by T. Kawakami (1910), B. Hayata (1917, as vol. 6 (supplement) of his *Icones*) and S. Sasaki (1928) and annotated enumerations by Matsumura and Hayata (1906) and Genkei Masamune (1936; see below). Taiwan plants were also incorporated into some Japanese manuals of the period.

Several decades and a change of government were, however, to pass before a modern descriptive flora was produced. A challenge expressed by C. G. G. J. van Steenis in the 1960s was taken up at National Taiwan University and the result was the six-volume *Flora of Taiwan* (Li *et al.*, 1975–79). In 1987–90 this work was revised with publication commencing in 1993. The original edition of the *Flora of Taiwan* incorporates in its final volume a checklist similar to that of Sasaki. There is also a nomenclator by Tsai-i Yang (1982).

EDITORIAL COMMITTEE OF THE FLORA OF TAIWAN (HUANG TSENG-CHIENG, gen. ed.), 1993– . *Flora of Taiwan/Tai-wan chih wu chih*. 2nd edn. Vols. 1– . Illus. (some col.), end-paper maps. Taipei: The Committee (for the National Science Council of the Republic of China). (Based on HUI-LIN LI *et al.* (eds.), 1975–79. *Flora of Taiwan*. 6 vols. Illus., end-paper maps. Taipei: Epoch.)

An illustrated, briefly descriptive, documentary flora of vascular plants; includes keys to all taxa, Chinese transcriptions of names, synonymy with newer references and citations (thus not duplicating the first edition), local and extralimital distribution, selected *exsiccatae* with localities, indication of habitat, text-figures, numerous color photographs of selected species (grouped in the back of each volume) and notes; index to all botanical names in each volume. Volume 1 contains the plan of the work and chapters on physical features, geology, climate, soils, tectonic history, origin and evolution of the flora, and floristics, phytogeography, chorology and vegetation (references, pp. 17–18). Volume 6 in the original edition contains a lengthy comprehensive bibliography and a systematic checklist of vascular plants as well as addenda, statistical tables, and complete indices. [Families are arranged following the 12th edition of the Englerian *Syllabus*, as in the original edition, and volumes are divided in the same fashion: pteridophytes and gymnosperms in vol. 1, angiosperms in the remaining volumes. Of the current edition, four volumes had been published as of 1999, covering all vascular plants except monocotyledons (expected in vol. 5).][136]

HAYATA, B., 1911–21. *Icones plantarum formosanarum nec non et contributiones ad floram Formosanam*. 10 vols., supplement. Illus. Taihoku: Government of Formosa, Bureau of Productive Industry. Continued as Y. YAMAMOTO, 1925–32. *Supplementa iconum plantarum formosanarum*. Fasc. 1–5. Illus. Taihoku: Department of Forestry, [Government] Research Institute.

The *Icones* comprise for the most part illustrated revisions of families and genera of vascular plants represented in Taiwan with keys to species, synonymy and references, literature citations, indication of local and extralimital range, taxonomic commentary, and notes on habitat, special features, etc.; checklist (vol. 6, supplement) and indices (vol. 10). Also included are a number of descriptions of new taxa from the Ryukyu Islands, the Bonin Islands, Hainan and Fujian. Volume 10 also contains (pp. 97–233) a detailed exposition of the author's 'dynamic' system of higher-plant phylogenetic classification.

MASAMUNE, G., 1936. *Saishu Taiwan shokubutsu so-mokuroku* [*Short flora of Formosa. An enumeration of the higher cryptogamic and phanerogamic plants hitherto known from the island of Formosa*]. [vi], 410 pp., frontisp. Taihoku: Editorial Department, 'Kudoa'.

Systematic list of native, naturalized and commonly cultivated vascular plants; includes concise indication of synonymy, references and citations, scientific and Japanese formal and vernacular names, and indication of local range; bibliography and indices to generic and family names and their equivalents at end. The brief introductory section includes statistical tables of the flora.

YANG, TSAI-I, 1982. *A list of plants in Taiwan*. xiii, 1281, 351 pp., 2 maps, folding table. Taipei: Natural Publishing Co. (Previous editions, 1969 (as *Nomenclature of plants in Taiwan*) and 1974.)

Systematic nomenclator-checklist (5998 species accounted for in 2066 genera and 244 families) of native and introduced vascular plants (the latter depicted in italics); entries include accepted scientific names, synonyms, and names in Chinese (including Hakka and Amoy dialects), Japanese (in Hiragana script), English, and the Austronesian languages of the interior (the last-named in Roman script, with the name of the tribe concerned). The indices cover respectively family and generic names, Chinese names (arranged by character-strokes), Japanese names and English names. There is a brief introductory part; statistics appear in a folding table.

Woody plants

The works given below, both well illustrated, were preceded by the two editions of Ryôzo Kanehira's *Formosan trees* (1917, 1936; a post-World War II reprint of the second edition, in reduced format, has also been seen). Both current standard works are now somewhat out of date; revised nomenclature may be found in LIAO JIH-CHANG, 1993. *A list of scientific names of woody plants in Taiwan.* 6, 212 pp. Taipei: Forestry Department, National Taiwan University.

LI, HUI-LIN, 1963. *Woody flora of Taiwan.* 992 pp., 371 text-figs., maps (end-papers). Philadelphia: Morris Arboretum; Narberth: Livingston Publ. Co. (Also unofficially in Taiwan, no date.)

Briefly descriptive flora of woody plants; includes keys to all taxa, synonymy and references, generalized indication of local range with citation of selected *exsiccatae* and summary of extralimital distribution, excluded species, taxonomic commentary and other miscellaneous notes; list of references (post-1936) and new taxa and index to all botanical names. The introductory section includes remarks on physical features, climate, soils, vegetation, forests, botanical exploration, floristics and phytogeography.

LIU TANG-SHUI, 1960–61. *Illustrations of native and introduced ligneous plants of Taiwan.* 2 vols. 1388 pp., 1109 text-figs., 20 pls. (some col., incl. maps). Taipei: National Taiwan University.

Atlas of woody plants, without keys; includes descriptive text in Chinese (one species per page), synonymy and references, Chinese and English vernacular names, and extensive notes on local range, habitat, special features, properties, uses, etc.; indices to all botanical and vernacular names at end of vol. 2.

Partial work: Lanyu (Botel Tobago)

Lanyu (Botel Tobago or Orchid Island) lies due east of Taiwan's South Cape but features a flora more closely related to that of the Philippines.

ANONYMOUS, n.d. *Lan yü chih wu* (*Plants of Lanyu*). [ii], 170 pp., col. pls. Taipei: Taiwan Provincial Dept. of Education.

Includes a checklist of known vascular plants, with names in Chinese and Latin; the work otherwise comprises color photographs of selected species with descriptive text. References (p. 169) and a character-stroke index to Chinese names are also included.

887

Hainan

Area: 34000 km^2. Vascular flora: *c.* 3500 species (Chun *et al.*, 1964–77, vol. 4, p. 522). A total of 2666 species, including 237 pteridophytes, were listed by Masamune (1943; see below). – Long administratively part of Guangdong, Hainan is now a separate province.

The plant life of this tropical/subtropical island in the South China Sea exhibits a number of floristic links with Southeast Asia and Malesia within an essentially southern Chinese framework. Three main accounts, all listed below, have covered the flora: *An enumeration of Hainan plants* (1927) by E. D. Merrill; *Kainan-to shokubutsu-shi* (1943) by Genkei Masamune; and the four-volume *Hai nan zhi wu zhi* (1964–77) by Woon-yung Chun and his associates in Guangzhou.

CHUN WOON-YOUNG *et al.*, 1964–77. *Hai nan zhi wu zhi/ Flora hainanica*. 4 vols. 1272 text-figs., col. map (vol. 4). Guangzhou: Ke xue chu ban she (Academia Sinica Press).

Briefly descriptive, well-illustrated flora of vascular plants, with keys to genera and species, synonymy, references and citations, Chinese transcriptions and local names, generalized indication of local and extralimital range, taxonomic commentary, and notes on habitat, occurrence, uses, etc.; complete indices in each volume. In volume 4 (pp. 521–530) is an account of the main features of the flora and vegetation types (with map) as well as a key to families, addenda, and complete general indices of botanical and Chinese names.[137]

MASAMUNE, G., 1943. *Kainan-to shokubutsu-shi/ Flora kainantensis*. xv, 443 pp., 2 maps. Taihoku: Taiwan Sotokufu Gaijabu. (Reprinted with new preface, 1975, Tokyo: Inoue Book Co.) Additions in G. MASAMUNE and Y. SYOZI, 1950–51. Florae novae kainantensis, I–II. *Acta Phytotax. Geobot.* **12**: 199–203; **14**: 87–90.

A largely compiled systematic enumeration of vascular plants, without keys; includes relevant synonymy, references and citations, Japanese transliterations of names, indication of distribution (with some localities), summary of extralimital range, and indices to families and genera and to Japanese names. An introductory section gives accounts of special features of the flora, phytogeography, vegetation formations, and the history of botanical investigation as well as statistics and a list of references.

MERRILL, E. D., 1927. An enumeration of Hainan plants. *Lingnan Sci. J.* **5**: 1–186.

Systematic enumeration of vascular plants, with descriptions of new taxa; limited synonymy, with occasional references; citation of *exsiccatae*, with localities; general indication of extralimital range; taxonomic commentary and notes on habitat, etc.; no index. The introductory section includes remarks on phytogeography as well as a list of references. [Though now quite incomplete, this early work provided a basis for the later elaboration of *Flora hainanica* (its author having been one of Merrill's local collaborators).]

888

South China Sea Islands

Area and vascular flora: no data seen but the extent of both is small. – Here are gathered three archipelagoes (and a large bank) in the South China Sea, all consisting of low islands, cays, sandbanks, shoals and reefs: the Pratas (Dong-sha) Islands, the Paracel Islands (Xi-sha), Macclesfield Bank (Chung-sha) and the Spratly (Nan-sha) Islands. All are claimed by China but, save for the Pratas group (definitively Chinese), the islands have been spoken for at different times by no fewer than six other countries. The native land plant life consists almost entirely of widely distributed oceanic species.[138]

I. Pratas group (Dong-sha)

Included here are the Vereker Banks. Salmon steak-shaped Pratas Island, the only land mass, rises to 50 m in an area of 1.74 km² but not surprisingly has a small flora although *Pisonia* trees were present. Useful old accounts include C. COLLINGWOOD, 1868. *Rambles of a naturalist on the shores and waters of the China Sea*. xiii, 445 pp., 3 pls., 7 figs. London; and H. F. HANCE, 1871. Note on *Portulaca psammotropha*. *J. Bot.* **9**: 201–202. [In the first work Pratas is covered on pp. 22–34; in the second there is a species list including, among others, the given *Portulaca*.]

HUANG, T.-S. *et al.*, 1994. The flora of Tungshatao (Pratas Island). *Taiwania* **39**: 27–53, map.

Enumeration of native (72 species), naturalized and commonly cultivated vascular plants (no ferns) with botanical and Chinese names, references, citations, indication of habitat and distribution, and voucher *exsiccatae*; references at end. The general part includes a description of the vegetation (heavily disturbed in past) and its formations. [A total of 110 species were recorded.]

II. Paracel group (Xi-sha) and Macclesfield Bank (Chung-sha)

The Paracels form a fairly compact group in the South China Sea south of Hainan and east of central

Vietnam. Macclesfield Bank, not hospitable to land plants, lies to their east. There has been a great increase in their known flora, largely through human introduction, since publication of *The vegetation of the Paracel Islands* by Chang Hung-ta (1948; in *Sunyatsenia* 7: 75–88, with map). In addition to the list by Chen (1983; see below) there is also ANONYMOUS, 1977. (*Plants and vegetation on Xisha Islands of China.*) 127 pp.

CHEN PANG-YU *et al.*, 1983. (An enumeration of plants from Xisha Qundao (Islands) of China.) *Acta Bot. Austro-Sinica* 1: 129–157.

Systematic enumeration of native and introduced species; includes synonymy with references and citations and localities with *exsiccatae*. For additions, see XING FU-WU *et al.*, 1994. (Additions to the flora of Xisha Islands, South China.) *Ibid.*, 9: 40–46.

III. Spratly group (Nan-sha, Truong-sa, Kalayaan)

These widely scattered low islands and cays, with their numerous associated reefs, shoals and banks, are situated on an extensive marine rise west of Palawan and north of Borneo. They have long been considered a serious hazard for shipping but feature potentially valuable underground and marine resources. The largest is Itu-aba (36 ha). References to vegetation and flora are inevitably scattered, and only in 1996 did a consolidated account (not seen) appear: XING FU-WU and T. WU (eds.), 1996. (*Flora of Nansha Islands and their neighboring islands.*) 375 pp. Beijing: China Ocean Press. The total extent of the flora is not yet established, but with more thorough investigation it appears to be rather greater than what was known before World War II.

GAGNEPAIN, F., 1934. Quelques plantes des îlots de la Mer de Chine. *Bull. Mus. Hist. Nat. (Paris)*, II, 6: 286.

Comprises a list of vascular plants collected from North Danger, Loaita, Itu-aba and Spratly (Storm) Islands as well as parts of the Paracels during the course of three fisheries surveys in 1930–33. Only 11 species were recorded.

HUANG, T.-S. *et al.*, 1994. The flora of Taipingtao (Aba Itu Island). *Taiwania* 39: 1–26, map.

Enumeration of native (81 species), naturalized and commonly cultivated vascular plants (including 3 pteridophytes) with botanical and Chinese names, references, citations, indication of habitat and distribution, and voucher *exsiccatae*; references at end. The general part includes a description of the vegetation along with an analysis of the flora and an account of phytogeography. [The area of the island is given as 48 ha. The authors' claim to be first with a botanical inventory is in error; however, rather more plants were recorded than in 1930–33.]

XING FU-WU *et al.*, 1994. (Study on the flora and vegetation of Nansha Islands, China.) *Guihaia* 14: 151–156.

Includes a checklist (with limited synonymy) of species with islands from which reported along with an account of the vegetation. [A figure of 48 species (native, established and cultivated) is given. Most natives are widespread 'strand' plants; those asterisked are considered as introduced.]

Region (Superregion) 89

Southeastern Asia

Area: 1 948 552 km^2. Vascular flora: 20 000–25 000 species. – Within this region as here circumscribed are the countries of the former Indochina (Vietnam, Laos and Cambodia), Thailand, Burma, and the Andaman and Nicobar Islands. For the Paracels and Spratlys, see **Region 88**.

The progress of botanical exploration in Southeast Asia has been, and for the present continues to be, fragmented due to geography, past colonial history, and political and other developments. No single modern flora covers the area as a whole, and the present array of past and current national and subregional floras is very heterogeneous. Furthermore, historically it has been overshadowed by India, China and Greater Malesia.

Early botanical explorations were few due to travel difficulties and restrictions and limited foreign contacts. Until well into the nineteenth century, the only significant contribution was *Flora cochinchinensis* (1790) by the Portuguese botanist João de Loureiro, long court physician at Hué in that part of modern

Vietnam known then as 'Cochinchina'. More detailed botanical exploration began, however, with increased trade and colonial penetration after 1800, particularly in Lower Burma from Arakhan to Pegu, Martaban and Tennaserim and in the Andaman and Nicobar Islands, chiefly through the work of Nathaniel Wallich, William Griffith, Johann Wilhelm Helfer, and, after them, Sulpiz Kurz. The remainder of the region received relatively little attention until after mid-century, although occasional visits were made to coastal areas including that of George Finlayson. Exploration then extended to Upper Burma and, with the work of Julien Harmand and Clovis Thorel in the Mekong Basin and Louis Pierre in Cambodia and southern Vietnam, to Indochina. In the last two decades of the century attention was directed to Tonkin and Annam and, in Burma, the Shan States and Kachin, with much of the serious exploration being accomplished only after World War I. Indeed, the Truong Son (Chaine Annamitique) in Indochina and some parts of northern Burma were seriously explored botanically only after 1920; one collector, Frank Kingdon Ward, moreover made significant new geographical observations. In Thailand, detailed botanical survey began after 1900 through the work of Arthur F. G. Kerr and other collectors (on behalf of the government) and Johannes Schmidt (in the southeast), since continued by Thai, Danish and other botanists. Some contributions were made in southern Thailand by botanists from the Straits Settlements and Malay States. By the 1930s botanical centers had been established in Rangoon, Bangkok, Saigon (now Ho Chi Minh City) and Hanoi.

World War II, though not directly affecting part of the region, nonetheless brought with it grave disruption. Revival of botanical work afterwards was uneven. In Burma and Indochina, after a moderate revival following the war, collecting gradually wound down in the 1950s owing to political turbulence, guerrilla activities and finally war. From about 1960, however, there was renewed activity in northern Vietnam where Vietnamese, Russian, and eastern European botanists were active. In Thailand, the establishment of the forestry research branch herbarium under Tem Smitinand initiated another cycle, which included substantial contributions from Danish and Dutch botanists under cooperative programmes and the establishment of a flora programme. In 1972, a station of the Botanical Survey of India was opened at Port

Blair, Andaman Islands, leading to a new collecting programme in the two Indian island groups.

The first modern general floras for any part of the region are those of Kurz. These include *Contributions towards a knowledge of the Burmese flora* (1874–77, not as such completed) and *Forest flora of British Burma* (1877). His early death was a serious loss to Burmese botany, though his *Contributions* were completed and published by William Theobald in the third edition of Mason's *Burma*. Apart from coverage in *Flora of British India* (see **810–40**) and other Indian works, no general works have appeared save the government checklist (first issued in 1912 and last revised in 1961) and some students' handbooks. More fortunate have been Thailand and the Indochina subregion. There, the more or less provisional *Flore générale de l'Indochine* (1907–51) and *Flora siamensis enumeratio* (1925–62) are gradually being supplanted respectively by the more definitive *Flore du Cambodge, du Laos, et du Viêt-Nam* (1960–), a French project like its predecessor, and *Flora of Thailand* (1970–), a Thai–Danish–Dutch project. The progress of the two current floras, while not rapid, is marked by mutual consultation; moreover, for both works several family treatments (or partial treatments) have been prepared by the same individual or individuals. The *Flore générale* was initiated at an early stage relative to the level of exploration in Indochina and, like the *Flora of Panama* (see **237**) and *Flore du Congo Belge* (see **560**), it came under attack for being too 'premature' and insufficiently critical. Other important works include Pierre's *Flore forestière de la Cochinchine* (1881–99 and 1907, not completed), *Cây-co miên-nam Viet-Nam*, a students' manual by Pham Hoàng Hô and Nguyên Van Duong (1960; revised 1970–72 and again in the 1990s), and *Forest flora of the Andaman Islands* (1923) by a forest officer, Charles E. Parkinson.

At the present time, the region remains unevenly explored, ranging from poorly to well known, the former being particularly true of northern and eastern Myanmar, much of Laos, and parts of the Truong Son (Chaine Annamitique) in Vietnam and Laos. Documentation also is uneven: as with the Mediterranean, the Himalaya, Mexico and Central America, and the West Indies, a modern synoptic enumeration is much to be desired. A level of coordination comparable to that which has existed for several decades for Greater Malesia has, however, yet to develop; particularly, the continuing 'isolation' of Myanmar remains a serious problem.

Progress

No overall review is available; however, passing references occur in *Short history of the phytography of Malaysian vascular plants* by H. C. D. De Wit (see **Superregion 91–93, Progress**). The situation as of the early 1970s is sketchily surveyed in P. LEGRIS, 1974. Vegetation and floristic composition of humid tropical continental Asia. In UNESCO, *Natural resources of humid tropical Asia*, pp. 217–238 (Natural resources research 12). Paris. Other pertinent remarks appear in the general surveys of tropical floristics by Prance (1978) and Campbell and Hammond (1989; for both, see **General references**).

For Indochina alone the main account has been J. E. VIDAL, 1979. Outline of ecology and vegetation of the Indochinese Peninsula. In K. Larsen and L. B. Holm-Nielsen (eds.), *Tropical botany*, pp. 109–123. London. An earlier, somewhat sketchy review is M. L. TARDIEU-BLOT, 1957. L'oeuvre botanique de la France en Indochine. In Pacific Science Association, *Proceedings of the Eighth Pacific Science Congress* (Manila, 1953), vol. 4, pp. 545–553. Manila. Some historical material is also included in the *Tome préliminaire* (1944) of *Flore générale de l'Indochine*. For Thailand, there are two useful earlier reviews: M. JACOBS, 1962. Reliquiae kerrianae. *Blumea* 11: 427–493; and K. LARSEN, 1979. Exploration of the flora of Thailand. In K. Larsen and L. B. Holm-Nielsen (eds.), *Tropical botany*, pp. 125–133. London.

The history of phytography in Myanmar – where since the end of the 1950s there has been relatively little botanical work – as well as in the Andamans and Nicobars is covered in passing in the reviews of South Asian work by Burkill (1965), Biswas (1943) and Santapau (1958) (see under **Superregion 81–84, Progress**). Separate reports on the Andamans and Nicobars include the papers by K. Thothathri and P. Balakrishnan in BOTANICAL SURVEY OF INDIA, 1977. *All-India symposium on floristic studies in India: present status and future strategies* (Howrah, 1977), *Abstracts*, pp. 18–19. Howrah, West Bengal, and M. K. VASUDEVA RAO, 1983 (1984). Early contributors to the botany of Andaman and Nicobar Islands. In *Hundred years of forestry in Andamans*, pp. 89–94. Port Blair, Andaman Islands.

Ongoing developments have also since 1947 been regularly reported in *Flora Malesiana Bulletin*.

Bibliographies. General bibliographies as for Division 8.

Regional bibliographies

ALLIED GEOGRAPHICAL SECTION, 1944. *An annotated bibliography of the Southwest Pacific and adjacent areas*. Vol. 3: *Malaya, Thailand, Indo-China*. [v], 256 pp., map. [Not restricted to botany or biology.]

DEPARTMENT OF EDUCATION, JAPAN, 1942. *Tōa kyō-ei-ken sigenkagaku bunken-mokuroku* [*Bibliographic index for the study of the natural resources of the Greater East Asia Co-Prosperity Sphere*]. Vol. 2: [*French Indo-China and Thailand*]. 253, 81 pp. Tokyo. [In Japanese, apart from references in other languages; not annotated.]

REED, C. F., 1969. *Bibliography to floras of Southeast Asia*. 191 pp., map. Baltimore: The author. [Entries arranged by author, without annotations. Evidently rather hastily compiled.]

SMITHSONIAN INSTITUTION, 1969. *A bibliography of the botany of South East Asia*. 161 pp. Washington. [Entries arranged by author, without annotations.]

Indices. General indices as for Division 8.

Regional index

STEENIS, C. G. G. J. VAN *et al.* (eds.), 1948– . *Flora Malesiana Bulletin*, 1– . Leiden: Flora Malesiana Foundation (later Rijksherbarium). (Mimeographed.) [At present issued bi-annually; the autumn installment contains a substantial bibliographic section which rather thoroughly covers ongoing new literature on Southeast Asia. For fuller description, see **Superregion 91–93**.]

890

Region in general

No overall general treatments exist. For Poaceae, however, notice may be taken of the following: M. LAZARIDES, 1980. *The tropical grasses of Southeast Asia (excluding bamboos)*. 225 pp. Vaduz,

Liechtenstein: Cramer/Gantner. (Phanerogamarum monographiae 12.)

891

'Indochina' (in general)

This term is here used in a geographical sense to account for works covering the whole of Vietnam, Laos, Cambodia, and the lower Mekong Basin in general; it thus also encompasses the former territory of French Indochina. The regional flora begun under the direction of H. Lecomte, *Flore générale de l'Indochine*, was premature – given the state of botanical exploration then existing in the subregion – and is now very incomplete. It moreover has had a lesser reputation than *Flora of British India* or most of the other 'Kew floras' of the period. Its successor, *Flore du Cambodge, du Laos, et du Viêt-Nam*, is more critical and possesses a better basis but is progressing only relatively slowly. Other literature on the area is rather scattered and for this reference should be made to standard bibliographies. A helpful semi-popular introductory work, with key references to more technical literature (useful also for Thailand), is J. E. VIDAL, 1997. *Paysages végétaux et plantes de la Péninsule indochinoise*. 245 pp., illus. (mostly col.). Paris: Karthala.[139]

Bibliographies

See also **Region 89**, **Regional index** (VAN STEENIS *et al.*).

PÉTELOT, A., 1955. *Bibliographie botanique de l'Indochine*. 102 pp. (Arch. Rech. Agron. Cambodge Laos Viêtnam 24). Saigon. Continued as J. E. VIDAL, 1972. Bibliographie botanique indochinoise de 1955 à 1969. *Bull. Soc. Études Indochinoises* **47**(4): [1], 657–748, [1]; J. E. VIDAL, Y. VIDAL and PHAM HOÀNG HÔ, 1988. *Bibliographie botanique indochinoise de 1970 à 1985*. 132 pp. Paris: Laboratoire de Phanérogamie, Muséum National d'Histoire Naturelle/Association de Botanique Tropicale; and J. E. VIDAL, H. FALAISE, PHAN KÊ LÔC and NGUYEN THI KY, 1994. *Bibliographie botanique indochinoise de 1986 à 1993*. 105 pp. Paris. (Documents pour la *Flore du Cambodge, du Laos, et du Viêtnam*.) [The original work includes itemization of monographs and revisions by authors under major plant groups; its coverage runs through 1954. The three subsequent installments comprise additions to Pételot's bibliography for the period 1955–93 along with amendments, the entries being arranged by author with subject and systematic indices. Russian titles have been translated into English and Vietnamese into French.]

AUBRÉVILLE, A. *et al.* (eds.), 1960– . *Flore du Cambodge, du Laos, et du Viêt-Nam* (from fasc. 23 as *Flore du Cambodge, du Laos, et du Viêtnam*). Fasc. 1– . Illus. Paris: Laboratoire de Phanérogamie, Muséum National d'Histoire Naturelle.

Large-scale descriptive 'research' flora of vascular plants, published in fascicles (each with one or more families); family treatments, similar in format to the *Flore générale de l'Indo-Chine* (see below), include keys to genera and species, synonymy, references and citations, typification, localities with citations of *exsiccatae* as well as general indication of local and extralimital distribution, taxonomic commentary, and notes on habitat, phenology, ecology, uses, etc. (the latter to a rather greater extent than in the earlier work), and complete indices. [Essentially a revised version of the *Flore générale* (see below); as of writing (1999), 29 fascicles have been published, the latest in 1997 (with no. 28, on gymnosperms, in 1996).]

LECOMTE, H. and H. HUMBERT (eds.), 1907–51. *Flore générale de l'Indo-Chine*. 7 vols., *Tome préliminaire* and *Supplément* (in 9). Illus. Paris: Masson (later Laboratoire de Phanérogamie, Muséum National d'Histoire Naturelle).

Comprehensive descriptive flora of vascular plants, with keys to genera and species, full synonymy, references and citations, vernacular names, indication of *exsiccatae* with localities, generalized indication of local and extralimital distribution, taxonomic commentary, and notes on uses, special features, etc.; index to genera in each volume. In the *Tome préliminaire* (1944) are chapters on physical features, climate, geology, vegetation and forests, botanical exploration (with maps), and a cyclopedia of collectors and contributors to the main work as well as a bibliography, general keys to families, and complete indices to genera and to vernacular names. The *Supplément* (1938–51) comprises in 10 parts additions and corrections to the original work (not, however, for all families). [The limits of the work encompass Vietnam, Laos, Cambodia, eastern Thailand (Mekong Basin), and Hainan. Most of it is now more of historical and specialist rather than practical interest.]

892

Vietnam

Area: 329 566 km^2. Vascular flora: 10 500 species (Hô, 1991–93), with an estimated likely total of *c.* 12 000 (Dao Van Tien and Phan Kê Lôc in PD; see **General references**). Phan Kê Lôc has recently made a more precise count of 9628 native species in 2010 genera and 291 families (the last following the Kew classification) as part of a full statistical survey of the flora.[140]

The first flora of any part of Vietnam is *Flora cochinchinensis* (1790) by João de Loureiro. More or less continuous botanical exploration began following the imposition of French rule, firstly in the south (in the 1860s, working from a re-founded Saigon – now part of Ho Chih Minh City – and its botanical garden, established in 1865) and later in the north (by the 1880s). The first major French work – unfortunately never completed – was *Flore forestière de la Cochinchine* (1881–99) by Louis Pierre, the first incumbent at the Saigon garden whose brief also included agriculture and forestry. The four decades following Pierre's tenure, from 1880 to 1918, saw a considerable expansion of geographical exploration and botanical work. The latter was distinguished by the collections of Bon, Balansa, Debeaux, d'Alliezette, Eberhardt, Harmand, Lecomte and Finet, Thorel and Chevalier. These formed the principal – though sketchy – basis upon which the *Flore générale de l'Indochine* was initially founded. Interwar collectors, rather fewer in number, included Clemens, Evrard, Poilane, Pételot and Tsang. Scientific units, including herbaria, and a university (in Hanoi) were established after 1919 but the actual task of flora-writing was concentrated in Paris and local research after 1925 became oriented to economic interests. Only after World War II did there arise more locally based floristic activities, in part made possible by the development of additional universities and institutes (including, from 1960, the present National Centre for Science and Technology in Hanoi).

The major independent work is now the third edition of *Cây-co Viêt-Nam* by Pham Hoàng Hô (1991–93; see below). This first appeared in 1960 as a one-volume students' manual for southern Vietnam under the title *Cây-co miên-nam Viêt-Nam*. An identification manual was also published in 1961 in the north under the authorship of Lê Kha Kê, Vu Van Chuyên and Thái Van Trùng; a later version by the same authors is the six-volume *Cây-co thuông thây o Viêt-Nam* (*Common plants of Viet Nam*) (1969–76; noted below under **Partial works**). Preparation of a new national flora has also been undertaken; in 1997 its first installment, an account of families and genera, was published through the National Centre for Science and Technology and its Institute for Ecology and Biological Resources. This same body also has a particular interest in the flora of Tây Nguyên (the Central Highlands) in southern Vietnam, floristically a key area (as indicated by Maurice Schmid in his *Végétation du Vietnam: Le Massif Sud-Annamitique* (1974)). The main forest trees were documented firstly in 1918–19 by Chevalier (in *Bulletin Économique de l'Indochine*) and then in *Cây Gô rùng Viêt Nam* (1971–88). This latter was revised with additional species and issued in English in 1996 as *Vietnam Forest Trees*.

Families and genera

NGUYÊN TIÊN BÂN (coord.), 1997. *Câm nang tra cúu và nhân biêt các ho thuc vât hat kín (Magnoliophyta, Angiospermae) o Viêt Nam / Handbook to reference and identification of the families of Angiospermae plants in Vietnam.* 532 pp., 6 pls. Hanoi: Nhà Xuât Ban Nông Nghiêp (Agriculture Publishing House). [A handbook to genera and families of flowering plants in Vietnam. Pp. 5–79 comprise family diagnoses (265 in all) with botanical and Vietnamese names and a few details, while on pp. 80–161 are spot-character multi-access identification tables (*in lieu* of traditional analytical keys). Pp. 162–454 feature a dictionary of genera with vernacular names, actual or estimated number of species, indication of type species, reductions, and corresponding families. The work concludes with an index to vernacular names with botanical equivalents, a table of families with corresponding botanical treatments in five standard works (*Flore générale de l'Indo-Chine, Flore du Cambodge, du Laos, et du Viêt-Nâm, Cây-co Viêt-Nam, Iconographia cormophytorum sinicarum,* and *Flora reipublicae popularis sinicae*), illustrations of common botanical terms (pp. 526–528), and principal references (pp. 529–530).]

AVERYANOV, L. V. *et al.*, 1990– . Конспект сосудистых растений флоры Вьетнама (*Konspekt sosudistykh rastenij flory V'etnama*) / *Trich yêu thu'c vât có mach thu'c vât chí Viêt Nam / Vascular plants synopsis of the Vietnamese flora.* Vols. 1– . Leningrad: 'Nauka' (vol. 1); St. Petersburg: 'Mir i Sem'ja-95' (vol. 2).

Systematic enumeration with synonymy, references and citations, types and type localities, distribution by provinces and localities, and (here and there) critical notes; index to all botanical names at end. The introductory part includes a map and table of names (in Russian, Vietnamese and English) of administrative units (which also include the Paracels (**888**) and Spratlys (**889**)). [Vol. 1 covers Orchidaceae, Amaranthaceae and Chenopodiaceae; vol. 2 (1996), miscellaneous families (as listed on the outside back cover).]

PHAM HOÀNG HÔ, 1991–93. *Cây-co Viêt-Nam* (*An illustrated flora of Vietnam*). 3 vols. in 6. Illus. Montreal: The author. (Original edn., 1960, Saigon, as *Cây-co miên-nam Viêt-Nam*; 2nd edn. in 2 vols., 1970–72, Saigon.)

Illustrated atlas-flora of vascular plants with all species (some 10 500) individually and consecutively numbered; includes keys to families, genera and species, descriptive text, limited synonymy, vernacular names, and notes on habitat, local range, occurrence, phenology, etc. including some diagnostic details in English; indices to genera, families and vernacular names in each volume. An introductory section (vol. 1) incorporates an illustrated organography. [The work is laid out along the lines of Coste's *Flore de France* (**651**) with a small figure for each species.][141]

Partial works

In addition to the following, see LÊ KHÀ KÊ, VU VAN CHUYÊN and THÁI VAN TRÙNG, 1969–76. *Cây co thuòng thây o Viêt-Nam* (*Common plants of Viet Nam*). 6 vols. Hanoi; and PHAM HOÀNG HÔ, 1972. (*Common plants of South Viet Nam*). Saigon: Lua Thieng.

NGUYÊN TIÊN BÂN, TRÂN DÌNH DAI and PHAN KÊ LÔC (eds.), 1984. *Danh luc thuc vât Tây Nguyên* (*Florae taynguyenensis enumeratio*). 235, [1] pp. Hanoi: Viên Khoa Hoc Viêt Nam [Vietnam Scientific Research Organization], Viên Sinh Vât Hoc [Institute of Biology].

Enumeration of seed plants (3201 species), the families alphabetically arranged within the major groups; includes accepted names, limited synonymy, vernacular names, provinces and localities, and (where necessary) notes; references (pp. 232–235). [Covers the Tây Nguyên or Central Highlands provinces from Lam Dong northwards through Dac Lac, Gia Lai and Kon Tum. Many species are simply numbered.]

Woody plants

For an earlier treatment of principal forest trees (limited to the northern provinces), see *Premier inventaire des bois at autres produits forestiers du Tonkin* by A. Chevalier (1918–19; in *Bull. Écon. Indochine* **20**: 497–524, 724–884; **21**: 495–552;

reprinted separately, 1919: 227 pp. Hanoi/Haiphong). The dendrological part is in the second installment.

VU VAN DUNG (ed.), 1996. *Vietnam Forest Trees.* vii, 788 pp. Hanoi: Agricultural Publishing House (for Forest Inventory and Planning Institute, Viêt Nam). (Based on FOREST INVENTORY AND PLANNING INSTITUTE, VIET NAM, 1971–88. *Cây gô rùng Viêt Nam* (*Essences forestières du Viêt Nam*). 7 parts. Hanoi: Nhà Xuât Ban Nông Nghiêp (Agricultural Publishing House).)

Dendrology of 763 native and exotic species, without keys and featuring one species per page; treatments include scientific and Vietnamese names, a botanical description, a quarter-page illustration, and notes on distribution, ecology, soils, phenology, growth and regeneration, and properties and uses. The botanical part is followed by illustrations of basic botanical details, alphabetical synopsis, indices to scientific and vernacular names, and (p. 788) key literature. The preface and acknowledgments are brief.[142]

PIERRE, J. B. L., 1881–99, 1907. *Flore forestière de la Cochinchine.* Vols. 1–5 (fasc. 1–26). 800 pp., 400 pls. Paris: Doin. (Reprinted in 1 vol., Lehre, Germany: Cramer.)

A sumptuous atlas-flora of woody plants, comprising finely executed and produced lithographic plates (by E. Delpy) with accompanying descriptive text; the latter includes extensive botanical descriptions, full synonymy with references, indication of *exsiccatae* with localities and general summary of distribution, and fairly detailed taxonomic commentaries. Complete general revisions are presented for some genera, notably *Garcinia*. The text and plates occupy fasc. 1–25; fasc. 26 (1907) includes a preface and a complete index to botanical names. [Never completed.][143]

Pteridophytes

There has so far been little coverage of pteridophytes since their publication in the *Flore générale*. Polypodiaceae have been documented in VU NGUYÊN TU, 1981. (Conspectus familiae Polypodiaceae Bercht. et J. Presl florae Vietnami.) *Nov. Sist. Vyss̆. Rast.* **18**: 5–50. Tu later also published 'A preliminary study of pteridophytes in Vietnam' (1987, in *Tap Chi Sinh Vât Hoc (Biol. J.)* **9**(2): 22–27). This latter accounts for all 132 known genera; comparisons are made with the account by Tardieu-Blot and Christensen in the *Flore générale* (121 genera and, for Vietnam, 669 species).

893

Cambodia (Kampuchea)

See also **891** (all works); **892** (PIERRE). – Area: 181 940 km². Vascular flora: no data available (PD; see **General references**).

Serious botanical exploration of Cambodia began in the 1860s, in particular through the work of Louis Pierre who had become the first director of the botanical garden in Saigon in nearby Cochinchina (southern Vietnam). Subsequent collecting, for example by Auguste Chevalier and his associate Poilane, was largely connected with activities in Vietnam until the end of French rule, and then under autochthonous (especially the University of Cambodia) and outside auspices until 1970.

The only known separate account, briefly described below, is of timber trees.

Woody plants (including forest trees)

BÉJAUD, M. and M. L. CONRARD, 1932. *Essences forestières du Cambodge*. 4 vols. 830 pls. Phnom Penh: Service Forestier, Cambodge. (Possibly mimeographed.)

A dendrological work with botanical, ecological and wood descriptions and notes on properties, uses and potential. Volume 1 comprises the text (484 pp.); vols. 2–4 the plates. [Not seen; bibliographic information and précis partly supplied by J. Vidal.][144]

894

Laos

See also **891** (all works). – Area: 236 725 km². Vascular flora: no data available (PD; see **General references**).

The initial botanical penetration of Laos took place as part of the French explorations of the Mekong referred to under Region 89. Thereafter, collecting progressed in tandem with the rest of French Indochina, but evidently at a relatively low level. Among the more extensive recent collections are those of Jules Vidal, active in the country from 1948 to 1954. Nevertheless, considerable parts remain poorly explored botanically, and no separate general floras are available. A useful introduction to the plant life of the area, however, is J. VIDAL, 1956–60. *La végétation du Laos*. 2 parts. 120, 462 pp., 37 pls., 6 maps (some col.) (Trav. Lab. Forest. Toulouse V/1, vol. 1, art. 3). Toulouse.

895

Myanmar (Burma)

See also **810–40** (BRANDIS; HOOKER). – Area: 678 031 km². Vascular flora: 7000 angiosperms, including about 200 not native (Hundley and Chit Ko Ko, 1961); this figure is likely to be too low. Nath Nair (1963, 1: 17) estimated the total as over 10 000; a more recent estimate is 14 000 (Davis *et al.*, 1994–97, 3: 5; see **General references**). Myanmar reached approximately its present limits in 1885 with the extension of British rule over Upper Burma, Kachin and the Shan States. Arakan, Lower Burma, Martaban and Tenasserim had become British territory in two stages respectively in 1826 and 1852. The country until 1937 was governed as part of India; from then until independence in 1948 it was a separate British dependency.

The history of botany in Myanmar was until well into the twentieth century closely tied with that of South Asia (Superregion 81–84); details can be gleaned from the accounts of I. H. Burkill, R. Desmond and others. Leading early and mid-nineteenth century contributors included Nathaniel Wallich, Wilhelm Helfer, William Griffith, C. Parish, and particularly Sulpiz Kurz. A first checklist was written by the British-born American missionary-scholar Francis Mason as *Flora burmanica* (1851, Tavoy). But it was the interests of the teak and other forests which from the beginning drove much of the botanical exploration in the country, and would continue to do so for a century or more. Kurz, the country's leading nineteenth-century botanist, was himself in the Forest Service prior to his association with the Indian Botanic Garden. His major works appeared in the 1870s: a general enumeration (1874–77) and the still-standard *Forest flora of British Burma* (1877). After Kurz's premature death a collaborator of Mason, the resident encyclopedist William Theobald, gathered up the still-unfinished enumeration and incorporated it into the second volume of the third edition of Mason's *Burma* (1883). Although long overlooked, this is, with Kurz's *Forest flora*, the most definitive of the older works. Dietrich Brandis, later head of the Indian Forest Service, was from 1856 to 1864 also in Myanmar; from his own experience as well as that of Kurz he included the country in his *Indian trees* (1906; see **810–40**). Brandis's work made possible the first forest-botanical checklist by John Henry Lace (1912; revised edition by A. Rodger, 1923). The latter edition accounted for 2927 species, with Rodger also noting the 'slight foundations' of Burmese forest botany as then known and calling for a full botanical survey. In 1927, Department of Agriculture staff in Mandalay produced a separate checklist (*A classified list of the plants of Burma*, by A. M. Sawyer and Daw Nyun) – apparently with no reference to the works of Lace and Rodger.

Exploration of the northern border areas after 1885, activities on the part of the Botanical Survey of India, and after 1900 the work of the celebrated 'plant hunters' and sponsored expeditions would, however, add considerably – if nowhere near enough – to knowledge of the flora. Most of the earlier collections were written up in India or the United Kingdom. A key work was *On a collection of plants from Upper Burma and the Shan States* by the later Shimla botanist Henry Collett with W. B. Hemsley (1889; in *J. Linn. Soc., Bot.* **28**: 1–150). Many novelties and notes from twentieth-century collections were written up by Indian forest botanist C. E. C. Fischer and others; these mainly appeared as 'Contributions to the flora of Burma' over 18 installments (1926–40; in *Bulletin of Miscellaneous Information* (Kew)). The vast, mostly northern collections of Reginald Farrer, Frank Kingdon Ward and others made after World War I were largely also worked up in the United Kingdom, but those collected in the 1930s by W. Dickason and by Kingdon Ward on the Vernay-Cutting Expedition were studied in the United States. A report by E. D. Merrill on the latter appeared as *The Upper Burma plants collected by Capt. F. Kingdon Ward on the Vernay-Cutting expedition* (1938–39; in *Brittonia* **4**: 20–188); of the former, however, only the pteridophytes appeared as a unit (see below under the subheading).

After World War II, Kingdon Ward returned to collecting in Myanmar, making a number of trips up through 1956. His work was taken up and continued for some years through to 1962 by Jimmy Keenan with support from the Royal Botanic Garden, Edinburgh. In the forestry service, H. G. Hundley and Chit Ko Ko produced a revision of Lace's checklist in 1961; this remains the most recent of its kind. University teaching in botany was also developed and support works written; among the latter were manuals for identification of families prepared by Dewan Mohinder Nath Nair. Nath also prepared a catalogue of Burmese plants; this, however, remains unpublished.[145] Vegetation has been covered in a variety of works, recently by A. S. Rao in *Ecology and Biogeography of India* (1974) by M. S. Mani.

Since 1962, however, the country has been largely closed to outside exploration; resident activity has also greatly declined. Considerable parts remain botanically not at all or poorly known; moreover, much of the collecting done until then, particularly after World War II, has not been reported upon or only haphazardly so. Of all of tropical Asia, it 'has had the smallest proportion of its flora collected' (Prance). A revival of floristic work is unlikely without political liberalization, although in recent years some controlled biological exploration has been resumed.

Keys to families

NATH NAIR, D. M., 1962 (1963). *A numbered analytical dichotomous key to the families of Burmese flowering plants, with a glossary of botanical terms.* vi, 273 pp. Rangoon (Yangon): Rangoon University Press. Accompanied by *idem*, 1963. *The families of Burmese flowering plants.* 2 vols. 221, 196 pp. Illus. Rangoon: Department of Botany and Agriculture, Rangoon University. (Mimeographed.) [Both works are students' manuals. The *Numbered key* is an analytical key to 231 angiosperm families (pp. 20–152) preceded by instructions for use and followed (pp. 155–273) by an extensive but unillustrated glossary. The *Families* is a handbook with family accounts including descriptions, floral diagrams, indication of distribution, diagnostic features, statistics, and notes on species of economic, horticultural or evolutionary importance; indices at end of each volume. The arrangement of families follows the 1959 version of the Hutchinson system; for Burma 263 are accounted for. Bibliographies appear on pp. 206–208 of vol. 1 and pp. 180–183 of vol. 2.]

HUNDLEY, H. G. and U CHIT KO KO, 1961. *List of trees, shrubs, herbs and principal climbers, etc., recorded from Burma.* 3rd edn. xiv, 532, v pp. Rangoon: Superintendent of Government Printing and Stationery. (1st edn., 1912; 2nd edn., 1923.)

Systematically arranged list (in tabular form) with botanical and vernacular names, fairly extensive synonymy, relatively detailed indication of local range, and brief notes on habitat; lexica of vernacular names with botanical equivalents and index to generic and family names at end. The general part includes a glossary and a short bibliography.[146]

KURZ, S., 1874–77. Contributions towards a knowledge of the Burmese flora. Parts 1–4. *J. Asiat.*

Soc. Bengal **42**/2: 39–141; **44**/2: 128–190; **45**/2: 204–310; **46**/2: 49–258.

An enumeration of the seed plants of Burma, with non-dichotomous keys to genera and species together with family conspectuses; includes full synonymy (with references), general indication of range, with indication of some *exsiccatae* and localities, taxonomic commentary, and notes on habitat, phenology, etc.; no separate indices. Owing to the death of the author the work was left incomplete; eight parts in all were planned, of which the four published cover families from Ranunculaceae to Apocynaceae (Bentham and Hooker system).[147]

THEOBALD, W., 1883. Botany. In F. Mason, *Burma, its people and productions* (ed. 3, by W. Theobald), vol. 2. xv, 787 pp., pp. i–xxxvi (blank). Hertford: Stephen Austin (for the Chief Commissioner of British Burma).

Systematic enumeration of all plants and fungi (5043 species, with the arrangement of Le Maout and Decaisne being followed for flowering plants); includes 'synoptic' keys, indication of habit, diagnostic features, vernacular names, and notes. Appendices A–C (pp. 672–729) include addenda, in-depth anecdotes on certain species, notes on woods, a lexicon of vernacular names, and a glossary; indices to all botanical names follow (as well as a subject index to the work as a whole). General remarks on the flora and forest vegetation appear on pp. 1–14. [Compiled from a variety of sources, as acknowledged in the preface.][148]

Partial work: Shan States

NATH NAIR, D. M., 1960. Botanical survey of the southern Shan States. In Burma Research Society, *Burma Research Society Fiftieth Anniversary Publications*, 1: 161–418, figs. 1–12, 2 maps. [Rangoon]: Distributed by Rangoon University Press. (Reprinted separately.)

Systematic enumeration of vascular plants, with limited synonymy, localities, citation of *exsiccatae*, vernacular names, and notes on habitat, uses, etc.; indices at end. An introductory section covers physical features, climate, floristics, botanical exploration, etc. [Based largely on local field work carried out during the late 1950s.]

Woody plants

KURZ, S., 1877. *Forest flora of British Burma*. 2 vols. Calcutta: Government Printer. (Reprinted 1974, Dehra Dun: Bishen Singh Mahendra Pal Singh.)

Briefly descriptive treatment (covering about 2000 species) of woody plants, with non-dichotomous keys to genera and species; limited synonymy, with references; vernacular names; generalized indication of local range; abbreviated notes on habitat, habit, phenology, timbers, etc.; indices to all botanical and vernacular names in vol. 2. The introductory section includes remarks on physical features, geology, forests, etc., together with a synopsis of families. [The work is most useful in Tenasserim, Pegu and Arakan; coverage of the Shan States, Upper Burma and Kachin is relatively poor.]

Pteridophytes

DICKASON, F. G., 1946. The ferns of Burma. *Ohio J. Sci.* **46**: 109–142, map.

Systematic enumeration of ferns (460 species) with localities and *exsiccatae*; new records are especially indicated. The account proper is preceded by a brief introduction and key to genera and concludes with an index to all botanical names and list of references.

896

Thailand (Siam)

Area: 514 000 km^2. Vascular flora: *c.* 12 000 species, but with relatively low endemism. – The unit corresponds to the present limits of the Kingdom of Thailand.[149]

The history of botanical work to the present time is well summarized by Tem Smitinand in his chapter in *Floristic inventory of tropical countries* (1989; see **General references**) and only the salient points are given here. Until after 1900, except for parts of the Mekong Basin and in Tenasserim, Thailand remained botanically poorly known. The Danish missionary-physician to India, J. G. Koenig, was the first to collect in 1778–79, but his herbarium was lost and his *Chloris siamensis* did not appear until relatively recently. The link with Denmark in the natural sciences – active to the present day – was renewed in 1868 with a collecting trip by C. A. Feilberg; but the best-known of that country's botanists from the period before 1905 is Johannes Schmidt who undertook an intensive study of the southeastern island of Ko Chang in 1899–1900. Schmidt's florula appeared in 10 parts in *Botanisk Tidsskrift* **24**–**32** (*passim*) from 1901 to 1916. An initial national enumeration, *Liste des plantes connues du Siam* by F. N. Williams, appeared in 1904–05; this opened a long and continuing British interest. In the first four decades of this century, C. C. Hosseus and H. B. G.

Garrett collected in the north, while A. F. G. Kerr (from 1920 government botanist in the Department of Agriculture and Commerce) travelled extensively in the Kingdom and with others built up the collections on which William G. Craib's *Flora siamensis enumeratio* (1925–62) was largely based. Craib and others authored many associated contributions in the Kew *Bulletin of Miscellaneous Information* from 1911 to 1940. Botanists from Singapore also collected, mainly in the far south; their significant contribution was *The flora of lower Siam and an account of a botanical expedition to Lower Siam* by H. N. Ridley (1911; in *J. Straits Branch Roy. Asiat. Soc.* **59**: 15–234).

From 1946, much additional collecting has been carried out, with an important impetus furnished by the joint Thai–Danish botanical programme initiated in 1958 with an initial expedition and the establishment of a series of contributions (1961–69; cited below). From this time, precursory materials have also been published in the *Botany Bulletins* of the Thai Forest Department. By the late 1960s there was thought to exist a sufficient basis for a descriptive *Flora of Thailand*. Publication began in 1970 but until recent years progress has been relatively slow. Additional collecting and contributions have been carried out through some of the universities.

Bibliography

WALKER, E. H., 1952. A contribution toward a bibliography of Thai botany. *Nat. Hist. Bull. Siam Soc.* **15**: 27–88. Continued in B. HANSEN, 1973. Bibliography of Thai botany. *Ibid.*, **24**: 319–408. [Titles through 1969.]

CRAIB, W. G. *et al.* (eds.), 1925–62. *Flora siamensis enumeratio*. Vols. 1–2; 3, parts 1–3. Bangkok: The Siam Society.

Systematic enumeration of seed plants, without descriptions; includes synonymy with references and citations, vernacular names, detailed summary of local range with *exsiccatae*, general indication of extralimital range, and some taxonomic commentary; indices to generic and family names. The introductory section gives an outline of the general features of the flora, a discussion of local vernacular names, and lists of major and minor references. [Not completed; covers families from Ranunculaceae to Gesneriaceae (Bentham and Hooker system). Now partly superseded by *Flora of Thailand*.]

SMITINAND, T., K. LARSEN *et al.* (eds.), 1970– . *Flora of Thailand*. In several volumes. Illus. Bangkok: Applied Scientific Research Corporation of Thailand (later Thailand Institute of Scientific and Technological Research) (vols. 2 and 3(1), 1970–81); Forest Herbarium, Royal Forest Department (vols. 3(2–4), 4(1–2), 5 and 6(1–4), 1984–).

Briefly descriptive, well-annotated serial flora of vascular plants; includes keys to genera and species, full synonymy (with references and citations), typification, vernacular names, rather detailed indication of local range, generalized summary of extralimital distribution, and notes on habitat, ecology, phenology, special features, uses, etc.; representative illustrations; indices. The volumes are published in parts, with each containing one or more families as ready. The interpretation of local distribution is aided by maps on the back covers (of earlier issues). [As of writing (1999) the whole of volumes 2, 3 (wholly on pteridophytes) and 5, the first two parts of volume 4 (Fabaceae in part), and parts 1–4 of volume 6 have been published; other parts are in press. No systematic sequence is followed but certain volumes have been, or will be, devoted to discrete major taxa. Volume 1 is to comprise a general introduction along with a key to families.]

SUVATTI, C., 1978. *Flora of Thailand*. 2 vols. 1503 pp. Bangkok: Royal Institute.

A compilation (in both Thai and English) of notes, derived largely from literature, on listed non-vascular and vascular plants and fungi organized under species with a strong accent on special features, uses and properties; bibliography (pp. 1205–1237) and indices at end. The order is that of the Engler and Prantl system, with species continuously numbered within each of its four major divisions. [Not critical, and with many omissions.][150]

Partial works

The most extensive and useful of these is perhaps the series by Kai Larsen and collaborators, published as precursors to the *Flora of Thailand*.

LARSEN, K. *et al.*, 1961–69. Studies in the flora of Thailand. 59 parts. *Dansk Bot. Ark.* **20**: 1–275; **23**: 1–540; **27**(1): 1–107.

Comprises local revisions of a large number of families, based principally on collections by mixed Thai–Danish botanical parties, with extensive annotations and descriptions of new taxa; complete index to genera in part 59.

897

Preparis and the Coco Islands

Area: no data seen. Vascular flora: 307 species (Prain, 1892). – In contrast to the Indian-administered Andaman Islands (see below), these islands, which lie between them and the Irrawaddy Delta (in Myanmar), are controlled by the latter country. Prain's work was based on visits to the Coco Islands in 1889 and 1890, but Preparis was not then examined. The former group was found botanically to be related to the Andamans.

PRAIN, D., 1892. The vegetation of the Coco group. *J. Asiat. Soc. Bengal* **60**/2: 283–406. (Reprinted 1894 in the author's *Memoirs and memoranda, chiefly botanical.*)

Incorporates an annotated list of 358 species (of which 307 are vascular plants) with indication of local and extralimital distribution. An extensive introductory section accounts for physical features, vegetation, and characteristic plants, while following the main list are cultivated, adventive and naturalized species (pp. 367–372), a table of distribution, and a consideration of the marine and terrestrial littoral flora (all as part of a lengthy consideration of dispersal and dispersal-classes with the aim of deducing the origin of the whole flora; some 70 species were considered truly 'remnant').

898

Andaman Islands

See also **810–40** (BOTANICAL SURVEY OF INDIA; BRANDIS; HOOKER); **911** (KING and GAMBLE). – Area: 6408 km². Vascular flora (Andamans and Nicobars): 1416 flowering plant species listed by Vasudeva Rao (1986), of which 187 endemic. Kurz in 1870 had accounted for 669 species of phanerogams.[151] The Andamans are here considered also to include Narcondam and Barren Islands, which lie some way to the east; these were separately explored by David Prain in 1891 and reported on in D. PRAIN, 1893. On the flora of Narcondam and Barren Island. *J. Asiat. Soc. Bengal* **62**/2: 39–86, 2 maps. For Preparis and the Coco Islands, see **897**.

Botanical exploration of the Andamans began at the end of the eighteenth century, shortly after establishment of what is now Port Blair in South Andaman, but was for long intermittent; indeed, Port Blair itself has had a continuous existence only from 1858. In that year a penal colony was established, which operated until 1945. Prior to 1903, the main collectors were Wilhelm Helfer, Sulpiz Kurz, David Prain and Charles Gilbert Rogers, with C. E. Parkinson following in the years to 1921. The fine forests early attracted attention and were a primary reason for Kurz's visit in 1866; his collections and observations formed the basis for his *Report on the vegetation of the Andaman Islands* (1867; revised edn., 1870). In this work, which also took account of the results of Helfer's ill-fated survey of 1839, Kurz called attention to the (for India) unusually pristine flora, likening it in significance to such islands as St. Helena and the Galápagos. Forestry activities, however, began in the 1870s; in 1903 there appeared forest conservator Rogers' *Preliminary list* and, in 1923, conservator C. E. Parkinson's forest flora (both accounted for below).

Inevitably, however, without resident botanists and with numerous islands the flora remained imperfectly known. This was finally recognized in 1972 with establishment of a 'circle' of the Botanical Survey of India in Port Blair. Since then knowledge of the flora, as well as awareness of its fragility, has considerably increased. A new checklist appeared in 1986 and – too late to be included here – the Botanical Survey of India in 1999 announced the first volume of a new flora for both the Andaman and Nicobar Islands, edited by P. K. Hajra, P. S. Rao and V. Mudgal.[152]

Bibliography

SALDANHA, C. J., 1988. *A select bibliography on the Andaman and Nicobar Islands for an environmental impact assessment.* 107 pp. Bangalore: Centre for Taxonomic Studies, St. Joseph's College. Complemented by *idem*, 1990. *A select bibliography on the Andaman and Nicobar Islands for an environmental impact assessment: Appendix 1.* 18 pp. Bangalore. [The main work encompasses over 1000 items dating to the late eighteenth century in a wide range of fields. Part I (pp. 1–57) is the author listing; part II (pp. 58–107) is classified, with flora on pp. 93–99 and forestry on pp. 100–102. The appendix is of similar layout, with flora and forestry on pp. 13–14. However, although coverage is extensive there are some serious omissions.]

ROGERS, C. G. (with J. S. GAMBLE), 1903. *A preliminary list of the plants of the Andaman Islands.* ii,

51 pp. Port Blair, Andaman Islands: Chief Commissioner's Press.

Tabular systematic list of genera and species of seed plants arranged by families, with abbreviated notes on habitat and status. [Compiled for the author by Gamble from the records in *Flora of British India* and other sources, in particular Rogers' own collections. Now obsolete, but contains habit information not given in Vasudeva Rao's list.]

VASUDEVA RAO, M. K., 1986. A preliminary report on angiosperms of Andaman and Nicobar Islands. *J. Econ. Taxon. Bot.* 8(1): 107–184.

The main part of this work is a systematically arranged tabular checklist with indication of presence or absence in the islands and for mainland India and beyond India; literature records as well as collections have been alike accounted for. Two further lists follow, one for non-indigenous species (names only, with limited synonymy) and the other covering doubtful taxa (mainly from Helfer's imperfectly localized collections, at least some of which actually were from Tenasserim and the Mergui Archipelago in Myanmar). The introductory part encompasses physical features, geology, soil, climate, etc. along with the basis for and plan of the list, while an analysis and (pp. 173–184) an annotated bibliography conclude the work.

Woody plants

PARKINSON, C. E., 1923. *A forest flora of the Andaman Islands.* v, v, xiii, 325 pp., frontisp., 6 pls. Simla: Superintendent, Government Central Press. (Reprinted 1972, Dehra Dun: Bishen Singh Mahendra Pal Singh; 1984, Dehra Dun: R. P. Singh Gahlot for International Book Distributors.)

Descriptive field-manual of woody plants, with non-dichotomous, partly artificial keys to all taxa; literature citations; fairly detailed indication of local range; vernacular names; some taxonomic commentary; notes on habitat, special features, phenology, uses, etc.; index to all botanical and vernacular names. The introductory section includes remarks on physical features, climate, vegetation and forests, and botanical exploration as well as a list of major references and a synopsis of families; an appendix incorporates lists of plants in special categories.

899

Nicobar Islands

See also **810–40** (BOTANICAL SURVEY OF INDIA; BRANDIS; HOOKER); **898** (VASUDEVA RAO); **911** (KING and GAMBLE). – Area: 1841 km². Vascular flora: 624 species (Kurz, 1876); this is surely too low. The group includes the islands from Car Nicobar south to Little and Great Nicobar. Its flora shows more affinities with Malesia (especially Sumatra) than does that of the Andamans.

The Nicobars were initially claimed for Denmark in 1745 and thus became a primary objective of that country's 'Galathea' expedition of 1845–47. In early 1846 that party became the first to undertake serious exploration in the group. In 1869 the islands passed to British India and soon afterwards Sulpiz Kurz undertook a botanical survey. His results were embodied in his *Sketch* (1876; see below), still the most recent separate account. Subsequent collecting was very patchy until establishment in 1972 of the Andaman and Nicobar Circle of the Botanical Survey of India (see **898** above). Since then, considerable additional information has become available, but no synthesis is yet available apart from Vasudeva Rao's joint checklist (see **898**). However, a flora of Great Nicobar Island by B. K. Sinha was announced in 1999, but too late to be included here.

Bibliography
See **898** (SALDANHA).

KURZ, S., 1876. A sketch of the vegetation of the Nicobar Islands. *J. Asiat. Soc. Bengal* 45/2: 105–164, pls. 12–13.

Includes a systematic enumeration of vascular plants (624 species) with descriptions of new taxa, limited synonymy, general indication of local range, citation of some *exsiccatae* with localities, taxonomic commentary, and notes on habitat, frequency, etc.; no separate index. An introduction incorporates remarks on the geology, vegetation and forests of the islands.

Notes

1 Fyson's flora was designed and written above all for those who made the annual 'migration' to the well-known and popular 'hill stations' of southern peninsular India as named in the title. The original version contains some technical material, including citations of *exsiccatae*, omitted from its successor.

2 This *Atlas* is in the present author's opinion truly significant among works of its kind. It is indicative of what can be achieved from long-term, consistently supported, and multi-faceted biodiversity research programmes, in contrast to the results of too many projects of the last two decades. It is indeed fortunate that it covers an area as biogeographically important (and under serious threat) as the Western Ghats. A logical further step – acknowledged in a preface by François Houllier – is the creation of a full floristic and ecological information system; such would be comparable to 'FloraBase' in Western Australia (see Chapter 3).

3 An English version of this work, with a selection of 100 species and botanical and cultivation notes, appeared as *Alpine flowers of Japan* (1938, Tokyo). It includes an ecological introduction.

4 Takeda's *Alpine flora of Japan in color* strongly resembles the Hegi–Merxmüller *Alpenflora* (**603/IV**) and was based on many earlier publications by the author and others. The work of Shimizu is a manual, similar to others in the Hoikusha series (see **851**).

5 This work encompasses the Changdu district in Xizang (**768**), the Aba, Garzê and Liangshan districts in Sichuan (**877**), and the Lijiang, Dêqên, Nujiang and Dali districts in Yunnan (**881**). It centers on the 'Tibetan Marches' (**878**) of the present work, with an eastern limit just east of Ta-tsien-lu.

6 The preface contains pertinent remarks on the value of floras.

7 For its first volume, Drury's work drew much from Wight and Arnott's *Prodromus*; the second and third effectively covered new ground, requiring 'a far greater amount of labour' (from the preface to vol. 2).

8 Hooker to J. F. Duthie, 24 January 1901; in L. Huxley, *Life and Letters of Sir Joseph Dalton Hooker*, 2: 387 (1918), and R. Desmond, *Sir Joseph Dalton Hooker: traveller and plant collector*, p. 265 (1999).

9 The regional floras were, however, officially undertakings of the various provincial governments and not directly sponsored by the Botanical Survey. While proposals and advice could, and did, emanate from outside, formal agreements were drawn up between authors and responsible provincial departments. See, for example, D.

D. Sundararaj, 1958. A few notes on the preparation and publication of Gamble's *Flora of the Presidency of Madras. J. Bombay Nat. Hist. Soc.* **55**: 238–242.

10 Possible events contributing to the initial reduction of the Survey were the completion of *Flora of British India*, financial constraints related to the Boer War (1899–1902), changing botanical fashions, and a lack of real commitment to the enterprise. Few new expeditions were undertaken and with abolition of specific grants from 1901 contributions by provincial staff also declined as posts were abolished or their holders moved to other duties. With a limited headquarters staff sanctioned in 1910 it nevertheless long remained the only all-Indian government botanical organization, the Indian Botanic Garden and its herbarium remaining (respectively until 1963 and 1957) the responsibility *de jure* of the Government of Bengal. In 1911 the Industrial Section of the Indian Museum, responsible for economic botany, was fully absorbed into the Survey and after 1914 the cinchona bark industry, previously under the Botanic Garden, became a major responsibility. Probably in consequence of the India Act of 1935, which granted greater autonomy to provinces, the directorship of the Survey was separated from the post of Superintendent of the Royal Botanic Garden with effect from 1937; it was, however, not filled until after 1947 and in the interim the Survey had but two staff! See N. L. Bor in *Proc. Linn. Soc.* **160**: 64–66 (1948); K. Biswas, *Thirtieth Annual General Meeting, Botanical Society of Bengal: Presidental Address.* 23 pp. Calcutta, n.d. [1949]; E. K. Janaki Ammal, 1954. The Botanical Survey of India – a retrospect. *Sci. Cult.* **19**: 322–328.

11 From this acquisition and their own continuing activities from late in the nineteenth century has developed the second largest herbarium in India, after Sibpur. Also active through most of the twentieth century has been the Blatter Herbarium in Bombay.

12 As observed by R. E. Holttum; see K. M. Matthew, 1989. Plant taxonomy in India: present state and future tasks. *J. Swamy Bot. Club* **6**(1–2): 27–32.

13 Initial ideas for this reorganization had been put forward as early as 1919; however, its implementation may have drawn on the example set by the Soviet Union. The first director following the reorganization was Fr. Santapau. Biswas, *op.cit.*; J. C. Sen Gupta, 1959. Botanical Survey of India: its past, present and future. *Bull. Bot. Surv. India* **1**(1): 9–29.

14 Although Fr. Santapau had proposed greater involvement by the BSI at district level, in these years district floras were largely the province of interested groups in tertiary institutions. See H. Santapau, *Mem. Indian Bot. Soc.* **1**: 117–121 (1958) and K. Subramanyam and M. P. Nayar, *Bull. Bot. Surv. India* **13**: 147–151 (1971).

15 S. K. Jain in the first installment of *Fascicles of the Flora of India* (1978); see *Fl. Males. Bull.* **32**: 3203 (1979).

16 Work by universities, whose botanists have contributed several district floras, has also been encompassed by the BSI programme. Allocation of a district flora project to a university circle has been based on needs, interests and the presence of an institution in or near the given area. Financial support was also made available. Publication is either independent or through BSI. See Anonymous, 1984. *Botanical Survey of India*. 63 pp. Howrah.

17 Some family accounts, such as a treatment of Papaveraceae by H. S. Debnath and M. P. Nayar (1986), have appeared separately. Progress is being documented in the *Reports* of the Botanical Survey.

18 K. M. Matthew, personal communication, 1995. The strong orientation towards floristics in the Survey and related programmes has allowed little opportunity for original work.

19 Much assistance in boundary delimitation has been gained from J. E. Schwartsberg (ed.), 1978. *A historical atlas of South Asia*. Chicago: University of Chicago Press.

20 For more on *Hortus malabaricus* see K. S. Manilal, 1980. *Botany and history of* Hortus malabaricus. Balkema. The work of clergymen and missionaries is reviewed in R. R. Stewart, 1982. Missionaries and clergymen as botanists in India and Pakistan. *Taxon* **31**: 57–64.

21 *Taxon* **28**: 168 (1979).

22 Fascicle 20 (1990) contains at its end a list of families or parts thereof (31 in all) published in fascicles 1–19. A 21st fascicle appeared in 1995 (devoted to the Indigoferae in Fabaceae) and a 22nd in 1996 (of which the largest taxon among six families was Campanulaceae–Campanuloideae). Preparation of this work was initially under the direction of S. K. Jain, then-director of BSI, and after his retirement, his successor M. P. Nayar. The 20th fascicle seems to have been long delayed in publication, as no citations after 1982 are given.

23 Four lists of additions and corrections to *Flora of British India* have been produced under the Botanical Survey of India, viz.: C. C. Calder *et al.*, 1926. *List of species and genera of Indian phanerogams not included in Sir J. D. Hooker's 'Flora of British India'*. ii, 157 pp. (Rec. Bot. Surv. India 11(1)). Calcutta. (Reprinted 1978, Dehra Dun); B. A. Razi, 1959. *A second list of species and genera of Indian phanerogams not included in J. D. Hooker's 'Flora of British India'*. ii, 56 pp. (*ibid.*, 18(1)); M. P. Nayar and K. Ramamurthy, 1976. Third list of species and genera of Indian phanerogams not included in J. D. Hooker's *Flora of British India* (excluding Bangladesh, Burma, Ceylon, Malayan Peninsula and Pakistan). *Bull. Bot. Surv. India* **15**: 204–234; and M. P. Nayar and S. Karthikeyan, 1981. Fourth list of species and genera of Indian phanerogams not included in J. D. Hooker's *The flora of British India* (excluding Bangladesh, Burma, Sri Lanka, Malayan Peninsula and Pakistan). *Rec. Bot. Surv. India* 21(2): 129–152. Other lists are cited by NAITHANI.

24 The *Florae indicae enumeratio* forms part of a work programme instituted in 1988 (called their 'Action Plan'). It is a product of a 'National Data Base' held by BSI. Some 1376 additional species of monocotyledons have been recorded for present-day India since publication of *Flora of British India* and, similarly, 400 Asteraceae additional to the 608 recorded by Clarke and Hooker.

25 The account of Asteraceae was the work of a team at the BSI Northern Circle, Dehra Dun.

26 The need for a new handbook in place of retreads was also stressed by R. E. Holttum in his review of the Chandra and Kaur *Guide* (*Kew Bull.* **44**: 548–550 (1989)). There is also a critical but rambling review of recent taxonomic work in the subcontinent (with many reductions and other changes): C. R. Fraser-Jenkins, 1997. *'New species syndrome' in Indian pteridology and the ferns of Nepal*. x, 403, [1] pp. Dehra Dun: International Book Distributors.

27 For an alternative revision of fern genera, see the 1984 paper by C. R. Fraser-Jenkins listed under the subheading.

28 Stewart in his supplement believed that with relatively little revision Parker's *Forest flora* could effectively serve all of Pakistan as 'the best thing which has been published'.

29 The area of the 'Upper Gangetic Plain' quoted is taken from Raizada *et al.*, 'Grasses of upper Gangetic Plain', I (1961; see main text under Duthie).

30 Botanical progress was reviewed, with an extensive list of references, by N. C. Nair and V. J. Nair in *Bull. Bot. Surv. India* **19**: 25–32 (1977 (1979)). More than 280 taxa had by then been added to the flora since 1925.

31 Duthie was until retirement in 1903 based at the old East India Company garden at Saharanpur, in the northern plains north of Delhi towards Dehra Dun, and its herbarium (now at the Forest Research Institute, Dehra Dun). That he did not complete the grasses for his *Flora* is ironic: in the 1880s he published several contributions on the family including a regional checklist (1883). For a fuller account of Poöideae, see M. B. Raizada *et al.*, 1983. *Poöideae of the upper Gangetic Plain*. 123 pp., text-fig., 14 pls. Dehra Dun.

32 Floristic work has been reviewed in S. Sharma, 1981. Floristic studies in Rajasthan: in retrospect and prospect. *J. Econ. Taxon. Bot.* **1**: 55–75.

33 Puri's work is marred by many inconsistencies, according to Gupta and Bhandari (*Ann. Arid Zone* **4**: 236–238 (1965)).

34 The state of botanical knowledge was first reviewed, and suggestions for progress made, in C. E. Hewetson, 1951. Preparation of a flora for Madhya Pradesh and the central parts of the Indian Union. *J. Bombay Nat. Hist. Soc.* **50**: 431–433. Among his recommendations was the formation of a state botanical survey. [Hewetson was then Conservator of Forests in the southeastern district of Bastar.] For progress to the mid-1970s, see G. Sen Gupta, 1977. A resumé of botanical explorations and floristic studies in the Central Indian state of Madhya Pradesh. *Bull. Bot. Surv. India* **19**: 71–88; pp. 82–88 comprise a bibliography. A map prepared by the Botanical Survey of India showing areas explored in the state to 1970 appears on p. 2 in Oommachan's *Flora of Bhopal*; only Surguja (in former Chota Nagpur) and parts of Bundelkhand were then considered effectively known (more than 60 percent explored).

35 This work cannot be wholly critical; for example, Araliaceae have been omitted although species of *Schefflera* are known from Pachmarhi in the Satpura Range (indeed, one is accounted for by Mukherjee in his *Flora of Pachmarhi and Bori Reserves*) and in Shadol district (according to Panigrahi and Murti in their *Flora of Bilaspur District*) and some odoriferous *Polyscias* cultivars surely are cultivated. The absence of taxonomic commentary is also to be regretted. Nevertheless, it fills a long-standing gap in floristic coverage of India at state and regional level.

36 A figure of 1800 species (in 803 genera) is also furnished by G. L. Shah and A. R. Menon in *BSI Symposium Abstracts* (1977): 11.

37 BSI-WC on its re-formation assumed control of the Cooke-Talbot herbarium, established in 1880 and a basis for their regional floras.

38 Cooke's work is in general less concise than the other contemporary regional Indian floras.

39 The 904 monocotyledons included here are more than double the 431 recorded in the older flora.

40 The flora project was initiated in 1964 in connection with a survey of medicinal plants.

41 Rao noted that in particular the territories outside of Goa proper had botanically been almost untouched.

42 Progress in floristic studies has been reviewed by P. N. Rao, V. S. Raju and R. S. Rao, 1981. Floristic research work in Andhra Pradesh. *Bull. Bot. Surv. India* **23**: 90–95. J. Joseph (see **827**: **Kerala**) noted that at least in the Madras Herbarium at Coimbatore representation of material from the state was poor by comparison with present-day Tamil Nadu.

43 This flora shows signs of compilation, and should be used with caution.

44 There was, however, no 'hill station' here.

45 Progress to 1977 has been reviewed in B. A. Razi in *BSI Symposium Abstracts* (1977): 16.

46 This flora was prepared at a then-cost of US$75 000, met by surplus agricultural funds made available under US PL-480. It was thought to account for some 75 percent of the vascular flora of present-day Karnataka.

47 This is a little smaller than Indiana (U.S.A.) which, however, has 2265 native, naturalized and adventive species. The districts of Bidar, Gulbarga and Raichur were formerly part of the Nizam's Dominions (**824**), Bijapur was in the former Bombay Presidency, Bellary was in the Madras Presidency, and the remainder in Mysore State. Only the first three thus were not covered in the Bombay or Madras Presidency floras; however, only small areas had been extensively collected.

48 See also K. S. Manilal, 1980. *The botany of* Hortus malabaricus. Balkema.

49 J. Joseph in *BSI Symposium Abstracts* (1977): 15–16; the situation was considered comparable to that in Andhra Pradesh. Even at the long-established Madras Herbarium in Coimbatore both states were in its collections poorly represented in comparison with Tamil Nadu. Similar points were also raised in the report by C. E. Ridsdale and A. J. G. H. Kostermans on their 1975 Rijksherbarium field trip in Kerala and Tamil Nadu (*Fl. Males. Bull.* **30**: 2759–2766).

50 C. E. Ridsdale, personal communication, *c.* 1979.

51 This book regrettably suffers from many typographical errors.

52 Cf. D. Sunderaj, *J. Bombay Nat. Hist. Soc.* **55**(2): 238–242 (1958).

53 Fr. Matthew and his circle have adapted the college at Shembanganur as an environmental study center, the Anglade Institute. With its historical and modern collections, it has served as a base for their flora of the Palni Hills (see **802**).

54 Some details respecting preparation of this work are furnished by D. Sundaraj in *J. Bombay Nat. Hist. Soc.* **55**(2): 238–242 (1958).

55 This remarkable series effectively continues a tradition earlier espoused by Wight, Beddome and (in Bombay) Talbot. The author's preface emphasizes the necessity for a contemporary basis for any new floristic undertaking and the importance of monographic studies. A substantial collecting programme was undertaken, with 30 000 numbers obtained over five years. Plans exist for a future *Flora of Central Tamilnadu* combining this work and the author's Palnis flora (see **802**).

56 For some observations on botanical work in Sri Lanka in the first half of the twentieth century, including that of Willis and Alston, see J. Ewan, 1983. Louis Cutter Wheeler (1910–1980) in California and Ceylon. *Taxon* **32**: 545–548.

57 As in so many books of its period, the paper has deteriorated badly.

58 The history of the project has been briefly discussed by Fosberg in the introductory part of the first series of the *Revised handbook* (1973). Further advice has been supplied by Seymour H. Sohmer and Derek Clayton.

59 The first series appeared as B. A. Abeywickrama and M. D. Dassanayake (eds.), 1973–77. *A revised handbook to the flora of Ceylon*. 1st edn. Vol. 1, parts 1–2. Illus. Peradeniya: University of Ceylon. This was an illustrated descriptive treatment in the style of *Flora Malesiana* and inclusive of keys, synonymy, vernacular names, *exsiccatae*, distribution, critical remarks and other notes. All family accounts published therein have appeared in the second edition following further revision.

60 The work, completed by Hooker after the death of Trimen in 1896, formed part of the 'Kew Series' of British colonial floras.

61 Most of the photographs were made from freshly collected specimens.

62 See Chauhan *et al.* (1996, under **847**, **Partial works**).

63 In 1992 Chota Nagpur attracted attention owing to a desire by tribal peoples in the area for a separate state of Jharkhand, which would be formed from parts of Bihar and adjacent states (and correspond functionally to the former Eastern States Agency).

64 Preparation of this work was associated with 'Project Tiger' in the 1970s and 1980s.

65 The district florulas also formed part of the plan for a state flora. Those as yet unpublished are fully listed by Naskar in his flora of the Lower Ganges Delta.

66 Work in Assam has been reviewed by A. S. Rao in *BSI Symposium Abstracts* (1977): 9. The Assam forest herbarium at Shillong, established in 1870 and a main basis for the *Flora of Assam*, is now part of BSI-EC.

67 The Tokyo group has since 1986 published a *Newsletter of Himalayan Botany*, with 20 issues and cumulative index to 1996.

68 Three volumes of contributions in this series have been published, respectively in 1988, 1991 and 1999 (the first two as *Bull. Univ. Tokyo Mus.* 31 and 34). The first includes an enumeration of Nepalese pteridophytes by K. Iwatsuki (see **844**).

69 For a review of plant collecting in Kashmir, see G. N. Javeid, 1971. History of plant exploration in Kashmir. *Kashmir Sci.* 8: 51–64.

70 *Flora simlensis*, perhaps not surprisingly, is reminiscent of contemporaneous English county floras. Moreover, the flora itself has a temperate, Eurasian facies. Perhaps nowhere else in India was such a conjunction possible.

71 The work has long served as a students' manual for classes at the Forestry College, Dehra Dun but, although several times reprinted (most recently in 1969), it has never again been revised. Publication of the second edition occasioned an extensive, partly philosophical review by the Dehra Dun forest botanist and lecturer Robert S. Hole, still worthy of consultation (*Indian Forester* 37: 537–552 (1911)).

72 Source libraries utilized were all in Asia or Japan.

73 The *Bulletin* series encompasses other floristic works but these by and large are outside our scope.

74 This work was based mostly on herbarium specimens.

75 The Flora of Bhutan project was initially described by Grierson and Long in *Notes Roy. Bot. Gard. Edinburgh* 36: 139–141 (1978). The enumerative style of the Nepal flora was not adopted as the work had to be more 'self-contained', given a limited stock of collections and literature *in situ*. Funds have recently been secured to complete publication (David Long, personal communication, 1999).

76 Parts of Kingdon Ward's collections have yet to be fully worked up.

77 For progress to the mid-1970s, see C. L. Malhotra and P. K. Hajra in *BSI Symposium Abstracts* (1977): 10.

78 The standard of this work seems mediocre, at least in Araliaceae where not even all species are accounted for.

79 The 'Abor Expedition' of 1911–12 was a so-called 'punitive' expedition consequent to a tribal attack in the present Siang district. For a general account, see A. Hamilton, 1912. *In Abor jungles*. London.

80 Miquel also prepared an enumerative *Flora japonica* (1870), the first (and only) part of his *Catalogus musei botanici lugduno-batavi*.

81 The *Chikin-shô* shows little or no European influence but the illustrations are comparable to those in John Parkinson's works.

82 *Honzo-zufu* was again reissued in 1986 with commentary by Siro Kitamura. The 1874 edition of *Somoku-zusetsu*, in 20 volumes, was revised by Yosiwo Tanaka and Paul Savatier, the latter furnishing the Latin names; the plants were arranged according to the Linnaean classes. The edition of 1907–12 was the work of Tomitaro Makino. Only the section on herbaceous plants ever saw print; the woody plants exist only in manuscript and the sections on grasses and vascular cryptogams were never written.

83 Koyama's paper is complemented by T. Koyama, n.d. *A summary bibliography of the flora of Japan*. Unpublished typescript; copy in Kew Library.

84 For early Japanese botany, the most complete treatments in English are M. Shirai Kotaro, 1926. A brief history of botany in old Japan. In Third Pan-Pacific Science Congress, *Scientific Japan, past and present*, pp. 213–227. Tokyo, and H. H. Bartlett and H. Shohara, 1961. Japanese botany during the period of wood-block printing. *Asa Gray Bull.*, N.S. 3: 295–561 (reissued 1961, Los Angeles: Dawson's Book Shop). The growth of the

natural sciences in general is documented in M. Shirai Kotaro, 1908. *Nippon hakubutsugaku nempyo* [*A chronological table of natural history in Japan*] and *idem*, 1934. *Nihon hakubutsugaku nempyo* (with a further revision in 1943; a MS translation of this last is at the Hunt Institute for Botanical Documentation). Of particular importance were the foundations in 1877 of Tokyo University and the Tokyo (now National) Science Museum.

85 The coverage of early Japanese works in this bibliography does not match the some 1000 titles and editions from the 1730–1840 period now known. Moreover, no specialized union-catalogue of these and other works in the indigenous tradition in collections outside Japan has been published. See G. S. Daniels, 1970. *Survey report of botanical libraries of the Far East*. Pittsburgh: Hunt Institute for Botanical Documentation. (Unpublished; copy in Kew Library.)

86 This well-produced though conservative work was conceived as a successor to the 1965 English translation of Ohwi's flora.

87 The plan of this work (and its predecessors) has evidently had a considerable influence elsewhere in eastern Asia for it has, with variations, been followed elsewhere in the region.

88 The work may be compared with those of Salmon (New Zealand) and Rodríguez, Quesada and Matthei (Chile) – but the Japanese tree flora is richer! The order of plates may have been dependent upon the availability of material for illustration, as with *Endemic flora of Tasmania*.

89 For literature, see M. M. J. van Balgooy, *Plant-geography of the Pacific*, 2nd edn. (1971; see **001**).

90 I thank R. Govaerts (Kew) for making available a copy of the revised edition, and also K. Woolliams, formerly of the Waimea Arboretum, Haleiwa, Oahu, Hawaii, for a copy of the 1978 version.

91 The distribution maps are moreover inaccurate and do not necessarily reflect the current situation (R. Govaerts, personal communication, 1999). The presence of goats, invasive plants (as well as introduced pollinators), and other factors have, however, certainly contributed to serious decline in many species since settlement.

92 For historical background, see E. H. Walker, 1952. A botanical mission to Okinawa and the southern Ryukyus. *Asa Gray Bull.*, N.S. 1: 225–244.

93 Whether such a large and bulky work is appropriate for a group of relatively small islands, and then only an administratively determined portion thereof, is arguable. Extension of the work to include the whole of the Ryukyus would have been desirable but perhaps officially not possible.

94 The contributions of Palibin and other Russian scientists were surveyed in V. A. Zaičikov, 1948. Вклад русских ученых в исследование Кореи (Vklad russkikh učenykh v issledovanie Korei). *Voprosy Geogr.* 8: 37–60.

95 Cf. Y. B. Suh, C. W. Park and B. Y. Sun, 1997. Biodiversity assessment of Korean vascular plants: current status in the nation. In B. H. Lee, J. C. Choe and H. Y. Han (eds.), *Taxonomy and biodiversity in East Asia: proceedings of international conference on taxonomy and biodiversity conservation in East Asia* (Seoul, 1997), pp. 126–136. Seoul. (KIBIO (Korea Research Institute of Bioscience and Biotechnology) Series 2.)

96 K. C. Oh *in litt.*, 1966.

97 Indeed, data for the Republic of Korea in the eighth edition (1990) of *Herbaria of the World* (Index Herbariorum, I) indicates that all existing herbaria date from 1952–53. Any resources which may have survived World War II may have been lost or removed during the Korean War, a point made also by Sr. Alice Goode in the introduction to her 1956 bibliography (see main text).

98 For a detailed survey (in Korean) of the history of plant systematics in Korea (covering all groups), see Y. H. Chŏng (ed.), 1986. (*An outline of the taxonomic history of Korean plants.*) vi, 404 pp., portr., frontisp. Seoul: Academy Books. It includes a dedication to T. H. Chŏng. Vascular plants are the subject of the first chapter. A more recent overview (in English) of the state of knowledge is in Suh *et al.* (*op.cit.*).

99 This work is somewhat comparable to Sargent's *Silva of North America* (**100**) which may indeed have served as a model.

100 The *Plants in danger* figure of 30 000 species came from Yü (1979; see **Progress**). This coincides with that given by Wu and Raven in their introduction to volume 17 of *Flora of China* (1994).

101 The *Nan-fang ts'ao-mu chuang* (literally 'Plants of the southern regions') was translated into English by Hui-lin Li and published as *Nan-fang ts'ao-mu chuang: a fourth-century flora of Southeast Asia* (1979, Hong Kong: Chinese University Press).

102 The work was also very influential in Japan. A first copy arrived through Nagasaki in 1607, and an edition in Japanese appeared in 1637. In northern Europe, an edition of 1735 was owned by Linnaeus and called by him 'Caput oculus' after the last two ideographs; it is still in the library of the Linnean Society of London. A first English edition of its botanical content appeared in Shanghai in 1911 as *Chinese Materia Medica: Vegetable Kingdom*.

103 The works of Joseph Needham and James Duke do, however, represent considerable advances. Much was also accomplished by G. W. Groff while at Canton Christian College (later Lingnan University) in the early twentieth century; his files are in the Pennsylvania State University Library.

104 A thorough review of this era has been presented by Bretschneider (1898). For a bio-bibliography and anthology of that eminent Sinologist, see Hartmut Walravens (comp.), 1983. *Emil Bretschneider, russischer Gesandtschafarzt, Geograph und Erforscher der chinesischen Botanik: eine Bibliographie*. vi, 42 pp., illus. Hamburg: C. Bell. (Han-pao Tung-ya shu-chi mu-lu 22.)

105 Considerable areas of the country were, however, more or less neglected by outside botanists. In the late 1930s H. H. Hu wrote that 'a vague belief seems to exist among western botanists that there is little to be expected in botanical exploration in South-Eastern China, and that the botanical wealth of Southern China has already been exhausted by earlier botanical collections. It is left to Chinese botanists to make fresh contributions from these regions and the malaria-ridden tropical jungles of Southern Yunnan'. See H. H. Hu, 1938. Recent progress in botanical exploration in China. *J. Roy. Hort. Soc.* **63**: 381–389.

106 On the surveys and related results see Hu (1938), Li (1944) and Tsiang (1950).

107 In 1930 only six significant herbaria were in existence. However, after World War II many more were established, reaching by 1988 a total of 102 despite some amalgamations and transfers, especially in the years after 1949. By 1998, 318 had been registered, with 22 having 150 000 specimens or more. Nevertheless, national collection density remains significantly lower than in much of Europe or North America. See Kit Tan and Li Yun Chang, 1988. A guide to the location and contents of herbaria in China. *Notes Roy. Bot. Gard. Edinburgh* **45**: 471–479; and Ma Jin-shuang and Liu Quan-ru, 1998. The present situation and prospects of plant taxonomy in China. *Taxon* **47**: 67–74.

108 The existence of this journal as part of the Academy's 'stable' represents official recognition of the size and importance of the national flora and, along with *Flora reipublicae popularis sinicae*, a commitment to its documentation. Australia furnishes a contemporary parallel, where *Australian Journal of Systematic Botany* is likewise one of a suite of journals published by that country's national research organization (CSIRO).

109 The major works are listed by Ma and Liu (1998; see **Superregion 86–88, Progress**).

110 The joint *Flora of China* project includes the issue of an irregular newsletter.

111 Compilation of the original work was carried out by Hemsley at the Kew Herbarium with the assistance of a dozen collaborators and financial aid from grants. Forbes merely acted as a zealous promoter, although he himself had collected extensively in China in 1857–74 and 1877–82.

112 From the project's initiation until 1966 only three parts were published, although others were ready at the onset of the 'Great Proletarian Cultural Revolution' which forced suspension of publication for almost a decade. Some activity, however, continued in the interim, and with the fall of the 'Gang of Four' came renewal: in 1977–78 alone five parts appeared, and publication since has been rapid. Upwards of 270 botanists throughout the country have been in one way or another engaged in the project, organized in cadres under the control of senior workers and guided by a 10-man editorial committee. Its initial leaders were Sung-shu Chien and Woon-yung Chun. Notable contributors of family treatments so far have included R. C. Ching (pteridophytes), W. P. Fang (Aceraceae and others), Y. Tsiang (Apocynales), and T. T. Yü (Rosaceae). External reports on the progress of the work have been given by F. A. Stafleu in *Taxon* **23**: 198–199 (1974), Hu (1975; see **Progress** above), and most thoroughly (including a detailed list of families with actual or probable collaborators) by Bartholomew (1979; see **Progress**). As of 1999 the work is very far advanced though not yet completed (contrary to earlier projections). There will be in all 125 volumes or part-volumes, of which about 90 or so have been published. – For a time in the early 1950s, a privately supported Chinese flora project was pursued in the United States. A large card file was developed, but only a 'pilot' fascicle was ever published: Arnold Arboretum of Harvard University, 1955. *Flora of China*. Family 153, Malvaceae, by Shiu-ying Hu. Illus. Cambridge, Mass. Dr. Hu's revision of Compositae, also undertaken for this project, later appeared as S. Y. Hu, 1965–69. *The Compositae of China*. 704 pp. Taipei. (Reprinted from *Quart. J. Taiwan Mus.* **18**(1)–**22**(2) *passim*.) This latter treatment remains the main modern reference for Chinese Compositae, a *Flora* treatment not yet having been published.

113 The description of this work, not seen by the author, was compiled in 1979 from information and sample pages from vol. 1 sent by courtesy of Messrs. Eugene Wu, Cambridge, Mass., and Reuben Frodin, now of Hanover, New Hampshire. The work furnishes a useful model for an 'amplified' generic flora of a kind which might be practicable in countries with large floras and limited scientific resources.

114 The citation and description of this work, not seen by the author, are based on Walker (under **Regional bibliographies**) and on notes supplied by Dr. Joseph Needham, Cambridge.

115 The *Iconographia* was compiled largely by more junior botanists in Peking and other centers (Bartholomew, 1979; see **Progress**). It evidently also provided a significant and strongly practically oriented focus for taxonomic botany in China in the later years of the

Cultural Revolution. A successor work, *Higher plants of China*, commenced publication in 1999.

116 For numerically arranged catalogues of collections from this and later expeditions by Forrest, see L. Diels, 1912–13. Plantae chinenses forrestianae. *Ibid.*, **7**: 1–411 (first expedition, nos. 1–5099); and Royal Botanic Garden, Edinburgh, 1925–30. Plantae chinenses forrestianae. *Ibid.*, **14** (1925): 75–393 (fifth expedition, 1921–23, nos. 19334–23258); **17** (1930): 1–406 (fourth expedition, 1917–20, nos. 13599–19333).

117 The 10-year delay in publication is unexplained; but in 1963 its author was 71 years old.

118 Chinese publications, mainly of the 1950s, were reviewed by A. Baranov in *Taxon* **30**: 731–732 (1981); Kitagawa's 1979 *Neo-lineamenta* similarly in *ibid.*, **32**: 153–154 (1983). Baranov regarded Kitagawa's work as representing the most effective synthesis of the Chinese, Japanese and Russian floristic traditions in Northeast China.

119 Liu's woody flora does not give a publication date on its title-page or colophon. The bibliographies of Walker, the Chinese Botanical Society, and Chen *et al.* (see **Superregion 86–88, Supraregional bibliographies**) all give 1955 but Ma (1989; see also there) gives 1959. The colophon shows a first proof date of February 1955 and a second proof date of March 1958. Unless the latter is a misprint, Ma's argument should be accepted.

120 Description originally prepared in part with the aid of material kindly supplied by Eugene Wu, Cambridge, Mass., U.S.A.

121 The author was an 'old Manchurian hand'.

122 The Basin is north of Xian and west of the Huang Ho; to the west lies the Liu-p'an Shan and on the north is the Pai-yu Shan. Within the area is Yenan (in Shaanxi), the World War II stronghold of Mao Ze-dong and his army.

123 Vols. 1–2 will cover bryophytes, vol. 3, pteridophytes and gymnosperms, and vols. 4–11, angiosperms.

124 According to Ma and Liu (1998; see **Superregion 86–88, Progress**), there have been no funds allocated for research towards a provincial flora.

125 According to Ma and Liu (1998; see **Superregion 86–88, Progress**) no funds have ever been made available for research towards a provincial flora, let alone its publication.

126 Ma and Liu (1998; see **Superregion 86–88, Progress**) have indicated that the text for a provincial flora was completed in the late 1980s but no funds have been made available for publication.

127 Modern Ningxia does not correspond to the former Inner Mongolian province.

128 Most of the text on angiosperms is ready (O. William Borrell to the author, June 1997).

129 According to Ma and Liu (1998; see **Superregion 86–88, Progress**), research had also commenced towards a provincial flora but was not yet completed due to insufficient funds.

130 In a review in *Taxon* **31**: 601–602 (1982), W. R. Philipson indicated that the work was relatively critical when compared with *Iconographia cormophytorum sinicorum*. Fang's treatment of his specialty, Aceraceae, appears in vol. 1.

131 The *Catalogue*, although carelessly prepared (cf. E. H. M. Cox in *Plant hunting in China*, 1945), furnishes the most complete record of collections in the province by the missionary-priests active from the 1870s.

132 The work resembles, and was possibly modeled on, the Forbes and Hemsley *Index*.

133 Research towards a new descriptive flora was pursued for several years in the 1970s and 1980s (S. Thrower, personal communication) but appears now to be dormant.

134 The work was officially the first of the 'Kew Series' of colonial floras initiated by W. J. Hooker around 1860; the slightly earlier *Flora capensis* and *Flora of the British West Indian islands* were 'co-opted'.

135 Vol. 1 (revised edn.) published 1991; vol. 2, 1985; vol. 3, 1987; vol. 4, 1989; vol. 5, 1993; vol. 6, 1996. Some taxonomy and nomenclature is not up to date; for example, in the Araliaceae (vol. 4) *Schefflera kwangsiensis* is still used although already in the 1970s reduced to *S. leucantha* (Vietnam).

136 The work is in a larger format than its predecessor and with more white space, but a conservative style has prevailed in both editions. No karyotypes and little ecological or biological information are included. As in Walker's Okinawan flora (**855**) and other works of that 'school' of mid-twentieth century floristic writers on eastern Asia, the emphasis is on description, taxonomic references, geographical distribution and critical commentary. Some taxonomic and floristic additions and changes have not been accounted for (as partially admitted by the editors in their preface to volume 3).

137 The work represents the culmination of botanical studies on the island begun in the 1920s by the senior author and others. Publication after volume 2 (1965) was interrupted by the Cultural Revolution and volume 3 appeared only in 1974.

138 For general issues, see D. Heinzig, 1976. *Disputed islands in the South China Sea: Paracels, Spratlys, Pratas, Macclesfield Bank*. 46, 12 pp. Wiesbaden: Harrassowitz, and Peter Kien-hong Yu, 1988. *A study of the Pratas, Macclesfield Bank, Paracels, and Spratlys in the South China Sea*. 52 pp., maps. Taipei: Tzeng Brothers Publications (Critical issues in Asian studies monograph series).

139 Some revised statistics on the Indochina flora, reflecting changes brought about by modern systematic work in comparison with family treatments in the *Flore générale*, are given in J. E. Vidal, 1964. Endémisime végétal et systématique en Indochine. *Comptes Rend. Soc. Biogéogr.* (Paris) **41**: 153–159.

140 Phan Kê Lôc, 1998. On the systematic structure of the Vietnamese flora. In Zhang Ao-luo and Wu Su-gong (eds.), *Floristic characteristics and diversity of East Asian plants; proceedings of the first international symposium* (Kunming, 1996), pp. 120–129. Beijing: China Higher Education Press; Berlin: Springer.

141 Earlier editions of this work, limited to southern Vietnam, relied heavily on literature and whatever collections were then available in Ho Chi Minh City. For the third edition the author worked also in Paris over a period of six years. He also expanded its coverage to the whole country.

142 This work, despite its faults (mainly typographical errors and some outdated nomenclature and taxonomy) is to be commended as the only modern dendrology for Southeast Asia north of Malaysia.

143 The *Flore forestière* was abandoned on account of the author's death in 1905 and perhaps also because Henri Lecomte, by then head of the Laboratoire de Phanérogamie at the Muséum National d'Histoire Naturelle in Paris, was more interested in a full flora of Indochina. All the published and unpublished drawings are now filed in the herbarium of the Laboratoire.

144 The work is now extremely rare.

145 Information from Nath Nair, *Numbered analytical dichotomous key*, p. iii.

146 Earlier editions of this work were limited to woody plants.

147 Kurz's work was soon after absorbed into Theobald's edition of Mason's *Burma* (1883).

148 For additional information on this work as well as on Mason and Theobald, see D. J. Mabberley, 1985. William Theobald (1829–1908), unwitting reformer of botanical nomenclature? *Taxon* **34**: 152–156. Its stricter application of priority was contrary to common usage at the period; for this reason (among possible others) it was ignored and 'disappeared'.

149 K. Larsen, Exploration of the flora of Thailand. In K. Larsen and L. B. Holm-Nielsen (eds.), *Tropical botany* (1979; see **Progress** under Region 89).

150 This compilation, by a zoologist, is hardly worth the paper it is printed on.

151 An estimate of 2200 vascular species from the Andamans and Nicobars together was given by K. Thothathri and N. P. Balakrishnan in *BSI Symposium Abstracts* (1977; see **Progress** under Region 89). Vasudeva Rao (1986; see main text) has shown that only 28 percent of species are shared with the Nicobar Islands.

152 Further historical details appear in Thothathri and Balakrishnan, *op. cit.*, and M. K. Vasudeva Rao, 1983 (1984). Early contributors to the botany of Andaman and Nicobar Islands. In *Hundred years of forestry in Andamans* (see **Progress** under Region 89). It is worthy of note that 550 species – nearly a third of the vascular flora – were first recorded only after 1972.

Division 9

• • •

Greater Malesia and Oceania

The great 'Horn of Plenty', the cornucopia of our Malaysian Flora, which was opened by van Rheede and by Rumphius shall still flow for a long time. I wish great joy to those that shall have the privilege to examine its contents.

L. G. M. Baas Becking, *Flora Malesiana*, I, 4: iii (1948).

It is not merely new species and rarities about which we want to learn, but the real occurrence of widespread plants. It would be useful to draw up lists of at least 100 species . . . which botanical explorers should know in order to record their distribution.

E. J. H. Corner, *Pacific Science Information Bulletin* 24(3/4): 19 (1972).

In the end, a practical taxonomy must rest on a limited number of characters which are observable without very elaborate equipment, which is one reason why uninformed academic botanists regard it as unscientific. I hope and believe that this Flora is producing new contributions to that end.

R. E. Holttum, *Flora Malesiana*, II, 1: (17) (1982).

Progress in taxonomy and biology in general is closely linked with identification. To teach botany (and ecology in particular) in a country without a practical Flora must be a thankless task.

P. H. Davis and V. H. Heywood, *Principles of angiosperm taxonomy*: 266 (1963).

This almost entirely insular division comprises Peninsular Malaysia and all islands east and south of the Nicobar Islands, the Paracel and Spratly Islands, Botel Tobago (off the south coast of Taiwan), and the Bonin (Ogasawara) and Volcano (Kazan) Islands and north of the continent of Australia (with the Torres Strait and Ashmore and Cartier Islands), Lord Howe and Norfolk Islands, and the Kermadecs. The outer limits are marked by the Hawaiian chain and by Rapa Nui (Easter) and Sala-y-Gómez Islands. Australia, Tasmania and New Zealand are in Division 4, and the eastern Pacific islands from Guadalupe to Juan Fernandez are in Division 0 (as Region 01). The division as a whole largely corresponds to the northern half (with the inclusion of New Caledonia) of the 'Pacific region' as delimited and subdivided by van Balgooy (1971; see below).

Two superregions are recognized, based upon biogeographical, bio-historical and bibliographical criteria: Greater Malesia (Superregion 91–93) and Oceania (Superregion 94–99). The dividing line lies east and north of Papuasia (including the Solomon Islands). The individual regions are arranged roughly in a west-to-east configuration, with delimitation on practical and biogeographical criteria. In Oceania, the richest regions come first, followed by those containing most of the low islands. The Hawaiian Islands, with their high endemism and greater percentage of American elements, are placed last. The whole of the division is covered in the series of maps (accompanied by descriptive notes and copious literature citations) published in *Pacific plant areas* (Steenis and van Balgooy, 1963–93; **001**). The most thorough recent introduction to floristic plant geography in the division, with emphasis on Oceania, remains M. M. J. van Balgooy, 1971. *Plant-geography of the Pacific*. 222 pp., maps, tables (Blumea, Suppl. 6). Leiden.

General bibliographies. Bay, 1910; Blake and Atwood, 1942; Frodin, 1964, 1984; Goodale, 1879; Holden and Wycoff, 1911–14; Jackson, 1881; Pritzel,

1871–77; Rehder, 1911; Sachet and Fosberg, 1955, 1971; USDA, 1958. See also under the superregions.

General indices. BA, 1926– ; BotA, 1918–26; BC, 1879–1944; BS, 1940– ; CSP, 1800–1900; EB, 1959–98; FB, 1931– ; IBBT, 1963–69; ICSL, 1901–14; JBJ, 1873–1939; KR, 1971– ; NN, 1879–1943; RŽ, 1954– . See also under the superregions.

Conspectus

903

'Alpine' (and montane) regions

Since the initiation of extensive studies by C. G.
G. J. van Steenis in the 1930s, the mountains of Malesia
– for the greater part geologically relatively young –
have received much attention from botanists; several
significant works, including floras, have appeared. Van
Steenis's initial floristic analysis (1934–36) was followed by two major works: firstly his own *Mountain
flora of Java* (1972) and then Pieter van Royen's *Alpine
flora of New Guinea* (1979–83). Only the latter is a
proper flora in the sense of this *Guide*, but as the others
are of basic importance the opportunity has been taken
to list them here.

STEENIS, C. G. G. J. VAN, 1934–36. On the origin
of the Malaysian mountain flora, I–III. *Bull. Jard. Bot.
Buitenz.*, III, **13**: 135–262, 289–417; **14**: 56–72, 10 figs.
(incl. maps), 2 folding maps, table.

Pages 155–260 in the first part comprise a checklist of bryophyte and vascular plant genera occurring
above 1000 m in the mountains of Malesia, the
Philippines and Papuasia and possessing temperate
affinities, arranged within families. It includes more or
less detailed information on species present, distribution, localities, and altitudinal range. The remainder of
the work is phytogeographical.

Java

STEENIS, C. G. G. J. VAN (with A. HAMZAH and
Md. TOHA), 1972. *The mountain flora of Java*. ix, 90
pp., frontisp., 2 text-figs., 71 halftones, 7 col. pls.
Leiden: Brill.

Handsomely produced atlas of colored illustrations (by Hamzah and Toha) of a wide range of the
more conspicuous members of the Javanese mountain
flora, accompanied by extensive descriptive text (by van
Steenis); the latter includes botanical details, local and
extralimital range, and notes on ecology, biology,
special features, related species, etc., as well as limited
taxonomic commentary. An index is also provided. The
general part (80 pp.) covers the geography, environmental factors, and biota of the mountains as well as
vegetation, plant formations, phytogeography, geobotanical history, the 'mountain mass elevation effect'
question, flower biology, and introduced plants.

New Guinea

ROYEN, P. VAN, 1979–83. *The alpine flora of New
Guinea*. 4 vols. lxviii, 3160 pp., *c*. 993 text-figs., 218 pls.,
frontisps. Vaduz, Liechtenstein: Cramer/Gantner.

Volumes 2–4 (reproduced by offset lithography
direct from the typescript) constitute a copiously illustrated documentary vascular flora of the high-mountain
areas of New Guinea above *c*. 3000 m; included are keys
to all taxa, generally lengthy descriptions, synonymy
with references and citations, vernacular names,
typification, indication of internal and (where appropriate) extralimital range with citation of *exsiccatae*, taxonomic commentary, brief notes on habitat, altitudinal

zonation, phenology, etc., and indices in each volume (with a general index at the end of the work). Volume 2 (1979) covers gymnosperms and monocotyledons (including a lengthy treatment of the Orchidaceae); vols. 3 and 4 are devoted to the dicotyledons save for an enumeration of high-altitude ferns. Some of the treatments, including that on ferns, are (at least in part) by specialists. Volume 1 (typeset and fairly extensively illustrated) contains a miscellany of chapters (by various authors) on physical features, geology, climate, soils, general ecology, vegetation formations and communities, vegetation history, and the origin, affinities and distribution of the high-altitude flora along with chapters (by the senior author) on general geography, botanical exploration, and 'languages and native names'.[1]

Annotated checklist of hydrophytes with synonymy, references, literature citations, and indication of distribution and habitat; index. Mangroves and seagrasses are also included.

STEENIS, C. G. G. J. VAN, 1932. Die Pteridophyten und Phanerogamen der deutschen limnologischen Sunda-Expedition. *Arch. Hydrobiol. Suppl.*, II, **3**: 231–387, 36 pls., 8 figs., 4 tables.

Illustrated systematic account with descriptions of novelties and extensive discussion of special features, distribution, ecology, and biology of plants accounted for; no keys. [Like post-World War II works in the United States, it encompasses hydrophytes and helophytes. Geographically it relates mainly to Java and Sumatra.]

908

Wetlands

Until recent decades, the aquatic and marsh macrophytes of Malesia and Oceania had received relatively little attention. An early major study was that of van Steenis (1932) on the collections of the German Limnological Sunda Expedition. Subsequent to World War II, Philippine aquatics were documented in 1967 and those of Papua New Guinea in 1985.

LEACH, G. and P. OSBORNE, 1985. *Freshwater plants of Papua New Guinea*. xv, 254 pp., 59 text-figs. (incl. 4 maps), 32 col. photographs. Port Moresby: University of Papua New Guinea Press.

More or less briefly descriptive flora of hydrophytic macrophytes; includes keys, significant synonymy, local distribution (with *exsiccatae*, mainly those housed in Papua New Guinea), commentaries on habitat, taxonomy, ecology, biology and uses, and representative illustrations; glossary but no index. Family and generic headings include descriptions and significant references. The introductory portion includes background remarks, especially on habitats, practical notes, references (pp. 11–12), abbreviations and symbols, a key to families, and a synoptic checklist of included species (pp. 22–24).

MENDOZA, D. R. and R. M. DEL ROSARIO, 1967. *Philippine aquatic flowering plants and ferns*. 53 pp. (Museum Publication 1). Manila: National Museum of the Philippines.

909

Oceans, islands, reefs and the littoral

While numerous atoll florulas have been published in *Atoll Research Bulletin* and elsewhere, few if any publications have concerned themselves more generally with systematic records of higher plants of the Malesian and Pacific littoral since Schimper's *Die indomalayische Strandflora* (1891) and its supplement by Booberg (1933; for both, see **009**). A useful popularly oriented introduction, however, is W. A. WHISTLER, 1992. *Flowers of the Pacific island seashore: a guide to the littoral plants of Hawai'i, Tahiti, Samoa, Tonga, Cook Islands, Fiji, and Micronesia*. [vi], 154 pp., col. photographs. Honolulu: Isle Botanica. This work, a geographically expanded version of the author's 1980 *Coastal flowers of the tropical Pacific*, covers some 120 species from Micronesia through Fiji to southeastern Polynesia and Hawaii and moreover includes a survey of the main vegetation formations. [A new checklist for the South China Sea littoral by I. M. Turner, Xing Fu-wu and R. T. Corlett appeared in 2000 in *Raffles Bulletin of Zoology, Supplement* 8.]

Superregion 91–93

Greater Malesia

Area: over 3 000 000 km^2. Vascular flora: at least 42 000 species (*Plant diversity in Malesia*, III: 233; see **Progress** (*Flora Malesiana*) below) – more than three-fifths as many again as the 25 000 estimated by van Steenis at the launch of the flora project in the late 1940s (Van Steenis, 1947, 1948; see **Progress**). All works treating Indonesia (the former Netherlands East Indies) as a whole, including *Flora Malesiana*, are here placed under **910–30**. Works on the constituent parts of Malaysia are placed respectively under **911** and **917**; those on the Philippines appear at **925**; while floras and other works for Papua New Guinea are given at **930** and those on the Solomon Islands at **938**. The system of units and regions adopted here for Malesia is based on geographical and biological criteria; political boundaries are largely arbitrary.

The fabled East Indies or Malay Archipelago, the greatest island empire in the world, was in 1857 given the name 'Malesia' by Heinrich Zollinger. This Swiss botanist-explorer, active in the southern part of the archipelago in the 1840s and 1850s, at the same time proposed botanical subdivisions within it.[2] It is here considered to extend from the northern end of present-day Peninsular Malaysia (botanically a part of this area rather than continental Asia) and the tip of Sumatra south and east in a flood of islands, large and small, all the way to San Cristobal in the southwestern Pacific. It includes among others Bangka, Java, Borneo, the Anambas and Natunas, Sulawesi (Celebes), the Philippines, Maluku (the Moluccas or Spice Islands), New Guinea, the Bismarcks, and the Solomons. This is similar to the delimitation adopted for *Flora Malesiana* by van Steenis, except that here the whole of Papuasia is included (the Louisiades and the Solomons being essentially biogeographic extensions of New Guinea) and that the Torres Strait Islands, save those nearest New Guinea, are as a group referred to eastern Australia (Region 43).[3]

The long, illustrious history of Malesian botanical exploration and floristic writing, involving hundreds of people and many institutions over more than four centuries, has been fairly well digested by H. C. D. de Wit in his *Short history of the phytography of Malaysian vascular plants* (1949; in volume 4 of *Flora Malesiana*). That essay is further supported by *Malaysian plant collectors and collections* by M. J. van Steenis-Kruseman (1950), the first volume of the *Flora*. Both contributions are thoroughly documented, and supplements to the latter appeared in 1958 and 1974.[4] Progress and literature since 1947 have been chronicled in *Flora Malesiana Bulletin*. Only a summary of events as they relate to the production of the current standard floras can be given here.

The era of botanical 'discovery' in Malesia begins in Java with the work of Jacob Bontius but the greatest early writer was unquestionably George E. Rumpf (referred to here in its Latinized form, Rumphius), resident in the region from 1653 to 1702. The latter's *Herbarium amboinense* was prepared at Ambon in southern Maluku, then the center of the spice trade, from about 1662 to 1697, with the *Auctuarium* (supplement) by 1702; for security and other reasons, however, publication was delayed until the 1740s following editorial work by Johannes Burman at Amsterdam (the supplementary *Auctuarium* followed in 1755). Its significance for the Malesian flora, however, was apparently not effectively realized by Linnaeus, who 'bypassed' it (save for one of his theses, published in 1754 – the first of several efforts towards its interpretation), and (due to involuntary deficiencies in the plates and the almost complete lack of voucher specimens) not adequately 'interpreted' until 1917. For Maluku, in the absence of any modern treatment, it remains a 'standard' work (there are also many references to plants elsewhere in Malesia and even beyond). Other early contributions were made by Kamel (Philippines), N. Burman (*Flora indica*, 1768), and (from India) Henrik van Rheede, but of extensive botanical exploration there was little until the last third of the eighteenth century.

In 1778, there was established in present-day Jakarta the *Bataviaasch Genootschap van Kunsten en Wetenschappen*, the first local learned society, with house, library, museum (forerunner of the present Muzium Pusat), and herbarium. Early contributors included Radermacher and von Wurmb, but after their deaths in the early 1780s the bottom dropped out of Dutch botanical work. From then until 1817 the major contributions were made by outsiders, partly through the *Bataviaasch Genootschap*. Even then, the island interiors, save Java, were scarcely visited; most collecting was confined to a limited number of areas surrounding ports of call. The Spanish Malaspina expedition spent a considerable time in the Philippines in 1792, but most other contributions were by naturalists from the famous British and French exploring expeditions of the era, as well as by the North American naturalist Thomas Horsfield, resident in Java for over 16 years.

It was Horsfield, following upon the Frenchmen Deschamps and Leschenault, who in extensive travels in Java first effectively explored the interior of any part of Malesia. Although his planned *Flora javana* was not realized in full, the large collections and many illustrations of the Pennsylvanian, from 1811 attached to Stamford Raffles' 'court', were later partly worked up for *Plantae javanicae rariores* (1838–52) by Bennett and Brown and in the 1850s loaned to Miquel for his *Flora indiae batavae*. Penetration of the interiors of the Malay Peninsula and the other large islands only gradually followed, in Indonesia often only in the wake of extension of direct rule by the Dutch. Korthals made extensive collections in Sumatra and southeastern Borneo in the 1830s, Mt. Kinabalu was first ascended by Hugh Low in the 1850s, and Odoardo Beccari was the first to explore the interior of Sarawak (in the 1860s) and the mountains of New Guinea (1872 and 1875).

The reforms of Raffles, which included plans for a central botanical garden, strongly influenced the new Dutch administration which in 1816 took control from the British. Under Governor-General van der Capellen, the present Botanic Garden at Bogor was established in 1817 by Reinwardt, who had arrived with Capellen. In 1820 the Indies Natural Sciences Commission came into being; over the next 30 years it conducted, with marked loss of life, extensive exploration and land surveys in many parts of modern Indonesia. Between them, these organizations assumed most of the biological work of the *Bataviaasch Genootschap*. Very considerable botanical collections resulted, which are for the most part at Leiden and (from 1844) Bogor. Significant contributors included Reinwardt, Blume, Teysmann, Hasskarl, Kuhl, van Hasselt, Kurz, Zipelius, Korthals, Junghuhn and Zollinger. Outside Dutch territory, Jack and Wallich collected in present-day Peninsular Malaysia and Singapore in the 1820s, Cuming operated in the Philippines (and the Malay Peninsula, Singapore and Sumatra) in 1834–38 and Griffith in Malacca in

1841–42 and 1844–45. Their collections also included many from Penang, the first of the British 'Straits Settlements'. Many parts of the superregion were also visited by the great voyages of discovery, exploration and mercantile diplomacy which were mounted in the decades following Waterloo, chiefly by Britain and France but also including the United States Exploring Expedition.

The first major floristic work on the Dutch Indies to appear after 1815 was Carl Blume's *Bijdragen tot de flora van Nederlandsch Indië* (1825–27). Although mainly covering Javan plants and not properly a flora, it is fundamental to almost all later work on the superregion. Contemporary works from parts not under Dutch rule included *Flora de Filipinas* by Manuel Blanco (1837; 2nd edn., 1845; 3rd edn., with colored plates, 1877–83), written in almost total isolation and originally published under royal decree, and William Jack's *Descriptions of Malayan Plants* (1822), sadly limited by the author's early death and the loss shortly afterwards of many of his collections. This last work also encompassed southern Sumatra, Bengkulu continuing under British rule until 1824. No other key floras appeared until after 1859, save for Blume's incomplete *Enumeratio plantarum Javae* (1827–28) and the 'prestigious' *Flora Javae* (1828–51, 1858), and two florulas of Timor, respectively by Decaisne (1834) and Spanoghe/Schlechtendahl (1841).

Increasing interest in the Malesian flora and a growing xenophobia on the part of Blume, from 1829 to 1862 director of the Rijksherbarium (at first in Brussels, but due to the Belgian revolution moved to Leiden in 1830), gave rise by the mid-nineteenth century to alternative initiatives towards a first synthesis of the flora. The strongest impulse came from Friedrich A. W. Miquel, then at Amsterdam where the botanical garden had been well stocked with Malesian plants. Experience with the preparation of *Plantae junghuhnianae* (1851–57) led Miquel to the conception of an overall flora for Malesia, *Flora van Nederlandsch Indië* (*Flora indiae batavae*). Initiated in 1854 with a grant-in-aid from the Ministry of Colonies, this work, inspired also by *Flora brasiliensis* and the Candollean *Prodromus*, appeared between 1855 and 1859. It covered all Malesia (save the Philippines and eastern New Guinea) and prompted Zollinger's phytogeographical proposals. A more or less critical compilation of what was then known and available to the author, but less refined than J. D. Hooker and T. Thomson's *Flora indica* and the future 'Kew floras', this flora represents the major Dutch contribution to the great group of synthetic floras written or begun in Europe in the mid-nineteenth century. However, it was prepared without effective support from the Rijksherbarium, direct access to the collections being yet refused by Blume although by government order loans had to be made. The work was thus deprived of its most important potential foundation, and, although the key collections of Horsfield, Junghuhn and Zollinger, as well as those held by Miquel himself (at Utrecht), Teysmann (at Bogor), and others were utilized freely, it perforce had to rely heavily on published sources. It also lacks keys. The enthusiastic Miquel was as much concerned with speed and service as with painstaking accuracy. Never a visitor to the tropics, the author only gradually became aware of how incomplete and imperfect his work would be, a defect particularly noticeable in the flora's first supplement, the uncritical *Prodromus florae sumatranae* (1860–61).

With Miquel's accession to the directorship of the Rijksherbarium in 1862, however, the many unstudied collections which came to light there were critically analyzed and written up in a series of studies in the four-volume *Annales Musei Botanici Lugduno-Batavi* (1863–69), which with the *Prodromus florae sumatranae* may be considered an extension of the original *Flora*. Even so, all these works are now known to have covered rather less than a third of the presently known vascular flora of the Malesian superregion. The task of general compilation had become too great for one man: even in the *Annales* Miquel was assisted by others, among them one of his two systematics students and only effective successor, Rudolph Scheffer. Apart from Boerlage's *Handleiding*, no successor to Miquel's syntheses was to appear until the mid-twentieth century and the initiation of *Flora Malesiana*.

Contemporaneously with *Flora indiae batavae* there appeared a survey – published in both Dutch and German – of the plant geography of Malesia by Heinrich Zollinger (1857). This included a botanical subdivision of the archipelago which fundamentally is sound but which has long been overshadowed by the more 'showy' zoological subdivisions of Huxley (1868) and especially Wallace (1860, 1876), whose 'Line' has acquired lasting popular fame although conceptually it is imperfect.[5] For plants, however, Zollinger – not of course very aware of the peculiarities of the Papuasian flora, then almost unknown – stressed the overall unity

of the flora of Malesia as distinct from that of Australia or mainland Asia north of the Isthmus of Kra. He had earlier been the first to propose the writing of a general Malesian flora, a project which, however, he himself would never realize.

Miquel's interest in the Malesian flora continued to the end of his life, culminating in his *Illustrations de la flore de l'Archipel Indien* (1870–71, with 37 plates by Ver Huell and Kouwels). After 1871, however, there was a notable diminution in direct Dutch contributions until well beyond 1900. This was evidently due to an onset of provincialism as well as a lack of students and growing interest in other areas of botany. In spite of Miquel's final words, Malesian floristics did not again become a major activity of the Rijksherbarium until the second quarter of the twentieth century. In the archipelago, the period from 1850 (the end of the Natural Sciences Commission and the publication of Junghuhn's definitive *Java* of 1850–54) to 1880 (the beginning of Melchior Treub's directorate at Lands Plantentuin) was comparably quiescent, with little work in systematics until 1868 when Miquel's student Rudolph Scheffer became director at Bogor and revived local phytographic studies (including in 1876 a significant enumeration of the known New Guinea flora). Nevertheless, much collecting and observation were accomplished in island interiors at this time and miscellaneous reports published. Notable contributors were Johannes Teysmann and Odoardo Beccari who, apart from visits to many other islands, were the initiators of serious botanical exploration in New Guinea, by 1870 the 'last frontier'. They were shortly afterwards followed by a large number of collectors in eastern Papuasia, most of them working at the behest of Ferdinand von Mueller in Melbourne. In the Philippines, local phytography in the Philippines was from 1876 revived through the work of the Spanish forest botanist Sebastian Vidal y Soler but upon his death in 1889 work again lapsed until after 1900. Continuous resident botanical work in the Malay Peninsula also had its beginnings in the 1870s.

The period from 1880 until World War I was one of the most active in Malesian floristic exploration and phytography, associated as it was with markedly increased colonial development and economic activities. It was also the first in which work was to a large extent conducted not from metropolitan but from local institutions (e.g., Bogor, Penang, Perak, Manila, Singapore) or peripheral centers (e.g., Calcutta and

Taipei). Each large local center had its dominant personality: Treub at Bogor, Ridley at Singapore, and (from 1902) Merrill at Manila. Treub was not himself a floristic botanist like the other two but made good use of whomever became available, Herbarium Bogoriense playing a key role in his increasingly renowned 'Institut Botanique de Buitenzorg'. As had earlier been the case at Calcutta, a small professional taxonomic staff gradually came into being there, followed by Manila after 1902 and Singapore soon after. On the other hand, nearly all descriptive work on northeastern New Guinea was accomplished in Berlin, and on British New Guinea in Australia and Britain. Many organized biological survey and collecting expeditions, often part of general exploration and 'contact', were made throughout Malesia, aided by improving administrative control and communications, and by the end of the period a floristic picture much improved over that available to Zollinger had emerged, especially with regard to Borneo (where some major botanical surveys had been carried out, notably on the Nieuwenhuis expeditions and by Hubert Winkler), Sulawesi (where especially S. H. Koorders and the Sarasin cousins were active) and Papuasia (with several large expeditions active between 1900 and 1915, with more to come between 1920 and 1926). More accessible areas were not neglected: very much was accomplished in Java by Koorders and Backer (not least in their respective floristic *voorloopers*), and for Singapore and Peninsular Malaysia Ridley (1917) reviewed progress in natural history, noting that 9000 species of plants were already known – 'a very large proportion of the flora, and enough to base at least some deductions as to [its] origin and history' – but 'practically a sample collection, not a complete one'.

Most floristic work and publication in this period was area-centered, occasioned by the sheer size of the vascular flora in relation to the limited scientific manpower available. Gone were the days when one or two men could cover the whole flora of Malesia, but at the same time the modern tradition of international interdependence had not yet developed. The fashion for large, synthetic floras had by 1900 also decayed, in part to be succeeded by 'contributions'. Rapid colonial development demanded 'results' and, coupled with the then-prevailing vertical national–imperial structure of activities, evidently led to an acceptance of local work as expedient, with the consequence that, as in neighboring Australia (particularly after von Mueller's time)

and Southeast Asia, names proliferated and species concepts and critical standards in published work varied widely. There also came to be a belief that the richness of the Malesian flora was 'inexhaustible' – leading later to a retrospectively not too fanciful claim by E. D. Merrill that the total was around 45 000 species – and that every difference had to be 'optimistically' exploited in print. A large, fragmented, and ever more complex 'literature', intelligible only to specialists, was thus created.

These constraints doubtless influenced Treub in advising against a replacement for Miquel's flora, and guiding taxonomic work at Bogor into more *ad hoc* activities: studies of economic groups, the flora of Java (with special attention to West Java), descriptions of novelties, etc. Almost the only overall work of the period on phanerogams was the generic flora, *Handleiding tot de kennis der flora van Nederlandsch-Indië* (1890–1900), by Boerlage. Publication of this work, which included some lists of species but was largely based upon Bentham and Hooker's *Genera plantarum*, was aided by the Dutch colonial ministry but owing to the author's death at Ternate in North Maluku it was not completed. No successor has been published, current professional opinion among Malesian specialists being skeptical of the value of such works, and its usage is limited to those with a knowledge of Dutch. However, at the end of the period there appeared the first general treatises on pteridophytes, all by van Alderwerelt van Rosenburgh: *Malayan ferns* (1908), *Malayan fern allies* (1915), and a *Supplement* (1917). The works of both Boerlage and Alderwerelt van Rosenburgh covered Malesia as a whole, but the former omitted eastern New Guinea. On the other hand, the locally oriented, *ad hoc* activities of the various botanical centers, both within and without Malesia, resulted in a number of key area floras and related works, including some series of 'contributions'. Most are still more or less current, lacking effective successors although often of limited practical value, and are here described under their appropriate headings. Some of these were continued or, like Ridley's *Flora of the Malay Peninsula*, did not appear until after World War I, but all were based upon work initiated before 1914.

The collections made before World War I by the various botanical centers and by locally resident botanists were increasingly supplemented by large contributions from the developing forestry services (beginning in Java and the Philippines and extending later to other parts of Indonesia, the British territories in the Malay Peninsula and Borneo, and eastern New Guinea), and by visiting phytographers, among them many Central Europeans. One of the most traveled of the latter was the German Otto Warburg in 1885–89 who, in underlining the overall floristic unity of his 'Monsunia', argued for the essentially Malesian character of Papuasia.

World War I, while by and large not directly affecting Malesia, nevertheless caused marked disruptions which, accompanied by postwar economic recession and other developments, were to alter significantly the general patterns of Malesian botanical work, a process also influenced by conceptual and methodological changes in biology. In Indonesia, with stimuli from other agencies around 1917 and the appointment of W. M. Docters van Leeuwen as director at Bogor in 1918, increased emphasis was given to floristic synthesis and revisionary studies in Herbarium Bogoriense. This policy was continued until 1942, despite severe difficulties caused by the Depression of the 1930s. In addition to many family revisions, published from 1923 as a series *Contributions à l'étude de la flore des Indes Néerlandaises*, new collecting (particularly by botanical field parties) was carried out in most islands, large and small. Collecting also continued in other parts of Malesia, notably New Guinea. To these contributions of botanical centers were added those of the forestry services and of many visiting and resident individuals, amateur and professional. Much of the interwar effort in certain areas was accomplished in one way or another under American auspices, largely 'coordinated' by Merrill (firstly in Manila, later in the United States) as 'headquarters' botanist, with assistance from Bartlett on collections from Americans in Sumatra. Direct or indirect sponsorship also came from other 'metropolitan' countries. A notable feature of the period was an increase in international cooperation and division of labor, and communication through such media as the periodic Pacific Science Congresses, begun in 1920. By 1941 enough new work had been accomplished to make feasible a more definitive assessment of Malesian plant geography and floristics; this was to find various expressions including in particular van Steenis's 1934–36 series on the origin of the Malesian mountain flora and, as will be indicated below, thoughts towards a new general flora.

By contrast, new area floras from the period

between the world wars were few. While external factors doubtless played their role, the wisdom of adding further to a stock of, at best, partially critical works in a superregion then generally sketchily known botanically, and with an inadequate critical foundation – *Flora indiae batavae* being in this respect inferior to *Flora australiensis* and *Flora of British India* – came to be questioned. Merrill in 1915 considered that only the environs of Singapore, Manila, Jakarta and Bogor were by Northern Hemisphere standards sufficiently well known to enable passable local florulas to be written. Most contributions were essentially reports on places of botanical interest with annotated checklists. The only larger local works based on specific initiatives were more or less specialized: *Geïllustreerd Handboek der Javaansche Theeonkruiden* by Backer and van Slooten (1924) on tea plantation weeds, *Onkruidflora der Javasche Suikerrietgronden* by Backer (1928–34; atlas completed in 1973) covering sugar plantation weeds, *Mountain flora of Java* by van Steenis (completed by 1945 but not published until 1972), and, finally, *Wayside trees of Malaya* by Corner (1940; revised editions in 1952 and 1988). The last was a true 'field flora' and remains in regular use.

It was in the difficult years of the 1930s that the idea of a new Malesian flora gradually took root in the minds of certain botanists at Bogor and elsewhere. One of them, Hermann J. Lam, became director at Leiden in 1933 and subsequently, in developing the Rijksherbarium into a significant systematics center, brought it back into the mainstream of Malesian botany. Others included Danser, later at Groningen, and van Steenis, at Bogor until 1946. Experience with the *Contributions*, which by 1941 extended to 34 numbers with detailed coverage of some 2000 species, had suggested that they were not suitable in the wider context envisaged, being laboriously conceived and aimed mainly at specialists. A more practical 'handbook', *Flora Malesiana*, was first publicly advocated by van Steenis (1938). After considerable discussions as to its format and geographical scope – particularly whether it would be restricted to the Dutch possessions or extended to the whole region – it was officially adopted as a project in 1940 under L. G. M. Baas Becking, recently appointed to the Bogor directorship. However, the Japanese invasion postponed the provisional launch of the *Flora* until 1948, giving time, though, for preparation of much background material. Formal launching took place in 1950 with the creation of the Flora Malesiana Foundation, headed by van Steenis as general editor. An accompanying newsletter, *Flora Malesiana Bulletin*, begin life in 1947.

For the last five decades, *Flora Malesiana* has dominated progress in Malesian botany; by 2000 some 7720 species of non-orchidaceous spermatophytes and 1288 pteridophytes had been treated, covering about 26 percent of the vascular flora. Additional precursory studies extend these figures by 10 percent or so. At the rate of production – relatively constant from the launch of the project – prevailing at the end of the 1980s, 'completion' would, even within the original estimate of 25000 seed plant species, have not taken place until 2150; since then it has become recognized that the seed plant flora is some one-third higher. The situation as of the late 1970s was described by Kalkman and Vink (1978, 1979) and van Steenis (1979), and in 1989 – on the occasion of the first Flora Malesiana Symposium three years after the death of van Steenis – an open review of *Flora Malesiana* was conducted which has led to modifications in philosophy, management and style if not yet also an effective information system. Notable among these changes has been a move away from semi-monography (cf. Geesink, 1990). Its published historical, philosophical and bibliographical foundation nevertheless remains without peer among modern tropical floras.[6]

Some area works, mainly forest floras, have continued to be published, notably in Malaysia and Papua New Guinea. Among these may be mentioned *Flora of Malaya* (1953– , currently dormant), *Tree flora of Malaya* (1972–89), *Trees of Sabah* (1976–80, not completed) and related works on East Malaysian Dipterocarpaceae, *Handbooks of the flora of Papua New Guinea* (1978–), and *Tree flora of Sabah and Sarawak* (1995–). A flora of Mt. Kinabalu began publication in 1994, while an account of the plants of Brunei appeared early in 1997. There is also a two-volume handbook to the Singapore flora. In Indonesia, the only major area work to date is *Flora of Java* (1963–68), published in the Netherlands. Throughout the region, there are also checklists for sites of special interest such as Pulau Tioman off the east coast of the Peninsula and the intermontane Bulolo-Watut Basin in Morobe Province in Papua New Guinea.

Works for school and university use remain few. Important early steps were *Flora voor de scholen in Indonesië* of van Steenis, first published in Dutch in 1949 and in Indonesian in 1975, and *Orders and fami-*

lies of Malayan seed plants by Hsuan Keng (1969; 2nd edn., 1978). In New Guinea, Robert Johns compiled several works respectively on New Guinea pteridophytes, monocotyledons and dicotyledons (none of them, however, being as yet completed); these have been joined by florulas for Kairiru Island off the north coast (1989) and Motupore Island near Port Moresby (1996).

At the present time, the superregion is on the average sketchily to moderately well explored, with a basis large enough for a meaningful critical general flora but less complete than Central America and well behind South Asia or much of tropical Africa. Only two units have been relatively well studied: Java and Peninsular Malaysia. The least-collected major areas are western New Guinea (Irian Jaya), Sulawesi, and parts of Sumatra and the provinces of Kalimantan in Borneo. It is believed, however, that on average 98 percent of all vascular species have now been collected at least once.

Progress

The best overall introduction to the flora, vegetation, plant geography, patterns of field work, and the organization and development of Malesian botany remains M. JACOBS, 1974. Botanical panorama of the Malesian archipelago (vascular plants). In UNESCO, *Natural resources of humid tropical Asia*, pp. 263–294, illus., maps, table. Paris (Natural resources research 12). For a detailed but somewhat fragmented historical survey, see H. C. D. DE WIT, 1949. Short history of the phytography of Malaysian vascular plants. In C. G. G. J. van Steenis (ed.), *Flora Malesiana*, I, 4: lxx–clxi, illus. Supplementary material has appeared since 1947 in *Flora Malesiana Bulletin*, and for individual parts of the region some separate historical accounts also exist.[7] The state of exploration to 1950 was reviewed and suggestions for future work furnished by van Steenis, also in *Flora Malesiana* (vol. 1, 1950). Later assessments are in E. J. H. CORNER (comp.), 1972. Urgent exploration needs: Pacific floras. *Pacific Sci. Inform. Bull.* 24(3/4): 17–27, and in PRANCE (1978; see **General references**). Reviews for individual islands or other areas have also been published.[8] Much information on the work of the garden and institutes at Bogor (Buitenzorg) appears in P. HONIG and F. VERDOORN, 1945. *Science and scientists in the Netherlands Indies.* New York. Institutional organization at the end of the 1970s is described in C. KALKMAN and W. VINK, 1979. Report on a visit to centres of systematic botany in Southeast Asia, September–October 1978. 18 pp. Leiden. Area reports also exist.[9]

Flora Malesiana: For an early conception of probable natural geographical limits, see H. ZOLLINGER, 1857. Ueber den Begriff und Umfang einer 'Flora Malesiana'. *Viertejahrsschr. Naturf. Ges. Zürich* **2**: 317–349, map; also in Dutch as *idem*, 1857. Over het begrip en den omvang eener 'Flora Malesiana'. *Natuurk. Tijdschr. Ned. Ind.* **13**: 293–322. The genesis, development and advocacy of the *Flora* are documented in C. G. G. J. VAN STEENIS, 1938. Recent progress and prospects in the study of the Malaysian flora. *Chron. Bot.* **4**: 392–397; *idem*, 1947. Doel, opzet en omvang der Flora Malesiana. *Chron. Nat.* **103**: 67–70; *idem*, 1948. Introduction. *Flora Malesiana*, I, 4: v–xii; *idem*, 1949. De Flora Malesiana en haar beteekenis voor de Nederlandsche botanici. *Vakbl. Biol.* **29**: 24–33; *idem*, 1951. Flora Malesiana: present and prospects. *Taxon* **1**: 21–24. The limits as adopted are based on studies reported in C. G. G. J. VAN STEENIS, 1948. Hoofdlijnen van de plantengeografie van de Indische Archipel op grond van de verspreiding der phanerogamen-geslachten. *Tijdschr. Kon. Ned. Aardrijksk. Genootsch.* **65**: 193–208; addition of the Louisiade and Solomon Islands represents a later development. A quarter-century on, progress was documented by M. JACOBS, 1973. De Flora Malesiana en haar intellectuele bekoring. *Ibid.*, **53**: 287–289, map; C. KALKMAN and W. VINK, 1978. General information on 'Flora Malesiana'. 11 pp. [Leiden]; and C. G. G. J. VAN STEENIS, 1979. The Rijksherbarium and its contribution to the knowledge of the tropical Asiatic flora. *Blumea* **25**: 57–77. The reassessments of the project at the end of the 1980s appeared as R. GEESINK, 1990. The general progress of *Flora Malesiana*. In P. Baas, K. Kalkman and R. Geesink (eds.), *The plant diversity of Malesia*, pp. 11–16. Dordrecht; and ANONYMOUS, 1990. *The future of Flora Malesiana*. 60 pp. Leiden (Fl. Males. Bull., Special Vol. 1). Progress has also been reported in *Plant diversity in Malesia*, III (1997 (1998), Kew), especially on pp. 231–246.

General bibliographies. Bay, 1910; Blake and Atwood, 1942; Frodin, 1964, 1984; Goodale, 1879; Holden and Wycoff, 1911–14; Jackson, 1881; Pritzel, 1871–77; Rehder, 1911; Sachet and Fosberg, 1955, 1971; USDA, 1958.

Supraregional bibliographies

ALLIED GEOGRAPHICAL SECTION, SOUTH-WEST PACIFIC AREA, 1944. *An annotated bibliography of the southwest Pacific and adjacent areas.* 3 vols. Maps. N.p. (Reprinted in 1990s.) [Annotated lists, not primarily botanical. Volume 1 covers the present area of East Malaysia, Indonesia, and the Philippines; vol. 2, Papua New Guinea, the British Solomon Islands, the New Hebrides, and Micronesia; and vol. 3, Malaya, Thailand, Indochina, and parts of eastern Asia. Based on Australian library holdings.]

DEPARTMENT OF EDUCATION, JAPAN, 1942–44. *Toa kyō-ei-ken sigenkagaku bunken mokuroku* [*Bibliographic index for the study of the natural resources of the Greater East Asia Co-Prosperity Sphere*]. Vols. 1, 3, 4, 6. Tokyo. [In Japanese, apart from references in other languages; titles arranged by subject areas, but not annotated. Volume 1 (1942) covers New Guinea (botany, pp. 20–61); vol. 3 (1942), the Philippines (botany, pp. 35–155); vol. 4 (1943), the Malay Peninsula (botany, pp. 38–108); and vol. 6 (1944), the (Indonesian and Malaysian) East Indies (botany, pp. 1–206, with pp. 1–144 on phanerogams and pp. 145–157 on pteridophytes).]

STEENIS, C. G. G. J. VAN, 1955. Annotated selected bibliography. In *idem* (ed.), *Flora Malesiana*, I, 5: i–cxliv. Groningen: Noordhoff. (Reprinted separately.) [Provides detailed coverage, except for the Solomon Islands; briefly annotated. Arranged by areas as well as by families (and genera).]

General indices. BA, 1926– ; BotA, 1918–26; BC, 1879–1944; BS, 1940– ; CSP, 1800–1900; EB, 1959–98; FB, 1931– ; IBBT, 1963–69; ICSL, 1901–14; JBJ, 1873–1939; KR, 1971– ; NN, 1879–1943; RŽ, 1954– .

Supraregional indices

STEENIS, C. G. G. J. VAN *et al.* (eds.), 1948– . *Flora Malesiana Bulletin*, 1– . Leiden: Flora Malesiana Foundation. (Mimeographed; later offset-printed.) [Annual or semi-annual in different periods; contains a substantial bibliographic section classified by major plant groups with entries arranged by author and usually tersely annotated. Larger works may be separately reviewed. Covers very thoroughly a much wider area than Malesia, extending to the whole of Australasia, the Pacific, and monsoon Asia to the Indus and Japan.]

910–30

Superregion in general

Flora Malesiana, announced in 1948 with a prospectus and formally commenced in the same year, is the only modern descriptive treatment on the higher plants of the superregion. Its two current series, I (Spermatophyta except Orchidaceae) and II (Pteridophyta), have as of 2000 covered some 25 percent (7720) and 28 percent (1288) respectively of the present estimates for their respective groups of 30 500 and 4500 species. A later associated series, *Orchid Monographs*, covers to date for Malesia 200, or 3 percent, of an estimated 6500 species.[10] Revisionary studies made and in a state of readiness for the work account for another 5 percent, while other more or less recent revisions, or parts thereof, account for a further 13 percent, giving a rough total of 40 percent of the species featuring more or less modern and moderately practical coverage. The other general works on Malesia, *Flora indiae batavae* by Miquel and the *Handleiding* by Boerlage, and the 1908–17 trio of works on ferns and fern-allies by van Alderwerelt van Rosenburgh, are now largely of historico-documentary interest. For the tree flora, only one overall work is available: the somewhat forbidding *Geslachtstabellen* or generic keys by Endert (1928; 2nd edn., 1953; English edn., 1956). With respect to orchids, the only current floristic treatments relate to individual islands or land masses; Peninsular Malaysia, Java, Borneo and the Philippines are fortunate with post-1980 coverage. Keng's introductory *Orders and families of Malayan seed plants* is listed under **911**.

Dictionary

STEENIS, C. G. G. J. VAN, 1987. *Checklist of generic names in Malesian botany (spermatophytes)*. 162 pp. Leiden: Flora Malesiana Foundation. [A taxonomically validated list of accepted families and genera, with intercalated synonymy and indication of biological and nomenclatural status. An extensive general part gives 'guiding lines' and includes sources, statistics (2382 indigenous and 237 naturalized alien genera) and phytogeographic data with special reference to endemic, Australasian and Pacific elements.][11]

BOERLAGE, J. G., 1890–1900. *Handleiding tot de kennis der flora van Nederlandsch-Indië*. Vols. 1–3(1). Leiden: Brill.

Descriptive generic flora of seed plants (in style resembling the *Genera plantarum* of Bentham and Hooker); includes synoptic keys, summaries of sections within genera, and taxonomic commentary. In some instances species recorded for the region are listed at the end of the family, with brief indication of distribution. [Not completed, owing to the author's untimely death in Ternate in 1900; covers dicotyledons and gymnosperms.]

MIQUEL, F. A. W., 1855–59. *Flora van Nederlandsch Indië/Flora indiae batavae*. 3 vols. (in 4). 44 pls., 2 maps, tables. Amsterdam: van der Post.

Descriptive flora, with synoptic keys to species, of the seed plants of the former Netherlands East Indies with inclusion of those from adjacent regions such as the Malay Peninsula and the Philippines; species entries include synonymy, with references and citations, vernacular names (where known), general indication of internal range, with some citations given for less frequently encountered species; taxonomic commentary, and notes on habitat, uses, etc.; index to species at end of each volume. The appendices in volume 3 include statistics of the flora and a summary of the phytogeography of Java. For the *Supplementum*, wholly concerned with Sumatra, see **913**. [The work, to a large extent compiled – as with Paul Standley and Middle America, speed was a characteristic of Miquel – is laid out in a format resembling that of the 'Kew Series' of British colonial floras. The botanical text is in Latin, the commentary in Dutch. Species found entirely outside the Netherlands East Indies are printed in smaller type.][12]

STEENIS, C. G. G. J. VAN, C. KALKMAN and W. J. J. O. DE WILDE (eds.), 1948– . *Flora Malesiana*, I (Spermatophyta), 1– . Illus., maps. Jakarta, Groningen (later Leiden, Groningen, Alphen a/d Rijn): Noordhoff (later Wolters-Noordhoff; Sijthoff & Nordhoff); The Hague (later Dordrecht): Nijhoff; Dordrecht: Kluwer; (from 1992) Leiden: Flora Malesiana Foundation. Complemented by R. E. HOLTTUM and H. NOOTEBOOM (eds.), 1959– . *Flora Malesiana*, II (Pteridophyta), 1– . Illus., maps. Leiden (later Groningen, Alphen a/d Rijn, The Hague, Dordrecht); (from 1991) Leiden.

Comprehensive descriptive floras respectively of seed plants and pteridophytes; included are keys to genera and species, rather detailed synonymy (with references and many citations), vernacular names (with provenance), generalized indication of internal and extralimital distribution (with a number of maps), taxonomic commentary, and extensive notes on habitat, phenology, special features, ecology, phytochemistry, etc. (the latter aspects with particular emphasis at family and generic levels, as in the *Pflanzenfamilien* and *Flora Neotropica*), and numerous drawings and photographs of representative species; index to all botanical names as well as addenda and corrigenda in each volume of revisions. The introductory sections in volumes 4 and 5 of series I contain chapters on the scope and basis for the work, general botanical considerations affecting species delimitation, and the long essays 'Specific and infraspecific delimitation' by van Steenis (vol. 5, pp. clxvii–ccxxxiv) and 'Short history of the phytography of Malaysian vascular plants' by H. C. D. de Wit (vol. 4, pp. lxx–clxi) as well as the 'Annotated selected bibliography' (see above). Volume 1 in series I is largely given over to a detailed (and widely acclaimed) cyclopedia of collectors by Mme. M. J. van Steenis-Kruseman (with supplements in vols. 4 and 7). In the first volume of series II may be found a bibliography of pteridological works on Malesia published since 1934 (i.e., subsequent to the third supplement of Christensen's *Index filicum*). [In both series, one or more families are published in a fascicle, but no systematic order is followed (an index to those published appears in the cover of the latest fascicle of each series). In series I, volumes 1, 4, 5, 6, 7, 8, 9, 10, 11, 12 and 13 have as of writing (1999) been published, covering 7385 species; perhaps 40 will now be required. Volumes 2 and 3 were originally projected to cover, respectively, Malesian vegetation and phytogeography, but at present no firm plans exist for their publication; priority is being given to family revisions. In series II, of the five or six volumes now projected, volume 1 (with five fascicles) was completed in 1982 and volume 2 is in progress, with two fascicles published.][13]

Forest trees

The only work which covers more than a single island or island group is the key to genera by Endert (1928, 1953; English edn., 1956). In recent years, selected species have been treated in the PROSEA Handbooks series (especially series 5), but in the usually accepted sense these are not floras.

Given the importance of the timber industry in Malesia, national, state and provincial-level coverage in the superregion not surprisingly is extensive (if uneven). With a history of nearly eight decades of research, knowledge in Malaysia is becoming relatively well consolidated; there is a fair idea of the number and extent of tree species and several

good publications. In Papua New Guinea and the Philippines, it was partly advanced by the 1980s (though with relatively little activity in the latter in recent decades, a situation into which Papua New Guinea more recently has fallen).

The size of Indonesia and the variable state of knowledge of the flora at the time of formation of its Forest Service (in 1913), however, inevitably necessitated other approaches. S. H. Koorders made extensive surveys from 1890 onwards, mainly in Java but sometimes to other islands; with Th. Valeton he produced *Bijdragen tot de Kennis der Boomsoorten van Java* (1894–1914; atlas, 1913–18). This remains a basic work. From World War I, increasing official interest in the 'outer islands' necessitated improved knowledge of their forests and forest flora; many surveys were thus undertaken by forest officers and assistants. Under the direction of successive Service botanists F. Endert, whose key to genera (see below) first appeared in 1928, and F. H. Hildebrand, mimeographed (later offset-printed), alphabetically arranged one-line enumerations of forest tree species, or *Daftar-daftar nama pohon-pohonan*, were introduced. These were compiled from 1940 onwards (save for World War II and the years immediately following) in the botany section of the Forest Research Institute, Bogor (Lembaga Penelitian Hutan; now the Forest and Nature Conservation Research and Development Centre), and cover every residency (later province) or, in some cases, their individual divisions (*afdelingen* or *kabupaten-kabupaten*). Largely internally generated documents, they feature comparatively little control from other sources and therefore cannot in botanical terms be considered critical or reliable. They remained a charge of Hildebrand until 1954; subsequent reissues by Soewanda Among Prawira and others, though labeled as revisions, were largely unchanged save for the addition of generally good, simply executed illustrations.[14]

In the 1980s, grants through the then-Agency for Agricultural Research and Development in Indonesia enabled extensive revision and reorganization of the *Daftar-daftar nama*.[15] The number of separate lists was reduced to six, one for each major unit in Indonesia except Java (thought to be effectively covered by the *Flora of Java*). Those for Kalimantan (in three volumes) and Irian Jaya (in one) were additionally designed respectively to cover the whole of Borneo and New Guinea. The new checklists generally cover all the species in any genus containing one or more of marketable dimensions (at least 20 m tall and 35 cm diameter at breast height, the latter measure representing a reduction of 5 cm from the minimum in the earlier lists). Where good background data (recent monographs, revisions, well-named collections) are wanting, however, coverage in some families or genera (e.g., Lauraceae, *Eugenia* sensu latiss., *Garcinia*) is skeletal or absent and reference to the earlier lists may become necessary. Certain other information in those lists, such as more detailed distribution, number of collections and durability classes, has moreover been omitted in the new versions. The *Daftar-daftar nama* are therefore retained in the *Guide*. A full *Tree flora of Indonesia* remains a long-term goal of the Indonesian Forestry Department; a recent manual for Kalimantan Tengah represents a beginning.[16]

Dipterocarpaceae, a family of major importance to the timber industry, has often been documented separately. The older works of Symington and Ashton are accounted for respectively under Peninsular Malaysia (**911**) and Borneo (**917**), but for convenience I aggregate here for mention the recent series of seven *Manuals of Dipterocarps for Foresters* by M. F. Newman, P. F. Burgess and T. C. Whitmore (1995–98) published by the Royal Botanic Garden, Edinburgh, and the Center for International Forestry Research, Jakarta. They include *Singapore* (1995); *Philippines* (1996); *Borneo Island, Light Hardwoods* (1996); *Sumatra, Light Hardwoods* (1996); *Borneo Island, Medium and Heavy Hardwoods* (1998); *Sumatra, Medium and Heavy Hardwoods* (1998); and *Java to New Guinea* (1998). All feature organographies, glossaries, descriptions, indication of distribution and habitat, illustrations, maps, commentary, and references; they also are furnished with one or more 3.5 in. floppy disks. The volumes together cover all of Malesia except Peninsular Malaysia, for which Symington's manual was thought not to require a successor. [In 1999 a CD-ROM consolidating the material in all seven printed manuals, and also including the species of Peninsular Malaysia which they did not cover, was released.]

ENDERT, F. H., 1928. *Geslachtstabellen voor Nederlandsch-Indische boomsoorten naar vegetatieve kenmerken*. 242 pp. (Meded. Bosbouwproefstat. 20). Buitenzorg. (2nd edn., 1953, Bogor, revised by F. H. Hildebrand.) English edn.: *idem*, 1956. *Key to the tree genera in Indonesia*. 2nd edn., transl. R. D. Hoogland. [iii], 78, [9] pp. Canberra: Division of Land Research and Regional Survey, CSIRO, Australia. (Mimeographed.) [Comprises artificial keys, based almost entirely on bark and vegetative features, to the genera of Indonesian trees attaining at least 10 m in height, a bole of 2 m and a diameter at breast height of 40 cm index. Malaysia, New Guinea as a whole, and the Philippines are only imperfectly covered through lack of knowledge or exclusion. Effective use of these keys requires considerable practice and knowledge of characters and terminology.]

Pteridophytes
For series II of *Flora Malesiana*, see that entry above. The following works are now chiefly of scholarly and historical interest.

ALDERWERELT VAN ROSENBURGH, C. R. W. K. VAN, 1908. *Malayan ferns. Handbook to the determination of the ferns of the Malayan islands*. xl, 899, 11 pp. Batavia: Department of Agriculture, Netherlands India. Complemented by *idem*, 1915. *Malayan fern allies. Handbook to the determination of the fern allies of the Malayan islands*. xvi, 261, 1 pp. Batavia:

Department of Agriculture, Industry and Commerce; both continued as *idem*, 1917. *Malayan ferns and fern allies. Supplement 1*. 577, 73 pp. Batavia.

These three works together form a briefly descriptive comprehensive treatment, with partly dichotomous keys to species and genera, of the pteridophytes of Malesia; entries include synonymy (with references and citations), generalized indication of internal distribution, short taxonomic commentaries, and other notes. Indices to all botanical names appear at the end of each volume. [The work, although by now very out of date, remains an important reference given that series II of *Flora Malesiana* is still far from complete.]

Region 91

Western and Southern Malesia

This region comprises the Malay Peninsula, Sumatra, Palawan (with Calamian), and Borneo together with all associated island chains and groups, which together make up Western Malesia ('Sundaland' of many authors), and Java and the Lesser Sunda Islands as far as the Timor arc, which constitute Southern Malesia. Save for the inclusion of the last-named, necessary on phytogeographical grounds, the region is bounded on the east by the Asian continental shelf and its associated biogeographical line, Huxley's line (as modified by Simpson).[17]

While the region as a whole is served by *Flora Malesiana* and (to varying degrees) other works on the Malesian superregion, several works are now available for individual units. Of these, peninsular Malaysia, Singapore and Java are the best provided for, owing to their long histories as centers of development and scientific activity. Sumatra has one and Borneo two general works, neither very recent. Phytogeographic surveys exist for Nusa Tenggara (Lesser Sunda Islands) and the Anambas and Natuna Islands, a rather elderly florula is available for Timor, and there is a recent florula (with but limited circulation) for the Manggarai district in Flores; in general, however, the smaller islands and groups in the region have only fragmentary separate coverage.

Available major works on Peninsular Malaysia and Singapore include, apart from the earlier *Materials for a flora of the Malay Peninsula* by King and Gamble (1889–1936) and Ridley (1907–08) and the uncritical

Flora of the Malay Peninsula by Ridley (1922–25), the 'revised' *Flora of Malaya* (1953–71), covering ferns, grasses and orchids; the *Tree flora of Malaya* (1972–89); and the famous, unconventional *Wayside trees of Malaya* by Corner (1940; 2nd edn., 1952; 3rd edn., 1988). For student use, reference should also be made to *Orders and families of Malayan seed plants* by Keng (1969; later edns., 1978, 1983). A number of more local florulas and checklists are also available, as well as a much-appreciated work on mangrove trees widely used outside as well as inside the Peninsula.

For Java, the major works are the notorious *Exkursionsflora von Java* by Koorders (1911–12; atlas, 1913–37, not completed) and the critical but mannered *Flora of Java* by Backer and Bakhuizen van den Brink, Jr. (1963–68). The woody plants are covered in the 13-part *Bijdragen tot de kennis der boomsoorten van Java* by Koorders and Valeton (1894–1914; atlas, 1913–18). An especially attractive album is *Mountain flora of Java* by van Steenis, Hamzah and Toha (1972; see **903**). For school use, there is *Flora untuk sekolah di Indonesia* by van Steenis (1975; based on the Dutch version of 1949).

With regard to other units, Borneo features two general checklists, *A bibliographic enumeration of Bornean plants* by Merrill (1921) and the less critical *Enumeratio phanerogamarum borneanum* and its companion *Enumeratio pteridophytarum borneanum*, both by Genkei Masamune (1942, 1945). At a more local level (mainly in the Malaysian states and Brunei) are several works, notable recent additions being *Plants of Mt. Kinabalu* (1992–), *A checklist of the flowering plants and gymnosperms of Brunei Darussalam* (1996), and *Tree flora of Sabah and Sarawak* (1995–). In Sumatra coverage is much less satisfactory, with the only general work being *Prodromus florae sumatranae* by Miquel (1860–61). The virtual absence of botanical centers therein has limited the production of partial florulas. Nusa Tenggara and Timor are sketchily covered in *Prodromus florae timorensis*, a checklist by Britten *et al.* (1885), a paper on woody plants by Meijer Drees (1950), *Die Flora der Manggarai* [Flores] by Fr. Schmutz produced in the 1970s and 1980s, and a treatment of ferns by Posthumus (1944). Many 'collection reports' for areas of greater or lesser extent as well as relatively detailed phytogeographic surveys are also available.

Attention should also be drawn to the tree lists which between them cover every part of the region. The most complete of these are *Pocket check-list of timber trees*, covering Peninsular Malaysia and origi-nally compiled by Wyatt-Smith in the 1950s, and *A checklist of the trees of Sarawak* by J. A. R. Anderson (1980). Pocket lists are also available for Brunei and Sabah. In the Indonesian islands, uncritical mimeo-graphed lists in A4 format have been available in a series originally begun by the Forest Research Institute, Bogor, in 1940; most lists, each in general cor-responding to one province, have reached their third issues, generally without significant revision. Sumatra (with the Western Islands, the Riau and Lingga groups, Banka and Billiton) is covered in nine such lists, Java in three, Kalimantan (Indonesian Borneo) in five, and Nusa Tenggara (with Timor) in two. The current ver-sions are all published in *Laporan-laporanan Lembaga Penelitian Hutan* (Reports, Forest Research Institute). For many areas, they constitute the only recent floristic lists of any kind. They have been partly replaced by a new series, *Tree species lists of Indonesia*, covering in the present region Sumatra, Nusa Tenggara and Borneo.

The level of collecting somewhat directly varies with the level of organized documentation. Peninsular Malaysia (with Singapore) and Java (with Madura) are well ahead, and indeed are among the best-known areas in the tropics. Elsewhere, the overall average is low although certain areas such as North Sumatra, Sabah, parts of Sarawak, and Timor have received considerable atten-tion. The situation as of the mid-1980s has been surveyed by Ashton, van Balgooy, and de Wilde in Campbell and Hammond (1989; see **General references**).

Bibliographies. General and supraregional bibliog-raphies as for Superregion 91–93. For a useful intro-duction to the main botanical literature on Region 91, see J. GAUDET and B. C. STONE, 1968. Plant life of Malaysia. *Quart. Rev. Biol.* **43**: 306–310.

Indices. General and supraregional indices as for Superregion 91–93.

911

Peninsular Malaysia (Malaya, Tanah Melayu) and Singapore

Area: 132 124 km^2. Vascular flora (phanerogams): 8000–8500 species (Keng; see below). Ridley in the 1920s had accounted for 6766 species. – Encompasses the Malay Peninsula, Penang (Pinang) and Singapore along with smaller offshore islands including the botanically important Langkawi and Tioman groups.

Initial knowledge of Peninsular Malaysia and Singapore was achieved in relation to the formation or acquisition by the British of the various Straits Settlements (Penang, Malacca and Singapore) and sponsorship of botanical work by the East India Company. This work was furthered in mid-century in particular by William Griffith; in addition, formation of the Botanic Gardens in Singapore in 1859 furnished a new base. Serious botanical survey of the interior, however, began with the extension of British protection to the sultanates. The results of these explorations, along with earlier records, formed the basis for *Materials for a flora of the Malay Peninsula*, initiated in 1889. In the decades leading up to World War II, the Singapore Botanic Gardens (along with, from the 1920s, the Forest Research Institute of Malaya) took the lead in further exploration. A plan to shift the Gardens' herbarium to Kuala Lumpur, however, was not pursued. Both institutions have produced notable work, continuing to the present at what is now the Forest Research Institute of Malaysia (FRIM). Knowledge of the very rich flora is now relatively advanced, although specialist exploration and study continue to yield new information.

The several available works together provide relatively thorough coverage, although treatment of much of its non-woody element is not particularly recent and moreover partly uncritical. Ian Turner at Singapore produced a new Peninsular checklist in 1996–97. For an introduction to area floristics and phytogeography, see H. KENG, 1970. Size and affinities of the flora of the Malay Peninsula. *J. Trop. Geog.* 31: 43–56.

Peninsular Malaysia and Singapore are in addition to 'professional' works relatively well supplied with semi-popular offerings. Corner's *Wayside trees of Malaya* (see below) is written as far as possible in 'lay' language and is well illustrated. A complementary introduction to the shrubby, lianous and herbaceous flora is M. R. HENDERSON, 1949–54. *Malayan wild flowers*. 2 vols. Illus. Kuala Lumpur: Malayan Nature Society. (Vol. 1 originally published in 3 parts as *Malayan Nature J.* 4(3–4), 1949; 6(1), 1950; 6(2), 1951; reprinted in 1 vol., 1951; 2nd reprint, 1959. Whole work reprinted 1974 by the Society.) 1974 reprint complemented by B. C. STONE [1974]. *Malayan wild flowers: Appendix*. 27 pp. Kuala Lumpur.

Bibliographies
See **Superregion 91–93, Supraregional bibliographies** (ALLIED GEOGRAPHICAL SECTION, vol. 3; DEPARTMENT OF EDUCATION, JAPAN, vol. 4 (botany, pp. 38–108)).

Keys to families and genera
KENG, H., 1983. *Orders and families of Malayan seed plants.* 3rd edn. xli, 441 pp., 211 text-figs., frontisp. Singapore: Singapore University Press. (Reissued 1987 with revisions. 1st edn., 1969, Kuala Lumpur; 2nd edn., 1978, Singapore.) [Students' introductory manual, with analytical keys to orders, families and some genera of seed plants in Peninsular Malaysia and Singapore (and, by and large, for Western and Southern Malesia in general) accompanied by descriptive notes, illustrations, a glossary, and list of references; index. A most useful work.]

I. General and Peninsular (Semananjung) Malaysia

HOLTTUM, R. E. (vol. 3 coordinated by H. B. GILLILAND), 1953–71. *A revised flora of Malaya.* Vols. 1–3. Illus. Singapore: Government Printer. (3rd edn. of vol. 1, 1968; 2nd edn. of vol. 2, 1968.)

Well-illustrated descriptive systematic treatments incorporating keys to genera and species, synonymy (with references and principal citations), generalized indication of local and extralimital range, taxonomic commentary, and notes on habitat, occurrence, affinities, special botanical features, etc.; indices to all botanical names in each volume. In each volume is also an introductory section with chapters on general morphology, ecology, systematics, etc., of the group concerned. Of this series, only three volumes have appeared, viz.: vol. 1, *Orchids* (by Holttum); vol. 2, *Ferns* (by Holttum), and vol. 3, *Grasses* (mainly by H. B. Gilliland).[18]

KING, G. and J. S. GAMBLE (continued by A. GAGE), 1889–1915, 1936. *Materials for a flora of the Malay Peninsula.* 5 vols. (in 26 parts). Calcutta. (Reprinted with separate pagination from various issues of *J. Asiat. Soc. Bengal* 58/II–75.) Complemented by H. N. RIDLEY, 1907–08. *Materials for a flora of the Malay Peninsula: Monocotyledons.* 3 vols. Singapore: Government Printer.

These two works together comprise, in flora form, a systematic series of detailed revisions of angiosperm families occurring in Peninsular Malaysia and Singapore; included are keys to genera and species, full synonymy with references and citations, general indication of local and extralimital range (with citation of *exsiccatae* for less well-known species), and indices to all botanical names in each volume. The King and Gamble volumes, though limited to dicotyledons, also encompass the Andaman and Nicobar Islands (**898, 899**). The Urticaceae sensu latiss. and part of the Euphorbiaceae were never completed.[19]

RIDLEY, H. N., 1922–25. *Flora of the Malay Peninsula.* 5 vols. Illus. London (later Ashford, Kent): Reeve. (Reprinted 1968, Amsterdam: Asher.)

Briefly descriptive flora of vascular plants; includes non-dichotomous keys to species, synonymy with indication of some references, vernacular names, generalized summary of local and extralimital range, taxonomic commentary, notes on habitat, dispersal, etc., and indices to botanical and vernacular names. The introductory section has an account of botanical exploration, while volume 5 incorporates a supplement. [Based in large part upon the *Materials* (see preceding entry).][20]

TURNER, I. M., 1995 (1996–97). A catalogue of the vascular plants of Malaya. *Gardens Bull. Singapore* **47**(1–2): 1–346, 347–757.

Annotated enumeration with indication of sources, habit, habitat, geographical status, and distribution by states (and sometimes more limited areas). All taxa are alphabetically arranged and numerically coded. The enumeration proper is preceded by a general introduction covering background and the plan of the work and a list of accepted families and genera, while at the end are an extensive bibliography (pp. 657–702) along with indices to genera and to synonyms. [As of 1999 complete; encompasses Peninsular Malaysia and Singapore. A statistical synopsis is unfortunately wanting.]

Partial work: limestone hills

CHIN, S. C., 1977–83. The limestone hill flora of Malaya, I–IV. *Gardens Bull. Singapore* **30**: 165–219; **32**: 64–203; **35**: 137–190; **36**: 31–91.

Keyed enumeration with descriptions for the 'exclusives' or limestone-dependent taxa and notes on habitat, frequency, etc.; preceded by a general part covering environment, history of studies on the limestone flora, floristics, statistics (1216 species in 582 genera), and categories (257 species, or 21.1 percent, are 'exclusives'; other categories include 'preferred', 'indifferent', and 'strangers'). Part IV contains references (pp. 75–77) and a list of *exsiccatae* (at end).

Woody plants (including forest trees)

Mention may also be made here of *Tree flora of Pasoh Forest* by K. M. Kochummen (1997, Kuala Lumpur: Forest Research Institute of Malaysia. (Malayan Forest Rec. 44.)). This includes keys and a selection of color photographs.

CORNER, E. J. H., 1988. *Wayside trees of Malaya.* 2 vols. 260 text-figs., 236 pls. (halftones), maps (end-papers). Kuala Lumpur: Malayan Nature Society. (1st edn., 1940, Singapore: Government Printing Office; 2nd edn., 1952, with slight additions and alterations.)

Illustrated descriptive layman's treatment of more commonly encountered trees, both native and introduced, in Peninsular Malaysia and Singapore; includes (pp. 61–103) artificial keys to families (and some genera) and, within the text, keys to genera and species, Malay and English vernacular names, generalized indication of local and extralimital range (with many notes on particular occurrences), taxonomic commentary, and numerous observations on habitat, phenology, special botanical and biological features, dispersal, uses, etc.; indices to vernacular names and to families and genera at end. In the introductory section are chapters on terminology, hints to identification, tree form and growth, phenological patterns, vegetation formations, trees of local interest (arranged by state), and railway-trees; references are given on pp. 781–785 following a short appendix on Hallé-Oldeman tree models (with key).[21]

WHITMORE, T. C. and F. S. P. NG (eds.), 1972–89. *Tree flora of Malaya.* 4 vols. Illus. (Malayan Forest Rec. 26). Kuala Lumpur: Longman. Complemented by C. F. SYMINGTON 1943 (A.N. 2503). *Foresters' manual of dipterocarps.* xliii, 244 pp., 114 text-figs. (Malayan Forest Rec. 16). Kuala Lumpur. (Reprinted 1976, Kuala Lumpur: University of Malaya Press, with added halftones.)

The *Tree flora* is an illustrated, briefly descriptive treatment of native, naturalized and commonly cultivated trees, except for Dipterocarpaceae; included are keys to genera and species, limited synonymy and essential references, vernacular names, generalized indication of local and extralimital range, notes on habitat, ecology, occurrence, frequency, etc., and complete indices at the end of each volume. Family and generic headings include citations of significant taxonomic works as well as additional notes on classification, biology, timber, properties and uses, etc. An introductory section (vol. 1) includes a synopsis of tree families, a general description of a tree and its important features (for students), and easily recognized trees; however, there is no general key to tree families. As in *Flora Malesiana* and the *Revised handbook to the flora of Ceylon* a systematic sequence has not been followed. Volumes 1–2 were edited by T. C. Whitmore; vols. 3–4 by F. S. P. Ng. The associated manual on Dipterocarpaceae, in its genre long considered a classic, is a well-illustrated systematic treatment, with keys, substantial descriptions, and summary of local range.[22]

WYATT-SMITH, J., 1999. *Pocket check-list of timber trees.* 4th edn., revised by K. M. Kochummen. xiii, 367 pp., 32 figs. (24 col.) (Malayan Forest Rec. 17). Kepong. (1st edn., 1952; 2nd edn., 1965; 3rd edn., 1979.)

The first and second parts of this manual comprise tabular checklists respectively of dipterocarp and non-dipterocarp trees, arranged according to 'standard' Malay vernacular names and including botanical names, alternative vernacular names, trade and timber names, size class, and distribution and habitat. Part 3 covers vegetative characters useful for identification (with which all the illustrations are associated), while parts 4 and 5 comprise field-keys respectively for dipterocarps and non-dipterocarps (with among the latter a distinction between mangrove and estuarine trees and those from inland; there are also descriptions of non-dipterocarp genera). At the end of the work are a lexicon of botanical names with vernacular equivalents and a glossary. The color photographs depict buttress, bole and bark types.

Pteridophytes

The standard treatment has long been that by Holttum in the second volume of *Flora of Malaya* (see above). A recent, well-produced complementary work, covering 392 species and varieties of ferns alone (nearly 80 percent of those recorded) is A. G. PIGGOTT, 1988. *Ferns of Malaysia in colour.* xi, 458 pp., 1363 col. photographs. Kuala Lumpur: Tropical Press.

II. Singapore

While all works on Peninsular Malaysia to date have also covered Singapore, local interest has naturally given rise to separate accounts. The first such was *The flora of Singapore* by H. N. Ridley (1900; in *J. Straits Branch Roy. Asiat. Soc.* 33: 27–196). A 'Flora of Syonan' was compiled during World War II at the Botanic Gardens but never published. From the 1970s there was a renewal of local botanical work in the face of massive urban development. Several publications have resulted or are planned. These include a preliminary enumeration, students' manuals and a definitive descriptive flora. Preparation of the last is underway (H. Keng and Hugh Tan, personal communication).[23] The flora is now quite thoroughly documented, and its dynamics in the face of changing land use and environment can be tracked. Some 2400 native and naturalized vascular plants have been recorded for the city-state; many are now no longer present. For the preliminary enumeration, see H. KENG, 1973–87. Annotated list of seed plants of Singapore, I–XI. *Gardens Bull. Singapore* **26**: 233–237; **27**: 67–83, 247–266; **28**: 237–258; **31**:

84–113; **33**: 329–367; **35**: 83–103; **36**: 103–124; **38**: 149–174; **39**: 67–95; **40**: 113–132. A basic summary appears as I. M. TURNER, H. T. TAN and K. S. CHUA, 1990. A checklist of the native and naturalized vascular plants of the Republic of Singapore. *J. Singapore Natl. Acad. Sci.* **18/19**: 58–88. The students' manuals are described below.

KENG, H., 1990. *The concise flora of Singapore: gymnosperms and dicotyledons.* 364 pp., illus. Singapore: Singapore University Press. Continued as H. KENG, S. C. CHIN and H. T. W. TAN, 1998. *The concise flora of Singapore, II: Monocotyledons.* xix, 215 pp., 293 text-figs. Singapore. Complemented by A. JOHNSON, 1977. *The ferns of Singapore Island.* 140 pp., illus. Singapore.

The *Concise flora* is a well-illustrated, annotated enumeration of nearly 1300 native and naturalized and over 520 commonly cultivated species, the families systematically arranged; includes keys to families and genera, brief family descriptions and, for species, notes on distribution, status, occurrence, etc. *Ferns* is an illustrated descriptive account of pteridophytes with keys, habitat and occurrence, and other notes.

912

Western Islands of Sumatra

See **913** (MIQUEL; SOEWANDA and TANTRA, 1973 (Aceh), 1973 (Sumatera Utara), 1973 (Bengkulu), 1974 (Sumatera Barat); WHITMORE and TANTRA, 1986). – Area: 1845 km² (Simeuluë and associated islands); 318 km² (Banyak group); 4064 km² (Nias); 1201 km² (Batu group); 3900 km² (Siberut and Sipora); 1330 km² (the Pagais); 457 km² (Enggano and P. Mega). Vascular flora: no data.

The Western Islands, which together form a distinct chain running from Simeuluë in the north through Enggano in the south, have rarely been the subject of separate floristic works; indeed, even by the mid-twentieth century some were still botanically not or but very poorly known. Only the Mentawais feature an (incomplete) checklist, one of the results of a primarily zoological expedition to Siberut and Sipora conducted in 1924 by the then-Raffles Museum in Singapore: H. N. RIDLEY, 1926. The flora of the Mentawi Islands. *Bull. Misc. Inform.* (Kew) **1926**: 57–94 (as part of *Spolia mentawiensia*, coordinated by

C. Boden Kloss). Data on tree species appear also in the respective Bogor tree species lists for the respective mainland provinces to which the islands are administratively attached, as indicated above.

913

Sumatra (Sumatera)

Area: 435 000 km². Vascular flora: 8000–10 000 species (de Wilde in Campbell and Hammond (1989, p. 106; see **General references**)). – Sumatra here comprises the main island along with immediately adjacent smaller islands such as P. Weh (off Banda Aceh), P. Musala (off Sibolga in Tapanuli), the Krakatoa group and those off the coast of Bengkalis in Riau. For the Western Islands, see **912**; for the Riau and Lingga groups, Banka and Belitung, see **914**; for the Anambas and Natuna Islands, see **915**.

'Sumatra still wants its florist', wrote the naturalist and armchair traveler Thomas Pennant in 1800 in the fourth volume of his *Outlines of the Globe.* Miquel's *Prodromus* (see below), completed in 1861 and reflecting little more than 40 years of exploration and publication, was manifestly an imperfect work. Subsequent collection checklists and other floristic papers on Sumatra (along with the Western Islands, the Riau and Lingga groups, Banka and Belitung) are widely scattered and with one exception – the series of tree species lists produced by the Forest Research Institute at Bogor – do not fall into or comprise any distinctive series. For more specific details, therefore, reference should perforce be made to standard bibliographies and indices. Useful introductions may be found in E. D. MERRILL *et al.*, 1934. *An enumeration of plants collected in Sumatra by W. N. and C. M. Bangham.* 178 pp., 13 pls. (Contr. Arnold Arbor. 8). Jamaica Plain, Mass., and A. J. WHITTEN, S. J. DAMANIK, JAZANUL ANWAR and NAZARUDDIN HISYAM, 1987. *The ecology of Sumatra.* 2nd edn. xx, [iv], 583 pp., illus. (part col.). Yogyakarta: Gadjah Mada University Press. (The ecology of Indonesia series I.) It has to be said, however, that in general the island remains quite undercollected in spite of significant past efforts in certain areas, and organized documentation poor; Pennant's maxim still holds.

MIQUEL, F. A. W., 1860–61. *Flora van Nederlandsch Indië, eerste bijvoegsel: Sumatra, zijne planten-* *wereld en hare voortbrengselen/ Florae indiae batavae, supplementum primum: Prodromus florae sumatranae.* xxiv, 656 pp., 4 pls. Amsterdam/Utrecht: van der Post; Leipzig: Fleischer. (Reissued 1862 as *Sumatra, zijne plantenwereld en hare voortbrengselen.*) German edn. as *idem*, 1862. *Sumatra, seine Pflanzenwelt und deren Erzeugnisse.* xxi, 656 pp., 4 pls. Amsterdam/Utrecht, Leipzig.

Pages 104–276 of this work comprise a systematic enumeration of the known seed plants of Sumatra (including Bangka and the Riau Archipelago), with synonymy, references and essential citations, brief indication of local range, and vernacular names. The latter part of the work consists of descriptions and notes on new and critical Sumatran taxa not included in the original 1855–59 flora; this is followed by an additional supplement (pp. 618–626) and an index to all botanical names. The introductory section includes a list of references and a lengthy description of physical features, climate, vegetation, agriculture, etc. [Apart from botanical descriptions (in Latin) the text is respectively in Dutch or German.][24]

Woody plants (including forest trees)

The arrangement of the *Daftar nama* with respect to Sumatra is by provinces from north to south.

WHITMORE, T. C. and I. G. M. TANTRA (eds.), 1986. *Tree flora of Indonesia: check list for Sumatra.* xi, 381 pp., illus. Bogor: Forest Research and Development Centre.

Part 1 of this four-part work comprises a concisely annotated checklist, arranged alphabetically by families and genera with the names of smaller trees enclosed in parentheses; includes abbreviated citations of key literature, indication of maximum size, habitat and overall distribution, and occasional *exsiccatae.* Part 2 is a lexicon of vernacular and trade names with their botanical equivalents; the names have been taken from the *Daftar nama pohon-pohonan* and *Flora Malesiana.* Part 3 comprises keys to Dipterocarpaceae (by P. S. Ashton), with all species but one illustrated in part 4 (which also includes figures of some important non-dipterocarp species). At the beginning of part 1 are lists of families and genera with statistics.

Provincial checklists (Daftar nama)

SOEWANDA AMONG PRAWIRA, R. and I. G. M. TANTRA (comp.), 1973. *Daftar nama pohon-pohonan: Aceh.* Revised edn. Illus. (Lap. Lemb. Penilit. Hutan 179). Bogor. (Mimeographed. Originally published 1941, Buitenzorg, as F. H. HILDEBRAND (comp.), *Lijst van boomnamen van Atjeh en onderhorigheden* (Bosbouwproefstation, Serie Boomnamenlijsten 8); revised 1950, Bogor, as *idem*, *Daftar nama pohon-pohonan: Atjeh-Simalur.* [i], 70 pp. (Lap. Balai Penjelidik. Kehut. 32; also designated as Seri Daftar nama pohon-pohonan 23).)

SOEWANDA AMONG PRAWIRA, R. and I. G. M. TANTRA (comp.), 1973. *Daftar nama pohon-pohonan: Sumatera Utara.* Revised edn. 123 pp., 57 pls. (Lap. Lemb. Penilit. Hutan 171). Bogor. (Mimeographed. Preceded by separate reports for the former residencies respectively of Sumatra East Coast and Tapanuli: F. H. HILDEBRAND (comp.), 1949. *Lijst van boomsoorten verzameld in Sumatera Timur.* [iv], 50 pp. (Rapp. Bosbouwproefsta. 9; also designated as Serie Boomnamenlijsten 15). Buitenzorg, and *idem*, 1950. *Daftar nama pohon-pohonan: Tapanuli* [former Residency of Tapanuli]. [i], 43 pp. (Lap. Balai Penjelidik. Kehut. 29; also designated as Seri Daftar nama pohon-pohonan 22). Bogor; former revised 1952 as *idem, Daftar nama pohon-pohonan Sumatera-Timur.* 66 pp. (Lap. Balai Penjelidik. Kehut. 56; also designated as Seri Daftar nama pohon-pohonan 33); latter revised 1954 as *idem, Daftar nama pohon-pohonan: Tapanuli.* 50 pp. (Lap. Balai Penjelidik. Kehut. 67; also designated as Seri Daftar nama pohon-pohonan 43).)

SOEWANDA AMONG PRAWIRA, R. and I. G. M. TANTRA (comp.), 1974. *Daftar nama pohon-pohonan: Sumatera Barat.* Revised edn. 64 pp., 46 pls. (Lap. Lemb. Penelit. Hutan 187). Bogor. (Mimeographed. Originally published as F. H. HILDEBRAND (comp.), 1950. *Daftar nama pohon-pohonan: Sumatera-Barat.* [vi], 60 pp. (Lap. Balai Penjelidik. Kehut. 26; also designated as Seri Daftar nama pohon-pohonan 21). Bogor; revised 1953 as *idem, Daftar nama pohon-pohonan: Sumatera-Barat.* [i], 69 pp. (Lap. Balai Penjelidik. Kehut. 64; also designated as Seri Daftar nama pohon-pohonan 41).)

SOEWANDA AMONG PRAWIRA, R. (comp.), 1970. *Daftar nama pohon-pohonan: Riau.* Revised edn. Illus. (Lap. Lemb. Penelit. Hutan 106). Bogor. (Mimeographed. Originally published as F. H. HILDEBRAND (comp.), 1949. *Lijst van boomsoorten verzameld in Riau, Bengkalis en Indragiri.* [iv], 67 pp. (Rapp. Bosbouwproefsta. 11; also designated as Serie Boomnamenlijsten 16). Buitenzorg; revised in two numbers, 1953, Bogor, as *idem, Daftar nama pohon-pohonan: Bengkalis* [northern Riau]. 28 pp. (Lap. Balai Penjelidik. Kehut. 59; also designated as Seri Daftar nama pohon-pohonan 36), and *idem, Daftar nama pohon-pohonan: Riau dan Inderagiri* [southern Riau]. 69 pp. (Lap. Balai Penjelidik. Kehut. 61; also designated as Seri Daftar nama pohon-pohonan 38).) [The 1970 revision represents an amalgamation of the 1953 numbers.]

SOEWANDA AMONG PRAWIRA, R. and I. G. M. TANTRA (comp.), 1974. *Daftar nama pohon-pohonan: Jambi.* Revised edn. 26 pp., 29 pls. (Lap. Lemb. Penelit. Hutan 185). Bogor. (Mimeographed. Originally published 1949, Buitenzorg, as AFDELING BOSBOTANIE, *Lijst van boomsoorten*

verzameld in Djambi-Sumatra. [iv], 14 pp. (Rapp. Bosbouwproefsta. 8; also designated as Serie Boomnamenlijsten 14), revised 1954, Bogor, as BAGIAN BOTANI (comp.), *Daftar pohon-pohonan Djambi.* 24 pp. (Lap. Balai Penjelidik. Kehut. 8a; also designated as Seri Daftar nama pohon-pohonan 14a).)

SOEWANDA AMONG PRAWIRA, R. and I. G. M. TANTRA (comp.), 1973. *Daftar nama pohon-pohonan: Bengkulu.* Revised edn. 60 pp., 42 pls. (Lap. Lemb. Penelit. Hutan 159). Bogor. (Mimeographed. Originally published 1949, Buitenzorg, as F. H. HILDEBRAND (comp.), *Lijst van boomsoorten verzameld in Benkuelen.* [iv], 47 pp. (Rapp. Bosbouwproefsta. 22; also designated as Serie Boomnamenlijsten 20).)

SOEWANDA AMONG PRAWIRA, R. and I. G. M. TANTRA (comp.), 1972. *Daftar nama pohon-pohonan: Palembang (Sumatera Selatan).* Revised edn. 91, 4 pp., 60 pls. (Lap. Lemb. Penelit. Hutan 141). Bogor. (Mimeographed. Originally published 1949, Buitenzorg, as F. H. HILDEBRAND (comp.), *Lijst van boomsoorten verzameld in Palembang.* [iv], 78 pp. (Rapp. Bosbouwproefsta. 19; also designated as Serie Boomnamenlijsten 18).)

SOEWANDA AMONG PRAWIRA, R. and I. G. M. TANTRA (comp.), 1972. *Daftar nama pohon-pohonan: Lampung.* Revised edn. 29, 2 pp., 40 pls. (Lap. Lemb. Penelit. Hutan 143). Bogor. (Mimeographed. Originally published 1949, Buitenzorg, as F. H. HILDEBRAND (comp.), *Lijst van boomsoorten verzameld in de Lampungse districten.* [iv], 18 pp. (Rapp. Bosbouwproefsta. 20; also designated as Serie Boomnamenlijsten 19); revised 1954, Bogor, as *idem, Daftar nama pohon-pohonan: Lampung.* 27 pp. (Lap. Balai Penjelidik. Kehut. 65; also designated as Seri Daftar nama pohon-pohonan 42).)

914

Islands from the Riau group to Belitung

See also **913** (MIQUEL; SOEWANDA, 1970 (Riau); SOEWANDA and TANTRA, 1972 (Palembang)). – Area: 5316 km² (Riau and Lingga Archipelagoes); 11941 km² (Bangka and neighboring islands); 4833 km² (Belitung and associated islands). Vascular flora: no data. These islands are administratively part of two Sumatran provinces: Riau (Riau (Rhiow) and Lingga Archipelagoes) and Sumatera Selatan (Bangka (Banka) and Belitung (Billiton)).

No separate recent coverage is available. For the

Riau (Rhiow) and Lingga Archipelagoes, works on Peninsular Malaysia and Singapore (**911**) may be found useful. Bangka, Belitung and associated islets were formerly a separate residency and thus featured independent treatment in the Bogor Forest Research Institute tree species lists during their first 'cycle' of publication (see **913** for listing). Little else of value exists except perhaps the following, which includes a vegetation survey (by the author) and an annotated checklist of 959 species: S. KURZ, 1864. Korte schets der vegetatie van het eiland Bangka. *Natuurk. Tijdschr. Ned.-Indië* 27: 142–258.

Woody plants (including forest trees)
The only separate coverage of the woody flora is in an uncritical tree species checklist from the Indonesian Forestry Research Institute.

HILDEBRAND, F. H. (comp.), 1952. *Daftar nama pohon-pohonan: Bangka dan Billiton*. 48 pp. (Lap. Balai Penjelidik. Kehut. 57; also designated as Seri Daftar nama pohon-pohonan 34). Bogor. (Mimeographed. Originally published 1949, Buitenzorg, as *idem*, *Lijst van boomsoorten verzameld in Bangka en Billiton*. [iv], 29 pp. (Rapp. Bosbouwproefsta. 12; also designated as Serie Boomnamenlijsten 17).)

915
Anambas and Natuna Islands

See **913** (SOEWANDA, 1970 (Riau); WHITMORE and TANTRA, 1986). Works for Peninsular Malaysia and Singapore (**911**) will also be found useful, along with those on Borneo (**917**). – Area: 2480 km². Vascular flora: no data. The Anambas and Natuna Islands lie at the southern end of the South China Sea between Peninsular Malaysia and Borneo and represent partly drowned mountains on a shallow continental shelf. Administratively they are attached to Riau Province in Sumatra.

Unfortunately, no full checklist is available for either of the groups. The most useful source of information, with lists of collections preceded by some general remarks on the flora and its affinities, is C. G. G. J. VAN STEENIS, 1932. Botanical results of a trip to the Anambas and Natoena Islands. *Bull. Jard. Bot. Buitenz.*, III, **12**: 151–211, 11 figs. Other coverage, though very haphazard and unreliable, is provided in

the Forest Research Institute tree species list for Riau and its successor for all of Sumatra (see **913** for coverage).

916
Palawan (with Calamian)

See also **925** (MERRILL). – Area: 14896 km². Vascular flora: no data. Along with Palawan are here included the more northerly Calamian Island and its neighbors as well as the Cagayan Sulu group and other associated small islands.

As is well known, biogeographically the Palawan group is intermediate between Western Malesia and the Philippines although rather nearer the former. However, no separate checklist is available and other floristic papers are few and scattered. Records as far as known by the 1920s are included with all others from the Philippines in Merrill's *Enumeration*. A more recent, but unpublished, account is D. D. SOEJARTO, J. C. REGALADO, D. A. MADULID and C. E. RIDSDALE, 1995. *Preliminary checklist of the flowering plants of Palawan*. Unpaged, map. [Chicago, Ill.] This comprises an alphabetically arranged checklist with references, synonymy, types, citations to the *Enumeration* (where applicable), and localities with *exsiccatae* and their places of deposit.[25]

917
Borneo (Kalimantan)

Area: 734000 km². Vascular flora: 10000–15000 species (Ashton in Campbell and Hammond (1989; p. 94; see **General references**)). Merrill estimated 10000–11000 species, with 5250 recorded as of 1926. – Borneo (Kalimantan) here includes the Sultanate of Brunei Darussalam, the Malaysian states of Sabah and Sarawak, and the four provinces of Indonesian Borneo along with the large offshore islands of Balambangan, Bangui and Laut.

Botanical survey of the third largest island in the world commenced with the arrival of the Dutch Natural Sciences Commission team in the southeast in

1839. Penetration of what is now Sarawak and Sabah began relatively soon afterwards, with the first ascent of Mt. Kinabalu made in 1851. The early records were incorporated into Miquel's works. A first local base was established with the formation of the Sarawak Museum in Kuching in 1883. Elmer D. Merrill's interest began around 1915 with his bibliography; this was followed in 1921 by the *Enumeration*, with additions to 1926. A revised enumeration by Genkei Masamune appeared in 1942, with the pteridophytes in 1945.

Both these works, however, are now out of date; the latter is moreover to a large extent based on information from herbarium labels. Subsequent taxonomic literature pertaining to Bornean plants is quite scattered; reference should be made to standard bibliographies and indices. With the growth of the timber industry in the decades following World War II particular attention has been paid to the woody flora, and several publications have resulted. A definitive tree flora for Sabah and Sarawak was initiated in 1992; two volumes have appeared at the time of writing (1998). Other vascular plants remain less well known (apart from Nepenthaceae and Orchidaceae); however, advances are represented in accounts of the flora of Brunei (1996) and Mt. Kinabalu (begun in 1992). Some ideas of the distribution of diversity within the island have begun to emerge, but caution should be exercised; apparent patterns may be mere artifacts. More intensive, targeted collecting may greatly increase known diversity for a given area, as has been shown by the Brunei and Kinabalu flora projects. Comparable levels have yet to be achieved in the Indonesian provinces, although some research foci now exist.

Bibliography

MERRILL, E. D., 1915. A contribution to the bibliography of the botany of Borneo. *Sarawak Mus. J.* **2**: 99–136. [For additions to 1921, see the author's *Bibliographic enumeration*, pp. 2–6 (see below).]

MASAMUNE, G., 1942. *Enumeratio phanerogamarum bornearum.* 739 pp., map. Taihoku: Taiwan Sotukufu Gaijabu. Complemented by *idem*, 1945. *Enumeratio pteridophytarum bornearum.* ii, 124 pp. Taihoku.

Systematic enumerations respectively of known Bornean seed plants and pteridophytes, including synonymy, references and principal citations, Japanese scientific names, indication of local range (with citation

of some localities), and indices to generic and family names. The introductory section of the 1942 volume gives an account of botanical exploration, notes on special features of the flora, statistics, and a short list of references; in the 1945 volume is a discussion of the fern flora. These compilations are somewhat uncritical, and should be used with caution. [Description prepared in part from material supplied by Professor H. Hara, Tokyo.]

MERRILL, E. D., 1921. *A bibliographic enumeration of Bornean plants.* 637 pp. (J. Straits Branch Roy. Asiat. Soc., Special Number). Singapore. Supplemented in *idem*, 1922–26. Additions . . . [I]–II. *Philipp. J. Sci.* **21**: 515–534; **30**: 79–87.

Systematic enumeration of seed plants; includes synonymy with references, literature citations, generalized indication of local and extralimital range (with localities and citations of *exsiccatae* for less well-known species), limited taxonomic commentary, and index to all botanical names. The introductory section incorporates a summary of general features of the flora and vegetation along with an account of botanical exploration in Borneo; there are also additions to the author's 1915 bibliography (see above).[26]

Woody plants (including forest trees)

Since 1940 a considerable literature has appeared on the tree flora of Borneo. The only work covering the whole island is, however, the Kalimantan volume of *Tree flora of Indonesia* (see below). *Tree flora of Sabah and Sarawak* is a descriptive account for the two Malaysian states; it also appears immediately below. Other works refer to single polities or parts thereof and are here grouped under four subheadings: I. Sabah; II. Brunei; III. Sarawak; and IV. Indonesian Borneo (Kalimantan). The last is in turn divided into four tertiary headings, one for each province.

SOEPADMO, E., K. M. WONG and L. G. SAW (eds.), 1995– . *Tree flora of Sabah and Sarawak.* Vols. 1– . Illus. Kepong: Forest Research Institute of Malaysia (for the Forest Departments of Sabah and Sarawak and the Forest Research Institute of Malaysia).

Illustrated descriptive flora with bracketed keys (based as far as possible on vegetative features), synonymy with references and citations, types, vernacular names, and notes on distribution, ecology, silviculture, timber structure and properties, and uses; glossary and indices to botanical and vernacular names in each volume. Keys to genera also include non-tree taxa, their ultimate leads being annotated. The first volume contains the background to the project, a brief history of botanical work in Borneo through 1980 (pp. xxi–xlii), and an account of biogeography and ecology (by P. S. Ashton). [Full inclusion of a species is based on a trunk diameter of 30 cm or more at breast height, with 'commercial' timber and forest-product species attracting the most complete treatment. Lesser species are, as in *Tree flora of Malaya*, covered only in the keys. Eight volumes have been planned for publication covering 2829 species; as of 1999, two had appeared, accounting for 54 families with these in each being alphabetically arranged.][27]

WHITMORE, T. C., I. G. M. TANTRA and U. SUTISNA (eds.), 1989–90. *Tree flora of Indonesia: check list for Kalimantan.* 3 vols. Bogor: Forest Research and Development Centre.

Part 1 of this four-part work, encompassing the first two volumes, comprises a concisely annotated checklist, arranged alphabetically by families and genera with the names of smaller trees enclosed in parentheses; includes abbreviated citations of key literature, indication of maximum size, habitat and overall distribution, and occasional *exsiccatae*. Volume 3 encompasses the three remaining parts. Part 2 is a lexicon of vernacular and trade names with their botanical equivalents; the names have been taken from the *Daftar nama pohon-pohonan* and *Flora Malesiana*. Part 3 comprises keys to Dipterocarpaceae (by P. S. Ashton); all these species bar one are illustrated in part 4 (which also includes figures of important non-dipterocarp species). At the beginning of part 1 are lists of families and genera with statistics.

I. Sabah (North Borneo)

Mt. Kinabalu and its surroundings were the principal attraction for botanists from 1851 onwards; a first summary of the mountain flora, by Otto Stapf, appeared as *On the flora of Mount Kinabalu, in North Borneo* (1896, in *Trans. Linn. Soc. London*, II, Bot. **4**: 69–263). A second account is that of Lilian Gibbs (1914). The remainder of the state began to receive more attention only with formation of the North Borneo Forestry Department around 1918. For some years strong links were maintained with the Philippines.[28] Since World War II there has been substantial collecting throughout the state. Preliminary accounts of parts of the tree flora were produced in the 1950s and 1960s; a first definitive work was *Dipterocarps of Sabah* (1964; see below). *Trees of Sabah* was initiated in 1976 for non-dipterocarp species; however, both will be succeeded by *Tree flora of Sabah and Sarawak* (see above).

Partial works

BEAMAN, J. H. *et al.*, 1992– . *The plants of Mt. Kinabalu*, 1– . Col. illus. Kew: Royal Botanic Gardens.

Annotated enumeration with synonymy, references, indication of distribution, altitudinal range and habitat, and

localities with *exsiccatae*; bibliography, list of *exsiccatae*, glossary, appendix and index in each part. The general parts are extensive with sections on methodology, geography, vegetation and land use, collections, distribution, associations, special features of the plants concerned, and keys down to genera. [Vol. 1 covers pteridophytes; vol. 2 (1993), Orchidaceae; vol. 3 (1998), other monocotyledons. Preparation of the rest of the work is underway.]

Forest trees

COCKBURN, P. F. *et al.*, 1976–80. *Trees of Sabah*. Vols. 1–2. xv, 261 pp., 54 text-figs., 16 pls., maps (end-papers) (Sabah Forest Rec. 10). Kuching, Sarawak: Borneo Literature Bureau (vol. 1); Dewan Pustaka dan Bahasa, East Malaysian Branch (vol. 2). Complemented by W. MEIJER and G. H. S. WOOD, 1964. *Dipterocarps of Sabah*. 344 pp., 59 text-figs., 30 pls., map (Sabah Forest Rec. 5). Kuching.

The first work, covering 45 families, is an illustrated descriptive dendrological treatment of non-dipterocarp species; it includes keys, descriptions, synonymy, local vernacular and trade names, indication of local distribution (but without citations of *exsiccatae*), some critical remarks, and notes on ecology, special features, uses, etc.; complete indices at end of each volume. Sources of information, etc., are given in the general part of vol. 1. The treatment of Dipterocarpaceae is similar, but with the addition of *exsiccatae*.[29]

FOX, J. E. D., 1970. *Preferred check-list of Sabah trees*. 65 pp. (Sabah Forest Rec. 7). Kuching: Borneo Literature Bureau (for Sabah Forest Department).

Tabular checklist in pocket-sized format; includes notes on tree size, local range, and ecology and a list of vernacular names, with botanical equivalents.

II. Brunei Darussalam

Area: 5765 km^2. Vascular flora: *c.* 4300 species of seed plants (M. J. E. Coode, personal communication). – Forest trees were surveyed in the 1950s and from 1989 to 1996 a botanical survey was undertaken by the Forests Department with the assistance of the Royal Botanic Gardens, Kew. The result of the latter programme was a definitive enumeration, published in 1996. For Dipterocarpaceae, see also P. S. ASHTON, 1964. *A manual of the dipterocarp trees of Brunei State.* xii, 242 pp., 58 pls., 20 text-figs. Oxford: Oxford University Press. Some of the non-dipterocarp trees were covered in the unfinished *Manual of the non-dipterocarp trees of Sarawak* by Ashton and in *Forest trees of Sarawak and Brunei* by Browne (for both, see below under **917/III**).

COODE, M. J. E. *et al.* (eds.), 1996. *A checklist of the flowering plants and gymnosperms of Brunei Darussalam.* 477 pp., 16 col. pls., 10 col. figs. (inside front cover), 3 portrs., maps (frontisp. and inside back cover), portr. [N.p.]: Ministry of Industry and Primary Resources, Brunei Darussalam.

Tightly formatted enumeration of seed plants with local vernacular names (where known), occasional synonyms, and indication of habit, habitat, vegetation type(s), substrate, and altitudinal range as well as localities with *exsiccatae* and wider distribution (if applicable); list of orchids under cultivation at Leiden as of 1996; diagnoses and descriptions of new taxa; index to all botanical names. The brief introductory part gives the background to and plan of the work.

Forest trees

HASAN BIN PUKUL and P. S. ASHTON, 1966. *A checklist of Brunei trees.* 132 pp. Brunei: Government of Brunei (distributed by Borneo Literature Bureau, Kuching). (Reissued 1988.)

Tabular checklist of non-dipterocarp trees, arranged according to vernacular names (with botanical equivalents); includes notes on local range, ecology, etc., and a list of botanical names with vernacular equivalents.

III. Sarawak

The lure of this state, first formed in 1841 as a 'buffer' between Brunei and Pontiniak (West Kalimantan), attracted many collectors from the mid-nineteenth century, notably Odoardo Beccari. A further stimulus came with the advent of the Sarawak Museum in 1883 and, somewhat later, its curator G. D. Haviland; its collections, along with others, formed a basis for Merrill's *Enumeration.* The Forest Department was established in 1921; in 1955 its herbarium and that of the Museum were combined.

The 1955 account by Browne of the trees, the first such for Borneo, is highly selective. P. S. Ashton with others later undertook an expanded version; however, only two volumes were ever published: one an extension to the manual of Brunei Dipterocarpaceae, the other covering some families of non-dipterocarp trees (for both, see ASHTON below).

Forest trees

ANDERSON, J. A. R., 1980. *A check list of the trees of Sarawak.* [vi], 364 pp. Kuching: Dewan Bahasa dan Pustaka, Sarawak Branch (for Sarawak Forest Department).

Annotated checklist of over 2500 species of trees and treelets 'known to have a straight woody stem, even . . . a few feet high' (p. 6), in 101 families, divided into dipterocarp (pp. 110–133) and non-dipterocarp (pp. 133–349) portions; each entry includes a code number, vernacular name(s) (including those standardized in forestry), maximum girth, and distribution, occurrence and known forest types. The list proper is preceded by indices to genera and families (pp. 91–109) and vernacular names (pp. 12–90). The introductory section gives the background to the work (completed in 1973) and its plan (with definitions and abbreviations, including forest types (pp. 9–11)). There is a special concise list of trees of peat-swamp forests (pp. 350–364); all species here, however, appear in the main checklist. A revision is in preparation. Trees with a more fastigate, shrublike habit, as well as sub-shrubs, epiphytes and lianas (among them all species of *Rhododendron* and most in *Vaccinium*), are omitted.

ASHTON, P. S., 1988. *Manual of the non-dipterocarp trees of Sarawak.* Vol. 2. xix, 490 pp., text-figs., 52 halftone pls. Kuching: Dewan Bahasa dan Pustaka, Sarawak Branch (for Sarawak Forest Department). Complemented by *idem*, 1968. *A manual of the dipterocarp trees of Brunei State and Sarawak: Supplement.* viii, 129 pp., 23 pls., 15 text-figs. Kuching: Sarawak Forest Department.

The first work is an illustrated, amply descriptive botanical treatment (covering both Sarawak and Brunei) encompassing commercial and other tree species; it includes keys to genera and species (including forest-keys), synonymy, references and citations, generalized indication of local and extralimital distribution, documentation of *exsiccatae* (as represented in the Sarawak Herbarium), habitats, uses, and critical commentary. Derivations of generic and species names are also presented, and at the end is an index to scientific names. The treatment of Dipterocarpaceae, designed to complement the author's earlier *Manual of the dipterocarp trees of Brunei State* (1964; see **917/II**), is similar in format. [The manual for non-dipterocarp trees was not continued; it is being succeeded by *Tree flora of Sabah and Sarawak.* For vernacular names, reference must be made to ANDERSON.][30]

BROWNE, F. G., 1955. *Forest trees of Sarawak and Brunei.* 369 pp. Kuching: Sarawak Government Printer.

Illustrated, quite detailed descriptive dendrological treatment of principal tree species; includes notes on distribution, special features, properties, uses, etc., and an index. [Partly superseded, notably for the Dipterocarpaceae.]

IV. Indonesian Borneo

Control by the Dutch of Kalimantan did not begin until after 1825; the first scientific survey was undertaken only in the 1830s. Over the next century, other surveys and expeditions were from time to time mounted, aided by the spread of government control. Forest surveys followed from the early twentieth century onwards; however, as in Sumatra local research resources were nowhere established. Some improvement has ensued with establishment of universities in the major centers but more particularly the recent development of the research station of Wanariset in East Kalimantan.

Until recently the only organized floristic documentation was that embodied in the relevant issues of the tree species lists published for the different parts of Indonesia by the Forest Research Institute (Lembaga Penelitian Hutan), Bogor (for background, see supraregional introduction). Six numbers are available for the four provinces (of which Kalimantan Timur, in the east, and the central Kalimantan Tengah each had two forest districts); however, South Kalimantan (Banjarmasin and Hulu Sungai) saw no reissue after 1953 and the Sampit region (in western Kalimantan Tengah) no coverage since 1941. In 1989 all were succeeded by the Kalimantan volumes of the new series *Tree flora of Indonesia* (see above). As they contain some information not in the later work, however, they are retained here. Their order of citation follows the provinces from west to east.

Kalimantan Barat

SOEWANDA AMONG PRAWIRA, R. (comp.), 1978. *Daftar nama pohon-pohonan: Kalimantan Barat.* 3rd revised edn. 107 pp., illus. (Lap. Lemb. Penelit. Hutan 265). Bogor. (Mimeographed. Originally published 1941, Buitenzorg, as F. H. HILDEBRAND (comp.), *Boomnamenlijst: Residentie Westerafdeling van Borneo* (Bosbouwproefstation, Serie Boomnamenlijsten 10); 1st revision, 1952, Bogor, as *idem, Daftar nama pohon-pohonan: Kalimantan-Barat.* 82 pp. (Lap. Balai Penjelidik. Kehut. 54; also designated as Seri Daftar nama pohon-pohonan 31); 2nd revision, 1970, as R. SOEWANDA AMONG PRAWIRA *et al., Daftar nama pohon-pohonan: Kalimantan Barat.* 99 pp. (Lap. Lemb. Penelit. Hutan 107).)

Kalimantan Tengah

Organized in 1957, this province encompasses the Sampit (Kotawaringin) and Kapuas-Barito districts of the former Dutch residency of South and East Borneo. For clarity, the titles below are arranged chronologically rather than alphabetically as is usual in the *Guide*.

HILDEBRAND, F. H. (comp.), 1941. *Boomnamenlijst van Sampit (Zuid-Oost-Borneo).* 15 pp. (Bosbouwproefstation, Serie Boomnamenlijsten 7). Buitenzorg. (Mimeographed.) [Apparently never revised.]

SOEWANDA AMONG PRAWIRA, R. (comp.), 1971. *Daftar nama pohon-pohonan: Kapuas-Barito (Kalimantan Tengah-Selatan).* Revised edn. 90 pp., 90 pls. (Lap. Lemb. Penelit. Hutan 121). Bogor. (Mimeographed. Originally published 1949, Buitenzorg, as AFDELING BOSBOTANIE, *Lijst van boomsoorten verzameld in de afdeling Kapoeas-Barito, Zuid-Borneo.* [iv], 61 pp. (Rapp. Bosbouwproefsta. 3; also designated as Serie Boomnamenlijsten 12); revised 1953, Bogor, as F. H. HILDEBRAND, *Daftar nama pohon-pohonan: Kapuas-Barito (Kalimantan-Selatan).* 79 pp. (Lap. Balai Penjelidik. Kehut. 62; also designated as Seri Daftar nama pohon-pohonan 39).)

ARGENT, G., A. SARIDAN, E. J. F. CAMPBELL and P. WILKIE (eds.), n.d. [1998]. *Manual of the larger and more important non-dipterocarp trees of Central Kalimantan, Indonesia.* 2 vols. xxvi, 685, [20] pp., 202 pls., illus., map. Samarinda: Forest Research Institute, [Forestry Research and Development Agency, Indonesia].

Descriptive account with keys, occasional synonyms, and notes on distribution, habitat, vernacular names, illustrations and uses, the families and genera alphabetically arranged; list of authors, illustrations (with illustrators), bibliography (pp. 668–670), abbreviations, glossary (with figures) and index to genera at end. The general part is practical: on the use of keys, tree names, and making of specimens; it is followed by a list of families covered (p. xi), illustration captions (pp. xii–xviii) and the main key.

Kalimantan Selatan

Includes the Banjarmasin and Hulu Sungei districts along with Pulau Laut. Before 1957 the province also included Kalimantan Tengah. No revisions of the 1953 list have been seen. For secondary tree species, see **Kalimantan Timur**.

HILDEBRAND, F. H. (comp.), 1953. *Daftar nama pohon-pohonan: Bandjarmasin-Hulu Sungai (Kalimantan-Tenggara).* 64 pp. (Lap. Balai Penjelidik. Kehut. 63; also designated as Seri Daftar nama pohon-pohonan 40). Bogor. (Mimeographed. Originally published 1949, Buitenzorg, as AFDELING BOSBOTANIE, *Lijst van boomsoorten verzameld in de afdelingen Bandjarmasin en Hoeloe Soengei, Zuid-Oost-Borneo.* [iv], 48 pp. (Rapp. Bosbouwproefsta. 5; also designated as Serie Boomnamenlijsten 13).)

Kalimantan Timur

Coverage for this large province in the *Daftar-daftar nama* appeared in two parts (corresponding to its two forest districts): Bulungan and Berau (north of the Kutai Basin and north of the so-called 'W' Range, thus east of G. Murud in Sarawak and south of Sabah) and Samarinda (the Kutai Basin, the Mahakam valley, and the oil region of Balikpapan, all south of the 'W' Range). The southern forest district is now also covered by *Trees of the Balikpapan-Samarinda area* (see below). In addition, note should be taken of *Checklist of*

Secondary Forest Trees in East and South Kalimantan by P. J. A. Keßler, K. Sidayasa, Ambriansyah and A. Zainal (1995, Wageningen, as *Tropenbos Documents* 8), which is complementary to the two first authors' 1994 work.

KEßLER, P. J. A. and K. SIDIYASA, 1994. *Trees of the Balikpapan-Samarinda area, East Kalimantan, Indonesia.* 446 pp. (incl. 197 pls.), map. Wageningen: Tropenbos Foundation. (Tropenbos Series 7.)

Illustrated dendrology covering some 280 species with local vernacular and trade names, literature references, descriptions, and notes on habitat, frequency, distribution, representative *exsiccatae*, and taxonomic notes; indices to botanical and vernacular names (pp. 235–241, preceding the plates). The introductory part covers the plan and scope of the work along with keys for identification and a glossary; references, p. 33.

SOEWANDA AMONG PRAWIRA, R. (comp.), 1974. *Daftar nama pohon-pohonan: Bulungan dan Berau (Kalimantan Timur).* Revised edn. [ii], 89 pp., 80 pls., map (Lap. Lemb. Penelit. Hutan 196). Bogor. (Mimeographed. Originally published 1941, Buitenzorg, as F. H. HILDEBRAND (comp.), *Boomnamenlijst: afdeeling Boeloengan en Beraoe (Zuid- en Oost-Borneo).* (Bosbouwproefstation, Serie Boomnamenlijsten 9); revised 1952, Bogor, as *idem, Daftar nama pohon-pohonan: Bulungan dan Berau (Kalimantan–Timur).* 83, [2] pp. (Lap. Balai Penjelidik. Kehut. 55; also designated as Seri Daftar nama pohon-pohonan 32).)

SOEWANDA AMONG PRAWIRA, R. and I. G. M. TANTRA (comp.), 1971. *Daftar nama pohon-pohonan: Samarinda (Kalimantan Timur).* Revised edn. [ii], 118, 2 pp., 61 pls. (Lap. Lemb. Penelit. Hutan 123). Bogor. (Mimeographed. Originally published 1949, Buitenzorg, as AFDELING BOSBOTANIE (comp.), *Lijst van boomsoorten verzameld in Samarinda (Oost-Borneo).* [iv], 76 pp. (Rapp. Bosbouwproefsta. 2; also designated as Serie Boomnamenlijsten 11); revised 1952, Bogor, as F. H. HILDEBRAND, *Daftar nama pohon-pohonan: Samarinda (Kalimantan–Timur).* 107 pp. (Lap. Balai Penjelidik. Kehut. 58; also designated as Seri Daftar nama pohon-pohonan 35).)

918

Java (with Madura)

Area (inclusive of associated islands): 132 319 km². Vascular flora: 4911 species, of which 4598 native (Backer and Bakhuizen, 1963–68; see below). The associated islands of Krakatoa, Bawean and Kangean are included here. For Christmas Island, see **046**.

Along with Maluku, Java was one of the earliest parts of Malesia to be botanized, but detailed exploration and documentation began only late in the eighteenth century with the work of Arnold Radermacher (a founder of the Batavian Society and author of a first checklist (*Naamlijst der planten, die gevonden worden op het eiland Java mit beschrijving van eenige nieuwe geslagten en soorten*, 1780–82)). Major surveys followed from the beginning of the nineteenth century; notable were those of Horsfield, the Natural Sciences Commission (Kuhl, van Hasselt, Korthals and Junghuhn), Zollinger, Koorders, and Backer. The mountains also received close attention, particularly from van Steenis in the 1920s and 1930s, as also did the weed flora. In spite of a number of attempts, however, no complete and critical flora was to appear in print until the 1960s. Conflicting demands, false starts, rivalries, a bitter controversy, varying institutional policies and support, and social and political changes marked the intervening period of nearly two centuries. The flora – or what is left of it – is now in general well documented, with a number of works of varying scope and style; however, an *accessible* work is needed.

BACKER, C. A. and R. C. BAKHUIZEN VAN DEN BRINK, JR., 1963–68. *Flora of Java.* 3 vols. 32 halftones, portr., 2 maps. Groningen: Noordhoff (later Wolters-Noordhoff). Complemented by C. A. BACKER and O. POSTHUMUS, 1939. *Varenflora voor Java.* 370 pp., 90 text-figs. Buitenzorg: s'Lands Plantentuin.

The first of this complementary pair of works is a concise, largely unillustrated descriptive manual of seed plants (6100 species, of which over 4000 are native), with species treatments in the form of keys; these treatments incorporate limited synonymy and abbreviated indication of local range, altitudinal zonation, habitat and phenology, with taxonomic commentary (and some literature references) in footnotes. A complete index appears in vol. 3. In the introductory section (vol. 1) is a biography of Backer (with list of publications) and a brief account of the preparation and editing of the work, along with a somewhat forbidding key to families. Volume 2 features an illustrated chapter on the vegetation and phytogeography of Java (by C. G. G. J. van Steenis and A. F. Schippers-Lammertse). The companion work on the pteridophytes (in Dutch) is similar in format but includes figures of representative species; it is preceded by an introductory section with chapters on organography, fern communities and fern geography, a glossary, and a list of references, as well as general keys to all genera.[31]

KOORDERS, S. H., 1911–12. *Exkursionsflora von Java, umfassend die Blütenpflanzen.* Vols. 1–3. xxiv, 2528 pp., 20 pls. (1 col.), 139 text-figs., 4 maps. Jena: Fischer. Complemented by *idem*, 1913–37. *Exkursionsflora von Java, umfassend die Blütenpflanzen.* Vol. 4 (*Atlas*). Pp. 1–688, 865–1020, pls. 1–976, 1166–1313. Jena.

The main work is a sparsely illustrated manual-key to native, naturalized and commonly cultivated seed plants (5067 species), with family and generic descriptions of variable length and, for montane species, a separate, more detailed accounting. Diagnoses in the key leads include limited synonymy, references, localities and Javan range, altitudinal zonation, vernacular names, status, and terse notes on habit, habitat, biology, etc.; entries on montane species include additional synonymy (with citations), *exsiccatae*, localities, altitudes, etc., as well as extra-Javan distribution. Literature citations also appear in family and generic headings, and partial indices are given for each volume (addenda, corrigenda, and complete index in vol. 3). Arrangement follows the Englerian system, with use of Dalla Torre and Harms numbers. The never-completed *Atlas* comprises 1110 plates with large figures of most species in families Cycadaceae through Ranunculaceae as well as part of the Leguminosae.[32]

Partial work

STEENIS, C. G. G. J. VAN *et al.*, 1975. *Flora untuk sekolah di Indonesia.* Transl. at Universitas Gadja Mada, Yogyakarta. 495 pp., 46 text-figs. Jakarta-Pusat: Pradnya Paramita. Dutch edn.: *idem*, 1949. *Flora voor de scholen in Indonesië.* Jakarta. (2nd edn., 1951.)

Concise illustrated students' manual-key of commonly encountered native, naturalized and cultivated lowland plants; includes vernacular names and notes on habit, distribution, phenology, special features, etc.; index. Covers about 400 species. The Indonesian translation, which remains in print, is based on the 1949 version and is in a slightly larger format than the original.

Woody plants

Java was the first part of Malesia to have good documentation of its woody flora, first undertaken under the Treub regime at the Bogor Botanic Gardens and published as the *Bijdragen*. The next treatment, based on some further botanical exploration, formed part of the Forest Research Institute series of tree species lists (for background, see supraregional introduction). The first edition appeared in one number in 1950; the 1976–77 reissue, however, is in three parts corresponding to the long-standing basic geographical and administrative divisions of Java and Madura. In view of *Flora of Java*, it was thought not necessary to undertake a further revision for *Tree species lists of Indonesia*. The Java lists are distinguished from those for the 'Outer Provinces' by lesser minimum tree size limits.

KOORDERS, S. H. and T. VALETON, 1894–1914. *Bijdragen tot de kennis der boomsoorten van (op) Java.* 13 parts (parts 1–10 in *Meded. Lands Plantentuin* 11, 14, 16, 17, 33, 40, 42, 59, 61, 68; parts 11–13 in *Meded. Dept. Landb. Ned.-Indië* 2, 10, 18). Batavia, The Hague. Complemented by S. H. KOORDERS, 1913–18. *Atlas der Baumarten von Java (im Anschluss an die 'Bijdragen tot de kennis der boomsoorten van Java').* 16 parts in 4 vols. 800 pls. Leiden.

The *Bijdragen*, with text in both Dutch and Latin, comprise a thoroughly documented, copiously descriptive flora of woody plants, comprising detailed family revisions with keys to genera and species; treatments include full synonymy (including references), literature citations, vernacular names, generalized indication of local and extralimital range (with some indication of *exsiccatae* and specific localities, notably with reference to the 'Herbarium Koordersianum', now in the Herbarium Bogoriense), taxonomic commentary, and notes on habitat, phenology and silviculture; general index to all botanical and vernacular names in part 13. Part 1 includes a short introduction to the work. The *Atlas* comprises finely executed lithographic plates with short captions fully cross-referenced to the *Bijdragen*; vernacular names are included as well as a brief synonymy, and the specimens used for the drawings have been cited. [The *Bijdragen*, largely worked up at present-day Bogor with perforce limited reference to collections in Holland or elsewhere, was for its day an outstanding work.]

SOEWANDA AMONG PRAWIRA, R. (comp.), 1976. *Daftar nama pohon-pohonan [di] Java-Madura*, I: *Java Barat* (West Java). 124 pp., 51 pls. (Lap. Lemb. Penelit. Hutan 219). Bogor; *idem*, 1977. *Daftar nama pohon-pohonan [di] Java-Madura*, III: *Java Tengah* (Central Java). 131 pp., 54 pls. (*ibid.*, 244); and *idem*, 1977. *Daftar nama pohon-pohonan [di] Java-Madura*, III: *Java Timur* (East Java). 143 pp., 5 pls. (*ibid.*, 253). (All parts mimeographed. Preceded by F. H. HILDEBRAND (comp.), 1950. *Daftar nama pohon-pohonan: Djawa-Madura.* [i], 171 pp. (Lap. Balai Penjelidik. Kehut. 35; also designated as Seri Daftar nama pohon-pohonan 24). Bogor; revised and expanded 1951 as *idem*, *Daftar nama pohon-pohonan: Djawa-Madura.* [i], 183 pp. (Lap. Balai Penjelidik. Kehut. 50; also designated as Seri Daftar nama pohon-pohonan 30).)

Briefly annotated checklists of tree species (1280 in all) known to attain a height of 10 m and a diameter at breast height of 15 cm, arranged alphabetically by families and genera and including some synonymy, vernacular names, information on distribution, altitudinal range, wood properties, and (variously) notes on habitat, frequency and special features. Some introduced trees have also been included. Alphabetical lists of vernacular names (Indonesian, Sundanese, Javanese, Madurese, Dutch and English) with species reference numbers are also given. [The 1976–77 issues are, however, only slightly changed from that of 1951.][33]

919

Lesser Sunda Islands (Nusa Tenggara, Bali and East Timor)

Area: 88 488 km[2]. Vascular flora: no data seen. – This unit comprises the island arc east of Java, from Bali to the Damar group, Alor and Timor.

Timor was the first island to receive any amount of attention, the town of Kupang at its western end being frequented by exploring expeditions from the mid-eighteenth century. Knowledge of its flora was first consolidated by Joseph Decaisne in *Description d'un herbier de l'île de Timor* (1834, in *Nouv. Ann. Mus. Natl. Hist. Nat.* **3**; reprinted 1835 as *Herbarii timorensis descriptio*); further treatments have been *Prodromus florae timorensis* by J. B. Spanoghe (1841, in *Linnaea* **15**) and *Prodromus florae timorensis* by James Britten and others (1885; see below). The last-named encompassed in particular Forbes's collections from East Timor, an area hitherto scarcely documented.[34] Knowledge of other islands developed more slowly, but just sufficiently to enable preparation by Posthumus (1944; see below) of a consolidated enumeration of the ferns together with a summary review of botanical work. An analysis of the vascular flora followed a decade later, as C. KALKMAN, 1955. A plant-geographical analysis of the Lesser Sunda Islands. *Acta Bot. Neerl.* **4**: 200–225. Tree species lists also became available for most areas by the mid-twentieth century, with Meijer Drees additionally producing an account of woodland trees and shrubs in relation to potential management.

In the following decades substantial collections were made in central Flores by the missionary-priests Schmutz and Verheijen. From 1977 to 1986 Fr. Schmutz put together a five-part *Die Flora van Manggarai* which represented a considerable advance

in knowledge. Their work suggested that while endemism was relatively low and, as Kalkman had shown, the flora was primarily an extension of that of Java, the floristic transition from West to East Malesia was not as one-sided or sudden as had been supposed; some distinct East Malesian elements appeared in Flores.[35] Preparation of a properly documented separate flora or enumeration would now be worthwhile, such is the phytogeographic interest of the whole chain.

A useful general introduction to ecology and biodiversity of the islands is K. A. MONK, Y. DE FRETES and G. REKSODIHARJO-LILLEY, 1997. *The ecology of Nusa Tenggara and Maluku*. xvii, 966 pp., illus., maps. Hong Kong: Periplus. (The ecology of Indonesia series V.)

Woody plants (including forest trees)

The tree species lists for forest districts issued by the Forest Research Institute in Bogor have recently been succeeded by a volume in their recent series *Tree flora of Indonesia*. The earlier lists, however, contain information not retained in their successor and so are accounted for below under individual island groups.

WHITMORE, T. C., I. G. M. TANTRA and U. SUTISNA, 1989. *Tree flora of Indonesia: check list for Bali, Nusa Tenggara and Timor*. vii, [ii], 119 pp., illus. Bogor: Forest Research and Development Centre.

Part 1 of this four-part work comprises a concisely annotated checklist, arranged alphabetically by families and genera with the names of smaller trees enclosed in parentheses; includes abbreviated citations of key literature, indication of maximum size, habitat and overall distribution, and occasional *exsiccatae*. Part 2 is a lexicon of vernacular and trade names with their botanical equivalents; the names have been taken from the *Daftar nama pohon-pohonan* and *Flora Malesiana*. Part 3 comprises a key to Dipterocarpaceae (by P. S. Ashton), and part 4 figures of 12 key timber species including the three known dipterocarps. At the beginning of part 1 are lists of families and genera with statistics.

Bali and Lombok

SOEWANDA AMONG PRAWIRA, R. and I. G. M. TANTRA (comp.), 1972. *Daftar nama pohon-pohonan: Bali dan Lombok*. 24, 2 pp., 41 pls. (Lap. Lemb. Penelit. Hutan 145). Bogor. (Mimeographed. Preceded by F. H. HILDEBRAND (comp.), 1940. *Boomnamenlijst van Bali en Lombok*. (Bosbouwproefstation, Serie Boomnamenlijsten 2). Buitenzorg.)

Uncritical, sparsely annotated checklist, with each entry usually on one line; includes illustrations of key timber species. Species are arranged both by families and genera and by vernacular names.

Islands from Sumbawa to Wetar (including Flores)

A florula for the Manggarai area in central Flores was produced and distributed between 1977 and 1986. No other florulas are available. The woody plants of eastern Sumbawa are included in Meijer Drees' study of woodlands (see **Timor** below).

SCHMUTZ, E., Fr., [1977–86]. *Die Flora der Manggarai.* Heft 1–5. [N.p.] (Reproduced from typescript.)

A collection of notes organized by families, genera and species with numerous field and other observations, vernacular names, and localities (both near and as far as Timor) with *exsiccatae*. Within each part families and genera are alphabetically arranged but there is no complete index. Additions and corrections appear in parts 2 and 5 and field notes for collections from no. 4380 upwards appear also in the last part.

Timor and associated islands

Included here are Timor, Roti and Semau. No full account of these botanically interesting islands has appeared since 1885. However, the trees alone were further documented by Hildebrand in 1940 and 1953 in the *Daftar-daftar nama* and the woody plants of the woodland savanna by Meijer Drees in two works in 1950.

BRITTEN, J. *et al.*, 1885. Prodromus florae timorensis; compiled in the Botanical Department of the British Museum. In H. O. Forbes, *A naturalist's wanderings in the Eastern Archipelago*, pp. 497–523. London: Sampson Low; New York: Harper.

Briefly annotated systematic list of vascular plants known from Timor (as appendix VI); includes descriptions of new taxa, localities with some *exsiccatae* (the latter mostly of the author's collections), occasional taxonomic commentary, and brief notes on habitat, previous records, etc.; no separate index. A short introductory section includes a sketch of prior botanical exploration in the island (prepared by Forbes with assistance from H. N. Ridley).

HILDEBRAND, F. H. (comp.), 1953. *Daftar nama pohon-pohonan: Timor (Kepulauan Sunda Ketjil).* 51 pp. (Lap. Balai Penjelidik. Kehut. 60; also designated as Seri Daftar nama pohon-pohonan 37). Bogor. (Mimeographed. Originally issued 1940, Buitenzorg, as *idem, Boomnamenlijst van Timor en onderhorigheden* (Bosbouwproefstation, Serie Boomnamenlijsten 3).)

Uncritical, sparsely annotated checklist arranged alphabetically by families and genera. The botanical checklist is preceded by one arranged by vernacular names. [No revisions have been seen.]

MEIJER DREES, E., 1950. *Distribution, ecology and silvicultural possibilities of the trees and shrubs from the savanna-forest region in eastern Sumbawa and Timor.* 146 pp. (Pengum. Balai Penjelidik. Kehut. 33). Bogor. Accompanied by *idem*, 1950, Lesser Sunda Islands (woody plants, other islands).

1950. *Daftar nama-nama pohon dan perdu, Pulau Timor/ Lijst van boom- en struiknamen van het eiland Timor.* 61 pp. (Lap. Balai Penjelidik. Kehut. 33). Bogor. (Mimeographed; also published as Seri pemeriksaan tumbuh-tumbuhan 1.)

The main work includes in its section 6 an account of woodland trees and shrubs from Timor and eastern Sumbawa (both within the range of savanna and woodland species of *Eucalyptus*), with short descriptions and relatively extensive notes on distribution, ecology and silviculture. Families and genera are arranged alphabetically. The general part (sections 1–5) covers suitability of species for silvicultural (plantation) purposes, explanatory notes, and vegetatiological units, while at the end are an Indonesian and Dutch summaries, references, and an index to botanical names. The *Daftar nama-nama* covers scientific and vernacular names, arranged by families and genera as well as by their names in four different dialects (all arranged in a single sequence).

Pteridophytes

POSTHUMUS, O., [1944]. *The ferns of the Lesser Sunda Islands.* Buitenzorg. (*Ann. Bot. Gard. Buitenzorg*, vol. hors série: 35–113.) Buitenzorg.

Systematic enumeration of known ferns, with synonymy (including references) and citations, detailed indication of local range (with listing of specimens seen), taxonomic commentary, and notes on habitat; no separate index. The introductory section includes an account of previous botanical work in the area along with remarks on climate, etc., of representative localities.

Region 92

Central Malesia (including the Philippines)

In this region are included Sulawesi (Celebes), the Philippines (except for Palawan), and the Moluccas (Maluku), as well as all associated islands. This region, along with the Lesser Sunda Islands (on botanical evidence here included in Region 91), has been called 'Wallacea' by a number of authors and by George Gaylord Simpson an 'unstable zone' – a kind of biological 'no man's land' between the Oriental and Australasian zoogeographic regions, traversed by 'Weber's Line of Faunal Balance' as redefined by Ernst Mayr.[36] The western and eastern limits of this region are here by and large conventionally defined by the respective continental shelves, save for the Lesser

Sundas as already noted. Zoological and botanical evidence both suggest that Region 92 is not uniform but is marked by various states of biogeographic transition, a concept advanced by Simpson and subscribed to here. The limits or units in this region have therefore been fairly finely drawn.

Although the region is as a whole served by *Flora Malesiana* and, to varying degrees, other works on the Greater Malesian superregion, separate documentation for the various units here recognized is very uneven, a situation to some extent paralleled by the current level of botanical exploration (as well as by the existence of well-founded research centers). Much of the available literature is now comparatively old.

The best, though not most recently, documented unit certainly is the Philippines. Here, *An enumeration of Philippine flowering plants* by E. D. Merrill (1923–26) has long been a standard. With this may be associated *Fern flora of the Philippines* by E. B. Copeland (1958–60). A very good local flora is *Flora of Manila* by Merrill (1912; reprinted 1974). However, forest-botanical literature by comparison with Malaysia or Indonesia is scanty; this reflects the relatively limited development of forest botany as a research field in the country in spite of the example set by Sebastian Vidal in the nineteenth century. Of parts of the Philippines designated by separate headings, the isolated Batan Islands to the north were treated by Sumihiko Hatusima in his *An enumeration of the plants of Batan Islands, northern Philippines* (1966) but for the Sulu Archipelago no separate accounts exist. Major forest trees were documented and illustrated in vol. 3 of the University of the Philippines' *Guide to Philippine flora and fauna* (1986) and wetland plants by Mendoza and del Rosario in 1967 (see **908**).

Current floristic coverage of Sulawesi and Maluku remains very scanty. For the former, there remains little else but *Enumeratio specierum phanerogamarum Minahassae* by S. H. Koorders (1898; additions 1902, 1918–22) and *Die Farnflora von Celebes* by Hermann Christ (1898; additions 1904). Of outlying groups, no separate accounts exist for the Sanghie Islands, while for the Talauds there is one relatively substantial collection report (Holthuis and Lam, 1942). For Maluku, herein divided into three subunits, the only large-scale coverage remains the classic *Herbarium amboinense* by Georg Eberhard Rumphius (1741–50; *Auctuarium*, 1755), a work made effectively more useful through its interpretation by Merrill (1917; a revised

list of modern botanical names, by H. C. D. de Wit, appeared in 1959). The islands of southeastern Maluku were sketchily covered by Hemsley (1885), while the northern islands have no separate lists.

As in Western and Southern Malesia, the Indonesian islands were all covered by mimeographed lists of forest tree species in A4 format, part of a series initiated by the Forest Research Institute, Bogor, in 1940. Two lists cover Sulawesi and one deals with North and South Maluku. These have been supplanted by two volumes in the *Tree flora of Indonesia* series.

Sulawesi and Maluku are, along with Irian Jaya (western New Guinea), among the least-explored large areas in Greater Malesia. In Sulawesi, only the northeastern part of Minahasa and the southwestern peninsula have had more than a few collectors, although the situation elsewhere is improving with much effort focused on declared and potential natural reserves. Maluku is unevenly known although overall collecting is at a somewhat higher level. North Maluku in particular is very poorly known and documented.

Bibliographies. General and supraregional bibliographies as for Superregion 91–93.

Indices. General and supraregional indices as for Superregion 91–93.

921

Sulawesi (Celebes)

With the main island are here included the adjacent islands of Salajar, the Butung group and the Banggai Archipelago. The Sula Islands have been grouped with the North Moluccas. It is regrettable that for an area of great phytogeographic interest only scattered trip reports and checklists are available; easily the most substantial of these is Koorders' report on the northeastern Minahasa district (see below). There is also an enumeration of pteridophytes by Hermann Christ (1898) as well as tree species lists, the latest version being that of Whitmore *et al.* (1989).

Good, though now aging, general introductions to Sulawesian floristics and plant geography, both with many references, are embodied in two papers by Hermann Lam: H. J. LAM, 1942. Notes on the historical phytogeography of Celebes. *Blumea* 5: 600–640; and *idem*, 1950. Proeve eener plantengeografie van Celebes. *Tijdschr. Kon. Ned. Aardrijksk. Genootsch.* 67: 566–604, 3 text-figs., 7 halftones. A good recent general reference on ecology and biodiversity is A. J. WHITTEN, M. MUSTAFA and G. HENDERSON, 1987. *The ecology of Sulawesi*. Illus., maps. Yogyakarta: Gadjah Mada University Press. (The ecology of Indonesia series IV.) There are several contributions in more specialized biogeographical collections.[37]

Partial works: Minahassa

KOORDERS, S. H., 1898. *Verslag eener botanische diensreis door de Minahasa, tevens eerste overzicht der Flora van N. O. Celebes*. xxvi, 716 pp., 17 pls. (incl. maps) (Meded. Lands Plantentuin 19). Batavia/The Hague. Continued as *idem*, 1902–04. [1.]–3. Nachtrag zu meiner 'Enumeratio specierum phanerogamarum Minahassae'. *Natuurk. Tijdschr. Ned.-Indië* 61: 250–261; 63: 76–89, 90–99; and *idem*, 1918–22. *Supplementum op het eerste overzicht der Flora van N. O. Celebes*, 1(1–2), 2/3. 50, 121 pp.; 13, 127 pls. Batavia: The author (part 1(1)); Buitenzorg: Mme. A. Koorders-Schumacher (parts 1(2), 2/3). (Part 1(2) also published as *Bull. Jard. Bot. Buitenz.*, III, 2: 242–260, pls. 11–13 (1920).)

Part III (pp. 253–645) of the main work comprises the author's *Enumeratio specierum phanerogamarum Minahassae*, an annotated systematic account of the seed plants known from northeastern Celebes (for pteridophytes, see below under Christ's *Die Farnflora von Celebes*); this enumeration includes limited synonymy, vernacular names, brief descriptive notes, detailed indication of local range, remarks on habitats, uses, etc., and an index to vernacular names. The introductory section to the whole work (pp. i–xxvi) includes chapters on physical features, geography and botanical exploration, while part I (pp. 1–110) describes Koorders' itinerary and part II (pp. 111–252) includes an account of useful plants.

Woody plants (including forest trees)

Included here are the tree species lists produced by the Forest Research Institute, Bogor (for background, see supraregional introduction). That by Whitmore *et al.* (1989) is a partial successor to those produced initially by Hildebrand in 1940.

SOEWANDA AMONG PRAWIRA, R. (comp.), 1972. *Daftar nama pohon-pohonan: Sulawesi Selatan, Tenggara dan sekitarnja* [South and Southeast Celebes and dependencies]. Revised edn. 113, 4 pp., 55 pls. (Lap. Lemb. Penelit. Hutan 151). Bogor. (Mimeographed. Originally issued 1940, Buitenzorg, as F. H. HILDEBRAND (comp.), *Boomnamenlijst van Selebes en onderhorigheden* (Bosbouwproefstation, Serie Boomnamenlijsten 4); revised 1950, Bogor, as *idem*, *Daftar nama pohon-pohonan: Selebes*. [i], 105 pp. (Lap. Balai Penjelidik. Kehut. 43; also designated as Seri Daftar nama pohon-pohonan 26).)

SOEWANDA AMONG PRAWIRA, R. (comp.), 1972. *Daftar nama pohon-pohonan: Sulawesi Tengah dan Utara* (Central and North Celebes). Revised edn. Illus. (Lap. Lemb. Penelit. Hutan 156). Bogor. (Mimeographed. Originally issued 1940, Buitenzorg, as F. H. HILDEBRAND (comp.), *Boomnamenlijst van Manado* (Bosbouwproefstation, Serie Boomnamenlijsten 1); revised 1951, Bogor, as *idem*, *Daftar nama pohon-pohonan: Manado*. [i], 53 pp. (Lap. Balai Penjelidik. Kehut. 44; also designated as Seri Daftar nama pohon-pohonan 27).)

Sparsely annotated, uncritical checklists of forest trees known to attain a diameter at breast height of 40 cm or more, arranged alphabetically by families and genera; each entry (usually on one line) includes a vernacular name (Malay, Buginese, Macassarese, Tobelo, Toraja, Minahassan, etc.), distribution (by subdistricts), number of collections (but no *exsiccata*-numbers), and durability class (on a scale of five). Each of the systematic lists proper is preceded by a list with all records, fully annotated, arranged alphabetically by vernacular names; at the end of each of the two treatments are lists of amended botanical names, a lexicon of Indonesian names with more important local vernacular equivalents, an alphabetical lexicon of genera with family equivalents, and an indexed album (in current issues) of full-page illustrations of key species depicting botanical features. No keys are included.

WHITMORE, T. C., I. G. M. TANTRA and U. SUTISNA, 1989. *Tree flora of Indonesia: check list for Sulawesi*. vii, [1], 204 pp., illus. Bogor: Forest Research and Development Centre.

Part 1 of this four-part work comprises a concise annotated checklist arranged alphabetically by families and genera with the names of smaller trees enclosed in parentheses. Entries include abbreviated citations of key literature, indication of maximum size, habitat and overall distribution, and occasional *exsiccatae*. Part 2 is a lexicon of vernacular and trade names with their botanical equivalents; the names have been taken from the *Daftar nama pohon-pohonan* and *Flora Malesiana*. Part 3 comprises a key to Dipterocarpaceae (by P. S. Ashton) covering all species from Sulawesi eastwards, and part 4 features full-page figures of 23 key timber species including five dipterocarps. At the beginning of part 1 are lists of families and genera with statistics (not, however, summarized).

Pteridophytes

Included here is the only flora covering all Sulawesi, a checklist of pteridophytes by Hermann Christ (1898; supplement, 1904). Both were based on collections by Warburg, Koorders and the Sarasin brothers along with others of earlier date in Herbarium Bogoriense.

CHRIST, H., 1898. Die Farnflora von Celebes. *Ann. Jard. Bot. Buitenz.* **15**: 73–186, pls. 13–17 (incl. map). Continued as *idem*, 1904. Zur Farnflora von Celebes II. *Ibid.*, **19**: 33–44.

The main work comprises a systematic enumeration of pteridophytes, with descriptions of some novelties and inclusion of synonyms, localities with *exsiccatae*, and notes on local and extralimital distribution, habitat, etc.; figures of selected species and map with collecting localities at end. The 1904 continuation is a supplement to the main work. [Produced in association with Koorders' *Enumeratio* but more comprehensive in coverage and sources.]

922

Sangihe Islands

This chain of partly active volcanic islands extends from the end of the Minahassa Peninsula in northeastern Celebes north towards the Philippines. No separate checklists appear to be available and, until recently, collections were few.

923

Sulu Archipelago

No separate checklists are available for this island chain stretching between the southern Philippines and eastern Sabah and including Jolo and Tawitawi as its largest islands. Records up to 1923 are, however, included in E. D. Merrill's *Enumeration of Philippine flowering plants* (**925**). Biogeographically the area is, like Palawan, intermediate between Borneo and the Philippines with, however, more links to the latter.

924

Batan Islands

This isolated small island group is located between the Philippines and Taiwan but biogeographically is most closely related to the former. There is one

relatively recent checklist, by Sumihiko Hatusima (see below).

HATUSIMA, S., 1966. An enumeration of the plants of Batan Islands, northern Philippines. *Mem. Fac. Agric. Kagoshima Univ.* **5**: 13–70, 10 halftones, map.

Systematic enumeration, based on the results of a staff–student field trip but incorporating earlier published records; includes synonymy, references, localities with *exsiccatae*, and notes on occurrence, habitat, special features, etc., followed by a discussion of floristics. Features some new species and many records additional to those in Merrill's *Enumeration of Philippine flowering plants* (**925**).

925

The Philippines (Pilipinas)

Area: 300000 km². Vascular flora: 8030 seed plant and 900 pteridophyte species (GB; see **General references**); these figures are slightly less than the 9100 vascular plants of Merrill (1926; see below). Exclusion of Palawan (**917**), the Sulu Archipelago (**923**) and the Batan Islands (**924**) will reduce this figure somewhat.[38] – Works relating to the Philippines as a whole are included here; for Palawan, the Sulu Archipelago and the Batan Islands alone, see their respective headings.

The major works on the Philippine flora prior to 1900 are the various editions of Fr. Manuel Blanco's *Flora de Filipinas* (1837; 2nd edn., 1845; 3rd edn. by A. Navés, C. Fernández-Villar and D. Vidal y Soler, 1877–80(–83)). The last edition (reissued in 1993) was produced as a connoisseur's work, with inclusion of 477 colored plates; it also included as its *Novissima Appendix* (vol. 4a) a revised account of the flora by Navés and Fernández-Villar, notorious for its many misapplied names but the only standard until the 1920s. Interpretation of Blanco's species continued to be a problem, however, and in 1918 Merrill, then government botanist, published *Species blancoanae* on the basis of observation and authenticating collections. Domingo Vidal y Soler's younger brother Sebastián, as royal forest botanist from 1879 until his death in 1889, produced several key works including a forest flora (1883; noted below) and *Revisión de plantas vasculares*

Filipinas (1886). Sadly, this progress was largely negated by civil unrest and war in the next decade, with most local resources lost by 1900.

The substantial amount of exploration and floristic and systematic work accomplished in the Philippines in the first quarter-century after 1900 was for the seed plants summarized, along with earlier work, by Elmer D. Merrill in his well-known *Enumeration* (1923–26). The ferns were revised in a descriptive manual by Edwin Copeland (1958–60). Merrill also produced a useful local flora of the Manila district (1912). However, apart from the first volume of a flora of Mt. Makiling by Juan Pancho, little else of substance has been published. A new initiative, the 'Flora of the Philippines Project', based in Manila and Honolulu, was mounted in 1989 with the aim of producing a new flora and database by the end of the 1990s. As of 1997 the 'writing-up' phase was pending, the plan having shifted from a 'monographic' flora to a 'revised enumeration' designed to appear both as hard copy and in electronic form.[39]

With regard to the forest flora, the lack of a strong tradition of forest botany compared with that in Malaysia, Indonesia and New Guinea is a possible reason for the paucity of significant manuals or checklists on the forest or tree flora comparable to those elsewhere in Asia or Malesia.[40] Until 1986, the only descriptive treatments were Sebastian Vidal's work of 1883 and the dendrologies of George Ahern (1901) and Harry Whitford (1911). In that year a selective dendrology appeared in the third volume of *Guide to Philippine flora and fauna*, a series produced by the Natural Resources Management Center of the Ministry of Natural Resources and the University of the Philippines.

Bibliographies
See also **Superregion 91–93, Supraregional bibliographies** (ALLIED GEOGRAPHICAL SECTION, vol. 1; DEPARTMENT OF EDUCATION, JAPAN, vol. 3 (botany, pp. 35–155)).

MADULID, D. A. and M. G. AGOO, 1992. *A bibliography of biodiversity research in the Philippines*, I: *Flora*. ix, 189 pp. Manila: National Museum of the Philippines. [An extension to Nemenzo's 1966 bibliography; this and other sources, pp. v–vi. Titles, in part annotated, are arranged by author within six sections: all items (pp. 1–85), floristics and taxonomy, forest resources and vegetation, plant conservation, general conservation, and regional accounts (therein further arranged by province). Emphasis is on literature since 1965

except for the last section which is fully retrospective; also, relevant items not in Nemenzo's work are included. Much unpublished work is also accounted for.]

MERRILL, E. D., 1926. Bibliography of Philippine botany. In his *Enumeration of Philippine flowering plants*, 4: 155–239. Manila. [Titles alphabetically arranged; unannotated.]

NEMENZO, C. A., 1969. *The flora and fauna of the Philippines, 1851–1966: an annotated bibliography*, I: *Plants*. 307 pp. (Nat. Appl. Sci. Bull. Univ. Philipp. 21). Quezon City. [1493 entries. Based on holdings available in Manila.]

MERRILL, E. D., 1922 (1923)–26. *An enumeration of Philippine flowering plants*. 4 parts. 3 text-figs., 6 maps (Publ. Bur. Sci. Philipp. 18). Manila: Department of Agriculture and Natural Resources, Philippine Islands. (Reprinted 1968, Amsterdam: Asher.)

Detailed systematic enumeration of seed plants; includes full synonymy (with references and literature citations), vernacular names, generalized indication of local and extralimital distribution, with citations of *exsiccatae* for lesser-known species, brief taxonomic commentary, and notes on habitat, etc., as recorded. Volume 4, apart from additions and corrections to the enumeration proper, is devoted to a general introduction, with chapters on physical features, climate, Philippine peoples and alphabets, origin of local vernacular names, botanical exploration and research, ecology, the biogeographical subdivisions of the archipelago, overall relationships of the flora and fauna, and geological correlations; a detailed bibliography (pp. 155–239) and complete indices to botanical and vernacular names are also provided. For additions, see E. QUISUMBING, 1930–44. New or interesting Philippine plants, I–II. *Philipp. J. Sci.* 41: 315–371, 28 figs., 3 pls.; 76: 37–56; and E. QUISUMBING and E. D. MERRILL, 1928–53. New Philippine plants, I–II. *Ibid.*, 37: 133–213, 4 pls.; 82: 323–339, 5 pls.

Partial works
Florulas exist for the Manila district and for Mt. Makiling near Los Baños, both in Luzon; the latter has so far not been completed.

MERRILL, E. D., 1912. *A flora of Manila*. 490 pp. (Publ. Bur. Sci. Philipp. 5). Manila: Department of Agriculture and Natural Resources, Philippine Islands. (Reprinted 1968, Lehre, Germany: Cramer; 1974, Manila: Bookmark.)

Descriptive manual of wild and more commonly cultivated vascular plants of an area – now largely built up – of about 100 km[2] around the city of Manila; includes analytical keys, essential synonymy, English and Tagalog vernacular names, indication of local and extralimital range, and notes on habitat, phenology, origin, cultivation, special features, etc.; indices to vernacular, generic and family names at end. The introductory section includes an introduction to descriptive botany together with a glossary. [Useful generally in inhabited lowland areas of the Philippines as it adequately covers the widespread secondary and adventive flora.][41]

PANCHO, J. V., 1983. *Vascular flora of Mount Makiling and vicinity (Luzon: Philippines)*. Part 1. iii, 476 pp., 147 textfigs. (Kalikasan, Suppl. 1). College, Laguna, Philippines.

Briefly descriptive illustrated flora with keys to genera and species, concise synonymy, references and citations, vernacular names (with languages), local and Philippines distribution, origins of non-native plants, citations of representative *exsiccatae*, and notes on habitat but no critical remarks; index. The general part encompasses the justification for the work along with sections on physical features, climate, soil, biota, floristics, basic vegetation formations, and botanical history; abbreviations and a key to families are also featured. [Five parts in all were planned to cover some 2038 species in an area of *c.* 200 km[2].][42]

Woody plants (including forest trees)

The best nineteenth-century work, unfortunately only to generic level, is *Sinopsis de familias y géneros de plantas leñosas de Filipinas* by Sebastian Vidal (1883, in 2 vols. respectively of text and atlas). The 100 atlas plates are well documented. From 1900 through the 1970s, the main dendrological works were G. P. AHERN, 1901. *Compilation of notes on the most important timber tree species of the Philippine Islands.* 112 pp., 43 col. pls. Manila; and H. N. WHITFORD, 1911. *The forests of the Philippines*, II: *The principal forest trees.* 113 pp., 103 pls. (Bureau of Forestry, Bull. 10(2)). Manila. The latter covered 106 principal species with 277 more meriting passing mention (from a total tree flora then estimated at 2500 species). These were succeeded respectively by an enumeration and a checklist: F. M. SALVOZA and M. LAGRIMAS, 1940–41. Check list of the trees of the Philippines, I–IV. *Philipp. J. For.* **3**: 477–548; **4**: 63–104, 191–218, 285–313 (never completed); and F. M. SALVOZA (as F. M. SALVOSA), 1963. *Lexicon of Philippine trees.* 136 pp. (Forest Products Res. Inst. Bull. 1). Los Baños, Laguna.[43] The first comparable successor to Whitford's work was G. SEEBER, H.-J. WEIDELT and V. S. BANAAG, 1979. *Dendrological characters of important forest trees from eastern Mindanao.* 440 pp., illus. (Schriftenreihe der GTZ 73). Eschborn: German Agency for Technical Cooperation. (A second edition, with nomenclatural changes (among others the substitution of *Eugenia* for *Syzygium*), appeared in 1985.)

The best current nationwide dendrology, an effective successor to that of Whitford, is that presented in *Guide to Philippine flora and fauna* (see below).

NATURAL RESOURCES MANAGEMENT CENTER, MINISTRY OF NATURAL RESOURCES, PHILIPPINES, 1986. *Guide to Philippine flora and fauna*, vols. 1–4. Illus. Manila (distributed by Goodwill Bookstore).

Volume 3 covers forest trees with sections respectively on dipterocarp (56) and non-dipterocarp (299) species (the latter inclusive of mangroves); for each species are given a description along with notes on habitat, distribution, frequency and economic importance as well as illustrations; references and a glossary at end. The arrangement of the non-dipterocarp trees is after Cronquist. Volume 2 includes a chapter on gymnosperms while vol. 4 has chapters on bamboo (26) and palm (97) species. Each volume is fully indexed. [Other parts of the first four volumes cover zoosporic fungi, seaweeds, endemic mosses, ferns, and grasses.][44]

Pteridophytes

Some fern species are covered in vol. 2 of *Guide to Philippine flora and fauna* (see above).

COPELAND, E. B., 1958–60(–61). *Fern flora of the Philippines.* 3 parts. iv, 557, [iv, iv] pp. (Monogr. Natl. Inst. Sci. Tech. 6). Manila: National Institute of Science and Technology, Philippines.

Briefly descriptive systematic treatment of true ferns; includes keys to genera and species, full synonymy (with references and citations), generalized indication of local and extralimital distribution (with citations of *exsiccatae* for less common species), and taxonomic commentary; index to all botanical names. The introductory section includes an account of fern geography along with an explanation of terminology.

926

Talaud Islands

Area: no data seen. – This small, non-volcanic group of islands lies between the Philippines and North Maluku. No general checklist is available, but the substantial report of a botanical survey of 1926 is accounted for below.

HOLTHUIS, L. B. and H. J. LAM, 1942. A first contribution to our knowledge of the flora of the Talaud Islands and Morotai. *Blumea* **5**: 93–256.

Systematic enumeration of collections made on a 1926 expedition, with descriptions of novelties, critical remarks, citations of *exsiccatae*, and notes on phytogeography, etc.

927

Maluku (The Moluccas): general

No floras or enumerations cover Maluku as a whole save for the one (actually two under one cover) of the tree species lists issued as part of the Indonesia-wide series produced by the Forest Research Institute, Bogor (for background, see under **913**) and its successor. Key partial works are here given under **928**.

Woody plants (including forest trees)
    SOEWANDA AMONG PRAWIRA, R. (comp.), 1975. *Daftar nama pohon-pohonan: Maluku Utara dan Selatan (North and South Moluccas)*. Revised edn. 130 pp., 61 pls. (Lap. Lemb. Penelit. Hutan 210). Bogor. (Mimeographed. Originally issued 1941, Buitenzorg, as F. H. HILDEBRAND (comp.), *Boomnamenlijst van Molukken* (Bosbouwproef-station, Serie Boomnamenlijsten 5); revised in two parts, 1951, Bogor, as *idem*, *Daftar nama pohon-pohonan Maluku-Utara*. [i], 53 pp. (Lap. Balai Penjelidik. Kehut. 45; also designated as Seri Daftar nama pohon-pohonan 28), and *idem*, *Daftar nama pohon-pohonan: Maluku-Selatan*. [i], 59 pp. (Lap. Balai Penjelidik. Kehut. 49; also designated as Seri Daftar nama pohon-pohonan 29).)[45]
    Uncritical, sparsely annotated checklists of forest trees respectively of Maluku Selatan and Maluku Utara known to attain a diameter at breast height of 40 cm or more, arranged alphabetically by families and genera; each entry (usually on one line) includes a vernacular name (Alfur, Malay, etc.), distribution (by subdistricts), number of collections (but not *exsiccata*-numbers), and durability class (on a scale of five). Each of the systematic lists proper is preceded by a list with all records, fully annotated, arranged alphabetically by vernacular names; at the end of each treatment are lists of amended botanical names, a lexicon of Indonesian names with more important local vernacular equivalents, and an alphabetical lexicon of genera with family equivalents. Following the separate lists is an indexed album of full-page illustrations of key species depicting botanical features. No keys are included.[46]
    WHITMORE, T. C., I. G. M. TANTRA and U. SUTISNA, 1989. *Tree flora of Indonesia: check list for Maluku*. 186 pp., illus. Bogor: Forest Research and Development Centre.
    Part 1 of this four-part work comprises a concise annotated checklist arranged alphabetically by families and genera with the names of smaller trees enclosed in parentheses; includes abbreviated citations of key literature, indication of maximum size, habitat and overall distribution, and occasional *exsiccatae*. Part 2 is a lexicon of vernacular and trade names with their botanical equivalents; the names have been taken from the *Daftar nama pohon-pohonan* and *Flora Malesiana*. Part 3 comprises a key to Dipterocarpaceae (by P. S. Ashton) covering all species from Sulawesi eastwards, and part 4 figures of 28 key timber species including eight dipterocarps. At the beginning of part 1 are lists of families and genera with statistics (not summarized).

928

Maluku: subdivisions

Under this heading are listed works referring to one or another of the three provinces into which Maluku currently is divided: Maluku Utara (North Moluccas), Maluku Tengah (Central Moluccas), and Maluku Tenggara (Southeastern Moluccas). The last two together constituted the former division of Maluku Selatan (South Moluccas).

I. Maluku Utara (North Moluccas)

This province comprises the northern Moluccas from Morotai in the north through Halmahera, Bacan and Obi to the Sula group, along with associated smaller islands such as Gebé, Ternate and Tidore. As befitting a part of the 'Spice Islands', botanical contacts have been prolonged; collecting, however, has been sporadic and even now some of the islands are not well known although efforts continue. The only separate checklist available is the section on Maluku Utara (pp. 67–126) in the 1975 tree species list by SOEWANDA (**927**).

II. Maluku Tengah (Central Moluccas)

The main islands of the present province of Maluku Tengah – the heart of what was formerly Maluku Selatan (South Moluccas) – are Buru, Seram (Ceram) and Ambon; also included are the Banda group and the southeastern islands stretching from the end of Seram out to Seram Laut. The only passably general work for this part of the 'Spice Islands' is the

'Golden Age' *Herbarium amboinense* by G. E. Rumphius. The only other overall checklist available is the section on Maluku Selatan in the 1975 tree species list by SOEWANDA (**927**).

RUMPHIUS, G. E., 1741–50. *Herbarium amboinense/Het amboinsche kruid-boek*. Transl. and ed. J. Burman. 6 vols. 669 pls., portr. Amsterdam: Uytwerf (and associates). (Reissued 1750.) Completed as *idem*, 1755. *Herbarium amboinense auctuarium*. 74, [20] pp., 30 pls. Amsterdam.

Amply descriptive, well-illustrated treatment (with parallel Latin and Dutch text) of vascular and some larger non-vascular plants (and corals, then thought to be partly plant-like); includes local vernacular names, indications of localities where observed or reported, and copious notes on life-form, natural history, special features, cultivation, properties, uses, folklore, etc. The author's preface in volume 1, dated 1690, explains the arrangement of the work (in 12 'books' or primary classes based on habit, practical criteria, local custom and other features) and its geographical coverage (not wholly restricted to Ambon). The *Auctuarium* includes additions as well as a complete index (the latter with some Linnaean binomials).[47]

III. Maluku Tenggara (Southeastern Moluccas)

The Southeastern Moluccas, biogeographically more or less intermediate between the Lesser Sunda Islands, the Central and Northern Moluccas, and Papuasia, includes the so-called 'Southwestern Islands' (Kepulauan Barat Daya) from Wetar to Damar and Leti to Babar, the Tanimbar (Timor Laut) group, the Kai (Ké or Kei) Islands, and (politically if not biogeographically) the Aru Islands (see also **931**). Only two organized floristic lists cover this area: the section on Maluku Selatan in the 1975 tree species list by SOEWANDA (see above under **927**) and Hemsley's enumeration, described below.

HEMSLEY, W. B., 1885. The south-eastern Moluccas. In W. Thomson and J. Murray (eds.), *Reports on the scientific results of the voyage of HMS Challenger during the years 1873–76, Botany*, 1(3): 101–226, pls. 64–65. London: HMSO. (Reprinted 1965, New York: Johnson.)

Includes among other writings an annotated enumeration of non-vascular and vascular plants from the Kai, Tenimber and Aru Islands as well as other parts of the Southeastern Moluccas; entries include descriptions of novelties, synonymy, references, brief indication of range, *exsiccatae*, critical remarks, and notes on occurrence, phytogeography, special features, cultivation, etc. No separate index is provided, but in an appendix is a discussion of Beccari's species from the area.[48]

Region 93

Papuasia (East Malesia)

Area: *c.* 912 000 km². Vascular flora: 15 000–17 000 species (Frodin, unpublished); other estimates range as high as 25 000. Good (1960, in *Bull. Brit. Mus. (Nat. Hist.), Bot.* **2**(8): 205–226) suggested (p. 221) a figure of 9000–9250 species of angiosperms for mainland New Guinea. – Papuasia, a biogeographical term first introduced by Otto Warburg in the 1890s, here includes the great island of New Guinea together with associated islands and extends eastwards to the end of the Solomon Islands group. Its western limit is defined by the continental shelf with which Lydekker's Line is associated. The bulk of the Torres Strait Islands, however, has been assigned to **439** as part of eastern Australia (**Region 43**). The region thus includes the Aru Islands, the Indonesian province of Irian Jaya, the independent state of Papua New Guinea (i.e., the former Territory of Papua and the Trust Territory of New Guinea, including Buka and Bougainville), and the state of Solomon Islands (formerly British Solomon Islands Protectorate) exclusive of the Santa Cruz group and associated islands (**951, 952**).

Of the general works on Malesia given under **910–30**, only *Flora Malesiana* covers Papuasia at all adequately, and even then its earlier volumes omit the Louisiades (**935**) and Solomons and the outer atolls (**938, 939**). At the time the *Flora Malesiana* project was being developed, these groups, then botanically very imperfectly known, were loosely thought to be of 'Pacific' affinity. Subsequent study has shown that their flora is essentially Papuasian, although in both groups unusual elements exist and many 'mainland' taxa are absent.

No complete comprehensive flora for Papuasia is yet available. A beginning has been made with publication of the first three volumes of the series *Handbooks of the flora of Papua New Guinea* (1978–), a project of the Division of Botany of the Papua New Guinea Forest Research Institute at Lae; as yet, rather less than 10 percent of the vascular flora has been covered, with current prospects for continuation poor. A forerunner from the same institution, terminated far from completion, was *Manual of the forest trees of Papua and New Guinea* (1964–70). While concentrating on Papua New Guinea, both works extend coverage to Irian Jaya (West New Guinea) and the Solomon Islands through utilization of literature records and collections available at Lae.

Apart from these two works, the existing literature comprises a bewildering variety of expedition checklists, series of 'contributions', partial floras and florulas, and works on special groups (including trees), as well as monographs and revisions, some not limited to Papuasia; this situation continues to the present. Coming to grips with it is almost as formidable a task as that which faced Lam in the 1930s, although bibliographic aids in the interim have improved.[49] To a very large extent, this reflects the confused but compelling bio-history of the region: different political/administrative regimes in space and time, the lack of local botanical centers until the 1940s, imperial rivalries, availability of resources and motivation, the disruptions of war, occupation and other events, and above all the magnetic attraction of the 'Last Unknown' for natural historians from many countries.[50] Collections and associated materials are thus widely scattered, in contrast to, say, Peninsular Malaysia or Indochina. Any synthesis, while sorely needed, would thus be a formidable undertaking, although it is a matter for argument whether it be better handled through the *Flora Malesiana* project or as a separate regional series like the *Handbooks* or even an 'interim enumeration' such as that of Merrill for the Philippines or Brako and Zarucchi for Peru, the last-named taking advantage of the three or four nomenclatural/bibliographic card files known to exist. An alternative possibility is production of a 'generic flora' along the lines of that now being written for Madagascar.

The most comprehensive group of contributions, although not covering all families, is the great *Beiträge zur Flora von Papuasien* (1912–42), comprising 150 numbers in 26 parts successively edited by Carl Lauterbach and Ludwig Diels. Containing many keyed regional revisions, it was largely completed by the time of the destruction of the greater part of the Berlin-Dahlem collections through bombing in 1943 (Friedrich Markgraf, personal communication). Along with the *Handbooks* it is treated here as a major regional work. Other leading contribution-series are the botanical volumes of the original *Nova Guinea* (1909–36), edited by August Pulle, and the 'Botanical results of the Archbold Expeditions', including its two major elements *Plantae papuanae archboldianae* by Elmer Merrill and Lily Perry (1939–53; the last three parts by Perry alone) and *Studies of Papuasian plants* by Albert C. Smith (1941–44). As practical contributions the Dutch and North American series are, however, by and large of lesser consequence (although both are for the most part separately indexed). Some contributions in *Nova Guinea*, a work largely limited to the present-day Irian Jaya province of Indonesia, were written by authors also contributing to the *Beiträge*. Paralleling these works, but undertaken a generation or more earlier, is the major Australian contribution, the 'ad hoc' *Descriptive notes on Papuan plants* by von Mueller (1875–90); this relates mainly to southern Papua New Guinea (the former British New Guinea).

Many other floristic treatments of the late nineteenth and early twentieth centuries, some of them relatively substantial or significant like those by Lilian Gibbs, Henry Ridley, Th. Valeton, Otto Warburg or Cyril T. White, also exist but as they are essentially collection reports they cannot be considered here. On the other hand, with the so far limited coverage furnished by modern general works, attention is given to some key florulas ranging from Karl Schumann's *Die Flora von Neu-Pommern* (1898) to Brother William Borrell's *Flora of Kairiru Island* (1989); a listing is also given for the two parts of *Flora der deutschen Schutzgebiete in der Südsee* by Schumann and Lauterbach (1901–05), the first general account for former German New Guinea (in the first part also including Micronesia and Western Samoa). The substantial *Alpine flora of New Guinea* by van Royen appears at **903** (under New Guinea).

The associated islands, although better covered than in 1942, are still imperfectly documented. Worthy of note are the accounts by the Kew botanist William Hemsley for the Aru Islands and the Admiralty group (1885), visited in the 1870s by the *Challenger* expedition, Wagner and Grether's 1948 treatment of pteridophytes in the Admiralties (perhaps the leading contribution based on collecting by servicemen during World War II), Don Foreman's 1972 checklist of the flora of Bougainville, and, finally, Father Gerhard Peekel's remarkable *Flora of the Bismarck Archipelago for naturalists* (1984), based on his final typescript *Illustrierte Flora des Bismarckarchipels für Naturfreunde*, saved from possible loss during World War II and completed in 1947. Two standard checklists exist for the Solomon Islands: Whitmore (1966) and Hancock and Henderson (1988), both accounted for below.

The major forest trees of Irian Jaya, Papua New Guinea and the Solomons have each been covered in one or more processed or printed publications of varying quality. Those for Irian Jaya and Papua New Guinea have been listed under respective subheadings within **930**, while those on the Solomon Islands are given under **938**. A full tree checklist – the first for the whole of New Guinea – appeared in 1997 in the *Tree flora of Indonesia* checklist series; this appears under **930**.

A rather imperfect set of keys to families and genera, with a bibliography of sources, was distributed by van Royen in 1959 and is accounted for below under the main heading. A projected successor has not so far been published.

Botanically, the region is at present on average about as well collected as Borneo but coverage is still quite uneven. Irian Jaya lags considerably behind Papua New Guinea and the Solomons and the southern fall of the central cordillera and parts of the foothill zone in the north are moreover somewhat neglected. Undercollected areas of more limited extent exist elsewhere, but much work in recent years has been carried out in the Louisiades, Bismarcks and Admiralties, making sketchy floristic checklists feasible. In the 1990s, much new collecting has been accomplished in Kepala Burung (the Vogelkop) and in the PT Freeport mining concession lands from G. Jaya to the south coast; additional work is also expected in the Cyclops Range near Jayapura.[51]

Bibliographies. General and supraregional bibliographies as for Superregion 91–93.

Regional bibliographies

See also **Superregion 91–93**, **Supraregional bibliographies** (ALLIED GEOGRAPHICAL SECTION, vol. 1; DEPARTMENT OF EDUCATION, JAPAN, vol. 1 (botany, pp. 20–61).

LAM, H. J., 1934. Materials towards a study of the flora of the island of New Guinea. *Blumea* 1: 115–159, 2 maps. [This fine survey includes *inter alia* a selection of more important references on New Guinea, as well as detailed indices to the family revisions in *Beiträge zur Flora von Papuasien* and *Nova Guinea, Botanique* as far as then available.]

SAULEI, S., 1996 (1997). A bibliography of the flora and vegetation of Papua New Guinea. *Papua New Guinea J. Agric., Forestry & Fisheries* 39(2): 20–168. [Titles arranged by author in eight primary sections; unannotated. Quite complete albeit with some bibliographic errors. Some unpublished material is included.]

Indices. General and supraregional indices as for Superregion 91–93.

930
Region in general (including mainland New Guinea)

The major works on the region, all with keys, are *Beiträge zur Flora von Papuasien* (1912–42), *Handbooks of the flora of Papua New Guinea* (1978–), and the never-completed *Manual of the forest trees of Papua and New Guinea* (1964–70). Key subsidiary works include *Flora der deutschen Schutzgebiete in der Südsee* (1900–05), *Nova Guinea, Botanique* (1909–36), and *Plantae papuanae archboldianae* (1939–53), here given under **Partial works**. These are followed by a selected group of specialized works on forest trees.

Keys to families and genera

ROYEN, P. VAN [1959]. *Keys to the families and genera of higher plants in New Guinea*. 3 parts. 227 pp. Leiden. (Mimeographed.) [Largely compiled, artificial analytical keys to the genera and families of seed plants in mainland New Guinea, with family descriptions; index. An appendix gives a useful list of source references, arranged by families. Now greatly in need of revision.][52]

JOHNS, R. J. and A. HAY (eds.), [1981]–84. *A students' guide to the monocotyledons of Papua New Guinea*. Parts 1–4 (part 4 by N. H. S. Howcroft). Illus. (Part 1 also published as *Training Manual for the Forestry*

College, vol. 13). Bulolo (part 1); [Lae] (parts 2–3); Lae: L. J. Brass Memorial Herbarium, Forestry Department, Papua New Guinea University of Technology (part 4). Accompanied by R. J. Johns, 1987–89. *The flowering plants of Papuasia*. Parts 1–3. Illus. Lae: Papua New Guinea University of Technology.

Students' manuals, with keys to genera and species, generic and family descriptions, lists of species, and notes on distribution, identification (with *exsiccatae* sometimes cited), ecology, and other features; many line drawings; references (only those in English) at end of each family treatment; no indices. An introduction and a page 'How to recognize a monocotyledon' precede the first of the monocotyledon parts. [In the monocotyledon series, part 1 includes treatments (all by Hay) of Stemonaceae, Dioscoreaceae, Taccaceae, Philydraceae, Philesiaceae, Pontederiaceae, Araceae and Hypoxidaceae. Part 2 covers Amaryllidaceae, Iridaceae, Cyperaceae, Restionaceae, Joinvilleaceae, Hanguanaceae, Haemodoraceae and Agavaceae (by Hay) and Corsiaceae (by Johns); part 3 is on Palmae (by Hay), and part 4 on part of the Orchidaceae (by Howcroft). In the series on dicotyledons, the order of families is that of Cronquist. Part 1 covers Magnoliidae; part 2, Dilleniidae; part 3, Caryophyllidae; particular features are references to key literature and indices to genera and families.][53]

Lauterbach, C. and L. Diels (eds.), 1912–42. Beiträge zur Flora von Papuasien, I–XXVI (in 150 parts). Illus. *Bot. Jahrb. Syst.* **49–72**, *passim*. (Separate reprints of at least some parts known.)

Comprises regional revisions and miscellaneous contributions (often with keys) on a wide range of families and genera; individual treatments, many by specialists, contain descriptions, synonymy (with references and citations), localities with *exsiccatae*, and notes on classification, habitat, special features, etc.; detailed indices in *Beiträge* VI and X.[54] [Many novelties have been described in this series, and much material also published in *Nova Guinea* (see below) incorporated. Some treatments are limited to former German New Guinea (exclusive of Micronesia and Nauru), but others deal with Papuasia as a whole and even beyond. General introductions to many families were written from a floristic point of view by the senior editor (who also in 1928–30 devoted some parts in the series to a floristically based vegetation survey). Some family revisions have been published elsewhere; the most notable is R. Schlechter, 1911–14. *Die Orchidaceen von Deutsch Neu-Guinea*. lxvi, 1079 pp. (Repert. Spec. Nov. Regni Veg. [Fedde], Beih. 1). Berlin: Fedde. (Reprinted 1974, Königstein/Ts., Germany: Koeltz. English translation, 1982, Melbourne: Australian Orchid Foundation.)][55]

Womersley, J. S. *et al.* (eds.), 1978– . *Handbooks of the flora of Papua New Guinea*. Vols. 1– . Illus., maps (one folding). Melbourne: Melbourne University Press (for the Papua New Guinea Government).

Comprises a random group of well-illustrated regional family revisions by various authors; each treatment includes keys to genera and species, descriptions, some synonymy, generalized indication of internal range (by 'district', i.e., province, in Papua New Guinea; or *kabupaten* (regency), in Irian Jaya) and extralimital distribution, critical remarks, notes on ecology, special features, biology, dispersal, uses, etc., and family/generic literature sources; illustrated glossary, list of higher plant families in Papua New Guinea (209 in all), selected references (pp. 273–274), and index to botanical names in each volume. The general part (vol. 1) includes technical notes and a brief introduction to the vegetation. [Although covering the whole of Papuasia as recognized here, emphasis in this work has been placed on Papua New Guinea. The three volumes so far published treat 608 species in 42 families (about 4 percent of all vascular species in the region); subsequent volumes will contain groups of additional families as they are ready, without observation of a systematic sequence. Volume 2, covering mostly small families but including Elaeocarpaceae and the Loranthaceae sensu stricto, was not released until well into 1982. Volume 3 appeared in 1995 – after a lengthy delay – and among other families includes parts of Araliaceae and Clusiaceae as well as the Proteaceae.][56]

Secondary and partial works

Merrill, E. D. and L. M. Perry (eds.), 1939–49. Plantae papuanae archboldianae, I–XVIII. *J. Arnold Arbor.* **20–30**, *passim*. Continued as L. M. Perry, 1949–53. Plantae papuanae archboldianae, XIX–XXI. *Ibid.*, **30**: 139–165; **32**: 369–389; **34**: 191–257. Accompanied by A. C. Smith, 1941–44. Studies of Papuasian plants, I–VI. *Ibid.*, **22**: 60–80, 231–252, 343–374, 497–528; **23**: 417–443; **25**: 104–298.

These sets of contributions – which together form the

bulk of the so-called 'Botanical results of the Archbold Expeditions' – comprise treatments of a majority of flowering plant families in Papuasia. The papers by Merrill and Perry mostly take the form of critical notes, descriptions of novelties, and new records with *exsiccatae* and localities; while those by Perry alone and Smith comprise full revisions of families and genera, sometimes (in Perry's treatments) also with keys but in all cases including descriptions, localities with *exsiccatae*, and critical notes. The last part of the Merrill and Perry series has an index to genera and families covered. [The principal basis were the collections of L. J. Brass made on the first three Archbold Expeditions of 1933–34, 1936–37, and 1938–39. Many treatments were limited to material available in the U.S.A.][57]

PULLE, A. A. *et al.*, 1909–36. Botanique. In H. A. Lorentz *et al.* (eds.), *Nova Guinea. Résultats de l'expédition scientifique néerlandaise de la Nouvelle Guinée*, 8, 12, 14, 18. Illus. Leiden: Brill.

Comprises miscellaneous contributions in many families and genera (but with few keys) based for the most part on botanical collections from West New Guinea (now the Indonesian province of Irian Jaya) made between 1907 and 1921. Most accounts consist principally of descriptions of new taxa along with miscellaneous records of known species and critical notes; of these, easily the most significant is the set of papers on Orchidaceae by J. J. Smith, accompanied by numerous delicately executed plates. Some papers are complementary to their counterparts by the same authors in *Beiträge zur Flora von Papuasien* (see above). No detailed index was, however, ever published. Additional contributions on these and later collections are found in *Nova Guinea*, N.S. 1–10 (1937–59) and *Nova Guinea, Botany* 1–24 (1960–66).[58]

Woody plants (including forest trees)

The tree flora of Papuasia is among the wonders of the tropical world. All that is available, however, is a heterogeneous collection of treatments of principal forest trees over the whole or its constituent parts, all but one comparatively limited in scope. Only in 1997 was a first full checklist published. The sole descriptive account is the set of manuals by van Royen (and others) of 1964–70 which have been treated below as a 'general' work. Whitmore's treatment of the 'big trees' of the Solomons (in his *Guide to the forests of the British Solomon Islands*) appears under **938**. The remaining works are accounted for under the subheadings for Irian Jaya and Papua New Guinea.

ROYEN, P. VAN *et al.*, 1964–70. *Manual of the forest trees of Papua and New Guinea*. 9 parts (in 6 fascicles). Illus., maps. Port Moresby, Papua New Guinea: Department of Forests, Territory of Papua and New Guinea (reprinted). (1st edn. of part 1, 1964; 2nd edn., 1970.)

Descriptive treatments (in part compiled) of individual families including forest trees (with emphasis on the territories now comprising Papua New Guinea and on the more important species), with keys, partial synonymy, local distribution (with some *exsiccatae*), and notes on taxonomy, habitat, life-form, wood, properties, uses, etc.; some vernacular as well as trade names included. [Now for the greater part superseded.][59]

WHITMORE, T. C., I. G. M. TANTRA and U. SUTISNA (eds.), 1996. *Tree flora of Indonesia: check list for Irian Jaya*. ix, 367 pp., illus. Bogor: Forest Research and Development Centre.

Part 1 of this four-part work comprises a concisely annotated checklist, arranged alphabetically by families and genera with the names of smaller trees enclosed in parentheses; includes abbreviated citations of key literature, indication of maximum size, habitat and overall distribution, and occasional *exsiccatae*. Part 2 is a lexicon of vernacular and trade names with their botanical equivalents; the names have been taken from the *Daftar nama pohon-pohonan* and *Flora Malesiana*. Part 3 comprises keys to Dipterocarpaceae (by P. S. Ashton), with most species illustrated in part 4 (which also includes figures of some important non-dipterocarp species). At the beginning of part 1 are lists of families and genera with numbers of species classed by relative size. [Coverage of certain families and genera is sketchy due to imperfect taxonomic knowledge; among these are Lauraceae (apart from *Cinnamomum*) and *Syzygium*.]

Pteridophytes

Included here is the not-yet-completed series of students' manuals by R. J. Johns and A. Bellamy, which despite their titles furnish coverage for all of Papuasia. They may be considered as 'gap-filling' but generally are not critical nor do they provide full and consistent coverage to species level. No continuation, however, has appeared since 1981.

JOHNS, R. J. and A. BELLAMY, 1979 (1980). *The ferns and fern allies of Papua New Guinea*. Parts 1–5 (in 1 fascicle). Illus., maps. [Bulolo, P.N.G.: Forestry College.] Continued as R. J. JOHNS, *The ferns and fern-allies of Papua New Guinea*. Parts 6–12 (in 1 fascicle). Illus., maps (Papua New Guinea University of Technology Research Reports 48–81). Lae, Papua New Guinea.

Students' manual, with keys to genera and species, short descriptions, synonymy, citations of *exsiccatae* (without localities), generalized indication of distribution and ecology, and (sometimes) taxonomic commentary; illustrations (figures and/or monochrome prints) of most taxa covered as well as dot distribution maps; literature references. [Part 1 comprises a brief introductory section and a synopsis of genera. Parts 2–12 cover Parkeriaceae, Matoniaceae, Cheiropleuraceae, Dipteridaceae, Ophioglossaceae, Marattiaceae (in part), Osmundaceae, Psilotaceae, Marsileaceae, Salviniaceae, and Azollaceae.][60]

I. Irian Jaya (Papua Barat)

No general or partial floras for Irian Jaya so far have been published. Work is, however, currently in progress on a database and florula for the northwestern Kepala Burung (Jazirah Doberai, Vogelkop) peninsula, and some preliminary materials have been circulated.

Forest trees

The Irian Jaya province of Indonesia, earlier known as Irian Barat (West Irian), until recently had its most extensive (but least critical) coverage of forest trees in one of the tree species lists produced by the Forest Research Institute of Indonesia at Bogor (for background details, see **913**). A successor in the *Tree flora of Indonesia* series appeared in 1997; as it covers the *whole* of New Guinea, however, it is described above under **930, Woody plants (including forest trees)**. Other works listed here are rather more detailed but cover fewer species.

HILDEBRAND, F. H., 1953. *Daftar nama pohon-pohonan: Irian.* 46 pp. (Lap. Balai Penjelidik. Kehut., hors série; also designated as Seri Daftar nama pohon-pohonan 35). Bogor. (Originally published 1941, Buitenzorg, as *idem*, *Boomnamenlijsten van Nieuw-Guinea* (Bosbouwproefstation, Serie Boomnamenlijsten 6); 1953 version reissued 1963 as *Daftar nama pohon-pohonan: Irian Barat* (Lap. Balai Penjelidik. Kehut. 91).)

Sparsely annotated, uncritical checklist of forest trees known to attain a diameter at breast height of 40 cm or more, alphabetically arranged by families and genera; each entry (usually on one line) includes a vernacular name, distribution (by subdistricts), number of collections (but not *exsiccata*-numbers), and durability class (on a scale of five). The systematic list is preceded by a list with the fully annotated records all arranged alphabetically by vernacular names; other lexica appear at the end. No keys are included. [The deficiencies of these lists have been discussed under Region 91, with an example under **927**.]

VERSTEEGH, C., 1971. *Key to the most important native timber trees of Irian Barat (Indonesia) based on field characters.* 63 pp., 7 pls. (Meded. Landbouwhoogesch. Wageningen 71-19). Wageningen.

Largely unillustrated dendrological treatment of 108 leading tree species recorded from the northern lowlands of Irian Jaya, with a glossary, key for identification, and standardized descriptive notes stressing habit, trunk and bark characters arranged under subheadings. A related Indonesian-language guide, with numerous colored illustrations of habit, bark and wood, is DIREKTORAT JENDERAL KEHUTANAN, INDONESIA, 1976. *Mengenal beberapa jenis kayu Irian Jaya.* Jilid (Vol.) 1. ix, 114 pp., col. illus., 2 maps. Jakarta. [Covers 70 species (or species-groups) of forest trees with colored illustrations of boles (with crowns) and woods and descriptive text; includes many vernacular names, but no keys.]

II. Papua New Guinea

In addition to the following, a modern illustrated florula for a small portion of the Port Moresby region – an offshore island of *c.* 13 ha – has been published: H. FORTUNE HOPKINS and J. I. MENZIES, 1995 (1996). *The flora of Motupore Island.* 212 pp., illus., 8 col. pls. University, N.C.D., Papua New Guinea: University of Papua New Guinea Press.

Partial works

SCHUMANN, K. and C. LAUTERBACH, 1901 (1900). *Die Flora der deutschen Schutzgebiete in der Südsee.* xvi, 613 pp., 22 pls., map. Leipzig: Borntraeger; and *idem*, 1905. *Nachtrage zur Flora der deutschen Schutzgebiete in der Südsee (mit Ausschluss Samoa's und der Karolinen).* 446 pp., 14 pls., portr. Leipzig. (Reprinted in 1 vol., 1976, Vaduz, Liechtenstein: Cramer (*apud* Gantner)).

The main work comprises a systematic enumeration of the then-known non-vascular and vascular plants of German New Guinea, Samoa and Micronesia, with descriptions of new taxa, detailed synonymy (with references and citations), vernacular names (without provenance), *exsiccatae* with localities, summary of overall range both within and outside the area, and miscellaneous notes; list of references; index to genera. The *Nachtrage* similarly documents collections made between 1899 and 1904, particularly those of Rudolf Schlechter. The introductory sections include chapters on the history of botanical collecting as well as short biographies of important collectors, while a complete general index to the whole work appears at the end. Encompasses 2208 species (1560 vascular), perhaps 15 percent of the total flora of Papuasia (exclusive of the Solomon Islands) but including many secondary and weed species. [The reprint is in a considerably reduced and in some respects more convenient format than the original volumes.]

STREIMANN, H., 1983. *The plants of the Upper Watut watershed of Papua New Guinea.* iv, 209 pp., map. [Canberra]: National Botanic Gardens, Department of Territories and Local Government, Australia.

Annotated checklist of lichenized fungi, bryophytes and vascular plants with indication of habitat(s), localities (for less frequently collected species), and *exsiccatae*. Pp. 187–194 comprise a list of type collections described from the area; pp. 195 to the end contain references. The introductory section includes a historical review of collecting and an outline of the plan and scope of the work. All known collections to 1972 are covered. [1789 phanerogams and 325 ferns and allies accounted for; no figure for the area covered is, however, given.]

Forest trees

As indicated in the previous edition, it remains worthy of note that the Division of Botany of the Department of Forests (since 1989 incorporated into the Papua New Guinea Forest Research Institute) has so far failed to complete a substantial forest flora or tree checklist. The principal contributions, largely covering the principal species, have come from the Papua New Guinea Forestry College, Bulolo. A predecessor there of Johns's *Common forest trees* (and largely incorporated within it) was J. J. HAVEL, 1975 (1976). *Forest botany*. 2 parts. Illus. (Training Manual for the Forestry College 3). Bulolo, Papua New Guinea.[61]

JOHNS, R. J., 1975–77. *Common forest trees of Papua New Guinea*. 13 parts. Illus. (Training Manual for the Forestry College 8). Bulolo, Papua New Guinea.

Copiously illustrated, largely compiled descriptive students' guide to more common forest tree species, but including many remarks on other trees and other sorts of plants found in the forest along with keys, ecological notes, etc.; short lists of references in each part; index (part 13). [Much of Havel's *Forest botany* (see above) is incorporated.]

WHITE, C. T., 1961. *Forest botany lectures*. Revised by K. J. White and E. E. Henty. 89 pp. [Port Moresby: Department of Forests, Territory of Papua and New Guinea.] (Mimeographed.)

The closest approach in Papua New Guinea to a forest flora, this is an introductory treatment in forest botany with families grouped into 10 'lectures' or sections; partial keys, descriptions (with emphasis on principal tree species), and miscellaneous biological and other notes are included, with an index to families and genera at end. [Originally written for use during World War II, but last revised in 1961 and now somewhat outdated. More recent works, however, lack its scope and readability.]

931

Aru Islands

See **928** (HEMSLEY). No separate checklists are available, though sporadic collections have been made over a long period. Although biologically Papuasian, the islands are administratively part of South Maluku Province (**928**).

932

Islands west of New Guinea

These comprise Misool, Salawati, Batanta, Waigeo, and associated smaller islands (including Gebé). No checklists are available except for parts of Waigeo: P. VAN ROYEN, 1960. The vegetation of some parts of Waigeo Island. *Nova Guinea, Bot.* **3**: 25–62, illus.

933

Islands in Cendrawasih (Geelvink) Bay

The main islands are Biak and Japen. No separate florulas or checklists are available.

934

Trobriand and Woodlark Islands

This unit covers the chain of islands from the Trobriands through Woodlark (Murua) to the Laughlans. Extensive collections have been made only since World War II, and no separate checklists are yet available. [The large, high d'Entrecasteaux Islands are here considered part of the mainland; see **930**.]

935

Louisiade Archipelago

This comprises the large islands of Misima, Tagula (Sudest) and Rossel along with many associated smaller islands and islets. As with the preceding unit, extensive collecting has only been carried out since World War II and no separate checklist is available at present. The islands are of considerable botanical interest, with certain links with the flora of New Caledonia, and here the important timber tree family

Dipterocarpaceae reaches its eastern limit. A separate florula or checklist would be highly desirable, particularly as they were not included in early volumes of *Flora Malesiana*.

936

Northern 'horstian' islands

Included here, from east to west, are the Kairiru group off Wewak, the Schouten Islands, Manam, Karkar, Long and Umboi. Biogeographically they are more or less distinct from the mainland. A florula was recently published for Kairiru.

Partial work

BORRELL, O. W., 1989. *An annotated checklist of the flora of Kairiru Island, New Guinea*. xii, 241 pp., illus., pls. (part col.), maps. Bulleen, Vic.: The author.

Illustrated descriptive treatment, without keys; includes limited synonymy, some English and vernacular names, distribution and vouchers, and index to all botanical names. The introduction includes some notes about the island, which rises to 940 m.

937

Bismarck Archipelago and Admiralty Islands

Included here are the Bismarck Archipelago proper (with the islands of New Britain, New Ireland, Lavongai (New Hanover) and many smaller islands) and the associated Admiralty Islands (from Manus and its neighbors west to Aua and Wuvulu). Apart from the scattered citations provided by general works on Papuasia (**930**), especially Schumann and Lauterbach's flora (under **Papua New Guinea**) and the *Beiträge*, no useful coverage is available save for the works described below.

I. Admiralty Islands

Two references, one very old, refer to the area but with much new collecting in recent years both are now very incomplete. The fern flora, however, has very useful keys for identification.

HEMSLEY, W. B., 1885. The Admiralty Islands. In W. Thomson and J. Murray (eds.), *Reports on the scientific results of the voyage of HMS* Challenger *during the years 1873–76, Botany*, 1(3): 227–275. London: HMSO. (Reprinted 1965, New York: Johnson.)

Includes an enumeration of plants known from the northwest of Manus (Nares Harbor) and other areas, based to a large extent on the collections of the expedition's naturalist, H. N. Moseley, with descriptions of novelties and incorporating synonymy, references, distribution (with *exsiccatae*), and notes on occurrence, dispersal, etc.; no separate index.

WAGNER, W. H., JR. and D. F. GRETHER, 1948. *The pteridophytes of the Admiralty Islands*. Berkeley, Calif. (*Univ. Calif. Publ. Bot.* 23(2): 17–110, pls. 5–25.)

Keyed, annotated enumeration of all known ferns and fern-allies of (mainly) Manus Island, with descriptions of new taxa; includes essential synonymy (with references), localities (with citations of *exsiccatae*), general indication of overall range, and notes on distinguishing features, taxonomy, habitat, frequency, life-form, special features, etc.; photographs of new or interesting species; no separate index. The introductory section gives a general description of Manus and its fern flora; species concepts are also discussed.[62]

II. Bismarck Archipelago

The available works are now very incomplete in the light of additional exploration since World War II, especially in the mountains.

PEEKEL, G., Fr., 1984 (1985). *Flora of the Bismarck Archipelago for naturalists*. Transl. E. E. Henty. [v], 638 pp., over 911 text-figs., map. Lae: Division of Botany, Office of Forests, Papua New Guinea.

Illustrated manual of vascular plants (mainly of lowland New Ireland and the the northeastern Gazelle Peninsula of New Britain), without keys; includes scientific names (both as used by Peekel and their modern equivalents) and their derivation, synonymy, vernacular names, local distribution (somewhat sketchily indicated) and notes on habit, habitat, etc. but no *exsiccatae*; synopsis of families, lists of medicinal, poisonous and useful plants, the coastal flora (pp. 592–595), drift fruits and seeds, a phenological calendar, summary of novelties originating from the author's collections (pp. 604–609), glossary, abbreviations of publications, and general index to all names (not always accurate) at end. The beginning of the work contains a foreword by the author and a biography by H. Sleumer.[63]

Partial work: Gazelle Peninsula

SCHUMANN, K. *et al.*, 1898. *Die Flora von Neu-Pommern*. Berlin. (*Notizbl. Bot. Gart. Berlin* **2**: 59–158, map.)

Annotated systematic enumeration of known non-vascular and vascular plants, with descriptions of new taxa; includes synonymy, citations of literature and *exsiccatae*, localities, and taxonomic commentary. The introductory section features notes on botanical exploration, physiography, and vegetational/floristic formations together with an index. [Coverage limited to the Gazelle Peninsula in northeastern New Britain, centering on Simpson Harbor and treating particularly the large collections of the German biologist Friedrich Dahl. All records are included in SCHUMANN and LAUTERBACH (**930/II**).][64]

938

Solomon Islands (including Bougainville)

Area: 28784 km^2 (state of Solomon Islands). Vascular flora: 3210 species (Hancock and Henderson, 1988; see below). – In this archipelago are the islands from Nissan, Buka and Bougainville to San Cristobal, along with Sikaiana, Rennell and Bellona. All of them except the first three now form part of the state of the Solomon Islands; Bougainville, Buka and Nissan are nominally part of Papua New Guinea. For the outer atolls, see **939**; for the Santa Cruz Islands and dependencies (comprising the Temotu province of Solomon Islands), see **951** and **952**.

Serious collecting in the Solomon Islands began in the 1880s with a further surge in the 1930s. Extensive forest surveys in the 1960s as well as an expedition organized by the Royal Society of London in 1965 added substantially to available collections, facilitating compilation of a first complete checklist (Whitmore, 1966; see **Woody plants** below). A complementary work for Bougainville was published in 1972. A new Solomons checklist appeared in 1988; this accounts for additional collections and moreover is better oriented to the herbaceous, useful and cultivated flora.

The Solomon Islands in the past were not included within the scope of *Flora Malesiana*, but more recent treatments have extended coverage therein (exclusive, however, of the Santa Cruz group and its dependencies, biogeographically quite distinct).

HANCOCK, I. R. and C. P. HENDERSON, 1988. *Flora of the Solomon Islands*. 203 pp. (Research Bull. 7). Honiara: Research Department, Ministry of Agriculture and Lands, Solomon Islands.

Tabular checklists of identical content, the genera and species alphabetically arranged in the first and by major taxon and family in the second; for each specific and infraspecific entry are included vernacular names (mainly Kwara'ae, continuing a usage established during the forest-botanical survey of the 1960s) and codes standing for status, habit and uses (explained in section 3 of the general part); there is, however, *no* breakdown of distribution by island or island group nor any citation of *exsiccatae*. Associated with the checklists in the special part is an alphabetically arranged conspectus of the families represented in the flora. The general part, in addition to its explanation of the fields in the checklist, includes in its other sections a description of vegetation formations, acknowledgments and (p. 19) a short list of references.

Woody plants (including forest trees)

For additional information on Bougainville and Buka, including a dendrological treatment of the major trees, see D. B. FOREMAN, 1972. *A check list of the vascular plants of Bougainville, with descriptions of some common forest trees*. 194 pp., illus., map (Bot. Bull. (Lae) 5). Lae: Division of Botany, Department of Forests, Papua New Guinea.

WHITMORE, T. C., 1966. *Guide to the forests of the British Solomon Islands*. 226 pp., 7 text-figs., maps (in endpapers). London: Oxford University Press.

The latter part of this work comprises an alphabetical list of all species of vascular plants recorded from the Solomon Islands (including Bougainville and Santa Cruz), with Kwara'ae vernacular names (where known) and references for those species originally described from the area. The remainder of the book is devoted mainly to an illustrated descriptive treatment of the more common large forest trees. The introductory section includes notes on vegetation formations and methods of tree description, with lists of special characteristics (milky sap, deciduousness, etc.). A list of references is found at the end.

939

Outer northeastern atolls

Included here are the atolls of Nuguria, Kilinailau, Tauu, Nukumanu and Ontong Java, all in Papua New Guinea except for the last-named (in Solomon Islands). Scattered collections have been made, but no checklists appear to be available. [As of 1992, 64 native species were recorded from Ontong Java, based on unpublished data from T. Bayliss Smith.]

Superregion 94–99

Oceania

This vast superregion – the real 'South Seas' – comprises all Pacific islands and island groups from western Micronesia, the Santa Cruz Islands and New Caledonia (along with the Coral Sea Islands) across to Rapa Nui (Easter Island) and Sala-y-Gómez and north to the Hawaiian Islands. It thus corresponds largely to the primary limits of coverage of Merrill's 1947 bibliography and from a biogeographical point of view to the Polynesian and Neocaledonian subregions of the Oriental Region of Thorne (1963).[65] Botanically its affinities are with Malesia, although in the Hawaiian Islands there is an appreciable tropical American element (Balgooy, 1971).[66]

Oceania has attracted much attention from botanists of many countries ever since the latter part of the eighteenth century, when the classical contributions of Philibert Commerson, Joseph Banks, Daniel Solander and the Forsters were made, and at the present time is relatively well – if unevenly – known. Good bibliographical coverage is also to hand in *A botanical bibliography of the islands of the Pacific* by E. D. Merrill (1947; subject index by E. H. Walker), *Flora Malesiana Bulletin*, and several more local or specialized cumulations including *Island bibliographies* (1955, 1971) by F. R. Fosberg and M.-H. Sachet.

This exceptional interest, however, has also led in time to serious fragmentation in botanical progress and floristic publication, with attempts at synthesis comparatively few and then more monographic than floristic. Broadly speaking, though, four stages in the development of the present level of botanical knowledge of Oceania can be recognized.

Until the 1840s, botanical exploration in the superregion was largely the province of the omnibus voyages of discovery and exploration, ranging from Bougainville and Cook to Aubert du Petit-Thouars and Wilkes. The majority carried naturalists, these often also acting as medical officers, and from their observations and collections a rough general picture of the flora was pieced together. There were, however, considerable gaps resulting from uneven geographical coverage, temporal and logistical limitations of many shore visits, and here and there local hostility. The 'high' islands were thus by mid-century still very imperfectly known, while the 'low' islands and atolls were for the most part ignored as their apparently 'monotonous' flora was thought to be just that – and would be so for another century: not until after World War II was this recognized as fallacious. Thus the period of primary floristic investigation could not be said to have closed by 1842. Nevertheless, syntheses were attempted from an early date. The first of these was from Georg Forster; his *Florulae insularum australium prodromus* (1786) was based in part on the collections of the *Resolution* and *Adventure* and arranged according to the Linnaean system. Its successor, half a century later, was *Bemerkungen über die Flora der Südseeinseln* by the Viennese botanist Stephan Endlicher (1836, in *Ann. Wiener Mus. Naturgesch.* 1: 129–190, pls. 13–16). Endlicher's work accounts for all then-published records, with authorities, from the *whole* of our superregion along with those from New Zealand and its outliers (Region 41) and was long a standard work. Nothing of comparable geographical scope has since appeared; Drake's *Illustrationes florae insularum maris Pacifici* covers a lesser area.[67]

The second stage, comprising the years from the 1840s until about 1920, was characterized largely by single island or island-group exploration: a process by and large related to political and economic penetration and, for most, eventual 'protection' or annexation. Contributions were also made by detailed coastal surveying and charting expeditions. A large proportion of collecting was accomplished by resident or visiting non-specialists, amateur or semi-professional, with their finds being sent to public and private herbaria in several 'metropolitan' countries, including Australia and New Zealand. Official biological expeditions were few, largely taking the form of resource surveys. Among the most important of the latter was the British Mission to Fiji of 1860–61, to which Berthold Seemann was attached as botanist. One of its results was his *Flora vitiensis* (1865–73), a key work also incorporating other collections, not only from Fiji but also from other South Pacific islands. Seemann thus added substantially to the overall foundation laid by Endlicher 30 years before. Other important contributions of the period were *Flora of the Hawaiian Islands* by William Hillebrand (1888), a synthesis based on the author's long experience in those islands and the first true manual for any part of the superregion; *Flore de la Polynésie française* by Emmanuel Drake (1893), *Die Flora von Samoa-Inseln* by Reinecke (1896–98; later incorporated in *Die Flora der deutschen Schutzgebiete in der Südsee* by Schumann and Lauterbach (1901; see **930**) which also consolidated what was then known of Micronesia; and *Catalogue des plantes phanérogames de la Nouvelle-Calédonie et dépendances* by André Guillaumin (1911). All were standard works for decades and some so remain.

Any 'primary' phase of exploration eventually comes to an end, and the work considered to mark this transition in the eastern half of the superregion (as far west as Fiji) is Éduard Drake's *Illustrationes florae insularum maris Pacifici* (1886–92). This was sumptuously produced in the tradition of Hooker's mid-century southern zone works. In the western half, there was, and probably could have been, no comparable successor to Endlicher's *Bemerkungen*. A comparable level of exploration was not reached until well into the twentieth century, with continued finds in New Caledonia by Schlechter, Compton, Le Rat and others in rich New Caledonia, the explorations of Morrison and others in present-day Vanuatu (the New Hebrides), and those of Georg Volkens, Augustin Krämer and Carl Ledermann in Micronesia. Attention also began to be drawn to the low islands and atolls through the work of Henry Guppy (centered on Fiji), the Australian Museum (especially in Tuvalu), the Bishop Museum in Marcus, Midway and Palmyra (the last leading to the fine illustrated *Palmyra Island* (1916) by Joseph Rock), and (after World War I) William A. Setchell (in central and southeastern Polynesia). However, not since Drake's *Illustrationes* have any supraregional synthetic floras or

enumerations been published, although such would be contemplated by Elmer D. Merrill after 1920.

The third stage recognized here comprises the period between the world wars, when Pacific biological exploration, developing on a larger and more independent scale with the aid of private wealth, was dominated by large American museums, notably the Bernice P. Bishop Museum in Honolulu. Established in 1889 (in a then-independent state), this museum, the first institution of its kind in Oceania, at first emphasized the Hawaiian Islands in its research programme; but with a change of direction after 1919 work expanded into nearly all other parts of Oceania, covering anthropology, botany and zoology. A rapid expansion of botanical and other collections ensued through the efforts of several expeditions and many resident and visiting individuals, either under institutional auspices or self-sponsored (the latter sometimes with the aid of small grants-in-aid). Collaborative work with scientists from Japan, the only other metropolitan power then significantly active in Oceania (mainly in Micronesia), was carried out in the 1930s. Development of the Museum botanical programme, which included the publication of island floras and other contributions, was from 1920 until the early 1950s guided by Merrill as consultant botanist. By 1941 the basis had been laid for what is now 'without doubt the largest and most valuable collection of Pacific plant materials in the world'.[68] Further substantial contributions arrived in the decennium after 1945 from the extensive American surveys of Micronesia and other sources.

A goodly number of still-current standard floras and related works were produced in the quarter-century from 1920 through 1945, and the basis laid for others which were not to appear until after World War II or have even yet not been fully realized.[69] The greater part was accomplished by botanical staff or 'associates' of the Bishop Museum and published in that institution's *Bulletin* or *Occasional Papers*. Among them were floras of the Northern Line, Howland and Baker Islands (1927) and of Samoa (1935–38) by Christophersen; the Hawaiian Leeward Islands (1931) by Christophersen and Caum; Rarotonga (1931) and Makatea (1934) by Wilder; a less than critical treatment of parts of southeastern Polynesia (1931–35) by Brown and Brown; some fern floras by Copeland (1929–38); and floras of Niue (1943) and the Manua Islands in Samoa (1945) by Yuncker. Under other auspices there appeared Skottsberg's flora of Easter Island (1922, with

ferns by Christensen who also treated Samoan pteridophytes for the Bishop Museum in 1943); Kanehira's
illustrated one-volume manual (1933) and enumeration
(1935) of the flora of the Japanese mandated islands in
Micronesia; Setchell's contributions on Tutuila and
Rose Atoll in Samoa (1924) and on Tahiti (1926);
Guillaumin's many contributions on New Caledonia
and the New Hebrides throughout the period; and,
finally, Degener's inimitable illustrated *Flora hawaiiensis* (1932–80). Most of these works were, however, documentary in nature, lacking keys and sometimes
descriptions, and some were little more than vehicles
for ethnobotanical information. Perhaps the most practical work was Kanehira's *Flora micronesica*, written in
Japanese but well illustrated and with keys later (1956)
made available in English. Coverage of the superregion
was thus greatly improved but remained patchy.

In 1928 Merrill (1929; see **Progress**) surveyed
the status of exploration in Oceania, noting that the
various islands were still very unevenly known, with
overall knowledge insufficient to warrant production of
an annotated enumeration along the lines of those on
Borneo (1921) and the Philippines (1922–26) and so
succeeding Endlicher's work. A call was made for an
organized Pacific botanical survey, with emphasis on
the 'high' islands, especially in the south-central sector
(Region 95) which then was considered very poorly
known and phytogeographically crucial. Towards this
end, as an extension of his Polynesian botanical bibliography (whose first version appeared in 1924, with new
editions in 1937 and 1947), Merrill began compilation
of a nomenclatural card file as a contribution towards a
future general enumeration, adding to this until the
end of the 1940s when retirement stopped work. The
enumeration project has never been resumed, but the
bibliographical materials were later utilized by Fosberg
and Sachet for their *Island bibliographies*.

This unrealized attempt at synthesis, which
would have resulted in a true successor to Endlicher's
Bemerkungen, is symptomatic of the more fragmented
pattern of floristic work which began to develop after
World War II and has largely prevailed since the early
1950s – the fourth of the developmental stages recognized here. Institutionally, this period was characterized by a marked decline in the level of Oceanian
botanical contributions from the Bishop Museum (due
to financial stringency and changes in priorities) and
the concomitant rise of two other activities: the
Washington-based atoll research programme (initiated

in the U.S. Office of Naval Research and later supported by the National Research Council and the
Smithsonian Institution in Washington, D.C.) and the
Pacific floristic mapping project at the Rijksherbarium
at Leiden. The latter (see **001**) was originally conceived
by Hermann J. Lam in 1939 and later strongly prosecuted by van Steenis and van Balgooy with their collaborators. The *Atoll Research Bulletin*, organ of the atoll
research programme, began publication in the 1950s
and continues to the present, while the first volume of
Pacific plant areas appeared in 1963 (followed by four
further volumes, respectively in 1966, 1975, 1984 and
1993). By contrast, the last floras published as Bishop
Museum Bulletins were *Flora of Ponape* (1952) by
Glassman and *Plants of Tonga* (1959) by Yuncker.
Other floristic projects initiated under the Museum's
interwar programme largely lapsed or continued under
alternative auspices.

The 'Pacific' botanical circle at the Smithsonian
Institution, until 1978 headed by Fosberg but in 1987
renewed with the appointment of Warren L. Wagner as
leader, was responsible for many florulas of low islands
and atolls in the Pacific (and other oceans), produced
either separately or as parts of larger studies and for the
most part published in the *Atoll Research Bulletin*. That
the plants of these low islands exhibit phytogeographical patterns paralleling those of the 'high' islands has
been a significant result of the atoll programme.[70]
Other work of the unit related to Pacific floristic botany
has included the preparation of *Flora of Micronesia*
(1975–), a flora of the Marquesas (1975–), an enumeration for the Northern Marianas (1975), and the only
completed portion of Grant's flora of the Society
Islands (1974) – the absence of which had been depre-
cated by Merrill in 1951 – all in *Smithsonian
Contributions to Botany*, and the tripartite *Island
bibliographies* (1955; supplement, 1971), covering
Micronesian botany, Pacific Islands vegetation, and low
islands and atolls in all oceans. In Hawaii, the Pacific
(now National) Tropical Botanical Garden published a
new checklist of Hawaiian native and introduced
flowering plants (1973) by Harold St. John, a revised
reprint of Joseph Rock's fine *Indigenous trees of Hawaii*
(1974), and from 1979 to 1991 (with an index in 1996)
the five volumes of *Flora vitiensis nova* by Albert C.
Smith. The Bishop Museum channeled its energies
into a new flora of Hawaii in succession to that of
Hillebrand; the result was the two-volume *Manual of
the flowering plants of Hawai'i* (1990) by Wagner, Derral

Herbst and Seymour Sohmer. From Guam there appeared *Flora of Guam* by Benjamin Stone (1970). A number of these floras and other works, however, represent projects begun in the interwar period; more recent American taxonomic research in the Pacific has, apart from the Hawaiian *Manual* and many small contributions in whole or in part by St. John, tended to be synthetic – likewise a continuation of an older tradition exemplified by the work of Rock, Sherff, Fosberg and others in their revisionary studies.

A renewed awareness of the Pacific Islands on the part of European metropolitan powers as well as New Zealand also took place, which in botanical terms has manifested itself in much new field work and publication, notably in New Caledonia, Fiji, the New Hebrides, and the Santa Cruz group as well as the adjacent Solomon Islands (**938**). Some of this work was conducted from already existing (Suva) or newly established (Honiara, Guam, Noumea) herbaria. Collections made from the early 1950s onwards have greatly improved botanical knowledge in New Caledonia and the south-central Pacific Islands (the latter comprising the 'Fijian region' of the many regional revisions of A. C. Smith and his collaborators), and a number of interesting discoveries have been made. Notable contributions to date have included the large-scale, documentary *Flore de la Nouvelle-Calédonie et dépendances* (1966–), and an annotated checklist, *Plants of the Fiji Islands* by John Parham (1964; 2nd edn., 1972). From New Zealand, Garth Brownlie revised the pteridophytes of several islands, notably New Caledonia and Fiji, and William Sykes has made substantial additions to Yuncker's Niue flora in *Contributions to the flora of Niue* (1970). A number of revisions of Pacific genera have resulted from work on *Flora Malesiana* at Leiden, and these and others have been used for *Pacific plant areas*. Guillaumin's earlier work was encapsulated in his ill-documented, partly compiled *Flore analytique et synoptique de la Nouvelle-Calédonie* (1948), now quite incomplete but at the time one of the few Oceanic identification manuals.

The state of knowledge at the beginning of the last third of the twentieth century was surveyed by Corner (1972) and Prance (1978), the latter based on an unpublished report by Fosberg written for the Thirteenth Pacific Science Congress in 1975. These represent the latest reports prepared under the aegis of the Standing Committee on Pacific Botany of the Pacific Science Association, a coordinating body for the Pacific (and Malesian) superregions. Several islands and groups are still considered inadequately explored and/or documented, and almost throughout there are marked threats from increasing human pressure and 'development'. Many raised limestone flat-topped islands have been comparatively neglected and require, or are receiving, urgent attention; among these are Henderson, Makatea and Nauru.[71] Nevertheless, the level of exploration as a whole compares favorably with the West Indies or the Mediterranean. With respect to a new general enumeration as conceived by Merrill, the considerable and more or less inevitable fragmentation of floristic work and collections, both in the past and more recently, will make very difficult the preparation of a new general enumeration, although Merrill's strictures are less applicable now than in 1928. The creation of an 'information center' should be encouraged.[72] Simple florulas for education purposes, such as Sheila Gowers' illustrated, locally produced treatment of common trees of Vanuatu (1976), continue to be a desideratum.

The most important recent developments in the last decade or so have been the completion of a *Manual of the flowering plants of Hawai'i*, a wish expressed in these pages in 1984, and the full realization of *Flora vitiensis nova*. Other initiatives include the renewal of work on the flora of southeastern Polynesia (at the Smithsonian Institution and in French centers, in the latter notably by Jacques Florence), continuation of *Flore de la Nouvelle-Calédonie* and *Flora of Micronesia*, and further collecting in many islands or island groups, some now with more or less serious threats to their flora. The Pitcairn Islands were in 1991–92 the subject of a multidisciplinary survey, the Sir Peter Scott Commemorative Expedition. Botanical research remains active in Honolulu, the National Tropical Botanical Garden (Lawai, Kauai, Hawaii), the University of Guam, the University of the South Pacific at Suva, and at the ORSTOM centers in Papeete and Nouméa.[73]

Progress
No overall historical review for the superregion has been seen. The state of knowledge in 1928 and some suggestions for advancement were given in E. D. MERRILL, 1929. Pacific botanical survey. In H. E. Gregory (comp.), *Annual report of the Director for 1928*, pp. 53–57 (Bernice P. Bishop Mus. Bull. 65). Honolulu. Later reviews include E. J. H. CORNER (comp.), 1972.

Urgent exploration needs: Pacific floras. *Pacific Sci. Inform. Bull.* **24**(3/4): 17–27, and an extract from an unpublished paper by Fosberg in the general survey by Prance (1978; see **General references**). Emergent reef surfaces in the Pacific and other parts of the world were reviewed most recently in 1985 (Fosberg, 1986, see **002**). Further details may be found in Bishop Museum annual reports and various proceedings of the Pacific Science Congresses (e.g., H. ST. JOHN, 1975. Floristic needs in the Pacific Basin – Polynesia. In *Proceedings of the Thirteenth Pacific Science Congress (Pacific Science Association)*, 1: 111–112. Vancouver).[74]

General bibliographies. Bay, 1910; Blake and Atwood, 1942; Frodin, 1964, 1984; Goodale, 1879; Holden and Wycoff, 1911–14; Jackson, 1881; Pritzel, 1871–77; Rehder, 1911; Sachet and Fosberg, 1955, 1971; USDA, 1958.

Supraregional bibliographies

 LEESON, I., 1954. *A bibliography of bibliographies of the South Pacific.* vii, 61 pp. London: Oxford University Press.

 MERRILL, E. D., 1947. *A botanical bibliography of the islands of the Pacific*; and E. H. WALKER, *A subject index to Elmer D. Merrill's 'A botanical bibliography of the islands of the Pacific'.* Pp. 1–404 (Contr. U.S. Natl. Herb. 30(1)). Washington. [Concisely annotated bibliography by Merrill, with 3850 entries; detailed subject index by Walker. In addition to all the islands in Superregion 94–99, full coverage is extended to Juan Fernández, the Kermadec group, and Norfolk and Lord Howe Islands. Succeeds *Bibliography of Polynesian botany* (1924) and *Polynesian botanical bibliography, 1773–1935* (1937) by E. D. Merrill.]

 STEENIS, C. G. G. J. VAN, 1955. Annotated selected bibliography. In *idem* (ed.), *Flora Malesiana*, I, 5: i–cxliv. Groningen: Noordhoff. (Reprinted separately.) [Provides a limited selection of more important floras.]

General indices. BA, 1926– ; BotA, 1918–26; BC, 1879–1944; BS, 1940– ; CSP, 1800–1900; EB, 1959–98; FB, 1931– ; IBBT, 1963–69; ICSL, 1901–14; JBJ, 1873–1939; KR, 1971– ; NN, 1879–1943; RŽ, 1954– .

Supraregional indices

 STEENIS, C. G. G. J. VAN *et al.* (eds.), 1948– . *Flora Malesiana Bulletin*, 1– . Leiden: Flora Malesiana

Foundation. (Initially mimeographed; since 1988 a desktop-publication). [Issued nowadays annually, this contains a substantial bibliographic section also very thoroughly covering the Pacific/Oceanic literature. A fuller description appears under Superregion 91–93.]

940–90
Superregion in general

Only two overall accounts, both enumerations with the latter limited to the central and eastern islands, have ever been compiled: *Bemerkungen über die Flora der Südseeinseln* by Stephan Endlicher (1836, in *Ann. Wiener Mus. Naturgesch.* 1: 129–190, pls. 13–16), now mainly of historical interest, and *Illustrationes florae insularum maris Pacifici* (1886–92; see below) by Éduard Drake, a large-scale work intended partly for the connoisseur. From the 1920s card files were developed by Merrill for a projected enumeration but nothing was ever published apart from bibliographies (see above). Recently there has appeared an illustrated popular guide to the lowland flora: W. A. WHISTLER, 1995. *Wayside plants of the islands: a guide to the lowland flora of the Pacific Islands including Hawai'i, Samoa, Tonga, Tahiti, Fiji, Guam and Belau.* 202 pp., col. illus. Honolulu: Isle Botanica. For a contemporary review of vegetation, see D. MUELLER-DOMBOIS and F. R. FOSBERG, 1998. *Vegetation of the tropical Pacific Islands.* xxvii, 733 pp., illus., maps. Berlin: Springer. (Vegetationsmonographien der einzelnen Grossräume IX.)

 DRAKE DEL CASTILLO, É., 1886–92. *Illustrationes florae insularum maris Pacifici.* 458 pp., 50 pls. Paris: Masson. (Reprinted 1977, Vaduz, Liechtenstein: Cramer.)

 Pages 103–408 of this work contain a systematic enumeration (covering 2189 species) of the known vascular plants of central and eastern Oceania, with synonymy, references and citations, indication of localities with *exsiccatae*, general summary of intra- and extra-limital distribution, and taxonomic commentary on individual species, genera and families; index to all botanical names at end. The lengthy introductory section (pp. 1–100) encompasses chapters on the physical characteristics of the islands, general features of the flora, floristics and phytogeography, and the progress of botanical exploration. At the end of the work is an atlas of 50 plates of representative species, each with accompanying descriptive text. [Encompasses the Fiji Islands, Tonga, Samoa, Niue, the Marquesas, the Societies, the Tuamotu Archipelago, and the Hawaiian Islands.]

Region 94

New Caledonia and dependencies

Area: 16750 km^2. Vascular flora: 3344 native and *c.* 1600 introduced species, with 76 percent endemism (Morat, 1994; see Note 77). – This region, almost all under French administration, encompasses the main island of New Caledonia ('La Grande Terre', including Kunié and Îles Belep), the Loyalty Islands, the islands from Walpole through Matthew and Hunter to Conway Reef, and the more westerly groups of Chesterfield and the Coral Sea Islands Territory (the last-named an Australian dependency).

'La Grande Terre' is distinguished by one of the most peculiar floras in the Pacific (and indeed the world), with high endemism, many evidently archaic taxonomic and morphological 'relicts', and quite mixed floristic affinities compressed within a relatively small area. Its biogeographic classification has been moot, though affinities with Australia and New Guinea are strongest; for practical and bio-historical reasons, though, it has here been placed along with its 'dependencies' in Division 9 as part of tropical Oceania.[75]

The first organized account of the New Caledonian flora is *Sertum austro-caledonicum* (1824–25) by J. J. H. de Labilliardière. Extensive collecting, however, began only in the 1850s following proclamation of French sovereignty. A stream of new discoveries followed over the next several decades, in the nineteenth century mainly from French workers and in the early twentieth from Rudolf Schlechter, Fritz Sarasin and J. Roux, Robert H. Compton and Albert U. Däniker. The latter period was, however, dominated by the work of André Guillaumin, whose two compilations, the *Catalogue des plantes phanérogames* (1911) and the *Flore analytique et synoptique* (1948), remain the only nominally complete accounts of the vascular flora.

Guillaumin also documented the flora in two long-running series: the *Matériaux*, containing family and generic revisions precursory to a general flora, and the *Contributions*, comprising more or less extensive lists of new collections, florulas of given localities or small islands, and miscellaneous matters worthy of record. Only the former have been indexed, and then only through 1945, just prior to publication of the *Flore*. Largely compiled, part uncritical, and without bibliographic aids or commentary, this last is in need of thorough revision or replacement (M. Schmid, personal communication).

The advent of André Aubréville as director of the Laboratoire de Phanérogamie at the Muséum National d'Histoire Naturelle in Paris brought about an expansion of the unit's interest in large-scale definitive floras for *la francophonie tropicale*. New Caledonia, with the added attraction of its unusual flora, was no exception, and so *Flore de la Nouvelle-Calédonie et dépendances* came into being. The first part, on Aubréville's specialty, the Sapotaceae, appeared in 1967; as of 1999, 21 parts covering some 45 families and other taxa or groups have been published. Although accounting for the large amount of new material and information which has become available since World War II, full realization of this enterprise will, as with *Flora Malesiana* and the new *Flora of Micronesia*, take several more decades and even then be a reference rather than a practical work. There is thus scope for preparation and publication of a concise 'interim' treatment along the lines of Hermann Merxmüller's *Prodromus einer Flora von Südwestafrika* (521).[76]

The tree flora features in two works: *Notice sur les bois de la Nouvelle-Calédonie* by H. Sébert and J. A. I. Pancher (1874; reprinted from installments in *Revue Maritime et Coloniale* 37–41 (1873–74)) and its somewhat uncritical successor, *Bois et forêts de la Nouvelle-Calédonie* by P. Sarlin (1954; cited below under **Forest trees**).

Of the associated entities here recognized, no properly published coverage exists for the undercollected Loyalties and only sketchy documentation is available for the eastern islands and for the Huon and Chesterfield groups. The Coral Sea Islands, the only part of the region not under French rule, are now documented in volume 50 of the *Flora of Australia* (1993). The eastern islands are small, 'high' volcanic entities, while those to the west are widely scattered low cays and atolls, some on the drowned 'Queensland Rise'.[77]

Bibliographies. General and supraregional bibliographies as for Superregion 94–99. For further details respecting Guillaumin's *Matériaux* and *Contributions*, see the bibliographies of Blake and Atwood (1942) and Merrill (1947).

Regional bibliography

O'REILLY, P., 1955. *Bibliographie méthodique, analytique et critique de la Nouvelle-Calédonie.* ix, 361 pp. Paris: Musée de l'Homme. (Société des Océanistes, Publ. 4.) [Annotated general bibliography; includes references in botany.]

Indices. General and supraregional indices as for Superregion 94–99. Additions to Guillaumin's *Contributions* from 1947 are listed in *Flora Malesiana Bulletin.*

940

Region in general

In addition to the *Catalogue* (1911) and the *Flore analytique et synoptique* (1948), both described below, Guillaumin was responsible for two major floristic and systematic series on the New Caledonian flora. These comprise the *Matériaux pour la flore de la Nouvelle-Calédonie* 1–85 (1914–44), which consists very largely of family revisions (indexed in *Bull. Soc. Bot. France* **92**: 76–77 (1946)), and *Contributions à la flore de la Nouvelle-Calédonie* 1–130 (1911–73), generally devoted to reports on collections, floristic lists and taxonomic notes. Other 'collection reports' exist, some of them quite substantial and worthy of inclusion here (under **Partial works**). Of the *Flore de la Nouvelle-Calédonie et dépendances*, the pteridophytes, gymnosperms and a number of angiosperm families (including certain key taxa such as the Orchidaceae, Myrtaceae–Leptospermoideae and Proteaceae) have been published, but the work as yet covers less than one-third of the native vascular flora. *Bois et forêts de la Nouvelle-Calédonie* by P. Sarlin (1954) remains the only current work on the tree flora.

AUBRÉVILLE, A. *et al.* (eds.), 1967– . *Flore de la Nouvelle-Calédonie et dépendances.* Fasc. 1– . Illus., maps. Paris: Muséum National d'Histoire Naturelle, Laboratoire de Phanérogamie. Accompanied by H. S. MACKEE, 1994. *Catalogue des plantes introduites et cul-*

tivées en Nouvelle-Calédonie. 2nd edn. Paris. (Flore de la Nouvelle-Calédonie, hors série.)

The *Flore* is a copiously illustrated, extensively descriptive 'research' flora, with each family treatment including keys to genera and species, full synonymy, localities with *exsiccatae* and (where applicable) a general statement of overall range, distribution maps, taxonomic commentary, and notes on habitat, distribution, frequency, biology, juvenile forms (bizarre blastogenic features are among the characteristic features in the flora) and special features of individual species. Each fascicle, which contains one or more higher taxa above the rank of genus, is separately indexed. The *Catalogue* by MacKee is an annotated enumeration of introduced and naturalized species. [As of 1999, 23 fascicles of the main work have appeared, of which fasc. 3 constitutes a detailed revision of the pteridophytes (by G. Brownlie), fasc. 4, the gymnosperms (by D. J. de Laubenfels) and fasc. 8, the Orchidaceae (by N. Hallé). Parts of the work, however, are already out of date.]

GUILLAUMIN, A., 1911. Catalogue des plantes phanérogames de la Nouvelle-Calédonie et dépendances (Îles des Pins et Loyalty). *Ann. Inst. Bot.-Géol. Colon. Marseille* **19**: 77–290, map. (Reprinted separately in 3 parts.)

Systematic enumeration of seed plants; includes synonymy, vernacular names, localities with *exsiccatae* and other sources, indication of useful plants, and an index to family and vernacular names. The introductory chapter includes a fairly extensive section on the history of local botanical exploration and an annotated list of collectors. The work furnishes an important summary of earlier exploration and research and serves as a starting-point for the author's extensive *Matériaux* and *Contributions* (referred to under the heading).

GUILLAUMIN, A., 1948. *Flore analytique et synoptique de la Nouvelle-Calédonie (Phanérogames).* 369 pp. Paris: Office de la Récherche Scientifique Coloniale.

Unannotated manual-key to seed plants; includes brief descriptions of the characteristics of each family and an index to generic and family names. The work is in essence a distillation from the author's *Matériaux* and *Contributions*; unfortunately, no bibliographic aids whatever to these or other sources were furnished.

Partial works

The two most extensive precursory series are, as already noted, Guillaumin's *Matériaux* and *Contributions* (the latter continuing almost up to the time of his death in 1974). They are not, however, described here as they are too numerous and scattered; moreover, those published before 1948 have been implicitly accounted for in his *Flore analytique et synoptique*. On the other hand, there also exist a number of significant more self-contained contributions, both by Guillaumin (on the Sarasin herbarium but especially on the several thousands of collections of Marcel Baumann, Hans Hürlimann and himself made in 1950–52 on the so-called 'Franco-Swiss' expedition) and by some non-French botanists (notably Schlechter, Sarasin, Roux and collaborators, Alfred Rendle and collaborators, Däniker, and Robert F. Thorne). The most extensive, those by Däniker (1932–43) and Guillaumin (1957–74), are accounted for below.

DÄNIKER, A. U., 1932–43. *Ergebnisse der Reise von Dr A. U. Däniker nach Neu-Caledonien und den Loyaltäts-Inseln*, 4: *Katalog der Pteridophyta und Embryophyta siphonogama*. 5 parts. 507 pp. (Vierteljahrsschr. Naturf. Ges. Zürich 77–78, Beibl. 19). Zurich. (Also published as *Mitt. Bot. Mus. Univ. Zürich* 142.)

Critical systematic enumeration of vascular plants in the author's collection of 1924–25; includes descriptions of new taxa, full synonymy (with references and citations), vernacular names, localities with *exsiccatae*, taxonomic commentary, and extensive notes on habitat, biology, etc.; index to generic and family names at end of work (part 5). The introductory section includes a general description of the several islands as well as the author's itinerary and an account of past botanical exploration.

GUILLAUMIN, A., 1957–74. *Résultats scientifiques de la mission franco-suisse de botanique en Nouvelle-Calédonie 1950–2*, I–V. 5 parts (Mém. Mus. Natl. Hist. Nat., II/B, Bot. 8(1), 8(3), 15(1), 15(2), 23). Paris.

Annotated lists of collections, in each part arranged systematically, with indication of localities (together with *exsiccatae*), descriptions of novelties, references, and some taxonomic commentary; keys (mainly as supplementation for the *Flora analytique*) presented for a few higher taxa (e.g., *Araucaria* in part 1; *Dysoxylum* in part 4); no index. The last two parts contain, along with additional material, extensive addenda and corrigenda to treatments in parts 1–3. Many treatments are by specialists. Part 5 (1974), the author's last contribution to New Caledonian botany, appeared posthumously.

Forest trees
In addition to the long-standard work of Sarlin (1954), the following may also be consulted: CENTRE TECHNIQUE FORESTIER TROPICAL, 1975. *Inventaire des resources forestières de la Nouvelle-Calédonie*. 227 pp. Nogent-sur-Marne. (Mimeographed.)

SARLIN, P., 1954. *Bois et forêts de la Nouvelle-Calédonie*. 303 pp., 131 pls. (incl. halftones), 3 folding maps

(Publ. CTFT 6). Nogent-sur-Marne, France: Centre Technique Forestier Tropical.

Part III of this work (pp. 81–212) comprises a crudely illustrated dendrological treatment of principal forest trees, with botanical descriptions, trade and vernacular names, occurrence, habitat, properties, structure of wood and its description, and provenances, with cross-referenced illustrations at the end of the work. The introduction (part I) covers background (including the problems posed by a lack of previous forest-botanical studies and collections, including wood samples), topography, climate, soils, biotic factors, land use and forest exploitation; in part II is a descriptive treatment of vegetation and the forests. Part IV gives conclusions along with a classification of actual and potential uses. [Of most value for its illustrations; otherwise an uncritical work, hastily prepared by a non-specialist without adequate assistance (M. Schmid, personal communication).]

945

Loyalty Islands

Area: 1981 km^2. Vascular flora: 550–600 native and naturalized species; 360 considered truly indigenous, with at the most 10 percent endemism (Schmid, 1967). – The islands comprise the more or less raised atolls of Ouvéa, Lifou and Maré and their associated islets.

The flora has usually been accounted for in overall works on New Caledonia, but still is considered undercollected. The only relatively recent separate survey is M. SCHMID, 1967. *Note sur la végétation des îles Loyalty*. 70 pp. Noumea: ORSTOM, Centre Nouméa. (Mimeographed.)

946

Islands east of New Caledonia

Area: 1.78 km^2 (Walpole, 125 ha; Hunter, 41 ha; Matthew, 12 ha). Vascular flora: Walpole, 70 native species, 28 naturalized or otherwise introduced; Matthew, 10 species. – The 'Eastern Dependencies' encompass Walpole, Matthew and Hunter Islands. Walpole is a distant outlier of the Loyalty chain, while Matthew and Hunter are isolated. Conway Reef (Ceva i Ra) is for convenience also included here.

Prior to 1972, no botanical collections had been made on the islands and even now no checklist for Hunter has been published. The paper below, the last of Guillaumin's 130 numbered *Contributions*, was thus a first for the Eastern Dependencies. For Walpole, a 4 km × 0.4–1 km raised *makatea* or 'coral plateau' to 80 m high, subsequent studies have been incorporated in a revised checklist by A. Renevier and J.-F. Cherrier.

GUILLAUMIN, A., 1973. Contributions à la flore de la Nouvelle-Calédonie, 130: Plantes des îles Walpole et Matthew. *Bull. Mus. Natl. Hist. Nat.* (Paris), III, **192** (Bot., 12): 180–183.

Includes unannotated lists of plants collected from Walpole (4 species) and Matthew (10 species). [For Walpole now superseded.]

RENEVIER, A. and J.-F. CHERRIER, 1991. *Flore et végétation de l'île de Walpole.* 21 pp., 3 text-figs. (maps) (Atoll Res. Bull. 351). Washington.

Includes a vouchered checklist with indication of status (pp. 14–19); references, p. 21. Geography, climate, soils and vegetation are also considered.

947

D'Entrecasteaux Reefs (with Huon Islands)

Area: 0.65 km². – The two d'Entrecasteaux Reefs lie to the northwest of 'La Grande Terre', beyond Îles Belep, and encompass Huon Islands. (on the northern reef) and Surprise, Fabre, Le Leizour and the Sand Islets (on the southern reef).

GUILLAUMIN, A. and J.-M. VEILLON, 1969. Plantes des archipels Huon et Chesterfield. *Bull. Mus. Natl. Hist. Nat.* (Paris), II, 41: 606–607.

Includes, for Surprise Island, unannotated lists of 25 species (based on a 1965 collection by Blanchon) and of 17 species (collected in 1876–77 by mariners from the *Curieux* and submitted to Montrouzier).

948

Islands of the Bellona Plateau (including the Chesterfield group)

Area: less than 1 km² (Long I., 22 ha). – The low islands of the Bellona Plateau form part of the Coral Sea Rise between New Caledonia and the Great Barrier Reef of Australia. Only the Chesterfield group, with 11 islets including the largest, Long Island, has been botanically studied. Other entities include Observatory Cay, Sable, and Bellona Reefs.

COHIC, F., 1959. *Report on a visit to the Chesterfield Islands, September 1957.* 11 pp., maps (Atoll Res. Bull. 63). Washington.

Pages 3–4 contain an unannotated list (encompassing 20 species) of vascular plants collected by the author from Long Island. The remainder of the report describes previous visits to the islands, the vegetation, and the terrestrial fauna; a list of references is appended. A second, alternative list of 10 species is given in GUILLAUMIN and VEILLON (1969; see **947**).[78]

949

Coral Sea Islands Territory (Australia)

Area: no data but miniscule. Vascular flora: *c.* 26 species, of which 5 naturalized and 2 unsubstantiated (Telford in Australian Biological Resources Study, 1993); composition likely to be changeable due to tropical storms. – The 46 scattered low cays of the Coral Sea Islands, an Australian 'External Territory' partly associated with the Coral Sea Rise, lie northeast of the Great Barrier Reef and generally west of the Chesterfields. Their 'center' is Willis Island with its weather station, established in 1922 and described by J. K. Davis in *Willis Island, a storm warning station in the Coral Sea* (1923, Melbourne: Critchley Parker). Davis included as his appendix B a list of seven plant species; however, more collecting has since taken place and the flora is now well documented in volume 50 of *Flora of Australia*.[79]

AUSTRALIAN BIOLOGICAL RESOURCES STUDY, 1993. *Flora of Australia,* 50: Oceanic Islands 2. xxvi, 606 pp., 97 figs. (part col. and incl. maps). Canberra: Australian Government Publishing Service.

Pages 47–51 comprise an introduction to the Coral Sea Islands (by Ian R. H. Telford), encompassing background information, vegetation formations and characteristic species (the latter grouped by synusiae), floristics, usage, history and collecting (here, on Wreck Reef, were lost the living and first dried set of Robert Brown's Australian collections), and references along with a key to families and a concisely annotated list of species with indication of their status and whether or not they were recorded elsewhere.[80] All species are further treated in the descriptive part – which follows the format of other *Flora* volumes – together with those from other islands or groups (Cocos (Keeling), **045**; Christmas, **046**; Ashmore and Cartier, **451**; Macquarie, **081**; and the McDonald (Heard) group, **082**).

Region 95

South-central Pacific Islands

Within the South-central Pacific Islands, sometimes also known as the 'Fijian region', are included the 'high' islands of eastern Melanesia and southwestern Polynesia, as well as many low islands and atolls. The boundaries are marked by the Santa Cruz Islands to the northwest, Samoa and Niue in the east, and a line running from south of Aneityum (in Vanuatu, the former New Hebrides) to south of the Minerva Reefs (south of Fiji). Conway Reef is included in Region 94 with New Caledonia and the Loyalties.

A heterogeneous assemblage of floras and enumerations more or less completely covers the region, but by current standards only one work can be seriously taken as definitive: *Flora vitiensis nova* by A. C. Smith (1979–91; general index, 1996) for Fiji in the eastern subregion. That country also features a comparatively recent, well-documented though unkeyed checklist, *Plants of the Fiji Islands* by a former government botanist, John Parham (2nd edn., 1972). Elsewhere, useful works – neither of them, however, with keys – exist for Niue (Yuncker, 1943, with a

supplementary work by Sykes, 1970) and Tonga (Yuncker, 1959); the last-named, however, does not effectively account for the country as a whole as some islands were then undercollected. Divided Samoa features a variety of accounts, none of them complete and the latest nearly 40 years old; parts of the group are also considered inadequately collected. In recent years, considerable work has been carried out in the group by Paul Cox and W. Arthur Whistler and several papers published. Documentation of the Wallis and Futuna Islands has been further improved through the checklists of St. John and Smith (1971) and Morat and Veillon (1985); both reflect renewed collecting in recent decades.

Coverage of the western subregion (Rotuma, the Tikopia group, the Reef Islands, Santa Cruz, and the New Hebrides) remains more fragmented and uneven, both in literature and in available collections, although much new field work has been carried out since 1960. The Tikopia and Santa Cruz groups, along with the Reef Islands, are part of the state of the Solomon Islands (**938**) and as such have been sketchily accounted for in Tim Whitmore's *Guide to the forests of the British Solomon Islands* (1966), with its scarcely annotated checklist, and in *Flora of the Solomon Islands* (1988) by I. R. Hancock and C. P. Henderson. For Vanuatu, the nominally complete enumeration by André Guillaumin, the *Compendium* (1948), is now virtually useless; other writings are very scattered apart from two tree guides (1976, 1992). Significant additions to the flora are moreover continuing as exploration has proceeded.[81] In Rotuma, unfortunately already much deforested by the 1930s, considerable collecting has been carried out but so far only the pteridophytes have enjoyed separate treatment (St. John, 1954); coverage must otherwise be sought in A. C. Smith's *Flora vitiensis nova* and John Parham's *Plants of the Fiji Islands*. Many families occurring in this western subregion have, however, been accounted for in regional revisions by Smith and his associates in preparation for the Fijian flora.

Bibliographies. General and supraregional bibliographies as for Superregion 94–99.

Regional bibliography
SNOW, P. A., 1969. *A bibliography of Fiji, Tonga and Rotuma.* 418 pp. Canberra. [Includes references in the natural sciences.]

Indices. General and supraregional indices as for Superregion 94–99.

950

Region in general

No general works are available; however, many family revisions relating to the region as a whole can be found in A. C. Smith's *Studies of Pacific Island plants*. Initiated in 1941 in preparation for what ultimately appeared as *Flora vitiensis nova* (see **955**) and in part written with or by other authors, this series, published in various professional journals, concluded in 1978 with its 34th installment.[82] The ruderal flora of lowland areas is pictorially introduced by W. A. Whistler in his *Wayside plants of the islands* (1995) while the eastern island groups (Fiji, Samoa and Tonga) are in addition covered by É. Drake del Castillo's *Illustrationes florae insularum maris Pacifici* (1886–92; for both, see **940–90**); the latter work, however, is now largely of historical interest.

951

Santa Cruz Islands

See also **938** (all works). – Area: 958 km². Vascular flora: no data. Included with the Santa Cruz group proper (Ndende or Santa Cruz, Utupua, Tinakula and Vanikoro) are the Taumako (Duff) Islands and the Reef Islands. For Tikopia, Anuta and Fatutaka (Mitre), see **952**. The group administratively is part of Solomon Islands, constituting its Temotu province.

No separate checklists are available, but known records up to the mid-1980s are, at least in theory, accounted for in the Solomon Islands checklists by Whitmore and by Hancock and Henderson (under **938**). The group is still considered undercollected in spite of considerable attention in the 1930s and again in the 1960s and early 1970s. Vanikoro, one of the larger islands, was once noted for its stands of kauri (*Agathis obtusa*).

952

Tikopia, Anuta and Fatutaka (Mitre)

See also **938** (all works). – Area and vascular flora: no data available. These three comparatively small islands form a loose group lying east of the Santa Cruz Islands. Like that archipelago, they are administratively part of Solomon Islands.

No separate checklists appear to be available; however, any records have in theory been accounted for in the Solomon Islands checklists by Whitmore and by Hancock and Henderson (under **938**). Moreover, relatively few collections have been made given a past history of limited access. A brief account of the flora of Tikopia, including a list of trees and shrubs seen, appears in P. V. KIRCH and D. E. YEN, 1982. *Tikopia: the prehistory and ecology of a Polynesian outlier.* xviii, 396 pp., 129 text-figs. (Bernice P. Bishop Mus. Bull. 238). Honolulu.

953

Vanuatu (New Hebrides)

Area: 14763 km². Vascular flora: 1000 species (Schmid, 1978, p. 658; see **General references**). – Vanuatu, from 1906 to 1980 an Anglo–French condominium, extends from the Torres Islands in the north through the Banks Group, Santo and Efate (Vaté) to Aneityum (Anatom). For Matthew and Hunter Islands, see **946**.

No satisfactory modern flora or checklist is available apart from the sketchy and now virtually useless *Compendium* by Guillaumin (1948), the first of its kind. This was essentially a summary of several earlier reports by him (given below). Some individual island florulas, however, have become available and are also accounted for here, while forest trees were treated in an atlas by Gowers (1976) and a guide by Wheatley (1992). Much new collecting has been accomplished, beginning in the 1960s, by both anglophone and francophone workers (seldom jointly!) but the country is still considered unevenly known. Available collections, particularly from before continuous participation by the ORSTOM center in New Caledonia and the establishment of a herbarium at Vila, are somewhat widely scattered abroad so that preparation of any more definitive consolidated accounts will require some international collaboration.

A useful introduction to the biota of Vanuatu (for flora, see the papers by W. L. Chew, E. J. H. Corner, A. Gillison and M. Schmid) is E. J. H. CORNER and K. E. LEE (coords.), 1975. Discussion on the results of the 1971 Royal Society expedition to the New Hebrides. *Philos. Trans. Roy. Soc.* (London), B, **272**: 267–486.

Bibliography

O'REILLY, P., 1958. *Bibliographie méthodique, analytique et critique des Nouvelles-Hébrides.* ix, 304 pp. Paris: Musée de l'Homme. (Société des Océanistes, Publ. 8.) [Annotated general bibliography; includes references in botany.]

GUILLAUMIN, A., 1947–48 (1948). Compendium de la flore phanérogamique des Nouvelles-Hébrides. *Ann. Inst. Bot.-Géol. Colon. Marseille* **55/56**: 5–58.

Briefly annotated systematic enumeration of seed plants, with limited synonymy; includes indication of local distribution according to islands with, where applicable, a summary of extralimital distribution, and occasional notes; index to families at end. The brief preface gives a list of the principal collectors in the islands up to World War II. A goodly number of entries are not identified beyond the genus. [The work was based largely on the following precursory papers (largely reports of collections, including those of S. Kajewski and A. de la Rüe): A. GUILLAUMIN, 1919–29. Contribution à la flore des Nouvelles-Hébrides, I–III. *Bull. Soc. Bot. France* **66**: 267–277, **74** (1927): 693–712, **76**: 298–303; and *idem*, 1931–33. Contribution to the flora of the New Hebrides. *J. Arnold Arbor.* **12**: 221–264, figs. 1–3 (incl. map), **13**: 1–29, 81–124, figs. 1–3, pl. 43, **14**: 53–61; and *idem*, 1935–37. Contribution à la flore des Nouvelles-Hébrides. *Bull. Soc. Bot. France* **82**: 316–354, map, and *Bull. Mus. Natl. Hist. Nat.* (Paris), II, **9**: 283–306, 1 fig.]

Partial works

A number of florulas exist now for individual islands; these should, however, not be regarded as critical or, in the case of those by Maurice Schmid, formally published (all these latter were mimeographically produced for relatively limited distribution). The Santo florula by Guillaumin is now very incomplete. Apart from those listed below, a florula of Tanna by Schmid was also circulated but no copy has been available for examination.

GUILLAUMIN, A., 1938. A florula of the island of Espiritu Santo, one of the New Hebrides. *J. Linn. Soc., Bot.* **51**: 547–566.

A checklist with introductory matter; based mainly on the results of the 1933–34 Baker expedition from the United Kingdom.

SCHMID, M., 1970. *Florule d'Anatom.* 53 pp. Nouméa: Centre ORSTOM; also *idem*, 1973. *Espèces de végétaux supérieurs observés à Vaté.* 42 pp.; *idem*, 1974. *Florule de Erromango.* 52 pp.; and *idem*, 1974. *Florule de Pentecôte.* 25 pp. (All mimeographed.)

Briefly annotated checklists, usually excluding Orchidaceae.

Woody plants (including forest trees)

GOWERS, S., 1976. *Some common trees of the New Hebrides and their vernacular names.* 189 pp., illus. Port Vila: Forestry Branch, Department of Agriculture, New Hebrides Condominium Government. (Mimeographed. Reprinted with corrections, 1978.)

Illustrated dendrological treatment of about 60 more common native, naturalized or commonly cultivated tree species, with large figures and accompanying botanical and timber descriptions, local vernacular and Bislama (Vanuatu pidgin) names, and notes on uses, special features, etc. An introductory section includes notes on the limits of the work, remarks on vernacular names, and a general key (pp. 1–14), while at the end are an illustrated glossary of terms and indices to vernacular and botanical names. Arrangement of tree species is alphabetical by genus.

WHEATLEY, J. I., 1992. *A guide to the common trees of Vanuatu.* 308 pp., illus., maps. Port Vila: Department of Forestry, Vanuatu.

Alphabetically arranged illustrated treatment, the text covering botanical features, phenology, dispersal, habitat and ecology, altitudinal range, distribution, uses and vernacular names; lexicon of ni-Vanuatu names (arranged by island and language); classes of uses; checklist of trees and shrubs (pp. 283–300) in tabular form with Bislama names, use codes, lifeform and notes; references (p. 301) and index to families and species at end. The introductory part includes an introduction to the flora and vegetation along with an illustrated descriptive glossary, list of families, list of islands and languages, and (pp. 27–34) a field-key.

954

Rotuma

See also **955** (PARHAM; SMITH); administratively the island is part of Fiji. – Area: no data, but the island is only 12×4 km. While a 'manual' for the ferns has been produced by St. John (1954), no separate work of recent date for the seed plants is available for this floristically most easterly outpost of 'Melanesia'. Substantial collections have, however, been made, particularly by St. John.[83]

Pteridophytes

ST. JOHN, H., 1954. Ferns of Rotuma Island, a descriptive manual. *Occas. Pap. Bernice P. Bishop Mus.* **21**(9): 161–208, 11 figs. (incl. map).

Briefly descriptive fern flora; includes keys to all taxa, full synonymy (with references and citations), localities with *exsiccatae* or sight records, taxonomic remarks, and notes on frequency, habitat, uses, etc.; list of references and index to genera at end. The introduction treats extensively geography, climate, geology, deforestation, the economy, features of the fern flora and its relationships, and general attributes of vernacular names; a short history of exploration is also included.

955

Fiji Islands

Area: 18 333 km². Vascular flora: 2591 species, of which 2130 native (812 endemic) and 461 introduced (Smith, 1996, in *Flora vitiensis nova: Comprehensive indices*; see below). – Included here are the islands of the Fijian Archipelago, corresponding to the limits of the state of Fiji (except Rotuma; see **954**).

The first significant collections from the islands were those of the United States Exploring Expedition around 1840 and Berthold Seemann in 1860. These and others formed the basis for the latter's *Flora vitiensis* (1865–73, London; reprinted 1977 by Cramer/Gantner, Vaduz), with its hand-colored plates drawn by the premier Victorian botanical artist, Walter Fitch. The importance of this work extended far beyond the limits of the archipelago, and remained definitive for many decades. The middle two quarters of the twentieth century were marked by a second wave of collect-

ing, initially sponsored through the Bishop Museum as part of its Pacific botanical programme; a local herbarium came into being in 1933. The largest contributor of this period was Albert C. Smith, who from the 1930s began work towards a new definitive flora. Key precursory works include *New plants from Fiji*, I–III (1930–32) by J. W. Gillespie, and *Fijian plant studies*, I–II (1936–42) and *Studies of Pacific Island plants*, 1–34 (1941–78) by A. C. Smith.[84]

Bibliography

A detailed chronological bibliography appears on pp. 414–454 of *Plants of the Fiji Islands* by John Parham (see below). Selected items from 1968 to 1988 appear in Ash and Vodonaivalu (1989; see Note 74).

PARHAM, J. W., 1972. *Plants of the Fiji Islands.* Revised edn. xv, 462, xxix pp., 104 figs., 1 col. pl. Suva: Government Printer. 1st edn., 1964.)

Annotated systematic enumeration of native, naturalized, adventive and commonly cultivated vascular plants; includes essential synonymy, vernacular names, generalized indication of local range (with more details for less common species), numerous literature citations, and an extensive chronologically arranged bibliography; glossary and general index to family, generic and vernacular names at end. The introductory section includes a brief history of botanical work in Fiji as well as general notes on physical features, flora and vegetation.

SMITH, A. C., 1979–91. *Flora vitiensis nova: a new flora of Fiji.* 5 vols. Illus. (part col.). Lawai, Kauai, Hawaii: Pacific Tropical Botanical Garden. Complemented by *idem*, 1996. *Flora vitiensis nova . . . Comprehensive indices* (comp. J. L. Leopold and M. Egan). 125 pp., end-paper maps. Lawai: National Tropical Botanical Garden.

Copiously annotated, briefly descriptive illustrated flora of native, naturalized, adventive and commonly cultivated seed plants; includes keys to all taxa, full synonymy (with references and citations), vernacular names, general indication of local range with citation of representative *exsiccatae*, extralimital distribution where appropriate, taxonomic commentary and typification with historical remarks, and notes on habitat, frequency, occurrence, special features, properties, local uses, etc. The introductory section (volume 1), with several scenic colored plates, gives the plan and genesis of the work (for which preliminary studies began in 1933 in continuation of work initiated by J. W. Gillespie in the late 1920s), physical features, climate, floristics, vegetation, and phytogeography of the islands, a discussion of the delimitation of taxa, and an account of botanical exploration. An addendum and preliminary general index appear in volume 5. The general index of 1996 also contains a floristic analysis, with statistics. [Within the major flowering plant groups (the monocotyledons preceding the dicotyledons), a linear version of the Takhtajan system (1969) is followed (save that Orchidaceae are in the last volume).][85]

Pteridophytes

Brownlie's account succeeds E. B. COPELAND, 1929. *Ferns of Fiji.* 105 pp., 5 pls. (Bernice P. Bishop Mus. Bull. 59). Honolulu.

BROWNLIE, G., 1977. *The pteridophyte flora of Fiji.* 397 pp., 44 pls., 3 maps (Beih. Nova Hedwigia 55). Vaduz, Liechtenstein: Cramer/Gantner.

Illustrated descriptive treatment of ferns and fern-allies; includes keys, synonymy (with references), typification of taxa, general indication of local and extralimital distribution with citation of *exsiccatae*, critical commentary, distinguishing features, and notes on habitat, etc.; complete botanical index at end. An introduction includes an account of exploration, sources of information, remarks on introduced and excluded species, technical notes, and a synoptic list of species; list of major references, pp. 7–8. The illustrations are of representative taxa.

956

Tonga

Area: 699 km². Vascular flora: 771 species, including introductions (Yuncker, 1959). – Comprises the island chain – and the Kingdom of Tonga – from Minerva Reefs in the south to Niuafo'ou and Tafahi in the north.

Despite the relatively modern flora by Yuncker, the islands are seen as botanically unevenly explored. Yuncker's work succeeded *The flora of the Tonga or Friendly Islands* by W. B. Hemsley (1894, published in *J. Linn. Soc., Bot.*), the first flora of the group. Another early account is *The flora of Vavau, one of the Tonga Islands* by I. H. Burkill (1901, published in the same journal).

YUNCKER, T. G., 1959. *Plants of Tonga.* 283 pp., 16 text-figs. (Bernice P. Bishop Mus. Bull. 220). Honolulu.

Systematic enumeration of known vascular plants, bryophytes and fungi, with very brief descriptions of all vascular species; includes essential synonymy, vernacular names, localities with citations of *exsiccatae* and relevant earlier literature, general indication of external range, and notes on habitat and frequency as well as status (native, naturalized, etc.); indices to all taxa and to vernacular names at end. The introduction includes a general description of the islands and their climate as well as an account of previous botanical work.

957

Niue

Area: 259 km². Vascular flora: 629 taxa, of which *c.* 175 indigenous (Sykes, 1970). – Along with Niue, a self-governing state in association with New Zealand, this unit includes Beveridge Reef.

The floras of Yuncker (1943) – the first separate work – and Sykes (1970) are complementary, and between them provide fairly thorough coverage.

SYKES, W. R., 1970. *Contributions to the flora of Niue.* 321 pp., 45 halftones (New Zealand Div. Sci. Indust. Res. Bull. 200). Wellington.

Systematic enumeration of vascular plants and bryophytes; includes descriptions for many species, limited synonymy, vernacular names, localities with *exsiccatae*, extensive taxonomic commentary and notes on habitat, special features, biology, etc.; brief glossary, list of references, and indices to all botanical and vernacular names at end. The introductory section gives general remarks on climate and vegetation, botanical exploration, and the results of the author's 1965 survey. [Intended as a supplement to Yuncker's flora (see below).]

YUNCKER, T. G., 1943. *The flora of Niue Island.* ii, 126 pp., 3 text-figs. (incl. map), 4 pls. (Bernice P. Bishop Mus. Bull. 178). Honolulu.

Briefly descriptive flora of vascular plants and bryophytes, without keys; includes vernacular names (Niuean and English), indication of local range (with some citations of *exsiccatae*) and extralimital distribution, critical taxonomic remarks, and notes on habitat, uses and status (native, naturalized, etc.); index to all taxa and to vernacular names. The introductory section deals with the geography, soils, climate, history and economy of the island as well as the general features of the flora and vegetation.

958

Samoa

Area: 3137 km^2 (Western Samoa, 2842 km^2; American Samoa exclusive of Swains I., 195 km^2). Vascular flora: American Samoa, 489 native and naturalized species. – Samoa as here delimited includes the islands from Savai'i in the west to Rose Atoll in the east. For Swains Island, politically part of American Samoa, see **974**. At present Savai'i and Upolu (and associated islets) comprise the independent state of Western Samoa while Tutuila (with its harbor of Pago Pago), the Manua Islands and Rose Atoll form the bulk of American Samoa.

Although botanical exploration has been extensive, both before and after partition in 1899, no fully retrospective modern flora with keys is available. The first account was *Die Flora der Samoainseln* (1896–98; see below) by Franz Reinecke, to which significant additions were made by Carl Lauterbach. Additional data notably were furnished by Karl Rechinger in the six parts of his *Botanische und zoologische Ergebnisse einer wissenschaftlichen Forschungsreise nach den Samoa-Inseln, dem Neu-Guinea Archipel, und den Salomonsinseln von März bis Dezember, 1905* (1907–15; index in its last part). These were succeeded by the Bishop Museum-sponsored accounts respectively of the flowering plants (by Erling Christophersen) and pteridophytes (by Carl Christensen); however, the former in particular was largely an account of collections made for the Museum after 1920, including those of the author in 1929 and 1931–32 (mostly from Western Samoa). There are several partial works, mostly for American Samoa; in addition, some key genera (*Meryta*, *Pandanus*, *Psychotria* and *Syzygium*) have been revised in recent years in addition to those covered in A. C. Smith's *Studies of Pacific Island plants* (see **950**). For environment and vegetation, see W. A. WHISTLER, 1992. Vegetation of Samoa and Tonga. *Pacific Sci.* **46**: 159–178.[86]

Bibliography

PEREIRA, J. A. (comp.), 1983. *A check list of selected material on Samoa*, 1: *General bibliography*. iii, 437 pp. Western Samoa: University of the South Pacific Extension Centre. [Botany, pp. 178–195.]

CHRISTOPHERSEN, E., 1935–38. *Flowering plants of Samoa*, [I–]II. 2 parts. 53 text-figs., 3 pls. (Bernice P. Bishop Mus. Bull. 128, 154). Honolulu. Complemented by C. CHRISTENSEN, 1943. *A revision of the Pteridophyta of Samoa*. 138 pp., 4 pls. (Bernice P. Bishop Mus. Bull. 177). Honolulu.

Flowering plants is a systematic enumeration of angiosperms (with some families contributed by specialists), without keys; includes descriptions of new taxa, synonymy (with references and citations), vernacular names, localities with *exsiccatae*, critical remarks and notes on habitat, local uses, etc.; list of references (in both parts) and indices to all botanical and vernacular names. Ethnobotanical data are extensive, but there is little direct reference to earlier German work or even the data of the United States Exploring Expedition. The complementary *Pteridophyta*, also an enumeration, includes some keys to species.

REINECKE, F., 1896–98. Die Flora der Samoainseln, [I–II]. *Bot. Jahrb. Syst.* **23**: 237–368, 8 text-figs., pls. 4–5; **25**: 578–708, 1 text-fig., pls. 8–13.

Systematic enumeration of then-known non-vascular and vascular plants of Samoa; includes descriptions of new taxa, limited synonymy, vernacular names, localities with *exsiccatae*, brief mention of overall distribution (if applicable), sometimes extensive taxonomic notes, and generalized observations on individual species; full index to all taxa in both parts at the end of part II. The introduction includes sections on physical features, geology, physiography and climate of the islands, as well as the general features of the flora. For additions, see C. LAUTERBACH, 1908. Beiträge zur Flora von Samoa-Inseln. *Bot. Jahrb. Syst.* **41**: 215–238.

American Samoa

AMERSON, A. B., JR., W. A. WHISTLER and T. D. SCHWANER, 1982. Accounts of flora and fauna. In R. C. Banks (ed.), *Wildlife and wildlife habitat of American Samoa*, part 2. ii, 151 pp., 26 text-figs. (mainly maps). Washington: U.S. Fish and Wildlife Service.

Includes on pp. 17–36 a checklist of vascular plants (489 species) with indication of habit, habitat, overall distribution and status as well as representative *exsiccatae* (mostly collections by W. A. Whistler). The list is within the general section, which concludes with references (pp. 77–79); the rest of the work features site data, with locations appearing in table 2 and the data itself in tables 4–44.

SETCHELL, W. A., 1924. *American Samoa.* vi, 275 pp., 57 text-figs., 37 pls. (Publ. Carnegie Inst. Washington 341 (Dept. of Marine Biology Pap. 20)). Washington.

Within this monograph is an enumeration (pp. 41–129) of non-vascular and vascular plants of Tutuila including all records up to 1924 with descriptions of new taxa, vernacular names, localities with citations of *exsiccatae*, much critical taxonomic discussion, and notes on habitats, biology, uses, special features, etc.; list of references and indices at end. The introduction deals with geography, physiography, geology, soils, climate, vegetation associations, and origin of the flora, as well as the history of botanical exploration. Part III of the work gives an account of the plants known from the very small Rose Atoll (0.7 km²).

YUNCKER, T. G., 1945. *Plants of the Manua Islands.* 73 pp., map (Bernice P. Bishop Mus. Bull. 184). Honolulu.

Systematic enumeration of known vascular plants and mosses; includes descriptions of new or little-known taxa, vernacular names, localities with *exsiccatae*, overall distribution, some taxonomic remarks, and indication of habitat; index to all botanical and vernacular names at end. The introduction includes a general description of the group (whose largest island is Tau, 44.2 km²). See also *idem*, 1946. Additions to the flora of the Manua Islands. *Occas. Pap. Bernice P. Bishop Mus.* 18(14): 207–209.

959

Wallis and Futuna (Horne) Islands

Area: 255 km². Vascular flora: 475 species, of which 292 indigenous (Morat and Veillon, 1985; see below); endemism is low (no genera, 7 species). – Included in this French territory are Wallis and the Futuna (Horne) Islands (the latter encompassing Futuna and Alofi).

Wallis and Futuna for long had botanically been insufficiently known and the list by St. John and Smith (1971) was avowedly provisional. This has now been succeeded by Morat and Veillon's 1985 list which takes into account additional collections.

Bibliography

O'REILLY, P., 1964. *Bibliographie des îles Wallis et Futuna.* 68 pp. Paris: Musée de l'Homme. (Société des Océanistes, Publ. 13.) [Annotated general bibliography; includes references in botany.]

MORAT, P. and J.-M. VEILLON, 1985. Contribution à la connaissance de la végétation et de la flore de Wallis et Futuna. *Bull. Mus. Natl. Hist. Nat.*, IV/B (Adansonia) 7: 259–329.

'Annexe I' (pp. 291–325) comprises an alphabetically arranged, tabular species list of vascular plants with separate columns for each of the three main islands incorporating presence, status, locality or localities, collections and 'biotopes'; introduced species are asterisked. The rest of the paper relates to vegetation, floristics and phytogeography.

ST. JOHN, H. and A. C. SMITH, 1971. The vascular plants of the Horne and Wallis Islands. *Pacific Sci.* 25: 313–348, 2 figs.

Systematic enumeration of known vascular plants, with descriptions of some new taxa; includes very limited synonymy, localities with *exsiccatae* and literature references, extralimital distribution, taxonomic commentary, and notes on habitat, uses, special features, etc.; list of references at end but no separate index. The introductory section includes notes on geography, previous botanical work in the islands, and a brief analysis of the vascular flora.

Region 96

Micronesia

Area: *c.* 2550 km². Vascular flora: 2227 species, of which 1228 indigenous (Fosberg *et al.*, 1979–87; see below under **960**). – Covering an immense expanse of the western Pacific, Micronesia covers all the islands of the former Trust Territory of the Pacific Islands as well as Guam, Marcus, Wake, Nauru, Banaba (Ocean) and Tungaru (the Gilbert Islands, now the main part of Kiribati). Useful maps, though politically now out of date, appear in Gressitt's introductory volume for *Insects of Micronesia*, Fosberg's *Vegetation of Micronesia* and in each number of the *Flora of Micronesia* (see below under **960**); our limits are as depicted by Fosberg except for the exclusion of the Mapia Islands (see **933**). Current nomenclature can be sought in the map *The New Pacific*, published by Pacific Geographic Maps, Honolulu, and reproduced in *Pacific* magazine.

Floristic knowledge of the region through 1974 was summarized by Fosberg and Sachet (1975). Until this time, there were available only two nominally complete general compendia, both by Ryôzo Kanehira and dating from before World War II: the illustrated *Flora micronesica* (1933), in Japanese but with the keys later translated into English, and the *Enumeration of Micronesian plants* (1935). Both works were based on comparatively limited primary resources and covered only the islands of the former Trust Territory. They did, however, attempt to account for earlier botanical knowledge, including the work of the exploring expeditions up to about 1840 (mainly in the Marianas, although Duperrey visited Kosrae in 1824) and subseqent Spanish, German and Japanese botanical surveys as well as the records in *Die Flora der deutschen Schutzgebiete in der Südsee* (1900) by Karl Schumann and Carl Lauterbach.[87] Successor works include the three *Geographical checklists* (1979–87) and *Flora of Micronesia* (1975–) by F. Raymond Fosberg and his associates. The latter, a definitive critical account but unfortunately without illustrations, is truly synthetic, being based not only on the results of all known field work in the Trust Territory and other Micronesian islands both before and after World War II but also on an unpublished revised enumeration by Takahide Hosokawa (a student of Kanehira), appropriate floras, monographs and revisions, and extensive critical studies by the authors and others. The former, a preliminary name list, appeared in three installments in *Micronesica* and was produced with the admission that progress on the *Flora* would be slow. Although naturally more up-to-date than Kanehira's list, it is less detailed, lacking citations of *exsiccatae* or commentary.

Coverage of individual units within the region is patchy. Guam, perhaps the most thoroughly collected, has two descriptive floras, respectively by Safford (1905) and Stone (1971); only the latter is accounted for here. For other islands, more or less extensively annotated enumerations have been published: Palau (Belau) (Otobed, 1977; Fosberg, 1980); Yap (Volkens, 1901); the Truk group (Hosokawa, 1937); Pohnpei (Glassman, 1952); the Northern Marianas (Hosokawa, 1934; Fosberg *et al.*, 1975); Marcus (Sakagami, 1961); Wake (Fosberg, 1959; Fosberg and Sachet, 1969); the north-

ern Marshalls (Fosberg, 1955); Kiribati (Volkens, 1903, also covering the Marshalls); Nauru (Thaman, 1994) and a number of the low Carolines, including Kapingamarangi. No florulas are available for Banaba or for Kosrae, both considered more or less undercollected. Also imperfectly known are some of the smaller Western Carolines.

Progress

FOSBERG, F. R. and M.-H. SACHET, 1975. Micronesia: status of floristic knowledge. In *Proceedings of the Thirteenth Pacific Science Congress* (Pacific Science Association), 1: 98. Vancouver. [Abstract only.]

SCHLECHTER, R., 1921. Die Orchidaceen von Mikronesien. *Bot. Jahrb. Syst.* 56: 434–501. [Includes a brief history of investigations to 1914 including Ledermann's tour. The influence of Volkens in stimulating collecting and observations by resident officials, missionaries and others is noted.]

Bibliographies. General and supraregional bibliographies as for Superregion 94–99.

Regional bibliographies

SACHET, M.-H. and F. R. FOSBERG, 1955. Annotated bibliography of Micronesian botany. In M.-H. Sachet and F. R. Fosberg (eds.), *Island bibliographies*, pp. 1–132 (Natl. Acad. Sci./Natl. Res. Council Publ. 335). Washington. Continued as *idem*, 1971. Supplement to Annotated bibliography of Micronesian botany. In M.-H. Sachet and F. R. Fosberg (eds.), *Island bibliographies, Supplement*, pp. 3–75 (*ibid.*, Supplement). Washington. [Both works comprise briefly annotated lists, with subject, geographical and systematic indices.]

UTINOMI, H., 1944. Bibliographia micronesica scientiae naturalis et cultis. In *Toa kyō-ei-ken sigenkagaku bunken mokuroku* [*Bibliographic index for the study of the natural resources of the Greater East Asia Co-Prosperity Sphere*], 5 (*Micronesia*). 208 pp. Tokyo: Department of Education, Japan. English edn.: *idem*, 1952. *Bibliography of Micronesia*. Transl. and ed. O. A. Bushnell *et al.* 157 pp. Honolulu: University of Hawaii Press. [Unannotated; botany, pp. 1–21 (pp. 3–16 in English edn.).]

Indices. General and supraregional indices as for Superregion 94–99.

960

Region in general

As already noted, the pre-war *Flora micronesica* and *Enumeration of Micronesian plants* and the more recent *Geographical checklists* are the only nominally complete floristic works. The first two deal largely with the former Trust Territory, thus only sketchily accounting for Nauru, Banaba, the Tungaru (Gilbert) group, Marcus and Wake. The Trust Territory islands (and Guam) were also nominally covered in *Flora der deutschen Schutzgebiete in der Südsee* (**930**). From its commencement in 1975 publication of *Flora of Micronesia* was slow; through 1999 only five fascicles had appeared, the most recent in 1993.[88] No continuation of the work appears now to be in prospect. Nomenclators for dicotyledons (1979) and other groups (1982–87) have appeared in the meantime.

Not much really useful background material has been published. No introductory section has yet appeared in *Flora of Micronesia*, and a *Botanical report on Micronesia* by Fosberg prepared for the U.S. Commercial Company (350 pp., 445 figs., 1946) was never as such published, save for its treatment of the vegetation (in a revised, more definitive form; see below). The best biologically oriented modern general and topographical work on the region, which includes a gazetteer, is J. L. GRESSITT, 1954. *Insects of Micronesia: introduction.* ix, 257 pp., 70 text-figs., map (Insects of Micronesia 1). Honolulu: Bishop Museum Press. The above-mentioned vegetatiological treatise, unfortunately not completed, appeared as F. R. Fosberg, 1960. *The vegetation of Micronesia, 1. General descriptions, the vegetation of the Marianas Islands, and a detailed consideration of the vegetation of Guam.* Pp. 1–75, 2 maps, pls. 1–40 (Bull. Amer. Mus. Nat. Hist. 119(1)). New York. [For a successor to the latter work, see MUELLER-DOMBOIS and FOSBERG, 1998 (under **940–90**).]

FOSBERG, F. R., M.-H. SACHET and R. L. OLIVER, 1975. *Flora of Micronesia.* Parts 1– (Smithsonian Contr. Bot. 20, *passim*). Washington.

Critical 'research' flora of native, naturalized and common introduced seed plants, issued in installments; each fascicle (covering one or more families) includes keys to genera and species, full synonymy (including references and citations), vernacular names, detailed indication of regional distribution (including *exsiccatae*), extralimital range, taxonomic commentary, notes on habitat, biology, ethnobotany, etc. (the last often extensive), and, at the end, references and index. No systematic order is followed in publication, although a notional arrangement in accordance with the traditional Englerian sequence was envisaged. [As of 1999, five fascicles had been published, the most recent in 1993.]

FOSBERG, F. R., M.-H. SACHET and R. L. OLIVER, 1979. A geographical checklist of the Micronesian Dicotyledoneae. *Micronesica* 15: 41–295, map. Continued as *idem*, 1982. A geographical checklist of the Micronesian Pteridophyta and Gymnospermae. *Ibid.*, **18**: 23–82; and *idem*, 1987. A geographical checklist of the Micronesian Monocotyledoneae. *Ibid.*, **20**: 19–129.

Systematic name lists covering native and introduced vascular plants, with synonymy and indication of status and islands of occurrence. A list of references and a family index are given at the end of each installment. Covers a total of 2227 species with 1228 considered to be native.

KANEHIRA, R., 1933. *Flora micronesica.* 8, 468, 37 pp., 211 text-figs., 21 pls., map. Tokyo: South Seas Bureau.

Illustrated, keyed descriptive treatment of trees and shrubs together with an enumeration of all known vascular plants; the latter inclusive of synonymy (with references), vernacular names, and local and extralimital range. The introductory section includes a general sketch of the climate, vegetation, and general features of the flora, an account of botanical exploration, statistics of the flora, and an analysis of the vegetation on different islands; an index to botanical names concludes the work. [Wholly in Japanese; for a free English rendition of the keys, see H. ST. JOHN, 1956. A translation of the keys in 'Flora Micronesica' of Ryôzo Kanehira (1933). Pacific Sci. 10: 96–102.]

KANEHIRA, R., 1935. Enumeration of Micronesian plants. *J. Dept. Agric. Kyushu Imp. Univ.* 4: 237–464, pl. 2 (map).

Systematic enumeration of vascular plants; includes synonymy (with references), vernacular names, localities with *exsiccatae*, and an index to all botanical names. The introductory section includes an account of botanical exploration. [Both this and the preceding work cover Guam as well as the Northern Marianas, Palau, Caroline and Marshall Islands (i.e., the then-Japanese Mandated Territory).]

961

Palau

Area: 494 km². Vascular flora: no data available. – The islands of Palau (Belau) extend from Ngarungl and Ngaiangl (Kayangel) in the north through Babeldaob (Babelthuap) and Koror to Tobi (and Helen Reef) in the south. Although geographically usually associated with the Caroline Islands, they are not, however, in the Federated States of Micronesia; Palau is at present independent 'in free association' with the United States.[89]

The 1980 checklist does not wholly supplant that of Demei Otobed; both are here included. The latter, prepared with assistance from Fosberg and others, was in part based on a local herbarium now at the Conservation Service in Koror.

FOSBERG, F. R. *et al.*, 1980. *Vascular plants of Palau with vernacular names.* ii, 43 pp. Washington, D.C.: Department of Botany, Smithsonian Institution.

Systematically arranged, unannotated checklist with (where known) equivalent English and Palauan vernacular names and indication of status (if exotic). An introductory section gives an indication of the scope and method of development of the work.

OTOBED, D. O., 1977. *Guide list of plants of the Palau Islands.* Revised edn. vii, 52 pp. Koror, Palau: Biology Laboratory, Trust Territory of the Pacific Islands. (Mimeographed; first issued 1961 with subsequent versions in 1967, 1971 and 1972.)

Part 1 of this unannotated work comprises a systematically arranged checklist of botanical names, with English and Palauan equivalents where known. Part 2 is a lexicon of local names with scientific equivalents. The introductory part encompasses a brief introduction and (pp. ii–vii) an index to families; a rather brief list of references concludes the work. Covers vascular plants along with some mosses and algae.[90]

962

Federated States of Micronesia (Caroline Islands)

Land area: 700 km². Vascular flora: no data available. – The Federated States of Micronesia, stretching through some 30 degrees of longitude, extend from Yap and Ngulu in the west to Kosrae in the east, with a southward extension to the isolated atoll of Kapingamarangi. They correspond to the Caroline Islands (without the Palau group). Among them are both 'high' and 'low' islands, with among the latter a wide range of atoll types. Formed in the 1980s from the dissolution of the Trust Territory of the Pacific Islands and currently independent in 'free association' with the United States, the country comprises four states: Yap, Chuuk (Truk), Pohnpei (Ponape) and Kosrae (Kusaie); its capital is at Kolonia on Pohnpei.

Floristic research and writing, surely influenced by the islands' physical situation and features as well as other factors, has been relatively local in character and, in the absence of a completed *Flora of Micronesia*, presents a highly fragmented picture. Documentation has also been affected by uneven botanical collecting and, sometimes, limited opportunities for effective coverage; several remote places have seldom been visited, and indeed until after World War II most of the low islands and atolls were largely neglected.[91]

As a reflection of geographical and other factors, entries are here grouped under seven subheadings.

I. Yap group

Includes Ngulu, Yap, Ulithi, Faïs and Sorol, together making up the 'monsoon' islands of Yap, the most westerly of the Federated States. For the remaining islands in Yap State, see subheading **VII** below.

Of particular interest (and covered in one of the works listed below) is Faïs, a phosphate-bearing raised atoll comparable to Makatea in the Tuamotus; it is 19 m high and 2.8 km² in area with reputedly a relatively rich but underexplored flora. Geologically it is associated with Babeldaob (in Palau) and the southern Marianas.

FOSBERG, F. R. and M. EVANS, 1969. *A collection of plants from Fais, Caroline Islands.* 15 pp. (Atoll Res. Bull. 133). Washington.

Enumeration of plants collected by the authors in 1965 during a 2½ hour visit from a 'field trip ship'; includes indication of status, local names, citation of *exsiccatae*, and ecological and biological notes. The list is preceded by general remarks on the vegetation. Covers 120 species (of which 59 not native), considered only a 'fraction' of the total.

Volkens, G., 1901. Die Vegetation der Karolinen, mit besonderer Berücksichtigung der von Yap. *Bot. Jahrb. Syst.* **31**: 412–477, pls. 11–14.

Includes a briefly annotated list, with vernacular names, of the vascular and some non-vascular plants known from Yap. The remainder of the work is taken up with a consideration of physical features, climate, geology, vegetation formations, etc., on this and other islands in the Carolines. [Based on field work in 1899–1900.][92]

II. Chuuk (Truk) Archipelago

Corresponds to the Chuuk (Truk) Archipelago, the principal part of Chuuk, the second of the Federated States.

Hosokawa, T., 1937. [A preliminary account of the phytogeographical study on Truk, Caroline.] *Bull. Biogeogr. Soc. Japan* 7: 171–255, figs. 1–51.

This mixed work (in Japanese apart from the enumeration proper, in English) includes a systematic enumeration of vascular plants, with vernacular names, indication of localities (with *exsiccatae*), earlier records, general occurrence, and habitat, life-form and special features. The introductory section gives accounts of physical features, climate and botanical exploration, while at the end are accounts of vegetation, floristics and phytogeography as well as lists of plants by life-forms. A summary (also in English) and list of references are also given.

III. Pohnpei (Ponape)

Area: 336 km². Vascular flora: 249 indigenous angiosperms (Glassman, 1952). – The main and only 'high' island in Pohnpei, the third of the Federated States.

Glassman, S. F., 1952. *The flora of Ponape*. iii, 152 pp., 19 pls. (Bernice P. Bishop Mus. Bull. 209). Honolulu.

Systematic enumeration of vascular plants; includes descriptions of new taxa, synonymy (with references and citations), indication of *exsiccatae* with localities, generalized summary of extralimital range, taxonomic commentary, and notes on uses; list of references and full index. The introductory section includes chapters on physical features, general attributes of the

flora and vegetation, botanical exploration, and agriculture and economic botany.

IV. Kosrae (Kusaie)

Area: 109 km². Vascular flora: no data. – Corresponds to the state of Kosrae, the most easterly of the Federated States.

No separate flora or checklist is available for the island. For a vegetatiological treatment of inland areas, see B. D. Maxwell, 1982. Floristic description of native upland forests on Kosrae, Eastern Caroline Islands. *Micronesica* 18: 109–120.

V. Kapingamarangi

This atoll, the most isolated of the Caroline Islands, lies due north of Nuguria (**939**) at about 1° N latitude. It is in the Federated States of Micronesia state of Pohnpei. As of 1992, 43 indigenous species were known.

Niering, W. A., 1956. *Bioecology of Kapingamarangi Atoll, Caroline Islands: terrestrial aspects*. iv, 32 pp., 33 text-figs. (incl. maps) (Atoll Res. Bull. 49). Washington.

Primarily ecological but includes a table of species of vascular plants, divided into trees, shrubs and herbs with each islet being separately considered; for each species are given notes on localities, frequency and mode of origin. A related synthesis is *idem*, 1963. Terrestrial ecology of Kapingamarangi Atoll, Caroline Islands. *Ecol. Monogr.* 33: 131–160.

VI. Eastern low islands and atolls (Kosrae and Pohnpei)

Included here are the islands from Pingelap to Nukuoro and Oroluk, all now in the state of Pohnpei. Florulas are available for Pingelap, Ant and Mokil.

Glassman, S. F., 1953. New plant records from the eastern Caroline Islands, with a comparative study of the plant names. *Pacific Sci.* 7: 291–311.

Includes reports on collections from Mokil, Ant and Pingelap, based on one-day visits, with annotated checklists of the higher (and lower) plants found in each island unit featuring local names, habitat, frequency of occurrence, and *exsiccatae*; linguistic comparison and list of references at end. For Pingelap see also H. St. John, 1948. Report on the flora of Pingelap Atoll, Caroline Islands, and observations on the vocabulary of the native inhabitants. *Ibid.*, **2**: 96–113. [57 taxa, of which 32 indigenous.]

VII. Central and western low islands and atolls (Chuuk and Yap)

Included here are all islands from Lukunor and Satawan west to Eauripik, passing through, among others, Namoluk (southeast of the Chuuk group), Satawal (west of the Chuuk group, in Yap State), and Namonuito and the Hall Islands (north and northwest of the Chuuk group). Separate florulas are available for those named. The Chuuk/Yap state boundary passes east of Satawal and Lamotrek, at 147° 30′ E.

Of particular concern in Marshall's study of Namoluk was the occurrence of severe hurricanes with accompanying ecosystem destruction.

FOSBERG, F. R., 1969. *Plants of Satawal Island, Caroline Islands.* 13 pp. (Atoll Res. Bull. 132). Washington.

Enumeration of vascular plants (203 taxa of which 46 introduced) collected on a two-day visit in 1965; includes indication of status, local names, *exsiccatae*, and notes on biology, ecology, etc. The list is preceded by a general account of the vegetation. [As of 1992, 50 species were considered indigenous.]

MARSHALL, M., 1975. *The natural history of Namoluk Atoll, eastern Caroline Islands.* ii, 53 pp., 13 pls. (Atoll Res. Bull. 189). Washington.

Comprises a general natural history and ethnobiological survey, with an account of the vascular flora including a list of 113 species; within the list local names, biology, uses, and voucher specimens are indicated. Twelve photographs of vegetation are appended.

STONE, B. C., 1959. Flora of Namonuito and the Hall Islands. *Pacific Sci.* **13**: 88–104.

Annotated enumeration of higher plants (94 species, with 52 indigenous); includes keys to species, localities (with citation of some exsiccatae), and notes on habitat, special features, cultivation, etc. [As of 1992, 41 species were considered indigenous on Namonuito.]

963

Marianas Islands (including Guam)

Area: Guam, 541 km^2; Northern Marianas, 476 km^2. Vascular flora: Guam, 930 species of which 330 certainly and 20 doubtfully native (Stone, 1970); Northern Marianas, 286 species (Fosberg *et al.*, 1975). – The Marianas comprise the islands from Guam north through Rota, Saipan and Pagan to Farallon de Pajaros. Guam, a United States possession, remains politically separate from the remaining islands, now incorporated as the Commonwealth of the Northern Marianas Islands (affiliated with the United States). Although a case could be made for subdivision of the Marianas along geological lines – the southern islands from Guam through Saipan to Farallon de Medinilla are essentially raised limestone, the remainder being volcanic (and some actively so) – for convenience the century-old political division is here followed.

Fosberg's work on Micronesian vegetation (see above under **960**) is most complete for the Marianas and especially Guam, and gives a good impression of the great changes which have occurred in the historical period, at least through the 1950s.

I. Guam

STONE, B. C., 1970. *The flora of Guam.* vi, 629 pp., 97 text-figs., 16 pls. (3 col.), 4 maps (Micronesica 6). Agaña.

Briefly descriptive flora of native, naturalized and commonly cultivated vascular plants; includes keys to all taxa, synonymy (with references and citations of pertinent literature), vernacular names, generalized indication of local, Marianan and extralimital range (with some citations of *exsiccatae*), short notes on habitat, special botanical features, uses, etc., and separate indices to vernacular (Chamorro and English) and all botanical names. The introductory section includes chapters on the history of Guam, botanical exploration, floristics and vegetation formations, phytogeography, forest and other plant resources, natural reserves, agriculture and economic botany, and suggestions for future work.

II. Northern Marianas

This subunit corresponds to the Commonwealth of the Northern Marianas. The below-cited list by Fosberg, Falanruw and Sachet, however, focuses on the volcanic islands from Anatahan to Farallon de Pajaros (hitherto for the most part poorly documented) and does **not** account for Rota, Aguiguan, Tinian or Saipan; the southernmost island included is Farallon de Medinilla.

FOSBERG, F. R., M. V. C. FALANRUW and M.-H. SACHET, 1975. Vascular flora of the northern Marianas Islands. 45 pp., 2 maps (Smithsonian Contr. Bot. 22). Washington.

Annotated enumeration of vascular plants; includes localities with *exsiccatae*, notes on habitat, special features, variability, wider distribution, etc., list of references and index to families. The introduction among other topics gives an account of previous work on the flora. No keys or synonymy are included. For additions, see *idem*, 1977. Additional records of vascular plants from the Northern Marianas Islands. *Micronesica* 13: 27–31.

HOSOKAWA, T., 1934. [Preliminary account of the vegetation of the Marianne Islands group.] *Bull. Biogeogr. Soc. Japan* 5: 124–172, figs. 1–9, pls. 10–14 (incl. map).

Includes (pp. 129–151) a systematic list (in Japanese save for the list proper and the tables) of vascular plants, with indication of presence or absence by island; tabular list of local distribution, with summary of extralimital range (if applicable). The remainder of the work consists of sections on physical features, climate, botanical collectors, vegetation and phytogeography; a summary and list of references is appended.

964

Okino-tori-shima (Parece Vela)

Area: less than 0.01 km^2. – This pair of emergent rocks, associated with Douglas Reef, lie in the Philippine Sea at 20° 25′ N by 136° 05′ E, well west of Farallon de Pajaros in the Marianas and southwest of Kazan Retto (Volcano Islands). They have been claimed by Japan since 1931; recent concern there over their possible erosion and submergence has led to plans to stabilize and 'enhance' them, so retaining for the nation the surrounding 200-mile 'economic zone'.[93] With each island at high tide no larger than a king-size bed, the presence of vascular plants is unlikely.

965

Minami-tori-shima (Marcus Island)

Area: 0.3 km^2. Vascular flora: no data. – This isolated raised atoll is a Japanese possession.

Sakagami's paper of 1961 (although primarily zoological) appears to provide the most recent information on the limited vascular flora. An earlier, still-useful work is W. A. BRYAN, 1903. A monograph of Marcus Island. *Occas. Pap. Bernice P. Bishop Mus.* 2: 77–124.

SAKAGAMI, S. F., 1961. An ecological perspective of Marcus Island, with special reference to land animals. *Pacific Sci.* 15: 82–104, illus., map.

Includes a list of plants as well as a description of the vegetation, with accompanying photographs and a map showing distribution of the plant formations. A historical sketch is also provided.

966

Wake Island

See also **998** (CHRISTOPHERSEN). – Area: 7.4 km^2. Vascular flora: 94 species, of which 20 indigenous (Fosberg and Sachet, 1969). Wake is a United States possession, at one time used as a way-station for air services to and from North America and East Asia and also as a military base.

FOSBERG, F. R., 1959. *Vegetation and flora of Wake Island*. 20 pp. (Atoll Res. Bull. 67). Washington.

Enumeration of the native, naturalized and commonly cultivated vascular plants and algae known from the atoll, with notes on the occurrence, frequency, habitats and general features of each species; also includes local range (with *exsiccatae* and literature records), some taxonomic commentary, and a list of references. The introductory section covers climate, soils and the main vegetation features. For further data, see F. R. FOSBERG and M.-H. SACHET, 1969. *Wake Island vegetation and flora, 1961–1963.* 15 pp. (Atoll Res. Bull. 123). Washington.

967

Marshall Islands

Area: 176 km². Vascular flora: no data. – The islands included here, a German colony from 1885 to 1914 and afterwards under Japanese and then United States administration, currently comprise the Republic of the Marshall Islands, a state in 'free association' with the U.S. They are more or less aggregated into two chains: Radak in the north and Ralik in the south.

The only overall work, now long out of date, is that by Volkens (1903). Several modern florulas of individual atolls or groups thereof have appeared; they are here accounted for under subheadings for the Radak and Ralik groups. Much of the work reported in these was carried out under U.S. military sponsorship.

VOLKENS, G., 1903. Die Flora der Marshallinseln. *Notizbl. Königl. Bot. Gart. Berlin* **4**: 83–91.

Briefly annotated list of the known vascular plants and fungi of the Marshall (and Gilbert) Islands; includes vernacular names, localities with *exsiccatae* (but without a generalized indication of local range), some descriptive remarks, and notes on uses. The introductory section includes some general comments on the vegetation.

I. Radak chain

Comprises the northern islands from Bikini and Pokak through Aur, Majuro and Arno to Mili and Narik.

FOSBERG, F. R., 1955. *Northern Marshalls expedition 1951–1952. Land biota; vascular plants.* 22 pp. (Atoll Res. Bull. 39). Washington.

Systematic enumeration of vascular plants, with indication of presence by atoll or islet (including citation of *exsiccatae*) and notes on habitat, occurrence, frequency, special features, etc.; vernacular names; critical remarks. Based mainly on military expeditions of

1951–52 but incorporating records from field trips in 1946 and 1950. See also *idem*, 1959. *Additional records of phanerogams from the northern Marshall Islands.* 9 pp. (Atoll Res. Bull. 68). Washington.

Individual atolls

ST. JOHN, H., 1951. Plant records from Aur Atoll and Majuro Atoll, Marshall Islands. *Pacific Sci.* **5**: 279–286, map. (Pacific plant studies 9.)

Systematic enumeration of vascular plants with essential synonymy, Marshallese vernacular names, *exsiccatae* or observations with localities and field notes, some taxonomic notes (particularly on *Canavalia sericea* A. Gray); literature. [Based on the first collections ever made on Aur and Majuro during a 1945 field trip by the author. The latter is now home to the capital of Marshall Islands.]

STONE, B. C. and H. ST. JOHN, 1960. *A brief field guide to the plants of Majuro, Marshall Islands.* Majuro, Marshall I.: Marshall Islands Intermediate School.

Provides a key to species, descriptive text, and many illustrations. [Not seen; cited from J. Lawyer *et al.* (unpubl.). *A guide to selected current literature on vascular plant floristics for the contiguous United States* (see **Division 1, Divisional bibliographies**).]

TAYLOR, W. R., 1950. *Plants of Bikini and other northern Marshall Islands.* 227 pp., 79 pls. (incl. map), frontisp. Ann Arbor: University of Michigan Press.

Copiously illustrated, briefly descriptive flora of nonvascular and vascular plants, without keys; limited synonymy; citation of *exsiccatae*, notes on ecology, special features, biology, cultivation, etc.; list of references at end. – Especially useful for marine algae.

II. Ralik chain

Comprises the southern islands from isolated Ujelang and Enewetak (Eniwetok) in the northwest through Kwajalein and Jaluit (the former German administrative center) to Ebon in the south. In contrast to the Radak chain, no separate general florula exists.

Individual atolls

See also J. O. LAMBERSON, 1987. Natural history of terrestrial vascular plants of Enewetak Atoll. In D. M. Devaney (ed.), *The natural history of Enewetak Atoll*, 2: *Biogeography and systematics*, pp. 17–35. Oak Ridge, Tenn.: U.S. Department of Energy, Office of Scientific and Technical Information (for Office of Energy Research, Office of Health and Environmental Research, Ecological Research Division).

FOSBERG, F. R. and M.-H. SACHET, 1962. *Vascular*

plants recorded from *Jaluit Atoll*. 39 pp. (Atoll Res. Bull. 92). Washington.

Annotated enumeration of vascular plants with descriptive remarks, vernacular names, citations, and indication of exsiccatae; historical notes; list of references.

ST. JOHN, H., 1960. Flora of Eniwetok Atoll. *Pacific Sci.* **14**: 313–336.

Amply descriptive flora of vascular plants (42 species); includes keys, synonymy, local range (with citation of *exsiccatae*), taxonomic commentary, and notes on distribution and status; list of references at end.

968

Tungaru (Gilbert) group

See also **967** (VOLKENS). Although the nation of Kiribati (independent since 1979) extends to include Banaba (**969** in part) and the Phoenix group and most of the Line Islands in the Central Pacific (Region 97), only the Tungaru (Gilbert) group, from Makin Meang in the north through Tarawa, Tabiteuea and Onotua to Arorae in the south, is accounted for here.

In addition to the works listed below, an unannotated plant list appears in C. M. WOODFORD, 1895. The Gilbert Islands. *Geogr. J.* (London) **6**: 325–350, map. Also of use are J. G. ALLERTON and D. R. HERBST, 1972–73. Report from the Gilbert and Ellice Islands. *Bull. Pacific Trop. Bot. Gard.* **2**(4): 63–68; **3**(1): 2–6; R. OVERY, I. POLUNIN and D. W. G. WIMBLATT, 1982. *Some plants of Kiribati: an illustrated list.* Col. photographs. Tarawa: National Library and Archives, Kiribati; and R. R. THAMAN, 1987. *Plants of Kiribati: a listing and analysis of vernacular names.* 42 pp. (Atoll Res. Bull. 296). Washington.

FOSBERG, F. R. and M.-H. SACHET, 1987. *Flora of the Gilbert Islands, Kiribati.* 33 pp. (Atoll Res. Bull. 295). Washington.

A largely compiled checklist of vascular plants, with 'recent' introductions asterisked; includes synonymy and occurrence by island unit but no other data.[94]

Individual atolls

LUOMALA, K., 1953. *Ethnobotany of the Gilbert Islands.* v, 129 pp., map (Bernice P. Bishop Mus. Bull. 213). Honolulu.

Includes a plant checklist for Tabiteuea Atoll, together with notes on earlier records for other atolls and islands in Kiribati proper. Other sections of the monograph cover physical features, soil, vegetation, etc.

MOUL, E. T., 1957. *Preliminary report on the flora of Onotua Atoll, Gilbert Islands.* ii, 48 pp., map (Atoll Res. Bull. 57). Washington.

An annotated enumeration of non-vascular and vascular plants, along with a description of the vegetation; includes localities (with *exsiccatae*) and notes on habitat, occurrence, biology, vernacular names, uses, etc.

969

Banaba (Ocean Island) and Nauru

Area: Banaba, 6.5 km^2; Nauru, 22 km^2. Vascular flora: Banaba, no data; Nauru, 493 species, of which 59 'possibly indigenous' (Thaman *et al.*, 1994). – Banaba is part of Kiribati; Nauru is independent.

These famous raised 'phosphate' islands were botanically long amongst the most poorly documented in Micronesia, unfortunately the more so because great disturbance to their vegetation has occurred in the present century with the extensive mining operations (on Banaba now discontinued and on Nauru all but exhausted), the effects of World War II, and (on Nauru) urbanization of the coastal belt. Past access was difficult through mining company policy as well as isolation; until recently only casual collections had been made and botanical references have been few and scattered. Publication of the Nauru account of Thaman *et al.* (1994) thus fills a significant gap; it is moreover based on substantial field work carried out by the authors and others from 1979 through 1988.

Older works of interest include a topographical account of the islands with references to the vegetation in A. F. ELLIS, 1936. *Ocean Island and Nauru.* 319 pp. Sydney: Angus & Robertson, and (for Nauru) notes on the vegetation in N. A. Burges, 1934. Nauru. *Sci. J.* (Sydney) **13**: 30–35.[95]

Partial work: Nauru

THAMAN, R. R., F. R. FOSBERG, H. I. MANNER and D. C. HASSALL, 1994. *The flora of Nauru: a compilation and analysis of the vegetation and flora of the equatorial Pacific Ocean island of Nauru.* vi, 223 pp. (Atoll Res. Bull. 392). Washington.

Descriptive account of vascular plants, alphabetically arranged by families with inclusion for each taxon of vernacular, English and scientific nomenclature, synonymy, status (and origin if not native), abundance or frequency, features, local distribution and habitat, indirect or direct uses, and sources including records or collections (the last category being coded, with explanation on pp. 32–33); references (pp. 209–215), statistical summary and ecological and ethnobotanical table at end. The general part (pp. 1–28) accounts for previous studies along with geography, history, the people, the economy, impacts on the flora, vegetation formations and indicator species, the flora in general (summary table, p. 15; list of indigenous or possibly indigenous species, pp. 16–18) and comparison with other island floras, and ethnobotany.

Region 97

Central Pacific Islands

This spread-out region includes all the low islands and atolls extending from Tuvalu (Ellice Islands) and Howland and Baker in the west towards the Line Islands in the east and the Northern Cook Islands and the Tokelau group (with Swains) in the south.

The great diffusion of the various islands and atolls, the perceived poverty and monotony of the vascular flora (at least partly due to the relatively low average rainfall in many groups), and administrative fragmentation and disinterest have doubtless contributed to the lack of any general flora or enumeration. The level of botanical survey is, not surprisingly, variable: from very well-known (Phoenix Islands, which include the one-time air way-station of Kanton Island) to poorly collected (most of Tuvalu). Most of the seven units here designated have, however, nominally complete checklists or florulas, or have one or more individual islands with such a work, here deemed 'representative'.

Units with overall coverage include the Howland and Baker Islands together with the Northern Line Islands (Christophersen, 1927) and the Tokelau Islands (Parham, 1971). Tuvalu is 'represented' by Funafuti (Maiden, 1904) and Nui (Woodroffe, 1985); the Phoenix Islands by Kanton (Degener and Gillaspy, 1955); the Northern Cook Islands by Manihiki

(Cranwell, 1933), and the Southern Line Islands by Caroline (now Millennium; Clapp and Sibley, 1971), Vostok (Fosberg, 1937; Clapp and Sibley, 1971) and Flint (St. John and Fosberg, 1937). In the Northern Line Islands, supplementary coverage is available for Kiritimati (Christmas; Chock and Hamilton, 1962) and Tabuaeran (Fanning) (St. John, 1974), along with that classic of atoll florulas, *Palmyra Island* (Rock, 1916). Other information is very scattered, and standard bibliographies should be consulted.

Bibliographies. General and supraregional bibliographies as for Superregion 94–99.

Indices. General and supraregional indices as for Superregion 94–99.

971

Tuvalu (Ellice Islands)

Area: 26 km^2. Vascular flora: 86 native and introduced species have been recorded for Nui Atoll (Woodroffe, 1985). – Nine atolls comprise present-day Tuvalu, Funafuti being the largest.

No flora or enumeration covering the whole country is available, and indeed botanical work has been sporadic. Of individual islands or atolls, florulas exist only for Funafuti and Nui. The former was one of the results of an Australian Museum (Sydney) atoll research expedition in the 1890s, while the latter arose from the work of a United Kingdom-sponsored land resources survey in the 1970s.[96]

Individual atolls
MAIDEN, J. H., 1904. The botany of Funafuti, Ellice group. *Proc. Linn. Soc. New South Wales* **29**: 539–556.

An enumeration of vascular plants, bryophytes and lichens; includes limited synonymy, vernacular names, generalized indication of distribution, and extensive notes on plant forms, habitat and uses. A brief introduction is attached, while at the end are remarks on dispersal methods and mechanisms.[97]

WOODROFFE, C. D., 1985. *Vegetation and flora of Nui Atoll, Tuvalu.* 18, 10 pp., 8½ halftone pls., 8 maps (Atoll Res. Bull. 283). Washington.

Includes a systematic list of native and introduced plants with vouchers or sight records; discussion and references. The introductory part includes a description of the atoll and its climate, soil and vegetation.

972

Howland and Baker Islands

These two islands, east of Tuvalu and north of the Phoenix group, are United States territory. The only floristic coverage is in *Vegetation of the Pacific Equatorial Islands* by Erling Christophersen (**976**).

973

Phoenix Islands

Area: 52.5 km². Vascular flora: 87 species, of which 28 are considered native (Fosberg and Stoddart, 1994). – The eight Phoenix Islands, now part of Kiribati (**968**), comprise a relatively compact group; the largest island is Kanton (Canton), a former air base and way-station for commercial flights between the United States and Australasia and the South Pacific.

A descriptive flora appeared in 1994. Published accounts also exist for Kanton and surveys have been made in other islands.

FOSBERG, F. R. and D. R. STODDART, 1994. *Flora of the Phoenix Islands, Central Pacific*. 60 pp., 2 maps (Atoll Res. Bull. 393). Washington.

Descriptive treatment with essential synonymy, commentary, citation of *exsiccatae* and references for each taxon (keyed to the bibliography, pp. 54–55); index at end. The introduction includes geographical details as well as descriptions of climate, environment, history and sources. No keys are included.

Individual islands

Only Kanton, inhabited from 1938 (on establishment of the air base) until 1968, has had a separate florula: O. DEGENER and E. GILLASPY, 1955. *Canton Island, South Pacific*. ii, 51 pp. (Atoll Res. Bull. 41). Washington. For additions, see O. DEGENER and I. DEGENER, 1959. *Canton Island, South Pacific (resurvey of 1958)*. 24 pp. (Atoll Res. Bull. 64). Washington. As of 1992, 24 native species were known.

974

Tokelau Islands (Tokelau group and Swains Island)

Area: Tokelau group, 10 km²; Swains I., 2.1 km². Vascular flora: Tokelau group, 40 species; Swains I., 95 species (including introductions). – Includes the three Tokelauan atolls of Atafu, Nukunonu, and Fakaofo (administratively a self-governing territory of New Zealand) along with Swains Island (from 1925 a part of American Samoa).

Tokelau group

PARHAM, B. E. V., 1971. The vegetation of the Tokelau Islands with special reference to the plants of Nukunonu Atoll. *New Zealand J. Bot.* 9: 576–609, 11 text-figs. (incl. maps), table.

Table 1 comprises an enumeration of known vascular plants of the Tokelau Islands (both indigenous and adventive), with indication of life-form and their occurrence elsewhere in the South Pacific. Separate lists are given for each of the three Tokelauan atolls, with local vernacular names and biogeographic commentary. The remainder of the paper deals with physical features, vegetation, and human influence, and ends with a list of references.

Swains Island

WHISTLER, W. A., 1983. *The flora and vegetation of Swains Island*. 25 pp., 10 text-figs. (Atoll Res. Bull. 262). Washington.

Includes an annotated systematic enumeration of vascular plants with vernacular names, indication of frequency, and habitat and biological notes along with *exsiccatae*; references at end. The introduction includes remarks on the history of botanical work and the vegetation.

975

Northern Cook Islands

Area: 26 km². Vascular flora: Suvorov, 23 native species (Woodroffe and Stoddart, 1992; see Note 98). – There are six atolls: the Danger Islands (Pukapuka), Nassau, Suvorov (Suwarrow), Manihiki, Rakahanga and Tongareva (Penrhyn). All are administratively part of the Cook Islands (**981**).

No general enumeration is available, and to date the only individual island florula is that by Cranwell (1933) for Manihiki.[98]

Individual atolls

CRANWELL, L., 1933. Flora of Manihiki, Cook group. *Rec. Auckland Inst. Mus.* 1: 169–171.

Briefly annotated list of vascular and non-vascular plants; includes limited synonymy, vernacular names and notes on habitats and life-forms. An introductory section incorporates general remarks on the atoll, and a list of references is attached.

976

Northern Line Islands

Vascular flora: 42 indigenous species, only 9 of them on all four subaerially habitable islands (exclusive of Jarvis; Wester, 1985 (see below)). – In the Northern Line Islands are, from north to south, Kingman Reef (at high tide completely submerged), Palmyra, Teraina (Washington), Tabuaeran (Fanning) and Kiritimati (Christmas) along with the more southwesterly, somewhat removed Jarvis Island. Teraina, Tabuaeran and Kiritimati are part of Kiribati; the others are United States possessions. Their individual forms show distinct differences, with in the main chain a gradual increase in altitude from north to south. The more northerly islands are for part of the year strongly influenced by the northern branch of the Pacific intertropical convergence zone; they thus enjoy relatively high rainfall, at Palmyra over 4000 mm/annum.

Christophersen's 1927 survey remains an important reference point for Wester's 1985 summary as well as for recent studies of Kiritimati, Tabuaeran and Teraina. The classic early florula for Palmyra by Rock (1916) has also been included.

CHRISTOPHERSEN, E., 1927. *Vegetation of the Pacific equatorial islands.* 79 pp., 13 text-figs., 7 pls. (Bernice P. Bishop Mus. Bull. 44). Honolulu.

Includes separate enumerations of the then-known vascular plants of Jarvis, Kiritimati, Tabuaeran, Teraina and Palmyra Islands as well as those of Howland and Baker Islands (**972**). Each list, accompanied also by general remarks on climate, soils, vegetation and land use, contains notes on life-form, habitat,

occurrence and biology for the species covered, as well as citations of exsiccatae and some taxonomic remarks. At the end of the monograph are notes on dispersal methods of the plants, introduced species, and a list of references; pp. 73–76 furthermore give a tabular summary of geographical distribution of species within the islands concerned.

WESTER, L., 1985. *Checklist of the vascular plants of the Northern Line Islands.* 38 pp. (Atoll Res. Bull. 287). Washington.

A systematic enumeration of vascular plants with critical review notes including alternative names, occurrence, frequency, habitat, and taxonomic commentary along with distribution and authorities. Asterisks indicate more or less recent introductions. The enumeration is preceded by introductory paragraphs on geography, history and plant collectors and a summary account of the flora, and pp. 34–38 encompass a list of references.

Individual atolls

In addition to the following, mention may be made here of L. WESTER, J. O. JUVIK and P. HOLTHUIS, 1992. *Vegetation history of Washington Island (Teraina), Northern Line Islands.* 50 pp., illus., maps (Atoll Res. Bull. 358). Washington.

CHOCK, A. K. and D. C. HAMILTON, Jr., 1962. *Plants of Christmas Island.* 7 pp. (Atoll Res. Bull. 90). Washington.

Briefly annotated list of vascular plants (41 species, of which some 35 were considered spontaneous) along with some fungi; includes citations of *exsiccatae*, references to earlier records, and occasional taxonomic notes with a bibliography at the end. The introductory section includes an account of previous collecting on the island.

ROCK, J. F., 1916. *Palmyra Island: with a description of its flora.* 53 pp., 20 halftones, map (Coll. Hawaii Publ., Bull. 4). Honolulu.

Copiously illustrated annotated enumeration of non-vascular and vascular plants; includes notes on their habitat, life-form and biology, synonymy (with references), and citations of exsiccatae with generalized indication of extralimital range. The introductory section gives a general description of the atoll, with remarks on the climate, vegetation, animal life, and history of human contact. [This florula, enhanced by superb photographs from its author, was long the best of its kind.]

ST. JOHN, H., 1974. Vascular flora of Fanning Island, Line Islands, Pacific Ocean. *Pacific Sci.* 28: 339–355.

Systematic enumeration, irrespective of status, of all reported vascular plants (indigenous species in bold face), with citation of *exsiccatae*. A total of 102 taxa are accounted for (20 indigenous, 28 adventive). The remainder of the paper comprises descriptions of new taxa of Pandanus. [As of 1992, 22 indigenous species were known.]

977

Southern Line Islands

Includes the widely spread-out Caroline (now Millennium), Vostok, Flint, Starbuck and Malden Islands (along with Filippo Reef), politically all now part of Kiribati. Small in size, some are influenced by cooler waters emerging from the Humboldt Current and thus, as at lower elevations in the Galápagos (017), the rainfall is scanty and unreliable. The range of spontaneous vascular plants may accordingly be meager. No overall treatment is available; however, florulas exist for the first three entities named, with that for Caroline (Kepler and Kepler, 1994) being particularly thorough with many good illustrations and maps.

Individual atolls

KEPLER, A. K. and C. B. KEPLER, 1994. *The natural history of Caroline Atoll, Southern Line Islands,* I: *History, physiography, botany, and islet descriptions.* xi, 225 pp., 57 textfigs. (incl. maps), 72 halftone pls. (Atoll Res. Bull. 397). Washington.

Includes (pp. 34–64) tabular lists and a descriptive enumeration, with detailed species maps, of the vascular flora; species accounts encompass synonymy, current and former distribution throughout the atoll, ecology and phenology. The enumeration is followed (pp. 64–69) by a general discussion on the size and composition of the flora (including the influence of geographical, habitat and other factors) and a valuable table of the sizes of known atoll floras throughout the Pacific. The initial part of the work is primarily geographical and historical, while the remainder (pp. 69–171) covers successions and ecology, with documentation of all individual islets or motus. [Succeeds the less critical enumeration (covering 35 species) in R. B. CLAPP and F. C. SIBLEY, 1971. *Notes on the vascular flora and terrestrial vertebrates of Caroline Atoll, southern Line Islands.* ii, 16 pp., 5 text-figs. (Atoll Res. Bull. 145). Washington.]

FOSBERG, F. R., 1937. Vegetation of Vostok Island, central Pacific. *Proc. Hawaiian Acad. Sci.* 11 (1935–36) (*Bernice P. Bishop Mus. Spec. Publ.* 30): 19.

Includes a 'list' of one non-vascular and two vascular plants (*Boerhaavia diffusa* and *Pisonia grandis*). A fuller account of the vegetation, based on a new survey in 1965 and reporting the presence of the same two vascular plants, is given in R. B. CLAPP and F. C. SIBLEY, 1971. *The vascular flora and terrestrial vertebrates of Vostok Island, South-Central Pacific.* 10 pp., illus., map (Atoll Res. Bull. 144). Washington.

ST. JOHN, H. and F. R. FOSBERG, 1937. *Vegetation of Flint Island, central Pacific.* 4 pp. (Occas. Pap. Bernice P. Bishop Mus. 12(24)). Honolulu.

Includes a list of native and more significant introduced vascular plants, with notes on their status (and method of introduction) and *exsiccatae*. A brief prefatory section features a description of the island and its extant vegetation.

Region 98

Islands of southeastern Polynesia

Area: 3200 km² (French Polynesia). Vascular flora: French Polynesia, 675 native flowering plant species, of which 500 endemic (Florence, 1997; see below). Pteridophytes are proportionally quite numerous, with 254 recorded – also for French Polynesia – by Robertson in 1952 (see below). – This region covers all the islands from the southern Cooks eastwards and southeastwards to Rapa Nui (Easter) and Sala-y-Gómez and northeastwards to the Marquesas, thus also encompassing the Societies, the Austral Islands, Rapa Iti (Rapa), the Tuamotu (Low) Archipelago (including Makatea), the Mangareva (Gambier) group and the Pitcairn Islands (including Oeno, Henderson and Ducie). The majority constitute French Polynesia; only the southern Cooks (Cook Islands), the Pitcairn Islands (Britain), and Rapa Nui/Sala-y-Gómez (Chile) are politically apart.

Because these islands and archipelagoes are so scattered and diverse, not only biotically but also politically, and botanical work long fragmented and episodic under a variety of auspices, no overall floras, apart from the three discussed below, have been published. Both local coverage and the level of botanical exploration vary widely; but with respect to the latter the region is generally less well covered than the Hawaiian Islands, many parts being still viewed as insufficiently well collected. In the last decade, however, a regional botanical center has been established in Papeete (Tahiti) under the auspices of ORSTOM and collecting and documentation programmes initiated. This is now bearing fruit in a new *Flore de la Polynésie Française* (1997–).

Only the last-named work, although not encompassing the entire region, can be considered definitive. Its predecessors, *Flore de la Polynésie française* by Emmanuel Drake (1893) and *Flora of southeastern Polynesia* by Elizabeth and Forrest Brown (1931–35), covered a more limited range of materials out of necessity or design. Drake worked with what was available in France and the Browns largely if not even wholly with Bishop Museum collections; these were perforce mutually distinct. The Browns' flora was furthermore severely criticized by Merrill in his farewell *aloha*, *The Botany of Cook's Voyages* (1954), as having been overambitious in conception and prepared without proper guidance. It moreover was, in the words of Marie-Hélène Sachet in the first part of her *Flora of the Marquesas* (**989**), characterized by 'erratic writing and nonexistent editing'.

Since these floras were written, there have been many critical revisionary studies, including those on *Cyrtandra* (Gesneriaceae) by George W. Gillett and on the Rubiaceae by Stephen P. Darwin and others. These constitute important contributions towards the new synthesis now being undertaken by Jacques Florence. Given, however, that his new *Flore de la Polynésie Française* may take some years to complete, a useful interim step might be the preparation of a keyed enumeration for the whole region, so enabling taxonomic problems to be pinpointed.

Of the several floras, florulas and enumerations for the nine units designated here (or for parts thereof), mostly dating from after 1920 when intensive botanical exploration expanded from a few readily accessible islands, the most critical are Sachet's Marquesian flora as well as the fragment of a Societies flora by Grant *et al.* (1974). Both represent complete reviews of available collections, literature, and unpublished data. It is regrettable that Grant's work was never completed, as Welsh's 1998 flora is not truly critical. Because it, along with the projected Raiatean flora of J. W. Moore, originated as a Bishop Museum project parallel to that of the Browns (the latter originally planned to cover only the Marquesas), the Societies (and, for similar reasons, the Cooks) were not encompassed in *Flora of southeastern Polynesia* although on phytogeographical grounds such might have been expected.

Organized contributions for smaller areas are fairly numerous. The southern Cooks are covered to a goodly extent in the florulas for Rarotonga (Wilder, 1931) and Aitutaki (Fosberg, 1975) as well as by a general survey of pteridophytes (Brownlie and Philipson, 1971), while the Societies have coverage for Raiatea (Moore, 1933, 1963 – neither proper florulas) and Tahiti (Nadéaud, 1873; Setchell, 1926 – the latter not a florula) with the pteridophytes as a whole treated by Copeland (1932). In the Tuamotus, separate coverage exists for Makatea (Wilder, 1934), Nukutipipi (Salvat and Salvat, 1992), Raroia (Doty *et al.*, 1954) and Rangiroa (Stoddart and Sachet, 1969). The Gambiers have been treated in an unsatisfactory paper by Huguenin (1974) and again by St. John (1988), while the four Pitcairn Islands have been best summarized by Florence *et al.* (1995). No separate coverage is available for the Australs and Rapa Iti save for a checklist by N. Hallé (1980; supplement, 1986). The depauperate flora of distant and now politically distinct Rapa Nui (Easter Island) has, however, been well documented in the papers by Skottsberg (1922, 1951), Christensen (1920) and Zizka (1991).

The pteridophytes for the region as a whole were last revised in 1931 by Elizabeth Brown and in 1938 by Copeland (who also published a revision for the Societies in 1932). More recent treatments are for individual islands or groups, published either separately or as parts of vascular or general florulas.

Bibliographies. General and supraregional bibliographies as for Division 9 and Superregion 94–99.

Regional bibliography

O'REILLY, P. and E. REITMAN, 1967. *Bibliographie de Tahiti et de la Polynésie française.* xvi, 1046 pp. Paris: Musée de l'Homme. (Société des Océanistes, Publ. 14.) [Annotated general bibliography; botany, pp. 248–288.]

Indices. General and supraregional indices as for Division 9 and Superregion 94–99.

980

Region in general

Neither of the two general floras accounted for below covers the entire region, and that by the Browns is in part not exhaustive. With their differing sources, they are, however, mutually somewhat complementary. Drake's flora (1893) covers those islands under French administration (thus excluding the southern Cooks, the Pitcairn group and Rapa Nui), while that by the Browns omits Rapa Nui, the southern Cooks and the Societies and gives only sketchy coverage beyond the Marquesas, its originally projected limits. It certainly did not meet the need for a critical general flora. Contributions towards such a goal have included a revision of the ferns by Copeland (1938; see below) and the first installment of a checklist (Robertson, 1952), which did not cover more than the ferns and fern-allies. In 1997 the first volume of a new *Flore de la Polynésie Française*, by Jacques Florence, was published.

BROWN, F. B. H. and E. W. BROWN, 1931–35. *Flora of southeastern Polynesia, I–III.* 3 vols. Illus. (Bernice P. Bishop Mus. Bull. 84, 89, 130). Honolulu.

Descriptive flora of native, naturalized and 'village' plants; includes keys to all taxa, occasional indication of synonymy (without references), some citations of standard floras and other works in the text, vernacular names, indication of localities (with *exsiccatae* if present in the Bishop Museum) and summary of overall range, sometimes extensive taxonomic commentary, and notes on sources of introduced species, local varieties of cultivated plants, special features, etc.; lists of references and complete indices at end of each part. The introductory sections deal with physical features, vegetation (rather sketchily), floristics, native agricultural systems, etc., in the Marquesas (other areas virtually neglected), as well as the itinerary of the Bayard Dominick expedition (which yielded the greater part of the materials, collected by the authors, on which the flora was based). [Most complete for the Marquesas, the work also includes records from the Tuamotus, the Austral Islands, Rapa Iti, and (sketchily) the Mangareva and Pitcairn groups but omits the Societies and Southern Cook Islands and does not extend to Rapa Nui. Forrest Brown was mainly responsible for parts I and III (on monocotyledons and dicotyledons respectively), and his wife Elizabeth Wuist Brown for part II (1931, on pteridophytes). For additions and revisions to part II, see COPELAND below under **Pteridophytes**.]

DRAKE DEL CASTILLO, É., 1893. *Flore de la Polynésie française.* xxiv, 352 pp., col. map. Paris: Masson.

Briefly descriptive flora of native, naturalized and 'village' vascular plants; includes keys to all taxa,

synonymy (with references and literature citations), vernacular names, localities with *exsiccatae* and summary of external distribution or origin, critical remarks, and brief habitat and other notes; index to all botanical names. An introductory section contains notes on physical features, floristics, etc. of the French territories in southeastern Polynesia (as well as the Wallis Islands), a brief history of botanical exploration, statistics of the flora and a list of references.[99]

FLORENCE, J., 1997– . *Flore de la Polynésie Française.* Vols. 1– . 393 pp., 52, 4 text-figs., 4 col. pls., 2 maps. Paris: Éditions ORSTOM. (Collection faune et flore tropicales 34.)

Descriptive flora of native and introduced plants with keys, synonymy, references and citations, types, vernacular names (where known), indication of phenology, internal and external distribution (the latter including the Cook and Pitcairn groups if a given French Polynesian species is there also represented), habitats and ecology, phenology, endemism and conservation status, taxonomic commentary (where necessary), and full-page figures of selected species; index to all vernacular and botanical names at end along with a general key to families (and certain cultivated plants not generally flowering in the area), illustrated glossary, and (pp. 325–393) list of specimens seen (arranged both by collectors and by taxa). The general part covers insular biogeography and floristics including remarks on dispersal.[100] The geography, vegetation and floristics of individual island groups follow, along with botanical exploration and the plan of the work (with references cited), and abbreviations and conventions. [The first volume covers Cannabaceae, Cecropiaceae, Euphorbiaceae, Moraceae, Piperaceae, Ulmaceae and Urticaceae (the first two represented only by introductions).]

ROBERTSON, R., 1952. Catalogue des plantes vasculaires de la Polynésie française. *Bull. Soc. Études Océanien.* (Papeete) 8: 371–406.

Introduction; systematic checklist (the species alphabetically arranged within the genera) with authorities, synonymy, distribution by archipelago and island, and probable origin if not native; no references to places of publication, citations or commentaries. Names used by Drake are asterisked but no infraspecific taxa are accounted for. Species not known to have been collected after 1900 or otherwise queried in the literature are treated as doubtful and not numbered. [Only the first part, on pteridophytes, completed; this covers 254 confirmed species.]

Pteridophytes

See also Brown and Brown (1931–35; especially part II, 1931, by E. W. Brown) and Robertson (1952) above.

COPELAND, E. B., 1938. Ferns of southeastern Polynesia. *Occas. Pap. Bernice P. Bishop Mus.* **14**(5): 45–101, 25 text-figs.

Introduction; enumeration with collection records, generalized overall distributions, descriptions of novelties, and commentary. Complements Brown and Brown (see above) and Copeland (1932; see **981**).[101]

981

Society Islands

The 14 Society Islands, high and low, extend from Motuone (Bellingshausen) in the west to Mehetia in the east. – Area: 1614 km^2. Vascular flora: c. 700 native species (Grant, 1974); 200 of these are ferns (PD; see **General references**). Welsh (1998) accepted just over 400 flowering plant species as native. Tetiaroa Atoll has 47 natives (Sachet and Fosberg, 1983).

Drake's obsolete *Flore de la Polynésie française* (see above under **980**) remains the only nominally comprehensive work and until 1997 was the last French-language flora. It was a successor to the first flora of the group, *Zephyritis taitensis* (1836–37) by J. B. A. Guillemin. Of the considerable botanical results of American expeditions and local field programmes (the latter principally by Martin L. Grant and John W. Moore) in the 1920s and 1930s, only a regrettably small proportion ever appeared in floristic form (Copeland, 1932, with additions in 1938; Grant *et al.*, 1974 (but largely written by 1936)). Miscellaneous contributions on individual islands have been published since 1920, but florulas projected for Tahiti and Raiatea were never realized.[102] A new flora was compiled by Stanley Welsh in 1998. Vegetation, floristics and phytogeography have been extensively treated in H. R. PAPY, (1951–)1954–55. *Tahiti et les îles voisines. La végétation des îles de la Société et de Makatéa (Océanie française)*. 2 fascicles. 386 pp., illus., map (Trav. Lab. Forest. Toulouse, V, 2, part 1(3)). Toulouse, and by F. R. FOSBERG in *Pacific Sci.* **46**: 232–250 (1992).

GRANT, M. L., F. R. FOSBERG and H. M. SMITH, 1974. *Partial flora of the Society Islands: Ericaceae to Apocynaceae*. vii, 85 pp. (Smithsonian Contr. Bot. 17). Washington.

Rambling systematic treatment of certain 'sympetalous' families; includes analytical keys to genera, species and lower taxa, moderately long to long descriptions, full synonymy (with references), vernacular names, localities (with *exsiccatae*), extralimital range (if applicable), critical remarks, and extensive (largely compiled) notes on ethnobotany as well as on habitats and associates; index to botanical names at end. The introductory section includes a biographical sketch of the senior author as well as chapters on geography, climate and ethnobotany, and sources for the work, while at the end is a bibliography and an account (by H. M. Smith) of botanical work in the area. [No continuation of this work was ever published.][103]

WELSH, S. L., 1998. *Flora societensis: a summary revision of the flowering plants of the Society Islands*. iv, 420 pp., map (in two copies, towards end-papers). Orem, Utah: E.P.S. (for the author).

Descriptive flora (covering just over 400 native species along with significant introductions) featuring keys, synonymy, references, rather full citation of types, vernacular names, representative *exsiccatae*, and commentary; generic entries with references to significant taxonomic literature; general references (pp. 381–384), glossary, lexicon of author abbreviations and complete index at end. The general part includes the basis for the work, including the author's own field research (between 1991 and 1996; he did not visit Tubuai Manu, Motuone (Bellingshausen) or Fenuaura (Scilly) and they were therefore not accounted for); there is also much discussion of introduced, useful and ornamental plants as well as geographical notes, a detailed statistical and analytical section, and general keys to families (pp. 10–17). [Much of this flora is compiled and there are many errors. A broader concept of species is here taken than in Florence's *Flore de la Polynésie Française* (as, for example, in *Glochidion*). No summary of taxonomic changes (new species, names or combinations) is presented.]

Pteridophytes

See also the author's *Ferns of southeastern Polynesia* (**980**).

COPELAND, E. B., 1932. *Pteridophytes of the Society Islands*. ii, 86 pp., 16 pls. (Bernice P. Bishop Mus. Bull. 93). Honolulu.

Briefly descriptive treatment of lower vascular plants; includes keys to all taxa, essential synonymy, localities with *exsiccatae* and general summary of extralimital range, more or less extensive critical commentary, and index. Based mainly on the collections of Grant in 1930–31 and other American sources, without full revision of the collections covered by Drake.

Partial works: Tahiti

The only separate florula, an annotated checklist, is J. NADEAUD, 1873. *Énumération des plantes indigènes de l'île de Tahiti*. v, 86 pp. Paris. All its records have been incorporated into Drake's general flora (**980**). It succeeded the already-mentioned *Zephyritis taitensis*. Many additions and critical notes appear in W. A. SETCHELL, 1926. Tahitian spermato-phytes collected by W. A. Setchell, C. B. Setchell and H. E. Parks. *Univ. Calif. Publ. Bot.* **12**: 143–213, pls. 23–26.

Partial works: other 'high' islands

Only Maupiti, the most westerly of the outer 'high' islands, is now covered by a full florula. John W. Moore's projected florula for Raiatea, perhaps the best-collected island, was never published although two precursory papers appeared, viz.: J. W. MOORE, 1933. *New and critical plants from Raiatea*. 53 pp. (Bernice P. Bishop Mus. Bull. 102). Honolulu; and *idem*, 1963. *Notes on Raiatean flowering plants, with descriptions of new species and varieties*. 26 pp., illus. (*ibid.*, 226).

FOSBERG, F. R. and M.-H. SACHET, 1987. *Flora of Maupiti, Society Islands*. 70 pp. (Atoll Res. Bull. 294). Washington.

A well-annotated enumeration covering 237 native, naturalized, adventive and commonly cultivated species with citation of sight records or *exsiccatae*, commentary and indication of habitat, habit, origin (if appropriate), etc. The introduction incorporates a description of the island and its vegetation with mention of indicator species and a discussion of processes.[104]

Partial works: low islands and atolls

Included here are, from west to east, Motuone (Bellingshausen), Fenuaura (Scilly), Mopihaa (Mopelia), Tubuai Manu (Tupai) and, close to Tahiti, Tetiaroa. No separate account is yet available for the first-named.

SACHET, M.-H., 1983. *Botanique de l'île de Tupai, Îles de la Société*. 35 pp., 18 halftones, map (Atoll Res. Bull. 276). Washington.

Introduction, geography, vegetation and botanical history (in French); plant list (in English) with provenances and for the Grant collections of 1930 the native 'districts' along with some vernacular names and critical commentary; notes on the fauna and list of references at end.

SACHET, M.-H., 1983. *Natural history of Mopelia Atoll, Society Islands*. 52 pp., 30 halftones, map (Atoll Res. Bull. 274). Washington.

Introduction with physical background, climate, soils, vegetation, and notes on the fauna and marine life; list of species of mollusks; plant list with brief annotations

including habitat, frequency and provenance (*exsiccatae* or observations); references, pp. 36–37. Includes reports from the Scilly and Bellingshausen atolls.

SACHET, M.-H., 1983 (1984). Végétation et flore terrestre de l'atoll de Scilly (Fenua Ura). *J. Soc. Océanistes* **39**: 29–34.

Includes sight records along with collections, the latter mostly by Quayle.

SACHET, M.-H. and F. R. FOSBERG, 1983. *An ecological reconnaissance of Tetiaroa Atoll, Society Islands*. 88 pp., 42 halftones, map (Atoll Res. Bull. 275). Washington.

Introduction, history of the atoll, physical background, climate, vegetation, present state of knowledge and ecological objectives, and needs, problems and recommendations; plant list (pp. 31–54) and geographical table (pp. 55–62). The plant list encompasses terrestrial lower as well as higher plants and includes vernacular names (Tahitian and English), localities, *exsiccatae*, occurrence, and habitat along with phytographical notes.[105]

982

Southern Cook Islands

Area: 219 km². Vascular flora: Rarotonga, 560 species (Wilder, 1931); many of these are, however, introduced. GB (see **General references**) gives a figure of 284 vascular plants (of which 100 are pteridophytes) for the whole of the Cook Islands. – Included here are the 'high' islands of Rarotonga and Mangaia, the part-*makatea* islands of Ma'uke, Miti'aro and Atiu, the 'almost-atoll' of Aitutaki, and some atolls (including, for geographical reasons, Palmerston which botanically is closer to the Northern Cooks (**975**)).

No general flora is available. The only plants treated for the southern Cooks as a whole to date are the pteridophytes (Brownlie and Philipson, 1971). Of individual islands, significant florulas exist only for Aitutaki (Fosberg, 1975) and Rarotonga (Cheeseman, 1903; Wilder, 1931). Although St. John (1975; see **Progress** under **Superregion 94–99**) indicated that in part the Cook Islands (apart from Rarotonga) were among the less well-known in the oceanic Pacific, considerable collecting was undertaken in 1974 by William R. Sykes and work is in progress towards an overall flora (W. R. Sykes, personal communication, 1981). A review by Sykes of available knowledge was included in Kinloch's bibliography (see below).

Bibliography

KINLOCH, D. I. (ed.) 1980. *Bibliography of research on the Cook Islands.* 164 pp. (New Zealand MAB Report 4). Lower Hutt, New Zealand: Soil Bureau, Department of Scientific and Industrial Research, New Zealand (for New Zealand National Commission for UNESCO). [Entries with detailed annotations; botany (pp. 9–67) by W. R. Sykes with, on pp. 11–17, a state-of-knowledge review.]

Pteridophytes

BROWNLIE, G. and W. R. PHILIPSON, 1971. Pteridophyta of the southern Cook Group. *Pacific Sci.* **25**: 502–511.

Annotated systematic enumeration covering 80 taxa, with synonymy, references, citations of *exsiccatae*, critical commentary, and notes on frequency, habitat, etc. An introductory section includes plant-geographical considerations. Incorporates collecting and observations made during the Cook Bicentenary expedition of 1969 and supersedes for these plants the works of Cheeseman and Wilder described below.

Partial works: Rarotonga

Cheeseman's flora was never effectively superseded by Wilder's differently oriented contribution. In the latter work the additions came largely from cultivated, adventive and naturalized species which by the end of the 1920s had become more numerous. Wilder's work is moreover by current standards considered only semi-critical. For pteridophytes, both works have been superseded by Brownlie and Philipson's account (see above). Other recent contributions are W. R. PHILIPSON, 1971. Floristics of Rarotonga. *Bull. Roy. Soc. New Zealand* **8**: 49–54; and D. R. STODDART and F. R. FOSBERG, 1972. *Reef islands of Rarotonga, with list of vascular flora.* 14 pp. (Atoll Res. Bull. 160). Washington.

CHEESEMAN, T. F., 1903. *The flora of Rarotonga, the chief island of the Cook group.* Pp. 261–313, pls. 31–35, map (Trans. Linn. Soc. London, II, Bot. 6(6)). London.

Annotated enumeration of vascular plants; includes Latin descriptions of novelties (these partly by W. B. Hemsley) and commentaries on habitats, biology, special features, uses, etc., with summaries of external distribution (where appropriate); not separately indexed. The introductory section gives a general account of the island of Rarotonga, with descriptions of the vegetation, noteworthy plants, crops, and introduced species as well as of geographical features, climate, geology, etc. Based largely upon a visit of three months in 1899 – shortly before formal annexation by Britain – the work accounts for 334 species (233 of them considered indigenous).

WILDER, G. P., 1931. *Flora of Rarotonga.* 113 pp., 3 text-figs., 8 pls. (Bernice P. Bishop Mus. Bull. 86). Honolulu.

Systematic enumeration, with a strong ethnobotanical orientation, of native and introduced (including many merely cultivated) vascular plants; includes brief diagnostic notes, vernacular names, remarks on habitat, status, frequency, phenology, uses, etc., and indication of local range (with for some less common species the mention of particular localities but without any citation of *exsiccatae* or references). The introductory section gives a general description and historical sketch of the island along with notes on general features of the flora, while the index includes cross-references to Cheeseman's flora (see above).

Partial works: other 'high' islands

Included here are Mangaia, Ma'uke, Miti'aro, Atiu and Aitutaki. Only the last-named has to date been the subject of a floristic treatment, part of an 'island monograph'. Botanically they remain imperfectly known.

FOSBERG, F. R., 1975. Vascular plants of Aitutaki. In D. R. Stoddart and P. E. Gibbs (eds.), *Almost-atoll of Aitutaki (Cook Islands)*, pp. 73–84 (Atoll Res. Bull. 190). Washington.

Briefly annotated systematic list, with citations of exsiccatae and localities along with indication of status, habitat, biology, etc. [The first florula for the island. Vegetation is described by Stoddart elsewhere in the monograph.]

Partial works: atolls

Included here are Takutea, the Hervey Islands (Manuae), and Palmerston. No florulas for any of these have been seen.

983

Tubuai (Austral) Islands

Area: 121 km². Vascular flora (including Rapa Iti and Morotiri): 758 native and other species (Hallé and Florence, 1986; see **984**). – Included here are Maria Atoll and the part-makatea islands of Rimatara, Rurutu, Tubuai and Raivavae. For Rapa Iti, see **984**.

Until 1980, the only overall floristic coverage was notionally that in BROWN and BROWN (**980**); it, however, contains but few pertinent records. This gap has now been filled after a fashion by N. Hallé (1980; see below); this at least accounts for the numerous collections made after 1930 (including, for example, those of the 1934 Mangarevan Expedition) which hitherto had been partly treated in scattered papers (many of them by Harold St. John, one of the two expedition botanists). Additions were published in 1983 and 1986, reflecting new field work in both the Australs proper and Rapa Iti.

The somewhat isolated Maria Atoll has been separately documented; see F. R. FOSBERG and H. ST. JOHN, 1952. Végétation et flore de l'atoll Maria, îles Australes. *Rev. Sci. Bourbonn.* (1951): 1–7.

HALLÉ, N., 1980. Les orchidées de Tubuaï (Archipel des Australes, Sud Polynésie); suivies d'un catalogue des plantes à fleurs et fougères des îles Australes. *Cah. Indo-Pacifique* 2(3–4): 69–130, pls. 1–12.

Pages 85–129 comprise an alphabetically arranged checklist of 627 species of vascular plants of the Austral Islands and Rapa Iti, with citations of *exsiccatae* arranged by island; bibliography, pp. 129–130. For additions, see *idem*, 1983. Végétation de l'île Rurutu et additions au catalogue de la flore des îles Australes. *Bull. Mus. Natl. Hist. Nat.*, IV/B (Adansonia) 5(2): 141–150, 2 text-figs.; and J. FLORENCE and N. HALLÉ, 1986. Suite du Catalogue des plantes à fleurs et fougères des îles Australes. In DIRECTION DES CENTRES D'EXPÉRIMENTATIONS NUCLÉAIRES, SERVICE MIXTE DE CONTRÔLE BIOLOGIQUE, FRANCE, *Rapa*, pp. 151–158 (see 984).

984

Rapa Iti and Morotiri (Bass Islets)

See also 983 (HALLÉ). – Area: Rapa Iti, 40 km^2; Morotiri, 0.2 km^2. Vascular flora: Rapa Iti, 152 species including 66 ferns, with 62 percent endemism (St. John, 1975; see Superregion 94–99). Included here are the extra-tropical reefless 'high' island of Rapa Iti (Rapa) and tiny Morotiri (the Bass Islets) to the southeast.

Although Rapa Iti is of great botanical (and also paleobotanical) interest, until recent years the only general coverage was nominally in regional floras (980). The results of the Mangarevan Expedition of the Bishop Museum, which stayed a month in 1934, have gradually appeared in scattered papers, many by Harold St. John in his series 'Pacific plant studies'. In 1980 all known records for Rapa Iti and Morotiri were included in Nicholas Hallé's enumeration for the Austral Islands (983); additions appeared in 1983 and 1986.

Morotiri is covered in a separate florula, based on

collections made in 1934: H. ST. JOHN, 1982. Marotiri, rock pinnacles in the South Pacific. *Occas. Pap. Bernice P. Bishop Mus.* 25(4): 1–4.

DIRECTION DES CENTRES D'EXPÉRIMENTATIONS NUCLÉAIRES, SERVICE MIXTE DE CONTRÔLE BIOLOGIQUE, FRANCE, 1986. Rapa. 236, [1] pp., illus., maps. N.p.

Multidisciplinary island monograph; plants are covered in the 'Suite du Catalogue des plantes à fleurs et fougères des îles Australes' (pp. 151–158) by J. Florence and N. Hallé (983, under HALLÉ) and 'Description de 10 espèces rares de plantes à fleurs de l'île de Rapa' by N. Hallé and J. Florence (pp. 129–149, 10 pls.). The latter is an iconography with text which incorporates revisions to the Browns' flora (980).

985

Tuamotu Archipelago (excluding Mangareva)

Includes the whole of the Tuamotus (except for Mangareva and Timoe). – Area: 842 km^2. Vascular flora: Makatea, 71 species including 6 ferns; other atolls, from 21 to 54 native species each.

Almost all islands in this large archipelago are low, with a small, generally monotonous vascular flora. Only Makatea, a raised, phosphate-bearing atoll, is relatively rich. For convenience, the works treated here are divided into two groups, respectively for Makatea and all the low islands. Few florulas have been published for the latter, but nominal general coverage is available in the regional floras (980); and in the Raroia florula by Doty *et al.* (1954; see below) a key to the species commonly expected in the low islands as a whole (by F. R. Fosberg) is provided.

Partial works: Makatea

The Wilder florula of 1934 is by current standards only semicritical and moreover is now incomplete. Informed opinion has also suggested that more collecting is needed. A contribution [not seen] towards a revised flora is J. FLORENCE, 1982. *Introduction a l'étude de la flore et de la végétation de l'île de Makatéa (Tuamotu)* (Mém. de stage, Rech. Bot. Poly. 69–93). ORSTOM, Papeete.

WILDER, G. P., 1934. *Flora of Makatea*. 49 pp., 5 pls., map (Bernice P. Bishop Mus. Bull. 120). Honolulu.

Systematic enumeration of native, naturalized and commonly cultivated vascular plants; includes vernacular names, generalized indication of local range (with but few *exsiccatae*) and brief diagnostic notes together with sometimes extensive remarks on habitat, biology, cultivation, uses, etc.; index only to genera and species. No synonymy or references are included. An introductory section includes notes on physical features, fauna, the vegetation and features of the flora as well as on the now-defunct phosphate extraction industry.

Partial works: low islands

Separate florulas appear to be available only for Rangiroa (near Makatea, in the western islands), Raroia (of *Kon-Tiki* fame) in the central islands, and Takapoto. The floras are small: Doty *et al.* (1954) recorded 51 species of native and aboriginally introduced vascular plants on Raroia, while Stoddart and Sachet (1969) listed 41 native species for Rangiroa. An additional treatment, unfortunately with its list of plants unannotated, is F. SALVAT and B. SALVAT, 1992. *Nukutipipi Atoll, Tuamotu Archipelago: geomorphology, land and marine flora and fauna and interrelationships.* 43 pp., illus., maps (Atoll Res. Bull. 357). Washington. This south-central atoll in the Duke of Gloucester group, 5 km² in extent and uninhabited by Polynesians during historical time (1767 C.E. to present), has only 21 native species.

DOTY, M. S. et al., 1954. *Floristics and plant ecology of Raroia atoll, Tuamotus.* 58 pp. (Atoll Res. Bull. 33). Washington.

Includes an enumeration (pp. 24–35) of native and aboriginally introduced seed plants (by M. Doty) with an appendix (pp. 36–41) on 'village' plants; includes vernacular names, citations of *exsiccatae* with localities, and notes on occurrence, habitat, biology, etc. Pteridophytes (by K. Wilson) appear on pp. 57–58, and treatments of some non-vascular plants are also given. Pages 14–23 feature a 'key to commonly expected Tuamotuan vascular plants' originally prepared by F. R. Fosberg.

SACHET, M.-H., 1983. *Takapoto Atoll, Tuamotu Archipelago: terrestrial vegetation and flora.* 46 pp., 9 halftones, map (Atoll Res. Bull. 277). Washington.

Incorporates a systematic checklist of plants from Takapoto and Manihi atolls with localities, *exsiccatae* and/or sight records, and ecological and biological notes (introduced species are asterisked); references, pp. 40–41.

STODDART, D. R. and M.-H. SACHET, 1969. *Reconnaissance geomorphology of Rangiroa Atoll, Tuamotu Archipelago, with list of vascular flora of Rangiroa.* 44 pp. (Atoll Res. Bull. 125). Washington.

Pages 33–44 feature an annotated checklist (by the second author) based mainly on a 1963 field trip; included are localities, citations of *exsiccatae*, and notes on habitat, occur-

rence, etc., along with introductory remarks. A total of 121 species was recorded but only some 41 were considered indigenous.

986
Mangareva group (Gambier Islands)

Area: 25 km². Vascular flora: 82 indigenous and endemic species, 65 adventive weeds, 101 cultivated ornamentals and 68 'esculents' were recorded by St. John (1988). A total of 41 indigenous species (including 12 pteridophytes) with 27 percent endemism were reported in Huguenin (1974). – The group as delimited here comprises the Mangareva archipelago proper: three 'high' islands surrounded by a reef and, to the southeast, the atoll of Temoe. Mangareva is noteworthy for 'megalithic' churches (the work of a zealous Roman Catholic missionary in the 1830s) and much grassland but little 'bush'. St. John (1975; see Superregion 94–99) noted that the natural vegetation was '98 per cent devastated'. St. John's florula incorporates the plants of the 1934 American expedition. Additional coverage may be found in the regional floras (**980**).

HUGUENIN, B., 1974. La végétation des îles Gambier. Relevé botanique des espèces introduites. *Cah. Pacifique* 18(2): 459–471.

Sparsely annotated checklist of introduced and native vascular plants, with indication of status, origin and local occurrence; no references. [Some 200 introductions alone were recorded during a 1966 field trip (upon which the checklist, part of a larger 'island monograph', was mainly based).]

ST. JOHN, H., 1988. *Census of the flora of the Gambier Islands, Polynesia.* 34 pp., map. Honolulu: The author. (Pacific plant studies 43.)

Systematic enumeration with vernacular names, localities and *exsiccatae*; some descriptions of novelties and, at end, an index to all names. Native species are asterisked (*) and endemics are indicated with a plus sign (+). The work of Huguenin is cited throughout, with other references on p. 28. An introductory section includes an account of previous botanical exploration, including the author's own visit in 1934.

987

Pitcairn Islands

Area: Pitcairn, 6.6 km²; Oeno, 0.62 km²; Henderson, 37.2 km²; Ducie, 0.74 km². Vascular flora: 105 native species in all (Pitcairn, 66 including 23 pteridophytes; Henderson, 63 including 9 pteridophytes; Oeno, 16 including 2 ferns; Ducie, 2 flowering plants) (Florence *et al.*, 1995; see below). – Here are included Pitcairn Island (of *Bounty* fame) as well as Oeno, Henderson (Elizabeth) and Ducie. They are structurally varied but politically united as the last British colony in the Pacific.

All the islands were recently documented by Florence *et al.* (1995), with a primary basis in the work of the Sir Peter Scott Commemorative Expedition of 1991–92. This undertaking followed a 1987 expedition from the Smithsonian Institution, whose botanical results appear as F. R. FOSBERG *et al.*, 1989. *New collections and notes on the plants of Henderson, Pitcairn, Oeno, and Ducie Islands.* 18 pp. (Atoll Res. Bull. 329). Washington. Prior coverage related to individual islands, with patchy coverage also furnished in the general flora of BROWN and BROWN (**980**).

FLORENCE, J., S. WALDREN and A. J. CHEPSTOW-LUSTY, 1995. The flora of the Pitcairn Islands: a review. *Biol. J. Linn. Soc.* **56**: 79–119, 1 fig.

Comprises individual enumerations for Ducie (p. 83), Henderson (pp. 84–97), Oeno (pp. 97–100) and Pitcairn (pp. 100–114) with two appendices respectively on pollen of the endemic seed plants of Henderson and the lichens of Henderson and Oeno. The floristic accounts include synonymy, citations in prior area literature, concise descriptive notes and vouchers (all from the Scott Expedition). Plants not seen on the expedition but previously documented, as well as records on sight but not collected by expedition participants, are gathered in a running format at the end of each island account. The introductory part includes statistics and notes on general features of the flora. [An expedition report more than a fully documented florula.]

Partial works: Pitcairn

The most extensive account of the history of botanical work on Pitcairn is in St. John (1987). Some collections housed in Australia, Fiji and perhaps elsewhere have, however, yet to be listed or re-examined. As elsewhere in southeastern Polynesia, the flora is remarkable for a high percentage of pteridophytes; their relationships are with Greater Malesia (Brownlie, 1961).

BROWNLIE, G., 1961. The pteridophyte flora of Pitcairn Island (Studies in Pacific ferns, IV). *Pacific Sci.* **15**: 297–300.

Enumeration of 20 species; includes synonymy, references, *exsiccatae*, critical remarks, and notes on habitat, frequency, etc.

MAIDEN, J. H., 1901. Notes on the botany of Pitcairn Island. *Rep. Meetings Australasian Assoc. Adv. Sci.* **8**: 262–271.

Briefly annotated list of vascular plants, including vernacular names and notes on local uses. An introduction includes remarks on botanical work in the island and a short list of references. Based mainly on collections by Miss R. A. Young (now at herb. NSW).

ST. JOHN, H., 1987. *An account of the flora of Pitcairn Island with new* Pandanus *species.* 65 pp., 5 unpaged pls. (2 figs., 3 halftones). Honolulu: The author. (Pacific plant studies 46.)

Annotated enumeration of vascular plants (arranged according to the Englerian system) with localities and citations of *exsiccatae* and/or published sources; indigenous and endemic species are specially indicated. A brief introduction covers physical and biological aspects along with remarks on vernacular names, a list of endemic taxa and a relatively extensive survey of botanical activity; at the end are references (pp. 58–59) and an index to vernacular (but not botanical) names.

Partial works: Oeno, Henderson and Ducie

Henderson (Elizabeth in older maps and literature), an uninhabited *makatea* or raised atoll, has a much richer flora than smaller and lower Oeno or more distant Ducie; its indigenous vegetation is also relatively intact due to remoteness and difficulty of access. The subject of a controversial prospective land deal in the early 1980s, it was afterwards 'saved' and is now on the World Heritage List. Ducie, 'at the end of the line', has, along with Vostok Island (**977**), one of the smallest island vascular floras in the world: *Argusia argentea* (Boraginaceae) and – since 1987 – *Pemphis acidula* (Lythraceae), both Pacific wides.[106] Much new information was collected by the Smithsonian and Scott expeditions (see main entry) including species not recorded in the florulas listed below.

FOSBERG, F. R., M.-H. SACHET and D. R. STODDART, 1983. *Henderson Island* (Southeastern Polynesia)*: summary of current knowledge.* [iv], 47, [vi] pp., 2 figs., 12 pls. (Atoll Res. Bull. 272). Washington.

General monograph and status report with in appendix 1 (pp. 29–33) a plant list (basically a revision of the 1962 work of St. John and Philipson, no further collections being available); references, pp. 41–47. [The report was prepared in response to the aforementioned threat of a land deal.]

St. John, H. and W. R. Philipson, 1960. List of the flora of Oeno Atoll. *Trans. Roy. Soc. New Zealand* 88(2): 401–403.

Systematic enumeration of 17 vascular plants and one alga, based largely on collections from the 1934 Mangarevan Expedition; includes synonymy, *exsiccatae*, and notes on habitats and local distribution.

St. John, H. and W. R. Philipson, 1962. An account of the flora of Henderson Island, south Pacific Ocean. *Trans. Roy. Soc. New Zealand, Bot.* 1(14): 175–194, 4 text-figs., 7 halftones.

Annotated, illustrated systematic enumeration of nonvascular and vascular plants with descriptions of new taxa, synonymy including references, and literature citations (mainly from the Browns' *Flora of southeastern Polynesia*), *exsiccatae*, critical remarks, and notes on habitat, occurrence, etc.; list of references at end. The introductory section includes descriptive remarks on the island and its vegetation as well as an account of previous botanical work.

988

Rapa Nui (Easter) and Sala-y-Gómez Islands

See also **390** (all works). – Area: 117 km^2. Vascular flora: 179 angiosperms, of which 30 native (with at least 1 extinct as well as 1 known only in cultivation); 141 species are considered to have been introduced of which 67 are now established (Zizka, 1991). There are also 16 species of pteridophytes (Looser, 1958).

Rapa Nui (Easter Island), a very isolated, extratropical island of volcanic origin renowned for its 'mysterious' megalithic statues, has been visited by many scientists including botanists. The first floristic account appeared as *Reseña botánica sobre la Isla de Pascua* (1913, publ. 1914) by Francisco Fuentes, based on work done on a 1911 expedition from Chile led by Walter Knoche. The Skottsbergs visited the island in 1917; their results were incorporated into Carl Skottsberg's well-known enumeration (1922; additions 1951), with the pteridophytes contributed by Carl Christensen (1920). The island, with vegetation already recorded as sparse at first contact in 1722, is now virtually treeless and the distinctive *toromiro (Sophora toromiro)*, last

collected on Thor Heyerdahl's 1955–56 expedition, has become extinct in the wild, surviving only in cultivation in Göteborg, Sweden (where Skottsberg long was head of the botanical institute and garden) and elsewhere. Much research has been accomplished in recent decades, including the collection and analysis of pollen samples and fossil remains; particularly interesting was the discovery of an extinct palm (now described as *Paschalococos disperta*). Many questions, however, remain with respect to the 'original' flora. For casual notes on the vascular plants (two or three species) of the more easterly Sala-y-Gómez, an island group consisting of little more than rocks rising to 29 m, see R. M. Norris, 1960. Sala-y-Gomez – lonely landfall. *Pacific Disc.* 13(6): 20–25.

Bibliography

Marticorena, C. 1992. *Bibliografía botánica taxonómica de Chile*. See 390.

Skottsberg, C. J. F., 1922. The phanerogams of Easter Island. In C. J. F. Skottsberg (ed.), *Natural history of Juan Fernandez and Easter Island*, 2 (Botany): 61–84, 2 text-figs., pls. 6–9. Uppsala: Almqvist & Wiksell. Complemented by C. Christensen and C. J. F. Skottsberg, 1920. The ferns of Easter Island. *Ibid.*: 49–53, 3 text-figs.; and C. J. F. Skottsberg, 1951. A supplement to the pteridophytes and phanerogams of Juan Fernandez and Easter Island. *Ibid.*: 763–792, 1 text-fig., pls. 55–57.

The initial contributions respectively include critical systematic enumerations of flowering plants and of ferns, with descriptions of novelties, synonymy (with references), *exsiccatae* with localities, summary of extralimital distribution (where applicable), and extensive notes on habitat, frequency, etc., as well as critical commentary. The introductory section of Skottsberg's contribution includes an account of botanical work on the island by the author and his predecessors. The additional information in the supplement is based on subsequent exploration.

Zizka, G., 1991. *Flowering plants of Easter Island*. 108 pp., 54 text-figs. (mostly col. photographs but including 2 maps) (Palmarum Hortus Francofurtensis, Wiss. Ber. 3). Frankfurt-am-Main.

An island botanical monograph including (pp. 30–95) a formal floristic treatment including keys, descriptions, synonymy with references and citations, vernacular names, indication of distribution, dispersal class, habitat and frequency, taxonomic and other notes, and *exsiccatae*; references, list of all *exsiccatae* and index to species names at end. The general part covers physical features, past and current botanical studies, features of the flora and vegetation, and remarks on phytogeography and conservation.

Pteridophytes
See also CHRISTENSEN and SKOTTSBERG above.
LOOSER, G., 1958. Los helechos de la Isla de Pascua. *Revista Univ. Católica* (Chile) 43: 39–64, 16 text-figs.

Illustrated enumeration of vascular cryptogams with keys, brief descriptions, synonymy, references, citations of *exsiccatae*, indication of local and overall range, and critical notes; general notes on ecology, references and index at end. The introductory part incorporates a family key.

989

Marquesas

Area: 1275 km². Vascular flora: *c.* 420 native species and as many or more naturalized, with 33 percent of the natives being pteridophytes (Wagner and Lorence, 1997, p. 222; see Note 107). – The Marquesas are a very distinctive group of 12 'high' islands in the north of southeastern Polynesia, exhibiting some floristic affinities with the Hawaiian Islands.

The main early works on the natural history of the group are by É. Jardin and were published in the 1850s and 1860s. These include *Essai d'une flore de l'archipel des Marquises* (1858). The next treatments are those of Drake and the Browns (**980**), the latter with a particular focus on these islands and incorporating considerable collections made for the Bishop Museum in the 1920s (particularly from the Bayard Dominick and Whitney expeditions). As already related, however, the Browns' work has proved unsatisfactory in several respects and, with a view to unifying the two main streams of data (respectively French and American) work towards a new flora was initiated by Marie-Hélène Sachet. Its first (and so far only) part appeared in 1975. Meanwhile, collecting was renewed in the 1950s and has continued more or less to the present. One key result was the discovery of a remarkable endemic, *Lebronnecia* (Malvaceae). Among more recent collectors have been Jacques Florence, who from 1982 made repeated trips preparatory to his *Flore de la*

Polynésie Française (1997– ; see **980**). In 1987 the Fosberg–Sachet project was renewed by Warren L. Wagner and David Lorence. Florence and Lorence have moreover written a general botanical review: J. FLORENCE and D. H. LORENCE, 1997. Introduction to the flora and vegetation of the Marquesas Islands. *Allertonia* 7: 226–237, map.[107]

SACHET, M.-H., 1975. *Flora of the Marquesas*. Part 1, *Ericaceae–Convolvulaceae*. iv, 34 pp., map (Smithsonian Contr. Bot. 23). Washington.

Detailed critical 'research' flora (including native, naturalized and 'village' plants), arranged according to the Englerian system; includes keys, long descriptions, extensive synonymy (with references and citations), vernacular names, indication of *exsiccatae* from all sources with localities, general summary of distribution, taxonomic commentary, and notes on habitat, biology, occurrence, frequency, local uses, etc.; no index. An introductory section covers the history of botanical collecting, the background to the work, and the logistical and other problems of preparing such a Pacific Islands flora. [Not completed.]

Region 99

Hawaiian Islands

Area: 16705 km² (exclusive of **999**). Vascular flora: 1935 native and naturalized flowering plant species, of which 1033 native with 850 endemic (based on Eldredge and Miller, 1997).[108] Some 233 species of pteridophytes (of which 201 native with 130 endemic) have also been recorded (based on Eldredge and Miller, 1995, 1997).[109] Hillebrand in 1888 had recorded 999 native and naturalized species of vascular plants, while in 1948 Fosberg suggested for Zimmermann (1948; noted below) a total of 2000 native species and varieties. More recently St. John (1973; see below) listed 2668 species and varieties (in 1442 species), and Degener (1975; in *Sida* **6**: 120–122) an idiosyncratic figure of 20000 to 30000 species. – The Hawaiian Islands here comprise the chain from the 'Big Island' of Hawaii northwest through the other main Hawaiian Islands (to Niihau) and on to the smaller (and in part extra-tropical) Leeward Islands, ending at Midway and Kure. Also included are Johnston Island and, for convenience, Luna.

The indigenous vascular flora of the 'high' islands is for its size about the most peculiar and isolated in the world, with very high endemism, and at the same time one of the most fragile, with a long official list of extinct, endangered and threatened vascular plant taxa. With many visiting and some resident collectors and naturalists almost from the time of outside discovery in 1778, a first understanding of the flora was achieved in somewhat over a century. This came through publication of what would become a classic work, *Flora of the Hawaiian Islands* (1888; reissued 1965, 1981) by W. H. Hillebrand.[110] However, another century would elapse before an effective successor appeared. The main intervening works were *Observations on the ferns and flowering plants of the Hawaiian Islands* (1897; in *Bull. Geol. & Nat. Hist. Surv. Minnesota* 9 (Bot. Ser. 2): 760–922, or *Minnesota Botanical Studies* 1) by A. A. Heller, largely a supplement to the Hillebrand flora but incorporating the results of its author's own extensive collecting, the lavishly illustrated, loose-leaf *Flora hawaiiensis* (1932–80) by O. Degener, and *List and summary of the flowering plants in the Hawaiian Islands* (1973) by H. St. John.

The two latter works are now obsolete (save for pteridophytes and, in *Flora hawaiiensis*, the illustrations). Degener's work, never completed, was notorious for its format (and intricate nomenclature), while St. John's *List* was little more than a nomenclator based partly upon work done before 1930 by E. H. Bryan, Jr. and padded out with scores of garden and other exotics not even part of the large adventive and naturalized component in the modern flora.[111] On the other hand, in the latter half of the twentieth century modern floristic reviews (as parts of island monographs) became available for Johnston and the various Leeward Islands; these supplement the enumerations of Christophersen published between the wars. For trees, the best work remains without doubt Rock's *Indigenous trees of Hawaii* (1913; reprinted 1974), the reprint being enhanced through revisions by Derral Herbst.[112]

In 1980, a change of botanical leadership took place at the Bernice P. Bishop Museum in Honolulu. High on the agenda was a new manual, hopefully in time for the centennial of Hillebrand's *Flora*. A 'project' was organized, with three main aims: a retrospective bibliography, a 'standard-list' and the manual. The bibliography appeared in 1988, the checklist (by C. Imada, W. L. Wagner and Herbst) in 1989, and the two-volume *Manual of the flowering plants of Hawai'i* by Wagner, Herbst and S. H. Sohmer in 1990. While earlier twentieth-century accounts credited Hawaii with anywhere from 1800 to a very fanciful 30000 species, the new *Manual* took a relatively conservative approach, leading to a total for native flowering plants not much higher than that of Hillebrand. Such a philosophy was rooted in a felt need to bring out a work accessible to conservationists and other users and at the same time to be a vehicle for more 'realistic' taxonomic concepts. Most work in the previous century or more had been rooted in pre-biosystematic ideologies which called for formal treatment of every observed variation. With origins in other parts of the world, such did not translate well to an isolated, then only imperfectly known insular flora and so gave rise to a seemingly impenetrable thicket of names. Nevertheless, with shifts in taxonomic fashions as well as perceptions of biodiversity opinions are subject to change.[113]

A companion work to the *Manual* for pteridophytes remains a desideratum. In addition, no account has effectively succeeded the following as a general introduction to the Hawaiian environment and biota: E. C. Zimmermann, 1948. *Insects of Hawaii*, 1: *Introduction*. xx, 206 pp., 52 text-figs. (incl. maps). Honolulu: University of Hawaii Press. This includes a chapter on the flora by F. R. Fosberg; within is a statistical table by families from which his above-quoted estimate of the vascular flora was taken.

Bibliographies. General and supraregional bibliographies as for Superregion 94–99.

Regional bibliography

Mill, S. W., D. P. Gowing, D. R. Herbst and W. L. Wagner, 1988. *Indexed bibliography on the flowering plants of Hawai'i*. vi, 214, [1] pp. Honolulu: University of Hawaii Press/Bishop Museum Press. (Bishop Museum Spec. Publ. 82.) [Introduction, including a historical survey of the literature, plan of the work, and definitions of subject categories (as used in the subject index); unannotated, numbered author list of 3257 titles with subject and plant name indices.]

Indices. General and supraregional indices as for Superregion 94–99.

990

Region in general

The publication in 1990 of *Manual of the flowering plants of Hawai'i* from the team of Warren Wagner, Derral Herbst and S. H. Sohmer provided, at least for seed plants, the first effective successor to Hillebrand's 1888 classic. A modern treatment of pteridophytes is, however, wanting, and for trees there has been no successor to Rock's *Indigenous trees of Hawaii* (1913).

Keys to families and genera

ST. JOHN, H. and F. R. FOSBERG, 1938–40. *Identification of Hawaiian plants*, I–II. 53, 47 pp. (Occas. Pap. Univ. Hawaii 36, 41). Honolulu. [Keys to and descriptions of families, with keys to genera; indices. Dicotyledoneae in part I; Monocotyledoneae and Gymnospermae in part II.]

DEGENER, O., 1932–80. *Flora hawaiiensis; or the new illustrated flora of the Hawaiian Islands*. Vols. 1–7. Illus. Honolulu (later Waialua, Oahu), Hawaii: The author. (Vols. 1–4 reprinted under one cover, 1946, Honolulu.)

This, for long the only major 'loose-leaf' flora in the world, is a profusely illustrated, copiously descriptive work covering spontaneous and commonly cultivated vascular plants, with keys to all taxa, full synonymy (including references and literature citations), vernacular names (Hawaiian and English), general indication of local and extralimital distribution, type localities, more or less extensive taxonomic commentary, and notes on status, ecology, occurrence, uses, special features, etc. The introductory leaves include historical matter as well as a glossary and index to families; successive 'temporary' indices to all botanical and vernacular names have also been provided. [Six volumes of 'leaves' were published by 1963, but subsequent progress was rather slower and by termination of the work vol. 7 remained only partly filled. The volumes serve merely as receptacles; no taxonomic progression in publication was ever followed although for sorting each leaf was numbered according to the Englerian sequence. The 'leaves' – comprising in all 1144 bibliographic units – were published with text on the front and (usually) an illustration on the reverse side for each taxon. The offset reprint of 1946, combining vols. 1–4

but retaining the loose-leaf format, was made following destruction of stock in a tidal wave. For a list of and index to the individual entities in this work, see S. W. MILL, W. L. WAGNER and D. R. HERBST, 1985. Bibliography of Otto and Isa Degener's Hawaiian floras. *Taxon* 34: 229–259.]

IMADA, C., W. L. WAGNER and D. R. HERBST, 1989. Checklist of native and naturalized flowering plants of Hawai'i. *Occas. Pap. Bernice P. Bishop Mus.* 29: 31–87.

Name list, with indication of distribution and status (both biological and for conservation); index to genera. An introductory section includes explanations of the text as well as statistics of the flora. [Supersedes the core of St. John's 1973 checklist; accounts for 1094 native species and non-nominate subordinate taxa, with 89 percent of the species endemic; 869 further taxa naturalized, of which 25 date from before 1778.]

ST. JOHN, H., 1973. *List and summary of the flowering plants in the Hawaiian Islands*. 519 pp. (Mem. Pacific Tropical Bot. Gard. 1). Lawai, Kauai, Hawaii.

Briefly annotated systematic list of the accepted botanical names, basionyms, and more important synonyms of all native, naturalized, adventive and cultivated flowering plants in the Hawaiian Islands; entries include year of publication, citations of major works on the Hawaiian flora in which the plant is mentioned, indication of status (and country of origin, if applicable), and vernacular names. An addendum, a list of new names and combinations, and a complete general index are also provided. The introductory section gives an explanation of the work as well as statistics of the flora (1442 indigenous species). [Now largely obsolete.][114]

WAGNER, W. L., D. R. HERBST and S. H. SOHMER, 1990. *Manual of the flowering plants of Hawaii*. 2 vols. xviii, 1853 pp., illus. (part col.). Honolulu: University of Hawaii Press/Bishop Museum Press. (Bishop Museum Spec. Publ. 83. Reissued 1999 with corrections and a supplementary section.)

Illustrated descriptive flora of native and naturalized seed plants; includes keys, essential synonymy, vernacular names, status, karyotype(s), habitat, local and (if applicable) extralimital distribution, occasional *exsiccatae* (for extinct, very rare and critical species), and taxonomic and other notes with, where known, dates of introduction or first collection of non-native species; references under genera to key treatments;

illustrated glossary, literature cited (pp. 1659–1703), addendum, (with new combinations in *Melicope* for species formerly in *Pelea* (Rutaceae)), illustration vouchers and other sources, and index to all vernacular and botanical names at end.

Woody plants (including trees)

For a non-technical introduction (in 8.5×11 in. format) to native trees and shrubs, see S. H. LAMB, 1981. *Native trees and shrubs of the Hawaiian Islands.* 159, [1] pp., illus. Santa Fé, New Mexico: Sunstone Press.

ROCK, J. F., 1913. *The indigenous trees of the Hawaiian Islands.* [viii], 518 pp., 215 pls. (halftones). Honolulu: The author. (Reprinted 1974 as *Indigenous trees of Hawaii.* xx, 548 pp. Rutland, Vt., and Tokyo: Tuttle (in association with Pacific Tropical Botanical Garden).)

Amply descriptive account of Hawaiian trees; includes non-dichotomous keys to genera and species and a general key to families, full synonymy (with references and literature citations), Hawaiian and English vernacular names, rather detailed indication of local and extralimital distribution, critical remarks, and extensive notes on habitat, field attributes, ecology, timber properties, and uses; complete indices to botanical and vernacular names at end. The introductory section includes an illustrated account of vegetation and floristic regions in the islands, as well as of forest zones. The excellent photographs, also by the author, are all from life or from freshly collected material. [This 'tree book' has no rivals in the Pacific and but few elsewhere in the tropics. The 1974 reprint includes a new introduction by Sherwin Carlquist and a detailed table of nomenclatural changes by Derral Herbst.]

Pteridophytes

A modern treatment of Hawaiian ferns and fern-allies is not yet available. Christensen's 1925 revision (see below) succeeded *A taxonomic study of the Pteridophyta of the Hawaiian Islands* by Winifred J. Robinson (1912–14, in *Bull. Torrey Bot. Club* **39–41**, *passim*); her keys were, however, in Christensen's eyes 'often misleading'. Naturalized species were first covered separately in W. H. WAGNER, Jr., 1950. Ferns naturalized in Hawaii. *Occas. Pap. Bernice P. Bishop Mus.* **20**: 95–121; 30 in all are now known. Recent research has been summarized in W. H. WAGNER, JR., F. S. WAGNER and T. FLYNN, 1995. Taxonomic notes on the pteridophytes of Hawaii. *Contr. Univ. Michigan Herb.* **20**: 241–260.

Christensen, C., 1925. *Revised list of Hawaiian Pteridophyta.* 30 pp. (Bernice P. Bishop Mus. Bull. 25). Honolulu.

A critical enumeration of 159 species with citations of earlier accounts (Hillebrand, Robinson) and indication of status (many species being endemic) and related species elsewhere; extended commentaries on some species (pp. 22–29).

and brief bibliography at end. The general part is introductory, without sections on physical features or floristics.

995

Hawaiian Leeward Islands

These include all islands and atolls west of the main islands from Nihoa west through Laysan and Midway to Kure. The only work covering the chain as a whole remains the 1931 enumeration of Christophersen and Caum, although, as indicated below, revised lists have in more recent years appeared for all islands. It is to be hoped that a consolidated botanical report can be published so as to provide a new reference point.

CHRISTOPHERSEN, E. and E. L. CAUM, 1931. *Vascular plants of the Leeward Islands, Hawaii.* 41 pp., 16 pls., 3 maps (Bernice P. Bishop Mus. Bull. 81). Honolulu.

Enumeration of known vascular plants of the islands from Nihoa in the east to Midway and Kure in the west; includes synonymy (with references), descriptions of new or critical taxa, localities with *exsiccatae*, general indication of overall range, and remarks on ecology and taxonomy, with a list of references at end. The introductory section incorporates chapters on the physical features, flora and vegetation of each island or island group.

Individual islands

During the 1960s and 1970s, detailed biological and habitat surveys under the general direction of R. B. Clapp were carried out on all the Leeward Islands (except Midway). Comprehensive reports were published in *Atoll Research Bulletin* and include modern floristic accounts with checklists. All are similar in style and content, providing accepted botanical names, locality and collection records, references, extensive ecological notes, and records of changes since earlier surveys. Only minimal citations are given here.

Nihoa: Clapp and Kridler, 1977 (*A.R.B.* 207).

Necker: Clapp, Kridler and Flett, 1977 (*A.R.B.* 206).

French Frigate Shoals: Amerson and Clapp, 1971 (*A.R.B.* 150).

Gardner Pinnacles: Clapp, 1972 (*A.R.B.* 163).

Laysan: Ely and Clapp, 1973 (*A.R.B.* 171).

Lisianski: Clapp and Wintz, 1975 (*A.R.B.* 186).

Pearl and Hermes Reef: Amerson, Clapp and Wintz, 1974 (*A.R.B.* 174).

Kure Atoll: Woodward, 1972 (*A.R.B.* 164).

The report on Midway Island was based on a 1954 resurvey; see I. A. NEFF and P. A. DUMONT, 1955. *A partial list of the plants of the Midway Islands.* 11 pp. (Atoll Res. Bull. 45). Washington. [Extensive changes to the flora had occurred as a result of World War II and other human activities.]

998

Johnston Atoll

Area: 2.5 km². Vascular flora: 3 species in 1923, 124 by 1976, with all additions the result of accidental or deliberate introduction. – For more than five decades the atoll was a United States military base. It is now about the most altered in the Pacific, physically and floristically. In 1923, the date of the survey upon which Christophersen's paper was based, it was still more or less in its pristine state. By the 1970s all natural vegetation had disappeared and the shapes of the islets on the atoll had been completely altered, their areas greatly increased through dredging.[115]

AMERSON, A. B., Jr. and P. C. SHELTON, 1976. *The natural history of Johnston Atoll, Central Pacific Ocean.* xx, 479 pp., illus., maps (Atoll Res. Bull. 192). Washington.

Pages 47–65 of this exhaustive topographic account cover plant life; incorporated is a tabular checklist on pp. 52–60 and bibliography. Appendices 2–6 (pp. 387–442) give detailed records of plant arrivals for each of the five islets.

CHRISTOPHERSEN, E., 1931. *Vascular plants of Johnston and Wake Islands.* 20 pp., 5 pls., 3 maps (Occas. Pap. Bernice P. Bishop Mus. 9(13)). Honolulu.

Separate enumerations of vascular plants respectively for Johnston and Wake Islands; each includes literature records, exsiccatae, and extensive notes on habitat, frequency, biology and taxonomy. The introduction contains remarks on the general features of the islands, their flora, and their vegetation.

999

Luna

No non-vascular or vascular plants have yet been found for study by explorers and exobiologists, despite costly explorations from 1969 to 1973.

Notes

1 The *Alpine flora*, like the *Handbooks of the flora of Papua New Guinea* (**930**), is more prestigious than practical. Its species concept moreover is often narrow; in the Myrsinaceae, for example, subsequent revision has relegated several of the species recognized to synonymy. For orchids it is unsatisfactory both in the field and in other respects (P. Cribb and T. Reeve, personal communications). The high posted price for the set has doubtless caused the work to be relatively rare in Irian Jaya and Papua New Guinea, save for a few libraries and interested individuals. Its execution is furthermore unsatisfactory: muddy photographs and uneven text in volume 1 and scores of typographical errors in volume 2 (necessitating a seven-page errata leaflet).

2 See H. Zollinger, 1857. Ueber den Begriff und Umfang einer 'Flora Malesiana'. *Vierteljahrsschr. Naturf. Ges. Zürich* **2**: 317–349, map; reviewed in H. J. Lam, 1937. On a forgotten floristic map of Malaysia (H. Zollinger 1857). *Blumea, Suppl.* **1**: 176–182, map.

3 For van Steenis's phytogeographic demarcation of Malesia, upon which were based the initial limits of *Flora Malesiana*, see C. G. G. J. van Steenis, 1948. Hoofdlijnen van de plantengeographie van de Indische Archipel op grond van de verspreiding der phanerogamen-geslachten. *Tijdschr. Kon. Ned. Aardrijksk. Genootsch.* **65**: 193–208.

4 They have been incorporated in the 1985 Koeltz reprint of the main work.

5 Among recent reviews of the famous biogeographical transition within Malesia from west to east are (for zoology) G. G. Simpson, 1977. Too many lines; the limits of the Oriental and Australian zoogeographic regions. *Proc. Amer. Philos. Soc.* **121**(2): 107–120, and (for botany) C. G. G. J. van Steenis, 1979. Plant geography of east Malesia. *Bot. J. Linn. Soc.* **79**(2): 97–178. Additional knowledge of the tectonic evolution of the region and the application of the methods of cladistic biogeography has allowed further insights on its biological history to be developed, most recently with reference to individual genera; see in particlar R. Hall and J. D. Holloway (eds.), 1998. *Biogeography and geological evolution of SE Asia.* ii, 417 pp., illus., maps. Leiden: Backhuys.

6 Financial and other constraints beginning in the 1980s have brought about changes to *Flora Malesiana*. These have included greater international representation in the membership of the Flora Malesiana Foundation, development of a wider circle of actual and potential contributors, 'prioritization', changes in format and publication

arrangements, and simplification of the taxonomic treatments. There was also a growing recognition that the time needed for 'completion' of the *Flora* would be much greater than originally supposed; the number of species to be covered had also increased by more than one-half from initial estimates.

7 For example D. G. Frodin, 1990. Explorers, institutions and outside influences: botany north of Thursday. In P. Short (ed.), *History of systematic botany in Australasia*, pp. 193–215. Melbourne: Australian Systematic Botany Society.

8 For example for all areas of natural history in Peninsular Malaysia and Singapore in H. N. Ridley, 1917. The scientific exploration of the Peninsula. *J. Straits Branch Roy. Asiat. Soc.* **75**: vii–xi; in papers by B. Tan and J. P. Rojo, P. S. Ashton, M. M. J. van Balgooy, W. J. J. O. de Wilde, E. F. de Vogel and P. F. Stevens in D. G. Campbell and H. D. Hammond (eds.), 1989. *Floristic inventory of tropical countries*, pp. 91–132. New York: The New York Botanical Garden (see **General references**); and D. G. Frodin, 1990. Botanical progress in Papuasia. In P. Baas, C. Kalkman and R. Geesink (eds.), *The plant diversity of Malesia*, pp. 235–247. Dordrecht: Kluwer. A progress review by M. M. J. van Balgooy, presented to the Thirteenth Pacific Science Congress in Vancouver in 1975, evidently never was published.

9 For example D. G. Frodin, 1985. Herbaria in Papua New Guinea and nearby areas. In S. H. Sohmer (ed.), *Forum on systematic resources in the Pacific*, pp. 54–62. Honolulu (Bernice P. Bishop Mus. Spec. Publ. 74); D. A. Madulid, 1985. Status of plant systematic collections in the Philippines. *Idem*, pp. 71–75; and A. Latiff Md. (ed.), 1991. *Status of herbaria in Malaysia*. Kuala Lumpur.

10 These figures are revised from those given in an undated brochure *The Flora Malesiana Project: Plant Diversity of Tropical Southeast Asia (Malesia)* issued in 1996 by the Flora Malesiana Foundation, Leiden. As already noted the species estimates have increased by at least one-half since the originally proposed figures of 25 000 spermatophytes and 3000 pteridophytes still current by the late 1970s (see Kalkman and Vink (1978) and van Steenis (1979) under **Progress** (p. 848) following the supraregional introduction). The estimates of coverage by precursory revisions, however, remain as taken from those papers.

11 The background to this work is given in its Introduction. It was originally to have formed part of vol. 3 of series I of *Flora Malesiana*.

12 Like the later *Flora Malesiana*, the inclusion of the Malay Peninsula, the Philippines, and eastern New Guinea (then still little known) in this work gives the work a biogeographic rather than political orientation. Its genesis and unusual circumstances with respect to preparation are covered in the general introduction to

910–30. Numerous additions are presented as family revisions and other notes in the author's *Annales musei botanici Lugduno-Batavi*. 4 vols. Leiden, 1863–69.

13 The conception and development of this flora project have been touched upon in the general introduction to **910–30**, where references are given. Species production to 1989 was charted in *The future of Flora Malesiana*, p. 2; progress has been maintained or increased since. It has, however, relied rather less on 'hired staff' and more on postgraduate students than some other large flora projects (R. Polhill in *op. cit.*, pp. 11–20).

14 Until project aid became available in the 1980s, limited staff and resources militated against extensive revision, however much it was desired (I. G. M. Tantra, personal communication).

15 Support for forestry research has since been transferred to a separate Forestry Research and Development Agency under the Ministry of Forestry.

16 Progress with tree taxonomy in Malesia has been briefly reviewed in Faridah Hanum Ibrahim and Lesmy Tipot, 1993. Status of tree taxonomy in Southeast Asia with special reference to Malaysia. In T. Whiffin *et al.* (eds.), *Proceedings of the symposium on tropical trees*, pp. 13–19 (BIOTROP Spec. Publ. 51). Bogor.

17 Simpson, 1977, p. 117.

18 The orchid volume of the *Flora* has been succeeded by G. Seidenfaden and J. J. Wood, 1992. *The orchids of Peninsular Malaysia and Singapore*. 779 pp., numerous text-figs., 48 col. pls. Fredensborg, Denmark: Olsen & Olsen (with the Royal Botanic Gardens, Kew, and the Botanic Gardens, Singapore).

19 For a review particularly of King and Gamble's work, including dates of publication, see F. S. P. Ng and M. Jacobs, 1983. A guide to King's *Materials for a flora of the Malayan Peninsula*. *Gardens Bull. Singapore* **36**: 177–185. Further details about the work may be had in Thistleton-Dyer (1906; see **General references**). Preparation of the Materials as an extension from the imperfect coverage in *Flora of British India* was initially advocated by Sir Hugh Low, British Resident in Perak in the 1880s and sponsor of the botanical exploration of that state (and adjacent Selangor). Its preparation and publication were subsidized by the Straits Settlements, Perak and Selangor following initiatives by Sir Cecil Clementi-Smith (who urged that coverage extend to the whole of the Peninsula along with Penang and Singapore). Although formally admitted as a 'Kew flora', most of the research and writing was carried out at Calcutta and Singapore. By World War I the *Materials*, arranged according to the Bentham and Hooker system, were largely completed and together with Ridley's complementary work on monocotyledons they became a primary basis for the latter's *Flora of the Malay*

Peninsula. The advent of the *Flora* proximately caused the Asiatic Society of Bengal effectively to terminate the *Materials*, with Andrew Gage's treatment of part of the Euphorbiaceae – the last to appear – being published only in 1936.

20 Ridley's *Flora* was reviewed critically by E. J. H. Corner in *J. Malay Branch Roy. Asiat. Soc.* **1933**: 42 and by C. F. Symington in *Bull. Misc. Inform.* (Kew) **1937**: 318–320. The work was compiled within a traditional structure and philosophy and is useful largely as a checklist, being valueless for field work (in spite of a claim by its author that it was 'so written and illustrated that [the public] will have no difficulty in identifying any plants that they meet with' (Ridley, 1917, *loc.cit.*)). Posterity has credited the work as showing signs of hasty compilation and often being uncritical; it remains, however, the last nominally complete flora.

21 *Wayside trees of Malaya* was as heterodox a work as its author, but as a tropical biological flora it has few if any equals. It developed as a consequence of Corner's exasperation with the 'traditional' type of flora embodied in Ridley's work.

22 *Tree flora of Malaya* remains available, and sets can be purchased locally; but at M$600 it is relatively dear and moreover apparently beyond the means of some local herbaria (personal observation, 1992; see also A. L. Lim in A. Latiff Md., 1991, p. 39). The photographs meant for Symington's manual were temporarily lost during World War II, only coming to light afterwards; they were incorporated in the reprint.

23 See also *Gardens Bull. Singapore* **44**: 1–2 (1992). Installments began to appear from vol. 46 (1995).

24 Many additions may be found in *idem*, 1863–70. *Annales musei botanici Lugduno-Batavi*. Vols. 1–4. Leiden.

25 This checklist also accounts for all collections made in conjunction with the U.S. National Cancer Institute screening surveys in the 1980s and 1990s as well as those made under Hilleshög auspices in the 1980s. Three versions have been circulated from 1994 through 1995.

26 This work was an important source for Masamune's enumerations.

27 The *Tree flora* has been supported by the Malaysian Government, the U.K. Overseas Development Administration (now Department for International Development) and the International Tropical Timbers Organisation. The project was originally scheduled to run for 10 years from 1992, but with the economic recession in Malaysia and the death early in 1999 of one of its principal contributors, K. M. Kochummen, its completion may be delayed. Workshops towards eventual flora treatments of difficult families such as Lauraceae continue, however, to be held. Such a 'collective' approach can only be commended.

28 An example of this collaboration was a checklist for Bangi Island, off the coast towards Palawan: E. D. Merrill, 1926. The flora of Banguey Island. *Philipp. J. Sci.* **29**: 341–427.

29 With the advent of *Tree flora of Sabah and Sarawak*, *Trees of Sabah* will not now be completed in its original form.

30 The 25 families included in vol. 2 (of which the largest is Guttiferae, or Clusiaceae) 'merely represent those which had been treated by the author at the termination of his service with the Sarawak Government, plus some others . . . he has found time to complete in the years following' (i.e., through 1973). Due perhaps in part to the vagaries of publication financing, there was a 15-year delay between completion of the manuscript and publication, rendering the book outdated on its appearance. No revision was essayed in the meantime, perhaps for want of time or opportunity; as a result such developments as the revision of *Calophyllum* by Peter Stevens are not accounted for. No other volumes have been published, including the projected introductory vol. 1. With the advent of *Tree flora of Sabah and Sarawak* its continuation is unlikely.

31 Like Koorders' *Exkursionsflora von Java* (following entry), the *Flora of Java* and the *Varenflora* were patterned after the 'excursion-floras' of Europe. They also recall, however, the author's earlier, never-completed *Schoolflora voor Java* (1911) in being more technically descriptive and highly mannered with actually less documentation and auxiliary information than in Koorders' work. A notable omission for such 'critical' works is a somewhat more complete synonymy. The absence of illustrations in the seed plant flora reportedly was intentional; the senior author believed them to have no place in a students' work. The Dutch version of the *Flora of Java* was completed by Backer with the assistance of others and reproduced in mimeographed form as *Beknopte flora van Java* (22 parts, 1942–61, Leiden: Rijksherbarium). This so-called 'emergency' edition was initially undertaken as a precautionary measure against possible destruction of the manuscript due to World War II.

32 This hastily compiled, uneven and rather uncritical work, designed after the manner of European 'excursion-floras' and originally oriented towards the mountain flora, has a few features not in its more critical successor: vernacular names, more detailed indication of local distribution, more synonymy and citations, and illustrations. When published, the work sparked off a long and bitter controversy, gaining it considerable notoriety (H. C. D. de Wit in *Flora Malesiana*, I, 4: cxxvi–cxxvii (1949) who labeled the work 'unfortunate'). Reasoned judgment has not on the whole been favorable.

According to de Wit, Koorders was originally contracted to write a mountain flora of Java but on his own account enlarged his brief. With the short time available, quality was sacrificed in favor of compilation and other short-cuts. Nevertheless, he had 'bested' Backer who had entertained the ambition to write *the* complete critical flora of Java (and to that effect had published some 'pre-cursory' works, including, in 1911, the first volume of his *Schoolflora voor Java*). The result was that until his death in 1963 Backer never forgave Koorders. Yet, were it not for strong representations by Hermann J. Lam (who in 1933 became director of the Rijksherbarium at the University of Leiden), Backer's own flora might never have been completely realized, even in its so-called 'emergency' edition. Two polemical tracts, with some-times colorful language, appeared shortly after publica-tion of the three text volumes of the *Exkursionsflora*: C. A. Backer, 1913. *Kritiek op de 'Exkursionsflora von Java' (bearbeitet von Dr S. H. Koorders)*. i, 67 pp. Weltevreden, Java: Visser (for the author); and S. H. Koorders, 1914. *Opmerkingen over eene Buitenzorgsche kritiek op mijne 'Exkursionsflora von Java'*. vii, 201 pp. Batavia: Kolff. The controversy, perhaps the most bitter and long-lasting in the whole history of floristic botany, highlights an almost insoluble equation in flora-writing: scholar-ship versus expediency, with time (and finance) as gov-erning factors. As already suggested, this question has in recent years also been faced by *Flora Malesiana*.

33 The lists for Java relied heavily on published floras including those described in the present work. They are thus in some respects more reliable than the 'Outer Provinces' lists which were largely dependent on Forest Service collections.

34 Little more would be accomplished during the remain-ing years of Portuguese rule. The only Portuguese floristic work seen for the period from 1885 to World War II is a series of notes: P. E. C. Saitos, 1934. *Aportamentos para o estudio da flora de Macau e de Timor*. 76 pp., 4 pp. addenda (loose insert), 27 halftones. [Lisbon: Jardim Colonial de Lisboa.] After 1945, however, some further serious investigation did take place under the auspices particularly of the then-recently organized Junta de Investigações Colonais (now IICT) in Lisbon; with respect to botany and forestry this was accomplished by Ruy Cinatti (see particularly his *Explorações botânicas em Timor* (1950) and *Reconhecimento preliminar das florestais no Timor Português* (1950)).

35 A bibliography appears in *Bot. J. Linn. Soc.* 79: 175–176 (1979) as part of van Steenis's review of East Malesian plant geography.

36 Simpson, 1977.

37 For biogeography, see T. C. Whitmore (ed.), 1981. *Wallace's Line and plate tectonics*. 91 pp., illus., maps (Oxford monographs on biogeography [1]). Oxford: Oxford University Press; and *idem*, 1987. *Biogeographical evolution of the Malay Archipelago*. 145 pp., illus., maps (Oxford monographs on biogeography 4). Oxford. Another perspective on Sulawesi may be found in R. I. Vane-Wright, 1991. Transcending the Wallace Line: do the western edges of the Australian region and the Australian plate coincide? In P. Y. Ladiges *et al.* (eds.), *Austral biogeography*, pp. 183–197. Melbourne: Com-monwealth Scientific and Industrial Research Organisa-tion, Australia (also published as *Austral. Syst. Bot.* 4(1)).

38 Many of the species described by Elmer and Merrill prior to 1926 were admittedly 'optimistic'. Following initiation of the *Flora Malesiana* project, successive family revisions have often produced reductions – some-times significant – in the number of species accepted (cf. van Steenis in *Flora Malesiana*, I, 5: ccxxii (1957)). This trend has continued to the present. In 1998, Domingo Madulid reported to the Fourth Flora Malesiana Symposium (Kuala Lumpur) that the number of flowering plant species had fallen below 8000. The largest reductions have occurred in tree genera with their best development at low to medium elevations, with *Cleistanthus* (Euphorbiaceae) the latest example (in *Blumea* 44 (1999)).

39 Activities of the Flora of the Philippines Project have been chronicled in *Philippine Flora Newsletter* (1991– ; 7 issues to 1994) and in D. A. Madulid and S. H. Sohmer, 1997 (1998). An update on the Flora of the Philippines project. In J. Dransfield, M. J. E. Coode and D. A. Simpson (eds.), *Plant diversity in Malesia*, III: 135–140. Kew: Royal Botanic Gardens. In the Philippines, also in the 1990s, there was established a national Biodiversity Information Center. Among its projects have been the creation of a floristic database derived in the first instance from Merrill's *Enumeration* and one on endemic species (*Philippine Biodiversity Information Center, Plants Unit, Newsletter* 3, July 1995).

40 The unification of most pure and applied research in a single Bureau of Science under the United States administration surely was a limiting factor. All Forestry Bureau collections were lodged there before distribu-tion; only those now outside the Philippines have sur-vived, with any unicates lost.

41 A *Flora of Manila* was prepared in part as a students' manual following the author's part-time academic appointment in 1909 to the new University of the Philippines. As the only such work in the Philippines it is still in use for this purpose although now sorely in need of revision.

42 An initial basis for this work was an unpublished type-script flora of Mt. Makiling compiled by A. D. E. Elmer early in the twentieth century.

43 The *Checklist* by Salvoza and Lagrimas, which follows the order of Merrill's *Enumeration*, ends with *Neolitsea* in the Lauraceae. What manuscript remained to be published was lost or destroyed during World War II (*Philipp. J. For.* 5 (1947)).

44 The work in all encompasses 12 volumes; vols. 5–12 are zoological. Some nomenclature is out of date.

45 In the 1975 version, the two lists of 1951 remain as discrete entities but share one collection of illustrations.

46 Among the defects of these lists is the lack of consideration of much more recent taxonomic work. For example, under Maluku Selatan there is listed *Tristania* sp. 15 which in the list of name changes (pp. 58–59) was amended to *Metrosideros nigroviridis*; but nowhere is there an indication of the further change to *Lindsayomyrtus brachyandrus*, the currently accepted name (which in 1990 had to undergo a further renaming, as *L. racemoides* (Greves) Craven; the reassembly of seemingly unconnected 'anomalies' may take time). Similar instances could be quoted. Nevertheless, these two lists, along with their successor, represent the sole organized floristic works covering all of Maluku.

47 The botanical names in this work (save for those few with Linnaean binomials in the general index) are pre-Linnaean. Several attempts, beginning with Linnaeus himself in 1754, were made to identify Rumphius's plants. The most extensive and best-founded is E. D. Merrill, 1917. *An interpretation of Rumphius's Herbarium amboinense.* 595 pp., 2 maps (Publ. Bur. Sci. Philipp. 9). Manila. This was brought further up to date in H. C. D. de Wit, 1959. A checklist to Rumphius's Herbarium amboinense. In *idem* (ed.), *Rumphius memorial volume*, pp. 339–460. Baarn: Hollandia. The secondary literature on *Herbarium amboinense* is considerable, but attention should be drawn to the following: A. Peeters, 1979. Nomenclature and classification in Rumphius's *Herbarium amboinense*. In R. F. Ellen and D. Reason (eds.), *Classifications in their social context*, pp. 145–166. London: Academic Press, and E. M. Beekman (ed.), 1981. *The Poison Tree: selected writings of Rumphius on the natural history of the Indies.* Amherst: University of Massachusetts Press (reissued 1993, Kuala Lumpur: Oxford University Press, as an Oxford in Asia Paperback).

48 The two plates actually appeared in 1884.

49 H. J. Lam in *Blumea* 1: 115–159 (1934). A new, though taxonomically unclassified, bibliography, is that of Simon Saulei (1996 (1997)). Both are fully cited in the main text.

50 The best overall review of exploration in English remains G. Souter, 1964. *New Guinea: the last unknown.* Sydney. An evaluation of the dynamics of biological exploration is found in D. G. Frodin, 1988. The natural world of New Guinea: hopes, realities and legacies. In R. MacLeod and P. F. Rehbock (eds.), *Nature in its greatest extent: Western science in the Pacific*, pp. 89–138. Honolulu: University of Hawaii Press.

51 Recent accounts of botanical progress include P. F. Stevens, 1990. New Guinea. In D. G. Campbell and H. D. Hammond (eds.), *Floristic inventory of tropical countries*, pp. 120–132. New York; D. G. Frodin, 1990 (*op.cit.*, Note 7) and B. J. Conn, 1994. Documentation of the flora of New Guinea. In C.-I. Peng and C.-H. Chou (eds.), *Biodiversity and terrestrial ecosystems*, pp. 123–156. Taipei. (Institute of Botany, Academia Sinica Monograph Series 14.)

52 A revised version has been in gestation for more than 20 years.

53 The parts on dicotyledons are largely compiled, while that on palms incorporates much original work. Both series provide coverage for some families at present not readily available elsewhere, but have had rather limited circulation (of the first part of the monocotyledon series, only 300 copies were printed).

54 Families in *Beiträge* I–XX have also been indexed by Lam (1934).

55 Up to the present, the *Beiträge* collectively remain the most important single contribution to Papuasian botany. With some exceptions they have had a more lasting value than the contributions in *Nova Guinea* or those issued as *Plantae papuanae archboldianae* (and associated 'Reports from the Archbold Expeditions').

56 A fourth volume is currently in preparation; like the third, it is under the editorship of B. J. Conn, Royal Botanic Gardens and National Herbarium, Sydney.

57 Other 'Botanical results of the Archbold Expeditions', mostly individual papers, appeared in both *Brittonia* and the *Journal of the Arnold Arboretum* through the mid-1950s.

58 Families in vols. 8, 12 and 14 of the original series have been indexed in Lam (1934).

59 The best treatment is the revised version of the Combretaceae (by M. Coode); others cover Anacardiaceae, Apocynaceae, Dipterocarpaceae, Eupomatiaceae, Himantandraceae, Magnoliaceae, Sapindaceae and Sterculiaceae. The series was discontinued in favor of the *Handbooks* and some family treatments prepared for it were therefore never published.

60 The synopsis of genera follows the system of Crabbe *et al.* (*Fern Gaz.* 11: 141–162 (1975)). The layout of this work is wasteful; judicious editing and printing reduction would have saved considerable space without any real loss in content. Nevertheless, what has been published can serve as a useful supplement to the outdated treatments by Brause (ferns), Herter (Lycopodiaceae) and Hieronymus (Selaginellaceae) in the *Beiträge* of

Lauterbach and Diels (see above) as well as that of Wagner and Grether for the Admiralty Islands (937) and general Malesian pteridophyte works (910–30).

61 Publication of part 2, an illustrated dendrological treatment, was delayed for 10 years.

62 This work is useful well beyond its nominal coverage (J. Croft, personal communication).

63 The translation is based on a typescript: G. Peekel, [1947]. *Illustrierte Flora des Bismarck-Archipels für Naturfreunde*, I–XII, Supplements I–II. Folios 1–40, 15–2016, 1–129, 1–12. Illus. (Reproduced on microfiche as Inter-Documentation Company (IDC) set B-8096.) The original is now housed in the library of the *Provinzialät* of the *Orden des Heiligen Jesu* in Hiltrup, Rhein-Westphalia, Germany. The author's collections, made between 1908 and 1940 and unfortunately to a goodly extent now lost, were for the most part determined by specialists in Germany and the results published in various botanical journals.

64 Friedrich Dahl was in 1896–97 resident at a 'biological station' in the area under the patronage of 'Queen' Emma Forsayth, a successful early planter and society hostess.

65 R. F. Thorne, 1963. Biotic distribution patterns in the tropical Pacific. In J. L. Gressitt (ed.), *Pacific basin biogeography*, pp. 311–350. Honolulu.

66 M. M. J. van Balgooy, 1971. *Plant-geography of the Pacific as based on a census of phanerogam genera*. 222 pp., maps (Blumea, Suppl. 6). Leiden.

67 Initially drawn, perhaps, to the Pacific flora by the collections and drawings brought by Ferdinand Bauer in 1826 on removing from England to Austria, Endlicher first wrote up the *Investigator* artist's plants from Norfolk Island as *Prodromus florae norfolkicae* (1833). Later he would write on the Australian flora.

68 R. W. Force, 1977. *Annual reports of the Director, Bernice P. Bishop Museum, 1971–75*, p. 6. Honolulu.

69 On the pre-war contributions, see Blake and Atwood in *Geographical guide*, 1: 11–12 (1942), and St. John in *Bernice P. Bishop Mus. Bull.* 133: 56 (1935).

70 F. R. Fosberg, 1976. Phytogeography of atolls and other coral islands. In *Proceedings of the Second International Coral Reef Symposium* (Brisbane, 1974), 1: 389–396. Brisbane.

71 *Ibid*. Floras of Henderson and Nauru have been published in recent years.

72 This, however, may be some time in realization; indeed, current circumstances at the Bishop Museum cannot be considered favorable.

73 For a recent account of Pacific biogeography, see D. R. Stoddart, 1992. Biogeography of the tropical Pacific. *Pacific Sci.* 46: 276–293. A 'typological' approach to classification is here adopted.

74 Some Pacific island groups were covered in *Floristic inventory of tropical countries* (1989; see **General references**): Fiji by J. Ash and S. Vodonaivalu (pp. 166–176), New Caledonia by K. Kendrick (pp. 177–180) and the Hawaiian Islands by D. Frame *et al*. (pp. 181–186).

75 The main issue has been the island's situation respective to the Oriental and Australasian floristic 'realms' or their derivatives. By Thorne (in *Pacific basin biogeography*, p. 326) New Caledonia was treated as a subregion of his Oriental Region, but van Balgooy (*Plant geography of the Pacific* (Leiden, 1971; cf. also *Blumea* 25: 81 (1979)) placed it in his Australian kingdom, 'albeit in a high hierarchical rank'. However, if a biogeography without boundaries be advocated, such speculation becomes mere sophistry.

76 Mention should also be made here of the very idiosyncratic *Systematik der Flora von Neu-Caledonien* by M. G. Baumann-Bodenheim (1988–), planned to occupy 17 *Bände* (each in one or more parts), of which *Bde*. 1, 3, 4, 5, 6, 7 and 14 have so far appeared, most recently in 1992. Of the documentary sections, only those on cryptogams (including pteridophytes) are now available. *Bde*. 4 and 5 are primarily photographic records, mostly from the author's tenure on the Franco-Swiss expedition. As of 1999, however, it seemed unlikely that any additional parts would be published.

77 General accounts on the floristics and plant geography of New Caledonia may be found in R. F. Thorne, 1965. Floristic relationships of New Caledonia. *Univ. Iowa Stud. Nat. Hist.* 20(7); J. D. Holloway, 1979. *A survey of the Lepidoptera, biogeography and ecology of New Caledonia*, pp. 1–14. The Hague: Junk; and P. Morat, 1994. La flore: caractéristiques et composition floristique des principales formations végétales. *Bois Forêts Trop.* 242: 7–30. For information on progress in floristic work, see Kendrick (1989; under **Progress** in **Superregion 94–99**).

78 Guillaumin and Veillon in their list appear to have overlooked Cohic's paper.

79 A further description of the islands is given by K. A. Hindwood in *Australian Natural History* 14(10): 305–311 (1964).

80 The Davis list was, unfortunately, here overlooked.

81 An example is *Schefflera cabalionii* P. Lowry, collected for the first time only in 1979 and described in 1989; it filled a 'gap' in the geographical distribution of its section (formerly the genus *Plerandra*) previously thought 'puzzling'.

82 References to most appear in the bibliographies of PARHAM (see 955) as well as of Ash and Vodonaivalu (1989; see Note 74).

83 There is also a treatment of the genus *Pandanus*, also by St. John. A separate florula would be desirable (Ash and Vodonaivalu, 1989; see Note 74).

84 Details of these precursory papers are in the bibliographies in Parham's checklist and the review by Ash and Vodonaivalu. The latter give a good general survey of the state of knowledge of the flora as of the late 1980s.

85 An interesting feature of the introduction is the author's discussion of the question of 'splitting' or 'lumping' of taxa in the context of Pacific botany. The often sweeping 'lumping' of the van Steenis 'school' as exemplified in *Flora Malesiana* is in particular strongly criticized. Such a cavalier approach arguably obscures the potentially differing evolutionary histories of plants in Oceania as opposed to their Malesian relatives (see also A. C. Smith, 1978. A precursor to a new flora of Fiji. *Allertonia* 1(6): 331–414, especially p. 332).

86 Another account of the flora is B. E. V. Parham, 1972. *Plants of Samoa.* 162 pp. (New Zealand Dept. Sci. Indust. Res., Inform. Ser. 85). Wellington.

87 Many further records from the remaining years of German administration, including those of Carl Ledermann in 1914, are documented in G. Volkens (ed.), 1914. Beiträge zur Flora von Mikronesien, I. *Bot. Jahrb. Syst.* 52: 1–18, and L. Diels (ed.), 1921–38. Beiträge zur Flora von Mikronesien und Polynesien, II–V. *Ibid.*, 56: 429–577; 59: 1–29; 63: 271–323; 69: 395–400. Orchids were covered by R. Schlechter in 1921 (for reference, see under **Progress**). Japanese collections to 1935 are documented in Kanehira's *Enumeration of Micronesian plants.*

88 Regrettably, they have not been published by the Smithsonian Institution as a distinct series available for general distribution.

89 Final negotiations on independence and association were long delayed on account of the country's constitutional ban on nuclear materials and perceptions of potential U.S. military needs.

90 I thank Demei Otobed for sending a copy of his work for annotation.

91 One of the most comprehensive general surveys ever attempted in the Carolines remains that made in 1908–10 by the German research vessel *Peiho*, which visited many atolls and low islands. Little attention, however, was paid to the terrestrial flora and only incidental collections were made by the ethnologist A. Krämer. A good opportunity for the collection of meaningful floristic data was thus lost, or perhaps at the time not realized.

92 Volkens was in the islands as a member of the German annexation commission following their transfer from Spain. His stay on Yap during 1900 lasted seven months due to a lack of opportunities for further travel.

93 For a recent account of Parece Vela, see *New York Times*, 4 January 1988, pp. A1, A5. It was recommended at the time that the Japanese Construction Ministry encase the islets with large steel and concrete blocks.

94 The authors note in their introduction a marked unevenness in collecting; only Tarawa, Onotua and Butaritari have received relatively detailed attention.

95 Neither work is cited by Thaman *et al.* (1994).

96 All the 'outer' islands were resurveyed in the 1980s with the aid of a team from the Department of Geography, Auckland University, but so far no separate botanical contribution has been published.

97 The full account of the Australian Museum expedition, the main source of the collections seen by Maiden, is contained in C. Hedley (ed.), 1896–1900. *The Atoll of Funafuti, Ellice Group.* ix, [3], 609 pp. (Austral. Mus. Mem. 3). Sydney. Hedley was a pioneer in serious study of Pacific insular biogeography.

98 There is, however, a biogeographical study of Suvorov: C. D. Woodroffe and D. R. Stoddart, 1992. *Substrate specificity and episodic catastrophe: constraints on the insular plant biogeography of Suwarrow Atoll, Northern Cook Islands.* 19 pp., 3 halftones, 2 maps (Atoll Res. Bull. 362). Washington. Some 23 native (list, p. 11) and 22 introduced species were recorded, and a florula was indicated as being then in preparation.

99 Much of this work presumably was based on the author's *Illustrationes florae insularum maris Pacifici* (1886–92; see **940–90**), and, like that work, utilized for its primary sources collections in France.

100 The Rubiaceae are the largest family and Euphorbiaceae the second largest, but Myrsinaceae and Urticaceae notably are relatively numerous in relation to the world average for these families, probably on account of endozoochory.

101 Most of the new material reported here was from the 1934 Mangarevan Expedition of the Bishop Museum (whose flowering plants were, however, reported on only haphazardly).

102 Referred to as 'in preparation' in *Bernice P. Bishop Mus. Bull.* 175: 15 (1942).

103 A draft flora of the Societies – possibly a continuation of that of Grant – was, according to S. L. Welsh in his *Flora societensis*, prepared by Fosberg and Sachet. It was, however, made available to him only in 1997, a few years after Fosberg's death.

104 The flora is thought to have been larger before the advent of man; the vegetation is now very heavily altered.

105 The atoll was at the time owned by the actor Marlon Brando. Investigations were commissioned in relation to possible development of parts of the plantation land.

106 Ducie was fully described in H. A. Rehder and J. E. Randall, 1975. *Ducie Atoll: its history, physiography and biota.* 40 pp., illus. (Atoll Res. Bull. 183). Washington. Two other plants, both herbaceous, were reported in 1922 but may have since been washed away in tropical storms.

107 For an account of the history of botanical work on the Marquesan flora, see W. L. Wagner and D. H. Lorence, 1997. Studies of Marquesan vascular plants: introduction. *Allertonia* 7: 221–225.

108 L. G. Eldredge and S. Miller, 1997. Numbers of Hawaiian species: supplement 2, including a review of freshwater invertebrates. *Occas. Pap. Bernice P. Bishop Mus.* **48**: 3–22.

109 L. G. Eldredge and S. Miller, 1995. How many species are there in Hawaii? *Occas. Pap. Bernice P. Bishop Mus.* **41**: 3–18; Eldredge and Miller, 1997 (*op. cit.*).

110 This remarkable manual, written in the style of the 'Kew floras', was realized following a 20-year residence in the then-kingdom in the 1850s and 1860s but published posthumously by W. F. Hillebrand following the author's death in 1886. The introduction, never completed, is unfortunately fragmentary. During his time in the islands, W. Hillebrand also contributed to the establishment of the present Foster Botanic Garden in Honolulu.

111 For nearly half a century the publication method of *Flora hawaiiensis* would remain 'unique among floras at present', but Blake and Atwood's hope of 'forever' inverted itself into a botanical Spandau. Neither flora nor prison, however, now continue: Degener died in 1986 and the prison was demolished in 1987 following the death of the last of its inmates – to be succeeded by a supermarket.

112 Also useful is M. C. Neal, 1965. *In gardens of Hawaii.* xix, 924 pp., illus. Honolulu: Bishop Museum Press. (Bernice P. Bishop Mus. Spec. Publ. 50. Originally published 1928 as *In Honolulu gardens*.) The work is currently under revision at the Museum as *In gardens of Hawaii II*.

113 For an account of the genesis and development of the *Manual*, see S. H. Sohmer, 1994. Conservation and the *Manual of the flowering plants of Hawai'i*: a sense of reality; or, Nine easy steps to producing a relevant flora. In C.-I. Peng and C.-H. Chou (eds.), *Biodiversity and terrestrial ecosystems*, pp. 43–51. Taipei. (Institute of Botany, Academia Sinica, Monograph Series 14.)

114 The list was based in the first instance on a manuscript compilation made before 1930 by E. H. Bryan, Jr.

115 From 1971 the atoll also was a storage dump for obsolete chemical and biological agents as well as other weapons; these included 25 000 drums of Agent Orange. Since 1990, however, all weapons and agents have been in process of destruction, unfortunately so far leaving dioxin and other residues. The atoll is now a U.S. National Wildlife Refuge and, following pressure from the City of Honolulu (administratively the responsible local government), the munitions disposal operation was due to conclude in 2000 with ecological restoration to follow (*Pacific Islands Monthly* **69**(5): 22–23 (1999)). The fate of the introduced plants would make an interesting study.

Appendix A

• • •

Major general bibliographies, indices and library catalogues covering world floristic literature

Introductory remarks

The following paragraphs are intended to serve as a background to an annotated enumeration of the principal general retrospective botanical bibliographies, indices and abstracting journals in print, as well as printed library catalogues most likely to be consulted in a search for floristic literature references – particularly in areas not effectively covered by regional source bibliographies and indices. Not included here are general taxonomic works, as these have been dealt with under Division 0 and moreover are described in greater or lesser detail in standard textbooks and other introductory accounts.

Records of floristic literature

With the movement of the last decade or so towards widespread use of on-line bibliographic search tools, it may properly be asked: why concern oneself with traditional print? The researcher primarily concerned with the description or analysis of structures and processes would by now probably find most, if not all, of his sources through computerized databases as well as by following citation trails from article to article. Relatively few sources would in the end be more than, say, 10 years old, and recourse to those bibliographies and indices (or parts thereof) available purely in print would rarely be necessary. On the other hand, floristic botany is not only spatially but also historically based, as is the literature of record in natural history, ethnobiology, uses of biota, evolution, and biodiversity in general.[1] Most on-line bibliographic databases do not at present go back earlier than the late 1960s (with full-text journals first appearing in the 1990s), and as yet there has not been a general movement in biology for retrospective conversion of indices, monographic bibliographies, or full texts even of the more important journals.[2] Some knowledge of the more traditional bibliographic media pertinent to the field – and their evolution – therefore remains essential.[3]

As indicated in Chapter 1 of Part I, floristic literature, including new monographic floras and enumerations, has with respect to reporting and indexing had a variable record over the course of the nineteenth and twentieth centuries. Of particular importance for current and retrospective searching have been works embodying a good geographical as well as subject organization – not to mention sufficient comprehensiveness. Coverage was probably at its best over the four decades or so through 1914, furnished by as many as six different services – all established during the great classification and documentation movement of the late nineteenth and early twentieth centuries.[4] The botany union card catalogue in Washington, D.C. was also commenced in this period, as was the *Index to American Botanical Literature*.[5] With respect to monographic works, the classic *Thesaurus* of Pritzel and Jackson's *Guide* (as well as Goodale's concise survey of 1879) were followed after 1900 by Bay's *Bibliographies of botany*, the Lloyd Library catalogue of floras by Holden and Wycoff, and Rehder's monumental *Bradley bibliography* along with the catalogues of such key libraries as Kew, the then-British Museum (Natural History), the Arnold Arboretum of Harvard University, and the Imperial Botanical Garden of St. Petersburg.[6] In the following three decades (the years of the two world wars excepted), the general situation remained relatively satisfactory; losses (such as the *International Catalogue of Scientific Literature*) were counterbalanced by gains (including *Botanical Abstracts* (later *Biological Abstracts*), the several abstracting serials of the Imperial Bureaux in the United Kingdom (now *CAB Abstracts* of CAB International), *Fortschritte der Botanik* (now *Progress in Botany*), and the completed *Bradley bibliography* and British Museum (Natural History) catalogue). Some notable regional monographs, such as Meisel's three-volume *Bibliography of American natural history: the pioneer century, 1769–1865* (1924–29) and Merrill and Walker's *Bibliography of Eastern Asiatic botany* (1938), were published; and towards the end of this period, the first volume of Blake and Atwood's *Geographical guide* appeared.

The least satisfactory period of the century from the point of view of retrospective searches is, however, that from 1950 through 1970 – a time also of great expansion in the sciences as a whole. All the Central European serials ceased publication by the end of World War II save for *Fortschritte der Botanik*, while the union card catalogue in Washington was closed in 1952.[7] Regional bibliographies – among them the *Index to American Botanical Literature, AETFAT Index* (begun in 1952) and the bibliographical pages in *Flora Malesiana Bulletin* (begun in 1947), as well as various monographic works – excepted, floristic and systematic botany is covered only by *Biological Abstracts* (from 1959 until 1993 without geographical indices or a distinct geographical search field), *Referativnyj Žurnal* (established in the 1950s along the lines of *Biological Abstracts*), and *Excerpta Botanica* (established in 1959), a partial successor to *Botanisches Zentralblatt*. From its inception in 1951 the journal *Taxon* included book notices and reviews, but it was in these decades not as comprehensive as had been *Naturae Novitates*. In 1961, the second volume of the *Geographical guide*, covering western Europe, was published, and in 1965 and 1967 respectively systematists in general benefited from two further works: *Biosystematic Literature* by Solbrig and Gadella and the first edition of *Taxonomic Literature* by Stafleu.

The final three decades of the century were a time of transition to electronic media, although as a medium for delivery of abstracting and indexing information print retained its primacy for at least the first 10–15 years or so. The addition of *Kew Record* in 1971, representing a union and expansion of certain published and unpublished regional indices prepared at the Royal Botanic Gardens, Kew, was a welcome development, which moreover coincided with an overall review of approaches to biological information presented as a series of lectures at the Linnean Society of London.[8] From 1969 Biosciences Information Service, the publishers of *Biological Abstracts*, furnished citations electronically as *BIOSIS Previews* in advance of print publication of full entries in *Biological Abstracts* (and *Bioresearch Index*, from 1980 *Biological Abstracts/ RRM*); abstracts themselves appeared in electronic form from 1976. The years 1976 to 1988 saw publication of 'die moderne Pritzel', the seven-volume second edition of *Taxonomic Literature* by Stafleu and Cowan. By 1986, fully electronic compilation and formatting brought greater currency to *Kew Record*, although its primary medium of delivery continued to be as conventional print.[9] Continuing advances in information technology led to the implementation in 1993 of significant reforms to *Biological Abstracts* and *Biological Abstracts/RRM*, including independent search fields for geopolitical units.[10]

Following the closure in 1998 of *Excerpta Botanica* – which never also adopted electronic search capabilities or output – there are now seven main services useful to floristic botany, two of them geographically limited. Among them are BIOSIS (*Biological Abstracts, Biological Abstracts/RRM*, and *BIOSIS Previews*, all also available electronically), *Flora Malesiana Bulletin* (available only in print), *Index to American Botanical Literature* (disseminated since 1995 from the New York Botanical Garden and from 1999 entirely electronic), *Kew Record* (also available on-line), PASCAL (*Bibliographie Internationale*, formerly *Bulletin Signalétique*; also available on-line), *Progress in Botany* (the former *Fortschritte der Botanik*), and *Referativnyj Žurnal*.[11] Among serials of a more applied orientation – useful for areas such as agronomy, forestry and ethnobotany – may be mentioned *Bibliography of Agriculture* (and its electronic equivalent, AGRICOLA), the abstracts service of the Food and Agriculture Organization of the United Nations (available electronically as AGRIS), and – arguably the most complete – *CAB Abstracts* from CABI Biosciences (also available electronically). The 'Reviews and Notices' section of *Taxon*, since 1987 under the leadership of Rudolf Schmid, now also furnishes extensive coverage. Much material of interest to floristic botany has moreover been consolidated into state, national and regional bibliographies, now quite numerous if not always easy to use; nearly all, however, have – at least up to the time of writing – largely remained print-oriented. To these latter have been added, as already mentioned, biodiversity information systems.[12] The publication in the mid-1980s of *Plants in danger* – along with the original edition of the present *Guide* – also represented major advances in information consolidation and facilitation of access to the literature and at the same time aided the development of new research programmes in biogeography and biodiversity.[13]

The present edition of this *Guide*, however, may be the last to be conceived largely in terms of a printed monograph. Any successor is likely to be based upon an updatable information system, with one-volume 'derivatives' issued in print (and CD-ROM) from time to time to meet a probable continuing need for 'elbow-books'. Within such a work, references to regional, national and state floras may be at least supplemented by hypertext links to floristic information systems; there similarly might be links to on-line bibliographies. To aid his or her work, a compiler ideally would have a custom search profile or profiles established with an individual service or an integrator such as DIALOG, BIDS or EINS.[14] Links would also have to be maintained with *Kew Record*, now the only indexing service specifically covering systematic botany on a global basis, and perhaps an Internet listserver established for informal communications. Nevertheless, any floras and related works selected would still have to be evaluated and annotated ('recensed') as in the past.[15]

Key sources for compilation of this Guide

When the first edition of *this Guide* was prepared, key guides to new material were the *Kew Current Awareness List*, issued by the Kew Library (in advance of *Kew Record*, then incurring a lag of up to four years), and the annual or biennial surveys of plant geography (in a wider sense) by E. J. Jäger in *Fortschritte der Botanik* (now entitled *Progress in Botany*). Other sources included reviews and notices in *Taxon* and other botanical journals as well as the annotated catalogues of two leading specialist book firms, Flück-Wirth (Teufen, Switzerland) and Koeltz (Königstein/Ts., Germany) and comparable trade outlets (including pre-publication leaflets).[16] Searches through national and regional bibliographies, as well as many library visits, were also undertaken, and much advice (along with notes on or extracts from specific items) was received from colleagues.

The *modus operandi* for the present edition has been similar to that of the first, with the advantage of the author's having been since 1993 a member of the Kew staff. The publication by IUCN of *Plants in danger* in 1986, with independently prepared bibliographies for all countries (and some major islands or island groups), not unnaturally revealed titles which had been overlooked for the original edition. Also in that year, *Kew Record* became a quarterly, in the process absorbing *Kew Current Awareness List*; and from the following year, under Rudolf Schmid (University of California, Berkeley), the 'Reviews and Notices' section of *Taxon* was reorganized and expanded. Another useful recent source has been the lists of acquisitions by the library of the Linnean Society of London, given in each issue of *The Linnean*, the Society's newsletter and proceedings initiated in 1985. For coverage of the many historically important but non-current floras – a feature of this edition – use was again made of classical and more contemporary world, regional and national bibliographies.

Floristic bibliography in the context of botanical and general bibliography (and textbooks of systematics)

The several primary sources herein considered in detail, both monographic and serial, are variously treated in an extensive range of secondary literature of greater or lesser scope – much of it arguably better known to the librarian or bibliographer than to the practising botanist. A concise account is here presented.

Among large, general reference works for libraries may be mentioned *A world bibliography of bibliographies* by the Voltaire scholar Theodore Besterman (4th edn., 5 vols., 1965–66, Lausanne), and *Les sources du travail bibliographique* compiled under the direction of L.-N. Malclès (3 vols., 1952–58, Geneva; reprinted 1965). The former is a massive empirical compilation in which full bibliographic details as well as (where necessary) collations, along with the number of entries accounted for, are out of its enormous range given for most of the source works described here, while the latter furnishes for each discipline an analytically arranged, annotated selection of monographs, key serials, and other works as well as bibliographies thought to characterize that discipline, so facilitating an overview.[17] Bibliographies of theses and dissertations through 1975 are covered in *A guide to theses and dissertations* by M. M. Reynolds (1975, Detroit).[18] For materials relating to botanical exploration, the best current overall work is *The history of natural history: an annotated bibliography* by G. D. R. Bridson (1994, New York).[19]

Several introductory texts covering the use of biological and botanical literature are also available, ranging in style and texture from the empirical to the analytical and aimed at students of biology or library science (and bibliography) or both. They are mentioned here as all of them give consideration to floristic literature along with related works on geographical botany and various kinds of bibliographies. Broadest in scope, and serving to place the biological sciences at a glance in relation to the whole range of human knowledge, is *Manuel de bibliographie* by L.-N. Malclès (1963, Paris; 2nd edn., 1969); in this work, chapters 27 and 29 respectively cover general biology and botany. More empirically oriented is *Bibliographies – subject and national* by R. L. Collison (1951, London; 3rd edn., 1968.) For biology as a whole, significant works include *Éléments d'un guide bibliographique du naturaliste* by F. Bourlière (1940–41, Mâcon/Paris); Путеводитель для биологов по библиографическим изданиям (*Putevoditel' dlja biologov po bibliografičeskim izdanijam*) by V. L. Levin, V. G. Levina and D. V. Lebedev (1978, Leningrad); *Smith's guide to the literature of the life sciences* by R. C. Smith, W. M. Reid and A. E. Luchsinger (9th edn., 1980, Minneapolis; 1st edn., 1942); *Guide to sources for agricultural and biological research* by J. R. Blanchard and L. Farrell (1981 (1982), Berkeley); *Biologische Fachliteratur* by G. Ewald (2nd edn., 1983, Stuttgart; 1st edn., 1973, as *Führer zur biologischen Fachliteratur*); *Using the biological literature: a practical guide* by E. B. Davis and D. Schmidt (1995, New York); and *Information sources in the life sciences* by H. V. Wyatt (4th edn., 1997, London; 1st and 2nd edns., 1966, 1971, as *The use of biological literature* by R. T. Bottle and H. V. Wyatt).[20] In botany, more or less recent works include *Guide to information sources in the botanical sciences* by E. B. Davis and D. Schmidt (1996, Englewood, Colo.; 1st edn. by Davis alone, 1987); *Botanical bibliographies* by L. H. Swift (1970, Minneapolis, Minn.; reprinted 1974, Königstein/Ts., Germany), and *Esbôco de un guia da literatura botânica* by C. T. Rizzini (1957, Rio de Janeiro).[21] The lists of key reference works given by S. F. Blake in the introductions of the two volumes of the *Geographical guide* should also be consulted.

Well-founded systematics textbooks containing more or less extensive lists of literature, with indication of general taxonomic works and varying selections of floras, include *Taxonomy of vascular plants* by G. H. M. Lawrence (1951, New York), *Taxonomy of flowering plants* by C. L. Porter (1967, San Francisco), *Vascular plant systematics* by A. E. Radford *et al.* (1974, New York), *Contemporary plant systematics* by D. W. Woodland (3rd edn., 2000, Berrien Springs, Mich.), and *Plant systematics: a phylogenetic approach* by W. S. Judd, C. S. Campbell, E. A. Kellogg and P. F. Stevens (1999, Sunderland, Mass.).

The only overall review of the structure of the information network catering for systematic biology remains a collection of articles of varying scope entitled *Storage and retrieval of biological information*, edited by P. I. Edwards (1971, in *Biol. J. Linn. Soc.* 3(3): 165–299). This was based on a series of lectures given at the Linnean Society of London in 1970.[22] A more recent, though brief, account is *The biological literature* by E. B. Davis and D. Schmidt (1998; in *Encyclopedia of Library and Information Science* (ed. A. Kent), 61: 1–16). Also serving as reviews are the already-mentioned texts by Bourlière, Ewald, Levin *et al.* and Wyatt. Rather more succinct but serving to place biological bibliography in relation to the art as a whole are the interpretative treatments of L. E. Bamber (for biology) and J. R. Blanchard (for agriculture) in *Bibliography: current state and future prospects* edited by R. B. Downs and F. B. Jenkins (1967, Urbana, Ill.) as well as L.-N. Malclès in her already-mentioned *Manuel de bibliographie*. Finally, there should be mentioned two more theoretical studies of the whole biological information system: *Die Bibliographie der Biologie* by H.-R. Simon (1977, Stuttgart), to which reference has been made in the introductory chapters of this book, and *The literature of the life sciences* by D. A. Kronick (1985, Philadelphia).

Summary

In concluding these remarks, it is perhaps worthy of note that relatively few, if any, analytico-critical surveys of the systematics information network have been written. The above-mentioned collection of Edwards was, as already noted, based on a series of lectures, while the work of Simon was conceived – as he himself noted – as only a beginning to serious research (and further development) on the biological information system, both in itself and towards improvement in its effectiveness. However, the initiative represented by the 1970 lecture series evidently did not lead to a continuing 'study group'; and Stafleu's review of the state of botanical bibliography and indexing at the 1975 International Botanical Congress in St. Petersburg suggested to the author at the time that, while activity was considerable and there was moreover in some quarters increased recognition of the importance in its own right of biological information work, little coordination among projects existed – with in some cases a consequent waste of effort. Put more simply, taxonomic botanists had not yet developed a stand on what they wanted in an information network. The ultimate goal of a world systematic and floristic information system, as first conceived by Leopold Nicotra of Messina in 1884 and strongly pushed by J. P. Lotsy and others after 1900[23] as well as by the Utrecht group from the 1930s,[24] yet seemed impossible, at least in the eyes of the Congress President.[25]

Developments in the systematics information system from 1975 to about 1992 remained relatively conventional in the face of economic and social as well as technical considerations. Bibliographies produced directly from databases, either independently or as parts of larger projects, remained for the most part relatively modest in scale until well into the 1980s.[26] While technological progress enabled *Kew Record* to be compiled on-line from 1986 (with retrospective conversion of the files for 1971–85), the major new contributions of the era continued to appear wholly in print. Among these were *Taxonomic Literature-2* (1976–88), the complete *Index nominum genericorum* (1979; supplement, 1986) and the first edition of the present book (1984, publ. 1985).[27] The production of major floras continued to be conventional, although computer-based techniques such as word-processing and databases were increasingly brought into use as microcomputers began their now-ubiquitous spread. Some floristic works, notably checklists, were by 1990 nevertheless being produced entirely from files derived from databases, and a number of floristic information systems were moreover under development.[28]

The introduction of hypertext links, mark-up language for files, browsers and other informational techniques in the 1980s and 1990s – together with the spread of the Internet as well as exponentially increasing capacity and a vastly expanded range of software – were to create entirely new possibilities. In 1985 the Taxonomic Databases Working Group (TDWG) was formed to promote integration of resources; among its initial (though long-term) goals was a unified plant reference system.[29] To this end, an international conference was held in 1990.[30] Continuing developments, notably the World Wide Web – formally launched in 1992 – as well as client–server technology, the Java language, and other features soon made a different vision possible, one enabling the integration of distributed resources as well as attractive presentation and comparative ease of navigation.[31]

It is now possible, more than at any time since before World War I, to foresee the effective implementation of an integrated, albeit distributed, taxonomic information system covering *inter alia* taxa, floristic regions and bibliographic data. The political climate for such a network is also improving, given the requirements of the Convention on Biological Diversity, the Convention on International Trade in Endangered Species, and other agreements as well as increasing public awareness. At the same time, there is an increased interest in retrospective capture of data not yet in computer-accessible form (M. Nesbitt, personal communication).[32] From the point of view of bibliography, the kinds of technical problems encountered in the 1970s – as noted in the original edition of this *Guide* – are, as forecast, now largely historical or can be overcome. With respect to the capture of floristic citations, useful initiatives – in rough order of priority – might include (1) a serious 'attack' on the poorly consolidated 1950–70 period, using a variety of printed and other sources; (2) conversion of the many national and state or provincial bibliographies (where not already incorporated into floristic information systems); (3) capture of the *Botany Subject Index*; (4) capture of the other monographic and serial sources described below as they relate to floristic and systematic botany and their allied areas (except for *Kew Record*);[33] and (5) capture of the major supraregional indices (already in progress on the part of *Index to American Botanical Literature*). In addition, consideration might be given to closer collaboration among bibliographic indices in the manner of the *International Plant Names Index*.[34] It should also be noted that a number of large botanical libraries can also be accessed on-line (via the Web or Telnet or both); some of these moreover now have full or substantial retrospective coverage.[35]

The works described in this appendix are given in three sections under the following headings:

General bibliographies

General indices (and abstracting journals)

Major library catalogues.

In the first two of these, each title is preceded by an abbreviated designation of (for general bibliographies) author(s) and date(s) of publication or (for general indices) the initials of the title and year(s) of coverage. These abbreviations correspond to those given under the subheadings 'General bibliographies' and 'General indices' at the beginning of each division and superregion in the *Guide* itself. Annotations have been based upon personal knowledge and examination as well as on the previously mentioned works of Besterman, Collison, Ewald, Malclès, Simon, Stafleu and Cowan, and others, *Bibliographies of botany* by J. Christian Bay (described below), *Published library catalogues* by R. L. Collison (1973, London: Mansell), and other sources (including articles directly or indirectly referring to particular works). The abbreviation '**BG**' under both 'General bibliographies' and 'General

indices' stands for 'bibliographic guide', where systematic reference to those of Ewald and Levin *et al.* has been made. Most of these works have also been discussed by Simon in *Die Bibliographie der Biologie* (1977, Stuttgart).[36]

General bibliographies

BAY, 1909.   Bay, J. C., 1909. *Bibliographies of botany*. [126 pp.] (*Progressus rei botanicae* 3(2): 331–456.) Jena. [Written as a precursor to a projected bibliography of botany for 1870 through 1899, this compilation, subtitled *A contribution toward a bibliotheca bibliographica*, includes annotated lists of 'General and comprehensive bibliographies' (heading 3) and 'National (regional) bibliographies' (heading 3a). Very good coverage of nineteenth-century (and earlier) source bibliographies, with full collations of some bibliographic periodicals up to 1908, is also furnished (heading 2). The work was compiled successively at the Missouri Botanical Garden, the Library of Congress, and the John Crerar Library in Chicago (of which last Bay was later head librarian). (The year 1909 is the correct date of publication, not 1910 as given in some references. His larger work regrettably never appeared.) **BG**: Levin *et al.*, 1978, p. 117.]

BLAKE and ATWOOD, 1942.   Blake, S. F. and A. C. Atwood, 1942. *Geographical guide to floras of the world*, 1: *Africa, Australia, North America, South America, and islands of the Atlantic, Pacific and Indian Ocean*. 336 pp. (USDA Misc. Publ. 401). Washington. (Reprinted 1963, 1967, New York: Hafner; 1974, Königstein/Ts., Germany: Koeltz.) [Comprehensive, briefly annotated critical bibliography of general, regional and local floras and enumerations, covering all parts of the world except Eurasia. Continued by Blake, 1961 (see below). Primary sources for the work included the USDA botany subject card catalogue (see below under USDA, 1958), Bay (see above), Holden and Wycoff, and Rehder (for both, see below), the Arnold Arboretum Library catalogue (see under **Major library catalogues**), and searches on shelves and in 'certain' periodicals, mainly in Washington, D.C., U.S.A. Total entries 3025 (Ewald, 1973). **BG**: Collison, 1968, p. 60; Ewald, 1973, p. 129; Levin *et al.*, 1978, p. 165.]

BLAKE, 1961.   Blake, S. F., 1961. *Geographical guide to floras of the world*, 2: *Western Europe*. 742 pp. (USDA Misc. Publ. 797). Washington. (Reprinted 1973, Königstein/Ts., Germany: Koeltz.) [A continuation of the preceding; covers western Europe from Scandinavia and Iceland to Italy but omits Germany and Austria. Based primarily upon the same sources as part 1 but with the addition of a number of European libraries, bibliographies and indices. No further continuation was published or even contemplated. 3757 primary and 3085 subsidiary entries (Blake). **BG**: Collison, 1968, p. 60; Ewald, 1973, p. 129; Levin *et al.*, 1978, p. 165.]

FRODIN, 1964, 1984. Frodin, D. G., 1964. *Guide to the standard floras of the world (arranged geographically)*. iv, 59 pp. Knoxville, Tenn.: Department of Botany, University of Tennessee; and *idem*, 1984 (publ. 1985). *Guide to standard floras of the world.* xx, 619 pp., 2 maps. Cambridge: Cambridge University Press. [The predecessors of the present work. The 1964 edition is a mimeographed classified list, limited to publications providing more or less complete coverage of vascular (or seed) plants for a given area. Its primary source was the botanical library of the Field Museum of Natural History, Chicago, supplemented by several other primary and secondary sources including the Lloyd Library, Cincinnati. 900 entries (author's estimate). The 1984 work differs from that presented here primarily in its lesser coverage of historical 'standards'. **BG**: Levin *et al.*, 1978, p. 165 (1964 version).]

GOODALE, 1879. Goodale, G. L., 1879. *The floras of different countries.* 12 pp. (Bibliographical Contributions of Harvard University Library 9). Cambridge, Mass. [Briefly annotated, geographically arranged list of the more important independently published floristic works of the period. Few smaller local floras are included. Locations (if held) within Harvard libraries are given. Largely compiled from Pritzel (see below) and Harvard sources. 400 entries (Besterman).]

HOLDEN and WYCOFF, 1911–14. Holden, W. and E. Wycoff, 1911–14. *Bibliography relating to the floras.* 513 pp. (Bibliographic Contributions from the Lloyd Library, Cincinnati, Ohio 1(2–13)). Cincinnati, Ohio. [Each number of this work comprises a self-contained, unannotated list, arranged by author, of the floristic literature of a given geographical area – either large or small – as known to the compilers (who were successively head librarians) with an indication of those actually held in the Lloyd Library. Holden was responsible for nos. 2–6, Wycoff for the remainder. (The geographical classification regrettably is less detailed than in most other works of the same genre. In the main, only independently published works are accounted for;

moreover, apart from Lloyd Library holdings, compilation was – as with the Goodale bibliography (see above) – second hand.) 6750 entries (Besterman).]

HULTÉN, 1958. Hultén, E., 1958. [Bibliography.] In *idem*, *The amphi-Atlantic plants*, pp. 298–330 (Kongl. Svenska. Vetenskapsakad. Handl., IV, 7(1)). Stockholm. (Whole work reprinted 1973, Königstein/Ts., Germany: Koeltz.) [Geographically arranged, unannotated list of floristic literature for almost the entire Holarctic region. An exceptional feature is that the areal coverage of nearly every work cited is shown on an accompanying map. Compiled in Stockholm from sources also used for *Index holmiensis* (see **001**).]

JACKSON, 1881. Jackson, B. Daydon, 1881. *Guide to the literature of botany.* xl, 626 pp. London: Longmans. (Index Society Publications 8.) (Reprinted 1964, New York: Hafner; 1974, Königstein/Ts., Germany: Koeltz.) [Compiled as an avocation by a then-'City man' of London, with a practically-oriented credo which aimed at being 'suggestive' rather than 'exhaustive', this subject-arranged but largely unannotated short-title bibliography contains in its sections 72 through 103 (pp. 221–405) a geographically arranged listing of independently published general, regional and local floristic works, with coverage through 1880. Many titles not in Pritzel (see below) are included. 9000 titles in all (Jackson) but Besterman gives a figure of 10000 (of which the present author estimates about one-third to be floristic). **BG**: Collison, 1968, p. 60; Ewald, 1973, pp. 133–134; Levin *et al.*, 1978, p. 112.]

PRITZEL, 1871–77. Pritzel, G. A., 1871–77. *Thesaurus literaturae botanicae.* iv, 577 pp. Leipzig: Brockhaus. (Reprinted 1950, Milan: Görlich; 1972, Königstein/Ts., Germany: Koeltz. 1st edn., 1847–51.) [The main body of this classic work has entries arranged by author, but its appendix comprises a short-title classified list arranged under subject and geographical headings. Under the latter (nos. 20–36) are found independently published regional, national and local floras and other floristic works, all accounted for more fully in the alphabetical division. This bibliography is essential for historical regional floristic literature, largely omitted from the present *Guide* save for passing references. Contains 10871 numbered entries covering some 15000 works (Pritzel). **BG**: Collison, 1968, p. 60; Ewald, 1973, pp. 135–136; Levin *et al.*, 1978, p. 112.][37]

REHDER, 1911–18.   Rehder, A., 1911–18. *The Bradley bibliography*. 5 vols. (Publ. Arnold Arbor. 3). Cambridge, Mass.: Riverside Press. (Reprinted 1976, Königstein/Ts., Germany: Koeltz.) [Volume 1 (*Dendrology*, part 1) includes among other subject areas geographically arranged listings of regional and local works on trees and on woody plants in general published before 1900, the closing date for the whole work. Volume 5 is an author and general subject index, with full details given for each entry (making cross-checking optional). (This very comprehensive work, the fruit of a large private benefaction, was prepared largely at the library of the Arnold Arboretum (then in Jamaica Plain, Mass., U.S.A.) as well as at other Boston-area libraries.) Covers in all 145000 entries (including the fully detailed author and subject indices) on almost 3900 pages, with 75000 titles alone in the two dendrological volumes (1 and 2) (Besterman). **BG**: Collison, 1968, p. 60; Levin et al. 1978, p. 144.]

SACHET and FOSBERG, 1955, 1971.   Sachet, M. H. and F. R. Fosberg, 1955. *Island bibliographies*. v, 577 pp.; and *idem*, 1971. *Island bibliographies: supplement*. ix, 427 pp. (Nat. Acad. Sci./Nat. Res. Council Publ. 335, 335[a]). Washington. [Each of the two parts of this work comprises three annotated bibliographies, each separately indexed. The first bibliography in both parts deals with Micronesian botany; the second with geography, physical features, etc., of coral atolls and other low islands in all oceans; and the third with vegetation on tropical Pacific islands of all kinds. A number of relatively obscure florulas and floristic lists are herein accounted for. 7500 entries in original work, of which 3500 comprise the section on Micronesian botany (Besterman).]

USDA, 1958.   United States Department of Agriculture, National Agricultural Library, 1958. *Botany subject index*. 15 vols. Boston: Microphotography Co. [G. K. Hall]. [A reproduction in book form of one of the greatest classified card catalogues covering systematic botany, the Botany Subject Catalogue in the USDA central library in Washington (now the National Agricultural Library at Beltsville, Md.). Begun in 1906 and continued until July 1952, this union catalogue has a worldwide scope with countless headings, including among others geographical units in great detail. Although not fully inclusive of entries from the *Index to American Botanical Literature* (see below), it is considered to be the most complete guide to the botanical literature of the first half of the twentieth century (a period otherwise imperfectly served in retrospective bibliography – especially between 1914 and 1926, at least for English-language materials). It was the principal source for the *Geographical guide* by Blake and Atwood (see above) and its discontinuance through lack of funds can only be regarded as short-sighted. The book version contains in each volume a table of contents. 315000 entries are included (Simon, p. 82). **BG**: Collison, 1973, p. 54; Levin *et al.*, 1978, p. 114.][38]

General indices (and abstracting journals)

Apart from *Biological Abstracts and Referativny Žurnal*, as of 2000 there is only one current service covering systematic and geographical botany; it does not include summaries. Mention should, however, be made of the substantial 'Reviews and Notices' section in the journal *Taxon*, since 1987 conducted by R. Schmid.

BA, 1926– .   *Biological Abstracts*. Vols. 1– . Baltimore (later Philadelphia), 1926– . URL: http://www.biosis.org/. [Large-scale classified abstracting journal, currently published semi-monthly in print, with at present five comprehensive indices: author, subject (KWIC-type), 'CROSS' (concept relation of subject specialties, now simply termed a concept index), 'biosystematic' (for groups of organisms), and (from 1974) generic-specific (for specific taxa), with cumulation semi-annual (and every five years on microform). Since 1965 BA has been accompanied by *Bioresearch Index* (in 1980 renamed *Biological Abstracts/RRM*) which extracts titles from books, book contents, annual reports, conference materials and reports, and review journals as well as reporting bibliographies and developments in nomenclature; its contents are largely additional to those in *Biological Abstracts* proper. Appearing monthly, this latter is indexed similarly to the parent work (but without long-term cumulations). From 1969 all titles have also been made available (as *BIOSIS Previews*) in electronic form, with retrospective searches possible; abstracts followed in mid-1976. The service is available through direct subscription or through a service provider. Special individual or group subject 'profiles' can be established to be searched regularly. Despite their extent, however, the two stablemates of BIOSIS (for BIOSciences Information Service) do not cover all biological literature. For floristic material it should be regarded largely as a secondary source unless used with a custom search profile. The termination in 1959 of geographical and conventional subject indices in favor of the more quickly produced contextual KWIC index was also disadvantageous for floristic botany; this situation remained until 1993 when a separate geopolitical search field was introduced. Supplementation from other indexing and abstracting journals is therefore necessary for some searches. **BG**: Collison, 1968, p. 59; Ewald, 1973, pp. 108–110; Levin *et al.*, 1978, pp. 31–35; also in most other standard works. Available on-line as *BIOSIS Previews* from 1969, with additional facilities (including abstracts) and media at later dates (e.g., on CD-ROM from 1982).][39]

BotA, 1918–26. *Botanical Abstracts*. Vols. 1–15. Baltimore, 1918–26. [Monthly abstracting journal in (usually) two volumes per year, with in each issue the abstracts arranged under broad subject headings; semi-annual cumulative author indices. In scope restricted to periodical and serial articles, initially all from American outlets. By 1926 coverage had risen to 9929 items per year (Besterman). In that year, it was absorbed into *Biological Abstracts*. Like its successor, however, it was considered by Blake and Atwood as a source of only secondary importance for compilation of their *Geographical guide*. **BG**: Collison, 1968, p. 61; Ewald, 1973, p. 111.][40]

BC, 1879–1944. *Botanisches Centralblatt* (from 1938 *Botanisches Zentralblatt*). Jahrgang 1–40 (Bände 1–142); N.S., Jg. 1–37 (Bde. 143–179, no. 8). Cassel, 1880–1901; Leiden, 1902–05; Jena/Dresden, 1906–45. (Suspended 1920–21.) [The first 22 years of this work, a key resource for botany particularly until just after the close of World War I, comprised, in quarterly volumes, a miscellany of material – not unlike the present-day journal *Taxon* – including reports of new books and periodical literature together with original articles and reviews. Growing professional pressure for a more properly organized abstracting/indexing journal gradually manifested itself on both sides of the Atlantic and led to a change of ownership and editorial reorganization in 1902. From then until 1919, publication was weekly, with three volumes per year (two containing notices or reviews, and the third – issued in parallel – devoted to 'Neue Literatur', a classified index in which only very short abstracts appeared with the titles). Following another reorganization, a 'new series' was begun in 1922, though its organization remained similar. Some 330000 items were accounted for, averaging 2500–3000 per year (Besterman) with its highest level attained during 1902–10. It covered monographs

as well as serial and periodical articles, and some cumulative indices were published, particularly in the *Neue Folge* or New Series of 1922–45. For their *Geographical guide*, Blake and Atwood viewed the *Zentralblatt* as a key secondary resource. **BG**: Collison, 1968, p. 61; Ewald, 1973, pp. 111–112; Levin *et al.*, 1978, pp. 115–116.][41]

BS, 1940– . *Bulletin Signalétique* (through 1955 *Bulletin Analytique*; from 1984 known as *Bibliographie Internationale*). Vols. 1– . Paris, 1940– . [Produced by the Centre National de la Recherche Scientifique, Institut de l'Information Scientifique et Technique (INIST), France, this comprises a monthly index serial for all fields of science with, from the 1950s, separate series for different subject areas (from 1969 division 370 (formerly 17), *Biologie et physiologie végétales* (BPV), accounted for floristic botany). The contents of each division have been subdivided in some detail, but as each entry is accompanied only by a brief précis it is less useful as an information source than *Biological Abstracts* or *Referativnyj Žurnal, Biologija*. Only periodical articles are covered. Titles appear in original languages with a French translation (from 1982 also in English). With respect to floristic botany it appears to have weaknesses similar to *Biological Abstracts*, but offsetting this for some time was the continued appearance of the botanical division under separate covers (abandoned in *Biological Abstracts* after 1959). As a 'newsletter', *Bulletin Signalétique* has some value but must be used with other sources for effective coverage of floristic botany. **BG**: Collison, 1968, pp. 47–48; Ewald, 1973, p. 112; Levin *et al.*, 1978, pp. 35–36 (BPV, pp. 109–110). Available on-line as PASCAL with coverage from 1973.]

CSP, 1800–1900. *Catalogue of Scientific Papers*. 22 vols. London (later Cambridge), 1867–1925. [A comprehensive index to periodical articles, published under the auspices of the Royal Society of London and covering all scientific fields from 1800 through 1900. Comprises four series and one supplementary volume (19 volumes in all) with citations arranged by author in each chronological series (1800–63; 1864–73; 1874–83; 1884–1900). Nearly 2.3 million titles were in all accounted for. Three further volumes of cumulative indices were published, but regrettably these cover only pure mathematics, mechanics and physics. For annual coverage of the years 1901–14, see ICSL below. **BG**: Collison, 1968, p. 46; Ewald, 1973, pp. 130–131, where defined as a monographic bibliography and with a discussion of the peculiarities of its system of title abbreviations; Levin *et al.*, 1978, pp. 55–56.][42]

EB, 1959–98. *Excerpta Botanica*, sectio A: *Taxonomia et chorologia*. Vols. 1–65. Stuttgart, 1959–98. [Begun through an agreement between Fischer, publishers of the erstwhile *Botanisches Zentralblatt*, and the International Association for Plant Taxonomy, this journal contained briefly annotated listings of new books and periodical articles in the field of systematics, including floristics (chorology); in the case of the latter these are classified under only seven geographical headings. Book reviews were also included. Published bimonthly (later monthly), with an additional cumulative annual (later semi-annual) index appearing later (sometimes tardily so); seven parts comprised a volume. Each subject area and geographical unit, however, was covered only once in any volume. Through 1963 it accounted for 1000–2000 titles per year (Besterman), with perhaps 3000 by the early 1980s. It appears to be most thorough for Eurasia but generally requires to be used with other sources. Multilingual (German, English, French). **BG**: Ewald, 1973, p. 116; Levin *et al.*, 1978, p. 109.][43]

FB, 1931– . *Fortschritte der Botanik* (now *Progress in Botany*). Vols. 1– . Berlin, 1932– . [Concise annual review of all fields of botany, organized under several major headings and prepared by specialists. Systematics and floristics of higher plants in earlier years usually comprised one major heading, which normally included lists of the more important new floristic works brought to notice in the year under review. In more recent years floristics has appeared separately, and from the early 1980s at intervals of two years or longer. Useful as a secondary information source, and corroborative in terms of assessment of relative importance, a key factor in determining the inclusion or not of a given work in a selective bibliography such as this *Guide*. Its botanical coverage rose from 1000 per annum to 2000 in 1962 (Besterman). **BG**: Ewald, 1973, p. 116; Levin *et al.*, 1978, p. 110.][44]

IBBT, 1963–69. *Index Bibliographique de Botanique Tropicale*. 9 vols. Paris, 1964–71. [Trimestrial annual compilation of literature relating to tropical botany, including tropical floristics. Each annual volume arranged by author, with a final part devoted to a cumulative subject index. Published under the auspices of the French Office de la Recherche Scientifique et Technique d'Outre-Mer but suspended in 1971 due to unspecified technical difficulties.]

ICSL, 1901–14. *International Catalogue of Scientific Literature*, section M: *Botany*. 14 vols. London, 1902–19. [A continuation of the *Catalogue of Scientific Papers* (see above), this journal, one of 17 of its kind covering all fields of science, comprised an annual index to new botanical literature in general, including that on floristics. It encompassed 82 582 items in all, with annual coverage ranging from 4728 to 7355 (Besterman); although covering both monographic and periodical literature it was to some, however, insufficiently comprehensive. It was utilized for the *Geographical guide* by Blake and Atwood together with other secondary sources. **BG**: Collison, 1968, p. 46; Ewald, 1973, pp. 119–120; Levin *et al.*, 1978, pp. 56–57 (section M, p. 116).][45]

JBJ, 1873–1939. *Just's Botanischer Jahresbericht*. Jahrgang 1–67. Berlin, 1874–1944. [Subtitled *Systematisch geordnetes Repertorium der botanischen Literatur aller Länder*, this comprised an annual analytical classified guide (in two major divisions, each with a number of subdivisions) to botanical literature in all fields. In earlier years, one subdivision (in part II) dealt with floristic (geographical) botany. This floristic subdivision comprised two sections: (α) for titles, arranged by author, and (β) for geographically arranged reviews of individual items in abstract form. At a later period floristics appeared in part I, divided into sections for Europe and beyond Europe. Each volume was furnished with complete author, subject and taxon indices. New taxa, with places of publication, were also listed for all major groups. Until World War I coverage had been anywhere from one to three years behind; afterwards, however, with more difficult circumstances and a rapidly rising volume of literature, compilation lagged still further and some of the last volumes were never completed. The *Jahresbericht* ceased publication during World War II. Total coverage had by then reached over 1.1 million items (Simon, 1977), with an annual increment ranging up to 19 500 (for 1911; Besterman does not provide figures for later years, but these appear to have fallen somewhat by the 1930s with the increasing restriction in coverage to systematic, floristic and comparative botany). The *Jahresbericht* served as one of the secondary sources for Blake and Atwood's *Geographical guide*. **BG**: Collison, 1968, p. 61; Ewald, 1973, pp. 121–122; Levin *et al.*, 1978, p. 115.][46]

KR, 1971– . *Kew Record of Taxonomic Literature*. Vols. 1– . London, 1974– . [Classified annual index to new taxa (through 1980) and new literature in systematics, floristics and related fields, arranged by subject and region. The geographical classification, based upon the system used in the Kew Herbarium, was initially exceptionally detailed but later was simplified. An index to authors appears at the end. With nearly 8000 citations per year (some, inevitably, replicated) by 1978, it is currently the most comprehensive indexing source in systematic and floristic botany (and from 1998 the only one still in print), and the single work in the field comparable to the long-established *Zoological Record*. **BG**: Levin *et al.*, 1978, p. 130.][47]

NN, 1879–1943. *Naturae Novitates*. Vols. 1–65. Berlin, 1879–1943. [Subtitled *Bibliographie neuer Erscheinungen aller Länder auf dem Gebiete der Naturgeschichte und der exacten Wissenschaften*, this bimonthly trade guide from the Friedländer publishing house reported the appearance of new books and other key contributions in biology and other sciences (including also journal issues, installments of serially published monographs, and independently published offprints); each volume was separately indexed. Floras enjoyed a distinct subheading. Coverage through 1912 ranged from 5000 to 9492 items per year, but after World War I it dropped to an average of 3500 until 1938; during World War II there was a further fall to less than half this figure (Besterman). It was moreover of much reduced importance to botany after World War I. **BG**: Ewald, 1973, p. 123; Levin *et al.*, 1978, p. 51.][48]

RŽ, 1954– . Реферативный журнал (*Referativnyj Žurnal*): (04) Биология (*Biologija*) and (04V) Ботанкиа (*Botanika*). Vols. [1] (1953)– . Moscow, 1954– . [Two of 61 different series published by the Russian (formerly Soviet) All-Union Institute for Scientific and Technical Information (VINITI), Moscow, *RŽ Biologija* (and *RŽ Botanika*) comprise monthly abstracting journals, as of 1978 classified into 17 divisions of which one (04), established from the beginning, deals with biology. Botany was covered within *Biologija* under the subseries (04V), but was physically separated in 1992. Coverage is worldwide, with some emphasis on literature from the Russian Federation and neighboring countries. Within each division, material is further subdivided, sometimes to a higher degree than in *Biological Abstracts*. Designation of abstracts is alphanumeric and coordinated with month of issue and principal subdivisions; each abstract is moreover provided with a UDC number. In 1972, 106000 items were accounted for. Subject and author indices are provided in an annual cumulation.

Generally speaking, *RŽ Biologija* was less comprehensive than *Biological Abstracts* but at the same time less markedly interdisciplinary or biomedically oriented. **BG**: Collison, 1968, pp. 47–48; Ewald, 1973, p. 125; Levin *et al.*, 1978, pp. 26–29.][49]

Major library catalogues

With the exception of the Herder/Klinge catalogue for St. Petersburg, only those catalogues containing botanical works listing a substantial number of floras in relation to the totality of entries and published in print in the twentieth century are given here. Nineteenth-century catalogues as a whole are well treated by Bay in his *Bibliographies of Botany* (see above). The twentieth-century catalogues listed here are all touched upon by Collison in his *Published library catalogues* to which reference was earlier made (see end of **Introductory remarks**). Of exceptional value for bibliographical control of older works, particularly in the final stages of preparation and correction of copy for the first edition and for the initial stages of this edition of the *Guide*, has been the U.S. Library of Congress *National Union Catalog: Pre-1956 Imprints* (1968–80, London: Mansell) and its supplements, although these are not here described. Since then, many major libraries have databased all or many of their holdings; a goodly proportion of them may now be searched freely with the aid of such Internet tools as the World Wide Web or Telnet.[50]

HERDER, F. E., 1886. Систематический каталог библиотеки имп. С.-Петербургского ботанического сада (*Sistematičeskij katalog biblioteki imp. S.-Peterburgskogo botaničeskogo sada*). xii, 510 pp. St. Petersburg. Continued as J. KLINGE, 1899. Систематический каталог библиотеки имп. С.-Петербургского ботанического сада с 1886–1896 гг. (*Sistematičeskij katalog biblioteki imp. S.-Peterburgskogo botaničeskogo sada s 1886–1896 gg.*). viii, 253 pp. St. Petersburg; and *idem*, 1899. Систематический каталог библиотеки имп. С.-Петербургского ботанического сада за 1898 г. (*Sistematičeskij katalog biblioteki imp. S.-Peterburgskogo botaničeskogo sada za 1898 g.*). 47 pp. St. Petersburg.

A classified printed catalogue, the original work and its two supplements covering some 14000 items. [This collection, now the core of the botanical library of the Komarov Botanical Institute of the Russian Academy of Sciences in St. Petersburg, is one of the most important of its kind in Europe and indeed the world, with much unique or rare material. It is essential for effective knowledge of the literature of central and northern Eurasia.]

ROYAL BOTANIC GARDENS, KEW, 1974. *Author catalogue of the Royal Botanic Gardens Library, Kew, England*. 5 vols. Boston: Hall. Complemented by *idem*, 1974. *Classified catalogue of the Royal Botanic Gardens Library, Kew, England*. 4 vols. Boston. [A successor to *idem*, 1899. *Catalogue of the library of the Royal Botanic Gardens, Kew*. 790 pp. (Bull. of Misc. Inform. (Kew), Addit. Ser. 3). London, and its *Supplement*, 1919.]

The 1974 catalogue, photo-reduced from cards, includes some 100 000 items in its author division and 64 680 items in its classified division. It supersedes the printed catalogue of 1899–1919, which covered author entries through 1915. [The Kew Library is one of the largest, if not *the* largest, botanical libraries in the world. Almost completely computerized, but to date available on-line only within the institution. The classification of floras follows the geographical arrangement used in the Herbarium.]

TUCKER, E. M., 1914–33. *Catalogue of the library of the Arnold Arboretum of Harvard University*. 3 vols. (Publ. Arnold Arbor. 6). Cambridge, Mass.

A printed catalogue with author and subject divisions. Volumes 1–2 cover holdings up through 1917. Volume 3 (authors only) comprises a supplement for 1917–33. A projected fourth volume to contain the complementary subject index for 1917–33 never materialized. [All holdings are now on-line, along with those of the Gray Herbarium and other Harvard botanical libraries, as part of Harvard's HOLLIS system; see http://hplus.harvard.edu/.]

UNITED STATES DEPARTMENT OF AGRICULTURE, 1967–70. *Dictionary catalog of the National Agricultural Library*, 1862–1965. 73 vols. New York: Rowman & Littlefield. Continued as *idem*, 1972–73. *Dictionary catalog of the National Agricultural Library*, 1966–70. 12 vols. Totowa, N.J.: Rowman & Littlefield.

The main work comprises a photo-reproduced card catalogue, with authors and subject entries in one sequence; in vol. 73 are translations of articles. The first supplement has authors in vols. 1–8 and subjects and translations in vols. 9–12. Since 1966 the *Dictionary Catalog* has also been updated by Rowman & Littlefield on a monthly basis, with one volume per year and cumulations every five years. Automation of cataloguing from 1970 and the concomitant creation of the AGRICOLA database have enabled supplements from vol. 6 (1971) onwards to be arranged on a broad subject basis, with inclusion of indices. [This library was the primary resource for the compilation of the *Geographical guide* of Blake and Atwood – its junior author was on the library staff – and in the 1970s and early 1980s served as a principal resource for compilation of *Taxonomic Literature*-2. It contains the largest single collection of botanical literature in the U.S.A. All records in the original *Dictionary catalog* (through 1965) have now been added to the on-line catalogue; see http://www.nal.usda.gov/.]

WOODWARD, B. B. and A. C. TOWNSEND (eds.), 1903–40. *British Museum (Natural History): catalogue of the books, manuscripts, maps and drawings*. 8 vols. London: Trustees of the British Museum. (Reprinted 1964, Weinheim, Germany: Cramer (in association with Wheldon & Wesley, Codicote near Hitchin, Herts., England); a paperback version in four volumes is also available.)

Printed catalogue, arranged solely by author (save some special categories) and published in two series (vols. 1–5, 1903–15 and 6–8, 1922–40). Coverage thus runs up to anywhere from 1920 through 1938 depending upon the letter of the alphabet. Botany is not separately treated. [Most holdings are now also catalogued on-line; see http://library.nhm.ac.uk/.]

Notes

1 See, for example, G. H. M. Lawrence, 1970. Botanical libraries and collections. In A. Kent and H. Lancour (eds.), *Encyclopedia of Library and Information Science*, 3: 104–112. Pittsburgh; and T. J. Delendick, 1990. Citation analysis of the literature of systematic botany: a preliminary survey. *J. Amer. Soc. Inform. Sci.* 41: 535–543. Delendick notes that a similar temporal distinction had been made in the 1960s by the historian of science Derek J. de Solla Price.

2 Exceptions include *CAB Abstracts*, with some files available back to 1939, and *Dissertation Abstracts International*, fully retrospective to 1861. Some key journals in biology are likely to become fully available on-line through JSTOR, an information consortium established in 1995 (http://www.jstor.ac.uk/about/). At the Royal Botanic Gardens, Kew, the entire plant anatomy and economic botany files, with records going back to the nineteenth century, are internally on-line.

3 There has inevitably been much discussion as to how far on-line information retrieval would displace the use of printed resources, particularly as its relative cost has declined. For the first edition of this book, following Ian McPhail ('Information resources for botanical gardens', *American Association of Botanic Gardens and Arboreta Bulletin* (July 1981): 90–95), it was suggested that 'it may be some little time, if ever, before the systematic botanical information system dispenses entirely with "traditional" resources and means of communication, despite hyperbolical forecasts by some writers of a "paperless" future'. Subsequent progress has been patchy, given available resources and perceived priorities as well as movements in information technology as a whole. One example is floristic information systems; though developed by the original FNA Program in North America in the 1960–70s and ESFEDS in Europe in the 1970–80s, their effective implementation did not really take place until after the emergence of mark-up language and the World Wide Web in the 1990s. Even now, they are still mostly at state or national level. Capture of herbarium collection information, first discussed in the late 1960s, has also developed in a patchy fashion, reflecting individual initiatives. Nevertheless, change is taking place; with respect to the bibliography of floristics a considerable advance will occur when *Kew Record* becomes generally available on the Web.

4 An important manifestation of this trend was the formation in 1895 of the Institut International de Bibliographie in Brussels by Henri La Fontaine and Paul Otlet. From this there evolved, after World War I, what is now the International Federation for Information and Documentation in The Hague.

5 A. C. Atwood, 1911. *Description of the comprehensive catalogue of botanical literature in the libraries of Washington*. 7 pp. (USDA Bureau of Plant Industry Circular 87). Washington.

6 Most available resources were reviewed in J. C. Bay, 1906. Contributions to the theory and history of botanical bibliography. *Bibliographical Society of America, Proceedings and Papers 1904–05* 1(1): 75–83. Bay regarded the best of the monographic works as being Pritzel's *Thesaurus* and, among library catalogues, that of the British Museum (Natural History). Among serial sources the then-weekly *Botanisches Centralblatt*, with its timely separate lists of new literature, was regarded as the most generally useful. 'What is needed for botany, by way of periodical lists of new literature, is a yearly conspectus, roughly classified, published promptly on the close of each year, and sold at a moderate price.'

7 A partial successor to the card catalogues was the USDA *Bibliography of Agriculture*, a serial begun in 1942 and from 1970 also available electronically as AGRICOLA (Agricultural OnLine Access).

8 Published under the editorship of P. I. Edwards in *Biological Journal of the Linnean Society* 3(3): 165–299 (1971).

9 At that time it absorbed the annual *AETFAT Index* as well as the very useful monthly *Kew Current Awareness List* and became a quarterly – at a much greater cost than the outlets it replaced.

10 For a short time in the 1970s there also appeared *Asher's Guide to Botanical Periodicals* (1974–75, Amsterdam). This was similar in concept, but for the field more complete, than *Current Contents*. Its production, however, came in an era already marked by rapidly rising costs and falling library budgets, and could not be sustained.

11 *Progress in Botany* is no longer as timely as a bibliographic source due to longer subdisciplinary review cycles and, more recently, it has become for new floras less useful due to changes in reviewers. *Kew Record* was available on-line outside the Royal Botanic Gardens only through the Bath Information and Documentation System (BIDS; see below) but as of September 2000 became more generally accessible.

12 See in particular D. L. Hawksworth, P. M. Kirk and S. Dextre Clarke (eds.), 1997. *Biodiversity information: needs and options*. x, 194 pp. Wallingford: CAB International. Of the several papers in this volume, those by V. P. Canhos, G. P. Manfio and D. A. L. Canhos (chapter 11, pp. 147–156) and J. R. Burley, P. R. Scott and A. W. Speedy (chapter 12, pp. 156–171) are perhaps the most apropos.

13 It should nevertheless be emphasized that, though somewhat diminished in more recent years, *Excerpta Botanica*'s network of collaborators around the world gave it certain advantages in covering materials with possibly only a limited circulation. *Kew Record* does not have a comparable system, relying mostly on acquisitions in major libraries in southeastern England.

14 DIALOG is a U.S. commercial information provider (established by Lockheed in 1972); EINS (European Information Network Services) is a consortium of European providers; BIDS (Bath Information and Documentation System) was established in 1991 in the first instance for United Kingdom academic clients.

15 All the developments described in this and preceding paragraphs not unnaturally relate to changes in general in science and technology as well as to social, political, military and other events. In the late nineteenth and early twentieth centuries Germany (and Central Europe) dominated bibliography as well as science in general; moreover, there was still a strong interest in maintaining connectivity among areas of knowledge. The severe disruption of World War I was followed by a time of transition; universality was supplanted by looser international federations as well as increasing specialization and greater national isolation. From World War II the United States

of America became dominant in bibliographic documentation as in most areas of the sciences, with France covering the francophone network through *Bulletin Signalétique* (now PASCAL), the Soviet Union contributing *Referativnyj Žurnal*, and the United Kingdom remaining a key player in certain areas where great historical strength remained (through, for example, the CABI group of abstracting serials, *Zoological Record* – now associated with BIOSIS – and *Kew Record*). Moreover, organismal botany (and zoology), still prominent in the late nineteenth century, had by the late twentieth become relatively small areas within a vastly more diverse, and generally more process-related, array of biological sciences. Only large comprehensive outlets such as BIOSIS – moreover without a history of disruption by war and national division – had a sufficient market as well as technological basis to be self-sustaining.

16 The catalogues (and, latterly, their on-line counterparts) of Flück-Wirth, Koeltz, and Natural History Book Service (NHBS), are the nearest modern counterpart of *Naturae Novitates* (1879–1944), dedicated to new monographic literature in the natural sciences and in its day much sought after.

17 For the benefit of biologists, all the biological entries in *A world bibliography of bibliographies* were separately reprinted in a handy small format as T. Besterman, 1971. *Biological sciences: a bibliography of bibliographies*. 471 pp. Totowa, N.J.: Rowman & Littlefield.

18 This last was reviewed by R. Schmid in *Taxon* **35**: 101–105 (1986). Theses and dissertations contain a wealth of unpublished or only partially published floristic and taxonomic data. Reynolds' work contains 2001 entries, of which the largest source is *Dissertation Abstracts International*.

19 Reviewed by R. Schmid in *ibid.*, **43**: 514 (1994).

20 The *Putevoditel'* of Levin *et al.* contains illuminating graphical tables showing the chronological ranges of many of the sources described in this Appendix. Bourlière's *Éléments*, with some 6300 entries arranged in geographical as well as systematic sections, remains one of the most comprehensive of its kind. Davis and Schmidt's *Practical guide* was noted by R. Schmid in *ibid.*, **46**: 618 (1997); plant sciences are covered therein on pp. 258–319.

21 Davis and Schmidt's *Guide* was noted by R. Schmid in *ibid.*, **48**: 881 (1999). Additional select lists cited in the original edition of this book were *Basic books for the library* by G. P. de Wolf (1970, in *Arnoldia* **30**: 107–113) and *Some botanical reference works* by P. I. Edwards (1971, in *Biol. J. Linn. Soc.* 3: 269–275).

22 A planned book based on these lectures (Phyllis Edwards, personal communication, 1981) did not, however, see publication. Interest in information handling in system-atics has nevertheless continued to be manifest in various ways. As this is written (February 2000) an announcement has been made for a meeting on 'Insect Information: from Linnaeus to the Internet' to be held in London in autumn 2000.

23 Through the 'Association Internationale des Botanistes'.

24 The 'Utrecht group' is defined here in particular as encompassing J. Lanjouw, F. Verdoorn and F. A. Stafleu and their associates there (and in Washington, D.C.).

25 A. L. Takhtajan, Presidential Address; in *Proceedings of the Twelfth International Botanical Congress* (1979, Leningrad: Secretariat of the Twelfth International Botanical Congress), pp. 61–62. In the previous edition of the present *Guide*, I noted that this compared unfavorably with anthropology, then well served by the Human Relations Area Files at New Haven (Connecticut, U.S.A.); it could be argued, however, that in spite of losses during World War II several major libraries, herbaria and archives together already formed a strong 'backbone' for an information system. Communication is being facilitated through such bodies as the Council for Botanical and Horticultural Libraries (in North America) and the Council for Botanical and Horticultural Libraries in Europe.

26 Problems were encountered, for example, with production of what was claimed to be the first botanical bibliography processed directly from a database, the Strandell Catalogue of Linnaeana (see G. H. M. Lawrence and R. W. Kiger in *Svenska Linnésällskap. Årsskr.* (1978, publ. 1979), pp. 276–295). In the end the Catalogue was never published, nor was another major Hunt Institute project, *Bibliographia Huntiana*.

27 Computerization of *Index Kewensis* was also begun in the 1980s with retrospective capture complete by 1990; a first CD-ROM edition appeared in 1992. *Index nominum genericorum* had earlier been partly issued in card form. As for the first edition of the present book, the typescript was produced manually, word-processing not becoming generally available in Papua New Guinea until after its submission in 1980.

28 See, for example, R. Allkin and F. A. Bisby (eds.), 1984. *Databases in systematics*. London: Academic Press (Systematics Association Spec. Vol. 26); and P. Frost-Olsen and L. B. Holm-Nielsen, 1986. *A brief introduction to the AAU-Flora of Ecuador information system*. Aarhus. (Reports from the Botanical Institute, University of Aarhus 14.)

29 F. A. Bisby, 1993. A global plant species information system (GPSIS): 'blue skies design' or tomorrow's workplan? In F. A. Bisby, G. F. Russell and R. J. Pankhurst (eds.), *Designs for a global plant species information system*, pp. 1–6. Oxford: Oxford University Press. (Systematics Association Spec. Vol. 48.)

30 At Delphi, Greece; see Bisby, Russell and Pankhurst, *op. cit.* One consequence was the formation of the International Organization for Plant Information in 1991.

31 For biodiversity information in general these developments are best covered in Hawksworth, Kirk and Dextre Clarke, *op. cit.* This reflected the situation in 1996–97; subsequent developments have largely been incremental.

32 For JSTOR, see Note 2. It appears likely – and should be welcomed – that individual initiatives of societies and other organizations will enable additional journals – such as, for example, *Taxon* – to be retrospectively 'captured' and made available on-line or as CD-ROMs.

33 As already noted, *Kew Record* is fully retrospective to its establishment in 1971. A Web client–server interface is soon to become generally available.

34 A collaborative venture, launched in 1997, of *Index Kewensis*, the *Gray Herbarium index*, and the *Australian plant name index*.

35 Among them are the libraries of the Missouri Botanical Garden and the New York Botanical Garden, the botanical libraries of Harvard University, the U.S. National Agricultural Library, and the libraries of the Natural History Museum (London). More details appear below under **Major library catalogues**.

36 A finer, somewhat historical distinction should perhaps be made among the types of services than merely separating them into abstracts and indices. Within the first-named category are actually two kinds of outlets: the modern abstracting journal, such as *Biological Abstracts*, and the 'recension' journal, such as *Excerpta Botanica*. For the former, abstracts (*Referate*) are nowadays supplied by authors through the journals in which their papers are published. For the latter, articles were summarized by collaborators in the form of a précis (*Recensio*). However, during the first half of the twentieth century this distinction was less marked; indeed, it is probably true to say that the modern abstracting journal has evolved from 'recension' antecedents. An additional class is represented by *Progress in Botany*: the subdiscipline-oriented 'annual review' with literature surveys – for our purposes more useful than similar series made up entirely of commissioned, usually more topical articles.

37 For a commemorative retrospective, see F. A. Stafleu, 1973. Pritzel and his *Thesaurus. Taxon* **22**: 119–130.

38 For an early description, see Atwood, 1911 (Note 5).

39 *BIOSIS Previews* – the advance citations – were initially distributed monthly on tape, with the abstracts themselves in turn also so distributed. The CD-ROM version (quarterly, through SilverPlatter) was inaugurated in 1982 (*BA/RRM* from 1989), and Telnet and World Wide Web links were opened respectively in 1985 and 1993. Individual search profiles were also introduced, with monthly distribution of materials by tape or disk possible

by 1983 for use on microcomputers. From 1927 through 1957, just under 775 000 abstracts were published by *BA*, reaching by the latter year an annual rate of 40 000. The rapid growth of the literature in the 1950s and 1960s made mechanization and, later, computerization imperative. When electronic searching was finally introduced in 1969, *BA* and *Bioresearch Index* respectively covered 135 010 and 85 000 citations. By 1980, the total was 290 000 items per year (165 000 in *BA*; 125 000 in *BA/RRM*); for 1999, it was estimated that 559 000 citations would be covered (350 000 in *BA*; 209 000 in *BA/RRM*). This last total should, however, be compared with the totals of 541 630 for 1993 and 561 106 for 1996. There have been numerous reviews and analyses of *Biological Abstracts*; only a few can be noted here. For zoological systematic literature it was claimed (E. O. Wiley, *Phylogenetic systematics*; 1981, New York) that *BA* is 'much less complete' than *Zoological Record*. A similar claim was made for ornithology by R. Mengel (in T. R. Buckman (ed.), *Bibliography and natural history*, pp. 121–130; 1966, Lawrence, Kans.). Its effectiveness for botanical systematic literature was most recently surveyed by Delendick (1990; see Note 1), though a comparison with *Kew Record* was not essayed. A further critical review, with many additional references, appears on pp. 166–172 of Simon's *Die Bibliographie der Biologie*. In the experience of the author, *BA* was relatively weak in its coverage of independently published works like floras and monographs; this corroborated Blake and Atwood's earlier observations. Indeed, it is probably better for general biology (and biomedicine) than for comparative biology. The first half-century of BIOSIS – a non-profit corporation – has been recorded in W. C. Steere, 1976. *Biological Abstracts/BIOSIS: the first fifty years*. New York: Plenum.

40 *Botanical Abstracts* was initiated by B. E. Livingston of Johns Hopkins University and others as a reaction to the suspension in 1917 in transmission to the United States of *Botanisches Centralblatt*, but was nonetheless continued after the close of World War I. An account of its genesis, and its relationship with the *Centralblatt*, can be found in F. Verdoorn, 1945. Farlow's interest in an international abstracting journal. *Farlowia* **2**(1): 71–82.

41 The *Zentralblatt* was by Bay (Note 6) considered to be the most timely, with its *Neue Literatur* section being of the greatest value. Its nearest effective modern equivalents (at least for taxonomic literature) have been the mimeographed *Current Awareness List* issued by the Library of the Royal Botanic Gardens, Kew, on a monthly basis from 1973 through 1985 and, since 1986, *Kew Record* as a quarterly. Although after its reorganization in 1902 under the newly formed 'Association Internationale des Botanistes', publication was transferred to the Netherlands, close links between that country and Germany in the years

before 1914 led to the return of the *Zentralblatt* in 1906 to a German publisher. This was to have unfortunate consequences in the difficult aftermath of World War I, as well as during hostilities when severe restrictions upon the serial's entry into the U.S.A. were imposed. It was the latter which finally unleashed latent interest in an American abstracting journal and in 1918 *Botanical Abstracts* came into being. In 1919 the Association (also responsible for the International Botanical Congresses) collapsed. Not until 1922 did it reappear, again under new auspices – this time the Deutsche Botanische Gesellschaft. However, the *Zentralblatt* of the interwar years was comparatively less international in scope, with a lower level of coverage than in the 1900s, and at the end of World War II it ceased publication (a partial successor was *Excerpta Botanica* (see below), also published by Fischer). Editorial policies and style of the *Zentralblatt* were reviewed in memorials by E. Bornet and J. P. Lotsy in *Bot. Zentralbl.* 23(89): 1–7 (1902). Some historical information appears in F. Verdoorn, 1945. *Plants and plant science in Latin America* (ed. F. Verdoorn), pp. xv–xxii. Waltham, Mass., as well as in the same author's article in *Farlowia* cited in Note 40. A more recent retrospective review is that by R. Schmid in *Taxon* 36: 101–105 (1987).

42 The continuous arrangement by author in each main series has been much criticized. Nevertheless, the work is considered as of inestimable value for the historian of science; it is certainly useful for tracing papers by nineteenth-century authors, particularly as the contemporary standard bibliographies all deal with independently published works. Indeed, the existence of the *Catalogue* was specifically acknowledged by Pritzel in the introduction to the second edition of his *Thesaurus*, and references to it appear throughout. The work was initiated by Act of Parliament in 1864 and continued until 1925 on completion of coverage through 1900. It was by then clear that available resources could no longer cope with a continuation of coverage beyond that date.

43 The journal, in its time well conceived but appearing increasingly old-fashioned in the face of changes in information technology in the 1980s and especially the 1990s, was discontinued in 1998 consequent to a fall in subscriptions to uneconomic levels. It moreover never was adapted to electronic dissemination. Its passing may, however, create a gap not entirely bridged by other, differently oriented services. It was reviewed by R. Schmid in *Taxon* 38: 70 (1989), with a rejoinder by K. Fægri (one of the editors) in *ibid.*, 39: 231 (1990). Among Schmid's criticisms were the tardiness of indices, a lack of keyword indices, brevity of the summaries, and high cost. To him it was reminiscent of *Botanisches Zentralblatt* and *Just's Botanischer Jahresbericht*. Fægri saw this as too severe a criticism, but admitted that it was Europe-centered and

that the apparent high cost was due to the fall of the dollar relative to the mark (a process which would continue well into the mid-1990s).

44 Jäger's reviews of floristics, with their associated bibliographies, have from 1980 through 1995 appeared at intervals of two or more years. A treatment of plant biodiversity by Porembski and Barthlott in vol. 61 (2000) is different in scope. A general review of *Progress in Botany* by R. Schmid appears in *Taxon* 36: 301–302 (1987); he notes that it evidently had been poorly known or underused in the U.S.A.

45 The background and organization for the ICSL, and its editorial and bureaux system, are given as prefaces to each annual volume, along with the classification of contents (which followed a code peculiar to the work). B. Daydon Jackson was disciplinary referee in botany for the ICSL board of control – the latter an international council established under the patronage of the Royal Society of London and directed by H. Foster Morley from a central office in London. Affiliated bureaux had been established in 32 countries; in Belgium its work was handled by La Fontaine and Otlet's Office International de Bibliographie. World War I was ultimately fatal to the enterprise. Publication was suspended in 1921 after the 14th annual issue had been completed (covering 1914). A conference was held in Brussels in 1922 regarding its future, but financial difficulties in the end prevented its resumption. Costs had become too high when expressed in terms of the 'depreciated currency of many . . . countries'. The *Annual Reports of the Smithsonian Institution* – home to the U.S. national office – record further attempts in the 1920s at revival, but these were ultimately unsuccessful (the U.S. office being disbanded in 1933). Bay (*op. cit.*; see Note 6) had regarded the botanical section as less satisfactory than either *Botanisches Centralblatt* or *Just's Botanischer Jahresbericht*.

46 From the beginning, much use was made by its compilers of the Botanical Museum and other special libraries in Berlin; from Jg. 11 (1885), when it received its permanent title upon transfer of the editorship from Leopold Just to Emil Koehne, its direction was linked with the Museum itself. In 1906 the editorship passed to the independent botanist Friedrich Fedde but a close association continued. The *Jahresbericht* was thus inevitably forced into suspension upon the destruction of the Museum library in March 1943, a year after Fedde's death. A successor to *Repertorium der periodischen botanischen Literatur* (produced from 1864 to 1873 as *Beiblätter* to the Regensburg journal *Flora*), the *Jahresbericht* had been for six decades or so from its foundation into the early 1930s the most complete and well-indexed guide to botanical literature available. Early in the twentieth century it was rated by W. G. Farlow at Harvard University (see Verdoorn in

Farlowia, in Note 40 under *Botanical Abstracts*) as being of better value than (although not as timely as) *Botanisches Centralblatt*; however, a contrary view had earlier been taken by J. C. Bay on the grounds of value for money (Bay, *op. cit.*). Over the years there was a steady reduction in the length and discursiveness of reviews, as well as some changes in organization of contents; moreover, by the late 1920s coverage of some general areas such as genetics, physiology and biochemistry had fallen away (most likely in favor of the more timely coverage of *Biological Abstracts* and *Berichte über die gesamte Biologie*, both formed in 1926; for the latter, see Simon, 1977, pp. 160–162). Some later section reports in the *Jahresbericht* were effectively classified lists of literature resembling the modern *Kew Record*; an example is the 1935 survey by K. Krause for higher-plant morphology and systematics, the last of its kind ever to appear (during 1943–44). The growing 'lag' noted by Simon – already seven or more years by 1935 – and the increasingly brief reviews may well have been at least partial inducements for Richard Wettstein to establish *Fortschritte der Botanik* (now *Progress in Botany*). A modern review of the *Jahresbericht* is given by R. Schmid in *Taxon* 36: 101–105 (1987). From 1906 Fedde introduced as an *Anhang* to the *Jahresbericht* the journal *Repertorium specierum novarum regni vegetabilis* (now known as *Feddes Repertorium*), for the publication (or republication) of taxonomic novelties – much as Walpers had attempted in the nineteenth century by way of supplementing the Candollean *Prodromus*.

47 Sources for *Kew Record* encompass the libraries of Kew Gardens, the Natural History Museum (London), the Oxford Forestry Institute (since the mid-1980s a part of the Plant Sciences Department of Oxford University) and the mycological branch of CABI Bioscience (at Egham, Surrey). For a time in the 1970s and 1980s, *Kew Record* showed a serious lag (3–4 years or more); moreover, the quality of the geographical classification changed for the worse consequent to a production change made starting with the 1976 number. The change

to a quarterly took place with the introduction of compilation into a computer-based textbase; earlier entries were retrospectively converted, this work being completed in 1993–94. A Web-based client interface was introduced in 1999 with a view towards more open general accessibility in the future. It may be worthy of note that as of the late 1970s, less than half of the some 3000 periodical titles scanned were also being covered by *Biological Abstracts*. 175 000 items are now to hand.

48 For a retrospective review, see Stafleu in *Taxon* 33: 152 (1984). A reissue on microfiche by Inter-Documentation Company appeared around the same time. The early years of *Naturae Novitates* overlapped another repertorium of new books and monographs, *Bibliotheca historico-naturalis, physiochemica et mathematica* (37 vols., 1851–87, Göttingen; published semi-annually). A more recent serial index of new monographic works – also covering botany – was *Recent publications in natural history* (1983–92, New York), issued through the American Museum of Natural History.

49 For a guide to *Referativnyj Žurnal*, see E. J. Copley, 1975. *A guide to* Referativnyj Žurnal. 3rd edn. 26 pp. London: British Library. *Novye Knigi SSSR* (available in English translation from 1964 for the biological sciences) may also be consulted for new botanical books from Russia and neighboring states. From 1994, *Referativnyj Žurnal* has also been available on microfiche.

50 It should be noted, however, that among major botanical libraries retrospective databasing has been to date less general in Europe than in North America. Important collections with full retrospective coverage in addition to those covered here include the Missouri Botanical Garden Library and the New York Botanical Garden Library. In addition, national libraries or collection networks, such as LOCIS (the Library of Congress Online Catalog), OPAC97 (British Library) or LIBRIS (the national network in Sweden based on the Swedish Royal Library), should be consulted for rare or unusual items, or for general bibliographic control.

Appendix B

• • •

Abbreviations of serials cited

Introductory remarks

The following lexicon lists abbreviations of serials and periodicals cited in the bibliographic entries along with the introductory digests in Part II of this work.

The standard for abbreviations largely follows that in *Botanico-periodicum-huntianum* (*B-P-H*), edited by G. H. M. Lawrence et al. (1968, Pittsburgh: Hunt Library), and its supplement, *Botanico-periodicum-huntianum supplementum* (*B-P-H/S*), edited by G. Büchheim *et al.* (1991, Pittsburgh: Hunt Institute for Botanical Documentation). Together the two volumes represent the most comprehensive currently available reference specific to the field. The original *B-P-H* was, however, not without errors and moreover in terms of the present work was distinctly incomplete, even accounting for the lapse of time from its publication through 1980. It had in fact originally been conceived and developed in relation to the retrospective *Bibliographia huntiana* project (a definitive critical bibliography of botanical literature from 1730 through 1840) as well as – to some extent – *Taxonomic Literature* by Frans Stafleu (1967, Zug). [For *B-P-H/S* coverage was greatly expanded, utilizing a variety of additional sources; there remain, nevertheless, a number of omissions.]

For the original version of these listings, therefore, other sources had to be consulted. These included: *British union-catalogue of periodicals* (BUCOP) and its *Supplement to 1960* (5 vols., 1955–62, London); BUCOP *New serial titles 1950–70, 1971–5*, and annual volumes from 1976 through 1980 (as well as *Subject guide to* New serial titles 1950–70); *Scientific serials in Australian libraries* (3 vols. in loose-leaf, 1959–61, with added and revised pages to 1979); *Union list of serials in New Zealand libraries* (2nd edn. in 3 vols., 1964, Wellington); *Union list of serials in the libraries of the United States and Canada* (3rd edn. in 5 vols., 1965, New York), *Serial publications of foreign governments 1815–1931* edited by W. Gregory (1932, New York; reprinted 1973, Millwood, N.Y.); and *World list of scientific periodicals* (4th edn. (1900–60) in 3 vols., 1963–65, London) and its supplements. Use was also made of some specialized periodical listings: *An index of state geological survey publications issued in series* by J. B. Corbin (1965, New York); *List of periodical publications in the Library, Royal Botanic Gardens, Kew* (1978, London); *Serial publications in the British Museum (Natural History) Library* (3rd edn. in 3 vols., 1980, London); and the 1979 and 1980 editions of the annual *Serial sources for the BIOSIS Data Base* (Philadelphia). The valuable lexica in Blake and Atwood's *Geographical guide to floras of the world*, Merrill and Walker's *A bibliography of Eastern Asiatic botany* (1938) and its *Supplement* (1960; for both see **860–80**), and Rehder's *Bibliography of cultivated trees and shrubs hardy in the cooler regions of the Northern Hemisphere* (1949, Jamaica Plain) were also consulted.

For the present edition use was made not only of *B-P-H/S* but also of the most recent version of the Natural History Museum (London) serials catalogue, *Serial titles held in the Department of Library and Information Services* (4th edn. in 5 vols., 1995, London); the 1999 edition of the annual *Serials in the British Library* (London); the 1999 edition of the annual *Current Serials Received* from the British Library Lending Division (Boston Spa); and recent issues of what is now called *BIOSIS Serial Sources* (Philadelphia). Other serials lists utilized included *Bibliothèque Centrale du Muséum National d'Histoire Naturelle (Paris): liste des periodiques en cours* (1989, Paris); *Catalogue des périodiques de la bibliothèque des Conservatoire et Jardin Botaniques de la Ville de Genève* by H. M. Burdet *et al.* (1980, Geneva) and its *Supplément 1980–1995* by F. Maiullari and H. M. Burdet (1995); and *Union list of scientific serials in Canadian libraries* (1977, Ottawa; 2nd edn., 1981). Two additional compilations worthy of mention here but now largely of historical interest are *A catalogue of scientific and technical periodicals, 1665–1895* by H. C. Bolton (1897, Washington, D.C.; reprinted 1965, New York), and *Catálogo universal de revistas de ciencias exactas, físicas y naturales* by E. Sparn (1920, Córdoba, Argentina, as *Revista de la Universidad Nacional de Córdoba* 6/7) and its *Supplement* (1922), the two latter covering journals through 1919. Also helpful was *Catalogue of periodical literature in the Lloyd Library, Cincinnati, Ohio* (1911, in their *Bibliographical*

Contributions). Several dates were moreover checked in the Kew Library and at the Natural History Museum.

It has to be said, however, that within the last 20 years these printed sources have to a considerable extent been supplanted by a new development: on-line searching over the Internet, initially using Telnet and, since the mid-1990s, also the World Wide Web. Full advantage was taken of this opportunity, although locating resources and building up a list of 'bookmarks' takes time. Printed serials catalogues do, however, retain their value in being for many purposes faster to use and – like that of the Natural History Museum Library – it is to be hoped that such will continue to be issued from time to time.

Only indefinite serials (and periodicals) as conventionally defined are accounted for here.[1] Expedition reports are not included, nor are multi-volumed symposium and congress proceedings; these are spelt out in full in the text. Also omitted are serially produced floras, although these are sometimes treated as (and a few, such as *Flora Neotropica*, are) indefinite serials. Conversely, some publications are here treated, or also referred to, as serials although, being more usually catalogued as monographs, they have not appeared in serial indices. More analysis of these – beyond our scope here – would be desirable.

The dates of commencement (and termination) have been given as far as available resources reasonably allow. In those instances for which no data have been available, the volume and year of publication concerned are given instead. Commentary has been given where deemed necessary; this includes notes on breaks in publication, title changes or successor serials.

This lexicon is not meant to be a scholarly work in itself, and the user is at all times advised to refer also to the various aforementioned standard works. It may be noted here, however, that a modern reference giving collations and 'phylogenies' for a large range of botanical and 'parabotanical' serials remains a desideratum. It would fulfill a wish of J. Christian Bay in remarks made in the introduction to his *Bibliographies of botany* (here described in Appendix A) as well as the example set by Jonas Dryander in his now two centuries-old *Catalogus bibliothecae historico-naturalis Josephi Banks* (1796–1800). Even in the massive second edition of *Taxonomic Literature* by Stafleu and Cowan (and its supplemental volumes) only a few serials are fully collated.

Note

1 W. H. Huff, 1967. Periodicals. In R. B. Downs and F. B. Jenkins (eds.), *Bibliography: current state and future trends*, pp. 62–83. Urbana: University of Illinois Press. See also D. A. Kronick, 1962. *A history of scientific and technical periodicals: the origins and development of the scientific and technological press*. New York: Scarecrow Press; and B. Houghton, 1975. *Scientific periodicals: their historical development, characteristics and control*. London: Clive Bingley.

Lexicon

Note: An asterisk (*) indicates a serial new for this edition.

*AAU Rep.: AAU reports. *See* Reports from the botanical institute of Aarhus University.

AIBS Education Review: American institute of biological sciences education review. Arlington, Va., 1972– .

*ANARE Res. Notes: Australian National Antarctic Research Expeditions, research notes. Kingston, Tasmania, 1982– .

Abh. Auslandsk. (Hamburg), Reihe C, Naturwiss.: Abhandlungen aus dem Gebiet der Auslandskunde. Reihe C: Naturwissenschaften. Hamburg, 1920– . [Hamburg University.]

Abh. Königl. Akad. Wiss. Berlin: Abhandlungen der Königlichen Akademie der Wissenschaften in Berlin, [Physikalischen-Mathematischen Klasse]. Berlin, 1815–1900.

Abh. Königl. Ges. Wiss. Göttingen: Abhandlungen der Königlichen Gesellschaft der Wissenschaften zu Göttingen. Göttingen, 1843–92.

*Abh. Ges. Wiss. Göttingen, Math.-Phys. Kl.: Abhandlungen der [Königlichen] Gesellschaft der Wissenschaften zu Göttingen, Mathematisch-Physikalische Klasse, Neue Folge. Göttingen, 1893– . [*Previously entitled* Commetationes societatis regiae scientiarum gottingensis (1778–1837); Abhandlungen der Königlichen Gesellschaft der Wissenschaften zu Göttingen (1843–92).]

Acad. Ci. Cuba, Ser. Biol.: Academia de Ciencias de Cuba, seria Biologia. Havana, 1967– .

Acta Amazonica: Acta amazonica. Manaus, Brazil, 1972– .

Acta Borealia, A, Sci.: Acta Borealia, A: Scientia. Tromsø, Norway, 1951– .

*Acta Bot. Austro-Sinica: Acta botanica austro-sinica. Guangzhou, 1983– .

*Acta Bot. Gall.: Acta botanica gallica. Paris, 1993– . [*Successor* to Bulletin de la société botanique de France, *without break in volumination*.]

*Acta Bot. Malacit.: Acta botanica malicitana. Málaga, 1975– .

*Acta Bot. Méx.: Acta botánica méxicana. Pátzcuaro, 1988– .

Acta Bot. Neerl.: Acta botanica neerlandica. Amsterdam, 1952–98. [*Subsequently merged with* Botanica Acta (*Germany*) *and from 1999 succeeded by* Plant biology.]

Acta Bot. Venez.: Acta botanica venezuelica. Caracas, 1965– .

Acta Horti Gothob.: Acta horti gothoburgensis. Göteborg, Sweden, 1924–66.

Acta Horti Petrop.: Acta horti petropolitani. *See* Trudy Imperatorskago S.-Peterburgskago botaničeskago sada.

Acta Phytotax. Barcinon.: Acta phytotaxonomica barcinonensia. Barcelona, 1968– .

Acta Phytotax. Geobot.: Acta phytotaxonomica et geobotanica. Kyoto, Japan, 1932– .

Acta Phytotax. Sin.: Acta phytotaxonomica sinica. Peking, 1951– .

Acta Soc. Fauna Fl. Fenn.: Acta societatis pro fauna et flora fennica. Helsinki, 1875– .

Acta Univ. Lund.: Acta universitatis lundensis [Lunds universitets årsskrift]. Lund, 1864–1904; N.S., 1905– . [*Inclusive of* Kongl. Svenska fysiografiska sällskapets handlingar.]

Adansonia, sér. 2: Adansonia, série 2. Paris, 1961–80.

Agric. Handb. USDA: Agriculture Handbooks, United States Department of Agriculture. Washington, 1950– .

Akad. Wiss. Wien, Math.-Naturwiss. Kl., Denkschr.: Akademie der Wissenschaften in Wien. Mathematisch-Naturwissenschaftliche Klasse. Denkschriften. Vienna, 1919–51.

Aliso: [El] Aliso. Claremont, Calif., 1948– .

*Allertonia: Allertonia. Lawai, Kauai, Hawaii, 1975– .

*Amer. Fern J.: American fern journal. Port Richmond, N.Y., etc., 1911– .

*Amer. J. Bot.: American journal of botany. Lancaster, Pa., etc., 1914– .

Amer. Midl. Naturalist: American midland naturalist. South Bend, Ind., 1909– .

Anais Junta Invest. Colon.: Anais do Junta de Investigações Colonais. *See next entry*.

Anais Junta Invest. Ultram.: Anais do Junta de Investigações do Ultramar. Lisbon, 1946–60.

Anales Jard. Bot. Madrid: Anales del jardin botánico de Madrid. Madrid, 1941– . [*From 1950–70 entitled* Anales del instituto botánico 'A.J. Cavanilles'.]

Anales Mus. Nac. Hist. Nat. Buenos Aires: Anales del museo nacional de historia natural de Buenos Aires. Buenos Aires, 1864–1931.

Anales Mus. Nac. Montevideo: Anales del museo nacional de Montevideo. Montevideo, 1894–1909; ser. 2, 1925– .

*Anales R. Acad. Farm. Madrid: Anales de la real academia de farmacia de Madrid. Madrid, 1940– . [*Preceded by* Anales de la academia nacional de farmacia de Madrid. Madrid, 1932–36.]

*Anales Soc. Cient. Argent.: Anales de la Sociedad científica argentina. Buenos Aires, 1876– .

Anales Univ. Chile: Anales de la universidad de Chile. Santiago, 1843– .

Ann. Arid Zone: Annals of arid zone. Jodhpur, Rajasthan, India, 1962– .

Ann. Bot. (London): Annals of botany. London, 1887– . [*Two successive series.*]

Ann. Bot. Gard. Buitenzorg: Annals of the botanic gardens, Buitenzorg. Buitenzorg (Bogor), Indonesia, 1941–52.

Ann. Carnegie Mus.: Annals of the Carnegie Museum. Lancaster, Pa., etc., 1901– .

Ann. Inst. Bot.-Géol. [*later* Mus.] Colon. Marseille: Annales de l'institut botanico-géologique colonial de Marseille (*later* Annales du musée colonial de Marseille). Paris, etc., 1893–1954. [*Title varies. Several successive series.*]

Ann. Jard. Bot. Buitenzorg: Annales du jardin botanique de Buitenzorg. Batavia (Jakarta), Leiden, 1876–1940. [*Succeeded* by Annales bogorienses.]

Ann. Kentucky [Soc.] Nat. Hist.: Annals of [the] Kentucky [society of] natural history. Louisville, Ky., 1941– . [Vol. 2, 1968.]

Ann. Missouri Bot. Gard.: Annals of the Missouri botanical garden. St. Louis, 1914– .

Ann. Mus. Goulandris: Επετηρις μουσειον Γουλανδρη/Annales musei Goulandris. Kifissia, Greece, 1973– .

Ann. Mus. Roy. Afrique Centr., Série-en-8°, Sci. Econ.: Annales de la musée royale d'Afrique centrale, série-en-octavo: sciences économiques (*initially as* Annales . . . série-en-octavo: sciences historiques et économiques). Tervuren, Belgium, 1947– .

*Ann. Naturh. Mus. Wien: Annalen des [k.k.] naturhistorischen [Hof]museums in Wien. Vienna, 1886– . [*Present title from 1918.*]

Ann. New York Acad. Sci.: Annals of the New York academy of sciences. New York, 1879– .

Ann. Roy. Bot. Gard. Peradeniya: Annals of the royal botanic gardens. Peradeniya (*later* Colombo), Ceylon, 1901–32. [*From vol. 9 (1924) alternatively titled (as one of several concurrent series) Ceylon journal of science, Sect. A, Botany. After 1932 wholly under that title.*]

Ann. Sci. Nat., Bot.: Annales des sciences naturelles, botanique. Paris, 1824–1991. [*Several successive series, with* Botanique *added to title from sér. 2 (1833); closing as sér. 13,* Botanique et biologie végétale.]

Ann. Soc. Sci. Nat. Charente-Infér.: Annales de la société de sciences naturelles de Charente-Inférieur. La Rochelle, France, 1854– .

Ann. Transvaal Mus.: Annals of the Transvaal museum. Pretoria, 1908– .

*Ann. Wiener Mus. Naturgesch.: Annalen des Wiener Museums der Naturgeschichte. Vienna, 1835–40.

Annual Rep. Div. Plant Industry, CSIRO, Australia: Annual reports, Division of plant industry, Commonwealth scientific and industrial research organisation of Australia. Canberra, 1960/61– .

Annual Rep. Geol. Surv. Arkansas: Annual reports of the Geological Survey of Arkansas. Little Rock, Ark., 1887–92.

Annual Rep. Geol. & Nat. Hist. Surv. Minnesota: Annual reports of the geological and natural history survey of Minnesota. St. Paul, Minn., 1873–99.

Annual Rep. Michigan Acad. Sci.: Annual reports of the Michigan academy of science, arts and letters. Lansing, Mich., 1894– .

Annual Rep. Missouri Bot. Gard.: Annual reports of the Missouri botanical garden. St. Louis, 1889–1912.

Annual Rep. New Jersey State Mus.: Annual reports of the New Jersey state museum. Trenton. [Reports for 1910, 1911.]

Annuario Reale Ist. Bot. Roma: Annuario del Reale istituto botanico di Roma. Rome, 1884–1907.

*Anuário Soc. Brot.: Anuário de Sociedade broteriana. Coimbra, Portugal, 1935– .

Arch. Bot. Biogeogr. Ital. [*or* Arch. Bot. (Forlì)]: Archivio botanico e biogeografico italiano. Forlì, Emilia-Romagna, Italy, 1925– . [*Title varies. Initially as* Archivio botanico per la sistematica, fitogeografia e genetica . . .; *from 1935–55 as* Archivio botanico.]

*Arch. Hydrobiol., Beih. Ergeb. Limnol.: Archiv für Hydrobiologie: Beihefte (Ergebnisse der Limnologie/Advances in limnology). Stuttgart, 1964– . [*Title varies; subtitle now exclusively in English.*]

Arch. Hydrobiol., Suppl.: Archiv für Hydrobiologie: Supplementbände. Stuttgart, 1911– .

*Arch. Nat. Hist.: Archives of natural history. London, 1981– . (*Previously* Journal of the society for the bibliography of natural history.)

*Arch. Naturk. Liv. Est. & Kurlands, II: Archiv für die Naturkunde Liv-, Est- und Kurland. Serie 2, Biologische Naturkunde. Dorpat (Tartu), 1859–1905. [*Continued as* Eesti loodusteaduse archiv. Tartu, 1931–61.]

Arch. Nauk Biol. Towarz. Nauk. Warszawsk.: Archiwum nauk biologicznych towarzystwa naukowego warszawskiego. Warsaw, 1921– .

Arch. Rech. Agron. Cambodge Laos Viêtnam: Archives des recherches agronomiques au Cambodge, au Laos et au Viêtnam. Saigon, 1951– .

*Arctica: Arctica. Leningrad, 1933–37. [*Succeeded by* Problemy arktiki.]

*Arct. & Alp. Res.: Arctic and alpine research. Boulder, Colo., 1969– .

Ark. Bot.: Arkiv für botanik. Stockholm, 1903–49; N.S., 1950– .

Arnoldia: Arnoldia. Jamaica Plain, Mass., 1941– .

Arq. Jard. Bot. Rio de Janeiro: Arquivos do jardim botânico do Rio de Janeiro. Rio de Janeiro, 1915– .

Arq. Mus. Nac. Rio de Janeiro: Arquivos [*initially* Archivos] do museu nacional do Rio de Janeiro. Rio de Janeiro, 1876– . [*Title varies.*]

*Arq. Serv. Florest.: Arquivos do servicio florestal. Rio de Janeiro, 1939– .

Atoll Res. Bull.: Atoll research bulletin. Washington, 1951– .

*Atti Accad. Naz. Lincei, Rendiconti Cl. Sci. Fis. Mat. Nat.: Atti della accademia nazionale dei lincei. Rendiconti, classe di scienze fisiche, matematiche e naturali. Serie 8. Rome, 1948– .

Atti Ist. Bot. 'Giovanni Briosi' (R. Univ. Pavia): Atti dell'istituto botanico 'Giovanni Briosi' e laboratorio crittogamico italiano della Reale università di Pavia. *See next entry.*

Atti Ist. Bot. Lab. Crittogam. Univ. Pavia: Atti dell'istituto botanico della [Reale] università di Pavia, [Reale] laboratorio crittogamico. Ser. I *et seq.*; Suppl. A–J (to ser. V). Milan, etc., 1874–78; 1888– (Suppl. A–J, Forlì, 1944–55). [*Ser. I* as Archivio triennale del laboratorio di botanica crittogamica; *Ser. II–VI as above with slight variations, notably as in the preceding entry.*]

*Atti Mus. Civico Storia Nat. Trieste: Atti del museo civico di storia naturale di Trieste. Trieste, 1852– .

Atti Soc. Acclim. Agric. Sicilia: (Giornale ed) Atti della società di acclimazione ed agricoltura in Sicilia. Palermo, 1860–91.

Austral. Conservation Found., Spec. Publ.: Australian conservation foundation, special publications. Canberra, 1968– .

Austral. J. Sci.: Australian journal of science. Sydney, 1938–70. [*Succeeded by* Search.]

Austral. Mus. Mem.: Australian museum memoirs. Sydney, 1851– .

Austral. Syst. Bot. Soc. Newsletter: Australian systematic botany society newsletter. Brisbane, etc., 1973– .

BSBI Conf. Rep.: Botanical society of the British Isles: conference reports. London, etc., 1949– .

*Bartonia: Bartonia. Philadelphia, 1908– .

Beih. Bot. Centralbl.: Beihefte zum Botanischen Centralblatt. Kassel, 1891–1944.

Beih. Nova Hedwigia: Beihefte zur Nova Hedwigia. Weinheim, Germany, etc., 1962– .

Ber. Deutsch. Bot. Ges.: Berichte der Deutschen Botanischen Gesellschaft. Berlin, 1883–1987. [*Continued as* Botanica acta *without change in volumination.*]

Bergens Mus. Skr.: Bergens museums skrifter. Bergen, 1878–1943.

Bernice P. Bishop Mus. Bull.: Bernice P. Bishop museum bulletin. Honolulu, Hawaii, 1922– .

Bibliogr. Bull. USDA: Bibliographical bulletin, United States Department of Agriculture. Washington, 1943–54.

Biblioth. Bot.: Bibliotheca botanica. Kassel, 1886– .

Biol. Conservation: Biological conservation. Barking, Essex, 1968– .

Biol. Sci. Ser. Western Illinois Univ.: Biological sciences series, Western Illinois University. Macomb, Ill. [Vol. 10, 1972.]

Biologia (Lahore): Biologia. Lahore, Pakistan, 1955– .

BioScience: BioScience. Washington, 1964– .

Biotropica: Biotropica. Pullman, Washington State, etc., 1969– .

*Bjull. Moskovsk. Obšč. Isp. Prir., [Otd. Biol.]: Bjulleten' Moskovskogo obščestva ispytatelej prirody, otdel biologičeskij. Moscow, 1829–86; N.S., 1887– . [*Title originally in French as* Bulletin de la Société Impériale des Naturalistes de Moscou; *present title (and subdivision) from 1922.*]

Blumea: Blumea. Leiden, Holland, 1934– .

*Blumea, Suppl.: Blumea, supplementary series. Leiden, Holland, 1937– .

*Bocconea: Bocconea. Palermo, 1991– .

*Boissiera: Boissiera. Geneva, 1936– .

*Bol. Agríc. Pecuár. Moçambique: Boletim agrícola e pecuária, Colónia de Moçambique. Lourenço Marques, 1928(–33).

Bol. Bot. Latinoamer.: Boletín botánico latinoamericano. Bogotá, 1978– .

*Bol. Bot. Univ. São Paulo: Boletim de botânica, universidade de São Paulo. São Paulo, Brazil, 1973– .

Bol. Inst. Centr. Biociências (Porto Alegre): Boletim do instituto central de biociências (*initially as* Boletim do instituto de ciências naturais) da universidade do Rio Grande do Sul. Porto Alegre, Rio Grande do Sul, 1955– .

Bol. Inst. Ci. Nat. Univ. Rio Grande do Sul: Boletim do instituto de ciências naturais da universidade do Rio Grande do Sul. *See* Boletim do instituto central de biociências da universidade do Rio Grande do Sul.

*Bol. Inst. Vasco da Gama: Boletim do instituto 'Vasco da Gama'. Nova Goa, etc., 1926–61. [*Succeeded by* Boletim do Instituto 'Menezes Bragança'.]

Bol. Minist. Agric. [Brazil]: Boletim do ministerio de agricultura. Rio de Janeiro, 1912– .

Bol. Minist. Agric., Serv. Florest. [Brazil]: Boletim do ministerio de agricultura, serviço florestal. Rio de Janeiro, 1929–31; 1956– .

Bol. Mus. Munic. Funchal: Boletim do museu municipal do Funchal. Funchal, Madeira, 1945– .

*Bol. Mus. Munic. Funchal, Suppl.: Boletim do museu municipal do Funchal. Supplemento. Funchal, Madeira, 1990– . [Vol. 1, 1990.]

Bol. Mus. Nac. Hist. Nat. (Santiago): Boletín del museo nacional de historia natural. Santiago, 1937– .

Bol. Mus. Nac. Rio de Janeiro: Boletim do museu nacional de Rio de Janeiro. Rio de Janeiro, 1923–41. [*Continued in several disciplinarily distinct 'new series', including* Botânica *from 1944.*]

*Bol. Pro-Cult. Reg.: Boletín de Pro-cultura regional, S.C.L. Mazatlán, Sinaloa, Mexico, 1929– .

*Bol. Soc. Argent. Bot.: Boletín de la Sociedad argentina de botánica. La Plata, 1945– .

*Bol. Soc. Biol. Concepción: Boletín de la Sociedad de biología de Concepción. Concepción, Chile, 1927– .

Bol. Soc. Brot.: Boletim da Sociedade broteriana. Coimbra, Portugal, 1880–1920; sér. II, 1922– .

*Boll. Accad. Gioenia Sci. Nat. Catania: Bollettino delle sedute accademia gioenia di scienze naturali. Catania, 1888– . [*Three successive series.*]

Bot. Arch.: Botanisches Archiv. Königsberg, Berlin, 1922–44.

Bot. Bull. (Lae): Botany bulletin, Department of forests, Papua New Guinea. Lae, Papua New Guinea, 1969– .

*Bot. Chron.: Βοτανικα χρονικα/Botanika chronika. Patras, Greece, 1981– .

*Bot. Helv.: Botanica helvetica. Basel/Geneva, 1891– . [*Present title from 1981; previously entitled* Berichte der Schweizerischen Botanischen Gesellschaft/Bulletin de la Société Botanique de Suisse.]

Bot. J. Linn. Soc.: Botanical journal of the Linnean society of London. London, 1969– . [*Previously entitled* Journal of the Linnean society, botany.]

Bot. Jahrb. Syst.: Botanische Jahrbücher für Systematik, Pflanzengeschichte und Pflanzengeographie. Leipzig (*later* Stuttgart), 1881– .

*Bot. Közlem.: Botanikai (*initially* Növénytani) közlemények. Budapest, 1902– .

Bot. Mag. (Tokyo): Botanical magazine. Tokyo, 1887– .

Bot. Mater. Gerb. Bot. Inst. Komarova Akad. Nauk SSSR: Botaničeskie materialy Gerbarija Botaničeskogo instituti imeni V. L. Komarova Akademii nauk SSSR. (Notulae systematicae ex herbario instituti botanici nomine V. L. Komarovi academiae scientiarum URSS.) Leningrad, 1919–26; 1931–63. [*Title varies. Succeeded by* Novosti sistematiki vysšikh rastenij.]

Bot. Notis.: Botaniska notiser. Lund, Sweden, 1841–1980. [*Succeeded by* Nordic journal of botany.]

*Bot. Tidsskr.: Botanisk tidsskrift. Copenhagen, 1866–1979. [*Succeeded by* Nordic journal of botany.]

*Bot. Zeit. (Berlin): Botanische zeitung. Berlin (*later* Leipzig), 1843–1892; 2. Abteilung, 1893–1910.

*Bot. Žurn. RBO: Botaničeskij Žurnal Russkago Botaničeskago Obščestva. Petrograd (*currently* Moscow/St. Petersburg), 1916– . (*Initially* Žurnal Russkago Botaničeskago Obščestva pri Imperatorskoj Akademii nauk, Petrograd; *title altered from 1917 but from 1924 to 1992 as* Botaničeskij Žurnal SSSR).

Bot. Žurn. SSSR: Botaničeskij Žurnal SSSR. *See* Botaničeskij Žurnal Russkago Botaničeskago Obščestva.

*Bradea: Bradea. Rio de Janeiro, 1969– .

Brittonia: Brittonia. New York, 1931– .

Brotéria (*later* Brotéria, Sér. Bot.): Brotéria. Lisbon, 1902–31.

Bul. Grăd. Muz. Bot. Univ. Cluj: Buletinul grădinii botanice i al muzeului botanic de la universitatea din Cluj. Cluj-Napoca, Romania, 1926–48.

Bull. Acad. Int. Géogr. Bot.: Bulletin de l'académie internationale de géographie botanique. Le Mans, France, 1899–1919.

*Bull. Agric. Exp. Sta. Puerto Rico (Río Piedras): Bulletin of the agricultural experiment station, Puerto Rico. Río Piedras, P.R., 1911– .

Bull. Agric. Fish. Dept. [Hong Kong]: Bulletin, Agriculture and fisheries department. Hong Kong, 1974– .

Bull. Biogeogr. Soc. Japan: Bulletin of the biogeographical society of Japan. Tokyo, 1929– .

Bull. Biol. Sci. Washington: Bulletin of the biological society of Washington. Washington, D.C., 1918– . [Vol. 2, 1972.]

Bull. Bot. Soc. Bengal: Bulletin of the botanical society of Bengal. Calcutta, 1947– .

Bull. Bot. Surv. India: Bulletin of the botanical survey of India. Calcutta, 1959– .

Bull. Brit. Antarctic Surv.: Bulletin of the British antarctic survey. London, 1963– .

Bull. Brit. Mus. (Nat. Hist.), Bot.: Bulletin of the British museum (natural history), botany (*later* Bulletin of the natural history museum, botany). London, 1951– .

*Bull. Buffalo Soc. Nat. Sci.: Bulletin of the Buffalo society of natural science. Buffalo, N.Y., 1874– .

Bull. Conservation Dept. Univ. Nebraska: Bulletin of the conservation department, University of Nebraska. Lincoln, Nebr., 1928– .

*Bull. Cranbrook Inst. Sci.: Bulletin of the Cranbrook institute of science. Bloomfield Hills, Mich., 1931– .

Bull. Dept. Medicinal Pl., Nepal: Bulletin of the department of medicinal plants, Nepal. Kathmandu, 1967– .

Bull. Fac. Sci. (Cairo Univ.): Bulletin of the faculty of science, Cairo University. Cairo, 1934– .

*Bull. For. Res. Inst. Chittagong, Pl. Tax. Ser.: Bulletin of the [Bangladesh] forest research institute (Chittagong), plant taxonomy series. Chittagong, 1980– .

Bull. Forests Dept. Western Australia: Bulletin of the forests department, Western Australia. Perth, Western Australia, 1919– .

Bull. Geol. & Nat. Hist. Surv. Minnesota: Bulletin of the geological and natural history survey of Minnesota. St. Paul, Minn., 1885– .

Bull. Herb. Boissier: Bulletin de l'herbier Boissier. Geneva, 1893–1908.

Bull. Inst. Jamaica, Sci. Ser.: Bulletin of the institute of Jamaica, science series. Kingston, Jamaica, 1940– .

Bull. Jard. Bot. Buitenzorg: Bulletin du jardin botanique de Buitenzorg (*from 1941* Bulletin of the botanical garden, Buitenzorg). Buitenzorg (Bogor), Indonesia, 1898–1950. [*Three consecutive series. From 1898 to 1905 entitled* Bulletin de l'institut botanique de Buitenzorg; *from 1906 to 1911 as* Bulletin du département de l'agriculture aux index néerlandaises. *Succeeded by* Reinwardtia.]

Bull. Jard. Bot. État: Bulletin du jardin botanique de l'État (*from 1938 also entitled* Bulletin van den (de) rijksplantentuin; *from 1969 retitled* Bulletin du jardin botanique national de Belgique/Bulletin van den nationale plantentuin van België). Brussels (*later* Meise), 1902–98. [*Continued as* Systematics and geography of plants *without change in volumination*.]

Bull. Jard. Bot. Natl. Belg.: Bulletin du jardin botanique national de Belgique/Bulletin van den nationale plantentuin van België. *See* Bulletin du jardin botanique de l'État/Bulletin van den rijksplantentuin.

Bull. Josselyn Bot. Soc. Maine: Bulletin of the Josselyn botanical society of Maine. Portland, Maine, 1907– .

Bull. Mauritius Inst.: Bulletin of the Mauritius institute. Port Louis, Mauritius, 1937– .

*Bull. Mens. Soc. Naturalistes Luxemb.: Bulletins mensuels de la société des naturalistes luxembourgois (*after 1949 titled* Bulletin de la société des naturalistes luxembourgois). Luxembourg, 1907– .

Bull. Misc. Inform. (Kew): Bulletin of miscellaneous information. Kew, England, 1887–1942. [*Succeeded by* Kew bulletin.]

Bull. Misc. Inform. (Kew), Addit. Ser.: Bulletin of miscellaneous information, Kew: additional series. London, 1898–1936. [*Succeeded by* Kew bulletin, additional series.]

*Bull. Misc. Inform. (Trinidad): Bulletin of miscellaneous information, royal botanic garden, Trinidad (*later* Bulletin of miscellaneous information, botanical department, Trinidad). Port-of-Spain, 1888–1908.

*Bull. Mus. Life Sci., Louisiana State Univ., Shreveport: Bulletin of the museum of life sciences, Louisiana State University at Shreveport. Shreveport, La., 1979– .

Bull. Mus. [Natl.] Hist. Nat. (Paris): Bulletin du Muséum [national] d'histoire naturelle. Paris, 1895–1970. [*Two

consecutive series. Succeeded by* Bulletin du Muséum national d'histoire naturelle, série III (*to 1978*) and série IV (*from 1979; for subseries B see* Bulletin du muséum national d'histoire naturelle, série IV/B, Botanique, biologie et écologie végétales, phytochimie).

Bull. Mus. Natl. Hist. Nat. (Paris), IV/B (Adansonia): Bulletin du muséum national d'histoire naturelle, série IV/B, Botanique, biologie et écologie végétales, phytochimie (*from 1981 subtitled* Adansonia: botanique, phytochimie). Paris, 1979–96. [*Continued as* Adansonia, série 3, *without change in volumination*.]

Bull. Nat. Hist. Soc. New Brunswick: Bulletin of the natural history society of New Brunswick. St. John, New Brunswick, 1882–1914.

Bull. Natl. Mus. Canada: Bulletin of the national museum of Canada. Ottawa, 1913–68. [*Includes several series, among them* Biological series *and* Contributions to botany.]

Bull. Natl. Sci. Mus. (Tokyo): Bulletin of the National science museum, Tokyo, Japan. Tokyo, 1939–54; N.S., 1954–74. [*From 1974 further new series initiated in separate subject areas*.]

*Bull. New Hampshire Acad. Sci.: Bulletin of the New Hampshire academy of science. Durham, 1940– .

Bull. New York Bot. Gard.: Bulletin of the New York botanical garden. Lancaster, Pa., etc., 1896–1932.

Bull. Ohio Biol. Surv.: Bulletin of the Ohio biological survey. Columbus, Ohio., 1913– .

Bull. Oklahoma Agric. Mech. Coll.: Bulletin of the Oklahoma agricultural and mechanical college (*from 1957* Oklahoma State University). Stillwater, Okla., ?1915– . [Vol. 43(21), 1946.]

Bull. Peking Soc. Nat. Hist.: Bulletin of the Peking society of natural history. Peking, 1926–30.

Bull. Raffles Mus. Singapore: Bulletin of the Raffles museum, Singapore. Singapore, 1928–70.

Bull. Roy. Soc. New Zealand: Bulletin of the royal society of New Zealand. Wellington, 1910–30; 1953– .

*Bull. S. Calif. Acad. Sci.: Bulletin of the Southern California academy of sciences. Los Angeles, 1902– .

Bull. Soc. Bot. France: Bulletin de la société botanique de France. Paris, 1854–1992. [*From 1978–79 in two series*, Actualités Botaniques *and* Lettres Botaniques; *from 1993 reunited under one cover and continued as* Acta botanica gallica *without change in volumination*.]

*Bull. Soc. Bot. Genève: Bulletin de la Société botanique de Genève. Geneva, Switzerland, 1909–52. [*Succeeded from 1970 by* Saussurea.]

Bull. Soc. Études Indochin.: Bulletin de la société d'études indochinoises (de Saigon). Saigon, etc., 1883– (N.S., 1926–).

Bull. Soc. Études Océanien. (Papeete): Bulletin de la société des études océaniennes. Papeete, Tahiti, 1917– .

Bull. Soc. Roy. Bot. Belgique: Bulletin de la société royale de botanique de Belgique. Brussels, 1862– . [*Title varies.*]

*Bull. Soc. Sci. Hist. Nat. Corse: Bulletin de la société des sciences historiques et naturelles de la Corse. Bastia, 1881–1938. [*Special numbers have, however, appeared since.*]

Bull. Soc. Sci. Nat. (Phys.) Maroc: Bulletin de la société des sciences naturelles (et physiques) du Maroc. Rabat, Morocco, 1921– .

*Bull. Sul Ross State Teachers' Coll.: Bulletin of the Sul Ross state teachers' college. Alpine, Tex. [Vol. 19(4), 1938.]

Bull. Techn. Sci. Serv., Minist. Agric. [Egypt]: Bulletin of the technical and scientific service, Ministry of agriculture, Egypt. Cairo, 1916– .

Bull. Torrey Bot. Club: Bulletin of the Torrey botanical club (*from 1997 retitled* Journal of the Torrey Botanical Society). Lancaster, Pa., etc., 1870– .

*Bull. Univ. Texas: Bulletin of the University of Texas. Austin, Tex., 1905–15.

Bull. Univ. Tokyo Mus.: Bulletin of the University of Tokyo museum (Sogo kenkyu shiryokan). Tokyo, 1970– .

Bull. Wellington Bot. Soc.: Bulletin of the Wellington botanical society. Wellington, N.Z., 1941– .

*Butlletí Inst. Catal. Hist. Nat.: Butlletí de la institució catalana d'història natural. Barcelona, 1901–49; 1974– .

*CNFRA (Paris): *See* Publications de la Comité national français des recherches antarctiques (CNFRA), biologie.

Cah. Indo-Pacifique: Cahiers de l'Indo–Pacifique. Paris, 1979–80.

Cah. Pacifique: Cahiers du Pacifique. Paris, 1958–78. [*Succeeded by* Cahiers de l'Indo-Pacifique.]

*Canad. Field–Nat.: Canadian field–naturalist. Ottawa, 1919– .

Candollea: Candollea. Geneva, 1922– .

Caribbean Forest.: Caribbean forester. Río Pedras, Puerto Rico, 1939–63.

Castanea: Castanea. Morgantown, W.Va., 1936– .

Ceiba: Ceiba. Tegucigalpa, Honduras, 1950– .

Ceylon J. Sci., Biol. Sci.: Ceylon journal of science. Biological sciences. Colombo, Sri Lanka, 1957– .

Chron. Bot.: Chronica botanica. Leiden (*later* Waltham, Mass.), 1935–53/54.

Chron. Nat.: Chronica naturae. Batavia (Jakarta), 1947–50. [*A continuation of* Natuurkundig tijdschrift voor Nederlandsch-Indië. *Subsequently retitled* Madjalah ilmu alam untuk Indonesia/Indonesian journal for natural science; *ceased publication 1957.*]

*Ci. Invest.: Ciencia e investigación. Buenos Aires, Argentina, 1945– .

Ci. & Naturalez. (Quito): Ciencia y naturaleza. Quito, 1957– .

Colecc. Científ. INTA: Colección científica, Instituto national de técnologia agropecuaria. Buenos Aires, 1959– .

Coll. Hawaii Publ. Bull.: College of Hawaii publications. Bulletins. Honolulu, Hawaii, 1911–16.

*Collect. New-York Hist. Soc.: Collections of the New-York historical society. New York, 1809–29; ser. 2, 1841–59.

Colloq. Int. CNRS: Colloques internationaux du Centre national de la recherche scientifique (CNRS), France. Paris, 1946– .

*Commonw. Forest. Rev.: Commonwealth forestry review. London, etc., 1922–98. [*Initially titled* Empire forestry; *from 1923 to 1945* Empire forestry journal; *from 1946 to 1962* Empire forestry review; *from 1962 to 1998 as given here. From 1999 succeeded by* International forestry review.]

*Comptes Rend. Soc. Biogéogr. (Paris): Comptes rendus sommaires des séances, Société de biogéographie. Paris, 1923– . [*From 1993 retitled* Biogeographica.]

Connecticut State Geol. Surv. Bull.: Connecticut state geological and natural history survey bulletin. Hartford, Conn., 1903– .

Contr. Arnold Arbor.: Contributions from the Arnold Arboretum, Harvard University. Jamaica Plain, Mass., 1932–38. [*Succeeded by* Sargentia.]

Contr. Biol. Lab. Chin. Assoc. Advancem. Sci., Sect. Bot.: Contributions from the biological laboratory of the Chinese association for the advancement of science. Section botany. Nanking (Nanjing), 1930–39.

Contr. Gray Herb., N.S.: Contributions of the Gray herbarium of Harvard University. New series. Cambridge, Mass., 1891–1984. [*Succeeded by* Harvard papers in botany.]

*Contr. Inst. Agricole Oka: Contributions de l'institut [agricole] d'Oka. Oka, Qué., 1911–35; ser. 2, 1948– . [*Title varies.*]

Contr. Inst. Bot. Univ. Montréal: Contributions de l'institut botanique de l'université de Montréal. Montreal, 1938– .

Contr. Inst. Ecuat. Ci. Nat.: Contribuciones del instituto ecuatoriano de ciencias naturales. Quito, <1950– .

Contr. New South Wales Natl. Herb.: Contributions from the New South Wales national herbarium. Sydney, 1939–74. [*Succeeded by* Telopea.]

Contr. New South Wales Natl. Herb., Flora Ser.: Contributions from the New South Wales national herbarium, flora series. Sydney, 1961–74.

Contr. Ocas. Mus. Hist. Nat. Colegio 'La Salle': Contribuciones ocasionales del museo de historia natural del colegio de 'La Salle'. Havana, 1944–60.

*Contr. Reed Herb.: Contributions from the Reed Herbarium. Baltimore, Md. [Vol. 30, 1986.]

*Contr. Sci. Nat. Hist. Mus. Los Angeles County:
Contributions in science, natural history museum of
Los Angeles County. Los Angeles, 1945– . [*Title varies.*]

Contr. U.S. Natl. Herb.: Contributions from the United
States national herbarium (Smithsonian Institution).
Washington, 1890–1974; 2000– .

Contr. Univ. Michigan Herb.: Contributions from the
University of Michigan herbarium. Ann Arbor, Mich.,
1939– .

Cornell Univ. Agric. Exp. Sta. Mem.: Cornell University
agricultural experiment station memoirs. Ithaca, N.Y.,
1913– .

*Courier Forschungsinst. Senckenb.: Courier,
Forschungsinstitut Senckenberg. Frankfurt, 1973– .

Črnog. Akad. Nauk Umjetn., Bibliografije: Črnogorska
akademija nauka i umjetnosti (ČANU), Bibliografije.
Titograd (Podgorica). [No. 1, 1980.]

Dansk Bot. Ark.: Dansk botanisk arkiv udgivet af Dansk
botanisk forening. Copenhagen, 1913–80. [*Absorbed into*
Opera botanica.]

Dir. Gen. Agric. [Iraq], Bull.: Directorate-general of
agriculture, Iraq. Bulletin. Baghdad, <1943–?60.
[*Succeeded by* Directorate-general of agricultural
research, Iraq. Technical bulletin.]

Dir. Gen. Agric. Res. [Iraq], Tech. Bull.: Directorate-
general of agricultural research, Iraq (Mudiryat al-
Buhuth Wa-al-Mashari al-Zira'iyah al-Ammah).
Technical bulletin. Baghdad, 1960– .

Ecol. Monogr.: Ecological monographs. Durham, N.C.,
1931– .

Ecos: Ecos – CSIRO environmental research. Melbourne,
1974– .

*Englera: Englera. Berlin, 1979– .

Estudos, Ensaios e Documentos, Junta Invest. Ultram.:
Estudos, ensaios e documentos, Junta de Investigações
do Ultramar, Portugal. Lisbon, 1950– . [*To at least vol.
160, 1995.*]

Études Bot. IEMVPT: Études botaniques, Institut d'élevage
et de médecine vétérinaire des pays tropicaux. Maisons-
Alfort, Val-de-Marne, France, 1972– .

FNA Rep.: Flora North America reports. *Abbreviated here as*
Fl. N. Amer. Rep.

*Feddes Repert.: Feddes repertorium. Berlin, 1944– . [*Title
varies. Previously entitled* Repertorium specierum
novarum regni vegetabilis.]

Fern Gaz.: [British] fern gazette. Kendal, etc., 1909– .

Field Studies: Field studies. London, 1959– .

Fieldiana, Bot.: Fieldiana, botany. Chicago, 1946–77; N.S.,
1979– . [*Previously entitled* Publications of the Field
museum of natural history, botanical series.]

*Fl. Butt.: Flora buttensis. Chico, Calif./St. Louis, Mo.,
1980–84.

Fl. Males. Bull.: Flora malesiana bulletin. Leiden, 1947– .

*Fl. N. Amer. Newsl.: Flora of North America newsletter. St.
Louis, 1987– .

Fl. N. Amer. Rep.: Flora North America reports.
Washington, D.C., 1967–78. [*84 numbers.*]

*Flora: Flora (oder allgemeine botanische Zeitung).
Regensburg, etc., 1818– . [*In 1965 divided into*
Abteilung A *and* Abteilung B, *the latter soon assuming the
original title.*]

*Florist. Rundbr.: Floristische Rundbriefe (*initially* Göttinger
floristische Rundbriefe). Göttingen, 1977– .

*Florist. Rundbr., Beih.: Floristische Rundbriefe, Beihefte.
Göttingen, 1990– .

Forest Bull. (Georgetown): Forest bulletin. Georgetown,
Guyana, 1948– .

Forest Bull. India: Forest bulletin, India. Calcutta, 1906–07;
1911– .

Forest Prod. Res. Inst. Bull. [Philippines]: Forest products
research institute bulletin. Los Baños, Laguna,
Philippines, 1963– .

Fortschr. Bot.: Fortschritte der Botanik. *See* Progress in
botany.

*Fragm. Flor. Geobot., Suppl.: Fragmenta floristica et
geobotanica. Supplementum. Kráków, 1991– .

Garcia de Orta: Garcia de Orta. Lisbon, 1953–71. [*Continued
in several disciplinarily distinct series, including, from
1973,* Botânica.]

*Garcia de Orta, Bot.: Garcia de Orta, série de botânica. *See*
Garcia de Orta.

*Gayana, Bot.: Gayana, botánica. Concepción, Chile, 1948– .

Geogr. J. (London): Geographical journal. London, 1893– .

Geogr. Rev. (New York): Geographical review. New York,
1916– .

*Glasn. Inst. Bot. Bašte Univ. Beograd: Glasnik instituta za
botaniku i botaničke bašte univerziteta u Beogradu.
Belgrade, 1959– .

*Glasn. Republ. Zavoda Zaštitu Prir. Prirodnjačk. Zbirke,
Titograd.: Glasnik republičkog zavoda za zaštitu
prirode i prirodnjačke zbirke (*later* prirodnjačkog
muzeja) u Titogradu (*later* Podgorica). Titograd
(Podgorica), Montenegro, 1968–91.

Glasn. Zemaljsk. Muz. Bosni Hercegovini: Glasnik
zemaljskog muzeja u Bosni i Hercegovini. Sarajevo,
1889–1937.

*God. Zborn. Prir.-Mat. Fak. Univ. Skoplje: Godišen zbornik
prirodno-matematički fakultet na univerzitetot Skoplje.
Skopje, 1948– . [*Title varies.*]

Göteborgs Kungl. Vetensk. Vitterh. Samhälles Handl.:
Göteborgs Kungliga Vetenskaps- och Vitterhets-
Samhälles handlingar. Göteborg, 1778–1966. [*Series V,*

1928–40, *and Series VI, 1941–66, each divided into* Sect. A, Humanistiska skrifter, *and* Sect. B, Matematiska och naturvetenskapliga skrifter.]

*Gorteria: Gorteria. Leiden, 1961– .

*Great Basin Naturalist: Great Basin naturalist. Provo, Utah, 1939– .

Gulf Res. Rep.: Gulf research reports. Ocean Springs, Miss., 1961– .

*Handb. Brit. Columbia Prov. Mus.: Handbooks of the British Columbia provincial museum. Victoria, B.C., 1942– .

Handb. Peking Soc. Nat. Hist.: Handbooks of the Peking society of natural history. Peking, 1926– . [*Initially as* Educational series.]

*Harvard Pap. Bot.: Harvard papers in botany. Cambridge, Mass., 1989– .

Health Bulletin (India): Health bulletin, Medical department, (Government of) India. Calcutta, etc., 1916– .

*Hist. Rec. Austral. Sci.: Historical records of Australian science. Canberra, 1966– . [*Until 1980 entitled* Records of the Australian academy of science.]

[Hooker's] Icon. Pl.: [Hooker's] icones plantarum. London (*later* Kew), 1836–1990.

*Huntia: Huntia. Pittsburgh, 1964– .

Illinois State Mus. Sci. Pap. Ser.: Illinois state museum, scientific papers series. Springfield, Ill., 1940– .

Indian For. Rec.: Indian forest records. Calcutta, etc., 1907–34; N.S., Botany, 1937–64, 1980– . [Botany *is one of several concurrent series established after 1934.*]

Indian Forester: Indian forester. Allahabad, India, 1875– .

*Indian J. Forest., Addit. Ser.: Indian journal of forestry, additional series. Dehra Dun. [Vol. 2, 1981– .]

*Inform. Bot. Ital.: Informatore botanico italiano – bollettino della società botanica italiana. Florence, 1969– .

Iowa State Coll. J. Sci.: Iowa state college journal of science. Ames, Iowa, 1926– .

Izv. Bot. Inst. (Sofia): Izvestija na botaničeskaja institut, Bălgarska akademija na naukite. Sofia, Bulgaria, 1950–74. [*Succeeded by* Fitologija.]

*Izv. Imp. St-Peterburgsk. Bot. Sada: Izvestija imperatorskogo St-Peterburgskogo botaničeskogo sada. St Petersburg, 1901–32. [*Title varies. Succeeded by* Sovetskaja botanika.]

*Izv. Tsarsk. Prirodonauč. Inst. (Sofia): Izvestija na tsarskite prirodonaučni instituti v Sofija. Sofia, Bulgaria, 1928–43.

*J. Adelaide Bot. Gard.: Journal of the Adelaide botanic gardens. Adelaide, South Australia, 1975– .

J. Agric. Trop. Bot. Appl.: Journal d'agriculture tropicale (*later* traditionelle) et de botanique appliqué (JATBA). Paris, 1954– .

J. Appl. Ecol.: Journal of applied ecology. Oxford, 1964– .

J. Arnold Arbor.: Journal of the Arnold Arboretum of Harvard University. Jamaica Plain (*later* Cambridge), Mass., 1920–90. [*Absorbed into* Harvard papers in botany.]

*J. Arnold Arbor., Suppl. Ser.: Journal of the Arnold Arboretum of Harvard University, supplementary series. Cambridge, Mass., 1991. [*Only one volume published.*]

J. Asiat. Soc. Bengal: Journal of the Asiatic society of Bengal. Calcutta, 1834–1936. [*After 1864 divided into parts, with natural history in Part II.*]

J. Bombay Nat. Hist. Soc.: Journal of the Bombay natural history society. Bombay, 1886– .

J. Bot.: Journal of botany, British and foreign. London, 1863–1942.

*J. Chosen Nat. Hist. Soc.: Journal of the Chosen natural history society. Keijo, Korea, 1924–42.

J. Coll. Sci. Imp. Univ. Tokyo: Journal of the college of science, Imperial University of Tokyo. Tokyo, 1887–1925.

J. Coll. Sci. Sun Yat-sen Univ.: Journal of the college of science, Sun Yat-sen University. Canton (Guangzhou), 1928– .

J. Dept. Agric. Kyushu Imp. Univ.: Journal of the department of agriculture of the Kyushu Imperial University. Fukuoka, Japan, 1923–46.

*J. Econ. Taxon. Bot.: Journal of economic and taxonomic botany. Jodhpur, 1980– .

J. Fac. Agric. Hokkaido Univ.: Journal of the faculty of agriculture of the Hokkaido University. Sapporo, Japan, 1902– .

J. Forest. (Tsinan): Journal of forestry [Chinese forestry associations]. Tsinan, China, 1921.

*J. Forest. (Washington): Journal of forestry. Washington, D.C., 1917– .

J. Gen. Biol. (USSR): Journal of general biology, USSR. *See* Žurnal obščej biologii (SSSR).

J. Indian Bot. Soc.: Journal of the Indian botanical society. Madras, 1921– . [*First two volumes entitled* Journal of Indian botany.]

J. Linn. Soc., Bot.: Journal of the Linnean society, botany. London, 1865–1968. [*Subsequently retitled* Botanical journal of the Linnean society.]

J. Mysore Univ., B.: Journal of the Mysore University, series B. Mysore, India, 1940– .

*J. New York Bot. Gard.: Journal of the New York botanical garden. New York, N.Y., 1900–50.

J. & Proc. Roy. Soc. Western Australia: Journal and proceedings of the royal society of Western Australia. Perth, Australia, 1914– .

J. S. African Bot.: Journal of South African botany (later South African journal of botany). Kirstenbosch, Western Cape Province, 1935– .

J. S. African Bot., Suppl.: Journal of South African botany, supplementary volumes. Kirstenbosch, Western Cape Province, 1944– .

*J. Sci. Univ. Tehran: Journal of science, University of Tehran. Tehran. [Vol. 18, 1989. *Not to be confused with* Journal of science, Islamic Republic of Iran.]

J. Soc. Bibliogr. Nat. Hist.: Journal of the society for the bibliography of natural history. London, 1936–80. [*Continued as* Archives of natural history.]

J. Straits Branch Roy. Asiat. Soc.: Journal of the Straits branch of the Royal Asiatic Society. Singapore, 1878–1922. [*Includes a* Special Number, 1921. *Afterwards continued as* Journal of the Malayan branch, Royal Asiatic Society.]

*J. Swamy Bot. Club: Journal of the Swamy botanical club. [Vol. 6, 1989.]

J. Tennessee Acad. Sci.: Journal of the Tennessee academy of science. Nashville, Tenn., 1926– .

J. Trans. Victoria Inst. (London): Journal of the transactions of the Victoria institute. London, 1866– .

J. Trop. Geog.: Journal of tropical geography (*initially as* Malayan journal of tropical geography). Singapore, 1953– .

*J. Washington Acad. Sci.: Journal of the Washington academy of sciences. Baltimore, Md., etc., 1911– .

*J. Wuhan Bot. Res.: Journal of Wuhan botanical research. Wuhan, 1983– (*from 1985 retitled* Wuhan botanical research).

*Kakteen Sukk.: Kakteen und andere Sukkulenten. Berlin, 1937–38; 1949– .

Kashmir Sci.: Kashmir science. Srinagar, Kashmir, 1964– .

Kew Bull.: Kew bulletin. London (*later* Kew), 1946– .

Kew Bull. Addit. Ser.: Kew bulletin, additional series. London, 1958– .

Kirkia: Kirkia. Harare (Salisbury), Zimbabwe, 1960– .

*Kongel. Danske Vidensk. Selsk., Biol. Meddel.: Det kongelige danske videnskabernes selskab. Biologiske meddelelser. Copenhagen, 1917–71.

Kongel. Danske Vidensk. Selsk., Biol. Skr.: Det kongelige danske videnskabernes selskab. Biologiske skrifter. Copenhagen, 1939– .

Kongl. Svenska Fysiograf. Sällskap. Handl.: Kongl. svenska fysiografiska sällskapets handlingar. *See* Acta universitatis lundensis.

Kongl. Svenska Vetenskapsakad. Handl.: Kongl. svenska vetenskapsakademiens handlingar. Stockholm, 1855–1971. [*Several successive series.*]

Kwangsi Agric.: Kwangsi agriculture. Liuzhou, China, 1940– .

*Lagascalia: Lagascalia. Seville, 1971– .

Lap. Balai Penjelidik. Kehut.: Laporan-laporanan Balai penjelidikan kehutanan. *See* Laporan-laporanan Lembaga penelitian hutan.

Lap. Lembaga Penelit. Hutan: Laporan-laporanan Lembaga penelitian hutan (*initially* Rapports van het bosbouwproefstation, *then* Laporan-laporanan Balai penjelidikan kehutanan, *then under above title; afterwards* Laporan-laporanan Pusat penelitian dan pengembangan hutan (P3H)). Bogor (Buitenzorg), 1948–84. [*Continued as* Buletin penelitian hutan (Forest research bulletin) *without change in numeration*.]

Lav. Ist. Bot. Reale Univ. Modena: Lavori dell'istituto botanico della Reale università di Modena. Modena, Italy. [Vol. 3, 1932.]

*Lav. [Reale] Ist. Bot. Palermo: Lavori del reale istituto botanico di Palermo (e Bollettino del giardino coloniale) (*later* Lavori del [reale] istituto botanico di Palermo e del [reale] giardino coloniale di Palermo). Palermo, 1930–74. [*Continues with different voluminination as irregular collections of offprints from other journals.*]

*Lav. Soc. Ital. Biogeogr., N.S.: Lavori della Società italiana di biogeografia. Nuova serie. Forlì. [Vol. 3, 1972 (1973).]

Leafl. W. Bot.: Leaflets of western botany. San Francisco, 1932–66.

Lilloa: Lilloa. Tucumán, Argentina, 1937– .

Lingnan Sci. J.: Lingnan science journal. Canton, 1927–50.

*Linnaea: Linnaea. Berlin, 1826–82.

*Linnean: The linnean. London, 1985– .

*Linzer Biol. Beitr.: Linzer biologische Beiträge. Linz, 1975– .

Lloydia: Lloydia. Cincinnati, Ohio, 1938– . [*Now entitled* Journal of natural products.]

Louisiana Forest. Commiss. Bull.: Louisiana forestry commission bulletin. Baton Rouge, La., 1945– .

*Lunds Univ. Årsskr.: Lunds universitets årsskrift. *See* Acta universitatis lundensis.

Madroño: Madroño. Berkeley, Calif., 1916– .

Magyar Bot. Lapok: Magyar botanikai lapok. Budapest, 1902–34.

Malayan Forest Rec.: Malayan forest records. Singapore and Calcutta (*now* Kuala Lumpur), 1922– .

Malayan Nat. J.: Malayan nature journal. Kuala Lumpur, 1940– .

Malpighia: Malpighia. Messina, Italy, 1887–1937.

Maria Moors Cabot Foundation, Spec. Publ.: Maria Moors Cabot Foundation, special publications. Petersham, Mass., 1947– .

Maryland Weather Serv., Spec. Publ.: Maryland weather service, special publications. Baltimore, 1899–1910.

Mater. Pozn. Fauny Fl. SSSR, Otd. Bot.: Materialy k poznaniju fauny i flory SSSR, izdavaemye Moskovskim obščestvom ispytatelej prirody. Otdel botaničeskij. Moscow, 1940– .

*Meddel. Grønland: Meddeleser om Grønland. Copenhagen, 1879– .

Meded. Bosbouwproefstat.: Mededeelingen van het bosbouwproefstation. Buitenzorg (Bogor), Indonesia, 1915–50.

Meded. Bot. Mus. Herb. Rijks Univ. Utrecht: Mededeelingen van het botanisch museum en herbarium van de rijks universiteit te Utrecht. Utrecht, 1932–85. [*A collection chiefly of offprints from other journals or works also released independently. Succeeded by* Miscellaneous publications of the University of Utrecht herbarium, 1983–90.]

Meded. Dept. Landb. Ned.-Indie: Mededeelingen uitgaande van het departement van landbouw [in Nederlandsch-Indië]. Batavia (Jakarta), Indonesia, 1905–14. [*Previously entitled* Mededeelingen uit 's Lands plantentuin.]

Meded. Landbouwhogeschool (Wageningen): Mededeelingen van de landbouwh[o]ogeschool. Wageningen, 1908– . [*From 1984 entitled* Wageningen Agricultural University papers.]

Meded. Lands Plantentuin: Mededeelingen uit 's Lands plantentuin. Batavia (Jakarta), Indonesia, 1884–1904. [*Continued as* Mededeelingen uitgaande van het departement van landbouw in Nederlandsch-Indië.]

Meded. Rijks-Herb. Leiden: Mededeelingen van 's Rijks-Herbarium. Leiden, 1910–33. [*Succeeded by* Blumea.]

*Mem. Acad. Ci. Lisboa, Cl. Ci.: Memórias da academia das Ciências de Lisboa, Classe de Ciências. Lisbon, 1937– .

*Mem. Acad. Nac. Ci. 'Antonio Alzate': Memórias de la academia nacional de ciencias 'Antonio Alzate'. Mexico City, 1931– . [*Previously entitled* Memórias [y revista] de la Sociedad científica 'Antonio Alzate'.]

*Mem. Amer. Philos. Soc.: Memoirs of the American philosophical society. Philadelphia, Pa., 1935– .

Mem. Boston Soc. Nat. Hist.: Memoirs of the Boston society of natural history. Boston, Mass., 1862–1936.

Mem. Bot. Surv. S. Africa: Memoirs of the botanical survey of South Africa. Pretoria, 1919– .

Mem. Brooklyn Bot. Gard.: Memoirs of the Brooklyn botanical garden. Brooklyn, New York, 1918–36.

*Mem. Calif. Acad. Sci.: Memoirs of the California academy of sciences. San Francisco, 1868– .

Mem. Fac. Agric. Kagoshima Univ.: Memoirs of the faculty of agriculture, Kagoshima University. Kagoshima, Japan, 1952– .

*Mem. Indian Bot. Soc.: Memoirs of the Indian botanical society. Bangalore, 1958–63. [*Four volumes.*]

Mém. Inst. Égypte: Mémoires de l'institut d'Égypte. Cairo, 1862– .

Mém. Inst. Études Centrafr.: Mémoires de l'institut d'études centrafricaines. Brazzaville, Congo Republic, 1948– .

Mém. Inst. Franç. Afrique Noire: Mémoires de l'institut français d'Afrique noire. Dakar, Senegal, 1939– .

Mem. Inst. Invest. Agron. Moçambique: Memórias do instituto de investigação agronomia de Moçambique, Centro de Documentação Agrario. Maputo, Mozambique, 1966– .

Mém. Inst. Rech. Saharien. Univ. Alger: Mémoires de l'institut des recherches sahariennes de l'Université d'Alger. Algeria, 1954– .

Mem. Junta Invest. Ultram., II. sér.: Memórias do Junta de Investigações do Ultramar, séria II. Lisbon, 1958– .

Mém. Mus. Natl. Hist. Nat., sér. II/A (Zool.): Mémoires du muséum national d'histoire naturelle. Série II/A, Zoologie. Paris, 1950– .

Mém. Mus. Natl. Hist. Nat., sér. II/B (Bot.): Mémoires du muséum national d'histoire naturelle. Série II/B, Botanique. Paris, 1950– .

*Mem. Natl. Acad. Sci.: Memoirs of the [U. S.] national academy of sciences. Washington, D. C., 1866–1928.

Mem. New York Bot. Gard.: Memoirs of the New York botanical garden. New York, 1900– .

Mem. Pacific Tropical Bot. Gard.: Memoirs of the Pacific tropical botanical garden [*now* National tropical botanical garden]. Lawai, Kauai, Hawaii, 1973– .

Mem. Real Acad. Ci. Barcelona: Memorias de la real academia de ciencias y artes de Barcelona. Barcelona, 1892– .

*Mem. Reale Accad. Sci. (Bologna): Memorie della reale accademia delle scienze dell'istituto di Bologna. Bologna, 1850– . [*Several successive series (ser. 8, 1923–33); from 1907 additionally titled* Classe di scienze fisiche.]

*Mem. Reale Ist. Veneto Sci.: Memorie del reale istituto veneto di scienze, lettere ed arti. Venice, 1843–1940. [*Revived 1956 under a different title.*]

*Mem. San Diego Soc. Nat. Hist.: Memoirs of the San Diego society of natural history. San Diego, 1931– .

Mem. Sci. Soc. China: Memoirs of the science society of China. Shanghai, 1924–32.

Mem. Soc. Brot.: Memórias da Sociedade broteriana. Coimbra, Portugal, 1930–31, 1943– .

*Mem. Soc. Ci. Nat. "La Salle": Memórias de la Sociedad de ciencias naturales "La Salle". Caracas, Venezuela, 1941– .

*Mem. Soc. Cient. 'Antonio Alzate': Memórias [y revista] de la Sociedad científica 'Antonio Alzate'. Mexico City, 1887–1931. [*Title varies. Succeeded by* Memórias de la academia nacional de ciencias 'Antonio Alzate'.]

*Mém. Soc. Hist. Nat. Afrique Nord: Mémoires de la Société d'histoire naturelle de l'Afrique du Nord. Algiers, 1926–34; N.S., 1956–83.

Mém. Soc. Linn. Normandie: Mémoires de la société linnéenne de Normandie. Caen, Paris, 1824–28; 1835–1924.

Mém. Soc. Sci. Nat. Maroc: Mémoires de la société des

sciences naturelles du Maroc. Rabat, Morocco, 1921–52.

Mem. Torrey Bot. Club: Memoirs of the Torrey botanical club. New York, 1889– .

*Michigan Agr. Expt. Sta. Techn. Bull.: Michigan agricultural experiment station, technical bulletin. East Lansing, Mich., 1908–64.

Michigan Bot.: Michigan botanist. Ann Arbor, Mich., 1962– .

Micronesica: Micronesica. Agaña, Guam, 1964– .

*Milwaukee Public Mus. Contr. Biol. Geol.: Milwaukee public museum, contributions in biology and geology. Milwaukee, 1974– .

Milwaukee Public Mus., Publ. Bot.: Milwaukee public museum, publications in botany. Milwaukee, 1955–65. [*Three numbers.*]

Minnesota Geol. Surv. Rep., Bot. Ser.: Minnesota geological survey reports, botanical series. Minneapolis, Minn., 1892–1912.

Misc. Publ. USDA: Miscellaneous publications, United States Department of Agriculture. Washington, 1927– .

Misc. Rep. Res. Inst. Nat. Resources [Japan]: Miscellaneous reports of the research institute for natural resources. Tokyo, 1943– .

Mississippi State Geol. Surv. Bull.: Mississippi state geological survey bulletin. Jackson, Miss., 1907– .

Mitt. Bot. Staatssamml. München: Mitteilungen (aus) der botanischen Staatssammlung, München. Munich, 1950–91. [*From 1993 succeeded by* Sendtnera.]

Mitt. Inst. Allg. Bot. Hamburg: Mitteilungen aus dem Institut für allgemeine Botanik in Hamburg. Hamburg, 1916– .

Monde Pl.: Le monde des plantes. Le Mans (*later* Toulouse), 1891– .

Monogr. Alabama Geol. Surv.: Monographs of the Alabama geological survey. University, Ala., 1883–1945.

Monogr. Amer. Midl. Naturalist: Monographs of the American midland naturalist. South Bend, Ind., 1944– .

Monogr. Canad. Dept. Agric. Res. Br.: Monographs of the Canadian department of agriculture, research branch. Ottawa, <1966– .

Monogr. Inst. Estud. Pirenaicos: Monografías del instituto de estudios pirenaicos [del Consejo superior de investigaciones científicas, Spain]. Zaragoza, Spain, 1948– . [*Comprises several series, including a* ser. Botánica.]

Monogr. Philipp. Inst. Sci. Tech.: Monographs of the Philippine (national) institute of science and technology. Manila, 1951–63. [*Successor to* Publications of the Bureau of Science, Philippines.]

*Monogr. Rég. Inst. Rech. Sahar. Univ. Alger: Monographes régionales, Institut de recherches sahariennes, Université d'Alger. Algiers, 1953– .

*Monogr. Soc. Ci. Nat. "La Salle": Monografías de la sociedad de ciencias naturales "La Salle". Caracas. [Vol. 34, 1985.]

*Monogr. Syst. Bot. Missouri Bot. Gard.: Monographs in systematic botany from the Missouri Botanical Garden. St. Louis, Mo., 1978– .

*Monogr. Univ. Nac. Auton. Méx., Inst. Geofis.: Monografías del instituto de geofísica de la Universidad national autonómico de México. Mexico City, 1959– .

Muelleria: Muelleria. Melbourne, 1955– .

*Mus. Genève: Musées de Genève. Geneva, 1944– .

Mus. Nac. Costa Rica, Ser. Bot.: Museo nacional de Costa Rica, serie Botánica. San José, Costa Rica, 1937– .

*N.W. Nat.: North western naturalist. Arbroath, Scotland, etc., 1926– .

Nat. Appl. Sci. Bull. Univ. Philipp.: Natural and applied sciences bulletin, University of the Philippines. Manila, 1930– .

Nat. Hist. Bull. Siam Soc.: Natural history bulletin of the Siam society. Bangkok, 1947– .

*Nat. Hist. Notes (Jamaica): Natural history notes, natural history society of Jamaica. Kingston, 1941– .

Nat. Res.: Nature and resources. Paris, 1965– .

Natl. Acad. Sci. – Natl. Res. Council Publ.: National academy of sciences – National research council publications. Washington, D.C., 1951– .

Natl. Mus. Nat. Sci. Canada, Publ. Bot.: National museum of natural sciences of Canada, publications in botany. Ottawa, 1969– . [*Successor in part to* Bulletin of the national museum of Canada.]

*Natur & Landschaft: Natur und Landschaft. Lüneberg, 1953– .

Natura Mosana: Natura mosana. Liège, Belgium, 1948– .

Natural resources research: Natural resources research [UNESCO]. Paris, 1963– .

Naturaliste Canad.: Le naturaliste canadien. Québec, 1869–1993. [*Full index*, 1993.]

Naturaliste Malg.: Le naturaliste malgache. Tananarive, 1949–62.

Nature: Nature. London, 1869– .

Natuurk. Tijdschr. Ned.-Indië: Natuurkundig tijdschrift voor Nederlandsch-Indië. Batavia (Jakarta), Indonesia, 1850–1940. [*Subsequently retitled* Chronica naturae.]

*New Hampshire Agr. Expt. Sta. Bull.: New Hampshire agricultural experiment station bulletin. Durham, N.H., 1888– .

New York State Mus. Bull.: New York state museum bulletin. Albany, N.Y., 1887– .

New Zealand Div. Sci. Indust. Res. Bull.: New Zealand division of scientific and industrial research bulletins. Wellington, New Zealand, 1927– . [*Inclusive of* New Zealand division of scientific and industrial research, Cape expeditions science bulletins.]

*New Zealand Div. Sci. Indust. Res., Cape Exped. Sci. Bull.: New Zealand division of scientific and industrial research, Cape expeditions science bulletins. *See* New Zealand division of scientific and industrial research bulletins.

New Zealand J. Bot.: New Zealand journal of botany. Wellington, 1963– .

New Zealand J. Sci. (Wellington): New Zealand journal of science. Wellington, 1958– .

*Newsl. Nat. Hist.: Newsletter of natural history. Tokyo. [No. 9, 1992.]

*Nigerian J. Forest.: Nigerian journal of forestry. Ibadan, 1971– .

*Nordic J. Bot.: Nordic journal of botany. Copenhagen, 1980– .

*Northern Territory Bot. Bull.: Northern Territory botanical bulletin. Alice Springs (*later* Darwin), 1976– .

Northw. Sci.: Northwest science. Cheney, Washington State, 1927– .

Notes Roy. Bot. Gard. Edinburgh: Notes from the royal botanic garden, Edinburgh. Edinburgh, 1900–89. [*From 1990 continued as* Edinburgh journal of botany.]

Notizbl. [Königl.] Bot. Gart. Berlin [-Dahlem]: Notizblatt des [Königlichen] botanischen Gartens und Museums zu Berlin [-Dahlem]. Berlin, 1895–1945. [*Title varies. Succeeded by* Willdenowia.]

Notul. Syst. (Paris): Notulae systematicae. Paris, 1909–60. [*Succeeded by* Adansonia, série 2.]

Nouv. Arch. Mus. Hist. Nat.: Nouvelles archives du muséum [national] d'histoire naturelle. Paris, 1865–1914. [*Title varies.*]

Nov. Sist. Vysš. Rast.: Novosti sistematiki vysšikh rastenij. Leningrad, 1964– .

Nova Acta Regiae Soc. Sci. Upsal.: Nova acta regiae societatis scientiarum upsaliensis. Uppsala, 1773– . [*Several successive series*; ser. IV, 1905– .]

Nova Guinea: Nova Guinea. Leiden, 1909–36.

Nova Guinea, Bot.: Nova Guinea, botany. Leiden, 1960–66.

*Nova Hedwigia: Nova Hedwigia. Weinheim, etc., Germany, 1959– .

*Nova Hedwigia, Beih.: Nova Hedwigia, Beihefte. Lehre, etc., Germany, 1962– .

OPTIMA Newsl.: Organization for the Phyto-taxonomic Investigation of the Mediterranean Area (OPTIMA), newsletter. Geneva, etc., 1975– .

Occas. Pap. Bernice P. Bishop Mus.: Occasional papers of the Bernice P. Bishop museum of Polynesian ethnology and natural history. Honolulu, Hawaii, 1898– .

*Occas. Pap. British Columbia Prov. Mus.: Occasional papers of the British Columbia provincial museum. Victoria, B.C., 1939– .

Occas. Pap. Univ. Hawaii: Occasional papers of the University of Hawaii. Honolulu, Hawaii, 1923– .

Ohio J. Sci.: Ohio journal of science. Columbus, Ohio, 1901– .

Ohio State Univ. Bull.: Ohio State University bulletin. Columbus, 1896–1906; 1909– .

Oklahoma Agric. Exp. Sta. Bull.: Oklahoma agricultural experiment station bulletin. Stillwater, Okla., 1892– .

Oklahoma Agric. Mech. Coll. Bull.: Oklahoma agricultural and mechanical college, bulletin [biological (science) series]. Stillwater, Okla., 1950– .

Opera Bot.: Opera botanica a societate botanica lundensi. Lund, Sweden, 1953– .

Opera Bot., B: Opera botanica, series B [Flora of Ecuador]. Lund, 1973–75. [*Subsequently issued as an independent work.*]

Opera Lilloana: Opera lilloana. Tucumán, Argentina, 1957– .

Pacific Disc.: Pacific discovery. San Francisco, 1948– .

Pacific Insects, Monogr.: Pacific insects, monographs. Honolulu, Hawaii, 1961– .

Pacific Sci.: Pacific science. Honolulu, Hawaii, 1947– .

Pacific Sci. (Assoc.) Inform. Bull.: Pacific science association information bulletin. Honolulu, Hawaii, 1947– .

Pakistan J. Forest.: Pakistan journal of forestry. Upper Topa, Pakistan, 1951– .

Pakistan Syst.: Pakistan systematics. Rawalpindi, 1977– .

*Pamiętn. Fizyogr.: Pamiętnik fizyograficzny. Warsaw, 1881–1918.

*Papua New Guinea J. Agric. Forest. Fisheries: Papua New Guinea journal of agriculture, forestry and fisheries. Rabaul (*later* Port Moresby), 1936– . [*Title varies. Initially published as* New Guinea agricultural gazette; *later known as* Papua and New Guinea agricultural journal.]

Phil. Trans.: Philosophical transactions of the royal society of London. London, 1665–1886. [*Succeeded in part by* Philosophical transactions of the royal society of London. Series B.]

Phil. Trans., B.: Philosophical transactions of the royal society of London. Series B. London, 1887– .

Philipp. J. Sci.: Philippine journal of science. Manila, 1906– .

Phytologia: Phytologia. N.Y., etc., 1933– .

Phytologia, Mem.: Phytologia, memoirs. Plainfield, N.J., etc., 1980– .

Phyton (Horn): Phyton. Horn, Austria, 1948– .

Pl. Sci. Bull.: Plant science bulletin. Urbana, Ill., etc., 1955– .

Polar Rec.: Polar record. Cambridge, England, 1931– .

*Polar Res.: Polar research. Oslo, 1982– .

Popular Sci. Monthly: Popular science monthly. New York, 1872– .

Posebna Izd. Biol. Inst. & Zemaljsk. Muz. u Sarajevu:
Posebna izdanja biologii institut i zemaljskog museja
(Bosni Hercegovini) u Sarajevu. Sarajevo, 1950– .

Posebna Izd. Srpska [Kral.] Akad.: Posebna izdanja Srpska
[Kral'evsk] Akademia. Belgrade, 1890– .

Posebni Izd. Filoz. Fak. Univ. Skoplje: Posebni izdanija,
Filozofski fakultet na Universitetot Skoplje. Skopje,
1950– .

Prace Monogr. Komis. Fizjogr.: Prace monograficzne
komisji fizjograficznej. Cracow, Poland, 1925–31.

Preslia: Preslia. Prague, 1914– .

*Proc. Acad. Nat. Sci. Philadelphia: Proceedings of the
academy of natural sciences of Philadelphia.
Philadelphia, Pa., 1841– .

Proc. Amer. Philos. Soc.: Proceedings of the American
philosophical society. Philadelphia, Pa., 1838– .

*Proc. Arkansas Acad. Sci.: Proceedings of the Arkansas
academy of science. Fayetteville, Ark., 1941– .

Proc. Boston Soc. Nat. Hist.: Proceedings of the Boston
society of natural history. Boston, Mass., 1841–1942.

*Proc. Bot. Soc. British Isles: Proceedings of the botanical
society of the British Isles. Arbroath, Scotland, etc.,
1954–69. [*Partly absorbed into* Watsonia.]

Proc. Calif. Acad. Sci.: Proceedings of the California
academy of sciences. San Francisco, 1854– .

Proc. Hawaiian Acad. Sci.: Proceedings of the Hawaiian
academy of science. Honolulu, Hawaii, 1926– .

*Proc. Indiana Acad. Sci.: Proceedings of the Indiana
academy of science. Brookville, Ind., 1891– .

Proc. Iowa Acad. Sci.: Proceedings of the Iowa academy of
science. Des Moines, Iowa, 1887– .

*Proc. Isle of Man Nat. Hist. Antiq. Soc.: Proceedings of the
Isle of Man natural history and antiquarian society.
Douglas, 1906– .

*Proc. Linn. Soc. London: Proceedings of the Linnean
society of London. London, 1839–1968. [*Succeeded by*
Biological journal of the Linnean society *and (from
1985)* The Linnean.]

Proc. Linn. Soc. New South Wales: Proceedings of the
Linnean society of New South Wales. Sydney, 1875– .

*Proc. Louisiana Soc. Naturalists: Proceedings of the
Louisiana society of naturalists. New Orleans. [Volume
for 1897–99(1900).]

Proc. Natl. Acad. Sci. India, sect. B., Biol. Sci.: Proceedings
of the national academy of sciences of India. Section B.
Biological sciences. Allahabad, India, 1936– .

Proc. Nova Scotian Inst. Sci.: Proceedings of the Nova Scotian
institute of science. Halifax, Nova Scotia, 1863– .

Proc. Roy. Irish Acad.: Proceedings of the royal Irish
academy. Dublin, 1841– .

Proc. Roy. Soc. Arts Sci. Mauritius: Proceedings of the royal
society of arts and science of Mauritius. [New series.]
Port Louis, 1950– .

Proc. Roy. Soc. Queensland: Proceedings of the royal society
of Queensland. Brisbane, 1883– .

Proc. & Trans. Roy. Soc. Canada: Proceedings and
transactions of the royal society of Canada. *See*
Transactions of the royal society of Canada.

*Proc. Univ. Durham Philos. Soc.: Proceedings of the
University of Durham philosophical society. Durham,
1896–1963.

*Proc. Utah Acad. Sci.: Proceedings of the Utah academy of
science. Salt Lake City, 1929– .

Prog. Bot.: Progress in botany (*initially* Fortschritte der
Botanik). Berlin, 1932– .

Provancheria: Provancheria. Québec, 1966– .

Publ. Arnold Arbor.: Publications of the Arnold Arboretum.
Cambridge, Mass., 1891–1921(–1933).

Publ. Biol. (Jeddah): Publications in biology, King Abdulaziz
University. Jeddah, 1978– .

Publ. Bur. Sci. Philipp.: Publications (*later* Monographs) of
the bureau of science, Philippines. Manila, 1908–34.
[*Succeeded by* Monographs of the Philippine institute of
science and technology.]

*Publ. CNFRA, Biol.: Publications de la Comité national
français des recherches antarctiques (CNFRA),
biologie. Paris, 1962– . [*Title varies, with* 'pour les'
sometimes substituted for 'des'.]

Publ. CTFT: Publications du Centre technique de forestier
tropical (CTFT). Nogent-sur-Marne, 1950– .

Publ. Cairo Univ. Herb.: Publications of the Cairo
University herbarium. Cairo, 1968– .

Publ. Canada Dept. Agric. Res. Br.: Publications of the
Canada department of agriculture (*later* Agriculture
Canada), research branch. Ottawa, 1951– (*as unified
series; no. 846 onwards*).

Publ. Carnegie Inst. Wash.: Publications of the Carnegie
institution of Washington. Washington, 1902– .

Publ. Comitié étud. Hist. Sci. A.O.F.: Publications du
Comitié d'études historiques et scientifiques de
l'Afrique occidentale française. Dakar/Gorée, Sénégal,
1934–39 (series A, 1934–39; series B, 1935–39).

*Publ. Conservatoire Jard. Bot.: Publications *hors-série* du
conservatoire et jardin botaniques. Geneva. [Vol. 4,
1980; vol. 4b, 1995.]

*Publ. Conservatoire Jard. Bot., Sér. Doc.: Publications du
conservatoire et jardin botaniques, série documentaire.
[No. 17, 1985. *A 'popular' series.*]

*Publ. Dépt. Écol. (Pondichéry): Publications du
département d'écologie, Institut français de Pondichéry.
See Travaux de la section scientifique et technique,
Institut français de Pondichéry.

Publ. Field Columbian Mus., Bot. Ser.: Publications of the
Field Columbian museum. Botanical series. Chicago,
1895–1932. [*Subsequently retitled* Publications of the
Field museum of natural history, botanical series.]

Publ. Field Mus. Nat. Hist., Bot. Ser.: Publications of the Field museum of natural history (*later* Chicago natural history museum), botanical series. Chicago, 1930–44. [*Subsequently retitled* Fieldiana, Botany.]

Publ. Inst. Nac. Hielo Continental Patagonica: Publicaciones, Instituto nacional de hielo continental patagonica. Buenos Aires. [Vol. 8, 1965.]

Publ. Inst. Natl. Rech. Sci. Rwanda: Publications de l'institut national de recherches scientifiques du Rwanda. Butare, Rwanda, <1971– .

Publ. Natl. Sci. Res. Council, Iran: Publications of the national science research council, Iran. Tehran. [Vol. 21, 1978.]

Publ. Soc. Océan.: Publications de la société des océanistes. Paris, 1951.

Publ. Syst. Assoc.: Publications of the systematics association. London, 1940; 1953– .

Publ. Univ. Pretoria: Publications of the University of Pretoria. Pretoria, 1921– . [Natural science, 1936– ; new series, 1956– .]

*Publ. Wormsloe Found.: Publications of the Wormsloe foundation. Athens, Ga. [No. 18, 1988.]

*Qatar Univ. Sci. Bull.: Qatar University science bulletin (*from 1992* Qatar University science journal). Doha, Qatar, 1981– .

*Quart. J. Chinese Forest.: Quarterly journal of Chinese forestry. Taipei, 1967– .

Quart. J. Taiwan Mus.: Quarterly journal of the Taiwan museum. Taipei, 1948– .

Quart. Rev. Biol.: Quarterly review of biology. Baltimore, Md., etc., 1926– .

*Queensland Bot. Bull.: Queensland botany bulletin [new series]. Brisbane, 1982– .

Queensland Naturalist: Queensland naturalist. Brisbane, 1908– .

Rapp. Bosbouwproefsta.: Rapporten van het bosbouwproefstation. *See* Laporan-laporanan Lembaga penelitian hutan.

Razpr. Slovensk. Akad. Znan.: Razprave Slovenskie akademija znanosti in umjetnosti v Ljubljani. Ljubljana, 1950– . [*Comprises a number of concurrent series, including* ser. IV, 1951– .]

Rec. Auckland Inst. Mus.: Records of the Auckland institute and museum. Auckland, 1930– .

Rec. Bot. Surv. India: Records of the botanical survey of India. Calcutta, 1893– .

Rec. Domin. Mus.: Records of the Dominion museum. Wellington, New Zealand, 1942– . [*Includes also a* Miscellaneous Series.]

Recueil Trav. Bot. Néerl.: Recueil des travaux botaniques néerlandais. Nijmegen, Netherlands, 1904–50. [*Succeeded by* Acta botanica neerlandica.]

Regnum Veg.: Regnum vegetabile. Utrecht, etc., 1953– .

Rendiconti Seminarii Fac. Sci. Reale Univ. Cagliari: Rendiconti seminarii (Rendiconti del seminario) della facoltà di scienze di reale università di Cagliari. Cagliari, Sardinia, Italy, 1931– .

Rep. Austral. Natl. Antarctic Res. Exped.: Reports of the Australian national antarctic research expeditions. Melbourne, 1948– .

*Rep. Bot. Inst. Aarhus Univ.: Reports from the botanical institute of Aarhus University (*from 1988* AAU reports). Aarhus, Denmark, 1976– .

*Rep. Bot. Soc. Exch. Club Brit. Isles: Reports, Botanical society and exchange club of the British Isles (*initially* Reports, Botanical exchange club and society of the British Isles). Manchester, etc., 1880–1948. [*Title varies. Succeeded by* Year book of the botanical society of the British Isles *and, later,* Proceedings of the botanical society of the British Isles *as well as* Watsonia.]

*Rep. Canad. Arctic Exped.: Reports of the Canadian arctic expedition, 1913–18. Ottawa, 1919–28.

Rep. Inst. Sci. Res. Manchukuo: Report of the institute of scientific research, Manchukuo. Hsinking, Manchuria (= Changchun, China), 1936–40.

Rep. Meetings Australas. Assoc. Advancem. Sci.: Reports of meetings of the Australasian association for the advancement of science. Sydney, 1888–1954.

Rep. State Biol. Surv. Kansas: Reports of the state biological survey of Kansas. [Lawrence], Kans. <1976– .

Repert. Spec. Nov. Regni Veg. [Fedde]: Repertorium specierum novarum regni vegetabilis (ed. F. Fedde). Berlin, 1905–42. [*Subsequently retitled* Feddes repertorium.]

Repert. Spec. Nov. Regni Veg. [Fedde], Beih.: Repertorium specierum novarum regni vegetabilis (ed. F. Fedde). Beihefte. Berlin, 1911–44. [*Subsequently retitled* Feddes repertorium, Beihefte.]

*Res. Bull. Panjab Univ., N.S.: Research bulletin of the Panjab University of science, new series. Hoshiarpur, 1954– .

*Rev. Écol. (Terre Vie): Revue d'écologie (terre et vie). Paris, 1980– . [*From 1931 to 1935 and 1947 to 1979 published as* Terre et vie.]

Rev. Int. Bot. Appl. Agric. Trop.: Revue international de botanique appliquée et d'agriculture tropicale. Paris, 1921–53. [*Succeeded by* Journal d'agriculture tropicale et de botanique appliquée (JATBA).]

*Rev. Roum. Biol., Bot.: Revue roumaine de biologie, série de botanique (*later* biologie végétale). Bucharest, 1964– .

*Rev. Sci. Bourbonn.: Revue scientifique du Bourbonnais et du centre de la France. Moulins, 1888–1914; 1921– . [*Not voluminated.*]

*Rev. Valdôtaine Hist. Nat.: Revue valdôtaine d'histoire naturelle. Aoste, 1972– .

Revista Acad. Col. Ci. Exact.: Revista de la academia colombiana de ciencias exactas, fisicas y naturales correspondiente de la española. Bogotá, 1936– .

*Revista Agric. (Cochabamba): Revista de agricultura, Cochabamba, Bolivia. Cochabamba, 1943– .

Revista Agron. (Rio Grande do Sul): Revista agronómica. Porto Alegre, Brazil, 1937– .

*Revista Chilena Hist. Nat.: Revista chilena de historia natural. Valparaiso, 1897–1946.

Revista Fac. Agron. (Maracay): Revista de la facultad de agronomía [universidad central de Venezuela]. Maracay, Venezuela, 1952– .

Revista Fac. Ci. Univ. Lisboa, sér. 2, C (Ci. Nat.): Revista da faculdade de ciências, universidade de Lisboa. Séria 2, C (ciências naturais). Lisbon, 1950– .

Revista Fac. Nac. Agron., Medellín: Revista, facultad nacional de agronomia, universidad de Antioquia, Medellín. Medellín, 1939– .

Revista Mus. La Plata, Secc. Bot.: Revista del museo de La Plata. Sección botánica. La Plata, Argentina, 1936– .

Revista Mus. Nac. (Lima): Revista del museo nacional. Lima, Peru, 1932– .

Revista Mus. Prov. Ci. Nat. Córdoba: Revista del museo provincial de ciencias naturales de Córdoba. Córdoba, Argentina. [Vol. 3, 1938.]

*Revista Pedagog.: Revista de pedagogie [Romania]. [Vol. 5(3), 1935.]

Revista Soc. Colomb. Ci. Nat.: Revista (*initially* Boletín) de la sociedad colombiana de ciencias naturales. Bogotá, 1913–31.

Revista Sudamer. Bot.: Revista sudamericana de botánica. Montevideo, 1934–56.

*Revista Univ. (Cuzco): Revista universitaria, Cuzco, Peru. Cuzco, 1912– .

Revista Univ. (Santiago): Revista universitaria, Universidad católica de Chile. Santiago, 1915–68; N.S. 1978– .

Revista Univ. Nac. Córdoba: Revista de la Universidad nacional de Córdoba. Córdoba, Argentina, 1914– .

Rhodesian Agric. J.: Rhodesian agricultural journal. Salisbury (*now* Harare), Zimbabwe, 1902– .

Rhodora: Rhodora. Lancaster, Pa., etc., 1899– .

Rodriguésia: Rodriguésia. Rio de Janeiro, 1935– .

S. African Biol. Soc., Pamphl.: South African biological society, pamphlets. Pretoria, 1931–59.

*S. African J. Antarct. Res.: South African journal of antarctic research. Pretoria, 1971– .

S. African J. Sci.: South African journal of science (*initially* Reports of the annual meetings of the South African association for the advancement of science). Cape Town, 1903– .

Sabah Forest Rec.: Sabah forest records (*initially* North Borneo forest records). Sandakan, Sabah (*later also* Kuching, Sarawak), 1938– .

Sarawak Mus. J.: Sarawak museum journal. Kuching, Sarawak, 1911– .

Sargentia: Sargentia. Jamaica Plain, Mass., 1942–49.

Sarracenia: Sarracenia. Montreal, 1959– .

Saudi Biol. Soc. Publ.: Saudi biological society: publications. N.p., <1979– .

*Schriftenreihe GTZ: Schriftenreihe der GTZ ([Deutsche] Gesellschaft für technischen Zusammenarbeit). Eschborn, 1972– . [Hors-série *issues also exist.*]

Sci. Bull. Brigham Young Univ., Biol. Ser.: Science bulletin of the Brigham Young University, biological series. Provo, Utah, 1955– .

*Sci. Cult.: Science and culture. Calcutta, 1935– .

Sci. J. (Sydney): Science journal. [University of Sydney science association.] Sydney, 1917–38.

Sci. Rep. Kanazawa Univ., Biol.: Science reports of the Kanazawa University; biology. Kanazawa, Japan, 1951– .

*Scient. Rep. Australas. Antarct. Exped. 1911–1914: Scientific reports of the Australasian antarctic expedition, 1911–1914. Adelaide, 1916–48.

Scient. Rep. Brit. Antarctic Surv.: Scientific reports of the British antarctic survey. London, etc., 1953– .

*Scripta Geobotanica: Scripta geobotanica. Göttingen, 1970– .

Search: Search. Sydney, 1971– .

Selbyana: Selbyana. Sarasota, Fla., 1975– .

Sellowia: Sellowia (*initially* Anais botânicos). Itajaí, Brazil, 1949– .

Sida: Sida; contributions to botany. Dallas (*later* Fort Worth), Tex., 1962– .

*Sida, Bot. Misc.: Sida, botanical miscellany. Dallas (*later* Fort Worth), Tex., 1987– .

*Sitzungsber. Naturf.-Ges. Dorpat: Sitzungsberichte der Naturforscher-Gesellschaft zu Dorpat (*from 1881* Sitzungsberichte der Naturforscher-Gesellschaft bei der Universität Dorpat; *after 1892* Sitzungsberichte der Naturforscher-Gesellschaft bei der Universität Jurjew; *after 1896 also entitled* Protokoly obščestva estestvoispytatelej pri Imperatorskom Jur'evskom Universitete). Dorpat (Jurjew; = Tartu), 1861–78; 1881–1916. [*From 1925 to 1943 continued as* Tartu ulikooli juures oleva loodusuurijati seltsi kirjatööd (*with title also in German or Latin*). Tartu; *and from 1955 onwards by* [Eesti] loodusuurijate seltsi aastaraamat. Tallinn.]

Skr. Norsk Polarinst.: Skrifter Norsk polarinstitutt. Oslo, 1948– .

Skr. Svalbard Ishavet: Skrifter om Svalbard og Ishavet. Oslo, 1927–40. [*Continued as* Skrifter Norsk polarinstitutt, *q.v.*]

Slovenská Vlastiv.: Slovenská vlastiveda. (Slovenské akadémia vied a umení.) Bratislava, 1943–48.

Smithsonian Contr. Bot.: Smithsonian contributions to botany. Washington, 1969– .

*Smithsonian Misc. Collect.: Smithsonian miscellaneous collections. Washington, 1862– .

*Sommerfeltia: Sommerfeltia. Oslo, 1985– .

South Australian Woods & Forests Dept. Bull.: South Australian woods and forests department, bulletin. Adelaide, 1928– .

Southwestern Louisiana J.: Southwestern Louisiana journal. Lafayette, La., 1951– .

*Sovetsk. Bot.: Sovetskja botanika. Moscow/Leningrad, 1933–47. [*Successor to* Izvestija imperatorskogo St-Peterburgskogo botaničeskogo sada *and, finally,* Izvestija botaničeskogo sada Akademii nauk SSSR. *After 1947 absorbed into* Botaničeskij Žurnal.]

Special Number Educ. Misc. Hunan: Special number of the educational miscellany of Hunan. [Changsha], Hunan. [Vol. 4, 1924.]

Special Publ. Brit. Columbia Forest Serv.: Special publications of the British Columbia forest service [B-series]. Victoria, B.C., <1949– .

Special Publ. Brit. Columbia Prov. Mus. Nat. Hist.: Special publications of the British Columbia provincial museum of natural history and anthropology. Vancouver, 1947– .

*Special Publ. Carnegie Mus. Nat. Hist.: Special publications of the Carnegie museum of natural history. Pittsburgh, Pa. [No. 14, 1989.]

State Biol. Surv. Kansas, Techn. Publ.: State biological survey of Kans.: technical publications. [Lawrence], Kans., <1976– .

Stud. Nat. Hist. Iowa Univ.: Studies in natural history, Iowa University (*initially* Bulletin of Iowa University's laboratories of natural history). Iowa City, Iowa, 1888– .

Sudan Forests Bull.: Sudan forests bulletins. Khartoum, 1958– .

Sudan Notes & Rec.: Sudan notes and records. Khartoum, 1918– .

Sunyatsenia: Sunyatsenia. Canton, China, 1930–38.

Svensk Bot. Tidskr.: Svensk botanisk tidskrift. Stockholm, 1907– .

*Syllogeus: Syllogeus. Ottawa, 1972– .

Symb. Bot. Upsal.: Symbolae botanicae upsalienses. Uppsala, Sweden, 1932– .

Syst. Bot.: Systematic botany. Tallahassee, Florida, 1976– .

TC Ankara Üniv. Fen Fakult., Yayınları: Türkiye cumhuriyet Ankara universitesi, fen fakültesi. Yayınları. Ankara, 1948– (Botanik, 1952–).

TÜRDOK, Bibliogr. Ser.: TÜRDOK, Bibliografiya seriya. Ankara. [Vol. 5, 1972.]

*Taiwania: Taiwania. Taipei, 1948– .

*Tap Chi Sinh Vât Hoc (Biol. J.): Tap Chi Sinh Vât Hoc (Biological journal). Hanoi, 1978– . [*Sometime prior to 1987 retitled* Tap Chi Sinh Hoc.]

*Tasman. Nat.: Tasmanian naturalist. Hobart, 1907– . [*Publication not continuous.*]

Taxodium: Taxodium. [Vol. 1, 1943.]

Taxon: Taxon. Utrecht (*later* Berlin*; now* Vienna), 1951– .

*Techn. Bull. Agric. Canada Res. Br.: Technical bulletin, Agriculture Canada research branch. Ottawa. [No. 1986-3E. *Not to be confused with* Technical bulletin, Department of agriculture, Canada. Ottawa, 1935–50.]

Techn. Bull. Lafayette Nat. Hist. Mus.: Technical bulletin of the Lafayette natural history museum. Lafayette, La., 1969– .

Techn. Bull. Univ. Brit. Columbia Bot. Gard.: Technical bulletin of the University of British Columbia botanical garden. Vancouver, 1972– .

*Techn. Bull. Univ. Florida Agric. Exp. Sta.: Technical bulletins of the University of Florida agricultural experiment station. Gainesville, Fla., 1910– .

Techn. Rep. Meteorol. Climatol. Arid Regions, Inst. Atmos. Phys., Univ. Arizona: Technical reports, meteorology and climatology of arid regions, Institute of atmospheric physics of the University of Arizona. Tucson, 1956– .

Texas Agric. Expt. Sta. Publ.: Texas agricultural experiment station, publications. College Station, Tex. [No. MP-1399, 1978; MP-1655, 1990.]

Tijdschr. Kon. Ned. Aardrijksk. Genootsch.: Tijdschrift van het koninklijk nederlandsch aardrijkskundig genootschap. Amsterdam, 1874– .

*Tijdschr. Natuurl. Gesch. Physiol.: Tijdschrift voor natuurlijke geschiedenis en physiologie. Amsterdam, Netherlands, 1834–45.

Trab. Soc. Ci. Nat. "La Salle": Trabajos de la sociedad de ciencias naturales "La Salle". Caracas. [Vol. 20, 1960.]

Trans. Acad. Sci. St. Louis: Transactions of the academy of science of St. Louis. St. Louis, 1860– .

Trans. [& Proc.] Bot. Soc. Edinburgh: Transactions and proceedings of the botanical society of Edinburgh. Edinburgh, 1844–1990. [*Subsequently retitled* Botanical journal of Scotland *without change in volumination.*]

Trans. Kentucky Acad. Sci.: Transactions of the Kentucky academy of science. Lexington, Ky., 1914– .

*Trans. Linn. Soc. London: Transactions of the Linnean society of London. London, 1791–1875. [*Succeeded by* Transactions of the Linnean society of London, Series 2, botany *and* Series 2, zoology.]

Trans. Linn. Soc. London, II, Bot.: Transactions of the Linnean society of London. Ser. 2, botany. London, 1875–1922.

Trans. Linn. Soc. London, II, Zool.: Transactions of the Linnean society of London. Ser. 2, zoology. London, 1875–1936.

Trans. Nat. Hist. Soc. Taiwan: Transactions of the natural history society of Taiwan. Taihoku (Taipei), Taiwan, 1914– .

Trans. Oceanogr. Inst. Moscow: Transactions of the oceanographic institute of Moscow. *See* Trudy Gosudarstvennogo okeanografičeskogo instituta.

Trans. & Proc. New Zealand Inst.: Transactions and proceedings of the New Zealand institute. Wellington, New Zealand, 1868–1933. [*Succeeded by* Transactions and proceedings of the royal society of New Zealand.]

Trans. & Proc. Roy. Soc. New Zealand: Transactions and proceedings of the royal society of New Zealand. Wellington, New Zealand, 1934–52. [*Succeeded by* Transactions of the royal society of New Zealand.]

Trans. & Proc. Roy. Soc. South Australia: Transactions and proceedings of the royal society of South Australia. Adelaide, 1877– .

*Trans. Roy. Canad. Inst.: Transactions of the royal Canadian institute. Toronto, 1889– .

*Trans. Roy. Soc. Canada: Transactions of the royal society of Canada (*initially* Proceedings and transactions of the royal society of Canada). Montreal, 1882– . [*Title varies. Three successive series.*]

Trans. Roy. Soc. Edinburgh: Transactions of the royal society of Edinburgh. Edinburgh, 1788– .

Trans. Roy. Soc. New Zealand: Transactions of the royal society of New Zealand. Wellington, New Zealand, 1953–60. [*From 1961 divided into series; for* Botany *see next entry.*]

Trans. Roy. Soc. New Zealand, Bot.: Transactions of the royal society of New Zealand. Botany. Wellington, New Zealand, 1961–68.

Trans. Roy. Soc. South Africa: Transactions of the royal society of South Africa. Cape Town, 1908– .

Trans. Wisconsin Acad. Sci.: Transactions of the Wisconsin academy of sciences, arts and letters. Madison, Wis., 1870– .

Trav. Inst. Sci. Chérifien, Sér. Bot.: Travaux de l'institut scientifique chérifien (*from vol. 34 (1984)* Travaux de l'institut scientifique, Université Mohammed V). Série botanique. Tangier (*later* Rabat), 1952– .

Trav. Lab. Forest. Toulouse: Travaux du laboratoire forestier de Toulouse. Toulouse, 1928– . (*Comprises several concurrent series.*)

*Trav. Sect. Sci. Tech. Inst. Franç. Pondichéry: Travaux de la section scientifique et technique, Institut français de Pondichéry. Pondicherry, 1957– . [*From 1991 retitled* Publications du département d'écologie, Institut français de Pondichéry, *without change in volumination.*]

Tree species lists, [Buitenzorg]: Tree species lists. [Forest Research Institute, Buitenzorg.] Buitenzorg (Bogor), Indonesia, 1940–41.

Trop. Woods: Tropical woods. New Haven, Conn., 1925–60.

Trudy Bot. Inst. Akad. Nauk Tadžiksk. SSR: Trudy Botaničeskogo institut Akademii nauk Tadžikskoj SSR. Dushanbe, Moscow/Leningrad, 1962– .

Trudy Bot. Inst. AN SSSR: Trudy Botaničeskogo institut Akademii nauk SSSR. Moscow/Leningrad, 1933–92. [*Comprises several concurrent series. Succeeded by* Trudy Bot. Inst. RAN.]

*Trudy Bot. Inst. RAN: Trudy Botaničeskogo institut Rossijiskej Akademii nauk. St. Petersburg, 1993– .

Trudy Bot. Muz. Imp. Akad. Nauk: Trudy Botaničeskago muzeja Imperatorskej Akademii nauk. St. Petersburg (Leningrad), 1902–32. [*After 1916* 'Imperatorskej' *was replaced by* 'Rossijskej'.]

Trudy Dal'nevostočnogo Fil. 'Komarova', ser. Bot.: Trudy Dal'nevostočnogo filial imeni V. L. Komarova. Serija botaničeskaja. Moscow/Leningrad, 1956– .

Trudy Glavn. Bot. Sada (SSSR): Trudy Glavnago botaničeskago sada (SSSR). *See* Trudy Imperatorskago S.-Peterburgskago botaničeskago sada.

Trudy Gosud. Nikitsk. Bot. Sada: Trudy Gosudarstvennago Nikitskago botaničeskago sada. Yalta, Ukraine, 1934– . [*Slight change of title after 1953.*]

Trudy Gosud. Okeanogr. Inst. (Moscow): Trudy Gosudarstvennogo okeanografičeskogo instituta. Moscow, 1931–34; 1947– .

Trudy Imp. S.-Peterburgsk. Bot. Sada: Trudy Imperatorskago S.-Peterburgskago botaničeskago sada (*also with Latin title* Acta horti petropolitani). St. Petersburg, 1871–1930. [*From 1915 title changed to* Trudy Glavnago botaničeskago sada.]

Trudy Inst. Biol. Akad. Nauk SSSR Ural'sk Fil.: Trudy Instituta biologii, Akademija nauk SSSR, Ural'skij filial. Moscow/Leningrad, 1948– .

Trudy Kazakhstansk. Fil. Akad. Nauk SSSR: Trudy Kazakhstanskij filial, Akademija nauk SSSR. Moscow/Leningrad, 1937–41.

Trudy Mongol'sk. Komiss.: Trudy Mongol'skoj komissii. Moscow/Leningrad, 1928– .

Trudy Poljarn. Komiss.: Trudy Poljarnoj komissii. Moscow/Leningrad, 1930–37.

Trudy Rostovsk. Obl. Biol. Obšč.: Trudy Rostovskogo oblastnogo biologičeskogo obščestva. Rostov, Russian SFSR, 1937–40.

Trudy Severn. Bazy AN SSSR: Trudy Severnoj bazy (Akademii nauk SSSR). Moscow/Leningrad. [Vol. 2, 1937.]

Trudy Tiflissk. Bot. Sada: Trudy Tiflisskago botaničeskago sada (*later* Trudy Tbilsskago botaničeskago sada). Tbilisi (Tiflis), 1895–1917; 1920–34; 1938–49.

Tuatara: Tuatara. Wellington, New Zealand, 1947– .

*Tulane Stud. Zool. Bot.: Tulane studies in zoology and botany. New Orleans, La., 1969– .

U.S. Forest Surv. Res. Pap.: United States forest service: research papers. Washington, D.C., etc., 1964– . [*Encompasses several concurrent series issued by regional centers, each with separate numeration.*]

*Uchen. Zap. Krasnojarsk. Pedagog. Inst.: Uchenje [Uchenye] zapiski Krasnojarskii pedagogičeskii institut. Krasnojarsk. [Vol. 24(4), 1963.]

Uitgaven Natuurw. Studiekring Suriname Ned. Antillen: Uitgaven van de natuurwetenschapplijke studiekring voor Suriname en de Nederlandse Antillen (*initially* Uitgaven van de . . . studiekring voor Suriname en Curaçao). The Hague (*later* Utrecht), 1946– .

Uitgaven Natuurw. Werkgroep Ned. Antillen: Uitgaven van de 'Natuurwetenschappelijke werkgroep Nederlandse Antillen'. Willemstad, Curaçao, 1951–71.

*Ukrajins'k. Bot. Žurn.: Ukrajins'kyj botaničnyj žurnal. Kiev, 1921–29; [N.S.], 1940– . [*From 1940 to 1955 as* Botaničnyj žurnal (Kiev)*; present title from 1956.*]

Univ. Calif. Publ. Bot.: University of California publications in botany. Berkeley, etc., Calif., 1902– .

*Univ. Colorado Stud., Ser. Biol.: University of Colorado studies, series in biology. Boulder, Colo., 1950–70. [*Separated from* University of Colorado studies. Boulder, 1902–57.]

*Univ. Tokyo Mus., Material Reports: University of Tokyo museum, material reports. Tokyo, 1976– .

Vakbl. Biol.: Vakblad voor biologen. Amsterdam, 1919– .

Verh. Kon. Ned. Akad. Wetensch., II. reeks, Afd. Natuurk.: Verhandelingen der Koninklijke Nederlandse akademie van wetenschappen, tweede reeks (*until 1951* tweede sectie), afdeling Natuurkunde. Amsterdam, 1892– .

*Verh. Zool.-Bot. Ges. Österreich: Verhandlungen der [k.k.] zoologisch-botanischen Gesellschaft in Wien (*from 1978* Österreich). Vienna, 1852– . [*Title varies.*]

Vermont Agric. Exp. Sta. Bull.: Vermont state agricultural college, agricultural experiment station, bulletin. Burlington, Vt., 1887– .

*Veröff. Geobot. Inst. ETH, Stiftung Rübel, Zürich: Veröffentlichungen des Geobotanischen Institutes der Eidgenössisches Technisches Hochschule in Zürich, Stiftung Rübel. Zürich, 1924– . [*Title varies; institute part of ETH only from 1959.*]

*Veröff. Joachim Jungius-Ges. Wiss. Hamburg: Veröffentlichungen, Joachim Jungius-Gesellschaft der Wissenschaften, Hamburg. Göttingen. [Vol. 43, 1980.]

*Victorian Hist. Mag.: Victorian historical magazine (*from 1975* Victorian historical journal). Melbourne, 1911– .

Victorian Naturalist: Victorian naturalist. Melbourne, 1884–.

*Vidensk. Meddel. Dansk. Naturhist. Foren. Kjøbenhavn: Videnskabelige meddelelser fra Dansk naturhistorisk forening i Kjøbenhavn. Copenhagen, 1849– .

Vierteljahrsschr. Naturf. Ges. Zürich: Vierteljahrsschrift der Naturforschenden Gesellschaft in Zürich. Zürich, 1856– .

Vierteljahrsschr. Naturf. Ges. Zürich, Beibl.: Vierteljahrsschrift der Naturforschenden Gesellschaft in Zürich. Beiblätter. Zürich, 1923– .

Virginia Agric. Exp. Sta. Tech. Bull.: Virginia agricultural experiment station. Technical bulletin. Blacksburg, 1915–69.

*Virginia Agric. Ext. Serv. Bull.: Virginia [polytechnic institute and state university] agricultural extension service, bulletin. Blacksburg, 1915– . [*Title varies.*]

W. Virginia Univ. Bull.: West Virginia university bulletins. Morganstown, W.Va., 1888– .

*Watsonia: Watsonia. Arbroath, Scotland, etc., 1949– .

Webbia: Webbia. Florence, 1905– .

Willdenowia: Willdenowia. Berlin, 1953– .

Wiss. Mitt. Bosnien-Herzegowina: Wissenschaftliche Mitt(h)eilungen aus Bosnien und der Herzegowina. Vienna, 1893–1916.

Wyoming Agric. Exp. Sta. Bull.: Wyoming agricultural experiment station. Bulletin. Laramie, Wyo., 1891– .

Wyoming Agric. Exp. Sta. Res. J.: Wyoming agricultural experiment station. Research journal. Laramie, Wyo., 1966– .

Zambia Forest Res. Bull.: Zambia forest research bulletin (*initially* Research bulletin, Forest department, Northern Rhodesia). Ndola, 1960– .

Zap. Imp. Akad. Nauk, Fiz.-Mat. Otd.: Zapiski Imperatorskoj akademii nauk po Fiziko-matematičeskomu otdeleniju. Ser. 8. St. Petersburg, 1894–1916.

Zap. Južno-Ussurijsk. Otd. Gosud. Russk. Geogr. Obšč.: Zapiski Južno-Ussurijskogo otdela Gosudarstvennogo russkogo geografičeskogo obščestva. Vladivostok, Russian SFSR, 1927–29.

Žurn. Obščej Biol. (SSSR): Žurnal obščej biologii (Journal of general biology), SSSR (USSR). Moscow, 1940– .

Addenda in proof

The following came to the author's notice too late for inclusion in the main text or, if published in 2000, represent significant additions to the literature. All are based on personal observation (*) or have been taken from listings in *Taxon* or recent catalogues from Koeltz Scientific Books (nos. 886, 889 and 905, 2000) as well as other sources. A superscript bullet (•) indicates new works; all others represent continuations, revisions, translations or issues in alternative media. 'a' indicates the left-hand column of a page, 'b' the right-hand column.

Conventions and abbreviations
  p. 91a: KR (*Kew Record of Taxonomic Literature*). See **Appendix A** below.

Division 0
  pp. 95b, 96a, and Note 5: *Gray Herbarium index* and *Index Kewensis*. The on-line combination of these indices (together with the *Australian plant names index*), the *International Plant Names Index* (IPNI), went 'public' during 2000 (http://www.ipni.org/).
  p. 104a: •TURNER, I. M., XING FUWU and R. C. CORLETT, 2000. An annotated check-list of the vascular plants of the South China Sea and its shores. *Raffles Bulletin of Zoology, Supplement* 8: 23–116. [Tabular checklist; complements Schimper's *Die indo-malayische Strandflora*.]*
  p. 119b: •CRONK, Q. C. B., 2000. *The endemic flora of St Helena*. 119 pp., 25 pls., 30 halftones. Oswestry, Salop.: Anthony Nelson. [Descriptions, commentary and portraits.]*
  p. 120a: •VÁLKA ALVES, R. J., 1998. *Ilha de Trindade e arquipélago Martin Vaz: um ensaio geobotânico*. Rio de Janeiro. [Includes a systematic account covering 55 spp. (20 pteridophytes, 35 angiosperms), making the work the modern definitive treatment for the higher flora. (The work therefore should have a full listing rather than merely a passing mention as given.)]*

p. 128b: *Svalbards flora* (RØNNING). 3rd Norwegian edn., 1996; also published in English as O. I. RØNNING, 1996. *The flora of Svalbard*. 184 pp., 77 text-figs. (part col.), col. maps (end-papers), map (in text). Oslo: Norsk Polarinstitutt. (Polarhåndbok 10.)*

p. 131a: *Flora of the Russian Arctic* (TOLMATCHEV). Vol. 3, 2000. Berlin: Cramer/ Borntraeger. [Corresponds to fascicles 5–6 (Salicaceae–Ranunculaceae) of the Russian edn.]

Division 1

p. 164b: *CHADDE, S. W., 1998. *A Great Lakes wetland flora*. x, 569 pp., illus. [A field-guide; regionally complements the larger works of Fassett/Crow/ Hellquist. – Koeltz 905.]

p. 174b: *DOUGLAS, G. W., G. B. STRALEY, D. MEIDINGER and J. POJAR (eds.), 1998– . *Illustrated flora of British Columbia*. Vols. 1– . Victoria, B.C.: Ministry of Environment, Lands and Parks/Ministry of Forests, British Columbia. [3 vols. as of 2000.]*

p. 194a: *RHOADS, A. F. and T. A. BLOCK, 2000. *The plants of Pennsylvania: an illustrated manual*. vii, 1061 pp., 2645 text-figs. (by A. Anisko). Philadelphia: University of Pennsylvania Press. [Concisely descriptive manual with keys and numerous illustrations; over 3000 spp. – Further information from http://www.upenn.edu/pennpress/.]

p. 236a: *Trees and shrubs of Trans-Pecos Texas* (POWELL). Revised as A. M. POWELL, 1998. *Trees and shrubs of the Trans-Pecos and adjacent areas*. xviii, 498 pp., 398 text-figs., 7 halftones, 1 map. Austin: University of Texas Press. [447 species.]*

Division 2

p. 258b: *Flora Neotropica*. See **Division 3** below.

p. 275b: *DIX, M. A. and M. W. DIX, 2000. *Orchids of Guatemala: a revised annotated checklist*. iii, 62 pp. (Monogr. Syst. Bot. Missouri Bot. Gard. 78). St. Louis. [734 taxa; complementary to *Orchids of Guatemala* (AMES/CORRELL).]

p. 279a: *STEVENS, W. D., C. ULLOA ULLOA, A. POOL and O. M. MONTIEL (eds.), 2000. *Flora de Nicaragua*. 3 vols. (Monogr. Syst. Bot. Missouri Bot. Gard. 80). St. Louis. [The first critical national flora, covering native and introduced seed plants.]

p. 288b: *Flora de la República de Cuba*, serie A. Fascicles 3 and 4, 2000. Königstein/Ts.: Koeltz.

p. 289b: *Flora of the Cayman Islands* (PROCTOR). For additions, see G. R. PROCTOR, 1996. Additions and corrections to "Flora of the Cayman Islands". *Kew*

Bull. **51**: 483–507. [74 species added; one deleted; total non-cultivated species now 674, a 12 percent increase.]

p. 304a: *Flora of Trinidad and Tobago*. For palms (not so far in the *Flora*) see L. H. BAILEY, 1947. Indigenous palms of Trinidad. *Gentes Herbarum* **7**: 351–445, illus.* For pteridophytes, a revision by C. D. Adams is in progress (*idem*, personal communication, 2000).

Division 3

p. 317a: *Flora Neotropica*. Monographs 78 (Cladoniaceae), 79 (Buddlejaceae), 2000. New York.

p. 330a: *Flora of the Venezuelan Guayana*. Vol. 6, 2000 (Liliaceae–Myrsinaceae). St. Louis.

p. 349a: *ALVES CAMARGOS, J. A., C. M. CZARNESKI *et al.*, 1996. *Catálogo de árvores do Brasil*. 888 pp., illus. Brasília: Instituto Brasileiro do Meio Ambiente e dos Recursos Naturais, Laboratorio de Produtos Florestais, Brasil. [Primary list by vernacular names, with (pp. 601ff.) a lexicon by botanical names; no keys.]*

Division 4

p. 386b: *Flora of New Zealand*. Vol. 5, *Grasses* [Gramineae], 2000, by E. Edgar and H. E. Connor. lxxxii, 650 pp., illus. Lincoln, New Zealand: Manaaki Whenua [Landcare Research] Press. [Twenty years' study required for this account of 188 native and 226 naturalized species, which completes the series begun in 1961.]*

p. 392a: The Australian vascular flora is now estimated to have anywhere from 20 000 to 25 000 species (ABRS, *Biologue* **23**: 6 (2000)).

p. 395b: *Flora of Australia*. Vol. 17A, 2000 (Proteaceae 2, by R. O. Makinson).

p. 422b: *Flore des Mascareignes*. Families 27–30bis, 1997 (gymnosperms); 69–79, 1997 (Meliaceae–Connaraceae); 81–89, 1997 (Rosaceae–Callitrichaceae); 149–152, 1998 (Aristolochiaceae–Monimiaceae).*

p. 425b: *WISE, R., 1998. *A fragile Eden: portraits of the endemic plants of the granitic Seychelles*. xvi, 216 pp., col. pls. Princeton, N.J.: Princeton University Press. [Portraits with descriptions and commentaries.]*

Division 5

p. 451a: *Genera of Southern African flowering plants* (DYER). Succeeded by O. A. LEISTNER (ed.), 2000. *Seed plants of southern Africa: families and genera*.

775 pp. (Strelitzia 10). Pretoria: National Botanical Institute, South Africa. [In fulfillment of Dyer's expressed wish for a successor to the *Genera* by 2000.]*

p. 451b: *Flora of Southern Africa*. Vol. 5(1), 2000 (Aloeaceae 1: *Aloe*); vol. 28(1), 2000 (Convolvulaceae). The work is now under the general editorship of G. Germishuizen.*

p. 453b. *Plants of the Cape flora* (BOND/GOLDBLATT). As presaged in Note 20 (p. 511), this has been succeeded by P. GOLDBLATT and J. MANNING, *Cape plants: a conspectus of the Cape flora of South Africa*, with publication (as Strelitzia 9) announced for 2000 by the National Botanical Institute of South Africa.

p. 460a: *Flora Zambesiaca*. Vol. 3(6), 2000 (Leguminosae in part).*

p. 485b: *Flore du Gabon*. Part 5bis, 1999 (Gramineae, supplement).

p. 509b: *FENNANE, MD., MD. IBN TATTOU, J. MATHEZ, A. OUYAHYA and J. EL OUALIDI (eds.), 2000. *Flore pratique du Maroc*. Vol. 1. xv, 558 pp., illus. Rabat-Agdal: Institut Scientifique, Université Mohammed V (as Trav. Inst. Sci., Sér. Bot. 36). [Pteridophytes, gymnosperms and dicotyledons Lauraceae–Neuradaceae.]

Division 6

p. 525b: *Flora Europaea*. CD-ROM edn., 2000. Cambridge. [Interactive keys and glossary added. – Koeltz 905 as well as Cambridge University Press, personal communication.]

p. 533b: *Atlas der Alpenflora* (DEUTSCHER UND ÖSTERREICHISCHER ALPENVEREIN). 2nd German (Palla) edn. reissued 1987 (Koeltz 905).

p. 542b: *Atlas clasificatorio de la flora de España peninsular y balear* (GARCIA ROLLÁN). Vol. 1 reissued 1999 with corrections.

p. 545b: *Flora dels Països Catalans*. Vol. 4 (of 4) in press as of 1999 (*OPTIMA Newsl.* 34).

p. 564a: *Flora e Shqipërisë/Flore de l'Albanie*. Vol. 4 (of 4), 2000. [Compositae and monocotyledons. – Koeltz 905.]

p. 569a: *Flora of Turkey*, vol. 11. See **Division 7** below.

p. 573a: *Illustrierte Flora von Mitteleuropa*, 2. Aufl. Bd. 4(2C), Lfg. A, pp. 1–108 (*Rosa*), 2000. [An extension of Bd. 4(2B); first revision of *Rosa* since its 1923 appearance in the 1. Auflage. The other genera to be covered in Bd. 4(2C) are *Potentilla* and *Sibbaldia*.]*

p. 575a: *TASENKEVICH, L., 1998. *Flora of the Carpathians: checklist of the native vascular plant species*. xiii, 609 pp. [3808 species and infraspecific taxa. May be seen as a partial realization of Dostál's Carpatho-Pannonian flora proposal (Division 6, Note 95). – Koeltz 905.]

p. 579a: *Mala flora Slovenije* (MARTINČIČ/SUŠNIK). Revised as A. MARTINČIČ (ed.), 1999. *Mala flora Slovenije: ključ za določanje praprotnic in semenk*. 3rd edn., 845 pp., 605 text-figs. Ljubljana. [2979 native and naturalized vascular species.]*

p. 587b: *Flora von Deutschland und angrenzender Länder* (SCHMEIL/FITSCHEN). 91. Aufl., 2000, revised by K. Senghas and S. Seybold. Wiebelsheim: Quelle & Mayer.

p. 606b: *JONSELL, B. (ed.), 2000. *Flora nordica*. Vol. 1. Stockholm: Bergius Foundation, Royal Swedish Academy of Sciences. [The first of five planned volumes; covers pteridophytes, gymnosperms and Salicaceae–Polygonaceae. Families arranged essentially as in *Flora Europaea*.]*

p. 615a: *Taimede välimääraja* (KASK *et al.*). 2nd edn, 1975. 423 pp.

p. 616a: *GAVRILOVA, G. and V. SULCS, 1999. *Latvijas vaskularo augu flora: taksonu saraksts/Flora of Latvian vascular plants: list of taxa*. 136 pp. Riga: [Latvian Academy Press]. [*Euro+Med PlantBase Newsletter* 2.]

p. 617b: *GUDZINSKAS, Z., 1999. *Lietuvos induociai augalai* [Vascular plants of Lithuania]. Vilnius. [*Euro+Med PlantBase Newsletter* 2.]

p. 624b: *Flora of Russia: the European part and bordering regions* (English edn. of *Flora evropejskoj časti SSSR/Flora vostočnoj Evropy*). Vol. 3, 2000. [Vols. 1 and 2, 1999. Contents correspond to the volumes of the Russian edition, without revision.]

p. 632a: *MOSYAKIN, S. L. and M. M. FEDORONCHUK, 1999. *Vascular plants of the Ukraine: a nomenclatural checklist*. 345 pp. Kiev. [In English; covers *c.* 5000 native and established species. – Koeltz 886; cf. also Note 204.]

p. 632b: *Opredelitel' vysšikh rastenij Ukrainy* (PROKUDIN). Reissued as YU. N. PROKUDIN *et al.*, 1999. (*Determination key to vascular plants of the Ukraine*.) 544 pp., 570 text-figs. Kiev. [In Ukrainian, according to Koeltz 905.]

Division 7

p. 662b: *Flora of Siberia* (English edn. of *Flora Sibiri*). Vol. 1, 2000. Enfield, N.H./Plymouth, England: Science Publishers. [Contents correspond to the volumes of the Russian edition, without revision.]*

p. 680a: *Flora Armenii*. Vol. 10, 2000. [Monocotyledons except Poaceae. – Koeltz 905.]

p. 688b: *Materialy k bibliografii po flore i rastitel'nosti Tsentral'noj Azii* (LEBEDEV). Appears as *Bibliography for flora and vegetation of Central Asia* on pp. 114–188 of *Plants of Central Asia* (see below).*

p. 689a: *Plants of Central Asia* (English edn. of *Plantae asiae centralis*). Vol. 1, 1999; vol. 2, 2000. Enfield, N.H./Plymouth, England: Science Publishers. [Contents correspond to the fascicles of the Russian edition, without revision.]*

p. 690a: *Manual of the vascular plants of Mongolia* (English edn. of *Opredelitel' sosudistykh rastenij Mongolii* (GRUBOV, 1982)). Announced for 2000 in 2 vols.; publishers as above. [See also Koeltz 905.]

p. 691b: *Flora xinjiangensis*. Vol. 5, 1999 (Compositae).

p. 697b: *Flora of Turkey and the east Aegean islands*. Vol. 11, 2000, by A. Güner, N. Özhatay, T. Ekim and K. H. C. Baser (eds.) with assistance from I. C. Hedge. 600 pp. Edinburgh: Edinburgh University Press. [Accounts for 400 new species or new records for Turkey; also indices to chromosome numbers and phytochemical data. – Information from I. C. Hedge.]

p. 698b: *VINEY, D., 1994–96. An illustrated flora of North Cyprus. 2 vols. Illus. Königstein/Ts.: Koeltz (vol. 1), Vaduz: Gantner (vol. 2). [Vol. 1, vascular plants exclusive of grasses, sedges and pteridophytes; vol. 2, grasses, sedges and pteridophytes. Covers a large proportion of the total flora of Cyprus.]*

p. 700a: *FRAGMAN, O., U. PLITMAN, D. HELLER and A. SHMIDA, 1999. Checklist and ecological data-base of the flora of Israel and its surroundings*. Jerusalem. [Tabular checklist, covering in all 2944 native and established higher plant species. 2444 are in Israel proper with 2138 thought truly native.]*

p. 704a: *CHAUDHARY, S. A. (ed.), 1999. Flora of the kingdom of Saudi Arabia illustrated*. Vol. 1. 691 pp., 341 pls. (line drawings). Riyadh: National Herbarium, National Agriculture and Water Research Center, Ministry of Agriculture and Water, Saudi Arabia; and *idem*, 1989. *Grasses of Saudi Arabia*. iii, 465, 63 pp., illus. Riyadh. [The *Flora* so far encompasses pteridophytes, gymnosperms, and 'earlier' dicotyledons from Annonaceae through Mimosaceae. The volume on grasses is complementary. Both are intended to succeed the floras of Migahid.]*

p. 709a and Note 69: *Flora iranica*, no. 174 (**Papilionaceae III**, *Astragalus*). Two or three further

parts expected, with part 2 due at end of 2000 or early 2001 (*Taxon* 49: 869 (2000)).

Division 8

p. 723b: *LI HENG, GUO HUIJUN and DAO ZHILING (eds.), 2000. Flora of Gaoligong Mountains*. 1350 pp., illus., maps. Beijing: Science Press. [Covers 4303 seed plants (434 strictly endemic) in an area of 111 000 km^2 on the Yunnan side of this geologically complex and rainy 'hotspot' range (the remainder being in the Kachin state of Myanmar).]*

p. 731b: *Fascicles of flora of India*. To fasc. 24, 1999.

p. 772a: *PRESS, J. R., K. K. SHRESTHA and D. A. SUTTON, 2000. Annotated checklist of the flowering plants of Nepal*. ix, 430 pp. London: The Natural History Museum. [Includes distribution, types, life-forms and synonym index; 6200 species. Complements the *Enumeration* of HARA *et al.* (1978–82).]*

p. 793b: *Flora of China*. Vol. 24, 2000 (Flagellariaceae–Marantaceae).*

p. 793b: *Flora of China, Illustrations*. Vols. 15 and 18, 2000.*

p. 794a: *Higher plants of China/Zhong guo gian den zhi wu*. Illus., maps. Qingdao (Tsingtao). Vol. 3, 2000 (gymnosperms and Magnoliaceae–Daphniphyllaceae); vol. 9, 1999 (Loganiaceae–Lamiaceae). [Planned to cover about 17 000 species, this large-format publication will be a functional successor to *Iconographia cormophytorum sinicarum* (1972–83).]*

p. 799b: *Flora Loess-Plateaus Sinicae*. Vol. 1, 2000.

p. 803a: *Henan zhi wi zhi*. Vol. 4, 1998 (Tiliaceae–Compositae).

p. 803a: *Flora Shanxiensis*. Vol. 3, 2000 (Tropaeolaceae–Boraginaceae).

p. 809b: *Flora sichuanica*. Vol. 15, 2000 (Gentianaceae, Menyanthaceae).

p. 827b: *GARDNER, S., P. SIDISUNTHORN and V. ANUSARNSUNTHORN, 2000. A field guide to forest trees of northern Thailand*. xiv, 545 pp., illus. (part col.). Bangkok: Kobfai Publishing Project (distributed by Asia Books). [Lavishly illustrated field-guide.]*

p. 828a: *HAJRA, P. K., P. S. N. RAO and V. N. MUDGAL (eds.), 1999. Flora of Andaman and Nicobar Islands*. Vol. 1. Illus. Calcutta: Botanical Survey of India. (Flora of India, series 2.) [Ranunculaceae–Combretaceae. One further volume expected.]*

p. 829b: *SINHA, B. K., 1999. Flora of Great*

Nicobar Island. v, 525 pp., illus. Calcutta: Botanical Survey of India.*

Division 9

p. 850a: *Flora Malesiana*, series I: Vol. 14, 2000 (Myristicaceae). [Covers 335 species, bringing the total of seed plants covered since 1948 to 7720.]*

p. 859b: *Tree flora of Sabah and Sarawak*. Vol. 3, 2000.*

p. 921b: *Manual of the flowering plants of Hawaii* (WAGNER *et al.*). [Published as Bernice P. Bishop Museum Special Publication 97, the 2nd edn. features a new preface and in vol. 2 a supplement of 64 pp. following the original index.]

Appendix A

pp. 943b (Note 11) and 947b (Note 47): *Kew Record of Taxonomic Literature*. Publicly available on-line from summer 2000 (http://www.rbgkew.org.uk/kr/KRHomeExt.html). [*VISTA* (Kew) 4, 25 September 2000; *Taxon* 49: 831 (2000).]

Appendix B

p. 949a: *Catalogue de Périodiques* (BIBLIOTHÈQUE CENTRALE DU MUSÉUM NATIONAL D'HISTOIRE NATURELLE). Revised edn., 1995, as *Catalogue de Périodiques 1995*. Paris. [Also available on-line.]*

Geographical index

The reference numbers in the following index stand for the geographical unit headings (**000–999**) as well as the designations used for major divisions (**D0–D9**), superregions (numbers prefixed by **SR**) and regions (**R01–R99**). These all appear in the text in bold sans-serif along with the name of the polity or other unit in question, and should enable a desired heading to be found more quickly than otherwise.

As explained in Chapter 1 of the General Part, each number basically consists of three digits: the first, or third-order digit, represents the appropriate major geographical division; the second, or second-order digit, represents the geographical region or, if it be 0, denotes that the category is of very broad scope or is physiographically or synusially founded; and the third, or first-order digit, represents the basic geographical unit or, if it be 0, that the said category is of regional scope (and is used for regional floras, etc.). Some numbers are connected by a dash; these generally represent superregions or superregional floras and are inclusive from the first to the last given region. A few regions are linked by a slash (/); these are 'double' regions wherein a significant number of units has to be recognized without there being a basis for the designation of normal regions.

For the first edition, an attempt was made to account for a number of synonymous geographical names, with the guidance in particular of the geographical scheme given in the entomological informatic system proposal of Travis *et al.* (1962).[1] However, 'colonial' and other generally superseded geographical names have for the most part been omitted. Use of a good world or historical atlas is recommended in cases of omission or doubt. For the current edition, necessary nomenclatural changes as well as additions have been made; additionally, note has been taken of the Taxonomic Databases Working Group (TDWG) *World geographical scheme* (Hollis and Brummitt, 1992).[2]

References

1 TRAVIS, B. V., H. H. CASWELL, JR., W. B. ROWAN, H. STARCKE and C. H. ROSS, 1962. *Classification and coding system for compilations from the world literature on insects and other arthropods that affect the health and comfort of man.* 259 pp., illus., map (United States Army, Quartermaster Research and Engineering Center, Technical Report ES-4). Natick, Mass.

2 HOLLIS, S. and R. K. BRUMMITT, 1992. *World geographical scheme for recording plant distributions.* ix, 105 pp., maps. Pittsburgh, Pa.: Hunt Institute for Botanical Documentation (for the International Working Group on Taxonomic Databases for Plant Sciences). (Plant Taxonomic Database Standards, 2: version 1.0.) [A revision, also to be published through the Hunt Institute, was in press as of late 2000.]

Author index

The following index accounts for all authors cited in Part II of this book, save for those only mentioned in passing in heading and other commentaries as well as those responsible for such reference items as general bibliographies, general indices and major library catalogues. The latter items are fully covered in Appendix A, wherein each class is alphabetically arranged (the bibliographies and catalogues by author(s), the indices (including abstracting works) by title).

No attempt has been made to index Part I, as it is basically an extended introduction to the *Guide* proper, which constitutes Part II.

The year following the author's name is the first (or only) date of publication. Numerals in **bold** type refer to geographical headings (see Geographical index for full explanation). Page numbers appear in *italic*. Entries in *italic* represent those which appear in running or introductory text only; entries in Roman are those with a full annotation.

The format of the index is modeled on that used by Blake and Atwood for the first volume of their *Geographical guide* (1942) as well as by Blake alone for the second volume of that work (1961). Vital statistics of authors, however, have here been omitted for practical reasons and because there is now a greater range of taxonomic reference works containing this kind of information than was the case in 1939.

The principal source for authors' dates is now *Authors of plant names*, edited by R. K. Brummitt and E. Powell (1992, Kew: Royal Botanic Gardens). For further information on authors whose works were (first) published before about 1941, users are advised in the first instance to consult *Taxonomic literature-2* (*TL-2* for short) by F. A. Stafleu and R. S. Cowan (1976–88, Utrecht; supplement, 1992–2000, Königstein/Ts., Germany). The great majority of floras and their kin published or begun before World War II are there treated in bibliographical detail and with recensional and other pertinent information (but without a précis and commentary on the contents as is the case in the present work). It should be noted, however, that *TL-2* – particularly in the original volumes – treats only works which can for at least some copies be demonstrated as having been independently published and/or distributed.

More 'recent' authors (who, as before 1941, may have also been, or acted only as, editors) are not so singularly treated, and a considerable variety of sources, too numerous to detail here, must therefore be consulted for further information. The introductory parts of each volume of *TL-2*, though, collectively account for much the greatest part of this range, thus providing a starting-point for further studies on any given author and his works. Also useful here is *The history of natural history: an annotated bibliography* by G. D. R. Bridson (1994, New York: Garland Publishing).

And in such indexes, although small pricks
To their subsequent volumes, there is seen
The baby figure of the giant mass
Of things to come at large.

Shakespeare, *Troilus and Cressida*, Act 1, Scene 3.

ABEYESUNDERE, L. A. J.
  1939, Sri Lanka (woody plants) **829**, *757*
ABEYESUNDERE, L. A. J. and ROSAYRO, R. A. DE
  1939, Sri Lanka (woody plants) **829**, *757*
ABEYWICKRAMA, B. A.
  1959, Sri Lanka **829**, *756*
  1978, Sri Lanka (pteridophytes) **829**, *757*
ABRAMS, L.
  1923–60, Pacific Coast United States (in general) **190**, *241*
ABRAMS, L. and FERRIS, R. S.
  1923–60, Pacific Coast United States (in general) **190**, *241*
ACADÉMIE MALGACHE
  1931–35, Madagascar and associated islands **460**, *420*
ACEBES GINOVÉS, J. R.
  1978, Salvage Islands **023**, *113*
ACEVEDO RODRÍGUEZ, P.
  1996, Virgin Islands (St. John, American V. I.) **257**, *294*
ACEVES DE LA MORA, J. L.
  1988, Oaxaca (bibliography) **227**, *272*
ACOCKS, J. P. H.
  1988, Southern Africa **510**, *450*
ACOSTA-SOLÍS, M.
  1968, Colombia and Ecuador (progress) **R31**, *333*
  1968, Ecuador **329**, *337*

ADAM, J.-G.
  1971–83, 'Old mountains', Africa (Nimba Massif) **502**, *444*
  1980–81, 'Old mountains', Africa (Loma Massif) **502**, *445*
ADAMS, C. D.
  1972, Jamaica **253**, *290*
  1983, Maldive Islands **042**, *122*
ADAMS, J.
  1928–30, Canada (bibliography) **R12/13**, *170*
  1930, Canada (bibliography). *See* ADAMS, J., 1928–30.
  1932, Canada (bibliography). *See* ADAMS, J., 1928–30.
  1936, Canada (bibliography). *See* ADAMS, J., 1928–30.
ADAMS, J. and NORWELL, M. H.
  1932, Canada (bibliography). *See* ADAMS, J., 1928–30.
  1936, Canada (bibliography). *See* ADAMS, J., 1928–30.
ADAMS, N. M.
  1963, New Zealand (woody plants) **410**, *388*
  1995, Alpine and upper montane zones (New Zealand) **403/I**, *383*
ADAMSON, R. S.
  1950, Former Cape Province (Cape Peninsula) **512**, *454*
ADAMSON, R. S. and SALTER, T. M.
  1950, Former Cape Province (Cape Peninsula) **512**, *454*
ADEM, J. *et al.*
  1960, Revillagigedo Islands **013**, *107*
ADILOV (ODILOV), T.
  1988, Uzbekistan (partial works) **753**, *686*
ADLER, W.
  1994, Austria **644**, *580*
ADLER, W., OSWALD, K. and FISCHER, R.
  1994, Austria **644**, *580*
ADSERSEN, J.
  1987, Galápagos Islands **017**, *109*
ÆLIJEV, A. R.
  1961–70, Azerbaijan (woody plants) **748**, *681*
AEDO, C.
  1999, Benin Islands (Bioko and Pagalu) **579**, *490*
  1999, Equatorial Guinea **574**, *486*
AEDO, C., TELLERÍA, Mª. T. and VELAYOS, M.
  1999, Benin Islands (Bioko and Pagalu) **579**, *490*
  1999, Equatorial Guinea **574**, *486*
AESCHIMANN, D.
  1986, Alpine regions (C Europe, Alps). *See* LANDOLT, E., 1992.
  1994, Switzerland. *See* BINZ, A., 1990.
  1996, Switzerland **649**, *588*
AESCHIMANN, D. and BURDET, H. M.
  1994, Switzerland. *See* BINZ, A., 1990.
AESCHIMANN, D. and HEITZ, C.
  1996, Switzerland **649**, *588*

HARA, H., STEARN, W. T., CHATER, A. O. and WILLIAMS, L. H. J.
  1978–82, Nepal **844**, *771*
HARCOME, G. F.
  1982, British Columbia (family keys, dicotyledons) **124**, *174*
  1982, British Columbia (family keys, monocotyledons). *See* RAFIQ, M., HARCOME, G. F. and OGILVIE, R. T., 1982 (family keys, dicotyledons).
HARDEN, G.
  1990–93, New South Wales **432**, *403*
HARDIN, J. W.
  1982, The Carolinas (bibliography) **162**, *214*
HARGER, E. B. *et al.*
  1931, New England (Connecticut). See GRAVES, C. B. *et al.*, 1910.
HARIDASAN, K.
  1985–87, Old Assam Region (Meghalaya, woody plants) **837/II**, *764*
HARIDASAN, K. and RAO, R. R.
  1985–87, Old Assam Region (Meghalaya, woody plants) **837/II**, *764*
HARKER-USECHE, M. A.
  1990, Colombia (pteridophytes) **321**, *335*
HARLEY, R. M.
  1980, Bahia **359**, *354*
  1986, Bahia (partial works) **359**, *354*
HARLEY, R. M. and MAYO, S. J.
  1980, Bahia **359**, *354*
HARLEY, R. M. and SIMMONS, N. A.
  1986, Bahia (partial works) **359**, *354*
HARLEY, W. J.
  1955, Liberia (pteridophytes) **585/I**, *496*
HARLING, G.
  1973– , Ecuador **329**, *338*
HARLING, G., SPARRE, B. *et al.*
  1973– , Ecuador **329**, *338*
HARLOW, W. M.
  1942, Northeastern and North Central United States (woody plants) **140**, *186*
HARPER, R. M.
  1928, Alabama (woody plants) **166**, *219*
HARRAR, E. S.
  1946, Southeastern United States (woody plants) **160**, *212*
HARRAR, E. S. and HARRAR, J. G.
  1946, Southeastern United States (woody plants) **160**, *212*
HARRAR, J. G.
  1946, Southeastern United States (woody plants) **160**, *212*
HARRINGTON, H. D.
  1964, Colorado **184**, *235*

HARRIS, T. Y.
  1970, Alpine and upper montane regions (Australia) **403/II**, *383*
HARRISON, W.
  1975, Bahama Archipelago (bibliography) **R24**, *286*
HARSHBERGER, J. W.
  1911, North America (progress) **D1**, *153*
HARTLEY, E.
  1965, Montana (bibliography) **181**, *233*
HARTLEY, T. G.
  1966, Midwest subregion **151**, *200*
HARTMANN, H.
  1966, Alpine and upper montane regions (Karakoram) **703/IV**, *654*
HARVEY, W. H.
  1859(1860)–65, Southern Africa **510**, *451*
HARVEY, W. H. and SONDER, O. W.
  1859(1860)–65, Southern Africa **510**, *451*
HARVILL, A. M., JR.
  1970, Virginia (partial works, spring flora) **147**, *197*
  1992, Virginia (distribution maps) **147**, *197*
HASAN BIN PUKUL
  1966, Borneo (Brunei) **917/II**, *861*
HASLAM, S. M.
  1975, Wetlands (NW European Islands) **608**, *538*
  1977, Malta **627**, *555*
  1978, World – river plants **006**, *102*
  1987, World – river plants **006**, *102*
HASLAM, S. M., SELL, P. D. and WOLSELEY, P. A.
  1977, Malta **627**, *555*
HASLAM, S. M., SINKER, C. and WOLSELEY, P. A.
  1975, Wetlands (NW European Islands) **608**, *538*
HASSALL, D. C.
  1994, Banaba and Nauru **969**, *905*
HASSIB, M.
  1956, Egypt **591**, *503*
HASSLER, E.
  1917, Paraguay. *See* CHODAT, R., 1898–1907.
HATCH, S. L.
  1990, Texas **171**, *224*
HATCH, S. L., GANDHI, K. N. and BROWN, L. E.
  1990, Texas **171**, *224*
HATUSIMA, S.
  1966, Batan Islands **924**, *869*
  1971, Nansei-shoto **856**, *782*
  1975, Nansei-shoto. *See* HATUSIMA, S., 1971.
  1994, Nansei-shoto (southern islands) **856**, *782*
HATUSIMA, S. and AMANO, T.
  1994, Nansei-shoto (southern islands) **856**, *782*
HAUMAN, L.
  1934, Belgium (bibliography) **656**, *595*
  1984, Argentina (families and genera) **380**, *366*

MOUTERDE, P., Fr.
    1965, Syria and Lebanon. See MOUTERDE, P., Fr.,
        1966–83(–84).
    1966–83(–84), Syria and Lebanon **774**, *699*
MUELLER, F.
    1864, Chatham Islands **414**, *390*
MUENSCHER, W. C.
    1944, Wetlands (North America, United States) **108**,
        164
    1950, Northeastern and North Central United States
        (woody plants) **140**, *186*
MUKHERJEE, A.
    1988, Sikkim and Darjiling (partial works) **845**, *773*
MUKHERJEE, A. K.
    1984, Madhya Pradesh (district/local works, Satpura
        Ranges) **819**, *740*
MULDASEV, A. A.
    1988–89, Eastern Russia (partial works, Bashkir
        Republic) **699**, *636*
MÜLLER, T.
    1994, Germany. See OBERDORFER, E., 1994.
MULLIN, J. M.
    1991, Outer Hebrides **666**, *605*
MUÑOZ PIZARRO, C.
    1966, Chile **390**, *373*
MUÑOZ, J. DE D.
    1982, Argentine 'Mesopotamia' (Entre Rios, woody
        plants) **381**, *368*
MUNZ, P. A.
    1959, California **195**, *245*
    1968, California. See MUNZ, P. A. and KECK, D. D.,
        1959.
    1974, Southern California **197**, *246*
MUNZ, P. A. and KECK, D. D.
    1959, California **195**, *245*
MURATA, G.
    1971–79, Japan. See KITAMURA, S. *et al.*, 1957–64.
MURILLO-PULIDO, M. T.
    1990, Colombia (pteridophytes) **321**, *335*
MURILLO-PULIDO, M. T. and HARKER-USECHE, M. A.
    1990, Colombia (pteridophytes) **321**, *335*
MURPHY, R. C.
    1952, Territorio Colón (Venezuela) **283**, *302*
MURRELL, G.
    1996– , Western European Is. **660**, *602*
MURTI, S. K.
    1989, Madhya Pradesh (district/local works, Bilaspur)
        819, *741*
MUSCHLER, R.
    1912, Egypt **591**, *503*
MUŠEGJAN, A. M.
    1962–66, Kazakhstan (woody plants) **751**, *684*

MUSEO BOTÁNICO, FACULDAD DE CIENCIAS EXACTAS,
    FÍSICAS Y NATURALES, UNIVERSIDAD NACIONAL DE
    CÓRDOBA
    1994– , Central Argentina (in general) **385**, *370*
MUSEO NACIONAL DE HISTORIA NATURAL, URUGUAY
    1958– , Uruguay **375**, *363*
MUSTAFA, M.
    1987, Sulawesi **921**, *868*
MUTIS, J. C.
    1954– , Colombia **321**, *334*
MWASUMBI, L. B.
    1976, Tanzania (pteridophytes) **531**, *469*
MYERS, R. M.
    1972, Illinois **154**, *204*
MYERS, W. S.
    1937, Oklahoma **172**, *225*
MYLKIDZHANJAN, JA. I.
    1951, Armenia (woody plants) **747**, *680*

NABIEV, M. M.
    1988, Uzbekistan (partial works) **753**, *686*
NABIL EL-HADIDI, M. *et al.*
    1980– , Egypt. See TÄCKHOLM, V., TÄCKHOLM, G. and
        DRAR, M., 1941–69.
NADEAUD, J.
    1873, Society Islands (Tahiti) **981**, *913*
NAIK, V. N.
    1979, Maharashtra (district/local works, Osmanabad)
        822, *746*
NAIR, K. K. N.
    1986–87, Tamil Nadu (partial/local works, Courtallum)
        828, *755*
NAIR, K. K. N. and NAYAR, M. P.
    1986–87, Tamil Nadu (partial/local works, Courtallum)
        828, *755*
NAIR, N. C.
    1977, Himachal Pradesh (district/local works, Bashahr
        Himalaya) **842**, *769*
    1978, East Punjab, Harayana, Chandigarh and Delhi **814**,
        736
    1983–89, Tamil Nadu **828**, *754*
    1991, Tamil Nadu (partial/local works, Coimbatore) **828**,
        755
NAIR, N. C., HENRY, A. N., KUMARI, G. R., CHITHRA, V. and
    BALAKRISHNAN, N. P.
    1983–89, Tamil Nadu **828**, *754*
NAIR, V. J.
    1988, Karnataka (district/local works, Cannanore) **827**,
        753
NAITHANI, B. D.
    1984–85, Uttarkhand (Kumaon, partial works (Chamoli))
        843/II, *770*